THE PRINCIPAL FUNCTIONAL GROUPS OF ORGANIC CHEM

	Example	Acceptable Name(s) of Example	Characteristic Reaction Type
Hydrocarbons			
Alkanes	CH_3CH_3	Ethane	Free-radical substitution of hydrogen by halogen
Alkenes	$H_2C{=}CH_2$	Ethene or ethylene	Electrophilic addition to double bond
Alkynes	$HC{\equiv}CH$	Ethyne or acetylene	Electrophilic addition to triple bond
Dienes	$H_2C{=}CHCH{=}CH_2$	1,3-Butadiene	Electrophilic addition to double bonds
Arenes	(benzene ring)	Benzene	Electrophilic aromatic substitution
Halogen-substituted derivatives of hydrocarbons			
Alkyl halides	CH_3CH_2Cl	Chloroethane or ethyl chloride	Nucleophilic substitution; elimination
Alkenyl halides	$H_2C{=}CHCl$	Chloroethene or vinyl chloride	Electrophilic addition to double bond; elimination
Aryl halides	C_6H_5Cl	Chlorobenzene	Electrophilic aromatic substitution; nucleophilic aromatic substitution
Oxygen-containing organic compounds			
Alcohols	CH_3CH_2OH	Ethanol or ethyl alcohol	Dehydration; conversion to alkyl halides; esterification
Phenols	C_6H_5OH	Phenol	Electrophilic aromatic substitution
Ethers	$CH_3CH_2OCH_2CH_3$	Ethoxyethane or diethyl ether	Cleavage by hydrogen halides
Epoxides	$H_2C{-}CH_2$ bridged by O	Epoxyethane or ethylene oxide or oxirane	Nucleophilic ring opening
Aldehydes	$CH_3\overset{O}{\overset{\Vert}{C}}H$	Ethanal or acetal-dehyde	Nucleophilic addition to carbonyl group
Ketones	$CH_3\overset{O}{\overset{\Vert}{C}}CH_3$	2-Propanone or acetone	Nucleophilic addition to carbonyl group
Carboxylic acids	$CH_3\overset{O}{\overset{\Vert}{C}}OH$	Ethanoic acid or acetic acid	Ionization of carboxyl; esterification

THE PRINCIPAL FUNCTIONAL GROUPS OF ORGANIC CHEMISTRY

	Example	*Acceptable Name(s) of Example*	*Characteristic Reaction Type*
Carboxylic acid derivatives			
Acyl halides	$CH_3C(=O)Cl$	Ethanoyl chloride or acetyl chloride	Nucleophilic acyl substitution
Acid anhydrides	$CH_3C(=O)OC(=O)CH_3$	Ethanoic anhydride or acetic anhydride	Nucleophilic acyl substitution
Esters	$CH_3C(=O)OCH_2CH_3$	Ethyl ethanoate or ethyl acetate	Nucleophilic acyl substitution
Amides	$CH_3C(=O)NHCH_3$	*N*-Methylethanamide or *N*-methylacetamide	Nucleophilic acyl substitution
Nitrogen-containing organic compounds			
Amines	$CH_3CH_2NH_2$	Ethanamine or ethylamine	Nitrogen acts as a base or as a nucleophile
Nitriles	$CH_3C{\equiv}N$	Ethanenitrile or acetonitrile	Nucleophilic addition to carbon–nitrogen triple bond
Nitro compounds	$C_6H_5NO_2$	Nitrobenzene	Reduction of nitro group to amine
Sulfur-containing organic compounds			
Thiols	CH_3CH_2SH	Ethanethiol	Oxidation to a sulfenic, sulfinic, or sulfonic acid or to a disulfide
Sulfides	$CH_3CH_2SCH_2CH_3$	Diethyl sulfide	Alkylation to a sulfonium salt; oxidation to a sulfoxide or sulfone

Organic Chemistry

ELEVENTH EDITION

Francis A. Carey
University of Virginia

Robert M. Giuliano
Villanova University

Neil T. Allison
University of Arkansas

Susan L. Bane
Binghamton University

ORGANIC CHEMISTRY, ELEVENTH EDITION

Published by McGraw-Hill Education, 2 Penn Plaza, New York, NY 10121.

This book is printed on acid-free paper.

1 2 3 4 5 6 7 8 9 LWI 21 20 19

ISBN 978-1-260-14892-3
MHID 1-260-14892-0

Portfolio Manager: Michelle Hentz
Product Developers: Mary E. Hurley & Megan Platt
Marketing Manager: Tamara Hodge
Content Project Managers: Laura Bies, Rachael Hillebrand & Sandra Schnee
Buyer: Sandy Ludovissy
Design: Daivd W. Hash
Content Licensing Specialists: Lorraine Buczek
Cover Image: ©Ella Maru Studio
Compositor: Aptara, Inc.

Library of Congress Cataloging-in-Publication Data

Names: Carey, Francis A., 1937- author. | Giuliano, Robert M., 1954- author. | Allison, Neil T. (Neil Thomas), 1953- author. | Tuttle, Susan L. Bane, author.
Title: Organic chemistry / Francis A. Carey (University of Virginia), Robert M. Giuliano (Villanova University), Neil T. Allison (University of Arkansas), Susan L. Bane Tuttle (Binghamton University).
Description: Eleventh edition. | New York, NY: McGraw-Hill Education, 2018. | Includes index.
Identifiers: LCCN 2018024902| ISBN 9781260148923 (alk. paper) | ISBN 1260148920 (alk. paper)
Subjects: LCSH: Chemistry, Organic. | Chemistry, Organic—Textbooks
Classification: LCC QD251.3 .C37 2018 | DDC 547—dc23 LC record available at https://lccn.loc.gov/2018024902

mheducation.com/highered

Each of the eleven editions of this text has benefited from the individual and collective contributions of the staff at McGraw-Hill. They are the ones who make it all possible. We appreciate their professionalism and thank them for their continuing support.

About the Authors

Before **Frank Carey** retired in 2000, his career teaching chemistry was spent entirely at the University of Virginia.

In addition to this text, he is coauthor (with Robert C. Atkins) of *Organic Chemistry: A Brief Course* and (with Richard J. Sundberg) of *Advanced Organic Chemistry,* a two-volume treatment designed for graduate students and advanced undergraduates.

Frank and his wife Jill are the parents of Andy, Bob, and Bill and the grandparents of Riyad, Ava, Juliana, Miles, Wynne, and Michael.

Robert M. Giuliano was born in Altoona, Pennsylvania, and attended Penn State (B.S. in chemistry) and the University of Virginia (Ph.D., under the direction of Francis Carey). Following postdoctoral studies with Bert Fraser-Reid at the University of Maryland, he joined the chemistry department faculty of Villanova University in 1982, where he is currently Professor. His research interests are in synthetic organic and carbohydrate chemistry.

Bob and his wife Margot, an elementary school teacher he met while attending UVa, are the parents of Michael, Ellen, and Christopher and the grandparents of Carina, Aurelia, Serafina, Lucia, and Francesca.

Neil T. Allison was born in Athens, Georgia, and attended Georgia College (B.S., 1975, in chemistry) and the University of Florida (Ph.D., 1978, under the direction of W. M. Jones). Following postdoctoral studies with Emanuel Vogel at the University of Cologne, Germany, and Peter Vollhardt at the University of California, Berkeley, he joined the faculty of the Department of Chemistry and Biochemistry, University of Arkansas in 1980. His research interests are in physical organometallic chemistry and physical organic chemistry.

Neil and his wife Amelia met while attending GC, and are the parents of Betsy, Joseph, and Alyse and the grandparents of Beau.

Susan L. Bane was raised in Spartanburg, South Carolina, and attended Davidson College (B.S., 1980, in chemistry) and Vanderbilt University (Ph.D., 1983, in biochemistry under the direction of J. David Puett and Robley C. Williams, Jr.). Following postdoctoral studies in bioorganic chemistry with Timothy L. Macdonald at the University of Virginia, she joined the faculty of the Department of Chemistry of Binghamton University, State University of New York, in 1985. She is currently Professor of Chemistry and director of the Biochemistry Program. Her research interests are in bioorganic and biophysical chemistry.

Susan is married to David Tuttle and is the mother of Bryant, Lauren, and Lesley.

Brief Contents

Contents

CHAPTER 4

Chirality 134

CHAPTER 5

Alcohols and Alkyl Halides: Introduction to Reaction Mechanisms 172

CHAPTER 6

Nucleophilic Substitution 210

CHAPTER 7

Structure and Preparation of Alkenes: Elimination Reactions 244

CHAPTER 8

Addition Reactions of Alkenes 288

CHAPTER 9

Alkynes 330

CHAPTER 10

Introduction to Free Radicals 356

CHAPTER 11

Conjugation in Alkadienes and Allylic Systems 384

CHAPTER 12

Arenes and Aromaticity 426

CHAPTER 13

Electrophilic and Nucleophilic Aromatic Substitution 476

CHAPTER 14

Spectroscopy 532

CHAPTER 15

Organometallic Compounds 600

CHAPTER 16

Alcohols, Diols, and Thiols 638

CHAPTER 17

Ethers, Epoxides, and Sulfides 676

CHAPTER 18

Aldehydes and Ketones: Nucleophilic Addition to the Carbonyl Group 714

CHAPTER 22

Amines 890

CHAPTER 23

Carbohydrates 946

CHAPTER 24

Lipids 996

CHAPTER **25**

Amino Acids, Peptides, and Proteins 1036

CHAPTER **26**

Nucleosides, Nucleotides, and Nucleic Acids 1098

CHAPTER **27**

Synthetic Polymers 1140

List of Important Features

Mechanisms

Tables

Boxed Essays

Descriptive Passage and Interpretive Problems

Preface

Overview

"There is a close analogy between organic chemistry in its relation to biochemistry and pure mathematics in its relation to physics."

Sir Robert Robinson

This quote from Sir Robert Robinson exemplifies two broad goals in the creation of the eleventh edition of Francis Carey's organic chemistry textbook. We want our students to have a deeper understanding of the physical concepts that underlie organic chemistry, and we want them to have a broader knowledge of the role of organic chemistry in biological systems. The Carey team now includes two new coauthors for the eleventh edition, Neil Allison, and Susan Bane, that have expertise in physical organic chemistry and in biochemistry. Significant changes in the areas of structure and mechanism and in bioorganic chemistry have been incorporated into the eleventh edition. These and other changes are highlighted below.

Mechanism

The text is organized according to functional groups—structural units within a molecule that are most closely identified with characteristic properties. Reaction mechanisms are emphasized early and often in an effort to develop the student's ability to see similarities in reactivity across the diverse range of functional groups encountered in organic chemistry. Mechanisms are developed from observations; thus, reactions are normally presented first, followed by their mechanism.

In order to maintain consistency with what our students have already learned, this text presents multistep mechanisms in the same way as most general chemistry textbooks—that is, as a series of *elementary steps*. Additionally, we provide a brief comment about how each step contributes to the overall mechanism. Section 1.11 "Curved Arrows, Arrow Pushing, and Chemical Reactions" provides the student with an early introduction to the notational system employed in all of the mechanistic discussions in the text.

Numerous reaction mechanisms are accompanied by potential energy diagrams. Section 5.8 "Reaction of Alcohols with Hydrogen Halides: The S_N1 Mechanism" shows how the potential energy diagrams for three elementary steps are combined to give the diagram for the overall reaction.

Enhanced Graphics

The teaching of organic chemistry has especially benefited as powerful modeling and graphics software has become routinely available. Computer-generated molecular models and electrostatic potential maps were integrated into the third edition of this text and their number has increased in succeeding editions; also seeing increasing use are molecular orbital theory and the role of orbital interactions in chemical reactivity.

Coverage of Biochemical Topics

From its earliest editions, four chapters have been included on biochemical topics and updated to cover topics of recent interest.

- Chapter 23 Carbohydrates
- Chapter 24 Lipids
- Chapter 25 Amino Acids, Peptides, and Proteins
- Chapter 26 Nucleosides, Nucleotides, and Nucleic Acids

Generous and Effective Use of Tables

Annotated summary tables have been a staple of *Organic Chemistry* since the first edition. Some tables review reactions from earlier chapters, others the reactions or concepts of a current chapter. Still other tables walk the reader step-by-step through skill builders and concepts unique to organic chemistry. Well received by students and faculty alike, these summary tables remain one of the text's strengths.

Problems

- ▶ Problem-solving strategies and skills are emphasized throughout. Understanding is progressively reinforced by problems that appear within topic sections.
- ▶ For many problems, sample solutions are given, including examples of handwritten solutions from the authors.
- ▶ The text now contains more than 1400 problems, many of which contain multiple parts. End-of-chapter problems are now organized to conform to the primary topic areas of each chapter.

Pedagogy

- ▶ A list of tables, mechanisms, boxed features, and Descriptive Passages and Interpretive Questions is included in the front matter as a quick reference to these important learning tools in each chapter.
- ▶ Each chapter begins with an opener that is meant to capture the reader's attention. Chemistry that is highlighted in the opener is relevant to chemistry that is included in the chapter.
- ▶ End-of-Chapter Summaries highlight and consolidate all of the important concepts and reactions within a chapter.

Audience

Organic Chemistry is designed to meet the needs of the "mainstream," two-semester undergraduate organic chemistry course. From the beginning and with each new edition, we have remained grounded in some fundamental notions. These include important issues concerning the intended audience. Is the topic appropriate for them with respect to their interests, aspirations, and experience? Just as important is the need to present an accurate picture of the present state of organic chemistry. How do we know what we know? What makes organic chemistry worth knowing? Where are we now? Where are we headed?

Descriptive Passages and Interpretive Problems

Many organic chemistry students later take standardized pre-professional examinations composed of problems derived from a descriptive passage; this text includes comparable passages and problems to familiarize students with this testing style.

Thus, *every* chapter concludes with a self-contained *Descriptive Passage and Interpretive Problems* unit that complements the chapter's content while emulating the "MCAT style." These 27 passages—listed on page xxii—are accompanied by more than 100 total multiple-choice problems.

The passages focus on a wide range of topics—from structure, synthesis, mechanism, and natural products. They provide instructors with numerous opportunities to customize their own organic chemistry course, while giving students practice in combining new information with what they have already learned.

A Student-Focused Revision

For the eleventh edition, real student data points and input, derived from thousands of our LearnSmart users, were used to guide the revision. LearnSmart Heat Maps provided a quick visual snapshot of usage of portions of the text and the relative difficulty students experienced in mastering the content.

This process was used to direct many of the revisions for this new edition. Of course, many updates have also been made according to changing scientific data, based on current events, and so forth. The following "What's New" summary lists the more major additions and refinements.

What's New

General Revisions

- Several new chapter openers have been created for this edition. Chapter openers are designed to peak student interest and understanding of the importance of the chapter's concepts.
- The inclusion of kcal and Angstrom units was added for consistency throughout the text, problems, and figures to aid student understanding.
- Color has been added and revised for consistency in many areas to help students better understand three-dimensional structure, stereochemistry, and reactions.
- New sample Problems and illustrations have also been added throughout the new edition to clarify topics and enhance the student learning experience.

Chapter-Specific Revisions

- A new section 2.10 Bonding in Water and Ammonia: Hybridization of Oxygen and Nitrogen was added in Chapter 2.
- In the stereochemistry chapter, enantiomeric ratio has been introduced in Section 4.4.
- New art was added to Chapters 6, 11, and 12 that reinforces the S_N2 reaction. Figure 6.2 includes electrostatic potential maps of hydroxide and methyl bromide, and Figure 6.3 shows a revised figure of the molecular orbital description. The revised art in Sections 11.2 and 12.9, for allyl and benzyl MO systems, ties together alkyl halides, S_N2, and MO theory.
- In Chapter 7, Section 7.2 was revised to introduce π molecular orbitals of ethene so that when students study π molecular orbitals in allyl intermediates and butadiene in Chapter 10, the topic should come as a natural extension. These revisions are intended to aid and reinforce student understanding of the theories of chemical bonding. The structure and reactivity concept of using sterically hindered bases giving the less substituted alkene in higher yields was added to the E2 discussion in Sections 7.14 and 7.15.
- Electrostatic potential maps that support carbanion stability trends were added in Figure 9.4 of the alkyne chapter.
- Section 11.15 includes a new boxed essay, "Pericyclic Reactions in Chemical Biology."
- A new section 15.9 Carbenes and Carbenoids and the mechanism of cyclopropanation have been added in Chapter 15, as it draws a useful comparison to the mechanism of peracid epoxidation. Also, the discussion of carbine structure complements material on carbocations, radicals, and anions elsewhere in the textbook. A catalytic cycle has been added to enhance the material on palladium-catalyzed cross-couplings. A new Descriptive Passage and Interpretive Problems, "Allylindium Reagents," has been added at the end of Chapter 15, keeping with the theme of sustainable organic chemistry.
- Section 19.16 includes a new boxed essay, "Enzymatic Decarboxylation of a β-Keto Acid."
- Chapter 21 includes two new sections: 21.1 Aldehyde, Ketone, and Ester Enolates and 21.8 Some Chemical and Stereochemical Consequences of Enolization. This new organization improves topic flow, allowing instructors to cover the aldol reaction during the first lecture on enolates, and adds to the emphasis on enolates, the main topic of the chapter. A new Descriptive Passage and Interpretive Problems on the Knoevenagel reaction has also been added to Chapter 21.

- Chapter 23 Phenols was eliminated for the eleventh edition and its contents redistributed as follows, enhancing the topics in other chapters:
  - Phenol acidity has been moved to Chapter 1, along with new problems, to enhance the discussion of the use of resonance in providing stability to the conjugate base.
  - Electrophilic aromatic substitution of phenols, along with new problems, is now covered in Chapter 13.
  - Cleavage of aryl ethers by hydrogen halides is covered in chapter 17 and new problems have also been added there.
- The coverage of biochemistry has been extensively updated and revised for the eleventh edition:
  - Chapter 23 includes new text and problems on reducing sugars and updated material on oligosaccharide synthesis and glycobiology. Section 23.20 includes a new boxed essay, "Oligosaccharides in Infectious Disease."
  - Chapter 24 includes updated coverage and illustrations for liposomes used in drug delivery and lipid rafts and membrane organization.
  - Chapter 25 has been reorganized and updated. Changes include moving the presentation of biochemical reactions of amino acids to later in the chapter (Section 25.21), after a revised presentation of enzymology (Section 25.20). Protein sequencing has been revised and organized to allow instructors greater flexibility with this material. Protein sequencing using mass spectrometry is added in a new section (Section 25.12 Mass Spectrometry of Peptides and Proteins). Chemical synthesis of proteins emphasizes the importance of solid-phase synthesis over solution methods, and orthogonal amine protecting groups are introduced (Sections 25.13–25.17).
  - Chapter 26 includes revised coverage and new illustrations for purines and pyrimidines; the importance of tautomers rather than aromaticity is emphasized. The bioenergetics sections (Sections 26.4–26.5) have been revised for clarity. Distinctions in the chemical properties of RNA and DNA are detailed in a new section (Section 26.7 Phosphoric Acid Esters). Updated coverage of RNA includes important new discoveries such as RNAi (Section 26.11 Ribonucleic Acids). New illustrations and an example of next-generation sequencing have been added to Section 26.17 Recombinant DNA Technology.
  - Chapter 27 also includes a new essay on Bakelite and the historical development of polymers.

Instructor Resources

Presentation Tools

Accessed from the Instructor Resources in the Connect Library, Presentation Tools contains photos, artwork, and Lecture PowerPoints that can be used to create customized lectures, visually enhanced tests and quizzes, compelling course websites, or attractive printed support materials. All assets are copyrighted by McGraw-Hill Higher Education, but can be used by instructors for classroom purposes. The visual resources in this collection include:

- **Art** Full-color digital files of all illustrations in the book can be readily incorporated into lecture presentations, exams, or custom-made classroom materials. In addition, all files are pre-inserted into PowerPoint slides for ease of lecture preparation.
- **Photos** The photo collection contains digital files of photographs from the text, which can be reproduced for multiple classroom uses.
- **PowerPoint® Lecture Outlines** Ready-made presentations that combine art and lecture notes are provided for each chapter of the text.

Also accessed through your textbook's Instructor Resources in the Connect Library are:

- **Classroom Response System Questions** (Clicker Questions) Nearly 600 questions covering the content of the *Organic Chemistry* text are available on the *Organic Chemistry* site for use with any classroom response system.
- **Animations** covering the most important mechanisms for *Organic Chemistry* are provided.

Test Bank

A test bank with over 1300 questions is available with the eleventh edition. The Test Bank is available in a test-generating software, as Word files, and is assignable through Connect to quickly create customized exams.

Student Resources

Solutions Manual

The Student Solutions Manual provides step-by-step solutions guiding the student through the reasoning behind each problem in the text. There is also a self-test section at the end of each chapter that is designed to assess the student's mastery of the material.

Schaum's Outline of Organic Chemistry

This helpful study aid provides students with hundreds of solved and supplementary problems for the organic chemistry course.

McGraw Hill Education **connect**®

Students—study more efficiently, retain more and achieve better outcomes. Instructors—focus on what you love—teaching.

SUCCESSFUL SEMESTERS INCLUDE CONNECT

FOR INSTRUCTORS

You're in the driver's seat.

Want to build your own course? No problem. Prefer to use our turnkey, prebuilt course? Easy. Want to make changes throughout the semester? Sure. And you'll save time with Connect's auto-grading too.

65%

Less Time Grading

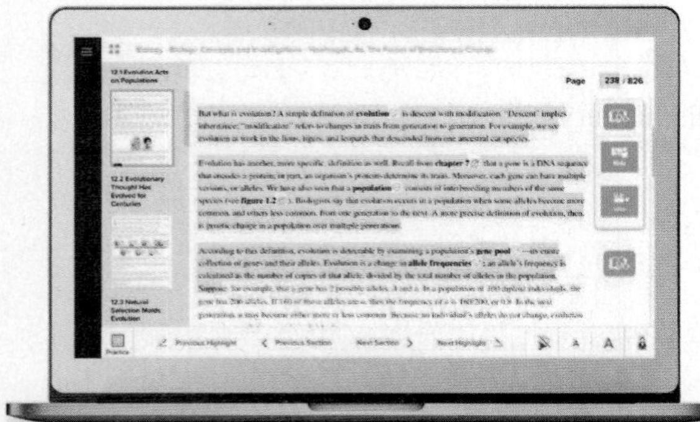

They'll thank you for it.

Adaptive study resources like SmartBook® help your students be better prepared in less time. You can transform your class time from dull definitions to dynamic debates. Hear from your peers about the benefits of Connect at **www.mheducation.com/highered/connect**

Make it simple, make it affordable.

Connect makes it easy with seamless integration using any of the major Learning Management Systems—Blackboard®, Canvas, and D2L, among others—to let you organize your course in one convenient location. Give your students access to digital materials at a discount with our inclusive access program. Ask your McGraw-Hill representative for more information.

©Hill Street Studios/Tobin Rogers/Blend Images LLC

Solutions for your challenges.

A product isn't a solution. Real solutions are affordable, reliable, and come with training and ongoing support when you need it and how you want it. Our Customer Experience Group can also help you troubleshoot tech problems—although Connect's 99% uptime means you might not need to call them. See for yourself at **status.mheducation.com**

For Students

Effective, efficient studying.

Connect helps you be more productive with your study time and get better grades using tools like SmartBook, which highlights key concepts and creates a personalized study plan. Connect sets you up for success, so you walk into class with confidence and walk out with better grades.

©Shutterstock/wavebreakmedia

"I really liked this app—it made it easy to study when you don't have your textbook in front of you."

—Jordan Cunningham,
Eastern Washington University

Study anytime, anywhere.

Download the free ReadAnywhere app and access your online eBook when it's convenient, even if you're offline. And since the app automatically syncs with your eBook in Connect, all of your notes are available every time you open it. Find out more at **www.mheducation.com/readanywhere**

No surprises.

The Connect Calendar and Reports tools keep you on track with the work you need to get done and your assignment scores. Life gets busy; Connect tools help you keep learning through it all.

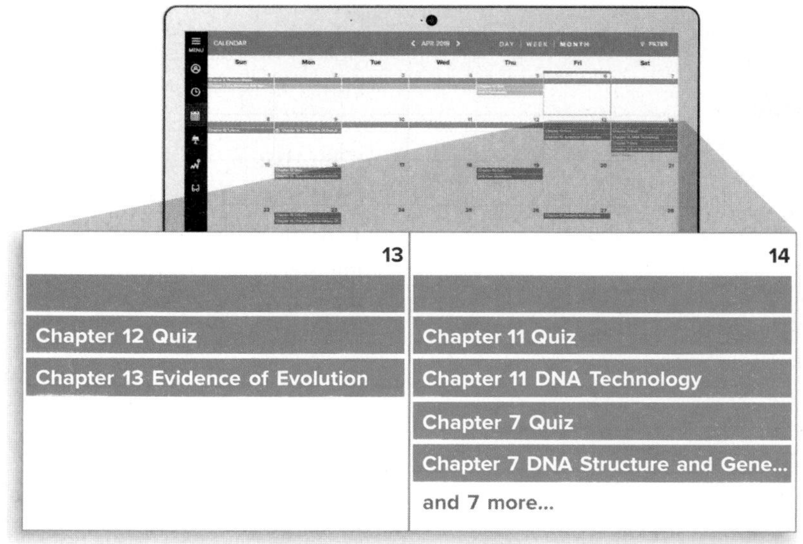

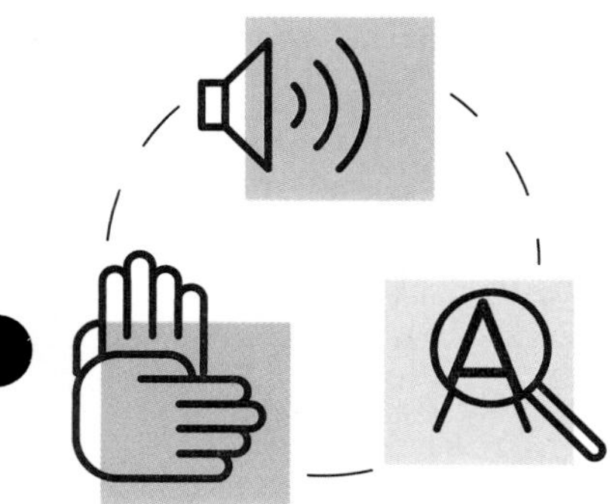

Learning for everyone.

McGraw-Hill works directly with Accessibility Services Departments and faculty to meet the learning needs of all students. Please contact your Accessibility Services office and ask them to email accessibility@mheducation.com, or visit **www.mheducation.com/accessibility** for more information.

ACKNOWLEDGEMENTS

Special thanks to the authors of the Student Solutions Manual, Neil Allison, Robert Giuliano, and Susan Bane Tuttle, who had a monumental task in updating the manual for this edition. The authors also acknowledge the generosity of Sigma-Aldrich for providing almost all of the 300-MHz NMR spectra.

Reviewers

Hundreds of teachers of organic chemistry have reviewed this text in its various editions. Our thanks to all of them.

The addition of SmartBook to the McGraw-Hill digital offerings has been invaluable. Thank you to the individuals who gave their time and talent to develop SmartBook for *Organic Chemistry*:

Tania Houjeiry, *Clemson University*

David Jones, *St. David's School in Raleigh, NC*

Margaret Ruth Leslie, *Kent State University*

Brooke Van Horn, *College of Charleston*

Organic Chemistry is also complemented by the exemplary digital products in Connect. We are extremely appreciative for the talents of the following individuals who played important roles in the authoring and content development for our digital products.

Ned B. Bowden, *University of Iowa*

Philip A. Brown, *North Carolina State*

William E. Crowe, *Louisiana State University*

Kimi Hatton, *George Mason University*

Ed Hilinski, *Florida State University*

T. Keith Hollis, *Mississippi State University*

Jennifer A. Irvin, *Texas State University*

Phil Janowicz, *California State University – Fullerton*

Margaret Ruth Leslie, *Kent State University*

Michael Lewis, *Saint Louis University*

James M. Salvador, *University of Texas at El Paso*

Buchang Shi, *Eastern Kentucky University*

Brooke A. Van Horn, *College of Charleston*

Organic Chemistry

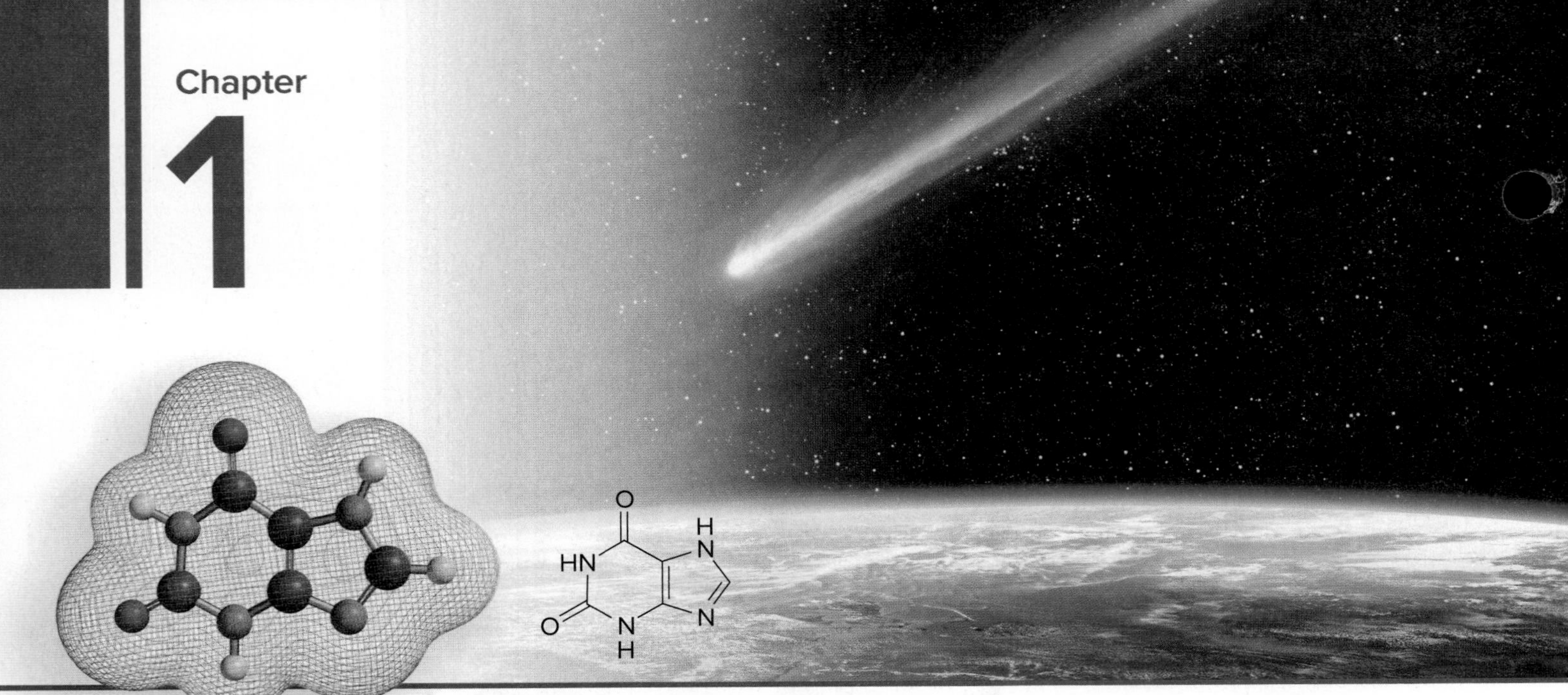

A class of meteors that strikes our planet is classified as "carbonaceous". They bring organic compounds with them—even compounds as complicated as the DNA base xanthine.
©Paul Fleet/Getty Images

CHAPTER OUTLINE

Structure Determines Properties

An emerging field of study called "Big History" is nothing if not ambitious. It could also be called "The History of Everything" and covers the period from when the universe was formed in a "Big Bang" 13 billion years ago until the present. During the first few seconds of time, protons and electrons combined to give hydrogen and helium atoms. Then, some 4.7 billion years ago, local gravitational forces caused the region of the universe around a particular star to assume that unique feature we call our solar system. Earth was located at the proper distance from the sun to allow the spontaneous formation of molecules, assemblies of molecules, replication of molecular assemblies, and so on. The earliest living things appeared about 4 billion years ago, the earliest ape-like animals 10 million years ago, and modern humans less than 100,000 years ago.

A Neanderthal looking at the lights in the night sky, for example, is not science. But it can lead to science—an organized body of knowledge based on theory and experiment. Astronomy is regarded as the oldest science but was limited to studying the night sky until the telescope was invented in 1608 and pointed skyward by Galileo a year later. Chemists, however, like to call their own discipline "The Central Science" because of its relationship to physics on one side and biology on the other. In the same way, among the various subdisciplines of chemistry, organic chemistry can be said to be central. It adapts the physical principles that underlie the content of general chemistry courses to the relationships between structure and properties of compounds based on carbon—the most versatile of all the elements.

This chapter begins your training toward understanding the relationship between structure and properties by reviewing the fundamentals of the Lewis approach to molecular structure and bonding and describes the various graphical ways molecular structures are presented. Principles of acid–base chemistry—emphasized in a quantitative way in introductory chemistry courses—are revisited qualitatively as a tool for introducing the effect of structure on properties. This structure/property relationship is what makes organic chemistry important. The same atom (carbon) is common to many structural types, countless compounds with different properties, and much variation in the degree to which a particular property is expressed. What is equally remarkable is the degree to which a relatively small group of principles suffice to connect the structure of a substance to its properties.

1.1 Atoms, Electrons, and Orbitals

Before discussing structure and bonding in *molecules,* let's first review some fundamentals of *atomic* structure. Each element is characterized by a unique **atomic number *Z*,*** which is equal to the number of protons in its nucleus. A neutral atom has equal numbers of protons, which are positively charged, and electrons, which are negatively charged.

Electrons were believed to be particles from the time of their discovery in 1897 until 1924, when the French physicist Louis de Broglie suggested that they have wavelike properties as well. Two years later Erwin Schrödinger took the next step and calculated the energy of an electron in a hydrogen atom by using equations that treated the electron as if it were a wave. Instead of a single energy, Schrödinger obtained a series of energies, each of which corresponded to a different mathematical description of the electron wave. These mathematical descriptions are called **wave functions** and are symbolized by the Greek letter ψ (psi).

According to the Heisenberg uncertainty principle, we can't tell exactly where an electron is, but we can tell where it is most likely to be. The probability of finding an electron at a particular spot relative to an atom's nucleus is given by the square of the wave function (ψ^2) at that point. Figure 1.1 illustrates the probability of finding an electron at various points in the lowest energy (most stable) state of a hydrogen atom. The darker the color in a region, the higher the probability. The probability of finding an electron at a particular point is greatest near the nucleus and decreases with increasing distance from the nucleus but never becomes zero.

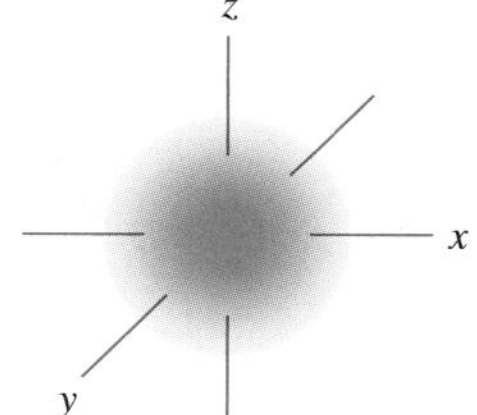

Figure 1.1

Probability distribution (ψ^2) for an electron in a 1s orbital.

Wave functions are also called **orbitals.** For convenience, chemists use the term "orbital" in several different ways. A drawing such as Figure 1.1 is often said to represent an orbital. We will see other kinds of drawings in this chapter, and use the word "orbital" to describe them too.

Orbitals are described by specifying their size, shape, and directional properties. Spherically symmetrical ones such as shown in Figure 1.1 are called *s orbitals.* The letter *s* is preceded by the **principal quantum number** n (n = 1, 2, 3, etc.), which specifies the **shell** and is related to the energy of the orbital. An electron in a 1*s* orbital is likely to be found closer to the nucleus, is lower in energy, and is more strongly held than an electron in a 2*s* orbital.

Instead of probability distributions, it is more common to represent orbitals by their **boundary surfaces,** as shown in Figure 1.2 for the 1*s* and 2*s* orbitals. The region enclosed by a boundary surface is arbitrary but is customarily the volume where the probability of finding an electron is high—on the order of 90–95%. Like the probability distribution plot from which it is derived, a picture of a boundary surface is usually described as a drawing of an orbital.

*A glossary of the terms shown in boldface may be found immediately before the index at the back of the book.

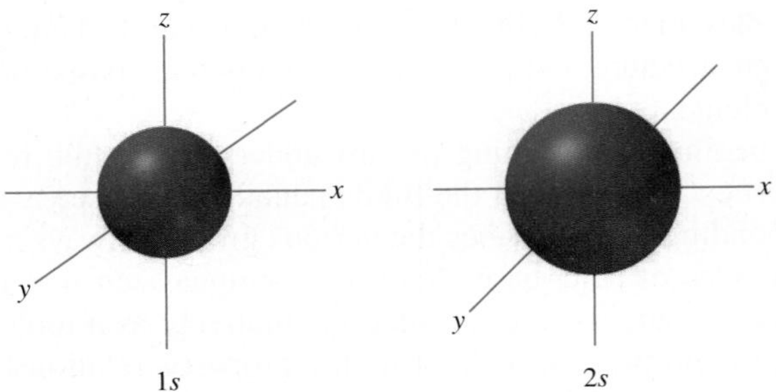

Figure 1.2

Boundary surfaces of a 1s orbital and a 2s orbital.

A hydrogen atom ($Z = 1$) has one electron; a helium atom ($Z = 2$) has two. The single electron of hydrogen occupies a 1*s* orbital, as do the two electrons of helium. We write their electron configurations as

$$\text{Hydrogen: } 1s^1 \quad \text{Helium: } 1s^2$$

In addition to being negatively charged, electrons possess the property of **spin.** The **spin quantum number** of an electron can have a value of either $+\frac{1}{2}$ or $-\frac{1}{2}$. According to the **Pauli exclusion principle,** two electrons may occupy the same orbital only when they have opposite, or "paired," spins. For this reason, no orbital can contain more than two electrons. Because two electrons fill the 1*s* orbital, the third electron in lithium ($Z = 3$) must occupy an orbital of higher energy. After 1*s*, the next higher energy orbital is 2*s*. The third electron in lithium therefore occupies the 2*s* orbital, and the electron configuration of lithium is

$$\text{Lithium: } 1s^2 2s^1$$

A complete periodic table of the elements is presented at the back of the book.

The **period** (or **row**) of the periodic table in which an element appears corresponds to the principal quantum number of the highest numbered occupied orbital ($n = 1$ in the case of hydrogen and helium). Hydrogen and helium are first-row elements; lithium ($n = 2$) is a second-row element.

With beryllium ($Z = 4$), the 2*s* level becomes filled and, beginning with boron ($Z = 5$), the next orbitals to be occupied are $2p_x$, $2p_y$, and $2p_z$. These three orbitals (Figure 1.3) are of equal energy and are characterized by boundary surfaces that are usually described as "dumbbell-shaped." The axes of the three 2*p* orbitals are at right angles to one another. Each orbital consists of two "lobes," represented in Figure 1.3 by regions of different colors. Regions of a single orbital, in this case, each 2*p* orbital, may be separated by **nodal surfaces** where the wave function changes sign and the probability of finding an electron is zero.

Other methods are also used to contrast the regions of an orbital where the signs of the wave function are different. Some mark one lobe of a *p* orbital + and the other −. Others shade one lobe and leave the other blank. When this level of detail isn't necessary, no differentiation is made between the two lobes.

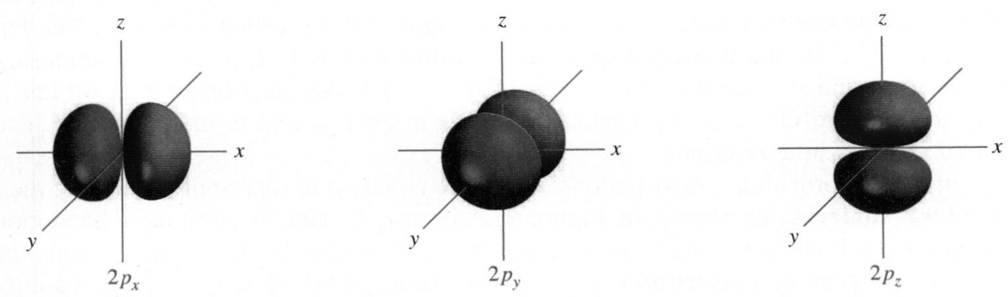

Figure 1.3

Boundary surfaces of the 2*p* orbitals. The wave function changes sign at the nucleus. The two halves of each orbital are indicated by different colors. The *yz*-plane is a nodal surface for the $2p_x$ orbital. The probability of finding a $2p_x$ electron in the *yz*-plane is zero. Analogously, the *xz*-plane is a nodal surface for the $2p_y$ orbital, and the *xy*-plane is a nodal surface for the $2p_z$ orbital.

Organic Chemistry: The Early Days

Eighteenth-century chemists regarded their science as being composed of two branches. One dealt with substances obtained from natural or living sources and was called *organic chemistry;* the other dealt with materials from nonliving matter—minerals and the like—and was called *inorganic chemistry.* Over time, combustion analysis established that the compounds derived from natural sources contained carbon, and a new definition of organic chemistry emerged: *Organic chemistry is the study of carbon compounds.* This is the definition we still use today.

As the eighteenth century gave way to the nineteenth, many scientists still subscribed to a doctrine known as *vitalism,* which held that living systems possessed a "vital force" that was absent in nonliving systems. Substances derived from natural sources (organic) were thought to be fundamentally different from inorganic ones. It was believed that inorganic compounds could be synthesized in the laboratory, but organic compounds could not—at least not from inorganic materials.

In 1823, Friedrich Wöhler, after completing medical studies in Germany, spent a year in Stockholm studying under one of the world's foremost chemists of the time, Jöns Jacob Berzelius. Wöhler subsequently went on to have a distinguished independent career, spending most of it at the University of Göttingen. He is best remembered for a brief paper he published in 1828 in which he noted that, on evaporating an aqueous solution of ammonium cyanate, he obtained "colorless, clear crystals often more than an inch long," which were not ammonium cyanate but were instead urea.

$$\underset{\text{(inorganic)}}{\underset{\text{Ammonium cyanate}}{NH_4OCN}} \longrightarrow \underset{\text{(organic)}}{\underset{\text{Urea}}{O{=}C(NH_2)_2}}$$

This transformation was remarkable at the time because an *inorganic* salt, ammonium cyanate, was converted to urea, a known *organic* substance earlier isolated from urine. It is now recognized as a significant early step toward overturning the philosophy of vitalism. Although Wöhler himself made no extravagant claims concerning the relationship of his discovery to vitalist theory, the die was cast, and over the next generation organic chemistry outgrew vitalism. What particularly seemed to excite Wöhler and Berzelius had very little to do with vitalism. Berzelius was interested in cases in which two clearly different materials had the same elemental composition, and he invented the word *isomers* to apply to them. Wöhler's observation that an inorganic compound (ammonium cyanate) of molecular formula CH_4N_2O could be transformed into an organic compound (urea) of the same molecular formula had an important bearing on the concept of isomerism.

From the concept of isomerism we can trace the origins of the *structural theory*—the idea that a specific arrangement of atoms uniquely defines a substance. Ammonium cyanate and urea are different compounds because they have different structures.

Three mid-nineteenth-century scientists, August Kekulé, Archibald S. Couper, and Alexander M. Butlerov, stand out for separately proposing the elements of the structural theory. The essential features of Kekulé's theory, developed and presented while he taught at Heidelberg in 1858, were that carbon normally formed four bonds and had the capacity to bond to other carbons so as to form long chains. Isomers were possible because the same elemental composition (say, the CH_4N_2O molecular formula common to both ammonium cyanate and urea) accommodates more than one pattern of atoms and bonds. Shortly thereafter, Couper, a Scot working at the École de Médecine in Paris, and Butlerov, a Russian chemist at the University of Kazan, proposed similar theories.

In the late nineteenth and early twentieth centuries, major discoveries about atoms and electrons placed theories of molecular structure and bonding on a more secure, physics-based foundation. Several of these are described at the beginning of this section.

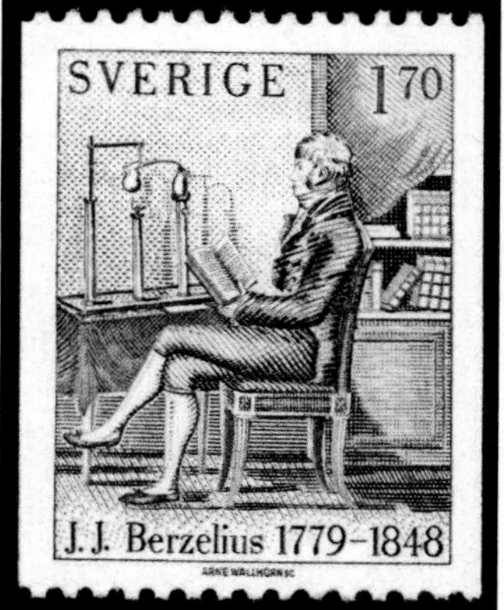

(Left & top right): ©David Tietz/Editorial Image, LLC; *(bottom right):* ©H.S. Photos/Alamy Stock Photo

The electron configurations of the first 12 elements, hydrogen through magnesium, are given in Table 1.1. In filling the $2p$ orbitals, notice that each is singly occupied before any one is doubly occupied. This general principle for orbitals of equal energy is known as **Hund's rule.** Of particular importance in Table 1.1 are *hydrogen, carbon, nitrogen,* and *oxygen.* Countless organic compounds contain nitrogen, oxygen, or both in addition to carbon, the essential element of organic chemistry. Most of them also contain hydrogen.

It is often convenient to speak of the **valence electrons** of an atom. These are the outermost electrons, the ones most likely to be involved in chemical bonding and reactions. For second-row elements these are the $2s$ and $2p$ electrons. Because four orbitals ($2s$, $2p_x$, $2p_y$, $2p_z$) are involved, the maximum number of electrons in the **valence shell**

TABLE 1.1 Electron Configurations of the First Twelve Elements of the Periodic Table

Element	Atomic number Z	Number of electrons in indicated orbital					
		$1s$	$2s$	$2p_x$	$2p_y$	$2p_z$	$3s$
Hydrogen	1	1					
Helium	2	2					
Lithium	3	2	1				
Beryllium	4	2	2				
Boron	5	2	2	1			
Carbon	6	2	2	1	1		
Nitrogen	7	2	2	1	1	1	
Oxygen	8	2	2	2	1	1	
Fluorine	9	2	2	2	2	1	
Neon	10	2	2	2	2	2	
Sodium	11	2	2	2	2	2	1
Magnesium	12	2	2	2	2	2	2

of any second-row element is 8. Neon, with all its 2*s* and 2*p* orbitals doubly occupied, has eight valence electrons and completes the second row of the periodic table. For **main-group elements,** the number of valence electrons is equal to its group number in the periodic table.

Detailed solutions to all of the problems are found in the *Student Solutions Manual* along with a brief discussion and advice on how to do problems of the same type.

Problem 1.1

How many electrons does carbon have? How many are valence electrons? What third-row element has the same number of valence electrons as carbon?

Once the 2*s* and 2*p* orbitals are filled, the next level is the 3*s*, followed by the $3p_x$, $3p_y$, and $3p_z$ orbitals. Electrons in these orbitals are farther from the nucleus than those in the 2*s* and 2*p* orbitals and are of higher energy.

In-chapter problems that contain multiple parts are accompanied by a sample solution to part (a).

Problem 1.2

Referring to the periodic table as needed, write electron configurations for all the elements in the third period.

Sample Solution The third period begins with sodium and ends with argon. The atomic number Z of sodium is 11, and so a sodium atom has 11 electrons. The maximum number of electrons in the 1*s*, 2*s*, and 2*p* orbitals is ten, and so the eleventh electron of sodium occupies a 3*s* orbital. The electron configuration of sodium is $1s^22s^22p_x^{\,2}2p_y^{\,2}2p_z^{\,2}3s^1$.

Neon, in the second period, and argon, in the third, have eight electrons in their valence shell; they are said to have a complete **octet** of electrons. Helium, neon, and argon belong to the class of elements known as **noble gases** or **rare gases.** The noble gases are characterized by an extremely stable "closed-shell" electron configuration and are very unreactive.

Structure determines properties and the properties of atoms depend on atomic structure. All of an element's protons are in its nucleus, but the element's electrons are distributed among orbitals of various energy and distance from the nucleus. More than anything else, we look at its electron configuration when we wish to understand how an element behaves. The next section illustrates this with a brief review of ionic bonding.

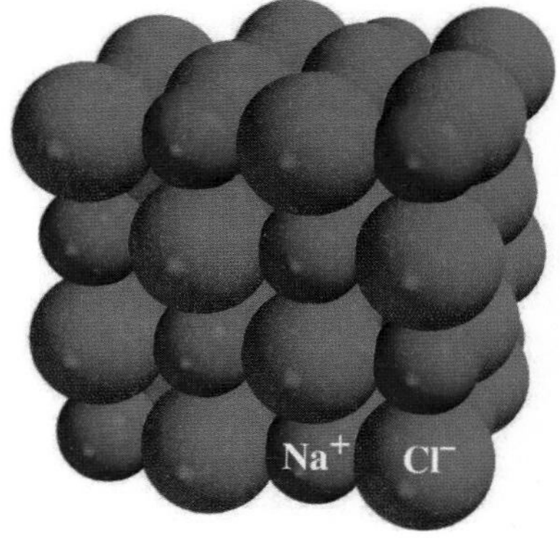

Figure 1.4

An ionic bond is the force of attraction between oppositely charged ions. Each Na^+ ion in the crystal lattice of solid NaCl is involved in ionic bonding to each of six surrounding Cl^- ions and vice versa. The smaller spheres are Na^+ and the larger spheres are Cl^-.

1.2 Ionic Bonds

Atoms combine with one another to give **compounds** having properties different from the atoms they contain. The attractive force between atoms in a compound is a **chemical bond.** One type of chemical bond, called an **ionic bond,** is the force of attraction between oppositely charged species **(ions)** (Figure 1.4). Positively charged ions are referred to as **cations;** negatively charged ions are **anions.**

Whether an element is the source of the cation or anion in an ionic bond depends on several factors, for which the periodic table can serve as a guide. In forming ionic compounds, elements at the left of the periodic table typically lose electrons, giving a cation that has the same electron configuration as the preceding noble gas. Loss of an electron from sodium, for example, yields Na^+, which has the same electron configuration as neon.

$$\underset{\substack{\text{Sodium atom} \\ 1s^22s^22p^63s^1}}{\text{Na}(g)} \longrightarrow \underset{\substack{\text{Sodium ion} \\ 1s^22s^22p^6}}{\text{Na}^+(g)} + \underset{\text{Electron}}{e^-}$$

[The symbol (*g*) indicates that the species is present in the gas phase.]

Problem 1.3

Species that have the same number of electrons are described as *isoelectronic*. What +2 ion is isoelectronic with Na^+? What −2 ion?

A large amount of energy, called the **ionization energy,** must be transferred to any atom to dislodge an electron. The ionization energy of sodium, for example, is 496 kJ/mol (119 kcal/mol). Processes that absorb energy are said to be **endothermic.** Compared with other elements, sodium and its relatives in group 1A have relatively low ionization energies. In general, ionization energy increases across a row in the periodic table.

The SI (*Système International d'Unités*) unit of energy is the *joule* (J). An older unit is the *calorie* (cal). Many chemists still express energy changes in units of kilocalories per mole (1 kcal/mol = 4.184 kJ/mol).

Elements at the right of the periodic table tend to gain electrons to reach the electron configuration of the next higher noble gas. Adding an electron to chlorine, for example, gives the anion Cl^-, which has the same closed-shell electron configuration as the noble gas argon.

$$\underset{\substack{\text{Chlorine atom} \\ 1s^22s^22p^63s^23p^5}}{\text{Cl}(g)} + \underset{\text{Electron}}{e^-} \longrightarrow \underset{\substack{\text{Chloride ion} \\ 1s^22s^22p^63s^23p^6}}{\text{Cl}^-(g)}$$

Problem 1.4

Which of the following ions possess a noble gas electron configuration?

(a) K^+
(b) He^+
(c) H^-
(d) O^-
(e) F^-
(f) Ca^{2+}

Sample Solution (a) Potassium has atomic number 19, and so a potassium atom has 19 electrons. The ion K^+, therefore, has 18 electrons, the same as the noble gas argon. The electron configurations of both K^+ and Ar are $1s^22s^22p^63s^23p^6$.

Energy is released when a chlorine atom captures an electron. Energy-releasing reactions are described as **exothermic,** and the energy change for an exothermic process has a negative sign. The energy change for addition of an electron to an atom is referred to as its **electron affinity** and is −349 kJ/mol (−83.4 kcal/mol) for chlorine.

We can use the ionization energy of sodium and the electron affinity of chlorine to calculate the energy change for the reaction:

$$\underset{\text{Sodium atom}}{Na(g)} + \underset{\text{Chlorine atom}}{Cl(g)} \longrightarrow \underset{\text{Sodium ion}}{Na^+(g)} + \underset{\text{Chloride ion}}{Cl^-(g)}$$

Were we to simply add the ionization energy of 496 kJ/mol (119 kcal/mol) for sodium and the electron affinity of −349 kJ/mol (−83.4 kcal/mol) for chlorine, we would conclude that the overall process is endothermic by +147 kJ/mol (+35 kcal/mol). The energy liberated by adding an electron to chlorine is insufficient to override the energy required to remove an electron from sodium. This analysis, however, fails to consider the force of attraction between the oppositely charged ions Na^+ and Cl^-, as expressed in terms of the energy released in the formation of solid NaCl from the separated gas-phase ions:

$$\underset{\text{Sodium ion}}{Na^+(g)} + \underset{\text{Chloride ion}}{Cl^-(g)} \longrightarrow \underset{\text{Sodium chloride}}{NaCl(s)}$$

This *lattice energy* is 787 kJ/mol (188 kcal/mol) and is more than sufficient to make the overall process for formation of sodium chloride from the elements exothermic. Forces between charged particles are called **electrostatic,** or **Coulombic,** and constitute an ionic bond when they are attractive.

Ionic bonding was proposed by the German physicist Walther Kossel in 1916 in order to explain the ability of substances such as molten sodium chloride to conduct an electric current. He was the son of Albrecht Kossel, winner of the 1910 Nobel Prize in Physiology or Medicine for early studies of nucleic acids.

Problem 1.5

What is the electron configuration of C^+? Of C^-? Does either one of these ions have a noble gas (closed-shell) electron configuration?

Ionic bonds are very common in *inorganic* compounds, but rare in *organic* ones. The ionization energy of carbon is too large and the electron affinity too small for carbon to realistically form a C^{4+} or C^{4-} ion. What kinds of bonds, then, link carbon to other elements in millions of organic compounds? Instead of losing or gaining electrons, carbon *shares* electrons with other elements (including other carbon atoms) to give what are called covalent bonds.

1.3 Covalent Bonds, Lewis Formulas, and the Octet Rule

Gilbert Newton Lewis has been called the greatest American chemist.

The **covalent,** or **shared electron pair,** model of chemical bonding was first suggested by G. N. Lewis of the University of California in 1916. Lewis proposed that a *sharing* of two electrons by two hydrogen atoms permits each one to have a stable closed-shell electron configuration analogous to that of helium.

H· ·H	H:H
Two hydrogen atoms, each with a single electron	Hydrogen molecule: covalent bonding by way of a shared electron pair

The amount of energy required to dissociate a hydrogen molecule H_2 to two separate hydrogen atoms is its **bond dissociation enthalpy.** For H_2 it is quite large, amounting to +435 kJ/mol (+104 kcal/mol). The main contributor to the strength of the covalent bond in H_2 is the increased Coulombic force exerted on its two electrons. Each electron in H_2 "feels" the attractive force of two nuclei, rather than one as it would in an isolated hydrogen atom.

Only the electrons in an atom's valence shell are involved in covalent bonding. Fluorine, for example, has nine electrons, but only seven are in its valence shell. Pairing a valence electron of one fluorine atom with one of a second fluorine gives a fluorine molecule (F_2) in which each fluorine has eight valence electrons and an electron configuration equivalent to that of the noble gas neon. Shared electrons count toward satisfying the octet of both atoms.

:F̤̈·   ·F̤̈:

Two fluorine atoms, each with seven electrons in its valence shell

:F̤̈:F̤̈:

Fluorine molecule: covalent bonding by way of a shared electron pair

The six valence electrons of each fluorine that are not involved in bonding comprise three **unshared pairs.**

Unshared pairs are also called *lone pairs.*

Structural formulas such as those just shown for H_2 and F_2 where electrons are represented as dots are called **Lewis formulas,** or **Lewis structures.** It is usually more convenient to represent shared electron-pair bonds as lines and to sometimes omit electron pairs.

The Lewis model limits second-row elements (Li, Be, B, C, N, O, F, Ne) to a total of eight electrons (shared plus unshared) in their valence shells. Hydrogen is limited to two. Most of the elements that we'll encounter in this text obey the **octet rule:** *In forming compounds they gain, lose, or share electrons to achieve a stable electron configuration characterized by eight valence electrons.* When the octet rule is satisfied for carbon, nitrogen, oxygen, and fluorine, each has an electron configuration analogous to that of the noble gas neon. The Lewis formulas of methane (CH_4), ammonia (NH_3), water (H_2O), and hydrogen fluoride (HF) given in Table 1.2 illustrate the octet rule.

With four valence electrons, carbon normally forms four covalent bonds as shown in Table 1.2 for CH_4. In addition to C—H bonds, most organic compounds contain covalent C—C bonds. Ethane (C_2H_6) is an example.

Combine two carbons and six hydrogens: H· ·Ċ· ·Ċ· ·H (each C with an H· above and below) to write a Lewis structure for ethane:

```
 H H                H   H
H:C:C:H     or      |   |
 H H             H—C—C—H
                    |   |
                    H   H
```

TABLE 1.2 Lewis Formulas of Methane, Ammonia, Water, and Hydrogen Fluoride

Compound	Atom	Number of valence electrons in atom	Atom and sufficient number of hydrogen atoms to complete octet	Lewis formula: Dot	Lewis formula: Line
Methane	Carbon	4	H· ·Ċ· ·H (H above and below)	H:C:H (H above and below)	H—C—H (H above and below)
Ammonia	Nitrogen	5	H· ·N̈· ·H (H below)	H:N̈:H (H below)	H—N̈—H (H below)
Water	Oxygen	6	H· ·Ö· ·H	H:Ö:H	H—Ö—H
Hydrogen fluoride	Fluorine	7	H· ·F̤̈:	H:F̤̈:	H—F̤̈:

Problem 1.6

Write Lewis formulas, including unshared pairs, for each of the following. Carbon has four bonds in each compound.

(a) Propane (C_3H_8) (c) Methyl fluoride (CH_3F)

(b) Methanol (CH_4O) (d) Ethyl fluoride (C_2H_5F)

Sample Solution (a) The Lewis formula of propane is analogous to that of ethane but the chain has three carbons instead of two.

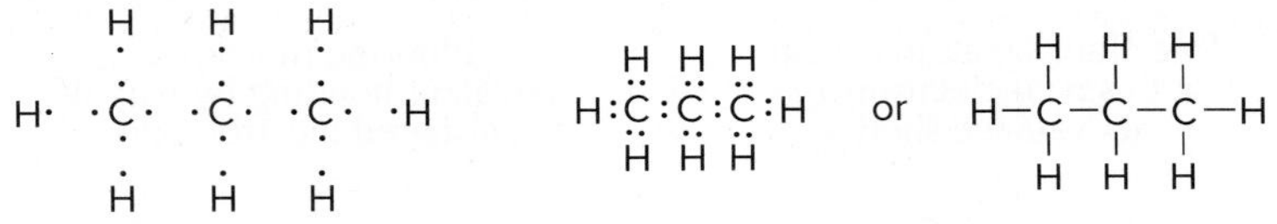

H· ·C· ·C· ·C· ·H (each carbon with H above and below) H:C:C:C:H or H—C—C—C—H (each carbon bearing H above and below)

Combine three carbons and eight hydrogens to write a Lewis formula for propane

The ten covalent bonds in the Lewis formula shown account for 20 valence electrons, which is the same as that calculated from the molecular formula (C_3H_8). The eight hydrogens of C_3H_8 contribute 1 electron each and the three carbons 4 each, for a total of 20 (8 from the hydrogens and 12 from the carbons). Therefore, all the valence electrons are in covalent bonds; propane has no unshared pairs.

Lewis's concept of shared electron-pair bonds allows for four-electron double bonds and six-electron triple bonds. Ethylene (C_2H_4) has 12 valence electrons, which can be distributed as follows:

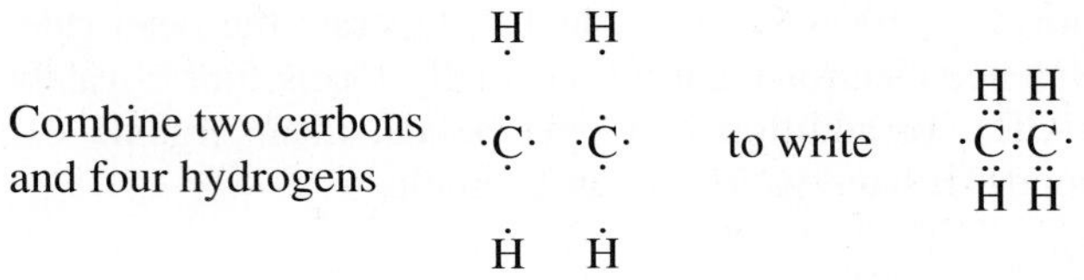

Combine two carbons and four hydrogens ·C· ·C· (with H above and below each) to write ·C:C· (with H above and below each)

The structural formula produced has a single bond between the carbons and seven electrons around each. By pairing the unshared electron of one carbon with its counterpart of the other carbon, a **double bond** results and the octet rule is satisfied for both carbons.

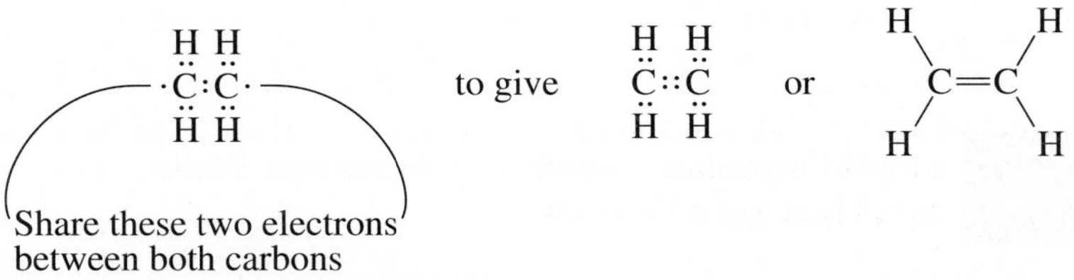

·C:C· (with H above and below each) to give C::C (with H above and below each) or $H_2C{=}CH_2$

Share these two electrons between both carbons

Likewise, the ten valence electrons of acetylene (C_2H_2) can be arranged in a structural formula that satisfies the octet rule when six of them are shared in a **triple bond** between the carbons.

H:C:::C:H or H—C≡C—H

Carbon dioxide (CO_2) has two carbon–oxygen double bonds, thus satisfying the octet rule for both carbon and oxygen.

:Ö::C::Ö: or :Ö=C=Ö:

Problem 1.7

All of the hydrogens are bonded to carbon in both of the following. Write a Lewis formula that satisfies the octet rule for each.

(a) Formaldehyde (CH_2O) (b) Hydrogen cyanide (HCN)

Sample Solution (a) Formaldehyde has 12 valence electrons: 4 from carbon, 2 from two hydrogens, and 6 from oxygen. Connect carbon to oxygen and both hydrogens by covalent bonds.

Combine H· ·Ċ· ·H and ·Ö· to give H:C:O (with H on carbon above and below; carbon and oxygen each bearing one unpaired electron)

Pair the unpaired electron on carbon with the unpaired electron on oxygen to give a carbon–oxygen double bond. The resulting structural formula satisfies the octet rule.

Share these two electrons between carbon and oxygen — to give H₂C::Ö or $H_2C{=}O$

1.4 Polar Covalent Bonds, Electronegativity, and Bond Dipoles

Electrons in covalent bonds are not necessarily shared equally by the two atoms that they connect. If one atom has a greater tendency to attract electrons toward itself than the other, the electron distribution is *polarized,* and the bond is described as **polar covalent.** The tendency of an atom to attract the electrons in a covalent bond toward itself defines its **electronegativity.** An electronegative element attracts electrons; an electropositive one donates them.

Hydrogen fluoride, for example, has a polar covalent bond. Fluorine is more electronegative than hydrogen and pulls the electrons in the H—F bond toward itself, giving fluorine a partial negative charge and hydrogen a partial positive charge. Two ways of representing the polarization in HF are

$^{\delta+}$H—F$^{\delta-}$ (The symbols δ+ and δ− indicate partial positive and partial negative charge, respectively)

$\overset{\longleftrightarrow}{\text{H—F}}$ (The symbol ⟷ represents the direction of polarization of electrons in the H—F bond)

A third way of illustrating the electron polarization in HF is graphically, by way of an **electrostatic potential map,** which uses the colors of the rainbow to show the charge distribution. Blue through red tracks regions of greater positive charge to greater negative charge. (For more details, see the boxed essay *Electrostatic Potential Maps* in this section.)

Positively charged region of molecule

Negatively charged region of molecule

Contrast the electrostatic potential map of HF with those of H_2 and F_2.

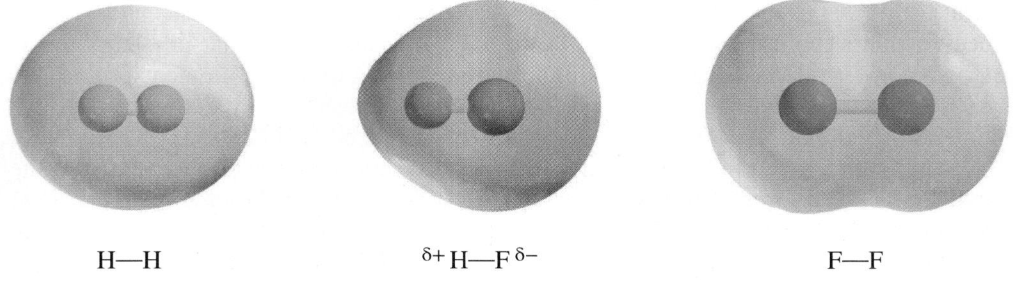

The covalent bond in H_2 joins two hydrogen atoms. Because the bonded atoms are identical, so are their electronegativities. There is no polarization of the electron distribution, the H—H bond is nonpolar, and a neutral yellow-green color dominates the electrostatic potential map. Likewise, the F—F bond in F_2 is nonpolar and its electrostatic potential map resembles that of H_2. The covalent bond in HF, on the other hand, unites two atoms of different electronegativity, and the electron distribution is very polarized. Blue is the dominant color near the positively polarized hydrogen, and red the dominant color near the negatively polarized fluorine.

Linus Pauling (1901–1994) was born in Portland, Oregon, and was educated at Oregon State University and at the California Institute of Technology, where he earned a Ph.D. in chemistry in 1925. In addition to research in bonding theory, Pauling studied the structure of proteins and was awarded the Nobel Prize in Chemistry for that work in 1954. Pauling won a second Nobel Prize (the Peace Prize) in 1962 for his efforts to limit the testing of nuclear weapons. He was one of only four scientists to have won two Nobel Prizes. The first double winner was a woman. Can you name her?

The most commonly used electronegativity scale was devised by Linus Pauling. Table 1.3 keys Pauling's electronegativity values to the periodic table.

Electronegativity *increases* from left to right across a row in the periodic table. Of the second-row elements, the most electronegative is fluorine, the least electronegative is lithium. Electronegativity *decreases* going down a column. Of the halogens, fluorine is the most electronegative, then chlorine, then bromine, then iodine. Indeed, fluorine is the most electronegative of all the elements; oxygen is second.

In general, the greater the electronegativity difference between two elements, the more polar the bond between them.

TABLE 1.3 Selected Values from the Pauling Electronegativity Scale

	Group number						
Period	**1A**	**2A**	**3A**	**4A**	**5A**	**6A**	**7A**
1	H 2.1						
2	Li 1.0	Be 1.5	B 2.0	C 2.5	N 3.0	O 3.5	F 4.0
3	Na 0.9	Mg 1.2	Al 1.5	Si 1.8	P 2.1	S 2.5	Cl 3.0
4	K 0.8	Ca 1.0					Br 2.8
5							I 2.5

Problem 1.8

In which of the compounds CH_4, NH_3, H_2O, SiH_4, or H_2S is $\delta+$ for hydrogen the greatest? In which one does hydrogen bear a partial negative charge?

Table 1.4 compares the polarity of various bond types according to their **bond dipole moments.** A dipole exists whenever opposite charges are separated from each other, and a **dipole moment μ** is the product of the amount of the charge *e* multiplied by the distance *d* between the centers of charge.

$$\mu = e \times d$$

The debye unit is named in honor of Peter Debye, a Dutch scientist who did important work in many areas of chemistry and physics and was awarded the Nobel Prize in Chemistry in 1936.

Because the charge on an electron is 4.80×10^{-10} electrostatic units (esu) and the distances within a molecule typically fall in the 10^{-8} cm range, molecular dipole moments are on the order of 10^{-18} esu·cm. To simplify the reporting of dipole moments, this value of 10^{-18} esu·cm is defined as a **debye, D.** Thus the experimentally determined dipole moment of hydrogen fluoride, 1.7×10^{-18} esu·cm, is stated as 1.7 D.

The bond dipoles in Table 1.4 depend on the difference in electronegativity of the bonded atoms and on the bond distance. The polarity of a C—H bond is relatively low; substantially less than a C—O bond, for example. Don't lose sight of an even more important difference between a C—H bond and a C—O bond, and that is the *direction*

TABLE 1.4 Selected Bond Dipole Moments

Bond*	Dipole moment, D	Bond*	Dipole moment, D
H—F	1.7	C—F	1.4
H—Cl	1.1	C—O	0.7
H—Br	0.8	C—N	0.4
H—I	0.4	C=O	2.4
H—C	0.3	C=N	1.4
H—N	1.3	C≡N	3.6
H—O	1.5		

*The direction of the dipole moment is toward the more electronegative atom. In the listed examples, hydrogen and carbon are the positive ends of the dipoles. Carbon is the negative end of the dipole associated with the C—H bond.

Electrostatic Potential Maps

All of the material in this text, and most of chemistry generally, can be understood on the basis of what physicists call the *electromagnetic force*. Its major principle is that opposite charges attract and like charges repel. A good way to connect structure to properties such as chemical reactivity is to find the positive part of one molecule and the negative part of another. Most of the time, these will be the reactive sites.

Imagine that you bring a positive charge toward a molecule. The interaction between that positive charge and some point in the molecule will be attractive if the point is negatively charged, repulsive if it is positively charged, and the strength of the interaction will depend on the magnitude of the charge. Computational methods make it possible to calculate and map these interactions. It is convenient to display this map using the colors of the rainbow from red to blue. Red is the negative (electron-rich) end and blue is the positive (electron-poor) end.

The electrostatic potential map of hydrogen fluoride (HF) was shown in the preceding section and is repeated here. Compare it with the electrostatic potential map of lithium hydride (LiH).

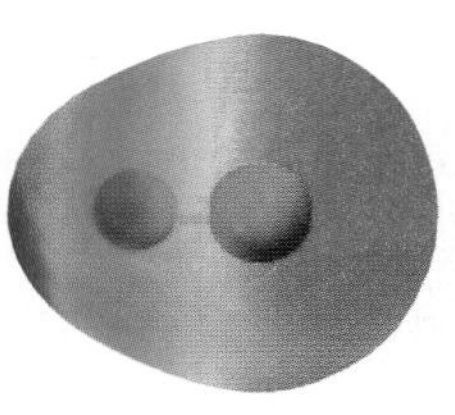

H—F

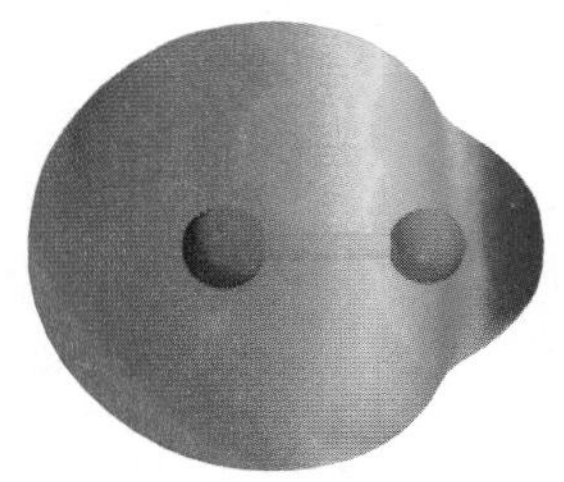

H—Li

The H—F bond is polarized so that hydrogen is partially positive (blue) and fluorine partially negative (red). Because hydrogen is more electronegative than lithium, the H—Li bond is polarized in the opposite sense, making hydrogen partially negative (red) and lithium partially positive (blue).

We will use electrostatic potential maps often to illustrate charge distribution in both organic and inorganic molecules. However, we need to offer one cautionary note. Electrostatic potential mapping within a single molecule is fine, but we need to be careful when comparing maps of different molecules. The reason for this is that the entire red-to-blue palette is used to map the electrostatic potential regardless of whether the charge difference is large or small. This is apparent in the H—F and H—Li electrostatic potential maps just shown. If, as shown in the following map, we use the same range for H—F that was used for H—Li we see that H is green instead of blue and the red of F is less intense.

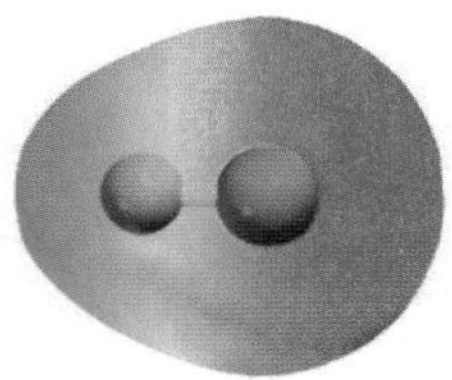

H—F

Thus, electrostatic potential maps can give an exaggerated picture of the charge distribution when the entire palette of colors is used. In most cases, that won't matter to us inasmuch as we are mostly concerned with the distribution within a single molecule. When we want to compare trends in a series of molecules, we'll use a common scale and will point that out. For example, the electrostatic potentials of H_2, F_2, and HF that were compared on page 11 were mapped using the same color scale.

of the dipole moment. In a C—H bond the electrons are drawn away from H, toward C. In a C—O bond, electrons are drawn from C toward O. As we'll see in later chapters, the kinds of reactions that a substance undergoes can often be related to the size and direction of key bond dipoles.

Problem 1.9

Indicate the direction of the dipole for the following bonds using the symbol $\longleftrightarrow$ and δ+, δ− notation.

H—O H—N C—O C=O C—N C=N C≡N

1.5 Formal Charge

Lewis formulas frequently contain atoms that bear a positive or negative charge. If the molecule as a whole is neutral, the sum of its positive charges must equal the sum of its negative charges. An example is nitromethane CH_3NO_2.

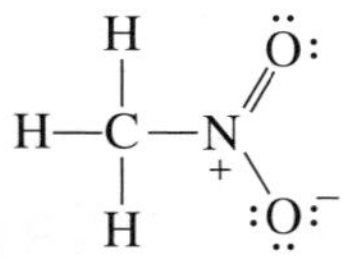

As written, the Lewis formula for nitromethane shows one of the oxygens doubly bonded to nitrogen while the other is singly bonded. The octet rule is satisfied for nitrogen, carbon, and both oxygens. Carbon, the three hydrogens, and the doubly bonded oxygen are uncharged, but nitrogen bears a charge of +1 and the singly bonded oxygen a charge of −1. These charges are called formal charges and are required for the Lewis formula of nitromethane to be complete.

Formal charges correspond to the difference between the number of valence electrons in the neutral free atom and the number of valence electrons in its bonded state. The number of electrons in the neutral free atom is the same as the atom's group number in the periodic table. To determine the electron count of an atom in a Lewis formula, we add the total number of electrons in unshared pairs to one-half the number of electrons in bonded pairs. It's important to note that counting electrons for the purpose of assigning formal charge differs from counting electrons to see if the octet rule is satisfied. A second-row element has a complete octet if the sum of all the electrons around it, shared and unshared, is eight. When counting the electrons to assign formal charge, half the number of electrons in covalent bonds are assigned to each atom.

It will always be true that a covalently bonded hydrogen has no formal charge (formal charge = 0).

Figure 1.5 applies this procedure to the calculation of formal charges in nitromethane. Starting with the three hydrogens, we see that each is associated with two electrons, giving each an electron count of $\frac{1}{2}(2) = 1$. Because a neutral hydrogen atom

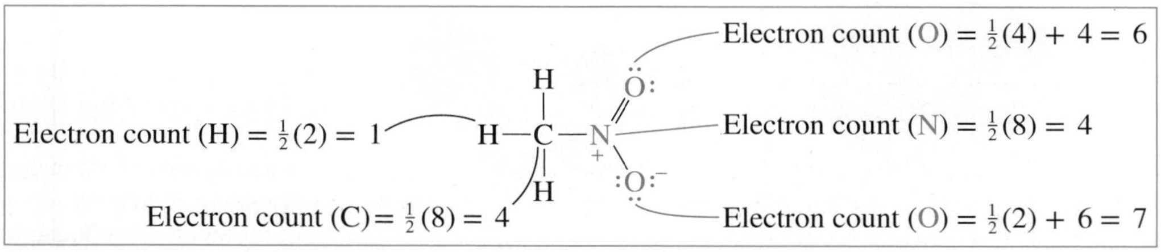

Electron count (O) = $\frac{1}{2}(4) + 4 = 6$

Electron count (H) = $\frac{1}{2}(2) = 1$

Electron count (N) = $\frac{1}{2}(8) = 4$

Electron count (C) = $\frac{1}{2}(8) = 4$

Electron count (O) = $\frac{1}{2}(2) + 6 = 7$

Figure 1.5

Counting electrons in nitric acid. The electron count of each atom is equal to half the number of electrons it shares in covalent bonds plus the number of electrons in its own unshared pairs.

has one electron, the hydrogens of nitromethane have no formal charge. Similarly for carbon, the electron count is $\frac{1}{2}(8) = 4$, which is the number of electrons on a neutral carbon atom, so carbon has no formal charge in nitromethane.

Moving to nitrogen, we see that it has four covalent bonds, so its electron count is $\frac{1}{2}(8) = 4$, which is one less than the number of valence electrons of a nitrogen atom; therefore, its formal charge is +1. The doubly bonded oxygen has an electron count of six (four electrons from the two unshared pairs + two from the double bond). An electron count of six is the same as that of a neutral oxygen's valence electrons; therefore, the doubly bonded oxygen has no formal charge. The singly bonded oxygen, however, has an electron count of seven: six for the three nonbonded pairs plus one for the single bond to nitrogen. This total is one more than the number of valence electrons of a neutral oxygen, so the formal charge is −1.

It will always be true that a nitrogen with four covalent bonds has a formal charge of +1. (A nitrogen with four covalent bonds cannot have unshared pairs, according to the octet rule.)

It will always be true that an oxygen with two covalent bonds and two unshared pairs has no formal charge.

It will always be true that an oxygen with one covalent bond and three unshared pairs has a formal charge of −1.

Problem 1.10

Why is the formula shown for nitromethane incorrect?

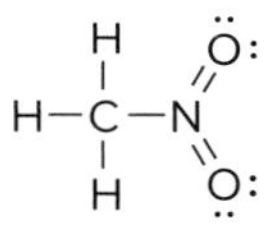

Problem 1.11

The following inorganic species will be encountered in this text. Calculate the formal charge on each of the atoms in the Lewis formulas given.

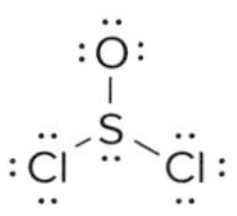

(a) Thionyl chloride 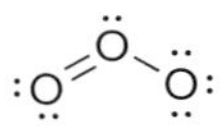 (b) Ozone 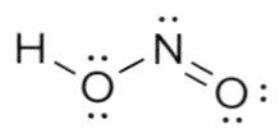 (c) Nitrous acid

Sample Solution (a) The formal charge is the difference between the number of valence electrons in the neutral atom and the electron count in the Lewis formula. (The number of valence electrons is the same as the group number in the periodic table for the main-group elements.)

	Valence electrons of neutral atom	Electron count	Formal charge
Sulfur:	6	$\frac{1}{2}(6) + 2 = 5$	+1
Oxygen:	6	$\frac{1}{2}(2) + 6 = 7$	−1
Chlorine:	7	$\frac{1}{2}(2) + 6 = 7$	0

The formal charges are shown in the Lewis formula of thionyl chloride as

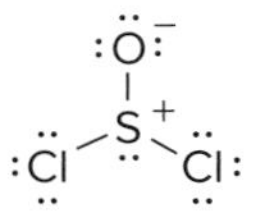

The method described for calculating formal charge has been one of reasoning through a series of logical steps. It can be reduced to the following equation:

$$\text{Formal charge} = \text{Group number in periodic table} - \text{Electron count}$$

where

$$\text{Electron count} = \tfrac{1}{2}(\text{Number of shared electrons}) + \text{Number of unshared electrons}$$

So far we've only considered neutral molecules—those in which the sums of the positive and negative formal charges were equal. With ions, these sums will not be equal. Ammonium cation and borohydride anion, for example, are ions with net charges of +1 and −1, respectively. Nitrogen has a formal charge of +1 in ammonium ion, and boron has a formal charge of −1 in borohydride. None of the hydrogens in the Lewis formulas shown for these ions bears a formal charge.

```
     H               H
     |               |
 H—N⁺—H          H—B⁻—H
     |               |
     H               H
Ammonium ion    Borohydride ion
```

Formal charges are based on Lewis formulas in which electrons are considered to be shared equally between covalently bonded atoms. Actually, polarization of N—H bonds in ammonium ion and of B—H bonds in borohydride leads to some transfer of positive and negative charge, respectively, to the hydrogens.

Problem 1.12

Calculate the formal charge on each nitrogen in the following Lewis formula (azide ion) and the net charge on the species.

$:\ddot{N}{=}N{=}\ddot{N}:$

Determining formal charges on individual atoms in Lewis formulas is an important element in good "electron bookkeeping." So much of organic chemistry can be made more understandable by keeping track of electrons that it is worth taking some time at the beginning to become proficient at the seemingly simple task of counting them.

1.6 Structural Formulas of Organic Molecules: Isomers

Most organic compounds are more complicated than the examples we've seen so far and require a more systematic approach to writing structural formulas for them. The approach outlined in Table 1.5 begins (step 1) with the **molecular formula** that tells us which atoms and how many of each are present in the compound. From the molecular formula we calculate the number of valence electrons (step 2).

In step 3 we set out a partial structure that shows the order in which the atoms are connected. This is called the **connectivity** of the molecule and is almost always determined by experiment. Most of the time carbon has four bonds, nitrogen has three, and oxygen two. It frequently happens in organic chemistry that two or more different compounds have the same molecular formula, but different connectivities. Ethanol and dimethyl ether—the examples shown in the table—are different compounds with different properties, yet have the same molecular formula (C_2H_6O). Ethanol is a liquid with a boiling point of 78°C. Dimethyl ether is a gas at room temperature; its boiling point is −24°C.

Different compounds that have the same molecular formula are classified as **isomers.** Isomers can be either **constitutional isomers** (differ in connectivity) or **stereoisomers** (differ in arrangement of atoms in space). Constitutional isomers are also sometimes called **structural isomers.** Ethanol and dimethyl ether are constitutional isomers of each other. Stereoisomers will be introduced in Section 3.11.

The framework of covalent bonds revealed by the connectivity information accounts for 16 of the 20 valence electrons in C_2H_6O (step 4). The remaining four valence electrons are assigned to each oxygen as two unshared pairs in step 5 to complete the Lewis formulas of ethanol and dimethyl ether.

The suffix *-mer* in the word "isomer" is derived from the Greek word *meros,* meaning "part," "share," or "portion." The prefix *iso-* is also from Greek (*isos,* meaning "the same"). Thus isomers are different molecules that have the same parts (elemental composition).

TABLE 1.5 A Systematic Approach to Writing Lewis Formulas

Step	Illustration
1. The molecular formula is determined experimentally.	Ethanol and dimethyl ether both have the molecular formula C_2H_6O.
2. Based on the molecular formula, count the number of valence electrons.	In C_2H_6O, each hydrogen contributes 1 valence electron, each carbon contributes 4, and oxygen contributes 6 for a total of 20.
3. Given the connectivity, connect bonded atoms by a shared electron-pair bond (:) represented by a dash (—).	Oxygen and the two carbons are connected in the order CCO in ethanol and COC in dimethyl ether. The connectivity and the fact that carbon normally has four bonds in neutral molecules allow us to place the hydrogens of ethanol and dimethyl ether. [Structural formulas: Ethanol; Dimethyl ether]
4. Count the number of electrons in the bonds (twice the number of bonds), and subtract this from the total number of valence electrons to give the number of electrons that remain to be added.	The structural formulas in step 3 contain eight bonds, accounting for 16 electrons. Because C_2H_6O contains 20 valence electrons, 4 more are needed.
5. Add electrons in pairs so that as many atoms as possible have eight electrons. It is usually best to begin with the most electronegative atom. (Hydrogen is limited to two electrons.) *Under no circumstances can a second-row element such as C, N, or O have more than eight valence electrons.*	Both carbons already have complete octets in the structures illustrated in step 3. The remaining four electrons are added to each oxygen as two unshared pairs to complete its octet. The Lewis structures are: [Lewis structures: Ethanol; Dimethyl ether]
6. If one or more atoms (excluding hydrogens) have fewer than eight electrons, use an unshared pair from an adjacent atom to form a double or triple bond to complete the octet. Use one double bond for each deficiency of two electrons to complete the octet for each atom.	All the carbon and oxygen atoms in the structural formulas of ethanol and dimethyl ether have complete octets. No double bonds are needed.
7. Calculate formal charges.	None of the atoms in the Lewis formulas shown in step 5 bears a formal charge.

Problem 1.13

Write structural formulas for all the constitutional isomers that have the given molecular formula.

(a) C_2H_7N (b) C_3H_7Cl (c) C_3H_8O

Sample Solution (a) The molecular formula C_2H_7N requires 20 valence electrons. Two carbons contribute a total of eight, nitrogen contributes five, and seven hydrogens contribute a total of seven. Nitrogen and two carbons can be connected in the order CCN or CNC. Assuming four bonds to each carbon and three to nitrogen, we write these connectivities as

—C—C—N— and —C—N—C—

continued

Place a hydrogen on each of the seven available bonds of each framework.

```
   H  H  H                H     H
   |  |  |                |     |
H—C—C—N—H      and      H—C—N—C—H
   |  |                   |  |  |
   H  H                   H  H  H
```

The nine bonds in each structural formula account for 18 electrons. Add an unshared pair to each nitrogen to complete its octet and give a total of 20 valence electrons as required by the molecular formula.

```
   H  H  H                H     H
   |  |  |                |  ·· |
H—C—C—N—H      and      H—C—N—C—H
   |  |  ··               |  |  |
   H  H                   H  H  H
```

These two are constitutional isomers.

Now let's consider a slightly more complex molecule, methyl nitrite, in which we have to include multiple bonds when writing the Lewis structure (step 6). Methyl nitrite has the molecular formula CH_3NO_2 and is a constitutional isomer of nitromethane (Section 1.5). All the hydrogens are attached to the carbon, and the order of atom connections is CONO. First count the number of valence electrons.

```
   H
   |
H—C—O—N—O
   |
   H
```

Each hydrogen contributes 1 valence electron, carbon 4, nitrogen 5, and each oxygen 6, for a total of 24. The partial structure shown contains 6 bonds equivalent to 12 electrons, so another 12 electrons must be added. Add these 12 electrons in pairs to oxygen and nitrogen. Both oxygens will end up with eight electrons, but nitrogen, because it is less electronegative, will have only six.

```
   H
   |   ··  ··  ··
H—C—O—N—O:
   |   ··      ··
   H
```

All atoms have complete octets except nitrogen, which has a deficiency of two electrons. Use an electron pair from the terminal oxygen to form a double bond to nitrogen to complete its octet. The resulting structure is the best (most stable) for methyl nitrite. All atoms (except hydrogen) have eight electrons in their valence shell.

```
   H
   |   ··  ··  ··
H—C—O—N=O:
   |   ··
   H
```

You may wonder why the electron pair from the terminal oxygen and not the oxygen in the middle was used to form the double bond to nitrogen in the Lewis structure. The structure that would result from using the electron pair on the middle oxygen has a separation of a positive and a negative charge. Although the resulting Lewis structure satisfies the octet rule, its charge separation makes it less stable than the other. When two or more Lewis structures satisfy the octet rule, we need to know how to choose the one that best represents the true structure. The following section describes a procedure for doing this.

```
   H
   |   +   ··  ··
H—C—O=N—O:⁻
   |   ··      ··
   H
```

Problem 1.14

Nitrosomethane and formaldoxime both have the molecular formula CH_3NO and the connectivity CNO. All of the hydrogens are bonded to carbon in nitrosomethane. In formaldoxime, two of the hydrogens are bonded to carbon and one to oxygen. Write Lewis formulas for (a) nitrosomethane and (b) formaldoxime that satisfy the octet rule and are free of charge separation.

Sample Solution

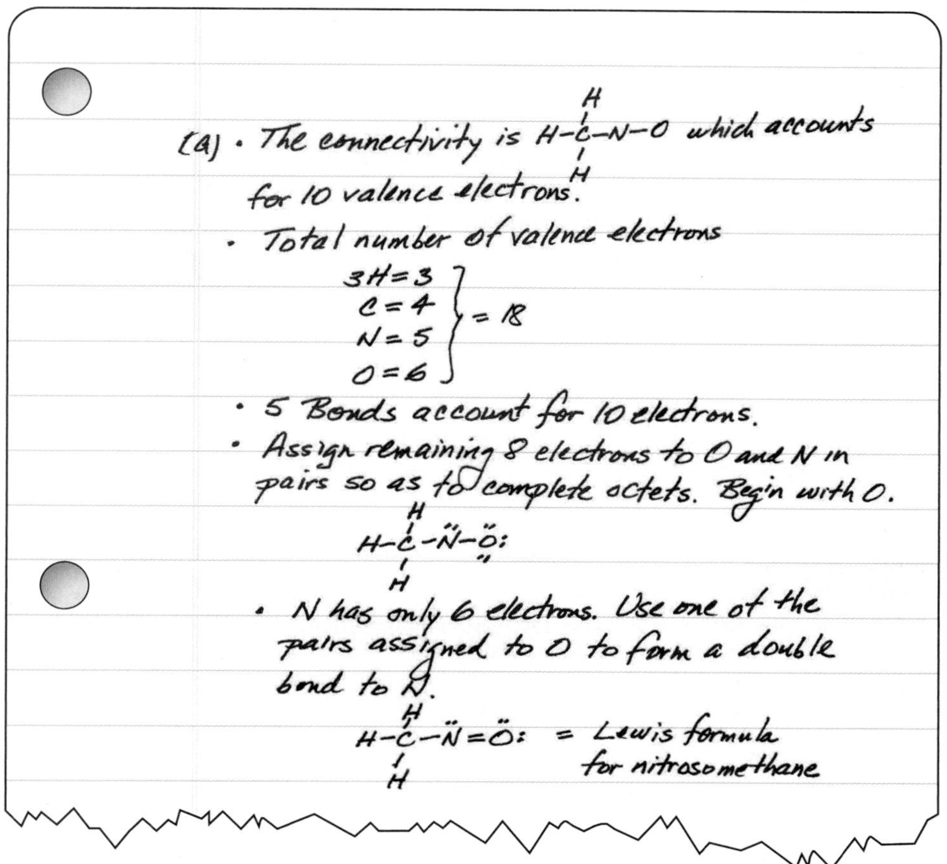

(a) • The connectivity is H–C–N–O which accounts for 10 valence electrons.
• Total number of valence electrons
3H = 3
C = 4
N = 5
O = 6
= 18
• 5 Bonds account for 10 electrons.
• Assign remaining 8 electrons to O and N in pairs so as to complete octets. Begin with O.
• N has only 6 electrons. Use one of the pairs assigned to O to form a double bond to N.
= Lewis formula for nitrosomethane

As illustrated for diethyl ether, chemists often find condensed formulas and line formulas in which carbon chains are represented as a zigzag collection of bonds to be more convenient than Lewis formulas.

$CH_3CH_2\ddot{\underset{\cdot\cdot}{O}}CH_2CH_3$ or $(CH_3CH_2)_2\ddot{O}:$

Condensed formulas

Bond-line formula

In a **condensed formula,** we omit the bonds altogether. Atoms and their attached hydrogens are grouped and written in sequence; subscripts indicate the number of identical groups attached to a particular atom. **Bond-line formulas** are formulas in which labels for individual carbons are omitted and hydrogens attached to carbon are shown only when necessary for clarity. **Heteroatoms**—atoms other than carbon or hydrogen—are shown explicitly as are hydrogens attached to them. Unshared electron pairs are shown when necessary, but are often omitted.

$CH_3CH_2CH_2CH_2OH$ simplifies to OH

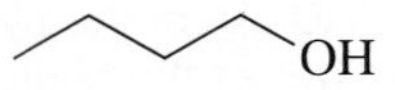

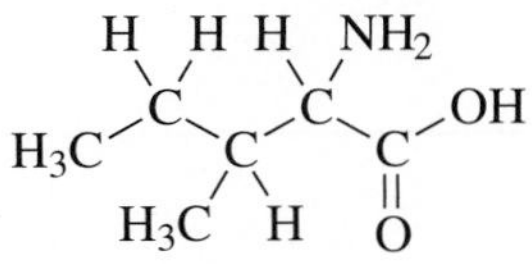

simplifies to

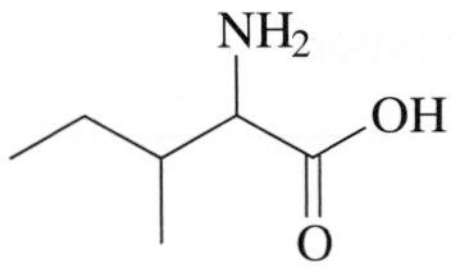

The structural language of organic chemistry has been developed so that complex molecules can be described in a clear, yet economical way. A molecule as complex as cholesterol can be drawn rapidly in a bond-line formula, while drawing even a condensed formula would require a prohibitive amount of time.

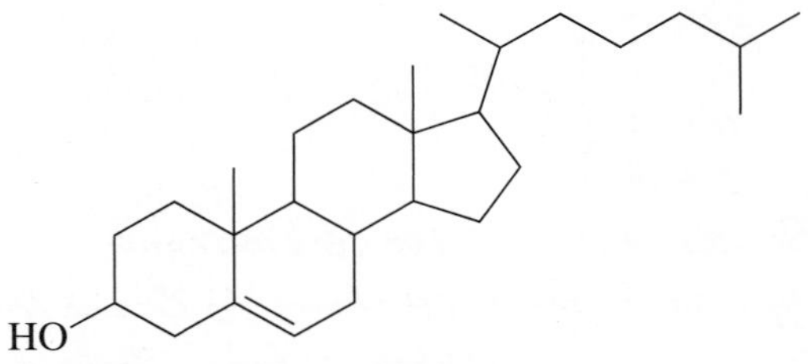

Cholesterol

Problem 1.15

Expand the bond-line formulas of the amino acid cysteine and the neurotransmitter serotonin to show all the unshared electron pairs. Molecular formulas of organic compounds are customarily presented in the fashion $C_aH_bX_cY_d$. Carbon and hydrogen are cited first, followed by the other atoms in alphabetical order. What are the molecular formulas of cysteine and serotonin?

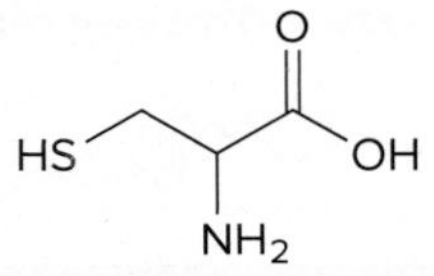

Cysteine

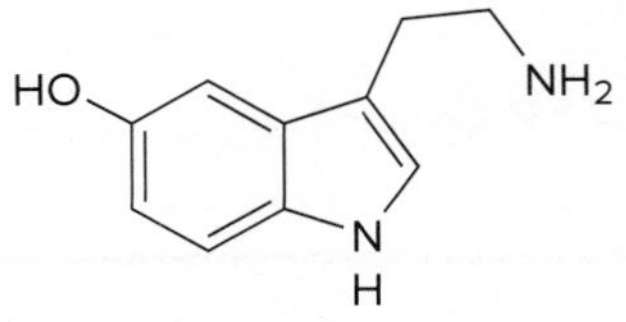

Serotonin

1.7 Resonance and Curved Arrows

Ozone occurs naturally in large quantities in the upper atmosphere where it screens the surface of the Earth from much of the sun's ultraviolet rays.

Sometimes more than one Lewis formula can be written for a molecule, especially if the molecule contains a double or triple bond. A simple example is ozone (O_3), for which we can write

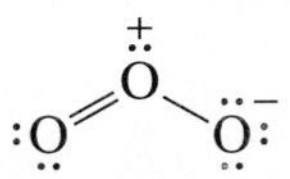

We will express bond distances in both picometers (pm) and angstroms, a unit commonly used by organic chemists. Picometer is an SI unit (1 pm = 10^{-12} m). To convert pm to angstrom units (1 Å = 10^{-10} m), divide by 100.

This Lewis formula, however, is inconsistent with the experimentally determined structure. On the basis of the Lewis formula, we would expect ozone to have two different O—O bond lengths, one of them similar to the O—O single bond distance of 147 pm (1.47 Å) in hydrogen peroxide (HO—OH) and the other similar to the 121 pm (1.21 Å) double bond distance in O_2. In fact, both bond distances are the same (128 pm, 1.28 Å)—somewhat shorter than a single bond, somewhat longer than a double bond. *The structure of ozone requires that the central oxygen must be identically bonded to both terminal oxygens.*

An electrostatic potential map shows the equivalence of the two terminal oxygens. Notice, too, that the central oxygen is blue (positively charged) and both terminal oxygens are red (negatively charged).

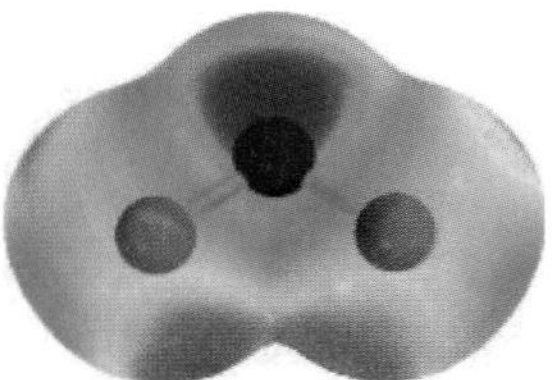

To deal with circumstances such as the bonding in ozone, yet retain Lewis formulas as a useful tool for representing molecular structure, the notion of **resonance** was developed. According to the resonance concept, when two or more Lewis formulas that *differ only in the distribution of electrons* can be written for a molecule, no single Lewis formula is sufficient to describe the true electron distribution. The true structure is said to be a **resonance hybrid** of the various Lewis formulas, called **contributing structures,** that can be written for the molecule. In the case of ozone, the two Lewis formulas are equivalent and contribute equally to the resonance hybrid. We use a double-headed arrow to signify resonance and read it to mean that the Lewis formulas shown contribute to, but do not separately describe, the electron distribution in the molecule.

$$:\ddot{O}=\overset{+}{\ddot{O}}-\ddot{\underset{..}{O}}:^{-} \longleftrightarrow {}^{-}:\ddot{\underset{..}{O}}-\overset{+}{\ddot{O}}=\ddot{O}:$$

Resonance attempts to correct a fundamental defect in Lewis formulas. Lewis formulas show electrons as being **localized;** they either are shared between two atoms in a covalent bond or are unshared electrons belonging to a single atom. In reality, electrons distribute themselves in the way that leads to their most stable arrangement. This means that a pair of electrons can be **delocalized,** or shared by several nuclei. In the case of ozone, resonance attempts to show the delocalization of four electrons (an unshared pair of one oxygen plus two of the electrons in the double bond) over the three oxygens.

It is important to remember that the double-headed resonance arrow does *not* indicate a *process* in which contributing Lewis formulas interconvert. Ozone, for example, has a *single* structure; it does not oscillate back and forth between two contributors. An average of the two Lewis formulas is sometimes drawn using a dashed line to represent a "partial" bond. In the dashed-line notation the central oxygen is linked to the other two by bonds that are halfway between a single bond and a double bond, and the terminal oxygens each bear one half of a unit negative charge. The structure below represents the resonance hybrid for ozone.

Resonance is indicated by a double-headed arrow $\longleftrightarrow$; equilibria are described by two arrows $\rightleftharpoons$.

$$-\tfrac{1}{2}\,\ddot{\underset{..}{O}} \cdots \overset{+}{\ddot{O}} \cdots \ddot{\underset{..}{O}}\,-\tfrac{1}{2}$$

Dashed-line notation

Writing the various Lewis formulas that contribute to a resonance hybrid can be made easier by using **curved arrows** to keep track of delocalized electrons. We can convert one Lewis formula of ozone to another by moving electron pairs as shown:

The main use of curved arrows is to show electron flow in chemical reactions and will be described in Section 1.11. The technique is also known as "arrow pushing" and is attributed to the English chemist Sir Robert Robinson.

Move electron pairs as shown by curved arrows $\quad {}^{-}:\ddot{\underset{..}{O}}-\overset{+}{\ddot{O}}=\ddot{\underset{..}{O}}: \quad$ to convert one Lewis formula to another $\quad :\ddot{\underset{..}{O}}=\overset{+}{\ddot{O}}-\ddot{\underset{..}{O}}:^{-}$

Curved arrows show the origin and destination of a pair of electrons. In the case of ozone, one arrow begins at an unshared pair and becomes the second half of a double bond. The other begins at a double bond and becomes an unshared pair of the other oxygen.

Problem 1.16

All of the bonds in the carbonate ion (CO_3^{2-}) are between C and O. Write Lewis formulas for the major resonance contributors, and use curved arrows to show their relationship. Apply the resonance concept to explain why all of the C—O bond distances in carbonate are equal.

In most cases, the various resonance structures of a molecule are not equivalent and do not contribute equally to the resonance hybrid. The electron distribution in the molecule resembles that of its major contributor more closely than any of its alternative resonance structures. Therefore, it is important that we develop some generalizations concerning the factors that make one resonance form more important (more stable) than another. Table 1.6 outlines the structural features that alert us to situations when resonance needs to be considered and lists criteria for evaluating the relative importance of the contributing structures.

TABLE 1.6 Introduction to the Rules of Resonance

Rule	Illustration
I. *When can resonance be considered?*	
1. The connectivity must be the same in all contributing structures; only the electron positions may vary among the various contributing structures. Atom movement is not allowed.	The Lewis formulas A and B are *not* resonance forms of the same compound. They are *isomers* (different compounds with the same molecular formula). 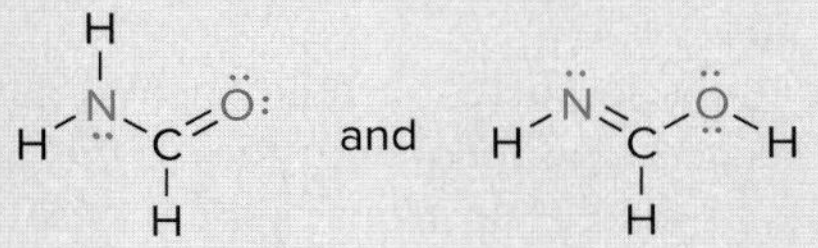 [Structures A and B] The Lewis formulas A, C, and D are resonance forms of a single compound. 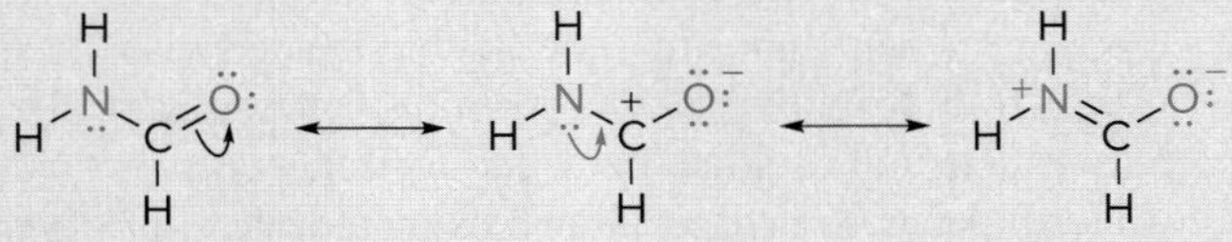 [Structures A ⟷ C ⟷ D]
2. Each contributing structure must have the same number of electrons and the same *net* charge. The formal charges of individual atoms may vary among the various Lewis formulas.	Structures A, C, and D (preceding example) all have 18 valence electrons and a net charge of 0, even though they differ in respect to formal charges on individual atoms. Structure E has 20 valence electrons and a net charge of −2. It is not a resonance structure of A, C, or D. 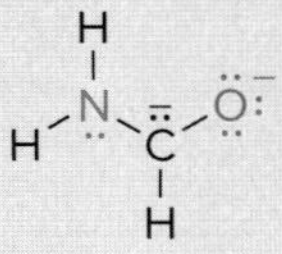 [Structure E]
3. Each contributing structure must have the same number of *unpaired* electrons.	Structural formula F has the same atomic positions and the same number of electrons as A, C, and D, but is not a resonance form of any of them. F has two unpaired electrons; all the electrons in A, C, and D are paired. 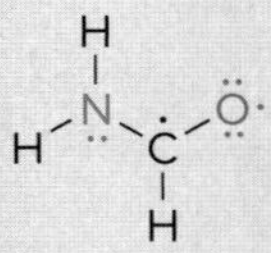 [Structure F]

TABLE 1.6 Introduction to the Rules of Resonance (*Continued*)

Rule	Illustration
4. Contributing structures in which the octet rule is exceeded for second-row elements make no contribution. (The octet rule may be exceeded for elements beyond the second row.)	Lewis formulas G and H are resonance contributors to the structure of nitromethane. Structural formula I is not a permissible Lewis formula because it has ten electrons around nitrogen. 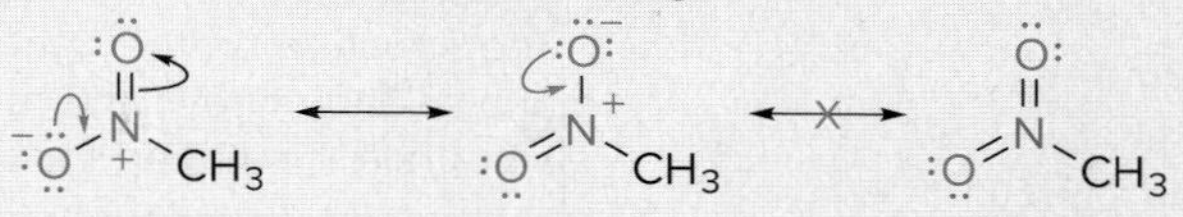 G ⟷ H ⟷ (not permitted) I
II. *Which resonance form contributes more?*	
5. As long as the octet rule is not exceeded for second-row elements, the contributing structure with the greater number of covalent bonds contributes more to the resonance hybrid. Maximizing the number of bonds and satisfying the octet rule normally go hand in hand. This rule is more important than rules 6 and 7.	Of the two Lewis formulas for formaldehyde, the major contributor J has one more bond than the minor contributor K. 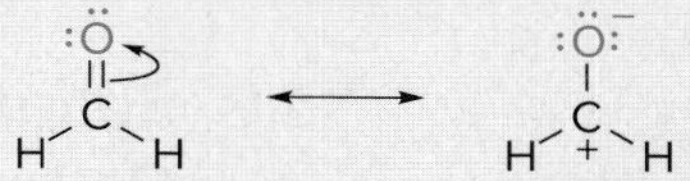 J (Major contributor) ⟷ K (Minor contributor)
6. When two or more structures satisfy the octet rule, the major contributor is the one with the smallest separation of oppositely charged atoms.	The two structures L and M for methyl nitrite have the same number of bonds, but L is the major contributor because it lacks the separation of positive and negative charge that characterizes M. 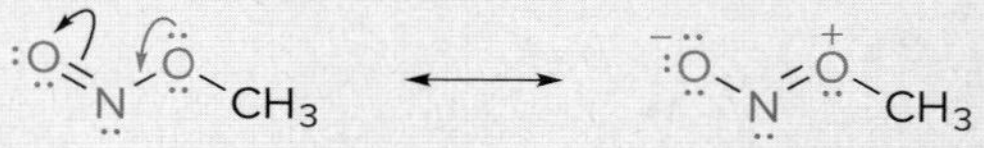 L (Major contributor) ⟷ M (Minor contributor)
7. Among structural formulas that satisfy the octet rule and in which one or more atoms bear a formal charge, the major contributor is the one in which the negative charge resides on the most electronegative atom, and the positive charge on the least electronegative element.	The major contributing structure for cyanate ion is N because the negative charge is on its oxygen. 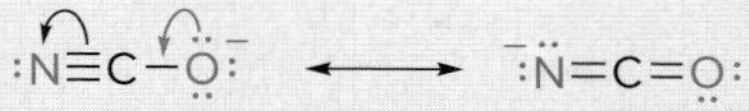 :N≡C—Ö:⁻ ⟷ ⁻:N=C=O: N (Major contributor) O (Minor contributor) In O the negative charge is on nitrogen. Oxygen is more electronegative than nitrogen and can better support a negative charge.
III. *What is the effect of resonance?*	
8. Electron delocalization stabilizes a molecule. Resonance is a way of showing electron delocalization. Therefore, the true electron distribution is more stable than any of the contributing structures. The degree of stabilization is greatest when the contributing structures are of equal stability.	Structures P, Q, and R for carbonate ion are equivalent and contribute equally to the electron distribution. The true structure of carbonate ion is a hybrid of P, Q, and R and is more stable than any of them. 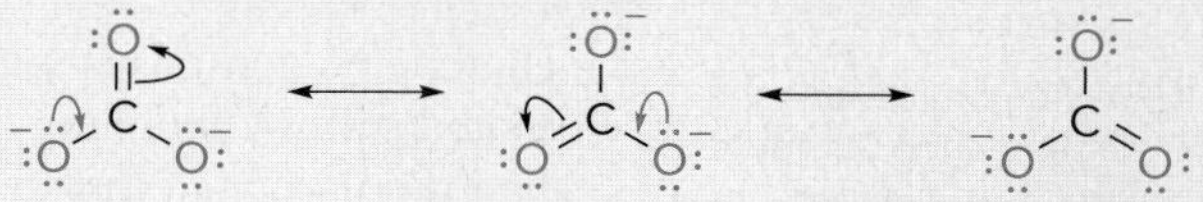 P ⟷ Q ⟷ R

Problem 1.17

Write the resonance structure obtained by moving electrons as indicated by the curved arrows. Compare the stabilities of the two Lewis formulas according to the guidelines in Table 1.6. Are the two structures equally stable, or is one more stable than the other? Why?

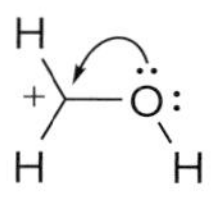

(a) 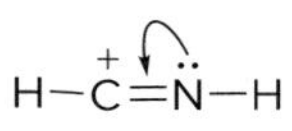 (b) H—C=N—H (c) 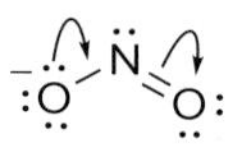 (d)

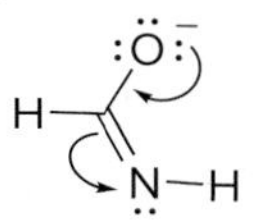

continued

Sample Solution (a) The curved arrow shows how we move an unshared electron pair assigned to oxygen so that it becomes shared by carbon and oxygen. This converts a single bond to a double bond and leads to a formal charge of +1 on oxygen.

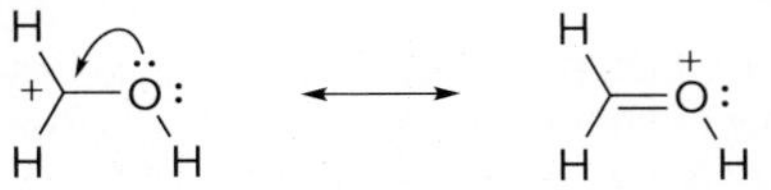

The structure on the right is more stable because it has one more covalent bond than the original structure. Carbon did not have an octet of electrons in the original structure, but the octet rule is satisfied for both carbon and oxygen in the new structure.

It is good chemical practice to represent molecules by their most stable contributing structure. However, the ability to write alternative resonance forms and to assess their relative contributions can provide insight into both molecular structure and chemical behavior.

1.8 Sulfur and Phosphorus-Containing Organic Compounds and the Octet Rule

Applying the Lewis rules to compounds that contain a *third-row element* such as sulfur is sometimes complicated by a conflict between minimizing charge separation and following the octet rule. Consider the two structural formulas **A** and **B** for dimethyl sulfoxide:

A **B**

According to resonance, **A** and **B** are contributing structures, and the actual structure is a hybrid of both. The octet rule favors **A,** but maximizing bonding and eliminating charge separation favor **B.** The justification for explicitly considering **B** is that sulfur has vacant 3*d* orbitals that permit it to accommodate more than eight electrons in its valence shell.

The situation is even more pronounced in dimethyl sulfone in which structural formula **C** has 8 electrons in sulfur's valence shell, **D** has 10, and **E** has 12.

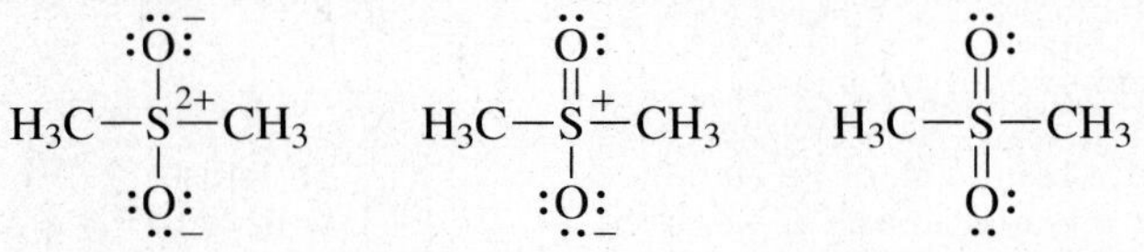

C **D** **E**

There is no consensus regarding which Lewis formula is the major contributor in these and related sulfur-containing compounds. The IUPAC recommends writing double bonds rather than dipolar single bonds; that is, **B** for dimethyl sulfoxide and **E** for dimethyl sulfone.

The 2008 IUPAC Recommendations "Graphical Representation Standards for Chemical Structure Diagrams" can be accessed at http://www.iupac.org/publications/pac/80/2/0277/. For more on the IUPAC, see the boxed essay "What's in a Name? Organic Nomenclature" in Chapter 2.

Similarly, compounds with four atoms or groups bonded to phosphorus can be represented by contributing structures of the type **F** and **G** shown for trimethylphosphine oxide.

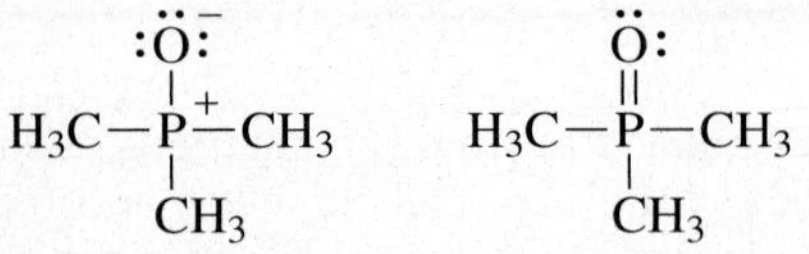

F **G**

Phosphorus shares 8 electrons in **F,** 10 in **G.** The octet rule favors **F;** involvement of 3*d* orbitals allows **G.** As with sulfur-containing compounds, the IUPAC recommends **G,** but both formulas have been used. Many biochemically important compounds—adenosine triphosphate (ATP), for example—are *phosphates* and can be written with either a P=O or $^{+}P—O^{-}$ unit. The same recommendation applies to them; the double-bonded structure is preferred.

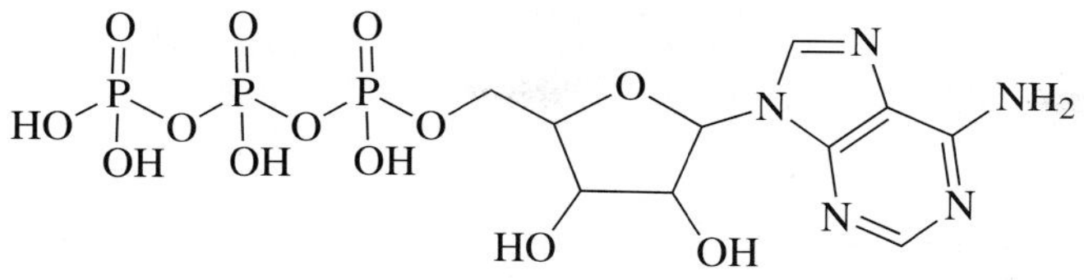

Adenosine triphosphate

Problem 1.18

Listing the atoms in the order CHNOP, what is the molecular formula of ATP? (You can check your answer by entering *adenosine triphosphate* in your web browser.) Show the location of all its unshared electron pairs. How many are there?

Before leaving this introduction to bonding in sulfur and phosphorus compounds, we should emphasize that the only valence orbitals available to second-row elements (Li, Be, B, C, N, O, F, Ne) are 2*s* and 2*p,* and the octet rule cannot be exceeded for them.

Problem 1.19

Of the four structural formulas shown, three are permissible and one is not. Which one is not a permissible structure? Why?

$(CH_3)_3N{=}CH_2$ $\quad (CH_3)_3\overset{+}{N}—\overset{-}{\underset{\cdot\cdot}{C}}H_2$ $\quad (CH_3)_3P{=}CH_2$ $\quad (CH_3)_3\overset{+}{P}—\overset{-}{\underset{\cdot\cdot}{C}}H_2$

1.9 Molecular Geometries

So far we have emphasized structure in terms of "electron bookkeeping." We now turn our attention to molecular geometry and will see how we can begin to connect the three-dimensional shape of a molecule to its Lewis formula. Table 1.7 lists some simple compounds illustrating the geometries that will be seen most often in our study of organic chemistry.

As shown in Table 1.7, methane (CH_4) has a tetrahedral geometry. Each hydrogen occupies a corner of the tetrahedron with the carbon at its center. The table also shows a common method of representing three-dimensionality through the use of different bond styles. A solid wedge (▬) stands for a bond that projects toward you, a "hashed" wedge (┅) for one that points away from you, and a simple line (—) for a bond that lies in the plane of the paper.

The tetrahedral geometry of methane is often explained with the **valence shell electron-pair repulsion (VSEPR) model.** The VSEPR model rests on the idea that an electron pair, either a bonded pair or an unshared pair, associated with a particular atom will be as far away from the atom's other electron pairs as possible. Thus, a tetrahedral geometry permits the four bonds of methane to be maximally separated and is characterized by H—C—H angles of 109.5°, a value referred to as the **tetrahedral angle.**

Water, ammonia, and methane share the common feature of an approximately tetrahedral arrangement of four electron pairs. Because we describe the shape of a molecule according to the positions of its atoms only rather than by the orientation of its electron pairs, water is said to be *bent,* and ammonia is *trigonal pyramidal.*

Although reservations have been expressed concerning VSEPR as an *explanation* for molecular geometries, it remains a useful *tool* for predicting the shapes of organic compounds.

The H—O—H angle in water (105°) and the H—N—H angles in ammonia (107°) are slightly smaller than the tetrahedral angle. These bond-angle contractions are easily accommodated by VSEPR by reasoning that bonded pairs take up less space than unshared pairs. A bonded pair feels the attractive force of two nuclei and is held more

TABLE 1.7 VSEPR and Molecular Geometry

Compound	Structural formula	Repulsive electron pairs	Arrangement of repulsive electron pairs	Molecular shape	Molecular model
Methane (CH_4)	109.5°, H, H, C—H, H; 109.5°, 109.5°, 109.5°	Carbon has four bonded pairs	Tetrahedral	**Tetrahedral**	
Water (H_2O)	105°, H, H, Ö:	Oxygen has two bonded pairs + two unshared pairs	Tetrahedral	**Bent**	
Ammonia (NH_3)	107°, H, H, N:, H	Nitrogen has three bonded pairs + one unshared pair	Tetrahedral	**Trigonal pyramidal**	
Boron trifluoride (BF_3)	120°, :F̤, B—F̤:, :F̤	Boron has three bonded pairs	Trigonal planar	**Trigonal planar**	
Formaldehyde (H_2CO)	H, H, C=Ö:	Carbon has two bonded pairs + one double bond, which is counted as one bonded pair	Trigonal planar	**Trigonal planar**	
Carbon dioxide (CO_2)	180°, :Ö=C=Ö:	Carbon has two double bonds, which are counted as two bonded pairs	Linear	**Linear**	

tightly than an unshared pair localized on a single atom. Thus, repulsive forces increase in the order:

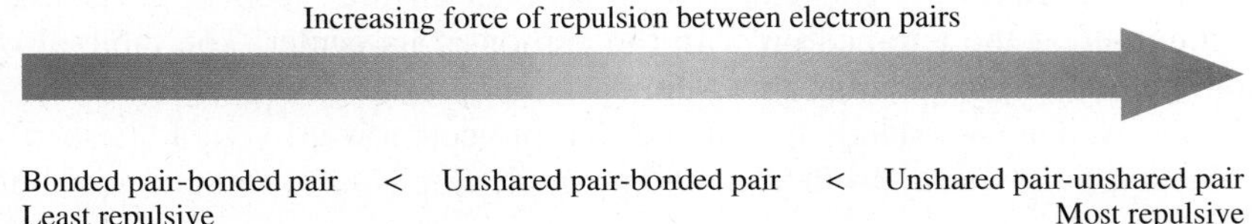

Repulsions among the four bonded pairs of methane give the normal tetrahedral angle of 109.5°. Repulsions among the unshared pair of nitrogen in ammonia and the three bonded pairs cause the bonded pair-bonded pair H—N—H angles to be smaller than 109.5°. In water, a larger repulsive force exists because of two unshared pairs, and the H—O—H angle is compressed further to 105°.

Boron trifluoride is a *trigonal planar* molecule. There are six electrons, two for each B—F bond, associated with the valence shell of boron. These three bonded pairs are farthest apart when they are coplanar, with F—B—F bond angles of 120°.

Problem 1.20

Sodium borohydride, $NaBH_4$, has an ionic bond between Na^+ and the anion BH_4^-. What are the H—B—H angles in borohydride anion?

Molecular Models and Modeling

We can gain a clearer idea about the features that affect structure and reactivity when we examine the three-dimensional shape of a molecule, using either a physical model or a graphical one. Physical models are tangible objects and first appeared on the chemistry scene in the nineteenth century. They proved their worth in two of the pioneering scientific achievements of the mid-twentieth century—Pauling's protein α-helix and the Watson–Crick DNA double helix. But physical models are limited to information about overall shape, angles, and distances and have given way to computer graphics rendering of models in twenty-first-century chemistry, biochemistry, and molecular biology.

At its lowest level, computer graphics substitutes for a physical molecular modeling kit. It is a simple matter to assemble atoms into a specified molecule, then display it in a variety of orientations and formats. Three of these formats are illustrated for methane in Figure 1.6. The most familiar are ball-and-stick models (Figure 1.6*b*), which direct attention to both the atoms and the bonds that connect them. Framework models (Figure 1.6*a*) and space-filling models (Figure 1.6*c*) represent opposite extremes. Framework models emphasize a molecule's bonds while ignoring the sizes of the atoms. Space-filling models emphasize the volume occupied by individual atoms at the cost of a clear depiction of the bonds; they are most useful in those cases where we wish to examine the overall molecular shape and to assess how closely nonbonded atoms approach each other.

Collections such as the Protein Data Bank (PDB) are freely available on the Internet along with viewers for manipulating the models. As of 2015 the PDB contains experimentally obtained structural data for more than 105,000 "biological macromolecular structures"—mainly proteins—and serves as a resource for scientists seeking to understand the structure and function of important biomolecules. Figure 1.7 shows a model of human insulin in a display option in which the two chains are shown as ribbons of different colors.

Computational chemistry takes model making to a yet higher level. Most modeling software also incorporates programs that identify the most stable geometry of a molecule by calculating the energies of possible candidate structures. More than this, the electron distribution in a molecule can be calculated and displayed as described in the boxed essay *Electrostatic Potential Maps* earlier in this chapter.

Molecular models of various types are used throughout this text. Their number and variety of applications testify to their importance in communicating the principles and applications of molecular structure in organic chemistry.

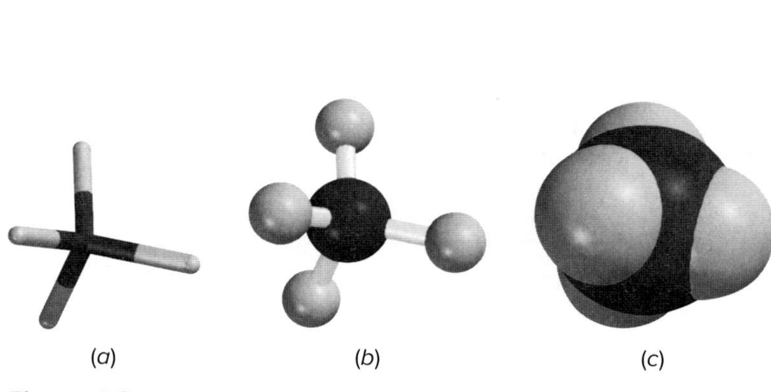

Figure 1.6

Molecular models of methane (CH_4). (*a*) Framework models show the bonds connecting the atoms but not the atoms themselves. (*b*) Ball-and-stick models show the atoms as spheres and the bonds as sticks. (*c*) Space-filling models portray overall molecular size; the radius of each sphere approximates the van der Waals radius of the atom.

Figure 1.7

A ribbon model of the two strands of human insulin. The model may be accessed, viewed, and downloaded in various formats by entering 2KJJ as the PDB ID at http://www.rcsb.org/pdb/home/home.do.

Source: From coordinates deposited with the Protein Data Bank, PDB ID: 2KJJ. Q. X. Hua, M. A. Weiss, "Dynamics of Insulin Probed by 1H NMR. Amide Proton Exchange. Anomalous Flexibility of the Receptor-Binding Surface." http://www.rcsb.org/pdb/home/home.do.

A multiple bond (double or triple) is treated as a unit in the VSEPR model. Formaldehyde is a trigonal planar molecule in which the electrons of the double bond and those of the two single bonds are maximally separated. A linear arrangement of atoms in carbon dioxide allows the electrons in one double bond to be as far away as possible from the electrons in the other double bond.

Problem 1.21

Specify the shape of the following:

H—C≡N:	H_4N^+	:N̈=N̈⁺=N̈:	CO_3^{2-}
Hydrogen cyanide (a)	Ammonium ion (b)	Azide ion (c)	Carbonate ion (d)

Sample Solution (a) The structure shown accounts for all the electrons in hydrogen cyanide. No unshared electron pairs are associated with carbon, and so the structure is determined by maximizing the separation between its single bond to hydrogen and the triple bond to nitrogen. Hydrogen cyanide is a *linear* molecule.

1.10 Molecular Dipole Moments

We can combine our knowledge of molecular geometry with a feel for the polarity of chemical bonds to predict whether a molecule has a dipole moment or not. The **molecular dipole moment** is the resultant of all of the individual bond dipole moments of a substance. Some molecules, such as carbon dioxide, have polar bonds, but lack a dipole moment because their geometry causes the individual C=O bond dipoles to cancel.

:Ö=C=Ö: Dipole moment = 0 D

Carbon dioxide

Carbon tetrachloride, with four polar C—Cl bonds and a tetrahedral shape, has no net dipole moment, because the result of the four bond dipoles, as shown in Figure 1.8, is zero. Dichloromethane, on the other hand, has a dipole moment of 1.62 D. The C—H bond dipoles reinforce the C—Cl bond dipoles.

Problem 1.22

Which of the following compounds would you expect to have a dipole moment? If the molecule has a dipole moment, specify its direction.

(a) BF_3
(b) H_2O
(c) CH_4
(d) CH_3Cl
(e) CH_2O
(f) HCN

Sample Solution (a) As we saw in Table 1.7, boron trifluoride is planar with 120° bond angles. Although each boron–fluorine bond is polar, their combined effects cancel and the molecule has no dipole moment.

$\mu = 0$ D

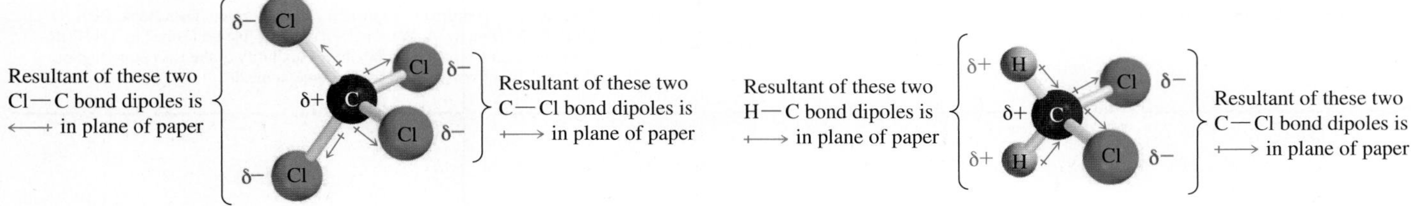

(*a*) There is a mutual cancellation of individual bond dipoles in carbon tetrachloride. It has no dipole moment.

(*b*) The H—C bond dipoles reinforce the C—Cl bond moment in dichloromethane. The molecule has a dipole moment of 1.62 D.

Figure 1.8

Contribution of individual bond dipole moments to the molecular dipole moments of (*a*) carbon tetrachloride (CCl_4) and (*b*) dichloromethane (CH_2Cl_2).

The opening paragraphs of this chapter emphasized that the connection between structure and properties is central to understanding organic chemistry. We have just seen one such connection. From the Lewis formula of a molecule, we can use electronegativity to tell us about the polarity of bonds and combine that with VSEPR to predict whether the molecule has a dipole moment. In the next several sections we'll see a connection between structure and *chemical reactivity* as we review acids and bases.

1.11 Curved Arrows, Arrow Pushing, and Chemical Reactions

In Section 1.7 we introduced curved arrows as a tool for systematically converting one resonance contributor to another. Their more common use is to track electron flow in chemical reactions. The remainder of this chapter introduces acid–base chemistry and illustrates how curved-arrow notation enhances our understanding of chemical reactions by focusing on electron movement.

There are two kinds of curved arrows. A double-barbed arrow (↷) shows the movement of a *pair* of electrons, either a bonded pair or a lone pair. A single-barbed, or fishhook, arrow (↷) shows the movement of *one* electron. For now, we'll concern ourselves only with reactions that involve electron pairs and focus on double-barbed arrows.

We'll start with some simple examples—reactions involving only one electron pair. Suppose the molecule A—B dissociates to cation A^+ and anion B^-. A chemical equation for this ionization could be written as

$$AB \longrightarrow A^+ + B^-$$

Alternatively, we could write

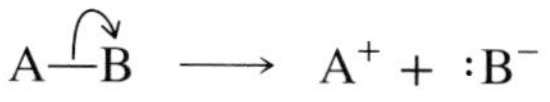

$$A{-}B \longrightarrow A^+ + :B^-$$

The reaction is the same but the second equation provides more information by including the bond that is broken during ionization and showing the flow of electrons. The curved arrow begins where the electrons are originally—in the bond—and points to atom B as their destination where they become an unshared pair of the anion B^-.

Dissociations of this type are common in organic chemistry and will be encountered frequently as we proceed through the text. In many cases, the species A^+ has its positive charge on carbon and is referred to as a **carbocation.** Dissociation of an alkyl bromide, for example, involves breaking a C—Br bond with the two electrons in that bond becoming an unshared pair of bromide ion.

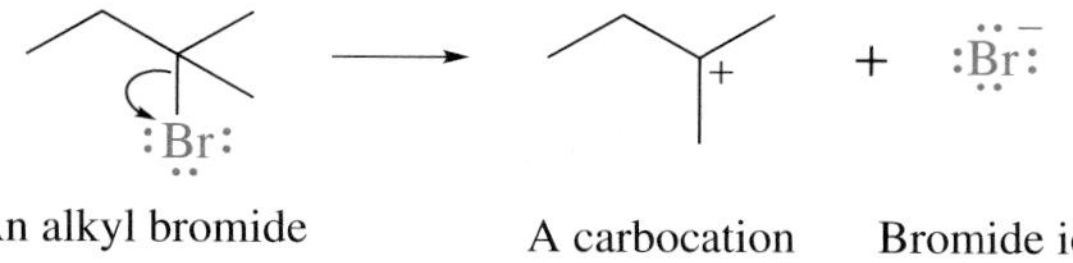

:Br: → + + :Br:⁻

An alkyl bromide — A carbocation — Bromide ion

Charge is conserved, as it must be in all reactions. Here, the reactant is uncharged, and the net charge on the products is 0. In a conceptually related dissociation, a net charge of +1 is conserved when a positively charged reactant dissociates to a carbocation and a neutral molecule.

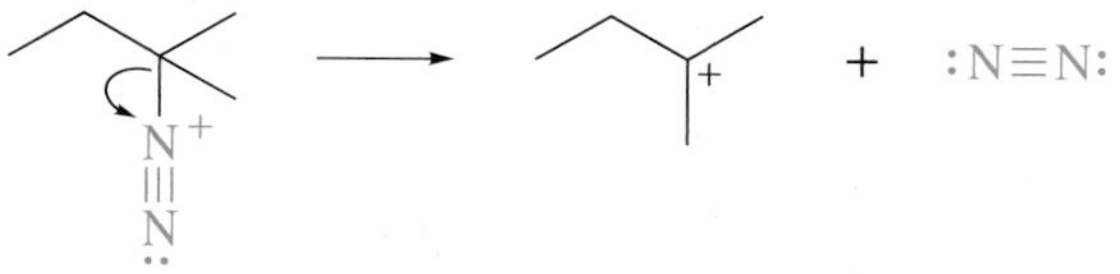

N⁺≡N: → + + :N≡N:

A diazonium ion — A carbocation — Nitrogen

These alkyl bromide and diazonium ion dissociations are discussed in detail in Sections 7.18 and 22.15, respectively.

Problem 1.23

Using the curved arrow to guide your reasoning, show the products of the following dissociations. Include formal charges and unshared electron pairs. Check your answers to ensure that charge is conserved.

(a) 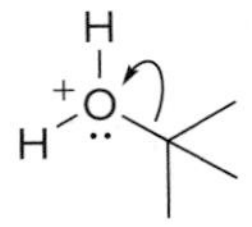 (b)

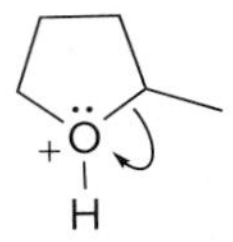

Sample Solution (a) The curved arrow tells us that the C—O bond breaks and the pair of electrons in that bond becomes an unshared electron pair of oxygen.

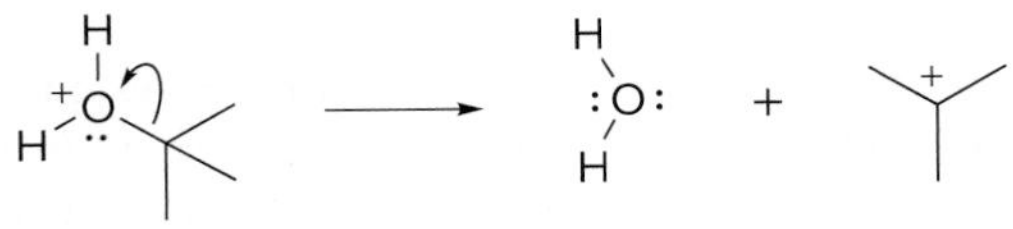

Water is one product of the reaction. The organic species produced is a cation. Its central carbon has six electrons in its valence shell and a formal charge of +1. Charge is conserved in the reaction. The net charge on both the left and right side of the equation is +1.

The reverse of a dissociation is a combination, such as the formation of a covalent bond between a cation A^+ and an anion $:B^-$.

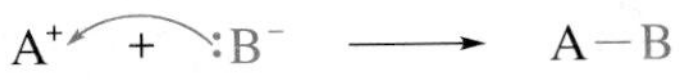

$$A^+ + :B^- \longrightarrow A-B$$

Here the tail of the curved arrow begins at the middle of the unshared electron pair of $:B^-$ and the head points to the location of the new bond—in this case the open space just before A^+. *Electrons flow from sites of higher electron density to lower.* The unshared electron pair of $:B^-$ becomes the shared pair in the A—B bond.

Problem 1.24

Write equations, including curved arrows, describing the reverse reactions of Problem 1.23.

Sample Solution (a) First write the equation for the reverse process. Next, use a curved arrow to show that the electron pair in the C—O bond in the product originates as an unshared electron pair of oxygen in water.

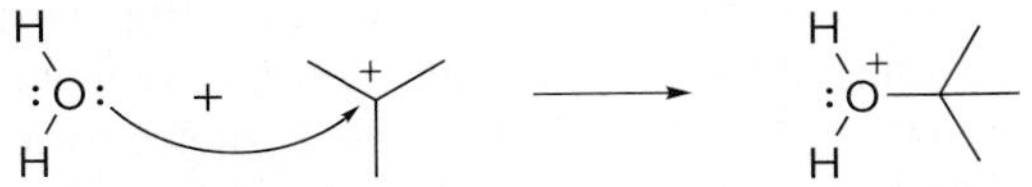

Many reactions combine bond making with bond breaking and require more than one curved arrow.

$$^-A: + B-C \longrightarrow A-B + :C^-$$

Note that the electron counts and, therefore, the formal charges of A and C, but not B, change. An example is a reaction that will be discussed in detail in Section 6.3.

$$H\ddot{O}:^- + H_3C-\ddot{Br}: \longrightarrow H\ddot{O}-CH_3 + :\ddot{Br}:^-$$

An unshared electron pair of a negatively charged oxygen becomes a shared electron pair in a C—O bond. Again, notice that electrons flow from electron-rich to electron-poor sites. Hydroxide ion is negatively charged and, therefore, electron-rich while the carbon of H_3CBr is partially positive because of the polarization of the C—Br bond (Section 1.4).

A very common process is the transfer of a proton from one atom to another as in the reaction that occurs when hydrogen bromide dissolves in water.

$$H_2\ddot{O}: + H{-}\ddot{Br}: \longrightarrow H_2\overset{+}{O}{-}H + :\ddot{Br}:^-$$

Numerous other proton-transfer reactions will appear in the remainder of this chapter.

Curved-arrow notation is also applied to reactions in which double and triple bonds are made or broken. Only one component (one electron pair) of the double or triple bond is involved. Examples include:

$$(CH_3)_2C{=}CH_2 + H{-}\ddot{Br}: \longrightarrow (CH_3)_2\overset{+}{C}{-}CH_3 + :\ddot{Br}:^-$$

$$H_3N: + H_2C{=}\ddot{O}: \longrightarrow H_3\overset{+}{N}{-}CH_2{-}\ddot{O}:^-$$

Problem 1.25

Reactions of the type shown are an important part of Chapter 21. Follow the arrows to predict the products. Show formal charges and include all unshared electron pairs.

$$H\ddot{O}:^- + H{-}CH_2{-}CH{=}\ddot{O}: \longrightarrow$$

Before we conclude this section, we should emphasize an important point:

- *Resist the temptation to use curved arrows to show the movement of atoms*. Curved arrows always show *electron* flow.

Although our eyes are drawn to the atoms when we look at a chemical equation, following the electrons provides a clearer understanding of how reactants become products.

1.12 Acids and Bases: The Brønsted–Lowry View

Acids and bases are a big part of organic chemistry, but the emphasis is much different from what you may be familiar with from your general chemistry course. Most of the attention in general chemistry is given to numerical calculations: pH, percent ionization, buffer problems, and so on. Some of this returns in organic chemistry, but mostly we are concerned with acids and bases as reactants, products, and catalysts in chemical reactions. We'll start by reviewing some general ideas about acids and bases.

According to the theory proposed by Svante Arrhenius, a Swedish chemist and winner of the 1903 Nobel Prize in Chemistry, an acid is a substance that ionizes to give protons when dissolved in water; a base ionizes to give hydroxide ions.

A more general theory of acids and bases was devised by Johannes Brønsted (Denmark) and Thomas M. Lowry (England) in 1923. In the Brønsted–Lowry approach, an acid is a **proton donor,** and a base is a **proton acceptor.** The reaction that occurs between an acid and a base is *proton transfer*.

$$\underset{\text{Base}}{B:} + \underset{\text{Acid}}{H{-}A} \rightleftharpoons \underset{\text{Conjugate acid}}{\overset{+}{B}{-}H} + \underset{\text{Conjugate base}}{:A^-}$$

In the equation shown, the base uses an unshared pair of electrons to remove a proton from an acid. The base is converted to its **conjugate acid,** and the acid is converted to its **conjugate base.** A base and its conjugate acid always differ by a single proton. Likewise, an acid and its conjugate base always differ by a single proton.

In the Brønsted–Lowry view, an acid doesn't dissociate in water; it transfers a proton to water. Water acts as a base.

$$H_2\ddot{O}: + H{-}A \rightleftharpoons H_2\overset{+}{O}{-}H + :A^-$$

Water (base) Acid Conjugate acid of water Conjugate base

The systematic name for the conjugate acid of water (H_3O^+) is **oxonium ion.** Its common name is **hydronium ion.**

Problem 1.26

Write an equation for proton transfer from hydrogen chloride (HCl) to

(a) Ammonia ($:NH_3$)
(b) Trimethylamine [$(CH_3)_3N:$]

Identify the acid, base, conjugate acid, and conjugate base and use curved arrows to track electron movement.

Sample Solution (a) We are told that a proton is transferred from HCl to $:NH_3$. Therefore, HCl is the Brønsted acid and $:NH_3$ is the Brønsted base.

$$H_3N: + H{-}\ddot{Cl}: \rightleftharpoons H_3\overset{+}{N}{-}H + :\ddot{Cl}:^-$$

Ammonia (base) Hydrogen chloride (acid) Ammonium ion (conjugate acid) Chloride ion (conjugate base)

The strength of an acid is measured by its **acidity constant** K_a defined as

$$K_a = \frac{[H_3O^+][:A^-]}{[HA]}$$

Even though water is a reactant (a Brønsted base), its concentration does not appear in the expression for K_a because it is the solvent. The convention for equilibrium constant expressions is to omit concentration terms for pure solids, liquids, and solvents.

Water can also be a Brønsted acid, donating a proton to a base. Sodium amide ($NaNH_2$), for example, is a source of the strongly basic amide ion, which reacts with water to give ammonia.

$$H_2\ddot{N}:^- + H{-}\ddot{O}{-}H \rightleftharpoons H_2\ddot{N}{-}H + ^-:\ddot{O}{-}H$$

Amide ion (base) Water (acid) Ammonia (conjugate acid) Hydroxide ion (conjugate base)

Problem 1.27

Potassium hydride (KH) is a source of the strongly basic hydride ion ($:H^-$).

Using curved arrows to track electron movement, write an equation for the reaction of hydride ion with water. What is the conjugate acid of hydride ion?

A convenient way to express the strength of an acid is by its **pK_a**, defined as

$$pK_a = -\log_{10}K_a$$

Thus, acetic acid with $K_a = 1.8 \times 10^{-5}$ has a pK_a of 4.7. The advantage of pK_a over K_a is that it avoids exponentials. You are probably more familiar with K_a, but most organic chemists and biochemists use pK_a. It is a good idea to be comfortable with both systems, so you should practice converting K_a to pK_a and vice versa.

Problem 1.28

Salicylic acid, the starting material for the preparation of aspirin, has a K_a of 1.06×10^{-3}. What is its pK_a?

Problem 1.29

Hydrogen cyanide (HCN) has a pK_a of 9.1. What is its K_a?

Table 1.8 lists a number of acids, their acidity constants, and their conjugate bases. The list is more extensive than we need at this point, but we will return to it repeatedly throughout the text as new aspects of acid–base behavior are introduced. The table is organized so that acid strength decreases from top to bottom. Conversely, the strength of the conjugate base increases from top to bottom. Thus, *the stronger the acid, the weaker its conjugate base. The stronger the base, the weaker its conjugate acid.*

The Brønsted–Lowry approach involving conjugate relationships between acids and bases makes a separate basicity constant K_b unnecessary. Rather than having separate tables listing K_a for acids and K_b for bases, the usual practice is to give only K_a or pK_a as was done in Table 1.8. Assessing relative basicities requires only that we remember that the weaker the acid, the stronger the conjugate base and find the appropriate acid–base pair in the table.

Problem 1.30

Which is the stronger base in each of the following pairs? (*Note:* This information will prove useful when you get to Chapter 9.)

(a) Sodium ethoxide ($NaOCH_2CH_3$) or sodium amide ($NaNH_2$)
(b) Sodium acetylide ($NaC{\equiv}CH$) or sodium amide ($NaNH_2$)
(c) Sodium acetylide ($NaC{\equiv}CH$) or sodium ethoxide ($NaOCH_2CH_3$)

Sample Solution (a) $NaOCH_2CH_3$ contains the ions Na^+ and $CH_3CH_2O^-$. $NaNH_2$ contains the ions Na^+ and H_2N^-. $CH_3CH_2O^-$ is the conjugate base of ethanol; H_2N^- is the conjugate base of ammonia.

Base	$CH_3CH_2O^-$	H_2N^-
Conjugate acid	CH_3CH_2OH	NH_3
pK_a of conjugate acid	16	36

The conjugate acid of $CH_3CH_2O^-$ is stronger than the conjugate acid of H_2N^-. Therefore, H_2N^- is a stronger base than $CH_3CH_2O^-$.

TABLE 1.8 Acidity Constants (pK_a) of Acids

Acid	pK_a	Formula	Conjugate base	Discussed in Section
Hydrogen iodide	−10.4	HI	I^-	1.13; 1.14
Hydrogen bromide	−5.8	HBr	Br^-	1.13; 1.14
Sulfuric acid	−4.8	$HOSO_3H$	$^-OSO_3H$	1.14
Hydrogen chloride	−3.9	HCl	Cl^-	1.13; 1.14
Protonated ethanol	−2.4	$CH_3CH_2\overset{+}{O}H_2$	CH_3CH_2OH	1.14
Hydronium ion*	−1.7	$H_3\overset{+}{O}$	H_2O	1.14
Nitric acid	−1.4	$HONO_2$	$^-ONO_2$	1.13
Hydrogen sulfate ion	2.0	$^-OSO_3H$	$^-OSO_3^-$	1.14
Hydrogen fluoride	3.1	HF	F^-	1.13
Anilinium ion	4.6	$C_6H_5\overset{+}{N}H_3$	$C_6H_5NH_2$	22.4
Acetic acid	4.7	$CH_3C(=O)OH$	$CH_3C(=O)O^-$	1.13; 1.14; 19.5
Pyridinium ion	5.2	$C_5H_5\overset{+}{N}—H$	C_5H_5N	12.22; 22.4
Carbonic acid	6.4	$HOCO_2H$	$^-OCO_2H$	19.9
Hydrogen sulfide	7.0	H_2S	HS^-	7.19
2,4-Pentanedione	9	$CH_3C(=O)CH_2C(=O)CH_3$	$CH_3C(=O)\overset{-}{C}HC(=O)CH_3$	21.1
Hydrogen cyanide	9.1	$HC\equiv N$	$^-C\equiv N$	7.19
Ammonium ion	9.3	$^+NH_4$	NH_3	22.4
Glycine	9.6	$H_3\overset{+}{N}CH_2CO_2^-$	$H_2NCH_2CO_2^-$	26.3
Phenol	10	C_6H_5OH	$C_6H_5O^-$	1.14
Hydrogen carbonate ion	10.2	$HOCO_2^-$	$^-OCO_2^-$	19.9
Methanethiol	10.7	CH_3SH	CH_3S^-	16.12
Dimethylammonium ion	10.7	$(CH_3)_2\overset{+}{N}H_2$	$(CH_3)_2NH$	22.4

*For acid–base reactions in which water is the solvent, the pK_a of H_3O^+ is zero and the pK_a of H_2O is 14.

TABLE 1.8 Acidity Constants (pK_a) of Acids (*Continued*)

Acid	pK_a	Formula	Conjugate base	Discussed in Section
Ethyl acetoacetate	11			21.1
Piperidinium ion	11.2			22.4
Diethyl malonate	13			21.1
Methanol	15.2	CH_3OH	CH_3O^-	1.13; 15.4
2-Methylpropanal	15.5			21.1
Water*	15.7	H_2O	HO^-	1.13
Ethanol	16	CH_3CH_2OH	$CH_3CH_2O^-$	1.13
Cyclopentadiene	16			12.20
Isopropyl alcohol	17			1.13
tert-Butyl alcohol	18			1.13
Acetone	19			21.1
Ethyl acetate	25.6			21.1
Acetylene	26	$HC{\equiv}CH$	$^-C{\equiv}CH$	9.5
Hydrogen	35	$H{-}H$	H^-	21.8
Ammonia	36	NH_3	H_2N^-	1.13; 1.14; 15.4
Diisopropylamine	36			21.1

continued

*For acid–base reactions in which water is the solvent, the pK_a of H_3O^+ is zero and the pK_a of H_2O is 14.

TABLE 1.8 Acidity Constants (pK_a) of Acids (*Continued*)

Acid	pK_a	Formula	Conjugate base	Discussed in Section
Benzene	43	C_6H_5—H	$C_6H_5:^-$	15.4; 15.5
Ethylene	45	$H_2C{=}CH$—H	$H_2C{=}CH:^-$	9.4; 9.5
Methane	60	H_3C—H	$H_3C:^-$	1.13; 9.4; 9.5; 15.4
Ethane	62	CH_3CH_2—H	$CH_3\ddot{C}H_2^-$	15.5

Web collections of pK_a data include that of H. Reich (University of Wisconsin) at https://www.chem.wisc.edu/areas/reich/pkatable/

*For acid–base reactions in which water is the solvent, the pK_a of H_3O^+ is zero and the pK_a of H_2O is 14.

1.13 How Structure Affects Acid Strength

In this section we'll introduce some generalizations that will permit us to connect molecular structure with acidity in related compounds. The main ways in which structure affects acidity in solution depend on:

1. The strength of the bond to the atom from which the proton is lost
2. The electronegativity of the atom from which the proton is lost
3. Electron delocalization in the conjugate base

Bond Strength. The effect of bond strength is easy to see by comparing the acidities of the hydrogen halides.

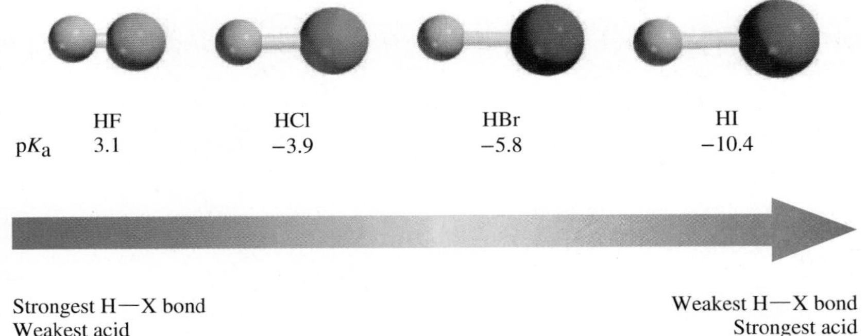

In general, bond strength decreases going down a group in the periodic table. As the halogen X becomes larger, the H—X bond becomes longer and weaker and acid strength increases. This is the dominant factor in the series HCl, HBr, HI and also contributes to the relative weakness of HF.

With HF, a second factor concerns the high charge-to-size ratio of F^-. Other things being equal, processes that give ions in which the electric charge is constrained to a small volume are less favorable than processes in which the charge is more spread out. The strong H—F bond and the high charge-to-size ratio of F^- combine to make HF the weakest acid of the hydrogen halides.

Because of the conjugate relationship between acidity and basicity, the strongest acid (HI) has the weakest conjugate base (I^-), and the weakest acid (HF) has the strongest conjugate base (F^-).

Problem 1.31

Which is the stronger acid, H_2O or H_2S? Which is the stronger base, HO^- or HS^-? Check your predictions against the data in Table 1.8.

Electronegativity. The effect of electronegativity on acidity is evident in the following series involving bonds between hydrogen and the second-row elements C, N, O, and F.

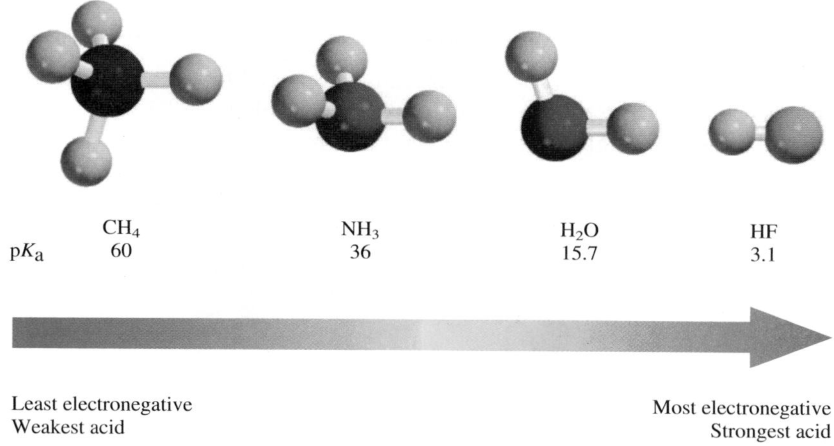

As the atom (A) to which H is bonded becomes more electronegative, the polarization $^{\delta+}H—A^{\delta-}$ becomes more pronounced and the equilibrium constant K_a for proton transfer increases.

$$H_2\ddot{O}: + H—A \rightleftharpoons H_2\overset{+}{\ddot{O}}—H + :A^-$$

Bond strength is more important than electronegativity when comparing elements in the same group of the periodic table as the pK_a's for the hydrogen halides show. Fluorine is the most electronegative and iodine the least electronegative of the halogens, but HF is the weakest acid while HI is the strongest. Electronegativity is the more important factor when comparing elements in the same row of the periodic table.

Problem 1.32

Try to do this problem without consulting Table 1.8.

(a) Which is the stronger acid: $(CH_3)_3\overset{+}{N}H$ or $(CH_3)_2\overset{+}{\ddot{O}}H$?

(b) Which is the stronger base: $(CH_3)_3N:$ or $(CH_3)_2\ddot{O}:$?

Sample Solution (a) The ionizable proton is bonded to N in $(CH_3)_3\overset{+}{N}H$ and to O in $(CH_3)_2\overset{+}{\ddot{O}}H$

$$H—\overset{+}{N}(CH_3)_3 \qquad H—\overset{+}{\ddot{O}}(CH_3)_2$$

Nitrogen and oxygen are in the same row of the periodic table, so their relative electronegativities are the determining factor. Oxygen is more electronegative than nitrogen; therefore, $(CH_3)_2\overset{+}{\ddot{O}}H$ is a stronger acid than $(CH_3)_3\overset{+}{N}H$.

In many acids, the acidic proton is bonded to oxygen. Such compounds can be considered as derivatives of water. Among organic compounds, the ones most closely related to water are alcohols. Most alcohols are somewhat weaker acids than water.

	HO—H	CH_3O—H	CH_3CH_2O—H	$(CH_3)_2CHO$—H	$(CH_3)_3CO$—H
	Water	Methanol	Ethanol	Isopropyl alcohol	*tert*-Butyl alcohol
pK_a	15.7	15.2	16	17	18

Problem 1.33

Which is a stronger base, ethoxide ($CH_3CH_2\ddot{O}:^-$) or *tert*-butoxide [$(CH_3)_3C\ddot{O}:^-$]?

Electronegative atoms in a molecule can affect acidity even when they are not directly bonded to the ionizable proton. Compare ethanol (CH_3CH_2OH) with a related compound in which a CF_3 group replaces the CH_3 group.

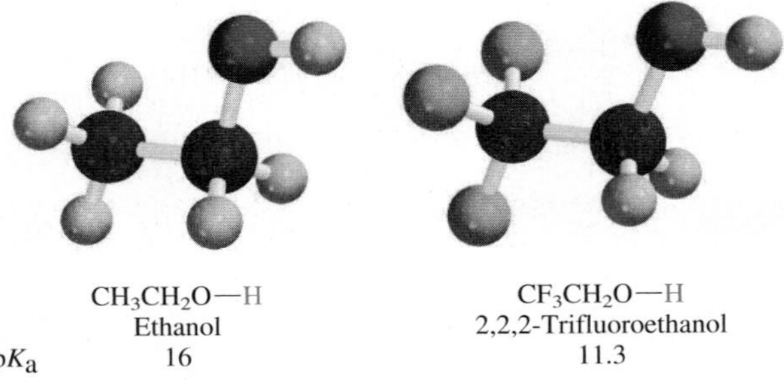

	CH_3CH_2O—H	CF_3CH_2O—H
	Ethanol	2,2,2-Trifluoroethanol
pK_a	16	11.3

We see that the substitution of C—H bonds by C—F increases the acidity of the O—H proton by 4.7 pK_a units, which corresponds to a difference of $10^{4.7}$ in K_a. The simplest explanation for this enhanced acidity is that the electronegative fluorines attract electrons and that this attraction is transmitted through the bonds, increasing the positive character of the O—H proton.

$$F_3C\text{—}CH_2\text{—}O\text{—}H^{\delta+}$$

The greater positive character, hence the increased acidity, of the O—H proton of 2,2,2-trifluoroethanol can be seen in the electrostatic potential maps displayed in Figure 1.9.

We can also explain the greater acidity of CF_3CH_2OH relative to CH_3CH_2OH by referring to the equations for their ionization.

$$X_3CCH_2\ddot{O}\text{—}H + :\ddot{O}H_2 \rightleftharpoons X_3CCH_2\ddot{O}:^- + H\text{—}\overset{+}{O}H_2$$

X = H: Ethanol
X = F: 2,2,2-Trifluoroethanol

X = H: Conjugate base of ethanol
X = F: Conjugate base of 2,2,2-trifluoroethanol

Figure 1.9

Electrostatic potential maps of ethanol and 2,2,2-trifluoroethanol. As indicated by the more blue, less green color in the region near the OH proton in 2,2,2-trifluoroethanol, this proton bears a greater degree of positive charge and is more acidic than the OH proton in ethanol. The color scale is the same in both maps.

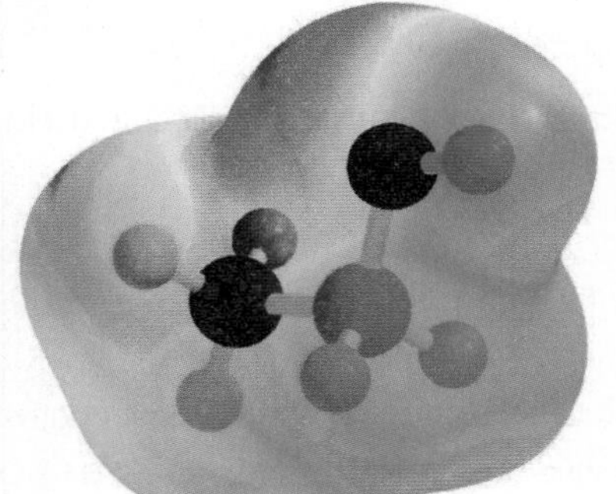

Ethanol (CH_3CH_2OH)

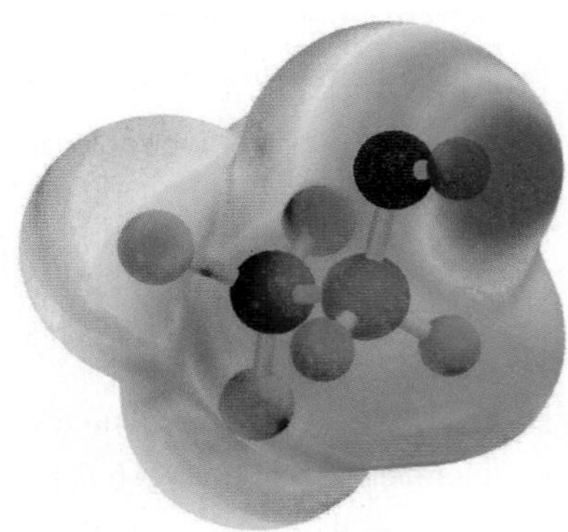

2,2,2-Trifluoroethanol (CF_3CH_2OH)

The conjugate base of 2,2,2-trifluoroethanol, the anion $CF_3CH_2O^-$, is stabilized by its three fluorines, which attract electrons from the negatively charged oxygen, dispersing the negative charge. Because of this stabilization, the equilibrium for ionization of CF_3CH_2OH lies farther to the right than that of CH_3CH_2OH.

Structural effects that are transmitted through bonds are called **inductive effects.** A substituent *induces* a polarization in the bonds between it and some remote site.

The same kind of inductive effects that make CF_3CH_2OH a stronger acid than CH_3CH_2OH makes the trifluoro derivative of acetic acid more than 4 pK_a units stronger than acetic acid.

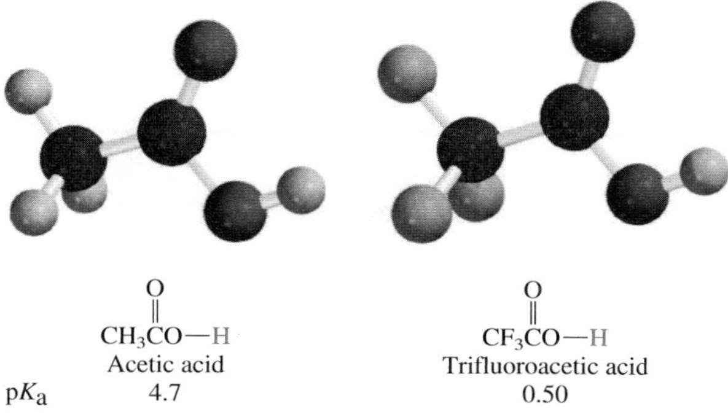

	$CH_3\overset{O}{\overset{\|}{C}}O{-}H$	$CF_3\overset{O}{\overset{\|}{C}}O{-}H$
	Acetic acid	Trifluoroacetic acid
pK_a	4.7	0.50

Problem 1.34

Hypochlorous and hypobromous acid (HOCl and HOBr) are weak acids. Write chemical equations for the ionization of each in water and predict which one is the stronger acid.

Inductive effects depend on the electronegativity of the substituent and the number of bonds between it and the affected site. As the number of bonds between the two units increases, the inductive effect decreases. For instance, the two butanoic acids differ in the placement of chlorines by two bonds and differ by more than 1.5 pK_a units.

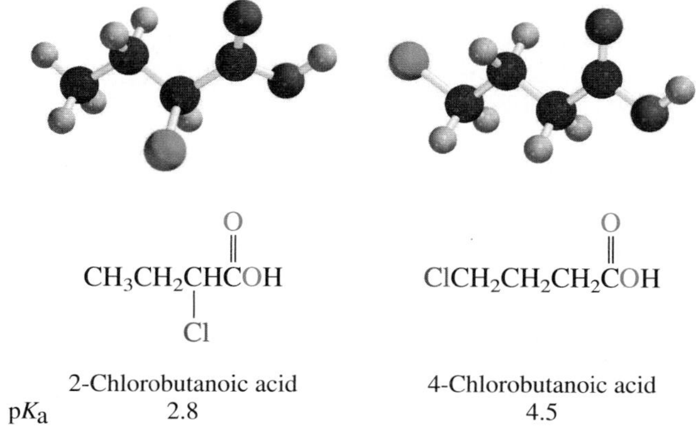

	$CH_3CH_2\underset{Cl}{\underset{\|}{C}}H\overset{O}{\overset{\|}{C}}OH$	$ClCH_2CH_2CH_2\overset{O}{\overset{\|}{C}}OH$
	2-Chlorobutanoic acid	4-Chlorobutanoic acid
pK_a	2.8	4.5

Electron Delocalization in the Conjugate Base. With a pK_a of −1.4, nitric acid is almost completely ionized in water. If we look at the Lewis formula of nitric acid in light of what we have said about inductive effects, we can see why. The N atom in nitric acid is not only electronegative in its own right, but bears a formal charge of +1, which enhances its ability to attract electrons away from the —OH group. But inductive effects are only part of the story. When nitric acid transfers its proton to water, nitrate ion is produced.

$$(O{=})(^-O{-})\overset{+}{N}{-}\ddot{O}{-}H + H{-}\ddot{O}{:}{-}H \rightleftharpoons (O{=})(^-O{-})\overset{+}{N}{-}\ddot{O}{:}^- + H{-}\overset{+}{O}(H){-}H$$

Nitric acid (acid)	Water (base)	Nitrate ion (conjugate base)	Hydronium ion (conjugate acid)

Nitrate ion is stabilized by electron delocalization, which we can represent in terms of resonance between three equivalent contributing structures:

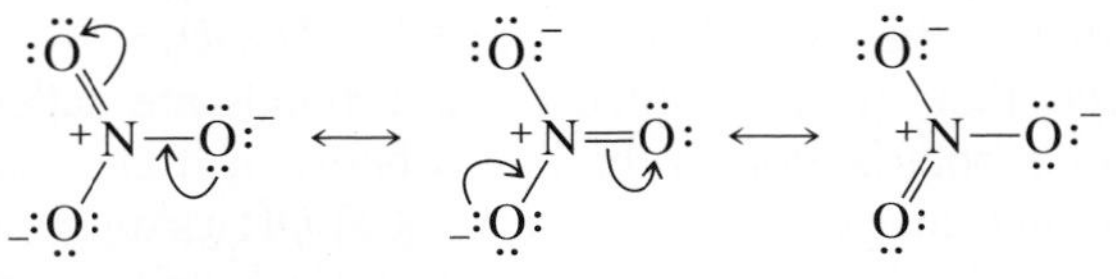

The negative charge is shared equally by all three oxygens. Stabilization of nitrate ion by electron delocalization increases the equilibrium constant for its formation.

Problem 1.35

What is the average formal charge on each oxygen in nitrate ion?

Acetic acid, with a pK_a of 4.7, derives its acidity from the electron delocalization of its conjugate base acetate ion. Acetic acid is over 11 pK_a units lower than ethanol.

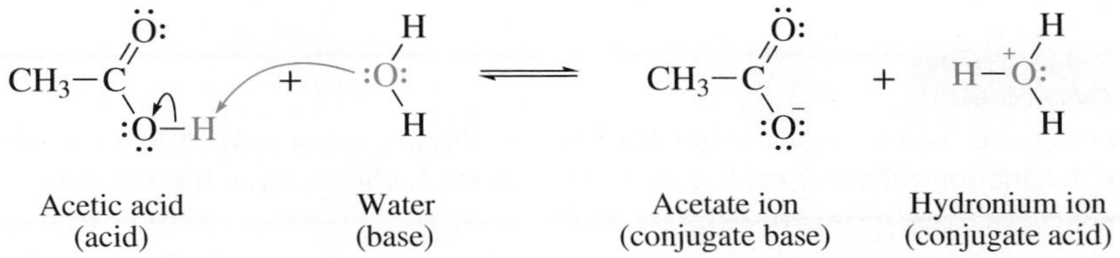

Acetic acid (acid) + Water (base) ⇌ Acetate ion (conjugate base) + Hydronium ion (conjugate acid)

The two equivalent resonance contributors with the negative charge shared equally on both oxygen atoms stabilize this structure.

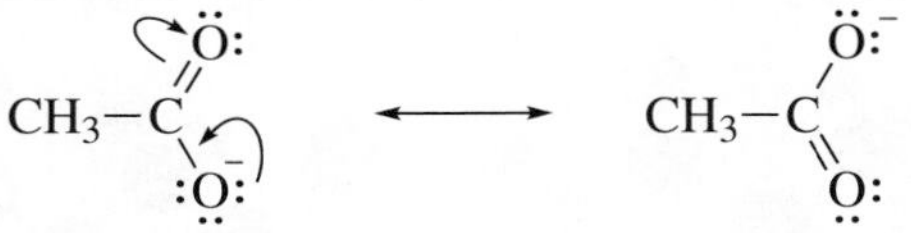

Problem 1.36

Is 3-chlorobutanoic acid more or less acidic than 2-chlorobutanoic acid? Explain your reasoning.

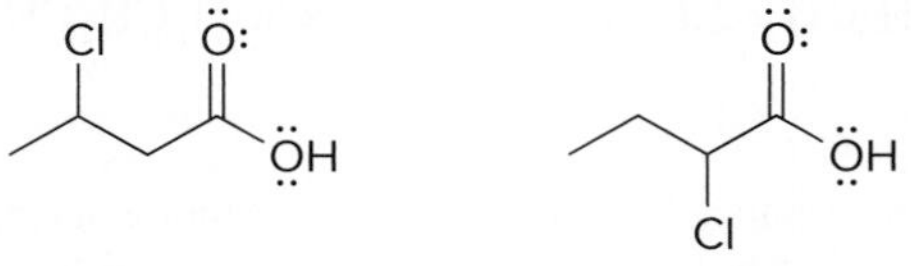

3-Chlorobutanoic acid 2-Chlorobutanoic acid

Problem 1.37

Show by writing appropriate resonance structures that the two compounds shown form the same conjugate base on ionization. Which atom in the conjugate base, O or S, bears the greater share of negative charge?

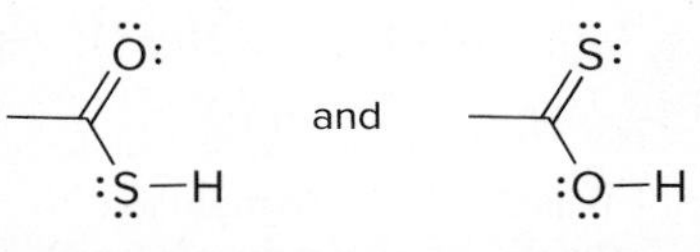

and

Electron delocalization in the conjugate base is also responsible for the increased acidity of phenols compared to alcohols.

$$C_6H_5OH + H_2O \rightleftharpoons C_6H_5O^- + H_3O^+ \qquad pK_a = 10$$

This electron delocalization is represented by resonance among the various contributing structures:

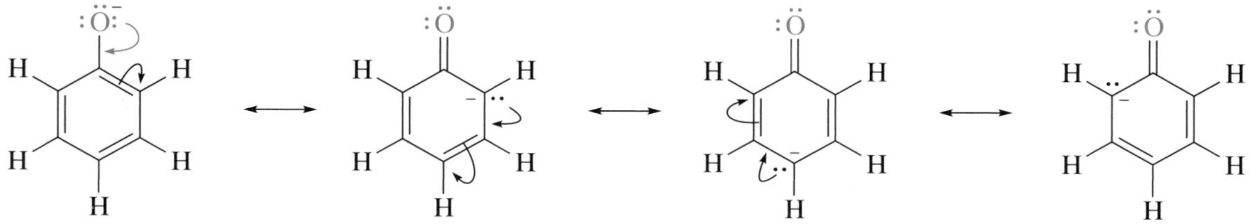

Problem 1.38

Rank the following phenols in order of increasing acidity. Explain your reasoning.

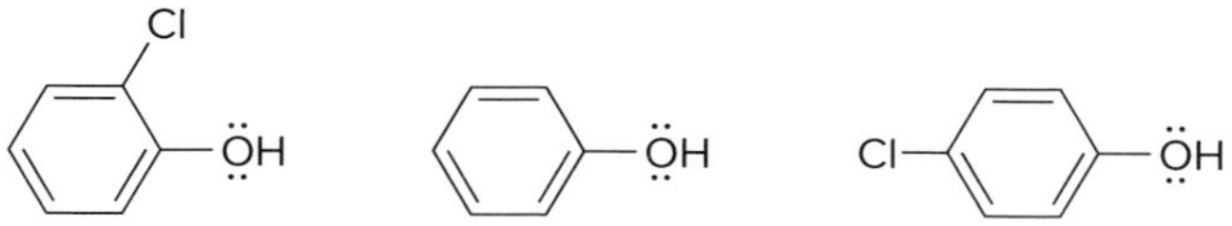

Cl (ortho-chlorophenol) ÖH   phenol ÖH   Cl— (para-chlorophenol) ÖH

Organic chemistry involves a good bit of reasoning by analogy and looking for trends. At the beginning of this section we listed three ways that structure can affect acidity. The last two—electronegativity of the atom from which the proton is lost, and electron delocalization in the conjugate base—are both related to the stability of the conjugate base. A useful trend emerges: *factors that stabilize the conjugate base increase the acidity of the parent acid.*

1.14 Acid–Base Equilibria

In any proton-transfer reaction:

$$\text{Acid} + \text{Base} \rightleftharpoons \text{Conjugate acid} + \text{Conjugate base}$$

we are concerned with the question of whether the position of equilibrium lies to the side of products or reactants. There is an easy way to determine this. The reaction proceeds in the direction that converts the stronger acid and the stronger base to the weaker acid and the weaker base.

$$\text{Stronger acid} + \text{Stronger base} \xrightleftharpoons{K>1} \text{Weaker acid} + \text{Weaker base}$$

This generalization can be stated even more simply. *The reaction will be favorable when the stronger acid is on the left and the weaker acid is on the right.* The equilibrium favors dissociation of the stronger acid.

Consider first the case of adding a strong acid such as HBr to water. The equation for the Brønsted acid–base reaction that occurs between them is:

$$H_2O + HBr \rightleftharpoons H_3O^+ + Br^-$$

Water	Hydrogen bromide	Hydronium ion	Bromide ion
	$pK_a = -5.8$	$pK_a = 0$	
	stronger acid	weaker acid	

For acid–base reactions in which water is the solvent, the pK_a of $H_3O^+ = 0$. See Table 1.8.

We identify the acid on the left and the acid on the right and compare their pK_a's to decide which is stronger. (Remember, the more negative the pK_a, the stronger the acid.)

The acid on the left is HBr, which has a pK_a of −5.8. The acid on the right is H_3O^+, which has a pK_a of 0. The stronger acid (HBr) is on the left and the weaker acid (H_3O^+) is on the right, so the position of equilibrium lies to the right. The equilibrium constant K_{eq} for an acid–base reaction is given by the ratio of the K_a of the reactant acid to the K_a of the product acid.

$$K_{eq} = \frac{K_a \text{ of reactant acid}}{K_a \text{ of product acid}}$$

Since $10^{-pK_a} = K_a$, we rewrite the expression as:

$$K_{eq} = \frac{10^{-pK_a} \text{ of reactant acid}}{10^{-pK_a} \text{ of product acid}}$$

and substitute the pK_a values of HBr and H_3O^+ to calculate K_{eq}.

$$K_{eq} = \frac{10^{5.8}}{10^{0}}$$

This equilibrium constant is so large that we consider HBr to be completely ionized in water.

Compare the reaction of HBr with water to that of acetic acid with water.

Water + Acetic acid (pK_a = 4.7, weaker acid) ⇌ Hydronium ion (pK_a = 0, stronger acid) + Acetate ion

Here, the weaker acid (acetic acid) is on the left and the stronger acid (hydronium ion) is on the right. The equilibrium constant $K_{eq} = 10^{-4.7}$, and the position of equilibrium lies far to the left.

Problem 1.39

What is the equilibrium constant for the following acid–base reactions?

(a) ammonia and acetic acid
(b) fluoride ion and acetic acid
(c) ethanol and hydrobromic acid

Sample Solution (a) Always start with an equation for an acid–base reaction. Ammonia is a Brønsted base and accepts a proton from the —OH group of acetic acid. Ammonia is converted to its conjugate acid, and acetic acid to its conjugate base.

Ammonia + Acetic acid (pK_a = 4.7, stronger acid) ⇌ Ammonium ion (pK_a = 9.3, weaker acid) + Acetate ion

From their respective pK_a's, we see that acetic acid is a much stronger acid than ammonium ion. Therefore, the equilibrium lies to the right. The equilibrium constant for the process is

$$K_{eq} = \frac{10^{-pK_a} \text{ of acetic acid (reactant)}}{10^{-pK_a} \text{ of ammonium ion (product)}} = \frac{10^{-4.7}}{10^{-9.3}} = 10^{4.6}$$

An unexpected fact emerges by working through this exercise. We see that although acetic acid is a weak acid and ammonia is a weak base, the acid–base reaction between them is virtually complete.

Two important points come from using relative pK_a's to analyze acid–base equilibria:

1. They permit clear-cut distinctions between strong and weak acids and bases. *A strong acid is one that is stronger than* H_3O^+. Conversely, *a weak acid* is one that is weaker than H_3O^+.
   Example: The pK_a's for the first and second ionizations of sulfuric acid are −4.8 and 2.0, respectively. Sulfuric acid ($HOSO_2OH$) is a strong acid; hydrogen sulfate ion ($HOSO_2O^-$) is a weak acid.
   A strong base is one that is stronger than HO^-.
   Example: A common misconception is that the conjugate base of a weak acid is strong. This is sometimes, but not always, true. It is true, for example, for ammonia, which is a very weak acid (pK_a 36). Its conjugate base amide ion (H_2N^-) is a much stronger base than HO^-. It is not true, however, for acetic acid; both acetic acid and its conjugate base acetate ion are weak. The conjugate base of a weak acid will be strong only when the acid is a weaker acid than water.
2. The strongest acid present in significant amounts at equilibrium after a strong acid is dissolved in water is H_3O^+. The strongest acid present in significant amounts when a weak acid is dissolved in water is the weak acid itself.
   Example: $[H_3O^+]$ = 1.0 M in a 1.0 M aqueous solution of HBr. The concentration of undissociated HBr molecules is near zero. $[H_3O^+]$ = 0.004 M in a 1.0 M aqueous solution of acetic acid. The concentration of undissociated acetic acid molecules is near 1.0 M. Likewise, HO^- is the strongest base that can be present in significant quantities in aqueous solution.

Problem 1.40

Rank the following in order of decreasing concentration in a solution prepared by dissolving 1.0 mol of sulfuric acid in enough water to give 1.0 L of solution. (It is not necessary to do any calculations.)

$$H_2SO_4,\ HSO_4^-,\ SO_4^{2-},\ H_3O^+$$

Analyzing acid–base reactions according to the Brønsted–Lowry picture provides yet another benefit. Table 1.8, which lists acids according to their strength in descending order along with their conjugate bases, can be used to predict the direction of proton transfer. Acid–base reactions in which a proton is transferred from an acid to a base that lies below it in the table have favorable equilibrium constants. Proton transfers from an acid to a base that lies above it in the table are unfavorable. Thus, the equilibrium constant for proton transfer from phenol to hydroxide ion is greater than 1, but that for proton transfer from phenol to hydrogen carbonate ion is less than 1.

Phenol + Hydroxide ion $\xrightarrow{K > 1}$ Phenoxide ion + Water

Phenol + Hydrogen carbonate ion $\xleftarrow{K < 1}$ Phenoxide ion + Carbonic acid

Hydroxide ion lies below phenol in Table 1.8; hydrogen carbonate ion lies above phenol. The practical consequence of the reactions shown is that NaOH is a strong enough base to convert phenol to phenoxide ion, but $NaHCO_3$ is not.

Problem 1.41

Verify that the position of equilibrium for the reaction between phenol and hydroxide ion lies to the right by comparing the pK_a of the acid on the left to the acid on the right. Which acid is stronger? Do the same for the reaction of phenol with hydrogen carbonate ion.

1.15 Acids and Bases: The Lewis View

The same G. N. Lewis who gave us electron dot formulas also suggested a way to classify acids and bases that is more general than the Brønsted–Lowry approach. Where Brønsted and Lowry viewed acids and bases as donors and acceptors of protons (positively charged), Lewis took the opposite view and focused on electron pairs (negatively charged). According to Lewis, *an acid is an electron-pair acceptor, and a base is an electron-pair donor.*

$$A^+ + :B^- \rightleftharpoons A{-}B$$

Lewis acid Lewis base

An unshared pair of electrons from the Lewis base is used to form a covalent bond between the Lewis acid and the Lewis base. The Lewis acid and the Lewis base are shown as ions in the equation, but they need not be. If both are neutral molecules, the corresponding equation becomes:

$$A + :B \rightleftharpoons \overset{-}{A}{-}\overset{+}{B}$$

Lewis acid Lewis base

We can illustrate this latter case by the reaction:

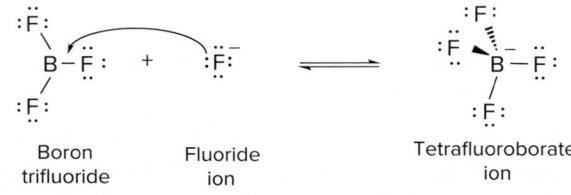

$$BF_3 + (CH_3CH_2)_2\ddot{O}: \rightleftharpoons F_3\overset{-}{B}{-}\overset{+}{O}(CH_2CH_3)_2$$

Boron trifluoride (Lewis acid) Diethyl ether (Lewis base) "Boron trifluoride etherate" (Lewis acid/Lewis base complex)

Verify that the formal charges on boron and oxygen in "boron trifluoride etherate" are correct.

The product of this reaction, a **Lewis acid/Lewis base complex** called informally "boron trifluoride etherate," may look unusual but it is a stable species with properties different from those of the reactants. Its boiling point (126°C), for example, is much higher than that of boron trifluoride—a gas with a boiling point of −100°C—and diethyl ether, a liquid that boils at 34°C.

Problem 1.42

Write an equation for the Lewis acid/Lewis base reaction between boron trifluoride and each of the following. Use curved arrows to track the flow of electrons and show formal charges if present.

(a) Fluoride ion
(b) Dimethyl sulfide [$(CH_3)_2S$]
(c) Trimethylamine [$(CH_3)_3N$]

Sample Solution (a) Fluoride ion has 8 electrons (4 pairs) in its valence shell. It acts as a Lewis base and uses one pair to bond to boron in BF_3.

$$BF_3 + :\ddot{F}:^- \rightleftharpoons BF_4^-$$

Boron trifluoride Fluoride ion Tetrafluoroborate ion

The Lewis acid/Lewis base idea also includes certain **substitution** reactions in which one atom or group replaces another.

$$HO^- + H_3C{-}Br \rightleftharpoons HO{-}CH_3 + Br^-$$

Hydroxide ion (Lewis base) | Bromomethane (Lewis acid) | Methanol | Bromide ion

The carbon atom in bromomethane can accept an electron pair if its covalent bond with bromine breaks with both electrons in that bond becoming an unshared pair of bromide ion. Thus, bromomethane acts as a Lewis acid in this reaction.

Notice the similarity of the preceding reaction to one that is more familiar to us.

$$HO^- + H{-}Br \rightleftharpoons HO{-}H + Br^-$$

Hydroxide ion (Lewis base) | Hydrogen bromide (Lewis acid) | Water | Bromide ion

The two reactions are analogous and demonstrate that the reaction between hydroxide ion and hydrogen bromide is simultaneously a Brønsted–Lowry acid–base reaction and a Lewis acid/Lewis base reaction. *Brønsted–Lowry acid–base reactions constitute a subcategory of Lewis acid/Lewis base reactions.*

Many important biochemical reactions involve Lewis acid/Lewis base chemistry. Carbon dioxide is rapidly converted to hydrogen carbonate ion in the presence of the enzyme *carbonic anhydrase*.

$$HO^- + O{=}C{=}O \xrightarrow[]{\text{carbonic anhydrase}} HO{-}C({=}O){-}O^-$$

Hydroxide ion (Lewis base) | Carbon dioxide (Lewis acid) | Hydrogen carbonate ion

Recall that the carbon atom of carbon dioxide bears a partial positive charge because of the electron-attracting power of its attached oxygens. When hydroxide ion (the Lewis base) bonds to this positively polarized carbon, a pair of electrons in the carbon–oxygen double bond leaves carbon to become an unshared pair of oxygen.

Lewis bases use an unshared pair to form a bond to some other atom and are also referred to as **nucleophiles** ("nucleus seekers"). Conversely, Lewis acids are **electrophiles** ("electron seekers"). We will use these terms hundreds of times throughout the remaining chapters.

Examine the detailed table of contents. What chapters include terms related to *nucleophile* or *electrophile* in their title?

1.16 SUMMARY

Section 1.1 A review of some fundamental knowledge about atoms and electrons leads to a discussion of **wave functions, orbitals,** and the **electron configurations** of atoms. Neutral atoms have as many electrons as the number of protons in the nucleus. These electrons occupy orbitals in order of increasing energy, with no more than two electrons in any one orbital. The most frequently encountered atomic orbitals in this text are *s* orbitals (spherically symmetrical) and *p* orbitals ("dumbbell-shaped").

Boundary surface of a carbon 2*s* orbital

Boundary surface of a carbon 2*p* orbital

Section 1.2 An **ionic bond** is the force of electrostatic attraction between two oppositely charged ions. Atoms at the upper right of the periodic table, especially fluorine and oxygen, tend to gain electrons to form anions. Elements toward the left of the periodic table, especially metals such as sodium, tend to lose electrons to form cations. Ionic bonds in which carbon is the cation or anion are rare.

Section 1.3 The most common kind of bonding involving carbon is **covalent bonding:** the sharing of a pair of electrons between two atoms. **Lewis formulas** are written on the basis of the **octet rule,** which limits second-row elements to no more than eight electrons in their valence shells. In most of its compounds, carbon has four bonds. Many organic compounds have **double** or **triple bonds** to carbon. Four electrons are involved in a double bond, six in a triple bond.

```
    H   H                 H       H
    |   |                  \     /
H—C—C—Ö—H                  C=C           H—C≡C—H
    |   |                  /     \
    H   H                 H       H
```

Each carbon has four bonds in ethyl alcohol; oxygen and each carbon are surrounded by eight electrons.

Ethylene has a carbon–carbon double bond.

Acetylene has a carbon–carbon triple bond.

Section 1.4 When two atoms that differ in **electronegativity** are covalently bonded, the electrons in the bond are drawn toward the more electronegative element.

$$\overset{\delta+}{C}\overset{\longrightarrow}{—}\overset{\delta-}{F}$$

The electrons in a carbon–fluorine bond are drawn away from carbon, toward fluorine.

Section 1.5 Counting electrons and assessing charge distribution in molecules is essential to understanding how structure affects properties. A particular atom in a Lewis formula may be neutral, positively charged, or negatively charged. The **formal charge** of an atom in the Lewis formula of a molecule can be calculated by comparing its electron count with that of the neutral atom itself.

$$\text{Formal charge} = \text{Group number in periodic table} - \text{Electron count}$$

where

$$\text{Electron count} = \tfrac{1}{2}(\text{Number of shared electrons}) + \text{Number of unshared electrons}$$

Section 1.6 Table 1.5 in this section sets forth the procedure to be followed in writing Lewis formulas for organic molecules. It begins with experimentally determined information: the **molecular formula** and the **connectivity** (order in which the atoms are connected).

```
    H  :O:
    |   ‖
H—C—C—Ö—H
    |
    H
```

The Lewis formula of acetic acid

Different compounds that have the same molecular formula are called **isomers.** If they are different because their atoms are connected in a different order, they are called **constitutional isomers.**

```
:Ö        H      H
  \\     /        \
    C—N:           C=N̈
  /      \        /    \
H         H      H     :Ö—H
```

Formamide (*left*) and formaldoxime (*right*) are constitutional isomers; both have the same molecular formula (CH_3NO), but the atoms are connected in a different order.

Condensed formulas and line formulas are used to economize the drawing of organic structures.

$CH_3CH_2CH_2CH_2OH$
or
$CH_3(CH_2)_2CH_2OH$

condensed formula

line formula

Section 1.7 Many molecules can be represented by two or more Lewis formulas that differ only in the placement of electrons. In such cases the electrons are delocalized, and the real electron distribution is a hybrid of the **contributing structures.** The rules of resonance are summarized in Table 1.6.

Two Lewis structures (resonance contributors) of formamide; the atoms are connected in the same order, but the arrangement of the electrons is different.

Section 1.8 The octet rule can be exceeded for second-row elements. Resonance contributors in which sulfur contains 10 or 12 electrons in its valence shell are permissible, as are phosphorus compounds with 10 electrons. Familiar examples include sulfuric acid and phosphoric acid.

Section 1.9 The shapes of molecules can often be predicted on the basis of **valence shell electron-pair repulsions.** A tetrahedral arrangement gives the maximum separation of four electron pairs (*left*); a trigonal planar arrangement is best for three electron pairs (*center*), and a linear arrangement for two electron pairs (*right*).

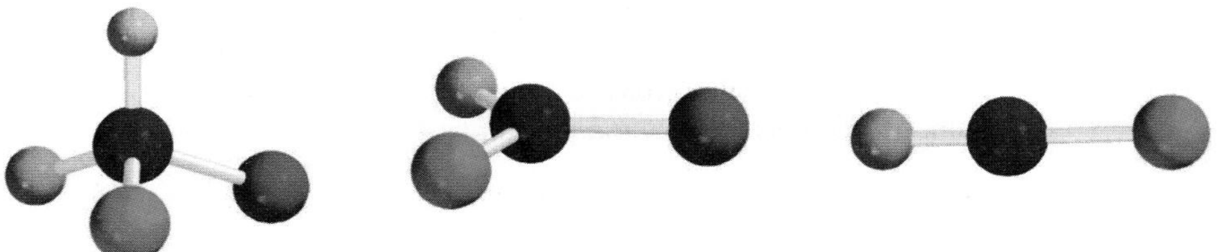

Section 1.10 Knowing the shape of a molecule and the polarity of its various bonds allows the presence or absence of a **molecular dipole moment** and its direction to be predicted.

Both water and carbon dioxide have polar bonds, but water has a dipole moment while carbon dioxide does not.

Section 1.11 Curved arrows increase the amount of information provided by a chemical equation by showing the flow of electrons associated with bond making and bond breaking. In the process:

$$:\ddot{Br}-CH_3 \quad :NH_3 \longrightarrow :\ddot{Br}:^- + H_3C-\overset{+}{N}H_3$$

an electron pair of nitrogen becomes the pair of electrons in a C—N bond. The C—Br bond breaks, with the pair of electrons in that bond becoming an unshared pair of bromide ion.

Section 1.12 According to the Brønsted–Lowry definitions, an acid is a **proton donor** and a base is a **proton acceptor.**

$$\text{B:} + \text{H—A} \rightleftharpoons \overset{+}{\text{B}}\text{—H} + \text{:A}^-$$

Base Acid Conjugate acid Conjugate base

The strength of an acid is given by its equilibrium constant K_a for ionization in aqueous solution:

$$K_a = \frac{[H_3O^+][:A^-]}{[HA]}$$

or more conveniently by its pK_a:

$$pK_a = -\log_{10}K_a$$

Section 1.13 The strength of an acid depends on the atom to which the proton is bonded. Two important factors are the strength of the H—X bond and the electronegativity of X. Bond strength is more important for atoms in the same group of the periodic table; electronegativity is more important for atoms in the same row. Electronegative atoms elsewhere in the molecule can increase the acidity by **inductive effects.**

Electron **delocalization** in the conjugate base, usually expressed via resonance between contributing Lewis formulas, increases acidity by stabilizing the conjugate base.

Section 1.14 The position of equilibrium in an acid–base reaction lies to the side of the weaker acid.

$$\text{Stronger acid} + \text{Stronger base} \xrightleftharpoons{K > 1} \text{Weaker acid} + \text{Weaker base}$$

This is a very useful relationship. You should practice writing equations according to the Brønsted–Lowry definitions of acids and bases and familiarize yourself with Table 1.8, which gives the pK_a's of various Brønsted acids.

Section 1.15 The Lewis definitions of acids and bases provide for a more general view of acid–base reactions than the Brønsted–Lowry picture. A **Lewis acid** is an electron-pair acceptor. A **Lewis base** is an electron-pair donor. The Lewis approach incorporates the Brønsted–Lowry approach as a subcategory in which the atom that accepts the electron pair in the Lewis acid is a hydrogen.

PROBLEMS

Structural Formulas

1.43 Write a Lewis formula for each of the following organic molecules:

(a) C_2H_3Cl (vinyl chloride: starting material for the preparation of PVC plastics)

(b) $C_2HBrClF_3$ (halothane: a nonflammable inhalation anesthetic; all three fluorines are bonded to the same carbon)

(c) $C_2Cl_2F_4$ (Freon 114: formerly used as a refrigerant and as an aerosol propellant; each carbon bears one chlorine)

1.44 Write structural formulas for all the constitutionally isomeric compounds having the given molecular formula.

(a) C_4H_{10} (c) $C_2H_4Cl_2$ (e) C_3H_9N

(b) C_5H_{12} (d) C_4H_9Br

1.45 Write structural formulas for all the constitutional isomers of

(a) C_3H_8 (b) C_3H_6 (c) C_3H_4

1.46 Write structural formulas for all the constitutional isomers of molecular formula C_3H_6O that contain

(a) Only single bonds (b) One double bond

1.47 Expand the following structural representations so as to more clearly show all the atoms and any unshared electron pairs. What are their molecular formulas? Are any of them isomers?

(a) 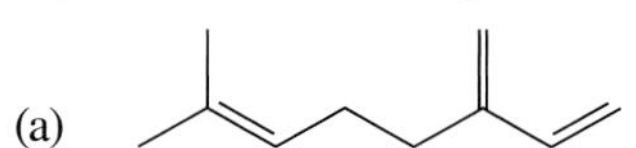

Occurs in bay and verbena oil

(d)

Found in Roquefort cheese

(b) 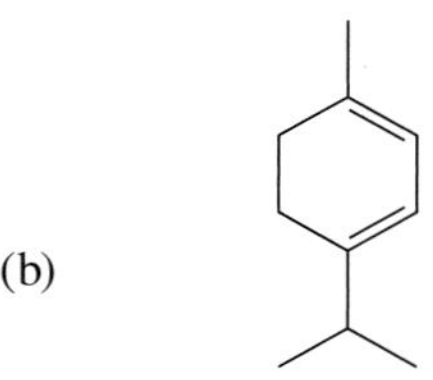

Pleasant-smelling substance found in marjoram oil

(e) 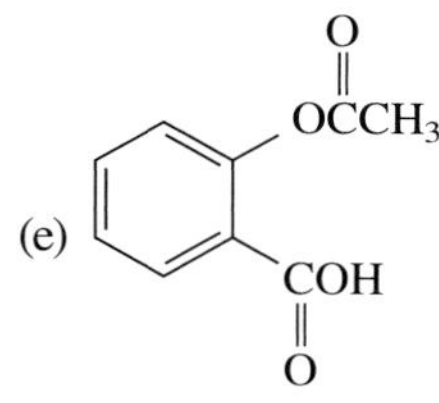

Aspirin

(c)

Present in oil of cloves

(f) 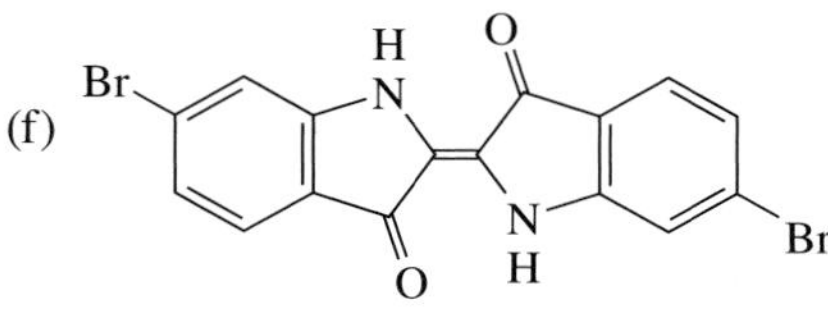

Tyrian purple: a purple dye extracted from a species of Mediterranean sea snail

Formal Charge and Resonance

1.48 Each of the following species will be encountered at some point in this text. They all have the same number of electrons binding the same number of atoms and the same arrangement of bonds; they are *isoelectronic*. Specify which atoms, if any, bear a formal charge in the Lewis formula given and the net charge for each species.

(a) :N≡N: (c) :C≡C: (e) :C≡O:

(b) :C≡N: (d) :N≡O:

1.49 Consider Lewis formulas A, B, and C:

$$\underset{A}{H_2\ddot{C}—N≡N:} \qquad \underset{B}{H_2C=N=\ddot{N}:} \qquad \underset{C}{H_2C—\ddot{N}=\ddot{N}:}$$

(a) Are A, B, and C constitutional isomers, or are they resonance contributors?

(b) Which have a negatively charged carbon?

(c) Which have a positively charged carbon?

(d) Which have a positively charged nitrogen?

(e) Which have a negatively charged nitrogen?

(f) What is the net charge on each?

(g) Which is a more stable structure, A or B? Why?

(h) Which is a more stable structure, B or C? Why?

(i) What is the CNN geometry in each according to VSEPR?

1.50 In each of the following pairs, determine whether the two represent resonance contributors of a single species or depict different substances. If two structures are not resonance contributors, explain why.

(a) $:\ddot{\underset{..}{N}}—N≡N:$ and $:\ddot{N}=N=\ddot{N}:$

(b) $:\ddot{\underset{..}{N}}—N≡N:$ and $:\ddot{N}—N=\ddot{N}:$

(c) $:\ddot{\underset{..}{N}}—N≡N:$ and $:\ddot{\underset{..}{N}}—\ddot{\underset{..}{N}}—\ddot{\underset{..}{N}}:$

1.51 (a) Which one of the following is *not* a permissible contributing structure? Why?

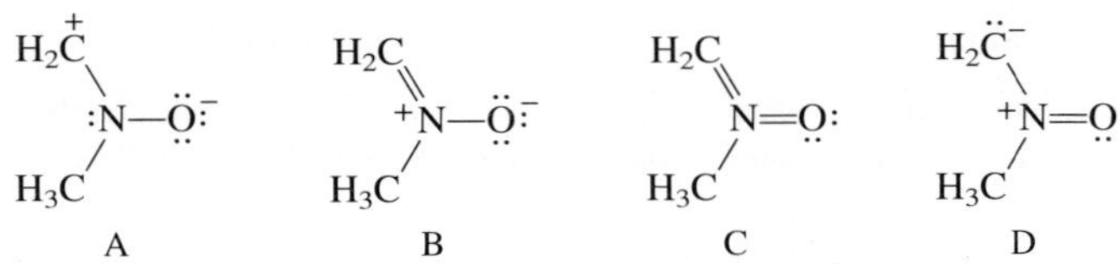

(b) Rank the three remaining structures in order of their contribution to the resonance hybrid. Explain your reasoning.

(c) Using curved arrows, show the electron movement that connects the three resonance contributors.

1.52 Of two possible structures A and B for the conjugate acid of guanidine, the more stable is the one that is better stabilized by electron delocalization. Which one is it? Write resonance structures showing this electron delocalization.

:NH
‖
$H_2\ddot{N}$—C—$\ddot{N}H_2$
Guanidine

:NH
‖
$H_2\ddot{N}$—C—$\overset{+}{N}H_3$
A

$^{+}NH_2$
‖
$H_2\ddot{N}$—C—$\ddot{N}H_2$
B

1.53 Write a more stable contributing structure for each of the following. Use curved arrows to show how to transform the original Lewis formula to the new one. Be sure to specify formal charges, if any.

(a) $H_3C—\ddot{N}=\overset{+}{N}:$

(b) H—C(—:Ö:⁻)(=$\overset{+}{O}$—H)

(c) $H_2\overset{+}{C}—\overset{-}{\ddot{C}}H_2$

(d) $H_2\overset{+}{C}—CH=CH—\overset{-}{\ddot{C}}H_2$

(e) $H_2\overset{+}{C}—CH=CH—\ddot{O}:^{-}$

(f) $H_2\overset{-}{\ddot{C}}$—C(=Ö:)—H

(g) $H—\overset{+}{C}=\ddot{O}:$

(h) $H_2\overset{+}{C}—\ddot{O}H$

(i) $H_2\overset{-}{\ddot{C}}—\ddot{N}=\overset{+}{N}H_2$

Dipole Moment

1.54 The connectivity of carbon oxysulfide is OCS.

(a) Write a Lewis formula for carbon oxysulfide that satisfies the octet rule.

(b) What is the molecular geometry according to VSEPR?

(c) Does carbon oxysulfide have a dipole moment? If so, what is its direction?

1.55 For each of the following molecules that contain polar covalent bonds, indicate the positive and negative ends of the dipole, using the symbol ⟷. Refer to Table 1.3 as needed.

(a) HCl
(b) HI
(c) H_2O
(d) HOCl

1.56 The compounds FCl and ICl have dipole moments μ that are similar in magnitude (0.9 and 0.7 D, respectively) but opposite in direction. In one compound, chlorine is the positive end of the dipole; in the other it is the negative end. Specify the direction of the dipole moment in each compound, and explain your reasoning.

1.57 Which compound in each of the following pairs would you expect to have the greater dipole moment μ? Why?

(a) HF or HCl
(b) HF or BF_3
(c) $(CH_3)_3CH$ or $(CH_3)_3CCl$
(d) $CHCl_3$ or CCl_3F
(e) CH_3NH_2 or CH_3OH
(f) CH_3NH_2 or CH_3NO_2

Acids and Bases

1.58 With a pK_a of 11.6, hydrogen peroxide is a stronger acid than water. Why?

1.59 The structure of montelukast, an antiasthma drug, is shown here.

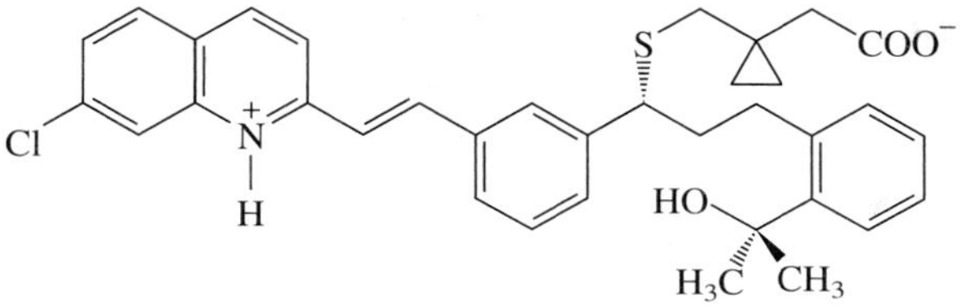

(a) Use Table 1.8 to identify the most acidic and most basic sites in the molecule. (Although you won't find an exact match in structure, make a prediction based on analogy with similar groups in simpler molecules.)

(b) Write the structure of the product formed by treating montelukast with one equivalent of sodium hydroxide.

(c) Write the structure of the product formed by treating montelukast with one equivalent of HCl.

1.60 (a) One acid has a pK_a of 2, the other has a pK_a of 8. What is the ratio of their K_a's?

(b) Two acids differ by a factor of 10,000 in their K_a's. If the pK_a of the weaker acid is 5, what is the pK_a of the stronger acid?

1.61 Calculate K_a for each of the following acids, given its pK_a. Rank the compounds in order of decreasing acidity.

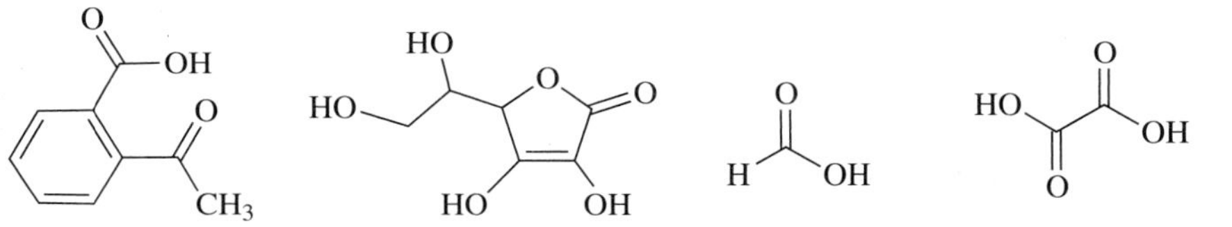

Aspirin
pK_a = 3.48

Vitamin C
pK_a = 4.17
(ascorbic acid)

Formic acid
pK_a = 3.75
(present in sting of ants)

Oxalic acid
pK_a = 1.19
(poisonous substance in certain berries)

1.62 Rank the following in order of decreasing acidity. Although none of these specific structures appear in Table 1.8, you can use analogous structures in the table to guide your reasoning.

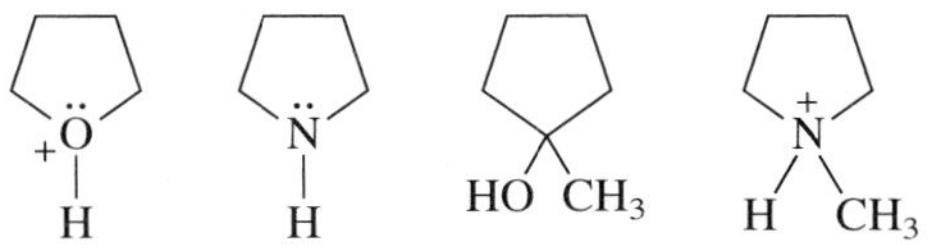

1.63 Rank the following in order of decreasing basicity. As in the preceding problem, Table 1.8 should prove helpful.

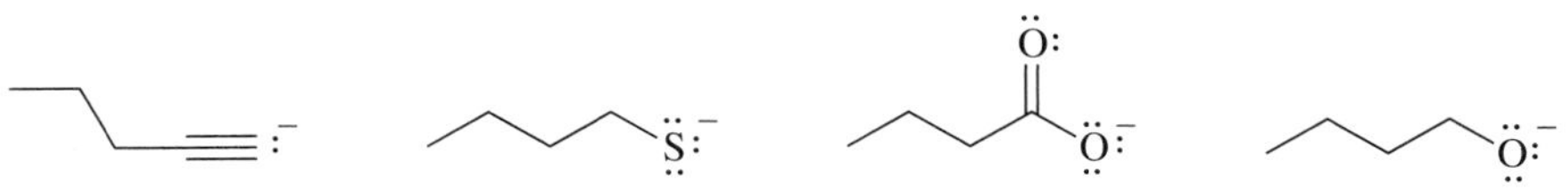

1.64 Consider 1.0 M aqueous solutions of each of the following. Which solution is more basic?

(a) Sodium cyanide (NaCN) or sodium fluoride (NaF)

(b) Sodium carbonate (Na_2CO_3) or sodium acetate ($CH_3\overset{O}{\overset{\|}{C}}ONa$)

(c) Sodium sulfate (Na_2SO_4) or sodium methanethiolate ($NaSCH_3$)

1.65 Write an equation for the Brønsted–Lowry acid–base reaction that occurs when each of the following acids reacts with water. Show all unshared electron pairs and formal charges, and use curved arrows to track electron movement.

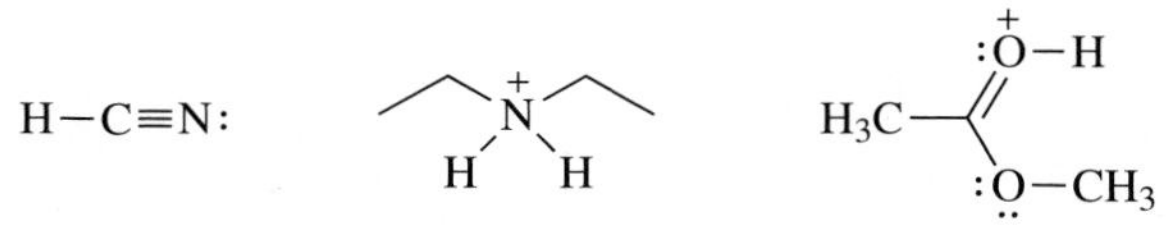

(a) H−C≡N: (b) [diethylammonium ion, $\overset{+}{N}$ with H, H]   (c) H_3C–C(=$\overset{+}{\ddot{O}}$–H)(–$\ddot{O}$–CH_3)

1.66 Write an equation for the Brønsted–Lowry acid–base reaction that occurs when each of the following bases reacts with water. Show all unshared electron pairs and formal charges, and use curved arrows to track electron movement.

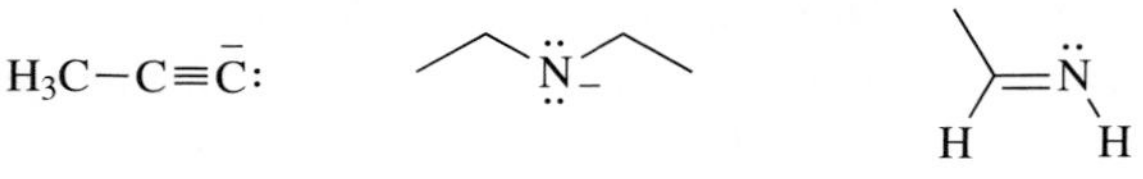

(a) $H_3C-C\equiv\bar{C}:$ (b) [diethylamide anion, $\ddot{N}^-$]   (c) [imine =$\ddot{N}$, with H, H]

1.67 All of the substances shown in the following acid–base reactions are found in Table 1.8, and the equilibrium lies to the right in each case. Following the curved arrows, complete each equation to show the products formed. Identify the acid, base, conjugate acid, and conjugate base. Calculate the equilibrium constant for each reaction.

(a) 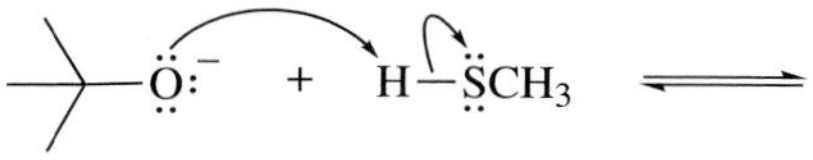 [tert-butoxide, $-\ddot{O}:^-$] + H−$\ddot{S}CH_3$ ⇌

(b) 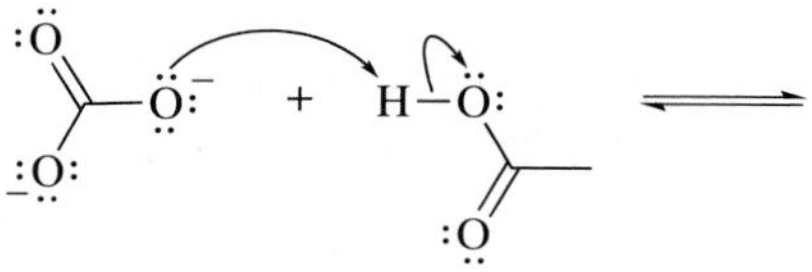 [carbonate-type anion with :Ö and :Ö:⁻, $-\ddot{O}:^-$] + H−$\ddot{O}$: [acetic acid, :O] ⇌

(c) 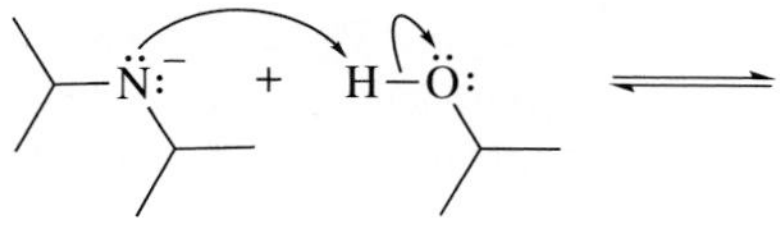 [diisopropylamide, $\ddot{N}:^-$] + H−$\ddot{O}$: [isopropanol] ⇌

(d) 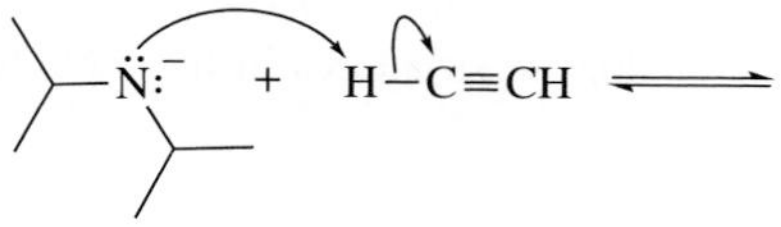 [diisopropylamide, $\ddot{N}:^-$] + H−C≡CH ⇌

1.68 Each of the following acid–base reactions involves substances found in Table 1.8. Use the pK_a data in the table to help you predict the products of the reactions. Use curved arrows to show electron flow. Predict whether the equilibrium lies to the left or to the right and calculate the equilibrium constant for each reaction.

(a) HC≡CH + $^-:\ddot{N}H_2$ ⇌

(b) HC≡CH + $^-:\ddot{O}CH_3$ ⇌

(c) 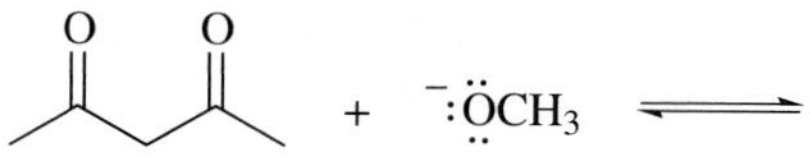 [2,4-pentanedione, O O] + $^-:\ddot{O}CH_3$ ⇌

(d) 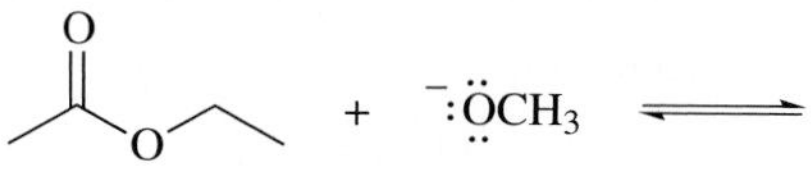 [ethyl acetate, O, O] + $^-:\ddot{O}CH_3$ ⇌

(e) 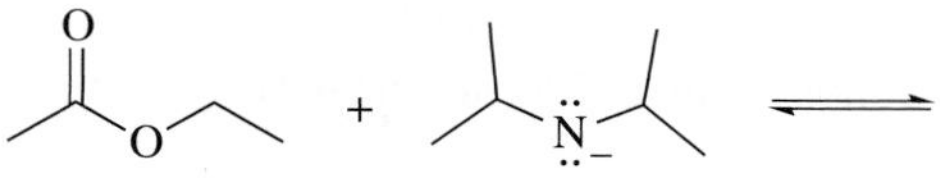 [ethyl acetate, O, O] + [diisopropylamide, $\ddot{N}^-$] ⇌

1.69 With a pK_a of 1.2, squaric acid is unusually acidic for a compound containing only C, H, and O.

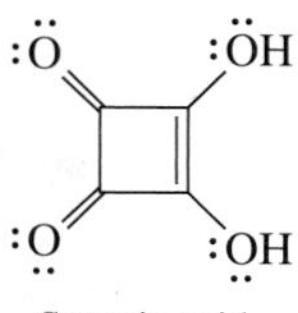

Squaric acid

Write a Lewis formula for the conjugate base of squaric acid and, using curved arrows, show how the negative charge is shared by two oxygens.

1.70 What are the products of the following reaction based on the electron flow represented by the curved arrows? Which compound is the Lewis acid? Which is the Lewis base?

$$(H_2N)_2C{=}S \; + \; H_3C{-}Br \longrightarrow$$

Descriptive Passage and Interpretive Problems 1

Amide Lewis Structural Formulas

Lewis formulas are the major means by which structural information is communicated in organic chemistry. These structural formulas show the atoms, bonds, location of unshared pairs, and formal charges.

Two or more Lewis formulas, differing only in the placement of electrons, can often be written for a single compound. In such cases the separate structures represented by the Lewis formulas are said to be in *resonance,* and the true electron distribution is a *hybrid* of the electron distributions of the *contributing structures.*

The amide function is an important structural unit in peptides and proteins. Formamide, represented by the Lewis structure shown, is the simplest amide. It is a planar molecule with a dipole moment of 3.7 D. Lewis structures I–IV represent species that bear some relationship to the Lewis structure for formamide.

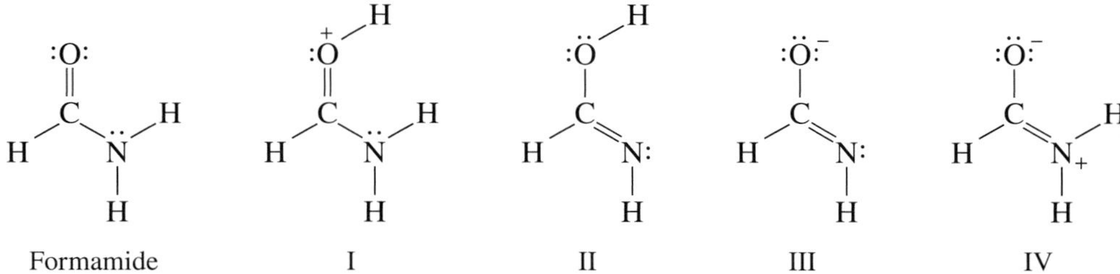

Formamide I II III IV

1.71 Formamide is a planar molecule. According to VSEPR, does the structural formula given for formamide satisfy this requirement?

A. Yes

B. No

1.72 Which Lewis formula is both planar according to VSEPR and a resonance contributor of formamide?

A. I C. III

B. II D. IV

1.73 According to VSEPR, which Lewis formula has a pyramidal arrangement of bonds to nitrogen?

A. I C. III

B. II D. IV

1.74 Which Lewis formula is a constitutional isomer of formamide?

A. I C. III

B. II D. IV

1.75 Which Lewis formula is a conjugate acid of formamide?

A. I C. III

B. II D. IV

1.76 Which Lewis formula is a conjugate base of formamide?

A. I C. III

B. II D. IV

Chapter

2

H H
H—C—C—H
H H

CHAPTER OUTLINE

In our solar system, only Earth and Titan, a moon of Saturn, are known to possess lakes. The spacecraft *Cassini* arrived at Saturn in 2004 and stayed operational until September 2017. *Cassini* and its European-built probe *Huygens,* which landed on Titan's surface in 2005, revealed that instead of lakes of water, Titan's low temperature of −179°C allows for lakes of methane and ethane. The scientific community continues to discuss the origins of hydrocarbons on Titan.

(Left): Source: NASA/JPL-Caltech/USGS; *(right):* Source: NASA/JPL-Caltech/Space Science Institute

Alkanes and Cycloalkanes: Introduction to Hydrocarbons

This chapter continues the connection between structure and properties begun in Chapter 1. In it we focus on the simplest organic compounds—those that contain only carbon and hydrogen, called *hydrocarbons.* These compounds occupy a key position in the organic chemical landscape. Their framework of carbon–carbon bonds provides the scaffolding on which more reactive *functional groups* are attached. We'll introduce functional groups in a preliminary way in this chapter but will have much more to say about them beginning in Chapter 5.

By focusing on hydrocarbons, we'll expand our picture of bonding by introducing two approaches that grew out of the idea that electrons can be described as waves: the *valence bond* and *molecular orbital* models. In particular, one aspect of the valence bond model, called *orbital hybridization,* will be emphasized.

A major portion of this chapter deals with how we name organic compounds. The system used throughout the world is based on a set of rules for naming hydrocarbons, then extending the rules to include compounds that contain functional groups.

2.1 Classes of Hydrocarbons

Hydrocarbons are divided into two main classes: aliphatic and aromatic. This classification dates from the nineteenth century, when organic chemistry was devoted almost entirely to the study of materials from natural sources, and terms were coined that reflected a substance's origin. Two sources were fats and oils, and the word *aliphatic* was derived from the Greek word *aleiphar* meaning "fat." Aromatic hydrocarbons, irrespective of their own odor, were typically obtained by chemical treatment of pleasant-smelling plant extracts.

Aliphatic hydrocarbons include three major groups: *alkanes, alkenes,* and *alkynes.* **Alkanes** are hydrocarbons in which all the bonds are single bonds, **alkenes** contain at least one carbon–carbon double bond, and **alkynes** contain at least one carbon–carbon triple bond. Examples of the three classes of aliphatic hydrocarbons are the two-carbon compounds *ethane, ethylene,* and *acetylene*. Another name for aromatic hydrocarbons is **arenes.** The most important aromatic hydrocarbon is *benzene*.

$H_3C{-}CH_3$ — Ethane (alkane)

$H_2C{=}CH_2$ — Ethylene (alkene)

$H{-}C{\equiv}C{-}H$ — Acetylene (alkyne)

C_6H_6 — Benzene (arene)

Different properties in these hydrocarbons are the result of the different types of bonding involving carbon. The shared electron pair, or Lewis model of chemical bonding described in Section 1.3, does not account for all of the differences. In the following sections, we will consider two additional bonding theories: the valence bond model and molecular orbital model.

2.2 Electron Waves and Chemical Bonds

G. N. Lewis proposed his shared electron-pair model of bonding in 1916, almost a decade before Louis de Broglie's theory of wave–particle duality. De Broglie's radically different view of an electron, and Erwin Schrödinger's success in using wave equations to calculate the energy of an electron in a hydrogen *atom,* encouraged the belief that bonding in *molecules* could be explained on the basis of interactions between electron waves. This thinking produced two widely used theories of chemical bonding; one is called the *valence bond model,* the other the *molecular orbital model.*

> De Broglie's and Schrödinger's contributions to our present understanding of electrons were described in Section 1.1.

Before we describe these theories in the context of organic molecules, let's first think about bonding between two hydrogen atoms in the most fundamental terms. We'll begin with two hydrogen atoms that are far apart and see what happens as the distance between them decreases. The forces involved are electron–electron (−−) repulsions, nucleus–nucleus (++) repulsions, and electron–nucleus (−+) attractions. All of these forces *increase* as the distance between the two hydrogens *decreases*. Because the electrons are so mobile, however, they can choreograph their motions so as to minimize their mutual repulsion while maximizing their attractive forces with the protons. Thus, as shown in Figure 2.1, a net, albeit weak, attractive force exists between the two hydrogens even when the atoms are far apart. This interaction becomes stronger as the two atoms approach each other—the electron of each hydrogen increasingly feels the attractive force of two protons rather than one, the total energy decreases, and the system becomes more stable. A potential energy minimum is reached when the separation between the nuclei reaches 74 pm (0.74 Å), which corresponds to the H—H bond length in H_2. At distances shorter than this, nucleus–nucleus and electron–electron repulsions dominate, and the system becomes less stable.

> All of the forces in chemistry, except for nuclear chemistry, are electrical. Opposite charges attract; like charges repel. This simple fact can take you a long way.

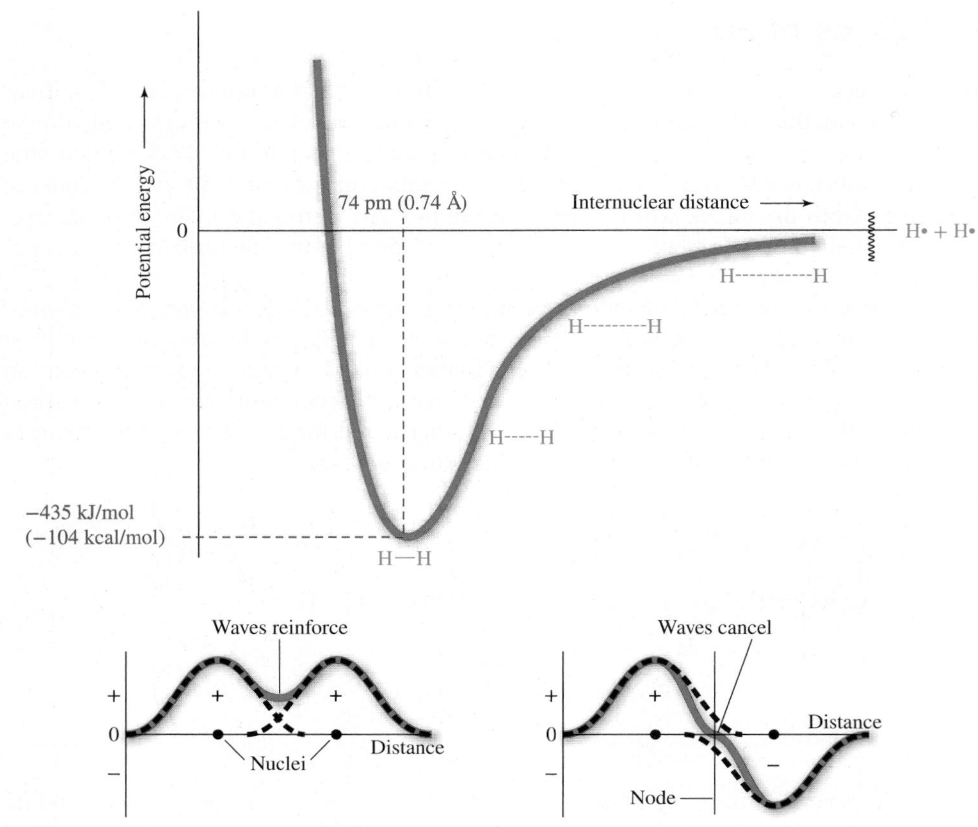

Figure 2.1

Plot of potential energy versus distance for two hydrogen atoms. At long distances, there is a weak attractive force. As the distance decreases, the potential energy decreases, and the system becomes more stable because each electron now "feels" the attractive force of two protons rather than one. The lowest energy state corresponds to a separation of 74 pm (0.74 Å), which is the normal bond distance in H_2. At shorter distances, nucleus–nucleus and electron–electron repulsions are greater than electron–nucleus attractions, and the system becomes less stable.

Figure 2.2

Interference between waves. (*a*) Constructive interference occurs when two waves combine in phase with each other. The amplitude of the resulting wave at each point is the sum of the amplitudes of the original waves. (*b*) Destructive interference decreases the amplitude when two waves are out of phase with each other.

(*a*) Amplitudes of wave functions added

(*b*) Amplitudes of wave functions subtracted

Valence bond and molecular orbital theory both incorporate the wave description of an atom's electrons into this picture of H_2, but in somewhat different ways. Both assume that electron waves behave like more familiar waves, such as sound and light waves. One important property of waves is called interference in physics. *Constructive interference* occurs when two waves combine so as to reinforce each other (in phase); *destructive interference* occurs when they oppose each other (out of phase) (Figure 2.2).

Recall from Section 1.1 that electron waves in atoms are characterized by their wave function, which is the same as an orbital. For an electron in the most stable state of a hydrogen atom, for example, this state is defined by the 1*s* wave function and is often called the 1*s* orbital. The *valence bond* model bases the connection between two atoms on the overlap between half-filled orbitals of the two atoms. The *molecular orbital* model assembles a set of molecular orbitals by combining the atomic orbitals of *all* of the atoms in the molecule.

For a molecule as simple as H_2, valence bond and molecular orbital theory produce very similar pictures. The next two sections describe these two approaches.

2.3 Bonding in H_2: The Valence Bond Model

The characteristic feature of **valence bond theory** is that it pictures a covalent bond between two atoms in terms of an in-phase overlap of a half-filled orbital of one atom with a half-filled orbital of the other, illustrated for the case of H_2 in Figure 2.3. Two hydrogen atoms, each containing an electron in a 1*s* orbital, combine so that their orbitals overlap to give a new orbital associated with both of them. In-phase orbital overlap (constructive interference) increases the probability of finding an electron in the region between the two nuclei where it feels the attractive force of both of them. Electrostatic potential maps show this build-up of electron density in the region between two hydrogen atoms as they approach each other closely enough for their orbitals to overlap.

Figure 2.3

Valence bond picture of bonding in H_2 as illustrated by electrostatic potential maps. The 1s orbitals of two hydrogen atoms overlap to give an orbital that contains both electrons of an H_2 molecule.

(*a*) The 1*s* orbitals of two separated hydrogen atoms, sufficiently far apart so that essentially no interaction takes place between them. Each electron is associated with only a single proton.

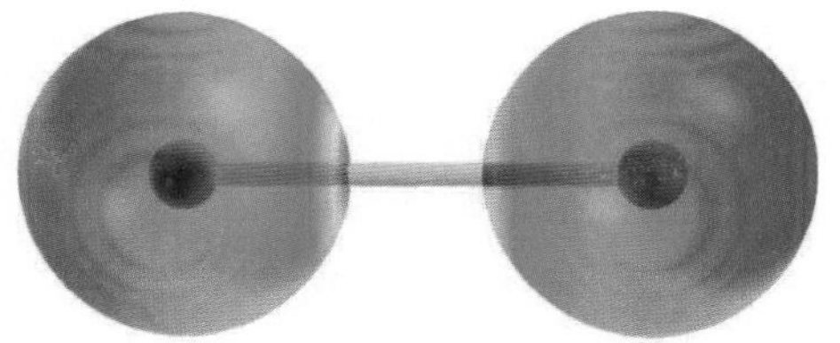

(*b*) As the hydrogen atoms approach each other, their 1*s* orbitals begin to overlap and each electron begins to feel the attractive force of both protons.

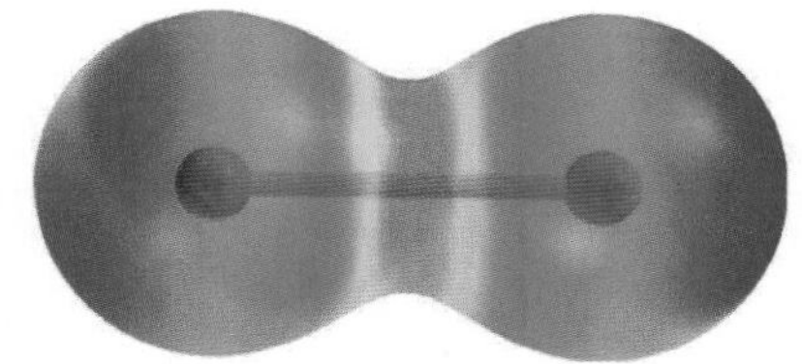

(*c*) The hydrogen atoms are close enough so that appreciable overlap of the two 1*s* orbitals occurs. The concentration of electron density in the region between the two protons is more readily apparent.

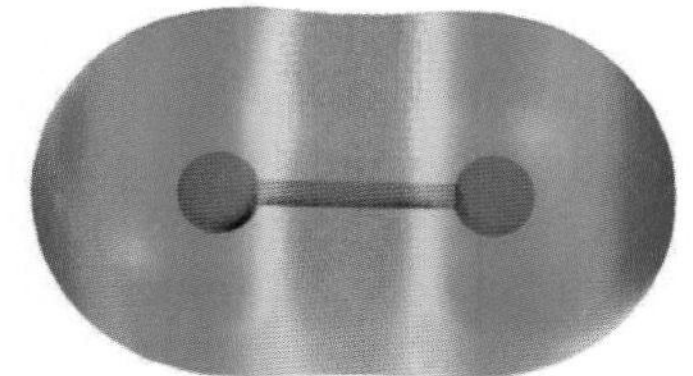

(*d*) A molecule of H_2. The center-to-center distance between the hydrogen atoms is 74 pm (0.74 Å). The two individual 1*s* orbitals have been replaced by a new orbital that encompasses both hydrogens and contains both electrons. The electron density is greatest in the region between the two hydrogens.

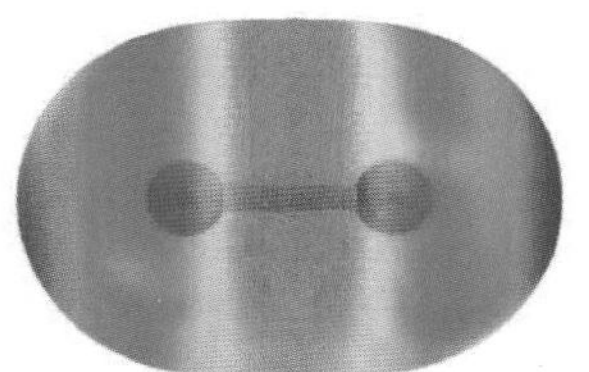

A bond in which the orbitals overlap along a line connecting the atoms (the *internuclear axis*) is called a **sigma (σ) bond.** The electron distribution in a σ bond is cylindrically symmetrical; were we to slice through a σ bond perpendicular to the internuclear axis, its cross section would appear as a circle. Another way to see the shape of the electron distribution is to view the molecule end-on.

Orbitals overlap along a line connecting the two atoms

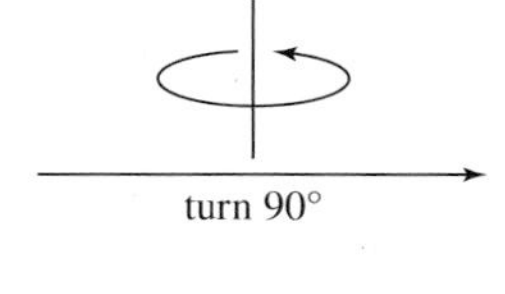

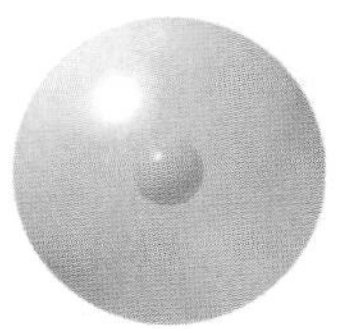

Circular electron distribution when viewing down the H—H bond

We will use the valence bond approach extensively in our discussion of organic molecules and expand on it shortly. First though, let's introduce the molecular orbital method to see how it uses the 1*s* orbitals of two hydrogen atoms to generate the orbitals of an H_2 molecule.

2.4 Bonding in H_2: The Molecular Orbital Model

The molecular orbital theory of chemical bonding rests on the notion that, as electrons in atoms occupy *atomic orbitals,* electrons in molecules occupy *molecular orbitals*. Just as our first task in writing the electron configuration of an atom is to identify the atomic orbitals that are available to it, so too must we first describe the orbitals available to a molecule. In the molecular orbital method this is done by representing molecular orbitals

(*a*) Add the 1*s* wave functions of two hydrogen atoms to generate a bonding molecular orbital (σ) of H_2. There is a high probability of finding both electrons in the region between the two nuclei.

(*b*) Subtract the 1*s* wave function of one hydrogen atom from the other to generate an antibonding molecular orbital (σ*) of H_2. There is a nodal plane where there is a zero probability of finding the electrons in the region between the two nuclei.

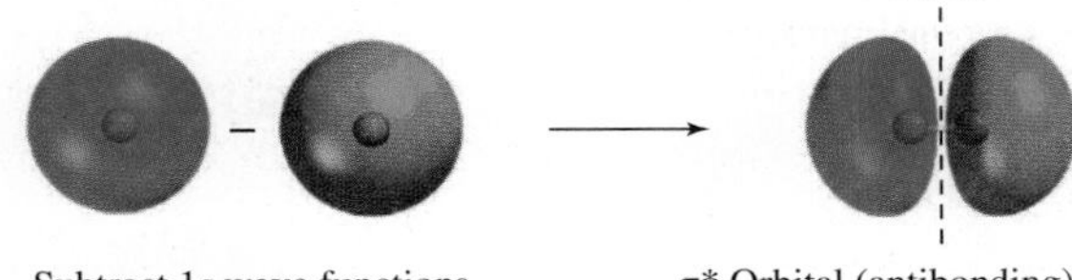

Figure 2.4

Generation of σ and σ* molecular orbitals of H_2 by combining 1s orbitals of two hydrogen atoms.

as combinations of atomic orbitals, the *linear combination of atomic orbitals-molecular orbital* (LCAO-MO) method.

Two molecular orbitals (MOs) of H_2 are generated by combining the 1*s* atomic orbitals (AOs) of two hydrogen atoms. In one combination, the two wave functions are added; in the other they are subtracted. The two new orbitals that are produced are portrayed in Figure 2.4. The additive combination generates a **bonding orbital;** the subtractive combination generates an **antibonding orbital.** Both the bonding and antibonding orbitals have σ symmetry, meaning that they are symmetrical with respect to the internuclear axis. The two are differentiated by calling the bonding orbital σ and the antibonding orbital σ* ("sigma star"). The bonding orbital is characterized by a region of high electron probability between the two atoms, whereas the antibonding orbital has a nodal plane between them.

A molecular orbital diagram for H_2 is shown in Figure 2.5. The customary format shows the starting AOs at the left and right sides and the MOs in the middle. It must always be true that *the number of MOs is the same as the number of AOs that combine to produce them.* Thus, when the 1*s* AOs of two hydrogen atoms combine, two MOs result. The bonding MO (σ) is lower in energy and the antibonding MO (σ*) higher in energy than either of the original 1*s* orbitals.

When assigning electrons to MOs, the same rules apply as for writing electron configurations of atoms. Electrons fill the MOs in order of increasing orbital energy, and the maximum number of electrons in any orbital is two. Both electrons of H_2 occupy the bonding orbital and have opposite spins, and both are held more strongly than they would be in separated hydrogen atoms. There are no electrons in the antibonding orbital.

For a molecule as simple as H_2, it is hard to see much difference between the valence bond and molecular orbital methods. The most important differences appear in molecules with more than two atoms. In those cases, the valence bond method continues to view a molecule as a collection of bonds between connected atoms. The molecular orbital method, however, leads to a picture in which the same electron can be associated with many, or even all, of the atoms in a molecule. We'll have more to say about the

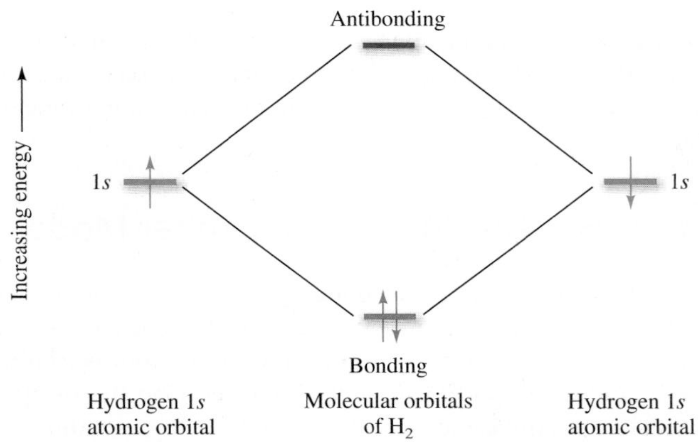

Figure 2.5

Two molecular orbitals (MOs) are generated by combining two hydrogen 1s atomic orbitals (AOs). The bonding MO is lower in energy than either of the AOs that combine to produce it. The antibonding MO is of higher energy than either AO. Each arrow indicates one electron, and the electron spins are opposite in sign. Both electrons of H_2 occupy the bonding MO.

similarities and differences in valence bond and molecular orbital theory as we continue to develop their principles, beginning with the simplest alkanes: methane, ethane, and propane.

Problem 2.1

Construct a diagram similar to Figure 2.5 for diatomic helium (He_2). Why is helium monatomic instead of diatomic?

2.5 Introduction to Alkanes: Methane, Ethane, and Propane

Alkanes have the general molecular formula C_nH_{2n+2}. The simplest one, methane (CH_4), is also the most abundant. Large amounts are present in our atmosphere, in the ground, and in the oceans. Methane has been found on Mars, Jupiter, Saturn, Uranus, Neptune, and Pluto, on Halley's Comet, even in the atmosphere of a planet in a distant solar system. About 2–8% of the atmosphere of Titan, Saturn's largest moon, is methane.

Ethane (C_2H_6: CH_3CH_3) and propane (C_3H_8: $CH_3CH_2CH_3$) are second and third, respectively, to methane in many ways. Ethane is the alkane next to methane in structural simplicity, followed by propane. Ethane (≈10%) is the second and propane (≈5%) the third most abundant component of natural gas, which is ≈75% methane. Natural gas is colorless and nearly odorless, as are methane, ethane, and propane. The characteristic odor of the natural gas we use for heating our homes and cooking comes from trace amounts of unpleasant-smelling sulfur-containing compounds, called thiols, that are deliberately added to it to warn us of potentially dangerous leaks.

Methane is the lowest boiling alkane, followed by ethane, then propane.

	CH_4	CH_3CH_3	$CH_3CH_2CH_3$
	Methane	Ethane	Propane
Boiling point:	−160°C	−89°C	−42°C

Boiling points cited in this text are at 1 atm (760 mm Hg) unless otherwise stated.

All the alkanes with four or fewer carbons are gases at room temperature and atmospheric pressure. With the highest boiling point of the three, propane is the easiest one to liquefy. We are all familiar with "propane tanks." These are steel containers in which a propane-rich mixture of hydrocarbons called *liquefied petroleum gas* (LPG) is maintained in a liquid state under high pressure as a convenient clean-burning fuel.

It is generally true that as the number of carbon atoms increases, so does the boiling point. The C_{70}-alkane heptacontane [$CH_3(CH_2)_{68}CH_3$] boils at 653°C, and its C_{100} analog hectane at 715°C.

The structural features of methane, ethane, and propane are summarized in Figure 2.6. All of the carbon atoms have four bonds, all of the bonds are single bonds, and the bond angles are close to tetrahedral. In the next section we'll see how to adapt the valence bond model to accommodate the observed structures.

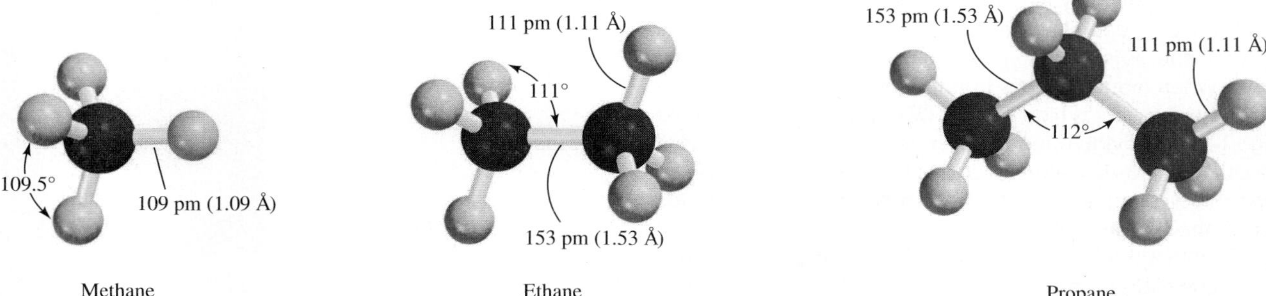

Figure 2.6

Structures of methane, ethane, and propane showing bond distances and bond angles.

2.6 sp^3 Hybridization and Bonding in Methane

Before we describe the bonding in methane, it is worth emphasizing that bonding theories attempt to describe a molecule on the basis of its component atoms; bonding theories do not attempt to explain *how* bonds form. The world's methane does *not* come from the reaction of carbon atoms with hydrogen atoms; it comes from biological processes.

Methane and the Biosphere

One of the things that environmental scientists do is to keep track of important elements in the biosphere—in what form do these elements normally occur, to what are they transformed, and how are they returned to their normal state? Careful studies have given clear, although complicated, pictures of the "nitrogen cycle," the "sulfur cycle," and the "phosphorus cycle," for example. The "carbon cycle" begins and ends with atmospheric carbon dioxide. It can be represented in an abbreviated form as:

$$CO_2 + H_2O + \text{energy} \underset{\text{respiration}}{\overset{\text{photosynthesis}}{\rightleftharpoons}} \text{carbohydrates}$$

carbohydrates → naturally occurring substances of numerous types

naturally occurring substances of numerous types $\xrightarrow{\text{respiration}}$ $CO_2 + H_2O + \text{energy}$

Methane is one of literally millions of compounds in the carbon cycle, but one of the most abundant. It is formed when carbon-containing compounds decompose in the absence of air (*anaerobic* conditions). The organisms that bring this about are called *methanoarchaea*. Cells can be divided into three types: *archaea, bacteria,* and *eukarya*. Methanoarchaea convert carbon-containing compounds, including carbon dioxide and acetic acid, to methane. Virtually anywhere water contacts organic matter in the absence of air is a suitable place for methanoarchaea to thrive—at the bottom of ponds, bogs, rice fields, even on the ocean floor. They live inside termites and grass-eating animals; one source quotes 20 L/day as the methane output of a large cow.

The scale on which the world's methanoarchaea churn out methane, estimated to be 10^{11}–10^{12} lb/year, is enormous. About 10% of this amount makes its way into the atmosphere, but most of the rest simply ends up completing the carbon cycle. It exits the anaerobic environment where it was formed and enters the aerobic world where it is eventually converted to carbon dioxide. But not all of it. Much of the world's methane lies trapped beneath the Earth's surface. *Firedamp*, an explosion hazard to coal miners, is mostly methane, as is the *natural gas* that accompanies petroleum deposits. When methane leaks from petroleum under the ocean floor and the pressure is high enough (50 atm) and the water cold enough (4°C), individual methane molecules become trapped inside clusters of 6–18 water molecules as *methane clathrates* or *methane hydrates* (Figure 2.7). Aggregates of these hydrates remain at the bottom of the ocean in what looks like a lump of dirty ice, ice that burns (Figure 2.8). Far from being mere curiosities, methane hydrates are potential sources of energy on a scale greater than that of all the known oil reserves combined. The extraction of methane from hydrates has been demonstrated on a small scale, and estimates suggest some modest contribution to the global energy supply by 2020.

Methane hydrates contributed to the 2010 environmental disaster in the Gulf of Mexico in an unexpected and important way. Because the hydrates are stable only under the extreme conditions of pressure and temperature found in the deep ocean, their effect on the methods used to repair damage to the oil rigs proved difficult to anticipate and their ice-like properties interfered with attempts to cap the flow of oil in its early stages.

In a different vein, environmental scientists are looking into the possibility that methane hydrates contributed to a major global warming event that occurred 55 million years ago, lasted 40,000 years, and raised the temperature of the Earth some 5°C. They speculate that a modest warming of the oceans encouraged the dissociation of hydrates, releasing methane into the atmosphere. Methane is a potent greenhouse gas, and the resulting greenhouse effect raised the temperature of the Earth. This, in turn, caused more methane to be released from the oceans into the atmosphere, causing more global warming. Eventually a new, warmer equilibrium state was reached.

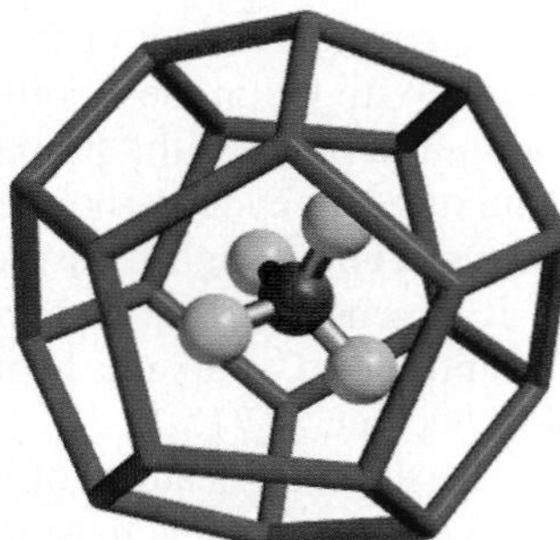

Figure 2.7

In a hydrate a molecule of methane is surrounded by a cage of hydrogen-bonded water molecules. The cages are of various sizes; the one shown here is based on a dodecahedron. Each vertex corresponds to one water molecule, and the lines between them represent hydrogen bonds (O—H·····O).

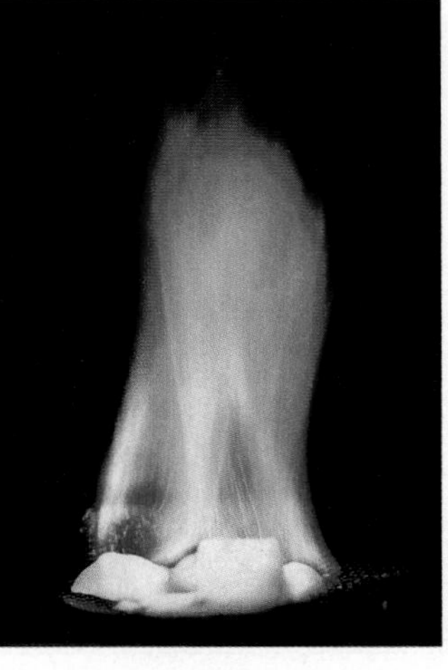

Figure 2.8

Methane burning as it is released from a clathrate.

Source: Photo by J. Pinkston and L. Stern, U.S. Geological Survey

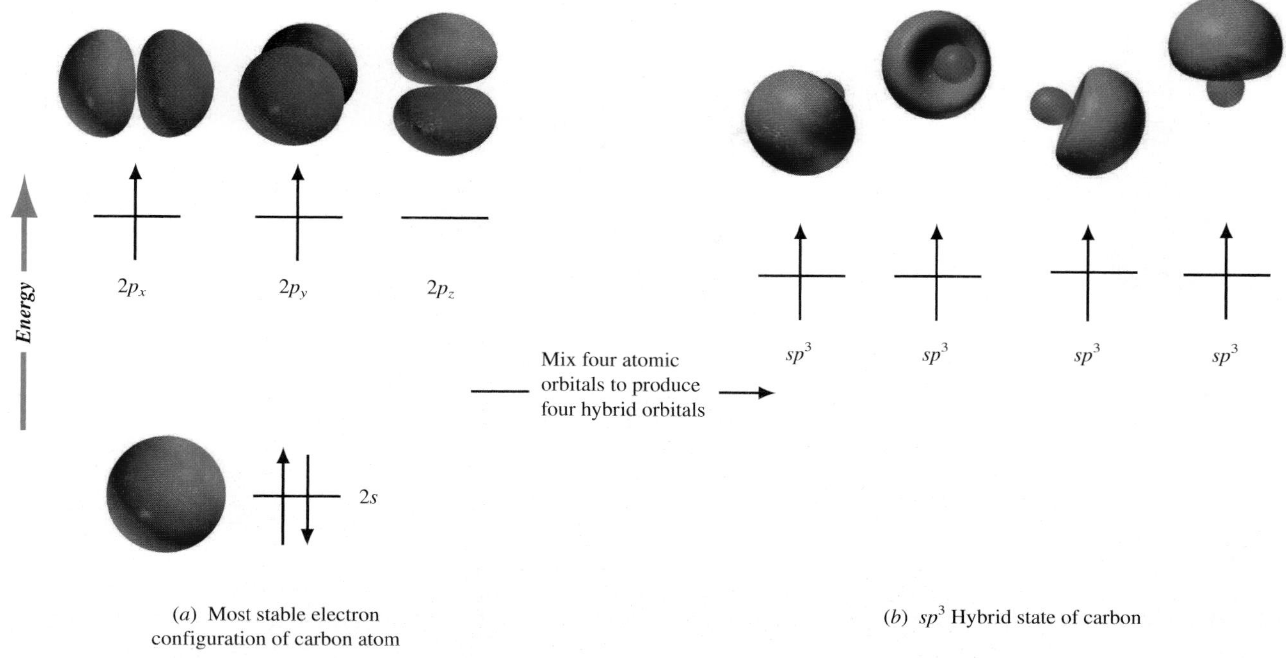

Figure 2.9

sp^3 Hybridization. (*a*) Electron configuration of carbon in its most stable state. (*b*) Mixing the σ orbital with the three *p* orbitals generates four sp^3 hybrid orbitals. The four sp^3 hybrid orbitals are of equal energy; therefore, the four valence electrons are distributed evenly among them. The axes of the four sp^3 orbitals are directed toward the corners of a tetrahedron.

The boxed essay *Methane and the Biosphere* tells you more about the origins of methane and other organic compounds.

We *begin* with the experimentally determined three-dimensional structure of a molecule, *then* propose bonding models that are consistent with the structure. We do not claim that the observed structure is a result of the bonding model. Indeed, there may be two or more equally satisfactory models. Structures are facts; bonding models are theories that we use to try to understand the facts.

A vexing puzzle in the early days of valence bond theory concerned the fact that methane is CH_4 and that the four bonds to carbon are directed toward the corners of a tetrahedron. Valence bond theory is based on the in-phase overlap of half-filled orbitals of the connected atoms. But with an electron configuration of $1s^22s^22p_x^12p_y^1$ carbon has only two half-filled orbitals (Figure 2.9*a*). How, then, can it have four bonds?

In the 1930s Linus Pauling offered an ingenious solution to this puzzle. He suggested that the electron configuration of a carbon bonded to other atoms need not be the same as that of a free carbon atom. By mixing ("hybridizing") the 2*s*, $2p_x$, $2p_y$, and $2p_z$ orbitals, four new orbitals are obtained (Figure 2.9*b*). These four new orbitals are called **sp^3 hybrid orbitals** because they come from one *s* orbital and three *p* orbitals. Each sp^3 hybrid orbital has 25% *s* character and 75% *p* character. Among their most important features are the following:

1. *All four* sp^3 *orbitals are of equal energy*. Therefore, according to Hund's rule (Section 1.1) the four valence electrons of carbon are distributed equally among them, making four half-filled orbitals available for bonding.
2. *The axes of the* sp^3 *orbitals point toward the corners of a tetrahedron*. Therefore, sp^3 hybridization of carbon is consistent with the tetrahedral structure of methane. Each C—H bond is a σ bond in which a half-filled 1*s* orbital of hydrogen overlaps with a half-filled sp^3 orbital of carbon along a line drawn between them (Figure 2.10).

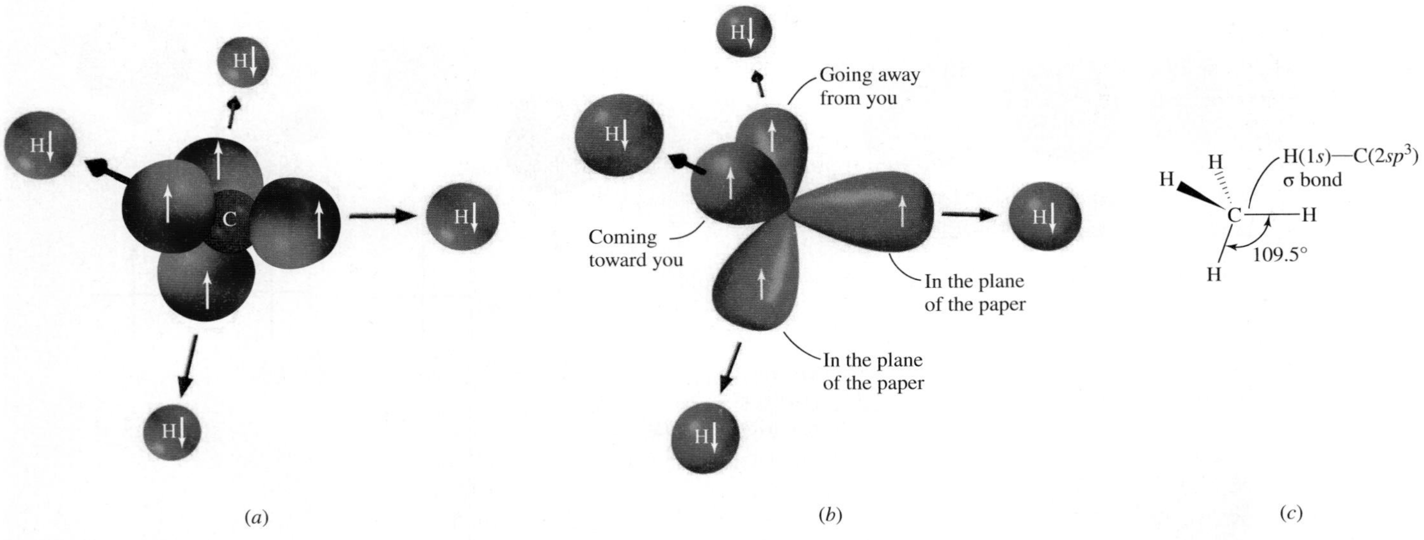

Figure 2.10

Only the major orbital lobe of each sp^3 orbital is shown in (*a*) and in a commonly utilized stylized version (*b*). Each half-filled sp^3 orbital overlaps with a half-filled hydrogen 1s orbital along a line between them giving a tetrahedral arrangement of four σ bonds. (*c*) All bond angles are 109.5°.

3. *σ Bonds involving* sp^3 *hybrid orbitals of carbon are stronger than those involving unhybridized* 2s *or* 2p *orbitals.* Each sp^3 hybrid orbital has two lobes of unequal size, making the electron density greater on one side of the nucleus than the other. In a C—H σ bond, it is the larger lobe of a carbon sp^3 orbital that overlaps with a hydrogen 1*s* orbital. This concentrates the electron density in the region between the two atoms.

The orbital hybridization model accounts for carbon having four bonds rather than two, the bonds are stronger than they would be in the absence of hybridization, and they are arranged in a tetrahedral fashion around carbon.

2.7 Bonding in Ethane

The orbital hybridization model of covalent bonding is readily extended to carbon–carbon bonds. As Figure 2.11 illustrates, ethane is described in terms of a carbon–carbon σ bond joining two CH_3 **(methyl)** groups. Each methyl group consists of an sp^3-hybridized carbon attached to three hydrogens by sp^3–1*s* σ bonds. Overlap of the remaining half-filled sp^3 orbital of one carbon with that of the other generates a σ bond between them. Here is a third kind of σ bond, one that has as its basis the overlap of two half-filled sp^3-hybridized orbitals. *In general, you can expect that carbon will be* sp^3*-hybridized when it is directly bonded to four atoms.*

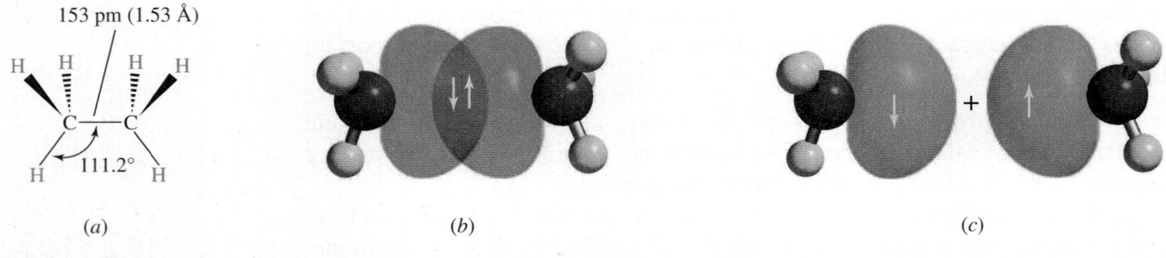

Figure 2.11

(*a*) The carbons in ethane are tetrahedral. The C—C bond length is 153 pm (1.53 Å). (*b*) The C—C σ bond of ethane is viewed as a combination of two half-filled sp^3 orbitals (*c*).

Problem 2.2

Describe the bonding in propane according to the orbital hybridization model.

In the next few sections we'll examine the application of the valence bond-orbital hybridization model to alkenes and alkynes, then return to other aspects of alkanes in Section 2.12. We'll begin with ethylene.

2.8 sp^2 Hybridization and Bonding in Ethylene

Ethylene is planar with bond angles close to 120° (Figure 2.12); therefore, some hybridization state other than sp^3 is required. The hybridization scheme is determined by the number of atoms to which carbon is directly attached. In sp^3 hybridization, four atoms are attached to carbon by σ bonds, and so four equivalent sp^3 hybrid orbitals are required. In ethylene, three atoms are attached to each carbon, so three equivalent hybrid orbitals are needed. As shown in Figure 2.13, these three orbitals are generated by mixing the

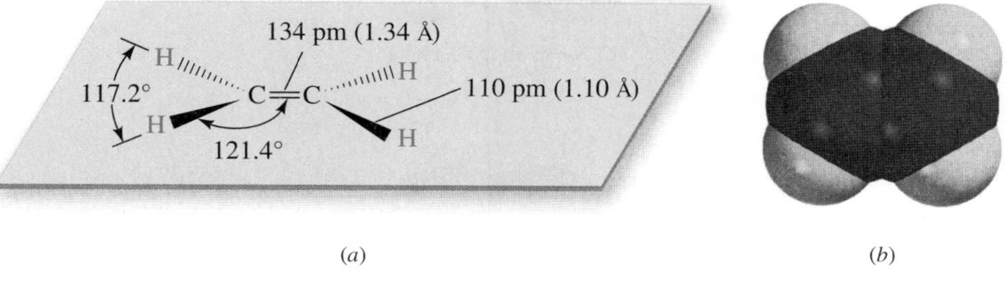

Figure 2.12

(*a*) All the atoms of ethylene lie in the same plane, the bond angles are close to 120°, and the carbon–carbon bond distance is significantly shorter than that of ethane. (*b*) A space-filling model of ethylene.

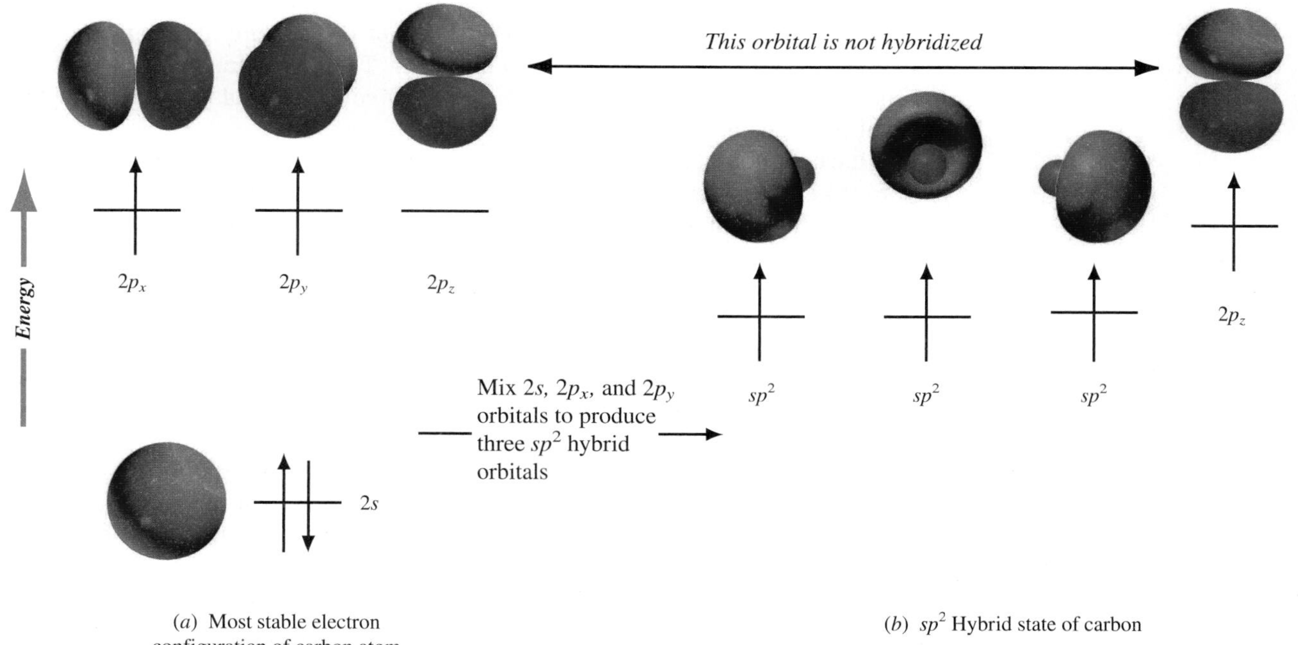

Figure 2.13

sp^2 Hybridization. (*a*) Electron configuration of carbon in its most stable state. (*b*) Mixing the *s* orbital with two of the three *p* orbitals generates three sp^2 hybrid orbitals and leaves one of the 2*p* orbitals untouched. The axes of the three sp^2 orbitals lie in the same plane and make angles of 120° with one another.

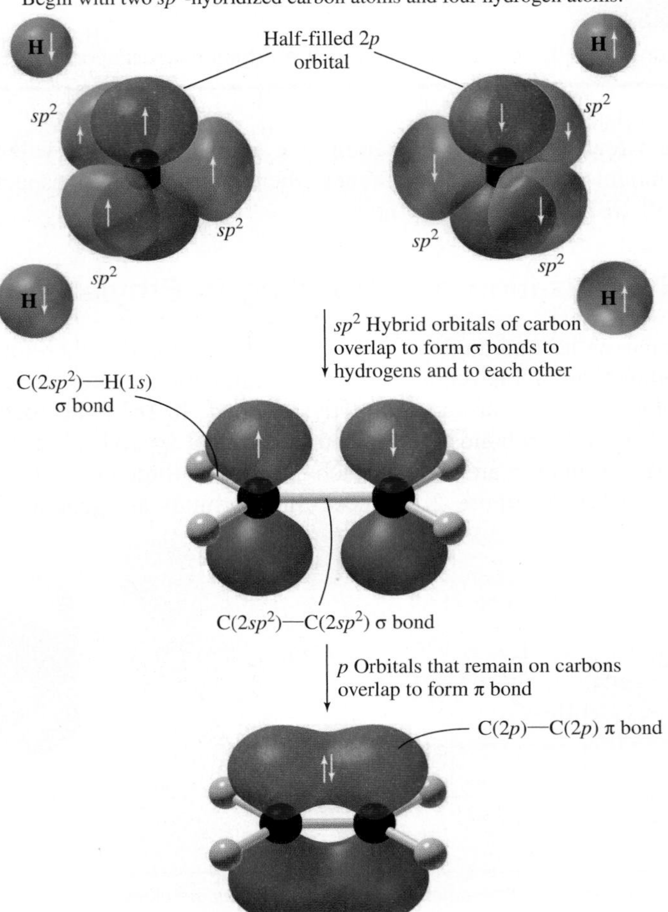
Begin with two sp^2-hybridized carbon atoms and four hydrogen atoms:
Half-filled $2p$ orbital
H
sp^2
sp^2 Hybrid orbitals of carbon overlap to form σ bonds to hydrogens and to each other
C($2sp^2$)—H($1s$) σ bond
C($2sp^2$)—C($2sp^2$) σ bond
p Orbitals that remain on carbons overlap to form π bond
C($2p$)—C($2p$) π bond

Figure 2.14

The carbon–carbon double bond in ethylene has a σ component and a π component. The σ component arises from overlap of sp^2-hybridized orbitals along the internuclear axis. The π component results from a side-by-side overlap of $2p$ orbitals.

carbon $2s$ orbital with *two* of the $2p$ orbitals and are called **sp^2 hybrid orbitals.** One of the $2p$ orbitals is left unhybridized. The three sp^2 orbitals are of equal energy; each has one-third s character and two-thirds p character. Their axes are coplanar, and each has a shape much like that of an sp^3 orbital. The three sp^2 orbitals and the unhybridized p orbital each contain one electron.

Each carbon of ethylene uses two of its sp^2 hybrid orbitals to form σ bonds to two hydrogen atoms, as illustrated in the first part of Figure 2.14. The remaining sp^2 orbitals, one on each carbon, overlap along the internuclear axis to give a σ bond connecting the two carbons.

Each carbon atom still has, at this point, an unhybridized $2p$ orbital available for bonding. These two half-filled $2p$ orbitals have their axes perpendicular to the framework of σ bonds of the molecule and overlap in a side-by-side manner to give a **pi (π) bond.** The carbon–carbon double bond of ethylene is viewed as a combination of a σ bond plus a π bond. The additional increment of bonding makes a carbon–carbon double bond both stronger and shorter than a carbon–carbon single bond.

Electrons in a π bond are called **π electrons.** The probability of finding a π electron is highest in the region above and below the plane of the molecule. The plane of the molecule corresponds to a nodal plane, where the probability of finding a π electron is zero.

In general, you can expect that carbon will be sp^2*-hybridized when it is directly bonded to three atoms in a neutral molecule.*

Problem 2.3

Identify the orbital overlaps of all of the bonds in propene ($H_2C{=}CHCH_3$) and classify them as σ or π as appropriate.

2.9 *sp* Hybridization and Bonding in Acetylene

One more hybridization scheme is important in organic chemistry. It is called ***sp* hybridization** and applies when carbon is directly bonded to two atoms, as in acetylene. The structure of acetylene is shown in Figure 2.15 along with its bond distances and bond angles. Its most prominent feature is its linear geometry.

Because each carbon in acetylene is bonded to two other atoms, the orbital hybridization model requires each carbon to have two equivalent orbitals available for σ bonds as outlined in Figure 2.16. According to this model the carbon 2*s* orbital and one of its 2*p* orbitals combine to generate two *sp* hybrid orbitals, each of which has 50% *s* character and 50% *p* character. These two *sp* orbitals share a common axis, but their major lobes are oriented at an angle of 180° to each other. Two of the original 2*p* orbitals remain unhybridized.

As portrayed in Figure 2.17, the two carbons of acetylene are connected to each other by a 2*sp*–2*sp* σ bond, and each is attached to a hydrogen substituent by a 2*sp*–1*s* σ bond. The unhybridized 2*p* orbitals on one carbon overlap with their counterparts on

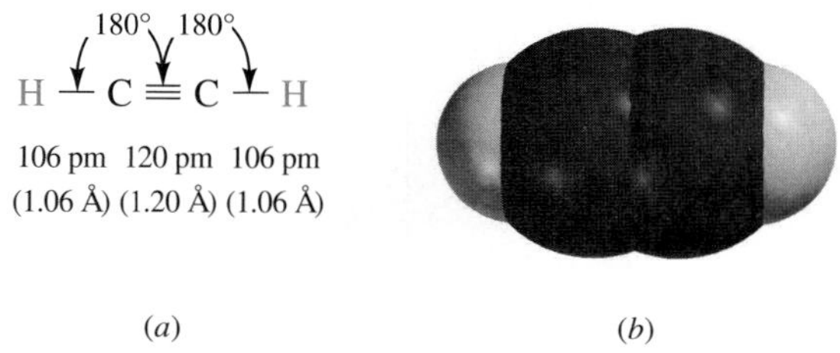

Figure 2.15

Acetylene is a linear molecule as indicated in (*a*) the structural formula and (*b*) a space-filling model.

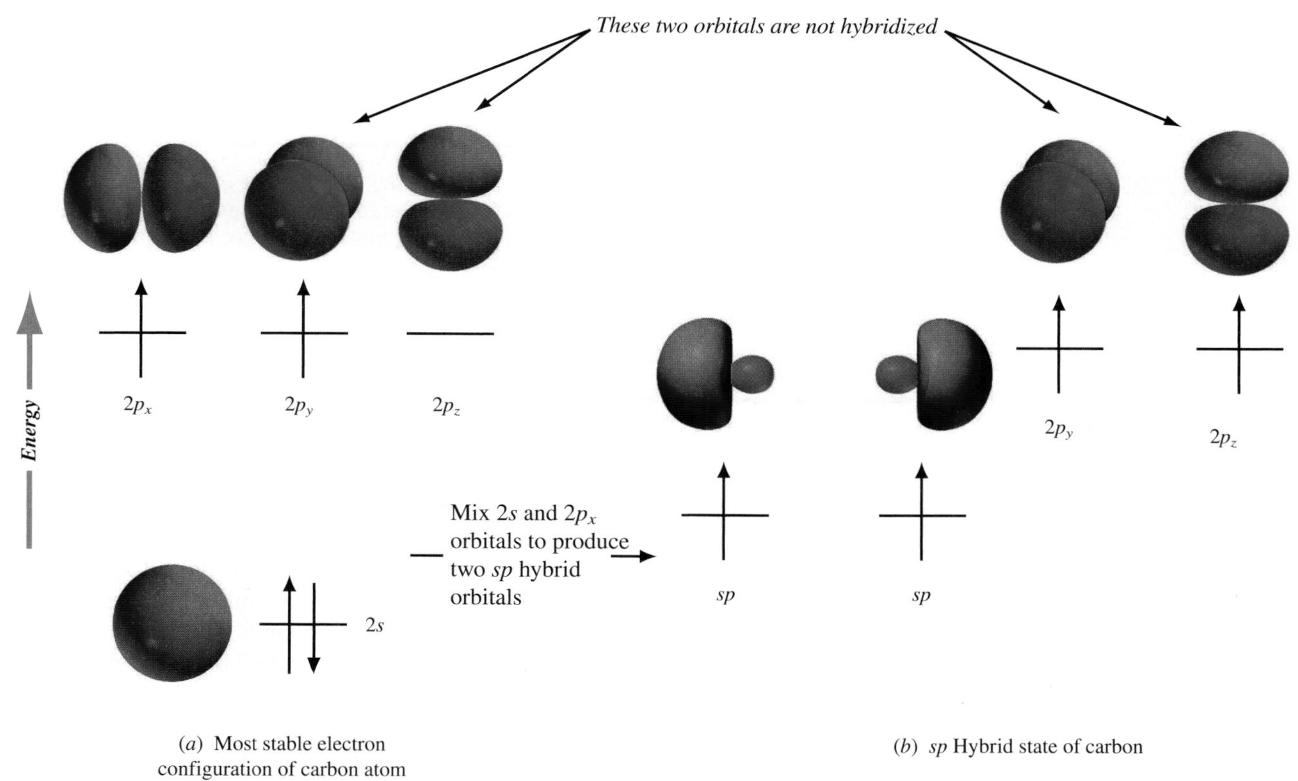

Figure 2.16

sp Hybridization. (*a*) Electron configuration of carbon in its most stable state. (*b*) Mixing the *s* orbital with one of the three *p* orbitals generates two *sp* hybrid orbitals and leaves two of the 2*p* orbitals untouched. The axes of the two *sp* orbitals make an angle of 180° with each other.

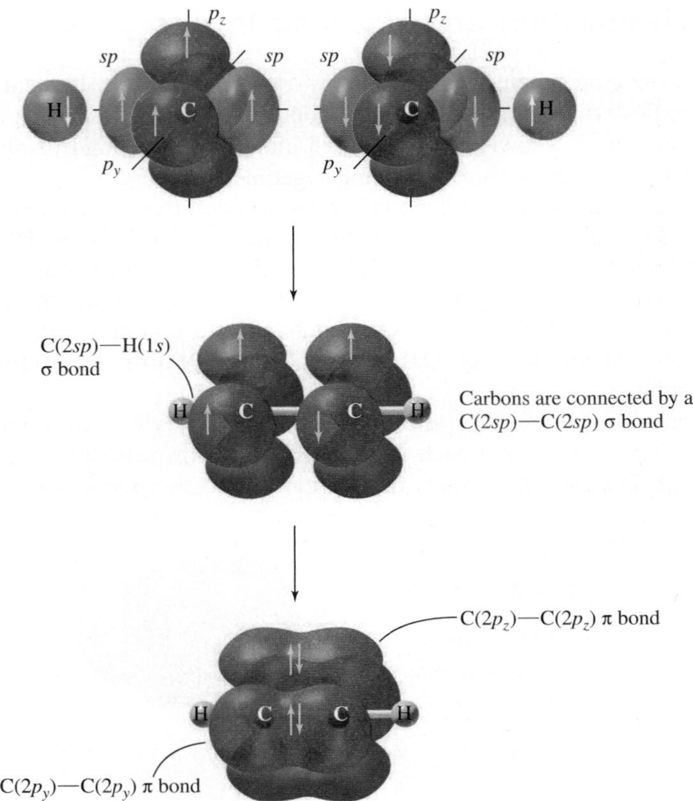

Figure 2.17

Bonding in acetylene based on *sp* hybridization of carbon. The carbon–carbon triple bond is viewed as consisting of one σ bond and two π bonds.

the other to form two π bonds. The carbon–carbon triple bond in acetylene is viewed as a multiple bond of the $\sigma + \pi + \pi$ type.

In general, you can expect that carbon will be sp*-hybridized when it is directly bonded to two atoms in a neutral molecule.*

Problem 2.4

The hydrocarbon shown, called *vinylacetylene,* is used in the synthesis of neoprene, a synthetic rubber. Identify the orbital overlaps involved in the indicated bond. How many σ bonds are there in vinylacetylene? How many π bonds?

$$H_2C{=}CH{-}C{\equiv}CH$$

(arrow indicates the CH—C single bond)

2.10 Bonding in Water and Ammonia: Hybridization of Oxygen and Nitrogen

The valence bond model and the accompanying idea of orbital hybridization have applications beyond those of bonding in hydrocarbons. A very simple extension to inorganic chemistry describes the bonding to the nitrogen of ammonia and the oxygen of water.

Like the carbon of methane, nitrogen is surrounded by four electron pairs in ammonia—three bonded pairs and an unshared pair. Oxygen in water has two bonded pairs and two unshared pairs. As we saw when discussing the tetrahedral geometry of CH_4, the pyramidal geometry of NH_3, and the bent geometry of H_2O in Section 1.9, all three are consistent with a tetrahedral arrangement of electron pairs. By associating a tetrahedral arrangement of electron pairs with sp^3 orbital hybridization, we see that CH_4, NH_3, and H_2O share a common valence bond description, differing only in the number of their valence electrons. C, N, and O are each sp^3-hybridized, with σ bonds connecting them to four, three, and two hydrogens, respectively. The unshared pair of nitrogen in

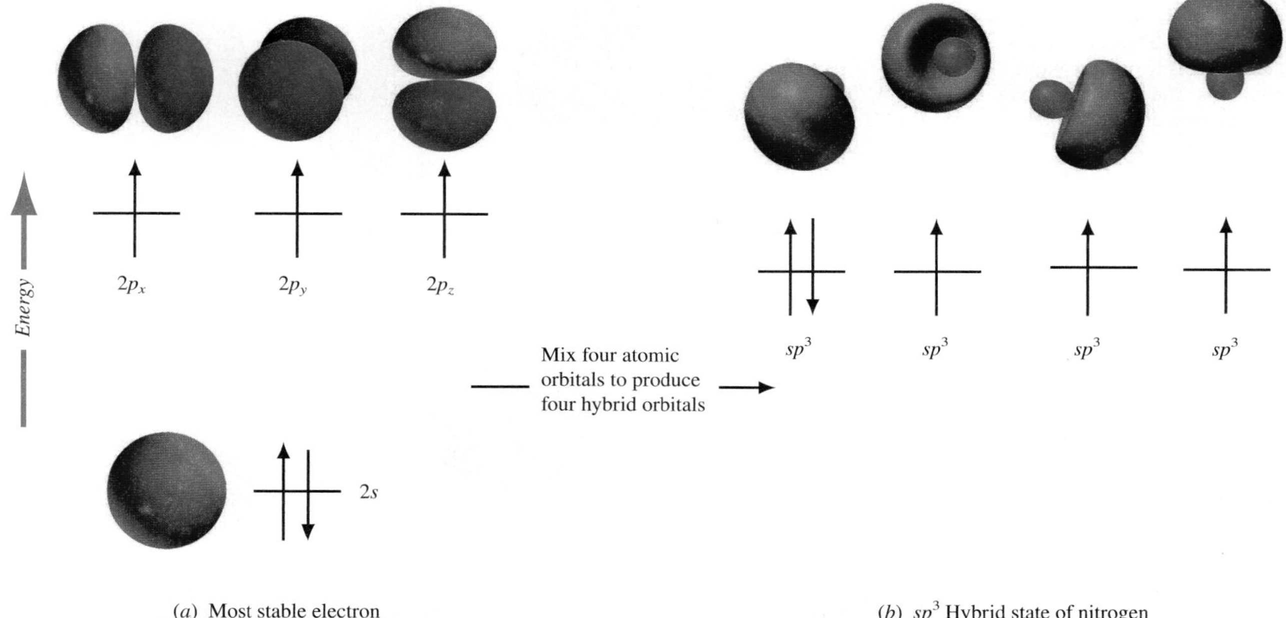

Figure 2.18

sp^3 Hybridization of nitrogen. (*a*) Electron configuration of nitrogen in its most stable state. (*b*) Mixing the *s* orbital with the three *p* orbitals generates four sp^3 hybrid orbitals. The four sp^3 hybrid orbitals are of equal energy. One sp^3 orbital contains a pair of electrons; the other three each contain a single electron.

NH_3 occupies an sp^3-hybridized orbital. The two unshared pairs of oxygen in H_2O occupy two sp^3-hybridized orbitals.

Methane | Ammonia (sp^3) | Water (sp^3)

As noted in Section 1.9, the H—N—H angle in ammonia (107°) and the H—O—H angle in water (105°) are slightly smaller than the tetrahedral angle of 109.5° because of the larger volume required by unshared pairs.

Figure 2.18 illustrates sp^3 hybridization of nitrogen in ammonia. Except for the fact that nitrogen has one more valence electron than carbon, the diagram is identical to that shown for methane in Figure 2.9.

Problem 2.5

Construct an orbital diagram to show the hybrid orbitals of sp^3-hybridized oxygen in water.

Problem 2.6

Two contributing structures A and B that satisfy the octet rule can be written for formamide:

A: $H_2\ddot{N}$–CH=$\ddot{O}$:   B: $H_2\overset{+}{N}$=CH–$\ddot{O}$:$^-$

(a) What is the hybridization of carbon and nitrogen in each?

(b) Formamide is planar or nearly so. Which structural formula better fits this fact?

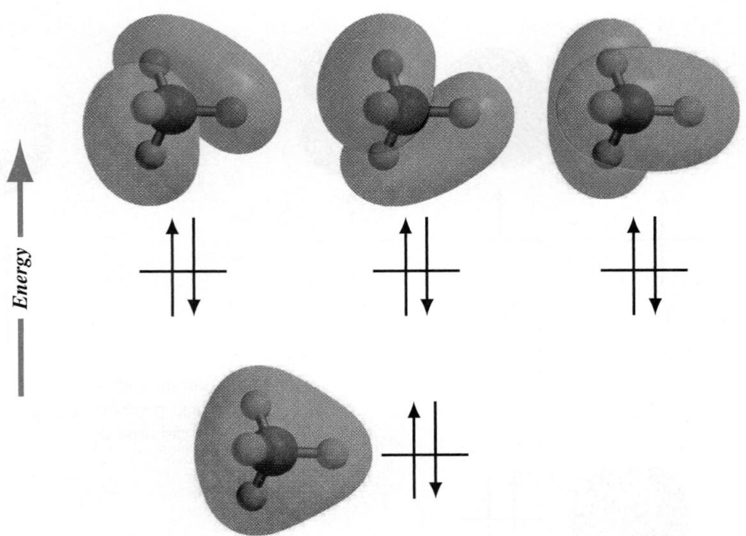

Figure 2.19

Bonding molecular orbitals of methane. Each orbital contains two of the eight valence electrons. The carbon 1s orbital and its two electrons are not shown.

2.11 Molecular Orbitals and Bonding in Methane

Compared to the Lewis and orbital hybridization models, molecular orbital theory is the least intuitive and requires the most training, background, and experience to apply. We have so far *discussed* molecular orbital theory only in the context of bonding in H_2 but have *used* the results of molecular orbital theory without acknowledging it. Electrostatic potential maps, for example, are obtained by molecular orbital calculations. You will see other results of molecular orbital theory often in this text, but the theory itself will be developed only as needed.

We saw in Section 2.6 that the valence bond model for bonding in methane rests on the overlap of a hydrogen 1*s* orbital with an sp^3-hybridized orbital of carbon. The pair of electrons in each of the four σ bonds is delocalized, but only between two atoms—carbon and the attached hydrogen. According to molecular orbital theory, as illustrated in Figure 2.19, the bonding electrons in methane are more delocalized in that each of the four bonding orbitals involves carbon and all of the hydrogens and each contains two of the eight valence electrons. The lowest-energy molecular orbital in the figure has no nodes; each of the other three has one node. There are an equal number of antibonding orbitals, all of which are vacant; these plus the carbon 1*s* orbital and its two electrons are not involved in bonding and are not shown in the figure.

Which theory of chemical bonding is best: Lewis, valence bond, or molecular orbital? The answer is that organic chemists use all three, depending on the situation. The Lewis rules are straightforward and most familiar. Valence bond theory, especially when coupled with the concept of orbital hybridization, enhances the information content of Lewis formulas by distinguishing among various types of atoms, electrons, and bonds. Molecular orbital theory, although the least intuitive of the three methods, can provide insights into structure and reactivity that the Lewis and valence bond models cannot. All three theories are used by chemists with the choice being determined according to which one seems most appropriate.

2.12 Isomeric Alkanes: The Butanes

Methane is the only alkane of molecular formula CH_4, ethane the only one that is C_2H_6, and propane the only one that is C_3H_8. Beginning with C_4H_{10}, however, constitutional isomers (Section 1.6) are possible; two alkanes have this particular molecular formula. In one, called ***n*-butane,** four carbons are joined in a continuous chain. The *n* in *n*-butane stands for "normal" and means that the carbon chain is unbranched. The second isomer has a branched carbon chain and is called **isobutane.**

	$CH_3CH_2CH_2CH_3$	CH_3CHCH_3 (with CH_3 on the middle carbon) or $(CH_3)_3CH$
	n-Butane	Isobutane
Boiling point:	−0.4°C	−10.2°C
Melting point:	−139°C	−160.9°C

As noted in Section 2.7, CH_3 is called a *methyl* group. In addition to having methyl groups at both ends, *n*-butane contains two CH_2, or **methylene** groups. Isobutane contains three methyl groups bonded to a CH unit. The CH unit is called a **methine** group.

n-Butane and isobutane have the same molecular formula but differ in connectivity. They are *constitutional isomers* of each other and have different properties. Both are gases at room temperature, but *n*-butane boils almost 10°C higher than isobutane and has a melting point that is over 20°C higher.

Bonding in *n*-butane and isobutane continues the theme begun with methane, ethane, and propane. All of the carbon atoms are sp^3-hybridized, all of the bonds are σ bonds, and the bond angles at carbon are close to tetrahedral. This generalization holds for all alkanes regardless of the number of carbons they have.

"Butane" lighters contain about 5% *n*-butane and 95% isobutane in a sealed container. The pressure produced by the two compounds (about 3 atm) is enough to keep them in the liquid state until opening a small valve emits a fine stream of the vaporized mixture across a spark, which ignites it.

2.13 Higher *n*-Alkanes

n-Pentane and *n*-hexane are *n*-alkanes possessing five and six carbon atoms, respectively.

$CH_3CH_2CH_2CH_2CH_3$	$CH_3CH_2CH_2CH_2CH_2CH_3$
n-Pentane	*n*-Hexane

The condensed formulas can be abbreviated by indicating within parentheses the number of methylene groups in the chain. Thus, *n*-pentane may be written as $CH_3(CH_2)_3CH_3$ and *n*-hexane as $CH_3(CH_2)_4CH_3$. This shortcut is especially convenient with longer-chain alkanes. The laboratory synthesis of the "ultralong" alkane $CH_3(CH_2)_{388}CH_3$ was achieved in 1985; imagine trying to write its structural formula in anything other than an abbreviated way!

Problem 2.7

An *n*-alkane of molecular formula $C_{28}H_{58}$ has been isolated from a certain fossil plant. Write a condensed structural formula for this alkane.

n-Alkanes have the general formula $CH_3(CH_2)_xCH_3$ and constitute a **homologous series** of compounds. A homologous series is one in which successive members differ by a —CH_2— group.

Problem 2.8

Much of the communication between insects involves chemical messengers called *pheromones*. A species of cockroach secretes a substance from its mandibular glands that alerts other cockroaches to its presence and causes them to congregate. One of the principal components of this *aggregation pheromone* is the alkane shown. Give the molecular formula of this substance, and represent it by a condensed formula.

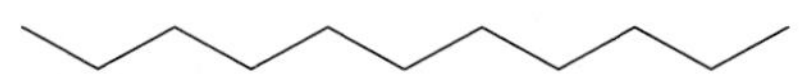

2.14 The C_5H_{12} Isomers

Three isomeric alkanes have the molecular formula C_5H_{12}. The unbranched isomer is *n*-pentane. The isomer with a single methyl branch is called *isopentane*. The third isomer has a three-carbon chain with two methyl branches and is called *neopentane*.

n-Pentane: $CH_3CH_2CH_2CH_2CH_3$ or $CH_3(CH_2)_3CH_3$ or [bond-line structure]

Isopentane: $CH_3CHCH_2CH_3$ (with CH_3 on C-2) or $(CH_3)_2CHCH_2CH_3$ or [bond-line structure]

Neopentane: CH_3CCH_3 (with CH_3 above and below C-2) or $(CH_3)_4C$ or [bond-line structure]

Table 2.1 lists the number of possible alkane isomers according to the number of carbon atoms they contain. As the table shows, the number of isomers increases enormously with the number of carbon atoms and raises two important questions:

1. How can we tell when we have written all the possible isomers corresponding to a particular molecular formula?
2. How can we name alkanes so that each one has a unique name?

The answer to the first question is that you cannot easily calculate the number of isomers. The data in Table 2.1 were determined by a mathematician who concluded that no simple expression can calculate the number of isomers. The best way to ensure that you have written all the isomers of a particular molecular formula is to work systematically, beginning with the unbranched chain and then shortening it while adding branches one by one. It is essential that you be able to recognize when two different-looking structural formulas are actually the same molecule written in different ways. The key

TABLE 2.1 The Number of Constitutionally Isomeric Alkanes of Particular Molecular Formulas

Molecular formula	Number of constitutional isomers
CH_4	1
C_2H_6	1
C_3H_8	1
C_4H_{10}	2
C_5H_{12}	3
C_6H_{14}	5
C_7H_{16}	9
C_8H_{18}	18
C_9H_{20}	35
$C_{10}H_{22}$	75
$C_{15}H_{32}$	4,347
$C_{20}H_{42}$	366,319
$C_{40}H_{82}$	62,491,178,805,831

point is the *connectivity* of the carbon chain. For example, the following structural formulas do *not* represent different compounds; they are just a portion of the many ways we could write a structural formula for isopentane. Each one has a continuous chain of four carbons with a methyl branch located one carbon from the end of the chain, and all represent the same compound.

$CH_3CHCH_2CH_3$ (with CH_3 below C-2) $CH_3CHCH_2CH_3$ (with CH_3 above C-2) $CH_3CH_2CHCH_3$ (with CH_3 below C-3) $CH_3CH_2CHCH_3$ (with CH_3 above C-3) $CHCH_2CH_3$ (with CH_3 above and below C-1)

Problem 2.9

Write condensed and bond-line formulas for the five isomeric C_6H_{14} alkanes.

Sample Solution

- The unbranched isomer is

$CH_3CH_2CH_2CH_2CH_2CH_3$ = (bond-line zigzag)

- Shortening the chain by one carbon and adding a CH_3 branch gives two more isomers.

$CH_3CHCH_2CH_2CH_3$ (with CH_3 on C-2) and $CH_3CH_2CHCH_2CH_3$ (with CH_3 on C-3)

- A four-carbon chain can have two CH_3 branches on the same carbon or on adjacent carbons.

$CH_3C(CH_3)_2-CH_2CH_3$ and $CH_3CH(CH_3)-CH(CH_3)CH_3$

The answer to the second question—how to provide a name that is unique to a particular structure—is presented in the following section. It is worth noting, however, that being able to name compounds in a *systematic* way is a great help in deciding whether two structural formulas represent isomers or are the same compound written in two different ways. By following a precise set of rules, you will always get the same systematic name for a compound, regardless of how it is written. Conversely, two different compounds will always have different names.

2.15 IUPAC Nomenclature of Unbranched Alkanes

In the preceding section we saw that the three C_5H_{12} isomers all incorporate "pentane" in their names and are differentiated by the prefixes "*n*-," "iso," and "neo." Extending this approach to alkanes beyond C_5H_{12} fails because we run out of descriptive prefixes before all the isomers have unique names. As difficult as it would be to invent different names for the 18 constitutional isomers of C_8H_{18}, for example, it would be even harder to remember which structure corresponded to which name. For this and other reasons, organic chemists developed systematic ways to name compounds based on their structure. The most widely used approach is called the **IUPAC rules;** *IUPAC* stands for the International Union of Pure and Applied Chemistry. (See the boxed essay *What's in a Name? Organic Nomenclature.*)

Alkane names form the foundation of the IUPAC system; more complicated compounds are viewed as being derived from alkanes. The IUPAC names assigned to unbranched alkanes are shown in Table 2.2. Methane, ethane, propane, and butane are retained for CH_4, CH_3CH_3, $CH_3CH_2CH_3$, and $CH_3CH_2CH_2CH_3$, respectively. Thereafter, the number of carbon atoms in the chain is specified by a Greek prefix preceding the suffix *-ane,* which identifies the compound as a member of the alkane family. Notice that the prefix *n*- is not part of the IUPAC system. The IUPAC name for $CH_3CH_2CH_2CH_3$ is butane, not *n*-butane.

Figure 2.20

Worker bees build the hive with an alkane-containing wax secreted from their abdominal glands.

Problem 2.10

Refer to Table 2.2 as needed to answer the following questions:

(a) Beeswax (Figure 2.20) contains 8–9% hentriacontane. Write a condensed structural formula for hentriacontane.

(b) Octacosane has been found to be present in a certain fossil plant. Write a condensed structural formula for octacosane.

(c) What is the IUPAC name of the alkane described in Problem 2.8 as a component of the cockroach aggregation pheromone?

Sample Solution (a) Note in Table 2.2 that hentriacontane has 31 carbon atoms. All the alkanes in Table 2.2 have unbranched carbon chains. Hentriacontane has the condensed structural formula $CH_3(CH_2)_{29}CH_3$.

TABLE 2.2 IUPAC Names of Unbranched Alkanes

Number of carbon atoms	Name	Number of carbon atoms	Name	Number of carbon atoms	Name
1	Methane	11	Undecane	21	Henicosane
2	Ethane	12	Dodecane	22	Docosane
3	Propane	13	Tridecane	23	Tricosane
4	Butane	14	Tetradecane	24	Tetracosane
5	Pentane	15	Pentadecane	30	Triacontane
6	Hexane	16	Hexadecane	31	Hentriacontane
7	Heptane	17	Heptadecane	32	Dotriacontane
8	Octane	18	Octadecane	40	Tetracontane
9	Nonane	19	Nonadecane	50	Pentacontane
10	Decane	20	Icosane	100	Hectane

In Problem 2.9 you were asked to write structural formulas for the five isomeric alkanes of molecular formula C_6H_{14}. In the next section you will see how the IUPAC rules generate a unique name for each isomer.

2.16 Applying the IUPAC Rules: The Names of the C_6H_{14} Isomers

We can present and illustrate the most important of the IUPAC rules for alkane nomenclature by naming the five C_6H_{14} isomers. By definition (see Table 2.2), the unbranched C_6H_{14} isomer is hexane.

$CH_3CH_2CH_2CH_2CH_2CH_3$ or

IUPAC name: hexane

The IUPAC rules name branched alkanes as *substituted derivatives* of the unbranched parent alkanes listed in Table 2.2. Consider the C_6H_{14} isomer represented by the structure

$CH_3CHCH_2CH_2CH_3$ (with CH_3 on C-2) or

Step 1

Pick out the *longest continuous carbon chain,* and find the IUPAC name in Table 2.2 that corresponds to the unbranched alkane having that number of carbons. This is the parent alkane from which the IUPAC name is to be derived.

In this case, the longest continuous chain has *five* carbon atoms; the compound is named as a derivative of pentane. The key word here is *continuous*. It does not matter whether the carbon skeleton is drawn in an extended straight-chain form or in one with many bends and turns. All that matters is the number of carbons linked together in an uninterrupted sequence.

Step 2

Identify the substituent groups attached to the parent.

The parent pentane chain bears a methyl (CH_3) group as a substituent.

Step 3

Number the longest continuous chain in the direction that gives the lowest number to the substituent at the first point of branching.

The numbering scheme

$\overset{1}{C}H_3\overset{2}{C}H\overset{3}{C}H_2\overset{4}{C}H_2\overset{5}{C}H_3$ (with CH_3 on C-2) is equivalent to $CH_3\overset{2}{C}H\overset{3}{C}H_2\overset{4}{C}H_2\overset{5}{C}H_3$ (with $\overset{1}{C}H_3$ on C-2)

Both count five carbon atoms in their longest continuous chain and bear a methyl group as a substituent at the second carbon. An alternative numbering sequence that begins at the other end of the chain is incorrect:

$\overset{5}{C}H_3\overset{4}{C}H\overset{3}{C}H_2\overset{2}{C}H_2\overset{1}{C}H_3$ (with CH_3 on C-4) (methyl group attached to C-4)

Step 4

Write the name of the compound. The parent alkane is the last part of the name and is preceded by the names of the substituents and their numerical locations (**locants**). Hyphens separate the locants from the words.

$CH_3CHCH_2CH_2CH_3$ (with CH_3 on C-2) or

IUPAC name: 2-methylpentane

What's in a Name? Organic Nomenclature

Systematic Names and Common Names *Systematic names are derived according to a prescribed set of rules, common names are not.*

Many compounds are better known by **common names** than by their **systematic names.**

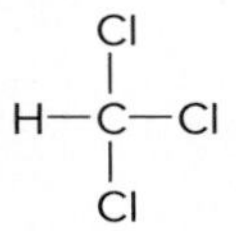

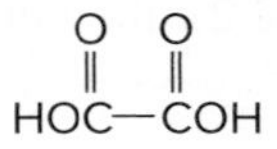

Common name: Chloroform — Oxalic acid

Systematic name: Trichloromethane — Ethanedioic acid

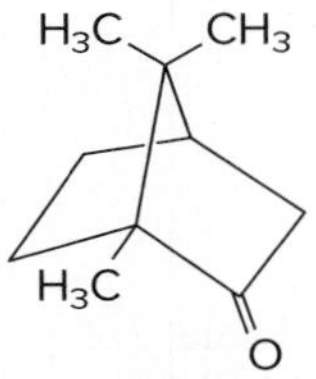

Common name: Camphor

Systematic name: 1,7,7-Trimethylbicyclo[2.2.1]heptan-2-one

Common names, despite their familiarity in certain cases, suffer serious limitations compared with systematic ones. The number of known compounds (more than 50 million) far exceeds our capacity to give each one a unique common name, and most common names are difficult to connect directly to a structural formula. A systematic approach *based on structure* not only conveys structural information, but also generates a unique name for each structural variation.

Evolution of the IUPAC Rules *A single compound can have several acceptable systematic names but no two compounds can have the same name.*

As early as 1787 with the French publication of *Méthode de nomenclature chimique,* chemists suggested guidelines for naming compounds according to chemical composition. Their proposals were more suited to inorganic compounds than organic ones, and it was not until the 1830s that comparable changes appeared in organic chemistry. Later (1892), a group of prominent chemists met in Geneva, Switzerland, where they formulated the principles on which our present system of organic nomenclature is based.

During the twentieth century, what we now know as the *International Union of Pure and Applied Chemistry* (IUPAC) carried out major revisions and extensions of organic nomenclature culminating in the IUPAC Rules of 1979 and 1993. Both versions are accessible via the Internet and are the basis for the names used in this text.* As of this writing, the 2013 rules are available only in print form.

Our practice will be to name compounds in the manner of most active chemists and to use nomenclature as a tool to advance our understanding of organic chemistry.

Other Nomenclatures Chemical Abstracts Service, a division of the American Chemical Society, surveys all the world's leading scientific journals and publishes brief abstracts of their chemistry papers. *Chemical Abstracts* nomenclature has evolved in a direction geared to computerized literature searches and, although once similar to IUPAC, it is now much different. In general, it is easier to make the mental connection between a structure and its IUPAC name than its *Chemical Abstracts* name.

The **generic name** of a drug is not derived from systematic nomenclature. The group responsible for most generic names in the United States is the U.S. Adopted Names (USAN) Council, a private organization founded by the American Medical Association, the American Pharmacists Association, and the U.S. Pharmacopeial Convention.

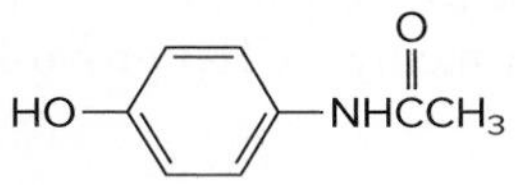

USAN Acetaminophen
INN Paracetamol
IUPAC *N*-(4-Hydroxyphenyl)acetamide

Tylenol

The USAN name is recognized as the official name by the U.S. Food and Drug Administration. International Nonproprietary Names (INN) are generic names as designated by the World Health Organization.

*The 1979 and 1993 IUPAC rules may be accessed at http://www.acdlabs.com/iupac/nomenclature.

The same sequence of four steps gives the IUPAC name for the isomer that has its methyl group attached to the middle carbon of the five-carbon chain.

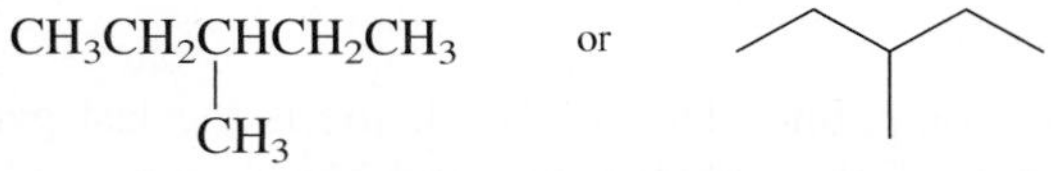

$$CH_3CH_2CH(CH_3)CH_2CH_3$$ or

IUPAC name: 3-methylpentane

Both remaining C_6H_{14} isomers have two methyl groups as substituents on a four-carbon chain. Thus the parent chain is butane. When the same substituent appears more than once, use the multiplying prefixes *di-, tri-, tetra-,* and so on. A separate locant is used for each substituent, and the locants are separated from each other by commas and from the words by hyphens.

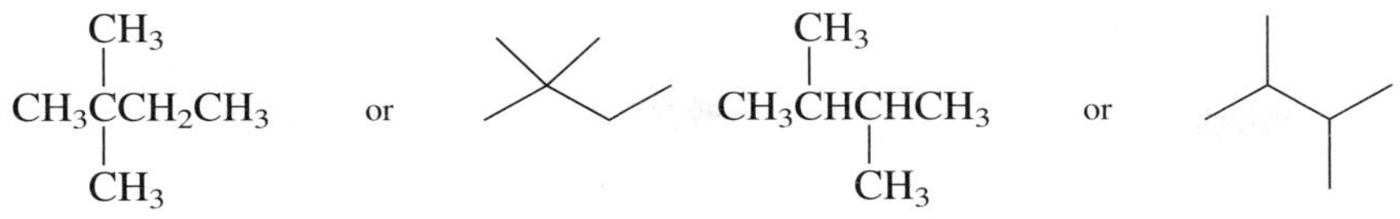

$$\begin{array}{c} CH_3 \\ | \\ CH_3CCH_2CH_3 \\ | \\ CH_3 \end{array}$$ or [bond-line structure]

IUPAC name: 2,2-dimethylbutane

$$\begin{array}{c} CH_3 \\ | \\ CH_3CHCHCH_3 \\ | \\ CH_3 \end{array}$$ or [bond-line structure]

IUPAC name: 2,3-dimethylbutane

Problem 2.11

Phytane is the common name of a naturally occurring alkane produced by the alga *Spirogyra* and is a constituent of petroleum. The IUPAC name for phytane is 2,6,10,14-tetramethylhexadecane. Write a bond-line formula for phytane. What is its molecular formula?

Problem 2.12

Give the IUPAC names for

(a) The isomers of C_4H_{10}

(b) The isomers of C_5H_{12}

(c) 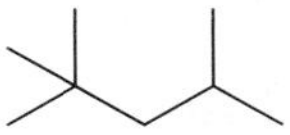 [bond-line structure]

(d) 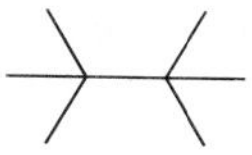 [bond-line structure]

Sample Solution (a) There are two C_4H_{10} isomers. Butane (see Table 2.2) is the IUPAC name for the isomer that has an unbranched carbon chain. The other isomer has three carbons in its longest continuous chain with a methyl branch at the central carbon; its IUPAC name is 2-methylpropane.

$CH_3CH_2CH_2CH_3$

IUPAC name: butane

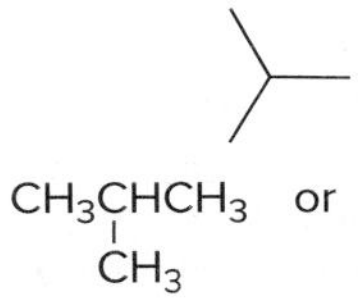

$$\begin{array}{c} CH_3CHCH_3 \\ | \\ CH_3 \end{array}$$ or $(CH_3)_3CH$

IUPAC name: 2-methylpropane

So far, the only branched alkanes that we've named have methyl groups attached to the main chain. What about groups other than CH_3? What do we call these groups, and how do we name alkanes that contain them?

2.17 Alkyl Groups

An **alkyl group** lacks one of the hydrogens of an alkane. A methyl group ($—CH_3$) is an alkyl group derived from methane (CH_4). Unbranched alkyl groups in which the point of attachment is at the end of the chain are named in IUPAC nomenclature by replacing the *-ane* endings of Table 2.2 by *-yl*.

$CH_3CH_2—$	$CH_3(CH_2)_5CH_2—$	$CH_3(CH_2)_{16}CH_2—$
Ethyl group	**Heptyl** group	**Octadecyl** group

The dash at the end of the chain represents a potential point of attachment for some other atom or group.

Carbon atoms are classified according to their degree of substitution by other carbons. A **primary** carbon is *directly* attached to one other carbon. Similarly, a **secondary** carbon is directly attached to two other carbons, a **tertiary** carbon to three, and a **quaternary** carbon to four. Alkyl groups are designated as primary, secondary, or tertiary according to the degree of substitution of the carbon at the potential point of attachment.

```
    H      Primary          C      Secondary        C      Tertiary
    |   ↙ carbon            |   ↙ carbon            |   ↙ carbon
 C—C—                    C—C—                    C—C—
    |                       |                       |
    H                       H                       C
```

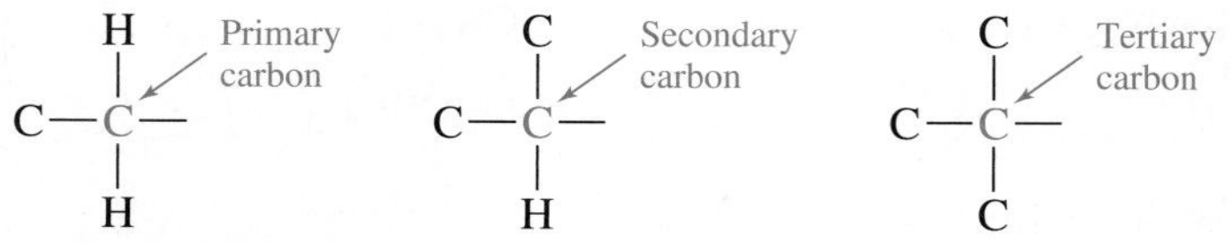

Primary alkyl group — Secondary alkyl group — Tertiary alkyl group

Ethyl (CH_3CH_2—), heptyl [$CH_3(CH_2)_5CH_2$—], and octadecyl [$CH_3(CH_2)_{16}CH_2$—] are examples of primary alkyl groups.

Branched alkyl groups are named by using the longest continuous chain *that begins at the point of attachment* as the parent. Thus, the systematic names of the two C_3H_7 alkyl groups are propyl and 1-methylethyl. Both are better known by their common names, *n*-propyl and isopropyl, respectively.

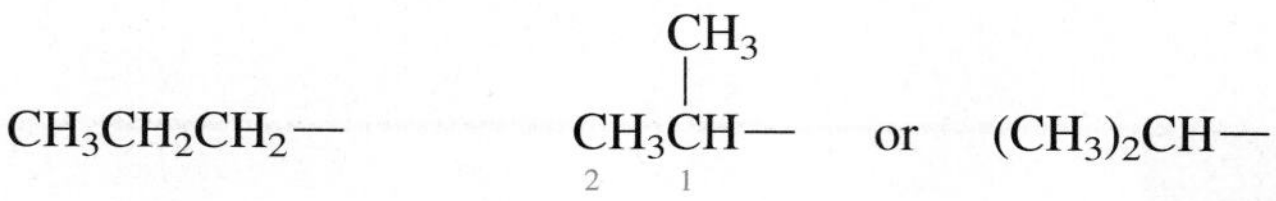

$CH_3CH_2CH_2$—

Propyl group (common name: ***n*-propyl**)

$CH_3\underset{2}{C}H(CH_3)$— (numbered 2, 1) or $(CH_3)_2CH$—

1-Methylethyl group (common name: **isopropyl**)

An isopropyl group is a *secondary* alkyl group. Its point of attachment is to a secondary carbon atom, one that is directly bonded to two other carbons.

The C_4H_9 alkyl groups may be derived either from the unbranched carbon skeleton of butane or from the branched carbon skeleton of isobutane. Those derived from butane are the butyl (*n*-butyl) group and the 1-methylpropyl (*sec*-butyl) group.

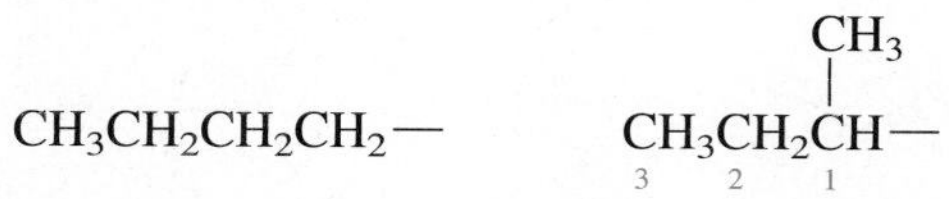

$CH_3CH_2CH_2CH_2$—

Butyl group (common name: ***n*-butyl**)

$CH_3CH_2CH(CH_3)$— (numbered 3, 2, 1)

1-Methylpropyl group (common name: ***sec*-butyl**)

Those derived from isobutane are the 2-methylpropyl (isobutyl) group and the 1,1-dimethylethyl (*tert*-butyl) group. Isobutyl is a primary alkyl group because its potential point of attachment is to a primary carbon. *tert*-Butyl is a tertiary alkyl group because its potential point of attachment is to a tertiary carbon.

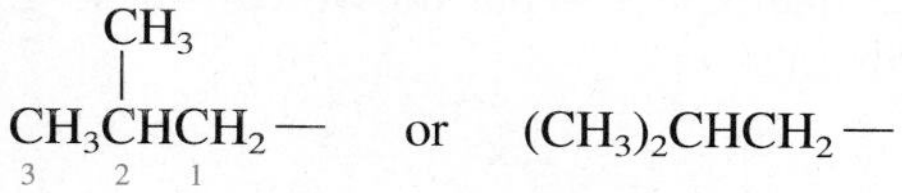

$CH_3CH(CH_3)CH_2$— (numbered 3, 2, 1) or $(CH_3)_2CHCH_2$—

2-Methylpropyl group (common name: **isobutyl**)

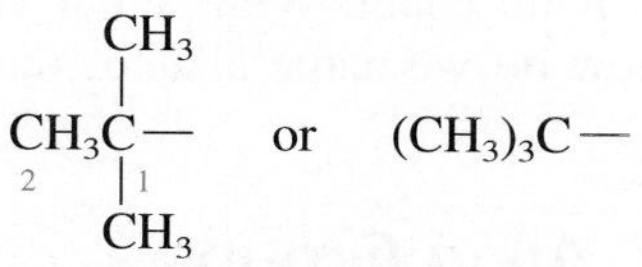

$CH_3C(CH_3)_2$— (numbered 2, 1) or $(CH_3)_3C$—

1,1-Dimethylethyl group (common name: ***tert*-butyl**)

Problem 2.13

Give the structures and IUPAC names of all the C_5H_{11} alkyl groups, and identify them as primary, secondary, or tertiary, as appropriate.

Sample Solution Consider the alkyl group having the same carbon skeleton as $(CH_3)_4C$. All the hydrogens are equivalent; replacing any one of them by a potential point of attachment is the same as replacing any of the others.

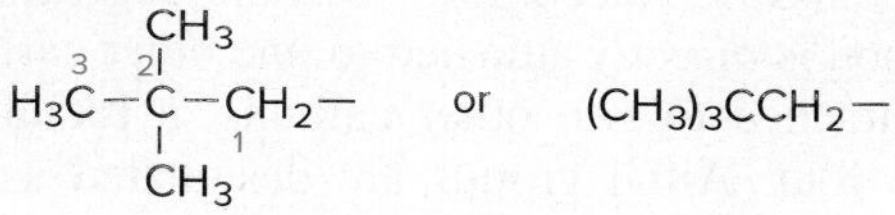

$H_3C—C(CH_3)_2—CH_2$— (numbered 3, 2, 1) or $(CH_3)_3CCH_2$—

Numbering always begins at the point of attachment and continues through the longest continuous chain. In this case the chain is three carbons and there are two methyl groups at C-2. The IUPAC name of this alkyl group is *2,2-dimethylpropyl*. (The common name for this group is *neopentyl*.) It is a *primary* alkyl group because the carbon that bears the potential point of attachment (C-1) is itself directly bonded to one other carbon.

In addition to methyl and ethyl groups, *n*-propyl, isopropyl, *n*-butyl, *sec*-butyl, isobutyl, *tert*-butyl, and neopentyl groups will appear often throughout this text. You should be able to recognize these groups on sight and to give their structures when needed.

The names and structures of the most frequently encountered alkyl groups are given on the inside back cover.

2.18 IUPAC Names of Highly Branched Alkanes

By combining the basic principles of IUPAC notation with the names of the various alkyl groups, we can develop systematic names for highly branched alkanes. We'll start with the following alkane, name it, then increase its complexity by successively adding methyl groups at various positions.

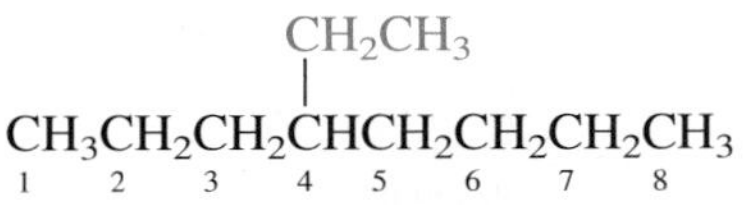

or

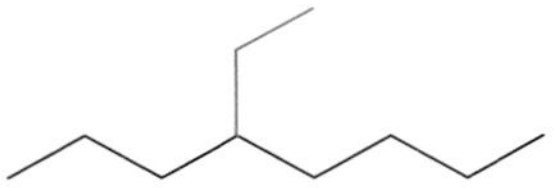

As numbered on the structural formula, the longest continuous chain contains eight carbons, and so the compound is named as a derivative of octane. Numbering begins at the end nearest the branch, and so the ethyl substituent is located at C-4, and the name of the alkane is *4-ethyloctane*.

What happens to the IUPAC name when a methyl group replaces one of the hydrogens at C-3?

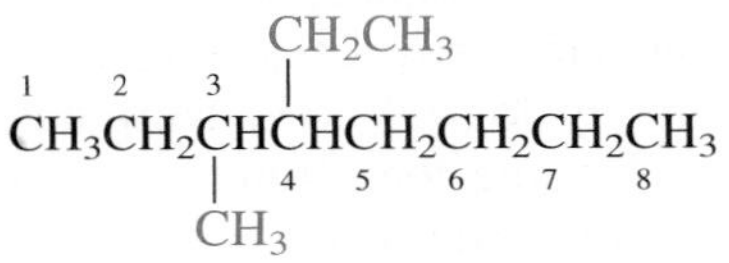

or

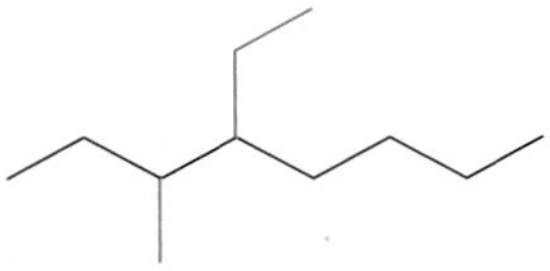

The compound becomes an octane derivative that bears a C-3 methyl group and a C-4 ethyl group. *When two or more different substituents are present, they are listed in alphabetical order in the name*. The IUPAC name for this compound is *4-ethyl-3-methyloctane*.

Replicating prefixes such as *di-*, *tri-*, and *tetra-* (Section 2.16) are used as needed but are ignored when alphabetizing. Adding a second methyl group to the original structure, at C-5, for example, converts it to *4-ethyl-3,5-dimethyloctane*.

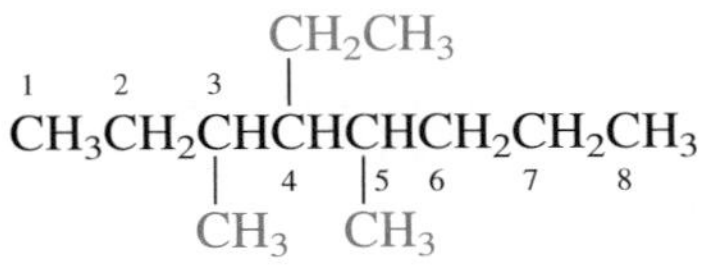

or

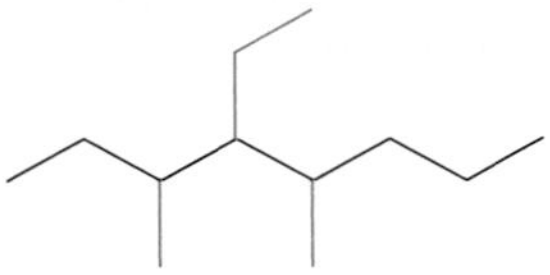

Italicized prefixes such as *sec-* and *tert-* are ignored when alphabetizing except when they are compared with each other. *tert*-Butyl precedes isobutyl, and *sec*-butyl precedes *tert*-butyl.

Problem 2.14

Give an acceptable IUPAC name for each of the following alkanes:

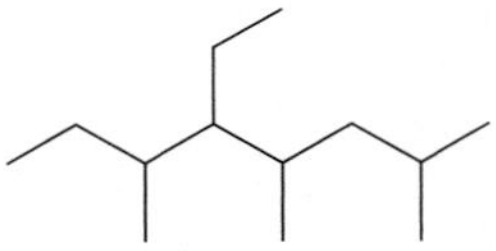

(a)

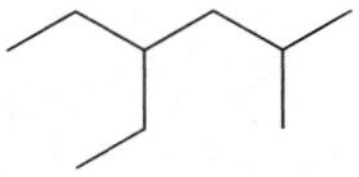

(b)

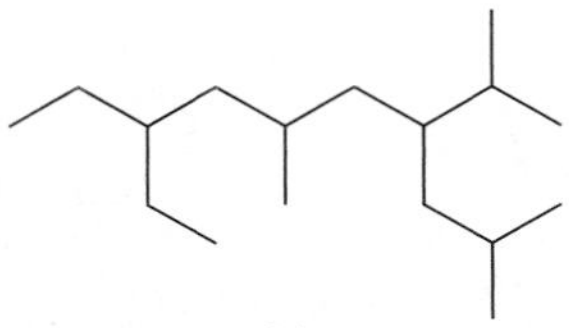

(c)

continued

Sample Solution (a) This problem extends the preceding discussion by adding a third methyl group to 4-ethyl-3,5-dimethyloctane, the compound just described. It is, therefore, an *ethyltrimethyloctane*. Notice, however, that the numbering sequence needs to be changed in order to adhere to the rule of numbering from the end of the chain nearest the first branch. When numbered properly, this compound has a methyl group at C-2 as its first-appearing substituent.

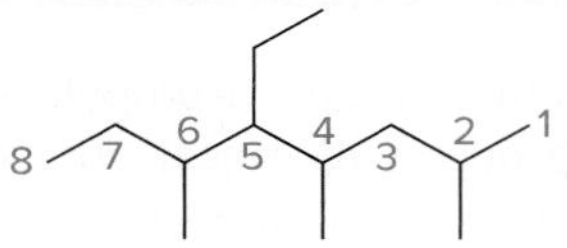

5-Ethyl-2,4,6-trimethyloctane

An additional feature of IUPAC nomenclature that concerns the direction of numbering is the "first point of difference" rule. Consider the two directions in which the following alkane may be numbered:

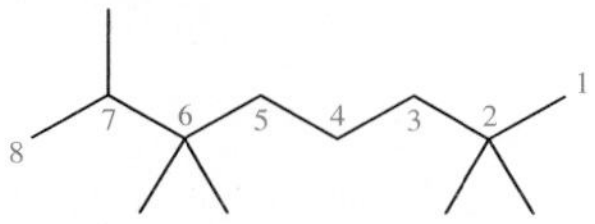

2,2,6,6,7-Pentamethyloctane (correct)

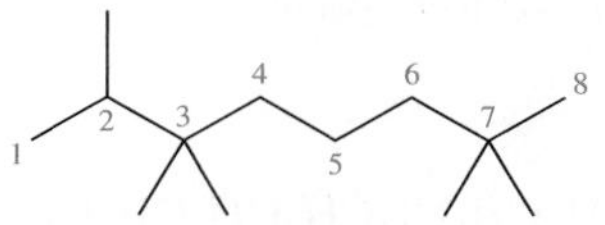

2,3,3,7,7-Pentamethyloctane (incorrect!)

When deciding on the proper direction, a point of difference occurs when one order gives a lower locant than another. Thus, although 2 is the first locant in both numbering schemes, the tie is broken at the second locant, and the rule favors 2,2,6,6,7, which has 2 as its second locant, whereas 3 is the second locant in 2,3,3,7,7.

Finally, when the same group of locants are generated from two different numbering directions, choose the direction that gives the lower number to the substituent that appears first in the name. (Remember, substituents are listed alphabetically.)

The IUPAC nomenclature system is inherently logical and incorporates healthy elements of common sense into its rules. Granted, some long, funny-looking, hard-to-pronounce names are generated. Once one knows the code (rules of grammar) though, it becomes a simple matter to convert those long names to unique structural formulas.

Tabular summaries of the IUPAC rules for alkane and alkyl group nomenclature appear in Tables 2.4 and 2.5 on pages 91–92.

2.19 Cycloalkane Nomenclature

Cycloalkanes are one class of *alicyclic* (*aliphatic cyclic*) hydrocarbons.

Cycloalkanes are alkanes that contain a ring of three or more carbons. They are frequently encountered in organic chemistry and are characterized by the molecular formula C_nH_{2n}. They are named by adding the prefix *cyclo-* to the name of the unbranched alkane with the same number of carbons as the ring.

Cyclopropane 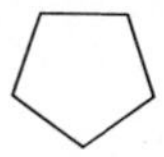 Cyclopentane 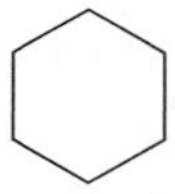 Cyclohexane 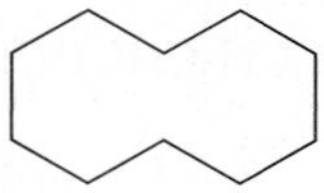 Cyclodecane

Substituents are identified in the usual way. Their positions are specified by numbering the carbon atoms of the ring in the direction that gives the lowest number to the substituents at the first point of difference.

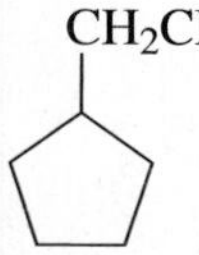

CH_2CH_3 or

Ethylcyclopentane

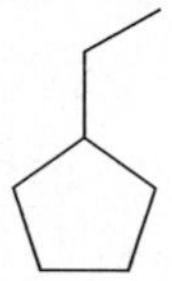

H_3C CH_3 CH_2CH_3 or

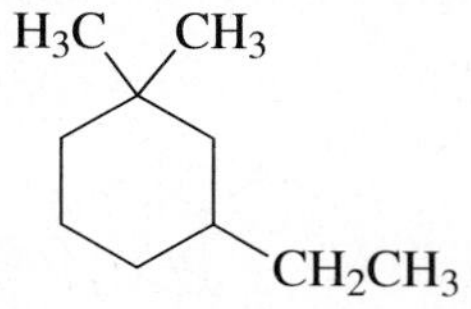

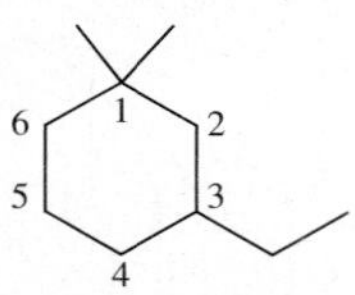

3-Ethyl-1,1-dimethylcyclohexane
(not 1-ethyl-3,3-dimethylcyclohexane, because first point of difference rule requires 1,1,3 substitution rather than 1,3,3)

When the ring contains fewer carbon atoms than an alkyl group attached to it, the compound is named as an alkane, and the ring is treated as a *cycloalkyl* substituent:

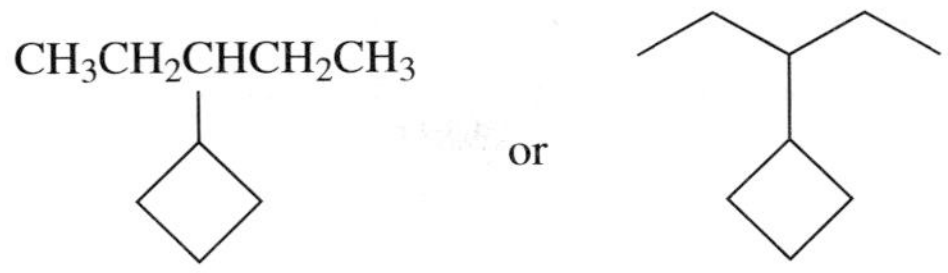

$CH_3CH_2CHCH_2CH_3$ or

3-Cyclobutylpentane

Problem 2.15

Name each of the following compounds:

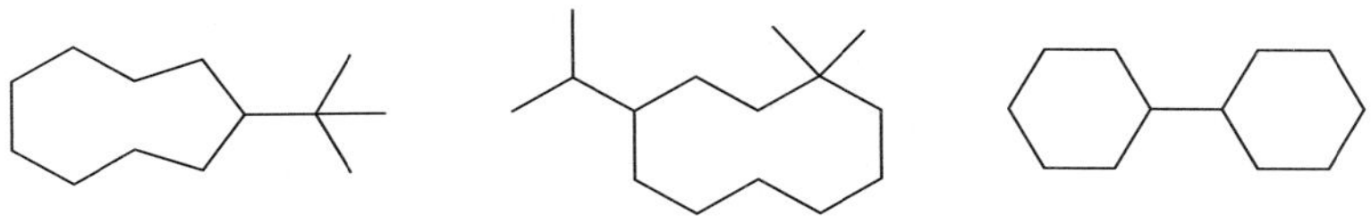

(a) (b) (c)

Sample Solution (a) The molecule has a *tert*-butyl group bonded to a nine-membered cycloalkane. It is *tert*-butylcyclononane. Alternatively, the *tert*-butyl group could be named systematically as a 1,1-dimethylethyl group, and the compound would then be named (1,1-dimethylethyl)cyclononane. Parentheses are used when necessary to avoid ambiguity. In this case the parentheses alert the reader that the locants 1,1 refer to substituents on the alkyl group and not to ring positions.

2.20 Introduction to Functional Groups

Much of the rich chemistry of organic compounds is associated with the presence of structural units called **functional groups** attached to or otherwise incorporated within the carbon framework, and it is standard practice to classify a compound according to its functional group. Although we will defer a more extensive presentation until Section 5.1, it is useful at this point to introduce the role of functional groups as an organizational tool.

The IUPAC defines a functional group as an "atom, or a group of atoms that has similar chemical properties whenever it occurs in different compounds."

A commonly used shorthand represents a generic alkyl group by the symbol R and a functionalized derivative of it as RX, where X is a functional group. Thus, a reaction in which X is replaced by some other group Y is written as:

$$RX + Y \longrightarrow RY + X$$

Among functional groups, hydroxyl (X = OH) and halogens, particularly chlorine (X = Cl) and bromine (X = Br), are especially frequently encountered.

Table 5.1 lists the functional groups encountered most often in this text.

A widely used method called **functional class nomenclature** for compounds that bear a functional group is to name the alkyl group followed by the functional group as a separate word.

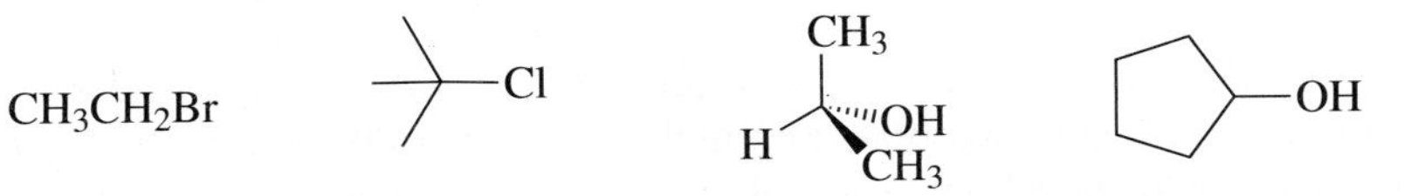

CH_3CH_2Br — Ethyl bromide

tert-Butyl chloride

Isopropyl alcohol

Cyclopentyl alcohol

2.21 Sources of Alkanes and Cycloalkanes

As noted earlier, natural gas is mostly methane but also contains ethane and propane, along with smaller amounts of other low-molecular-weight alkanes. Natural gas is often found associated with petroleum deposits. Petroleum is a liquid mixture containing hundreds of substances, including approximately 150 hydrocarbons, roughly half

The word *petroleum* is derived from the Latin words for "rock" (*petra*) and "oil" (*oleum*).

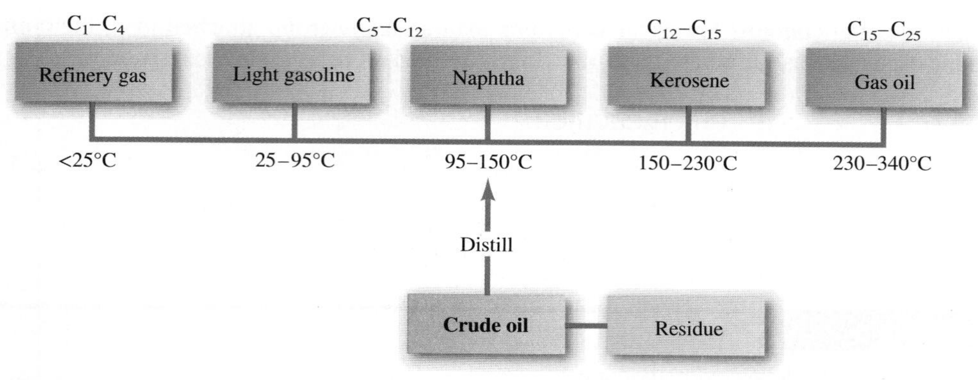

Figure 2.21

Distillation of crude oil yields a series of volatile fractions having the names indicated, along with a nonvolatile residue. The number of carbon atoms that characterize the hydrocarbons in each fraction is approximate.

of which are alkanes or cycloalkanes. Distillation of crude oil gives a number of volatile fractions, which by custom have the names given in Figure 2.21. High-boiling fractions such as kerosene and gas oil find wide use as fuels for diesel engines and furnaces, and the nonvolatile residue can be processed to give lubricating oil, greases, petroleum jelly, paraffin wax, and asphalt. Jet fuels are obtained from the naphtha–kerosene fractions.

Although both are closely linked in our minds and by our own experience, the petroleum industry predated the automobile industry by half a century. The first oil well, drilled in Titusville, Pennsylvania, by Edwin Drake in 1859, provided "rock oil," as it was then called, on a large scale. This was quickly followed by the development of a process to "refine" it so as to produce kerosene. As a fuel for oil lamps, kerosene burned with a bright, clean flame and soon replaced the vastly more expensive whale oil then in use (Figure 2.22). Other oil fields were discovered, and uses for other petroleum products were found—illuminating city streets with gas lights, heating homes with oil, and powering locomotives. There were oil refineries long before there were automobiles. By the time the first Model T rolled off Henry Ford's assembly line in 1908, John D. Rockefeller's Standard Oil holdings had already made him one of the half-dozen wealthiest people in the world.

Figure 2.22

The earliest major use for petroleum was as a fuel for oil lamps.
©J. Lekavicius/Shutterstock

Modern petroleum **refining** involves more than distillation, however, and includes two major additional operations:

1. **Cracking.** The more volatile, lower-molecular-weight hydrocarbons are useful as automotive fuels and as a source of petrochemicals. Cracking increases the proportion of these hydrocarbons at the expense of higher-molecular-weight ones by processes that involve the cleavage of carbon–carbon bonds induced by heat (*thermal cracking*) or with the aid of certain catalysts (*catalytic cracking*).
2. **Reforming.** The physical properties of the crude oil fractions known as *light gasoline* and *naphtha* (Figure 2.21) are appropriate for use as a motor fuel, but their ignition characteristics in high-compression automobile engines are poor and give rise to preignition, or "knocking." Reforming converts the hydrocarbons in petroleum to aromatic hydrocarbons and highly branched alkanes, both of which show less tendency for knocking than unbranched alkanes and cycloalkanes.

The tendency of a gasoline to cause "knocking" in an engine is given by its octane number. The lower the octane number, the greater the tendency. The two standards are heptane (assigned a value of 0) and "isooctane" (2,2,4-trimethylpentane, which is assigned a value of 100). The octane number of a gasoline is equal to the percentage of isooctane in a mixture of isooctane and heptane that has the same tendency to cause knocking as that sample of gasoline.

Petroleum is not the only place where alkanes occur naturally. Solid *n*-alkanes, especially those with relatively long chains, have a waxy constituency and coat the outer surface of many living things where they help prevent the loss of water. Pentacosane [$CH_3(CH_2)_{23}CH_3$] is present in the waxy outer layer of most insects. Hentriacontane [$CH_3(CH_2)_{29}CH_3$] is a component of beeswax (see Problem 2.10) as well as the wax that coats the leaves of tobacco, peach trees, pea plants, and numerous others. The C_{23}, C_{25}, C_{27}, C_{29}, and C_{31} *n*-alkanes have been identified in the surface coating of the eggs of honeybee queens.

Cyclopentane and cyclohexane are present in petroleum, but as a rule, hydrocarbons based on cycloalkane frameworks rarely occur naturally. An exception is a group of more than 200 hydrocarbons called *hopanes,* related to the parent having the carbon skeleton shown.

Hopane

Hopanes were first found in petroleum and geological sediments, later as components of certain bacterial cell membranes. Although present in small amounts, hopanes are so widespread that they rank among the most abundant natural products on Earth.

Problem 2.16

What is the molecular formula of hopane?

2.22 Physical Properties of Alkanes and Cycloalkanes

Boiling Point. As we have seen earlier in this chapter, methane, ethane, propane, and butane are gases at room temperature. The unbranched alkanes pentane (C_5H_{12}) through heptadecane ($C_{17}H_{36}$) are liquids, whereas higher homologs are solids. As shown in Figure 2.23, the boiling points of unbranched alkanes increase with the number of carbon atoms and the boiling points of 2-methyl-branched alkanes are lower than those of the unbranched isomer. By exploring at the molecular level the reasons for the increase in boiling point with the number of carbons and the difference in boiling point between branched and unbranched alkanes, we can continue to connect structure with properties.

A substance exists as a liquid rather than a gas because attractive forces between molecules **(intermolecular attractive forces)** are greater in the liquid. Attractive forces

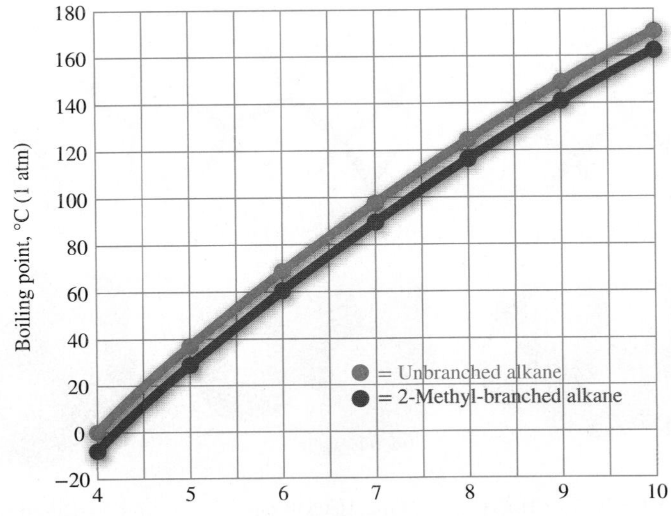

Figure 2.23

Boiling points of unbranched alkanes and their 2-methyl-branched isomers.

between neutral species (atoms or molecules, but not ions) are referred to as **van der Waals forces** and may be of three types:

Induced-dipole/induced-dipole attractive forces are often called *London forces,* or *dispersion forces.*

1. dipole–dipole (including hydrogen bonding)
2. dipole/induced-dipole
3. induced-dipole/induced-dipole

These forces are electrical in nature, and in order to vaporize a substance, enough energy must be added to overcome them. Most alkanes have no measurable dipole moment, and therefore the only van der Waals force to be considered is the **induced-dipole/induced-dipole attractive force.**

It might seem that two nearby molecules A and B of the same nonpolar substance would be unaffected by each other.

In fact, the electric fields of both A and B are dynamic and fluctuate in a complementary way that results in a temporary dipole moment and a weak attraction between them.

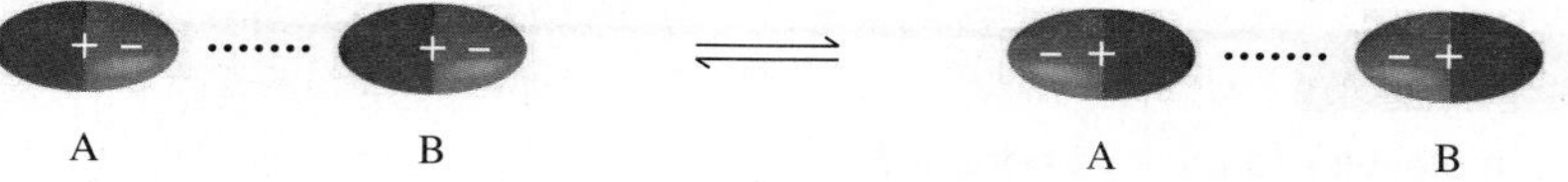

Extended assemblies of induced-dipole/induced-dipole attractions can accumulate to give substantial intermolecular attractive forces. An alkane with a higher molecular weight has more atoms and electrons and, therefore, more opportunities for intermolecular attractions and a higher boiling point than one with a lower molecular weight.

As noted earlier in this section, branched alkanes have lower boiling points than their unbranched isomers. Isomers have the same number of atoms and electrons, but a molecule of a branched alkane has a smaller surface area than an unbranched one. The extended shape of an unbranched alkane permits more points of contact for intermolecular associations. Compare the boiling points of pentane (36°C) and its isomers 2-methylbutane (28°C) and 2,2-dimethylpropane (bp 9°C).

The shapes of these isomers are evident in the space-filling models depicted in Figure 2.24. Pentane has the most extended structure and the largest surface area available for "sticking" to other molecules by way of induced-dipole/induced-dipole attractive forces; it has the highest boiling point. 2,2-Dimethylpropane has the most compact, most spherical structure, engages in the fewest induced-dipole/induced-dipole attractions, and has the lowest boiling point.

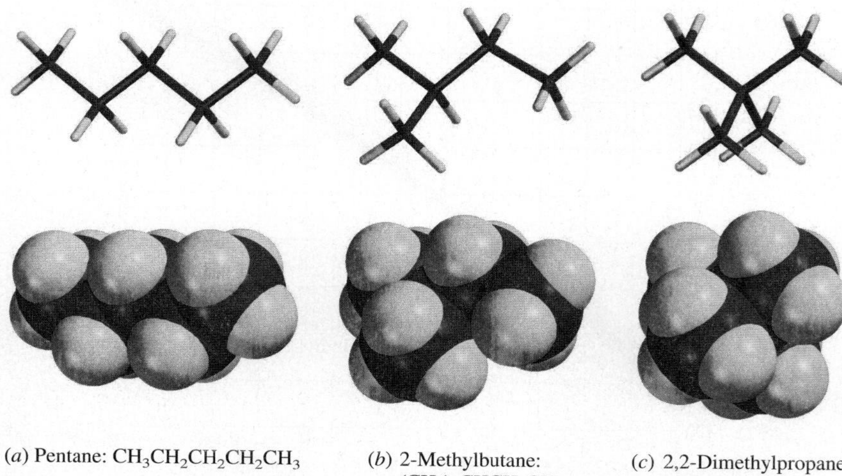

(*a*) Pentane: $CH_3CH_2CH_2CH_2CH_3$

(*b*) 2-Methylbutane: $(CH_3)_2CHCH_2CH_3$

(*c*) 2,2-Dimethylpropane: $(CH_3)_4C$

Figure 2.24

Tube (*top*) and space-filling (*bottom*) models of (*a*) pentane, (*b*) 2-methylbutane, and (*c*) 2,2-dimethylpropane. The most branched isomer, 2,2-dimethylpropane, has the most compact, most spherical three-dimensional shape.

Induced-dipole/induced-dipole attractions are very weak forces individually, but a typical organic substance can participate in so many of them that they are collectively the most important of all the contributors to intermolecular attraction in the liquid state. They are the only forces of attraction possible between nonpolar molecules such as alkanes.

Problem 2.17

Match the boiling points with the appropriate alkanes.
Alkanes: octane, 2-methylheptane, 2,2,3,3-tetramethylbutane, nonane
Boiling points (°C, 1 atm): 106, 116, 126, 151

Cyclopentane has a higher boiling point (49.3°C) than pentane (36°C), indicating greater forces of association in the cyclic alkane than occur in the straight-chain alkane.

Melting Point. Solid alkanes are soft, generally low-melting materials. The forces responsible for holding the crystal together are the same induced-dipole/induced-dipole interactions that operate between molecules in the liquid, but the degree of organization is greater in the solid phase. By measuring the distances between the atoms of one molecule and its neighbor in the crystal, it is possible to specify a distance of closest approach characteristic of an atom called its **van der Waals radius.** In space-filling molecular models, such as those of pentane, 2-methylbutane, and 2,2-dimethylpropane shown in Figure 2.24, the radius of each sphere corresponds to the van der Waals radius of the atom it represents. The van der Waals radius for hydrogen is 120 pm (1.20 Å). When two alkane molecules are brought together so that a hydrogen of one molecule is within 240 pm (2.40 Å) of a hydrogen of the other, the balance between electron–nucleus attractions versus electron–electron and nucleus–nucleus repulsions is most favorable. Closer approach is resisted by a strong increase in repulsive forces.

Solubility in Water. A familiar physical property of alkanes is contained in the adage "oil and water don't mix." Alkanes—indeed all hydrocarbons—are virtually insoluble in water. In order for a hydrocarbon to dissolve in water, the framework of hydrogen bonds between water molecules would become more ordered in the region around each molecule of the dissolved hydrocarbon. This increase in order, which corresponds to a decrease in entropy, signals a process that can be favorable only if it is reasonably exothermic. Such is not the case here. The hydrogen bonding among water molecules is too strong to be disrupted by nonpolar hydrocarbons. Being insoluble, and with densities in the 0.6–0.8 g/mL range, alkanes float on the surface of water. The exclusion of nonpolar molecules, such as alkanes, from water is called the **hydrophobic effect.**

2.23 Chemical Properties: Combustion of Alkanes

An older name for alkanes is **paraffin hydrocarbons.** *Paraffin* is derived from the Latin words *parum affinis* ("with little affinity") and testifies to the low level of reactivity of alkanes.

Although essentially inert in acid–base reactions, alkanes do participate in oxidation–reduction reactions as the compound that undergoes oxidation. Burning in air **(combustion)** is the best known and most important example. Combustion of hydrocarbons is exothermic and gives carbon dioxide and water as the products.

Alkanes are so unreactive that George A. Olah of the University of Southern California was awarded the 1994 Nobel Prize in Chemistry in part for developing novel substances that do react with alkanes.

$$\underset{\text{Methane}}{CH_4} + \underset{\text{Oxygen}}{2O_2} \longrightarrow \underset{\text{Carbon dioxide}}{CO_2} + \underset{\text{Water}}{2H_2O} \qquad \Delta H° = -891 \text{ kJ/mol } (-212.9 \text{ kcal/mol})$$

$$\underset{\text{2-Methylbutane}}{(CH_3)_2CHCH_2CH_3} + \underset{\text{Oxygen}}{8O_2} \longrightarrow \underset{\text{Carbon dioxide}}{5CO_2} + \underset{\text{Water}}{6H_2O} \qquad \Delta H° = -3529 \text{ kJ/mol } (-843.4 \text{ kcal/mol})$$

Problem 2.18

Write a balanced chemical equation for the combustion of cyclohexane.

TABLE 2.3 Heats of Combustion ($-\Delta H°$) of Representative Alkanes

Compound	Formula	$-\Delta H°$ kJ/mol	kcal/mol
Unbranched alkanes			
Hexane	$CH_3(CH_2)_4CH_3$	4,195	1002.6
Heptane	$CH_3(CH_2)_5CH_3$	4,853	1160.0
Octane	$CH_3(CH_2)_6CH_3$	5,512	1317.4
Nonane	$CH_3(CH_2)_7CH_3$	6,171	1474.9
Decane	$CH_3(CH_2)_8CH_3$	6,830	1632.4
Undecane	$CH_3(CH_2)_9CH_3$	7,488	1789.7
Dodecane	$CH_3(CH_2)_{10}CH_3$	8,147	1947.2
Hexadecane	$CH_3(CH_2)_{14}CH_3$	10,781	2576.8
2-Methyl-branched alkanes			
2-Methylpentane	$(CH_3)_2CHCH_2CH_2CH_3$	4,187	1000.8
2-Methylhexane	$(CH_3)_2CH(CH_2)_3CH_3$	4,847	1158.4
2-Methylheptane	$(CH_3)_2CH(CH_2)_4CH_3$	5,505	1315.8

The heat released on combustion of a substance is called its **heat of combustion** and is equal to $-\Delta H°$ for the reaction. By convention

$$\Delta H° = H°_{\text{products}} - H°_{\text{reactants}}$$

where $H°$ is the heat content, or **enthalpy,** of a compound in its standard state, that is, the gas, pure liquid, or crystalline solid at a pressure of 1 atm. In an exothermic process the enthalpy of the products is less than that of the starting materials, and $\Delta H°$ is a negative number.

Table 2.3 lists the heats of combustion of several alkanes. Unbranched alkanes have slightly higher heats of combustion than their 2-methyl-branched isomers, but the most important factor is the number of carbons. The unbranched alkanes and the 2-methyl-branched alkanes constitute two separate *homologous series* (Section 2.13) in which there is a regular increase of about 659 kJ/mol (157.4 kcal/mol) in the heat of combustion for each additional CH_2 group.

Problem 2.19

Using the data in Table 2.3, estimate the heat of combustion of

(a) 2-Methylnonane
(b) Eicosane

Sample Solution (a) The last entry for the group of 2-methylalkanes in the table is 2-methylheptane. Its heat of combustion is 5505 kJ/mol (1315.8 kcal/mol). Because 2-methylnonane has two more methylene groups than 2-methylheptane, its heat of combustion is 2 × 659 kJ/mol (or 2 × 157.4 kcal/mol) higher.

$$\text{Heat of combustion of 2-methylnonane} = 5505 + 2(659) = 6823 \text{ kJ/mol}$$
$$(1315.8 + 2(157.4) = 1630.6 \text{ kcal/mol})$$

Heats of combustion can be used to compare the relative stability of isomeric hydrocarbons. They tell us not only which isomer is more stable than another, but also

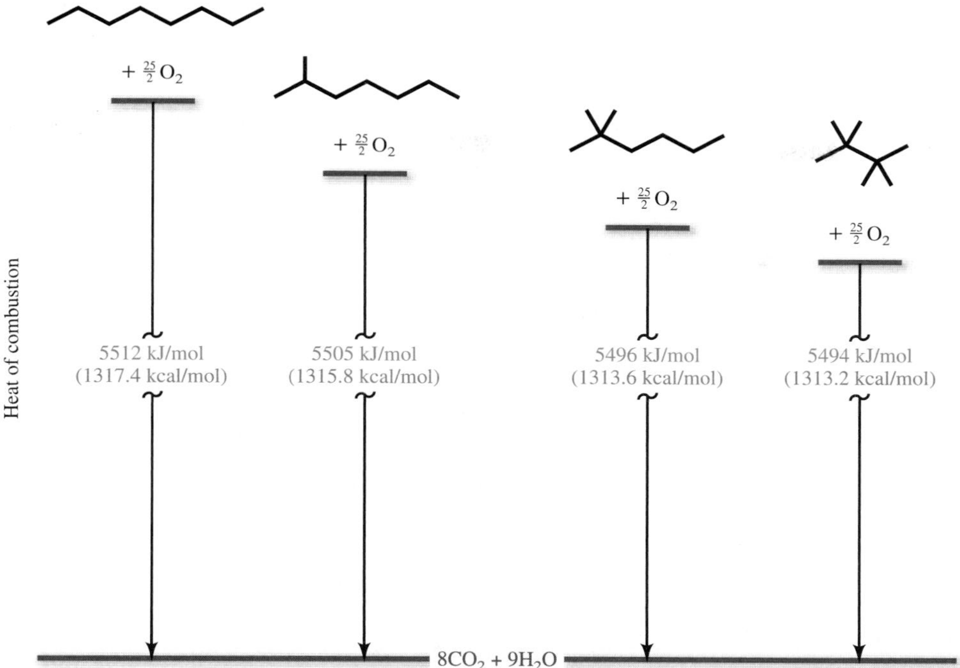

Figure 2.25

Energy diagram comparing heats of combustion of isomeric C_8H_{18} alkanes. The most branched isomer is the most stable; the unbranched isomer has the highest heat of combustion and is the least stable.

by how much. Figure 2.25 compares the heats of combustion of several C_8H_{18} isomers on a *potential energy diagram.* **Potential energy** is comparable with enthalpy; it is the energy a molecule has exclusive of its kinetic energy. These C_8H_{18} isomers all undergo combustion to the same final state according to the equation:

$$C_8H_{18} + \frac{25}{2}O_2 \longrightarrow 8CO_2 + 9H_2O$$

therefore, the differences in their heats of combustion translate directly to differences in their potential energies. *When comparing isomers, the one with the lowest potential energy (in this case, the smallest heat of combustion) is the most stable.* Among the C_8H_{18} alkanes, the most highly branched isomer, 2,2,3,3-tetramethylbutane, is the most stable, and the unbranched isomer octane is the least stable. It is generally true for alkanes that a more branched isomer is more stable than a less branched one.

The small differences in stability between branched and unbranched alkanes result from an interplay between attractive and repulsive forces *within* a molecule (**intramolecular forces**). These forces are nucleus–nucleus repulsions, electron–electron repulsions, and nucleus–electron attractions, the same set of fundamental forces we met when talking about chemical bonding (Section 2.2) plus the van der Waals forces between molecules (Section 2.22). When the energy associated with these interactions is calculated for all of the nuclei and electrons within a molecule, it is found that the attractive forces increase more than the repulsive forces as the structure becomes more compact. Sometimes, though, two atoms in a molecule are held too closely together. We'll explore the consequences of that in Chapter 3.

Problem 2.20

Without consulting Table 2.3, arrange the following compounds in order of decreasing heat of combustion: pentane, 2-methylbutane, 2,2-dimethylpropane, hexane.

Thermochemistry

Thermochemistry is the study of the heat changes that accompany chemical processes. It has a long history dating back to the work of the French chemist Antoine Laurent Lavoisier in the late eighteenth century. Thermochemistry provides quantitative information that complements the qualitative description of a chemical reaction and can help us understand why some reactions occur and others do not. It is of value when assessing the relative value of various materials as fuels, when comparing the stability of isomers, or when determining the practicality of a particular reaction. In the field of bioenergetics, thermochemical information is applied to the task of sorting out how living systems use chemical reactions to store and use the energy that originates in the sun.

By allowing compounds to react in a calorimeter, it is possible to measure the heat evolved in an exothermic reaction or the heat absorbed in an endothermic one. Thousands of reactions have been studied to produce a rich library of thermochemical data. These data take the form of heats of reaction and correspond to the value of the standard enthalpy change $\Delta H°$ for a particular reaction of a particular substance.

In this section you have seen how heats of combustion can be used to determine relative stabilities of isomeric alkanes. In later sections we shall expand our scope to include the experimentally determined heats of certain other reactions, to see how $\Delta H°$ values from various sources can aid our understanding of structure and reactivity.

The **standard heat of formation** (ΔH_f°) is the enthalpy change for formation of one mole of a compound directly from its elements, and is one type of heat of reaction. In cases such as the formation of CO_2 or H_2O from the combustion of carbon or hydrogen, respectively, the heat of formation of a substance can be measured directly. In most other cases, heats of formation are not measured experimentally but are calculated from the measured heats of other reactions. Consider, for example, the heat of formation of methane. The reaction that defines the formation of methane from its elements,

$$\underset{\text{Carbon}}{\text{C (graphite)}} + \underset{\text{Hydrogen}}{2H_2(g)} \longrightarrow \underset{\text{Methane}}{CH_4(g)}$$

can be expressed as the sum of three reactions:

$$\begin{array}{llll}
(1)\ \text{C (graphite)} + O_2(g) & \longrightarrow CO_2(g) & \Delta H° = -393\text{ kJ/mol} & (-93.9\text{ kcal/mol}) \\
(2)\ 2H_2(g) + O_2(g) & \longrightarrow 2H_2O(l) & \Delta H° = -572\text{ kJ/mol} & (-136.7\text{ kcal/mol}) \\
(3)\ CO_2(g) + 2H_2O(l) & \longrightarrow CH_4(g) + 2O_2(g) & \Delta H° = +890\text{ kJ/mol} & (+212.7\text{ kcal/mol}) \\
\hline
\text{C (graphite)} + 2H_2(g) & \longrightarrow CH_4(g) & \Delta H° = -75\text{ kJ/mol} & (-17.9\text{ kcal/mol})
\end{array}$$

Equations (1) and (2) are the heats of formation of one mole of carbon dioxide and two moles of water, respectively. Equation (3) is the reverse of the combustion of methane, and so the heat of reaction is equal to the heat of combustion but opposite in sign. The sum of equations (1)–(3) is the enthalpy change for formation of one mole of methane from its elements. Thus, $\Delta H_f^\circ = -75$ kJ/mol (−17.9 kcal/mol).

The heats of formation of most organic compounds are derived from heats of reaction by arithmetic manipulations similar to that shown. Chemists find a table of ΔH_f° values to be convenient because it replaces many separate tables of $\Delta H°$ values for individual reaction types and permits $\Delta H°$ to be calculated for any reaction, real or imaginary, for which the heats of formation of reactants and products are available.

Problem 2.21

Given the standard enthalpies of formation (ΔH_f°) of +53.3 kJ/mol (+12.7 kcal/mol) for cyclopropane and −123.4 kJ/mol (−29.5 kcal/mol) for cyclohexane, calculate $\Delta H°$ for the reaction:

2   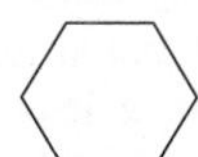

2.24 Oxidation–Reduction in Organic Chemistry

As we have just seen, the reaction of alkanes with oxygen to give carbon dioxide and water is called *combustion*. A more fundamental classification of reaction types places it in the **oxidation–reduction** category. To understand why, let's review some principles of oxidation–reduction, beginning with the **oxidation number** (also known as **oxidation state**).

There are a variety of methods for calculating oxidation numbers. In compounds that contain a single carbon, such as methane (CH_4) and carbon dioxide (CO_2), the oxidation number of carbon can be calculated from the molecular formula. For neutral molecules the algebraic sum of all the oxidation numbers must equal zero. Assuming, as is customary, that the oxidation state of hydrogen is +1, the oxidation state of carbon in CH_4 then is calculated to be −4. Similarly, assuming an oxidation state of −2 for oxygen, carbon is +4 in H_2CO_3 and CO_2.

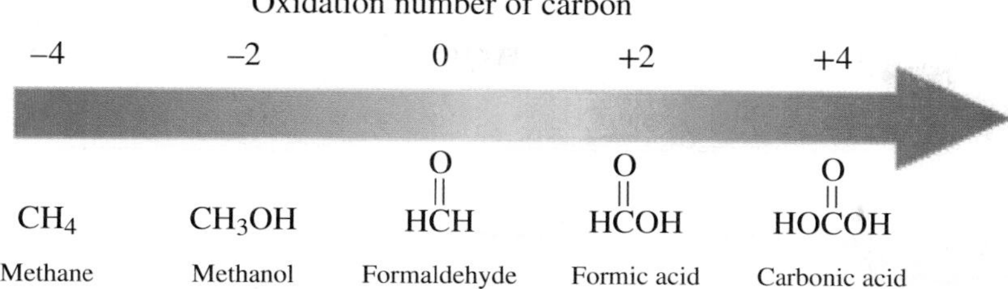

The carbon in methane has the lowest oxidation number (−4) of any of the compounds shown and contains carbon in its most *reduced* form. Carbon dioxide and carbonic acid have the highest oxidation numbers (+4) for carbon, corresponding to its most *oxidized* state. When methane or any alkane undergoes combustion to form carbon dioxide, carbon is oxidized and oxygen is reduced.

A useful generalization is the following:

Oxidation of carbon corresponds to an increase in the number of bonds between carbon and oxygen or to a decrease in the number of carbon–hydrogen bonds. Conversely, reduction corresponds to an increase in the number of carbon–hydrogen bonds or to a decrease in the number of carbon–oxygen bonds.

Each successive increase in oxidation state increases the number of bonds between carbon and oxygen and decreases the number of carbon–hydrogen bonds. Methane has four C—H bonds and no C—O bonds; carbonic acid and carbon dioxide both have four C—O bonds and no C—H bonds.

Among the various classes of hydrocarbons, alkanes contain carbon in its most reduced state, and alkynes contain carbon in its most oxidized state.

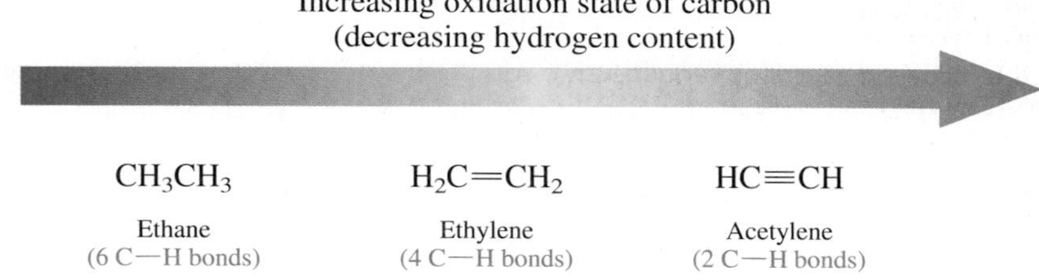

CH_3CH_3	$H_2C{=}CH_2$	$HC{\equiv}CH$
Ethane (6 C—H bonds)	Ethylene (4 C—H bonds)	Acetylene (2 C—H bonds)

We can extend the generalization by recognizing that the pattern of oxidation states is not limited to increasing oxygen or hydrogen content. Any element *more electronegative* than carbon will have the same effect on oxidation number as oxygen. Thus, the oxidation numbers of carbon in CH_3Cl and CH_3OH are the same (−2). The reaction of chlorine with methane (to be discussed in Sections 10.2–10.3) involves *oxidation* at carbon.

$$\underset{\text{Methane}}{CH_4} + \underset{\text{Chlorine}}{Cl_2} \longrightarrow \underset{\text{Chloromethane}}{CH_3Cl} + \underset{\text{Hydrogen chloride}}{HCl}$$

Any element *less electronegative* than carbon will have the same effect on oxidation number as hydrogen. Thus, the oxidation numbers of carbon in CH_3Li and CH_4 are the same (−4), and the reaction of CH_3Cl with lithium (to be discussed in Section 15.3) involves *reduction* at carbon.

$$\underset{\text{Chloromethane}}{CH_3Cl} + \underset{\text{Lithium}}{2Li} \longrightarrow \underset{\text{Methyllithium}}{CH_3Li} + \underset{\text{Lithium chloride}}{LiCl}$$

The oxidation number of carbon decreases from −2 in CH_3Cl to −4 in CH_3Li.

The generalization illustrated by the preceding examples can be expressed in terms broad enough to cover these reactions and many others, as follows: *Oxidation of carbon occurs when a bond between a carbon and an atom that is less electronegative than carbon is replaced by a bond to an atom that is more electronegative than carbon. The reverse process is reduction.*

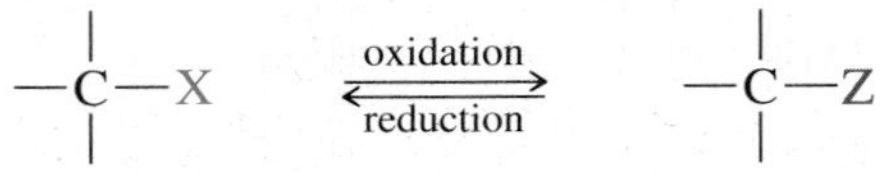

$$-\overset{|}{\underset{|}{C}}-X \underset{\text{reduction}}{\overset{\text{oxidation}}{\rightleftharpoons}} -\overset{|}{\underset{|}{C}}-Z$$

X is less electronegative than carbon

Z is more electronegative than carbon

Problem 2.22

Both of the following reactions will be encountered later in this text. One is oxidation–reduction, the other is not. Which is which?

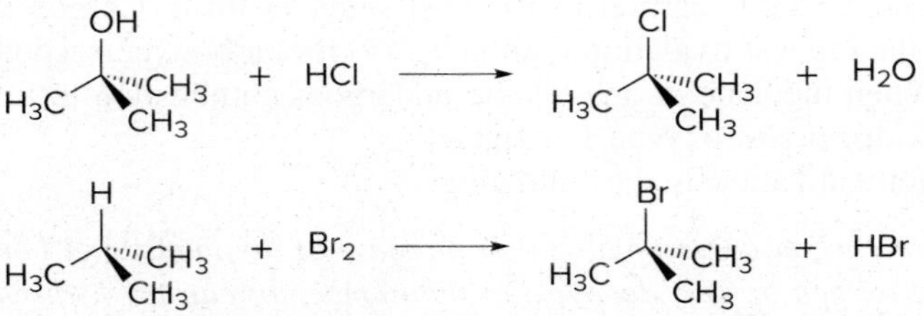

$(CH_3)_3COH + HCl \longrightarrow (CH_3)_3CCl + H_2O$

$(CH_3)_3CH + Br_2 \longrightarrow (CH_3)_3CBr + HBr$

Many, indeed most, organic compounds contain carbon in more than one oxidation state. Consider ethanol (CH_3CH_2OH), for example. One carbon is connected to three hydrogens; the other carbon to two hydrogens and one oxygen. Although we could calculate the actual oxidation numbers, we rarely need to. Most of the time we are only concerned with whether a particular reaction is an oxidation or a reduction. The ability to recognize when oxidation or reduction occurs is of value when deciding on the kind of reactant with which an organic molecule must be treated to convert it into a desired product.

Problem 2.23

Which of the following reactions requires an oxidizing agent, a reducing agent, or neither?

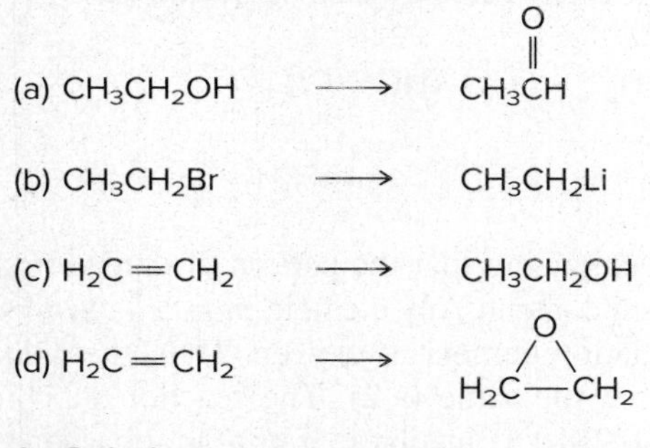

(a) $CH_3CH_2OH \longrightarrow CH_3\overset{O}{\overset{\|}{C}}H$

(b) $CH_3CH_2Br \longrightarrow CH_3CH_2Li$

(c) $H_2C{=}CH_2 \longrightarrow CH_3CH_2OH$

(d) $H_2C{=}CH_2 \longrightarrow$ $H_2C—CH_2$ (epoxide ring, with O bonded to both carbons)

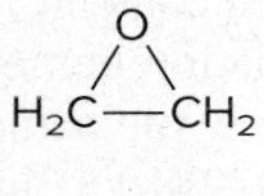

Sample Solution (a) The CH_3 carbon is unchanged in the reaction; however, the carbon of CH_2OH now has two bonds to oxygen. Therefore, the reaction requires an oxidizing agent.

2.25 SUMMARY

Section 2.1 The classes of hydrocarbons are **alkanes, alkenes, alkynes,** and **arenes.** Alkanes are hydrocarbons in which all of the bonds are *single* bonds and are characterized by the molecular formula C_nH_{2n+2}.

Section 2.2 Two theories of bonding, valence bond and molecular orbital theory, are based on the wave nature of an electron. Constructive interference between the electron wave of one atom and that of another gives a region between the two atoms in which the probability of sharing an electron is high—a bond.

Section 2.3 In valence bond theory a covalent bond is described in terms of in-phase overlap of a half-filled orbital of one atom with a half-filled orbital of another. When applied to bonding in H_2, the orbitals involved are the 1*s* orbitals of two hydrogen atoms and the bond is a σ bond.

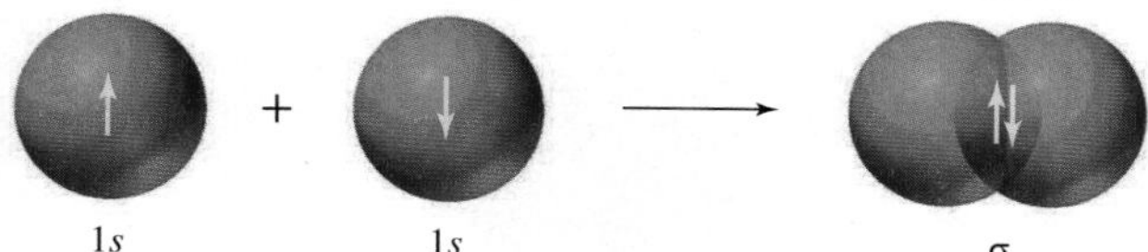

Section 2.4 In molecular orbital theory, the molecular orbitals (MOs) are approximated by combining the atomic orbitals (AOs) of all of the atoms in a molecule. The number of MOs must equal the number of AOs that are combined.

Section 2.5 The first three alkanes are methane (CH_4), ethane (CH_3CH_3), and propane ($CH_3CH_2CH_3$).

Section 2.6 Bonding in methane is most often described by an orbital hybridization model, which is a modified form of valence bond theory. Four equivalent sp^3 hybrid orbitals of carbon are generated by mixing the 2*s*, $2p_x$, $2p_y$, and $2p_z$ orbitals. In-phase overlap of each half-filled sp^3 hybrid orbital with a half-filled hydrogen 1*s* orbital gives a σ bond.

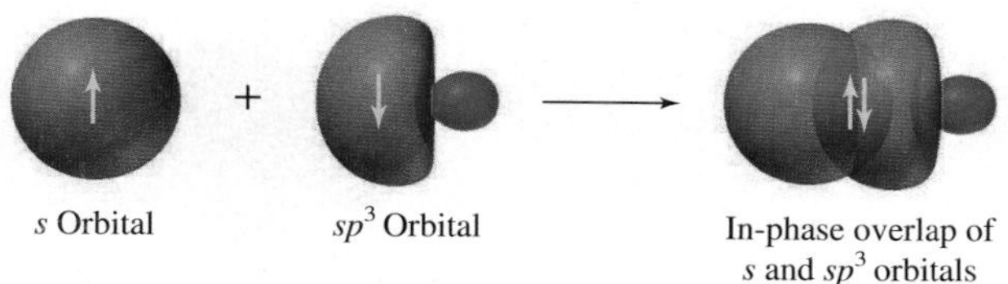

Section 2.7 The carbon–carbon bond in ethane is a σ bond viewed as an overlap of a half-filled sp^3 orbital of one carbon with a half-filled sp^3 orbital of another.

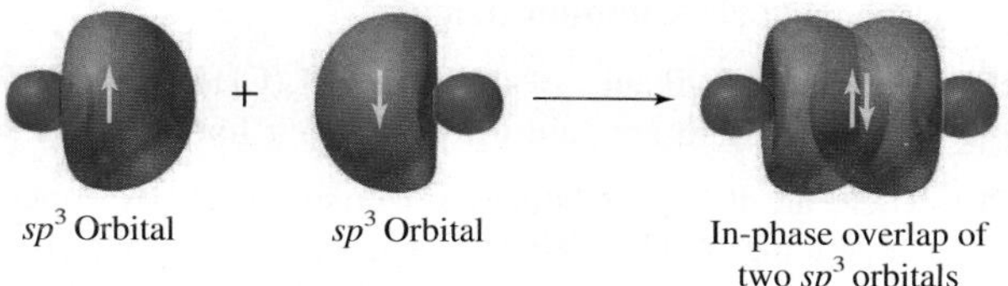

Section 2.8 Carbon is ***sp*²-hybridized** in ethylene, and the double bond has a σ component and a π component. The sp^2 hybridization state is derived by mixing the 2*s* and two of the three 2*p* orbitals. Three equivalent sp^2 orbitals result, and their axes are coplanar. Overlap of a half-filled sp^2 orbital of one carbon with a half-filled sp^2 orbital of another gives a σ bond between the two carbons. Each carbon still has one unhybridized *p* orbital available for bonding, and "side-by-side" overlap of half-filled *p* orbitals of adjacent carbons gives a π bond between them.

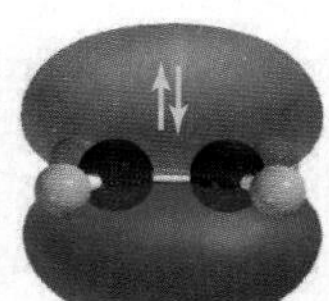

The π bond in ethylene is generated by overlap of half-filled *p* orbitals of adjacent carbons.

Section 2.9 Carbon is ***sp*-hybridized** in acetylene, and the triple bond is of the σ + π + π type. The 2*s* orbital and one of the 2*p* orbitals combine to give two equivalent *sp* orbitals that have their axes in a straight line. A σ bond between the two carbons is supplemented by two π bonds formed by overlap of the remaining half-filled *p* orbitals.

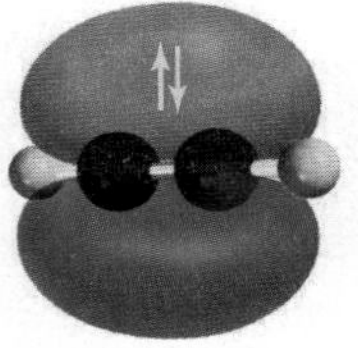
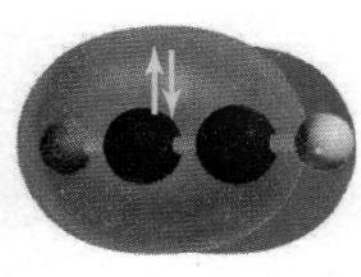

The triple bond of acetylene has a σ bond component and two π bonds; the two π bonds are shown here and are perpendicular to each other.

Section 2.10 The hybridization model for carbon can also be extended to oxygen and nitrogen. Oxygen and nitrogen are sp^3-hybridized in water and ammonia, respectively.

Section 2.11 Instead of being localized in a bond between two atoms, electron pairs are more delocalized in molecular orbital theory. The lowest energy bonding molecular orbital in butane and other alkanes, for example, encompasses all of the atoms. Higher-energy orbitals have one or more nodal surfaces where the probability of finding an electron is zero.

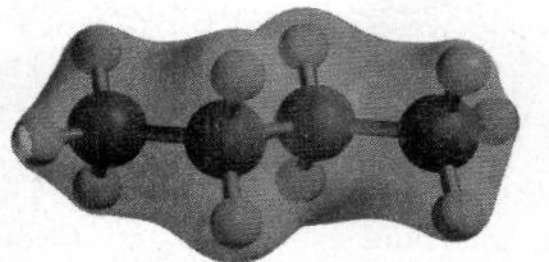

Lowest-energy molecular orbital of butane

Section 2.12 Two constitutionally isomeric alkanes have the molecular formula C_4H_{10}. One has an unbranched chain ($CH_3CH_2CH_2CH_3$) and is called ***n*-butane;** the other has a branched chain [$(CH_3)_3CH$] and is called **isobutane.** Both *n*-butane and isobutane are **common names.**

Section 2.13 Unbranched alkanes of the type $CH_3(CH_2)_xCH_3$ are often referred to as *n*-alkanes, and are said to belong to a **homologous series.**

Section 2.14 There are three constitutional isomers of C_5H_{12}: *n*-pentane ($CH_3CH_2CH_2CH_2CH_3$), isopentane [$(CH_3)_2CHCH_2CH_3$], and neopentane [$(CH_3)_4C$].

Sections 2.15–2.19 A single alkane may have several different names; a name may be a common name, or it may be a *systematic name* developed by a well-defined set of rules. The most widely used system is **IUPAC** nomenclature. Table 2.4 summarizes the rules for alkanes and cycloalkanes. Table 2.5 gives the rules for naming alkyl groups.

Section 2.20 Much of organic chemistry is concerned with structural units called **functional groups** incorporated into their structure. Alcohols and alkyl halides are introduced in this section as representative classes of compounds that contain a functional group.

Section 2.21 Natural gas is an abundant source of methane, ethane, and propane. Petroleum is a liquid mixture of many hydrocarbons, including alkanes. Alkanes also occur naturally in the waxy coating of leaves and fruits.

Section 2.22 Alkanes and cycloalkanes are nonpolar and insoluble in water. The forces of attraction between alkane molecules are **induced-dipole/induced-dipole** attractive forces. The boiling points of alkanes increase as the number of carbon atoms increases. Branched alkanes have lower boiling points than their unbranched isomers. There is a limit to how closely two atoms can approach each other, which is given by the sum of their **van der Waals radii.**

Section 2.23 Alkanes and cycloalkanes burn in air to give carbon dioxide, water, and heat. This process is called **combustion.**

$$\underset{\text{2-Methylbutane}}{(CH_3)_2CHCH_2CH_3} + \underset{\text{Oxygen}}{8O_2} \longrightarrow \underset{\text{Carbon dioxide}}{5CO_2} + \underset{\text{Water}}{6H_2O}$$

$$\Delta H° = -3529 \text{ kJ/mol } (-843.4 \text{ kcal/mol})$$

The heat evolved on burning an alkane increases with the number of carbon atoms. The relative stability of isomers may be determined by comparing their respective **heats of combustion.** The more stable of two isomers has the lower heat of combustion.

Section 2.24 Combustion of alkanes is an example of **oxidation–reduction.** Although it is possible to calculate oxidation numbers of carbon in organic molecules, it is more convenient to regard oxidation of an organic substance as an increase in its oxygen content or a decrease in its hydrogen content.

TABLE 2.4 Summary of IUPAC Nomenclature of Alkanes and Cycloalkanes

Rule	Example
A. Alkanes **1.** Find the longest continuous chain of carbon atoms, and assign a parent name to the compound corresponding to the IUPAC name of the unbranched alkane having the same number of carbons.	The longest continuous chain in the alkane shown is six carbons. 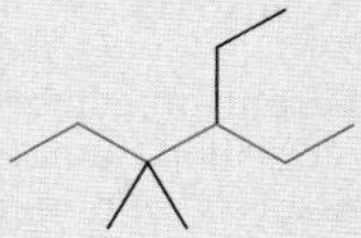 This alkane is named as a derivative of hexane.
2. List the substituents attached to the longest continuous chain in alphabetical order. Use the prefixes *di-*, *tri-*, *tetra-*, and so on, when the same substituent appears more than once. Ignore these prefixes when alphabetizing.	The alkane bears two methyl groups and an ethyl group. It is an *ethyldimethylhexane*. 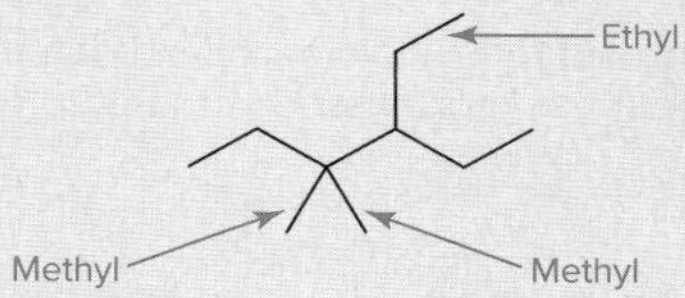 Ethyl Methyl Methyl
3. Number from the end of the chain in the direction that gives the lower locant to a substituent at the first point of difference.	When numbering from left to right, the substituents appear at carbons 3, 3, and 4. When numbering from right to left, the locants are 3, 4, and 4; therefore, number from left to right. 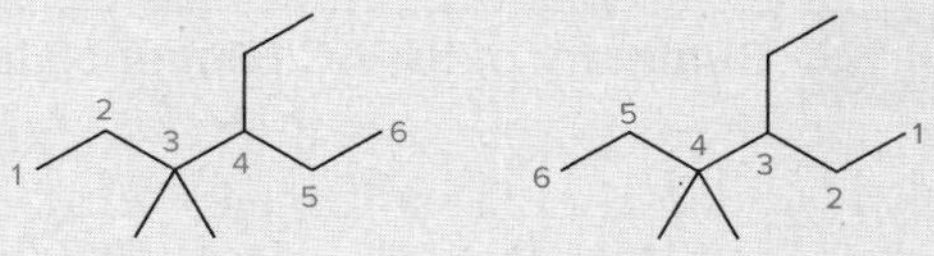 Correct Incorrect The correct name is *4-ethyl-3,3-dimethylhexane*.
4. When two different numbering schemes give equivalent sets of locants, choose the direction that gives the lower locant to the group that appears first in the name.	In the following example, the substituents are located at carbons 3 and 4 regardless of the direction in which the chain is numbered. 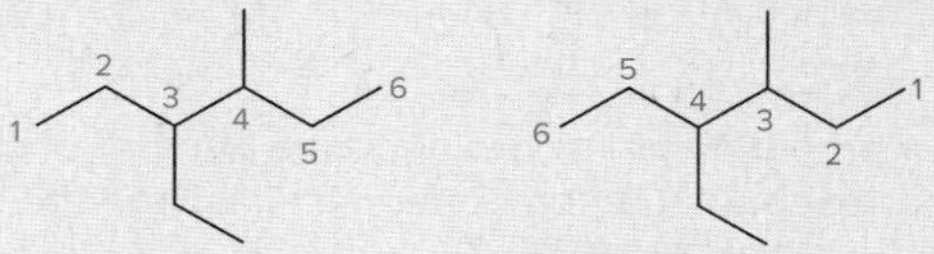 Correct Incorrect Ethyl precedes methyl in the name; therefore, *3-ethyl-4-methylhexane* is correct.

continued

TABLE 2.4 Summary of IUPAC Nomenclature of Alkanes and Cycloalkanes (*Continued*)

Rule	Example
5. When two chains are of equal length, choose the one with the greater number of substituents as the parent. (Although this requires naming more substituents, the substituents have simpler names.)	Two different chains contain five carbons in the alkane: Correct / Incorrect. The correct name is *3-ethyl-2-methylpentane* (disubstituted chain), rather than 3-isopropylpentane (monosubstituted chain).
B. Cycloalkanes **1.** Count the number of carbons in the ring, and assign a parent name to the cycloalkane corresponding to the IUPAC name of the unbranched cycloalkane having the same number of carbons.	The compound shown contains five carbons in its ring. (cyclopentane)–$CH(CH_3)_2$ It is named as a derivative of *cyclopentane*.
2. Name the alkyl group, and append it as a prefix to the cycloalkane. No locant is needed if the compound is a monosubstituted cycloalkane. It is understood that the alkyl group is attached to C-1.	The previous compound is *isopropylcyclopentane*. Alternatively, the alkyl group can be named according to the rules summarized in Table 2.5, whereupon the name becomes *(1-methylethyl)cyclopentane*. Parentheses are used to set off the name of the alkyl group as needed to avoid ambiguity.
3. When two or more different substituents are present, list them in alphabetical order, and number the ring in the direction that gives the lower number at the first point of difference.	The compound shown is *1,1-diethyl-4-hexylcyclooctane.* (CH_3CH_2, CH_3CH_2 at C-1; $CH_2CH_2CH_2CH_2CH_2CH_3$ at C-4)
4. Name the compound as a cycloalkyl-substituted alkane if the substituent has more carbons than the ring.	(cyclopentane)–$CH_2CH_2CH_2CH_2CH_3$ is *pentylcyclopentane*; (cyclopentane)–$CH_2CH_2CH_2CH_2CH_2CH_3$ is *1-cyclopentylhexane*

TABLE 2.5 Summary of IUPAC Nomenclature of Alkyl Groups

Rule	Example
1. Number the carbon atoms beginning at the point of attachment, proceeding in the direction that follows the longest continuous chain.	The longest continuous chain that begins at the point of attachment in the group shown contains six carbons. $CH_3CH_2CH_2\overset{1\,|\,2}{C}(CH_3)CH_2\overset{3}{C}H(CH_3)\overset{4}{C}H_2\overset{5}{C}H_2\overset{6}{C}H_3$
2. Assign a parent name according to the number of carbons in the corresponding unbranched alkane. Drop the ending *-ane* and replace it with *-yl*.	The alkyl group shown in step 1 is named as a substituted *hexyl* group.
3. List the substituents attached to the basis group in alphabetical order using replicating prefixes when necessary.	The alkyl group in step 1 is a *dimethylpropylhexyl* group.
4. Locate the substituents according to the numbering of the main chain described in step 1.	The alkyl group is a *1,3-dimethyl-1-propylhexyl* group.

PROBLEMS

Structure and Bonding

2.24 The general molecular formula for alkanes is C_nH_{2n+2}. What is the general molecular formula for:

(a) Cycloalkanes

(b) Alkenes

(c) Alkynes

(d) Cyclic hydrocarbons that contain one double bond

2.25 A certain hydrocarbon has a molecular formula of C_5H_8. Which of the following is *not* a structural possibility for this hydrocarbon?

(a) It is a cycloalkane.

(b) It contains one ring and one double bond.

(c) It contains two double bonds and no rings.

(d) It is an alkyne.

2.26 Which of the hydrocarbons in each of the following groups are isomers?

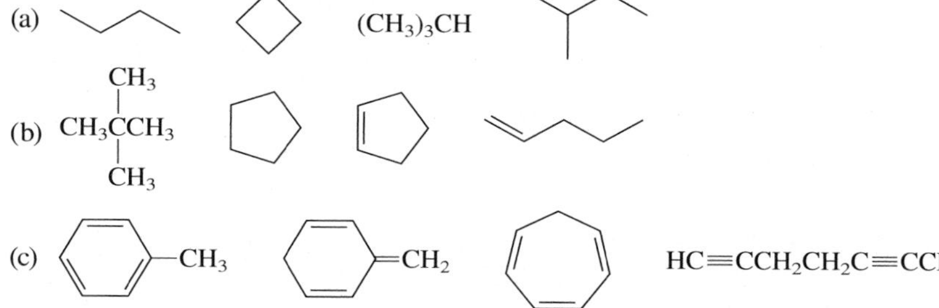

2.27 Write the structural formula of a compound of molecular formula $C_4H_8Cl_2$ in which

(a) All the carbons belong to methylene groups

(b) None of the carbons belong to methylene groups

2.28 What is the hybridization of each carbon in $CH_3CH{=}CHC{\equiv}CH$? What are the CCC bond angles?

2.29 Of the overlaps between an *s* and a *p* orbital as shown in the illustration, one is bonding, one is antibonding, and the third is nonbonding (neither bonding nor antibonding). Which orbital overlap corresponds to which interaction? Why?

2.30 Does the overlap of two *p* orbitals in the fashion shown correspond to a σ bond or to a π bond? Explain.

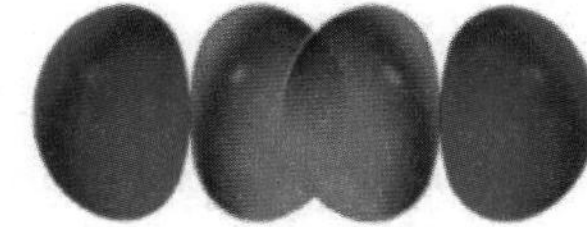

2.31 Pheromones are chemical compounds that animals, especially insects, use to signal others of the same species. Female tiger moths, for example, signify their presence to male moths this way. The sex attractant is a 2-methyl-branched alkane having a molecular weight of 254. What is its structure?

2.32 Aphids secrete an alarm pheromone having the structure shown. What is its molecular formula? Classify each carbon according to its hybridization state.

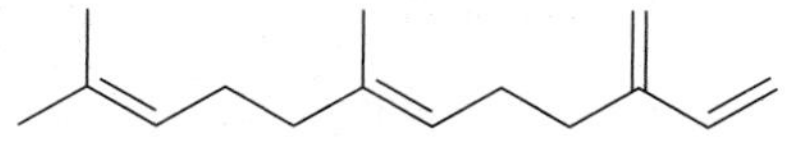

Nomenclature

2.33 All the parts of this problem refer to the alkane having the carbon skeleton shown.

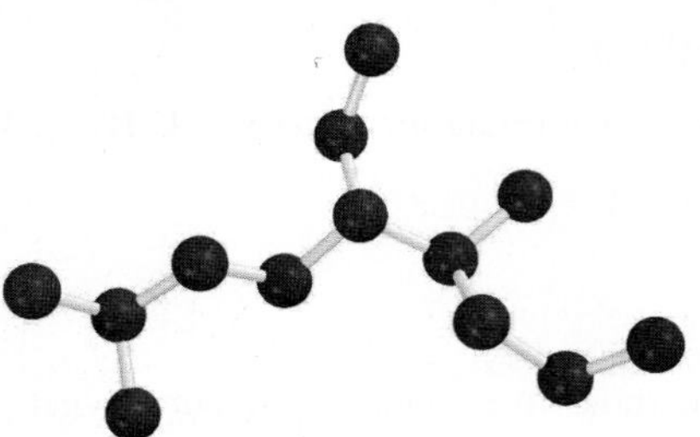

(a) What is the molecular formula of this alkane?

(b) What is its IUPAC name?

(c) How many methyl groups are present in this alkane? Methylene groups? Methine groups?

(d) How many carbon atoms are primary? Secondary? Tertiary? Quaternary?

2.34 *Pristane* is an alkane that is present to the extent of about 14% in shark liver oil. Its IUPAC name is 2,6,10,14-tetramethylpentadecane. Write its structural formula.

2.35 Write a structural formula for each of the following compounds:

(a) 6-Isopropyl-2,3-dimethylnonane

(b) 4-*tert*-Butyl-3-methylheptane

(c) 4-Isobutyl-1,1-dimethylcyclohexane

(d) *sec*-Butylcycloheptane

(e) Cyclobutylcyclopentane

2.36 Write structural formulas and give the IUPAC names for the nine alkanes that have the molecular formula C_7H_{16}.

2.37 From among the 18 constitutional isomers of C_8H_{18}, write structural formulas, and give the IUPAC names for those that are named as derivatives of

(a) Heptane (b) Hexane (c) Pentane (d) Butane

2.38 Give the IUPAC name for each of the following compounds:

(a) $CH_3(CH_2)_{25}CH_3$

(b) $(CH_3)_2CHCH_2(CH_2)_{14}CH_3$

(c) $(CH_3CH_2)_3CCH(CH_2CH_3)_2$

(d)

(e)

(f)

2.39 Using the method outlined in Section 2.17, give an IUPAC name for each of the following alkyl groups, and classify each one as primary, secondary, or tertiary:

(a) $CH_3(CH_2)_{10}CH_2—$

(b) $—CH_2CH_2\underset{|}{C}HCH_2CH_2CH_3$ with CH_2CH_3 on the branched carbon

(c) $—C(CH_2CH_3)_3$

(d) $—CHCH_2CH_2CH_3$ with a cyclopropyl group on CH

(e) cyclohexyl$—CH_2CH_2—$

(f) cyclohexyl$—CH—$ with CH_3 on CH

2.40 It has been suggested that the names of alkyl groups be derived from the alkane having the same carbon chain as the alkyl group. The *-e* ending of that alkane is replaced by *-yl,* and the chain is numbered from the end that gives the carbon at the point of attachment its lower number. This number immediately precedes the *-yl* ending and is bracketed by hyphens.

$\overset{5}{C}H_3\overset{4}{C}H\overset{3}{C}H_2\overset{2}{C}H_2\overset{1}{C}H_2—$ with CH_3 on C-4
4-Methylpentan-1-yl

$\overset{1}{C}H_3\overset{2}{C}\overset{3}{C}H_2\overset{4}{C}H_2\overset{5}{C}H_3$ with CH_3 on C-2 (point of attachment at C-2)
2-Methylpentan-2-yl

Name the C_4H_9 alkyl groups according to this system.

Reactions

2.41 Write a balanced chemical equation for the combustion of each of the following compounds:

(a) Decane

(b) Cyclodecane

(c) Methylcyclononane

(d) Cyclopentylcyclopentane

2.42 The heats of combustion of methane and butane are 890 kJ/mol (212.8 kcal/mol) and 2876 kJ/mol (687.4 kcal/mol), respectively. When used as a fuel, would methane or butane generate more heat for the same mass of gas? Which would generate more heat for the same volume of gas?

2.43 In each of the following groups of compounds, identify the one with the largest heat of combustion and the one with the smallest. (Try to do this problem without consulting Table 2.3.)

(a) Hexane, heptane, octane

(b) 2-Methylpropane, pentane, 2-methylbutane

(c) 2-Methylbutane, 2-methylpentane, 2,2-dimethylpropane

(d) Pentane, 3-methylpentane, 3,3-dimethylpentane

(e) Ethylcyclopentane, ethylcyclohexane, ethylcycloheptane

2.44 (a) Given $\Delta H°$ for the reaction

$$H_2(g) + \tfrac{1}{2}O_2(g) \longrightarrow H_2O(l) \qquad \Delta H° = -286 \text{ kJ/mol } (-68.4 \text{ kcal/mol})$$

along with the information that the heat of combustion of ethane is 1560 kJ/mol (−372.8 kcal/mol) and that of ethylene is 1410 kJ/mol (337.0 kcal/mol), calculate $\Delta H°$ for the hydrogenation of ethylene:

$$H_2C{=}CH_2(g) + H_2(g) \longrightarrow CH_3CH_3(g)$$

(b) If the heat of combustion of acetylene is 1300 kJ/mol (310.7 kcal/mol), what is the value of $\Delta H°$ for its hydrogenation to ethylene? To ethane?

(c) What is the value of $\Delta H°$ for the hypothetical reaction

$$2H_2C{=}CH_2(g) \longrightarrow CH_3CH_3(g) + HC{\equiv}CH(g)$$

2.45 We have seen in this chapter that, among isomeric alkanes, the unbranched isomer is the least stable and has the highest boiling point; the most branched isomer is the most stable and has the lowest boiling point. Does this mean that one alkane boils lower than another *because* it is more stable? Explain.

2.46 Higher octane gasoline typically contains a greater proportion of branched alkanes relative to unbranched ones. Are branched alkanes better fuels because they give off more energy on combustion? Explain.

2.47 The reaction shown is important in the industrial preparation of dichlorodimethylsilane for eventual conversion to silicone polymers.

$$2CH_3Cl + Si \longrightarrow (CH_3)_2SiCl_2$$

(a) Is carbon oxidized, reduced, or neither in this reaction?

(b) On the basis of the molecular model of $(CH_3)_2SiCl_2$, deduce the hybridization state of silicon in this compound. What is the principal quantum number n of the silicon s and p orbitals that are hybridized?

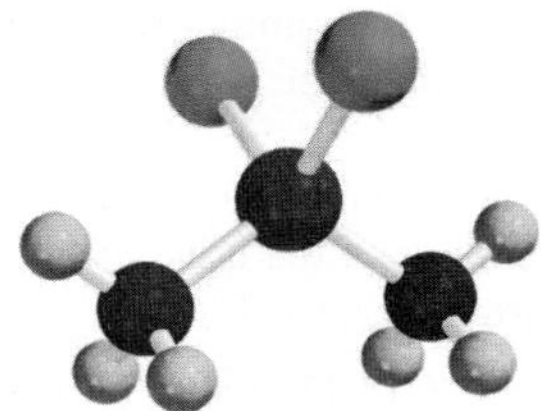

2.48 Alkanes spontaneously burst into flame in the presence of elemental fluorine. The reaction that takes place between pentane and F_2 gives CF_4 and HF as the only products.

(a) Write a balanced equation for this reaction.

(b) Is carbon oxidized, is it reduced, or does it undergo no change in oxidation state in this reaction?

2.49 Which atoms in the following reaction undergo changes in their oxidation state? Which atom is oxidized? Which one is reduced?

$$2CH_3CH_2OH + 2Na \longrightarrow 2CH_3CH_2ONa + H_2$$

2.50 Compound A undergoes the following reactions:

Compound A

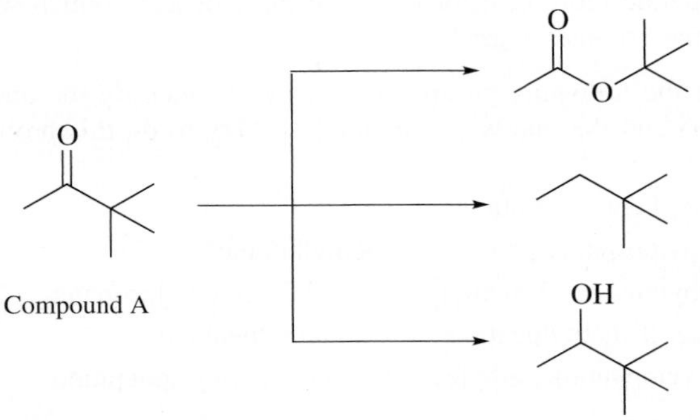

(a) Which of the reactions shown require(s) an oxidizing agent?

(b) Which of the reactions shown require(s) a reducing agent?

2.51 Each of the following reactions will be encountered at some point in this text. Classify each one according to whether the organic reactant is oxidized or reduced in the process.

(a) $CH_3C{\equiv}CH + 2Na + 2NH_3 \longrightarrow CH_3CH{=}CH_2 + 2NaNH_2$

(b) 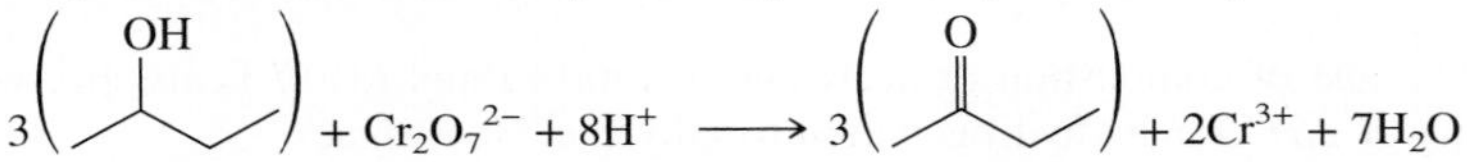 3 (OH) $+ Cr_2O_7^{2-} + 8H^+ \longrightarrow$ 3 (O) $+ 2Cr^{3+} + 7H_2O$

(c) $HOCH_2CH_2OH + HIO_4 \longrightarrow 2H_2C{=}O + HIO_3 + H_2O$

(d) 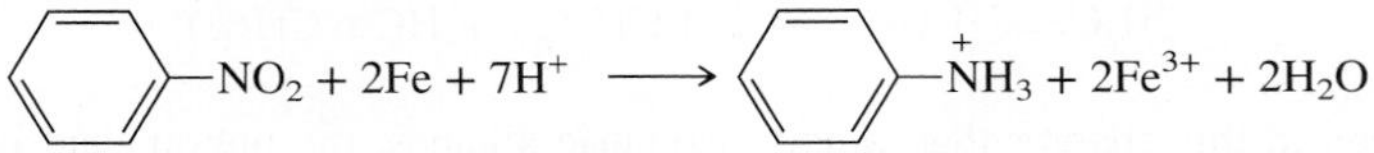 $-NO_2 + 2Fe + 7H^+ \longrightarrow$ $-\overset{+}{N}H_3 + 2Fe^{3+} + 2H_2O$

(e) 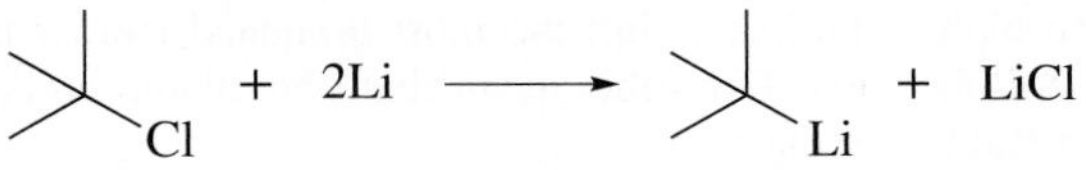 Cl + 2Li $\longrightarrow$ Li + LiCl

(f)  O + Cl_2 $\longrightarrow$ O, Cl + HCl

Descriptive Passage and Interpretive Problems 2

Some Biochemical Reactions of Alkanes

Alkanes occur naturally in places other than petroleum deposits—in insects, for example. The waxy alkanes dispersed in its cuticle help protect an insect from dehydration. Some insects use volatile alkanes to defend themselves or communicate with others of the same species. Alkanes even serve as starting materials that the insect converts to other biologically important substances.

The major biosynthetic pathway leading to alkanes is by enzyme-catalyzed decarboxylation (loss of CO_2) of fatty acids, compounds of the type $CH_3(CH_2)_nCO_2H$ in which n is an even number and the chain has 14 or more carbons.

$$CH_3(CH_2)_nCO_2H \longrightarrow CH_3(CH_2)_{n-1}CH_3 + CO_2$$

Biochemical conversion of alkanes to other substances normally begins with oxidation.

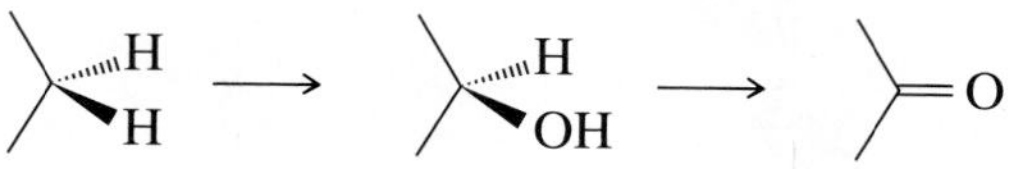

In addition to alkanes, the oxidation of drugs and other substances occurs mainly in the liver and is catalyzed by the enzyme cytochrome P-450. Molecular oxygen and nicotinamide adenine dinucleotide (NAD) are also required.

Oxidation by microorganisms has been extensively studied and is often selective for certain kinds of C—H bonds. The fungus *Pseudomonas oleovorans,* for example, oxidizes the CH_3 groups at the end of the carbon chain of 4-methyloctane faster than the CH_3 branch and faster than the CH_2 and CH units within the chain.

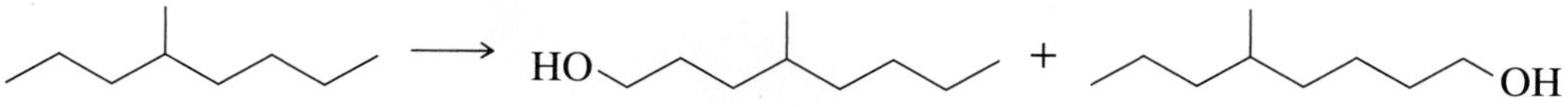

2.52 Tridecane $[CH_3(CH_2)_{11}CH_3]$ is a major component of the repellent that the stink bug *Piezodorus guildinii* releases from its scent glands when attacked. What fatty acid gives tridecane on decarboxylation?

A. $CH_3(CH_2)_{10}CO_2H$
B. $CH_3(CH_2)_{11}CO_2H$
C. $CH_3(CH_2)_{12}CO_2H$
D. $CH_3(CH_2)_{13}CO_2H$

2.53 Assuming a selectivity analogous to that observed in the microbiological oxidation of 4-methyloctane by *Pseudomonas oleovorans,* which of the following is expected to give two constitutionally isomeric alcohols on oxidation?

A. Heptane
B. 3-Methylheptane
C. 4-Methylheptane
D. 4,4-Dimethylheptane

2.54 Female German cockroaches convert the alkane shown to a substance that attracts males.

$$CH_3CH_2\underset{CH_3}{\underset{|}{C}}H(CH_2)_7\underset{CH_3}{\underset{|}{C}}H(CH_2)_{16}CH_2CH_3$$

Oxidation at C-2 of the alkane gives the sex attractant, which has a molecular formula $C_{31}H_{62}O$ and the same carbon skeleton as the alkane. What is the structure of the sex attractant?

A. $CH_3\overset{OH}{\overset{|}{C}}H\underset{CH_3}{\underset{|}{C}}H(CH_2)_7\underset{CH_3}{\underset{|}{C}}H(CH_2)_{16}CH_2CH_3$

B. $CH_3CH_2\underset{CH_3}{\underset{|}{C}}H(CH_2)_7\underset{CH_3}{\underset{|}{C}}H(CH_2)_{16}\overset{OH}{\overset{|}{C}}HCH_3$

C. $CH_3\overset{O}{\overset{\|}{C}}\underset{CH_3}{\underset{|}{C}}H(CH_2)_7\underset{CH_3}{\underset{|}{C}}H(CH_2)_{16}CH_2CH_3$

D. $CH_3CH_2\underset{CH_3}{\underset{|}{C}}H(CH_2)_7\underset{CH_3}{\underset{|}{C}}H(CH_2)_{16}\overset{O}{\overset{\|}{C}}CH_3$

2.55 Biological oxidation of the hydrocarbon adamantane by the fungus *Absidia glauca* gives a mixture of two alcohols.

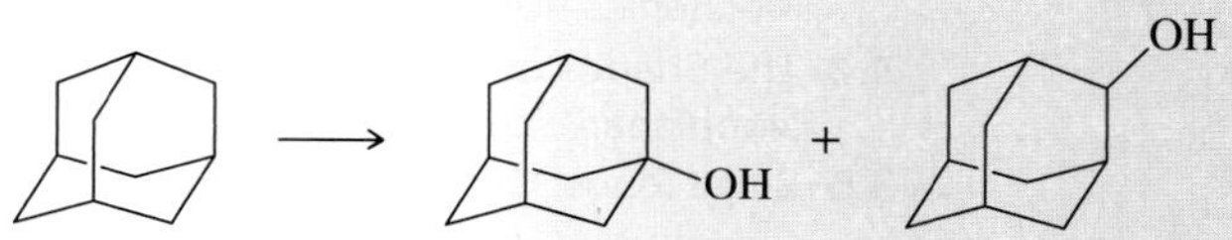

Adamantane → Major + Minor

Classify the carbon in adamantane that is oxidized in forming the major product.

A. Primary
B. Secondary
C. Tertiary
D. Quaternary

Chapter

3

CHAPTER OUTLINE

Carbon's unrivaled ability to form bonds to itself can produce not only chains and rings, but also compact frameworks of many rings. Diamonds represent the ultimate elaboration of the pattern introduced by the progression of structural formulas shown. The three most complicated structures, plus many analogous but even larger ones, are all found in petroleum.

Alkanes and Cycloalkanes: Conformations and cis–trans Stereoisomers

Hydrogen peroxide is formed in the cells of plants and animals but is toxic to them. Consequently, living systems have developed mechanisms to rid themselves of hydrogen peroxide, usually by enzyme-catalyzed reduction to water. An understanding of how reactions take place, be they in living systems or in test tubes, begins with a thorough knowledge of the structure of the reactants, products, and catalysts. Even a molecule as simple as hydrogen peroxide (four atoms!) may be structurally more complicated than you think. Suppose we wanted to write the structural formula for H_2O_2 in enough detail to show the positions of the atoms relative to one another. We could write two different planar geometries A and B that differ by a 180° rotation about the O—O bond. We could also write an infinite number of nonplanar structures, of which C is but one example, that differ from one another by tiny increments of rotation about the O—O bond.

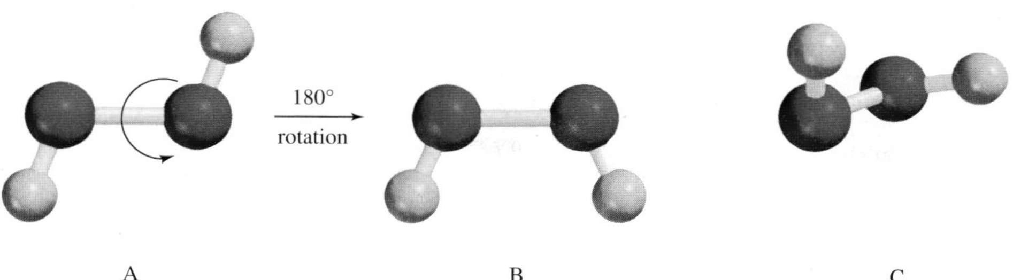

Structures A, B, and C represent different **conformations** of hydrogen peroxide. *Conformations are different spatial arrangements of a molecule that are generated by rotation about single bonds.* Although we can't tell from simply looking at these structures, we now know from experimental studies that all are in rapid equilibrium and that C with an H—O—O—H angle of about 114.5° is the most stable conformation.

Conformational analysis is the study of how conformational factors affect the structure of a molecule and its properties. In this chapter we'll examine the conformations of various alkanes and cycloalkanes, focusing most of our attention on three of them: *ethane, butane,* and *cyclohexane.* You will see that even simple organic molecules can exist in many conformations. Conformational analysis will help us to visualize organic molecules as three-dimensional objects and to better understand their structure and properties.

3.1 Conformational Analysis of Ethane

Ethane is the simplest hydrocarbon that can have distinct conformations. Two, the **staggered conformation** and the **eclipsed conformation,** deserve special mention and are illustrated with molecular models in Figure 3.1.

In the staggered conformation, each C—H bond of one carbon bisects an H—C—H angle of the other carbon. In the eclipsed conformation, each C—H bond of one carbon is aligned with a C—H bond of the other carbon.

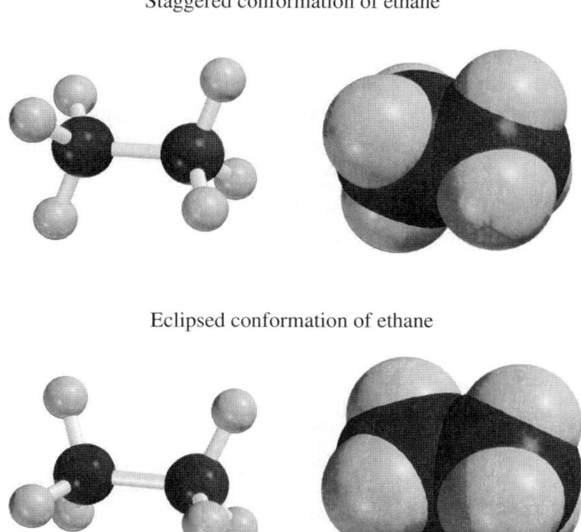

Figure 3.1

The staggered and eclipsed conformations of ethane shown as ball-and-spoke models (*left*) and as space-filling models (*right*).

Sawhorses are beams with four legs that are used in pairs to support a plank for sawing, or in other uses as a support structure or road marker.

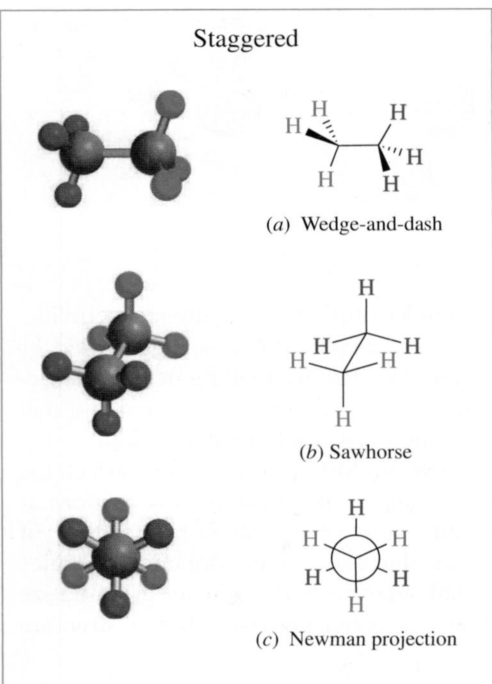

Figure 3.2

Molecular models and graphical representations of the staggered conformation of ethane.

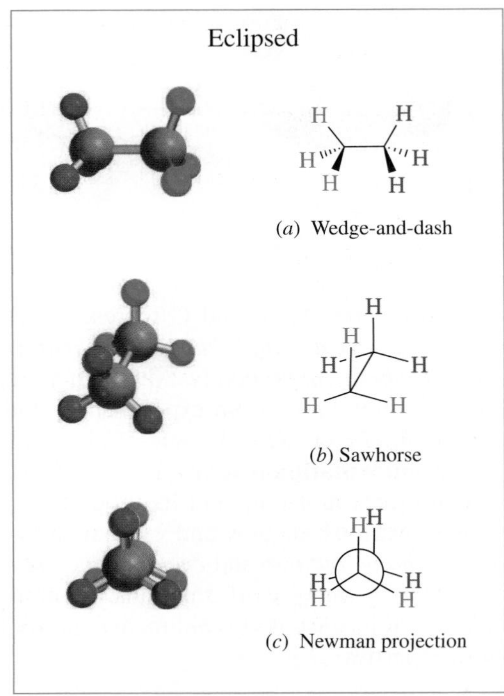

Figure 3.3

Molecular models and graphical representations of the eclipsed conformation of ethane.

The staggered and eclipsed conformations interconvert by rotation about the C—C bond, and do so very rapidly. We'll see just how rapidly later in this section.

Among the various ways in which the staggered and eclipsed forms are portrayed, wedge-and-dash, sawhorse, and Newman projection drawings are especially useful. These are shown for the staggered conformation of ethane in Figure 3.2 and for the eclipsed conformation in Figure 3.3.

Newman projections were devised by Professor Melvin S. Newman of Ohio State University.

We used *wedge-and-dash* drawings in earlier chapters, and so Figures 3.2*a* and 3.3*a* are familiar to us. A *sawhorse* drawing (Figures 3.2*b* and 3.3*b*) shows the conformation of a molecule without having to resort to different styles of bonds. In a *Newman projection* (Figures 3.2*c* and 3.3*c*), we sight down the C—C bond, and represent the front carbon by a point and the back carbon by a circle. Each carbon has three other bonds that are placed symmetrically around it.

Figures 3.2 and 3.3 illustrate the spatial relationship between bonds on adjacent carbons. Each H—C—C—H unit in ethane is characterized by a *torsion angle* or *dihedral angle,* which is the angle between the H—C—C plane and the C—C—H plane. The torsion angle is easily seen in a Newman projection of ethane as the angle between C—H bonds of adjacent carbons.

0° HH — Torsion angle = 0° Eclipsed

H 60° H — Torsion angle = 60° Gauche

H 180° H — Torsion angle = 180° Anti

Eclipsed bonds are characterized by a torsion angle of 0°. When the torsion angle is approximately 60° the spatial relationship is **gauche;** and when it is 180° it is **anti.**

Staggered conformations have only gauche or anti relationships between bonds on adjacent atoms.

Problem 3.1

Identify the alkanes corresponding to each of the drawings shown.

(*a*) Newman projection (*b*) Sawhorse (*c*) and (*d*) Alternative wedge-and-dash representations (*e*) Bond-line

Sample Solution (a) The Newman projection of this alkane resembles that of ethane, except one of the hydrogens has been replaced by a methyl group. The drawing is a Newman projection of propane, $CH_3CH_2CH_3$.

Of the two conformations of ethane, the staggered is 12 kJ/mol (2.9 kcal/mol) more stable than the eclipsed. The staggered conformation is the most stable conformation, the eclipsed is the least stable conformation. Two main explanations have been offered for this difference. One holds that repulsions between bonds on adjacent atoms *destabilize* the eclipsed conformation. The other suggests that better electron delocalization *stabilizes* the staggered conformation. Both effects contribute to the preference for the staggered conformation.

The ethane conformations in which the torsion angles between adjacent bonds are other than 60° are said to have **torsional strain.** Eclipsed bonds produce the most torsional strain; staggered bonds none. Because three pairs of eclipsed bonds are responsible for 12 kJ/mol (2.9 kcal/mol) of torsional strain in ethane, it is reasonable to assign an "energy cost" of 4 kJ/mol (1 kcal/mol) to each pair. In this chapter we'll learn of additional sources of strain in molecules, which together with torsional strain comprise **steric strain.**

Steric is derived from the Greek word *stereos* for "solid" and refers to the three-dimensional or spatial aspects of chemistry.

In principle, ethane has an infinite number of conformations that differ by only tiny increments in their torsion angles. Not only is the staggered conformation more stable than the eclipsed, it also is the most stable of all the conformations; the eclipsed is the least stable. Figure 3.4 shows how the potential energy of ethane changes for a 360° rotation

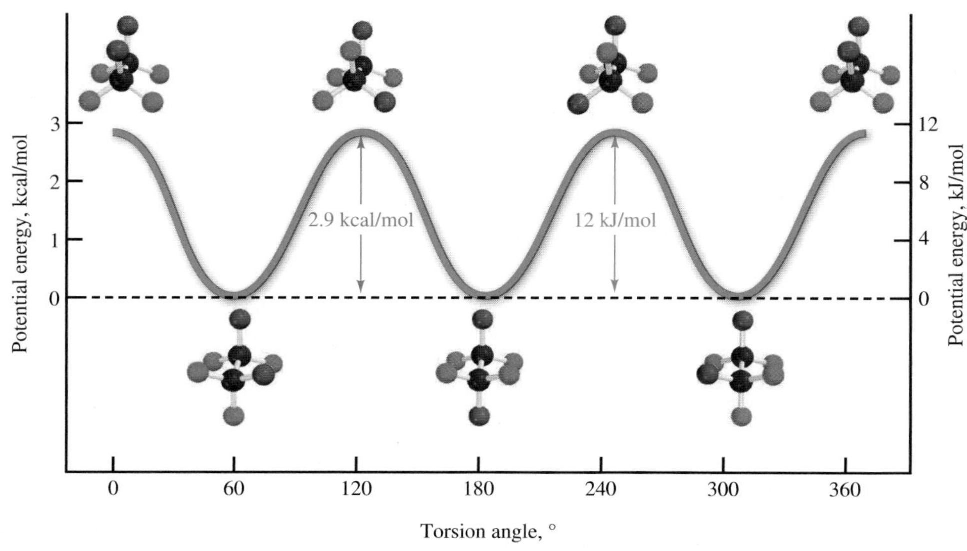

Figure 3.4

Potential energy diagram for rotation about the carbon–carbon bond in ethane. Two of the hydrogens are shown in red and four in green so as to indicate more clearly the bond rotation.

about the carbon–carbon bond. Three equivalent eclipsed conformations and three equivalent staggered conformations occur during the 360° rotation; the eclipsed conformations appear at the highest points on the curve (*potential energy maxima*), the staggered ones at the lowest (*potential energy minima*). Conformations that correspond to potential energy minima are called **conformers.**

At any instant, almost all of the molecules are in staggered conformations; hardly any are in eclipsed conformations.

Problem 3.2

Find the conformations in Figure 3.4 in which the hydrogens marked in red are (a) gauche and (b) anti.

Diagrams such as Figure 3.4 help us understand how the potential energy of a system changes during a process. The process can be as simple as the one described here—rotation about a carbon–carbon bond. Or it might be more complicated—a chemical reaction, for example. We will see applications of potential energy diagrams to a variety of processes throughout the text.

Let's focus our attention on a portion of Figure 3.4. The region that lies between a torsion angle of 60° and 180° tracks the conversion of one staggered conformer of ethane to the next one. Both conformers are equivalent and equal in energy, but for one to get to the next, it must first pass through an eclipsed conformation and needs to gain 12 kJ/mol (2.9 kcal/mol) of energy to reach it. This amount of energy is the **activation energy (E_a)** for the process. Molecules must become energized in order to undergo a chemical reaction or, as in this case, to undergo rotation about a carbon–carbon bond. Kinetic (thermal) energy is absorbed by a molecule from collisions with other molecules and is transformed into potential energy. When the potential energy exceeds E_a, the unstable arrangement of atoms that exists at that instant relaxes to a more stable structure, giving off its excess potential energy in collisions with other molecules or with the walls of a container. The point of maximum potential energy encountered by the reactants as they proceed to products is called the **transition state.** The eclipsed conformation is the transition state for the conversion of one staggered conformation of ethane to another.

The structure that exists at the transition state is sometimes referred to as the *transition structure* or the *activated complex.*

Rotation about carbon–carbon bonds is one of the fastest processes in chemistry. Among the ways that we can describe the rate of a process is by its *half-life,* which is the length of time it takes for one half of the molecules to have reacted. It takes less than 10^{-6} s for half of the molecules in a sample of ethane to have gone from one staggered conformation to another at 25°C.

A second way is by citing the experimentally determined **rate constant *k*,** which is related to the energy of activation by the **Arrhenius equation:**

$$k = Ae^{-E_a/RT}$$

where A is a frequency factor related to the collision rate and geometry. The $e^{-E_a/RT}$ term is the probability that a collision will result in reaction, T is the temperature in kelvins, and R is a constant (8.314 kJ/K · mol or 1.987×10^{-3} kcal/K · mol). E_a is calculated by comparing reaction rates as a function of temperature. Raising the temperature decreases E_a/RT and increases $Ae^{-E_a/RT}$, thereby increasing k. Small increases in E_a result in large decreases in rate. Quantitative studies of reaction rates are grouped under the general term **kinetics** and provide the basis for many of the structure–reactivity relationships that we will see in this and later chapters.

As shown in Figure 3.5, most of the molecules in a sample have energies that are clustered around some average value. Only molecules with a potential energy greater than E_a, however, are able to go over the transition state and proceed on. The number of these molecules is given by the shaded areas under the curve in Figure 3.5. The energy distribution curve flattens out at higher temperatures, and a greater proportion of molecules have energies in excess of E_a at T_2 (higher) than at T_1 (lower). *The effect of temperature is quite pronounced; an increase of only 10°C produces a two- to threefold increase in the rate of a typical chemical process.*

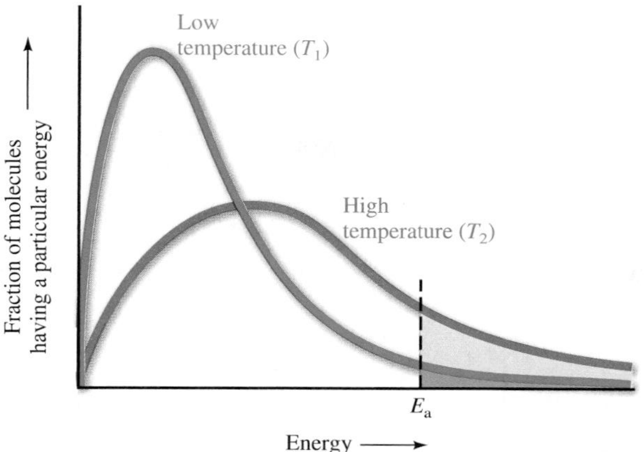

Figure 3.5

Distribution of energies. The number of molecules with energy greater than E_a at temperature T_1 is shown as the darker green-shaded area. At some higher temperature T_2, the curve is flatter, and more molecules have energies in excess of E_a.

3.2 Conformational Analysis of Butane

The next alkane that we will examine is butane. In particular, we consider conformations related by rotation about the bond between the middle two carbons. Unlike ethane, in which the staggered conformations are equivalent, butane has two different staggered conformations, as shown in Figure 3.6. The methyl groups of butane are gauche to each other in one, anti in the other. Both conformations are staggered, so are free of torsional strain, but two of the methyl hydrogens of the gauche conformation lie within 210 pm (2.10 Å) of each other. This distance is less than the sum of their van der Waals radii (240 pm; 2.40 Å), and there is a repulsive force between the two hydrogens. The destabilization of a molecule that results when two of its atoms are too close to each other is called **van der Waals strain,** or **steric hindrance,** and contributes to the total steric strain. In the case of butane, van der Waals strain makes the gauche conformation approximately 3.8 kJ/mol (0.9 kcal/mol) less stable than the anti.

Figure 3.7 illustrates the potential energy relationships among the various conformations of butane about the central carbon–carbon bond. The staggered conformations are more stable than the eclipsed. At any instant, almost all the molecules exist in staggered conformations, and more are present in the anti conformation than in the gauche. The point of maximum potential energy lies some 21 kJ/mol (5 kcal/mol) above the anti conformation. The total strain in this structure is approximately equally divided between the torsional strain associated with three pairs of eclipsed bonds that was determined for

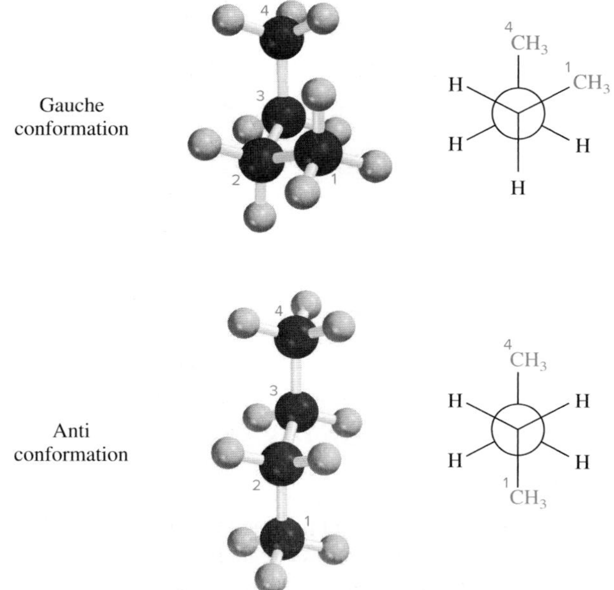

Figure 3.6

The gauche and anti conformations of butane shown as ball-and-spoke models (*left*) and as Newman projections (*right*). The gauche conformation is less stable than the anti because of the van der Waals strain between the methyl groups. Sighting along the C(2)—C(3) bond produces the perspective in the Newman projection.

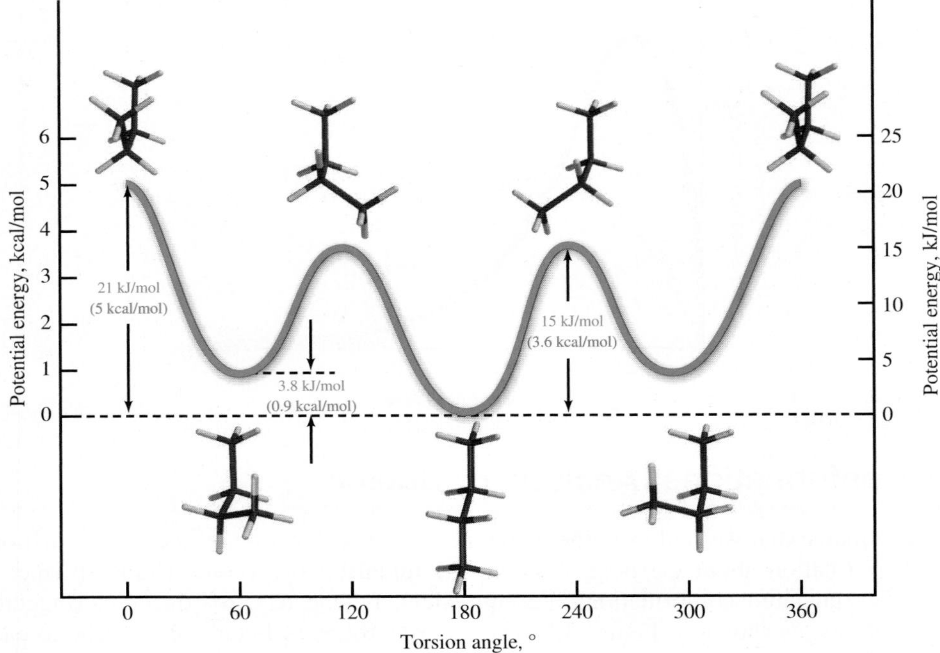

Figure 3.7

Potential energy diagram for rotation about the central carbon–carbon bond in butane.

ethane in Figure 3.4 (12 kJ/mol; 2.9 kcal/mol) and the van der Waals strain between the eclipsed methyl groups.

Problem 3.3

Sketch sawhorse and Newman projections for the staggered conformations of butane in Figure 3.7. Identify conformations with gauche and anti methyl groups.

Problem 3.4

Sketch a potential energy diagram for rotation about a carbon–carbon bond in propane. Identify each potential energy maximum and minimum with a structural formula that shows the conformation of propane at that point. Does your diagram more closely resemble that of ethane or of butane? Would you expect the activation energy for bond rotation in propane to be more than or less than that of ethane? Of butane?

Problem 3.5

Acetylcholine is a neurotransmitter in the central nervous system in humans. Sighting along the C—C bond, draw Newman projection formulas for the anti and gauche conformations of acetylcholine.

$$(CH_3)_3\overset{+}{N}CH_2CH_2OC(=O)CH_3$$

3.3 Conformations of Higher Alkanes

Higher alkanes having unbranched carbon chains are, like butane, most stable in their all-anti conformations. The energy difference between gauche and anti conformations is similar to that of butane, and appreciable quantities of the gauche conformation are present in liquid alkanes at 25°C. In depicting the conformations of higher alkanes it is often more helpful

Computational Chemistry: Molecular Mechanics and Quantum Mechanics

Molecular mechanics is a method for calculating the energy of a molecule by comparing selected structural features with those of "unstrained" standards. It makes no attempt to explain why the van der Waals radius of hydrogen is 120 pm (1.20 Å), why the bond angles in methane are 109.5°, why the C—C bond distance in ethane is 153 pm (1.53 Å), or why the staggered conformation of ethane is 12 kJ/mol (2.9 kcal/mol) more stable than the eclipsed. Instead, it uses these and other experimentally determined values as benchmarks to create a "force field" to which the features of other substances are compared. If we assume that there are certain ideal values for bond angles, bond distances, and so on, it follows that deviations from them will destabilize a particular structure. The resulting increase in potential energy is referred to as the *steric energy* (E_{steric}) of the structure. Other terms include *strain energy* and *steric strain.*

Arithmetically, the steric energy of a structure can be separated into several components:

$$E_{steric} = E_{bond\ stretch} + E_{angle\ bend} + E_{torsion} + E_{nonbonded}$$

where $E_{bond\ stretch}$ is the strain that results when bond distances are distorted from their ideal values, $E_{angle\ bend}$ from the expansion or contraction of bond angles, $E_{torsion}$ from deviation of torsion angles from their stable relationship, and $E_{nonbonded}$ from attractive or repulsive forces between atoms that aren't bonded to one another. It often happens that the shape of a molecule causes two atoms to be close in space, even though they may be separated by many bonds. Although van der Waals forces in alkanes are weakly attractive at most distances, two atoms that are closer than the sum of their van der Waals radii experience repulsive forces that can dominate the $E_{nonbonded}$ term. This resulting destabilization is called *van der Waals strain.* Another frequently encountered nonbonded interaction is *Coulombic,* which is the attractive force between oppositely charged atoms or the repulsive force between atoms of like charge. In molecular mechanics, each component of strain is separately described by a mathematical expression developed and refined so that it gives solutions that match experimental observations for reference molecules. A computer-driven *steric energy minimization* routine searches for the combination of bond angles, distances, torsion angles, and nonbonded interactions that has the lowest total strain.

Consider the rotation about the C(2)—C(3) bond in butane discussed in Section 3.2 and its potential energy diagram shown in Figure 3.7. As calculated by the molecular mechanics force field, MM3, E_{steric} for the anti and gauche conformations of butane are 14.5 and 18.4 kJ/mol (3.46 and 4.40 kcal/mol), respectively. The 3.9 kJ/mol (0.94 kcal/mol) difference between these values is in good agreement with the experimentally determined value of 3.8 kJ/mol (0.9 kcal/mol) cited in the opening paragraph of Section 3.2.

Problem 3.6

As calculated by molecular mechanics, E_{steric} for the methyl eclipsed conformation of butane is 36.9 kJ/mol (8.8 kcal/mol). On the basis of this and the E_{steric} values just cited, calculate the activation energy for rotation about the C(2)—C(3) bond.

Quantum mechanical calculations are much different and are based on the Schrödinger equation (Section 1.1). Instead of treating molecules as collections of atoms and bonds, quantum mechanics focuses on nuclei and electrons and treats electrons as waves. The energy of a chemical species is determined as the sum of the attractive (nucleus–electron) and repulsive (nucleus–nucleus and electron–electron) forces plus the kinetic energies of the electrons and the nuclei. Minimizing the total energy gives a series of solutions called *wave functions,* which are equivalent to orbitals. Calculations based on quantum mechanics are generally referred to as *molecular orbital (MO) calculations*.

The computing requirements are so much greater for MO calculations than for molecular mechanics that MO calculations were once considered a specialty area. That is no longer true and it is now a routine matter to carry out MO calculations on personal computers. Steric-energy minimization by molecular mechanics is increasingly seen as a preliminary step prior to carrying out an MO calculation. A molecule is constructed, its geometry minimized by molecular mechanics, then MO methods are used to calculate energies, geometries, and other properties.

Our first encounter with the results of computational methods is often visual. Figure 3.8*a* shows a ball-and-stick model of the methyl–methyl eclipsed conformation of butane as it would appear before doing any calculations. Either a molecular mechanics or MO calculation can produce, among other renderings, the space-filling model of butane shown in Figure 3.8*b,* which reveals the close contact between hydrogens that contributes to van der Waals strain in the eclipsed conformation. However, only an MO calculation can generate an electrostatic potential map, which we see in Figure 3.8*c* as an overlay of charge distribution on a van der Waals surface.

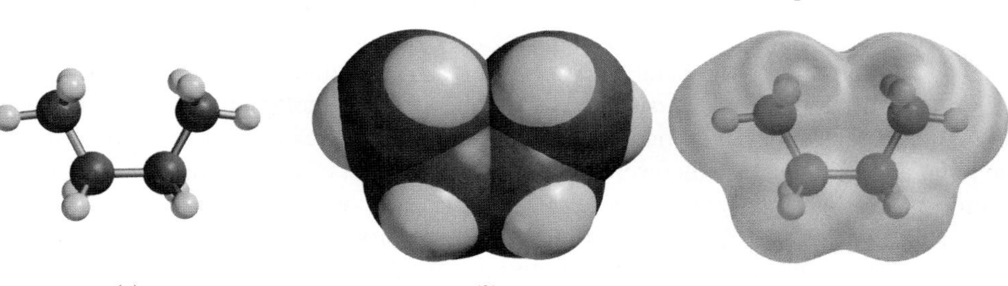

Figure 3.8

The methyl–methyl eclipsed conformation of butane: (*a*) ball-and-stick model, (*b*) space-filling model, and (c) electrostatic potential map. The molecule itself is the same size in each.

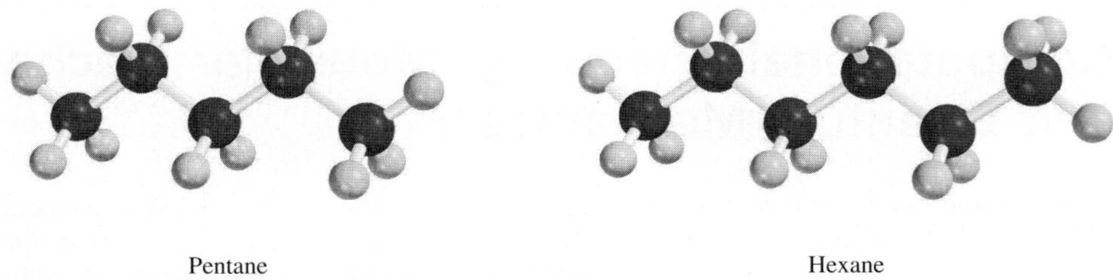

Figure 3.9

Ball-and-spoke models of pentane and hexane in their all-anti (zigzag) conformations.

to look at them from the side rather than end-on as in a Newman projection. Viewed from this perspective, the most stable conformations of pentane and hexane have their carbon "backbones" arranged in a zigzag fashion, as shown in Figure 3.9. All the bonds are staggered, and the chains are characterized by anti arrangements of C—C—C—C units.

3.4 The Shapes of Cycloalkanes: Planar or Nonplanar?

During the nineteenth century it was widely believed—incorrectly, as we'll see—that cycloalkane rings are planar. A leading advocate of this view was the German chemist Adolf von Baeyer. He noted that compounds containing rings other than those based on cyclopentane and cyclohexane were rarely encountered naturally and were difficult to synthesize. Baeyer connected both observations with cycloalkane stability, which he suggested was related to how closely the internal angles of planar rings match the tetrahedral value of 109.5°. For example, the 60° bond angle of cyclopropane and the 90° bond angles of a planar cyclobutane ring are much smaller than the tetrahedral angle of 109.5°. Baeyer suggested that three- and four-membered rings suffer from what we now call angle strain. **Angle strain** is the strain a molecule has because one or more of its bond angles deviate from the ideal value; in the case of alkanes the ideal value is 109.5°.

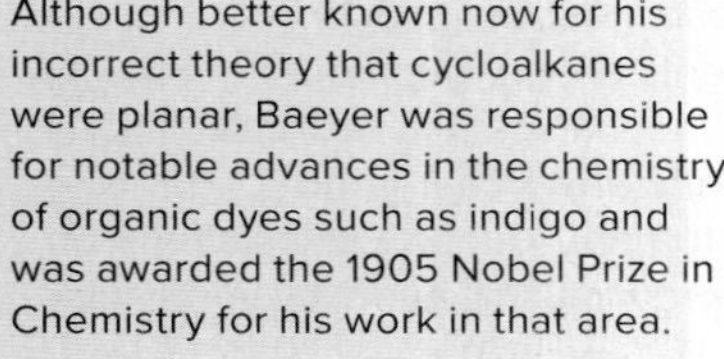
Although better known now for his incorrect theory that cycloalkanes were planar, Baeyer was responsible for notable advances in the chemistry of organic dyes such as indigo and was awarded the 1905 Nobel Prize in Chemistry for his work in that area.

According to Baeyer, cyclopentane should be the most stable of all the cycloalkanes because the ring angles of a planar pentagon, 108°, are closer to the tetrahedral angle than those of any other cycloalkane. A prediction of the *Baeyer strain theory* is that the cycloalkanes beyond cyclopentane should become increasingly strained and correspondingly less stable. The angles of a regular hexagon are 120°, and the angles of larger polygons deviate more and more from the ideal tetrahedral angle.

Problems with the Baeyer strain theory become apparent when we use heats of combustion (Table 3.1) to probe the relative energies of cycloalkanes. The most important column in the table is the heat of combustion per methylene (CH_2) group. Because all of the cycloalkanes have molecular formulas of the type C_nH_{2n}, dividing the heat of combustion by n allows direct comparison of ring size and potential energy. Cyclopropane has the highest heat of combustion per methylene group, which is consistent with the idea that its potential energy is raised by angle strain. Cyclobutane has less angle strain at each of its carbon atoms and a lower heat of combustion per methylene group. Cyclopentane, as expected, has a lower value still. Notice, however, that contrary to the prediction of the Baeyer strain theory, cyclohexane has a *smaller* heat of combustion per methylene group than cyclopentane. If angle strain were greater in cyclohexane than in cyclopentane, the opposite would have been observed.

Furthermore, the heats of combustion per methylene group of the very large rings are all about the same and similar to that of cyclopentane and cyclohexane. Rather than rising because of increasing angle strain in large rings, the heat of combustion per methylene group remains constant at approximately 659 kJ/mol (157.4 kcal/mol), the value cited in Section 2.23 as the difference between successive members of a homologous series of alkanes. We conclude, therefore, that the bond angles of large cycloalkanes are not much different from the bond angles of alkanes themselves. The prediction of the Baeyer strain theory that angle strain increases steadily with ring size is contradicted by experimental fact.

The Baeyer strain theory is useful to us in identifying angle strain as a destabilizing effect. Its fundamental flaw is its assumption that the rings of cycloalkanes are planar. *With the exception of cyclopropane, cycloalkanes are nonplanar.* Sections 3.5–3.13 describe the shapes of cycloalkanes. We'll begin with cyclopropane.

TABLE 3.1 Heats of Combustion ($-\Delta H°$) of Cycloalkanes

Cycloalkane	Number of CH_2 groups	Heat of combustion		Heat of combustion per CH_2 group	
		kJ/mol	kcal/mol	kJ/ mol	kcal/mol
Cyclopropane	3	2,091	499.8	697	166.6
Cyclobutane	4	2,745	656.1	686	164.0
Cyclopentane	5	3,320	793.4	664	158.7
Cyclohexane	6	3,953	944.7	659	157.5
Cycloheptane	7	4,637	1108.4	662	158.3
Cyclooctane	8	5,310	1269.2	664	158.7
Cyclononane	9	5,981	1429.6	665	158.8
Cyclodecane	10	6,639	1586.8	664	158.7
Cycloundecane	11	7,293	1743.2	663	158.5
Cyclododecane	12	7,922	1893.4	660	157.8
Cyclotetradecane	14	9,234	2207.0	660	157.6
Cyclohexadecane	16	10,548	2521.0	659	157.6

3.5 Small Rings: Cyclopropane and Cyclobutane

Conformational analysis is far simpler in cyclopropane than in any other cycloalkane. Cyclopropane's three carbon atoms are, of geometric necessity, coplanar, and rotation about its carbon–carbon bonds is impossible. You saw in Section 3.4 how angle strain in cyclopropane leads to an abnormally large heat of combustion. Let's now look at cyclopropane in more detail to see how our orbital hybridization bonding model may be adapted to molecules of unusual geometry.

Strong sp^3–sp^3 σ bonds are not possible for cyclopropane, because the 60° bond angles of the ring do not permit the orbitals to be properly aligned for effective overlap (Figure 3.10). The less effective overlap that does occur leads to what chemists refer to as "bent" bonds. The electron density in the carbon–carbon bonds of cyclopropane does not lie along the internuclear axis but is distributed along an arc between the two carbon atoms. The ring bonds of cyclopropane are weaker than other carbon–carbon σ bonds.

In keeping with the "bent-bond" description of Figure 3.10, the carbon–carbon bond distance in cyclopropane (151 pm; 1.51 Å) is slightly shorter than that of ethane (153 pm; 1.53 Å) and cyclohexane (154 pm; 1.54 Å).

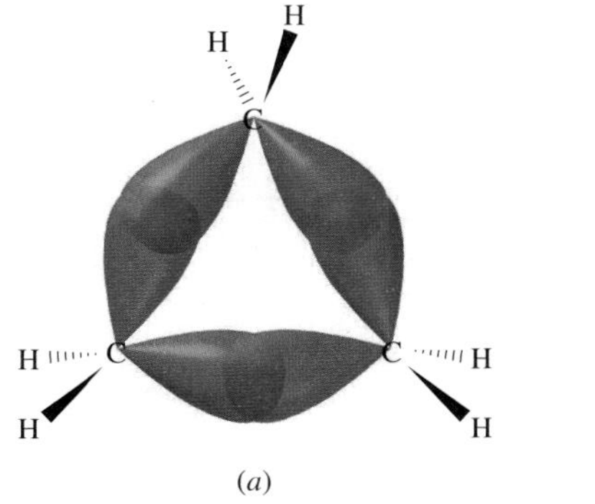

(*a*)

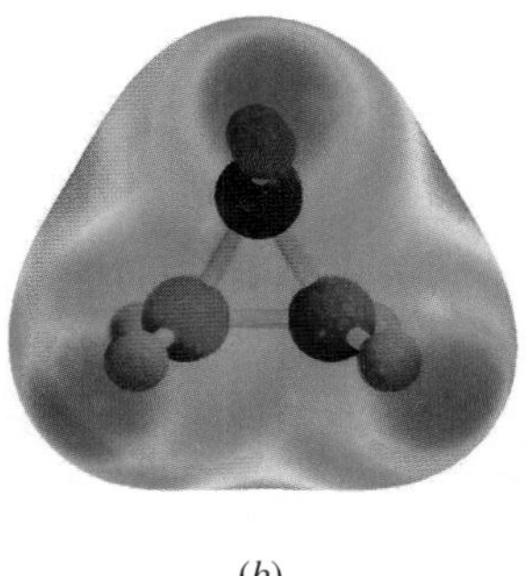

(*b*)

Figure 3.10

"Bent bonds" in cyclopropane. (*a*) The orbitals involved in carbon–carbon bond formation overlap in a region that is displaced from the internuclear axis. (*b*) The three areas of greatest negative electrostatic potential (red) correspond to those predicted by the bent-bond description.

In addition to angle strain, cyclopropane is destabilized by torsional strain. Each C—H bond of cyclopropane is eclipsed with two others.

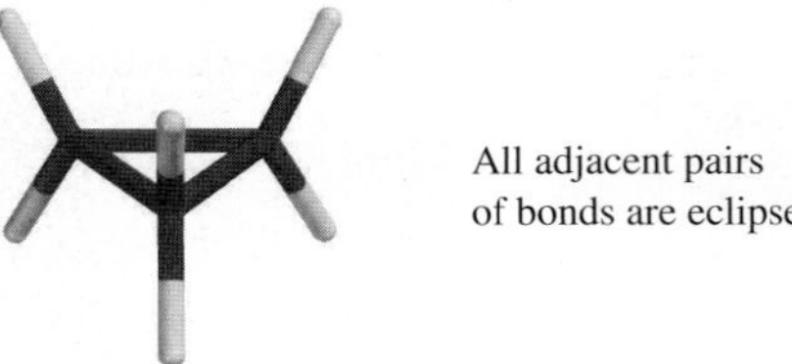

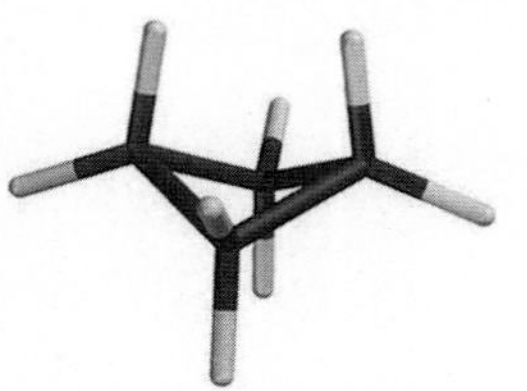

Figure 3.11

Nonplanar ("puckered") conformation of cyclobutane. The nonplanar conformation reduces the eclipsing of bonds on adjacent carbons that characterizes the planar conformation.

Cyclobutane not only has less angle strain than cyclopropane but also can reduce the torsional strain that goes with a planar geometry by adopting the nonplanar "puckered" conformation shown in Figure 3.11, in which adjacent C—H bonds are twisted away from one another. A fully staggered arrangement in cyclobutane is not possible, but eclipsing interactions are decreased in the puckered form.

Problem 3.7

From Table 3.1, cyclohexane has the lowest heat of combustion per CH_2 group of 659 kJ/mol (157.5 kcal/mol). This represents a CH_2 group in a cyclic ring with negligible strain. Calculate the total strain energy for (a) the three CH_2 groups of cyclopropane and (b) the four CH_2 groups of cyclobutane.

Sample Solution (a) The heat of combustion for one cyclopropane CH_2 group is 697 kJ/mol (166.6 kcal/mol). Subtracting this value from the cyclohexane value gives the strain energy for one of the cyclopropane CH_2 groups: 697 − 659 = 38 kJ/mol (166.6 − 157.5 = 9.1 kcal/mol). Cyclopropane, with three CH_2 groups, has a total strain energy of three times that amount: 38 kJ/mol × 3 = 114 kJ/mol (9.1 kcal/mol × 3 = 27.3 kcal/mol).

Problem 3.8

The heats of combustion of ethylcyclopropane and methylcyclobutane have been measured as 3352 and 3384 kJ/mol (801.2 and 808.8 kcal/mol). Assign the correct heat of combustion to each isomer.

3.6 Cyclopentane

Neighboring C—H bonds are eclipsed in any planar cycloalkane. Thus all planar conformations are destabilized by torsional strain.

Angle strain in the planar conformation of cyclopentane is relatively small because the 108° angles of a regular pentagon are not much different from the normal 109.5° bond angles of sp^3-hybridized carbon. The torsional strain, however, is substantial, because five bonds are eclipsed on the top face of the ring, and another set of five are eclipsed on the bottom face (Figure 3.12*a*). Some, but not all, of this torsional strain is relieved in nonplanar conformations. Two nonplanar conformations of cyclopentane, the **envelope** (Figure 3.12*b*) and the **half-chair** (Figure 3.12*c*), are of similar energy.

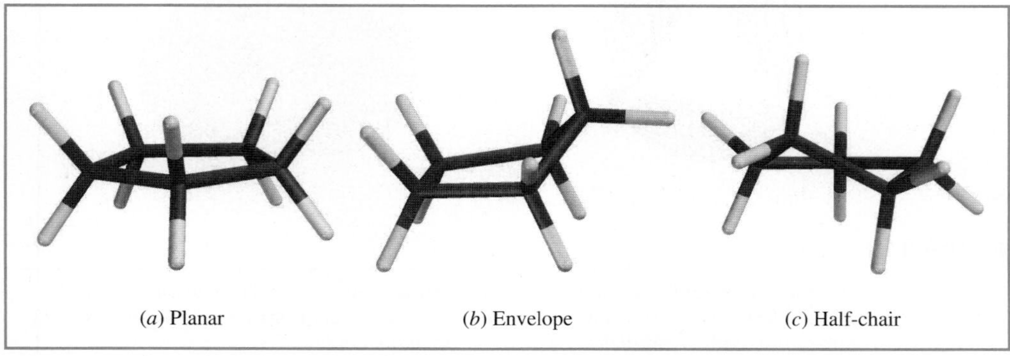

Figure 3.12

The (*a*) planar, (*b*) envelope, and (*c*) half-chair conformations of cyclopentane.

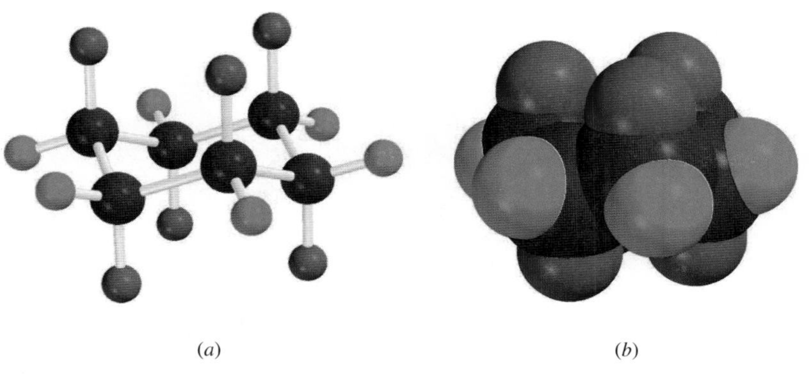

Figure 3.13

(*a*) A ball-and-stick model and (*b*) a space-filling model of the chair conformation of cyclohexane.

Chair cyclohexane bears some resemblance to a chaise lounge.
©Floortje/iStock/Getty Images

In the envelope conformation four of the carbon atoms are coplanar. The fifth carbon is out of the plane of the other four. There are three coplanar carbons in the half-chair conformation, with one carbon atom displaced above that plane and another below it. In both the envelope and the half-chair conformations, in-plane and out-of-plane carbons exchange positions rapidly. Equilibration between conformations of cyclopentane is very fast and occurs at rates similar to that of rotation about the carbon–carbon bond of ethane.

3.7 Conformations of Cyclohexane

Experimental evidence indicating that six-membered rings are nonplanar began to accumulate in the 1920s. Eventually, Odd Hassel of the University of Oslo established that the most stable conformation of cyclohexane has the shape shown in Figure 3.13. This is called the **chair** conformation. With C—C—C bond angles of 111°, the chair conformation is nearly free of angle strain. All of its bonds are staggered, making it free of torsional strain as well. The staggered arrangement of bonds in the chair conformation of cyclohexane is apparent in a Newman-style projection.

Hassel shared the 1969 Nobel Prize in Chemistry with Sir Derek Barton of Imperial College (London). Barton demonstrated how Hassel's structural results could be extended to an analysis of conformational effects on chemical reactivity.

H H H CH_2 H H CH_2 H H H

Staggered arrangement of bonds in chair conformation of cyclohexane

The cyclohexane chair is best viewed from the side-on perspective, which is useful for describing its conformational properties. You may draw chair cyclohexane in this pespective using different techniques, but your final drawing must have the following features. Bonds that are across the ring from each other are parallel, as indicated for the pairs of red, green, and blue bonds in the following drawing. Notice also that the bonds shown in red are drawn with longer lines to show the side-on perspective. In reality, all of the C—C bonds of cyclohexane are of the same length. Bonds are slanted as indicated. Although not planar, the cyclohexane ring should be level with respect to carbons 2 and 4 and carbons 1 and 5. The side-on perspective of cyclohexane is sometimes depicted with wedge bonds for C-1 to C-2 and C-3 to C-4 and a bold line for C-2 to C-3.

1 6 5 3 2 4

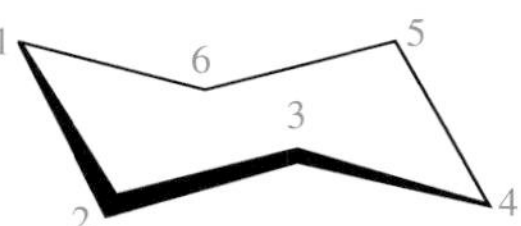

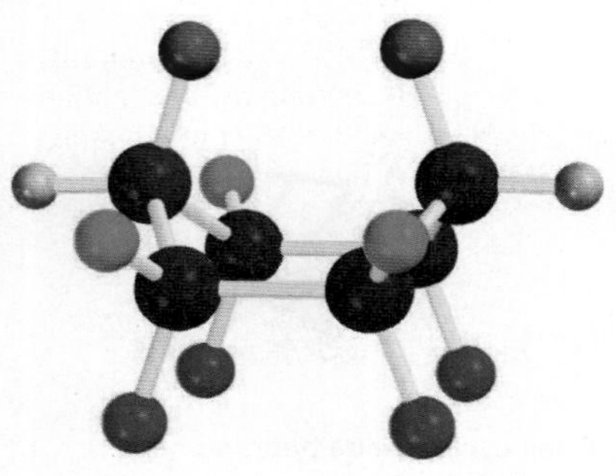

(*a*)

(*b*)

Figure 3.14

(*a*) A ball-and-spoke model and (*b*) a space-filling model of the boat conformation of cyclohexane. Torsional strain from eclipsed bonds and van der Waals strain involving the "flagpole" hydrogens (red) make the boat less stable than the chair.

(*a*) (*b*)

Figure 3.15

(*a*) The boat and (*b*) skew boat conformations of cyclohexane. Some of the torsional strain in the boat is relieved by rotation about C—C bonds in going to the twist conformation. This motion also causes the flagpole hydrogens to move away from one another, reducing the van der Waals strain between them.

A second, but much less stable, nonplanar conformation called the **boat** is shown in Figure 3.14. Like the chair, the boat conformation has bond angles that are approximately tetrahedral and is relatively free of angle strain. It is, however, destabilized by the torsional strain associated with eclipsed bonds on four of its carbons. The close approach of the two "flagpole" hydrogens shown in Figure 3.14 contributes a small amount of van der Waals strain as well. Both sources of strain are reduced by rotation about the carbon–carbon bond to give the slightly more stable **twist boat,** or **skew boat,** conformation (Figure 3.15).

The various conformations of cyclohexane are in rapid equilibrium with one another, but at any moment almost all of the molecules exist in the chair conformation. Less than 5 molecules per 100,000 are present in the skew boat conformation at 25°C. Thus, the discussion of cyclohexane conformational analysis that follows focuses exclusively on the chair conformation.

3.8 Axial and Equatorial Bonds in Cyclohexane

One of the most important findings to come from conformational studies of cyclohexane is that its 12 hydrogen atoms can be divided into two groups, as shown in Figure 3.16. Six of the hydrogens, called **axial** hydrogens, have their bonds parallel to a vertical axis that passes through the ring's center. These axial bonds alternately are directed up and down on adjacent carbons. The second set of six hydrogens, called **equatorial** hydrogens, are located approximately along the equator of the molecule. Notice that the four bonds to each carbon are arranged tetrahedrally, consistent with an sp^3 hybridization of carbon.

The conformational features of six-membered rings are fundamental to organic chemistry, so it is essential that you have a clear understanding of the directional properties of axial and equatorial bonds and be able to represent them accurately. Figure 3.17 offers some guidance.

Figure 3.16

Axial and equatorial bonds in cyclohexane.

Axial C—H bonds

Equatorial C—H bonds

Axial and equatorial bonds together

(1) Begin with the chair conformation of cyclohexane.

(2) Draw the axial bonds before the equatorial ones, alternating their direction on adjacent atoms. Always start by placing an axial bond "up" on the uppermost carbon or "down" on the lowest carbon.

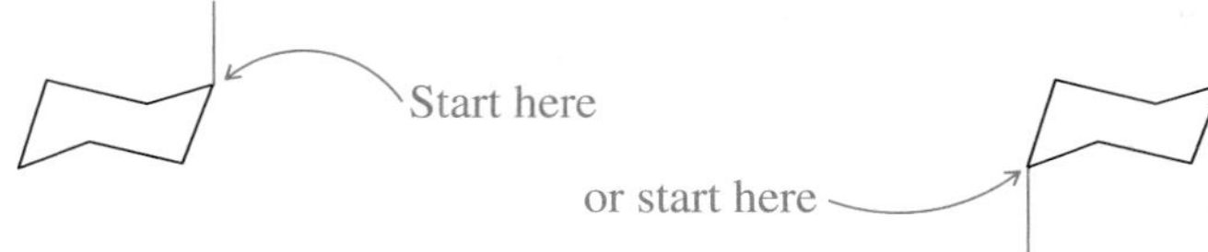

Then alternate to give

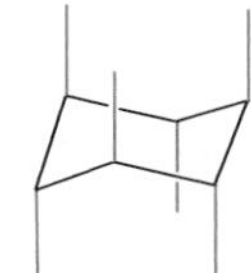

in which all the axial bonds are parallel to one another

(3) Place the equatorial bonds so as to approximate a tetrahedral arrangement of the bonds to each carbon. The equatorial bond of each carbon should be parallel to the ring bonds of its two nearest neighbor carbons.

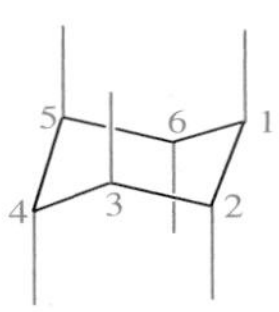

Place equatorial bond at C-1 so that it is parallel to the bonds between C-2 and C-3 and between C-5 and C-6.

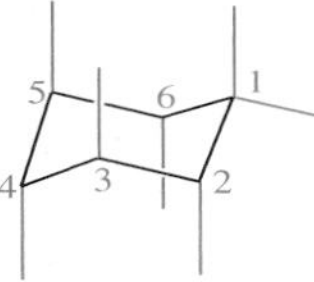

Following this pattern gives the complete set of equatorial bonds.

(4) Practice drawing cyclohexane chairs oriented in either direction.

and

Figure 3.17

A guide to representing the orientations of the bonds in the chair conformation of cyclohexane.

It is no accident that sections of our chair cyclohexane drawings resemble sawhorse projections of staggered conformations of alkanes. The same spatial relationships seen in alkanes carry over to substituents on a six-membered ring. In the structure

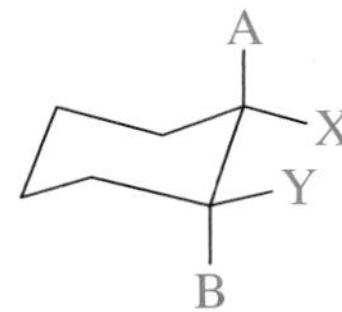

(The substituted carbons have the spatial arrangement shown)

substituents A and B are anti to each other, and the other relationships—A and Y, X and Y, and X and B—are gauche.

Problem 3.9

Given the following partial structure, add a substituent X to C-1 so that it satisfies the indicated stereochemical requirement. What is the A—C—C—X torsion (dihedral) angle in each?

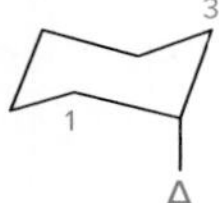

(a) Anti to A

(b) Gauche to A

(c) Anti to C-3

(d) Gauche to C-3

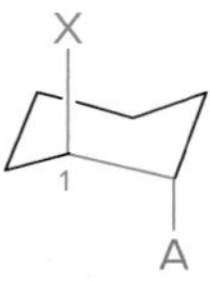

Sample Solution (a) In order to be anti to A, substituent X must be axial. The blue lines in the drawing show the A—C—C—X torsion angle to be 180°.

3.9 Conformational Inversion in Cyclohexane

We have seen that alkanes are not locked into a single conformation. Rotation about their C—C bonds occurs rapidly, interconverting anti and gauche conformations. Cyclohexane, too, is conformationally mobile. Through a process known as **ring inversion,** or **chair–chair interconversion,** one chair conformation is converted to another.

A potential energy diagram for chair–chair interconversion in cyclohexane is shown in Figure 3.18. In the first step, the chair is converted to a twist conformation. In this step, cyclohexane passes through a higher-energy half-chair conformation. The twist is converted to an alternate twist, via the boat conformation that is only 4–8 kJ/mol (1–2 kcal/mol) higher in energy. The second twist then proceeds to the inverted chair via another half-chair conformation. The twist conformations are *intermediates* in the process of ring inversion. Unlike a transition state, an **intermediate** is not a potential energy maximum but is a local minimum on the potential energy profile. The half-chair

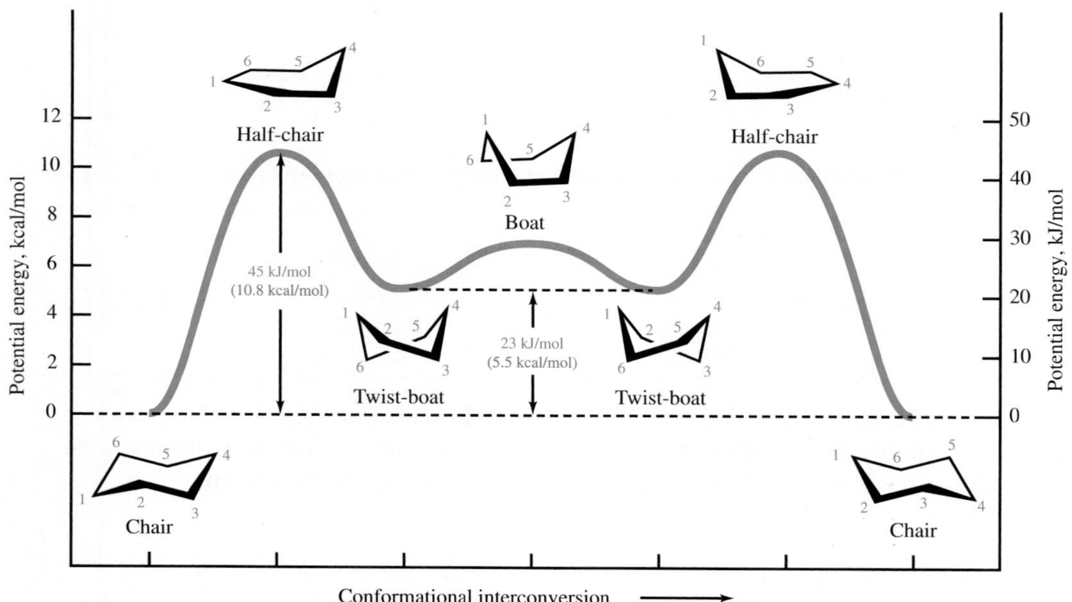

Figure 3.18

Energy diagram for ring inversion in cyclohexane. The energy of activation is the difference in energy between the chair and half-chair conformations. The twist conformations are intermediates. The boat and half-chair conformations are transition states.

conformations are highest in energy because they have the most eclipsing interactions. The difference in energy between the chair and half-chair conformations is the activation energy for the chair–chair interconversion, which is 45 kJ/mol (10.8 kcal/mol). It is a very rapid process with a half-life of 10^{-5} s at 25°C.

The most important result of ring inversion is that any substituent that is axial in the original chair conformation becomes equatorial in the ring-inverted form and vice versa.

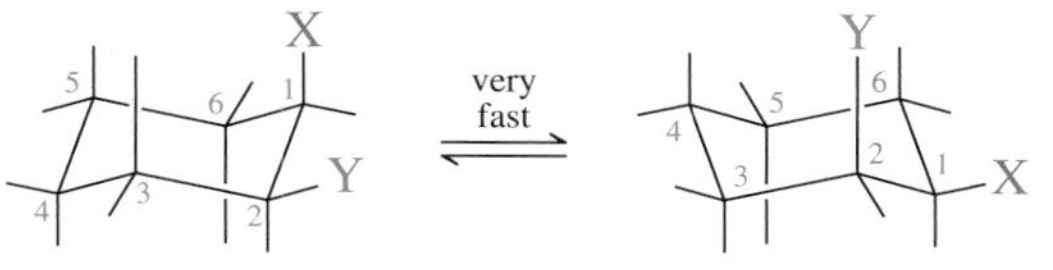

X axial; Y equatorial ⇌ (very fast) X equatorial; Y axial

The consequences of this point are developed for a number of monosubstituted cyclohexane derivatives in the following section, beginning with methylcyclohexane.

3.10 Conformational Analysis of Monosubstituted Cyclohexanes

Ring inversion in methylcyclohexane differs from that of cyclohexane in that the two chair conformations are not equivalent. In one chair the methyl group is axial; in the other it is equatorial. At room temperature approximately 95% of the molecules of methylcyclohexane are in the chair conformation that has an equatorial methyl group, whereas only 5% of the molecules have an axial methyl group.

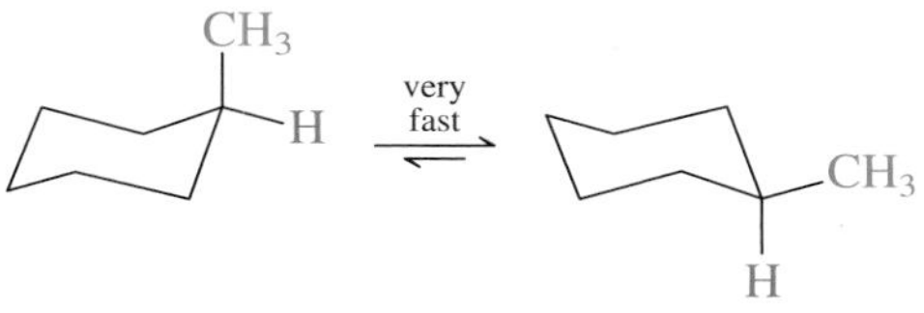

5% ⇌ (very fast) 95%

When two conformations of a molecule are in equilibrium with each other, the one with the lower free energy predominates. Why is equatorial methylcyclohexane more stable than axial methylcyclohexane?

See the boxed essay *Enthalpy, Free Energy, and Equilibrium Constant* accompanying this section for a discussion of these relationships.

A methyl group is less crowded when it is equatorial than when it is axial. One of the hydrogens of an axial methyl group is within 190–200 pm (1.90–2.00 Å) of the axial hydrogens at C-3 and C-5. This distance is less than the sum of the van der Waals radii of two hydrogens (240 pm; 2.40 Å) and causes van der Waals strain in the axial conformation. When the methyl group is equatorial, it experiences no significant crowding.

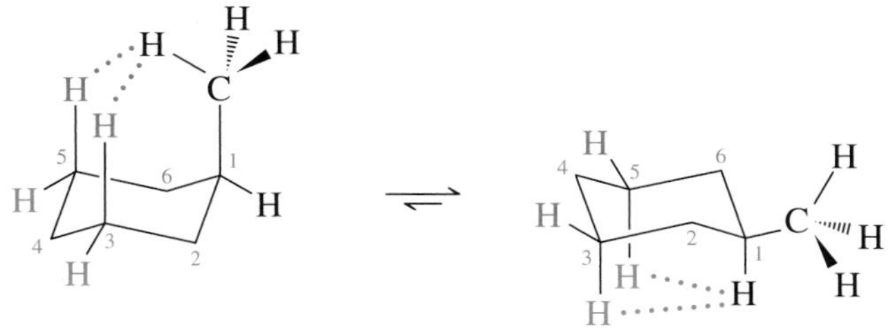

Van der Waals strain between hydrogen of axial CH_3 and axial hydrogens at C-3 and C-5

Negligible van der Waals strain between hydrogen at C-1 and axial hydrogens at C-3 and C-5

The greater stability of an equatorial methyl group, compared with an axial one, is another example of a *steric effect* (Section 3.2). An axial substituent is said to be crowded because of **1,3-diaxial repulsions** between itself and the other two axial substituents located on the same side of the ring.

Problem 3.10

The following questions relate to a cyclohexane ring in the chair conformation shown.

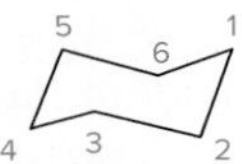

(a) Is a methyl group at C-6 that is "down" axial or equatorial?

(b) Is a methyl group that is "up" at C-1 more or less stable than a methyl group that is up at C-4?

(c) Place a methyl group at C-3 in its most stable orientation. Is it up or down?

Sample Solution (a) First indicate the directional properties of the bonds to the ring carbons. A substituent is down if it is below the other substituent on the same carbon atom. A methyl group that is down at C-6 is therefore axial.

We can relate the conformational preference for an equatorial methyl group in methylcyclohexane to the conformation of butane. The red bonds in the following structural formulas trace paths through four carbons, beginning at an equatorial methyl group. The zigzag arrangement described by each path mimics the anti conformation of butane.

When the methyl group is axial, each path mimics the gauche conformation of butane.

The preference for an equatorial methyl group in methylcyclohexane is therefore analogous to the preference for the anti conformation in butane. Two gauche butane-like units are present in axial methylcyclohexane that are absent in equatorial methylcyclohexane. As we saw earlier in Figure 3.7, the anti conformation of butane is 3.8 kJ/mol (0.9 kcal/mol) lower in energy than the gauche. Therefore, the energy difference between the equatorial and axial conformations of methylcyclohexane should be twice that, or 7.6 kJ/mol (1.8 kcal/mol). The experimentally measured difference of 7.2 kJ/mol (1.74 kcal/mol) gives us confidence that the same factors that govern the conformations of noncyclic compounds also apply to cyclic ones. What we call 1,3-diaxial repulsions in substituted cyclohexanes are really the same as van der Waals strain in the gauche conformations of alkanes.

Other substituted cyclohexanes are similar to methylcyclohexane. Two chair conformations exist in rapid equilibrium, and the one in which the substituent is equatorial is more stable. The relative amounts of the two conformations depend on the effective size of the substituent. The size of a substituent, in the context of cyclohexane conformations, is related to the degree of branching at the atom connected to the ring. A single atom, such as a halogen, does not take up much space, and its preference for an equatorial orientation is less than that of a methyl group.

40% 60%

The halogens F, Cl, Br, and I do not differ much in their preference for the equatorial position. As the atomic radius increases in the order F < Cl < Br < I, so does the carbon–halogen bond distance, and the two effects tend to cancel.

Enthalpy, Free Energy, and Equilibrium Constant

One of the fundamental equations of thermodynamics concerns systems at equilibrium and relates the equilibrium constant K to the difference in **standard free energy** ($\Delta G°$) between the products and the reactants.

$$\Delta G° = G°_{products} - G°_{reactants} = -RT \ln K$$

where T is the absolute temperature in kelvins and the constant R equals 8.314 J/mol · K (1.99 cal/mol · K).

For the equilibrium between the axial and equatorial conformations of a monosubstituted cyclohexane,

the equilibrium constant is given by the expression

$$K = \frac{[\text{products}]}{[\text{reactants}]}$$

Inserting the appropriate values for R, T (298 K), and K gives the values of $\Delta G°$ listed in the following table for the various substituents discussed in Section 3.10.

The relationship between $\Delta G°$ and K is plotted in Figure 3.19. A larger value of K is associated with a more negative $\Delta G°$.

Free energy and enthalpy are related by the expression

$$\Delta G° = \Delta H° - T\Delta S°$$

where $\Delta S°$ is the difference in *entropy* between the products and reactants. A positive $\Delta S°$ is accompanied by an increase in the disorder of a system. A positive $\Delta S°$ leads to a $\Delta G°$ that is more negative than $\Delta H°$ and a larger K than expected on the basis of enthalpy considerations alone. Conversely, a negative $\Delta S°$ gives a smaller K than expected. In the case of conformational equilibration between the chair forms of a substituted cyclohexane, $\Delta S°$ is close to zero and $\Delta G°$ and $\Delta H°$ are approximately equal.

				$\Delta G°_{298\text{ K}}$	
Substituent X	**Percent axial**	**Percent equatorial**	**K**	**kJ/mol**	**kcal/mol**
$-CH_3$	5	95	19	−7.2	−1.74
$-CH_2CH_3$	4.6	95.4	21	−7.5	−1.79
$-CH(CH_3)_2$	3	97	38	−9.0	−2.15
$-C(CH_3)_3$	<0.01	>99.99	>9999	−22.8	−5.5
—F	40	60	1.5	−1.0	−0.25
—Cl	29	71	2.4	−2.2	−0.52
—OH	17	83	4.8	−3.9	−0.93
$-NH_2$	9	91	11	−5.9	−1.4

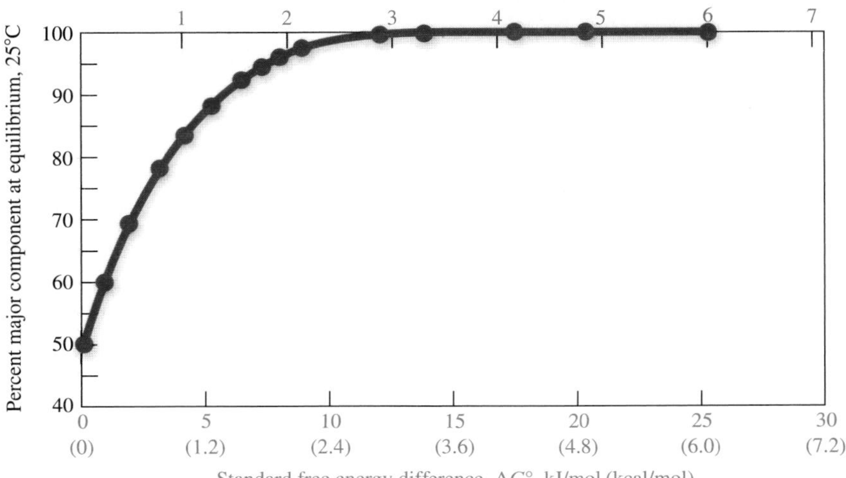

Figure 3.19

Distribution of two products at equilibrium at 25°C as a function of the standard free energy difference ($\Delta G°$) between them.

Highly branched groups such as *tert*-butyl are commonly described as "bulky."

A branched alkyl group such as isopropyl exhibits a slightly greater preference for the equatorial orientation than does methyl, but a *tert*-butyl group is so large that *tert*-butylcyclohexane exists almost entirely in the conformation in which the group is equatorial. The amount of axial *tert*-butylcyclohexane present is too small to measure.

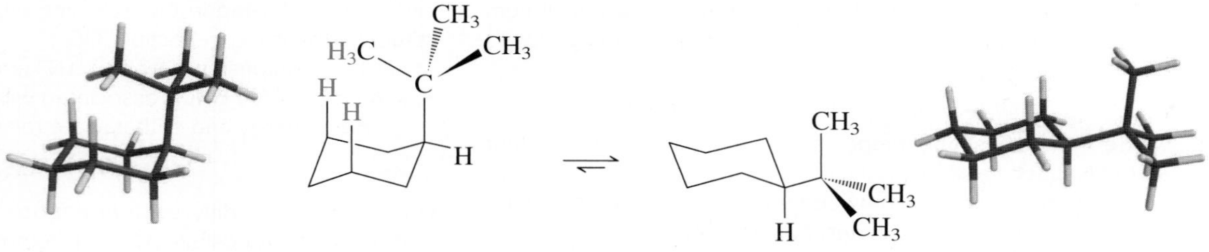

Less than 0.01%
(Serious 1,3-diaxial repulsions involving *tert*-butyl group)

Greater than 99.99%
(Decreased van der Waals strain)

Problem 3.11

Draw the most stable conformation of 1-*tert*-butyl-1-methylcyclohexane.

3.11 Disubstituted Cycloalkanes: cis–trans Stereoisomers

When a cycloalkane bears two substituents on different carbons—methyl groups, for example—these substituents may be on the same or on opposite sides of the ring. When substituents are on the same side, we say they are **cis** to each other; if they are on opposite sides, they are **trans** to each other. Both terms come from the Latin, in which cis means "on this side" and trans means "across."

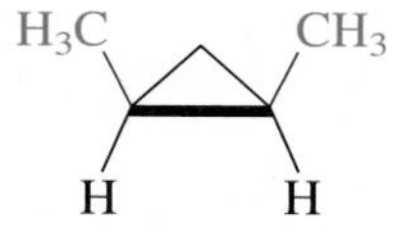

cis-1,2-Dimethylcyclopropane

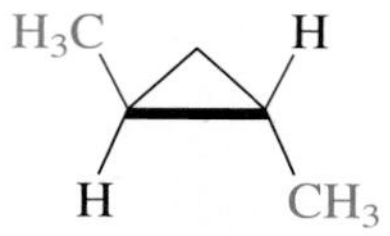

trans-1,2-Dimethylcyclopropane

Problem 3.12

Exclusive of compounds with double bonds, four hydrocarbons are *constitutional* isomers of *cis*- and *trans*-1,2-dimethylcyclopropane. Identify these compounds.

Stereochemistry is the term applied to the three-dimensional aspects of molecular structure and reactivity.

The cis and trans forms of 1,2-dimethylcyclopropane are stereoisomers. **Stereoisomers** are isomers that have their atoms bonded in the same order—that is, they have the same constitution, but they differ in the arrangement of atoms in space. You learned in Section 2.23 that constitutional isomers could differ in stability. What about stereoisomers?

We can measure the energy difference between *cis*- and *trans*-1,2-dimethylcyclopropane by comparing their heats of combustion. As illustrated in Figure 3.20, the difference in their heats of combustion is a direct measure of the difference in their energies. Because the heat of combustion of *trans*-1,2-dimethylcyclopropane is 5 kJ/mol (1.2 kcal/mol) less than that of its cis stereoisomer, it follows that *trans*-1,2-dimethylcyclopropane is 5 kJ/mol (1.2 kcal/mol) more stable than *cis*-1,2-dimethylcyclopropane.

In this case, the relationship between stability and stereochemistry is easily explained on the basis of van der Waals strain. The methyl groups on the same side of the ring in *cis*-1,2-dimethylcyclopropane crowd each other and increase the potential

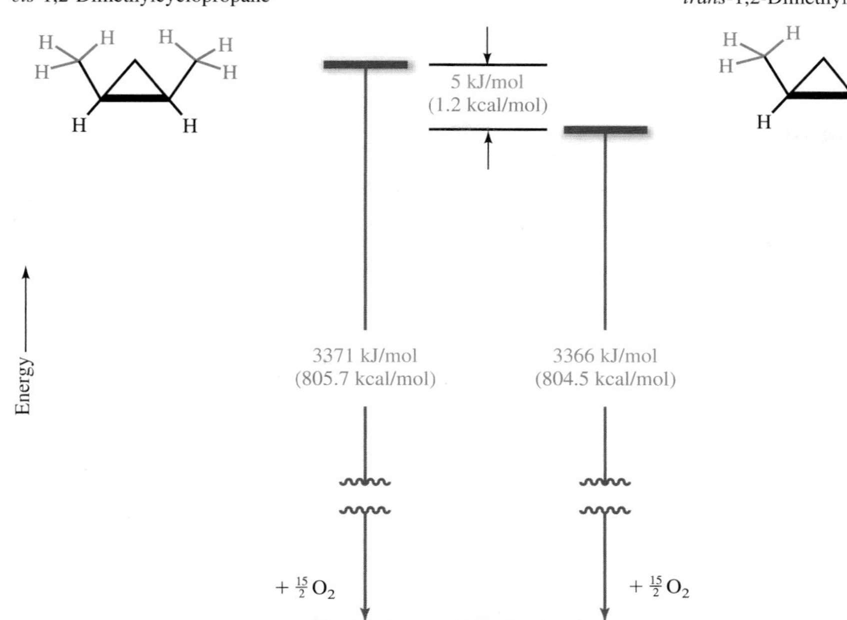

Figure 3.20

The enthalpy difference between *cis*- and *trans*-1,2-dimethylcyclopropane can be determined from their heats of combustion. Van der Waals strain between methyl groups on the same side of the ring makes the cis stereoisomer less stable than the trans.

energy of this stereoisomer. Steric hindrance between methyl groups is absent in *trans*-1,2-dimethylcyclopropane.

Problem 3.13

Chrysanthemic acid, from the chrysanthemum flower, is a naturally occurring insecticide, with the constitution indicated. Draw the structures of the cis and trans stereoisomers of chrysanthemic acid.

HO_2C — $CH{=}C(CH_3)_2$

H_3C CH_3

Disubstituted cyclopropanes exemplify one of the simplest cases involving stability differences between stereoisomers. A three-membered ring has no conformational mobility, so cannot reduce the van der Waals strain between cis substituents on adjacent carbons without introducing other strain. The situation is different in disubstituted derivatives of cyclohexane.

3.12 Conformational Analysis of Disubstituted Cyclohexanes

We'll begin with *cis*- and *trans*-1,4-dimethylcyclohexane.

H_3C CH_3 H H H_3C H H CH_3

cis-1,4-Dimethylcyclohexane *trans*-1,4-Dimethylcyclohexane

Wedges fail to show conformation, and it's important to remember that the rings of *cis*- and *trans*-1,2-dimethylcyclohexane exist in a chair conformation. This fact must be taken into consideration when evaluating the relative stabilities of the stereoisomers.

TABLE 3.2 Heats of Combustion of Isomeric Dimethylcyclohexanes

Compound	Orientation of methyl groups in most stable conformation	Heat of combustion		Difference in heat of combustion		More stable stereoisomer
		kJ/mol	kcal/mol	kJ/mol	kcal/mol	
cis-1,2-Dimethylcyclohexane	Axial–equatorial	5263	1257.8			
trans-1,2-Dimethylcyclohexane	Diequatorial	5255	1255.9	8	1.9	trans
cis-1,3-Dimethylcyclohexane	Diequatorial	5250	1254.8			
trans-1,3-Dimethylcyclohexane	Axial–equatorial	5258	1256.7	8	1.9	cis
cis-1,4-Dimethylcyclohexane	Axial–equatorial	5258	1256.7			
trans-1,4-Dimethylcyclohexane	Diequatorial	5250	1254.8	8	1.9	trans

Their heats of combustion (Table 3.2) reveal that *trans*-1,4-dimethylcyclohexane is 8 kJ/mol (1.9 kcal/mol) more stable than the cis stereoisomer. It is unrealistic to believe that van der Waals strain between cis substituents is responsible, because the methyl groups are too far away from each other. To understand why *trans*-1,4-dimethylcyclohexane is more stable than *cis*-1,4-dimethylcyclohexane, we need to examine each stereoisomer in its most stable conformation.

cis-1,4-Dimethylcyclohexane can adopt either of two equivalent chair conformations, *each having one axial methyl group and one equatorial methyl group.* The two are in rapid equilibrium with each other by ring inversion. The equatorial methyl group becomes axial, and the axial methyl group becomes equatorial.

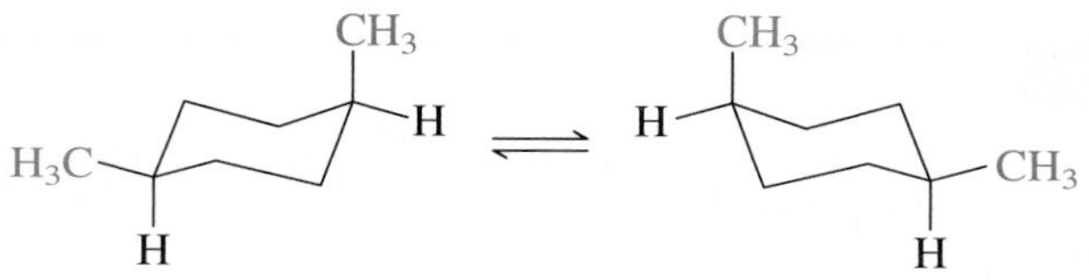

CH_3 H H_3C H ⇌ CH_3 H CH_3 H

(One methyl group is axial, the other equatorial) (One methyl group is axial, the other equatorial)

(Both methyl groups are up)

cis-1,4-Dimethylcyclohexane

The methyl groups are cis because both are up relative to the hydrogen present at each carbon. If both methyl groups were down, they would still be cis to each other. Notice that ring inversion does not alter the cis relationship between the methyl groups. Nor does it alter their up-versus-down quality; substituents that are up in one conformation remain up in the ring inverted form.

The most stable conformation of trans-1,4-dimethylcyclohexane has both methyl groups in equatorial orientations. The two chair conformations of *trans*-1,4-dimethylcyclohexane are not equivalent. One has two equatorial methyl groups; the other, two axial methyl groups.

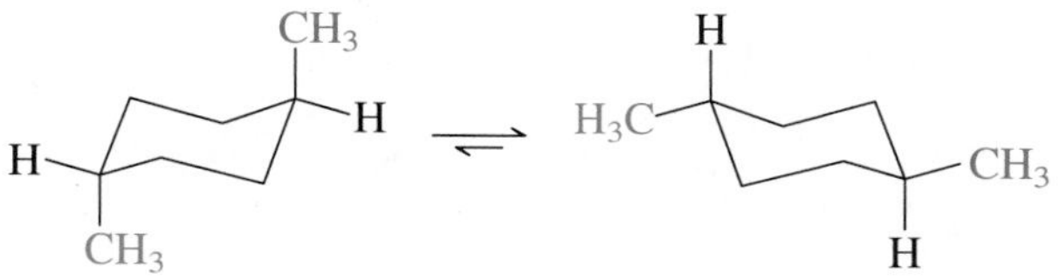

CH_3 H H CH_3 ⇌ H H_3C CH_3 H

(Both methyl groups are axial: less stable chair conformation) (Both methyl groups are equatorial: more stable chair conformation)

(One methyl group is up, the other down)

trans-1,4-Dimethylcyclohexane

The more stable chair—the one with both methyl groups equatorial—is adopted by most of the *trans*-1,4-dimethylcyclohexane molecules.

trans-1,4-Dimethylcyclohexane is more stable than *cis*-1,4-dimethylcyclohexane because both of the methyl groups are equatorial in its most stable conformation. One methyl group must be axial in the cis stereoisomer. Remember, it is a general rule that any substituent is more stable in an equatorial orientation than in an axial one. It is worth pointing out that the 8 kJ/mol (1.9 kcal/mol) energy difference between *cis*- and *trans*-1,4-dimethylcyclohexane is nearly the same as the energy difference between the axial and equatorial conformations of methylcyclohexane. There is a simple reason for this: in both instances the less stable structure has one axial methyl group, and the 8 kJ/mol (1.9 kcal/mol) energy difference can be considered the "energy cost" of having a methyl group in an axial rather than an equatorial orientation.

Like the 1,4-dimethyl derivatives, *trans*-1,2-dimethylcyclohexane has a lower heat of combustion (see Table 3.2) and is more stable than *cis*-1,2-dimethylcyclohexane. The cis stereoisomer has two chair conformations of equal energy, each containing one axial and one equatorial methyl group.

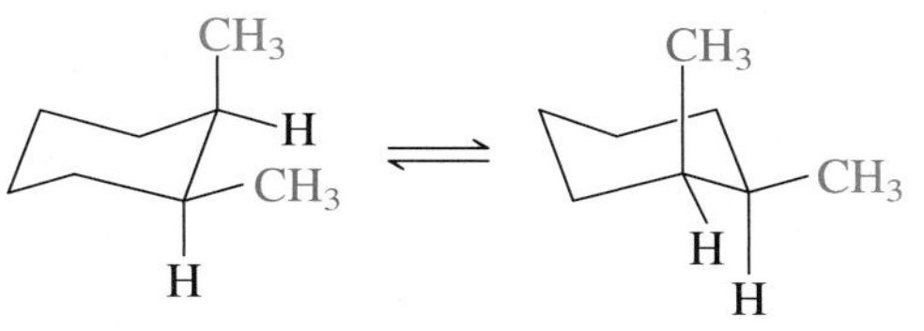

CH_3 H CH_3 H ⇌ CH_3 CH_3 H H

cis-1,2-Dimethylcyclohexane

Both methyl groups are equatorial in the most stable conformation of *trans*-1,2-dimethylcyclohexane.

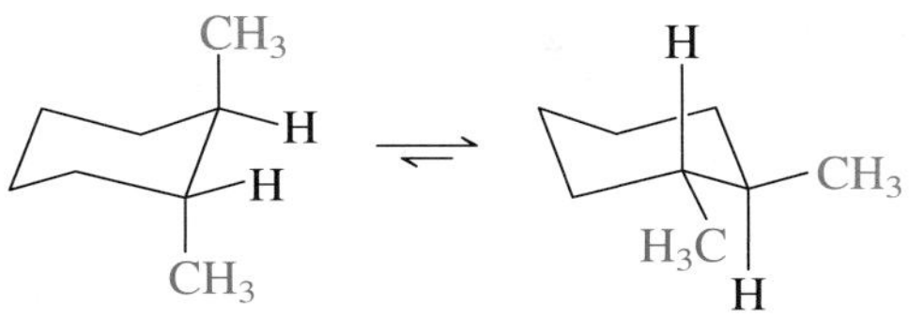

CH_3 H H CH_3 ⇌ H CH_3 H_3C H

(Both methyl groups are axial: less stable chair conformation)

(Both methyl groups are equatorial: more stable chair conformation)

trans-1,2-Dimethylcyclohexane

As in the 1,4-dimethylcyclohexanes, the 8 kJ/mol (1.9 kcal/mol) energy difference between the more stable (trans) and the less stable (cis) stereoisomer is attributed to the strain associated with the presence of an axial methyl group in the cis stereoisomer.

Probably the most interesting observation in Table 3.2 concerns the 1,3-dimethylcyclohexanes. Unlike the 1,2- and 1,4-dimethylcyclohexanes, in which the trans stereoisomer is more stable than the cis, we find that *cis*-1,3-dimethylcyclohexane is 8 kJ/mol (1.9 kcal/mol) more stable than *trans*-1,3-dimethylcyclohexane. Why?

The most stable conformation of *cis*-1,3-dimethylcyclohexane has both methyl groups equatorial.

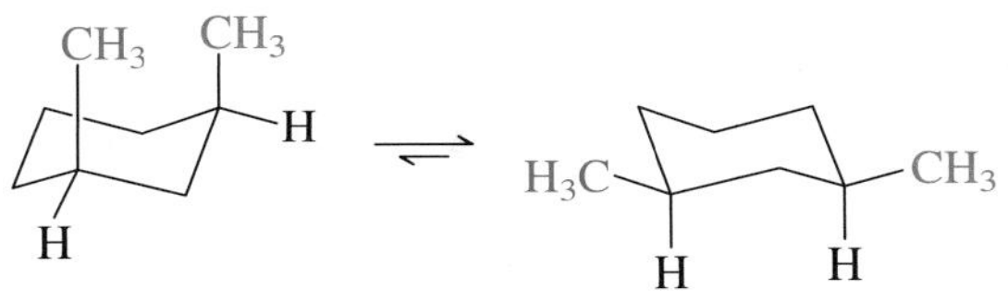

CH_3 CH_3 H H ⇌ H_3C CH_3 H H

(Both methyl groups are axial: less stable chair conformation)

(Both methyl groups are equatorial: more stable chair conformation)

cis-1,3-Dimethylcyclohexane

The two chair conformations of *trans*-1,3-dimethylcyclohexane are equivalent to each other. Both contain one axial and one equatorial methyl group.

(One methyl group is axial, the other equatorial) ⇌ (One methyl group is axial, the other equatorial)

trans-1,3-Dimethylcyclohexane

Thus the trans stereoisomer, with one axial methyl group, is less stable than *cis*-1,3-dimethylcyclohexane where both methyl groups are equatorial.

Problem 3.14

The following questions relate to a cyclohexane ring in the chair conformation shown.

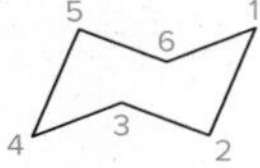

(a) Draw the more stable 1,4-dimethylcyclohexane with methyl substituents at C-3 and C-6. Are the methyl substituents cis or trans?

(b) Draw a 1,2-dimethycyclohexane using positions C-1 and C-2 with both methyl substituents in equatorial positions. What is the spatial relationship of the methyl groups?

(c) Place one methyl axial at C-2 and one equatorial at C-5. Are these methyl substituents cis or trans?

Sample Solution (a) For 1,4-dimethylcyclohexane, placing methyl substituents in equatorial positions on C-3 and C-6 gives the lower energy cyclohexane.

Labeling the methyl substituents and hydrogens on each carbon reveals that one methyl group is up and one down. These methyl groups are trans.

Problem 3.15

Based on what you know about disubstituted cyclohexanes, which of the following two stereoisomeric 1,3,5-trimethylcyclohexanes would you expect to be more stable?

cis-1,3,5-Trimethylcyclohexane

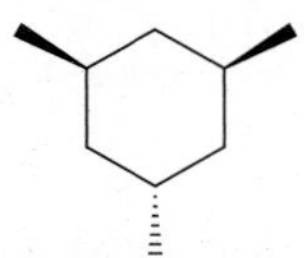

trans-1,3,5-Trimethylcyclohexane

If a disubstituted cyclohexane has two different substituents, then the most stable conformation is the chair that has the larger substituent in an equatorial orientation. This is most apparent when one of the substituents is a bulky group such as *tert*-butyl. Thus, the most stable conformation of *cis*-1-*tert*-butyl-2-methylcyclohexane has an equatorial *tert*-butyl group and an axial methyl group.

(Less stable conformation: larger group is axial) (More stable conformation: larger group is equatorial)

cis-1-*tert*-Butyl-2-methylcyclohexane

Problem 3.16

Write structural formulas for the most stable conformation of each of the following compounds:

(a) *trans*-1-*tert*-Butyl-3-methylcyclohexane
(b) *cis*-1-*tert*-Butyl-3-methylcyclohexane
(c) *trans*-1-*tert*-Butyl-4-methylcyclohexane
(d) *cis*-1-*tert*-Butyl-4-methylcyclohexane

Sample Solution

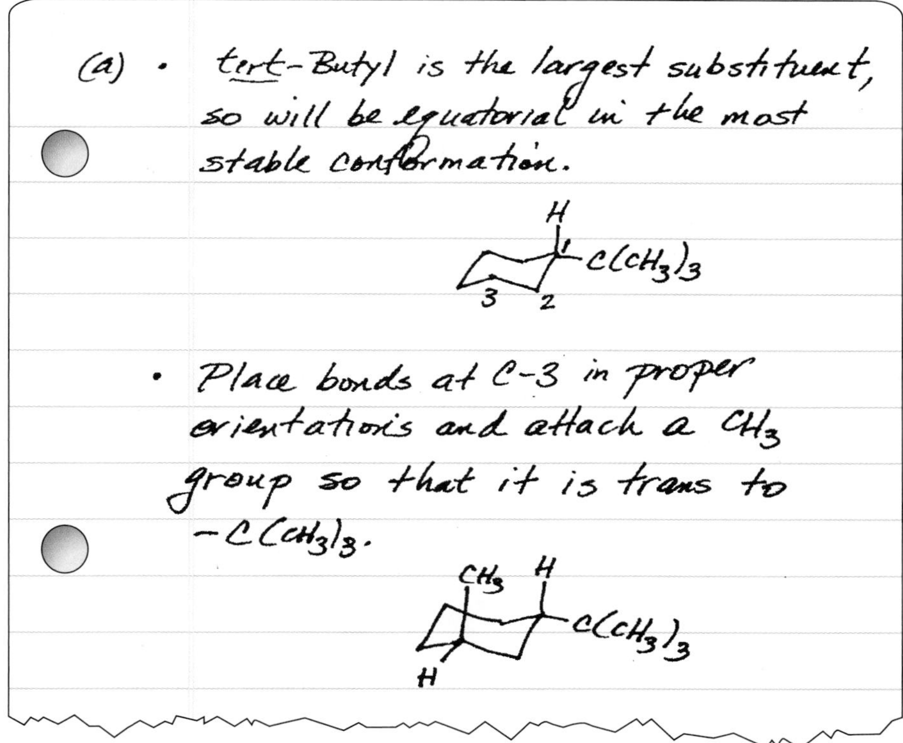

(a) • tert-Butyl is the largest substituent, so will be equatorial in the most stable conformation.
H
C(CH3)3
1
3 2
• Place bonds at C-3 in proper orientations and attach a CH3 group so that it is trans to -C(CH3)3.
CH3 H
C(CH3)3
H

Cyclohexane rings that bear *tert*-butyl substituents are examples of conformationally biased molecules. A *tert*-butyl group has such a pronounced preference for the equatorial orientation that it will strongly bias the equilibrium to favor such conformations. This does not mean that ring inversion does not occur, however. Ring inversion does occur, but at any instant only a tiny fraction of the molecules exist in conformations having axial *tert*-butyl groups. It is not strictly correct to say that *tert*-butylcyclohexane and its derivatives are "locked" into a single conformation; conformations related by ring inversion are in rapid equilibrium with one another, but the distribution between them strongly favors those in which the *tert*-butyl group is equatorial.

3.13 Medium and Large Rings

The largest known cycloalkane has a ring with 288 carbons.

Beginning with cycloheptane, which has four conformations of similar energy, conformational analysis of cycloalkanes becomes more complicated. The same fundamental principles apply to medium and large rings as apply to smaller ones—but there are more atoms, more bonds, and more conformational possibilities.

3.14 Polycyclic Ring Systems

Polycyclic compounds are those that contain more than one ring. The IUPAC classifies polycyclic structures according to the minimum number of bond cleavages required to generate a noncyclic structure. The structure is *bicyclic* if two bond disconnections yield an open-chain structure, *tricyclic* if three, *tetracyclic* if four, and so on. Adamantane, a naturally occurring hydrocarbon found in petroleum, for example, is tricyclic because three bond cleavages are needed before an open-chain structure results.

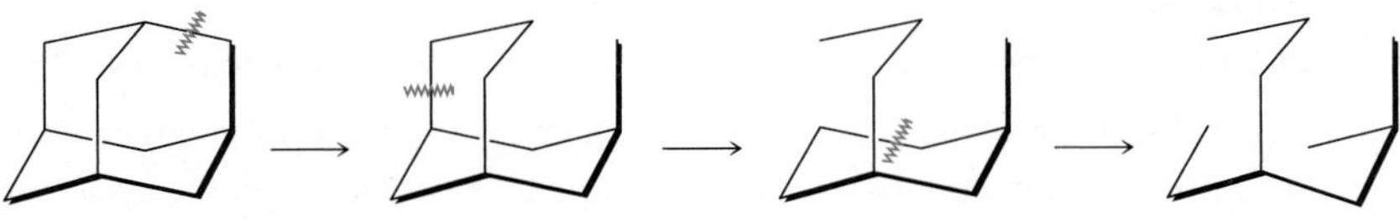

The correct number of rings may be determined by different sets of disconnections, and the final open-chain structure need not be the same for different sets. All that matters is finding the minimum number of disconnections.

Problem 3.17

Cubane (C_4H_8) is the common name of the polycyclic hydrocarbon shown. As its name implies, its structure is that of a cube. How many rings are present in cubane according to the bond-disconnection rule?

In addition to classifying polycyclic compounds according to the number of rings they contain, we classify them with respect to the way in which the rings are joined. In a **spiro** compound, one atom is common to two rings.

The simplest spiro alkane is *spiro[2.2]pentane,* a molecular model of which illustrates an interesting structural feature of spiro compounds. The two rings lie at right angles to each other.

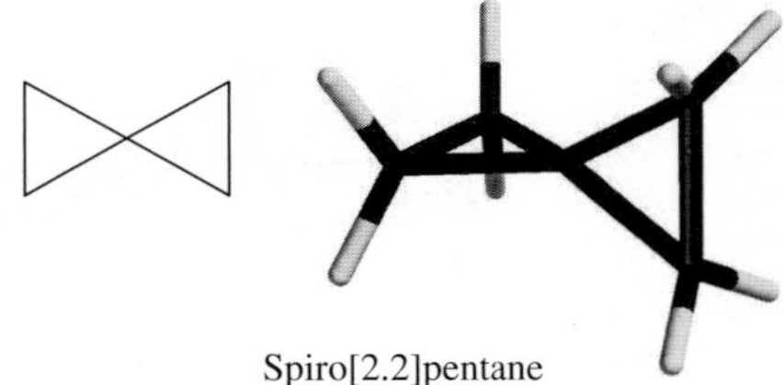
Spiro[2.2]pentane

The IUPAC names of spiro alkanes take the form *spiro[number.number]alkane.* The *alkane* suffix is simply the name of the unbranched alkane having the same number of carbons as those in the two rings. The numbers inside the brackets are, in ascending order, the number of carbons unique to each ring. Thus, eight carbons make up the two rings of spiro[3.4]octane; the spiro carbon is bridged by three carbons of one ring and four carbons of the other.

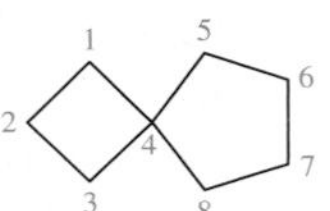

Spiro[3.4]octane

When substituents are present, numbering begins in the smaller ring adjacent to the spiro carbon and proceeds consecutively around the smaller ring away from the spiro carbon, through it, then around the larger ring. As with alkanes, the direction is chosen so as to give the lower locant at the first point of difference, substituents are listed in alphabetical order, and the locants and substituents appear first in the name.

Problem 3.18

Vetiver, a soothing oil popular in aromatherapy (Figure 3.21), contains β-vetivone, which can be viewed as a derivative of compound A. What is the IUPAC name of A?

β-Vetivone Compound A

Figure 3.21

Vetiver grass is the source of vetiver oil. ©Pungem/iStock/Getty Images

In a **bridged** compound, two atoms are common to two or more rings. *Camphene,* a naturally occurring hydrocarbon obtained from pine oil, is a representative bridged bicyclic hydrocarbon. It is convenient to regard camphene as a six-membered ring (indicated by the blue bonds in the following structure) in which the two carbons designated by asterisks (*) are bridged by a CH_2 group. The two designated carbons are known as *bridgehead* carbons.

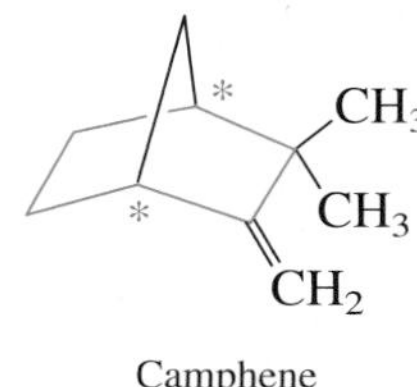

Camphene

Problem 3.19

Use the bond-cleavage criterion to verify that camphene is bicyclic.

Bridged bicyclic alkanes are named in the manner *bicyclo[number.number.number] alkane.* As illustrated for bicyclo[3.2.1]octane, the parent alkane is the one with the same number of carbons as the total in the bicyclic skeleton.

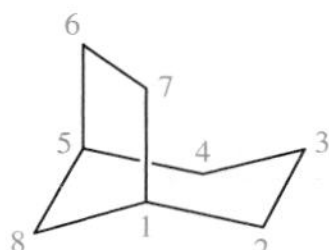

Bicyclo[3.2.1]octane

The bracketed numbers identify the number of carbons in the three bridges in descending order. Numbering begins at a bridgehead position and proceeds consecutively in the direction of the largest bridge and continues through the next largest. The atoms in the smallest bridge are numbered last.

Problem 3.20

Write structural formulas for each of the following bicyclic hydrocarbons:

(a) Bicyclo[2.2.1]heptane
(b) 1,7,7-Trimethylbicyclo[2.2.1]heptane
(c) Bicyclo[3.1.1]heptane

continued

Sample Solution (a) The bicyclo[2.2.1]heptane ring system is one of the most frequently encountered bicyclic structural types. It contains seven carbon atoms, as indicated by the suffix *-heptane*. The bridging groups contain two, two, and one carbon, respectively.

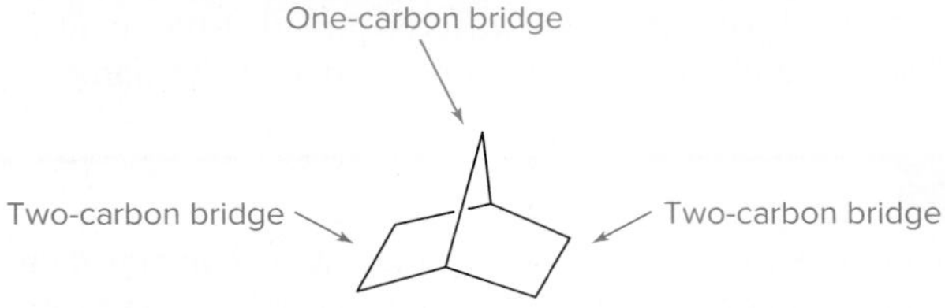

One-carbon bridge

Two-carbon bridge

Two-carbon bridge

Bicyclo[2.2.1]heptane

Many compounds contain rings that share a common side. Such compounds are normally referred to as *fused-ring* compounds, but for classification and naming purposes they are placed in the "bridged" category. The bridge in these cases is the common side and is given a value of zero atoms. The two stereoisomeric bicyclo[4.4.0]decanes, called *cis-* and *trans*-decalin, are important examples.

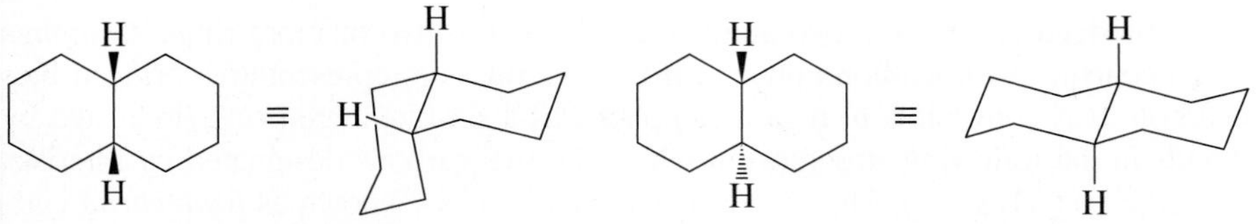

cis-Bicyclo[4.4.0]decane (*cis*-decalin)

trans-Bicyclo[4.4.0]decane (*trans*-decalin)

The hydrogen atoms at the ring junctions are on the same side in *cis*-decalin and on opposite sides in *trans*-decalin. Both rings adopt the chair conformation in each stereoisomer.

Decalin ring systems appear as structural units in a large number of naturally occurring substances, particularly the steroids. Cholic acid, for example, a steroid present in bile that promotes digestion, incorporates *cis*-decalin and *trans*-decalin units into a rather complex *tetracyclic* structure.

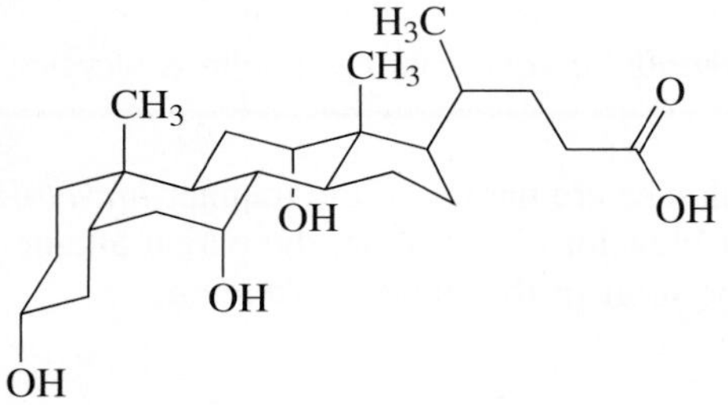

Cholic acid

Problem 3.21

Geosmin is a natural product that smells like dirt. It is produced by several microorganisms and can be obtained from beet extracts. Complete the following decalin ring skeleton, placing the substituents of geosmin in their proper orientations.

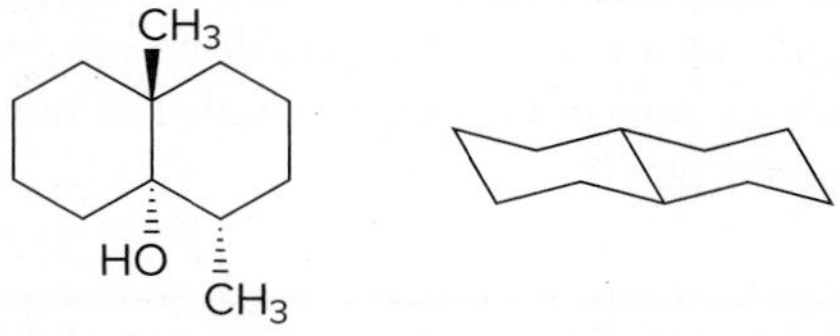

3.15 Heterocyclic Compounds

Not all cyclic compounds are hydrocarbons. Many substances include an atom other than carbon, called a *heteroatom* (Section 1.6), as part of a ring. A ring that contains at least one heteroatom is called a *heterocycle,* and a substance based on a heterocyclic ring is a **heterocyclic compound.** Each of the following heterocyclic ring systems will be encountered in this text:

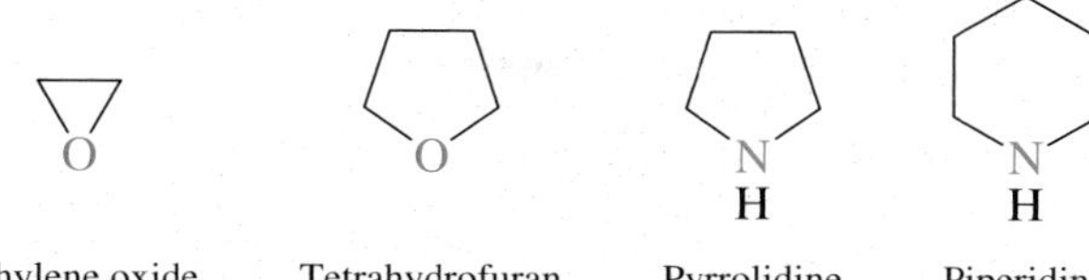

The names cited are common names, which have been in widespread use for a long time and are acceptable in IUPAC nomenclature.

The shapes of heterocyclic rings are very much like those of their all-carbon analogs. Thus, six-membered heterocycles such as piperidine exist in a chair conformation analogous to cyclohexane.

The hydrogen attached to nitrogen can be either axial or equatorial, and both chair conformations are approximately equal in stability.

Problem 3.22

Draw what you would expect to be the most stable conformation of the piperidine derivative in which the hydrogen bonded to nitrogen has been replaced by methyl.

Sulfur-containing heterocycles are also common. Compounds in which sulfur is the heteroatom in three-, four-, five-, and six-membered rings, as well as larger rings, are all well known. Two interesting heterocyclic compounds that contain sulfur–sulfur bonds are *lipoic acid* and *lenthionine.*

Lipoic acid: a growth factor required by a variety of different organisms

Lenthionine: contributes to the odor of shiitake mushrooms

Cyclic structures also exist in inorganic chemistry. The most stable form of elemental sulfur is an eight-membered ring of sulfur atoms.

Many heterocyclic systems contain double bonds and are related to arenes. The most important representatives of this class are introduced in Sections 12.21 and 12.22.

3.16 SUMMARY

In this chapter we explored the three-dimensional shapes of alkanes and cycloalkanes. The most important point to be taken from the chapter is that a molecule adopts the shape that minimizes its total strain. The sources of strain in alkanes and cycloalkanes are:

1. ***Bond length distortion:*** destabilization of a molecule that results when one or more of its bond distances are different from the normal values
2. ***Angle strain:*** destabilization that results from distortion of bond angles from their normal values
3. ***Torsional strain:*** destabilization that results when bonds on adjacent atoms are not staggered
4. ***Van der Waals strain:*** destabilization that results when atoms or groups on nonadjacent atoms are too close to one another

The various spatial arrangements available to a molecule by rotation about single bonds are called **conformations,** and **conformational analysis** is the study of the differences in stability and properties of the individual conformations. Rotation about carbon–carbon single bonds is normally very fast, occurring hundreds of thousands of times per second at room temperature. Molecules are rarely frozen into a single conformation but engage in rapid equilibration among the conformations that are energetically accessible.

Section 3.1 The most stable conformation of ethane is the **staggered** conformation. It is approximately 12 kJ/mol (2.9 kcal/mol) more stable than the **eclipsed,** which is the least stable conformation.

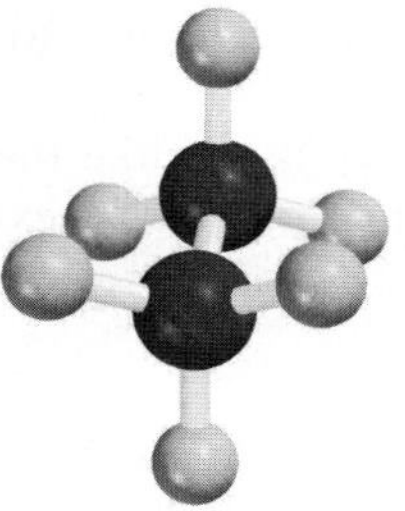

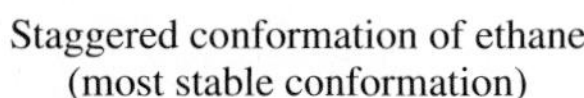

Staggered conformation of ethane (most stable conformation)

Eclipsed conformation of ethane (least stable conformation)

The difference in energy between the two results from two effects: a destabilization of the eclipsed conformation due to electron–electron repulsion in aligned bonds, and a stabilization of the staggered conformation due to better electron delocalization. At any instant, almost all the molecules of ethane reside in the staggered conformation.

Section 3.2 The two staggered conformations of butane are not equivalent. The **anti** conformation is more stable than the **gauche.**

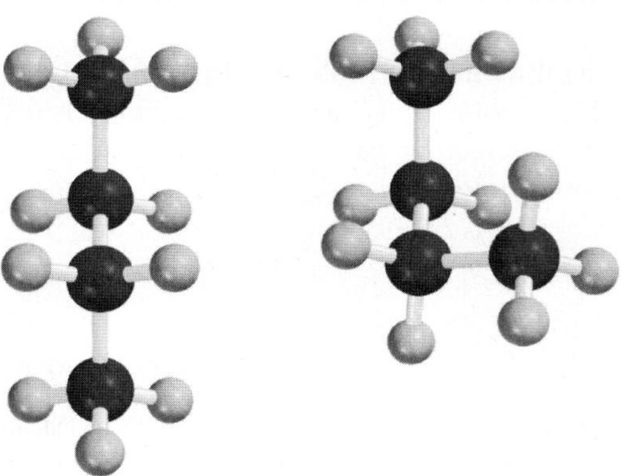

Anti conformation of butane Gauche conformation of butane

Neither conformation suffers torsional strain, because each has a staggered arrangement of bonds. The gauche conformation is less stable because of van der Waals strain involving the methyl groups.

Section 3.3 Higher alkanes adopt a zigzag conformation of the carbon chain in which all the bonds are staggered.

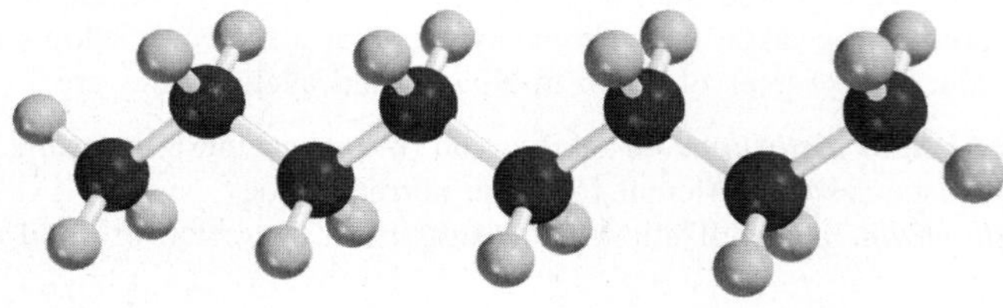

Octane

Section 3.4 At one time the rings of all cycloalkanes were believed to be planar. It was expected that cyclopentane would be the least strained cycloalkane because the angles of a regular pentagon (108°) are closest to the tetrahedral angle of 109.5°. Heats of combustion established that this is not so. With the exception of cyclopropane, the rings of all cycloalkanes are nonplanar.

Section 3.5 Cyclopropane is planar and destabilized by angle strain and torsional strain. Cyclobutane is nonplanar and less strained than cyclopropane.

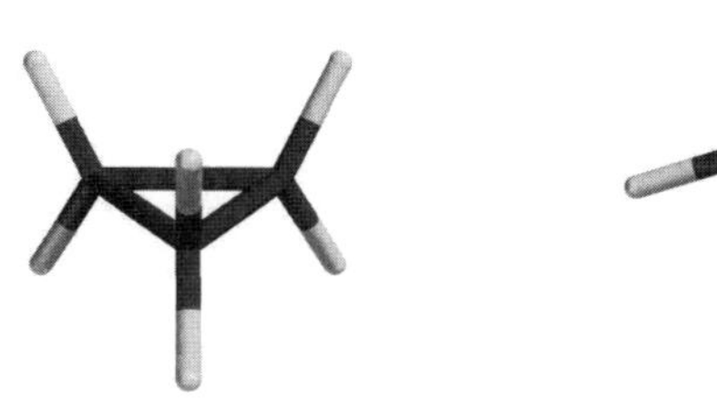

Cyclopropane

Cyclobutane

Section 3.6 Cyclopentane has two nonplanar conformations that are of similar stability: the **envelope** and the **half-chair.**

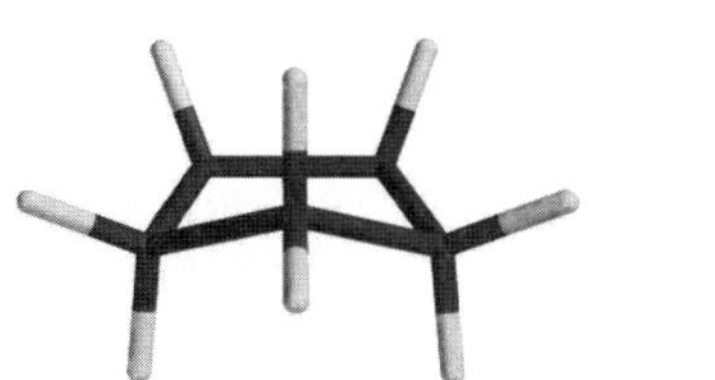

Envelope conformation of cyclopentane

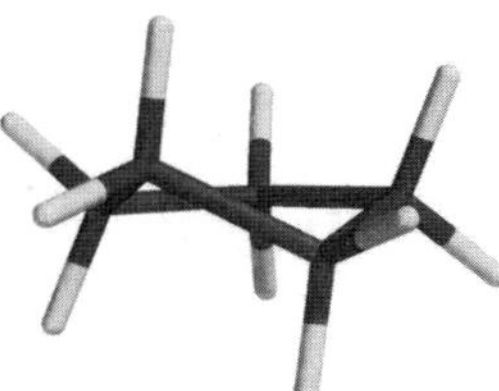

Half-chair conformation of cyclopentane

Section 3.7 Three conformations of cyclohexane have approximately tetrahedral angles at carbon: the chair, the boat, and the twist. The chair is by far the most stable; it is free of torsional strain, but the others are not. When a cyclohexane ring is present in a compound, it almost always adopts a chair conformation.

Chair Twist Boat

Section 3.8 The C—H bonds in the chair conformation of cyclohexane are not all equivalent but are divided into two sets of six each, called **axial** and **equatorial.**

Axial bonds to H in cyclohexane

Equatorial bonds to H in cyclohexane

Section 3.9 Conformational inversion is rapid in cyclohexane and causes all axial bonds to become equatorial and vice versa. As a result, a monosubstituted derivative of cyclohexane adopts the chair conformation in which the substituent is equatorial. *No bonds are made or broken in this process.*

Section 3.10 A substituent is less crowded and more stable when it is equatorial than when it is axial on a cyclohexane ring. Ring inversion of a monosubstituted cyclohexane allows the substituent to become equatorial.

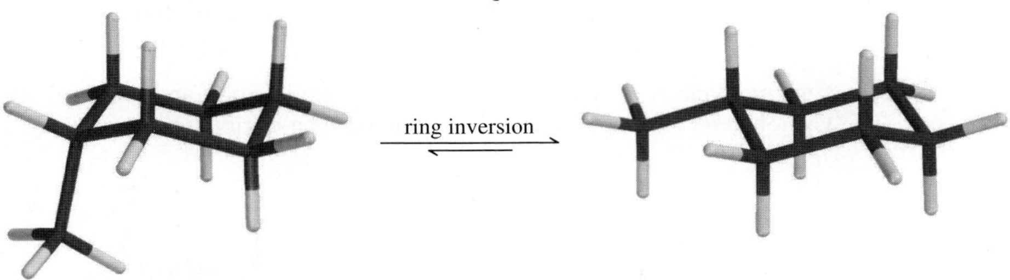

Methyl group axial (less stable) | Methyl group equatorial (more stable)

Branched substituents, especially *tert*-butyl, have an increased preference for the equatorial position.

Sections 3.11–3.12 **Stereoisomers** are isomers that have the same constitution but differ in the arrangement of atoms in space. *cis*-1,3-Dimethylcyclohexane and *trans*-1,3-dimethylcyclohexane are stereoisomers. The cis isomer is more stable than the trans.

Most stable conformation of *cis*-1,3-dimethylcyclohexane (no axial methyl groups)

Most stable conformation of *trans*-1,3-dimethylcyclohexane (one axial methyl group)

Section 3.13 Higher cycloalkanes have angles at carbon that are close to tetrahedral and are sufficiently flexible to adopt conformations that reduce their torsional strain. They tend to be populated by several different conformations of similar stability.

Section 3.14 Cyclic hydrocarbons can contain more than one ring. **Spiro** hydrocarbons are characterized by the presence of a single carbon that is common to two rings. Bicyclic alkanes contain two rings that share two or more atoms.

Section 3.15 Substances that contain one or more atoms other than carbon as part of a ring are called **heterocyclic** compounds. Rings in which the heteroatom is oxygen, nitrogen, or sulfur rank as both the most common and the most important.

6-Aminopenicillanic acid (bicyclic and heterocyclic)

PROBLEMS

Nomenclature and Terminology

3.23 Give the IUPAC name of each of the following:

(a) (b) (c)

3.24 Draw Newman projections for the gauche and anti conformations of 1,2-dichloroethane ($ClCH_2CH_2Cl$).

3.25 Identify all atoms that are (a) anti and (b) gauche to bromine in the conformation shown for $CH_3CH_2CH_2Br$.

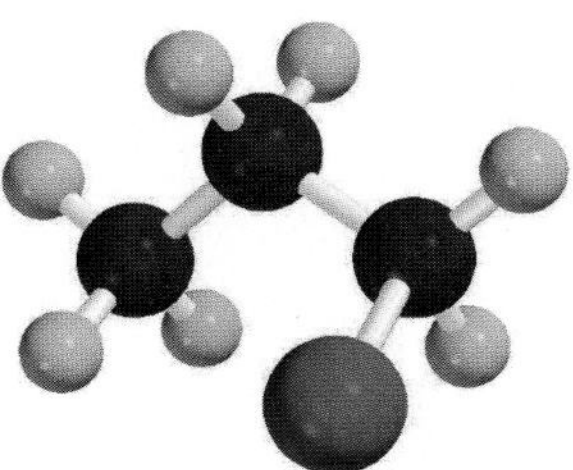

3.26 Excluding compounds that contain methyl or ethyl groups, write structural formulas for all the bicyclic isomers of (a) C_5H_8 and (b) C_6H_{10}.

3.27 Biological oxidation of hydrocarbons is a commonly observed process.

(a) To what class of hydrocarbons does the reactant in the following equation belong? What is its IUPAC name?

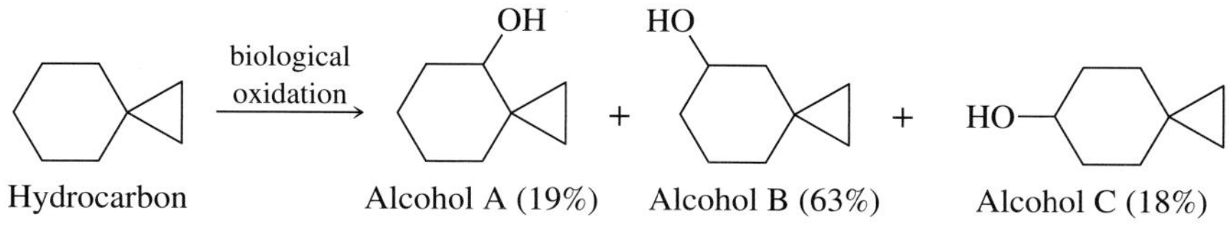

Hydrocarbon —biological oxidation→ Alcohol A (19%) + Alcohol B (63%) + Alcohol C (18%)

(b) Identify by IUPAC locant the carbon that is oxidized in the formation of each product.

(c) How are alcohols A, B, and C related? Are they constitutional isomers or stereoisomers?

3.28 A typical steroid skeleton is shown along with the numbering scheme used for this class of compounds. Specify in each case whether the designated substituent is axial or equatorial.

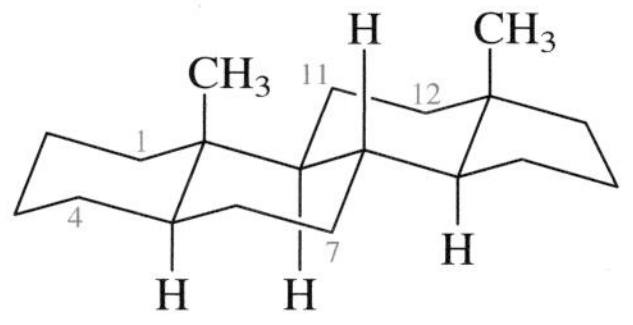

(a) Substituent at C-1 cis to the methyl groups

(b) Substituent at C-4 cis to the methyl groups

(c) Substituent at C-7 trans to the methyl groups

(d) Substituent at C-11 trans to the methyl groups

(e) Substituent at C-12 cis to the methyl groups

3.29 Repeat the preceding problem for the stereoisomeric steroidal skeleton having a cis ring fusion between the first two rings.

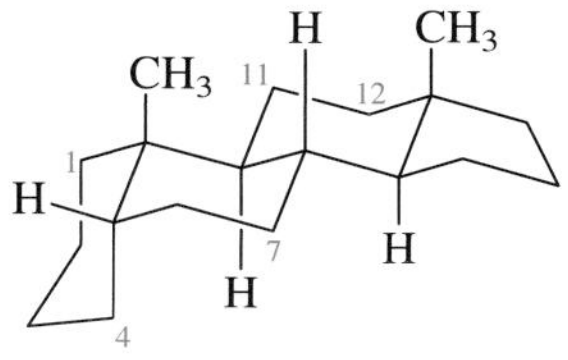

Constitutional Isomers, Stereoisomers, and Conformers

3.30 Determine whether the two structures in each of the following pairs represent *constitutional isomers,* different *conformations* of the same compound, or *stereoisomers* that cannot be interconverted by rotation about single bonds.

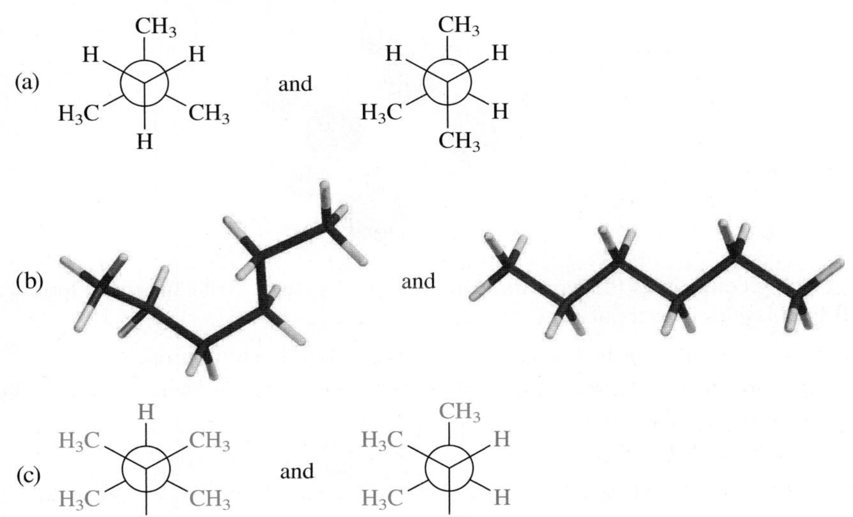

(d) *cis*-1,2-Dimethylcyclopentane and *trans*-1,3-dimethylcyclopentane

(e) CH_3CH_2, CH_3 and H_3C, CH_2CH_3

(f) CH_3CH_2, CH_3 and CH_3CH_2, CH_3

(g) CH_3, H and H, CH_3

3.31 Select the compounds in each group that are isomers and specify whether they are constitutional isomers or stereoisomers.

(a) $(CH_3)_3CCH_2CH_2CH_3$ $\quad$ $(CH_3)_3CCH_2CH_2CH_2CH_3$ $\quad$ $(CH_3)_2CHCH(CH_3)CH_2CH_3$

(b)

(c) H_3C, CH_3

(d) H_3C, CH_3 $\quad$ CH_3, H_3C $\quad$ H_3C, H_3C

(e) CH_3, CH_3 $\quad$ CH_3, CH_3 $\quad$ H_3C, CH_3

3.32 Oxidation of 4-*tert*-butylthiane proceeds according to the equation shown, but the resulting sulfoxide is a mixture of two isomers. Explain by writing appropriate sructural formulas.

3.33 The following are representations of two forms of glucose. The six-membered ring is known to exist in a chair conformation in each form. Draw clear representations of the most stable conformation of each. Are they two different conformations of the same molecule, or are they stereoisomers that cannot be interconverted by rotation about single bonds? Which substituents (if any) occupy axial sites?

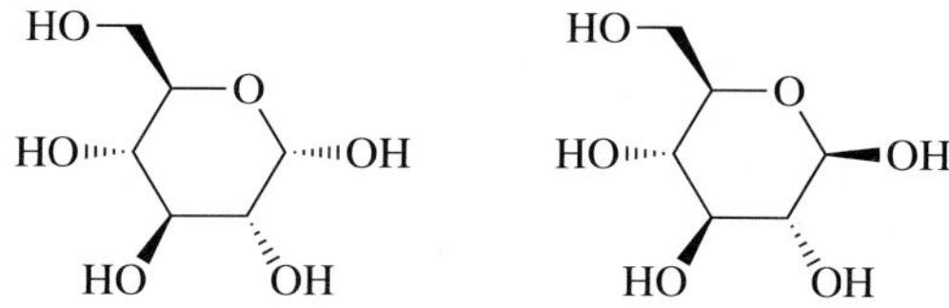

Conformations: Relative Stability

3.34 Draw (a) a Newman projection of the most stable conformation sighting down the C-3—C-4 bond and (b) a bond-line depiction of 2,2,5,5-tetramethylhexane.

3.35 Write a structural formula for the most stable chair conformation of each of the following compounds:

(a) *cis*-1-Isopropyl-3-methylcyclohexane

(b) *trans*-1-Isopropyl-3-methylcyclohexane

(c) *cis*-1-*tert*-Butyl-4-ethylcyclohexane

(d) *cis*-1,1,3,4-Tetramethylcyclohexane

(e)

3.36 Sight down the C-2—C-3 bond, and draw Newman projection formulas for the

(a) Most stable conformation of 2,2-dimethylbutane

(b) Two most stable conformations of 2-methylbutane

(c) Two most stable conformations of 2,3-dimethylbutane

3.37 One of the staggered conformations of 2-methylbutane in Problem 3.36b is more stable than the other. Which one is more stable? Why?

3.38 Sketch an approximate potential energy diagram for rotation about the carbon–carbon bond in 2,2-dimethylpropane similar to that shown in Figures 3.4 and 3.7. Does the form of the potential energy curve of 2,2-dimethylpropane more closely resemble that of ethane or that of butane?

3.39 Repeat Problem 3.38 for the case of 2-methylbutane.

3.40 Even though the methyl group occupies an equatorial site, the conformation shown is not the most stable one for methylcyclohexane. Explain.

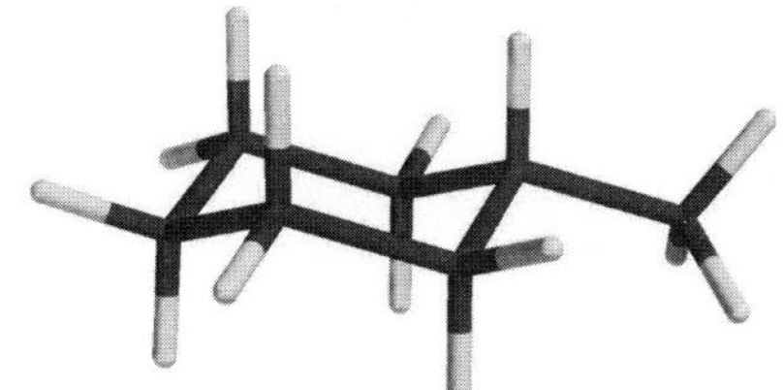

3.41 Which do you expect to be the more stable conformation of *cis*-1,3-dimethylcyclobutane, A or B? Why?

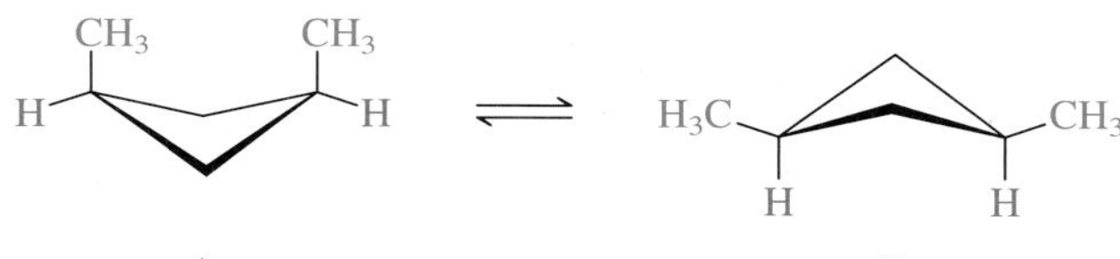

Stereoisomers: Relative Stability

3.42 Arrange the trimethylcyclohexane isomers shown in order of decreasing stability.

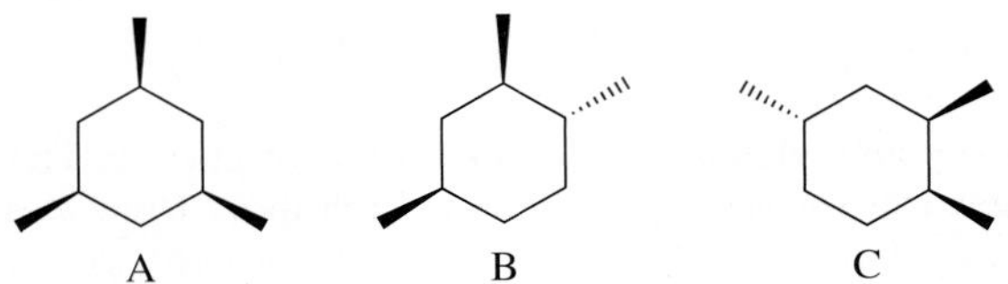

3.43 Identify the more stable stereoisomer in each of the following pairs, and give the reason for your choice:

(a) *cis*- or *trans*-1-Isopropyl-2-methylcyclohexane

(b) *cis*- or *trans*-1-Isopropyl-3-methylcyclohexane

(c) *cis*- or *trans*-1-Isopropyl-4-methylcyclohexane

(d) H_3C, CH_3, CH_3 or H_3C, CH_3, CH_3

(e) H_3C, CH_3, CH_3 or H_3C, CH_3, CH_3

(f) CH_3, CH_3 or CH_3, H_3C

3.44 One stereoisomer of 1,1,3,5-tetramethylcyclohexane is 15 kJ/mol (3.7 kcal/mol) less stable than the other. Indicate which isomer is the less stable, and identify the reason for its decreased stability.

3.45 One of the following two stereoisomers is 20 kJ/mol (4.9 kcal/mol) less stable than the other. Indicate which isomer is the less stable, and identify the reason for its decreased stability.

Properties

3.46 In each of the following groups of compounds, identify the one with the largest heat of combustion and the one with the smallest. In which cases can a comparison of heats of combustion be used to assess relative stability?

(a) Cyclopropane, cyclobutane, cyclopentane

(b) *cis*-1,2-Dimethylcyclopentane, methylcyclohexane, 1,1,2,2-tetramethylcyclopropane

(c)

(d)

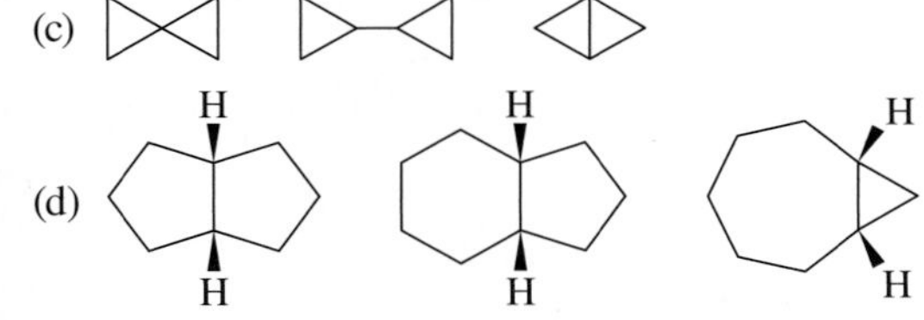

3.47 The heats of combustion of the more and less stable stereoisomers of the 1,2-, 1,3-, and 1,4-dimethylcyclohexanes are given here. The values are higher for the 1,2-dimethylcyclohexanes than for the 1,3- and 1,4-isomers. Suggest an explanation.

Dimethylcyclohexane	1,2	1,3	1,4
Heats of combustion, kJ/mol (kcal/mol):			
More stable stereoisomer	5255 (1255.9)	5250 (1254.8)	5250 (1254.8)
Less stable stereoisomer	5263 (1257.8)	5258 (1256.7)	5258 (1257.7)

3.48 The measured dipole moment of $ClCH_2CH_2Cl$ is 1.12 D. Which one of the following statements about 1,2-dichloroethane is false?

(1) It may exist entirely in the anti conformation.

(2) It may exist entirely in the gauche conformation.

(3) It may exist as a mixture of anti and gauche conformations.

3.49 Which one of the $C_2H_3Cl_3$ isomers has the largest dipole moment?

3.50 Which one of the dichlorocyclohexane isomers has the smallest dipole moment?

Descriptive Passage and Interpretive Problems 3

Cyclic Forms of Carbohydrates

Five- and six-membered ring structures are common in carbohydrates and are often in equilibrium with each other. The five-membered ring structures are called furanose forms; the six-membered ring structures are pyranose forms. D-Ribose, especially in its β-furanose form, is a familiar carbohydrate.

D-Ribose: β-Furanose form ⇌ β-Pyranose form

3.51 The β-furanose and β-pyranose forms of D-ribose are:

A. Conformational isomers
B. Constitutional isomers
C. Resonance forms
D. Stereoisomers

3.52 What is the orientation of the OH groups at C-2 and C-3 in the β-pyranose form of D-ribose?

A. Both are axial.
B. Both are equatorial.
C. C-2 is axial; C-3 is equatorial.
D. C-2 is equatorial; C-3 is axial.

3.53 The OH groups at C-2 and C-3 in the β-pyranose form of D-ribose are:

A. Cis and gauche
B. Cis and anti
C. Trans and gauche
D. Trans and anti

3.54 All of the OH groups of the β-pyranose form of D-xylose are equatorial. Which of the following is the β-furanose form of D-xylose?

A. B. C. D.

3.55 The carbohydrate shown here is a component of a drug used in veterinary medicine. Which is its most stable pyranose conformation?

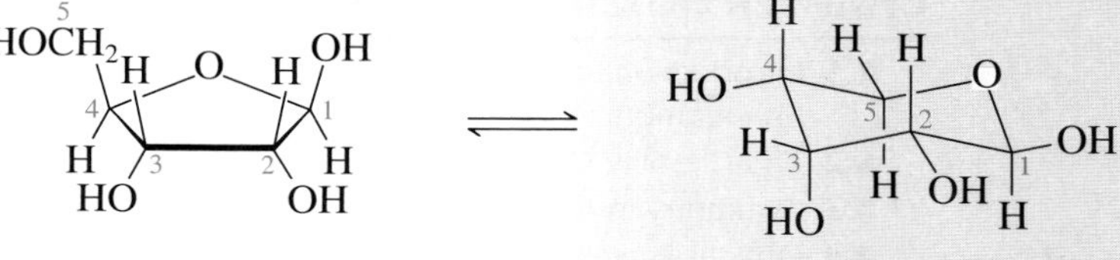

A. B. C. D.

3.56 What are the O—C(1)—C(2)—O and O—C(2)—C(3)—O torsion (dihedral) angles in the β-pyranose form of D-ribose?

A. 60° and 180°, respectively
B. 180° and 60°, respectively
C. Both are 60°.
D. Both are 180°.

Chapter

4

Bromochlorofluoromethane molecules come in right- and left-handed versions. The left and right horns of the markhor goats of Central Asia are chiral and enantiomeric. One horn traces a clockwise path, whereas the other traces a counterclockwise path.

CHAPTER OUTLINE

Chirality

In Chapter 3 we saw that rotation about a single bond in a molecule could generate a different *conformation*. In this chapter we explore a more subtle kind of isomerism, one of mirror-image *configurational* relationships. Its foundations were laid by Louis Pasteur's 1848 discovery of mirror-image tartaric acid crystals and his subsequent suggestion that the spatial arrangement of its atoms was responsible. Later, in 1874, Jacobus van't Hoff* and Joseph Achille Le Bel each proposed that the four bonds to carbon were directed toward the corners of a tetrahedron.

Our major objectives in this chapter are to describe the fundamental principles and properties of molecules as three-dimensional objects and to become familiar with stereochemical terms and notation. A full understanding of organic and biological chemistry requires an awareness of the spatial requirements for interaction between molecules; this chapter provides the basis for that understanding.

4.1 Introduction to Chirality: Enantiomers

Because the structure of a substance determines its properties, it is useful to remind ourselves of the factors that contribute to molecular structure. It begins with the atoms present and their

*Van't Hoff received the first Nobel Prize in Chemistry in 1901 for his work in chemical dynamics and osmotic pressure—two topics far removed from stereochemistry.

ratios in the form of an empirical formula, relates this to a molecular weight, and expresses the result as a molecular formula.

Elemental Composition --------→ Empirical Formula --------→ Molecular Formula

From the molecular formula, a variety of experimental data leads to a molecule's **constitution**—that is, the order in which the atoms are connected. Next, the three-dimensional spatial arrangement of substituent groups defines the **configuration** of specific atoms or structural units.

Molecular Formula --------→ Constitution/Connectivity --------→ Configuration

We already have experience with configuration in the case of cis- and trans-disubstituted cycloalkanes of the kind seen in Sections 3.11–3.12.

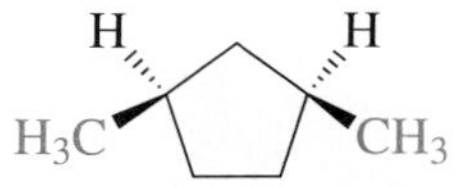

cis-1,3-Dimethylcyclopentane 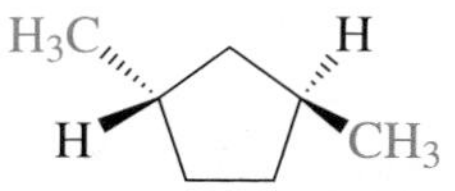 *trans*-1,3-Dimethylcyclopentane

Here, the cis and trans prefixes describe **relative configuration.** In the cis stereoisomer, both methyl groups lie on the same side of the cyclopentane ring. In the trans, they are on opposite sides.

All of these structural factors contribute to the various **conformations** available to a molecule by rotation about single bonds (Sections 3.1–3.10).

Configuration --------→ Conformation

A full understanding, however, requires that we expand the concept of configuration so as to include **chirality.** Chirality is derived from the Greek word *cheir,* for "hand," and was coined by William Thomson (Lord Kelvin) in 1894 who defined it this way:

> *I call any geometrical figure, or group of points, chiral, and say it has chirality, if its image in a plane mirror, ideally realized, cannot be brought to coincide with itself.*

Applying Thomson's term to chemistry, we say that *a molecule is chiral if its two mirror-image forms are not superimposable in three dimensions.* Thus, it is entirely appropriate to speak of the "handedness" of molecules. The opposite of chiral is **achiral.** A molecule that *is* superimposable on its mirror image is achiral.

Thomson was one of history's greatest scientists, responsible for fundamental contributions in numerous areas. His work toward determining the absolute zero of temperature was honored by naming it the Kelvin scale.

In organic chemistry, chirality most often occurs in molecules that contain a carbon that is attached to four different groups. An example is bromochlorofluoromethane (BrClFCH).

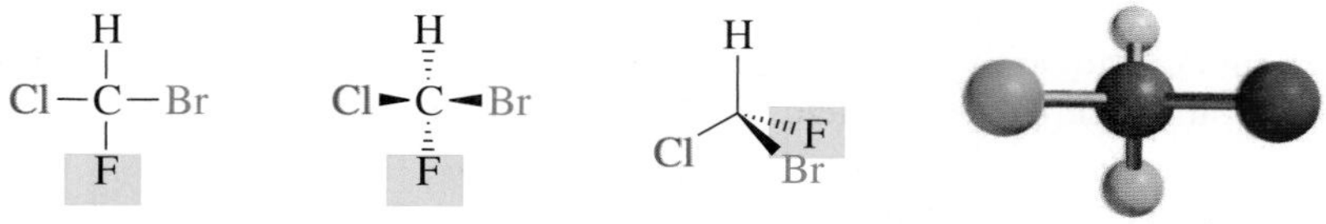

Bromochlorofluoromethane

As shown in Figure 4.1, the two mirror images of bromochlorofluoromethane cannot be superimposed on each other. *Because the two mirror images of bromochlorofluoromethane are not superimposable, BrClFCH is chiral.*

The mirror images of bromochlorofluoromethane have the same constitution. That is, the atoms are connected in the same order. But they differ in the arrangement of their atoms in space; they are stereoisomers. Stereoisomers that are related as an object and its nonsuperimposable mirror image are classified as **enantiomers.** The word *enantiomer* describes a particular relationship between two objects. One cannot look at a single molecule in isolation and ask if it is an enantiomer any more than one can look at an

In 1989 two chemists at the Polytechnic Institute of New York University described a method for the preparation of BrClFCH that is predominantly one enantiomer.

Figure 4.1

A molecule with four different groups attached to a single carbon is chiral. Its two mirror-image forms are not superimposable.

(*a*) Structures A and B are mirror-image representations of bromochlorofluoromethane (BrClFCH).

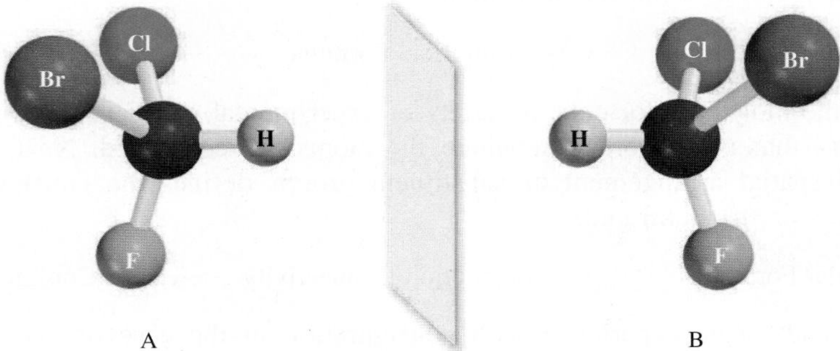

(*b*) To test for superimposability, reorient B by turning it 180°.

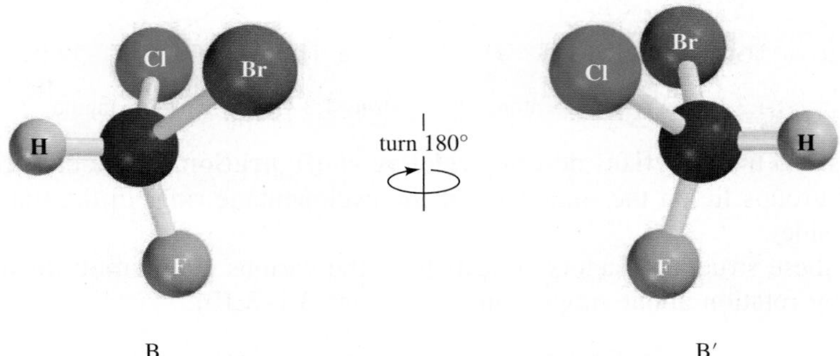

(*c*) Compare A and B′. The two do not match. A and B′ cannot be superimposed on each other. Bromochlorofluoromethane is therefore a chiral molecule. The two mirror-image forms are called enantiomers.

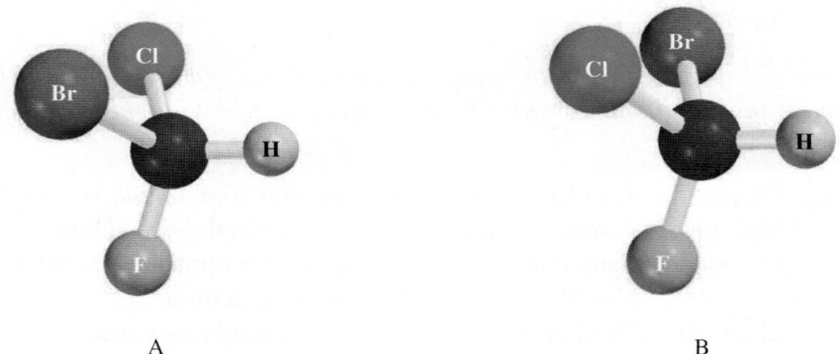

individual human being and ask, "Is that person a cousin?" Furthermore, just as an object has one, and only one, mirror image, a chiral molecule can have one, and only one, enantiomer.

Notice in Figure 4.1*c*, where the two enantiomers of bromochlorofluoromethane are similarly oriented, that the difference between them corresponds to an interchange of the positions of bromine and chlorine (A compared to B′). It will generally be true for species of the type C(w,x,y,z), where w, x, y, and z are different atoms or groups, that an exchange of two of them converts a structure to its enantiomer, but an exchange of three returns the original structure.

Consider next a molecule such as chlorodifluoromethane (ClF_2CH), in which two of the atoms attached to carbon are the same. Figure 4.2 shows two molecular models of ClF_2CH drawn so as to be mirror images. As is evident from these drawings, it is a simple matter to merge the two models so that all the atoms match. *Because mirror-image representations of chlorodifluoromethane are superimposable on each other, ClF_2CH is achiral.*

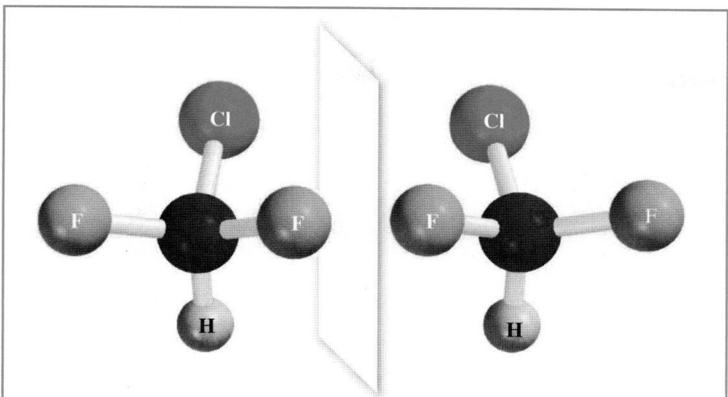

Figure 4.2

Mirror-image forms of chlorodifluoromethane are superimposable on each other. Chlorodifluoromethane is achiral.

The surest test for chirality is a careful examination of mirror-image forms for superimposability. Working with models provides the best practice in dealing with molecules as three-dimensional objects and is strongly recommended.

4.2 The Chirality Center

As we've just seen, molecules of the general type

are chiral when *w, x, y,* and *z* are different. The IUPAC recommends that a tetrahedral carbon attached to four different atoms or groups be called a **chirality center,** which is the term that we will use. Several earlier terms, including *asymmetric center, asymmetric carbon, chiral center, stereogenic center,* and *stereocenter,* are still widely used.

The IUPAC recommendations for stereochemical terms can be viewed at www.chem.qmul.ac.uk/iupac/stereo

Noting the presence of one (but not more than one) chirality center is a simple, rapid way to determine if a molecule is chiral. For example, C-2 is a chirality center in 2-butanol; it bears H, OH, CH_3, and CH_3CH_2 as its four different groups. By way of contrast, none of the carbon atoms bear four different groups in the achiral alcohol 2-propanol.

How we name alcohols and alkyl halides systematically is described in Sections 5.2–5.3.

H
$H_3C—C—CH_2CH_3$
OH

2-Butanol
Chiral; four different groups at C-2

H
$H_3C—C—CH_3$
OH

2-Propanol
Achiral; two of the groups at C-2 are the same

Problem 4.1

Examine the following for chirality centers:

Br

(a) 2-Bromopentane

Br

(b) 3-Bromopentane

Br

(c) 1-Bromo-2-methylbutane

Br

(d) 2-Bromo-2-methylbutane

Sample Solution A carbon with four different groups attached to it is a chirality center. (a) In 2-bromopentane, C-2 satisfies this requirement. (b) None of the carbons in 3-bromopentane has four different substituents, and so none of its atoms is a chirality center.

©Pixtal/SuperStock

Linalool is one of the half-dozen most abundant of the more than 90 organic compounds that remain after evaporating the water from orange juice.

Molecules with chirality centers are very common, both as naturally occurring substances and as the products of chemical synthesis. (Carbons that are part of a double bond or a triple bond can't be chirality centers.)

$$CH_3CH_2CH_2-\overset{CH_3}{\underset{CH_2CH_3}{C}}-CH_2CH_2CH_2CH_3$$

4-Ethyl-4-methyloctane
(a chiral alkane)

$$(CH_3)_2C{=}CHCH_2CH_2-\overset{CH_3}{\underset{OH}{C}}-CH{=}CH_2$$

Linalool
(a pleasant-smelling oil obtained from oranges)

A carbon atom in a ring can be a chirality center if it bears two different groups and the path traced around the ring from that carbon in one direction is different from that traced in the other. The carbon atom that bears the methyl group in 1,2-epoxypropane, for example, is a chirality center. The sequence of groups is O—CH_2 as one proceeds clockwise around the ring from that atom, but is H_2C—O in the counterclockwise direction. Similarly, C-4 is a chirality center in limonene.

H_2C—$CHCH_3$ (bridged by O)

1-2-Epoxypropane

Limonene
(a constituent of lemon oil)

Problem 4.2

Identify the chirality centers, if any, in

(a) 2-Cyclopentenol and 3-Cyclopentenol

(b) 1,1,2-Trimethylcyclobutane and 1,1,3-Trimethylcyclobutane

Sample Solution (a) The hydroxyl-bearing carbon in 2-cyclopentenol is a chirality center. There is no chirality center in 3-cyclopentenol, because the sequence of atoms 1 → 2 → 3 → 4 → 5 is equivalent regardless of whether one proceeds clockwise or counterclockwise.

2-Cyclopentenol

3-Cyclopentenol
(does not have a chirality center)

Even isotopes qualify as different substituents at a chirality center. The stereochemistry of biological oxidation of a derivative of ethane that is chiral because of deuterium (D = ^{2}H) and tritium (T = ^{3}H) atoms at carbon, has been studied and shown to proceed as follows:

$$\text{CHTD}(CH_3) \xrightarrow{\text{biological oxidation}} \text{C(OH)TD}(CH_3)$$

The stereochemical relationship between the reactant and the product, revealed by the isotopic labeling, shows that oxygen becomes bonded to carbon on the same side from

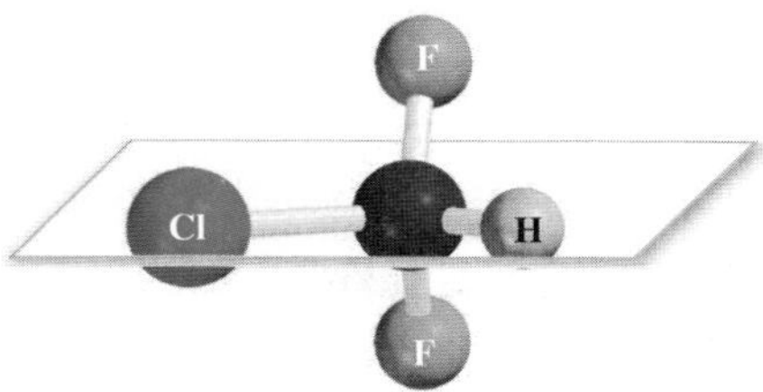

Figure 4.3

A plane of symmetry defined by the atoms H—C—Cl divides chlorodifluoromethane into two mirror-image halves. Note that the Cl and H atoms lie within the plane and reflect upon themselves.

which H is lost. As you will see in this and the chapters to come, determining the three-dimensional aspects of a chemical or biochemical transformation can be a subtle, yet powerful, tool for increasing our understanding of how these reactions occur.

One final, very important point: *Everything we have said in this section concerns molecules that have one and only one chirality center; molecules with more than one chirality center may or may not be chiral.* Molecules that have more than one chirality center will be discussed in Sections 4.10–4.12.

4.3 Symmetry in Achiral Structures

Certain structural features can help us determine whether a molecule is chiral or achiral. For example, a molecule that has a *plane of symmetry* or a *center of symmetry* is superimposable on its mirror image and is achiral.

A **plane of symmetry** bisects a molecule so that one half of the molecule is the mirror image of the other half. The achiral molecule chlorodifluoromethane, for example, has the plane of symmetry shown in Figure 4.3.

Problem 4.3

Locate any planes of symmetry in each of the following compounds. Which of the compounds are chiral? Which are achiral?

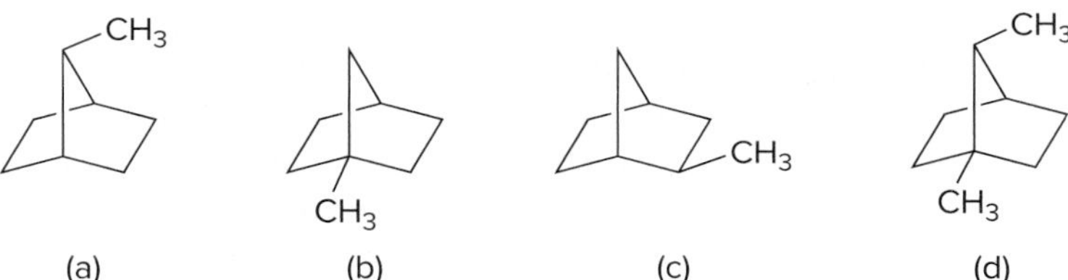

Sample Solution (a) A plane of symmetry passes through the methyl group, C-7, the midpoint of the C-2—C-3 bond, and the midpoint of the C-5—C-6 bond. The compound is achiral.

A point in the center of a molecule is a **center of symmetry** if any line drawn from it to some element of the structure will, when extended an equal distance in the opposite direction, encounter an identical element. *trans*-1,3-Cyclobutanediol has a plane of symmetry as well as a center of symmetry. The center of symmetry is the center of the molecule. A line starting at one of the hydroxyl groups and drawn through the center of the molecule encounters the equidistant hydroxyl group on the opposite side. Mirror images A and B are superimposable, and *trans*-1,3-cyclobutanediol is achiral.

A B

In IUPAC nomenclature, the *-anol* ending of an alcohol is replaced by *-anediol* in compounds with two —OH groups. Separate locants are used for each —OH group.

Problem 4.4

(a) Where is the plane of symmetry in *trans*-1,3-cyclobutanediol?

(b) Does *cis*-1,3-cyclobutanediol possess a center of symmetry? A plane of symmetry? Is it chiral or achiral?

Planes of symmetry are easier to identify and more common than centers of symmetry. Because either one is sufficient to make a molecule achiral, look first for a plane of symmetry. A molecule without a plane or center of symmetry is *likely* to be chiral, but the superimposability test must be applied to be certain.

4.4 Optical Activity

The phenomenon of optical activity was discovered by the French physicist Jean-Baptiste Biot in 1815.

The experimental facts that led van't Hoff and Le Bel to propose that molecules having the same constitution could differ in the arrangement of their atoms in space concerned the physical property of optical activity. **Optical activity** is the ability of a chiral substance to rotate the plane of plane-polarized light and is measured using an instrument called a **polarimeter** (Figure 4.4).

The light used to measure optical activity has two properties: it consists of a single wavelength and it is plane-polarized. The wavelength used most often is 589 nm (called the *D line*), which corresponds to the yellow light produced by a sodium lamp. Except for giving off light of a single wavelength, a sodium lamp is like any other lamp in that its light is unpolarized, meaning that the plane of its electric field vector can have any orientation along the line of travel. A beam of unpolarized light is transformed to plane-polarized light by passing it through a polarizing filter, which removes all the waves except those that have their electric field vector in the same plane. This plane-polarized light now passes through the sample tube containing the substance to be examined, either in the liquid phase or as a solution in a suitable solvent (usually water, ethanol, or chloroform). The sample is "optically active" if it rotates the plane of polarized light. The direction and magnitude of rotation are measured using a second polarizing filter (the "analyzer") and cited as α, the observed rotation.

To be optically active, the sample must contain a chiral substance and one enantiomer must be present in excess of the other. A substance that does not rotate the plane of polarized light is said to be optically inactive. *All achiral substances are optically inactive.*

What causes optical rotation? The plane of polarization of a light wave undergoes a minute rotation when it encounters a chiral molecule. Enantiomeric forms of a chiral molecule cause a rotation of the plane of polarization in exactly equal amounts but in opposite directions. A solution containing equal quantities of enantiomers therefore

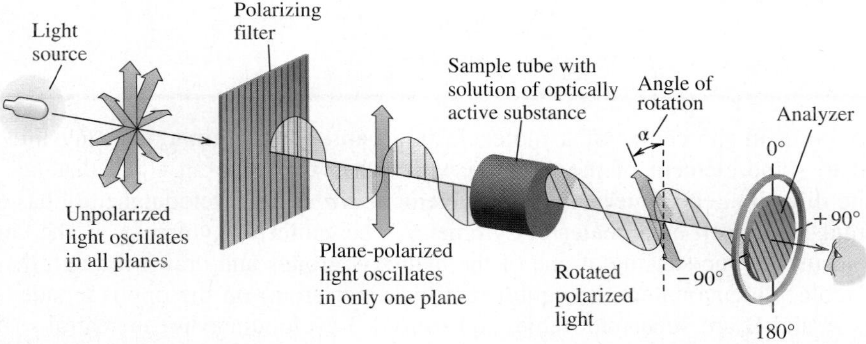

Figure 4.4

The sodium lamp emits light moving in all planes. When the light passes through the first polarizing filter, only one plane emerges. The plane-polarized beam enters the sample compartment, which contains a solution enriched in one of the enantiomers of a chiral substance. The plane rotates as it passes through the solution. A second polarizing filter (called the analyzer) is attached to a movable ring calibrated in degrees that is used to measure the angle of rotation α.

exhibits no net rotation because all the tiny increments of clockwise rotation produced by molecules of one "handedness" are canceled by an equal number of increments of counterclockwise rotation produced by molecules of the opposite handedness.

Mixtures containing equal quantities of enantiomers are called **racemic mixtures.** *Racemic mixtures are optically inactive.* Conversely, when one enantiomer is present in excess, a net rotation of the plane of polarization is observed. When all the molecules are of the same handedness, the substance is **homochiral** or **enantiopure.** In common practice, when we have a mixture of enantiomers where one isomer is present in larger amounts, percent **enantiomeric excess** (ee) or **enantiomeric ratio** (er) is used to describe the mixture, where:

$$\text{ee} = (\%\ \text{Major enantiomer}) - (\%\ \text{Minor enantiomer})$$

and

$$\text{er} = (\%\ \text{Major enantiomer}):(\%\ \text{Minor enantiomer})$$

For example, a mixture of enantiomers comprised of 80% one isomer and 20% of the other, has an 80:20 enantiomeric ratio. The excess of the major isomer above the minor isomer gives the enantiomeric excess, ee = 80 − 20 = 60.

Problem 4.5

A sample of the chiral molecule limonene has an enantiomeric ratio (er) of 97.5:2.5. What is the percent enantiomeric excess (ee)?

Rotation of the plane of polarized light in the clockwise sense is taken as positive (+), and rotation in the counterclockwise sense as negative (−). Older terms for these are *dextrorotatory* and *levorotatory,* from the Latin prefixes *dexter-* ("to the right") and *laevus-* ("to the left"), respectively. At one time, the symbols *d* and *l* were used to distinguish between enantiomers. Thus, the dextrorotatory enantiomer of 2-butanol was called *d*-2-butanol, and the levorotatory form *l*-2-butanol; a racemic mixture of the two was referred to as *dl*-2-butanol. Current custom uses algebraic signs instead, as in (+)-2-butanol, (−)-2-butanol, and (±)-2-butanol, respectively.

The observed rotation α of an optically pure substance depends on how many molecules the light beam encounters. A filled polarimeter tube twice the length of another produces twice the observed rotation, as does a solution twice as concentrated. To account for the effects of path length and concentration, chemists use the term **specific rotation,** given the symbol [α] and calculated from the observed rotation according to the expression

$$[\alpha] = \frac{100\alpha}{cl}$$

where c is the concentration of the sample in grams per 100 mL of solution, and l is the length of the polarimeter tube in decimeters. (One decimeter is 10 cm.)

If concentration is expressed as grams per milliliter of solution instead of grams per 100 mL, an equivalent expression is

$$[\alpha] = \frac{\alpha}{cl}$$

Specific rotation is a physical property of a substance, just as melting point, boiling point, density, and solubility are. For example, the lactic acid obtained from milk is exclusively a single enantiomer. We cite its specific rotation in the form $[\alpha]_D^{25} = +3.8°$. The temperature in degrees Celsius and the wavelength of light at which the measurement was made are indicated as superscripts and subscripts, respectively. The subscript D stands for the so-called "D-line" of sodium, which has a wavelength of 589 nm. Optical purity corresponds to enantiomeric excess and is calculated from the specific rotation:

$$\text{ee} = \text{Optical purity} = \frac{\text{specific rotation of sample}}{\text{specific rotation of pure enantiomer}} \times 100$$

Problem 4.6

(a) Cholesterol isolated from natural sources is enantiopure. The observed rotation of a 0.3-g sample of cholesterol in 15 mL of chloroform solution contained in a 10-cm polarimeter tube is −0.78°. Calculate the specific rotation of cholesterol.

continued

(b) A sample of synthetic cholesterol consisting entirely of (+)-cholesterol was mixed with some natural (−)-cholesterol. The specific rotation of the mixture was −13°. What fraction of the mixture was (+)-cholesterol?

It is convenient to distinguish between enantiomers by prefixing the sign of rotation to the name of the substance. For example, we refer to one of the enantiomers of 2-butanol as (+)-2-butanol and the other as (−)-2-butanol. Optically pure (+)-2-butanol has a specific rotation $[\alpha]_D^{27}$ of +13.5°; optically pure (−)-2-butanol has an exactly opposite specific rotation $[\alpha]_D^{27}$ of −13.5°.

4.5 Absolute and Relative Configuration

The three-dimensional spatial arrangement of substituents at a chirality center is its **absolute configuration.** Neither the sign nor the magnitude of rotation by itself can tell us the absolute configuration of a substance. Thus, one of the following structures is (+)-2-butanol and the other is (−)-2-butanol, but without additional information we can't tell which is which.

In several places throughout the chapter we will use red and blue frames to call attention to structures that are enantiomeric.

CH_3CH_2, H, C—OH, H_3C

H, CH_2CH_3, HO—C, CH_3

Although no absolute configuration was known for any substance until the mid-twentieth century, organic chemists had experimentally determined the configurations of thousands of compounds relative to one another (their **relative configurations**) through chemical interconversion. To illustrate, consider a reaction that doesn't involve any of the bonds directly attached to the chirality center.

For an introduction to the reaction of hydrogen with alkenes, see Section 8.1.

3-Buten-2-ol $[\alpha]_D^{27}$ +33.2° + H_2 (Hydrogen) $\xrightarrow{Pd}$ 2-Butanol $[\alpha]_D^{27}$ +13.5°

The fact that reactant and product have the same sign of rotation when they have the same relative configuration is established by the experiment; it could not have been predicted in advance of the experiment. Note also that the experiment does not tell us what the arrangement of bonds is; it tells us only that the relative configuration is the same when both compounds are dextrorotatory.

Compounds that have the same relative configuration can have optical rotations of opposite sign. For example, treatment of (−)-2-methyl-1-butanol with hydrogen bromide converts it to (+)-1-bromo-2-methylbutane.

The reaction of alcohols with hydrogen halides is one of the main topics in Chapter 5.

2-Methyl-1-butanol $[\alpha]_D^{25}$ −5.8° + HBr (Hydrogen bromide) $\longrightarrow$ 1-Bromo-2-methylbutane $[\alpha]_D^{25}$ +4.0° + H_2O (Water)

This reaction does not involve any of the bonds to the chirality center, and so both the starting alcohol (−) and the product bromide (+) have the same relative configuration.

An elaborate network connecting signs of rotation and relative configurations was developed that included the most important compounds of organic and biological chemistry. When, in 1951, the absolute configuration of (+)-tartaric acid was determined, the absolute configurations of all the compounds whose configurations had been related to it stood revealed as well. Thus, returning to the pair of 2-butanol enantiomers that began this section, their absolute configurations are now known to be as shown.

(+)-2-Butanol (−)-2-Butanol

Problem 4.7

Does the molecular model shown represent (+)-2-butanol or (−)-2-butanol?

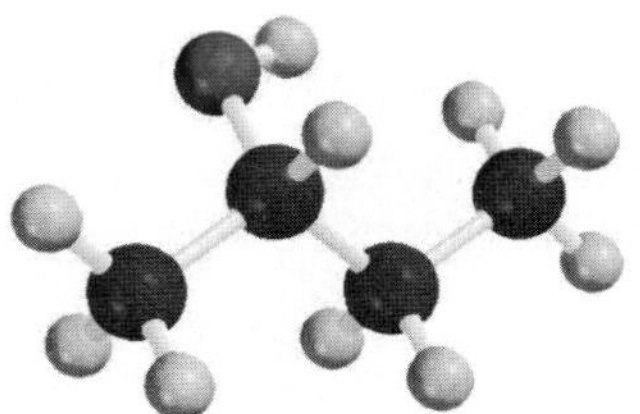

4.6 Cahn–Ingold–Prelog *R,S* Notation

Shortly after the problem of absolute configuration was settled, a method to specify configuration symbolically as an alternative to drawings was developed by R. S. Cahn, Christopher Ingold, and Vladimir Prelog. Their approach contains two elements: rules for ranking substituents at a chirality center and a protocol for viewing their spatial orientation. As an example, consider the enantiomers **A** and **B** of bromochlorofluoromethane.

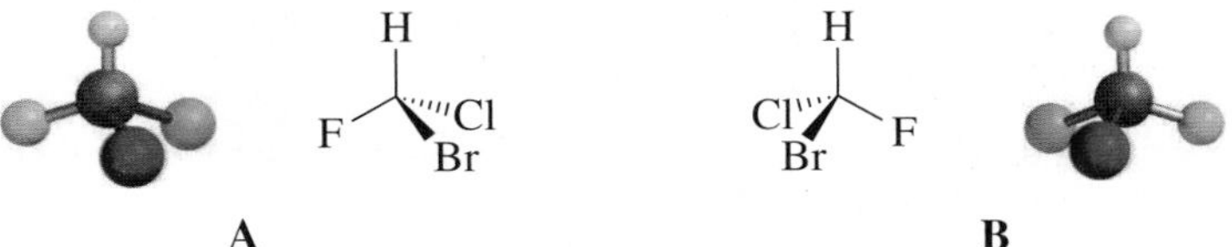

The **Cahn–Ingold–Prelog system** ranks substituents at the chirality center in order of decreasing atomic number. Here the order is Br > Cl > F > H. The next step is to orient the molecules so that the bond to the lowest ranked substituent—in this case H—points away from us and determine if the sequence of decreasing priority of the other three substituents is clockwise or counterclockwise.

If this path is clockwise as in **A,** the configuration is *R* (Latin "*rectus*" "right," "correct"). If it is counterclockwise as in **B,** the configuration is *S* (Latin "*sinister*" "left").

It has been experimentally determined that (*R*)-bromochlorofluoromethane is levorotatory (−). Therefore, (*S*)-bromochlorofluoromethane is dextrorotatory (+). When the connection between configuration and sign of rotation is known, it is common practice to incorporate both into the name of the compound. Thus, **A** is (*R*)-(−)-bromochlorofluoromethane and **B** is (*S*)-(+)-bromochlorofluoromethane.

In most organic compounds, of course, the groups attached to a chirality center are more complicated than single atoms, and the Cahn–Ingold–Prelog system incorporates a series of rules to accommodate this fact. These **sequence rules** are described in Table 4.1.

TABLE 4.1 The Cahn–Ingold–Prelog Sequence Rules

Rule	Example
1. Higher atomic number takes precedence over lower.	As described for the case of bromochlorofluoromethane, the sequence priority is Br > Cl > F > H.
2. When two atoms directly attached to the chirality center are the same, compare the atoms attached to them on the basis of their atomic numbers. Precedence is determined at the first point of difference.	The substituents at the chirality center in 2-chlorobutane are —Cl, —H, $—CH_3$, and $—CH_2CH_3$. 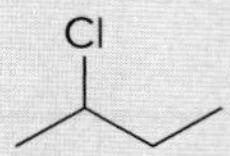 Cl Ethyl [—C(C,H,H) outranks methyl [—C(H,H,H)]. In order of decreasing priority, the substituents at the chirality center are: $—Cl > —CH_2CH_3 > CH_3 > H$
3. Work outward from the point of attachment, comparing all the atoms attached to a particular atom before proceeding further along the chain.	The substituents at the chirality center in 1-chloro-3,4-dimethylpentane are $—CH_2CH_2Cl$, —H, $—CH_3$, and $—CH(CH_3)_2$. 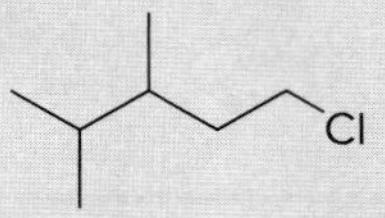 Cl Isopropyl $[—CH(CH_3)_2]$ counts as [—C(C,C,H)] and outranks $—CH_2CH_2Cl$, which counts as [—C(C,H,H)].
4. When working outward from the point of attachment, always evaluate substituent atoms one by one, never as a group.	The substituents at the chirality center in 1-chloro-2,3,3-trimethylpentane are $—CH_2Cl$, —H, $—CH_3$, and $—C(CH_3)_3$. 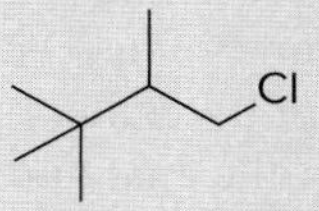 Cl $—CH_2Cl$ counts as [—C(Cl,H,H)] and outranks *tert*-butyl $[—C(CH_3)_3]$, which is treated as [—C(C,C,C)].
5. An atom that is multiply bonded to another atom is considered to be replicated as a substituent on that atom.	The substituents at the chirality center in 2-methyl-1-pentene are $—CH_2CH_3$, —H, $—CH_3$, and $—CH{=}CH_2$. 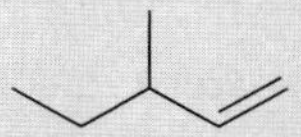 $—CH{=}CH_2$ counts as [—C(C,C,H)] and outranks $—CH_2CH_3$, which is treated as [—C(C,H,H)].

Problem 4.8

Assign absolute configurations as *R* or *S* to each of the following compounds:

(a) 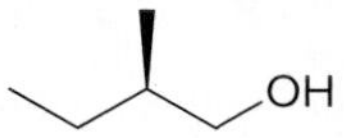 OH
(+)-2-Methyl-1-butanol

(b) 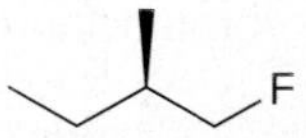 F
(+)-1-Fluoro-2-methylbutane

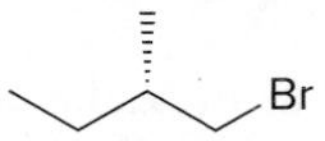

(c) Br
(+)-1-Bromo-2-methylbutane

(d) 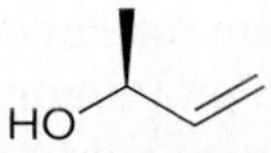 HO
(+)-3-Buten-2-ol

Sample Solution (a) The highest ranking substituent at the chirality center of 2-methyl-1-butanol is CH_2OH; the lowest is H. Of the remaining two, ethyl outranks methyl.

Order of precedence: $CH_2OH > CH_3CH_2 > CH_3 > H$

The lowest ranking group (hydrogen) points away from us in the drawing. The three highest ranking groups trace a clockwise path from $CH_2OH \rightarrow CH_3CH_2 \rightarrow CH_3$.

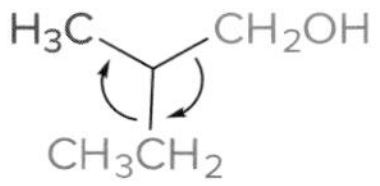

H_3C CH_2OH CH_3CH_2

This compound therefore has the *R* configuration. It is (*R*)-(+)-2-methyl-1-butanol.

Compounds in which a chirality center is part of a ring are handled in an analogous fashion. To determine, for example, whether the configuration of (+)-4-methylcyclohexene is *R* or *S*, treat the right- and left-hand paths around the ring as if they were independent groups.

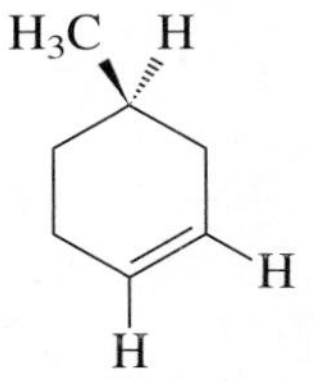

H_3C H H H is treated as Lower priority path H_2C H_2C H_3C H CH_2 C—H C C C Higher priority path

(+)-4-Methylcyclohexene

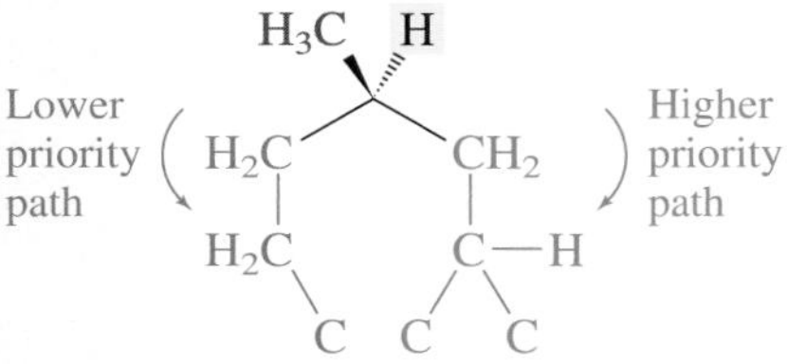

With the lowest ranked group (hydrogen) directed away from us, we see that the order of decreasing sequence rule precedence is *clockwise*. The absolute configuration is *R*.

Problem 4.9

Draw three-dimensional representations of

(a) The *R* enantiomer of

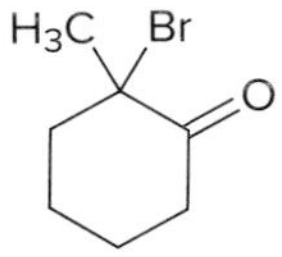

H_3C Br O

(b) The *S* enantiomer of

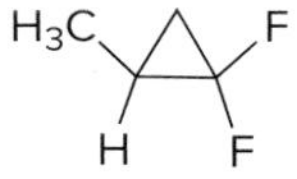

H_3C F H F

Sample Solution (a) The chirality center is the one that bears the bromine. In order of decreasing precedence, the substituents attached to the chirality center are

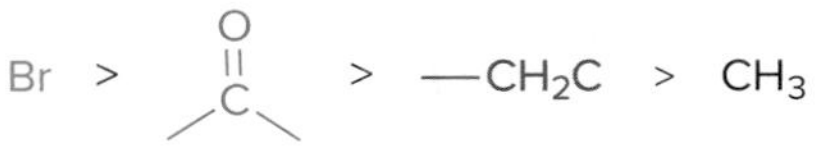

Br > C=O > $—CH_2C$ > CH_3

When the lowest ranked substituent (the methyl group) is away from us, the order of decreasing precedence of the remaining groups must appear in a clockwise sense in the *R* enantiomer.

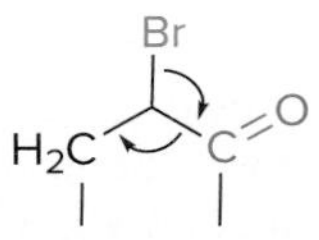

Br H_2C C=O which leads to the structure Br CH_3 O

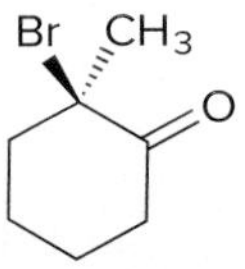

(*R*)-2-Bromo-2-methylcyclohexanone

The Cahn–Ingold–Prelog system is the standard method of stereochemical notation. It replaced an older system based on analogies to specified reference compounds that used the prefixes D and L, a system that is still used for carbohydrates and amino acids. We will use D and L notation when we get to Chapters 23–26, but won't need it until then.

Homochirality and Symmetry Breaking

The classic work of Louis Pasteur in 1848 showed that an optically inactive substance then known as "racemic acid" found in grapes is a 1:1 mixture of (+)- and (−)-tartaric acids.

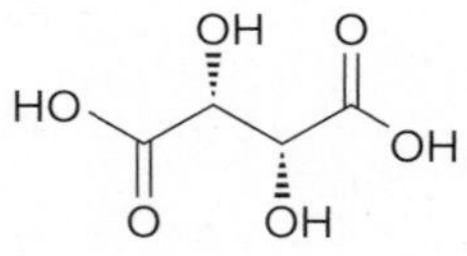

(+)-Tartaric acid 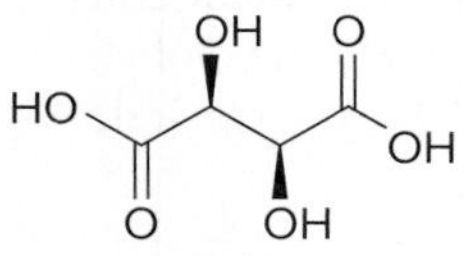 (−)-Tartaric acid

Although Pasteur's discovery was transformative in respect to the progress of science, the tartaric acid case turns out to be an exceptional one. Naturally occurring chiral compounds are almost always homochiral—their biosynthesis provides only a single enantiomer. Lemons contain only (*R*)-(+)-limonene, and apples only (*S*)-(−)-malic acid. Only the *S* enantiomer of methionine, never the *R*, is one of the amino acid building blocks of peptides and proteins.

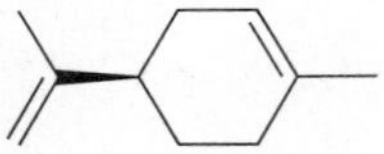

(*R*)-(+)-Limonene 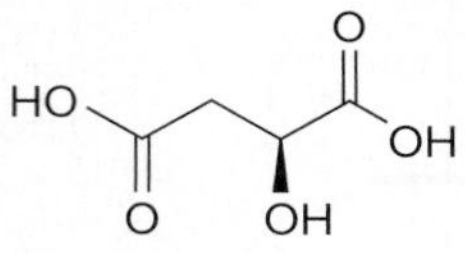 (*S*)-(−)-Malic acid 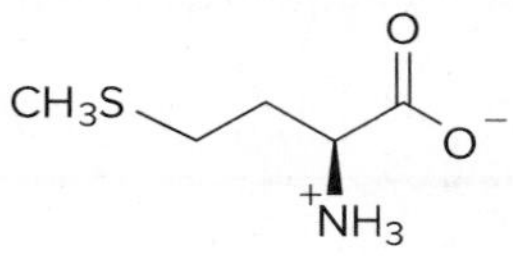 (*S*)-(−)-Methionine

There are, however, examples where each enantiomer of a natural product occurs to the exclusion of the other. The *R*-(−) enantiomer of 5-methylheptan-3-one is present in the male bristle worm *Platynereis dumerilii*, for example, and the *S*-(+) enantiomer in the female.

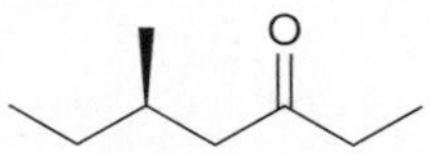

(*R*)-(−)-5-Methylheptan-3-one 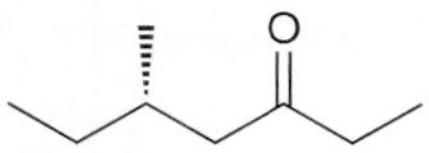 (*S*)-(+)-5-Methylheptan-3-one

How molecular homochirality came to dominate the natural world to the degree it does remains one of the great unanswered, and perhaps unanswerable, questions of science. The main problem is known as "symmetry breaking," especially as it applies to what origins-of-life theories term the "last universal ancestor" (LUA). Without going into detail, the LUA is the most recent organism from which all living things on Earth have descended, where "recent" encloses a time period between now and 3.5 billion years ago. Symmetry breaking is more fundamental in that it simply recognizes that our world and its mirror image are equally likely in the absence of some event or force that favors one enantiomer of the LUA or one of its descendants over all others.

What event? What force?

In one scenario, the event involves the seeding of Earth with extraterrestrial homochiral organisms or compounds. Proponents of this theory point to the presence of a large number of amino acids, including some that are enriched in one enantiomer, in a meteorite that fell in Murchison, Australia, in 1969.

In terms of forces, circularly polarized ultraviolet light—a type of radiation associated with neutron stars—is the most favored candidate (Figure 4.5). Numerous experiments in which racemic mixtures of chiral substances were irradiated with circularly polarized light resulted in enrichment of one enantiomer because of preferential destruction of the other.

Research directed toward finding symmetry-breaking mechanisms for homochiral generation is as fundamental as science can be, but standing in the way of even a modest degree of progress is the time window through which one is obliged to look. Most theories rely on a principle that once generated in a population, homochirality will be amplified by natural selection to the point that competing stereoisomers vanish.

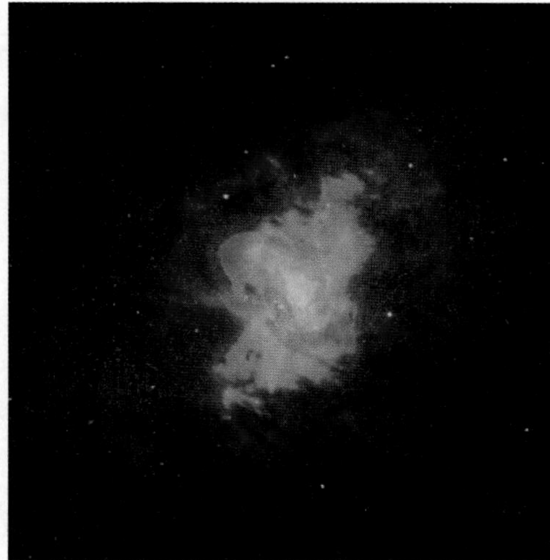

Figure 4.5

A neutron star lies at the center of the Crab Nebula. Radiation from a neutron star has been proposed as the polarized light source that served as a "symmetry breaker" in theories concerning the origin of homochirality.

Source: NASA

4.7 Fischer Projections

Stereochemistry deals with the three-dimensional arrangement of a molecule's atoms, and we have attempted to show stereochemistry with wedge-and-dash drawings and computer-generated models. It is possible, however, to convey stereochemical information in an abbreviated form using a method devised by the German chemist Emil Fischer.

Fischer was the foremost organic chemist of the late nineteenth century. He won the 1902 Nobel Prize in Chemistry for his pioneering work in carbohydrate and protein chemistry.

Let's return to bromochlorofluoromethane as a simple example of a chiral molecule. The two enantiomers of BrClFCH are shown as ball-and-spoke models, as wedge-and-dash drawings, and as **Fischer projections** in Figure 4.6. Fischer projections are always generated the same way: the molecule is oriented so that the vertical bonds at the chirality center are directed away from you and the horizontal bonds point toward you. A projection of the bonds onto the page is a cross. The chirality center lies at the center of the cross but is not explicitly shown.

It is customary to orient the molecule so that the carbon chain is vertical with the lowest numbered carbon at the top as shown for the Fischer projection of (*R*)-2-butanol.

The Fischer projection HO–(CH$_3$ top, CH$_2$CH$_3$ bottom)–H corresponds to HO–C–H (CH$_3$ top, CH$_2$CH$_3$ bottom)

(*R*)-2-Butanol

To verify that the Fischer projection has the *R* configuration at its chirality center, rotate the three-dimensional representation so that the lowest-ranked atom (H) points away from you. Be careful to maintain the proper stereochemical relationships during the operation.

HO–C–H (CH$_3$ top, CH$_2$CH$_3$ bottom) —rotate 180° around vertical axis→ H–C–OH (CH$_3$ top, CH$_2$CH$_3$ bottom)

With H pointing away from us, we can see that the order of decreasing precedence OH > CH$_2$CH$_3$ > CH$_3$ traces a clockwise path, verifying the configuration as *R*.

H–C–OH (CH$_3$ top, CH$_2$CH$_3$ bottom) H$_3$C, OH, CH$_2$CH$_3$

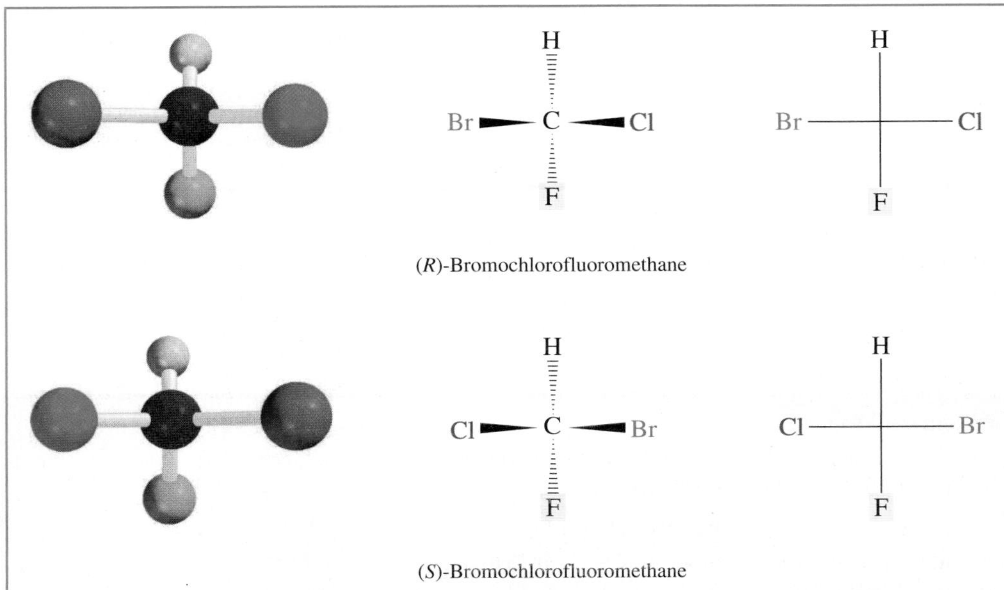

Figure 4.6

Ball-and-spoke models (*left*), wedge-and-dash drawings (*center*), and Fischer projections (*right*) of the *R* and *S* enantiomers of bromochlorofluoromethane.

Problem 4.10

What is the absolute configuration (*R* or *S*) of the compounds represented by the Fischer projections shown here?

(a) Fischer projection: top CH_2OH; left H; right OH; bottom CH_2CH_3

(b) Fischer projection: top CH=O; left HO; right H; bottom CH_2OH

Sample Solution

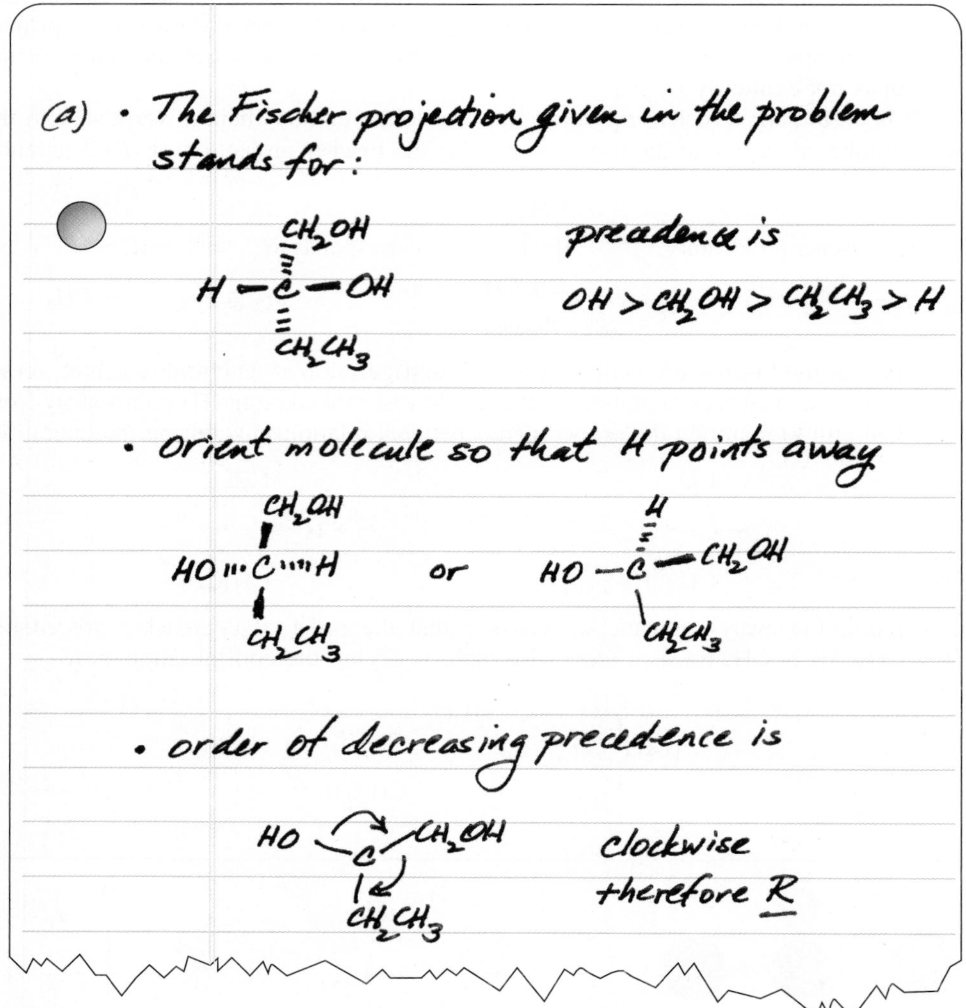

(a) • The Fischer projection given in the problem stands for:
CH2OH
H—C—OH
CH2CH3
precedence is
OH > CH2OH > CH2CH3 > H
• orient molecule so that H points away
CH2OH
HO C H
CH2CH3
or
H
HO—C—CH2OH
CH2CH3
• order of decreasing precedence is
HO
C
CH2OH
CH2CH3
clockwise
therefore R

As you work with Fischer projections, you may notice that some routine structural changes lead to predictable outcomes—outcomes that may reduce the number of manipulations you need to do to solve stereochemistry problems. Instead of listing these shortcuts, Problem 4.11 invites you to discover some of them for yourself.

Problem 4.11

Using the Fischer projection of (*R*)-2-butanol shown, explain how each of the following affects the configuration of the chirality center.

(a) Switching the positions of H and OH.

(b) Switching the positions of CH_3 and CH_2CH_3.

(c) Switching the positions of three groups.

Fischer projection: top CH_3; left HO; right H; bottom CH_2CH_3

(d) Switching H with OH, and CH_3 with CH_2CH_3.
(e) Rotating the Fischer projection 180° about an axis perpendicular to the page.

Sample Solution (a) Exchanging the positions of H and OH in the Fischer projection of (*R*)-2-butanol converts it to the mirror-image Fischer projection. The configuration of the chirality center goes from *R* to *S*.

CH_3 / HO—|—H / CH_2CH_3 —exchange the positions of H and OH→ CH_3 / H—|—OH / CH_2CH_3

(*R*)-2-Butanol (*S*)-2-Butanol

Switching the positions of two groups in a Fischer projection reverses the configuration of the chirality center.

We mentioned in Section 4.6 that the D,L system of stereochemical notation, while outdated for most purposes, is still widely used for carbohydrates and amino acids. Likewise, Fischer projections find their major application in these same two families of compounds.

4.8 Properties of Enantiomers

The usual physical properties such as density, melting point, and boiling point are identical for both enantiomers of a chiral compound.

Enantiomers can have striking differences, however, in properties that depend on the arrangement of atoms in space. Take, for example, the enantiomeric forms of carvone. (*R*)-(−)-Carvone is the principal component of spearmint oil. Its enantiomer, (*S*)-(+)-carvone, is the principal component of caraway seed oil. The two enantiomers do not smell the same; each has its own characteristic odor.

Spearmint leaves
©DAJ/Getty Images

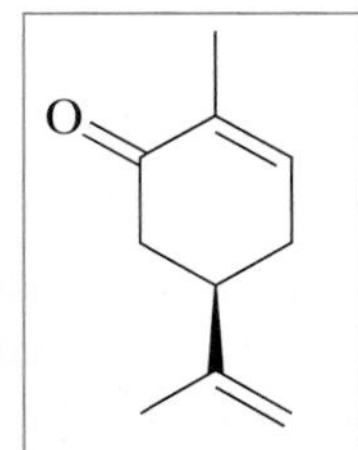

(*R*)-(−)-Carvone (from spearmint oil)

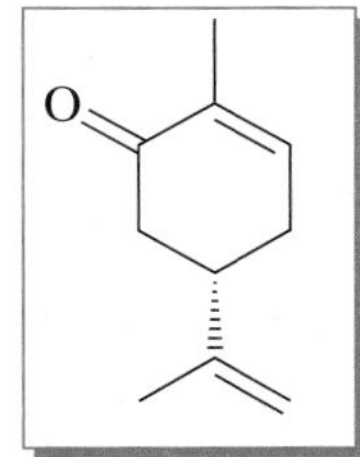

(*S*)-(+)-Carvone (from caraway seed oil)

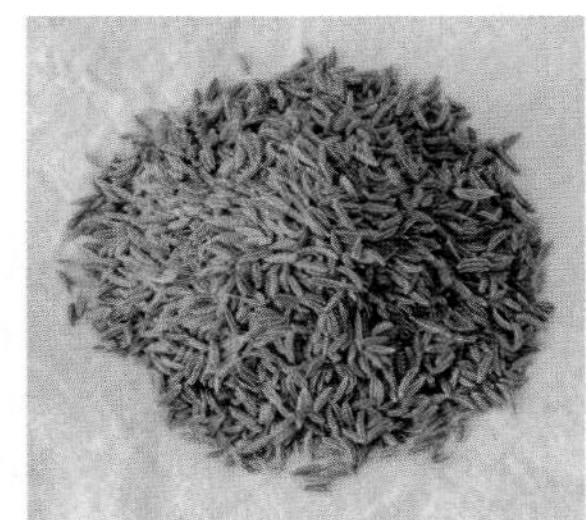
Caraway seeds
©McGraw-Hill Education/Elite Images

The difference in odor between (*R*)- and (*S*)-carvone results from their different behavior toward receptor sites in the nose. It is believed that volatile molecules occupy only those odor receptors that have the proper shape to accommodate them. Because the receptor sites are themselves chiral, one enantiomer may fit one kind of receptor while the other enantiomer fits a different kind. An analogy that can be drawn is to hands and gloves. Your left hand and your right hand are enantiomers. You can place your left hand into a left glove but not into a right one. The receptor (the glove) can accommodate one enantiomer of a chiral object (your hand) but not the other.

The term *chiral recognition* refers to a process in which some chiral receptor or reagent interacts selectively with one of the enantiomers of a chiral molecule. Very high levels of chiral recognition are common in biological processes. (−)-Nicotine, for example, is much more toxic than (+)-nicotine, and (+)-adrenaline is more active than (−)-adrenaline in constricting blood vessels. (−)-Thyroxine, an amino acid of the thyroid gland that speeds up metabolism, is one of the most widely used of all prescription drugs—about 10 million people in the United States take (−)-thyroxine on a daily basis. Its enantiomer, (+)-thyroxine, has none of the metabolism-regulating effects, but was formerly given to heart patients to lower their cholesterol levels.

Problem 4.12

Assign appropriate *R,S* symbols to the chirality centers in (−)-nicotine, (−)-adrenaline, and (−)-thyroxine.

(−)-Nicotine (−)-Adrenaline (−)-Thyroxine

4.9 The Chirality Axis

We have, so far, restricted our discussion of chiral molecules to those that contain a chirality center. Although these are the most common, they are not the only kinds of chiral molecules. A second group consists of molecules that contain a **chirality axis**—an axis about which a set of atoms or groups is arranged so that the spatial arrangement is not superimposable on its mirror image. We can think of two enantiomers characterized by a chirality axis as being analogous to a left-handed screw and a right-handed screw.

Among molecules with a chirality axis, substituted derivatives of biaryls have received much attention. **Biaryls** are compounds in which two aromatic rings are joined by a single bond: biphenyl and 1,1′-binaphthyl, for example.

Biphenyl 1,1′-Binaphthyl

Although their individual rings are flat, the molecules themselves are not. Rotation about the single bond connecting the two rings in biphenyl reduces the steric strain between nearby hydrogens of one ring (red) and those of the other (green). This rotation makes the "twisted" conformation more stable than one in which all of the atoms lie in the same plane.

The experimentally measured angle between the two rings of biphenyl in the gas phase is 44°.

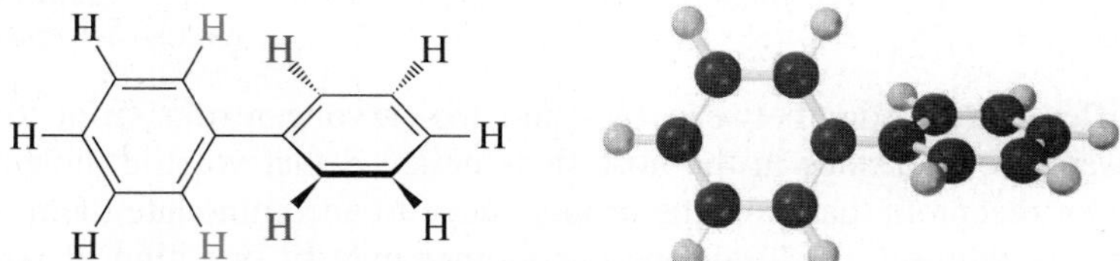

Nonplanar "twisted" conformation of biphenyl

Rotation about the bond joining the two rings is very fast in biphenyl, about the same as in ethane, but is slowed when the carbons adjacent to the ones joining the two rings bear groups other than hydrogen.

If the substituents are large enough, the steric strain that accompanies their moving past each other during rotation about the single bond can decrease the rate of equilibration

Chiral Drugs

A recent estimate places the number of prescription and over-the-counter drugs marketed throughout the world at more than 2000. Approximately one third of these are either naturally occurring substances themselves or are prepared by chemical modification of natural products. Most of the drugs derived from natural sources are chiral and are almost always obtained as a single enantiomer rather than as a racemic mixture. Not so with the over 500 chiral substances represented among the more than 1300 drugs that are the products of synthetic organic chemistry. Until recently, such substances were, with few exceptions, prepared, sold, and administered as racemic mixtures even though the desired therapeutic activity resided in only one of the enantiomers. Spurred by a number of factors ranging from safety and efficacy to synthetic methodology and economics, this practice is undergoing rapid change as more and more chiral synthetic drugs become available in enantiomerically pure form.

Because of the high degree of chiral recognition inherent in most biological processes (Section 4.8), it is unlikely that both enantiomers of a chiral drug will exhibit the same level, or even the same kind, of effect. At one extreme, one enantiomer has the desired effect, and the other exhibits no biological activity at all. In this case, which is relatively rare, the racemic form is simply a drug that is 50% pure and contains 50% "inert ingredients." Real cases are more complicated. For example, the *S* enantiomer is responsible for the pain-relieving properties of ibuprofen, normally sold as a racemic mixture. The 50% of racemic ibuprofen that is the *R* enantiomer is not completely wasted, however, because enzyme-catalyzed reactions in our body convert much of it to active (*S*)-ibuprofen.

Ibuprofen

A much more serious drawback to using chiral drugs as racemic mixtures is illustrated by thalidomide, briefly employed as a sedative and antinausea drug in Europe during the period 1959–1962. The desired properties are those of (*R*)-thalidomide. (*S*)-Thalidomide, however, has a very different spectrum of biological activity and was shown to be responsible for over 2000 cases of serious birth defects in children born to women who took it while pregnant.

Thalidomide

Basic research aimed at controlling the stereochemistry of chemical reactions has led to novel methods for the synthesis of chiral molecules in enantiomerically pure form. Aspects of this work were recognized with the award of the 2001 Nobel Prize in Chemistry to William S. Knowles (Monsanto), Ryoji Noyori (Nagoya University), and K. Barry Sharpless (The Scripps Research Institute). Most major pharmaceutical companies are examining their existing drugs to see which are the best candidates for synthesis as single enantiomers and, when preparing a new drug, design its synthesis so as to provide only the desired enantiomer. One incentive to developing enantiomerically pure versions of existing drugs, called a "chiral switch," is that the novel production methods they require may make them eligible for extended patent protection.

Problem 4.13

Find the chirality center in the molecular model of thalidomide shown above and identify its configuration as *R* or *S*.

so much that it becomes possible to isolate the two conformations under normal laboratory conditions.

When A $\neq$ B, and X $\neq$ Y, the two conformations are nonsuperimposable mirror images of each other; that is, they are enantiomers. The bond connecting the two rings lies along a chirality axis.

Chirality axis when A $\neq$ B and X $\neq$ Y

The first compound demonstrated to be chiral because of restricted rotation about a single bond was 6,6′-dinitrobiphenyl-2,2′-dicarboxylic acid in 1922.

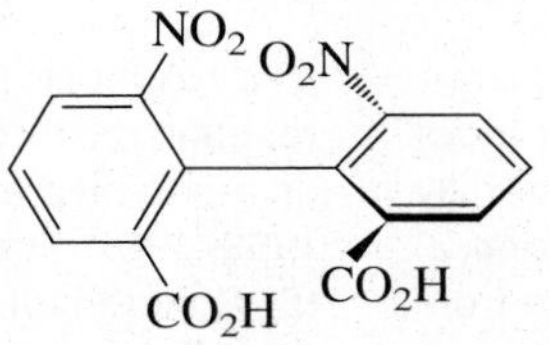

(+)-6,6′-Dinitrobiphenyl-2,2′-dicarboxylic acid $[\alpha]_D^{29}$ +127° (methanol)

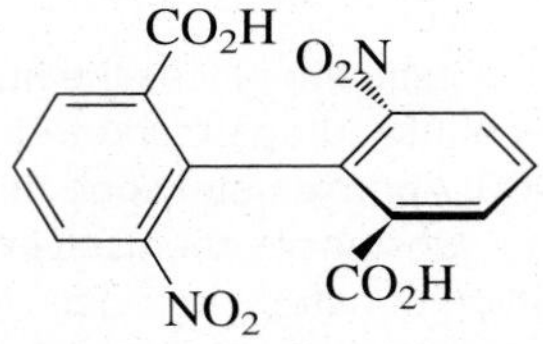

(−)-6,6′-Dinitrobiphenyl-2,2′-dicarboxylic acid $[\alpha]_D^{29}$ −127° (methanol)

Problem 4.14

The 3,3′-5,5′ isomer of the compound just shown has a chirality axis, but its separation into isolable enantiomers would be extremely difficult. Why?

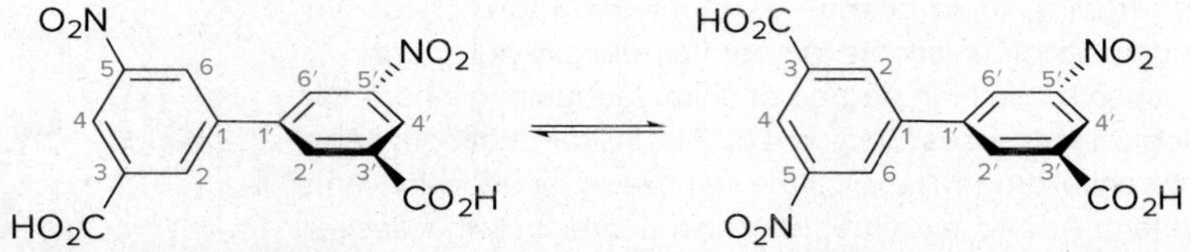

Chemists don't agree on the minimum energy barrier for bond rotation that allows isolation of enantiomeric atropisomers at room temperature, but it is on the order of 100 kJ/mol (24 kcal/mol). Recall that the activation energy for rotation about C—C single bonds in alkanes is about 12 kJ/mol (3 kcal/mol).

Structures such as chiral biaryls, which are related by rotation about a single bond yet are capable of independent existence, are sometimes called **atropisomers,** from the Greek *a* meaning "not" and *tropos* meaning "turn." They represent a subcategory of conformers.

Derivatives of 1,1′-binaphthyl exhibit atropisomerism, due to hindered rotation about the single bond that connects the two naphthalene rings. A commercially important application of chiral binaphthyls is based on a substituted derivative known as BINAP, a component of a hydrogenation catalyst. In this catalyst, ruthenium is bound by the two phosphorus atoms present on the groups attached to the naphthalene rings.

BINAP is an abbreviation for 2,2′-bis(diphenylphosphino)-1,1′-binaphthyl.

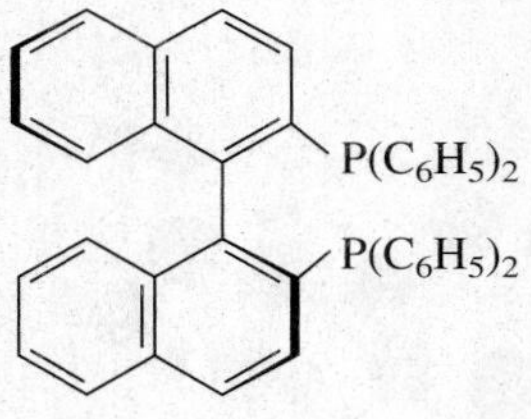

(*S*)-(−)-BINAP

We will explore the use of the ruthenium BINAP catalysts in the synthesis of chiral drugs in Section 15.13.

4.10 Chiral Molecules with Two Chirality Centers

When a molecule contains two chirality centers, as does 2,3-dihydroxybutanoic acid, how many stereoisomers are possible?

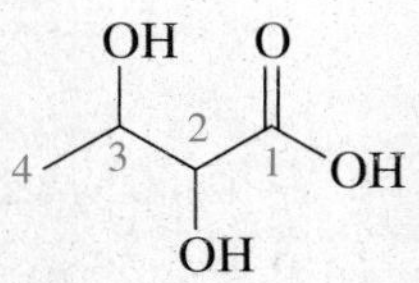

2,3-Dihydroxybutanoic acid

We can use straightforward reasoning to come up with the answer. The absolute configuration at C-2 may be *R* or *S*. Likewise, C-3 may have either the *R* or the *S* configuration. The four possible combinations of these two chirality centers are

(2*R*,3*R*)	(stereoisomer I)	(2*S*,3*S*)	(stereoisomer II)
(2*R*,3*S*)	(stereoisomer III)	(2*S*,3*R*)	(stereoisomer IV)

Figure 4.7 presents structural formulas for these four stereoisomers. Stereoisomers I and II are enantiomers of each other; the enantiomer of (*R*,*R*) is (*S*,*S*). Likewise stereoisomers III and IV are enantiomers of each other, the enantiomer of (*R*,*S*) being (*S*,*R*).

Stereoisomer I is not a mirror image of III or IV, so it is not an enantiomer of either one. Stereoisomers that are not related as an object and its mirror image are called diastereomers; **diastereomers** *are stereoisomers that are not mirror images.* Thus, stereoisomer I is a diastereomer of III and a diastereomer of IV. Similarly, II is a diastereomer of III and IV.

To convert a molecule with two chirality centers to its enantiomer, the configuration at *both* centers must be changed. Reversing the configuration at only one chirality center converts it to a diastereomer.

Enantiomers must have equal and opposite specific rotations. Diastereomers can have different rotations, with respect to both sign and magnitude. Thus, as Figure 4.7 shows, the (2*R*,3*R*) and (2*S*,3*S*) enantiomers (I and II) have specific rotations that are equal in magnitude but opposite in sign. The (2*R*,3*S*) and (2*S*,3*R*) enantiomers (III and IV) likewise have specific rotations that are equal to each other but opposite in sign. The magnitudes of rotation of I and II are different, however, from those of their diastereomers III and IV.

In writing Fischer projections of molecules with two chirality centers, the molecule is arranged in an *eclipsed* conformation for projection onto the page, as shown in Figure 4.8. Again, horizontal lines in the projection represent bonds coming toward you; vertical lines represent bonds pointing away.

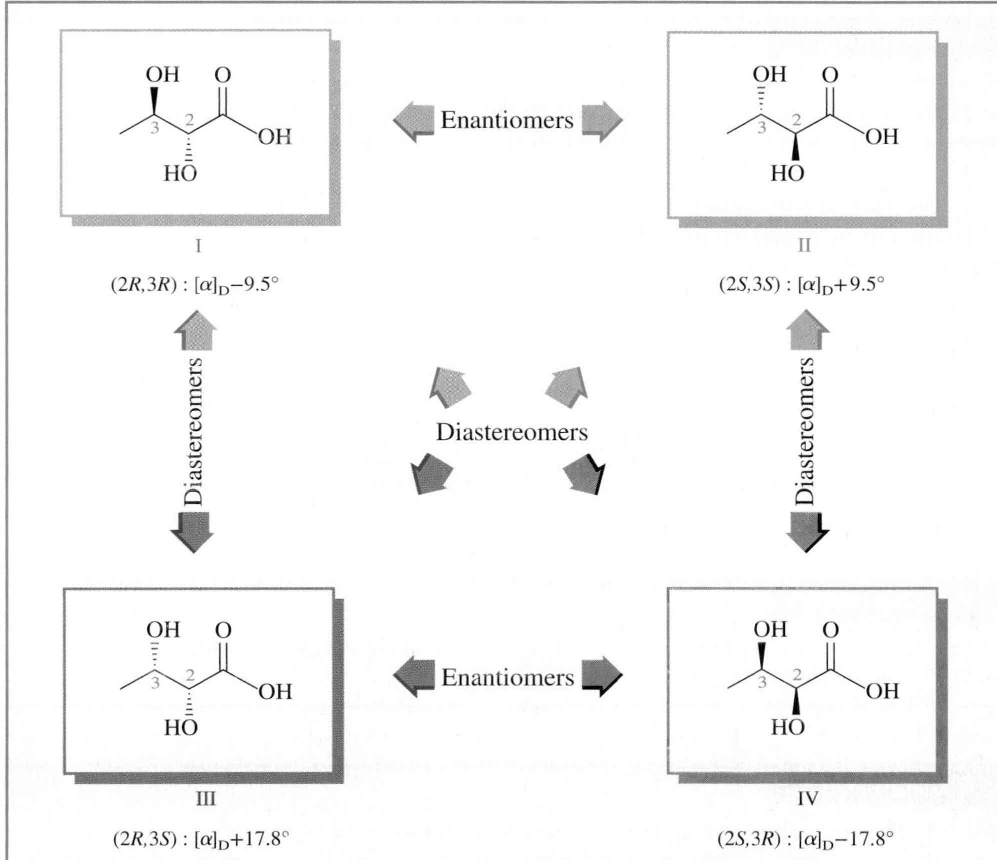

Figure 4.7

Stereoisomeric 2,3-dihydroxybutanoic acids. Stereoisomers I and II are enantiomers. Stereoisomers III and IV are enantiomers. All other relationships are diastereomeric (see text).

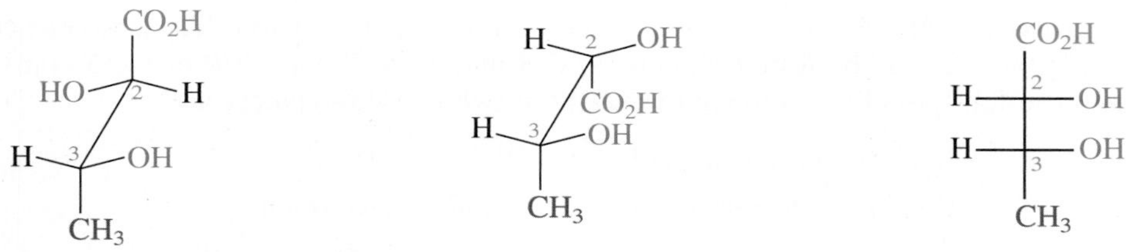

(*a*) (*b*) (*c*)

Figure 4.8

Representations of (2*R*,3*R*)-dihydroxybutanoic acid. (*a*) The staggered conformation is the most stable, but is not properly arranged to show stereochemistry as a Fischer projection. (*b*) Rotation about the C-2—C-3 bond gives the eclipsed conformation, and projection of the eclipsed conformation onto the page gives (c) a correct Fischer projection.

When the carbon chain is vertical and like substituents are on the same side of the Fischer projection, the molecule is described as the **erythro** diastereomer. When like substituents are on opposite sides of the Fischer projection, the molecule is described as the **threo** diastereomer. Thus, as seen in the Fischer projections of the stereoisomeric 2,3-dihydroxybutanoic acids, compounds I and II are erythro stereoisomers and III and IV are threo.

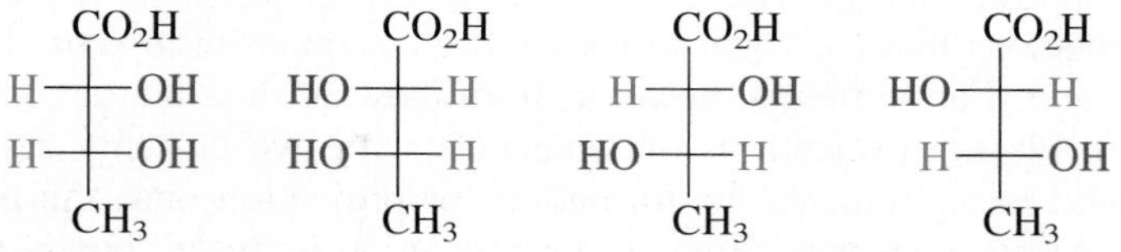

I	II	III	IV
erythro	erythro	threo	threo

Problem 4.15

Assign the *R* or *S* configuration to the chirality centers in the four isomeric 2,3-dihydroxybutanoic acids shown in the preceding Fischer projections. Consult Figure 4.7 to check your answers.

Because diastereomers are not mirror images of each other, they can have quite different physical and chemical properties. For example, the (2*R*,3*R*) stereoisomer of 3-amino-2-butanol is a liquid, but the (2*R*,3*S*) diastereomer is a crystalline solid.

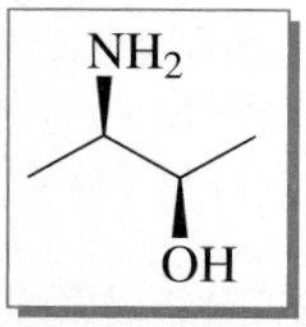

(2*R*,3*R*)-3-Amino-2-butanol (liquid)

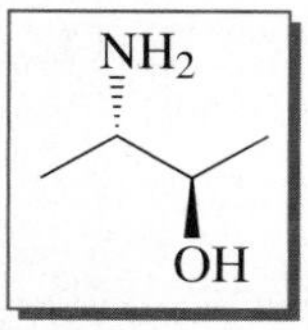

(2*R*,3*S*)-3-Amino-2-butanol (solid, mp 49°C)

Problem 4.16

Draw Fischer projections of the four stereoisomeric 3-amino-2-butanols, and label each erythro or threo as appropriate.

Problem 4.17

One other stereoisomer of 3-amino-2-butanol is a crystalline solid. Which one?

The situation is the same when the two chirality centers are present in a ring. There are four stereoisomeric 1-bromo-2-chlorocyclopropanes: a pair of enantiomers in which the halogens are trans and a pair in which they are cis. The cis compounds are diastereomers of the trans.

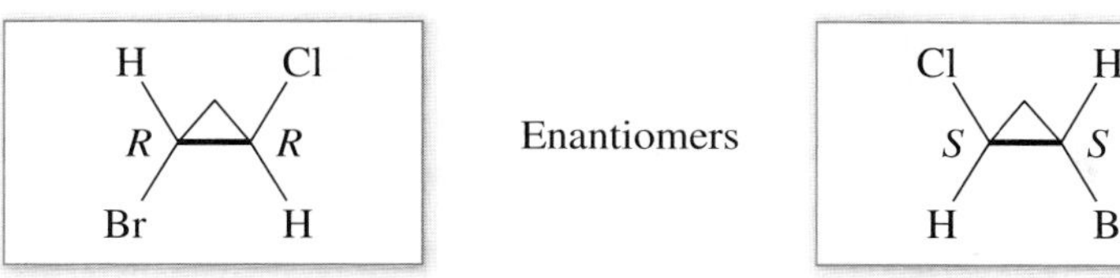

(1*R*,2*R*)-1-Bromo-2-chlorocyclopropane (1*S*,2*S*)-1-Bromo-2-chlorocyclopropane

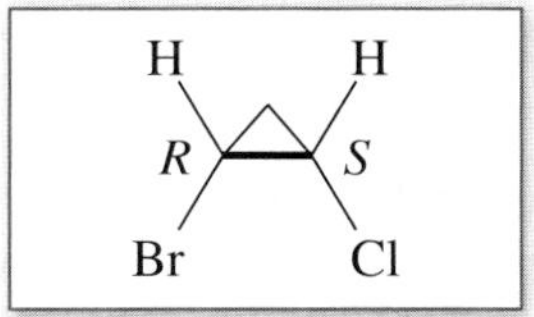

Enantiomers

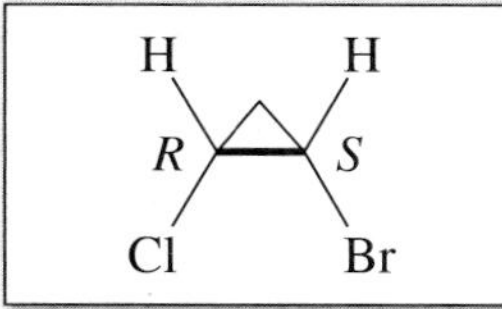

(1*R*,2*S*)-1-Bromo-2-chlorocyclopropane (1*S*,2*R*)-1-Bromo-2-chlorocyclopropane

A good thing to remember is that the cis and trans isomers of a particular compound are diastereomers of each other.

4.11 Achiral Molecules with Two Chirality Centers

Now think about a molecule, such as 2,3-butanediol, which has two chirality centers that are equivalently substituted.

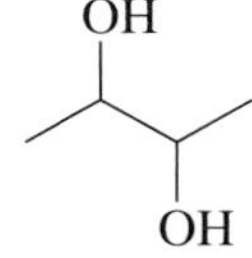

2,3-Butanediol

Only *three,* not four, stereoisomeric 2,3-butanediols are possible. These three are shown in Figure 4.9. The (2*R*,3*R*) and (2*S*,3*S*) forms are enantiomers and have equal and opposite optical rotations. A third combination of chirality centers, (2*R*,3*S*), however, gives an *achiral* structure that is superimposable on its (2*S*,3*R*) mirror image. Because it is achiral, this third stereoisomer is *optically inactive.* We call achiral molecules that have chirality centers **meso forms.** The meso form in Figure 4.9 is known as *meso*-2,3-butanediol.

One way to demonstrate that *meso*-2,3-butanediol is achiral is to recognize that its eclipsed conformation has a plane of symmetry that passes through and is perpendicular to the C-2—C-3 bond, as illustrated in Figure 4.10*a*. The anti conformation is achiral as well. As Figure 4.10*b* shows, this conformation is characterized by a center of symmetry at the midpoint of the C-2—C-3 bond.

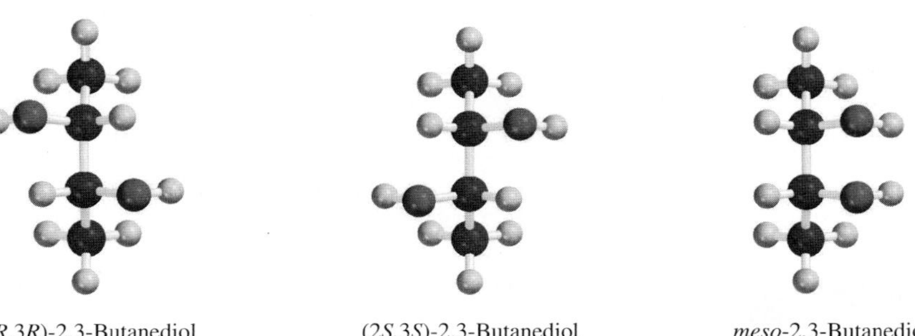

(2*R*,3*R*)-2,3-Butanediol (2*S*,3*S*)-2,3-Butanediol *meso*-2,3-Butanediol

(*a*) (*b*) (*c*)

Figure 4.9

Stereoisomeric 2,3-butanediols shown in their eclipsed conformations for convenience. Stereoisomers (*a*) and (*b*) are enantiomers. Structure (*c*) is a diastereomer of (*a*) and (*b*), and is achiral. It is called *meso*-2,3-butanediol.

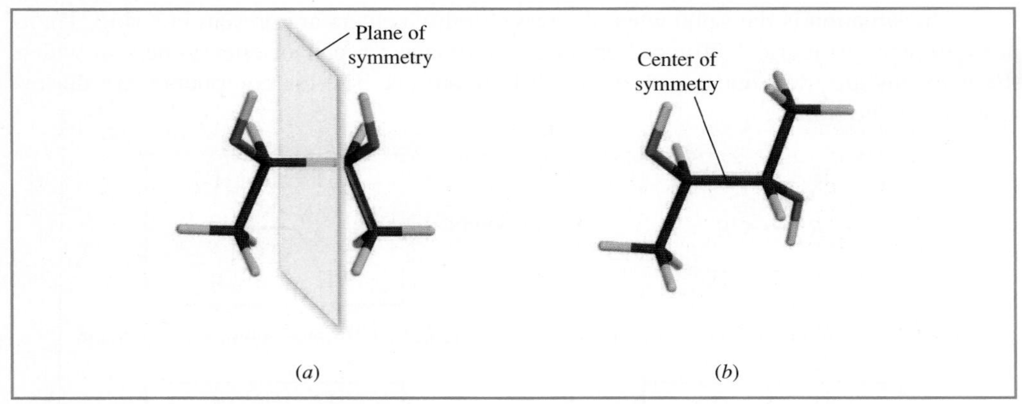

Figure 4.10

(*a*) The eclipsed conformation of *meso*-2,3-butanediol has a plane of symmetry. (*b*) The anti conformation of *meso*-2,3-butanediol has a center of symmetry.

Fischer projections can help us identify meso forms. Of the three stereoisomeric 2,3-butanediols, notice that only in the meso stereoisomer does a dashed line through the center of the Fischer projection divide the molecule into two mirror-image halves.

In the same way that a Fischer formula is a projection of the eclipsed conformation onto the page, the line drawn through its center is a projection of the plane of symmetry that is present in the eclipsed conformation of *meso*-2,3-butanediol.

(2*R*,3*R*)-2,3-Butanediol (2*S*,3*S*)-2,3-Butanediol *meso*-2,3-Butanediol

When using Fischer projections for this purpose, however, be sure to remember what three-dimensional objects they stand for. One should not, for example, test for superimposition of the two chiral stereoisomers by a procedure that involves moving any part of a Fischer projection out of the plane of the paper in any step.

Problem 4.18

A meso stereoisomer is possible for one of the following compounds. Which one?

Turning to cyclic compounds, we see that there are only three, not four, stereoisomeric 1,2-dibromocyclopropanes. Of these, two are enantiomeric *trans*-1,2-dibromocyclopropanes. The cis diastereomer is a meso form; it has a plane of symmetry.

(1*R*,2*R*)-1,2-Dibromocyclopropane (1*S*,2*S*)-1,2-Dibromocyclopropane *cis*-1,2-Dibromocyclopropane

Problem 4.19

One of the stereoisomers of 1,3-dimethylcyclohexane is a meso form. Which one?

Chirality of Disubstituted Cyclohexanes

Disubstituted cyclohexanes present us with a challenging exercise in stereochemistry. Consider the seven possible dichlorocyclohexanes: 1,1-; *cis*- and *trans*-1,2-; *cis*- and *trans*-1,3-; and *cis*- and *trans*-1,4-. Which are chiral? Which are achiral?

Four isomers—the ones that are achiral because they have a plane of symmetry—are relatively easy to identify:

Achiral Dichlorocyclohexanes

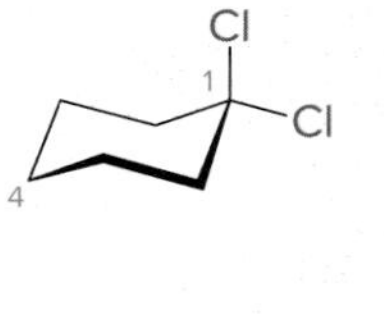

1,1
(plane of symmetry through C-1 and C-4)

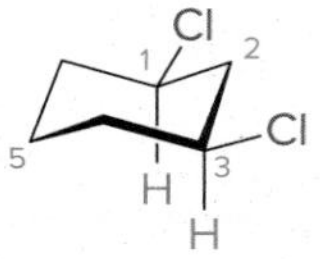

cis-1,3
(plane of symmetry through C-2 and C-5)

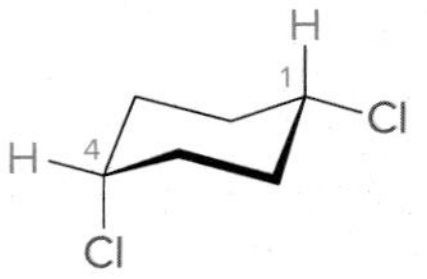

cis-1,4
(plane of symmetry through C-1 and C-4)

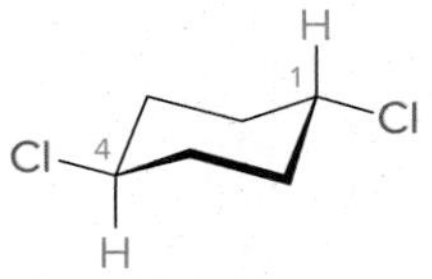

trans-1,4
(plane of symmetry through C-1 and C-4)

The remaining three isomers are chiral:

Chiral Dichlorocyclohexanes

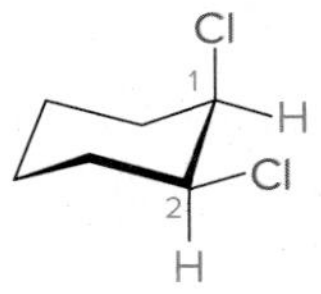

cis-1,2

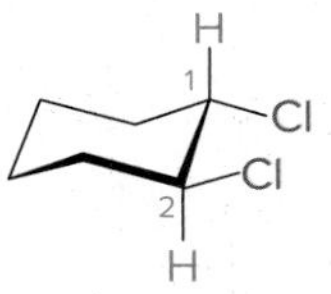

trans-1,2

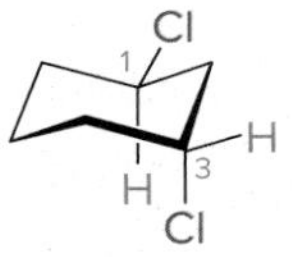

trans-1,3

Among all the isomers, *cis*-1,2-dichlorocyclohexane is unique in that the ring-inverting process typical of cyclohexane derivatives converts it to its enantiomer.

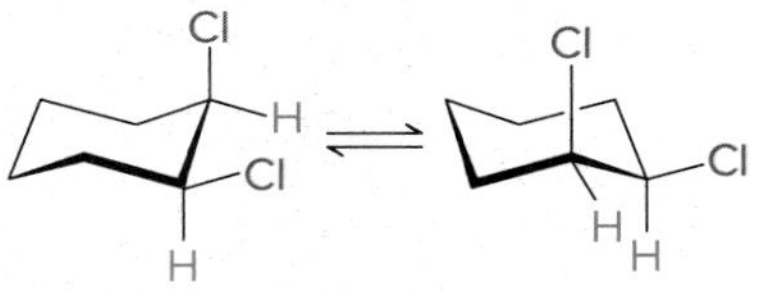

A ⇌ A′ which is equivalent to A′

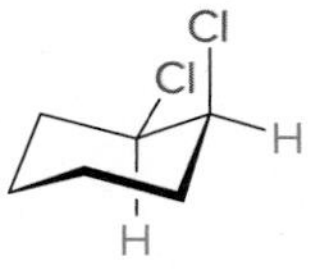

Structures A and A′ are nonsuperimposable mirror images of each other. Thus, although *cis*-1,2-dichlorocyclohexane is chiral, it is optically inactive when chair–chair interconversion occurs. Such interconversion is rapid at room temperature and converts optically active A to a racemic mixture of A and A′. Because A and A′ are enantiomers interconvertible by a conformational change, they are sometimes referred to as **conformational enantiomers.**

The same kind of spontaneous racemization occurs for any cis-1,2 disubstituted cyclohexane in which both substituents are the same. Because such compounds are chiral, it is incorrect to speak of them as meso compounds, which are achiral molecules that have chirality centers. Rapid chair–chair interconversion, however, converts them to a 1:1 mixture of enantiomers, and this mixture is optically inactive.

4.12 Molecules with Multiple Chirality Centers

Many naturally occurring compounds contain several chirality centers. By an analysis similar to that described for the case of two chirality centers, it can be shown that the maximum number of stereoisomers for a particular constitution is 2^n, where n is equal to the number of chirality centers.

Problem 4.20

Using *R* and *S* descriptors, write all the possible combinations for a molecule with three chirality centers.

When two or more of a molecule's chirality centers are equivalently substituted, meso forms are possible, and the number of stereoisomers is then less than 2^n. Thus, 2^n represents the *maximum* number of stereoisomers for a molecule containing n chirality centers.

The best examples of substances with multiple chirality centers are the *carbohydrates*. One class of carbohydrates, called *aldohexoses,* has the constitution:

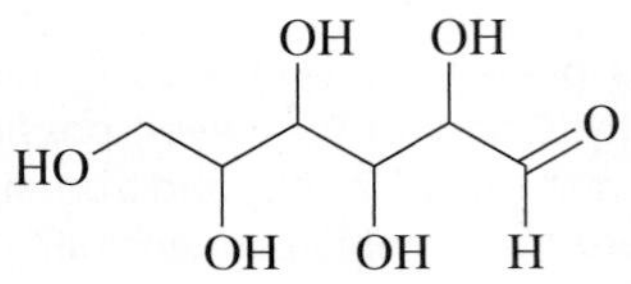

An aldohexose

Because there are four chirality centers and no possibility of meso forms, there are 2^4, or 16, stereoisomeric aldohexoses. All 16 are known, having been isolated either as natural products or as the products of chemical synthesis.

Problem 4.21

A second category of six-carbon carbohydrates, called *ketohexoses,* has the constitution shown. How many stereoisomeric 2-ketohexoses are possible?

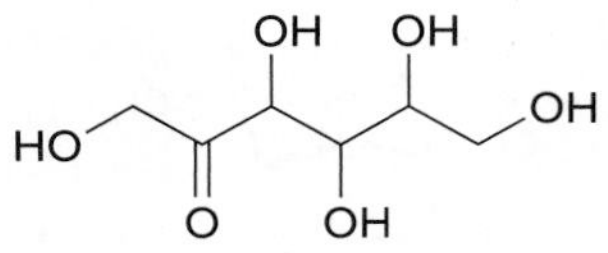

A 2-ketohexose

Steroids are another class of natural products with multiple chirality centers. One such compound is *cholic acid,* which can be obtained from bile. Its structural formula is given in Figure 4.11. Cholic acid has 11 chirality centers, and so a total (including cholic acid) of 2^{11}, or 2048, stereoisomers have this constitution. Of these 2048 stereoisomers, how many are diastereomers of cholic acid? Remember! Diastereomers are stereoisomers that are not enantiomers, and any object can have only one mirror image. Therefore, of the 2048 stereoisomers, one is cholic acid, one is its enantiomer, and the other 2046 are diastereomers of cholic acid. Only a small fraction of these compounds are known, and (+)-cholic acid is the only one ever isolated from natural sources!

Eleven chirality centers may seem like a lot, but it is nowhere close to a world record. It is a modest number when compared with the more than 100 chirality centers typical for most small proteins and the billions of chirality centers present in human DNA.

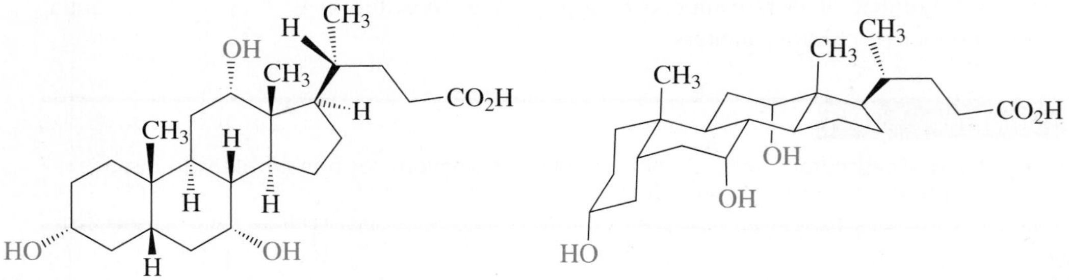

Figure 4.11

Cholic acid. Its 11 chirality centers are those carbons at which stereochemistry is indicated in the structural drawing at the left. The drawing at the right more clearly shows the overall shape of the molecule.

4.13 Resolution of Enantiomers

The separation of a racemic mixture into its enantiomeric components is termed **resolution.** The first resolution, that of tartaric acid, was carried out by Louis Pasteur in 1848. Tartaric acid is a byproduct of wine making and is almost always found as its dextrorotatory 2*R*,3*R* stereoisomer, shown here in a perspective drawing and in a Fischer projection.

OH
HO_2C CO_2H
OH

CO_2H
H—|—OH
HO—|—H
CO_2H

(2*R*,3*R*)-Tartaric acid (mp 170°C, $[\alpha]_D$ +12°)

Problem 4.22

There are two other stereoisomeric tartaric acids. Write their Fischer projections, and specify the configuration at their chirality centers.

Occasionally, an optically inactive sample of tartaric acid was obtained. Pasteur noticed that the sodium ammonium salt of optically inactive tartaric acid was a mixture of two mirror-image crystal forms. With microscope and tweezers, Pasteur carefully separated the two. He found that one kind of crystal (in aqueous solution) was dextrorotatory, whereas the mirror-image crystals rotated the plane of polarized light an equal amount but were levorotatory.

Although Pasteur was unable to provide a structural explanation—that had to wait for van't Hoff and Le Bel a quarter of a century later—he correctly deduced that the enantiomeric quality of the crystals was the result of enantiomeric molecules. The rare form of tartaric acid was optically inactive because it contained equal amounts of (+)-tartaric acid and (−)-tartaric acid. It had earlier been called *racemic acid* (from Latin *racemus,* meaning "a bunch of grapes"), a name that subsequently gave rise to our present term for an equal mixture of enantiomers.

Problem 4.23

Could the unusual, optically inactive form of tartaric acid studied by Pasteur have been *meso*-tartaric acid?

Pasteur's technique of separating enantiomers not only is laborious but requires that the crystals of the enantiomers be distinguishable. This happens very rarely. Consequently, alternative and more general approaches for resolving enantiomers have been developed. Most are based on a strategy of temporarily converting the enantiomers of a racemic mixture to diastereomeric derivatives, separating these diastereomers, then regenerating the enantiomeric starting materials.

Figure 4.12 illustrates this strategy. Say we have a mixture of enantiomers, which, for simplicity, we label as C(+) and C(−). Assume that C(+) and C(−) bear some functional group that can combine with a reagent P to yield adducts C(+)-P and C(−)-P. Now, if reagent P is chiral, and if only a single enantiomer of P, say, P(+), is added to a racemic mixture of C(+) and C(−), as shown in the first step of Figure 4.12, then the products of the reaction are C(+)-P(+) and C(−)-P(+). These products are not mirror images; they are diastereomers. Diastereomers can have different physical properties, which can serve as a means of separating them. The mixture of diastereomers is separated, usually by recrystallization from a suitable solvent. In the last step, an appropriate chemical transformation liberates the enantiomers and restores the resolving agent.

Whenever possible, the chemical reactions involved in the formation of diastereomers and their conversion to separate enantiomers are simple acid–base reactions. For

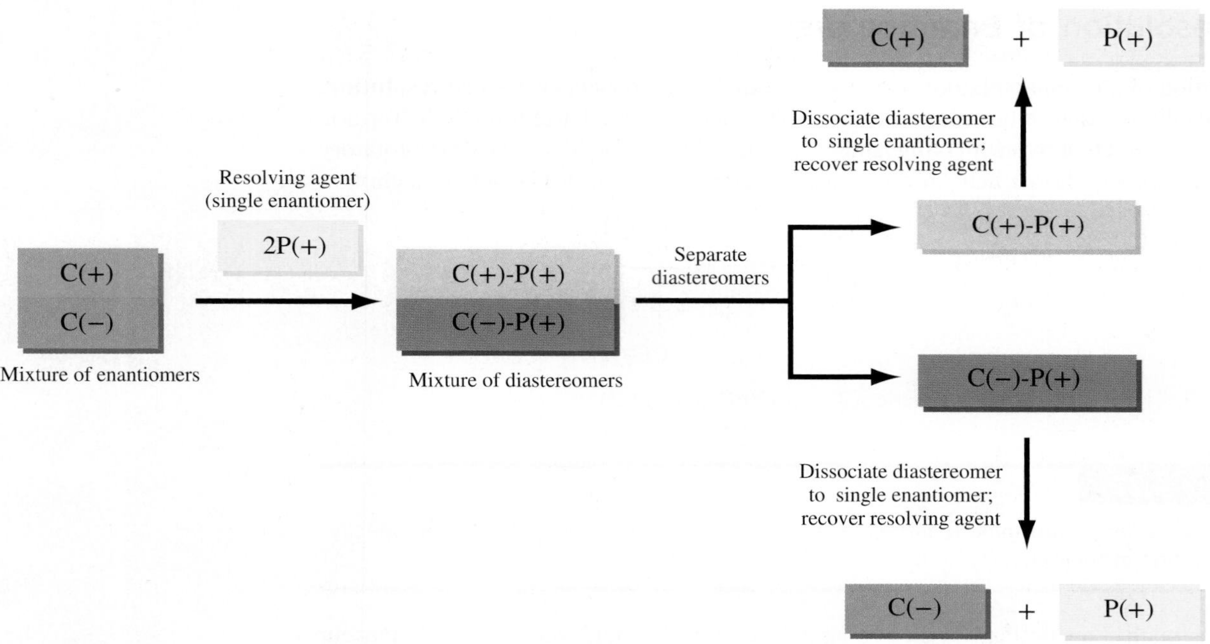

Figure 4.12

The general procedure for resolving a chiral substance into its enantiomers. Reaction with a single enantiomer of a chiral resolving agent P(+) converts the racemic mixture of enantiomers C(+) and C(−) to a mixture of diastereomers C(+)-P(+) and C(−)-P(+). The mixture of diastereomers is separated—by fractional crystallization, for example. A chemical reaction is then carried out to convert diastereomer C(+)-P(+) to C(+) and the resolving agent P(+). Likewise, diastereomer C(−)-P(+) is converted to C(−) and P(+). C(+) has been separated from C(−), and the resolving agent P(+) can be recovered for further use.

©Pixtal/age fotostock

Most resolving agents are isolated as single enantiomers from natural sources. *S*-(−)-Malic acid is obtained from apples.

example, naturally occurring (*S*)-(−)-malic acid is often used to resolve amines such as 1-phenylethylamine. Amines are bases, and malic acid is an acid. Proton transfer from (*S*)-(−)-malic acid to a racemic mixture of (*R*)- and (*S*)-1-phenylethylamine gives a mixture of diastereomeric salts.

$$C_6H_5CH(CH_3)NH_2 + HO_2CCH_2CH(OH)CO_2H \longrightarrow C_6H_5CH(CH_3)\overset{+}{N}H_3 \quad HO_2CCH_2CH(OH)CO_2^-$$

1-Phenylethylamine (racemic mixture) | (*S*)-(−)-Malic acid (resolving agent) | 1-Phenylethylammonium (*S*)-malate (mixture of diastereomeric salts)

The diastereomeric salts are separated and the individual enantiomers of the amine liberated by treatment with a base:

$$C_6H_5CH(CH_3)\overset{+}{N}H_3 \quad HO_2CCH_2CH(OH)CO_2^- + 2OH^- \longrightarrow$$

1-Phenylethylammonium (*S*)-malate (a single diastereomer) | Hydroxide ion

$$C_6H_5CH(CH_3)NH_2 + {}^-O_2CCH_2CH(OH)CO_2^- + 2H_2O$$

1-Phenylethylamine (a single enantiomer) | (*S*)-(−)-Malic acid (recovered resolving agent) | Water

Problem 4.24

In the resolution of 1-phenylethylamine using (*S*)-(−)-malic acid, the compound obtained by recrystallization of the mixture of diastereomeric salts is (*R*)-1-phenylethylammonium (*S*)-malate. The other component of the mixture is more soluble and remains in solution. What is the configuration of the more soluble salt?

This method is widely used for the resolution of chiral amines and carboxylic acids. Analogous methods based on the formation and separation of diastereomers have been developed for other functional groups; the precise approach depends on the kind of chemical reactivity associated with the functional groups present in the molecule.

As the experimental tools for biochemical transformations have become more powerful and procedures for carrying out these transformations in the laboratory more routine, the application of biochemical processes to mainstream organic chemical tasks including the production of enantiomerically pure chiral molecules has grown.

Another approach, called **kinetic resolution,** depends on the different rates of reaction of two enantiomers with a chiral reagent. A very effective form of kinetic resolution uses enzymes as chiral biocatalysts to selectively bring about the reaction of one enantiomer of a racemic mixture (**enzymatic resolution**). *Lipases,* or *esterases*—enzymes that catalyze ester hydrolysis and formation—have been successfully used in many kinetic resolutions. In a representative procedure, one enantiomer of an ester undergoes hydrolysis and the other is left unchanged.

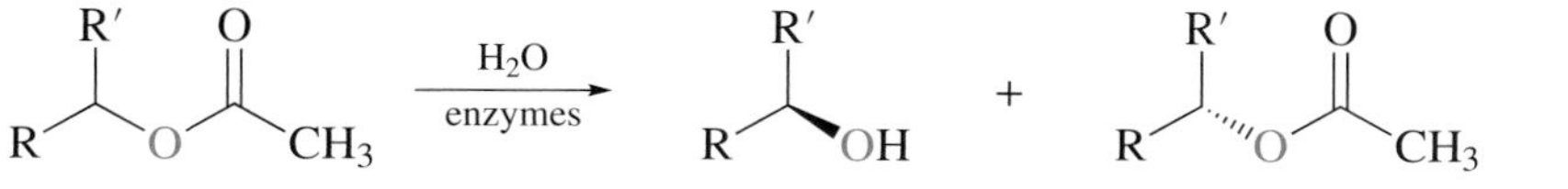

Acetate ester of racemic alcohol → (H_2O, enzymes) → Enantiomerically enriched alcohol + Enantiomerically enriched acetate ester (unhydrolyzed)

This procedure has been applied to the preparation of a key intermediate in the industrial synthesis of *diltiazem,* a drug used to treat hypertension, angina, and arrhythmia. In this case, the racemic reactant is a methyl ester, and lipase-catalyzed hydrolysis selectively converts the undesired enantiomer to its corresponding carboxylic acid, leaving behind the unhydrolyzed ester in greater than 99% enantiomeric excess.

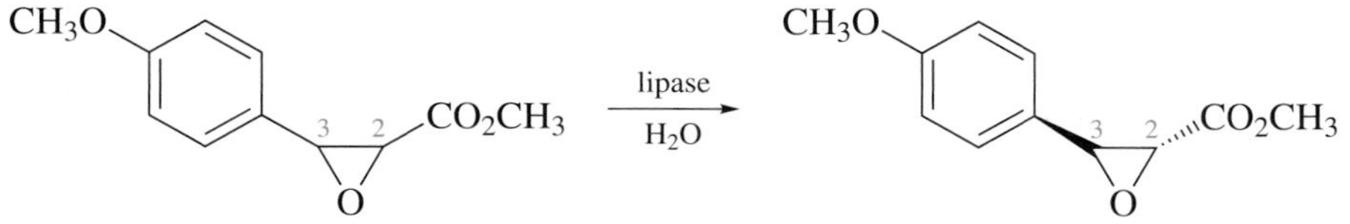

Racemic mixture of 2*R*,3*S* and 2*S*,3*R* → (lipase, H_2O) → Only 2*R*,3*S*

Enzymatic resolution, like other methods based on biocatalysis, may lead to more environmentally benign, or "green," processes for preparing useful intermediates on a commercial scale.

4.14 Chirality Centers Other Than Carbon

Atoms other than carbon may also be chirality centers. Silicon, like carbon, has a tetrahedral arrangement of bonds when it bears four substituents. Unlike carbon, there are no naturally occurring chiral organosilicon compounds to provide a ready source for stereochemical studies. Beginning in the 1960s, however, Leo H. Sommer and his students at Penn State began a systematic study of organosilicon reaction mechanisms made possible by their successful resolution of the enantiomers of 1-naphthylphenylmethylsilane.

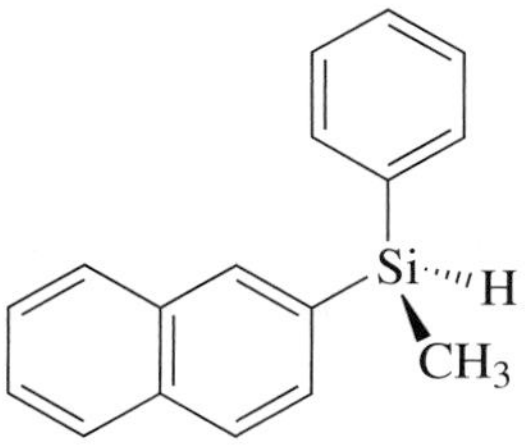

(*R*)-(+)-1-Naphthylphenylmethylsilane

Trigonal pyramidal molecules are chiral if the central atom bears three different groups. If one is to resolve substances of this type, however, the pyramidal inversion that interconverts enantiomers must be slow at room temperature. Pyramidal inversion at nitrogen is so fast that attempts to resolve chiral amines fail because of their rapid racemization.

very fast

Phosphorus is in the same group of the periodic table as nitrogen, and tricoordinate phosphorus compounds (phosphines), like amines, are trigonal pyramidal. Phosphines, however, undergo pyramidal inversion much more slowly than amines, and a number of optically active phosphines have been prepared.

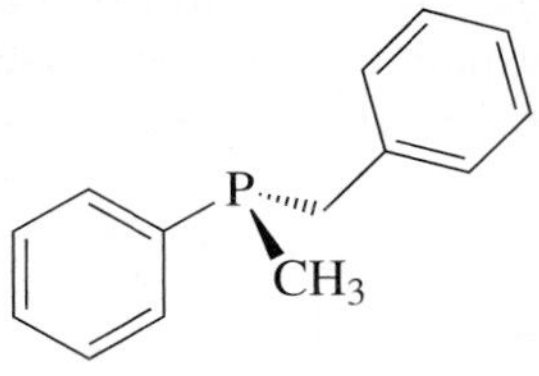

(*S*)-(+)-Benzylmethylphenylphosphine

Problem 4.25

When applying Cahn–Ingold–Prelog *R,S* stereochemical notation to phosphines, the unshared electron pair of phosphorus is taken to be the lowest ranked substituent. Use this information to verify that (+)-benzylmethylphenylphosphine (shown above) has the *S* configuration.

Tricoordinate sulfur compounds are pyramidal and chiral when sulfur bears three different groups. The most common examples are sulfoxides. *Alliin,* which occurs naturally in garlic, has two chirality centers; the one at carbon has the *R* configuration; the one at sulfur is *S*. *Omeprazole,* which is a racemic mixture, is widely used for treatment of acid reflux and has been joined in the marketplace by an enantiopure *S* version, which is appropriately named *Esomeprazole*.

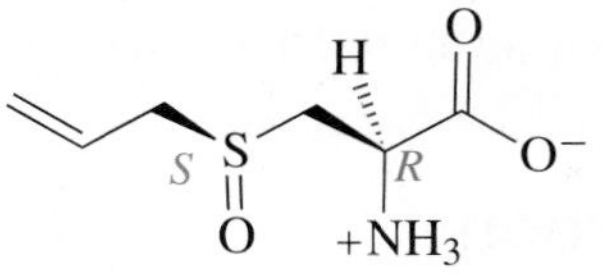

Alliin

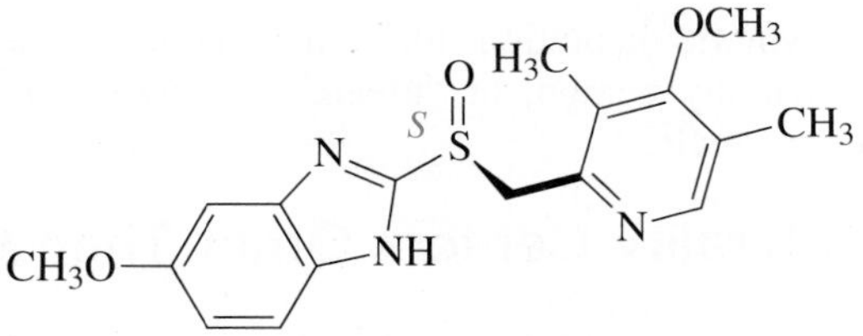

Esomeprazole

4.15 SUMMARY

Stereochemistry is chemistry in three dimensions. At its most fundamental level, its concern is molecular structure; at another level, it is chemical reactivity. Table 4.2 summarizes some of its basic definitions.

Section 4.1 A molecule is **chiral** if it cannot be superimposed on its mirror image. *Nonsuperimposable mirror images* are **enantiomers** of one another. Molecules in which mirror images are superimposable are achiral.

1-Chloropentane achiral

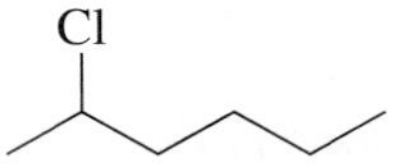

2-Chloropentane chiral

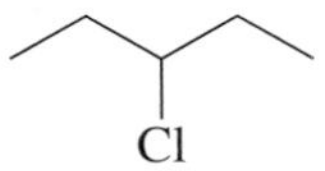

3-Chloropentane achiral

Section 4.2 The most common kind of chiral molecule contains a carbon atom that bears four different atoms or groups. Such an atom is called a **chirality center.** Table 4.2 shows the enantiomers of 2-chlorobutane. C-2 is a chirality center in 2-chlorobutane.

Section 4.3 A molecule that has a plane of symmetry or a center of symmetry is achiral. *cis*-4-Methylcyclohexanol (Table 4.2) has a plane of symmetry that bisects the molecule into two mirror-image halves and is achiral. The same can be said for *trans*-4-methylcyclohexanol.

TABLE 4.2 Classification of Isomers

Definition	Example
Isomers are different compounds that have the same molecular formula. They may be either constitutional isomers or stereoisomers.	
1. *Constitutional isomers* are isomers that differ in the order in which their atoms are connected.	Three constitutionally isomeric compounds have the molecular formula C_3H_8O: OH OH O 1-Propanol 2-Propanol Ethyl methyl ether
2. *Stereoisomers* are isomers that have the same constitution but differ in the arrangement of their atoms in space.	
(a) *Enantiomers* are stereoisomers that are related as an object and its nonsuperimposable mirror image.	The two enantiomeric forms of 2-chlorobutane are Cl and Cl (*R*)-(−)-2-Chlorobutane (*S*)-(+)-2-Chlorobutane
(b) *Diastereomers* are stereoisomers that are not mirror images.	The cis and trans isomers of 4-methylcyclohexanol are stereoisomers, but they are not related as an object and its mirror image; they are diastereomers. CH_3 HO and HO CH_3 *cis*-4-Methylcyclohexanol *trans*-4-Methylcyclohexanol

Section 4.4 **Optical activity,** or the degree to which a substance rotates the plane of polarized light, is a physical property used to characterize chiral substances. Enantiomers have equal and opposite optical rotations. To be optically active a substance must be chiral, and one enantiomer must be present in excess of the other. A **racemic mixture** is optically inactive and contains equal quantities of enantiomers. When enantiomers are present in unequal amounts, **enantiomeric excess** or **enantiomeric ratio** is used to specify the purity of the enantiomeric mixtures. Enantiomeric excess corresponds to **optical purity.**

Section 4.5 **Relative configuration** compares the arrangement of atoms in space to some reference. The prefix *cis* in *cis*-4-methylcyclohexanol, for example, describes relative configuration by referencing the orientation of the CH_3 group to the OH. **Absolute configuration** is an exact description of the arrangement of atoms in space.

Section 4.6 Absolute configuration in chiral molecules is best specified using the prefixes *R* and *S* of the Cahn–Ingold–Prelog notational system. Substituents at a chirality center are ranked in order of decreasing precedence. If the three highest ranked substituents trace a clockwise path (highest → second highest → third highest) when the lowest ranked substituent is held away from you, the configuration is *R*. If the path is counterclockwise, the configuration is *S*. Table 4.2 shows the *R* and *S* enantiomers of 2-chlorobutane.

Section 4.7 A **Fischer projection** shows how a molecule would look if its bonds were projected onto a flat surface. Horizontal lines represent bonds pointing toward you; vertical lines represent bonds pointing away from you. The projection is normally drawn so that the carbon chain is vertical, with the lowest numbered carbon at the top.

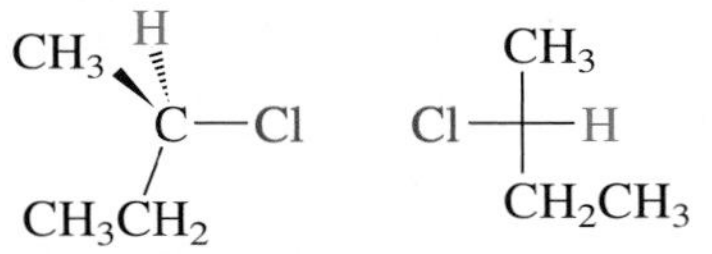

(*R*)-2-Chlorobutane 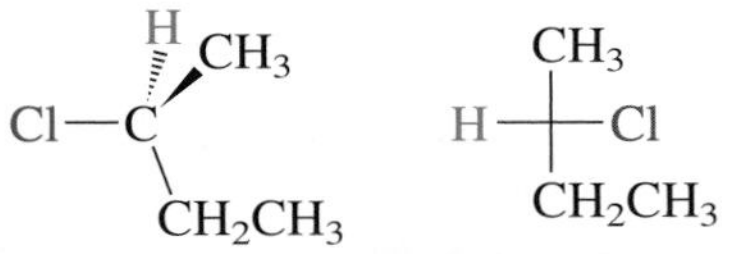 (*S*)-2-Chlorobutane

Section 4.8 Both enantiomers of the same substance are identical in most of their physical properties. The most prominent differences are biological ones, such as taste and odor, in which the substance interacts with a chiral receptor site. Enantiomers also have important consequences in medicine, in which the two enantiomeric forms of a drug can have much different effects on a patient.

Section 4.9 Molecules without chirality centers can be chiral. Biphenyls that are substituted can exhibit an **axis of chirality.** When $A \neq B$, and $X \neq Y$, the two conformations are nonsuperimposable mirror images of each other; that is, they are enantiomers. The bond connecting the two rings lies along a chirality axis.

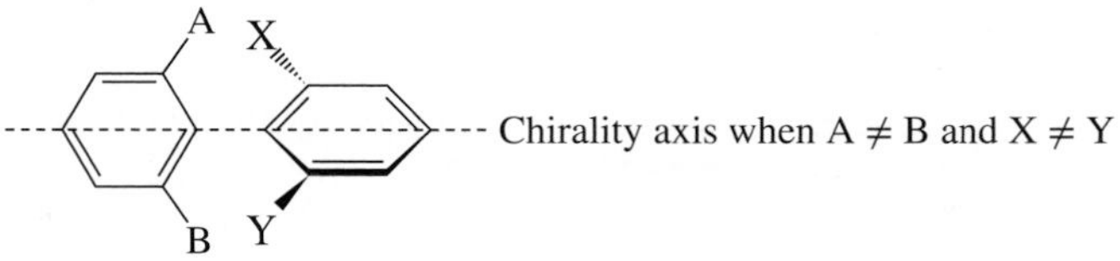

Chirality axis when $A \neq B$ and $X \neq Y$

Section 4.10 When a molecule has two chirality centers and these two chirality centers are not equivalent, four stereoisomers are possible.

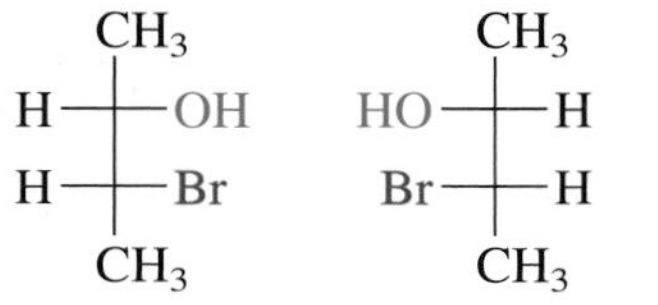

Enantiomers of *erythro*-3-bromo-2-butanol 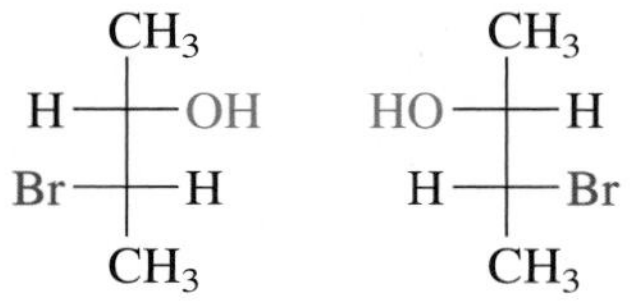 Enantiomers of *threo*-3-bromo-2-butanol

Stereoisomers that are not mirror images are classified as **diastereomers.** Each enantiomer of *erythro*-3-bromo-2-butanol is a diastereomer of each enantiomer of *threo*-3-bromo-2-butanol.

Section 4.11 Achiral molecules that contain chirality centers are called **meso forms.** Meso forms typically contain (but are not limited to) two equivalently substituted chirality centers. They are optically inactive.

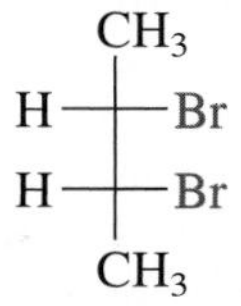

meso-2,3-Dibromobutane 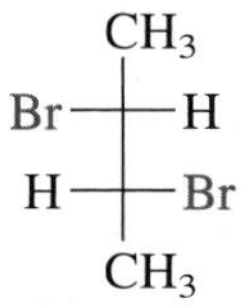 (2*R*,3*R*)-2,3-Dibromobutane 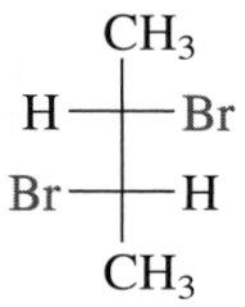 (2*S*,3*S*)-2,3-Dibromobutane

Section 4.12 For a particular constitution, the maximum number of stereoisomers is 2^n, where n is the number of chirality centers. The number of stereoisomers is reduced to less than 2^n when there are meso forms.

Section 4.13 **Resolution** is the separation of a racemic mixture into its enantiomers. It is normally carried out by converting the mixture of enantiomers to a mixture of diastereomers, separating the diastereomers, then regenerating the enantiomers.

Section 4.14 Atoms other than carbon can be chirality centers. Examples include those based on tetracoordinate silicon and tricoordinate sulfur or phosphorus as the chirality center. In principle, tricoordinate nitrogen can be a chirality center in compounds of the type N(x,y,z), where x, y, and z are different, but inversion of the nitrogen pyramid is so fast that racemization occurs virtually instantly at room temperature.

PROBLEMS

Molecular Chirality

4.26 Which of the isomeric alcohols having the molecular formula $C_6H_{14}O$ are chiral? Which are achiral?

4.27 Including stereoisomers, write structural formulas for all of the compounds that are trichloro derivatives of

(a) Cyclobutane

(b) Cyclopentane

Which are chiral? Which are achiral?

4.28 In each of the following pairs of compounds one is chiral and the other is achiral. Identify each compound as chiral or achiral, as appropriate.

(a) Cl, OH, OH and HO, OH, Cl

(b) Br and Br

(c) CH_3, H_2N—H, H—NH_2, CH_3 and CH_3, H—NH_2, H—NH_2, CH_3

(d) Cl and Cl

(e) OH, OH and OH, OH

(f) CH_3, OH and H_3C, OH

(g) O and O

4.29 Compare 2,3-pentanediol and 2,4-pentanediol with respect to the number of stereoisomers possible for each. Which ones are chiral? Which are achiral?

OH, OH — 2,3-Pentanediol

OH, OH — 2,4-Pentanediol

4.30 Of the isomers shown, which are chiral? Which ones are constitutional isomers of each other? Stereoisomers? Enantiomers? Diastereomers?

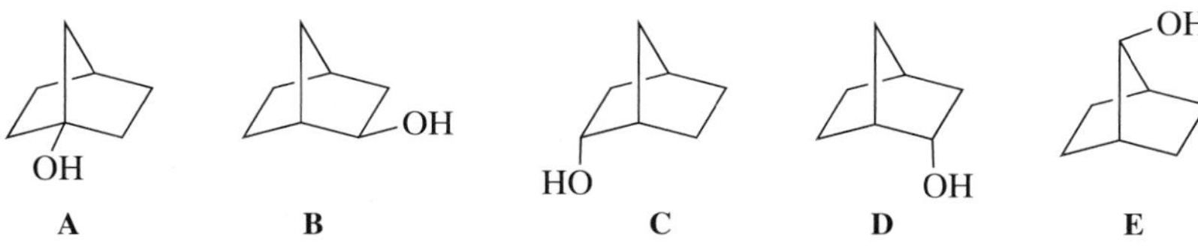

4.31 *Diltiazem* is prescribed to treat hypertension, and *simvastatin* is a cholesterol-lowering drug. Locate the chirality centers in each.

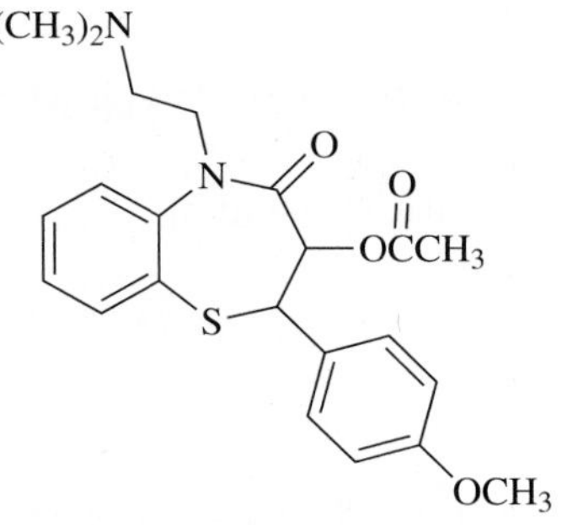

Diltiazem

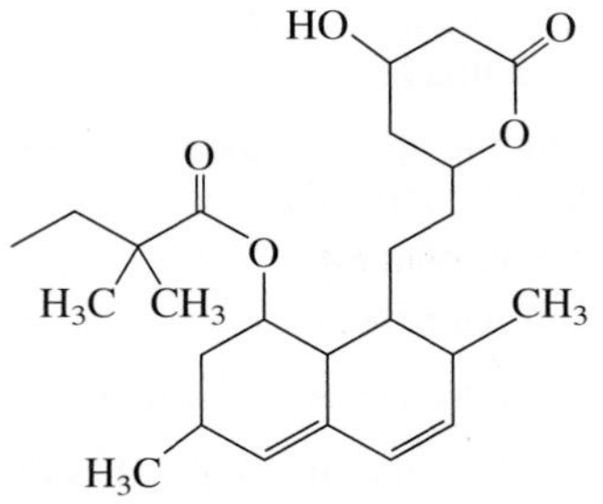

Simvastatin

4.32 Among compounds (a)–(d), identify those that have a chirality axis.

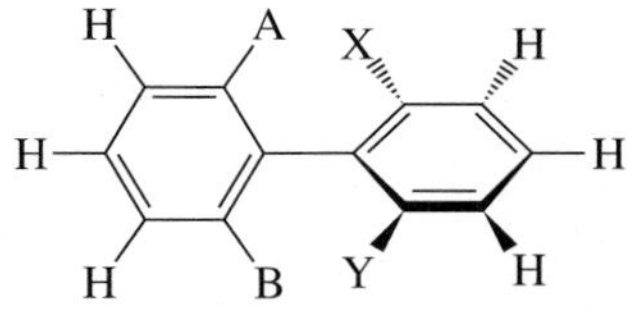

	A	B	X	Y
(a)	$(CH_3)_3C$—	H—	H—	H—
(b)	$(CH_3)_3C$—	$(CH_3)_3C$—	H—	H—
(c)	$(CH_3)_3C$—	H—	$(CH_3)_3C$—	H—
(d)	$(CH_3)_3C$—	$(CH_3)_3C$—	$(CH_3)_3C$—	$(CH_3)_3C$—

R,S-Configurational Notation

4.33 The absolute configuration of (−)-bromochlorofluoromethane is *R*. Which of the following is (are) (−)-BrClFCH?

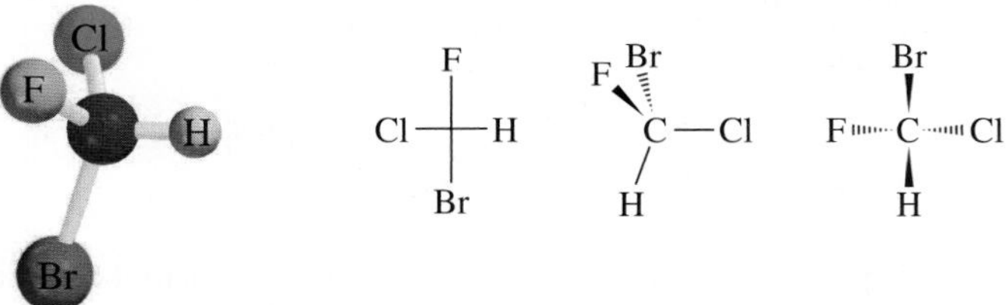

4.34 A subrule of the Cahn–Ingold–Prelog system specifies that higher mass number takes precedence over lower when distinguishing between isotopes.

(a) Determine the absolute configurations of the reactant and product in the biological oxidation of isotopically labeled ethane described in Section 4.2.

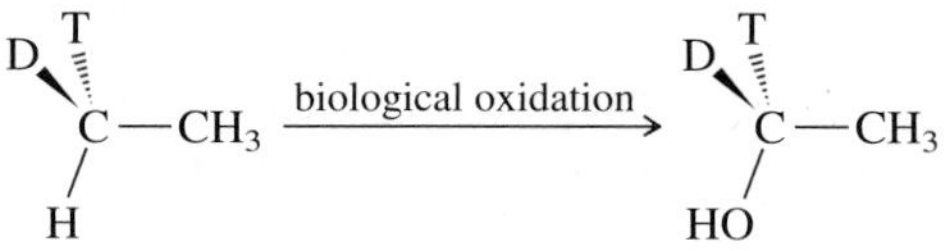

(b) Because OH becomes bonded to carbon at the same side from which H is lost, the oxidation proceeds with retention of configuration. Compare this fact with the *R* and *S* configurations you determined in part (a) and reconcile any apparent conflicts.

4.35 Specify the configuration of the chirality center as *R* or *S* in each of the following.

(a) (−)-2-Octanol

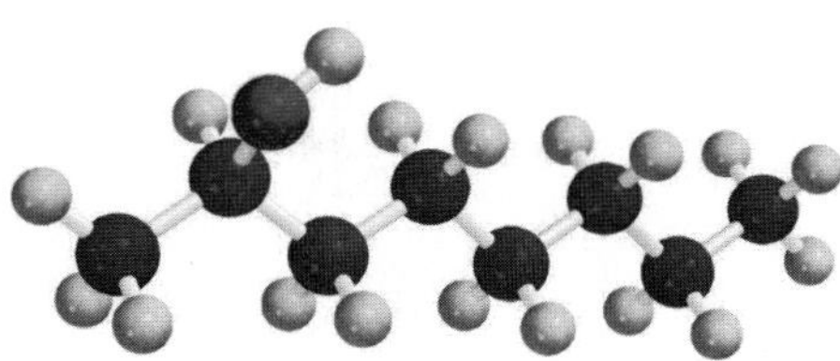

(b) Monosodium L-glutamate (only this stereoisomer is a flavor-enhancing agent)

O O

$^-$O O$^-$ Na+

+NH$_3$

4.36 The name *cis*-3-bromocyclohexanol correctly describes the constitution and *relative* stereochemistry of the compound shown. The molecule, however, is chiral so if we wish to distinguish between it and its enantiomer we need to specify its *absolute* configuration using *R,S* notation. Which of the four possibilities is correct?

Br OH

(1*R*,3*R*)-3-bromocyclohexanol	(1*S*,3*S*)-bromocyclohexanol
(1*R*,3*S*)-3-bromocyclohexanol	(1*S*,3*R*)-3-bromocyclohexanol

4.37 The antiparkinson drug *droxidopa* has the structural formula shown with configurations at C-2 and C-3 of *S* and *R,* respectively. Add appropriate wedges and/or dashes to show the stereochemistry.

OH, HO, 1, 2, 3, CO_2^-, +NH$_3$, HO

Structural Relationships

4.38 Identify the relationship in each of the following pairs. Do the drawings represent constitutional isomers or stereoisomers, or are they just different ways of drawing the same compound? If they are stereoisomers, are they enantiomers or diastereomers?

(a) H, CH_3, C—Br, CH_3CH_2 and H, Br, C—CH_3, CH_3CH_2

(b) CH_2OH, H—OH, CH_2OH and CH_2OH, HO—H, CH_2OH

(c) OH and OH

(d) HO, CH_2OH and HO, CH_2OH

(e) H, CH_3, H, CH_3 and CH_3, H, H_3C, H

(f) 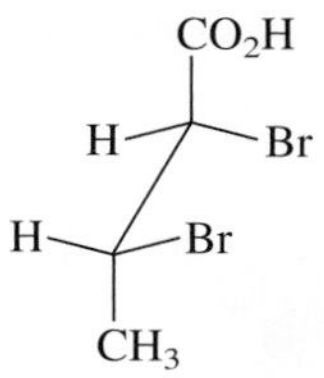 [structure] and [structure]

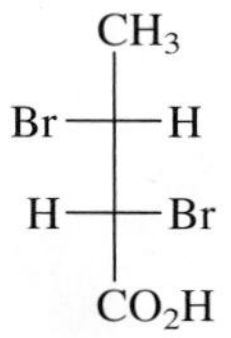

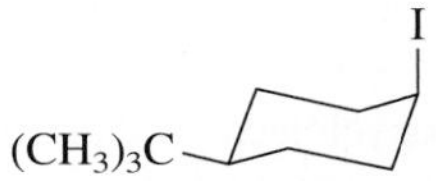

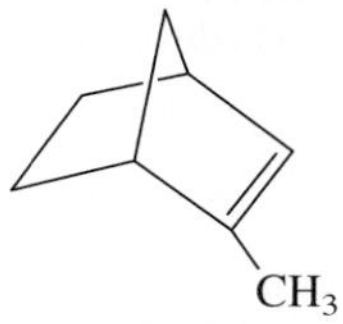

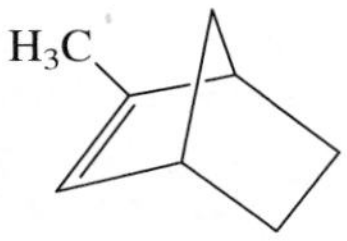

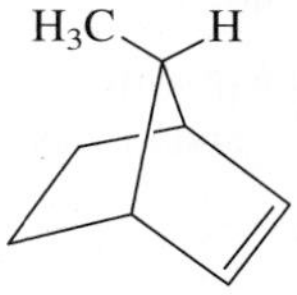

(g) 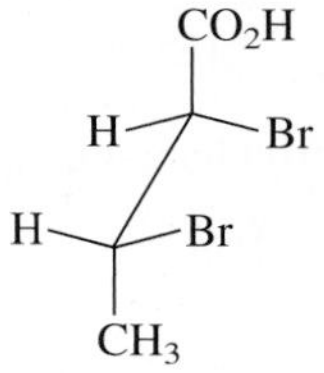 [structure] and [structure]

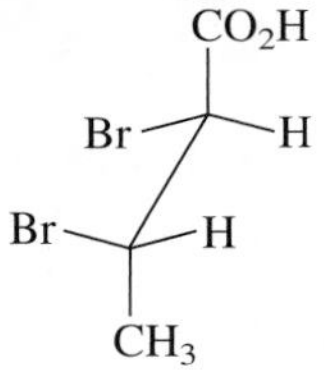

(h) 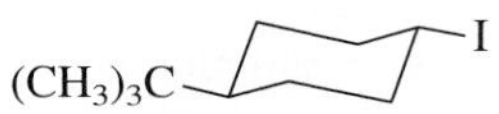 $(CH_3)_3C$–cyclohexane–I and $(CH_3)_3C$–cyclohexane–I

(i) [structure] and [structure]

(j) [structure] and [structure]

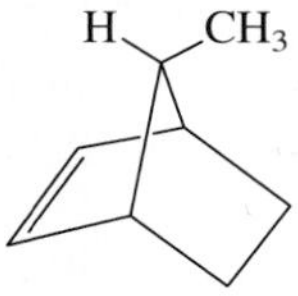

4.39 *Muscarine* is a poisonous substance present in the mushroom *Amanita muscaria*. Its structure is represented by the constitution shown here.

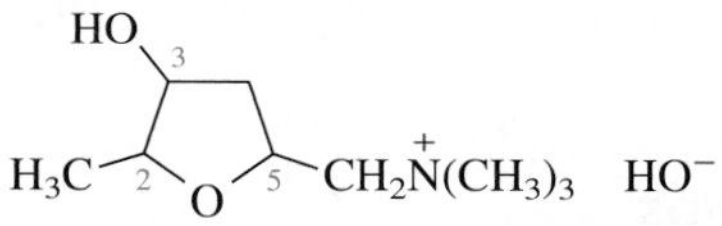

HO, H_3C, O, $CH_2\overset{+}{N}(CH_3)_3$ HO^-

(a) Including muscarine, how many stereoisomers have this constitution?

(b) One of the substituents on the ring of muscarine is trans to the other two. How many of the stereoisomers satisfy this requirement?

(c) Muscarine has the configuration 2*S*,3*R*,5*S*. Write a structural formula of muscarine showing its correct stereochemistry.

4.40 (−)-Menthol, used to flavor various foods, is the most stable stereoisomer of 2-isopropyl-5-methylcyclohexanol and has the *R* configuration at the hydroxyl-substituted carbon.

(a) Draw the most stable conformation of (−)-menthol. Is the hydroxyl group cis or trans to the isopropyl group?

(b) (+)-Isomenthol has the same constitution as (−)-menthol. The configurations at C-1 and C-2 of (+)-isomenthol are the opposite of the corresponding chirality centers of (−)-menthol. Write the preferred conformation of (+)-isomenthol.

Optical Activity

4.41 A certain natural product having $[\alpha]_D$ + 40.3° was isolated. Two very different structures were independently proposed for this compound. Which one do you think is more likely to be correct? Why?

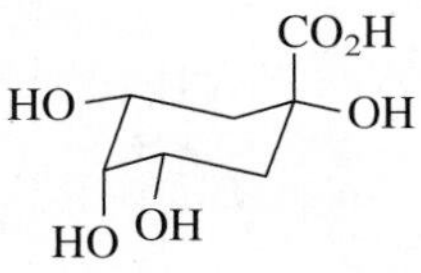

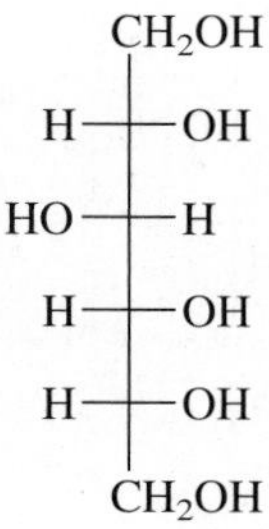

[structures]

4.42 One of the principal substances obtained from archaea (one of the oldest forms of life on Earth) is derived from a 40-carbon diol. Given the fact that this diol is optically active, is it compound A or is it compound B?

Compound A

Compound B

4.43 (a) An aqueous solution containing 10 g of optically pure fructose was diluted to 500 mL with water and placed in a polarimeter tube 20 cm long. The measured rotation was −5.20°. Calculate the specific rotation of fructose.

(b) If this solution were mixed with 500 mL of a solution containing 5 g of racemic fructose, what would be the specific rotation of the resulting fructose mixture? What would be its optical purity?

4.44 A kinetic resolution (Section 4.13) of *N*-succinylphenylalanine was carried out by treatment with the enzyme succinylase. The phenylalanine that was obtained had the *R* configuration at its chirality center and was obtained in 87.6% enantiomeric excess (ee).

(a) Draw the structure of the major phenylalanine product and the recovered *N*-succinylphenylalanine to show stereochemistry at their chirality centers.

(b) What is the enantiomeric ratio (er) of the phenylalanine that is produced in this reaction?

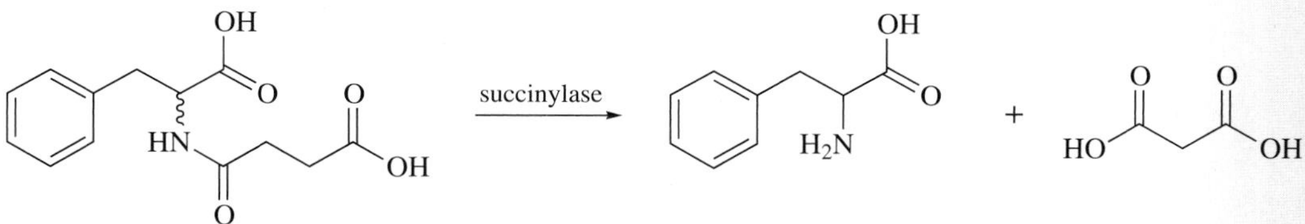

N-Succinylphenylalanine (racemic)   Phenylalanine   Succinic acid

+ *N*-Succinyl-L-phenylalanine

Descriptive Passage and Interpretive Problems 4

Prochirality

Consider two chemical changes: one occurring at a tetrahedral sp^3 carbon C(x,x,y,z), the other at a trigonal sp^2 carbon C(x,y,z), where x, y, and z are different atoms or groups attached to C. Each reactant is achiral; both are converted to the chiral product C(w,x,y,z). In the first case, w replaces one of the x atoms or groups; in the other, w adds to the trigonal carbon.

Achiral   Chiral   Achiral

Both transformations convert C in each achiral reactant to a chirality center in the product. The two achiral reactants are classified as **prochiral.** C is a **prochirality center** in C(x,x,y,z) and has two **prochiral faces** in C(x,y,z).

In achiral molecules with tetrahedral prochirality centers, substitution of one of the two x groups by w gives the enantiomer of the product that results from substitution of the other. The two x groups occupy mirror-image sites and are **enantiotopic.**

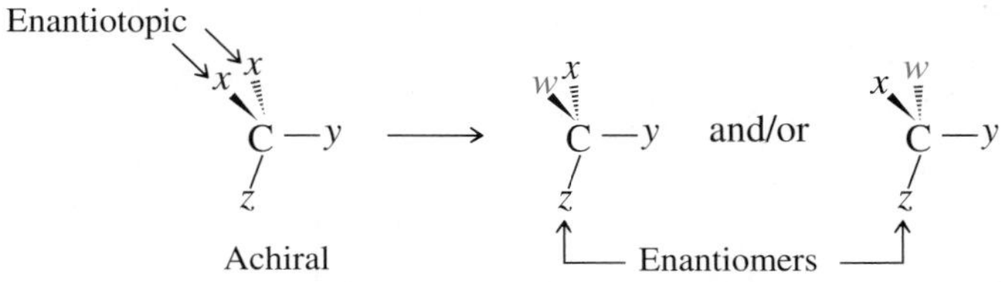

Enantiotopic

Achiral → and/or

Enantiomers

Enantiotopic groups are designated as *pro-R* or *pro-S* by a modification of Cahn–Ingold–Prelog notation. One is assigned a higher priority than the other without disturbing the priorities of the remaining groups, and the *R,S* configuration of the resulting chirality center is determined in the usual way. If it is *R,* the group assigned the higher rank is *pro-R*. If *S,* this group is *pro-S*. Ethanol and citric acid illustrate the application of this notation to two prochiral molecules.

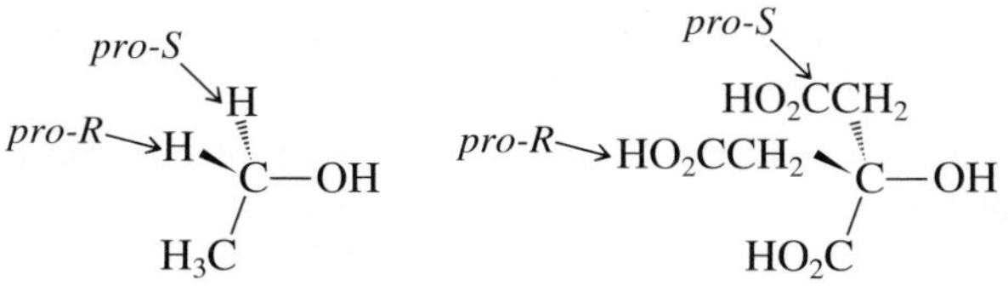

pro-S *pro-R* Ethanol *pro-S* *pro-R* Citric acid

Citric acid played a major role in the development of the concept of prochirality. Its two CH_2CO_2H chains groups behave differently in a key step of the Krebs cycle, so differently that some wondered whether citric acid itself were really involved. Alexander Ogston (Oxford) provided the answer in 1948 when he pointed out that the two CH_2CO_2H groups are differentiated when citric acid interacts with the chiral environment of an enzyme.

The two prochiral faces of a trigonal atom C(x,y,z) are enantiotopic and designated *Re* and *Si* according to whether x, y, and z trace a clockwise (*Re*) or counterclockwise (*Si*) path in order of decreasing Cahn–Ingold–Prelog precedence. An acetaldehyde molecule that lies in the plane of the paper, for example, presents either the *Re* or *Si* face according to how it is oriented.

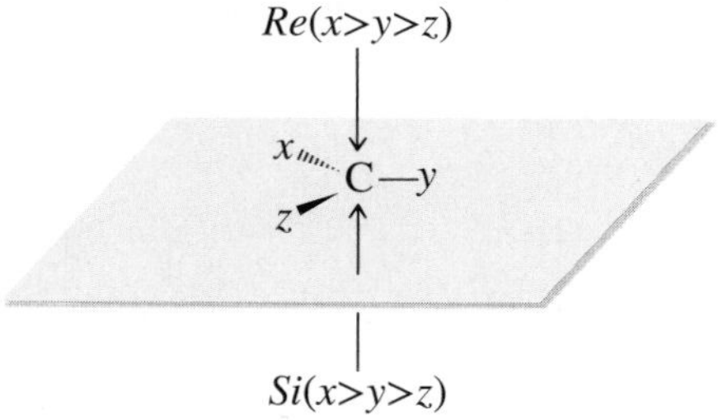

$Re(x>y>z)$

$Si(x>y>z)$

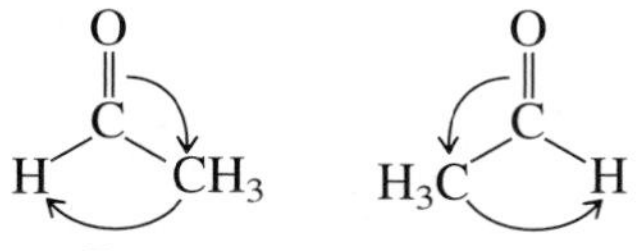

Re Acetaldehyde *Si*

The stereochemical aspects of many enzyme-catalyzed reactions have been determined. The enzyme *alcohol dehydrogenase* catalyzes the oxidation of ethanol to acetaldehyde by removing the *pro-R* hydrogen (abbreviated as H_R). When the same enzyme catalyzes the reduction of acetaldehyde to ethanol, hydrogen is transferred to the *Re* face.

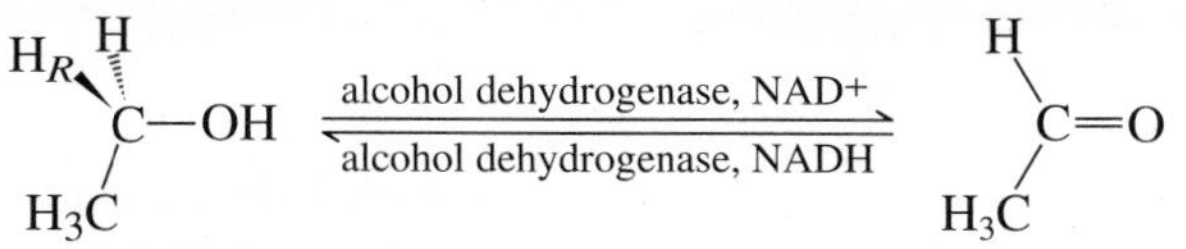

alcohol dehydrogenase, NAD+ / alcohol dehydrogenase, NADH

Ethanol Acetaldehyde

4.45 Which molecule is prochiral?

A. Ethane
B. Propane
C. Butane
D. Cyclopropane

4.46 How many of the carbons in 2-methylpentane [$(CH_3)_2CHCH_2CH_2CH_3$] are prochirality centers?

A. One
B. Two
C. Three
D. Four

4.47 What are the *pro-R* and *pro-S* designations for the enantiotopic hydrogens in 1-propanol?

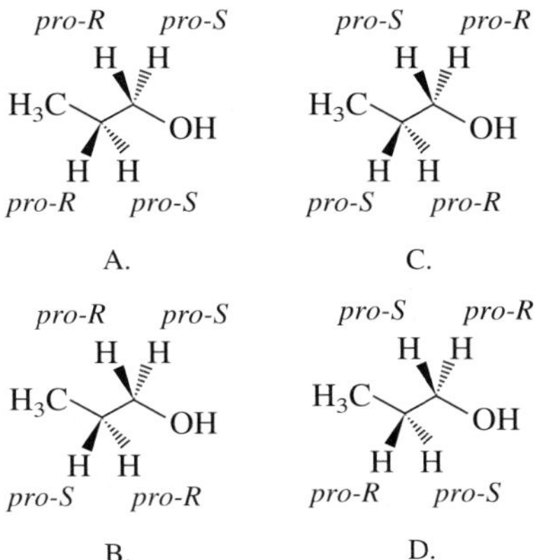

4.48 The enzyme fumarase catalyzes the addition of water to the double bond of fumaric acid.

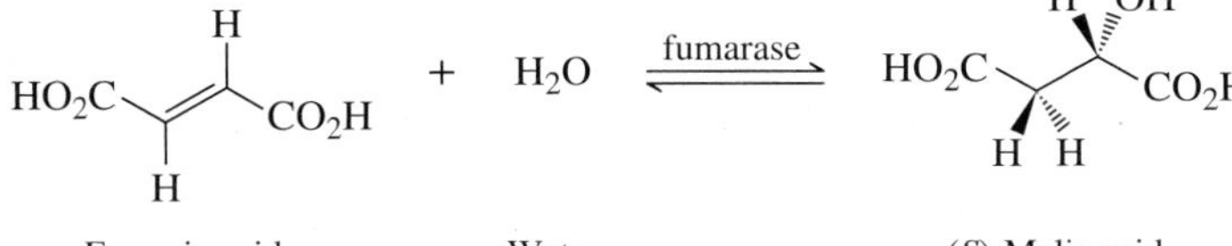

The —OH group and the *pro-R* hydrogen of the CH_2 group of (*S*)-(−)malic acid come from water. What stereochemical pathway describes the addition of water to the double bond?

A. syn Addition
B. anti Addition

4.49 To which prochiral face of the double bond of fumaric acid does the —OH group add to in the fumarase-catalyzed hydration of fumaric acid described in the preceding problem?

A. *Re*
B. *Si*

4.50 A method for the stereoselective synthesis of chiral epoxides gave the product shown in high enantiomeric excess. To which faces of the doubly bonded carbons is oxygen transferred?

A. *Re Re*
B. *Re Si*
C. *Si Si*
D. *Si Re*

4.51 When the achiral dione shown (below) was incubated in water with baker's yeast, reduction of one of the C=O groups occurred to give a single stereoisomer of the product. This product corresponded to hydrogen transfer to the *Re* face of the *pro-R* carbonyl group. Which product is this?

Achiral dione

A.
B.
C.
D.

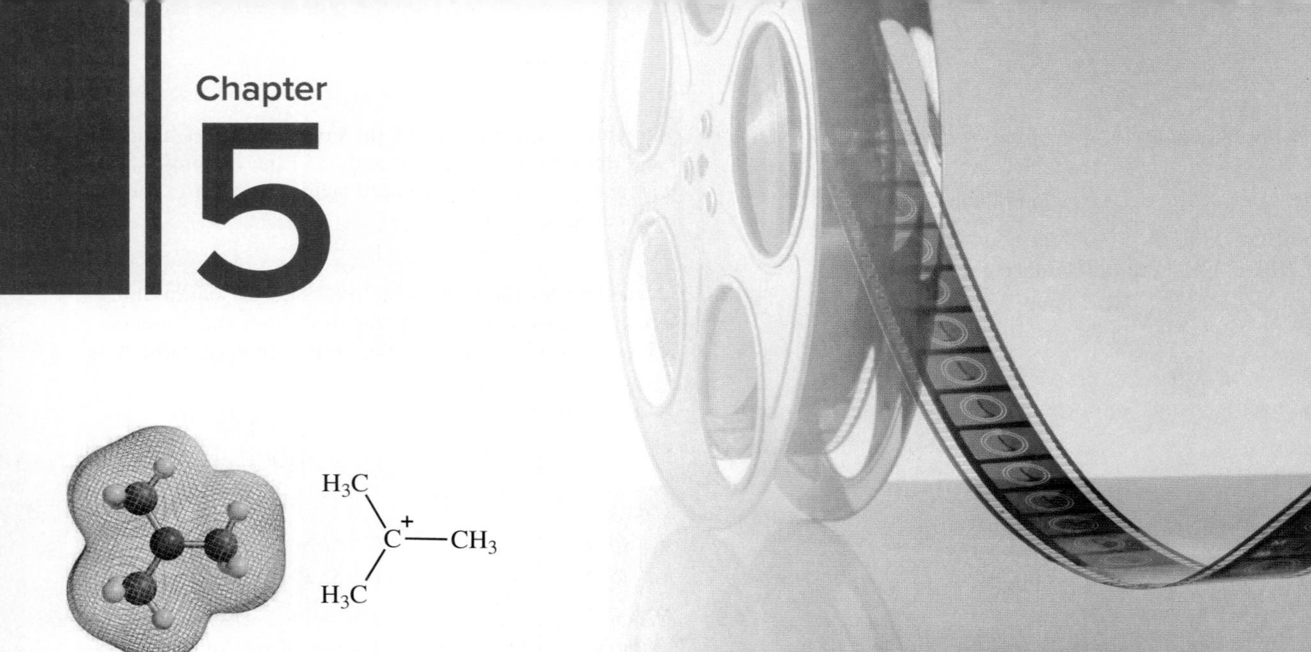

CHAPTER OUTLINE

As a motion picture tells a story, so too a mechanism tells us how a chemical reaction takes place. If we could look at a reaction "frame-by-frame," as we can with a film reel, we would observe intermediates and transition states that appear during it.

Alcohols and Alkyl Halides: Introduction to Reaction Mechanisms

Our first four chapters established some fundamental principles concerning the *structure* of organic molecules and introduced the connection between structure and *reactivity* with a review of acid–base reactions in Sections 1.12–1.15. In the present chapter, we'll explore the relation between structure and reactivity in more detail by developing two concepts: *functional groups* and *reaction mechanisms.* A **functional group** is the atom or group in a molecule most responsible for the reaction the compound undergoes under a prescribed set of conditions. *How* the structure of the reactant is transformed to that of the product is what we mean by the reaction **mechanism.**

Organic compounds are grouped into families according to the functional groups they contain. Two of the most important families are **alcohols** and **alkyl halides;** both of which are versatile starting materials for preparing numerous other families and *will appear in virtually all of the remaining chapters of this text.*

The major portion of the present chapter concerns the conversion of alcohols to alkyl halides by reaction with hydrogen halides:

$$\underset{\text{Alcohol}}{R{-}OH} + \underset{\text{Hydrogen halide}}{H{-}X} \longrightarrow \underset{\text{Alkyl halide}}{R{-}X} + \underset{\text{Water}}{H{-}OH}$$

It is convenient in equations such as this to represent generic alcohols and alkyl halides as ROH and RX, respectively, where **"R"** stands for an alkyl group. In addition to convenience, this notation lets us focus more clearly on the functional-group transformation; the OH functional group of an alcohol is replaced by a halogen, such as chlorine (X = Cl) or bromine (X = Br).

While developing the connections between structure, reaction, and mechanism, we will also extend the fundamentals of IUPAC nomenclature to functional-group families, beginning with alcohols and alkyl halides.

5.1 Functional Groups

The families of hydrocarbons—*alkanes, alkenes, alkynes,* and *arenes*—were introduced in Section 2.1. The double bond is a functional group in an alkene, the triple bond a functional group in an alkyne, and the benzene ring itself is a functional group in an arene. Alkanes (RH) are not considered to have a functional group, although as we'll see in later chapters, reactions that replace a hydrogen atom can take place. In general though, hydrogen atoms of alkanes are relatively unreactive and any other group attached to the hydrocarbon framework will be the functional group.

Table 5.1 lists the major families of organic compounds covered in this text and their functional groups.

Problem 5.1

(a) Write a structural formula for a sulfide having the molecular formula C_3H_8S.

(b) What two thiols have the molecular formula C_3H_8S?

Sample Solution (a) According to Table 5.1, sulfides have the general formula RSR and the Rs may be the same or different. The only possible connectivity for a sulfide with three carbons is C—S—C—C. Therefore, the sulfide is $CH_3SCH_2CH_3$.

Problem 5.2

A footnote to Table 5.1 states "The example given is a *primary* amine (RNH_2). *Secondary* amines have the general structure R_2NH; *tertiary* amines are R_3N." Eight constitutionally isomeric amines have the molecular formula $C_4H_{11}N$. Write their structural formulas and classify each as a primary, secondary, or tertiary amine as appropriate.

We have already touched on some of these functional-group families in our discussion of acids and bases. We have seen that alcohols resemble water in pK_a and that carboxylic acids, although weak acids, are stronger acids than alcohols. Carboxylic acids belong to one of the most important classes of organic compounds—those that contain carbonyl groups (C=O). They and other carbonyl-containing compounds rank among the most abundant and biologically significant naturally occurring substances. In this chapter we focus our attention on two classes of organic compounds listed in Table 5.1: alkyl halides and alcohols.

Carbonyl group chemistry is discussed in a block of four chapters (Chapters 18–21).

Problem 5.3

Many compounds contain more than one functional group. Elenolic acid is obtained from olive oil and contains three carbonyl groups. Classify each type according to Table 5.1. Identify the most acidic proton in elenolic acid and use Table 1.8 to estimate its pK_a.

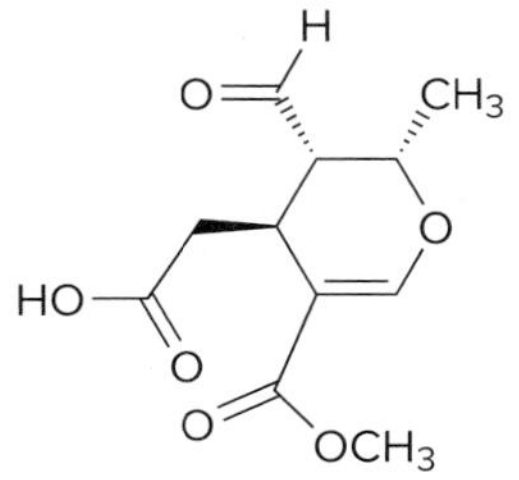

H
O
CH_3
O
HO
O
O
OCH_3

TABLE 5.1 Functional Groups in Some Important Classes of Organic Compounds

Class	Generalized abbreviation*	Representative example	Name of example[†]
Alcohol	ROH	CH_3CH_2OH	Ethanol
Alkyl halide	RCl	CH_3CH_2Cl	Chloroethane
Amine[‡]	RNH_2	$CH_3CH_2NH_2$	Ethanamine
Epoxide	$R_2C{-}CR_2$ (bridged by O)	$H_2C{-}CH_2$ (bridged by O)	Oxirane
Ether	ROR	$CH_3CH_2OCH_2CH_3$	Diethyl ether
Nitrile	RC≡N	$CH_3CH_2C{\equiv}N$	Propanenitrile
Nitroalkane	RNO_2	$CH_3CH_2NO_2$	Nitroethane
Sulfide	RSR	CH_3SCH_3	Dimethyl sulfide
Thiol	RSH	CH_3CH_2SH	Ethanethiol
Aldehyde	$\overset{O}{\overset{\|\|}{R C H}}$	CH_3CHO	Ethanal
Ketone	$\overset{O}{\overset{\|\|}{R C R}}$	$CH_3COCH_2CH_3$	2-Butanone
Carboxylic acid	$\overset{O}{\overset{\|\|}{R C OH}}$	CH_3COOH	Ethanoic acid
Carboxylic acid derivatives			
Acyl halide	$\overset{O}{\overset{\|\|}{R C X}}$	CH_3COCl	Ethanoyl chloride
Acid anhydride	RCOCR (both C=O)	$CH_3COOCOCH_3$	Ethanoic anhydride
Ester	$\overset{O}{\overset{\|\|}{R C OR}}$	$CH_3COOCH_2CH_3$	Ethyl ethanoate
Amide	$\overset{O}{\overset{\|\|}{R C NR_2}}$	CH_3CONH_2	Ethanamide

*When more than one R group is present, the groups may be the same or different.

[†] Most compounds have more than one acceptable name.

[‡] The example given is a *primary* amine (RNH_2). *Secondary* amines have the general structure R_2NH; *tertiary* amines are R_3N.

5.2 IUPAC Nomenclature of Alkyl Halides

The IUPAC rules permit certain common alkyl group names to be used. These include *n*-propyl, isopropyl, *n*-butyl, *sec*-butyl, isobutyl, *tert*-butyl, and neopentyl (Section 2.17).

According to the IUPAC rules, alkyl halides may be named in two different ways, called *functional class* nomenclature and *substitutive* nomenclature. In **functional class nomenclature** the alkyl group and the halide *(fluoride, chloride, bromide,* or *iodide)* are designated as separate words. The alkyl group is named on the basis of its longest continuous chain beginning at the carbon to which the halogen is attached.

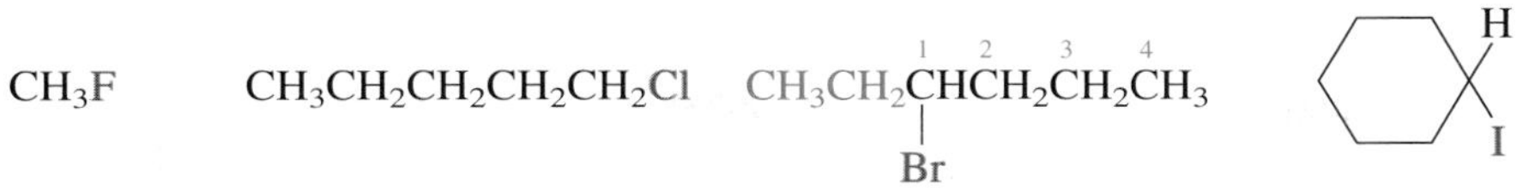

CH_3F	$CH_3CH_2CH_2CH_2CH_2Cl$	$CH_3CH_2\overset{1}{C}H(Br)\overset{2}{C}H_2\overset{3}{C}H_2\overset{4}{C}H_3$	(cyclohexane ring bearing H and I)
Methyl fluoride	Pentyl chloride	1-Ethylbutyl bromide	Cyclohexyl iodide

Substitutive nomenclature of alkyl halides treats the halogen as a *halo (fluoro-, chloro-, bromo-,* or *iodo-) substituent* on an alkane chain. The carbon chain is numbered in the direction that gives the substituted carbon the lower number.

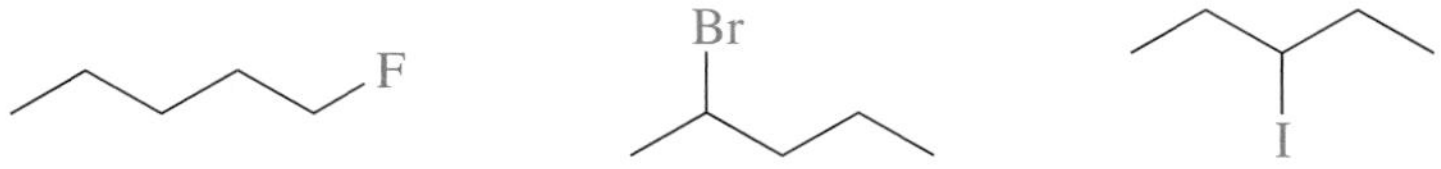

1-Fluoropentane 2-Bromopentane 3-Iodopentane

When the carbon chain bears both a halogen and an alkyl substituent, the two are considered of equal rank, and the chain is numbered so as to give the lower number to the substituent nearer the end of the chain.

5-Chloro-2-methylheptane 2-Chloro-5-methylheptane

Problem 5.4

Write structural formulas and give the functional class and substitutive names of all the isomeric alkyl chlorides that have the molecular formula C_4H_9Cl.

Substitutive names are preferred, but functional class names are sometimes more convenient or more familiar and are often encountered in organic chemistry.

Functional class names are part of the IUPAC system; they are not "common names."

5.3 IUPAC Nomenclature of Alcohols

Functional class names of alcohols are derived by naming the alkyl group that bears the hydroxyl substituent (—OH), then adding *alcohol* as a separate word. The chain is always numbered beginning at the carbon to which the hydroxyl group is attached.

Substitutive names of alcohols are developed by identifying the longest continuous chain that bears the hydroxyl group and replacing the *-e* ending of the corresponding alkane by *-ol*. The position of the hydroxyl group is indicated by number, choosing the sequence that assigns the lower locant to the carbon that bears the hydroxyl group.

The 1993 IUPAC recommendations alter the substitutive names of alcohols by bracketing the numerical locant for the substituted carbon with hyphens and placing it immediately before the *-ol* ending.

	CH_3CH_2OH	(OH on C-2 of hexane chain)	(HO on C-2 of 2-methylpentane)
Functional class name:	Ethyl alcohol	1-Methylpentyl alcohol	1,1-Dimethylbutyl alcohol
Substitutive names:	Ethanol	2-Hexanol Hexan-2-ol	2-Methyl-2-pentanol 2-Methylpentan-2-ol

Several alcohols are commonplace substances, well known by common names that reflect their origin (wood alcohol, grain alcohol) or use (rubbing alcohol). Wood alcohol is *methanol* (methyl alcohol, CH_3OH), grain alcohol is *ethanol* (ethyl alcohol, CH_3CH_2OH), and rubbing alcohol is *2-propanol* [isopropyl alcohol, $(CH_3)_2CHOH$].

Hydroxyl groups take precedence over ("outrank") alkyl groups and halogens in determining the direction in which a carbon chain is numbered. The OH group is assumed to be attached to C-1 of a cyclic alcohol.

6-Methyl-3-heptanol 6-Methylheptan-3-ol	*trans*-2-Methylcyclopentanol *trans*-2-Methylcyclopentan-1-ol	3-Fluoro-1-propanol 3-Fluoropropan-1-ol

Problem 5.5

Write structural formulas, and give the functional class and substitutive names of all the isomeric alcohols that have the molecular formula $C_4H_{10}O$.

5.4 Classes of Alcohols and Alkyl Halides

Alcohols and alkyl halides are classified as primary, secondary, or tertiary according to the degree of substitution of the carbon that bears the functional group (Section 2.17). Thus, *primary alcohols* and *primary alkyl halides* are compounds of the type RCH_2G (where G is the functional group), *secondary alcohols* and *secondary alkyl halides* are R_2CHG, and *tertiary alcohols* and *tertiary alkyl halides* are R_3CG.

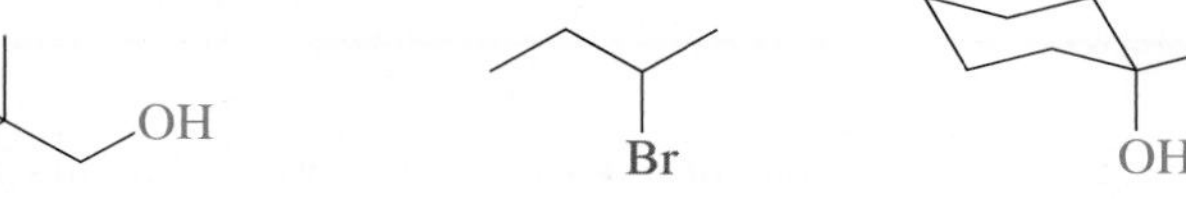

2,2-Dimethyl-1-propanol (a primary alcohol)	2-Bromobutane (a secondary alkyl halide)	1-Methylcyclohexanol (a tertiary alcohol)	2-Chloro-2-methylpentane (a tertiary alkyl halide)

Problem 5.6

Classify the isomeric $C_4H_{10}O$ alcohols as being primary, secondary, or tertiary.

Many of the properties of alcohols and alkyl halides are affected by whether their functional groups are attached to primary, secondary, or tertiary carbons. We will see a number of cases in which a functional group attached to a primary carbon is more reactive than one attached to a secondary or tertiary carbon, as well as other cases in which the reverse is true.

5.5 Bonding in Alcohols and Alkyl Halides

The carbon that bears the functional group is sp^3-hybridized in alcohols and alkyl halides. Figure 5.1 illustrates bonding in methanol. The bond angles at carbon are approximately tetrahedral, as is the C—O—H angle. A similar orbital hybridization model applies to alkyl halides, with the halogen connected to sp^3-hybridized carbon by a σ bond. Carbon–halogen bond distances in alkyl halides increase in the order C—F (140 pm; 1.40 Å) < C—Cl (179 pm; 1.79 Å) < C—Br (197 pm; 1.97 Å) < C—I (216 pm; 2.16 Å).

Carbon–oxygen and carbon–halogen bonds are polar covalent bonds, and carbon bears a partial positive charge in alcohols ($^{\delta+}C—O^{\delta-}$) and in alkyl halides ($^{\delta+}C—X^{\delta-}$).

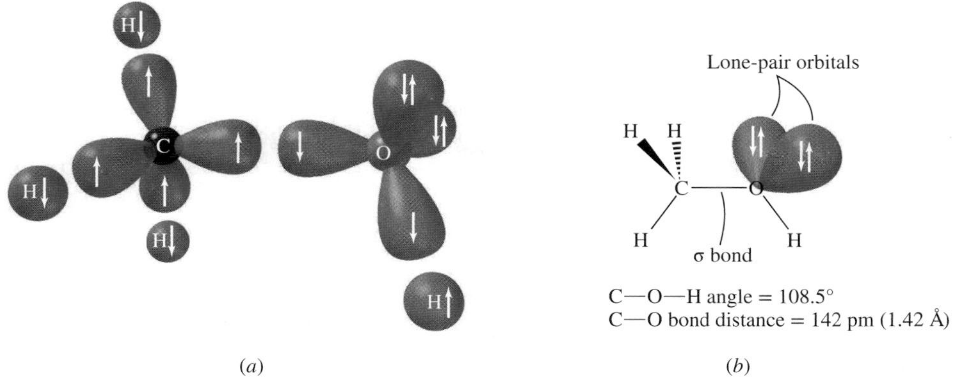

Figure 5.1

Orbital hybridization model of bonding in methanol. (*a*) The orbitals used in bonding are the 1*s* orbital of hydrogen and sp^3-hybridized orbitals of carbon and oxygen. (*b*) The bond angles at carbon and oxygen are close to tetrahedral, and the carbon–oxygen σ bond is about 10 pm (0.10 Å) shorter than a carbon–carbon single bond.

Alcohols and alkyl halides are polar molecules. The dipole moments of methanol and chloromethane are very similar to each other and to water.

$H-O-H$	H_3C-O-H	H_3C-Cl
Water	Methanol	Chloromethane
(μ = 1.8 D)	(μ = 1.7 D)	(μ = 1.9 D)

Problem 5.7

Bromine is less electronegative than chlorine, yet methyl bromide and methyl chloride have very similar dipole moments. Why?

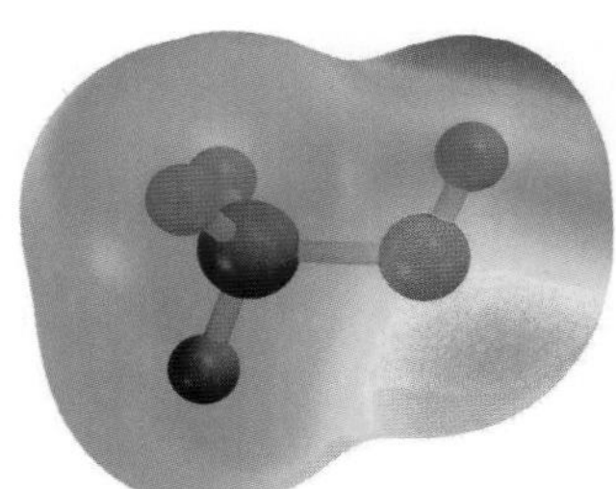

Methanol (CH_3OH)

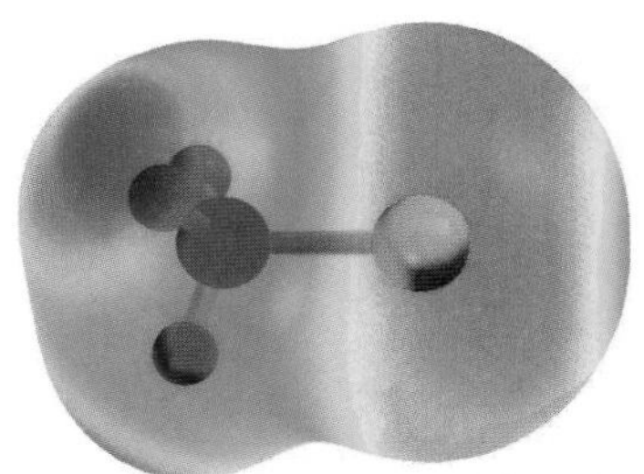

Chloromethane (CH_3Cl)

Figure 5.2

Electrostatic potential maps of methanol and chloromethane. The electrostatic potential is most negative near oxygen in methanol and near chlorine in chloromethane. The most positive region is near the O—H proton in methanol and near the methyl group in chloromethane.

Figure 5.2 maps the electrostatic potential in methanol and chloromethane. Both are similar in that the sites of highest negative potential (red) are near the electronegative atoms: oxygen and chlorine. The polarization of the bonds to oxygen and chlorine, as well as their unshared electron pairs, contribute to the concentration of negative charge on these atoms.

Relatively simple notions of attractive forces between opposite charges are sufficient to account for many of the properties of chemical substances. You will find it helpful to keep the polarity of carbon–oxygen and carbon–halogen bonds in mind as we develop the properties of alcohols and alkyl halides in later sections.

5.6 Physical Properties of Alcohols and Alkyl Halides: Intermolecular Forces

Boiling Point. When describing the effect of alkane structure on boiling point in Section 2.22, we pointed out that van der Waals attractive forces between neutral molecules are of three types.

1. Induced-dipole/induced-dipole forces (dispersion forces; London forces)
2. Dipole/induced-dipole forces
3. Dipole–dipole forces

Induced-dipole/induced-dipole forces are the only intermolecular attractive forces available to nonpolar molecules such as alkanes and are important in polar molecules as well. In addition, polar molecules also engage in dipole–dipole and dipole/induced-dipole

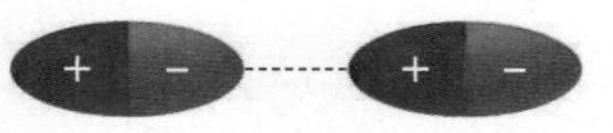

Figure 5.3

A dipole–dipole attractive force. Two molecules of a polar substance associate so that the positively polarized region of one and the negatively polarized region of the other attract each other.

attractions. The **dipole–dipole attractive force** is easiest to visualize and is illustrated in Figure 5.3. Two molecules of a polar substance experience a mutual attraction between the positively polarized region of one molecule and the negatively polarized region of the other. The **dipole/induced-dipole force** combines features of both the induced-dipole/induced-dipole and dipole–dipole attractive forces. A polar region of one molecule alters the electron distribution in a nonpolar region of another in a direction that produces an attractive force between them.

We can gain a sense of the relative importance of these intermolecular forces by considering three compounds similar in size and shape: the alkane propane, the alkyl halide fluoroethane, and the alcohol ethanol. Both of the polar compounds, ethanol and fluoroethane, have higher boiling points than the nonpolar one, propane. We attribute this to a combination of dipole/induced-dipole and dipole–dipole attractive forces that are present in the liquid states of ethanol and fluoroethane, but absent in propane.

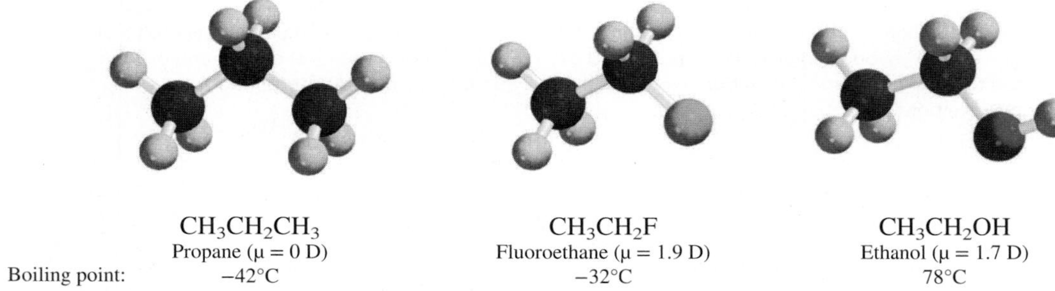

The most striking difference, however, is that despite the similarity in their dipole moments, ethanol has a much higher boiling point than fluoroethane. This suggests that the attractive forces in ethanol are unusually strong. They are an example of a special type of dipole–dipole attraction called **hydrogen bonding** and involve, in this case, the positively polarized proton of the —OH group of one ethanol molecule with the negatively polarized oxygen of another. The oxygen of the —OH group of alcohols serves as a hydrogen bond *acceptor*, while the hydrogen attached to the oxygen serves as a hydrogen bond *donor*. Having both hydrogen bond acceptor and donor capability in the same molecule creates a strong network among ethanol molecules in the liquid phase.

$$:\ddot{O}{-}\overset{\delta+}{H}\cdots\cdots\overset{\delta-}{:\ddot{O}:}{-}H$$

Figure 5.4 shows the association of two ethanol molecules to form a hydrogen-bonded complex. The proton in the hydrogen bond (O—H---O) is not shared equally between the two oxygens, but is closer to and more strongly bonded to one oxygen than the other. Typical hydrogen bond strengths are on the order of 20 kJ/mol (about 5 kcal/mol), making them some 15–20 times weaker than most covalent bonds. Extended networks of hydrogen bonds are broken when individual ethanol molecules escape from the liquid to the vapor phase, but the covalent bonds remain intact.

Among organic compounds, hydrogen bonding involves only OH or NH protons, as in:

O—H---O O—H---N N—H---O N—H---N

Hydrogen bonds between —OH groups are stronger than those between —NH groups, as a comparison of the boiling points of water (H_2O, 100°C) and ammonia (NH_3, −33°C) demonstrates.

The hydrogen must be bonded to a strongly electronegative element in order for the bond to be polar enough to support hydrogen bonding. Therefore, C—H groups do not participate in hydrogen bonds.

Problem 5.8

Write structural formulas for all the isomers of (C_3H_8O). One of these is a gas at 25°C. Which one? Why?

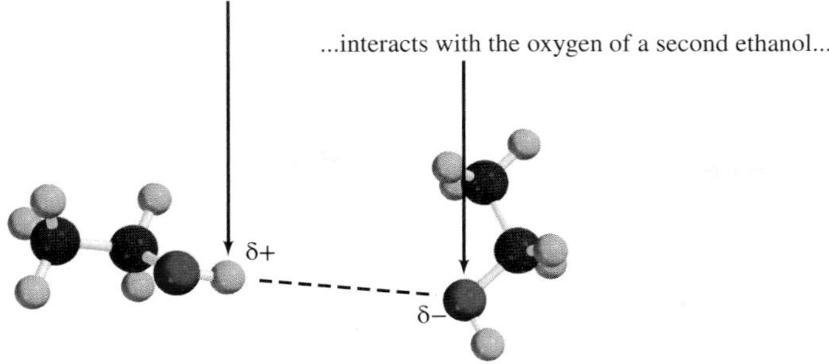
An OH proton of one ethanol molecule...
...interacts with the oxygen of a second ethanol...
δ+
δ−
...to create a hydrogen bond between the two molecules.

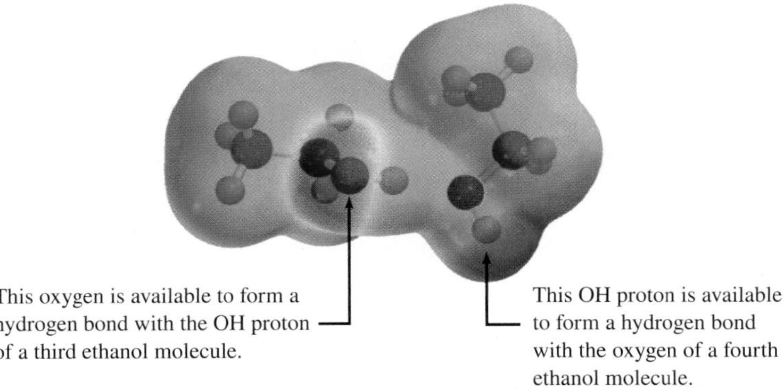
This oxygen is available to form a hydrogen bond with the OH proton of a third ethanol molecule.
This OH proton is available to form a hydrogen bond with the oxygen of a fourth ethanol molecule.

Figure 5.4

Hydrogen bonding in ethanol involves the oxygen of one molecule and the proton of the —OH group of another. A network of hydrogen-bonded complexes composed of many molecules characterizes the liquid phase of ethanol.

More than other dipole–dipole attractions, intermolecular hydrogen bonds are strong enough to impose a relatively high degree of structural order on systems in which they occur. We'll see, in Chapters 25 and 26, that the three-dimensional structures adopted by proteins and nucleic acids, the organic chemicals of life, are strongly influenced by hydrogen bonds.

Table 5.2 lists the boiling points of some representative alkyl halides and alcohols. When comparing the boiling points of related compounds as a function of the *alkyl group,* we find that the boiling point increases with the number of carbon atoms, as it does with alkanes.

The importance of hydrogen bonding in alcohols is evident in the last column of the table where it can be seen that the boiling points of alcohols are consistently higher than the corresponding alkyl fluoride, chloride, or bromide.

Among alkyl halides, the boiling point increases with increasing size of the halogen; alkyl fluorides have the lowest boiling points, alkyl iodides the highest. Induced-dipole/induced-dipole attractive forces are mainly responsible and are favored when the

TABLE 5.2 Boiling Points of Some Alkyl Halides and Alcohols

Name of alkyl group	Formula	Substituent X and boiling point, °C (1 atm)				
		X = F	X = Cl	X = Br	X = I	X = OH
Methyl	CH_3X	−78	−24	3	42	65
Ethyl	CH_3CH_2X	−32	12	38	72	78
Propyl	$CH_3CH_2CH_2X$	−3	47	71	103	97
Pentyl	$CH_3(CH_2)_3CH_2X$	65	108	129	157	138
Hexyl	$CH_3(CH_2)_4CH_2X$	92	134	155	180	157

electron cloud around an atom is easily distorted. This property of an atom is its **polarizability** and is more pronounced when the electrons are farther from the nucleus (iodine) than when they are closer (fluorine). Thus, induced-dipole/induced-dipole attractions are strongest in alkyl iodides, weakest in alkyl fluorides, and the boiling points of alkyl halides reflect this.

The boiling points of the chlorinated derivatives of methane increase with the number of chlorine atoms because the induced-dipole/induced-dipole attractive forces increase with each replacement of hydrogen by chlorine.

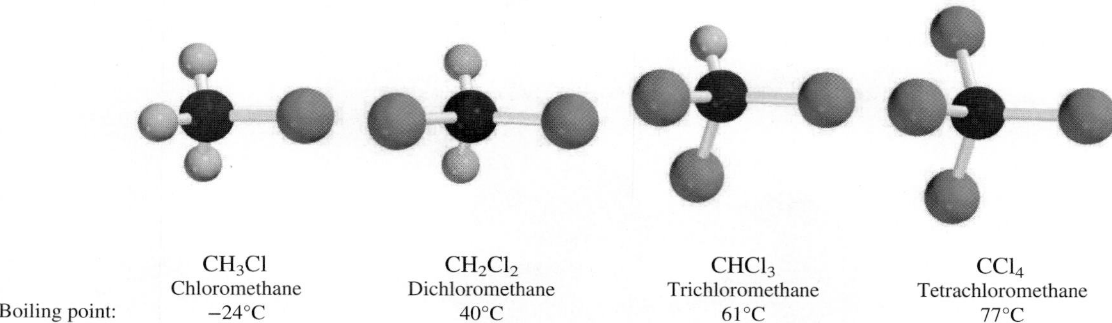

	CH_3Cl	CH_2Cl_2	$CHCl_3$	CCl_4
	Chloromethane	Dichloromethane	Trichloromethane	Tetrachloromethane
Boiling point:	−24°C	40°C	61°C	77°C

Fluorine is unique among the halogens in that increasing the number of fluorines does not lead to higher and higher boiling points.

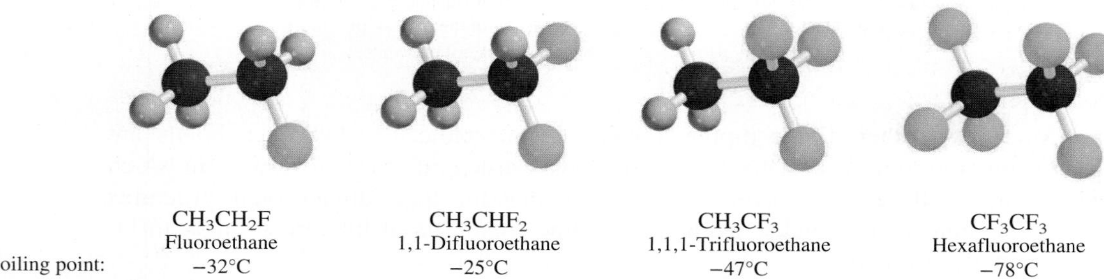

	CH_3CH_2F	CH_3CHF_2	CH_3CF_3	CF_3CF_3
	Fluoroethane	1,1-Difluoroethane	1,1,1-Trifluoroethane	Hexafluoroethane
Boiling point:	−32°C	−25°C	−47°C	−78°C

These boiling points illustrate why we should do away with the notion that boiling points always increase with increasing molecular weight.

Thus, although the difluoride CH_3CHF_2 boils at a higher temperature than CH_3CH_2F, the trifluoride CH_3CF_3 boils at a lower temperature than either of them. Even more striking is the observation that the hexafluoride CF_3CF_3 is the lowest boiling of any of the fluorinated derivatives of ethane. Its boiling point is, in fact, only 11°C higher than that of ethane itself. The reason for this has to do with the very low polarizability of fluorine and a decrease in induced-dipole/induced-dipole forces that accompanies the incorporation of fluorine substituents into a molecule. Their weak intermolecular attractive forces give fluorinated hydrocarbons certain desirable physical properties such as that found in the "no stick" *Teflon* coating of frying pans. Teflon is a *polymer* (Sections 10.8 and 27.2) made up of long chains of —CF_2CF_2— units.

Solubility in Water. Alkyl halides and alcohols differ markedly from one another in their solubility in water. All alkyl halides are insoluble in water, but low-molecular-weight alcohols (methyl, ethyl, *n*-propyl, and isopropyl) are soluble in all proportions. Their ability to participate in intermolecular hydrogen bonding not only affects the boiling points of alcohols, but also enhances their water solubility. Hydrogen-bonded networks of the type shown in Figure 5.5, in which alcohol and water molecules associate with one another, replace the alcohol–alcohol and water–water hydrogen-bonded networks present in the pure substances.

Higher alcohols become more "hydrocarbon-like" and less water-soluble. 1-Octanol, for example, dissolves to the extent of only 1 mL in 2000 mL of water. As the alkyl chain gets longer, the hydrophobic effect (Section 2.22) becomes more important, to the point that it, more than hydrogen bonding, governs the solubility of alcohols.

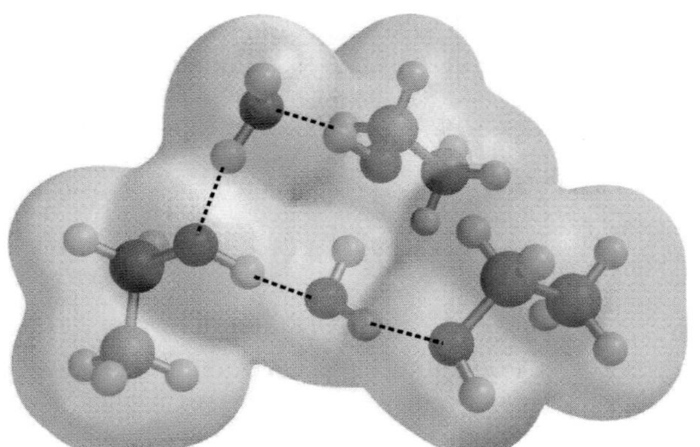

Figure 5.5

Hydrogen bonding between molecules of ethanol and water.

Density. Alkyl fluorides and chlorides are less dense, and alkyl bromides and iodides more dense, than water.

	$CH_3(CH_2)_6CH_2F$	$CH_3(CH_2)_6CH_2Cl$	$CH_3(CH_2)_6CH_2Br$	$CH_3(CH_2)_6CH_2I$
Density (20°C):	0.80 g/mL	0.89 g/mL	1.12 g/mL	1.34 g/mL

Because alkyl halides are insoluble in water, a mixture of an alkyl halide and water separates into two layers. When the alkyl halide is a fluoride or chloride, it is the upper layer and water is the lower. The situation is reversed when the alkyl halide is a bromide or an iodide. In these cases the alkyl halide is the lower layer. Polyhalogenation increases the density. The compounds CH_2Cl_2, $CHCl_3$, and CCl_4, for example, are all more dense than water.

All liquid alcohols have densities of approximately 0.8 g/mL and are less dense than water.

5.7 Preparation of Alkyl Halides from Alcohols and Hydrogen Halides

Much of what organic chemists do is directed toward practical goals. Chemists in the pharmaceutical industry synthesize new compounds as potential drugs. Agricultural chemicals designed to increase crop yields include organic compounds used for weed control, insecticides, and fungicides. Among the "building block" molecules used as starting materials to prepare new substances, alcohols and alkyl halides are especially valuable.

By knowing how to prepare alkyl halides, we can better appreciate the material in later chapters, where alkyl halides figure prominently in key functional-group transformations. *Just as important, the preparation of alkyl halides will serve to introduce some fundamental principles of reaction mechanisms.* We'll begin with the preparation of alkyl halides from alcohols by reaction with hydrogen halides according to the general equation:

$$\underset{\text{Alcohol}}{R{-}OH} + \underset{\text{Hydrogen halide}}{H{-}X} \longrightarrow \underset{\text{Alkyl halide}}{R{-}X} + \underset{\text{Water}}{H{-}OH}$$

The reaction shown is an example of a **substitution.** A halogen, usually chlorine or bromine, replaces a hydroxyl group on carbon.

The order of reactivity of the hydrogen halides parallels their acidity: HI > HBr > HCl > HF. Hydrogen iodide is used infrequently, however, and the reaction of alcohols with hydrogen fluoride is not a useful method for the preparation of alkyl fluorides.

Among the various classes of alcohols, tertiary alcohols are the most reactive and primary alcohols the least.

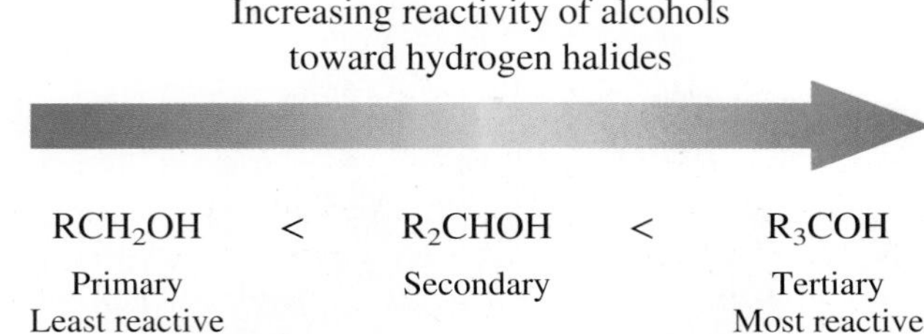

RCH_2OH < R_2CHOH < R_3COH

Primary, Least reactive — Secondary — Tertiary, Most reactive

Tertiary alcohols are converted to alkyl chlorides in high yield within minutes on reaction with hydrogen chloride at room temperature and below.

2-Methyl-2-propanol (*tert*-butyl alcohol) + HCl (Hydrogen chloride) $\xrightarrow{25°C}$ 2-Chloro-2-methylpropane (*tert*-butyl chloride) (78–88%) + H_2O (Water)

The efficiency of a synthetic transformation is normally expressed as a *percent yield,* or percentage of the theoretical yield. *Theoretical yield* is the amount of product that could be formed if the reaction proceeded to completion and did not lead to any products other than those given in the equation.

Secondary and primary alcohols do not react with HCl at rates fast enough to make the preparation of the corresponding alkyl chlorides a method of practical value. Therefore, the more reactive hydrogen halide HBr is used; even then, elevated temperatures are required to increase the rate of reaction.

Cyclohexanol + HBr (Hydrogen bromide) $\xrightarrow{80–100°C}$ Bromocyclohexane (73%) + H_2O (Water)

$CH_3(CH_2)_5CH_2OH$ + HBr $\xrightarrow{120°C}$ $CH_3(CH_2)_5CH_2Br$ + H_2O

1-Heptanol — Hydrogen bromide — 1-Bromoheptane (87–90%) — Water

The same kind of transformation may be carried out by heating an alcohol with sodium bromide and sulfuric acid.

1-Butanol $\xrightarrow[\text{heat}]{NaBr,\ H_2SO_4}$ 1-Bromobutane (90%)

We'll often write chemical equations in the abbreviated form just shown, in which reagents, especially inorganic ones, are not included in the body of the equation but instead are indicated over the arrow. Inorganic products—in this case, water—are usually omitted.

Problem 5.9

Write chemical equations for the reaction that takes place between each of the following pairs of reactants:

(a) 2-Butanol and hydrogen bromide
(b) 3-Ethyl-3-pentanol and hydrogen chloride
(c) 1-Tetradecanol and hydrogen bromide

Sample Solution (a) An alcohol and a hydrogen halide react to form an alkyl halide and water. In this case 2-bromobutane was isolated in 73% yield.

2-Butanol + HBr (Hydrogen bromide) ⟶ 2-Bromobutane + H_2O (Water)

5.8 Reaction of Alcohols with Hydrogen Halides: The S_N1 Mechanism

The reaction of an alcohol with a hydrogen halide is a **substitution.** A halogen, usually chlorine or bromine, replaces a hydroxyl group as a substituent on carbon. In addition to knowing what the reactants and products of a chemical reaction are, it is useful to evaluate the energy relationships between them **(thermodynamics)** and the pathway by which reactants become products **(mechanism)**.

With respect to thermodynamics, we've already seen how experimentally determined heats of combustion provide quantitative information concerning the relative stability of constitutional isomers (see Section 2.23) and of stereoisomers (see Section 3.11). These, along with a variety of other studies, have provided a library of thermochemical data from which the enthalpy change for formation of a particular compound from its component elements under defined conditions can be calculated. The resulting compilation of **standard enthalpy of formation** (ΔH_f°) values can then be used to calculate the enthalpy change ΔH° for a particular reaction such as that between *tert*-butyl alcohol and hydrogen chloride.

$(CH_3)_3COH(l)$	+	$HCl(g)$	$\longrightarrow$	$(CH_3)_3CCl(l)$	+	$H_2O(l)$
tert-Butyl alcohol		Hydrogen chloride		*tert*-Butyl chloride		Water
$\Delta H_f^\circ = -359$ kJ/mol (−85.8 kcal/mol)		$\Delta H_f^\circ = -92$ kJ/mol (−22.0 kcal/mol)		$\Delta H_f^\circ = -211$ kJ/mol (−50.4 kcal/mol)		$\Delta H_f^\circ = -286$ kJ/mol (−68.4 kcal/mol)

Subtracting the standard enthalpies of formation of the reactants from those of the products gives the standard enthalpy change for the reaction.

$$\Delta H^\circ_{\text{reaction}} = (-211 \text{ kJ/mol} - 286 \text{ kJ/mol}) - (-359 \text{ kJ/mol} - 92 \text{ kJ/mol}) = -46 \text{ kJ/mol}$$

$$(-50.4 \text{ kcal/mol} - 68.4 \text{ kcal/mol}) - (-85.8 \text{ kcal/mol} - 22.0 \text{ kcal/mol}) = -11 \text{ kcal/mol}$$

Energy is transferred to the surroundings, the products are of lower energy than the reactants, and the conversion of *tert*-butyl alcohol to *tert*-butyl chloride under the conditions shown is calculated to be exothermic.

For the corresponding reaction using aqueous hydrochloric acid, the standard enthalpy of formation of HCl(*aq*) is −167 kJ/mol (−39.9 kcal/mol), which makes $\Delta H^\circ_{\text{reaction}} = +29$ kJ/mol (+6.9 kcal/mol). Although this would seem to indicate that the reaction of *tert*-butyl alcohol with hydrochloric acid might not be suitable as a method for making *tert*-butyl chloride, we need to remember that concentrated hydrochloric acid is 12 *M* in HCl versus 1 *M* for the standard state used in calculations. The high HCl concentration causes the position of equilibrium to shift to the side of products and makes the reaction of *tert*-butyl alcohol with concentrated hydrochloric acid an effective method for preparing *tert*-butyl chloride.

In developing a mechanism for any reaction, we combine some basic principles of chemical reactivity with experimental observations to deduce the most likely sequence of steps. A mechanism can never be proven correct but is our best present assessment of how a reaction proceeds. If new experimental data appear that conflict with the proposed mechanism, the mechanism must be modified to accommodate them. In the absence of conflicting data, our confidence grows that our proposed mechanism is likely to be correct. The generally accepted mechanism for the reaction of *tert*-butyl alcohol with hydrogen chloride is presented as a series of three equations in Mechanism 5.1.

Problem 5.10

Adapt Mechanism 5.1 so that it applies to the preparation of *tert*-butyl chloride using aqueous hydrochloric acid according to the following equation.

$$(CH_3)_3COH + H_3O^+ + Cl^- \longrightarrow (CH_3)_3CCl + 2H_2O$$

Show each elementary step and write the structural formulas in a bond-line format showing curved arrows and all unshared electron pairs.

Mechanism 5.1

Formation of *tert*-Butyl Chloride from *tert*-Butyl Alcohol and Hydrogen Chloride

THE OVERALL REACTION:

THE MECHANISM:

Step 1: Protonation of *tert*-butyl alcohol to give an alkyloxonium ion:

Step 2: Dissociation of *tert*-butyloxonium ion to give a carbocation:

Step 3: Capture of *tert*-butyl cation by chloride ion:

Each equation in Mechanism 5.1 represents a single **elementary step,** meaning that it involves only one transition state. A particular reaction might proceed by way of a single elementary step, in which it is described as a **concerted reaction,** or by a series of elementary steps as in Mechanism 5.1. To be valid a proposed mechanism must meet a number of criteria, one of which is that the sum of the equations for the elementary steps must correspond to the equation for the overall reaction. Before we examine each step in detail, you should verify that the process in Mechanism 5.1 satisfies this requirement.

Step 1: Proton Transfer

We saw in Chapter 1, especially in Table 1.8, that alcohols resemble water in respect to their Brønsted acidity (ability to donate a proton *from oxygen*). They also resemble water in their Brønsted basicity (ability to accept a proton *on oxygen*). Just as proton transfer to a water molecule gives oxonium ion (hydronium ion, H_3O^+), proton transfer to an alcohol gives an **alkyloxonium ion** ($ROH_2{}^+$).

Recall from Section 1.11 that curved arrows indicate the *movement of electrons* in chemical reactions.

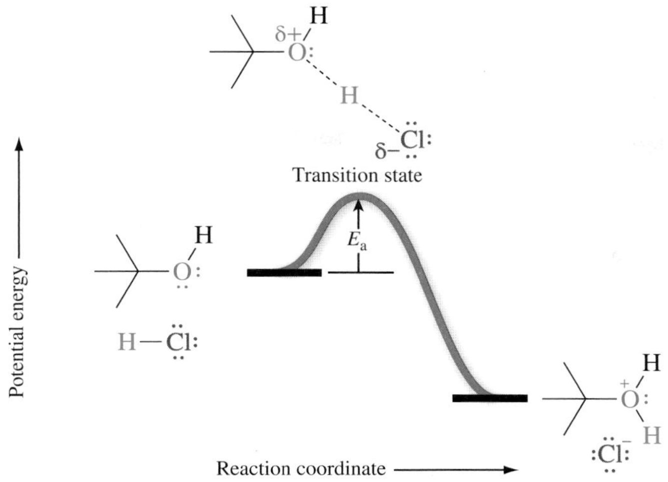

Figure 5.6

Potential energy diagram for proton transfer from hydrogen chloride to *tert*-butyl alcohol (step 1 of Mechanism 5.1).

Furthermore, a strong acid such as HCl that ionizes completely when dissolved in water, also ionizes completely when dissolved in an alcohol. Many important reactions of alcohols involve strong acids either as reactants or as catalysts. In all these reactions the first step is formation of an alkyloxonium ion by proton transfer from the acid to the alcohol.

The **molecularity** of an elementary step is given by the number of species that undergo a chemical change in that step. Transfer of a proton from hydrogen chloride to *tert*-butyl alcohol is **bimolecular** because two molecules [HCl and $(CH_3)_3COH$] undergo chemical change.

The *tert*-butyloxonium ion formed in step 1 is an **intermediate.** It was not one of the initial reactants, nor is it formed as one of the final products. Rather it is formed in one elementary step, consumed in another, and lies on the pathway from reactants to products.

Potential energy diagrams of the kind introduced in Section 3.1 are especially useful when applied to reaction mechanisms. One for proton transfer from hydrogen chloride to *tert*-butyl alcohol is shown in Figure 5.6. The potential energy of the system is plotted against the "reaction coordinate," which is a measure of the degree to which the reacting molecules have progressed on their way to products. Several aspects of the diagram are worth noting:

- Because this is an elementary step, it involves a single transition state.
- Proton transfers from strong acids to water and alcohols rank among the most rapid chemical processes and occur almost as fast as the molecules collide with one another. Thus, the height of the energy barrier E_a for proton transfer must be quite low.
- The step is known to be exothermic, so the products are placed lower in energy than the reactants.

The 1967 Nobel Prize in Chemistry was shared by Manfred Eigen, a German chemist who developed novel methods for measuring the rates of very fast reactions such as proton transfers.

The concerted nature of proton transfer contributes to its rapid rate. The energy cost of breaking the H—Cl bond is partially offset by the energy released in forming the new bond between the transferred proton and the oxygen of the alcohol. Thus, the activation energy is far less than it would be for a hypothetical two-step process in which the H—Cl bond breaks first, followed by bond formation between H^+ and the alcohol.

The species present at the transition state is not a stable structure and cannot be isolated or examined directly. Its structure is assumed to be one in which the proton being transferred is partially bonded to both chlorine and oxygen simultaneously, although not necessarily to the same extent.

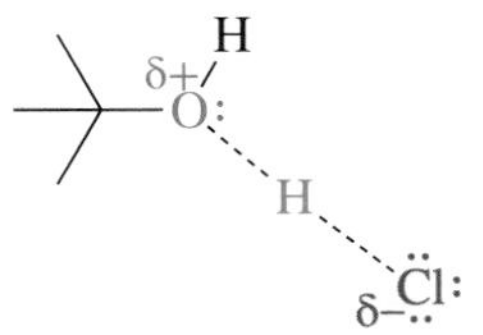

Dashed lines in transition-state structures represent *partial* bonds, that is, bonds in the process of being made or broken.

Hammond made his proposal in 1955 while at Iowa State University. He later did pioneering work in organic photochemistry at CalTech.

Inferring the structure at the transition state on the basis of the reactants and products of the elementary step in which it is involved is a time-honored practice in organic chemistry. Speaking specifically of transition states, George S. Hammond suggested that *if two states are similar in energy, they are similar in structure.* This rationale is known as **Hammond's postulate.** One of its corollaries is that the structure of a transition state more closely resembles the immediately preceding or following state to which it is closer in energy. In the case of the exothermic proton transfer in Figure 5.6, the transition state is closer in energy to the reactants and so resembles them more closely than it does the products of this step. We often call this an "early" transition state. The next step of this mechanism will provide us with an example of a "late" transition state.

Step 2: Carbocation Formation

In the second elementary step of Mechanism 5.1, the alkyloxonium ion dissociates to a molecule of water and a **carbocation,** an ion that contains a positively charged carbon.

slow

tert-Butyloxonium ion — *tert*-Butyl cation — Water

One way to name carbocations in the IUPAC system is to add the word "cation" to the name of the alkyl group.

Only one species, *tert*-butyloxonium ion, undergoes a chemical change in this step. Therefore, the step is **unimolecular.**

Like *tert*-butyloxonium ion, *tert*-butyl cation is an intermediate along the reaction pathway. It is, however, a relatively unstable species and its formation by dissociation of the alkyloxonium ion is endothermic. Step 2 is the slowest step in the mechanism and has the highest activation energy. Figure 5.7 shows a potential energy diagram for this step.

- Because this step is endothermic, the products of it are placed higher in energy than the reactants.

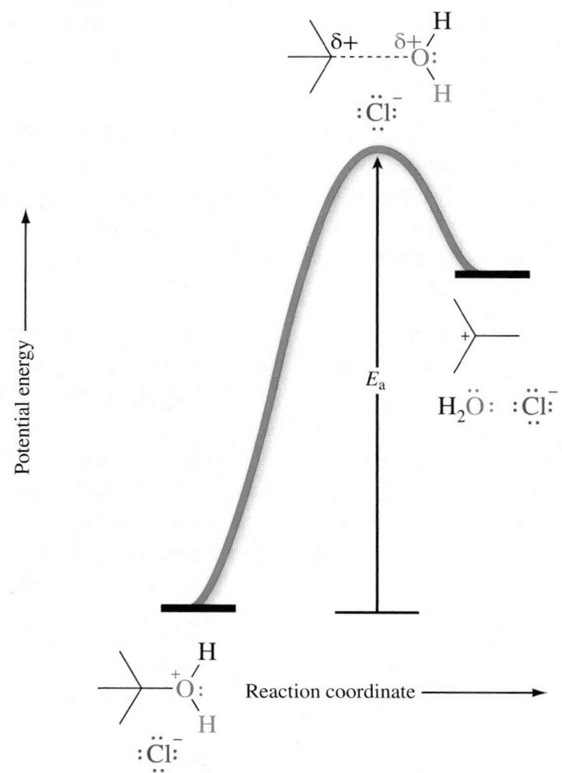

Figure 5.7

Potential energy diagram for dissociation of *tert*-butyloxonium ion to *tert*-butyl cation (step 2 of Mechanism 5.1).

- The transition state is closer in energy to the carbocation (*tert*-butyl cation), so, according to Hammond's postulate, its structure more closely resembles the carbocation than it resembles *tert*-butyloxonium ion. The transition state has considerable "carbocation character," meaning that a significant degree of positive charge has developed at carbon, and its hybridization is closer to sp^2 than sp^3.

δ+ ·········· δ+ O: (H)(H)

There is ample evidence from a variety of sources that carbocations are intermediates in many chemical reactions but are almost always too unstable to isolate. The simplest reason for the instability of carbocations is that the positively charged carbon has only six electrons in its valence shell—the octet rule is not satisfied for the positively charged carbon.

The properties of *tert*-butyl cation can be understood by focusing on its structure shown in Figure 5.8. With only six valence electrons, which are distributed among three coplanar σ bonds, the positively charged carbon is sp^2-hybridized. The unhybridized 2*p* orbital that remains on the positively charged carbon contains no electrons.

The positive charge on carbon and the vacant *p* orbital combine to make carbocations strongly **electrophilic** ("electron-loving" or "electron-seeking"). Electrophiles are Lewis acids (Section 1.15). They are electron-pair acceptors and react with Lewis bases (electron-pair donors). Step 3, which follows and completes the mechanism, is a Lewis acid/Lewis base reaction. We'll return to carbocations to describe them in more detail in Section 5.9.

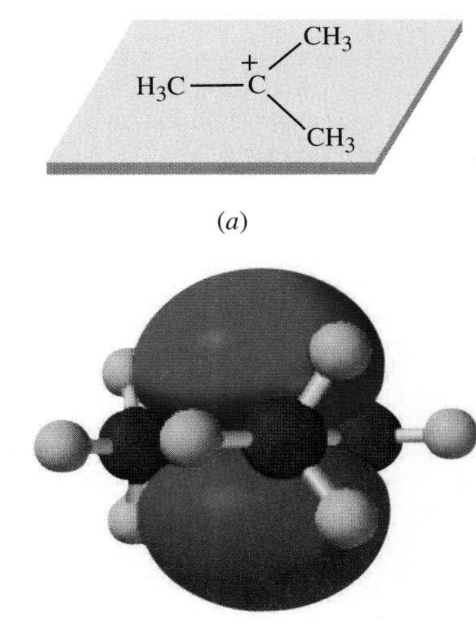

Figure 5.8

tert-Butyl cation. (*a*) The positively charged carbon is sp^2-hybridized. Each methyl group is attached to the positively charged carbon by a σ bond, and these three bonds lie in the same plane. (*b*) The sp^2-hybridized carbon has an empty 2*p* orbital, the axis of which is perpendicular to the plane of the carbon atoms.

Step 3: Reaction of tert-Butyl Cation with Chloride Ion

The Lewis bases that react with electrophiles are called **nucleophiles** ("nucleus seekers"). They have an unshared electron pair that they can use in covalent bond formation. The nucleophile in step 3 of Mechanism 5.1 is chloride ion.

tert-Butyl cation (electrophile)	+	:Cl:⁻ Chloride ion (nucleophile)	$\xrightarrow{\text{fast}}$	*tert*-Butyl chloride

Step 3 is bimolecular because two species, the carbocation and chloride ion, react together. Figure 5.9 is a potential energy diagram for this step, and Figure 5.10 shows the orbitals involved in C—Cl bond formation.

- The step is exothermic; it leads from the carbocation intermediate to the stable isolated products of the reaction.
- The activation energy for this step is small, and bond formation between a positive ion and a negative ion occurs rapidly.
- The transition state for this step involves partial bond formation between *tert*-butyl cation and chloride ion.

δ+ ·········· :Cl: δ−

Having seen how Mechanism 5.1 for the reaction of *tert*-butyl alcohol with hydrogen chloride can be supplemented with potential energy diagrams for its three elementary steps, we'll complete the picture by combining these diagrams into one that covers the entire process.

The composite diagram (Figure 5.11) has three peaks and two valleys. The peaks correspond to transition states, one for each of the three elementary steps. The valleys correspond to the reactive intermediates—*tert*-butyloxonium ion and *tert*-butyl cation—species formed in one step and consumed in another. The transition state for formation

Figure 5.9

Potential energy diagram for reaction of *tert*-butyl cation with chloride anion (step 3 of Mechanism 5.1).

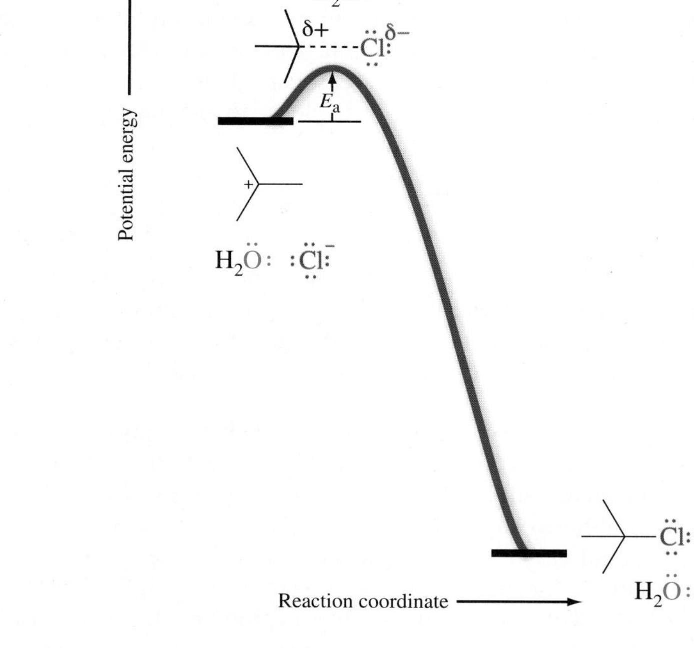

Figure 5.10

Combination of *tert*-butyl cation and chloride anion to give *tert*-butyl chloride. In-phase overlap between a vacant *p* orbital of $(CH_3)_3C^+$ and a filled *p* orbital of Cl^- gives a C—Cl σ bond.

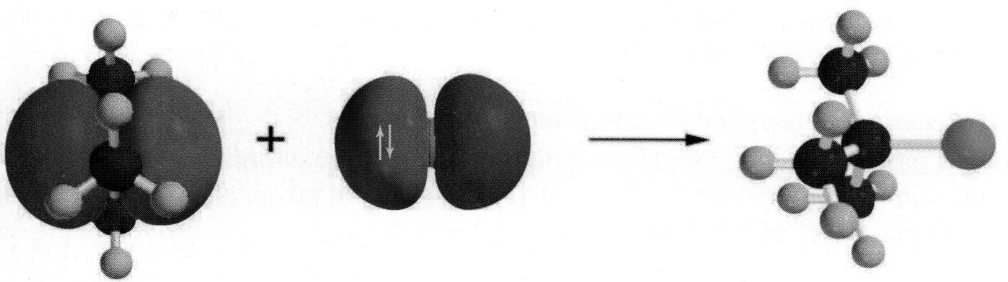

of *tert*-butyl cation from the oxonium ion is the point of highest energy on the diagram, which makes this elementary step the slowest of the three. It is called the **rate-determining** step, and the overall reaction can proceed no faster than the rate of this, its slowest step.

Substitution reactions, of which the reaction of alcohols with hydrogen halides is but one example, will be discussed in more detail in Chapter 6. There, we will make extensive use of a notation originally introduced by Sir Christopher Ingold. Ingold proposed the symbol, S_N, to stand for *substitution nucleophilic,* to be followed by the number *1* or *2* according to whether the rate-determining step is unimolecular or bimolecular. The reaction of *tert*-butyl alcohol with hydrogen chloride, for example, is said to follow an **S_N1 mechanism** because its slow step (dissociation of *tert*-butyloxonium ion) is unimolecular. Only the alkyloxonium ion undergoes a chemical change in this step.

Problem 5.11

Assume the reaction

Cyclohexanol (cyclohexyl–OH) + HBr (Hydrogen bromide) ⟶ Bromocyclohexane (cyclohexyl–Br) + H_2O (Water)

follows an S_N1 mechanism, and write a chemical equation for the rate-determining step. Use curved arrows to show the flow of electrons.

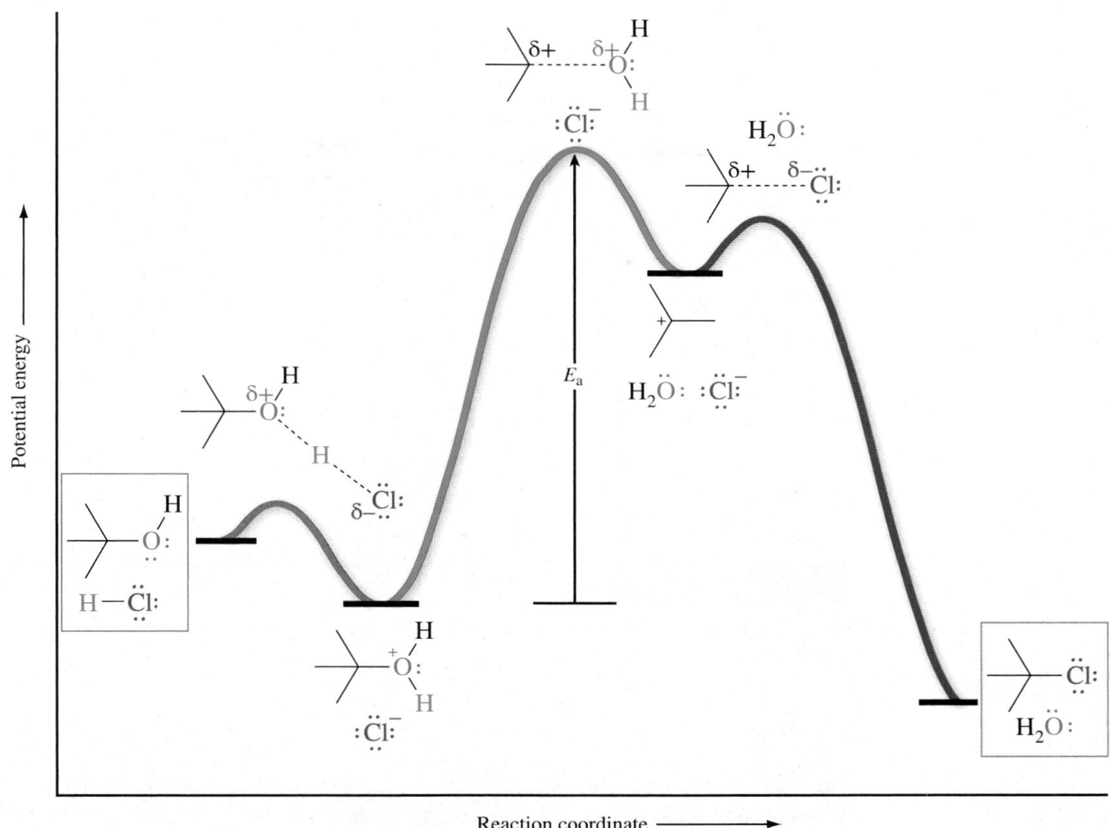

Figure 5.11

Potential energy diagram for the reaction of *tert*-butyl alcohol and hydrogen chloride according to the S_N1 mechanism (Mechanism 5.1).

5.9 Structure, Bonding, and Stability of Carbocations

As we have just seen, the rate-determining step in the reaction of *tert*-butyl alcohol with hydrogen chloride is formation of the carbocation $(CH_3)_3C^+$. Convincing evidence from a variety of sources tells us that carbocations can exist, but are relatively unstable. When carbocations are involved in chemical reactions, it is as reactive intermediates, formed slowly in one step and consumed rapidly in another.

Numerous other studies have shown that *alkyl groups directly attached to the positively charged carbon stabilize a carbocation.* Figure 5.12 illustrates this generalization for CH_3^+, $CH_3CH_2^+$, $(CH_3)_2CH^+$, and $(CH_3)_3C^+$. Among this group, CH_3^+ is the least stable and $(CH_3)_3C^+$ the most stable.

Carbocations are classified according to the degree of substitution at the positively charged carbon. The positive charge is on a primary carbon in $CH_3CH_2^+$, a secondary carbon in $(CH_3)_2CH^+$, and a tertiary carbon in $(CH_3)_3C^+$. Ethyl cation is a primary carbocation, isopropyl cation a secondary carbocation, and *tert*-butyl cation a tertiary carbocation.

As carbocations go, CH_3^+ is particularly unstable, and its existence as an intermediate in chemical reactions has never been demonstrated. Most primary carbocations, although more stable than CH_3^+, are still too unstable to be involved as intermediates in chemical reactions. The threshold of stability is reached with secondary carbocations. Many reactions, including the reaction of secondary alcohols with hydrogen halides, are believed to involve secondary carbocations. The evidence in support of tertiary carbocation intermediates is stronger yet.

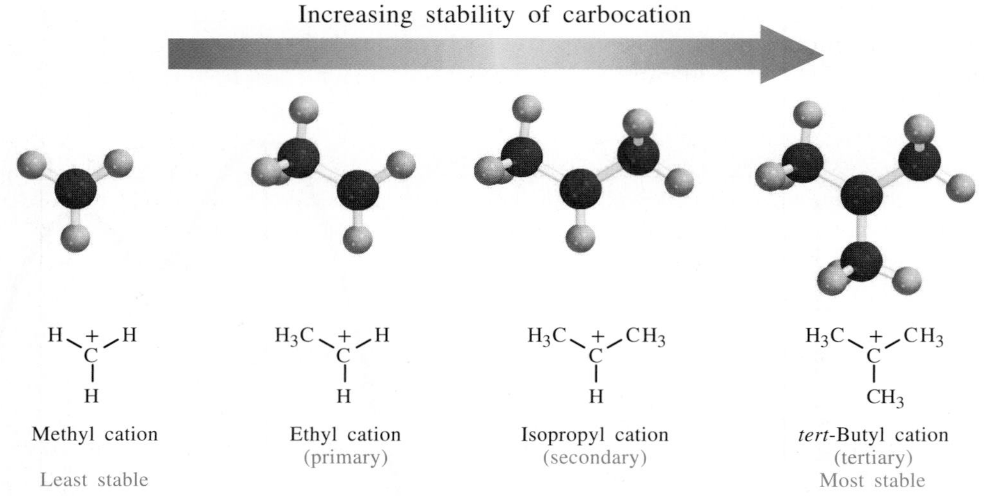

Figure 5.12

The order of carbocation stability is methyl < primary < secondary < tertiary. Alkyl groups that are directly attached to the positively charged carbon stabilize carbocations.

Problem 5.12

Carbocations are key intermediates in petroleum refining. Of particular importance is one having the carbon skeleton shown.

How many different $C_8H_{17}^+$ carbocations have this carbon skeleton? Write a bond-line formula for each and classify the carbocation as primary, secondary, or tertiary. The most stable of them corresponds to the intermediate in petroleum refining. Which one is it?

Alkyl groups stabilize carbocations by releasing electron density to the positively charged carbon, thereby dispersing the positive charge. Figure 5.13 illustrates this effect by comparing the electrostatic potential maps of CH_3^+, $CH_3CH_2^+$, $(CH_3)_2CH^+$, and $(CH_3)_3C^+$. The decreased intensity of the blue color reflects the greater dispersal of positive charge as the number of methyl groups on the positively charged carbon increases.

Dispersal of positive charge goes hand in hand with delocalization of electrons. The redistribution of negative charge—the electrons—is responsible for spreading out the positive charge. There are two main ways that methyl and other alkyl groups act as electron sources to stabilize carbocations:

- Inductive effect (by polarization of σ bonds)
- Hyperconjugation (by delocalization of electrons in σ bonds)

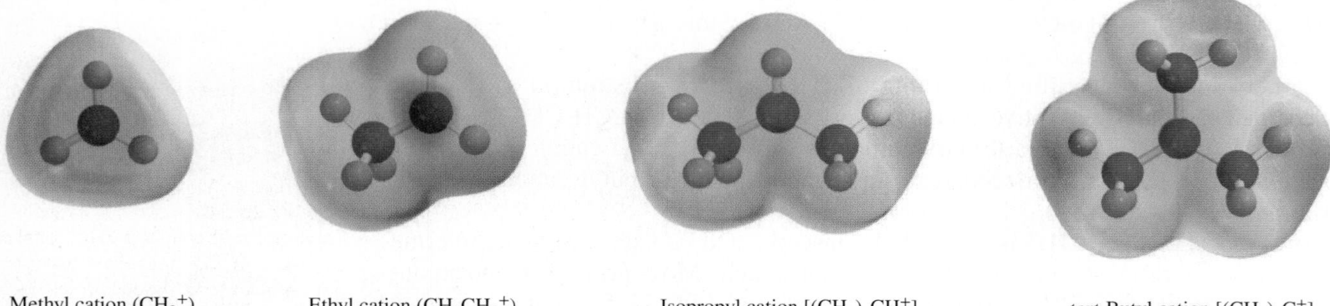

Figure 5.13

Electrostatic potential maps of carbocations. The positive charge (blue) is most concentrated in CH_3^+ and most spread out in $(CH_3)_3C^+$. (The electrostatic potentials were mapped on the same scale to allow direct comparison.)

Recall from Section 1.13 that an inductive effect is an electron-donating-or-withdrawing effect of a substituent that is transmitted by the polarization of σ bonds. As illustrated for $CH_3CH_2^+$ in Figure 5.14, the positively charged carbon draws the electrons in its σ bonds toward itself and away from the atoms attached to it. Electrons in a C—C bond are more polarizable than those in a C—H bond, so replacing hydrogens by alkyl groups reduces the net charge on the positively charged carbon. Alkyl groups are electron-releasing substituents with respect to their inductive effect. The more alkyl groups that are directly attached to the positively charged carbon, the more stable the carbocation.

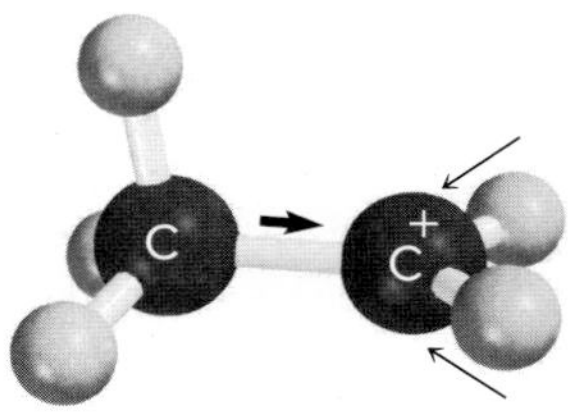

Figure 5.14

The charge in ethyl cation is stabilized by polarization of the electron distribution in the σ bonds to the positively charged carbon atom. Alkyl groups release electrons better than hydrogen. Arrows indicate direction of electron release.

Problem 5.13

Which would you expect to be more stable: $(CH_3)_3C^+$ or $(CF_3)_3C^+$? Why?

Hyperconjugation refers to the delocalization of electrons in a σ bond through a system of overlapping orbitals. Its application to carbocations such as $CH_3CH_2^+$ can be described in terms of valence bond and molecular orbital models.

The valence bond approach to hyperconjugation in $CH_3CH_2^+$ is illustrated in Figure 5.15*a*. Overlap of an orbital associated with one of the C—H σ bonds of the methyl group with the vacant *p* orbital of the positively charged carbon gives an extended orbital that encompasses both and permits the electrons in the σ bond to be shared by both carbons. The positive charge is dispersed, and the delocalized electrons feel the attractive force of both carbons.

A molecular orbital approach parallels the valence bond model. One of the filled bonding MOs of $CH_3CH_2^+$ (Figure 5.15*b*) combines a portion of the 2*p* orbital of the positively charged carbon with orbitals associated with the CH_3 group. The pair of electrons in this MO are shared by the CH_3 group and by the positively charged carbon.

When applying hyperconjugation to carbocations more complicated than $CH_3CH_2^+$, it is helpful to keep track of the various bonds. Begin with the positively charged carbon and label the three bonds originating from it with the Greek letter α. Proceed along the chain, labeling the bonds extending from the next carbon β, those from the next carbon γ, and so on.

$$\overset{\gamma}{-}\underset{\gamma}{\overset{\gamma}{C}}\overset{}{\underset{\beta}{-}}\underset{\beta}{\overset{\beta}{C}}\overset{\alpha}{-}\overset{+}{C}\begin{matrix}\alpha\\ \alpha\end{matrix}$$

Only electrons in bonds that are β to the positively charged carbon can stabilize a carbocation by hyperconjugation. Moreover, it doesn't matter whether hydrogen or another carbon is at the far end of the β bond; stabilization by hyperconjugation will still operate. The key point is that electrons in bonds that are β to the positively charged carbon are more stabilizing than electrons in an α $^+$C—H bond. Thus, successive replacement of first one, then two, then three hydrogens of CH_3^+ by alkyl groups increases the opportunities for hyperconjugation, which is consistent with the observed order of carbocation stability: $CH_3^+ < CH_3CH_2^+ < (CH_3)_2CH^+ < (CH_3)_3C^+$.

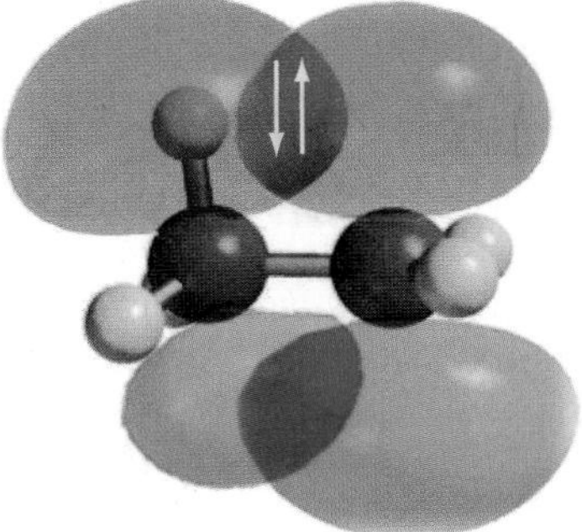

(*a*) Valence bond

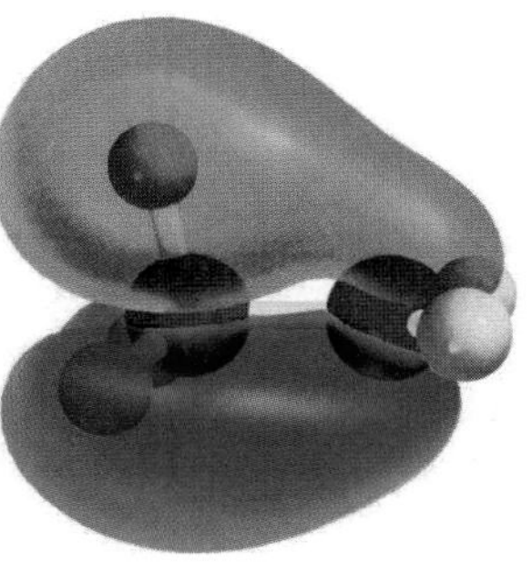

(*b*) Molecular orbital

Figure 5.15

Two views of the stabilization of $CH_3CH_2^+$ by hyperconjugation. (*a*) *Valence bond:* Overlap of the vacant 2*p* orbital of the positively charged carbon with the σ orbital of a C—H bond delocalizes the σ electrons and disperses the positive charge. (*b*) *Molecular orbital:* One of the molecular orbitals of $CH_3CH_2^+$ encompasses both the CH_3 group and the positively charged carbon; it is a bonding MO and contains two electrons.

Problem 5.14

For the general case of R = any alkyl group, how many bonded pairs of electrons are involved in stabilizing R_3C^+ by hyperconjugation? How many in R_2CH^+? In RCH_2^+?

We will see numerous reactions that involve carbocation intermediates as we proceed through the text, so it is important to understand how their structure determines their properties.

5.10 Effect of Alcohol Structure on Reaction Rate

For a proposed reaction mechanism to be valid, the sum of its elementary steps must equal the equation for the overall reaction and the mechanism must be consistent with all experimental observations. The S_N1 process set forth in Mechanism 5.1 satisfies the first criterion. What about the second?

One important experimental fact is that the rate of reaction of alcohols with hydrogen halides increases in the order primary < secondary < tertiary. This reactivity order parallels the carbocation stability order and is readily accommodated by the mechanism we have outlined.

The rate-determining step in the S_N1 mechanism is dissociation of the alkyloxonium ion to the carbocation.

$$\text{C}-\overset{+}{\ddot{\text{O}}}\text{H}_2 \longrightarrow \overset{\delta+}{\text{C}}\text{----}\overset{\delta+}{\ddot{\text{O}}}\text{H}_2 \longrightarrow \text{C}^+ + \text{:}\ddot{\text{O}}\text{H}_2$$

Alkyloxonium ion — Transition state — Carbocation — Water

The rate of this step is proportional to the concentration of the alkyloxonium ion:

$$\text{Rate} = k\left[\text{C}-\overset{+}{\ddot{\text{O}}}\text{H}_2\right]$$

The rate of any chemical reaction increases with increasing temperature. Thus the value of k for a reaction is not constant, but increases as the temperature increases.

where k is a constant of proportionality called the *rate constant.* The value of k is related to the activation energy for alkyloxonium ion dissociation and is different for different alkyloxonium ions. A low activation energy implies a large value of k and a rapid rate of alkyloxonium ion dissociation. Conversely, a large activation energy is characterized by a small k for dissociation and a slow rate.

The transition state is closer in energy to the carbocation and, according to Hammond's postulate, more closely resembles it than the alkyloxonium ion. Thus, structural features that stabilize carbocations stabilize transition states leading to them. It follows, therefore, that alkyloxonium ions derived from tertiary alcohols have a lower energy of activation for dissociation and are converted to their corresponding carbocations faster than those derived from secondary and primary alcohols. Simply put: *more stable carbocations are formed faster than less stable ones.* Figure 5.16 expresses this principle via a series of potential energy diagrams.

The S_N1 mechanism is generally accepted to be correct for the reaction of tertiary and most secondary alcohols with hydrogen halides. It is almost certainly *not* correct for methyl alcohol and primary alcohols because methyl and primary carbocations are believed to be much too unstable, and the activation energies for their formation much too high, for them to be reasonably involved. Before describing the mechanism by which methyl and primary alcohols are converted to their corresponding halides, we'll complete our discussion of the S_N1 mechanism by examining its stereochemical aspects in Section 5.11 and some novel properties of carbocations in Section 5.12.

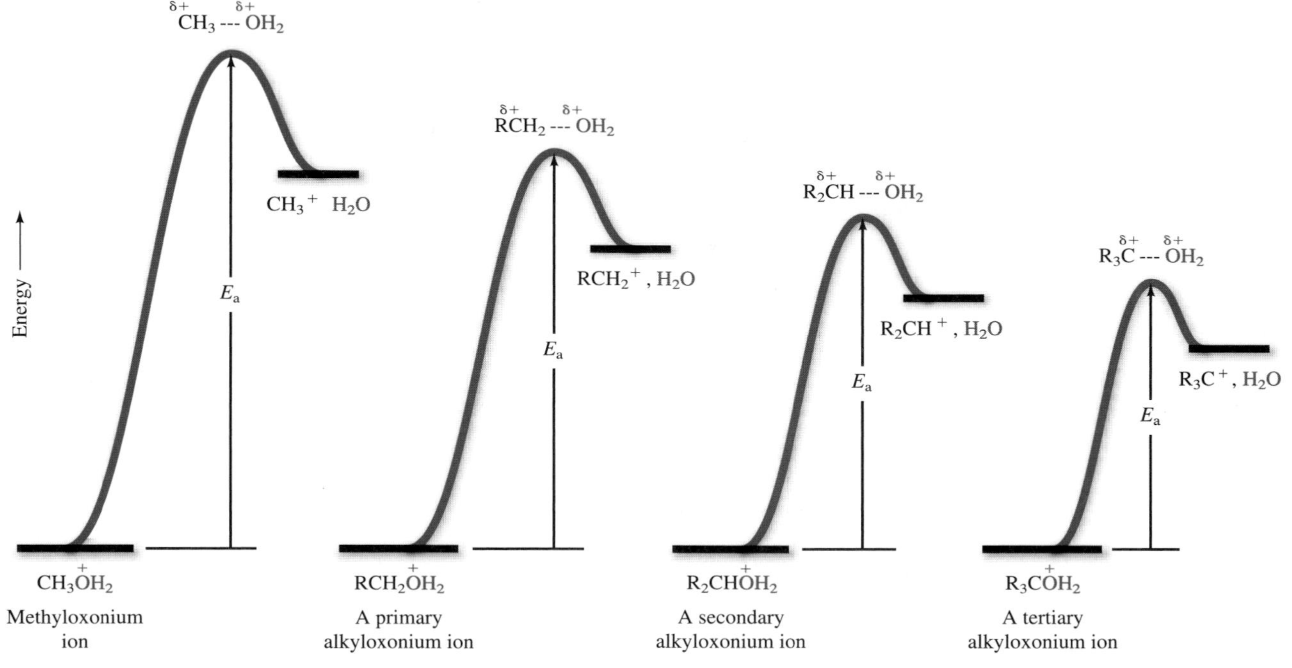

Figure 5.16

Energies of activation for formation of carbocations from alkyloxonium ions of methyl, primary, secondary, and tertiary alcohols.

5.11 Stereochemistry and the S_N1 Mechanism

When studying reactions that are believed to involve carbocations as intermediates, it is common to test this proposal by assessing the stereochemical relationship between the organic reactant and its product. Almost all the studies of the reactions of hydrogen halides with optically active alcohols have involved secondary alcohols, the case of 2-butanol being typical.

OH → HCl(*g*) → Cl

2-Butanol — 84% *S*; 16% *R*

2-Chlorobutane (96% yield) — 70% *R*; 30% *S*

Even with a reactant that is not enantiopure, the data tell a clear story. The change in configuration at the chirality center from 84% *S* in the alcohol to 70% *R* in the product is an example of a reaction that proceeds with predominant **inversion of configuration** at the chirality center. A modified version of the S_N1 mechanism is generally accepted to account for the observed stereochemistry. According to this model, dissociation of the oxonium ion intermediate to a carbocation and water remains the rate-determining step. The two of them, however, retain a close spatial orientation that favors approach of chloride from the side opposite the original C—O bond.

slow → fast → fast

With a water molecule shielding one face of the carbocation, the opposite face is more exposed to the medium and chloride bonds preferentially to that face.

Were this the only possibility, configurational inversion would be 100% and is sometimes observed. At the other extreme, the rate at which halide bonds to one face of the carbocation is the same as the rate at which it bonds to the other and the product is an optically inactive racemic mixture.

Problem 5.15

Two stereoisomeric bromides are formed in the reaction of hydrogen bromide with the 4-methyl-2-hexanol stereoisomer shown.

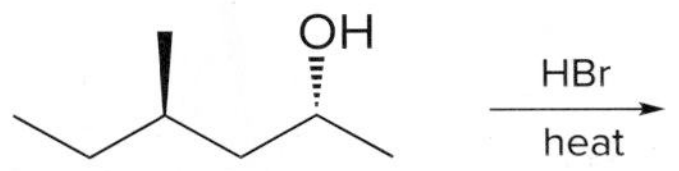

Write structural formulas showing their stereochemistry, identify the configuration of all chirality centers as *R* or *S*, and predict which stereoisomer is the major product.

In a slightly different example involving relative stereochemistry in an achiral tertiary alcohol with hydrogen chloride, both the cis and trans stereoisomers of 4-*tert*-butyl-1-methylcyclohexanol give the same two products with identical yields.

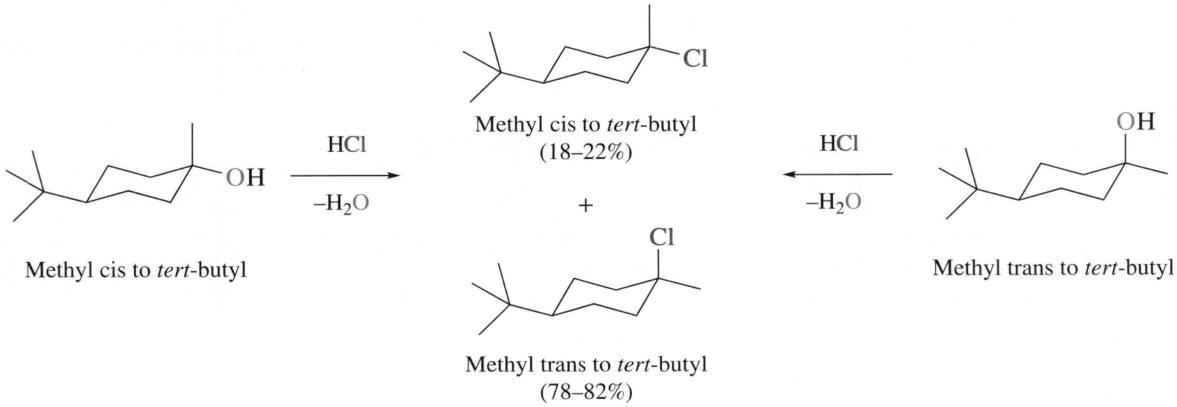

The identical yields from the different alcohols indicates chloride reacts with a common intermediate carbocation, void of any shielding from the water molecule.

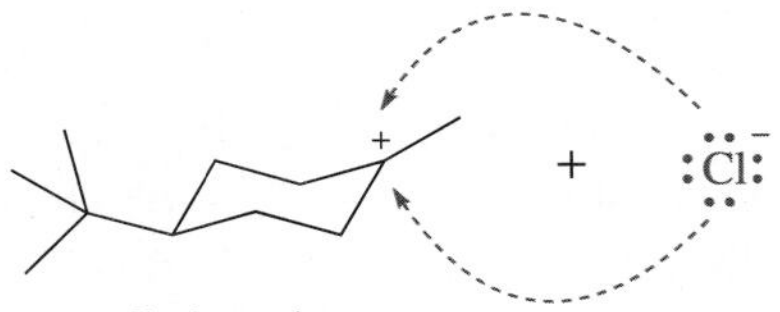

A carbocation that is not shielded by a water molecule may be expected to give equal amounts of the stereoisomeric 4-*tert*-butyl-1-methylcyclohexyl halides instead of the approximately 80:20 ratio observed. The reason for the different yields is not because of shielding of the water molecule—instead, structural features of the cyclohexyl cation ring allow for the formation of the more stable cyclohexane product with the chlorine axial. Since these two products are related as cis and trans isomers, they are diastereomers. Because one diastereomer is formed in higher yield, this is a **diastereoselective** reaction.

Reactions involving carbocation intermediates can sometimes give products having a carbon skeleton different from the starting alcohol. When and how this occurs is the topic of the next section.

5.12 Carbocation Rearrangements

A reaction is said to have proceeded with **rearrangement** when an atom or group in the reactant migrates from the atom to which it is attached and becomes bonded to another.

In the example shown:

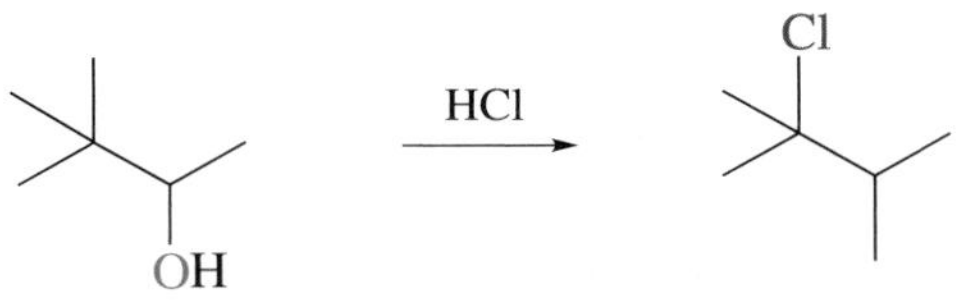

HCl

3,3-Dimethyl-2-butanol → 2-Chloro-2,3-dimethylbutane (83%)

instead of chlorine simply replacing the hydroxyl group at C-2, one of the methyl groups at C-3 has moved to C-2 and its original place at C-3 is taken by the incoming chlorine.

The generally accepted explanation for this rearrangement is outlined in Mechanism 5.2. It extends the S_N1 mechanism of Section 5.8 by introducing a new reaction path for carbocations. Not only can a carbocation react with a halide ion, it can also rearrange to a more stable carbocation prior to capture by the nucleophile.

Similar reactions called Wagner–Meerwein rearrangements were discovered over one hundred years ago. The mechanistic explanation is credited to Frank Whitmore of Penn State who carried out a systematic study of rearrangements during the 1930s.

Mechanism 5.2

Carbocation Rearrangement in the Reaction of 3,3-Dimethyl-2-butanol with Hydrogen Chloride

THE OVERALL REACTION:

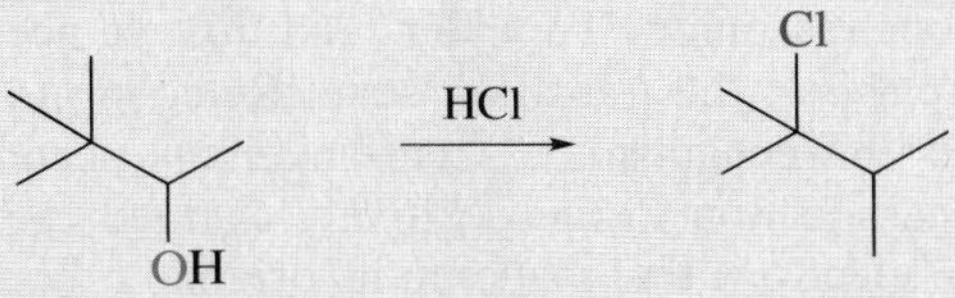

HCl

3,3-Dimethyl-2-butanol → 2-Chloro-2,3-dimethylbutane

THE MECHANISM:

Step 1: This is a proton-transfer reaction. Hydrogen chloride is the proton donor (Brønsted acid) and the alcohol is the proton acceptor (Brønsted base).

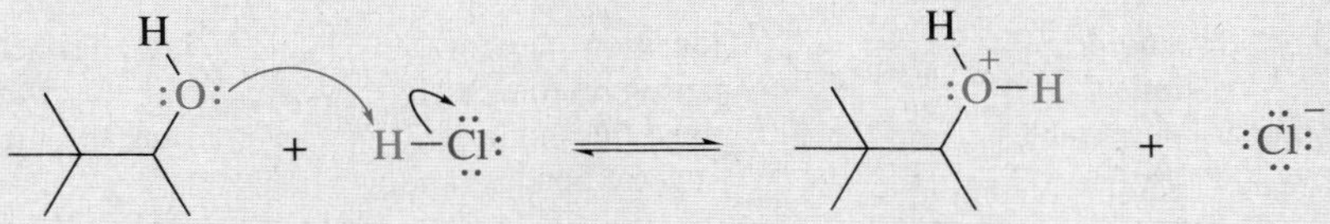

3,3-Dimethyl-2-butanol + Hydrogen chloride ⇌ 1,2,2-Trimethylpropyloxonium ion + Chloride ion

Step 2: The oxonium ion dissociates to a carbocation and water.

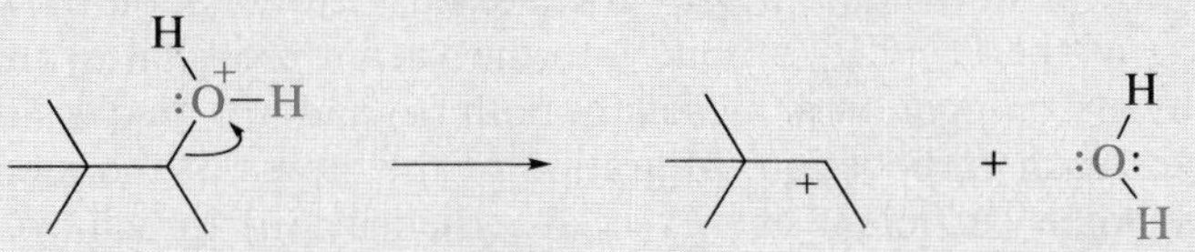

1,2,2-Trimethylpropyloxonium ion → 1,2,2-Trimethylpropyl cation + Water

continued

Step 3: The secondary carbocation formed in step 2 rearranges to a more stable tertiary carbocation. One of the methyl groups at C-3 migrates with its electron pair to C-2.

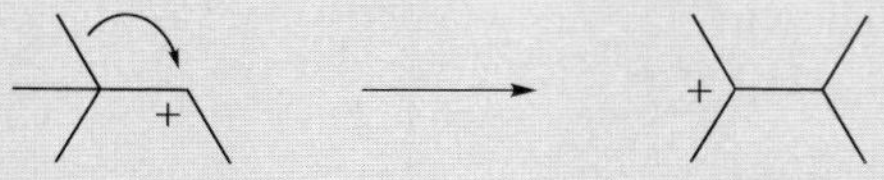

1,2,2-Trimethylpropyl cation → 1,1,2-Trimethylpropyl cation

Step 4: Chloride ion acts as a nucleophile and bonds to the positively charged carbon.

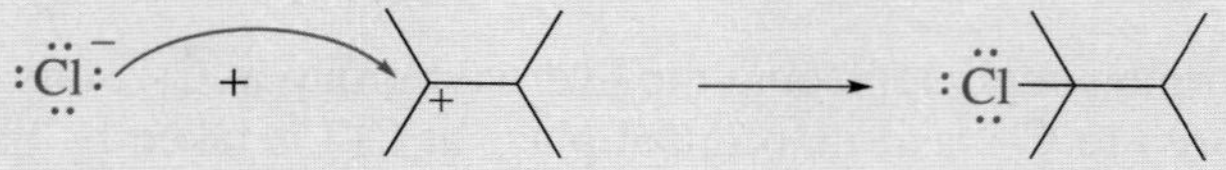

Chloride ion + 1,1,2-Trimethylpropyl cation → 2-Chloro-2,3-dimethylbutane

Why do carbocations rearrange? The answer is straightforward once we recall that tertiary carbocations are more stable than secondary ones (Section 5.9); rearrangement of a secondary to a tertiary carbocation is energetically favorable. As shown in Mechanism 5.2, the carbocation that is formed first in the dehydration of 3,3-dimethyl-2-butanol is secondary; the rearranged carbocation is tertiary. Rearrangement occurs, and the alkyl halide comes from the tertiary carbocation.

How do carbocations rearrange? To understand this we need to examine the structural change that takes place at the transition state. Referring to the initial (secondary) carbocation intermediate in Mechanism 5.2, rearrangement occurs when a methyl group shifts from C-2 of the carbocation to the positively charged carbon. The methyl group migrates with the pair of electrons that made up its original σ bond to C-2. In the curved arrow notation for this methyl migration, the arrow shows the movement of both the methyl group and the electrons in the σ bond.

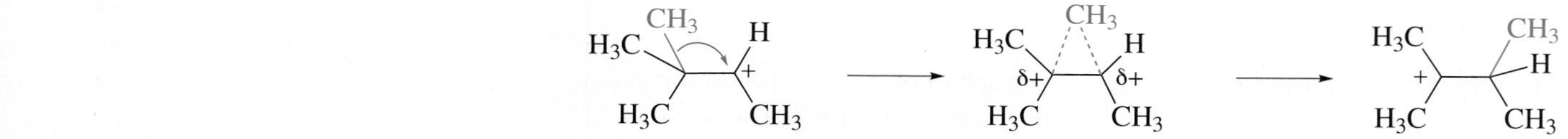

1,2,2-Trimethylpropyl cation (secondary, less stable) → Transition state for methyl migration (dashed lines indicate partial bonds) → 1,1,2-Trimethylpropyl cation (tertiary, more stable)

At the transition state for rearrangement, the methyl group is partially bonded both to its point of origin and to the carbon that will be its destination.

This rearrangement is shown in orbital terms in Figure 5.17. The relevant orbitals of the secondary carbocation are shown in structure (*a*), those of the transition state for rearrangement in (*b*), and those of the tertiary carbocation in (*c*). Delocalization of the electrons of the C—CH_3 σ bond into the vacant *p* orbital of the positively charged carbon by hyperconjugation is present in both (*a*) and (*c*), requires no activation energy, and stabilizes each carbocation. Migration of the atoms of the methyl group, however, occurs only when sufficient energy is absorbed by (*a*) to achieve the transition state (*b*). The activation energy is modest, and carbocation rearrangements are normally quite fast.

Many carbocation rearrangements involve migration of a hydrogen and are called **hydride shifts.** The same requirements apply to hydride shifts as to alkyl group migrations; they proceed in the direction that leads to a more stable carbocation; the origin

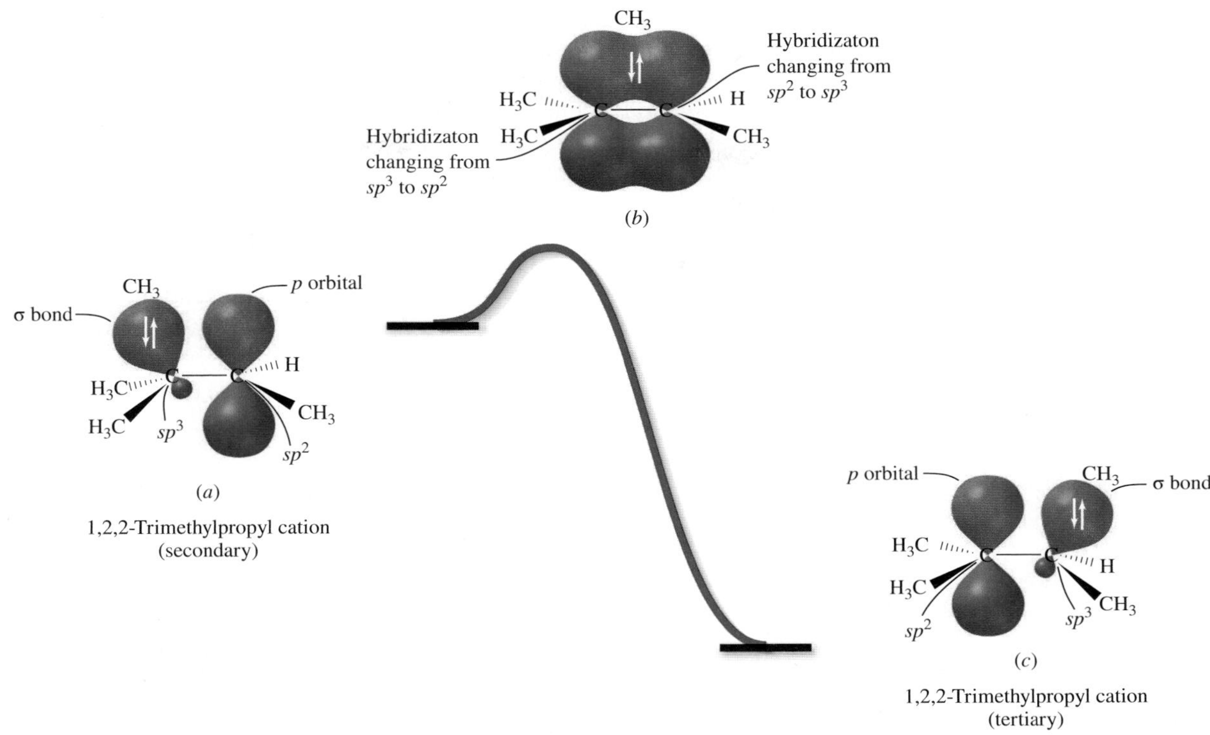

Figure 5.17

Methyl migration in 1,2,2-trimethylpropyl cation. Structure (*a*) is the initial secondary carbocation; structure (*b*) is the transition state for methyl migration; and structure (c) is the final tertiary carbocation.

and destination of the migrating hydrogen are adjacent carbons, one of which must be positively charged; and the hydrogen migrates with a pair of electrons.

OH — HCl → Cl

2-Methyl-3-pentanol → 2-Chloro-2-methylpentane (89%)

In this example, a secondary carbocation is formed initially then converted to a more stable tertiary one by migration of a hydrogen with its pair of electrons.

2-Methyl-3-pentanol —HCl→ 2-Methyl-3-pentyl cation → 2-Methyl-2-pentyl cation —Cl^-→ 2-Chloro-2-methylpentane

5.13 Reaction of Methyl and Primary Alcohols with Hydrogen Halides: The S_N2 Mechanism

Unlike tertiary and secondary carbocations, methyl and primary carbocations are too high in energy to be intermediates in chemical reactions. However, methyl and primary alcohols are converted, albeit rather slowly, to alkyl halides on treatment with hydrogen halides. Therefore, they must follow a different mechanism, one that avoids carbocation intermediates. This alternative process is outlined in Mechanism 5.3 for the reaction of 1-heptanol with hydrogen bromide.

The first step of this new mechanism is exactly the same as that seen earlier for the reaction of *tert*-butyl alcohol with hydrogen chloride—formation of an alkyloxonium ion by proton transfer from the hydrogen halide to the alcohol. Like the earlier example, this is a rapid, reversible Brønsted acid–base reaction.

The major difference between the two mechanisms is the second step. The second step in the reaction of *tert*-butyl alcohol with hydrogen chloride is the unimolecular dissociation of *tert*-butyloxonium ion to *tert*-butyl cation and water. Heptyloxonium ion, however, instead of dissociating to an unstable primary carbocation, reacts differently. It reacts with bromide ion, which acts as a nucleophile. We can represent the transition state of this displacement as:

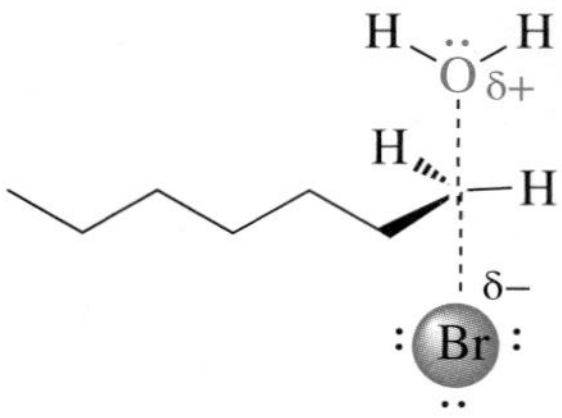

Transition state for step 2

Bromide ion forms a bond to the primary carbon by "pushing off" a water molecule. This step is bimolecular because it involves both bromide and heptyloxonium ion. Step 2 is slower than the proton transfer in step 1, so it is rate-determining. Using Ingold's terminology, we classify nucleophilic substitutions that have a bimolecular rate-determining step by the mechanistic symbol **S_N2.** In the present case, the S_N2 mechanism is favored because the reaction site, being a primary carbon, is less crowded than that of a secondary or tertiary carbon. Further, because primary carbocations are much less stable than secondary and tertiary ones, an S_N1 mechanism is too slow to compete with S_N2.

Mechanism 5.3

Formation of 1-Bromoheptane from 1-Heptanol and Hydrogen Bromide

THE OVERALL REACTION:

1-Heptanol + Hydrogen bromide (HBr) → 1-Bromoheptane + Water (H_2O)

THE MECHANISM:

Step 1: Proton transfer from hydrogen bromide to 1-heptanol to give the corresponding alkyloxonium ion:

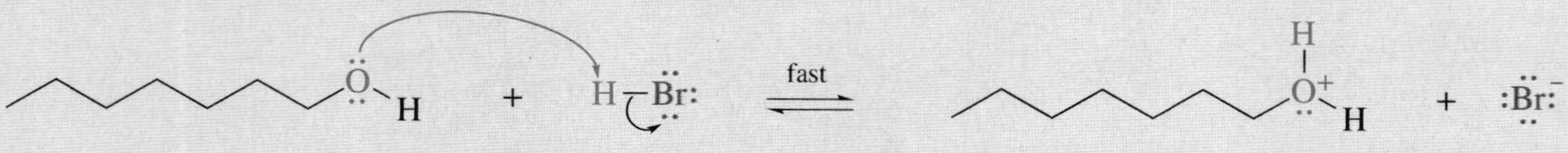

1-Heptanol + Hydrogen bromide ⇌ (fast) Heptyloxonium ion + Bromide ion

Step 2: Displacement of water from the alkyloxonium ion by bromide:

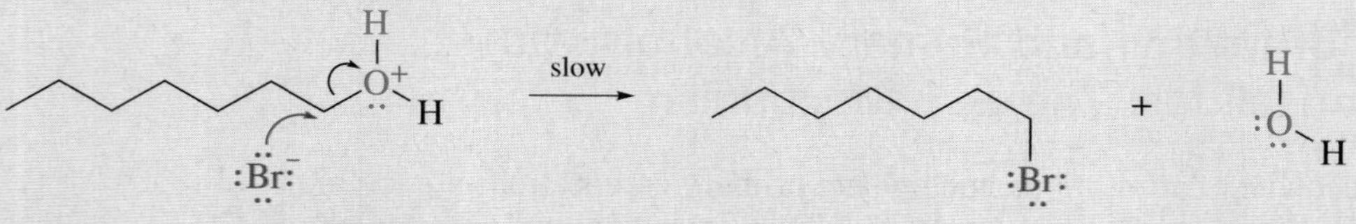

Heptyloxonium ion + bromide ion → (slow) 1-Bromoheptane + Water

Problem 5.16

Sketch a potential energy diagram for the reaction of 1-heptanol with hydrogen bromide, paying careful attention to the positioning and structures of the intermediates and transition states.

It is important to note that although methyl and primary alcohols react with hydrogen halides by a mechanism that involves fewer steps than the corresponding reactions of secondary and tertiary alcohols, fewer steps do not translate to faster reaction rates. Remember, the observed order of reactivity of alcohols with hydrogen halides is tertiary > secondary > primary. Reaction rate is governed by the activation energy of the slowest step, regardless of how many steps there are.

We described the effect of temperature on reaction rates in Section 3.1 and will examine concentration effects beginning in Section 6.3. These and other studies provide additional information that can be used to determine reaction mechanisms and a deeper understanding of how reactions occur.

5.14 Other Methods for Converting Alcohols to Alkyl Halides

Alkyl halides are such useful starting materials for preparing other functional-group types that chemists have developed several alternative methods for converting alcohols to them. Of these, those based on thionyl chloride ($SOCl_2$) and the phosphorus trihalides PCl_3 and PBr_3, bear special mention.

Thionyl chloride reacts with alcohols to give alkyl chlorides.

$$\underset{\text{Alcohol}}{\text{ROH}} + \underset{\substack{\text{Thionyl}\\\text{chloride}}}{\text{SOCl}_2} \longrightarrow \underset{\substack{\text{Alkyl}\\\text{chloride}}}{\text{RCl}} + \underset{\substack{\text{Sulfur}\\\text{dioxide}}}{\text{SO}_2} + \underset{\substack{\text{Hydrogen}\\\text{chloride}}}{\text{HCl}}$$

The reactions are typically carried out in the presence of an amine such as pyridine or triethylamine. In such cases, the amine acts as both a Brønsted base and a catalyst.

$$\underset{\text{Alcohol}}{\text{ROH}} + \underset{\substack{\text{Thionyl}\\\text{chloride}}}{\text{SOCl}_2} + \underset{\text{Triethylamine}}{(\text{CH}_3\text{CH}_2)_3\text{N}} \longrightarrow \underset{\substack{\text{Alkyl}\\\text{chloride}}}{\text{RCl}} + \underset{\substack{\text{Sulfur}\\\text{dioxide}}}{\text{SO}_2} + \underset{\substack{\text{Triethylammonium}\\\text{chloride}}}{(\text{CH}_3\text{CH}_2)_3\overset{+}{\text{N}}\text{H}\ \text{Cl}^-}$$

2-Ethyl-1-butanol $\xrightarrow[\text{pyridine}]{SOCl_2}$ 1-Chloro-2-ethylbutane (82%)

Inversion of configuration is normally observed when optically active alcohols are used.

(*S*)-3,7-Dimethyl-3-octanol $\xrightarrow[\text{triethylamine}]{SOCl_2}$ (*R*)-3-Chloro-3,7-dimethyloctane (54%)

The reaction mechanism differs according to whether the alcohol is primary, secondary, or tertiary and whether it is carried out in the presence of an amine or not. When an amine is present, we can regard it as proceeding in two stages, each of which involves more than one elementary step. In the first stage, reaction with thionyl chloride converts the alcohol to a chlorosulfite. Pyridine, shown above the arrow in the equation, acts as a catalyst in one reaction and as a weak base in another. Triethylamine can serve the same function.

Alcohol + Thionyl chloride → (pyridine) Alkyl chlorosulfite + Chloride ion

In the second stage, the alkyl halide is formed by reaction of the chlorosulfite with chloride ion. In the presence of pyridine or trimethylamine as a catalyst, two separate reactions characterize this stage.

Alkyl chlorosulfite + chloride ion → Alkyl halide + Sulfur dioxide + Chloride ion

If the starting alcohol is primary, (R′═H), carbon–chlorine bond formation and carbon–oxygen bond-breaking occur in the same step by an S_N2 process as shown. If the starting alcohol is secondary or tertiary, an S_N1 process in which a carbon–oxygen bond-breaking step precedes carbon–chlorine bond-making is involved.

Problem 5.17

For the reaction of a primary alcohol RCH_2OH with thionyl chloride in pyridine, use curved arrows to show how the last intermediate can be transformed to RCH_2Cl by reacting with chloride ion in a single step. Two additional compounds are formed. What are they?

Phosphorus trichloride and phosphorus tribromide react with alcohols to give alkyl chlorides and bromides, respectively.

$$3ROH + PX_3 \longrightarrow 3RX + P(OH)_3$$

Alcohol, Phosphorus trihalide, Alkyl halide, Phosphorous acid

2-Methyl-1-propanol →(PBr_3) 1-Bromo-2-methylpropane (55–60%)

Cyclopentanol →(PBr_3) Cyclopentyl bromide (78–84%)

An early step in the mechanism of phosphorus trihalides with alcohols is the formation of a series of intermediates of the type $ROPX_2$, $(RO)_2PX$, $(RO)_3P$, which react with the nucleophilic halide ion in subsequent steps. Chiral alcohols are converted to the corresponding alkyl halides with predominant inversion of configuration.

5.15 Sulfonates as Alkyl Halide Surrogates

As pointed out at the beginning of this chapter, alkyl halides are commonly used to prepare other functional-group families of organic compounds. On exploring the preparation of alkyl halides themselves, we emphasized the role of alcohols as their precursors and saw that the stereochemical integrity of the alcohol is not always completely maintained in the process. Usually an optically active alcohol gives an alkyl halide with net, but incomplete and varying, inversion of configuration at its chirality center. Moreover, we'll see in Chapter 6 that the customary reaction conditions invite the formation of an alkyl halide mixture containing isomers having carbon skeletons different from the original alcohol.

Organic chemists have devised alternatives to minimize these shortcomings by using alkyl sulfonates instead of alkyl halides. Sulfonates undergo many of the same kinds of reactions as alkyl halides but have the advantage that their preparation involves the oxygen of the alcohol, not the carbon to which the oxygen is attached.

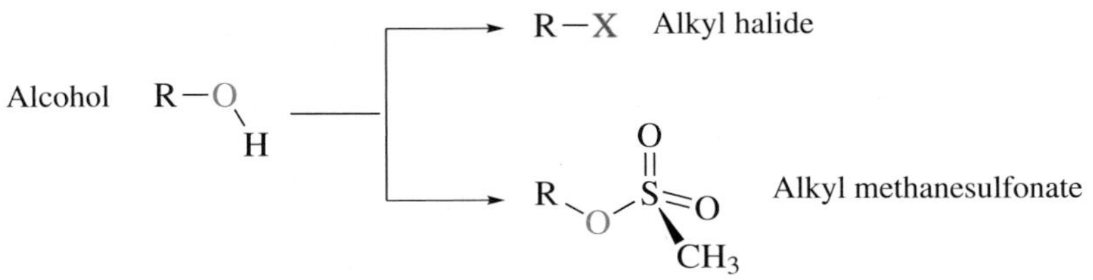

Alcohol R—O—H → R—X Alkyl halide

Alcohol R—O—H → R—O—$S(=O)_2$—CH_3 Alkyl methanesulfonate

Thus, both the carbon skeleton and stereochemistry of the original alcohol are maintained on converting it to a sulfonate. Alkyl sulfonates are prepared by the reaction of an alcohol with one of a number of sulfonyl chlorides including the following:

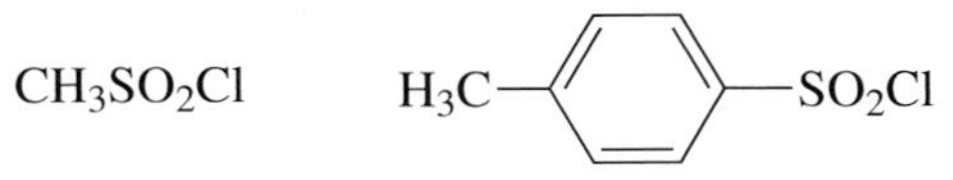

CH_3SO_2Cl — Methanesulfonyl chloride

H_3C—C_6H_4—SO_2Cl — *p*-Toluenesulfonyl chloride

The reactions are normally carried out in the presence of an amine such as trimethylamine or pyridine. The first example demonstrates the maintenance of the carbon skeleton during the process. The second shows that the stereochemical configuration at the chirality center is retained. We will discuss these reactions again in Section 6.10.

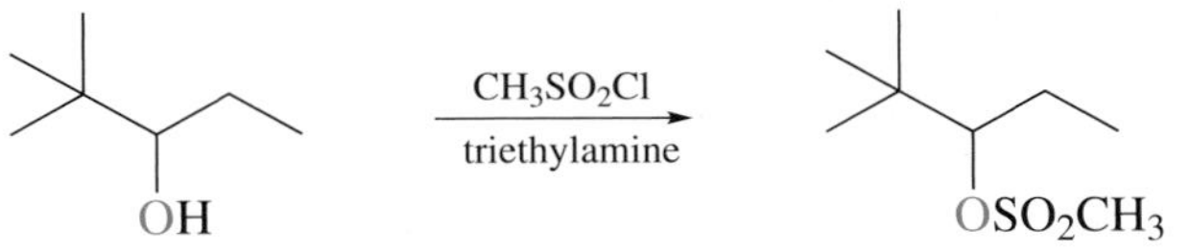

2,2-Dimethyl-3-pentanol (OH) $\xrightarrow[\text{triethylamine}]{CH_3SO_2Cl}$ 1-Ethyl-2,2-dimethylpropyl methanesulfonate (90%) (OSO_2CH_3)

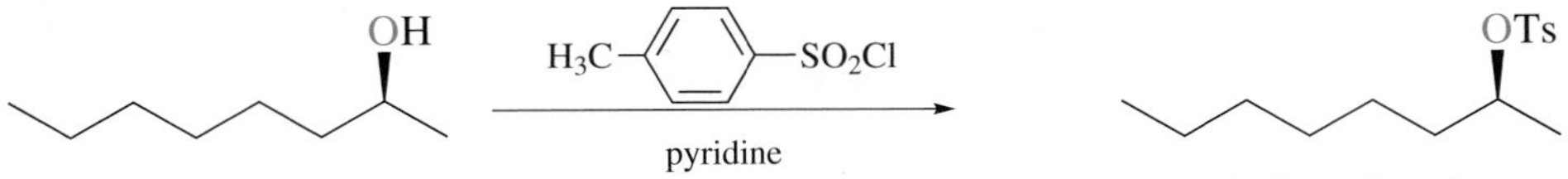

(*S*)-(+)-2-Octanol (enantiopure) (OH) $\xrightarrow[\text{pyridine}]{H_3C-C_6H_4-SO_2Cl}$ (*S*)-(+)-1-Methylheptyl *p*-toluenesulfonate (enantiopure) (OTs)

p-Toluenesulfonates are commonly referred to as "tosylates" and their compounds written as ROTs. Likewise, methanesulfonates are called "mesylates" and written as ROMs.

Problem 5.18

Write a chemical equation describing the preparation of *cis*-4-*tert*-butylcyclohexyl methanesulfonate.

5.16 SUMMARY

Chemical reactivity and functional-group transformations involving the conversion of alcohols to alkyl halides comprise the main themes of this chapter.

Section 5.1 **Functional groups** are the structural units responsible for the characteristic reactions of a molecule. The hydrocarbon chain to which a functional group is attached can often be considered as a supporting framework. The most common functional groups characterize the families of organic compounds listed on first pages of the text.

Section 5.2 Alcohols and alkyl halides may be named using either **substitutive** or **functional class** IUPAC nomenclature. In substitutive nomenclature alkyl halides are named as halogen derivatives of alkanes. The parent is the longest continuous chain that bears the halogen substituent, and in the absence of other substituents the chain is numbered from the direction that gives the lower number to the carbon that bears the halogen. The functional class names of alkyl halides begin with the name of the alkyl group and end with the halide as a separate word.

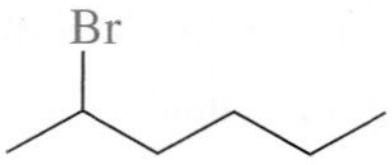

Br

Substitutive IUPAC name: 2-Bromohexane
Functional class IUPAC name: 1-Methylpentyl bromide

Section 5.3 The substitutive names of alcohols are derived by replacing the *-e* ending of an alkane with *-ol.* The longest chain containing the OH group becomes the basis for the name. Functional class names of alcohols begin with the name of the alkyl group and end in the word *alcohol.*

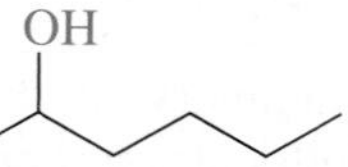

OH

Substitutive IUPAC name: 2-Hexanol or hexan-2-ol
Functional class IUPAC name: 1-Methylpentyl alcohol

Section 5.4 Alcohols (X = OH) and alkyl halides (X = F, Cl, Br, or I) are classified as primary, secondary, or tertiary according to the degree of substitution at the carbon that bears the functional group.

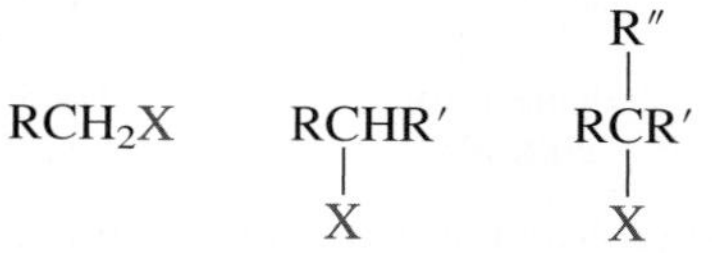

RCH_2X	$RCHR'$ with X on C	RCR' with R″ and X on C
Primary	Secondary	Tertiary

Section 5.5 The halogens (especially fluorine and chlorine) and oxygen are more electronegative than carbon, and the carbon–halogen bond in alkyl halides and the carbon–oxygen bond in alcohols are polar. Carbon is the positive end of the dipole and halogen or oxygen the negative end.

Cl   Cl δ– δ+

Section 5.6 Dipole/induced-dipole and dipole–dipole attractive forces make alcohols higher boiling than alkanes of similar molecular size. The attractive force between —OH groups is called **hydrogen bonding.**

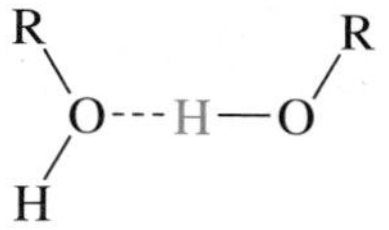

R R
O---H—O
H

Hydrogen bonding between the hydroxyl group of an alcohol and water makes the water-solubility of alcohols greater than that of hydrocarbons. Low-molecular-weight alcohols [CH_3OH, CH_3CH_2OH, $CH_3CH_2CH_2OH$, and $(CH_3)_2CHOH$] are soluble in water in all proportions. Alkyl halides are insoluble in water.

Section 5.7 See Table 5.3.

Section 5.8 Secondary and tertiary alcohols react with hydrogen halides by an S_N1 mechanism that involves formation of a carbocation intermediate in the rate-determining step.

1. $ROH + HX \overset{\text{fast}}{\rightleftharpoons} R\overset{+}{O}H_2 + X^-$
   Alcohol, Hydrogen halide, Alkyloxonium ion, Halide anion

2. $R\overset{+}{O}H_2 \xrightarrow{\text{slow}} R^+ + H_2O$
   Alkyloxonium ion, Carbocation, Water

3. $R^+ + X^- \xrightarrow{\text{fast}} RX$
   Carbocation, Halide ion, Alkyl halide

TABLE 5.3 Conversions of Alcohols to Alkyl Halides and Sulfonates

Reaction (section) and comments	General equation and specific example(s)
Reactions of alcohols with hydrogen halides (Section 5.7) Alcohols react with hydrogen halides to yield alkyl halides. The reaction is useful as a synthesis of alkyl halides. The reactivity of hydrogen halides decreases in the order HI > HBr > HCl > HF. Alcohol reactivity decreases in the order tertiary > secondary > primary.	$ROH + HX \longrightarrow RX + H_2O$ Alcohol, Hydrogen halide, Alkyl halide, Water 1-Methylcyclopentanol $\xrightarrow{HCl}$ 1-Chloro-1-methylcyclopentane (96%)
Reaction of alcohols with thionyl chloride (Section 5.14) Thionyl chloride converts alcohols to alkyl chlorides. As in the reaction of chiral alcohols with hydrogen halides, the reaction proceeds with net, but varying degrees of inversion of configuration.	$ROH + SOCl_2 \longrightarrow RCl + SO_2 + HCl$ Alcohol, Thionyl chloride, Alkyl chloride, Sulfur dioxide, Hydrogen chloride 1-Pentanol $\xrightarrow[\text{pyridine}]{SOCl_2}$ 1-Chloropentane (80%)
Reaction of alcohols with phosphorus tribromide (Section 5.14) As an alternative to converting alcohols to alkyl bromides with hydrogen bromide, the inorganic reagent phosphorus tribromide is sometimes used. Net, but incomplete, inversion of configuration is normally observed with chiral alcohols.	$3ROH + PBr_3 \longrightarrow 3RBr + H_3PO_3$ Alcohol, Phosphorus tribromide, Alkyl bromide, Phosphorous acid 2-Pentanol $\xrightarrow{PBr_3}$ 2-Bromopentane (67%)
Preparation of alkyl sulfonates (Section 5.15) Alkyl sulfonates undergo certain reactions analogous to those of alkyl halides and are often used instead of them in synthesis. They are prepared by the reaction of an alcohol with a sulfonyl chloride.	$ROH + R'SO_2Cl \xrightarrow{\text{pyridine}} ROSO_2R'$ Alcohol, A sulfonyl chloride, Alkyl sulfonate $CH_3(CH_2)_8CH_2OH + H_3C{-}C_6H_4{-}SO_2Cl$ 1-Decanol, *p*-Toluenesulfonyl chloride $\xrightarrow{\text{pyridine}} CH_3(CH_2)_8CH_2OSO_2{-}C_6H_4{-}CH_3$ Decyl *p*-toluenesulfonate (98%)

Section 5.9 Carbocations contain a positively charged carbon with only three atoms or groups attached to it. This carbon is sp^2-hybridized and has a vacant $2p$ orbital.

Carbocations are stabilized by alkyl substituents attached directly to the positively charged carbon. Alkyl groups are *electron-releasing* substituents. Stability increases in the order

(least stable) $CH_3^+ < RCH_2^+ < R_2CH^+ < R_3C^+$ (most stable)

Carbocations are strong **electrophiles** (Lewis acids) and react with **nucleophiles** (Lewis bases).

Section 5.10 The rate at which alcohols are converted to alkyl halides depends on the rate of carbocation formation: tertiary alcohols are most reactive; primary alcohols do not react. A low activation energy gives a large reaction rate.

Section 5.11 The reaction of optically active alcohols with hydrogen halides normally proceeds with predominant inversion of configuration accompanied by varying degrees of racemization. A reaction that yields one diastereomer over another is called a **diastereoselective** reaction.

Section 5.12 A complicating feature of reactions that involve carbocation intermediates is their capacity to undergo rearrangement.

Section 5.13 Primary alcohols and methanol do not react with hydrogen halides by way of carbocation intermediates. The nucleophilic species (Br^- for example) attacks the alkyloxonium ion and displaces a water molecule from carbon in a bimolecular step. This step is rate-determining, and the mechanism is S_N2.

Section 5.13 See Table 5.3.

Section 5.14 See Table 5.3.

Section 5.15 See Table 5.3.

PROBLEMS

Structure and Nomenclature

5.19 Write structural formulas for each of the following alcohols and alkyl halides:

(a) Cyclobutanol
(b) *sec*-Butyl alcohol
(c) 3-Heptanol
(d) *trans*-2-Chlorocyclopentanol
(e) 2,6-Dichloro-4-methyl-4-octanol
(f) *trans*-4-*tert*-Butylcyclohexanol
(g) 1-Cyclopropylethanol
(h) 2-Cyclopropylethanol

5.20 Name each of the following compounds according to substitutive IUPAC nomenclature:

(a) $(CH_3)_2CHCH_2CH_2CH_2Br$
(b) $(CH_3)_2CHCH_2CH_2CH_2OH$
(c) Cl_3CCH_2Br
(d) $Cl_2CHCHBr$ (with Cl on the CH carbon)
(e) CF_3CH_2OH
(f)

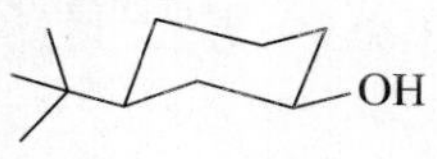

(g)

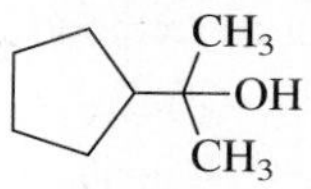

(h)

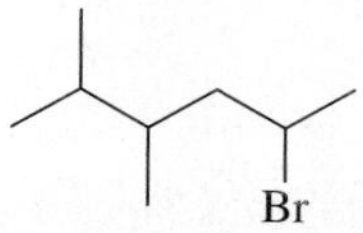

(i)

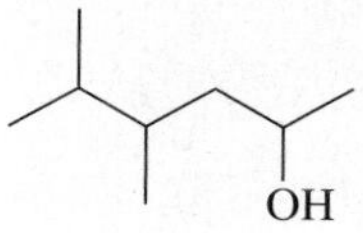

5.21 Each of the following is a *functional class name* developed according to the 1993 IUPAC recommendations. Alkyl group names of this type are derived by naming the longest continuous chain that includes the point of attachment, numbering in the direction so as to give the substituted carbon the lower number. The *-e* ending of the corresponding alkane is replaced by *-yl,* which is preceded by the number corresponding to the substituted carbon bracketed by hyphens. Write a structural formula for each alkyl halide.

(a) 6-Methylheptan-3-yl chloride

(b) 2,2-Dimethylpentan-3-yl bromide

(c) 3,3-Dimethylcyclopentan-1-yl alcohol

5.22 Write structural formulas for all the constitutionally isomeric alcohols of molecular formula $C_5H_{12}O$. Assign a substitutive and a functional class name to each one, and specify whether it is a primary, secondary, or tertiary alcohol.

5.23 A hydroxyl group is a somewhat "smaller" substituent on a six-membered ring than is a methyl group. That is, the preference of a hydroxyl group for the equatorial orientation is less pronounced than that of a methyl group. Given this information, write structural formulas for all the isomeric methylcyclohexanols, showing each one in its most stable conformation. Give the substitutive IUPAC name for each isomer.

Functional Groups

5.24 *Epichlorohydrin* is the common name of an industrial chemical used as a component in epoxy cement. The molecular formula of epichlorohydrin is C_3H_5ClO. Epichlorohydrin has an epoxide functional group; it does not have a methyl group. Write a structural formula for epichlorohydrin.

5.25 (a) Complete the structure of the pain-relieving drug *ibuprofen* on the basis of the fact that ibuprofen is a carboxylic acid that has the molecular formula $C_{13}H_{18}O_2$, X is an isobutyl group, and Y is a methyl group.

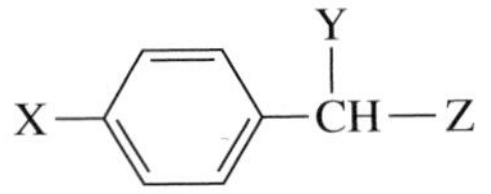

Y
X—(benzene ring)—CH—Z

(b) *Mandelonitrile* may be obtained from peach flowers. Derive its structure from the template in part (a) given that X is hydrogen, Y is the functional group that characterizes alcohols, and Z characterizes nitriles.

5.26 *Isoamyl acetate* is the common name of the substance most responsible for the characteristic odor of bananas. Write a structural formula for isoamyl acetate, given the information that it is an ester in which the carbonyl group bears a methyl substituent and there is a 3-methylbutyl group attached to one of the oxygens.

5.27 *n-Butyl mercaptan* is the common name of a foul-smelling substance obtained from skunk spray. It is a thiol of the type RX, where R is an *n*-butyl group and X is the functional group that characterizes a thiol. Write a structural formula for this substance.

5.28 Some of the most important organic compounds in biochemistry are the *α-amino acids,* represented by the general formula shown.

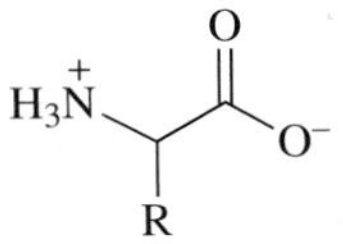

$H_3\overset{+}{N}$—CH(R)—C(=O)—O^-

Write structural formulas for the following α-amino acids.

(a) Alanine (R = methyl)

(b) Valine (R = isopropyl)

(c) Leucine (R = isobutyl)

(d) Isoleucine (R = *sec*-butyl)

(e) Serine (R = XCH_2, where X is the functional group that characterizes alcohols)

(f) Cysteine (R = XCH_2, where X is the functional group that characterizes thiols)

(g) Aspartic acid (R = XCH_2, where X is the functional group that characterizes carboxylic acids)

5.29 The compound *zoapatanol* was isolated from the leaves of a Mexican plant. Classify each oxygen in zoapatanol according to the functional group to which it belongs. If an oxygen is part of an alcohol, classify the alcohol as primary, secondary, or tertiary.

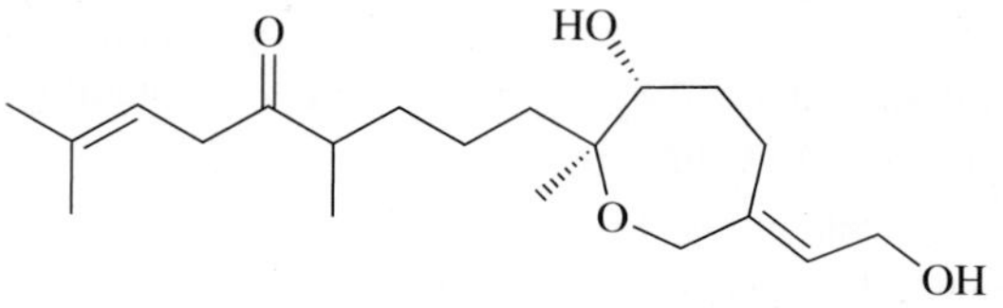

5.30 Consult Table 5.1 and classify each nitrogen-containing functional group in the anesthetic *lidocaine* according to whether it is an amide, or a primary, secondary, or tertiary amine.

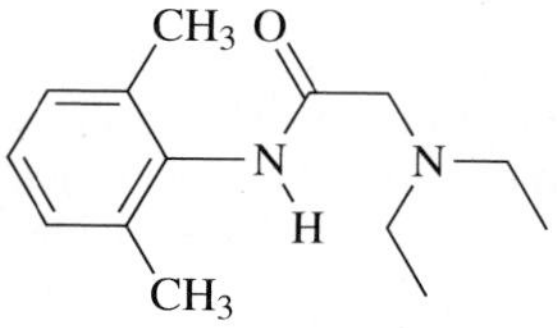

5.31 *Uscharidin* is a natural product present in milkweed. It has the structure shown. Locate all of the following in uscharidin:

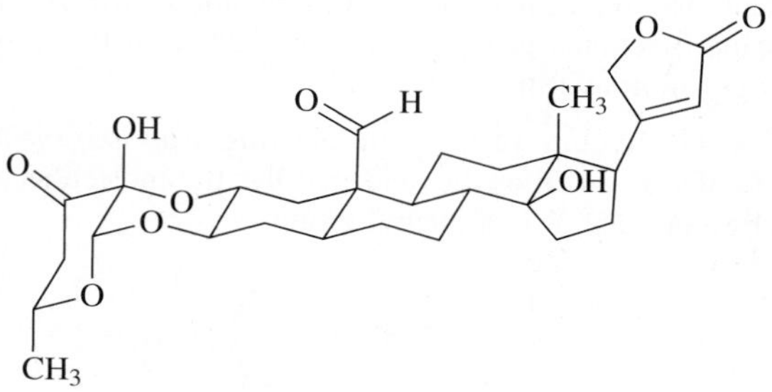

(a) Alcohol, aldehyde, ketone, and ester functional groups
(b) Methylene groups
(c) Primary carbons

Reactions and Mechanisms

5.32 Write a chemical equation for the reaction of 1-butanol with each of the following:
(a) Sodium amide ($NaNH_2$)
(b) Hydrogen bromide, heat
(c) Sodium bromide, sulfuric acid, heat
(d) Phosphorus tribromide
(e) Thionyl chloride
(f) Methanesulfonyl chloride, pyridine

5.33 Each of the following reactions has been described in the chemical literature and involves an organic reactant somewhat more complex than those we have encountered so far. Nevertheless, on the basis of the topics covered in this chapter, you should be able to write the structure of the principal organic product of each reaction.

(a) 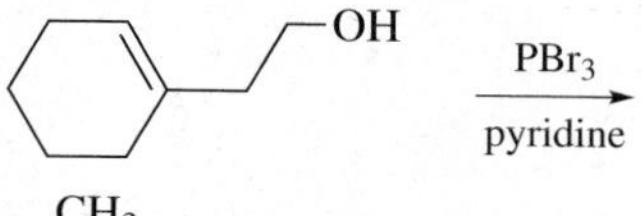 $\xrightarrow[\text{pyridine}]{PBr_3}$

(b) 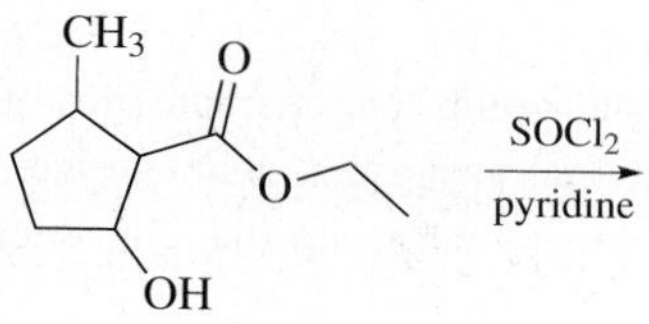 $\xrightarrow[\text{pyridine}]{SOCl_2}$

(c) [2-bromophenyl]–C(CH$_3$)$_2$OH $\xrightarrow{HCl}$

(d) HOCH$_2$CH$_2$–C$_6$H$_4$–CH$_2$CH$_2$OH + 2HBr $\xrightarrow{heat}$

(e) (CH$_3$O)$_3$C$_6$H$_2$–CH$_2$CH$_2$CH$_2$CH$_2$OH + CH$_3$–C$_6$H$_4$–SO$_2$Cl $\xrightarrow{pyridine}$

5.34 Select the compound in each of the following pairs that will be converted to the corresponding alkyl bromide more rapidly on being treated with hydrogen bromide. Explain the reason for your choice.

(a) 1-Butanol or 2-butanol

(b) 2-Methyl-1-butanol or 2-butanol

(c) 2-Methyl-2-butanol or 2-butanol

(d) 2-Methylbutane or 2-butanol

(e) 1-Methylcyclopentanol or cyclohexanol

(f) 1-Methylcyclopentanol or *trans*-2-methylcyclopentanol

(g) 1-Cyclopentylethanol or 1-ethylcyclopentanol

5.35 On reaction with hydrogen chloride, one of the trimethylcyclohexanols shown gives a single product, the other gives a mixture of two stereoisomers. Explain.

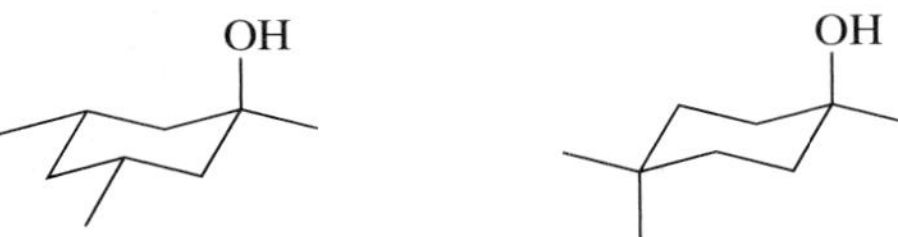

5.36 Which alcohol is converted to the corresponding chloride at the fastest rate on reaction with concentrated hydrochloric acid? Which one reacts at the slowest rate?

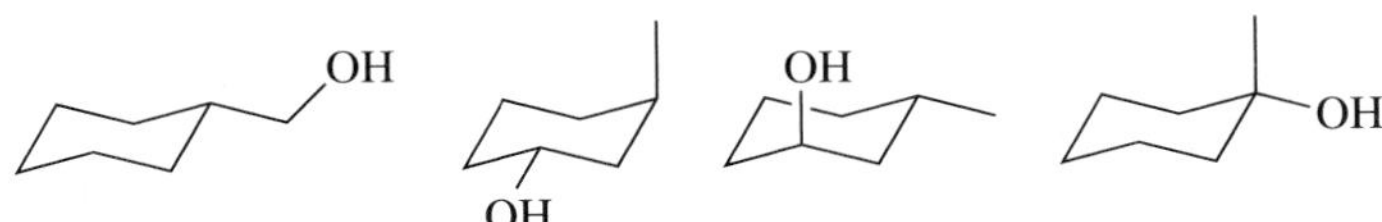

5.37 The compound shown was used as a building block in the synthesis of an HIV protease inhibitor. Write an equation for the preparation of this compound from a suitable alcohol and a reactant of your choice.

O= [lactone ring, O] –CH$_2$OSO$_2$CH$_3$

5.38 Alcohols in which the hydroxyl group occupies a "bridgehead" position such as bicyclo[2.2.1]heptan-1-ol are relatively unreactive toward hydrogen halides. Why?

5.39 Each of the following chiral alcohols is shown in a different format.

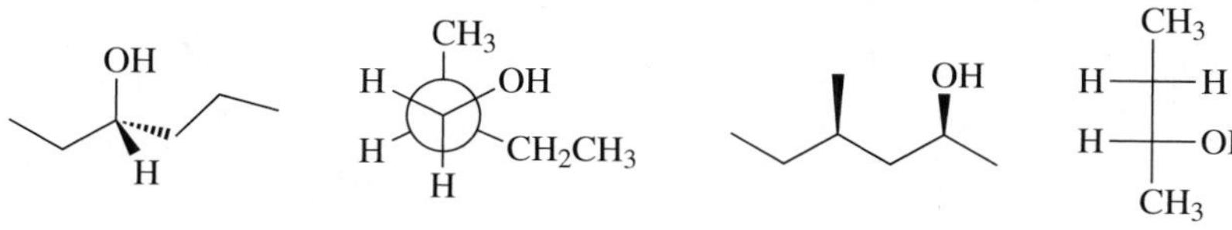

(a) (b) (c) (d)

Draw a structural formula for the alkyl bromide formed as the major product from each alcohol on reaction with hydrogen bromide. Use the same format for the alkyl bromide as the original alcohol. Give the IUPAC name for each alkyl bromide including its stereochemistry according to the Cahn-Ingold-Prelog *R,S* system.

5.40 (a) Assuming that the rate-determining elementary step in the reaction of *trans*-4-methylcyclohexanol with hydrogen bromide is unimolecular, write an equation for this step. Use curved arrows to show electron flow.

(b) Two stereoisomers of 1-bromo-4-methylcyclohexane are formed when *trans*-4-methylcyclohexanol reacts with hydrogen bromide. Write a separate equation for the elementary step that gives each stereoisomer.

5.41 The reaction of 3-*tert*-butyl-3-pentanol with hydrochloric acid gave the products shown. Write appropriate equations explaining the formation of each.

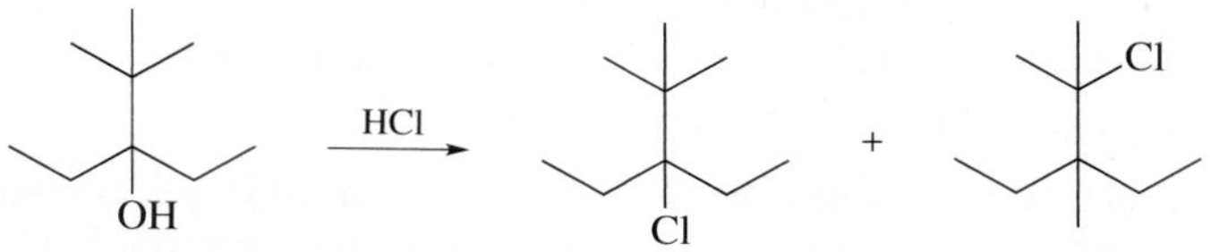

5.42 Bromomethylcycloheptane has been prepared in 92% yield by the reaction shown. Write a stepwise mechanism and use curved arrows to show electron flow. The reaction was carried out in water, so use H_3O^+ as the proton donor in your mechanism. Is the rate-determining step unimolecular (S_N1) or bimolecular (S_N2)?

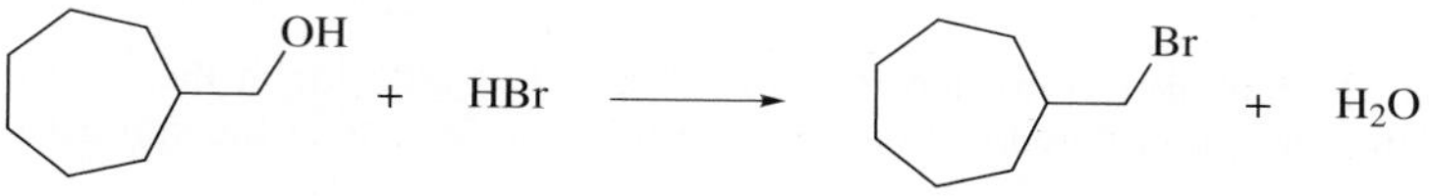

$$+ \quad HBr \longrightarrow \quad + \quad H_2O$$

5.43 Although useful in agriculture as a soil fumigant, methyl bromide is an ozone-depleting chemical, and its production is being phased out. The industrial preparation of methyl bromide is from methanol, by reaction with hydrogen bromide. Write a mechanism for this reaction and classify it as S_N1 or S_N2.

Descriptive Passage and Interpretive Problems 5

More About Potential Energy Diagrams

Chapter 6 will describe *elimination* reactions and their mechanisms. In one example, heating *tert*-butyl bromide in ethanol gives the alkene 2-methylpropene by a two-step mechanism:

Step 1: $(CH_3)_3C{-}\ddot{\underset{..}{Br}}: \rightleftharpoons (CH_3)_2\overset{+}{C}{-}CH_3 + :\ddot{\underset{..}{Br}}:^-$

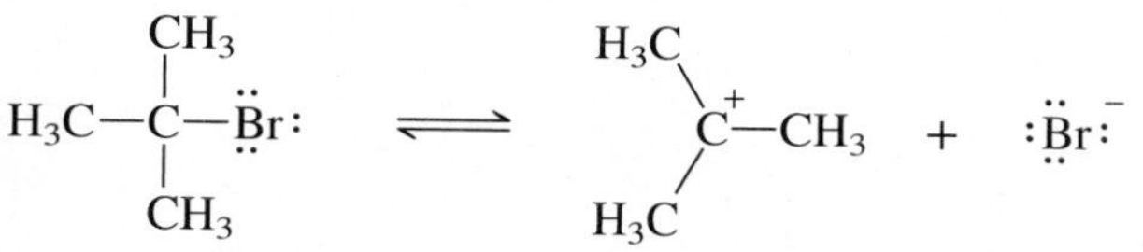

tert-Butyl bromide — *tert*-Butyl cation — Bromide ion

Step 2: $(CH_3)_2\overset{+}{C}{-}CH_3 + CH_3CH_2{-}\ddot{\underset{..}{O}}H \longrightarrow (CH_3)_2C{=}CH_2 + CH_3CH_2{-}\overset{+}{O}H_2$

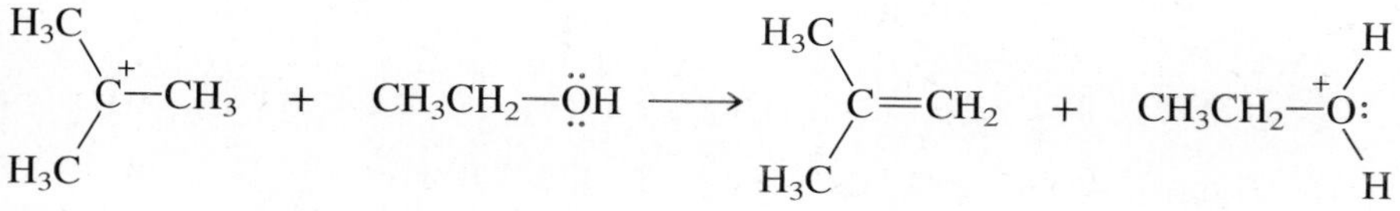

tert-Butyl cation — Ethanol — 2-Methylpropene — Ethyloxonium ion

A potential energy diagram for the reaction provides additional information to complement the mechanism expressed in the equations for the two elementary steps.

The energy relationships in the diagram are not only useful in their own right, but also aid in understanding the structural changes occurring at the transition state. Hammond's postulate tells us that if two states occur consecutively, the closer they are in energy, the more similar they are in structure.

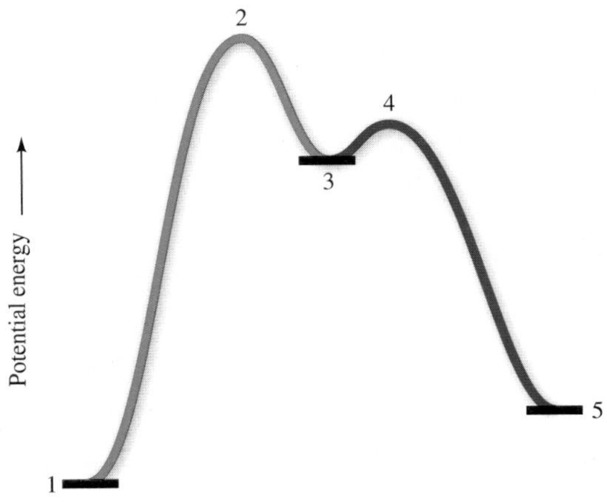

5.44 Ethanol is:

A. A catalyst
B. A reactive intermediate
C. A Brønsted acid
D. A Brønsted base

5.45 According to the potential energy diagram, the overall reaction is:

A. Endothermic
B. Exothermic

5.46 Classify the elementary steps in the mechanism according to their molecularity.

A. Step 1 is unimolecular; step 2 is bimolecular.
B. Step 1 is bimolecular; step 2 is unimolecular.
C. Both steps are unimolecular.
D. Both steps are bimolecular.

5.47 Classify states 2–4 in the potential energy diagram.

A. 2, 3, and 4 are transition states.
B. 2, 3, and 4 are reactive intermediates.
C. 2 and 4 are transition states; 3 is a reactive intermediate.
D. 2 and 4 are reactive intermediates; 3 is a transition state.

5.48 According to the diagram, the activation energy of the slow step is given by the energy difference between states

A. 1 and 2
B. 2 and 3
C. 3 and 4
D. 1 and 5

5.49 What best describes the species at the rate-determining transition state?

A. $(CH_3)_2(H_3C)C^{\delta+}$-----$:\ddot{Br}:^{\delta-}$ (H₃C—C(CH₃)₂ with δ+ on C, partial bond to Br with δ−)

B. $(H_3C)_2\overset{+}{C}—CH_3$

C. $(H_3C)_2C^{\delta+}$=CH_2---H---$:\ddot{O}^{\delta+}$—CH_2CH_3 (O also bonded to H)

D. $^{\delta-}:\ddot{Br}:$ --- $H_3C—C(H_3C)$=CH_2---H---$:\ddot{O}^{\delta+}$—CH_2CH_3 (O also bonded to H)

5.50 By applying Hammond's postulate to the potential energy diagram for this reaction, we can say that:

A. The structure of 2 is more carbocation-like than 4.
B. The structure of 2 is less carbocation-like than 4.
C. The structure of 2 resembles 1 more than it resembles 3.
D. The structure of 4 resembles 5 more than it resembles 3.

Chapter

6

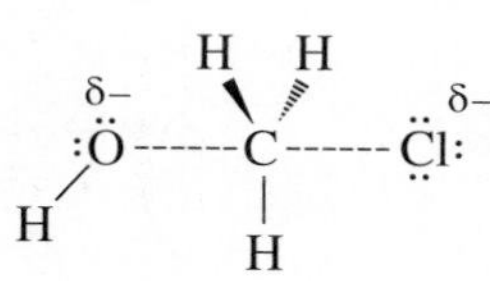

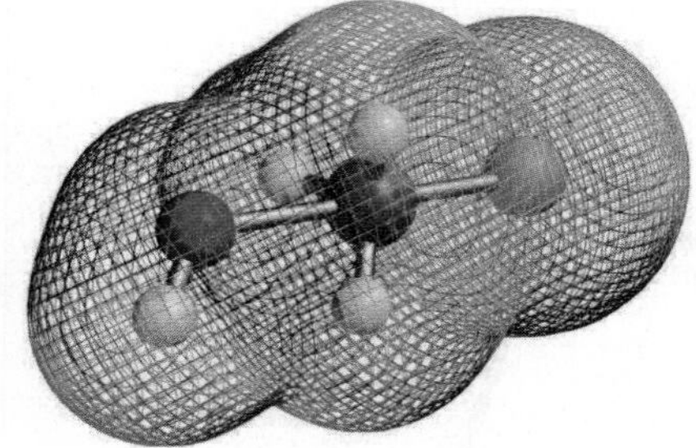

This electrostatic potential map is of the transition state for the reaction of hydroxide ion with chloromethane. The tetrahedral arrangement of bonds inverts like an umbrella in a storm during the reaction.

CHAPTER OUTLINE

Nucleophilic Substitution

Nucleophilic substitution was introduced in Chapter 5 with the reaction of alcohols with hydrogen halides to form alkyl halides. Now we'll see how alkyl halides themselves can be converted by nucleophilic substitution to other classes of organic compounds. Like Chapter 5, the present chapter has a mechanistic emphasis designed to achieve a practical result. By understanding the mechanisms by which alkyl halides undergo nucleophilic substitution, we can choose experimental conditions best suited to carrying out a particular functional-group transformation. The difference between a successful reaction that leads cleanly to a desired product and one that fails is often a subtle one. Mechanistic analysis helps us to appreciate these subtleties and use them to our advantage.

6.1 Functional-Group Transformation by Nucleophilic Substitution

We've learned in Sections 5.8–5.13 that the initial intermediate in the reaction of alcohols with hydrogen halides is the conjugate acid of the alcohol and that, once formed, two mechanistic pathways are available to it. In one, called **S_N2** for *substitution nucleophilic bimolecular,* rate-determining step displacement of a water molecule by a halide anion gives an alkyl halide.

$$\text{R-}\ddot{O}\text{-H} \xrightleftharpoons[\text{fast}]{\text{HX}} \text{R-}\overset{+}{O}\text{H}_2 \xrightarrow[\text{slow}]{:\ddot{X}:^-} \text{RX} + H_2O$$

Alcohol — Alkyloxonium ion — Alkyl halide — Water

In the other, called **S_N1** for *substitution nucleophilic unimolecular,* the rate-determining step in alkyl halide formation is dissociation of the alkyloxonium ion to give a carbocation that is then captured by halide.

$$\text{R-}\ddot{O}\text{-H} \xrightleftharpoons[\text{fast}]{\text{HX}} \text{R-}\overset{+}{O}\text{H}_2 \xrightleftharpoons[\text{slow}]{-H_2O} R^+ \xrightarrow[\text{fast}]{:\ddot{X}:^-} \text{RX}$$

Alcohol — Alkyloxonium ion — Carbocation — Alkyl halide

In the present chapter, alkyl halides feature prominently as *starting materials* for preparing a large variety of other classes of organic compounds by reactions of the S_N2 type:

$$M^+ \ {}^-Y: + R-\ddot{X}: \longrightarrow R-Y + M^+ \ :\ddot{X}:^-$$

Nucleophilic reagent — Alkyl halide — Product of nucleophilic substitution — Metal halide

M^+ in the nucleophilic reagent is typically Li^+, Na^+, or K^+. Frequently encountered nucleophilic reagents include

MOH (a metal *hydroxide,* a source of the nucleophilic anion $H\ddot{O}:^-$)

MOR (a metal *alkoxide,* a source of the nucleophilic anion $R\ddot{O}:^-$)

MOCR (O double-bonded to C) (a metal *carboxylate,* a source of the nucleophilic anion $RC(=O)-\ddot{O}:^-$)

MSH (a metal *hydrogen sulfide,* a source of the nucleophilic anion $H\ddot{S}:^-$)

MCN (a metal *cyanide,* a source of the nucleophilic anion $^-:C{\equiv}N:$)

MN_3 (a metal *azide,* a source of the nucleophilic anion $:\overset{-}{N}{=}\overset{+}{N}{=}\overset{-}{N}:$)

Table 6.1 illustrates an application of each of these to a functional-group transformation. The anionic portion of the salt substitutes for the halogen of an alkyl halide. The metal cation portion becomes a lithium, sodium, or potassium halide.

Notice that all the examples in Table 6.1 involve *alkyl* halides, that is, compounds in which the halogen is attached to an sp^3-hybridized carbon. Alkenyl halides and aryl halides, compounds in which the halogen is attached to sp^2-hybridized carbons, are essentially

Alkenyl halides are also referred to as *vinylic halides*.

TABLE 6.1 Functional-Group Transformation via Nucleophilic Substitution

Nucleophile and comments	General equation and specific example
Hydroxide ion: The oxygen atom of hydroxide is nucleophilic and replaces the halogen of an alkyl halide. The product is an *alcohol.*	$H\ddot{O}:^- + R-\ddot{X}: \longrightarrow H\ddot{O}R + :\ddot{X}:^-$ (Hydroxide ion + Alkyl halide → Alcohol + Halide ion) NaOH + Ethyl bromide (CH₃CH₂Br) —dimethyl sulfoxide→ Ethanol (CH₃CH₂OH) Sodium hydroxide, Ethyl bromide, Ethanol

continued

TABLE 6.1 Functional-Group Transformation via Nucleophilic Substitution (*Continued*)

Nucleophile and comments	General equation and specific example
Alkoxide ion: The oxygen atom of a metal alkoxide is nucleophilic and replaces the halogen of an alkyl halide. The product is an *ether*.	$RO^- + R'-X \longrightarrow ROR' + X^-$ Alkoxide ion + Alkyl halide → Ether + Halide ion Sodium isobutoxide + Ethyl bromide —(isobutyl alcohol)→ Ethyl isobutyl ether (66%)
Carboxylate ion: An ester is formed when the negatively charged oxygen of a carboxylate replaces the halogen of an alkyl halide.	$RCO_2^- + R'-X \longrightarrow RCOR' + X^-$ (RC(=O)O⁻ → RC(=O)OR') Carboxylate ion + Alkyl halide → Ester + Halide ion $CH_3(CH_2)_{16}COK + CH_3CH_2I \xrightarrow[\text{water}]{\text{acetone}} CH_3(CH_2)_{16}COCH_2CH_3$ Potassium octadecanoate + Ethyl iodide → Ethyl octadecanoate (95%)
Hydrogen sulfide ion: Using hydrogen sulfide as a nucleophile permits the conversion of alkyl halides to *thiols*.	$HS^- + R'-X \longrightarrow HSR' + X^-$ Hydrogen sulfide ion + Alkyl halide → Thiol + Halide ion $KSH + CH_3CH(Br)(CH_2)_6CH_3 \xrightarrow[\text{water}]{\text{ethanol}} CH_3CH(SH)(CH_2)_6CH_3$ Potassium hydrogen sulfide + 2-Bromononane → 2-Nonanethiol (74%)
Cyanide ion: The negatively charged carbon of cyanide is the site of its nucleophilic character. Cyanide reacts with alkyl halides to extend a carbon chain by forming an *alkyl cyanide* or *nitrile*.	$:N{\equiv}C^- + R'-X \longrightarrow :N{\equiv}CR' + X^-$ Cyanide ion + Alkyl halide → Alkyl cyanide + Halide ion NaCN + Cyclopentyl chloride —(dimethyl sulfoxide)→ Cyclopentyl cyanide (70%) Sodium cyanide + Cyclopentyl chloride → Cyclopentyl cyanide (70%)
Azide ion: Sodium azide makes carbon–nitrogen bonds by converting an alkyl halide to an *alkyl azide*.	$^-N{=}\overset{+}{N}{=}N^- + R'-X \longrightarrow {}^-N{=}\overset{+}{N}{=}NR' + X^-$ Azide ion + Alkyl halide → Alkyl azide + Halide ion NaN_3 + Pentyl iodide —(1-propanol, water)→ Pentyl azide (52%) Sodium azide + Pentyl iodide → Pentyl azide (52%)
Iodide ion: Alkyl chlorides and bromides are converted to *alkyl iodides* by treatment with sodium iodide in acetone. NaI is soluble in acetone, but the NaCl or NaBr that is formed is not and crystallizes from the reaction mixture, making the reaction irreversible.	$I^- + R'-X \longrightarrow IR' + X^-$ Iodide ion + Alkyl chloride or bromide → Alkyl iodide + Chloride or bromide ion NaI + 2-Bromopropane —(acetone)→ 2-Iodopropane (63%) Sodium iodide + 2-Bromopropane → 2-Iodopropane (63%)

unreactive under these conditions, and the principles to be developed in this chapter do not apply to them.

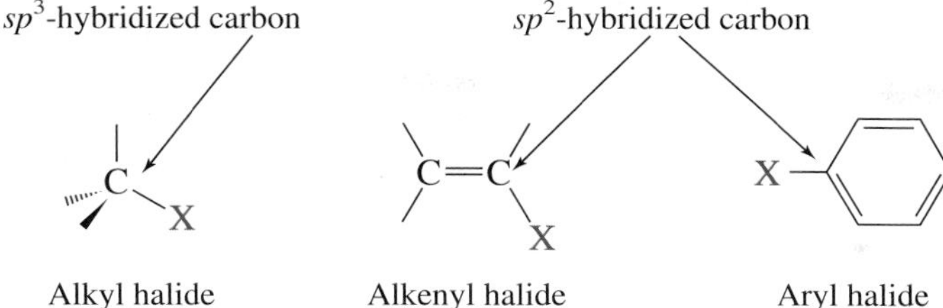

To ensure that reaction occurs in homogeneous solution, solvents are chosen that dissolve both the alkyl halide and the ionic salt. Alkyl halides are soluble in organic solvents, but the salts often are not. Inorganic salts are soluble in water, but alkyl halides are not. Mixed solvents such as ethanol–water mixtures that can dissolve both the alkyl halide and the nucleophile are frequently used. Many salts, as well as most alkyl halides, possess significant solubility in dimethyl sulfoxide (DMSO) or *N,N*-dimethylformamide (DMF), which makes them good solvents for carrying out nucleophilic substitution reactions (Section 6.9).

Problem 6.1

Write a structural formula for the principal organic product formed in the reaction of methyl bromide with each of the following compounds:

(a) NaOH (sodium hydroxide)
(b) $KOCH_2CH_3$ (potassium ethoxide)
(c) $NaOC(=O)-C_6H_5$ (sodium benzoate)
(d) LiN_3 (lithium azide)
(e) KCN (potassium cyanide)
(f) NaSH (sodium hydrogen sulfide)
(g) NaI (sodium iodide)

Sample Solution (a) The nucleophile in sodium hydroxide is the negatively charged hydroxide ion. The reaction that occurs is nucleophilic substitution of bromide by hydroxide. The product is methyl alcohol.

$$\underset{\text{Hydroxide ion (nucleophile)}}{HO^{-}} + \underset{\text{Methyl bromide (substrate)}}{H_3C-Br} \longrightarrow \underset{\text{Methyl alcohol (product)}}{H_3C-OH} + \underset{\text{Bromide ion (leaving group)}}{Br^{-}}$$

With Table 6.1 as background, you can begin to see how useful alkyl halides are in synthetic organic chemistry. Their ease of preparation from alcohols makes them readily available as starting materials for the preparation of other functionally substituted organic compounds. The range of compounds that can be prepared by nucleophilic substitution reactions of alkyl halides is quite large; the examples shown in Table 6.1 illustrate only a few of them. Numerous other examples will be added to the list in this and subsequent chapters.

6.2 Relative Reactivity of Halide Leaving Groups

Among alkyl halides, alkyl iodides undergo nucleophilic substitution at the fastest rate, alkyl fluorides the slowest.

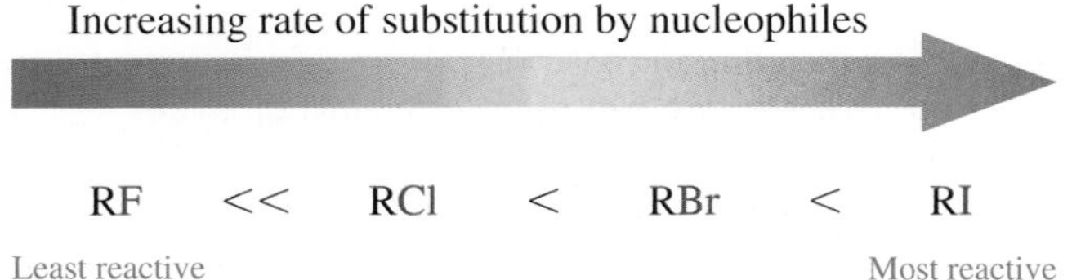

Alkyl iodides are several times more reactive than alkyl bromides and from 50 to 100 times more reactive than alkyl chlorides. Alkyl fluorides are several thousand times less reactive than alkyl chlorides and are rarely used in nucleophilic substitutions. These reactivity differences can be related to (1) the carbon–halogen bond strength and (2) the basicity of the halide anion. Alkyl iodides have the weakest carbon–halogen bond and require the lowest activation energy to break; alkyl fluorides have the strongest carbon–halogen bond and require the highest activation energy. Regarding basicity of the halide leaving group, iodide is the weakest base, fluoride the strongest. It is generally true that the less basic the leaving group, the smaller the energy requirement for cleaving its bond to carbon and the faster the rate.

Problem 6.2

1-Bromo-3-chloropropane reacts with one molar equivalent of sodium cyanide in aqueous ethanol to give a single organic product. What is this product?

6.3 The S_N2 Mechanism of Nucleophilic Substitution

Much of what we know about the mechanisms of nucleophilic substitution is due to studies carried out by Sir Christopher Ingold and Edward D. Hughes at University College, London, in the 1930s using kinetics and stereochemistry as their key experimental tools. Indeed, it is these studies of substitution in alkyl halides that are the foundation upon which our earlier discussion of the reactions of alcohols with hydrogen halides in Chapter 5 rested.

Kinetics. Because the rate of nucleophilic substitution of an alkyl halide—be it methyl, primary, secondary, or tertiary—depends on the leaving group (I > Br > Cl > F), the carbon–halogen bond must break in the slow (rate-determining) step, and the concentration of the alkyl halide must appear in the rate law. What about the nucleophile? Hughes and Ingold found that many nucleophilic substitutions, such as the hydrolysis of methyl bromide in base:

$$\underset{\text{Methyl bromide}}{CH_3\ddot{Br}:} + \underset{\text{Hydroxide ion}}{H\ddot{O}:^-} \longrightarrow \underset{\text{Methyl alcohol}}{CH_3\ddot{O}H} + \underset{\text{Bromide ion}}{:\ddot{Br}:^-}$$

obey a *second-order* rate law, first order in the alkyl halide and first order in the nucleophile.

$$\text{Rate} = k[CH_3Br][HO^-]$$

They concluded that the rate-determining step is *bimolecular* and proposed the **substitution nucleophilic bimolecular** (S_N2) mechanism shown as an equation in Mechanism 6.1 and as a potential energy diagram in Figure 6.1.

The Hughes–Ingold S_N2 mechanism is a one-step concerted process in which both the alkyl halide and the nucleophile are involved at the transition state. Cleavage of the bond between carbon and the leaving group is assisted by formation of a bond between carbon and the nucleophile. Carbon is partially bonded to both the incoming nucleophile and the departing halide at the transition state. Progress is made toward the transition state as the nucleophile begins to share a pair of its electrons with carbon and the halide ion leaves, taking with it the pair of electrons in its bond to carbon.

Problem 6.3

Is the two-step sequence depicted in the following equations consistent with the second-order kinetic behavior observed for the hydrolysis of methyl bromide?

$$H_3C-\ddot{Br}: \xrightarrow{\text{slow}} H_3C^+ + :\ddot{Br}:^-$$

$$H_3C^+ + :\ddot{O}^--H \xrightarrow{\text{fast}} H_3C-\ddot{O}-H$$

Mechanism 6.1

The S_N2 Mechanism of Nucleophilic Substitution

THE OVERALL REACTION:

$$CH_3Br + HO^- \longrightarrow CH_3OH + Br^-$$

Methyl bromide — Hydroxide ion — Methyl alcohol — Bromide ion

THE MECHANISM: The reaction proceeds in a single step. Hydroxide ion acts as a nucleophile. While the C—Br bond is breaking, the C—O bond is forming.

$$H{-}\ddot{O}{:}^- + H_3C{-}\ddot{Br}{:} \longrightarrow H{-}\ddot{O}{-}CH_3 + {:}\ddot{Br}{:}^-$$

Hydroxide ion — Methyl bromide — Methyl alcohol — Bromide ion

THE TRANSITION STATE: Hydroxide ion attacks carbon from the side opposite the C—Br bond.

$$\overset{\delta-}{H{-}\ddot{O}}\text{------}C(H)(H)(H)\text{------}\overset{\delta-}{\ddot{Br}{:}}$$

Carbon is partially bonded to both hydroxide and bromide. Oxygen and bromine, both partially negative, share the charge. The arrangement of bonds undergoes tetrahedral inversion from \C— to —C/ as the reaction progresses.

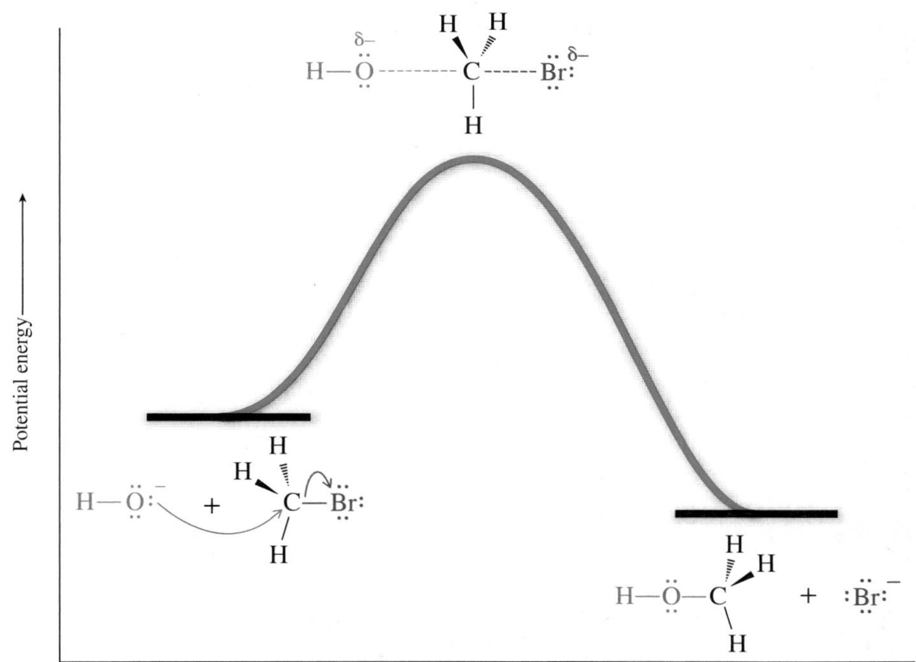

Figure 6.1

Potential energy diagram for the reaction of methyl bromide with hydroxide ion by the S_N2 mechanism.

Stereochemistry. The diagram for the transition state in Mechanism 6.1 and Figure 6.1 for the reaction of methyl bromide with hydroxide illustrates a key stereochemical feature of the S_N2 mechanism. *The nucleophile attacks carbon from the side opposite the bond to the leaving group.* Figure 6.2, which maps the electrostatic potential of methyl bromide and hydroxide, supports this fact, where the highest negative potential (red) of

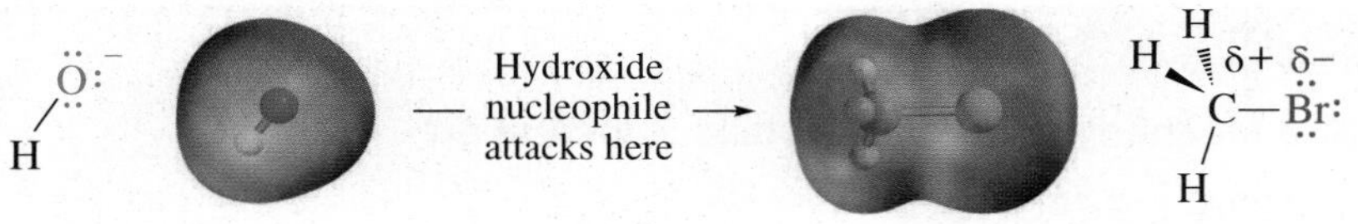

Hydroxide nucleophile attacks here

Figure 6.2

Electrostatic potential maps of hydroxide and methyl bromide. Hydroxide is most negative (red) at the oxygen. The most positive region of methyl bromide (blue) resides on the methyl group, opposite the side of the bromine.

the oxygen atom in hydroxide attacks the positive (blue) carbon of methyl bromide, opposite to the bromide.

Another way of expressing the same point, especially when substitution occurs at a chirality center, is that *S_N2 reactions proceed with* ***inversion of configuration*** *at the carbon that bears the leaving group*. The tetrahedral arrangement of bonds in the reactant is converted to an inverted tetrahedral arrangement in the product.

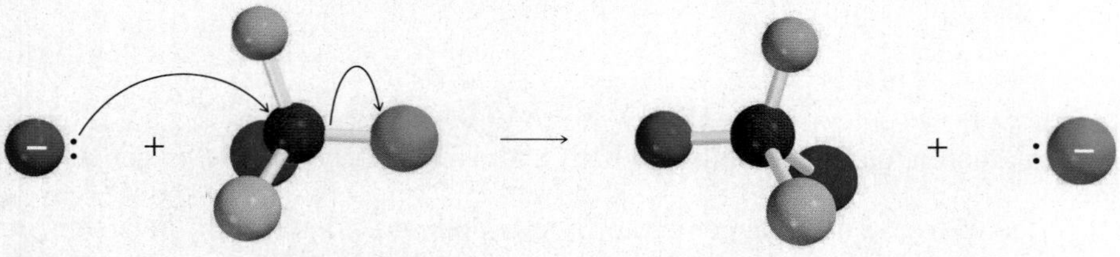

Nucleophile   Alkyl halide S_N2 product   Leaving group

This generalization comes from studies of nucleophilic substitutions of optically active alkyl halides such as that shown in the equation:

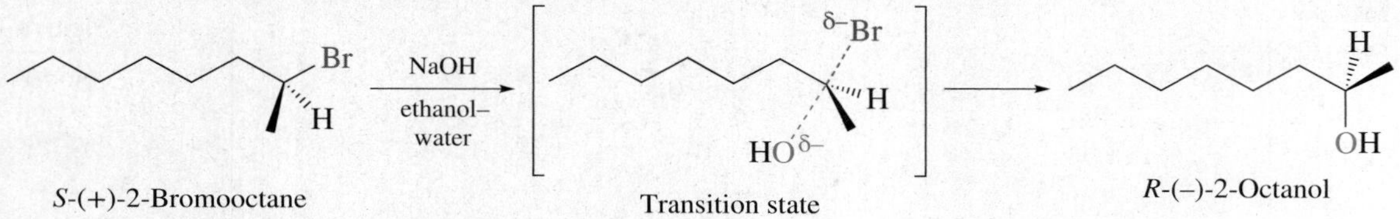

NaOH, ethanol–water

S-(+)-2-Bromooctane   Transition state   *R*-(−)-2-Octanol

Problem 6.4

The Fischer projection for (+)-2-bromooctane is shown. Write the Fischer projection of the (−)-2-octanol formed from it by the S_N2 mechanism.

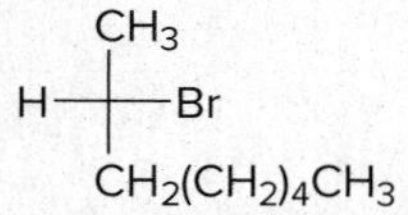

CH_3 (top); H (left); Br (right); $CH_2(CH_2)_4CH_3$ (bottom)

Problem 6.5

Would you expect the 2-octanol formed by S_N2 hydrolysis of (−)-2-bromooctane to be optically active? If so, what will be its absolute configuration and sign of rotation? What about the 2-octanol formed by hydrolysis of racemic 2-bromooctane?

Countless experiments have confirmed that substitution by the S_N2 mechanism is stereospecific and suggest that there exists a *stereoelectronic* requirement for the nucleophile to approach carbon from the side opposite the bond to the leaving group. The results of molecular orbital calculations help us understand why.

When a nucleophile such as hydroxide ion reacts with methyl bromide, electrons flow from the highest occupied molecular orbital (HOMO) of HO^- to the lowest unoccupied molecular orbital (LUMO) of CH_3Br. Directing our attention to the LUMO of CH_3Br, we find three main regions where the HOMO of the nucleophile can overlap with the LUMO. One of these—the blue region shown at the right—can be ignored because it is associated only with Br, and nucleophilic attack from that direction does not produce a C—O bond.

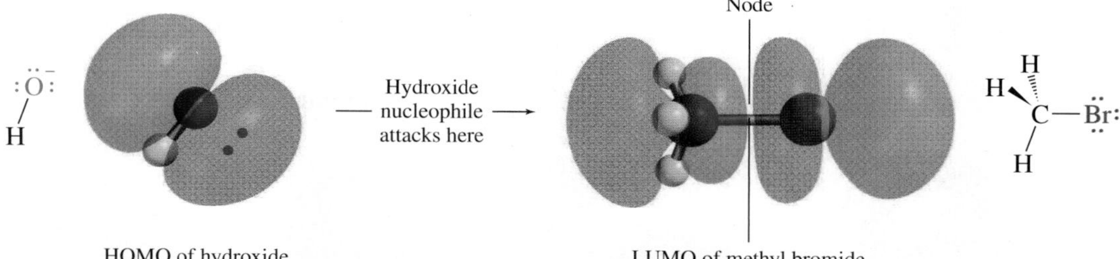

The region between carbon and bromine contains a nodal surface; therefore, no net bonding results from its overlap with the HOMO of HO^-. The remaining possibility, *which is also the one that coincides with experimental observation,* is overlap of the HOMO of HO^- with the LUMO of CH_3Br in the region opposite the C—Br bond. It involves a major region of the LUMO, avoids a node, and gives a C—O bond with inversion of configuration at carbon.

The S_N2 mechanism is believed to describe most substitutions in which simple primary and secondary alkyl halides react with negatively charged nucleophiles. *All the examples that introduced nucleophilic substitution in Table 6.1 proceed by the S_N2 mechanism* (or a mechanism very much like S_N2—remember, mechanisms can never be established with certainty but represent only our best present explanations of experimental observations).

Problem 6.6

Sketch the structure of the S_N2 transition state for the following reaction taken from Table 6.1. Na^+ is a spectator ion and can be omitted from the transition state.

$$(CH_3)_2CHBr + NaI \xrightarrow{\text{acetone}} (CH_3)_2CHI + NaBr$$

We saw in Section 6.2 that the rate of nucleophilic substitution depends strongly on the leaving group—alkyl iodides are the most reactive, alkyl fluorides the least. In the next section, we'll see that the structure of the alkyl group can have an even larger effect.

6.4 Steric Effects and S_N2 Reaction Rates

There are very large differences in the rates at which the various kinds of alkyl halides—methyl, primary, secondary, or tertiary—undergo nucleophilic substitution. For the reaction:

$$\underset{\text{Alkyl bromide}}{RBr} + \underset{\text{Lithium iodide}}{LiI} \xrightarrow{\text{acetone}} \underset{\text{Alkyl iodide}}{RI} + \underset{\text{Lithium bromide}}{LiBr}$$

the rates of nucleophilic substitution of a series of alkyl bromides differ by a factor of over 10^6.

Increasing relative reactivity toward S_N2 substitution
(RBr + LiI in acetone, 25°C)

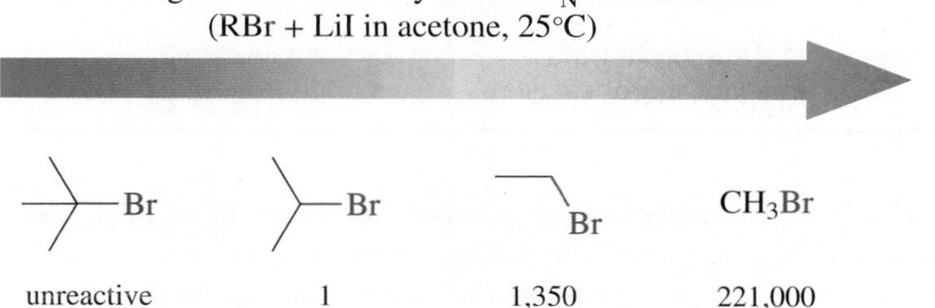

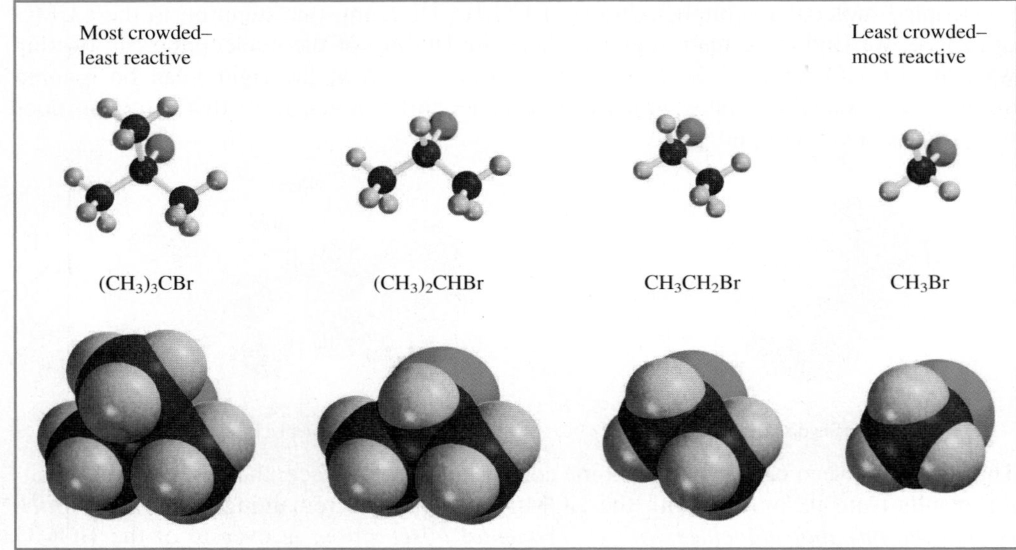

Figure 6.3

Ball-and-stick (*top*) and space-filling (*bottom*) models of alkyl bromides, showing how substituents shield the carbon atom that bears the leaving group from attack by a nucleophile. The nucleophile must attack from the side opposite the bond to the leaving group.

The very large rate differences between methyl, ethyl, isopropyl, and *tert*-butyl bromides reflect the **steric hindrance** each offers to nucleophilic attack. The nucleophile must approach the alkyl halide from the side opposite the bond to the leaving group, and, as illustrated in Figure 6.3, this approach is hindered by alkyl substituents on the carbon that is being attacked. The three hydrogens of methyl bromide offer little resistance to approach of the nucleophile, and a rapid reaction occurs. Replacing one of the hydrogens by a methyl group somewhat shields the carbon from approach of the nucleophile and causes ethyl bromide to be less reactive than methyl bromide. Replacing all three hydrogens by methyl groups almost completely blocks approach to the tertiary carbon of $(CH_3)_3CBr$ and shuts down bimolecular nucleophilic substitution.

In general, S_N2 reactions of alkyl halides show the following dependence of rate on structure: CH_3X > primary > secondary > tertiary.

Problem 6.7

Identify the compound in each of the following pairs that reacts with sodium iodide in acetone at the faster rate:

(a) 1-Chlorohexane or cyclohexyl chloride
(b) 1-Bromopentane or 3-bromopentane
(c) 2-Chloropentane or 2-fluoropentane
(d) 2-Bromo-2-methylhexane or 2-bromo-5-methylhexane
(e) 2-Bromopropane or 1-bromodecane

Sample Solution (a) Compare the structures of the two chlorides. 1-Chlorohexane is a primary alkyl chloride; cyclohexyl chloride is secondary. Primary alkyl halides are less crowded at the site of substitution than secondary ones and react faster in substitution by the S_N2 mechanism. 1-Chlorohexane is more reactive.

Cl Cl

1-Chlorohexane (primary, more reactive) Cyclohexyl chloride (secondary, less reactive)

Alkyl groups at the carbon atom *adjacent* to the point of nucleophilic attack also decrease the rate of the S_N2 reaction. Taking ethyl bromide as the standard and successively replacing its C-2 hydrogens by methyl groups, we see that each additional methyl group decreases the rate of displacement of bromide by iodide. When C-2 is completely substituted by methyl groups, as it is in neopentyl bromide [$(CH_3)_3CCH_2Br$], we see the unusual case of a primary alkyl halide that is practically inert to substitution by the S_N2 mechanism because of steric hindrance.

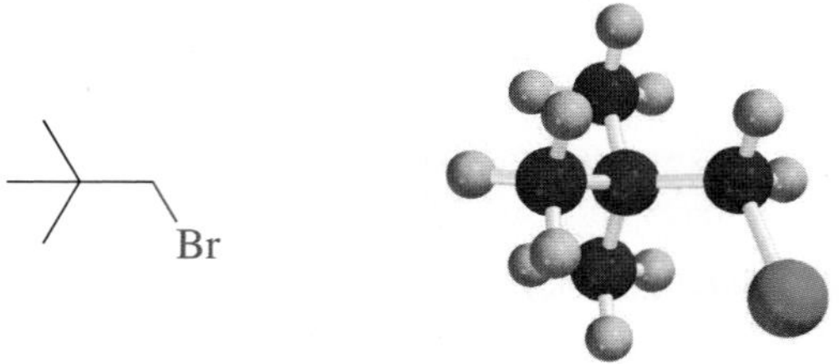

Neopentyl bromide
(1-Bromo-2,2-dimethylpropane)

Increasing relative reactivity toward S_N2 substitution
(RBr + LiI in acetone, 25°C)

Br	Br	Br	Br
0.00002	0.036	0.8	1

Problem 6.8

The first step in the synthesis of the antimalarial drug *primaquine* is summarized in the equation shown. What is the structure of the product?

NK + Br ... Br ⟶ $C_{13}H_{14}BrNO_2$

6.5 Nucleophiles and Nucleophilicity

The Lewis base that acts as the nucleophile often is, but need not always be, an anion. Neutral Lewis bases such as amines (R_3N:), phosphines (R_3P:), and sulfides ($R_2\ddot{S}$:) can also serve as nucleophiles.

$$(H_3C)_2\ddot{S}: + CH_3-\ddot{I}: \longrightarrow (H_3C)_2\overset{+}{S}-CH_3 \quad :\ddot{I}:^-$$

Dimethyl sulfide — Methyl iodide — Trimethylsulfonium iodide

Other common examples of substitutions involving neutral nucleophiles include **solvolysis** reactions—substitutions where the nucleophile is the solvent in which the reaction is carried out. Solvolysis in water (*hydrolysis*) converts an alkyl halide to an alcohol.

$$RX + 2H_2O \longrightarrow ROH + H_3O^+ + X^-$$

Alkyl halide — Water — Alcohol — Hydronium ion — Halide ion

The reaction occurs in two steps. The first yields an alkyloxonium ion by nucleophilic substitution and is rate-determining. The second gives the alcohol by proton transfer—a rapid Brønsted acid–base reaction.

$$H_2\ddot{O}: + R-\ddot{X}: \xrightarrow{\text{slow}} :\ddot{X}:^- + H_2\overset{+}{O}-R \xrightarrow[\text{fast}]{H_2\ddot{O}:} H\ddot{O}-R + H_3\overset{+}{O}:$$

Water | Alkyl halide | Halide ion | Alkoxonium ion | Alcohol | Hydronium ion

Analogous reactions take place in other solvents that, like water, contain an —OH group. Solvolysis in methanol (*methanolysis*) gives a methyl ether.

$$CH_3\ddot{O}H + R-\ddot{X}: \xrightarrow{\text{slow}} :\ddot{X}:^- + H_3C(H)\overset{+}{O}-R \xrightarrow[\text{fast}]{CH_3\ddot{O}H} H_3C\ddot{O}-R + CH_3\overset{+}{O}H_2$$

Methanol | Alkyl halide | Halide ion | Alkylmethyloxonium ion | Alkyl methyl ether | Methyloxonium ion

Because attack by the nucleophile is the rate-determining step of the S_N2 mechanism, the rate of substitution varies from nucleophile to nucleophile. Nucleophilic strength, or **nucleophilicity,** is a measure of how fast a Lewis base displaces a leaving group from a suitable substrate. Table 6.2 compares the rate at which various Lewis bases react with methyl iodide in methanol, relative to methanol as the standard nucleophile.

As long as the nucleophilic atom is the same, the more basic the nucleophile, the more reactive it is. An alkoxide ion (RO^-) is more basic and more nucleophilic than a carboxylate ion (RCO_2^-).

$$R-\ddot{O}:^- \quad \text{is more nucleophilic than} \quad RC(=O)-\ddot{O}:^-$$

Stronger base
Conjugate acid is ROH: $pK_a = 16$

Weaker base
Conjugate acid is RCO_2H: $pK_a = 5$

The connection between basicity and nucleophilicity holds when comparing atoms in the *same row* of the periodic table. Thus, HO^- is more basic and more nucleophilic than F^-, and NH_3 is more basic and more nucleophilic than H_2O. *It does not hold when proceeding down a column in the periodic table.* In that case, polarizability involving the distortion of the electron density surrounding an atom or ion comes into play. The more easily distorted the electron distribution, the more nucleophilic the atom or ion. Among the halide ions, for example, I^- is the least basic but the most nucleophilic, F^- the most basic but the least nucleophilic. In the same vein, phosphines (R_3P) are more nucleophilic than amines (R_3N), and thioethers (R_2S) more nucleophilic than their oxygen counterparts (R_2O).

TABLE 6.2 Nucleophilicity of Some Common Nucleophiles

Reactivity class	Nucleophile	Relative reactivity*
Very good nucleophiles	I^-, HS^-, RS^-	$>10^5$
Good nucleophiles	Br^-, HO^-, RO^-, CN^-, N_3^-	10^4
Fair nucleophiles	NH_3, Cl^-, F^-, RCO_2^-	10^3
Weak nucleophiles	H_2O, ROH	1
Very weak nucleophiles	RCO_2H	10^{-2}

*Relative reactivity is *k*(nucleophile)/*k*(methanol) for typical S_N2 reactions and is approximate. Data pertain to methanol as the solvent.

Another factor, likely the one most responsible for the inverse relationship between basicity and nucleophilicity among the halide ions, is the degree to which the ions are solvated by ion–dipole forces of the type illustrated in Figure 6.4. Smaller anions, because of their high charge-to-size ratio, are more strongly solvated than larger ones. In order to act as a nucleophile, the halide must shed some of the solvent molecules that surround it. Among the halide anions, ion–dipole forces are strongest for F^- and weakest for I^-. Thus, the nucleophilicity of F^- is suppressed more than that of Cl^-, Cl^- more than Br^-, and Br^- more than I^-. Similarly, HO^- is smaller, more solvated, and less nucleophilic than HS^-. The importance of solvation in reducing the nucleophilicity of small anions more than larger ones can be seen in the fact that, when measured in the gas phase where solvation forces don't exist, the order of halide nucleophilicity reverses and does track basicity: $F^- > Cl^- > Br^- > I^-$.

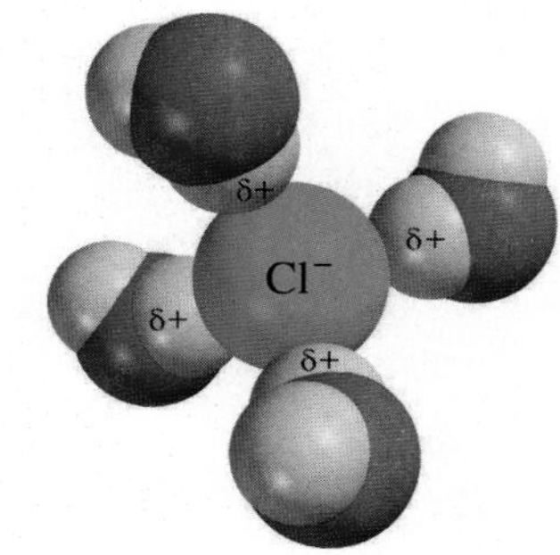

Figure 6.4

Solvation of a chloride ion by water.

Enzyme-Catalyzed Nucleophilic Substitutions of Alkyl Halides

Nucleophilic substitution is one of a variety of mechanisms by which living systems detoxify halogenated organic compounds introduced into the environment. Enzymes that catalyze these reactions are known as *haloalkane dehalogenases*. The hydrolysis of 1,2-dichloroethane to 2-chloroethanol, for example, is a biological nucleophilic substitution catalyzed by the dehalogenase shown in Figure 6.5.

Cl–CH₂CH₂–Cl $\xrightarrow[\text{dehalogenase}]{H_2O}$ Cl–CH₂CH₂–OH

This haloalkane dehalogenase is believed to act by *covalent catalysis* using one of its side-chain carboxylates to displace chloride by an S_N2 mechanism.

Enzyme–C(=O)O⁻ + ClCH₂CH₂Cl $\xrightarrow[-Cl^-]{S_N2}$ Enzyme–C(=O)O–CH₂CH₂Cl

The product of nucleophilic substitution then reacts with water, restoring the enzyme to its original state and giving the observed products of the reaction.

Enzyme–C(=O)O–CH₂CH₂Cl $\xrightarrow[\text{several steps}]{H_2O}$ Enzyme–C(=O)O⁻ + HO–CH₂CH₂–Cl

Both stages of the mechanism are faster than the hydrolysis of 1,2-dichloroethane in the absence of the enzyme.

Enzyme-catalyzed hydrolysis of racemic 2-chloropropanoic acid is a key step in the large-scale preparation (2000 tons per year!) of (*S*)-2-chloropropanoic acid used in the production of agricultural chemicals.

Racemic 2-chloropropanoic acid $\xrightarrow{H_2O,\ \text{dehalogenase}}$ (*S*)-2-Chloropropanoic acid + (*S*)-Lactic acid

In this enzymatic resolution, the dehalogenase enzyme catalyzes the hydrolysis of the *R*-enantiomer of 2-chloropropanoic acid to (*S*)-lactic acid. The desired (*S*)-2-chloropropanoic acid is unaffected and recovered in a nearly enantiomerically pure state.

Some of the most common biological S_N2 reactions involve attack at methyl groups, especially the methyl group of *S*-adenosylmethionine. Examples of these will be given in Chapter 17.

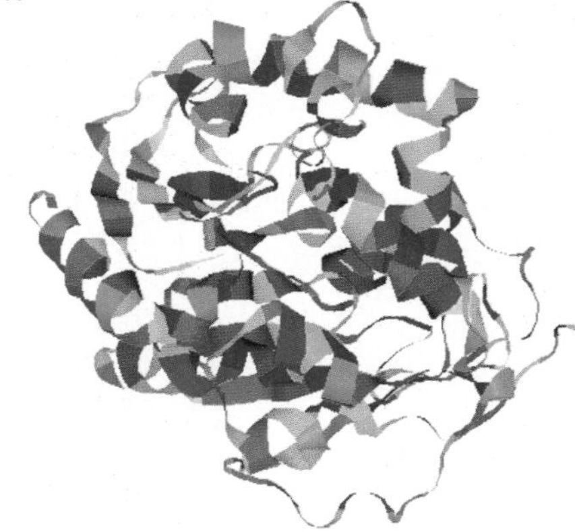

Figure 6.5

A ribbon diagram of the dehalogenase enzyme that catalyzes the hydrolysis of 1,2-dichloroethane. The progression of amino acids along the chain is indicated by a color change. The nucleophilic carboxylate group is near the center of the diagram.

When comparing species that have the same nucleophilic atom, a negatively charged nucleophile is more reactive than a neutral one.

$R—\ddot{\underset{..}{O}}:^-$	is more nucleophilic than	$R—\ddot{\underset{..}{O}}—H$
Alkoxide ion		Alcohol
$RC(=O)—\ddot{\underset{..}{O}}:^-$	is more nucleophilic than	$R—C(=O)—\ddot{\underset{..}{O}}—H$
Carboxylate ion		Carboxylic acid

6.6 The S_N1 Mechanism of Nucleophilic Substitution

Having seen that tertiary alkyl halides are practically inert to substitution by the S_N2 mechanism because of steric hindrance, we might wonder whether they undergo nucleophilic substitution at all. They do, but by a variant of the mechanism introduced in Section 5.8.

In their studies of reaction kinetics, Hughes and Ingold observed that the hydrolysis of *tert*-butyl bromide follows a *first-order* rate law:

$$\underset{tert\text{-Butyl bromide}}{(CH_3)_3CBr} + \underset{\text{Water}}{2H_2O} \longrightarrow \underset{tert\text{-Butyl alcohol}}{(CH_3)_3COH} + \underset{\text{Hydronium ion}}{H_3O^+} + \underset{\text{Bromide ion}}{Br^-}$$

$$\text{Rate} = k(CH_3)_3CBr$$

The reaction rate depends only on the concentration of *tert*-butyl bromide. Just as they interpreted a second-order rate law in terms of a bimolecular rate-determining step, Hughes and Ingold saw first-order kinetics as evidence for a *unimolecular* rate-determining step involving only the alkyl halide while independent of both the concentration and identity of the nucleophile. Like the S_N1 mechanism for the reaction of alcohols with hydrogen halides (Section 5.8), this pathway is characterized by the formation of a carbocation in the rate-determining step.

The S_N1 mechanism for the hydrolysis of *tert*-butyl bromide is presented as a series of elementary steps in Mechanism 6.2 and as a potential energy diagram in Figure 6.6.

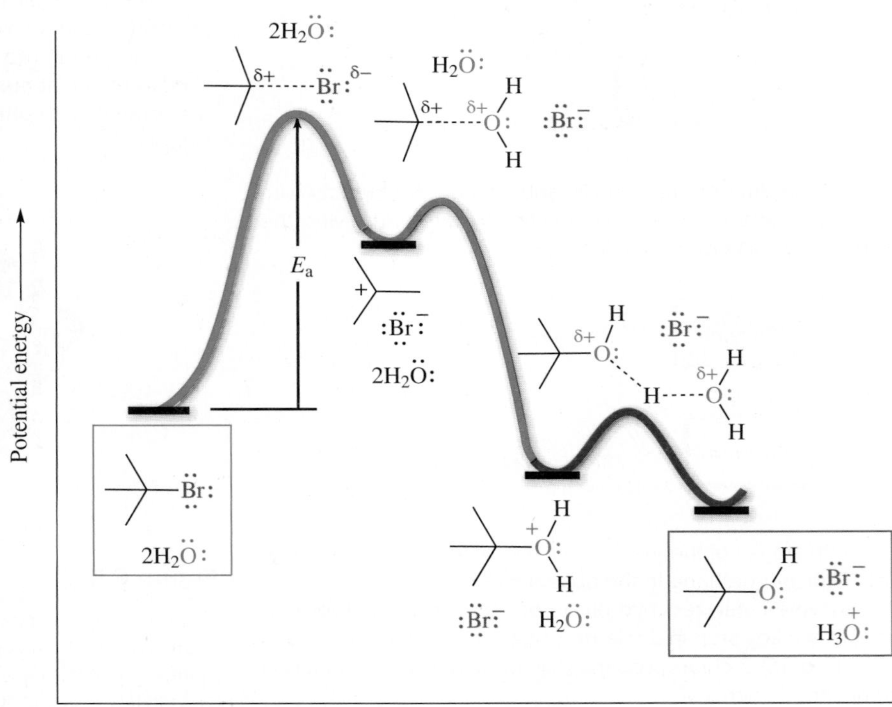

Figure 6.6

Energy diagram illustrating the S_N1 mechanism for hydrolysis of *tert*-butyl bromide.

Mechanism 6.2

The S_N1 Mechanism of Nucleophilic Substitution

THE OVERALL REACTION:

$$(CH_3)_3C{-}Br + 2H_2O \longrightarrow (CH_3)_3C{-}OH + H_3O^+ + Br^-$$

tert-Butyl bromide — Water — *tert*-Butyl alcohol — Hydronium ion — Bromide ion

THE MECHANISM:

Step 1: The alkyl halide dissociates to a carbocation and a halide ion.

$$(CH_3)_3C{-}\ddot{Br}: \xrightarrow{\text{slow}} (CH_3)_3C^+ + :\ddot{Br}:^-$$

tert-Butyl bromide — *tert*-Butyl cation — Bromide ion

Step 2: The carbocation formed in step 1 reacts rapidly with water, which acts as a nucleophile. This step completes the nucleophilic substitution stage of the mechanism and yields an alkyloxonium ion.

$$(CH_3)_3C^+ + H_2\ddot{O}: \longrightarrow (CH_3)_3C{-}\overset{+}{O}H_2$$

tert-Butyl cation — Water — *tert*-Butyloxonium ion

Step 3: This step is a fast acid–base reaction that follows the nucleophilic substitution. Water acts as a base to remove a proton from the alkyloxonium ion to give the observed product of the reaction, *tert*-butyl alcohol.

$$(CH_3)_3C{-}\overset{+}{O}H_2 + H_2\ddot{O}: \longrightarrow (CH_3)_3C{-}\ddot{O}H + H_3O^+$$

tert-Butyloxonium ion — Water — *tert*-Butyl alcohol — Hydronium ion

The key step is the first: a rate-determining unimolecular ionization of the alkyl halide to give a carbocation and a halide ion. Following this, capture of the carbocation by a water molecule acting as a nucleophile gives an alkyloxonium ion, which is then deprotonated by a second water molecule acting as a Brønsted base to complete the process.

In order to compare S_N1 rates in a range of alkyl halides, experimental conditions of low nucleophilicity such as solvolysis (Section 6.5) are chosen so as to suppress competition from S_N2. Under these conditions, the structure/reactivity trend among alkyl halides is exactly opposite to the S_N2 profile.

Increasing relative reactivity toward solvolysis
(aqueous formic acid)

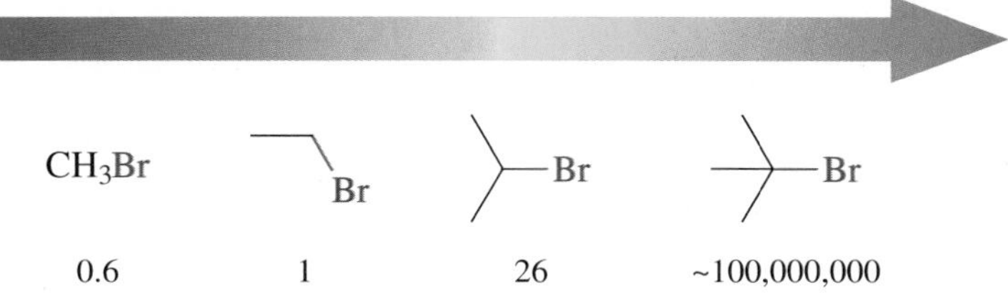

S_N1 reactivity: methyl < primary < secondary < tertiary

We have seen a similar trend in the reaction of alcohols with hydrogen halides (Section 5.10); the more stable the carbocation, the faster it is formed, and the faster the reaction rate. Methyl and primary carbocations are so high in energy that they are unlikely intermediates in nucleophilic substitutions. Although methyl and ethyl bromide undergo hydrolysis under the conditions just described, substitution probably takes place by an S_N2 process in which water is the nucleophile.

In general, methyl and primary alkyl halides never react by the S_N1 mechanism; tertiary alkyl halides never react by S_N2.

Secondary alkyl halides occupy a borderline region in which the nature of the nucleophile is the main determining factor in respect to the mechanism. Secondary alkyl halides usually react with good nucleophiles by the S_N2 mechanism, and with weak nucleophiles by S_N1.

Problem 6.9

Identify the compound in each of the following pairs that reacts at the faster rate in an S_N1 reaction:

(a) Isopropyl bromide or isobutyl bromide
(b) Cyclopentyl iodide or 1-methylcyclopentyl iodide
(c) Cyclopentyl bromide or 1-bromo-2,2-dimethylpropane
(d) *tert*-Butyl chloride or *tert*-butyl iodide

Sample Solution (a) Isopropyl bromide, $(CH_3)_2CHBr$, is a secondary alkyl halide, whereas isobutyl bromide, $(CH_3)_2CHCH_2Br$, is primary. Because the rate-determining step in an S_N1 reaction is carbocation formation and secondary carbocations are more stable than primary ones, isopropyl bromide is more reactive than isobutyl bromide in nucleophilic substitution by the S_N1 mechanism.

Problem 6.10

Numerous studies of their solvolysis reactions (S_N1) have established the approximate rates of nucleophilic substitution in bicyclic compounds A and B relative to their *tert*-butyl counterpart, where X is a halide or sulfonate leaving group.

X X X

tert-Butyl: 1 A: 10^{-6} B: 10^{-14}

Suggest a reasonable explanation for the very large spread in reaction rates.

6.7 Stereochemistry of S_N1 Reactions

Although S_N2 reactions are stereospecific and proceed with inversion of configuration at carbon, the situation is not as clear-cut for S_N1. When the leaving group departs from a chirality center of an optically active halide, the positively charged carbon that results is sp^2-hybridized and cannot be a chirality center. The three bonds to that carbon define a plane of symmetry.

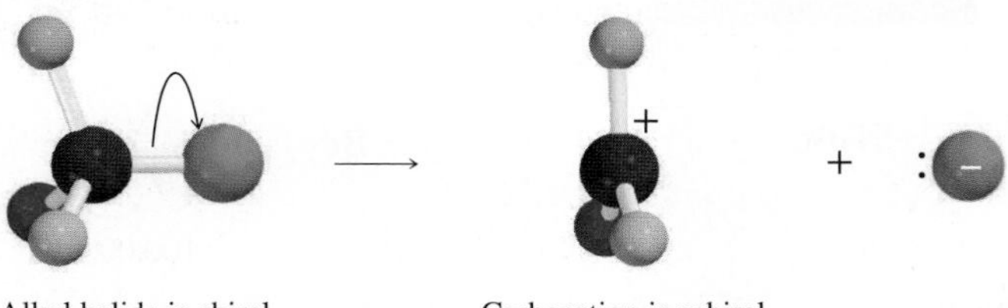

Alkyl halide is chiral Carbocation is achiral

If a nucleophile can approach each face of the carbocation equally well, substitution by the S_N1 mechanism should give a 1:1 mixture of enantiomers irrespective of whether the starting alkyl halide is *R*, *S*, or racemic. S_N1 reactions of alkyl halides should give racemic products from optically active starting materials.

But they rarely do. Methanolysis of the tertiary alkyl halide (*R*)-3-chloro-3,7-dimethyloctane, which almost certainly proceeds by an S_N1 mechanism, takes place with a high degree of inversion of configuration.

Cl — CH_3OH → OCH_3 + OCH_3

(*R*)-6-Chloro-2,6-dimethyloctane → (*S*)-2,6-Dimethyl-6-methoxyoctane (89%) + (*R*)-2,6-Dimethyl-6-methoxyoctane (11%)

Similarly, hydrolysis of (*R*)-2-bromooctane follows a first-order rate law and yields 2-octanol with 66% net inversion of configuration.

Br, H — H_2O, ethanol → H, OH + OH, H

(*R*)-2-Bromooctane → (*S*)-2-Octanol (83%) + (*R*)-2-Octanol (17%)

Partial but not complete loss of optical activity in S_N1 reactions is explained as shown in Figure 6.7. The key feature of this mechanism is that when the carbocation is formed, it is not completely free of the leaving group. Although ionization is complete, the leaving group has not yet diffused very far away from the carbon to which it was attached and partially blocks approach of the nucleophile from that direction. Nucleophilic attack on this species, called an *ion pair,* occurs faster from the side opposite the leaving group. Once the leaving group has diffused away, however, both faces of the carbocation are equally accessible to nucleophiles.

The stereochemistry of S_N1 substitution depends on the relative rates of competing processes—attack by the nucleophile on the ion pair versus separation of the ions. Consequently, the observed stereochemistry varies considerably according to the alkyl halide, nucleophile, and experimental conditions. Some give predominant, but incomplete, inversion of configuration. Others give products that are almost entirely racemic.

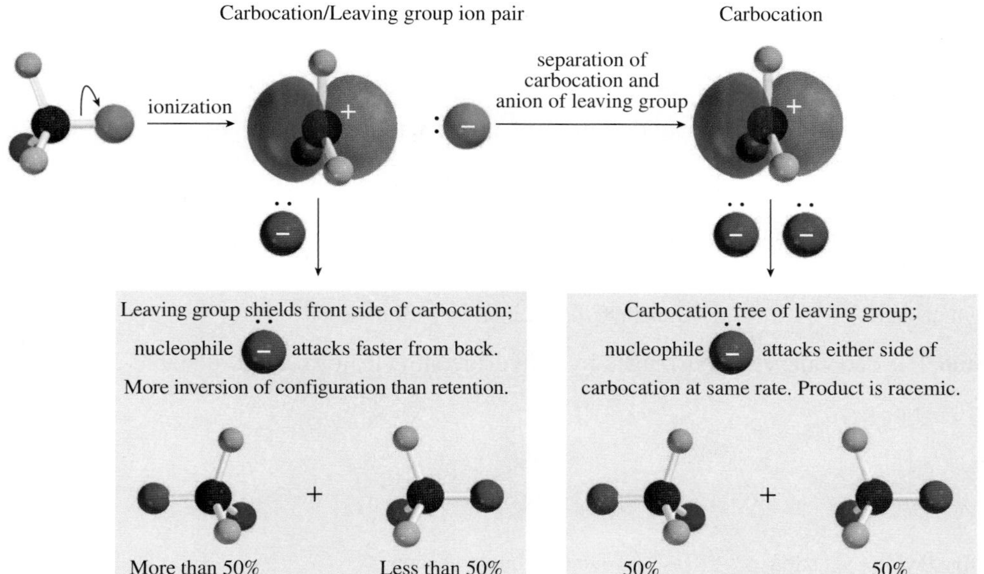

Carbocation/Leaving group ion pair
Carbocation
ionization
separation of carbocation and anion of leaving group
Leaving group shields front side of carbocation; nucleophile attacks faster from back. More inversion of configuration than retention.
Carbocation free of leaving group; nucleophile attacks either side of carbocation at same rate. Product is racemic.
More than 50%
Less than 50%
50%
50%

Figure 6.7

S_N1 stereochemistry. The carbocation formed by ionization of an alkyl halide is shielded on its "front" side by the leaving group. The nucleophile attacks this carbocation–halide ion pair faster from the less shielded "back" side and the product is formed with net inversion of configuration. In a process that competes with nucleophilic attack on the ion pair, the leaving group diffuses away from the carbocation. The nucleophile attacks the carbocation at the same rate from either side to give equal amounts of enantiomers.

Problem 6.11

What two stereoisomeric substitution products would you expect to isolate from the hydrolysis of *cis*-1,4-dimethylcyclohexyl bromide? From hydrolysis of *trans*-1,4-dimethylcyclohexyl bromide?

6.8 Carbocation Rearrangements in S_N1 Reactions

Additional evidence for carbocation intermediates in certain nucleophilic substitutions comes from observing rearrangements of the kind described in Section 5.12. For example, hydrolysis of the secondary alkyl bromide 2-bromo-3-methylbutane yields the rearranged tertiary alcohol 2-methyl-2-butanol as the only substitution product.

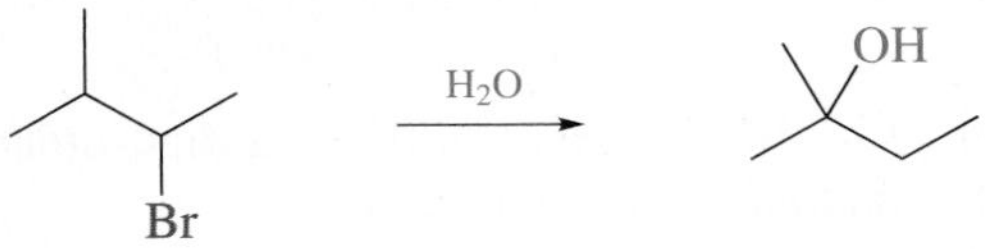

Br $\xrightarrow{H_2O}$ OH

2-Bromo-3-methylbutane 2-Methyl-2-butanol (93%)

Mechanism 6.3 for this reaction assumes rate-determining ionization of the alkyl halide (step 1), followed by a hydride shift that converts a secondary carbocation to a more stable tertiary one (step 2). The tertiary carbocation then reacts with water to yield the observed product (steps 3 and 4).

Problem 6.12

Why does the carbocation intermediate in the hydrolysis of 2-bromo-3-methylbutane rearrange by way of a hydride shift rather than a methyl shift?

Mechanism 6.3

Carbocation Rearrangement in the S_N1 Hydrolysis of 2-Bromo-3-methylbutane

THE OVERALL REACTION:

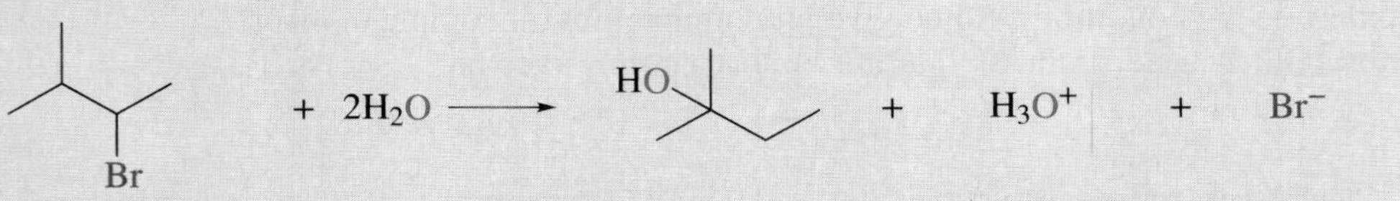

Br + $2H_2O$ ⟶ HO + H_3O^+ + Br^-

2-Bromo-3-methylbutane Water 2-Methyl-2-butanol Hydronium ion Bromide ion

THE MECHANISM:

Step 1: The alkyl halide ionizes to give a carbocation and bromide ion. This is the rate-determining step.

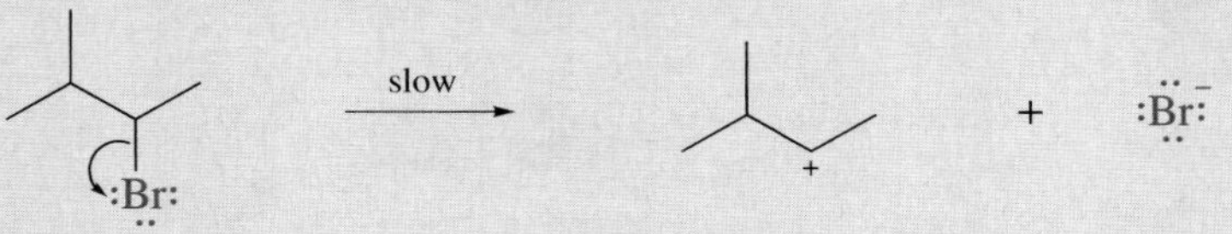

:Br: $\xrightarrow{\text{slow}}$ + + :Br:$^-$

2-Bromo-3-methylbutane 1,2-Dimethylpropyl cation Bromide ion

Step 2: The carbocation formed in step 1 is secondary; it rearranges by a hydride shift to form a more stable tertiary carbocation.

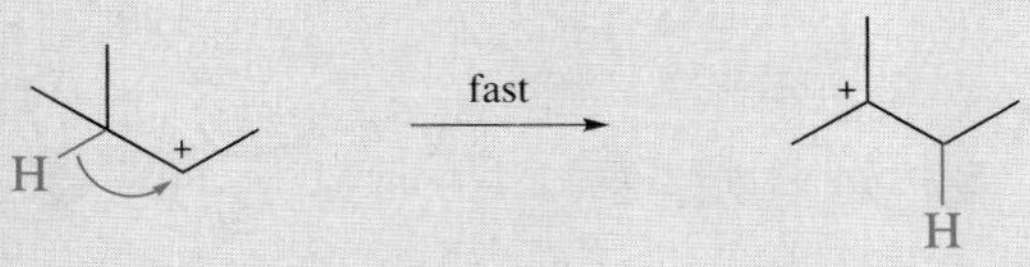

H + $\xrightarrow{\text{fast}}$ + H

1,2-Dimethylpropyl cation 1,1-Dimethylpropyl cation

Step 3: The tertiary carbocation is attacked by water acting as a nucleophile.

1,1-Dimethylpropyl cation + Water $\xrightarrow{\text{fast}}$ 1,1-Dimethylpropyloxonium ion

Step 4: Proton transfer from the alkyloxonium ion to water completes the process.

1,1-Dimethylpropyloxonium ion + Water $\xrightarrow{\text{fast}}$ 2-Methyl-2-butanol + Hydronium ion

Rearrangements, when they do occur, are taken as evidence for carbocation intermediates and point to the S_N1 mechanism as the reaction pathway. Rearrangements are never observed in S_N2 reactions of alkyl halides.

6.9 Effect of Solvent on the Rate of Nucleophilic Substitution

The major effect of the solvent is on the *rate* of nucleophilic substitution, not on what the products are. Thus, we need to consider two related questions:

1. What properties of the *solvent* influence the rate most?
2. How does the rate-determining step of the *mechanism* respond to the properties of the solvent?

We begin by looking at the solvents commonly employed in nucleophilic substitutions, then proceed to examine how these properties affect the S_N1 and S_N2 mechanisms. Because these mechanisms are so different from each other, we discuss each one separately.

Classes of Solvents. Table 6.3 lists a number of solvents in which nucleophilic substitutions are carried out and classifies them according to two criteria: whether they are *protic* or *aprotic,* and *polar* or *nonpolar.*

Protic solvents are those that are capable of hydrogen-bonding interactions. Most have —OH groups, as do the examples in Table 6.3 (water, formic acid, methanol, and acetic acid). The **aprotic** solvents in the table (dimethyl sulfoxide, *N,N*-dimethylformamide, and acetonitrile) lack —OH groups.

Unlike protic and aprotic, which constitute an "either-or" pair, polar and nonpolar belong to a continuous gradation with no sharply defined boundary separating them.

The polarity of a solvent is related to its **dielectric constant (ε),** which is a measure of the ability of a material to moderate the force of attraction between oppositely charged particles. The standard dielectric is a vacuum, assigned a value ε of exactly 1, to which the polarities of other materials are then compared. The higher the dielectric constant ε, the better the medium is able to support separated positively and negatively charged species. Solvents with high dielectric constants are classified as **polar** solvents; those with low dielectric constants are **nonpolar.**

Problem 6.13

Diethyl ether ($CH_3CH_2OCH_2CH_3$) has a dielectric constant of 4. What best describes its solvent properties: polar protic, nonpolar protic, polar aprotic, or nonpolar aprotic?

TABLE 6.3 Properties of Some Solvents Used in Nucleophilic Substitution

Solvent	Structural formula	Protic or Aprotic	Dielectric constant ε*	Polarity
Water	H_2O	Protic	78	Most polar
Formic acid	$HC(=O)OH$	Protic	58	
Dimethyl sulfoxide	$(CH_3)_2\overset{+}{S}-\overset{-}{O}$	Aprotic	49	
Acetonitrile	$CH_3C\equiv N$	Aprotic	37	
N,N-Dimethylformamide	$(CH_3)_2NC(=O)H$	Aprotic	37	
Methanol	CH_3OH	Protic	33	
Acetic acid	$CH_3C(=O)OH$	Protic	6	Least polar

*Dielectric constants are approximate and temperature-dependent.

Solvent Effects on the Rate of Substitution by the S_N2 Mechanism. Polar solvents are required in typical bimolecular substitutions because ionic substances, such as the sodium and potassium salts cited earlier in Table 6.1, are not sufficiently soluble in nonpolar solvents to give a high enough concentration of the nucleophile to allow the reaction to occur at a rapid rate. Other than the requirement that the solvent be polar enough to dissolve ionic compounds, however, the effect of solvent polarity on the rate of S_N2 reactions is small. What is more important is whether the polar solvent is protic or aprotic. Protic solvents such as water, formic acid, methanol, and acetic acid all have —OH groups that allow them to form hydrogen bonds to anionic nucleophiles.

$$\underset{\text{Solvent}}{R\overset{\delta-}{O}-\overset{\delta+}{H}} + \underset{\text{Nucleophile}}{:Y^-} \longrightarrow \underset{\text{Hydrogen-bonded complex}}{R\overset{\delta-}{O}-H\text{---}\overset{\delta-}{Y}}$$

This clustering of *protic* solvent molecules (*solvation*) around an anion suppresses its nucleophilicity and retards the rate of bimolecular substitution.

Aprotic solvents, on the other hand, lack —OH groups and do not solvate anions very strongly, leaving the anions much more able to express their nucleophilic character. Table 6.4 compares the second-order rate constants k for S_N2 substitution of 1-bromobutane by azide ion (a good nucleophile) in several polar aprotic solvents with the corresponding k's for the much slower reactions in polar protic solvents.

$$\underset{\text{Butyl bromide}}{CH_3CH_2CH_2CH_2Br} + \underset{\text{Azide ion}}{N_3^-} \longrightarrow \underset{\text{Butyl azide}}{CH_3CH_2CH_2CH_2N_3} + \underset{\text{Bromide ion}}{Br^-}$$

Problem 6.14

Unlike protic solvents, which solvate anions, polar aprotic solvents form complexes with cations better than with anions. Use a dashed line to show the interaction between dimethyl sulfoxide [$(CH_3)_2\overset{+}{\ddot{S}}-\ddot{\underset{..}{O}}:^-$] with a cation, using sodium azide (NaN_3) as the source of the cation.

TABLE 6.4 Relative Rate of S_N2 Displacement of 1-Bromobutane by Azide in Various Solvents*

Solvent	Structural formula	Dielectric constant ε	Type of solvent	Relative rate
Methanol	CH_3OH	33	Polar protic	1
Water	H_2O	78	Polar protic	7
Dimethyl sulfoxide	$(CH_3)_2\overset{+}{S}-\overset{-}{O}$	49	Polar aprotic	1300
N,N-Dimethylformamide	$(CH_3)_2NCH{=}O$	37	Polar aprotic	2800
Acetonitrile	$CH_3C{\equiv}N$	37	Polar aprotic	5000

*Ratio of second-order rate constant for substitution in indicated solvent to that for substitution in methanol at 25°C.

The large rate enhancements observed for bimolecular nucleophilic substitutions in polar aprotic solvents offer advantages in synthesis. One example is the preparation of alkyl cyanides (nitriles) by the reaction of sodium cyanide with alkyl halides:

$$\text{Hexyl halide (X)} + \text{NaCN} \longrightarrow \text{Hexyl cyanide (CN)} + \text{NaX}$$

Hexyl halide · Sodium cyanide · Hexyl cyanide · Sodium halide

When the reaction was carried out in aqueous methanol as the solvent, hexyl bromide was converted to hexyl cyanide in 71% yield. Although this is perfectly acceptable for a synthetic reaction, it required heating for a period of over *20 hours*. Changing the solvent to dimethyl sulfoxide increased the reaction rate to the extent that the less reactive (and less expensive) hexyl chloride could be used and the reaction was complete (91% yield) in only *20 minutes!*

The *rate* at which reactions occur can be important in the laboratory, and understanding how solvents affect rate is of practical value. As we proceed through the text, however, and see how nucleophilic substitution is applied to a variety of functional-group transformations, be aware that the nature of both the substrate and the nucleophile, more than anything else, determines what *product* is formed.

Solvent Effects on the Rate of Substitution by the S_N1 Mechanism. Table 6.5 gives the relative rate of solvolysis of *tert*-butyl chloride in several protic solvents listed in order of increasing dielectric constant. As the table illustrates, the rate of solvolysis of *tert*-butyl chloride (which is equal to its rate of ionization) increases dramatically as the solvent becomes more polar.

TABLE 6.5 Relative Rate of S_N1 Solvolysis of *tert*-Butyl Chloride as a Function of Solvent Polarity*

Solvent	Dielectric constant ε	Relative rate
Acetic acid	6	1
Methanol	33	4
Formic acid	58	5,000
Water	78	150,000

*Ratio of first-order rate constant for solvolysis in indicated solvent to that for solvolysis in acetic acid at 25°C.

Figure 6.8

A polar solvent stabilizes the transition state of an S_N1 reaction and increases its rate.

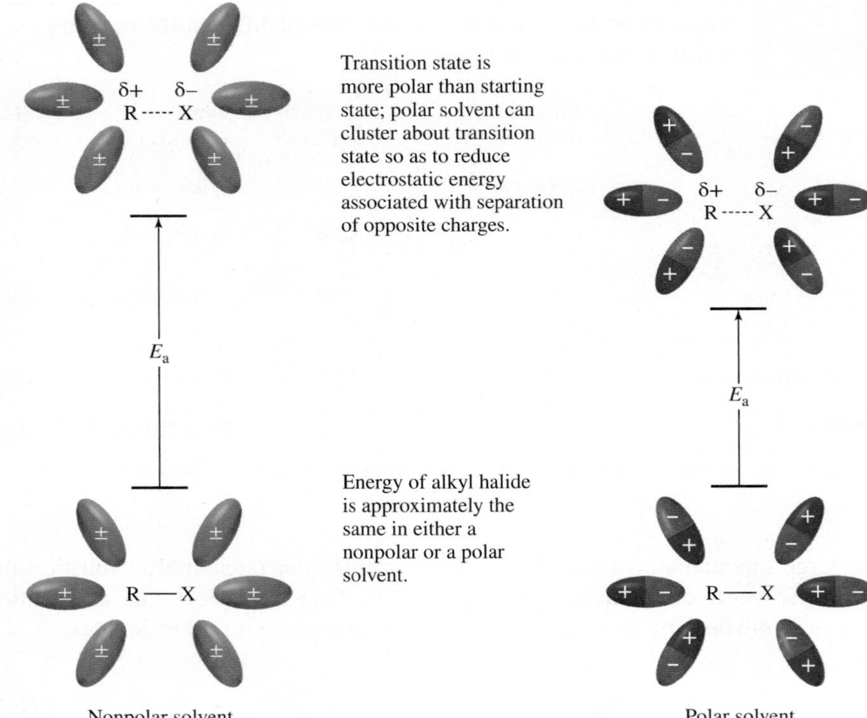
δ+ δ−
R----- X
Transition state is more polar than starting state; polar solvent can cluster about transition state so as to reduce electrostatic energy associated with separation of opposite charges.
δ+ δ−
R----- X
E_a
E_a
Energy of alkyl halide is approximately the same in either a nonpolar or a polar solvent.
R—X
R—X
Nonpolar solvent
Polar solvent

According to the S_N1 mechanism, a molecule of an alkyl halide ionizes to a positively charged carbocation and a negatively charged halide ion in the rate-determining step. As the alkyl halide approaches the transition state for this step, positive charge develops on the carbon and negative charge on the halogen. The effects of a nonpolar and a polar solvent on the energy of the transition state are contrasted in Figure 6.8. Polar and nonpolar solvents are similar in their interaction with the starting alkyl halide, but differ markedly in how they stabilize the transition state. A solvent with a low dielectric constant has little effect on the energy of the transition state, whereas one with a high dielectric constant stabilizes the charge-separated transition state, lowers the activation energy, and increases the rate of the reaction.

If the solvent, like those listed in Table 6.4, is protic, stabilization of the transition state is even more pronounced because of the hydrogen bonding that develops as the leaving group becomes negatively charged.

6.10 Nucleophilic Substitution of Alkyl Sulfonates

A few other classes of organic compounds undergo nucleophilic substitution reactions analogous to those of alkyl halides; the most important of these are sulfonates.

Sulfonic acids such as methanesulfonic acid and *p*-toluenesulfonic acid are strong acids, comparable in acidity with sulfuric acid.

$$H_3C-S(=O)_2-OH \qquad H_3C-C_6H_4-S(=O)_2-OH$$

Methanesulfonic acid　　*p*-Toluenesulfonic acid

TABLE 6.6 Approximate Relative Leaving-Group Abilities*

Leaving group	Relative rate	Conjugate acid of leaving group	pK_a of conjugate acid
F^-	10^{-5}	HF	3.1
Cl^-	10^0	HCl	−3.9
Br^-	10^1	HBr	−5.8
I^-	10^2	HI	−10.4
H_2O	10^1	H_3O^+	−1.7
$CH_3SO_2O^-$	10^4	CH_3SO_2OH	−2.6
TsO^-	10^5	TsOH	−2.8
$CF_3SO_2O^-$	10^8	CF_3SO_2OH	−6.0

*Values are approximate and vary according to substrate.

Alkyl sulfonates are derivatives of sulfonic acids prepared by treating an alcohol with the appropriate sulfonyl chloride, usually in the presence of pyridine.

$$CH_3CH_2OH + H_3C-C_6H_4-SO_2Cl \xrightarrow{\text{pyridine}} CH_3CH_2OSO_2-C_6H_4-CH_3$$

Ethanol — *p*-Toluenesulfonyl chloride — Ethyl *p*-toluenesulfonate (72%)

Alkyl sulfonates resemble alkyl halides in their ability to undergo nucleophilic substitution. Those used most frequently are the *p*-toluenesulfonates, commonly known as *tosylates* and abbreviated as ROTs.

(3-Cyclopentenyl)methyl *p*-toluenesulfonate (OTs) $\xrightarrow[\text{ethanol–water}]{\text{KCN}}$ (4-Cyanomethyl)cyclopentene (86%) (CN)

As shown in Table 6.6, alkyl tosylates undergo nucleophilic substitution at rates that are even faster than those of the corresponding iodides. Iodide is the weakest base and the best leaving group among the halide anions. Similarly, sulfonate ions rank among the least basic of the oxygen-containing leaving groups. The weaker the base, the better the leaving group. Trifluoromethanesulfonate (*triflate,* $CF_3SO_2O^-$) is a much weaker base than *p*-toluenesulfonate and is the best leaving group in the table.

Notice that strongly basic leaving groups are absent from Table 6.6. In general, any species that has pK_a greater than about 2 for its conjugate acid cannot be a leaving group in a nucleophilic substitution. Thus, hydroxide (HO^-) is far too strong a base to be displaced from an alcohol (ROH), and alcohols themselves do not undergo nucleophilic substitution. In strongly acidic media, alcohols are protonated to give alkyloxonium ions, and these do undergo nucleophilic substitution, because the leaving group is a weakly basic water molecule. S_N2 reactions are most favorable when a more basic nucleophile displaces a less basic leaving group.

Because halides are poorer leaving groups than *p*-toluenesulfonate, alkyl *p*-toluenesulfonates can be converted to alkyl halides by S_N2 reactions involving chloride, bromide, or iodide as the nucleophile.

sec-Butyl p-toluenesulfonate (OTs) + NaBr (Sodium bromide) $\xrightarrow{\text{DMSO}}$ sec-Butyl bromide (82%) + NaOTs (Sodium p-toluenesulfonate)

sec-Butyl *p*-toluenesulfonate | Sodium bromide | *sec*-Butyl bromide (82%) | Sodium *p*-toluenesulfonate

Problem 6.15

Write a chemical equation showing the preparation of octadecyl *p*-toluenesulfonate.

Problem 6.16

Write equations showing the reaction of octadecyl *p*-toluenesulfonate with each of the following reagents:

(a) Potassium acetate ($KOCCH_3$, with C=O)
(b) Potassium iodide (KI)
(c) Potassium cyanide (KCN)
(d) Potassium hydrogen sulfide (KSH)
(e) Sodium butanethiolate ($NaSCH_2CH_2CH_2CH_3$)

Sample Solution All these reactions of octadecyl *p*-toluenesulfonate have been reported in the chemical literature, and all proceed in synthetically useful yield. You should begin by identifying the nucleophile in each of the parts to this problem. The nucleophile replaces the *p*-toluenesulfonate leaving group in an S_N2 reaction. In part (a) the nucleophile is acetate ion, and the product of nucleophilic substitution is octadecyl acetate.

$$CH_3C(=O)O^- + H_2C(-OTs)(CH_2)_{16}CH_3 \longrightarrow CH_3C(=O)OCH_2(CH_2)_{16}CH_3 + {}^-OTs$$

Acetate ion | Octadecyl tosylate | Octadecyl acetate | Tosylate ion

An advantage that sulfonates have over alkyl halides is that their preparation from alcohols does not involve any of the bonds to carbon. The alcohol oxygen becomes the oxygen that connects the alkyl group to the sulfonyl group. Thus, the stereochemical configuration of a sulfonate is the same as that of the alcohol from which it was prepared. If we wish to study the stereochemistry of nucleophilic substitution in an optically active substrate, for example, we know that a tosylate will have the same configuration and the same optical purity as the alcohol from which it was prepared.

(*S*)-(+)-2-Octanol $\xrightarrow[\text{pyridine}]{p\text{-toluenesulfonyl chloride}}$ (*S*)-(+)-1-Methylheptyl *p*-toluenesulfonate

(*S*)-(+)-2-Octanol
$[\alpha]_D^{25}$ +9.9°
(optically pure)

(*S*)-(+)-1-Methylheptyl *p*-toluenesulfonate
$[\alpha]_D^{25}$ +7.9°
(optically pure)

The same cannot be said about reactions with alkyl halides. The conversion of optically active 2-octanol to the corresponding halide *does* involve a bond to the chirality center, and so the optical purity and absolute configuration of the alkyl halide need to be independently established.

The mechanisms by which sulfonates undergo nucleophilic substitution are the same as those of alkyl halides. Inversion of configuration is observed in S_N2 reactions of alkyl sulfonates and predominant inversion accompanied by racemization in S_N1 processes.

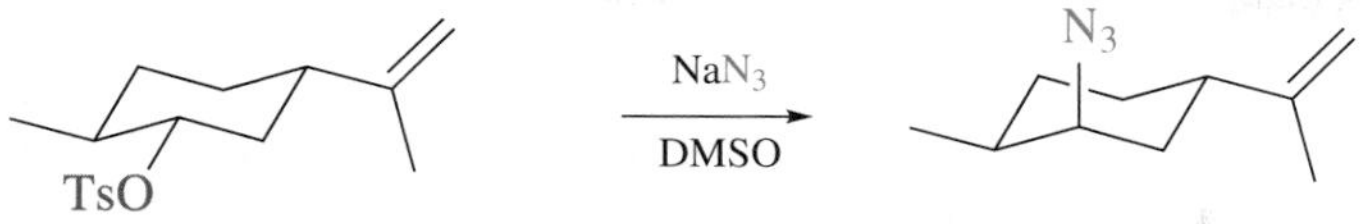

TsO —(NaN_3, DMSO)→ N_3

(1*S*, 2*S*, 5*S*)-5-Isopropenyl-2-methylcyclohexyl *p*-toluenesulfonate

(1*R*, 2*S*, 5*S*)-5-Isopropenyl-2-methylcyclohexyl azide (76%)

Problem 6.17

The hydrolysis of sulfonates of 2-octanol is stereospecific and proceeds with complete inversion of configuration. Write a structural formula that shows the stereochemistry of the 2-octanol formed by hydrolysis of an optically pure sample of (*S*)-(+)-1-methylheptyl *p*-toluenesulfonate, identify the product as *R* or *S*, and deduce its specific rotation.

6.11 Introduction to Organic Synthesis: Retrosynthetic Analysis

Retrosynthetic analysis is one component of a formal system of synthetic planning developed by E. J. Corey (Harvard). Corey received the 1990 Nobel Prize in Chemistry for his achievements in synthetic organic chemistry.

An important concern to chemists is synthesis, the challenge of preparing a particular compound in an economical way with confidence that the method chosen will lead to the desired structure. In this section we introduce the topic of synthesis, emphasizing the need for systematic planning to find the best sequence of steps to prepare a desired product (the target molecule).

Two critical features of synthetic planning are to *always use reactions that you know will work* and to *reason backward from the target to the starting material.* A way to represent this backward reasoning is called **retrosynthetic analysis,** characterized by an arrow of the type ⟹ pointing from the product of a synthetic step toward the reactant.

Suppose you wanted to prepare (*R*)-2-butanethiol from (*S*)-2-butanol.

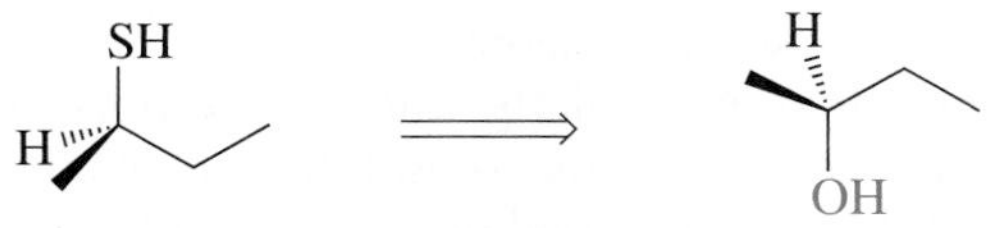

SH, H ⟹ H, OH

(*R*)-2-Butanethiol (*S*)-2-Butanol

Begin by asking the question "From what can I prepare the target molecule in a single step?" Table 6.1 tells us that alkyl halides react with HS^- by nucleophilic substitution to give thiols. Therefore, we are tempted to consider the retrosynthesis shown.

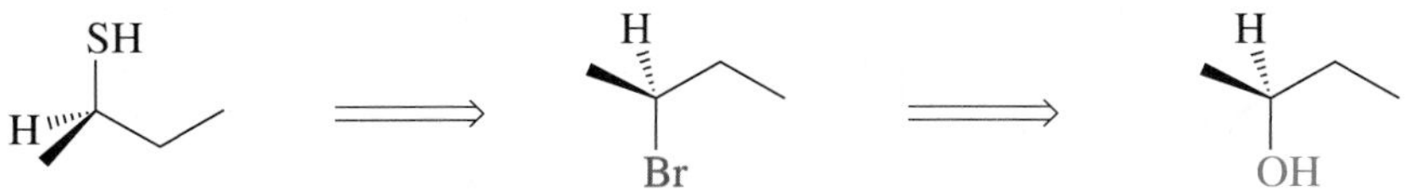

SH, H ⟹ H, Br ⟹ H, OH

(*R*)-2-Butanethiol (*S*)-2-Bromobutane (*S*)-2-Butanol

The flaw in this plan is stereochemical in that it requires that (*S*)-2-butanol be converted to (*S*)-2-bromobutane in order to get the correct enantiomer of the desired thiol. We know, however, that the reaction of alcohols with hydrogen halides proceeds with inversion of configuration (Section 5.11). Therefore, we need an alternative plan—one that takes into account both the functionality and configuration at C-2. One based on a sulfonate leaving group satisfies both considerations.

(*R*)-2-Butanethiol ⟹ (*S*)-1-Methylpropyl *p*-toluenesulfonate ⟹ (*S*)-2-Butanol

A synthesis based on this plan is straightforward and satisfies considerations of both connectivity and stereochemistry.

(*S*)-2-Butanol —*p*-toluenesulfonyl chloride, pyridine→ (*S*)-1-Methylpropyl *p*-toluenesulfonate —NaSH→ (*R*)-2-Butanethiol

Problem 6.18

Use retrosynthetic analysis to devise a synthesis of 2-bromo-3-methylbutane. You will find it useful to refer to Sections 6.3 and 6.10.

Chemists are often presented with the task of preparing a target molecule from an unspecified starting material. Usually there will be several possibilities and choosing the best one includes considering a number of factors—availability of the starting materials, cost, scale, and disposal of hazardous waste, among others. As we proceed through the text and develop a larger inventory of functional-group transformations and methods for forming carbon–carbon bonds, our ability to evaluate alternative synthetic plans will increase. In most cases the best synthetic plan is the one with the fewest steps.

6.12 Substitution versus Elimination: A Look Ahead

Early in this chapter, Table 6.1 listed a number of examples in which alkyl halides were converted to other functional-group classes by nucleophilic substitution. As the chapter progressed, important aspects of nucleophilic substitution were developed from a mechanistic perspective based on the S_N1/S_N2 concepts of Hughes and Ingold. As we close the chapter, we return to the use of nucleophilic substitution as a synthetic method for functional-group transformation, particularly with regard to its major limitation.

The major limitation to nucleophilic substitution as applied to both alkyl halides and sulfonates is that similar reaction conditions can lead to **elimination** as well as, or even instead of, substitution.

Alkyl halide or sulfonate + nucleophile ($+ Y^-$) —nucleophilic substitution→ (substitution product) $+ X^-$

—elimination→ (alkene) $+ H{-}Y + X^-$

Thus, a particular combination of reactants can give a mixture containing the products of elimination and/or substitution depending on a number of factors. These factors will be explored in detail in Chapter 7. For the present chapter, however, nucleophilic substitution has predominated over elimination in all of the reactions discussed so far and will in the end-of-chapter problems as well.

6.13 SUMMARY

Section 6.1 Nucleophilic substitution is one of the main methods for functional-group transformations. Examples of synthetically useful nucleophilic substitutions were given in Table 6.1. It is a good idea to return to that table and review its entries now that the details of nucleophilic substitution have been covered.

Sections 6.2–6.9 These sections show how a variety of experimental observations led to the proposal of the S_N1 and the S_N2 mechanisms for nucleophilic substitution. Summary Table 6.7 integrates the material in these sections.

TABLE 6.7 Comparison of S_N1 and S_N2 Mechanisms of Nucleophilic Substitution in Alkyl Halides

	S_N1	S_N2
Characteristics of mechanism	Two elementary steps: Step 1: $R—\ddot{X}: \rightleftharpoons R^+ + :\ddot{X}:^-$ Step 2: $R^+ + :Nu^- \longrightarrow R—Nu$ Ionization of alkyl halide (step 1) is rate-determining. (Section 6.6)	Single step: $^-Nu: \; R—\ddot{X}: \longrightarrow Nu—R + :\ddot{X}:^-$ Nucleophile displaces leaving group; bonding to the incoming nucleophile accompanies cleavage of the bond to the leaving group. (Section 6.3)
Rate-determining transition state	$^{\delta+}R---\ddot{X}:^{\delta-}$ (Section 6.6)	$^{\delta-}Nu---R---\ddot{X}:^{\delta-}$ (Section 6.3)
Molecularity	Unimolecular (Section 6.6)	Bimolecular (Section 6.3)
Kinetics and rate law	First order: Rate = k[alkyl halide] (Section 6.6)	Second order: Rate = k[alkyl halide][nucleophile] (Section 6.3)
Relative reactivity of halide leaving groups	RI > RBr > RCl >> RF (Section 6.2)	RI > RBr > RCl >> RF (Section 6.2)
Effect of structure on rate	$R_3CX > R_2CHX > RCH_2X > CH_3X$ Rate is governed by stability of carbocation that is formed in ionization step. Tertiary alkyl halides can react only by the S_N1 mechanism; they never react by the S_N2 mechanism. (Section 6.6)	$CH_3X > RCH_2X > R_2CHX > R_3CX$ Rate is governed by steric effects (crowding in transition state). Methyl and primary alkyl halides can react only by the S_N2 mechanism; they never react by the S_N1 mechanism. (Section 6.6)
Effect of nucleophile on rate	Rate of substitution is independent of both concentration and nature of nucleophile. Nucleophile does not participate until after rate-determining step. (Section 6.6)	Rate depends on both nature of nucleophile and its concentration. (Sections 6.3 and 6.5)
Effect of solvent on rate	Rate increases with increasing polarity of solvent as measured by its dielectric constant ε. (Section 6.9)	Polar aprotic solvents give fastest rates of substitution; solvation of $Nu:^-$ is minimal and nucleophilicity is greatest. (Section 6.9)
Stereochemistry	Not stereospecific: racemization accompanies inversion when leaving group is located at a chirality center. (Section 6.7)	Stereospecific: 100% inversion of configuration at reaction site. Nucleophile attacks carbon from side opposite bond to leaving group. (Section 6.3)
Potential for rearrangements	Carbocation intermediate capable of rearrangement. (Section 6.8)	No carbocation intermediate; no rearrangement.

Section 6.10 Nucleophilic substitution can occur with leaving groups other than halide. Alkyl *p*-toluenesulfonates (*tosylates*), which are prepared from alcohols by reaction with *p*-toluenesulfonyl chloride, are often used.

$$\text{ROH} + \text{H}_3\text{C}-\text{C}_6\text{H}_4-\text{S}(=\text{O})_2-\text{Cl} \xrightarrow{\text{pyridine}} \text{ROS}(=\text{O})_2-\text{C}_6\text{H}_4-\text{CH}_3 \ (\text{ROTs})$$

Alcohol — *p*-Toluenesulfonyl chloride — Alkyl *p*-toluenesulfonate (alkyl tosylate)

In its ability to act as a leaving group, *p*-toluenesulfonate is even more reactive than iodide.

$$\bar{\text{N}}\text{u}: \; \text{R}-\text{OTs} \longrightarrow \text{Nu}-\text{R} + {}^{-}\text{OTs}$$

Nucleophile — Alkyl *p*-toluenesulfonate — Substitution product — *p*-Toluenesulfonate ion

Section 6.11 Retrosynthetic analysis can suggest a synthetic transformation by disconnecting a bond to a functional group and considering how that group can be introduced into the carbon chain by nucleophilic substitution.

$$\text{R} \,\vdots\, \text{Y} \Longrightarrow \text{R}-\text{X} + \text{Y}^-$$

Section 6.12 When nucleophilic substitution is used for synthesis, the competition between substitution and elimination must favor substitution. The factors that influence this competition will be described in Chapter 7.

PROBLEMS

Predict the Products

6.19 Write the structure of the major organic product from the reaction of 1-bromopropane with each of the following:

(a) Sodium iodide in acetone

(b) Sodium acetate ($CH_3\overset{O}{\overset{\|}{C}}ONa$) in acetic acid

(c) Sodium ethoxide in ethanol

(d) Sodium cyanide in dimethyl sulfoxide

(e) Sodium azide in aqueous ethanol

(f) Sodium hydrogen sulfide in ethanol

(g) Sodium methanethiolate ($NaSCH_3$) in ethanol

6.20 Each of the following nucleophilic substitution reactions has been reported in the chemical literature. Many of them involve reactants that are somewhat more complex than those we have dealt with to this point. Nevertheless, you should be able to predict the product by analogy to what you know about nucleophilic substitution in simple systems.

(a) $\text{BrCH}_2\text{C(=O)CH}_2\text{CH}_2\text{CH}_3 \xrightarrow[\text{acetone}]{\text{NaI}}$

(b) $\text{O}_2\text{N}-\text{C}_6\text{H}_4-\text{CH}_2\text{Cl} \xrightarrow[\text{acetic acid}]{\text{CH}_3\overset{O}{\overset{\|}{C}}\text{ONa}}$

(c) 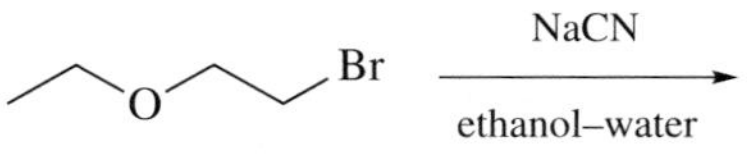 [structure: ethoxyethyl bromide] $\xrightarrow[\text{ethanol–water}]{NaCN}$

(d) 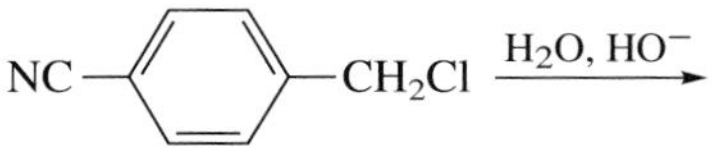 NC–C₆H₄–CH_2Cl $\xrightarrow{H_2O,\ HO^-}$

(e) 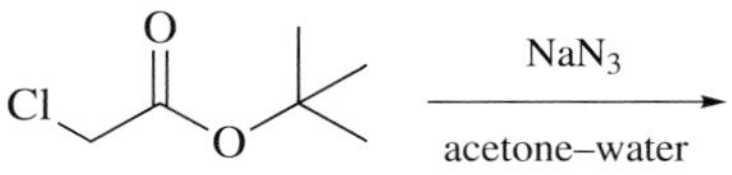 [structure: tert-butyl chloroacetate] $\xrightarrow[\text{acetone–water}]{NaN_3}$

(f) 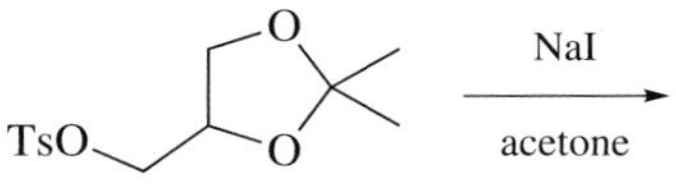 [structure: TsO–CH₂–dioxolane] $\xrightarrow[\text{acetone}]{NaI}$

(g) 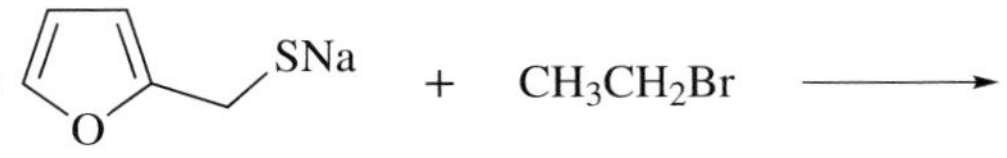 [structure: furylmethyl–SNa] + CH_3CH_2Br $\longrightarrow$

(h) [structure: trimethoxyphenylbutanol, OCH_3, CH_3O, CH_3O, OH] $\xrightarrow[\text{2. LiI, acetone}]{\text{1. TsCl, pyridine}}$

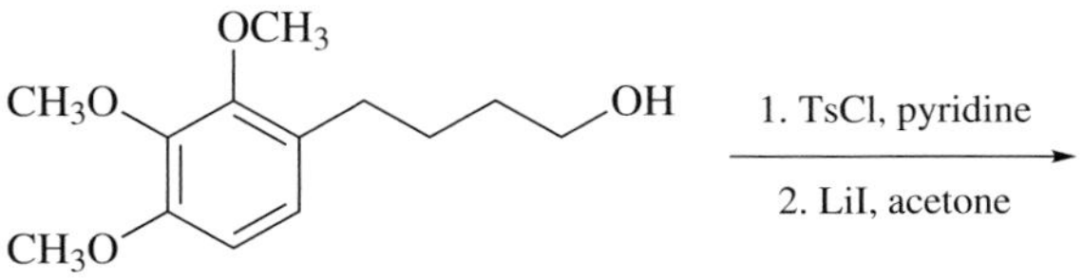

6.21 Both of the following reactions involve nucleophilic substitution. The product of reaction (a) is an isomer of the product of reaction (b). What kind of isomer? By what mechanism does nucleophilic substitution occur? Write the structural formula of the product of each reaction.

(a) 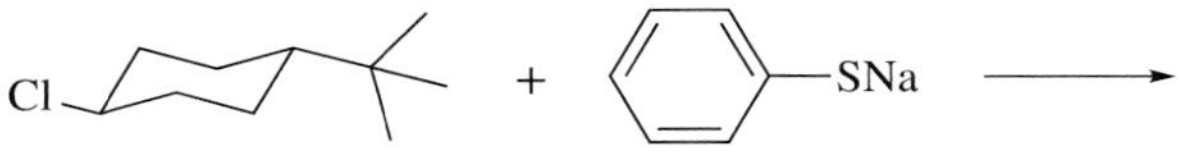 [structure: Cl-substituted tert-butylcyclohexane] + C_6H_5SNa $\longrightarrow$

(b) 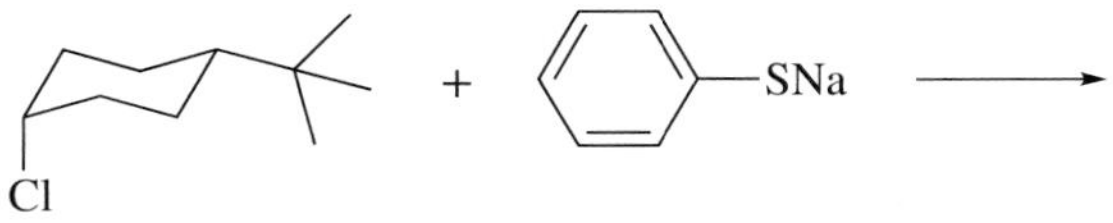 [structure: Cl-substituted tert-butylcyclohexane (axial Cl)] + C_6H_5SNa $\longrightarrow$

6.22 Identify the product in each of the following reactions:

(a) 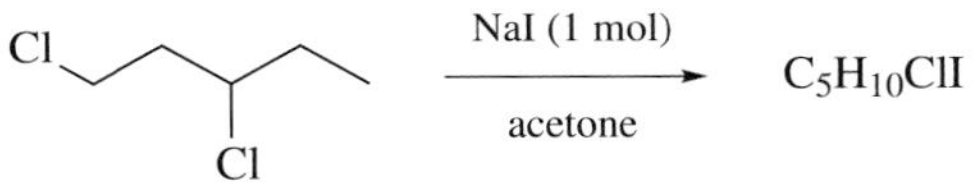 [structure: 1,3-dichloropentane] $\xrightarrow[\text{acetone}]{\text{NaI (1 mol)}}$ $C_5H_{10}ClI$

(b) $BrCH_2CH_2Br$ + $NaSCH_2CH_2SNa$ $\longrightarrow$ $C_4H_8S_2$

(c) $ClCH_2CH_2CH_2CH_2Cl$ + Na_2S $\longrightarrow$ C_4H_8S

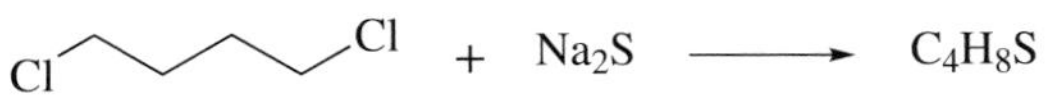

6.23 The compound KSCN is a source of *thiocyanate* ion.

(a) Write the two most stable Lewis structures for thiocyanate ion and identify the atom in each that bears a formal charge of −1.

(b) Two constitutionally isomeric products of molecular formula C_5H_9NS were isolated in a combined yield of 87% in the reaction shown. (DMF stands for *N,N*-dimethylformamide, a polar aprotic solvent.) Suggest reasonable structures for these two compounds.

$$CH_3CH_2CH_2CH_2Br \xrightarrow[DMF]{KSCN}$$

6.24 Sodium nitrite ($NaNO_2$) reacted with 2-iodooctane to give a mixture of two constitutionally isomeric compounds of molecular formula $C_8H_{17}NO_2$ in a combined yield of 88%. Suggest reasonable structures for these two isomers.

6.25 Reaction of ethyl iodide with triethylamine [$(CH_3CH_2)_3N$:] yields a crystalline compound $C_8H_{20}NI$ in high yield. This compound is soluble in polar solvents such as water but insoluble in nonpolar ones such as diethyl ether. It does not melt below about 200°C. Suggest a reasonable structure for this product.

Rate and Mechanism

6.26 There is an overall 29-fold difference in reactivity of 1-chlorohexane, 2-chlorohexane, and 3-chlorohexane toward potassium iodide in acetone.

(a) Which one is the most reactive? Why?

(b) Two of the isomers differ by only a factor of 2 in reactivity. Which two are these? Which one is the more reactive? Why?

6.27 In each of the following, indicate which reaction will occur faster. Explain your reasoning.

(a) $CH_3CH_2CH_2CH_2Br$ or $CH_3CH_2CH_2CH_2I$ with sodium cyanide in dimethyl sulfoxide

(b) 1-Chloro-2-methylbutane or 1-chloropentane with sodium iodide in acetone

(c) Hexyl chloride or cyclohexyl chloride with sodium azide in aqueous ethanol

(d) Solvolysis of 1-bromo-2,2-dimethylpropane or *tert*-butyl bromide in ethanol

(e) Solvolysis of isobutyl bromide or *sec*-butyl bromide in aqueous formic acid

(f) Reaction of 1-chlorobutane with sodium acetate in acetic acid or with sodium methoxide in methanol

(g) Reaction of 1-chlorobutane with sodium azide or sodium *p*-toluenesulfonate in aqueous ethanol

6.28 The reaction of 2,2-dimethyl-1-propanol with HBr is very slow and gives 2-bromo-2-methylbutane as the major product.

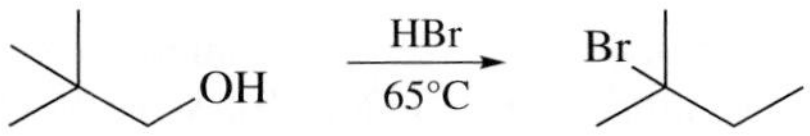

OH —HBr, 65°C→ Br

Give a mechanistic explanation for these observations.

6.29 If the temperature is not kept below 25°C during the reaction of primary alcohols with *p*-toluenesulfonyl chloride in pyridine, it is sometimes observed that the isolated product is not the desired alkyl *p*-toluenesulfonate but is instead the corresponding alkyl chloride. Suggest a mechanistic explanation for this observation.

6.30 The reaction of cyclopentyl bromide with sodium cyanide to give cyclopentyl cyanide

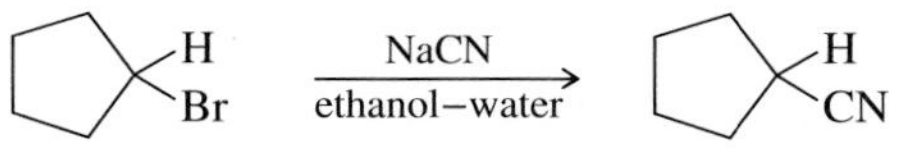

H, Br —NaCN, ethanol–water→ H, CN

Cyclopentyl bromide — Cyclopentyl cyanide

proceeds faster if a small amount of sodium iodide is added to the reaction mixture. Can you suggest a reasonable mechanism to explain the catalytic function of sodium iodide?

6.31 In a classic experiment, Edward D. Hughes (a colleague of Ingold's at University College, London) studied the rate of racemization of 2-iodooctane by sodium iodide in acetone and compared it with the rate of incorporation of radioactive iodine into 2-iodooctane.

$$RI + [I^*]^- \longrightarrow RI^* + I^-$$

(I* = radioactive iodine)

How will the rate of racemization compare with the rate of incorporation of radioactivity if

(a) Each act of exchange proceeds stereospecifically with retention of configuration?

(b) Each act of exchange proceeds stereospecifically with inversion of configuration?

(c) Each act of exchange proceeds in a stereorandom manner, in which retention and inversion of configuration are equally likely?

6.32 Give the mechanistic symbols (S_N1, S_N2) that are most consistent with each of the following statements:

(a) Methyl halides react with sodium ethoxide in ethanol only by this mechanism.

(b) Unhindered primary halides react with sodium ethoxide in ethanol mainly by this mechanism.

(c) The substitution product obtained by solvolysis of *tert*-butyl bromide in ethanol arises by this mechanism.

(d) Reactions proceeding by this mechanism are stereospecific.

(e) Reactions proceeding by this mechanism involve carbocation intermediates.

(f) This mechanism is most likely to have been involved when the products are found to have a different carbon skeleton from the substrate.

(g) Alkyl iodides react faster than alkyl bromides in reactions that proceed by these mechanisms.

Stereochemistry

6.33 Write an equation, clearly showing the stereochemistry of the starting material and the product, for the reaction of (*S*)-1-bromo-2-methylbutane with sodium iodide in acetone. What is the configuration (*R* or *S*) of the product?

6.34 Give the structures, including stereochemistry, of compounds A and B in the following sequence of reactions:

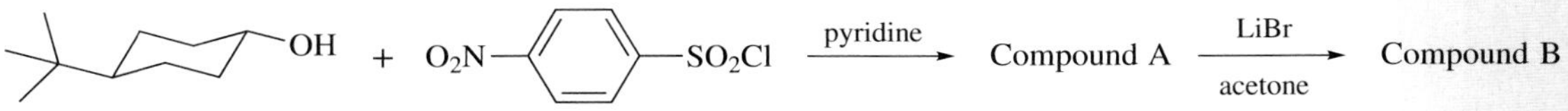

[4-tert-butylcyclohexanol] OH + O_2N–[benzene ring]–SO_2Cl $\xrightarrow{\text{pyridine}}$ Compound A $\xrightarrow[\text{acetone}]{\text{LiBr}}$ Compound B

6.35 Based on what we know about nucleophiles and leaving groups, we suspect that the reaction of (*R*)-2-chlorobutane with sodium iodide in acetone would not be useful as a synthesis of (*S*)-2-iodobutane. Explain.

6.36 Optically pure (*S*)-(+)-2-butanol was converted to its methanesulfonate according to the reaction shown.

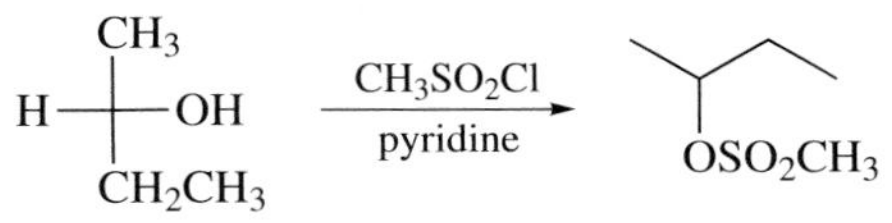

Fischer projection: CH_3 (top), H (left), OH (right), CH_2CH_3 (bottom) $\xrightarrow[\text{pyridine}]{CH_3SO_2Cl}$ [sec-butyl skeleton] OSO_2CH_3

(a) Write the Fischer projection of the *sec*-butyl methanesulfonate formed in this reaction.

(b) The *sec*-butyl methanesulfonate in part (a) was treated with $NaSCH_2CH_3$ to give a product having an optical rotation α_D of −25°. Write the Fischer projection of this product. By what mechanism is it formed? What is its absolute configuration (*R* or *S*)?

(c) When treated with PBr_3, optically pure (*S*)-(+)-2-butanol gave 2-bromobutane having an optical rotation α_D = −38°. This bromide was then allowed to react with $NaSCH_2CH_3$ to give a product having an optical rotation α_D of +23°. Write the Fischer projection for (−)-2-bromobutane and specify its configuration as *R* or *S*. Does the reaction of 2-butanol with PBr_3 proceed with predominant inversion or retention of configuration?

(d) What is the optical rotation of optically pure 2-bromobutane?

Synthesis

6.37 Outline an efficient synthesis of each of the following compounds from the indicated starting material and any necessary organic or inorganic reagents:

(a) Isobutyl iodide from isobutyl chloride

(b) Isopropyl azide from isopropyl alcohol

(c) (*S*)-*sec*-Butyl azide from (*R*)-*sec*-butyl alcohol

(d) (*S*)-$CH_3CH_2CHCH_3$ (with SH on the CH) from (*R*)-*sec*-butyl alcohol

6.38 The sex pheromone (*matsuone*) of a parasitic insect (*Matsucoccus*) that infests pine trees was prepared in a multistep synthesis from (−)-citronellol by way of the nitrile shown.

OH CN

(*S*)-(−)-Citronellol (*S*)-4,8-Dimethyl-7-nonenenitrile

(a) Relate the nitrile to (−)-citronellol by a retrosynthetic analysis.

(b) Convert your retrosynthesis to a synthesis, showing appropriate reagents for each step.

6.39 Suggest a reasonable series of synthetic transformations for converting *trans*-2-methylcyclopentanol to *cis*-2-methylcyclopentyl acetate.

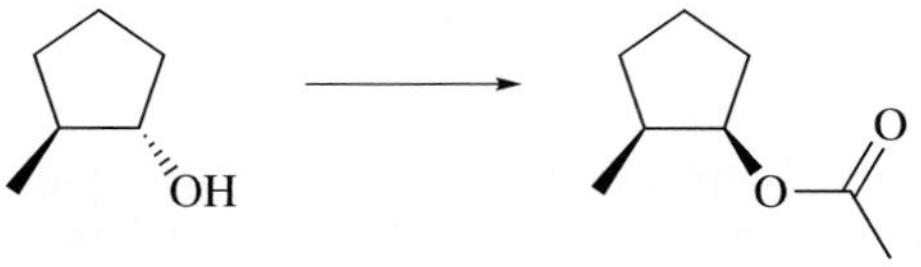

OH → O O

6.40 The preparation of the serotonin reuptake inhibitor, fluparoxan, is accomplished in four synthetic steps.

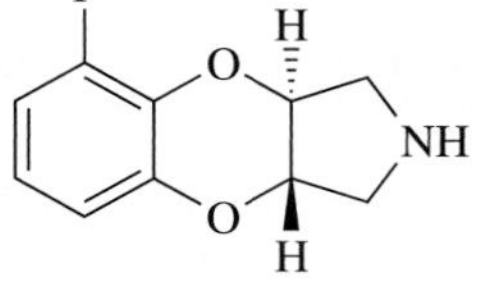

F O H NH O H

Fluparoxan

A retrosynthesis for the formation of the second ring in this synthesis, a tetraether, is shown. Propose a synthetic path for the formation of the tetraether from the starting fluorocatechol and butanediol.

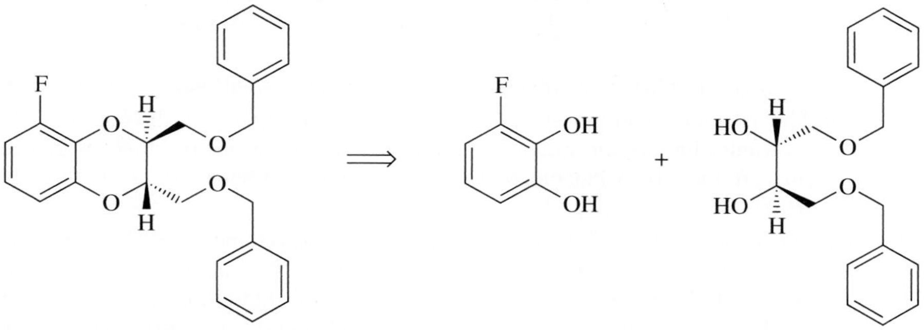

F O H O O O H ⟹ F OH OH + HO H O HO H O

Tetraether 3-Fluorocatechol Butanediol

6.41 Tomelukast has been examined as a leukotriene receptor antagonist. Its synthesis can be accomplished in five steps from readily available starting materials.

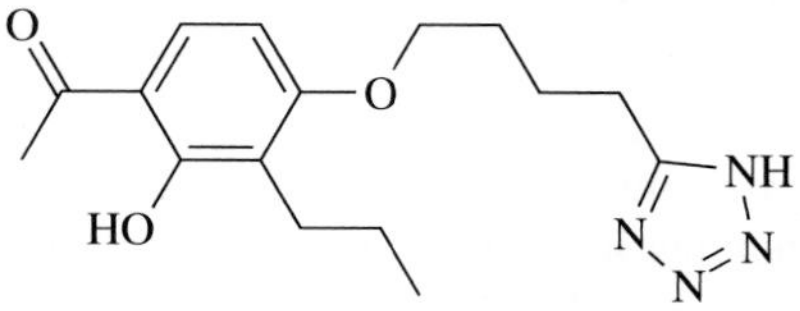

O O HO NH N N N

Tomelukast

A retrosynthetic analysis for the formation of the penultimate compound in the synthesis of Tomelukast is shown below. Devise a synthesis for the preparation of the indicated compound. Use any other reagents needed for the conversion.

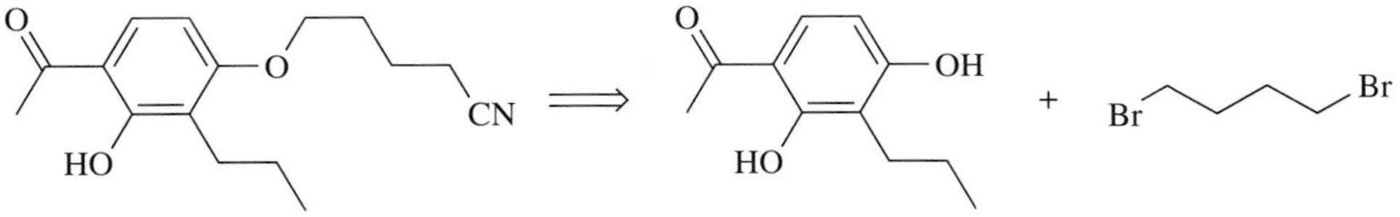

O HO O CN ⟹ O HO OH + Br Br

Descriptive Passage and Interpretive Problems 6

Nucleophilic Substitution

These problems differ from those in earlier chapters in that they directly test your knowledge of core material rather than using a descriptive passage to extend the material or introduce new ideas. The number of factors that contribute to nucleophilic substitution can be daunting. The really major ones, though, are few and readily applied to specific reactions by using the S_N1 and S_N2 mechanisms to guide your analysis.

6.42 Which compound undergoes substitution by the S_N1 mechanism at the fastest rate?

Br Br Br Br

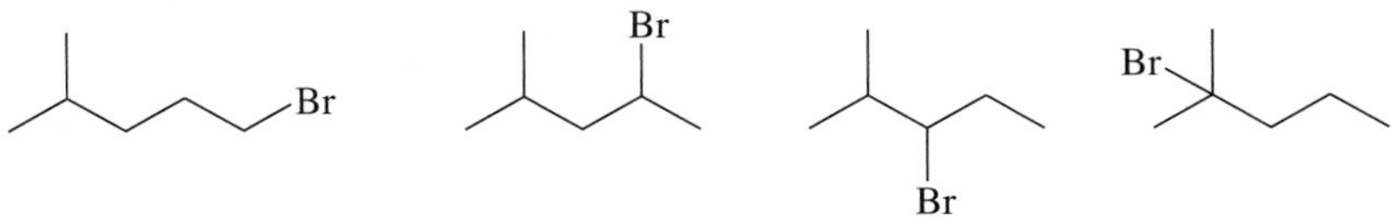

A. B. C. D.

6.43 Which compound undergoes substitution by the S_N2 mechanism at the fastest rate?

Br Br Br Br

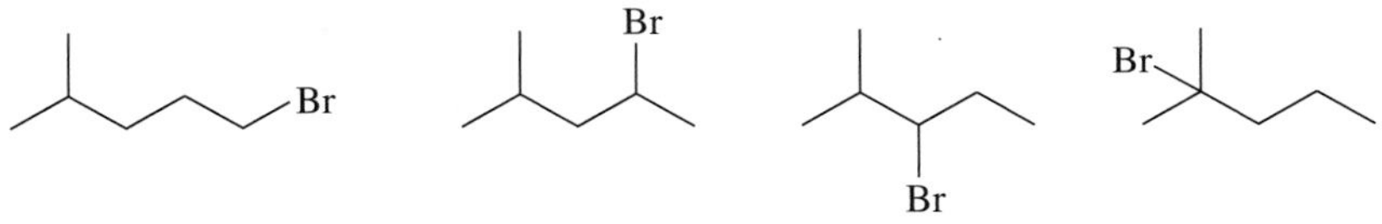

A. B. C. D.

6.44 Which reaction takes place at the fastest rate?

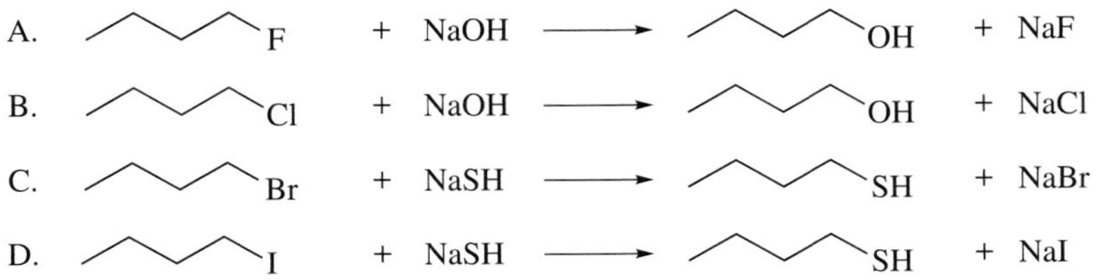

A. F + NaOH ⟶ OH + NaF

B. Cl + NaOH ⟶ OH + NaCl

C. Br + NaSH ⟶ SH + NaBr

D. I + NaSH ⟶ SH + NaI

6.45 What is the major product of the reaction shown?

CH_3 / H—C—Br / H_3C—C—H / CH_2CH_3 $\xrightarrow[\text{DMSO, 25°C}]{\text{NaCN}}$

A. CH_3; H—C—CN; H_3C—C—H; CH_2CH_3

B. CH_3; NC—C—H; H_3C—C—H; CH_2CH_3

C. CH_3; H—C—CN; H—C—CH_3; CH_2CH_3

D. CH_3; NC—C—H; H—C—CH_3; CH_2CH_3

6.46 What are reactant X and product Y in the following sequence of reactions?

Reactant X $\xrightarrow[\text{pyridine}]{H_3C-C_6H_4-SO_2Cl}$ [H_3C, H; cyclopentyl; $O-SO_2-C_6H_4-CH_3$, H] $\xrightarrow[\text{ethanol–water}]{NaN_3}$ Product Y

	Reactant X	Product Y
A.	H_3C, H; OH, H	H_3C, H; N_3, H
B.	H_3C, H; OH, H	H_3C, H; H, N_3
C.	H_3C, H; H, OH	H_3C, H; H, N_3
D.	H_3C, H; H, OH	H_3C, H; N_3, H

6.47 Trimethyloxonium tetrafluoroborate reacts with methanol (CH_3OH) to give dimethyl ether (CH_3OCH_3). Which equation, including the curved arrows, best represents the rate-determining step in the mechanism?

$$(H_3C)_2\overset{+}{O}{-}CH_3 \quad BF_4^-$$

Trimethyloxonium tetrafluoroborate

A. $H_3C{-}\overset{+}{O}(CH_3)_2 \longrightarrow H_3C^+ + :O(CH_3)_2$

B. $H_3C{-}\ddot{O}{-}H + H_3C^+ \longrightarrow H_3C{-}\overset{+}{O}(CH_3){-}H$

C. $H_3C{-}\ddot{O}{-}H + H_3C{-}\overset{+}{O}(CH_3)_2 \longrightarrow H_3C{-}\overset{+}{O}(CH_3){-}H + :O(CH_3)_2$

D. $H_3C{-}\ddot{O}^- + H_3C{-}\overset{+}{O}(CH_3)_2 \longrightarrow H_3C{-}\ddot{O}{-}CH_3 + :O(CH_3)_2$

CHAPTER OUTLINE

Squalene ($C_{30}H_{50}$) is a hydrocarbon with six carbon–carbon double bonds. It is present in plants and animals and, as will be seen in Chapter 24, is the biosynthetic precursor to steroids such as cholesterol.

Structure and Preparation of Alkenes: Elimination Reactions

Alkenes are hydrocarbons that contain a carbon–carbon double bond. A carbon–carbon double bond is both an important structural unit and an important functional group in organic chemistry. The shape of an organic molecule is influenced by its presence, and the double bond is the site of most of the chemical reactions that alkenes undergo.

This chapter is the first of two dealing with alkenes; it describes their structure, bonding, and preparation. Chapter 8 examines their chemical reactions.

7.1 Alkene Nomenclature

We give alkenes IUPAC names by replacing the *-ane* ending of the corresponding alkane with *-ene*. The two simplest alkenes are ethene and propene. Both are also well known by their common names *ethylene* and *propylene*.

$H_2C{=}CH_2$	$CH_3CH{=}CH_2$
IUPAC name: **ethene**	IUPAC name: **propene**
Common name: ethylene	Common name: propylene

The alkene corresponding to the longest continuous chain that includes the double bond is considered the parent, and the chain is numbered in the direction that gives the doubly bonded carbons their lower numbers. The locant (or numerical position) of only one of the doubly bonded carbons is specified in the name; it is understood that the other doubly bonded carbon must follow in sequence. According to various versions of the IUPAC rules, the locant may precede the parent chain or the *-ene* suffix. Both versions are widely used. *Carbon–carbon double bonds take precedence over alkyl groups and halogens in determining the main carbon chain and the direction in which it is numbered.*

$H_2C{=}CHCH_2CH_3$ (numbered 1 2 3 4) — 1-Butene or But-1-ene

$CH_3CHCH{=}CH_2$ with CH_3 on C-3 (numbered 4 3 2 1) — 3-Methyl-1-butene or 3-Methylbut-1-ene

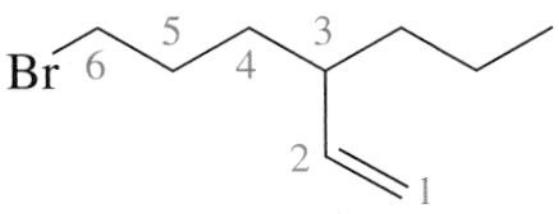

Br (chain numbered 6, 5, 4, 3, 2, 1) — 6-Bromo-3-propyl-1-hexene or 6-Bromo-3-propylhex-1-ene

Hydroxyl groups, however, outrank the double bond, and a chain that contains both an —OH group and a double bond is numbered in the direction that gives the carbon attached to the —OH group the lower number. Compounds that contain both a double bond and a hydroxyl group combine the suffixes *-en* + *-ol* to signify that both functional groups are present.

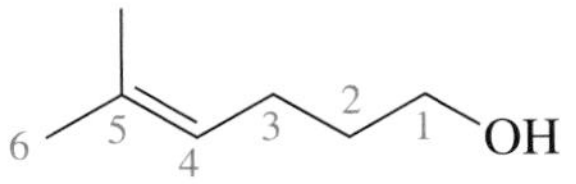

OH (chain numbered 6, 5, 4, 3, 2, 1) — 5-Methyl-4-hexen-1-ol or 5-Methylhex-4-en-1-ol

Problem 7.1

Name each of the following using IUPAC nomenclature:

(a) $(CH_3)_2C{=}C(CH_3)_2$

(b) $(CH_3)_3CCH{=}CH_2$

(c)

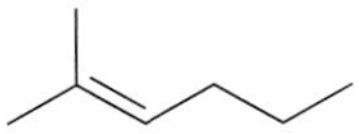

(d) 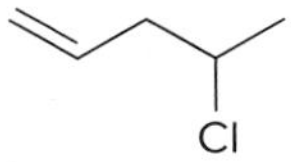 Cl

(e) 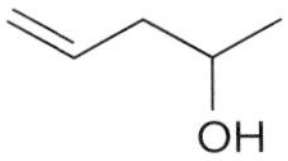 OH

Sample Solution (a) The longest continuous chain in this alkene contains four carbon atoms. The double bond is between C-2 and C-3, and so it is named as a derivative of 2-butene.

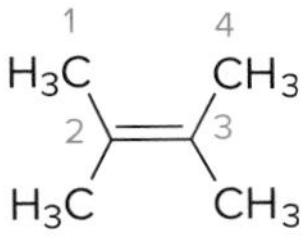

H_3C, CH_3, H_3C, CH_3 (numbered 1, 2, 3, 4) — 2,3-Dimethyl-2-butene or 2,3-Dimethylbut-2-ene

The two methyl groups are substituents attached to C-2 and C-3 of the main chain.

The common names of three frequently encountered *alkenyl* groups—**vinyl, allyl,** and **isopropenyl**—are acceptable in the IUPAC system.

$H_2C{=}CH—$ (Vinyl) as in $H_2C{=}CHCl$ or Cl (Vinyl chloride)

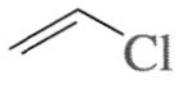

$H_2C{=}CHCH_2—$ (Allyl) as in $H_2C{=}CHCH_2OH$ or OH (Allyl alcohol)

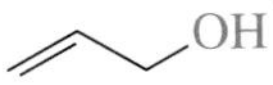

$H_2C{=}C(CH_3)—$ (Isopropenyl) as in $H_2C{=}C(CH_3)Cl$ or Cl (Isopropenyl chloride)

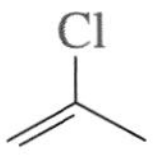

Vinyl chloride is an industrial chemical produced in large amounts (10^{10} lb/year in the United States) and is used in the preparation of poly(vinyl chloride). Poly(vinyl chloride), often called simply *vinyl*, has many applications, including siding for houses, wall coverings, and PVC piping.

When a CH_2 group is doubly bonded to a ring, the prefix *methylene* is added to the name of the ring.

$=CH_2$

Methylenecyclohexane

Cycloalkenes and their derivatives are named by adapting cycloalkane terminology to the principles of alkene nomenclature.

Cyclopentene | 1-Methylcyclohexene | 3-Chlorocycloheptene (not 1-chloro-2-cycloheptene)

No locants are needed in the absence of substituents; it is understood that the double bond connects C-1 and C-2. Substituted cycloalkenes are numbered beginning with the double bond, proceeding through it, and continuing in sequence around the ring. The direction is chosen so as to give the lower of two possible numbers to the substituent.

Problem 7.2

Write structural formulas and give the IUPAC names of all the monochloro-substituted derivatives of cyclopentene. Include enantiomers and assign *R*- and *S*-configurations as appropriate. Which isomers are allylic chlorides? Which are vinylic?

7.2 Structure and Bonding in Alkenes

The structure of ethylene and the orbital hybridization model for its double bond were presented in Section 2.8 and are briefly reviewed in Figure 7.1. Ethylene is planar, each carbon is sp^2-hybridized, and the double bond has a σ component and a π component. The σ component arises from overlap of sp^2 hybrid orbitals along a line connecting the two carbons, the π component via a "side-by-side" overlap of two p orbitals. Regions of high π-electron density are present above and below the plane of the molecule. Most of the reactions of ethylene and other alkenes involve these π electrons.

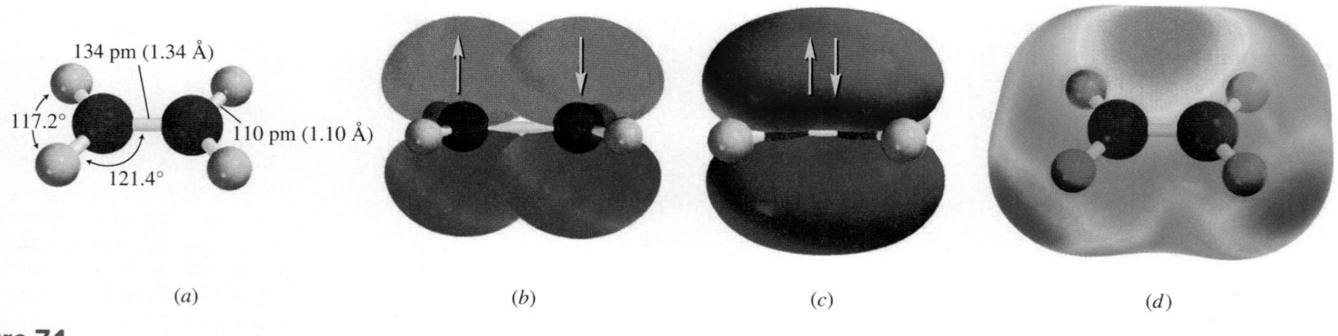

Figure 7.1

(*a*) The planar framework of σ bonds in ethylene showing bond distances and angles. (*b*) and (*c*) The half-filled p orbitals of two sp^2-hybridized carbons overlap to produce a π bond. (*d*) The electrostatic potential map shows a region of high negative potential due to the π electrons above and below the plane of the atoms.

Ethylene

Ethylene was known to chemists in the eighteenth century and isolated in pure form in 1795. An early name for ethylene was *gaz oléfiant* (French for "oil-forming gas"), to describe the fact that an oily liquid product is formed when two gases—ethylene and chlorine—react with each other.

$$\underset{\substack{\text{Ethylene}\\ \text{(bp: }-104^\circ\text{C)}}}{H_2C{=}CH_2} + \underset{\substack{\text{Chlorine}\\ \text{(bp: }-34^\circ\text{C)}}}{Cl_2} \longrightarrow \underset{\substack{\text{1,2-Dichloroethane}\\ \text{(bp: }83^\circ\text{C)}}}{ClCH_2CH_2Cl}$$

The term *gaz oléfiant* was the forerunner of the general term *olefin,* formerly used as the name of the class of compounds we now call *alkenes.*

Ethylene occurs naturally in small amounts as a plant hormone. It is formed in a complex series of steps from a compound containing a cyclopropane ring:

$$\underset{\substack{\text{1-Amino-}\\ \text{cyclopropane-}\\ \text{carboxylic acid}}}{\text{(cyclopropane with } \overset{+}{N}H_3 \text{ and } CO_2^-)} \xrightarrow{\text{several steps}} \underset{\text{Ethylene}}{H_2C{=}CH_2} + \text{other products}$$

Even minute amounts of ethylene can stimulate the ripening of fruits, and the rate of ripening increases with the concentration of ethylene. This property is used to advantage in the marketing of bananas. Bananas are picked green in the tropics, kept green by being stored with adequate ventilation to limit the amount of ethylene present, and then induced to ripen at their destination by passing ethylene over the fruit.

Ethylene is the cornerstone of the world's mammoth petrochemical industry and is produced in vast quantities. In a typical year the amount of ethylene produced in the United States (5×10^{10} lb) exceeds the combined weight of all of its people. In one process, ethane from natural gas is heated to bring about its dissociation into ethylene and hydrogen:

$$\underset{\text{Ethane}}{CH_3CH_3} \xrightarrow{750^\circ\text{C}} \underset{\text{Ethylene}}{H_2C{=}CH_2} + \underset{\text{Hydrogen}}{H_2}$$

This **dehydrogenation** is simultaneously both a source of ethylene and one of the methods by which hydrogen is prepared on an industrial scale. Most of this hydrogen is subsequently used to reduce nitrogen to ammonia for the preparation of fertilizer.

Similarly, dehydrogenation of propane gives propene:

$$\underset{\text{Propane}}{CH_3CH_2CH_3} \xrightarrow{750^\circ\text{C}} \underset{\text{Propene}}{H_2C{=}CHCH_3} + \underset{\text{Hydrogen}}{H_2}$$

Propene is the second most important petrochemical and is produced on a scale about half that of ethylene.

Almost any hydrocarbon can serve as a starting material for production of ethylene and propene. Cracking of petroleum (Section 2.21) gives ethylene and propene by processes involving cleavage of carbon–carbon bonds of higher-molecular-weight hydrocarbons. An area of current research interest is directed toward finding catalytic methods for converting methane from natural gas to ethylene.

The major uses of ethylene and propene are as starting materials for the preparation of polyethylene and polypropylene plastics, fibers, and films. These and other applications will be described in Chapter 8.

Molecular orbital (MO) theory offers an alternative to the valence bond theory for understanding the structure of alkenes. Recall from Section 2.4 that the number of molecular orbitals is equal to the number of atomic orbitals (AOs) that combine to form them. When we combine the two 2*p* AOs, one from each of the sp^2-hybridized carbons of ethene, the two π MOs are formed as shown in Figure 7.2. The **highest occupied molecular orbital,** or **HOMO** (π_1), is doubly occupied, whereas the **lowest unoccupied molecular orbital,** or **LUMO** (π_2), is vacant. Since both lobes above and both lobes below the sigma bond of the HOMO are the same, π bonding occurs.

On the basis of their bond-dissociation enthalpies, the C═C bond in ethylene is stronger than the C—C single bond in ethane, but it is not twice as strong.

$$\underset{\text{Ethylene}}{H_2C{=}CH_2} \longrightarrow \underset{\text{Methylene}}{2\,\ddot{C}H_2} \quad \Delta H^\circ = +730 \text{ kJ/mol (172 kcal/mol)}$$

$$\underset{\text{Ethane}}{H_3C{-}CH_3} \longrightarrow \underset{\text{Methyl}}{2\,\dot{C}H_3} \quad \Delta H^\circ = +375 \text{ kJ/mol (90 kcal/mol)}$$

While it is not possible to apportion the C═C bond energy of ethylene between its σ and π components, the data suggest that the π bond is weaker than the σ bond.

Figure 7.2

Two molecular orbitals (MOs) are generated by combining two 2*p* atomic orbitals (AOs). The bonding MO is lower in energy than either of the AOs that combine to produce it. The antibonding MO is higher energy than either AO.

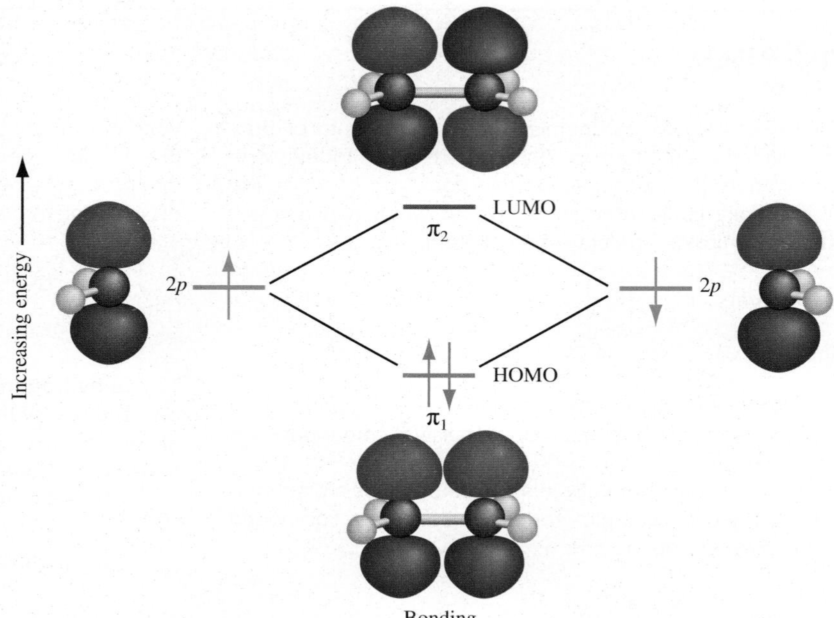

There are two different types of carbon–carbon bonds in propene, $CH_3CH{=}CH_2$. The double bond is of the $\sigma + \pi$ type, and the bond to the methyl group is a σ bond formed by sp^3–sp^2 overlap.

sp^3-Hybridized carbon

sp^2-Hybridized carbon

C—C bond length = 150 pm (1.50 Å)
C=C bond length = 134 pm (1.34 Å)

Problem 7.3

How many carbon atoms are sp^2-hybridized in the alkene shown? How many are sp^3-hybridized? How many bonds are of the sp^2–sp^3 type? How many are of the sp^3–sp^3 type?

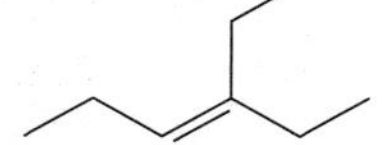

7.3 Isomerism in Alkenes

Although ethylene is the only two-carbon alkene, and propene the only three-carbon alkene, there are *four* isomeric alkenes of molecular formula C_4H_8:

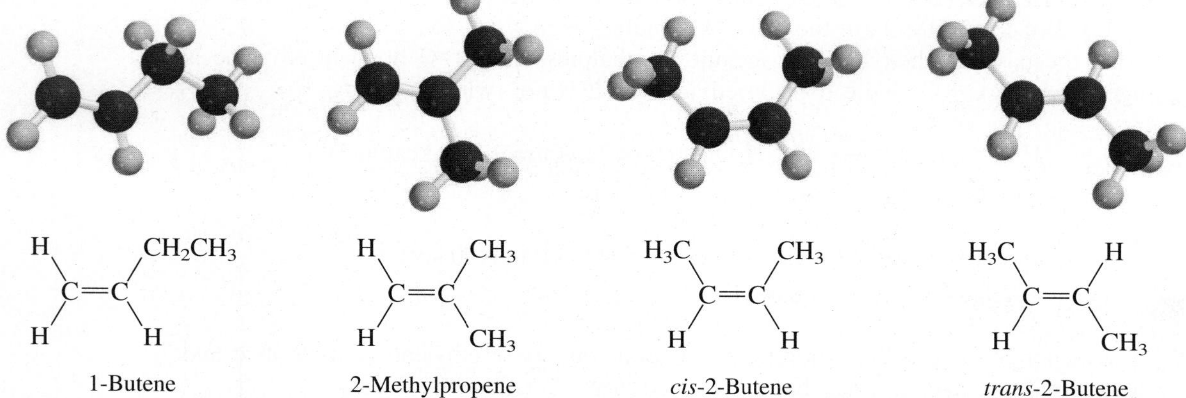

1-Butene has an unbranched carbon chain with a double bond between C-1 and C-2. It is a constitutional isomer of the other three. Similarly, 2-methylpropene, with a branched carbon chain, is a constitutional isomer of the other three.

The pair of isomers designated *cis*- and *trans*-2-butene are *stereoisomers* of each other. They have the same constitution; but the cis isomer has both of its methyl groups on the same side of the double bond, while the methyl groups in the trans isomer are on opposite sides.

Cis–trans stereoisomerism in alkenes is not possible when one of the doubly bonded carbons bears two identical substituents. Thus, neither 1-butene nor 2-methylpropene can have stereoisomers.

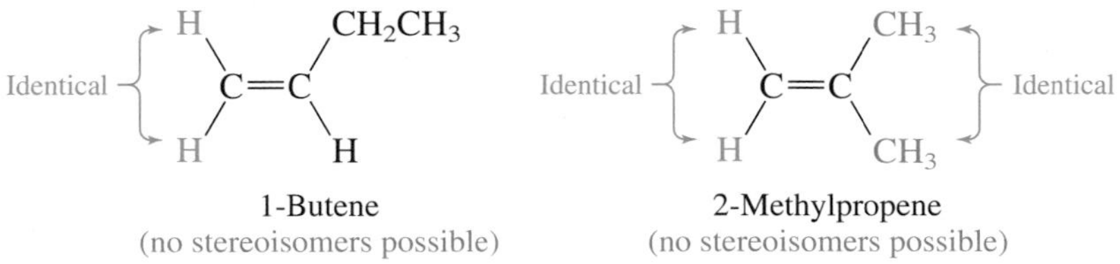

Identical; H, H, CH_2CH_3, H; C=C

1-Butene
(no stereoisomers possible)

Identical; H, H, CH_3, CH_3; C=C; Identical

2-Methylpropene
(no stereoisomers possible)

Problem 7.4

How many alkenes have the molecular formula C_5H_{10}? Write their structures and give their IUPAC names. Specify the configuration of stereoisomers as cis or trans as appropriate.

In principle, *cis*-2-butene and *trans*-2-butene may be interconverted by rotation about the C-2—C-3 double bond. However, unlike rotation about single bonds, which is quite fast, rotation about double bonds is restricted. Interconversion of the cis and trans isomers of 2-butene has an activation energy that is 10–15 times greater than that for rotation about the single bond of an alkane and does *not* occur under normal circumstances.

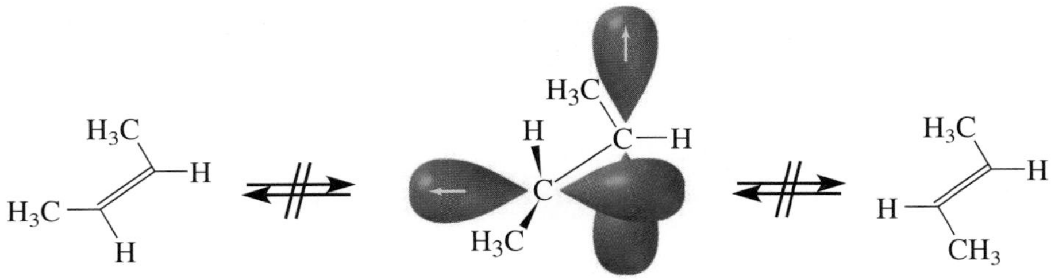

H_3C, H_3C, H, H; H_3C, H, C—H, H_3C, C; H_3C, H, H, CH_3

cis-2-Butene
(stable)

Transition state for rotation about C=C
(high energy: perpendicular *p* orbitals
eliminate π bonding)

trans-2-Butene
(stable)

π-Bonding in *cis*- and *trans*-2-butene is strong because of the favorable parallel alignment of the *p* orbitals at C-2 and C-3. Interconverting the two stereoisomers, however, requires these *p* orbitals to be at right angles to each other, eliminates their overlap, and breaks the π component of the double bond.

Problem 7.5

Are *cis*-2-hexene and *trans*-3-hexene stereoisomers? Explain.

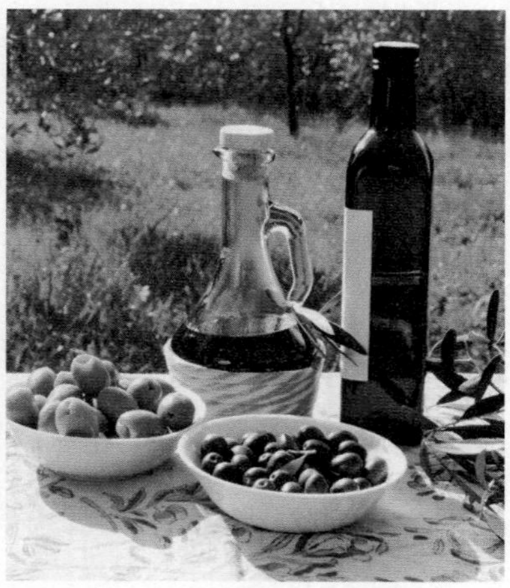
Oleic acid is prepared from olive oil.

Cinnamaldehyde gives cinnamon its flavor.

7.4 Naming Stereoisomeric Alkenes by the *E–Z* Notational System

When the groups on either end of a double bond are the same or are structurally similar to each other, it is a simple matter to describe the configuration of the double bond as cis or trans. Oleic acid, for example, has a cis double bond. Cinnamaldehyde has a trans double bond.

Oleic acid   Cinnamaldehyde

Problem 7.6

Female houseflies attract males by sending a chemical signal known as a *pheromone*. The substance emitted by the female housefly that attracts the male has been identified as *cis*-9-tricosene, $C_{23}H_{46}$. Write a structural formula, including stereochemistry, for this compound.

The terms *cis* and *trans* are ambiguous, however, when it is not obvious which substituent on one carbon is "similar" or "analogous" to a reference substituent on the other. An unambiguous system for specifying double-bond stereochemistry has been adopted by the IUPAC based on the Cahn–Ingold–Prelog sequence rules (Section 4.6). When atoms or groups of higher sequence-rule precedence are on the same side of the double bond, the double bond has the *Z* configuration, where *Z* stands for the German word ***zusammen,*** meaning "together." When they are on opposite sides, the configuration is ***E,*** standing for the German word ***entgegen,*** meaning "opposite." The *E* or *Z* descriptor is placed between parentheses and precedes the rest of the name.

Higher → Cl, Br ← Higher; C=C; Lower → H, F ← Lower

Z configuration
Higher ranked substituents (Cl and Br) are on same side of double bond

Higher → Cl, F ← Lower; C=C; Lower → H, Br ← Higher

E configuration
Higher ranked substituents (Cl and Br) are on opposite sides of double bond

Problem 7.7

Determine the configuration of each of the following as *Z* or *E* as appropriate:

OH (a)   F (b)   OH (c)   (d)

Sample Solution (a) One of the doubly bonded carbons bears a methyl group and a hydrogen. According to the Cahn–Ingold–Prelog sequence rules, methyl outranks hydrogen. The other carbon atom of the double bond bears a methyl and a $—CH_2OH$ group. The $—CH_2OH$ group is of higher priority than methyl.

Higher (C) ⟶ H_3C (C=C) CH_2OH ⟵ Higher C(O,H,H)

Lower (H) ⟶ H CH_3 ⟵ Lower C(H,H,H)

Higher ranked groups are on the same side of the double bond; the configuration is *Z*.

Problem 7.8

Give the IUPAC name of each of the compounds in Problem 7.7, including *E* or *Z* as appropriate.

Sample Solution (a) The longest continuous chain has 4 carbons and bears a hydroxyl group at C-1 and a methyl group at C-2. Its double bond connects C-2 and C-3 and has the *Z* configuration. Its IUPAC name is (*Z*)-2-methyl-2-buten-1-ol.

A molecule that has both chirality centers and double bonds incorporates opportunities for stereoisomerism from both structural units. For example, the configuration of the chirality center in 3-penten-2-ol may be either *R* or *S*, and the double bond may be either *E* or *Z*. Thus, there are four stereoisomers of 3-penten-2-ol.

OH OH OH OH

(2*R*,3*E*)-3-Penten-2-ol (2*S*,3*E*)-3-Penten-2-ol (2*R*,3*Z*)-3-Penten-2-ol (2*S*,3*Z*)-3-Penten-2-ol

The relationship of the (2*R*,3*E*) stereoisomer to the others is that it is the enantiomer of (2*S*,3*E*)-3-penten-2-ol and is a diastereomer of the (2*R*,3*Z*) and (2*S*,3*Z*) isomers.

7.5 Physical Properties of Alkenes

Alkenes resemble alkanes in most of their physical properties. The lower-molecular-weight alkenes through C_4H_8 are gases at room temperature and atmospheric pressure.

The dipole moments of most alkenes are quite small. Among the C_4H_8 isomers, 1-butene, *cis*-2-butene, and 2-methylpropene have dipole moments in the 0.3–0.5 D range; *trans*-2-butene has no dipole moment. Nevertheless, we can learn some things about alkenes by looking at the effect of substituents on dipole moments.

Experimental measurements of dipole moments give size, but not direction. We normally deduce the overall direction by examining the individual bond dipoles. With alkenes the basic question concerns the alkyl groups attached to C═C. *Does an alkyl group donate electrons to or withdraw electrons from a double bond?* This question can be approached by comparing the effect of an alkyl group, methyl for example, with other substituents.

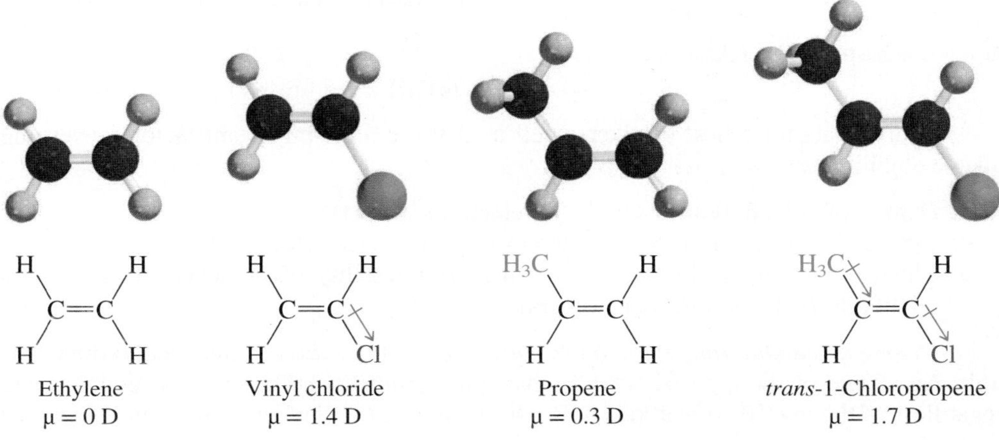

Ethylene $\mu = 0$ D | Vinyl chloride $\mu = 1.4$ D | Propene $\mu = 0.3$ D | *trans*-1-Chloropropene $\mu = 1.7$ D

Ethylene has no dipole moment. Replacing one of its hydrogens by an electron-attracting chlorine atom gives vinyl chloride, which has a dipole moment of 1.4 D. The effect is much smaller when one of the hydrogens of ethylene is replaced by methyl; propene has a dipole moment of only 0.3 D. Now place CH_3 and Cl trans to each other on the double bond. If methyl releases electrons better than H, then the dipole moment of *trans*-$CH_3CH{=}CHCl$ should be larger than that of $H_2C{=}CHCl$, because the effects of CH_3 and Cl reinforce each other. If methyl is electron attracting, the opposite should occur, and the dipole moment of *trans*-$CH_3CH{=}CHCl$ will be smaller than 1.4 D. In fact, the dipole moment of *trans*-$CH_3CH{=}CHCl$ is larger than that of $H_2C{=}CHCl$, indicating that a methyl group acts as an electron-donating substituent on the double bond.

Problem 7.9

Arrange the following in order of increasing dipole moment.

(*E*)-1-Chloropropene 1,1-Dichloro-2-methylpropene (*E*)-2,3-Dichloro-2-butene

A methyl group releases electrons to an attached double bond in much the same way that it releases electrons to an sp^2-hybridized carbon of a carbocation—by an inductive effect and by hyperconjugation. These effects are less pronounced compared to carbocations since a severely electron-deficient carbocation center is not involved.

Other alkyl groups resemble methyl in respect to their ability to stabilize double bonds by hyperconjugation. We'll see examples of this substituent effect in Section 7.6.

7.6 Relative Stabilities of Alkenes

We have seen how heats of combustion can be used to compare the stabilities of isomeric alkanes (Section 2.23) and dimethylcyclohexanes (Section 3.12). When a similar analysis of heats of combustion data is applied to the four alkenes of molecular formula C_4H_8, we find that l-butene is the least stable isomer and 2-methylpropene the most stable. Of the pair of stereoisomeric 2-butenes, *trans*-2-butene is more stable than *cis*-.

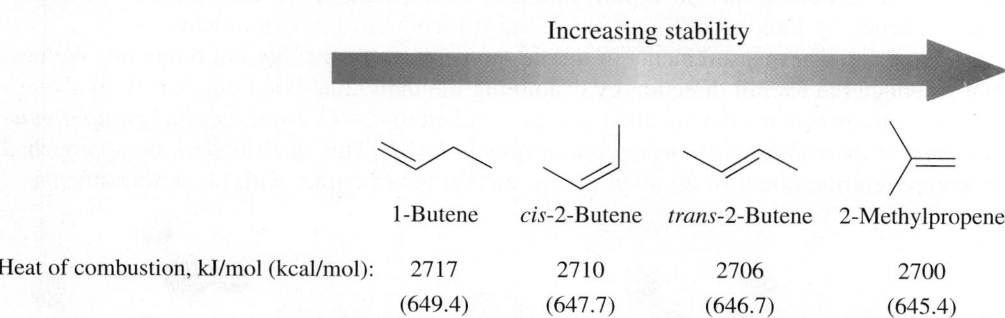

Similar data for a host of alkenes tell us that the most important factors governing alkene stability are:

1. Degree of substitution of C=C (an electronic effect)
2. van der Waals strain in the cis stereoisomer (a steric effect)
3. Chain branching (analogous to the increased stability of branched alkane chains relative to their unbranched isomers)

Degree of substitution refers to the number of carbons *directly* attached to the C=C unit. An alkene of the type $RCH{=}CH_2$ has a *monosubstituted* or *terminal* double bond regardless of the number of carbons in R. Disubstituted, *trisubstituted,* and *tetrasubstituted* double bonds have two, three, and four carbon atoms, respectively, directly attached to C=C. Among the C_4H_8 isomeric alkenes, only 1-butene has a monosubstituted double bond; the other three have disubstituted double bonds and are, as measured by their heats of combustion, more stable than 1-butene.

Problem 7.10

Write structural formulas and give the IUPAC names for all the alkenes of molecular formula C_6H_{12} that contain a trisubstituted double bond. (Don't forget to include stereoisomers.)

Like the *sp*²-hybridized carbons of carbocations and free radicals, the *sp*²-hybridized carbons of double bonds are electron attracting, and alkenes are stabilized by substituents that release electrons to these carbons. As we saw in the preceding section, alkyl groups are better electron-releasing substituents than hydrogen and are, therefore, better able to stabilize an alkene.

In general, alkenes with more highly substituted double bonds are more stable than isomers with less substituted double bonds.

Problem 7.11

The heats of combustion are known for all 17 isomeric C_6H_{12} alkenes, with the most stable being 4006 kJ/mol (957.5 kcal/mol) and the least stable being 4034 kJ/mol (964.2 kcal/mol). Which is the least stable and which is the most stable?

An effect that results when two or more atoms or groups interact so as to alter the electron distribution in a molecule is called an **electronic effect.** The greater stability of more highly substituted alkenes is an example of an electronic effect.

Van der Waals strain in alkenes is a **steric effect** most commonly associated with repulsive forces between substituents that are cis to each other and is reflected in the observation that the heat of combustion of *cis*-2-butene is 4 kJ/mol (1.0 kcal/mol) greater than *trans*-2-butene. The source of this difference is illustrated in the space-filling models of Figure 7.3, where it can be seen that the methyl groups crowd each other in *cis,* but not *trans*-2-butene.

Van der Waals strain, or steric effect, is often expressed by the notation used below for *cis*-2-butene's C-1 and C-4 hydrogens.

In general, trans alkenes are more stable than their cis stereoisomers.

Problem 7.12

Arrange the following alkenes in order of decreasing stability: 1-pentene; (*E*)-2-pentene; (*Z*)-2-pentene; 2-methyl-2-butene.

A more dramatic example of this difference is observed between stereoisomeric alkenes with bulky alkyl groups as substituents on the double bond. The heat of combustion of the cis stereoisomer of 2,2,5,5-tetramethyl-3-hexene, for example, is 44 kJ/mol (10.5 kcal/mol) higher than that of the trans because of van der Waals strain between cis *tert*-butyl groups!

The common names of these alkenes are *cis*- and *trans*-di-*tert*-butylethylene. In cases such as this the common names are more convenient than the IUPAC names because they are more readily associated with molecular structure.

cis-2,2,5,5-Tetramethyl-3-hexene
(less stable)

trans-2,2,5,5-Tetramethyl-3-hexene
(more stable)

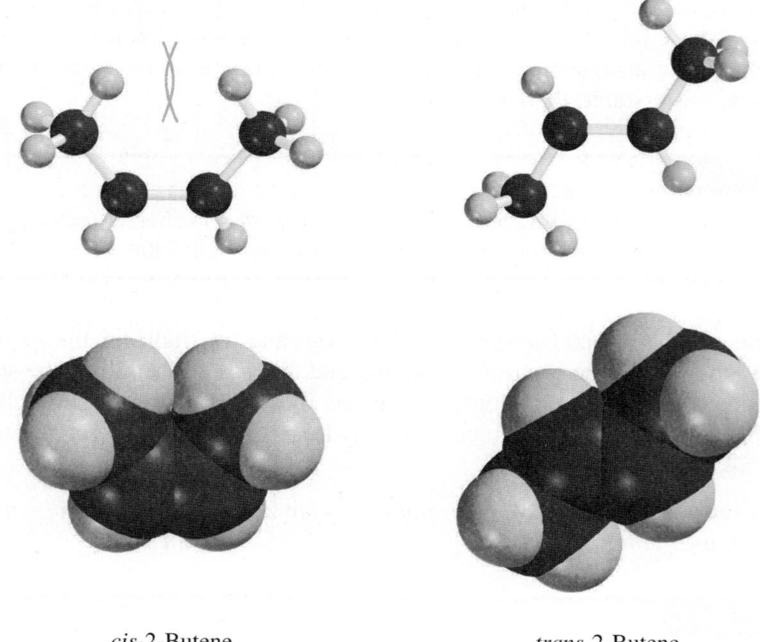

Figure 7.3

Ball-and-stick and space-filling models of *cis*- and *trans*-2-butene. The space-filling model shows the serious van der Waals strain between two of the hydrogens in *cis*-2-butene. The molecule adjusts by expanding those bond angles that increase the separation between the crowded atoms. The combination of angle strain and van der Waals strain makes *cis*-2-butene less stable than *trans*-2-butene.

Problem 7.13

Despite numerous attempts, the alkene 3,4-di-*tert*-butyl-2,2,5,5-tetramethyl-3-hexene has never been synthesized. Can you explain why?

Chain branching was seen earlier (Section 2.23) to have a stabilizing effect on alkanes. The same is true of carbon chains that include a double bond. Of the three disubstituted C_4H_8 alkenes, the branched isomer $(CH_3)_2C{=}CH_2$ is more stable than either *cis*- or *trans*-$CH_3CH{=}CHCH_3$.

In general, alkenes with branched chains are more stable than unbranched isomers. This effect is usually less important than the degree of substitution or stereochemistry of the double bond.

Problem 7.14

Write structural formulas for the six isomeric alkenes of molecular formula C_5H_{10} and arrange them in order of increasing stability (smaller heat of combustion, more negative ΔH_f°).

7.7 Cycloalkenes

Double bonds are accommodated by rings of all sizes. The smallest cycloalkene, cyclopropene, was first synthesized in 1922. A cyclopropene ring is present in sterculic acid, a substance derived from the oil present in the seeds of a tree (*Sterculia foetida*) that grows in the Philippines and Indonesia.

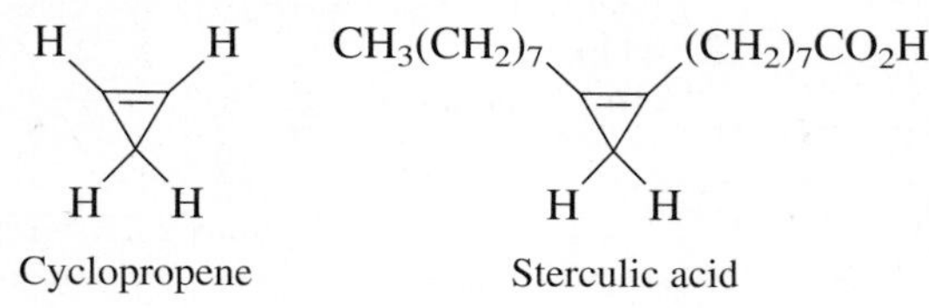

Fruit, seeds, and leaves of *Sterculia foetida*.

As we saw in Section 3.5, cyclopropane is destabilized by angle strain because its 60° bond angles are much smaller than the normal 109.5° angles associated with sp^3-hybridized carbon. Cyclopropene is even more strained because of the distortion of the bond angles at its doubly bonded carbons from their normal sp^2-hybridization value of 120°. Cyclobutene has less angle strain than cyclopropene, and the angle strain in cyclopentene, cyclohexene, and higher cycloalkenes is negligible.

The presence of the double bond in cycloalkenes affects the conformation of the ring. The conformation of cyclohexene is a half-chair, with carbons 1, 2, 3, and 6 in the same plane, and carbons 4 and 5 above and below the plane. Substituents at carbons 3 and 6 are tilted from their usual axial and equatorial orientations and are referred to as *pseudoaxial* and *pseudoequatorial.* Conversion to the alternative half-chair occurs readily, with an energy barrier of 22.2 kJ/mol (5.3 kcal/mol), which is about one half that required for chair-to-chair interconversion in cyclohexane.

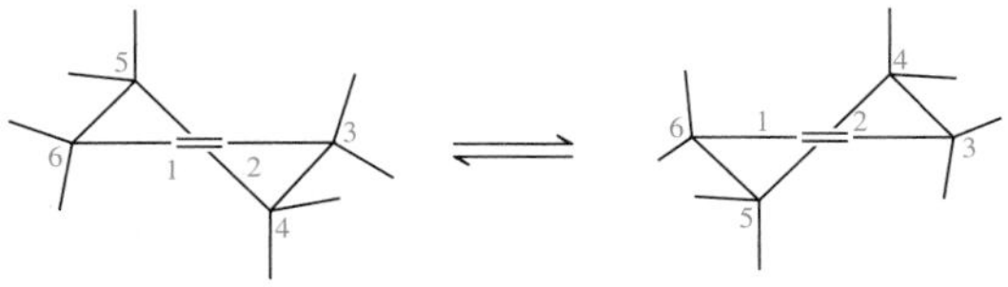

So far we have represented cycloalkenes by structural formulas in which the double bonds are of the cis configuration. If the ring is large enough, however, a trans stereoisomer is also possible. The smallest trans cycloalkene that is stable enough to be isolated and stored in a normal way is *trans*-cyclooctene.

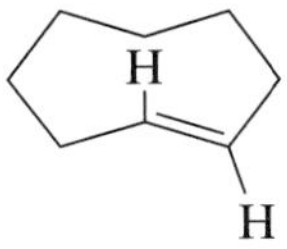

Energy difference =
39 kJ/mol (9.2 kcal/mol)

(*E*)-Cyclooctene
(*trans*-cyclooctene)
Less stable

(*Z*)-Cyclooctene
(*cis*-cyclooctene)
More stable

trans-Cycloheptene has been prepared and studied at low temperature (−90°C) but is too reactive to be isolated and stored at room temperature. Evidence has also been presented for the fleeting existence of the even more strained *trans*-cyclohexene as a reactive intermediate in certain reactions.

Problem 7.15

Place a double bond in the carbon skeleton shown so as to represent

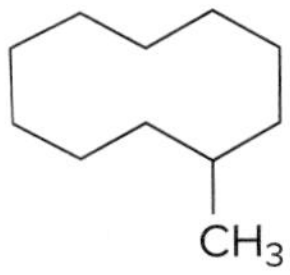

(a) (*Z*)-1-Methylcyclodecene
(b) (*E*)-1-Methylcyclodecene
(c) (*Z*)-3-Methylcyclodecene
(d) (*E*)-3-Methylcyclodecene
(e) (*Z*)-5-Methylcyclodecene
(f) (*E*)-5-Methylcyclodecene

Sample Solution (a) and (b) Because the methyl group must be at C-1, there are only two possible places to put the double bond:

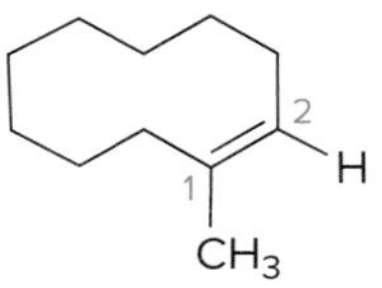

(*Z*)-1-Methylcyclodecene

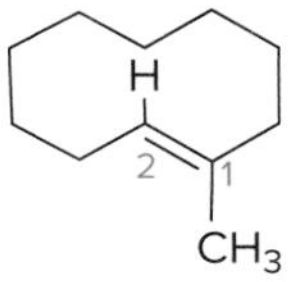

(*E*)-1-Methylcyclodecene

In the *Z* stereoisomer the two lower priority substituents—the methyl group and the hydrogen—are on the same side of the double bond. In the *E* stereoisomer these substituents are on opposite sides of the double bond. The ring carbons are the higher ranking substituents at each end of the double bond.

Because larger rings have more carbons with which to span the ends of a double bond, the strain associated with a trans cycloalkene decreases with increasing ring size. The strain eventually disappears when a 12-membered ring is reached and *cis*- and *trans*-cyclododecene are of approximately equal stability. When the rings are larger than 12-membered, trans cycloalkenes are more stable than cis. In these cases, the ring is large enough and flexible enough that it is energetically similar to a noncyclic cis alkene.

7.8 Preparation of Alkenes: Elimination Reactions

The rest of this chapter describes how alkenes are prepared by elimination—that is, reactions of the type:

$$X-\overset{|}{\underset{|}{C}}{}^{\alpha}-\overset{|}{\underset{|}{C}}{}^{\beta}-Y \longrightarrow \;\rangle C{=}C\langle\; + X-Y$$

Alkene formation requires that X and Y be substituents on adjacent carbon atoms. By making X the reference atom and identifying the carbon attached to it as the α carbon, we see that atom Y is a substituent on the β carbon. Carbons succeedingly more remote from the reference atom are designated γ, δ, and so on. Only β elimination reactions will be discussed in this chapter. **β Eliminations** are also known as *1,2 eliminations.*

You are already familiar with one type of β elimination, having seen in Section 7.2 that ethylene and propene are prepared on an industrial scale by the high-temperature *dehydrogenation* of ethane and propane. Both reactions involve β elimination of H_2.

$$\underset{\text{Ethane}}{CH_3CH_3} \xrightarrow{750°C} \underset{\text{Ethylene}}{H_2C{=}CH_2} + \underset{\text{Hydrogen}}{H_2}$$

$$\underset{\text{Propane}}{CH_3CH_2CH_3} \xrightarrow{750°C} \underset{\text{Propene}}{CH_3CH{=}CH_2} + \underset{\text{Hydrogen}}{H_2}$$

Many reactions classified as dehydrogenations occur within the cells of living systems at 25°C. H_2 is not one of the products, however. Instead, the hydrogens are lost in separate steps of an enzyme-catalyzed process.

$$\underset{\text{Succinic acid}}{HOOCCH_2CH_2COOH} \xrightarrow{\text{succinate dehydrogenase}} \underset{\text{Fumaric acid}}{HOOCCH{=}CHCOOH}$$

Dehydrogenation of alkanes is not a practical *laboratory* synthesis for the vast majority of alkenes. The principal methods by which alkenes are prepared in the laboratory are two other β eliminations: the *dehydration* of alcohols and the *dehydrohalogenation* of alkyl halides. A discussion of these two methods makes up the remainder of this chapter.

7.9 Dehydration of Alcohols

In the **dehydration** of alcohols, H and OH are lost from adjacent carbons. An acid catalyst is necessary. Before dehydrogenation of ethane became the dominant method, ethylene was prepared by heating ethyl alcohol with sulfuric acid.

$$\underset{\text{Ethyl alcohol}}{CH_3CH_2OH} \xrightarrow[160°C]{H_2SO_4} \underset{\text{Ethylene}}{H_2C{=}CH_2} + \underset{\text{Water}}{H_2O}$$

Other alcohols behave similarly. Secondary alcohols undergo elimination at lower temperatures than primary alcohols, and tertiary alcohols at lower temperatures than secondary.

OH
H_2SO_4, 140°C → + H_2O

Cyclohexanol → Cyclohexene (79–87%) + Water

OH
H_2SO_4, heat → + H_2O

tert-Butyl alcohol → 2-Methylpropene (82%) + Water

Reaction conditions, such as the acid used and the temperature, are chosen to maximize the formation of alkene by elimination. Sulfuric acid (H_2SO_4) and phosphoric acid (H_3PO_4) are the acids most frequently used in alcohol dehydrations. Potassium hydrogen sulfate ($KHSO_4$) is also often used.

HSO_4^- and H_3PO_4 are very similar in acid strength. Both are much weaker than H_2SO_4, which is a strong acid.

Problem 7.16

Identify the alkene obtained on dehydration of each of the following alcohols:

(a) 3-Ethyl-3-pentanol
(b) 1-Propanol
(c) 2-Propanol
(d) 2,3,3-Trimethyl-2-butanol

Sample Solution (a) The hydrogen and the hydroxyl are lost from adjacent carbons in the dehydration of 3-ethyl-3-pentanol.

OH, α, β, β, β
H^+ → + H_2O

3-Ethyl-3-pentanol → 3-Ethyl-2-pentene + Water

The hydroxyl group is lost from a carbon that bears three equivalent ethyl substituents. β elimination can occur in any one of three equivalent directions to give the same alkene, 3-ethyl-2-pentene.

Some biochemical processes involve alcohol dehydration as a key step. An example is the conversion of 3-dehydroquinic acid to 3-dehydroshikimic acid.

HO CO_2H, O, OH, OH
3-dehydroquinate dehydratase →
CO_2H, O, OH, OH + H_2O

3-Dehydroquinic acid → 3-Dehydroshikimic acid + Water

This reaction is catalyzed by a *dehydratase* enzyme and is one step along a complex pathway by which plants convert glucose to the amino acid tyrosine.

7.10 Regioselectivity in Alcohol Dehydration: The Zaitsev Rule

Except for the biochemical example just cited, the structures of all of the alcohols in Section 7.9 were such that each one could give only a single alkene by β elimination. What about elimination in alcohols such as 2-methyl-2-butanol, in which dehydration can occur in two different directions to give alkenes that are constitutional isomers? Here, a double bond can be generated between C-1 and C-2 or between C-2 and C-3. Both processes occur but not nearly to the same extent. Under the usual reaction conditions 2-methyl-2-butene is the major product, and 2-methyl-1-butene the minor one.

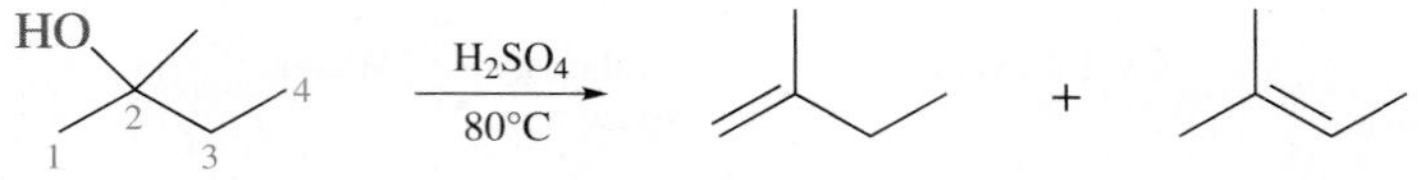

$$\text{2-Methyl-2-butanol} \xrightarrow[80°C]{H_2SO_4} \text{2-Methyl-1-butene (10\%)} + \text{2-Methyl-2-butene (90\%)}$$

Dehydration of this alcohol is selective in respect to its *direction.* Elimination occurs in the direction that leads to the double bond between C-2 and C-3 more than between C-2 and C-1. Reactions that can proceed in more than one direction, but in which one direction is preferred, are said to be **regioselective.**

Although Russian, Zaitsev published most of his work in German scientific journals, where his name was transliterated as *Saytzeff.* The spelling used here (*Zaitsev*) corresponds to the currently preferred style.

In 1875, Alexander M. Zaitsev of the University of Kazan (Russia) set forth a generalization describing the regioselectivity of β eliminations. **Zaitsev's rule** summarizes the results of numerous experiments in which alkene mixtures were produced by β elimination. In its original form, Zaitsev's rule stated that *the alkene formed in greatest amount is the one that corresponds to removal of the hydrogen from the β carbon having the fewest hydrogens.*

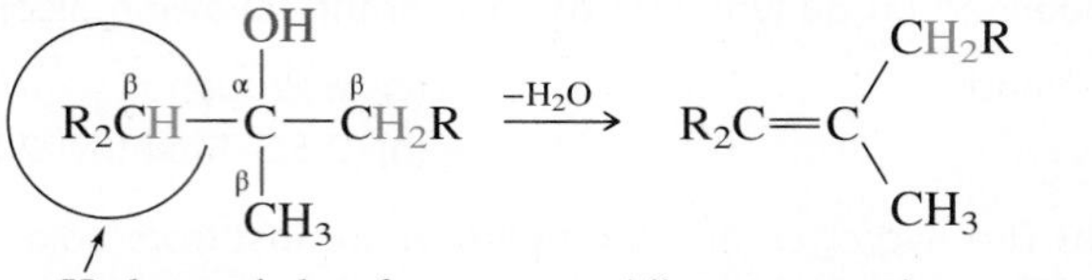

$$R_2CH-C(OH)(CH_3)-CH_2R \xrightarrow{-H_2O} R_2C=C(CH_2R)(CH_3)$$

Hydrogen is lost from β carbon having the fewest attached hydrogens

Alkene present in greatest amount in product

Zaitsev's rule as applied to the acid-catalyzed dehydration of alcohols is now more often expressed in a different way: *β elimination reactions of alcohols yield the most highly substituted alkene as the major product.* Because, as was discussed in Section 7.6, the most highly substituted alkene is also normally the most stable one, Zaitsev's rule is sometimes expressed as a preference for *predominant formation of the most stable alkene that could arise by β elimination.*

Problem 7.17

Each of the following alcohols has been subjected to acid-catalyzed dehydration and yields a mixture of two isomeric alkenes. Identify the two alkenes in each case, and predict which one is the major product on the basis of the Zaitsev rule.

(a) $(CH_3)_2C(OH)CH(CH_3)_2$ (b) 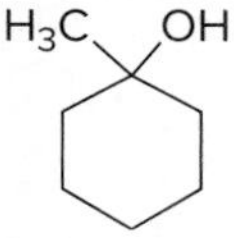 1-methylcyclohexanol (H_3C, OH) (c) 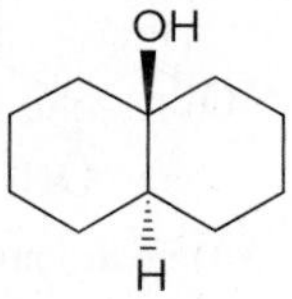 decalinol (OH, H)

Sample Solution (a) Dehydration of 2,3-dimethyl-2-butanol can lead to either 2,3-dimethyl-1-butene by removal of a C-1 hydrogen or to 2,3-dimethyl-2-butene by removal of a C-3 hydrogen.

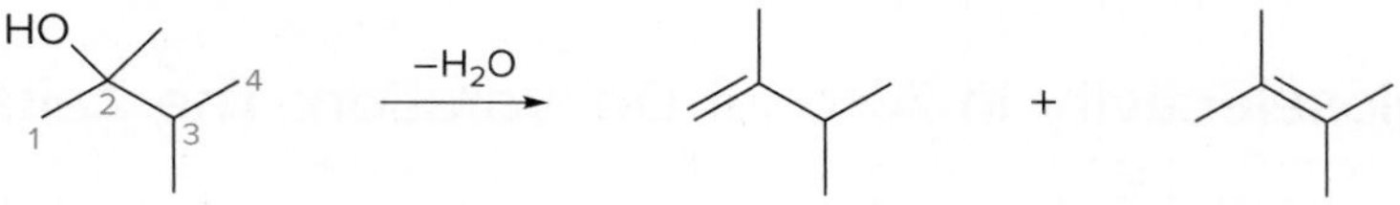

$$\text{2,3-Dimethyl-2-butanol} \xrightarrow{-H_2O} \text{2,3-Dimethyl-1-butene (minor product)} + \text{2,3-Dimethyl-2-butene (major product)}$$

The major product is 2,3-dimethyl-2-butene. It has a tetrasubstituted double bond and is more stable than 2,3-dimethyl-1-butene, which has a disubstituted double bond. The major alkene arises by loss of a hydrogen from the β carbon that has fewer attached hydrogens (C-3) rather than from the β carbon that has the greater number of hydrogens (C-1).

7.11 Stereoselectivity in Alcohol Dehydration

In addition to being regioselective, alcohol dehydrations are stereoselective. A **stereoselective** reaction is one in which a single starting material can yield two or more stereoisomeric products, but gives one of them in greater amounts than any other. Alcohol dehydrations tend to produce the more stable stereoisomer of an alkene. Dehydration of 3-pentanol, for example, yields a mixture of *trans*-2-pentene and *cis*-2-pentene in which the more stable trans stereoisomer predominates.

OH

3-Pentanol $\xrightarrow[\text{heat}]{H_2SO_4}$ *cis*-2-Pentene (25%) (minor product) + *trans*-2-Pentene (75%) (major product)

Problem 7.18

What three alkenes are formed in the acid-catalyzed dehydration of 2-pentanol?

7.12 The E1 and E2 Mechanisms of Alcohol Dehydration

We saw in Chapter 6 that nucleophilic substitution mechanisms were classified by Hughes and Ingold as S_N1 or S_N2 according to whether the rate-determining elementary step is unimolecular or bimolecular. They adopted similar terminology for elimination reactions via the labels **E1** and **E2** standing for **elimination unimolecular** and **elimination bimolecular,** respectively.

The dehydration of alcohols resembles the reaction of alcohols with hydrogen halides (Section 5.7) in two important ways:

1. Both reactions are promoted by acids.
2. The relative reactivity of alcohols increases in the order primary < secondary < tertiary.

These common features suggest that carbocations are key intermediates in alcohol dehydrations, just as they are in the reaction of alcohols with hydrogen halides. Mechanism 7.1 portrays a three-step process for the acid-catalyzed dehydration of *tert*-butyl alcohol. Steps 1 and 2 describe the generation of *tert*-butyl cation by a process similar to that which led to its formation as an intermediate in the reaction of *tert*-butyl alcohol with hydrogen chloride.

Like the reaction of *tert*-butyl alcohol with hydrogen chloride, step 2 in which *tert*-butyloxonium ion dissociates to $(CH_3)_3C^+$ and water, is rate-determining. Because the rate-determining step is unimolecular, the overall dehydration process is referred to as a *unimolecular elimination* and given the symbol E1.

Step 3 is an acid–base reaction in which the carbocation acts as a Brønsted acid, transferring a proton to a Brønsted base (water). This is the property of carbocations that is of the most significance to elimination reactions. Carbocations are strong acids; they are the conjugate acids of alkenes and readily lose a proton to form alkenes. Even weak bases such as water are sufficiently basic to abstract a proton from a carbocation.

Step 3 in Mechanism 7.1 shows water as the base that abstracts a proton from the carbocation. Other Brønsted bases present in the reaction mixture that can function in the same way include *tert*-butyl alcohol and hydrogen sulfate ion.

Mechanism 7.1

The E1 Mechanism for Acid-Catalyzed Dehydration of *tert*-Butyl Alcohol

THE OVERALL REACTION:

$$(CH_3)_3COH \xrightarrow[\text{heat}]{H_2SO_4} (CH_3)_2C{=}CH_2 + H_2O$$

tert-Butyl alcohol — 2-Methylpropene — Water

THE MECHANISM:

Step 1: Protonation of *tert*-butyl alcohol:

tert-Butyl alcohol + Hydronium ion ⇌ (fast) *tert*-Butyloxonium ion + Water

Step 2: Dissociation of *tert*-butyloxonium ion to a carbocation and water:

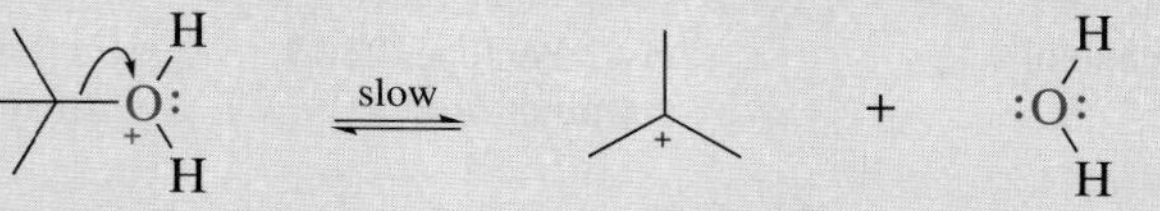

tert-Butyloxonium ion ⇌ (slow) *tert*-Butyl cation + Water

Step 3: Deprotonation of *tert*-butyl cation:

tert-Butyl cation + Water ⇌ (fast) 2-Methylpropene + Hydronium ion

Problem 7.19

Write a structural formula for the carbocation intermediate formed in the dehydration of each of the alcohols in Problem 7.17 (Section 7.10). Using curved arrows, show how each carbocation is deprotonated by water to give a mixture of alkenes.

Sample Solution (a) The carbon that bears the hydroxyl group in the starting alcohol is the one that becomes positively charged in the carbocation.

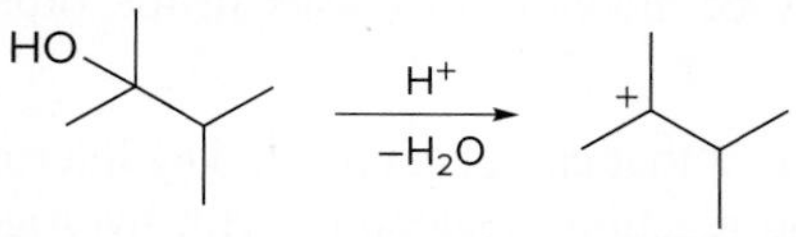

$$\xrightarrow[-H_2O]{H^+}$$

Water may remove a proton from either C-1 or C-3 of this carbocation. Loss of a proton from C-1 yields the minor product 2,3-dimethyl-1-butene. (This alkene has a disubstituted double bond.)

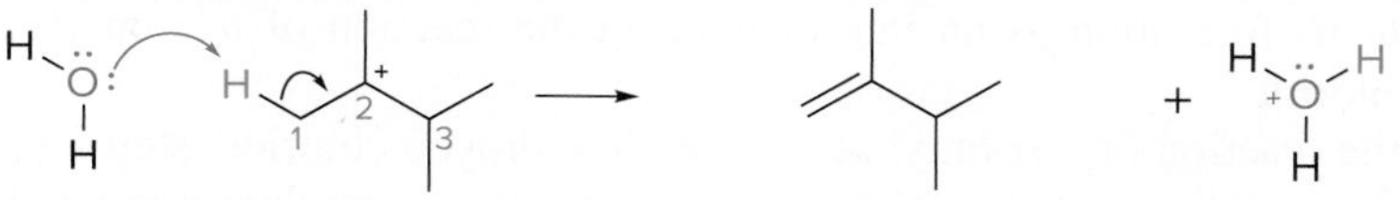

2,3-Dimethyl-1-butene

Loss of a proton from C-3 yields the major product 2,3-dimethyl-2-butene. (This alkene has a tetrasubstituted double bond.)

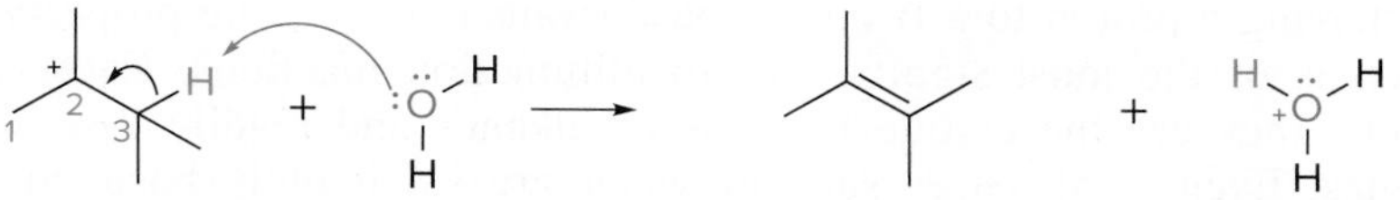

2,3-Dimethyl-2-butene

As noted earlier (Section 5.9), primary carbocations are too high in energy to be intermediates in most chemical reactions. If primary alcohols don't form primary carbocations, then how do they undergo elimination? A modification of our general mechanism for alcohol dehydration offers a reasonable explanation. For primary alcohols it is believed that a proton is lost from the alkyloxonium ion in the same step in which carbon–oxygen bond cleavage takes place. For example, the rate-determining step in the sulfuric acid–catalyzed dehydration of ethanol may be represented as:

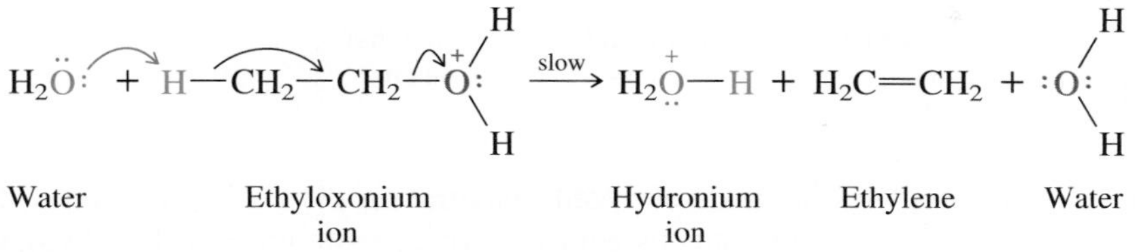

$$H_2\ddot{O}: + H{-}CH_2{-}CH_2{-}\overset{+}{O}H_2 \xrightarrow{\text{slow}} H_2\overset{+}{O}{-}H + H_2C{=}CH_2 + :\ddot{O}H_2$$

Water | Ethyloxonium ion | Hydronium ion | Ethylene | Water

Because the rate-determining step involves two molecules—the alkyloxonium ion and water—the overall reaction is classified as a **bimolecular elimination** and given the symbol E2.

Like tertiary alcohols, secondary alcohols normally undergo dehydration by way of carbocation intermediates.

Chapters 5 and 6 introduced carbocations as intermediates in S_N1 reactions and showed that a less stable carbocation could rearrange to a more stable one prior to reacting with a nucleophile. In the present chapter, we've seen that a carbocation can lose a proton to form an alkene according to an E1 mechanism. Section 7.13 describes how carbocation rearrangements affect the outcome of E1 reactions in a manner similar to that of their S_N1 relatives.

7.13 Rearrangements in Alcohol Dehydration

Some alcohols undergo dehydration to yield alkenes having carbon skeletons different from the starting alcohols. In the example shown, only the least abundant of the three alkenes in the isolated product has the same carbon skeleton as the original alcohol. A **rearrangement** of the carbon skeleton has occurred during the formation of the two most abundant alkenes.

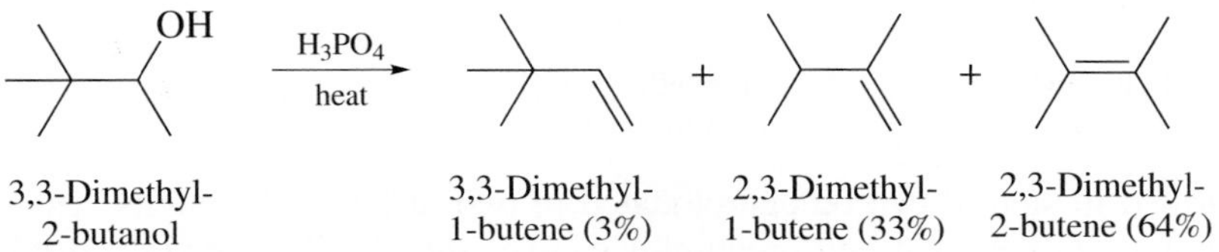

$\xrightarrow[\text{heat}]{H_3PO_4}$

3,3-Dimethyl-2-butanol → 3,3-Dimethyl-1-butene (3%) + 2,3-Dimethyl-1-butene (33%) + 2,3-Dimethyl-2-butene (64%)

The generally accepted explanation for this rearrangement is outlined in Mechanism 7.2. It extends the E1 mechanism for acid-catalyzed alcohol dehydration by combining it with the tendency of carbocation intermediates to rearrange. Not only can a carbocation give an alkene by deprotonation, it can also rearrange to a more stable carbocation that becomes the source of the constitutionally isomeric alkenes.

Similar reactions called Wagner–Meerwein rearrangements were discovered over one hundred years ago. The mechanistic explanation is credited to Frank Whitmore of Penn State who carried out a systematic study of rearrangements during the 1930s.

Problem 7.20

The alkene mixture obtained on dehydration of 2,2-dimethylcyclohexanol contains appreciable amounts of 1,2-dimethylcyclohexene. Give a mechanistic explanation for the formation of this product.

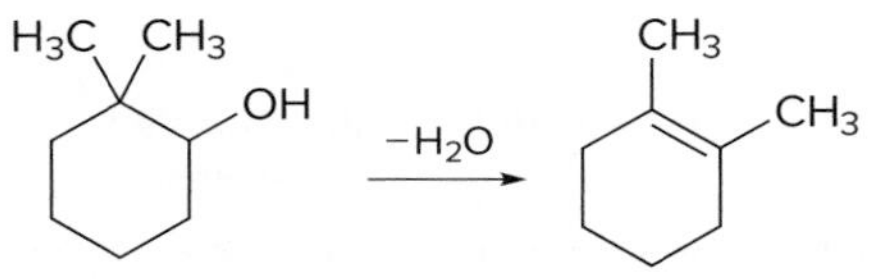

H_3C CH_3 OH $\xrightarrow{-H_2O}$ CH_3 CH_3

Mechanism 7.2

Carbocation Rearrangement in Dehydration of 3,3-Dimethyl-2-butanol

THE OVERALL REACTION:

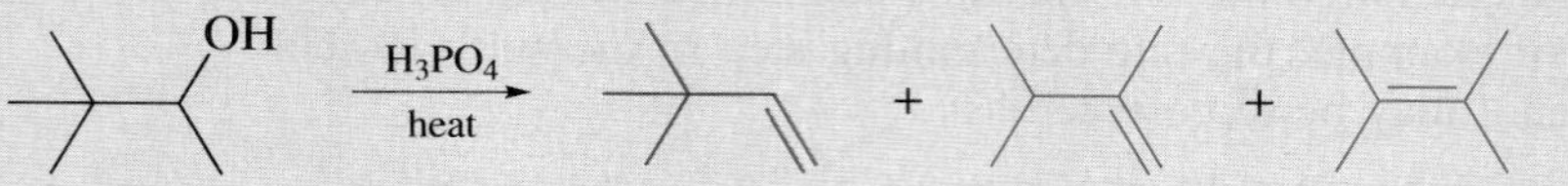

OH, H_3PO_4, heat

3,3-Dimethyl-2-butanol → 3,3-Dimethyl-1-butene + 2,3-Dimethyl-1-butene + 2,3-Dimethyl-2-butene

THE MECHANISM:

Steps 1 and 2: These are analogous to the first two steps in the acid-catalyzed dehydration of *tert*-butyl alcohol described in Mechanism 7.1. The alcohol is protonated in aqueous acid to give an oxonium ion that dissociates to a carbocation and water.

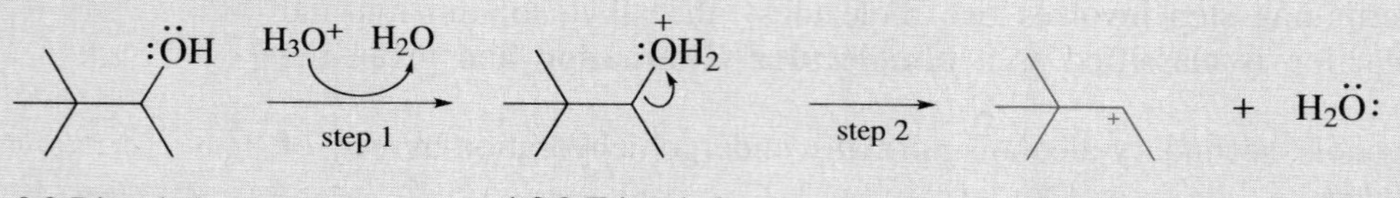

:ÖH, H_3O^+ H_2O, step 1, $\overset{+}{:}OH_2$, step 2, + $H_2\ddot{O}$:

3,3-Dimethyl-2-butanol → 1,2,2-Trimethylpropyloxonium ion → 1,2,2-Trimethylpropyl cation + Water

Steps 3 and 3′: The carbocation formed in step 2 can do two things. It can give an alkene by transferring a proton to a Brønsted base such as water present in the reaction mixture (step 3), or it can rearrange (step 3′). Because alkenes with a rearranged carbon skeleton predominate in the product, we conclude that step 3′ is faster than step 3.

Step 3:

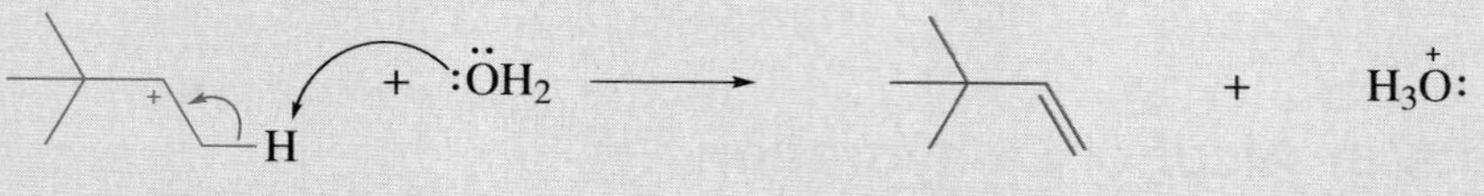

H + :ÖH₂ → + $H_3\overset{+}{\ddot{O}}$:

1,2,2-Trimethylpropyl cation + Water → 3,3-Dimethyl-1-butene + Hydronium ion

Step 3′: Rearrangement by methyl migration is driven by the conversion of a less stable secondary carbocation to a more stable tertiary one.

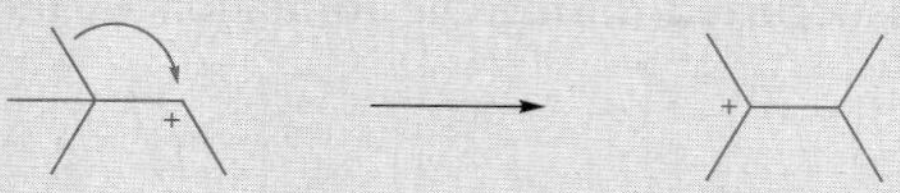

1,2,2-Trimethylpropyl cation (secondary) → 1,1,2-Trimethylpropyl cation (tertiary)

Steps 4 and 4′: The tertiary carbocation formed in step 3′ can be deprotonated in two different directions. The major pathway (step 4′) gives a tetrasubstituted double bond and predominates over step 4, which gives a disubstituted double bond.

Step 4:

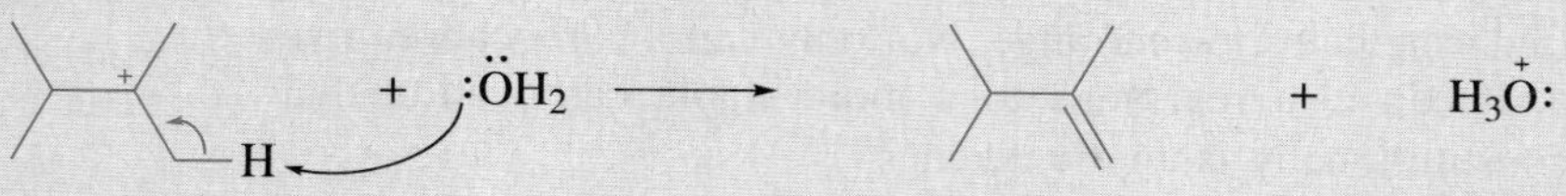

H + :ÖH₂ → + $H_3\overset{+}{\ddot{O}}$:

1,1,2-Trimethylpropyl cation + Water → 2,3-Dimethyl-1-butene + Hydronium ion

Step 4′:

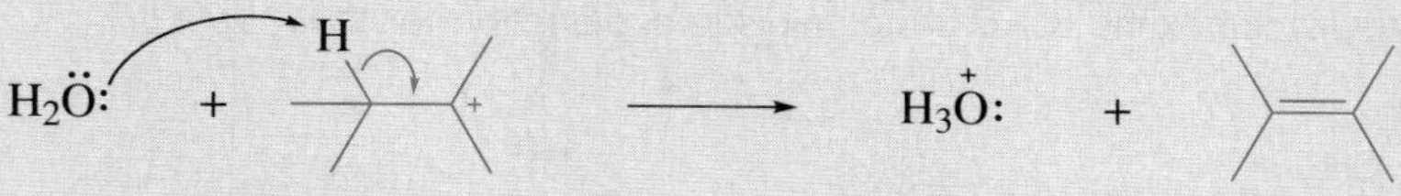

$H_2\ddot{O}$: + H → $H_3\overset{+}{\ddot{O}}$: +

Water + 1,1,2-Trimethylpropyl cation → Hydronium ion + 2,3-Dimethyl-2-butene

In Mechanism 7.2, Step 3′, a methyl group migrates to give a more stable carbocation, however, alkyl groups other than methyl can also migrate to a positively charged carbon.

Many carbocation rearrangements involve hydride shifts (Section 5.12) and proceed in the direction that leads to a more stable carbocation.

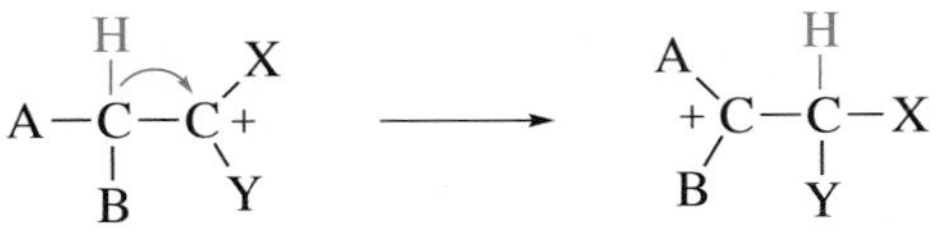

$$\text{A}-\overset{\text{H}}{\underset{\text{B}}{\text{C}}}-\overset{+}{\text{C}}(\text{X})(\text{Y}) \longrightarrow \text{A}(\text{B})\overset{+}{\text{C}}-\overset{\text{H}}{\underset{\text{Y}}{\text{C}}}-\text{X} \quad \text{Hydride shift}$$

Hydride shifts often occur during the dehydration of primary alcohols. Thus, although 1-butene might be expected to be the only alkene formed on dehydration of 1-butanol, it is in fact accompanied by a mixture of *cis*- and *trans*-2-butene.

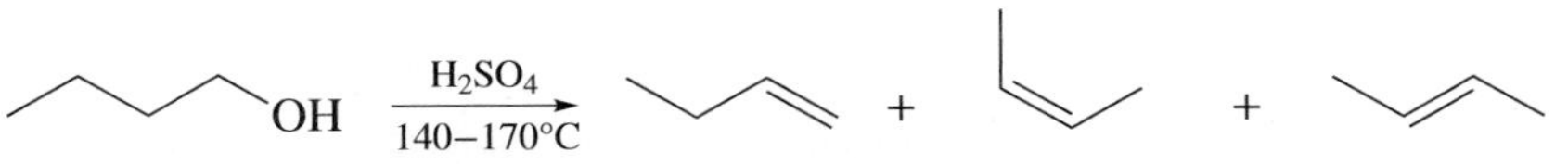

1-Butanol $\xrightarrow[140-170°C]{H_2SO_4}$ 1-Butene (12%) + *cis*-2-Butene (32%) + *trans*-2-Butene (56%)

The formation of three alkenes in the dehydration of 1-butanol begins with protonation of the hydroxyl group as shown in step 1 of Mechanism 7.3. Because 1-butanol is a primary alcohol, it can give 1-butene by an E2 process in which a proton at C-2 of butyloxonium ion is removed while a water molecule departs from C-1 (step 2). The cis and trans stereoisomers of 2-butene, however, are formed from the secondary carbocation that arises via a hydride shift from C-3 to C-2, which accompanies loss of water from the oxonium ion (step 2′).

Mechanism 7.3

Hydride Shift in Dehydration of 1-Butanol

THE OVERALL REACTION:

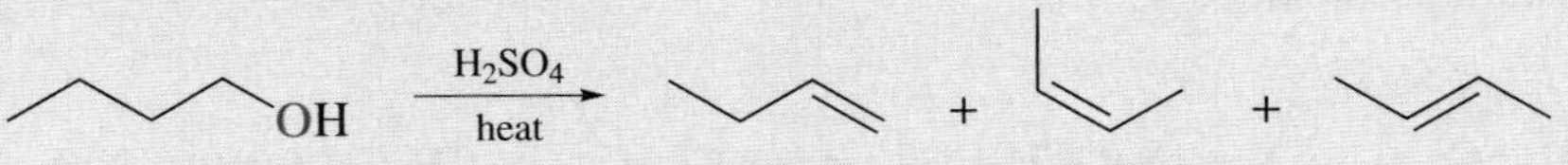

1-Butanol $\xrightarrow[heat]{H_2SO_4}$ 1-Butene + *cis*-2-Butene + *trans*-2-Butene

THE MECHANISM:

Step 1: Protonation of the alcohol gives the corresponding alkyloxonium ion.

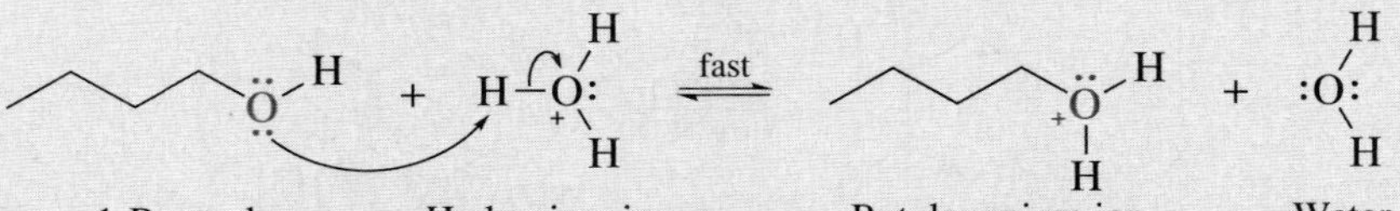

1-Butanol + Hydronium ion (H_3O^+) $\underset{}{\overset{\text{fast}}{\rightleftharpoons}}$ Butyloxonium ion + Water

Step 2: One reaction available to butyloxonium ion leads to 1-butene by an E2 pathway.

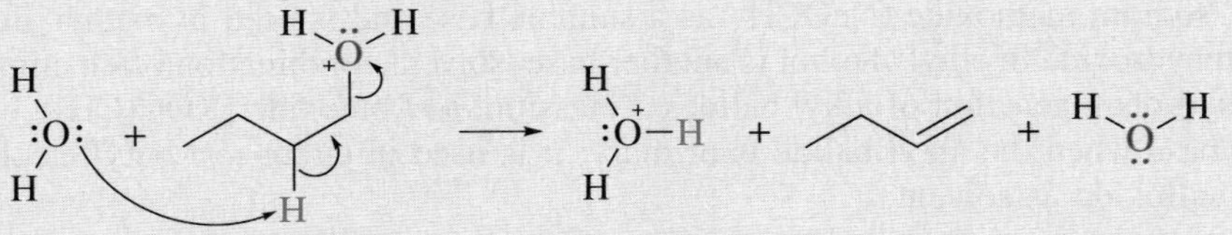

Water + Butyloxonium ion $\longrightarrow$ Hydronium ion + 1-Butene + Water

continued

Step 2′: Alternatively, butyloxonium ion can form a secondary carbocation by loss of water accompanied by a hydride shift.

Butyloxonium ion → Transition state → *sec*-Butyl cation + Water

Step 3: The secondary carbocation formed in step 2′ can give 1-butene or a mixture of *cis*-2-butene and *trans*-2-butene, depending on which proton is removed.

sec-Butyl cation → 1-Butene + *cis*-2-Butene + *trans*-2-Butene

This concludes discussion of our second functional-group transformation involving *alcohols:* the first was the conversion of alcohols to alkyl halides (Chapter 5), and the second the conversion of alcohols to alkenes. In the remaining sections of the chapter the conversion of *alkyl halides* to alkenes by dehydrohalogenation is described.

7.14 Dehydrohalogenation of Alkyl Halides

Dehydrohalogenation is the loss of a hydrogen and a halogen from an alkyl halide. It is one of the most useful methods for preparing alkenes by β elimination. When applied to the preparation of alkenes, the reaction is carried out in the presence of a strong base, such as sodium ethoxide.

$$\text{H}-\text{C}-\text{C}-\text{X} + \text{NaOCH}_2\text{CH}_3 \longrightarrow \text{C}=\text{C} + \text{CH}_3\text{CH}_2\text{OH} + \text{NaX}$$

Alkyl halide, Sodium ethoxide, Alkene, Ethyl alcohol, Sodium halide

$$\text{Cyclohexyl chloride} \xrightarrow[\text{CH}_3\text{CH}_2\text{OH, 55°C}]{\text{NaOCH}_2\text{CH}_3} \text{Cyclohexene (100\%)}$$

Dimethyl sulfoxide (DMSO) has the structure $(CH_3)_2\overset{+}{\ddot{S}}-\ddot{O}:^-$. It is a relatively inexpensive solvent, obtained as a byproduct in paper manufacture.

Similarly, sodium methoxide ($NaOCH_3$) is a suitable base and is used in methyl alcohol. Potassium hydroxide in ethyl alcohol is another base–solvent combination often employed in the dehydrohalogenation of alkyl halides. Potassium *tert*-butoxide [$KOC(CH_3)_3$] is the preferred base when the alkyl halide is primary; it is used in either *tert*-butyl alcohol or dimethyl sulfoxide as solvent.

$$\underset{\text{1-Chlorooctadecane}}{\text{CH}_3(\text{CH}_2)_{15}\text{CH}_2\text{CH}_2\text{Cl}} \xrightarrow[\text{DMSO, 25°C}]{\text{KOC(CH}_3)_3} \underset{\text{1-Octadecene (86\%)}}{\text{CH}_3(\text{CH}_2)_{15}\text{CH}=\text{CH}_2}$$

When using smaller bases like hydroxide, methoxide, and ethoxide, the regioselectivity of dehydrohalogenation of alkyl halides follows the Zaitsev rule; β elimination predominates in the direction that leads to the more highly substituted alkene.

$KOCH_2CH_3$ / $HOCH_2CH_3$, 70°C

2-Bromo-2-methylbutane → 2-Methyl-1-butene (29%) + 2-Methyl-2-butene (71%)

However, with more sterically hindered bases, like *tert*-butoxide, the less substituted alkene is formed in higher yields. Section 7.15 discusses this reaction in more detail.

$KOC(CH_3)_3$ / $HOC(CH_3)_3$, 70°C

2-Bromo-2-methylbutane → 2-Methyl-1-butene (73%) + 2-Methyl-2-butene (27%)

Problem 7.21

Write the structures of all the alkenes that can be formed by dehydrohalogenation of each of the following alkyl halides with sodium ethoxide. Apply the Zaitsev rule to predict the alkene formed in the greatest amount in each case.

(a) 2-Bromo-2,3-dimethylbutane
(b) *tert*-Butyl chloride
(c) 3-Bromo-3-ethylpentane
(d) 2-Bromo-3-methylbutane
(e) 1-Bromo-3-methylbutane
(f) 1-Iodo-1-methylcyclohexane

Sample Solution (a) First analyze the structure of 2-bromo-2,3-dimethylbutane with respect to the number of possible β elimination pathways.

Br; 1, 2, 3, 4 — The two possible alkenes are → 2,3-Dimethyl-1-butene (minor product) and 2,3-Dimethyl-2-butene (major product)

The major product, predicted on the basis of Zaitsev's rule, is 2,3-dimethyl-2-butene. It has a tetrasubstituted double bond. The minor alkene has a disubstituted double bond.

In addition to being regioselective, dehydrohalogenation of alkyl halides is stereoselective and favors formation of the more stable stereoisomer. Usually, as in the case of 5-bromononane, the trans (or *E*) alkene is formed in greater amounts than its cis (or *Z*) stereoisomer.

$KOCH_2CH_3$ / CH_3CH_2OH

5-Bromononane (Br) → *cis*-4-Nonene (23%) + *trans*-4-Nonene (77%)

Problem 7.22

Write structural formulas for all the alkenes that can be formed in the reaction of 2-bromobutane with potassium ethoxide ($KOCH_2CH_3$).

Dehydrohalogenation of cycloalkyl halides leads exclusively to cis cycloalkenes when the ring has fewer than ten carbons. As the ring becomes larger, it can accommodate either a cis or a trans double bond, and large-ring cycloalkyl halides give mixtures of cis and trans cycloalkenes.

Br

$\xrightarrow[CH_3CH_2OH]{KOCH_2CH_3}$ +

Bromocyclodecane

cis-Cyclodecene
(*Z*)-cyclodecene
(85%)

trans-Cyclodecene
(*E*)-cyclodecene
(15%)

7.15 The E2 Mechanism of Dehydrohalogenation of Alkyl Halides

We have seen that nucleophilic substitution mechanisms were classified by Hughes and Ingold as S_N1 or S_N2 according to whether the rate-determining elementary step is unimolecular or bimolecular. They adopted similar terminology for elimination reactions via the labels E1 and E2 standing for **elimination unimolecular** and **elimination bimolecular.** The next several sections focus on eliminations that follow the E2 mechanism because these are the ones with the greater synthetic application. Only rarely does a synthetic plan rest on carrying out an elimination under conditions that favor E1.

Like S_N2, E2 is a one-step mechanism. The equation for the general reaction, when supplemented by curved arrows to show the electron flow, also describes the mechanism.

$$\text{Base}:^- + \text{H}-\text{CR}_2-\text{CR}_2-\ddot{\text{X}}: \longrightarrow \text{Base}-\text{H} + \text{R}_2\text{C}=\text{CR}_2 + :\ddot{\text{X}}:^-$$

Kinetic studies show that the rate of elimination of an alkyl halide on treatment with a strong base is directly proportional to the concentration of both the base and the alkyl halide.

$$\text{Rate} = k[\text{Alkyl halide}][\text{Base}]$$

The rate constant k (Section 5.10) depends on the alkyl halide and the base among other experimental variables (temperature, solvent, etc.). The larger the rate constant, the more reactive the alkyl halide. The value of k for the formation of cyclohexene from cyclohexyl *bromide* and sodium ethoxide, for example, is over 60 times larger than that of cyclohexyl *chloride*. Among the halogens, iodide is the best **leaving group** in dehydrohalogenation, fluoride the poorest. Fluoride is such a poor leaving group that alkyl fluorides are rarely used to prepare alkenes.

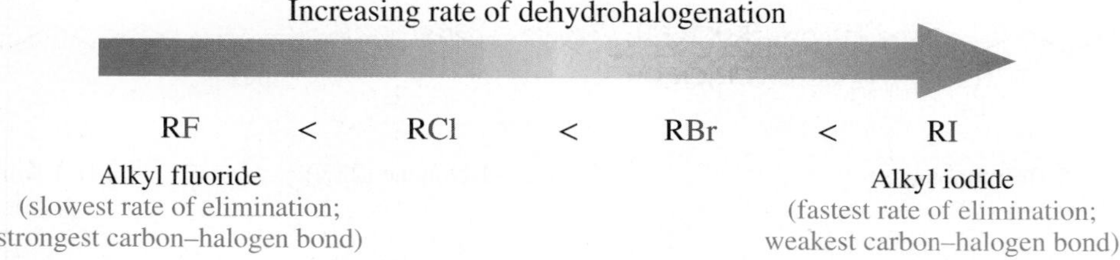

For leaving-group effects in nucleophilic substitution, see Section 6.2.

This trend in leaving-group behavior correlates with certain other properties, such as decreasing carbon–halogen bond strength and decreasing basicity of the halide leaving group. Among the alkyl halides, the C—I bond is the weakest, C—F the strongest. Among halide ions, I^- is the weakest base, F^- the strongest.

Mechanism 7.4

The E2 Mechanism of 1-Chlorooctadecane

THE OVERALL REACTION:

$$(CH_3)_3CO^-K^+ + CH_3(CH_2)_{15}CH_2CH_2Cl \longrightarrow (CH_3)_3COH + CH_3(CH_2)_{15}CH{=}CH_2 + KCl$$

Potassium *tert*-butoxide; 1-Chlorooctadecane; *tert*-Butyl alcohol; 1-Chlorooctadecene; Potassium chloride

THE MECHANISM: The reaction takes place in a single step in which the strong base *tert*-butoxide abstracts a proton from C-2 of the alkyl halide concurrent with loss of chloride from C-1. We can omit writing K^+ in the equation because it appears on both sides of the equation (a "spectator ion").

$$(CH_3)_3C\ddot{O}:^- + CH_3(CH_2)_{15}CH(H){-}CH_2{-}\ddot{Cl}: \longrightarrow (CH_3)_3C\ddot{O}{-}H + CH_3(CH_2)_{15}CH{=}CH_2 + :\ddot{Cl}:^-$$

tert-Butoxide ion; 1-Chlorooctadecane; *tert*-Butyl alcohol; 1-Chlorooctadecene; Chloride ion

THE TRANSITION STATE: *tert*-Butoxide ion attacks the hydrogen on the carbon adjacent to the carbon that bears the chlorine. Oxygen is partially bonded to hydrogen, and the C—H and C—Cl bond breaks while the π bond forms. The oxygen and chlorine, both partially negative, share the charge.

$\delta-$ O----H; H; H; H; C===C; $CH_3(CH_2)_{14}CH_2$; :Cl: $\delta-$

Problem 7.23

A study of the hydrolysis behavior of chlorofluorocarbons (CFCs) carried out by the U.S. Environmental Protection Agency found that 1,2-dichloro-1,1,2-trifluoroethane ($ClF_2C{-}CHClF$) underwent dehydrohalogenation on treatment with aqueous sodium hydroxide. Suggest a reasonable structure for the product of this reaction.

Four key elements

1. Base---H bond making
2. C---H bond breaking
3. C==C π bond development
4. C---X bond breaking

all contribute to the structure of the transition state.

Mechanism 7.4 shows the E2 mechanism for the reaction of 1-chlorooctadecane with potassium *tert*-butoxide presented in the preceding section. The bimolecular transition state is characterized by partial bonds between *tert*-butoxide and one of the hydrogens at C-2 of 2-chlorooctadecane, a partial double bond between C-1 and C-2, and a partial bond between C-1 and chlorine. Figure 7.4 is a potential energy diagram for a simpler reaction (ethyl chloride + hydroxide ion) that also illustrates the orbital interactions involved.

Problem 7.24

Use curved arrows to illustrate the electron flow in the chlorofluorocarbon dehydrohalogenation of Problem 7.23.

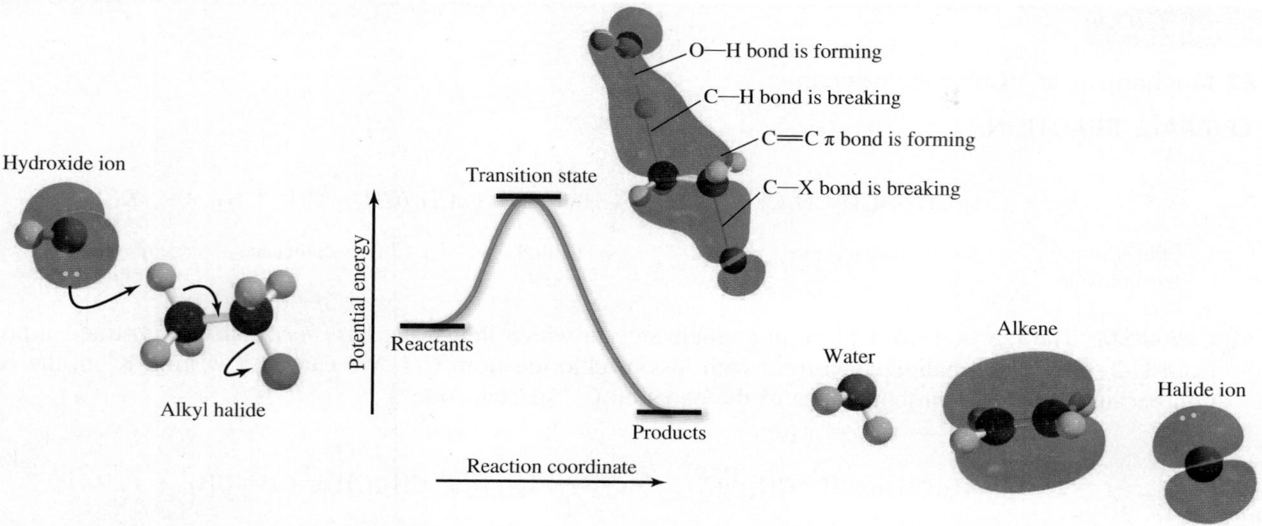

Figure 7.4

Potential energy diagram for E2 elimination of ethyl chloride.

Two aspects of dehydrohalogenation, both based on the stabilization of double bonds by alkyl groups, are accommodated by the E2 mechanism. As shown earlier (Section 7.14), the reaction with smaller bases like ethoxide:

2-Bromo-2-methylbutane $\xrightarrow[CH_3CH_2OH,\ 70°C]{KOCH_2CH_3}$ 2-Methyl-1-butene (29%) Disubstituted double bond + 2-Methyl-2-butene (71%) Trisubstituted double bond

is regioselective and follows the Zaitsev rule in that β elimination gives greater amounts of the more stable isomer—that is, the one with the more substituted double bond. Because alkyl groups stabilize double bonds, it is reasonable that they also stabilize a partially formed π bond in the transition state.

Transition state for formation of 2-methyl-2-butene ($:\ddot{Br}:^{\delta-}$; H- -$\ddot{O}^{\delta-}CH_2CH_3$) is lower energy than Transition state for formation of 2-methyl-1-butene ($:\ddot{Br}:^{\delta-}$; $CH_3CH_2\ddot{O}^{\delta-}$---H)

Problem 7.25

Predict the major product of the reaction shown.

(cyclohexane with Br) $\xrightarrow[ethanol]{NaOCH_2CH_3}$

Studies have shown that steric effects can change the product distribution in E2 reactions, even allowing for the less stable alkene to be formed as the major product. In contrast to the reaction of 2-bromo-2-methylbutane with potassium ethoxide, reacting the same haloalkane with the sterically hindered base, potassium *tert*-butoxide, 2-methyl-1-butene is formed as the major product. An elimination reaction that gives the less stable alkene as the major product is referred to as a Hofmann elimination. Other Hofmann elimination reactions are discussed in Section 22.13.

Br

$KOC(CH_3)_3$ / $HOC(CH_3)_3$, 70°C

+

2-Bromo-2-methylbutane

2-Methyl-1-butene (73%) disubstituted double bond

2-Methyl-2-butene (27%) trisubstituted double bond

With hindered bases, reactions with hydrogens on more substituted carbons, such as secondary versus primary, are slower. This is due to steric hindrance, which raises the energy of the transition state. The result is a lower yield of the more thermodynamically stable alkene product.

$\delta-$:Br: ... secondary ... H---O $\delta-$

is higher energy than

primary ... $\delta-$:Br: ... $\delta-$ O---H

Steric effect raises energy of transition state for 2-methyl-2-butene

Transition state for formation of 2-methyl-1-butene

Regardless of the choice of base, partial double-bond character in the transition state also contributes to the fact that the rate of elimination is fastest for tertiary alkyl halides, slowest for primary halides.

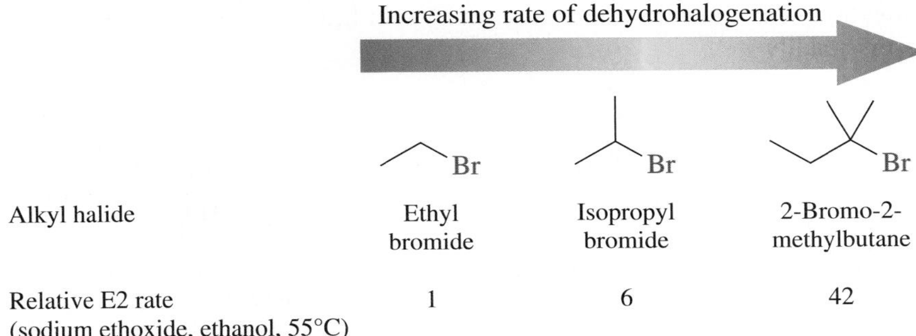

The two regioisomeric alkenes formed via the E2 transition state of the tertiary halide 2-bromo-2-methylbutane are both more substituted than the alkenes (ethylene and propene) formed from ethyl and isopropyl bromide, respectively.

The E2 mechanism is followed whenever an alkyl halide—be it primary, secondary, or tertiary—undergoes elimination in the presence of a strong base. If a strong base is absent, or present in very low concentration, elimination can sometimes still occur by a unimolecular mechanism (E1). The E1 mechanism for dehydrohalogenation will be described in Section 7.18.

7.16 Anti Elimination in E2 Reactions: Stereoelectronic Effects

Further insight into the E2 mechanism comes from stereochemical studies. One such experiment compares the rates of elimination of the cis and trans isomers of 4-*tert*-butylcyclohexyl bromide.

Br

$KOC(CH_3)_3$ / $(CH_3)_3COH$ → ← $KOC(CH_3)_3$ / $(CH_3)_3COH$

Br

cis-4-*tert*-Butylcyclohexyl bromide

4-*tert*-Butylcyclohexene

trans-4-*tert*-Butylcyclohexyl bromide

Although both stereoisomers yield 4-*tert*-butylcyclohexene as the only alkene, they do so at quite different rates. The cis isomer reacts over 500 times faster than the trans.

The difference in reaction rate results from different degrees of π bond development in the E2 transition state. Since π overlap of *p* orbitals requires their axes to be parallel, π bond formation is best achieved when the four atoms of the H—C—C—X unit lie in the same plane at the transition state. The two conformations that permit this are termed *syn coplanar* and *anti coplanar.*

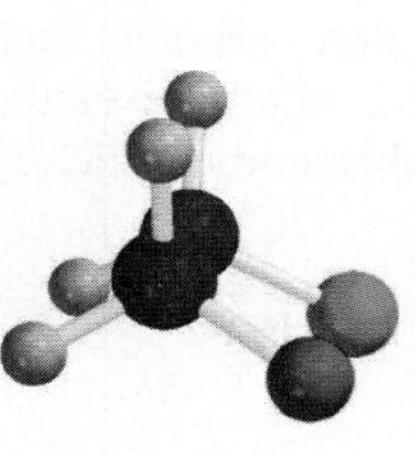

Syn coplanar
Eclipsed conformation
C—H and C—X bonds aligned

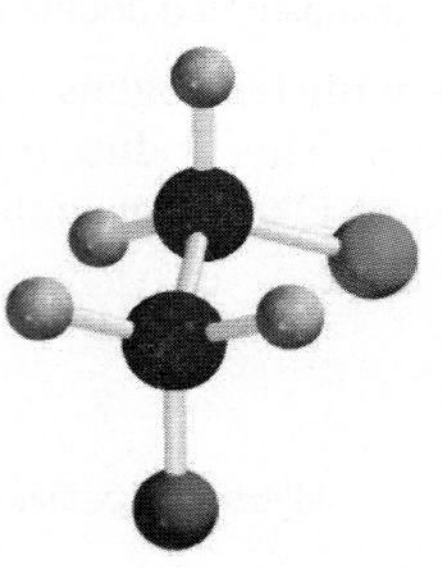

Gauche
Staggered conformation
C—H and C—X bonds not aligned

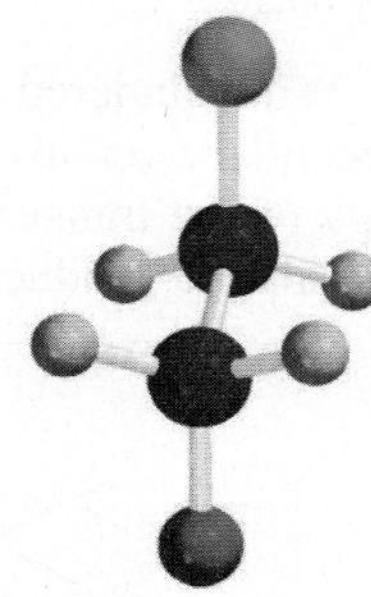

Anti coplanar
Staggered conformation
C—H and C—X bonds aligned

Because adjacent bonds are eclipsed when the H—C—C—X unit is syn coplanar, a transition state with this geometry is less stable than one that has an anti coplanar relationship between the proton and the leaving group.

Bromine is axial and anti coplanar to two axial hydrogens in the most stable conformation of *cis*-4-*tert*-butylcyclohexyl bromide and has the proper geometry for ready E2 elimination. The transition state is reached with little increase in strain, and elimination occurs readily.

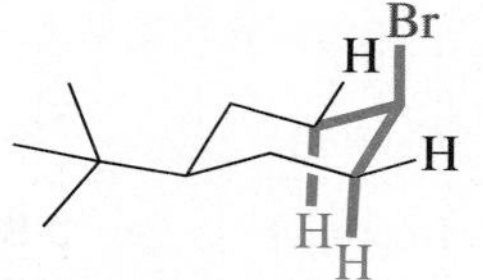

cis-4-*tert*-Butylcyclohexyl bromide
(faster E2 rate:
H and Br are anti coplanar)

H
Br
H
H
H

trans-4-*tert*-Butylcyclohexyl bromide
(slower E2 rate:
(no H atoms anti to Br)

In its most stable conformation, the trans stereoisomer has no β hydrogens anti to Br; all four are gauche. Strain increases significantly in going to the E2 transition state, and the rate of elimination is slower than for the cis stereoisomer.

Problem 7.26

Which stereoisomer do you predict will undergo elimination on treatment with sodium ethoxide in ethanol at the faster rate?

Br

H_3C CH_3

A

or

Br

H_3C CH_3

B

Effects on rate or equilibrium that arise because one spatial arrangement of electrons (or orbitals or bonds) is more stable than another are called **stereoelectronic effects.** We saw an important example of a stereoelectronic requirement in the S_N2 mechanism of nucleophilic substitution (Section 6.3) where the incoming nucleophile bonds to carbon at the transition state from the side opposite the bond to the leaving group. Likewise, *there is a stereoelectronic preference for the anti coplanar arrangement of proton and leaving group in elimination by the E2 mechanism.* Just as inversion of configuration characterizes the S_N2 mechanism, anti elimination characterizes the E2 mechanism. Although coplanarity of the *p* orbitals is the best geometry for the E2 process, modest deviations from it can be tolerated at the cost of a decrease in reaction rate.

The stereoelectronic preference for an anti coplanar arrangement of the H—C—C—X unit in the E2 mechanism, as illustrated in Figure 7.4, is also reflected in the preference for formation of trans rather than cis alkenes.

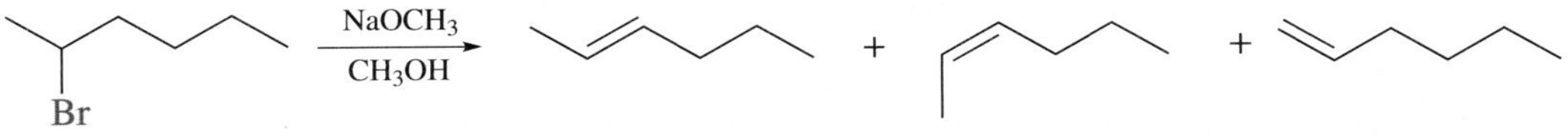

NaOCH$_3$ / CH$_3$OH

2-Bromohexane → *trans*-2-Hexene (54%) + *cis*-2-Hexene (18%) + 1-Hexene (28%)

Anti elimination from the more stable staggered conformation gives the major product.

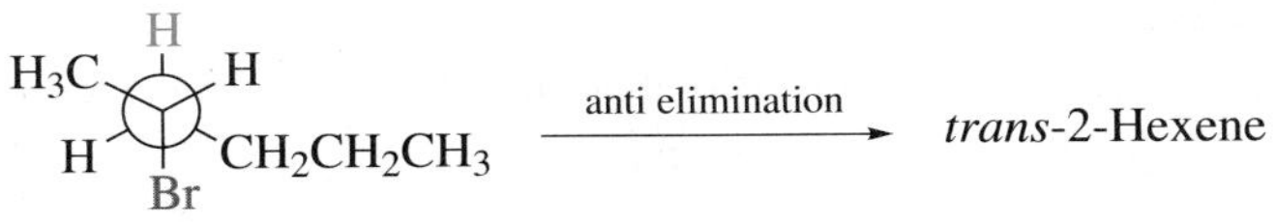

H, H$_3$C, H, H, CH$_2$CH$_2$CH$_3$, Br — anti elimination → *trans*-2-Hexene

More stable conformation

Anti elimination from the less stable staggered conformation gives the minor product.

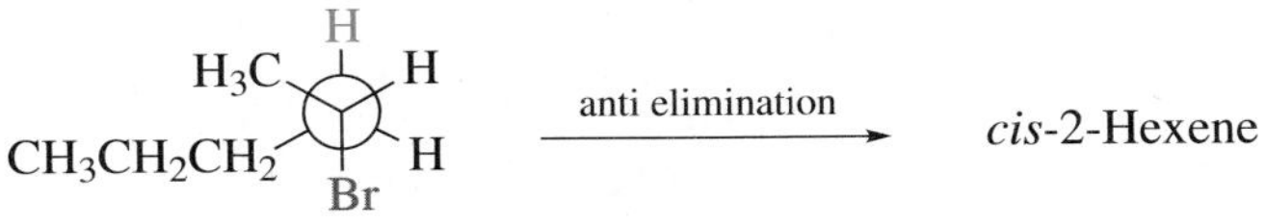

H, H$_3$C, H, CH$_3$CH$_2$CH$_2$, H, Br — anti elimination → *cis*-2-Hexene

Less stable conformation

Not only is this conformation less populated than the other, but van der Waals repulsions between the CH_3 and $CH_3CH_2CH_2$ groups increase in going to the transition state, which raises the energy of the activated complex, increases E_a, and decreases the reaction rate.

The preferential formation of *trans*-2-hexene from 2-bromohexane is an example of a **stereoselective reaction,** a reaction that can give two or more stereoisomeric products but gives one of them preferentially (Section 7.11). **Stereospecific reactions** are those in which stereoisomeric reactants yield products that are stereoisomers of each other. Terms such as "addition to the less-hindered side" describe stereoselectivity; "inversion of configuration" and "anti-elimination" describe stereospecificity.

7.17 Isotope Effects and the E2 Mechanism

The E2 mechanism as outlined in the preceding two sections receives support from studies of the rates of dehydrohalogenation of alkyl halides that contain deuterium (D = ^{2}H) instead of protium (^{1}H) at the β carbon. The fundamental *kinds* of reactions a substance undergoes are the same regardless of which isotope is present, but the reaction *rates* can be different.

A C—D bond is ≈ 12 kJ/mol (2.9 kcal/mol) stronger than a C—H bond, making the activation energy for breaking a C—D bond slightly greater than that of an analogous C—H bond. Consequently, the rate constant *k* for an elementary step in which a C—D bond breaks is smaller than for a C—H bond. This difference in rate is expressed as a ratio of the respective rate constants (k_H/k_D) and is a type of **kinetic isotope effect.** Because it compares ^{2}H to ^{1}H, it is also referred to as a **deuterium isotope effect.**

Typical deuterium isotope effects for reactions in which C—H bond breaking is rate-determining lie in the range k_H/k_D = 3–8. If the C—H bond breaks after the rate-determining step, the overall reaction rate is affected only slightly and k_H/k_D = 1–2. *Thus, measuring the deuterium isotope effect can tell us if a C—H bond breaks in the rate-determining step.*

According to the E2 mechanism for dehydrohalogenation, a base removes a proton from the β carbon in the same step as the halide is lost. This step, indeed it is the only step in the mechanism, is rate-determining. Therefore, elimination by the E2 mechanism should exhibit a deuterium isotope effect. This prediction was tested by comparing the rate of elimination in the reaction:

$$(CD_3)_2CHBr \xrightarrow[CH_3CH_2OH]{NaOCH_2CH_3} CD_2{=}C(H)CD_3$$

with that of $(CH_3)_2CHBr$. The measured value was $k_H/k_D = 6.7$, consistent with the idea that the β hydrogen is removed by the base in the rate-determining step, not after it.

Problem 7.27

Choose the compound in the following pairs that undergoes E2 elimination at the faster rate.

(a) $CH_3CH_2CH_2CD_2Br$ or $CH_3CH_2CD_2CH_2Br$

(b) $CH_3C(CH_3)(D)CH_2Br$ or $CH_3C(CH_3)(H)CD_2Br$

(c) $CD_3CD_2C(CD_3)(H)CH_2Br$ or $CH_3CH_2C(CH_3)(D)CH_2Br$

Sample Solution (a) A double bond is formed between C-1 and C-2 when either of the two compounds undergoes elimination. Bromine is lost from C-1, and H (or D) is lost from C-2. A C—H bond breaks faster than a C—D bond; therefore, E2 elimination is faster in $CH_3CH_2CH_2CD_2Br$ than in $CH_3CH_2CD_2CH_2Br$.

The size of an isotope effect depends on the ratio of the atomic masses of the isotopes; thus, those that result from replacing ^1H by ^2H or ^3H (tritium) are easiest to measure. This, plus the additional facts that most organic compounds contain hydrogen and many reactions involve breaking C—H bonds, have made rate studies involving hydrogen isotopes much more common than those of other elements.

In later chapters we'll see several additional examples of reactions in which deuterium isotope effects were measured in order to test proposed mechanisms.

7.18 The E1 Mechanism of Dehydrohalogenation of Alkyl Halides

The E2 mechanism is a concerted process in which the carbon–hydrogen and carbon–halogen bonds both break in the same elementary step. What if these bonds break in separate steps?

One possibility is the two-step process of Mechanism 7.5, in which the carbon–halogen bond breaks first to give a carbocation intermediate, followed by deprotonation of the carbocation in a second step.

The alkyl halide, in this case 2-bromo-2-methylbutane, ionizes to a carbocation and a halide anion by a heterolytic cleavage of the carbon–halogen bond. Like the dissociation of an alkyloxonium ion to a carbocation, this step is rate-determining. Because the rate-determining step is unimolecular—it involves only the alkyl halide and not the base—it is an E1 mechanism.

Mechanism 7.5

The E1 Mechanism for Dehydrohalogenation of 2-Bromo-2-methylbutane

THE OVERALL REACTION:

2-Bromo-2-methylbutane $\xrightarrow[\text{heat}]{CH_3CH_2OH}$ 2-Methyl-1-butene (25%) + 2-Methyl-2-butene (75%)

THE MECHANISM:

Step 1: *Ionization* The alkyl halide dissociates by heterolytic cleavage of the carbon–halogen bond. The products are a carbocation and a halide ion. This is the rate-determining step.

2-Bromo-2-methylbutane $\xrightarrow{\text{slow}}$ 1,1-Dimethylpropyl cation + Bromide ion ($:\ddot{Br}:^-$)

Step 2: *Deprotonation* Ethanol acts as a Brønsted base to remove a proton from the carbocation to give the two alkene products. Zaitsev's rule is followed, and the regioisomer with the more highly substituted double bond predominates.

Ethanol ($CH_3CH_2\ddot{O}H$) + 1,1-Dimethylpropyl cation $\xrightarrow{\text{fast}}$ Ethyloxonium ion ($CH_3CH_2\overset{+}{O}H_2$) + 2-Methyl-1-butene

Ethanol ($CH_3CH_2\ddot{O}H$) + 1,1-Dimethylpropyl cation $\xrightarrow{\text{faster}}$ Ethyloxonium ion ($CH_3CH_2\overset{+}{O}H_2$) + 2-Methyl-2-butene

Typically, elimination by the E1 mechanism is observed only for tertiary and some secondary alkyl halides, and then only when the base is weak or in low concentration. Unlike eliminations that follow an E2 pathway and exhibit second-order kinetic behavior:

$$\text{Rate} = k[\text{alkyl halide}][\text{base}]$$

those that follow an E1 mechanism obey a first-order rate law.

$$\text{Rate} = k[\text{alkyl halide}]$$

The reactivity order parallels the ease of carbocation formation.

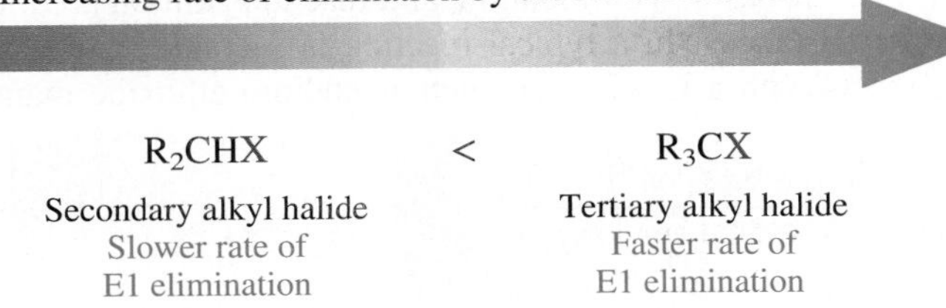

Because the carbon–halogen bond breaks in the slow step, the rate of the reaction depends on the leaving group. Alkyl iodides have the weakest carbon–halogen bond and are the most reactive; alkyl fluorides have the strongest carbon–halogen bond and are the least reactive.

Problem 7.28

Based on the E1 mechanism shown for it in Mechanism 7.5, would you expect elimination in 2-bromo-2-methylbutane to exhibit a deuterium isotope effect?

The most common examples of E1 eliminations are those carried out in the absence of added base. In the example cited in Mechanism 7.5, the base that abstracts the proton from the carbocation intermediate is a very weak one; it is a molecule of the solvent, ethyl alcohol. At even modest concentrations of strong base, elimination by the E2 mechanism is much faster than E1 elimination.

There is a strong similarity between the process shown in Mechanism 7.5 and the one shown for alcohol dehydration in Mechanism 7.1. The main difference between the dehydration of 2-methyl-2-butanol and the dehydrohalogenation of 2-bromo-2-methylbutane is the source of the carbocation. With alcohols, it is the corresponding alkyloxonium ion that dissociates to form the carbocation. Alkyl halides ionize directly to the carbocation.

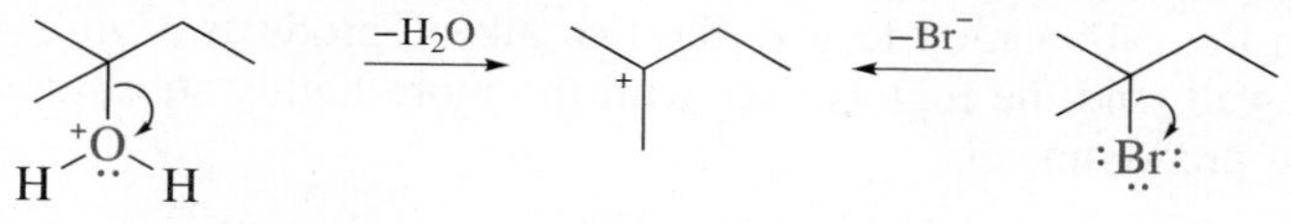

$-H_2O$ $-Br^-$

Alkyloxonium ion   Carbocation   Alkyl halide

Like alcohol dehydrations, E1 reactions of alkyl halides can be accompanied by carbocation rearrangements. Eliminations by the E2 mechanism, on the other hand, normally proceed without rearrangement. Consequently, if one wishes to prepare an alkene from an alkyl halide, conditions favorable to E2 elimination should be chosen. In practice this simply means carrying out the reaction in the presence of a strong base.

7.19 Substitution and Elimination as Competing Reactions

In this and the preceding chapter we have seen that a Lewis base can react with an alkyl halide by either substitution or elimination.

β elimination → $C{=}C$ + H—Y + X^-

$H-C-C(X)- + Y^-$

nucleophilic substitution → $H-C-C(Y)- + X^-$

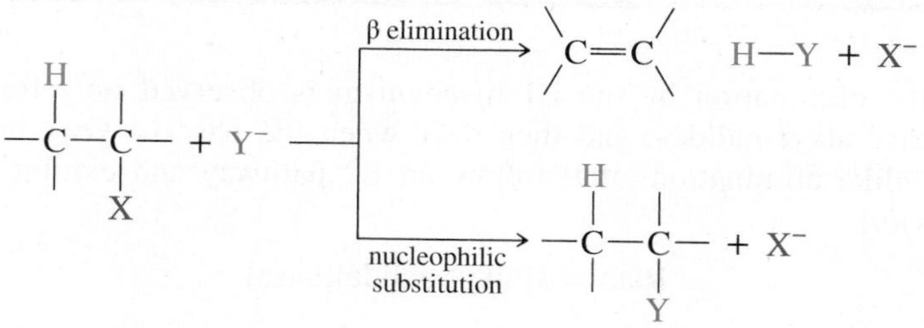

Substitution can take place by the S_N1 or the S_N2 mechanism, elimination by E1 or E2.

How can we predict whether substitution or elimination will predominate? The two most important factors are the *structure of the alkyl halide* and the *basicity of the anion*. It is useful to approach the question from the premise that the characteristic reaction of alkyl halides with Lewis bases is elimination, and that substitution predominates only under certain special circumstances. In a typical reaction, a secondary alkyl halide such as isopropyl bromide reacts with a Lewis base such as sodium ethoxide mainly by elimination:

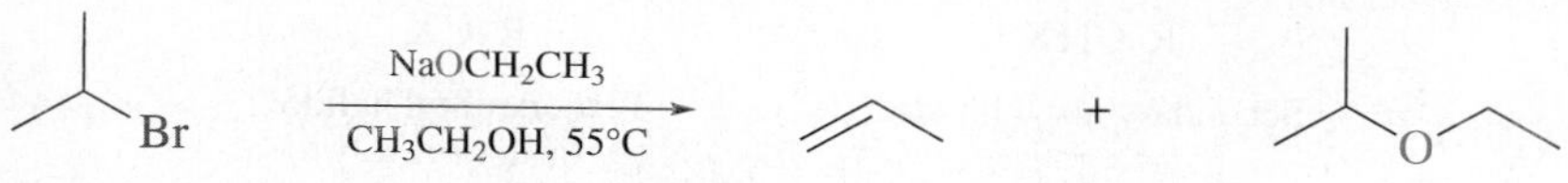

$NaOCH_2CH_3$ / CH_3CH_2OH, 55°C

Isopropyl bromide → Propene (87%) + Ethyl isopropyl ether (13%)

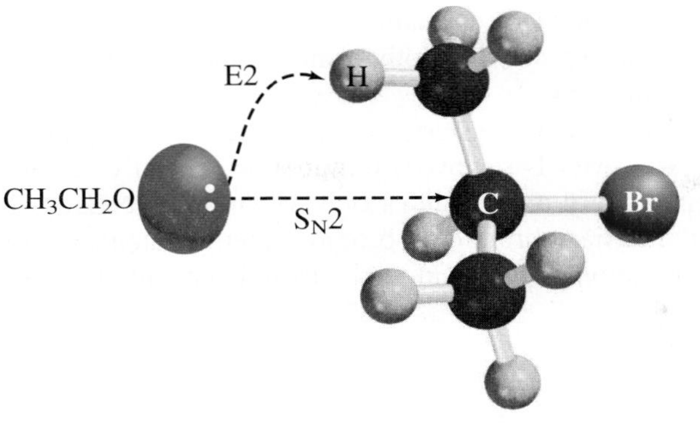

Figure 7.5

When a Lewis base reacts with an alkyl halide, either substitution or elimination can occur. Substitution (S_N2) occurs when the Lewis base acts as a nucleophile and attacks carbon to displace bromide. Elimination (E2) occurs when the Lewis base abstracts a proton from the β carbon. The alkyl halide shown is isopropyl bromide, and elimination (E2) predominates over substitution with alkoxide bases.

Figure 7.5 illustrates the close relationship between the E2 and S_N2 pathways for this case, and the results cited in the preceding equation show that E2 is faster than S_N2 when a secondary alkyl halide reacts with a strong base.

As crowding at the carbon that bears the leaving group decreases, the rate of nucleophilic substitution becomes faster than the rate of elimination. A low level of steric hindrance to approach of the nucleophile is one of the special circumstances that permit substitution to predominate, and primary alkyl halides react with alkoxide bases by an S_N2 mechanism in preference to E2.

Propyl bromide $\xrightarrow[CH_3CH_2OH,\ 55°C]{NaOCH_2CH_3}$ Propene (9%) + Ethyl propyl ether (91%)

If, however, the base itself is crowded, such as potassium *tert*-butoxide, even primary alkyl halides undergo elimination rather than substitution (87%).

$$CH_3(CH_2)_{15}CH_2CH_2Br \xrightarrow[(CH_3)_3COH,\ 40°C]{KOC(CH_3)_3} CH_3(CH_2)_{15}CH{=}CH_2 + CH_3(CH_2)_{15}CH_2CH_2OC(CH_3)_3$$

1-Bromooctadecane → 1-Octadecene (87%) + *tert*-Butyl octadecyl ether (13%)

A second factor that can tip the balance in favor of substitution is weak basicity of the nucleophile. Nucleophiles that are less basic than hydroxide react with both primary and secondary alkyl halides to give the product of nucleophilic substitution in high yield. To illustrate, cyanide ion is much less basic than hydroxide and reacts with 2-chlorooctane to give the corresponding alkyl cyanide as the major product.

2-Chlorooctane $\xrightarrow[DMSO]{KCN}$ 2-Cyanooctane (70%)

Cyanide is a weaker base than hydroxide because its conjugate acid HCN (pK_a 9.1) is a stronger acid than water (pK_a 15.7).

Azide ion (N_3^-) is an even weaker base than cyanide. It is a good nucleophile and reacts with secondary alkyl halides mainly by substitution.

Cyclohexyl iodide $\xrightarrow{NaN_3}$ Cyclohexyl azide (75%)

The conjugate acid of azide ion is called *hydrazoic acid* (HN_3). It has a pK_a of 4.6, and so is similar to acetic acid in its acidity.

Hydrogen sulfide (pK_a 7.0) is a stronger acid than water (pK_a 15.7). Therefore, HS^- is a much weaker base than HO^-.

Hydrogen sulfide ion HS^-, and anions of the type RS^-, are substantially less basic than hydroxide ion and react with both primary and secondary alkyl halides to give mainly substitution products.

Tertiary alkyl halides are so sterically hindered to nucleophilic attack that the presence of any anionic Lewis base favors elimination. Usually substitution predominates over elimination in tertiary alkyl halides only when anionic Lewis bases are absent. In the solvolysis of the tertiary bromide 2-bromo-2-methylbutane, for example, the ratio of substitution to elimination is 64:36 in pure ethanol but falls to 1:99 in the presence of 2 M sodium ethoxide.

Br; ethanol, 25°C; OCH_2CH_3

2-Bromo-2-methylbutane → 2-Ethoxy-2-methylbutane (Major product in absence of sodium ethoxide) + 2-Methyl-2-butene + 2-Methyl-1-butene (Alkene mixture is major product in presence of sodium ethoxide)

The substitution product in this case is formed by an S_N1 mechanism in both the presence and absence of sodium ethoxide. The alkenes are formed by an E1 mechanism in the absence of sodium ethoxide and by a combination of E2 (major) and E1 (minor) in its presence.

Problem 7.29

Predict the major organic product of each of the following reactions:

(a) Cyclohexyl bromide and potassium ethoxide
(b) Ethyl bromide and potassium cyclohexanolate
(c) *sec*-Butyl bromide solvolysis in methanol
(d) *sec*-Butyl bromide solvolysis in methanol containing 2 M sodium methoxide

Sample Solution (a) Cyclohexyl bromide is a secondary halide and reacts with alkoxide bases by elimination rather than substitution. The major organic products are cyclohexene and ethanol.

Cyclohexyl bromide + Potassium ethoxide ($KOCH_2CH_3$) → Cyclohexene + Ethanol (CH_3CH_2OH)

Regardless of the alkyl halide, raising the temperature increases the rate of both substitution and elimination. The rate of elimination, however, usually increases faster than substitution, so that at higher temperatures the proportion of elimination products increases at the expense of substitution products.

As a practical matter, if an anti coplanar arrangement of a proton and a suitable leaving group is structurally accessible, elimination can always be made to occur quantitatively. Strong bases, especially bulky ones such as *tert*-butoxide ion, react even with primary alkyl halides by an E2 process at elevated temperatures. The more difficult task is to find conditions that promote substitution. In general, the best approach is to choose conditions that favor the S_N2 mechanism—an unhindered substrate, a good nucleophile that is not strongly basic, and the lowest practical temperature consistent with reasonable reaction rates.

Problem 7.30

A standard method for the synthesis of ethers is an S_N2 reaction between an alkoxide and an alkyl halide.

$$R-\ddot{O}:^- + R'-\ddot{X}: \longrightarrow R-\ddot{O}-R' + :\ddot{X}:^-$$

Show possible combinations of alkoxide and alkyl halide for the preparation of the following ethers. Which of these can be prepared effectively by this method?

(a) $CH_3CH_2CH(CH_3)OCH(CH_3)_2$ (b) $CH_3CH_2CH_2OCH(CH_3)_2$

Sample Solution

(a) • There are two possible combinations of alkoxide and alkyl halide (shown as bromide).

$CH_3CH_2CH(CH_3)-\ddot{O}:^-$ + $BrCH(CH_3)_2$

$CH_3CH_2CH(CH_3)-Br$ + $^-:\ddot{O}-CH(CH_3)_2$

• In both cases the alkyl halide is secondary. Secondary alkyl halides undergo elimination (E2) rather than substitution (S_N2) with alkoxide bases.

Functional-group transformations that rely on substitution by the S_N1 mechanism are not as generally applicable as those of the S_N2 type. Hindered substrates are prone to elimination, and rearrangement is possible when carbocations are involved. Only in cases in which elimination is impossible are S_N1 reactions used in functional-group transformations.

7.20 Elimination Reactions of Sulfonates

Everything that we have said about E1 and E2 reactions of alkyl halides also applies to alkyl sulfonates. The following example is representative in respect to regioselectivity and stereoselectivity.

2-Hexyl *p*-toluenesulfonate $\xrightarrow[\text{CH}_3\text{OH, 60°C}]{\text{NaOCH}_3}$ *trans*-2-Hexene (43%) + *cis*-2-Hexene (23%) + 1-Hexene (34%)

The major products with the more substituted double bond have a combined yield of 66%, and the more stable trans stereoisomer of 2-hexene is formed in greater amounts than the less stable cis isomer.

7.21 SUMMARY

Section 7.1 Alkenes and cycloalkenes contain carbon–carbon double bonds. According to IUPAC nomenclature, alkenes are named by substituting *-ene* for the *-ane* suffix of the alkane that has the same number of carbon atoms as the longest continuous chain that includes the double bond. The chain is numbered in the direction that gives the lower number to the first-appearing carbon of the double bond. The double bond takes precedence over alkyl groups and halogens in dictating the direction of numbering, but is outranked by a hydroxyl group.

3-Ethyl-2-pentene 3-Bromocyclopentene 3-Buten-1-ol

Section 7.2 Bonding in alkenes is described according to an sp^2 orbital hybridization model. The double bond unites two sp^2-hybridized carbon atoms and is made of a σ component and a π component. The σ bond arises by overlap of an sp^2 hybrid orbital on each carbon. The π bond is weaker than the σ bond and results from a side-by-side overlap of half-filled *p* orbitals. By combining two 2*p* orbitals, the highest occupied molecular orbital (HOMO) and the lowest unoccupied molecular orbital (LUMO) are formed.

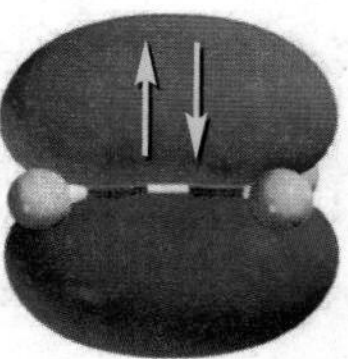

Sections 7.3–7.4 Isomeric alkenes may be either **constitutional isomers** or **stereoisomers.** There is a sizable barrier to rotation about a carbon–carbon double bond, which corresponds to the energy required to break the π component of the double bond. Stereoisomeric alkenes do not interconvert under normal conditions. Their configurations are described according to two notational systems. One system adds the prefix *cis-* to the name of the alkene when similar substituents are on the same side of the double bond and the prefix *trans-* when they are on opposite sides. The other ranks substituents according to a system of rules based on atomic number. The prefix *Z* is used for alkenes that have higher ranked substituents on the same side of the double bond; the prefix *E* is used when higher ranked substituents are on opposite sides.

cis-2-Pentene [(*Z*)-2-pentene] *trans*-2-Pentene [(*E*)-2-pentene]

Section 7.5 Alkenes are nonpolar. Alkyl substituents donate electrons to an sp^2-hybridized carbon to which they are attached slightly better than hydrogen does.

Section 7.6 Electron release from alkyl substituents stabilizes a double bond. In general, the order of alkene stability is:

1. Tetrasubstituted alkenes ($R_2C{=}CR_2$) are the most stable.
2. Trisubstituted alkenes ($R_2C{=}CHR$) are next.
3. Among disubstituted alkenes, *trans*-RCH=CHR is normally more stable than *cis*-RCH=CHR. Exceptions are cycloalkenes, cis cycloalkenes being more stable than trans when the ring contains fewer than 12 carbons.
4. Monosubstituted alkenes ($RCH{=}CH_2$) have a more stabilized double bond than ethylene (unsubstituted) but are less stable than disubstituted alkenes.

The greater stability of more highly substituted double bonds is an example of an **electronic effect.** The decreased stability that results from van der Waals strain between cis substituents is an example of a **steric effect.**

Section 7.7 Cycloalkenes that have trans double bonds in rings smaller than 12 members are less stable than their cis stereoisomers. *trans*-Cyclooctene can be isolated and stored at room temperature, but *trans*-cycloheptene is not stable above −30°C.

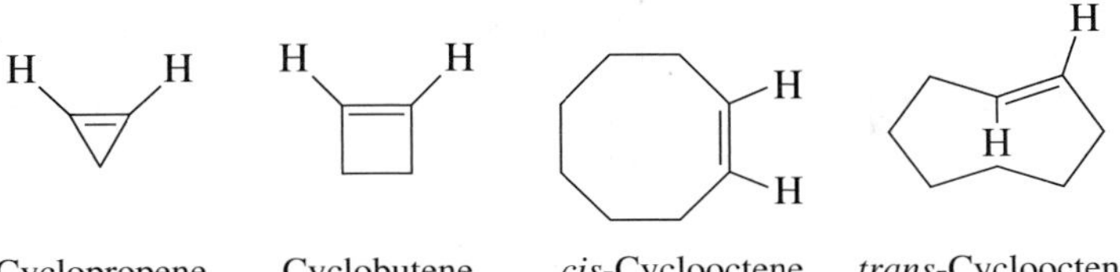

Cyclopropene Cyclobutene *cis*-Cyclooctene *trans*-Cyclooctene

Section 7.8 Alkenes are prepared by **β elimination** of alcohols and alkyl halides. These reactions are summarized with examples in Table 7.1. In both cases, β elimination proceeds in the direction that yields the more highly substituted double bond (**Zaitsev's rule**).

Sections 7.9–7.11 See Table 7.1.

Section 7.12 Secondary and tertiary alcohols undergo **dehydration** by an E1 mechanism involving carbocation intermediates.

Step 1 $R_2CH{-}CR'_2(\ddot{O}H) \underset{\text{fast}}{\overset{H_3O^+}{\rightleftharpoons}} R_2CH{-}CR'_2(\overset{+}{O}H_2)$

Alcohol Alkyloxonium ion

Step 2 $R_2CH{-}CR'_2(\overset{+}{O}H_2) \overset{\text{slow}}{\rightleftharpoons} R_2CH{-}\overset{+}{C}R'_2 + H_2\ddot{O}:$

Alkyloxonium ion Carbocation Water

Step 3 $H_2\ddot{O}: + R_2C(H){-}\overset{+}{C}R'_2 \xrightarrow{\text{fast}} R_2C{=}CR'_2 + H_3\overset{+}{O}:$

Water Carbocation Alkene Hydronium ion

Primary alcohols do not dehydrate as readily as secondary or tertiary alcohols, and their dehydration does not involve a primary carbocation. A proton is lost from the β carbon in the same step in which carbon–oxygen bond cleavage occurs. The mechanism is E2.

Section 7.13 Alkene synthesis via alcohol dehydration is sometimes accompanied by carbocation **rearrangement.** A less stable carbocation can rearrange to a more stable one by an alkyl group migration or by a hydride shift, opening the possibility for alkene formation from two different carbocations.

Secondary carbocation → Tertiary carbocation

(G is a migrating group; it may be either a hydrogen or an alkyl group)

Section 7.14 See Table 7.1.

Section 7.15 **Dehydrohalogenation** of alkyl halides by alkoxide bases is not complicated by rearrangements, because carbocations are not intermediates. The mechanism is

TABLE 7.1 Preparation of Alkenes by Elimination Reactions of Alcohols and Alkyl Halides

Reaction (section) and comments	General equation and specific example
Dehydration of alcohols (Sections 7.9–7.13) Dehydration requires an acid catalyst; the order of reactivity of alcohols is tertiary > secondary > primary. Elimination is regioselective and proceeds in the direction that gives the most highly substituted double bond. When stereoisomeric alkenes are possible, the more stable one is formed in greater amounts. An E1 (elimination unimolecular) mechanism via a carbocation intermediate is followed with secondary and tertiary alcohols. Primary alcohols react by an E2 (elimination bimolecular) mechanism. Sometimes elimination is accompanied by rearrangement.	$R_2CHCR'_2$ (with OH) $\xrightarrow{H^+}$ $R_2C{=}CR'_2$ + H_2O Alcohol → Alkene + Water 2-Methyl-2-hexanol $\xrightarrow[80°C]{H_2SO_4}$ 2-Methyl-1-hexene (19%) + 2-Methyl-2-hexene (81%)
Dehydrohalogenation of alkyl halides (Sections 7.14–7.16) Strong bases cause a proton and a halide to be lost from adjacent carbons of an alkyl halide to yield an alkene. When using small strong bases, regioselectivity is in accord with the Zaitsev rule. When using sterically hindered strong bases, like *tert*-butoxide, Hofmann elimination is observed. The order of halide reactivity is I > Br > Cl > F. A concerted E2 reaction pathway is followed; carbocations are not involved, and rearrangements do not occur. An anti coplanar arrangement of the proton being removed and the halide being lost characterizes the transition state.	$R_2CHCR'_2$ (with X) + $:B^-$ → $R_2C{=}CR'_2$ + H—B + X^- Alkyl halide + Base → Alkene + Conjugate acid of base + Halide 1-Chloro-1-methylcyclohexane $\xrightarrow[\text{ethanol, 100°C}]{KOCH_2CH_3}$ Methylenecyclohexane (6%) + 1-Methylcyclohexene (94%)

E2. It is a concerted process in which the base abstracts a proton from the β carbon while the bond between the halogen and the α carbon undergoes heterolytic cleavage.

$$\bar{B}: \; H{-}C{-}C{-}X \longrightarrow [{}^{\delta-}B\cdots H\cdots C{\cdots}C\cdots X_{\delta-}] \longrightarrow B{-}H + C{=}C + X:^-$$

Transition state

Section 7.16 The preceding equation shows the proton H and the halogen X in the *anti coplanar* relationship that is required for elimination by the E2 mechanism.

Section 7.17 A β C—D bond is broken more slowly in the E2 dehydrohalogenation of alkyl halides than a β C—H bond. The ratio of the rate constants k_H/k_D is a measure of the **deuterium isotope effect** and has a value in the range 3–8 when a carbon–hydrogen bond breaks in the rate-determining step of a reaction.

Section 7.18 In the absence of a strong base, alkyl halides eliminate by an E1 mechanism. Rate-determining ionization of the alkyl halide to a carbocation is followed by deprotonation of the carbocation.

Step 1 $R_2CH{-}CR'_2X \underset{slow}{\overset{-:\ddot{X}:^-}{\rightleftharpoons}} R_2CH{-}\overset{+}{C}R'_2$

Alkyl halide — Carbocation

Step 2 $R_2C(H){-}\overset{+}{C}R'_2 + \text{base}: \xrightarrow[fast]{-H^+} R_2C{=}CR'_2 + (\text{base}{-}H)^+$

Carbocation — Alkene

Section 7.19 The competition between substitution and elimination is influenced by the structure of the substrate (alkyl halide or sulfonate) and the size and basicity of the nucleophile/base. Unhindered substrates react with good nucleophiles that are not strongly basic by the S_N2 mechanism. Elimination is normally favored in the reactions of secondary or tertiary substrates with strong bases, or even primary substrates with strong, bulky bases.

Section 7.20 Alkyl sulfonates undergo elimination reactions under the same conditions as alkyl halides with similar outcomes.

PROBLEMS

Structure and Nomenclature

7.31 Write structural formulas for each of the following:

(a) 1-Heptene
(b) 3-Ethyl-2-pentene
(c) *cis*-3-Octene
(d) *trans*-1,4-Dichloro-2-butene
(e) (*Z*)-3-Methyl-2-hexene
(f) (*E*)-3-Chloro-2-hexene
(g) 1-Bromo-3-methylcyclohexene
(h) 1-Bromo-6-methylcyclohexene
(i) 4-Methyl-4-penten-2-ol
(j) Vinylcycloheptane
(k) 1,1-Diallylcyclopropane
(l) *trans*-1-Isopropenyl-3-methylcyclohexane

7.32 Write a structural formula and give two acceptable IUPAC names for each alkene of molecular formula C_7H_{14} that has a *tetrasubstituted* double bond.

7.33 Give an IUPAC name for each of the following compounds:

(a) $(CH_3CH_2)_2C{=}CHCH_3$
(b) $(CH_3CH_2)_2C{=}C(CH_2CH_3)_2$
(c) $(CH_3)_3CCH{=}CCl_2$
(d)

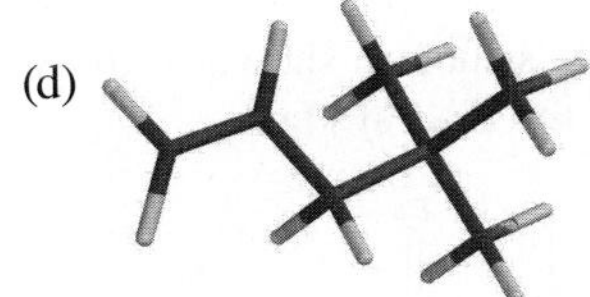

(e)

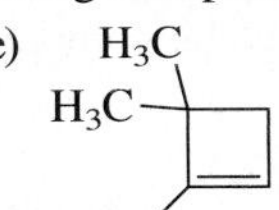

(f) H, H

(g)

7.34 (a) A hydrocarbon isolated from fish oil and from plankton was identified as 2,6,10,14-tetramethyl-2-pentadecene. Write its structure.

(b) Alkyl isothiocyanates are compounds of the type RN=C=S. Write a structural formula for *allyl isothiocyanate,* a pungent-smelling compound isolated from mustard.

(c) Grandisol is one component of the sex attractant of the boll weevil. Write a structural formula for grandisol given that R in the structure shown is an isopropenyl group.

$HOCH_2CH_2$, R, H_3C, H

7.35 *Multifidene* is a sperm-cell-attracting substance released by the female of a species of brown algae. It has the constitution shown.

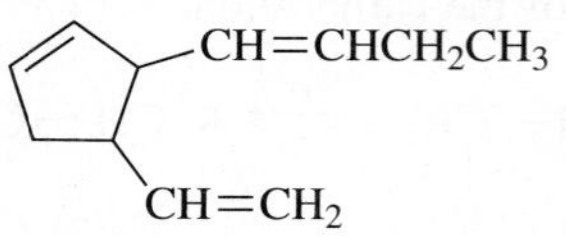

$CH{=}CHCH_2CH_3$

$CH{=}CH_2$

Assuming that the double bond in the five-membered ring of all of the isomers is cis:

(a) How many stereoisomers are represented by this constitution?

(b) If the substituents on the five-membered ring are cis to each other, how many stereoisomers are represented by this constitution?

(c) If the butenyl side chain has the Z configuration of its double bond, how many stereoisomers are possible?

(d) Draw stereochemically accurate representations of all the stereoisomers that satisfy the structural requirements just described.

(e) How are these stereoisomers related? Are they enantiomers or diastereomers?

7.36 *Sphingosine* is a component of membrane lipids, including those found in nerve and muscle cells. How many stereoisomers are possible?

NH_2 OH OH

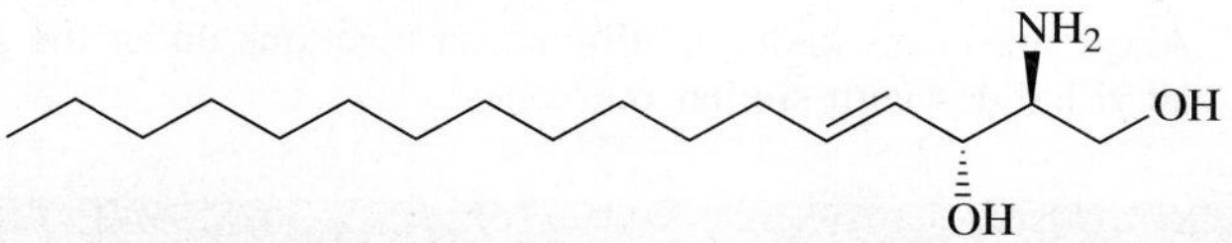

7.37 Write a bond-line formula for each of the following naturally occurring compounds, clearly showing their stereochemistry.

(a) (E)-6-Nonen-1-ol: the sex attractant of the Mediterranean fruit fly.

(b) Geraniol: a hydrocarbon with a rose-like odor present in the fragrant oil of many plants (including geranium flowers). It is the E isomer of

$$(CH_3)_2C{=}CHCH_2CH_2\underset{\displaystyle CH_3}{\underset{|}{C}}{=}CHCH_2OH$$

(c) Nerol: a stereoisomer of geraniol found in neroli and lemongrass oil.

(d) The worm in apples is the larval stage of the codling moth. The sex attractant of the male moth is the $2Z,6E$ stereoisomer of the compound shown.

$$CH_3CH_2CH_2\underset{\displaystyle CH_3}{\underset{|}{C}}{=}CHCH_2CH_2\underset{\displaystyle CH_2CH_3}{\underset{|}{C}}{=}CHCH_2OH$$

(e) The E stereoisomer of the compound is the sex pheromone of the honeybee.

$$CH_3\overset{\displaystyle O}{\overset{\|}{C}}(CH_2)_4CH_2CH{=}CHCO_2H$$

(f) A growth hormone from the cecropia moth has the structure shown. Express the stereochemistry of the double bonds according to the E–Z system.

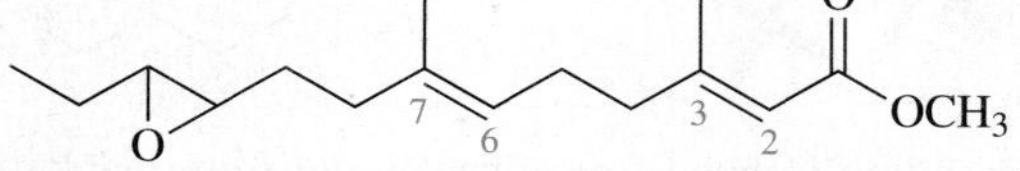

O 7 6 3 2 O OCH_3

7.38 Match each alkene with the appropriate heat of combustion:

Heats of combustion (kJ/mol): 5293; 4658; 4650; 4638; 4632

Heats of combustion (kcal/mol): 1264.9; 1113.4; 1111.4; 1108.6; 1107.1

(a) 1-Heptene

(b) 2,4-Dimethyl-1-pentene

(c) 2,4-Dimethyl-2-pentene

(d) (Z)-4,4-Dimethyl-2-pentene

(e) 2,4,4-Trimethyl-2-pentene

7.39 Choose the more stable alkene in each of the following pairs. Explain your reasoning.

(a) 1-Methylcyclohexene or 3-methylcyclohexene

(b) Isopropenylcyclopentane or allylcyclopentane

(c) 

Bicyclo[4.2.0]oct-7-ene   Bicyclo[4.2.0]oct-3-ene

(d) (*Z*)-Cyclononene or (*E*)-cyclononene

(e) (*Z*)-Cyclooctadecene or (*E*)-cyclooctadecene

7.40 a. Suggest an explanation for the fact that 1-methylcyclopropene is some 42 kJ/mol (10 kcal/mol) less stable than methylenecyclopropane.

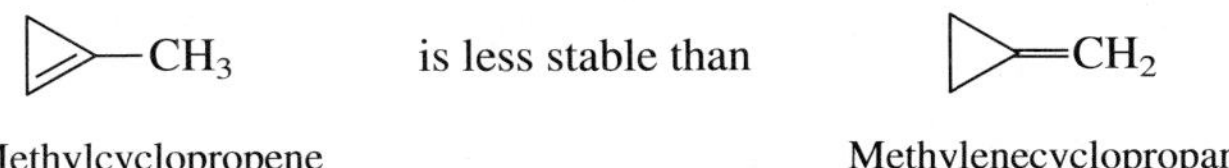

1-Methylcyclopropene   Methylenecyclopropane

b. On the basis of your answer to part (a), compare the expected stability of 3-methylcyclopropene with that of 1-methylcyclopropene and that of methylenecyclopropane.

Reactions

7.41 How many alkenes would you expect to be formed from each of the following alkyl bromides under conditions of E2 elimination? Identify the alkenes in each case.

(a) 1-Bromohexane

(b) 2-Bromohexane

(c) 3-Bromohexane

(d) 2-Bromo-2-methylpentane

(e) 2-Bromo-3-methylpentane

(f) 3-Bromo-2-methylpentane

(g) 3-Bromo-3-methylpentane

(h) 3-Bromo-2,2-dimethylbutane

7.42 Write structural formulas for all the alkene products that could reasonably be formed from each of the following compounds under the indicated reaction conditions. Where more than one alkene is produced, specify the one that is the major product.

(a) 1-Bromo-3,3-dimethylbutane (potassium *tert*-butoxide, *tert*-butyl alcohol, 100°C)

(b) 1-Methylcyclopentyl chloride (sodium ethoxide, ethanol, 70°C)

(c) 3-Methyl-3-pentanol (sulfuric acid, 80°C)

(d) 2,3-Dimethyl-2-butanol (phosphoric acid, 120°C)

(e) 3-Iodo-2,4-dimethylpentane (sodium ethoxide, ethanol, 70°C)

(f) 2,4-Dimethyl-3-pentanol (sulfuric acid, 120°C)

7.43 Choose the compound of molecular formula $C_7H_{13}Br$ that gives each alkene shown as the *exclusive* product of E2 elimination.

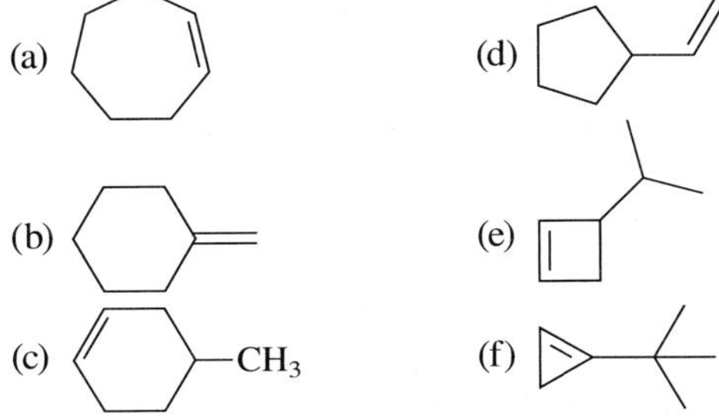

(d)

(e)

(f)

7.44 Give the structures of two different alkyl bromides both of which yield the indicated alkene as the *exclusive* product of E2 elimination.

(a) $CH_3CH{=}CH_2$

(b) $(CH_3)_2C{=}CH_2$

(c) $BrCH{=}CBr_2$

(d) CH_3 CH_3

7.45 Predict the major organic product of each of the following reactions.

(a) 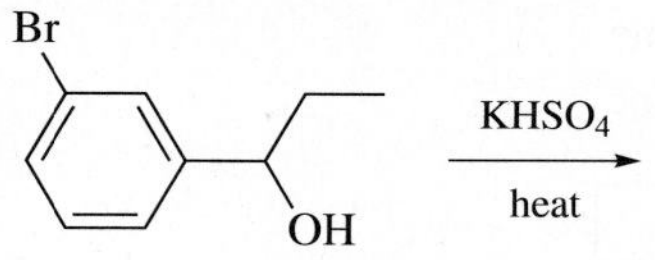 [structure: 1-(2-bromophenyl)propan-1-ol] $\xrightarrow[\text{heat}]{KHSO_4}$

(b) $ICH_2CH(OCH_2CH_3)_2 \xrightarrow[(CH_3)_3COH,\ heat]{KOC(CH_3)_3}$

(c) 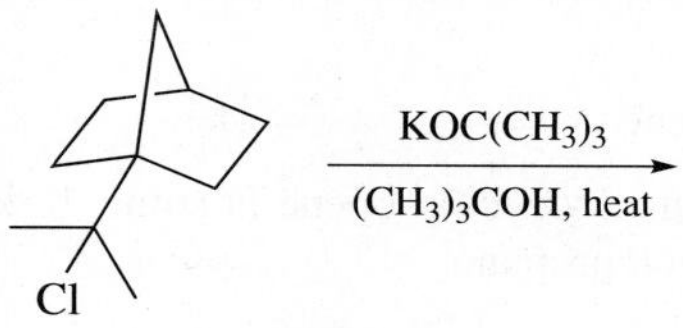 [structure, Cl] $\xrightarrow[(CH_3)_3COH,\ heat]{KOC(CH_3)_3}$

(d) 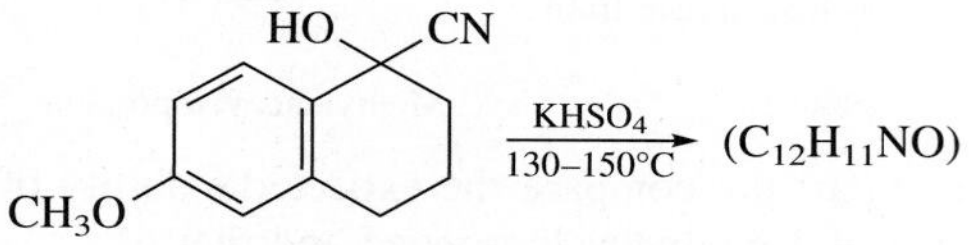 [structure: HO, CN, CH_3O] $\xrightarrow[130\text{–}150°C]{KHSO_4}$ $(C_{12}H_{11}NO)$

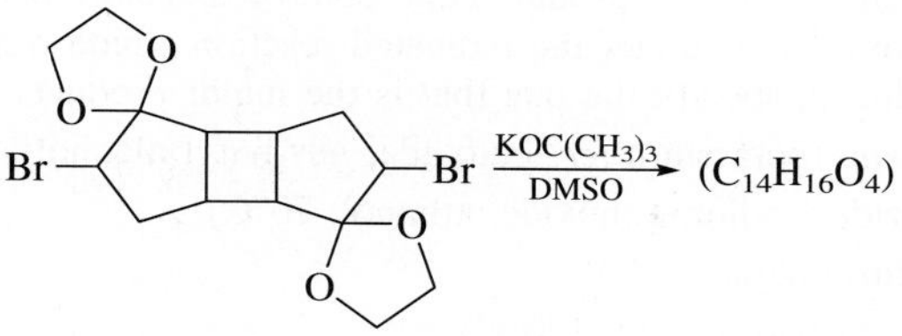

(e) 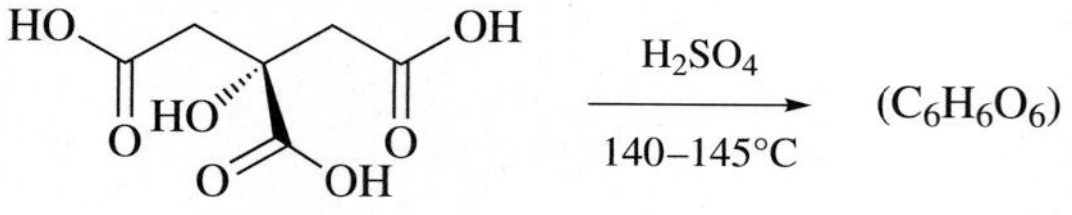 [structure: Citric acid] $\xrightarrow[140\text{–}145°C]{H_2SO_4}$ $(C_6H_6O_6)$

Citric acid

(f) 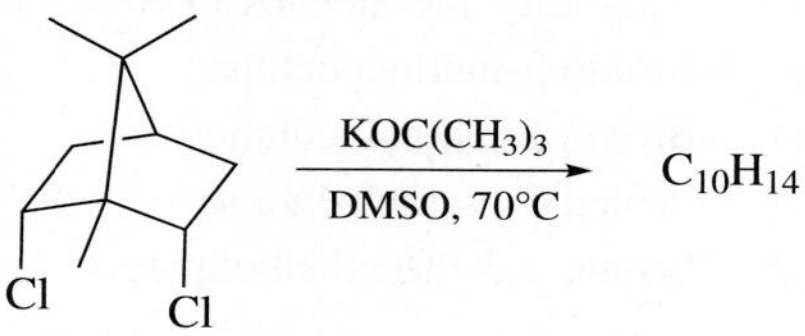 [structure, Cl, Cl] $\xrightarrow[DMSO,\ 70°C]{KOC(CH_3)_3}$ $C_{10}H_{14}$

(g) [structure, Br, Br] $\xrightarrow[DMSO]{KOC(CH_3)_3}$ $(C_{14}H_{16}O_4)$

(h) [structure, Br] $\xrightarrow[(CH_3)_3COH,\ 76°C]{KOC(CH_3)_3}$ (C_6H_{12})

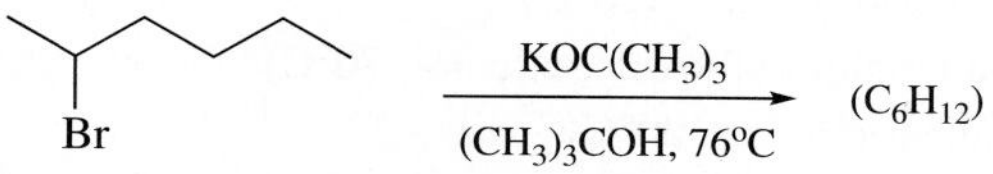

7.46 The following reaction sequence is described as an introductory organic chemistry laboratory experiment in the *Journal of Chemical Education,* vol. 78:1676–1678 (2001). Write structural formulas for compound A and compound B.

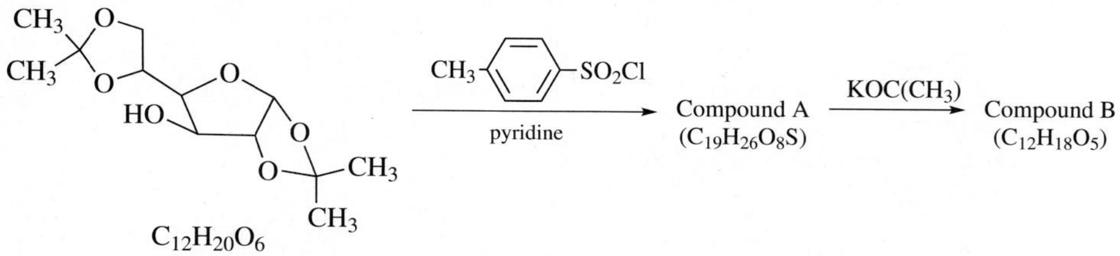

[structure: $C_{12}H_{20}O_6$] $\xrightarrow[\text{pyridine}]{CH_3\text{–}C_6H_4\text{–}SO_2Cl}$ Compound A $(C_{19}H_{26}O_8S)$ $\xrightarrow{KOC(CH_3)}$ Compound B $(C_{12}H_{18}O_5)$

7.47 Solvolysis of 2-bromo-2-methylbutane in acetic acid containing sodium acetate gives three organic products. What are they?

Mechanisms

7.48 The rate of the reaction

$$(CH_3)_3\ CCl + NaSCH_2CH_3 \rightarrow (CH_3)_2C{=}CH_2 + CH_3CH_2SH + NaCl$$

is first-order in $(CH_3)_3CCl$ and first-order in $NaSCH_2CH_3$. Give the symbol (E1 or E2) for the most reasonable mechanism, and use curved arrows to show the flow of electrons.

7.49 Menthyl chloride and neomenthyl chloride have the structures shown. One of these stereoisomers undergoes elimination on treatment with sodium ethoxide in ethanol much more readily than the other. Which reacts faster, menthyl chloride or neomenthyl chloride? Why?

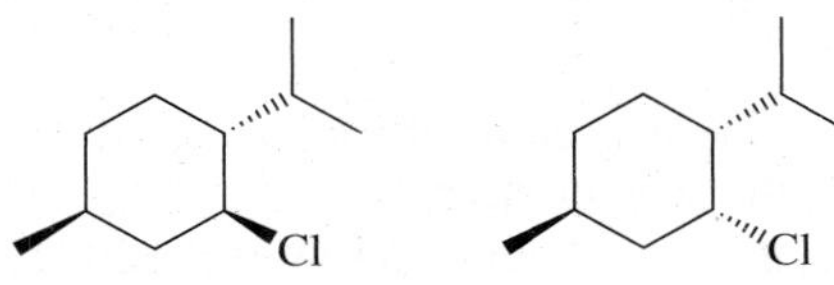

Menthyl chloride Neomenthyl chloride

7.50 Draw a Newman projection for the conformation adopted by 2-bromo-2,4,4-trimethylpentane in a reaction proceeding by the E2 mechanism. Assume the regioselectivity is consistent with the Zaitsev rule.

7.51 You have available 2,2-dimethylcyclopentanol (**A**) and 2-bromo-1,1-dimethylcyclopentane (**B**) and wish to prepare 3,3-dimethylcyclopentene (**C**). Which would you choose as the more suitable reactant, **A** or **B**, and with what would you treat it?

H_3C CH_3 — OH (**A**) H_3C CH_3 — Br (**B**) H_3C CH_3 (**C**)

7.52 In the acid-catalyzed dehydration of 2-methyl-1-propanol, what carbocation would be formed if a hydride shift accompanied cleavage of the carbon–oxygen bond in the alkyloxonium ion? What ion would be formed as a result of a methyl shift? Which pathway do you think will predominate, a hydride shift or a methyl shift?

7.53 Write a sequence of steps depicting the mechanisms of each of the following reactions. Use curved arrows to show electron flow.

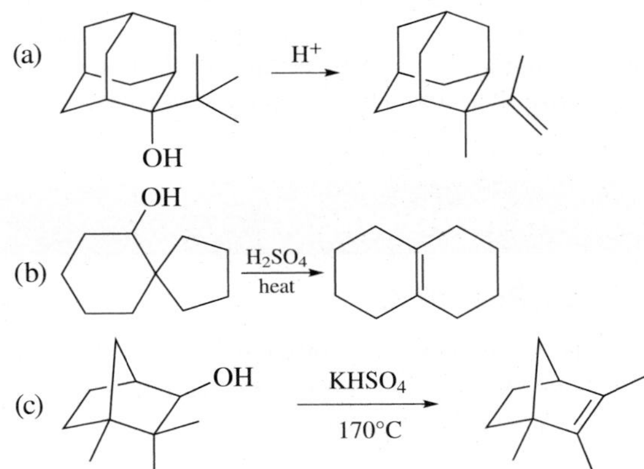

7.54 In Problem 7.20 (Section 7.13) we saw that acid-catalyzed dehydration of 2,2-dimethylcyclohexanol afforded 1,2-dimethylcyclohexene. To explain this product we must write a mechanism for the reaction in which a methyl shift transforms a secondary carbocation to a tertiary one. Another product of the dehydration of 2,2-dimethylcyclohexanol is isopropylidenecyclopentane. Write a mechanism to rationalize its formation, using curved arrows to show the flow of electrons.

OH $\xrightarrow[\text{heat}]{H^+}$ +

2,2-Dimethylcyclohexanol 1,2-Dimethylcyclohexene Isopropylidenecyclopentane

7.55 Acid-catalyzed dehydration of 2,2-dimethyl-1-hexanol gave a number of isomeric alkenes including 2-methyl-2-heptene as shown in the following equation.

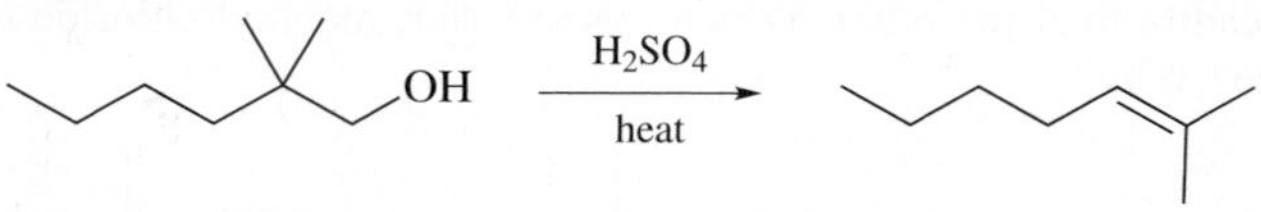

$$\text{OH} \xrightarrow[\text{heat}]{H_2SO_4}$$

(a) Write a stepwise mechanism for the formation of 2-methyl-2-heptene, using curved arrows to show the flow of electrons.

(b) What other alkenes do you think are formed in this reaction?

7.56 The ratio of elimination to substitution is exactly the same (26% elimination) for 2-bromo-2-methylbutane and 2-iodo-2-methylbutane in 80% ethanol/20% water at 25°C.

(a) By what mechanism does substitution most likely occur in these compounds under these conditions?

(b) By what mechanism does elimination most likely occur in these compounds under these conditions?

(c) Which one of the two alkyl halides undergoes substitution faster?

(d) Which one of the two alkyl halides undergoes elimination faster?

(e) What two substitution products are formed from each?

(f) What two elimination products are formed from each?

(g) Why do you suppose the ratio of elimination to substitution is the same for the two alkyl halides?

7.57 The reactant shown, having an axial *p*-toluenesulfonate in its most stable conformation, undergoes predominant elimination on reaction with sodium azide.

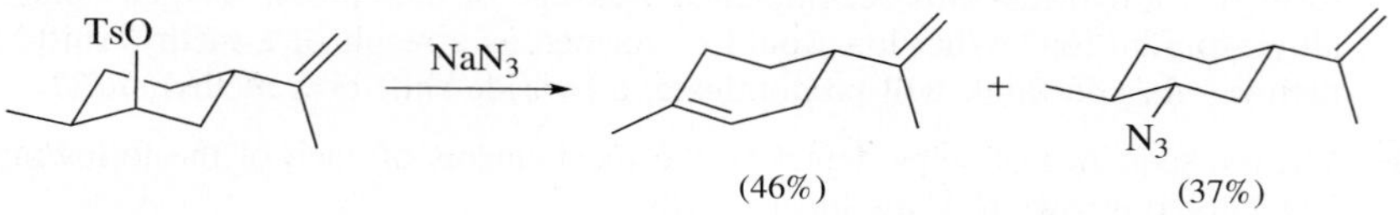

TsO $\xrightarrow{NaN_3}$ (46%) + N_3 (37%)

Its diastereomer having an equatorial *p*-toluenesulfonate gives predominant substitution (76%). Give the structure of the resulting azide and suggest an explanation for the difference in reactivity between the two diastereomeric *p*-toluenesulfonates.

Descriptive Passage and Interpretive Problems 7

A Mechanistic Preview of Addition Reactions

The following flow chart connects three of the reactions we have discussed that involve carbocation intermediates. *Each arrow may represent more than one elementary step in a mechanism.*

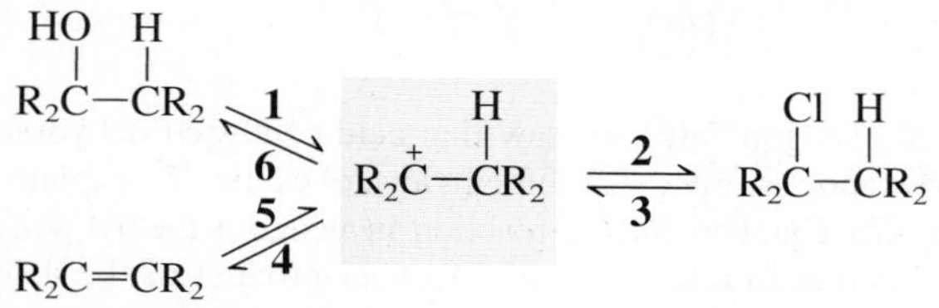

$$R_2\overset{HO}{\overset{|}{C}}-\overset{H}{\overset{|}{C}}R_2 \underset{6}{\overset{1}{\rightleftharpoons}} R_2\overset{+}{C}-\overset{H}{\overset{|}{C}}R_2 \underset{3}{\overset{2}{\rightleftharpoons}} R_2\overset{Cl}{\overset{|}{C}}-\overset{H}{\overset{|}{C}}R_2$$

$$R_2C{=}CR_2 \underset{4}{\overset{5}{\rightleftharpoons}} R_2\overset{+}{C}-\overset{H}{\overset{|}{C}}R_2$$

Arrows **1** and **2** summarize the conversion of alcohols to alkyl halides, **3** and **4** the dehydrohalogenation of an alkyl halide to an alkene by the E1 mechanism, and **1** and **4** the formation of an alkene by dehydration of an alcohol.

The reaction indicated by arrow **5** constitutes a major focus of the next chapter. There we will explore reactions that give overall *addition* to the double bond by way of carbocation intermediates. One such process converts alkenes to alkyl halides (**5** + **2**), another converts alkenes to alcohols (**5** + **6**).

7.58 Based on the S_N1 mechanism for the reaction of tertiary alcohols with HCl as summarized in arrows **1** and **2,** which arrow(s) represent(s) more than one elementary step?

A. Arrow **1** C. Both **1** and **2**
B. Arrow **2** D. Neither **1** nor **2**

7.59 Based on the E1 mechanism for the acid-catalyzed dehydration of a tertiary alcohol as summarized in arrows **1** and **4,** which arrow(s) represent(s) more than one elementary step?

A. Arrow **1** C. Both **1** and **4**
B. Arrow **4** D. Neither **1** nor **4**

7.60 Based on the E1 mechanism for the conversion of a tertiary alkyl chloride to an alkene as summarized in arrows **3** and **4,** which arrow(s) represent(s) more than one elementary step?

A. Arrow **3** C. Both **3** and **4**
B. Arrow **4** D. Neither **3** nor **4**

7.61 Based on the E1 mechanism for the conversion of a tertiary alkyl chloride to an alkene as summarized in arrows **3** and **4,** which arrow(s) correspond(s) to exothermic processes?

A. Arrow **3**
B. Arrow **4**
C. Both **3** and **4**
D. Neither **3** nor **4**

7.62 What term best describes the relationship between an alkene and a carbocation?

A. Isomers
B. Resonance contributors
C. Alkene is conjugate acid of carbocation
D. Alkene is conjugate base of carbocation

7.63 The overall equation for the addition of HCl to alkenes is:

$$R_2C{=}CR_2 \;+\; HCl \longrightarrow R_2\overset{\overset{\textstyle Cl}{|}}{C}{-}\overset{\overset{\textstyle H}{|}}{C}R_2$$

If the transition state for proton transfer from HCl to the alkene (arrow **5**) resembles a carbocation and this step is rate-determining, what should be the effect of alkene structure on the rate of the overall reaction?

	Fastest rate		Slowest rate
A.	$H_2C{=}CH_2$	$CH_3CH{=}CHCH_3$	$(CH_3)_2C{=}C(CH_3)_2$
B.	$CH_3CH{=}CHCH_3$	$(CH_3)_2C{=}C(CH_3)_2$	$H_2C{=}CH_2$
C.	$CH_3CH{=}CHCH_3$	$H_2C{=}CH_2$	$(CH_3)_2C{=}C(CH_3)_2$
D.	$(CH_3)_2C{=}C(CH_3)_2$	$CH_3CH{=}CHCH_3$	$H_2C{=}CH_2$

7.64 For the addition of HCl to alkenes according to the general equation given in the preceding problem, assume the mechanism involves rate-determining formation of the more stable carbocation (arrow **5**) and predict the alkyl chloride formed by reaction of HCl with $(CH_3)_2C{=}CH_2$.

A. $(CH_3)_2CHCH_2Cl$
B. $(CH_3)_3CCl$

7.65 *Zaitsev's rule* was presented in this chapter. In the next chapter we will introduce *Markovnikov's rule,* which is related to arrow **5.** To which arrow does *Zaitsev's rule* most closely relate from a mechanistic perspective?

A. **1** C. **3**
B. **2** D. **4**

Chapter

8

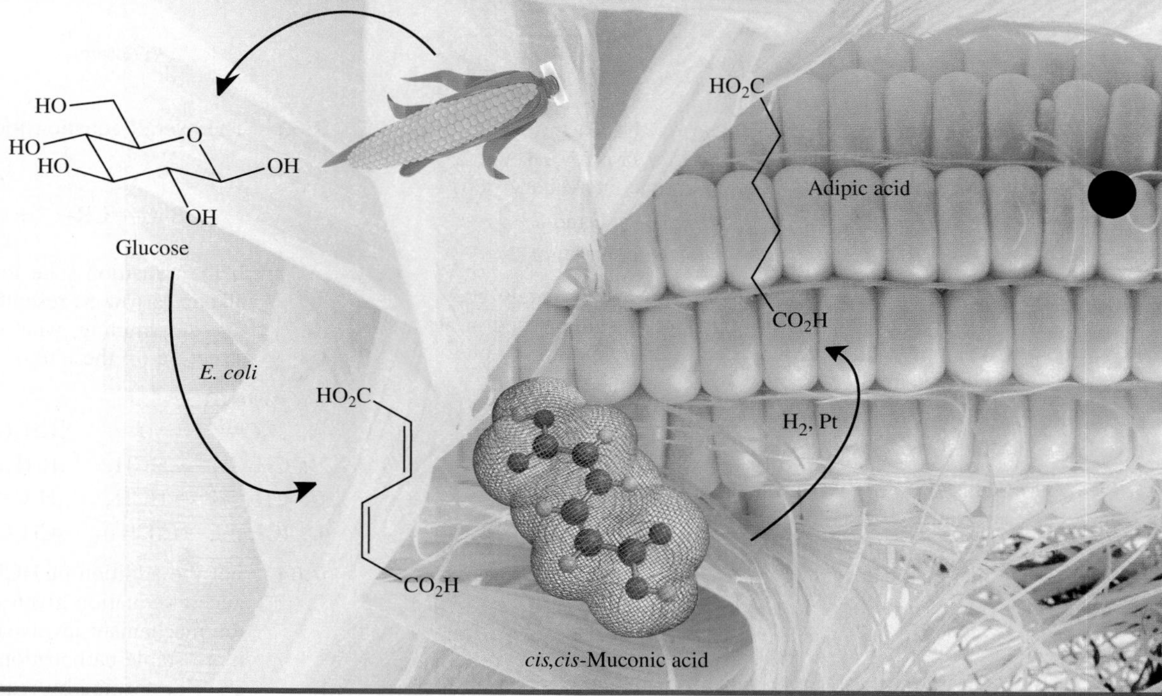

CHAPTER OUTLINE

Petroleum-derived benzene is the source of the six carbon atoms of adipic acid, an industrial chemical used to make nylon. An alternative process has been developed that uses genetically engineered strains of the bacterium *Escherichia coli* to convert glucose, a renewable resource obtained from cornstarch, to *cis,cis*-muconic acid. Subsequent hydrogenation gives adipic acid. This chapter is about reactions that involve addition to double bonds and begins with hydrogenation.

Addition Reactions of Alkenes

Now that we're familiar with the structure and preparation of alkenes, let's look at their chemical reactions, the most characteristic of which is **addition** to the double bond according to the general equation:

$$\mathrm{A{-}B} + \mathrm{C{=}C} \longrightarrow \mathrm{A{-}\overset{|}{\underset{|}{C}}{-}\overset{|}{\underset{|}{C}}{-}B}$$

The range of compounds represented as A–B in this equation offers a wealth of opportunity for converting alkenes to a number of other structural types.

Alkenes are commonly described as **unsaturated hydrocarbons** because they have the capacity to react with substances that add to them. Alkanes, on the other hand, are **saturated hydrocarbons** and are incapable of undergoing addition reactions.

8.1 Hydrogenation of Alkenes

The relationship between reactants and products in addition reactions can be illustrated by the *hydrogenation* of alkenes to yield alkanes. **Hydrogenation** is the addition of H_2 to a multiple bond, as illustrated in the conversion of ethylene to ethane.

$$\underset{\text{Ethylene}}{H_2C \overset{\pi}{=} CH_2} + \underset{\text{Hydrogen}}{H \overset{\sigma}{—} H} \xrightarrow{\text{Pt, Pd, Ni, or Rh}} \underset{\text{Ethane}}{H_3C—CH_3} \qquad \Delta H^\circ = -136 \text{ kJ/mol } (-32.6 \text{ kcal/mol})$$

The French chemist Paul Sabatier received the 1912 Nobel Prize in Chemistry for his discovery that finely divided nickel is an effective hydrogenation catalyst.

Hydrogenation of all alkenes is exothermic, so is characterized by a negative sign for ΔH°. The heat given off is called the **heat of hydrogenation** and cited without a sign. In other words, heat of hydrogenation = $-\Delta H^\circ$.

The uncatalyzed addition of hydrogen to an alkene, although exothermic, is too slow to ever occur, but is readily accomplished in the presence of certain finely divided metal catalysts, such as platinum, palladium, nickel, and rhodium. *Catalytic hydrogenation* is normally rapid at room temperature, and the alkane is the only product.

$$\underset{\text{2-Methyl-2-butene}}{(CH_3)_2C{=}CHCH_3} + \underset{\text{Hydrogen}}{H_2} \xrightarrow{\text{Pt}} \underset{\text{2-Methylbutane (100\%)}}{(CH_3)_2CHCH_2CH_3}$$

$$\underset{\text{5,5-Dimethyl(methylene)cyclononane}}{} + \underset{\text{Hydrogen}}{H_2} \xrightarrow{\text{Pt}} \underset{\text{1,1,5-Trimethylcyclononane (73\%)}}{}$$

5,5-Dimethyl(methylene)cyclononane (H$_3$C, CH$_3$, =CH$_2$) + H$_2$ (Hydrogen) $\xrightarrow{\text{Pt}}$ 1,1,5-Trimethylcyclononane (73%) (H$_3$C, CH$_3$, CH$_3$, H)

Problem 8.1

What three alkenes yield 2-methylbutane on catalytic hydrogenation?

The solvent used in catalytic hydrogenation is chosen for its ability to dissolve the alkene and is typically ethanol, hexane, or acetic acid. The metal catalysts are insoluble in these solvents (or, indeed, in any solvent). Two phases, the solution and the metal, are present, and the reaction takes place at the interface between them. Reactions involving a substance in one phase with a different substance in a second phase are called **heterogeneous reactions.** A number of organometallic compounds have been developed for catalytic hydrogenation under homogeneous conditions and are described in Chapter 15.

Catalytic hydrogenation of an alkene is believed to proceed by the series of steps shown in Mechanism 8.1. The addition of hydrogen to the alkene is very slow in the absence of a metal catalyst, meaning that any uncatalyzed mechanism must have a very high activation energy. The metal catalyst accelerates the rate of hydrogenation by providing an alternative pathway that involves a sequence of several low activation energy steps.

8.2 Stereochemistry of Alkene Hydrogenation

Two stereochemical aspects—*stereospecificity* and *stereoselectivity*—attend catalytic hydrogenation. Recall that a stereospecific reaction is one in which stereoisomeric starting materials give stereoisomeric products. Previous examples were the requirement of inversion of configuration in substitution by the S_N2 mechanism (Section 6.3) and an anti relationship between the proton lost and the leaving group in elimination by the E2 mechanism (Section 7.16). Catalytic hydrogenation is a stereospecific **syn addition;** both hydrogens add to the same face of the double bond. Its counterpart—**anti addition**—would be characterized by addition to opposite faces of a double bond.

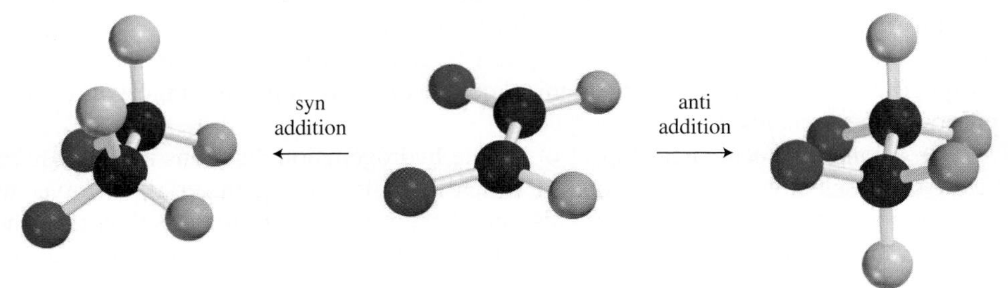

Elimination reactions that proceed by the E2 mechanism (see Sections 7.15–7.16) are stereospecific in that they require an anti relationship between the proton and leaving group.

Mechanism 8.1

Hydrogenation of Alkenes

Step 1: Hydrogen molecules react with metal atoms at the catalyst surface. The relatively strong hydrogen–hydrogen σ bond is broken and replaced by two weak metal–hydrogen bonds.

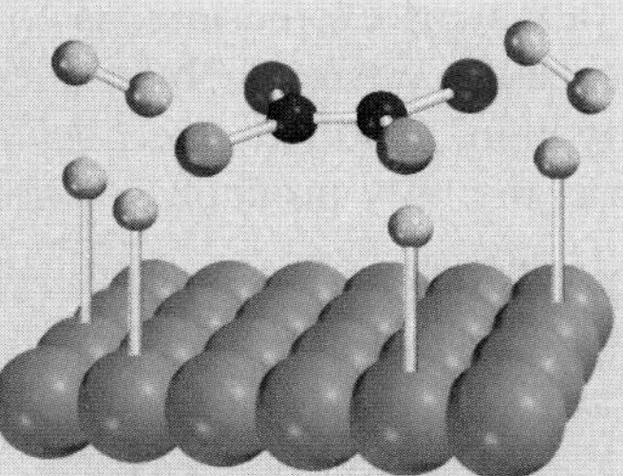

Step 2: The alkene reacts with the metal catalyst. The π component of the double bond between the two carbons is replaced by two relatively weak carbon–metal σ bonds.

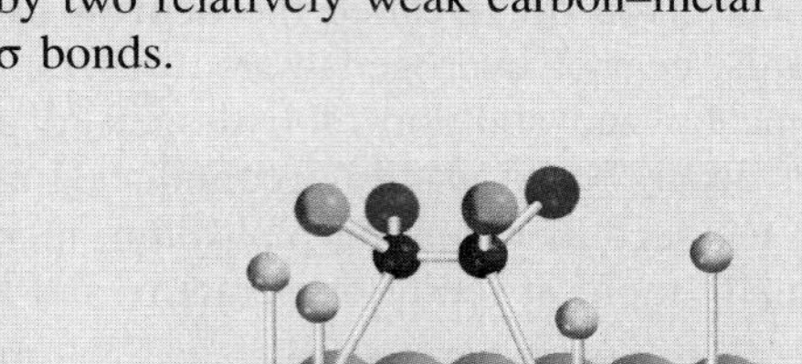

Step 3: A hydrogen atom is transferred from the catalyst surface to one of the carbons of the double bond.

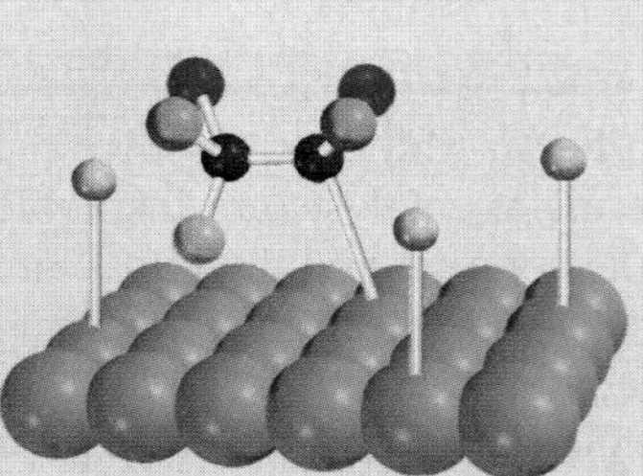

Step 4: The second hydrogen atom is transferred, forming the alkane. The sites on the catalyst surface at which the reaction occurred are free to accept additional hydrogen and alkene molecules.

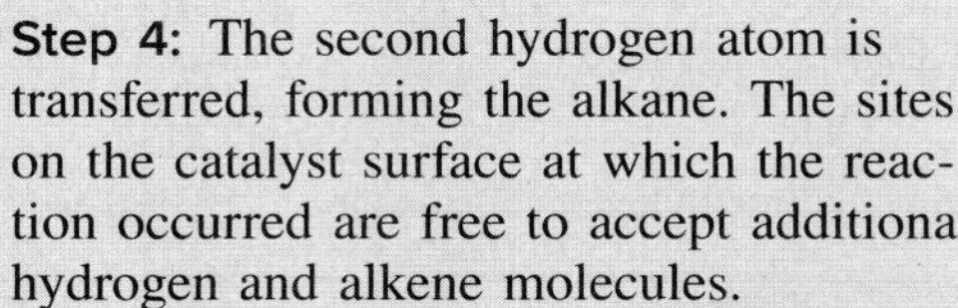

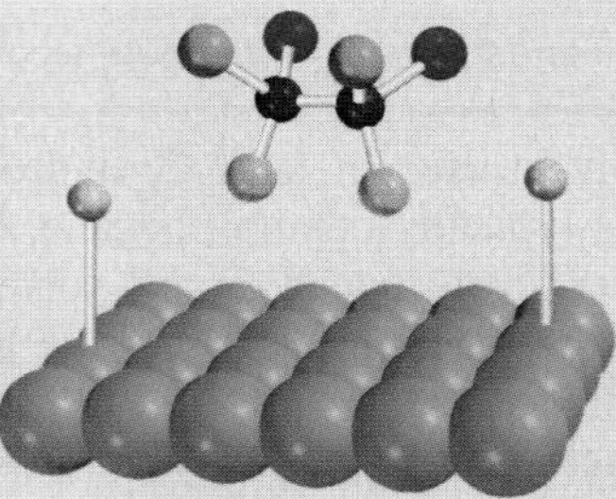

Experimental support for syn addition can be found in the hydrogenation of the cyclohexene derivative shown where the product of syn addition is formed exclusively, even though it is the less stable stereoisomer.

CO_2CH_3 / CO_2CH_3 (cyclohexene) $\xrightarrow[Pt]{H_2}$ CO_2CH_3, H, CO_2CH_3, H (cyclohexane) + H, CO_2CH_3, CO_2CH_3, H (cyclohexane)

Dimethyl cyclohexene-1,2-dicarboxylate — Product of syn addition (100%) — Product of anti addition (not formed)

Unlike catalytic hydrogenation, not all additions to alkenes are syn. As we proceed in this chapter, we'll see some that take place by anti addition, and others that are not stereospecific. Identifying the stereochemical course of a reaction is an important element in proposing a mechanism.

The second stereochemical aspect of alkene hydrogenation concerns its *stereoselectivity*. A stereoselective reaction is one in which a single starting material can give two or more stereoisomeric products but yields one of them in greater amounts than the other (or even to the exclusion of the other). Recall from Section 7.11 that the acid-catalyzed dehydration of alcohols is stereoselective in that it favors the formation of the more stable stereochemistry of the alkene double bond. In catalytic hydrogenation, stereoselectivity is associated with a different factor—the direction from which hydrogen atoms are transferred from the catalyst to the double bond. In the example shown:

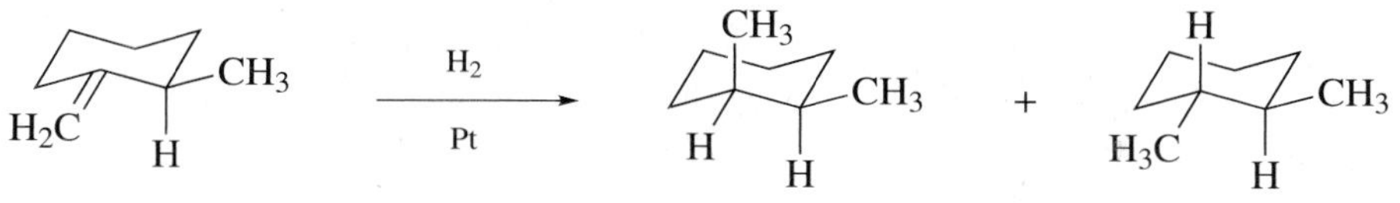

CH_3 H_2C H $\xrightarrow[Pt]{H_2}$ CH_3 CH_3 H H + H CH_3 H_3C H

2-Methyl(methylene)-cyclohexane — *cis*-1,2-Dimethyl-cyclohexane (68%) — *trans*-1,2-Dimethyl-cyclohexane (32%)

the major product is the cis stereoisomer of the product. The reason for this is that the face of the double bond that is opposite the C-2 methyl group in the alkene is less hindered and better able to contact the catalyst surface. Therefore, hydrogen is transferred predominantly to that face. We customarily describe the stereochemistry of alkene hydrogenation as proceeding by syn addition of hydrogen to the less hindered face of the double bond. Reactions that discriminate between nonequivalent sides or faces of a reactant are common in organic chemistry and are examples of steric effects on *reactivity.* Previously we saw steric effects on *stability* in the case of cis and trans stereoisomers of substituted cycloalkanes (Sections 3.11–3.12) and alkenes (Sections 7.6–7.7).

Problem 8.2

Could the amounts of stereoisomeric 1,2-dimethylcyclohexanes formed in the preceding equation reflect their relative stabilities?

Problem 8.3

Catalytic hydrogenation of α-pinene (a constituent of turpentine) is 100% stereoselective and gives only compound A. Explain using the molecular model of α-pinene to guide your reasoning.

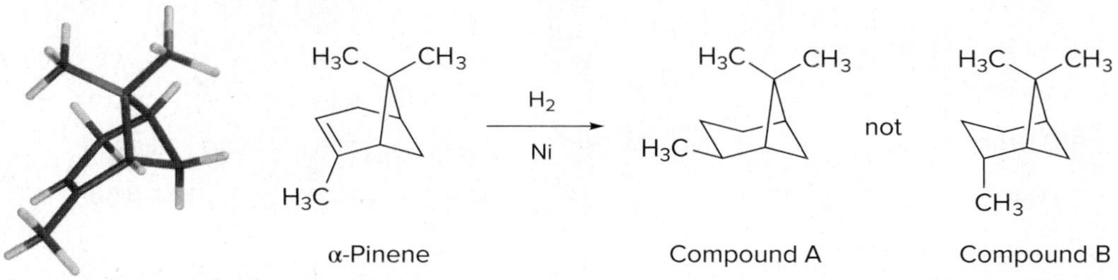

H_3C CH_3 H_3C $\xrightarrow[Ni]{H_2}$ H_3C CH_3 H_3C not H_3C CH_3 CH_3

α-Pinene — Compound A — Compound B

8.3 Heats of Hydrogenation

In much the same way as heats of combustion, heats of hydrogenation are used to compare the relative stabilities of alkenes. Both methods measure the differences in the energy of *isomers* by converting them to a product or products common to all. Catalytic hydrogenation of 1-butene, *cis*-2-butene, or *trans*-2-butene yields the same product—butane. As Figure 8.1 shows, the measured heats of hydrogenation reveal that *trans*-2-butene is 4 kJ/mol (1.0 kcal/mol) lower in energy than *cis*-2-butene and that *cis*-2-butene is 7 kJ/mol (1.7 kcal/mol) lower in energy than 1-butene.

Heats of hydrogenation can be used to *estimate* the stability of double bonds as structural units, even in alkenes that are not isomers. Table 8.1 lists the heats of hydrogenation for a representative collection of alkenes.

Remember that a catalyst affects the rate of a reaction but not the energy relationships between reactants and products. Thus, the heat of hydrogenation of a particular alkene is the same irrespective of what catalyst is used.

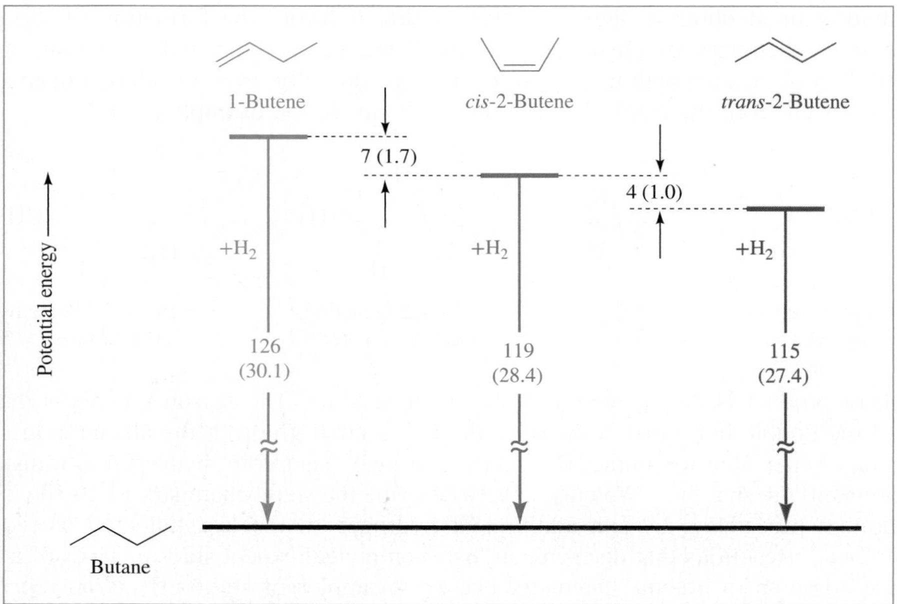

Figure 8.1

Heats of hydrogenation of butane isomers in kJ/mol (kcal/mol).

TABLE 8.1 Heats of Hydrogenation of Some Alkenes

Alkene	Structure	Heat of hydrogenation kJ/mol	Heat of hydrogenation kcal/mol
Ethylene	$H_2C{=}CH_2$	136	32.6
Monosubstituted alkenes			
Propene		125	29.9
1-Butene		126	30.1
1-Hexene		126	30.2
Cis-disubstituted alkenes			
cis-2-Butene		119	28.4
cis-2-Pentene		117	28.1
Trans-disubstituted alkenes			
trans-2-Butene		115	27.4
trans-2-Pentene		114	27.2
Trisubstituted alkenes			
2-Methyl-2-pentene		112	26.7
Tetrasubstituted alkenes			
2,3-Dimethyl-2-butene		110	26.4

The pattern of alkene stability determined from heats of hydrogenation parallels exactly the pattern deduced from heats of combustion.

Decreasing heat of hydrogenation and increasing stability of the double bond

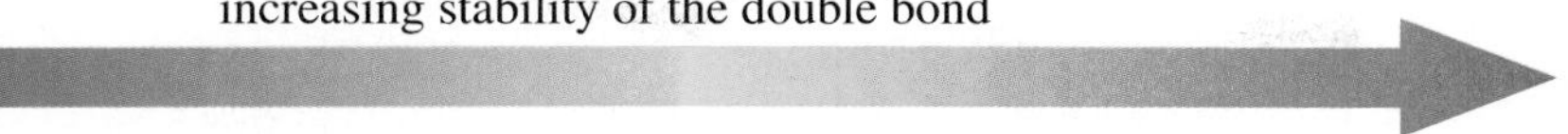

$H_2C{=}CH_2$	$RCH{=}CH_2$	$RCH{=}CHR$	$R_2C{=}CHR$	$R_2C{=}CR_2$
Ethylene	Monosubstituted	Disubstituted	Trisubstituted	Tetrasubstituted

Ethylene, which has no alkyl substituents to stabilize its double bond, has the highest heat of hydrogenation. Alkenes that are similar in structure to one another have similar heats of hydrogenation. For example, the heats of hydrogenation of the monosubstituted alkenes propene, 1-butene, and 1-hexene are almost identical. Cis-disubstituted alkenes have lower heats of hydrogenation than monosubstituted alkenes but higher heats of hydrogenation than their more stable trans stereoisomers. Alkenes with trisubstituted double bonds have lower heats of hydrogenation than disubstituted alkenes, and tetrasubstituted alkenes have the lowest heats of hydrogenation.

Problem 8.4

Match each alkene of Problem 8.1 with its correct heat of hydrogenation.

Heats of hydrogenation in kJ/mol (kcal/mol): 112 (26.7); 118 (28.2); 126 (30.2)

8.4 Electrophilic Addition of Hydrogen Halides to Alkenes

Addition to the double bond is the most characteristic chemical property of alkenes and in most cases the reactant, unlike H_2, is a polar molecule such as a hydrogen halide.

$$\text{C}{=}\text{C} \quad + \quad {}^{\delta+}\text{H}{-}\text{X}^{\delta-} \longrightarrow \text{H}{-}\text{C}{-}\text{C}{-}\text{X}$$

Alkene — Hydrogen halide — Alkyl halide

The reaction is classified as an **electrophilic addition,** a term analogous to **nucleophilic substitution** where the first word characterizes the attacking reagent and the second describes what happens to the organic reactant. In general, an electrophile is a Lewis acid in that it acts as an electron-pair acceptor in a reaction based on the alkene acting as an electron-pair donor.

The electrostatic potential maps in Figure 8.2 illustrate the complementary distribution of charge in hydrogen chloride and ethylene. The proton of hydrogen chloride is positively polarized (electrophilic), and the region of greatest negative character in the alkene is where the π electrons are—above and below the plane of the bonds to the sp^2-hybridized carbons. During the reaction, π electrons flow from the alkene toward the proton of the hydrogen halide. More highly substituted double bonds are more "electron-rich" and react faster than less substituted ones.

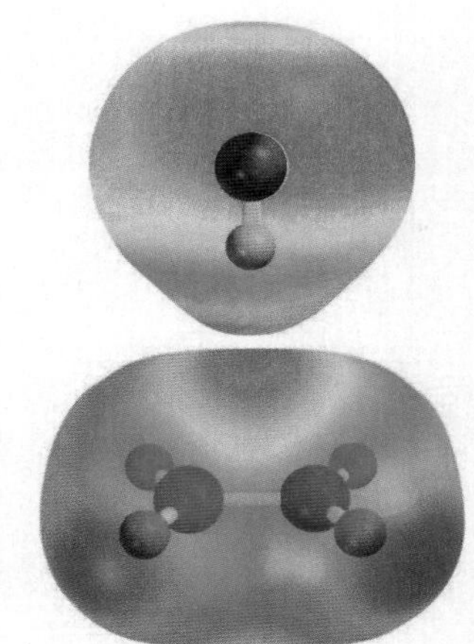

Figure 8.2

Electrostatic potential maps of HCl and ethylene. When the two react, the interaction is between the electron-rich site (red) of ethylene and electron-poor region (blue) of HCl. The electron-rich region of ethylene is associated with the π electrons of the double bond, and H is the electron-poor atom of HCl.

Increasing rate of reaction of alkene with hydrogen halides

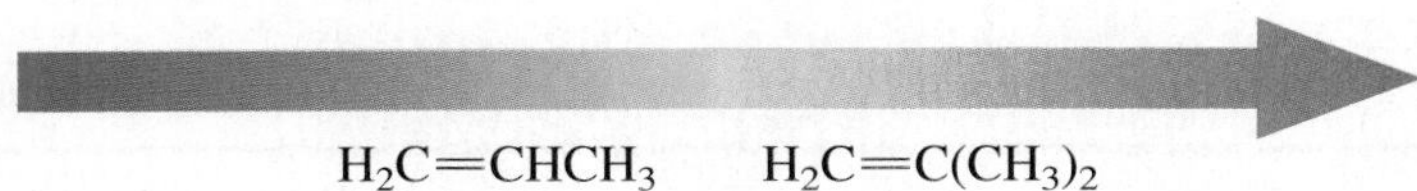

$H_2C{=}CHCH_3$	$H_2C{=}C(CH_3)_2$
Slower rate of addition; less substituted alkene	Faster rate of addition; more substituted alkene

Among the hydrogen halides, reactivity increases with acid strength; thus, hydrogen iodide reacts at the fastest rate, whereas hydrogen fluoride reacts at the slowest.

Increasing rate of addition of hydrogen halides to alkenes

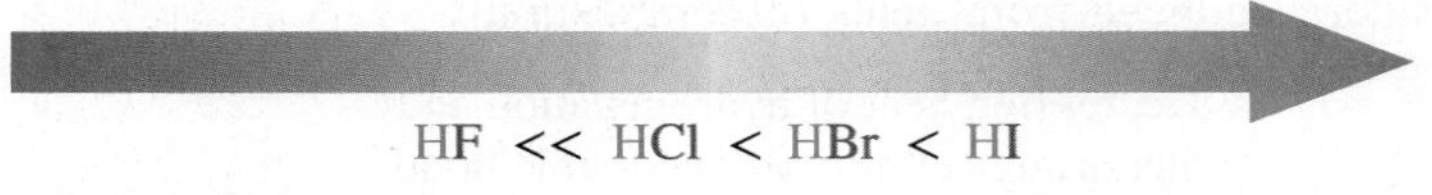

HF << HCl < HBr < HI

Slowest rate of addition; weakest acid — Fastest rate of addition; strongest acid

Recall from Section 7.10 that a regioselective reaction is one that can produce two (or more) constitutional isomers from a single reactant, but gives one in greater amounts than the other. A regiospecific reaction is one that is 100% regioselective.

As shown in the following examples, hydrogen halide addition to alkenes can be highly regioselective, even regiospecific. In both cases, two constitutionally isomeric alkyl halides can be formed by addition to the double bond, but one is formed in preference to the other.

1-Butene + HBr (Hydrogen bromide) $\xrightarrow{\text{acetic acid}}$ 2-Bromobutane (only product, 80% yield) not 1-Bromobutane (not formed)

1-Methylcyclopentene + HCl (Hydrogen chloride) $\xrightarrow{0°C}$ 1-Chloro-1-methylcyclopentane (only product, 100% yield) not 1-Chloro-2-methylcyclopentane (not formed)

Observations such as these prompted Vladimir Markovnikov, a colleague of Alexander Zaitsev at the University of Kazan (Russia), to offer a generalization in 1870. According to what is now known as **Markovnikov's rule,** *when an unsymmetrically substituted alkene reacts with a hydrogen halide, the hydrogen adds to the carbon that has the greater number of hydrogens, and the halogen adds to the carbon that has fewer hydrogens.*

Problem 8.5

Use Markovnikov's rule to predict the major organic product formed in the reaction of hydrogen chloride with each of the following:

(a) 2-Methyl-2-butene

(b) *cis*-2-Butene

(c) 2-Methyl-1-butene

(d) CH_3CH= (cyclohexylidene)

Sample Solution (a) Hydrogen chloride adds to the double bond of 2-methyl-2-butene in accordance with Markovnikov's rule. The proton adds to the carbon that has one attached hydrogen, chlorine to the carbon that has none.

Chloride bonds to this carbon — A proton bonds to this carbon

2-Methyl-2-butene → 2-Chloro-2-methylbutane

Like Zaitsev's rule (Section 7.10), Markovnikov's rule collects experimental observations into a form that allows us to predict the outcome of future experiments. To understand its basis we need to look at the mechanism by which these reactions take place.

Mechanism 8.2 outlines the two-step sequence for the electrophilic addition of hydrogen bromide to 2-methylpropene.

$$(CH_3)_2C{=}CH_2 \quad + \quad HBr \xrightarrow{\text{acetic acid}} (CH_3)_3CBr$$

2-Methylpropene; Hydrogen bromide; *tert*-Butyl bromide (only product, 90% yield)

Mechanism 8.2

Electrophilic Addition of Hydrogen Bromide to 2-Methylpropene

THE OVERALL REACTION:

2-Methylpropene + HBr → *tert*-Butyl bromide (—Br)

2-Methylpropene; Hydrogen bromide; *tert*-Butyl bromide

THE MECHANISM:

Step 1: This is the rate-determining step and is bimolecular. Protonation of the double bond occurs in the direction that gives the more stable of two possible carbocations. In this case the carbocation is tertiary. Protonation of C-2 would have given a less stable secondary carbocation.

2-Methylpropene + H—Br: $\xrightarrow{\text{slow}}$ (+)—H + :Br:⁻

2-Methylpropene; Hydrogen bromide; *tert*-Butyl cation; Bromide ion

Step 2: This step is the combination of a cation (Lewis acid, electrophile) with an anion (Lewis base, nucleophile) and occurs rapidly.

tert-Butyl cation + :Br:⁻ $\xrightarrow{\text{fast}}$ —Br:

tert-Butyl cation; Bromide ion; *tert*-Butyl bromide

The first step is rate-determining protonation of the double bond by the hydrogen halide, forming a carbocation. The regioselectivity of addition is set in this step and is controlled by the relative stabilities of the two possible carbocations.

(a) *Addition according to Markovnikov's rule:*

H—Br: ⇌ (+) H; :Br:⁻ → :Br: H

Tertiary carbocation; Observed product

(b) *Addition opposite to Markovnikov's rule:*

H—Br: → (+) H; :Br:⁻ → Br H

Primary carbocation; Not formed

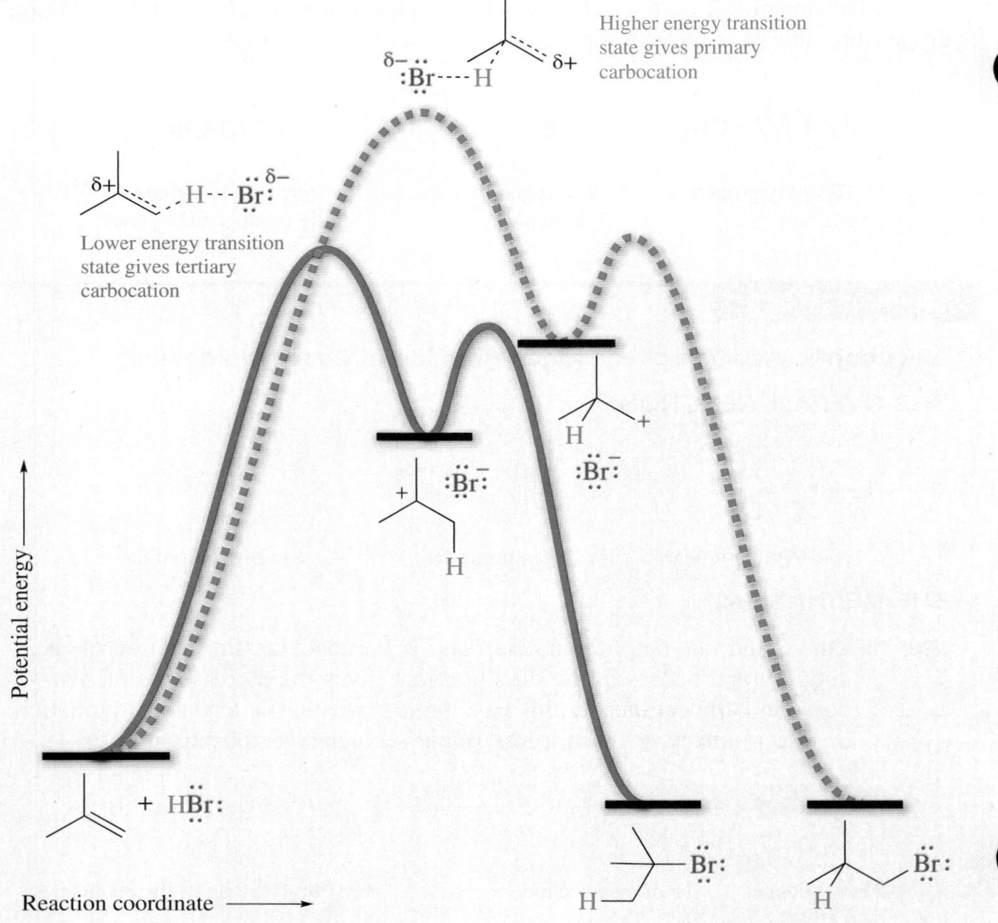

Figure 8.3

Energy diagram comparing addition of hydrogen bromide to 2-methylpropene according to Markovnikov's rule (solid red curve) and opposite to it (dashed blue curve). E_a is less, and the reaction is faster for the reaction that proceeds via the more stable tertiary carbocation.

Figure 8.3 compares potential energy diagrams for these two competing modes of addition. According to Hammond's postulate (Section 5.8), the transition state for protonation of the double bond resembles the carbocation more than the alkene, and E_a for formation of the more stable carbocation (tertiary) is less than that for formation of the less stable carbocation (primary). The major product is derived from the carbocation that is formed faster, and the energy difference between a primary and a tertiary carbocation is so great and their rates of formation so different that essentially all of the product is derived from the tertiary carbocation.

Problem 8.6

Give a structural formula for the carbocation intermediate that leads to the major product in each of the reactions of Problem 8.5.

Sample Solution (a) Protonation of the double bond of 2-methyl-2-butene can give a tertiary carbocation or a secondary carbocation.

1 2 3 4 H

2-Methyl-2-butene

faster — protonate C-3 → Tertiary carbocation (more stable)

slower — protonate C-2 → Secondary carbocation (less stable)

The product of the reaction is derived from the more stable carbocation—in this case, it is a tertiary carbocation that is formed more rapidly than a secondary one.

Rules, Laws, Theories, and the Scientific Method

As we have just seen, Markovnikov's rule can be expressed in two ways:

1. When a hydrogen halide adds to an alkene, hydrogen adds to the carbon of the alkene that has the greater number of hydrogens attached to it, and the halogen to the carbon that has the fewer hydrogens.
2. When a hydrogen halide adds to an alkene, protonation of the double bond occurs in the direction that gives the more stable carbocation.

The first of these statements is close to the way Vladimir Markovnikov expressed it in 1870; the second is the way we usually phrase it now. These two statements differ in an important way—a way that is related to the **scientific method.**

Adherence to the scientific method is what defines science. The scientific method has four major elements: observation, law, theory, and hypothesis.

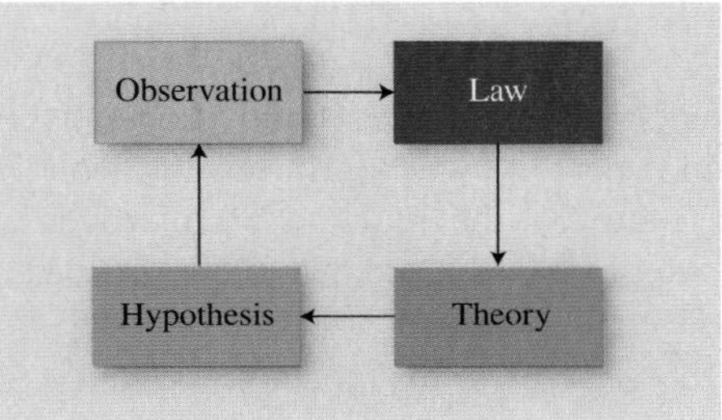

Most *observations* in chemistry come from experiments. If we do enough experiments we may see a pattern running through our observations. A *law* is a mathematical (the law of gravity) or verbal (the law of diminishing returns) description of that pattern. Establishing a law can lead to the framing of a *rule* that lets us predict the results of future experiments. This is what the 1870 version of Markovnikov's rule is: a statement based on experimental observations that has predictive value. The same can be said of Zaitsev's rule for elimination reactions.

A *theory* is our best present interpretation of why things happen the way they do. The modern version of Markovnikov's rule, which is based on mechanistic reasoning and carbocation stability, recasts the rule in terms of theoretical ideas. Mechanisms, and explanations grounded in them, belong to the theory part of the scientific method.

It is worth remembering that a theory can never be proven correct. It can only be proven incorrect, incomplete, or inadequate. Thus, theories are always being tested and refined. As important as anything else in the scientific method is the *testable hypothesis.* Once a theory is proposed, experiments are designed to test its validity. If the results are consistent with the theory, our belief in its soundness is strengthened. If the results conflict with it, the theory is flawed and must be modified. Section 8.5 describes some observations that support the theory that carbocations are intermediates in the addition of hydrogen halides to alkenes.

Regioselectivity is an important consideration in the addition of hydrogen halides to alkenes. What about stereoselectivity? Addition of hydrogen bromide to 1-butene, *cis*-2-butene, or *trans*-2-butene, all of which are achiral, yields the same chiral product, 2-bromobutane. The product, however, is not optically active because it is racemic. It is composed of equal amounts of (*R*)- and (*S*)-2-bromobutane because both are formed at equal rates regardless of whether the starting alkene is 1-butene, *cis*-2-butene, or *trans*-2-butene.

1-Butene (achiral) or *cis*-2-Butene (achiral) or *trans*-2-Butene (achiral) $\xrightarrow[\text{(achiral)}]{\text{HBr}}$ 2-Bromobutane (chiral, but racemic)

The carbocation intermediate has a plane of symmetry and is achiral. It reacts with bromide ion with equal probability from either side to give a 1:1 mixture of (*R*)- and (*S*)-2-bromobutane (Figure 8.4). This reaction illustrates a general principle, that *optically active products cannot be formed from an optically inactive starting material unless at least one optically active reactant, reagent, or catalyst is used.* This principle holds regardless of the mechanism.

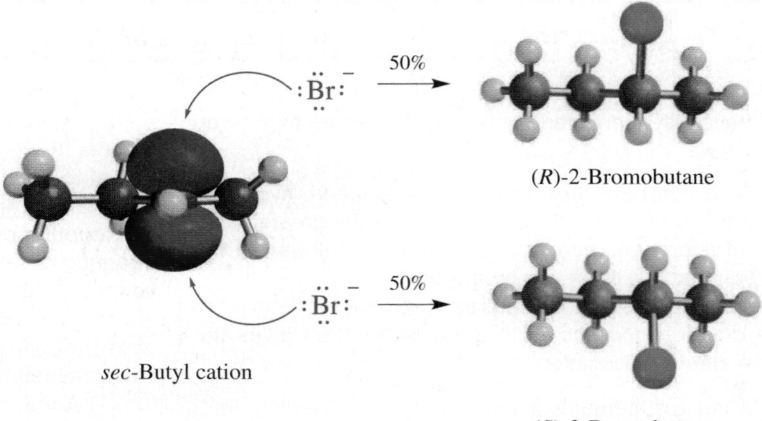

Figure 8.4

Addition of HBr to 1-butene, *cis*-2-butene, or *trans*-2-butene proceeds by way of an achiral carbocation intermediate, which reacts with bromide ion to give racemic 2-bromobutane.

8.5 Carbocation Rearrangements in Hydrogen Halide Addition to Alkenes

Our belief that carbocations are intermediates in the addition of hydrogen halides to alkenes is strengthened by the fact that rearrangements of the kind seen in alcohol dehydrations (Section 7.13) sometimes occur. For example, the reaction of hydrogen chloride with 3-methyl-1-butene is expected to produce 2-chloro-3-methylbutane. Instead, a mixture of 2-chloro-3-methylbutane and 2-chloro-2-methylbutane results.

3-Methyl-1-butene $\xrightarrow[0°C]{HCl}$ 2-Chloro-3-methylbutane (40%) + 2-Chloro-2-methylbutane (60%)

Addition begins in the usual way, by protonation of the double bond to give, in this case, a secondary carbocation.

Hydrogen chloride + 3-Methyl-1-butene $\xrightarrow{-:\ddot{Cl}:^-}$ 1,2-Dimethylpropyl cation (secondary) $\xrightarrow{\text{hydride shift}}$ 1,1-Dimethylpropyl cation (tertiary)

This carbocation can be captured by chloride to give 2-chloro-3-methylbutane (40%) or it can rearrange by way of a hydride shift to give a tertiary carbocation. The tertiary carbocation reacts with chloride ion to give 2-chloro-2-methylbutane (60%).

Problem 8.7

Addition of hydrogen chloride to 3,3-dimethyl-1-butene gives a mixture of two isomeric chlorides in approximately equal amounts. Suggest reasonable structures for these two compounds, and offer a mechanistic explanation for their formation.

8.6 Acid-Catalyzed Hydration of Alkenes

Analogous to the conversion of alkenes to alkyl halides by electrophilic addition of hydrogen halides across the double bond, acid-catalyzed addition of water gives alcohols.

$$\mathrm{C{=}C} + \mathrm{HOH} \xrightarrow{H^+} \mathrm{H{-}C{-}C{-}OH}$$

Alkene Water Alcohol

Markovnikov's rule is followed.

2-Methyl-2-butene $\xrightarrow{50\%\ H_2SO_4/H_2O}$ 2-Methyl-2-butanol (90%)

Mechanism 8.3 extends the general principles of electrophilic addition to acid-catalyzed hydration. In the first step of the mechanism, proton transfer converts the alkene to a carbocation, which then reacts with a molecule of water in step 2. The alkyloxonium ion formed in this step is the conjugate acid of the ultimate alcohol and yields it in step 3 while regenerating the acid catalyst.

Mechanism 8.3

Acid-Catalyzed Hydration of 2-Methylpropene

THE OVERALL REACTION:

2-Methylpropene + Water $\xrightarrow{H_3O^+}$ *tert*-Butyl alcohol

THE MECHANISM:

Step 1: Protonation of the carbon–carbon double bond in the direction that leads to the more stable carbocation:

2-Methylpropene + Hydronium ion $\underset{}{\overset{\text{slow}}{\rightleftharpoons}}$ *tert*-Butyl cation + Water

Step 2: Water acts as a nucleophile to capture *tert*-butyl cation:

tert-Butyl cation + Water $\overset{\text{fast}}{\rightleftharpoons}$ *tert*-Butyloxonium ion

Step 3: Deprotonation of *tert*-butyloxonium ion. Water acts as a Brønsted base:

tert-Butyloxonium ion + Water $\overset{\text{fast}}{\rightleftharpoons}$ *tert*-Butyl alcohol + Hydronium ion

Problem 8.8

Instead of the three-step process of Mechanism 8.3, the following two-step mechanism might be considered:

1. $(CH_3)_2C{=}CH_2 + H_3O^+ \xrightarrow{\text{slow}} (CH_3)_3C^+ + H_2O$
2. $(CH_3)_3C^+ + HO^- \xrightarrow{\text{fast}} (CH_3)_3COH$

This mechanism cannot be correct! What is its fundamental flaw?

The notion that carbocation formation is rate-determining follows from our previous experience with reactions that involve carbocation intermediates and by observing how the reaction rate is affected by the structure of the alkene. Alkenes that yield more stable carbocations react faster than those that yield less stable ones.

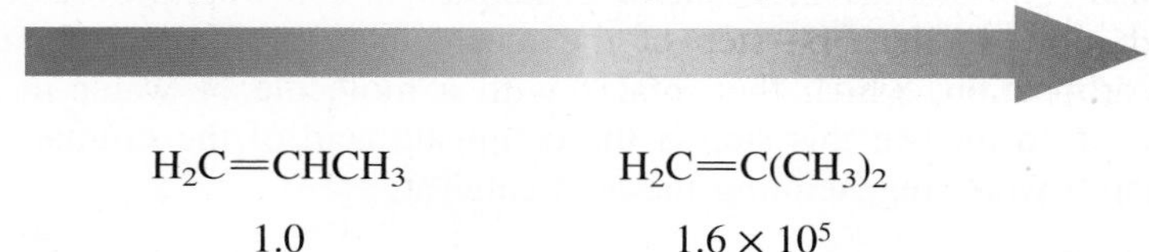

$H_2C{=}CHCH_3$	$H_2C{=}C(CH_3)_2$
1.0	1.6×10^5

Protonation of 2-methylpropene, the most reactive, gives a tertiary carbocation. The more stable the carbocation, the faster its rate of formation and the faster the overall reaction rate.

Problem 8.9

The rates of hydration of the two alkenes shown differ by a factor of over 7000 at 25°C. Which isomer is the more reactive? Why?

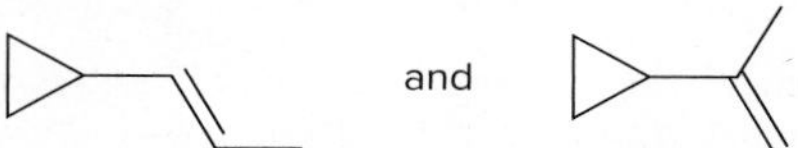

You may have noticed that the acid-catalyzed hydration of an alkene and the acid-catalyzed dehydration of an alcohol are the reverse of each other. For example:

2-Methylpropene + H_2O (Water) $\underset{}{\overset{H_3O^+}{\rightleftharpoons}}$ *tert*-Butyl alcohol (OH)

An important principle, called **microscopic reversibility,** connects the mechanisms of the forward and reverse reactions. It states that *in any equilibrium, the sequence of intermediates and transition states encountered as reactants proceed to products in one direction must also be encountered, and in precisely the reverse order, in the opposite direction.* Just as the reaction is reversible with respect to reactants and products, so too is each tiny increment of progress along the mechanistic pathway. Once we know the mechanism for the forward reaction, we also know the intermediates and transition states for its reverse. In particular, the three-step mechanism for the acid-catalyzed hydration of 2-methylpropene shown in Mechanism 8.3 is the reverse of that for the acid-catalyzed dehydration of *tert*-butyl alcohol in Mechanism 7.1.

It would be a good idea to verify the statement in the last sentence of this paragraph by revisiting Mechanisms 7.1 and 8.3.

Problem 8.10

Is the electrophilic addition of hydrogen chloride to 2-methylpropene the reverse of the E1 or E2 elimination of *tert*-butyl chloride?

Reaction mechanisms help us understand the "how" of reversible reactions, but not the "how much." To gain an appreciation for the factors that influence equilibria in addition reactions we need to expand on some ideas introduced when we discussed acid–base reactions in Chapter 1 and conformational equilibria in Chapter 3.

8.7 Thermodynamics of Addition–Elimination Equilibria

We have seen that both the forward and reverse reactions represented by the hydration–dehydration equilibrium are useful synthetic methods.

$$\mathrm{C{=}C} + \mathrm{H_2O} \overset{\mathrm{H^+}}{\rightleftharpoons} \mathrm{H{-}C{-}C{-}OH}$$

Alkene Water Alcohol

We can prepare alcohols from alkenes, and alkenes from alcohols, but how do we control the position of equilibrium so as to maximize the yield of the compound we want?

The qualitative reasoning expressed in **Le Châtelier's principle** is a helpful guide: *a system at equilibrium adjusts so as to minimize any stress applied to it.* For hydration–dehydration equilibria, the key stress factor is the water concentration. Adding water to a hydration–dehydration equilibrium mixture causes the system to respond by consuming water. More alkene is converted to alcohol, and the position of equilibrium shifts to the right. When we prepare an alcohol from an alkene, we use a reaction medium in which the molar concentration of water is high—dilute sulfuric acid, for example.

On the other hand, alkene formation is favored when the concentration of water is kept low. The system responds to the absence of water by causing more alcohol molecules to dehydrate, forming more alkene. The amount of water in the reaction mixture is kept low by using concentrated acids as catalysts. Distilling the reaction mixture is an effective way of removing water as it is formed, causing the equilibrium to shift to the left. If the alkene is low-boiling, it can also be removed by distillation. This offers the additional benefit of protecting the alkene from acid-catalyzed isomerization after it is formed.

Le Châtelier's principle helps us predict *qualitatively* how an equilibrium will respond to changes in experimental conditions. For a *quantitative* understanding, we need to examine reactions from a thermodynamic perspective.

At constant temperature and pressure, the direction in which a reaction proceeds—that is, the direction in which it is **spontaneous**—is the one that leads to a decrease in **free energy (G):**

$$\Delta G = G_{\text{products}} - G_{\text{reactants}} \qquad \text{spontaneous when } \Delta G < 0$$

Free energy is also called "Gibbs free energy." The official term is **Gibbs energy,** in honor of the nineteenth-century American physicist J. Willard Gibbs.

The free energy of the reactants and products depends on what they are and how much of each is present. Although G is always positive, ΔG can be positive or negative. If only the reactants are present at the beginning, $G_{\text{reactants}}$ has some value but G_{products} is zero; therefore, ΔG is negative and the reaction is spontaneous in the direction written. As the reaction proceeds, $G_{\text{reactants}}$ decreases while G_{products} increases until both are equal and $\Delta G = 0$. At this point the system is at equilibrium. Both the forward and reverse reactions continue to take place, but at equal rates.

Because reactions are carried out under a variety of conditions, it is convenient to define a *standard state* for substances and experimental conditions. The standard state is the form (solid, liquid, or gas) assumed by the pure substance at 1 atm pressure. For substances in aqueous solution, the standard-state concentration is 1 M. Standard-state values are designated by a superscript ° following the thermodynamic symbol as in $\Delta G°$.

For a reversible reaction

$$a\mathrm{A} + b\mathrm{B} \rightleftharpoons c\mathrm{C} + d\mathrm{D}$$

the relationship between ΔG and $\Delta G°$ is

$$\Delta G = \Delta G° + RT \ln \frac{[\mathrm{C}]^c[\mathrm{D}]^d}{[\mathrm{A}]^a[\mathrm{B}]^b}$$

where $R = 8.314$ J/(mol·K) or 1.99 cal/(mol·K) and T is the Kelvin temperature. At equilibrium $\Delta G = 0$, and $\frac{[C]^c[D]^d}{[A]^a[B]^b}$ becomes the equilibrium constant K. Substituting these values in the preceding equation and rearranging, we get

$$\Delta G^\circ = -RT \ln K$$

Reactions for which the sign of ΔG° is negative are **exergonic;** those for which ΔG° is positive are **endergonic.** Exergonic reactions have an equilibrium constant greater than 1; endergonic reactions have equilibrium constants less than 1.

Free energy has both an enthalpy (H) and an entropy (S) component.

$$G = H - TS$$

At constant temperature, $\Delta G^\circ = \Delta H^\circ - T\Delta S^\circ$

For the hydration of 2-methylpropene, the standard-state thermodynamic values are given beside the equation.

$$(CH_3)_2C{=}CH_2(g) + H_2O(\ell) \rightleftharpoons (CH_3)_3COH(\ell)$$

$\Delta G^\circ = -5.4$ kJ/mol	(−1.3 kcal/mol)	Exergonic
$\Delta H^\circ = -52.7$ kJ/mol	(−12.6 kcal/mol)	Exothermic
$\Delta S^\circ = -0.16$ kJ/K·mol	(−0.038 kcal/K·mol)	Entropy decreases

The negative sign for ΔG° tells us the reaction is exergonic. From the relationship

$$\Delta G^\circ = -RT \ln K$$

we can calculate the equilibrium constant at 25°C as $K = 9$.

Problem 8.11

You can calculate the equilibrium constant for the dehydration of $(CH_3)_3COH$ (the reverse of the preceding reaction) by reversing the sign of ΔG° in the expression $\Delta G^\circ = -RT \ln K$, but there is an easier way. Do you know what it is? What is K for the dehydration of $(CH_3)_3COH$?

The ΔH° term is dominated by bond strength. A negative sign for ΔH° almost always means that bonding is stronger in the products than in the reactants. Stronger bonding reduces the free energy of the products and contributes to a more negative ΔG°. Such is the normal case for addition reactions. Hydrogenation, hydration, and hydrogen halide additions to alkenes, for example, are all characterized by negative values for ΔH°.

The ΔS° term is a measure of the increase or decrease in the order of a system. A more ordered system has less entropy and is less probable than a disordered one. The main factors that influence ΔS° in a chemical reaction are the number of moles of material on each side of the balanced equation and their physical state. The liquid phase of a substance has more entropy (less order) than the solid, and the gas phase has much more entropy than the liquid. Entropy increases when more molecules are formed at the expense of fewer ones, as for example in elimination reactions. Conversely, addition reactions convert more molecules to fewer ones and are characterized by a negative sign for ΔS°.

The negative signs for both ΔH° and ΔS° in typical addition reactions of alkenes cause the competition between addition and elimination to be strongly temperature-dependent. Addition is favored at low temperatures, elimination at high temperatures. The economically important hydrogenation–dehydrogenation equilibrium that connects ethylene and ethane illustrates this.

$$\underset{\text{Ethylene}}{H_2C{=}CH_2(g)} + \underset{\text{Hydrogen}}{H_2(g)} \rightleftharpoons \underset{\text{Ethane}}{CH_3CH_3(g)}$$

Hydrogenation of ethylene converts two gas molecules on the left to one gas molecule on the right, leading to a decrease in entropy. The hydrogenation is sufficiently exothermic and $\Delta H°$ sufficiently negative, however, that the equilibrium lies far to the right over a relatively wide temperature range.

Very high temperatures—typically in excess of 750°C—reverse the equilibrium. At these temperatures, the $-T\Delta S°$ term in

$$\Delta G° = \Delta H° - T\Delta S°$$

becomes so positive that it eventually overwhelms $\Delta H°$ in magnitude, and the equilibrium shifts to the left. In spite of the fact that *dehydrogenation* is very endothermic, billions of pounds of ethylene are produced from ethane each year by this process.

Problem 8.12

Does the presence or absence of a catalyst such as finely divided platinum, palladium, or nickel affect the equilibrium constant for the ethylene–ethane conversion?

Problem 8.13

The gas phase reaction of ethanol with hydrogen bromide can occur by either elimination or substitution.

$$CH_3CH_2OH(g) \underset{}{\overset{HBr}{\rightleftharpoons}} H_2C{=}CH_2(g) + H_2O(g)$$

$$CH_3CH_2OH(g) + HBr(g) \rightleftharpoons CH_3CH_2Br(g) + H_2O(g)$$

Which product, ethylene or ethyl bromide, will increase relative to the other as the temperature is raised? Why?

8.8 Hydroboration–Oxidation of Alkenes

Acid-catalyzed hydration converts alkenes to alcohols according to Markovnikov's rule. Frequently, however, one needs an alcohol having a structure that corresponds to hydration of an alkene with a regioselectivity opposite to that of Markovnikov's rule. The conversion of 1-decene to 1-decanol is an example.

$$\underset{\text{1-Decene}}{CH_3(CH_2)_7CH{=}CH_2} \longrightarrow \underset{\text{1-Decanol}}{CH_3(CH_2)_7\overset{H}{\overset{|}{C}}H{-}\overset{OH}{\overset{|}{C}}H_2}$$

The synthetic method used to accomplish this is an indirect one known as **hydroboration–oxidation.** *Hydroboration* is a reaction in which a boron hydride, a compound of the type R_2BH, adds to a carbon–carbon π bond. A carbon–hydrogen bond and a carbon–boron bond result.

$$\underset{\text{Alkene}}{>C{=}C<} + \underset{\text{Boron hydride}}{R_2B{-}H} \longrightarrow \underset{\text{Organoborane}}{H{-}\overset{|}{\underset{|}{C}}{-}\overset{|}{\underset{|}{C}}{-}BR_2}$$

Hydroboration–oxidation was developed by Professor Herbert C. Brown [Nobel Prize in Chemistry (1979)] as part of a broad program designed to apply boron-containing reagents to organic chemical synthesis.

Following hydroboration, the organoborane is oxidized by treatment with hydrogen peroxide in aqueous base. This is the *oxidation* stage of the sequence; hydrogen peroxide is the oxidizing agent, and the organoborane is converted to an alcohol.

With sodium hydroxide as the base, boron of the alkylborane is converted to the water-soluble and easily removed sodium salt of boric acid.

$$\underset{\text{Organoborane}}{H{-}\overset{|}{\underset{|}{C}}{-}\overset{|}{\underset{|}{C}}{-}BR_2} + \underset{\text{Hydrogen peroxide}}{3H_2O_2} + \underset{\text{Hydroxide ion}}{HO^-} \longrightarrow \underset{\text{Alcohol}}{H{-}\overset{|}{\underset{|}{C}}{-}\overset{|}{\underset{|}{C}}{-}OH} + \underset{\text{Alcohol}}{2ROH} + \underset{\text{Borate ion}}{B(OH)_4^-}$$

Hydroboration–oxidation leads to the overall hydration of an alkene. Notice, however, that water is not a reactant. The hydrogen that becomes bonded to carbon comes from the organoborane, and the hydroxyl group from hydrogen peroxide.

With this as introduction, let us now look at the individual steps in more detail for the case of hydroboration–oxidation of 1-decene. A boron hydride that is often used is *diborane* (B_2H_6). Diborane adds to 1-decene to give tridecylborane according to the balanced equation:

Diglyme, shown above the arrow in the equation, is the solvent in this example. Diglyme is an acronym for *di*ethylene *gly*col di*me*thyl ether, and its structure is $CH_3OCH_2CH_2OCH_2CH_2OCH_3$.

$$\underset{\text{1-Decene}}{6CH_3(CH_2)_7CH{=}CH_2} + \underset{\text{Diborane}}{B_2H_6} \xrightarrow{\text{diglyme}} \underset{\text{Tridecylborane}}{2[CH_3(CH_2)_7CH_2CH_2]_3B}$$

There is a regioselective preference for boron to bond to the less substituted carbon of the double bond. Thus, the hydrogen atoms of diborane add to C-2 of 1-decene, and boron to C-1. Oxidation of tridecylborane gives 1-decanol. The net result is the conversion of an alkene to an alcohol with a regioselectivity opposite to that of acid-catalyzed hydration.

$$\underset{\text{Tridecylborane}}{[CH_3(CH_2)_7CH_2CH_2]_3B} \xrightarrow[\text{NaOH}]{H_2O_2} \underset{\text{1-Decanol}}{CH_3(CH_2)_7CH_2CH_2OH}$$

Customarily, we combine the two stages, hydroboration and oxidation, in a single equation with the operations numbered sequentially above and below the arrow.

$$\underset{\text{1-Decene}}{CH_3(CH_2)_7CH{=}CH_2} \xrightarrow[\text{2. } H_2O_2,\ HO^-]{\text{1. } B_2H_6,\text{ diglyme}} \underset{\text{1-Decanol (93\%)}}{CH_3(CH_2)_7CH_2CH_2OH}$$

$H_3\bar{B}{-}\overset{+}{O}$: (tetrahydrofuran ring)

Borane–tetrahydrofuran complex

A more convenient hydroborating agent is the borane–tetrahydrofuran complex ($H_3B{\cdot}THF$). It is very reactive, adding to alkenes within minutes at 0°C, and is used in tetrahydrofuran as the solvent.

2-Methyl-2-butene $\xrightarrow[\text{2. } H_2O_2,\ HO^-]{\text{1. } H_3B\cdot THF}$ 3-Methyl-2-butanol (98%)

OH

Carbocation intermediates are not involved in hydroboration–oxidation. Hydration of double bonds takes place without rearrangement, even in alkenes as highly branched as the following:

(*E*)-2,2,5,5-Tetramethyl-3-hexene $\xrightarrow[\text{2. } H_2O_2,\ HO^-]{\text{1. } B_2H_6,\text{ diglyme}}$ 2,2,5,5-Tetramethyl-3-hexanol (82%)

OH

Problem 8.14

Write the structure of the major organic product obtained by hydroboration–oxidation of each of the following alkenes:

(a) 2-Methylpropene
(b) *cis*-2-Butene
(c) (cyclobutane ring)$=CH_2$
(d) Cyclopentene
(e) 3-Ethyl-2-pentene
(f) 3-Ethyl-1-pentene

Sample Solution (a) In hydroboration–oxidation, H and OH are introduced with a regioselectivity opposite to that of Markovnikov's rule. In the case of 2-methylpropene, this leads to 2-methyl-1-propanol as the product.

1. hydroboration
2. oxidation

2-Methylpropene → 2-Methyl-1-propanol

Hydrogen becomes bonded to the carbon that has the fewer hydrogens, hydroxyl to the carbon that has the greater number of hydrogens.

Both operations, hydroboration and oxidation, are stereospecific and lead to syn addition of H and OH to the double bond.

1. B_2H_6, diglyme
2. H_2O_2, HO^-

1-Methylcyclopentene → *trans*-2-Methylcyclopentanol (only product, 86% yield)

Problem 8.15

Hydroboration–oxidation of α-pinene, like its catalytic hydrogenation (Problem 8.3), is stereoselective. Addition takes place at the less hindered face of the double bond, and a single alcohol is produced in high yield (89%). Suggest a reasonable structure for this alcohol.

8.9 Mechanism of Hydroboration–Oxidation

The regioselectivity and syn stereospecificity of hydroboration–oxidation, coupled with a knowledge of the chemical properties of alkenes and boranes, contribute to our understanding of the reaction mechanism.

In order to simplify our presentation, we'll regard the hydroborating agent as if it were borane (BH_3) itself rather than B_2H_6 or the borane–tetrahydrofuran complex. BH_3 is electrophilic; it has a vacant 2*p* orbital that interacts with the π electron pair of the alkene as shown in step 1 of Mechanism 8.4. The product of this step is an unstable intermediate called a *π complex* in which boron and the two carbons of the double bond are joined by a three-center, two-electron bond. Structures of this type in which boron and two other atoms share two electrons are frequently encountered in boron chemistry. Each of the two carbons of the π complex has a small positive charge, whereas boron is slightly negative. This negative character of boron assists one of its hydrogens to migrate with a pair of electrons (a hydride shift) from boron to carbon as shown in step 2, yielding a stable alkylborane. The carbon–boron bond and the carbon–hydrogen bond are formed on the same face of the alkene in a stereospecific syn addition.

Borane (BH_3) does not exist as such at room temperature and atmospheric pressure. Two molecules of BH_3 combine to give diborane (B_2H_6), which is the more stable form.

Step 1 is consistent with the regioselectivity of hydroboration. Boron, with its attached substituents, is more sterically demanding than hydrogen, and bonds to the less crowded carbon of the double bond; hydrogen bonds to the more crowded one. Electronic effects are believed to be less important than steric ones, but point in the same direction. Hydrogen is transferred with a pair of electrons to the carbon atom that bears more of the positive charge in the π complex, namely, the one that bears the methyl group.

The oxidation stage (Mechanism 8.5) of hydroboration–oxidation begins with the formation of the conjugate base of hydrogen peroxide in step 1, followed by its bonding to boron in step 2. The resulting intermediate expels hydroxide with migration of the alkyl group from boron to oxygen in step 3. It is in this step that the critical C—O bond is formed. The stereochemical orientation of this new bond is the same as that of the original C—B bond, thereby maintaining the syn stereochemistry

Mechanism 8.4

Hydroboration of 1-Methylcyclopentene

THE OVERALL REACTION:

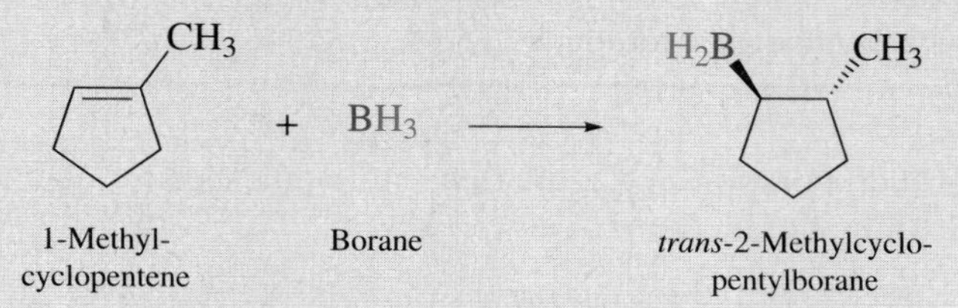

1-Methylcyclopentene | Borane | *trans*-2-Methylcyclopentylborane

THE MECHANISM:

Step 1: A molecule of borane (BH_3) attacks the alkene. Electrons flow from the π orbital of the alkene to the 2*p* orbital of boron. A π complex is formed.

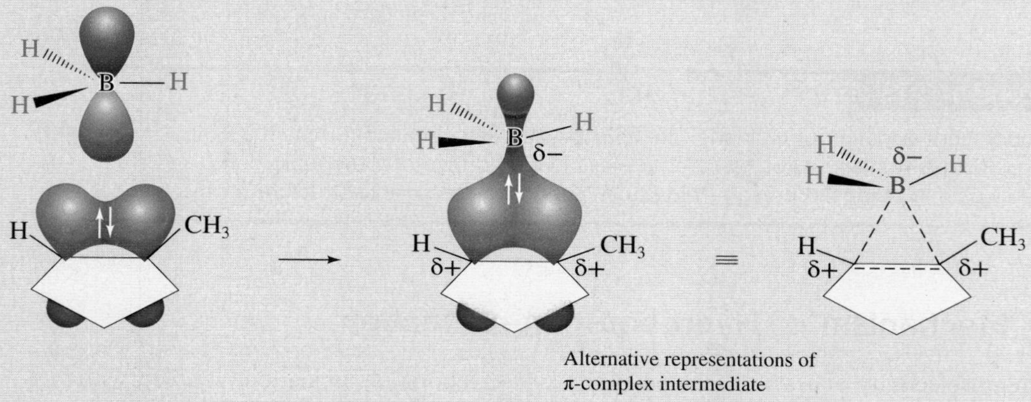

Alternative representations of π-complex intermediate

Step 2: The π complex rearranges to an organoborane. Hydrogen migrates from boron to carbon, carrying with it the two electrons in its bond to boron.

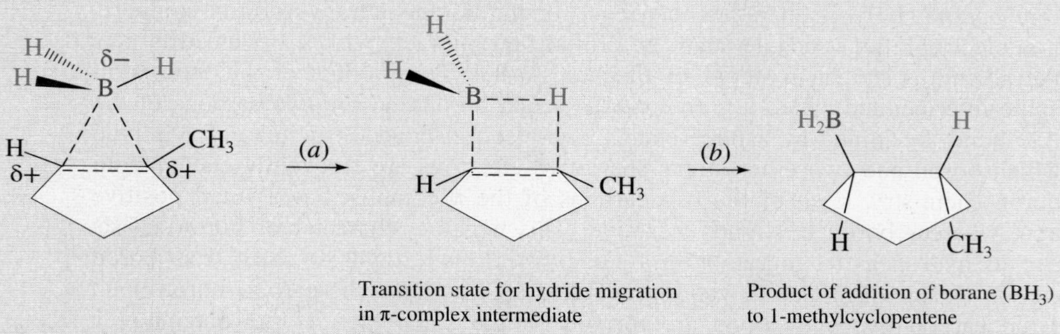

Transition state for hydride migration in π-complex intermediate

Product of addition of borane (BH_3) to 1-methylcyclopentene

of the hydroboration stage. Migration of the alkyl group from boron to oxygen occurs with *retention of configuration* at carbon. The alkoxyborane intermediate formed in step 3 undergoes subsequent base-promoted oxygen–boron bond cleavage in step 4 to give the alcohol product.

The mechanistic complexity of hydroboration–oxidation stands in contrast to the simplicity with which these reactions are carried out experimentally. Both the hydroboration and oxidation steps are extremely rapid and are performed at room temperature with conventional laboratory equipment. Ease of operation, along with the fact that hydroboration–oxidation leads to syn hydration of alkenes with a regioselectivity opposite to Markovnikov's rule, makes this procedure one of great value to the synthetic chemist.

Mechanism 8.5

Oxidation of an Organoborane

THE OVERALL REACTION:

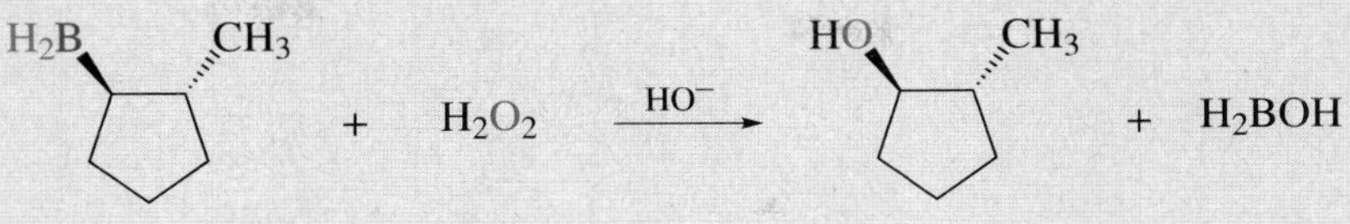

H_2B ... CH_3 (cyclopentane) + H_2O_2 $\xrightarrow{HO^-}$ HO ... CH_3 (cyclopentane) + H_2BOH

trans-2-Methylcyclopentylborane — Hydrogen peroxide — *trans*-2-Methylcyclopentanol — Hydroxyborane

THE MECHANISM:

Step 1: Hydrogen peroxide is converted to its anion in basic solution:

$$H—\ddot{O}—\ddot{O}—H + {}^-:\ddot{O}H \rightleftharpoons H—\ddot{O}—\ddot{O}:^- + H—\ddot{O}—H$$

Hydrogen peroxide — Hydroxide ion — Hydroperoxide ion — Water

Step 2: Anion of hydrogen peroxide acts as a nucleophile, attacking boron and forming an oxygen–boron bond:

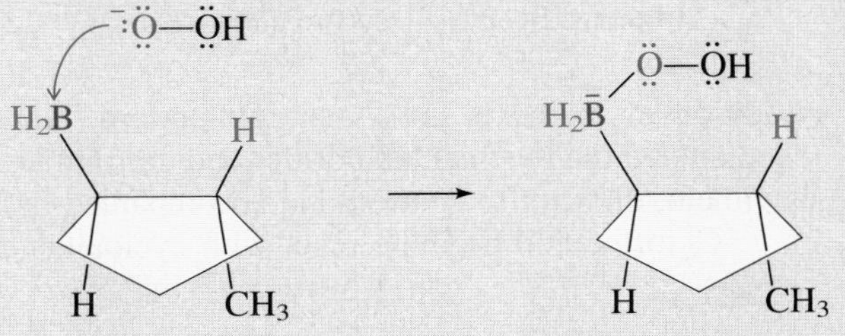

${}^-:\ddot{O}—\ddot{O}H$, H_2B, H, H, CH_3 → $\bar{H_2B}$—$\ddot{O}—\ddot{O}H$, H, H, CH_3

Organoborane intermediate from hydroboration of 1-methylcyclopentene

Step 3: Carbon migrates from boron to oxygen, displacing hydroxide ion. Carbon migrates with the pair of electrons in the carbon–boron bond; these become the electrons in the carbon–oxygen bond.

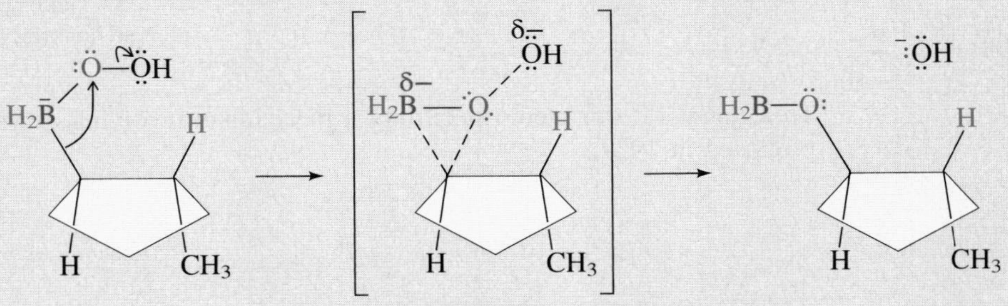

$H_2\bar{B}$—$:\ddot{O}—\ddot{O}H$, H, H, CH_3 → [δ− H_2B—O—(δ− OH), H, H, CH_3] → $H_2B—\ddot{O}:$, H, H, CH_3 + ${}^-:\ddot{O}H$

Transition state for migration of carbon from boron to oxygen — Alkoxyborane

Step 4: Hydrolysis cleaves the boron–oxygen bond, yielding the alcohol:

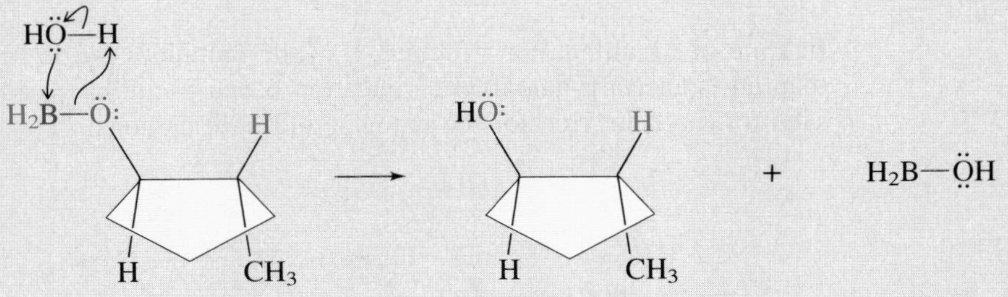

$H\ddot{O}—H$, $H_2B—\ddot{O}:$, H, H, CH_3 → $H\ddot{O}:$, H, H, CH_3 + $H_2B—\ddot{O}H$

Alkoxyborane — *trans*-2-Methylcyclopentanol

8.10 Addition of Halogens to Alkenes

Halogens react rapidly with alkenes by electrophilic addition. The products are called **vicinal dihalides,** meaning that the halogen atoms are attached to adjacent carbons.

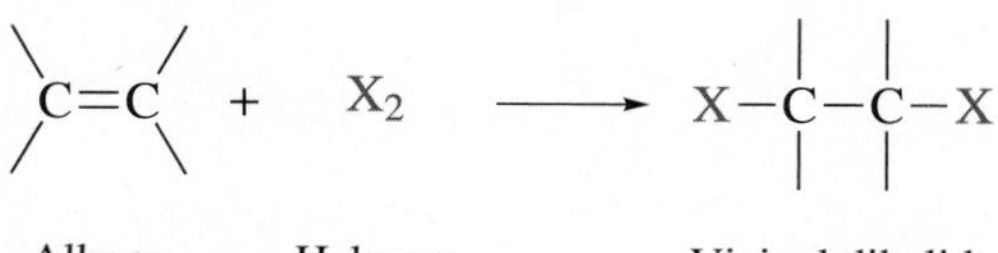

Alkene Halogen Vicinal dihalide

Like the word *vicinity, vicinal* comes from the Latin *vicinalis,* which means "neighboring."

Addition of chlorine or bromine takes place rapidly at room temperature and below in a variety of solvents, including acetic acid, carbon tetrachloride, chloroform, and dichloromethane.

Br_2 ($CHCl_3$, 0°C)

4-Methyl-2-pentene + Bromine → 2,3-Dibromo-4-methylpentane (100%)

Addition of iodine is not as straightforward, and vicinal diiodides are less commonly encountered than vicinal dichlorides and dibromides. The reaction of fluorine with alkenes is violent, difficult to control, and accompanied by substitution of hydrogens by fluorine.

Chlorine and bromine react with cycloalkenes by stereospecific anti addition.

Cl_2 ($CHCl_3$, −60°C)

Cyclooctene + Chlorine → *trans*-1,2-Dichlorocyclooctane (73% yield; none of the cis stereoisomer is formed)

To account for this stereospecificity a mechanism involving a cyclic **halonium ion** was proposed in 1937.

Alkene + halogen → Cyclic halonium ion + halide ion → Vicinal dihalide

In spite of its unfamiliar structure, a cyclic halonium ion is believed to be more stable than an isomeric β-haloalkyl carbocation because, unlike a carbocation, a halonium ion satisfies the octet rule for the halogen and both carbons.

Cyclic halonium ion β-Haloalkyl carbocation

Evidence in support of this mechanism has been obtained by the isolation of a stable bromonium ion by the reaction:

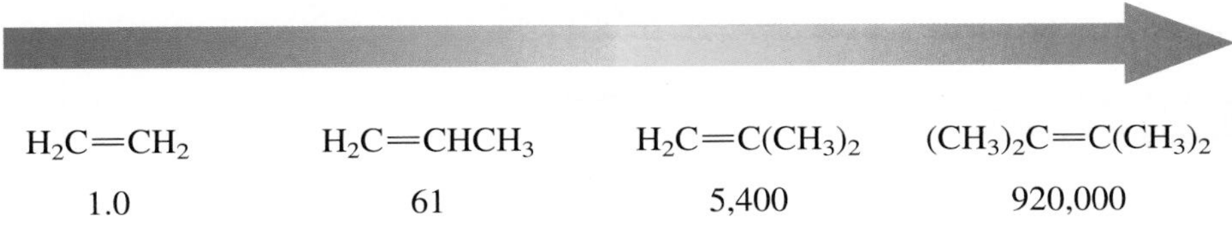

The crowded environment enclosing the three-membered ring prevents the next step in the mechanism and prevents formation of a vicinal dibromide.

The trend in relative rates as a function of alkene structure is consistent with a rate-determining step in which electrons flow from the alkene to the halogen.

Relative rate of bromine addition (25°C)

$H_2C{=}CH_2$	$H_2C{=}CHCH_3$	$H_2C{=}C(CH_3)_2$	$(CH_3)_2C{=}C(CH_3)_2$
1.0	61	5,400	920,000

Alkyl groups on the double bond release electrons, stabilize the transition state for the rate-determining step, and increase the reaction rate. The much greater reactivity of $(CH_3)_2C{=}C(CH_3)_2$ compared with $H_2C{=}C(CH_3)_2$ indicates that both of the carbons of the double bond participate in this stabilization.

Transition state for bromonium ion formation from an alkene and bromine

δ+ C ··· Br ··· Br δ−

Mechanism 8.6 describes the bromonium ion mechanism for the reaction of cyclopentene with bromine.

Mechanism 8.6

Bromine Addition to Cyclopentene

THE OVERALL REACTION:

Cyclopentene + Br_2 $\xrightarrow{CHCl_3}$ *trans*-1,2-Dibromocyclopentane (80%)

Cyclopentene Bromine *trans*-1,2-Dibromocyclopentane (80%)

THE MECHANISM:

Step 1: Bromine acts as an electrophile and reacts with cyclopentene to form a cyclic bromonium ion. This is the rate-determining step.

Cyclopentene + Bromine $\xrightarrow{slow}$ Bromonium ion intermediate + Bromide ion

continued

Step 2: Bromide ion acts as a nucleophile, forming a bond to one of the carbons of the bromonium ion and displacing the positively charged bromine from that carbon. Because substitutions of this type normally occur with the nucleophile approaching carbon from the side opposite the bond that is broken, the two bromine atoms are trans to one another in the product.

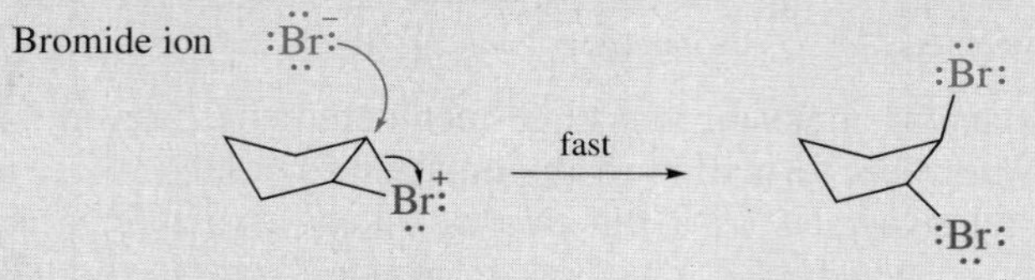

Bromide ion :Br:⁻ ; Bromonium ion intermediate —fast→ *trans*-1,2-Dibromocyclopentane

Problem 8.16

Arrange the compounds 2-methyl-1-butene, 2-methyl-2-butene, and 3-methyl-1-butene in order of decreasing reactivity toward bromine.

When bromine adds to (*Z*)- or (*E*)-2-butene, the product 2,3-dibromobutane contains two equivalently substituted chirality centers:

$$CH_3CH{=}CHCH_3 \xrightarrow{Br_2} CH_3\underset{Br}{C}H\underset{Br}{C}HCH_3$$

(*Z*)- or (*E*)-2-butene → 2,3-Dibromobutane

Three stereoisomers are possible: a pair of enantiomers and a meso form.

Two factors combine to determine which stereoisomers are actually formed in the reaction:

1. The (*E*)- or (*Z*)-configuration of the starting alkene
2. The anti stereochemistry of addition

Figure 8.5 shows the stereochemical differences associated with anti addition of bromine to (*E*)- and (*Z*)-2-butene, respectively. The trans alkene (*E*)-2-butene yields only *meso*-2,3-dibromobutane, but the cis alkene (*Z*)-2-butene gives a racemic mixture of (2*R*,3*R*)- and (2*S*,3*S*)-2,3-dibromobutane.

(*a*) Anti addition of Br_2 to (*E*)-2-butene gives *meso*-2,3-dibromobutane.

meso ← Br_2 50% — (*E*)-2-butene — Br_2 50% → *meso*

(*b*) Anti addition of Br_2 to (*Z*)-2-butene gives equal amounts of (2*R*,3*R*)- and (2*S*,3*S*)-2,3-dibromobutane.

(2*R*,3*R*) ← Br_2 50% — (*Z*)-2-butene — Br_2 50% → (2*S*,3*S*)

Figure 8.5

Addition of Br_2 to (*E*)- and (*Z*)-2-butene is stereospecific. Stereoisomeric products are formed from stereoisomeric reactants.

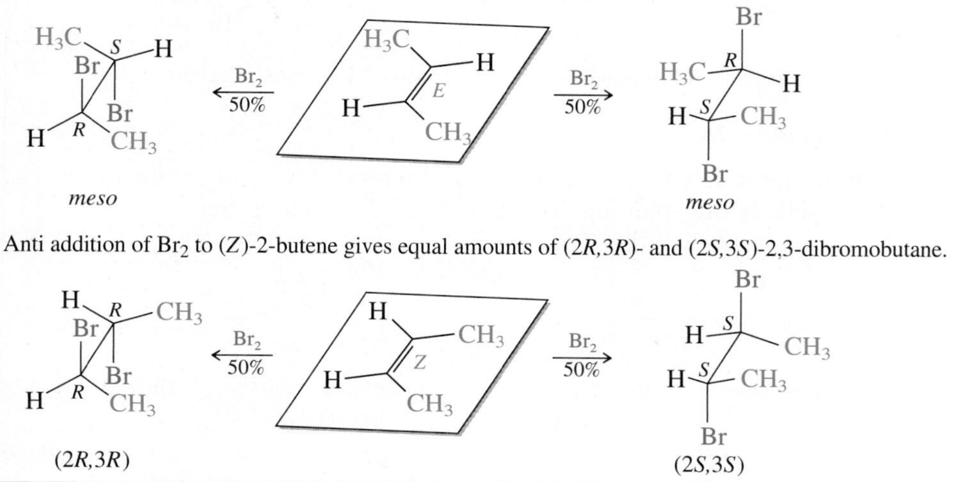

Bromine addition to alkenes is a **stereospecific reaction,** a reaction in which stereoisomeric starting materials yield products that are stereoisomers of each other. In this case the starting materials, in separate reactions, are the *E* and *Z* stereoisomers of 2-butene. The chiral dibromides formed from (*Z*)-2-butene are stereoisomers (diastereomers) of the meso dibromide from (*E*)-2-butene.

In a related reaction, chlorine and bromine react with alkenes in aqueous solution to give **vicinal halohydrins,** compounds that have a halogen and a hydroxyl group on adjacent carbons.

$$H_2C{=}CH_2 + Br_2 \xrightarrow{H_2O} HOCH_2CH_2Br$$

Ethylene — Bromine — 2-Bromoethanol (70%)

Like the acid-catalyzed hydration of alkenes and the reaction of alkenes with hydrogen halides, halohydrin formation is regioselective. The halogen bonds to the less substituted carbon of the double bond and hydroxyl to the more substituted carbon.

Br_2, H_2O

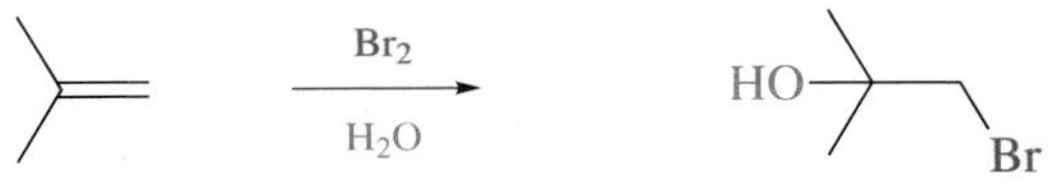

2-Methylpropene → 1-Bromo-2-methyl-2-propanol (77%)

The generally accepted mechanism for this reaction is modeled after the formation of vicinal dihalides. It begins with formation of a cyclic bromonium ion, followed by nucleophilic attack of water from the side opposite the carbon–halogen bond. The transition state for bromonium ion ring opening has some of the character of a carbocation; therefore, the reaction proceeds by breaking the bond between bromine and the more substituted carbon.

More stable transition state; positive charge shared by tertiary carbon

Less stable transition state; positive charge shared by primary carbon

Vicinal halohydrin formation also resembles that of vicinal dihalides in that addition is stereospecific and anti.

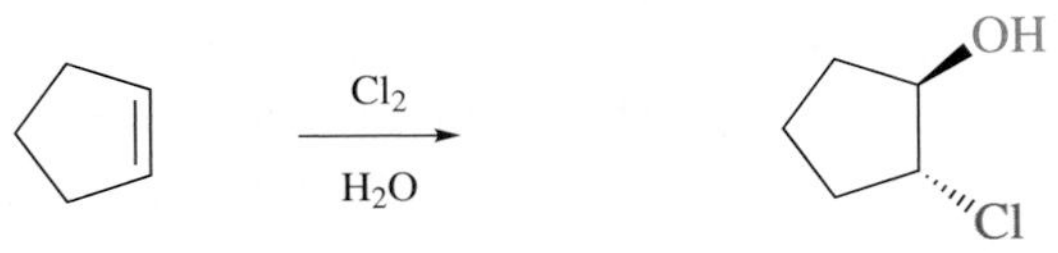

Cl_2, H_2O

Cyclopentene → *trans*-2-Chlorocyclopentanol (52–56% yield; cis isomer not formed)

A mechanism involving approach by a water molecule from the side opposite the carbon–chlorine bond of a cyclic chloronium ion accounts for the observed stereochemistry.

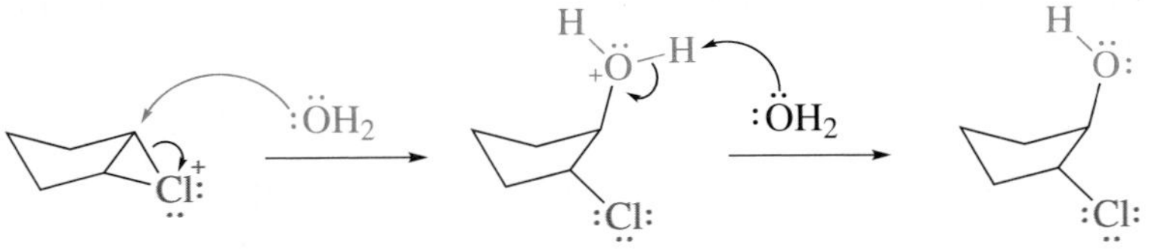

Chloronium ion intermediate → *trans*-2-Chlorocyclopentanol

Problem 8.17

Give the structure of the product formed when each of the following alkenes reacts with bromine in water:

(a) 2-Methyl-1-butene
(b) 2-Methyl-2-butene
(c) 3-Methyl-1-butene
(d) 1-Methylcyclopentene

Sample Solution (a) The hydroxyl group becomes bonded to the more substituted carbon of the double bond, and bromine bonds to the less substituted one.

2-Methyl-1-butene + Br_2 (Bromine) $\xrightarrow{H_2O}$ 1-Bromo-2-methyl-2-butanol (OH, Br)

8.11 Epoxidation of Alkenes

We have seen two reactions in which alkenes react with electrophilic reagents via cyclic transition states (hydroboration) or intermediates (halonium ions). In this section we'll introduce a reaction that gives a cyclic product—a three-membered oxygen-containing ring called an **epoxide.**

A second method for naming epoxides in the IUPAC system is described in Section 17.1.

Substitutive IUPAC nomenclature treats epoxides as *epoxy* derivatives of alkane parents; the *epoxy*-prefix is listed in alphabetical order like other substituents. Some industrial chemicals have common names formed by adding the word "oxide" to the name of the alkene.

$H_2C{-}CH_2$ (O bridging) — Epoxyethane (ethylene oxide)

$H_2C{-}CHCH_3$ (O bridging) — 1,2-Epoxypropane (propylene oxide)

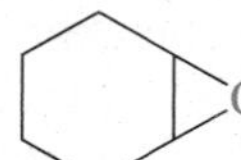

1,2-Epoxycyclohexane

$(H_3C)_2C{-}CH(CH_3)$ (O bridging) — 2,3-Epoxy-2-methylbutane

Countless numbers of naturally occurring substances are epoxides. *Disparlure,* the sex attractant of the female gypsy moth, is but one example.

Disparlure

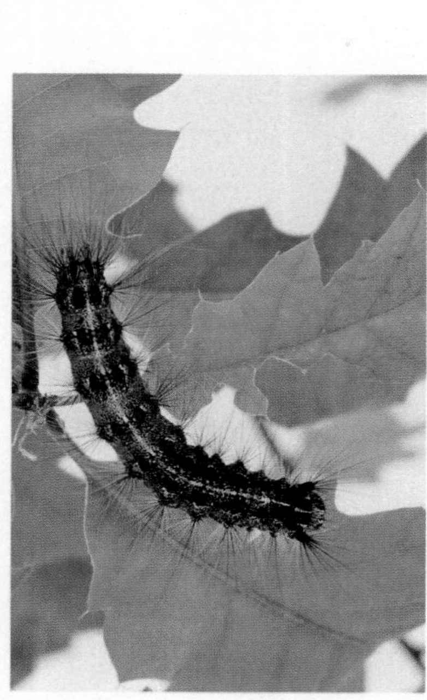

Gypsy moths were accidentally introduced into United States forests around 1869 in Medford, Massachusetts. They have become persistent pests throughout the Northeast and Middle Atlantic states, defoliating millions of acres of woodlands.
Source: Photo by Scott Bauer, USDA-ARS

In one strategy designed to control the spread of the gypsy moth, infested areas are sprayed with synthetic disparlure. With the sex attractant everywhere, male gypsy moths become hopelessly confused as to the actual location of individual females. Many otherwise fertile female gypsy moths then live out their lives without producing hungry gypsy moth caterpillars.

Problem 8.18

Give the substitutive IUPAC name, including stereochemistry, for disparlure.

Epoxides are very easy to prepare via the reaction of an alkene with a peroxy acid. This process is known as **epoxidation.**

$$\underset{\text{Alkene}}{>C{=}C<} + \underset{\text{Peroxy acid}}{RC(=O)OOH} \longrightarrow \underset{\text{Epoxide}}{>C{-}C<\ (\text{O bridging})} + \underset{\text{Carboxylic acid}}{RC(=O)OH}$$

A commonly used peroxy acid is peroxyacetic acid (CH_3CO_2OH). Peroxyacetic acid is normally used in acetic acid as the solvent, but epoxidation reactions tolerate a variety of solvents and are often carried out in dichloromethane or chloroform.

$$H_2C{=}CH(CH_2)_9CH_3 + CH_3\overset{O}{\overset{\|}{C}}OOH \longrightarrow H_2C\underset{O}{-}CH(CH_2)_9CH_3 + CH_3\overset{O}{\overset{\|}{C}}OH$$

1-Dodecene; Peroxyacetic acid; 1,2-Epoxydodecane (52%); Acetic acid

$$\text{Cyclooctene} + CH_3\overset{O}{\overset{\|}{C}}OOH \longrightarrow \text{1,2-Epoxycyclooctane} + CH_3\overset{O}{\overset{\|}{C}}OH$$

Cyclooctene; Peroxyacetic acid; 1,2-Epoxycyclooctane (86%); Acetic acid

Epoxidation of alkenes with peroxy acids is a syn addition to the double bond. Substituents that are cis to each other in the alkene remain cis in the epoxide; substituents that are trans in the alkene remain trans in the epoxide.

Problem 8.19

Give the structure of the alkene, including stereochemistry, that you would choose as the starting material in a preparation of synthetic disparlure.

Electron-releasing substitutents on the double bond increase the rate of epoxidation, which suggests that the peroxy acid acts as an electrophile toward the alkene.

Relative rate of epoxidation (peroxyacetic acid, 25°C)

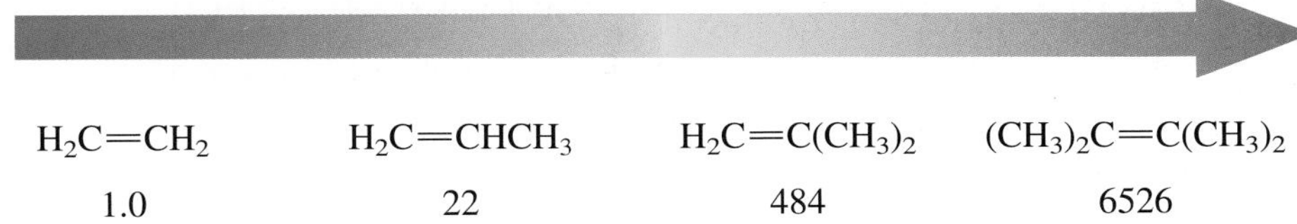

$H_2C{=}CH_2$	$H_2C{=}CHCH_3$	$H_2C{=}C(CH_3)_2$	$(CH_3)_2C{=}C(CH_3)_2$
1.0	22	484	6526

Alkene epoxidation is believed to occur in a single bimolecular step as shown in Mechanism 8.7.

Mechanism 8.7

Epoxidation of Bicyclo[2.2.1]-2-heptene

THE REACTION AND THE MECHANISM:

Oxygen is transferred from the peroxy acid to the less crowded (upper) face of the alkene.

Bicyclo[2.2.1]-2-heptene + Peroxyacetic acid ⟶ 2,3-Epoxybicyclo-[2.2.1]heptane + Acetic acid

continued

The various bonding changes occur in the same transition state.

8.12 Ozonolysis of Alkenes

Ozone (O_3), the triatomic form of oxygen, is a polar molecule ($\mu = 0.5$ D) that can be represented as a hybrid of its two most stable Lewis structures.

It is a powerful electrophile and reacts with alkenes to cleave the double bond, forming an **ozonide.** Ozonides undergo hydrolysis in water, giving carbonyl compounds.

$$\text{Alkene} \xrightarrow{O_3} \text{Ozonide} \xrightarrow{H_2O} \text{Two carbonyl compounds} + H_2O_2 \text{ (Hydrogen peroxide)}$$

Alkene Ozone Ozonide Two carbonyl compounds Hydrogen peroxide

Because hydrogen peroxide is a product of ozonide hydrolysis and has the potential to oxidize the products, the second half of this two-stage **ozonolysis** sequence is carried out in the presence of a reducing agent, usually zinc or dimethyl sulfide.

$$CH_3(CH_2)_5CH{=}CH_2 \xrightarrow[2.\ (CH_3)_2S]{1.\ O_3,\ CH_3OH} CH_3(CH_2)_5\overset{O}{\overset{\|}{C}}H + H\overset{O}{\overset{\|}{C}}H$$

1-Octene Heptanal (75%) Formaldehyde

$$\text{2-Methyl-1-hexene} \xrightarrow[2.\ H_2O,\ Zn]{1.\ O_3} \text{2-Hexanone (60\%)} + \text{Formaldehyde}$$

2-Methyl-1-hexene 2-Hexanone (60%) Formaldehyde

The types of carbonyl compounds that result are determined by the substituents on the doubly bonded carbons. Formaldehyde, aldehydes, or ketones are possible, depending on whether a particular carbon is attached to two hydrogens, a hydrogen and an alkyl group, or two alkyl groups respectively. Thus, the $=CH_2$ unit in each of the preceding examples gives formaldehyde. The remaining seven carbons of 1-octene are incorporated into the aldehyde heptanal, and the remaining six carbons of 2-methyl-1-hexene into the ketone 2-hexanone.

Problem 8.20

1-Methylcyclopentene gives a single compound ($C_6H_{10}O_2$) on ozonolysis. What is it?

Ozonolysis has both synthetic and analytical applications in organic chemistry. In synthesis, ozonolysis of alkenes provides a method for the preparation of aldehydes and ketones. When the objective is analytical, the products of ozonolysis are isolated and identified, thereby allowing the structure of the alkene to be deduced. In one such example, an alkene having the molecular formula C_8H_{16} was obtained from a chemical reaction and gave acetone and 2,2-dimethylpropanal as the products.

$$C_8H_{16} \xrightarrow[2.\ H_2O,\ Zn]{1.\ O_3} (CH_3)_2C{=}O \quad + \quad (CH_3)_3CCH{=}O$$

Acetone 2,2-Dimethylpropanal

Together, these two products contain all eight carbons of the starting alkene. The two carbonyl carbons correspond to those that were doubly bonded in the original alkene. Therefore, one of the doubly bonded carbons bears two methyl substituents; the other bears a hydrogen and a *tert*-butyl group and identifies the alkene as 2,4,4-trimethyl-2-pentene.

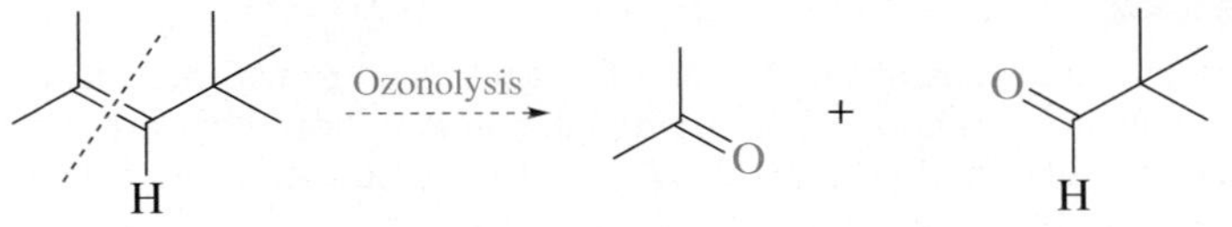

Ozonolysis

Acetone 2,2-Dimethylpropanal

Problem 8.21

The same reaction that gave 2,4,4-trimethyl-2-pentene also yielded an isomeric alkene. This second alkene produced formaldehyde and 4,4-dimethyl-2-pentanone on ozonolysis. Identify this alkene.

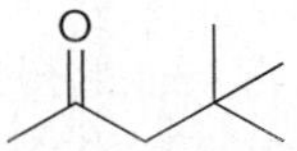

4,4-Dimethyl-2-pentanone

8.13 Enantioselective Addition to Alkenes

The preceding sections have introduced several stereochemical aspects of addition to double bonds. This section extends our discussion by focusing on additions in which an achiral reactant is converted to a chiral product. Countless examples of reactions of this type exist—a simple one being the hydration of the double bond of *cis*- or *trans*-2-butene. Under either the usual acid-catalyzed conditions or hydroboration–oxidation, 2-butanol is formed as a racemic mixture.

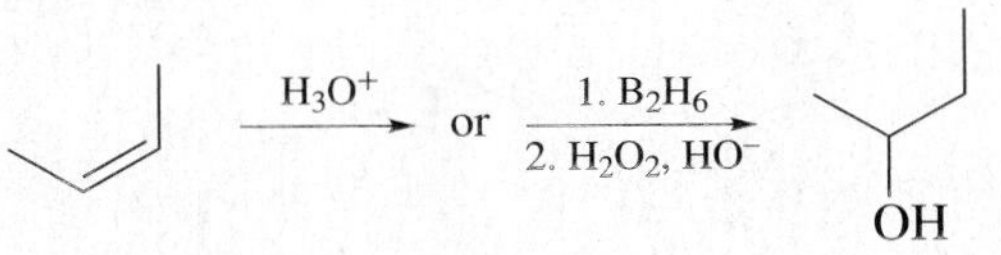

H_3O^+ or 1. B_2H_6 2. H_2O_2, HO^-

cis-2-Butene 2-Butanol

If, instead, certain chiral and enantiomerically enriched hydroborating agents are used instead of diborane itself, the 2-butanol that results is itself enriched in one of its enantiomers. Reactions such as this are said to be **enantioselective.**

Most biochemical reactions are catalyzed by enzymes and many require coenzymes to carry out their function. All enzymes and most coenzymes are chiral and enantiopure and provide an asymmetric environment for chemical reactions to take place. Enzyme-catalyzed reactions are enantioselective and occur with such a high level of stereoselectivity as to form a single enantiomer of a chiral substance exclusively.

One example is the biosynthesis of (*S*)-(−)-malic acid from fumaric acid in apples and other fruits.

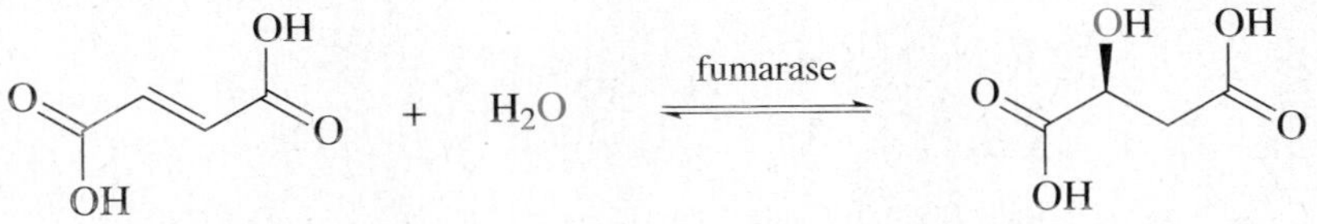

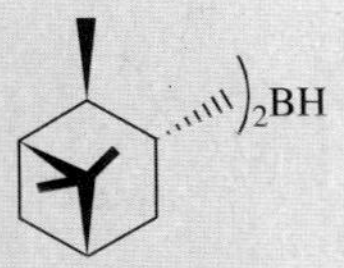

$(Ipc)_2BH$

Hydroboration–oxidation of *cis*-2-butene with the chiral reagent (−)-diisopinocamphenylborane gives (*R*)-2-butanol with 99% e.e.

+ H_2O fumarase

Fumaric acid Water (*S*)-(−)-Malic acid

From the equation, we see that the chemical reaction is hydration of a double bond. Neither fumaric acid nor water is chiral, but only the *S* enantiomer of malic acid is produced. The only source of chirality is the enzyme *fumarase*. According to the equation, we also see that the reaction is reversible—hydration of a double bond in one direction, dehydration in the other—and that fumarase is a catalyst for both processes. We should also point out the cis isomer of fumaric acid, called *maleic acid,* does not react under these or analogous conditions nor is there an enzyme-catalyzed biosynthesis that converts fumaric or maleic acid directly to (*R*)-(+)-malic acid.

Problem 8.22

The enzyme aconitase catalyzes the hydration of aconitic acid to two products: citric acid and isocitric acid. Isocitric acid is optically active; citric acid is not. What are the respective constitutions of isocitric acid and citric acid? Why isn't citric acid optically active?

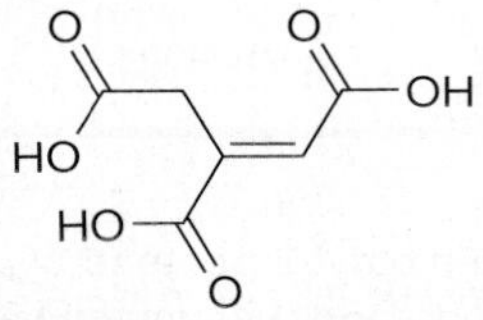

O O
OH
HO
HO
O

In a second example of an enzyme-catalyzed addition to an alkene, an enzyme of the class known as *monooxygenases* catalyzes the epoxidation of one of the double bonds of β-carotene.

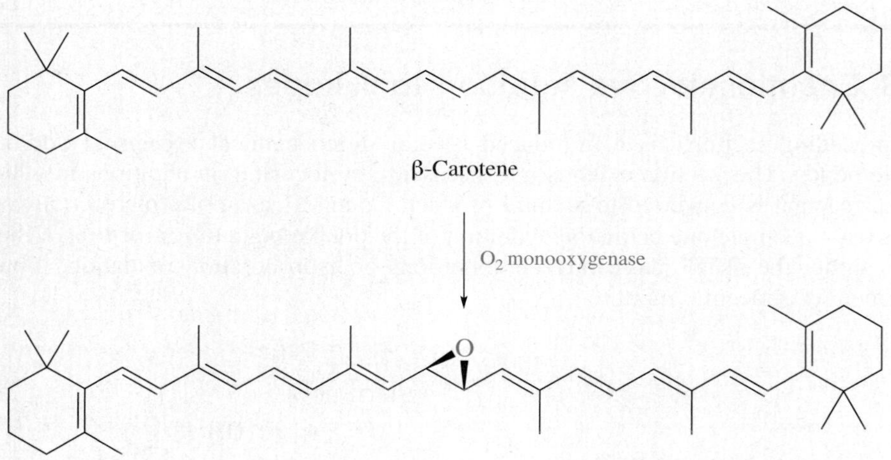

β-Carotene

O_2 monooxygenase

O

In addition to being stereoselective, the epoxidation is regioselective; of the 11 double bonds in β-carotene, only one of them is epoxidized.

8.14 Retrosynthetic Analysis and Alkene Intermediates

Chapters 6 and 7 focus on the preparation and reactions of alkenes, but are not only about alkenes. The two chapters also concern the compounds that lead to alkenes and the products of the reactions of alkenes. Suppose you wanted to prepare 1,2-epoxycyclohexane, given cyclohexanol as the starting material. We represent this retrosynthetically as

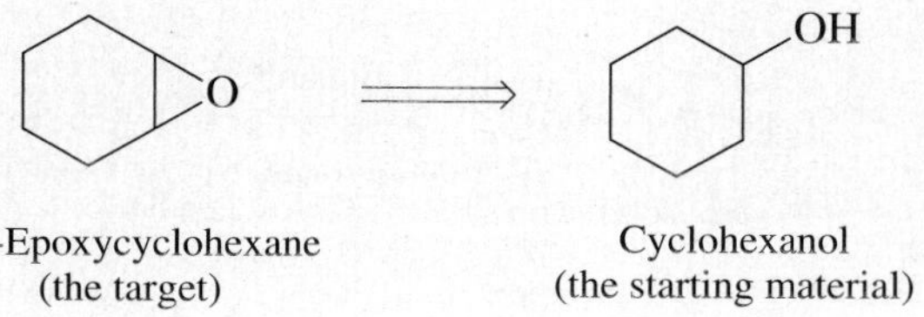

O ⟹ OH

1,2-Epoxycyclohexane (the target)

Cyclohexanol (the starting material)

However, we know of no reactions that convert alcohols directly to epoxides. Because we do know that epoxides are prepared from alkenes, we expand our retrosynthesis to reflect that.

1,2-Epoxycyclohexane $\Longrightarrow$ Cyclohexene $\Longrightarrow$ Cyclohexanol

Recognizing that cyclohexene can be prepared by acid-catalyzed dehydration of cyclohexanol, we write a suitable synthesis in the forward direction complete with the necessary reagents.

Cyclohexanol $\xrightarrow[\text{heat}]{H_2SO_4}$ Cyclohexene $\xrightarrow{CH_3C(O)OOH}$ 1,2-Epoxycyclohexane

Problem 8.23

Suggest a reasonable synthesis of *trans*-2-bromocyclohexanol from cyclohexyl bromide. Indicate appropriate reagents over the reaction arrow.

Now consider a slightly more difficult, but more realistic case—one in which the target is the nitrile $(CH_3)_2CHCH_2CN$ and the starting material is not specified other than to say it must be an alkene with five carbons or fewer.

Begin by recalling that the only reaction we have seen that yields a compound of the type RCN is a substitution of the S_N2 type. Thus, we consider a partial sequence of the type:

$(CH_3)_2CHCH_2CN \Longrightarrow (CH_3)_2CHCH_2Br \Longrightarrow (CH_3)_2CHCH_2OH$

Conversion of 1-bromo-2-methylpropane to the nitrile target is a reasonable last step, but successful preparation of the bromide is doubtful because of the possibility of rearrangement in the reaction of 2-methyl-1-propanol with hydrogen bromide. A safer choice is to use a sulfonate instead of a bromide.

$(CH_3)_2CHCH_2CN \Longrightarrow (CH_3)_2CHCH_2OSO_2CH_3 \Longrightarrow (CH_3)_2CHCH_2OH$

The problem specifies that the starting material be an alkene with five or fewer carbons. The most obvious choice is 2-methylpropene.

$(CH_3)_2CHCH_2CN \Longrightarrow (CH_3)_2CHCH_2OSO_2CH_3 \Longrightarrow (CH_3)_2CHCH_2OH \Longrightarrow (CH_3)_2C{=}CH_2$

We write the synthesis in the forward direction by choosing the appropriate reagents and begin by converting 2-methylpropene to 2-methyl-1-propanol by hydration of the double bond regioselectively by hydroboration–oxidation. Converting the alcohol to the methanesulfonate, followed by nucleophilic substitution of methanesulfonate by cyanide, completes the synthesis.

2-Methylpropene $\xrightarrow[\text{2. } H_2O_2,\ HO^-]{\text{1. } B_2H_6}$ 2-Methyl-1-propanol $\xrightarrow[\text{pyridine}]{CH_3SO_2Cl}$ 2-Methyl-1-propyl methanesulfonate $\xrightarrow{KCN}$ 3-Methylbutanenitrile

Problem 8.24

Fill in the missing compounds in the partial retrosynthesis shown and devise a synthesis showing all necessary reagents based on your retrosynthesis.

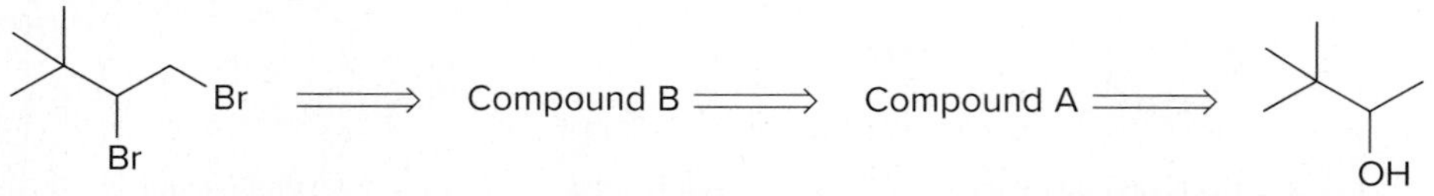

Br Br ⟹ Compound B ⟹ Compound A ⟹ OH

8.15 SUMMARY

Alkenes are **unsaturated hydrocarbons** and react with substances that add to the double bond. This chapter surveys the kinds of substances that react with alkenes, the mechanisms by which the reactions occur, and their synthetic applications.

Sections 8.1–8.3 See Table 8.2. Aspects of addition reactions are introduced in these sections as they apply to the catalytic hydrogenation of alkenes.

Sections 8.4–8.5 See Table 8.2. The mechanism of electrophilic addition is outlined for the reaction of hydrogen halides with alkenes. Carbocations are intermediates.

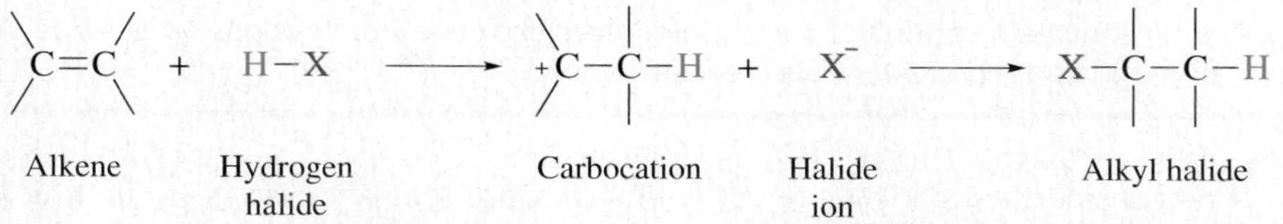

$$\text{C=C} + \text{H-X} \longrightarrow {}^{+}\text{C-C-H} + \text{X}^{-} \longrightarrow \text{X-C-C-H}$$

Alkene — Hydrogen halide — Carbocation — Halide ion — Alkyl halide

Addition is regioselective because protonation of the double bond occurs in the direction that gives the more stable of two possible carbocations.

Section 8.6 See Table 8.2. Acid-catalyzed addition of water to the double bond of an alkene gives an alcohol. The mechanism is analogous to that for electrophilic addition of hydrogen halides and is the reverse of that for acid-catalyzed dehydration of alcohols.

TABLE 8.2 Addition Reactions of Alkenes

Reaction (section) and Comments	General Equation and Specific Example
Catalytic hydrogenation (Sections 8.1–8.3) Alkenes react with hydrogen in the presence of a platinum, palladium, rhodium, or nickel catalyst to form the corresponding alkane. Both hydrogens add to the same face of the double bond (syn addition). Heats of hydrogenation can be used to compare the relative stability of various double-bond types.	$R_2C{=}CR_2$ (Alkene) + H_2 (Hydrogen) $\xrightarrow{\text{Pt, Pd, Rh, or Ni}}$ R_2CHCHR_2 (Alkane) 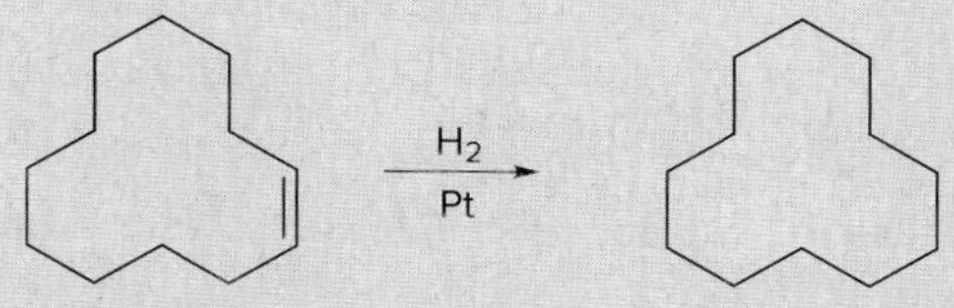 *cis*-Cyclododecene $\xrightarrow[\text{Pt}]{H_2}$ Cyclododecane (100%)
Addition of hydrogen halides (Sections 8.4–8.5) A proton and a halogen add to the double bond of an alkene to yield an alkyl halide. Addition proceeds in accordance with Markovnikov's rule: hydrogen adds to the carbon that has the greater number of hydrogens, halide to the carbon that has the fewer hydrogens. The regioselectivity is controlled by the relative stability of the two possible carbocation intermediates. Because the reaction involves carbocations, rearrangement is possible.	$RCH{=}CR'_2$ (Alkene) + HX (Hydrogen halide) $\longrightarrow$ $RCH_2{-}CR'_2(X)$ (Alkyl halide)  Methylenecyclohexane ($=CH_2$) + HCl (Hydrogen chloride) $\longrightarrow$ 1-Chloro-1-methylcyclohexane (75–80%) (CH_3, Cl)

TABLE 8.2 Addition Reactions of Alkenes (*Continued*)

Reaction (section) and Comments	General Equation and Specific Example
Acid-catalyzed hydration (Section 8.6) Addition of water to the double bond of an alkene takes place according to Markovnikov's rule in aqueous acid. A carbocation is an intermediate and is captured by a molecule of water acting as a nucleophile. Rearrangements are possible.	$RCH{=}CR'_2 + H_2O \xrightarrow{H^+} RCH_2{-}C(OH)R'_2$ Alkene, Water, Alcohol 2-Methylpropene $\xrightarrow{50\%\ H_2SO_4/H_2O}$ *tert*-Butyl alcohol (55–58%)
Hydroboration–oxidation (Sections 8.8–8.9) This two-step sequence converts alkenes to alcohols with a regioselectivity opposite to Markovnikov's rule. Addition of H and OH is stereospecific and syn. The reaction involves electrophilic addition of a boron hydride to the double bond, followed by oxidation of the intermediate organoborane with hydrogen peroxides. Carbocations are not intermediates and rearrangements do not occur.	$RCH{=}CR'_2 \xrightarrow[2.\ H_2O_2,\ HO^-]{1.\ B_2H_6,\ \text{diglyme}} RCH(OH){-}CHR'_2$ Alkene, Alcohol 4-Methyl-1-pentene $\xrightarrow[2.\ H_2O_2,\ HO^-]{1.\ H_3B\cdot THF}$ 4-Methyl-1-pentanol (80%)
Addition of Halogens (Section 8.10) Reactions with Br_2 or Cl_2 are the most common and yield vicinal dihalides except when the reaction is carried out in water. In water, the product is a vicinal halohydrin. The reactions involve a cyclic halonium ion intermediate and are stereospecific (anti addition). Halohydrin formation is regiospecific; the halogen bonds to the carbon of C=C that has the greater number of hydrogens.	$R_2C{=}CR'_2 + X_2 \longrightarrow R_2C(X){-}C(X)R'_2$ Alkene, Halogen, Vicinal dihalide $RCH{=}CR'_2 + X_2 + H_2O \longrightarrow RCH(X){-}C(OH)R'_2 + HX$ Alkene, Halogen, Water, Vicinal halohydrin, Hydrogen halide 1-Hexene $\xrightarrow{Br_2}$ 1,2-Dibromohexane (100%) Methylenecyclohexane ($=CH_2$) $\xrightarrow[H_2O]{Br_2}$ (1-Bromomethyl)cyclohexanol (89%) (CH_2Br, OH)
Epoxidation (Section 8.11) Peroxy acids transfer oxygen to the double bond of alkenes to yield epoxides. Addition is stereospecific and syn.	$R_2C{=}CR_2 + R'COOH$ (peroxy acid, $R'C(=O)OOH$) $\longrightarrow R_2C{-}CR_2$ (bridged by O) $+ R'COH$ ($R'C(=O)OH$) Alkene, Peroxy acid, Vicinal dihalide, Carboxylic acid 1-Methylcycloheptene $\xrightarrow{\text{peroxyacetic acid}}$ 1,2-Epoxy-1-methylcycloheptane (65%)

Section 8.7 Addition and elimination reactions are often reversible, and proceed spontaneously in the direction in which the free energy G decreases. The reaction is at equilibrium when $\Delta G = 0$. Free energy is related to enthalpy (H) and entropy (S) by the equations

$$G = H - TS \qquad \text{and} \qquad \Delta G = \Delta H - T\Delta S$$

The standard free energy change ΔG° is related to the equilibrium constant K by the equation

$$\Delta G^\circ = -RT \ln K$$

Sections 8.8–8.9 See Table 8.2. Hydroboration–oxidation is a synthetically useful and mechanistically novel method for converting alkenes to alcohols.

Section 8.10 See Table 8.2. Bromine and chlorine react with alkenes to give cyclic halonium ions, which react further to give vicinal dihalides. In aqueous solution, the product is a vicinal bromohydrin.

Section 8.11 See Table 8.2. Peroxy acids are a source of electrophilic oxygen and convert alkenes to epoxides.

Section 8.12 Alkenes are cleaved to carbonyl compounds by **ozonolysis.** This reaction is useful for both synthesis (preparation of aldehydes, ketones, or carboxylic acids) and analysis. When applied to analysis, the carbonyl compounds are isolated and identified, allowing the substituents attached to the double bond to be deduced.

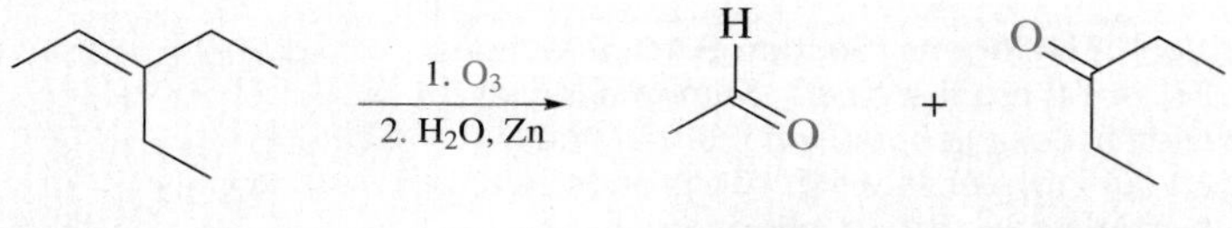

3-Ethyl-2-pentene $\xrightarrow[\text{2. } H_2O\text{, Zn}]{\text{1. } O_3}$ Acetaldehyde + 3-Pentanone

Section 8.13 Enzyme-catalyzed additions to alkenes occur enantioselectively and regioselectively. Steroid biosynthesis, for example, begins with the epoxidation of squalene according to the reaction shown.

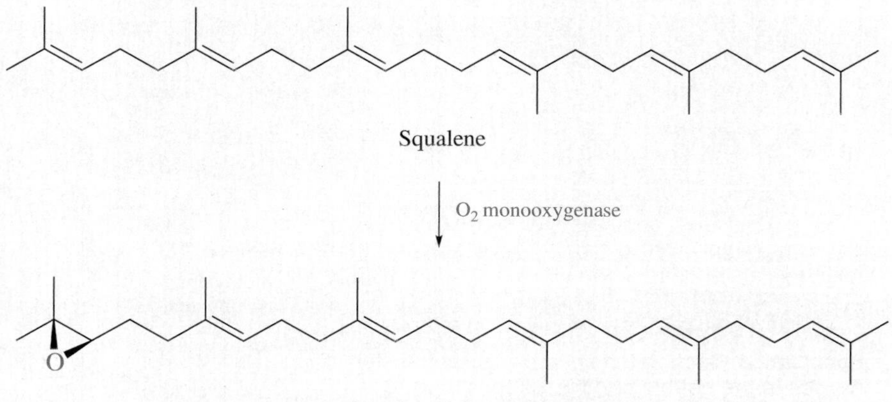

Squalene

O_2 monooxygenase

Squalene 2,3-epoxide

Section 8.14 Eliminations that lead to alkenes, followed by addition to the double bond, can constitute a key intermediate stage in synthesis. The synthesis of 3-bromo-2-butanol from 2-bromobutane, for example, can only be carried out by sacrificing the initial bromine and replacing it in a subsequent operation. The retrosynthesis:

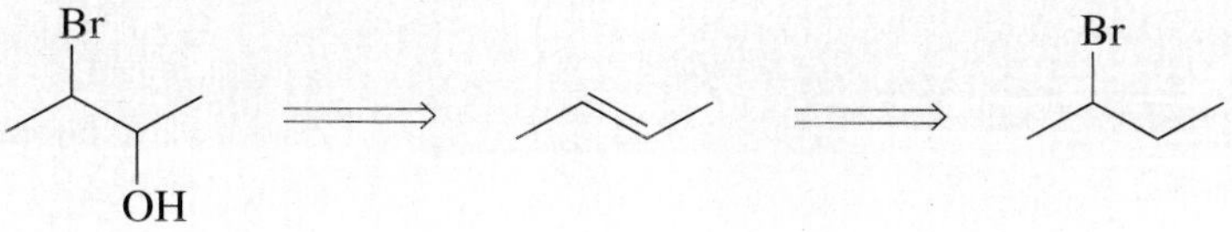

Br, OH ⟹ ⟹ Br

suggests the following:

Br — ethanol, $NaOCH_2CH_3$ → — Br_2, H_2O → Br, OH

2-Bromobutane *trans*-2-Butene 3-Bromo-2-butanol

PROBLEMS

Reactions of Alkenes

8.25 (a) How many alkenes yield 2,2,3,4,4-pentamethylpentane on catalytic hydrogenation?
(b) How many yield 2,3-dimethylbutane?
(c) How many yield methylcyclobutane?

8.26 Compound A undergoes catalytic hydrogenation much faster than does compound B. Why?

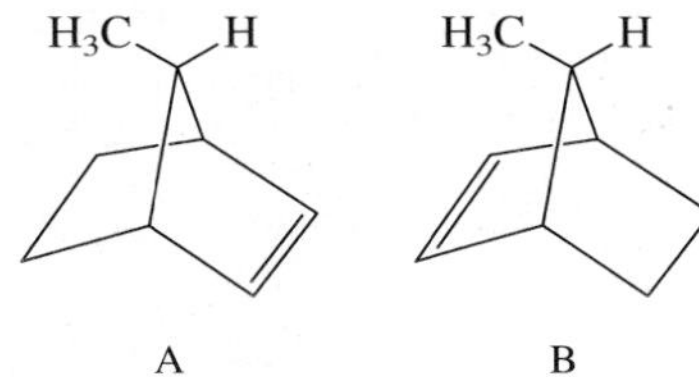

8.27 Catalytic hydrogenation of 1,4-dimethylcyclopentene yields a mixture of two products. Identify them. One of them is formed in much greater amounts than the other (observed ratio = 10:1). Which one is the major product?

8.28 Write the structure of the major organic product formed in the reaction of 1-pentene with each of the following:
(a) Hydrogen chloride
(b) Dilute sulfuric acid
(c) Diborane in diglyme, followed by basic hydrogen peroxide
(d) Bromine in carbon tetrachloride
(e) Bromine in water
(f) Peroxyacetic acid
(g) Ozone
(h) Product of part (g) treated with zinc and water
(i) Product of part (g) treated with dimethyl sulfide $(CH_3)_2S$

8.29 Repeat Problem 8.28 for 2-methyl-2-butene.

8.30 Repeat Problem 8.28 for 1-methylcyclohexene.

8.31 All the following reactions have been reported in the chemical literature. Give the structure of the principal organic product in each case.

(a) $\xrightarrow{HBr}$

(b) $\xrightarrow[2.\ H_2O_2,\ HO^-]{1.\ B_2H_6}$

(c) $\xrightarrow[2.\ H_2O_2,\ HO^-]{1.\ B_2H_6}$

(d) 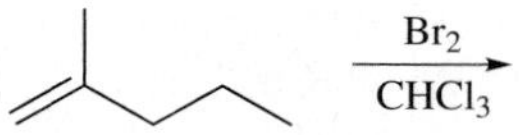 $\xrightarrow[CHCl_3]{Br_2}$

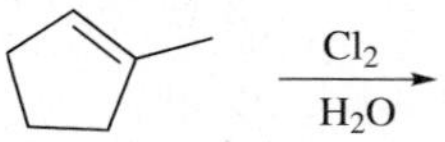

(e) 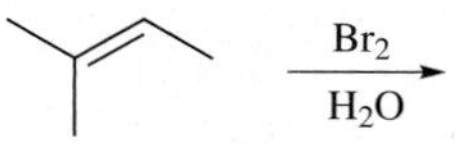 $\xrightarrow[H_2O]{Br_2}$

(f) $\xrightarrow[H_2O]{Cl_2}$

(g) $\xrightarrow{CH_3CO_2OH}$

(h) $\xrightarrow[2.\ H_2O]{1.\ O_3}$

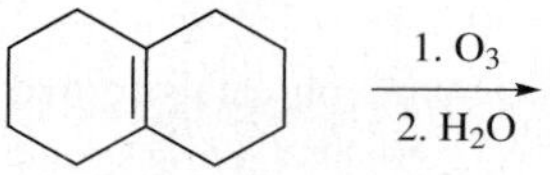

8.32 A single epoxide was isolated in 79–84% yield in the following reaction. Was this epoxide A or B? Explain your reasoning.

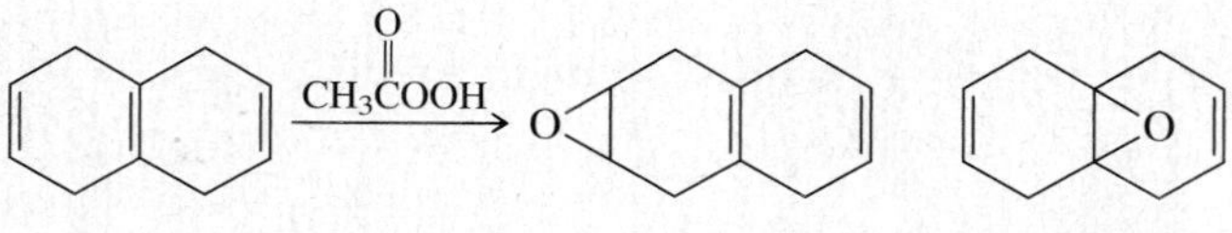

$\xrightarrow{CH_3COOH}$ (O over the carbonyl carbon)

A B

Stereochemistry

8.33 What two stereoisomeric alkanes are formed on catalytic hydrogenation of (*E*)-3-methyl-2-hexene? What are the relative amounts of each?

8.34 Two alkenes undergo hydrogenation to yield a mixture of *cis*- and *trans*-1,4-dimethylcyclohexane. Which two are these? A third, however, gives only *cis*-1,4-dimethylcyclohexane. What compound is this?

8.35 On catalytic hydrogenation over a rhodium catalyst, the compound shown gave a mixture containing *cis*-1-*tert*-butyl-4-methylcyclohexane (88%) and *trans*-1-*tert*-butyl-4-methylcyclohexane (12%). With this stereochemical result in mind, consider the reactions in (a) and (b).

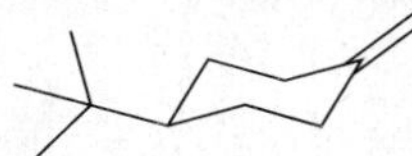

(a) What two products are formed in the epoxidation of this compound? Which one do you think will predominate?

(b) What two products are formed in the hydroboration–oxidation of this compound? Which one do you think will predominate?

8.36 Hydrogenation of 3-carene is, in principle, capable of yielding two stereoisomeric products. Write their structures. Only one of them was actually obtained on catalytic hydrogenation over platinum. Which one do you think is formed? Why?

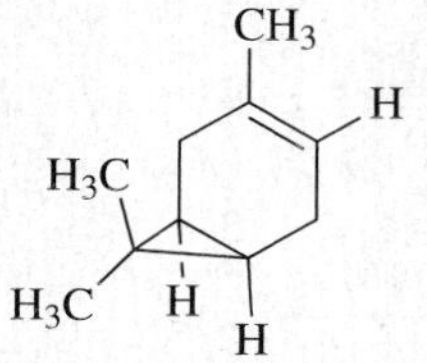

CH_3 H H_3C H_3C H H

8.37 When enantiopure 2,3-dimethyl-2-pentanol was subjected to dehydration, a mixture of two alkenes was obtained. Hydrogenation of this alkene mixture gave 2,3-dimethylpentane which was 50% optically pure. What were the two alkenes formed in the elimination reaction, and what were the relative amounts of each?

8.38 When (*R*)-3-buten-2-ol is treated with a peroxy acid, two stereoisomeric epoxides are formed in a 60:40 ratio. The minor stereoisomer has the structure shown.

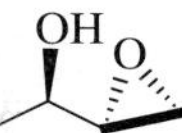

(a) What is the structure of the major isomer?

(b) What is the relationship between the two epoxides? Are they enantiomers or diastereomers?

(c) What four stereoisomeric products are formed when racemic 3-buten-2-ol is epoxidized under the same conditions? How much of each stereoisomer is formed?

8.39 Consider the ozonolysis of *trans*-4,5-dimethylcyclohexene having the configuration shown.

H_3C
H_3C

Structures A, B, and C are Fischer projections of three stereoisomeric forms of the reaction product.

A	B	C
CH=O	CH=O	CH=O
H—H	H—H	H—H
H_3C—H	H—CH_3	H—CH_3
H—CH_3	H_3C—H	H—CH_3
H—H	H—H	H—H
CH=O	CH=O	CH=O

(a) Which, if any, of the compounds A, B, and C are chiral?

(b) What product is formed in the reaction?

(c) What product would be formed if the methyl groups were cis to each other in the starting alkene?

Thermochemistry

8.40 1-Butene has a higher heat of hydrogenation than 2,3-dimethyl-2-butene. Which has the higher heat of combustion? Explain.

8.41 Match the following alkenes with the appropriate heats of hydrogenation:

A B C D E

Heats of hydrogenation in kJ/mol (kcal/mol): 151 (36.2); 122 (29.3); 114 (27.3); 111 (26.5); 105 (25.1).

8.42 The heats of reaction were measured for addition of HBr to *cis*- and *trans*-2-butene.

$$CH_3CH{=}CHCH_3 + HBr \longrightarrow CH_3CH_2CHCH_3 \text{ (Br on C-2)}$$

cis-2-butene: $\Delta H° = -77$ kJ/mol (-18.4 kcal/mol)

trans-2-butene: $\Delta H° = -72$ kJ/mol (-17.3 kcal/mol)

Use these data to calculate the energy difference between *cis*- and *trans*-2-butene. How does this energy difference compare to that based on heats of hydrogenation (Section 8.3) and heats of combustion (Section 7.6)?

8.43 Complete the following table by adding + and − signs to the $\Delta H°$ and $\Delta S°$ columns so as to correspond to the effect of temperature on a reversible reaction.

	Sign of	
Reaction is	**$\Delta H°$**	**$\Delta S°$**
(a) Exergonic at all temperatures		
(b) Exergonic at low temperature; endergonic at high temperature		
(c) Endergonic at all temperatures		
(d) Endergonic at low temperature; exergonic at high temperature		

8.44 Match the heats of hydrogenation—107 kJ/mol (25.6 kcal/mol), 114.5 kJ/mol (27.3 kcal/mol), and 119 kJ/mol (28.4 kcal/mol)—with the appropriate C_7H_{12} isomer.

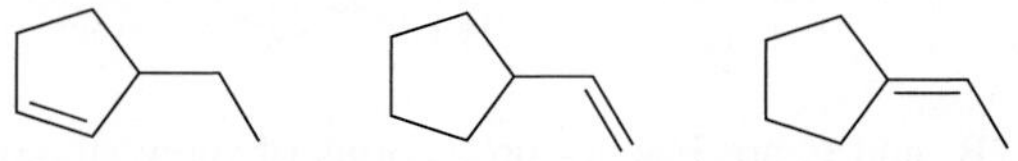

8.45 The iodination of ethylene at 25°C is characterized by the thermodynamic values shown.

$$H_2C{=}CH_2(g) + I_2(g) \rightleftharpoons ICH_2CH_2I(g)$$

$$\Delta H° = -48 \text{ kJ/mol } (11.5 \text{ kcal/mol}); \ \Delta S° = -0.13 \text{ kJ/K·mol } (0.31 \text{ kcal/K·mol})$$

(a) Calculate $\Delta G°$ and K at 25°C.
(b) Is the reaction exergonic or endergonic at 25°C?
(c) What happens to K as the temperature is raised?

Synthesis

8.46 Specify reagents suitable for converting 3-ethyl-2-pentene to each of the following:
(a) 2,3-Dibromo-3-ethylpentane
(b) 3-Chloro-3-ethylpentane
(c) 3-Ethyl-3-pentanol
(d) 3-Ethyl-2-pentanol
(e) 2,3-Epoxy-3-ethylpentane
(f) 3-Ethylpentane

8.47 (a) Which primary alcohol of molecular formula $C_5H_{12}O$ cannot be prepared from an alkene by hydroboration–oxidation? Why?
(b) Write equations describing the preparation of three isomeric primary alcohols of molecular formula $C_5H_{12}O$ from alkenes.
(c) Write equations describing the preparation of the tertiary alcohol of molecular formula $C_5H_{12}O$ by acid-catalyzed hydration of two different alkenes.

8.48 Identify compound A in the retrosynthesis shown and use this information to design a synthesis of the target molecule showing all necessary reagents.

8.49 Identify compounds A and B in the retrosynthesis shown and use this information to design a synthesis of the desired nitrile showing all necessary reagents.

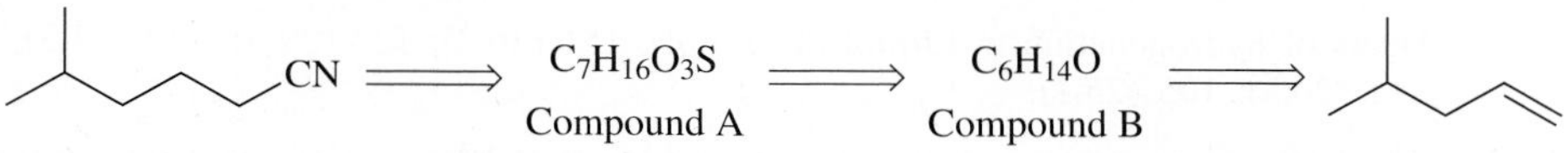

8.50 Apply retrosynthetic analysis to guide the preparation of each of the following compounds from the indicated starting material, then write out the synthesis showing the necessary reagents.
(a) 1-Propanol from 2-propanol
(b) 1,2-Dibromopropane from 2-bromopropane
(c) 1-Bromo-2-propanol from 2-propanol
(d) 1-Bromo-2-methyl-2-propanol from *tert*-butyl bromide

(e) 1,2-Epoxypropane from 2-propanol
(f) *tert*-Butyl alcohol from isobutyl alcohol
(g) *tert*-Butyl iodide from isobutyl iodide
(h) *trans*-2-Chlorocyclohexanol from cyclohexyl chloride

Structure Determination

8.51 On being heated with a solution of sodium ethoxide in ethanol, compound A ($C_7H_{15}Br$) yielded a mixture of two alkenes B and C, each having the molecular formula C_7H_{14}. Catalytic hydrogenation of the major isomer B or the minor isomer C gave only 3-ethylpentane. Suggest structures for compounds A, B, and C consistent with these observations.

8.52 Compound A ($C_7H_{15}Br$) is not a primary alkyl bromide. It yields a single alkene (compound B) on being heated with sodium ethoxide in ethanol. Hydrogenation of compound B yields 2,4-dimethylpentane. Identify compounds A and B.

8.53 Compounds A and B are isomers of molecular formula $C_9H_{19}Br$. Both yield the same alkene C as the exclusive product of elimination on being treated with potassium *tert*-butoxide in dimethyl sulfoxide. Hydrogenation of alkene C gives 2,3,3,4-tetramethylpentane. What are the structures of compounds A and B and alkene C?

8.54 Alcohol A ($C_{10}H_{18}O$) is converted to a mixture of alkenes B and C on being heated with potassium hydrogen sulfate ($KHSO_4$). Catalytic hydrogenation of B and C yields the same product. Assuming that dehydration of alcohol A proceeds without rearrangement, deduce the structures of alcohol A and alkene C.

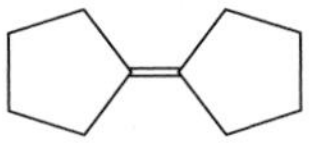

Compound B

8.55 A mixture of three alkenes (A, B, and C) was obtained by dehydration of 1,2-dimethylcyclohexanol. The composition of the mixture was A (3%), B (31%), and C (66%). Catalytic hydrogenation of A, B, or C gave 1,2-dimethylcyclohexane. The three alkenes can be equilibrated by heating with sulfuric acid to give a mixture containing A (0%), B (15%), and C (85%). Identify A, B, and C.

8.56 Reaction of 3,3-dimethyl-1-butene with hydrogen iodide yields two compounds A and B, each having the molecular formula $C_6H_{13}I$, in the ratio A:B = 90:10. Compound A, on being heated with potassium hydroxide in *n*-propyl alcohol, gives only 3,3-dimethyl-1-butene. Compound B undergoes elimination under these conditions to give 2,3-dimethyl-2-butene as the major product. Suggest structures for compounds A and B, and write a reasonable mechanism for the formation of each.

8.57 Dehydration of 2,2,3,4,4-pentamethyl-3-pentanol gave two alkenes A and B. Ozonolysis of the lower boiling alkene A gave formaldehyde ($H_2C{=}O$) and 2,2,4,4-tetramethyl-3-pentanone. Ozonolysis of B gave formaldehyde and 3,3,4,4-tetramethyl-2-pentanone. Identify A and B, and suggest an explanation for the formation of B in the dehydration reaction.

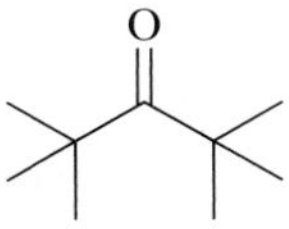

2,2,4,4-Tetramethyl-3-pentanone 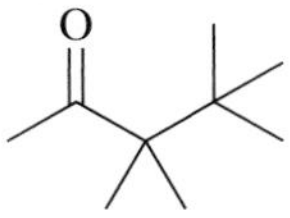 3,3,4,4-Tetramethyl-2-pentanone

8.58 Compound A ($C_7H_{13}Br$) is a tertiary bromide. On treatment with sodium ethoxide in ethanol, A is converted to B (C_7H_{12}). Ozonolysis of B gives the compound shown as the only product. Deduce the structures of A and B. What is the symbol for the reaction mechanism by which A is converted to B under the reaction conditions?

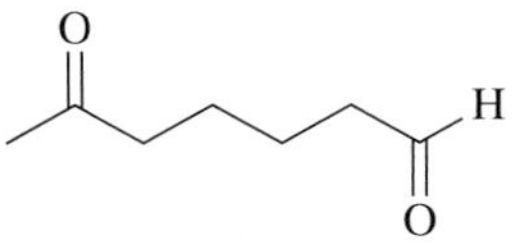

8.59 East Indian sandalwood oil contains a hydrocarbon given the name *santene* (C_9H_{14}). Ozonolysis of santene gives the compound shown. What is the structure of santene?

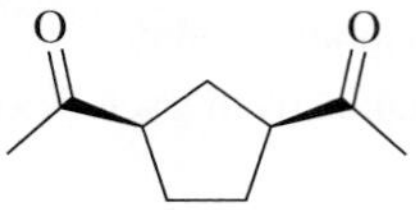

8.60 *Sabinene* and *Δ³-carene* are isomeric natural products with the molecular formula $C_{10}H_{16}$. (a) Ozonolysis of sabinene followed by hydrolysis in the presence of zinc gives compound A. What is the structure of sabinene? What other compound is formed on ozonolysis? (b) Ozonolysis of Δ^3-carene gives compound B. What is the structure of Δ^3-carene?

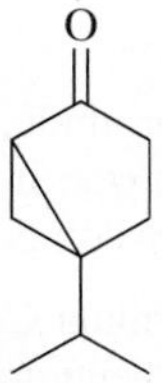

Compound A  Compound B

8.61 The sex attractant by which the female housefly attracts the male has the molecular formula $C_{23}H_{46}$. Catalytic hydrogenation yields an alkane of molecular formula $C_{23}H_{48}$. Ozonolysis yields

$$CH_3(CH_2)_7\overset{O}{\overset{\|}{C}}H \quad \text{and} \quad CH_3(CH_2)_{12}\overset{O}{\overset{\|}{C}}H$$

What is the structure of the housefly sex attractant?

8.62 A certain compound of molecular formula $C_{19}H_{38}$ was isolated from fish oil and from plankton. On hydrogenation it gave 2,6,10,14-tetramethylpentadecane. Ozonolysis gave $(CH_3)_2C{=}O$ and a 16-carbon aldehyde. What is the structure of the natural product? What is the structure of the aldehyde?

8.63 The sex attractant of the female arctiid moth contains, among other components, a compound of molecular formula $C_{21}H_{40}$ that yields

$$CH_3(CH_2)_{10}\overset{O}{\overset{\|}{C}}H \quad CH_3(CH_2)_4\overset{O}{\overset{\|}{C}}H \quad \text{and} \quad H\overset{O}{\overset{\|}{C}}CH_2\overset{O}{\overset{\|}{C}}H$$

on ozonolysis. What is the constitution of this material?

Mechanism

8.64 Suggest reasonable mechanisms for each of the following reactions. Use curved arrows to show electron flow.

(a) 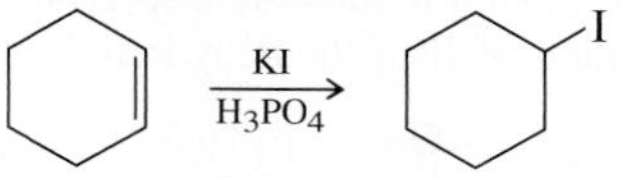 cyclohexene $\xrightarrow[H_3PO_4]{KI}$ iodocyclohexane

(b) 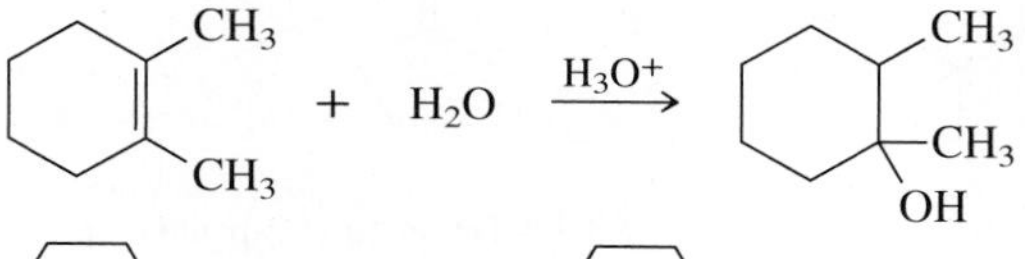 1,2-dimethylcyclohexene (CH₃, CH₃) + H_2O $\xrightarrow{H_3O^+}$ 1,2-dimethylcyclohexanol (CH₃, CH₃, OH) (mixture of stereoisomers)

(c) 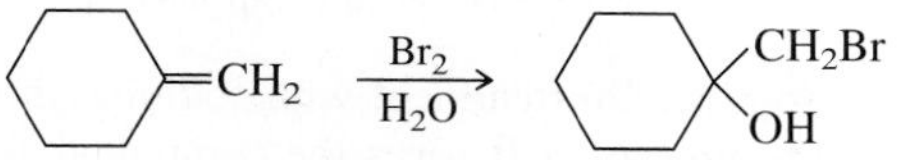 methylenecyclohexane (=CH₂) $\xrightarrow[H_2O]{Br_2}$ 1-(bromomethyl)cyclohexanol (CH₂Br, OH)

8.65 On the basis of the mechanism of acid-catalyzed hydration, can you suggest why the reaction

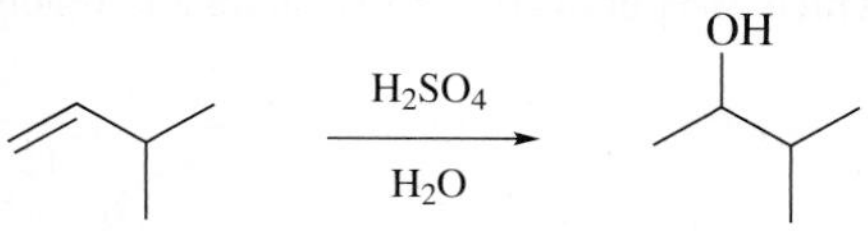

3-methyl-1-butene $\xrightarrow[H_2O]{H_2SO_4}$ 3-methyl-2-butanol (OH)

would probably *not* be a good method for the synthesis of 3-methyl-2-butanol?

8.66 As a method for the preparation of alkenes, a weakness in the acid-catalyzed dehydration of alcohols is that the initially formed alkene (or mixture of alkenes) sometimes isomerizes under the conditions of its formation. Write a stepwise mechanism for the reaction:

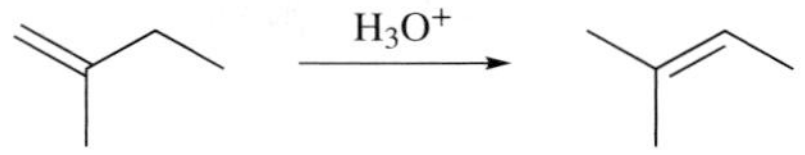

H_3O^+

8.67 Which of the following is the most reasonable structure for the product of the reaction of 4-*tert*-butyl-1-methylcyclohexene with bromine in methanol? Explain your reasoning.

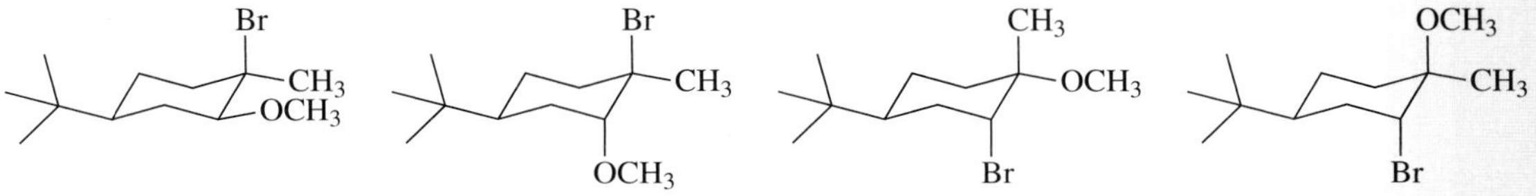

A: Br, CH_3, OCH_3 B: Br, CH_3, OCH_3 C: CH_3, OCH_3, Br D: OCH_3, CH_3, Br

8.68 The following reaction was performed as part of a research program sponsored by the National Institutes of Health to develop therapeutic agents for the treatment of cocaine addiction. Using what you have seen about the reactions of halogens with alkenes, propose a mechanism for this process.

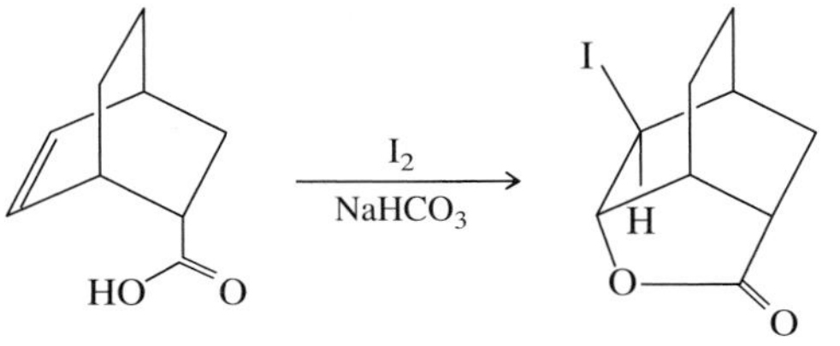

HO O — I_2, $NaHCO_3$ → I, H, O, O

8.69 How could you prepare *cis*-2-methylcyclopentyl acetate from 1-methylcyclopentanol?

Descriptive Passage and Interpretive Problems 8

Oxymercuration

Concerns about mercury's toxicity have led to decreased use of mercury-based reagents in synthetic organic chemistry. Alternatives exist for many of the transformations formerly carried out with mercury compounds while carrying much less risk. The chemistry of several of the reactions, however, is sufficiently interesting to examine here.

Among the synthetically useful reactions of Hg(II) salts with organic compounds, the most familiar is a two-stage procedure for alkene hydration called **oxymercuration–demercuration.** Its application in the conversion of 3,3-dimethyl-1-butene to 3,3-dimethyl-2-butanol illustrates the procedure.

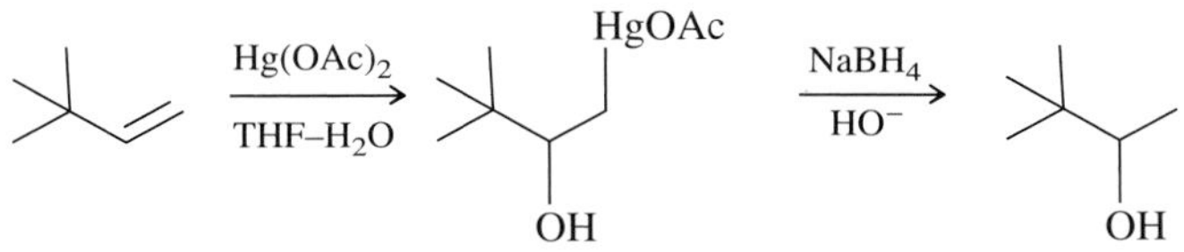

$Hg(OAc)_2$, THF–H_2O → HgOAc, OH → $NaBH_4$, HO^- → OH

Oxymercuration stage Demercuration stage

The reaction is performed in two operations, the first of which is oxymercuration. In this stage the alkene is treated with mercury(II) acetate [$Hg(O_2CCH_3)_2$, abbreviated as $Hg(OAc)_2$]. Mercury(II) acetate is a source of the electrophile ^+HgOAc, which bonds to C-1 of the alkene. The oxygen of water, one of the components in the THF–H_2O solvent mixture, bonds to C-2. The demercuration operation uses sodium borohydride ($NaBH_4$, a reducing agent) to convert C—Hg to C—H.

From the overall reaction, we see that oxymercuration–demercuration

1. accomplishes hydration of the double bond in accordance with Markovnikov's rule, and
2. carbocation rearrangements do not occur.

Additional information from stereochemical studies with other alkenes has established that

3. anti addition of HgOAc and OH characterizes the oxymercuration stage, and
4. the replacement of HgOAc by H in the demercuration stage is not stereospecific.

The structure of the intermediate in oxymercuration has received much attention and can be approached by considering what is likely to happen when the electrophile $^+$HgOAc reacts with the double bond of an alkene.

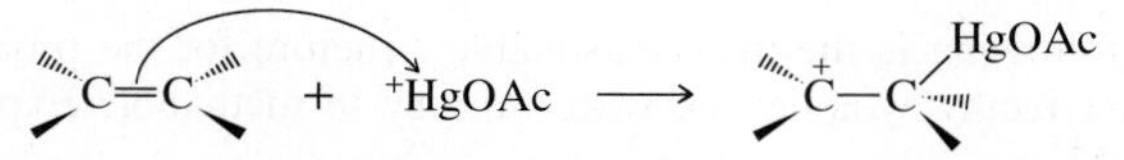

C=C + $^+$HgOAc ⟶ $\overset{+}{C}$—C—HgOAc

Recall from Section 5.9 that electrons in bonds that are β to a positively charged carbon stabilize a carbocation by hyperconjugation.

$\overset{+}{C}$—C(α)—H(β) $\overset{+}{C}$—C(α)—CH$_3$(β) $\overset{+}{C}$—C(α)—HgOAc(β)

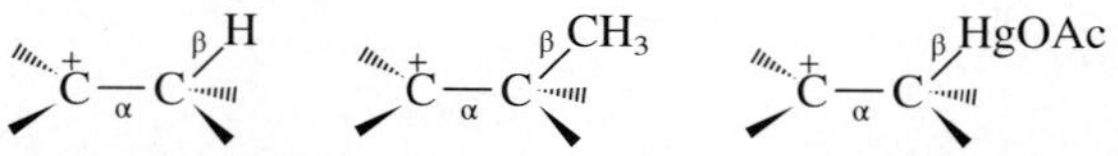

The electrons in a C—Hg σ bond are more loosely held than C—H or C—C electrons, making stabilization by hyperconjugation more effective for β-C—Hg than for β-C—H or β-C—C. Hyperconjugative stabilization of the intermediate in oxymercuration is normally shown using dashed lines to represent partial bonds. The intermediate is referred to as a "bridged" *mercurinium ion.*

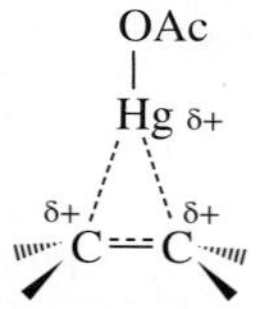

OAc—Hg δ+ bridged to δ+C==Cδ+

The problems that follow explore various synthetic aspects of oxymercuration–demercuration. Experimental procedures sometimes vary depending on the particular transformation. The source of the electrophile may be a mercury(II) salt other than $Hg(OAc)_2$, the nucleophile may be other than H_2O, and the reaction may be intramolecular rather than intermolecular.

8.70 Oxymercuration–demercuration of 1-methylcyclopentene gives which of the following products?

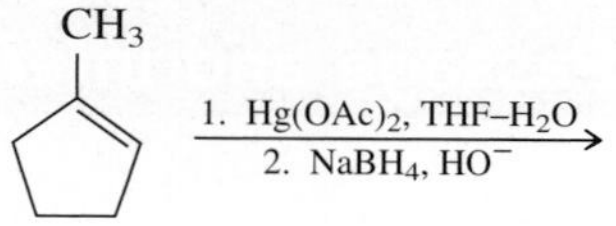

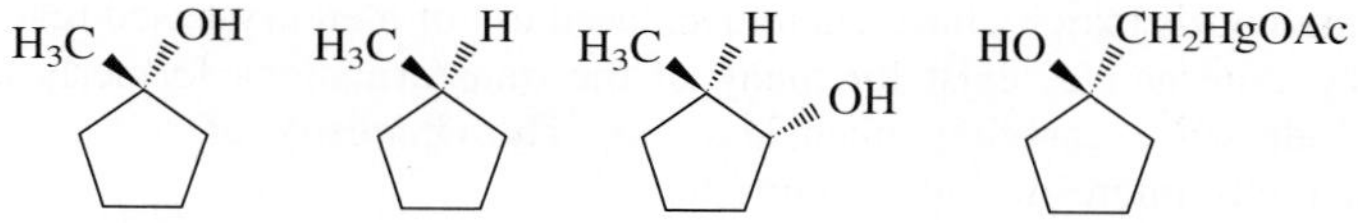

1-methylcyclopentene (CH$_3$) → 1. $Hg(OAc)_2$, THF–H_2O; 2. $NaBH_4$, HO^-

A. H$_3$C, OH B. H$_3$C, H C. H$_3$C, H, OH D. HO, CH$_2$HgOAc

8.71 Which alkene would be expected to give the following alcohol by oxymercuration–demercuration?

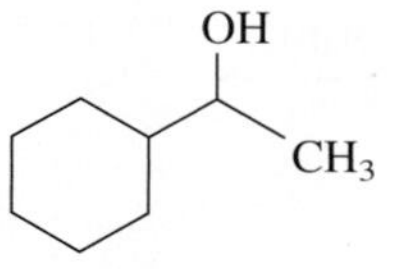

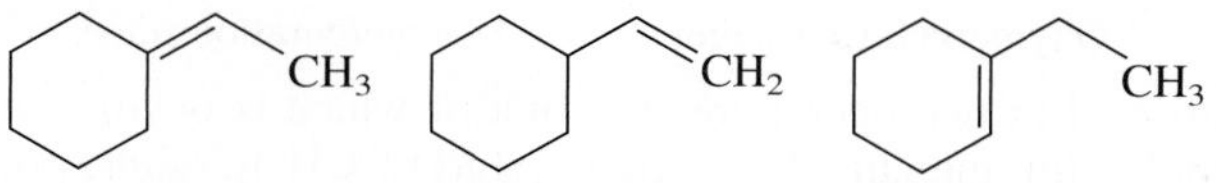

OH, CH$_3$ (1-cyclohexylethanol)

A. CH$_3$ B. CH$_2$ C. CH$_3$

8.72 Given that 2-methyl-1-pentene undergoes oxymercuration–demercuration approximately 35 times faster than 2-methyl-2-pentene, predict the major product from oxymercuration–demercuration of limonene.

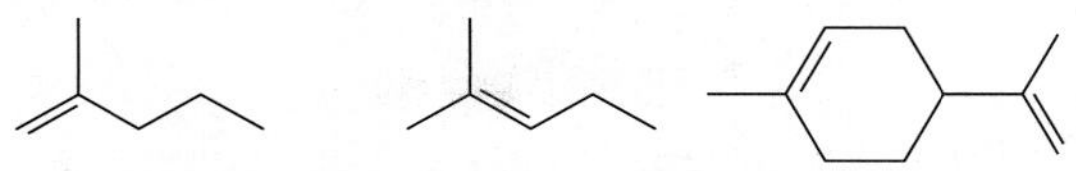

2-Methyl-1-pentene 2-Methyl-2-pentene Limonene

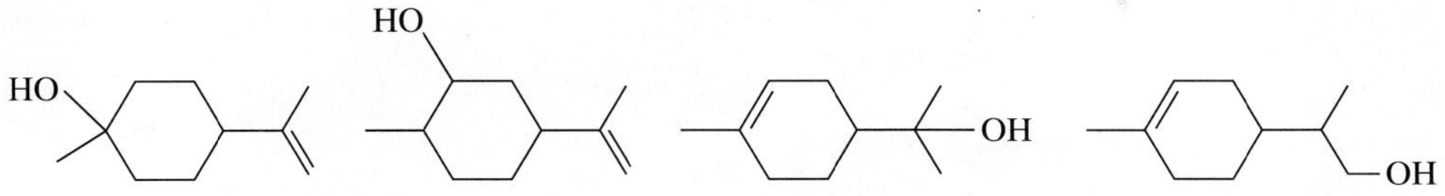

A. B. C. D.

8.73 In a procedure called solvomercuration–demercuration an alkene is treated with $Hg(OAc)_2$ or $Hg(OCCF_3)_2$ (with C=O) in an alcohol solvent rather than in the THF–H_2O mixture used in oxymercuration. The oxygen of the alcohol solvent reacts with the mercurinium ion during solvomercuration. What is the product of the following solvomercuration–demercuration?

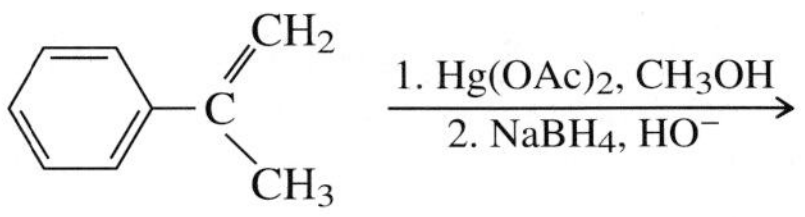

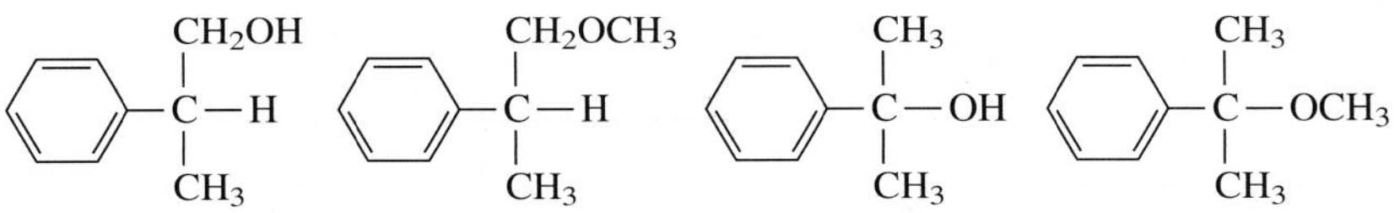

$C_6H_5C(=CH_2)CH_3$ → 1. $Hg(OAc)_2$, CH_3OH; 2. $NaBH_4$, HO^-

A. $C_6H_5CH(CH_2OH)CH_3$
B. $C_6H_5CH(CH_2OCH_3)CH_3$
C. $C_6H_5C(OH)(CH_3)_2$
D. $C_6H_5C(OCH_3)(CH_3)_2$

8.74 From among the same product choice as Problem 8.73, which one is the major product of the following reaction?

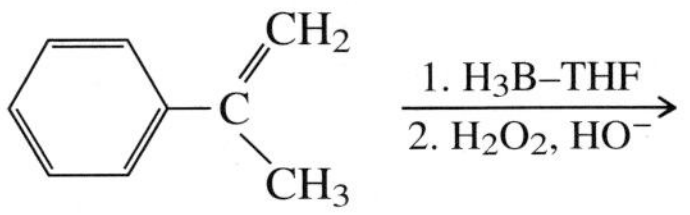

$C_6H_5C(=CH_2)CH_3$ → 1. H_3B–THF; 2. H_2O_2, HO^-

8.75 Oxymercuration–demercuration of allyl alcohol gives 1,2-propanediol.

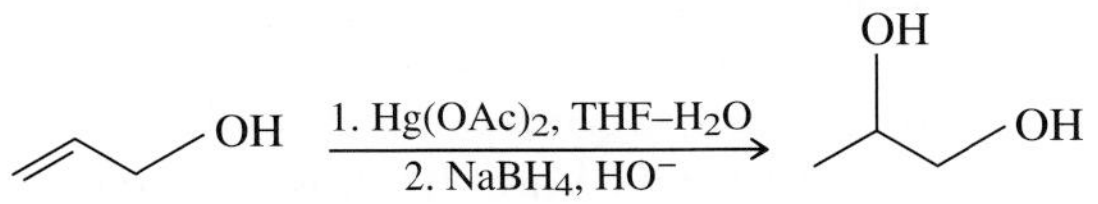

allyl alcohol → 1. $Hg(OAc)_2$, THF–H_2O; 2. $NaBH_4$, HO^- → 1,2-propanediol

Under the same conditions, however, 4-penten-1-ol yields a compound having the molecular formula $C_5H_{10}O$.

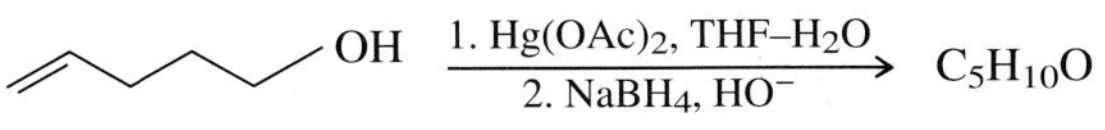

4-penten-1-ol → 1. $Hg(OAc)_2$, THF–H_2O; 2. $NaBH_4$, HO^- → $C_5H_{10}O$

What is the most reasonable structure for the product of this reaction?

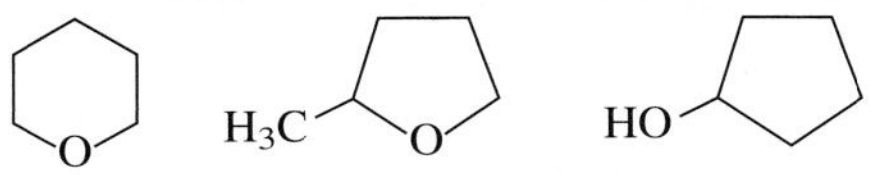

A. (tetrahydropyran) B. (H_3C-substituted tetrahydrofuran) C. (HO-cyclopentane)

CHAPTER OUTLINE

The brightly colored poison dart frogs of Central and South America store toxic substances such as the acetylenic alkaloid histrionicotoxin within their bodies to deter attacks by other animals.

Alkynes

Hydrocarbons that contain a carbon–carbon triple bond are called **alkynes.** Noncyclic alkynes have the molecular formula C_nH_{2n-2}. *Acetylene* (HC≡CH) is the simplest alkyne. We call compounds that have their triple bond at the end of a carbon chain (RC≡CH) *monosubstituted,* or **terminal, alkynes.** Disubstituted alkynes (RC≡CR′) have *internal* triple bonds. You will see in this chapter that a carbon–carbon triple bond is a functional group, reacting with many of the same reagents that react with the double bonds of alkenes.

The most distinctive aspect of the chemistry of acetylene and terminal alkynes is their acidity. As a class, compounds of the type RC≡CH are the most acidic of all hydrocarbons. The structural reasons for this property, as well as the ways in which it is used to advantage in chemical synthesis, are important elements of this chapter.

9.1 Sources of Alkynes

Acetylene was discovered in 1836 but did not command much attention until its large-scale preparation from calcium carbide near the end of the nineteenth century stimulated interest in industrial applications. In the first stage of that synthesis, limestone and coke, a material rich in elemental carbon obtained from coal, are heated in an electric furnace to form calcium carbide.

This reaction was accidentally discovered in 1892 by the Canadian inventor Thomas L. Willson while looking for a method to make aluminum.

$$\underset{\substack{\text{Calcium oxide} \\ \text{(from limestone)}}}{CaO} + \underset{\substack{\text{Carbon} \\ \text{(from coke)}}}{3C} \xrightarrow{1800\text{–}2100°C} \underset{\text{Calcium carbide}}{CaC_2} + \underset{\text{Carbon monoxide}}{CO}$$

Calcium carbide is the calcium salt of the doubly negative carbide ion ($:\bar{C}{\equiv}\bar{C}:$). Carbide ion is strongly basic and reacts with water to form acetylene:

$$\underset{\text{Calcium carbide}}{Ca^{2+}\left[:C{\equiv}C:\right]^{2-}} + \underset{\text{Water}}{2H_2O} \longrightarrow \underset{\text{Calcium hydroxide}}{Ca(OH)_2} + \underset{\text{Acetylene}}{HC{\equiv}CH}$$

Problem 9.1

Use curved arrows to show how calcium carbide reacts with water to give acetylene.

Beginning in the mid-twentieth century, alternative methods of acetylene production became practical. One is the dehydrogenation of ethylene.

$$\underset{\text{Ethylene}}{H_2C{=}CH_2} \overset{\text{heat}}{\rightleftharpoons} \underset{\text{Acetylene}}{HC{\equiv}CH} + \underset{\text{Hydrogen}}{H_2}$$

The reaction is endothermic, and the equilibrium favors ethylene at low temperatures but shifts to favor acetylene above 1150°C. Indeed, at very high temperatures most hydrocarbons, even methane, are converted to acetylene. Acetylene has value not only by itself but also as a starting material from which higher alkynes are prepared.

More than 1000 natural products contain carbon–carbon triple bonds. Many, such as stearolic acid and tariric acid, are **fatty acids**—carboxylic acids with unbranched chains of 12–20 carbon atoms—or are derived from them.

$$\underset{\text{Stearolic acid}}{CH_3(CH_2)_7C{\equiv}C(CH_2)_7\overset{O}{\overset{\|}{C}}OH} \qquad \underset{\text{Tariric acid}}{CH_3(CH_2)_{10}C{\equiv}C(CH_2)_4\overset{O}{\overset{\|}{C}}OH}$$

A major biosynthetic route to acetylenic fatty acids in certain flowering plants involves oxidation of analogous compounds with carbon–carbon double bonds, and is catalyzed by enzymes of the *desaturase* class known as *acetyleneases*. Crepenynic acid,

O
OH

Crepenynic acid

which is formed by oxidation of the fatty acid linoleic acid, is one example.

O
OH

Linoleic acid

Cultures of the bacterium *Micromonospora chersina* produce dynemicin A, a purple substance characterized by a novel structure containing a double bond and two triple bonds in a ten-membered ring (an *enediyne*). Dynemicin A has attracted interest because of its ability to cleave DNA by a novel mechanism, which may lead to the development of anticancer drugs that are based on the enediyne structure.

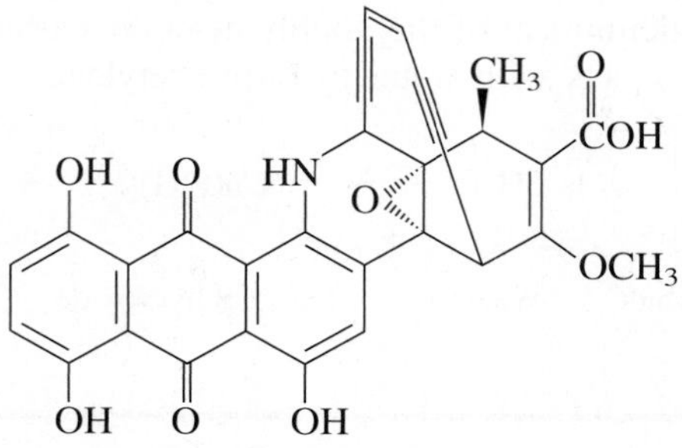

Dynemicin A

Diacetylene ($HC{\equiv}C{-}C{\equiv}CH$) has been identified as a component of the hydrocarbon-rich atmospheres of Uranus, Neptune, and Pluto. It is also present in the atmospheres of Titan and Triton, satellites of Saturn and Neptune, respectively. Most surprisingly, 2009 brought the report of the discovery of diacetylene on our own moon.

9.2 Nomenclature

In naming alkynes the usual IUPAC rules for hydrocarbons are followed, and the suffix *-ane* is replaced by *-yne*. Both acetylene and ethyne are acceptable IUPAC names for $HC{\equiv}CH$. The position of the triple bond along the chain is specified by number in a manner analogous to alkene nomenclature.

$HC{\equiv}CCH_3$	$HC{\equiv}CCH_2CH_3$	$CH_3C{\equiv}CCH_3$	$(CH_3)_3CC{\equiv}CCH_3$
Propyne or Prop-1-yne	1-Butyne or But-1-yne	2-Butyne or But-2-yne	4,4-Dimethyl-2-pentyne or 4,4-Dimethylpent-2-yne

Problem 9.2

Write structural formulas and give the IUPAC names for all the alkynes of molecular formula C_5H_8.

If a compound contains both a double bond and a triple bond, the chain is numbered so as to give the first multiple bond the lowest number, irrespective of whether it is a double bond or a triple bond. Ties are broken in favor of the double bond. An *en* suffix for the double bond precedes *yne* and is separated from it by the *yne* locant. Thus, the compound vinylacetylene $H_2C{=}CH{-}C{\equiv}CH$ is named but-1-en-3-yne according to the latest IUPAC rules.

Vinylacetylene is an industrial chemical used in the preparation of neoprene.

When the $-C{\equiv}CH$ group is named as a substituent, it is designated as an *ethynyl* group.

9.3 Physical Properties of Alkynes

Alkynes resemble alkanes and alkenes in their physical properties. They share with these other hydrocarbons the properties of low density and low water-solubility and have boiling points similar to those of alkanes.

9.4 Structure and Bonding in Alkynes: *sp* Hybridization

Acetylene is linear, with a carbon–carbon bond distance of 120 pm (1.20 Å) and carbon–hydrogen bond distances of 106 pm (1.06 Å).

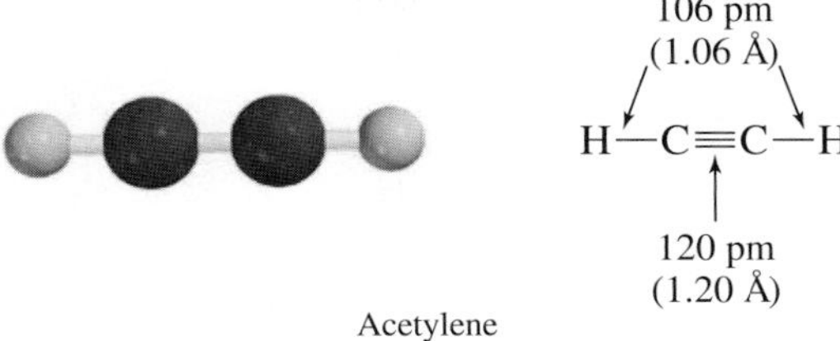

Acetylene

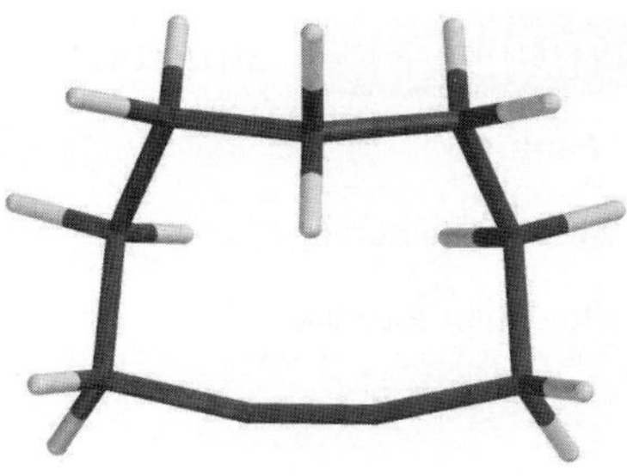

Figure 9.1

Molecular model of cyclononyne showing bending of the bond angles associated with the triply bonded carbons. This model closely matches the structure determined experimentally. Notice how the staggering of bonds on adjacent atoms governs the overall shape of the ring.

Linear geometries characterize the H—C≡C—C and C—C≡C—C units of terminal and internal triple bonds, respectively, as well. This linear geometry is responsible for the relatively small number of known *cycloalkynes*. Figure 9.1 shows a molecular model of cyclononyne in which the bending of the C—C≡C—C unit is clearly evident. Angle strain destabilizes cycloalkynes to the extent that cyclononyne is the smallest one that is stable enough to be stored for long periods. The next smaller one, cyclooctyne, has been isolated, but is relatively reactive and polymerizes on standing.

An *sp* hybridization model for the carbon–carbon triple bond was developed in Section 2.9 and is reviewed for acetylene in Figure 9.2. Figure 9.3 compares the electrostatic potential maps of ethylene and acetylene and shows how the two π bonds in acetylene cause a band of high electron density to encircle the molecule.

Table 9.1 compares some structural features of alkanes, alkenes, and alkynes. As we progress through the series in the order ethane → ethylene → acetylene:

1. The geometry at carbon changes from tetrahedral → trigonal planar → linear.
2. The C—C and C—H bonds become shorter and stronger.
3. The acidity of the C—H bonds increases.

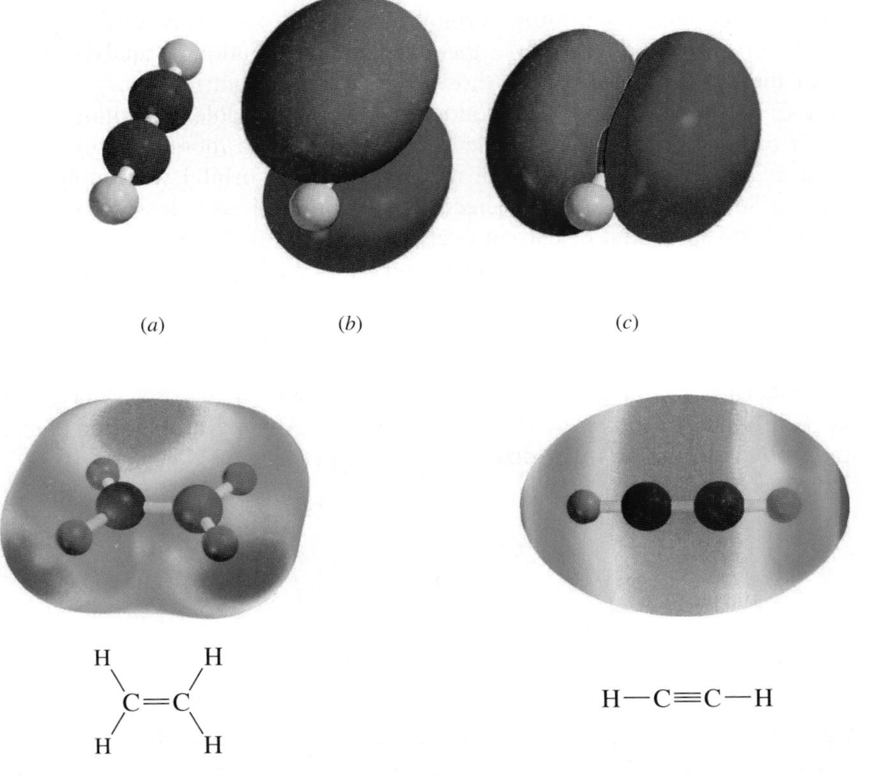

Figure 9.2

The carbon atoms of acetylene are connected by a σ + π + π triple bond. (*a*) Both carbon atoms are *sp*-hybridized, and each is bonded to a hydrogen by a σ bond. The two π bonds are perpendicular to each other and are shown separately in (*b*) and (*c*).

Ethylene

Acetylene

Figure 9.3

Electrostatic potential maps of ethylene and acetylene. The region of highest negative charge (red) is associated with the π bonds and lies between the two carbons in both. This electron-rich region is above and below the plane of the molecule in ethylene. Because acetylene has two π bonds, a band of high electron density encircles the molecule.

TABLE 9.1 Structural Features of Ethane, Ethylene, and Acetylene

Feature	Ethane	Ethylene	Acetylene
Systematic name	Ethane	Ethene	Ethyne
Molecular formula	C_2H_6	C_2H_4	C_2H_2
Structural formula	$H_3C{-}CH_3$	$H_2C{=}CH_2$	$H{-}C{\equiv}C{-}H$
C—C bond distance, pm (Å)	153 (1.53)	134 (1.34)	120 (1.20)
C—H bond distance, pm (Å)	111 (1.11)	110 (1.10)	106 (1.06)
H—C—C bond angles	111.0°	121.4°	180°
C—C bond dissociation enthalpy kJ/mol (kcal/mol)	375 (90)	720 (172)	961 (230)
C—H bond dissociation enthalpy, kJ/mol (kcal/mol)	421 (100.5)	464 (111)	547 (131)
Hybridization of carbon	sp^3	sp^2	sp
s character in C—H bonds	25%	33%	50%
Approximate pK_a	62	45	26

All of these trends can be accommodated by the orbital hybridization model. The bond angles are characteristic for the sp^3, sp^2, and sp hybridization states of carbon and don't require additional comment. The bond distances, bond strengths, and acidities are related to the *s* character in the orbitals used for bonding. *s* Character is the fraction of the hybrid orbital contributed by an *s* orbital. Thus, an sp^3 orbital has one quarter *s* character and three quarters *p,* an sp^2 orbital has one third *s* and two thirds *p,* and an *sp* orbital one half *s* and one half *p.* We then use this information to analyze how various qualities of the hybrid orbital reflect those of its *s* and *p* contributors.

Take C—H bond distance and bond strength, for example. Recalling that an electron in a 2*s* orbital is, on average, closer to the nucleus and more strongly held than an electron in a 2*p* orbital, it follows that an electron in an orbital with more *s* character will be more strongly held than an electron in an orbital with less *s* character. Thus, when a half-filled *sp* orbital of carbon overlaps with a half-filled hydrogen 1*s* orbital to give a C—H σ bond, the bond is stronger and shorter than one between hydrogen and sp^2-hybridized carbon. Similar reasoning holds for the shorter C—C bond distance of acetylene compared with ethylene, although here the additional π bond in acetylene is also a factor.

The hybridization model for bonding in acetylene is depicted in Figure 2.17.

The pattern is repeated in higher alkynes as shown when comparing propyne and propene. The bonds to the *sp*-hybridized carbons of propyne are shorter than the corresponding bonds to the sp^2-hybridized carbons of propene.

$H{-}C{\equiv}C{-}CH_3$: H—C 106 pm (1.06 Å); C≡C 121 pm (1.21 Å); C—CH₃ 146 pm (1.46 Å)

Propyne

$H_2C{=}CH{-}CH_3$: C—H 108 pm (1.08 Å); C=C 134 pm (1.34 Å); C—CH₃ 151 pm (1.51 Å)

Propene

A good way to think about the effect of the *s* character is to associate it with electronegativity. As its *s* character increases, so does a carbon's electronegativity (the electrons in the bond involving that orbital are closer to carbon). The hydrogens in C—H bonds behave as if they are attached to an increasingly more electronegative carbon in the series ethane → ethylene → acetylene.

Problem 9.3

How do bond distances and bond strengths change with electronegativity in the series NH_3, H_2O, and HF?

The property that most separates acetylene from ethane and ethylene is its acidity. It, too, can be explained on the basis of the greater electronegativity of *sp*-hybridized carbon compared with sp^3 and sp^2.

9.5 Acidity of Acetylene and Terminal Alkynes

The C—H bonds of hydrocarbons show little tendency to ionize, and alkanes, alkenes, and alkynes are all very weak acids. The acid-dissociation constant K_a for methane, for example, is too small to be measured directly but is estimated to be about 10^{-60} (pK_a 60).

$$H_3C\text{—}H + :\ddot{O}H_2 \rightleftharpoons H_3C:^- + H\text{—}\overset{+}{\ddot{O}}H_2$$

Methane (pK_a = 60)	Water	Methanide ion (a *carbanion*)	Hydronium ion (pK_a = −1.7)

The conjugate base of a hydrocarbon is called a **carbanion.** It is an anion in which the negative charge is borne by carbon. Because it is derived from a very weak acid, a carbanion such as $^-{:}CH_3$ is an exceptionally strong base.

Using the relationship from the preceding section that the electronegativity of carbon increases with its *s* character ($sp^3 < sp^2 < sp$), the order of hydrocarbon acidity is seen to increase with increasing *s* character of carbon.

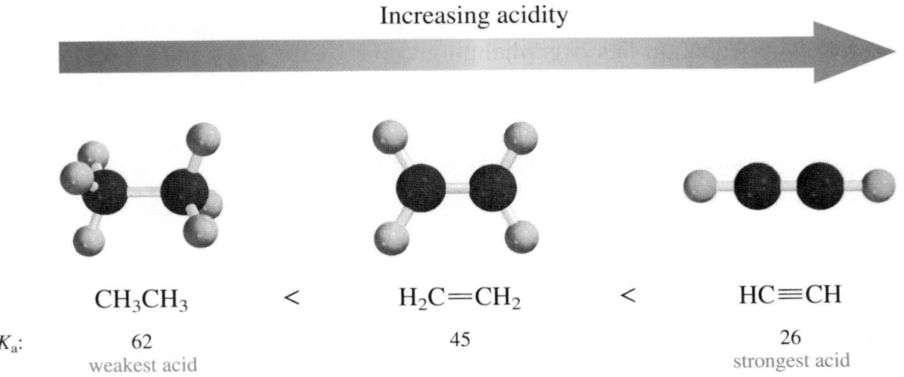

Ionization of acetylene gives acetylide ion in which the unshared electron pair occupies an orbital with 50% *s* character.

$$H\text{—}C{\equiv}C\text{—}H + :\ddot{O}H_2 \rightleftharpoons H\text{—}C{\equiv}C:^- (sp) + H\text{—}\overset{+}{\ddot{O}}H_2$$

Acetylene (pK_a = 26)	Water	Acetylide ion	Hydronium ion (pK_a = −1.7)

In the corresponding ionizations of ethylene and ethane, the unshared pair occupies an orbital with 33% (sp^2) and 25% (sp^3) *s* character, respectively. The greater % *s* character

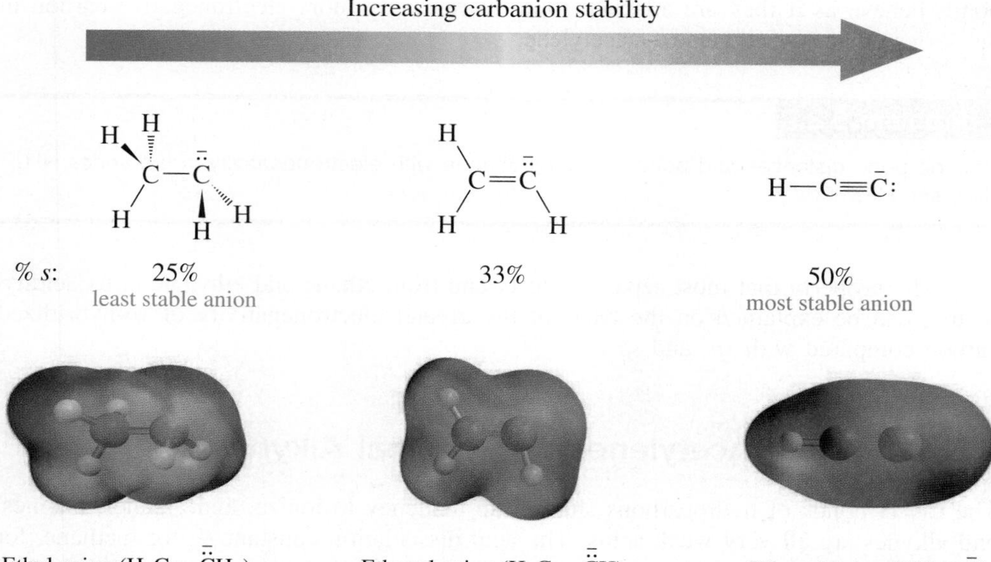

Figure 9.4

Electrostatic potential maps of carbanions. The negative charge (red) is most intense on the ethyl anion with the lowest % *s* character. The ethynyl anion has the highest % *s* character, which places the negative charge closest to the positively charged nucleus. (The electrostatic potentials were mapped on the same scale to allow direct comparison.)

helps to stabilize the anion charge by allowing the negative charge to be held closer to the positively charged nucleus. Figure 9.4 illustrates the effect of *s* character on the anion by comparing the conjugate bases of ethane, ethylene, and acetylene.

Terminal alkynes (RC≡CH) resemble acetylene in acidity.

$$(CH_3)_3CC{\equiv}CH \qquad pK_a = 25.5$$

3,3-Dimethyl-1-butyne

Although acetylene and terminal alkynes are far stronger acids than other hydrocarbons, we must remember that they are, nevertheless, very weak acids—much weaker than water and alcohols, for example. Hydroxide ion is too weak a base to convert acetylene to its anion in meaningful amounts. The position of the equilibrium described by the following equation lies overwhelmingly to the left:

$$H{-}C{\equiv}C{-}H + {:}\ddot{O}H^- \rightleftharpoons H{-}C{\equiv}C{:}^- + H{-}\ddot{O}H$$

Acetylene (weaker acid) $pK_a = 26$	Hydroxide ion (weaker base)	Acetylide ion (stronger base)	Water (stronger acid) $pK_a = 15.7$

Because acetylene is a far weaker acid than water and alcohols, these substances are not suitable solvents for reactions involving acetylide ions. Acetylide is instantly converted to acetylene by proton transfer from compounds that contain —OH groups.

Amide ion, however, is a much stronger base than acetylide ion and converts acetylene to its conjugate base quantitatively.

$$H{-}C{\equiv}C{-}H + {:}\ddot{N}H_2^- \rightleftharpoons H{-}C{\equiv}C{:}^- + H{-}\ddot{N}H_2$$

Acetylene (stronger acid) $pK_a = 26$	Amide ion (stronger base)	Acetylide ion (weaker base)	Ammonia (weaker acid) $pK_a = 36$

Solutions of sodium acetylide (HC≡CNa) may be prepared by adding *sodium amide* ($NaNH_2$) to acetylene in liquid ammonia as the solvent. Terminal alkynes react similarly to give species of the type RC≡CNa.

Problem 9.4

Complete each of the following equations to show the conjugate acid and the conjugate base formed by proton transfer between the indicated species. Use curved arrows to show the flow of electrons, and specify whether the position of equilibrium lies to the side of reactants or products.

(a) $CH_3C{\equiv}CH \quad + \quad {}^{-}{:}\ddot{O}CH_3 \quad \rightleftharpoons$

(b) $HC{\equiv}CH \quad + \quad H_2\ddot{C}CH_3 \quad \rightleftharpoons$

(c) $H_2C{=}CH_2 \quad + \quad {}^{-}{:}\ddot{N}H_2 \quad \rightleftharpoons$

(d) $CH_3C{\equiv}CCH_2OH \quad + \quad {}^{-}{:}\ddot{N}H_2 \quad \rightleftharpoons$

Sample Solution (a) The equation representing the acid–base reaction between propyne and methoxide ion is:

$$CH_3C{\equiv}C{-}H \quad + \quad {}^{-}{:}\ddot{O}CH_3 \quad \rightleftharpoons \quad CH_3C{\equiv}C{:}^{-} \quad + \quad H{-}\ddot{O}CH_3$$

Propyne (weaker acid; $pK_a = 26$) — Methoxide ion (weaker base) — Propynide ion (stronger base) — Methanol (stronger acid; $pK_a = 16$)

Alcohols are stronger acids than acetylene, and so the position of equilibrium lies to the left. Methoxide ion is not a strong enough base to remove a proton from acetylene.

Anions of acetylene and terminal alkynes are nucleophilic and react with methyl and primary alkyl halides to form carbon–carbon bonds by nucleophilic substitution. They are such strong bases, however, that they react with secondary and tertiary alkyl halides by elimination.

9.6 Preparation of Alkynes by Alkylation of Acetylene and Terminal Alkynes

Organic synthesis makes use of two major reaction types:

1. Carbon–carbon bond-forming reactions
2. Functional-group transformations

Both strategies are applied to the preparation of alkynes. In this section we shall see how to prepare alkynes by carbon–carbon bond-forming reactions. By attaching alkyl groups to acetylene **(alkylation),** more complex alkynes can be prepared.

$$H{-}C{\equiv}C{-}H \longrightarrow R{-}C{\equiv}C{-}H \longrightarrow R{-}C{\equiv}C{-}R'$$

Acetylene — Monosubstituted or terminal alkyne — Disubstituted derivative of acetylene

Alkylation of acetylene involves a sequence of two separate operations. In the first, acetylene is converted to its conjugate base by treatment with sodium amide.

$$HC{\equiv}CH \; + \; NaNH_2 \longrightarrow HC{\equiv}CNa \; + \; NH_3$$

Acetylene — Sodium amide — Sodium acetylide — Ammonia

Next, an alkyl halide is added to the solution of sodium acetylide. Acetylide ion acts as a nucleophile, displacing halide from carbon and forming a new carbon–carbon bond. Substitution occurs by an S_N2 mechanism.

$$HC{\equiv}CNa \; + \; RX \longrightarrow HC{\equiv}CR \; + \; NaX \quad \text{via} \quad HC{\equiv}C{:}^{-} \;\; R{-}X$$

Sodium acetylide — Alkyl halide — Alkyne — Sodium halide

Note the similarity of this reaction to that of an alkyl halide with cyanide ion.

The synthetic sequence is normally carried out in liquid ammonia, diethyl ether, or tetrahydrofuran as the solvent.

NaC≡CH + 1-Bromobutane $\xrightarrow{NH_3}$ 1-Hexyne (70–77%)

Sodium acetylide | 1-Bromobutane | 1-Hexyne (70–77%)

An analogous sequence starting with terminal alkynes (RC≡CH) yields alkynes of the type RC≡CR′.

4-Methyl-1-pentyne $\xrightarrow[NH_3]{NaNH_2}$ Sodium 4-methylpentynylide ($:^-$ Na^+) $\xrightarrow{CH_3Br}$ 5-Methyl-2-hexyne (81%)

Dialkylation of acetylene can be achieved by carrying out the sequence twice.

HC≡CH (Acetylene) $\xrightarrow[2.\ CH_3CH_2Br]{1.\ NaNH_2,\ NH_3}$ CH$_3$CH$_2$C≡CH $\xrightarrow[2.\ CH_3Br]{1.\ NaNH_2,\ NH_3}$ CH$_3$CH$_2$C≡C—CH$_3$

Problem 9.5

Outline efficient syntheses of each of the following alkynes from acetylene and any necessary organic or inorganic reagents:

(a) 1-Heptyne
(b) 2-Heptyne
(c) 3-Heptyne

Sample Solution (a) An examination of the structural formula of 1-heptyne reveals it to have a pentyl group attached to an acetylene unit. Alkylation of acetylene, by way of its anion, with a pentyl halide is a suitable synthetic route to 1-heptyne.

$$HC{\equiv}CH \xrightarrow[NH_3]{NaNH_2} HC{\equiv}CNa \xrightarrow{CH_3CH_2CH_2CH_2CH_2Br} HC{\equiv}CCH_2CH_2CH_2CH_2CH_3$$

Acetylene | Sodium acetylide | 1-Heptyne

The major limitation to this reaction is that synthetically acceptable yields are obtained only with methyl and primary alkyl halides. Acetylide anions are very basic, much more basic than hydroxide, for example, and react with secondary and tertiary alkyl halides by elimination.

H—C≡C:$^-$ + H—CH$_2$C(CH$_3$)$_2$—Br: $\xrightarrow{E2}$ H—C≡C—H + CH$_2$=C(CH$_3$)$_2$ + :Br:$^-$

Acetylide | *tert*-Butyl bromide | Acetylene | 2-Methylpropene | Bromide

Problem 9.6

Which of the alkynes of molecular formula C_5H_8 can be prepared in good yield by alkylation or dialkylation of acetylene? Explain why the preparation of the other C_5H_8 isomers would not be practical.

The following section describes an alternative strategy for alkyne synthesis by a two-fold elimination procedure.

9.7 Preparation of Alkynes by Elimination Reactions

Just as it is possible to prepare alkenes by dehydrohalogenation of alkyl halides, so may alkynes be prepared by a *double dehydrohalogenation* of dihaloalkanes. The dihalide may be a **geminal dihalide,** one in which both halogens are on the same carbon, or it may be a **vicinal dihalide,** one in which the halogens are on adjacent carbons.

Geminal dihalide or Vicinal dihalide $\xrightarrow{2NaNH_2}$ R—≡—R′ (Alkyne) + $2NH_3$ (Ammonia) + 2NaX (Sodium halide)

The most frequent applications of these procedures lie in the preparation of terminal alkynes. Because the terminal alkyne product is acidic enough to transfer a proton to amide anion, one equivalent of base in addition to the two equivalents required for double dehydrohalogenation is needed. Adding water or acid after the reaction is complete converts the sodium salt to the corresponding alkyne.

Double dehydrohalogenation of a geminal dihalide

1,1-Dichloro-3,3-dimethylbutane $\xrightarrow[NH_3]{3NaNH_2}$ Sodium salt of alkyne product (not isolated) (≡:⁻ Na^+) $\xrightarrow{H_2O}$ 3,3-Dimethyl-1-butyne (56–60%)

Double dehydrohalogenation of a vicinal dihalide

1,2-Dibromodecane $\xrightarrow[NH_3]{3NaNH_2}$ $CH_3(CH_2)_7C{\equiv}C{:}^-\ Na^+$ Sodium salt of alkyne product (not isolated) $\xrightarrow{H_2O}$ $CH_3(CH_2)_7C{\equiv}CH$ 1-Decyne (54%)

Double dehydrohalogenation to form terminal alkynes may also be carried out by heating geminal and vicinal dihalides with potassium *tert*-butoxide in dimethyl sulfoxide.

Problem 9.7

Give the structures of three isomeric dibromides that could be used as starting materials for the preparation of 3,3-dimethyl-1-butyne.

Because vicinal dihalides are prepared by addition of chlorine or bromine to alkenes (Section 8.10), alkenes, especially terminal alkenes, can serve as starting materials for the preparation of alkynes as shown in the following example:

3-Methyl-1-butene $\xrightarrow{Br_2}$ 1,2-Dibromo-3-methylbutane $\xrightarrow[2.\ H_2O]{1.\ NaNH_2,\ NH_3}$ 3-Methyl-1-butyne (52%)

Problem 9.8

Show, by writing an appropriate series of equations, how you could prepare propyne from each of the following compounds as starting materials. You may use any necessary organic or inorganic reagents.

(a) 2-Propanol
(b) 1-Propanol
(c) Isopropyl bromide
(d) 1,1-Dichloroethane
(e) Ethyl alcohol

Sample Solution (a) Because we know that we can convert propene to propyne by the sequence of reactions

Propene $\xrightarrow{Br_2}$ 1,2-Dibromopropane $\xrightarrow{\text{1. } NaNH_2, NH_3;\ \text{2. } H_2O}$ Propyne

all that remains to completely describe the synthesis is to show the preparation of propene from 2-propanol. Acid-catalyzed dehydration is suitable.

2-Propanol $\xrightarrow[\text{heat}]{H_2SO_4}$ Propene

9.8 Reactions of Alkynes

We have already discussed one important chemical property of alkynes, the acidity of acetylene and terminal alkynes. In the remaining sections of this chapter several other reactions of alkynes will be explored. Most of them will be similar to reactions of alkenes. Like alkenes, alkynes undergo addition reactions. We'll begin with a reaction familiar to us from our study of alkenes, namely, catalytic hydrogenation.

9.9 Hydrogenation of Alkynes

The conditions for hydrogenation of alkynes are similar to those employed for alkenes. In the presence of finely divided platinum, palladium, nickel, or rhodium, two molar equivalents of hydrogen add to the triple bond of an alkyne to yield an alkane.

$$\underset{\text{Alkyne}}{RC{\equiv}CR'} + \underset{\text{Hydrogen}}{2H_2} \xrightarrow{\text{Pt, Pd, Ni, or Rh}} \underset{\text{Alkane}}{RCH_2CH_2R'}$$

4-Methyl-1-hexyne + $2H_2$ (Hydrogen) $\xrightarrow{Ni}$ 3-Methylhexane (77%)

Problem 9.9

Write a series of equations showing how you could prepare octane from acetylene and any necessary organic and inorganic reagents.

The heat of hydrogenation of an alkyne is greater than twice the heat of hydrogenation of an alkene. When two moles of hydrogen add to an alkyne, addition of the first mole (triple bond → double bond) is more exothermic than the second (double bond → single bond).

Substituents affect the heats of hydrogenation of alkynes in the same way they affect alkenes. Compare the heats of hydrogenation of 1-butyne and 2-butyne, both of which give butane on taking up two moles of H_2.

	$CH_3CH_2C{\equiv}CH$	$CH_3C{\equiv}CCH_3$
	1-Butyne	2-Butyne
Heat of hydrogenation:	292 kJ/mol (69.9 kcal/mol)	275 kJ/mol (65.6 kcal/mol)

The internal triple bond of 2-butyne is stabilized relative to the terminal triple bond of 1-butyne. Alkyl groups release electrons to *sp*-hybridized carbon, stabilizing the alkyne and decreasing the heat of hydrogenation.

Like the hydrogenation of alkenes, hydrogenation of alkynes is a syn addition; cis alkenes are intermediates in the hydrogenation of alkynes to alkanes.

$$RC{\equiv}CR' \xrightarrow[\text{catalyst}]{H_2} \underset{\text{cis Alkene}}{(R)(H)C{=}C(R')(H)} \xrightarrow[\text{catalyst}]{H_2} RCH_2CH_2R'$$

Alkyne — cis Alkene — Alkane

The fact that cis alkenes are intermediates in the hydrogenation of alkynes suggests that partial hydrogenation of an alkyne would provide a method for preparing:

1. Alkenes from alkynes, and
2. cis Alkenes free of their trans stereoisomers

Both objectives are met with special hydrogenation catalysts. The most frequently used one is the **Lindlar catalyst,** a palladium on calcium carbonate combination to which lead acetate and quinoline have been added. Lead acetate and quinoline partially deactivate ("poison") the catalyst, making it a poor catalyst for alkene hydrogenation while retaining its ability to catalyze the addition of H_2 to the triple bond.

The structure of quinoline is shown on page 461. In subsequent equations, we will simply use the term *Lindlar Pd* to stand for all of the components of the Lindlar catalyst.

$$\text{1-Ethynylcyclohexanol} + H_2 \xrightarrow[\text{lead acetate, quinoline}]{Pd/CaCO_3} \text{1-Vinylcyclohexanol}$$

1-Ethynylcyclohexanol + Hydrogen → 1-Vinylcyclohexanol (90–95%)

Hydrogenation of alkynes with internal triple bonds gives cis alkenes.

$$\text{5-Decyne} \xrightarrow[\text{Lindlar Pd}]{H_2} \textit{cis}\text{-5-Decene (87\%)}$$

5-Decyne → *cis*-5-Decene (87%)

Problem 9.10

Write a series of equations showing how to prepare *cis*-5-decene from acetylene and 1-bromobutane as the source of all its carbons, using any necessary organic or inorganic reagents. (*Hint:* You may find it helpful to review Section 9.6.)

continued

Sample Solution

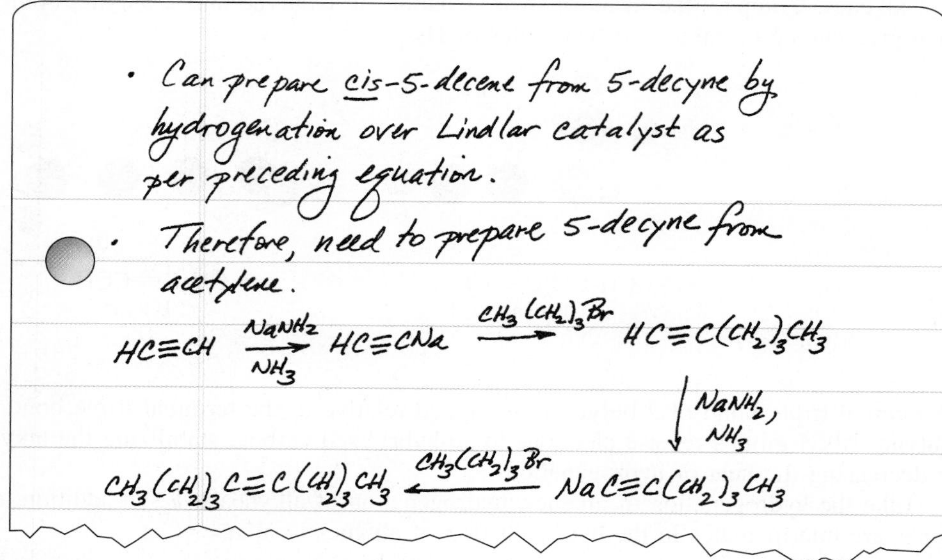

• Can prepare cis-5-decene from 5-decyne by hydrogenation over Lindlar catalyst as per preceding equation.
• Therefore, need to prepare 5-decyne from acetylene.
HC≡CH
NaNH2 / NH3
HC≡CNa
CH3(CH2)3Br
HC≡C(CH2)3CH3
NaNH2, NH3
CH3(CH2)3C≡C(CH2)3CH3
CH3(CH2)3Br
NaC≡C(CH2)3CH3

Hydrogenation of alkynes to alkenes using the Lindlar catalyst is attractive because it sidesteps the regioselectivity and stereoselectivity issues that accompany the dehydration of alcohols and dehydrohalogenation of alkyl halides. In terms of regioselectivity, the position of the double bond is never in doubt—it appears in the carbon chain at exactly the same place where the triple bond was. In terms of stereoselectivity, only the cis alkene forms. Recall that dehydration and dehydrohalogenation normally give a cis–trans mixture in which the cis isomer is the minor product.

9.10 Addition of Hydrogen Halides to Alkynes

Alkynes react with many of the same electrophilic reagents that add to the carbon–carbon double bond of alkenes. Hydrogen halides, for example, add to alkynes to form alkenyl halides.

$$RC{\equiv}CR' + HX \longrightarrow RCH{=}CR'(X)$$

Alkyne | Hydrogen halide | Alkenyl halide

The regioselectivity of addition follows Markovnikov's rule. A proton adds to the carbon that has the greater number of hydrogens, and halide adds to the carbon with the fewer hydrogens.

1-Hexyne + HBr ⟶ 2-Bromo-1-hexene (60%)

1-Hexyne | Hydrogen bromide | 2-Bromo-1-hexene (60%)

When formulating a mechanism for the reaction of alkynes with hydrogen halides, we could propose a process analogous to that of electrophilic addition to alkenes in which the first step is formation of a carbocation and is rate-determining. According to such a mechanism, the second step would be nucleophilic capture of the carbocation by a halide ion.

R—C≡C—H + H—$\ddot{X}$: —slow→ R—$\overset{+}{C}$=CH₂ + :$\ddot{X}$:⁻ —fast→ R(X)C=CH₂

Alkyne | Hydrogen halide | Alkenyl cation | Halide ion | Alkenyl halide

Figure 9.5

Curved-arrow notation for electrophilic addition of a hydrogen halide HX to an alkyne.

Evidence from a variety of sources, however, indicates that alkenyl cations (also called *vinylic cations*) are much less stable than simple alkyl cations, and their involvement in these additions has been questioned.

Kinetic studies suggest a transition state involving two molecules of the hydrogen halide and one of the alkyne. Figure 9.5 depicts a one-step termolecular process that avoids the formation of a very unstable alkenyl cation intermediate by invoking nucleophilic participation by the halogen at an early stage. Nevertheless, because Markovnikov's rule is observed, it seems likely that some degree of positive character develops at carbon and controls the regioselectivity of addition.

In the presence of excess hydrogen halide, geminal dihalides are formed by sequential addition of two molecules of hydrogen halide to the carbon–carbon triple bond.

$$\underset{\text{Alkyne}}{RC{\equiv}CR'} \xrightarrow{HX} \underset{\text{Alkenyl halide}}{RCH{=}C(X)R'} \xrightarrow{HX} \underset{\text{Geminal dihalide}}{RCH_2CX_2R'}$$

The second mole of hydrogen halide adds to the initially formed alkenyl halide in accordance with Markovnikov's rule. Overall, both protons become bonded to the same carbon and both halogens to the adjacent carbon.

3-Hexyne + 2HF (Hydrogen fluoride) → 3,3-Difluorohexane (76%)

Problem 9.11

Design a synthesis of 1,1-dichloroethane from each of the following:

(a) Ethylene
(b) Vinyl chloride ($H_2C{=}CHCl$)
(c) 1,1-Dibromoethane

Sample Solution (a) Reasoning backward, we recognize 1,1-dichloroethane as the product of addition of two molecules of hydrogen chloride to acetylene. Thus, the synthesis requires converting ethylene to acetylene as a key feature. As described in Section 9.7, this may be accomplished by conversion of ethylene to a vicinal dihalide, followed by double dehydrohalogenation. A suitable synthesis based on this analysis is as shown:

$$\underset{\text{Ethylene}}{H_2C{=}CH_2} \xrightarrow{Br_2} \underset{\text{1,2-Dibromoethane}}{BrCH_2CH_2Br} \longrightarrow \underset{\text{Acetylene}}{HC{\equiv}CH} \xrightarrow{2HCl} \underset{\text{1,1-Dichloroethane}}{CH_3CHCl_2}$$

9.11 Hydration of Alkynes

By analogy to the hydration of alkenes, hydration of an alkyne would be expected to yield an alcohol. This alcohol, however, would be a special kind, called an **enol,** one in which the —OH group is attached to a carbon–carbon double bond. Except for the enol derived by hydration of acetylene itself, the enols formed by hydration of alkynes rapidly isomerize to ketones under conditions of their formation.

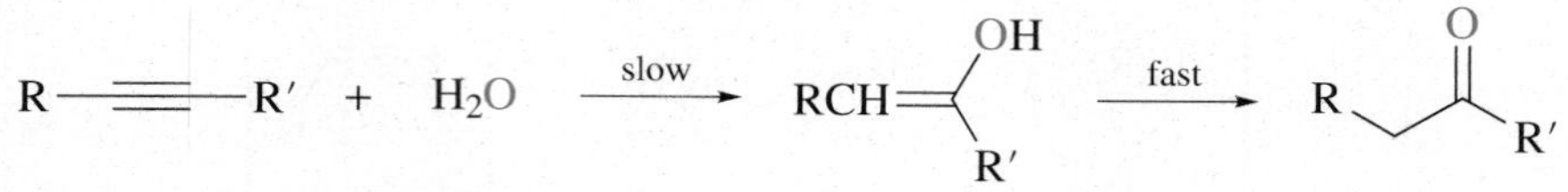

R—≡—R′ + H_2O —slow→ RCH=C(OH)R′ —fast→ $RCH_2C(=O)R'$

Alkyne · Water · Enol (not isolated) · Ketone

The ketone is called the **keto** form, and the keto ⇆ enol equilibration is referred to as *keto–enol isomerism* or *keto–enol tautomerism.* **Tautomers** are constitutional isomers that equilibrate by migration of an atom or group, and their equilibration is called **tautomerism.** Keto–enol isomerism involves the sequence of proton transfers shown in Mechanism 9.1.

Mechanism 9.1

Conversion of an Enol to a Ketone

THE OVERALL REACTION:

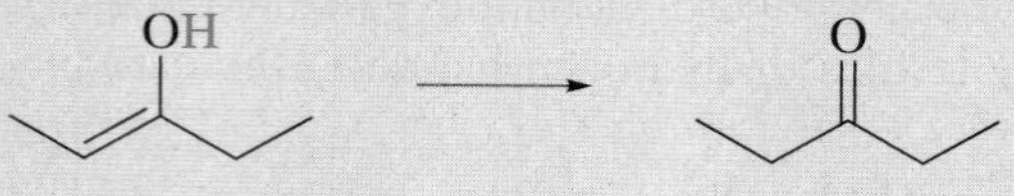

Enol → Ketone (or aldehyde)

THE MECHANISM:

Step 1: In aqueous acid, the first step is proton transfer to the carbon–carbon double bond.

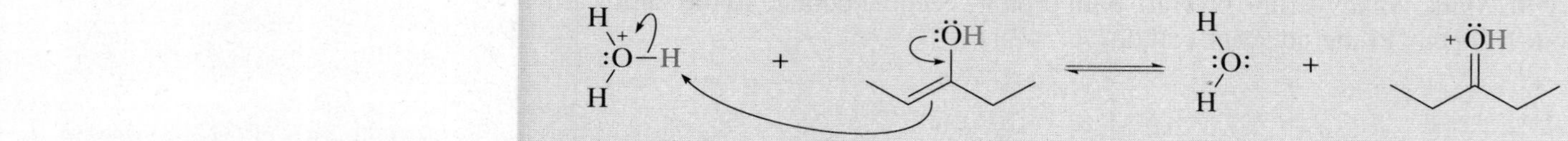

Hydronium ion + Enol ⇌ Water + Conjugate acid of ketone (or aldehyde)

Step 2: The conjugate acid of the ketone transfers a proton from oxygen to a water molecule, yielding a ketone.

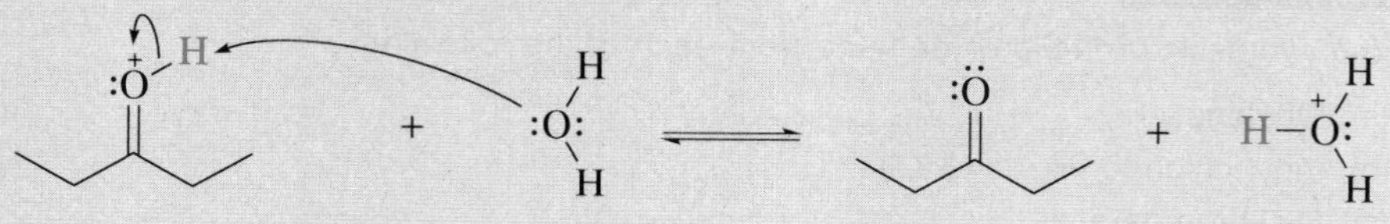

Conjugate acid of ketone (or aldehyde) + Water ⇌ Ketone (or aldehyde) + Hydronium ion

The first step, protonation of the double bond of the enol, is analogous to the protonation of the double bond of an alkene. It takes place more readily, however, because the carbocation formed in this step is stabilized by resonance involving delocalization of a lone pair of oxygen.

A ⟷ B

Of the two contributing structures, A satisfies the octet rule for both carbon and oxygen but B has only six electrons around its positively charged carbon.

Problem 9.12

Give the structure of the enol formed by hydration of 2-butyne, and write a series of equations showing its conversion to its corresponding ketone isomer.

In general, ketones are more stable than their enol isomers and are the products actually isolated when alkynes undergo acid-catalyzed hydration. The standard method for alkyne hydration employs aqueous sulfuric acid as the reaction medium and mercury(II) sulfate or mercury(II) oxide as a catalyst.

4-Octyne + H_2O (Water) $\xrightarrow{H_3O^+, Hg^{2+}}$ 4-Octanone (89%)

For previous examples of the reaction of mercury(II) salts with multiple bonds, see "Descriptive Passage and Interpretive Problems 8: Oxymercuration."

Hydration of alkynes follows Markovnikov's rule; terminal alkynes yield methyl-substituted ketones.

1-Octyne + H_2O (Water) $\xrightarrow[HgSO_4]{H_2SO_4}$ 2-Octanone (91%)

Problem 9.13

Show by a series of equations how you could prepare 2-octanone from acetylene and any necessary organic or inorganic reagents. How could you prepare 4-octanone?

9.12 Addition of Halogens to Alkynes

Alkynes react with chlorine and bromine to yield tetrahaloalkanes. Two molecules of the halogen add to the triple bond.

$$RC{\equiv}CR' + 2X_2 \longrightarrow RCX_2{-}CX_2R'$$

Alkyne; Halogen (chlorine or bromine); Tetrahaloalkane

Propyne + $2Cl_2$ (Chlorine) $\longrightarrow$ 1,1,2,2-Tetrachloropropane (63%)

Some Things That Can Be Made from Acetylene . . . But Aren't

Acetylene had several uses around the time of World War I, primarily because it burned with a hot, luminous flame. The oxyacetylene torch and automobile and bicycle headlamps made by the Prest-O-Lite Company are representative of this period.

In an attempt to find a route to acetylene other than from calcium carbide, Prest-O-Lite sponsored research carried out by George O. Curme at Pittsburgh's Mellon Institute. Curme's research, which was directed toward converting the gases produced during petroleum refining to acetylene, led to methods better suited for making ethylene than acetylene. Viewed from our present perspective, Curme's petroleum-based route to ethylene ranks as a major discovery. It wasn't at the time though, because ethylene had virtually no uses before the 1920s. Curme's second great contribution was the research he carried out to see what useful products he could make from ethylene. The first was ethylene glycol, which became Prestone antifreeze. Others followed, and now ethylene is clearly the most important industrial organic chemical—perhaps the most important of all industrial chemicals.

What about acetylene? Based on the reactions described in this chapter we can write the following equations, all of which lead to useful compounds.

$HC\equiv CH$ + HCl → Vinyl chloride

$HC\equiv CH$ + HCN → Acrylonitrile

$HC\equiv CH$ + H_2O → Acetaldehyde

$HC\equiv CH$ + CH_3CO_2H → Vinyl acetate

In fact, very little of each of these products is made from acetylene. Ethylene is the starting material for the preparation of vinyl chloride, vinyl acetate, and acetaldehyde. Propene is the starting material for acrylonitrile.

Economics dictate the choice of alkene in each case. Acetylene, because of the high energy cost of preparing it, is much more expensive than ethylene and propene. At present, acetylene is used as a starting material only in those few countries where local coal versus petroleum prices favor it. Ethylene comes from petroleum, acetylene can be made from coal. In time, as petroleum becomes increasingly expensive, acetylene-based syntheses may become competitive with ethylene-based ones.

A dihaloalkene is an intermediate and is the isolated product when the alkyne and the halogen are present in equimolar amounts. The stereochemistry of addition is anti.

3-Hexyne + Br_2 (Bromine) → (*E*)-3,4-Dibromo-3-hexene (90%)

9.13 Ozonolysis of Alkynes

Carboxylic acids are produced when alkynes are subjected to ozonolysis.

$$RC\equiv CR' \xrightarrow[2.\ H_2O]{1.\ O_3} R\overset{O}{\overset{\|}{C}}OH + HO\overset{O}{\overset{\|}{C}}R'$$

Recall that when carbonic acid is formed as a reaction product, it dissociates to carbon dioxide and water.

1-Hexyne $\xrightarrow[2.\ H_2O]{1.\ O_3}$ Pentanoic acid (51%) + Carbonic acid

As was the case in the reaction of ozone with alkenes (Section 8.12), ozonolysis of alkynes is sometimes used as a tool in structure determination. By identifying the carboxylic acids produced, we can deduce the structure of the alkyne. As with many other chemical methods of structure determination, however, it has been superseded by spectroscopic methods such as those to be described in Chapter 14.

Problem 9.14

A certain hydrocarbon had the molecular formula $C_{16}H_{26}$ and contained two triple bonds. Ozonolysis gave $CH_3(CH_2)_4CO_2H$ and $HO_2CCH_2CH_2CO_2H$ as the only products. Suggest a reasonable structure for this hydrocarbon.

9.14 Alkynes in Synthesis and Retrosynthesis

Acetylene occupies a useful position in organic synthesis in that it can be applied to C—C bond formation by reaction of its conjugate base with alkylating agents such as alkyl halides and sulfonates. The chain-extended product retains the triple bond and can undergo subsequent addition reactions to give other classes of organic compounds or be subjected to a second alkylation. The following illustrates the retrosynthetic approach and incorporates a chain extension plus two functional-group transformations.

Example: Outline a synthesis of 1,2-epoxybutane using ethyl bromide and acetylene as sources for all the carbon atoms.

We determine the last step in the synthesis by recognizing that the most common route to epoxides is via the reaction of alkenes with peroxy acids.

$$\text{1,2-epoxybutane} \Longrightarrow \text{1-butene}$$

The problem now is to prepare 1-butene from ethyl bromide and acetylene, a process that clearly requires C—C bond formation. We can do this in two operations, shown retrosynthetically as:

$$\text{1-butene} \Longrightarrow \text{1-butyne} \Longrightarrow CH_3CH_2Br + HC{\equiv}C{:}^-$$

Based on this analysis, the synthesis becomes:

$$HC{\equiv}CH \xrightarrow[NH_3]{NaNH_2} HC{\equiv}CNa \xrightarrow{CH_3CH_2Br} \text{1-butyne} \xrightarrow{H_2,\ \text{Lindlar Pd}} \text{1-butene} \xrightarrow{CH_3CO_2OH} \text{1,2-epoxybutane}$$

Problem 9.15

Outline a synthesis of (*Z*)-$H_2C{=}CHCH_2CH_2CH{=}CHCH_3$ from propyne, organic compounds with four carbons or fewer, and any necessary inorganic reagents.

9.15 SUMMARY

Section 9.1 **Alkynes** are hydrocarbons that contain a carbon–carbon *triple bond.* Simple alkynes having no other functional groups or rings have the general formula C_nH_{2n-2}. Acetylene is the simplest alkyne.

Section 9.2 Alkynes are named in much the same way as alkenes, using the suffix *-yne* instead of *-ene.*

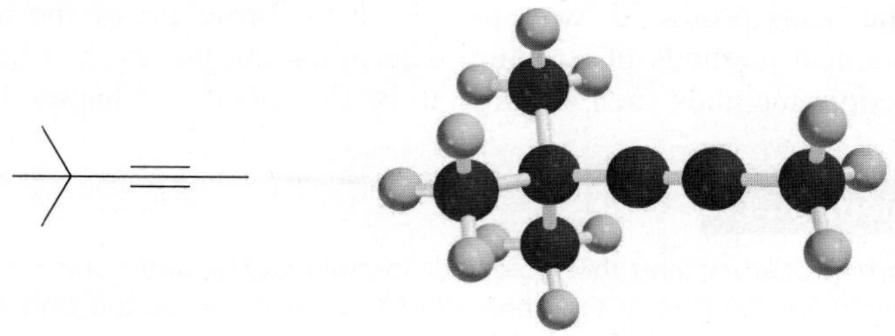

4,4-Dimethyl-2-pentyne

Section 9.3 The physical properties (boiling point, solubility in water, dipole moment) of alkynes resemble those of alkanes and alkenes.

Section 9.4 Acetylene is linear and alkynes have a linear geometry of their X—C≡C—Y units. The carbon–carbon triple bond in alkynes is composed of a σ and two π components.

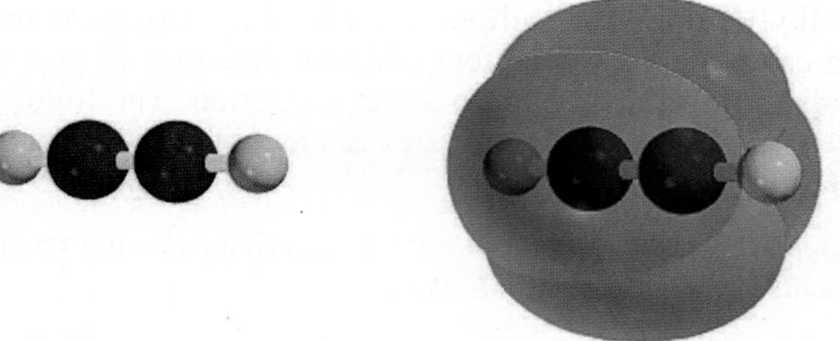

The triple-bonded carbons are *sp*-hybridized. The σ component of the triple bond contains two electrons in an orbital generated by the overlap of *sp*-hybridized orbitals on adjacent carbons. Each of these carbons also has two 2*p* orbitals, which overlap in pairs so as to give two π orbitals, each of which contains two electrons.

Section 9.5 Acetylene and terminal alkynes are more *acidic* than other hydrocarbons. They have pK_as of approximately 26, compared with about 45 for alkenes and about 60 for alkanes. Sodium amide is a strong enough base to remove a proton from acetylene or a terminal alkyne, but sodium hydroxide is not.

$$\underset{\text{1-Butyne}}{CH_3CH_2C{\equiv}CH} + \underset{\text{Sodium amide}}{NaNH_2} \longrightarrow \underset{\text{Sodium 1-butynide}}{CH_3CH_2C{\equiv}CNa} + \underset{\text{Ammonia}}{NH_3}$$

Sections 9.6–9.7 Table 9.2 summarizes the methods for preparing alkynes.

Section 9.8 Like alkenes, alkynes undergo addition reactions.

Section 9.9 Hydrogenation of alkynes in the presence of the customary metal catalysts consumes two molar equivalents of H_2 to give alkanes.

$2H_2$, Pt

Cyclodecyne → Cyclodecane (71%)

Special catalysts allow hydrogenation to be halted at the alkene stage. Lindlar palladium is often used. Hydrogenation occurs with syn stereochemistry to yield a cis alkene.

H_2 / Lindlar Pd

2-Heptyne → *cis*-2-Heptene (59%)

TABLE 9.2 Preparation of Alkynes

Reaction (section) and comments	General equation and specific example
Alkylation of acetylene and terminal alkynes (Section 9.6) The acidity of acetylene and terminal alkynes permits them to be converted to their conjugate bases on treatment with sodium amide. These anions are good nucleophiles and react with methyl and primary alkyl halides to form carbon–carbon bonds. Secondary and tertiary alkyl halides cannot be used, because they yield only elimination products under these conditions.	$RC{\equiv}CH + NaNH_2 \longrightarrow RC{\equiv}CNa + NH_3$ (Alkyne; Sodium amide; Sodium alkynide; Ammonia) $RC{\equiv}CNa + R'CH_2X \longrightarrow RC{\equiv}CCH_2R' + NaX$ (Sodium alkynide; Primary alkyl halide; Alkyne; Sodium halide) 3,3-Dimethyl-1-butyne → (1. $NaNH_2$, NH_3; 2. CH_3I) → 4,4-Dimethyl-2-pentyne (96%)
Double dehydrohalogenation of geminal dihalides (Section 9.7) An E2 elimination reaction of a geminal dihalide yields an alkenyl halide. If a strong enough base is used, sodium amide, for example, a second elimination step follows the first and the alkenyl halide is converted to an alkyne.	$RCH_2CX_2R' + 2NaNH_2 \longrightarrow RC{\equiv}CR' + 2NaX + 2NH_3$ (Geminal dihalide; Sodium amide; Alkyne; Sodium halide; Ammonia) 1,1-Dichloro-3,3-dimethylbutane → (1. $3NaNH_2$, NH_3; 2. H_2O) → 3,3-Dimethyl-1-butyne (56–60%)
Double dehydrohalogenation of vicinal dihalides (Section 9.7) Dihalides in which the halogens are on adjacent carbons undergo two elimination processes analogous to those of geminal dihalides.	$RCHX{-}CHXR' + 2NaNH_2 \longrightarrow RC{\equiv}CR' + 2NaX + 2NH_3$ (Vicinal dihalide; Sodium amide; Alkyne; Sodium halide; Ammonia) 1,2-Dibromobutane → (1. $3NaNH_2$, NH_3; 2. H_2O) → 1-Butyne (78–85%)

Section 9.10 Hydrogen halides add to alkynes in accordance with Markovnikov's rule to give alkenyl halides. In the presence of 2 mol of hydrogen halide, a second addition occurs to give a geminal dihalide.

Propyne + 2HBr (Hydrogen bromide) → 2,2-Dibromopropane (100%)

Section 9.11 Hydration of alkynes in the presence of Hg^{2+} salts yields ketones by way of an unstable enol intermediate. The enol arises by addition of water to the double bond according to Markovnikov's rule.

1-Hexyne + H_2O (Water) → (H_2SO_4, $HgSO_4$) → 2-Hexanone (80%)

Section 9.12 Addition of 1 mol of chlorine or bromine to an alkyne yields a trans dihaloalkene. A tetrahalide is formed on addition of a second equivalent of the halogen.

Propyne + $2Cl_2$ (Chlorine) ⟶ 1,1,2,2-Tetrachloropropane (63%)

Section 9.13 Carbon–carbon triple bonds can be cleaved by ozonolysis. The cleavage products are carboxylic acids.

2-Hexyne $\xrightarrow[2.\ H_2O]{1.\ O_3}$ Butanoic acid + Acetic acid

Section 9.14 Alkylation of acetylene followed by functional-group transformations involving the triple bond of the resulting alkyne can be applied to the synthesis of a wide range of organic compounds.

PROBLEMS

Structure and Nomenclature

9.16 Provide the IUPAC name for each of the following alkynes:

(a) $CH_3CH_2CH_2C{\equiv}CH$

(b) $CH_3CH_2C{\equiv}CCH_3$

(c) $CH_3C{\equiv}CCHCH(CH_3)_2$ with CH_3 on the CH

(d) cyclopropyl$-CH_2CH_2CH_2C{\equiv}CH$

(e)

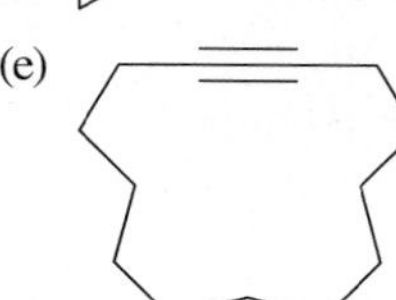

(f)

(g)

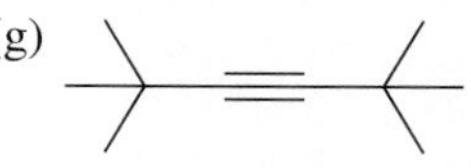

9.17 Write a structural formula for each of the following:

(a) 1-Octyne

(b) 2-Octyne

(c) 3-Octyne

(d) 4-Octyne

(e) 2,5-Dimethyl-3-hexyne

(f) 4-Ethyl-1-hexyne

(g) Ethynylcyclohexane

(h) 3-Ethyl-3-methyl-1-pentyne

9.18 All compounds in Problem 9.17 are isomers except one. Which one?

9.19 Oropheic acid is the common name of a naturally occurring acetylenic carboxylic acid having the molecular formula $C_{18}H_{22}O_2$. Its systematic name is 17-octadecene-9,11,13-triynoic acid. What is its structural formula?

Reactions

9.20 Write structural formulas for all the alkynes of molecular formula C_8H_{14} that yield 3-ethylhexane on catalytic hydrogenation.

9.21 An unknown acetylenic amino acid obtained from the seed of a tropical fruit has the molecular formula $C_7H_{11}NO_2$. On catalytic hydrogenation over platinum, this amino acid yielded homoleucine (an amino acid of known structure shown here) as the only product. What is the structure of the unknown amino acid?

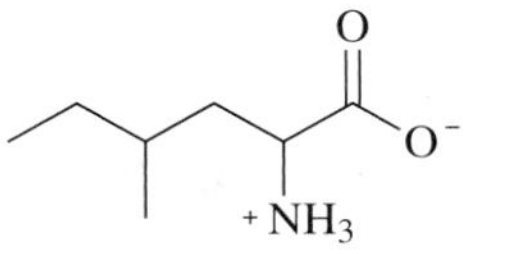

Homoleucine

9.22 Oleic acid and stearic acid are naturally occurring compounds, which can be isolated from various fats and oils. In the laboratory, each can be prepared by hydrogenation of a compound known as *stearolic acid,* which has the formula $CH_3(CH_2)_7C{\equiv}C(CH_2)_7CO_2H$. Oleic acid is obtained by hydrogenation of stearolic acid over Lindlar palladium; stearic acid is obtained by hydrogenation over platinum. What are the structures of oleic acid and stearic acid?

9.23 The alkane formed by hydrogenation of (*S*)-4-methyl-1-hexyne is optically active, but the one formed by hydrogenation of (*S*)-3-methyl-1-pentyne is not. Explain. Would you expect the products of hydrogenation of these two compounds in the presence of Lindlar palladium to be optically active?

9.24 Write the structure of the major organic product isolated from the reaction of 1-hexyne with

(a) Hydrogen (2 mol), platinum
(b) Hydrogen (1 mol), Lindlar palladium
(c) Sodium amide in liquid ammonia
(d) Product in part (c) treated with 1-bromobutane
(e) Product in part (c) treated with *tert*-butyl bromide
(f) Hydrogen chloride (1 mol)
(g) Hydrogen chloride (2 mol)
(h) Chlorine (1 mol)
(i) Chlorine (2 mol)
(j) Aqueous sulfuric acid, mercury(II) sulfate
(k) Ozone followed by hydrolysis

9.25 Write the structure of the major organic product isolated from the reaction of 3-hexyne with

(a) Hydrogen (2 mol), platinum
(b) Hydrogen (1 mol), Lindlar palladium
(c) Hydrogen chloride (1 mol)
(d) Hydrogen chloride (2 mol)
(e) Chlorine (1 mol)
(f) Chlorine (2 mol)
(g) Aqueous sulfuric acid, mercury(II) sulfate
(h) Ozone followed by hydrolysis

9.26 When 2-heptyne was treated with aqueous sulfuric acid containing mercury(II) sulfate, two products, each having the molecular formula $C_7H_{14}O$, were obtained in approximately equal amounts. What are these two compounds?

9.27 All the following reactions have been described in the chemical literature and proceed in good yield. In some cases the reactants are more complicated than those we have so far encountered. Nevertheless, on the basis of what you have already learned, you should be able to predict the principal product in each case.

(a) 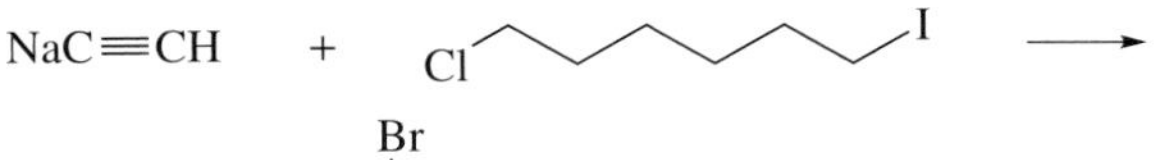 $NaC{\equiv}CH$ + Cl–(CH₂)₆–I ⟶

(b) 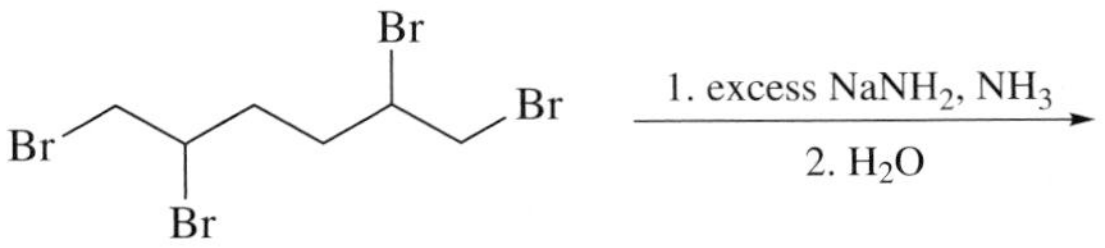 Br, Br, Br, Br (tetrabromide) $\xrightarrow[\text{2. } H_2O]{\text{1. excess } NaNH_2,\ NH_3}$

(c)  cyclopropyl–C(CH₃)(Cl)–Cl $\xrightarrow[\text{heat}]{KOC(CH_3)_3,\ DMSO}$

(d) 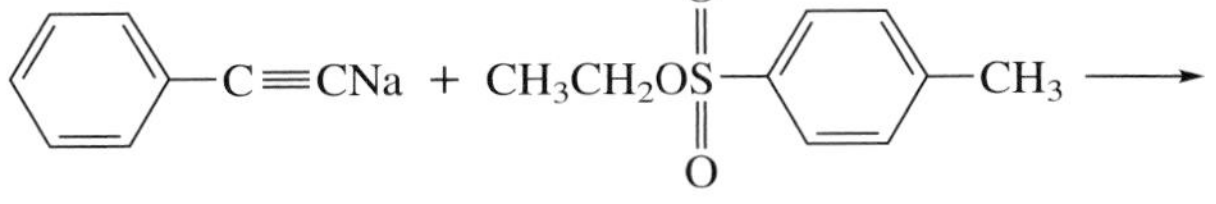 $C_6H_5C{\equiv}CNa + CH_3CH_2OSO_2C_6H_4CH_3$ ⟶

(e) Cyclodecyne $\xrightarrow[\text{2. } H_2O]{\text{1. } O_3}$

(f) 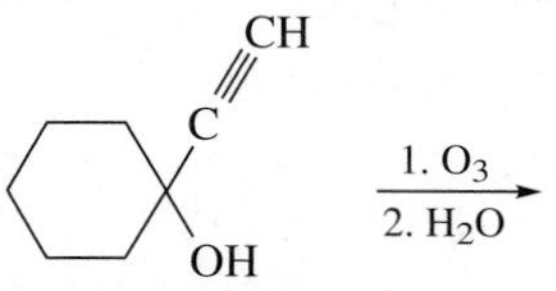 1-ethynylcyclohexanol (CH≡C– and OH on cyclohexane) $\xrightarrow[\text{2. } H_2O]{\text{1. } O_3}$

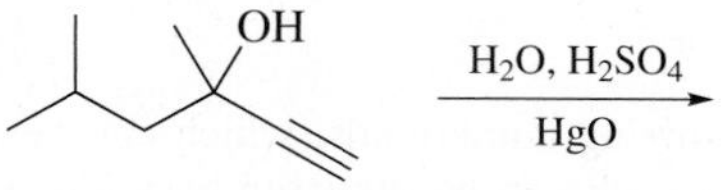

(g) (alkynyl tertiary alcohol with OH) $\xrightarrow[HgO]{H_2O,\ H_2SO_4}$

(h) (tetrahydropyranyl ether, $O(CH_2)_8Cl$) + $NaC{\equiv}CCH_2CH_2CH_2CH_3$ $\longrightarrow$

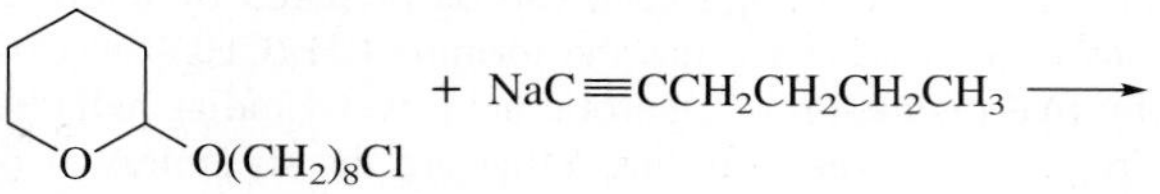

(i) Product of part (h) $\xrightarrow[\text{Lindlar Pd}]{H_2}$

9.28 Compound A has the molecular formula $C_{14}H_{25}Br$ and was obtained by reaction of sodium acetylide with 1,12-dibromododecane. On treatment of compound A with sodium amide, it was converted to compound B ($C_{14}H_{24}$). Ozonolysis of compound B gave the diacid $HO_2C(CH_2)_{12}CO_2H$. Catalytic hydrogenation of compound B over Lindlar palladium gave compound C ($C_{14}H_{26}$), and hydrogenation over platinum gave compound D ($C_{14}H_{28}$). C yielded $O{=}CH(CH_2)_{12}CH{=}O$ on ozonolysis. Assign structures to compounds A through D so as to be consistent with the observed transformations.

Synthesis

9.29 When 1,2-dibromodecane was treated with potassium hydroxide in aqueous ethanol, it yielded a mixture of three isomeric compounds of molecular formula $C_{10}H_{19}Br$. Each of these compounds was converted to 1-decyne on reaction with sodium amide in dimethyl sulfoxide. Identify these three compounds.

9.30 Show by writing appropriate chemical equations how each of the following compounds could be converted to 1-hexyne:

(a) 1,1-Dichlorohexane
(b) 1-Hexene
(c) Acetylene
(d) 1-Iodohexane

9.31 Show by writing appropriate chemical equations how each of the following compounds could be converted to 3-hexyne:

(a) 1-Butene (b) 1,1-Dichlorobutane (c) Acetylene

9.32 Diphenylacetylene can be synthesized by the double dehydrohalogenation of 1,2-dibromo-1,2-diphenylethene. The sequence starting from (*E*)-1,2-diphenylethene consists of bromination to give the dibromide, followed by dehydrohalogenation to give a vinylic bromide, then a second dehydrohalogenation to give diphenylacetylene.

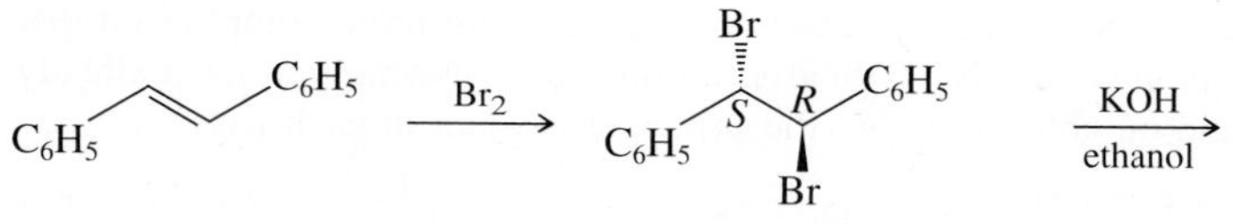

$C_6H_5CH{=}CHC_6H_5$ $\xrightarrow{Br_2}$ $C_6H_5CHBr{-}CHBrC_6H_5$ (S, R) $\xrightarrow[\text{ethanol}]{KOH}$

(*E*)-1,2-Diphenylethene

meso-1,2-Dibromo-1,2-diphenylethane

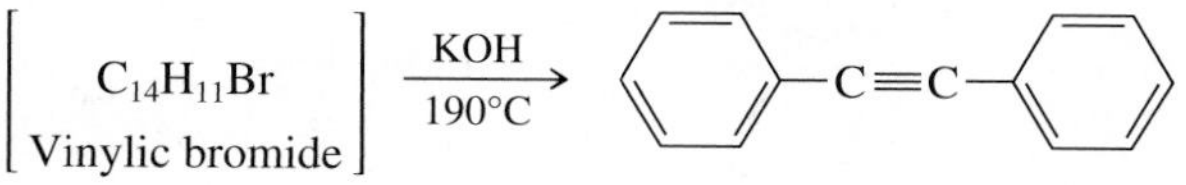

[$C_{14}H_{11}Br$ Vinylic bromide] $\xrightarrow[190°C]{KOH}$ $C_6H_5{-}C{\equiv}C{-}C_6H_5$

Diphenylacetylene

(a) What is the structure, including stereochemistry, of the vinylic bromide?

(b) If the sequence starts with (*Z*)-1,2-dibromo-1,2-diphenylethene, what is (are) the structure(s) of the intermediate dibromide(s)? What is the structure of the vinylic bromide?

9.33 (*Z*)-9-Tricosene [(*Z*)-$CH_3(CH_2)_7CH{=}CH(CH_2)_{12}CH_3$] is the sex pheromone of the female housefly. Synthetic (*Z*)-9-tricosene is used as bait to lure male flies to traps that contain insecticide. Using acetylene and alcohols of your choice as starting materials, along with any necessary inorganic reagents, show how you could prepare (*Z*)-9-tricosene.

9.34 The ketone 2-heptanone has been identified as contributing to the odor of a number of dairy products, including condensed milk and cheddar cheese. Describe a synthesis of 2-heptanone from acetylene and any necessary organic or inorganic reagents.

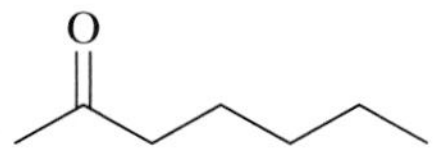

2-Heptanone

9.35 Assume that you need to prepare 4-methyl-2-pentyne and discover that the only alkynes on hand are acetylene and propyne. You also have available methyl iodide, isopropyl bromide, and 1,1-dichloro-3-methylbutane. Which of these compounds would you choose in order to perform your synthesis, and how would you carry it out?

9.36 Show by writing a suitable series of equations how you could prepare each of the following compounds from the designated starting materials and any necessary organic or inorganic reagents:

(a) 2,2-Dibromopropane from 1,1-dibromopropane
(b) 2,2-Dibromopropane from 1,2-dibromopropane
(c) 1,1,2,2-Tetrachloropropane from 1,2-dichloropropane
(d) 2,2-Diiodobutane from acetylene and ethyl bromide
(e) 1-Hexene from 1-butene and acetylene
(f) Decane from 1-butene and acetylene
(g) Cyclopentadecyne from cyclopentadecene
(h) 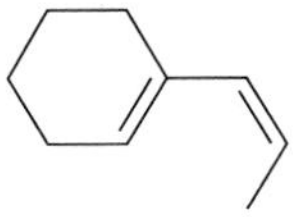 from 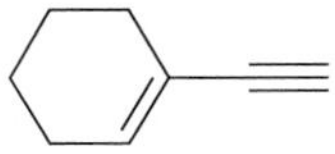

Mechanism

9.37 Alkynes undergo hydroboration to give alkenylboranes, which can be oxidized to give carbonyl compounds with hydrogen peroxide. The net result of the two-step sequence is hydration, which gives aldehydes from terminal alkynes.

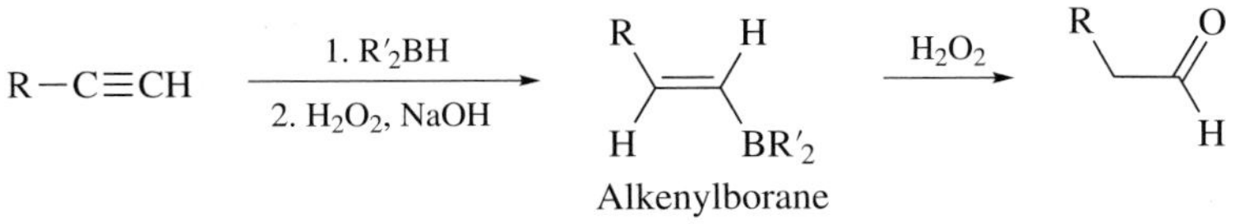

The oxidation step involves an enol intermediate. Using Mechanism 9.1 as a guide, write the structure of the enol that is formed in the conversion of 1-hexyne to hexanal.

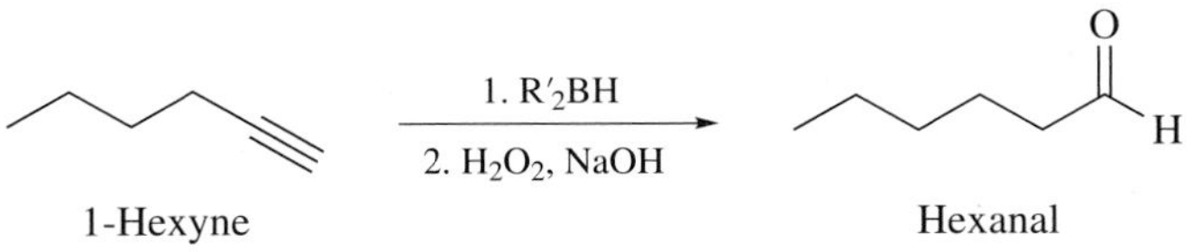

Descriptive Passage and Interpretive Problems 9

Thinking Mechanistically About Alkynes

The preparation and properties of alkynes extend some topics explored in earlier chapters:

- Alkynes can be prepared by elimination reactions related to the E2 dehydrohalogenation of alkyl halides used to prepare alkenes.
- Alkynes can be prepared by S_N2 reactions in which a nucleophile of the type $RC{\equiv}C{:}^-$ reacts with a primary alkyl halide.
- Alkynes undergo addition reactions, especially electrophilic addition, with many of the same compounds that add to alkenes.

The greater *s* character of *sp* hybrid orbitals compared with sp^3 and sp^2 gives alkynes certain properties beyond those seen in alkanes and alkenes. It is convenient to think of *sp*-hybridized carbon as more electronegative than its sp^2 or sp^3 counterparts.

- The $\equiv$C—H unit of an alkyne is more acidic than a C—H unit of an alkene or alkane, allowing acetylene and terminal alkynes to be converted to their conjugate bases $\equiv C{:}^-$ by $NaNH_2$.
- Unlike alkenes, alkynes are reduced by metals, especially Li, Na, and K.
- Unlike alkenes, alkynes can undergo nucleophilic as well as electrophilic addition.

—C≡C— with E^+ ; —C≡C— with $\bar{N}u{:}$

E^+ is an electrophile $Nu{:}^-$ is a nucleophile

Problems 9.38–9.42 emphasize mechanistic reasoning. By thinking mechanistically you reduce the need to memorize facts while increasing your ability to analyze and understand new material. Nucleophilic addition to alkynes, for example, is not covered in this chapter but is the focus of Problem 9.42, which can be solved by thinking mechanistically.

9.38 Which of the following best describes what happens in the first step in the mechanism of the reaction shown?

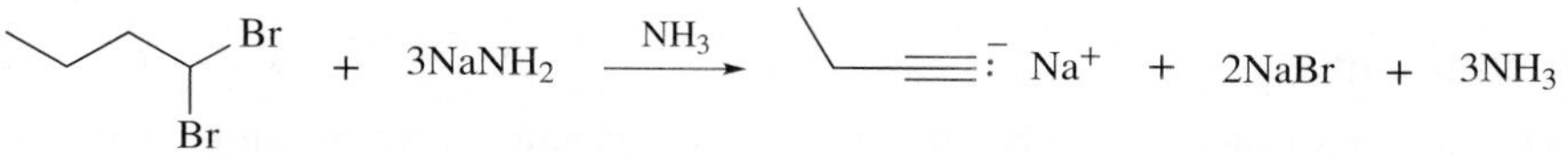

Br, Br (geminal dibromide) + $3NaNH_2$ $\xrightarrow{NH_3}$ ≡:$^-$ Na^+ + 2NaBr + $3NH_3$

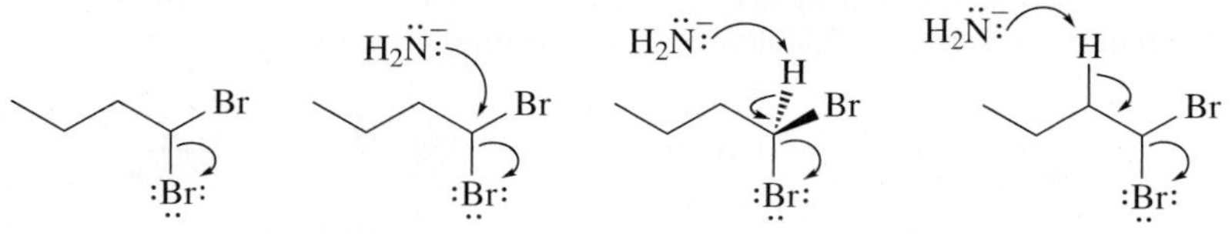

A. :Br: Br B. $H_2\ddot{N}{:}^-$ Br :Br: C. $H_2\ddot{N}{:}^-$ H Br :Br: D. $H_2\ddot{N}{:}^-$ H Br :Br:

9.39 Which of the following best describes what happens in the first step in the mechanism of the hydrogen–deuterium exchange reaction shown?

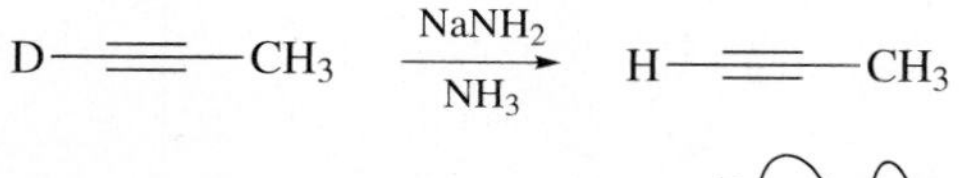

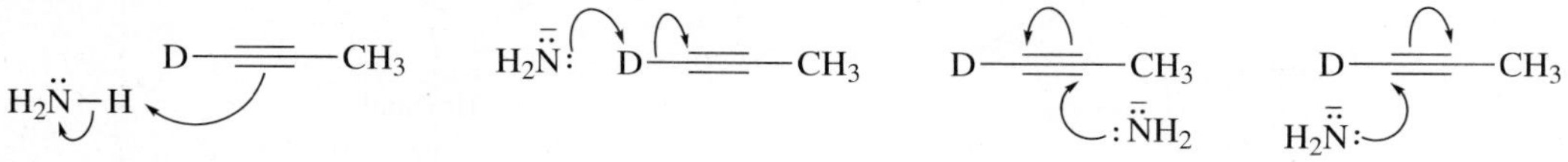

D—≡—CH_3 $\xrightarrow[NH_3]{NaNH_2}$ H—≡—CH_3

A. D—≡—CH_3 $H_2\ddot{N}$—H B. $H_2\ddot{N}{:}$ D—≡—CH_3 C. D—≡—CH_3 $:\bar{N}H_2$ D. D—≡—CH_3 $H_2\bar{N}{:}$

9.40 Electrophilic addition of fluorosulfonic acid (FSO_2OH) to propyne proceeds by way of a very unstable vinyl cation intermediate. What is the most reasonable structure, including geometry, of this intermediate? (*Hint:* Use VSEPR to deduce the geometry.)

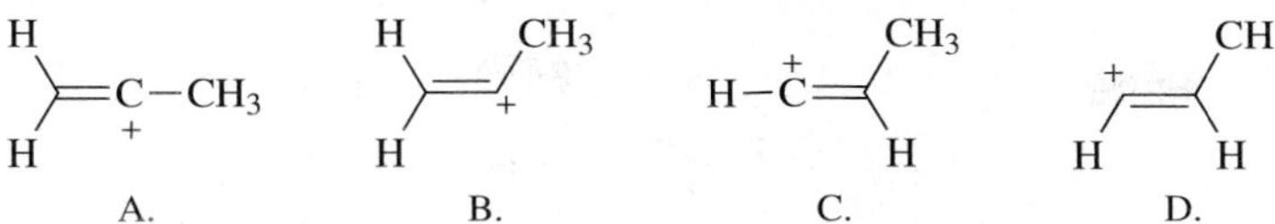

9.41 Rates of Br_2 addition were measured for a series of alkynes, giving the data shown.

Relative rate of Br_2 addition

$HC{\equiv}CH$	$CH_3C{\equiv}CH$	$CH_3C{\equiv}CCH_3$	$CH_3C{\equiv}CC(CH_3)_3$
1.0	13.4	120	558

Assuming that Br_2 addition to alkynes proceeds through rate-determining formation of a cyclic bromonium ion, what generalizations can you make about the structure of the rate-determining transition state?

Positive charge development at carbons in original triple bond	More important effect of substituents on triple bond
A. One carbon only	Electron donation
B. One carbon only	Steric hindrance
C. Both carbons	Electron donation
D. Both carbons	Steric hindrance

9.42 Nucleophilic addition can occur with alkynes that bear strong electron-attracting substituents such as CF_3 on the triple bond. Predict the product of nucleophilic addition of CH_3OD to 3,3,3-trifluoropropyne. The stereochemistry of addition is anti, and the first step in the mechanism is bond formation between CH_3O^- and one of the carbons of the triple bond.

$$HC{\equiv}CCF_3 \xrightarrow[CH_3OD]{NaOCH_3}$$

A. H, CF_3 / CH_3O, D

B. H, D / CH_3O, CF_3

C. H, CF_3 / D, OCH_3

D. D, CF_3 / H, OCH_3

Chapter

10

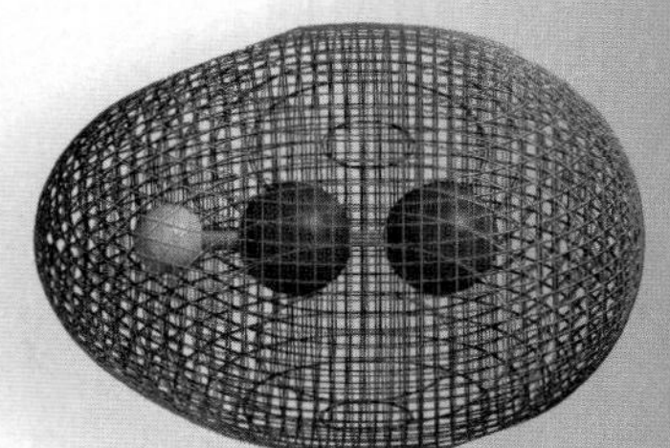

$H—C\equiv C\cdot$

CHAPTER OUTLINE

Stars that are well along in their lifetime are described as "evolved." While using up their nuclear fuel, these evolved stars produce a stellar wind that carries chemical substances from the interior to the envelope surrounding the surface. Most of these substances have very simple molecular formulas, but unusual structures. C_2H, for example, cannot be represented by a structural formula in which all of its electron spins are paired. It belongs to a class of compounds we call "free radicals."
Source: NASA

Introduction to Free Radicals

There is a pronounced chemical bias in favor of compounds that have an even number of electrons—electron *pairs* and the *octet* rule testify to this. Our familiarity with paired electrons doesn't mean, however, that compounds with an odd number of electrons don't exist. They do, in both inorganic and organic chemistry and must, of necessity, contain at least one unpaired electron. Two of the oxides of nitrogen furnish us with inorganic examples.

$\cdot\ddot{N}=\ddot{O}:$ Nitrogen monoxide

$:O=\overset{\cdot +}{N}-\ddot{O}:^{-}$ Nitrogen dioxide

Although known for hundreds of years, nitrogen monoxide has only recently been found to be an important biochemical messenger and moderator of so many biological processes that it's fair to ask "Which ones is it not involved in?" Nitrogen dioxide produced by combustion of various fuels, especially gasoline, contributes significantly to air pollution in cities.

Species that contain unpaired electrons are called **free radicals.** Some, such as nitrogen monoxide and dioxide, are stable species. Others, and these are the ones of interest in this chapter, occur as intermediates in chemical reactions—formed in one step of a mechanism and consumed in another. We'll begin with a description of structure and bonding in carbon-centered free radicals, then introduce reactions of alkanes, alkenes, and alkynes that involve them as intermediates.

For more on the role of NO in physiology, see the boxed essay *Oh NO! it's Inorganic!* in Chapter 25.

10.1 Structure, Bonding, and Stability of Alkyl Radicals

Alkyl radicals are characterized by the presence of a carbon with three bonds and are classified as primary, secondary, or tertiary according to the number of carbon atoms directly attached to the one that bears the unpaired electron.

$H_3C\cdot$	$RCH_2\cdot$	$R_2CH\cdot$	$R_3C\cdot$
Methyl radical	Primary radical	Secondary radical	Tertiary radical

Two possibilities can be considered for bonding in methyl radical. According to the model shown in Figure 10.1*a,* carbon is sp^2-hybridized and the unpaired electron occupies a $2p$ orbital. Alternatively, as shown in Figure 10.1*b,* carbon is sp^3-hybridized, and one of its orbitals contains only one electron. Of the two extremes, experimental studies indicate that the planar sp^2 model describes the bonding in alkyl radicals better than the pyramidal sp^3 model. Methyl radical is planar, and more highly substituted radicals such as *tert*-butyl are flattened pyramids closer in shape to that expected for sp^2-hybridized carbon than for sp^3.

Free radicals, like carbocations, are stabilized by substituents such as alkyl groups that can donate electrons to the unfilled orbital by hyperconjugation, where the unpaired electron, plus electrons in σ bonds that are β to the radical site, are delocalized. This delocalization is illustrated in Figure 10.2, which compares the calculated spin density in methyl and ethyl radical. **Spin density** is a measure of the unpaired electron density at a particular point in a molecule—it tells us where the unpaired electron is most likely to be. In the case of methyl radical, which cannot be stabilized by hyperconjugation, the spin density is concentrated on a single atom, carbon. In ethyl radical, hyperconjugation allows the spin density to be shared by the sp^2-hybridized carbon plus the three hydrogens of the methyl group.

Stabilization of carbocations by hyperconjugation was described in Section 5.9.

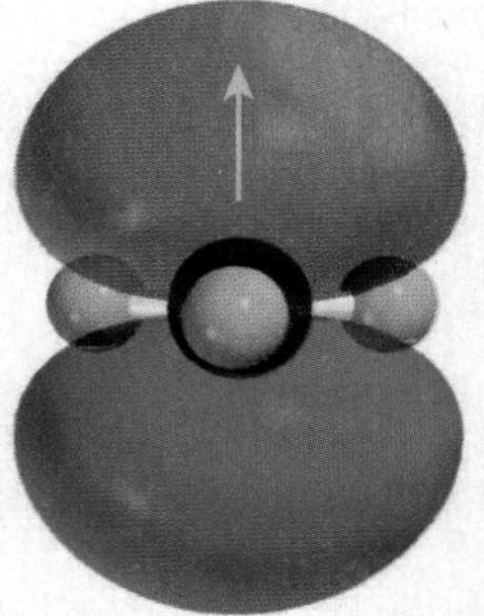

(*a*)
Planar CH_3
Carbon is sp^2-hybridized (120° bond angles). Unpaired electron is in $2p$ orbital.

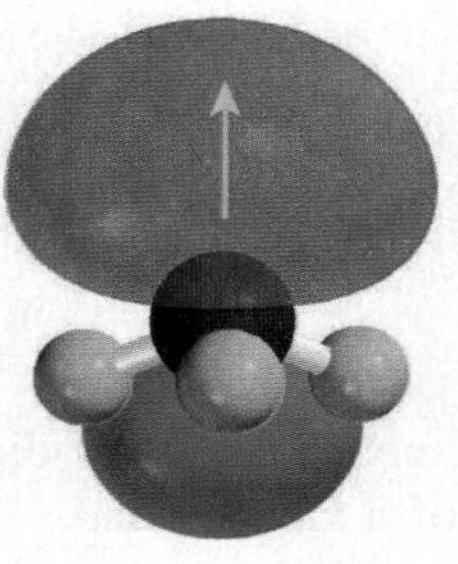

(*b*)
Pyramidal CH_3
Carbon is sp^3-hybridized (109.5° bond angles). Unpaired electron is in sp^3-hybridized orbital.

Figure 10.1

Bonding in methyl radical. Model (*a*) is more consistent with experimental observations.

Figure 10.2

The calculated spin density (yellow) in methyl and ethyl radical. (*a*) The unpaired electron in methyl radical is localized in a *p* orbital of sp^2-hybridized carbon. (*b*) The unpaired electron in ethyl radical is shared by the sp^2-hybridized carbon and by the hydrogens of the CH_3 group.

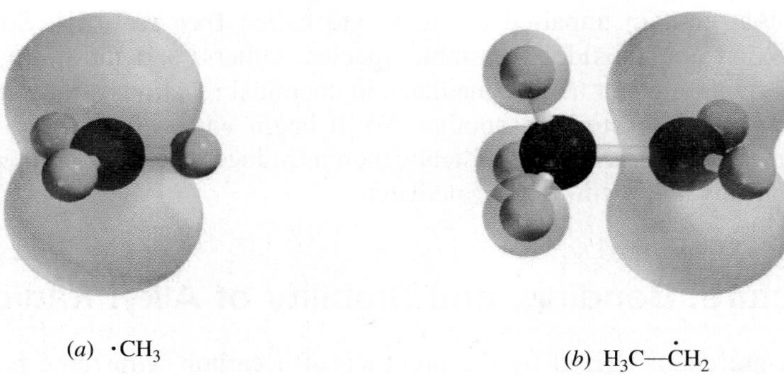

(*a*) $\cdot CH_3$ (*b*) $H_3C—\dot{C}H_2$

More highly substituted radicals are more stable than less highly substituted ones because they have more electron pairs β to the radical site. The order of free-radical stability parallels that of carbocations.

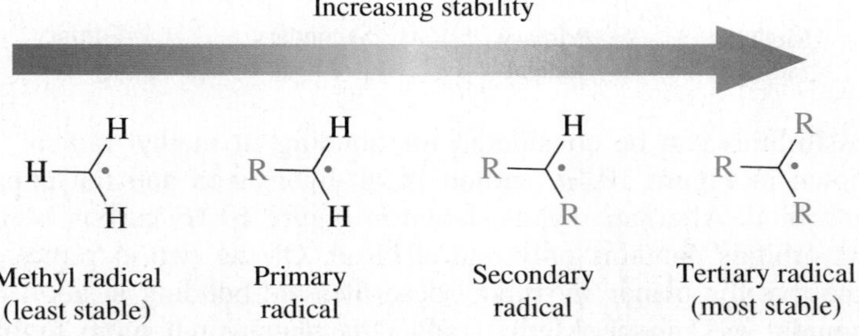

Problem 10.1

Write a line formula for all the free radicals that have the formula C_5H_{11} and classify each as primary, secondary, or tertiary. Which one is the most stable?

Some of the evidence indicating that alkyl substituents stabilize free radicals comes from comparing C—H bond strengths in alkanes. A covalent bond can be broken in two ways. In a **homolytic cleavage** a bond between two atoms is broken so that each atom retains one of the electrons in the bond. Conversely, both electrons are retained by one of the atoms in a **heterolytic cleavage.**

$$X{:}Y \longrightarrow X\cdot + \cdot Y \qquad\qquad X{:}Y \longrightarrow X^+ + :Y^-$$

Homolytic bond cleavage Heterolytic bond cleavage

We can assess the relative stability of alkyl radicals by measuring the enthalpy change ($\Delta H°$) for the homolytic cleavage of a C—H bond in an alkane:

$$R{-}H \longrightarrow R\cdot + \cdot H$$

The more stable the radical, the lower the energy required to generate it by homolytic cleavage of a C—H bond. This energy is called the **bond dissociation enthalpy (*D*).** Table 10.1 gives some examples.

As the table indicates, C—H bond dissociation enthalpies in alkanes are approximately 400–440 kJ/mol (95–105 kcal/mol). Cleaving the $H—CH_3$ bond in methane gives methyl radical and requires 439 kJ/mol (105 kcal/mol). The dissociation enthalpy of the $H—CH_2CH_3$ bond in ethane, which gives a primary radical, is somewhat less (421 kJ/mol, or 100.5 kcal/mol) and is consistent with the notion that ethyl radical (primary) is more stable than methyl.

TABLE 10.1 Some Bond Dissociation Enthalpies*

Bond	Bond dissociation enthalpy (*D*) kJ/mol	kcal/mol	Bond	Bond dissociation enthalpy (*D*) kJ/mol	kcal/mol
Diatomic molecules					
H—H	436	104	H—F	571	136
F—F	159	38	H—Cl	432	103
Cl—Cl	243	58	H—Br	366	87.5
Br—Br	193	46	H—I	298	71
I—I	151	36			
Alkanes					
CH_3—H	439	105	CH_3—CH_3	375	90
CH_3CH_2—H	421	100.5	CH_3CH_2—CH_3	369	88
$CH_3CH_2CH_2$—H	423	101			
$(CH_3)_2CH$—H	413	99			
$(CH_3)_2CHCH_2$—H	422	101	$(CH_3)_2CH$—CH_3	370	88
$(CH_3)_3C$—H	400	95	$(CH_3)_3C$—CH_3	362	86
Alkyl halides					
CH_3—F	459	110	$(CH_3)_2CH$—Cl	355	85
CH_3—Cl	351	84	$(CH_3)_2CH$—Br	297	72
CH_3—Br	292	70			
CH_3—I	238	57			
CH_3CH_2—Cl	350	83	$(CH_3)_3C$—Cl	349	83
$CH_3CH_2CH_2$—Cl	354	85	$(CH_3)_3C$—Br	292	69

*Bond dissociation enthalpies refer to the bond indicated in each structural formula and were calculated from standard enthalpy of formation values as recorded in the NIST Standard Reference Database Number 69, http://webbook.nist.gov/chemistry/.

The dissociation enthalpy of the terminal C—H bond in propane is almost the same as that of ethane. Like ethyl, propyl is a primary free radical.

Propane (H) → Propyl radical (primary) + H• (Hydrogen atom) $\Delta H° = +423$ kJ/mol (101 kcal/mol)

Note, however, that Table 10.1 includes two entries for propane. The second entry corresponds to the cleavage of a bond to one of the hydrogens of the methylene group. It requires slightly less energy to break a C—H bond in the methylene group than in the methyl group.

Propane (H) → Isopropyl radical (secondary) + H• (Hydrogen atom) $\Delta H° = +413$ kJ/mol (99 kcal/mol)

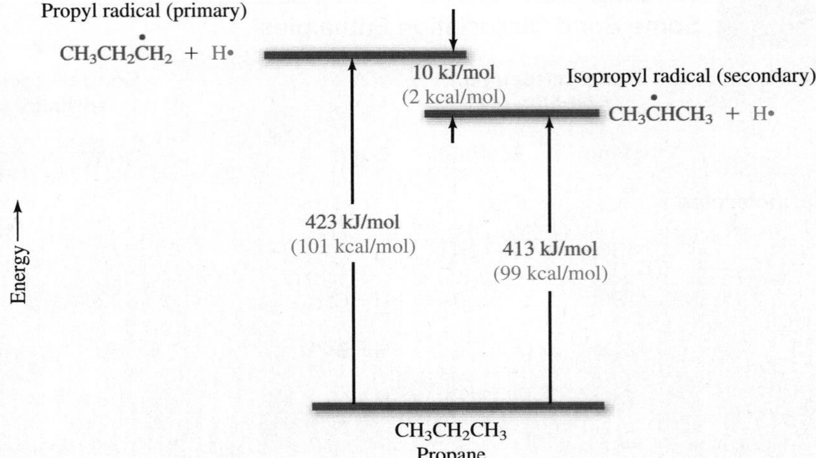

Figure 10.3

The bond dissociation enthalpies of methylene and methyl C—H bonds in propane reveal a difference in stabilities between two isomeric free radicals. The secondary radical is more stable than the primary.

Because the starting material (propane) and one of the products (H•) are the same in both processes, the difference in bond dissociation enthalpies is equal to the energy difference between a propyl radical (primary) and an isopropyl radical (secondary). As depicted in Figure 10.3, the secondary radical is 10 kJ/mol (2 kcal/mol) more stable than the primary radical.

Similarly, by comparing the bond dissociation enthalpies of the two different types of C—H bonds in 2-methylpropane, we see that a tertiary radical is 22 kJ/mol (6 kcal/mol) more stable than a primary radical.

2-Methylpropane ⟶ Isobutyl radical (primary) + H• Hydrogen atom $\Delta H° = +422$ kJ/mol (101 kcal/mol)

2-Methylpropane ⟶ *tert*-Butyl radical (tertiary) + H• Hydrogen atom $\Delta H° = +400$ kJ/mol (95 kcal/mol)

Problem 10.2

The C—Cl bond dissociation enthalpies of propyl and isopropyl chloride are the same within experimental error (see Table 10.1). However, it is incorrect to conclude that the data indicate equal stabilities of propyl and isopropyl radical. Why? Why are the bond dissociation enthalpies of propane a better indicator of the free-radical stabilities?

Like carbocations, most free radicals are exceedingly reactive species—too reactive to be isolated but capable of being formed as intermediates in chemical reactions. Methyl radical, as we shall see in Section 10.2, is an intermediate in the chlorination of methane.

10.2 Halogenation of Alkanes

Alkanes react with halogens according to the equation:

$$\underset{\text{Alkane}}{R\text{—}H} + \underset{\text{Halogen}}{X_2} \longrightarrow \underset{\text{Alkyl halide}}{R\text{—}X} + \underset{\text{Hydrogen halide}}{H\text{—}X}$$

The alkane is said to undergo *fluorination, chlorination, bromination,* or *iodination* according to whether X_2 is F_2, Cl_2, Br_2, or I_2, respectively. The general term is **halogenation.** Chlorination and bromination are the most widely used.

The reactivity of the halogens decreases in the order $F_2 > Cl_2 > Br_2 > I_2$. Fluorine is an extremely aggressive oxidizing agent, and its reaction with alkanes is strongly exothermic and difficult to control. Chlorination of alkanes is less exothermic than fluorination, and bromination less exothermic than chlorination. Iodine is unique among the halogens in that its reaction with alkanes is endothermic; consequently, alkyl iodides are never prepared by iodination of alkanes.

Volume 11 of *Organic Reactions,* an annual series that reviews reactions of interest to organic chemists, contains the statement "Most organic compounds burn or explode when brought in contact with fluorine."

Problem 10.3

Use the data in Table 10.1 to calculate $\Delta H°$ for the iodination of methane.

From Bond Enthalpies to Heats of Reaction

You have seen that measurements of heats of reaction, such as heats of combustion, can provide quantitative information concerning the relative stability of constitutional isomers (Section 2.23) and stereoisomers (Section 3.11). The boxed essay in Section 2.23 described how heats of reaction can be manipulated arithmetically to generate heats of formation ($\Delta H_f°$) for many molecules. The following material shows how two different sources of thermochemical information, heats of formation and bond dissociation enthalpies (see Table 10.1), can reveal whether a particular reaction is exothermic or endothermic and by how much.

Consider the chlorination of methane to chloromethane. The heats of formation of the reactants and products appear beneath the equation. These heats of formation for the chemical compounds are taken from published tabulations; the heat of formation of chlorine is zero, as it is for all elements.

$$CH_4 + Cl_2 \longrightarrow CH_3Cl + HCl$$

	CH_4	Cl_2	CH_3Cl	HCl
$\Delta H_f°$ kJ/mol:	−74.8	0	−83.7	−92.3
(kcal/mol)	(−17.9)	(0)	(−20.0)	(−22.1)

The overall heat of reaction is given by

$$\Delta H° = \Sigma \text{ (heats of formation of products)} - \Sigma \text{ (heats of formation of reactants)}$$

$$\Delta H° \text{ (kJ/mol)} = (-83.7 - 92.3) - (-74.8) = -101.2 \text{ kJ/mol}$$

$$\Delta H° \text{ (kcal/mol)} = (-20.0 - 22.1) - (-17.9) = -24.2 \text{ kcal/mol}$$

Thus, the chlorination of methane is calculated to be exothermic on the basis of heat of formation data.

The same conclusion is reached using bond dissociation enthalpies. The following equation shows the bond dissociation enthalpies of the reactants and products taken from Table 10.1:

$$CH_4 + Cl_2 \longrightarrow CH_3Cl + HCl$$

	CH_4	Cl_2	CH_3Cl	HCl
BDE kJ/mol:	439	243	351	432
(kcal/mol)	(105)	(58)	(84)	(103)

Because stronger bonds are formed at the expense of weaker ones, the reaction is exothermic and

$$\Delta H° = \Sigma\text{(BDE of bonds broken)} - \Sigma\text{(BDE of bonds formed)}$$

$$\Delta H° \text{ (kJ/mol)} = (439 + 243) - (351 + 432) = -101 \text{ kJ/mol}$$

$$\Delta H° \text{ (kcal/mol)} = (105 + 58) - (84 + 103) = -24 \text{ kcal/mol}$$

This value is in good agreement with that obtained from heats of formation.

Compare chlorination of methane with iodination. The relevant bond dissociation enthalpies are given in the equation.

$$CH_4 + I_2 \longrightarrow CH_3I + HI$$

	CH_4	I_2	CH_3I	HI
BDE kJ/mol:	439	151	238	298
(kcal/mol)	(105)	(36)	(57)	(71)

$$\Delta H° = \Sigma\text{(BDE of bonds broken)} - \Sigma\text{(BDE of bonds formed)}$$

$$\Delta H° \text{ (kJ/mol)} = (439 + 151) - (238 + 298) = +54 \text{ kJ/mol}$$

$$\Delta H° \text{ (kcal/mol)} = (105 + 36) - (57 + 71) = +13 \text{ kcal/mol}$$

A positive value for $\Delta H°$ signifies an **endothermic** reaction. The reactants are more stable than the products, and so iodination of alkanes is not a feasible reaction. You would not want to attempt the preparation of iodomethane by iodination of methane.

A similar analysis for fluorination of methane gives $\Delta H° = -432$ kJ/mol (−103 kcal/mol) for its heat of reaction. Fluorination of methane is about four times as exothermic as chlorination. A reaction this exothermic, if it also occurs at a rapid rate, can proceed with explosive violence.

Bromination of methane is exothermic, but less so than chlorination. The value calculated from bond dissociation enthalpies is $\Delta H° = -26$ kJ/mol (−6.2 kcal/mol). Although bromination of methane is energetically favorable, economic considerations cause most of the methyl bromide prepared commercially to be made from methanol by reaction with hydrogen bromide.

Chlorination of methane provides approximately one third of the annual U.S. production of chloromethane. The reaction of methanol with hydrogen chloride is the major synthetic method for the preparation of chloromethane.

Chlorination of methane is carried out in the gas phase on an industrial scale to give a mixture of chloromethane (CH_3Cl), dichloromethane (CH_2Cl_2), trichloromethane ($CHCl_3$), and tetrachloromethane (CCl_4).

$$\underset{\text{Methane}}{CH_4} + \underset{\text{Chlorine}}{Cl_2} \xrightarrow{400\text{–}440°C} \underset{\substack{\text{Chloromethane} \\ \text{(bp −24°C)}}}{CH_3Cl} + \underset{\substack{\text{Hydrogen} \\ \text{chloride}}}{HCl}$$

$$\underset{\text{Chloromethane}}{CH_3Cl} + \underset{\text{Chlorine}}{Cl_2} \xrightarrow{400\text{–}440°C} \underset{\substack{\text{Dichloromethane} \\ \text{(bp 40°C)}}}{CH_2Cl_2} + \underset{\substack{\text{Hydrogen} \\ \text{chloride}}}{HCl}$$

$$\underset{\text{Dichloromethane}}{CH_2Cl_2} + \underset{\text{Chlorine}}{Cl_2} \xrightarrow{400\text{–}440°C} \underset{\substack{\text{Trichloromethane} \\ \text{(bp 61°C)}}}{CHCl_3} + \underset{\substack{\text{Hydrogen} \\ \text{chloride}}}{HCl}$$

$$\underset{\text{Trichloromethane}}{CHCl_3} + \underset{\text{Chlorine}}{Cl_2} \xrightarrow{400\text{–}440°C} \underset{\substack{\text{Tetrachloromethane} \\ \text{(bp 77°C)}}}{CCl_4} + \underset{\substack{\text{Hydrogen} \\ \text{chloride}}}{HCl}$$

One of the chief uses of chloromethane is as a starting material from which silicone polymers are made. Dichloromethane is widely used as a paint stripper. Trichloromethane (chloroform) was once used as an inhalation anesthetic, but its toxicity caused it to be replaced by safer materials many years ago. Tetrachloromethane is the starting material for chlorofluorocarbons (CFCs), at one time widely used as refrigerant gases. Most of the world's industrialized nations have agreed to phase out all uses of CFCs because these compounds have been implicated in atmospheric processes that degrade the Earth's ozone layer.

Dichloromethane, trichloromethane, and tetrachloromethane are widely known by their common names methylene chloride, chloroform, and carbon tetrachloride, respectively.

The mechanism of free-radical chlorination, to be presented in Section 10.3, is fundamentally different from the mechanism by which alcohols react with hydrogen halides. Alcohols are converted to alkyl halides in reactions involving ionic (or "polar") intermediates—alkyloxonium ions and carbocations. The intermediates in the chlorination of methane and other alkanes are free radicals, not ions.

10.3 Mechanism of Methane Chlorination

Mechanism 10.1 describes the sequence of steps in the generally accepted mechanism for the chlorination of methane. The reaction is normally carried out in the gas phase at high temperature (400–440°C). Although free-radical chlorination of methane is strongly exothermic, energy must be put into the system to initiate the reaction. This energy goes into breaking the weakest bond in the system, which, as we see from the bond dissociation enthalpy data in Table 10.1, is the Cl—Cl bond with a bond dissociation enthalpy of 243 kJ/mol (58 kcal/mol). The step in which Cl—Cl bond homolysis occurs is called the **initiation step.**

The bond dissociation enthalpy of the other reactant, methane, is much higher. It is 439 kJ/mol (105 kcal/mol).

Each chlorine atom formed in the initiation step has seven valence electrons and is very reactive. Once formed, it abstracts a hydrogen atom from methane as shown in step 2 in Mechanism 10.1. Hydrogen chloride, one of the isolated products from the overall reaction, is formed in this step. A methyl radical is also formed, which then reacts with a molecule of Cl_2 in step 3 giving chloromethane, the other product of the overall reaction, along with a chlorine atom, which cycles back to step 2, and the process repeats. Steps 2 and 3 are called **propagation steps** and, when added together, reproduce the net equation. Because one initiation step can result in a great many propagation cycles, the overall process is called a free-radical **chain reaction.**

Problem 10.4

Write equations for the initiation and propagation steps for the formation of dichloromethane by free-radical chlorination of chloromethane.

Mechanism 10.1

Free-Radical Chlorination of Methane

THE OVERALL REACTION:

$$CH_4 + Cl_2 \longrightarrow CH_3Cl + HCl$$

Methane, Chlorine, Chloromethane, Hydrogen chloride

THE MECHANISM:

(*a*) Initiation

Step 1: Dissociation of a chlorine molecule into two chlorine atoms:

$$:\ddot{\underset{..}{Cl}}-\ddot{\underset{..}{Cl}}: \longrightarrow 2[:\ddot{\underset{..}{Cl}}\cdot]$$

Chlorine molecule — Two chlorine atoms

(*b*) Chain propagation

Step 2: Hydrogen atom abstraction from methane by a chlorine atom:

$$:\ddot{\underset{..}{Cl}}\cdot + H-CH_3 \longrightarrow :\ddot{\underset{..}{Cl}}-H + \cdot CH_3$$

Chlorine atom, Methane, Hydrogen chloride, Methyl radical

Step 3: Reaction of methyl radical with molecular chlorine:

$$:\ddot{\underset{..}{Cl}}-\ddot{\underset{..}{Cl}}: + \cdot CH_3 \longrightarrow :\ddot{\underset{..}{Cl}}\cdot + :\ddot{\underset{..}{Cl}}-CH_3$$

Chlorine molecule, Methyl radical, Chlorine atom, Chloromethane

Steps 2 and 3 then repeat many times.

In practice, side reactions intervene to reduce the efficiency of the propagation steps. The chain sequence is interrupted whenever two odd-electron species combine to give an even electron product. Reactions of this type are called **chain-terminating steps.** Chain-terminating steps in the chlorination of methane include:

Combination of a methyl radical with a chlorine atom

$$\dot{C}H_3 \quad \cdot\ddot{\underset{..}{Cl}}: \longrightarrow CH_3-\ddot{\underset{..}{Cl}}:$$

Methyl radical, Chlorine atom, Chloromethane

Combination of two methyl radicals

$$\dot{C}H_3 \quad \dot{C}H_3 \longrightarrow CH_3-CH_3$$

Two methyl radicals, Ethane

Combination of two chlorine atoms

$$:\ddot{\underset{..}{Cl}}\cdot \quad \cdot\ddot{\underset{..}{Cl}}: \longrightarrow :\ddot{\underset{..}{Cl}}-\ddot{\underset{..}{Cl}}:$$

Two chlorine atoms, Chlorine molecule

Termination steps are, in general, less likely to occur than the propagation steps. Each of the termination steps requires two free radicals to encounter each other in a medium that contains far greater quantities of other materials (methane and molecular Cl_2) with which they can react. Although some chloromethane undoubtedly arises via combination of methyl radicals with chlorine atoms, most of it is formed by the propagation sequence shown in Mechanism 10.1.

10.4 Halogenation of Higher Alkanes

Like the chlorination of methane, chlorination of ethane is carried out on an industrial scale as a high-temperature gas-phase reaction.

$$\underset{\text{Ethane}}{CH_3CH_3} + \underset{\text{Chlorine}}{Cl_2} \xrightarrow{420°C} \underset{\substack{\text{Chloroethane (78\%)} \\ \text{(ethyl chloride)}}}{CH_3CH_2Cl} + \underset{\text{Hydrogen chloride}}{HCl}$$

Problem 10.5

Chlorination of ethane yields, in addition to ethyl chloride, a mixture of two isomeric dichlorides. What are the structures of these two dichlorides?

Reactions that occur when light energy—usually visible or ultraviolet—is absorbed by a molecule are called **photochemical reactions** irrespective of their mechanism. In laboratory-scale syntheses, it is often convenient to carry out free-radical halogenations photochemically at room temperature.

Photochemical energy is indicated by writing "light" or "$h\nu$" above or below the arrow. The symbol $h\nu$ is equal to the energy of a light photon and will be discussed in more detail in Section 14.1.

Cyclobutane + Cl_2 $\xrightarrow{h\nu}$ Chlorocyclobutane (73%) (cyclobutyl chloride) + HCl (Hydrogen chloride)

Cyclobutane Chlorine Chlorocyclobutane (73%) (cyclobutyl chloride) Hydrogen chloride

The three examples described so far—chlorination of methane, ethane, and cyclobutane—share the common feature that each can give only a *single* monochloro derivative. Chlorination of alkanes in which the hydrogens are not all equivalent is more complicated in that a mixture of every possible monochloro derivative is formed, as illustrated for the chlorination of butane:

The percentages cited in the accompanying equation reflect the composition of the monochloride fraction of the product mixture rather than the isolated yield of each component.

Butane $\xrightarrow[h\nu,\ 35°C]{Cl_2}$ 1-Chlorobutane (28%) (*n*-butyl chloride) + 2-Chlorobutane (72%) (*sec*-butyl chloride)

Constitutionally isomeric products arise because a chlorine atom may abstract a hydrogen atom from either a methyl or a methylene group in the propagation step.

Butane (H) + $\cdot\ddot{Cl}:$ (Chlorine atom) ⟶ Butyl radical + H—Cl: (Hydrogen chloride)

Butane (H) + $\cdot\ddot{Cl}:$ (Chlorine atom) ⟶ *sec*-Butyl radical + H—Cl: (Hydrogen chloride)

The resulting free radicals react with chlorine to give the corresponding alkyl chlorides. Butyl radical gives only 1-chlorobutane; *sec*-butyl radical gives only 2-chlorobutane.

Butyl radical + Chlorine → 1-Chlorobutane (butyl chloride) + Chlorine atom

sec-Butyl radical + Chlorine → 2-Chlorobutane (*sec*-butyl chloride) + Chlorine atom

If every collision of a chlorine atom with a butane molecule resulted in hydrogen atom abstraction, the butyl/*sec*-butyl radical ratio and, therefore, the 1-chloro/2-chlorobutane ratio, would be given by the relative numbers of hydrogens in the two equivalent methyl groups of butane (six) compared with those in the two equivalent methylene groups (four). The product distribution *expected* on this basis would be 60% 1-chlorobutane and 40% 2-chlorobutane. The *experimentally observed* product distribution, however, is much different: 28% 1-chlorobutane and 72% 2-chlorobutane. *sec*-Butyl radical is therefore formed in greater amounts, and butyl radical in lesser amounts, than expected on a statistical basis.

Mechanistically, this behavior stems from the greater stability of secondary compared with primary free radicals. The transition state for the step in which a chlorine atom abstracts a hydrogen from carbon has free-radical character at carbon.

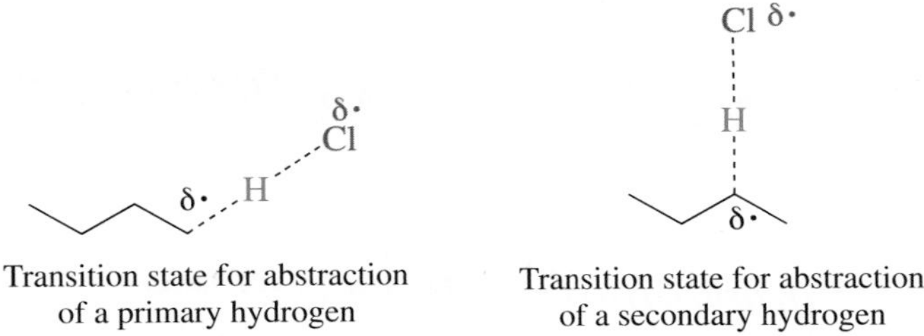

Transition state for abstraction of a primary hydrogen

Transition state for abstraction of a secondary hydrogen

A secondary hydrogen is abstracted faster than a primary hydrogen because the transition state with secondary radical character is of lower energy than the one with primary radical character. The same factors that stabilize a secondary radical stabilize a transition state with secondary radical character more than one with primary radical character and cause a hydrogen atom to be abstracted from a CH_2 group faster than one from a CH_3 group. We can calculate how much faster a *single* secondary hydrogen is abstracted compared with a *single* primary hydrogen from the experimentally observed product distribution.

$$\frac{72\%\ \text{2-chlorobutane}}{28\%\ \text{1-chlorobutane}} = \frac{\text{rate of secondary H abstraction} \times 4\ \text{secondary hydrogens}}{\text{rate of primary H abstraction} \times 6\ \text{primary hydrogens}}$$

$$\frac{\text{Rate of secondary H abstraction}}{\text{Rate of primary H abstraction}} = \frac{72}{28} \times \frac{6}{4} = \frac{3.9}{1}$$

A single secondary hydrogen in butane is abstracted by a chlorine atom 3.9 times faster than a single primary hydrogen.

Problem 10.6

Assuming the relative rate of secondary to primary hydrogen atom abstraction to be the same in the chlorination of propane as it is in that of butane, calculate the relative amounts of propyl chloride and isopropyl chloride obtained in the free-radical chlorination of propane.

A similar study of the chlorination of 2-methylpropane established that a tertiary hydrogen is removed 5.2 times faster than each primary hydrogen.

2-Methylpropane $\xrightarrow[h\nu,\ 35°C]{Cl_2}$ 1-Chloro-2-methylpropane (63%) (isobutyl chloride) + 2-Chloro-2-methylpropane (37%) (*tert*-butyl chloride)

Problem 10.7

Do the arithmetic involved in converting the preceding product composition to a 5.2:1 ratio in the rate of abstraction of a tertiary versus a primary hydrogen in 2-methylpropane by a chlorine atom.

Problem 10.8

How many constitutionally isomeric monochlorination products are possible from each of the following?

(a) 2-Methylpentane
(b) 3-Methylpentane
(c) 2,2-Dimethylbutane
(d) 2,3-Dimethylbutane

Sample Solution

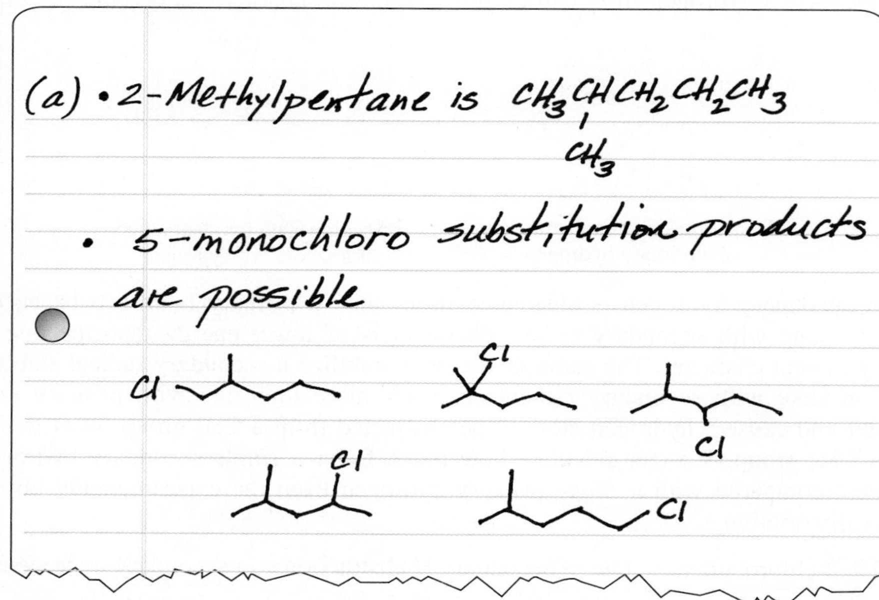
(a) • 2-Methylpentane is CH3CHCH2CH2CH3
CH3
• 5-monochloro substitution products
are possible
Cl
Cl
Cl
Cl
Cl

In summary, the chlorination of alkanes is not very selective. The various kinds of hydrogens present in a molecule (tertiary, secondary, and primary) differ by only a factor of 5 in the relative rate at which each reacts with a chlorine atom.

	R_3CH >	R_2CH_2 >	RCH_3
	(tertiary)	(secondary)	(primary)
Relative rate (chlorination)	5.2	3.9	1

Bromine reacts with alkanes by a free-radical chain mechanism analogous to that of chlorine. There is an important difference between chlorination and bromination, however.

Bromination is highly selective for substitution of *tertiary hydrogens.* The spread in reactivity among primary, secondary, and tertiary hydrogens is greater than 10^3.

	R_3CH >	R_2CH_2 >	RCH_3
	(tertiary)	(secondary)	(primary)
Relative rate (bromination)	1640	82	1

In practice, this means that when an alkane contains primary, secondary, and tertiary hydrogens, it is usually only the tertiary hydrogen that is replaced by bromine.

2-Methylpentane $\xrightarrow[hv,\ 60°C]{Br_2}$ 2-Bromo-2-methylpentane (76% isolated yield)

The percentage cited in this reaction is the isolated yield of purified product. Isomeric bromides constitute only a tiny fraction of the product.

We can understand why bromination is more selective than chlorination by using bond dissociation enthalpies (Table 10.1) to calculate the energy changes for the propagation step in which each halogen atom abstracts a hydrogen from ethane.

$$CH_3CH_2{-}H + \cdot\ddot{Cl}: \longrightarrow CH_3\dot{C}H_2 + H{-}\ddot{Cl}: \quad \Delta H° = -11\ \text{kJ/mol}\ (-2.5\ \text{kcal/mol})$$

Ethane, Chlorine atom, Ethyl radical, Hydrogen chloride

$$CH_3CH_2{-}H + \cdot\ddot{Br}: \longrightarrow CH_3\dot{C}H_2 + H{-}\ddot{Br}: \quad \Delta H° = +54\ \text{kJ/mol}\ (+13\ \text{kcal/mol})$$

Ethane, Bromine atom, Ethyl radical, Hydrogen bromide

The alkyl radical-forming step is exothermic for chlorination and endothermic for bromination. Applying Hammond's postulate to these elementary steps, we conclude that alkyl radical character is more highly developed in the transition state for abstraction of hydrogen by a bromine atom than by a chlorine atom. Thus, bromination is more sensitive to the stability of the free-radical intermediate than chlorination and more selective.

Problem 10.9

Give the structure of the major organic product formed by free-radical bromination of each of the following:

(a) Methylcyclopentane
(b) 2,2,4-Trimethylpentane
(c) 1-Isopropyl-1-methylcyclopentane

Sample Solution (a) Write the structure of the starting hydrocarbon, and identify any tertiary hydrogens that are present. The only tertiary hydrogen in methylcyclopentane is the one attached to C-1. This is the one replaced by bromine.

Methylcyclopentane $\xrightarrow[\text{light}]{Br_2}$ 1-Bromo-1-methylcyclopentane

The difference in selectivity between chlorination and bromination of alkanes needs to be kept in mind when one wishes to prepare an alkyl halide from an alkane:

1. *Chlorination* of an alkane yields every possible monochloride, so is used only when all the hydrogens in an alkane are equivalent.
2. *Bromination* of alkanes is highly regioselective for replacing tertiary hydrogens, so is mainly used to prepare tertiary alkyl bromides.

In Section 10.5 we'll see a second free-radical method for preparing alkyl halides from hydrocarbons—the addition of hydrogen bromide to *alkenes* under conditions different from those involving electrophilic addition in Section 8.4.

10.5 Free-Radical Addition of Hydrogen Bromide to Alkenes and Alkynes

The regioselectivity of addition of hydrogen bromide to alkenes puzzled chemists for a long time. In contrast to the HCl and HI additions to alkenes that faithfully obeyed Markovnikov's rule, HBr sometimes added in accordance with the rule, while at other times, seemingly under the same conditions, it added opposite to it. After hundreds of experiments during the period 1929–1933, Morris Kharasch and his students at the University of Chicago found that Markovnikov's rule was followed when peroxides were carefully excluded from the reaction mixture, but addition occurred opposite to the rule when peroxides were intentionally added.

1-Butene + HBr (Hydrogen bromide) —no peroxides→ 2-Bromobutane (only product; 90% yield)

1-Butene + HBr (Hydrogen bromide) —peroxides→ 1-Bromobutane (only product; 95% yield)

Kharasch called this the **peroxide effect** and proposed that the difference in regioselectivity was due to a peroxide-induced change in mechanism. The conventional electrophilic addition pathway via a carbocation is responsible for Markovnikov addition and operates in the absence of peroxides; the other mechanism involves a free-radical intermediate (Mechanism 10.2).

Problem 10.10

Kharasch's earliest studies in this area were carried out in collaboration with graduate student Frank R. Mayo. Mayo performed over 400 experiments in which allyl bromide (3-bromo-1-propene) was treated with hydrogen bromide under a variety of conditions, and determined the distribution of the "normal" and "abnormal" products formed during the reaction. What two products were formed? Which is the product of addition in accordance with Markovnikov's rule? Which one corresponds to addition opposite to the rule?

Like free-radical chlorination of methane described earlier in this chapter, the free-radical addition of hydrogen bromide to 1-butene (Mechanism 10.2) is characterized by *initiation* and *chain propagation* stages. The initiation stage, however, involves two steps rather than one, and it is this "extra" step that accounts for the role of peroxides. Peroxides are *initiators;* they are not incorporated into the product but act as a source of radicals necessary to get the chain reaction started.

The regioselectivity of *electrophilic* addition of HBr to alkenes is controlled by the tendency of a proton to add to the double bond to produce the more stable *carbocation*. Under *free-radical* conditions the regioselectivity is governed by addition of a bromine atom to give the more stable alkyl *radical*. For the case of 1-butene, electrophilic addition involves a secondary carbocation, free-radical addition involves a secondary free radical.

Electrophilic addition:

1-Butene —HBr→ 2-Bromobutane via 1-Methylpropyl cation

Mechanism 10.2

Free-Radical Addition of Hydrogen Bromide to 1-Butene

THE OVERALL REACTION:

1-Butene + Hydrogen bromide (HBr) $\xrightarrow[\text{light or heat}]{\text{ROOR}}$ 1-Bromobutane

THE MECHANISM:

(*a*) Initiation

Step 1: The weak O—O bond of the peroxide undergoes homolytic dissociation to give two alkoxy radicals.

RÖ—ÖR (Peroxide) $\xrightarrow{\text{light or heat}}$ RÖ· + ·ÖR (Two alkoxy radicals)

Step 2: An alkoxy radical abstracts a hydrogen atom from hydrogen bromide, generating a bromine atom and setting the stage for a chain reaction.

RÖ· (Alkoxy radical) + H—Br: (Hydrogen bromide) ⟶ RÖ—H (Alcohol) + ·Br: (Bromine atom)

(*b*) Chain propagation

Step 3: The regiochemistry of addition is set in this step. Bromine bonds to C-1 of 1-butene to give a secondary radical. If it had bonded to C-2, a less stable primary radical would have resulted.

1-Butene (carbons numbered 4, 3, 2, 1) + ·Br: (Bromine atom) ⟶ 1-(Bromomethyl)propyl radical

Step 4: The radical produced in step 3 abstracts a hydrogen atom from hydrogen bromide giving the product 1-bromobutane. This hydrogen abstraction also generates a bromine atom, which reacts with another molecule of alkene as in step 3.

:Br—H (Hydrogen bromide) + 1-(Bromomethyl)propyl radical ⟶ ·Br: (Bromine atom) + 1-Bromobutane

Steps 3 and 4 repeat many times unless interrupted by chain termination steps.

Free-radical addition:

1-Butene —HBr / peroxides→ 1-Bromobutane via 1-(Bromomethyl)propyl radical

Problem 10.11

Problem 8.5 asked you to predict the major organic product for addition of HCl to each of the following alkenes. Do the same for the addition of HBr to these alkenes, comparing the products formed in the absence of peroxides and in their presence.

(a) 2-Methyl-2-butene

(b) *cis*-2-Butene

(c) 2-Methyl-1-butene

(d)

Sample Solution (a) The addition of hydrogen bromide in the absence of peroxides exhibits a regioselectivity just like that of hydrogen chloride addition in that Markovnikov's rule is followed.

2-Methyl-2-butene + Hydrogen bromide (HBr) —no peroxides→ 2-Bromo-2-methylbutane

Under free-radical conditions in the presence of peroxides, however, addition takes place with a regioselectivity opposite to Markovnikov's rule.

2-Methyl-2-butene + Hydrogen bromide (HBr) —peroxides→ 2-Bromo-3-methylbutane

Free-radical addition of hydrogen bromide to alkenes can also be initiated photochemically, either with or without added peroxides.

Methylenecyclopentane + Hydrogen bromide (HBr) —$h\nu$→ Bromomethylcyclopentane (60%)

Although the possibility of having two different reaction paths available to an alkene and hydrogen bromide may seem like a complication, it can be an advantage in organic synthesis. It is often possible to regioselectively prepare either of two different alkyl bromides by choosing reaction conditions that favor electrophilic addition or free-radical addition of hydrogen bromide.

Problem 10.12

Electrophilic addition of HBr to $H_2C{=}CHCH(CH_3)_2$ gives a mixture of two constitutional isomers A and B. Only B is formed, however, when $CH_3CH{=}C(CH_3)_2$ reacts with HBr in the presence of peroxides. Identify A and B and explain your reasoning.

Hydrogen bromide (but not hydrogen chloride or hydrogen iodide) adds to alkynes by a free-radical mechanism when peroxides are present in the reaction mixture. As in the free-radical addition of hydrogen bromide to alkenes, the regioselectivity is opposite to Markovnikov's rule.

1-Hexyne + HBr (Hydrogen bromide) $\xrightarrow{\text{peroxides}}$ (*E* + *Z*)-1-Bromo-1-hexene (79%)

10.6 Metal–Ammonia Reduction of Alkynes

A useful alternative to catalytic partial hydrogenation for converting alkynes to alkenes is reduction by a Group 1 metal (lithium, sodium, or potassium) in liquid ammonia. The unique feature of metal–ammonia reduction is that it converts alkynes to trans alkenes, whereas catalytic hydrogenation yields cis. Thus, from the same alkyne one can prepare either a cis or a trans alkene by choosing the appropriate reaction conditions.

3-Hexyne $\xrightarrow[NH_3]{Na}$ *trans*-3-Hexene (82%)

Problem 10.13

Suggest an efficient synthesis of *trans*-2-heptene from propyne and any necessary organic or inorganic reagents.

The stereochemistry of metal–ammonia reduction of alkynes differs from that of catalytic hydrogenation because the mechanisms of the two reactions are different. The mechanism of hydrogenation of alkynes (Section 9.9) is similar to that of catalytic hydrogenation of alkenes (Section 8.2). Metal–ammonia reduction of alkynes is outlined in Mechanism 10.3.

The mechanism includes two single-electron transfers (steps 1 and 3) and two proton transfers (steps 2 and 4). Experimental evidence indicates that step 2 is rate-determining, and that the (*E*)- and (*Z*)-alkenyl radicals formed in this step interconvert rapidly.

(*Z*)-Alkenyl radical (less stable) ⇌ (*E*)-Alkenyl radical (more stable)

Reduction of these alkenyl radicals (step 3) gives a mixture of the (*E*)- and (*Z*)-alkenyl anions in which the more stable *E* stereoisomer predominates. Unlike the corresponding alkenyl radicals, the (*E*)- and (*Z*)-alkenyl anions are configurationally stable under the reaction conditions and yield an *E*/*Z* ratio of alkenes in step 4 that reflects the *E*/*Z* ratio of the alkenyl anions formed in step 3.

Mechanism 10.3

Sodium–Ammonia Reduction of an Alkyne

THE OVERALL REACTION:

On dissolving in liquid ammonia, sodium atoms dissociate into sodium ions and electrons, both of which are solvated by ammonia. To reflect this, the solvated electrons are represented in the equation as $e^-(am)$.

$$\mathrm{RC{\equiv}CR'} + 2e^-(am) + 2\mathrm{NH_3} \longrightarrow (E)\text{-RCH=CHR}' + 2\mathrm{NH_2^-}$$

Alkyne — Electrons — Ammonia — (*E*)-Alkene — Amide ion

THE MECHANISM:

Step 1: *Electron transfer.* An electron adds to one of the triply bonded carbons to give an anion radical.

$$\mathrm{RC{\equiv}CR'} + e^-(am) \underset{}{\overset{\text{fast}}{\rightleftharpoons}} \mathrm{R\dot{C}{=}\ddot{C}R'}$$

Alkyne — Electron — Anion radical

Step 2: *Proton transfer.* The anion radical formed in the first step is strongly basic and abstracts a proton from ammonia. This is believed to be the rate-determining step. The alkenyl radical that results is a mixture of rapidly equilibrating *E* and *Z* stereoisomers.

$$\mathrm{R\dot{C}{=}\ddot{C}R'} + \mathrm{H{-}\ddot{N}H_2} \xrightarrow{\text{slow}} \mathrm{R\dot{C}{=}CHR'} + \mathrm{:\ddot{N}H_2^-}$$

Anion radical — Ammonia — Alkenyl radical (*E*/*Z* mixture) — Amide ion

Step 3: *Electron transfer.* The alkenyl radical reacts with a solvated electron to give a vinyl anion. The more stable *E*-alkenyl anion predominates and *E*–*Z* equilibration is slow.

$$\mathrm{R\dot{C}{=}CHR'} + e^-(am) \xrightarrow{\text{fast}} \mathrm{R\ddot{C}{=}CHR'}$$

Alkenyl radical (*E*/*Z* mixture) — Electron — Alkenyl anion (mainly *E*)

Step 4: *Proton transfer.* The alkenyl anion abstracts a proton from ammonia to form the alkene. The *E*/*Z* ratio of the product reflects the *E*/*Z* ratio of the alkenyl anion.

$$\mathrm{H_2\ddot{N}{-}H} + \mathrm{R\ddot{C}{=}CHR'} \xrightarrow{\text{fast}} \mathrm{RCH{=}CHR'} + \mathrm{H_2\ddot{N}:^-}$$

Ammonia — Alkenyl anion (mainly *E*) — Alkene (mainly *E*) — Amide ion

10.7 Free Radicals and Retrosynthesis of Alkyl Halides

Consider the synthesis of the primary alkyl bromide 1-bromo-2,3,3-trimethylbutane from the alkane shown.

2,2,3-Trimethylbutane → 1-Bromo-2,3,3-trimethylbutane

The only method we have learned so far for introducing a **functional group** on to an alkane skeleton is free-radical halogenation. Direct bromination, however, would introduce bromine at the tertiary carbon of 2,2,3-trimethylbutane. Therefore, an indirect approach such as the following is required.

1-Bromo-2,3,3-trimethylbutane ⟹ 2,3,3-Trimethyl-1-butene ⟹ 2-Bromo-2,3,3-trimethylbutane ⟹ 2,2,3-Trimethylbutane

Written in the forward direction with the appropriate reagents requires two free-radical reactions—the first synthetic step introduces functionality via direct bromination. The second step is an E2 elimination to give an alkene, while the third step provides the desired regioselectivity.

2,2,3-Trimethylbutane $\xrightarrow[h\nu]{Br_2}$ 2-Bromo-2,3,3-trimethylbutane $\xrightarrow[CH_3CH_2OH]{NaOCH_2CH_3}$ 2,3,3-Trimethyl-1-butene $\xrightarrow[h\nu]{HBr}$ 1-Bromo-2,3,3-trimethylbutane

Problem 10.14

Replacing HBr by HI in the third step of the synthesis shown would not provide a suitable synthesis for 1-iodo-2,3,3-trimethylbutane. Why not? How could you extend the synthesis shown by an additional step to give the corresponding iodide?

10.8 Free-Radical Polymerization of Alkenes

The boxed essay *Ethylene and Propene: The Most Important Industrial Organic Chemicals* summarizes the main uses of these two alkenes, especially their polymerization to *polyethylene* and *polypropylene,* respectively. Of the methods used to prepare polyethylene, the oldest involves free radicals and is carried out by heating ethylene under pressure in the presence of oxygen or a peroxide initiator.

$$n(H_2C{=}CH_2) \xrightarrow[O_2 \text{ or peroxides}]{200°C,\ 2000 \text{ atm}} -CH_2-CH_2-(CH_2-CH_2)_{n-2}-CH_2-CH_2-$$

Ethylene — Polyethylene

In this reaction, n can have a value of thousands.

Mechanism 10.4 shows the steps in the free-radical polymerization of ethylene. Dissociation of a peroxide initiates the process in step 1. The resulting peroxy radical adds to the carbon–carbon double bond in step 2, giving a new radical, which then adds to a second molecule of ethylene in step 3. The carbon–carbon bond-forming process in step 3 can be repeated thousands of times to give long carbon chains. In spite of the *-ene* ending to its

Mechanism 10.4

Free-Radical Polymerization of Ethylene

THE OVERALL REACTION:

$$n(\mathrm{H_2C{=}CH_2}) \longrightarrow \mathrm{+CH_2CH_2+}_n$$

Ethylene → Polyethylene

THE MECHANISM:

Step 1: Homolytic dissociation of a peroxide produces alkoxy radicals that serve as free-radical initiators:

$$\mathrm{R\ddot{O}{-}\ddot{O}R} \longrightarrow \mathrm{R\ddot{O}\cdot} + \mathrm{\cdot\ddot{O}R}$$

Peroxide → Two alkoxy radicals

Step 2: An alkoxy radical adds to the carbon–carbon double bond:

$$\mathrm{R\ddot{O}\cdot} + \mathrm{H_2C{=}CH_2} \longrightarrow \mathrm{R\ddot{O}{-}CH_2{-}\dot{C}H_2}$$

Alkoxy radical + Ethylene → 2-Alkoxyethyl radical

Step 3: The radical produced in step 2 adds to a second molecule of ethylene:

$$\mathrm{R\ddot{O}{-}CH_2{-}\dot{C}H_2} + \mathrm{H_2C{=}CH_2} \longrightarrow \mathrm{R\ddot{O}{-}CH_2{-}CH_2{-}CH_2{-}\dot{C}H_2}$$

2-Alkoxyethyl radical + Ethylene → 4-Alkoxybutyl radical

The radical formed in step 3 then adds to a third molecule of ethylene, and the process continues, forming a long chain of methylene groups.

name, polyethylene is much more closely related to alkanes than to alkenes. It is simply a long chain of CH_2 groups bearing at its ends an alkoxy group (from the initiator) or a carbon–carbon double bond.

The properties that make polyethylene so useful come from its alkane-like structure. Except for the ends of the chain, which make up only a tiny portion of the molecule, polyethylene has no functional groups so is almost completely inert to most substances with which it comes in contact.

Teflon is made in a similar way by free-radical polymerization of tetrafluoroethylene.

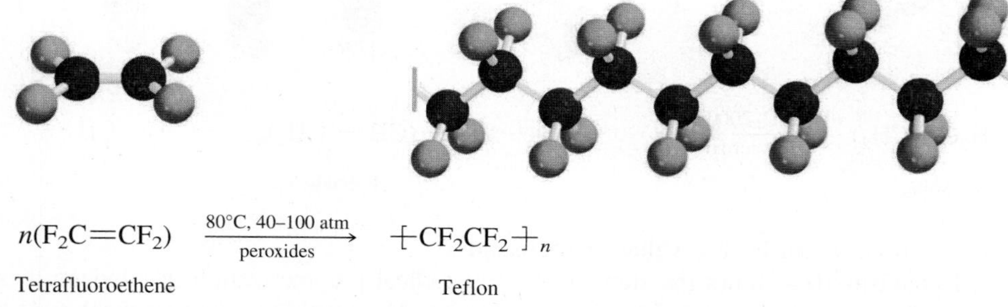

$$n(\mathrm{F_2C{=}CF_2}) \xrightarrow[\text{peroxides}]{80^\circ\mathrm{C},\ 40\text{–}100\ \mathrm{atm}} \mathrm{+CF_2CF_2+}_n$$

Tetrafluoroethene → Teflon

Carbon–fluorine bonds are quite strong (slightly stronger than C—H bonds), and like polyethylene, Teflon is a very stable, inert material. It is known by its "nonstick" surface, which can be understood by comparing it with polyethylene. The high electronegativity of fluorine makes C—F bonds less polarizable than C—H bonds, causing the induced-dipole/induced-dipole attractive forces to be weaker than in polyethylene and the surface to be slicker.

Problem 10.15

The materials shown in Table 10.2 are classified as *vinyl polymers* because the starting material, the *monomer,* contains a carbon–carbon double bond. Super Glue sticks because of its ready conversion to the vinyl polymer shown. What is the monomer?

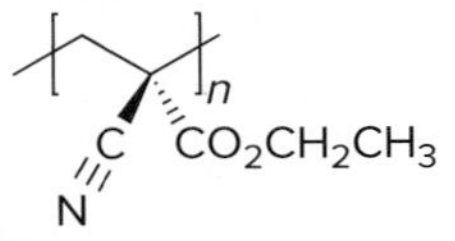

$$\left[\text{—}CH_2\text{—}C(C{\equiv}N)(CO_2CH_2CH_3)\text{—}\right]_n$$

A large number of compounds with carbon–carbon double bonds have been polymerized to yield materials with useful properties. Some of the more familiar ones are listed in Table 10.2. Not all are effectively polymerized under free-radical conditions, and much research has been carried out to develop alternative methods. The most notable of these, **coordination polymerization,** employs transition metal catalysts and is used to prepare polypropylene. Coordination polymerization is described in Sections 15.15 and 27.7.

Ethylene and Propene: The Most Important Industrial Organic Chemicals

Having examined the properties of alkenes and introduced the elements of polymers and polymerization, let's now look at some commercial applications of ethylene and propene.

Ethylene We discussed ethylene production in an earlier boxed essay (Section 7.2), where it was pointed out that the output of the U.S. petrochemical industry exceeds 5×10^{10} lb/year. Approximately 90% of this material is used for the preparation of four compounds (polyethylene, ethylene oxide, vinyl chloride, and styrene), with polymerization to polyethylene accounting for half the total. Vinyl chloride and styrene are polymerized to give poly(vinyl chloride) and polystyrene, respectively. Ethylene oxide is a starting material for the preparation of ethylene glycol for use as an antifreeze in automobile radiators and in the production of polyester fibers.

Propene The major use of propene is in the production of polypropylene. Two other propene-derived organic chemicals, acrylonitrile and propylene oxide, are also starting materials for polymer synthesis. Acrylonitrile is used to make acrylic fibers, and propylene oxide is one component in the preparation of *polyurethane* polymers. Cumene itself has no direct uses but rather serves as the starting material in a process that yields two valuable industrial chemicals: acetone and phenol.

We have not indicated the reagents employed in the reactions by which ethylene and propene are converted to the compounds shown. Because of patent requirements, different companies often use different processes. Although the processes may be different, they share the common characteristic of being extremely efficient. The industrial chemist faces the challenge of producing valuable materials, at low cost. Success in the industrial environment requires both an understanding of chemistry and an appreciation of the economics associated with alternative procedures.

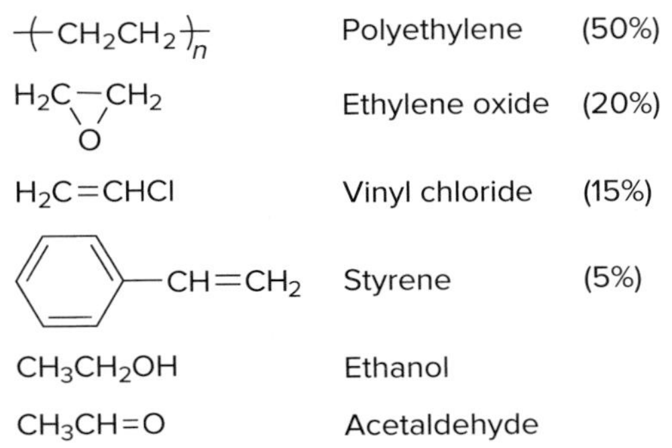

$\left[CH_2CH_2\right]_n$	Polyethylene	(50%)
$H_2C\text{—}CH_2$ (epoxide, O bridge)	Ethylene oxide	(20%)
$H_2C{=}CHCl$	Vinyl chloride	(15%)
$C_6H_5\text{—}CH{=}CH_2$	Styrene	(5%)
CH_3CH_2OH	Ethanol	
$CH_3CH{=}O$	Acetaldehyde	

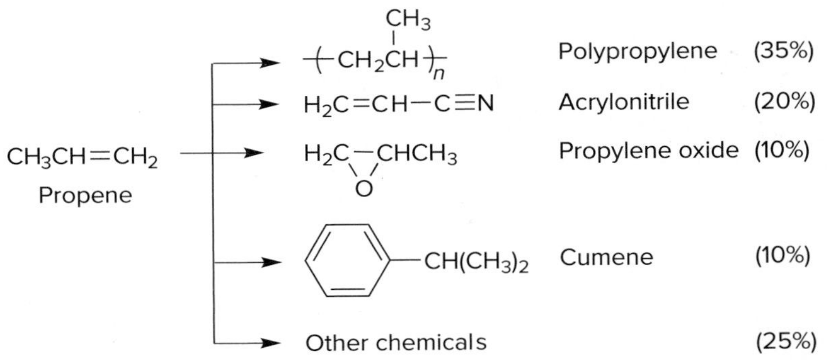

$CH_3CH{=}CH_2$ (Propene) →

$\left[CH_2CH(CH_3)\right]_n$	Polypropylene	(35%)
$H_2C{=}CH\text{—}C{\equiv}N$	Acrylonitrile	(20%)
$H_2C\text{—}CHCH_3$ (epoxide, O bridge)	Propylene oxide	(10%)
$C_6H_5\text{—}CH(CH_3)_2$	Cumene	(10%)
Other chemicals		(25%)

TABLE 10.2 Some Compounds with Carbon–Carbon Double Bonds Used to Prepare Polymers

A. Alkenes of the type $H_2C{=}CH{-}X$ used to form polymers of the type $\left(CH_2{-}\underset{X}{\underset{|}{CH}}\right)_n$

Compound	Structure	—X in polymer	Application
Ethylene	$H_2C{=}CH_2$	$-H$	Polyethylene films as packaging material; "plastic" squeeze bottles are molded from high-density polyethylene.
Propene	$H_2C{=}CH{-}CH_3$	$-CH_3$	Polypropylene fibers for use in carpets and automobile tires; consumer items (luggage, appliances, etc.); packaging material.
Styrene	$H_2C{=}CH{-}C_6H_5$ (phenyl ring)	$-C_6H_5$ (phenyl ring)	Polystyrene packaging, housewares, luggage, radio and television cabinets.
Vinyl chloride	$H_2C{=}CH{-}Cl$	$-Cl$	Poly(vinyl chloride) (PVC) has replaced leather in many of its applications; PVC tubes and pipes are often used in place of copper.
Acrylonitrile	$H_2C{=}CH{-}C{\equiv}N$	$-C{\equiv}N$	Wool substitute in sweaters, blankets, etc.

B. Alkenes of the type $H_2C{=}CX_2$ used to form polymers of the type $\left(CH_2{-}CX_2\right)_n$

Compound	Structure	X in polymer	Application
1,1-Dichloroethene (vinylidene chloride)	$H_2C{=}CCl_2$	Cl	Saran used as air- and watertight packaging film.
2-Methylpropene	$H_2C{=}C(CH_3)_2$	CH_3	Polyisobutylene is component of "butyl rubber," one of earliest synthetic rubber substitutes.

C. Others

Compound	Structure	Polymer	Application
Tetrafluoroethene	$F_2C{=}CF_2$	$\left(CF_2{-}CF_2\right)_n$ (Teflon)	Nonstick coating for cooking utensils; bearings, gaskets, and fittings.
Methyl methacrylate	$H_2C{=}\underset{CH_3}{\underset{|}{C}}CO_2CH_3$	$\left(CH_2{-}\underset{CH_3}{\underset{|}{\overset{CO_2CH_3}{\overset{|}{C}}}}\right)_n$	When cast in sheets, is transparent; used as glass substitute (Lucite, Plexiglas).
2-Methyl-1,3-butadiene	$H_2C{=}\underset{CH_3}{\underset{|}{C}}CH{=}CH_2$	$\left(CH_2\underset{CH_3}{\underset{|}{C}}{=}CH{-}CH_2\right)_n$ (Polyisoprene)	Synthetic rubber.

10.9 SUMMARY

Section 10.1 Alkyl radicals are neutral species in which one of the carbons has an unpaired electron and three substituents. Methyl radical is planar and sp^2-hybridized with its unpaired electron occupying a $2p$ orbital.

Like carbocations, free radicals are stabilized by alkyl substituents. Tertiary alkyl radicals are more stable than secondary, secondary are more stable than primary, and primary radicals more stable than methyl.

Section 10.2 Alkanes react with halogens by substitution of a halogen for a hydrogen on the alkane.

$$\underset{\text{Alkane}}{RH} + \underset{\text{Halogen}}{X_2} \longrightarrow \underset{\text{Alkyl halide}}{RX} + \underset{\text{Hydrogen halide}}{HX}$$

The reactivity of the halogens decreases in the order $F_2 > Cl_2 > Br_2 > I_2$. The ease of replacing a hydrogen decreases in the order tertiary > secondary > primary > methyl.

Section 10.3 Chlorination of methane, and halogenation of alkanes generally, proceed by way of free-radical intermediates. Elementary steps 1 through 3 describe the mechanism.

1. (initiation step) $$\underset{\text{Halogen molecule}}{X_2} \longrightarrow \underset{\text{Two halogen atoms}}{2X\cdot}$$

2. (propagation step) $$\underset{\text{Alkane}}{RH} + \underset{\text{Halogen atom}}{X\cdot} \longrightarrow \underset{\text{Alkyl radical}}{R\cdot} + \underset{\text{Hydrogen halide}}{HX}$$

3. (propagation step) $$\underset{\text{Alkyl radical}}{R\cdot} + \underset{\text{Halogen molecule}}{X_2} \longrightarrow \underset{\text{Alkyl halide}}{RX} + \underset{\text{Halogen atom}}{\cdot X}$$

Section 10.4 Among alkanes, tertiary hydrogens are replaced faster than secondary, and secondary faster than primary. Chlorination is not very selective and is used only when all the hydrogens of the alkane are equivalent. Bromination is highly selective, replacing tertiary hydrogens much more readily than secondary or primary ones.

2,2,3-Trimethylbutane $\xrightarrow[\text{light}]{Br_2}$ 2-Bromo-2,3,3-trimethylbutane (80%)

Section 10.5 Hydrogen bromide is unique among the hydrogen halides in that it can add to alkenes by either electrophilic or free-radical addition. Under photochemical conditions or in the presence of peroxides, free-radical addition is observed, and HBr adds to the double bond with a regioselectivity opposite to that of Markovnikov's rule.

$$\text{Methylenecyclopentane} + \text{HBr (Hydrogen bromide)} \xrightarrow{h\nu} \text{Bromomethylcyclopentane (60\%)}$$

Section 10.6 Group 1 metals—sodium is usually employed—in liquid ammonia as the solvent convert alkynes to trans alkenes. The reaction proceeds by a four-step sequence in which electron-transfer and proton-transfer steps alternate.

$$\text{2-Hexyne} \xrightarrow[NH_3]{Na} \textit{trans}\text{-2-Hexene (69\%)}$$

Section 10.7 Free-radical reactions often provide alternative routes to useful compounds when planning a synthesis. In one example, free-radical halogenation offers a method to attach a functional group to an otherwise unreactive alkane framework. In another, free-radical addition of hydrogen bromide to alkenes occurs with a regioselectivity opposite to that observed under more conventional ionic conditions.

Section 10.8 In their polymerization, many individual alkene molecules combine to give a high-molecular-weight product. Several economically important polymers such as polystyrene and low-density polyethylene are prepared by free-radical processes.

PROBLEMS

Structure and Bonding

10.16 Carbon–carbon bond dissociation enthalpies have been measured for many alkanes. Without referring to Table 10.1, identify the alkane in each of the following pairs that has the lower carbon–carbon bond-dissociation enthalpy, and explain the reason for your choice.

(a) Ethane or propane
(b) Propane or 2-methylpropane
(c) 2-Methylpropane or 2,2-dimethylpropane
(d) Cyclobutane or cyclopentane

10.17 (a) Use the bond dissociation enthalpy data in Table 10.1 to calculate $\Delta H°$ for the reaction of methane with a bromine atom.

(b) The activation energy for this reaction is 76 kJ/mol (18.3 kcal/mol). Sketch a potential energy diagram for it, labeling reactants, product, and the transition state. Does the transition state more closely resemble reactants or products?

10.18 Use the bond dissociation enthalpy data in Table 10.1 to calculate $\Delta H°$ for (a) the monofluorination and (b) the monoiodination of methane. What is the main reason the values are so different?

Reactions

10.19 Write the structure of the major organic product formed in the reaction of each of the following with hydrogen bromide in the absence of peroxides and in their presence.

(a) 1-Pentene

(b) 2-Methyl-2-butene

(c) 1-Methylcyclohexene

10.20 (a) Excluding enantiomers, free-radical chlorination of bicyclo[2.2.1]heptane yields four monochloro derivatives but bicyclo[2.2.2]octane gives only two. Draw structural formulas for each.

Bicyclo[2.2.1]heptane Bicyclo[2.2.2]octane

(b) Which of the compounds in your answer to part (a) are chiral? Draw the structural formula of the enantiomer of each chiral monochloro derivative.

10.21 What is the product of each of the following reactions?

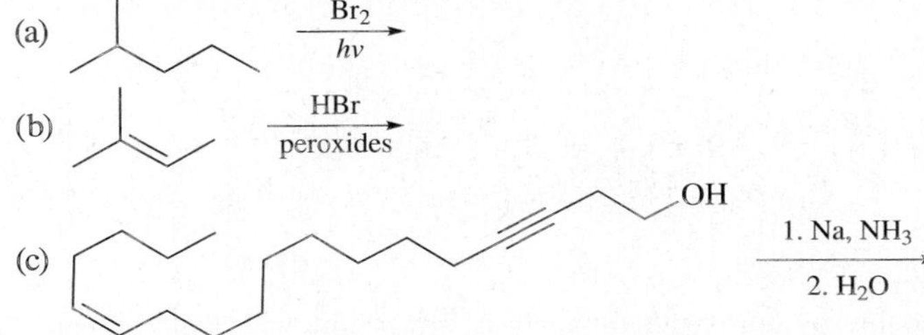

10.22 Including stereoisomers, how many compounds are likely to be formed on free-radical addition of HBr to *cis*-2-pentene? Write a stereochemically accurate formula for each and specify the configuration at each chirality center as *R* or *S*.

10.23 Photochemical chlorination of 1,2-dibromoethane gives a mixture of two stereoisomers of $C_2H_3Br_2Cl$. Write their structural formulas. Are they enantiomers or diastereomers? Are they chiral or achiral? Are they formed in equal amounts?

10.24 Photochemical chlorination of $(CH_3)_3CCH_2C(CH_3)_3$ gave a mixture of two monochlorides in a 4:1 ratio. The structures of these two products were assigned on the basis of their S_N1 hydrolysis rates in aqueous ethanol. The major product (compound A) underwent hydrolysis much more slowly than the minor one (compound B). Deduce the structures of compounds A and B and discuss the factors that influence their distribution.

10.25 Compound A (C_6H_{14}) gives three different monochlorides on photochemical chlorination. One of these monochlorides is inert to E2 elimination. The other two yield the same alkene B (C_6H_{12}) as the only product on being heated with potassium *tert*-butoxide in *tert*-butyl alcohol. Identify compound A, the three monochlorides, and alkene B.

10.26 In both of the following exercises, assume that all the methylene groups in the alkane are equally reactive as sites of free-radical chlorination.

(a) Photochemical chlorination of heptane gave a mixture of monochlorides containing 15% 1-chloroheptane. What other monochlorides are present? Estimate the percentage of each of these additional $C_7H_{15}Cl$ isomers in the monochloride fraction.

(b) Photochemical chlorination of dodecane gave a monochloride fraction containing 19% 2-chlorododecane. Estimate the percentage of 1-chlorododecane present in that fraction.

10.27 Photochemical chlorination of 2,2,4-trimethylpentane gives four isomeric monochlorides.

(a) Write structural formulas for these four isomers.

(b) The two primary chlorides make up 65% of the monochloride fraction. Assuming that all the primary hydrogens in 2,2,4-trimethylpentane are equally reactive, estimate the percentage of each of the two primary chlorides in the product mixture.

10.28 Photochemical chlorination of pentane gave a mixture of three constitutionally isomeric monochlorides. The principal monochloride constituted 46% of the total, and the remaining 54% was approximately a 1:1 mixture of the other two isomers. Write structural formulas for the three monochloride isomers and specify which one was formed in greatest amount. (Recall that a secondary hydrogen is abstracted three times faster by a chlorine atom than a primary hydrogen.)

Synthesis

10.29 Outline a synthesis of each of the following compounds from isopropyl alcohol. A compound prepared in one part can be used as a reactant in another. (*Hint:* Which of the compounds shown can serve as a starting material to all the others?)

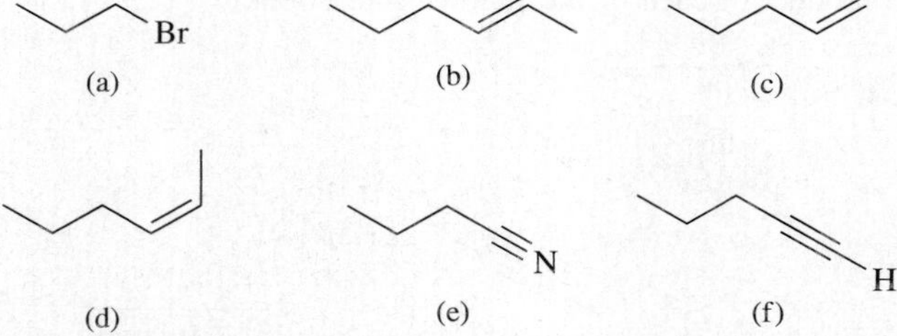

10.30 Guiding your reasoning by retrosynthetic analysis, show how you could prepare each of the following compounds from the given starting material and any necessary organic or inorganic reagents. All require more than one synthetic step.

(a) Cyclopentyl iodide from cyclopentane

(b) 1-Bromo-2-methylpropane from 2-bromo-2-methylpropane

(c) *meso*-2,3-Dibromobutane from 2-butyne

(d) 1-Heptene from 1-bromopentane

(e) *cis*-2-Hexene from 1,2-dibromopentane

(f) Butyl methyl ether ($CH_3CH_2CH_2CH_2OCH_3$) from 1-butene

(g) 

10.31 (*Z*)-9-Tricosene [(*Z*)-$CH_3(CH_2)_7CH{=}CH(CH_2)_{12}CH_3$] is the sex pheromone of the female housefly. Synthetic (*Z*)-9-tricosene is used as bait to lure male flies to traps that contain insecticide. Using acetylene and alcohols of your choice as starting materials, along with any necessary inorganic reagents, show how you could prepare (*Z*)-9-tricosene.

Mechanism

10.32 Suggest a reasonable mechanism for the following reaction. Use curved arrows to show electron flow.

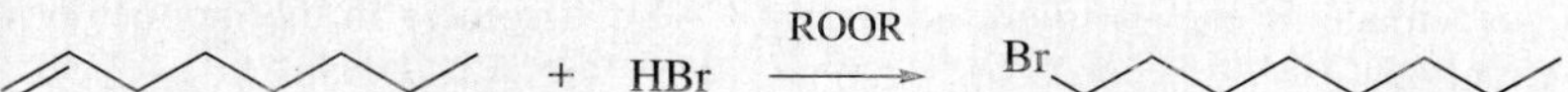

10.33 Cyclopropyl chloride has been prepared by the free-radical chlorination of cyclopropane. Write a stepwise mechanism for this reaction.

Descriptive Passage and Interpretive Problems 10

Free-Radical Reduction of Alkyl Halides

Earlier chapters have introduced several methods for preparing alkyl halides and demonstrated their value in preparing other classes of organic compounds. A thioether, for example, cannot be prepared from an alcohol directly but is readily obtained by first converting the alcohol to an alkyl halide.

$$\text{C}_6\text{H}_{11}\text{–OH} \xrightarrow{\text{HBr}} \text{C}_6\text{H}_{11}\text{–Br} \xrightarrow{\text{NaSCH}_3} \text{C}_6\text{H}_{11}\text{–SCH}_3$$

In this chapter we learned that another source of alkyl halides is free-radical halogenation of alkanes. This opens another reasonable route to the same synthetic target.

$$\text{C}_6\text{H}_{12} \xrightarrow[h\nu]{\text{Br}_2} \text{C}_6\text{H}_{11}\text{–Br} \xrightarrow{\text{NaSCH}_3} \text{C}_6\text{H}_{11}\text{–SCH}_3$$

Sometimes, we face the opposite situation: removing a halogen from carbon and replacing it with hydrogen. We might look at this in terms of nucleophilic substitution and consider sodium hydride (NaH), for example, as a source of hydride ion ($:H^-$) in an S_N2 reaction.

$$\text{C}_6\text{H}_{11}\text{–Br} \xrightarrow{\text{NaH?}} \text{C}_6\text{H}_{12}$$

Reactions of this type, however, usually fail. Sodium hydride is not only too basic (think E2 instead of S_N2) but is also incompatible with the solvents customarily used in nucleophilic substitutions.

An approach that does work is to use hydrides of metals that are less electropositive than sodium; tributyltin hydride [$(CH_3CH_2CH_2CH_2)_3SnH$, for example.

$$R{-}X + (CH_3CH_2CH_2CH_2)_3SnH \longrightarrow R{-}H + (CH_3CH_2CH_2CH_2)_3SnX$$

Tin, with an electronegativity of 1.8 compared to 0.9 for sodium, is closer in electronegativity to hydrogen (2.1). Thus, the Sn–H bond is not very polar and hydrogen is transferred as $\cdot H$ rather than $:H^-$. The reaction proceeds by a free-radical mechanism in which the propagation steps are

$$R{-}\ddot{\underset{..}{X}}: + \cdot Sn(CH_2CH_2CH_2CH_3)_3 \longrightarrow R\cdot + :\ddot{\underset{..}{X}}{-}Sn(CH_2CH_2CH_2CH_3)_3$$

$$R\cdot + H{-}Sn(CH_2CH_2CH_2CH_3)_3 \longrightarrow R{-}H + \cdot Sn(CH_2CH_2CH_2CH_3)_3$$

Tin hydride reduction of alkyl halides suffers from the fact that the reagent not only is used in stoichiometric amounts but organotin compounds present hazardous waste problems due to their toxicity.

In an alternative method, sodium borohydride ($NaBH_4$)—a reducing agent commonly used in a number of synthetic procedures—serves as the source of hydrogen atoms in a process catalyzed by indium chloride ($InCl_3$).

$$\text{C}_6\text{H}_5\text{CH}_2\text{CH}_2\text{CH}_2\text{Br} \xrightarrow[\text{InCl}_3]{\text{NaBH}_4} \text{C}_6\text{H}_5\text{CH}_2\text{CH}_2\text{CH}_3$$

1-Bromo-3-phenylpropane → 1-Phenylpropane (95%)

The actual reducing agent is $HInCl_2$ formed by the reaction of $NaBH_4$ with $InCl_3$. A free-radical mechanism involving the following propagation steps has been suggested.

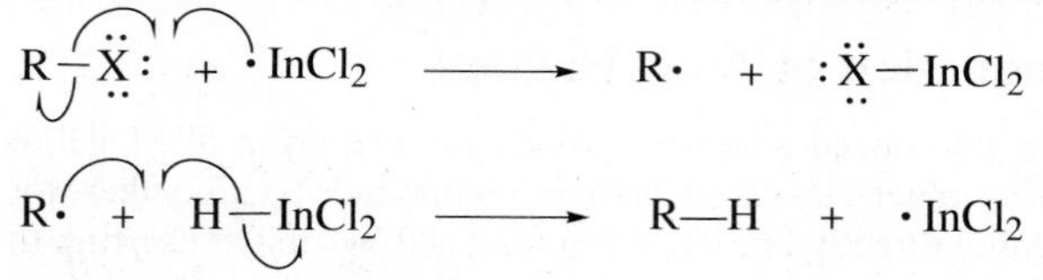

$$R{-}\ddot{\underset{\cdot\cdot}{X}}: + \cdot InCl_2 \longrightarrow R\cdot + :\ddot{\underset{\cdot\cdot}{X}}{-}InCl_2$$

$$R\cdot + H{-}InCl_2 \longrightarrow R{-}H + \cdot InCl_2$$

10.34 What is the product of the reaction shown?

1. HBr, peroxides
2. Tributyltin hydride, heat

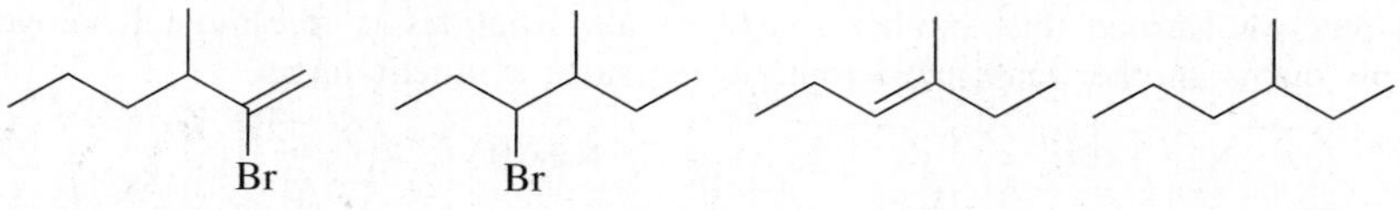

Br Br

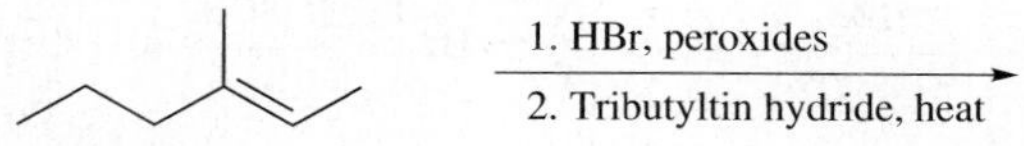

A. B. C. D.

10.35 The species formed in the reaction of an alkyl halide with $HInCl_2$ is also synthetically useful in that it has the capacity to add to double bonds. What is the product of the reaction shown?

I + O OCH$_3$ $\xrightarrow[InCl_3]{NaBH_4}$

O OCH$_3$ O OCH$_3$ O OCH$_3$ O OCH$_3$

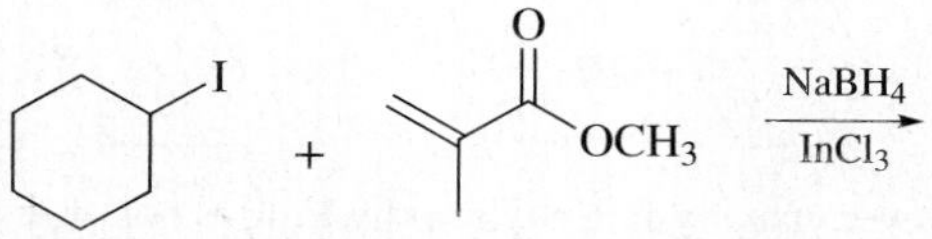

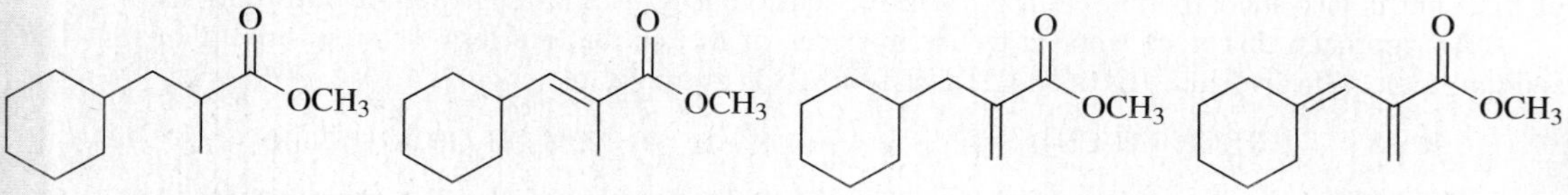

A. B. C. D.

10.36 In an intramolecular analog to the reaction shown in Problem 10.35, intermediate X in the following reaction undergoes cyclization to form the product shown. What is the structure of X?

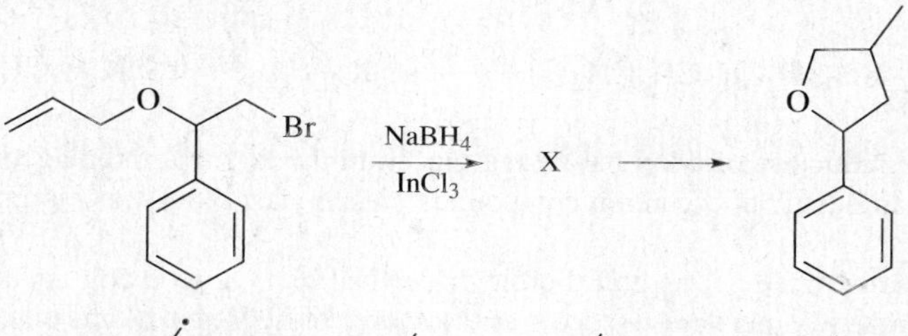

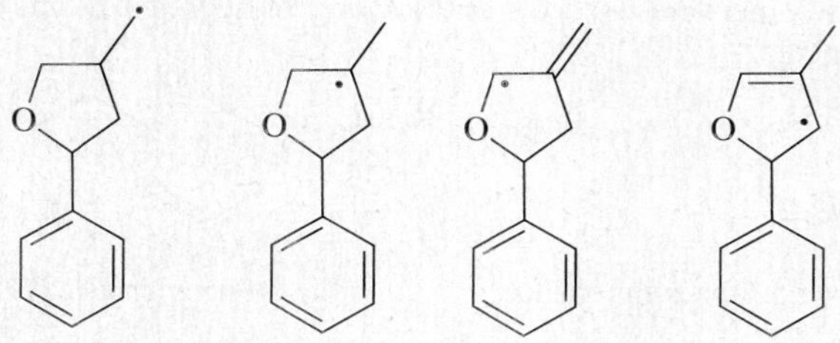

O Br $\xrightarrow[InCl_3]{NaBH_4}$ X $\longrightarrow$ O

O O O O

A. B. C. D.

10.37 Reduction of 6-bromo-1-hexene with tributyltin hydride gives a mixture of 1-methylcyclopentane (major) and cyclohexane (minor). Which of the following statements about this reaction are true?

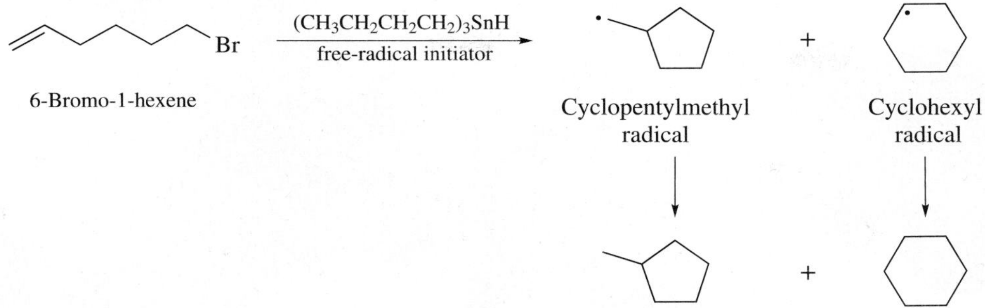

A. Cyclopentylmethyl radical is formed faster than cyclohexyl radical.
B. Cyclopentylmethyl radical is more stable than cyclohexyl radical.
C. Both (a) and (b) are true.

10.38 What combination of alkene and alkyl halide would you treat with tributyltin hydride in order to prepare the compound shown by free-radical addition?

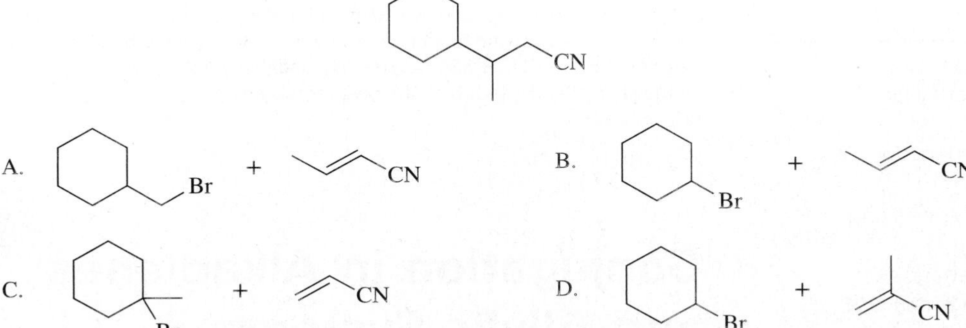

Chapter 11

CHAPTER OUTLINE

Allyl is derived from the botanical name for garlic (*Allium sativum*). Over a century ago it was found that the major component obtained by distilling garlic oil is $H_2C{=}CHCH_2SSCH_2CH{=}CH_2$, and the word *allyl* was coined for the $H_2C{=}CHCH_2$ group on the basis of this origin.

Conjugation in Alkadienes and Allylic Systems

Not all the properties of alkenes are revealed by focusing exclusively on the functional-group behavior of the double bond. A double bond can affect the properties of a second functional unit to which it is directly attached. It can be a substituent, for example, on a positively charged carbon in an **allylic carbocation,** on a carbon that bears an unpaired electron in an **allylic free radical,** on a negatively charged carbon in an **allylic anion,** or it can be a substituent on a second double bond in a **conjugated diene.**

Allylic carbocation　　Allylic free radical

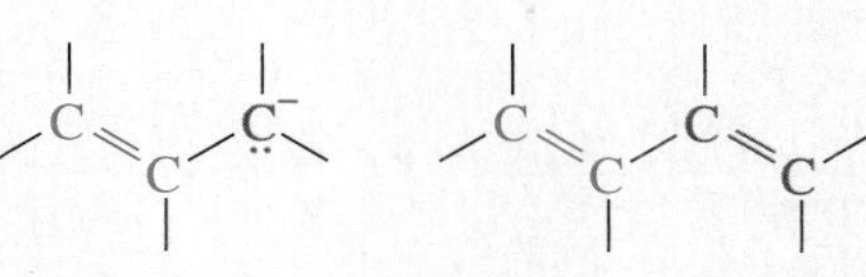

Allylic anion　　Conjugated diene

Conjugare is a Latin verb meaning "to link or yoke together," and allylic carbocations, allylic free radicals, allylic anions, and conjugated dienes are all examples of **conjugated systems.** In this chapter we'll see how conjugation permits two functional units within a molecule to display a kind of reactivity that is qualitatively different from that of either unit alone.

11.1 The Allyl Group

Allyl is both a common name and a permissible IUPAC name for the $H_2C{=}CHCH_2$ group. Its derivatives are better known by their functional class IUPAC names than by their substitutive ones:

Allyl alcohol (2-propen-1-ol)

Allyl chloride (3-chloro-1-propene)

Allylamine (2-propen-1-amine)

The sp^3-hybridized carbon of an allyl group is termed an **allylic carbon,** and atoms or groups attached to it are allylic substituents.

Problem 11.1

α-Terpineol is a pleasant-smelling oil obtained from pine. How many allylic hydrogens does it have? Is the hydroxyl group allylic?

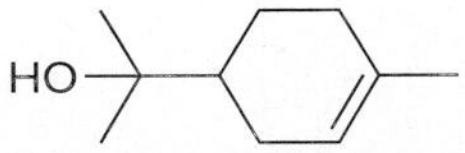

According to the number of its electrons, an allyl unit can be a positively charged carbocation ($H_2C{=}CHCH_2^+$), a neutral free radical ($H_2C{=}CHCH_2\cdot$), or a negatively charged carbanion ($H_2C{=}CHCH_2{:}^-$). Each is stabilized by delocalization involving the π electrons in the double bond. The positive charge, negative charge, or unpaired electron is shared by the two carbons at opposite ends of the allyl group.

Allyl cation: $H_2C{=}CH{-}\overset{+}{C}H_2 \longleftrightarrow H_2\overset{+}{C}{-}CH{=}CH_2$ or

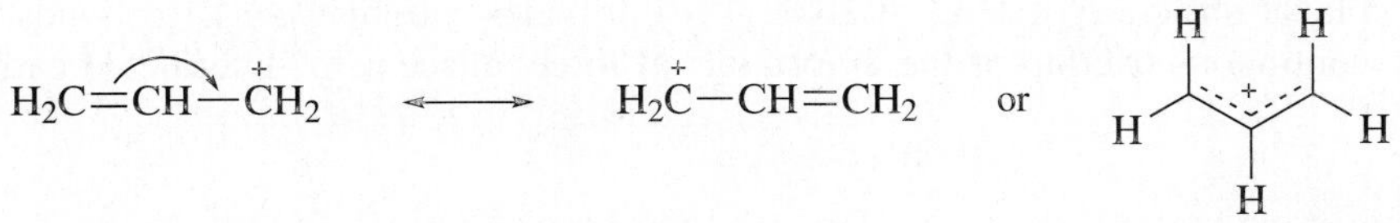

Allyl radical: $H_2C{=}CH{-}\dot{C}H_2 \longleftrightarrow H_2\dot{C}{-}CH{=}CH_2$ or

Allyl anion: $H_2C{=}CH{-}\overset{-}{\ddot{C}}H_2 \longleftrightarrow H_2\overset{-}{\ddot{C}}{-}CH{=}CH_2$ or

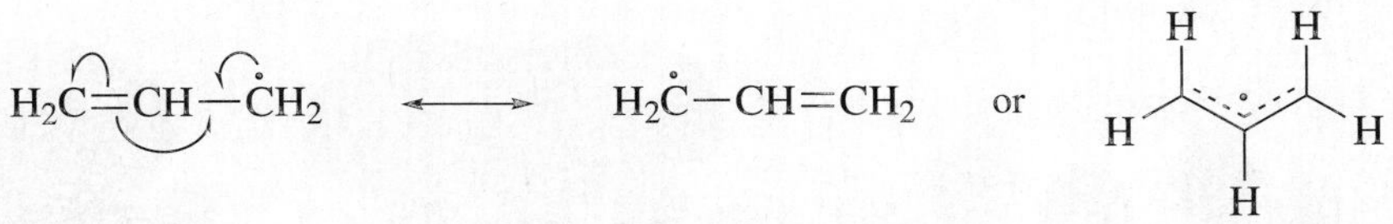

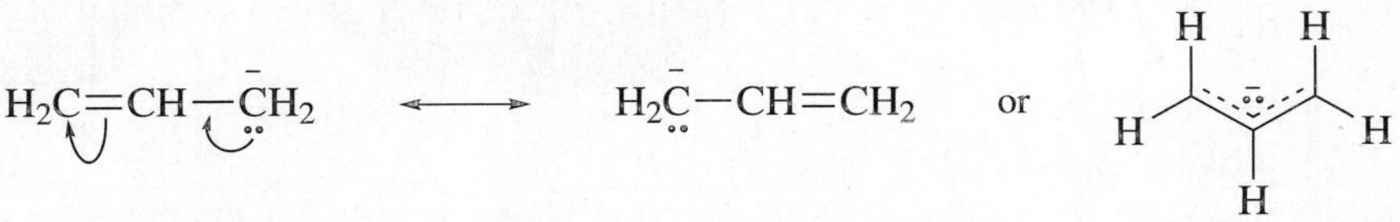

Another way to indicate this electron delocalization is via the dotted-line structures shown. It is important, however, to recognize that the +, $\overset{-}{\cdot\cdot}$, or · above the middle of the dashed line applies to the unit as a whole and is shared only by its end carbons.

In allylic species that are not symmetrically substituted, the two resonance structures are not equivalent and do not contribute equally to the hybrid.

or

Major contributor

Such an allylic carbocation more closely resembles a tertiary carbocation than a primary one in terms of its stability.

Problem 11.2

Write a second resonance contributor for each of the following. Is the charge or unpaired electron shared equally by both allylic carbons? If not, which one bears more of the charge or unpaired electron?

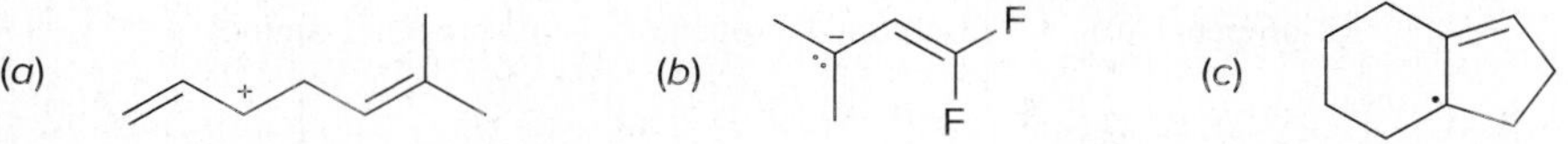

Sample Solution (a) First, identify the allylic unit by picking out the C=C—C$^+$ sequence. Of the two double bonds in this structure, only the one at the left is part of C=C—C$^+$. The double bond at the right is separated from the positively charged carbon by a CH_2 group, so is not conjugated to it. Move electrons in pairs from the double bond toward the positively charged carbon to generate a second resonance structure.

2
1 3

The two contributing structures are not equivalent; therefore, the positive charge is not shared equally between C-1 and C-3. C-1 is a primary carbon, C-3 is secondary. More of the positive charge resides on C-3 than on C-1. The original structure (left) contributes more to the resonance hybrid than the other (right).

Figure 11.1 displays a valence bond description of bonding in allyl cation. The planar structure of $H_2C{=}CHCH_2^+$ (*a*) provides a framework of σ bonds that allows for continuous overlap of the 2*p* orbitals of three adjacent sp^2-hybridized carbons (*b* and *c*).

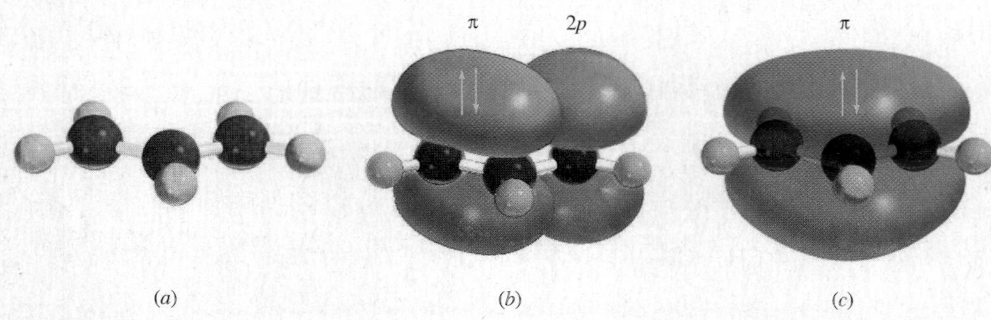

Figure 11.1

Bonding in allyl cation. (*a*) All of the atoms of $H_2C{=}CHCH_2^+$ lie in the same plane and each carbon is sp^2-hybridized. (*b*) The alignment of the π component of the double bond and the vacant *p* orbital permits overlap between them. (*c*) A π orbital encompasses all three carbons of $H_2C{=}CHCH_2^+$. The two electrons in this orbital are delocalized over three carbons.

Until now, we have only seen π orbitals involving two carbons. Conjugated systems are characterized by extended π orbitals that encompass three or more atoms.

Although satisfactory for allyl cation, Figure 11.1 is insufficient for species with more than two π electrons because the π orbital in (*c*) can accommodate only two electrons. Molecular orbital (MO) theory, however, offers an alternative to resonance and valence bond theory for understanding the structure and reactions of not only allylic cations, but radicals (three π electrons) and anions (four π electrons) as well. Recalling from Section 7.2 that the number of molecular orbitals is equal to the number of atomic orbitals (AOs) that combine to form them, we combine the three 2*p* AOs, one from each of the three sp^2-hybridized carbons of allyl, into the system of three π MOs shown in Figure 11.2.

The lowest-energy orbital π_1 is doubly occupied in all three species; whereas π_2 is vacant in allyl cation, singly occupied in allyl radical, and doubly occupied in allyl anion. When $H_2C{=}CHCH_2^+$ reacts with a nucleophile, electrons flow from the nucleophile to the **lowest unoccupied molecular orbital,** or **LUMO,** which in this case is π_2. Because π_2 is characterized by a node at C-2, only C-1 and C-3 are available for bonding of a nucleophile to allyl cation. At the other extreme, electrons flow *from* the **highest occupied molecular orbital,** or **HOMO** (π_2), of allyl anion when it bonds to an electrophile. Again, only C-1 or C-3 can participate in bond formation because of the node at C-2. The results are similar for bond formation in allyl radical.

Erich Hückel was a German physical chemist first known for his collaboration with Peter Debye in developing what remains the most widely accepted theory of electrolyte solutions, then later for his application of molecular orbital theory to conjugated hydrocarbons, especially aromatic hydrocarbons (Chapter 12).

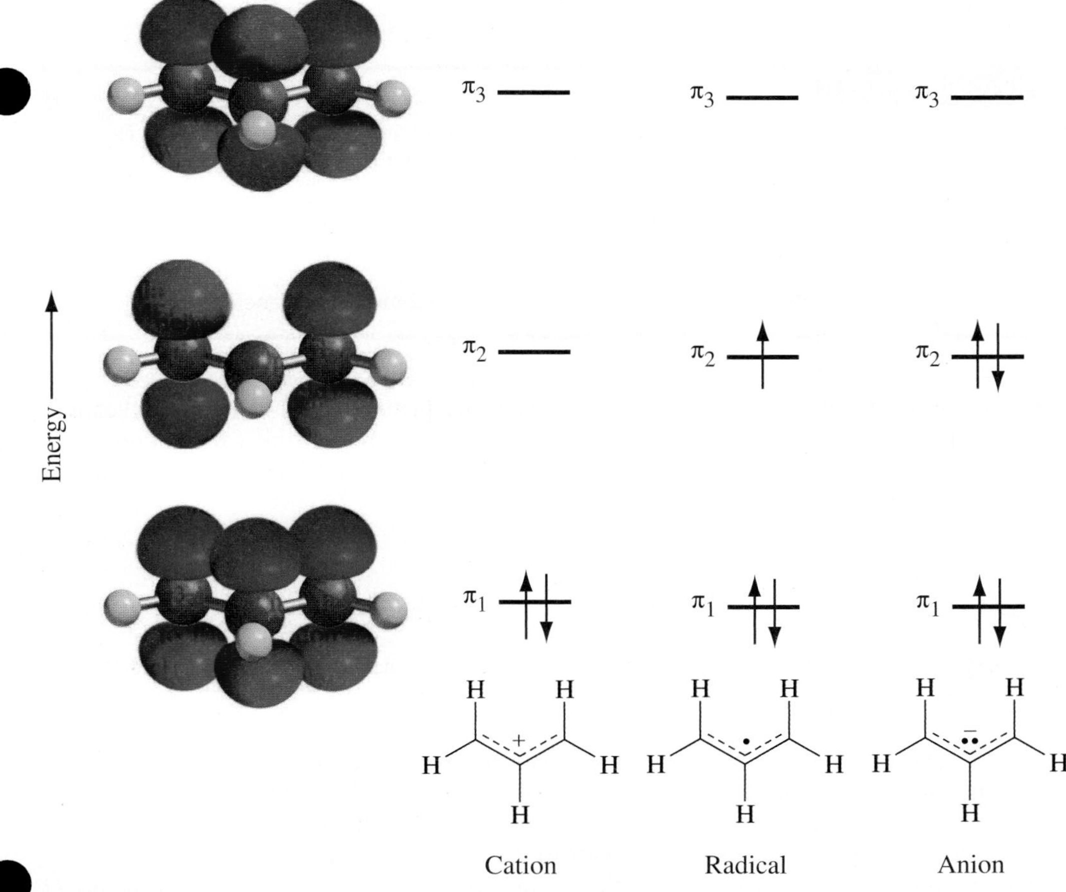

Figure 11.2

The π molecular orbitals of allyl cation, radical, and anion. Allyl cation has two π electrons, allyl radical has three, and allyl anion has four.

11.2 S_N1 and S_N2 Reactions of Allylic Halides

Much of our understanding about conjugation effects in allylic systems comes from rate and product studies of nucleophilic substitution, especially those that take place by the S_N1 mechanism. Allylic halides react faster than their nonallylic counterparts in both S_N1 and S_N2 reactions, but for different reasons. S_N1 reactions will be described first, followed later in this section by S_N2.

Relative S_N1 Rates Under S_N1 conditions such as solvolysis in ethanol, the tertiary allylic chloride 3-chloro-3-methyl-1-butene reacts over 100 times faster than *tert*-butyl chloride. Both reactions follow a first-order rate law, and their relative rates reflect the greater stability of $[(CH_3)_2C{=}CH{=}CH_2]^+$ compared with $(CH_3)_3C^+$.

Faster rate: $k_{rel} = 123$ — 3-Chloro-3-methyl-1-butene $\xrightarrow{-[:\ddot{Cl}:^-]}$ 1,1-Dimethylallyl cation

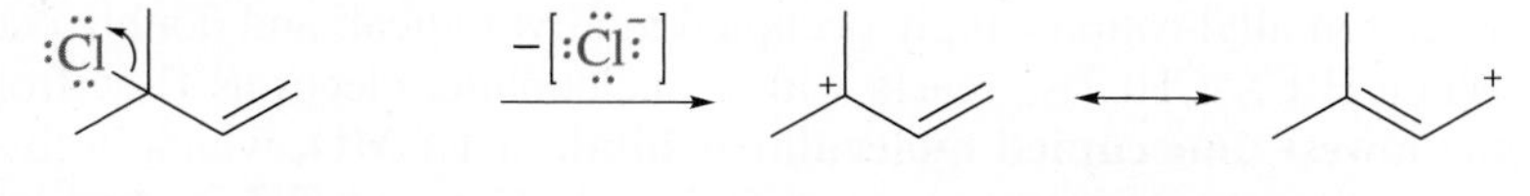

Slower rate: $k_{rel} = 1.0$ — *tert*-Butyl chloride $\xrightarrow{-[:\ddot{Cl}:^-]}$ *tert*-Butyl cation

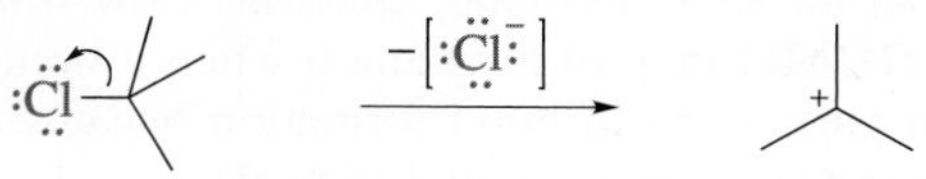

Allylic carbocations are more stable than simple alkyl cations and a vinyl group ($H_2C{=}CH$) is a better carbocation-stabilizing substituent than methyl (CH_3). Although $H_2C{=}CHCH_2^+$, for example, is a primary carbocation, it is about as stable as a typical secondary carbocation such as $(CH_3)_2CH^+$.

Problem 11.3

The two compounds shown differ by a factor of 60 in their first-order rate constants for hydrolysis in 50% ethanol water at 45°C. Which is more reactive? Why?

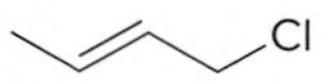

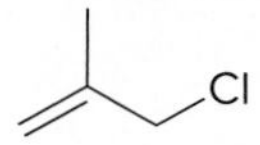

trans-1-Chloro-2-butene 3-Chloro-2-methylpropene

S_N1 Reaction Products According to the resonance picture for 1,1-dimethylallyl cation, the positive charge is shared by a tertiary and a primary carbon. If this carbocation reacts with a nucleophile, to which carbon does the nucleophile bond? The answer is *both,* but with a regioselective preference for the tertiary carbon.

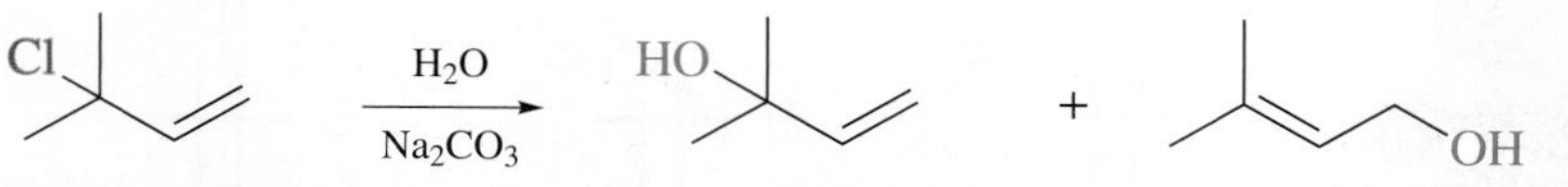

3-Chloro-3-methyl-1-butene $\xrightarrow[Na_2CO_3]{H_2O}$ 2-Methyl-3-buten-2-ol (85%) + 3-Methyl-2-buten-1-ol (15%) via 1,1-Dimethylallyl cation

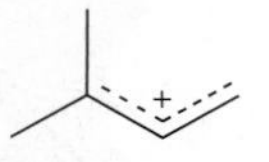

Mechanism 11.1 applies the S_N1 mechanism to this hydrolysis. Its key features are carbocation formation in step 1 and bonding of the nucleophile (water) to the carbocation in step 2. The oxygen of water can bond to either end of the allylic unit, but does so at different rates. The oxygen of water bonds to the carbon that carries more of the positive charge, giving the tertiary alcohol as the major product. Figure 11.3 illustrates the effect that methyl substitutents have on the allyl cation by comparing the electrostatic potential maps of allyl cation, 1-methylallyl cation, and 1,1-dimethylallyl cation. The minor product, a primary alcohol, results when the oxygen of water bonds to the primary carbon.

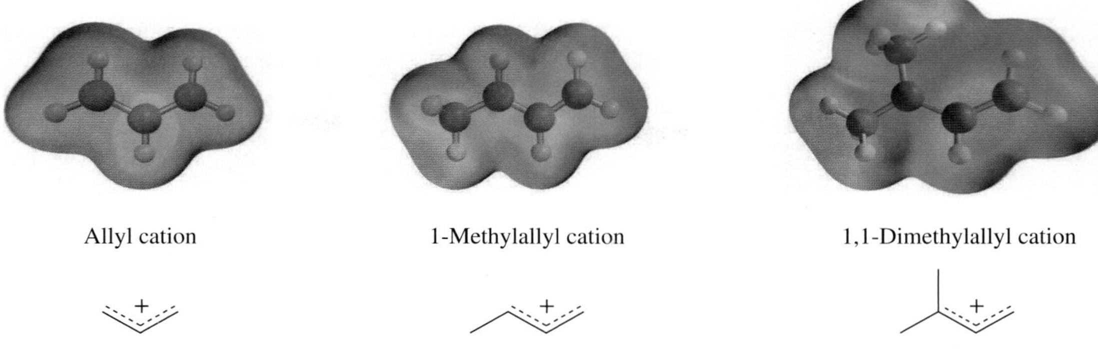

Figure 11.3

Electrostatic potential maps of allyl cations. The positive charge on allyl cation is evenly distributed on the terminal carbons. The positive charge of allyl cations with methyl substituents is concentrated on the more highly substituted terminal allyl carbon. (The electrostatic potentials were mapped on the same scale to allow direct comparison.)

Mechanism 11.1

S_N1 Hydrolysis of an Allylic Halide

THE OVERALL REACTION:

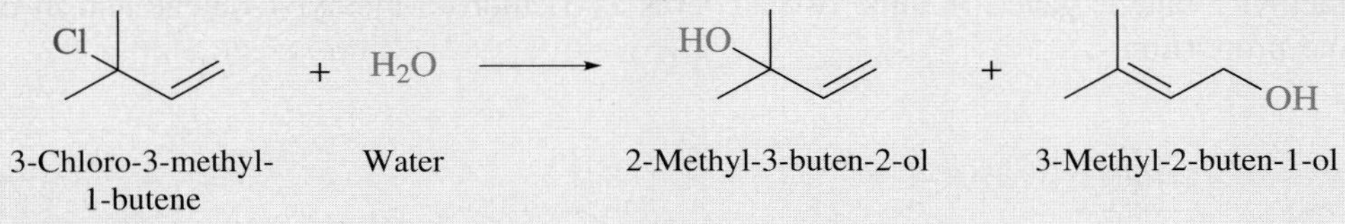

3-Chloro-3-methyl-1-butene + Water (H_2O) ⟶ 2-Methyl-3-buten-2-ol + 3-Methyl-2-buten-1-ol

THE MECHANISM:

Step 1: The alkyl halide ionizes to give a carbocation. This step is rate-determining and gives a delocalized carbocation.

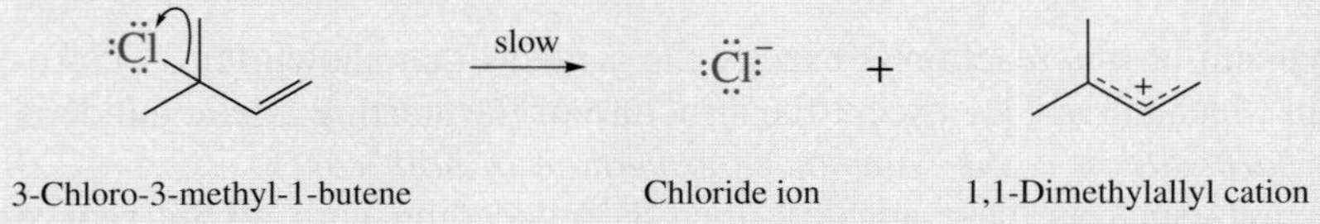

3-Chloro-3-methyl-1-butene —slow→ Chloride ion + 1,1-Dimethylallyl cation

Step 2: The carbocation (shown as its major contributor) reacts with water. Water acts as a nucleophile; its oxygen can bond to either the tertiary carbon (a) or the primary carbon (b).

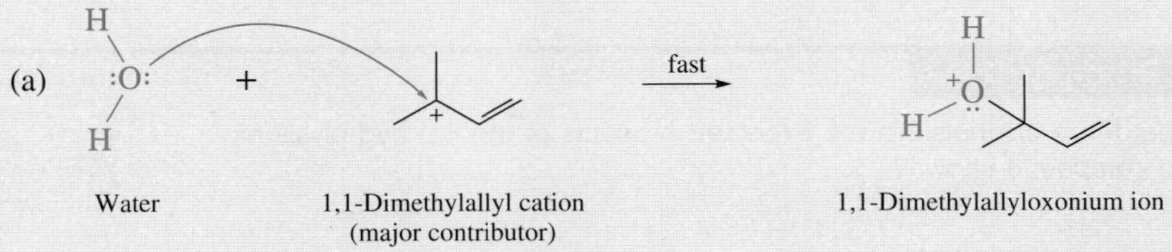

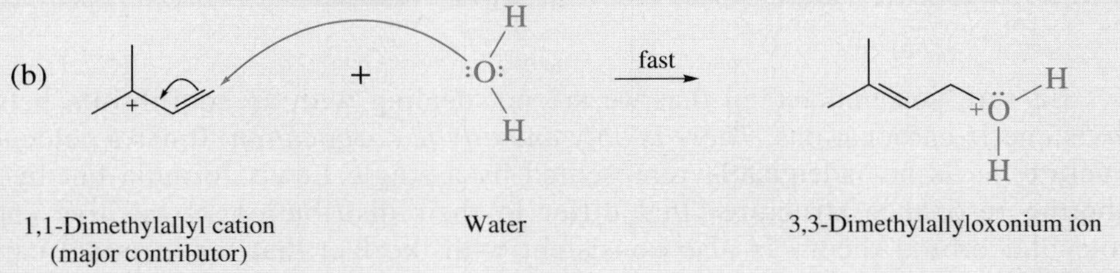

(a) Water + 1,1-Dimethylallyl cation (major contributor) —fast→ 1,1-Dimethylallyloxonium ion

(b) 1,1-Dimethylallyl cation (major contributor) + Water —fast→ 3,3-Dimethylallyloxonium ion

continued

Step 3: The alkyloxonium ions formed in step 2 are converted to the corresponding alcohols by proton transfer. Water is the proton acceptor.

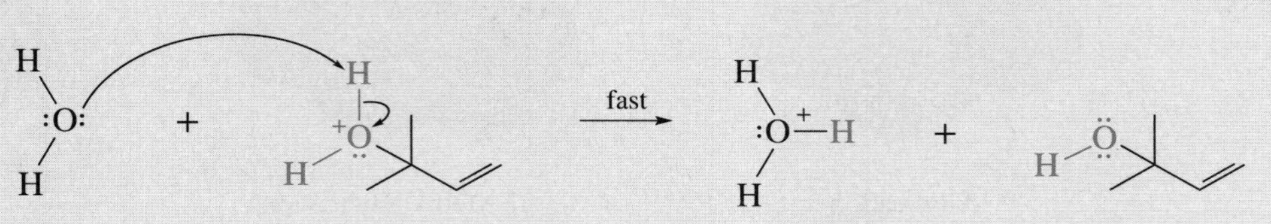

Water + 1,1-Dimethylallyloxonium ion $\xrightarrow{\text{fast}}$ Hydronium ion + 2-Methyl-3-buten-2-ol (major product)

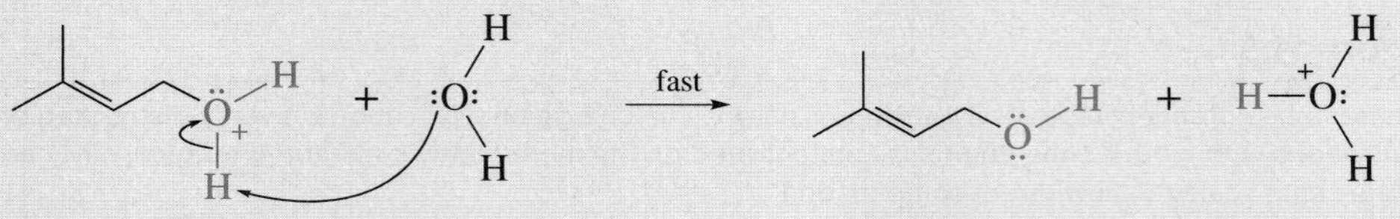

3,3-Dimethylallyloxonium ion + Water $\xrightarrow{\text{fast}}$ 3-Methyl-2-buten-1-ol (minor product) + Hydronium ion

In a parallel experiment, hydrolysis of the isomeric primary allylic chloride 1-chloro-3-methyl-2-butene gave the same two alcohols as 3-chloro-3-methyl-1-butene and in the same proportion.

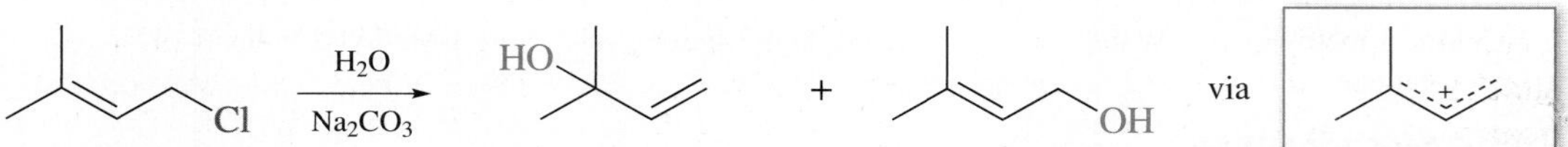

1-Chloro-3-methyl-2-butene $\xrightarrow[Na_2CO_3]{H_2O}$ 2-Methyl-3-buten-2-ol (85%) + 3-Methyl-2-buten-1-ol (15%) via 1,1-Dimethylallyl cation

The mechanism of this reaction is exactly the same as that shown for 3-chloro-3-methyl-1-butene in Mechanism 11.1 except the structure of the starting allylic halide is different. *The same carbocation is the common intermediate in both cases.*

Reactions such as these are described as proceeding with **allylic rearrangement.** They differ from the carbocation rearrangements of earlier chapters in that the latter involve structural changes resulting from atom or group migrations. Changes in *electron* positions are responsible for allylic rearrangements.

Problem 11.4

What three alcohols are the expected products of the S_N1 hydrolysis of the compound shown?

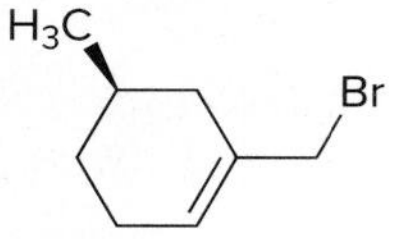

H_3C ... Br

Be sure you understand that we are not dealing with an equilibrium between two isomeric carbocations. *There is only one allylic carbocation.* It has a delocalized structure, so is not adequately represented by a single Lewis formula but by contributing resonance structures that differ in their distribution of positive charge. Molecular orbital theory is also consistent with the fact that the terminal carbons of the allyl unit are the main sites of interaction with an incoming nucleophile because, as we saw in Figure 11.2, these are the only ones that contribute a *p* orbital to the LUMO.

Problem 11.5

From among the following compounds, choose the two that yield the same carbocation on ionization.

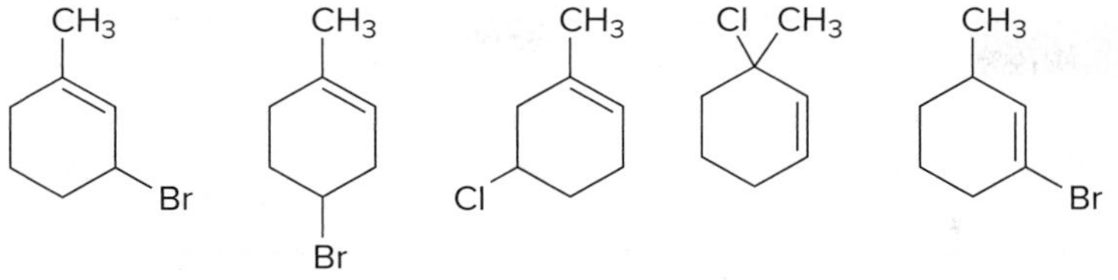

CH3 Br | CH3 Br | CH3 Cl | Cl CH3 | CH3 Br

Relative S_N2 Rates Like their S_N1 counterparts, S_N2 reactions of allylic halides take place more rapidly than the corresponding reactions of similar alkyl halides. The relative rate profile for a group of alkyl and allylic halides shows two significant trends.

Relative second-order rate constants for reaction with sodium ethoxide in ethanol; 45°C

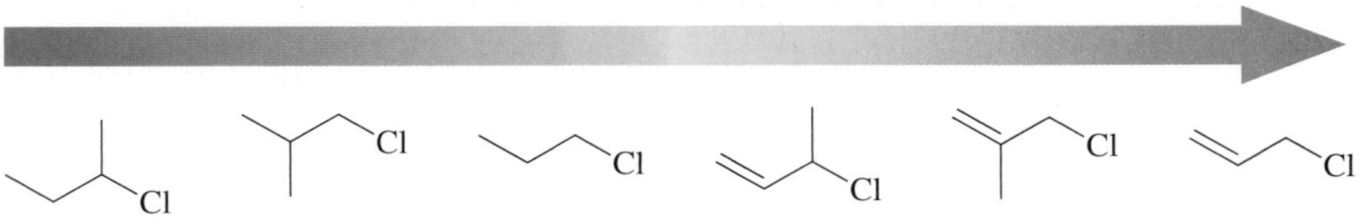

Cl 0.19 | Cl 0.24 | Cl 2.7 | Cl 4.9 | Cl 89 | Cl 100

The three halides that react at the fastest rates are all allylic; the three slowest are not. Within each group, the typical S_N2 order (primary faster than secondary) is observed.

The greater S_N2 reactivity of allylic halides results from a combination of two effects: steric and electronic. Sterically, a CH_2Cl group is less crowded and more reactive when it is attached to the sp^2-hybridized carbon of a vinyl group compared with being attached to the sp^3-hybridized carbon of an alkyl group. Electronically, molecular orbital treatments such as seen for the S_N2 mechanism in Section 6.3 are readily adapted to allyl chloride. According to that picture, electrons flow from the HOMO of the nucleophile to the LUMO of the alkyl halide.

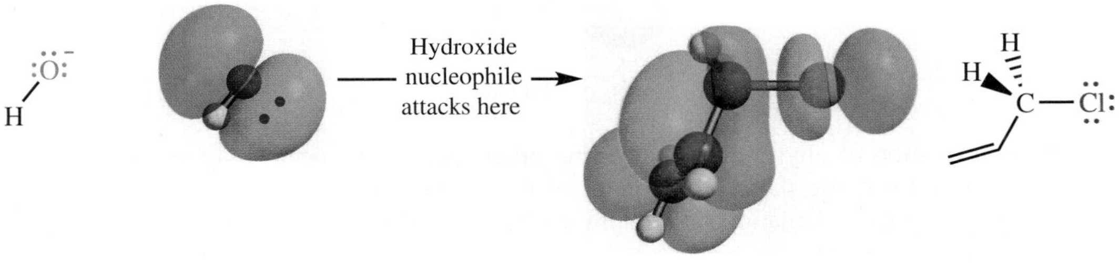

Hydroxide nucleophile attacks here

HOMO of hydroxide LUMO of allyl chloride

Because the LUMO of allyl chloride extends over all three carbons of the allyl group, it allows for greater electron delocalization than the corresponding LUMO of 1-chloropropane, a lower activation energy, and a faster rate of reaction.

S_N2 Reaction Products Typical S_N2 displacements occur when primary and unhindered secondary allylic halides react with good nucleophiles.

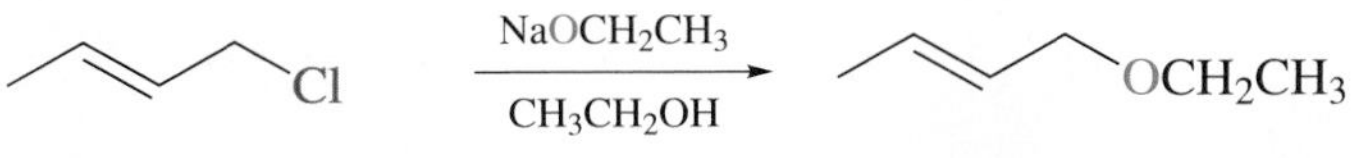

$$\text{Cl} \xrightarrow[\text{CH}_3\text{CH}_2\text{OH}]{\text{NaOCH}_2\text{CH}_3} \text{OCH}_2\text{CH}_3$$

trans-1-Chloro-2-butene → *trans*-1-Ethoxy-2-butene (only product; 82% yield)

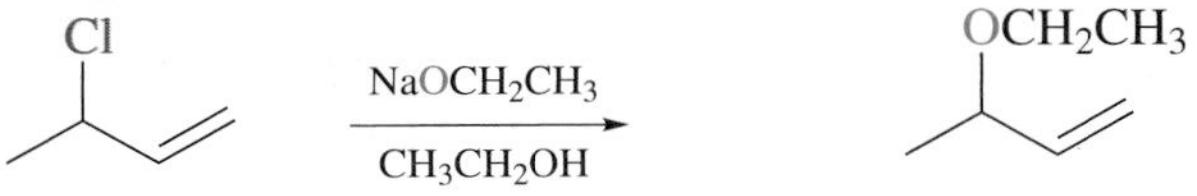

$$\text{Cl} \xrightarrow[\text{CH}_3\text{CH}_2\text{OH}]{\text{NaOCH}_2\text{CH}_3} \text{OCH}_2\text{CH}_3$$

3-Chloro-1-butene → 3-Ethoxy-1-butene (only substitution product; 53% yield)

At low concentrations of sodium ethoxide, S_N1 reactions compete with S_N2, and a mixture of direct displacement and allylic rearrangement products results.

Problem 11.6

As indicated in the preceding equations, the yield of the substitution product was much better in the reaction of *trans*-1-chloro-2-butene than 3-chloro-1-butene. Can you suggest a reason why?

11.3 Allylic Free-Radical Halogenation

As we have seen in Section 11.1, allyl radical is stabilized by electron delocalization expressed as resonance between contributing Lewis structures

$$H_2C{=}CH{-}\dot{C}H_2 \longleftrightarrow H_2\dot{C}{-}CH{=}CH_2 \quad \text{or}$$

or by the π-electron molecular orbital approximation. Both show the unpaired electron equally distributed between C-1 and C-3. In the π-electron approximation this unpaired electron singly occupies π_2, which is characterized by a node at C-2.

π_2 of allyl radical

Spin density is a measure of the unpaired electron density at an atom and was introduced in Section 10.1.

Bond formation of allyl radical with some other species can occur only at either end of the allyl unit because these are the sites of the greatest unpaired electron probability. A molecular orbital calculation of the spin density of allyl radical reinforces these interpretations (Figure 11.4).

Delocalization of the unpaired electron stabilizes allylic radicals and causes reactions that generate them to proceed more readily than those that give simple alkyl radicals.

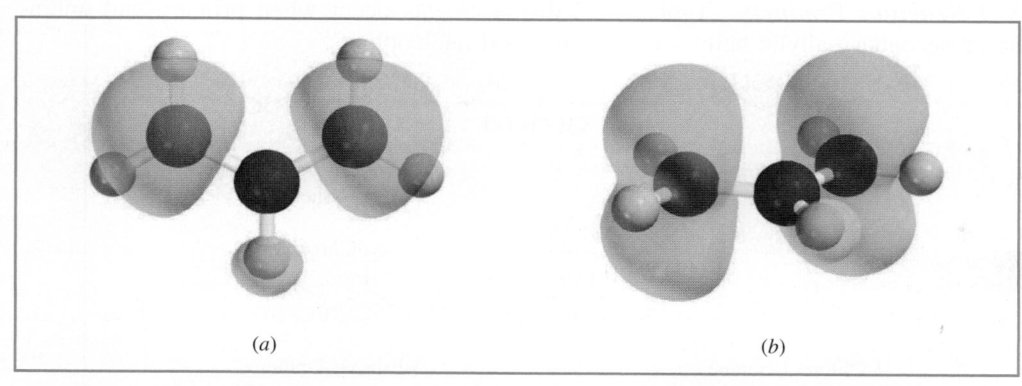

Figure 11.4

(*a*) The spin density (yellow) in allyl radical is equally divided between the two allylic carbons. There is a much smaller spin density at the C-2 hydrogen. (*b*) The odd electron is in an orbital that is part of the allylic π system.

Compare, for example, the bond dissociation enthalpies of the primary C—H bonds of propane and propene:

Propane → Propyl radical + Hydrogen atom (H·) $\Delta H° = +423$ kJ/mol (+101 kcal/mol)

Propene → Allyl radical + Hydrogen atom (H·) $\Delta H° = +368$ kJ/mol (+88 kcal/mol)

Breaking an allylic C—H bond in propene requires 55 kJ/mol (13 kcal/mol) less energy than a bond to a primary hydrogen in propane. Allyl radical is stabilized by π-electron delocalization; propyl radical is not.

Problem 11.7

Rank the C—H bonds of *trans*-3-hexene in order of decreasing bond dissociation enthalpy.

The greater stability of allylic radicals relative to their alkyl counterparts suggests that free-radical halogenation of alkenes should be both feasible and regioselective for the allylic position. Although, as we have already seen, the typical reaction of alkenes with halogens at room temperature and below is *electrophilic addition* to the double bond, *free-radical substitution* is favored at high temperature. The industrial-scale preparation of allyl chloride involves heating propene and chlorine at 300–500°C.

Propene + Chlorine (Cl_2) $\xrightarrow{500°C}$ Allyl chloride (80–85%) + Hydrogen chloride (HCl)

The reaction proceeds by the free-radical chain mechanism shown in Mechanism 11.2.

In the laboratory, allylic brominations are normally carried out using one of a number of specialized reagents such as *N*-bromosuccinimide. Small amounts of peroxides are sometimes added as free-radical initiators.

Cyclohexene + *N*-Bromosuccinimide (NBS) $\xrightarrow[CCl_4]{heat}$ 3-Bromocyclohexene (82–87%) + Succinimide

N-Bromosuccinimide provides a low concentration of molecular bromine, which reacts with alkenes by a mechanism analogous to that of other free-radical halogenations.

Problem 11.8

Assume that *N*-bromosuccinimide serves as a source of Br_2, and write equations for the propagation steps in the formation of 3-bromocyclohexene by allylic bromination of cyclohexene.

Mechanism 11.2

Allylic Chlorination of Propene

THE OVERALL REACTION:

$$H_2C{=}CHCH_3 + Cl_2 \xrightarrow{500°C} H_2C{=}CHCH_2Cl + HCl$$

Propene, Chlorine, Allyl chloride, Hydrogen chloride

THE MECHANISM:

Initiation step: A chlorine molecule dissociates to two atoms.

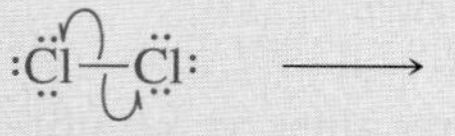

$$:\ddot{Cl}{-}\ddot{Cl}: \longrightarrow :\ddot{Cl}\cdot + \cdot\ddot{Cl}:$$

Chlorine → Chlorine atoms

Propagation steps: In the first propagation step a chlorine atom abstracts a hydrogen atom from the allylic carbon of propene forming allyl radical.

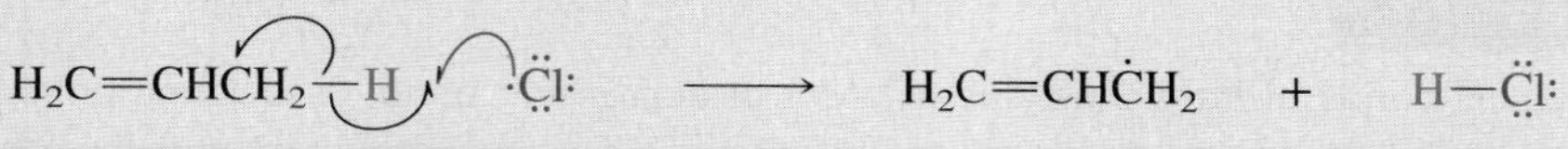

$$H_2C{=}CHCH_2{-}H \;\; \cdot\ddot{Cl}: \longrightarrow H_2C{=}CH\dot{C}H_2 + H{-}\ddot{Cl}:$$

Propene, Chlorine atom, Allyl radical, Hydrogen chloride

The allyl radical formed in the first propagation step reacts with Cl_2 to form allyl chloride.

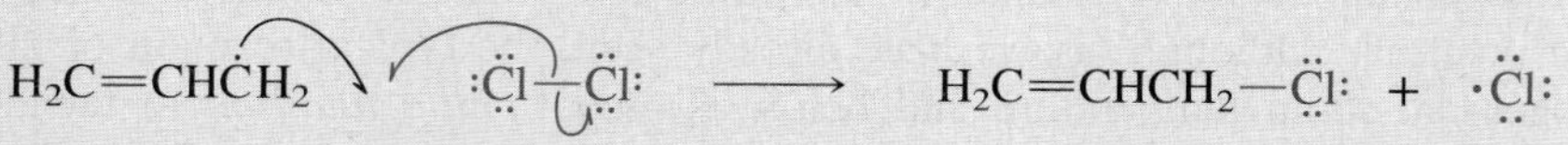

$$H_2C{=}CH\dot{C}H_2 \;\; :\ddot{Cl}{-}\ddot{Cl}: \longrightarrow H_2C{=}CHCH_2{-}\ddot{Cl}: + \cdot\ddot{Cl}:$$

Allyl radical, Chlorine, Allyl chloride, Chlorine atom

The chlorine atom generated in this propagation step then abstracts a hydrogen atom from another molecule of propene and the two propagation steps repeat over and over again.

Although allylic bromination and chlorination offer methods for attaching a reactive functional group to a hydrocarbon framework, we need to be aware of two important limitations. For allylic halogenation to be effective in a particular synthesis:

1. All the allylic hydrogens in the starting alkene must be equivalent, and
2. Both resonance forms of the allylic radical must be equivalent.

In the two examples cited so far, the chlorination of propene and the bromination of cyclohexene, both requirements are met.

All the allylic hydrogens of propene are equivalent. $H_2C{=}CH{-}CH_3$

The two resonance forms of allyl radical are equivalent. $H_2C{=}CH{-}\dot{C}H_2 \longleftrightarrow H_2\dot{C}{-}CH{=}CH_2$

All the allylic hydrogens of cyclohexene are equivalent.

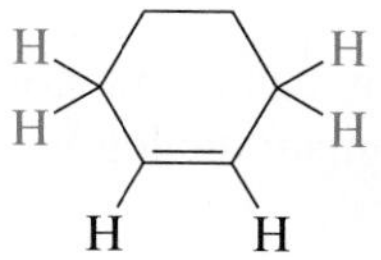

The two resonance forms of 2-cyclohexenyl radical are equivalent.

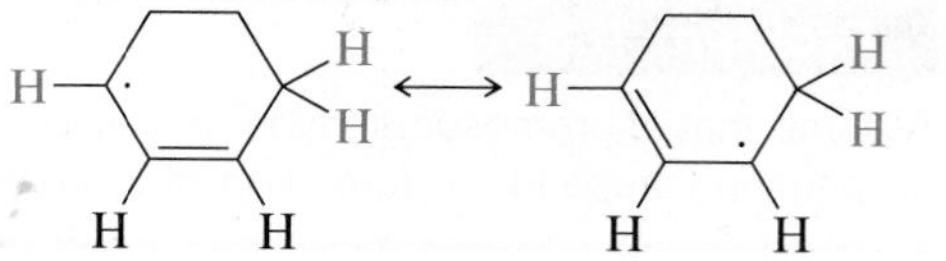

Unless both criteria are met, mixtures of constitutionally isomeric allylic halides result. The resonance forms of the allylic radical intermediate in the bromination of 1-octene, for example, are not equivalent and give both 3-bromo-1-octene and 1-bromo-2-octene, the latter as a mixture of cis and trans isomers.

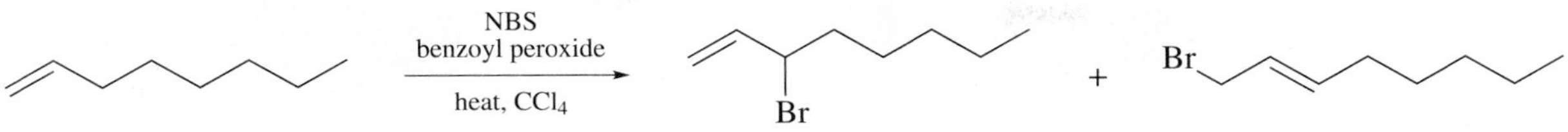

1-Octene —(NBS, benzoyl peroxide; heat, CCl_4)→ 3-Bromo-1-octene (17%) + 1-Bromo-2-octene (83%) (cis + trans)

via:

Problem 11.9

Evaluate 2,3,3-trimethyl-1-butene as a candidate for free-radical bromination. How many allylic bromides would you expect to result from its treatment with *N*-bromosuccinimide?

11.4 Allylic Anions

Like allyl cation and allyl radical, allyl anion is planar and stabilized by electron delocalization. The unshared pair plus the two π electrons of the double bond are shared by the three carbons of the allyl unit. This delocalization can be expressed in resonance terms

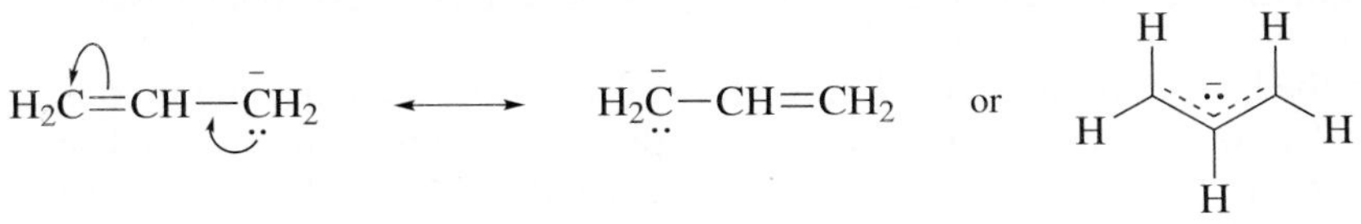

$$H_2C{=}CH{-}\ddot{\bar{C}}H_2 \longleftrightarrow H_2\ddot{\bar{C}}{-}CH{=}CH_2 \quad \text{or}$$

or by molecular orbital methods. According to the π-electron approximation, the four π electrons of allyl anion are distributed in pairs between the two bonding orbitals π_1 and π_2 as shown earlier in Figure 11.2. The electrons in the HOMO π_2 interact equally with C-1 and C-3. Thus, the negative charge is shared equally by these two carbons, and both are equivalent reactive sites.

The extent to which electron delocalization stabilizes allyl anion can be assessed by comparing the pK_as of propane and propene.

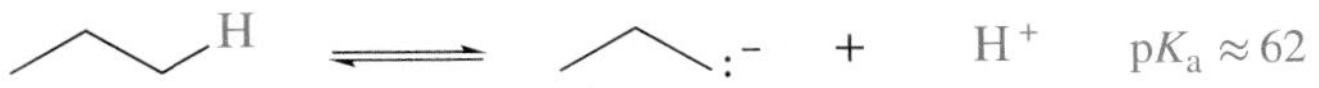

Propane ⇌ Propyl anion + H^+ (Proton) p$K_a \approx 62$

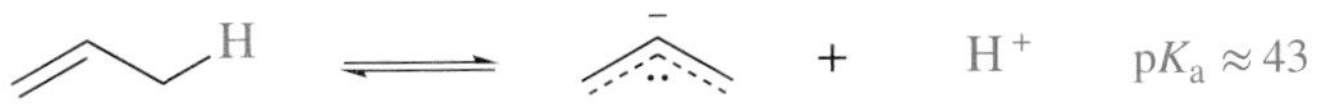

Propene ⇌ Allyl anion + H^+ (Proton) p$K_a \approx 43$

Although the pK_a values of hydrocarbons are subject to a fair degree of uncertainty and vary according to how they are measured, all of the methods agree that allyl anion is a much weaker base than propyl anion.

The electrostatic potential maps in Figure 11.5 illustrate the greater dispersal of negative charge in allyl anion versus propyl anion. In addition to the stabilization that results from electron delocalization of the π electrons, C-2 of propene is sp^2-hybridized, which increases the acidity of the allylic hydrogens by an electron-withdrawing inductive effect.

Recall from Section 9.5 that the electronegativity of carbon increases with increasing s character. *sp*-Hybridized carbon is more electronegative than sp^2, which is more electronegative than sp^3.

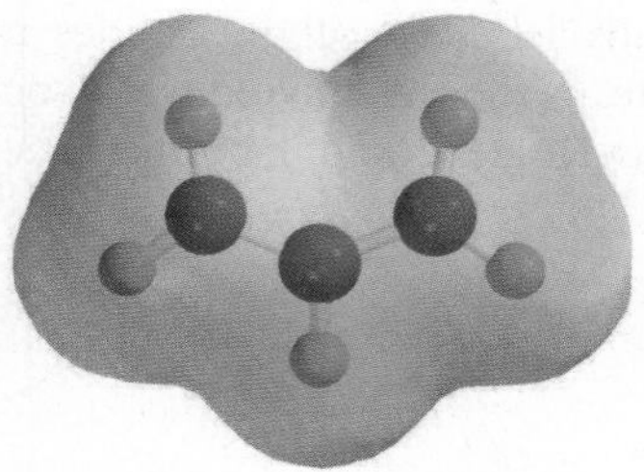

Allyl anion

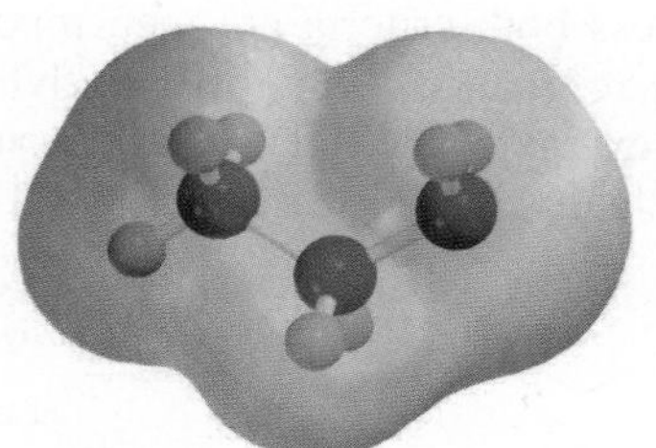

Propyl anion

Figure 11.5

Electrostatic potential maps for allyl and propyl anion. The charge is dispersed in allyl and shared equally by C-1 and C-3. The charge is localized at C-1 in propyl. The color scale is the same for both maps.

Problem 11.10

After heating a solution of allyl *tert*-butyl sulfide and sodium ethoxide in ethanol for several hours, *tert*-butyl propenyl sulfide was isolated in 66% yield. Suggest a stepwise mechanism for this isomerization. Which has the smaller pK_a (is the stronger acid), the reactant or the product?

$NaOCH_2CH_3$ / CH_3CH_2OH

Allyl *tert*-butyl sulfide → *tert*-Butyl propenyl sulfide

11.5 Classes of Dienes: Conjugated and Otherwise

As described in Sections 11.1–11.4, allylic carbocations, radicals, and anions are conjugated π-electron systems involved as intermediates in chemical reactions. The remaining sections of this chapter focus on stable molecules, especially hydrocarbons called **conjugated dienes,** which contain two C═C units joined by a single bond as in C═C—C═C. It begins by comparing their structure and stability to **isolated dienes,** in which the two C═C units are separated from each other by one or more sp^3-hybridized carbons, and to **cumulated dienes,** or **cumulenes** (also called **allenes**), in which two C═C units share a single carbon (C═C═C).

(*E*)-1,3-Pentadiene (conjugated)

1,4-Pentadiene (isolated)

$H_2C{=}C{=}CHCH_2CH_3$
1,2-Pentadiene (cumulated)

Problem 11.11

Many naturally occurring substances contain several carbon–carbon double bonds: some isolated, some conjugated, and some cumulated. Identify the types of carbon–carbon double bonds found in each of the following substances:

(a) β-Springene (a scent substance from the dorsal gland of springboks)

(b) Cembrene (occurs in pine resin)

(c) The sex attractant of the male dried-bean beetle

$CH_3(CH_2)_6CH_2CH{=}C{=}CH$–CH═CH–$CO_2CH_3$ (with H on each alkene carbon)

Sample Solution (a) β-Springene has three isolated double bonds and a pair of conjugated double bonds:

Conjugated double bonds

Isolated double bonds

Isolated double bonds are separated from other double bonds by at least one sp^3-hybridized carbon. Conjugated double bonds are joined by a single bond.

Alkadienes are named according to the IUPAC rules by replacing the *-ane* ending of an alkane with *-adiene* and locating the position of each double bond by number. Compounds with three carbon–carbon double bonds are called *alkatrienes* and named accordingly, those with four double bonds are *alkatetraenes,* and so on.

11.6 Relative Stabilities of Dienes

Which is the most stable arrangement of double bonds in an alkadiene: isolated, conjugated, or cumulated?

As we have seen before (Section 8.3), the relative stabilities of alkenes can be assessed from their heats of hydrogenation. Figure 11.6 compares these values for the isolated diene 1,4-pentadiene and its conjugated isomer (E)-1,3-pentadiene. The figure shows that an isolated pair of double bonds behaves much like two independent alkene units. The measured heat of hydrogenation of the two double bonds in 1,4-pentadiene is 252 kJ/mol (60.2 kcal/mol), exactly twice the heat of hydrogenation of 1-pentene. Furthermore, the heat evolved on hydrogenation of each double bond must be 126 kJ/mol (30.1 kcal/mol) because 1-pentene is an intermediate in the hydrogenation of 1,4-pentadiene to pentane.

By the same reasoning, hydrogenation of the terminal double bond in the conjugated diene (E)-1,3-pentadiene releases only 111 kJ/mol (26.5 kcal/mol) when it is hydrogenated to (E)-2-pentene. Hydrogenation of the terminal double bond in the conjugated

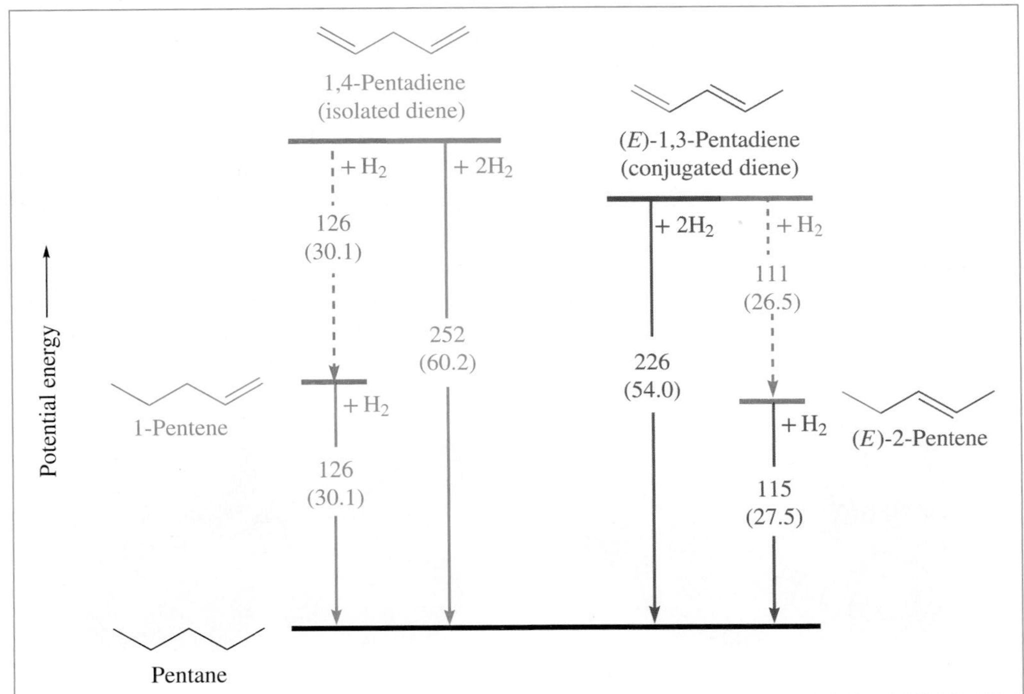

Figure 11.6

Heats of hydrogenation in kJ/mol (kcal/mol) are used to assess the stabilities of isolated versus conjugated double bonds. Comparing the measured heats of hydrogenation (*solid lines*) of the four compounds shown gives the values shown by the *dashed lines* for the heats of hydrogenation of the terminal double bond of 1,4-pentadiene and (E)-1,3-pentadiene. A conjugated double bond is approximately 15 kJ/mol (3.6 kcal/mol) more stable than an isolated double bond.

diene evolves 15 kJ/mol (3.6 kcal/mol) less heat than hydrogenation of a terminal double bond in the diene with isolated double bonds. *A conjugated double bond is 15 kJ/mol (3.6 kcal/mol) more stable than an isolated double bond.* This increased stability due to conjugation is the **delocalization energy, resonance energy,** or **conjugation energy.**

The cumulated double bonds of an allenic system are of relatively high energy. The heat of hydrogenation of allene is more than twice that of propene.

$$\underset{\text{Allene}}{H_2C{=}C{=}CH_2} + \underset{\text{Hydrogen}}{2H_2} \longrightarrow \underset{\text{Propane}}{CH_3CH_2CH_3} \qquad \Delta H^\circ = -295 \text{ kJ/mol } (-70.5 \text{ kcal/mol})$$

$$\underset{\text{Propene}}{CH_3CH{=}CH_2} + \underset{\text{Hydrogen}}{H_2} \longrightarrow \underset{\text{Propane}}{CH_3CH_2CH_3} \qquad \Delta H^\circ = -125 \text{ kJ/mol } (-29.9 \text{ kcal/mol})$$

Problem 11.12

Another way in which energies of isomers may be compared is by their heats of combustion. Match the heat of combustion with the appropriate diene.

Dienes:	1,2-Pentadiene, (*E*)-1,3-pentadiene, 1,4-pentadiene
Heats of combustion:	3186 kJ/mol (761 kcal/mol), 3217 kJ/mol (769 kcal/mol), 3251 kJ/mol (776 kcal/mol)

Thus, the order of alkadiene stability decreases in the order: conjugated diene (most stable) → isolated diene → cumulated diene (least stable). To understand this ranking, we need to look at structure and bonding in alkadienes in more detail.

11.7 Bonding in Conjugated Dienes

At 146 pm (1.46 Å) the C-2—C-3 distance in 1,3-butadiene is relatively short for a carbon–carbon single bond. This is most reasonably seen as a hybridization effect. In ethane both carbons are sp^3-hybridized and are separated by a distance of 153 pm (1.53 Å). The carbon–carbon single bond in propene unites sp^3- and sp^2-hybridized carbons and is shorter than that of ethane. Both C-2 and C-3 are sp^2-hybridized in 1,3-butadiene, and a decrease in bond distance between them reflects the tendency of carbon to attract electrons more strongly as its *s* character increases.

sp^3 sp^3	sp^3 sp^2	sp^2 sp^2
$H_3C{-}CH_3$	$H_3C{-}CH{=}CH_2$	$H_2C{=}CH{-}CH{=}CH_2$
153 pm (1.53 Å)	151 pm (1.51 Å)	146 pm (1.46 Å)

The factor most responsible for the increased stability of conjugated double bonds is the greater delocalization of their π electrons compared with the π electrons of isolated double bonds. As shown in Figure 11.7*a*, the four *p* orbitals of the conjugated diene 1,3-pentadiene combine to give a continuous π system, which allows each of the π electrons to interact with all four carbons. Continuous overlap of all four *p* orbitals of 1,4-pentadiene (Figure 11.7*b*) is not possible because a CH_2 group separates the two alkene units and their π electrons are no more delocalized than they are in ethylene.

(*a*) Conjugated double bonds in 1,3-pentadiene

(*b*) Isolated double bonds in 1,4-pentadiene

Figure 11.7

(*a*) In a conjugated diene, overlap of adjacent *p* orbitals gives an extended π system encompassing four carbons. (*b*) Isolated double bonds are separated from one another by one or more sp^3-hybridized carbons and cannot overlap to give an extended π orbital.

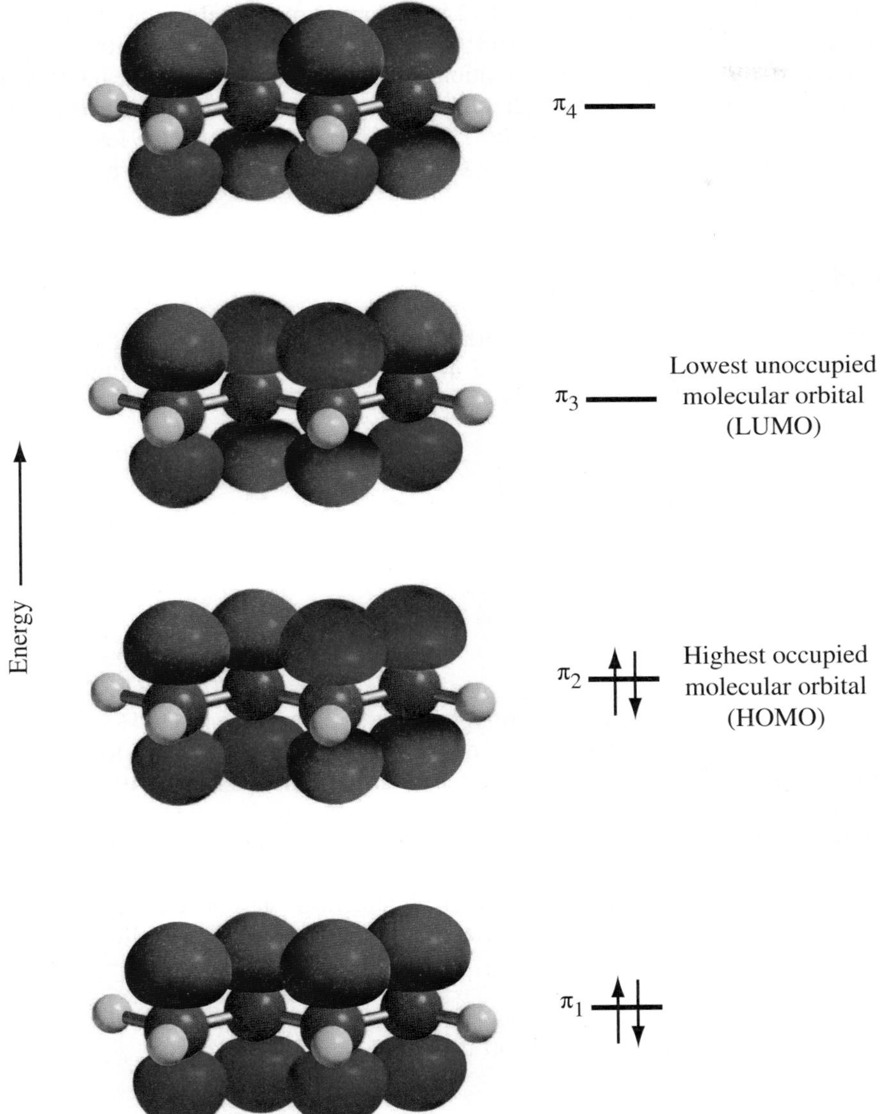

Figure 11.8

The four π molecular orbitals of 1,3-butadiene. The two lower molecular orbitals are both bonding and each is doubly occupied. The electrons in π_1 and in π_2 are delocalized over all four carbons. The highest occupied orbital is the HOMO and the lowest unoccupied molecular orbital is the LUMO.

A more detailed molecular orbital picture of electron delocalization is shown for the case of 1,3-butadiene in Figure 11.8. The four *p* AOs combine to give four π MOs, two of which (π_1 and π_2) are bonding; each of these π orbitals contains two electrons. The two antibonding orbitals are unoccupied.

Electron delocalization in 1,3-butadiene is most effective when all four carbons lie in the same plane. Two conformations, called *s*-cis and *s*-trans, permit this coplanarity.

s-Trans conformation of 1,3-butadiene ⇌ Transition state ⇌ *s*-Cis conformation of 1,3-butadiene

(The letter *s* in *s*-cis and *s*-trans refers to conformations around the single bond.)

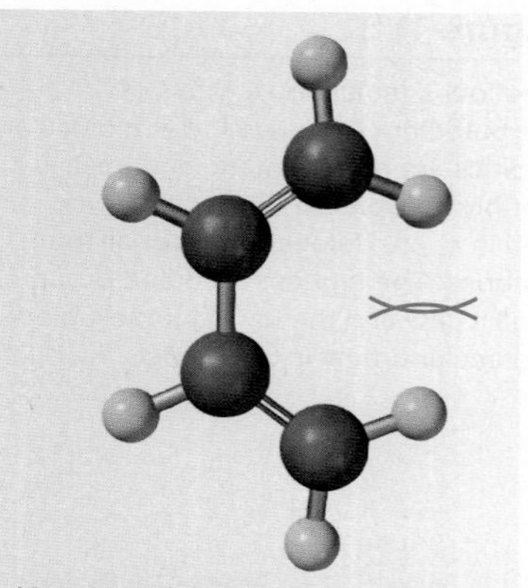

Van der Waals strain between interior hydrogens increases strain energy of the s-cis conformation of 1,3-butadiene.

The *s*-trans conformation is 12 kJ/mol (2.8 kcal/mol) more stable than *s*-cis, with van der Waals strain between the C-1 and C-4 "interior" hydrogens contributing to the decreased stability of the *s*-cis conformation. The two conformations are interconvertible by rotation about the C-2—C-3 single bond with an activation energy for the *s*-trans → *s*-cis conversion of 25 kJ/mol (6 kcal/mol). This energy cost reflects the loss of π-electron delocalization in going from a coplanar C=C—C=C arrangement to a nonplanar one at the transition state.

11.8 Bonding in Allenes

The three carbons of allene lie in a straight line, with relatively short carbon–carbon bond distances of 131 pm (1.31 Å). The middle carbon, because it has two π bonds, is *sp*-hybridized. The end carbons of allene are sp^2-hybridized.

sp sp^2

$118.4°$ $H_2C=C=CH_2$

108 pm (1.08 Å) 131 pm (1.31 Å)

Allene

As Figure 11.9 illustrates, allene is nonplanar; the plane of one HCH unit is perpendicular to the plane of the other. Figure 11.9 also shows the reason for this unusual geometry. The 2*p* orbital of each of the terminal carbons overlaps with a different 2*p* orbital of the central carbon. Because the 2*p* orbitals of the central carbon are perpendicular to each other, the perpendicular nature of the two HCH units follows naturally.

The nonplanarity of allenes has an interesting stereochemical consequence. 1,3-Disubstituted allenes are chiral; they are not superimposable on their mirror images.

(*a*) Planes defined by H(C-1)H and H(C-3)H are mutually perpendicular.

(*b*) The *p* orbital of C-1 and one of the *p* orbitals of C-2 can overlap so as to participate in π bonding.

(*c*) The *p* orbital of C-3 and one of the *p* orbitals of C-2 can overlap so as to participate in a second π orbital perpendicular to the one in (*b*).

(*d*) Allene is a nonplanar molecule characterized by a linear carbon chain and two mutually perpendicular π bonds.

Figure 11.9

Bonding and geometry in 1,2-propadiene (allene). The green and yellow colors are meant to differentiate the orbitals.

Even an allene as simple as 2,3-pentadiene ($CH_3CH{=}C{=}CHCH_3$) has been obtained as separate enantiomers.

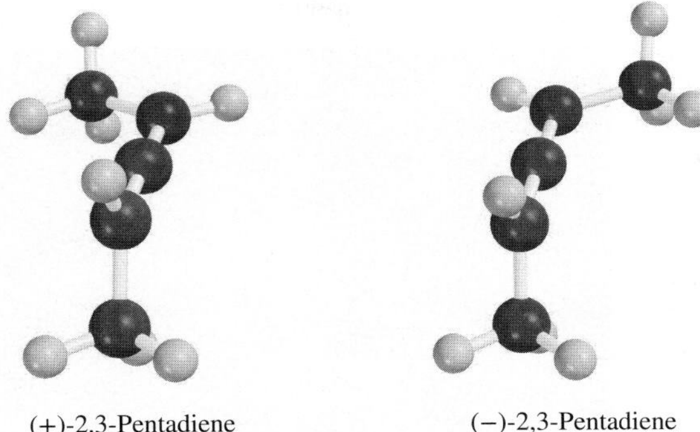

(+)-2,3-Pentadiene (−)-2,3-Pentadiene

The Cahn–Ingold–Prelog *R–S* notation has been extended to include molecules with a chirality axis. See the article by Mak in the November 2004 issue of the *Journal of Chemical Education* for a brief discussion of assigning *R* or *S* to chiral molecules that do not contain a chirality center.

The enantiomers shown are related as a right-hand and left-hand screw, respectively.

Chiral allenes are another example of molecules that are chiral, but do not contain a chirality center. Like the chiral biaryl derivatives that were described in Section 4.9, chiral allenes contain an axis of chirality. The axis of chirality in 2,3-pentadiene is a line passing through the three carbons of the allene unit (carbons 2, 3, and 4).

Problem 11.13

Is 2-methyl-2,3-pentadiene chiral? What about 2-chloro-2,3-pentadiene?

11.9 Preparation of Dienes

The use of 1,3-butadiene in the preparation of synthetic rubber is discussed in the boxed essay *Diene Polymers* that appears on page 402.

The conjugated diene 1,3-butadiene is used in the manufacture of synthetic rubber for automobile tires and is prepared on an industrial scale in vast quantities. In the presence of a suitable catalyst, butane undergoes thermal dehydrogenation to yield 1,3-butadiene.

$$CH_3CH_2CH_2CH_3 \xrightarrow[\text{chromia–alumina}]{590\text{–}675°C} H_2C{=}CHCH{=}CH_2 + 2H_2$$

Laboratory syntheses of conjugated dienes involve elimination reactions of unsaturated alcohols and alkyl halides. In the two examples that follow, the conjugated diene is produced in high yield even though an isolated diene is also possible.

3-Methyl-5-hexen-3-ol (OH) $\xrightarrow[88\%]{\text{KHSO}_4\text{, heat}}$ 4-Methyl-1,3-hexadiene $\xleftarrow[78\%]{\text{KOH, heat}}$ 4-Bromo-4-methyl-1-hexene (Br)

As we saw in Chapter 7, dehydrations and dehydrohalogenations are typically regioselective in the direction that leads to the most stable double bond. Conjugated dienes are more stable than isolated dienes and are formed faster via a lower-energy transition state.

Problem 11.14

What dienes containing isolated double bonds are capable of being formed, but are not observed, in the two preceding equations describing elimination in 3-methyl-5-hexen-3-ol and 4-bromo-4-methyl-1-hexene?

Diene Polymers

We begin with two trees, both cultivated on plantations in Southeast Asia. One, *Hevea brasiliensis,* is a source of natural rubber and was imported from Brazil in the nineteenth century. The other, *Isonandra gutta,* is native to Sumatra, Java, and Borneo and gives a latex from which gutta-percha is obtained.

Some 500 years ago during Columbus's second voyage to what are now the Americas, he and his crew saw children playing with balls made from the latex of trees that grew there. Later, Joseph Priestley called this material "rubber" to describe its ability to erase pencil marks by rubbing, and in 1823 Charles Macintosh demonstrated how rubber could be used to make waterproof coats and shoes. Shortly thereafter Michael Faraday determined an empirical formula of C_5H_8 for rubber. It was eventually determined that rubber is a polymer of 2-methyl-1,3-butadiene.

2-Methyl-1,3-butadiene
(common name: *isoprene*)

The structure of rubber corresponds to 1,4 addition of several thousand isoprene units to one another:

All the double bonds in rubber have the *Z* configuration.

Gutta-percha is a different polymer of isoprene. Its chains are shorter than those of natural rubber and have *E* double bonds.

Gutta-percha is flexible when heated, but is harder and more durable than rubber at room temperature. It was, at one time, the material of choice for golf ball covers. Gutta-percha's main claim to fame though lies out of sight on the floors of the world's oceans. The first global communication network—the telegraph—relied on insulated copper wire to connect senders and receivers. Gutta-percha proved so superior to natural rubber in resisting deterioration, especially underwater, that it coated the thousands of miles of insulated telegraph cable that connected most of the countries of the world by the close of the nineteenth century.

In natural rubber the attractive forces between neighboring polymer chains are relatively weak, and there is little overall structural order. The chains slide easily past one another when stretched and return, in time, to their disordered state when the distorting force is removed. The ability of a substance to recover its original shape after distortion is its *elasticity.* The elasticity of natural rubber is satisfactory only within a limited temperature range; it is too rigid when cold and too sticky when warm to be very useful. Rubber's elasticity is improved by *vulcanization,* a process discovered by Charles Goodyear in 1839. When natural rubber is heated with sulfur, a chemical reaction occurs in which neighboring polyisoprene chains become connected through covalent bonds to sulfur. Although these sulfur "bridges" permit only limited movement of one chain with respect to another,

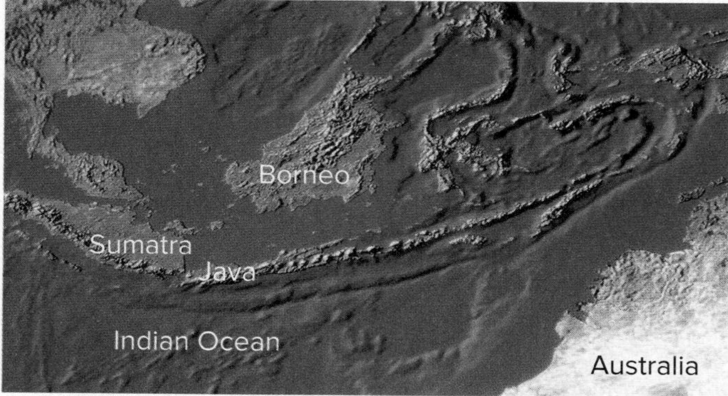

their presence ensures that the rubber will snap back to its original shape once the distorting force is removed.

As the demand for rubber increased, so did the chemical industry's efforts to prepare a synthetic substitute. One of the first **elastomers** (a synthetic polymer that possesses elasticity) to find a commercial niche was *neoprene,* discovered by chemists at Du Pont in 1931. Neoprene is produced by free-radical polymerization of 2-chloro-1,3-butadiene and has the greatest variety of applications of any elastomer. Some uses include electrical insulation, conveyer belts, hoses, and weather balloons.

2-Chloro-1,3-butadiene → Neoprene

The elastomer produced in greatest amount is *styrene-butadiene rubber* (SBR). Annually, just under 10^9 lb of SBR is produced in the United States, most of which is used in automobile tires. As its name suggests, SBR is prepared from styrene and 1,3-butadiene. It is an example of a **copolymer,** a polymer assembled from two or more different monomers. Free-radical polymerization of a mixture of styrene and 1,3-butadiene gives SBR.

1,3-Butadiene + Styrene →

Styrene-butadiene rubber

Coordination polymerization of isoprene using Ziegler–Natta catalyst systems (Section 15.15) gives a material similar in properties to natural rubber, as does polymerization of 1,3-butadiene. Poly(1,3-butadiene) is produced in about two thirds the quantity of SBR each year. It, too, finds its principal use in tires.

11.10 Addition of Hydrogen Halides to Conjugated Dienes

Electrophilic addition is the characteristic reaction of alkenes, and conjugated dienes undergo addition with the same electrophiles that react with alkenes, and by similar mechanisms. Hydrogen chloride, for example, adds to the diene unit of 1,3-cyclopentadiene to give 3-chlorocyclopentene. Mechanism 11.3 is analogous to the electrophilic addition of HCl to alkenes.

1,3-Cyclopentadiene + HCl ⟶ 3-Chlorocyclopentene (70–90%)

As with alkenes, the regioselectivity of electrophilic addition to conjugated dienes is governed by the stability of the resulting carbocation. Protonation of a conjugated diene always occurs at the end of the diene unit because an allylic carbocation results.

1,3-Cyclopentadiene + H—Cl: ⟶ Resonance contributors of 2-cyclopentenyl cation

Mechanism 11.3

Addition of Hydrogen Chloride to 1,3-Cyclopentadiene

THE OVERALL REACTION:

1,3-Cyclopentadiene + HCl (Hydrogen chloride) ⟶ 3-Chlorocyclopentene

THE MECHANISM:

Step 1: A proton is transferred from HCl to a carbon at the end of the diene system to give an allylic carbocation.

1,3-Cyclopentadiene + Hydrogen chloride ⟶ 2-Cyclopentenyl cation + Chloride ion

Step 2: Chloride ion acts as a nucleophile and bonds to the positively charged carbon of the carbocation.

2-Cyclopentenyl cation + Chloride ion ⟶ 3-Chlorocyclopentene

Problem 11.15

Carbons 1 and 4 of 1,3-cyclopentadiene are equivalent and give the same carbocation on protonation. Likewise, carbons 2 and 3 are equivalent. Write the structure of the carbocation formed by protonation of C-2 or C-3 to verify that it is not allylic and therefore not as stable as the one formed by protonation of C-1 or C-4.

Both resonance contributors of the allylic carbocation from 1,3-cyclopentadiene are equivalent and attack by chloride at either of the carbons that share the positive charge gives the same product, 3-chlorocyclopentene.

Such is not the case with 1,3-butadiene. Protonation of the diene is still regiospecific for the end carbon, but the two resonance forms of the resulting allylic carbocation are not equivalent.

1,3-Butadiene + Hydrogen bromide (H—Br:) ⟶ 1-Methylallyl cation (two resonance forms ⟷)

Consequently, a mixture of two regioisomeric allylic bromides is formed when HBr adds to 1,3-butadiene.

1,3-Butadiene —(HBr, −80°C)⟶ 3-Bromo-1-butene (81%) + (*E* + *Z*)-1-Bromo-2-butene (19%)

Both products are formed from the same allylic carbocation. The major product corresponds to addition of a proton to C-1 of 1,3-butadiene and bromine to C-2. This mode of addition is called **1,2 addition.** The minor product has its proton and bromide at C-1 and C-4, respectively, and is formed by **1,4 addition.**

At −80°C the product from 1,2 addition predominates because it is formed faster than the 1,4-addition product. The product distribution is governed by **kinetic control.**

At room temperature, a much different product ratio is observed. Under these conditions the 1,4-addition product predominates.

1,3-Butadiene —(HBr, room temperature)⟶ 3-Bromo-1-butene (44%) (1,2-addition) + (*E* + *Z*)-1-Bromo-2-butene (56%) (1,4-addition)

To understand why temperature affects the product composition, an important fact must be added. *The 1,2- and 1,4-addition products interconvert at elevated temperature in the presence of hydrogen bromide.*

1,2-Addition product formed faster ⇌ allylic cation + :Br:⁻ ⇌ 1,4-Addition product more stable

At 45°C, for example, interconversion is rapid and gives an equilibrium mixture containing 85% of the 1,4-addition product and 15% of the 1,2 product. This demonstrates that the 1,4 product is more stable, presumably because it has a disubstituted double bond, whereas the double bond in the 1,2 product is monosubstituted.

When addition occurs under conditions in which the products can equilibrate, the composition of the reaction mixture no longer reflects their relative rates of formation but tends to reflect their *relative stabilities.* Reactions of this type are governed by **thermodynamic control.**

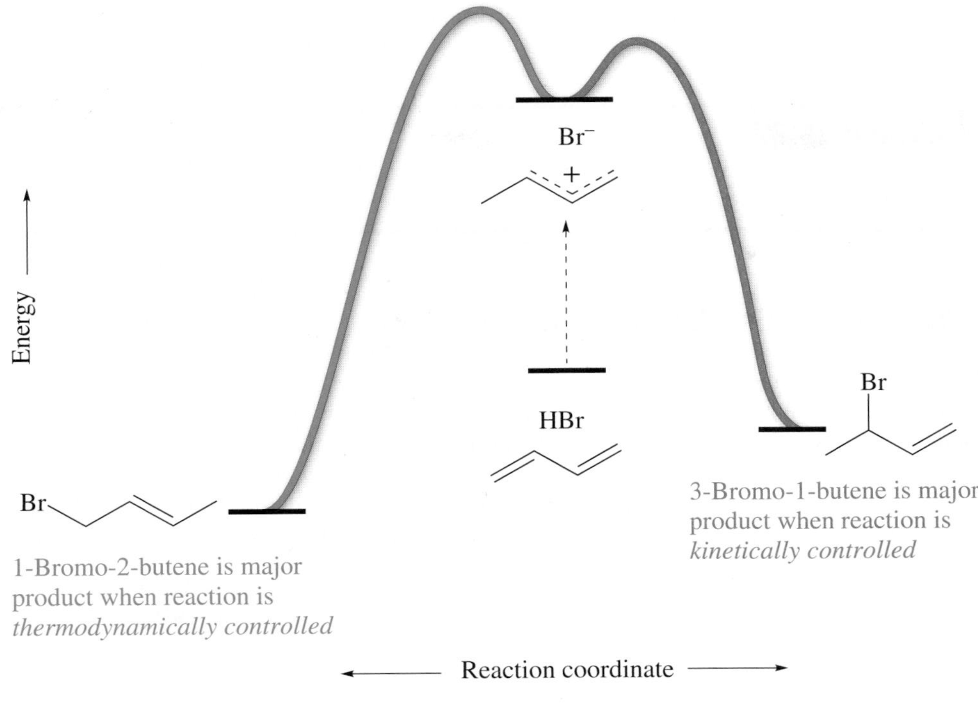
Br⁻
+
HBr
Br
Br
3-Bromo-1-butene is major product when reaction is *kinetically controlled*
1-Bromo-2-butene is major product when reaction is *thermodynamically controlled*
Energy
Reaction coordinate

Figure 11.10

3-Bromo-1-butene (red) is formed faster than 1-bromo-2-butene (blue) through a lower-energy transition state in the reaction of 1,3-butadiene with hydrogen bromide. At equilibrium, 1-bromo-2-butene predominates because it is the more stable isomer.

The energy diagram of Figure 11.10 illustrates kinetic and thermodynamic control in the addition of hydrogen bromide to 1,3-butadiene. At low temperature, addition takes place irreversibly. Isomerization is slow because insufficient thermal energy is available to permit the products to surmount the energy barrier for ionization. At higher temperatures isomerization is possible, and the more stable product predominates.

Dehydrohalogenation of alkyl halides using the sterically hindered base *tert*-butoxide, discussed in Section 7.14, was an earlier example of a reaction that gives kinetically controlled product distributions.

Before leaving this section, we should point out that the numbers in the terms 1,2 and 1,4 addition refer to carbons within the C═C—C═C structural unit wherever it may be in the molecule and not to the IUPAC numbering. For example, 1,2 and 1,4 addition to 2,4-hexadiene would involve the carbons shown.

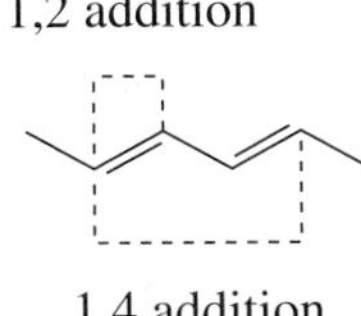
1,2 addition
1,4 addition

Problem 11.16

Write structural formulas for the products of 1,2 and 1,4 addition of hydrogen chloride to 2,4-hexadiene.

11.11 Halogen Addition to Dienes

Mixtures of 1,2- and 1,4-addition products are obtained when 1,3-butadiene reacts with chlorine or bromine.

$$\text{1,3-Butadiene} + \underset{\text{Bromine}}{Br_2} \xrightarrow{CHCl_3} \text{3,4-Dibromo-1-butene (37\%)} + (E)\text{-1,4-Dibromo-2-butene (63\%)}$$

The tendency for 1,4 addition is pronounced, and *E* double bonds are generated almost exclusively.

Problem 11.17

Exclusive of stereoisomers, how many products are possible in the electrophilic addition of 1 mol of bromine to 2-methyl-1,3-butadiene?

11.12 The Diels–Alder Reaction

We've already mentioned the value of carbon–carbon bond-forming reactions in organic synthesis. Imagine how useful it would be to have a reaction in which *two* carbon–carbon bonds are formed in a single operation simply by combining two compounds without having to add acids, bases, or other catalysts. For developing such a reaction, Otto Diels and Kurt Alder of the University of Kiel (Germany) shared the 1950 Nobel Prize in Chemistry. The **Diels–Alder reaction** is the conjugate addition of an alkene to a diene.

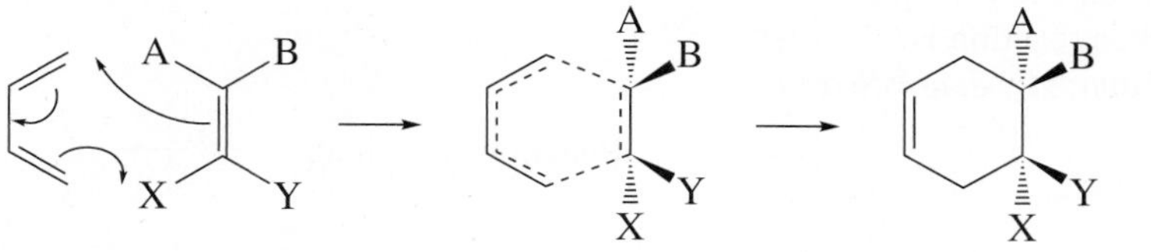

Diene + dienophile → Transition state for cycloaddition → Diels–Alder adduct

The alkene that adds to the diene is called the **dienophile** ("diene seeker"). The reaction is classified as a **cycloaddition,** and the product contains a cyclohexene ring.

The reaction occurs in a single step, without an intermediate, by a mechanism in which six atoms undergo bonding changes in the same transition state by way of cyclic reorganization of their π electrons. Concerted (one-step) reactions such as the Diels–Alder cycloaddition that proceed through a cyclic transition state are called **pericyclic reactions.**

Effect of Substituents on the Reactivity of the Dienophile The simplest of all Diels–Alder reactions, cycloaddition of ethylene to 1,3-butadiene, has a high activation energy and a low reaction rate, so does not proceed readily. However, electron-withdrawing substituents such as C=O and C≡N, when attached directly to the double bond, activate the dienophile toward cycloaddition. Acrolein ($H_2C{=}CHCH{=}O$), for example, reacts with 1,3-butadiene to give a high yield of the Diels–Alder adduct at a modest temperature.

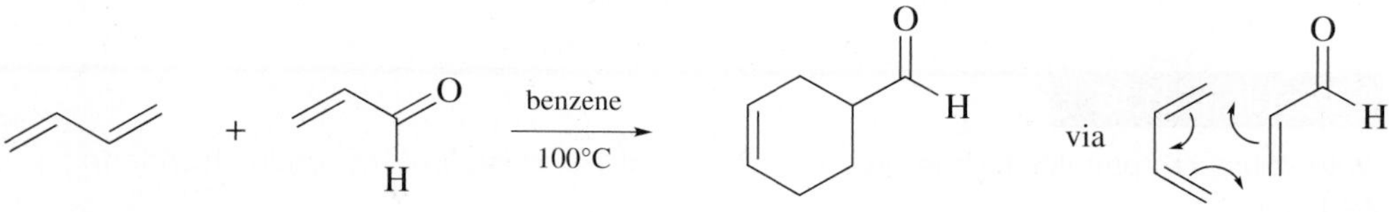

1,3-Butadiene + Acrolein $\xrightarrow[100°C]{\text{benzene}}$ Cyclohexene-4-carboxaldehyde (100%) via

Diethyl fumarate and maleic anhydride have two C=O functions on their double bond and are more reactive than acrolein; tetracyanoethylene is even more reactive.

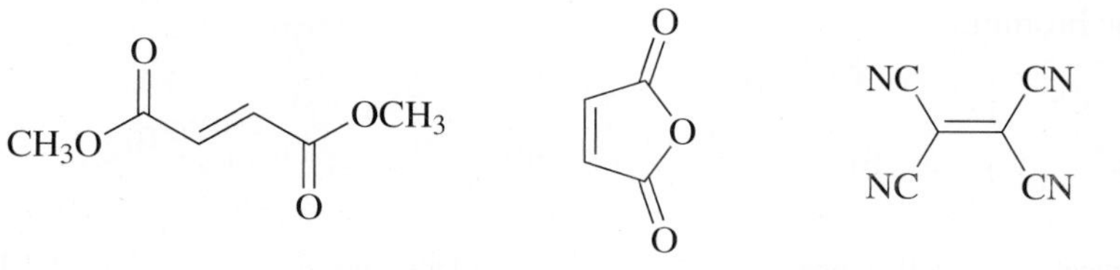

Dimethyl fumarate Maleic anhydride Tetracyanoethylene

The product of a Diels–Alder reaction always contains one more ring than the reactants. Maleic anhydride already contains one ring, so the product of its addition to 2-methyl-1,3-butadiene has two.

2-Methyl-1,3-butadiene + Maleic anhydride $\xrightarrow[100°C]{\text{benzene}}$ 1-Methylcyclohexene-4,5-dicarboxylic anhydride (100%) via

Problem 11.18

Dicarbonyl compounds such as quinones are reactive dienophiles.

(a) 1,4-Benzoquinone reacts with 2-chloro-1,3-butadiene to give a single product $C_{10}H_9ClO_2$ in 95% yield. Write a structural formula for this product.

(b) 2-Cyano-1,4-benzoquinone undergoes a Diels–Alder reaction with 1,3-butadiene to give a single product $C_{11}H_9NO_2$ in 84% yield. What is its structure?

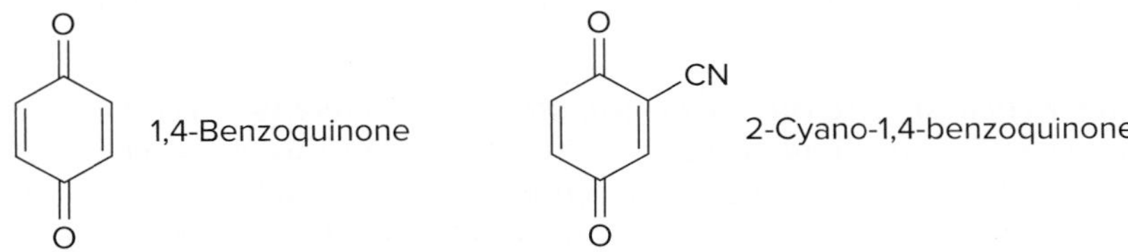

Sample Solution

(a)

- 2-Chloro-1,3-butadiene is $H_2C{=}C(Cl){-}CH{=}CH_2$
- $H_2C{=}C(Cl){-}CH{=}CH_2$ =
- reaction is

$C_{10}H_9ClO_2$

Conformational Effects on the Reactivity of the Diene The diene must be able to adopt the *s*-cis conformation in order for cycloaddition to occur. We saw in Section 11.7 that the *s*-cis conformation of 1,3-butadiene is 12 kJ/mol (2.8 kcal/mol) less stable than the *s*-trans form. This is a relatively small energy difference, so 1,3-butadiene is reactive in the Diels–Alder reaction. Dienes that cannot readily adopt the *s*-cis conformation are less reactive. For example, 4-methyl-1,3-pentadiene is a thousand times less reactive in the Diels–Alder reaction than *trans*-1,3-pentadiene because its *s*-cis conformation is destabilized by the steric effect imposed by the additional methyl group.

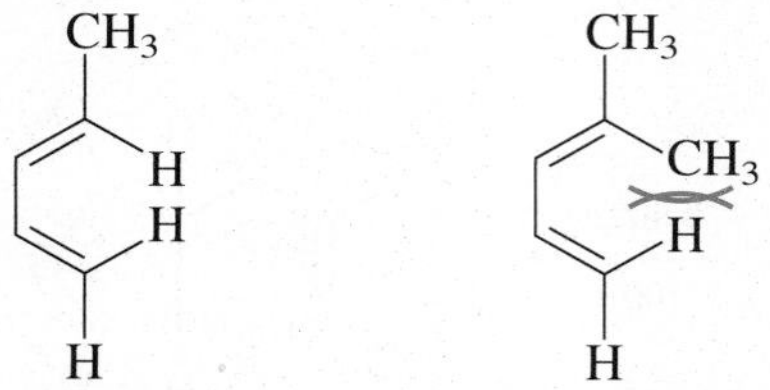

trans-1,3-Pentadiene 4-Methyl-1,3-pentadiene

Problem 11.19

2,3-Di-*tert*-butyl-1,3-butadiene is extremely unreactive in Diels–Alder reactions. Explain.

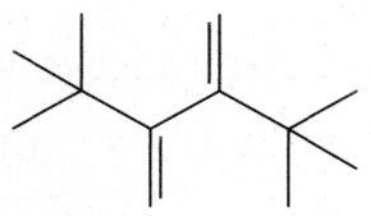

Cyclic conjugated dienes such as 1,3-cyclopentadiene are often used as the diene component in Diels–Alder reactions and are relatively reactive because the *s*-cis geometry is built into their structure.

Diels–Alder Reactions Are Stereospecific and Stereoselective The following pair of related cycloadditions combine to illustrate two important aspects of Diels–Alder reactions. First, recall that a stereospecific reaction is one in which stereoisomeric starting materials yield stereoisomeric products. In the example, 1,3-cyclopentadiene reacts with the two stereoisomeric dienophiles, dimethyl fumarate and dimethyl maleate. Dimethyl fumarate has a trans double bond; dimethyl maleate's double bond is cis.

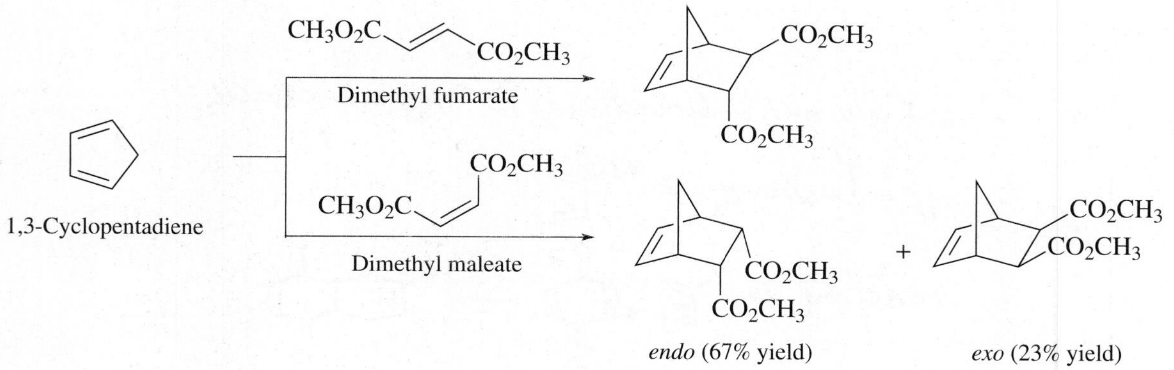

1,3-Cyclopentadiene; Dimethyl fumarate; Dimethyl maleate; *endo* (67% yield) + *exo* (23% yield)

Both reactions are stereospecific. The trans relationship between the —CO_2CH_3 substituents in dimethyl fumarate is retained in the cycloaddition product, and the cis relationship between them is retained in the product from dimethyl maleate. The customary terminology in bicyclic systems describes a substituent as exo if it is oriented toward the smaller bridge, endo if it is oriented away from it. Thus, one —CO_2CH_3 group is exo and the other endo in the dimethyl maleate Diels–Alder adduct, while both are endo in the major product from dimethyl fumarate, exo in the minor product.

In addition to being stereospecific, there is a stereoselective preference favoring the formation of endo-Diels–Alder adducts from cyclic dienes as is evident in the observed 3:1 *endo–exo* ratio in the reaction of dimethyl maleate with 1,3-cyclopentadiene shown in the preceding equation. Such results are common and are the basis of an empirical rule called the **endo rule** or **Alder rule,** which holds that for a dienophile that can yield two diastereoisomeric Diels–Alder adducts, the adduct with an endo orientation of the

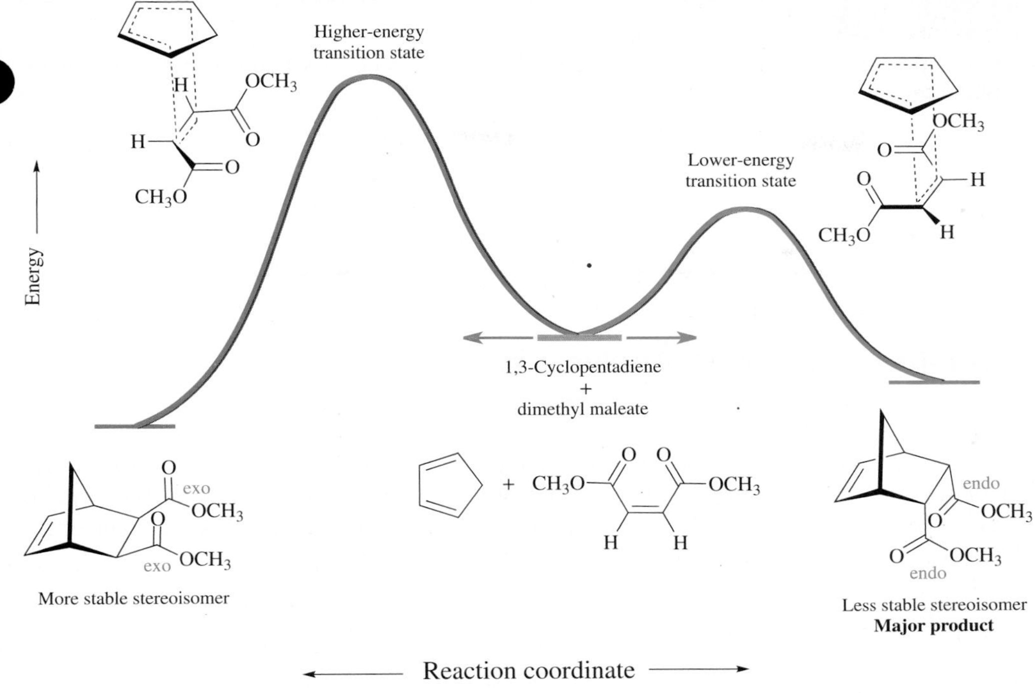

Figure 11.11

The less stable stereoisomer is formed preferentially in the Diels–Alder addition of dimethyl maleate to 1,3-cyclopentadiene because the endo transition state is lower in energy than the exo.

substituent predominates. Figure 11.11 illustrates the energy relationships involved in endo versus exo addition for the reaction of dimethyl maleate with 1,3-cyclopentadiene.

Problem 11.20

Methyl acrylate ($H_2C{=}CHCO_2CH_3$) reacts with 1,3-cyclopentadiene to give a mixture of two products. Write structural formulas for both and predict which one predominates.

11.13 Intramolecular Diels–Alder Reactions

Not only is the Diels–Alder reaction itself synthetically useful, it has also stimulated explorations of variants on its theme. The one most directly related to the traditional Diels–Alder reaction is an **intramolecular** analog in which both the diene and dienophile reside in the same molecule.

Intramolecular cycloaddition generates *two* new rings in a single operation. One of the new rings is, as in the intermolecular reaction, a cyclohexene; the other is typically five- or six-membered. The following example illustrates the formation of a bicyclic alkene from a noncyclic triene.

3-Methyl-1,3,9-decatriene —160°C→ 1-Methylbicyclo[4.4.0]-dec-1-ene (95%) via

Problem 11.21

The compound shown undergoes an intramolecular Diels–Alder reaction at room temperature. What is the structure of the product?

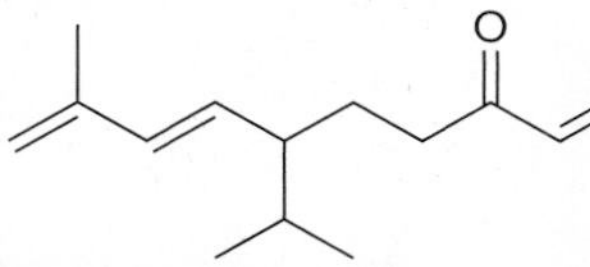

O

11.14 Retrosynthetic Analysis and the Diels–Alder Reaction

Diels–Alder reactions are widely used for making carbon–carbon bonds, and retrosynthetic analysis can reveal opportunities for their application. If a synthetic target contains a cyclohexene ring, start with the double bond and use curved arrows to disconnect the bonds to be formed in the sought-for cycloaddition. For example:

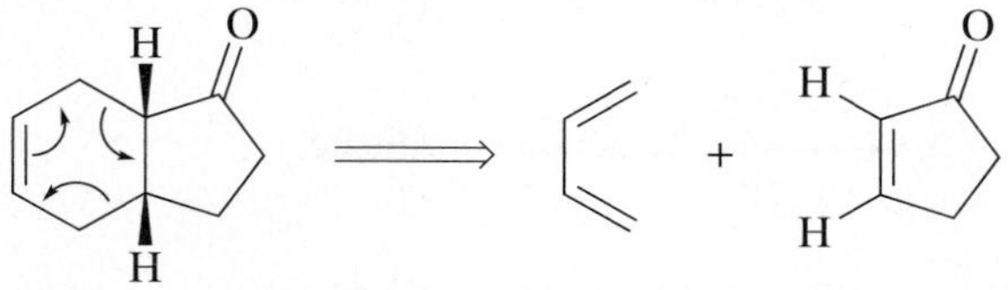

H O H O H + H H

In deciding whether the cycloaddition revealed by the disconnection is feasible or not, examine the alkene to make sure that it is a reactive dienophile. Usually this means that the double bond bears an electron-attracting group, especially C=O or C≡N.

Problem 11.22

What diene and dienophile could you use to prepare the compound shown?

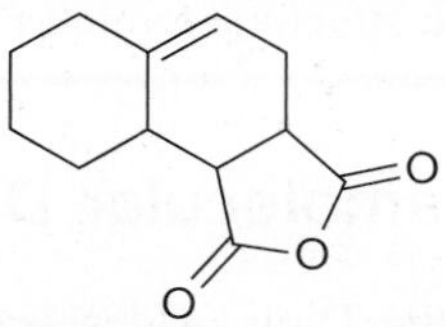

O O O

The number of reaction types that we have covered to this point is extensive enough to allow the construction of relatively complicated products from simple, readily available starting materials. Thus, recognizing that the bicyclo[2.2.2]octene derivative shown is accessible by a Diels–Alder reaction leads us back to cyclohexanol as the source of six of its carbons.

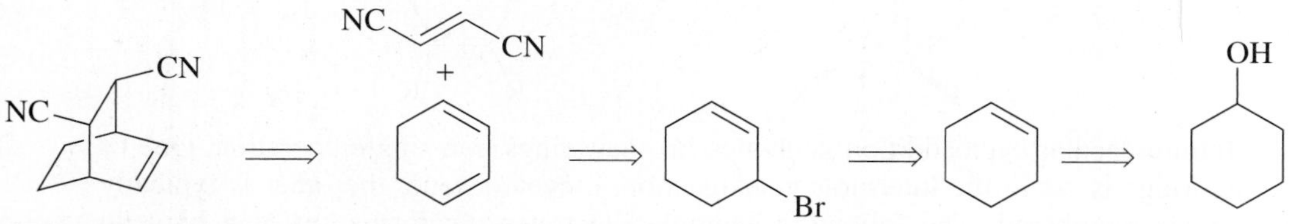

NC CN NC CN + Br OH

5,6-Dicyanobicyclo-[2.2.2]oct-2-ene

Cyclohexanol

Problem 11.23

Write equations in the synthetic direction for the preparation of 5,6-dicyanobicyclo[2.2.2]oct-2-ene from cyclohexanol and any necessary organic or inorganic reagents.

11.15 Molecular Orbital Analysis of the Diels–Alder Reaction

The conventional curved-arrow description of the Diels–Alder reaction shows a one-step mechanism involving a cyclic reorganization of six electrons—four from the diene plus two from the dienophile—in the transition state.

Diene + alkene Transition state Cyclohexene

Such a concerted process is more consistent with the stereochemical observations than one in which the two new σ bonds are formed in separate steps. Substituents that are cis in the dienophile remain cis in the product; those that are trans in the dienophile remain trans in the product.

Since all of the bonds are formed in the same step and only π electrons are involved, we can use the Hückel π-electron approximation to explore the process from a molecular orbital perspective and need examine only those orbitals of the reactants that are directly involved in bond formation. These are called the **frontier orbitals** and are usually the HOMO of one reactant and the LUMO of another. For the Diels–Alder reaction they are the HOMO of the diene and the LUMO of the dienophile and are as shown in Figure 11.12. The choice of this HOMO–LUMO combination is made to be consistent with the experimental fact that electron-withdrawing groups on the dienophile increase its reactivity and suggest that electrons flow from the diene to the dienophile. Notice that the symmetry properties of these two orbitals are such as to permit the in-phase overlap necessary for σ bond formation between the diene and dienophile. In MO terms, the Diels–Alder reaction is classified as **symmetry-allowed.**

Contrast the Diels–Alder reaction with one that looks similar, the cycloaddition of one ethylene molecule to another to give cyclobutane.

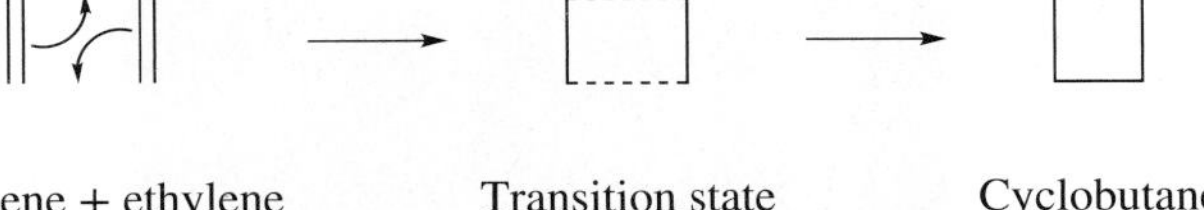

Cycloadditions of alkenes are rare and likely proceed in a stepwise fashion rather than by the concerted process implied in the equation. Figure 11.13 shows the interaction

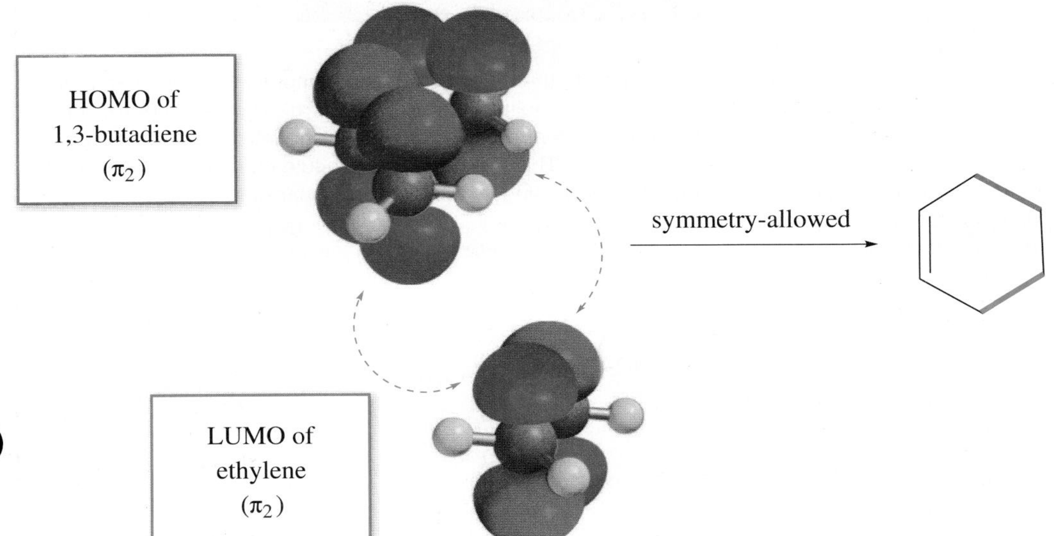

Figure 11.12

The LUMO of ethylene and the HOMO of 1,3-butadiene have the proper symmetry to allow cycloaddition. The two molecules approach each other in parallel planes, and electrons flow from the HOMO of 1,3-butadiene to the LUMO of ethylene. σ Bonds form when orbitals of the same symmetry (blue-to-blue and red-to-red) overlap. The molecular orbitals for ethylene were discussed in Section 7.2.

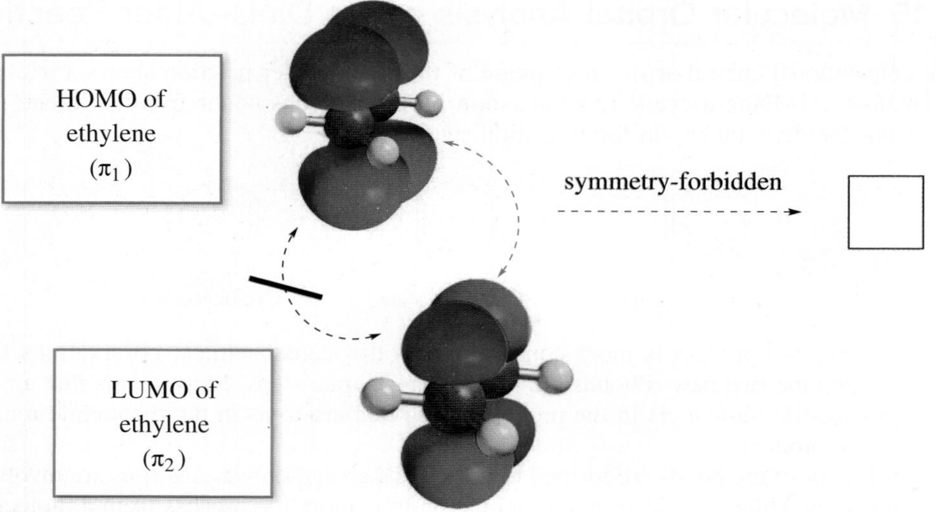

Figure 11.13

The HOMO and the LUMO lack the proper symmetry to allow for concerted cycloaddition of two ethylene molecules. The orbital mismatch precludes σ bond formation between the carbon of one and the carbon of the other.

Pericyclic Reactions in Chemical Biology

Although pericyclic reactions are widely used in synthetic chemistry, there are very few pericyclic reactions that naturally occur in living organisms. Chemists and chemical biologists have taken advantage of the rarity of these reactions to perform "bioorthogonal" chemical reactions. Bioorthogonal chemical reactions occur selectively in the presence of the plethora of reactive functional groups in biological systems, ideally under conditions that are compatible with biomolecules (aqueous solution, temperatures 37°C and below, near neutral pH). These reactions have found broad applications in biology and medicine, including basic research in cells and in living organisms, preparation of antibody–drug conjugates, and for medical imaging.

One of the most successful and widely used bioorthogonal reactions involves an azide–alkyne cycloaddition reaction. The copper-catalyzed version of this reaction was introduced by Nobel Laureate K. Barry Sharpless, who coined the term "click" for reactions that are simple, high yielding, and environmentally benign. Like the Diels–Alder reaction, the azide–alkyne "click" reaction is a six-electron process. The four-electron component is the azide, and the two-electron component comes from one π bond of the alkyne. This class of reactions is discussed in greater detail in the Descriptive Passage at the end of this chapter.

Azide

$:N \equiv \overset{+}{N} - \ddot{\overset{-}{N}} - R$ + $R' - C \equiv CH$ $\longrightarrow$ Product (+ Isomer)

Alkyne

Product

Neither of these functional groups occurs naturally in biopolymers, so chemical biologists continue to develop methods to incorporate them into the target biomolecule. In one example, a normal metabolic pathway in an organism is co-opted into accepting an "unnatural" substrate that contains one of the functional groups. This modified biomolecule can then be observed and identified using its bioorthogonal partner.

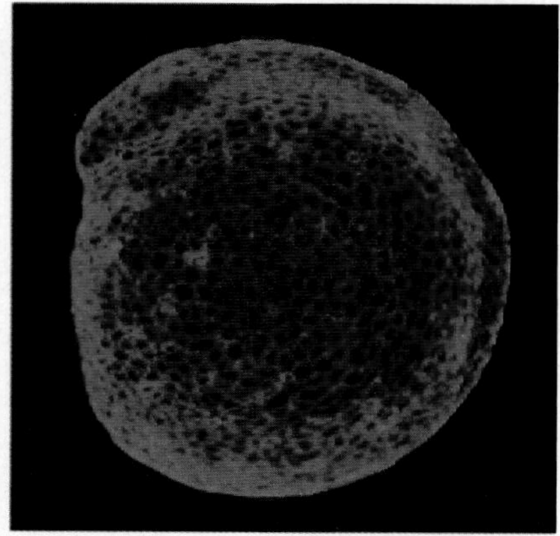

A zebrafish embryo was fed with the unnatural monosaccharide GalNAz, followed by an alkyne-containing fluorescent probe.

Courtesy of PNAS. Article "Visualizing enveloping layer glycans during zebrafish early embryogenesis" by J.M. Baskin et al. Published online 2010 May 20.

Natural monosaccharide — GalNAc

Unnatural monosaccharide — GalNAz (N_3)

between the HOMO of one ethylene molecule and the LUMO of another. The carbons that are to become σ-bonded to each other experience an antibonding interaction during cycloaddition, which raises the activation energy. The reaction is **symmetry-forbidden.** Reaction, were it to occur, would take place slowly and by a mechanism in which the two new σ bonds are formed in separate steps rather than one involving a single transition state.

Focusing on HOMO–LUMO interactions can aid our understanding of many organic reactions. Its early development is attributed to Professor Kenichi Fukui (Kyoto), and its application to Diels–Alder reactions constitutes but one part of the *Woodward–Hoffmann rules* proposed by Professors R. B. Woodward (Harvard) and Roald Hoffmann (Cornell).

Fukui and Hoffmann shared the 1981 Nobel Prize in Chemistry. Woodward had already won a Nobel Prize in 1965 for his achievements in the synthesis of complex organic compounds, and only his death in 1979 prevented him from being recognized a second time.

11.16 The Cope and Claisen Rearrangements

As noted in Section 11.12, reactions that proceed in a single step by way of a cyclic transition state are called *pericyclic reactions*. Diels–Alder cycloadditions are one example of a pericyclic reaction; the **Cope rearrangement** is another. In general terms, the Cope rearrangement is the one-step thermal transformation of one 1,5-diene unit to another by way of a cyclic transition state.

Arthur C. Cope (M.I.T.) was one of the most respected figures in American organic chemistry during the middle part of the twentieth century.

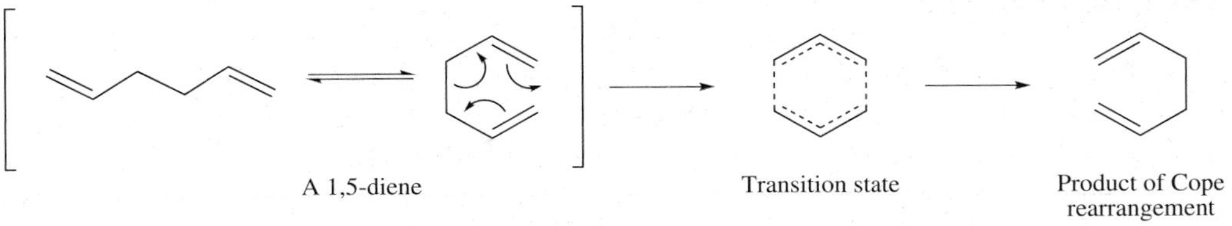

A 1,5-diene — Transition state — Product of Cope rearrangement

The Cope rearrangement shown of 1,5-hexadiene is a special case in which the reactant and the product are the same. More interesting are those in which they are different. A particularly striking example is the case of *cis*-1,2-divinylcyclopropane, which, on attempted preparation, spontaneously rearranges to 1,4-cycloheptadiene even at −40°C.

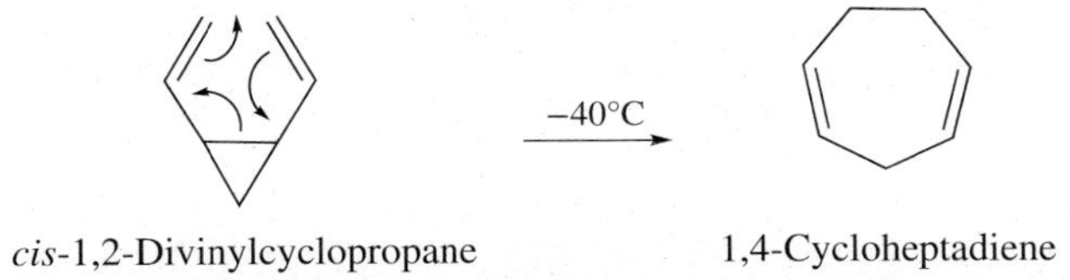

−40°C

cis-1,2-Divinylcyclopropane — 1,4-Cycloheptadiene

The low activation energy for this process results from the relief of angle strain as the cyclopropane ring undergoes partial cleavage in the transition state. Higher temperatures are required for relatively unstrained compounds that incorporate a 1,5-diene unit.

Problem 11.24

A Cope rearrangement of a divinylcyclopropane derivative was the final step in a laboratory synthesis of (*R*)-(−)-dictyopterene C, a constituent of Pacific seaweed. Deduce the structure, including stereochemistry of the reactant.

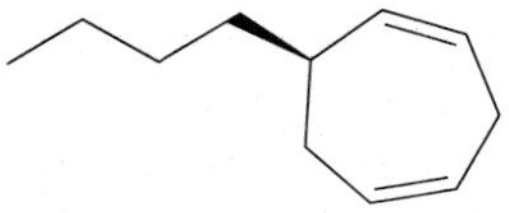

(*R*)-(−)-Dictyopterene C

Ludwig Claisen was a German chemist whose career spanned the end of the nineteenth century and the beginning of the twentieth. His name is associated with two other reactions in addition to the Claisen rearrangement. They are the *Claisen–Schmidt reaction* (Section 21.4) and the *Claisen condensation* (Section 21.4).

A significant number of structural types undergo reactions analogous to the Cope rearrangement of 1,5-dienes. In one—called the **Claisen rearrangement**—allyl vinyl ethers are converted to carbonyl compounds (aldehydes or ketones depending on substituents).

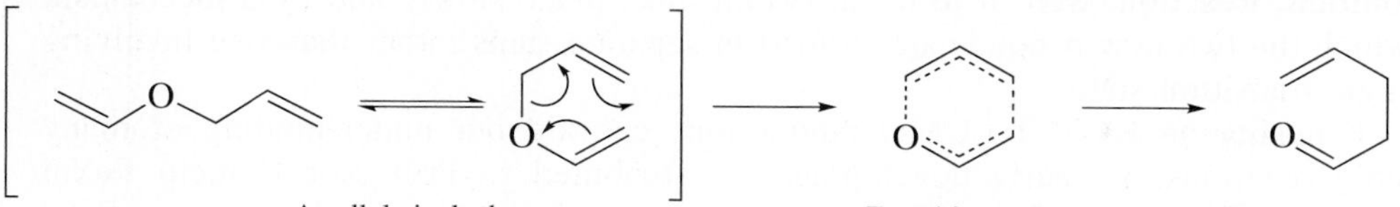

An allyl vinyl ether — Transition state — Product of Claisen rearrangement

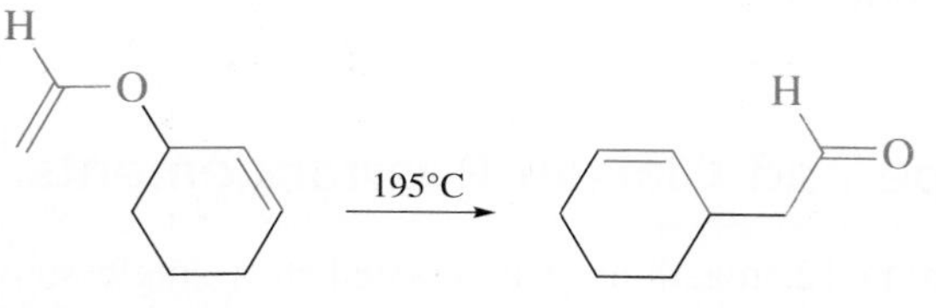

Cyclohex-3-enyl vinyl ether $\xrightarrow{195°C}$ Cyclohex-2-enylacetaldehyde (94%)

Problem 11.25

What is the product of the following reaction?

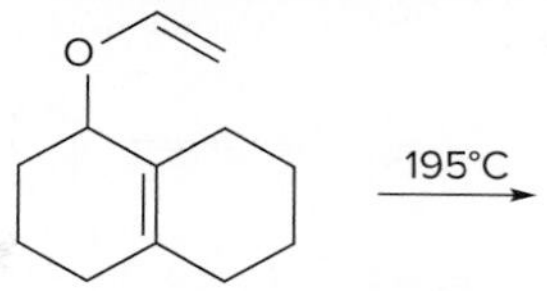

195°C

Thermodynamically, Cope and Claisen rearrangements differ in an important way. In its simplest terms, the pattern of bond types (C—C single and C=C double) is the same in both the reactant and product of a Cope rearrangement—one 1,5-diene is converted to an isomeric 1,5-diene—and $\Delta H_{reaction}$ is usually small. The Claisen rearrangement of an allyl vinyl ether, on the other hand, converts an O—C—O unit in the reactant to C=O at a cost of converting one of its C=C units to C—C. Inasmuch as the bond energy of C=O is normally more than twice that of C—O while the opposite holds for C=C versus C—C, Claisen rearrangements lead to stronger net bonding and tend to be exothermic.

11.17 SUMMARY

This chapter focused on the effect of a carbon–carbon double bond as a stabilizing substituent on a positively charged carbon in an **allylic carbocation,** on a carbon bearing an odd electron in an **allylic free radical,** on a negatively charged carbon in an **allylic anion,** and on a second double bond in a **conjugated diene.**

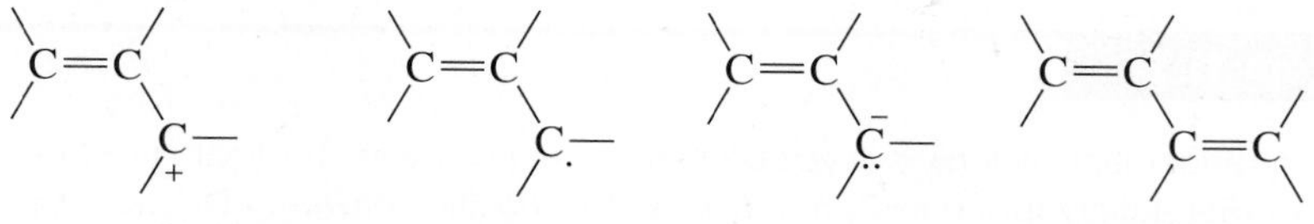

Allylic carbocation — Allylic radical — Allylic anion — Conjugated diene

Section 11.1 **Allyl** is the common name of the parent group $H_2C{=}CHCH_2—$ and is an acceptable name in IUPAC nomenclature.

Section 11.2 Allylic halides are more reactive than simple alkyl halides in both S_N1 and S_N2 reactions. The allylic carbocation intermediates in S_N1 reactions have their positive charge shared between the two end carbons of the allylic system. The products of nucleophilic substitution may be formed with the same pattern of bonds as the starting allylic halide or with *allylic rearrangement.*

Cl; Na_2CO_3, H_2O → OH + OH

3-Chloro-1-butene — 3-Buten-2-ol (65%) — 2-Buten-1-ol (35%)

Substitution by the S_N2 mechanism does not involve allylic rearrangement.

Cl; NaN_3, CH_3OH, H_2O → N_3

3-Chloro-1-heptene — 3-Azido-1-heptene (78%)

Section 11.3 Alkenes react with *N*-bromosuccinimide (NBS) to give allylic bromides by a free-radical mechanism. The reaction is used for synthetic purposes only when the two resonance forms of the allylic radical are equivalent. Otherwise a mixture of isomeric allylic bromides is produced.

NBS, CCl_4, heat → Br

Cyclodecene — 3-Bromocyclodecene (56%)

Section 11.4 Stabilization of allylic anions by electron delocalization causes an allylic hydrogen to be more acidic ($pK_a \approx 43$) than a hydrogen in an alkane ($pK_a \approx 62$).

Section 11.5 Dienes are classified as having **isolated, conjugated,** or **cumulated** double bonds.

$=C=CH_2$

Isolated — Conjugated — Cumulated

Section 11.6 Conjugated dienes are more stable than isolated dienes, and cumulated dienes are the least stable of all.

Section 11.7 Conjugated dienes are stabilized by electron delocalization to the extent of 12–16 kJ/mol (3–4 kcal/mol). Overlap of the *p* orbitals of four adjacent sp^2-hybridized carbons in a conjugated diene gives an extended π system through which the electrons are delocalized.

The two most stable conformations of conjugated dienes are the *s*-cis and *s*-trans. The *s*-trans conformation is normally more stable than the *s*-cis. Both conformations are planar, which allows the *p* orbitals to overlap to give an extended π system.

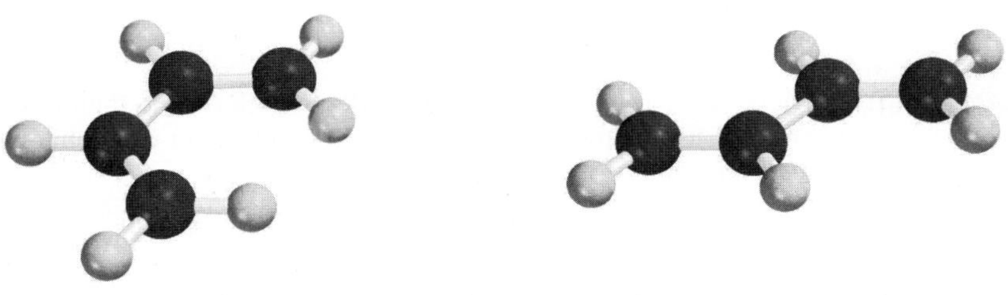

s-cis — *s*-trans

Section 11.8 1,2-Propadiene ($H_2C{=}C{=}CH_2$), also called **allene,** is the simplest cumulated diene. The two π bonds in an allene share an *sp*-hybridized carbon and are at right angles to each other. Certain allenes such as 2,3-pentadiene ($CH_3CH{=}C{=}CHCH_3$) possess a *chirality axis* and are chiral.

Section 11.9 Dehydration and dehydrohalogenation are commonly used methods for preparing dienes. Elimination is regioselective and gives a conjugated diene rather than an isolated or cumulated one.

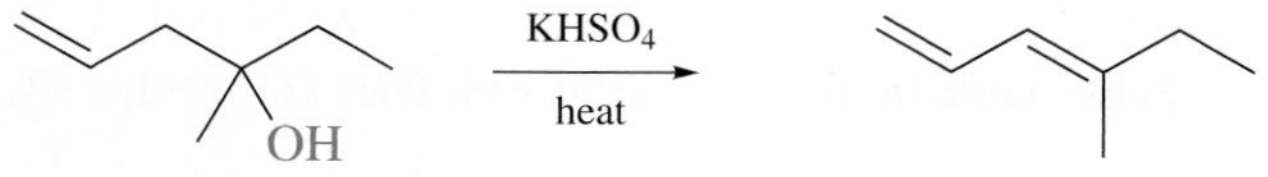

3-Methyl-5-hexen-3-ol —($KHSO_4$, heat)→ 4-Methyl-1,3-hexadiene (88%)

Section 11.10 Protonation at the terminal carbon of a conjugated diene system gives an allylic carbocation that can be captured by the halide nucleophile at either of the two sites that share the positive charge. Nucleophilic attack at the carbon adjacent to the one that is protonated gives the product of *1,2 addition*. Capture at the other site gives the product of *1,4 addition*.

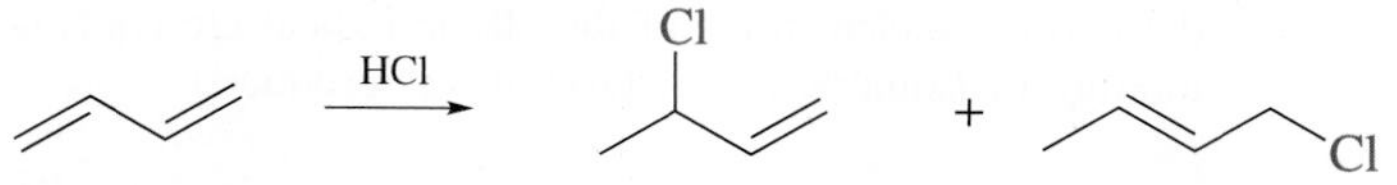

1,3-Butadiene —(HCl)→ 3-Chloro-1-butene (78%) + 1-Chloro-2-butene (22%)

Section 11.11 1,4-Addition predominates when Cl_2 and Br_2 add to conjugated dienes.

Section 11.12 The *Diels–Alder* reaction is the conjugate addition of an alkene (the *dienophile*) to a conjugated diene. It is concerted and stereospecific; substituents that are cis to each other on the dienophile remain cis in the product.

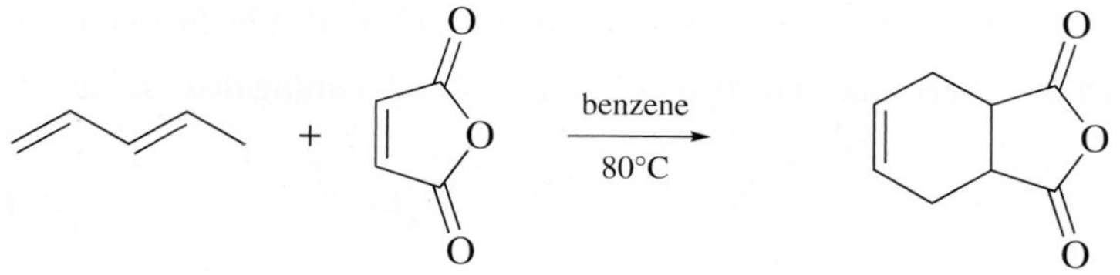

trans-1,3-Pentadiene + Maleic anhydride —(benzene, 80°C)→ 3-Methylcyclohexene-4,5-dicarboxylic anhydride (81%)

When endo and exo stereoisomers are possible, cycloaddition is stereoselective and favors formation of the endo isomer.

Section 11.13 A molecule that contains three double bonds, two of which are conjugated, can undergo an intramolecular Diels–Alder reaction.

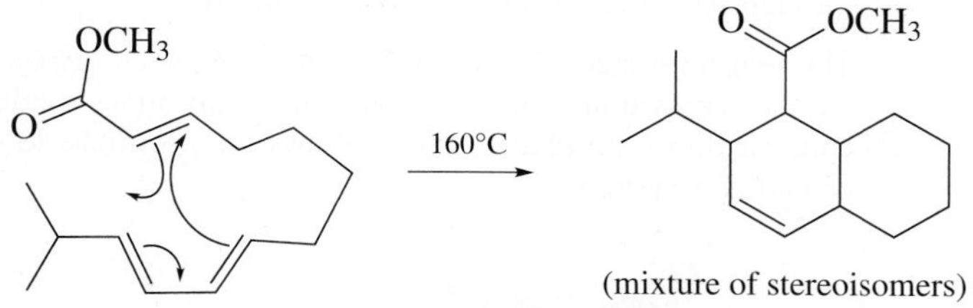

OCH3, O —(160°C)→ O, OCH3 (mixture of stereoisomers)

Section 11.14 Diels–Alder routes to cyclohexene derivatives can be found by retrosynthetic analysis using an approach in which the diene and dienophile are revealed by a disconnection of the type:

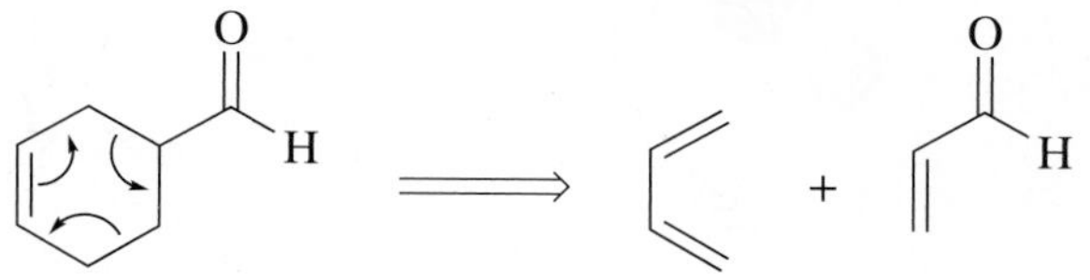

O, H ⟹ + O, H

Section 11.15 The Diels–Alder reaction is believed to proceed in a single step. Bonding changes in the transition state can be understood by examining the nodal properties of the highest occupied molecular orbital (HOMO) of the diene and the lowest unoccupied molecular orbital (LUMO) of the dienophile.

Section 11.16 Compounds that incorporate a 1,5-diene unit can undergo an intramolecular reaction that converts it to a different 1,5-diene. This reaction is called the *Cope rearrangement.*

heat

2-(2-Methylallyl)-methylenecyclopentane → 1-(3-Methyl-3-butenyl)-cyclopentene

In a *Claisen rearrangement,* an allyl vinyl ether undergoes an analogous reaction to give a product containing a C=C and a C=O group in a 1,5 relationship.

heat

3-(3,5-Dimethylhex-1-enyl) 1-propenyl ether → 2,5,7-Trimethyl-4-octenal

PROBLEMS

Structure and Nomenclature

11.26 Write structural formulas for each of the following:

(a) 3,4-Octadiene

(b) (*E,E*)-3,5-Octadiene

(c) (*Z,Z*)-1,3-Cyclooctadiene

(d) (*Z,Z*)-1,4-Cyclooctadiene

(e) (*E,E*)-1,5-Cyclooctadiene

(f) (2*E*,4*Z*,6*E*)-2,4,6-Octatriene

(g) 5-Allyl-1,3-cyclopentadiene

(h) *trans*-1,2-Divinylcyclopropane

(i) 2,4-Dimethyl-1,3-pentadiene

11.27 Give an acceptable IUPAC name for each of the following compounds:

(a) $H_2C{=}CH(CH_2)_5CH{=}CH_2$

(b)

(c) $(H_2C{=}CH)_3CH$

(d)

(e)

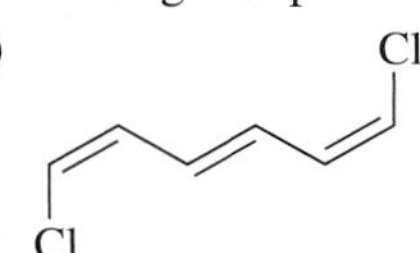

(f) $H_2C{=}C{=}CHCH{=}CHCH_3$

(g)

(h)

11.28 A certain species of grasshopper secretes an allenic substance of molecular formula $C_{13}H_{20}O_3$ that acts as an ant repellent. The carbon skeleton and location of various substituents in this substance are indicated in the partial structure shown. Complete the structure, adding double bonds where appropriate.

11.29 Which of the following are chiral?
(a) 2-Methyl-2,3-hexadiene
(b) 4-Methyl-2,3-hexadiene
(c) 2,4-Dimethyl-2,3-pentadiene

11.30 (a) Describe the molecular geometry expected for 1,2,3-butatriene ($H_2C{=}C{=}C{=}CH_2$).
(b) Two stereoisomers are expected for 2,3,4-hexatriene ($CH_3CH{=}C{=}C{=}CHCH_3$). What should be the relationship between these two stereoisomers?

11.31 The pK_a of $CH_3CH{=}CHCH{=}CH_2$ has been estimated to be about 35. Give the structure of its conjugate base and use resonance to show the sharing of negative charge among the various carbons.

Reactions

11.32 (a) What compound of molecular formula C_6H_{10} gives 2,3-dimethylbutane on catalytic hydrogenation over platinum?
(b) What two compounds of molecular formula $C_{11}H_{20}$ give 2,2,6,6-tetramethylheptane on catalytic hydrogenation over platinum?

11.33 Write structural formulas for all the
(a) Conjugated dienes
(b) Isolated dienes
(c) Cumulated dienes

that give 2,4-dimethylpentane on catalytic hydrogenation.

11.34 Give the structure, exclusive of stereochemistry, of the principal organic product formed on reaction of 2,3-dimethyl-1,3-butadiene with each of the following:
(a) 2 mol H_2, platinum catalyst
(b) 1 mol HCl (product of 1,2-addition)
(c) 1 mol HCl (product of 1,4-addition)
(d) 1 mol Br_2 (product of 1,2-addition)
(e) 1 mol Br_2 (product of 1,4-addition)
(f) 2 mol Br_2
(g)

11.35 Repeat the previous problem for the reactions of 1,3-cyclohexadiene.

11.36 Bromination of 1,5-cyclooctadiene with *N*-bromosuccinimide (NBS) gives a mixture of two constitutional isomers of $C_8H_{11}Br$. Suggest reasonable structures for these two isomers.

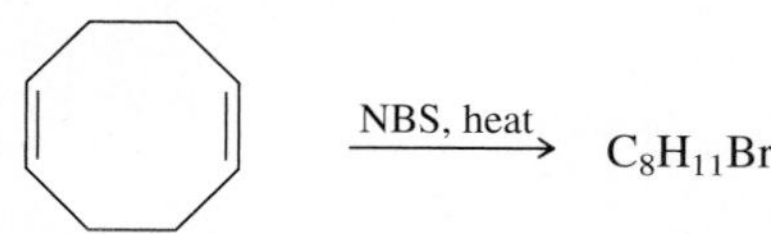

1,5-Cyclooctadiene

11.37 Identify the more reactive dienophile in each of the following pairs.

(a) 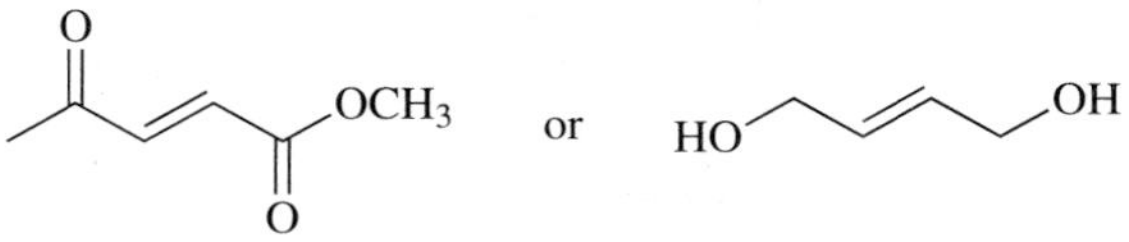 [structure] or [structure: HO, OH]

(b) 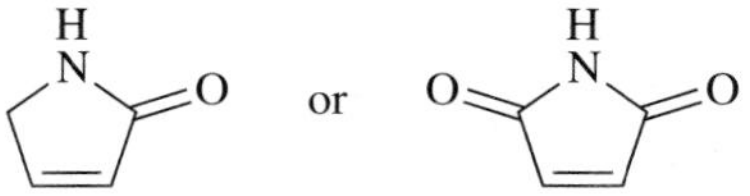 [structure] or [structure]

(c) 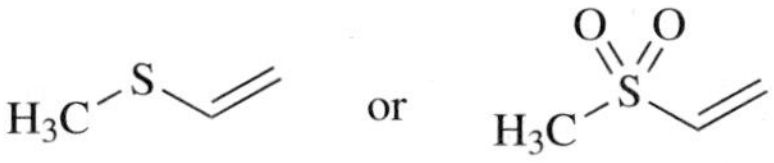 H_3C–S–CH=CH$_2$ or H_3C–SO_2–CH=CH$_2$

11.38 Which of the following two dienes can undergo the Diels–Alder reaction? Explain.

11.39 Two constitutional isomers of molecular formula $C_8H_{12}O$ are formed in the following reaction. Ignoring stereochemistry, suggest reasonable structures for these Diels–Alder adducts.

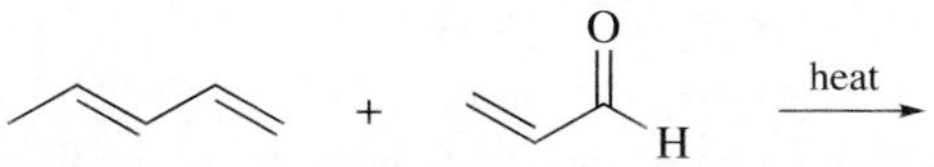

[diene] + [aldehyde] $\xrightarrow{\text{heat}}$

11.40 The two stereoisomers of cinnamic acid ($C_6H_5CH{=}CHCO_2H$) each give a different product on reaction with 1,3-butadiene. Write an equation for each reaction.

11.41 Acetylene resembles ethylene in that, although being a poor dienophile itself, it adds to dienes when its triple bond bears electron-attracting substituents. The reaction of 1,3-butadiene with diethyl acetylenedicarboxylate gives a Diels–Alder adduct in 98% yield. What is that product?

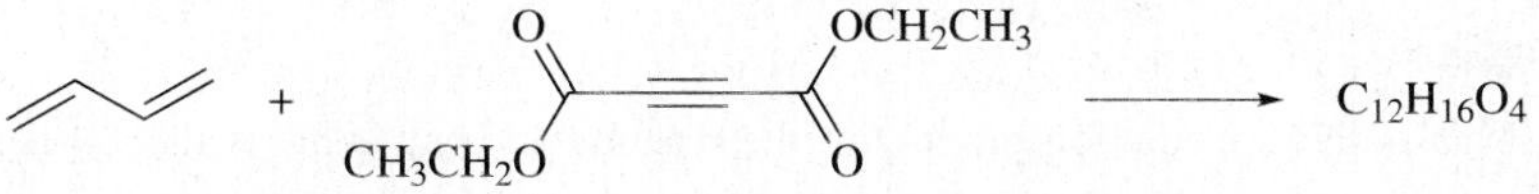

[1,3-butadiene] + CH_3CH_2O–C(=O)–C≡C–C(=O)–OCH_2CH_3 ⟶ $C_{12}H_{16}O_4$

11.42 Each of the following compounds was used in natural product syntheses where they undergo a Claisen rearrangement. What is the product of each?

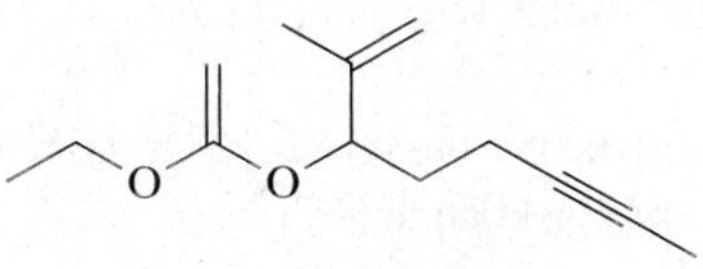

(a) 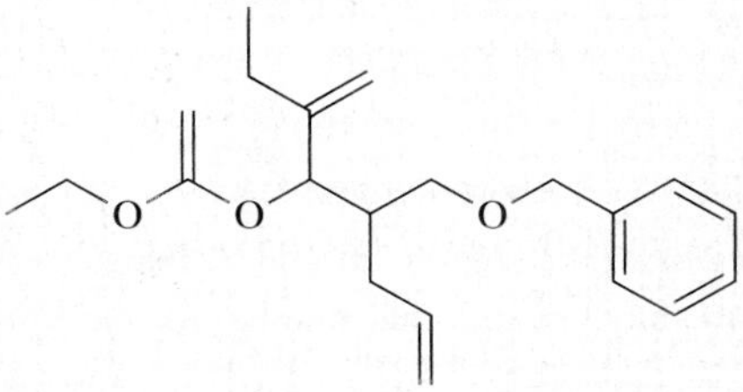 (b)

11.43 Heating the compound shown at 220°C gave a mixture of two stereoisomeric Claisen rearrangement products in approximately equal amounts. Write a structural formula for each.

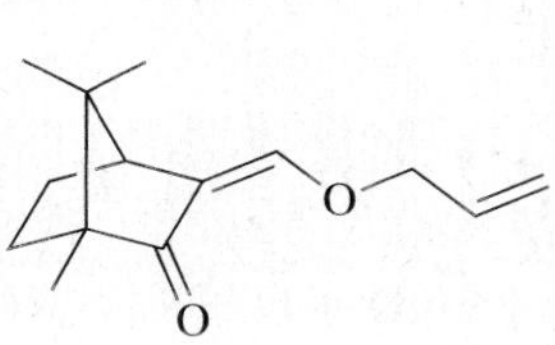

11.44 Allene can be converted to a trimer (compound A) of molecular formula C_9H_{12}. Compound A reacts with dimethyl acetylenedicarboxylate to give compound B. Deduce the structure of compound A.

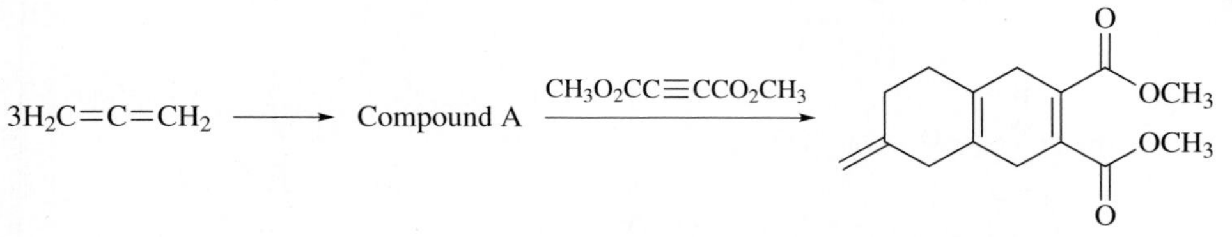

$3H_2C{=}C{=}CH_2$ ⟶ Compound A $\xrightarrow{CH_3O_2CC{\equiv}CCO_2CH_3}$

O, OCH$_3$, OCH$_3$, O

Compound B

11.45 Predict the constitution of the expected Diels–Alder adduct formed from the following combinations of dienes and dienophiles.

(a) 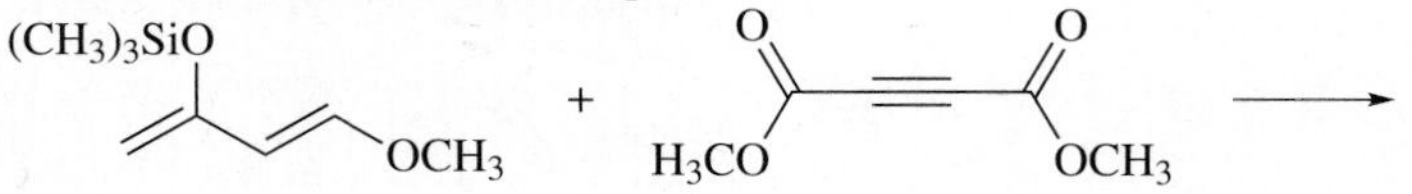 $(CH_3)_3SiO$ … OCH$_3$ + O, O, H$_3$CO, OCH$_3$ ⟶

(b) 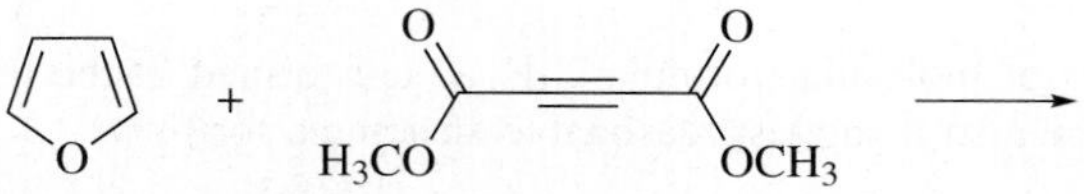 O + O, O, H$_3$CO, OCH$_3$ ⟶

(c) 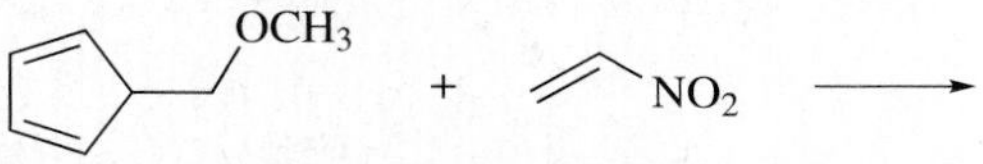 OCH$_3$ + NO$_2$ ⟶

11.46 Each of the following compounds undergoes an intramolecular Diels–Alder reaction. What is the product in each case?

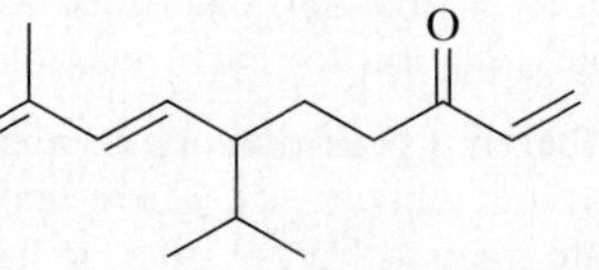

O (a)

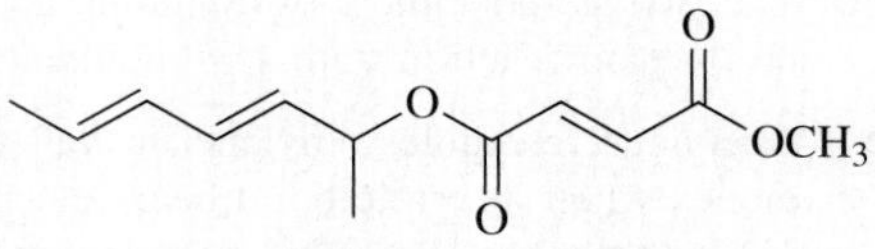

O, O, OCH$_3$, O (b)

Synthesis

11.47 What is the 1,2-addition product of the reaction shown? What is the 1,4-addition product?

$\xrightarrow{HCl}$

11.48 Show how to prepare each of the following compounds from propene and any necessary organic or inorganic reagents:

(a) Allyl bromide
(b) 1,2-Dibromopropane
(c) 1,3-Dibromopropane
(d) 1-Bromo-2-chloropropane
(e) 1,2,3-Tribromopropane
(f) Allyl alcohol
(g) Pent-1-en-4-yne ($H_2C{=}CHCH_2C{\equiv}CH$)
(h) 1,4-Pentadiene

11.49 Suggest reagents suitable for carrying out each step in the following synthetic sequence:

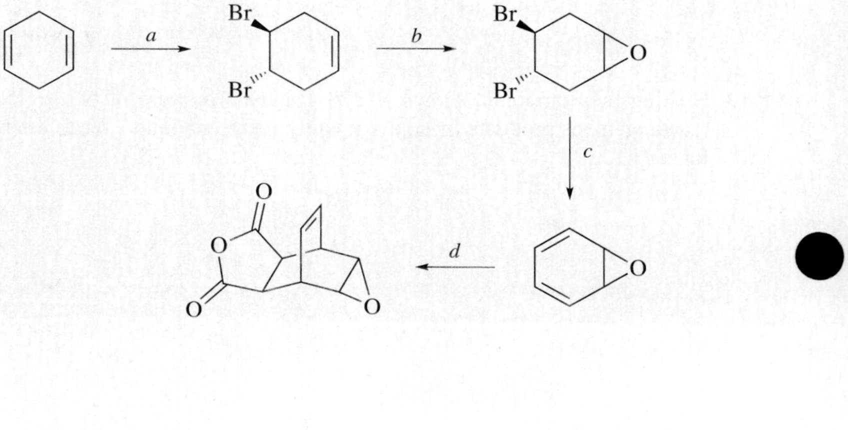

$\xrightarrow{a}$ Br, Br $\xrightarrow{b}$ Br, Br, O

$\downarrow c$

O, O, O, O $\xleftarrow{d}$ O

11.50 Each of the following compounds was prepared by a Diels–Alder reaction. Identify the diene and dienophile for each.

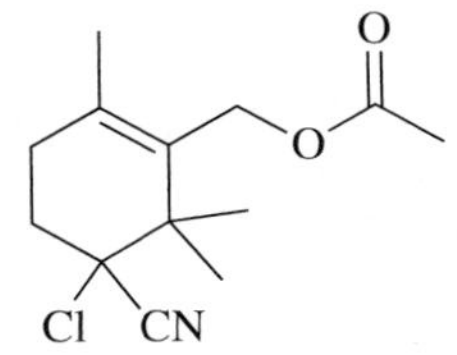

(a) 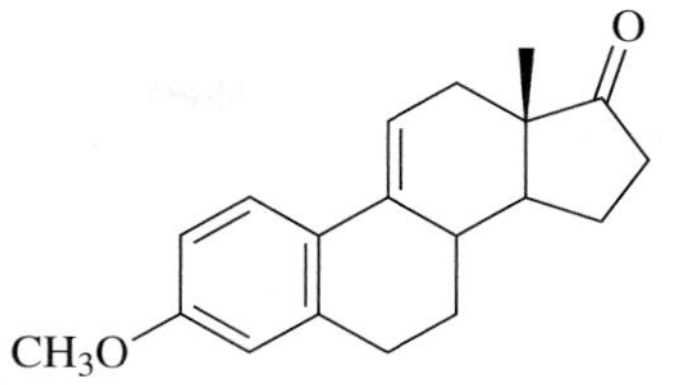 (b) 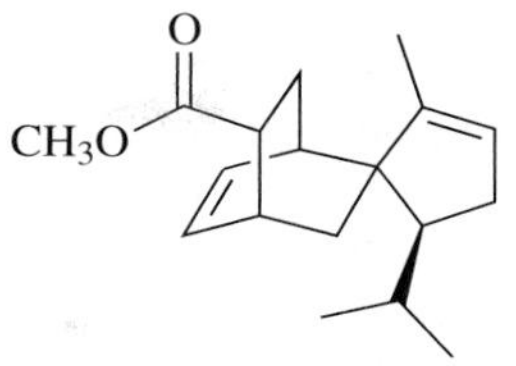 (c)

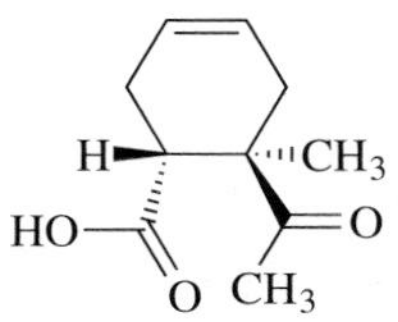

(d) 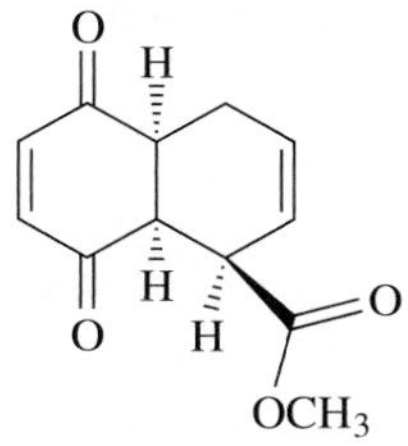 (e)

Mechanism

11.51 The following reaction gives only the product indicated. By what mechanism does this reaction most likely occur?

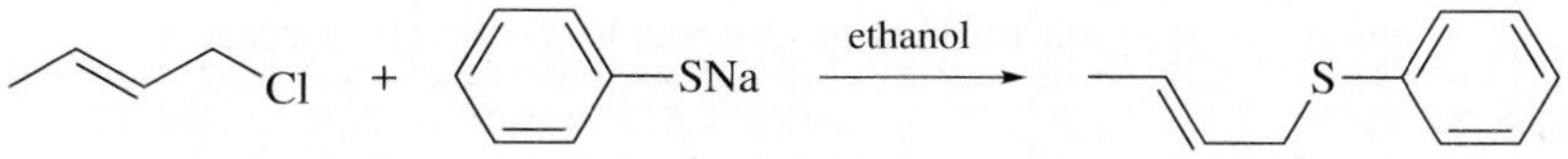

ethanol

11.52 Compound A was converted to compound B by the sequence shown. It is likely that the first step in this sequence gives two isomeric products having the molecular formula $C_{10}H_{15}BrO_2$, and both of these give compound B in the second step.

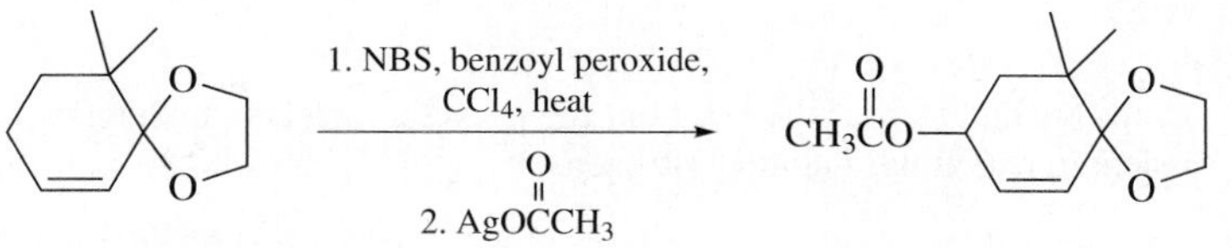

1. NBS, benzoyl peroxide, CCl_4, heat

2. $AgOCCH_3$

Compound A Compound B

(a) Suggest reasonable structures for the two isomers of $C_{10}H_{15}BrO_2$ formed in the first step.

(b) Give a mechanistic explanation why both isomers give the same product in the second step.

11.53 Suggest reasonable explanations for each of the following observations:

(a) The first-order rate constant for the solvolysis of $(CH_3)_2C{=}CHCH_2Cl$ in ethanol is over 6000 times greater than that of allyl chloride (25°C).

(b) After a solution of 3-buten-2-ol in aqueous sulfuric acid had been allowed to stand for 1 week, it was found to contain both 3-buten-2-ol and 2-buten-1-ol.

(c) Treatment of $CH_3CH{=}CHCH_2OH$ with hydrogen bromide gave a mixture of 1-bromo-2-butene and 3-bromo-1-butene.

(d) Treatment of 3-buten-2-ol with hydrogen bromide gave the same mixture of bromides as in part (c).

(e) The major product in parts (c) and (d) was 1-bromo-2-butene.

11.54 2-Chloro-1,3-butadiene (chloroprene) is the monomer from which the elastomer *neoprene* is prepared. 2-Chloro-1,3-butadiene is the thermodynamically controlled product formed by addition of hydrogen chloride to vinylacetylene ($H_2C{=}CHC{\equiv}CH$). The principal product under conditions of kinetic control is the allenic chloride 4-chloro-1,2-butadiene. Suggest a mechanism to account for the formation of each product.

11.55 Refer to the molecular orbital diagrams of allyl cation (Figure 11.2), ethylene, and 1,3-butadiene (Figure 11.12) to decide which of the following cycloaddition reactions are allowed and which are forbidden according to the Woodward–Hoffmann rules.

(a) 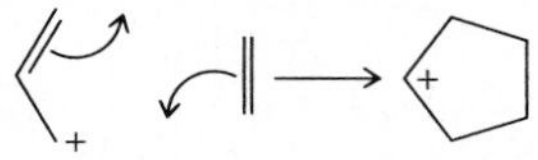 (b)

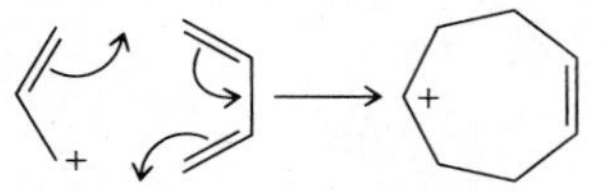

11.56 A second successful and popular pericyclic bioorthogonal chemical reaction is the tetrazene-strained alkene reaction.

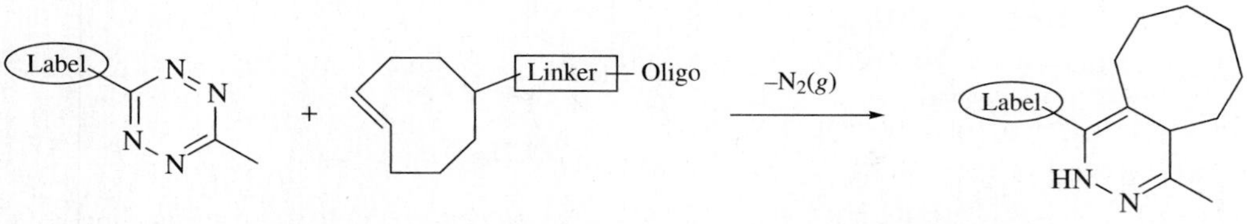

Label Linker Oligo $-N_2(g)$

Tetrazine Cyclooctene Cyclic alkene

The overall process is more complicated than the azide–alkyne bioorthogonal reaction because it involves two sequential pericyclic reactions followed by a rearrangement. Deduce the electron flow in each step of the reaction given the hints below.

a. In the first step of the reaction, the tetrazene and the *trans*-cyclooctene undergo a Diels–Alder reaction to form an intermediate tricyclic molecule.

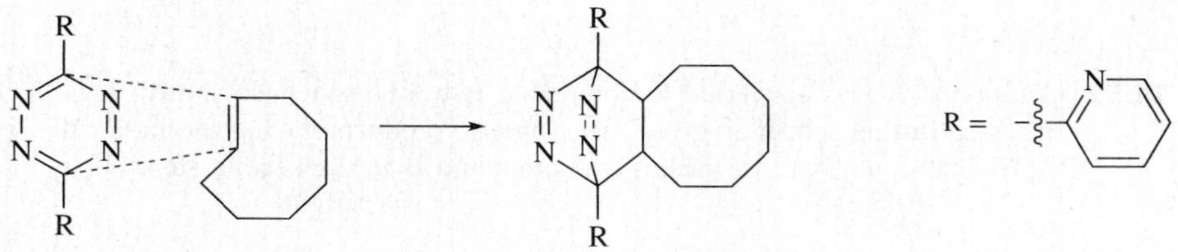

R =

b. In the second step of the reaction, the tricyclic molecule undergoes a retro-Diels–Alder reaction, expelling diatomic nitrogen.

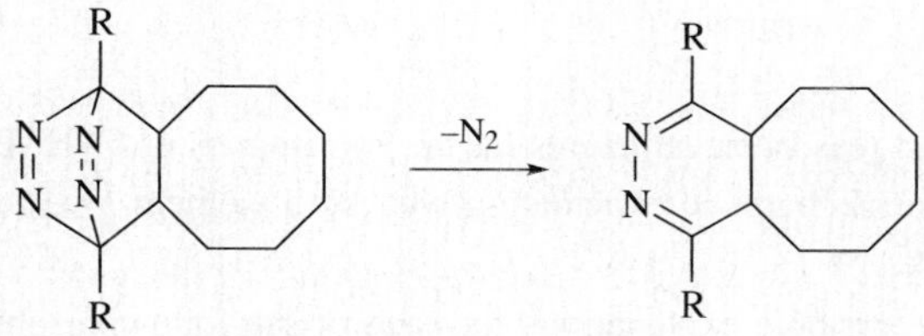

$-N_2$

c. The final product undergoes a rearrangement.

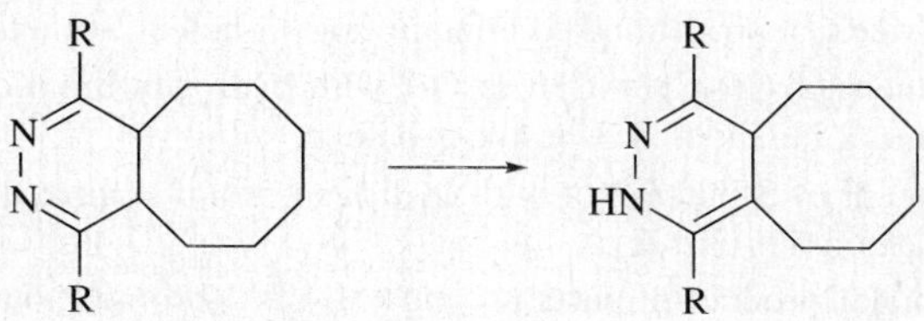

11.57 There are very few examples of naturally occurring pericyclic reactions in biology. One of the rare cases involves the enzyme chorismate mutase, which is an enzyme found in plants and bacteria as part of the biosynthesis of tyrosine and phenylalanine. The enzyme catalyzes the conversion of chorismate to prephenate via a Claisen rearrangement. Draw the structure of prephenate.

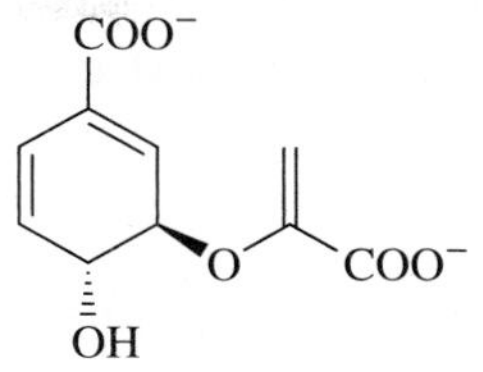

COO^- … O … COO^- … OH

Chorismate

Descriptive Passage and Interpretive Problems 11

1,3-Dipolar Cycloaddition

The Diels–Alder reaction is classified as a [4+2]-cycloaddition meaning that one reactant contributes four π electrons and the other two to a six π-electron transition state. The combination of a diene and an alkene, however, is not the only combination that satisfies the [4+2] standard. A second combination characterizes the class of reactions known as **1,3-dipolar cycloaddition,** represented generally as

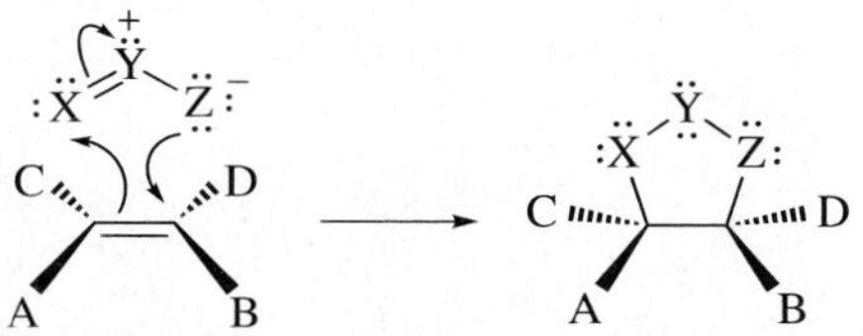

:X=Y⁺–Z:⁻ + C, D, A, B alkene → five-membered ring X–Y–Z with C, D, A, B

The alkene is the 2 π-electron component and is referred to in this context as a **dipolarophile.** The 4 π-electron component is any of a large number of species known as 1,3-dipoles that can be represented by a Lewis structure in which four electrons are delocalized over three atoms, one of which is positively charged and another negatively charged. Ozone is a familiar example. The ozonolysis of alkenes begins with the formation of an ozonide by cycloaddition.

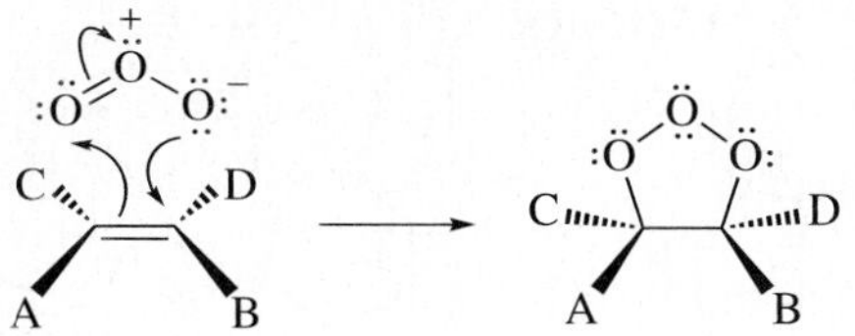

:O=O⁺–O:⁻ + C, D, A, B alkene → five-membered ring O–O–O with C, D, A, B

Other 1,3-dipolar compounds include

Alkyl azides $R-\bar{N}-\overset{+}{N}\equiv N: \longleftrightarrow R-N=\overset{+}{N}=N:^-$

Diazoalkanes $R_2\bar{C}-\overset{+}{N}\equiv N: \longleftrightarrow R_2C=\overset{+}{N}=N:^-$

Nitrile oxides $R-C\equiv\overset{+}{N}-O:^- \longleftrightarrow R-\bar{C}=\overset{+}{N}=O:$

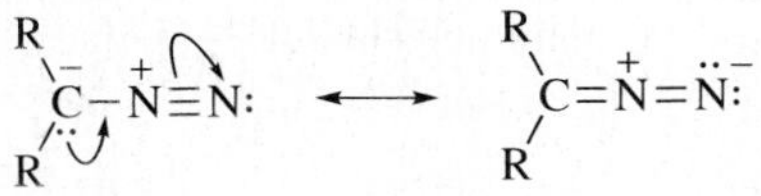

Nitrones $R_2\bar{C}-\overset{+}{N}(R)=O: \longleftrightarrow R_2C=\overset{+}{N}(R)-O:^-$

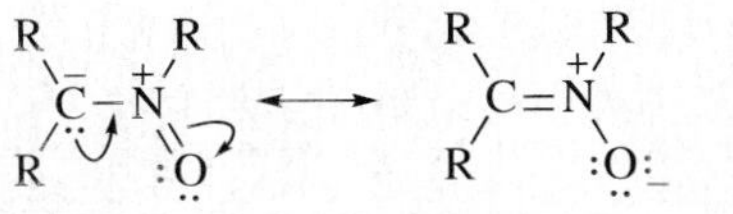

The resonance formulas show why we refer to such compounds as "1,3-dipolar." The positively charged atom remains unchanged between the pair of Lewis structures, but the negative charge is shared by atoms in a 1,3-relationship to each other.

1,3-Dipolar compounds have many synthetic applications. Not only are the initial products of the reaction of a 1,3-dipole with a dipolarophile useful in their own right, but their functionality can be used to advantage in subsequent transformations.

11.58 What is the 1,3-dipolar cycloaddition product formed in the reaction between diazomethane and maleic anhydride?

CH_2N_2

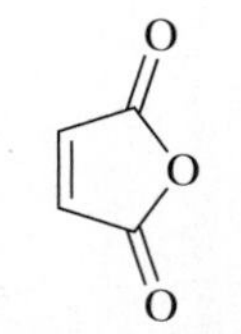

Diazomethane Maleic anhydride

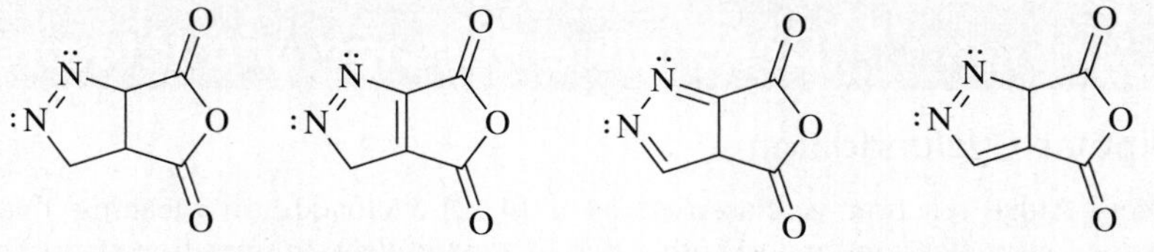

A. B. C. D.

11.59 What is the product of the following reaction?

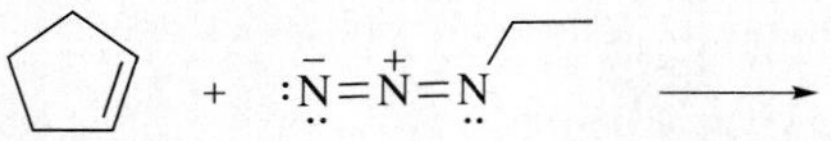

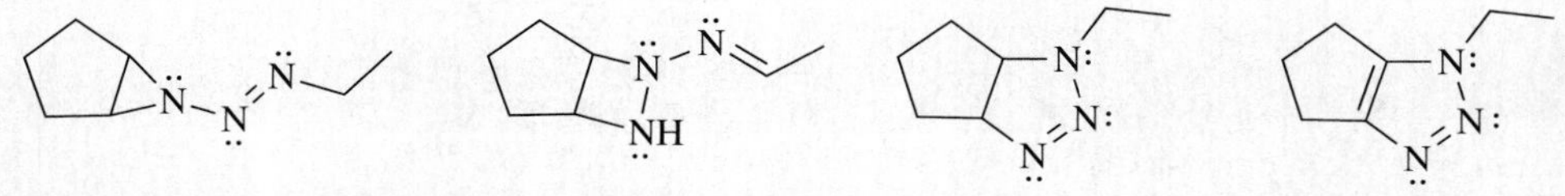

A. B. C. D.

11.60 Like alkenes, alkynes also undergo 1,3-dipolar cycloaddition. What is the product of the following reaction?

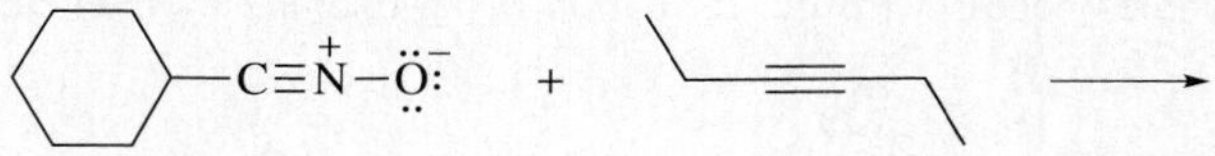

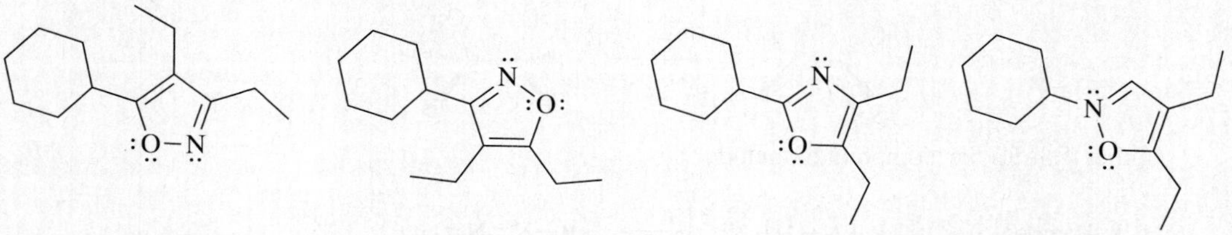

A. B. C. D.

11.61 The nitrone shown undergoes an intramolecular 1,3-dipolar cycloaddition. What is the product?

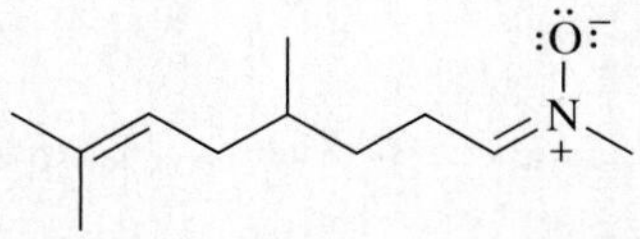

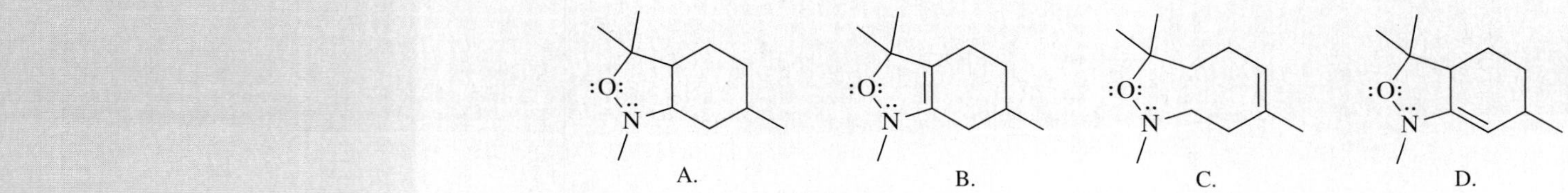

A. B. C. D.

11.62 In the second step of the mechanism of ozonolysis of alkenes, the ozonide described earlier undergoes a fragmentation–recombination process to give a 1,2,4-trioxolane.

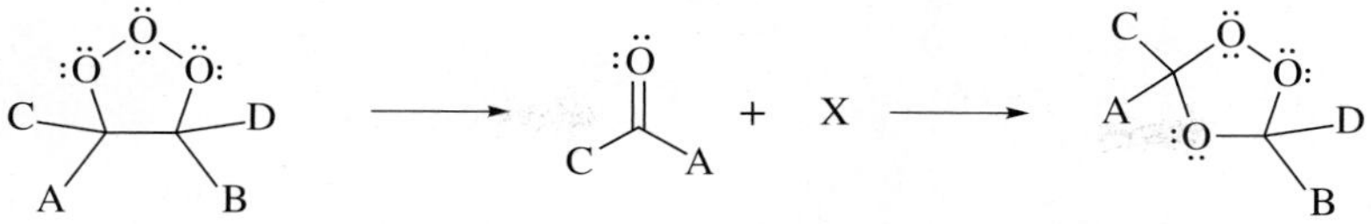

The fragmentation step is the reverse of a 1,3-dipolar cycloaddition and yields a carbonyl compound and a 1,3-dipolar compound X. The carbonyl compound and X then react to give the 1,2,4-trioxolane by 1,3-dipolar cycloaddition. What is the structure of X?

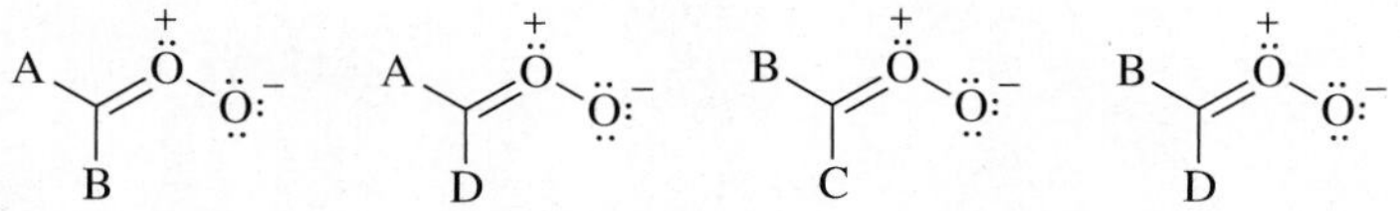

A. B. C. D.

Chapter

12

CHAPTER OUTLINE

Illuminating gas was used for lighting in nineteenth-century Europe, including the infamous chandelier in the Opéra de Paris, the setting for the *Phantom of the Opera*. Methane, ethylene, and hydrogen are the main components of illuminating gas, but other hydrocarbons are present in small amounts. One of these is benzene.

Arenes and Aromaticity

In this chapter and the next we extend our coverage of conjugated systems to include **arenes.** Arenes are hydrocarbons based on the benzene ring as a structural unit. Benzene, toluene, and naphthalene, for example, are arenes.

Benzene   Toluene   Naphthalene

One factor that makes conjugation in arenes special is its cyclic nature. A conjugated system that closes on itself can have properties that are much different from those of open-chain polyenes. Arenes are also referred to as aromatic hydrocarbons. Used in this sense, the word *aromatic* has nothing to do with odor but means instead that arenes are much more stable than we expect them to be based on their formulation as conjugated trienes. Our goal in this chapter is to develop an appreciation for the concept of **aromaticity**—to see what properties of benzene and its derivatives reflect its special stability and to explore the reasons for it. This chapter also examines the effect of a benzene ring as a substituent. The chapter following this one describes reactions that involve the ring itself.

Let's begin by tracing the history of benzene, its origin, and its structure. Many of the terms we use, including *aromaticity* itself, are of historical origin. We'll begin with the discovery of benzene.

12.1 Benzene

In 1825, Michael Faraday isolated a new hydrocarbon from illuminating gas, which he called "bicarburet of hydrogen." Nine years later Eilhardt Mitscherlich of the University of Berlin prepared the same substance by heating benzoic acid with lime and found it to be a hydrocarbon having the empirical formula C_nH_n.

Faraday is better known in chemistry for his laws of electrolysis and in physics for proposing the relationship between electric and magnetic fields and for demonstrating the principle of electromagnetic induction.

$$\underset{\text{Benzoic acid}}{C_6H_5CO_2H} + \underset{\text{Calcium oxide}}{CaO} \xrightarrow{\text{heat}} \underset{\text{Benzene}}{C_6H_6} + \underset{\text{Calcium carbonate}}{CaCO_3}$$

Eventually, because of its relationship to benzoic acid, this hydrocarbon came to be named *benzin,* then later *benzene,* the name by which it is known today.

Benzoic acid had been known for several hundred years by the time of Mitscherlich's experiment, having been prepared from *gum benzoin,* a pleasant-smelling resin used as incense and obtained from a tree native to Java. Similarly, toluene, the methyl derivative of benzene, takes its name from the South American *tolu* tree from which derivatives of it can be obtained.

Although benzene and toluene are not particularly fragrant compounds themselves, their origins in aromatic plant extracts led them and compounds related to them to be classified as *aromatic hydrocarbons.* Alkanes, alkenes, and alkynes belong to another class, the **aliphatic hydrocarbons.** The word *aliphatic* comes from the Greek *aleiphar* (meaning "oil" or "unguent") and was given to hydrocarbons that were obtained by the chemical degradation of fats.

Benzene was isolated from coal tar by August W. von Hofmann in 1845. Coal tar remained the primary source for the industrial production of benzene for many years, until petroleum-based technologies became competitive about 1950. Current production is about 6 million tons per year in the United States and almost 50 million tons worldwide. A substantial portion of this benzene is converted to styrene for use in the preparation of polystyrene plastics and films.

Toluene is also an important organic chemical. Like benzene, its early industrial production was from coal tar, but most of it now comes from petroleum.

12.2 The Structure of Benzene

Benzene puzzled the mid-nineteenth-century scientists who attempted to connect chemical properties to the still-novel concept of molecular structure. In spite of its C_6H_6 formula, benzene either failed to react with the compounds that added readily to alkenes and alkynes or, when it did, reacted by substitution instead. Reactions could be carried out that replaced one or more of its hydrogens, but benzene's six-carbon core remained, prompting early chemists to regard it as a unit. What was this unit and why was it "special"?

In 1866, only a few years after publishing his ideas concerning what we now recognize as the structural theory of organic chemistry, August Kekulé applied it to the structure of benzene. He based his reasoning on three premises:

1. Benzene is C_6H_6.
2. All the hydrogens of benzene are equivalent.
3. The structural theory requires that there be four bonds to each carbon.

In 1861, Johann Josef Loschmidt, who was later to become a professor at the University of Vienna, privately published a book containing a structural formula for benzene similar to the one Kekulé would propose five years later. Loschmidt's book reached few readers, and his ideas were not well known.

Kekulé proposed that the six carbon atoms of benzene were joined together in a ring. Four bonds to each carbon could be accommodated by a system of alternating single and double bonds with one hydrogen on each carbon.

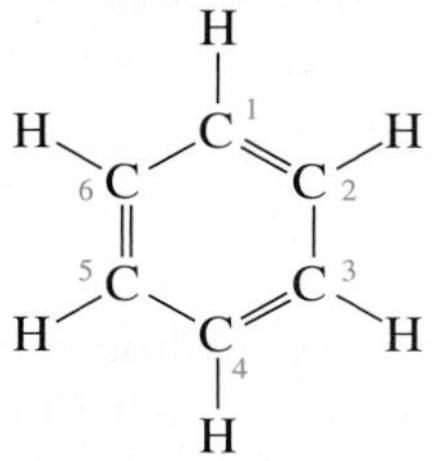

A flaw in the **Kekulé structure** for benzene was soon discovered. Kekulé's structure requires that 1,2- and 1,6-disubstitution create different compounds (isomers).

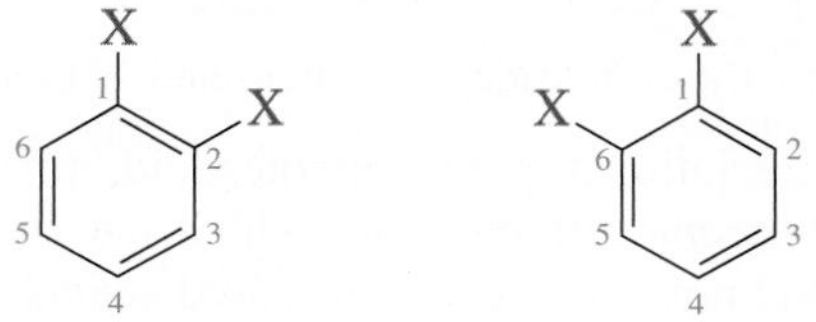

1,2-Disubstituted derivative of benzene

1,6-Disubstituted derivative of benzene

The two substituted carbons are connected by a double bond in one structure but by a single bond in the other. Because no such cases of isomerism in benzene derivatives were known, and none could be found, Kekulé suggested that two isomeric cyclohexatrienes could exist but interconverted too rapidly to be separated.

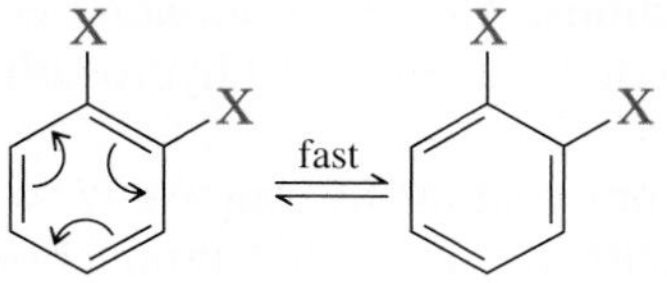

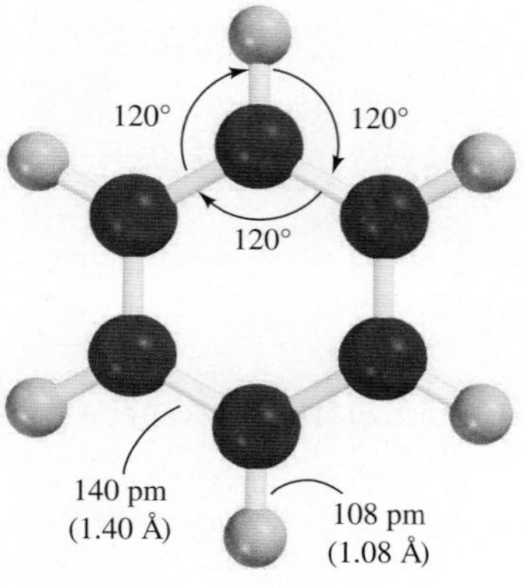

fast

120°

120°

120°

140 pm (1.40 Å)

108 pm (1.08 Å)

Figure 12.1

Bond distances and bond angles of benzene.

We now know that benzene is not cyclohexatriene, nor is it a pair of rapidly equilibrating isomers. Benzene is planar and its carbon skeleton has the shape of a regular hexagon. There is no evidence that it has alternating single and double bonds. As shown in Figure 12.1, all of the carbon–carbon bonds are the same length and the 120° bond angles correspond to perfect sp^2 hybridization. Moreover, the 140 pm (1.40 Å) bond distances in benzene are exactly midway between the typical sp^2–sp^2 single-bond distance of 146 pm (1.46 Å) and the sp^2–sp^2 double-bond distance of 134 pm (1.34 Å).

The two Kekulé structures for benzene have the same arrangement of atoms, but differ in the placement of electrons. Thus, they are resonance forms, and neither one by itself correctly describes the bonding in the actual molecule. As a hybrid of the two Kekulé structures, benzene is often represented by a hexagon containing an inscribed circle.

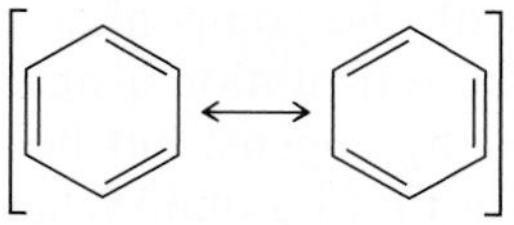

is equivalent to

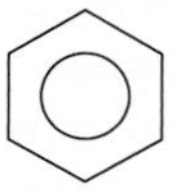

Robinson won the 1947 Nobel Prize in Chemistry for his studies of natural products. He may also have been the first to use curved arrows to track electron movement.

The circle-in-a-hexagon symbol was first suggested by the British chemist Sir Robert Robinson to represent the six delocalized π electrons of the three double bonds. Robinson's symbol is a convenient shorthand device, but Kekulé-type formulas are better for keeping track of electrons, especially in chemical reactions.

Problem 12.1

Write structural formulas for toluene ($C_6H_5CH_3$) and for benzoic acid ($C_6H_5CO_2H$) (a) as resonance hybrids of two Kekulé forms and (b) with the Robinson symbol.

Because the carbons that are singly bonded in one resonance form are doubly bonded in the other, the resonance description is consistent with the observed carbon–carbon bond distances in benzene. These distances not only are all identical but also are intermediate between typical single-bond and double-bond lengths.

We have come to associate electron delocalization with increased stability. On that basis alone, benzene ought to be stabilized. It differs from other conjugated systems that we have seen, however, in that its π electrons are delocalized over a *cyclic conjugated* system. Both Kekulé structures of benzene are of equal energy, and one of the principles of resonance theory is that stabilization is greatest when the contributing structures are of similar energy. Cyclic conjugation in benzene, then, leads to a greater stabilization than is observed in noncyclic conjugated trienes. How much greater can be estimated from heats of hydrogenation.

12.3 The Stability of Benzene

Hydrogenation of benzene and other arenes is more difficult than hydrogenation of alkenes and alkynes. Two of the more active catalysts are rhodium and platinum, and it is possible to hydrogenate arenes in the presence of these catalysts at room temperature and modest pressure. Benzene consumes three molar equivalents of hydrogen to give cyclohexane.

Benzene + $3H_2$ (Hydrogen, 2–3 atm pressure) $\xrightarrow[\text{acetic acid, 30°C}]{\text{Pt}}$ Cyclohexane (100%)

Nickel catalysts, although less expensive than rhodium and platinum, are also less active. Hydrogenation of arenes in the presence of nickel requires high temperatures (100–200°C) and pressures (100 atm).

The measured heat of hydrogenation of benzene to cyclohexane is the same regardless of the catalyst and is 208 kJ/mol (49.8 kcal/mol). To put this value into perspective, compare it with the heats of hydrogenation of cyclohexene and 1,3-cyclohexadiene, as shown in Figure 12.2. The most striking feature of Figure 12.2 is that *the heat of hydrogenation of*

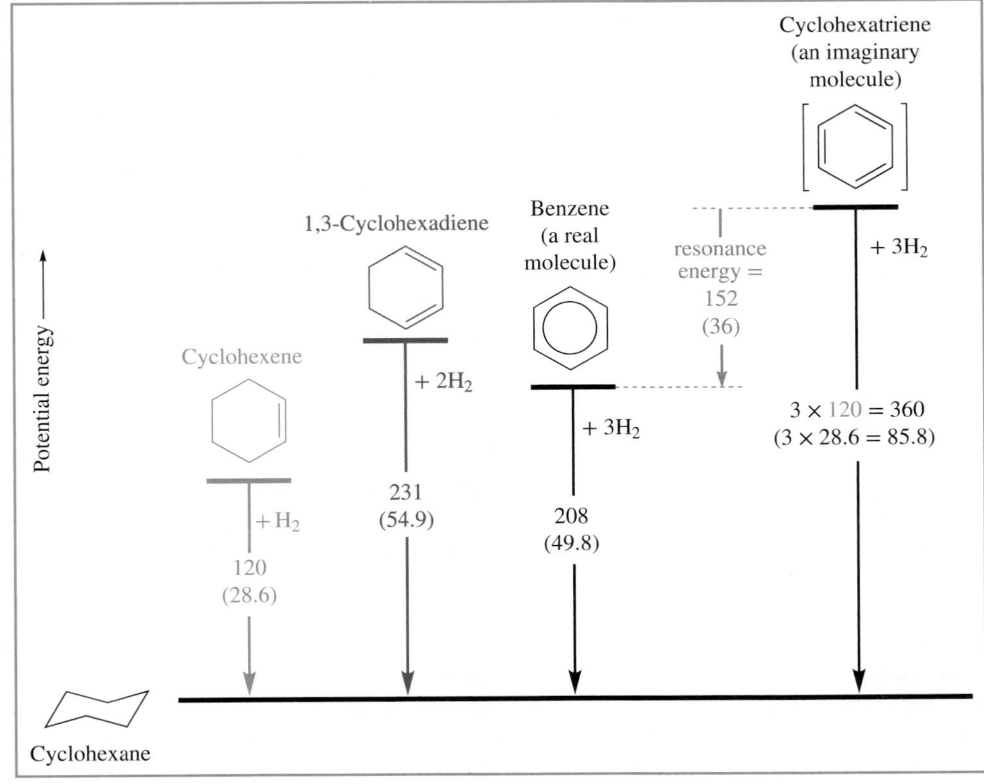

Figure 12.2

Heats of hydrogenation of cyclohexene, 1,3-cyclohexadiene, benzene, and a hypothetical 1,3,5-cyclohexatriene. Heats of hydrogenation are in kJ/mol (kcal/mol).

benzene, with three "double bonds," is less than the heat of hydrogenation of the two double bonds of 1,3-cyclohexadiene. The heat of hydrogenation of benzene is 152 kJ/mol (36 kcal/mol) *less* than expected for a hypothetical 1,3,5-cyclohexatriene with noninteracting double bonds. This is the **resonance energy** of benzene. It is a measure of how much more stable benzene is than would be predicted on the basis of its formulation as a pair of rapidly interconverting 1,3,5-cyclohexatrienes.

We reach a similar conclusion when comparing benzene with the noncyclic conjugated triene (*Z*)-1,3,5-hexatriene. Here we compare two real molecules, both conjugated trienes, but one is cyclic and the other is not. The heat of hydrogenation of (*Z*)-1,3,5-hexatriene is 337 kJ/mol (80.5 kcal/mol), a value which is 129 kJ/mol (30.7 kcal/mol) greater than that of benzene.

(*Z*)-1,3,5-Hexatriene	+	$3H_2$ Hydrogen	⟶	Hexane	$\Delta H° = -337$ kJ/mol (-80.5 kcal/mol)

The precise value of the resonance energy of benzene depends, as comparisons with 1,3,5-cyclohexatriene and (*Z*)-1,3,5-hexatriene illustrate, on the compound chosen as the reference. What is important is that the resonance energy of benzene is quite large, six to ten times that of a conjugated triene. It is this very large increment of resonance energy that places benzene and related compounds in a separate category that we call *aromatic*.

Problem 12.2

The heats of hydrogenation of cycloheptene and 1,3,5-cycloheptatriene are 110 kJ/mol (26.3 kcal/mol) and 305 kJ/mol (73.0 kcal/mol), respectively. In both cases cycloheptane is the product. What is the resonance energy of 1,3,5-cycloheptatriene? How does it compare with the resonance energy of benzene?

12.4 Bonding in Benzene

(*a*)

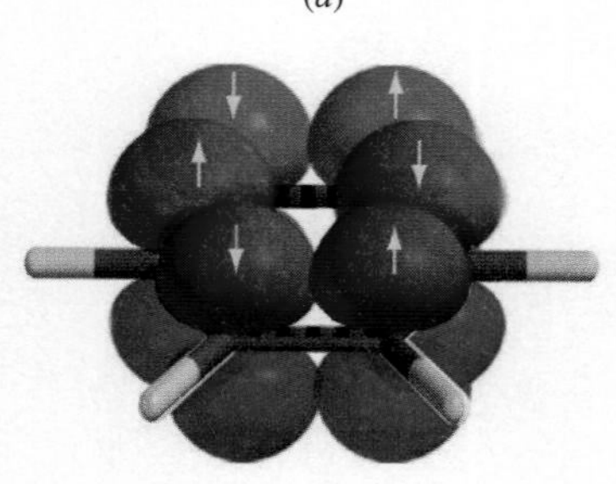

(*b*)

Figure 12.3

(*a*) The framework of bonds shown in the tube model of benzene are σ bonds. (*b*) Each carbon is sp^2-hybridized and has a half-filled 2*p* orbital perpendicular to the σ framework. Overlap of the 2*p* orbitals generates a π system encompassing the entire ring.

In the valence bond approach, the planar structure of benzene suggests sp^2 hybridization of carbon and the framework of σ bonds shown in Figure 12.3*a*. In addition to its three sp^2 orbitals, each carbon has a half-filled 2*p* orbital that can participate in π bonding by overlap with its counterpart on each of two adjacent carbons. Figure 12.3*b* shows the continuous π system that results from overlap of these orbitals and provides for delocalization of the π electrons over all six carbons.

The valence bond picture of benzene with six electrons in a delocalized π orbital is a useful, but superficial, one. Only two electrons can occupy a single orbital, be it an atomic orbital or a molecular orbital. The molecular orbital picture shown in Figure 12.4 does not suffer from this defect. We learned in Section 2.4 that when AOs combine to give MOs, the final number of MOs must equal the original number of AOs. Thus, the six 2*p* AOs of benzene combine to give six π MOs.

The orbitals in Figure 12.4 are arranged in order of increasing energy. Three orbitals are bonding; three are antibonding. Each of the three bonding MOs contains two electrons, accounting for the six π electrons of benzene. There are no electrons in the antibonding MOs. Benzene is said to have a **closed-shell** π electron configuration.

Recall that a wave function changes sign on passing through a nodal plane and is zero at a node (Section 1.1). In addition to the molecular plane, which contains a nodal surface common to all of them, five of the six π orbitals of benzene are characterized by nodal planes perpendicular to the molecule. The lowest-energy orbital π_1 has no such additional nodal surface; all of its *p* orbital interactions are bonding. The two other bonding orbitals π_2 and π_3 each have one nodal surface perpendicular to the molecule. The next three orbitals π_4^*, π_5^*, and π_6^* are antibonding and have, respectively, 2, 2, and 3 nodal planes perpendicular to the molecular plane. In the highest-energy orbital π_6^*, all interactions between adjacent *p* orbitals are antibonding.

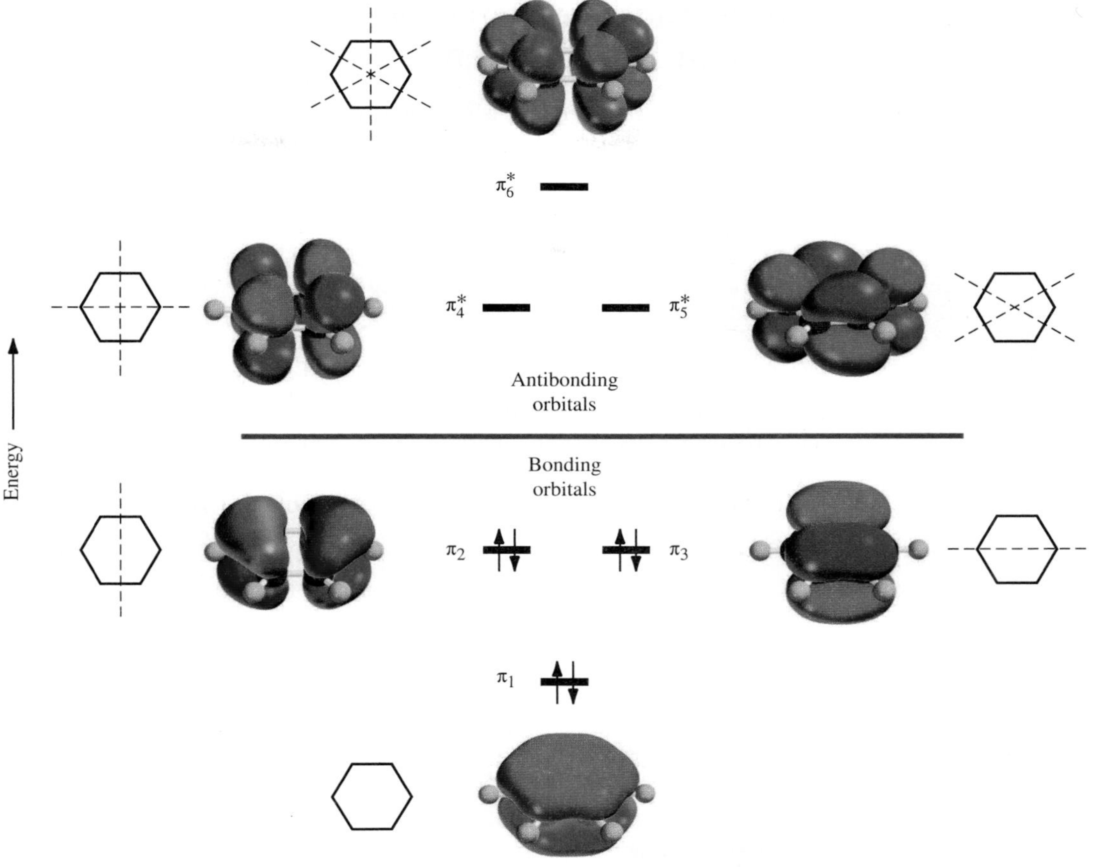

Figure 12.4

The π molecular orbitals of benzene arranged in order of increasing energy and showing nodal surfaces. The six π electrons of benzene occupy the three lowest-energy orbitals, all of which are bonding.

The pattern of orbital energies is different for benzene from the pattern it would have if the six π electrons were confined to three noninteracting double bonds. The delocalization provided by cyclic conjugation in benzene causes its π electrons to be held more strongly than they would be in the absence of cyclic conjugation. Stronger binding of its π electrons is the factor most responsible for the special stability—the aromaticity—of benzene.

But as the regions of high electron density above and below the plane of the ring in the electrostatic potential map (Figure 12.5) show, the π electrons are less strongly held than the electrons in the C—H bonds. In Chapter 13 we will see how this fact governs the characteristic chemical reactivity of benzene and its relatives.

Later in this chapter we'll explore the criteria for aromaticity in more detail to see how they apply to cyclic polyenes of different ring sizes. The next several sections introduce us to the chemistry of compounds that contain a benzene ring as a structural unit. We'll start with how we name them.

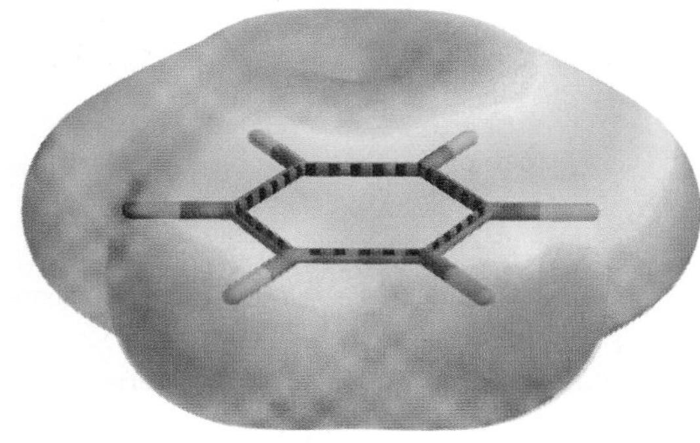

Figure 12.5

Electrostatic potential map of benzene. The red area in the center corresponds to the region above and below the plane of the ring where the π electrons are concentrated.

12.5 Substituted Derivatives of Benzene and Their Nomenclature

All compounds that contain a benzene ring are aromatic, and substituted derivatives of benzene make up the largest class of aromatic compounds. Many such compounds are named by attaching the name of the substituent as a prefix to *benzene.*

Br $C(CH_3)_3$ NO_2

Bromobenzene *tert*-Butylbenzene Nitrobenzene

Many simple monosubstituted derivatives of benzene have common names of long standing that have been retained in the IUPAC system. Table 12.1 lists some of the most important ones.

Dimethyl derivatives of benzene are called *xylenes.* There are three xylene isomers, the ortho (o)-, meta (m)-, and para (p)- substituted derivatives.

CH_3 CH_3 CH_3 CH_3 CH_3 CH_3

o-Xylene (1,2-dimethylbenzene) *m*-Xylene (1,3-dimethylbenzene) *p*-Xylene (1,4-dimethylbenzene)

TABLE 12.1 Names of Some Frequently Encountered Derivatives of Benzene

Structure	Systematic name	Common name*
O, H	Benzenecarbaldehyde	Benzaldehyde
O, OH	Benzenecarboxylic acid	Benzoic acid
	Vinylbenzene	Styrene
O	Methyl phenyl ketone	Acetophenone
OH	Benzenol	Phenol
OCH_3	Methoxybenzene	Anisole
NH_2	Benzenamine	Aniline

*These common names are acceptable in IUPAC nomenclature and are the names that will be used in this text.

The prefix **ortho** signifies a 1,2-disubstituted benzene ring, **meta** signifies 1,3 disubstitution, and **para** signifies 1,4 disubstitution. The prefixes *o, m,* and *p* can be used when a substance is named as a derivative of benzene or other parent, such as those in Table 12.1.

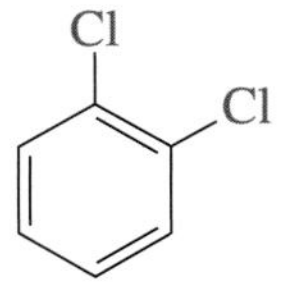

o-Dichlorobenzene (1,2-dichlorobenzene) 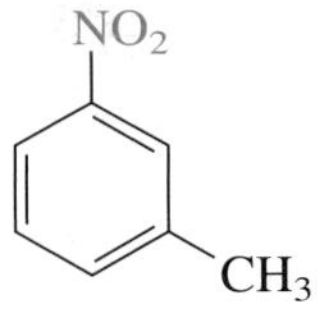 *m*-Nitrotoluene (3-nitrotoluene) 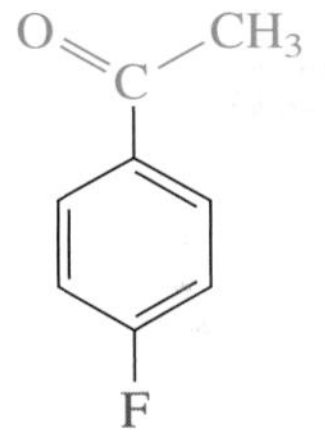 *p*-Fluoroacetophenone (4-fluoroacetophenone)

Problem 12.3

Write a structural formula for each of the following compounds:

(a) *o*-Ethylanisole (b) *m*-Chlorostyrene (c) *p*-Nitroaniline

Sample Solution (a) The parent compound in *o*-ethylanisole is anisole. Anisole, as shown in Table 12.1, has a methoxy (CH_3O—) substituent on the benzene ring. The ethyl group in *o*-ethylanisole is attached to the carbon adjacent to the one that bears the methoxy substituent.

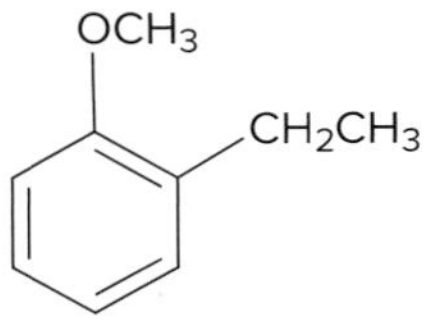

o-Ethylanisole

The *o, m,* and *p* prefixes are *not* used when three or more substituents are present on benzene; numerical locants must be used instead.

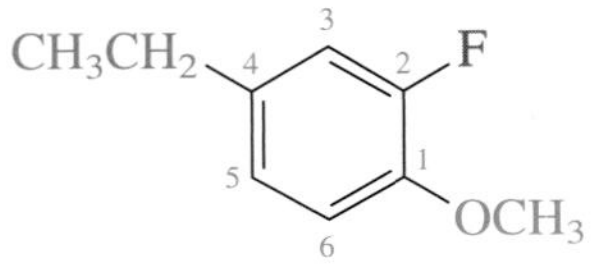

4-Ethyl-2-fluoroanisole 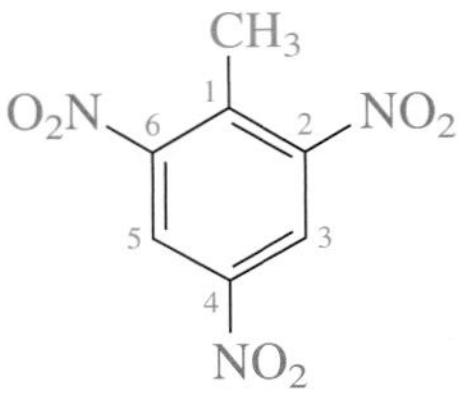 2,4,6-Trinitrotoluene 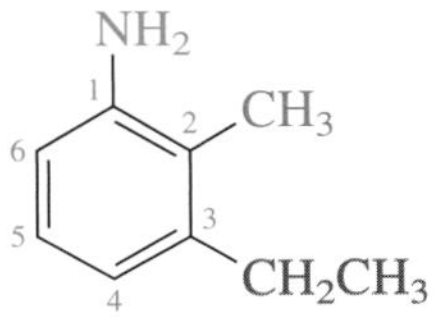 3-Ethyl-2-methylaniline

In these examples the name of the parent benzene derivative determines the carbon at which numbering begins: anisole has its methoxy group at C-1, toluene its methyl group at C-1, and aniline its amino group at C-1. The direction of numbering is chosen to give the next substituted position the lowest number irrespective of what substituent it bears. *The order of appearance of substituents in the name is alphabetical.* When no simple parent other than benzene is appropriate, positions are numbered so as to give the lowest locant at the first point of difference. Thus, each of the following examples is named as a 1,2,4-trisubstituted derivative of benzene rather than as a 1,3,4-derivative:

The "first point of difference" rule was introduced in Section 2.18.

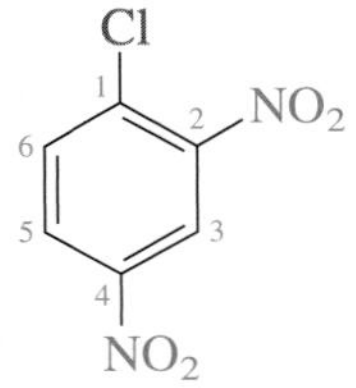

1-Chloro-2,4-dinitrobenzene 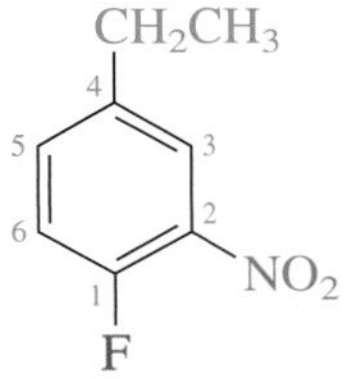 4-Ethyl-1-fluoro-2-nitrobenzene

When the benzene ring is named as a substituent, the word **phenyl** stands for C_6H_5—. Similarly, an arene named as a substituent is called an *aryl* group. A **benzyl group** is $C_6H_5CH_2$—.

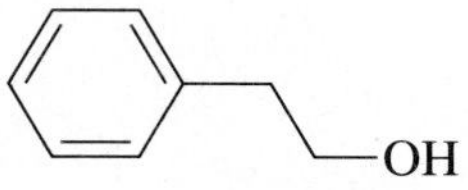

2-Phenylethanol 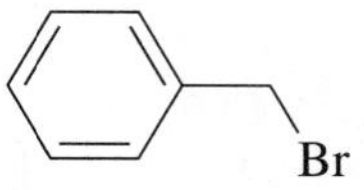   Benzyl bromide

Biphenyl is the accepted IUPAC name for the compound in which two benzene rings are connected by a single bond. In substituted biphenyls, each ring is numbered separately using primed and nonprimed numbers, beginning at the connection between the rings. If only one substituent is present, its position can be designated as being ortho, meta, or para to the other ring.

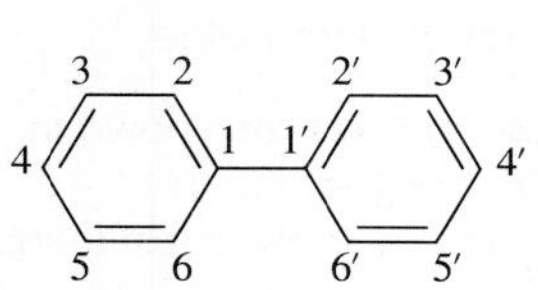

Biphenyl 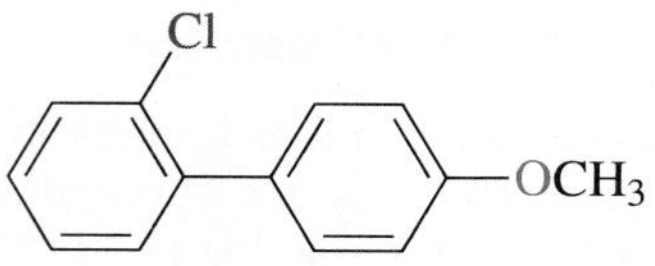   2-Chloro-4′-methoxybiphenyl 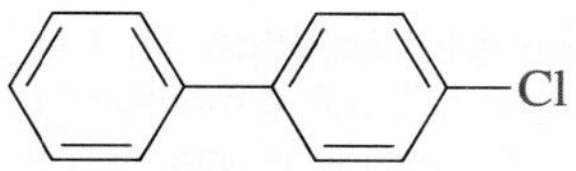   4-Chlorobiphenyl (*p*-Chlorobiphenyl)

Problem 12.4

Biphenyl is used as a fungicide, but some fungi are resistant and convert biphenyl to hydroxylated derivatives, one of which is 3,4,4′-trihydroxybiphenyl. Write a structural formula for this compound.

12.6 Polycyclic Aromatic Hydrocarbons

Naphthalene is a white crystalline solid melting at 80°C that sublimes readily. It has a characteristic odor and was formerly used as a moth repellent.

Members of a class of arenes called **polycyclic aromatic hydrocarbons** possess substantial resonance energies because each is a collection of benzene rings fused together.

Naphthalene, anthracene, and phenanthrene are the three simplest members of this class. They are all present in coal tar, a mixture of organic substances formed by heating coal at about 1000°C in the absence of air. Naphthalene is bicyclic and its two benzene rings share a common side. Anthracene and phenanthrene are both tricyclic aromatic hydrocarbons. Anthracene has three rings fused in a "linear" fashion; an "angular" fusion characterizes phenanthrene. The structural formulas of naphthalene, anthracene, and phenanthrene are shown along with the numbering system used to name their substituted derivatives:

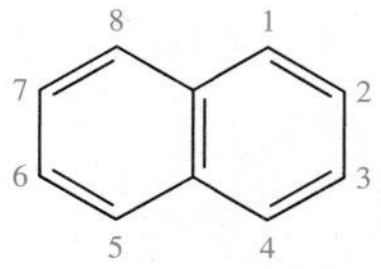

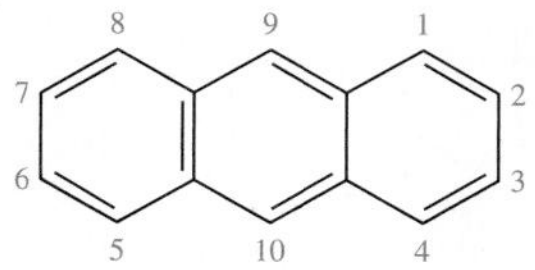

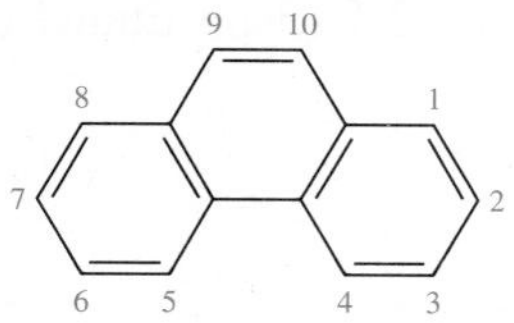

Arene:	Naphthalene	Anthracene	Phenanthrene
Resonance energy:	255 kJ/mol (61 kcal/mol)	347 kJ/mol (83 kcal/mol)	381 kJ/mol (91 kcal/mol)

Problem 12.5

How many monochloro derivatives of anthracene are possible? Write their structural formulas and give their IUPAC names.

In general, the most stable resonance contributor for a polycyclic aromatic hydrocarbon is the one with the greatest number of rings that correspond to Kekulé formulations of benzene. Naphthalene provides a fairly typical example:

Only left ring corresponds to Kekulé benzene. ⟷ **Most stable resonance form; major contributor** Both rings correspond to Kekulé benzene. ⟷ Only right ring corresponds to Kekulé benzene.

Notice that anthracene cannot be represented by any single Lewis structure in which all three rings correspond to Kekulé formulations of benzene, but phenanthrene can.

Problem 12.6

Chrysene is an aromatic hydrocarbon found in coal tar. Convert the molecular model to a Lewis structure in which all of the rings correspond to Kekulé formulas of benzene.

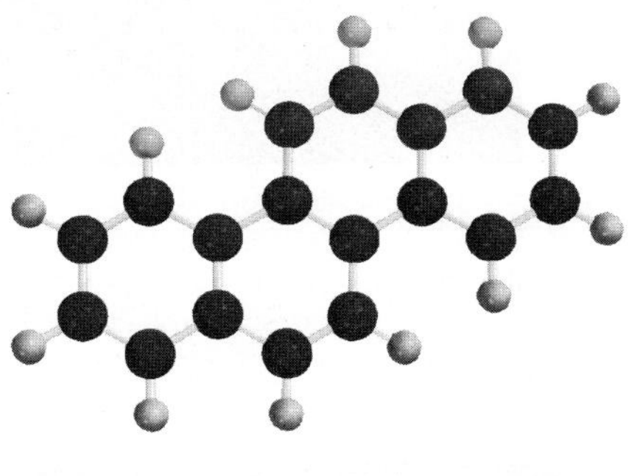

Chrysene

A large number of polycyclic aromatic hydrocarbons are known. Many have been synthesized in the laboratory, and several of the others are products of combustion. Benzo[a]pyrene, for example, is present in tobacco smoke, contaminates food cooked on barbecue grills, and collects in the soot of chimneys. Benzo[a]pyrene is a **carcinogen** (a cancer-causing substance). It is converted in the liver to a diol epoxide that can induce mutations leading to the uncontrolled growth of certain cells.

In 1775, the British surgeon Sir Percivall Pott suggested that scrotal cancer in chimney sweeps was caused by soot. This was the first proposal that cancer could be caused by chemicals present in the workplace.

Benzo[a]pyrene —oxidation in the liver→ 7,8-Dihydroxy-9,10-epoxy-7,8,9,10-tetrahydrobenzo[a]pyrene

O

HO

OH

Fullerenes, Nanotubes, and Graphene

In general, the term *nanoscale* applies to dimensions on the order of 1–100 nanometers (1 nm = 10^{-9} m), and one goal of *nanotechnology* is to develop useful nanoscale devices (*nanodevices*). Because typical covalent bonds range from 0.1 to 0.2 nm, chemical structures hold promise as candidates upon which to base nanodevices. Among them, much recent attention has been given to carbon-containing materials and even elemental carbon itself.

Until 1985, chemists recognized two elementary forms (*allotropes*) of carbon: diamond and graphite (Figure 12.6). Then, Professors Harold W. Kroto (University of Sussex), Robert F. Curl, and Richard E. Smalley (both of Rice University) reported that laser-induced evaporation of graphite gave a species with a molecular formula of C_{60} and proposed the spherical cluster of carbon atoms now called **buckminsterfullerene** (Figure 12.7) for it. Other closed carbon clusters, some larger than C_{60} and some smaller, were also formed in the experiment. These forms of carbon are now known as *fullerenes,* and Kroto, Smalley, and Curl were awarded the 1996 Nobel Prize in Chemistry for discovering them.

Research on fullerenes carried out at NEC Corporation (Japan) and at IBM (United States) led in 1991 to the isolation of fibrous clusters of *single-walled carbon nanotubes* (SWCNTs) (Figure 12.8). SWCNTs have since been joined by *multiwalled carbon nanotubes* (MWCNTs) (Figure 12.9) as well as nanotubes containing elements other than carbon.

CNTs are of interest because of their electrical and mechanical properties, and functionally modified ones are being examined in applications ranging from medical diagnosis and therapy to photovoltaic systems. Methods for adding functionality to CNTs include both covalent attachment of a reactive group and noncovalent coating of the outer surface of the CNT with a substance that itself bears a functional substituent.

Graphene is the most recent form of elemental carbon to emerge as a nanomaterial. As its name suggests, graphene is related to graphite in that graphite is an assembly of many graphene layers held together by van der Waals forces. The successful separation of single sheets of graphene was recognized with the award of the 2010 Nobel Prize in Physics to Andre Geim and Konstantin Novoselov of the University of Manchester. Much attention is being directed toward producing it on a scale that would make it available for use in novel materials and as a superior substitute for silicon in electronic devices.

Figure 12.6

Graphite is a form of elemental carbon composed of parallel layers of graphene.

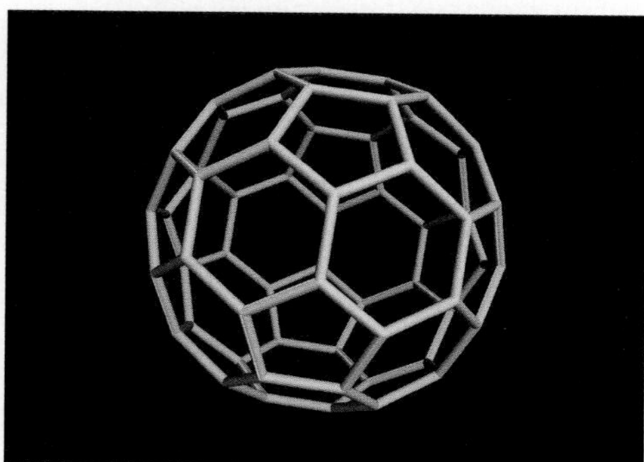

Figure 12.7

Buckminsterfullerene (C_{60}). All of the carbon atoms are equivalent and are sp^2-hybridized; each one simultaneously belongs to one five-membered ring and two benzene-like six-membered rings.
Courtesy of Dmitry Kazachkin, Temple University

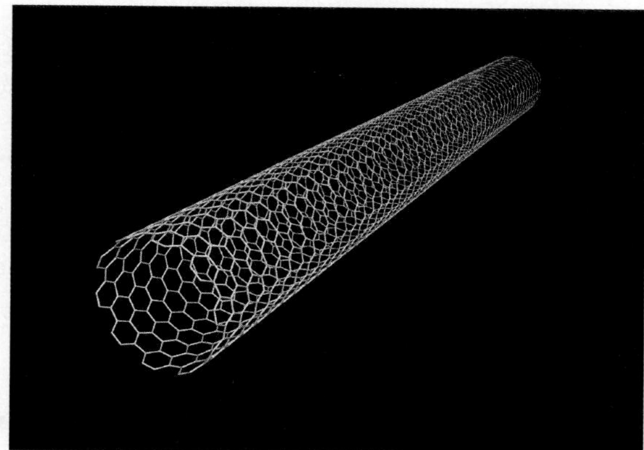

Figure 12.8

A single-walled carbon nanotube (SWCNT). SWCNTs can be regarded as a graphite sheet rolled into a cylinder.
Courtesy of Dmitry Kazachkin, Temple University

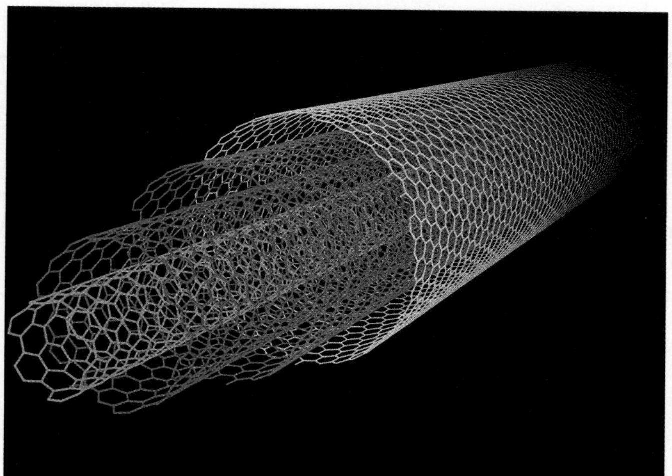

Figure 12.9

A multiwalled carbon nanotube (MWCNT).
Courtesy of Dmitry Kazachkin, Temple University

12.7 Physical Properties of Arenes

In general, arenes resemble other hydrocarbons in their physical properties. They are nonpolar, insoluble in water, and less dense than water. The boiling point of benzene is not much different from that of similar hydrocarbons and suggests that the aggregate intermolecular forces in the liquid phase are about equal for all of them.

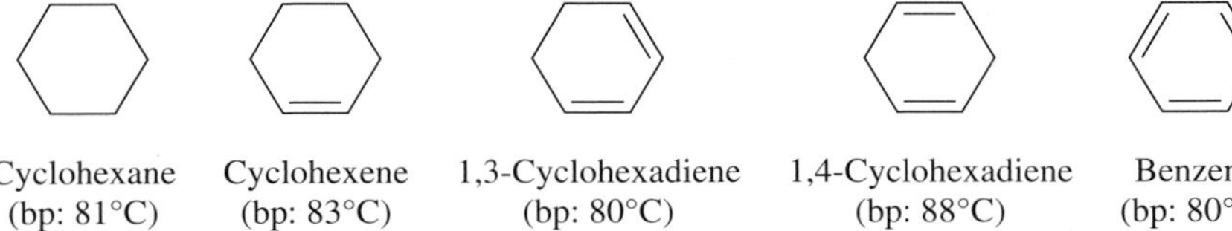

Cyclohexane (bp: 81°C) Cyclohexene (bp: 83°C) 1,3-Cyclohexadiene (bp: 80°C) 1,4-Cyclohexadiene (bp: 88°C) Benzene (bp: 80°C)

The intermolecular attractive forces in benzene have received considerable attention and prompted experimental and computational studies of *benzene dimer,* formed by noncovalent association between two benzene molecules. Figure 12.10 shows four geometries for this species. In (*a*) and (*b*) the rings are parallel to each other; in (*c*) and (*d*) they are perpendicular.

The parallel alignment of the rings in (*a*) and (*b*), referred to as π–π stacking, is destabilizing due to repulsion between the π electron systems of the two benzenes and subtracts from various stabilizing interactions such as intermolecular van der Waals attractive forces. Displacement of one ring versus the other reduces this electron–electron repulsion, making (*b*) more stable than (*a*). Both perpendicular geometries (*c*) and (*d*) avoid the repulsive π–π interaction and are more stable than the parallel dimers, with (*c*) slightly more stable than (*d*). The spread in energies among the four benzene dimers, however, is only a few kilojoules per mole, and other factors can tip the balance in favor of parallel versus perpendicular geometries when substituents are present.

In the solid phase, benzene molecules pack together in what is known as a "herringbone" pattern characterized by a combination of stacked, side-by-side, and perpendicular arrangements between molecules (Figure 12.11).

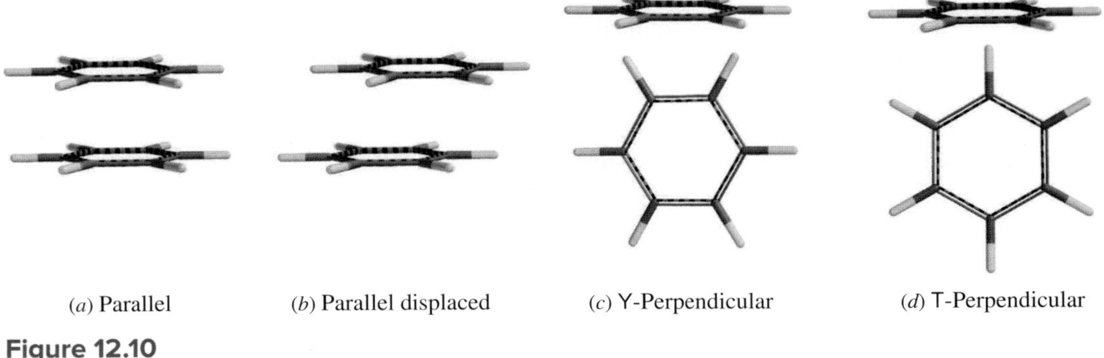

(*a*) Parallel (*b*) Parallel displaced (*c*) Y-Perpendicular (*d*) T-Perpendicular

Figure 12.10

Arrangement of benzene molecules in the dimer. All are close in energy, with (c) being the most stable according to electron diffraction studies.

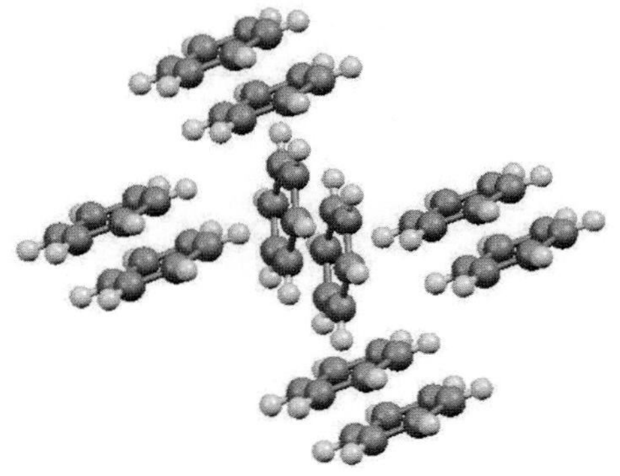

Figure 12.11

Packing of benzene molecules in the crystal.

Polycyclic aromatic compounds have significantly higher melting points than other hydrocarbons of similar size. The melting point of naphthalene is much higher than that of cyclodecane, and anthracene's is much higher than cyclotetradecane's.

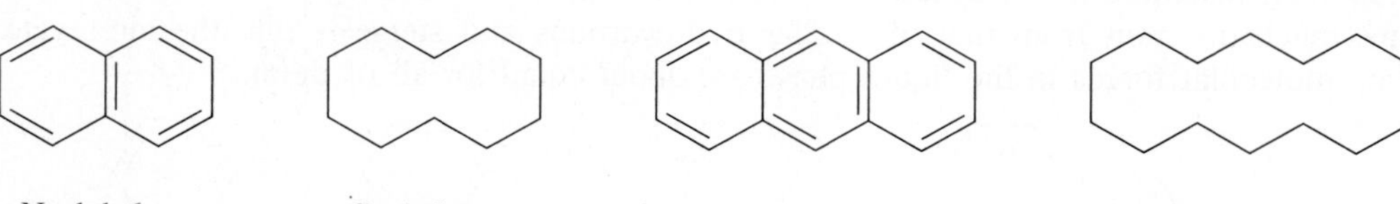

Naphthalene (mp: 80°C) Cyclodecane (mp: 10°C) Anthracene (mp: 218°C) Cyclotetradecane (mp: 56°C)

This reflects the increasing preference for parallel arrangements in the solid state of aromatic hydrocarbons larger than benzene. Intermolecular attractive forces like those in the perpendicular herringbone arrangement of benzene are outnumbered by van der Waals attractions in parallel displaced arrangements of polycyclic aromatics.

The stacking of graphene layers in graphite (see Figure 12.6 in the boxed essay *Fullerenes, Nanotubes, and Graphene*) is an example of such packing on a grand scale.

The three-dimensional shapes of proteins and nucleic acids can bring remote aromatic rings close enough in space so that interactions between them can contribute to the overall shape of the molecule. In general though, more familiar forces exert a greater influence than π–π stacking. We explore these forces in more detail in Chapters 25 and 26.

12.8 The Benzyl Group

A benzene ring can behave in two different ways in a chemical reaction; it can be a functional group in its own right or it can be a substituent that affects a reaction elsewhere in the molecule. In the most common case, this "elsewhere" is the carbon directly attached to the ring—the *benzylic* carbon.

Benzylic carbocations, radicals, and anions resemble their allylic counterparts in being conjugated systems stabilized by electron delocalization. This delocalization is describable in resonance, valence bond, and molecular orbital terms.

Resonance in benzylic carbocations:

Major contributor

Resonance in benzylic radicals:

Major contributor

Resonance in benzylic anions:

Major contributor

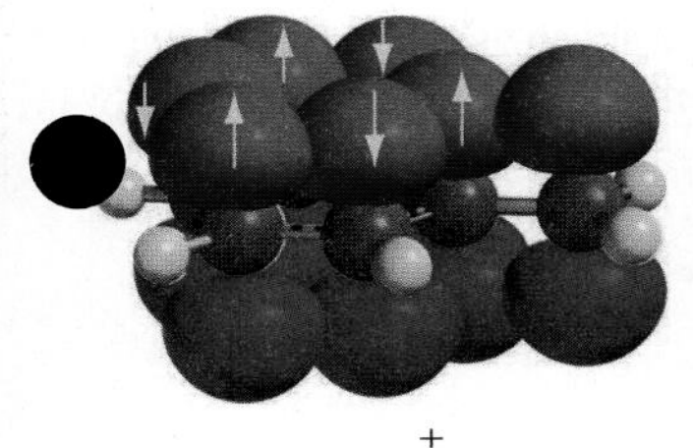
(*a*) $C_6H_5\overset{+}{C}H_2$

(*b*) $C_6H_5\dot{C}H_2$

(*c*) $C_6H_5\ddot{C}H_2^{-}$

Figure 12.12

Valence bond models of bonding in (*a*) benzyl cation, (*b*) benzyl radical, and (*c*) benzyl anion. Overlap of the 2*p* orbital of the benzylic carbon with the π system of the benzene ring creates an extended π system that stabilizes each one by electron delocalization.

Problem 12.7

Write a resonance contributor for each of the following in which the octet rule is followed for each atom other than hydrogen.

(a) $CH_3\ddot{O}-C_6H_4-\overset{+}{C}H_2$ (b) $O_2N^{+}-C_6H_4-\ddot{C}H_2^{-}$

Sample Solution (a) Start at an atom with only six electrons (the positively charged carbon) and move electrons in pairs toward it until the octet rule is satisfied. Add formal charges.

(a) $CH_3\ddot{O}-C_6H_4-\overset{+}{C}H_2 \longleftrightarrow CH_3\overset{+}{O}=C_6H_4=CH_2$

Valence Bond Description of Benzyl Cation, Radical, and Anion Figure 12.12 shows the valence bond approach to electron delocalization in benzyl cation, radical, and anion as the overlap of the π system of a benzene ring with, respectively, a vacant, half-filled, and filled 2*p* orbital of the benzylic carbon. According to this approach, benzyl cation (*a*) is stabilized by electron donation from the π system of the ring to the vacant 2*p* orbital of the benzylic carbon, thereby dispersing the positive charge and increasing the delocalization of the ring's π electrons. The charge dispersal contribution is absent in benzyl radical (*b*), but the stabilization due to electron delocalization remains and governs the rates of free-radical reactions of alkylbenzenes. Charge dispersal returns to join electron delocalization in stabilizing benzyl anion (*c*). In this case, the benzene ring acts as an electron-withdrawing group.

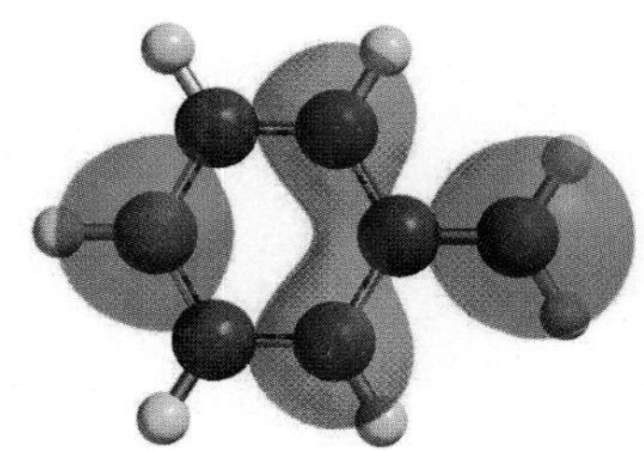

Figure 12.13

The lowest unoccupied molecular orbital (LUMO) of benzyl cation. The red and blue regions are associated with the atoms that contribute the largest share of the atomic orbitals that make up the LUMO. These atoms are the benzylic carbon and the carbons ortho and para to it.

Molecular Orbitals of Benzyl Cation, Radical, and Anion At the molecular orbital level, the most important MO for benzyl cation is the lowest unoccupied molecular orbital (LUMO), for benzyl radical it is the singly occupied molecular orbital (SOMO), and for benzyl anion the highest occupied molecular orbital (HOMO). Figure 12.13 shows the LUMO for benzyl cation where it is clear that the carbon 2*p* atomic orbitals that contribute the most are those of the benzylic carbon and the ring carbons that are ortho and para to it. These are the same carbons that, according to resonance, share the positive charge. The SOMO of benzyl radical and HOMO of benzyl anion are not shown, but are virtually identical to the LUMO of benzyl cation, in keeping with their respective resonance descriptions.

12.9 Nucleophilic Substitution in Benzylic Halides

Like allylic halides, benzylic halides undergo nucleophilic substitution, both S_N1 and S_N2, faster than simple alkyl halides and for similar reasons.

Relative S$_N$1 Rates Hydrolysis of the tertiary benzylic halide 2-chloro-2-phenylpropane occurs 620 times faster than hydrolysis of *tert*-butyl chloride under the same conditions (90% acetone–10% water at 25°C).

2-Chloro-2-phenylpropane
More reactive: k(rel) 620

tert-Butyl chloride
Less reactive: k(rel) 1

Because S$_N$1 rates reflect the activation energy for carbocation formation, we conclude that a phenyl substituent stabilizes a carbocation more than a methyl group does.

1-Methyl-1-phenylethyl cation is more stable than *tert*-Butyl cation

The electrostatic potential maps for the two carbocations (Figure 12.14) show the greater dispersal of positive charge in 1-methyl-1-phenylethyl cation compared with *tert*-butyl cation.

Problem 12.8

As measured by their first-order rate constants, the compound shown (R = CH_3) undergoes hydrolysis 26 times faster than 2-chloro-2-phenylpropane (R = H) in 90% acetone–10% water at 25°C. Offer a resonance explanation for this rate difference.

The rate-enhancing effect of phenyl substituents is cumulative; $(C_6H_5)_2CHCl$ undergoes S$_N$1 substitution faster than $C_6H_5CH_2Cl$, and triphenylmethyl perchlorate $[(C_6H_5)_3C^+ \ ClO_4^-]$ is a stable ionic compound that can be stored indefinitely.

S$_N$1 Reaction Products Unlike S$_N$1 reactions of allylic halides, dispersal of the charge in benzylic halides does not result in the nucleophile bonding to a carbon other than the one that had the leaving group. The ring's aromaticity is retained only if the nucleophile bonds to the benzylic carbon.

2-Chloro-2-phenylpropane $\xrightarrow{CH_3CH_2OH}$ 2-Ethoxy-2-phenylpropane (87%) via [cation]

Figure 12.14

The electrostatic potential maps show the positive charge is more dispersed in 1-methyl-1-phenylethyl cation (*a*) than in *tert*-butyl cation (*b*), where the blue color is concentrated at the tertiary carbon. The color range is the same for both models.

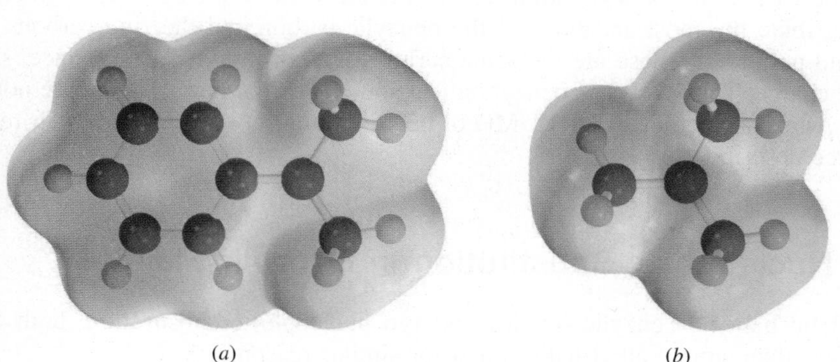

Relative S_N2 Rates Benzyl chloride undergoes S_N2 substitution with potassium iodide in acetone almost 200 times faster than propyl chloride and more than twice as fast as allyl chloride.

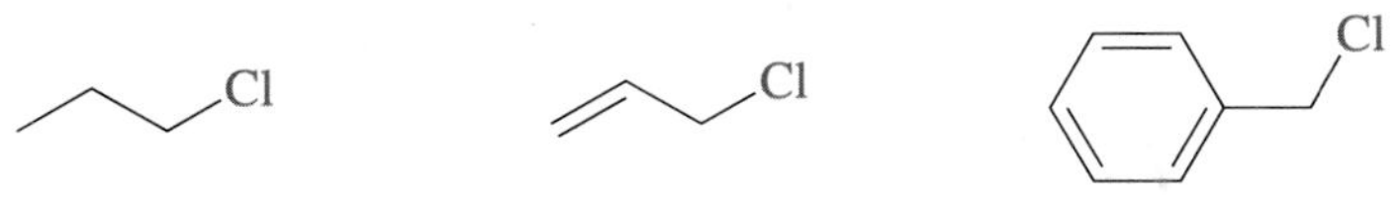

Propyl chloride
Least reactive: k(rel) 1.0

Allyl chloride
k(rel) 80

Benzyl chloride
Most reactive: k(rel) 197

As we saw for S_N2 reactions of alkyl halides (Section 6.3), and allylic halides (Section 11.2), the key interaction is between the HOMO of the nucleophile and the LUMO of the organic halide. In a benzylic halide, the LUMO encompasses both the CH_2Cl unit and the π system of the aromatic ring.

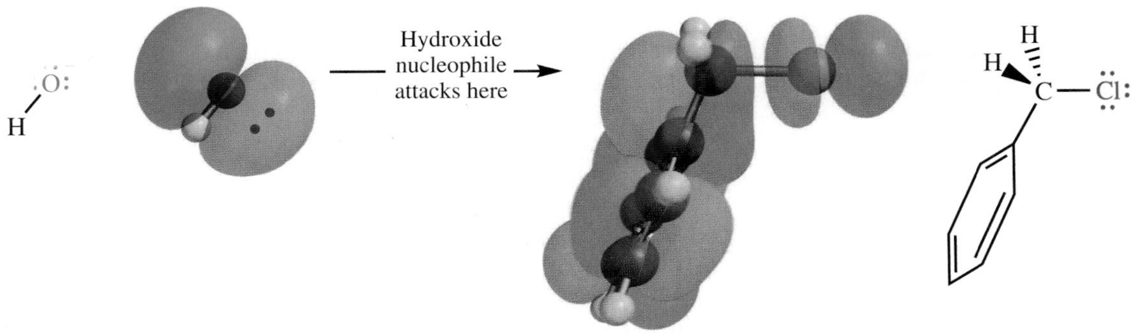

Hydroxide nucleophile attacks here

HOMO of hydroxide

LUMO of benzyl chloride

As electrons flow from the HOMO into the LUMO, they feel the attractive force not only of the benzylic carbon but of the ring carbons as well, which lowers the energy of the transition state and increases the reaction rate compared to allyl and propyl chloride.

S_N2 Reaction Products Primary benzylic halides are ideal substrates for S_N2 reactions. In addition to being very reactive, they are unable to undergo competing E2 elimination.

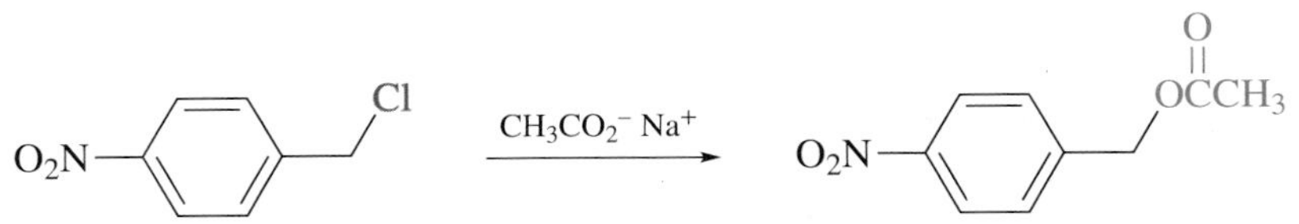

$$\text{p-Nitrobenzyl chloride} \xrightarrow{CH_3CO_2^-\ Na^+} \text{p-Nitrobenzyl acetate (78–82\%)}$$

p-Nitrobenzyl chloride

p-Nitrobenzyl acetate (78–82%)

Secondary and tertiary benzylic halides resemble secondary and tertiary alkyl halides in that they undergo substitution only when the nucleophile is weakly basic. If the nucleophile is a strong base such as sodium ethoxide, elimination by the E2 mechanism is faster than substitution.

Problem 12.9

Give the structure of the principal organic product formed on reaction of benzyl bromide with each of the following reagents:

(a) Sodium ethoxide
(b) Potassium *tert*-butoxide
(c) Sodium azide
(d) Sodium hydrogen sulfide
(e) Sodium iodide (in acetone)

continued

Sample Solution (a) Benzyl bromide is a primary bromide and undergoes S_N2 reactions readily. It has no hydrogens β to the leaving group and so cannot undergo elimination. Ethoxide ion acts as a nucleophile, displacing bromide and forming benzyl ethyl ether.

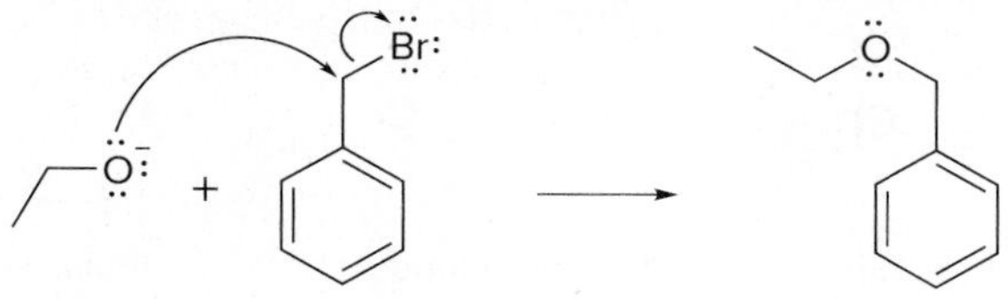

Ethoxide ion + Benzyl bromide → Benzyl ethyl ether

Triphenylmethyl Radical Yes, Hexaphenylethane No

An oft-told tale from organic chemistry's early days begins with a 1900 paper by Moses Gomberg of the University of Michigan entitled "An Instance of Trivalent Carbon: Triphenylmethyl." In an attempt to prepare hexaphenylethane by the reaction

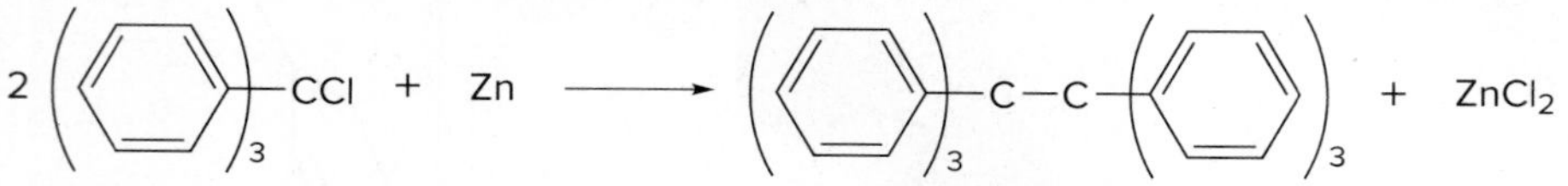

$$2\,(C_6H_5)_3CCl + Zn \longrightarrow (C_6H_5)_3C{-}C(C_6H_5)_3 + ZnCl_2$$

Triphenylmethyl chloride + Zinc → Hexaphenylethane + Zinc chloride

Gomberg found that the product was not hexaphenylethane but had properties more consistent with what we would now call triphenylmethyl radical [$(C_6H_5)_3C\cdot$] and suggested the two were in equilibrium with each other.

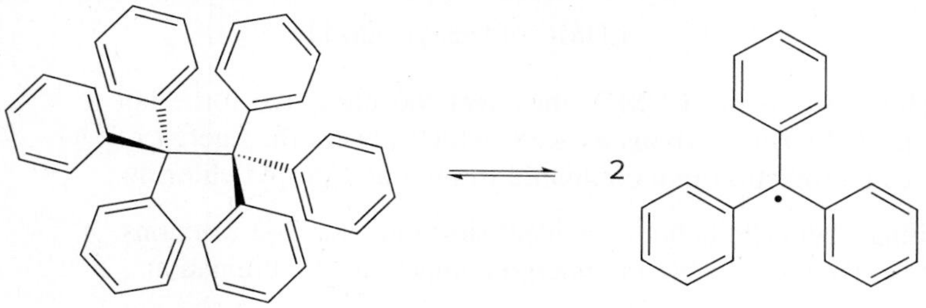

Hexaphenylethane ⇌ 2 Triphenylmethyl radical

Gomberg's paper received wide attention and stimulated alternative interpretations. One of these, suggested by Paul Jacobson, an official of a German chemical society and editor of its journal, raised the possible role of a species based on an alternative coupling of the two triphenylmethyl units.

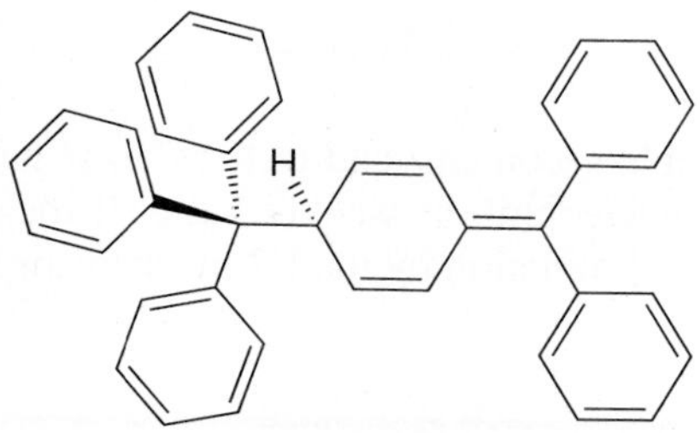

Jacobson's dimer

In spite of the fact that Gomberg himself grew to favor Jacobson's proposal, it became customary to regard hexaphenylethane as the product of the reaction and explain its properties in terms of an equilibrium between it and triphenylmethyl. The situation remained this way until 1968 when, on reinvestigating the reaction, it was conclusively determined that the structure proposed by Jacobson is the correct one.

It is now a routine matter to calculate relative energies of molecules. When compared this way, the Jacobson dimer is determined to be 62.9 kJ/mol (15.0 kcal/mol) more stable than hexaphenylethane. Although the Jacobson dimer has one less aromatic ring, the steric strain that would result from having six benzene rings on adjacent carbons is prohibitively high and it is likely that hexaphenylethane has never been prepared.

Problem 12.10

There is one more aspect to the story. It involves yet another dimer—one having the structure shown. It, too, was discovered by Gomberg who prepared it by treating what we now know is the Jacobson dimer with concentrated hydrochloric acid. Write a mechanism for this reaction. *Hint:* See Section 12.13.

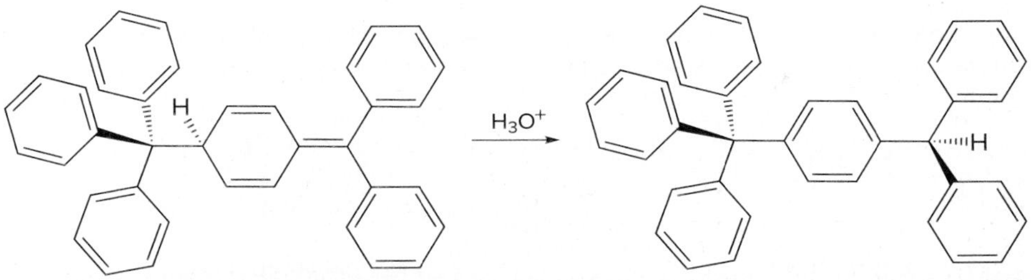

H_3O^+

12.10 Benzylic Free-Radical Halogenation

As measured by their bond-dissociation energies, benzylic C—H bonds are much weaker than allylic and alkyl C—H bonds.

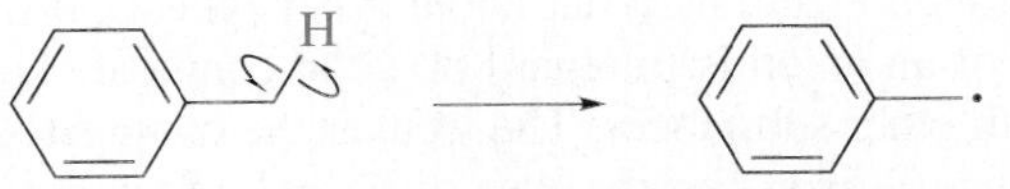

Toluene → Benzyl radical + H· $\Delta H° = 356$ kJ (85 kcal)

Propene → Allyl radical + H· $\Delta H° = 368$ kJ (88 kcal)

2-Methylpropane → *tert*-Butyl radical + H· $\Delta H° = 380$ kJ (91 kcal)

As in Section 11.3 where we counted the decreased bond-dissociation enthalpy in propene compared with 2-methylpropane as evidence for stabilization of allyl radical by delocalization of the unpaired electron, we attribute the decreased C—H bond strength in toluene versus propene as reflecting better electron delocalization in benzyl radical versus allyl.

The comparative ease with which a benzylic hydrogen is abstracted leads to a regioselective preference for substitution at the benzylic carbon in free-radical halogenations of alkylbenzenes. Bromination of alkylbenzenes using *N*-bromosuccinimide offers a convenient synthesis of benzylic bromides.

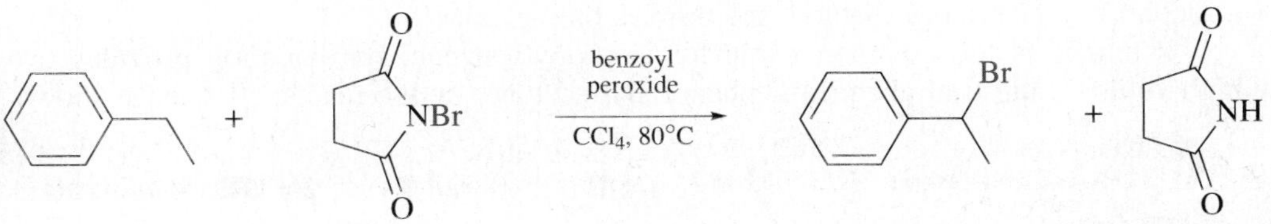

Ethylbenzene + *N*-Bromosuccinimide (NBS) →(benzoyl peroxide; CCl_4, 80°C) 1-Bromo-1-phenylethane (87%) + Succinimide

Problem 12.11

The reaction of *N*-bromosuccinimide with the following compounds has been reported in the chemical literature. Each compound yields a single product in 95% yield. Identify the product formed from each starting material.

(a) *p-tert*-Butyltoluene (b) 4-Methyl-3-nitroanisole

Sample Solution (a) The only benzylic hydrogens in *p-tert*-butyltoluene are those of the methyl group that is attached directly to the ring. Substitution occurs there to give *p-tert*-butylbenzyl bromide.

NBS; CCl_4, 80°C, free-radical initiator

p-tert-Butyltoluene → *p-tert*-Butylbenzyl bromide (Br)

12.11 Benzylic Anions

A benzylic *cation* is stabilized by delocalization of the π electrons of the ring into the vacant 2*p* orbital of the benzylic carbon. Similarly, the half-filled 2*p* orbital of a benzylic *radical* interacts with the π system of the ring to both increase the delocalization of the ring's electrons and allow delocalization of the unpaired electron. What about a benzylic *anion*?

One way to assess the stabilization of an anion is to regard it as the conjugate base of an acid and to compare the acid's pK_a with other substances. The weaker the conjugate base, the more strongly held is the unshared electron pair. For the case of benzyl anion, we compare toluene with other hydrocarbons—methane, diphenylmethane, and triphenylmethane.

Increasing acidity

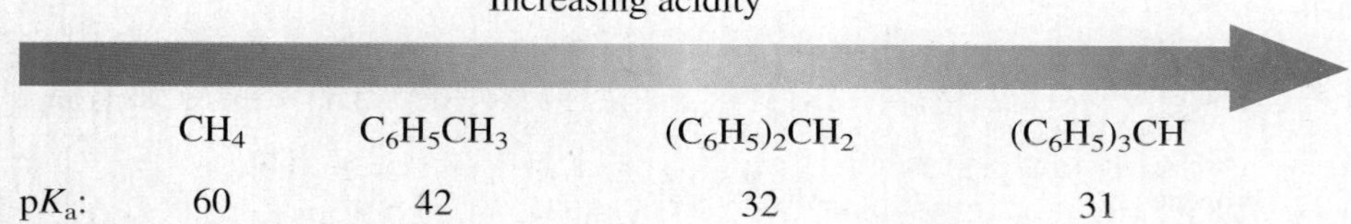

All four are very weak acids, but toluene is 18 pK_a units stronger than methane and its conjugate base is correspondingly that much weaker. Additional substitution of hydrogens by phenyl groups stabilizes the conjugate base and increases the acidity further.

Problem 12.12

Although we won't discuss amine basicity until Chapter 22, see if you can figure out which is the stronger base: *N*-methylaniline ($C_6H_5NHCH_3$) or benzylamine ($C_6H_5CH_2NH_2$). Explain your reasoning, supporting it with appropriate resonance contributors.

12.12 Oxidation of Alkylbenzenes

The term *oxidation* includes so many reaction types that space allows only a few to be included here, and those only briefly. Of these, one is a large-scale industrial synthesis, a second is a laboratory method, the third is biological.

Cumene is the common name for isopropylbenzene; its oxidation provides two high-volume industrial chemicals, phenol and acetone, by the reaction sequence shown.

Cumene $\xrightarrow{O_2}$ Cumene hydroperoxide (—OOH) $\xrightarrow{H_3O^+}$ Phenol (—OH) + Acetone (=O)

In the first step, oxygen abstracts a hydrogen atom from the benzylic carbon, setting the stage for a chain reaction that begins when the cumene hydroperoxy radical shown abstracts a benzylic hydrogen from a second molecule of cumene.

Cumene → Cumyl radical → Cumene hydroperoxy radical

> Show your understanding of the chain reaction, by writing an equation for the step in which cumyl hydroperoxy radical reacts with cumene.

Laboratory oxidations of alkylbenzenes typically employ inorganic oxidizing agents such as chromic acid (H_2CrO_4) or potassium permanganate ($KMnO_4$). Neither of these substances oxidize benzene or alkanes, but do oxidize alkylbenzenes containing at least one benzylic hydrogen to benzoic acid.

p-Isopropyltoluene $\xrightarrow[H_2O,\ H_2SO_4,\ heat]{Na_2Cr_2O_7}$ 1,4-Benzenedicarboxylic acid (45%)

> The combination of sodium dichromate and sulfuric acid is equivalent to chromic acid.

o-Chlorotoluene $\xrightarrow[2.\ HCl]{1.\ KMnO_4,\ H_2O}$ *o*-Chlorobenzoic acid (76–78%)

> The product of permanganate oxidation in step 1 is a carboxylate ion, so an acidification step follows in order to isolate the carboxylic acid.

Problem 12.13

(a) Chromic acid oxidation of 4-*tert*-butyl-1,2-dimethylbenzene yielded a single compound having the molecular formula $C_{12}H_{14}O_4$. What was this compound?

(b) What product is expected from chromic acid oxidation of 2,3-dihydroindene?

2,3-Dihydroindene

Sample Solution

(a)

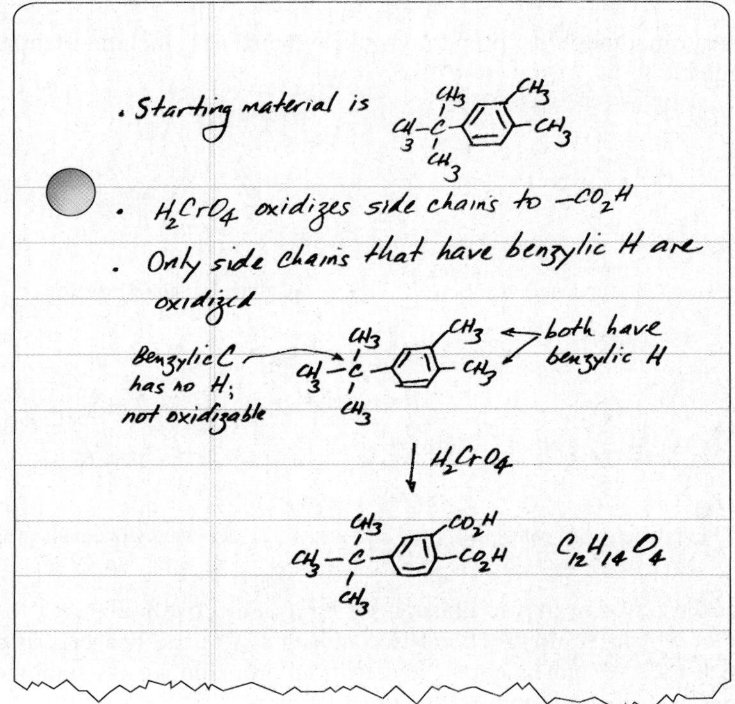

• Starting material is
• H2CrO4 oxidizes side chains to —CO2H
• Only side chains that have benzylic H are oxidized
Benzylic C has no H; not oxidizable
both have benzylic H
H2CrO4
C12H14O4

Side-chain oxidation of alkylbenzenes is important in certain metabolic processes. One way in which the body rids itself of foreign substances is by oxidation in the liver to compounds that are more easily excreted in the urine. Toluene, for example, is oxidized to benzoic acid and is eliminated rather readily.

$$C_6H_5CH_3 \xrightarrow[\text{(an enzyme in the liver)}]{O_2,\ \text{cytochrome P-450}} C_6H_5COH(=O)$$

Toluene → Benzoic acid

Benzene, with no alkyl side chain and no benzylic hydrogens, undergoes a different reaction under these conditions. Oxidation of the ring occurs to convert benzene to its epoxide.

$$\text{Benzene} \xrightarrow{O_2,\ \text{cytochrome P-450}} \text{Benzene oxide}$$

Benzene → Benzene oxide

Benzene oxide and compounds derived from it are carcinogenic and can react with DNA to induce mutations. This difference in the site of biological oxidation—ring versus side-chain—seems to be responsible for the fact that benzene is carcinogenic but toluene is not.

12.13 Alkenylbenzenes

Alkenylbenzenes are prepared by the various methods described in Chapter 7 for the preparation of alkenes: *dehydrogenation, dehydration,* and *dehydrohalogenation.*

Dehydrogenation of alkylbenzenes is not a convenient laboratory method but is used industrially to convert ethylbenzene to styrene for the preparation of polystyrene.

$$\text{Ethylbenzene} \xrightarrow[\text{ZnO}]{630°\text{C}} \text{Styrene} + H_2$$

Ethylbenzene → Styrene + Hydrogen

Laboratory methods for preparing alkenylbenzenes include dehydration and dehydrohalogenation.

$$\text{1-(}m\text{-Chlorophenyl)ethanol} \xrightarrow[\text{heat}]{KHSO_4} m\text{-Chlorostyrene (80–82\%)}$$

1-(*m*-Chlorophenyl)ethanol → *m*-Chlorostyrene (80–82%)

$$\text{2-Bromo-1-(}p\text{-methylphenyl)propane} \xrightarrow[CH_3CH_2OH,\ 50°\text{C}]{NaOCH_2CH_3} H_3C\text{-}C_6H_4\text{-}CH{=}CHCH_3$$

2-Bromo-1-(*p*-methylphenyl)propane → 1-*p*-Methylphenyl-1-propene (99%) (cis + trans)

The second of these two examples illustrates the regioselective preference for formation of the isomer in which the double bond is conjugated with the benzene ring. A hydrocarbon in which a C=C unit is conjugated with an aromatic ring is stabilized to about the same extent as a double bond in a conjugated diene.

The side-chain double bond is more reactive than the aromatic ring toward most electrophilic reagents. Many of the reactions of alkenes that were discussed in Chapter 8 find a parallel in the reactions of alkenylbenzenes. Thus, hydrogenation and halogen addition to a side-chain double bond can be achieved while leaving the ring unchanged.

Br, $CHCH_3$ —H_2, Pt→ Br

2-(*m*-Bromophenyl)-2-butene (cis + trans) → 2-(*m*-Bromophenyl)butane (92%)

—Br_2→ Br, Br

Styrene → 1,2-Dibromo-1-phenylethane (82%)

Problem 12.14

The cis and trans stereoisomers of 1,2-diphenylethylene give stereoisomeric products on addition of bromine. Draw a structural formula of each product. Which stereoisomer of 1,2-diphenylethylene gives a meso dibromide? Which one gives a chiral dibromide?

The regioselectivity of electrophilic addition is governed by the ability of an aromatic ring to stabilize an adjacent carbocation. This is clearly seen in the addition of hydrogen chloride to indene.

—HCl→ Cl

Indene → 1-Chloroindane (75–84%)

Only the benzylic chloride is formed because protonation of the double bond occurs in the direction that gives a carbocation that is both secondary and benzylic.

+ H—Cl: ⟶ + + :Cl:$^-$ ⟶ :Cl:

Protonation in the opposite direction also gives a secondary carbocation, but that carbocation is not benzylic and does not receive the extra increment of stabilization that its benzylic isomer does. The more stable benzylic carbocation is formed faster and is the one that determines the reaction product.

Problem 12.15

Each of the following reactions has been reported in the chemical literature and gives a single organic product in high yield. Write the structure of the product for each reaction.

continued

(a) 2-Phenylpropene + hydrogen chloride
(b) 2-Phenylpropene treated with diborane in tetrahydrofuran followed by oxidation with basic hydrogen peroxide
(c) Styrene + bromine in aqueous solution
(d) Styrene + peroxybenzoic acid (two organic products in this reaction; identify both by writing a balanced equation)

Sample Solution (a) Addition of hydrogen chloride to the double bond takes place by way of a tertiary benzylic carbocation.

2-Phenylpropene + Hydrogen chloride ⟶ [tertiary benzylic carbocation] + :Cl:⁻ ⟶ 2-Chloro-2-phenylpropane

In the presence of peroxides, hydrogen bromide adds to the double bond of styrene with a regioselectivity opposite to Markovnikov's rule. The reaction is a free-radical addition, and the regiochemistry is governed by preferential formation of the more stable radical as described in Section 10.5.

Styrene $\xrightarrow[\text{peroxides}]{\text{HBr}}$ 1-Bromo-2-phenylethane (only product) via [benzylic radical]

12.14 Polymerization of Styrene

As described in the boxed essay *Diene Polymers* in Chapter 11, most synthetic rubber is a copolymer of styrene and 1,3-butadiene.

The annual production of styrene in the United States is approximately 1.1×10^{10} lb, with about 65% of this output used to prepare polystyrene plastics and films. Styrofoam coffee cups are made from polystyrene. Polystyrene can also be produced in a form that is very strong and impact-resistant and is used widely in luggage, television and radio cabinets, and furniture.

Polymerization of styrene can be carried out under free-radical (Mechanism 12.1), cationic, anionic, or Ziegler–Natta conditions (see Section 15.15).

Mechanism 12.1

Free-Radical Polymerization of Styrene

Step 1: Polymerization of styrene usually employs a peroxide as an initiator. The peroxide dissociates on heating to produce two alkoxy radicals.

$$\text{RÖ–ÖR} \xrightarrow{\text{heat}} \text{RÖ·} + \text{·ÖR}$$

Peroxide Two alkoxy radicals

Step 2: The free radical produced in step 1 adds to the double bond of styrene. Addition occurs in the direction that produces a benzylic radical.

RO· + Styrene ⟶ A benzylic radical

Alkoxy radical Styrene A benzylic radical

Step 3: The benzylic radical produced in step 2 adds to a molecule of styrene. Again addition occurs in the direction that produces a benzylic radical.

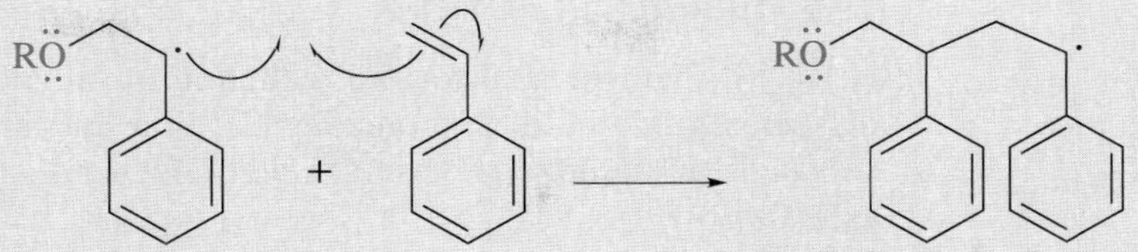

Benzylic radical from step 2 + Styrene → Chain-extended benzylic radical

Step 4: The radical produced in step 3 reacts with another styrene molecule, and the process repeats over and over to produce a long-chain polymer having phenyl substituents at every other carbon in the chain.

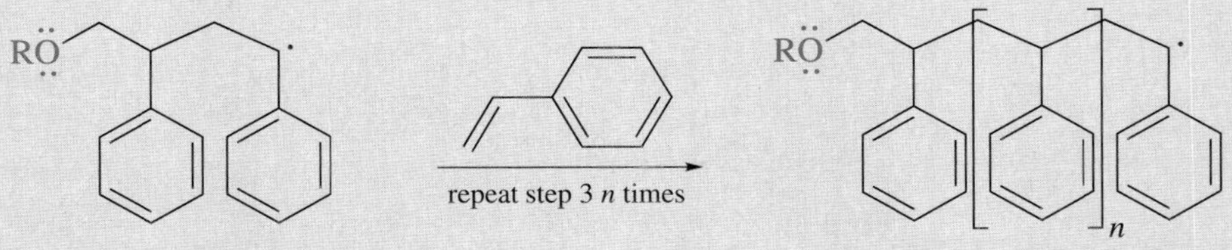

Benzylic radical from step 3 —(repeat step 3 *n* times)→ Growing polystyrene chain

12.15 The Birch Reduction

We saw in Section 10.6 that the combination of a Group 1 metal and liquid ammonia is a powerful reducing system capable of reducing alkynes to trans alkenes. In the presence of an alcohol, this same combination reduces arenes to *nonconjugated dienes*. Thus, treatment of benzene with sodium and methanol or ethanol in liquid ammonia converts it to 1,4-cyclohexadiene. Alkyl-substituted benzenes give 1,4-cyclohexadienes in which the alkyl group is a substituent on the double bond.

Metal–ammonia–alcohol reductions of aromatic rings are known as **Birch reductions,** after the Australian chemist Arthur J. Birch, who demonstrated their usefulness in organic synthesis.

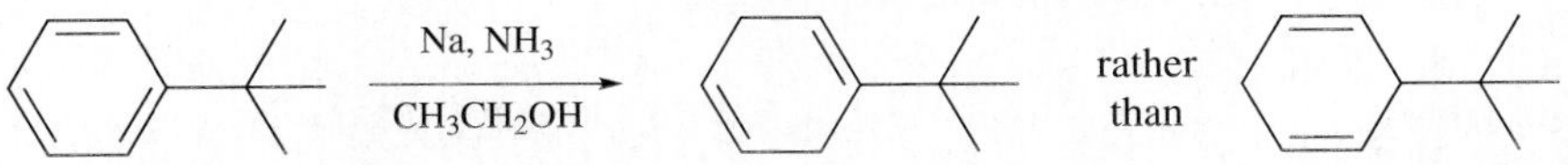

tert-Butylbenzene —(Na, NH_3 / CH_3CH_2OH)→ 1-*tert*-Butyl-1,4-cyclohexadiene (86%) rather than 3-*tert*-Butyl-1,4-cyclohexadiene

Problem 12.16

The regioselectivity of Birch reduction of alkoxy-substituted benzenes is the same as for alkylbenzenes. What did Arthur Birch isolate when he carried out the following reaction?

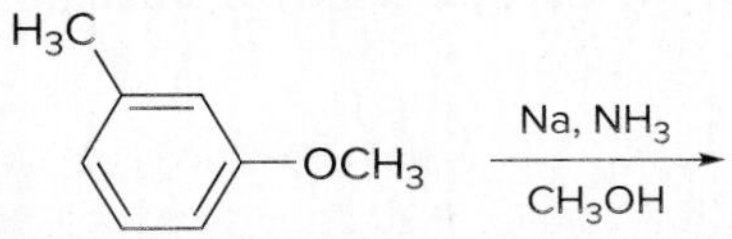

(H_3C, OCH_3-substituted benzene) —(Na, NH_3 / CH_3OH)→

The mechanism by which the Birch reduction of benzene takes place (Mechanism 12.2) is analogous to the mechanism for the metal–ammonia reduction of alkynes. It involves a sequence of four steps in which steps 1 and 3 are one-electron reductions and steps 2 and 4 are proton transfers from the alcohol.

Mechanism 12.2

The Birch Reduction

THE OVERALL REACTION:

As in Mechanism 10.3 for the reduction of alkynes by Group 1 metals, sodium dissociates in liquid ammonia to Na^+ ion and an electron, both of which are solvated by ammonia. These solvated electrons are represented in the equation as $e^-(am)$.

$$\text{Benzene} + 2e^-(am) + 2CH_3OH \xrightarrow{NH_3} \text{1,4-Cyclohexadiene} + 2CH_3O^-$$

Benzene   Electrons   Methanol   1,4-Cyclohexadiene   Methoxide ion

THE MECHANISM:

Step 1: *Electron transfer:* An electron adds to the LUMO of benzene to give an anion radical.

$$\text{Benzene} + e^-(am) \longrightarrow \text{Benzene anion radical}$$

Benzene   Electron   Benzene anion radical

Step 2: *Proton transfer:* The anion radical formed is strongly basic and abstracts a proton from methanol. As in the reduction of alkynes, this is believed to be the rate-determining step.

$$\text{Benzene anion radical} + H{-}\ddot{O}CH_3 \longrightarrow \text{Cyclohexadienyl radical} + :\ddot{O}CH_3$$

Benzene anion radical   Methanol   Cyclohexadienyl radical   Methoxide ion

Step 3: *Electron transfer:* The cyclohexadienyl radical produced in step 2 is converted to an anion by reaction with the solvated electron.

$$\text{Cyclohexadienyl radical} + e^-(am) \longrightarrow \text{Cyclohexadienyl anion}$$

Cyclohexadienyl radical   Electron   Cyclohexadienyl anion

Step 4: *Proton transfer:* Proton transfer from methanol to the anion gives 1,4-cyclohexadiene.

Cyclohexadienyl anion + $H-\ddot{O}CH_3$ ⟶ 1,4-Cyclohexadiene + $^{-}:\ddot{O}CH_3$

Cyclohexadienyl anion Methanol 1,4-Cyclohexadiene Methoxide ion

12.16 Benzylic Side Chains and Retrosynthetic Analysis

The relative stability of benzylic carbocations, radicals, and carbanions makes it possible to manipulate the side chains of aromatic rings. Functionalization at the benzylic position, for example, is readily accomplished by free-radical halogenation and provides access to the usual reactions (substitution, elimination) that we associate with alkyl halides.

To illustrate, consider the synthesis of (Z)-1-phenyl-2-butene. A major consideration—controlling the stereochemistry of the double bond—can be achieved by catalytic hydrogenation of the corresponding alkyne.

(Z)-1-Phenyl-2-butene ⟹ 1-Phenyl-2-butyne

The question then becomes one of preparing 1-phenyl-2-butyne. A standard method for the preparation of alkynes is the alkylation of acetylene and other alkynes. In the present case, a suitable combination is propyne and a benzylic halide. The benzylic halide can be prepared from toluene.

1-Phenyl-2-butyne ⟹ A benzyl halide ⟹ Toluene

This retrosynthesis suggests the following synthesis:

Toluene —(NBS; benzoyl peroxide, heat or light)→ Benzyl bromide —($NaC\equiv CCH_3$)→ 1-Phenyl-2-butyne —(H_2; Lindlar Pd)→ (Z)-1-Phenyl-2-butene

Problem 12.17

Use retrosynthetic analysis to describe the preparation of *trans*-2-phenylcyclopentanol from cyclopentylbenzene and write equations showing suitable reagents for the synthesis.

12.17 Cyclobutadiene and Cyclooctatetraene

During our discussion of benzene and its derivatives, it may have occurred to you that cyclobutadiene and cyclooctatetraene might be stabilized by cyclic π electron delocalization in a manner analogous to that of benzene.

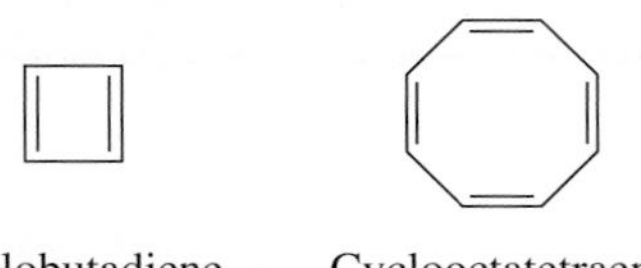

Willstätter's most important work, for which he won the 1915 Nobel Prize in Chemistry, was directed toward determining the structure of chlorophyll.

The same thought occurred to early chemists. However, the complete absence of naturally occurring compounds based on cyclobutadiene and cyclooctatetraene contrasted starkly with the abundance of compounds containing a benzene unit. Attempts to synthesize cyclobutadiene and cyclooctatetraene met with failure until 1911 when Richard Willstätter prepared cyclooctatetraene by a lengthy degradation of *pseudopelletierine,* a natural product obtained from the bark of the pomegranate tree. Today, cyclooctatetraene is prepared from acetylene in a reaction catalyzed by nickel cyanide.

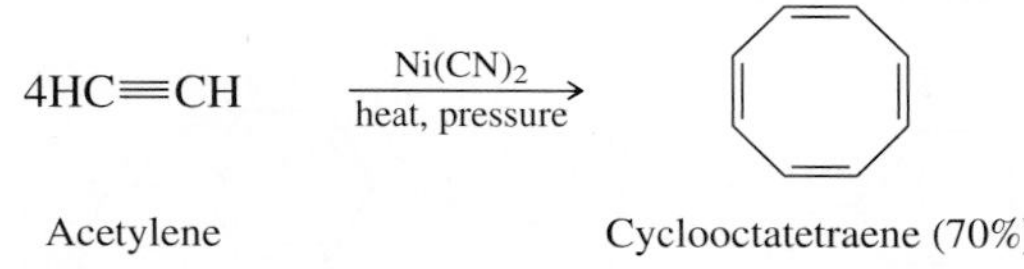

Pomegranate trees are best known for their large, juicy, seed-laden fruit.
©Valentyn Volkov/Getty Images

Cyclooctatetraene is relatively stable, but lacks the "special stability" of benzene. Unlike benzene, which we saw has a heat of hydrogenation that is 152 kJ/mol (36 kcal/mol) *less* than three times the heat of hydrogenation of cyclohexene, cyclooctatetraene's heat of hydrogenation is 26 kJ/mol (6 kcal/mol) *more* than four times that of *cis*-cyclooctene.

cis-Cyclooctene + Hydrogen (H_2) $\xrightarrow{\text{Pt}}$ Cyclooctane $\Delta H^\circ = -96$ kJ (-23 kcal)

Cyclooctatetraene + Hydrogen ($4H_2$) $\xrightarrow{\text{Pt}}$ Cyclooctane $\Delta H^\circ = -410$ kJ (-98 kcal)

Problem 12.18

Both cyclooctatetraene and styrene have the molecular formula C_8H_8 and undergo combustion according to the equation

$$C_8H_8 + 10O_2 \longrightarrow 8CO_2 + 4H_2O$$

The measured heats of combustion are 4393 and 4543 kJ/mol (1050 and 1086 kcal/mol). Which heat of combustion belongs to which compound?

Thermodynamically, cyclooctatetraene does not qualify as aromatic. Nor does its structure offer any possibility of the π electron delocalization responsible for aromaticity. As shown in Figure 12.15, cyclooctatetraene is *nonplanar* with four short and four long carbon–carbon bond distances. Cyclooctatetraene is satisfactorily represented by a single Lewis structure having alternating single and double bonds in a tub-shaped eight-membered ring. Experimental studies and theoretical calculations indicate that the structure of cyclooctatetraene shown in Figure 12.15 is about 75 kJ/mol (18 kcal/mol) more stable than the planar delocalized alternative. Cyclooctatetraene is not aromatic.

What about cyclobutadiene?

Cyclobutadiene escaped chemical characterization until the 1950s, when a variety of novel techniques succeeded in generating cyclobutadiene as a transient, reactive intermediate.

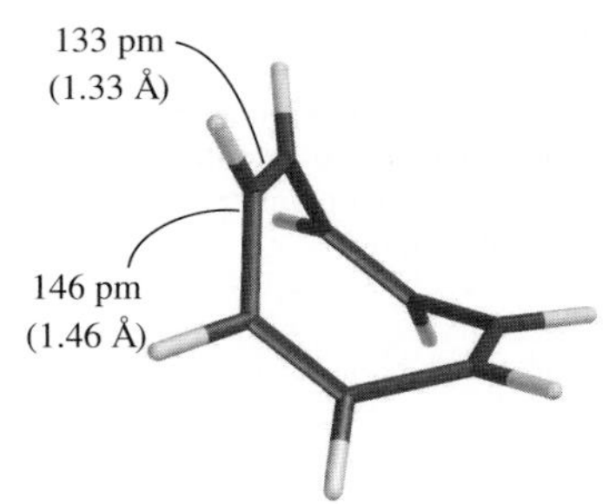

Figure 12.15

Molecular geometry of cyclooctatetraene. The ring is not planar, and the bond distances alternate between short double bonds and long single bonds.

Problem 12.19

One of the chemical properties that makes cyclobutadiene difficult to isolate is that it reacts readily with itself to give a dimer:

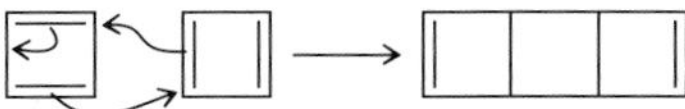

What reaction of dienes does this resemble?

Molecular orbital calculations of cyclobutadiene itself and experimentally measured bond distances of a stable, highly substituted derivative both reveal a pattern of alternating short and long bonds characteristic of a rectangular, rather than square, geometry.

H H $(CH_3)_3C$ $C(CH_3)_3$

135 pm (1.35 Å) 138 pm (1.38 Å)

H H $(CH_3)_3C$ CO_2CH_3

156 pm (1.56 Å) 151 pm (1.51 Å)

Cyclobutadiene Sterically hindered cyclobutadiene derivative

Experimental measurements place delocalized cyclobutadiene approximately 150 kJ/mol (36 kcal/mol) higher in energy than a structure with noninteracting double bonds; both square cyclobutadiene and planar cyclooctatetraene are *antiaromatic*. **Antiaromatic** molecules are *destabilized by delocalization of their π electrons* and cyclobutadiene and cyclooctatetraene adopt structures that minimize the delocalization of these electrons.

Cyclic conjugation, although necessary for aromaticity, is not sufficient for it. Some other factor or factors must contribute to the special stability of benzene and compounds based on the benzene ring. To understand these factors, let's return to the molecular orbital description of benzene.

12.18 Hückel's Rule

One of molecular orbital theory's early successes came in 1931 when Erich Hückel discovered an interesting pattern in the π orbital energy levels of benzene, cyclobutadiene, and cyclooctatetraene. By limiting his analysis to monocyclic conjugated polyenes and restricting the structures to planar geometries, Hückel found that whether a hydrocarbon of this type was aromatic depended on its number of π electrons. He set forth what we now call **Hückel's rule:**

Hückel was a German physical chemist. Before his theoretical studies of aromaticity, Hückel collaborated with Peter Debye in developing what remains the most widely accepted theory of electrolyte solutions.

> *Among planar, monocyclic, fully conjugated polyenes, only those possessing* $(4n + 2)$ *π electrons, where* n *is a whole number, will have special stability; that is, be aromatic.*

Thus, for this group of hydrocarbons, those with $(4n + 2) = 2, 6, 10, 14, \ldots$ π electrons will be aromatic. These values correspond to $(4n + 2)$ when $n = 0, 1, 2, 3, \ldots$

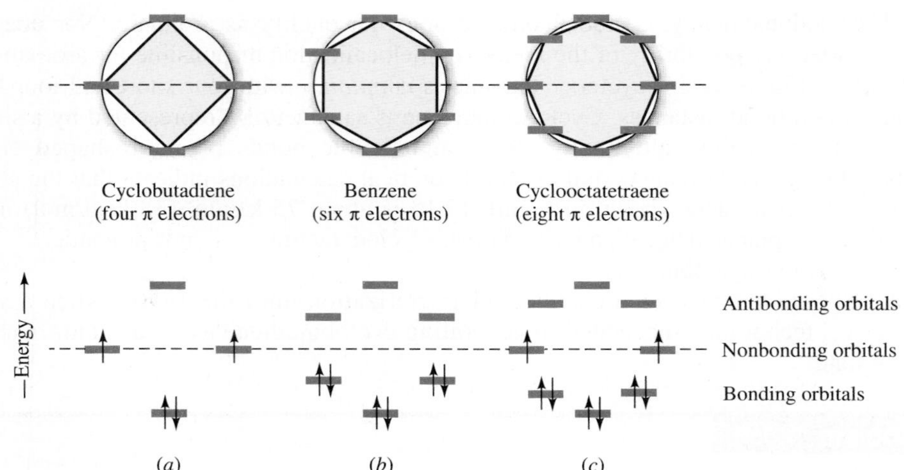

Figure 12.16

Frost's circle and the π molecular orbitals of (*a*) square cyclobutadiene, (*b*) benzene, and (*c*) planar cyclooctatetraene.

Hückel proposed his theory before ideas of antiaromaticity emerged. We now amplify his generalization by noting that among the hydrocarbons covered by Hückel's rule, those with $(4n)$ π electrons not only are not aromatic, they are antiaromatic.

Benzene, cyclobutadiene, and cyclooctatetraene provide clear examples of Hückel's rule. Benzene, with six π electrons, is a $(4n + 2)$ system and is predicted to be aromatic by the rule. Square cyclobutadiene and planar cyclooctatetraene are $4n$ systems with four and eight π electrons, respectively, and are antiaromatic.

The $(4n + 2)$ π electron standard follows from the pattern of orbital energies in monocyclic, completely conjugated polyenes. The π energy levels were shown for benzene earlier in Figure 12.4 and are repeated in Figure 12.16*b*. Figure 12.16*a* and 12.16*c* show the π energy levels for square cyclobutadiene and planar cyclooctatetraene, respectively.

The circle mnemonic was devised by Arthur A. Frost, a theoretical chemist at Northwestern University.

The diagrams in Figure 12.16 illustrate a simple method, called the **Frost circle,** for setting out the Hückel MOs of "planar, monocyclic, completely conjugated polyenes." By inscribing a polygon having the appropriate number of sides within a circle so that one of its corners lies at the bottom, the location of each of the polygon's corners defines a π electron energy level. A horizontal line drawn through the center of the circle separates the bonding and antibonding MOs; an orbital that lies directly on the line is nonbonding.

The pattern of orbital energies in Figure 12.16 provides a convincing explanation for why benzene is aromatic while square cyclobutadiene and planar cyclooctatetraene are not. We start by counting π electrons; cyclobutadiene has four, benzene six, and cyclooctatetraene has eight. These π electrons are assigned to MOs in accordance with the usual rules—lowest-energy orbitals first, a maximum of two electrons per orbital, and when two orbitals are of equal energy, each gets one electron before either orbital gets two (Hund's rule).

Benzene As seen earlier in Figure 12.4 (Section 12.4), the six π electrons of benzene are distributed in pairs among its three bonding π MOs, giving a closed-shell electron configuration. All the bonding orbitals are filled, and all the electron spins are paired.

Cyclobutadiene Square cyclobutadiene has one bonding π MO, two equal-energy nonbonding π MOs, and one antibonding π* MO. After the bonding MO is filled, the remaining two electrons are assigned to different nonbonding MOs in accordance with Hund's rule. This results in a species with two unpaired electrons—a diradical. In a square geometry, cyclobutadiene lacks a closed-shell electron configuration. It is not stabilized and, with two unpaired electrons, should be very reactive.

Cyclooctatetraene Six of the eight π electrons of planar cyclooctatetraene occupy three bonding orbitals. The remaining two π electrons occupy, one each, the two equal-energy nonbonding orbitals. Planar cyclooctatetraene should, like square cyclobutadiene, be a diradical.

An important conclusion to draw from the qualitative MO diagrams is that the customary geometry required for maximum π electron delocalization gives relatively unstable electron configurations for cyclobutadiene and cyclooctatetraene. Both escape to alternative geometries that have electron configurations which, although not aromatic, at least have all their electron spins paired. For cyclobutadiene the stable geometry is rectangular; for cyclooctatetraene it is tub-shaped.

Benzene's structure allows effective π electron conjugation and gives a closed-shell electron configuration. To understand why it also conveys special stability, we need to go one step further and compare the Hückel π MOs of benzene to those of a hypothetical "cyclohexatriene" with alternating single and double bonds. Without going into quantitative detail, we'll simply note that the occupied orbitals of a structure in which the π electrons are restricted to three noninteracting double bonds are of higher energy (less stable) than the occupied Hückel MOs of benzene.

Before looking at other applications of Hückel's rule, it is worth pointing out that its opening phrase, "Among planar, monocyclic, fully conjugated polyenes," does *not* mean that *only* "planar, monocyclic, fully conjugated polyenes" can be aromatic. It merely limits the rule to compounds of this type. There are countless aromatic compounds that are not monocyclic—naphthalene and related polycyclic aromatic hydrocarbons (Section 12.6), for example. All compounds based on benzene rings are aromatic. Cyclic conjugation *is* a requirement for aromaticity, however, and in those cases the conjugated system must contain $(4n + 2)$ π electrons. Cyclic conjugated systems with $4n$ π electrons are antiaromatic.

Problem 12.20

Give an explanation for each of the following observations:

(a) Compound A has six π electrons but is not aromatic.
(b) Compound B has six π electrons but is not aromatic.
(c) Compound C has 12 π electrons and is aromatic.

Compound A (C_7H_8) Compound B (C_7H_8) Compound C ($C_{12}H_{10}$)

Sample Solution (a) Cycloheptatriene (compound A) is not aromatic because, although it does contain six π electrons, its conjugated system of three double bonds does not close on itself—it lacks cyclic conjugation. The CH_2 group prevents cyclic delocalization of the π electrons.

In the next section we'll explore Hückel's rule for values of n greater than 1 to see how it can be extended beyond cyclobutadiene, benzene, and cyclooctatetraene.

12.19 Annulenes

The general term **annulene** refers to completely conjugated monocyclic hydrocarbons with more than six carbons. Cyclobutadiene and benzene retain their names, but higher members of the group are named **[*x*]annulene,** where *x* is the number of carbons in the ring. Thus, cyclooctatetraene becomes [8]annulene, cyclodecapentaene becomes [10]annulene, and so on.

Most of the synthetic work directed toward the higher annulenes was carried out by Franz Sondheimer and his students, first at Israel's Weizmann Institute and later at the University of London.

Problem 12.21

Use Frost's circle to construct orbital energy diagrams for (a) [10]annulene and (b) [12]annulene. Is either aromatic according to Hückel's rule?

Sample Solution (a) [10]Annulene is a ten-membered ring with five conjugated double bonds. Drawing a polygon with ten sides with its vertex pointing downward within a circle gives the orbital template. Place the orbitals at the positions where each vertex contacts the circle. The ten π electrons of [10]annulene satisfy the $(4n + 2)$ rule for $n = 2$ and occupy the five bonding orbitals in pairs. [10]Annulene is aromatic according to Hückel's rule.

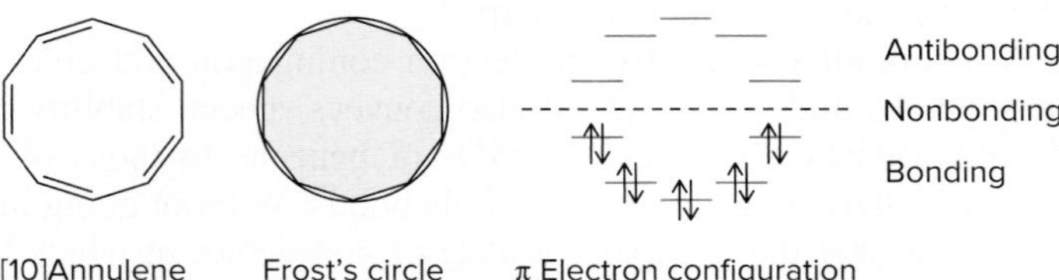

The size of each angle of a regular polygon is given by the expression

$$180° \times \frac{(\text{number of sides}) - 2}{(\text{number of sides})}$$

The prospect of observing aromatic character in conjugated polyenes having 10, 14, 18, and so on π electrons spurred efforts toward the synthesis of higher annulenes. A problem immediately arises in the case of the all-cis isomer of [10]annulene, the structure of which is shown in the preceding problem. Geometry requires a ten-sided regular polygon to have 144° bond angles; sp^2 hybridization at carbon requires 120° bond angles. Therefore, aromatic stabilization due to conjugation in all-*cis*-[10]annulene is opposed by the destabilizing angle strain at each of its carbon atoms. All-*cis*-[10]annulene has been prepared. It is highly reactive and lacks the "special stability" we associate with aromaticity.

A second isomer of [10]annulene (the cis, trans, cis, cis, trans stereoisomer) can have bond angles close to 120° but is destabilized by a close contact between two hydrogens directed toward the interior of the ring. To minimize the van der Waals strain between these hydrogens, the ring adopts a nonplanar geometry, which limits its ability to be stabilized by π electron delocalization. It, too, has been prepared and is not aromatic. Similarly, the next higher $(4n + 2)$ system, [14]annulene, is also somewhat destabilized by van der Waals strain, is nonplanar and not aromatic.

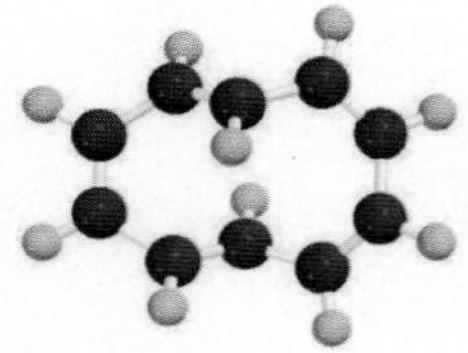

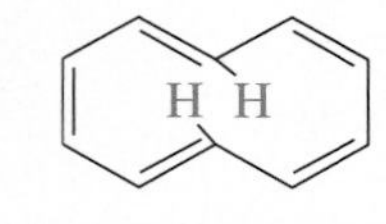

cis,trans,cis,cis,trans-[10]Annulene

Planar geometry required for aromaticity destabilized by van der Waals repulsions between indicated hydrogens

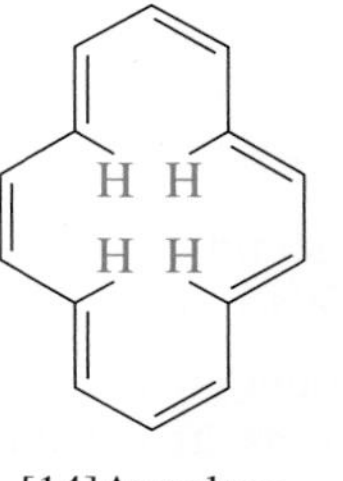

[14]Annulene

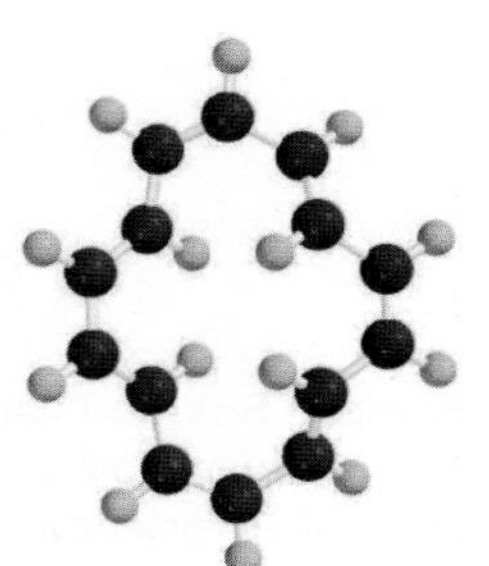

When the ring contains 18 carbon atoms, it is large enough to be planar while still allowing its interior hydrogens to be far enough apart so as to not interfere with one another. The [18]annulene shown is planar, or nearly so, and has all its carbon–carbon bond distances in the range 137–143 pm (1.37–1.43 Å), very much like those of benzene. Its resonance energy is estimated to be about 418 kJ/mol (100 kcal/mol). Although its structure and resonance energy attest to the validity of Hückel's rule, which predicts "special stability" for [18]annulene, its chemical reactivity does not. [18]Annulene behaves more like a polyene than like benzene in that it undergoes addition rather than substitution with bromine and forms a Diels–Alder adduct with maleic anhydride.

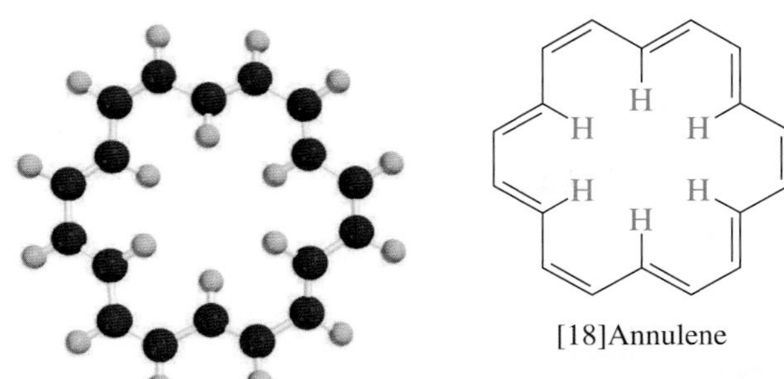

[18]Annulene

No serious repulsions among six interior hydrogens; molecule is planar and aromatic.

As noted earlier, planar annulenes with $4n$ π electrons are antiaromatic. A member of this group, [16]annulene, has been prepared. It is nonplanar and shows a pattern of alternating short (average 134 pm (1.34 Å)) and long (average 146 pm (1.46 Å)) bonds typical of a nonaromatic cyclic polyene.

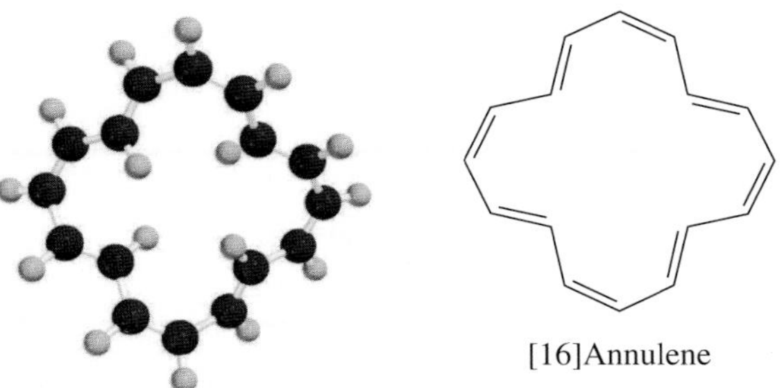
[16]Annulene

Problem 12.22

What does a comparison of the heats of combustion of benzene (3265 kJ/mol; 781 kcal/mol), cyclooctatetraene (4543 kJ/mol; 1086 kcal/mol), [16]annulene (9121 kJ/mol; 2182 kcal/mol), and [18]annulene (9806 kJ/mol; 2346 kcal/mol) reveal?

12.20 Aromatic Ions

Hückel realized that his molecular orbital analysis of conjugated systems could be extended beyond neutral hydrocarbons. He pointed out that cycloheptatrienyl cation, also called *tropylium ion,* contained a completely conjugated closed-shell six-π electron system analogous to that of benzene.

Benzene: completely conjugated, six π electrons delocalized over six carbons

Cycloheptatriene: lacks cyclic conjugation, interrupted by CH_2 group

Cycloheptatrienyl cation: completely conjugated, six π electrons delocalized over seven carbons

It is important to recognize the difference between the hydrocarbon cycloheptatriene and cycloheptatrienyl cation. The carbocation is aromatic; the hydrocarbon is not. Although cycloheptatriene has six π electrons in a conjugated system, the ends of the triene system are separated by an sp^3-hybridized carbon, which prevents continuous cyclic π electron delocalization.

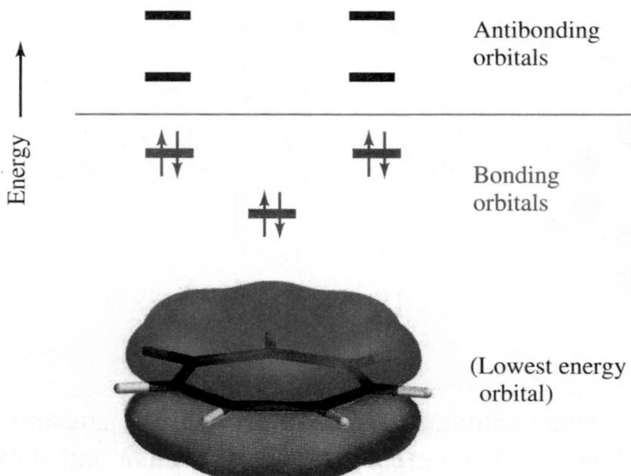

Figure 12.17

The π molecular orbitals of cycloheptatrienyl cation.

Figure 12.17 shows a molecular orbital diagram for cycloheptatrienyl cation. There are seven π MOs, three of which are bonding and contain the six π electrons of the cation. Cycloheptatrienyl cation is a Hückel ($4n + 2$) system and is an aromatic ion.

Problem 12.23

Show how you could adapt Frost's circle to generate the orbital energy level diagram shown in Figure 12.17 for cycloheptatrienyl cation.

Problem 12.24

Cycloheptatrienyl radical ($C_7H_7\cdot$) contains a cyclic, completely conjugated system of π electrons. Is it aromatic? Is it antiaromatic? Explain.

When we say cycloheptatriene is not aromatic but cycloheptatrienyl cation is, we are not comparing the stability of the two to each other. Cycloheptatriene is a stable hydrocarbon but does not possess the *special stability* required to be called *aromatic*. Cycloheptatrienyl cation, although aromatic, is still a carbocation and reasonably reactive toward nucleophiles. Its special stability does not imply a rock-like passivity, but rather a much greater ease of formation than expected on the basis of the Lewis structure drawn for it. A number of observations indicate that cycloheptatrienyl cation is far more stable than most other carbocations. To emphasize its aromatic nature, chemists often write the structure of cycloheptatrienyl cation in the Robinson circle-in-a-ring style.

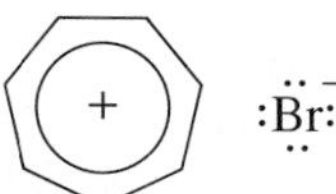

Tropylium bromide

For a novel cycloheptatriene derivative that occurs naturally, do an Internet search for *stipitatic acid*.

Tropylium bromide was first prepared, but not recognized as such, in 1891. The work was repeated in 1954, and the ionic properties of tropylium bromide were demonstrated. The ionic properties of tropylium bromide are apparent in its unusually high melting point (203°C), its solubility in water, and its complete lack of solubility in diethyl ether.

Problem 12.25

Write resonance structures for tropylium cation sufficient to show the delocalization of the positive charge over all seven carbons.

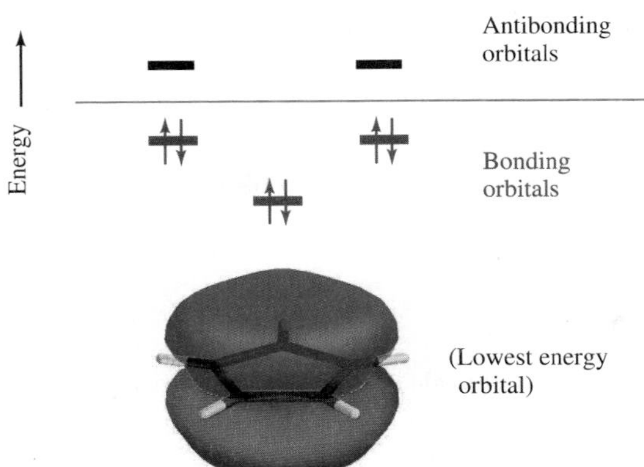

Figure 12.18

The π molecular orbitals of cyclopentadienyl anion.

The five-membered cyclopentadienyl system contrasts with cycloheptatrienyl. Cyclopentadienyl cation has four π electrons, is antiaromatic, very unstable, and very difficult to generate. Cyclopentadienyl anion, however, has six π electrons delocalized over five carbons and is aromatic.

Cyclopentadienyl cation   Cyclopentadienyl anion   is equivalent to

Figure 12.18 shows the MOs of cyclopentadienyl anion. Like benzene and cycloheptatrienyl cation, cyclopentadienyl anion has six π electrons and a closed-shell electron configuration.

Problem 12.26

Show how you could adapt Frost's circle to generate the orbital energy level diagram shown in Figure 12.18 for cyclopentadienyl anion.

The acidity of cyclopentadiene provides convincing evidence for the special stability of cyclopentadienyl anion.

Cyclopentadiene $pK_a = 16$ + Hydroxide ion ⇌ Cyclopentadienyl anion + Water $pK_a = 15.7$

With a pK_a of 16, cyclopentadiene is only a slightly weaker acid than water ($pK_a = 15.7$). It is much more acidic than other hydrocarbons—its K_a for ionization is 10^{10} times greater than acetylene, for example—because its conjugate base is aromatic and stabilized by electron delocalization.

Problem 12.27

Write resonance structures for cyclopentadienyl anion sufficient to show the delocalization of the negative charge over all five carbons.

There is a striking difference in the acidity of cyclopentadiene compared with cycloheptatriene. Cycloheptatriene has a pK_a of 36, which makes it 10^{20} times weaker in acid strength than cyclopentadiene.

Cycloheptatriene pK_a = 36 + Hydroxide ion ⇌ Cycloheptatrienyl anion + Water pK_a = 15.7

Even though resonance tells us that the negative charge in cycloheptatrienyl anion can be shared by all seven of its carbons, this delocalization offers little in the way of stabilization. Indeed with eight π electrons, cycloheptatrienyl anion is antiaromatic and relatively unstable.

Hückel's rule is now taken to apply to planar, monocyclic, completely conjugated systems generally, not just to neutral hydrocarbons.

A planar, monocyclic, continuous system of p orbitals possesses aromatic stability when it contains $(4n + 2)$ *π electrons.*

Other aromatic ions include cyclopropenyl cation (two π electrons) and cyclooctatetraene dianion (ten π electrons).

Cyclopropenyl cation

Cyclooctatetraene dianion

Here, we've taken liberties with the Robinson symbol. Instead of restricting it to a sextet of electrons, organic chemists use it as an all-purpose symbol for cyclic electron delocalization.

Problem 12.28

Is either of the following ions aromatic? Is either antiaromatic?

(a) Cyclononatetraenyl cation

(b) Cyclononatetraenide anion

Sample Solution (a) The crucial point is the number of π electrons in a cyclic conjugated system. If there are $(4n + 2)$ π electrons, the ion is aromatic. Electron counting is easiest if we write the ion as a single Lewis structure and remember that each double bond contributes two π electrons, a negatively charged carbon contributes two, and a positively charged carbon contributes none.

Cyclononatetraenyl cation has eight π electrons; it is *not aromatic.* It is antiaromatic ($4n = 8$; $n = 2$) if planar.

12.21 Heterocyclic Aromatic Compounds

Cyclic compounds that contain at least one atom other than carbon within their ring are called **heterocyclic compounds,** and those that possess aromatic stability are called **heterocyclic aromatic compounds.** Some representative heterocyclic aromatic compounds are *pyridine, pyrrole, furan,* and *thiophene.* The structures and the IUPAC numbering system used in naming their derivatives are shown. In their stability and chemical behavior, all these compounds resemble benzene more than they resemble alkenes.

Pyridine, pyrrole, and thiophene are present in coal tar. Furan is prepared from a substance called *furfural* obtained from corncobs.

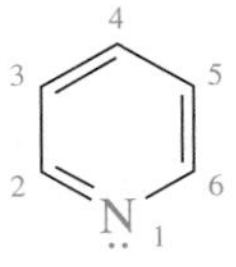

Pyridine 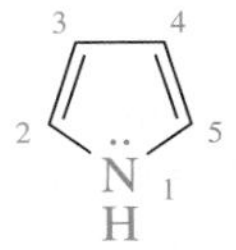 Pyrrole 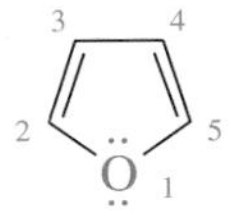 Furan 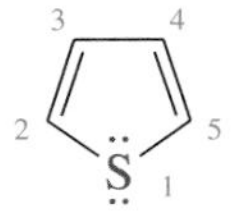 Thiophene

Heterocyclic aromatic compounds can be polycyclic as well. A benzene ring and a pyridine ring, for example, can share a common side in two different ways. One way gives a compound called *quinoline;* the other gives *isoquinoline.*

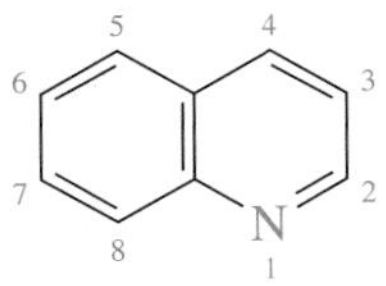

Quinoline 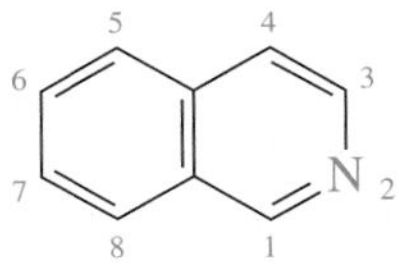 Isoquinoline

Analogous compounds derived by fusion of a benzene ring to a pyrrole, furan, or thiophene nucleus are called *indole, benzofuran,* and *benzothiophene.*

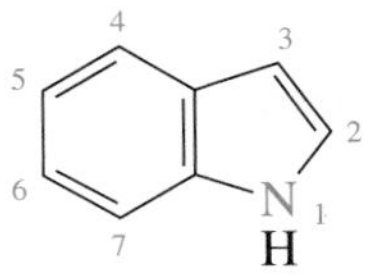

Indole 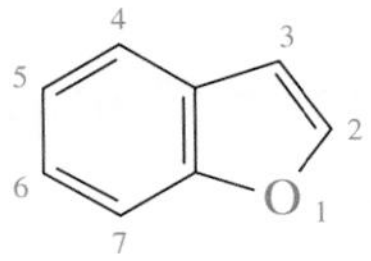 Benzofuran 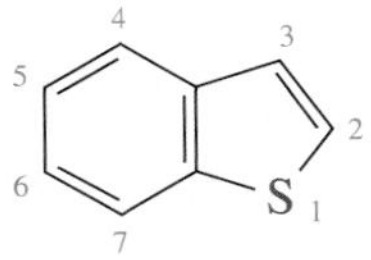 Benzothiophene

Problem 12.29

Unlike quinoline and isoquinoline, which are of comparable stability, the compounds indole and isoindole are quite different from each other. Which one is more stable? Explain the reason for your choice.

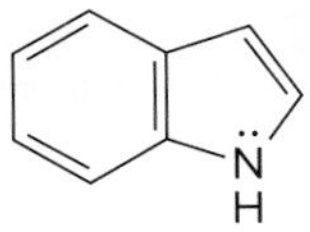

Indole 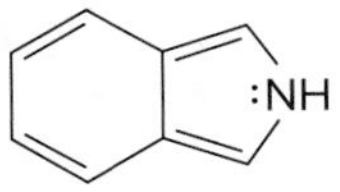 Isoindole

A large group of heterocyclic aromatic compounds are related to pyrrole by replacement of one of the ring carbons β to nitrogen by a second heteroatom. Compounds of this type are called *azoles.*

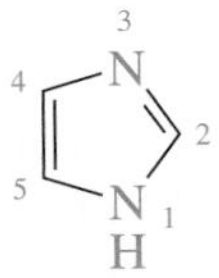

Imidazole 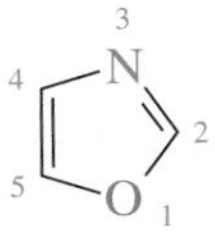 Oxazole 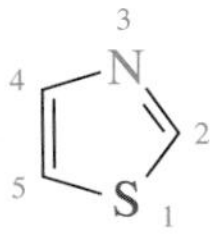 Thiazole

A widely prescribed drug for the treatment of gastric ulcers with the generic name *cimetidine* is a synthetic imidazole derivative. *Firefly luciferin* is a thiazole derivative that is the naturally occurring light-emitting substance present in fireflies.

Cimetidine Firefly luciferin

Firefly luciferin is an example of an azole that contains a benzene ring fused to the five-membered ring. Such structures are fairly common. Another example is *benzimidazole,* present as a structural unit in vitamin B_{12}. Some compounds related to benzimidazole include *purine* and its amino-substituted derivative *adenine,* one of the heterocyclic bases found in DNA and RNA (Chapter 26).

Benzimidazole Purine Adenine

Problem 12.30

Can you deduce the structural formulas of *benzoxazole* and *benzothiazole*?

12.22 Heterocyclic Aromatic Compounds and Hückel's Rule

Hückel's rule can be extended to heterocyclic aromatic compounds. A heteroatom such as oxygen or nitrogen can contribute either zero or two of its unshared electrons as needed to the π system so as to satisfy the $(4n + 2)$ π electron requirement.

The unshared pair in pyridine, for example, is not needed to satisfy the six π electron requirement for aromaticity, so is associated entirely with nitrogen and is not delocalized into the aromatic π system.

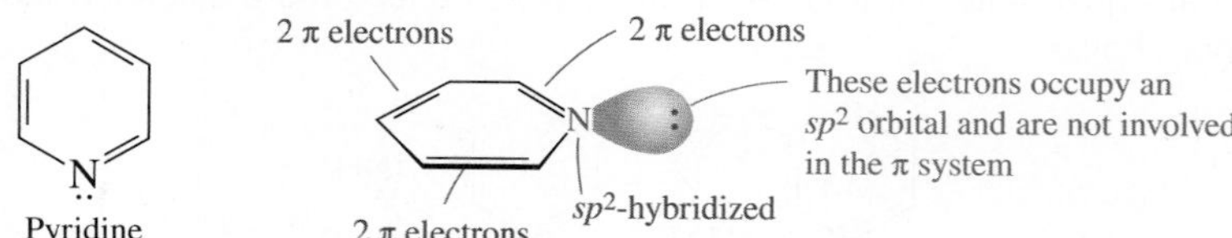

Pyridine

The unshared pair in the Lewis structure for pyrrole, on the other hand, must be added to the four π electrons of the two double bonds in order to meet the six π electron requirement.

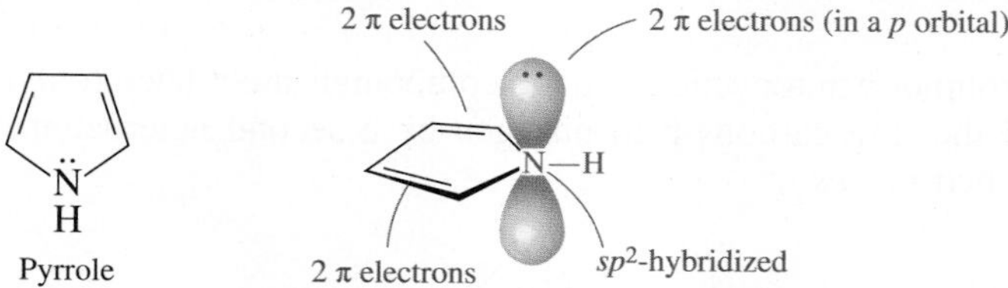

Pyrrole

In both pyridine and pyrrole the unshared electron pair occupies that orbital which provides the most stable structure. It is a different orbital in each case. In pyridine it is an sp^2-hybridized orbital localized on nitrogen. In pyrrole it is a *p* orbital of nitrogen that overlaps with the *p* orbitals of the ring carbons to give a delocalized π system.

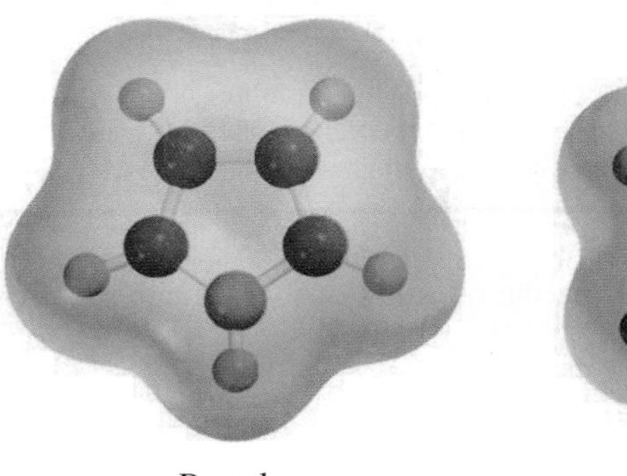

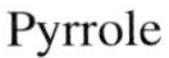

Pyrrole

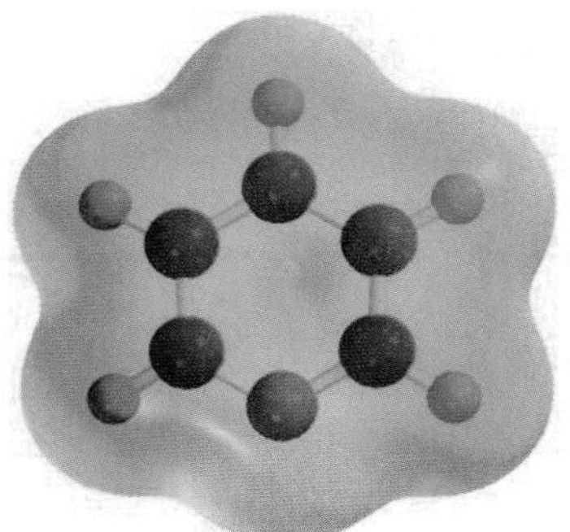

Pyridine

Figure 12.19

Electrostatic potential maps of pyridine and pyrrole. The color range is the same for both. In pyrrole the electron pair is delocalized into the π system of the ring. In pyridine the unshared electron pair is responsible for the concentration of electron density near nitrogen.

The electrostatic potential maps in Figure 12.19 show how pyridine and pyrrole differ with respect to their charge distribution. The unshared electron pair in pyridine gives rise to a region of high electron density (red) near nitrogen. A similar concentration of charge is absent in pyrrole because the corresponding electrons are delocalized among the five ring atoms.

The difference in bonding in pyridine and pyrrole is reflected in their properties. Although both are weak bases, pyridine is 10^7–10^9 times more basic than pyrrole. When pyridine acts as a Brønsted base, protonation of nitrogen converts an unshared pair (N:) to a bonded pair (N—H) while leaving the aromatic π system intact.

Pyridine + Water (H—O—H) $\underset{}{\overset{K = 10^{-10.5}}{\rightleftharpoons}}$ Pyridinium ion + Hydroxide ion

Weaker acid ($pK_a = 15.7$) — Stronger acid ($pK_a = 5.2$)

With pyrrole, however, the pair of electrons shown as an unshared pair in its Lewis formula is actually part of the aromatic π system. Were these two electrons to be involved in covalent bonding to a proton, all of the stabilization associated with aromaticity would be lost.

N: = N: but :N—H = N—H

Problem 12.31

Estimate the pK_a of the conjugate acid of pyrrole given that pyrrole is about 10^7–10^9 times less basic than pyridine and that the pK_a of the conjugate acid of pyridine is 5.2. Is the conjugate acid of pyridine strong or weak? What about the conjugate acid of pyrrole?

Imidazole is a heterocyclic aromatic compound with two nitrogens in a five-membered ring. One nitrogen has a pyridine-like unshared pair; the other has a pyrrole-like pair that is incorporated into the aromatic π system. Imidazole is somewhat more basic than pyridine. When imidazole acts as a Brønsted base, protonation of its pyridine-like nitrogen permits aromaticity to be retained by leaving the pyrrole-like nitrogen untouched.

Imidazole + Water (H—O—H) $\overset{K = 10^{-8.7}}{\rightleftharpoons}$ Imidazolium ion + Hydroxide ion

Weaker acid ($pK_a = 15.7$) — Stronger acid ($pK_a = 7$)

Problem 12.32

Refer to the structure of imidazolium ion in the preceding equation and write a second resonance contributor that obeys the octet rule and has its positive charge on the other nitrogen. Use curved arrows to show how you reorganized the electrons.

Turning to oxygen as a heteroatom, the question of two unshared pairs on the same atom arises.

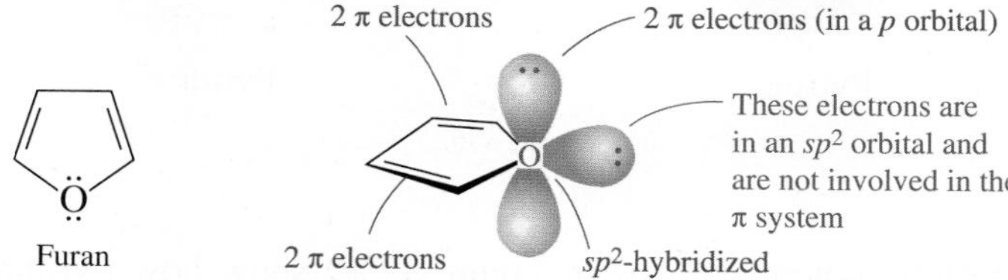

One pair is like the pair in pyrrole, occupying a *p* orbital and contributing two electrons to complete the six-π electron requirement for aromatic stabilization. The other electron pair in furan is an "extra" pair, not needed to satisfy the $4n + 2$ rule for aromaticity, and occupies an sp^2-hybridized orbital like the unshared pair in pyridine.

12.23 SUMMARY

Section 12.1 Benzene is the parent of a class of compounds called **arenes,** which are aromatic hydrocarbons.

Section 12.2 An important property of aromatic hydrocarbons is that they are much more stable and less reactive than other unsaturated compounds. The Kekulé formulas for benzene seem inconsistent with its low reactivity and with the fact that all of the C—C bonds in benzene are the same length (140 pm (1.40 Å)).

One explanation for the structure and stability of benzene and other arenes is based on resonance, according to which benzene is regarded as a hybrid of the two Kekulé structures.

Section 12.3 The extent to which benzene is more stable than either of the Kekulé structures is its **resonance energy,** which is estimated to be 152 kJ/mol (36 kcal/mol) from heats of hydrogenation data.

Section 12.4 According to the valence bond model, benzene has six π electrons, which are shared by all six sp^2-hybridized carbons. Regions of high π electron density are located above and below the plane of the ring.

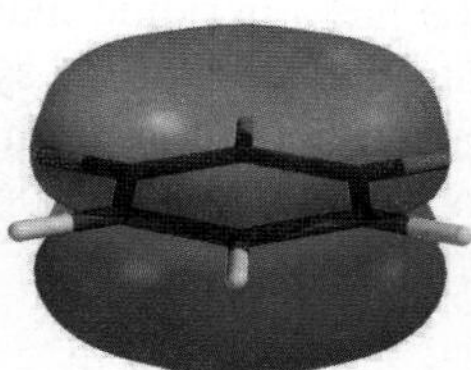

A molecular orbital description of benzene has three π orbitals that are bonding and three that are antibonding. Each of the bonding orbitals is fully occupied (two electrons each), and the antibonding orbitals are vacant.

Section 12.5 Many aromatic compounds are substituted derivatives of benzene and are named accordingly. Many others have names based on some other parent aromatic compound.

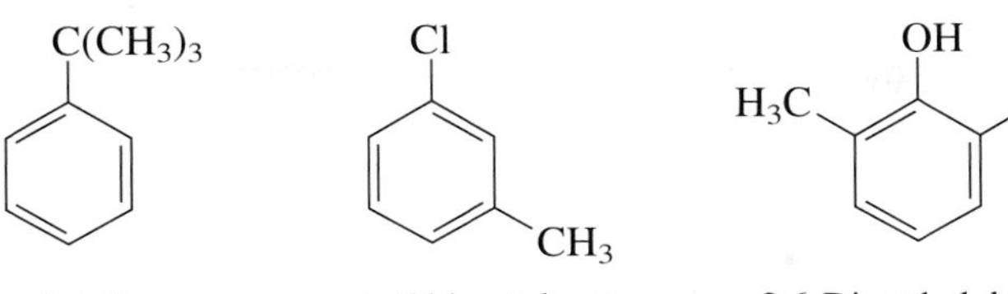

tert-Butylbenzene *m*-Chlorotoluene 2,6-Dimethylphenol

Section 12.6 **Polycyclic aromatic hydrocarbons,** of which anthracene is an example, contain two or more fused benzene rings.

Anthracene

Section 12.7 The physical properties of arenes resemble those of other hydrocarbons. Although weak, intermolecular attractive forces are somewhat stronger than those of other hydrocarbons of similar size.

Section 12.8 Chemical reactions of arenes can take place involving either the ring itself or a side chain. The most characteristic reactions proceed via benzylic carbocations, radicals, or anions.

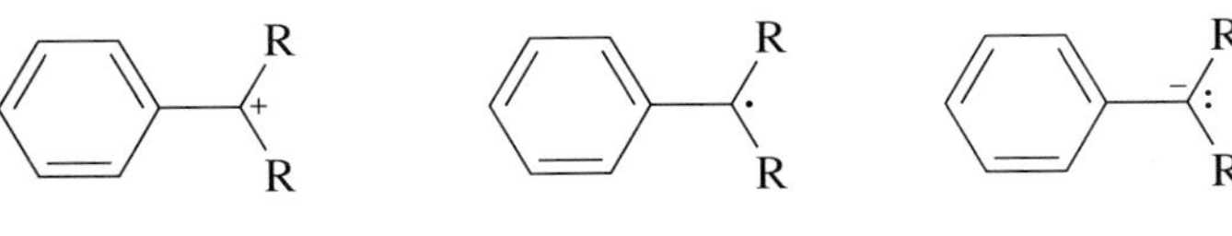

Benzylic carbocation Benzylic radical Benzylic anion

Sections 12.9–12.12 See Table 12.2.

Section 12.13 β eliminations that introduce double bonds that are conjugated to an aromatic ring occur readily.

OTs

KOH

ethanol, heat

+

cis-2-Phenylcyclohexyl *p*-toluenesulfonate

1-Phenylcyclohexene (87%)

3-Phenylcyclohexene (7%)

Examples of addition to these types of conjugated double bonds are shown in Table 12.2.

Section 12.14 Polystyrene is a widely used vinyl polymer prepared by the free-radical polymerization of styrene.

Polystyrene

TABLE 12.2 Reactions Involving Alkyl and Alkenyl Side Chains in Arenes and Arene Derivatives

Reaction (Section) and Comments	General Equation and Specific Example
Nucleophilic substitution (Section 12.9) S_N1 and S_N2 reactions of benzylic halides occur faster than the corresponding reactions of alkyl halides. Three nucleophilic substitutions are involved in the synthesis shown. The benzylic alcohol is converted to the corresponding chloride with HCl, then to the corresponding cyanide. Sodium iodide catalyzes the second reaction by converting the benzylic chloride to the more reactive iodide.	$ArCR_2LG \xrightarrow{\text{nucleophile}} ArCR_2Y$ LG = leaving group p-Methoxybenzyl alcohol $\xrightarrow[\text{2. NaCN, NaI(cat), acetone, heat}]{\text{1. HCl}}$ p-Methoxybenzyl cyanide (74–81%)
Free-radical halogenation (Section 12.10) Side-chain halogenation of alkylbenzenes is regiospecific for substitution at the benzylic carbon. Suitable reagents include elemental bromine and *N*-bromosuccinimide (NBS).	$ArCHR_2$ (Arene) $\xrightarrow[\text{benzoyl peroxide, } CCl_4\text{, 80°C}]{\text{NBS}}$ $ArCBrR_2$ (1-Arylalkyl bromide) *p*-Ethylnitrobenzene $\xrightarrow[CCl_4\text{, light}]{Br_2}$ 1-(*p*-Nitrophenyl)ethyl bromide (77%)
Acidity (Section 12.11) Alkylbenzenes are very weak acids but much stronger than alkanes. The pK_a of toluene, for example, is 42.	$ArCHR_2$ (Arylalkane) $\rightleftharpoons$ $Ar\ddot{C}R_2^-$ (Conjugate base) + H^+ (Proton) Toluene (weaker acid: pK_a = 42) + *tert*-Butoxide ion $\underset{K = 10^{-24}}{\rightleftharpoons}$ Benzyl anion + *tert*-Butyl alcohol (stronger acid: pK_a = 18)
Oxidation (Section 12.12) Oxidation of alkylbenzenes occurs at the benzylic carbon of the alkyl group and gives a benzoic acid derivative. Oxidizing agents include sodium or potassium dichromate in aqueous sulfuric acid. Potassium permanganate ($KMnO_4$) is also effective.	$ArCHR_2$ (Arene) $\xrightarrow{\text{oxidize}}$ $ArCO_2H$ (Arenecarboxylic acid) 2,4,6-Trinitrotoluene $\xrightarrow[H_2SO_4\text{, }H_2O\text{, heat}]{Na_2Cr_2O_7}$ 2,4,6-Trinitrobenzoic acid (57–69%)

TABLE 12.2 Reactions Involving Alkyl and Alkenyl Side Chains in Arenes and Arene Derivatives (*Continued*)

Reaction (Section) and Comments	General Equation and Specific Example
Hydrogenation (Section 12.13) Hydrogenation of aromatic rings is somewhat slower than hydrogenation of alkenes, and it is a simple matter to reduce the double bond of an unsaturated side chain in an arene while leaving the ring intact.	$ArCH{=}CR_2$ + H_2 $\xrightarrow{Pt}$ $ArCH_2CHR_2$ Alkenylarene, Hydrogen, Alkylarene 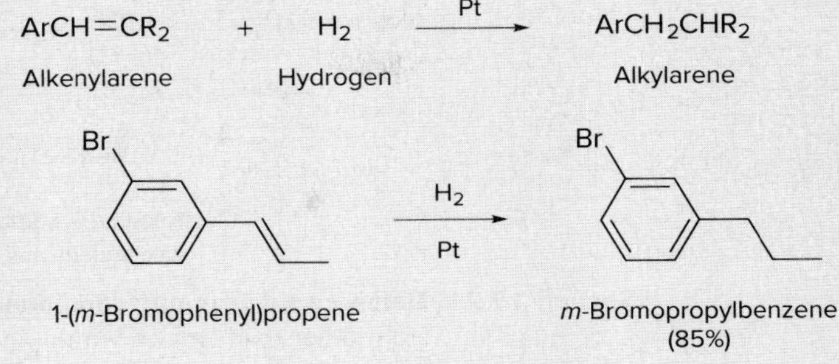 1-(*m*-Bromophenyl)propene $\xrightarrow[Pt]{H_2}$ *m*-Bromopropylbenzene (85%)
Electrophilic addition (Section 12.14) An aryl group stabilizes a benzylic carbocation and controls the regioselectivity of addition to a double bond involving the benzylic carbon. Markovnikov's rule is followed.	$ArCH{=}CH_2$ + $\overset{\delta+}{E}{-}\overset{\delta-}{Y}$ ⟶ $ArCHCH_2E$ (with Y on the benzylic carbon) Alkenylarene, Electrophile, Product of electrophilic addition Styrene $\xrightarrow{HBr}$ 1-Phenylethyl bromide (85%)

Section 12.15 An example of a reaction in which the ring itself reacts is the **Birch reduction.** The ring of an arene is reduced to a nonconjugated diene by treatment with a Group 1 metal (usually sodium) in liquid ammonia in the presence of an alcohol.

o-Xylene $\xrightarrow[CH_3OH]{Na,\ NH_3}$ 1,2-Dimethyl-1,4-cyclohexadiene (92%)

Section 12.16 The choice of synthetic routes to aromatic compounds is strongly influenced by considerations of the effect of the benzylic carbon on rate and regioselectivity.

Section 12.17 Although cyclic conjugation is a necessary requirement for aromaticity, this alone is not sufficient. If it were, cyclobutadiene and cyclooctatetraene would be aromatic. They are not.

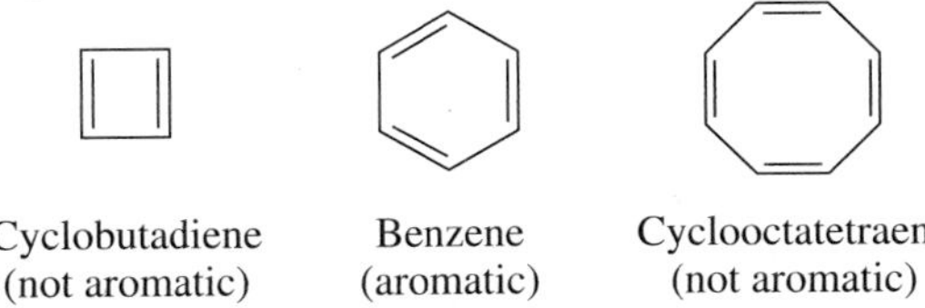

Cyclobutadiene (not aromatic) Benzene (aromatic) Cyclooctatetraene (not aromatic)

Section 12.18 An additional requirement for aromaticity is that the number of π electrons in conjugated, planar, monocyclic species must be equal to $4n + 2$, where n is an integer. This is called **Hückel's rule.** Benzene, with six π electrons, satisfies Hückel's rule for $n = 1$. Square cyclobutadiene (four π electrons) and planar cyclooctatetraene (eight π electrons) do not. Both are examples of systems with $4n$ π electrons and are **antiaromatic.**

Section 12.19 **Annulenes** are monocyclic, completely conjugated polyenes synthesized for the purpose of testing Hückel's rule. They are named by using a bracketed numerical prefix to indicate the number of carbons, followed by the word *annulene.* [4*n*]-Annulenes are characterized by rings with alternating short (double) and long (single) bonds and are *antiaromatic*. The expected aromaticity of $[4n + 2]$-annulenes is diminished by angle and van der Waals strain unless the ring contains 18 or more carbons.

Section 12.20 Species with six π electrons that possess "special stability" include certain ions, such as *cyclopentadienide* anion and *cycloheptatrienyl* cation.

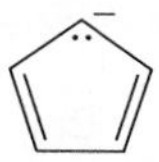

Cyclopentadienide anion (six π electrons)

Cycloheptatrienyl cation (six π electrons)

Section 12.21 **Heterocyclic aromatic compounds** are compounds that contain at least one atom other than carbon within an aromatic ring.

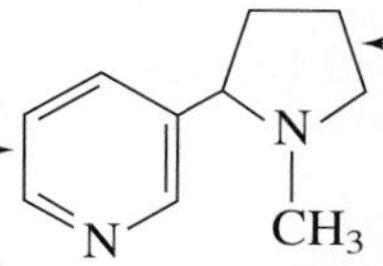

Ring is heterocyclic and aromatic.

Ring is heterocyclic but not aromatic.

CH_3

Nicotine

Section 12.22 Hückel's rule can be extended to heterocyclic aromatic compounds. Unshared electron pairs of the heteroatom may be used as π electrons as necessary to satisfy the $4n + 2$ rule.

PROBLEMS

Structure and Nomenclature

12.33 Write structural formulas and give the IUPAC names for all the isomers of $C_6H_5C_4H_9$ that contain a monosubstituted benzene ring.

12.34 Write a structural formula corresponding to each of the following:

(a) Allylbenzene
(b) (*E*)-1-Phenyl-1-butene
(c) (*Z*)-2-Phenyl-2-butene
(d) (*R*)-1-Phenylethanol
(e) *o*-Chlorobenzyl alcohol
(f) *p*-Chlorophenol
(g) 2-Nitrobenzenecarboxylic acid
(h) *p*-Diisopropylbenzene
(i) 2,4,6-Tribromoaniline
(j) *m*-Nitroacetophenone
(k) 4-Bromo-3-ethylstyrene

12.35 Using numerical locants and the names in Table 12.1 as a guide, give an acceptable IUPAC name for each of the following compounds:

(a) Estragole (principal component of wormwood oil)

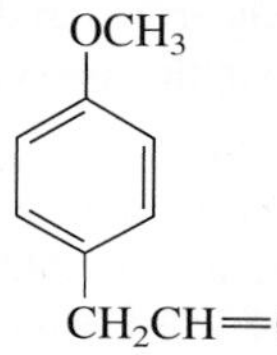

OCH_3

$CH_2CH{=}CH_2$

(b) Diosphenol (used in veterinary medicine to control parasites in animals)

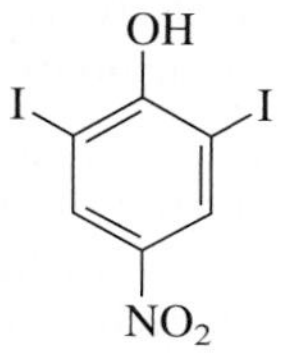

OH

I I

NO_2

(c) *m*-Xylidine (used in synthesis of lidocaine, a local anesthetic)

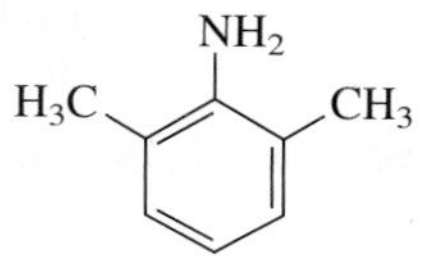

NH_2

H_3C CH_3

12.36 Write structural formulas and give acceptable names for all the isomeric

(a) Nitrotoluenes
(b) Dichlorobenzoic acids
(c) Tribromophenols
(d) Tetrafluorobenzenes
(e) Naphthalenecarboxylic acids

12.37 *Nitroxoline* is the generic name by which 5-nitro-8-hydroxyquinoline is sold as an antibacterial drug. Write its structural formula.

12.38 *Acridine* is a heterocyclic aromatic compound obtained from coal tar that is used in the synthesis of dyes. The molecular formula of acridine is $C_{13}H_9N$, and its ring system is analogous to that of anthracene except that one CH group has been replaced by N. The two most stable resonance structures of acridine are equivalent to each other, and both contain a pyridine-like structural unit. Write a structural formula for acridine.

Resonance, Aromaticity, and Mechanism

12.39 Each of the following may be represented by at least one alternative resonance structure in which all the six-membered rings correspond to Kekulé forms of benzene. Write such a resonance form for each.

(a) 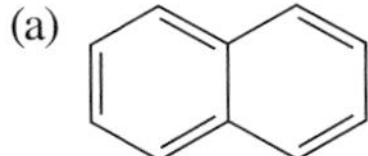
(b) 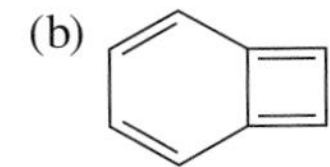
(c) 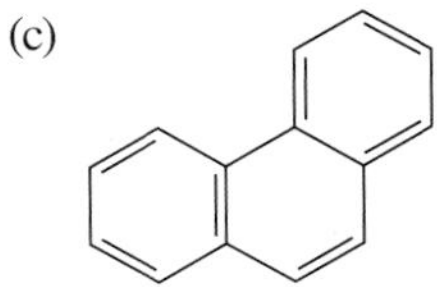

12.40 Cyclooctatetraene has two different tetramethyl derivatives with methyl groups on four adjacent carbon atoms. They are both completely conjugated and are not stereoisomers. Write their structures.

12.41 Evaluate each of the following processes applied to cyclooctatetraene, and decide whether the species formed is aromatic or not.

(a) Addition of one more π electron, to give $C_8H_8^-$
(b) Addition of two more π electrons, to give $C_8H_8^{2-}$
(c) Removal of one π electron, to give $C_8H_8^+$
(d) Removal of two π electrons, to give $C_8H_8^{2+}$

12.42 Evaluate each of the following processes applied to cyclononatetraene, and decide whether the species formed is aromatic or not.

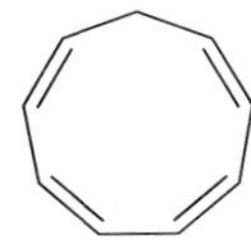
Cyclononatetraene

(a) Addition of one more π electron, to give $C_9H_{10}^-$
(b) Addition of two more π electrons, to give $C_9H_{10}^{2-}$
(c) Loss of H^+ from the sp^3-hybridized carbon
(d) Loss of H^+ from one of the sp^2-hybridized carbons

12.43 (a) Figure 12.20 is an electrostatic potential map of *calicene,* so named because its shape resembles a chalice (*calix* is the Latin word for "cup"). Both the electrostatic potential map and its calculated dipole moment (μ = 4.3 D) indicate that calicene is an unusually polar hydrocarbon. Which of the dipolar resonance forms, A or B, better corresponds to the electron distribution in the molecule? Why is this resonance form more important than the other?

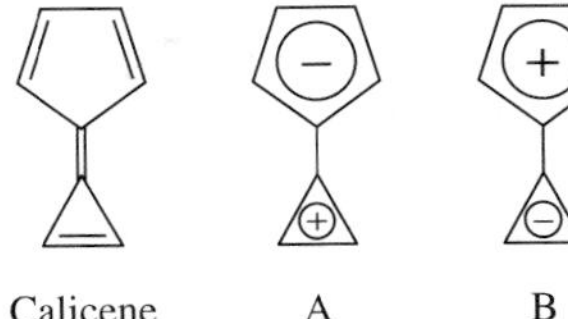

(b) Which one of the following should be stabilized by resonance to a greater extent? (*Hint:* Consider the reasonableness of dipolar resonance forms.)

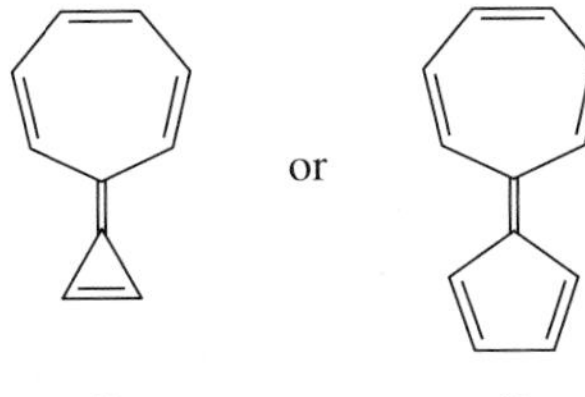

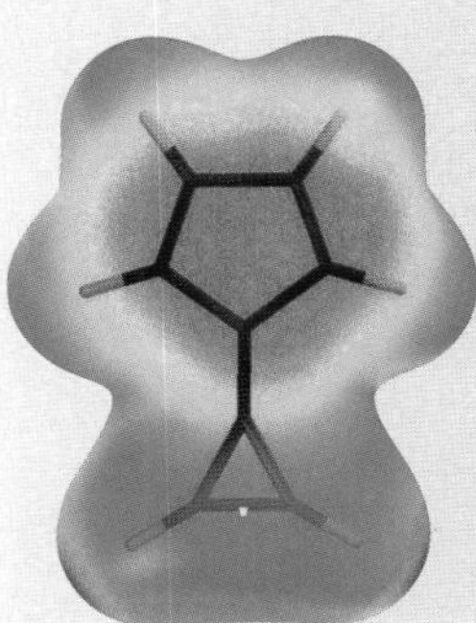

Figure 12.20

Electrostatic potential map of calicene (Problem 12.43).

12.44 Like calicene (Problem 12.43), the hydrocarbon azulene is planar and its electron distribution can be represented by a resonance contributor in which both rings satisfy Hückel's rule. Which is the major contributor, A or B?

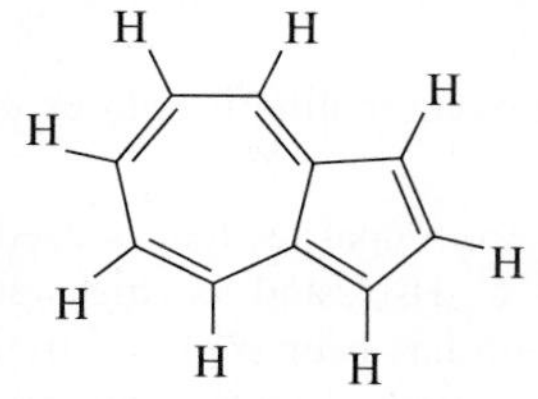

Azulene 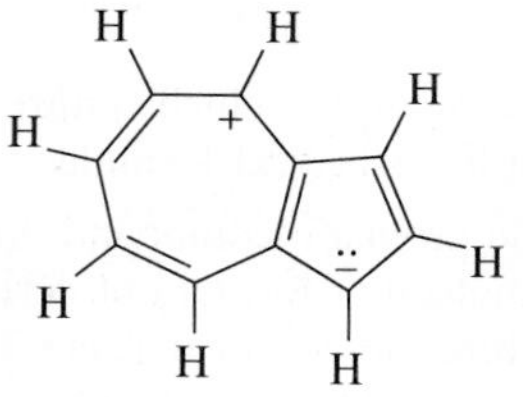 A 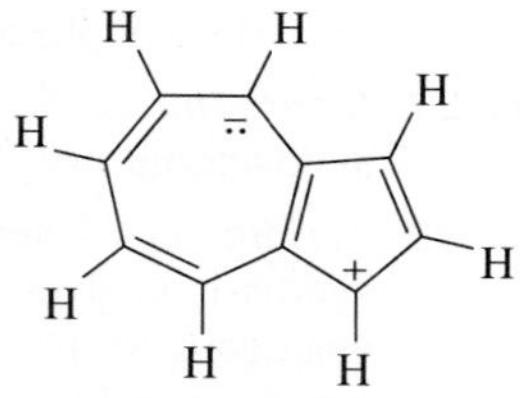 B

12.45 Classify each of the following molecules as aromatic or not, according to Hückel's rule. Are any antiaromatic?

(a) (c) (b) (d)

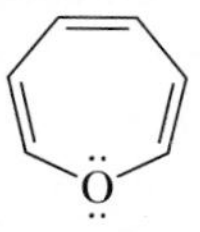

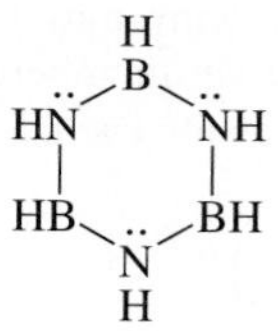

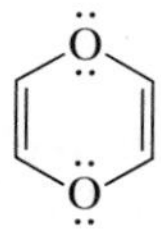

Reactions and Synthesis

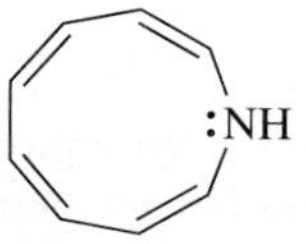

12.46 Bromine adds to the central ring of anthracene to give a 1,4-addition product. Write the structure of the product that would be formed if addition took place on one of the outer rings. By writing resonance structures for the product shown here and the one formed by addition to the outer ring, can you suggest why addition to the central ring is preferred?

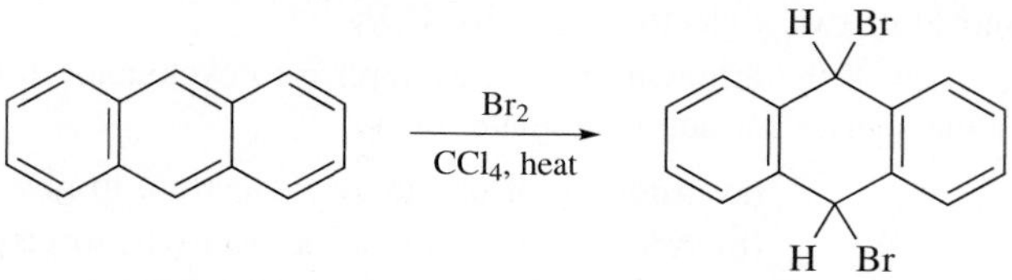

$\xrightarrow[CCl_4,\ heat]{Br_2}$

12.47 Anthracene undergoes a Diels–Alder reaction with maleic anhydride to give a cycloadduct with the formula $C_{18}H_{12}O_3$. What is its structure?

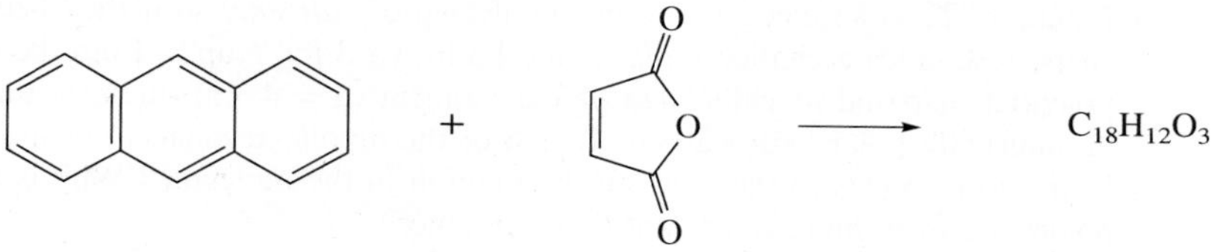

$+ \longrightarrow C_{18}H_{12}O_3$

12.48 As in the free-radical halogenation of alkanes, chlorination of alkylbenzenes is less selective than bromination. Given the relative rates per hydrogen for hydrogen atom abstraction from 1-phenylbutane by chlorine for the elementary step shown, calculate the percentage of 1-chloro-1-phenylbutane in the $C_{10}H_{13}Cl$ product.

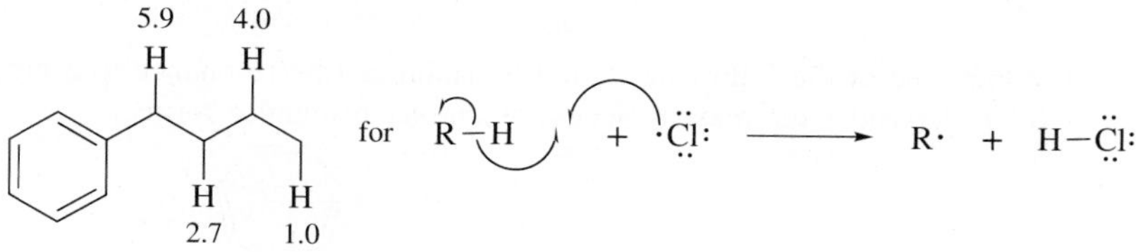

5.9 4.0 2.7 1.0

for $R{-}H + \cdot\ddot{Cl}: \longrightarrow R\cdot + H{-}\ddot{Cl}:$

12.49 Birch reduction of 2-methoxynaphthalene gave a mixture of two isomeric compounds, each having the molecular formula $C_{11}H_{14}O$. Suggest reasonable structures for these compounds.

12.50 The relative rates of reaction of ethane, toluene, and ethylbenzene with bromine atoms have been measured. The most reactive hydrocarbon undergoes hydrogen atom abstraction a million times faster than does the least reactive one. Arrange these hydrocarbons in order of decreasing reactivity.

12.51 Both 1,2-dihydronaphthalene and 1,4-dihydronaphthalene may be selectively hydrogenated to 1,2,3,4-tetrahydronaphthalene.

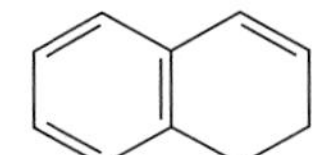
$\xrightarrow[\text{Pt}]{H_2}$
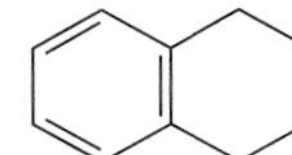
$\xleftarrow[\text{Pt}]{H_2}$
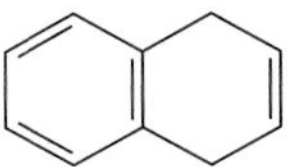

1,2-Dihydronaphthalene 1,2,3,4-Tetrahydronaphthalene 1,4-Dihydronaphthalene

One of these isomers has a heat of hydrogenation of 101 kJ/mol (24.1 kcal/mol), and the heat of hydrogenation of the other is 113 kJ/mol (27.1 kcal/mol). Match the heat of hydrogenation with the appropriate dihydronaphthalene.

12.52 Suggest an explanation for the observed order of S_N1 reactivity of the following compounds.

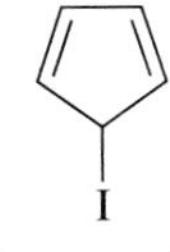
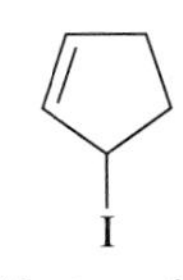

Least reactive Most reactive

12.53 A standard method for preparing sodium cyclopentadienide (C_5H_5Na) is by the reaction of cyclopentadiene with a solution of $NaNH_2$ in liquid ammonia. Write a net ionic equation for this reaction, identify the acid and the base, and use curved arrows to track the flow of electrons.

12.54 The same anion is formed by loss of the most acidic proton from 1-methyl-1,3-cyclopentadiene as from 5-methyl-1,3-cyclopentadiene. Explain.

12.55 Furan is less stabilized by resonance than benzene and undergoes a 1,4 addition of bromine to give an unstable dibromide $C_4H_4Br_2O$. What is the structure of this compound?

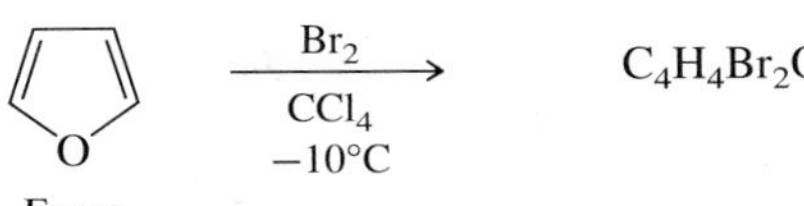

Furan

12.56 Give the structure of the expected product from the reaction of isopropylbenzene with

(a) Hydrogen (3 mol), Pt

(b) Sodium and ethanol in liquid ammonia

(c) Sodium dichromate, water, sulfuric acid, heat

(d) *N*-Bromosuccinimide in CCl_4, heat, benzoyl peroxide

(e) The product of part (d) treated with sodium ethoxide in ethanol

12.57 Each of the following reactions has been described in the chemical literature and gives a single organic product in good yield. Identify the product of each reaction.

(a) $\xrightarrow[\text{2. } H_2O_2,\ HO^-]{\text{1. } B_2H_6,\text{ diglyme}}$

(b) + H_2 (1 mol) $\xrightarrow{\text{Pt}}$

(c) $(C_6H_5)_2CH$—C6H4—CH_3 $\xrightarrow[CCl_4,\text{ light}]{\text{excess } Cl_2}$ $C_{20}H_{14}Cl_4$

(d) 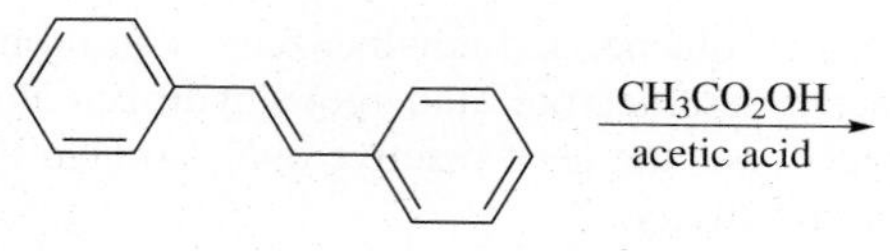 [structure] $\xrightarrow[\text{acetic acid}]{CH_3CO_2OH}$

(e) 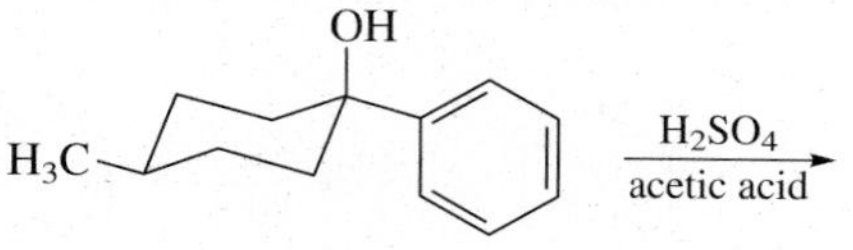 H_3C– [structure with OH] $\xrightarrow[\text{acetic acid}]{H_2SO_4}$

(f) 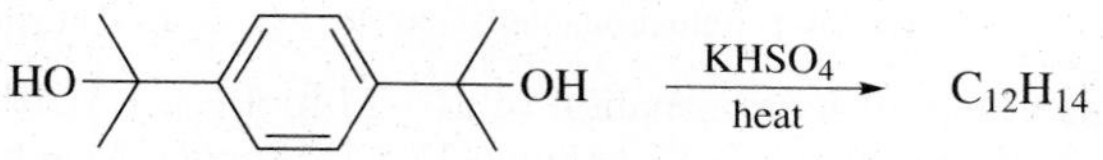 HO– [structure] –OH $\xrightarrow[\text{heat}]{KHSO_4}$ $C_{12}H_{14}$

(g) 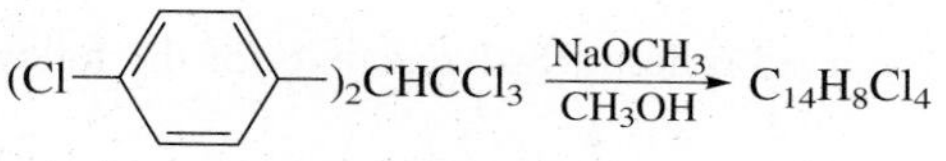 (Cl– [structure] –$)_2CHCCl_3$ $\xrightarrow[CH_3OH]{NaOCH_3}$ $C_{14}H_8Cl_4$

(DDT)

(h) 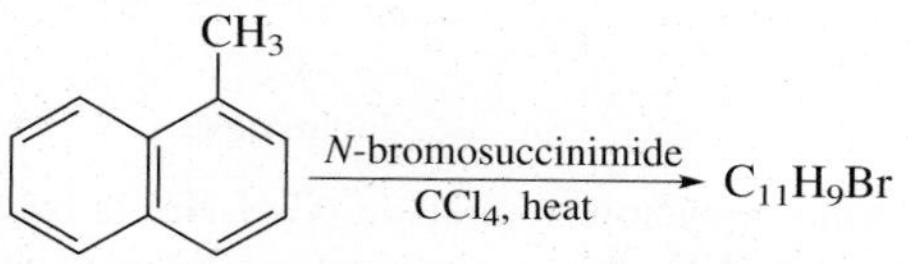 [structure with CH_3] $\xrightarrow[\text{CCl}_4\text{, heat}]{N\text{-bromosuccinimide}}$ $C_{11}H_9Br$

(i) 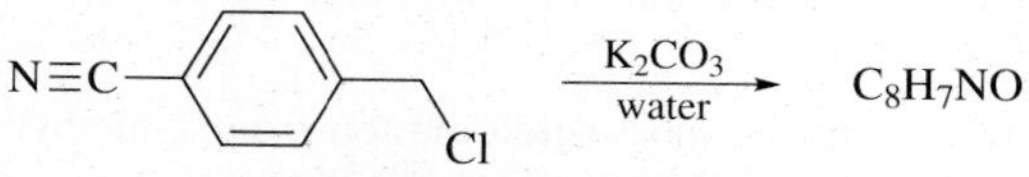 N≡C– [structure] –Cl $\xrightarrow[\text{water}]{K_2CO_3}$ C_8H_7NO

12.58 A certain compound A, when treated with *N*-bromosuccinimide and benzoyl peroxide under photochemical conditions in refluxing carbon tetrachloride, gave 3,4,5-tribromobenzyl bromide in excellent yield. Deduce the structure of compound A.

12.59 A compound was obtained from a natural product and had the molecular formula $C_{14}H_{20}O_3$. It contained three methoxy (—OCH_3) groups and a —$CH_2CH{=}C(CH_3)_2$ substituent. Oxidation with either chromic acid or potassium permanganate gave 2,3,5-trimethoxybenzoic acid. What is the structure of the compound?

12.60 Hydroboration–oxidation of (*E*)-2-(*p*-anisyl)-2-butene yielded an alcohol A, mp 60°C, in 72% yield. When the same reaction was performed on the *Z* alkene, an isomeric liquid alcohol B was obtained in 77% yield. Suggest reasonable structures for A and B, and describe the relationship between them.

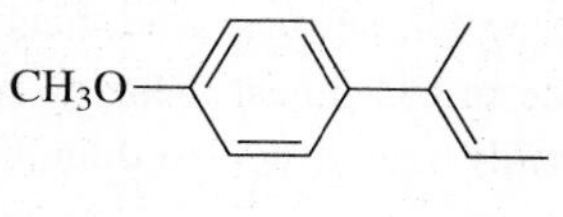

CH_3O– [structure]

(*E*)-2-(*p*-Anisyl)-2-butene

12.61 Suggest reagents suitable for carrying out each of the following conversions. In most cases more than one synthetic operation will be necessary.

(a)  C_6H_5 [structure] ⟶ Br, C_6H_5 [structure]

(b) Br, C_6H_5 [structure] ⟶ Br, C_6H_5, Br [structure]

(c) C_6H_5 [structure] ⟶ C_6H_5–≡

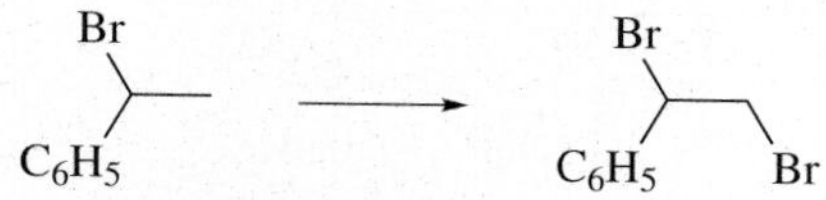

(d) 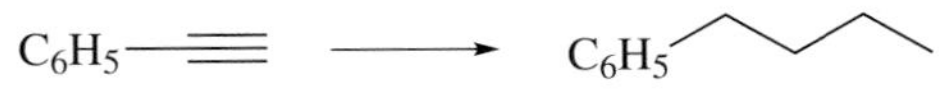 $C_6H_5-\equiv$ ⟶ C_6H_5-CH₂CH₂CH₂CH₃

(e)  $C_6H_5CH_2CH_2OH$ ⟶ $C_6H_5CH_2CH_2C\equiv CH$

(f)  $C_6H_5CH_2CH_2Br$ ⟶ $C_6H_5CH(OH)CH_2Br$

12.62 Pellagra is a disease caused by a deficiency of *niacin* ($C_6H_5NO_2$) in the diet. Niacin can be synthesized in the laboratory by the side-chain oxidation of 3-methylpyridine with chromic acid or potassium permanganate. Suggest a reasonable structure for niacin.

Descriptive Passage and Interpretive Problems 12

Substituent Effects on Reaction Rates and Equilibria

We have seen numerous examples of substituent effects on rates and equilibria of organic reactions and have developed a *qualitative* feel for steric and electronic effects as important factors. What about their *quantitative* treatment?

The first widely accepted method dealt with electronic effects and dates to the 1930s when Louis P. Hammett (Columbia University) developed an approach represented in the equations:

$$\log \frac{k}{k_0} = \sigma\rho \quad \text{and} \quad \log \frac{K}{K_0} = \sigma\rho$$

where k and k_0 are rate constants and K and K_0 are equilibrium constants. σ and ρ are experimentally determined constants characteristic of a substituent (σ) and a reaction (ρ). The standard substituent is H and is assigned $\sigma = 0$. The standard reaction, assigned a value of $\rho = 1.0$, is the ionization of meta- and para-substituted benzoic acids.

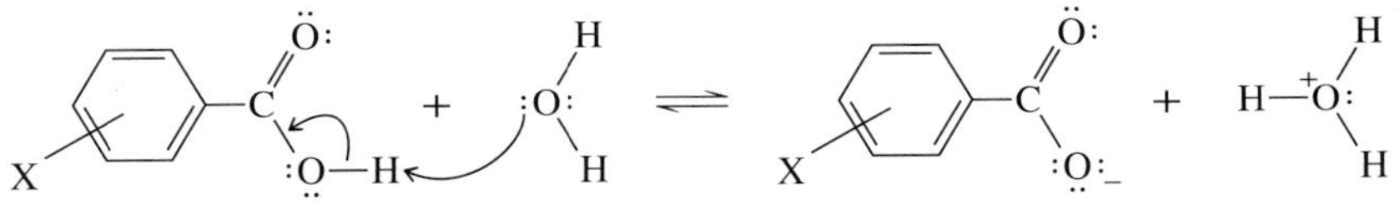

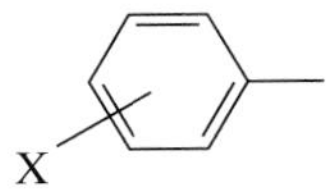

X-C_6H_4-C(=Ö:)-Ö-H + H-Ö:-H ⇌ X-C_6H_4-C(=Ö:)-Ö:$^-$ + $H-\overset{+}{O}H_2$

X-C_6H_4— is a meta- or para-substituted aryl group

When X is electron-withdrawing, the sign of σ is + and the acid is stronger than benzoic acid. Conversely, σ is negative for electron-releasing, acid-weakening groups. For individual substituents, σ differs according to whether it is meta or para to the reaction site. Experimental measures of K_a for a variety of substituted benzoic acids gave the following **substituent constants.** There are no σ_{ortho} values because of the possibility of steric interactions.

Substituent	$-NO_2$	$-CN$	$-CF_3$	$-Cl$	$-F$	$-OCH_3$	$-CH_3$
σ_{meta}	0.71	0.62	0.46	0.37	0.34	0.10	−0.06
σ_{para}	0.81	0.70	0.53	0.24	0.15	−0.12	−0.14

Example: Which of the following is the strongest acid? Which is the weakest?

m-Methylbenzoic acid | *p*-(Trifluoromethyl)benzoic acid
m-Fluorobenzoic acid | *p*-Methoxybenzoic acid

Answer: By definition: $\rho = +1.0$ for the ionization of substituted benzoic acids; therefore, the acid with the most positive value of σ for its substituent is strongest. Among the substituents *p*-F_3C with $\sigma = 0.53$ is the strongest electron-withdrawing group; therefore, *p*-(trifluoromethyl)benzoic acid is the strongest acid. *m*-Methylbenzoic acid is the weakest.

Experimentally determined substituent constants for more than 100 other groups have been determined and used to determine values of ρ for various reactions, resulting in a rich library of σ_{meta}, σ_{para}, and ρ values that can be applied to specific questions of reaction mechanisms, rates, and equilibria.

Example: ρ for the E2 elimination of a series of 2-arylethyl bromides with sodium ethoxide in ethanol is +2.1.

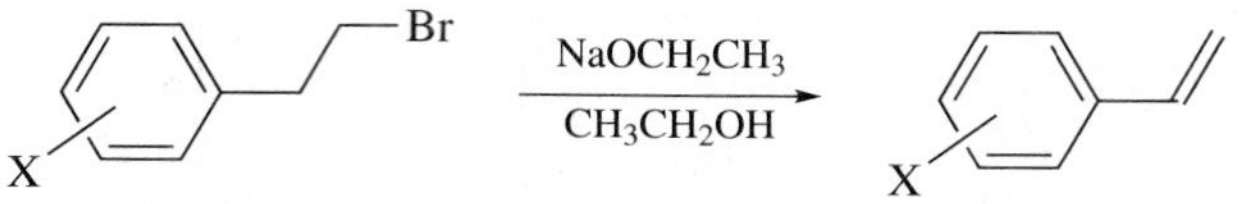

$NaOCH_2CH_3$ / CH_3CH_2OH

Which reacts at the faster rate: Compound A or Compound B?

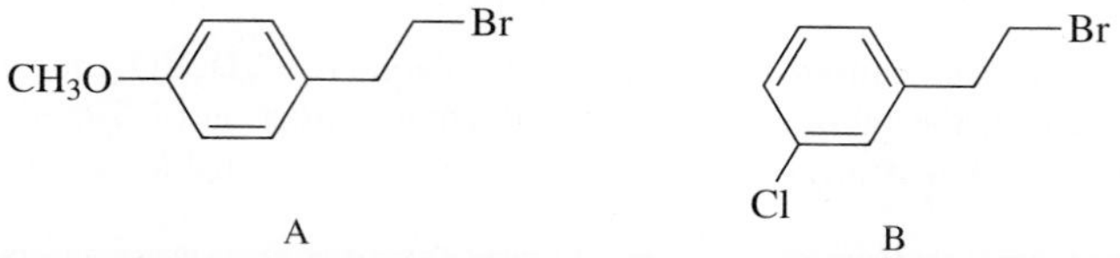

A (CH$_3$O–, Br) B (Cl, Br)

Answer: The positive value for ρ (+2.1) tells us that the reaction rate increases with increasingly positive values of σ. The substituent constant σ_{para} for —OCH_3 = −0.12 while σ_{meta} for Cl = 0.37; therefore, B reacts faster than A.

With respect to reaction mechanisms, the major application of the Hammett equation lies in using ρ as an indicator of the direction and degree of charge development in the transition state. In the E2 example just described (ρ = +2.1), we conclude that the substituted benzene ring is acting as an electron-withdrawing group and infer that C—H bond-breaking is more advanced than C—Br bond-breaking in the rate-determining transition state.

Historically, the Hammett equation provided the foundation for related applications, especially one developed by Robert Taft (Penn State, University of California Irvine). Taft adapted Hammett's approach so as to apply to aliphatic compounds by deriving an appropriate set of substituent constants (σ*) and including a term (δE_s) for steric effects where δ is a steric sensitivity factor for the reaction and E_s is a steric substituent constant.

$$\log \frac{k}{k_0} = \sigma^*\rho^* + \delta E_s \quad \text{and} \quad \log \frac{K}{K_0} = \sigma^*\rho^* + \delta E_s$$

Others, especially Marvin Charton (Pratt Institute), expanded the scope of what had come to be called *linear free-energy relationships* to more broadly based structure-activity correlations.

A potentially very important recent advance integrates stereochemistry into structure-activity correlations for the purpose of identifying reactants and catalysts best suited to produce chiral molecules with high levels of enantiomeric excess. As developed by Matthew Sigman (University of Utah), it applies relatively complex mathematical methods to identify the optimum experimental conditions for enantioselective synthesis. For example, in the reaction:

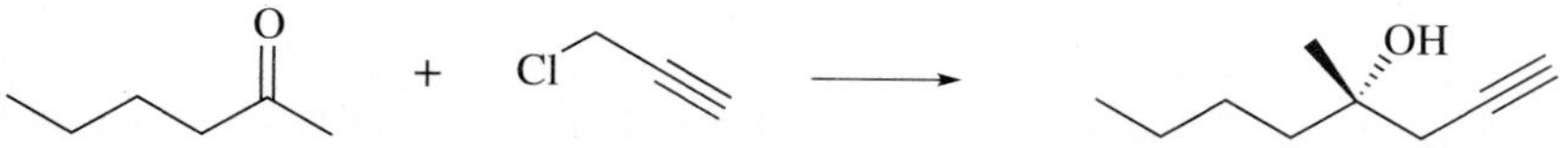

O + Cl → OH

70% yield; 70% enantiomeric excess

the reactants included not only the two shown in the equation, but also five other components in addition to an enantiomerically homogeneous catalyst chosen from a number of candidates.

12.63 When Ar is a m- or p-substituted phenyl group, the value of ρ for the following reaction is –5.1.

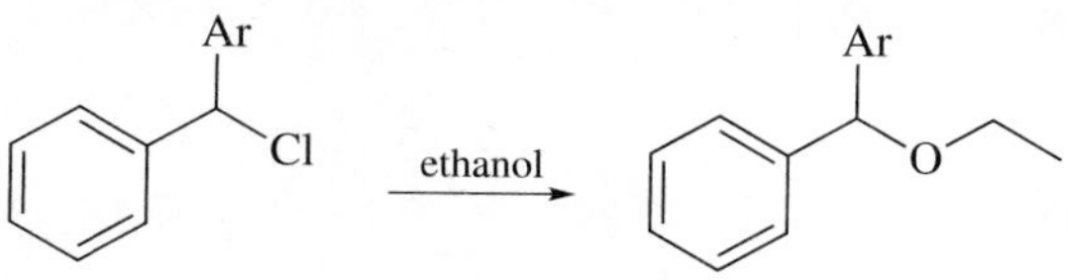

Ar, Cl → (ethanol) → Ar, O

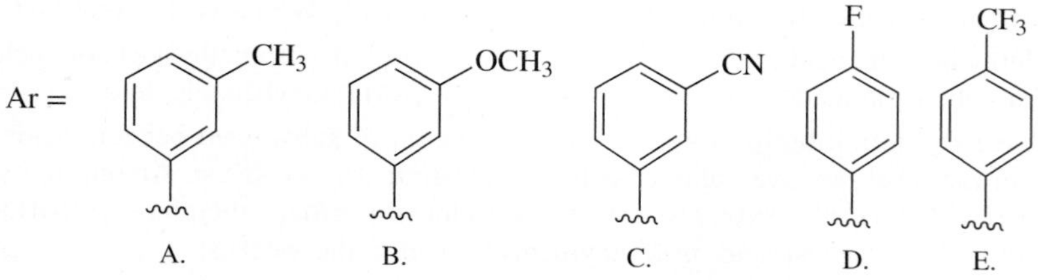

Ar = A. (CH_3) B. (OCH_3) C. (CN) D. (F) E. (CF_3)

(a) Is the mechanism S_N1 or S_N2?

(b) Which compound reacts at the fastest rate? (The substituent constants were given previously in this Descriptive Passage.)

(c) Which reacts at the slowest rate?

(d) What is the rate ratio $k_{\text{most reactive}}/k_{\text{least reactive}}$?

12.64 Nucleophilic substitution of the vinylic chloride shown follows an unusual two-step mechanism. The nucleophile adds to the double bond in the first step; chloride ion is expelled in the second.

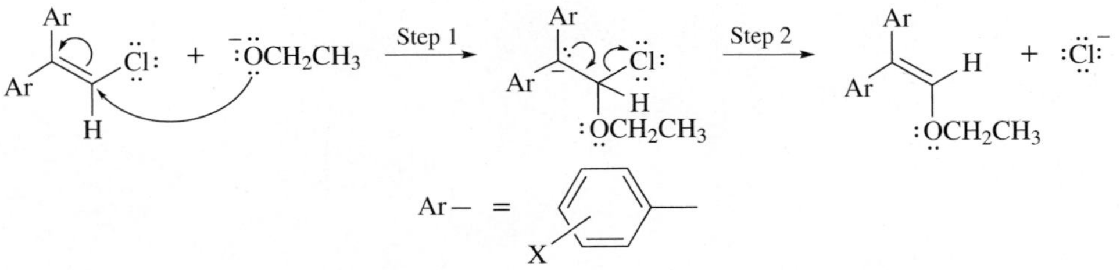

$$Ar_2C{=}CHCl + {}^-:\ddot{O}CH_2CH_3 \xrightarrow{\text{Step 1}} Ar_2\bar{C}{-}CH(Cl)(OCH_2CH_3) \xrightarrow{\text{Step 2}} Ar_2C{=}CH(OCH_2CH_3) + :\ddot{Cl}:^-$$

Ar— = X-substituted phenyl

The measured value of ρ for the overall reaction is reported to be +4.4. Which is the more reasonable choice for the rate-determining step in this reaction?

A. Step 1

B. Step 2

12.65 The value of ρ for the reaction of a series of benzylic chlorides with potassium iodide in acetone is +0.8.

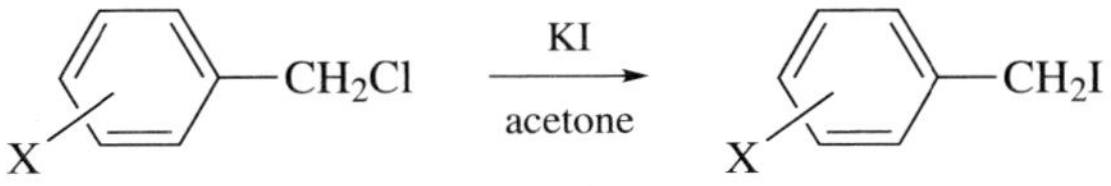

$$X{-}C_6H_4{-}CH_2Cl \xrightarrow[\text{acetone}]{\text{KI}} X{-}C_6H_4{-}CH_2I$$

(a) Is the mechanism S_N1 or S_N2?

From among the compounds shown:

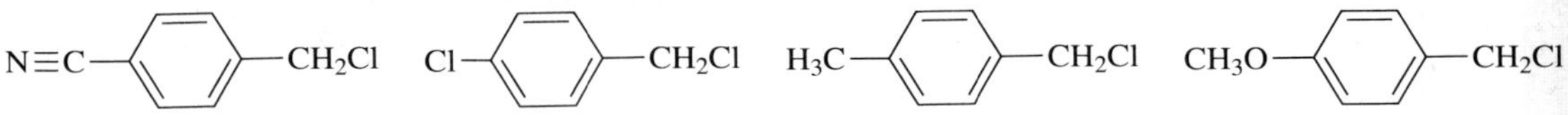

A. $N{\equiv}C{-}C_6H_4{-}CH_2Cl$ B. $Cl{-}C_6H_4{-}CH_2Cl$ C. $H_3C{-}C_6H_4{-}CH_2Cl$ D. $CH_3O{-}C_6H_4{-}CH_2Cl$

(b) Which compound reacts at the fastest rate?

(c) Which compound reacts at the slowest rate?

(d) What is the rate ratio $k_{\text{most reactive}}/k_{\text{least reactive}}$?

12.66 Transition states and ρ values are shown for E2 elimination in two series of compounds that differ in their leaving group. Based on their ρ values, in which series of reactants is there a greater degree of C—H bond-breaking at the transition state?

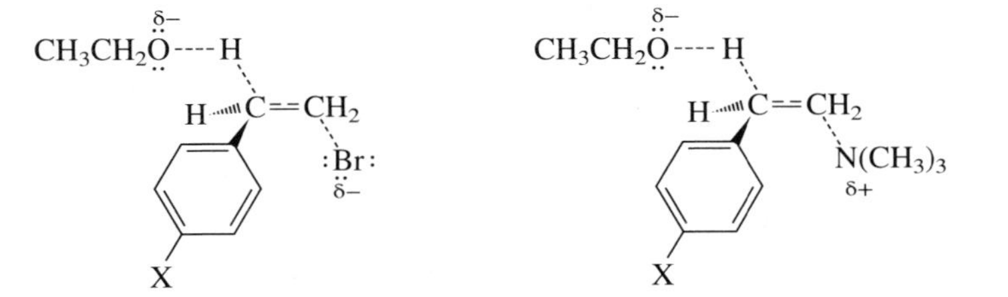

Reactant is $ArCH_2CH_2Br$; $\rho = +2.1$

Reactant is $ArCH_2CH_2\overset{+}{N}(CH_3)_3$; $\rho = +3.8$

A. $ArCH_2CH_2Br$

B. $ArCH_2CH_2\overset{+}{N}(CH_3)_3$

Chapter

13

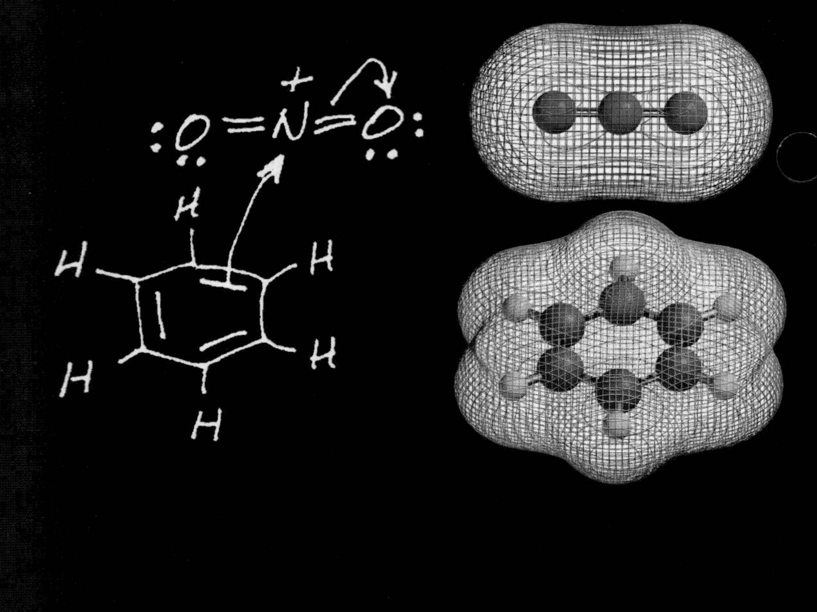

The blackboard shows the flow of electrons in the reaction of benzene with nitronium ion. The electrostatic potential maps show the complementarity between the π-electron system of benzene and the nitrogen of nitronium ion.

CHAPTER OUTLINE

Electrophilic and Nucleophilic Aromatic Substitution

In the preceding chapter the special stability of benzene was described, along with reactions in which an aromatic ring was present as a substituent. What about reactions that occur on the ring itself? What sort of reagents react with benzene and its derivatives, what products are formed, and by what mechanisms?

The largest and most important class of such reactions involves *electrophilic* reagents. We already have some experience with electrophiles, particularly with respect to their reaction with alkenes. Electrophilic reagents *add* to alkenes.

$$\mathrm{>C{=}C<} \;+\; \overset{\delta+}{\mathrm{E}}\!-\!\overset{\delta-}{\mathrm{Y}} \longrightarrow \mathrm{E{-}\overset{|}{\underset{|}{C}}{-}\overset{|}{\underset{|}{C}}{-}Y}$$

Alkene — Electrophilic reagent — Product of electrophilic addition

A different reaction occurs with arenes. *Substitution is observed instead of addition.* The electrophilic portion of the reagent replaces one of the hydrogens on the ring:

$$\mathrm{Ar{-}H} \;+\; \overset{\delta+}{\mathrm{E}}\!-\!\overset{\delta-}{\mathrm{Y}} \longrightarrow \mathrm{Ar{-}E} \;+\; \mathrm{H{-}Y}$$

Arene — Electrophilic reagent — Product of electrophilic aromatic substitution

This reaction is known as **electrophilic aromatic substitution.** It is one of the fundamental processes of organic chemistry and the major concern of this chapter.

What about nucleophilic substitution in aryl halides?

$$\underset{\text{Aryl halide}}{\text{Ar}-\ddot{\underset{..}{X}}:} + \underset{\text{Nucleophile}}{:\text{Nu}^-} \longrightarrow \underset{\text{Product of nucleophilic aromatic substitution}}{\text{Ar}-\text{Nu}} + \underset{\text{Halide ion}}{:\ddot{\underset{..}{X}}:^-}$$

In Section 6.1, we noted that aryl halides are normally much less reactive toward nucleophilic substitution than alkyl halides. In the present chapter we'll see examples of novel, useful, and mechanistically interesting **nucleophilic aromatic substitutions** and explore the structural features responsible for these reactions.

13.1 Representative Electrophilic Aromatic Substitution Reactions of Benzene

The scope of electrophilic aromatic substitution is quite large; both the aromatic compound and the electrophilic reagent are capable of wide variation. Indeed, it is this broad scope that makes electrophilic aromatic substitution so important. Electrophilic aromatic substitution is the main method by which substituted derivatives of benzene are prepared. We can gain a feeling for these reactions by examining the examples in Table 13.1. Each will be discussed in more detail in Sections 13.3 through 13.7. First, however, let us look at the general mechanism of electrophilic aromatic substitution.

TABLE 13.1 Representative Electrophilic Aromatic Substitution Reactions of Benzene

Reaction and comments	Equation
1. **Nitration** Warming benzene with a mixture of nitric acid and sulfuric acid gives nitrobenzene. A nitro group (—NO_2) replaces one of the ring hydrogens.	Benzene (C_6H_5—H) + $HONO_2$ (Nitric acid) $\xrightarrow[30\text{–}40°C]{H_2SO_4}$ Nitrobenzene (C_6H_5—NO_2) (95%) + H_2O (Water)
2. **Sulfonation** Treatment of benzene with hot concentrated sulfuric acid gives benzenesulfonic acid. A sulfonic acid group (—SO_2OH) replaces one of the ring hydrogens.	Benzene (C_6H_5—H) + $HOSO_2OH$ (Sulfuric acid) $\xrightarrow{\text{heat}}$ Benzenesulfonic acid (C_6H_5—SO_2OH) (100%) + H_2O (Water)
3. **Halogenation** Bromine reacts with benzene in the presence of iron(III) bromide as a catalyst to give bromobenzene. Chlorine reacts similarly in the presence of iron(III) chloride to give chlorobenzene.	Benzene (C_6H_5—H) + Br_2 (Bromine) $\xrightarrow{FeBr_3}$ Bromobenzene (C_6H_5—Br) (65–75%) + HBr (Hydrogen bromide)
4. **Friedel–Crafts alkylation** Alkyl halides react with benzene in the presence of aluminum chloride to yield alkylbenzenes.	Benzene (C_6H_5—H) + *tert*-Butyl chloride $\xrightarrow[0°C]{AlCl_3}$ *tert*-Butylbenzene (60%) + HCl (Hydrogen chloride)
5. **Friedel–Crafts acylation** An analogous reaction occurs when acyl halides react with benzene in the presence of aluminum chloride. The products are aryl ketones.	Benzene (C_6H_5—H) + Propanoyl chloride $\xrightarrow[40°C]{AlCl_3}$ 1-Phenyl-1-propanone (88%) + HCl (Hydrogen chloride)

13.2 Mechanistic Principles of Electrophilic Aromatic Substitution

Recall the general mechanism for electrophilic addition to alkenes:

slow → fast →

Alkene and electrophile | Carbocation and nucleophile | Product of electrophilic addition

The first step is rate-determining. In it a carbocation results when the pair of π electrons of the alkene is used to form a bond with the electrophile. The carbocation then undergoes rapid capture by some Lewis base present in the medium.

The first step in the reaction of electrophilic reagents with benzene is similar. An electrophile accepts an electron pair from the π system of benzene giving a carbocation.

slow → fast → + H−Y

Benzene and electrophile | Carbocation and anion | Product of electrophilic aromatic substitution

In the second step, this carbocation, called a **cyclohexadienyl cation, arenium ion,** or **σ-complex,** then undergoes deprotonation to restore the aromaticity of the ring. If the Lewis base (:Y^-) had acted as a nucleophile and bonded to carbon, the product would have been a nonaromatic cyclohexadiene derivative. Substitution occurs preferentially because a substantial driving force is present that favors rearomatization.

Figure 13.1 is a potential energy diagram describing the general mechanism of electrophilic aromatic substitution. In order to overcome the high activation energy that

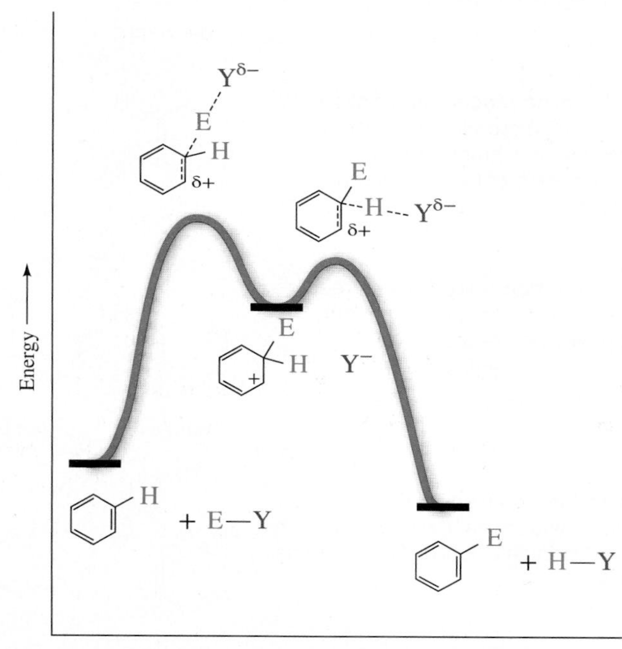

Figure 13.1

Potential energy diagram for electrophilic aromatic substitution.

characterizes the first step, the electrophile must be a reactive one. Many of the electrophilic reagents that react rapidly with alkenes do not react at all with benzene. Peroxy acids and diborane, for example, fall into this category. Others, such as bromine, react with benzene only in the presence of catalysts that increase their electrophilicity. The low level of reactivity of benzene toward electrophiles stems from the loss of aromaticity in the transition state for the rate-determining step.

Problem 13.1

Based on Hammond's postulate which holds that the closer two consecutive states in a reaction mechanism are in energy the closer they are in structure, does the structure of the transition state for formation of the carbocation intermediate more closely resemble benzene or cyclohexadienyl cation?

With this as background, we'll examine each of the electrophilic aromatic substitutions presented in Table 13.1 in more detail, especially with respect to the electrophile that reacts with benzene.

13.3 Nitration of Benzene

Having outlined the general mechanism for electrophilic aromatic substitution, we need only identify the specific electrophile in the nitration of benzene to have a fairly clear idea of how the reaction occurs.

$$C_6H_5H + HONO_2 \xrightarrow[30\text{–}40°C]{H_2SO_4} C_6H_5NO_2 + H_2O$$

Benzene | Nitric acid | Nitrobenzene (95%) | Water

The role of nitronium ion in the nitration of benzene was demonstrated by Sir Christopher Ingold—the same person who suggested the S_N1 and S_N2 mechanisms of nucleophilic substitution and who collaborated with Cahn and Prelog on the *R* and *S* notational system.

The electrophile (E^+) in this reaction is *nitronium ion* ($:\ddot{O}=\overset{+}{N}=\ddot{O}:$). The charge distribution in nitronium ion is evident both in its Lewis structure and in the electrostatic potential map of Figure 13.2. Nitronium ion is generated by the reaction of nitric acid with sulfuric acid, resulting in the protonation of nitric acid and loss of water:

$$HOSO_2O\text{–}H + HO\text{–}NO_2 \longrightarrow HOSO_2O^- + H_2\overset{+}{O}\text{–}NO_2 \longrightarrow H_2O + O=\overset{+}{N}=O$$

Sulfuric acid | Nitric acid | Hydrogen sulfate ion | Protonated nitric acid | Water | Nitronium ion

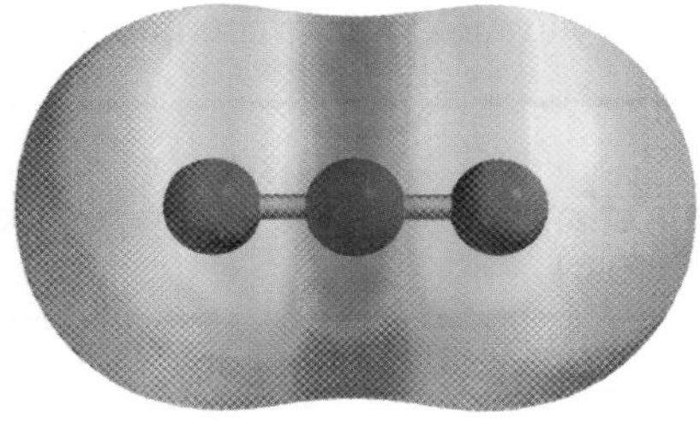

Figure 13.2

Electrostatic potential map of nitronium ion (NO_2^+). The positive charge is concentrated on nitrogen.

Mechanism 13.1

Nitration of Benzene

THE OVERALL REACTION:

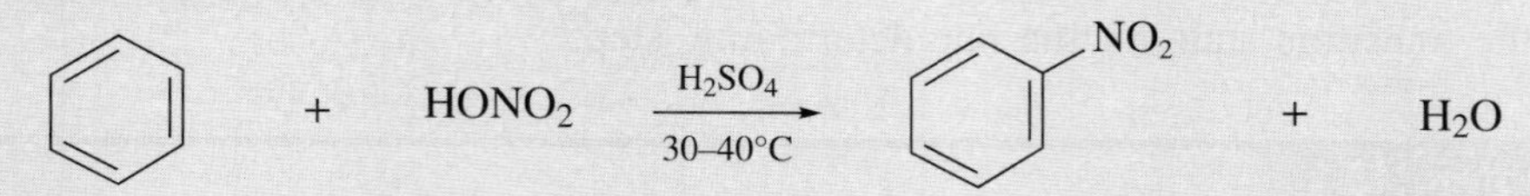

$$C_6H_6 + HONO_2 \xrightarrow[30\text{–}40°C]{H_2SO_4} C_6H_5NO_2 + H_2O$$

Benzene Nitric acid Nitrobenzene Water

THE MECHANISM:

Step 1: Nitronium ion is the active electrophile and reacts with the π system of the aromatic ring. This step is rate-determining. (The molecular model depicts the cyclohexadienyl cation intermediate.)

slow

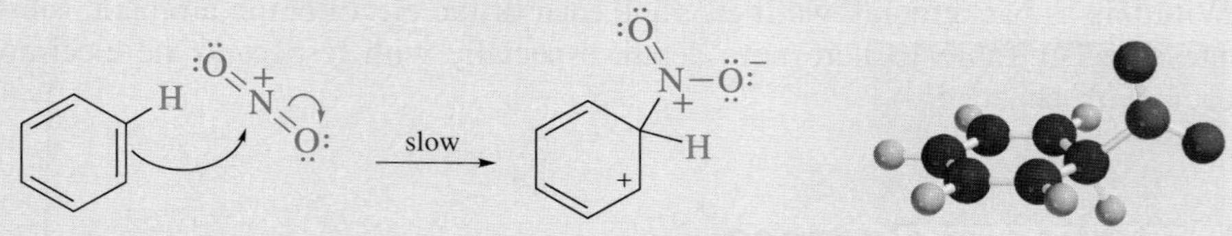

Benzene + nitronium ion → Cyclohexadienyl cation intermediate

Step 2: Deprotonation of the cyclohexadienyl cation restores the aromaticity of the ring.

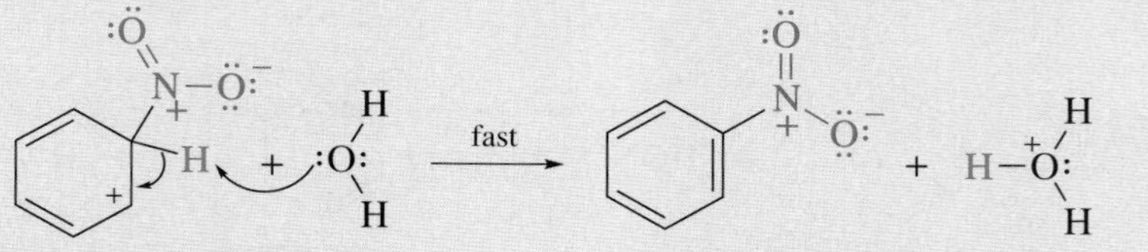

fast

Cyclohexadienyl cation intermediate + Water → Nitrobenzene + Hydronium ion

Mechanism 13.1 adapts the general mechanism of electrophilic aromatic substitution to the nitration of benzene. The first step is rate-determining; in it benzene reacts with nitronium ion to give the cyclohexadienyl cation intermediate. In the second step, the aromaticity of the ring is restored by loss of a proton from the cyclohexadienyl cation.

One way we know that step 1 is rate-determining is that nitration of benzene does not exhibit a deuterium isotope effect (Section 7.17). Loss of deuterium (D = ^2H) during nitration of C_6H_5D occurs at the same rate as loss of a single ^1H, which tells us that the C—D bond must break *after* the rate-determining step, not during it.

Nitration by electrophilic aromatic substitution is not limited to benzene alone, but is a general reaction of compounds that contain a benzene ring with at least one replaceable hydrogen. It would be a good idea to write out the answers to the following two problems to ensure that you understand the relationship of starting materials to products and the mechanism of aromatic nitration before continuing to the next section.

Problem 13.2

Nitration of 1,4-dimethylbenzene (*p*-xylene) gives a single product having the molecular formula $C_8H_9NO_2$ in high yield. What is this product?

Problem 13.3

Using $:\ddot{O}{=}\overset{+}{N}{=}\ddot{O}:$ as the electrophile, write a reasonable mechanism for the reaction given in Problem 13.2. Use curved arrows to show the flow of electrons.

13.4 Sulfonation of Benzene

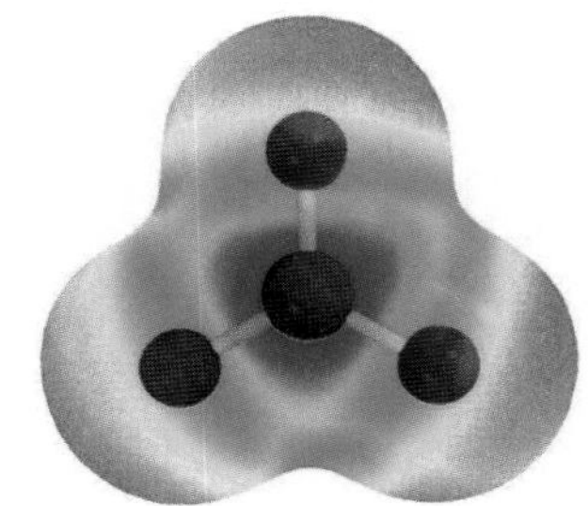

Figure 13.3

Electrostatic potential map of sulfur trioxide. The region of greatest positive charge surrounds sulfur.

The reaction of benzene with sulfuric acid to produce benzenesulfonic acid is reversible and can be driven to completion by several techniques. Removing the water formed in the reaction, for example, allows benzenesulfonic acid to be obtained in virtually quantitative yield.

$$C_6H_5\text{–}H + HOSO_2OH \xrightleftharpoons{\text{heat}} C_6H_5\text{–}SO_2OH + H_2O$$

Benzene | Sulfuric acid | Benzenesulfonic acid | Water

When a solution of sulfur trioxide in sulfuric acid is used as the sulfonating agent, the rate of sulfonation is much faster and the equilibrium is displaced entirely to the side of products.

$$C_6H_6 + SO_3 \xrightarrow{H_2SO_4} C_6H_5\text{–}SO_2OH$$

Benzene | Sulfur trioxide | Benzenesulfonic acid

Among the variety of electrophilic species present in concentrated sulfuric acid, sulfur trioxide (Figure 13.3) is probably the actual electrophile in aromatic sulfonation as shown in Mechanism 13.2.

Problem 13.4

On being heated with sulfur trioxide in sulfuric acid, 1,2,4,5-tetramethylbenzene was converted to a product of molecular formula $C_{10}H_{14}O_3S$ in 94% yield. Suggest a reasonable structure for this product.

Mechanism 13.2

Sulfonation of Benzene

THE OVERALL REACTION:

$$C_6H_6 + H_2SO_4 \longrightarrow C_6H_5\text{–}SO_2OH + H_2O$$

Benzene | Sulfuric acid | Benzenesulfonic acid | Water

THE MECHANISM:

Step 1: The active electrophile in the sulfonation of aromatic compounds is sulfur trioxide and reacts with benzene in the rate-determining step.

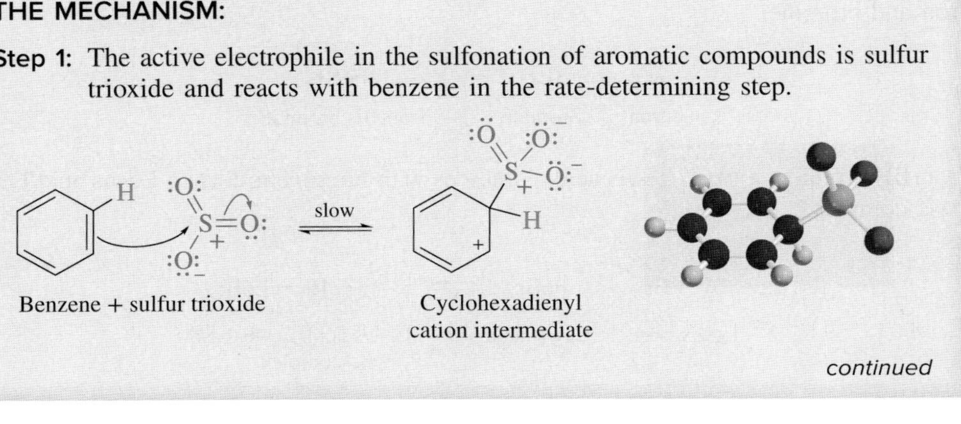

Benzene + sulfur trioxide | Cyclohexadienyl cation intermediate

continued

Step 2: A proton is abstracted from the sp^3-hybridized carbon of the intermediate to restore the aromaticity of the ring. The species shown that abstracts the proton is a hydrogen sulfate ion formed by ionization of sulfuric acid.

$$\text{Cyclohexadienyl cation intermediate} + {}^{-}\text{:}\ddot{\text{O}}\text{—SO}_2\text{OH} \underset{}{\overset{\text{fast}}{\rightleftharpoons}} \text{Benzenesulfonate ion} + \text{H—}\ddot{\text{O}}\text{—SO}_2\text{OH}$$

Cyclohexadienyl cation intermediate | Hydrogen sulfate ion | Benzenesulfonate ion | Sulfuric acid

Step 3: A rapid proton transfer from the oxygen of sulfuric acid to the oxygen of benzenesulfonate completes the process.

$$\text{Benzenesulfonate ion} + \text{H—}\ddot{\text{O}}\text{—SO}_2\text{OH} \overset{\text{fast}}{\rightleftharpoons} \text{Benzenesulfonic acid} + {}^{-}\text{:}\ddot{\text{O}}\text{—SO}_2\text{OH}$$

Benzenesulfonate ion | Sulfuric acid | Benzenesulfonic acid | Hydrogen sulfate ion

13.5 Halogenation of Benzene

According to the usual procedure for preparing bromobenzene, bromine is added to benzene in the presence of metallic iron (customarily a few carpet tacks) and the reaction mixture is heated.

$$C_6H_5H + Br_2 \xrightarrow[\text{heat}]{\text{Fe}} C_6H_5Br + HBr$$

Benzene | Bromine | Bromobenzene (65–75%) | Hydrogen bromide

Bromine, although it adds rapidly to alkenes, is too weak an electrophile to react at an appreciable rate with benzene. A catalyst that increases the electrophilic properties of bromine must be present. Somehow carpet tacks can do this. How?

Iron(III) bromide ($FeBr_3$) is also called *ferric bromide.*

The active catalyst is not iron itself but iron(III) bromide, formed by reaction of iron and bromine.

$$2Fe + 3Br_2 \longrightarrow 2FeBr_3$$

Iron | Bromine | Iron(III) bromide

Iron(III) bromide, a weak Lewis acid, combines with bromine to form a Lewis acid/Lewis base complex.

$$:\ddot{Br}—\ddot{Br}: + FeBr_3 \rightleftharpoons :\ddot{Br}—\overset{+}{\ddot{Br}}—\overset{-}{Fe}Br_3$$

Lewis base | Lewis acid | Lewis acid/Lewis base complex

Mechanism 13.3

Bromination of Benzene

THE OVERALL REACTION:

Benzene + Bromine (Br_2) —Fe→ Bromobenzene + Hydrogen bromide (HBr)

THE MECHANISM:

Step 1: The bromine–iron(III) bromide complex is the active electrophile that reacts with benzene. Two of the π electrons of benzene are used to form a bond to bromine and give a cyclohexadienyl cation intermediate. (The molecular model depicts the cyclohexadienyl cation intermediate.)

Benzene and bromine–iron(III) bromide complex —slow→ Cyclohexadienyl cation intermediate + Tetrabromoferrate ion

Step 2: Loss of a proton from the cyclohexadienyl cation yields bromobenzene.

Cyclohexadienyl cation intermediate + Tetrabromoferrate ion —fast→ Bromobenzene + Hydrogen bromide + Iron(III) bromide ($FeBr_3$)

Complexation of bromine with iron(III) bromide makes bromine more electrophilic, and it reacts with benzene to give a cyclohexadienyl intermediate as shown in step 1 of Mechanism 13.3. In step 2, as in nitration and sulfonation, loss of a proton from the cyclohexadienyl cation is rapid and gives the product of electrophilic aromatic substitution.

Only small quantities of iron(III) bromide are required. It is a catalyst for the bromination and, as Mechanism 13.3 indicates, is regenerated in the course of the reaction. We'll see later in this chapter that some aromatic compounds are much more reactive than benzene and react rapidly with bromine even in the absence of a catalyst.

Chlorination is carried out in a manner similar to bromination and follows an analogous mechanism to give aryl chlorides.

Iodination with I_2 is not very effective, but specialized reagents such as acetyl hypoiodite have been developed that provide a useful synthesis of aryl iodides.

Benzene + Acetyl hypoiodite → Iodobenzene (72%) + Acetic acid

Problem 13.5

Given the fact that iodine in acetyl hypoiodite is much more electrophilic than I_2, suggest a reasonable mechanism for the preceding reaction.

Biosynthetic Halogenation

Over 4000 natural products contain halogens. Some naturally occurring aryl halides include:

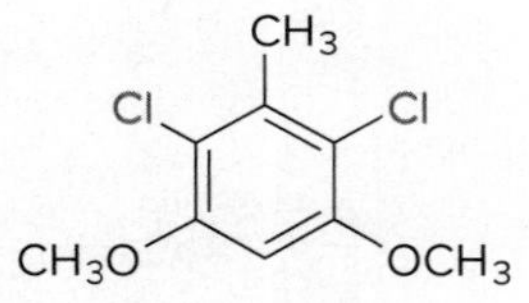

2,6-Dichloro-3,5-dimethoxytoluene: an antifungal compound isolated from lily plants

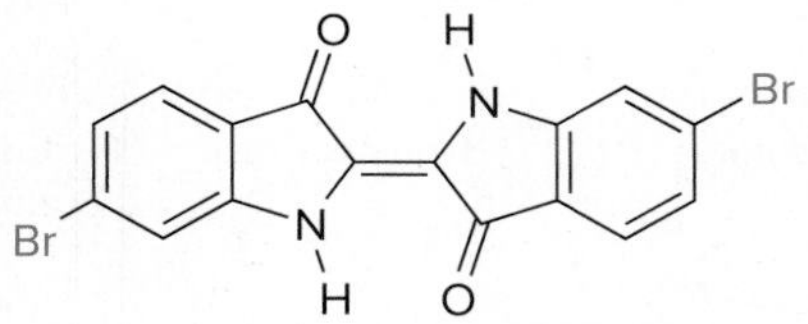

Dibromoindigo: main component of the dye known as Tyrian purple isolated from a species of Mediterranean sea snail and prized by ancient cultures

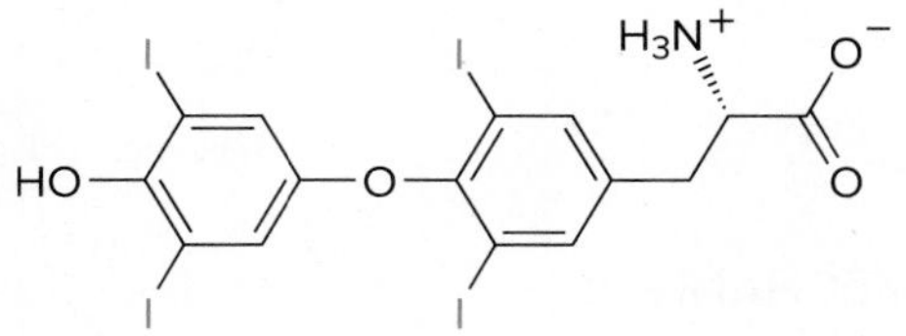

Thyroxine: a hormone of the thyroid gland; the (*S*) enantiomer is a widely used drug prescribed to increase metabolic rate

The presence of the halogen in these and in other halogenated natural products has a strong effect on their properties, and their biosynthetic origin was a scientific puzzle of longstanding. What are the biological halogenating agents, what enzymes catalyze the halogenation, and how do they do it? Recent studies have unlocked the answers to some of these questions.

Biosynthetic halogenation can occur through multiple pathways, but many halogenase enzymes use *electrophilic* halogenating species that are produced by oxidation of halide ions.

The biosynthesis of the antifungal antibiotic pyrrolnitrin begins with enzyme-catalyzed chlorination of the benzene ring of the amino acid tryptophan. The enzyme is a *halogenase,* and the electrophilic chlorinating agent is thought to be hypochlorous acid (HOCl), which is generated in a separate step. Although the oxidation state of chlorine in HOCl is +1, it alone is not electrophilic enough to chlorinate tryptophan but can be activated through hydrogen bonding with a nearby amino group found in the catalytic site of the enzyme. Loss of a proton to rearomatize the ring occurs from the σ complex and gives 7-chlorotryptophan, which undergoes subsequent conversion to pyrrolnitrin by way of additional enzyme-catalyzed processes, including a second biochemical chlorination.

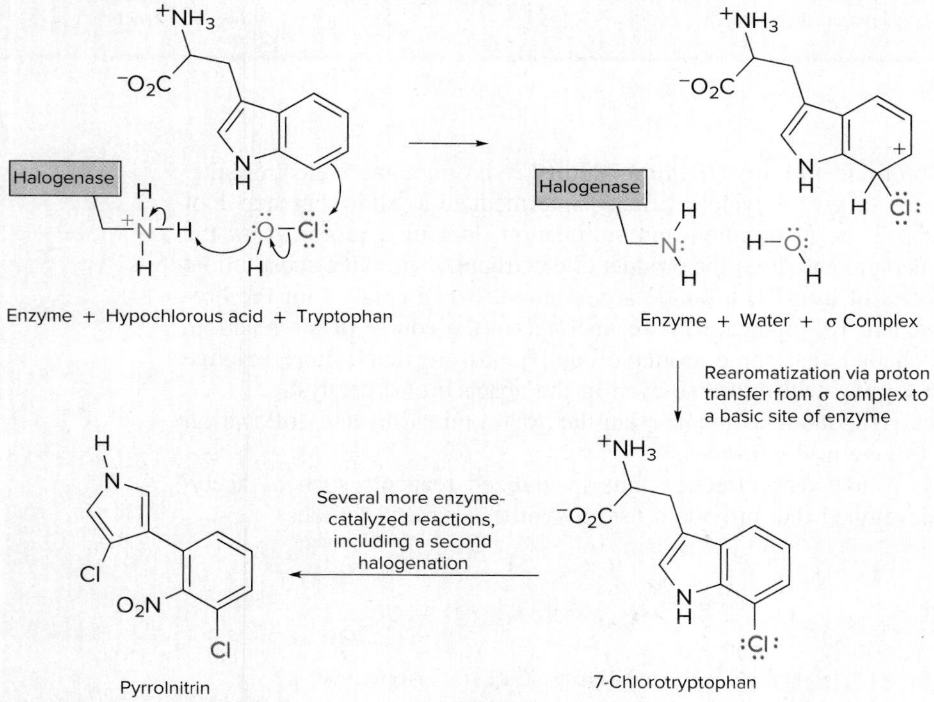

Enzyme + Hypochlorous acid + Tryptophan

Enzyme + Water + σ Complex

Rearomatization via proton transfer from σ complex to a basic site of enzyme

7-Chlorotryptophan

Several more enzyme-catalyzed reactions, including a second halogenation

Pyrrolnitrin

13.6 Friedel–Crafts Alkylation of Benzene

In a **Friedel–Crafts reaction,** alkyl halides react with benzene in the presence of aluminum chloride to give an alkylbenzene. It is one of the most useful synthetic methods in organic chemistry.

Benzene + *tert*-Butyl chloride $\xrightarrow[0°C]{AlCl_3}$ *tert*-Butylbenzene (60%) + Hydrogen chloride (HCl)

The reaction that bears their names was discovered in 1877 by Charles Friedel and James M. Crafts at the Sorbonne in Paris. Crafts later became president at M.I.T.

Although alkyl halides by themselves are insufficiently electrophilic to react with benzene, alkylation is catalyzed by aluminum chloride, which acts as a Lewis acid to convert secondary and tertiary alkyl halides to carbocations, which then alkylate the aromatic ring.

tert-Butyl chloride + Aluminum chloride ($AlCl_3$) ⟶ Lewis acid/Lewis base complex ⟶ *tert*-Butyl cation + Tetrachloroaluminate ion ($\bar{A}lCl_4$)

Mechanism 13.4 illustrates the reaction of benzene with *tert*-butyl cation (step 1) followed by formation of *tert*-butylbenzene by abstraction of a proton from the cyclohexadienyl cation intermediate (step 2).

Mechanism 13.4

Friedel–Crafts Alkylation

THE OVERALL REACTION:

Benzene + *tert*-Butyl chloride $\xrightarrow[0°C]{AlCl_3}$ *tert*-Butylbenzene + Hydrogen chloride (HCl)

THE MECHANISM:

Step 1: Once generated by the reaction of *tert*-butyl chloride and aluminum chloride, *tert*-butyl cation is attacked by the π electrons of benzene, and a carbon–carbon bond is formed. (The molecular model depicts the cyclohexadienyl cation intermediate.)

Benzene + *tert*-Butyl cation $\xrightarrow{\text{slow}}$ Cyclohexadienyl cation intermediate

Step 2: Loss of a proton from the cyclohexadienyl cation intermediate yields *tert*-butylbenzene.

Cyclohexadienyl cation intermediate + Tetrachloroaluminate ion ($:\ddot{C}l-\bar{A}lCl_3$) $\xrightarrow{\text{fast}}$ *tert*-Butylbenzene + Hydrogen chloride ($H-\ddot{C}l:$) + Aluminum chloride ($AlCl_3$)

Secondary alkyl halides react by a similar mechanism. Methyl and ethyl halides do not form carbocations under Friedel–Crafts conditions, but their aluminum chloride complexes contain highly polarized carbon–halogen bonds and these complexes do alkylate benzene.

$$H_3C-\overset{+}{\ddot{X}}-\bar{A}lX_3 \qquad CH_3CH_2-\overset{+}{\ddot{X}}-\bar{A}lX_3$$

Methyl halide/aluminum halide complex — Ethyl halide/aluminum halide complex

A drawback to Friedel–Crafts alkylation is that rearrangements can occur, especially with primary alkyl halides. For example, Friedel–Crafts alkylation of benzene with isobutyl chloride yields only *tert*-butylbenzene.

Other limitations to Friedel–Crafts reactions will be encountered in this chapter and are summarized in Table 13.4.

Benzene + Isobutyl chloride $\xrightarrow[0°C]{AlCl_3}$ *tert*-Butylbenzene + Hydrogen chloride (HCl)

Here, the electrophile is *tert*-butyl cation formed by a hydride migration that accompanies ionization of the carbon–chlorine bond.

We saw rearrangements involving hydride shifts earlier in Sections 6.8 and 7.13.

Isobutyl chloride/aluminum chloride complex $\longrightarrow$ *tert*-Butyl cation + $:\ddot{C}l-\bar{A}lCl_3$ Tetrachloroaluminate ion

Problem 13.6

Use curved arrows to show the formation of the cyclohexadienyl cation in the Friedel–Crafts alkylation of benzene with the electrophile formed from ethyl chloride and $AlCl_3$.

Benzene + $CH_3CH_2-\overset{+}{\ddot{C}l}-\bar{A}lCl_3$ $\longrightarrow$ cyclohexadienyl cation + $:\ddot{C}l-\bar{A}lCl_3$

Problem 13.7

In an attempt to prepare propylbenzene, a chemist alkylated benzene with 1-chloropropane and aluminum chloride. However, two isomeric hydrocarbons were obtained in a ratio of 2:1, the desired propylbenzene being the minor component. What do you think was the major product? How did it arise?

Because electrophilic aromatic substitution is simply another reaction available to a carbocation, other carbocation precursors can be used in place of alkyl halides. For example, alkenes, which are converted to carbocations by protonation, can be used to alkylate benzene.

Benzene + Cyclohexene $\xrightarrow{H_2SO_4}$ Cyclohexylbenzene (65–68%)

Problem 13.8

Write a reasonable mechanism for the formation of cyclohexylbenzene from the reaction of benzene, cyclohexene, and sulfuric acid.

Problem 13.9

tert-Butylbenzene can be prepared by alkylation of benzene using an alkene or an alcohol as the carbocation source. What alkene? What alcohol?

Alkenyl halides such as vinyl chloride ($H_2C{=}CHCl$) do *not* form carbocations on treatment with aluminum chloride and so cannot be used in Friedel–Crafts reactions. Thus, the industrial preparation of styrene from benzene and ethylene does not involve vinyl chloride but proceeds by way of ethylbenzene.

$$C_6H_6 + H_2C{=}CH_2 \xrightarrow{\text{HCl, AlCl}_3} C_6H_5{-}CH_2CH_3 \xrightarrow[\text{ZnO}]{630°\text{C}} C_6H_5{-}CH{=}CH_2$$

Benzene Ethylene Ethylbenzene Styrene

Dehydrogenation of alkylbenzenes, although useful in the industrial preparation of styrene, is not a general procedure and is not well suited to the laboratory preparation of alkenylbenzenes. In such cases an alkylbenzene is subjected to benzylic bromination (Section 12.10), and the resulting benzylic bromide is treated with base to effect dehydrohalogenation.

Problem 13.10

Outline a synthesis of 1-phenylcyclohexene from benzene and cyclohexene.

13.7 Friedel–Crafts Acylation of Benzene

Another version of the Friedel–Crafts reaction uses **acyl halides** instead of alkyl halides and yields aryl ketones.

$$C_6H_5{-}H + CH_3CH_2C({=}O)Cl \xrightarrow{\text{AlCl}_3} C_6H_5C({=}O)CH_2CH_3 + HCl$$

Benzene Propanoyl chloride 1-Phenyl-1-propanone (88%) Hydrogen chloride

An acyl group has the general formula $RC(=O)-$

The electrophile in a Friedel–Crafts acylation is an **acyl cation** (also referred to as an **acylium ion**) and is formed on reaction of acyl chlorides with aluminum chloride in much the same way as alkyl cations are formed from alkyl halides.

Electron delocalization in the acyl cation derived from propanoyl chloride is represented by the following two resonance contributors. Note that electron release from oxygen generates a contributing structure that both satisfies the octet rule and disperses the positive charge.

$$CH_3CH_2\overset{+}{C}{=}\ddot{O}: \longleftrightarrow CH_3CH_2C{\equiv}\overset{+}{O}:$$

Propanoyl cation

Unlike alkyl carbocations, *acyl cations do not rearrange.* An acyl cation is so strongly stabilized by electron delocalization that it is more stable than any other ion that would arise from it by a hydride or alkyl group shift.

The electrostatic potential map of propanoyl cation in Figure 13.4 illustrates the positive character of the acyl carbon, and it is this carbon that is the reactive site in electrophilic aromatic substitution (Mechanism 13.5).

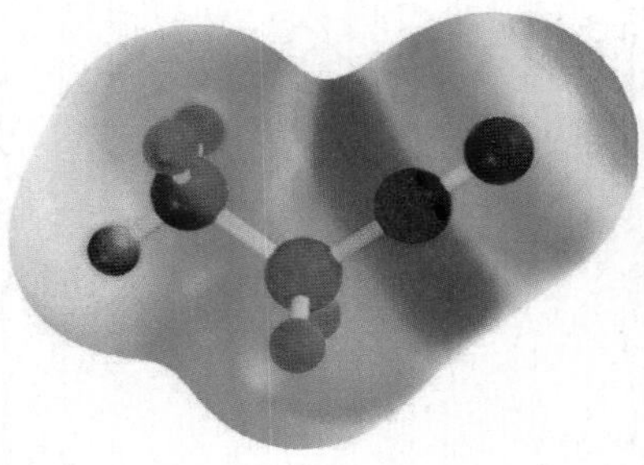

Figure 13.4

Electrostatic potential map of propanoyl cation. The region of greatest positive charge is associated with the carbon of the C=O group.

Mechanism 13.5

Friedel–Crafts Acylation

THE OVERALL REACTION:

THE MECHANISM:

Step 1: The acyl cation reacts with benzene. A pair of π electrons of benzene is used to form a covalent bond to the carbon of the acyl cation. (The molecular model depicts the cyclohexadienyl cation intermediate.)

Step 2: Aromaticity of the ring is restored when it loses a proton to give the aryl ketone.

Problem 13.11

The reaction shown gives a single product in 88% yield. What is that product?

Problem 13.12

What is the structure of the acylium ion that is formed in the reaction in Problem 13.11?

Acyl chlorides are readily prepared from carboxylic acids by reaction with thionyl chloride.

$$\underset{\text{Carboxylic acid}}{\text{RCOH}} + \underset{\text{Thionyl chloride}}{SOCl_2} \longrightarrow \underset{\text{Acyl chloride}}{\text{RCCl}} + \underset{\text{Sulfur dioxide}}{SO_2} + \underset{\text{Hydrogen chloride}}{\text{HCl}}$$

Carboxylic acid anhydrides, compounds of the type RCOCR, are also sources of acyl cations and, in the presence of aluminum chloride, acylate benzene. One acyl unit of an acid anhydride becomes attached to the benzene ring, and the other becomes part of a carboxylic acid.

Benzene + Acetic anhydride $\xrightarrow{AlCl_3}$ Acetophenone (76%) + Acetic acid

Acetophenone is one of the commonly encountered benzene derivatives listed in Table 12.1.

Problem 13.13

Succinic anhydride, the structure of which is shown, is a cyclic anhydride often used in Friedel–Crafts acylations. Give the structure of the product obtained when benzene is acylated with succinic anhydride in the presence of aluminum chloride.

13.8 Synthesis of Alkylbenzenes by Acylation–Reduction

Because acylation of an aromatic ring can be accomplished without rearrangement, it is frequently used as the first step in a procedure for the *alkylation* of aromatic compounds by *acylation–reduction.* As we saw in Section 13.6, Friedel–Crafts alkylation of benzene with primary alkyl halides normally yields products having rearranged alkyl groups. When preparing a compound of the type $ArCH_2R$, a two-step sequence is used in which the first step is a Friedel–Crafts acylation.

Benzene $\xrightarrow[AlCl_3]{\text{RCCl}}$ Aryl ketone (C_6H_5CR) $\xrightarrow{\text{reduction}}$ Alkylbenzene ($C_6H_5CH_2R$)

The second step is a reduction of the carbonyl group (C=O) to a methylene group (CH_2).

The most commonly used method for reducing an aryl ketone to an alkylbenzene employs a zinc–mercury amalgam in concentrated hydrochloric acid and is called the **Clemmensen reduction.** Zinc is the reducing agent.

An amalgam is a mixture or alloy of mercury with another metal. For many years silver amalgams were used in dental fillings.

The synthesis of butylbenzene illustrates the acylation–reduction sequence.

Benzene + Butanoyl chloride $\xrightarrow{AlCl_3}$ 1-Phenyl-1-butanone (86%) $\xrightarrow[HCl]{Zn(Hg)}$ Butylbenzene (73%)

Direct alkylation of benzene using 1-chlorobutane and aluminum chloride yields *sec*-butylbenzene by rearrangement and so cannot be used.

Problem 13.14

Using benzene and any necessary organic or inorganic reagents, suggest efficient syntheses of

(a) Isobutylbenzene, $C_6H_5CH_2CH(CH_3)_2$ (b) (2,2-Dimethylpropyl)benzene, $C_6H_5CH_2C(CH_3)_3$

Sample Solution (a) Friedel–Crafts alkylation of benzene with isobutyl chloride is not suitable, because it yields *tert*-butylbenzene by rearrangement (see Section 13.6).

Benzene + Isobutyl chloride $\xrightarrow{AlCl_3}$ *tert*-Butylbenzene (66%)

The two-step acylation–reduction sequence is required. Acylation of benzene puts the side chain on the ring with the correct carbon skeleton. Next, Clemmensen reduction converts the carbonyl group to a methylene group.

Benzene + 2-Methylpropanoyl chloride $\xrightarrow{AlCl_3}$ 2-Methyl-1-phenyl-propan-1-one (84%) $\xrightarrow{Zn(Hg), HCl}$ Isobutylbenzene (80%)

A second way to reduce aldehyde and ketone carbonyl groups is by **Wolff–Kishner reduction.** Heating an aldehyde or a ketone with hydrazine (H_2NNH_2) and sodium or potassium hydroxide in a high-boiling alcohol such as triethylene glycol (bp 287°C) converts the carbonyl to a CH_2 group.

1-Phenyl-1-propanone $\xrightarrow[\text{triethylene glycol, 175°C}]{H_2NNH_2, KOH}$ Propylbenzene (82%)

Both the Clemmensen and the Wolff–Kishner reductions convert an aldehyde or ketone carbonyl to a methylene group. Neither will reduce the carbonyl group of a carboxylic acid, nor are carbon–carbon double or triple bonds affected by these methods.

13.9 Rate and Regioselectivity in Electrophilic Aromatic Substitution

So far we've been concerned only with electrophilic substitution of benzene. Two important questions arise when we turn to substitution on rings that already bear at least one substituent:

1. What is the effect of a substituent on the *rate* of electrophilic aromatic substitution?
2. What is the effect of a substituent on the *regioselectivity* of electrophilic aromatic substitution?

To illustrate substituent effects on rate, consider the nitration of benzene, toluene, and (trifluoromethyl)benzene.

Increasing rate of nitration →

(Trifluoromethyl)benzene (least reactive) — Benzene — Toluene (most reactive)

The range of nitration rates among these three compounds is quite large; it covers a spread of approximately 1-millionfold. Toluene undergoes nitration some 20–25 times faster than benzene. Because toluene is more reactive than benzene, we say that a methyl group *activates* the ring toward electrophilic aromatic substitution. (Trifluoromethyl) benzene, on the other hand, undergoes nitration about 40,000 times more slowly than benzene. A trifluoromethyl group *deactivates* the ring toward electrophilic aromatic substitution. The structural factors responsible for these rate differences will be discussed in more detail beginning in the next section, but we can gain a clue in advance of that material from the three electrostatic potential maps in Figure 13.5, which illustrate how the aromatic rings of benzene and toluene are more "electron-rich" than the ring of (trifluoromethyl)benzene.

Just as there is a marked difference in how methyl and trifluoromethyl substituents affect the rate of electrophilic aromatic substitution, so too is there a marked difference in how they affect its regioselectivity.

Three products are possible from nitration of toluene: *o*-nitrotoluene, *m*-nitrotoluene, and *p*-nitrotoluene. All are formed, but not in equal amounts. Together, the ortho- and para-substituted isomers make up 97% of the product mixture; the meta only 3%.

Toluene $\xrightarrow[\text{acetic anhydride}]{HNO_3}$ *o*-Nitrotoluene (63%) + *m*-Nitrotoluene (3%) + *p*-Nitrotoluene (34%)

Because substitution in toluene occurs primarily at positions ortho and para to methyl, we say that *a methyl substituent is an* **ortho, para director.**

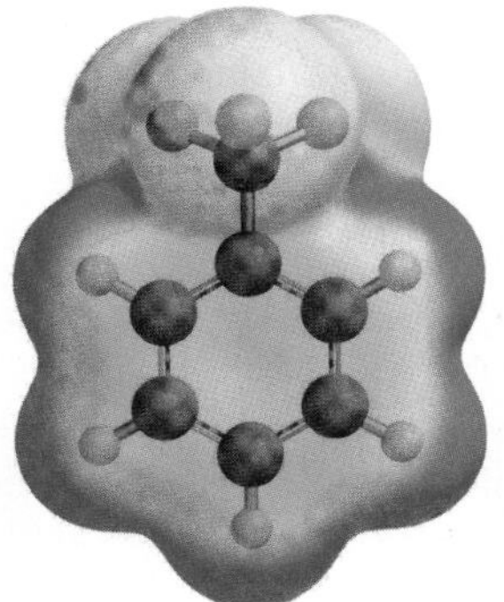

(Trifluoromethyl)benzene

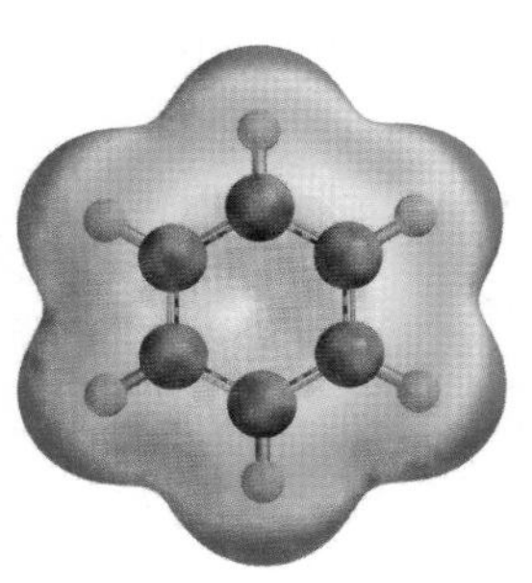

Benzene

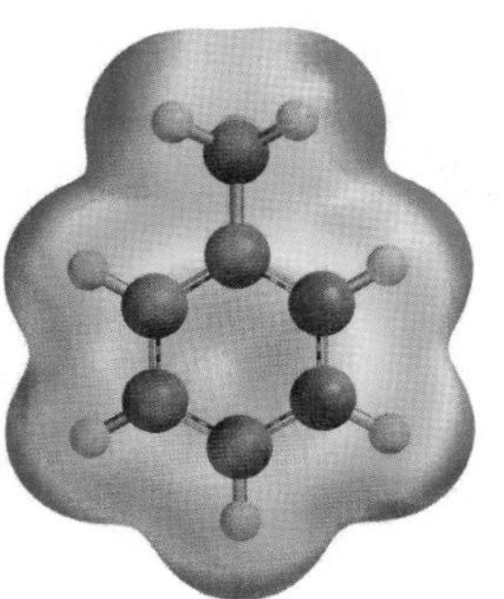

Toluene

Figure 13.5

Electrostatic potential maps of (trifluoromethyl)benzene, benzene, and toluene illustrating the decrease in π-electron density in the ring of (trifluoromethyl)benzene compared with benzene and toluene. The color range is the same for all three maps.

Nitration of (trifluoromethyl)benzene, on the other hand, yields almost exclusively *m*-nitro(trifluoromethyl)benzene (91%). The ortho- and para-substituted isomers are minor components of the reaction mixture.

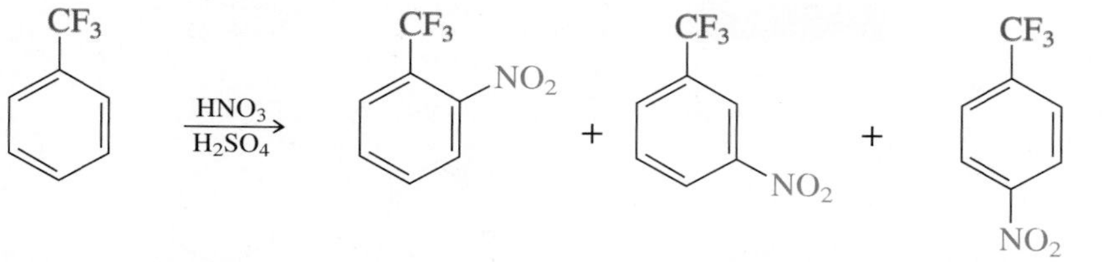

$$\text{(Trifluoromethyl)benzene} \xrightarrow[H_2SO_4]{HNO_3} \text{o-Nitro(trifluoromethyl)benzene (6\%)} + \text{m-Nitro(trifluoromethyl)benzene (91\%)} + \text{p-Nitro(trifluoromethyl)benzene (3\%)}$$

Because substitution in (trifluoromethyl)benzene occurs primarily at positions meta to the substituent, *a trifluoromethyl group is a* **meta director.**

The regioselectivity of substitution, like the rate, is strongly affected by the substituent. In the following several sections we will examine the relationship between the structure of the substituent and its effect on rate and regioselectivity of electrophilic aromatic substitution.

13.10 Rate and Regioselectivity in the Nitration of Toluene

Why is there such a marked difference between methyl and trifluoromethyl substituents in their influence on electrophilic aromatic substitution? Methyl is **activating** and ortho, para-directing; trifluoromethyl is **deactivating** and meta-directing. The first point to remember is that the regioselectivity of substitution is set once the cyclohexadienyl cation intermediate is formed. If we can explain why

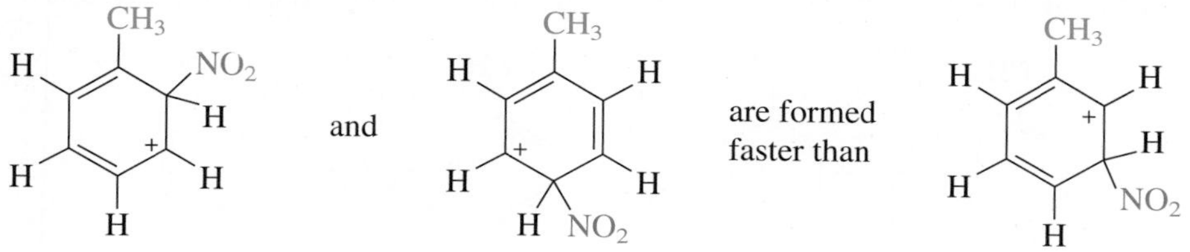

[ortho cyclohexadienyl cation] and [para cyclohexadienyl cation] are formed faster than [meta cyclohexadienyl cation]

in the rate-determining step, we will understand the reasons for the regioselectivity. A principle we have used before serves us well here: *a more stable carbocation is formed faster than a less stable one.* The most likely reason for the directing effect of a CH_3 group must be that the carbocations that give *o*- and *p*-nitrotoluene are more stable than the one that gives *m*-nitrotoluene.

One way to assess the relative stabilities of these carbocations is to examine electron delocalization in them using a resonance description. The cyclohexadienyl cations leading to *o*- and *p*-nitrotoluene have tertiary carbocation character. Each has a contributing structure in which the positive charge resides on the carbon that bears the methyl group.

Ortho nitration

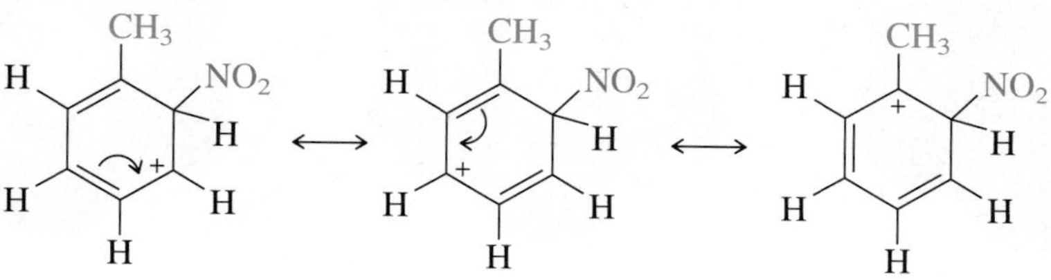

[three resonance structures of the ortho cyclohexadienyl cation, connected by ⟷]

This resonance contributor is a tertiary carbocation

Para nitration

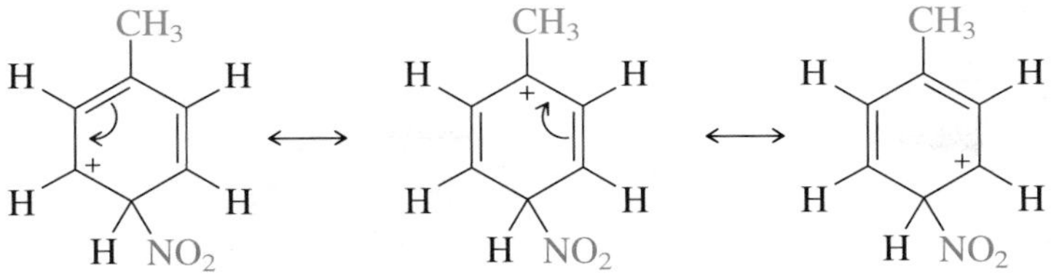

This resonance contributor is a tertiary carbocation

The three contributing resonance forms of the intermediate leading to meta substitution are all secondary carbocations.

Meta nitration

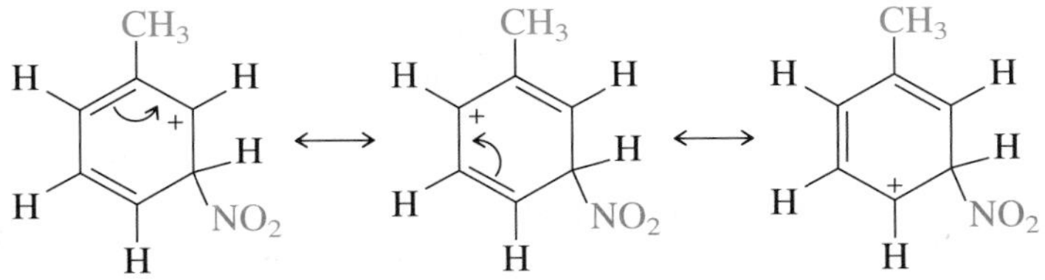

A methyl group is ortho, para-directing because the carbocations leading to *o*- and *p*-nitrotoluene are more stable and formed faster than the one leading to *m*-nitrotoluene. The greater stability of the carbocations for ortho and para substitution comes from their tertiary carbocation character. All of the contributing carbocation structures for meta substitution are secondary.

A methyl group is an activating substituent because it stabilizes the carbocation intermediate formed in the rate-determining step more than hydrogen does. Figure 13.6 compares the energies of activation for nitration at the various positions of toluene with each other and with benzene. Nitration of benzene has the highest activation energy, para-nitration of toluene the lowest.

Methyl is an **electron-releasing group** and activates *all* the available ring carbons toward electrophilic substitution. The ortho and para positions are activated more than meta. At 25°C, the relative rates of nitration at the various positions of toluene compared with a single carbon of benzene are:

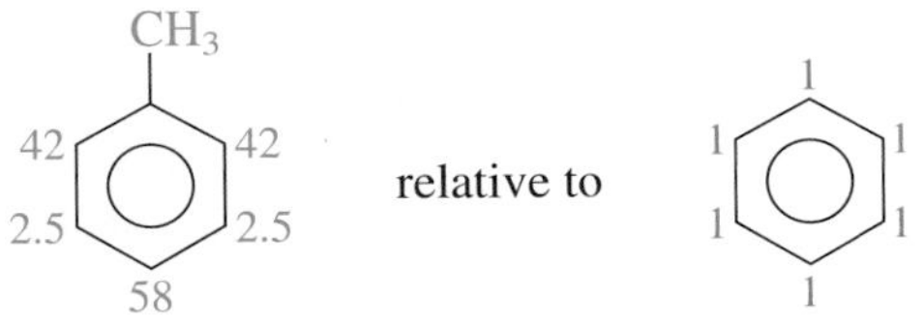

relative to

These relative rate data per position are experimentally determined and are known as **partial rate factors.** They offer a convenient way to express substituent effects in electrophilic aromatic substitutions.

The major influence of the methyl group is its *electronic* effect on carbocation stability. To a small extent, the methyl group sterically hinders the approach of the electrophile to the ortho positions, making substitution slightly slower at a single ortho carbon than at the para carbon. However, para substitution is at a statistical disadvantage because there are two equivalent ortho positions but only one para position.

Problem 13.15

The partial rate factors for nitration of *tert*-butylbenzene are as shown.

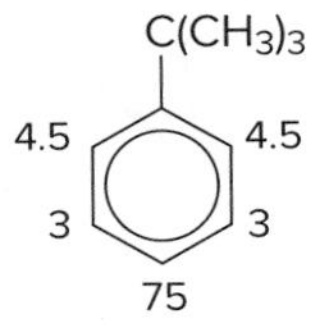

continued

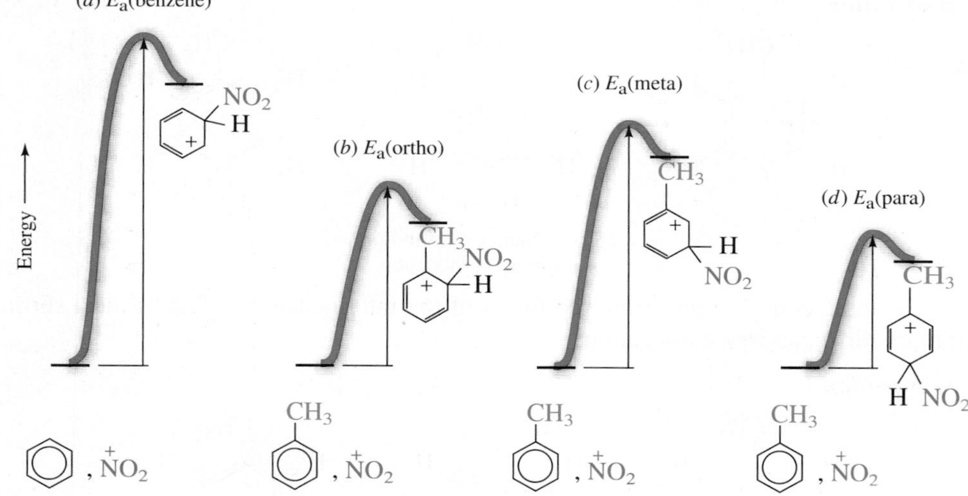

Figure 13.6

Comparative energy diagrams for reaction of nitronium ion with (*a*) benzene and at the (*b*) ortho, (*c*) meta, and (*d*) para positions of toluene. E_a (benzene) > E_a (meta) > E_a (ortho) > E_a (para).

(a) How reactive is *tert*-butylbenzene toward nitration compared with benzene?
(b) How reactive is *tert*-butylbenzene toward nitration compared with toluene?
(c) Predict the distribution among the various mononitration products of *tert*-butylbenzene.

Sample Solution (a) Benzene has six equivalent sites at which nitration can occur. Summing the individual relative rates of nitration at each position in *tert*-butylbenzene compared with benzene, we obtain

$$\frac{\textit{tert}\text{-Butylbenzene}}{\text{Benzene}} = \frac{2(4.5) + 2(3) + 75}{6(1)} = \frac{90}{6} = 15$$

tert-Butylbenzene undergoes nitration 15 times faster than benzene.

All alkyl groups, not just methyl, are electron-releasing, activating substituents and ortho, para directors. This is because any alkyl group, be it methyl, ethyl, isopropyl, *tert*-butyl, or any other, stabilizes a carbocation site to which it is directly attached. When R = alkyl,

and

are more stable than

where E^+ is any electrophile. All three structures are more stable for R = alkyl than for R = H and are formed faster.

13.11 Rate and Regioselectivity in the Nitration of (Trifluoromethyl)benzene

Recall from Section 1.13 that effects that are transmitted by the polarization of σ bonds are called *inductive effects*.

Turning now to electrophilic aromatic substitution in (trifluoromethyl)benzene, we consider the electronic properties of a trifluoromethyl group. Because of their high electronegativity the three fluorine atoms polarize the electron distribution in their σ bonds to carbon, so that carbon bears a partial positive charge.

Unlike a methyl group, which is slightly electron-releasing, trifluoromethyl is a powerful **electron-withdrawing group.** Consequently, CF_3 *destabilizes* a carbocation site to which it is attached.

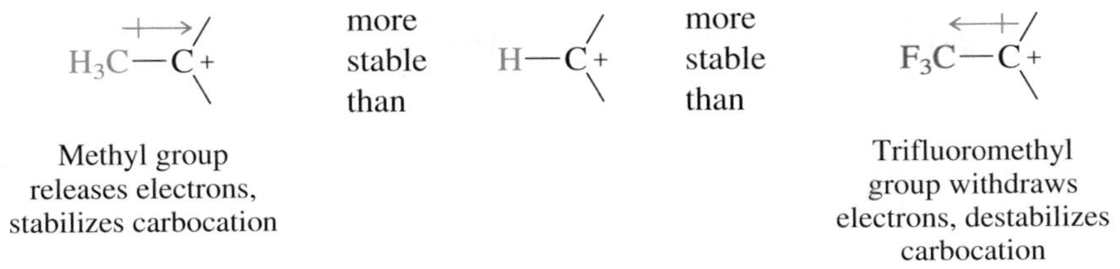

$H_3C—C^+$ more stable than $H—C^+$ more stable than $F_3C—C^+$

Methyl group releases electrons, stabilizes carbocation

Trifluoromethyl group withdraws electrons, destabilizes carbocation

When we examine the cyclohexadienyl cation intermediates involved in the nitration of (trifluoromethyl)benzene, we find that those leading to ortho and para substitution are strongly *destabilized.*

Ortho nitration

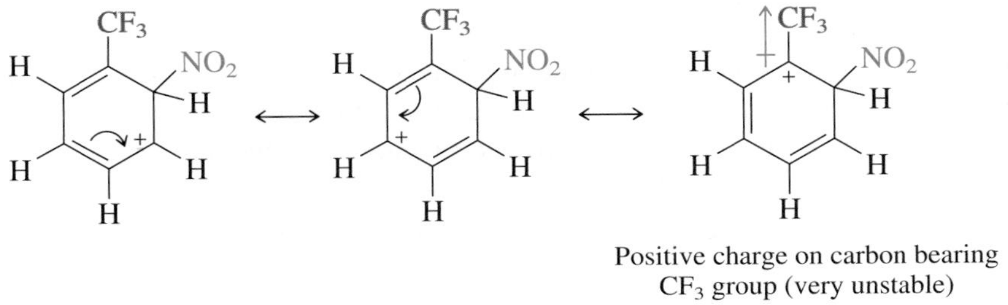

Positive charge on carbon bearing CF_3 group (very unstable)

Para nitration

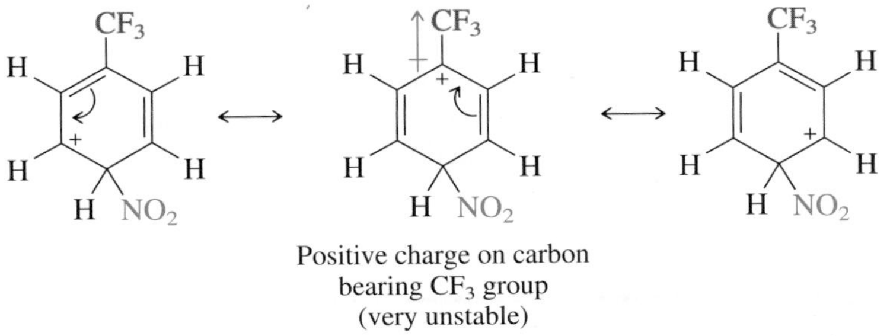

Positive charge on carbon bearing CF_3 group (very unstable)

None of the three major resonance contributors to the carbocation formed when the electrophile bonds to the meta position has a positive charge on the carbon bearing the $—CF_3$ group.

Meta nitration

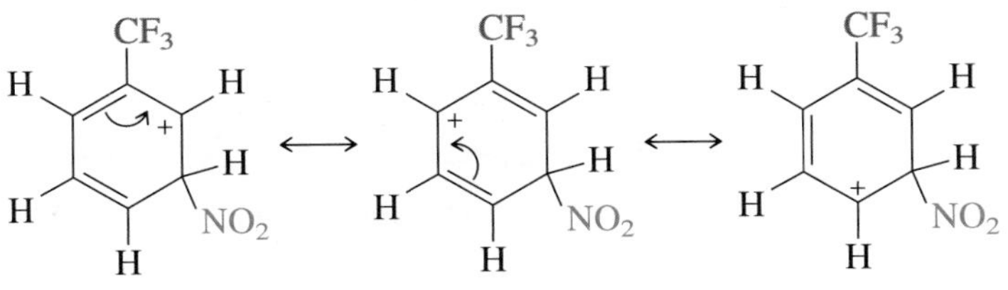

Bonding of NO_2^+ to the meta position gives a more stable intermediate than bonding at either the ortho or the para position, and so meta substitution predominates. Even the carbocation intermediate corresponding to meta nitration, however, is very unstable and is formed with difficulty. The trifluoromethyl group is only one bond farther removed from the positive charge here than it is in the ortho and para intermediates and so still exerts a significant, although somewhat diminished, destabilizing inductive effect.

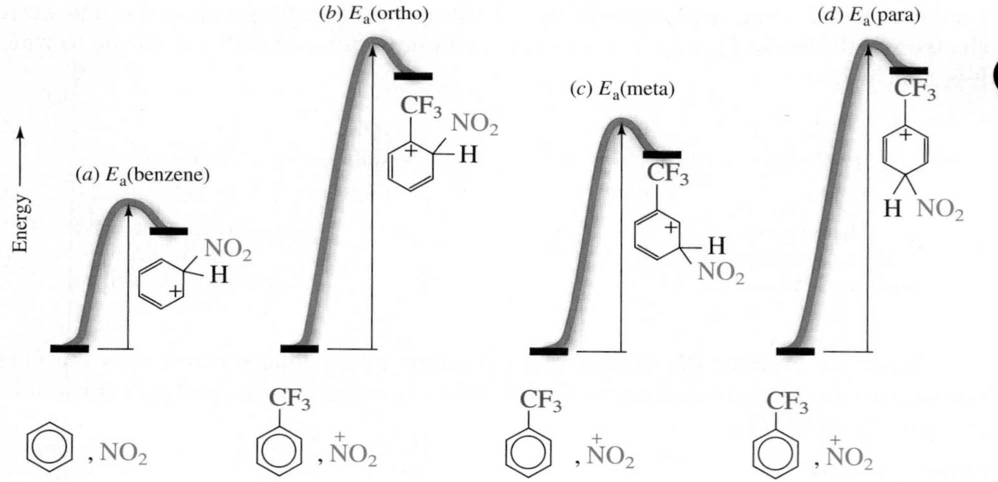

Figure 13.7

Comparative energy diagrams for nitronium ion attachment to (*a*) benzene and at the (*b*) ortho, (*c*) meta, and (*d*) para positions of (trifluoromethyl) benzene. E_a (ortho) > E_a (para) > E_a (meta) > E_a (benzene).

All the ring positions of (trifluoromethyl)benzene are deactivated compared with benzene. The meta position is simply deactivated *less* than the ortho and para positions. The partial rate factors for nitration of (trifluoromethyl)benzene are

CF_3 (ortho: 4.5×10^{-6}, 4.5×10^{-6}; meta: 67×10^{-6}, 67×10^{-6}; para: 4.5×10^{-6}) compared with benzene (1, 1, 1, 1, 1, 1)

Figure 13.7 compares the energy profile for nitration of benzene with those for the ortho, meta, and para positions of (trifluoromethyl)benzene. The presence of the electron-withdrawing trifluoromethyl group raises the activation energy at all the ring positions, but the increase is least for the meta position.

Problem 13.16

The compounds benzyl chloride ($C_6H_5CH_2Cl$), (dichloromethyl)benzene ($C_6H_5CHCl_2$), and (trichloromethyl)benzene ($C_6H_5CCl_3$) all undergo nitration more slowly than benzene. The proportion of m-nitro-substituted product is 4% in one, 34% in another, and 64% in another. Classify the substituents —CH_2Cl, —$CHCl_2$, and —CCl_3 according to each one's effect on rate and regioselectivity in electrophilic aromatic substitution.

13.12 Substituent Effects in Electrophilic Aromatic Substitution: Activating Substituents

Our analysis of substituent effects has so far centered on two groups: methyl and trifluoromethyl. We have seen that a methyl substituent is electron-releasing, activating, and ortho, para-directing. A trifluoromethyl group is strongly electron-withdrawing, deactivating, and meta-directing. What about other substituents?

Table 13.2 summarizes orientation and rate effects in electrophilic aromatic substitution for some frequently encountered substituents. It is arranged in order of decreasing activating power: the most strongly *activating* substituents are at the top, the most strongly

TABLE 13.2 Classification of Substituents in Electrophilic Aromatic Substitution Reactions

Effect on rate	Substituent		Effect on orientation
Very strongly activating	$-\ddot{N}H_2$	(amino)	Ortho, para-directing
	$-\ddot{N}HR$	(alkylamino)	
	$-\ddot{N}R_2$	(dialkylamino)	
	$-\ddot{O}H$	(hydroxyl)	
Strongly activating	$-\ddot{N}HC(=O)R$	(acylamino)	Ortho, para-directing
	$-\ddot{O}R$	(alkoxy)	
	$-\ddot{O}C(=O)R$	(acyloxy)	
Activating	$-R$	(alkyl)	Ortho, para-directing
	$-Ar$	(aryl)	
	$-CH{=}CR_2$	(alkenyl)	
Standard of comparison	$-H$	(hydrogen)	
Deactivating	$-X$ ($X = F, Cl, Br, I$)	(halogen)	Ortho, para-directing
	$-CH_2X$	(halomethyl)	
Strongly deactivating	$-C(=O)H$	(formyl)	Meta-directing
	$-C(=O)R$	(acyl)	
	$-C(=O)OH$	(carboxylic acid)	
	$-C(=O)OR$	(ester)	
	$-C(=O)Cl$	(acyl chloride)	
	$-C{\equiv}N$	(cyano)	
	$-SO_2OH$	(sulfonic acid)	
Very strongly deactivating	$-CF_3$	(trifluoromethyl)	Meta-directing
	$-NO_2$	(nitro)	

deactivating substituents are at the bottom. The main features of the table can be summarized as follows:

1. All activating substituents are ortho, para directors.
2. Halogen substituents are slightly deactivating but are ortho, para-directing.
3. Strongly deactivating substituents are meta directors.

Some of the most powerful activating substituents are those in which an oxygen atom is attached directly to the ring. These substituents include the hydroxyl group as well as alkoxy and acyloxy groups. All are ortho, para directors.

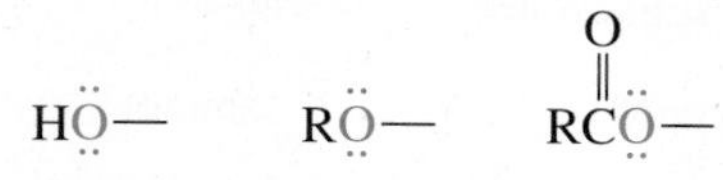

HÖ— Hydroxyl RÖ— Alkoxy RC(=O)Ö— Acyloxy

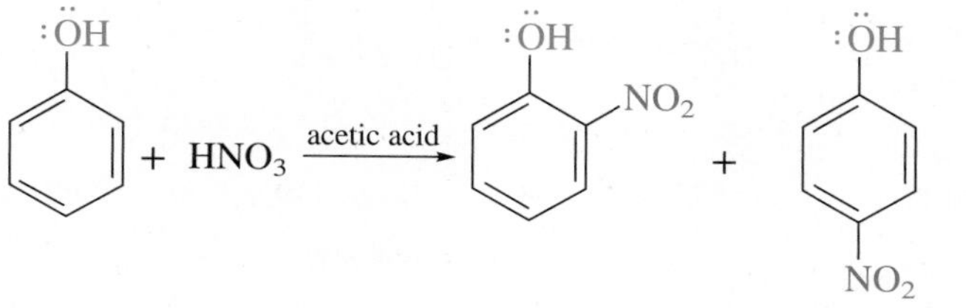

Phenol + HNO_3 $\xrightarrow{\text{acetic acid}}$ o-Nitrophenol (44%) + p-Nitrophenol (56%)

Phenol and anisole are among the commonly encountered benzene derivatives listed in Table 12.1.

Hydroxyl, alkoxy, and acyloxy groups activate the ring to such an extent that bromination occurs rapidly even in the absence of a catalyst.

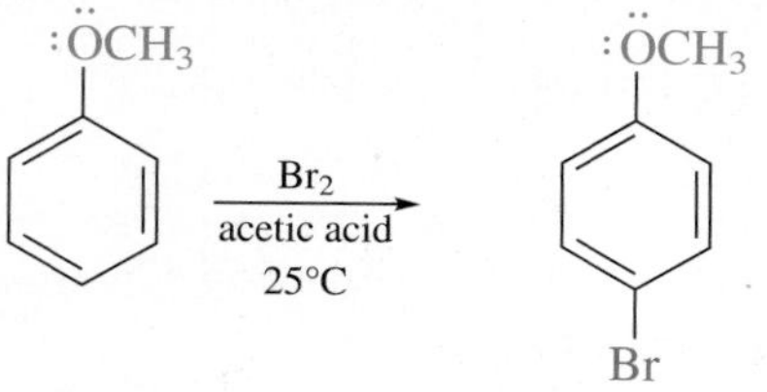

Anisole $\xrightarrow[\text{acetic acid, 25°C}]{Br_2}$ *p*-Bromoanisole (90%)

The *inductive* effect of hydroxyl and alkoxy groups, because of the electronegativity of oxygen, is to withdraw electrons and would seem to require that such substituents be deactivating. This electron-withdrawing inductive effect, however, is overcome by a much larger electron-releasing *resonance* effect involving the unshared electron pairs of oxygen. Bonding of the electrophile at positions ortho and para to a substituent of the type —ÖR gives a cation stabilized by delocalization of an unshared electron pair of oxygen into the π system of the ring.

Ortho attachment of E^+

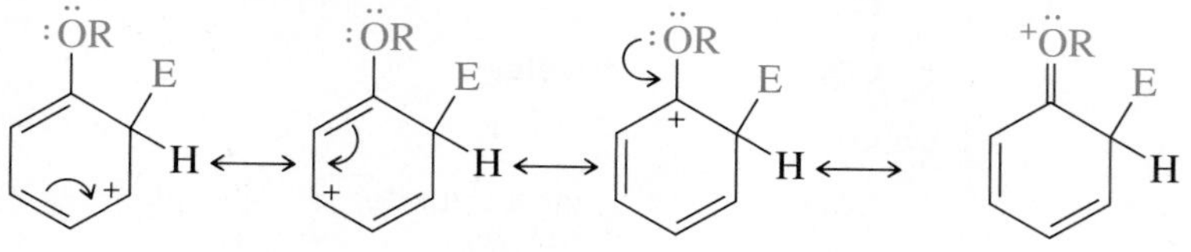

Most stable resonance contributor; oxygen and all carbons have octets of electrons

Para attachment of E^+

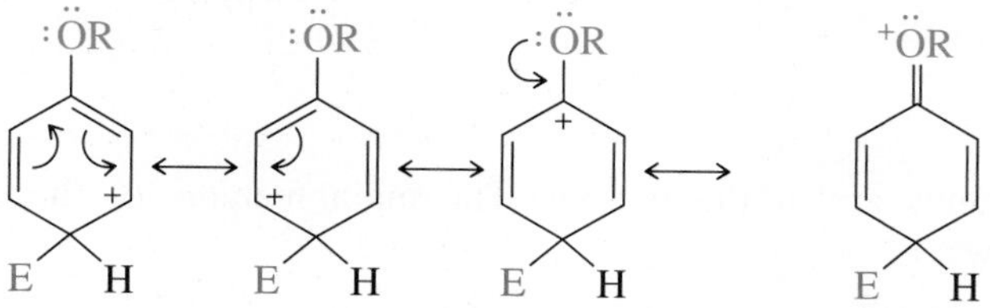

Most stable resonance contributor; oxygen and all carbons have octets of electrons

Oxygen-stabilized carbocations of this type are far more stable than tertiary carbocations. They are best represented by structures in which the positive charge is on oxygen because all the atoms then have octets of electrons. Their stability permits them to be formed rapidly, resulting in rates of electrophilic aromatic substitution that are much faster than that of benzene.

Meta attachment of E^+

The lone pair on oxygen cannot be directly involved in carbocation stabilization when the electrophile bonds to the meta carbon.

Oxygen lone pair cannot be used to stabilize positive charge in any of these structures; all have six electrons around positively charged carbon.

The greater stability of the carbocation intermediates arising from bonding of the electrophile to the ortho and para carbons compared with those at the carbon meta to oxygen explains the ortho, para-directing property of —ÖH, —ÖR, and —ÖC(O)R groups.

Monobromination of phenol occurs in organic solvents at low temperature and in the absence of a catalyst. In polar solvents such as water, however, it is difficult to limit the bromination of phenols to monosubstitution.

Problem 13.17

Bromination of phenol in water at 25°C gives a product with the formula $C_6H_3Br_3O$. What is its structure?

Nitrogen-containing substituents related to the amino group are even better electron-releasing groups and more strongly activating than the corresponding oxygen-containing substituents.

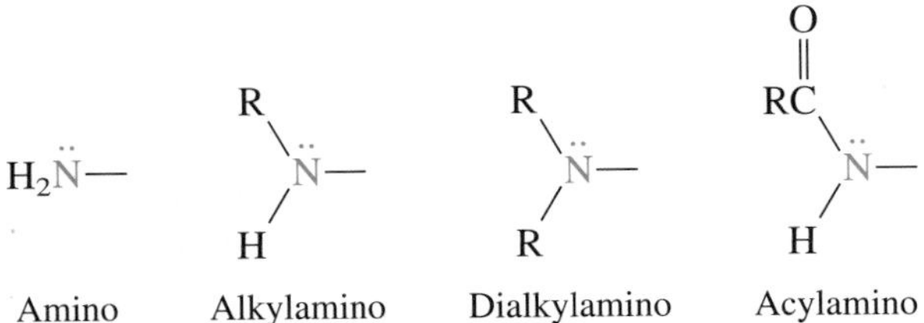

The nitrogen atom in each of these groups bears an electron pair that, like the unshared pairs of oxygen, stabilizes a carbocation to which it is attached. Nitrogen is less electronegative than oxygen, so is a better electron-pair donor and stabilizes the cyclohexadienyl cation intermediates in electrophilic aromatic substitution to an even greater degree.

Problem 13.18

Write structural formulas for the cyclohexadienyl cations formed from aniline ($C_6H_5NH_2$) during

(a) Ortho bromination (four resonance structures)
(b) Meta bromination (three resonance structures)
(c) Para bromination (four resonance structures)

Aniline and its derivatives are so reactive in electrophilic aromatic substitution that special strategies are usually necessary to carry out these reactions effectively. This topic is discussed in Section 22.14.

continued

Sample Solution (a) There are the customary three resonance contributors for the cyclohexadienyl cation plus a contributor (the most stable one) derived by delocalization of the nitrogen lone pair into the ring.

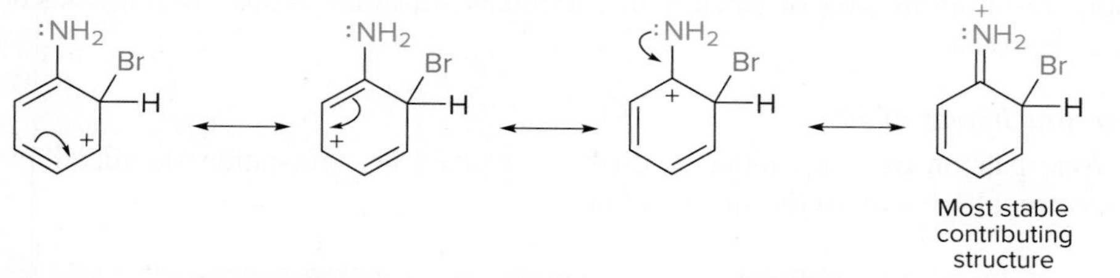

:NH2 Br H ⟷ :NH2 Br H ⟷ :NH2 Br H ⟷ $\overset{+}{N}H_2$ Br H

Most stable contributing structure

Alkyl groups are, as we saw when we discussed the nitration of toluene in Section 13.10, activating and ortho, para-directing substituents. Aryl and alkenyl substituents resemble alkyl groups in this respect; they too are activating and ortho, para-directing.

Problem 13.19

Treatment of biphenyl (see Section 12.5 to remind yourself of its structure) with a mixture of nitric acid and sulfuric acid gave two principal products both having the molecular formula $C_{12}H_9NO_2$. What are these two products?

The next group of substituents in Table 13.2 that we'll discuss are the ones near the bottom of the table, those that are meta-directing and strongly deactivating.

13.13 Substituent Effects in Electrophilic Aromatic Substitution: Strongly Deactivating Substituents

As Table 13.2 indicates, a number of substituents are *meta-directing and strongly deactivating*. We have already discussed one of these, the trifluoromethyl group. Several others have a carbonyl group attached directly to the aromatic ring.

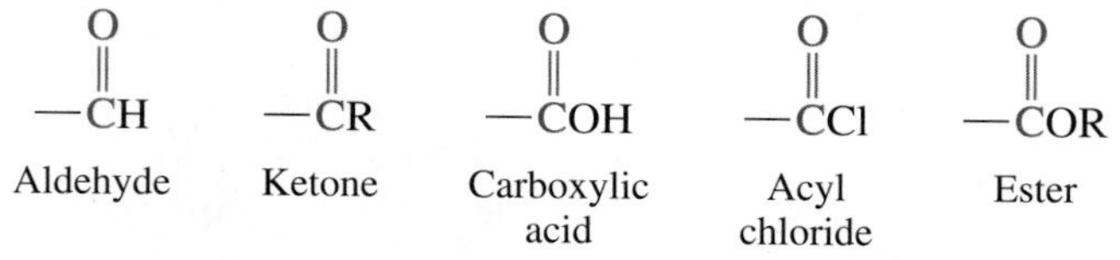

—CH(=O)	—CR(=O)	—COH(=O)	—CCl(=O)	—COR(=O)
Aldehyde	Ketone	Carboxylic acid	Acyl chloride	Ester

The behavior of aromatic aldehydes is typical. Nitration of benzaldehyde takes place several thousand times more slowly than that of benzene and yields *m*-nitrobenzaldehyde as the major product.

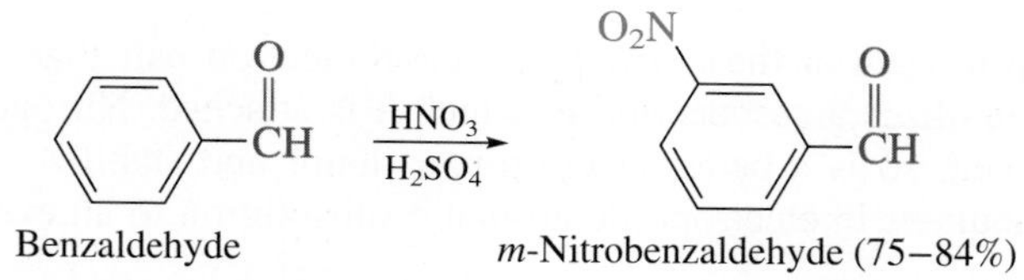

$$\text{C}_6\text{H}_5\text{CHO} \xrightarrow[H_2SO_4]{HNO_3} m\text{-}O_2N\text{C}_6\text{H}_4\text{CHO}$$

Benzaldehyde → *m*-Nitrobenzaldehyde (75–84%)

To understand the effect of a carbonyl group attached directly to the ring, consider its polarization. The electrons in the carbon–oxygen double bond are drawn toward oxygen and away from carbon, leaving the carbon attached to the ring with a partial positive charge.

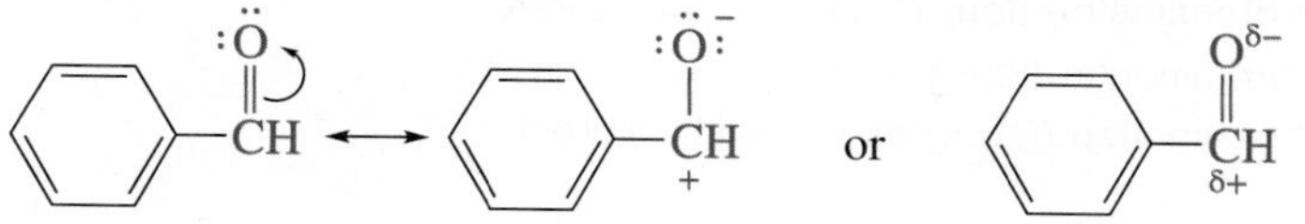

$$\text{C}_6\text{H}_5\text{CH}{=}\ddot{\text{O}}: \longleftrightarrow \text{C}_6\text{H}_5\overset{+}{\text{C}}\text{H}{-}\ddot{\text{O}}:^- \quad \text{or} \quad \text{C}_6\text{H}_5\overset{\delta+}{\text{C}}\text{H}{=}\text{O}^{\delta-}$$

Because the carbon atom attached to the ring is positively polarized, a carbonyl group is *strongly electron-withdrawing* and behaves in much the same way as a trifluoromethyl group to destabilize all the cyclohexadienyl cation intermediates in electrophilic aromatic substitution. Reaction at any ring position in benzaldehyde is slower than in benzene. The intermediates for ortho and para substitution are particularly unstable because each has a resonance contributor in which there is a positive charge on the carbon that bears the electron-withdrawing group. The intermediate for meta substitution avoids this unfavorable juxtaposition of positive charges, is not as unstable, and gives rise to most of the product. For the nitration of benzaldehyde:

Ortho nitration — $^{\delta+}CH{=}O^{\delta-}$, NO_2, H — Unstable because of adjacent positively polarized atoms

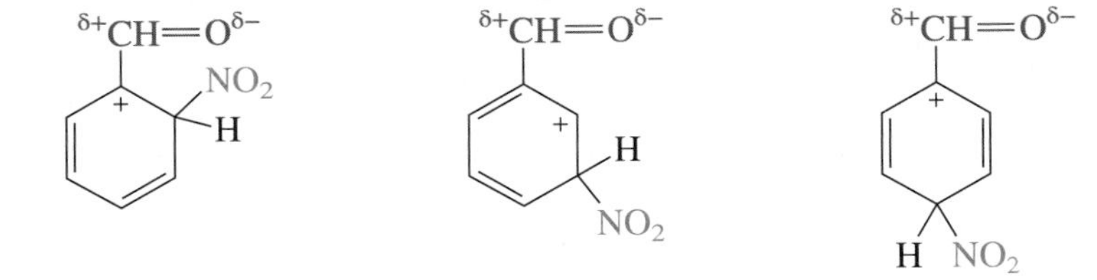

Meta nitration — $^{\delta+}CH{=}O^{\delta-}$, H, NO_2 — Positively polarized atoms not adjacent; most stable intermediate

Para nitration — $^{\delta+}CH{=}O^{\delta-}$, H NO_2 — Unstable because of adjacent positively polarized atoms

Problem 13.20

Each of the following reactions has been reported in the chemical literature, and the major organic product has been isolated in good yield. Write a structural formula for the product of each reaction.

(a) Treatment of benzoyl chloride ($C_6H_5\overset{O}{\overset{\|}{C}}Cl$) with chlorine and iron(III) chloride

(b) Treatment of methyl benzoate ($C_6H_5\overset{O}{\overset{\|}{C}}OCH_3$) with nitric acid and sulfuric acid

(c) Nitration of 1-phenyl-1-propanone ($C_6H_5\overset{O}{\overset{\|}{C}}CH_2CH_3$)

Sample Solution (a) Benzoyl chloride has a carbonyl group attached directly to the ring. A—C(O)Cl substituent is meta-directing. The combination of chlorine and iron(III) chloride introduces a chlorine onto the ring. The product is *m*-chlorobenzoyl chloride.

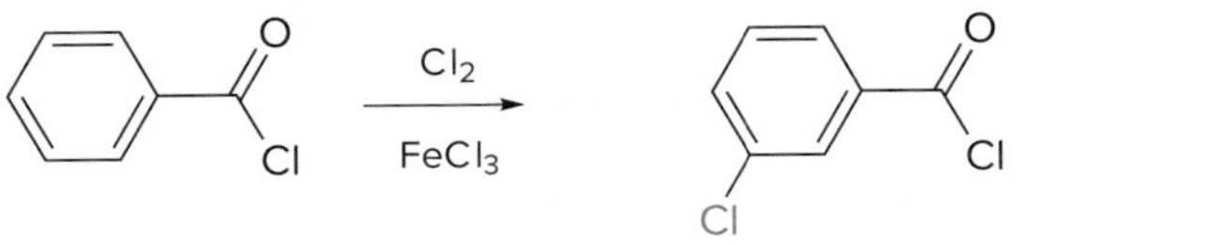

$$\text{Benzoyl chloride} \xrightarrow[FeCl_3]{Cl_2} m\text{-Chlorobenzoyl chloride (62\%)}$$

A cyano group is similar to a carbonyl for analogous reasons involving contributing resonance structures of the type shown for benzonitrile.

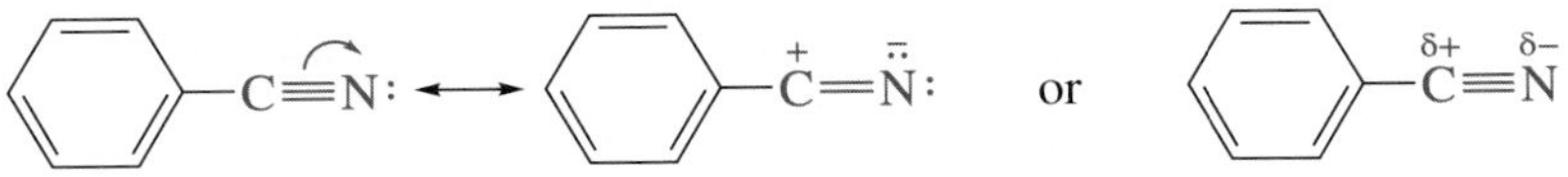

$$C_6H_5-C{\equiv}N: \longleftrightarrow C_6H_5-\overset{+}{C}{=}\ddot{N}:^{-} \quad \text{or} \quad C_6H_5-\overset{\delta+}{C}{\equiv}\overset{\delta-}{N}$$

Cyano groups are electron-withdrawing, deactivating, and meta-directing.

Sulfonic acid groups are electron-withdrawing because sulfur has a formal positive charge in several of its principal resonance contributors.

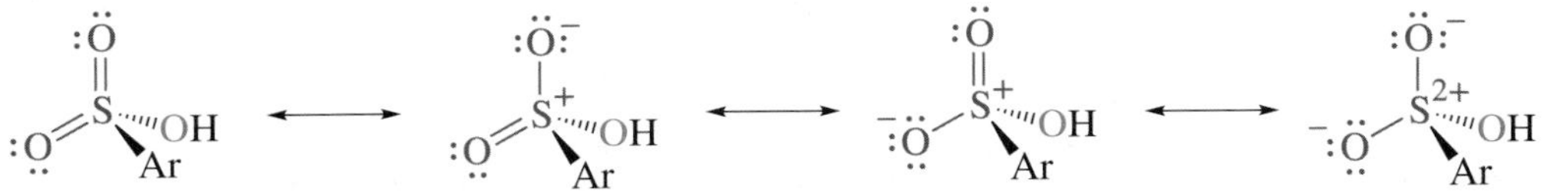

$$ArS(=O)_2OH \longleftrightarrow ArS^{+}(-O^{-})(=O)OH \longleftrightarrow ArS^{+}(=O)(-O^{-})OH \longleftrightarrow ArS^{2+}(-O^{-})_2OH$$

When benzene undergoes disulfonation, *m*-benzenedisulfonic acid is formed. The first sulfonic acid group to go on directs the second one meta to itself.

$$\text{Benzene} \xrightarrow[H_2SO_4]{SO_3} \text{Benzenesulfonic acid} \xrightarrow[H_2SO_4]{SO_3} m\text{-Benzenedisulfonic acid (90\%)}$$

The nitrogen atom of a nitro group bears a full positive charge in its two most stable contributing structures.

$$Ar-\overset{+}{N}(=O)-O^- \longleftrightarrow Ar-\overset{+}{N}(-O^-)=O$$

This makes the nitro group a powerful electron-withdrawing, deactivating substituent and a meta director.

$$\text{Nitrobenzene} \xrightarrow[Fe]{Br_2} m\text{-Bromonitrobenzene (60–75\%)}$$

Problem 13.21

Would you expect the substituent $-\overset{+}{N}(CH_3)_3$ to more closely resemble $-\ddot{N}(CH_3)_2$ or $-NO_2$ in its effect on rate and regioselectivity in electrophilic aromatic substitution? Why?

$-\ddot{N}(CH_3)_2$: unshared pair on N, strongly activating, ortho, para director

$-\overset{+}{N}(=O)-O^-$: positive charge on N, strongly deactivating, meta director

$-\overset{+}{N}(CH_3)_3$: positive charge on N. Therefore, resembles $-NO_2$ more than $-N(CH_3)_2$

Therefore, predict strongly deactivating, meta director

13.14 Substituent Effects in Electrophilic Aromatic Substitution: Halogens

Returning to Table 13.2, notice that *halogen substituents are ortho, para-directing, but deactivating*. Nitration of chlorobenzene is a typical example; its rate is some 30 times slower than the corresponding nitration of benzene, and the major products are *o*-chloronitrobenzene and *p*-chloronitrobenzene.

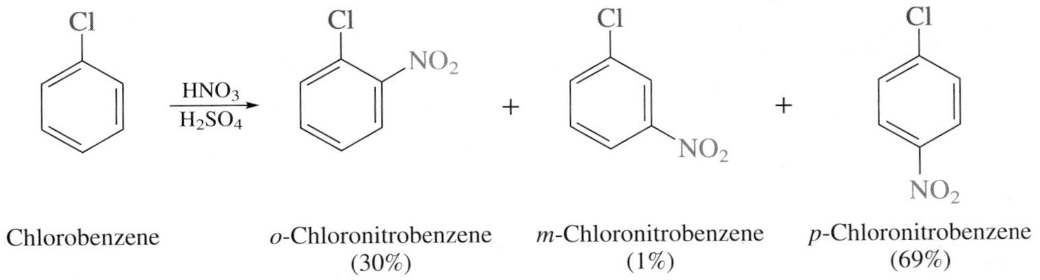

Chlorobenzene $\xrightarrow[H_2SO_4]{HNO_3}$ *o*-Chloronitrobenzene (30%) + *m*-Chloronitrobenzene (1%) + *p*-Chloronitrobenzene (69%)

Problem 13.22

Reaction of chlorobenzene with *p*-chlorobenzyl chloride and aluminum chloride gave a mixture of two products in good yield (76%). What were these two products?

Rate and product studies of electrophilic aromatic substitution in halobenzenes reveal a fairly consistent pattern of reactivity. The partial rate factors for chlorination show that, with one exception, all the ring positions of fluoro-, chloro-, and bromobenzene are deactivated. The exception is the para position of fluorobenzene, which is slightly more reactive than a single position of benzene.

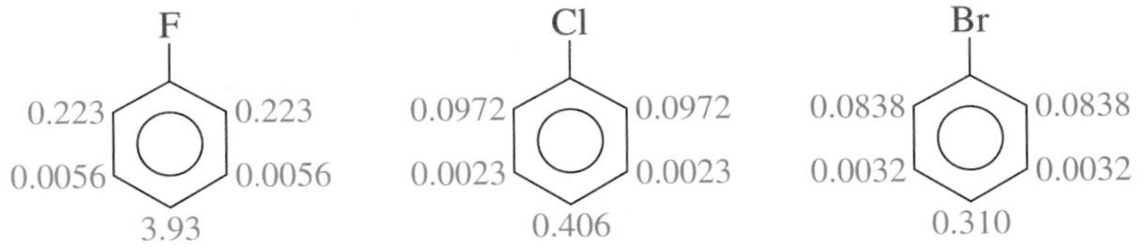

	F	Cl	Br
ortho	0.223	0.0972	0.0838
meta	0.0056	0.0023	0.0032
para	3.93	0.406	0.310

The range of reactivity is not large. Benzene undergoes chlorination only about 1.4 times faster than the most reactive of the group (fluorobenzene) and 14 times faster than the least reactive (bromobenzene). In each halobenzene the para position is the most reactive, followed by ortho.

Because we have come to associate activating substituents with ortho, para-directing effects and deactivating substituents with meta, the properties of halogen substituents appear on initial inspection to be unusual. The seeming inconsistency between regioselectivity and rate can be understood by analyzing the inductive and resonance effects of a halogen substituent.

Through its inductive effect, a halogen withdraws electrons from the ring by polarization of the σ framework. The effect is greatest for fluorine, least for iodine.

Inductive effect of halogen attracts electrons from ring

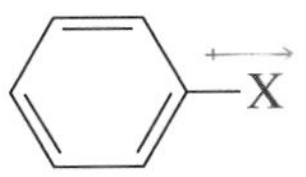

This polarization, in turn, causes the ring carbons to bind the π electrons more tightly and raises the activation energy for electrophilic aromatic substitution, and decreases the reaction rate. Figure 13.8 illustrates this effect by comparing the electrostatic potential maps of fluorobenzene and benzene.

Figure 13.8

Electrostatic potential maps of benzene and fluorobenzene. The high electronegativity of fluorine causes the π electrons of fluorobenzene to be more strongly held than those of benzene. This difference is reflected in the more pronounced red color associated with the π electrons of benzene. The color scale is the same for both models.

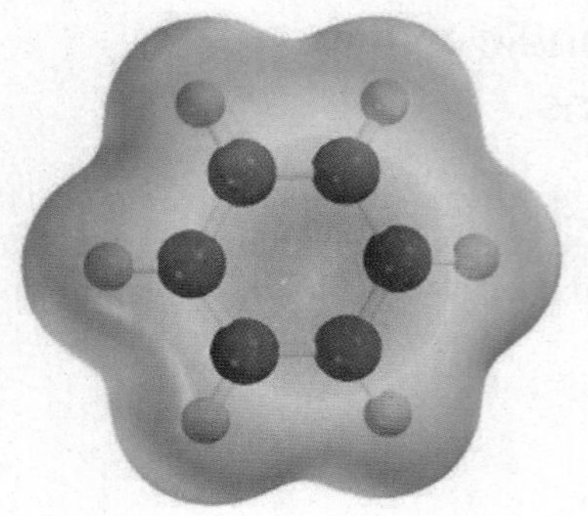

Benzene

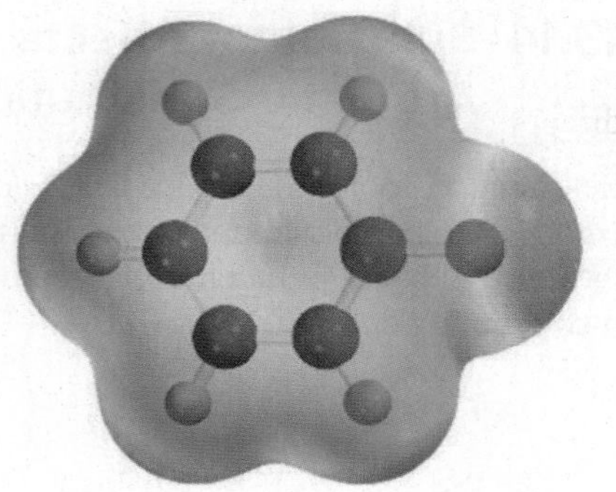

Fluorobenzene

Like $—\ddot{O}H$ and $—\ddot{N}H_2$ groups, however, halogen substituents possess unshared electron pairs that can be donated to a positively charged carbon. This electron donation into the π system stabilizes the intermediates for ortho and para substitution.

Ortho attachment of E^+ ***Para attachment of E^+***

Comparable stabilization of the intermediate for meta substitution is not possible. Thus, resonance involving their lone pairs causes halogens to be ortho, para-directing substituents.

The resonance effect is much greater for fluorine than for the other halogens. For resonance stabilization to be effective, the lone-pair *p* orbital of the substituent must overlap with the π system of the ring. The 2*p* orbital of fluorine is well suited for such overlap, but the 3*p* orbital of chlorine is not because of its more diffuse character and the longer C—Cl bond distance. The situation is even more pronounced for Br and I.

By stabilizing the cyclohexadienyl cation intermediate, lone-pair donation from fluorine counteracts the inductive effect to the extent that the rate of electrophilic aromatic substitution in fluorobenzene is usually only slightly less than that of benzene. With the other halogens, lone-pair donation is sufficient to make them ortho, para directors, but is less than that of fluorine.

13.15 Multiple Substituent Effects

When a benzene ring bears two or more substituents, both its reactivity and the site of further substitution can usually be predicted from the cumulative effects of its substituents.

Problems 13.2, 13.4, and 13.11 offer additional examples of reactions in which only a single product of electrophilic aromatic substitution is possible.

In the simplest cases all the available sites are equivalent, and substitution at any one of them gives the same product.

p-Xylene + Acetic anhydride $\xrightarrow{AlCl_3}$ 2,5-Dimethylacetophenone (99%)

Often the directing effects of substituents reinforce each other. Bromination of *p*-nitrotoluene, for example, takes place at the position that is ortho to the ortho, para-directing methyl group and meta to the meta-directing nitro group.

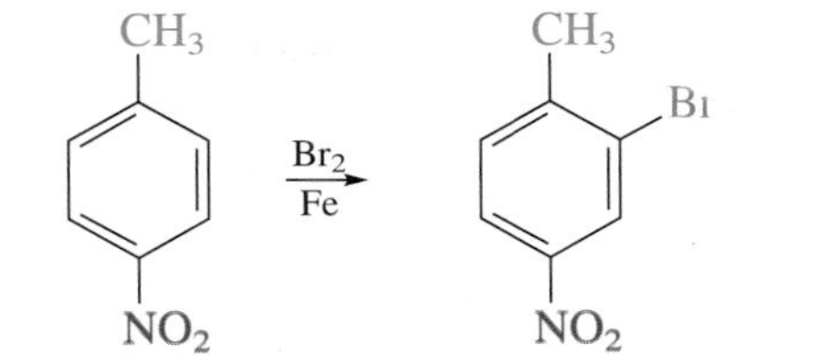

p-Nitrotoluene $\xrightarrow[\text{Fe}]{\text{Br}_2}$ 2-Bromo-4-nitrotoluene (86–90%)

In almost all cases, including most of those in which the directing effects of individual substituents oppose each other, *it is the more activating substituent that controls the regioselectivity of electrophilic aromatic substitution.* Thus, bromination occurs ortho to the *N*-methylamino group in 4-chloro-*N*-methylaniline because this group is a very powerful activating substituent while the chlorine is weakly deactivating.

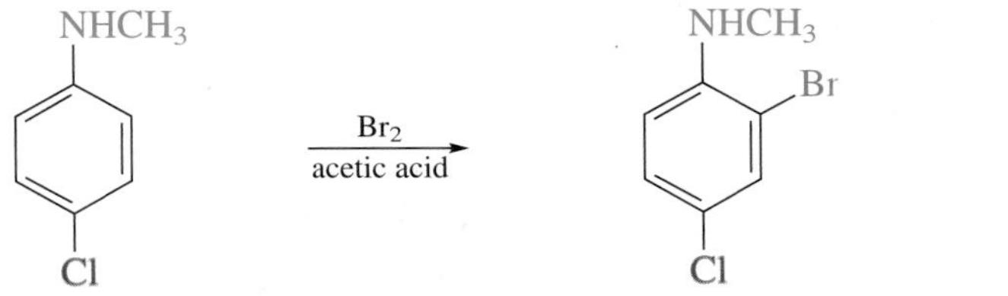

4-Chloro-*N*-methylaniline $\xrightarrow[\text{acetic acid}]{\text{Br}_2}$ 2-Bromo-4-chloro-*N*-methylaniline (87%)

Problem 13.23

The reactant in the preceding equation (4-chloro-*N*-methylaniline) is so reactive toward electrophilic aromatic substitution that no catalyst is necessary to bring about its bromination. Write a reasonable mechanism for the preceding reaction based on Br_2 as the electrophile.

Problem 13.24

Compound A is an intermediate in the synthesis of labetelol, a drug known as a "β-blocker" used to treat hypertension. What is the structure of compound A?

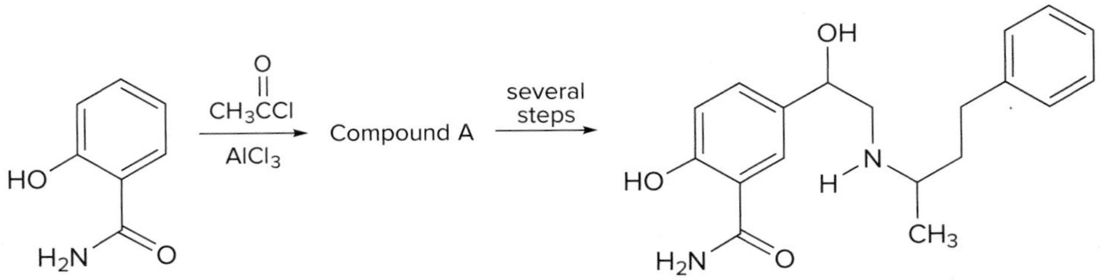

$\xrightarrow[\text{AlCl}_3]{\text{CH}_3\text{CCl (O)}}$ Compound A $\xrightarrow{\text{several steps}}$

Labetelol (a β-blocker)

When two positions are comparably activated by alkyl groups, substitution usually occurs at the less hindered site. Nitration of *p-tert*-butyltoluene takes place at positions ortho to the methyl group in preference to those ortho to the larger *tert*-butyl group. This is an example of a *steric effect.*

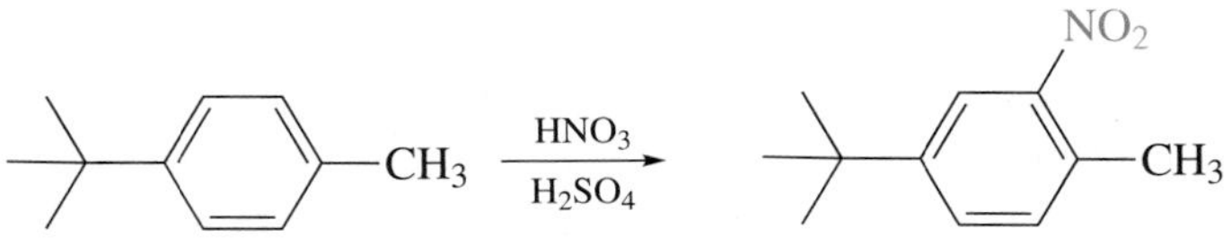

p-tert-Butyltoluene $\xrightarrow[\text{H}_2\text{SO}_4]{\text{HNO}_3}$ 4-*tert*-Butyl-2-nitrotoluene (88%)

Nitration of *m*-xylene is directed ortho to one methyl group and para to the other.

$$m\text{-Xylene} \xrightarrow[H_2SO_4]{HNO_3} \text{2,4-Dimethyl-1-nitrobenzene (98\%)}$$

The ortho position between the two methyl groups is less reactive because it is more sterically hindered.

Problem 13.25

Write the structure of the principal organic product obtained on nitration of each of the following:

(a) *p*-Methylbenzoic acid
(b) *m*-Dichlorobenzene
(c) *m*-Dinitrobenzene
(d) *p*-Methoxyacetophenone
(e) *p*-Methylanisole
(f) 4-hydroxy-3-methoxybenzaldehyde

Sample Solution (a) Of the two substituents in *p*-methylbenzoic acid, the methyl group is more activating and controls the regioselectivity of electrophilic aromatic substitution. The position para to the ortho, para-directing methyl group already bears a substituent (the carboxyl group), and so substitution occurs ortho to the methyl group. This position is meta to the m-directing carboxyl group, and the orienting properties of the two substituents reinforce each other. The product is 4-methyl-3-nitrobenzoic acid.

$$p\text{-Methylbenzoic acid} \xrightarrow[H_2SO_4]{HNO_3} \text{4-Methyl-3-nitrobenzoic acid}$$

An exception to the rule that regioselectivity is controlled by the most activating substituent occurs when the directing effects of alkyl groups and halogen substituents oppose each other. Alkyl groups and halogen substituents are weakly activating and weakly deactivating, respectively, and the difference between them is too small to allow a simple generalization.

13.16 Retrosynthetic Analysis and the Synthesis of Substituted Benzenes

Because the regioselectivity of electrophilic aromatic substitution is controlled by the directing effects of substituents already present on the ring, the synthesis of more highly substituted derivatives requires that careful thought be given to the *order* in which the reactions are performed. Retrosynthetic analysis provides a useful guide.

The analysis is often straightforward; one simply disconnects one of the substituents from the ring of the target molecule and examines the ring with respect to the directing properties of the remaining substituent. Consider *m*-bromoacetophenone:

Br^+ + (acetophenone) $\xLeftarrow{a}$ *m*-Bromoacetophenone $\xRightarrow{b}$ Bromobenzene + $CH_3C^+{=}O$

Bromine is an ortho, para director, acetyl a meta director. Reasoning backward from the target, disconnection *a* makes electrophilic bromination of acetophenone the last synthetic step; disconnection *b* makes Friedel–Crafts acylation of bromobenzene the last step. Of the two approaches, only bromination of acetophenone delivers the desired meta relationship of the two substituents and suggests the following synthesis.

Benzene → (acetic anhydride, $AlCl_3$) → Acetophenone (76–83%) → (Br_2, $AlCl_3$) → *m*-Bromoacetophenone (59%)

Aluminum chloride is a stronger Lewis acid than iron(III) bromide and has been used as a catalyst in electrophilic bromination when, as in the example shown, the aromatic ring bears a strongly deactivating substituent.

Conversely, when *p*-bromoacetophenone is the target, Friedel–Crafts acylation of bromobenzene is the last step (disconnection *b*)

Br^+ + (acetophenone) $\xLeftarrow{a}$ *p*-Bromoacetophenone $\xRightarrow{b}$ Bromobenzene + $CH_3C^+{=}O$

and the synthesis becomes:

Benzene → (Br_2, Fe) → Bromobenzene (65–75%) → (acetic anhydride, $AlCl_3$) → *p*-Bromoacetophenone (69–79%)

A less obvious example in which the success of a synthesis depends on the order of substituent placement on the ring is illustrated by the preparation of *m*-nitroacetophenone. Although both substituents are meta-directing, the only practical synthesis involves nitration of acetophenone (disconnection *a*).

$^+NO_2$ + (acetophenone) $\xLeftarrow{a}$ *m*-Nitroacetophenone $\xRightarrow{b}$ Nitrobenzene + $CH_3C^+{=}O$

Acylation must precede nitration because neither Friedel–Crafts acylation nor alkylation can be carried out successfully on strongly deactivated aromatics such as nitrobenzene.

Benzene —(acetic anhydride, $AlCl_3$)→ Acetophenone (76–83%) —(HNO_3, H_2SO_4)→ *m*-Nitroacetophenone (55%)

An aromatic ring more deactivated than a monohalobenzene cannot be alkylated or acylated under Friedel–Crafts conditions.

When the orientation of substituents in an aromatic compound precludes a synthesis as straightforward as the preceding ones, functional-group manipulation can be useful. *p*-Nitrobenzoic acid, for example, can't be prepared directly from either nitrobenzene or benzoic acid because each subsituent is a meta director. However, by recognizing that the carboxyl group is accessible via oxidation of a methyl group, we can use nitration of toluene to gain the correct regiochemistry.

p-Nitrobenzoic acid ⟹ *p*-Nitrotoluene ⟹ Toluene

With the strategy determined, we transform the retrosynthesis to a synthetic plan and include the appropriate reagents.

Toluene —(HNO_3, H_2SO_4)→ *p*-Nitrotoluene (separate from ortho isomer) —($Na_2Cr_2O_7$, H_2SO_4, H_2O)→ *p*-Nitrobenzoic acid

Problem 13.26

Many syntheses can involve several functional-group transformations. Identify compounds A–C in the retrosynthesis shown and suggest reagents for each synthetic step.

(2-methyl-2-phenyloxirane) ⟹ A ⟹ B ⟹ C ⟹ (benzene)

Problem 13.27

Use retrosynthetic analysis to devise a synthesis of *m*-chloroethylbenzene from benzene. Convert your retrosynthesis to a synthesis and show the necessary reagents for each step. (*Hint:* A Clemmensen or Wolff–Kishner reduction is necessary.)

13.17 Substitution in Naphthalene

Polycyclic aromatic hydrocarbons undergo electrophilic aromatic substitution when treated with the same reagents that react with benzene. In general, polycyclic aromatic hydrocarbons are more reactive than benzene. Most lack the symmetry of benzene, however, and mixtures of products may be formed even on monosubstitution. Among polycyclic aromatic hydrocarbons, we will discuss only naphthalene, and that only briefly.

Two sites are available for substitution in naphthalene: C-1 and C-2. The more reactive site of electrophilic attack is normally C-1.

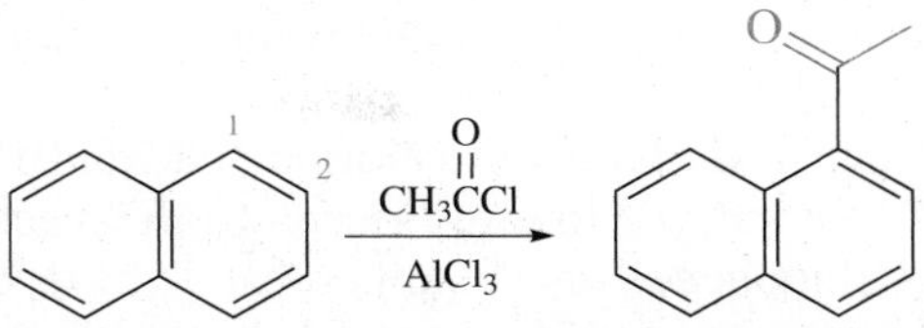

$$\text{Naphthalene} \xrightarrow[\text{AlCl}_3]{\text{CH}_3\overset{\text{O}}{\overset{\|}{\text{C}}}\text{Cl}} \text{1-Acetylnaphthalene (90\%)}$$

C-1 is more reactive because the intermediate formed when the electrophile bonds there is a relatively stable carbocation. A benzene-type pattern of bonds is retained in one ring, and the positive charge is delocalized by allylic resonance.

Attachment of E^+ to C-1

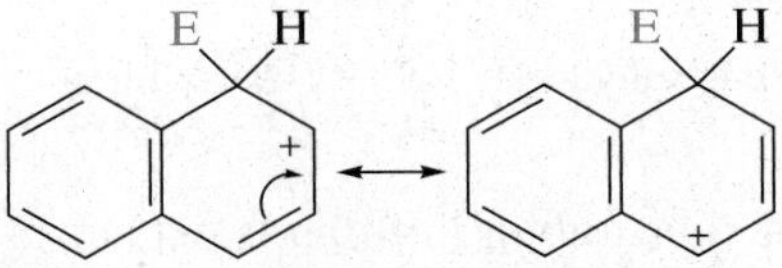

To involve allylic resonance in stabilizing the carbocation intermediate formed when the electrophile bonds to C-2, the benzene-like character of the other ring is sacrificed.

Attachment of E^+ to C-2

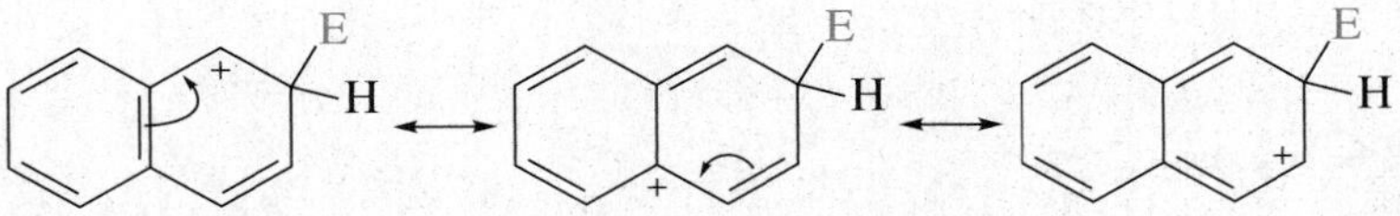

Problem 13.28

Sulfonation of naphthalene is reversible at elevated temperature. A different isomer of naphthalenesulfonic acid is the major product at 160°C than is the case at 0°C. Which isomer is the product of kinetic control? Which one is formed under conditions of thermodynamic control? Can you think of a reason why one isomer is more stable than the other?

13.18 Substitution in Heterocyclic Aromatic Compounds

Their great variety of structural types causes heterocyclic aromatic compounds to range from exceedingly reactive to practically inert toward electrophilic aromatic substitution.

Pyridine lies near one extreme in being far less reactive than benzene toward substitution by electrophilic reagents. In this respect it resembles strongly deactivated aromatic compounds such as nitrobenzene. It is incapable of being acylated or alkylated under Friedel–Crafts conditions, but can be sulfonated at high temperature. Electrophilic substitution in pyridine, when it does occur, takes place at C-3.

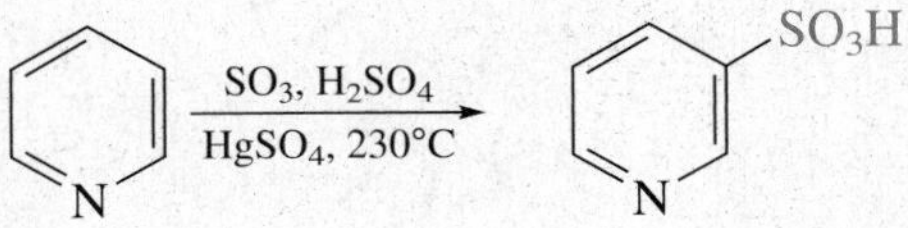

$$\text{Pyridine} \xrightarrow[\text{HgSO}_4,\ 230^\circ\text{C}]{\text{SO}_3,\ \text{H}_2\text{SO}_4} \text{Pyridine-3-sulfonic acid (71\%)}$$

One reason for the low reactivity of pyridine is that nitrogen is more electronegative than carbon, which causes the π electrons of pyridine to be held more tightly and raises the activation energy for bonding to an electrophile. Another is that the nitrogen of pyridine is protonated in sulfuric acid and the resulting pyridinium ion is even more deactivated than pyridine itself.

Lewis acid catalysts such as aluminum chloride and iron(III) halides also bond to nitrogen to strongly deactivate the ring toward Friedel–Crafts reactions and halogenation.

Pyrrole, furan, and thiophene, on the other hand, have electron-rich aromatic rings and are extremely reactive toward electrophilic aromatic substitution—more like phenol and aniline than benzene. Like benzene they have six π electrons, but these π electrons are delocalized over *five* atoms, not six, and are not held as strongly as those of benzene. Even when the ring atom is as electronegative as oxygen, substitution takes place readily.

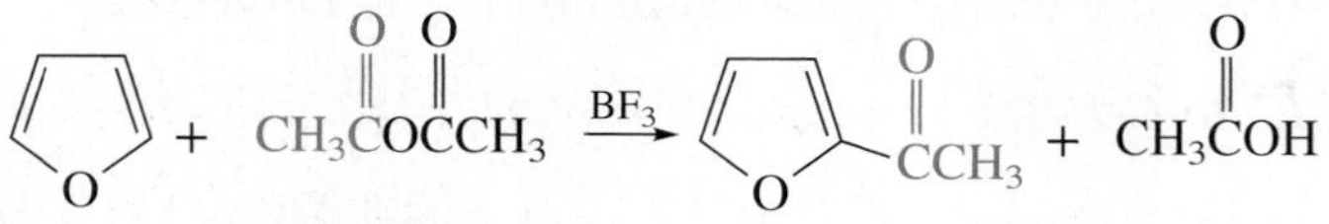

$$\text{Furan} + \underset{\text{Acetic anhydride}}{CH_3\overset{O}{\overset{\|}{C}}O\overset{O}{\overset{\|}{C}}CH_3} \xrightarrow{BF_3} \underset{\text{2-Acetylfuran (75–92\%)}}{\text{2-furyl}-\overset{O}{\overset{\|}{C}}CH_3} + \underset{\text{Acetic acid}}{CH_3\overset{O}{\overset{\|}{C}}OH}$$

The regioselectivity of substitution in furan is explained using a resonance description. When the electrophile bonds to C-2, the positive charge is shared by three atoms: C-3, C-5, and O.

Attachment of E⁺ to C-2

Carbocation *more stable;* positive charge shared by C-3, C-5, and O.

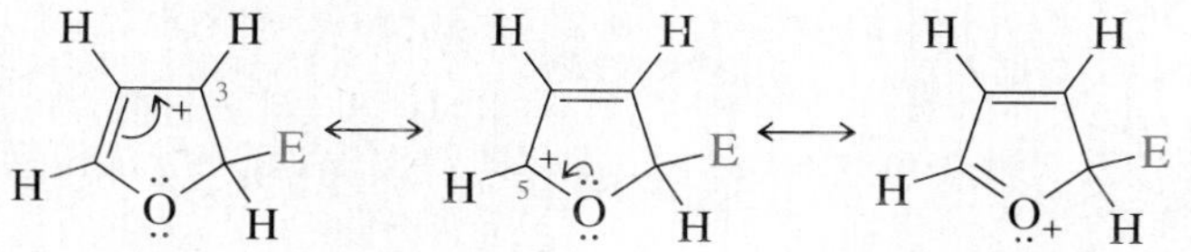

When the electrophile bonds to C-3, the positive charge is shared by only two atoms, C-2 and O, and the carbocation intermediate is less stable and formed more slowly.

Attachment of E⁺ to C-3

Carbocation *less stable;* positive charge shared by C-2 and O.

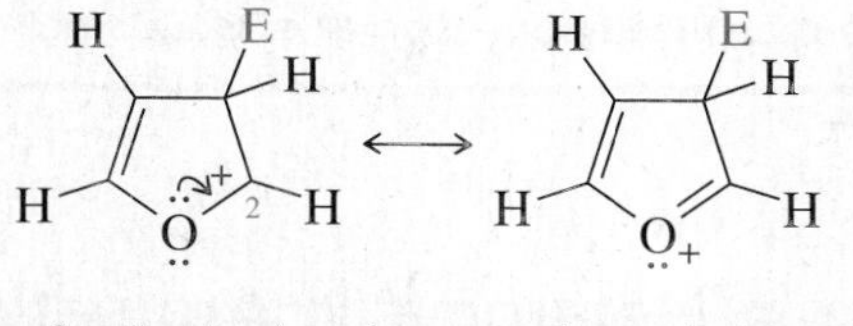

The regioselectivity of substitution in pyrrole and thiophene is like that of furan and for similar reasons.

Problem 13.29

Under acid-catalyzed conditions, the C-2 hydrogen of *N*-methylpyrrole is replaced by deuterium faster than the one at C-3 according to the equation:

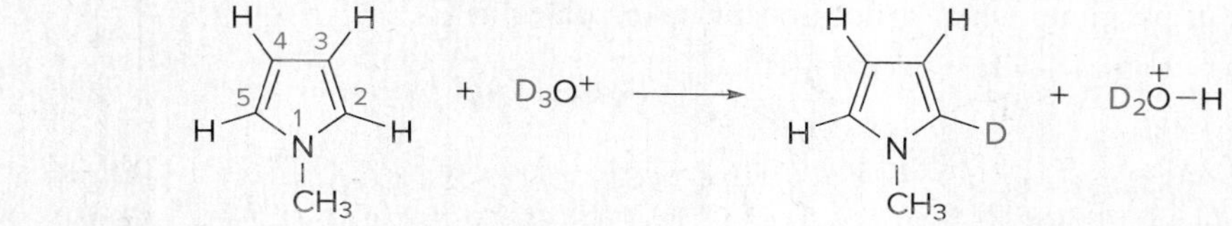

$$\text{N-methylpyrrole} + D_3O^+ \longrightarrow \text{2-deuterio-N-methylpyrrole} + D_2\overset{+}{O}-H$$

Suggest a reasonable mechanism for this reaction.

13.19 Nucleophilic Aromatic Substitution

We have seen numerous examples of *electrophilic* aromatic substitution in this chapter. What about *nucleophilic* aromatic substitutions: reactions in which a nucleophile displaces a halogen substituent from the ring?

In general, aryl halides are much less reactive than alkyl halides toward nucleophilic substitution. One reason for this is that breaking of the carbon–halogen bond is rate-determining in both the S_N1 and S_N2 mechanisms and the carbon–halogen bonds of aryl halides are stronger than those of alkyl halides. A second reason, illustrated in Figure 13.9, is that the optimal transition state geometry for S_N2 processes is inaccessible in aryl halides because the ring itself blocks approach of the nucleophile from the side opposite the carbon–halogen bond.

Aryl halides that bear electronegative substituents, most notably nitro, are an exception. These compounds do undergo nucleophilic substitution relatively readily. For example:

$$p\text{-Chloronitrobenzene} + \underset{\text{Sodium methoxide}}{NaOCH_3} \xrightarrow[85°C]{CH_3OH} \underset{(92\%)}{p\text{-Nitroanisole}} + \underset{\text{Sodium chloride}}{NaCl}$$

The position of the nitro group on the ring is important. Both *o*- and *p*-chloronitrobenzene react at similar rates, but *m*-chloronitrobenzene reacts thousands of times slower. The effect is cumulative; the more *o*- and *p*-nitro groups, the faster the rate.

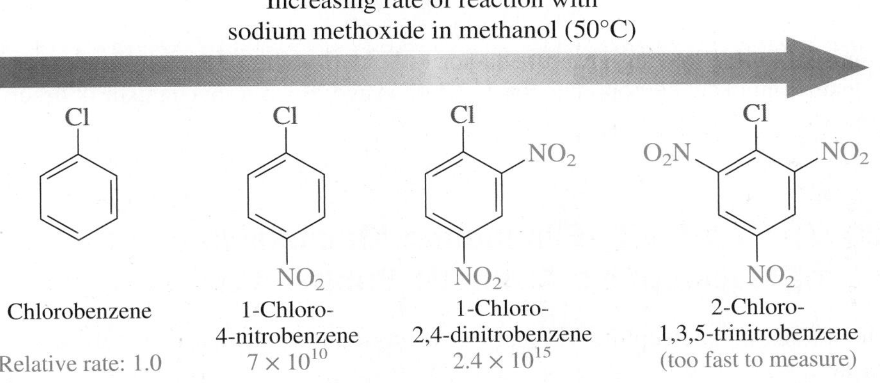

Problem 13.30

Write the structure of the expected product from the reaction of 1-chloro-2,4-dinitrobenzene with each of the following reagents:

(a) CH_3CH_2ONa (b) $C_6H_5CH_2SNa$ (c) CH_3NH_2

Sample Solution (a) Sodium ethoxide is a source of the nucleophile $CH_3CH_2O^-$, which displaces chloride from 1-chloro-2,4-dinitrobenzene.

$$\underset{\text{1-Chloro-2,4-dinitrobenzene}}{} + \underset{\text{Ethoxide ion}}{CH_3CH_2O^-} \longrightarrow \underset{\text{1-Ethoxy-2,4-dinitrobenzene}}{} + \underset{\text{Chloride ion}}{Cl^-}$$

(*a*) Hydroxide ion + methyl chloride

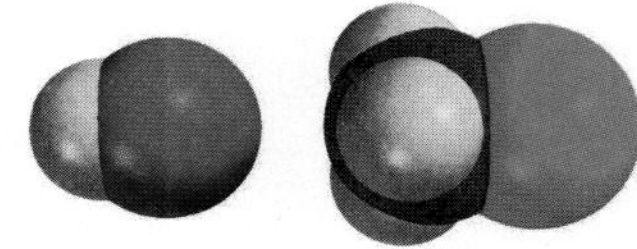

The nucleophile approaches carbon from the side opposite the bond to the leaving group.

(*b*) Hydroxide ion + chlorobenzene

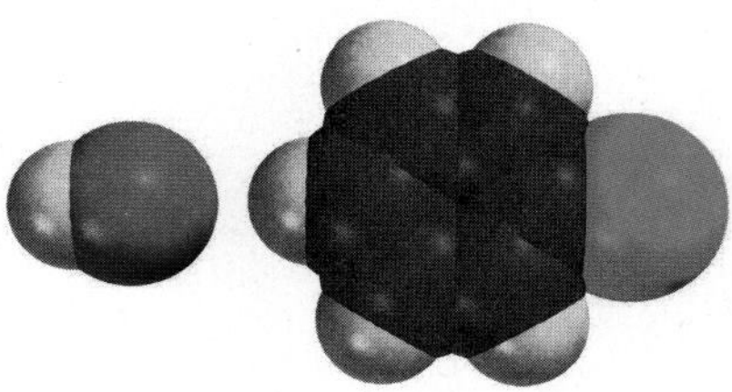

Approach of the nucleophile from opposite the leaving group is blocked by the ring.

Figure 13.9

Nucleophilic substitution by the S_N2 mechanism in aryl halides is blocked by the benzene ring.

In contrast to nucleophilic substitution in alkyl halides, where *alkyl* fluorides are exceedingly unreactive, *aryl* fluorides undergo nucleophilic substitution when the ring bears an o- or p-nitro group. The reaction of 1-fluoro-2,4-dinitrobenzene, known as *Sanger's reagent,* with amino acids takes place readily at room temperature and is the basis of a method used in protein structure determination.

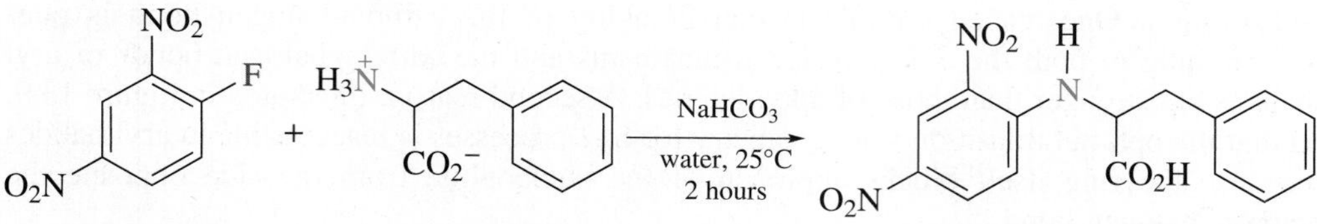

NO_2 F + $H_3\overset{+}{N}$ CO_2^- O_2N — $NaHCO_3$, water, 25°C, 2 hours → NO_2 H N CO_2H O_2N

1-Fluoro-2,4-dinitrobenzene + Phenylalanine → *N*-(2,4-Dinitrophenyl)phenylalanine (67%)

Indeed, the order of leaving-group reactivity in nucleophilic aromatic substitution is the opposite of that seen in aliphatic substitution. *Fluoride is the most reactive leaving group in nucleophilic aromatic substitution, iodide the least reactive.*

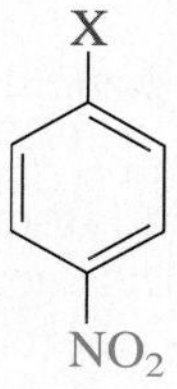

X, NO_2

Relative reactivity toward sodium methoxide in methanol (50°C):

X = F	312
X = Cl	1.0
X = Br	0.8
X = I	0.4

This unusual halide leaving-group behavior tells us that the rate-determining step in the mechanism cannot be carbon–halogen bond cleavage. Such a mechanism is described in the next section.

13.20 The Addition–Elimination Mechanism of Nucleophilic Aromatic Substitution

Kinetic studies of nucleophilic aromatic substitutions reveal that they follow a second-order rate law:

$$\text{Rate} = k[\text{Aryl halide}][\text{Nucleophile}]$$

which suggests a biomolecular rate-determining step. In this case, then, we look for a mechanism in which both the aryl halide and the nucleophile are involved in the slowest step.

This mechanism is sometimes called S_NAr (*substitution-nucleophilic-aromatic*).

The generally accepted mechanism for nucleophilic aromatic substitution in nitro-substituted aryl halides, illustrated for the reaction of *p*-fluoronitrobenzene with sodium methoxide, is outlined in Mechanism 13.6. It is a two-step **addition–elimination mechanism,** in which addition of the nucleophile to the aryl halide is followed by elimination of the halide leaving group. The mechanism is consistent with the following experimental observations:

1. *Kinetics:* Consistent with the observed second-order rate law, the rate-determining step (step 1) involves both the aryl halide and the nucleophile.
2. *Rate-enhancing effect of the nitro group:* The nucleophilic addition step is rate-determining because the aromatic character of the ring must be sacrificed to form the cyclohexadienyl anion intermediate. Only when the anionic intermediate is stabilized by the presence of a strong electron-withdrawing substituent ortho or para to the leaving group will the activation energy for its formation be low enough to

provide a reasonable reaction rate. We can illustrate the stabilization that a p-nitro group provides by examining the resonance structures for the cyclohexadienyl anion formed from methoxide and *p*-fluoronitrobenzene:

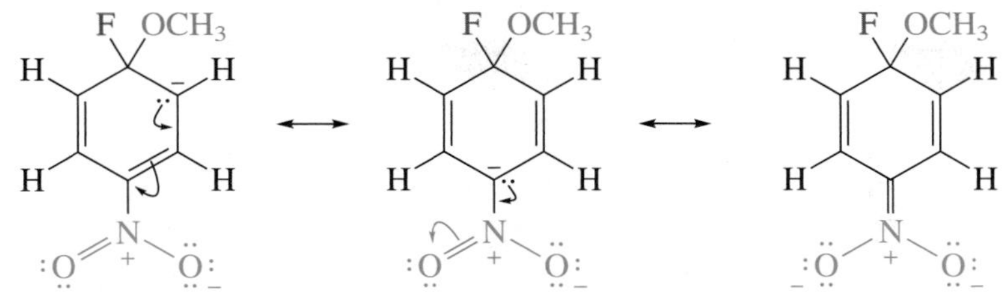

Most stable contributing structure; negative charge is on oxygen

Problem 13.31

Write the most stable contributing structure for the cyclohexadienyl anion formed by reaction of methoxide ion with *o*-fluoronitrobenzene.

m-Fluoronitrobenzene reacts with sodium methoxide 10^5 times more slowly than its ortho and para isomers. According to the resonance description, direct conjugation of the negatively charged carbon with the nitro group is not possible in the cyclohexadienyl anion intermediate from *m*-fluoronitrobenzene, and the decreased reaction rate reflects the decreased stabilization afforded this intermediate.

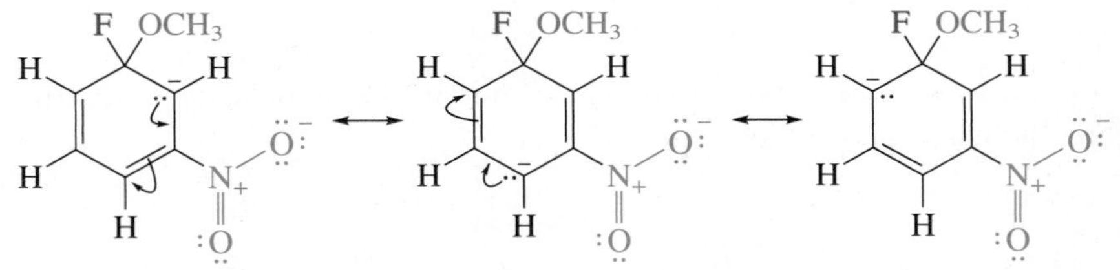

(Negative charge is restricted to carbon in all resonance contributors)

Problem 13.32

Reaction of 1,2,3-tribromo-5-nitrobenzene with sodium ethoxide in ethanol gave a single product, $C_8H_7Br_2NO_3$, in quantitative yield. Suggest a reasonable structure for this compound.

3. *Leaving-group effects:* Because aryl fluorides have the strongest carbon–halogen bond yet react fastest, the rate-determining step cannot involve carbon–halogen bond cleavage. According to Mechanism 13.6 the carbon–halogen bond breaks in the rapid elimination step that follows the rate-determining addition step. The unusually high reactivity of aryl fluorides arises because fluorine is the most electronegative of the halogens, and its greater ability to attract electrons increases the rate of formation of the cyclohexadienyl anion intermediate in the first step of the mechanism.

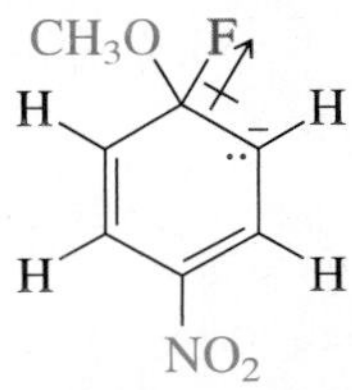

is more stable than

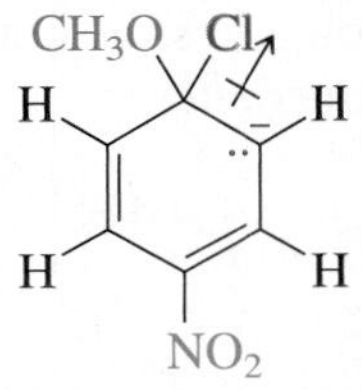

Fluorine stabilizes cyclohexadienyl anion by withdrawing electrons.

Chlorine is less electronegative than fluorine and does not stabilize cyclohexadienyl anion to as great an extent.

Mechanism 13.6

Nucleophilic Aromatic Substitution in *p*-Fluoronitrobenzene by the Addition–Elimination Mechanism

THE OVERALL REACTION:

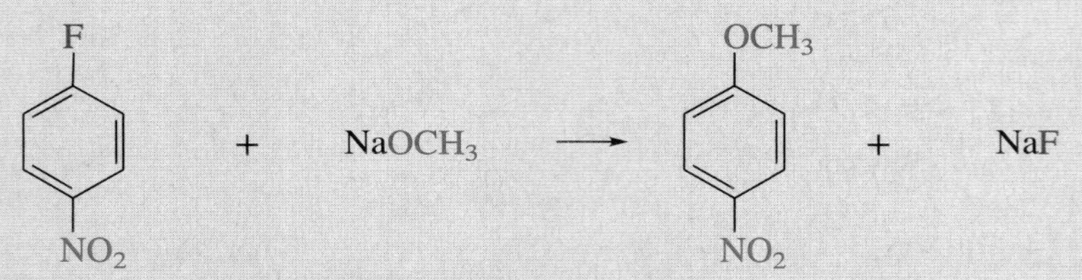

F, NO_2 + $NaOCH_3$ ⟶ OCH_3, NO_2 + NaF

p-Fluoronitrobenzene Sodium methoxide *p*-Nitroanisole Sodium fluoride

THE MECHANISM:

Step 1: Addition stage. The nucleophile, in this case methoxide ion, adds to the carbon atom that bears the leaving group to give a cyclohexadienyl anion intermediate.

$:\ddot{F}:$ H H H H NO_2 + $^{-}:\ddot{O}CH_3$ $\xrightarrow{\text{slow}}$ $:\ddot{F}:\ :\ddot{O}CH_3$ H H H H NO_2

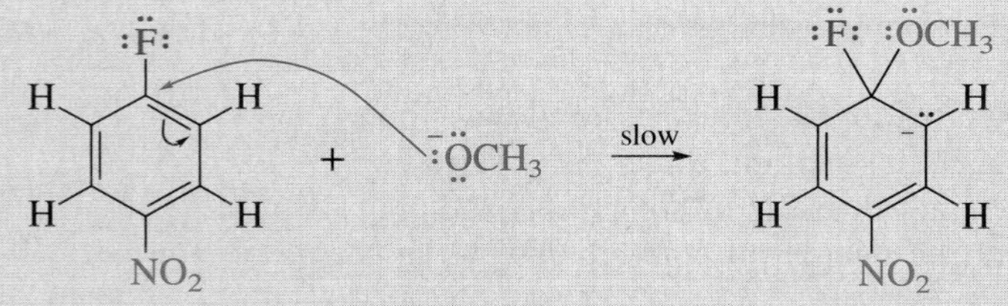

p-Fluoronitrobenzene Methoxide ion Cyclohexadienyl anion intermediate

Step 2: Elimination stage. Loss of halide from the cyclohexadienyl intermediate restores the aromaticity of the ring and gives the product of nucleophilic aromatic substitution.

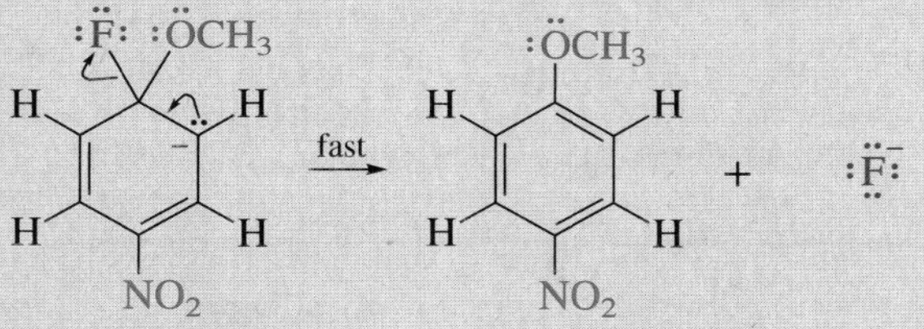

$:\ddot{F}:\ :\ddot{O}CH_3$ H H H H NO_2 $\xrightarrow{\text{fast}}$ $:\ddot{O}CH_3$ H H H H NO_2 + $:\ddot{F}:^{-}$

Cyclohexadienyl anion intermediate *p*-Nitroanisole Fluoride ion

Before leaving this mechanistic discussion, we should mention that the addition–elimination mechanism for nucleophilic aromatic substitution illustrates a principle worth remembering. The words *activating* and *deactivating* as applied to substituent effects in organic chemistry are without meaning when they stand alone. When we say that a group is activating or deactivating, we need to specify the reaction type that is being considered. A nitro group is a strongly *deactivating* substituent in *electrophilic* aromatic substitution, where it markedly destabilizes the key cyclohexadienyl cation intermediate:

NO_2 H E^+ $\xrightarrow{\text{very slow}}$ NO_2 + E H $\xrightarrow[\text{fast}]{-H^+}$ NO_2 E

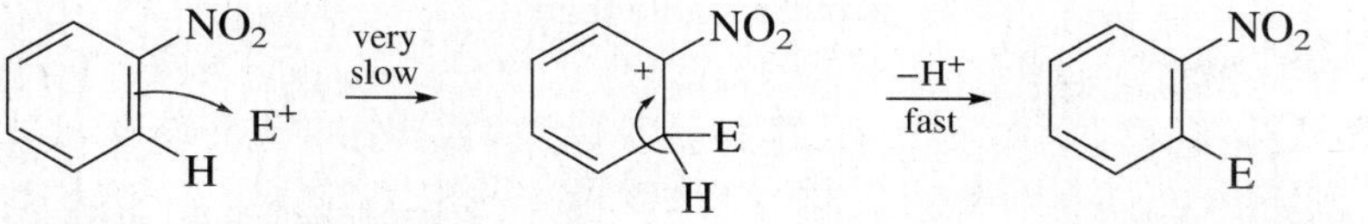

Nitrobenzene and an electrophile Cyclohexadienyl cation intermediate; nitro group is destabilizing Product of electrophilic aromatic substitution

A nitro group is a strongly *activating* substituent in *nucleophilic* aromatic substitution, where it stabilizes the key cyclohexadienyl anion intermediate:

NO_2 :Y⁻ slow addition → NO_2 Y X fast elimination → NO_2 Y + :X⁻

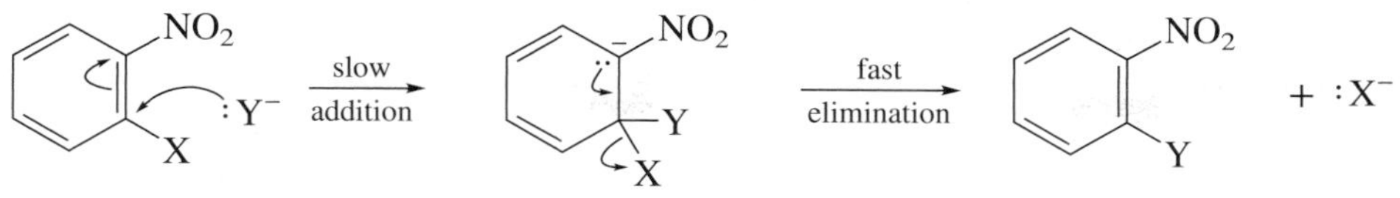

o-Halonitrobenzene (X = F, Cl, Br, or I) and a nucleophile

Cyclohexadienyl anion intermediate; nitro group is stabilizing

Product of nucleophilic aromatic substitution

A nitro group behaves the same way in both reactions: it attracts electrons. Reaction is retarded when electrons flow from the aromatic ring to the attacking species (electrophilic aromatic substitution). Reaction is facilitated when electrons flow from the attacking species to the aromatic ring (nucleophilic aromatic substitution). By being aware of the connection between reactivity and substituent effects, you will sharpen your appreciation of how chemical reactions occur.

13.21 Related Nucleophilic Aromatic Substitutions

The most common types of aryl halides in nucleophilic aromatic substitutions are those that bear o- or p-nitro substituents. Among other classes of reactive aryl halides, a few merit special consideration. One class includes highly fluorinated aromatic compounds such as hexafluorobenzene, which undergoes substitution of one of its fluorines on reaction with nucleophiles such as sodium methoxide.

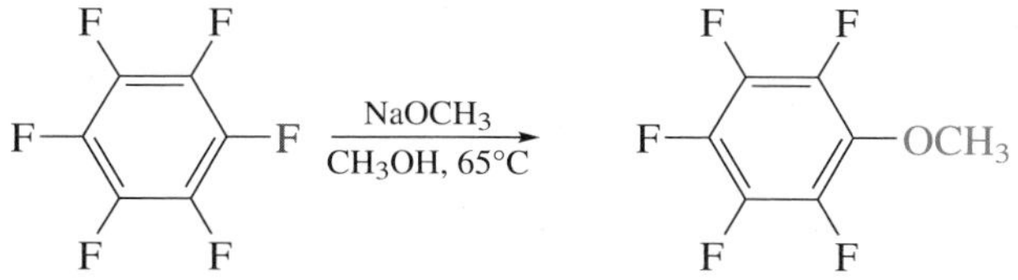

$$C_6F_6 \xrightarrow[CH_3OH,\ 65°C]{NaOCH_3} C_6F_5OCH_3$$

Hexafluorobenzene → 2,3,4,5,6-Pentafluoroanisole (72%)

Here it is the combined electron-attracting inductive effects of the six fluorine substituents that stabilize the cyclohexadienyl anion intermediate and permit the reaction to proceed so readily.

Problem 13.33

Write equations describing the addition–elimination mechanism for the reaction of hexafluorobenzene with sodium methoxide, clearly showing the structure of the rate-determining intermediate.

Halides derived from certain heterocyclic aromatic compounds are often quite reactive toward nucleophiles. 2-Chloropyridine, for example, reacts with sodium methoxide some 230 million times faster than chlorobenzene at 50°C.

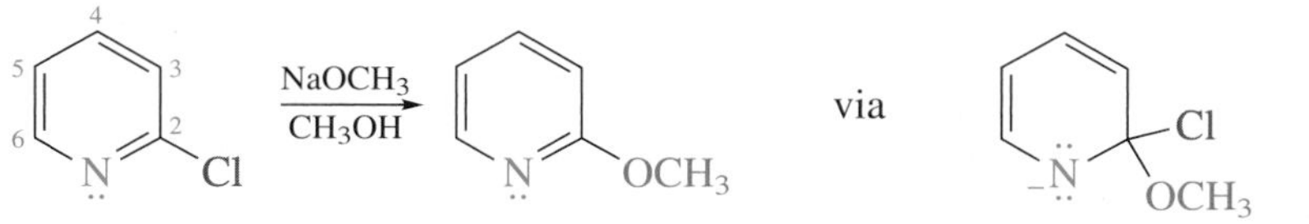

2-Chloropyridine $\xrightarrow[CH_3OH]{NaOCH_3}$ 2-Methoxypyridine via Anionic intermediate

Again, rapid reaction is attributed to the stability of the intermediate formed in the addition step. In contrast to chlorobenzene, where the negative charge of the intermediate must be borne by carbon, the anionic intermediate in the case of 2-chloropyridine has its negative charge on nitrogen. Because nitrogen is more electronegative than carbon, the intermediate is more stable and is formed faster than the one from chlorobenzene.

Problem 13.34

Offer an explanation for the observation that 4-chloropyridine is more reactive toward nucleophiles than 3-chloropyridine.

The reactivity of 2-chloropyridines and analogous compounds can be enhanced by the presence of strongly electron-withdrawing groups. In the following example, the chlorine leaving group is activated toward nucleophilic aromatic substitution by both of the ring nitrogens and is ortho to the electron-withdrawing cyano group as well. Substitution by ammonia takes place at 0°C to give a 96% yield of product.

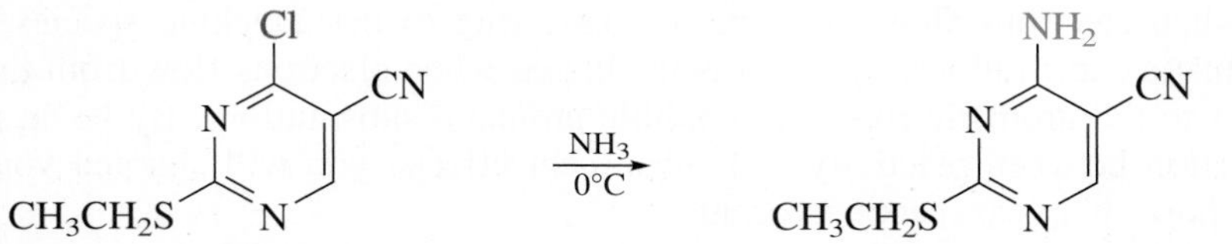

Cl, CN, N, CH_3CH_2S, N $\xrightarrow[0°C]{NH_3}$ NH_2, CN, N, CH_3CH_2S, N

4-Chloro-5-cyano-2-(thioethyl)pyrimidine → 4-Amino-5-cyano-2-(thioethyl)pyrimidine (96%)

Problem 13.35

Write contributing resonance structures to show how the negative charge in the intermediate in the preceding reaction is shared by three ring atoms and the nitrogen of the cyano group. Here is one of the contributing structures to get you started.

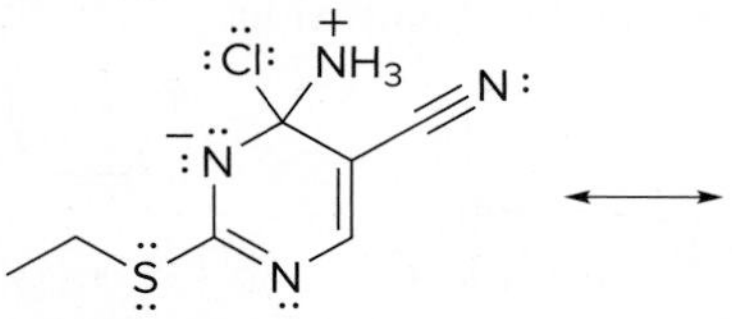

:Cl: $\overset{+}{N}H_3$ ≡N: $\overset{-}{N}$ S N ⟷

Very strong bases can bring about nucleophilic aromatic substitution by a mechanism other than the one we have been discussing. The intermediate in this other mechanism, outlined in the Descriptive Passage at the end of this chapter, may surprise you.

13.22 SUMMARY

Section 13.1 On reaction with electrophilic reagents, compounds that contain a benzene ring undergo **electrophilic aromatic substitution.** Table 13.1 in Section 13.1 and Table 13.3 in this summary give examples.

Section 13.2 The mechanism of electrophilic aromatic substitution involves two stages: bonding of the electrophile by the π electrons of the ring (slow, rate-determining), followed by rapid loss of a proton to restore the aromaticity of the ring.

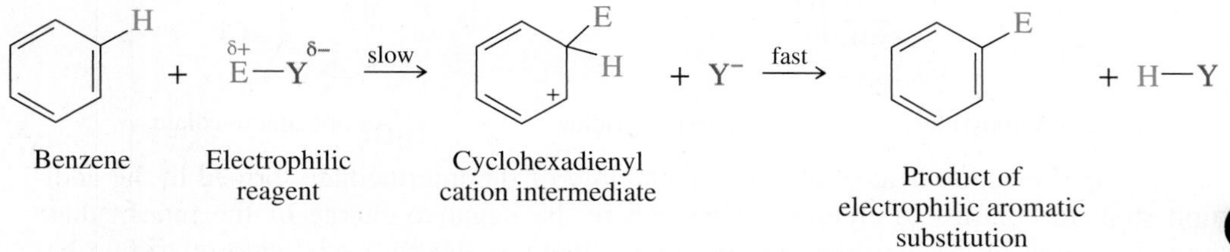

$$\text{Benzene (C}_6\text{H}_5\text{–H)} + \overset{\delta+}{E}—\overset{\delta-}{Y} \xrightarrow{\text{slow}} \text{Cyclohexadienyl cation intermediate} + Y^- \xrightarrow{\text{fast}} \text{Product of electrophilic aromatic substitution (C}_6\text{H}_5\text{–E)} + H—Y$$

Benzene — Electrophilic reagent — Cyclohexadienyl cation intermediate — Product of electrophilic aromatic substitution

Sections 13.3–13.5 See Table 13.3.

TABLE 13.3 Representative Electrophilic Aromatic Substitution Reactions

Reaction (section) and comments	General equation and specific example
Nitration (Section 13.3) The active electrophile in the nitration of benzene and its derivatives is nitronium cation (:Ö=N⁺=Ö:). It is generated by reaction of nitric acid and sulfuric acid. Very reactive arenes—those that bear strongly activating substituents—undergo nitration in nitric acid alone.	ArH (Arene) + HNO_3 (Nitric acid) $\xrightarrow{H_2SO_4}$ $ArNO_2$ (Nitroarene) + H_2O (Water) Fluorobenzene $\xrightarrow[H_2SO_4]{HNO_3}$ *p*-Fluoronitrobenzene (80%)
Sulfonation (Section 13.4) Sulfonic acids are formed when aromatic compounds are treated with sources of sulfur trioxide. These sources can be concentrated sulfuric acid (for very reactive arenes) or solutions of sulfur trioxide in sulfuric acid (for benzene and arenes less reactive than benzene).	ArH (Arene) + SO_3 (Sulfur trioxide) ⟶ $ArSO_3H$ (Arenesulfonic acid) 1,2,4,5-Tetramethylbenzene $\xrightarrow[H_2SO_4]{SO_3}$ 2,3,5,6-Tetramethylbenzenesulfonic acid (94%)
Halogenation (Section 13.5) Chlorination and bromination of arenes are carried out by treatment with the appropriate halogen in the presence of a Lewis acid catalyst. Very reactive arenes undergo halogenation in the absence of a catalyst.	ArH (Arene) + X_2 (Halogen) $\xrightarrow{FeX_3}$ ArX (Aryl halide) + HX (Hydrogen halide) Phenol $\xrightarrow[CS_2]{Br_2}$ *p*-Bromophenol (80–84%)
Friedel–Crafts alkylation (Section 13.6) Carbocations, usually generated from an alkyl halide and aluminum chloride, alkylate the aromatic ring. The arene must be at least as reactive as a halobenzene. Rearrangements can occur, especially with primary alkyl halides.	ArH (Arene) + RX (Alkyl halide) $\xrightarrow{AlCl_3}$ ArR (Alkylarene) + HX (Hydrogen halide) Benzene + Cyclopentyl bromide $\xrightarrow{AlCl_3}$ Cyclopentylbenzene (54%)
Friedel–Crafts acylation (Section 13.7) Acyl cations (acylium ions) generated by treating an acyl chloride or acid anhydride with aluminum chloride acylate aromatic rings to yield ketones. The arene must be at least as reactive as a halobenzene. Acyl cations are relatively stable, and do not rearrange.	ArH (Arene) + RCCl (O) (Acyl chloride) $\xrightarrow{AlCl_3}$ ArCR (O) (Ketone) + HCl (Hydrogen chloride) ArH (Arene) + RCOCR (O O) (Acid anhydride) $\xrightarrow{AlCl_3}$ ArCR (O) (Ketone) + RCOH (O) (Carboxylic acid) Anisole (CH_3O–C₆H₅) + acetic anhydride $\xrightarrow{AlCl_3}$ *p*-Methoxyacetophenone (90–94%)

TABLE 13.4 Limitations on Friedel–Crafts Reactions

1. The organic halide that reacts with the arene must be an alkyl halide (Section 13.6) or an acyl halide (Section 13.7).	*These will react with benzene under Friedel–Crafts conditions:* Alkyl halide; Benzylic halide (CH_2Cl); Acyl halide
Vinylic halides and aryl halides do not form carbocations under conditions of the Friedel–Crafts reaction and so cannot be used in place of an alkyl halide or an acyl halide.	*These will not react with benzene under Friedel–Crafts conditions:* Vinylic halide; Aryl halide
2. Rearrangement of alkyl groups can occur (Section 13.6).	Rearrangement is especially prevalent with primary alkyl halides of the type RCH_2CH_2X and R_2CHCH_2X. Aluminum chloride induces ionization with rearrangement to give a more stable carbocation. Benzylic halides and acyl halides do not rearrange.
3. Strongly deactivated aromatic rings do not undergo Friedel–Crafts alkylation or acylation (Section 13.16). Friedel–Crafts alkylations and acylations fail when applied to compounds of the following type, where EWG is a strongly electron-withdrawing group: (benzene–EWG)	*EWG:* —CF_3 —NO_2 —SO_3H —C≡N; —C(=O)H, —C(=O)R, —C(=O)OR, —C(=O)Cl
4. It is sometimes difficult to limit Friedel–Crafts alkylation to monoalkylation. Only monoacylation occurs during Friedel–Crafts acylation.	The first *alkyl* group that goes on makes the ring more reactive toward further substitution because alkyl groups are activating substituents. Monoacylation is possible because the first *acyl* group to go on is strongly electron-withdrawing and deactivates the ring toward further substitution.

Sections 13.6–13.7 See Tables 13.3 and 13.4.

Section 13.8 Friedel–Crafts acylation, followed by Clemmensen or Wolff–Kishner reduction, is a standard sequence used to introduce a primary alkyl group onto an aromatic ring.

1,2,4-Triethylbenzene $\xrightarrow[\text{AlCl}_3]{\text{CH}_3\text{CCl (O)}}$ 2,4,5-Triethylacetophenone (80%) $\xrightarrow[\text{triethylene glycol, heat}]{\text{H}_2\text{NNH}_2\text{, NaOH}}$ 1,2,4,5-Tetraethylbenzene (73%)

Section 13.9 Substituents on an aromatic ring can influence both the *rate* and *regioselectivity* of electrophilic aromatic substitution. Substituents are classified as *activating* or *deactivating* according to whether they cause electrophilic aromatic substitution to occur more rapidly or less rapidly than benzene. With respect to regioselectivity, substituents are either *ortho, para-directing* or *meta-directing.* A methyl group is activating and ortho, para-directing. A trifluoromethyl group is deactivating and meta-directing.

Sections 13.10–13.14 How substituents control rate and regioselectivity in electrophilic aromatic substitution results from their effect on carbocation stability. An electron-releasing substituent stabilizes the cyclohexadienyl cation intermediates leading to ortho and para substitution more than meta.

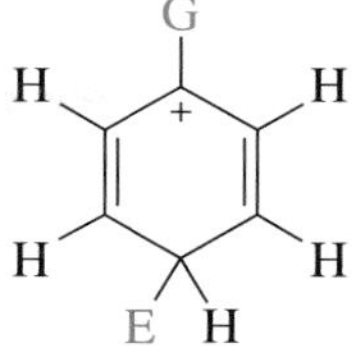

Stabilized when G is electron-releasing | Less stabilized when G is electron-releasing | Stabilized when G is electron-releasing

Conversely, an electron-withdrawing substituent destabilizes the cyclohexadienyl cations leading to ortho and para substitution more than meta. Thus, meta substitution predominates.

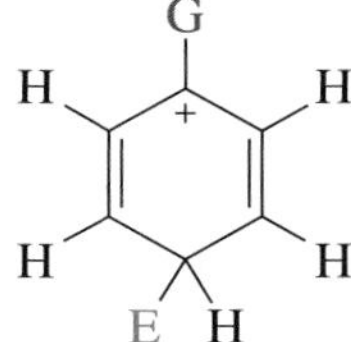

Destabilized when G is electron-withdrawing | Less destabilized when G is electron-withdrawing | Destabilized when G is electron-withdrawing

Substituents can be arranged into three major categories:

1. **Activating and ortho, para-directing:** These substituents stabilize the cyclohexadienyl cation formed in the rate-determining step. They include $-\ddot{N}R_2$, $-\ddot{\underset{..}{O}}R$, $-R$, $-Ar$, and related species. The most strongly activating members of this group are bonded to the ring by a nitrogen or oxygen atom that bears an unshared pair of electrons.
2. **Deactivating and ortho, para-directing:** The halogens are the most prominent members of this class. They withdraw electron density from all the ring positions by an inductive effect, making halobenzenes less reactive than benzene. Lone-pair electron donation stabilizes the cyclohexadienyl cation intermediates for ortho and para substitution more than those for meta substitution.
3. **Deactivating and meta-directing:** These substituents are strongly electron-withdrawing and destabilize carbocations. They include

$$-CH_3, \quad -\overset{O}{\overset{\|}{C}}R, \quad -C{\equiv}N, \quad -NO_2$$

and related species. All the ring positions are deactivated, but because the *meta* positions are deactivated less than the ortho and para, meta substitution is favored.

Section 13.15 When two or more substituents are present on a ring, the regioselectivity of electrophilic aromatic substitution is generally controlled by the directing effect of the more powerful *activating* substituent.

Section 13.16 The order in which substituents are introduced onto a benzene ring needs to be considered in order to prepare the desired isomer in a multistep synthesis.

Section 13.17 Polycyclic aromatic hydrocarbons undergo the same kind of electrophilic aromatic substitution reactions as benzene.

Section 13.18 Heterocyclic aromatic compounds may be more reactive or less reactive than benzene. Pyridine is much less reactive than benzene, but pyrrole, furan, and thiophene are more reactive. Pyridine undergoes substitution at C-3, whereas pyrrole, furan, and thiophene give mainly C-2-substituted products.

Section 13.19 Aryl halides are less reactive than alkyl halides in nucleophilic substitution reactions. Nucleophilic substitution in aryl halides is facilitated by the presence of a strong electron-withdrawing group, such as NO_2, ortho or para to the halogen.

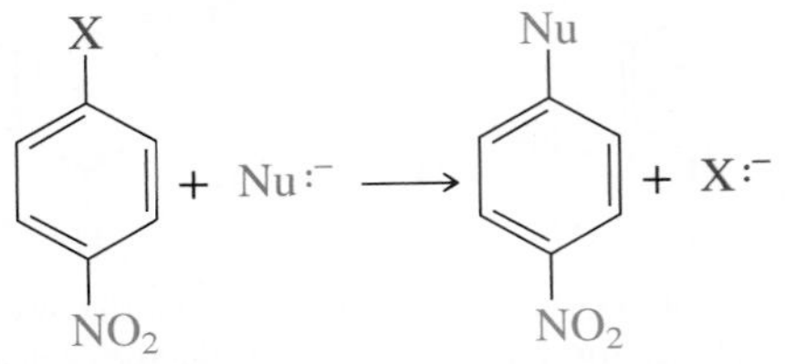

X, NO2 + Nu:⁻ ⟶ Nu, NO2 + X:⁻

In reactions of this type, fluoride is the best leaving group of the halogens and iodide the poorest.

Section 13.20 Nucleophilic aromatic substitutions of the type just shown follow an **addition–elimination mechanism.**

+ Nu:⁻ —slow addition stage→ —fast elimination stage→ + X:⁻

Nitro-substituted aryl halide

Cyclohexadienyl anion intermediate

Product of nucleophilic aromatic substitution

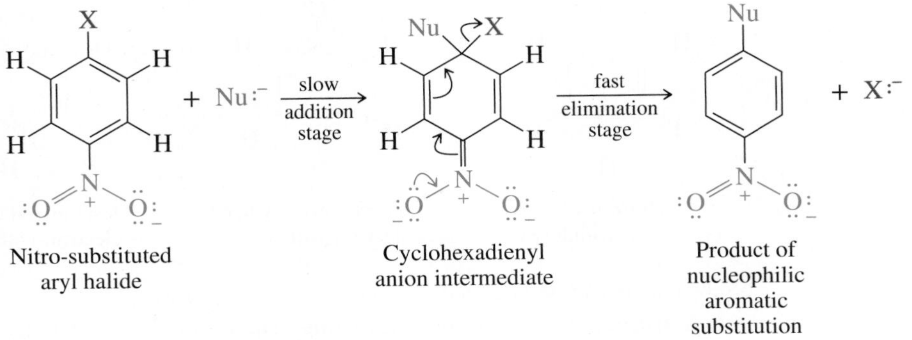

The rate-determining intermediate is a cyclohexadienyl anion and is stabilized by electron-withdrawing substituents.

Section 13.21 Other aryl halides that give stabilized anions can undergo nucleophilic aromatic substitution by the addition–elimination mechanism. Two examples are hexafluorobenzene and 2-chloropyridine.

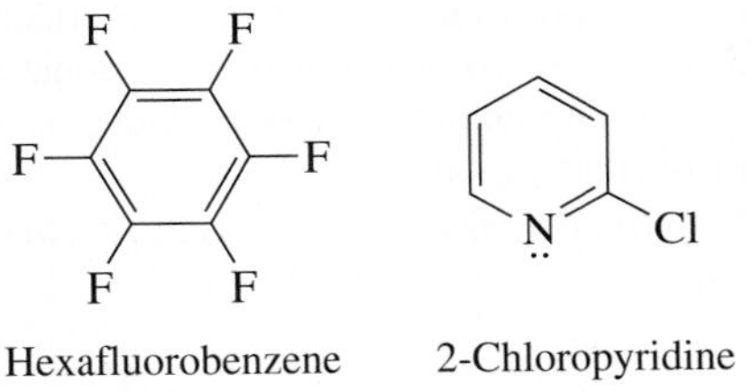

Hexafluorobenzene

2-Chloropyridine

PROBLEMS

Predict the Product

13.36 Write the structure of the organic product in each of the following reactions. If electrophilic aromatic substitution occurs, assume only monosubstitution.

(a) 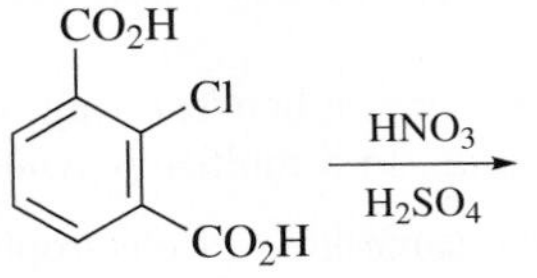 [2-chloroisophthalic acid: CO_2H, Cl, CO_2H] $\xrightarrow[H_2SO_4]{HNO_3}$

(b) 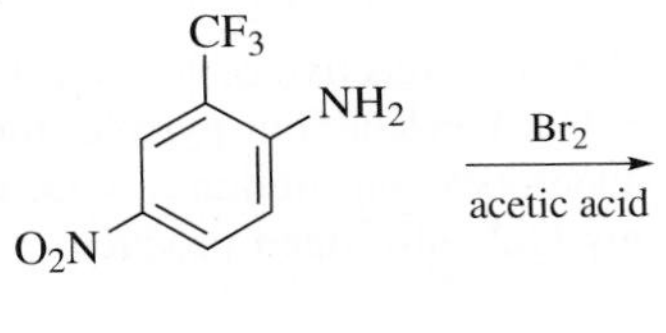 [CF_3, NH_2, O_2N substituted benzene] $\xrightarrow[\text{acetic acid}]{Br_2}$

(c) [biphenyl-4-ol] $\xrightarrow[\text{CHCl}_3]{\text{Br}_2}$

(d) [1-tert-butyl-3-isopropylbenzene] $\xrightarrow[\text{acetic acid}]{\text{HNO}_3}$

(e) [benzene] + [1-octene] $\xrightarrow{H_2SO_4}$

(f) [F, OCH_3 substituted benzene] + [acetic anhydride, O O O] $\xrightarrow{AlCl_3}$

(g) O_2N–[benzene]–[isopropyl] $\xrightarrow[H_2SO_4]{HNO_3}$

(h) [OCH_3, CH_3 substituted benzene] + [2-methylpropene] $\xrightarrow{H_2SO_4}$

(i) [benzyl-substituted dimethylphenol, OH] $\xrightarrow[\text{CHCl}_3]{\text{Br}_2}$

(j) [benzene]–F + [benzyl chloride, Cl] $\xrightarrow{AlCl_3}$

(k) Br–[benzene, NO_2]–Br + [piperidine, NH] $\longrightarrow$

(l) [O=C, HN, ethyl-substituted benzene] + [acetyl chloride, O, Cl] $\xrightarrow[CS_2]{AlCl_3}$

(m) [trimethylphenyl ketone, O] $\xrightarrow[\text{HCl}]{\text{Zn(Hg)}}$

(n) [thiophene-3-carboxylic acid, CO_2H, S] $\xrightarrow[\text{acetic acid}]{\text{Br}_2}$

(o) 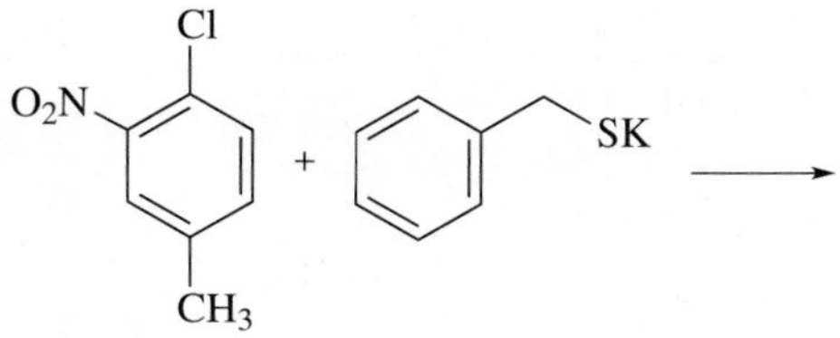 [structure: 1-chloro-2-nitro-4-methylbenzene: Cl, O_2N, CH_3] + [benzyl–SK] ⟶

(p) 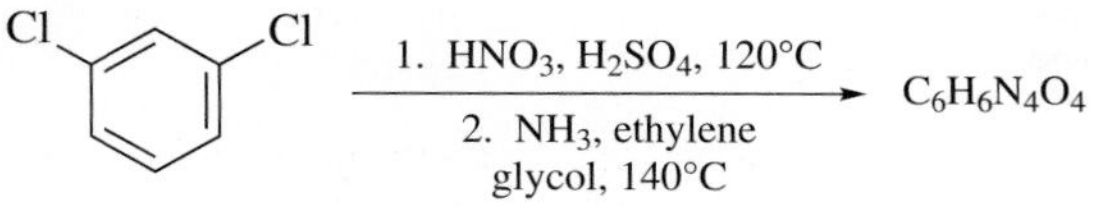 [1,3-dichlorobenzene: Cl, Cl] $\xrightarrow[\text{2. } NH_3\text{, ethylene glycol, 140°C}]{\text{1. } HNO_3\text{, } H_2SO_4\text{, 120°C}}$ $C_6H_6N_4O_4$

(q) 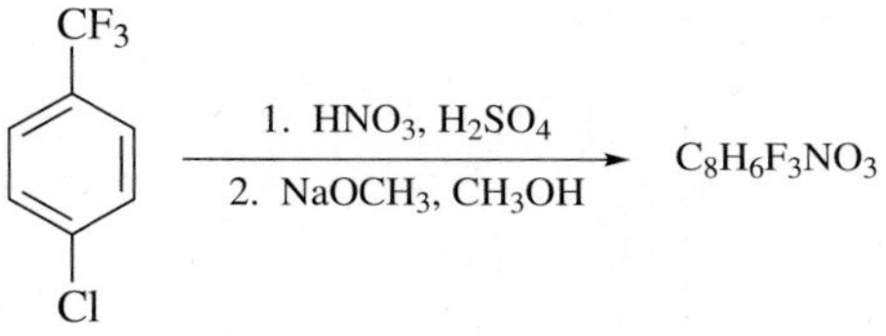 [4-chlorobenzotrifluoride: CF_3, Cl] $\xrightarrow[\text{2. } NaOCH_3\text{, } CH_3OH]{\text{1. } HNO_3\text{, } H_2SO_4}$ $C_8H_6F_3NO_3$

(r) 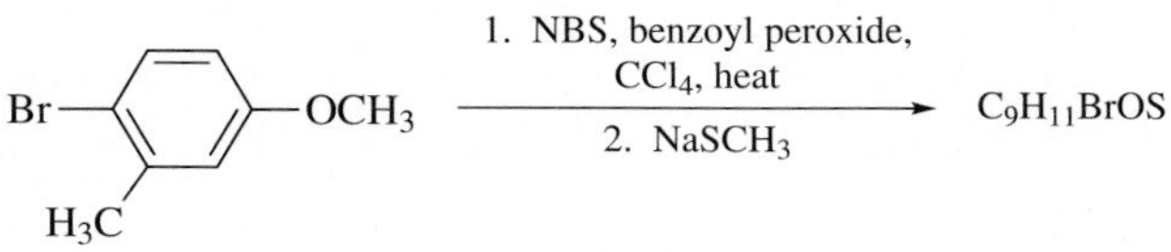 Br–[ring with H_3C]–OCH_3 $\xrightarrow[\text{2. } NaSCH_3]{\text{1. NBS, benzoyl peroxide, } CCl_4\text{, heat}}$ $C_9H_{11}BrOS$

13.37 Friedel–Crafts acylation of the individual isomers of xylene with acetyl chloride and aluminum chloride yields a single product, different for each xylene isomer, in high yield in each case. Write the structures of the products of acetylation of *o*-, *m*-, and *p*-xylene.

13.38 Reaction of benzanilide ($C_6H_5NH\overset{O}{\overset{\|}{C}}C_6H_5$) with chlorine in acetic acid yields a mixture of two monochloro derivatives. Suggest reasonable structures for these two isomers.

13.39 1,2,3,4,5-Pentafluoro-6-nitrobenzene reacts readily with sodium methoxide in methanol at room temperature to yield two major products, each having the molecular formula $C_7H_3F_4NO_3$. Suggest reasonable structures for these two compounds.

13.40 Each of the following reactions has been carried out under conditions such that disubstitution or trisubstitution occurred. Identify the principal organic product in each case.

(a) Nitration of *p*-chlorobenzoic acid (dinitration)
(b) Bromination of aniline (tribromination)
(c) Bromination of *o*-aminoacetophenone (dibromination)
(d) Bromination of *p*-nitrophenol (dibromination)
(e) Reaction of biphenyl with *tert*-butyl chloride and iron(III) chloride (dialkylation)
(f) Sulfonation of phenol (disulfonation)

13.41 The herbicide *trifluralin* is prepared by the following sequence of reactions. Identify compound A and deduce the structure of trifluralin.

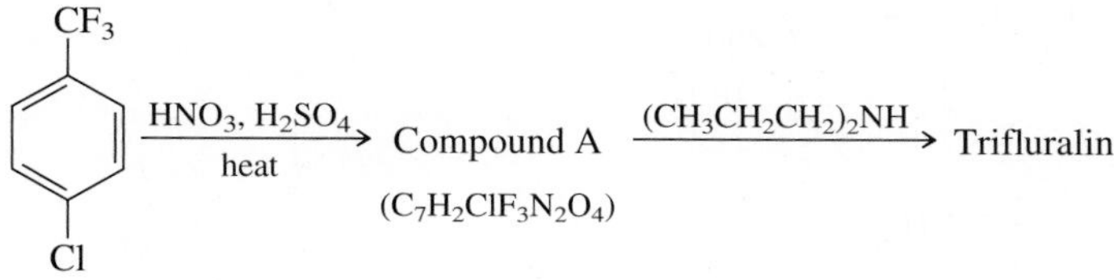

[4-chlorobenzotrifluoride: CF_3, Cl] $\xrightarrow[\text{heat}]{HNO_3\text{, } H_2SO_4}$ Compound A ($C_7H_2ClF_3N_2O_4$) $\xrightarrow{(CH_3CH_2CH_2)_2NH}$ Trifluralin

13.42 Each of the compounds indicated undergoes an intramolecular Friedel–Crafts acylation reaction to yield a cyclic ketone. Write the structure of the expected product in each case.

(a)

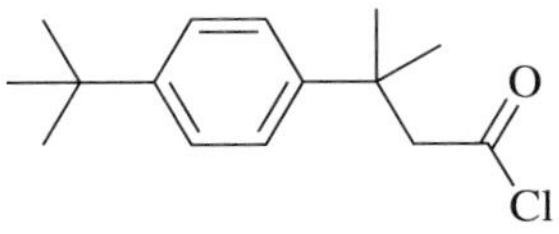

(b)

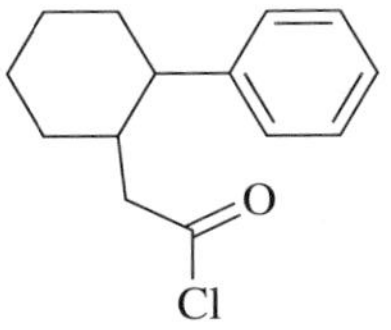

(c) 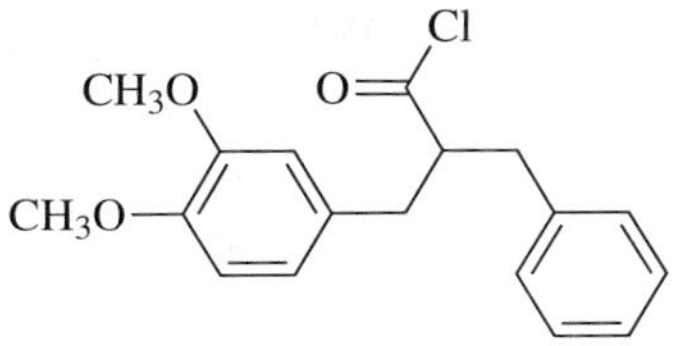 CH_3O, CH_3O, O, Cl

(d) 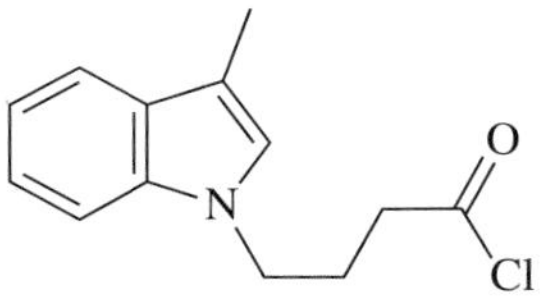 N, O, Cl

13.43 Treatment of the alcohol shown with sulfuric acid gave a tricyclic hydrocarbon of molecular formula $C_{16}H_{16}$ as the major organic product. Suggest a reasonable structure for this hydrocarbon.

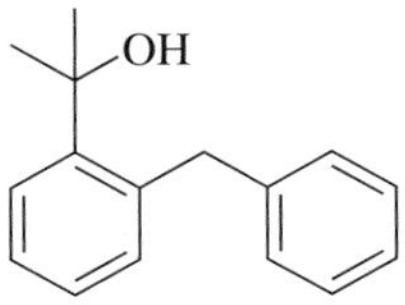

OH

13.44 When a dilute solution of 6-phenylhexanoyl chloride in carbon disulfide was slowly added (over a period of eight days!) to a suspension of aluminum chloride in the same solvent, it yielded a product A ($C_{12}H_{14}O$) in 67% yield. Oxidation of A gave benzene-1,2-dicarboxylic acid.

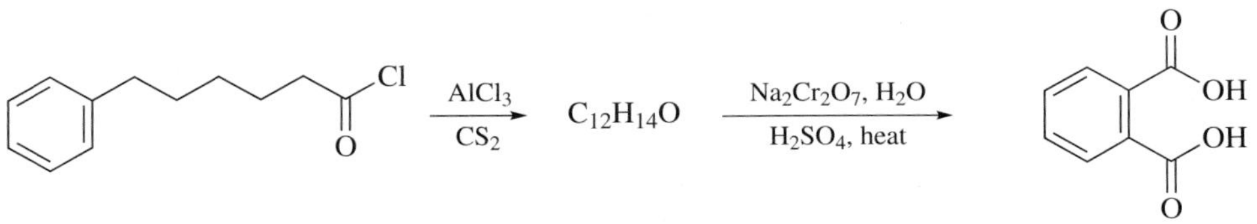

$$\text{6-Phenylhexanoyl chloride} \xrightarrow[CS_2]{AlCl_3} \underset{\text{Compound A}}{C_{12}H_{14}O} \xrightarrow[H_2SO_4,\ \text{heat}]{Na_2Cr_2O_7,\ H_2O} \text{Benzene-1,2-dicarboxylic acid}$$

Formulate a reasonable structure for compound A.

Reactivity

13.45 In each of the following pairs of compounds choose which one will react faster with the indicated reagent, and write a chemical equation for the faster reaction:

(a) Toluene or chlorobenzene with a mixture of nitric acid and sulfuric acid

(b) Fluorobenzene or (trifluoromethyl)benzene with benzyl chloride and aluminum chloride

(c) Methyl benzoate ($C_6H_5\overset{O}{\overset{\|}{C}}OCH_3$) or phenyl acetate ($C_6H_5O\overset{O}{\overset{\|}{C}}CH_3$) with bromine in acetic acid

(d) Acetanilide ($C_6H_5NH\overset{O}{\overset{\|}{C}}CH_3$) or nitrobenzene with sulfur trioxide in sulfuric acid

(e) *p*-Dimethylbenzene (*p*-xylene) or *p*-di-*tert*-butylbenzene with acetyl chloride and aluminum chloride

(f) Phenol or phenyl acetate with bromine

13.46 Choose the compound in each of the following pairs that reacts faster with sodium methoxide in methanol, and write a chemical equation for the faster reaction.

(a) Chlorobenzene or *o*-chloronitrobenzene

(b) *o*-Chloronitrobenzene or *m*-chloronitrobenzene

(c) 4-Chloro-3-nitroacetophenone or 4-chloro-3-nitrotoluene

(d) 2-Fluoro-1,3-dinitrobenzene or 1-fluoro-3,5-dinitrobenzene

(e) 1,4-Dibromo-2-nitrobenzene or 1-bromo-2,4-dinitrobenzene

13.47 Arrange the following five compounds in order of decreasing rate of bromination: benzene, toluene, *o*-xylene, *m*-xylene, 1,3,5-trimethylbenzene (the relative rates are 2×10^7, 5×10^4, 5×10^2, 60, and 1).

13.48 Of the groups shown, which is the most likely candidate for substituent X based on the partial rate factors for chlorination?

$—CF_3$ $—C(CH_3)_3$ $—Br$ $—SO_3H$ $—CH{=}O$

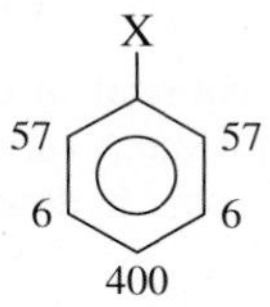

X

57 57

6 6

400

13.49 The partial rate factors for chlorination of biphenyl are as shown.

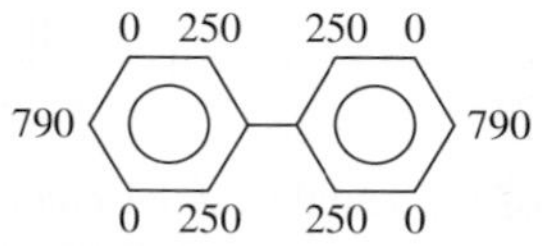

0 250 250 0

790 790

0 250 250 0

(a) What is the relative rate of chlorination of biphenyl compared with benzene?

(b) If, in a particular chlorination reaction, 10 g of *o*-chlorobiphenyl was formed, how much *p*-chlorobiphenyl would you expect to find?

13.50 Partial rate factors may be used to estimate product distributions in disubstituted benzene derivatives. The reactivity of a particular position in *o*-bromotoluene, for example, is given by the product of the partial rate factors for the corresponding position in toluene and bromobenzene. On the basis of the partial rate factor data given here for Friedel–Crafts acylation, predict the major product of the reaction of *o*-bromotoluene with acetyl chloride and aluminum chloride.

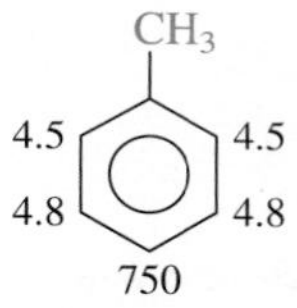

CH_3

4.5 4.5

4.8 4.8

750

Partial rate factors for reaction of toluene and bromobenzene with $CH_3\overset{O}{\overset{\|}{C}}Cl$, $AlCl_3$

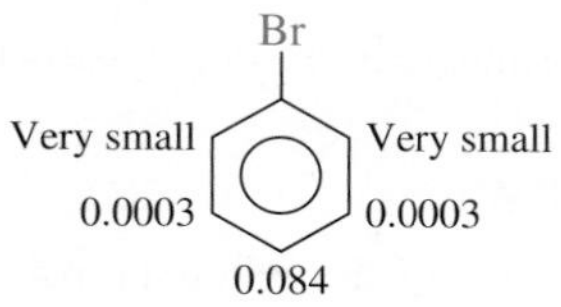

Br

Very small Very small

0.0003 0.0003

0.084

Synthesis

13.51 Give reagents suitable for carrying out each of the following reactions, and write the major organic products. If an ortho, para mixture is expected, show both. If the meta isomer is the expected major product, write only that isomer.

(a) Nitration of nitrobenzene

(b) Bromination of toluene

(c) Bromination of (trifluoromethyl)benzene

(d) Sulfonation of anisole

(e) Sulfonation of acetanilide ($C_6H_5NH\overset{O}{\overset{\|}{C}}CH_3$)

(f) Chlorination of bromobenzene

(g) Friedel–Crafts alkylation of anisole with benzyl chloride

(h) Friedel–Crafts acylation of benzene with benzoyl chloride (C_6H_5CCl)
(i) Nitration of the product from part (h)
(j) Clemmensen reduction of the product from part (h)
(k) Wolff–Kishner reduction of the product from part (h)

13.52 Write equations showing how to prepare each of the following from benzene or toluene and any necessary organic or inorganic reagents. If an ortho, para mixture is formed in any step of your synthesis, assume that you can separate the two isomers.

(a) Isopropylbenzene
(b) *p*-Isopropylbenzenesulfonic acid
(c) 2-Bromo-2-phenylpropane
(d) 4-*tert*-Butyl-2-nitrotoluene
(e) *m*-Chloroacetophenone
(f) *p*-Chloroacetophenone
(g) 3-Bromo-4-methylacetophenone
(h) 2-Bromo-4-ethyltoluene
(i) 3-Bromo-5-nitrobenzoic acid
(j) 2-Bromo-4-nitrobenzoic acid
(k) 1-Phenyloctane
(l) 1-Phenyl-1-octene
(m) 1-Phenyl-1-octyne
(n) 1,4-Di-*tert*-butyl-1,4-cyclohexadiene

13.53 Write equations showing how you could prepare each of the following from anisole and any necessary organic or inorganic reagents. If an ortho, para mixture is formed in any step of your synthesis, assume that you can separate the two isomers.

(a) *p*-Methoxybenzenesulfonic acid
(b) 2-Bromo-4-nitroanisole
(c) 4-Bromo-2-nitroanisole
(d) *p*-Methoxystyrene

13.54 Which is the best synthesis of the compound shown?

O_2N — O — NO_2

(a) — O — NO_2 $\xrightarrow[H_2SO_4]{HNO_3}$

(b) — ONa + F — (O_2N) — NO_2 $\longrightarrow$

(c) — F + NaO — (O_2N) — NO_2 $\longrightarrow$

13.55 What combination of acyl chloride or acid anhydride and arene would you choose to prepare each of the following compounds by Friedel–Crafts acylation?

(a)

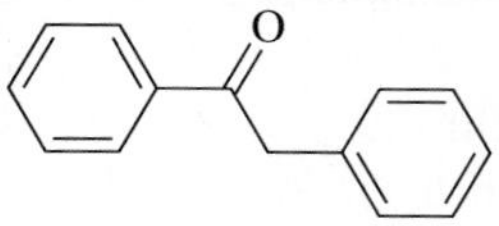

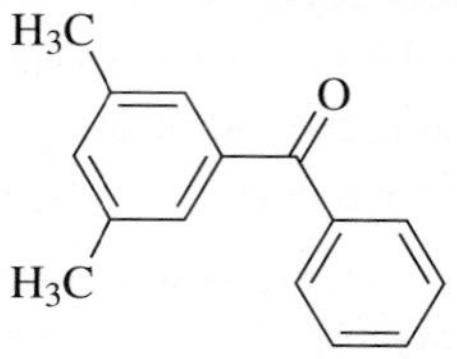

(b) H_3C ... CH_3 ... OH

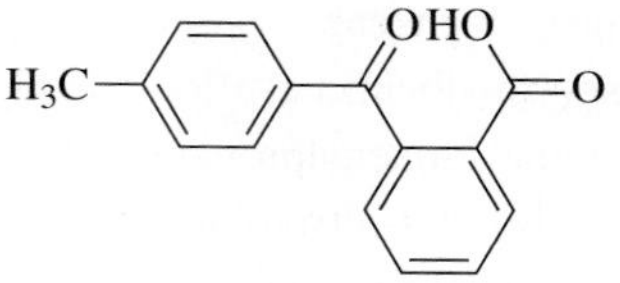

(c) O_2N

(d) H_3C ... H_3C

(e) H_3C ... O HO ... =O

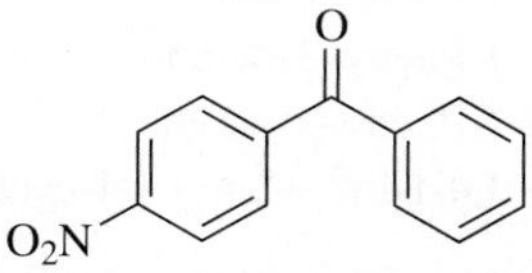

13.56 A standard synthetic sequence for building a six-membered cyclic ketone onto an existing aromatic ring is shown in outline as follows. Specify the reagents necessary for each step.

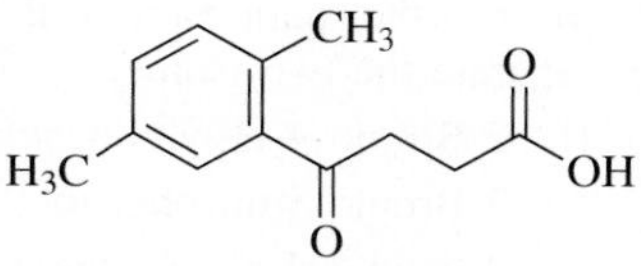

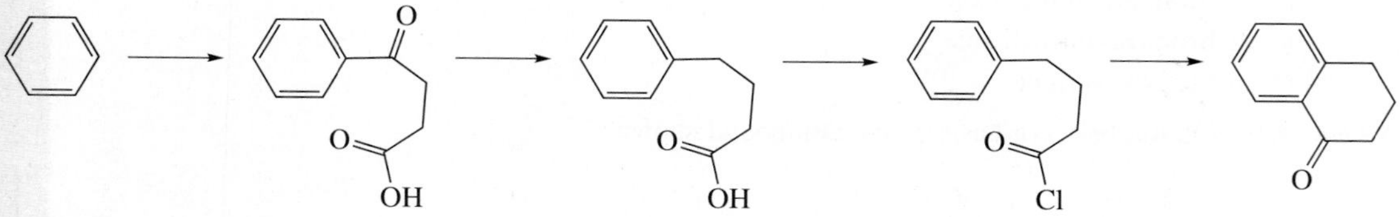

13.57 Suggest a suitable series of reactions for carrying out each of the following synthetic transformations:

(a) $CH(CH_3)_2$ to CO_2H ... SO_3H

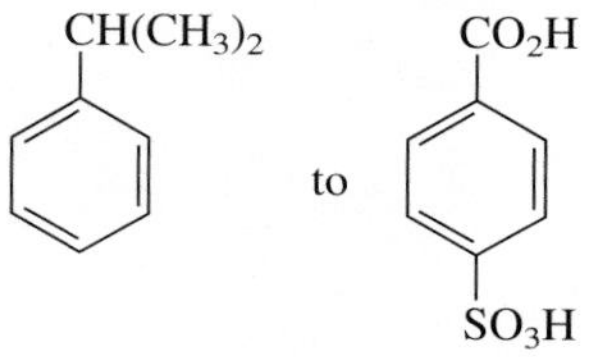

(b) CH_3 CH_3 to CO_2H CO_2H ... $C(CH_3)_3$

(c) to CH_3C(=O)

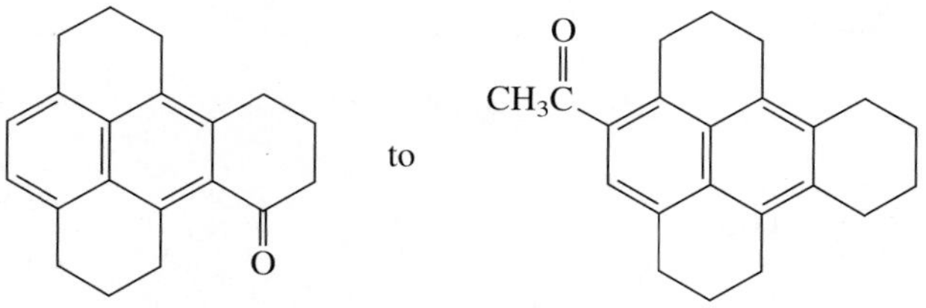

(d) OCH_3 OCH_3 to O_2N OCH_3 OCH_3 $C(CH_3)_3$

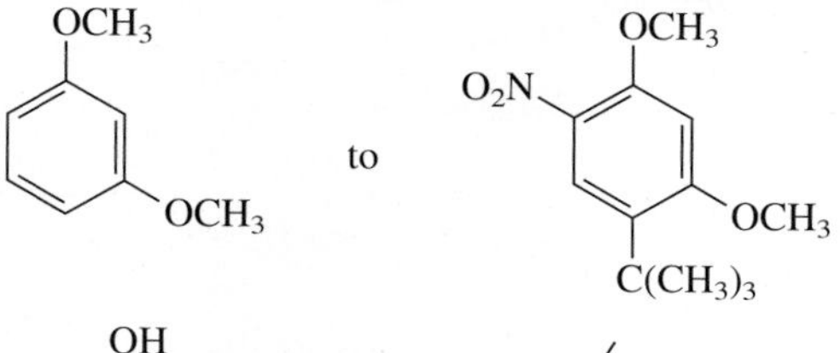

(e) OH to O ... NO_2

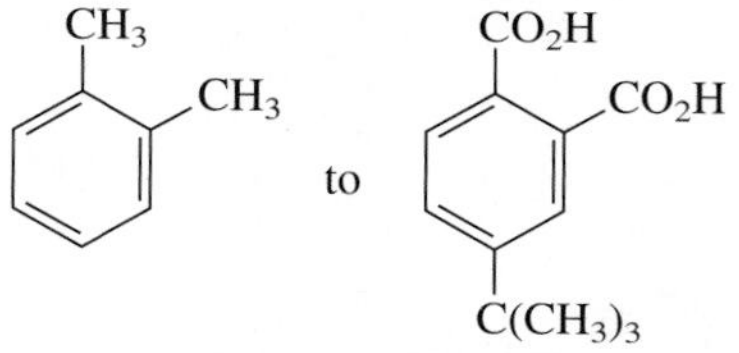

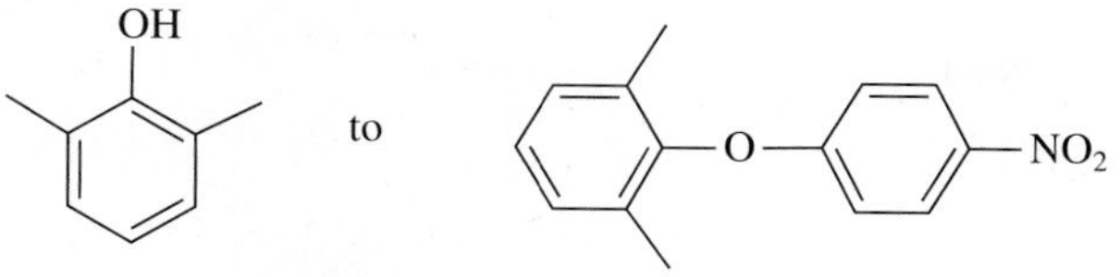

Mechanism

13.58 Write a structural formula for the most stable cyclohexadienyl cation intermediate formed in each of the following reactions. Is this intermediate more or less stable than the one formed from benzene?

(a) Bromination of *p*-xylene

(b) Chlorination of *m*-xylene

(c) Nitration of acetophenone

(d) Friedel–Crafts acylation of anisole with acetyl chloride ($CH_3\overset{O}{\overset{\|}{C}}Cl$)

(e) Nitration of isopropylbenzene

(f) Bromination of nitrobenzene

(g) Sulfonation of furan

(h) Bromination of pyridine

13.59 When 2-isopropyl-1,3,5-trimethylbenzene is heated with aluminum chloride (trace of HCl present) at 50°C, the major material present after 4 h is 1-isopropyl-2,4,5-trimethylbenzene. Suggest a reasonable mechanism for this isomerization.

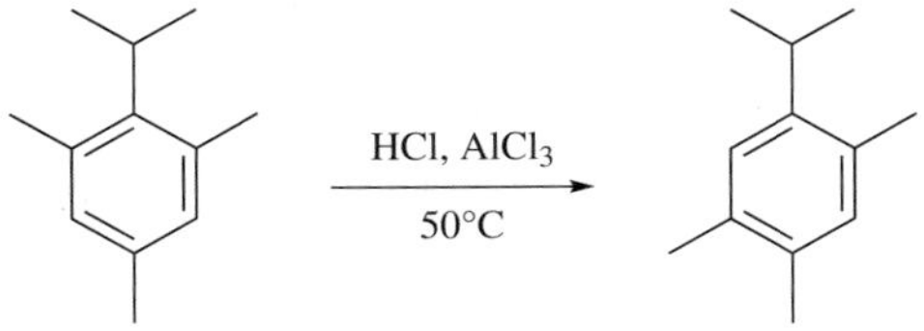

$\xrightarrow[50°C]{HCl,\ AlCl_3}$

13.60 In a general reaction known as the cyclohexadienone–phenol rearrangement, cyclohexadienones are converted to phenols under conditions of acid catalysis as shown in the following example. Write a mechanism for this reaction.

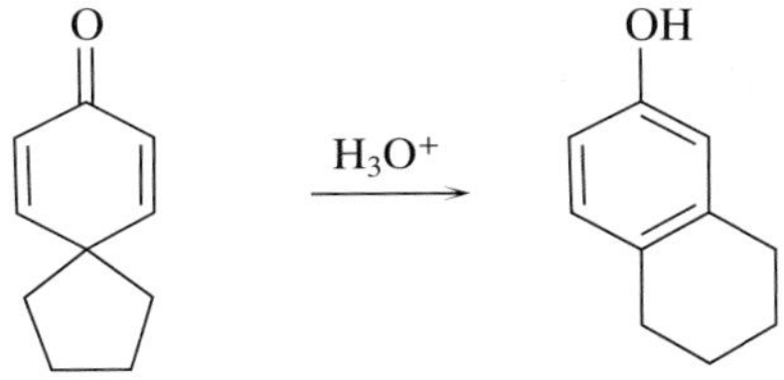

O $\xrightarrow{H_3O^+}$ OH

13.61 Trichloroisocyanuric acid (TCCA) is used as a swimming pool disinfectant and also can serve as an electrophilic chlorinating agent. Write a mechanism for the chlorination of anisole with TCCA.

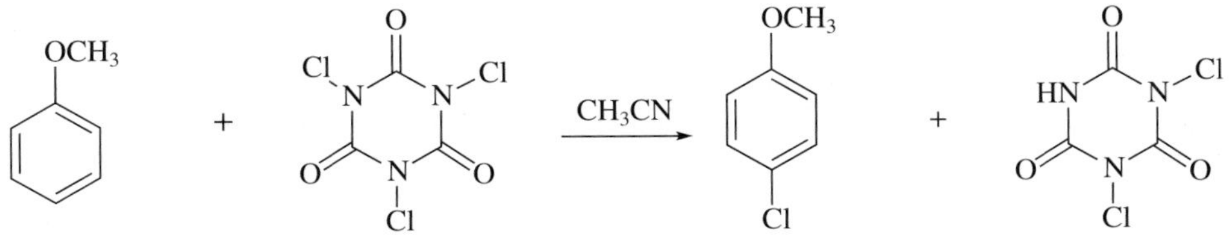

OCH_3 + Cl N O N Cl O N O Cl $\xrightarrow{CH_3CN}$ OCH_3 Cl + O HN N Cl O N O Cl

Anisole TCCA

13.62 Suggest a reasonable mechanism for the following reaction:

F F O OH F F F $\xrightarrow[heat]{K_2CO_3}$ F F O O F F

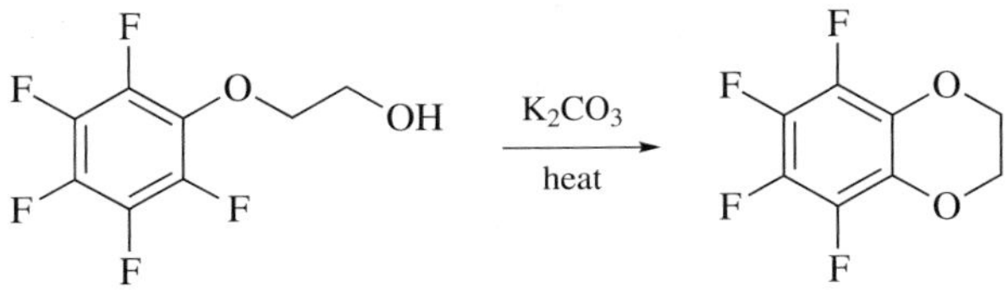

13.63 Reaction of hexamethylbenzene with methyl chloride and aluminum chloride gave a salt A, which, on being treated with aqueous sodium bicarbonate solution, yielded compound B. Suggest a mechanism for the conversion of hexamethylbenzene to B by correctly inferring the structure of A.

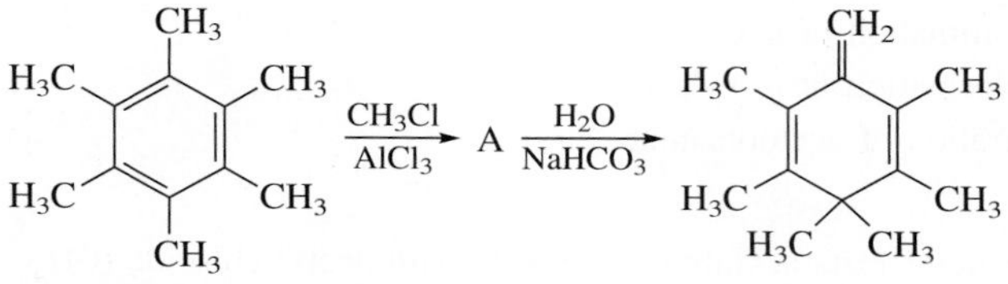

$$\text{Hexamethylbenzene} \xrightarrow[\text{AlCl}_3]{\text{CH}_3\text{Cl}} \text{A} \xrightarrow[\text{NaHCO}_3]{\text{H}_2\text{O}} \text{Compound B}$$

Hexamethylbenzene   Compound B

13.64 When styrene is heated with aqueous sulfuric acid, the two "styrene dimers" shown are the major products. Ignoring stereochemistry, suggest a reasonable mechanism for the formation of each isomer. Assume the proton donor in your mechanism is H_3O^+.

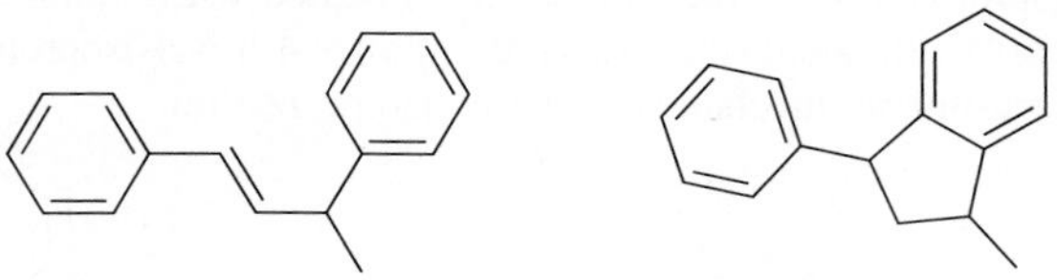

1,3-Diphenyl-1-butene   1-Methyl-3-phenylindan

Descriptive Passage and Interpretive Problems 13

Benzyne

Very strong bases such as sodium or potassium amide react readily with aryl halides, even those without electron-withdrawing substituents, to give products of nucleophilic substitution by the base. Substitution does not occur exclusively at the carbon with the halide, as shown for the following reaction of *o*-bromotoluene with sodium amide.

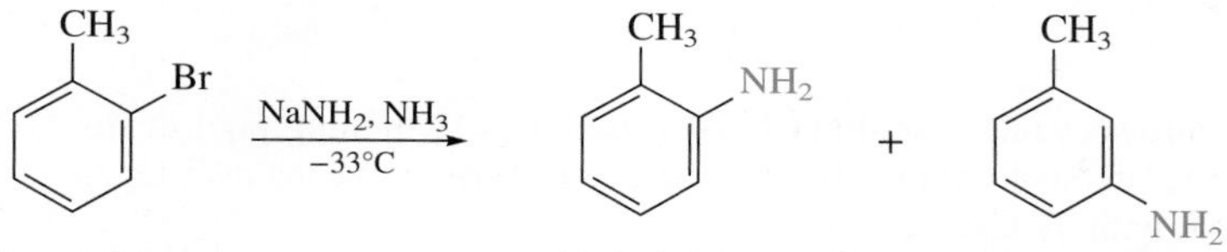

$$o\text{-Bromotoluene} \xrightarrow[-33^\circ\text{C}]{\text{NaNH}_2,\ \text{NH}_3} o\text{-Methylaniline} + m\text{-Methylaniline}$$

o-Bromotoluene   *o*-Methylaniline   *m*-Methylaniline

This reaction is inconsistent with substitution by an addition–elimination mechanism, because the nucleophile is not attached solely to the carbon from which the halide leaving group departed. An alternative mechanism was proposed on the basis of isotope experiments with ^{14}C-labeled chlorobenzene, in which the substitution product retained half of its label at C-1 and half at C-2.

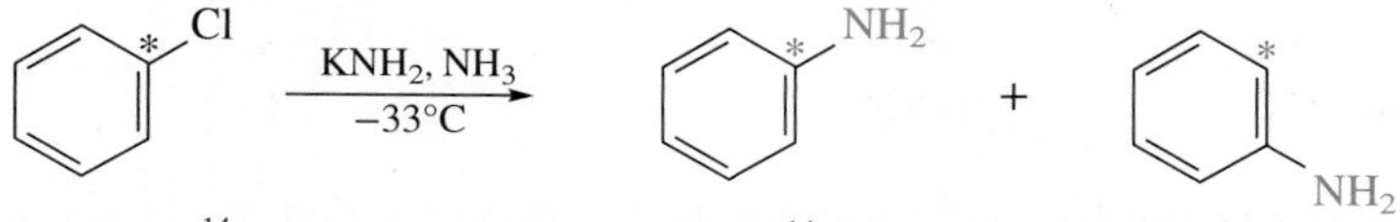

$$\text{Chlorobenzene-1-}^{14}\text{C} \xrightarrow[-33^\circ\text{C}]{\text{KNH}_2,\ \text{NH}_3} \text{Aniline-1-}^{14}\text{C} + \text{Aniline-2-}^{14}\text{C}$$

Chlorobenzene-1-^{14}C Aniline-1-^{14}C (48%)   Aniline-2-^{14}C (52%)

On the basis of the labeling experiment, an alternative mechanism was proposed for the substitution reaction of aryl halides with strong base—the elimination–addition mechanism. In the first step, the elimination stage, amide anion removes a proton from the carbon on the ring adjacent to the one with the halogen. The product is an unstable intermediate known as *benzyne*.

Chlorobenzene → Benzyne + $:NH_3$ + $:\ddot{C}l:^-$

In the second step, amide anion now acts as a nucleophile, adding to one of the carbons of the triple bond. This addition step gives an aryl anion.

Benzyne → Aryl anion

In the final step, the aryl anion abstracts a proton from ammonia to give aniline.

Aryl anion → Aniline + $:\bar{N}H_2$

The triple bond in benzyne is different from the usual triple bond of an alkyne. In benzyne, one of the π components of the triple bond results from *p–p* overlap and is part of the delocalized π system of the aromatic ring. The second π component, which results from overlapping adjacent sp^2-hybridized orbitals, lies in the plane of the ring and is not part of the aromatic π system, as shown in Figure 13.10.

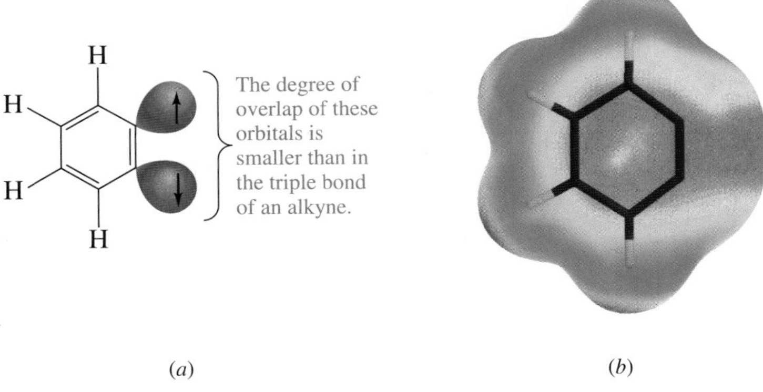
H
H
H
H
The degree of overlap of these orbitals is smaller than in the triple bond of an alkyne.
(a)
(b)

Figure 13.10

(*a*) The sp^2 orbitals in the plane of the ring in benzyne are not properly aligned for good overlap, and π bonding is weak. (*b*) The electrostatic potential map shows a region of high electron density associated with the "triple bond."

The intermediacy of benzyne in the elimination–addition mechanism for aryl halides accounts for the regioselectivity observed in the substitution reactions of labeled chlorobenzene and *o*-bromotoluene because both can give only a single aryne intermediate. Attack at either of the aryne carbons gives rise to the products.

The triple bond in benzyne is strained and is a dienophile in Diels–Alder reactions. Alternative methods exist for the generation of benzyne in cycloadditions and other synthetic applications. In the following example, *o*-bromofluorobenzene is treated with magnesium in tetrahydrofuran (THF). When carried out in the presence of cyclohexadiene, a Diels–Alder reaction occurs.

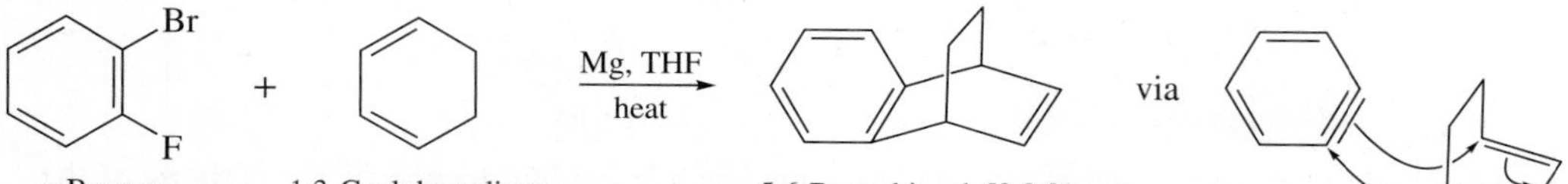

Br, F

o-Bromofluorobenzene + 1,3-Cyclohexadiene —(Mg, THF; heat)→ 5,6-Benzobicyclo[2.2.2]octa-2,5-diene (46%) via

13.65 Which of the following methylanilines can be formed by the reaction of *p*-bromotoluene with sodium amide in ammonia at −33°C?

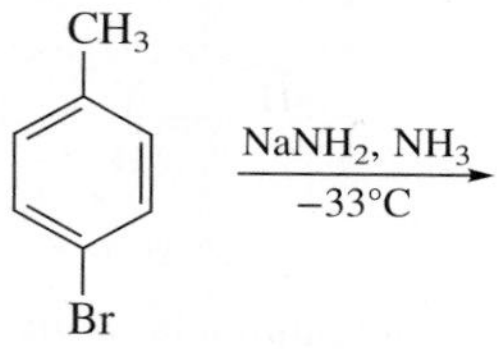

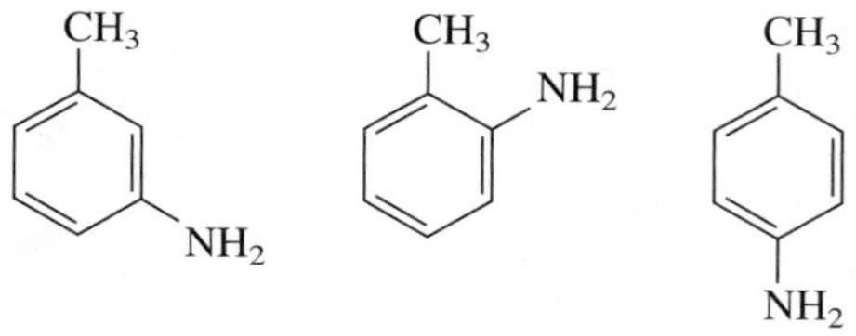

CH_3, Br —($NaNH_2$, NH_3; −33°C)→

I: CH_3, NH_2 II: CH_3, NH_2 III: CH_3, NH_2

A. I and II
B. II and III
C. I and III
D. I, II, and III

13.66 Which of the following methylanilines can be formed by the reaction of *m*-bromotoluene with sodium amide in ammonia at −33°C?

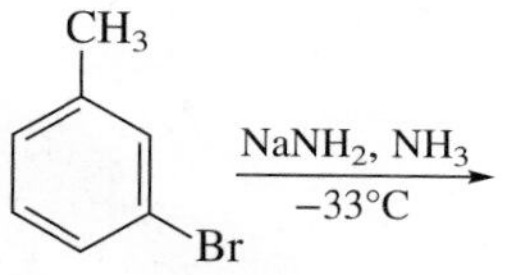

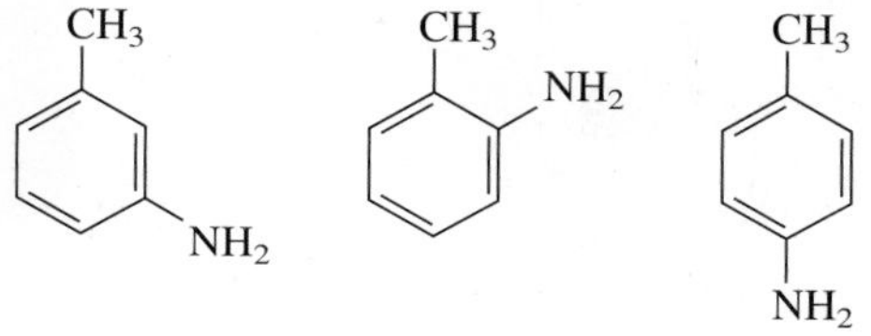

CH_3, Br —($NaNH_2$, NH_3; −33°C)→

I: CH_3, NH_2 II: CH_3, NH_2 III: CH_3, NH_2

A. I and II
B. II and III
C. I and III
D. I, II, and III

13.67 Which one of the following isomers of bromodimethylbenzene *cannot* undergo nucleophilic aromatic substitution by treatment with sodium amide in liquid ammonia?

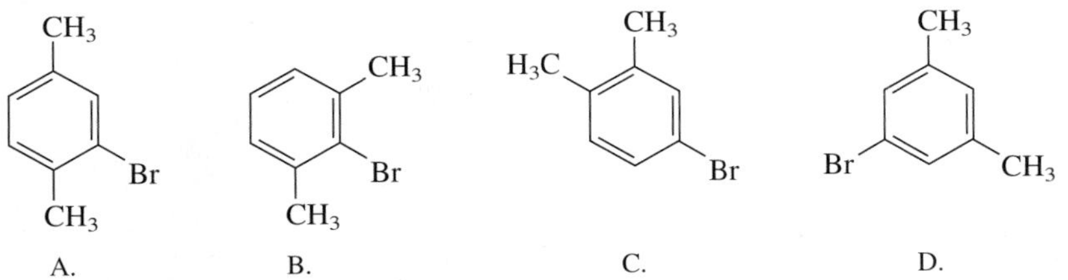

13.68 Two isomeric phenols are obtained in comparable amounts on hydrolysis of *p*-iodotoluene with 1 M sodium hydroxide at 300°C. What are the structures of each?

CH_3, I; 1 M NaOH, 300°C →

I	II	III	IV
CH_3, OH	CH_3, HO	CH_3, OH	OH, I

A. I and II

B. I and III

C. II and III

D. III and IV

Chapter 14

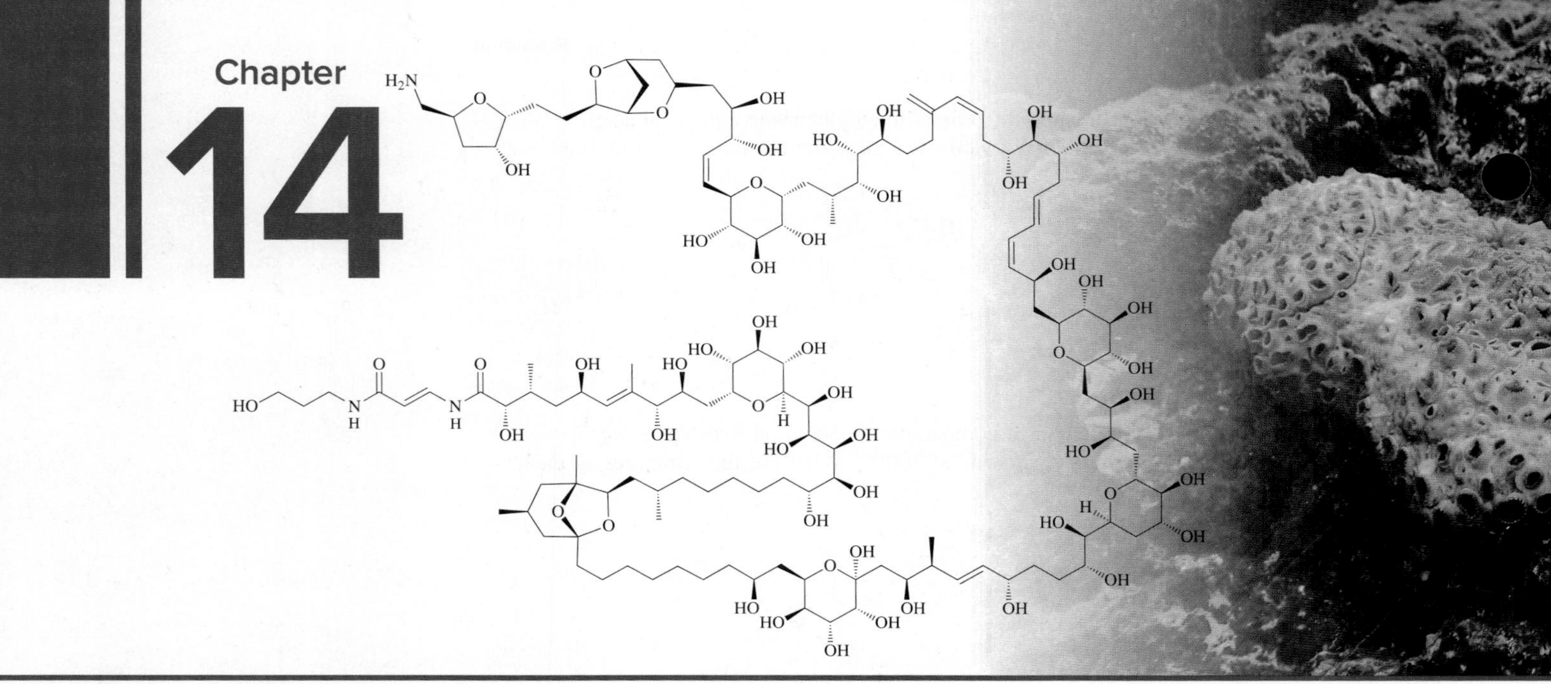

Palytoxin is one of the most toxic substances known. It is produced by a seaweed-like coral found in Hawaii known as *limu-make-o-Hana* and has the formula $C_{129}H_{223}N_3O_{54}$. NMR spectroscopy played a key role in the assignment of the structure of palytoxin.

Courtesy of James Davis Reimer

CHAPTER OUTLINE

Spectroscopy

Until the second half of the twentieth century, the structure of a substance—a newly discovered natural product, for example—was determined using information obtained from chemical reactions. This information included the identification of functional groups by chemical tests, along with the results of experiments in which the substance was broken down into smaller, more readily identifiable fragments. Typical of this approach is the demonstration of the presence of a double bond in an alkene by catalytic hydrogenation and determination of its location by ozonolysis. After considering all the available chemical evidence, the chemist proposed a candidate structure (or structures) consistent with the observations. Proof of structure was provided either by converting the substance to some already known compound or by an independent synthesis.

Qualitative tests and chemical degradation have given way to instrumental methods of structure determination. The main methods and the structural clues they provide are:

- **Nuclear magnetic resonance (NMR) spectroscopy,** which tells us about the carbon skeleton and the environments of the hydrogens attached to it.
- **Infrared (IR) spectroscopy,** which reveals the presence or signals the absence of key functional groups.

- **Ultraviolet-visible (UV-VIS) spectroscopy,** which probes the electron distribution, especially in molecules that have conjugated π-electron systems.
- **Mass spectrometry (MS),** which gives the molecular weight and formula, of both the molecule itself and various structural units within it.

As diverse as these techniques are, all of them are based on the absorption of energy by a molecule, and all measure how a molecule responds to that absorption. In describing these techniques our emphasis will be on their application to structure determination. We'll start with a brief discussion of electromagnetic radiation, which is the source of the energy that a molecule absorbs in NMR, IR, and UV-VIS spectroscopy. Mass spectrometry is unique in that, instead of electromagnetic radiation, its energy source is a stream of charged particles such as electrons.

14.1 Principles of Molecular Spectroscopy: Electromagnetic Radiation

Electromagnetic radiation, of which visible light is but one example, has the properties of both particles and waves. The particles are called **photons,** and each possesses an amount of energy referred to as a **quantum.** In 1900, the German physicist Max Planck proposed that the energy of a photon (E) is directly proportional to its **frequency** (ν).

"Modern" physics dates from Planck's proposal that energy is quantized, which set the stage for the development of quantum mechanics. Planck received the 1918 Nobel Prize in Physics.

$$E = h\nu$$

The SI units of frequency are reciprocal seconds (s^{-1}), given the name *hertz* and the symbol Hz in honor of the nineteenth-century physicist Heinrich R. Hertz. The constant of proportionality h is called **Planck's constant** and has the value

$$h = 6.63 \times 10^{-34}\ \text{J} \cdot \text{s}$$

Electromagnetic radiation travels at the speed of light ($c = 3.0 \times 10^8$ m/s), which is equal to the product of its frequency ν and its wavelength λ:

$$c = \nu\lambda$$

The range of photon energies is called the *electromagnetic spectrum* and is shown in Figure 14.1. Visible light occupies a very small region of the electromagnetic spectrum.

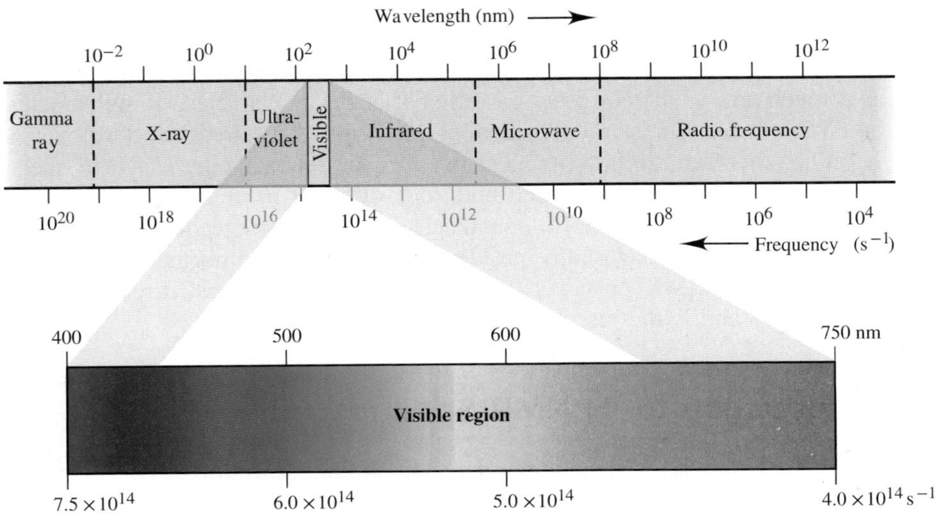

Figure 14.1

The electromagnetic spectrum.

1 nm = 10^{-9} m

It is characterized by wavelengths of 400 nm (violet) to 800 nm (red). When examining Figure 14.1 be sure to keep the following two relationships in mind:

1. *Frequency is inversely proportional to wavelength;* the greater the frequency, the shorter the wavelength.
2. *Energy is directly proportional to frequency;* electromagnetic radiation of higher frequency possesses more energy than radiation of lower frequency.

Gamma rays and X-rays are streams of very high energy photons. Radio waves are of relatively low energy. Ultraviolet radiation is of higher energy than the violet end of visible light. Infrared radiation is of lower energy than the red end of visible light. When a molecule is exposed to electromagnetic radiation, it may absorb a photon, increasing its energy by an amount equal to the energy of the photon. Molecules are highly selective with respect to the frequencies they absorb. Only photons of certain specific frequencies are absorbed by a molecule. The particular photon energies absorbed by a molecule depend on molecular structure and are measured with instruments called **spectrometers.** The data obtained are very sensitive indicators of molecular structure.

14.2 Principles of Molecular Spectroscopy: Quantized Energy States

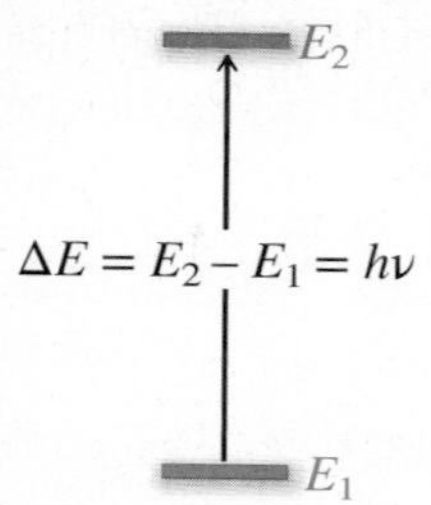

Figure 14.2

Two energy states of a molecule. Absorption of energy equal to $E_2 - E_1$ excites a molecule from its lower-energy state to the next higher state.

What determines whether electromagnetic radiation is absorbed by a molecule? The most important requirement is that the energy of the photon must equal the energy difference between two states, such as two nuclear spin states (NMR), two vibrational states (IR), or two electronic states (UV-VIS). In physics, the term for this is *resonance*—the transfer of energy between two objects that occurs when their frequencies are matched. In molecular spectroscopy, we are concerned with the transfer of energy from a photon to a molecule. Consider, for example, two energy states of a molecule designated E_1 and E_2 in Figure 14.2. The energy difference between them is $E_2 - E_1$, or ΔE. Unlike kinetic energy, which is continuous, meaning that all values of kinetic energy are available to a molecule, only certain energies are possible for electronic, vibrational, and nuclear spin states. These energy states are said to be **quantized.** More of the molecules exist in the lower-energy state E_1 than in the higher-energy state E_2. Excitation of a molecule from a lower state to a higher one requires the addition of an increment of energy equal to ΔE. Thus, when electromagnetic radiation strikes a molecule, only the frequency with energy equal to ΔE is absorbed.

Spectrometers are designed to measure the absorption of electromagnetic radiation by a sample. Basically, a spectrometer consists of a source of radiation, a compartment containing the sample through which the radiation passes, and a detector. The frequency of radiation is continuously varied, and its intensity at the detector is compared with that at the source. When the frequency is reached at which the sample absorbs radiation, the detector senses a decrease in intensity. The relation between frequency and absorption is plotted as a **spectrum,** which consists of a series of peaks at characteristic frequencies. Its interpretation can furnish structural information. Each type of spectroscopy developed independently of the others, and so the data format is different for each one. An NMR spectrum looks different from an IR spectrum, and both look different from a UV-VIS spectrum.

With this as background, we will now discuss spectroscopic techniques individually. NMR, IR, and UV-VIS spectroscopy provide complementary information, and all are useful. Among them, NMR provides the information that is most directly related to molecular structure and is the one we'll examine first.

14.3 Introduction to ^{1}H NMR Spectroscopy

Nuclear magnetic resonance spectroscopy depends on the absorption of energy when the nucleus of an atom is excited from its lowest-energy spin state to the next higher one. The nuclei of many elements can be studied by NMR, and the two elements that are the most common in organic molecules (carbon and hydrogen) have isotopes (^{1}H and ^{13}C) capable of giving NMR spectra that are rich in structural information. A proton nuclear magnetic resonance (^{1}H NMR) spectrum tells us about the environments of the various hydrogens in a molecule; a carbon-13 nuclear magnetic resonance (^{13}C NMR) spectrum does the same for the carbon atoms. Separately and together ^{1}H and ^{13}C NMR take us a long way toward determining a substance's molecular structure. We'll develop most of the general principles of NMR by discussing ^{1}H NMR, then extend them to ^{13}C NMR. The ^{13}C NMR discussion is shorter, not because it is less important than ^{1}H NMR, but because many of the same principles apply to both techniques.

Nuclear magnetic resonance of protons was first detected in 1946 by Edward Purcell (Harvard) and by Felix Bloch (Stanford). Purcell and Bloch shared the 1952 Nobel Prize in Physics.

Like an electron, a proton has two spin states with quantum numbers of $+\frac{1}{2}$ and $-\frac{1}{2}$. There is no difference in energy between these two nuclear spin states; a proton is just as likely to have a spin of $+\frac{1}{2}$ as $-\frac{1}{2}$. Absorption of electromagnetic radiation can occur only when the two spin states have different energies. A way to make them different is to place the sample in a magnetic field. A spinning proton behaves like a tiny bar magnet and has a magnetic moment associated with it (Figure 14.3). In the presence of an external magnetic field B_0, the spin state in which the magnetic moment of the nucleus is aligned with B_0 is lower in energy than the one in which it opposes B_0.

As shown in Figure 14.4, the energy difference between the two states is directly proportional to the strength of the applied field. Net absorption of electromagnetic radiation requires that the lower state be more highly populated than the higher one, and quite strong magnetic fields are required to achieve the separation necessary to give a detectable signal. A magnetic field of 4.7 T, which is about 100,000 times stronger than Earth's magnetic field, separates the two spin states of a proton by only 8×10^{-5} kJ/mol (1.9×10^{-5} kcal/mol). From Planck's equation $\Delta E = h\nu$, this energy gap corresponds to

The SI unit for magnetic field strength is the tesla (T), named after Nikola Tesla, a contemporary of Thomas Edison and who, like Edison, was an inventor of electrical devices.

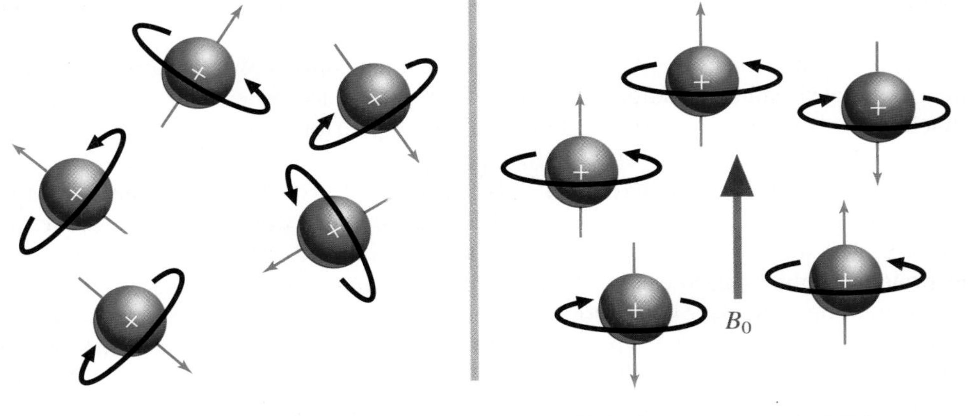

Figure 14.3

(*a*) In the absence of an external magnetic field, the nuclear spins of the protons are randomly oriented. (*b*) In the presence of an external magnetic field B_0, the nuclear spins are oriented so that the resulting nuclear magnetic moments are aligned either parallel or antiparallel to B_0. The lower energy orientation is the one parallel to B_0, and more nuclei have this orientation.

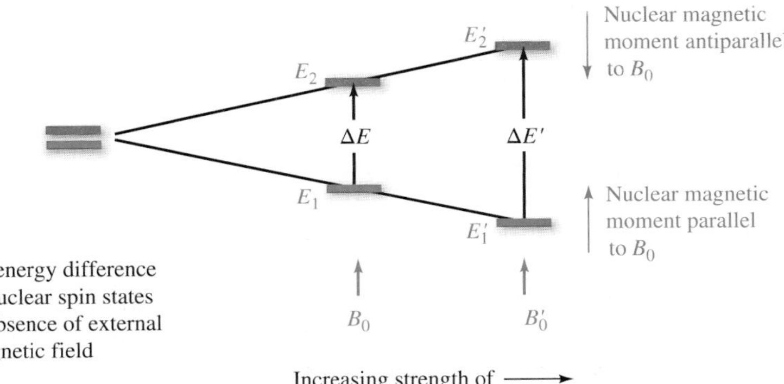

Figure 14.4

An external magnetic field causes the two nuclear spin states to have different energies. The difference in energy ΔE is proportional to the strength of the applied field.

radiation having a frequency of 2×10^8 Hz (200 MHz), which lies in the radiofrequency (rf) region of the electromagnetic spectrum (Figure 14.1).

Frequency of electromagnetic radiation (s^{-1} or Hz)	$\xrightarrow{\text{is proportional to}}$	Energy difference between nuclear spin states (kJ/mol or kcal/mol)	$\xrightarrow{\text{is proportional to}}$	Magnetic field (T)

Problem 14.1

Most of the NMR spectra in this text were recorded on a spectrometer having a field strength of 7 T (300 MHz for ^{1}H). The first generation of widely used NMR spectrometers were 60-MHz instruments. What was the magnetic field strength of these earlier spectrometers? What is the field strength of the 920-MHz instruments receiving increasing use?

The response of an atom to the strength of the external magnetic field is different for different elements, and for different isotopes of the same element. The resonance frequencies of most nuclei are sufficiently different that an NMR experiment is sensitive only to a particular isotope of a single element. The frequency for ^{1}H is 300 MHz at 7 T, but that for ^{13}C is 75.6 MHz. Thus, when recording the NMR spectrum of an organic compound, we see signals only for ^{1}H or ^{13}C, but not both; ^{1}H and ^{13}C NMR spectra are recorded in separate experiments with different instrument settings.

Problem 14.2

What will be the ^{13}C frequency of an NMR spectrometer that operates at 400 MHz for protons?

The essential features of an NMR spectrometer consist of a powerful magnet to align the nuclear spins, a radiofrequency (rf) transmitter as a source of energy to excite a nucleus from its lowest-energy state to the next higher one, and a way to monitor the absorption of rf radiation and display the spectrum.

NMR spectra are acquired using *pulsed Fourier-transform* nuclear magnetic resonance (FT-NMR) spectrometers (Figure 14.5). The sample is placed in a magnetic field and irradiated with a short, intense burst of rf radiation (the *pulse*), which excites *all* of the protons in the molecule at the same time. The magnetic field associated with the new orientation of nuclear spins induces an electrical signal in the receiver that decreases as the nuclei return to their original orientation. The resulting *free-induction decay* (FID) is a composite of the decay patterns of all of the protons in the molecule. The FID pattern is stored in a computer and converted into a spectrum by a mathematical process known as a *Fourier transform.* The pulse-relaxation sequence takes only about a second. The signal-to-noise ratio is enhanced by repeating the sequence many times, then averaging the data. Noise is random and averaging causes it to vanish; signals always appear at the same frequency and accumulate. All of the operations—the interval between pulses, collecting, storing, and averaging the data and converting it to a spectrum by a Fourier transform—are under computer control, which makes the actual recording of an FT-NMR spectrum a routine operation.

Richard R. Ernst of the Swiss Federal Institute of Technology won the 1991 Nobel Prize in Chemistry for devising pulse-relaxation NMR techniques.

14.4 Nuclear Shielding and ^{1}H Chemical Shifts

Our discussion so far has concerned ^{1}H nuclei in general without regard for the environments of individual protons in a molecule. Protons in a molecule are connected to other atoms—carbon, oxygen, nitrogen, and so on—by covalent bonds. The electrons in these bonds, indeed all the electrons in a molecule, affect the magnetic environment of the protons. Alone, a proton would feel the full strength of the external field, but a proton in an organic molecule responds to both the external field and any local fields within

1. Dissolve sample in deuterated chloroform ($CDCl_3$) and place in NMR tube.
2. Insert NMR tube into vertical cavity (bore) of the magnet.
3. Bore of magnet contains a probe that acts as a transmitter of radiofrequency (rf) pulses and receiver of signals from the sample. The transmitter is housed in a console along with other electronic equipment.
4. A short (5 μs), intense rf pulse is sent from the rf transmitter in the console to the probe. Absorption of rf energy tips the magnetic vector of the nuclei in the sample.
5. The magnetic field associated with the new orientation of the nuclei returns (relaxes) to the original state. Nuclei relax rapidly but at different rates that depend on their chemical environment. As the magnetic field changes, it generates an electrical impulse that is transmitted from the probe to a receiver in the console as a "free induction decay."
6. The pulse–relax sequence is repeated many times and the free-induction decay data stored in a computer in the console.
7. A mathematical operation called a Fourier transform carried out by the computer converts the amplitude-versus-time data of the free-induction decay to amplitude versus frequency and displays the resulting spectrum on the screen or prints it.

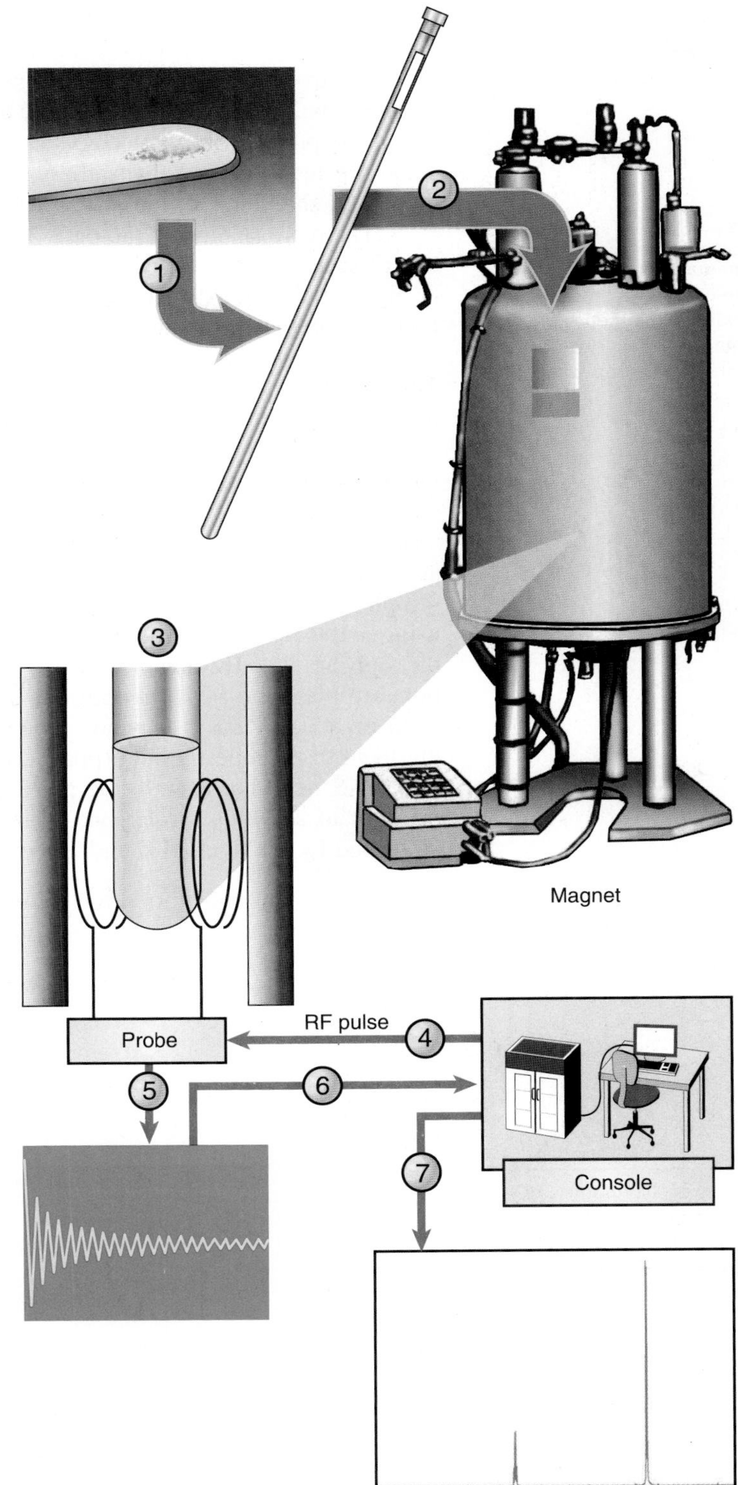

Figure 14.5

How an NMR spectrum is acquired using a pulsed Fourier-transform (FT) NMR spectrometer.

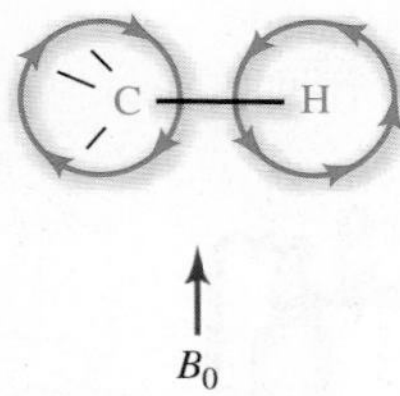

Figure 14.6

The induced magnetic field of the electrons in the carbon–hydrogen bond opposes the external magnetic field. The resulting magnetic field experienced by the proton and the carbon is slightly less than B_0.

the molecule. An external magnetic field affects the motion of the electrons in a molecule, inducing local fields characterized by lines of force that circulate in the *opposite* direction from the applied field (Figure 14.6). Thus, the net field felt by a proton in a molecule will always be less than the applied field, and the proton is said to be **shielded.** All of the protons of a molecule are shielded from the applied field by the electrons, but some are less shielded than others. The term *deshielded* is often used to describe this decreased shielding of one proton relative to another.

The more shielded a proton is, the greater must be the strength of the applied field in order to achieve resonance and produce a signal. A more shielded proton absorbs rf radiation at higher field strength (**upfield**) compared with one at lower field strength (**downfield**). Different protons give signals at different field strengths. *The dependence of the resonance position of a nucleus that results from its molecular environment is called its* **chemical shift.** This is where the real power of NMR lies. The chemical shifts of various protons in a molecule can be different and are characteristic of particular structural features.

Figure 14.7 shows the ^1H NMR spectrum of chloroform ($CHCl_3$) to illustrate how the terminology just developed applies to a real spectrum.

Instead of measuring chemical shifts in absolute terms, we measure them with respect to a standard—*tetramethylsilane* $(CH_3)_4Si$, abbreviated *TMS.* The protons of TMS are more shielded than those of most organic compounds, so all of the signals in a sample ordinarily appear at lower field than those of the TMS reference. When measured using a 100-MHz instrument, the signal for the proton in chloroform ($CHCl_3$), for example, appears 728 Hz downfield from the TMS signal. But because frequency is proportional to magnetic field strength, the same signal would appear 2184 Hz downfield from TMS on a 300-MHz instrument. We simplify the reporting of chemical shifts by converting them to parts per million (ppm) downfield from TMS, which is assigned a value of 0. The TMS need not actually be present in the sample, nor even appear in the spectrum in order to serve as a reference. When chemical shifts are reported this way, they are identified by the symbol δ and are independent of the magnetic field strength.

$$\frac{\text{Chemical shift } (\delta) = \text{position of signal} - \text{position of TMS peak}}{\text{spectrometer frequency}} \times 10^6$$

At 300 MHz the chemical shift for the proton in chloroform is:

$$\delta = \frac{2184 \text{ Hz} - 0 \text{ Hz}}{300 \times 10^6 \text{ Hz}} \times 10^6 = 7.28$$

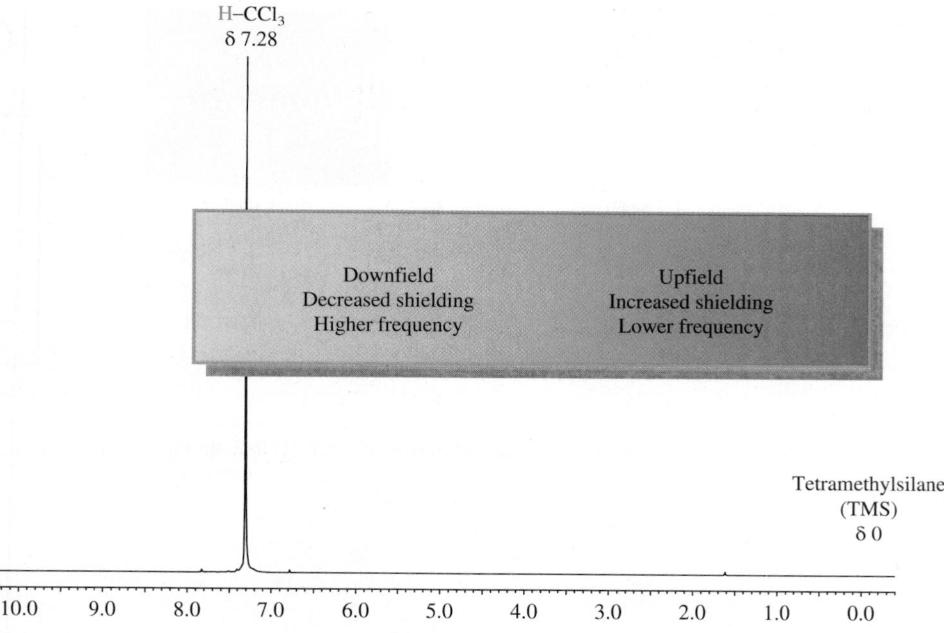

Figure 14.7

A ^1H NMR spectrum of chloroform ($CHCl_3$). Chemical shifts are measured along the *x*-axis in parts per million (ppm) from tetramethylsilane as the reference, which is assigned a value of zero.

Problem 14.3

The ^{1}H NMR signal for bromoform ($CHBr_3$) appears at 2065 Hz when recorded on a 300-MHz NMR spectrometer. (a) What is the chemical shift of this proton? (b) If the spectrum was recorded on a 400-MHz instrument, what would be the chemical shift of the $CHBr_3$ proton? (c) How many hertz downfield from TMS is the signal when recorded on a 400-MHz instrument?

NMR spectra are usually run in solution and, although chloroform is a good solvent for most organic compounds, it's rarely used because its own signal at δ 7.28 would be so intense that it would obscure signals in the sample. Because the magnetic properties of deuterium (D = ^{2}H) are different from those of ^{1}H, $CDCl_3$ gives no signals at all in a ^{1}H NMR spectrum and is the most commonly used solvent in ^{1}H NMR spectroscopy. Likewise, D_2O is used instead of H_2O for water-soluble substances such as carbohydrates.

14.5 Effects of Molecular Structure on ^{1}H Chemical Shifts

Nuclear magnetic resonance spectroscopy is such a powerful tool for structure determination because *protons in different environments experience different degrees of shielding and have different chemical shifts.* In compounds of the type CH_3X, for example, the shielding of the methyl protons increases as X becomes less electronegative.

Increased shielding of methyl protons
Decreasing electronegativity of atom attached to CH_3

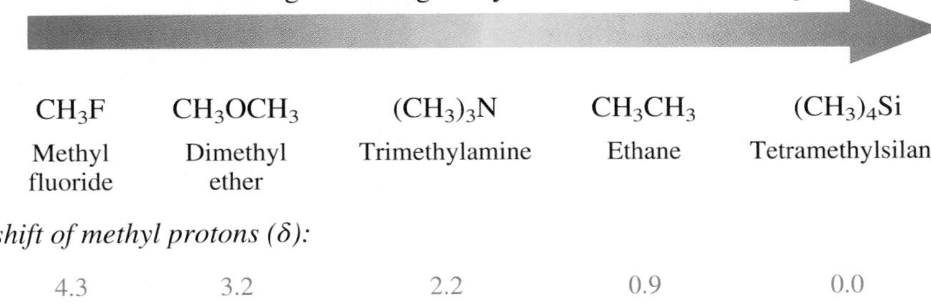

Inasmuch as the shielding is due to the electrons, it isn't surprising to find that the chemical shift depends on the degree to which X draws electrons away from the methyl group. A similar trend is seen in the methyl halides, in which the protons in CH_3F are the least shielded (δ 4.3) and those of CH_3I (δ 2.2) are the most.

Problem 14.3 in the preceding section was based on the chemical-shift difference between the proton in $CHCl_3$ and the proton in $CHBr_3$ and its relation to shielding.

The decreased shielding caused by electronegative substituents is primarily an inductive effect and, like other inductive effects, falls off rapidly as the number of bonds between the substituent and the proton increases. Compare the chemical shifts of the protons in propane and 1-nitropropane.

δ 0.9 δ 1.3 δ 0.9	δ 4.3 δ 2.0 δ 1.0
$H_3C—CH_2—CH_3$	$O_2N—CH_2—CH_2—CH_3$
Propane	1-Nitropropane

The strongly electron-withdrawing nitro group deshields the protons on C-1 by 3.4 ppm (δ 4.3–0.9). The effect is smaller on the protons at C-2 (0.7 ppm), and almost completely absent at C-3.

The deshielding effects of electronegative substituents are cumulative, as the chemical shifts for various chlorinated derivatives of methane indicate.

	CH_3Cl	CH_2Cl_2	$CHCl_3$
	Chloromethane	Dichloromethane	Trichloromethane
Chemical shift (δ):	3.1	5.3	7.3

Problem 14.4

Identify the most shielded and least shielded protons in

(a) 2-Bromobutane

(b) 1,1,2-Trichloropropane

(c) Tetrahydrofuran: 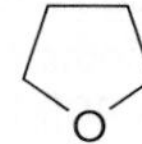

Sample Solution (a) Bromine is electronegative and will have its greatest electron-withdrawing effect on protons that are separated from it by the fewest bonds. Therefore, the proton at C-2 will be the least shielded, and those at C-4 the most shielded.

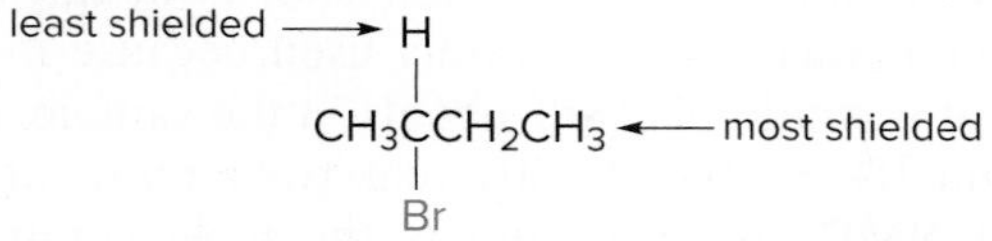

least shielded → H (on C-2)

$CH_3\overset{H}{\underset{Br}{C}}CH_2CH_3$ ← most shielded

The observed chemical shifts are δ 4.1 for the proton at C-2 and δ 1.1 for the protons at C-4. The protons at C-1 and C-3 appear in the range δ 1.7–2.0.

Figure 14.8 collects chemical-shift information for protons of various types. The major portion of the table concerns protons bonded to carbon. Within each type, methyl

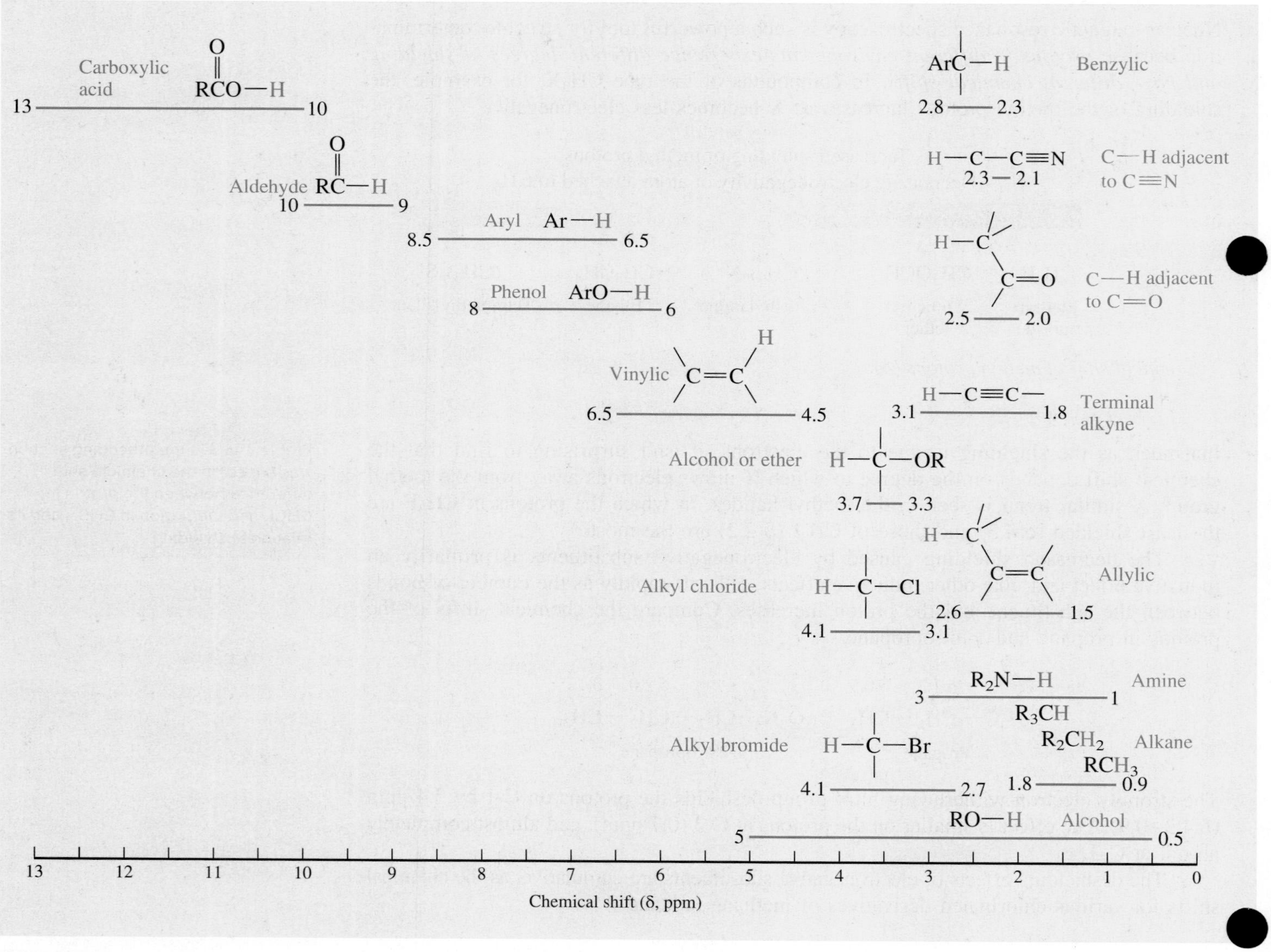

Carboxylic acid RCO—H (C=O): 13 — 10

Aldehyde RC—H (C=O): 10 — 9

Aryl Ar—H: 8.5 — 6.5

Phenol ArO—H: 8 — 6

Vinylic C=C—H: 6.5 — 4.5

Benzylic ArC—H: 2.8 — 2.3

C—H adjacent to C≡N, H—C—C≡N: 2.3 — 2.1

C—H adjacent to C=O, H—C—C=O: 2.5 — 2.0

Terminal alkyne H—C≡C—: 3.1 — 1.8

Alcohol or ether H—C—OR: 3.7 — 3.3

Allylic H—C—C=C: 2.6 — 1.5

Alkyl chloride H—C—Cl: 4.1 — 3.1

Amine R_2N—H: 3 — 1

Alkyl bromide H—C—Br: 4.1 — 2.7

Alkane R_3CH, R_2CH_2, RCH_3: 1.8 — 0.9

Alcohol RO—H: 5 — 0.5

13 12 11 10 9 8 7 6 5 4 3 2 1 0

Chemical shift (δ, ppm)

Figure 14.8

Approximate chemical-shift ranges for protons of various structural types in parts per million (δ) from tetramethylsilane. Protons in specific compounds may appear outside of the cited range depending on the shielding or deshielding effect of substituents. The chemical shifts of O—H and N—H protons depend on the conditions (solvent, temperature, concentration) under which the spectrum is recorded.

(CH_3) protons are more shielded than methylene (CH_2), and methylene protons are more shielded than methine (CH). The differences, however, are small.

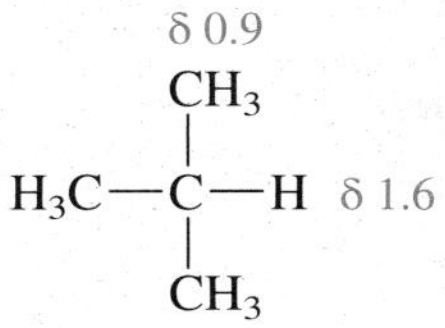

2-Methylpropane: $H_3C-C(CH_3)_2-H$; CH_3 δ 0.9, H δ 1.6

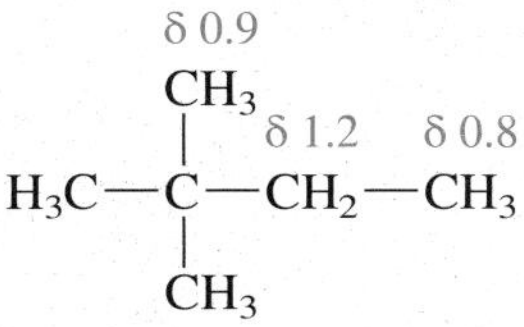

2,2-Dimethylbutane: $H_3C-C(CH_3)_2-CH_2-CH_3$; CH_3 δ 0.9, CH_2 δ 1.2, CH_3 δ 0.8

Given that the chemical shift of methane is δ 0.2, we attribute the decreased shielding of the protons of RCH_3, R_2CH_2, and R_3CH to the number of carbons attached to primary, secondary, and tertiary carbons, respectively. Carbon is more electronegative than hydrogen, so replacing the hydrogens of CH_4 by one, then two, then three carbons decreases the shielding of the remaining protons.

Likewise, the generalization that sp^2-hybridized carbon is more electronegative than sp^3-hybridized carbon is consistent with the decreased shielding of allylic and benzylic protons.

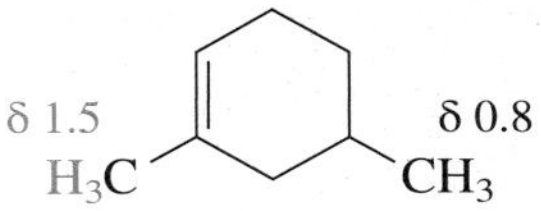

1,5-Dimethylcyclohexene: H_3C δ 1.5; CH_3 δ 0.8

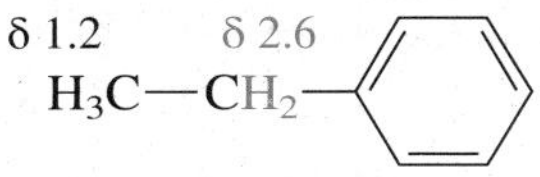

Ethylbenzene: $H_3C-CH_2-C_6H_5$; H_3C δ 1.2, CH_2 δ 2.6

Hydrogens that are directly attached to double bonds (vinylic protons) or to aromatic rings (aryl protons) are especially deshielded.

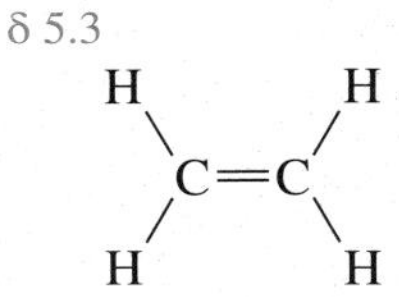

Ethylene: $H_2C=CH_2$ δ 5.3

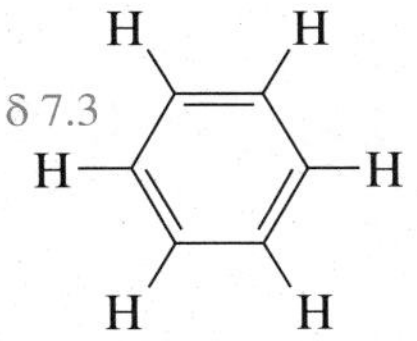

Benzene: δ 7.3

The main contributor to the deshielding of vinylic and aryl protons is the induced magnetic field associated with π electrons. We saw earlier in Section 14.4 that the local field resulting from electrons in a C—H σ bond opposes the applied field and shields a molecule's protons. The hydrogens of ethylene and benzene, however, lie in a region of the molecule where the induced magnetic field of the π electrons reinforces the applied field, *deshielding* the protons (Figure 14.9). In the case of benzene, this is described as a **ring current** effect that originates in the circulating π electrons. It has interesting consequences, some of which are described in the boxed essay *Ring Currents: Aromatic and Antiaromatic* in this chapter.

The induced field of C=C and aryl groups contributes to the deshielding of allylic and benzylic hydrogens.

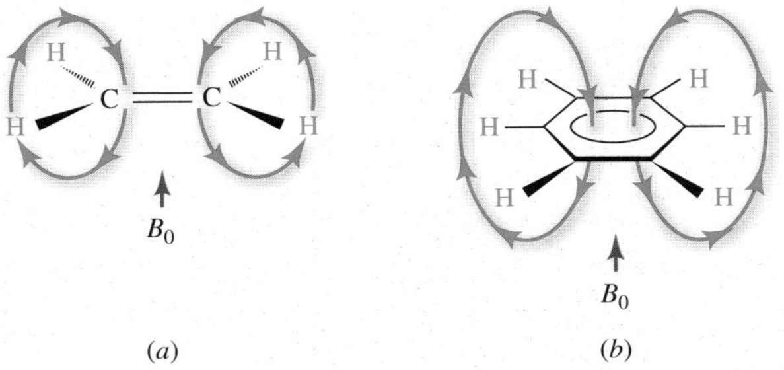

(*a*) (*b*)

Figure 14.9

The induced magnetic field of the π electrons of (*a*) ethylene and (*b*) benzene reinforces the applied field in the regions near vinyl and aryl protons and deshields them.

Problem 14.5

(a) Assign the chemical shifts δ 1.6, δ 2.2, and δ 4.8 to the appropriate protons of methylenecyclopentane.

$=CH_2$

Sample Solution

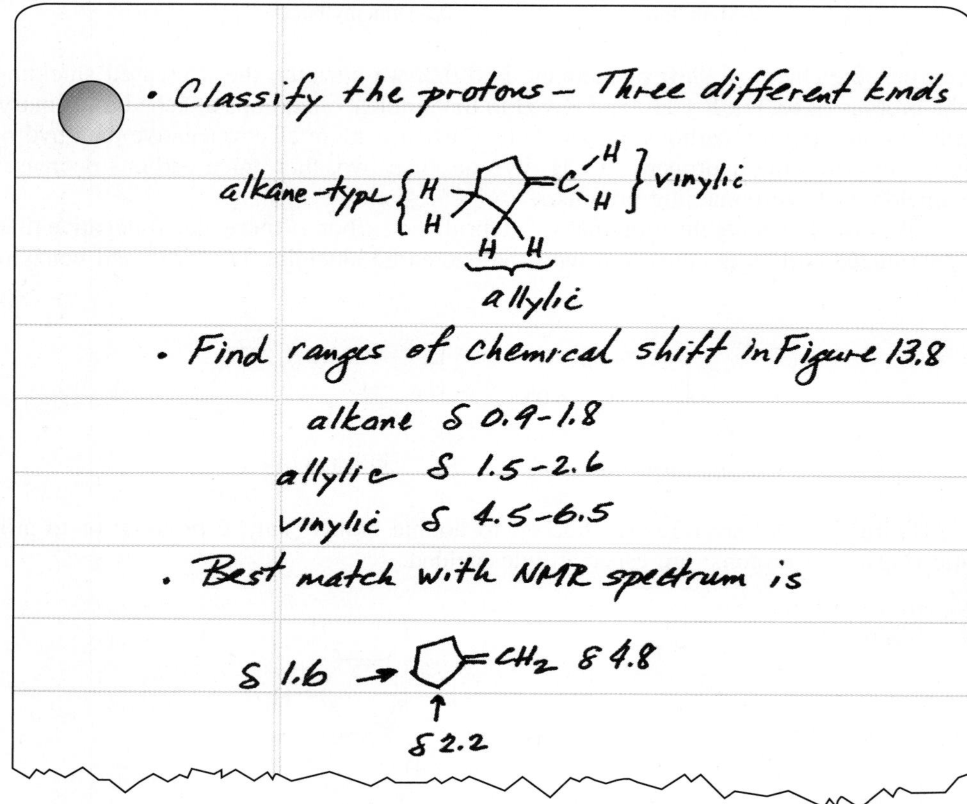

• Classify the protons – Three different kinds
alkane-type
vinylic
allylic
• Find ranges of chemical shift in Figure 13.8
alkane δ 0.9–1.8
allylic δ 1.5–2.6
vinylic δ 4.5–6.5
• Best match with NMR spectrum is
δ 1.6
CH₂ δ 4.8
δ 2.2

(b) Do the same for the chemical shifts δ 2.0, δ 5.1, and δ 7.2 of benzyl acetate.

CH_2OCCH_3 (with C=O)

Acetylenic hydrogens are unusual in that they are more shielded than we would expect for protons bonded to *sp*-hybridized carbon. This is because the π electrons circulate around the triple bond, not along it (Figure 14.10*a*). Therefore, the induced magnetic field is parallel to the long axis of the triple bond and shields the acetylenic proton (Figure 14.10*b*). Acetylenic protons typically have chemical shifts in the range δ 1.8–3.1.

δ 1.9

$H{-}C{\equiv}C{-}CH_2CH_2CH_2CH_3$

1-Hexyne

The induced field of a carbonyl group (C=O) deshields protons in much the same way that C=C does, and its oxygen makes it even more electron withdrawing.

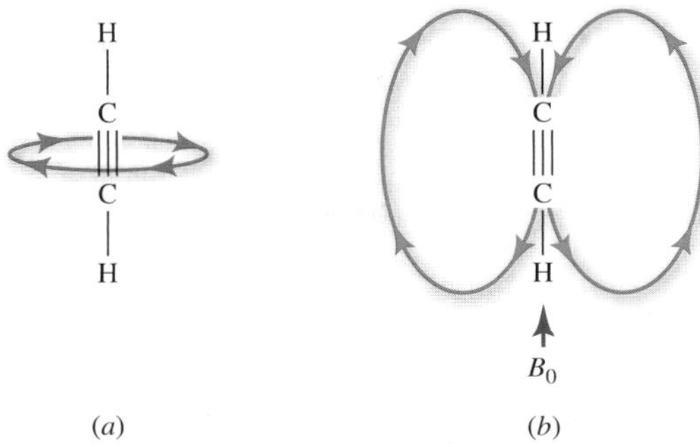

Figure 14.10

(*a*) The π electrons of acetylene circulate in a region surrounding the long axis of the molecule. (*b*) The induced magnetic field associated with the π electrons opposes the applied field and shields the protons.

Thus, protons attached to C═O in aldehydes are the least shielded of any protons bonded to carbon. They have chemical shifts in the range δ 9–10.

δ 1.1 H_3C—C(H, δ 2.4)(CH_3)—C(═O)—H δ 9.7

2-Methylpropanal

δ 1.4 CH_3CH_2O (δ 4.1)—C_6H_4 (H δ 6.9, H δ 7.8)—C(═O)—H δ 9.9

p-Ethoxybenzaldehyde

Protons on carbons adjacent to a carbonyl group are deshielded slightly more than allylic hydrogens.

Problem 14.6

Assign the chemical shifts δ 1.1, δ 1.7, δ 2.0, and δ 2.3 to the appropriate protons of 2-pentanone.

$CH_3\overset{O}{\overset{\|}{C}}CH_2CH_2CH_3$

The chemical shifts of O—H and N—H protons vary much more than those of protons bonded to carbon. This is because O—H and N—H groups can be involved in intermolecular hydrogen bonding, the extent of which depends on molecular structure, temperature, concentration, and solvent. Generally, an increase in hydrogen bonding decreases the shielding. This is especially evident in carboxylic acids. With δ values in the 10–12 ppm range, O—H protons of carboxylic acids are the least shielded of all of the protons in Figure 14.8. Hydrogen bonding in carboxylic acids is stronger than in most other classes of compounds that contain O—H groups.

Problem 14.7

Assign the chemical shifts δ 1.6, δ 4.0, δ 7.5, δ 8.2, and δ 12.0 to the appropriate protons of 2-(*p*-nitrophenyl)propanoic acid.

O_2N—C_6H_4—$CH(CH_3)\overset{O}{\overset{\|}{C}}OH$

Ring Currents: Aromatic and Antiaromatic

We saw in Chapter 12 that aromaticity reveals itself in various ways. Qualitatively, aromatic compounds are more stable and less reactive than alkenes. Quantitatively, their heats of hydrogenation are smaller than expected. Theory, especially Hückel's rule, furnishes a structural basis for aromaticity. Now let's examine some novel features of the NMR spectra of aromatic compounds.

We have mentioned that the protons in benzene appear at relatively low field because of deshielding by the magnetic field associated with the circulating π electrons. The amount of deshielding is sufficiently large—on the order of 2 ppm more than the corresponding effect in alkenes—that its presence is generally accepted as evidence for aromaticity. We speak of this deshielding as resulting from an *aromatic ring current.*

Something interesting happens when we go beyond benzene to apply the aromatic ring current test to annulenes.

[18]Annulene satisfies the Hückel $(4n + 2)$ π-electron rule for aromaticity, and many of its properties indicate aromaticity (Section 12.19). As shown in Figure 14.11*a*, [18]annulene contains two different kinds of protons; 12 lie on the ring's periphery ("outside"), and 6 reside near the middle of the molecule ("inside"). The 2:1 ratio of outside/inside protons makes it easy to assign the signals in the ^{1}H NMR spectrum. The outside protons have a chemical shift δ of 9.3 ppm, which makes them even less shielded than those of benzene. The six inside protons, on the other hand, have a *negative* chemical shift (δ–3.0), meaning that the signal for these protons appears at *higher field* (to the right) of the TMS peak. *The inside protons of [18]annulene are more than 12 ppm more shielded than the outside protons.*

As shown in Figure 14.11*a*, both the shielding of the inside protons and the deshielding of the outside ones result from the same aromatic ring current. When the molecule is placed in an external magnetic field B_0, its circulating π electrons produce their own magnetic field. This induced field opposes the applied field B_0 in the center of the molecule, shielding the inside protons. Because the induced magnetic field closes on itself, the outside protons lie in a region where the induced field reinforces B_0. The aromatic ring current in [18]annulene shields the 6 inside protons and deshields the 12 outside ones.

Exactly the opposite happens in [16]annulene (Figure 14.11*b*). Now it is the outside protons (δ 5.3) that are more shielded. The inside protons (δ 10.6) are less shielded than the outside ones and less shielded than the protons of both benzene and [18]annulene. This reversal of the shielding and deshielding regions in going from [18] to [16]annulene can only mean that the directions of their induced magnetic fields are reversed. Thus, [16]annulene, which has $4n$ π electrons and is antiaromatic, not only lacks an aromatic ring current, its π electrons produce exactly the opposite effect when placed in a magnetic field.

Score one for Hückel.

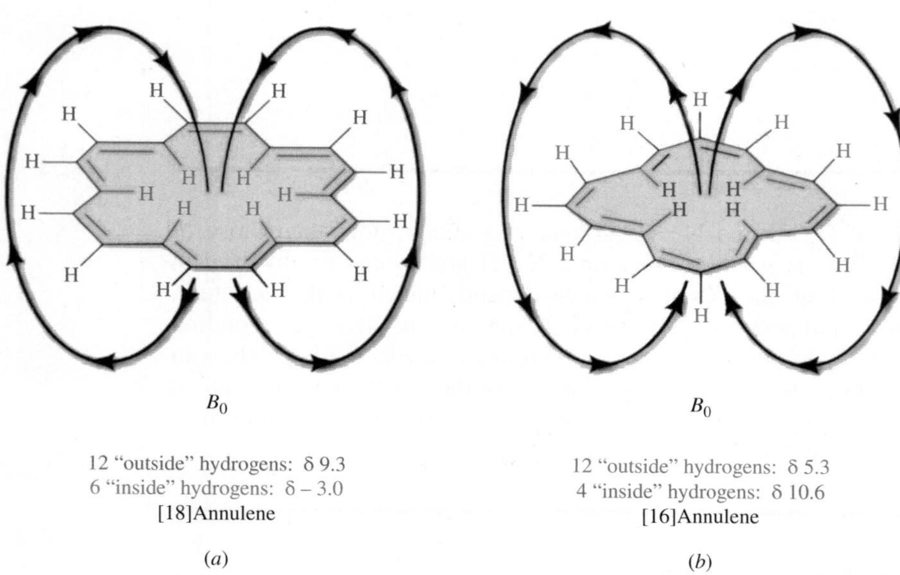

Figure 14.11

More shielded (red) and less shielded (blue) protons in (*a*) [18]annulene and (*b*) [16]annulene. The induced magnetic field associated with the aromatic ring current in [18]annulene shields the inside protons and deshields the outside protons. The opposite occurs in [16]annulene, which is antiaromatic.

It is common for several different kinds of protons to have similar chemical shifts. The range covered for ^1H chemical shifts is only 12 ppm, which is relatively small compared with (as we'll see) the 200-ppm range for ^{13}C chemical shifts. The ability of an NMR spectrometer to separate signals that have similar chemical shifts is termed its *resolving power* and is directly related to the magnetic field strength of the instrument. Even though the δ values of their chemical shifts in parts per million don't change, two signals that are closely spaced at 60 MHz become well separated at 300 MHz.

14.6 Interpreting ^1H NMR Spectra

In addition to suggesting what structural units might be present from the chemical shifts, an NMR spectrum provides other useful information.

1. *The number of signals,* which tells us how many different kinds of protons there are.
2. *The intensity of the signals* as measured by the area under each peak, which tells us the relative ratios of the different kinds of protons.
3. *The multiplicity, or splitting, of each signal,* which tells us how many protons are vicinal to the one giving the signal.

Recall (Section 9.7) that two atoms or groups are vicinal to each other if they are attached to adjacent carbons.

Protons that have different chemical shifts are said to be **chemical-shift-nonequivalent** (or **chemically nonequivalent**). A separate NMR signal is given for each chemical-shift-nonequivalent proton in a substance. Figure 14.12 shows the 300-MHz ^1H NMR spectrum of methoxyacetonitrile (CH_3OCH_2CN), a molecule with protons in two different environments. The three protons in the CH_3O group constitute one set, the two protons in the OCH_2CN group the other, and both can be assigned on the basis of their chemical shifts. The protons in the OCH_2CN group are connected to a carbon that bears two electronegative substituents (O and C≡N), are less shielded, and have a larger chemical shift (δ 4.2) than those of the CH_3O group (δ 3.4), which are attached to a carbon that bears only one electronegative atom (O).

Another way to assign the peaks is by comparing their intensities. The three equivalent protons of the CH_3O group give rise to a more intense peak than the two equivalent protons of the OCH_2CN group. This is clear by comparing the heights of the peaks in the spectrum in this case, but in general it is better to compare peak areas. This is done electronically at the time the NMR spectrum is recorded, and the **integrated areas** are displayed on the computer screen or printed out. Peak areas are proportional to the number of equivalent protons responsible for that signal.

It is important to remember that integration of peak areas gives relative, not absolute, proton counts. Thus, a 3:2 ratio of areas can, as in the case of CH_3OCH_2CN, correspond to a 3:2 ratio of protons. But in some other compound a 3:2 ratio of areas might correspond to a 6:4 or 9:6 ratio of protons.

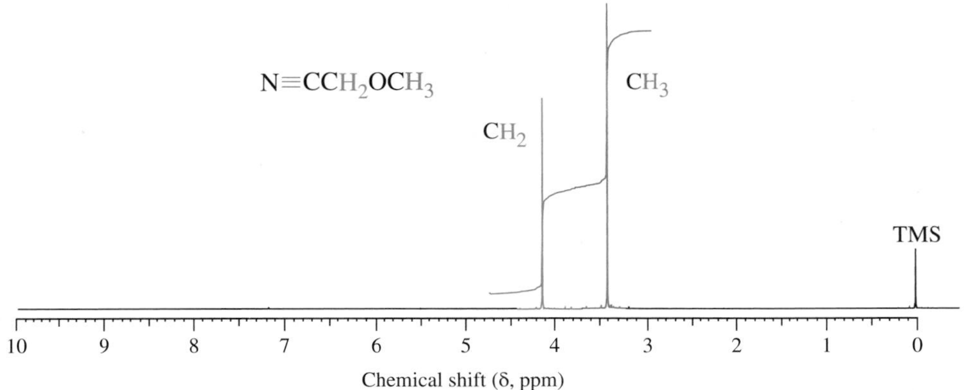

Figure 14.12

The 300-MHz ^1H NMR spectrum of methoxyacetonitrile (CH_3OCH_2CN).

Problem 14.8

The 300-MHz ^1H NMR spectrum of 1,4-dimethylbenzene looks exactly like that of CH_3OCH_2CN except the chemical shifts of the two peaks are δ 2.2 and δ 7.0. Assign the peaks to the appropriate protons of 1,4-dimethylbenzene.

Protons in equivalent environments have the same chemical shift. Often it is an easy matter to decide, simply by inspection, whether protons are equivalent or not. In more difficult cases, mentally replacing a proton in a molecule by a "test group" can help. We'll illustrate the procedure for a simple case—the protons of propane. To see if they have the same chemical shift, replace one of the methyl protons at C-1 by chlorine, then do the same thing for a proton at C-3. Both replacements give the same molecule, 1-chloropropane. Therefore, the methyl protons at C-1 are equivalent to those at C-3.

$CH_3CH_2CH_3$	$ClCH_2CH_2CH_3$	$CH_3CH_2CH_2Cl$
Propane	1-Chloropropane	1-Chloropropane

If the two structures produced by replacement of two different hydrogens in a molecule by a test group are the same, the hydrogens are chemically equivalent. Thus, the six methyl protons of propane are all chemically equivalent to one another and have the same chemical shift.

Replacement of either one of the methylene protons of propane generates 2-chloropropane. Both methylene protons are equivalent. Neither of them is equivalent to any of the methyl protons.

The ^1H NMR spectrum of propane contains two signals: one for the six equivalent methyl protons, the other for the pair of equivalent methylene protons.

Problem 14.9

How many signals would you expect to find in the ^1H NMR spectrum of each of the following compounds?

(a) 1-Bromobutane
(b) 1-Butanol
(c) Butane
(d) 1,4-Dibromobutane
(e) 2,2-Dibromobutane
(f) 2,2,3,3-Tetrabromobutane
(g) 1,1,4-Tribromobutane
(h) 1,1,1-Tribromobutane

Sample Solution (a) To test for chemical-shift equivalence, replace the protons at C-1, C-2, C-3, and C-4 of 1-bromobutane by some test group such as chlorine. Four constitutional isomers result:

1-Bromo-1-chlorobutane 1-Bromo-2-chlorobutane 1-Bromo-3-chlorobutane 1-Bromo-4-chlorobutane

Thus, separate signals will be seen for the protons at C-1, C-2, C-3, and C-4. Barring any accidental overlap, we expect to find four signals in the NMR spectrum of 1-bromobutane.

Chemical-shift nonequivalence can occur when two environments are stereochemically different. The two vinyl protons of 2-bromopropene have different chemical shifts.

2-Bromopropene

One of the vinyl protons is cis to bromine; the other trans. Replacing one of the vinyl protons by some test group, say, chlorine, gives the *Z* isomer of 2-bromo-1-chloropropene; replacing the other gives the *E* stereoisomer. The *E* and *Z* forms of 2-bromo-1-chloropropene are diastereomers. Protons that yield diastereomers on being replaced by some test group are *diastereotopic* and can have different chemical shifts. Because their environments are similar, however, the chemical-shift difference is usually small, and it sometimes happens that two diastereotopic protons accidentally have the same chemical shift. Recording the spectrum on a higher field NMR spectrometer is often helpful in resolving signals with similar chemical shifts.

Problem 14.10

How many signals would you expect to find in the ¹H NMR spectrum of each of the following compounds?

(a) Vinyl bromide
(b) 1,1-Dibromoethene
(c) *cis*-1,2-Dibromoethene
(d) *trans*-1,2-Dibromoethene
(e) Allyl bromide
(f) 2-Methyl-2-butene

Sample Solution (a) Each proton of vinyl bromide is unique and has a chemical shift different from the other two. The least shielded proton is attached to the carbon that bears the bromine. The pair of protons at C-2 are diastereotopic with respect to each other; one is cis to bromine and the other is trans to bromine. There are three proton signals in the NMR spectrum of vinyl bromide. Their observed chemical shifts are as indicated.

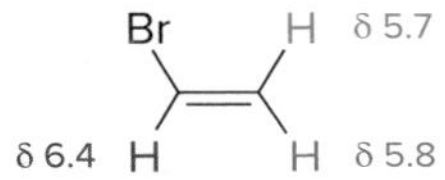

Br H δ 5.7
δ 6.4 H H δ 5.8

When enantiomers are generated by replacing first one proton and then another by a test group, the pair of protons are *enantiotopic* (Descriptive Passage and Interpretive Problems 4). The methylene protons at C-2 of 1-propanol, for example, are enantiotopic.

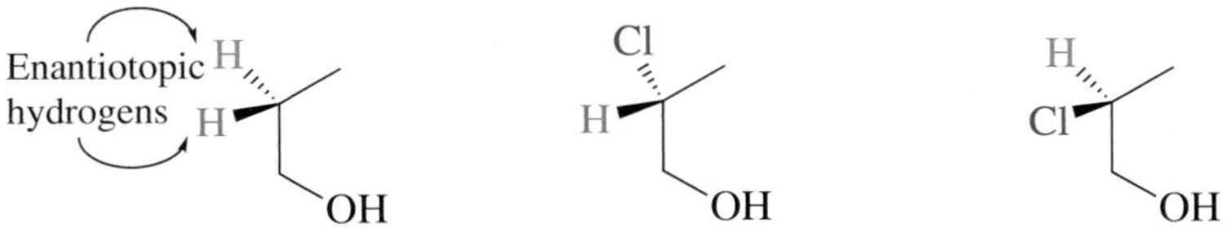

Enantiotopic hydrogens H H OH — 1-Propanol
Cl H OH — (*R*)-2-Chloro-1-propanol
H Cl OH — (*S*)-2-Chloro-1-propanol

Replacing one of these protons by chlorine as a test group gives (*R*)-2-chloro-1-propanol; replacing the other gives (*S*)-2-chloro-1-propanol. Enantiotopic protons have the same chemical shift, regardless of the field strength of the NMR spectrometer.

At the beginning of this section we noted that an NMR spectrum provides structural information based on chemical shift, the number of peaks, their relative areas, and the multiplicity, or splitting, of the peaks. We have discussed the first three of these features of ¹H NMR spectroscopy. Let's now turn our attention to peak splitting to see what kind of information it offers.

Enantiotopic protons can have different chemical shifts in an enantiomerically enriched chiral solvent. Because the customary solvent ($CDCl_3$) used in NMR measurements is achiral, this phenomenon is not observed in routine work.

14.7 Spin–Spin Splitting and ¹H NMR

The ¹H NMR spectrum of CH_3OCH_2CN (Figure 14.12) displayed in the preceding section is relatively simple because both signals are singlets; that is, each one consists of a single peak. It is quite common though to see a signal for a particular proton appear not as a singlet, but as a collection of peaks. The signal may be split into two peaks (a doublet), three peaks

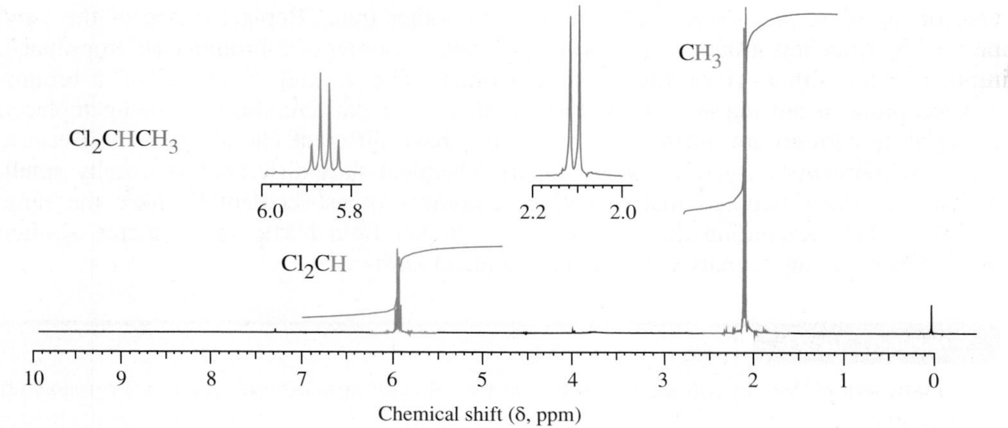

Figure 14.13

The 300-MHz ^{1}H NMR spectrum of 1,1-dichloroethane (Cl_2CHCH_3), showing the methine proton as a quartet and the methyl protons as a doublet. The peak multiplicities are seen more clearly in the scale-expanded insets.

(a triplet), four peaks (a quartet), or even more. Figure 14.13 shows the 300-MHz ^{1}H NMR spectrum of 1,1-dichloroethane (CH_3CHCl_2), which is characterized by a doublet centered at δ 2.1 for the methyl protons and a quartet at δ 5.9 for the methine proton.

Complex splitting patterns conform to an extension of the "$n + 1$" rule and will be discussed in Section 14.11.

The number of peaks into which the signal for a particular proton is split is called its **multiplicity.** For simple cases the rule that allows us to predict splitting in ^{1}H NMR spectroscopy is

$$\text{Multiplicity of signal for } H_a = n + 1$$

where n is equal to the number of equivalent protons that are vicinal to H_a. Two protons are vicinal to each other when they are bonded to adjacent atoms. Protons vicinal to H_a are separated from H_a by three bonds. The three methyl protons of 1,1-dichloroethane are equivalent and vicinal to the methine proton and split its signal into a quartet. The single methine proton, in turn, splits the methyl protons' signal into a doublet.

This proton splits the signal for the methyl protons into a doublet.

H—C(Cl)(Cl)—CH$_3$

These three protons split the signal for the methine proton into a quartet.

The physical basis for peak splitting in 1,1-dichloroethane can be explained with the aid of Figure 14.14, which examines how the chemical shift of the methyl protons is affected by the spin of the methine proton. There are two magnetic environments for the methyl protons: one in which the magnetic moment of the methine proton is parallel to the applied field, and the other in which it is antiparallel to it. When the magnetic moment of the methine proton is parallel to the applied field, it reinforces it. This decreases the shielding of the methyl protons and causes their signal to appear at slightly

Figure 14.14

The magnetic moments (blue arrows) of the two possible spin states of the methine proton affect the chemical shift of the methyl protons in 1,1-dichloroethane. When the magnetic moment is parallel to the external field B_0 (green arrow), it adds to the external field and a smaller B_0 is needed for resonance. When it is antiparallel to the external field, it subtracts from it and shields the methyl protons.

H—C(Cl)(Cl)—CH$_3$ (↑) B_0 H—C(Cl)(Cl)—CH$_3$ (↓)

Spin of methine proton reinforces B_0. Methyl signal appears at lower field (higher frequency).

Spin of methine proton shields methyl protons from B_0. Methyl signal appears at higher field (lower frequency).

lower field strength (higher frequency). Conversely, when the magnetic moment of the methine proton is antiparallel to the applied field, it opposes it and increases the shielding of the methyl protons. Instead of a single peak for the methyl protons, there are two of approximately equal intensity: one at slightly higher field than the "true" chemical shift, the other at slightly lower field.

Turning now to the methine proton, its signal is split by the methyl protons into a quartet. The same kind of analysis applies here and is outlined in Figure 14.15. The methine proton "sees" eight different combinations of nuclear spins for the methyl protons. In one combination, the magnetic moments of all three methyl protons reinforce the applied field. At the other extreme, the magnetic moments of all three methyl protons oppose the applied field. There are three combinations in which the magnetic moments of two methyl protons reinforce the applied field, whereas one opposes it. Finally, there are three combinations in which the magnetic moments of two methyl protons oppose the applied field and one reinforces it. These eight possible combinations give rise to four distinct peaks for the methine proton, with a ratio of intensities of 1:3:3:1.

We describe the observed splitting of NMR signals as **spin–spin splitting** and the physical basis for it as **spin–spin coupling.** It has its origin in the communication of nuclear spin information via the electrons in the bonds that intervene between the nuclei. Its effect is greatest when the number of bonds is small. Vicinal protons are separated by three bonds, and coupling between vicinal protons, as in 1,1-dichloroethane, is called **three-bond coupling,** or **vicinal coupling.** Four-bond couplings are weaker and not normally observable.

A very important characteristic of spin–spin splitting is that protons that have the same chemical shift do not split each other's signal. Ethane, for example, shows only a single sharp peak in its NMR spectrum. Even though there is a vicinal relationship between the protons of one methyl group and those of the other, they do not split each other's signal because they are equivalent.

Problem 14.11

Describe the appearance of the ^{1}H NMR spectrum of each of the following compounds. How many signals would you expect to find, and into how many peaks will each signal be split?

(a) 1,2-Dichloroethane
(b) 1,1,1-Trichloroethane
(c) 1,1,2-Trichloroethane
(d) 1,2,2-Trichloropropane
(e) 1,1,1,2-Tetrachloropropane

Sample Solution (a) All the protons of 1,2-dichloroethane ($ClCH_2CH_2Cl$) are chemically equivalent and have the same chemical shift. Protons that have the same chemical shift do not split each other's signal, and so the NMR spectrum of 1,2-dichloroethane consists of a single sharp peak.

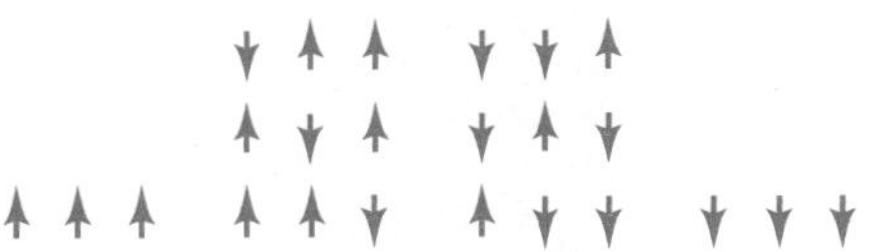

There are eight possible combinations of the nuclear spins of the three methyl protons in CH_3CHCl_2.

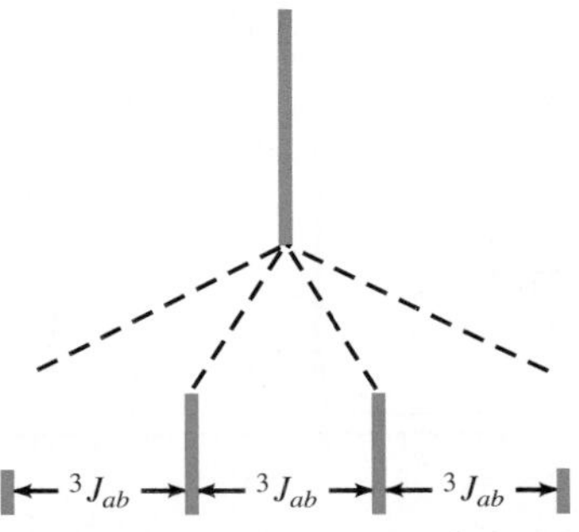

These eight combinations cause the signal of the $CHCl_2$ proton to be split into a quartet, in which the intensities of the peaks are in the ratio 1:3:3:1.

Figure 14.15

The methyl protons of 1,1-dichloroethane split the signal of the methine proton into a quartet.

Coupling of nuclear spins requires that the nuclei split each other's signal equally. The separation between the two halves of the methyl doublet in 1,1-dichloroethane is equal to the separation between any two adjacent peaks of the methine quartet. The extent to which two nuclei are coupled is given by the **coupling constant *J*** and in simple cases is equal to the separation between adjacent lines of the signal of a particular proton. The three-bond coupling constant $^3J_{ab}$ in 1,1-dichloroethane has a value of 7 Hz. *The size of the coupling constant is independent of the field strength;* the separation between adjacent peaks in 1,1-dichloroethane is 7 Hz, irrespective of whether the spectrum is recorded at 300 MHz or 500 MHz.

14.8 Splitting Patterns: The Ethyl Group

One of the most characteristic patterns of peaks is that of the ethyl group, represented in the NMR spectrum of ethyl bromide in Figure 14.16. In compounds of the type CH_3CH_2X, especially where X is an electronegative atom or group, such as bromine in ethyl bromide, the ethyl group appears as a *triplet–quartet pattern.* The signal for the methylene protons is split into a quartet by coupling with the three methyl protons. The signal for the methyl protons is a triplet because of vicinal coupling to the two protons of the adjacent methylene group.

$Br—CH_2—CH_3$

These two protons split the methyl signal into a triplet.

These three protons split the methylene signal into a quartet.

We have discussed in the preceding section why methyl groups split the signals due to vicinal protons into a quartet. Splitting by a methylene group gives a triplet corresponding to the spin combinations shown in Figure 14.17 for ethyl bromide. The relative intensities of the peaks of this triplet are 1:2:1.

Table 14.1 summarizes the splitting patterns and peak intensities expected for coupling to various numbers of protons.

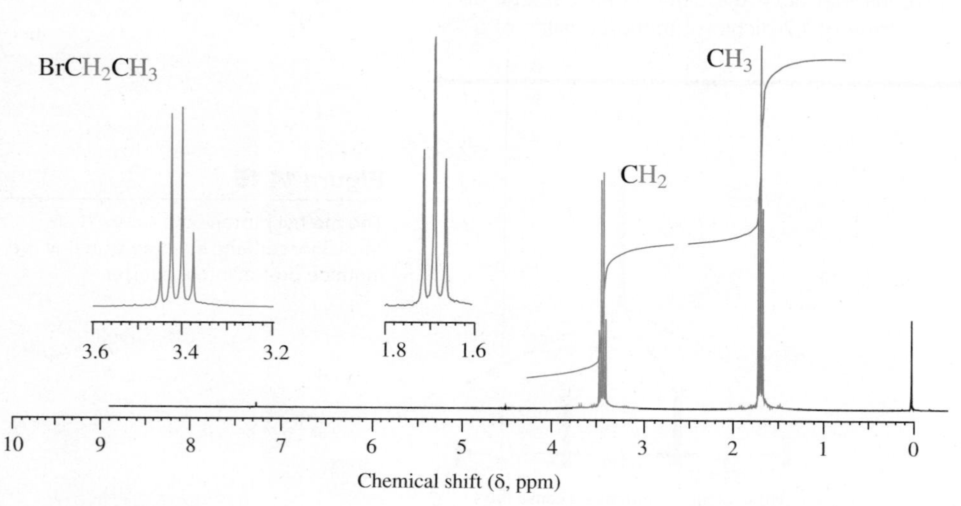

Figure 14.16

The 300-MHz ^{1}H NMR spectrum of ethyl bromide ($BrCH_2CH_3$), showing the characteristic triplet–quartet pattern of an ethyl group.

There are four possible combinations of the nuclear spins of the two methylene protons in CH_3CH_2Br.

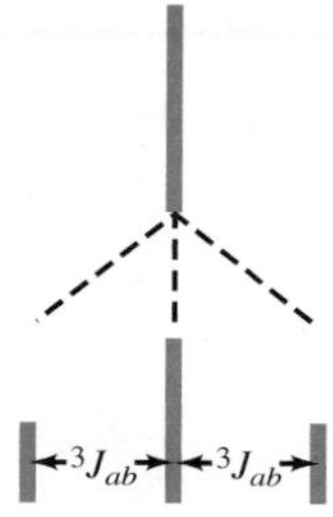

These four combinations cause the signal of the CH_3 protons to be split into a triplet, in which the intensities of the peaks are in the ratio 1:2:1.

Figure 14.17

The methylene protons of ethyl bromide split the signal of the methyl protons into a triplet.

TABLE 14.1 Splitting Patterns of Common Multiplets

Number of protons to which nucleus is equally coupled	Appearance of multiplet	Intensities of lines in multiplet
1	Doublet	1:1
2	Triplet	1:2:1
3	Quartet	1:3:3:1
4	Quintet	1:4:6:4:1
5	Sextet	1:5:10:10:5:1
6	Septet	1:6:15:20:15:6:1

The intensities correspond to the coefficients of a binomial expansion (Pascal's triangle).

Problem 14.12

Describe the appearance of the ^{1}H NMR spectrum of each of the following compounds. How many signals would you expect to find, and into how many peaks will each signal be split?

(a) $ClCH_2OCH_2CH_3$
(b) $CH_3CH_2OCH_3$
(c) $CH_3CH_2OCH_2CH_3$
(d) *p*-Diethylbenzene
(e) $ClCH_2CH_2OCH_2CH_3$

Sample Solution (a) Along with the triplet–quartet pattern of the ethyl group, the NMR spectrum of this compound will contain a singlet for the two protons of the chloromethyl group.

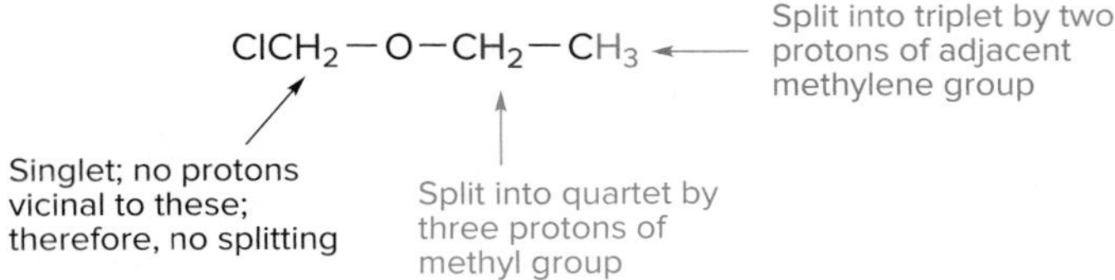

14.9 Splitting Patterns: The Isopropyl Group

The NMR spectrum of isopropyl chloride (Figure 14.18) illustrates the appearance of an isopropyl group. The signal for the six equivalent methyl protons at δ 1.5 is split into a doublet by the proton of the H—C—Cl unit. In turn, the H—C—Cl proton signal at δ 4.2

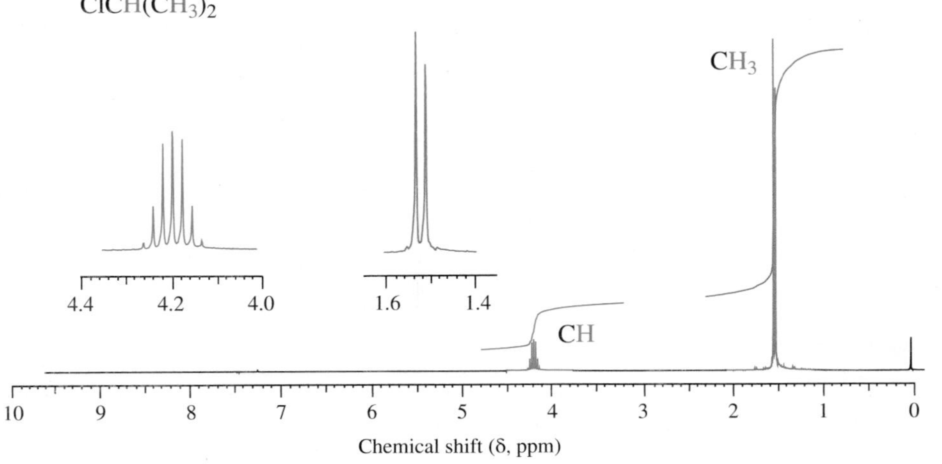

Figure 14.18

The 300-MHz ^{1}H NMR spectrum of isopropyl chloride, showing the doublet–septet pattern of an isopropyl group.

is split into a septet by the six methyl protons. A *doublet–septet* pattern is characteristic of an isopropyl group.

This proton splits the signal for the methyl protons into a doublet. H–C(Cl)(CH_3)(CH_3) These six protons split the methine signal into a septet.

14.10 Splitting Patterns: Pairs of Doublets

We often see splitting patterns in which the intensities of the individual peaks do not match those given in Table 14.1, but are distorted in that the signals for coupled protons "lean" toward each other. This leaning is a general phenomenon, but is most easily illustrated for the case of two nonequivalent vicinal protons as shown in Figure 14.19.

$$\text{W}-\underset{\text{H}_1}{\overset{\text{X}}{\text{C}}}-\underset{\text{H}_2}{\overset{\text{Y}}{\text{C}}}-\text{Z}$$

The appearance of the splitting pattern of protons 1 and 2 depends on their coupling constant J and the chemical-shift difference $\Delta\nu$ between them. When the ratio $\Delta\nu/J$ is large, two symmetrical 1:1 doublets are observed. We refer to this as the "AX" case, using two letters that are remote in the alphabet to stand for signals well removed from each other on the spectrum. Keeping the coupling constant the same while reducing $\Delta\nu$ leads to a steady decrease in the intensity of the outer two peaks with a simultaneous increase in the inner two as we progress from AX through AM to AB. At the extreme (A_2), the two protons have the same chemical shift, the outermost lines have disappeared, and no splitting is observed. Because of its appearance, it is easy to misinterpret an AB or AM pattern as a quartet, rather than the pair of skewed doublets it really is.

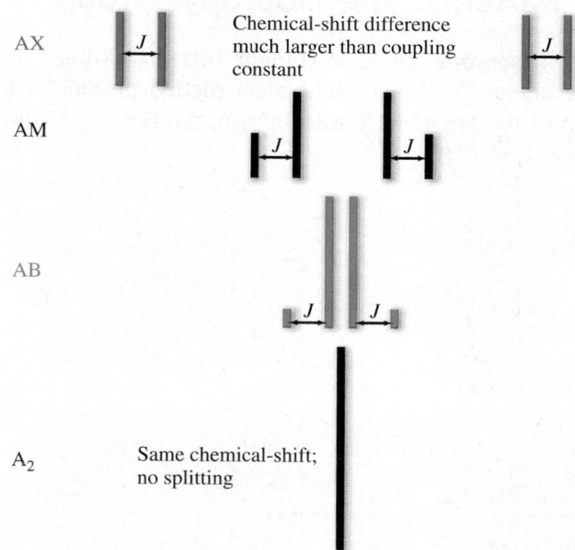

Figure 14.19

The appearance of the splitting pattern of two coupled protons depends on their coupling constant J and the chemical-shift difference $\Delta\nu$ between them. As the ratio $\Delta\nu/J$ decreases, the doublets become increasingly distorted. When the two protons have the same chemical shift, no splitting is observed.

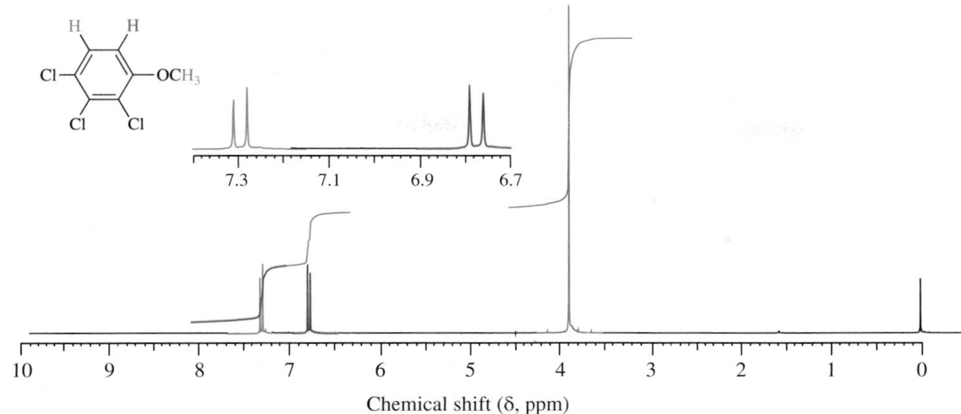

Figure 14.20

The 300-MHz ¹H NMR spectrum of 2,3,4-trichloroanisole, showing the splitting of the ring protons into a pair of doublets that "lean" toward each other.

A skewed pair of doublets is clearly visible in the ¹H NMR spectrum of 2,3,4-trichloroanisole (Figure 14.20). In addition to the singlet at δ 3.9 for the protons of the $—OCH_3$ group, we see doublets at δ 6.8 and δ 7.3 for the two protons of the aromatic ring.

Doublet δ 7.3 — H; H — Doublet δ 6.8; Cl; OCH_3 ← Singlet δ 3.9; Cl; Cl

2,3,4-Trichloroanisole

A pair of doublets frequently occurs with *geminal* protons (protons bonded to the same carbon). Geminal protons are separated by two bonds, and geminal coupling is referred to as *two-bond coupling* (^2J) in the same way that vicinal coupling is referred to as *three-bond coupling* (^3J). An example of geminal coupling is provided by the compound 1-chloro-1-cyanoethene, in which the two hydrogens appear as a pair of doublets. The splitting in each doublet is 2 Hz.

Doublet → H; Doublet → H; $^2J = 2$ Hz; C=C; Cl; CN

1-Chloro-1-cyanoethene

The protons in 1-chloro-1-cyanoethene are *diastereotopic* (Section 14.6). They are nonequivalent and have different chemical shifts. Remember, splitting can occur only between protons that have different chemical shifts.

Splitting due to geminal coupling is seen only in CH_2 groups and only when the two protons have different chemical shifts. All three protons of a methyl (CH_3) group are equivalent and cannot split one another's signal.

14.11 Complex Splitting Patterns

All the cases we've discussed so far have involved splitting of a proton signal by coupling to other protons that were equivalent to one another. Indeed, we have stated the splitting rule in terms of the multiplicity of a signal as being equal to $n + 1$, where n is equal to the number of equivalent protons to which the proton that gives the signal is coupled. What if all the vicinal protons are *not* equivalent?

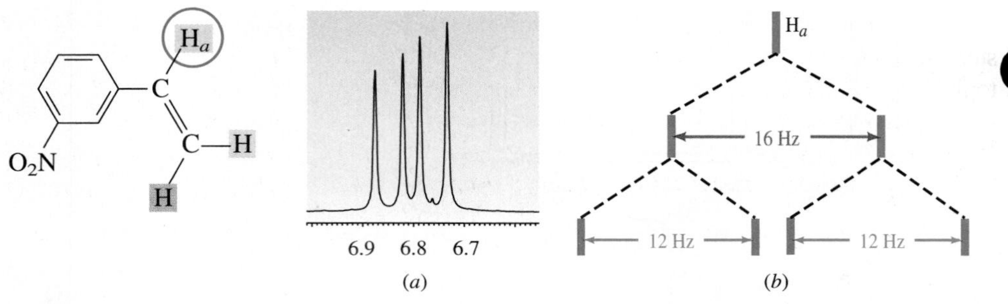

Figure 14.21

Splitting of a signal into a doublet of doublets by unequal coupling to two vicinal protons. (*a*) Appearance of the signal for the proton marked H_a in *m*-nitrostyrene as a set of four peaks. (*b*) Origin of these four peaks through successive splitting of the signal for H_a.

You will find it revealing to construct a splitting diagram similar to that of Figure 14.21 for the case in which the cis and trans H—C=C—H coupling constants are equal. Under those circumstances the four-line pattern simplifies to a triplet, as it should for a proton equally coupled to two vicinal protons.

Figure 14.21*a* shows the signal for the proton marked $ArCH_a{=}CH_2$ in *m*-nitrostyrene, which appears as a set of four peaks in the range δ 6.7–6.9. These four peaks are in fact a "doublet of doublets." The proton in question is *unequally coupled* to the two protons at the end of the vinyl side chain. The size of the vicinal coupling constant between protons trans to each other on a double bond is normally larger than that between cis protons. In this case the trans coupling constant is 16 Hz and the cis coupling constant is 12 Hz. Thus, as shown in Figure 14.21*b,* the signal for H_a is split into a doublet with a spacing of 16 Hz by one vicinal proton, and each line of this doublet is then split into another doublet with a spacing of 12 Hz.

The "$n + 1$ rule" should be amended to read: *When a proton* H_a *is coupled to* H_b, H_c, H_d, *etc., and* $J_{ab} \neq J_{ac}$, $\neq J_{ad}$, *etc., the original signal for* H_a is split into $n + 1$ *peaks by n* H_b *protons, each of these lines is further split into* $n + 1$ *peaks by n* H_c *protons, and each of these into* $n + 1$ *lines by n* H_d *protons, and so on.* Bear in mind that because of overlapping peaks, the number of lines actually observed can be less than that expected on the basis of the splitting rule.

Diastereotopic hydrogens can complicate the NMR spectra of chiral molecules, especially those in which there is a CH_2 group adjacent to a chirality center. Consider H_a and H_b in (*S*)-1,2-diphenyl-2-propanol, which has a chirality center at C-2 (Figure 14.22).

replace H_a with D

replace H_b with D

(1*S*,2*S*)

(1*R*,2*S*)

3.10 2.85

Chemical shift (δ, ppm)

(*S*)-1,2-Diphenyl-2-propanol

Diastereomers

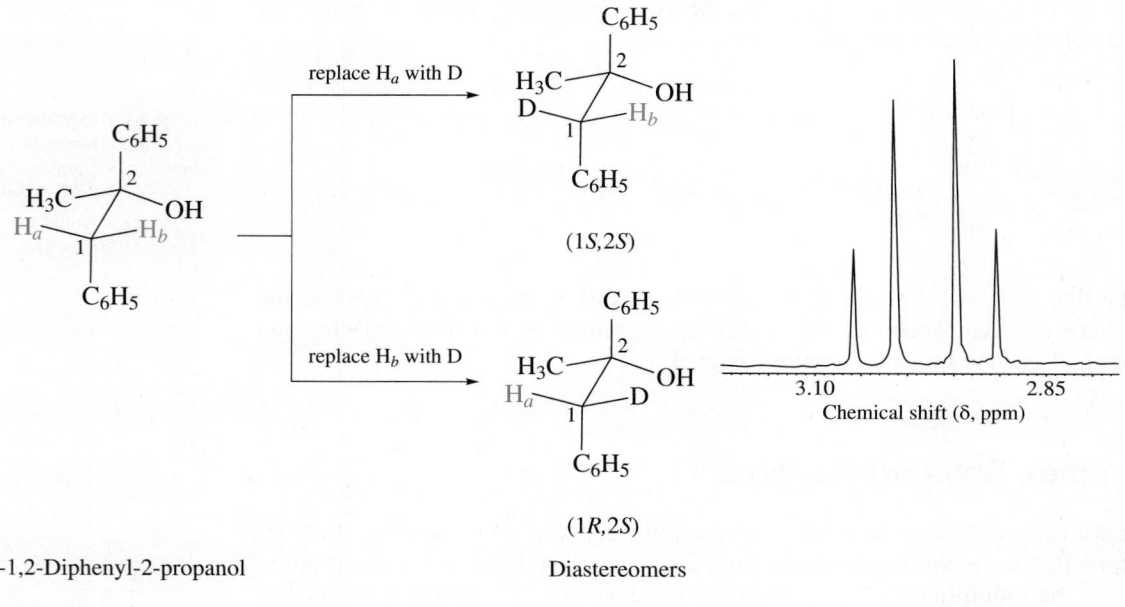

Figure 14.22

The methylene protons H_a and H_b of (*S*)-1,2-diphenyl-2-propanol are diastereotopic and appear as a pair of doublets in the 300-MHz ^1H NMR spectrum.

Replacement of H_a with a test group, for example, deuterium, gives a compound that is diastereomeric to the one that is generated by replacement of H_b. Recall that when diastereomers are produced by replacement of protons with test groups, the protons are diastereotopic and may have different chemical shifts (Section 14.6). In 1,2-diphenyl-2-propanol, H_a and H_b are diastereotopic, have different chemical shifts, and appear as a pair of doublets. They give an AM splitting pattern (Figure 14.19) because the chemical-shift differences between H_a and H_b are not much larger than the coupling constant.

Problem 14.13

Describe the splitting pattern expected for the proton at

(a) C-2 in (*Z*)-1,3-dichloropropene

(b) C-2 in $CH_3CH(Br)CH{=}O$

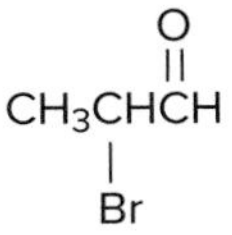

Sample Solution (a) The signal of the proton at C-2 is split into a doublet by coupling to the proton cis to it on the double bond, and each line of this doublet is split into a triplet by the two protons of the CH_2Cl group.

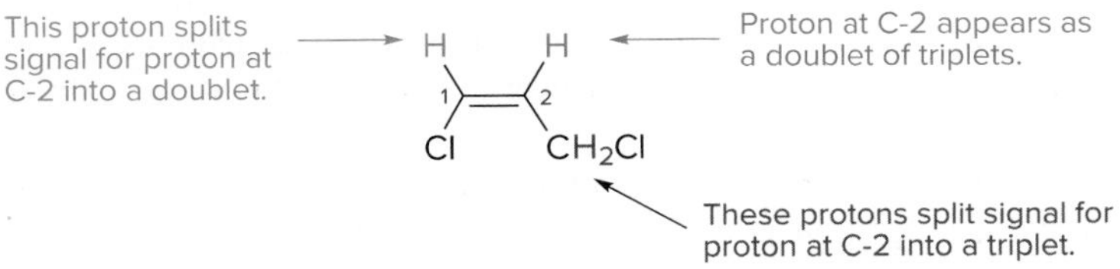

This proton splits signal for proton at C-2 into a doublet.

Proton at C-2 appears as a doublet of triplets.

These protons split signal for proton at C-2 into a triplet.

Problem 14.14

A portion of the ^{1}H NMR spectrum of the amino acid phenylalanine is shown in Figure 14.23. Why are eight lines observed for H_a and H_b?

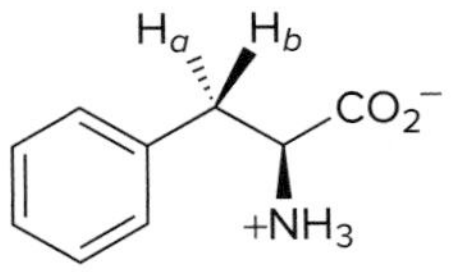

(*S*)-Phenylalanine

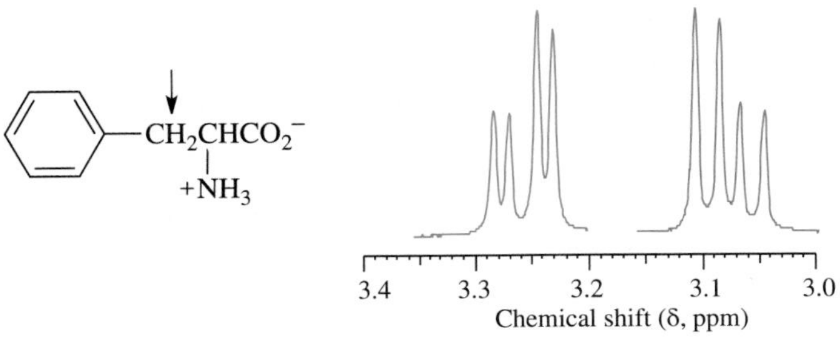

Figure 14.23

A portion of the 300-MHz ^{1}H NMR spectrum of phenylalanine for Problem 14.14 showing H_a and H_b.

14.12 ^{1}H NMR Spectra of Alcohols

We expect the hydroxyl proton of a primary alcohol RCH_2OH to be split into a triplet by vicinal coupling to the two protons of the CH_2 group. This is, in fact, exactly what we observe in the 300-MHz spectrum of benzyl alcohol shown in Figure 14.24. In reciprocal fashion, the OH proton splits the signal for the protons of the CH_2 group into a doublet. Often, however, splitting due to H—C—O—H coupling is not observed because of rapid exchange of OH protons between alcohol molecules, the rate of which increases with concentration and temperature and is subject to catalysis by acids and bases. Because higher field strength instruments allow more dilute solutions to be used, H—C—O—H splitting is more commonly seen in 300-MHz spectra than in earlier spectra run at 200 MHz and below.

Problem 14.15

Hydrogen bonding between the oxygen of dimethyl sulfoxide (DMSO) and the proton of an OH group is relatively strong. How could recording an NMR spectrum in DMSO instead of $CDCl_3$ be helpful in distinguishing between primary, secondary, and tertiary alcohols?

The chemical shift of the hydroxyl proton is variable, with a range of δ 0.5–5, depending on the solvent, the temperature at which the spectrum is recorded, and the concentration of the solution. The alcohol proton shifts to lower field in more concentrated solutions.

An easy way to verify that a particular signal belongs to a hydroxyl proton is to add D_2O. The hydroxyl proton is replaced by deuterium according to the equation:

$$RCH_2OH + D_2O \rightleftharpoons RCH_2OD + DOH$$

Deuterium does not give a signal under the conditions of ^{1}H NMR spectroscopy. Thus, replacement of a hydroxyl proton by deuterium leads to the disappearance of the OH peak of the alcohol. Protons bonded to nitrogen and sulfur also undergo exchange with D_2O. Those bound to carbon normally do not, which makes this a useful technique for assigning the proton resonances of OH, NH, and SH groups.

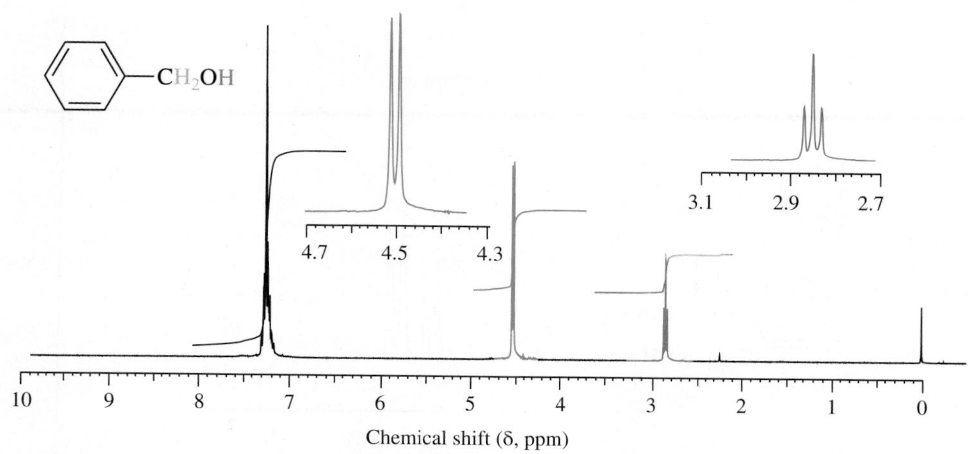

Figure 14.24

The 300-MHz ^{1}H NMR spectrum of benzyl alcohol. The hydroxyl proton and the methylene protons are vicinal and split each other's signal.

Magnetic Resonance Imaging (MRI)

It isn't often that someone goes to the emergency room because of a headache, and when the staff discovered that the man who did was due in court for sentencing the next day, some of them felt that there might not be anything wrong with him at all. There was.

The man's behavior toward the staff soon convinced them that he should be admitted, kept overnight, and seen by a neurologist the next day. After a preliminary examination, a magnetic resonance image, or MRI, was ordered which revealed a brain tumor. The tumor was located in the right frontal cortex, a portion of the brain known to be involved in controlling impulsive behavior.

The man had behaved normally until middle age; then his personality underwent significant changes, involving certain impulsive behaviors and criminal tendencies. These, as well as other behaviors, had not responded to drugs or counseling. Even though he had earned a master's degree, the man performed poorly on some simple mental tests and was unable to sketch the face of a clock or write a legible, coherent sentence.

Once the tumor was found, it was surgically removed. The man's ability to curb his impulses was restored, his mental, graphical, and writing skills improved to the normal range, and he successfully completed a rehabilitation program. About a year later though, the headaches and some of the earlier behaviors returned. When a new MRI showed that the tumor had regrown, it was removed and again the symptoms disappeared.

At a turning point in this man's life, an MRI made all the difference. MRI is NMR. The word *nuclear* is absent from the name to avoid confusion with nuclear medicine, which involves radioactive isotopes. MRI is noninvasive, requires no imaging or contrast agents, and is less damaging than X-rays. In the time since the first MRI of a living creature—a clam—was successfully obtained in the early 1970s, MRI has become a standard diagnostic tool. Two of its early developers, Paul Lauterbur (University of Illinois) and Peter Mansfield (University of Nottingham) were recognized with the 2003 Nobel Prize in Physiology or Medicine.

An MRI scanner is an NMR machine large enough to accommodate a human being, has a powerful magnet, operates in the pulse-FT mode, and detects protons—usually the protons in water and, to a lesser extent, lipids. The principles are the same as those of conventional FT-NMR spectroscopy but, because the goal is different, the way the data are collected and analyzed differs too. Some key features of MRI include:

1. A selective pulse is used in order to excite protons in a particular slice of the object to be imaged.
2. Unlike conventional NMR, the magnetic field in MRI is not uniform. A linear gradient is applied in addition to the static field so that the field strength varies as a function of position in the object but is precisely known. Because the frequency of a proton signal is directly proportional to the strength of the applied magnetic field, the measured resonance frequency is linearly related to the position in the magnetic field gradient.
3. Computer software carries out the essential task of reconstructing the 2D or 3D image from the NMR signals. The data are generally presented as a series of slices through the imaged object.
4. The intensity of the signal—its relative lightness or darkness in the image—depends on the concentration and spin relaxation times of the various protons. Spin relaxation time is the time it takes for the perturbed magnetization associated with a proton to return to its equilibrium value. The relaxation time is quite sensitive to the environment and is different for water in blood and various tissues.

New applications of nuclear magnetic resonance in biomedical science continue to appear. Functional MRI (fMRI) is an offshoot of MRI. Unlike MRI, which is used for diagnosis in a clinical setting, fMRI is a research tool that detects regions of the brain that are actively responding to stimuli. Increased brain activity is accompanied by an increase in blood flow to the region involved. This alters the ratio of oxygenated hemoglobin to its nonoxygenated counterpart. Because the two hemoglobins have different magnetic properties, the nuclear spin relaxation times of the protons in water are affected and can be studied by MRI.

14.13 NMR and Conformations

We know from Chapter 3 that the protons in cyclohexane exist in two different environments: axial and equatorial. The NMR spectrum of cyclohexane, however, shows only a single sharp peak at δ 1.4. All the protons of cyclohexane appear to be equivalent in the NMR spectrum. Why?

The answer is related to the very rapid rate of chair–chair interconversion in cyclohexane.

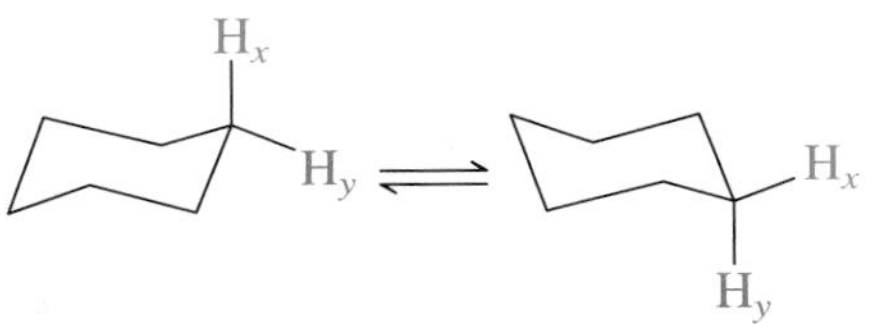

H_x H_y ⇌ H_x H_y

NMR is too slow to "see" the individual conformations of cyclohexane, but sees instead the *average* environment of the protons. Because chair–chair interconversion in cyclohexane converts each axial proton to an equatorial one and vice versa, the average environments of all the protons are the same. A single peak is observed that has a chemical shift midway between the true chemical shifts of the axial and the equatorial protons.

The rate of interconversion can be slowed down by lowering the temperature. At temperatures of about −100°C, separate signals are seen for the axial and equatorial protons of cyclohexane.

14.14 ^{13}C NMR Spectroscopy

We pointed out in Section 14.3 that both ^{1}H and ^{13}C are nuclei that can provide useful structural information when studied by NMR. Although a ^{1}H NMR spectrum helps us infer much about the carbon skeleton of a molecule, a ^{13}C NMR spectrum has the obvious advantage of probing the carbon skeleton directly. ^{13}C NMR spectroscopy is analogous to ^{1}H NMR in that the number of signals informs us about the number of different kinds of carbons, and their chemical shifts are related to particular chemical environments.

However, unlike ^{1}H, which is the most abundant of the hydrogen isotopes (99.985%), only 1.1% of the carbon atoms in a sample are ^{13}C. Moreover, the intensity of the signal produced by ^{13}C nuclei is far weaker than the signal produced by the same number of ^{1}H nuclei. In order for ^{13}C NMR to be a useful technique in structure determination, a vast increase in the signal-to-noise ratio is required. Pulsed FT-NMR provides for this, and its development was the critical breakthrough that led to ^{13}C NMR becoming the routine tool that it is today.

To orient ourselves in the information that ^{13}C NMR provides, let's compare the ^{1}H and ^{13}C NMR spectra of 1-chloropentane (Figures 14.25*a* and 14.25*b*, respectively). The ^{1}H NMR spectrum shows reasonably well-defined triplets for the protons of the CH_3 and CH_2Cl groups (δ 0.9 and 3.55, respectively). The signals for the six CH_2 protons at C-2, C-3, and C-4 of $CH_3CH_2CH_2CH_2CH_2Cl$, however, appear as two unresolved multiplets at δ 1.4 and 1.8.

The ^{13}C NMR spectrum, on the other hand, is very simple: *a separate, distinct peak is observed for each carbon.*

Notice, too, how well-separated these ^{13}C signals are: they cover a range of over 30 ppm, compared with less than 3 ppm for the proton signals of the same compound. In general, the window for proton signals in organic molecules is about 12 ppm; ^{13}C chemical shifts span a range of over 200 ppm. The greater spread of ^{13}C chemical shifts makes it easier to interpret the spectra.

Problem 14.16

How many signals would you expect to see in the ^{13}C NMR spectrum of each of the following compounds?

(a) Propylbenzene
(b) Isopropylbenzene
(c) 1,2,3-Trimethylbenzene
(d) 1,2,4-Trimethylbenzene
(e) 1,3,5-Trimethylbenzene

Sample Solution (a) The two ring carbons that are ortho to the propyl substituent are equivalent and so must have the same chemical shift. Similarly, the two ring carbons that are meta to the propyl group are equivalent to each other. The carbon atom para to the substituent is unique, as is the carbon that bears the substituent. Thus, there will be four signals for the ring carbons, designated *w, x, y,* and *z* in the structural formula. These four signals for the ring carbons added to those for the three nonequivalent carbons of the propyl group yield a total of seven signals.

x y
w z $CH_2CH_2CH_3$
x y

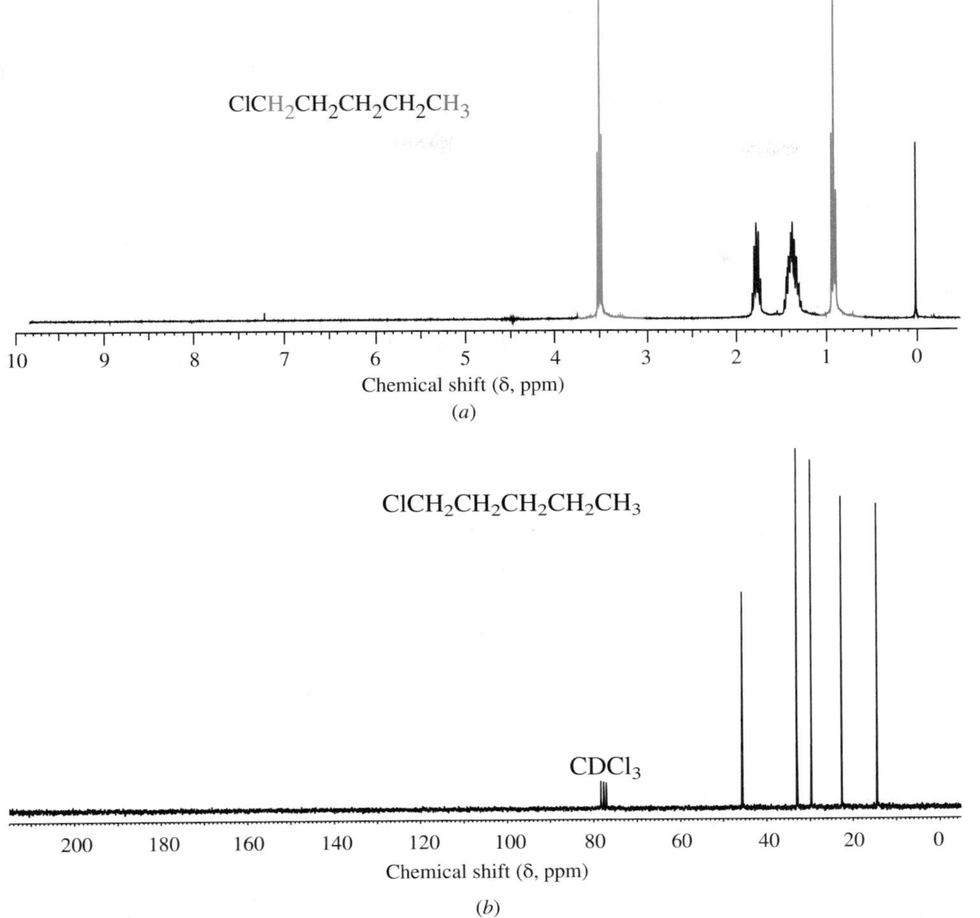

Figure 14.25

(*a*) The 300-MHz ^{1}H NMR spectrum and (*b*) the ^{13}C NMR spectrum of 1-chloropentane.

14.15 ^{13}C Chemical Shifts

Just as chemical shifts in ^{1}H NMR are measured relative to the *protons* of tetramethylsilane, chemical shifts in ^{13}C NMR are measured relative to the *carbons* of tetramethylsilane. Table 14.2 lists typical chemical-shift ranges for some representative types of carbon atoms.

In general, the factors that most affect ^{13}C chemical shifts are

1. The electronegativity of the groups attached to carbon
2. The hybridization of carbon

Substituent Effects Electronegative substituents affect ^{13}C chemical shifts in the same way as they affect ^{1}H chemical shifts, by withdrawing electrons. For ^{1}H NMR, recall that because carbon is more electronegative than hydrogen, the protons in methane (CH_4) are more shielded than primary hydrogens (RCH_3), primary hydrogens are more shielded than secondary (R_2CH_2), and secondary more shielded than tertiary (R_3CH). The same holds true for carbons in ^{13}C NMR, but the effects can be 10–20 times greater.

	$(CH_3)_4C$	$(CH_3)_3CH$	$CH_3CH_2CH_3$	CH_3CH_3	CH_4
Classification:	Quaternary	Tertiary	Secondary	Primary	
Chemical shift (δ), ppm:					
H		1.7	1.3	0.9	0.2
C	28	25	16	8	−2

TABLE 14.2 Chemical Shifts of Representative Carbons

Type of carbon	Chemical shift (δ) ppm*	Type of carbon	Chemical shift (δ) ppm*
Hydrocarbons		**Functionally substituted carbons**	
RCH_3	0–35	RCH_2Br	20–40
R_2CH_2	15–40	RCH_2Cl	25–50
R_3CH	25–50	RCH_2NH_2	35–50
R_4C	30–40	RCH_2OH and RCH_2OR	50–65
$RC{\equiv}CR$	65–90	$RC{\equiv}N$	110–125
$R_2C{=}CR_2$	100–150	$RC(=O)OH$ and $RC(=O)OR$	160–185
(benzene ring)	110–175	$RC(=O)H$ and $RC(=O)R$	190–220

*Approximate values relative to tetramethylsilane.

Likewise, for functionally substituted methyl groups:

	CH_3F	CH_3OH	CH_3NH_2	CH_4
Chemical shift (δ), ppm:				
H	4.3	3.4	2.5	0.2
C	75	50	27	−2

Figure 14.25 compared the appearance of the ^1H and ^{13}C NMR spectra of 1-chloropentane and drew attention to the fact each carbon gave a separate peak, well separated from the others. Let's now take a closer look at the ^{13}C NMR spectrum of 1-chloropentane with respect to assigning these peaks to individual carbons.

^{13}C chemical shift (δ), ppm: Cl–C(45)–C(33)–C(29)–C(22)–C(14)

The most obvious feature of these ^{13}C chemical shifts is that the closer the carbon is to the electronegative chlorine, the more deshielded it is. Peak assignments will not always be this easy, but the correspondence with electronegativity is so pronounced that *spectrum simulators* are available that allow reliable prediction of ^{13}C chemical shifts from structural formulas. These simulators are based on arithmetic formulas that combine experimentally derived chemical-shift increments for the various structural units within a molecule.

Problem 14.17

The ^{13}C NMR spectrum of 1-bromo-3-chloropropane contains peaks at δ 30, δ 35, and δ 43. Assign these signals to the appropriate carbons.

The effects of substituents on rate and orientation in electrophilic aromatic substitution described in Chapter 13 find parallels in their effect on the chemical shifts of aromatic ring carbons. For the group of compounds represented as

C_6H_5–G

the ^{13}C chemical shift of the para carbon is observed to correlate with the *o-p* versus *m*-directing effects of substituent G. Since shielding in NMR results from the local magnetic fields that accompany local electric fields, it follows that nuclear shielding—especially at carbons ortho or para to a substituent—increases as the substituent becomes more electron releasing or less electron withdrawing. The effect on chemical shift is quite pronounced and covers a range of over 16 ppm in the following group of *p*-substituted benzene derivatives.

Substituent

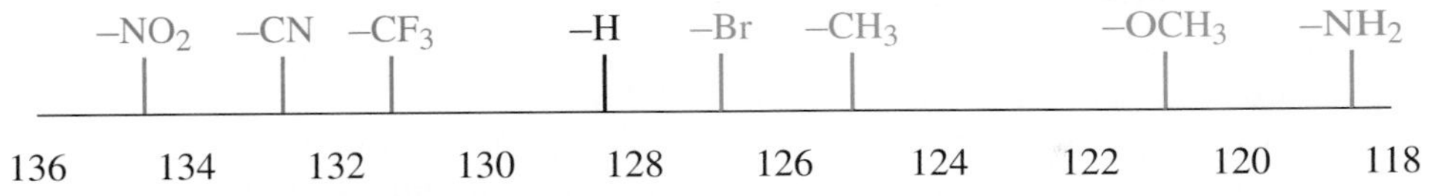

$-NO_2$	$-CN$	$-CF_3$	$-H$	$-Br$	$-CH_3$	$-OCH_3$	$-NH_2$

136 134 132 130 128 126 124 122 120 118

^{13}C Chemical shift of ring carbon para to substituent, ppm

Relative to H, deactivating *m*-directing substituents at C-1 deshield the carbon at C-4; activating o, p-directing substituents shield it. The halogen substituent in the group (bromine) behaves in much the same way as it does in electrophilic aromatic substitution, where, although weakly deactivating, it is an ortho, para director. In the case of ^{13}C chemical shifts, bromine increases the shielding (electron density) at the para position.

Hybridization Effects Here again, the effects are similar to those seen in ^{1}H NMR. As illustrated by 4-phenyl-1-butene, sp^3-hybridized carbons are more shielded than sp^2-hybridized ones.

^{13}C chemical shift (δ), ppm: 114 138 36 36 126–142

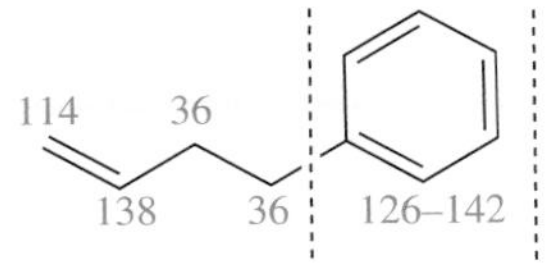

Of the sp^2-hybridized carbons, C-1 is the most shielded because it is bonded to only one other carbon. The least shielded carbon is the ring carbon to which the side chain is attached. It is the only sp^2-hybridized carbon connected to three other carbons.

Problem 14.18

Consider carbons *x*, *y*, and *z* in *p*-methylanisole. One has a chemical shift of δ 20, another has δ 55, and the third δ 157. Match the chemical shifts with the appropriate carbons.

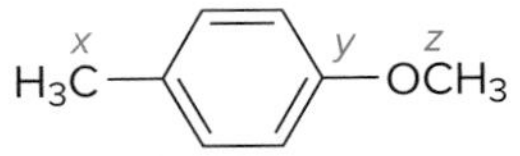

H_3C (*x*) — C_6H_4 — (*y*) OCH_3 (*z*)

Acetylenes are anomalous in ^{13}C, as in ^{1}H NMR. *sp*-Hybridized carbons are less shielded than sp^3-hybridized ones, but more shielded than sp^2-hybridized ones.

^{13}C chemical shift (δ), ppm: 68 84 22 20 13

Electronegativity and hybridization effects combine to make the carbon of a carbonyl group especially deshielded. Normally, the carbon of C=O is the least shielded one in a ^{13}C NMR spectrum.

^{13}C chemical shift (δ), ppm: 127–134 41 171 O 61 14

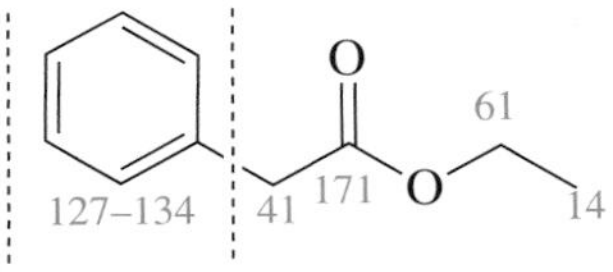

Problem 14.19

Which would you expect to be more shielded, the carbonyl carbon of an aldehyde or a ketone? Why?

We will have more to say about ^{13}C chemical shifts in later chapters when various families of compounds are discussed in more detail.

14.16 ^{13}C NMR and Peak Intensities

Two features that are fundamental to ^{1}H NMR spectroscopy—integrated areas and splitting patterns—are much less important in ^{13}C NMR.

Although it is a simple matter to integrate ^{13}C signals, it is rarely done because the observed ratios can be more misleading than helpful. The pulsed FT technique that is standard for ^{13}C NMR has the side effect of distorting the signal intensities, especially for carbons that lack attached hydrogens. Examine Figure 14.26, which shows the ^{13}C NMR spectrum of 3-methylphenol (*m*-cresol). Notice that, contrary to what we might expect for a compound with seven peaks for seven different carbons, the intensities of these peaks are not nearly the same. The two least intense signals, those at δ 140 and δ 157, correspond to carbons that lack attached hydrogens.

Problem 14.20

To which of the compounds of Problem 14.16 does the ^{13}C NMR spectrum of Figure 14.27 belong?

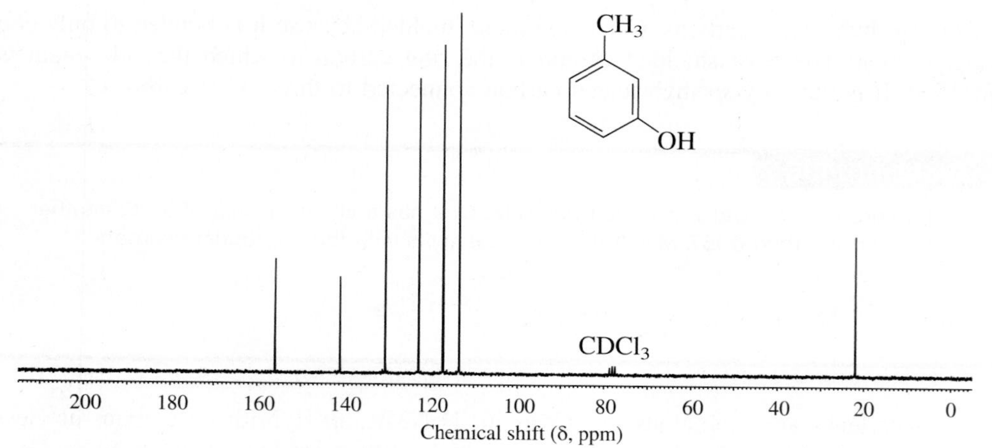

Figure 14.26

The ^{13}C NMR spectrum of *m*-cresol. Each of the seven carbons gives a separate peak. Integrating the spectrum would not provide useful information because the intensities of the peaks are so different, even though each one corresponds to a single carbon.

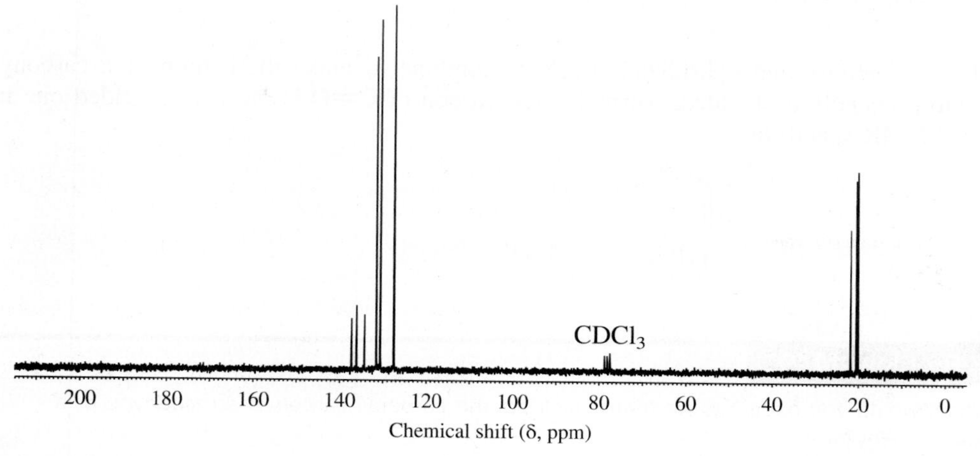

Figure 14.27

The ^{13}C NMR spectrum of the unknown compound of Problem 14.20.

14.17 ^{13}C—^{1}H Coupling

You have probably noticed another characteristic of ^{13}C NMR spectra—all of the peaks are singlets. With a spin of $\pm\frac{1}{2}$, a ^{13}C nucleus is subject to the same splitting rules that apply to ^{1}H, and we might expect to see splittings due to ^{13}C—^{13}C and ^{13}C—^{1}H couplings. We don't. Why?

The lack of splitting due to ^{13}C—^{13}C coupling is easy to understand. ^{13}C NMR spectra are measured on samples that contain ^{13}C at the "natural abundance" level. Only 1% of all the carbons in the sample are ^{13}C, and the probability that any molecule contains more than one ^{13}C atom is quite small.

Splitting due to ^{13}C—^{1}H coupling is absent for a different reason, one that has to do with the way the spectrum is run. Because a ^{13}C signal can be split not only by the protons to which it is directly attached, but also by protons separated from it by two, three, or even more bonds, the number of splittings might be so large as to make the spectrum too complicated to interpret. Thus, the spectrum is measured under conditions, called **broadband decoupling,** that suppress such splitting.

What we gain from broadband decoupling in terms of a simple-looking spectrum comes at the expense of some useful information. For example, being able to see splitting corresponding to one-bond ^{13}C—^{1}H coupling would immediately tell us the number of hydrogens directly attached to each carbon. The signal for a carbon with no attached hydrogens (a *quaternary* carbon) would be a singlet, the hydrogen of a CH group would split the carbon signal into a doublet, and the signals for the carbons of a CH_2 and a CH_3 group would appear as a triplet and a quartet, respectively. Although it is possible, with a technique called *off-resonance decoupling,* to observe such one-bond couplings, identifying a signal as belonging to a quaternary carbon or to the carbon of a CH, CH_2, or CH_3 group is normally done by a method called DEPT, which is described in the next section.

14.18 Using DEPT to Count Hydrogens

In general, a simple pulse FT-NMR experiment involves the following stages:

1. Equilibration of the nuclei between the lower and higher spin states under the influence of a magnetic field
2. Application of a radiofrequency pulse to give an excess of nuclei in the higher spin state
3. Acquisition of free-induction decay data during the time interval in which the equilibrium distribution of nuclear spins is restored
4. Mathematical manipulation (Fourier transform) of the data to plot a spectrum

The pulse sequence (stages 2–3) can be repeated hundreds of times to enhance the signal-to-noise ratio. The duration of time for stage 2 is on the order of milliseconds, and that for stage 3 is about 1 second.

Major advances in NMR have been made by using a second rf transmitter to irradiate the sample at some point during the sequence. There are several such techniques, of which we'll describe just one, called **distortionless enhancement of polarization transfer,** abbreviated as **DEPT.**

In the DEPT routine, a second transmitter excites ^{1}H, which affects the appearance of the ^{13}C spectrum. A typical DEPT experiment is illustrated for the case of 1-phenyl-1-pentanone in Figure 14.28. In addition to the normal spectrum shown in Figure 14.28*a,* four more spectra are run using prescribed pulse sequences. In one (Figure 14.28*b*), the signals for carbons of CH_3 and CH groups appear normally, whereas those for CH_2 groups are inverted and those for C without any attached hydrogens are nulled. In the others (not shown) different pulse sequences produce combinations of normal, nulled, and inverted peaks that allow assignments to be made to the various types of carbons with confidence.

Problem 14.21

DEPT spectra for a compound with the formula $C_6H_{12}O$ are shown in Figure 14.29. Assign a structure. (More than one answer is possible.)

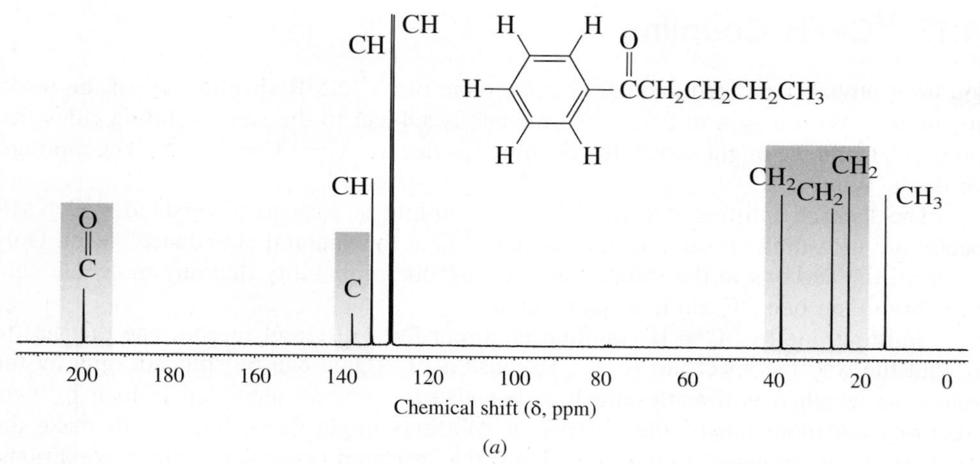

(*a*)

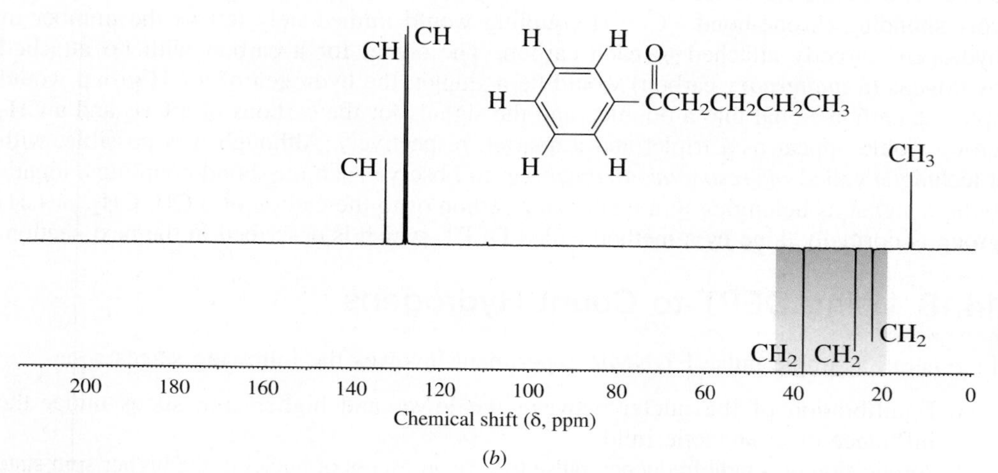

(*b*)

Figure 14.28

^{13}C NMR spectra of 1-phenyl-1-pentanone. (*a*) Normal spectrum. (*b*) DEPT spectrum recorded using a pulse sequence in which CH_3 and CH carbons appear as positive peaks, CH_2 carbons as negative peaks, and carbons without any attached hydrogens are nulled.

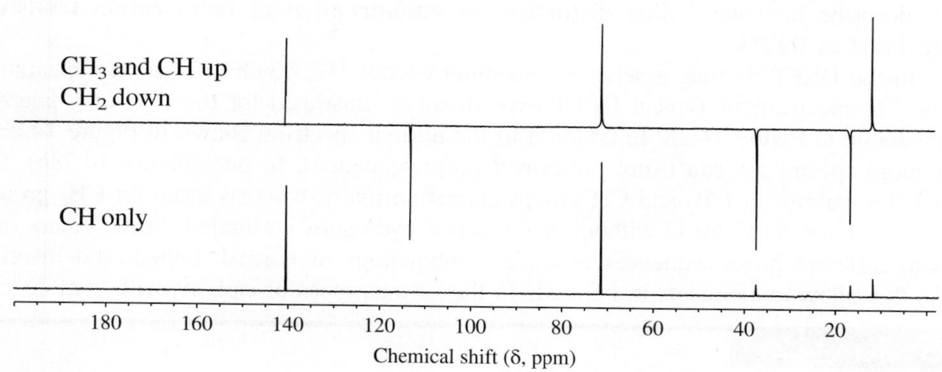

Figure 14.29

DEPT spectra for Problem 14.21.

14.19 2D NMR: COSY and HETCOR

The more information you can extract from an NMR spectrum, the better your chances at arriving at a unique structure. Like spin–spin splitting, which complicates the appearance of an ^{1}H NMR spectrum but provides additional information, 2D NMR looks more complicated than it is while making structure determination easier.

The key dimension in NMR is the frequency axis. All of the spectra we have seen so far are 1D spectra because they have only one frequency axis. In 2D NMR a standard pulse sequence adds a second frequency axis.

One kind of 2D NMR is called **COSY,** which stands for **correlated spectroscopy.** With a COSY spectrum you can determine by inspection which signals correspond to spin-coupled protons. Identifying coupling relationships is a valuable aid to establishing a molecule's *connectivity*.

Figure 14.30 is the COSY spectrum of 2-hexanone. Both the x- and y-axes are frequency axes expressed as chemical shifts. Displaying the 1D ^{1}H NMR spectrum of 2-hexanone along the x- and y-axes makes it easier to interpret the 2D information, which is the collection of contoured objects contained within the axes. To orient ourselves, first note that many of the contours lie along the diagonal that runs from the lower left to the upper right. This diagonal bisects the 2D NMR into two mirror-image halves. The off-diagonal contours are called *cross peaks* and contain the connectivity information we need.

Each cross peak has x and y coordinates. One coordinate corresponds to the chemical shift of a proton, the other to the chemical shift of a proton to which it is coupled. Because the diagonal splits the 2D spectrum in half, each cross peak is duplicated on the other side of the other diagonal with the same coordinates, except in reverse order. This redundancy means that we really need to examine only half of the cross peaks.

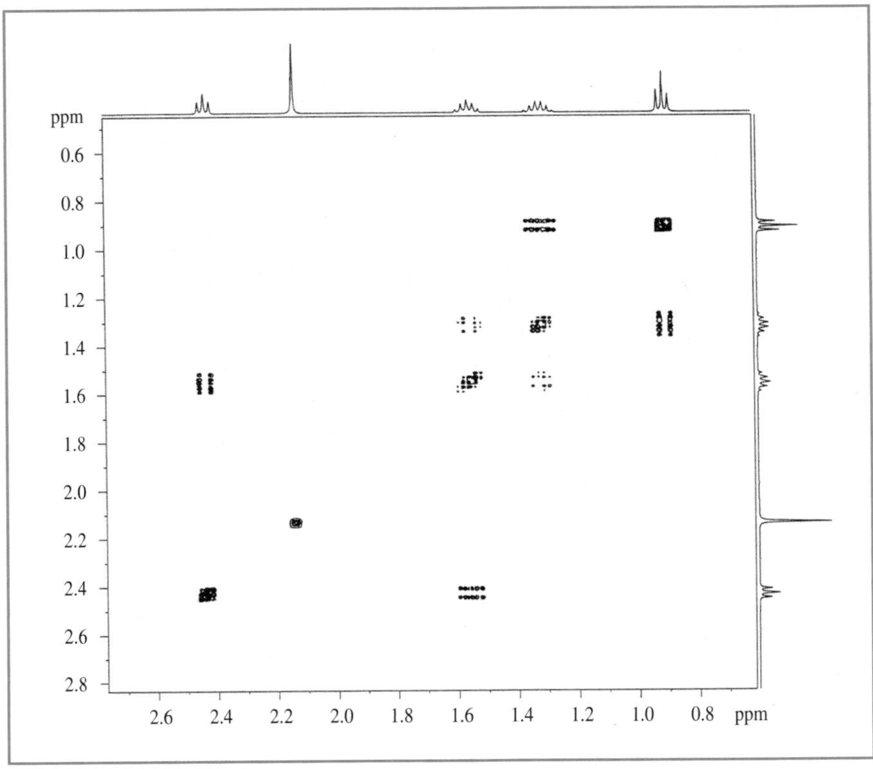

Figure 14.30

^{1}H-^{1}H COSY NMR spectrum of 2-hexanone.

To illustrate, start with the lowest field signal (δ 2.4) of 2-hexanone. We assign this signal, a triplet, to the protons at C-3 on the basis of its chemical shift and the splitting evident in the 1D spectrum.

$$CH_3\overset{\overset{\displaystyle O}{\|}}{C}\underset{\underset{\delta\ 2.4}{\uparrow}}{CH_2}CH_2CH_2CH_3$$

We look for cross peaks with the same x coordinate by drawing a vertical line from δ 2.4, finding a cross peak with a y coordinate of δ 1.6. *This means that the protons responsible for the signal at δ 2.4 are coupled to the ones at δ 1.6.* Therefore, the chemical shift of the C-4 protons is δ 1.6.

Now work from these C-4 protons. Drawing a vertical line from δ 1.6 on the x-axis finds two cross peaks. One cross peak simply confirms the coupling to the protons at C-3. The other has a y coordinate of δ 1.3 and, therefore, must correspond to the protons at C-5.

A vertical line drawn from δ 1.3 intersects the cross peaks at both δ 1.6 and δ 0.9. The former confirms the coupling of C-5 to C-4; the latter corresponds to the C-5 to C-6 coupling and identifies the signal at δ 0.9 as belonging to the protons at C-6.

Finally, a vertical line drawn from δ 2.1 intersects no cross peaks. The singlet at δ 2.1, as expected, is due to the protons at C-1, which are not coupled to any of the other protons in the molecule.

The complete connectivity and assignment of ^1H chemical shifts is

$$\underset{2.1}{H_3C}-\overset{\overset{\displaystyle O}{\|}}{C}-\underset{2.4}{CH_2}-\underset{1.6}{CH_2}-\underset{1.3}{CH_2}-\underset{0.9}{CH_3}$$

Although the 1D ^1H spectrum of 2-hexanone is simple enough to be interpreted directly, you can see that COSY offers one more tool we can call on in more complicated cases.

A second 2D NMR method called **HETCOR (heteronuclear chemical shift correlation)** is a type of COSY in which the two frequency axes are the chemical shifts for different nuclei, usually ^1H and ^{13}C. With HETCOR it is possible to relate a peak in a ^{13}C spectrum to the ^1H signal of the protons attached to that carbon. As we did with COSY, we'll use 2-hexanone to illustrate the technique.

The HETCOR spectrum of 2-hexanone is shown in Figure 14.31. It is considerably simpler than a COSY spectrum, lacking diagonal peaks and contoured cross peaks. Instead, we see objects that are approximately as tall as a ^1H signal is wide, and as wide as a ^{13}C signal. As with the COSY cross peaks, however, it is their coordinates that matter, not their size or shape. Interpreting the spectrum is straightforward. The ^{13}C peak at δ 30 correlates with the ^1H singlet at δ 2.1, which because of its multiplicity and chemical shift corresponds to the protons at C-1. Therefore, this ^{13}C peak can be assigned to C-1 of 2-hexanone. Repeating this procedure for the other carbons gives:

$$H_3C-\overset{\overset{\displaystyle O}{\|}}{C}-CH_2-CH_2-CH_2-CH_3$$

	H_3C	CH_2	CH_2	CH_2	CH_3
^1H chemical shift (δ), ppm:	2.1	2.4	1.6	1.3	0.9
^{13}C chemical shift (δ), ppm:	30	43	26	22	14

The chemical shift of the carbonyl carbon (δ 209) is not included because it has no attached hydrogens.

A number of 2D NMR techniques are available for a variety of purposes. They are especially valuable when attempting to determine the structure of complicated natural products and the conformations of biomolecules.

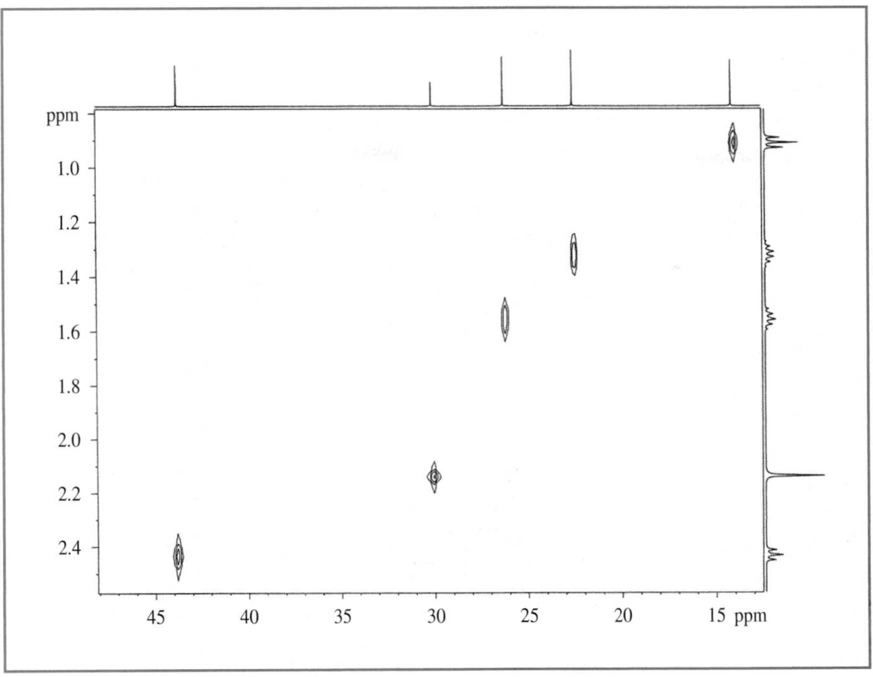

Figure 14.31

^1H-^{13}C HETCOR NMR spectrum of 2-hexanone.

14.20 Introduction to Infrared Spectroscopy

IR's earliest recognition came during World War II when it provided a key clue to the unusual β-lactam structure of the "miracle drug" penicillin.

Before the advent of NMR spectroscopy, infrared (IR) spectroscopy was the instrumental method most often applied to organic structure determination. Although NMR, in general, tells us more about the structure of an unknown compound, IR remains an important tool because of its usefulness in identifying the presence of certain *functional groups* within a molecule. Structural units, including functional groups, vibrate in characteristic ways and it is this sensitivity to *group vibrations* that is the basis of IR spectroscopy.

Among the ways a molecule responds to the absorption of energy is by vibrational motions such as the stretching and contracting of bonds and the opening and closing (bending) of bond angles. Vibrational motion and its energy are quantized. Only certain vibrational energy states are allowed.

We can visualize molecular vibrations by thinking of atoms and bonds as balls and springs.

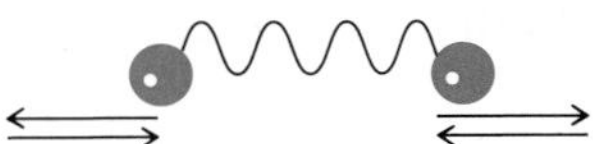

Zero-point energy is the term given to the energy of a molecule at absolute zero.

Even at the absolute zero of temperature, atoms in a molecule vibrate with respect to the bonds that connect them. At room temperature, the molecules are distributed among various vibrational energy states. *Frequency* is a property of the vibration and is related to the difference between vibrational energy states by $\Delta E = h\nu$ (Section 14.1). Promoting a molecule from a lower to a higher vibrational energy state increases the *amplitude* of the vibration.

For a sense of the variety of vibrational modes available to a molecule, consider a CH_2 group. Stretching and contracting the pair of C—H bonds can occur in two

Spectra by the Thousands

The best way to get good at interpreting spectra is by experience. Look at as many spectra and do as many spectroscopy problems as you can.

Among Web sites that offer spectroscopic problems, two stand out (Figure 14.32). One, called *WebSpectra,* was developed by Professor Craig A. Merlic (UCLA):

www.chem.ucla.edu/~webspectra/#Problems

The other is the *Organic Structure Elucidation* workbook, created by Professor Bradley D. Smith (Notre Dame):

https://www3.nd.edu/~smithgrp/structure/workbook.html

WebSpectra includes 75 problems. All the problems display the ^{1}H and ^{13}C spectra, several with DEPT or COSY enhancements. A number include IR spectra. *Organic Structure Elucidation* contains 64 problems, all with ^{1}H and ^{13}C NMR, IR, and mass spectra. The exercises in both *WebSpectra* and *Organic Structure Elucidation* are graded according to difficulty. Give them a try.

Vast numbers of NMR, IR, and mass spectra are freely accessible via the *Spectral Data Base System* (SDBS) maintained by the Japanese National Institute of Advanced Industrial Science and Technology at:

http://sdbs.db.aist.go.jp/sdbs/cgi-bin/cre_index.cgi

The SDBS contains 15,900 ^{1}H NMR, 14,200 ^{13}C NMR, 54,100 IR, and 25,000 mass spectra. Not only does the SDBS contain more spectra than anyone could possibly browse through, it incorporates some very useful search features. If you want spectra for a particular compound, entering the name of the compound calls up links to its spectra, which can then be displayed. If you don't know what the compound is, but know one or more of the following:

- Molecular formula
- ^{1}H or ^{13}C chemical shift of one or more peaks
- Mass number of mass spectra fragments

entering the values singly or in combination returns the names of the best matches in the database. You can then compare the spectra of these candidate compounds with the spectra of the sample to identify it.

As extensive as the SDBS is, don't be disappointed if the exact compound you are looking for is not there. There are, after all, millions of organic compounds. However, much of structure determination (and organic chemistry in general) is based on analogy. Finding the spectrum of a related compound can be almost as helpful as finding the one you really want.

These Web resources, in conjunction with the figures and problems in your text, afford a wealth of opportunities to gain practice and experience in what are now the standard techniques of structure determination.

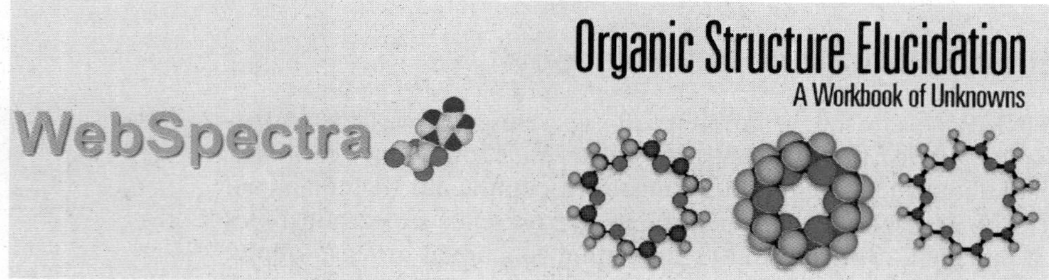

Figure 14.32

These two welcome screens open the door to almost 150 spectroscopy problems. The screens are used with permission of Professors Craig A. Merlic (*WebSpectra*) and Bradley D. Smith (*Organic Structure Elucidation*).

different ways. In the *symmetric* stretch, both C—H bonds stretch at the same time and contract at the same time. In the *antisymmetric* stretch, one C—H bond stretches while the other contracts.

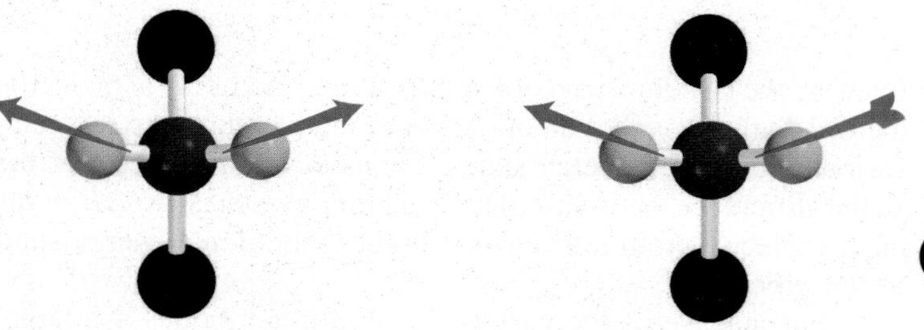

In addition to stretching vibrations, a CH_2 group can bend, and each bending mode has its own set of energy states.

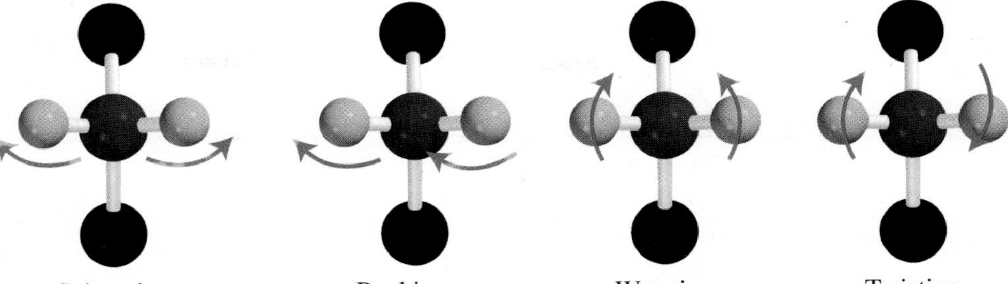

A molecule absorbs that portion of electromagnetic radiation having a frequency that matches the energy difference between two vibrational energy levels. This radiation lies in the infrared region of the electromagnetic spectrum (Figure 14.1). The wavelength λ of the infrared region that is the most useful for structure determination is 2.5–16 μm, where 1 μm = 10^{-6} m. Instead of wavelengths or SI units of frequency (s^{-1}), IR spectroscopy uses **wavenumbers,** which are equal to λ^{-1} and expressed in units of reciprocal centimeters (cm^{-1}). Thus, the region 2.5–16 μm corresponds to 4000–625 cm^{-1}. Wavenumbers are directly proportional to energy; 4000 cm^{-1} is the high-energy end of the scale for IR spectra, and 625 cm^{-1} is the low-energy end.

The energy difference between adjacent vibrational states is tens of thousands of times larger than what we saw for nuclear spin states in NMR.

Problem 14.22

Vibrational frequencies are sensitive to isotopic replacement. The O—H stretching frequency is near 3600 cm^{-1}, but that of O—D is about 2630 cm^{-1}. Which are closer in energy, two adjacent O—H or two adjacent O—D vibrational states?

Most molecules have many more vibrational modes than the ones just shown for a single CH_2 group. Some involve relatively simple structural units, others a substantial fraction of the atoms in a molecule. Thus, the infrared spectrum of each compound is unique, and superimposability of their IR spectra is convincing evidence that two substances are the same.

14.21 Infrared Spectra

IR spectra can be obtained regardless of the physical state of a sample—solid, liquid, gas, or dissolved in some solvent. If the sample is a liquid, a drop or two is placed between two sodium chloride disks, through which the IR beam is passed. Solids may be dissolved in a suitable solvent such as carbon tetrachloride or chloroform. More commonly, a solid sample is mixed with potassium bromide and the mixture pressed into a thin wafer, which is placed in the path of the IR beam. Newer instruments require little or no sample preparation. The present generation of IR spectrometers employs a technique known as attenuated total reflectance (ATR) coupled with FT data analysis. The whole range of vibrational states is sampled at once and transformed by Fourier analysis.

All IR spectra in this text were recorded without solvent using an ATR instrument.

Figure 14.33 orients us with respect to where we can expect to find IR absorptions for various structural units. Peaks in the range of 4000–1600 cm^{-1} are usually emphasized because this is the region in which the vibrations characteristic of particular functional groups are found. We'll look at some of these functional groups in more detail in Section 14.22. The region 1500–500 cm^{-1} is known as the **fingerprint region;** it is here that the pattern of peaks varies most from compound to compound.

An IR spectrum usually contains more peaks than we can assign, or even need to assign. We gain information by associating selected absorptions with particular structural units and functional groups, as well as noting what structural units can be excluded from consideration because a key peak that characterizes it is absent from the spectrum.

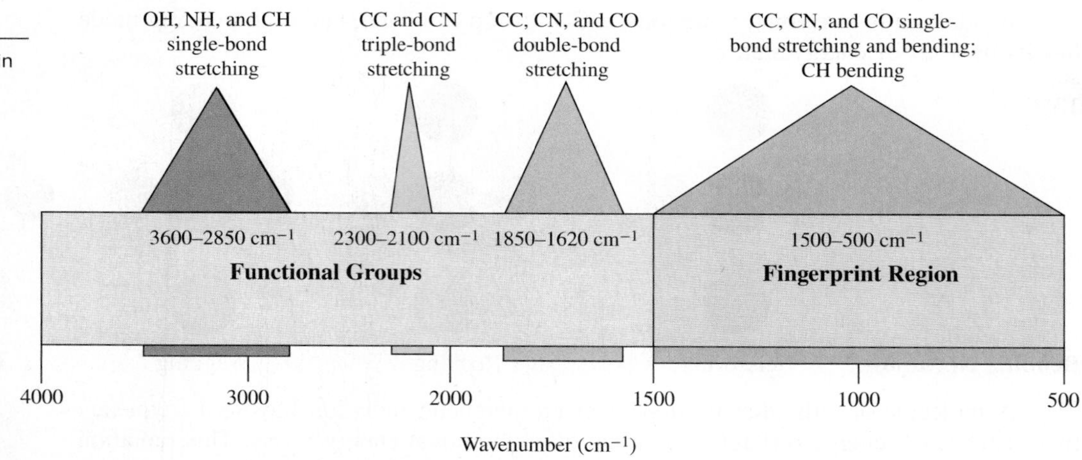

Figure 14.33

Structural units are commonly found in specific regions of the infrared spectrum.

Figure 14.34*a–d* shows the IR spectra of four hydrocarbons: hexane, 1-hexene, benzene, and hexylbenzene. Each spectrum consists of a series of absorption peaks of varying shape and intensity. Unlike NMR, in which intensities are related to the number of nuclei responsible for each signal, some IR vibrations give more intense peaks than others. To give an observable peak in the infrared, a vibration must produce a change in the molecular dipole moment, and peaks are usually more intense when they involve a bond between two atoms of different electronegativity. Consequently, C—C single-bond stretching vibrations normally give peaks of low intensity. The intensities of IR peaks are usually expressed in terms of percent transmittance (%T) and described as weak, medium, or strong.

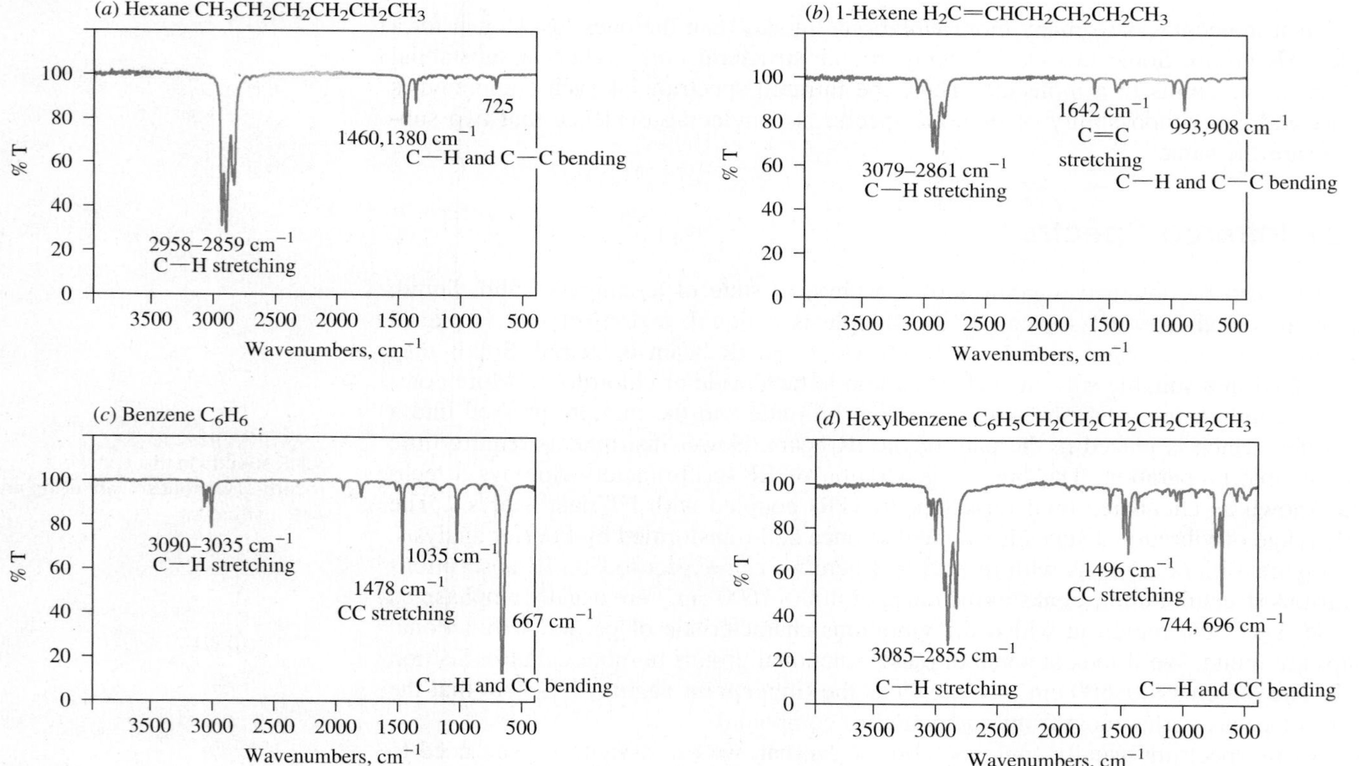

Figure 14.34

IR spectra of the hydrocarbons: (*a*) hexane, (*b*) 1-hexene, (*c*) benzene, and (*d*) hexylbenzene.

The IR spectrum of hexane (Figure 14.34*a*) is relatively simple, characterized by several peaks near 3000 cm^{-1} due to C—H stretching, along with weaker peaks at 1460, 1380, and 725 cm^{-1} from C—H and C—C bending.

Among the several ways in which the spectrum of the alkene 1-hexene (Figure 14.34*b*) differs from hexane, the most useful from the perspective of structure determination is found in the C—H stretching region. Although all the peaks for C—H stretching in *hexane* appear below 3000 cm^{-1}, *1-hexene* exhibits a peak at 3079 cm^{-1}. Peaks for C—H stretching above 3000 cm^{-1} are characteristic of hydrogens bonded to sp^2-hybridized carbon. The IR spectrum of 1-hexene also displays a weak peak at 1642 cm^{-1} corresponding to its C=C stretching vibration. The peaks at 993 and 908 cm^{-1} in the spectrum of 1-hexene, absent in the spectrum of hexane, are bending vibrations of the H_2C=C group.

All of the spectra in this and the next section are displayed on a common % T scale to better show how peak intensities differ among various groups.

Problem 14.23

Ethylene lacks a peak in its IR spectrum for C=C stretching. Why?

Benzene (Figure 14.34*c*) has *only* sp^2-hybridized carbons, and *all* of its peaks for C—H stretching lie above 3000 cm^{-1}. CC stretching gives a weak peak at 1478 cm^{-1}. The most intense peak in benzene (667 cm^{-1}) results from a vibration in which one of the C—H bonds bends out of the plane of the ring.

The hexylbenzene spectrum (Figure 14.34*d*) bears similarities to those of hexane and benzene. Peaks for C—H stretching are found both above and below 3000 cm^{-1} for sp^2 and sp^3 C—H stretching, respectively. The benzene ring is represented in the weak peak at 1496 cm^{-1}. The three peaks between 750 and 690 cm^{-1} include bending modes for the hexyl chain and the ring.

Rarely can the structure of a hydrocarbon ever be determined by IR alone. Figure 14.34 alerts us to the fact that most organic compounds give IR spectra in which many of the peaks are due to the carbon skeleton and its attached hydrogens. Chemists pay less attention to these peaks now that ^{1}H and ^{13}C NMR are available to gain the same information. What IR does best—identifying the presence or absence of functional groups—is described in the following section.

14.22 Characteristic Absorption Frequencies

Table 14.3 lists the **characteristic absorption frequencies** (in wavenumbers) for a variety of structural units found in organic compounds. Generally, absorptions above 1500 cm^{-1} for functional groups such as OH, C=O, and C≡N are the easiest to assign and provide the most useful information.

Some of these characteristic absorptions are reflected in the IR spectra of eight functional-group classes in Figure 14.35: alcohol, nitrile, carboxylic acid, ketone, ester, ether, amine, and amide. None of the specific compounds represented contains hydrogens bonded to sp^2-hybridized carbon, so all of the C—H absorbances lie below 3000 cm^{-1}. The compounds are related in that all have an unbranched six-carbon chain and, except for the peaks associated with the functional group, their spectra are similar, though not identical.

Problem 14.24

Which of the following is the most likely structure of the compound characterized by the IR spectrum shown in Figure 14.36?

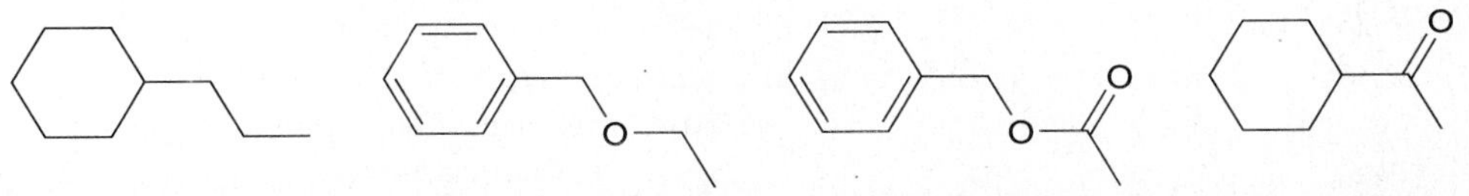

O O O

In later chapters, when families of compounds are discussed in detail, the IR frequencies associated with each type of functional group will be revisited.

(*a*) **Alcohols:** A broad peak at 3200–3400 cm^{-1} is characteristic of hydrogen-bonded OH groups. In dilute solution, hydrogen bonding is less, and a sharp second peak for "free" OH groups appears near 3600 cm^{-1}.

The peak at 1070 cm^{-1} lies in the range given in Table 14.3 (1025–1200 cm^{-1}) for C—O stretching and can be assigned to it.

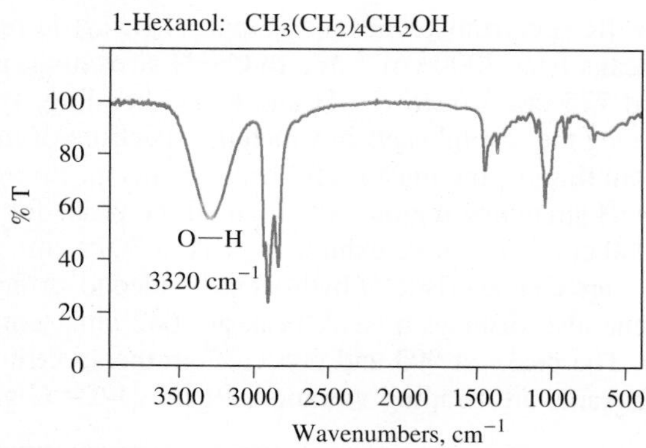

(*b*) **Nitriles:** The C≡N triple bond absorption is easily identifiable in the IR spectrum of a nitrile as a sharp peak of medium intensity at 2240–2280 cm^{-1}.

Very few other groups absorb in this region, the most notable being C≡C triple bonds (2100–2200 cm^{-1}).

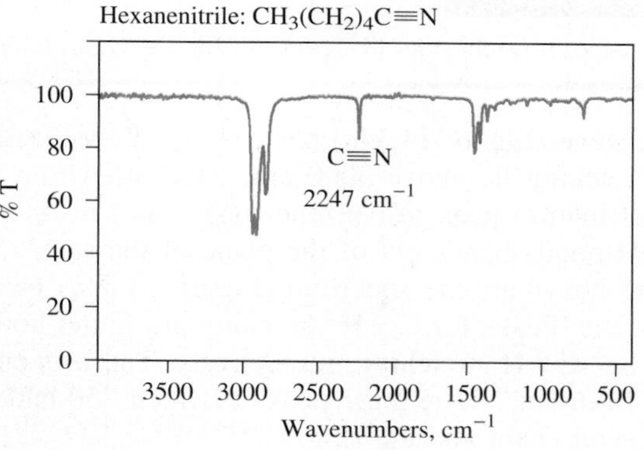

(*c*) **Carboxylic acids:** Carboxylic acids have two characteristic absorptions: a broad peak for O—H stretching in the range 2500–3600 cm^{-1} and a strong peak for C=O stretching at 1700–1725 cm^{-1}.

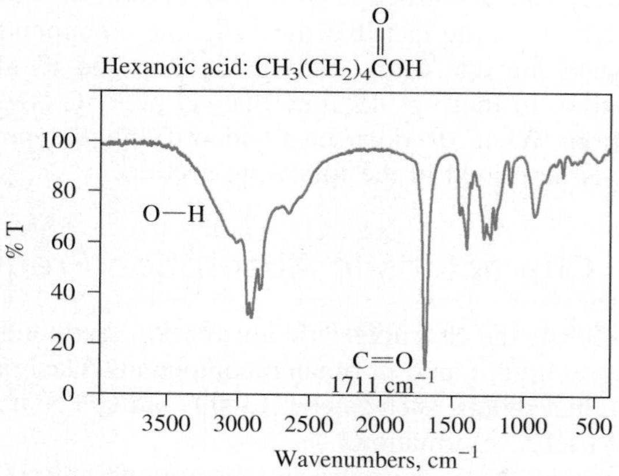

(*d*) **Aldehydes and ketones:** As in other carbonyl-containing compounds, the C=O stretching vibration gives the strongest peak in the IR spectra of aldehydes and ketones.

The C=O stretching frequencies of aldehydes are similar to those of ketones.

The C—H stretch of the CH=O group in aldehydes appears as a pair of bands in the range 2700–2900 cm^{-1}.

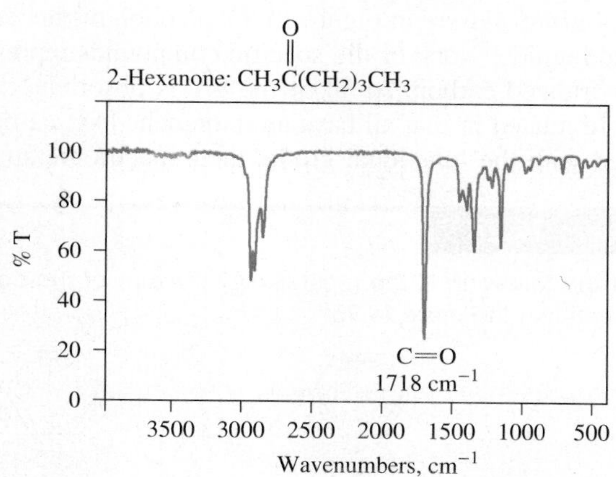

Figure 14.35

IR spectra of (*a*) 1-hexanol, (*b*) hexanenitrile, (*c*) hexanoic acid, (*d*) 2-hexanone, (*e*) methyl hexanoate, (*f*) dihexyl ether, (*g*) hexylamine, and (*h*) hexanamide.

(*Continued*)

(*e*) **Esters:** In addition to a strong C=O absorption (1730–1750 cm^{-1}), esters exhibit peaks for symmetric and antisymmetric C—O—C stretching at 1050–1300 cm^{-1}.

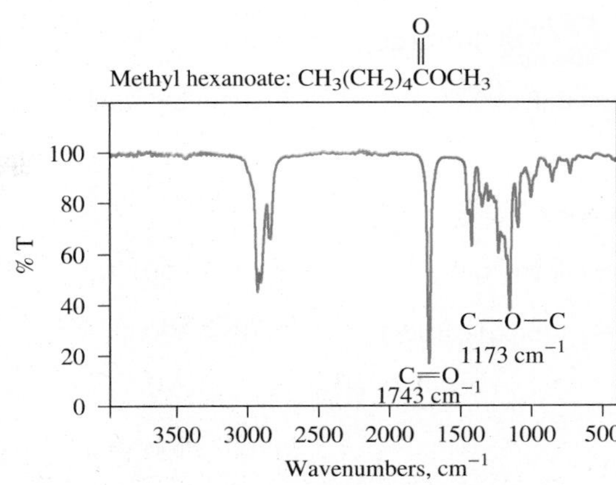

(*f*) **Ethers:** Peaks for C—O—C stretching in ethers appear in the range 1070–1150 cm^{-1}. Ethers of the type ROR′ where R and R′ are different have two peaks in this region.

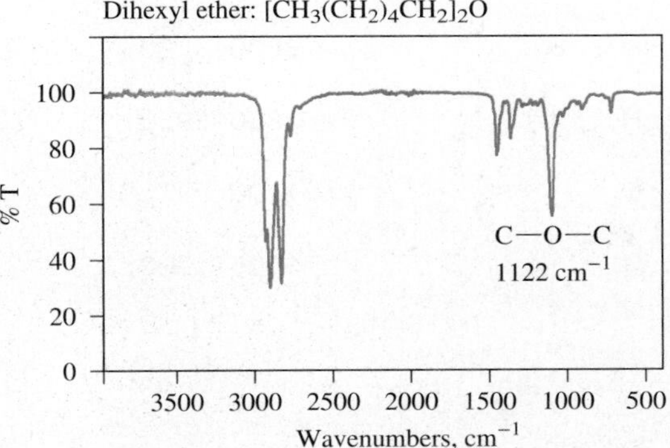

(*g*) **Amines:** Primary amines (RNH_2) have two peaks for the NH_2 group in the 3300–3500 cm^{-1} region, one for symmetric and the other for antisymmetric N—H stretching. Secondary amines (RNHR′) have only one peak (3310–3350 cm^{-1}).

An NH bending peak at 650–900 cm^{-1} occurs in both RH_2 and RNHR′. Primary amines also have an NH bending absorption at 1580–1650 cm^{-1}.

C—N stretching peaks are found at 1020–1250 cm^{-1}.

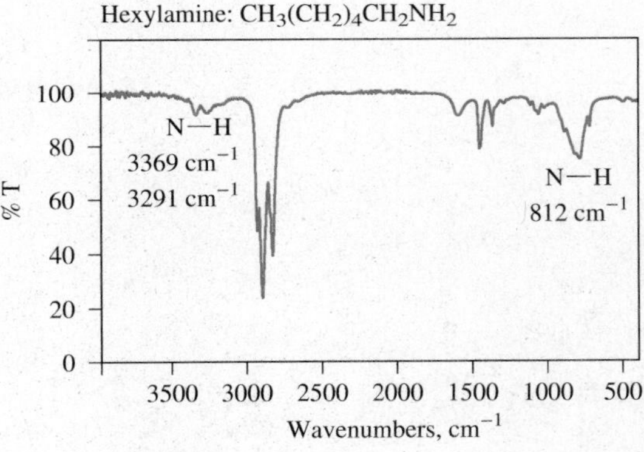

(*h*) **Amides:** Amides of the type $RC(O)NH_2$ have peaks for both symmetric and antisymmetric N—H stretching in the 3400–3150 cm^{-1} region.

The C=O absorption for amides appears at slightly lower frequency (1650–1700 cm^{-1}) than for ketones. Amides have a peak for NH_2 bending at a slightly lower frequency (1600–1650 cm^{-1}) than C=O.

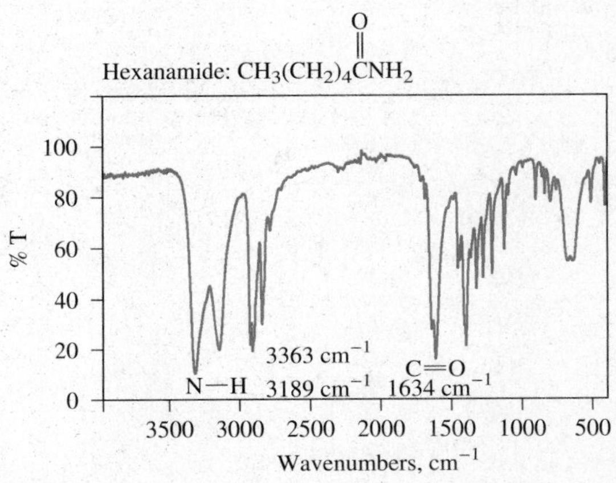

TABLE 14.3 Infrared Absorption Frequencies of Some Common Structural Units

Structural unit	Frequency, cm^{-1}	Structural unit	Frequency, cm^{-1}
	Stretching vibrations		
Single bonds		**Double bonds**	
—O—H (alcohols)	3200–3600		
—O—H (carboxylic acids)	2500–3600	>C=C<	1620–1680
>N—H	3350–3500	>C=O	
sp C—H	3310–3320	Aldehydes and ketones	1710–1750
sp^2 C—H	3000–3100	Carboxylic acids	1700–1725
sp^3 C—H	2850–2950	Acid anhydrides	1800–1850 and 1740–1790
sp^2 C—O	1200	Acyl halides	1770–1815
sp^3 C—O	1025–1200	Esters	1730–1750
		Amides	1680–1700
		Triple bonds	
		—C≡C—	2100–2200
		—C≡N	2240–2280
	Bending vibrations of diagnostic value		
Alkenes:		***Substituted derivatives of benzene:***	
$RCH=CH_2$	910, 990	Monosubstituted	730–770 and 690–710
$R_2C=CH_2$	890	Ortho-disubstituted	735–770
cis-RCH=CHR′	665–730	Meta-disubstituted	750–810 and 680–730
trans-RCH=CHR′	960–980	Para-disubstituted	790–840
$R_2C=CHR'$	790–840		

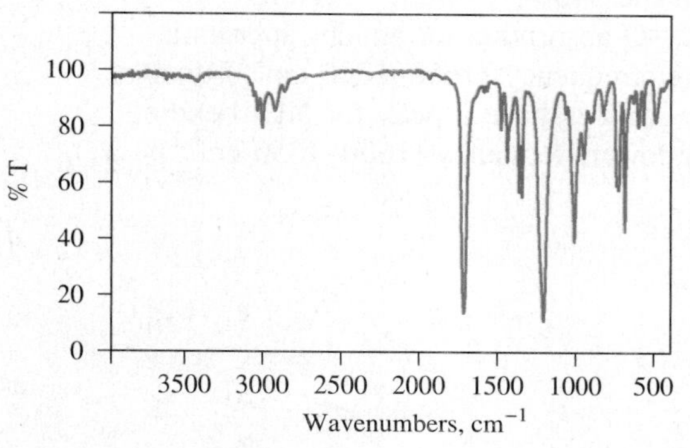

Figure 14.36

The IR spectrum of the unknown compound in Problem 14.24.

14.23 Ultraviolet-Visible Spectroscopy

The main application of UV-VIS spectroscopy, which depends on transitions between electronic energy levels, is in identifying conjugated π-electron systems.

Much greater energies separate electronic states than vibrational states. The energy required to promote an electron from one electronic state to the next lies in the visible and ultraviolet range of the electromagnetic spectrum (see Figure 14.1). We usually identify radiation in the UV-VIS range by its wavelength in nanometers. Thus, the visible region corresponds to 400–800 nm. Red light is the low-energy (long wavelength) end of the visible spectrum, violet light the high-energy (short wavelength) end. Ultraviolet light lies beyond the visible spectrum with wavelengths in the 200–400-nm range.

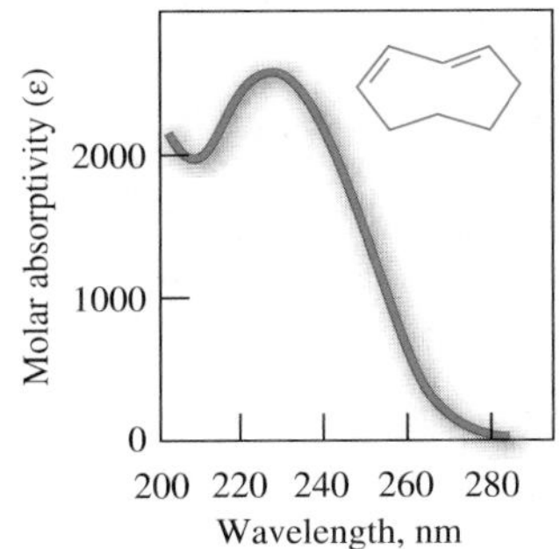

Figure 14.37

The UV spectrum of *cis,trans*-1,3-cyclooctadiene.

Figure 14.37 shows the UV spectrum of the conjugated diene *cis,trans*-1,3-cyclooctadiene, measured in ethanol as the solvent. As is typical of most UV spectra, the absorption is rather broad and is often spoken of as a "band" rather than a "peak." The wavelength at an absorption maximum is referred to as the λ_{max} of the band. For 1,3-cyclooctadiene, λ_{max} is 230 nm. In addition to λ_{max}, UV-VIS bands are characterized by their **absorbance** (*A*), which is a measure of how much of the radiation that passes through the sample is absorbed. To correct for concentration and path length effects, absorbance is converted to **molar absorptivity** (ϵ) by dividing it by the concentration *c* in moles per liter and the path length *l* in centimeters.

$$\epsilon = \frac{A}{c \cdot l}$$

Molar absorptivity, when measured at λ_{max}, is cited as ϵ_{max}. It is normally expressed without units. Both λ_{max} and ϵ_{max} are affected by the solvent, which is therefore included when reporting UV-VIS spectroscopic data. Thus, you might find a literature reference expressed in the form

cis,trans-1,3-Cyclooctadiene

$\lambda_{max}^{ethanol}$ 230 nm

$\epsilon_{max}^{ethanol}$ 2630

Figure 14.38 illustrates the transition between electronic energy states responsible for the 230-nm UV band of *cis,trans*-1,3-cyclooctadiene. Absorption of UV radiation excites an electron from the highest occupied molecular orbital (HOMO) to the lowest unoccupied molecular orbital (LUMO). In alkenes and polyenes, both the HOMO and LUMO are π type orbitals (rather than σ); the HOMO is the highest-energy π orbital and the LUMO is the lowest-energy π* orbital. Exciting one of the π electrons from a bonding π orbital to an antibonding π* orbital is referred to as a π → π* transition.

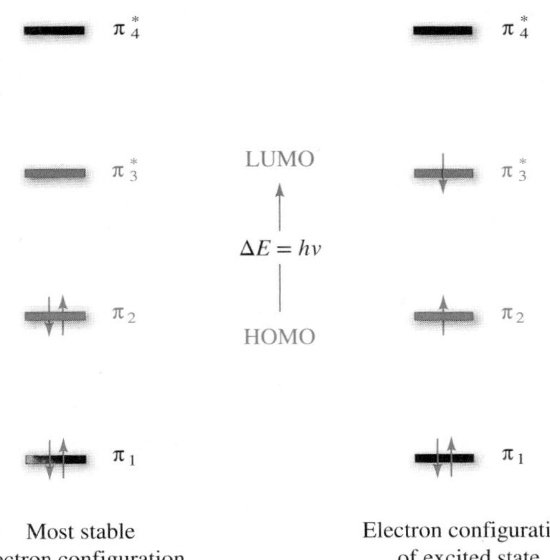

Figure 14.38

The π → π* transition in *cis,trans*-1,3-cyclooctadiene involves excitation of an electron from the highest occupied molecular orbital (HOMO) to the lowest unoccupied molecular orbital (LUMO).

Problem 14.25

λ_{max} for the $\pi \rightarrow \pi^*$ transition in ethylene is 170 nm. Is the HOMO–LUMO energy difference in ethylene greater than or less than that of *cis,trans*-1,3-cyclooctadiene (230 nm)?

The HOMO–LUMO energy gap and, consequently, λ_{max} for the $\pi \rightarrow \pi^*$ transition varies with the substituents on the double bonds. The data in Table 14.4 illustrate two substituent effects: adding methyl substituents to the double bond, and extending conjugation. Both cause λ_{max} to shift to longer wavelengths, but the effect of conjugation is the larger of the two. Based on data collected for many dienes it has been found that each methyl substituent on the double bonds causes a shift to longer wavelengths of about 5 nm, whereas extending the conjugation causes a shift of about 36 nm for each additional double bond.

Problem 14.26

Which one of the C_5H_8 isomers shown has its λ_{max} at the longest wavelength?

A striking example of the effect of conjugation on light absorption occurs in *lycopene,* one of the pigments in ripe tomatoes. Lycopene has a conjugated system of 11 double bonds and absorbs *visible light.* It has several UV-VIS bands, each characterized by a separate λ_{max}. Its longest wavelength absorption is at 505 nm. Note the inverse relationship between the color of a compound and the wavelength of light absorbed. Lycopene absorbs light in the blue region of the visible spectrum, yet appears red. The red color of lycopene is produced by the light that is not absorbed.

Lycopene

TABLE 14.4 Absorption Maxima of Some Representative Alkenes and Polyenes*

Compound	Structure	λ_{max} (nm)
Ethylene	$H_2C{=}CH_2$	170
2-Methylpropene	$H_2C{=}C(CH_3)_2$	188
1,3-Butadiene	$H_2C{=}CHCH{=}CH_2$	217
4-Methyl-1,3-pentadiene	$H_2C{=}CHCH{=}C(CH_3)_2$	234
2,5-Dimethyl-2,4-hexadiene	$(CH_3)_2C{=}CHCH{=}C(CH_3)_2$	241
(2*E*,4*E*,6*E*)-2,4,6-Octatriene	$CH_3CH{=}CHCH{=}CHCH{=}CHCH_3$	263
(2*E*,4*E*,6*E*,8*E*)-2,4,6,8-Decatetraene	$CH_3CH{=}CH(CH{=}CH)_2CH{=}CHCH_3$	299
(2*E*,4*E*,6*E*,8*E*,10*E*)-2,4,6,8,10-Dodecapentaene	$CH_3CH{=}CH(CH{=}CH)_3CH{=}CHCH_3$	326

*The value of λ_{max} refers to the longest wavelength $\pi \rightarrow \pi^*$ transition.

Many organic compounds such as lycopene are colored because their HOMO–LUMO energy gap is small enough that λ_{max} appears in the visible range of the spectrum. All that is required for a compound to be colored, however, is that it possess some absorption in the visible range. It often happens that a compound will have its λ_{max} in the UV region but that the peak is broad and extends into the visible. Absorption of the blue-to-violet components of visible light occurs, and the compound appears yellow.

A second type of absorption that is important in UV-VIS examination of organic compounds is the $n \rightarrow \pi^*$ transition of the carbonyl (C═O) group. One of the electrons in a lone-pair orbital of oxygen is excited to an antibonding orbital of the carbonyl group. The n in $n \rightarrow \pi^*$ identifies the electron as one of the nonbonded electrons of oxygen. This transition gives rise to relatively weak absorption peaks ($\epsilon_{max} < 100$) in the region 270–300 nm. The structural unit associated with an electronic transition in UV-VIS spectroscopy is called a **chromophore.** UV-visible spectroscopy has applications in biochemistry, where chromophores such as the heterocyclic bases found in nucleic acids and certain coenzymes involved in biochemical reactions can be studied.

Don't confuse the n in $n \rightarrow \pi^*$ with the n of Hückel's rule.

An important enzyme in biological electron transport called *cytochrome P450* gets its name from its UV absorption. The "P" stands for "pigment" because it is colored, and the "450" corresponds to the 450-nm absorption of one of its derivatives.

14.24 Mass Spectrometry

Mass spectrometry differs from the other instrumental methods discussed in this chapter in a fundamental way. It does not depend on the absorption of electromagnetic radiation but rather examines ions produced from a molecule in the gas phase. Several techniques have been developed for ionization in mass spectrometry. In one method, the molecule is bombarded with high-energy electrons. If an electron having an energy of about 10 electronvolts (10 eV = 230.5 kcal/mol) collides with an organic molecule, the energy transferred as a result of that collision is sufficient to dislodge one of the molecule's electrons.

$$\underset{\text{Molecule}}{\text{A:B}} + \underset{\text{Electron}}{e^-} \longrightarrow \underset{\text{Cation radical}}{\text{A}\overset{+}{\cdot}\text{B}} + \underset{\text{Two electrons}}{2e^-}$$

We say the molecule AB has been ionized by **electron impact.** The species that results, called the **molecular ion,** is positively charged and has an odd number of electrons—it is a **cation radical.** The molecular ion has the same mass (less the negligible mass of a single electron) as the molecule from which it is formed.

Although energies of about 10 eV are required, energies of about 70 eV are used. Electrons this energetic not only cause ionization of a molecule but also impart a large amount of energy to the molecular ion, enough energy to break chemical bonds. The molecular ion dissipates this excess energy by dissociating into smaller fragments. Dissociation of a cation radical produces a neutral fragment and a positively charged fragment.

$$\underset{\text{Cation radical}}{\text{A}\overset{+}{\cdot}\text{B}} \longrightarrow \underset{\text{Cation}}{\text{A}^+} + \underset{\text{Radical}}{\text{B}\cdot}$$

Ionization and fragmentation produce a mixture of particles, some neutral and some positively charged. To understand what follows, we need to examine the design of an electron-impact mass spectrometer, shown in Figure 14.39. The sample is bombarded with 70-eV electrons, and the resulting positively charged ions (the molecular ion as well as fragment ions) are directed into an analyzer tube surrounded by a magnet. This magnet deflects the ions from their original trajectory, causing them to adopt a circular path, the radius of which depends on their mass-to-charge ratio *(m/z).* Ions of small *m/z* are deflected more than those of larger *m/z*. By varying either the magnetic field strength or the degree to which the ions are accelerated on entering the analyzer, ions of a particular *m/z* can be selectively focused through a narrow slit onto a detector, where they are counted. Scanning all *m/z* values gives the distribution of positive ions, called a **mass spectrum,** characteristic of a particular compound.

Most mass spectrometers are capable of displaying the mass spectrum according to a number of different formats. Bar graphs on which relative intensity is plotted versus *m/z* are the most common. Figure 14.40 shows the mass spectrum of benzene in bar graph form.

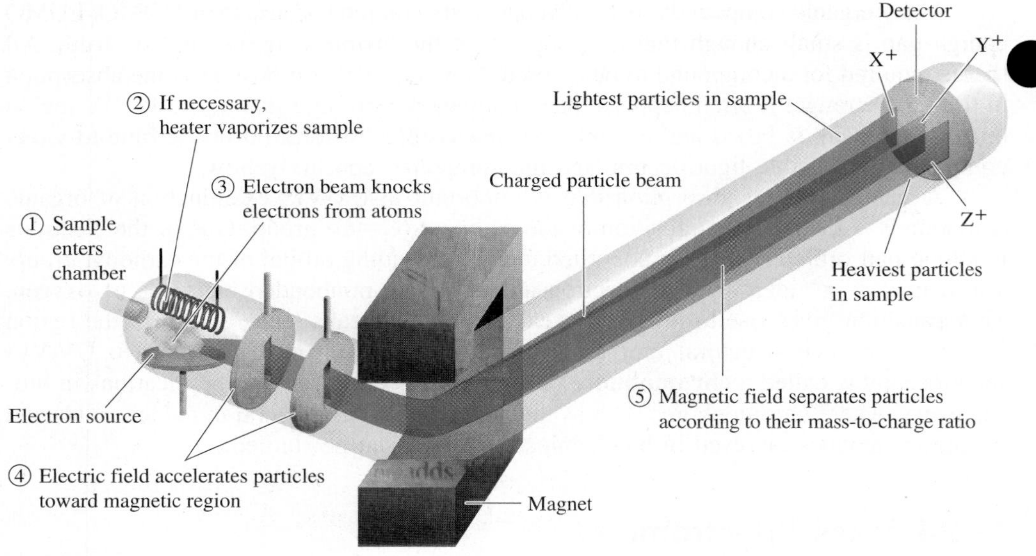

Figure 14.39

Diagram of a mass spectrometer. Only positive ions are detected. The cation X^+ has the lowest mass-to-charge ratio and its path is deflected most by the magnet. The cation Z^+ has the highest mass-to-charge ratio and its path is deflected least.

The mass spectrum of benzene is relatively simple and illustrates some of the information that mass spectrometry provides. The most intense peak in the mass spectrum is called the **base peak** and is assigned a relative intensity of 100. Ion abundances are proportional to peak intensities and are reported as intensities relative to the base peak. The base peak in the mass spectrum of benzene corresponds to the molecular ion (M^+) at m/z = 78.

Benzene + e^- ⟶ Molecular ion of benzene + $2e^-$

Benzene | Electron | Molecular ion of benzene | Two electrons

Benzene does not undergo extensive fragmentation; none of the fragment ions in its mass spectrum are as abundant as the molecular ion.

There is a small peak one mass unit higher than M^+ in the mass spectrum of benzene. What is the origin of this peak? What we see in Figure 14.40 as a single mass spectrum is actually a superposition of the spectra of three isotopically distinct benzenes. Most of the

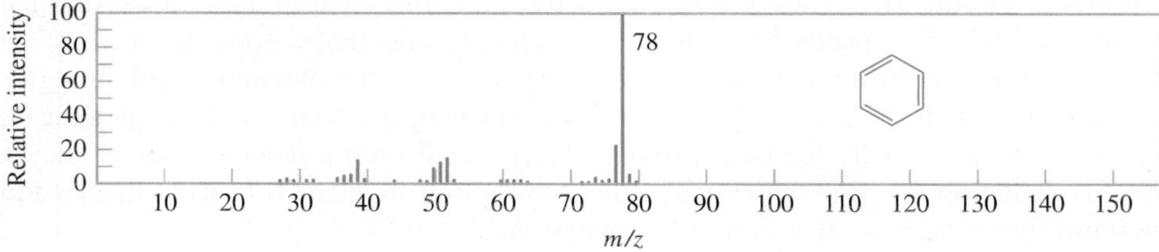

Figure 14.40

The mass spectrum of benzene. The peak at m/z = 78 corresponds to the C_6H_6 molecular ion.

benzene molecules contain only ^{12}C and ^{1}H and have a molecular mass of 78. Smaller proportions of benzene molecules contain ^{13}C in place of one of the ^{12}C atoms or ^{2}H in place of one of the protons. Both these species have a molecular mass of 79.

93.4% (all carbons are ^{12}C) Gives M^+ 78	6.5% (* = ^{13}C) Gives M^+ 79	0.1% (all carbons are ^{12}C) Gives M^+ 79

Not only the molecular ion peak but all the peaks in the mass spectrum of benzene are accompanied by a smaller peak one mass unit higher. Indeed, because all organic compounds contain carbon and most contain hydrogen, similar **isotopic clusters** will appear in the mass spectra of all organic compounds.

Isotopic clusters are especially apparent when atoms such as bromine and chlorine are present in an organic compound. The natural ratios of isotopes in these elements are

$$\frac{^{35}Cl}{^{37}Cl} = \frac{100}{32.7} \qquad \frac{^{79}Br}{^{81}Br} = \frac{100}{97.5}$$

Figure 14.41 presents the mass spectrum of chlorobenzene. There are two prominent molecular ion peaks, one at *m/z* 112 for $C_6H_5{}^{35}Cl$ and the other at *m/z* 114 for $C_6H_5{}^{37}Cl$. The peak at *m/z* 112 is three times as intense as the one at *m/z* 114.

Problem 14.27

Knowing what to look for with respect to isotopic clusters can aid in interpreting mass spectra. How many peaks would you expect to see for the molecular ion in each of the following compounds? At what *m/z* values would these peaks appear? (Disregard the small peaks due to ^{13}C and ^{2}H.)

(a) *p*-Dichlorobenzene
(b) *o*-Dichlorobenzene
(c) *p*-Dibromobenzene
(d) *p*-Bromochlorobenzene

Sample Solution (a) The two isotopes of chlorine are ^{35}Cl and ^{37}Cl. There will be three isotopically different forms of *p*-dichlorobenzene present. They have the structures shown as follows. Each one will give an M^+ peak at a different value of *m/z*.

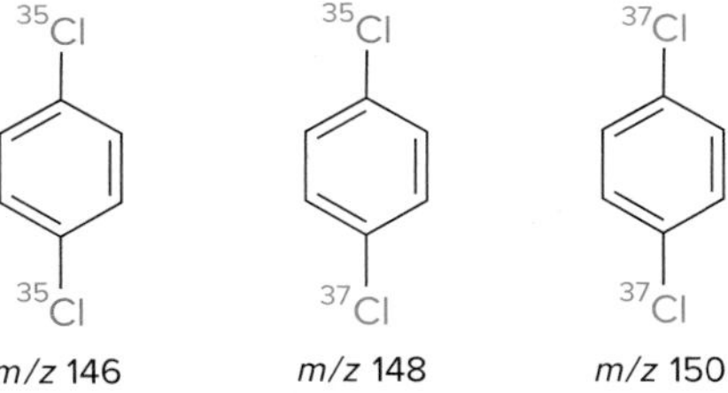

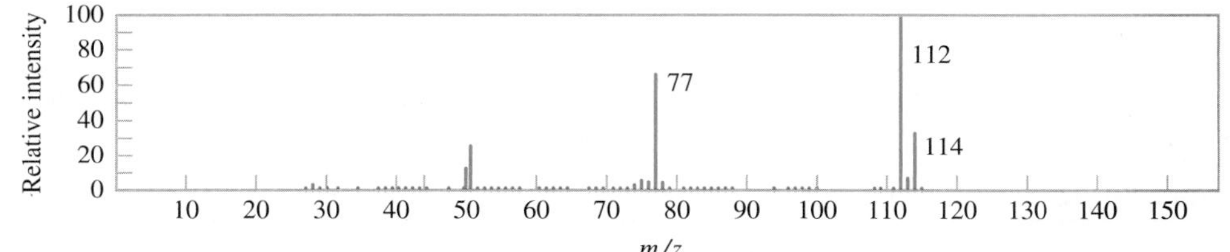

Figure 14.41

The mass spectrum of chlorobenzene.

Unlike the case of benzene, in which ionization involves loss of a π electron from the ring, electron-impact-induced ionization of chlorobenzene involves loss of an electron from an unshared pair of chlorine. The molecular ion then fragments by carbon–chlorine bond cleavage.

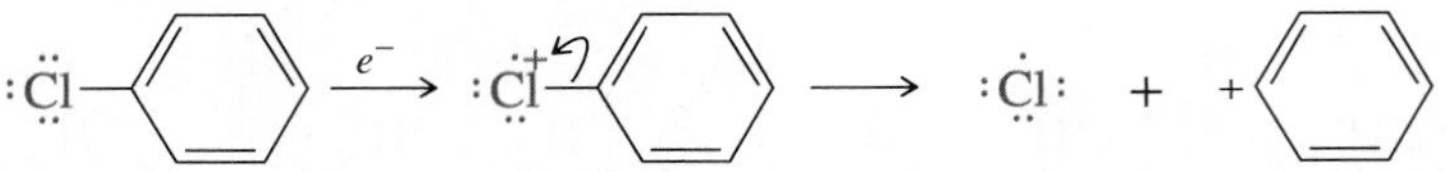
Chlorobenzene
Molecular ion of chlorobenzene
Chlorine atom
Phenyl cation m/z 77

The peak at m/z 77 in the mass spectrum of chlorobenzene in Figure 14.41 is attributed to this fragmentation. Because there is no peak of significant intensity two atomic mass units higher, we know that the cation responsible for the peak at m/z 77 cannot contain chlorine.

Some classes of compounds are so prone to fragmentation that the molecular ion peak is very weak. The base peak in most unbranched alkanes, for example, is m/z 43, which is followed by peaks of decreasing intensity at m/z values of 57, 71, 85, and so on. These peaks correspond to cleavage of each possible carbon–carbon bond in the molecule. This pattern is evident in the mass spectrum of decane, depicted in Figure 14.42. The points of cleavage are indicated in the following diagram:

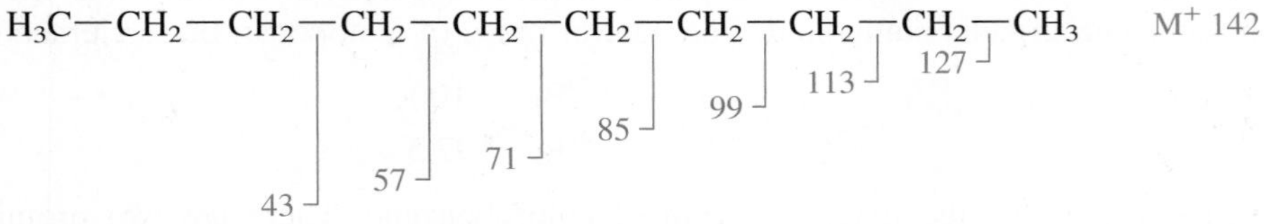
$H_3C-CH_2-CH_2-CH_2-CH_2-CH_2-CH_2-CH_2-CH_2-CH_3$
M^+ 142
127
113
99
85
71
57
43

Many fragmentations in mass spectrometry proceed so as to form a stable carbocation, and the principles that we have developed regarding carbocation stability apply. Alkylbenzenes of the type $C_6H_5CH_2R$ undergo cleavage of the bond to the benzylic carbon to give m/z 91 as the base peak. The mass spectrum in Figure 14.43 and the following fragmentation diagram illustrate this for propylbenzene.

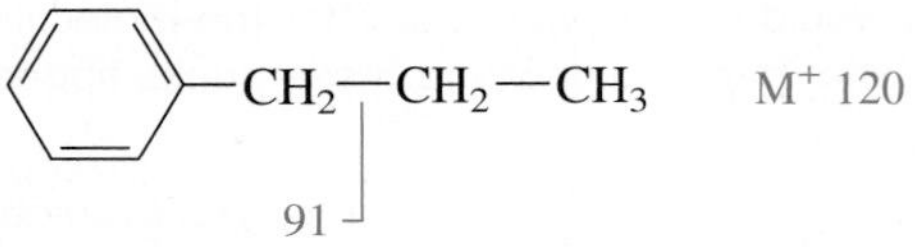
$-CH_2-CH_2-CH_3$
M^+ 120
91

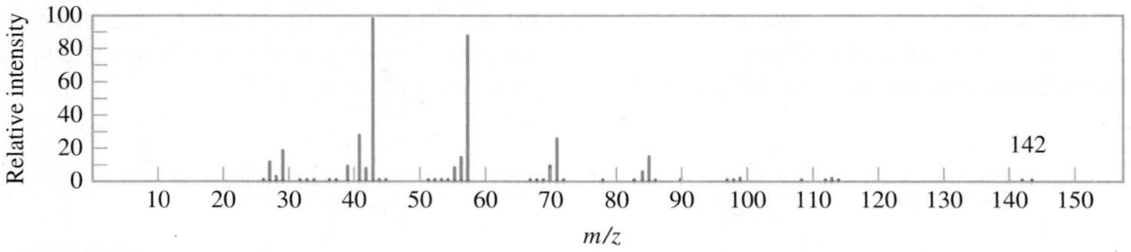
Relative intensity
142
m/z

Figure 14.42

The mass spectrum of decane. The peak for the molecular ion is extremely small. The most prominent peaks arise by fragmentation.

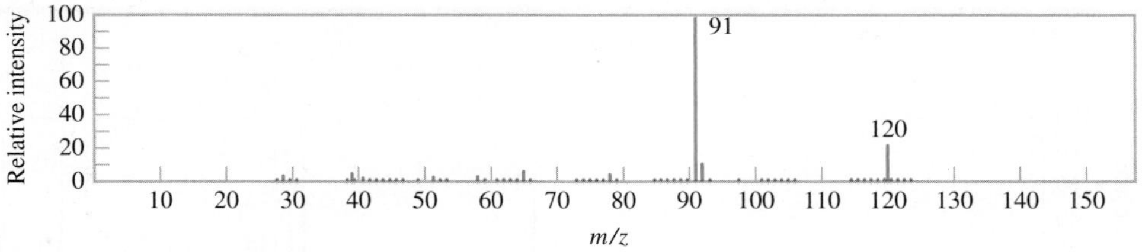
Relative intensity
91
120
m/z

Figure 14.43

The mass spectrum of propylbenzene. The most intense peak is $C_7H_7^+$.

Although this cleavage is probably driven by the stability of benzyl cation, evidence has been obtained suggesting that tropylium cation, formed by rearrangement of benzyl cation, is actually the species responsible for the peak.

The structure of tropylium cation is given in Section 12.20.

Problem 14.28

The base peak appears at *m/z* 105 for one of the following compounds and at *m/z* 119 for the other two. Match the compounds with the appropriate *m/z* values for their base peaks.

Problem 14.29

Mass spectra of 1-bromo-4-propylbenzene and (3-bromopropyl)benzene are shown in Figure 14.44. Match each spectrum to the appropriate compound. Write a structure for the ion that corresponds to the base peak in each spectrum.

Br

1-Bromo-4-propylbenzene

Br

(3-Bromopropyl)benzene

Understanding how molecules fragment upon electron impact permits a mass spectrum to be analyzed in sufficient detail to deduce the structure of an unknown compound. Thousands of compounds of known structure have been examined by mass spectrometry, and the fragmentation patterns that characterize different classes are well documented. As various groups are covered in subsequent chapters, aspects of their fragmentation behavior under conditions of electron impact will be described.

An alternate method of ionization is described in the boxed essay *Peptide Mapping and MALDI Mass Spectrometry* in Chapter 25.

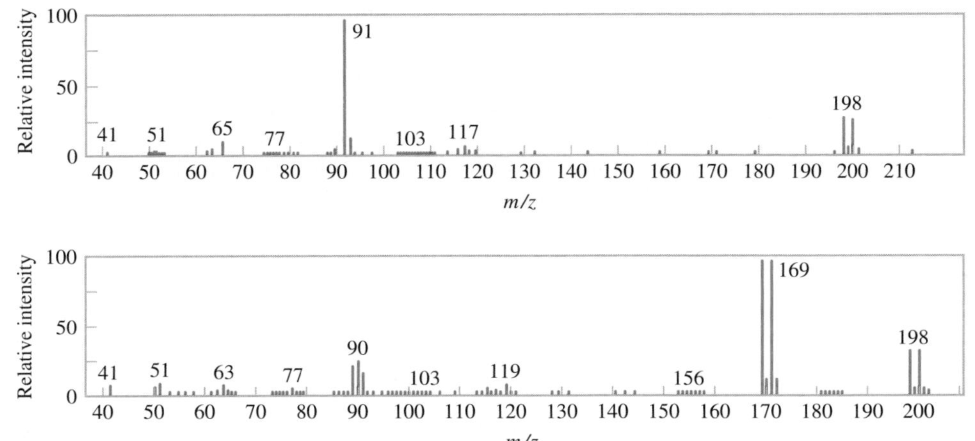

Figure 14.44

Mass spectra of 1-bromo-4-propylbenzene and (3-bromopropyl)benzene.

14.25 Molecular Formula as a Clue to Structure

As we have just seen, interpreting the fragmentation patterns in a mass spectrum in terms of a molecule's structural units makes mass spectrometry much more than just a tool for determining molecular weights. Nevertheless, even the molecular weight can provide more information than you might think.

A relatively simple example is the **nitrogen rule.** A molecule with an odd number of nitrogens has an odd molecular weight; a molecule with only C, H, and O or with an even number of nitrogens has an even molecular weight.

	Aniline (C_6H_7N)	*p*-Nitroaniline ($C_6H_6N_2O_2$)	2,4-Dinitroaniline ($C_6H_5N_3O_4$)
Molecular weight:	93	138	183

A second example concerns different compounds that have the same molecular weight, but different molecular formulas, such as heptane and cyclopropyl acetate.

Heptane (C_7H_{16}) Cyclopropyl acetate ($C_5H_8O_2$)

Because we normally round off molecular weights to whole numbers, both have a molecular weight of 100 and both have a peak for their molecular ion at *m/z* 100 in a typical mass spectrum. Recall, however, that mass spectra contain isotopic clusters that differ according to the isotopes present in each ion. Using the exact values for the major isotopes of C, H, and O, we calculate *exact masses* of *m/z* of 100.1253 and 100.0524 for the molecular ions of heptane (C_7H_{16}) and cyclopropyl acetate ($C_5H_8O_2$), respectively. As similar as these values are, it is possible to distinguish between them using a *high-resolution mass spectrometer.* This means that the exact mass of a molecular ion can usually be translated into a unique molecular formula.

You can't duplicate these molecular weights for C_7H_{16} and $C_5H_8O_2$ by using the atomic weights given in the periodic table. Those values are for the natural-abundance mixture of isotopes. The exact values are 12.00000 for ^{12}C, 1.00783 for ^{1}H, and 15.9949 for ^{16}O.

Once we have the molecular formula, it can provide information that limits the amount of trial-and-error structure writing we have to do. Consider, for example, heptane and its molecular formula of C_7H_{16}. We know immediately that the molecular formula belongs to an alkane because it corresponds to C_nH_{2n+2}.

What about a substance with the molecular formula C_7H_{14}? This compound cannot be an alkane but may be either a cycloalkane or an alkene, because both these classes of hydrocarbons correspond to the general molecular formula C_nH_{2n}. *Any time a ring or a double bond is present in an organic molecule, its molecular formula has two fewer hydrogen atoms than that of an alkane with the same number of carbons.*

The relationship between molecular formulas, multiple bonds, and rings is referred to as the **index of hydrogen deficiency** and can be expressed by the equation:

Other terms that mean the same thing as the index of hydrogen deficiency include *elements of unsaturation, sites of unsaturation,* and *the sum of double bonds and rings.*

$$\text{Index of hydrogen deficiency} = \tfrac{1}{2}(C_nH_{2n+2} - C_nH_x)$$

where C_nH_x is the molecular formula of the compound.

A molecule that has a molecular formula of C_7H_{14} has an index of hydrogen deficiency of 1:

$$\text{Index of hydrogen deficiency} = \tfrac{1}{2}(C_7H_{16} - C_7H_{14})$$

$$\text{Index of hydrogen deficiency} = \tfrac{1}{2}(2) = 1$$

Thus, the compound has one ring or one double bond. It can't have a triple bond.

A molecule of molecular formula C_7H_{12} has four fewer hydrogens than the corresponding alkane. It has an index of hydrogen deficiency of 2 and can have two rings, two double bonds, one ring and one double bond, or one triple bond.

What about substances other than hydrocarbons, 1-heptanol [$CH_3(CH_2)_5CH_2OH$], for example? Its molecular formula ($C_7H_{16}O$) contains the same carbon-to-hydrogen ratio as heptane and, like heptane, it has no double bonds or rings. Cyclopropyl acetate ($C_5H_8O_2$), the structure of which was given at the beginning of this section, has one ring and one double bond and an index of hydrogen deficiency of 2. *Oxygen atoms have no effect on the index of hydrogen deficiency.*

A halogen substituent, like hydrogen, is monovalent and when present in a molecular formula is treated as if it were hydrogen for counting purposes. If a nitrogen is present, one hydrogen is taken away from the formula. For example, $C_5H_{11}N$ is treated as C_5H_{10} when calculating the index of hydrogen deficiency.

How does one distinguish between rings and double bonds? This additional piece of information comes from catalytic hydrogenation experiments in which the amount of hydrogen consumed is measured exactly. Each of a molecule's double bonds consumes one molar equivalent of hydrogen, but rings are unaffected. For example, a substance with a hydrogen deficiency of 5 that takes up 3 mol of hydrogen must have two rings.

Problem 14.30

How many rings are present in each of the following compounds? Each consumes 2 mol of hydrogen on catalytic hydrogenation.

(a) $C_{10}H_{18}$
(b) C_8H_8
(c) $C_8H_8Cl_2$
(d) C_8H_8O
(e) $C_8H_{10}O_2$
(f) C_8H_9ClO
(g) C_3H_5N
(h) C_4H_5N

Sample Solution (a) The molecular formula $C_{10}H_{18}$ contains four fewer hydrogens than the alkane having the same number of carbon atoms ($C_{10}H_{22}$). Therefore, the index of hydrogen deficiency of this compound is 2. Because it consumes two molar equivalents of hydrogen on catalytic hydrogenation, it must have either a triple bond or two double bonds and no rings.

14.26 SUMMARY

Section 14.1 Structure determination in modern organic chemistry relies heavily on instrumental methods. Several of the most widely used ones depend on the absorption of electromagnetic radiation.

Section 14.2 Absorption of electromagnetic radiation causes a molecule to be excited from its most stable state (the *ground* state) to a higher-energy state (an *excited* state).

Spectroscopic method	*Transitions between*
Nuclear magnetic resonance	Spin states of an atom's nucleus
Infrared	Vibrational states
Ultraviolet-visible	Electronic states

Mass spectrometry is not based on absorption of electromagnetic radiation, but monitors what happens when a substance is ionized by collision with a high-energy electron.

^1H Nuclear Magnetic Resonance Spectroscopy

Section 14.3 In the presence of an external magnetic field, the $+\frac{1}{2}$ and $-\frac{1}{2}$ nuclear spin states of a proton have slightly different energies.

Section 14.4 The energy required to "flip" the spin of a proton from the lower-energy spin state to the higher state depends on the extent to which a nucleus is shielded from the external magnetic field by the molecule's electrons.

Section 14.5 Protons in different environments within a molecule have different **chemical shifts;** that is, they experience different degrees of shielding. Chemical shifts (δ) are reported in parts per million (ppm) from tetramethylsilane (TMS). Table 14.1 lists characteristic chemical shifts for various types of protons.

Section 14.6 In addition to *chemical shift,* a ^{1}H NMR spectrum provides structural information based on:

Number of signals, which tells how many different kinds of protons there are

Integrated areas, which tells the ratios of the various kinds of protons

Splitting pattern, which gives information about the number of protons that are within two or three bonds of the one giving the signal

Section 14.7 **Spin–spin splitting** of NMR signals results from coupling of the nuclear spins that are separated by two bonds (*geminal coupling*) or three bonds (*vicinal coupling*).

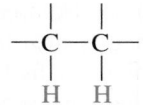

Geminal hydrogens are separated by two bonds

Vicinal hydrogens are separated by three bonds

In the simplest cases, the number of peaks into which a signal is split is equal to $n + 1$, where n is the number of protons to which the proton in question is coupled. *Protons that have the same chemical shift do not split each other's signal.*

Section 14.8 The methyl protons of an ethyl group appear as a *triplet* and the methylene protons as a *quartet* in compounds of the type CH_3CH_2X.

Section 14.9 The methyl protons of an isopropyl group appear as a *doublet* and the methine proton as a *septet* in compounds of the type $(CH_3)_2CHX$.

Section 14.10 A *pair of doublets* characterizes the signals for the protons of the type shown (where W, X, Y, and Z are not H or atoms that split H themselves).

X Y
W—C—C—Z
H H

Section 14.11 Complicated splitting patterns can result when a proton is unequally coupled to two or more protons that are different from one another.

Section 14.12 Splitting resulting from coupling to the O—H proton of alcohols is sometimes not observed, because the hydroxyl proton undergoes rapid intermolecular exchange with other alcohol molecules, which "decouples" it from other protons in the molecule.

Section 14.13 Many processes such as conformational changes take place faster than they can be detected by NMR. Consequently, NMR provides information about the *average* environment of a proton. For example, cyclohexane gives a single peak for its 12 protons even though, at any instant, 6 are axial and 6 are equatorial.

^{13}C *Nuclear Magnetic Resonance Spectroscopy*

Section 14.14 ^{13}C has a nuclear spin of $\pm\frac{1}{2}$ but only about 1% of all the carbons in a sample are ^{13}C. Nevertheless, high-quality ^{13}C NMR spectra can be obtained by pulse FT techniques and are a useful complement to ^{1}H NMR spectra.

Section 14.15 ^{13}C signals are more widely separated from one another than proton signals, and ^{13}C NMR spectra are relatively easy to interpret. Table 14.2 gives chemical-shift values for carbon in various environments.

Section 14.16 ^{13}C NMR spectra are rarely integrated because the pulse FT technique distorts the signal intensities.

Section 14.17 Carbon signals normally appear as singlets, but several techniques are available that allow one to distinguish among the various kinds of carbons shown.

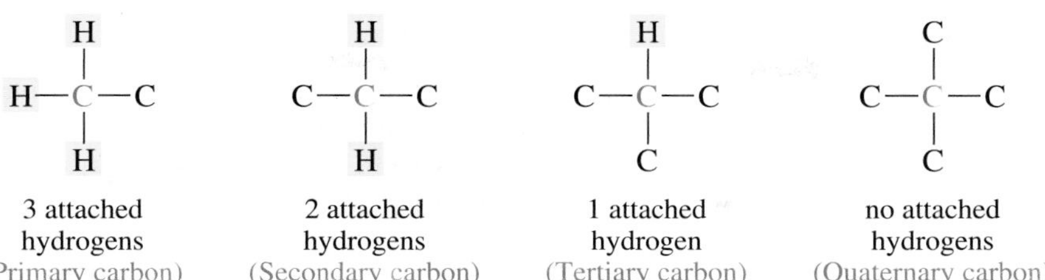

Section 14.18 One of the special techniques for distinguishing carbons according to the number of their attached hydrogens is called **DEPT.** A series of NMR measurements using different pulse sequences gives normal, nulled, and inverted peaks that allow assignment of primary, secondary, tertiary, and quaternary carbons.

Section 14.19 2D NMR techniques are enhancements that are sometimes useful in gaining additional structural information. A ^1H-^1H COSY spectrum reveals which protons are spin-coupled to other protons, which helps in determining connectivity. A HETCOR spectrum shows the C—H connections by correlating ^{13}C and ^1H chemical shifts.

Infrared Spectroscopy

Section 14.20 IR spectroscopy probes molecular structure by examining transitions between quantized vibrational energy levels using electromagnetic radiation in the 625–4000-cm^{-1} range, where cm^{-1} are units of **wavenumbers,** defined as λ^{-1}. Wavenumbers are proportional to frequency. The simplest vibration is the stretching of the bond between two atoms, but more complex vibrations can involve movement of many of a molecule's atoms.

Section 14.21 IR spectra are commonly regarded as consisting of a functional-group region (1500–4000 cm^{-1}) and a fingerprint region (500–1500 cm^{-1}). Included in the functional-group region are absorptions due to C—H stretching. In general, C—H stretching frequencies lie below 3000 cm^{-1} for sp^3-hybridized carbon and above 3000 cm^{-1} for sp^2. The fingerprint region is used less for determining structure than for verifying whether two compounds are identical or not.

Section 14.22 Functional-group identification is the main contribution of IR spectroscopy to organic chemistry. Various classes of compounds exhibit peaks at particular frequencies characteristic of the functional groups they contain (Table 14.3).

Ultraviolet-Visible Spectroscopy

Section 14.23 Transitions between electronic energy levels involving electromagnetic radiation in the 200–800-nm range form the basis of UV-VIS spectroscopy. The absorption peaks tend to be broad but are often useful in indicating the presence of particular π-electron systems within a molecule.

Mass Spectrometry

Section 14.24 Mass spectrometry exploits the information obtained when a molecule is ionized by electron impact and then dissociates to smaller fragments. Positive ions are separated and detected according to their mass-to-charge (m/z) ratio. By examining the fragments and by knowing how classes of molecules dissociate on electron impact, one can deduce the structure of a compound. Mass spectrometry is quite sensitive; as little as 10^{-9} g of compound is sufficient for analysis.

Section 14.25 A compound's molecular formula gives information about the number of double bonds and rings it contains and is a useful complement to spectroscopic methods of structure determination.

PROBLEMS

^{1}H NMR Spectroscopy

14.31 Each of the following compounds is characterized by a ^{1}H NMR spectrum that consists of only a single peak having the chemical shift indicated. Identify each compound.

(a) C_8H_{18}; δ 0.9
(b) C_5H_{10}; δ 1.5
(c) C_8H_8; δ 5.8
(d) C_4H_9Br; δ 1.8
(e) $C_2H_4Cl_2$; δ 3.7
(f) $C_2H_3Cl_3$; δ 2.7
(g) $C_5H_8Cl_4$; δ 3.7
(h) $C_{12}H_{18}$; δ 2.2
(i) $C_3H_6Br_2$; δ 2.6

14.32 Deduce the structure of each of the following compounds on the basis of their ^{1}H NMR spectra and molecular formulas:

(a) C_8H_{10}; δ 1.2 (triplet, 3H)
δ 2.6 (quartet, 2H)
δ 7.1 (broad singlet, 5H)

(b) $C_{10}H_{14}$; δ 1.3 (singlet, 9H)
δ 7.0 to 7.5 (multiplet, 5H)

(c) C_6H_{14}; δ 0.8 (doublet, 12H)
δ 1.4 (septet, 2H)

(d) C_6H_{12}; δ 0.9 (triplet, 3H)
δ 1.6 (singlet, 3H)
δ 1.7 (singlet, 3H)
δ 2.0 (quintet, 2H)
δ 5.1 (triplet, 1H)

(e) $C_4H_6Cl_4$; δ 3.9 (doublet, 4H)
δ 4.6 (triplet, 2H)

(f) $C_4H_6Cl_2$; δ 2.2 (singlet, 3H)
δ 4.1 (doublet, 2H)
δ 5.7 (triplet, 1H)

(g) C_3H_7ClO; δ 2.0 (quintet, 2H)
δ 2.8 (singlet, 1H)
δ 3.7 (triplet, 2H)
δ 3.8 (triplet, 2H)

(h) $C_{14}H_{14}$; δ 2.9 (singlet, 4H)
δ 7.1 (broad singlet, 10H)

14.33 From among the isomeric compounds of molecular formula C_4H_9Cl, choose the one having a ^{1}H NMR spectrum that

(a) Contains only a single peak
(b) Has several peaks including a doublet at δ 3.4
(c) Has several peaks including a triplet at δ 3.5
(d) Has several peaks including two distinct three-proton signals, one of them a triplet at δ 1.0 and the other a doublet at δ 1.5

14.34 The ^{1}H NMR spectrum of fluorene has signals at δ 3.8 and δ 7.2–7.7 in a 1:4 ratio. After heating with $NaOCH_3$ in CH_3OD at reflux for 15 minutes the signals at δ 7.2–7.7 remained, but the one at δ 3.8 had disappeared. Suggest an explanation and write a mechanism for this observation.

Fluorene

14.35 The vinyl proton region of the ^{1}H NMR spectrum of phenyl vinyl sulfoxide is shown in Figure 14.45. Construct a splitting diagram similar to the one in Figure 14.21 and label each of the coupling constants $J_{a,b}$, $J_{b,c}$, and $J_{a,c}$.

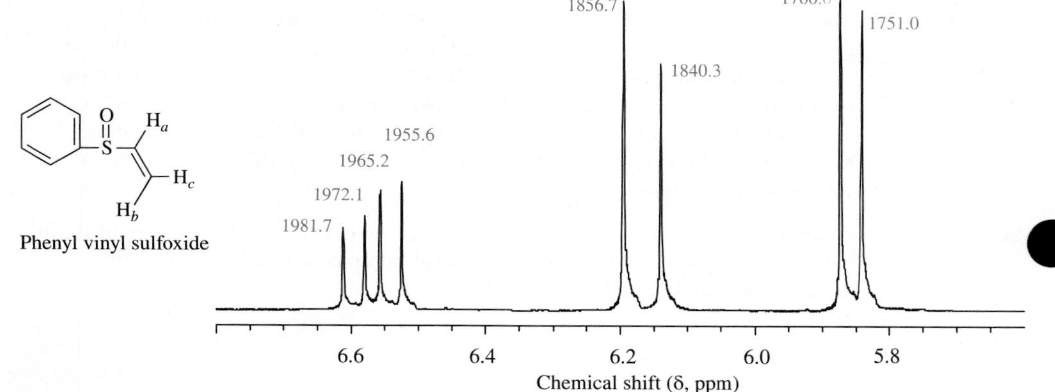

Figure 14.45

Vinyl proton region of the 300-MHz ^{1}H NMR spectrum of phenyl vinyl sulfoxide.

14.36 ^1H NMR spectra of four isomeric alcohols with formula $C_9H_{12}O$ are shown in Figure 14.46. Assign a structure for each alcohol and assign the peaks in each spectrum.

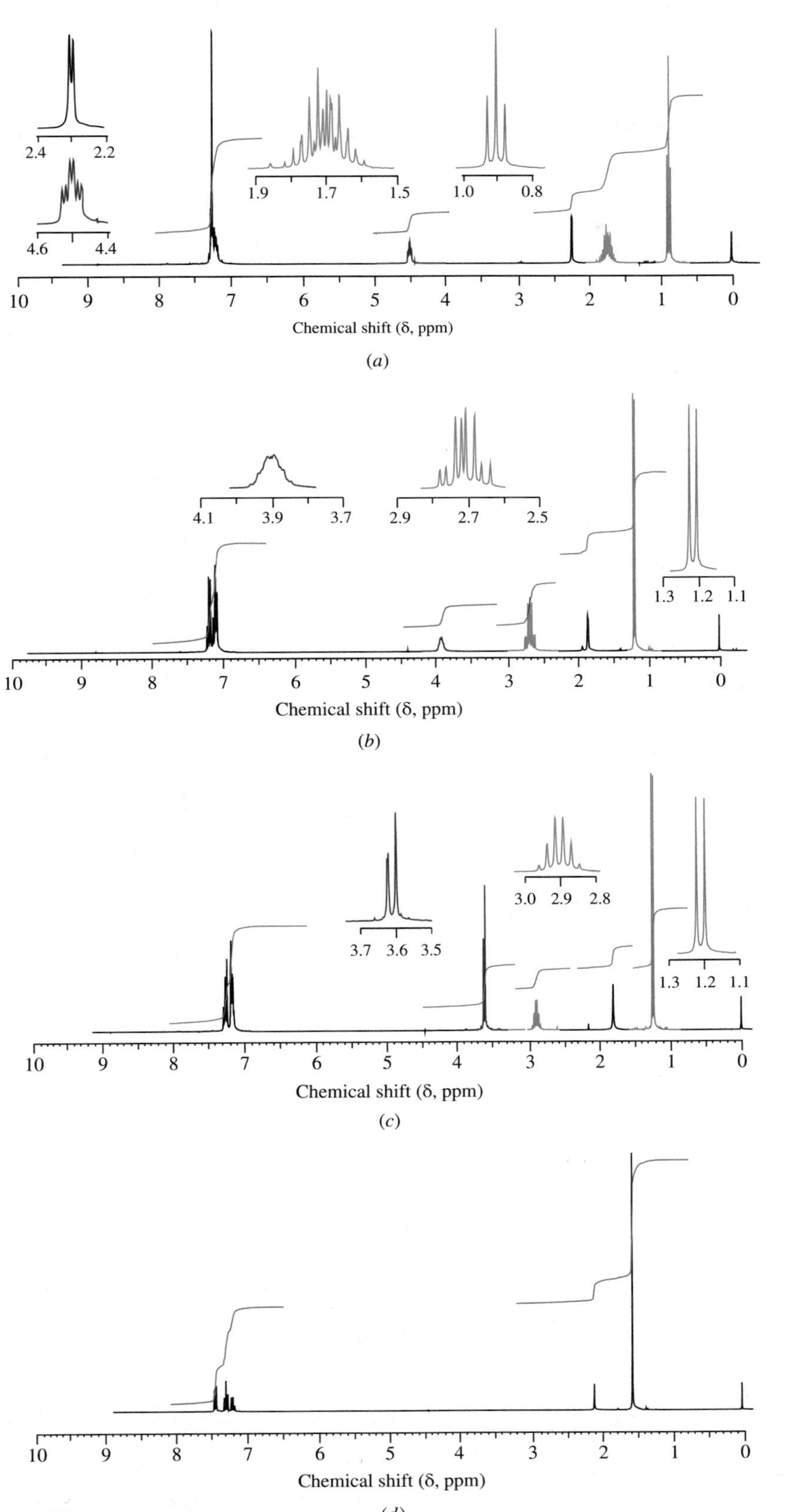

Figure 14.46

300-MHz ^1H NMR spectra of alcohols for Problem 14.36.

14.37 Which would you predict to be more shielded, the inner or outer protons of [24]annulene?

14.38 We noted in Section 14.13 that an NMR spectrum is an average spectrum of the conformations populated by a molecule. From the following data, estimate the percentages of axial and equatorial bromine present in bromocyclohexane.

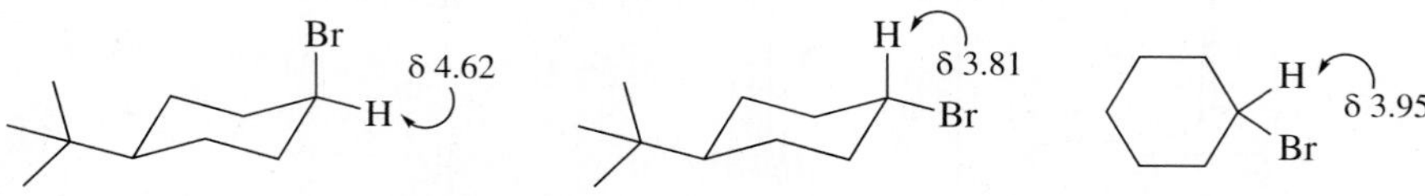

^{13}C NMR Spectroscopy

14.39 Identify each of the $C_4H_{10}O$ isomers on the basis of their ^{13}C NMR spectra:

(a) δ 18.9 (CH_3) (two carbons)
δ 30.8 (CH) (one carbon)
δ 69.4 (CH_2) (one carbon)

(b) δ 10.0 (CH_3)
δ 22.7 (CH_3)
δ 32.0 (CH_2)
δ 69.2 (CH)

(c) δ 31.2 (CH_3) (three carbons)
δ 68.9 (C) (one carbon)

14.40 A compound ($C_3H_7ClO_2$) exhibited three peaks in its ^{13}C NMR spectrum at δ 46.8 (CH_2), δ 63.5 (CH_2), and δ 72.0 (CH). Excluding compounds that have Cl and OH on the same carbon, which are unstable, what is the most reasonable structure for this compound?

14.41 Label nonequivalent carbons in the following compounds.

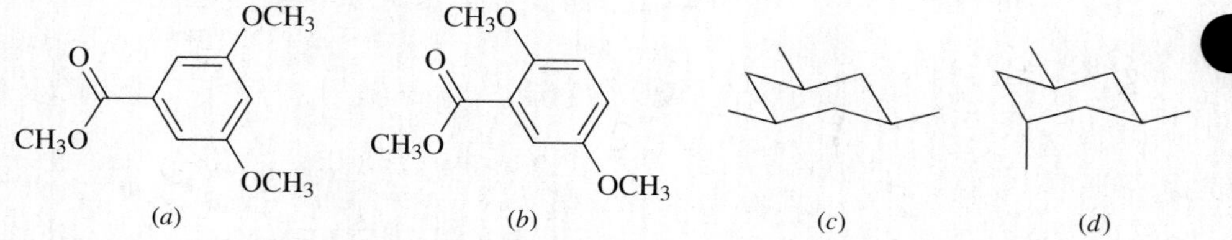

14.42 Compounds A and B are isomers of molecular formula $C_{10}H_{14}$. Identify each one on the basis of the ^{13}C NMR spectra presented in Figure 14.47.

Figure 14.47

The ^{13}C NMR spectrum of (*a*) compound A and (*b*) compound B, isomers of $C_{10}H_{14}$ (Problem 14.42).

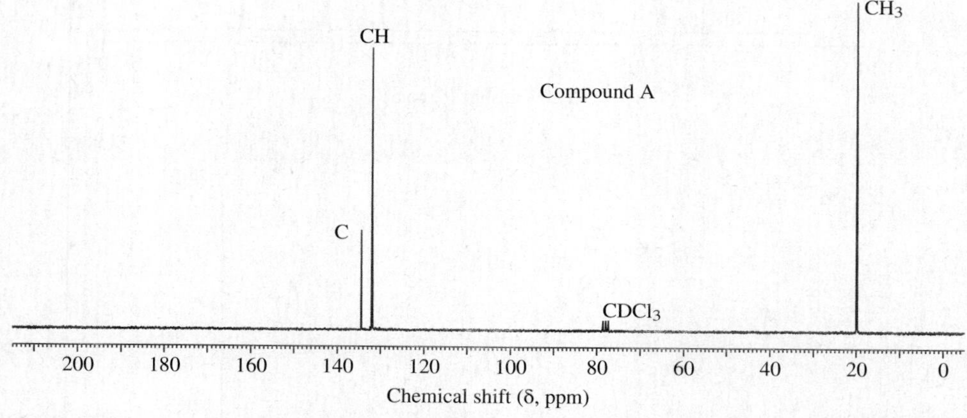

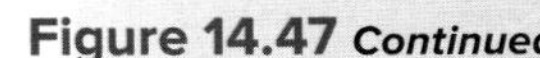
Figure 14.47 *Continued*

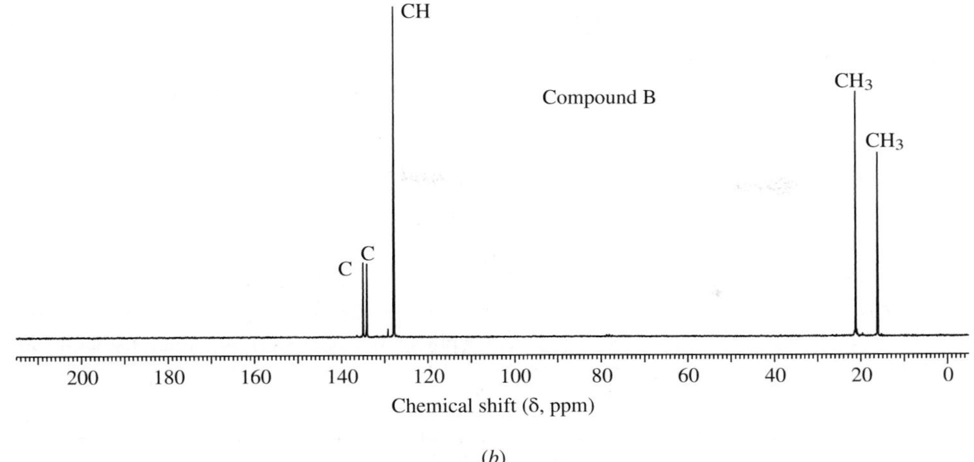

(*b*)

14.43 ^{13}C NMR spectra for four isomeric alkyl bromides with the formula $C_5H_{11}Br$ are shown in Figure 14.48. Multiplicities obtained from DEPT analysis are shown above each peak. Assign structures to each of the alkyl bromides and assign the peaks in each spectrum.

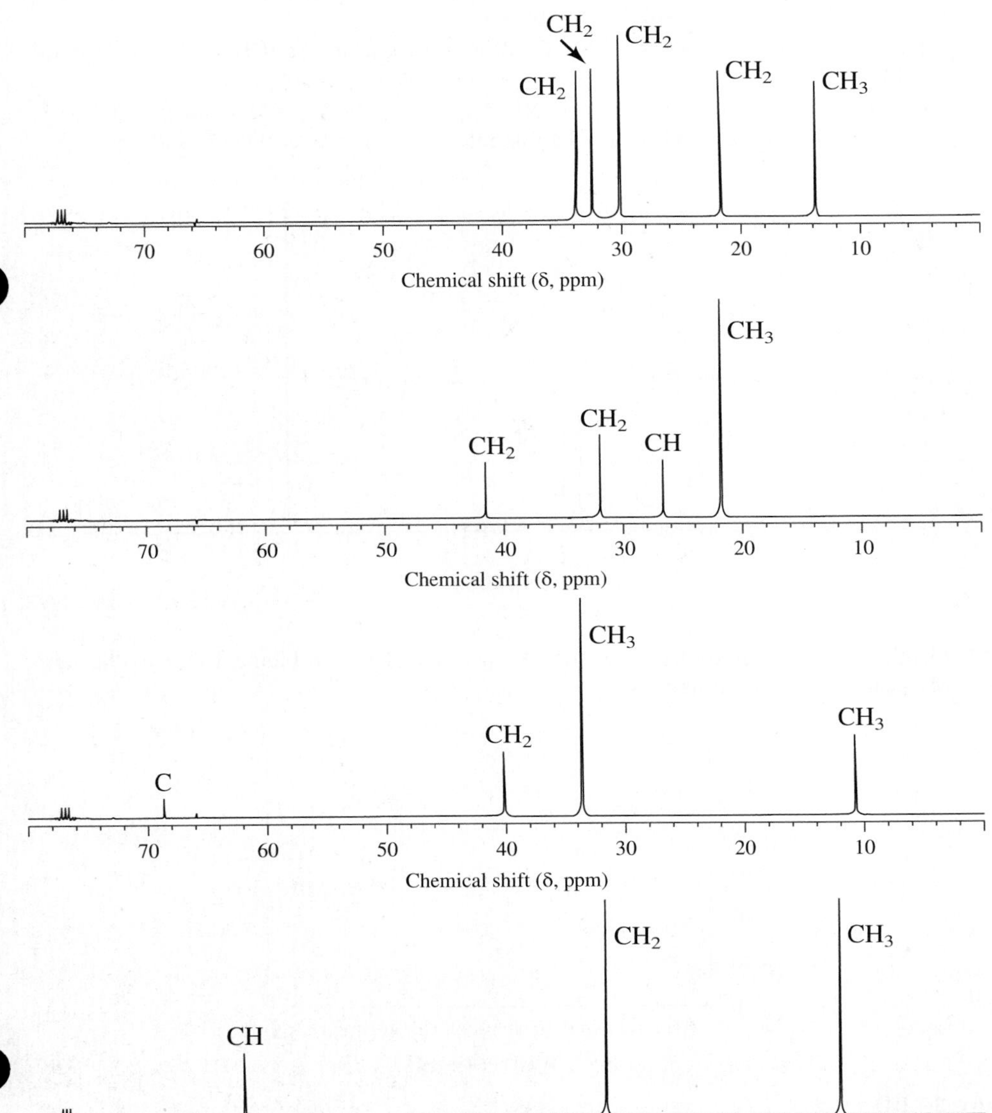

Figure 14.48

^{13}C NMR spectra for isomeric alkyl bromides in Problem 14.43.

^{19}F and ^{31}P NMR Spectroscopy

14.44 ^{19}F is the only isotope of fluorine that occurs naturally, and it has a nuclear spin of $\pm\frac{1}{2}$.

(a) Into how many peaks will the proton signal in the ^1H NMR spectrum of methyl fluoride be split?

(b) Into how many peaks will the fluorine signal in the ^{19}F NMR spectrum of methyl fluoride be split?

(c) The chemical shift of the protons in methyl fluoride is δ 4.3. Given that the geminal $^1H—^{19}F$ coupling constant is 45 Hz, specify the δ values at which peaks are observed in the proton spectrum of this compound at 300 MHz.

14.45 ^{31}P is the only phosphorus isotope present at natural abundance and has a nuclear spin of $\pm\frac{1}{2}$. The ^1H NMR spectrum of trimethyl phosphite, $(CH_3O)_3P$, exhibits a doublet for the methyl protons with a splitting of 12 Hz.

(a) Into how many peaks is the ^{31}P signal split?

(b) What is the difference in chemical shift (in hertz) between the lowest and highest field peaks of the ^{31}P multiplet?

Combined Spectra

14.46 Identify the C_3H_5Br isomers on the basis of the following information:

(a) Isomer A has the ^1H NMR spectrum shown in Figure 14.49.

(b) Isomer B has three peaks in its ^{13}C NMR spectrum: δ 32.6 (CH_2); 118.8 (CH_2); and 134.2 (CH).

(c) Isomer C has two peaks in its ^{13}C NMR spectrum: δ 12.0 (CH_2) and 16.8 (CH). The peak at lower field is only half as intense as the one at higher field.

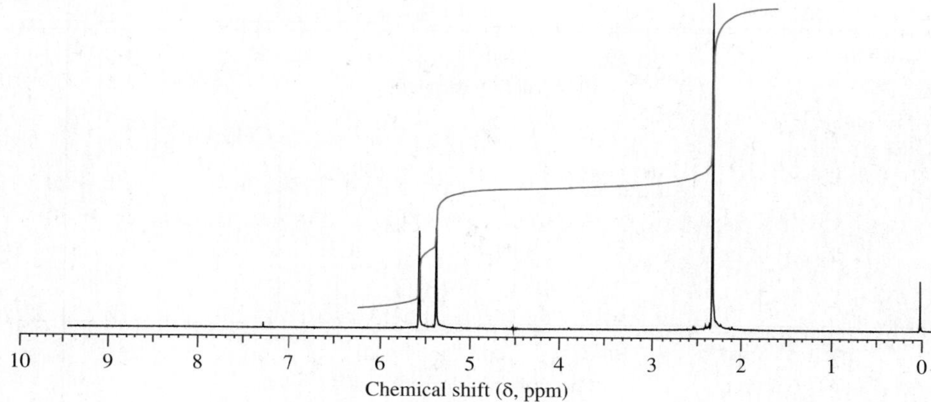

Figure 14.49

The 300-MHz ^1H NMR spectrum of isomer A (Problem 14.46a).

14.47 Identify the hydrocarbon that gives the IR spectrum shown in Figure 14.50 and has an M^+ peak at m/z 102 in its mass spectrum.

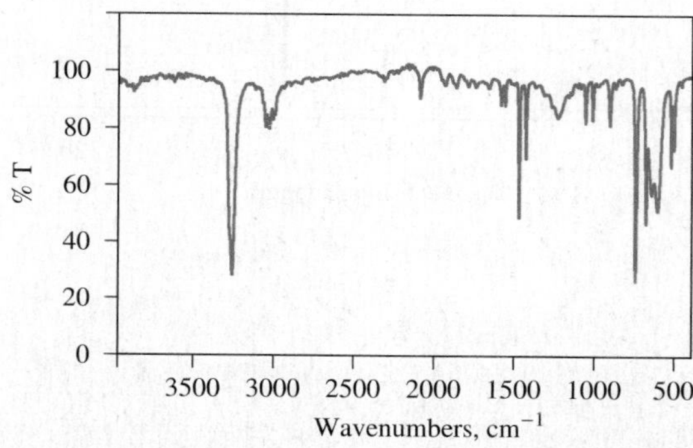

Figure 14.50

The IR spectrum of the hydrocarbon in Problem 14.47.

14.48 A compound ($C_8H_{10}O$) has the IR and ^1H NMR spectra presented in Figure 14.51. What is its structure?

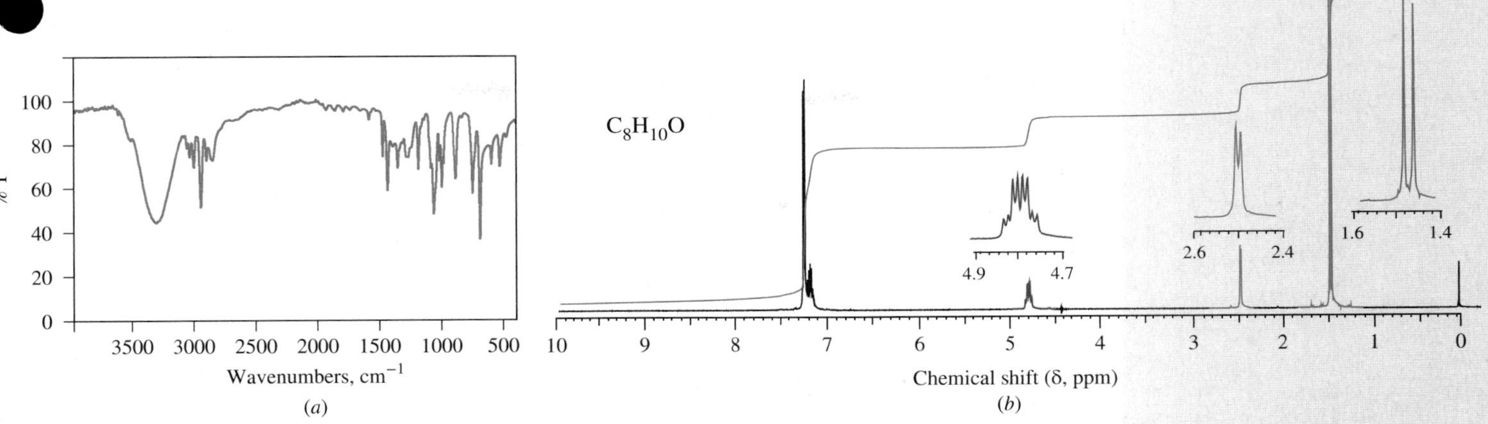

Figure 14.51

(*a*) IR and (*b*) 300-MHz ^1H NMR spectra of a compound $C_8H_{10}O$ (Problem 14.48).

14.49 Deduce the structure of a compound having the mass, IR, and ^1H NMR spectra presented in Figure 14.52.

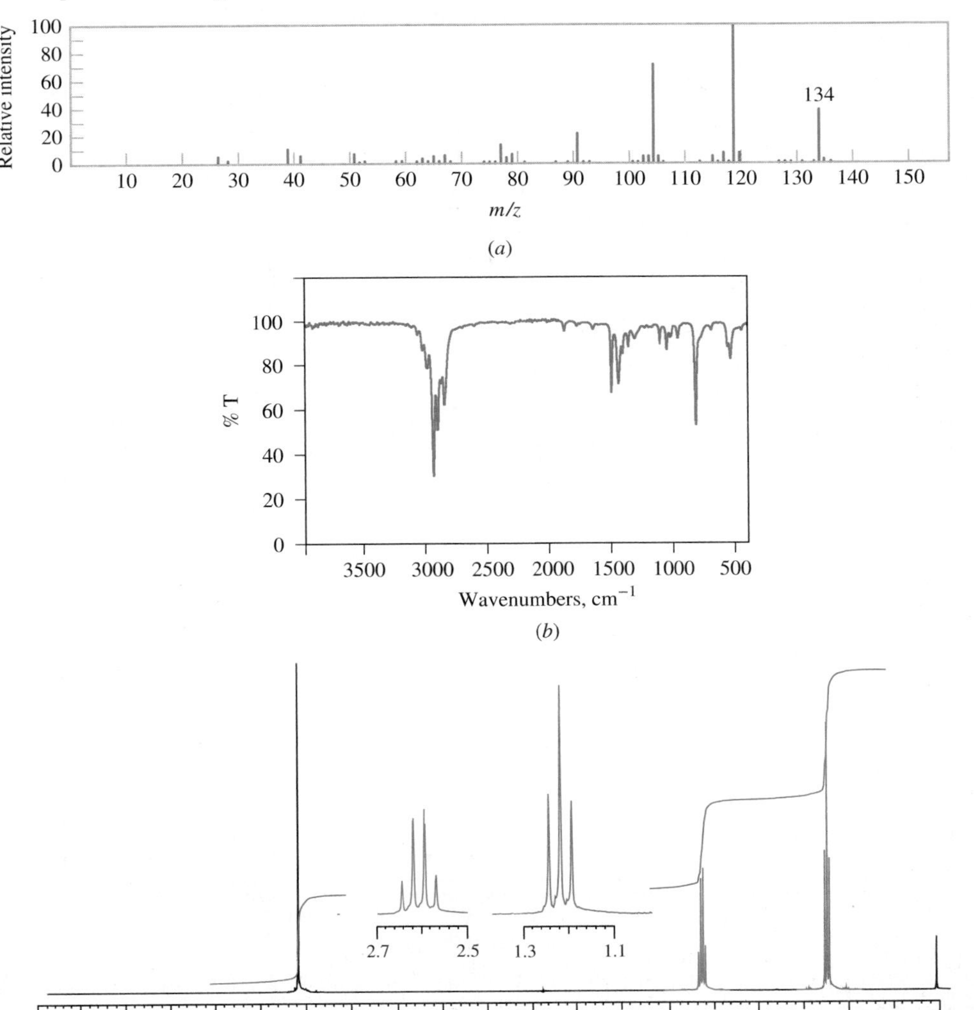

Figure 14.52

(*a*) Mass, (*b*) IR, and (*c*) 300-MHz ^1H NMR spectra of a compound (Problem 14.49).

14.50 Figure 14.53 presents IR, ^{1}H NMR, ^{13}C NMR, and mass spectra for a particular compound. What is it?

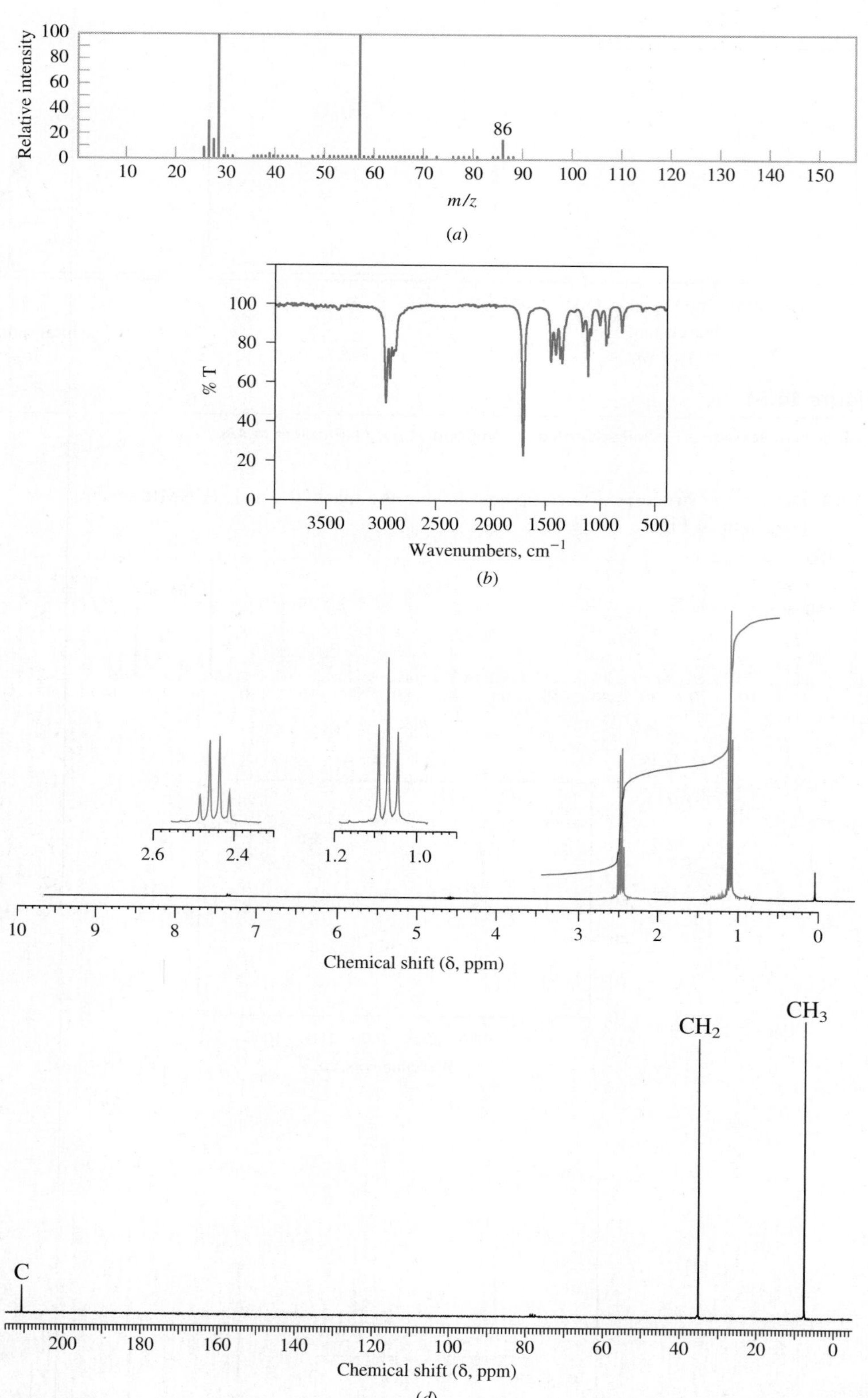

Figure 14.53

(a) Mass, (b) IR, (c) 300-MHz ^{1}H NMR, and (d) ^{13}C NMR spectra for the compound in Problem 14.50.

14.51 ^{1}H NMR, ^{13}C NMR, IR, and mass spectra are shown for a compound in Figure 14.54. Propose a structure and explain your answer based on spectral assignments.

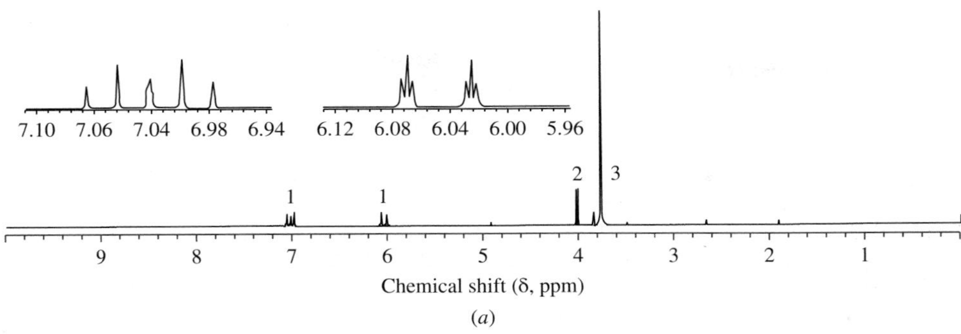

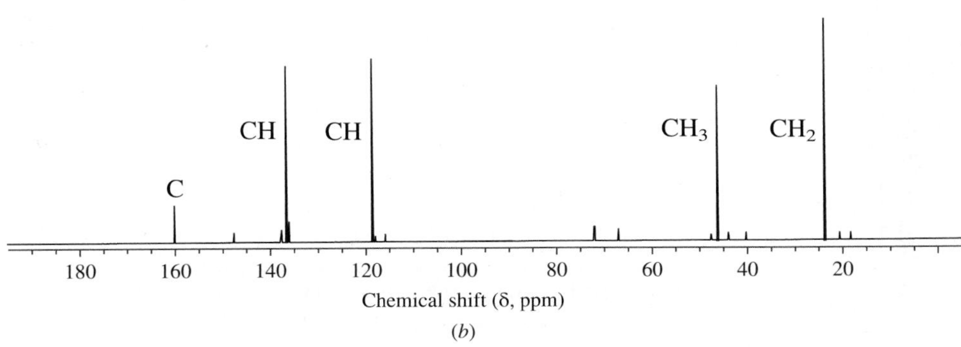

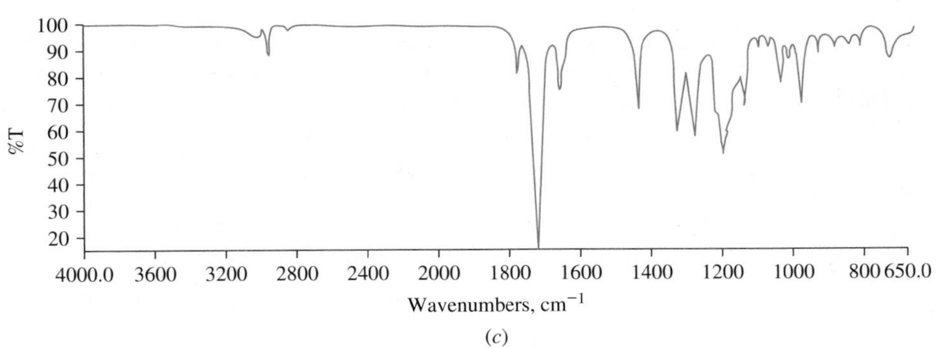

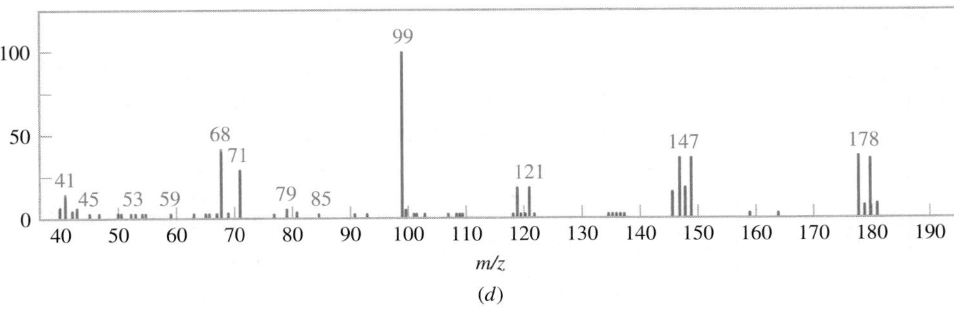

Figure 14.54

(*a*) ^{1}H NMR, (*b*) ^{13}C NMR, (*c*) IR, and (*d*) mass spectra (Problem 14.51).

14.52 ^{1}H NMR and IR spectra for a compound with the formula $C_7H_7NO_3$ are shown in Figure 14.55. Assign a structure and explain your reasoning.

Figure 14.55

(*a*) ^{1}H NMR and (*b*) IR spectra for Problem 14.52.

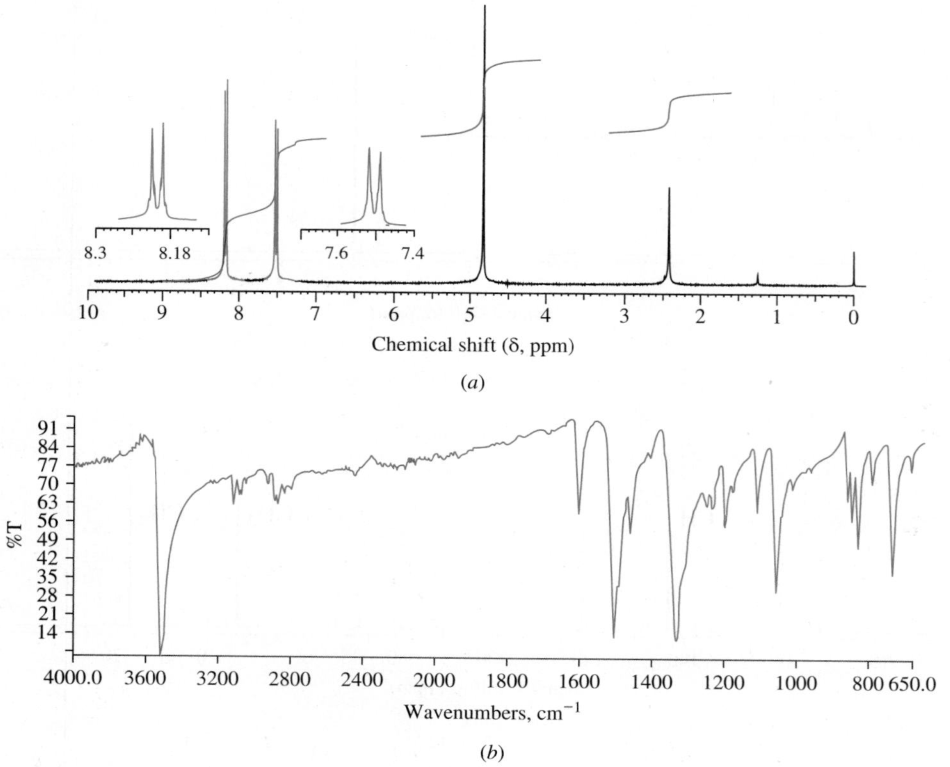

14.53 Friedel–Crafts alkylation of benzene with 1-chlorobutane gave a product for which the ^{1}H and ^{13}C NMR spectra are shown in Figure 14.56. The number of attached hydrogens from DEPT analysis are indicated on the ^{13}C NMR spectrum. Assign a structure to the product.

$$\text{benzene} + CH_3CH_2CH_2CH_2Cl \xrightarrow{AlCl_3} C_{10}H_{14}$$

Figure 14.56

^{1}H and ^{13}C NMR spectra for Problem 14.53.

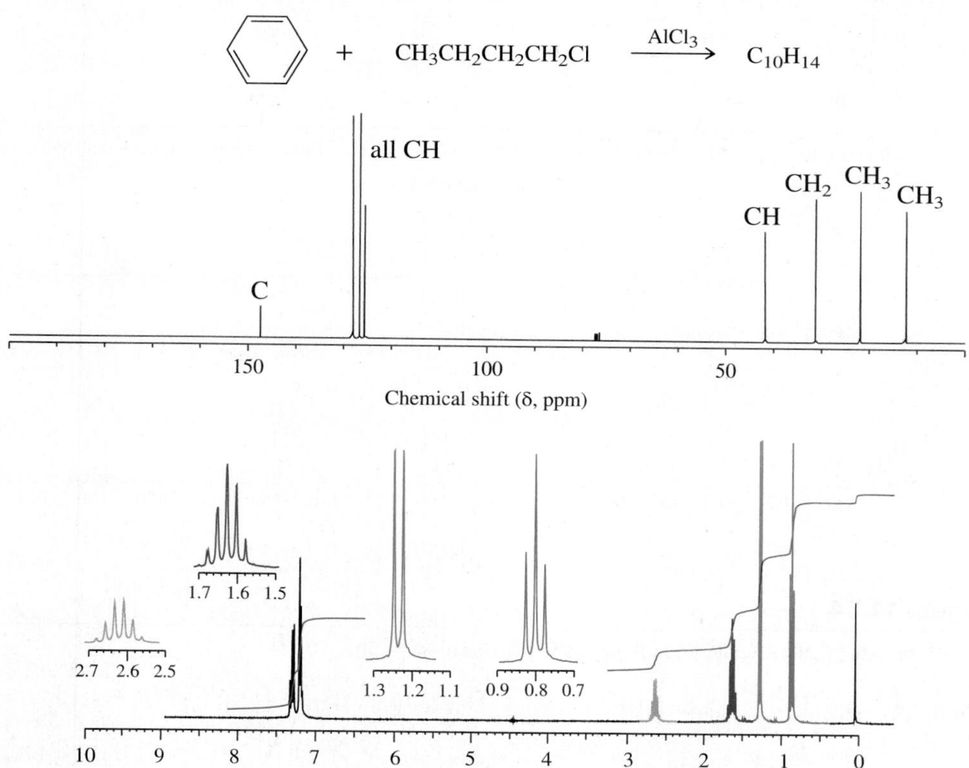

14.54 The following reaction produces an important chemical for polymer synthesis. Two equivalents of phenol and one equivalent of acetone react in the presence of acid to give $C_{15}H_{16}O_2$. ^{1}H and ^{13}C NMR spectra are shown in Figure 14.57. What is the structure of the product? (*Hints:* Two successive Freidel–Crafts alkylations occur in this process, the first involving protonated acetone as the electrophile. The peak at 9.2 ppm in the ^{1}H NMR spectrum disappears when the sample is treated with D_2O.)

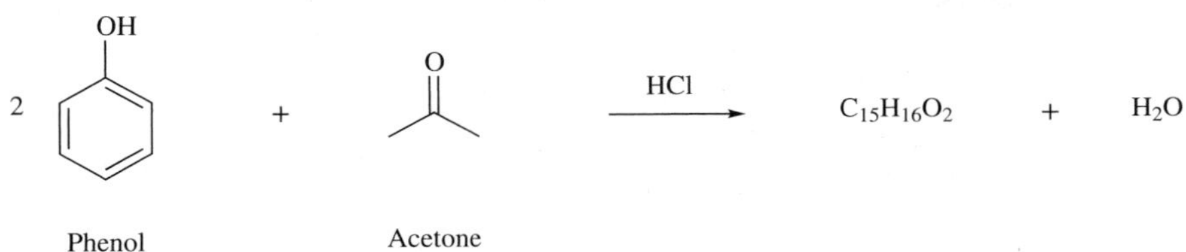

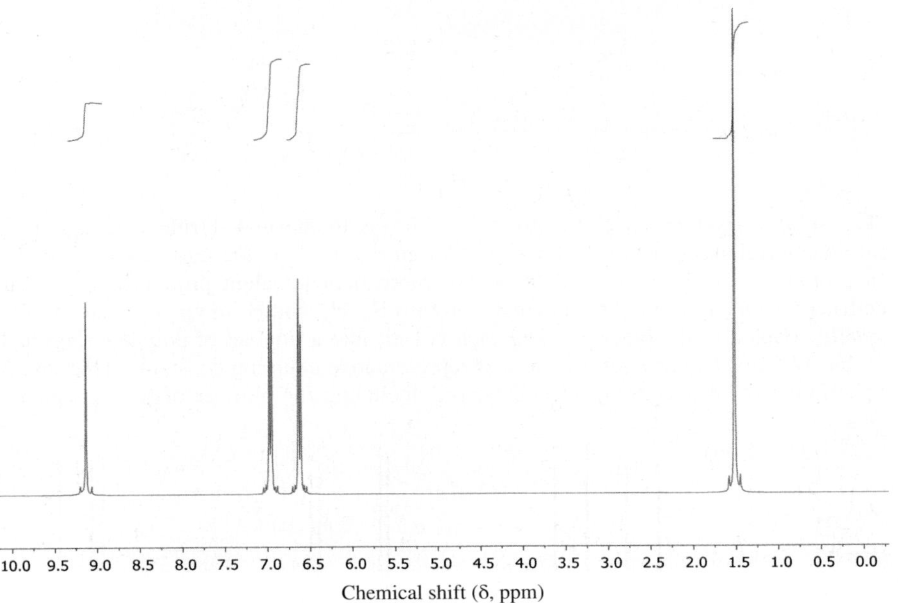

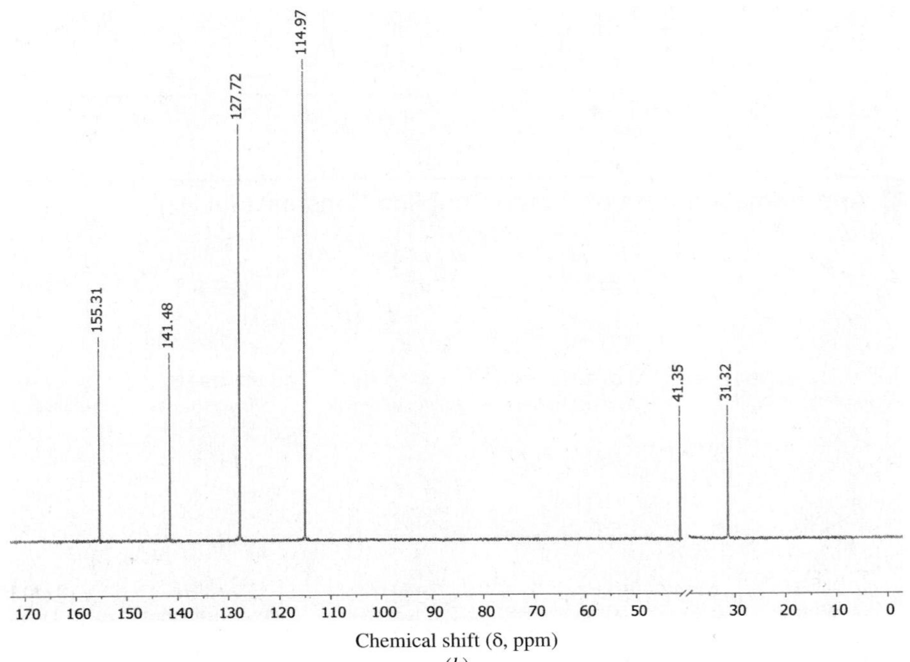

Figure 14.57

(*a*) ^{1}H and (*b*) ^{13}C NMR spectra for the compound in Problem 14.54.

Descriptive Passage and Interpretive Problems 14

More on Coupling Constants

As a result of the coupling of the nuclear spin of a proton with the spins of other protons, its ^{1}H NMR signal is often split into two or more smaller peaks. The chemical-shift difference in hertz between the individual peaks in the resulting multiplet can provide structural information and is governed by a coupling constant J, which in most cases can be determined directly from the spectrum. For example, the difference between any two adjacent lines in either the quartet or triplet in the ^{1}H NMR spectrum of ethyl bromide is 7.5 Hz and is cited as the vicinal, or three-bond (H—C—C—H), coupling constant (^3J).

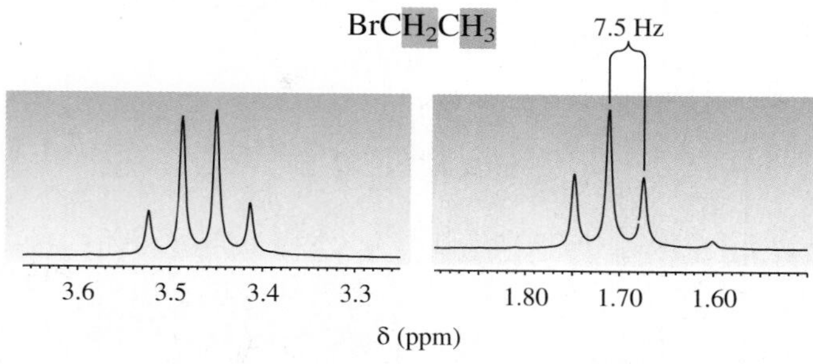

The splitting pattern for ethyl bromide conforms to the $n + 1$ rule, which states that n adjacent nonequivalent protons split the signal for an observed proton into $n + 1$ lines. However, when a proton is unequally coupled to two or more nonequivalent protons, the splittings are independent of each other. Each of the vinylic protons H_a, H_b, and H_c in vinyl acetate, for example, is unequally coupled to the other two and each is split into a doublet of doublets (Figure 14.58).

Table 14.5 gives ranges for a variety of representative coupling constants. Their exact value within the range is influenced by several factors, including the number of bonds separating the

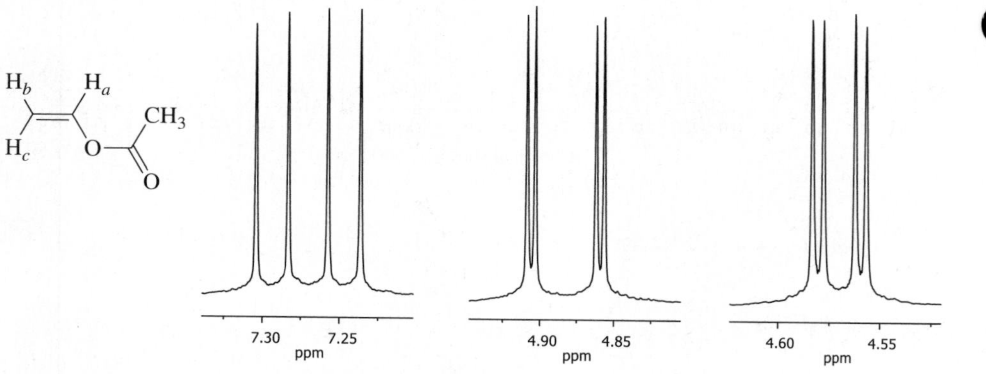

Figure 14.58

Each of the vinylic protons appears as a doublet of doublets in the 300-MHz ^{1}H NMR spectrum of vinyl acetate.

TABLE 14.5 Approximate Values of Proton Coupling Constants (in Hz)*

Type	J (Hz)
Vicinal in alkyl groups (H–C–C–H)	6–8 Hz
Geminal	9–15 Hz
Vinylic geminal	0–2 Hz
Allylic to vinylic	4–10 Hz
Vinylic trans	12–18 Hz
Vinylic cis	6–12 Hz
Ortho	6–10 Hz
Meta	1–3 Hz
Para	0–1 Hz
Diequatorial	2–3 Hz
Axial/equatorial	2–3 Hz
Diaxial	8–10 Hz

*Some ^{1}H coupling contants have negative J values, but this does not affect the appearance of the spectrum.

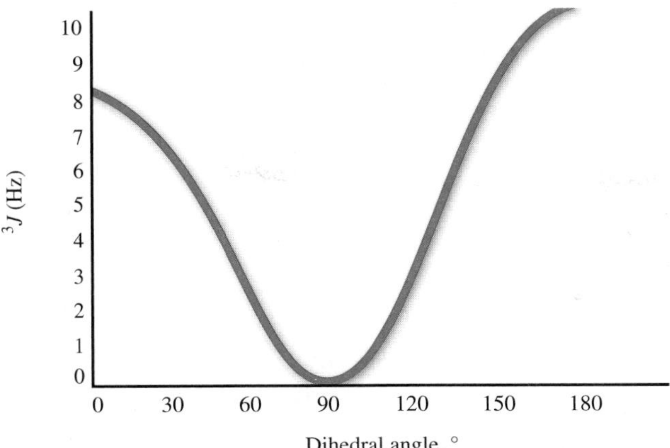

Figure 14.59

The Karplus relationship of vicinal ^{1}H coupling constant to dihedral angle of H—C—C—H.

spin-coupled protons, hybridization and electronegativity of attached atoms, bond and torsion angles, and the presence of π bonds.

As the table indicates, ^{1}H NMR can indicate whether two benzene protons are ortho, meta, or para to each other, whether two protons are cis, trans, or geminal on a double bond, or whether a vicinal pair on a cyclohexane ring is gauche or anti.

The relation of J to dihedral angle in a pair of vicinal protons was explored on a theoretical basis by Martin Karplus of Columbia University who calculated that ^3J is greatest when the H—C—C—H dihedral angle is 0° or 180° and smallest when the angle is 90° (Figure 14.59).

14.55 Refer to Figure 14.58 and Table 14.5 to assign chemical shifts for the vinylic protons H_a, H_b, and H_c in vinyl acetate.

(a) H_a: δ 4.57 H_b: δ 4.88 H_c: δ 7.28
(b) H_a: δ 4.88 H_b: δ 4.57 H_c: δ 7.28
(c) H_a: δ 7.28 H_b: δ 4.88 H_c: δ 4.57
(d) H_a: δ 7.28 H_b: δ 4.57 H_c: δ 4.88

14.56 Which one of the following statements *incorrectly* describes the expected coupling of the proton at C-1 in the stereoisomeric 4-*tert*-butylcyclohexanols?

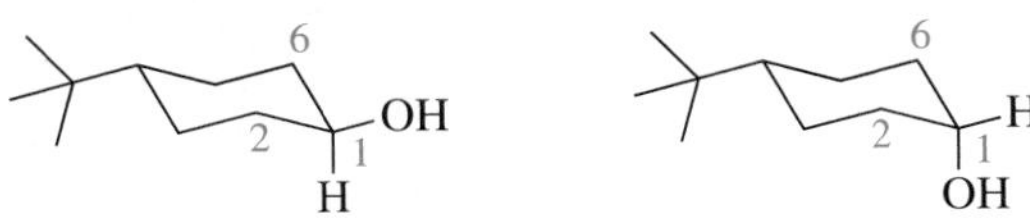

trans-4-*tert*-Butylcyclohexanol *cis*-4-*tert*-Butylcyclohexanol

In the trans isomer the coupling constant between:

(a) the proton at C-1 and the axial protons at C-2 and C-6 is 8 Hz.
(b) the proton at C-1 and the equatorial protons at C-2 and C-6 is 2 Hz.

In the cis isomer the coupling constant between:

(c) the proton at C-1 and the axial protons at C-2 and C-6 is 8 Hz.
(d) the proton at C-1 and the equatorial protons at C-2 and C-6 is 2 Hz.

14.57 Apiose is one of several naturally occurring carbohydrates characterized by a branched carbon chain and is conveniently isolated as the compound shown ("diacetone apiose"). Based on the observation that the protons at C-1 and C-2 have a coupling constant of 3.7 Hz, choose the correct statement.

Diacetone apiose

(a) The C-1 and C-2 protons are eclipsed.
(b) The H—C-1—C-2—H dihedral angle is in the range 30–60°.
(c) The H—C-1—C-2—H dihedral angle is in the range 145–165°.
(d) The C-1 and C-2 protons are anti.

14.58 The region of the ^{1}H NMR spectrum showing the signal for H_1 of a mixture of two isomers of glucose is shown in Figure 14.60. Which is the major isomer?
(a) Isomer A
(b) Isomer B

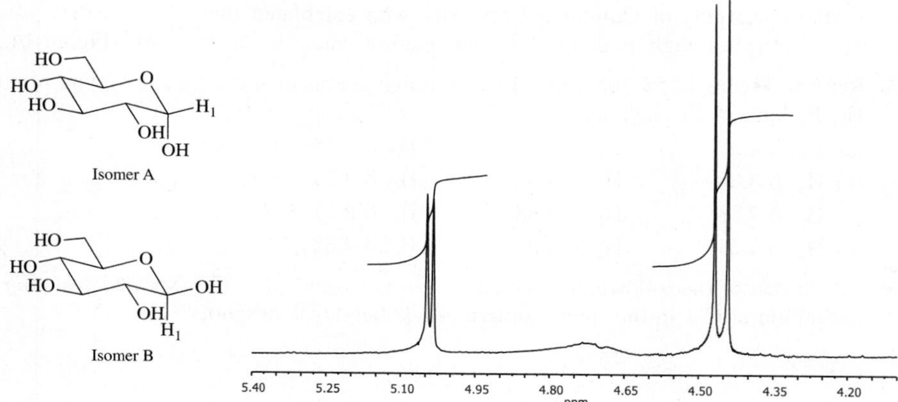

Figure 14.60

The portion of the 300-MHz spectrum where the signal for H_1 appears in a mixture of glucose stereoisomers A and B (Problem 14.58).

14.59 Figure 14.61 shows a portion of the ^1H NMR spectrum of 2,4-dibromoacetanilide. Which of the following corresponds to the chemical-shift assignments for the ring protons?

(a) H_a: δ 8.27 H_b: δ 7.68 H_c: δ 7.43
(b) H_a: δ 8.27 H_b: δ 7.43 H_c: δ 7.68
(c) H_a: δ 7.43 H_b: δ 8.27 H_c: δ 7.68
(d) H_a: δ 7.68 H_b: δ 7.43 H_c: δ 8.27

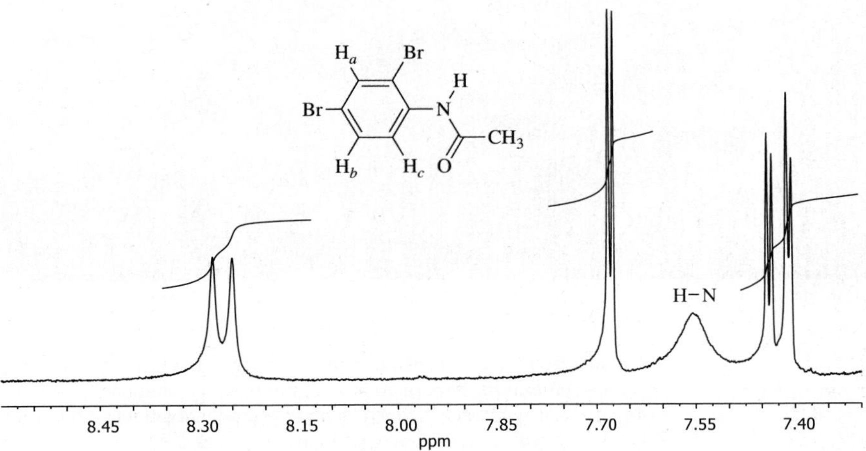

Figure 14.61

A portion of the 300-MHz ^1H NMR spectrum of 2,4-dibromoacetanilide (Problem 14.59).

Chapter

15

HO HO +NH3 O O−

CHAPTER OUTLINE

Parkinsonism results from a dopamine deficit in the brain that affects the "firing" of neurons. It responds to treatment with a chiral drug (L-dopa), one commercial synthesis of which involves the enantioselective organorhodium-catalyzed hydrogenation described in Section 15.13.

Organometallic Compounds

Organometallic compounds are *compounds that have a carbon–metal bond;* they occupy the place where organic and inorganic chemistry meet. You are already familiar with at least one organometallic compound, sodium acetylide ($NaC\equiv CH$), which has an ionic bond between carbon and sodium. But just because a compound contains both a metal and carbon isn't enough to classify it as organometallic. Like sodium acetylide, sodium methoxide ($NaOCH_3$) is an ionic compound. Unlike sodium acetylide, however, the negative charge in sodium methoxide resides on oxygen, not carbon.

$Na^+ \; :\bar{C}\equiv CH$	$Na^+ \; ^-\!:\ddot{O}CH_3$
Sodium acetylide (has a carbon-to-metal bond)	Sodium methoxide (does not have a carbon-to-metal bond)

The properties of organometallic compounds are much different from those of the other classes we have studied so far and differ among themselves according to the metal, its oxidation state, and the organic groups attached to the metal. Many organometallic compounds are sources of nucleophilic carbon, a quality that makes them especially valuable to the synthetic organic chemist who needs to make carbon–carbon bonds. For example, the preparation of alkynes by the reaction of sodium acetylide with alkyl halides (Section 9.6) depends on the presence of a negatively charged, nucleophilic carbon in acetylide ion. Conversely, certain other organometallic compounds behave as electrophiles.

A comprehensive treatment of organometallic chemistry would require a book of its own. In this chapter the preparation, properties, and usefulness of some of the most common organometallic reagents, those based on magnesium and lithium, are described in some detail. Other organometallic compounds derived from less familiar metals are introduced by highlighting some of their synthetic applications. We will also continue the story of alkene polymerization by introducing modern methods based on transition-metal catalysts.

15.1 Organometallic Nomenclature

Organometallic compounds of main-group metals are named as substituted derivatives of the metal. The metal is the parent, and the attached alkyl groups are identified by the appropriate prefix.

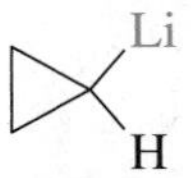

(cyclopropyl–Li, H)	$H_2C{=}CHNa$	$(CH_3CH_2)_2Mg$
Cyclopropyllithium	Vinylsodium	Diethylmagnesium

When the metal bears a substituent other than carbon, the substituent is treated as if it were an anion and named separately.

CH_3MgI	$(CH_3CH_2)_2AlCl$
Methylmagnesium iodide	Diethylaluminum chloride

Problem 15.1

Each of the following organometallic reagents will be encountered later in this chapter. Suggest a suitable name for each.

(a) 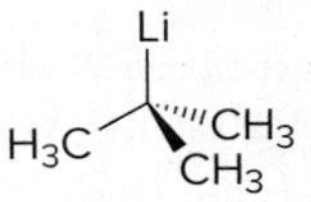 $(CH_3)_3CLi$ (Li attached to C bearing H_3C, CH_3, CH_3) (b) 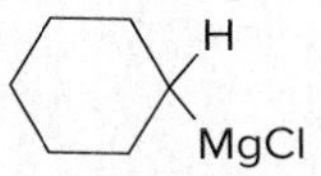 cyclohexyl with H and MgCl (c) ICH_2ZnI

Sample Solution (a) The metal lithium is considered the parent. The alkyl group to which it is bonded is *tert*-butyl, and so the name of this organometallic compound is *tert*-butyllithium. An alternative, equally correct name is 1,1-dimethylethyllithium.

An exception to this type of nomenclature is NaC≡CH, which is normally referred to as *sodium acetylide.* Both sodium acetylide and ethynylsodium are acceptable IUPAC names.

The second half of this chapter concentrates on organometallic complexes of transition metals. These complexes are normally named on the basis of the parent metal, with the attached groups (ligands) cited in alphabetical order preceding the metal. Their structural variety, however, requires a greater number of rules than is needed for our purposes and their nomenclature will be developed only to the degree necessary.

15.2 Carbon–Metal Bonds

With an electronegativity of 2.5 (Figure 15.1), carbon is neither strongly electropositive nor strongly electronegative. When carbon is bonded to an element more electronegative than itself, such as oxygen or chlorine, the electron distribution in the bond is polarized so that carbon is slightly positive and the more electronegative atom is slightly negative. Conversely, when carbon is bonded to a less electronegative element, such as a metal, the electrons in the bond are more strongly attracted toward carbon.

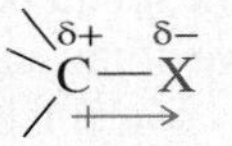

$\overset{\delta+}{C}-\overset{\delta-}{X}$	$\overset{\delta-}{C}-\overset{\delta+}{M}$
X is more electronegative than carbon	M is less electronegative than carbon

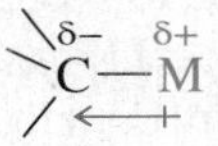

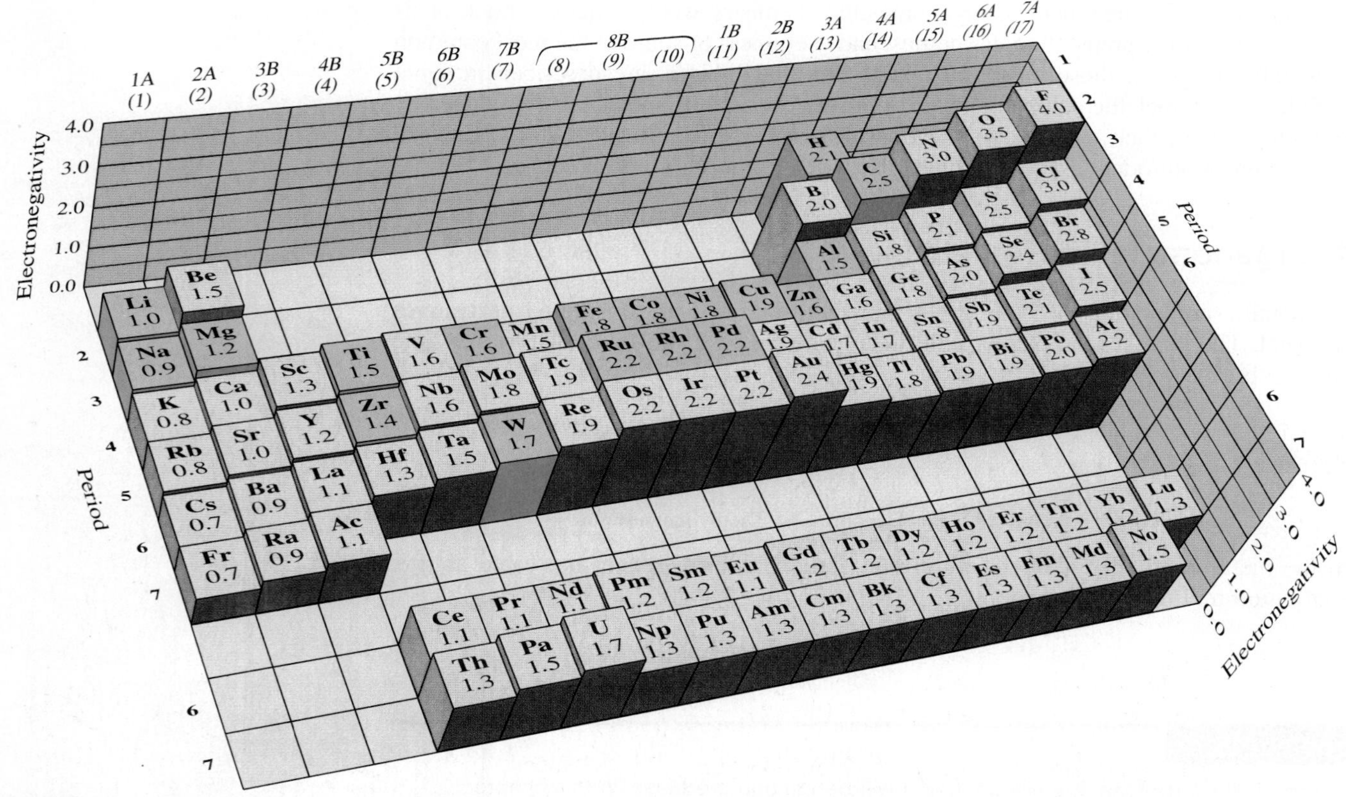

Figure 15.1

Electronegativities of the elements on the Pauling scale. The metals that appear in this chapter are shown in blue. Hydrogen and carbon are red.

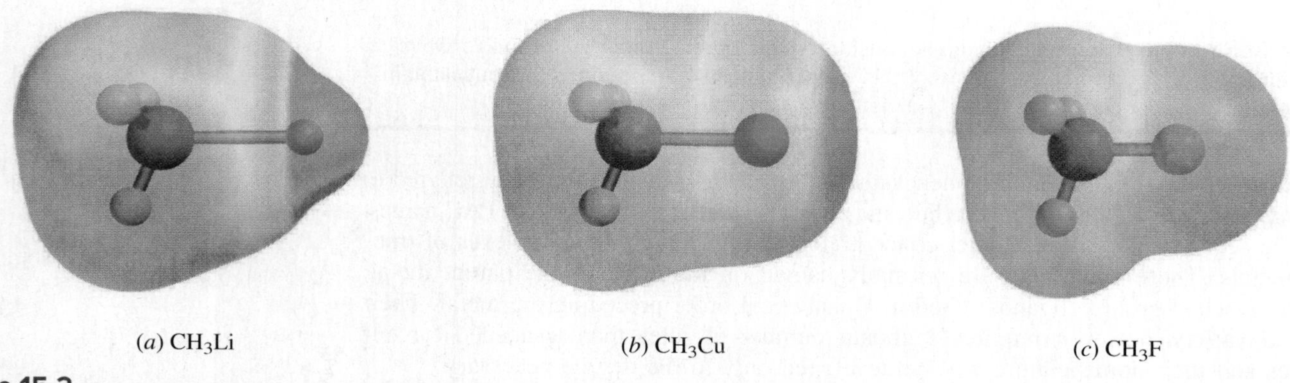

Figure 15.2

Electrostatic potential maps of (*a*) methyllithium, (*b*) methylcopper, and (*c*) methyl fluoride. The color range is the same in all three maps. The C—Li and C—F bonds are oppositely polarized. The C—Cu bond is the least polar.

This is especially true with Group 1A and 2A metals where the bonds range from ionic to polar covalent, depending on the metal and the nature of the organic group to which it is attached. Less electropositive metals such as copper have much less polarized covalent bonds to carbon (Figure 15.2).

Carbanions are the conjugate bases of hydrocarbons and were introduced in Section 9.5.

Many organometallic compounds have carbanionic character, and the ionic character of the carbon–metal bond becomes more pronounced as the metal becomes more electropositive. Organosodium and organopotassium compounds have ionic carbon–metal bonds; organolithium and organomagnesium compounds tend to have covalent, but rather polar, carbon–metal bonds with significant carbanionic character. *This carbanionic character makes these compounds especially useful as sources of nucleophilic carbon in organic synthesis.*

In general, carbon–metal bonds involving transition elements are not as polar as those of the elements in Groups 1A and 2A and exhibit less carbanionic character. The availability of *d* orbitals of transition elements, however, provides opportunities for novel and useful types of reactivity not available to main-group metals. The first part of this chapter deals with Group 1A and 2A organometallics, the second part with the transition-metal organometallics. Both emphasize synthetic applications.

15.3 Preparation of Organolithium and Organomagnesium Compounds

The most useful organometallic compounds of Groups 1A and 2A are those of lithium and magnesium. Organomagnesium compounds are called **Grignard reagents** after the French chemist Victor Grignard, who developed efficient methods for their preparation and demonstrated their value in synthesis. Grignard reagents and their lithium analogs are prepared by the reaction of the metal with an alkyl (primary, secondary, or tertiary), aryl, or vinyl halide, usually in diethyl ether as the solvent.

Grignard shared the 1912 Nobel Prize in Chemistry with Paul Sabatier, who showed that finely divided nickel is an effective hydrogenation catalyst.

$$\underset{\text{Organic halide}}{RX} + \underset{\text{Lithium}}{2Li} \longrightarrow \underset{\text{Organolithium}}{RLi} + \underset{\text{Lithium halide}}{LiX}$$

$$\underset{\textit{tert}\text{-Butyl chloride}}{(CH_3)_3CCl} + \underset{\text{Lithium}}{2Li} \xrightarrow[-30°C]{\text{diethyl ether}} \underset{\substack{\textit{tert}\text{-Butyllithium} \\ (75\%)}}{(CH_3)_3CLi} + \underset{\text{Lithium chloride}}{LiCl}$$

$$\underset{\text{Organic halide}}{RX} + \underset{\text{Magnesium}}{Mg} \longrightarrow \underset{\text{Organomagnesium halide}}{RMgX}$$

$$\underset{\text{Bromobenzene}}{C_6H_5Br} + \underset{\text{Magnesium}}{Mg} \xrightarrow[35°C]{\text{diethyl ether}} \underset{\substack{\text{Phenylmagnesium bromide} \\ (95\%)}}{C_6H_5MgBr}$$

The order of halide reactivity is I > Br > Cl > F, and alkyl halides are more reactive than aryl and vinyl halides. As shown in the preceding examples, the reactions are normally carried out in diethyl ether. When more vigorous reaction conditions are required, as in the preparation of vinylmagnesium chloride from vinyl chloride, the higher boiling solvent tetrahydrofuran is used.

$$\underset{\text{Vinyl chloride}}{CH_2{=}CHCl} \xrightarrow[\text{THF, 60°C}]{Mg} \underset{\substack{\text{Vinylmagnesium chloride} \\ (92\%)}}{CH_2{=}CHMgCl}$$

Recall the structure of tetrahydrofuran, abbreviated THF, from Section 3.15:

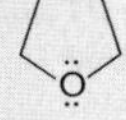

In all cases, *it is especially important that the solvent be anhydrous.* Even trace amounts of water or alcohols react with organolithium and organomagnesium reagents to form insoluble hydroxides or alkoxides that coat the surface of the metal and prevent it from reacting with the alkyl halide. Also, as we'll discuss in Section 15.4, organolithium and organomagnesium reagents are very strong bases and react instantly with water and alcohols to form hydrocarbons.

Problem 15.2

Write equations showing how you could prepare 2-methylpropylmagnesium bromide [$(CH_3)_2CHCH_2MgBr$] from 2-methylpropene and any necessary organic or inorganic reagents.

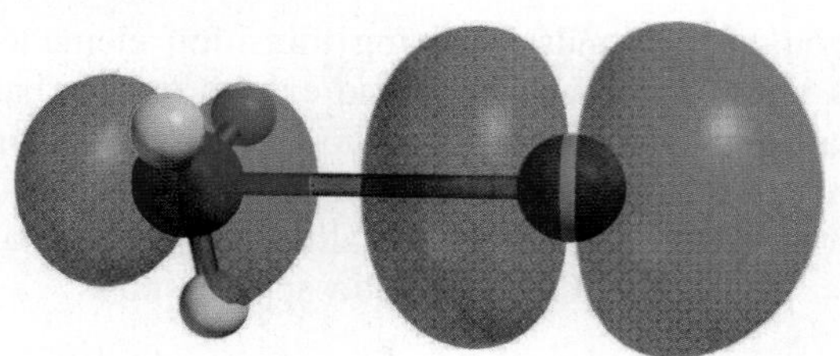

Figure 15.3

The highest occupied molecular orbital of the radical anion formed by donation of an electron to chloromethane is characterized by an antibonding interaction between carbon and chlorine.

The reaction of an organic halide with a metal is an oxidation–reduction in which the metal is the reducing agent. As shown in the following equations for the reactions of methyl chloride with lithium and with magnesium, a single-electron transfer from the metal converts methyl chloride to a *radical anion,* which then dissociates to a methyl radical and chloride ion. Bond formation between methyl radical and a metal species (·Li or Mg^+) follows.

$$\underset{\text{Methyl chloride}}{H_3C{-}\ddot{\underset{..}{Cl}}{:}} \xrightarrow{\cdot Li \;\; Li^+} \underset{\text{Radical anion}}{[H_3C{-}\ddot{\underset{..}{Cl}}{:}]^{\cdot -}} \xrightarrow{-{:}\ddot{\underset{..}{Cl}}{:}^-} \underset{\text{Methyl radical}}{H_3C\cdot} \xrightarrow{\cdot Li} \underset{\text{Methyllithium}}{H_3C{-}Li}$$

$$\underset{\text{Methyl chloride}}{H_3C{-}\ddot{\underset{..}{Cl}}{:}} \xrightarrow{Mg{:} \;\; \overset{+}{Mg}\cdot} \underset{\text{Radical anion}}{[H_3C{-}\ddot{\underset{..}{Cl}}{:}]^{\cdot -}} \xrightarrow{-{:}\ddot{\underset{..}{Cl}}{:}^-} \underset{\text{Methyl radical}}{H_3C\cdot} \xrightarrow[{:}\ddot{\underset{..}{Cl}}{:}^-]{\overset{+}{Mg}\cdot} \underset{\text{Methylmagnesium chloride}}{H_3C{-}Mg{-}\ddot{\underset{..}{Cl}}{:}}$$

We can understand the tendency for the radical anion $[H_3C{-}\dot{\underset{.}{Cl}}{:}]^{\cdot -}$ to dissociate to a methyl radical and chloride ion by referring to Figure 15.3, which illustrates the destabilizing antibonding interaction between carbon and chlorine in the highest occupied molecular orbital (HOMO) of the radical anion. This results in a much weaker bond and a C—Cl distance, calculated to be 324 pm (3.24 Å), that is longer than that of methyl chloride itself (181 pm, 1.81 Å).

The actual structures of organolithium and organomagnesium compounds are rarely monomeric as shown here; dimers are common, as well as higher aggregates, depending on the structure of the organometallic and the solvent.

Organolithium and organomagnesium compounds find their chief use in the preparation of alcohols by reaction with aldehydes and ketones. Before discussing these reactions, let us first examine their reactions with proton donors.

15.4 Organolithium and Organomagnesium Compounds as Brønsted Bases

Organolithium and organomagnesium compounds are stable species in aprotic solvents such as diethyl ether. They are strongly basic, however, and react instantly with proton donors even as weakly acidic as water and alcohols. A proton is transferred from the hydroxyl group to the negatively polarized carbon of the organometallic compound to form a hydrocarbon.

$$\underset{\text{stronger base}}{R{-}M} + \underset{\text{stronger acid}}{R'{-}\ddot{\underset{..}{O}}{-}H} \rightleftharpoons \underset{\text{weaker acid}}{R{-}H} + \underset{\text{weaker base}}{R'{-}\ddot{\underset{..}{O}}{:}^-\; M^+} \quad \textit{via} \quad \overset{\delta-}{R}(\overset{\delta+}{M}) \cdots \overset{\delta+}{H}{-}\ddot{\underset{..}{O}}R'$$

Because of their basicity organolithium compounds and Grignard reagents cannot be prepared or used in the presence of any material that bears an —OH group. Nor are they compatible with —NH or —SH groups, which can also convert an organolithium or organomagnesium compound to a hydrocarbon by proton transfer.

The carbon–metal bonds of organolithium and organomagnesium compounds have appreciable carbanionic character. Carbanions rank among the strongest bases that we'll see in this text. Their conjugate acids are hydrocarbons—very weak acids indeed, with pK_as in the 25–70 range.

$$\text{Carbanion (very strong base)} + H{-}\ddot{O}{:}{-}H \text{ (weak acid: p}K_a = 15.7) \longrightarrow \text{Hydrocarbon (weaker acid: p}K_a \approx 25\text{–}70) + {}^{-}{:}\ddot{O}{:}{-}H \text{ (Hydroxide ion (weaker base))}$$

Carbanion (very strong base) | Water (weak acid: pK_a = 15.7) | Hydrocarbon (weaker acid: pK_a ≈ 25–70) | Hydroxide ion (weaker base)

The basicity of Grignard and organolithium reagents has several applications. One is in the preparation of analogous organometallics of acetylene and terminal alkynes.

$$CH_3CH_2MgBr + HC{\equiv}CH \longrightarrow CH_3CH_3 + HC{\equiv}CMgBr$$

Ethylmagnesium bromide (stronger base) | Acetylene (stronger acid: pK_a = 26) | Ethane (weaker acid: pK_a = 62) | Ethynylmagnesium bromide (weaker base)

Another is in introducing deuterium at a specific position in a carbon chain.

$$CH_3CH{=}CHBr \xrightarrow[\text{THF}]{\text{Mg}} CH_3CH{=}CHMgBr \xrightarrow{D_2O} CH_3CH{=}CHD$$

1-Bromopropene | Propenylmagnesium bromide | 1-Deuteriopropene (70%)

A third is illustrated in the following problem.

Problem 15.3

Lithium diisopropylamide is often used as a strong base in organic synthesis and is prepared by an acid–base reaction between butyllithium and *N,N*-diisopropylamine. Complete the equation shown with appropriate structural formulas and use curved arrows to show the flow of electrons. What is the value of the equilibrium constant *K*?

Butyl anion + *N,N*-Diisopropylamine (pKa = 36) ⇌ Butane (pKa = 62) + *N,N*-Diisopropylamide ion

15.5 Synthesis of Alcohols Using Grignard and Organolithium Reagents

The main synthetic application of Grignard and organolithium reagents is their reaction with carbonyl-containing compounds to produce alcohols. Carbon–carbon bond formation is rapid and exothermic when Grignard and organolithium reagents react with an aldehyde or ketone.

$$\overset{\delta-}{:\ddot{O}}{=}\overset{\delta+}{C} + \overset{\delta-}{R}{-}\overset{\delta+}{MgX} \longrightarrow {}^{-}{:}\ddot{O}{-}C{-}R \; {}^{+}MgX \quad \text{usually written as} \quad R{-}C{-}OMgX$$

Aldehyde or ketone | Grignard reagent | Alkoxymagnesium halide

and

$$\overset{\delta-}{:\ddot{O}}{=}\overset{\delta+}{C} + \overset{\delta-}{R}{-}\overset{\delta+}{Li} \longrightarrow {}^{-}{:}\ddot{O}{-}C{-}R \; {}^{+}Li \quad \text{usually written as} \quad R{-}C{-}OLi$$

Aldehyde or ketone | Organolithium reagent | Lithium alkoxide

A carbonyl group is quite polar, and its carbon atom is electrophilic. Grignard and organolithium reagents are nucleophilic and add to carbonyl groups, forming a new carbon–carbon bond. This addition step leads to an alkoxymagnesium halide or a lithium alkoxide, which in the second stage of the synthesis is converted to the desired alcohol by adding aqueous acid.

$$R_3C{-}\ddot{O}{:}^{-} + H_3O^{+} \longrightarrow R_3C{-}\ddot{O}H + H_2\ddot{O}{:}$$

Lithium alkoxide Hydronium ion Alcohol Water

The type of alcohol produced depends on the carbonyl compound. Substituents present on the carbonyl group of an aldehyde or ketone stay there—they become substituents on the carbon that bears the hydroxyl group in the product. Thus, as shown in Table 15.1, formaldehyde reacts with Grignard reagents to yield primary alcohols, aldehydes yield secondary alcohols, and ketones yield tertiary alcohols. Analogous reactions take place with organolithium reagents.

Problem 15.4

Write the structure of the organic product of each of the following reactions.

(a) Benzaldehyde + vinyllithium (Li) $\xrightarrow[2.\ H_3O^+]{1.\ \text{diethyl ether}}$

(b) 2-Butanone + propylmagnesium bromide (MgBr) $\xrightarrow[2.\ H_3O^+]{1.\ \text{diethyl ether}}$

(c) $H_2C{=}O$ + ClMg–(3-methoxyphenyl, OCH_3) $\xrightarrow[2.\ H_3O^+]{1.\ \text{diethyl ether}}$

Sample Solution

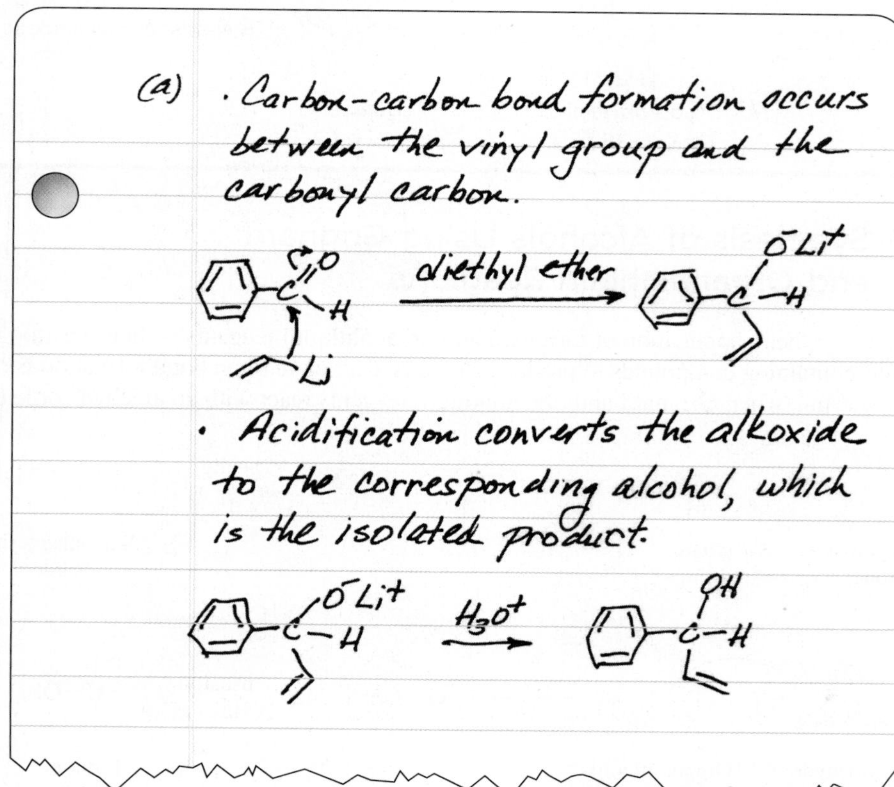

TABLE 15.1 Reactions of Grignard Reagents with Aldehydes and Ketones

Reaction and comments	General Equation and Specific Example
Reaction with formaldehyde Grignard reagents react with formaldehyde to give *primary alcohols* having one more carbon than the Grignard reagent.	RMgX (Grignard reagent) + $HCHO$ (Formaldehyde) —diethyl ether→ Primary alkoxymagnesium halide (R–CH_2–OMgX) —H_3O^+→ Primary alcohol (R–CH_2–OH) Cyclohexylmagnesium chloride + Formaldehyde —1. diethyl ether; 2. H_3O^+→ Cyclohexylmethanol (64–69%)
Reaction with aldehydes Grignard reagents react with aldehydes to give *secondary alcohols*.	RMgX (Grignard reagent) + R′CHO (Aldehyde) —diethyl ether→ Secondary alkoxymagnesium halide (RR′CH–OMgX) —H_3O^+→ Secondary alcohol (RR′CH–OH) Hexylmagnesium bromide + H_3CCHO, Ethanal (acetaldehyde) —1. diethyl ether; 2. H_3O^+→ 2-Octanol (84%)
Reaction with ketones Grignard reagents react with ketones to give *tertiary alcohols*.	RMgX (Grignard reagent) + R′COR″ (Ketone) —diethyl ether→ Tertiary alkoxymagnesium halide (RR′R″C–OMgX) —H_3O^+→ Tertiary alcohol (RR′R″C–OH) CH_3MgCl, Methylmagnesium chloride + Cyclopentanone —1. diethyl ether; 2. H_3O^+→ 1-Methylcyclopentanol (62%)

An ability to form carbon–carbon bonds is fundamental to organic synthesis. The addition of Grignard and organolithium reagents to aldehydes and ketones is one of the most frequently used reactions in synthetic organic chemistry. Not only does it permit the extension of carbon chains, but because the product is an alcohol, a wide variety of subsequent functional-group transformations is possible.

15.6 Synthesis of Acetylenic Alcohols

The first organometallic compounds we encountered were compounds of the type RC≡CNa obtained by treatment of terminal alkynes with sodium amide in liquid ammonia (Section 9.5):

$$\underset{\text{Terminal alkyne}}{RC{\equiv}CH} + \underset{\text{Sodium amide}}{NaNH_2} \xrightarrow[-33°C]{NH_3} \underset{\text{Sodium alkynide}}{RC{\equiv}CNa} + \underset{\text{Ammonia}}{NH_3}$$

These compounds are sources of the nucleophilic anion RC≡C:⁻, and their reaction with primary alkyl halides provides an effective synthesis of alkynes (Section 9.6). The nucleophilicity of acetylide anions is also evident in their reactions with aldehydes and ketones, which are entirely analogous to those of Grignard and organolithium reagents.

These reactions are normally carried out in liquid ammonia because that is the solvent in which the sodium salt of the alkyne is prepared.

$$\underset{\text{Sodium alkynide}}{RC{\equiv}CNa} + \underset{\text{Aldehyde or ketone}}{R'\overset{O}{\overset{\|}{C}}R''} \xrightarrow{NH_3} \underset{\text{Sodium salt of an alkynyl alcohol}}{RC{\equiv}C{-}\underset{R'}{\overset{R''}{C}}{-}ONa} \xrightarrow{H_3O^+} \underset{\text{Alkynyl alcohol}}{RC{\equiv}C\underset{R'}{\overset{R''}{C}}OH}$$

$$\underset{\text{Sodium acetylide}}{HC{\equiv}CNa} + \underset{\text{Cyclohexanone}}{\text{cyclohexanone}} \xrightarrow[2.\ H_3O^+]{1.\ NH_3} \underset{\text{1-Ethynylcyclohexanol (65–75\%)}}{\text{1-(HO)-1-(C{\equiv}CH)cyclohexane}}$$

As noted in Section 15.4, acetylenic Grignard reagents of the type RC≡CMgBr are prepared, not from an acetylenic halide, but by an acid–base reaction in which a Grignard reagent abstracts a proton from a terminal alkyne. Once formed, they react with aldehydes and ketones in the usual way.

$$\underset{\text{1-Hexynylmagnesium bromide}}{CH_3(CH_2)_3C{\equiv}CMgBr} + \underset{\text{Formaldehyde}}{H\overset{O}{\overset{\|}{C}}H} \xrightarrow[2.\ H_3O^+]{1.\ \text{diethyl ether}} \underset{\text{2-Heptyn-1-ol (82\%)}}{CH_3(CH_2)_3C{\equiv}CCH_2OH}$$

The corresponding acetylenic organolithium reagents are prepared by the reaction of terminal alkynes with methyllithium or butyllithium.

15.7 Retrosynthetic Analysis and Grignard and Organolithium Reagents

Constructing the desired carbon skeleton is a primary concern in synthetic organic chemistry, and we have already seen a number of methods for making carbon–carbon bonds in earlier chapters. The present section illustrates how to use retrosynthetic analysis to identify situations in which the reaction of a Grignard or organolithium reagent with an aldehyde or ketone can be used to advantage in building a carbon chain.

As illustrated earlier in Table 15.1, the main use of Grignard and organolithium reagents lies in the synthesis of alcohols. In such cases, the retrosynthesis focuses on the carbon that bears the hydroxyl group and begins by disconnecting one of its organic substituents as the corresponding anion.

$$\underset{R}{\overset{a\quad b}{C}}{-}OH \ (\text{disconnect C–R}) \Longrightarrow R{:}^- \quad \overset{b}{\underset{a}{C}}{=}O$$

This disconnection identifies the carbonyl component in the potential C—C bond-forming step, and R represents the group that adds to the carbonyl as a carbanion. For retrosynthetic purposes, Grignard and organolithium reagents (RMgX and RLi) are regarded as *synthetically equivalent* to a carbanion ($R:^-$). If the target is a primary alcohol (RCH_2OH), the carbonyl component is formaldehyde. If the target is a secondary alcohol, there are two retrosyntheses that differ in which R group is the carbanion equivalent and which remains attached to the carbonyl.

$R'{:}^-$ + RCH=O ⟸ R′(H)CH(OH)R ⟹ $R{:}^-$ + R′CH=O

Three disconnections are possible for tertiary alcohols.

There is often little advantage in choosing one route over another when preparing a particular target alcohol. For example, all three of the following combinations have been used to prepare the tertiary alcohol 2-phenyl-2-butanol, and each is effective.

CH_3MgI + 1-Phenyl-1-propanone $\xrightarrow[\text{2. } H_3O^+]{\text{1. diethyl ether}}$ 2-Phenyl-2-butanol

Methylmagnesium iodide — 1-Phenyl-1-propanone — 2-Phenyl-2-butanol

CH_3CH_2MgI + Acetophenone $\xrightarrow[\text{2. } H_3O^+]{\text{1. diethyl ether}}$ 2-Phenyl-2-butanol

Ethylmagnesium iodide — Acetophenone — 2-Phenyl-2-butanol

C_6H_5—MgBr + 2-Butanone $\xrightarrow[\text{2. } H_3O^+]{\text{1. diethyl ether}}$ 2-Phenyl-2-butanol

Phenylmagnesium bromide — 2-Butanone — 2-Phenyl-2-butanol

Problem 15.5

Use retrosynthetic analysis to develop a synthesis of 2-phenyl-2-butanol from benzene and 2-butyne as the source of all of the carbons and write a series of equations for the synthesis showing all necessary reagents.

All that has been said in this section applies with equal force to organolithium reagents. Grignard reagents are one source of nucleophilic carbon; organolithium reagents are another. Both have substantial carbanionic character in their carbon–metal bonds and undergo the same kinds of reactions with aldehydes and ketones.

15.8 An Organozinc Reagent for Cyclopropane Synthesis

Zinc reacts with alkyl halides in a manner similar to that of magnesium.

$$\underset{\text{Alkyl halide}}{RX} + \underset{\text{Zinc}}{Zn} \xrightarrow{\text{diethyl ether}} \underset{\text{Alkylzinc halide}}{RZnX}$$

Victor Grignard was led to study organomagnesium compounds because of earlier work he performed with organic derivatives of zinc.

Iodomethylzinc iodide is known as the *Simmons–Smith reagent,* after Howard E. Simmons and Ronald D. Smith of DuPont, who first described its use in the preparation of cyclopropanes.

Zinc is less electropositive than lithium and magnesium, and the carbon–zinc bond is less polar. Organozinc reagents are not nearly as reactive toward aldehydes and ketones as Grignard reagents and organolithium compounds.

An organozinc compound that occupies a special niche in organic synthesis is *iodomethylzinc iodide* (ICH_2ZnI). It is prepared by the reaction of zinc–copper couple [Zn(Cu)], zinc that has had its surface activated with a little copper, with diiodomethane in diethyl ether.

$$\underset{\text{Diiodomethane}}{CH_2I_2} + \underset{\text{Zinc}}{Zn} \xrightarrow[\text{Cu}]{\text{diethyl ether}} \underset{\text{Iodomethylzinc iodide}}{ICH_2ZnI}$$

Iodomethylzinc iodide is a useful reagent because it reacts with alkenes to give cyclopropanes. This reaction is called the **Simmons–Smith reaction** and is one of the few methods available for the synthesis of cyclopropanes. The reaction is *stereospecific.* Substituents that were cis in the alkene remain cis in the cyclopropane.

(*Z*)-3-Hexene $\xrightarrow[\text{Zn(Cu)}\\ \text{diethyl ether}]{CH_2I_2}$ *cis*-1,2-Diethylcyclopropane (34%)

(*E*)-3-Hexene $\xrightarrow[\text{Zn(Cu)}\\ \text{diethyl ether}]{CH_2I_2}$ *trans*-1,2-Diethylcyclopropane (15%)

Cyclopropanation of alkenes with the Simmons–Smith reagent bears some similarity to epoxidation. Both reactions are stereospecific cycloadditions, and iodomethylzinc iodide behaves, like peroxy acids, as a weak electrophile. Both cycloadditions take place faster with more highly substituted double bonds than less substituted ones, but are sensitive to steric hindrance in the alkene. These similarities are reflected in the mechanisms proposed for the two reactions shown in Mechanism 15.1. Both are believed to be concerted.

Mechanism 15.1

Similarities Between the Mechanisms of Reaction of an Alkene with Iodomethylzinc Iodide and a Peroxy Acid

Cyclopropanation

Alkene + Iodomethylzinc iodide → Transition state → A cyclopropane + Zinc iodide

Epoxidation

Alkene + Peroxy acid → Transition state → Epoxide + Carboxylic acid

Problem 15.6

Reaction of (*E*)-3-penten-2-ol with the Simmons–Smith reagent gave a mixture of two isomers having the molecular formula $C_6H_{12}O$. Suggest reasonable structures for these products.

A modified Simmons–Smith reaction has been used in the stereoselective synthesis of a naturally occurring substance called U-106305 containing six cyclopropane rings. In the synthesis, four of the six rings arise by Simmons–Smith-type cyclopropanation. The red lines in the structural formula identify the bonds to the CH_2 groups that are introduced in this way; the blue lines identify bonds that originated with the initial reactant.

HO ... OH $\xrightarrow[\text{5\% overall yield}]{\text{14 steps}}$ U-106305

Knowing the configuration of the starting diol to be (1*R*,2*R*) allowed the absolute configuration of all the chirality centers in the product to be established.

15.9 Carbenes and Carbenoids

Iodomethylzinc iodide is often referred to as a **carbenoid,** meaning that it resembles a carbene in its chemical reactions. Carbenes are neutral molecules in which one of the carbon atoms has six valence electrons. Such carbons are *divalent;* they are directly bonded to only two other atoms and have no multiple bonds. Iodomethylzinc iodide reacts as if it were a source of the carbene H—C̈—H.

Divalent carbon species first received attention with the work of the Swiss-American chemist J. U. Nef in the late nineteenth century; they were then largely ignored until the 1950s.

It is clear that free $:CH_2$ is not involved in the Simmons–Smith reaction, but there is substantial evidence to indicate that carbenes are formed as intermediates in certain other reactions that convert alkenes to cyclopropanes. The most studied examples of these reactions involve dichlorocarbene and dibromocarbene.

Dichlorocarbene Dibromocarbene

Carbenes are too reactive to be isolated and stored, but have been trapped in frozen argon for spectroscopic study at very low temperatures. Bonding in dihalocarbenes is based on sp^2 hybridization of carbon, shown for CCl_2 in Figure 15.4*a*. Two of carbon's sp^2 hybrid orbitals are involved in σ bonds to the halogen. The third sp^2 orbital contains the unshared

(*a*)

(*b*)

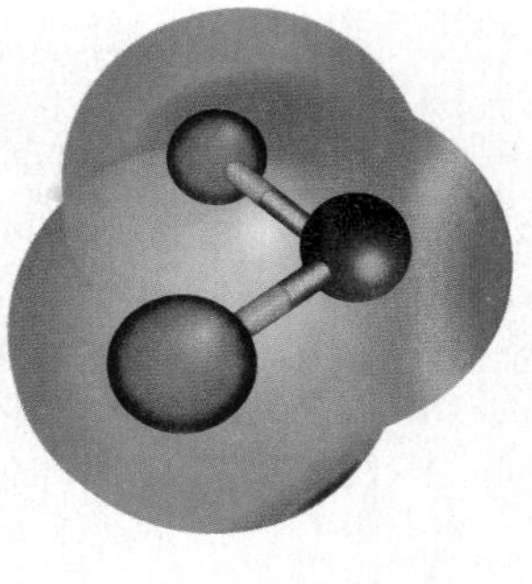
(*c*)

Figure 15.4

(*a*) The unshared electron pair occupies an sp^2-hybridized orbital in dichlorocarbene. (*b*) There are no electrons in the unhybridized *p* orbital. (*c*) An electrostatic potential map of dichlorocarbene shows negative charge is concentrated in the region of the unshared pair, whereas positive charge is concentrated above and below the carbon.

electron pair, and the unhybridized $2p$ orbital is vacant. The electrostatic potential map in Figure 15.4*b* illustrates this nicely with the highest negative character (red) concentrated in the region of the lone pair orbital, and the region of highest positive charge (blue) situated above and below the plane of the carbene.

15.10 Transition-Metal Organometallic Compounds

The transition elements occupy groups 3–11 of the periodic table and are characterized electronically as having a partially filled *d* subshell. A related but not equivalent term is ***d*-block element,** which includes all of these plus the elements in group 12. The first sequence of *d*-block elements begins in the fourth period with scandium, which has one 3*d* electron, and continues through zinc, which has ten. The second (4*d*) and third (5*d*) transition series are analogous.

3	4	5	6	7	8	9	10	11	12
Sc	Ti	V	Cr	Mn	Fe	Co	Ni	Cu	Zn

The group number (3–12) corresponds to the number of valence electrons, and the number of *d* electrons is two less than the group number. Manganese, for example, has seven valence electrons and five 3*d* electrons and is referred to as a d^5 element. Mn^{2+} is a d^3 ion, as is Fe^{3+}.

Problem 15.7

How many 4*d* electrons are there in Pd^{2+}? What +2 ion in the fourth period has the same number of 3*d* electrons? In the sixth period?

An already large and steadily increasing number of organometallic compounds of transition elements is known, and their applications in both laboratory and industrial processes have transformed synthetic organic chemistry. They not only make it possible to carry out reactions that are difficult to do otherwise but in many cases do so as catalysts rather than reactants. Before we look at the reactions of transition-metal organometallics in this and subsequent sections, some structural background is in order.

We'll start with nickel carbonyl, an intermediate in the purification of nickel first prepared over a hundred years ago. It is a neutral molecule (boiling point: 43°C) that forms spontaneously when carbon monoxide is passed over elemental nickel.

$$\underset{\text{Nickel}}{Ni} + \underset{\text{Carbon monoxide}}{4CO} \longrightarrow \underset{\text{Nickel carbonyl}}{Ni(CO)_4}$$

A structural unit attached to the metal is called a **ligand,** and can be an element, a compound (carbon monoxide in this case), or an ion; also, it can be organic or inorganic. Ligands are electron-pair donors (Lewis bases) and include the following plus many others.

$:H^-$	$:R^-$	$:\ddot{Cl}:^-$	$:NH_3$	$:P(C_6H_5)_3$	$:\bar{C}\equiv\overset{+}{O}:$	$:\bar{C}\equiv N:$
Hydride	Alkyl	Chloride	Ammonia	Triphenylphosphine	Carbon monoxide	Cyanide

Many transition-metal complexes obey the **18-electron rule,** which is to transition elements as the octet rule is to main-group elements. It states:

> *The number of ligands that can be attached to a transition metal are such that the sum of the electrons in the bonds to the metal plus the metal's valence electrons equals 18.*

Both the octet rule and the 18-electron rule connect stability to noble gas electron configurations. The noble gases in the second and third periods of the periodic table (Ne and Ar) have 8 valence electrons, those in the fourth and fifth (Kr and Xe) have 18.

Because the custom for writing structural formulas of transition-metal complexes omits electrons and formal charges, we need to first know the charge on the complex itself in order to apply the 18-electron rule. Take $Ni(CO)_4$, which is a neutral molecule, as an example. We count the valence electrons of nickel as given by its group number (10) and add to that the total number of electrons in Ni—C bonds (8) for a total of 18. The 18-electron rule is satisfied for $Ni(CO)_4$.

Problem 15.8

Like nickel, iron reacts with carbon monoxide to form a compound having the formula $M(CO)_n$ that obeys the 18-electron rule. What is the value of n in the formula $Fe(CO)_n$?

Transition-metal complexes that have an electron count less than 18 are described as *coordinatively unsaturated* and can take on additional ligands.

Not all ligands use just two electrons to bond to transition metals. Benzene uses its six π electrons in organometallic compounds such as (benzene)tricarbonylchromium.

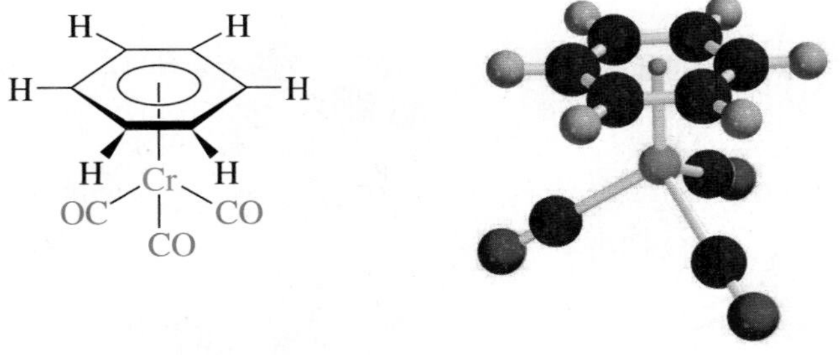

(Benzene)tricarbonylchromium

The complex is neutral, and chromium is a group 6 transition metal, so contributes six valence electrons. The three CO ligands contribute six more, and the six π electrons of benzene bring the total to 18.

Problem 15.9

What is the electron count of manganese in the cation shown? Is it coordinatively unsaturated?

$[(C_6H_6)Mn(CO)_3]^+$

OC Mn CO
CO

Like benzene, cyclopentadienyl anion is an aromatic six π-electron system (Section 12.20) and bonds to transition metals in a similar way. Ferrocene is the best known example, with a structure aptly described as a *sandwich.*

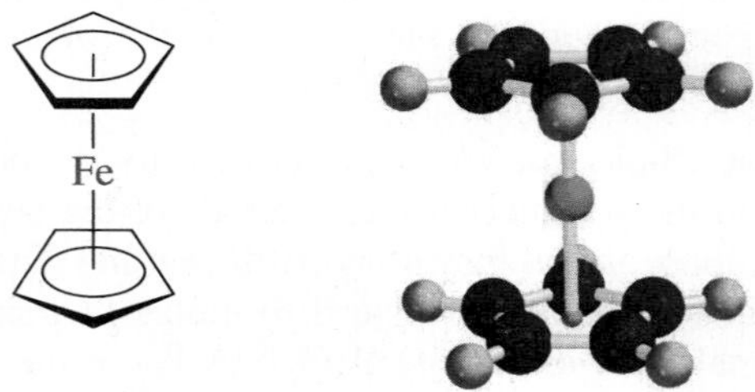

Ferrocene

Ferrocene is neutral, and each cyclopentadienyl ligand brings a charge of −1; therefore, the oxidation state of iron is +2. Iron is a group 8 transition metal, but being in the +2 oxidation state contributes six, rather than eight, electrons. These 6, plus 12 from the two cyclopentadienide ligands, bring the total to 18.

Problem 15.10

What is the oxidation state of manganese in the complex given in Problem 15.9?

Metallocenes, organometallic compounds that bear cyclopentadienyl ligands, are not only structurally interesting but many of them have useful applications as catalysts for industrial processes. Zirconium-based metallocenes, for example, are the most widely used catalysts for Ziegler–Natta polymerization of alkenes. We'll have more to say about them in Section 15.15 and Chapter 27.

Zirconocene dichloride

Problem 15.11

Zirconocene dichloride is neutral. What is the oxidation state and electron count of zirconium? Is zirconocene dichloride coordinatively unsaturated?

Naturally occurring compounds with carbon–metal bonds are very rare, coenzyme B_{12} being the best known example (see the boxed essay *An Organometallic Compound That Occurs Naturally: Coenzyme* B_{12} that accompanies this section).

In Sections 15.11 and 15.12, we'll describe some synthetic applications of organocopper and organopalladium compounds used to form C—C bonds between organic groups. The two types differ in that the organocopper compounds to be described are reagents present in stoichiometric amounts, whereas the organopalladium compounds are catalysts.

An Organometallic Compound That Occurs Naturally: Coenzyme B_{12}

Pernicious anemia is a disease characterized, as are all anemias, by a deficiency of red blood cells. Unlike ordinary anemia, pernicious anemia does not respond to treatment with sources of iron, and before effective treatments were developed, was often fatal. Injection of liver extracts was one such treatment, and in 1948 chemists succeeded in isolating the "anti-pernicious anemia factor" from beef liver as a red crystalline compound, which they called *vitamin B_{12}*. This compound had the formula $C_{63}H_{88}CoN_{14}O_{14}P$. Its complexity precluded structure determination by classical degradation techniques, and spectroscopic methods were too primitive to be of much help. The structure was solved by Dorothy Crowfoot Hodgkin of Oxford University in 1955 using X-ray diffraction techniques and is shown in Figure 15.5*a*. Structure determination by X-ray crystallography can be superficially considered as taking a photograph of a molecule with X-rays. It is a demanding task and earned Hodgkin the 1964 Nobel Prize in Chemistry. Modern structural studies by X-ray crystallography use computers to collect and analyze the diffraction data and take only a fraction of the time required years ago to solve the vitamin B_{12} structure.

The structure of vitamin B_{12} is interesting in that it contains a central cobalt atom that is surrounded by six atoms in an octahedral geometry. One substituent, the cyano (—CN) group, is what is known as an "artifact." It appears to be introduced into the molecule during the isolation process and leads to the synonym *cyanocobalamin* for vitamin B_{12}. This is the material used to treat pernicious anemia, but is not the form in which it exerts its activity. The biologically active substance is called *coenzyme B_{12}* and differs from vitamin B_{12} in the ligand attached to cobalt (Figure 15.5*b*). Coenzyme B_{12} was the first naturally occurring substance discovered to have a carbon–metal bond. In fact, coenzyme B_{12} was discovered before any compound containing an alkyl group σ-bonded to cobalt had ever been isolated in the laboratory!

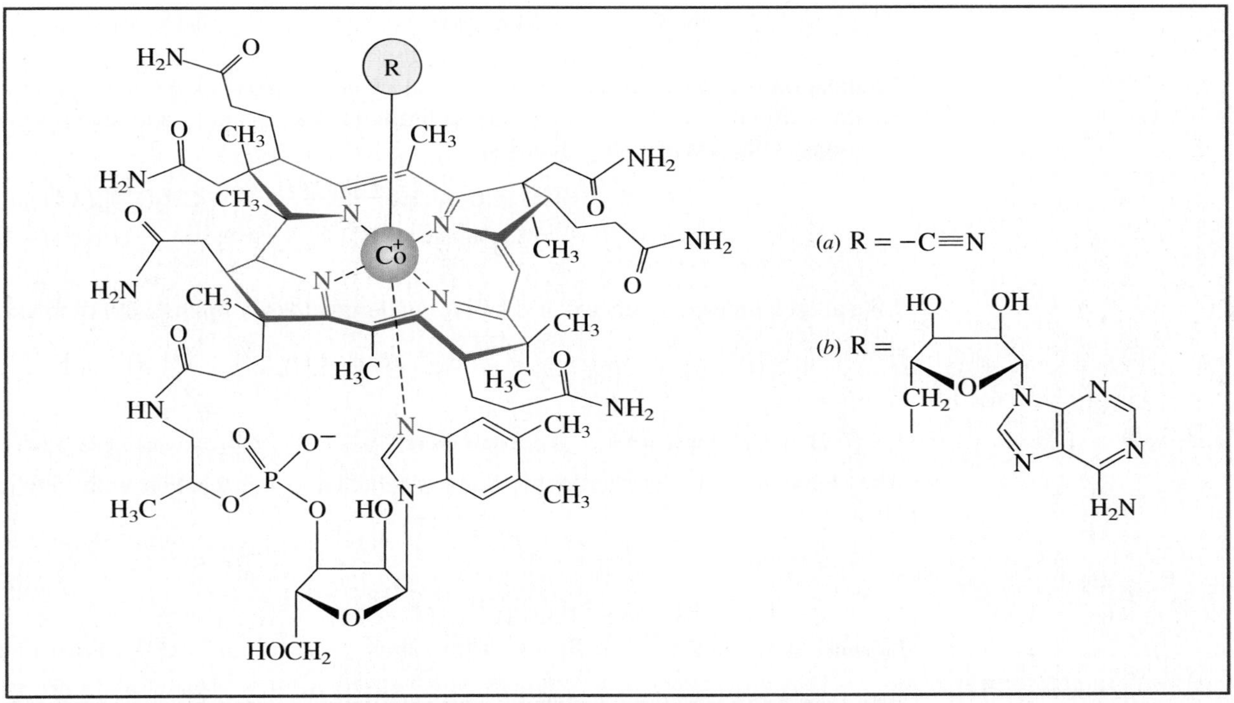

Figure 15.5

The structures of (*a*) vitamin B_{12} and (*b*) coenzyme B_{12}.

15.11 Organocopper Reagents

Early observations that certain Grignard reactions could be catalyzed by copper salts eventually led to systematic studies of organocopper compounds as reagents for carbon–carbon bond formation. Of these, lithium diorganocuprates known as **Gilman reagents** proved to be the most effective.

Henry Gilman, whose career at Iowa State spanned the period 1919–1975, was the first to prepare and study organocuprates.

Gilman reagents are prepared by the reaction of a copper(I) halide with two equivalents of an alkyl- or aryllithium in diethyl ether or tetrahydrofuran.

$$\underset{\substack{\text{Cu(I) halide}\\(X = Cl, Br, I)}}{CuX} \xrightarrow[\quad]{RLi \quad Li^+X^-} \underset{\text{Alkylcopper}}{R{-}Cu} \xrightarrow{RLi} \underset{\substack{\text{Lithium}\\\text{dialkylcuprate}}}{Li^+[R{-}Cu{-}R]^-}$$

Adding an alkyl halide to the solution of the lithium dialkylcuprate leads to carbon–carbon bond formation between the alkyl group of the halide and that of the cuprate.

$$\underset{\substack{\text{Lithium}\\\text{dialkylcuprate}}}{Li^+[R{-}Cu{-}R]^-} + \underset{\substack{\text{Alkyl}\\\text{halide}}}{R'{-}X} \longrightarrow \underset{\text{Alkane}}{R{-}R'} + \underset{\text{Alkylcopper}}{R{-}Cu} + \underset{\substack{\text{Lithium}\\\text{halide}}}{LiX}$$

The process is called *cross-coupling* when the groups that are joined from the two reactants are different. Methyl and primary alkyl halides, especially iodides, work best.

$$\underset{\substack{\text{Lithium}\\\text{dimethylcuprate}}}{(CH_3)_2CuLi} + \underset{\text{1-Iododecane}}{CH_3(CH_2)_8CH_2I} \xrightarrow[0°C]{\text{diethyl ether}} \underset{\text{Undecane (90\%)}}{CH_3(CH_2)_8CH_2CH_3}$$

Elimination becomes a problem with secondary and tertiary alkyl halides. Lithium diarylcuprates are prepared in the same way as lithium dialkylcuprates and undergo comparable reactions with primary alkyl halides.

$$\underset{\substack{\text{Lithium}\\\text{diphenylcuprate}}}{(C_6H_5)_2CuLi} + \underset{\text{1-Iodooctane}}{ICH_2(CH_2)_6CH_3} \xrightarrow{\text{diethyl ether}} \underset{\text{1-Phenyloctane (99\%)}}{C_6H_5CH_2(CH_2)_6CH_3}$$

Like alkyl halides, vinylic and aryl halides undergo cross-coupling with diorganocuprates.

$$\underset{\text{Lithium dibutylcuprate}}{(CH_3CH_2CH_2CH_2)_2CuLi} + \underset{\text{Iodobenzene}}{C_6H_5{-}I} \xrightarrow{\text{diethyl ether}} \underset{\text{Butylbenzene (75\%)}}{C_6H_5{-}CH_2CH_2CH_2CH_3}$$

The stereochemistry of the double bond is retained in the reaction with vinylic halides.

$$\underset{\text{Lithium diethylcuprate}}{(CH_3CH_2)_2CuLi} + \underset{(E)\text{-1-Bromo-2-phenylethene}}{C_6H_5CH{=}CHBr} \xrightarrow[-70°C]{\text{diethyl ether}} \underset{(E)\text{-1-Phenyl-1-butene (65\%)}}{C_6H_5CH{=}CHCH_2CH_3}$$

Problem 15.12

An antibacterial substance obtained from the bark of a flowering shrub (*plumeria*) has been synthesized by the reaction shown. What is its structure?

[Reaction scheme: chlorovinyl-substituted cyclopenta-fused pyranone bearing a CO_2CH_3 (O, OCH_3) group, Cl, O + di(1-propenyl)cuprate (Cu⁻) $\xrightarrow{\text{diethyl ether}}$]

The reactions of diorganocuprates with alkyl halides have characteristics that we've come to associate with the S_N2 mechanism:

Relative reactivity:	RI > RBr > RCl >> RF CH_3 > primary > secondary > tertiary
Substitution/Elimination:	Mainly substitution with primary alkyl halides; elimination with secondary and tertiary alkyl halides.
Stereochemistry:	Inversion of configuration at sp^3-hybridized carbon.

However, the fact that vinylic and aryl halides undergo similar reactions suggests that S_N2 cannot be the only possible mechanism. An alternative, and more likely, mechanism begins by regarding the dialkylcuprate as equivalent to the product of the reaction between Cu^+ and two methyl anions.

$$H_3\bar{C}: \;\; \overset{+}{Cu} \;\; :\bar{C}H_3 \longrightarrow [H_3C-Cu-CH_3]^-$$

Copper is a group 11 element, so Cu^+ contributes 10 valence electrons, and the two methyl anions contribute 4 more, for a total of 14. With fewer than 18 valence electrons, copper is coordinatively unsaturated and can accommodate additional ligands. When dimethylcuprate reacts with an alkyl halide RX, both R and X become ligands on copper.

$$[H_3C-Cu-CH_3]^- + R-X \xrightarrow[\text{addition}]{\text{oxidative}} \left[H_3C-\underset{X}{\overset{R}{Cu}}-CH_3\right]^-$$

Copper gains two electrons from the R—X bond, increasing its electron count from 14 to 16, and the 18-electron rule is not exceeded.

In spite of this increase in electron count, the reaction is classified as an *oxidative addition.* The reason for this seeming anomaly is that counting electrons for the purpose of the 18-electron rule differs from calculating oxidation state. Two new bonds, Cu—R and Cu—X, are formed by a process in which two electrons come from R—X and two *d* electrons from copper. We count all the electrons around an atom for the 18-electron rule, but when calculating oxidation number assign both electrons in a bond to the more electronegative partner, which is almost never the metal. Thus, copper is in the +1 oxidation state in R_2Cu^- and in the +3 oxidation state in the product of oxidative addition. Oxidative addition increases the electron count of a transition metal, but counts as a loss of two electrons when calculating oxidation state.

The counterpart of oxidative addition is *reductive elimination* and constitutes the next step in the reaction.

$$\left[H_3C-\underset{X}{\overset{R}{Cu}}-CH_3\right]^- \xrightarrow[\text{elimination}]{\text{reductive}} R-CH_3 + [H_3C-Cu-X]^-$$

Of the four electrons in the Cu—X and Cu—CH_3 bonds, two remain on copper after elimination and decrease its oxidation state from +3 to +1.

The sum of the oxidative addition and reductive elimination stages corresponds to the overall reaction.

$$\underset{\text{Lithium dialkylcuprate}}{R_2Cu^-} + \underset{\text{Alkyl halide}}{R'X} \xrightarrow[\text{addition}]{\text{oxidative}} \underset{\text{Intermediate}}{\left[R-\underset{R'}{\overset{X}{Cu}}-R\right]^-} \xrightarrow[\text{elimination}]{\text{reductive}} \underset{\text{Alkane}}{R-R'} + \underset{\text{Alkylcopper}}{RCu} + \underset{\text{Halide ion}}{X^-}$$

Problem 15.13

Use retrosynthetic analysis to devise a synthesis of each of the following from the indicated starting material and any necessary reagents.

(a) 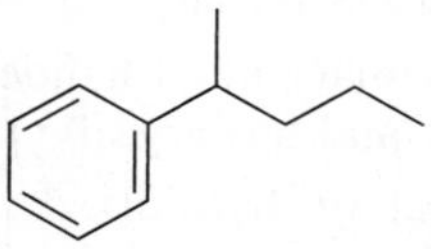 from

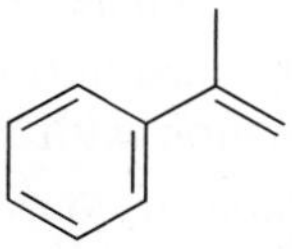

(b) 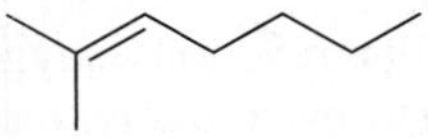 from $(CH_3)_2CHCHBr_2$

Sample Solution (a) The bond disconnection shown reveals the relationship between the carbon skeleton of the target and the given starting material.

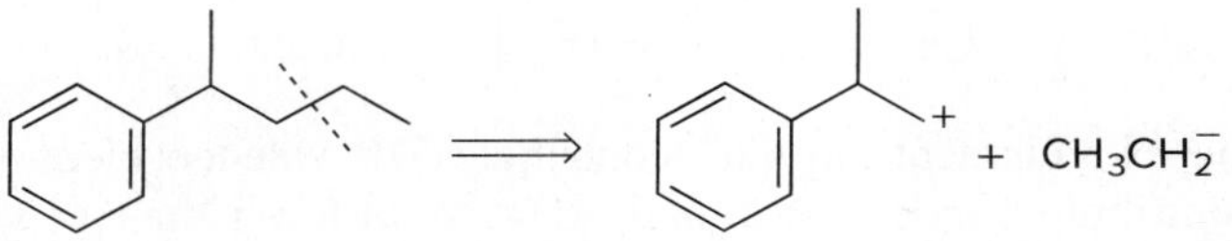

$\Longrightarrow$ + $CH_3CH_2^-$

Carbon–carbon bond formation between an ethyl anion equivalent and an electrophilic site on the side chain can be accomplished by the reaction between a primary alkyl bromide and lithium diethylcuprate.

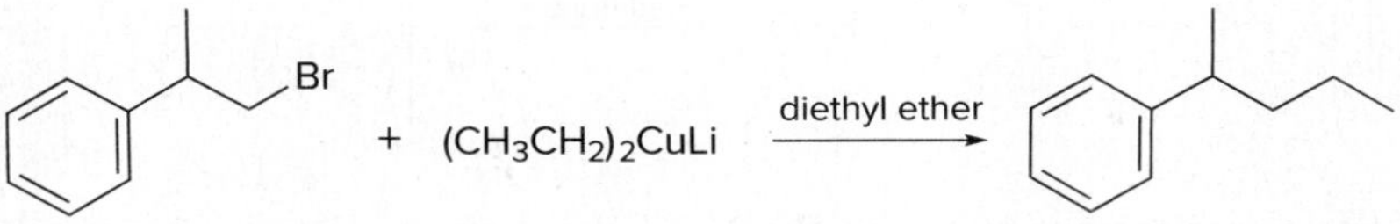

Br + $(CH_3CH_2)_2CuLi$ $\xrightarrow{\text{diethyl ether}}$

The primary bromide can be prepared by free-radical addition of HBr to the prescribed starting material.

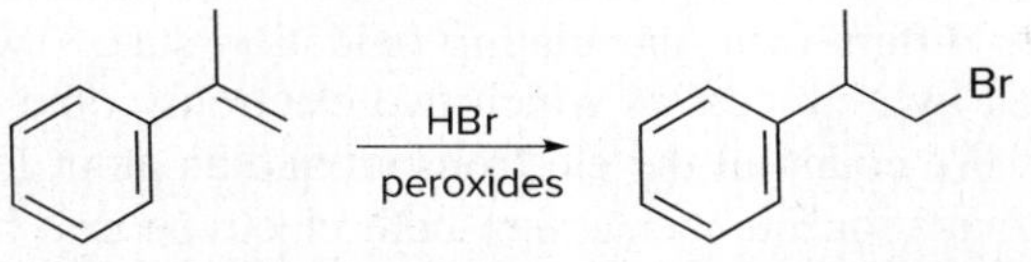

$\xrightarrow[\text{peroxides}]{\text{HBr}}$ Br

15.12 Palladium-Catalyzed Cross-Coupling

Four transition-metal catalyzed cross-coupling procedures, known separately as the Stille, Negishi, Suzuki-Miyaura, and Heck couplings, have emerged as powerful methods for making carbon–carbon bonds. Collectively, their most important qualities include their efficiency, tolerance of functionality elsewhere in the reacting molecules, versatility with respect to hybridization state, and the fact that palladium is used only in catalytic amounts. The four involve the Pd(0)-catalyzed reaction of a suitably functionalized organic group with an:

(a) Organotin reagent (Stille)

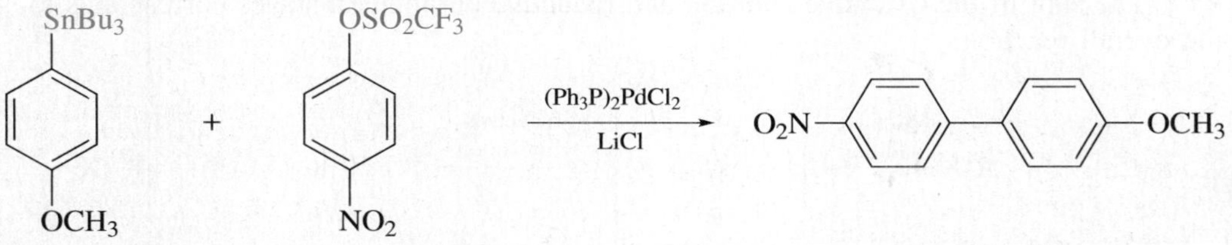

$SnBu_3$ / OCH_3 + OSO_2CF_3 / NO_2 $\xrightarrow[\text{LiCl}]{(Ph_3P)_2PdCl_2}$ O_2N–C_6H_4–C_6H_4–OCH_3

Tributyl(*p*-methoxyphenyl)stannane

p-Nitrophenyl trifluoromethanesulfonate

4-Methoxy-4′-nitrobiphenyl (48%)

(b) Organozinc reagent (Negishi)

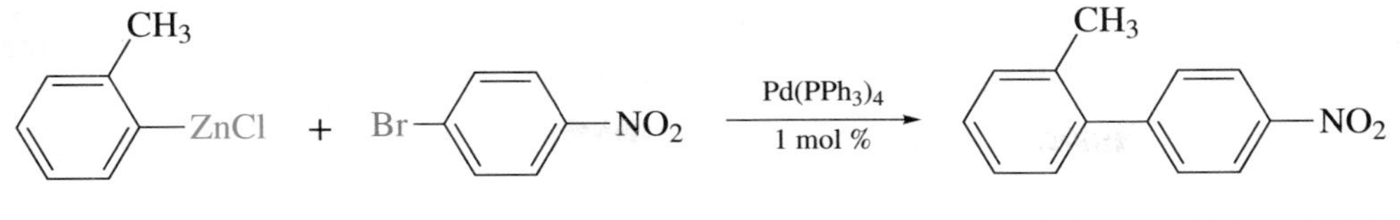

o-Tolylzinc chloride + 1-Bromo-4-nitrobenzene $\xrightarrow[\text{1 mol \%}]{Pd(PPh_3)_4}$ 2-Methyl-4′-nitrobiphenyl (78%)

(c) Organoboron reagent (Suzuki-Miyaura)

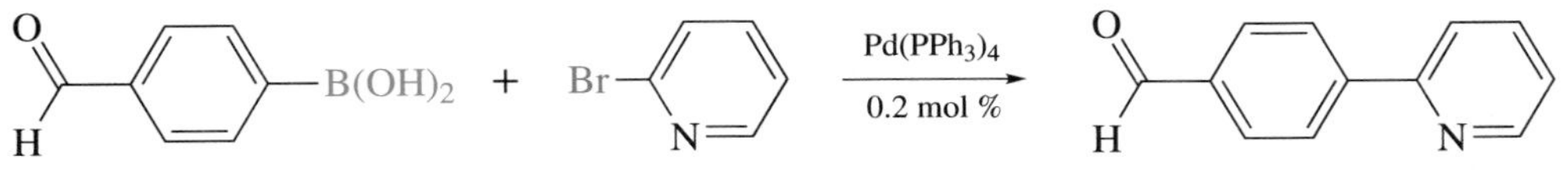

(*p*-Formylphenyl)boronic acid + 2-Bromopyridine $\xrightarrow[\text{0.2 mol \%}]{Pd(PPh_3)_4}$ 2-(*p*-Formylphenyl)pyridine (80%)

(d) Alkene (Heck)

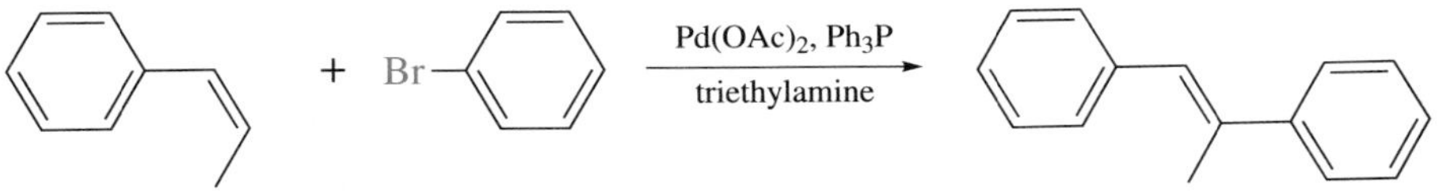

(*Z*)-1-Phenylpropene + Bromobenzene $\xrightarrow[\text{triethylamine}]{Pd(OAc)_2,\ Ph_3P}$ (*E*)-1,2-Diphenylpropene (79%)

The 2010 Nobel Prize in Chemistry was awarded to Richard F. Heck (University of Delaware), Ei-ichi Negishi (Purdue University), and Akira Suzuki-Miyaura (Hokkaido University) for their work on palladium-catalyzed cross-coupling reactions. Early and important contributions were made by John Stille (Colorado State University) before his untimely death in a 1989 plane crash.

As the examples illustrate, the synthesis of biaryls has received much attention. One reason is that they are difficult to synthesize, another is that they are often candidates for new drugs. The product in the Suzuki-Miyaura coupling, for example, is an intermediate in the synthesis of *atazanavir,* a protease inhibitor used in the treatment of HIV–AIDS. Numerous other classes of compounds are accessible by appropriate choices of reactants.

The reactions tolerate many functional groups, including OH, C═O, and NO_2, elsewhere in the molecule, and as shown in the Heck example, the stereochemistry of the double bond is retained when cross-coupling involves an sp^2-hybridized carbon.

It should also be noted that various sources of palladium are given in the examples. The active oxidation state is Pd(0) in all these reactions, even when the source is Pd^{2+} as in the Heck reaction example that uses palladium acetate.

Problem 15.14

Give the structure including stereochemistry of the product of each of the following reactions.

(a) 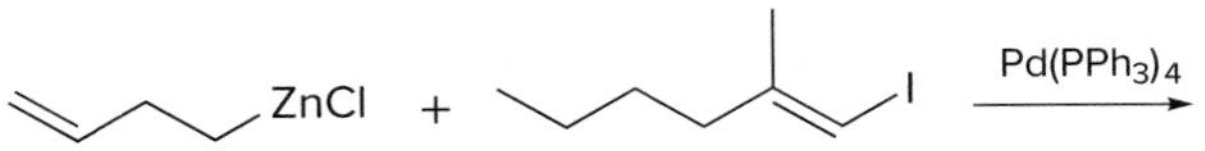 ZnCl + I $\xrightarrow{Pd(PPh_3)_4}$

(b) 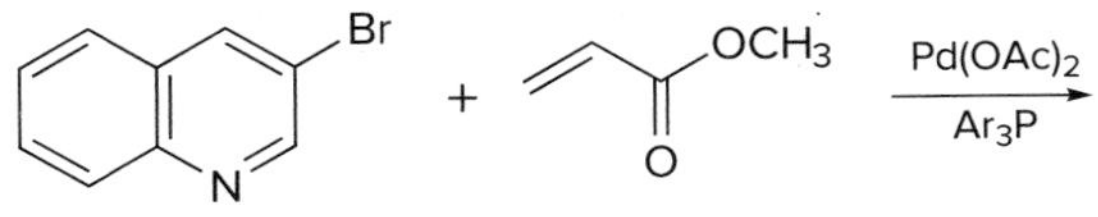 Br, N + OCH$_3$, O $\xrightarrow[Ar_3P]{Pd(OAc)_2}$

(c) 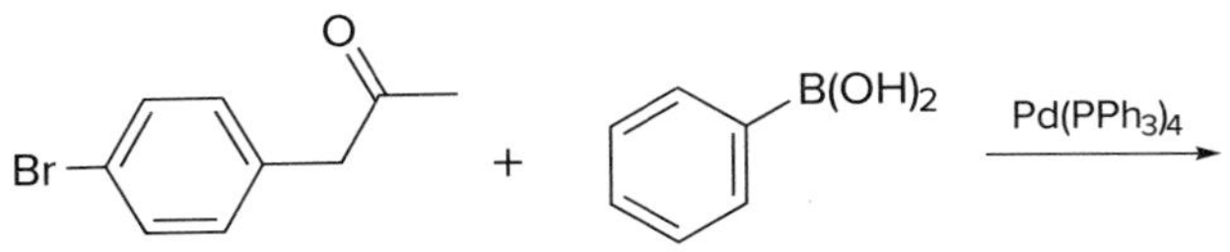 Br + B(OH)$_2$ $\xrightarrow{Pd(PPh_3)_4}$

Sample Solution (a) Palladium-catalyzed reactions of organozinc compounds with alkenyl halides give cross-coupling in which bond formation occurs between the carbon attached to zinc and the carbon attached to the halogen. The stereochemistry at the double bond of the iodoalkene is retained. The product is (*E*)-6-methyl-1,5-decadiene.

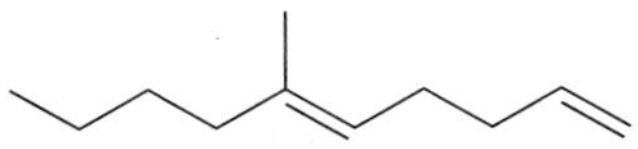

The mechanisms of palladium-catalyzed cross-couplings are complicated, but they all begin the same way—by oxidative addition of an organic halide to the catalyst (represented here as PdL_2, where L is a ligand). This is followed by transmetalation in the Stille, Negishi, and Suzuki-Miyaura methods through which palladium displaces another metal (Sn, Zn, or B) on the organometallic reactant. Reductive elimination creates a C—C bond between R and R′ and regenerates the catalyst.

$$RX + PdL_2 \xrightarrow{\text{oxidative addition}} L{-}Pd(R)(X){-}L \xrightarrow[\text{R}'\text{SnR}''_3,\ \text{R}'\text{ZnCl, or R}'\text{B(OR}'')_2]{\text{transmetalation}} L{-}Pd(R)(R'){-}L \xrightarrow{\text{reductive elimination}} R{-}R' + PdL_2$$

"Transmetalation" is sometimes spelled "transmetallation," and both forms are acceptable. The term was introduced by Henry Gilman, who spelled it with one "l."

The oxidation state of palladium is 0 in PdL_2, is +2 in the products of oxidative addition and transmetalation, and returns to 0 after reductive elimination.

Mechanistically, the Heck procedure can be summarized by the following series of steps.

$$L{-}Pd(R)(X){-}L \xrightarrow[-L]{=\!\!\diagup R'} L{-}Pd(R)(X){-}\|\!\!\diagdown R' \longrightarrow L{-}Pd(X){-}CH(CH_2R)R' \longrightarrow L{-}Pd(H)(X) + R{-}CH{=}CH{-}R'$$

π-Complex

Coordination of the π electrons of the double bond with palladium gives a π-complex, which rearranges by migration of the substituent R to the less substituted carbon of the double bond while palladium bonds to the other carbon. Dissociation of the complex gives the alkene. Other steps (not shown) restore the original form of the catalyst.

Catalytic processes are often represented by a catalytic cycle that shows the conversion of reactants to products and regeneration of the catalyst, in this case PdL_2. A catalytic cycle for the Suzuki-Miyaura coupling illustrated by reaction (c) earlier in this section is shown in Figure 15.6. The aryl boronic acid serves as the aryl group donor in the transmetalation step in this reaction.

Problem 15.15

Humulene is a naturally occurring hydrocarbon present in the seed cone of hops and has been synthesized several times. In one of these, the retrosynthetic strategy was based on the disconnection shown. Deduce the structure, including stereochemistry, of an allylic bromide capable of yielding humulene by an intramolecular Suzuki-Miyaura coupling in the last step in the synthesis. Represent the boron-containing unit as $—B(OH)_2$.

The scope of palladium-catalyzed cross-coupling has expanded beyond C—C bond formation to include C—O and C—N bond-forming methods.

3-Bromoquinoline + 2-Butanol $\xrightarrow{\text{Pd(0)}}$ *sec*-Butyl 3-quinolyl ether (88%)

$$\xrightarrow[\text{K}_2\text{CO}_3,\ \text{triethylamine}]{\text{Pd(PPh}_3)_4}$$

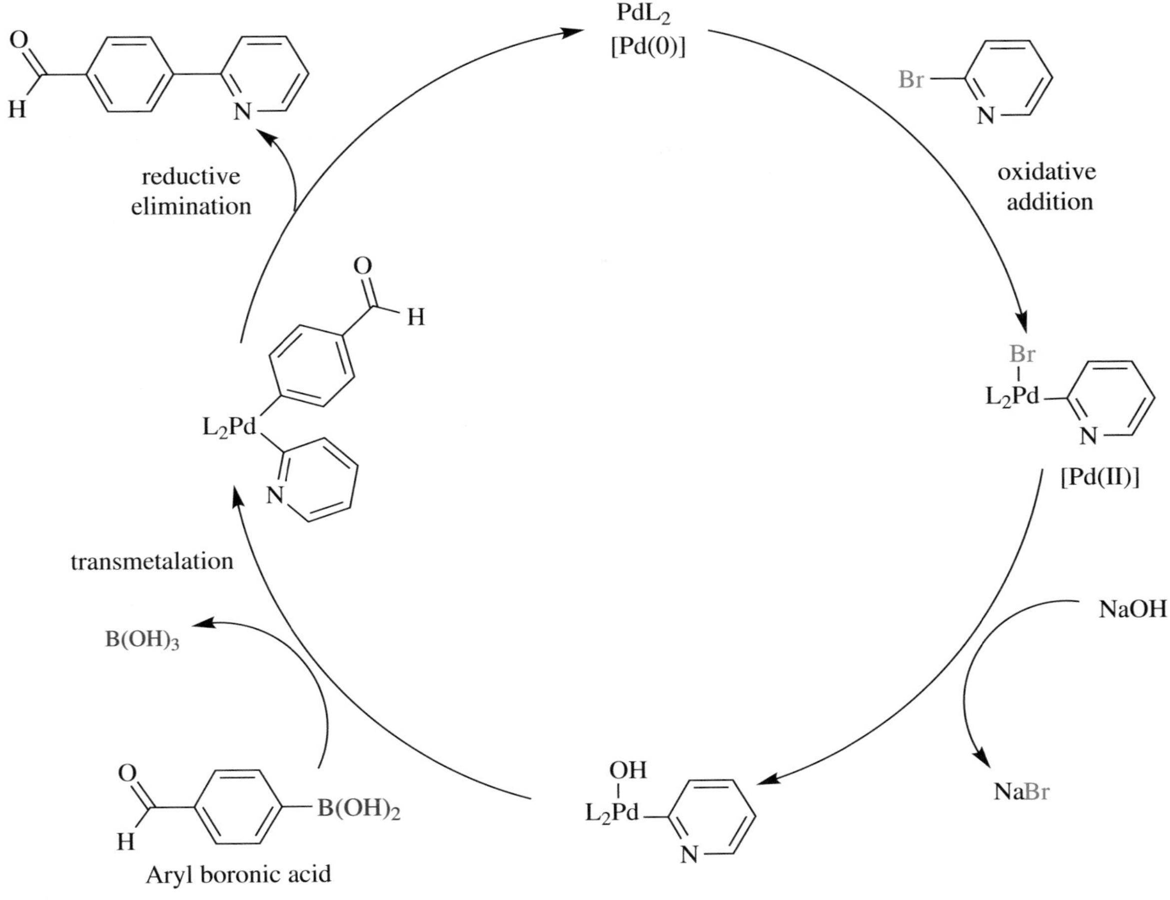

Figure 15.6

Catalytic cycle for the Suzuki-Miyaura coupling.

15.13 Homogeneous Catalytic Hydrogenation

We have seen numerous examples of the hydrogenation of alkenes catalyzed by various finely divided metals such as Ni, Pt, Pd, and Rh. In all those cases, the metal acted as a *heterogeneous catalyst,* present as a solid while the alkene was in solution. The idea of carrying out hydrogenations in homogeneous solution seems far-fetched inasmuch as no solvent is capable of simultaneously dissolving both metals and hydrocarbons. Nevertheless, there is a way to do it.

Rhodium is a good catalyst for alkene hydrogenation (Section 8.1), as are many of its complexes such as tris(triphenylphosphine)rhodium chloride (Wilkinson's catalyst).

Geoffrey Wilkinson (Imperial College, London) shared the 1973 Nobel Prize in Chemistry with Ernst O. Fischer (Munich) for their achievements in organometallic chemistry. In addition to his work on catalysts for homogeneous hydrogenation, Wilkinson collaborated on determining the structure of ferrocene as well as numerous other aspects of organometallic compounds.

$$(C_6H_5)_3P\text{—}\underset{\large P(C_6H_5)_3}{\overset{\large P(C_6H_5)_3}{Rh}}\text{—}Cl \quad = \quad [(C_6H_5)_3P]_3RhCl$$

Tris(triphenylphosphine)rhodium chloride

Like rhodium itself, Wilkinson's catalyst is an effective catalyst for alkene hydrogenation. Unlike rhodium metal, however, Wilkinson's catalyst is soluble in many organic solvents.

It is selective, reducing less-substituted double bonds faster than more-substituted ones and C=C in preference to C=O.

$$\text{Carvone} + \text{Hydrogen } (H_2) \xrightarrow[\text{benzene, 25°C, 1 atm}]{[(C_6H_5)_3P]_3RhCl} \text{Dihydrocarvone (90–94\%)}$$

Stereospecific syn addition is observed, and hydrogens are transferred to the less-hindered face of the double bond.

$$(Z)\text{-1-Methoxy-2-phenylethylene} + \text{Deuterium } (D_2) \xrightarrow{[(C_6H_5)_3P]_3RhCl} \textit{erythro}\text{-1,2-Dideuterio-1-methoxy-2-phenylethane}$$

Problem 15.16

Homogeneous catalytic hydrogenation of the compound shown gives two isomers in a 73:27 ratio. What are their structures, and which one is the major product?

The mechanism of hydrogenation in the presence of Wilkinson's catalyst begins with oxidative addition of hydrogen to the rhodium complex, with loss of one of the triphenylphosphine ligands.

$$\text{Tris(triphenylphosphine)rhodium chloride (Wilkinson's catalyst)} \xrightarrow{H_2} \text{Dihydride of Wilkinson's catalyst} \xrightarrow{-P(C_6H_5)_3} \text{Dihydride of bis(triphenylphosphine) complex}$$

Rhodium is in the +1 oxidation state in Wilkinson's catalyst, so is a d^8 ion. Its four ligands count for 8 more electrons, bringing the total number of valence electrons to 16. Addition of H_2 in the first step raises the number to 18. Loss of a triphenylphosphine ligand in the second step reduces the number of valence electrons to 16 and gives the active form of the catalyst, setting in motion the repeating cycle of four steps for alkene hydrogenation shown in Mechanism 15.2.

The effect that homogeneous transition-metal catalysis has had on stereoselective synthesis is especially impressive. Using chiral ligands, it is possible to control hydrogenation of double bonds so that new chirality centers have a particular configuration. The drug L-dopa, used to treat Parkinsonism, is prepared in multiton quantities by enantioselective hydrogenation catalyzed by an enantiomerically pure chiral rhodium complex.

$$\xrightarrow[\text{enantiomerically pure Rh catalyst}]{H_2} \xrightarrow[\text{2. neutralize}]{\text{1. } H_3O^+}$$

L-Dopa (100% yield; 95% enantiomeric excess)

Mechanism 15.2

Homogeneous Catalysis of Alkene Hydrogenation

THE OVERALL REACTION:

$$H_2C{=}CHCH_3 + H_2 \xrightarrow{[(C_6H_5)_3P]_3RhCl} CH_3CH_2CH_3$$

Propene Hydrogen Propane

THE MECHANISM:

Step 1: The active form of Wilkinson's catalyst described in the text is coordinatively unsaturated and reacts with the alkene to form an 18-electron complex.

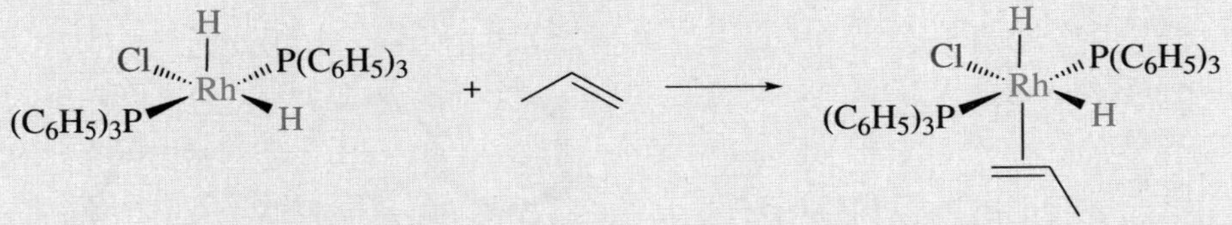

Dihydride of bis(triphenylphosphine) complex + Propene → Rhodium–propene complex

Step 2: The rhodium–propene complex rearranges. Rhodium bonds to the less-substituted carbon of the alkene while hydride migrates from rhodium to the more-substituted carbon.

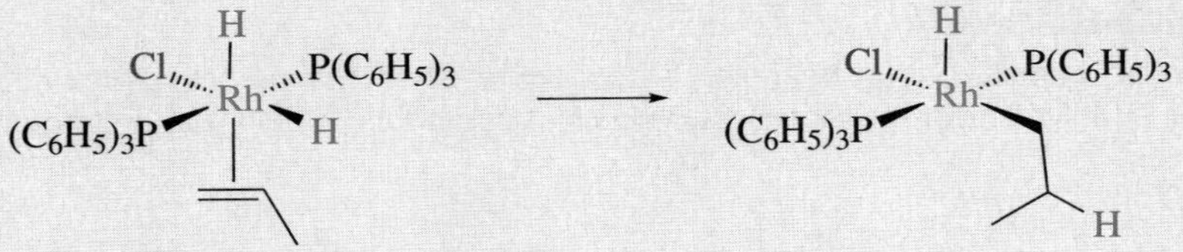

Rhodium–propene complex → Propylrhodium complex

Step 3: Hydride migrates from rhodium to carbon and the complex dissociates, releasing propane.

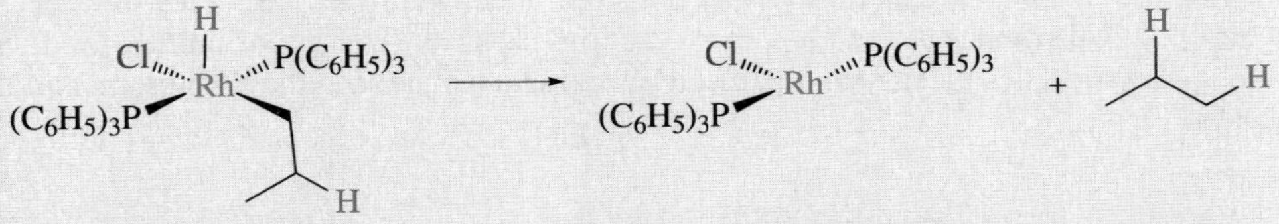

Propylrhodium complex → Bis(triphenylphosphine)rhodium chloride + Propane

Step 4: The complex formed in step 3 reacts with H_2 to restore the active form of the catalyst, which returns to step 1 to maintain a continuing cycle.

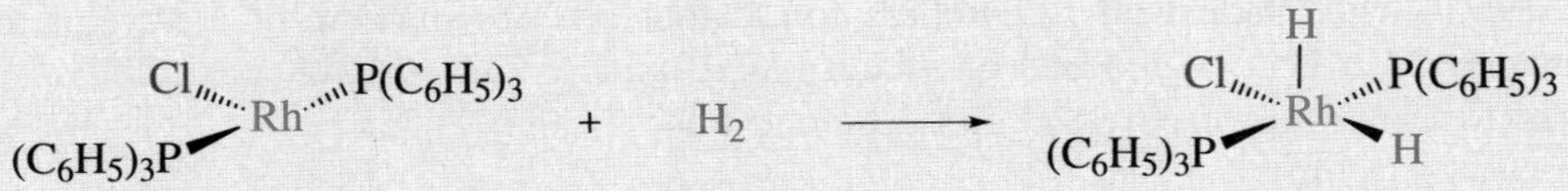

Bis(triphenylphosphine)rhodium chloride + H_2 (Hydrogen) → Dihydride of bis(triphenylphosphine) complex

The synthesis of L-dopa was one of the earliest of what has become an important advance in the pharmaceutical industry—the preparation and marketing of chiral drugs as single enantiomers (see the boxed essay, *Chiral Drugs,* in Chapter 4). William S. Knowles (Monsanto) and Ryoji Noyori (Nagoya, Japan) shared one half of the 2001 Nobel Prize in Chemistry for their independent work on enantioselective hydrogenations. Knowles devised and carried out the synthesis of L-dopa and Noyori developed a variety of chiral catalysts in which he varied both the metal and the ligands to achieve enantioselectivities approaching 100%.

BINAP is an abbreviation for 2,2′-bis(diphenylphosphino)-1,1′-binaphthyl.

Chirality is built into the catalysts by employing ligands with either chirality centers or axes. Noyori's widely used BINAP has a chirality axis, and crowding prevents interconversion of enantiomers by restricting rotation around the bond connecting the naphthalene rings. The metal, usually ruthenium, is held in place by the two phosphorus atoms (yellow) in a chiral environment. The steric demands in the cavity occupied by the metal in Ru-BINAP cause reaction to occur preferentially at one face of the double bond.

$P(C_6H_5)_2$

$P(C_6H_5)_2$

(*S*)-(−)-BINAP

Problem 15.17

The antiinflammatory drug *naproxen* is sold as its (*S*)-enantiomer. One large-scale synthesis uses a Ru-BINAP hydrogenation catalyst. What compound would you hydrogenate to prepare naproxen?

CH_3O O OH

A large number of enantioselective transition-metal catalysts have been developed, not just for hydrogenation but for other reactions as well. The opportunities for fine-tuning their properties by varying the metal, its oxidation state, and the ligands are almost limitless.

15.14 Olefin Metathesis

The word *metathesis* refers to an interchange, or transposition, of objects.

The 2005 Nobel Prize in Chemistry was jointly awarded to Robert H. Grubbs (Caltech), Yves Chauvin (French Petroleum Institute), and Richard R. Schrock (MIT) for establishing **olefin metathesis** as a reaction of synthetic versatility and contributing to an understanding of the mechanism of this novel process. Olefin metathesis first surfaced in the late 1950s when industrial researchers found that alkenes underwent a novel reaction when passed over a heated bed of mixed metal oxides. Propene, for example, was converted to a mixture of ethylene and 2-butene (cis + trans).

$$\underset{\text{Propene}}{2CH_3CH{=}CH_2} \overset{\text{catalyst}}{\rightleftharpoons} \underset{\text{Ethylene}}{H_2C{=}CH_2} + \underset{\textit{cis-} + \textit{trans-}\text{2-butene}}{CH_3CH{=}CHCH_3}$$

This same transformation was subsequently duplicated at lower temperatures by homogeneous transition-metal catalysis. An equilibrium is established, and the same mixture is obtained regardless of whether propene or a 1:1 mixture of ethylene and 2-butene is subjected to the reaction conditions. This type of olefin metathesis is called a *cross-metathesis.*

When cross-metathesis was first discovered, propene enjoyed only limited use and the reaction was viewed as a potential source of ethylene. Once methods were developed for the preparation of stereoregular polypropylene, however, propene became more valuable and cross-metathesis of ethylene and 2-butene now serves as a source of propene.

The relationship between reactants and products in cross-metathesis can be analyzed retrosynthetically by joining the double bonds in two reactant molecules by dotted lines, then disconnecting in the other direction.

$$CH_3CH{=}CH_2 + CH_3CH{=}CH_2 \Longrightarrow \begin{array}{c} CH_3CH\text{- - }CH_2 \\ \vdots \quad\quad \vdots \\ CH_3CH\text{- - }CH_2 \end{array} \Longleftarrow \begin{array}{c} CH_3CH \\ \| \\ CH_3CH \end{array} + \begin{array}{c} CH_2 \\ \| \\ CH_2 \end{array}$$

Two propene molecules — 2-Butene + Ethylene

Although this representation helps us relate products and reactants, *it is not related to the mechanism. Nothing containing a ring of four carbons is an intermediate in olefin cross-metathesis.*

Problem 15.18

What alkenes are formed from 2-pentene by olefin cross-metathesis?

The generally accepted mechanism for olefin cross-metathesis is outlined for the case of propene in Mechanism 15.3. The catalyst's structure is characterized by a carbon–metal double bond, and the metal is typically ruthenium (Ru), tungsten (W), or molybdenum (Mo).

One of the most widely used catalysts for olefin metathesis is the ruthenium complex shown. It is called *Grubbs' catalyst* and abbreviated $Cl_2(PCy_3)_2Ru{=}CHC_6H_5$.

$$Cl_2(PCy_3)_2Ru{=}CH{-}C_6H_5 \qquad \text{Cy = cyclohexyl}$$

Grubbs' catalyst

Olefin cross-metathesis is an intermolecular reaction between double bonds in separate molecules. Intramolecular metatheses in which two double bonds belong to the same molecule are also common and lead to ring formation. The process is called *ring-closing metathesis.*

$$o\text{-}(CH{=}CH_2)C_6H_4(OCH_2CH{=}CH_2) \xrightarrow[CH_2Cl_2,\ 25°C]{Cl_2(PCy_3)_2Ru{=}CHC_6H_5} \text{2H-1-Benzopyran} + H_2C{=}CH_2$$

Allyl *o*-vinylphenyl ether — 2H-1-Benzopyran (95%) — Ethylene

Although olefin metathesis is an equilibrium process, it can give high yields of the desired product when ethylene is formed as the other alkene. Being a gas, ethylene escapes from the reaction mixture, and the equilibrium shifts to the right in accordance with Le Châtelier's principle. Ring-closing metathesis has been widely and imaginatively applied to the synthesis of natural products. It occurs under mild conditions and tolerates the presence of numerous functional groups.

Mechanism 15.3

Olefin Cross-Metathesis

THE OVERALL REACTION:

$$2CH_3CH{=}CH_2 \underset{}{\overset{\text{catalyst}}{\rightleftharpoons}} H_2 + CH_3CH{=}CHCH_3$$

Propene — Ethylene — 2-Butene (cis + trans)

THE MECHANISM:

To simplify the presentation of the mechanism, the symbol (M) stands for the transition metal and its ligands. Steps have been omitted in which ligands leave or become attached to the metal; therefore, the number of ligands is not necessarily the same throughout a stage.

Stage 1: In this stage the sp^2-hybridized carbons of the alkene, with their attached groups, replace the benzylidene group of the catalyst. In the case of an unsymmetrical alkene such as propene, the two newly formed complexes (A and B) are different.

$$\text{M}{=}CHC_6H_5 + CH_3CH{=}CH_2 \longrightarrow \begin{array}{c}\text{M}{-}CHC_6H_5\\ |\quad\;\; |\\ CH_3CH{-}CH_2\end{array} + \begin{array}{c}\text{M}{-}CHC_6H_5\\ |\quad\;\; |\\ H_2C{-}CHCH_3\end{array}$$

$$\downarrow \qquad\qquad\qquad \downarrow$$

$$\begin{array}{c}\text{M}\\ \|\\ CH_3CH\end{array} + \begin{array}{c}CHC_6H_5\\ \|\\ CH_2\end{array} \qquad \begin{array}{c}\text{M}\\ \|\\ CH_2\end{array} + \begin{array}{c}CHC_6H_5\\ \|\\ CHCH_3\end{array}$$

A — B

Stage 2: *Complex A:* Propene adds to the double bond of the complex to give metallocyclobutanes C and D. Dissociation of C gives propene + A. Dissociation of D gives 2-butene + B.

$$\underset{\text{A}}{\text{M}{=}CHCH_3} + \underset{\text{Propene}}{CH_3CH{=}CH_2} \rightleftharpoons \underset{\text{C}}{\begin{array}{c}\text{M}{-}CHCH_3\\ |\quad\;\; |\\ CH_3CH{-}CH_2\end{array}} + \underset{\text{D}}{\begin{array}{c}\text{M}{-}CHCH_3\\ |\quad\;\; |\\ H_2C{-}CHCH_3\end{array}}$$

$$\downarrow \qquad\qquad\qquad \downarrow$$

$$\underset{\text{A}}{\begin{array}{c}\text{M}\\ \|\\ CH_3CH\end{array}} + \underset{\text{Propene}}{\begin{array}{c}CHCH_3\\ \|\\ CH_2\end{array}} \qquad \underset{\text{B}}{\begin{array}{c}\text{M}\\ \|\\ CH_2\end{array}} + \underset{\text{2-Butene}}{\begin{array}{c}CHCH_3\\ \|\\ CHCH_3\end{array}}$$

Complex B: Propene adds to the double bond of B to give metallocyclobutanes E and F. Dissociation of E gives ethylene + A. Dissociation of F gives propene + B.

$$\underset{\text{B}}{\text{M}{=}CH_2} + \underset{\text{Propene}}{CH_3CH{=}CH_2} \rightleftharpoons \underset{\text{E}}{\begin{array}{c}\text{M}{-}CH_2\\ |\quad\;\; |\\ CH_3CH{-}CH_2\end{array}} + \underset{\text{F}}{\begin{array}{c}\text{M}{-}CH_2\\ |\quad\;\; |\\ H_2C{-}CHCH_3\end{array}}$$

$$\downarrow \qquad\qquad\qquad \downarrow$$

$$\underset{\text{A}}{\begin{array}{c}\text{M}\\ \|\\ CH_3CH\end{array}} + \underset{\text{Ethylene}}{\begin{array}{c}CH_2\\ \|\\ CH_2\end{array}} \qquad \underset{\text{B}}{\begin{array}{c}\text{M}\\ \|\\ CH_2\end{array}} + \underset{\text{Propene}}{\begin{array}{c}CH_2\\ \|\\ CHCH_3\end{array}}$$

Stage 3: The two complexes A and B that react in stage 2 are also regenerated in the same stage. Thus, stage 3 is simply a repeat of stage 2 and the process continues.

Problem 15.19

The product of the following reaction was isolated in 99% yield. What is it?

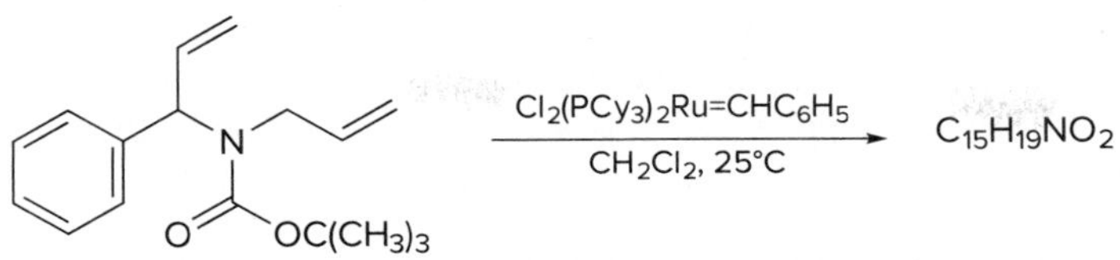

$$\xrightarrow[CH_2Cl_2,\ 25°C]{Cl_2(PCy_3)_2Ru{=}CHC_6H_5} C_{15}H_{19}NO_2$$

$OC(CH_3)_3$

Ring-opening metathesis is the converse of ring-closing metathesis and holds promise as a polymerization method. It is applied most often when ring opening is accompanied by relief of strain as in, for example, bicyclic alkenes.

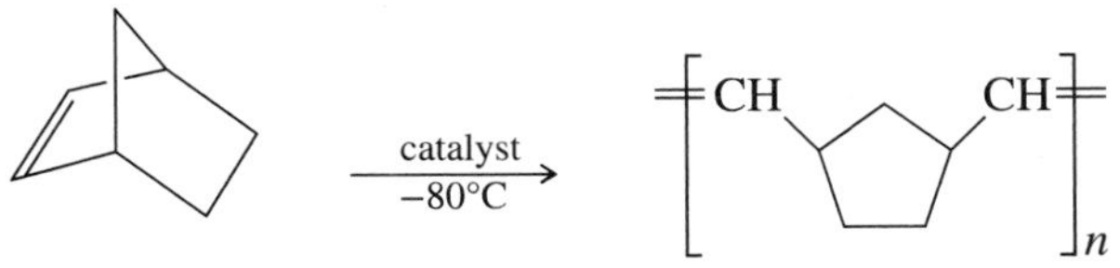

$$\xrightarrow[-80°C]{\text{catalyst}}$$

Bicyclo[2.2.1]hept-2-ene   Polynorbornene

Norbornene is a common name for bicyclo[2.2.1]hept-2-ene.

15.15 Ziegler–Natta Catalysis of Alkene Polymerization

We have already described the essentials of the free-radical polymerization of alkenes in earlier sections (10.8 and 12.14). In the present section we introduce a very different method—**coordination polymerization**—that has revolutionized the industry.

In the early 1950s, Karl Ziegler, then at the Max Planck Institute for Coal Research in Germany, was studying the use of aluminum compounds as catalysts for the oligomerization of ethylene.

$$nH_2C{=}CH_2 \xrightarrow{Al(CH_2CH_3)_3} CH_3CH_2(CH_2CH_2)_{n-2}CH{=}CH_2$$

Ethylene   Ethylene oligomers

Ziegler found that adding certain metals or their compounds to the reaction mixture led to the formation of ethylene oligomers with 6–18 carbons, but others promoted the formation of very long carbon chains giving polyethylene. Both were major discoveries. The 6–18 carbon ethylene oligomers constitute a class of industrial organic chemicals known as *linear α olefins* that are produced at a rate of 3×10^9 pounds/year in the United States. The Ziegler route to polyethylene is even more important because it occurs at modest temperatures and pressures and gives *high-density polyethylene,* which has properties superior to the low-density material formed by the free-radical polymerization described in Section 10.8.

Ziegler had a working relationship with the Italian chemical company Montecatini, for which Giulio Natta of the Milan Polytechnic Institute was a consultant. When Natta used Ziegler's catalyst to polymerize propene,

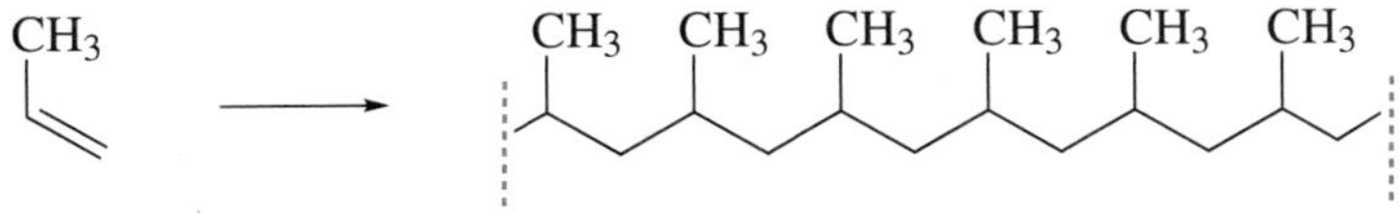

CH_3 → CH_3 CH_3 CH_3 CH_3 CH_3 CH_3

he discovered not only that the catalyst was effective but also that it gave mainly **isotactic** polypropylene.

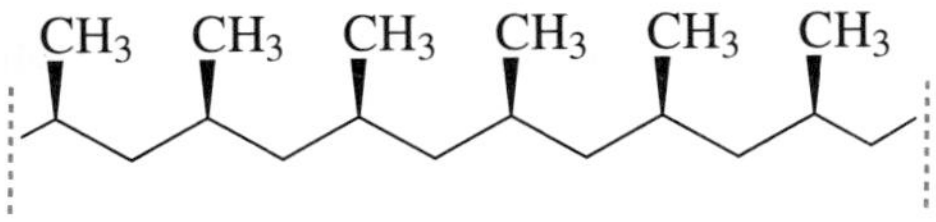

CH_3 CH_3 CH_3 CH_3 CH_3 CH_3 CH_3 CH_3 CH_3 CH_3 CH_3 CH_3

Isotactic polypropylene 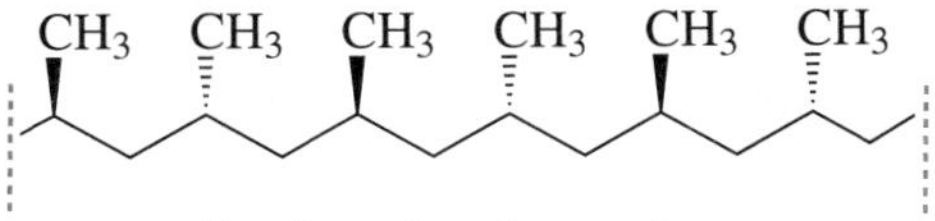   Syndiotactic polypropylene

Isotactic polypropylene is an example of a **stereoregular** polymer. All of its methyl groups are oriented in the same direction along the carbon chain. A second stereoregular form, classified as **syndiotactic,** has its methyl groups alternate front and back along the chain. When the orientation of the methyl groups is random, the polymer is not stereoregular and is referred to as **atactic.** Free-radical polymerization of propene gives atactic polypropylene. Isotactic polypropylene has a higher melting point than the atactic form and can be drawn into fibers or molded into hard, durable materials.

Most polypropylene products are made from isotactic polypropylene.

The earliest Ziegler–Natta catalysts were combinations of titanium tetrachloride ($TiCl_4$) and diethylaluminum chloride [$(CH_3CH_2)_2AlCl$], but these have given way to more effective zirconium-based metallocenes, the simplest of which is bis(cyclopentadienyl)-zirconium dichloride (Section 15.10).

Zirconium lies below titanium in the periodic table, so was an obvious choice in the search for other Ziegler–Natta catalysts.

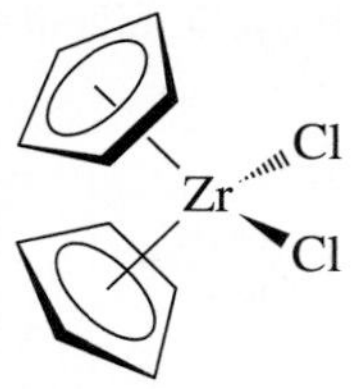

Bis(cyclopentadienyl)zirconium dichloride (Cp_2ZrCl_2)

Hundreds of analogs of Cp_2ZrCl_2 have been prepared and evaluated as catalysts for ethylene and propene polymerization. The structural modifications include replacing one or both of the cyclopentadienyl ligands by variously substituted cyclopentadienyl groups, linking the two rings with carbon chains, and so on. Some modifications give syndiotactic polypropylene, others give isotactic.

The metallocene catalyst is used in combination with a promoter, usually methylalumoxane (MAO).

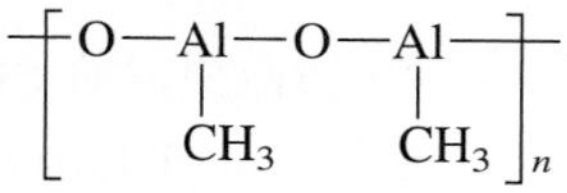

$$\left[-O-\underset{CH_3}{\underset{|}{Al}}-O-\underset{CH_3}{\underset{|}{Al}}- \right]_n$$

Methylalumoxane (MAO)

Mechanism 15.4 outlines ethylene polymerization in the presence of Cp_2ZrCl_2. Step 1 describes the purpose of the MAO promoter, which is to transfer a methyl group to the metallocene to convert it to its catalytically active form. This methyl group will be incorporated into the growing polymer chain—indeed, it will be the end from which the rest of the chain grows.

The active form of the catalyst, having one less ligand and being positively charged, acts as an electrophile toward ethylene in step 2.

With electrons flowing from ethylene to zirconium, the $Zr—CH_3$ bond weakens, the carbons of ethylene become positively polarized, and the methyl group migrates from zirconium to one of the carbons of ethylene. Cleavage of the $Zr—CH_3$ bond is accompanied by formation of a σ bond between zirconium and one of the carbons of ethylene in step 3. The product of this step is a chain-extended form of the active catalyst, ready to accept another ethylene ligand and repeat the chain-extending steps.

Before coordination polymerization was discovered by Ziegler and applied to propene by Natta, there was no polypropylene industry. Now, more than 10^{10} pounds of it are prepared each year in the United States. Ziegler and Natta shared the 1963 Nobel Prize in Chemistry: Ziegler for discovering novel catalytic systems for alkene polymerization and Natta for stereoregular polymerization. We'll see more about Ziegler–Natta polymerization in Chapter 27 when we examine the properties of synthetic polymers in more detail.

Mechanism 15.4

Polymerization of Ethylene in the Presence of Ziegler–Natta Catalyst

Step 1: Cp_2ZrCl_2 is converted to the active catalyst by reaction with the promoter methylalumoxane (MAO). A methyl group from MAO displaces one of the chlorine ligands of Cp_2ZrCl_2. The second chlorine is lost as chloride by ionization, giving a positively charged metallocene.

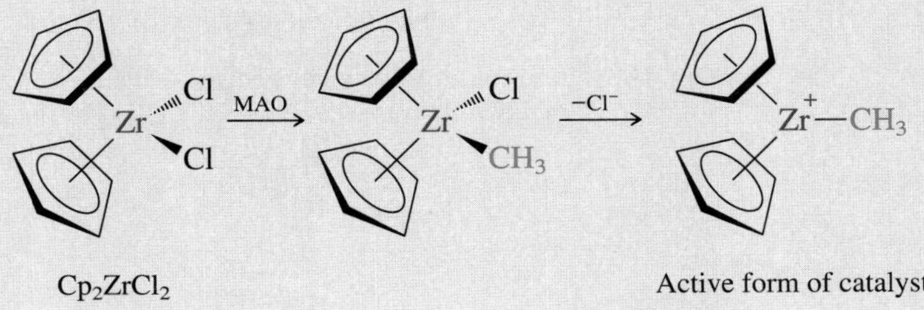

Cp_2ZrCl_2 $\xrightarrow{\text{MAO}}$ $Cp_2Zr(Cl)CH_3$ $\xrightarrow{-Cl^-}$ $Cp_2\overset{+}{Zr}-CH_3$

Cp_2ZrCl_2 Active form of catalyst

Step 2: Ethylene reacts with the active form of the catalyst. The two π electrons of ethylene are used to bind it as a ligand to zirconium.

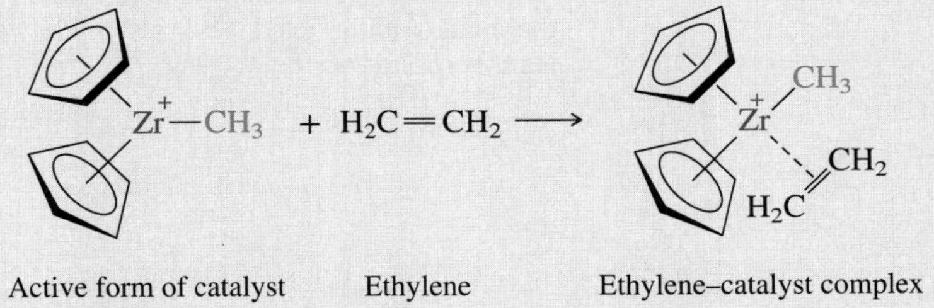

$Cp_2\overset{+}{Zr}-CH_3 + H_2C{=}CH_2 \longrightarrow$ Ethylene–catalyst complex

Active form of catalyst Ethylene Ethylene–catalyst complex

Step 3: The methyl group migrates from zirconium to one of the carbons of the ethylene ligand. At the same time, the π electrons of the ethylene ligand are used to form a σ bond between the other carbon and zirconium.

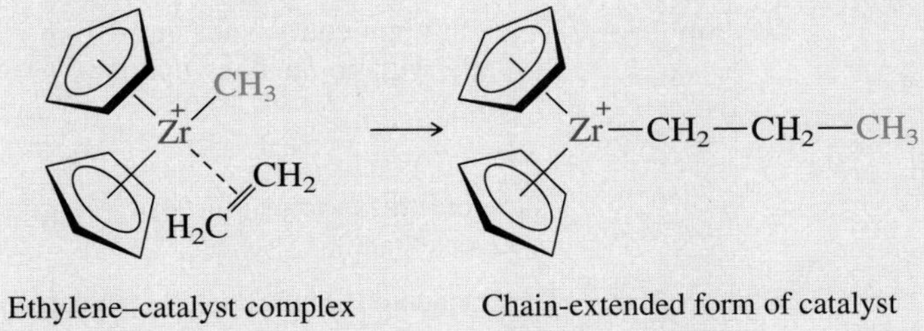

Ethylene–catalyst complex $\longrightarrow$ $Cp_2\overset{+}{Zr}-CH_2-CH_2-CH_3$

Ethylene–catalyst complex Chain-extended form of catalyst

Step 4: The catalyst now has a propyl group on zirconium instead of a methyl group. Repeating steps 2 and 3 converts the propyl group to a pentyl group, then a heptyl group, and so on. After thousands of repetitions, polyethylene results.

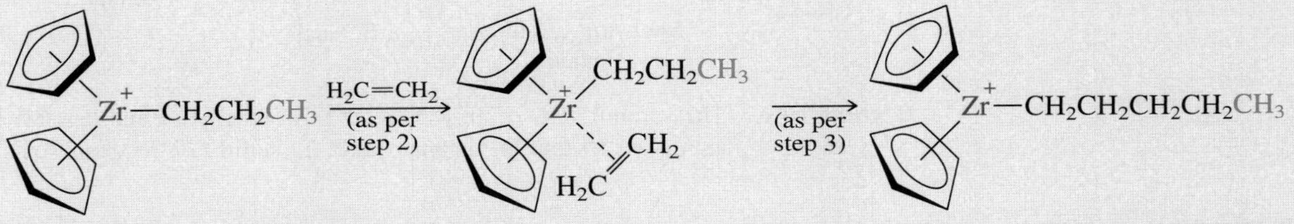

$Cp_2\overset{+}{Zr}-CH_2CH_2CH_3$ $\xrightarrow[\text{(as per step 2)}]{H_2C=CH_2}$ $Cp_2\overset{+}{Zr}(CH_2CH_2CH_3)(H_2C{=}CH_2)$ $\xrightarrow{\text{(as per step 3)}}$ $Cp_2\overset{+}{Zr}-CH_2CH_2CH_2CH_2CH_3$

15.16 SUMMARY

Section 15.1 Organometallic compounds contain a carbon–metal bond. Those derived from main-group metals are named as alkyl or aryl derivatives of the metal.

$CH_3CH_2CH_2CH_2Li$ or [bond-line structure] Li — Butyllithium

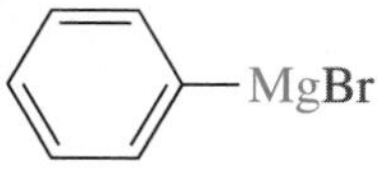

C_6H_5MgBr or [phenyl]—MgBr — Phenylmagnesium bromide

Section 15.2 Carbon is more electronegative than metals, and carbon–metal bonds are polarized so that carbon bears a partial to complete negative charge and the metal bears a partial to complete positive charge.

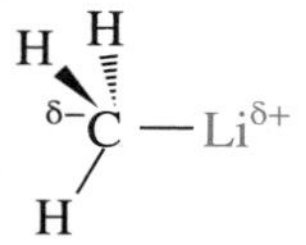

$H_3C^{\delta-}—Li^{\delta+}$ — Methyllithium has a polar covalent carbon–lithium bond.

$HC{\equiv}\bar{C}{:}\ Na^+$ — Sodium acetylide has an ionic bond between carbon and sodium.

Section 15.3 Organolithium compounds and Grignard reagents are prepared by reaction of the metal with an alkyl, aryl, or vinylic halide, usually in diethyl ether or tetrahydrofuran as the solvent.

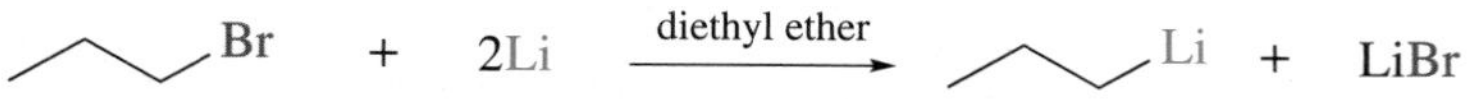

Propyl bromide + 2Li (Lithium) $\xrightarrow{\text{diethyl ether}}$ Propyllithium (78%) + LiBr (Lithium bromide)

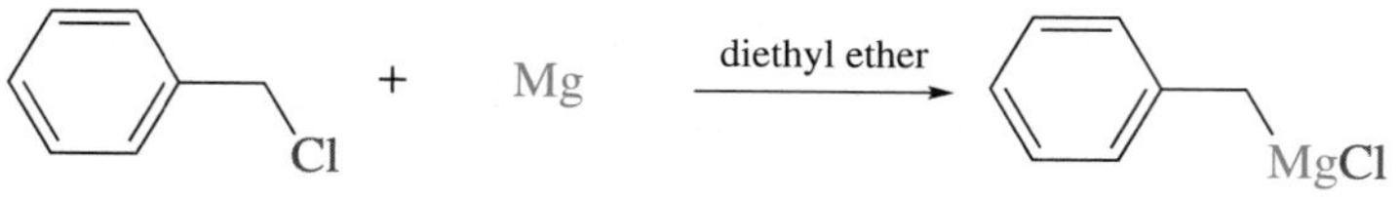

Benzyl chloride + Mg (Magnesium) $\xrightarrow{\text{diethyl ether}}$ Benzylmagnesium chloride (93%)

Section 15.4 Organolithium compounds and Grignard reagents are strong bases and react instantly with compounds that have —OH groups.

$$R—M + H—O—R' \longrightarrow R—H + M^+ \ {}^-O—R'$$

Therefore, these organometallic reagents cannot be used in solvents such as water or ethanol.

Section 15.5 The reaction of Grignard and organolithium reagents with carbonyl compounds is one of the most useful methods for making carbon–carbon bonds in synthetic organic chemistry. These reagents react with formaldehyde to yield primary alcohols, with aldehydes to give secondary alcohols, and with ketones to give tertiary alcohols.

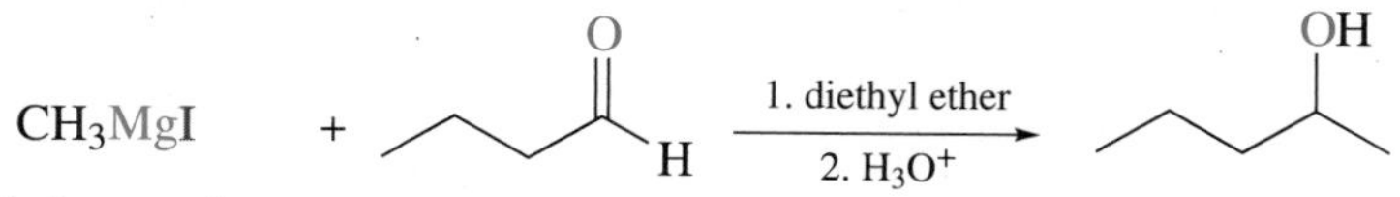

CH_3MgI (Methylmagnesium iodide) + Butanal $\xrightarrow[\text{2. } H_3O^+]{\text{1. diethyl ether}}$ 2-Pentanol (82%)

Section 15.6 The sodium salts of alkynes react with aldehydes and ketones in a manner analogous to that of Grignard and organolithium reagents.

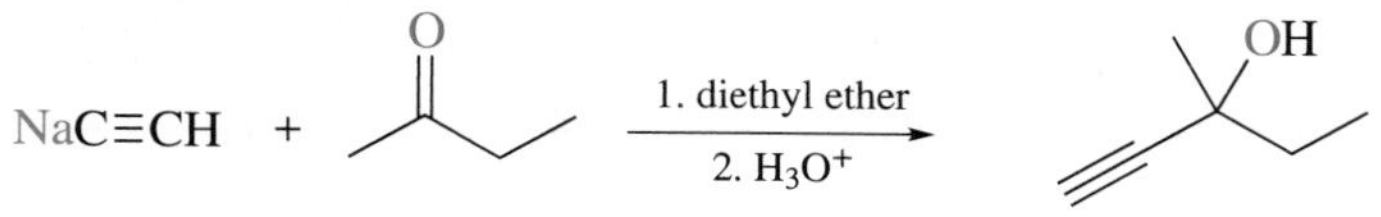

$NaC{\equiv}CH$ (Sodium acetylide) + 2-Butanone $\xrightarrow[\text{2. } H_3O^+]{\text{1. diethyl ether}}$ 3-Methyl-1-pentyn-3-ol (72%)

Section 15.7 Retrosynthetic analysis of alcohols via Grignard and organolithium reagents begins with a disconnection of one of the groups attached to the carbon that bears the oxygen. The detached group is viewed as synthetically equivalent to a carbanion, and the structural unit from which it is disconnected becomes the aldehyde or ketone component.

CH_3, OH ⟹ =O + $:\bar{C}H_3$

Section 15.8 Methylene transfer from iodomethylzinc iodide to alkene is called the *Simmons–Smith reaction* and converts alkenes to cyclopropanes.

CH_2I_2, Zn(Cu) / diethyl ether

2-Methyl-1-butene → 1-Ethyl-1-methylcyclopropane (79%)

Stereospecific syn addition of a CH_2 group to the double bond occurs.

Section 15.9 Carbenes are species that contain a *divalent carbon;* that is, a carbon with only two bond. Certain organometallic reagents resemble carbenes in their reactions and are referred to as **carbenoids.** Iodomethylzinc iodide is an example.

Section 15.10 Transition-metal complexes that contain one or more organic ligands offer a rich variety of structural types and reactivity. Organic ligands can be bonded to a metal by a σ bond or through its π system. The 18-electron rule is a guide to the number of ligands that may be attached to a particular metal.

Section 15.11 Lithium dialkylcuprates and diarylcuprates (R_2CuLi and Ar_2CuLi) are prepared by the reaction of a copper(I) salt with two equivalents of the corresponding organolithium reagent and undergo cross-coupling with primary alkyl halides and aryl and vinylic halides.

$(CH_3)_2CuLi$ + Cl (benzyl chloride) —diethyl ether→

Lithium dimethylcuprate, Benzyl chloride → Ethylbenzene (80%)

Section 15.12 Certain formulations of Pd(0) are catalysts for a number of useful carbon–carbon bond-forming processes represented by the general equation:

RX + PdL_2 —oxidative addition→ R–PdL_2–X —(R'_4Sn, R′ZnCl, $R'B(OR'')_2$; Transmetalation)→ R–PdL_2–R′ —reductive elimination→ R−R′ + PdL_2

The various methods are known as the Stille, Negishi, or Suzuki-Miyaura cross-couplings according to whether the organometallic component is a derivative of tin, zinc, or boron, respectively. The Heck reaction accomplishes the same transformation but uses an alkene as the reactant.

N≡C–C₆H₄–Br + ethyl acrylate (O, OCH_2CH_3) —$Pd(OAc)_2$, Ar_3P→ N≡C–C₆H₄–CH=CH–C(O)OCH_2CH_3

1-Bromo-4-cyanobenzene, Ethyl acrylate → Ethyl *p*-cyanocinnamate (70%)

Section 15.13 Organometallic compounds based on transition metals, especially rhodium and ruthenium, can catalyze the hydrogenation of alkenes under homogeneous conditions.

$$\text{Cinnamic acid} \xrightarrow[{[(C_6H_5)_3P]_3RhCl}]{H_2} \text{3-Phenylpropanoic acid (90\%)}$$

When a single enantiomer of a chiral catalyst is used, hydrogenations can be carried out with high enantioselectivity.

Section 15.14 The doubly bonded carbons of two alkenes exchange partners on treatment with transition-metal carbene complexes, especially those derived from ruthenium and tungsten.

$$2R_2C{=}CR'_2 \xrightarrow{\text{metallocarbene catalyst}} R_2C{=}CR_2 + R'_2C{=}CR'_2$$

Among other applications, **olefin metathesis** is useful in the synthesis of cyclic alkenes, the industrial preparation of propene, and in polymerization.

Section 15.15 Coordination polymerization of ethylene and propene has the biggest economic impact of any organic chemical process. Ziegler–Natta polymerization is carried out using catalysts derived from transition metals such as titanium and zirconium. π-Bonded and σ-bonded organometallic compounds are intermediates in coordination polymerization.

PROBLEMS

Preparation and Reactions of Main-Group Organometallic Compounds

15.20 Suggest appropriate methods for preparing each of the following organometallic compounds from the starting material of your choice.

(a) cyclopentyl–Li (b) tert-butyl–MgBr (c) $C_6H_5C{\equiv}C$–Li

15.21 Given the reactants in the preceding problem, write the structure of the principal organic product of each of the following.

(a) Cyclopentyllithium with formaldehyde in diethyl ether, followed by dilute acid.

(b) *tert*-Butylmagnesium bromide with benzaldehyde in diethyl ether, followed by dilute acid.

(c) Lithium phenylacetylide ($C_6H_5C{\equiv}CLi$) with cycloheptanone in diethyl ether, followed by dilute acid.

15.22 Predict the principal organic product of each of the following reactions:

(a) $C_6H_5\overset{O}{\overset{\|}{C}}C_6H_5 + NaC{\equiv}CH \xrightarrow[\text{2. } H_3O^+]{\text{1. liquid ammonia}}$

(b) adamantanone $+ CH_3CH_2Li \xrightarrow[\text{2. } H_3O^+]{\text{1. diethyl ether}}$

(c) 1-bromocyclopentene $\xrightarrow{\text{1. Mg, THF}}$ 2. $H\overset{O}{\overset{\|}{C}}H$ 3. H_3O^+

15.23 Addition of phenylmagnesium bromide to 4-*tert*-butylcyclohexanone gives two isomeric tertiary alcohols as products. Both alcohols yield the same alkene when subjected to acid-catalyzed dehydration. Suggest reasonable structures for these two alcohols.

O

4-*tert*-Butylcyclohexanone

Reactions of Transition-Metal Organometallic Compounds

15.24 Predict the principal organic product of each of these reactions involving transition-metal organic reagents.

(a) $\xrightarrow[\text{Zn(Cu)}\\ \text{diethyl ether}]{CH_2I_2}$

(b) $\xrightarrow[\text{Zn(Cu)}\\ \text{diethyl ether}]{CH_2I_2}$

(c) CH_3O ... —I + $LiCu(CH_3)_2$ ⟶

(d) O, O, S, O, O, CH_3 + $LiCu(CH_2CH_2CH_2CH_3)_2$ ⟶

(e) O, I, F, Br + O, OCH_2CH_3 $\xrightarrow[\text{5 mol \%}]{\text{palladium acetate}}$

15.25 Reaction of lithium diphenylcuprate with optically active 2-bromobutane yields 2-phenylbutane, with high net inversion of configuration. When the 2-bromobutane used has the absolute configuration shown, will the 2-phenylbutane formed have the (*R*)- or (*S*)-configuration?

Br

15.26 A different stereoisomer of 1-*tert*-butyl-4-methylcyclohexane was formed when lithium dimethylcuprate was allowed to react with each of the compounds shown.

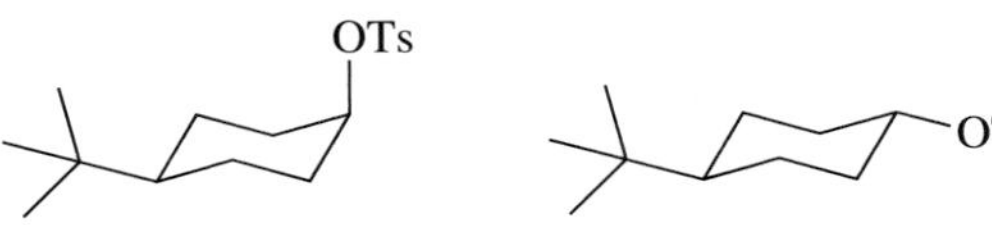

cis-4-*tert*-Butylcyclohexyl *p*-toluenesulfonate

trans-4-*tert*-Butylcyclohexyl *p*-toluenesulfonate

Give the structure of the product from each reactant. One reactant gave a higher yield of the substitution product than the other (36% versus 6%). Which one? What was the major product in each reaction?

Synthetic Applications of Organometallic Compounds

15.27 Using 1-bromobutane and any necessary organic or inorganic reagents, suggest efficient syntheses of each of the following alcohols:

(a) 1-Pentanol
(b) 2-Hexanol
(c) 1-Phenyl-1-pentanol
(d) 1-Butylcyclobutanol

15.28 Using phenyllithium and any necessary organic or inorganic reagents, suggest efficient syntheses of each of the following:

(a) Benzyl bromide
(b) 1-Bromo-1-phenylcyclohexane
(c) *trans*-2-Phenylcyclohexanol
(d) 2-Phenyl-1,3-butadiene

15.29 Apply retrosynthetic analysis to identify all the practical combinations of Grignard reagent and aldehyde or ketone that will give the required target.

(a) OH
(b) OCH_3 OH
(c) OH
(d) OH
(e) OH

15.30 A number of drugs are prepared by reactions in which carbon–carbon bond formation is the last step. Indicate what you believe would be a reasonable last step in the synthesis of each of the following:

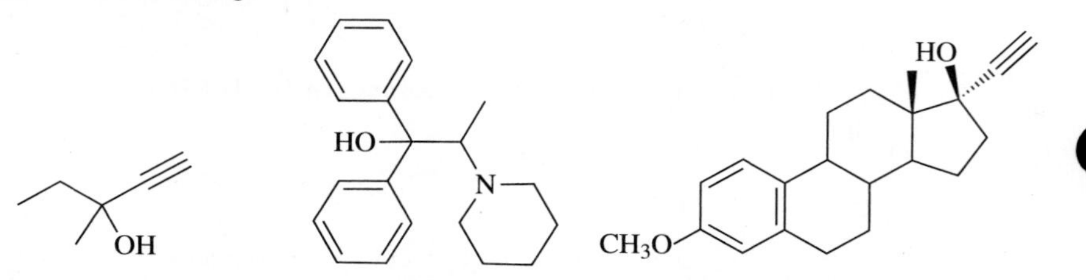

(a) Meparfynol: a mild hypnotic or sleep-inducing drug
(b) Diphepanol: a cough suppressant
(c) Mestranol: an estrogenic component in oral contraceptives

15.31 The following conversion was carried out in two steps, the first of which involved formation of a *p*-toluenesulfonate. Indicate the reagents for this step, and show how you could convert the *p*-toluenesulfonate to the desired product.

O OH —two steps→ O

15.32 Outline syntheses of (a) *cis*-1,2-diethylcyclopropane and (*b*) *trans*-1,2-diethylcyclopropane from compounds containing two or fewer carbons. (*Hint:* The last reaction in each case involves an organozinc reagent.)

Transition Metal-Catalyzed Reactions

15.33 (*S*)-(+)-Ibuprofen can be prepared by enantioselective hydrogenation. Give the structure of the $C_{13}H_{16}O_2$ isomer you would select as a candidate for this reaction.

O OH

(*S*)-(+)-Ibuprofent

15.34 Like other hydroborations, the reaction of alkynes with catecholborane is a syn addition and its regioselectivity is opposite to Markovnikov's rule.

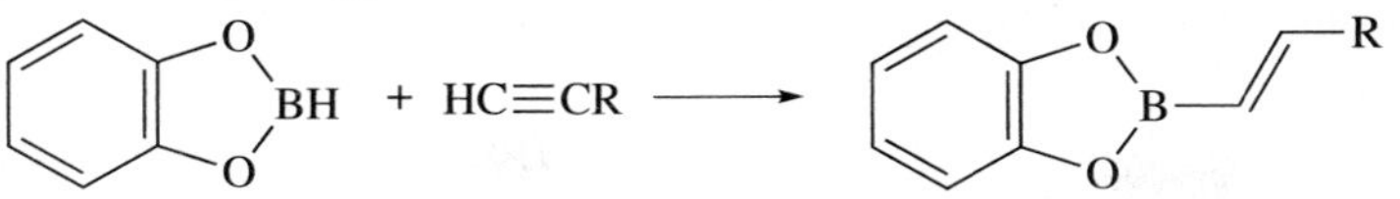

BH + HC≡CR ⟶ B—R

Use this fact to outline a synthesis of the compound shown from 1-hexyne and (Z)-$C_6H_5CH{=}CHBr$.

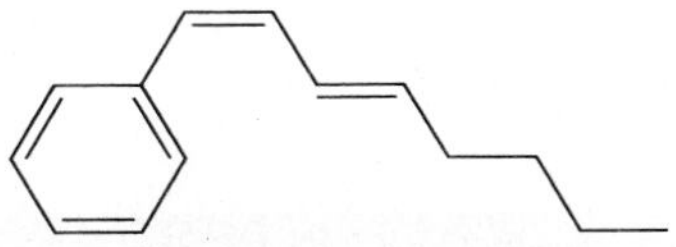

(1Z,3E)-1-Phenyl-1,3-octadiene

15.35 The sex attractant of the female silkworm has been synthesized by the reaction shown. What is its structure?

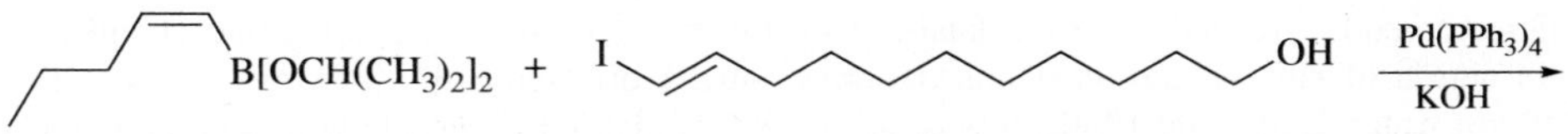

$B[OCH(CH_3)_2]_2$ + I ... OH $\xrightarrow[\text{KOH}]{\text{Pd(PPh}_3)_4}$

15.36 A compound having the molecular formula $C_{22}H_{32}O_2$ was isolated in 66% yield in the following reaction. Suggest a reasonable structure for this compound. What other organic compound is formed in this reaction?

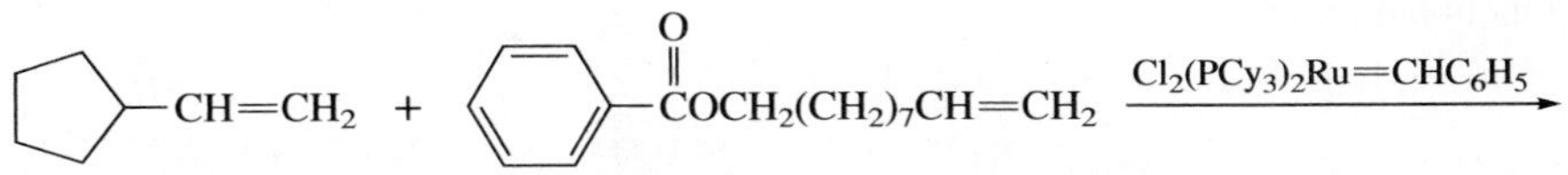

$-CH{=}CH_2$ + $-\overset{O}{\overset{\|}{C}}OCH_2(CH_2)_7CH{=}CH_2$ $\xrightarrow{Cl_2(PCy_3)_2Ru{=}CHC_6H_5}$

15.37 (a) *Exaltolide,* a musk substance, has been prepared by the reaction sequence shown. What is compound A?

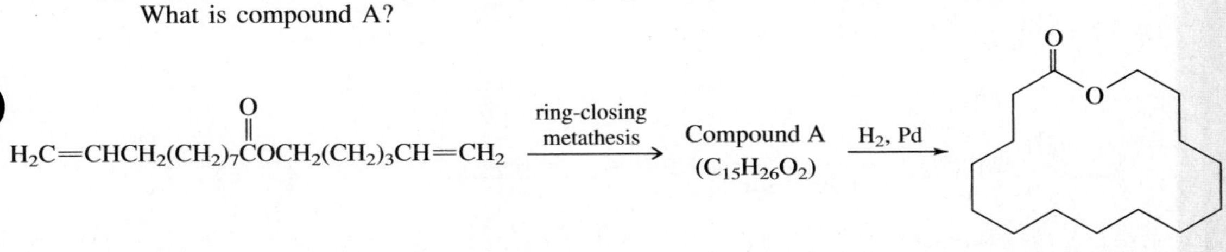

$H_2C{=}CHCH_2(CH_2)_7\overset{O}{\overset{\|}{C}}OCH_2(CH_2)_3CH{=}CH_2$ $\xrightarrow{\text{ring-closing metathesis}}$ Compound A ($C_{15}H_{26}O_2$) $\xrightarrow{H_2,\ Pd}$

Exaltolide

(b) An analogous sequence using $H_2C{=}CHCH_2(CH_2)_2\overset{O}{\overset{\|}{C}}OCH_2(CH_2)_8CH{=}CH_2$ as the reactant also gives Exaltolide. What is the product of ring-closing metathesis of this reactant?

15.38 On treatment with a Grubbs' olefin metathesis catalyst, the compound shown reacted with styrene to give a 95% yield of a product with the molecular formula $C_{25}H_{30}O_3$, which was later used in the synthesis of a metabolite isolated from a species of mollusk. Suggest a reasonable structure for the metathesis product.

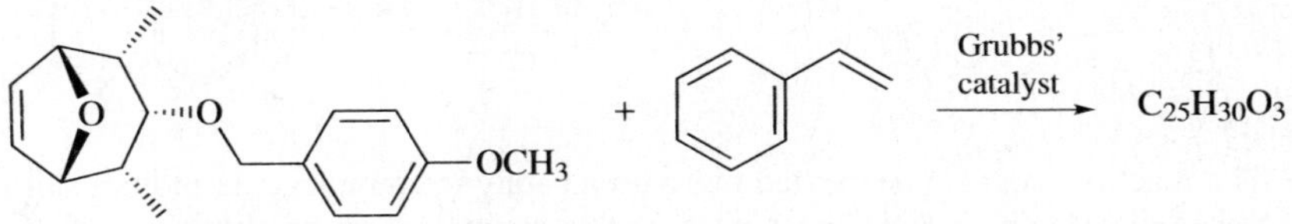

O ····O —OCH₃ + $\xrightarrow{\text{Grubbs' catalyst}}$ $C_{25}H_{30}O_3$

15.39 One synthetic advantage of olefin metathesis is that the catalyst tolerates a variety of functional groups in the reactant. In a synthesis of the antiinfluenza drug Tamiflu (oseltamivir), ring-closing metathesis was used to prepare the highly functionalized cyclohexene derivative shown. What was the reactant?

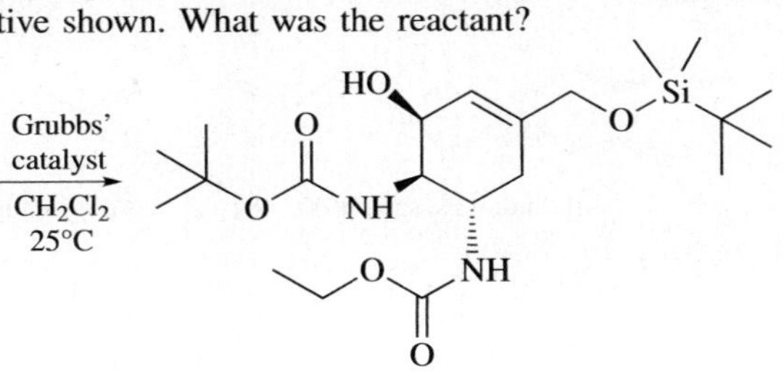

$\xrightarrow[\text{CH}_2\text{Cl}_2,\ 25°\text{C}]{\text{Grubbs' catalyst}}$ HO, O, Si, O, NH, NH, O, O

15.40 The following compound is a phytotoxin that can be synthesized by ring-closing metathesis, or RCM as it is commonly known. Two precursors based on RCM are possible. What are they?

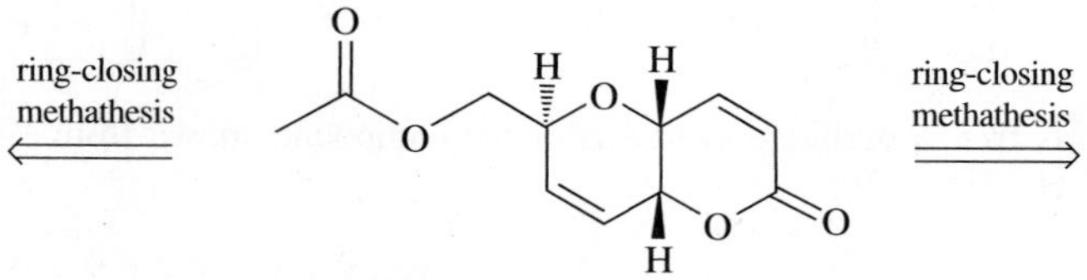

ring-closing methathesis ⟸ ⟹ ring-closing methathesis

Descriptive Passage and Interpretive Problems 15

Allylindium Reagents

Organometallic reagents are the basis of useful methods in synthetic organic chemistry for the formation of carbon–carbon bonds. However, many organometallic reagents are sensitive to air and/or water. In the late 1980s, it was discovered that carbonyl compounds can undergo *allylation,* the addition of an allyl group, with allyl halides in the presence of indium. While several solvents may also be used for this reaction, the use of water makes this procedure attractive from the standpoint of green, sustainable chemistry. Indium-mediated allylation is compatible with a variety of functional groups.

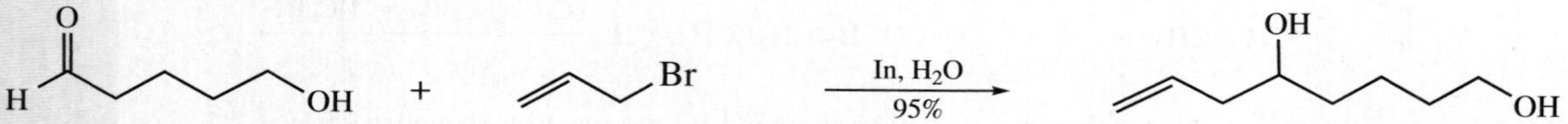

In, H_2O, 95%

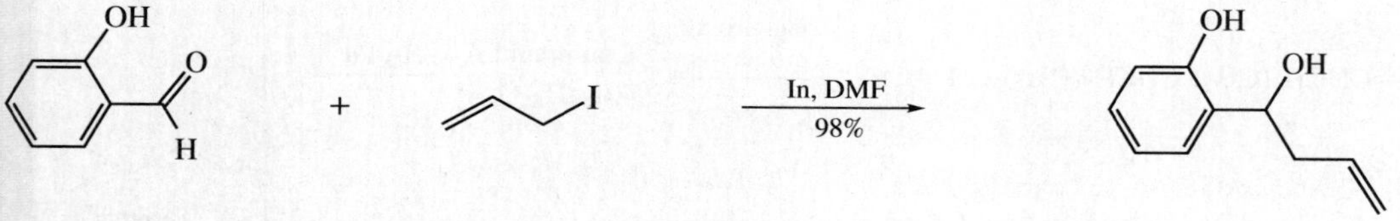

In, DMF, 98%

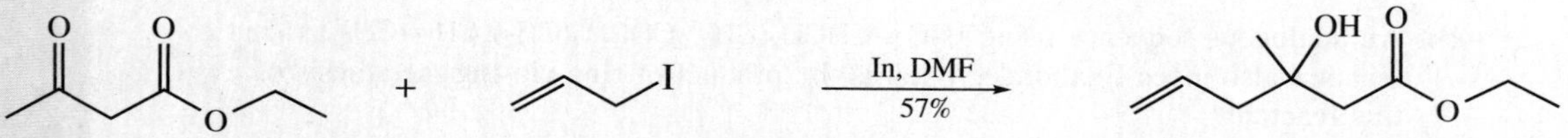

In, DMF, 57%

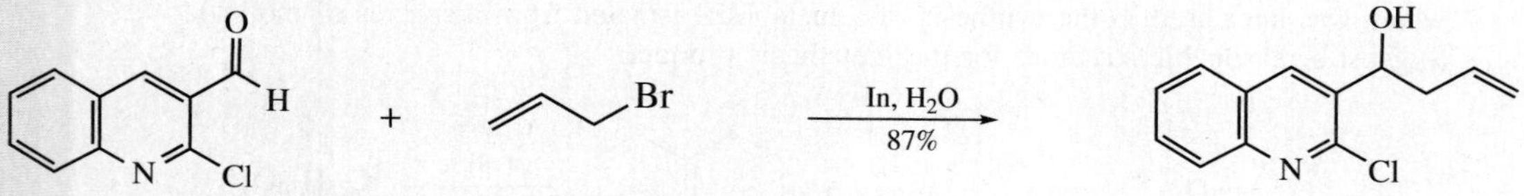

In, H_2O, 87%

The reaction can also be carried out with a catalytic amount of an indium salt such as $InCl_3$ and another metal (Zn or Al). The nature of the organoindium reagent depends on the starting reagents and the reaction conditions. An aggregate of 3 moles of alkyl halide, 2 moles of carbonyl compound, and 3 moles of indium (allylindium sesquihalide) and allylindium have both been proposed as intermediates.

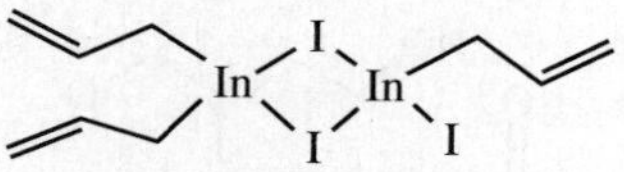

Allylindium sesquihalide 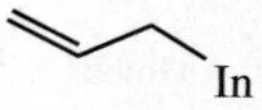 Allylindium

15.41 What is the product of the following reaction?

In, DMF

A. OH B. OH

C. HO D. OH

15.42 The following compound was synthesized from 2-phenylpropanal and which halide using indium allylation?

CO_2H OH ⟹ H O +

2-Phenylpropanal Halide

A. CO_2H Br B. Br CO_2H C. CO_2H Br D. Br CO_2H

15.43 Reactions that involve allylic intermediates are known to give products that result from rearrangement. The reaction shown below gives two products. What is the likely structure of the second one?

O H + Br In, H_2O → OH

A. OH B. OH C. OH D. OH

15.44 What is the product of the following three-step synthesis?

CH_3O O H I, In → Br, NaH → Grubbs' catalyst →

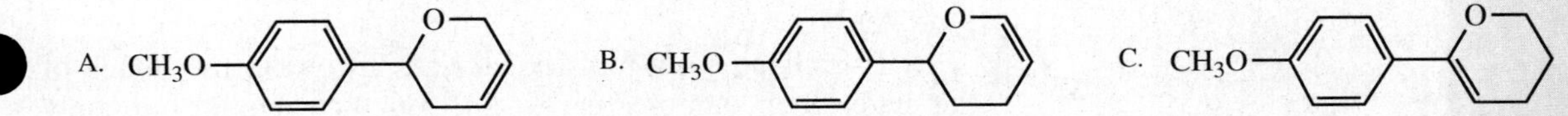

Chapter

16

This battery-powered electric car is equipped with a methanol fuel cell that recharges the battery while the vehicle is driven. The methanol in the fuel cell is not burned but converted electrochemically through the use of a catalyst to carbon dioxide and water. The process produces an electric current that charges the battery.
(Center): ©Serenergy; (Right): ©Car Culture® Collection/Getty Images

CHAPTER OUTLINE

Alcohols, Diols, and Thiols

The next several chapters deal with the chemistry of various oxygen-containing functional groups. The interplay of these important classes of compounds—alcohols, ethers, aldehydes, ketones, carboxylic acids, and derivatives of carboxylic acids—is fundamental to organic chemistry and biochemistry.

ROH	ROR′	RCH (C=O)	RCR′ (C=O)	RCOH (C=O)
Alcohol	Ether	Aldehyde	Ketone	Carboxylic acid

We'll start by discussing in more detail a class of compounds already familiar to us, *alcohols.* Alcohols were introduced in Chapter 5 and have appeared regularly since then. With this chapter we extend our knowledge of alcohols, particularly with respect to their relationship to carbonyl-containing compounds. In the course of studying alcohols, we shall also look at some relatives. **Diols** are alcohols in which two hydroxyl groups (—OH) are present; **thiols** are compounds that contain an —SH group. **Phenols** are compounds of the type ArOH.

This chapter is a transitional one. It ties together much of the material encountered earlier and sets the stage for our study of other oxygen-containing functional groups in the chapters that follow.

16.1 Sources of Alcohols

At one time, the major source of *methanol* was as a byproduct in the production of charcoal from wood—hence, the name *wood alcohol.* Now, most of the more than 10 billion lb of methanol used annually in the United States is synthetic, prepared by reduction of carbon monoxide with hydrogen. Carbon monoxide is normally made from methane.

$$\underset{\text{Carbon monoxide}}{CO} + \underset{\text{Hydrogen}}{2H_2} \xrightarrow[400°C]{ZnO/Cr_2O_3} \underset{\text{Methanol}}{CH_3OH}$$

The major use of methanol is in the preparation of formaldehyde as a starting material for various resins and plastics.

When vegetable matter ferments, its carbohydrates are converted to *ethanol* and carbon dioxide by enzymes present in yeast. Fermentation of barley produces beer; grapes give wine. The maximum ethanol content is on the order of 15%, because higher concentrations inactivate the enzymes, halting fermentation. Distillation of the fermentation broth gives "distilled spirits" of increased ethanol content. The characteristic flavors, odors, and colors of the various alcoholic beverages depend on both their origin and the way they are aged.

Synthetic ethanol and isopropyl alcohol are derived from petroleum by hydration of ethylene and propene, respectively.

Most alcohols of six carbons or fewer, as well as many higher alcohols, are commercially available at low cost. Some occur naturally; others are the products of efficient syntheses. Figure 16.1 presents the structures of a few naturally occurring alcohols. Table 16.1 summarizes the reactions encountered in earlier chapters that give alcohols and illustrates a thread that runs through the fabric of organic chemistry: *a reaction that is characteristic of one functional group often serves as a synthetic method for preparing another.*

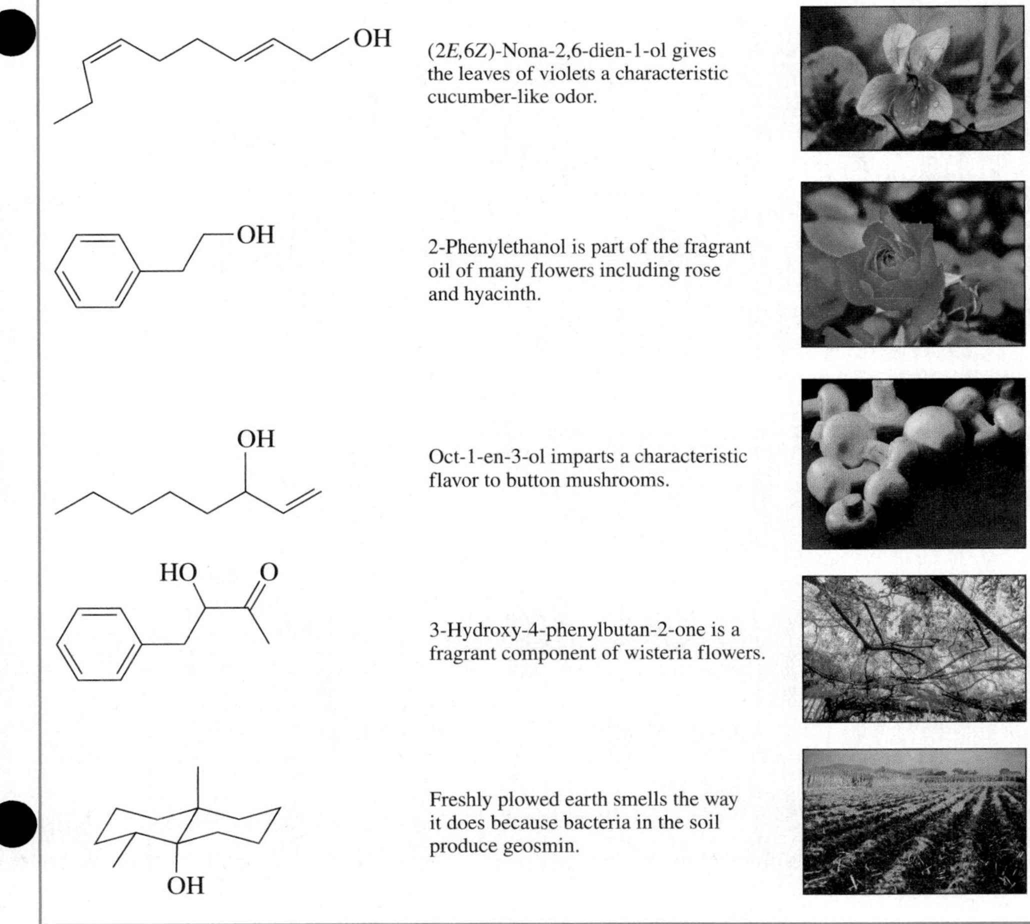

OH
(2E,6Z)-Nona-2,6-dien-1-ol gives the leaves of violets a characteristic cucumber-like odor.
OH
2-Phenylethanol is part of the fragrant oil of many flowers including rose and hyacinth.
OH
Oct-1-en-3-ol imparts a characteristic flavor to button mushrooms.
HO
O
3-Hydroxy-4-phenylbutan-2-one is a fragrant component of wisteria flowers.
Freshly plowed earth smells the way it does because bacteria in the soil produce geosmin.
OH

Figure 16.1

Several of the countless naturally occurring alcohols that stimulate our senses.

(Top): ©Brian Lasenby/Getty Images; (Top center): ©Hissyh2/Getty Images; (Center): ©Purestock/SuperStock; (Bottom center): ©Kit Leong/Shutterstock; (Bottom): ©Glow Images

TABLE 16.1 Reactions Discussed in Earlier Chapters That Yield Alcohols

Reaction (section) and comments	General equation and specific example
Acid-catalyzed hydration of alkenes (Section 8.6) Water adds to the double bond in accordance with Markovnikov's rule.	$R_2C{=}CR_2 + H_2O \xrightarrow{H^+} R_2C(OH)CHR_2$ Alkene Water Alcohol 2,3-Dimethyl-2-butene $\xrightarrow[H_2SO_4]{H_2O}$ 2,3-Dimethyl-2-butanol (90%)
Hydroboration–oxidation of alkenes (Section 8.8) H and OH add to the double bond with a regioselectivity opposite to Markovnikov's rule. Addition is syn, and rearrangements do not occur.	$R_2C{=}CR_2 \xrightarrow[2.\ H_2O_2,\ HO^-]{1.\ B_2H_6,\ \text{diglyme}} R_2C(OH)CHR_2$ Alkene Alcohol 1-Decene $\xrightarrow[2.\ H_2O_2,\ HO^-]{1.\ B_2H_6,\ \text{diglyme}}$ 1-Decanol (93%)
Hydrolysis of alkyl halides (Section 6.3) A reaction useful only with alkyl halides that do not undergo E2 elimination readily. It is rarely used for the synthesis of alcohols, since alkyl halides are normally prepared from alcohols.	$RX + HO^- \longrightarrow ROH + X^-$ Alkyl halide Hydroxide ion Alcohol Halide ion 2,4,6-Trimethylbenzyl chloride $\xrightarrow[\text{heat}]{H_2O,\ Ca(OH)_2}$ 2,4,6-Trimethylbenzyl alcohol (78%)
Reaction of Grignard and organolithium reagents with aldehydes and ketones (Section 15.5) A method that allows for alcohol preparation with formation of new carbon–carbon bonds. Primary, secondary, and tertiary alcohols are all accessible.	$RMgX \text{ or } RLi + R'C(=O)R'' \xrightarrow[2.\ H_3O^+]{1.\ \text{diethyl ether}} RR'R''C{-}OH$ Grignard or organolithium reagent Aldehyde or ketone Alcohol Cyclopentylmagnesium bromide + Formaldehyde ($HCHO$) $\xrightarrow[2.\ H_3O^+]{1.\ \text{diethyl ether}}$ Cyclopentylmethanol (62–64%) Butyllithium + Acetophenone $\xrightarrow[2.\ H_3O^+]{1.\ \text{diethyl ether}}$ 2-Phenyl-2-hexanol (67%)

16.2 Preparation of Alcohols by Reduction of Aldehydes and Ketones

The most obvious way to reduce an aldehyde or a ketone to an alcohol is by hydrogenation of the carbon–oxygen double bond. Like the hydrogenation of alkenes, the reaction is exothermic but exceedingly slow in the absence of a catalyst. Finely divided metals such as platinum, palladium, nickel, and ruthenium are effective catalysts for the hydrogenation of aldehydes and ketones. Aldehydes yield primary alcohols:

Recall from Section 2.24 that reduction corresponds to a decrease in the number of bonds between carbon and oxygen or an increase in the number of bonds between carbon and hydrogen (or both).

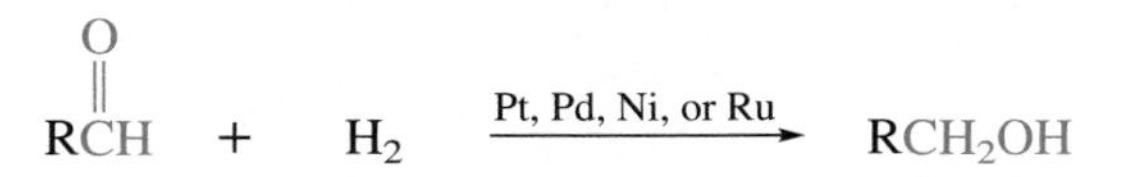

$$\underset{\text{Aldehyde}}{\overset{\displaystyle O}{\overset{\|}{\text{RCH}}}} + \underset{\text{Hydrogen}}{H_2} \xrightarrow{\text{Pt, Pd, Ni, or Ru}} \underset{\text{Primary alcohol}}{\text{RCH}_2\text{OH}}$$

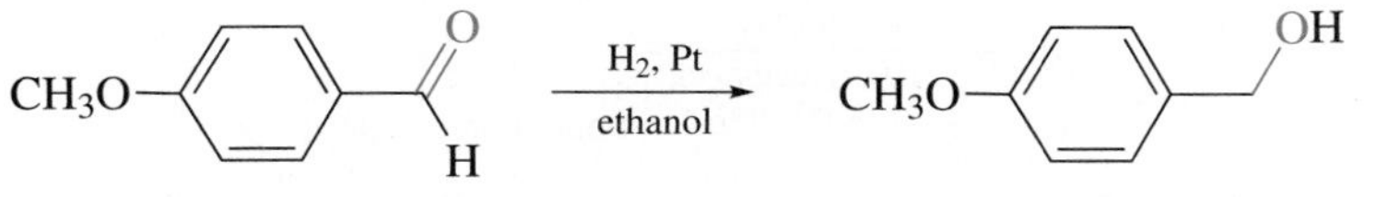

$$\underset{p\text{-Methoxybenzaldehyde}}{\text{CH}_3\text{O}-\text{C}_6\text{H}_4-\text{CHO}} \xrightarrow[\text{ethanol}]{\text{H}_2\text{, Pt}} \underset{p\text{-Methoxybenzyl alcohol (92\%)}}{\text{CH}_3\text{O}-\text{C}_6\text{H}_4-\text{CH}_2\text{OH}}$$

Ketones yield secondary alcohols:

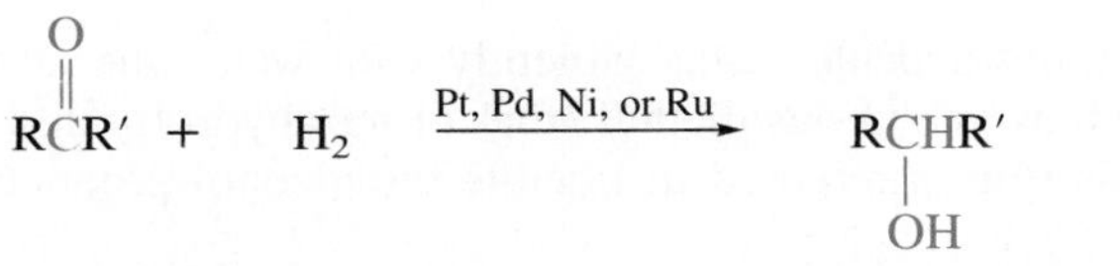

$$\underset{\text{Ketone}}{\overset{\displaystyle O}{\overset{\|}{\text{RCR}'}}} + \underset{\text{Hydrogen}}{H_2} \xrightarrow{\text{Pt, Pd, Ni, or Ru}} \underset{\text{Secondary alcohol}}{\underset{\displaystyle \text{OH}}{\underset{|}{\text{RCHR}'}}}$$

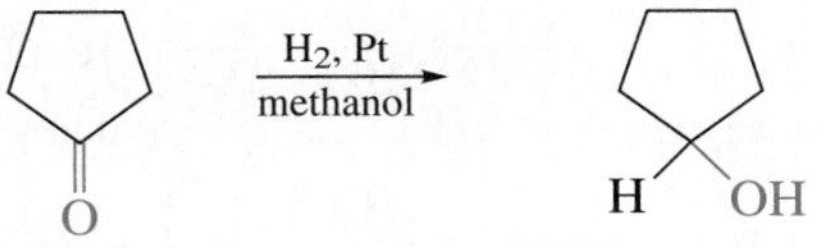

$$\underset{\text{Cyclopentanone}}{\text{C}_5\text{H}_8\text{O}} \xrightarrow[\text{methanol}]{\text{H}_2\text{, Pt}} \underset{\text{Cyclopentanol (93–95\%)}}{\text{C}_5\text{H}_9\text{OH}}$$

Problem 16.1

Which of the isomeric $C_4H_{10}O$ alcohols can be prepared by hydrogenation of aldehydes? Which can be prepared by hydrogenation of ketones? Which cannot be prepared by hydrogenation of a carbonyl compound?

For most laboratory-scale reductions of aldehydes and ketones, catalytic hydrogenation has been replaced by methods based on metal hydride reducing agents. The two most common reagents are sodium borohydride and lithium aluminum hydride.

$$\text{Na}^+ \left[\text{BH}_4\right]^- \qquad \text{Li}^+ \left[\text{AlH}_4\right]^-$$

Sodium borohydride ($NaBH_4$) Lithium aluminum hydride ($LiAlH_4$)

Sodium borohydride is especially easy to use, needing only to be added to an aqueous or alcoholic solution of an aldehyde or a ketone:

$$\underset{\text{Aldehyde}}{\text{RCH(=O)}} \xrightarrow[\text{water, methanol, or ethanol}]{\text{NaBH}_4} \underset{\text{Primary alcohol}}{\text{RCH}_2\text{OH}}$$

$$\underset{m\text{-Nitrobenzaldehyde}}{} \xrightarrow[\text{methanol}]{\text{NaBH}_4} \underset{m\text{-Nitrobenzyl alcohol (82\%)}}{}$$

$$\underset{\text{Ketone}}{\text{RCR}'\text{(=O)}} \xrightarrow[\text{water, methanol, or ethanol}]{\text{NaBH}_4} \underset{\text{Secondary alcohol}}{\text{RCH(OH)R}'}$$

$$\underset{\text{4,4-Dimethyl-2-pentanone}}{} \xrightarrow[\text{ethanol}]{\text{NaBH}_4} \underset{\text{4,4-Dimethyl-2-pentanol (85\%)}}{}$$

The same kinds of aprotic solvents are used for $LiAlH_4$ as for Grignard reagents.

Lithium aluminum hydride reacts violently with water and alcohols, so it must be used in solvents such as anhydrous diethyl ether or tetrahydrofuran. Following reduction, a separate hydrolysis step is required to liberate the alcohol product:

$$\underset{\text{Aldehyde}}{\text{RCH(=O)}} \xrightarrow[\text{2. H}_2\text{O}]{\text{1. LiAlH}_4\text{, diethyl ether}} \underset{\text{Primary alcohol}}{\text{RCH}_2\text{OH}}$$

$$\underset{\text{Heptanal}}{} \xrightarrow[\text{2. H}_2\text{O}]{\text{1. LiAlH}_4\text{, diethyl ether}} \underset{\text{1-Heptanol (86\%)}}{}$$

$$\underset{\text{Ketone}}{\text{RCR}'\text{(=O)}} \xrightarrow[\text{2. H}_2\text{O}]{\text{1. LiAlH}_4\text{, diethyl ether}} \underset{\text{Secondary alcohol}}{\text{RCH(OH)R}'}$$

$$\underset{\text{1,1-Diphenyl-2-propanone}}{} \xrightarrow[\text{2. H}_2\text{O}]{\text{1. LiAlH}_4\text{, diethyl ether}} \underset{\text{1,1-Diphenyl-2-propanol (84\%)}}{}$$

Sodium borohydride and lithium aluminum hydride react with carbonyl compounds in much the same way that Grignard reagents do, except that they function as *hydride donors* rather than as carbanion sources. The sodium borohydride reduction of an aldehyde or ketone can be outlined as:

Aldehyde or ketone + borohydride ion → Alkoxyborohydride $\xrightarrow{3R_2C{=}O}$ Tetraalkoxyborate $\xrightarrow[HO^-]{H_2O}$ 4 Alcohol + $(HO)_4\bar{B}$ Borate ion

Two points about the process bear special mention.

1. At no point is H_2 involved. The reducing agent is borohydride ion (BH_4^-).
2. In the reduction $R_2C{=}O \rightarrow R_2CHOH$, the hydrogen bonded to carbon comes from BH_4^-; the hydrogen on oxygen comes from an OH group of the solvent (water, methanol, or ethanol).

Problem 16.2

Sodium borodeuteride ($NaBD_4$) and lithium aluminum deuteride ($LiAlD_4$) are convenient reagents for introducing deuterium, the mass-2 isotope of hydrogen, into organic compounds. Write the structure of the organic product of the following reactions, clearly showing the position of all the deuterium atoms in each:

(a) $CH_3CHO \xrightarrow[H_2O]{NaBD_4}$

(b) $(CH_3)_2C{=}O \xrightarrow[CH_3OD]{NaBD_4}$

(c) $C_6H_5CHO \xrightarrow[CD_3OH]{NaBD_4}$

(d) $HCHO \xrightarrow[2.\ D_2O]{1.\ LiAlD_4,\ \text{diethyl ether}}$

Sample Solution (a) Sodium borodeuteride transfers deuterium to the carbonyl group of acetaldehyde, forming a C—D bond.

$CH_3CH{=}O + D{-}\bar{B}D_3 \longrightarrow CH_3CHD{-}O{-}\bar{B}D_3 \xrightarrow{3CH_3CH{=}O} CH_3CHD{-}O{-}\bar{B}(OCHDCH_3)_3$

Hydrolysis of $(CH_3CHDO)_4B^-$ in H_2O leads to the formation of ethanol, retaining the C—D bond formed in the preceding step while forming an O—H bond.

$CH_3CHD{-}O{-}\bar{B}(OCHDCH_3)_3 \xrightarrow{4H_2O} 4\,[CH_3CHDOH] + (HO)_4\bar{B}$

Ethanol-1-*d*

The mechanism of lithium aluminum hydride reduction of aldehydes and ketones is analogous to that of sodium borohydride except that the reduction and hydrolysis stages are independent operations. The reduction is carried out in diethyl ether, followed by a separate hydrolysis step when water is added to the reaction mixture.

$$4R_2C{=}O \xrightarrow[\text{diethyl ether}]{LiAlH_4} (R_2CHO)_4Al^- \xrightarrow{4H_2O} 4R_2CHOH + Al(OH)_4^-$$

Aldehyde or ketone → Tetraalkoxyaluminate → Alcohol

Neither sodium borohydride nor lithium aluminum hydride reduces isolated carbon–carbon double bonds. This makes possible the selective reduction of a carbonyl group in a molecule that contains both carbon–carbon and carbon–oxygen double bonds.

Catalytic hydrogenation would not be suitable for this transformation, because H_2 adds to carbon–carbon double bonds faster than it reduces carbonyl groups.

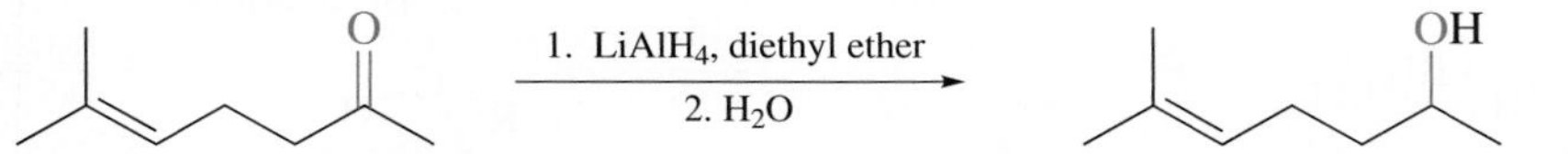

6-Methyl-5-hepten-2-one $\xrightarrow[\text{2. } H_2O]{\text{1. } LiAlH_4\text{, diethyl ether}}$ 6-Methyl-5-hepten-2-ol (90%)

16.3 Preparation of Alcohols by Reduction of Carboxylic Acids

Carboxylic acids are exceedingly difficult to reduce. Acetic acid, for example, is often used as a solvent in catalytic hydrogenations because it is inert under the reaction conditions. Lithium aluminum hydride is one of the few reducing agents capable of reducing a carboxylic acid to a primary alcohol.

$$\underset{\text{Carboxylic acid}}{RCOH} \xrightarrow[\text{2. } H_2O]{\text{1. } LiAlH_4\text{, diethyl ether}} \underset{\text{Primary alcohol}}{RCH_2OH}$$

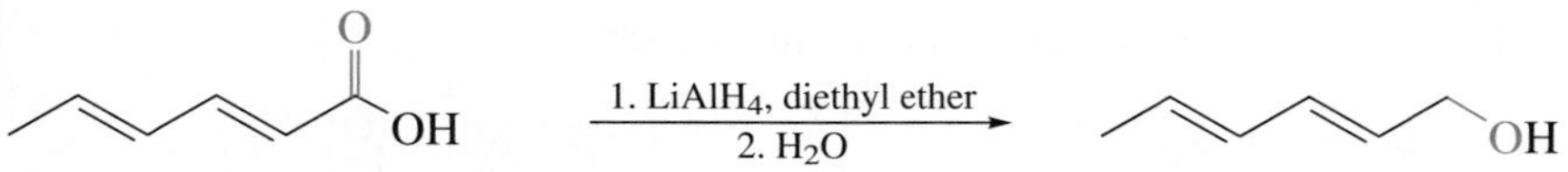

(2*E*,4*E*)-Hexa-2,4-dienoic acid $\xrightarrow[\text{2. } H_2O]{\text{1. } LiAlH_4\text{, diethyl ether}}$ (2*E*,4*E*)-Hexa-2,4-dien-l-ol (92%)

Esters can also be reduced to alcohols with lithium aluminum hydride. We will examine this reaction in detail in Section 20.11.

Sodium borohydride is not nearly as potent a hydride donor as lithium aluminum hydride and does not reduce carboxylic acids.

16.4 Preparation of Alcohols from Epoxides

Grignard reagents react with ethylene oxide to yield primary alcohols containing two more carbon atoms than the alkyl halide from which the organometallic compound was prepared.

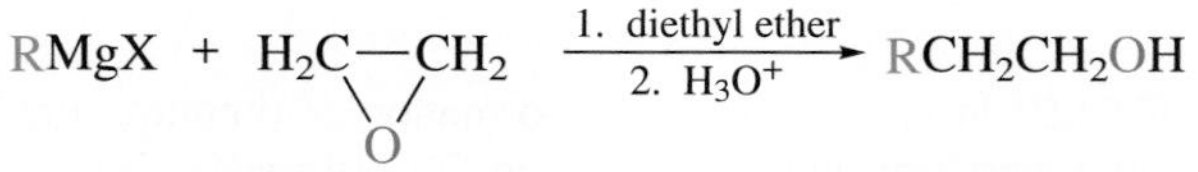

$$\underset{\text{Grignard reagent}}{RMgX} + \underset{\text{Ethylene oxide}}{H_2C\text{—}CH_2 \ (O)} \xrightarrow[\text{2. } H_3O^+]{\text{1. diethyl ether}} \underset{\text{Primary alcohol}}{RCH_2CH_2OH}$$

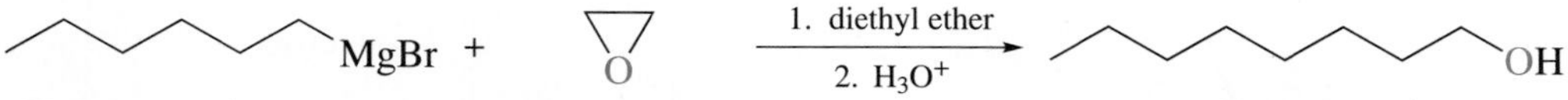

Hexylmagnesium bromide + Ethylene oxide $\xrightarrow[\text{2. } H_3O^+]{\text{1. diethyl ether}}$ 1-Octanol (71%)

Organolithium reagents react with epoxides in a similar manner.

Problem 16.3

Each of the following alcohols has been prepared by reaction of a Grignard reagent with ethylene oxide. Select the appropriate Grignard reagent in each case.

(a) 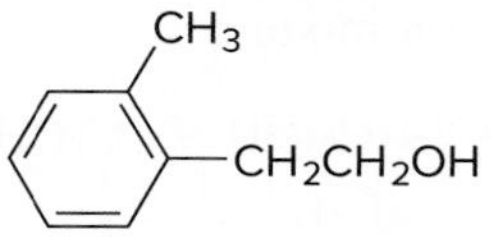 CH_3 / CH_2CH_2OH (2-methylphenyl ring)

(b)  OH

Sample Solution (a) Reaction with ethylene oxide results in the addition of a $—CH_2CH_2OH$ unit to the Grignard reagent. The Grignard reagent derived from *o*-bromotoluene (or *o*-chlorotoluene or *o*-iodotoluene) is appropriate here.

$$o\text{-}CH_3C_6H_4MgBr + \text{ethylene oxide} \xrightarrow[2.\ H_3O^+]{1.\ \text{diethyl ether}} o\text{-}CH_3C_6H_4CH_2CH_2OH$$

o-Methylphenylmagnesium bromide + Ethylene oxide → 2-(*o*-Methylphenyl)ethanol (66%)

Epoxide rings are readily opened with cleavage of the carbon–oxygen bond when attacked by nucleophiles. Grignard reagents and organolithium reagents react with ethylene oxide by serving as sources of nucleophilic carbon. The mechanism resembles an S_N2 reaction. Cleavage of the epoxide C—O bond is analogous to the cleavage of the bond between carbon and a leaving group.

$$\overset{\delta-}{R}—\overset{\delta+}{Mg}X + H_2C—CH_2\ (\text{epoxide O}) \longrightarrow R—CH_2—CH_2—\ddot{\underset{..}{O}}{:}^{-}\ \overset{+}{Mg}X \xrightarrow{H_3O^+} RCH_2CH_2OH$$

(may be written as RCH_2CH_2OMgX)

This kind of chemical reactivity of epoxides is rather general. Nucleophiles other than Grignard reagents react with epoxides, and epoxides more elaborate than ethylene oxide may be used. These features of epoxide chemistry will be discussed in Sections 17.11–17.12.

16.5 Preparation of Diols

Much of the chemistry of diols—compounds that bear two hydroxyl groups—is analogous to that of alcohols. Diols may be prepared, for example, from compounds that contain two carbonyl groups, using the same reducing agents employed in the preparation of alcohols. The following example shows the conversion of a dialdehyde to a diol by catalytic hydrogenation. Alternatively, the same transformation can be achieved by reduction with sodium borohydride or lithium aluminum hydride.

$$HC(=O)CH_2CH(CH_3)CH_2CH(=O)H \xrightarrow[\text{Ni, 125°C}]{H_2\ (100\ \text{atm})} HOCH_2CH_2CH(CH_3)CH_2CH_2OH$$

3-Methylpentanedial → 3-Methylpentane-1,5-diol (81–83%)

As can be seen in the preceding equation, the nomenclature of diols is similar to that of alcohols. The suffix *-diol* replaces *-ol,* and two locants, one for each hydroxyl group, are required. Note that the final *-e* of the parent alkane name is retained when the suffix begins with a consonant (*-diol*), but dropped when the suffix begins with a vowel (*-ol*).

Problem 16.4

Write an equation showing how 3-methylpentane-1,5-diol could be prepared from a dicarboxylic acid.

Vicinal diols are diols that have their hydroxyl groups on adjacent carbons. Two commonly encountered vicinal diols are ethane-1,2-diol and propane-1,2-diol.

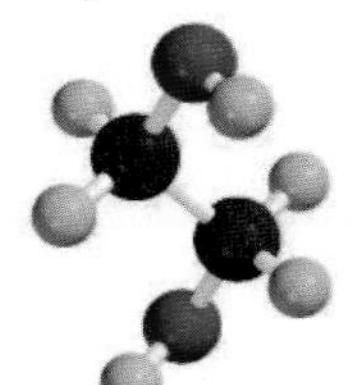

$HOCH_2CH_2OH$

Ethane-1,2-diol (ethylene glycol)

$CH_3CH(OH)CH_2OH$

Propane-1,2-diol (propylene glycol)

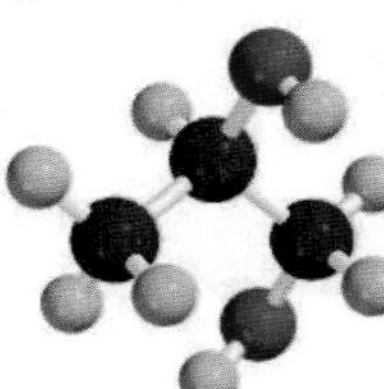

Ethylene glycol and propylene glycol are prepared industrially from the corresponding alkenes by way of their epoxides.

Ethylene glycol and *propylene glycol* are common names for these two diols and are acceptable IUPAC names. Aside from these two compounds, the IUPAC system does not use the word *glycol* for naming diols.

Vicinal diols are often prepared from alkenes using *osmium tetraoxide* (OsO_4). Osmium tetraoxide reacts rapidly with alkenes to give cyclic osmate esters.

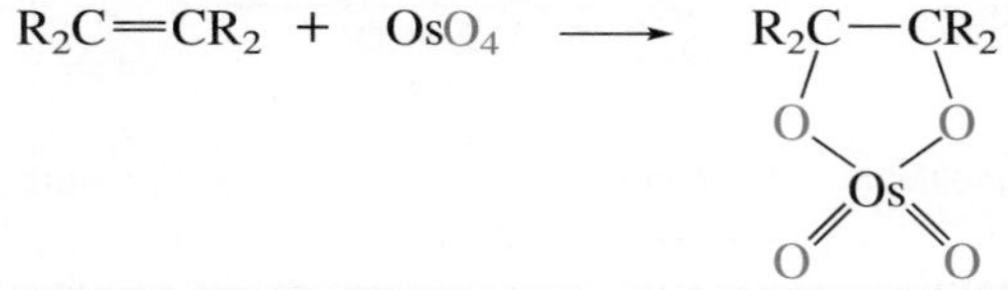

$$R_2C{=}CR_2 + OsO_4 \longrightarrow \text{Cyclic osmate ester}$$

Alkene Osmium tetraoxide Cyclic osmate ester

Osmate esters are fairly stable but are readily cleaved in the presence of an oxidizing agent such as *tert*-butyl hydroperoxide.

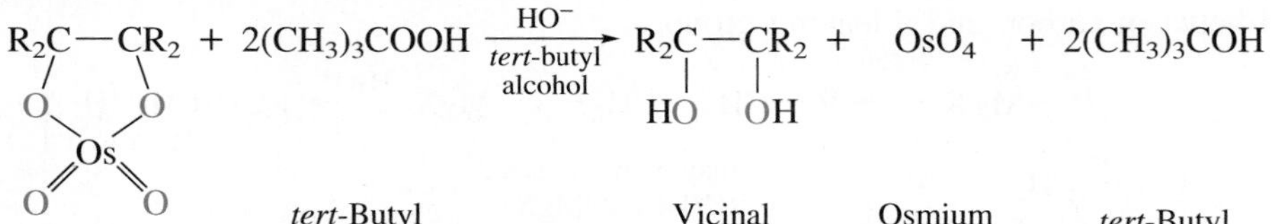

$$\text{Cyclic osmate ester} + 2(CH_3)_3COOH \xrightarrow[\text{tert-butyl alcohol}]{HO^-} R_2C(OH){-}C(OH)R_2 + OsO_4 + 2(CH_3)_3COH$$

tert-Butyl hydroperoxide; Vicinal diol; Osmium tetraoxide; *tert*-Butyl alcohol

Because osmium tetraoxide is regenerated in this step, alkenes can be converted to vicinal diols using only catalytic amounts of osmium tetraoxide in a single operation by simply allowing a solution of the alkene and *tert*-butyl hydroperoxide in *tert*-butyl alcohol containing a small amount of osmium tetraoxide and base to stand for several hours.

Overall, the reaction leads to addition of two hydroxyl groups to the double bond and is referred to as **dihydroxylation.** Both hydroxyl groups of the diol become attached to the same face of the double bond; *syn dihydroxylation of the alkene is observed.*

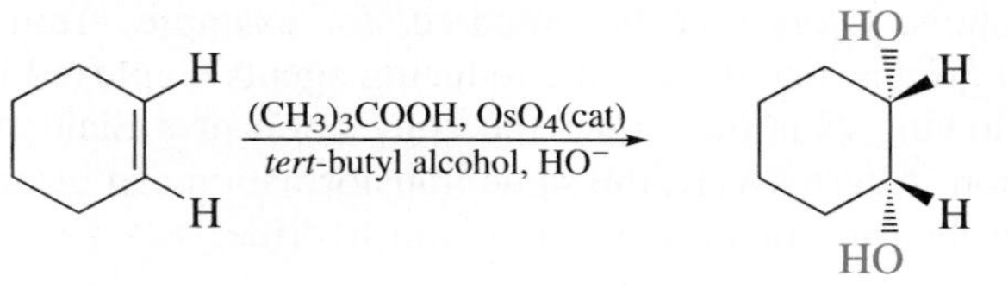

$$\text{Cyclohexene} \xrightarrow[\text{tert-butyl alcohol, } HO^-]{(CH_3)_3COOH,\ OsO_4(cat)} \textit{cis}\text{-1,2-Cyclohexanediol (62\%)}$$

Cyclohexene

cis-1,2-Cyclohexanediol (62%)

Problem 16.5

Give the structures, including stereochemistry, for the diols obtained by dihydroxylation of *cis*-2-butene and *trans*-2-butene.

Enantioselective dihydroxylation is one aspect of novel oxidation methods developed by K. Barry Sharpless (The Scripps Research Institute). Sharpless shared the 2001 Nobel Prize in Chemistry with William S. Knowles and Ryogi Noyori (Section 15.13).

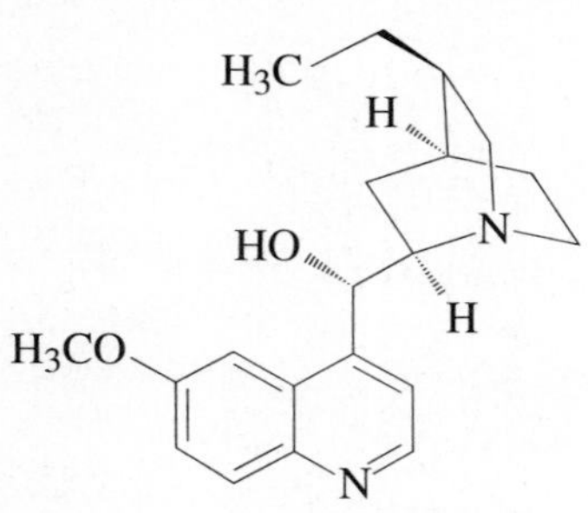

Osmium-catalyzed dihydroxylation of alkenes can be carried out with high enantioselectivity as illustrated by the following equation:

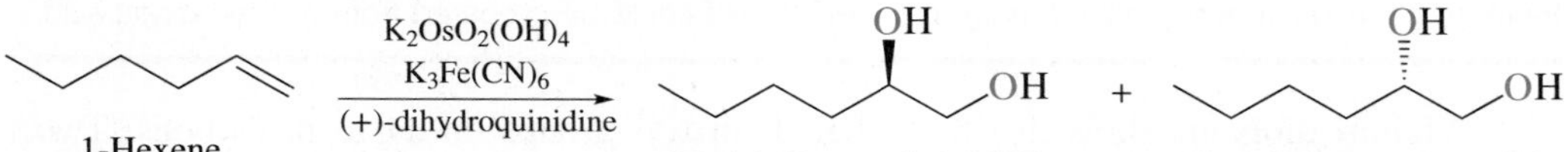

$$\text{1-Hexene} \xrightarrow[\text{(+)-dihydroquinidine}]{K_2OsO_2(OH)_4,\ K_3Fe(CN)_6} (R)\text{-1,2-Hexanediol (90\%)} + (S)\text{-1,2-Hexanediol (10\%)}$$

1-Hexene

(*R*)-1,2-Hexanediol (90%) (*S*)-1,2-Hexanediol (10%)

The chiral reactant in this example was the naturally occurring and readily available alkaloid (+)-dihydroquinidine.

(+)-dihydroquinidine

Problem 16.6

When *trans*-2-butene was subjected to enantioselective dihydroxylation, the 2,3-butanediol that was formed had the (*R*)-configuration at one carbon. What was the configuration at the other?

16.6 Reactions of Alcohols: A Review and a Preview

Alcohols are versatile starting materials for the preparation of a variety of organic functional groups. Several reactions of alcohols have already been seen in earlier chapters and are summarized in Table 16.2. The remaining sections of this chapter add to the list.

TABLE 16.2 Reactions of Alcohols Discussed in Earlier Chapters

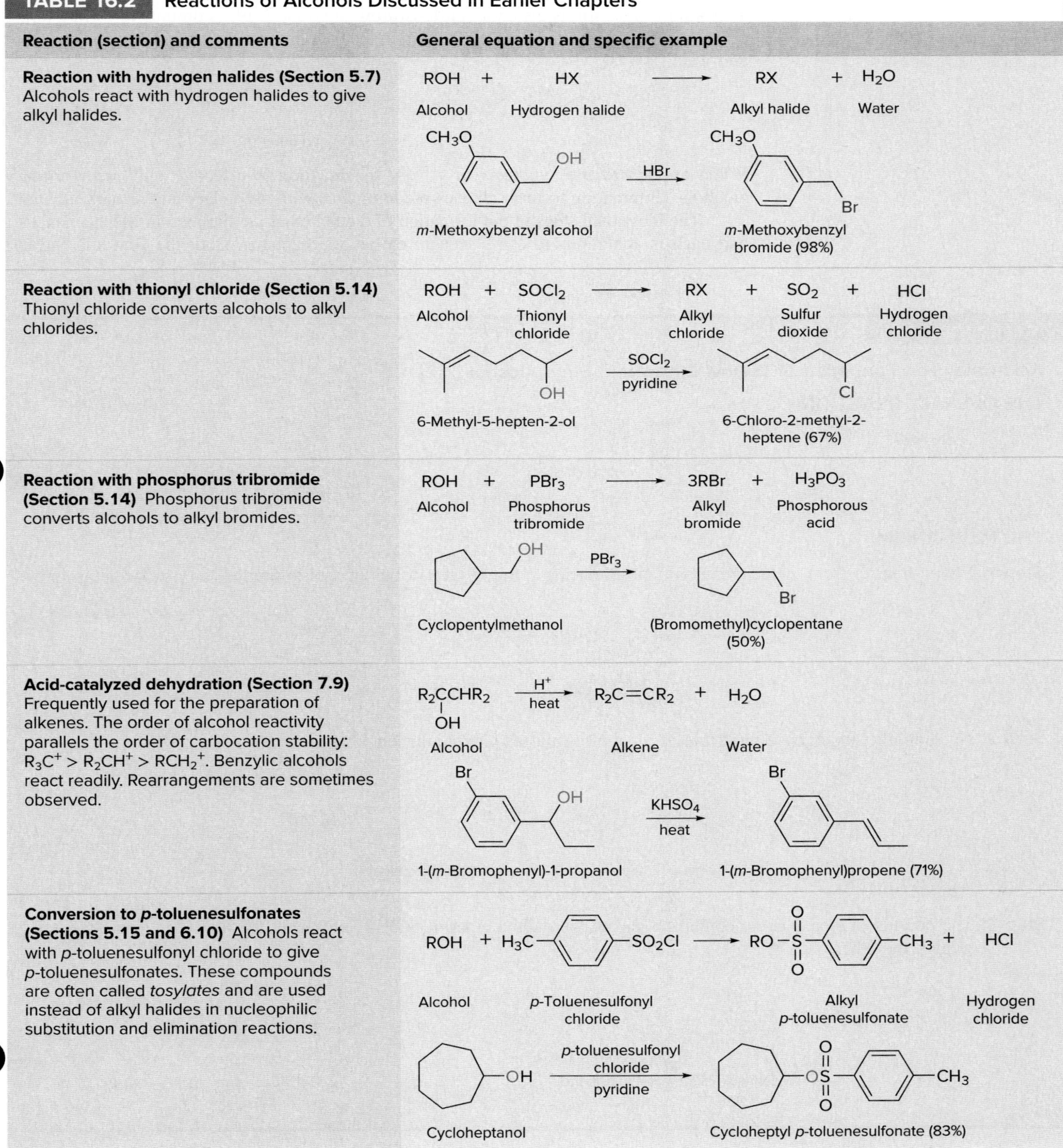

Reaction (section) and comments	General equation and specific example
Reaction with hydrogen halides (Section 5.7) Alcohols react with hydrogen halides to give alkyl halides.	ROH (Alcohol) + HX (Hydrogen halide) ⟶ RX (Alkyl halide) + H_2O (Water) *m*-Methoxybenzyl alcohol —HBr→ *m*-Methoxybenzyl bromide (98%)
Reaction with thionyl chloride (Section 5.14) Thionyl chloride converts alcohols to alkyl chlorides.	ROH (Alcohol) + $SOCl_2$ (Thionyl chloride) ⟶ RX (Alkyl chloride) + SO_2 (Sulfur dioxide) + HCl (Hydrogen chloride) 6-Methyl-5-hepten-2-ol —$SOCl_2$, pyridine→ 6-Chloro-2-methyl-2-heptene (67%)
Reaction with phosphorus tribromide (Section 5.14) Phosphorus tribromide converts alcohols to alkyl bromides.	ROH (Alcohol) + PBr_3 (Phosphorus tribromide) ⟶ 3RBr (Alkyl bromide) + H_3PO_3 (Phosphorous acid) Cyclopentylmethanol —PBr_3→ (Bromomethyl)cyclopentane (50%)
Acid-catalyzed dehydration (Section 7.9) Frequently used for the preparation of alkenes. The order of alcohol reactivity parallels the order of carbocation stability: $R_3C^+ > R_2CH^+ > RCH_2^+$. Benzylic alcohols react readily. Rearrangements are sometimes observed.	R_2CCHR_2 with OH (Alcohol) —H^+, heat→ $R_2C{=}CR_2$ (Alkene) + H_2O (Water) 1-(*m*-Bromophenyl)-1-propanol —$KHSO_4$, heat→ 1-(*m*-Bromophenyl)propene (71%)
Conversion to *p*-toluenesulfonates (Sections 5.15 and 6.10) Alcohols react with *p*-toluenesulfonyl chloride to give *p*-toluenesulfonates. These compounds are often called *tosylates* and are used instead of alkyl halides in nucleophilic substitution and elimination reactions.	ROH (Alcohol) + H_3C–C_6H_4–SO_2Cl (*p*-Toluenesulfonyl chloride) ⟶ RO–$S(=O)_2$–C_6H_4–CH_3 (Alkyl *p*-toluenesulfonate) + HCl (Hydrogen chloride) Cycloheptanol —*p*-toluenesulfonyl chloride, pyridine→ Cycloheptyl *p*-toluenesulfonate (83%)

16.7 Conversion of Alcohols to Ethers

Primary alcohols are converted to ethers on heating in the presence of an acid catalyst, usually sulfuric acid.

$$2RCH_2OH \xrightarrow{H^+, \text{ heat}} RCH_2OCH_2R + H_2O$$

Primary alcohol → Dialkyl ether + Water

This kind of reaction is called a **condensation**—two molecules combine to form a larger one plus some smaller molecule. Here, two alcohol molecules combine to give an ether and water.

$$2 \text{ 1-Butanol} \xrightarrow[130°C]{H_2SO_4} \text{Dibutyl ether (60\%)} + H_2O$$

1-Butanol → Dibutyl ether (60%) + Water

When applied to the synthesis of ethers, the reaction is effective only with primary alcohols. Elimination to form alkenes predominates with secondary and tertiary alcohols.

The individual steps in the formation of diethyl ether are outlined in Mechanism 16.1 and each is analogous to steps seen in earlier mechanisms. Both the first and the last

Mechanism 16.1

Acid-Catalyzed Formation of Diethyl Ether from Ethyl Alcohol

THE OVERALL REACTION:

$$2CH_3CH_2OH \xrightarrow[140°C]{H_2SO_4} CH_3CH_2OCH_2CH_3 + H_2O$$

Ethanol → Diethyl ether + Water

THE MECHANISM:

Step 1: Proton transfer from the acid catalyst (sulfuric acid) to the oxygen of the alcohol to produce an alkyloxonium ion.

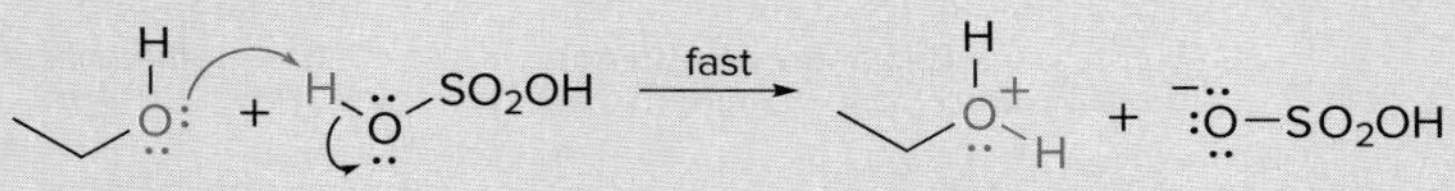

Ethanol + Sulfuric acid $\xrightarrow{\text{fast}}$ Ethyloxonium ion + Hydrogen sulfate ion ($^-$:O–SO_2OH)

Step 2: Nucleophilic attack by a molecule of alcohol on the alkyloxonium ion formed in step 1.

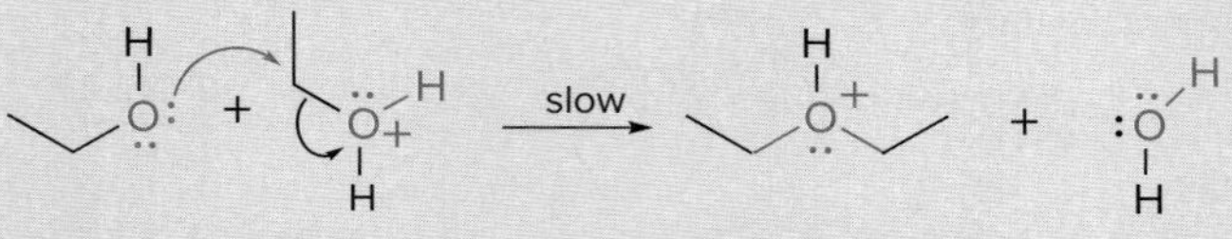

Ethanol + Ethyloxonium ion $\xrightarrow{\text{slow}}$ Diethyloxonium ion + Water

Step 3: The product of step 2 is the conjugate acid of the dialkyl ether. It is deprotonated in the final step of the process to give the ether.

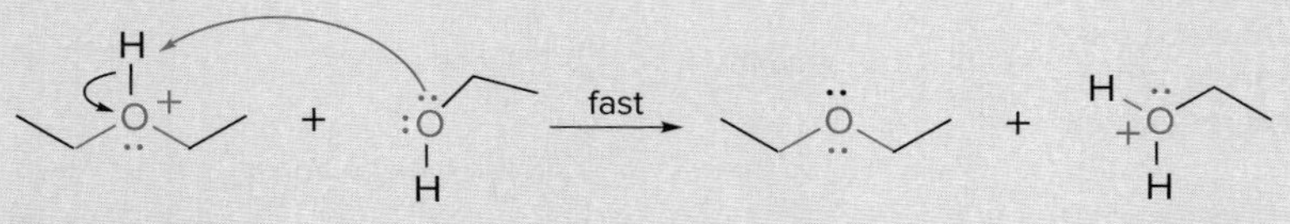

Diethyloxonium ion + Ethanol $\xrightarrow{\text{fast}}$ Diethyl ether + Ethyloxonium ion

steps are proton-transfers between oxygens. Reaction of a protonated alcohol with a nucleophile was encountered in the reaction of primary alcohols with hydrogen halides (Section 5.13), and the nucleophilic properties of alcohols were discussed in the context of solvolysis reactions (Section 6.5).

Diols react intramolecularly to form cyclic ethers when a five-membered or six-membered ring can result.

HO~~~~~OH $\xrightarrow[\text{heat}]{H_2SO_4}$ Oxane + H_2O

1,5-Pentanediol — Oxane (75%) — Water

Oxane is also called tetrahydropyran.

In these intramolecular ether-forming reactions, the alcohol may be primary, secondary, or tertiary.

Problem 16.7

On the basis of the acid-catalyzed formation of diethyl ether from ethanol in Mechanism 16.1, write a stepwise mechanism for the formation of oxane from 1,5-pentanediol.

16.8 Esterification

Acid-catalyzed condensation of an alcohol and a carboxylic acid yields an ester and water and is known as the **Fischer esterification.**

$$\underset{\text{Alcohol}}{ROH} + \underset{\text{Carboxylic acid}}{R'\overset{O}{\overset{\|}{C}}OH} \overset{H^+}{\rightleftharpoons} \underset{\text{Ester}}{R'\overset{O}{\overset{\|}{C}}OR} + \underset{\text{Water}}{H_2O}$$

Fischer esterification is reversible, and the position of equilibrium usually lies slightly to the side of products. For preparative purposes, the position of equilibrium can be made more favorable by using either the alcohol or the carboxylic acid in excess. In the following example, in which an excess of the alcohol was employed, the yield indicated is based on the carboxylic acid as the limiting reactant.

CH_3OH + C_6H_5COOH $\xrightarrow[\text{heat}]{H_2SO_4}$ $C_6H_5COOCH_3$ + H_2O

Methanol (0.6 mol) — Benzoic acid (0.1 mol) — Methyl benzoate (70% yield based on benzoic acid) — Water

Mechanism 19.1 shows the mechanism of this reaction.

Another way to shift the position of equilibrium to favor ester formation is to remove water from the reaction mixture by using benzene as a cosolvent and distilling the azeotropic mixture of benzene and water. This can be accomplished in the laboratory with a Dean–Stark trap.

An *azeotropic mixture* contains two or more substances that distill together at a constant boiling point. The benzene–water azeotrope contains 9% water and boils at 69°C.

sec-Butyl alcohol + Acetic acid $\xrightarrow[\text{heat}]{H^+\text{, benzene}}$ sec-Butyl acetate + H_2O

sec-Butyl alcohol (0.20 mol) — Acetic acid (0.25 mol) — *sec*-Butyl acetate (71% yield based on *sec*-butyl alcohol) — Water (codistills with benzene)

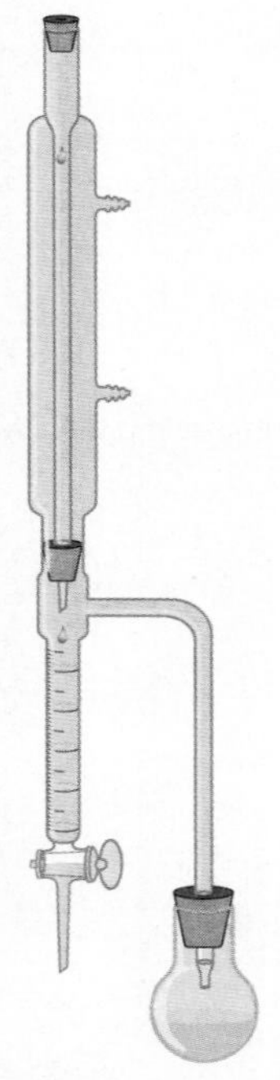

A reaction apparatus with a Dean–Stark trap. The water is denser than the ester–benzene mixture, and collects in the side arm of the trap.

Problem 16.8

Write the structure of the ester formed in each of the following reactions:

(a) $CH_3CH_2CH_2CH_2OH + CH_3CH_2CO_2H \xrightarrow[\text{heat}]{H_2SO_4}$

(b) $2CH_3OH + HO\overset{O}{\overset{\|}{C}}-C_6H_4-\overset{O}{\overset{\|}{C}}OH \xrightarrow[\text{heat}]{H_2SO_4} C_{10}H_{10}O_4$

Sample Solution (a) By analogy to the general equation and to the examples cited in this section, we can write the equation

$$CH_3CH_2CH_2CH_2OH + CH_3CH_2COOH \xrightarrow[\text{heat}]{H_2SO_4} CH_3CH_2COOCH_2CH_2CH_2CH_3 + H_2O$$

1-Butanol | Propanoic acid | Butyl propanoate | Water

As actually carried out in the laboratory, 3 mol of propanoic acid was used per mole of 1-butanol, and the desired ester was obtained in 78% yield.

Esters are also formed by the reaction of alcohols with acyl chlorides, usually in the presence of a weak base such as pyridine.

$$ROH + R'\overset{O}{\overset{\|}{C}}Cl \longrightarrow R'\overset{O}{\overset{\|}{C}}OR + HCl$$

Alcohol | Acyl chloride | Ester | Hydrogen chloride

$$(CH_3)_2CHCH_2OH + 3,5\text{-}(O_2N)_2C_6H_3COCl \xrightarrow{\text{pyridine}} 3,5\text{-}(O_2N)_2C_6H_3COOCH_2CH(CH_3)_2$$

Isobutyl alcohol | 3,5-Dinitrobenzoyl chloride | Isobutyl 3,5-dinitrobenzoate (86%)

Acid anhydrides react similarly to acyl chlorides.

$$ROH + R'\overset{O}{\overset{\|}{C}}O\overset{O}{\overset{\|}{C}}R' \longrightarrow R'\overset{O}{\overset{\|}{C}}OR + R'\overset{O}{\overset{\|}{C}}OH$$

Alcohol | Acid anhydride | Ester | Carboxylic acid

$$C_6H_5CH_2CH_2OH + F_3CCOOCOCF_3 \xrightarrow{\text{pyridine}} F_3CCOOCH_2CH_2C_6H_5 + F_3CCOOH$$

2-Phenylethanol | Trifluoroacetic anhydride | 2-Phenylethyl trifluoroacetate (83%) | Trifluoroacetic acid

The mechanisms of the Fischer esterification and the reactions of alcohols with acyl chlorides and acid anhydrides will be discussed in detail in Chapters 19 and 20 after some fundamental principles of carbonyl group reactivity have been developed. For the present, it is sufficient to point out that most of the reactions that convert alcohols to esters leave the C—O bond of the alcohol intact.

$$H-O-R \longrightarrow R'\overset{O}{\overset{\|}{C}}-O-R$$

This is the same oxygen that was attached to the R group in the starting alcohol.

The acyl group of the carboxylic acid, acyl chloride, or acid anhydride is transferred to the oxygen of the alcohol. This fact is most clearly evident in the esterification of chiral alcohols, where, because none of the bonds to the chirality center is broken in the process, *retention of configuration occurs.*

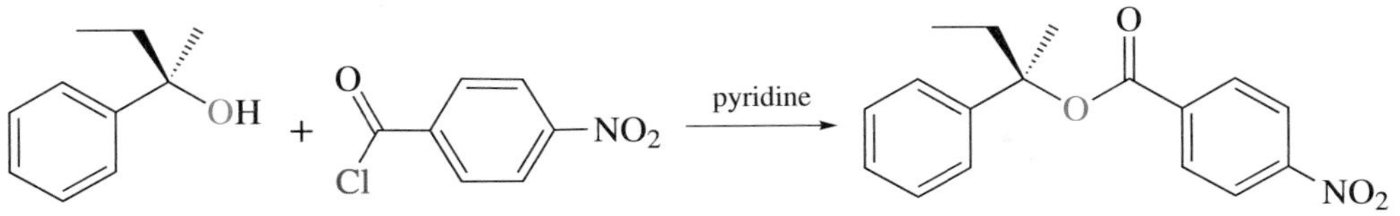

(*R*)-(+)-2-Phenyl-2-butanol + *p*-Nitrobenzoyl chloride $\xrightarrow{\text{pyridine}}$ (*R*)-(−)-1-Methyl-1-phenylpropyl *p*-nitrobenzoate (63%)

Problem 16.9

From what alcohol and acyl chloride can the following esters be synthesized? From what alcohol and acid anhydride?

(a) 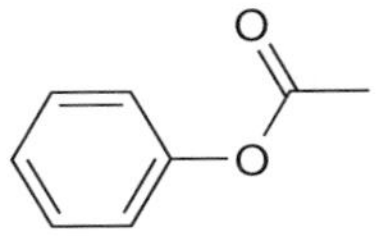 (b) 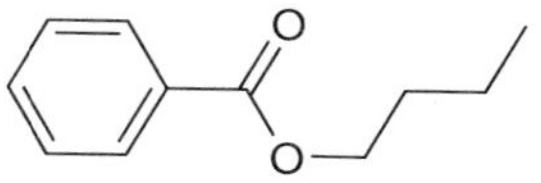 (c)

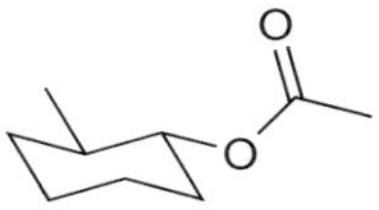

Sample Solution (a) The oxygen that has a single bond to the carbonyl carbon is the alcohol oxygen, and the carbonyl carbon is part of the acyl chloride or anhydride. The compound in part (a) is phenyl acetate, and it can be prepared from phenol and acetyl chloride, or acetic anhydride.

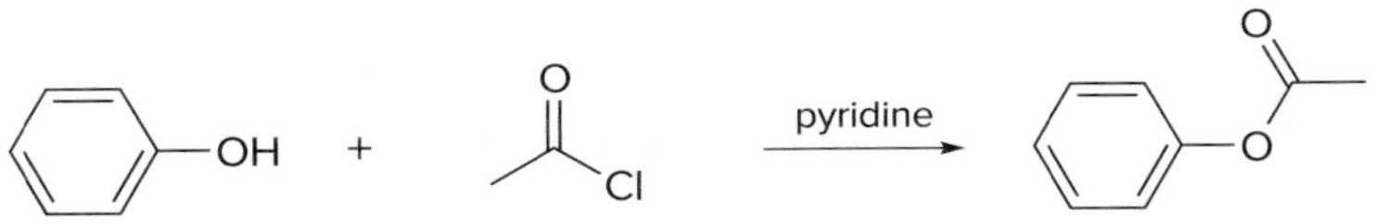

Phenol + Acetyl chloride $\xrightarrow{\text{pyridine}}$ Phenyl acetate

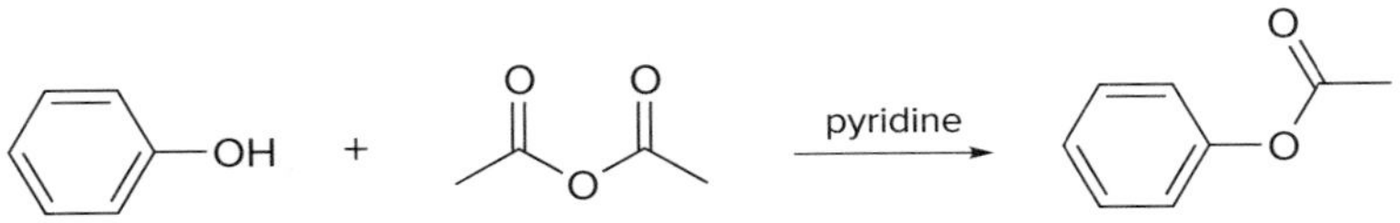

Phenol + Acetic anhydride $\xrightarrow{\text{pyridine}}$ Phenyl acetate

16.9 Oxidation of Alcohols

Oxidation of an alcohol yields a carbonyl compound. Whether the resulting carbonyl compound is an aldehyde, a ketone, or a carboxylic acid depends on the alcohol and on the oxidizing agent.

Primary alcohols are oxidized either to an aldehyde or to a carboxylic acid:

$$\underset{\text{Primary alcohol}}{RCH_2OH} \xrightarrow{\text{oxidize}} \underset{\text{Aldehyde}}{R\overset{O}{\overset{\|}{C}}H} \xrightarrow{\text{oxidize}} \underset{\text{Carboxylic acid}}{R\overset{O}{\overset{\|}{C}}OH}$$

Vigorous oxidation leads to the formation of a carboxylic acid, but a number of methods permit us to stop the oxidation at the intermediate aldehyde stage. The reagents most commonly used for oxidizing alcohols are based on high-oxidation-state transition metals, particularly chromium(VI).

Chromic acid (H_2CrO_4) is a good oxidizing agent and is formed when solutions containing chromate (CrO_4^{2-}) or dichromate ($Cr_2O_7^{2-}$) are acidified. Sometimes it is possible to obtain aldehydes in satisfactory yield before they are further oxidized, but in most cases carboxylic acids are the major products isolated on treatment of primary alcohols with chromic acid.

3-Fluoro-1-propanol $\xrightarrow[H_2SO_4,\ H_2O]{K_2Cr_2O_7}$ 3-Fluoropropanoic acid (74%)

Conditions that do permit the easy isolation of aldehydes in good yield by oxidation of primary alcohols employ various Cr(VI) species as the oxidant in *anhydrous* media. Two such reagents are pyridinium chlorochromate (PCC), $C_5H_5NH^+\ ClCrO_3^-$, and pyridinium dichromate (PDC), $(C_5H_5NH)_2^{2+}\ Cr_2O_7^{2-}$; both are used in dichloromethane.

1-Heptanol $\xrightarrow[CH_2Cl_2]{PCC}$ Heptanal (78%)

p-tert-Butylbenzyl alcohol $\xrightarrow[CH_2Cl_2]{PDC}$ *p-tert*-Butylbenzaldehyde (94%)

Secondary alcohols are oxidized to ketones by the same reagents that oxidize primary alcohols:

$$RCH(OH)R' \xrightarrow{\text{oxidize}} RC(=O)R'$$

Secondary alcohol → Ketone

Cyclohexanol $\xrightarrow[H_2SO_4,\ H_2O]{Na_2Cr_2O_7}$ Cyclohexanone (85%)

1-Octen-3-ol $\xrightarrow[CH_2Cl_2]{PDC}$ 1-Octen-3-one (80%)

Tertiary alcohols lack an H—C—O unit and are not as readily oxidized. When oxidation does occur (stronger oxidizing agents and/or higher temperatures), complex mixtures of products result.

Problem 16.10

Predict the principal organic product of each of the following reactions:

(a) $ClCH_2CH_2CH_2CH_2OH \xrightarrow[H_2SO_4,\ H_2O]{K_2Cr_2O_7}$

(b) $CH_3CH(OH)CH_2CH_2CH_2CH_2CH_2CH_3 \xrightarrow[H_2SO_4,\ H_2O]{Na_2Cr_2O_7}$

(c) $CH_3(CH_2)_6CH_2OH \xrightarrow[CH_2Cl_2]{PCC}$

Sample Solution (a) The reactant is a primary alcohol and so can be oxidized either to an aldehyde or to a carboxylic acid. Aldehydes are the major products only when the oxidation is carried out in anhydrous media. Carboxylic acids are formed when water is present. The reaction shown produced 4-chlorobutanoic acid in 56% yield.

4-Chloro-1-butanol $\xrightarrow[H_2SO_4,\ H_2O]{K_2Cr_2O_7}$ 4-Chlorobutanoic acid

The mechanism of chromic acid oxidation is complicated, but can be summarized as a combination of two stages. In the first, the alcohol and chromic acid react to give a chromate ester.

Alcohol + Chromic acid ⟶ Alkyl hydrogen chromate + Water (H_2O)

Next, the chromate ester undergoes a β elimination in which a proton is removed from carbon while the Cr—O bond breaks.

Water + Alkyl hydrogen chromate ⟶ Hydronium ion + Aldehyde or ketone + Hydrogen chromite ion

The second step is slower than the first as evidenced by the observation that $(CH_3)_2CHOH$ reacts almost seven times faster than $(CH_3)_2CDOH$. An H/D kinetic isotope effect this large is consistent with rate-determining carbon–hydrogen bond cleavage (Section 7.17).

As an alternative to chromium-based oxidants, chemists have developed other reagents for oxidizing alcohols, several of which are based on chlorodimethylsulfonium ion $[(CH_3)_2SCl^+]$. Most commonly, chlorodimethylsulfonium ion is generated under the reaction conditions by the reaction of dimethyl sulfoxide with oxalyl chloride.

$$(CH_3)_2S{=}O + ClC(=O)C(=O)Cl \xrightarrow[-78°C]{CH_2Cl_2} (CH_3)_2\overset{+}{S}{-}Cl + CO + CO_2 + Cl^-$$

Dimethyl sulfoxide + Oxalyl chloride ⟶ Chlorodimethylsulfonium ion + Carbon monoxide + Carbon dioxide + Chloride ion

The alcohol to be oxidized is then added to the solution of chlorodimethylsulfonium ion, followed by treatment with a weak base such as triethylamine. Primary alcohols yield aldehydes; secondary alcohols yield ketones.

Citronellol $\xrightarrow[2.\ (CH_3CH_2)_3N]{1.\ (CH_3)_2S=O,\ (COCl)_2,\ CH_2Cl_2,\ -50°C}$ Citronellal (83%)

Oxidations of alcohols with the DMSO–oxalyl chloride combination are known as Swern oxidations after Daniel Swern of Temple University, who developed this and related procedures.

Problem 16.11

The last intermediate in the oxidation of citronellol by dimethyl sulfoxide is believed to have the structure shown. Use curved arrows to describe its *unimolecular* dissociation to citronellal. What is the sulfur-containing product?

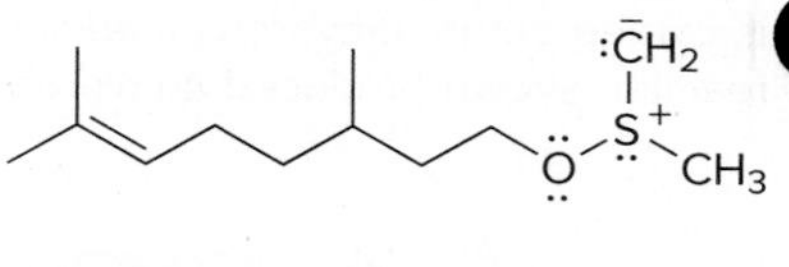

:$\bar{C}H_2$ S+ O CH_3

Sustainability and Organic Chemistry

In the 1970s, both the U.S. Environmental Protection Agency (EPA) and the United Nations Conference on the Human Environment independently addressed *sustainability*—the efficient and environmentally responsible use of our resources. Ideally, a "green" or "benign" chemical process should be efficient, based on renewable raw materials, produce minimum waste, use catalysts rather than stoichiometric reagents, avoid the use or formation of toxic or hazardous materials, require minimum energy, and yield a product that maximizes the incorporation of all materials.

These objectives have spurred research directed toward developing alternative synthetic methods. Take alcohol oxidation, for example. As described in Section 16.9, primary alcohols are converted to aldehydes by oxidation with pyridinium chlorochromate (PCC).

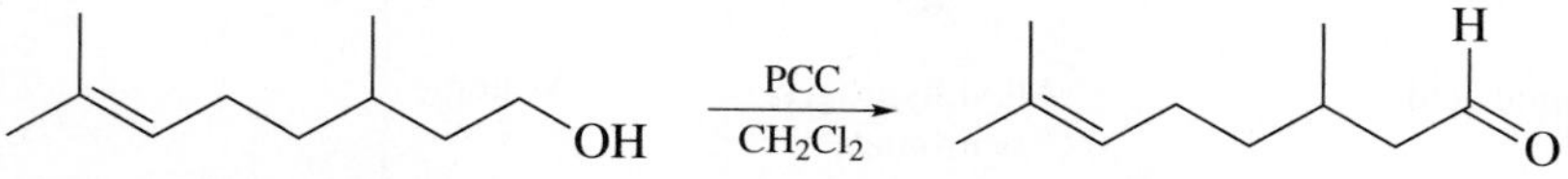

OH —PCC, CH_2Cl_2→ H O

3,7-Dimethyloct-6-en-1-ol 3,7-Dimethyloct-6-enal (82%)

While this oxidation proceeds in synthetically satisfactory yield, it is inefficient in terms of *atom economy.* PCC is used in stoichiometric amounts and none of its atoms are incorporated into the desired product. Moreover, the toxicity of chromium compounds introduces significant hazardous-waste disposal problems.

On the other hand, oxidation according to the equation:

$$RCH_2OH + NaOCl \longrightarrow RCH(=O) + NaCl + H_2O$$

Alcohol, Sodium hypochlorite, Aldehyde, Sodium chloride, Water

offers a more sustainable alternative in that it avoids toxicity problems (aqueous sodium hypochlorite is nothing more than household bleach), and the byproducts (water and sodium chloride) are benign. In practice, however, the reaction was not widely used until it was found that it could be catalyzed by the free-radical compound 2,2,6,6-tetramethylpiperidine-1-oxyl (TEMPO). The active catalyst is an oxoammonium cation formed by hypochlorite oxidation of TEMPO under the reaction conditions.

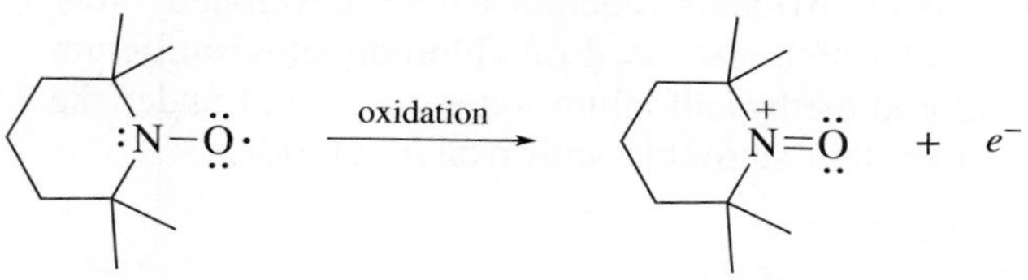

:N—Ö· —oxidation→ $\overset{+}{N}$=Ö + e^-

TEMPO An oxoammonium cation

TEMPO-catalyzed oxidations are replacing more familiar oxidation methods, especially in the pharmaceutical industry. An early step in the synthesis of an HIV protease inhibitor is the oxidation shown where the desired ketone is formed in 98% yield in the presence of 1 mol % of TEMPO.

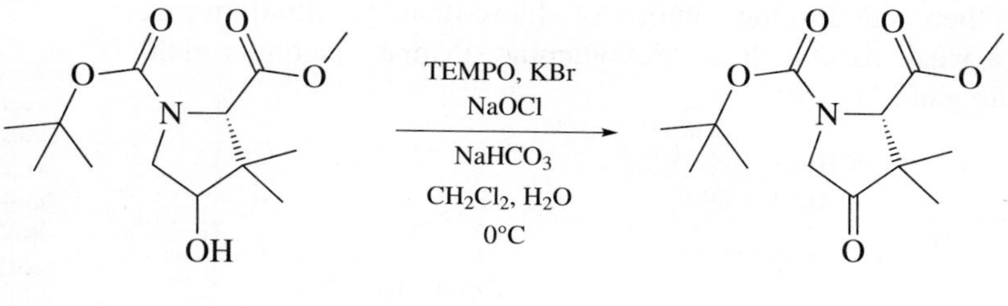

O O O N O OH —TEMPO, KBr, NaOCl / $NaHCO_3$, CH_2Cl_2, H_2O, 0°C→ O O O N O O

The 4-hydroxy derivative of TEMPO is used as the catalyst for the synthesis of a key aldehyde intermediate in the large-scale preparation of progesterone and corticosteroids.

4-hydroxy-TEMPO
KBr, NaOCl

Not only is the reaction itself green, but the raw material for the synthesis is a plant sterol obtained from soybean waste.

TEMPO-catalyzed oxidations can also be used to oxidize primary alcohols to carboxylic acids. In the example shown here, aerobic oxidation of the alcohol occurs in the presence of an Fe(III) salt and potassium chloride.

$Fe(NO_3)_3 \cdot 9H_2O$
TEMPO, KCl
O_2 or air
$ClCH_2CH_2Cl$

The mechanism of TEMPO-catalyzed oxidation depends on the particular experimental conditions but involves two stages. First, the alcohol undergoes nucleophilic addition to the oxoammonium ion:

and is followed by elimination of the species produced.

Subsequent steps with the oxidizing agent present in the reaction mixture (hypochlorite) restore the active form of the oxoammonium ion catalyst.

Because of their scale, the methods and practices of the chemical industry can have significant environmental effects. Fortunately, many of the qualities that characterize green chemistry are also the most desirable in economic terms. For example, two industrial chemicals—phenol and acetone—are produced in a process with high atom economy.

1. O_2, catalyst
2. H_2SO_4
3. H_2O

Isopropylbenzene is made from benzene and propene, both of which are readily available petrochemicals, O_2 is the ultimate green oxidizing agent, and the waste products are either benign (water) or easily managed (sulfuric acid).

On an even larger scale, all of the polymers listed in Table 10.2 and most of them in Chapter 27 are prepared by reactions that are efficient, use catalysts, produce minimum waste, and incorporate all of the atoms in the reactant into the product.

16.10 Biological Oxidation of Alcohols

Many biological processes involve oxidation of alcohols to carbonyl compounds or the reverse process, reduction of carbonyl compounds to alcohols. Ethanol, for example, is metabolized in the liver to acetaldehyde in a reaction catalyzed by the enzyme *alcohol dehydrogenase*.

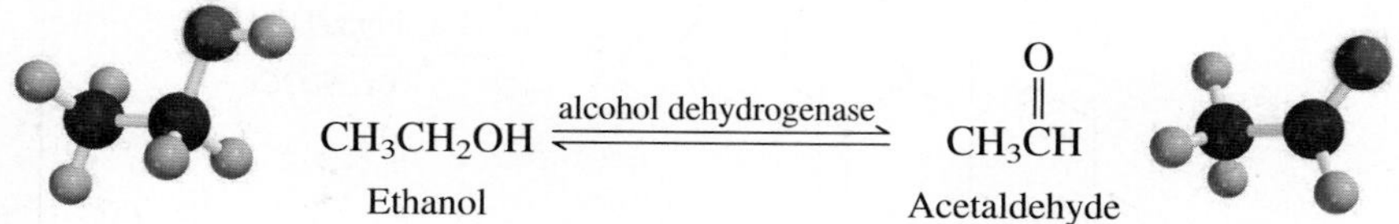

In addition to enzymes, biological oxidations require substances known as *coenzymes*. Coenzymes are organic molecules that, in concert with an enzyme, act on a substrate to bring about chemical change. Most vitamins are coenzymes. A coenzyme contains a functional group that is complementary to a functional group of the substrate; the enzyme catalyzes the interaction of these mutually complementary functional groups. If ethanol is oxidized, some other substance must be reduced. This other substance is the oxidized form of the coenzyme *nicotinamide adenine dinucleotide* (NAD). By representing the oxidized form as NAD^+ and the reduced form as NADH, the chemical equation for the biological oxidation of ethanol may be written:

$$\underset{\text{Ethanol}}{CH_3CH_2OH} + \underset{\text{Oxidized form of NAD coenzyme}}{NAD^+} \xrightleftharpoons{\text{alcohol dehydrogenase}} \underset{\text{Acetaldehyde}}{CH_3\overset{O}{\overset{\|}{C}}H} + \underset{\text{Reduced form of NAD coenzyme}}{NADH} + H^+$$

The structure of the oxidized form of nicotinamide adenine dinucleotide is shown in Figure 16.2. The only portion of the coenzyme that undergoes chemical change in the reaction is the substituted pyridine ring of the nicotinamide unit (framed in red in Figure 16.2). Representing the remainder of the coenzyme molecule by R, we track the flow of electrons in the oxidation of ethanol to acetaldehyde as:

NAD$^+$ + Ethanol + :Base ⟶ NADH + Acetaldehyde + H-Base

The key feature here is that hydrogen is transferred from ethanol to NAD^+ not as a proton (H^+), but as hydride ($:H^-$). The ability of ethanol to transfer hydride is enhanced by removal of the O—H proton by a basic site of the enzyme. Hydride is never free, but is transferred directly from ethanol to the positively charged pyridinium ring of NAD^+ to give NADH.

Figure 16.2

Structure of NAD^+, the oxidized form of the coenzyme nicotinamide adenine dinucleotide. The functional part of the coenzyme is framed in red.

Problem 16.12

The mechanism of enzymatic oxidation has been studied by isotopic labeling with the aid of deuterated derivatives of ethanol. Specify the number of deuterium atoms that you would expect to find attached to the dihydropyridine ring of NADH following enzymatic oxidation of each of the alcohols given:

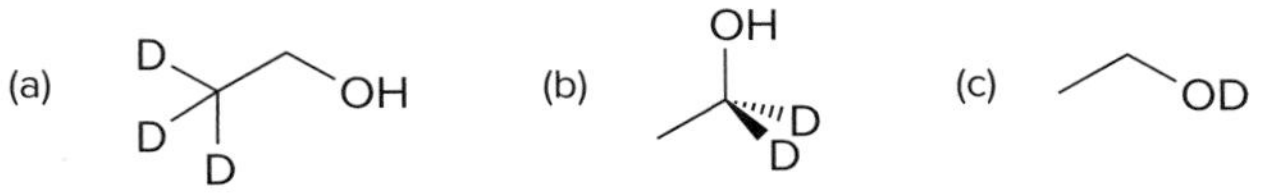

(a) (b) (c)

Sample Solution (a) The hydrogen that is transferred to the coenzyme comes from C-1 of ethanol. Therefore, the dihydropyridine ring will bear no deuterium atoms when CD_3CH_2OH is oxidized, because all the deuterium atoms of the alcohol are attached to C-2.

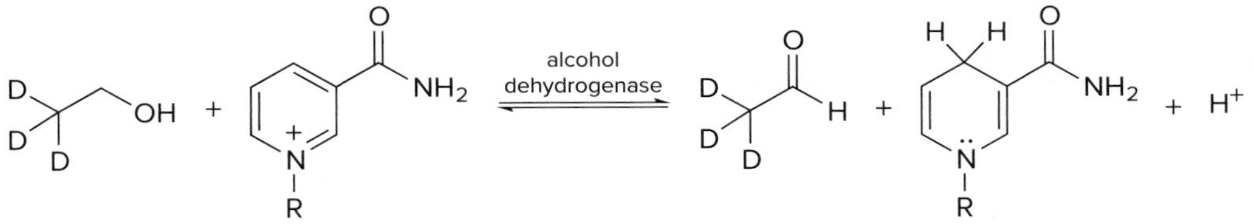

2,2,2,-Trideuterio-ethanol + NAD^+ ⇌ (alcohol dehydrogenase) 2,2,2,-Trideuterio-ethanal + NADH + H^+

The reverse reaction also occurs in living systems; NADH reduces acetaldehyde to ethanol in the presence of alcohol dehydrogenase. In this process, NADH serves as a hydride donor and is oxidized to NAD^+ while acetaldehyde is reduced.

The NAD^+–NADH coenzyme system is involved in a large number of biological oxidation–reductions. Another reaction similar to the ethanol–acetaldehyde conversion is the oxidation of lactic acid to pyruvic acid by NAD^+ and the enzyme *lactic acid dehydrogenase:*

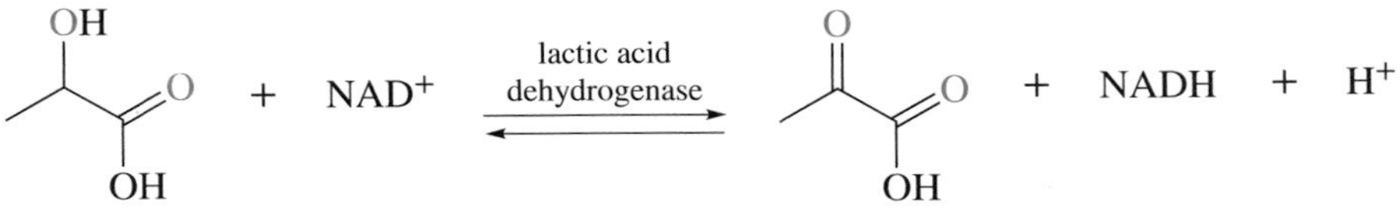

Lactic acid + NAD^+ ⇌ (lactic acid dehydrogenase) Pyruvic acid + NADH + H^+

We shall encounter other biological processes in which the $NAD^+ \rightleftharpoons$ NADH interconversion plays a prominent role in biological oxidation–reduction.

16.11 Oxidative Cleavage of Vicinal Diols

A reaction characteristic of vicinal diols is their oxidative cleavage on treatment with periodic acid (HIO_4). The carbon–carbon bond of the vicinal diol unit is broken and two carbonyl groups result. Periodic acid is reduced to iodic acid (HIO_3).

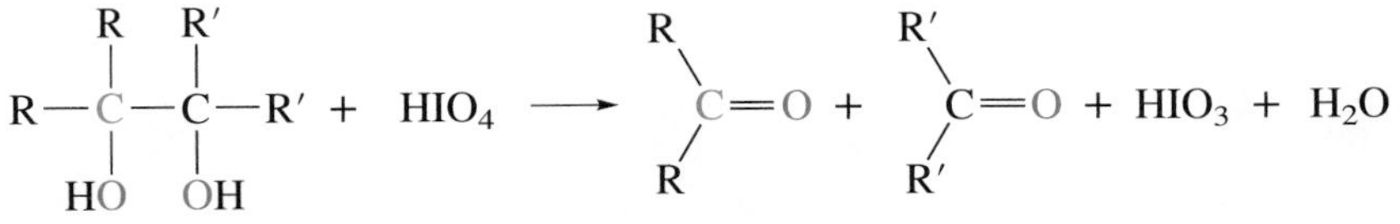

$$R_2C(OH){-}C(OH)R'_2 + HIO_4 \longrightarrow R_2C{=}O + R'_2C{=}O + HIO_3 + H_2O$$

Vicinal diol + Periodic acid → Aldehyde or ketone + Aldehyde or ketone + Iodic acid + Water

What is the oxidation state of iodine in HIO_4? In HIO_3?

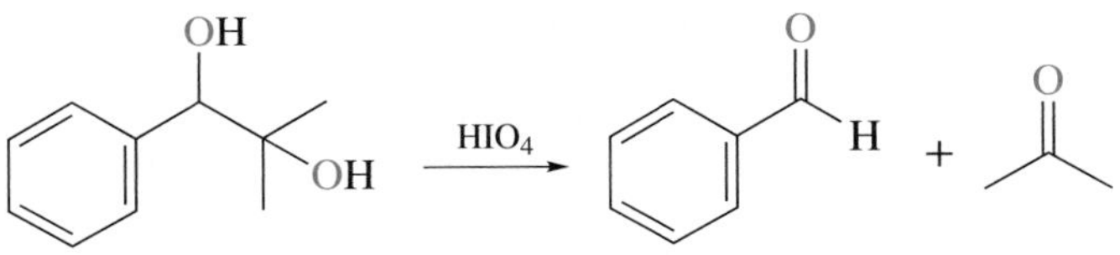

2-Methyl-1-phenyl-1,2-propanediol →(HIO_4) Benzaldehyde + Acetone

Can you remember what reaction of an alkene would give the same products as the periodic acid cleavage shown here?

This reaction occurs only when the hydroxyl groups are on adjacent carbons.

Problem 16.13

Predict the products formed on oxidation of each of the following with periodic acid:

(a) $HOCH_2CH_2OH$ (b) [structure with OH groups] (c) [structure with OH groups]

Sample Solution (a) The carbon–carbon bond of 1,2-ethanediol is cleaved by periodic acid to give two molecules of formaldehyde:

$$HOCH_2CH_2OH \xrightarrow{HIO_4} 2HCH(=O)$$

1,2-Ethanediol → Formaldehyde

Cyclic diols give dicarbonyl compounds. The reactions are faster when the hydroxyl groups are cis than when they are trans, but both stereoisomers are oxidized by periodic acid.

1,2-Cyclopentanediol (cis or trans) $\xrightarrow{HIO_4}$ Pentanedial

Periodic acid cleavage of vicinal diols is often used for analytical purposes as an aid in structure determination. By identifying the carbonyl compounds produced, the constitution of the starting diol may be deduced. This technique finds its widest application with carbohydrates and will be discussed more fully in Chapter 23.

16.12 Thiols

Sulfur lies just below oxygen in the periodic table, and many oxygen-containing organic compounds have sulfur analogs. The sulfur analogs of alcohols (ROH) are **thiols (RSH).** Thiols are given substitutive IUPAC names by appending the suffix *-thiol* to the name of the corresponding alkane, numbering the chain in the direction that gives the lower locant to the carbon that bears the —SH group. As with diols (Section 16.5), the final *-e* of the alkane name is retained. When the —SH group is named as a substituent, it is called a *mercapto,* or *sulfanyl,* group. It is also often referred to as a *sulfhydryl* group, but this is a generic term, not used in systematic nomenclature.

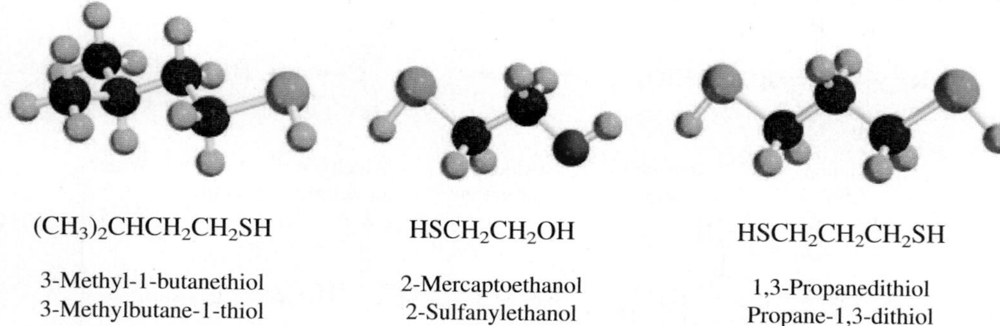

$(CH_3)_2CHCH_2CH_2SH$ — 3-Methyl-1-butanethiol, 3-Methylbutane-1-thiol

$HSCH_2CH_2OH$ — 2-Mercaptoethanol, 2-Sulfanylethanol

$HSCH_2CH_2CH_2SH$ — 1,3-Propanedithiol, Propane-1,3-dithiol

Thiols have a marked tendency to bond to mercury, and the word *mercaptan* comes from the Latin *mercurium captans,* which means "seizing mercury." The drug *dimercaprol* is used to treat mercury and lead poisoning; it is 2,3-disulfanyl-1-propanol.

At one time thiols were named **mercaptans.** Thus, CH_3CH_2SH was called "ethyl mercaptan" according to this system. This nomenclature was abandoned beginning with the 1965 revision of the IUPAC rules but is still sometimes encountered.

p-1-Menthene-8-thiol is a common name, not an IUPAC name.

The most obvious property of a low-molecular-weight thiol is its foul odor. Ethanethiol is added to natural gas so that leaks can be detected without special equipment—your nose is so sensitive that it can detect less than one part of ethanethiol in 10,000,000,000 parts of air! The odor of thiols weakens with the number of carbons, because both the volatility and the sulfur content decrease. 1-Dodecanethiol, for example, has only a faint odor. On the positive side, of the hundreds of substances that contribute to the aroma of freshly brewed coffee, the one most responsible for its characteristic odor is the thiol 2-(mercaptomethyl)furan. Likewise, the contribution of *p*-1-menthene-8-thiol to the taste and odor of freshly squeezed grapefruit juice far exceeds that of most of the more than 260 other volatile components so far identified.

3-(Mercaptomethyl)furan   *p*-1-Menthene-8-thiol

Problem 16.14

Two major components of a skunk's scent fluid are 3-methyl-1-butanethiol and *trans*-2-butene-1-thiol. Write structural formulas for each of these compounds.

Compare the boiling points of H_2S (−60°C) and H_2O (100°C).

Recall from Section 7.19 that the major pathway for reaction of *alkoxide* ions with secondary alkyl halides is E2, not S_N2.

The S—H bond is less polar than the O—H bond, as is evident in the electrostatic potential maps of Figure 16.3. The decreased polarity of the S—H bond, especially the decreased positive character of the proton, causes hydrogen bonding to be absent in thiols. Thus, methanethiol (CH_3SH) is a gas at room temperature (bp 6°C), whereas methanol (CH_3OH) is a liquid (bp 65°C).

In spite of S—H bonds being less polar than O—H bonds, thiols are stronger acids than alcohols. This is largely because S—H bonds are weaker than O—H bonds. We have seen that most alcohols have pK_as of 16–18. The corresponding value for a thiol is about 11. The significance of this difference is that a thiol can be quantitatively converted to its conjugate base (RS^-), called an **alkanethiolate,** by hydroxide. Consequently, thiols dissolve in aqueous base.

$$RS\text{—}H + {}^{-}OH \longrightarrow RS^{-} + H\text{—}OH$$

Alkanethiol (stronger acid) (pK_a = 11)	Hydroxide ion (stronger base)	Alkanethiolate ion (weaker base)	Water (weaker acid) (pK_a = 15.7)

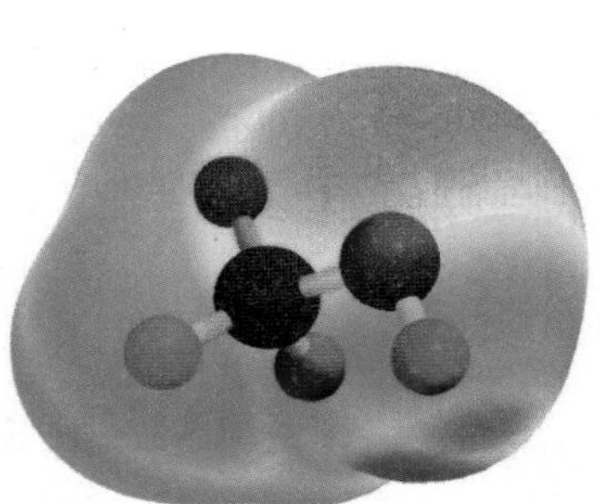

(*a*) Methanol (CH_3OH)

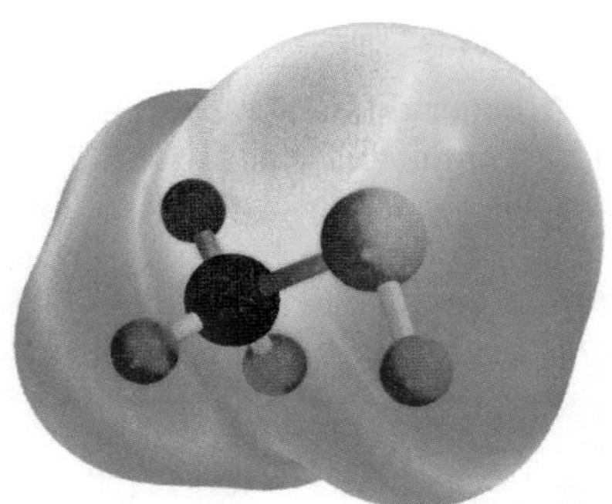

(*b*) Methanethiol (CH_3SH)

Figure 16.3

Electrostatic potential maps of (*a*) methanol and (*b*) methanethiol. The color scales were adjusted to be the same for both molecules to allow for direct comparison. The development of charge is more pronounced in the bluer color near the —OH proton in methanol than the —SH proton in methanethiol.

Alkanethiolate ions (RS^-) are weaker bases than alkoxide ions (RO^-), but they are powerful nucleophiles and undergo synthetically useful S_N2 reactions even with secondary alkyl halides.

3-Chlorocyclopentene $\xrightarrow[\text{THF}]{C_6H_5SNa}$ 3-Benzenesulfanylcyclopentene (75%)

Thiols themselves can be prepared by nucleophilic substitution using the conjugate base of H_2S.

1-Bromohexane $\xrightarrow[\text{ethanol}]{\text{KSH}}$ 1-Hexanethiol (67%)

Problem 16.15

Outline a synthesis of

(a) 1-Hexanethiol from 1-hexanol.

(b) (allyl phenyl sulfide) from allyl bromide

Sample Solution

(a)

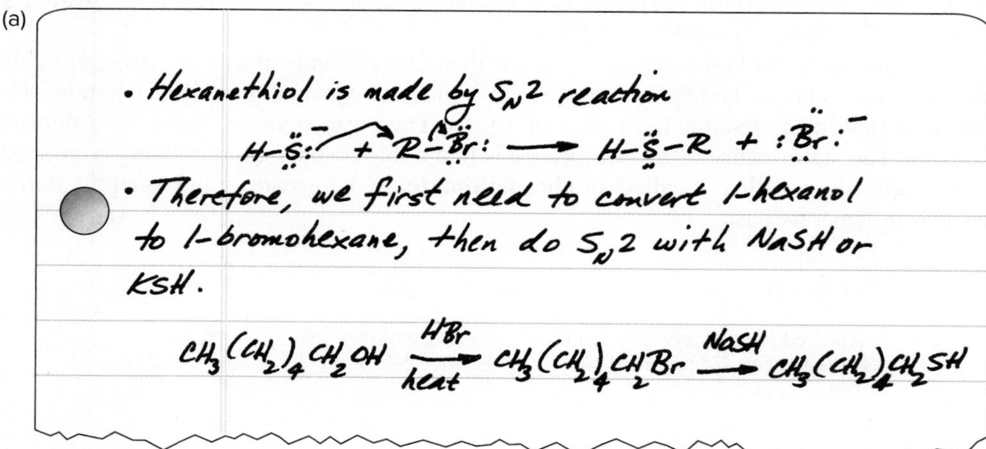

• Hexanethiol is made by S_N2 reaction
$H–\ddot{S}:^- + R–\ddot{Br}: \longrightarrow H–\ddot{S}–R + :\ddot{Br}:^-$
• Therefore, we first need to convert 1-hexanol to 1-bromohexane, then do S_N2 with NaSH or KSH.
$CH_3(CH_2)_4CH_2OH \xrightarrow[\text{heat}]{HBr} CH_3(CH_2)_4CH_2Br \xrightarrow{NaSH} CH_3(CH_2)_4CH_2SH$

A major difference between alcohols and thiols concerns their oxidation. We have seen earlier in this chapter that oxidation of alcohols produces carbonyl compounds. Analogous oxidation of thiols to compounds with C=S functions does *not* occur. Only sulfur is oxidized, not carbon, and compounds containing sulfur in various oxidation states are possible. These include a series of acids classified as *sulfenic, sulfinic,* and *sulfonic* according to the number of oxygens attached to sulfur.

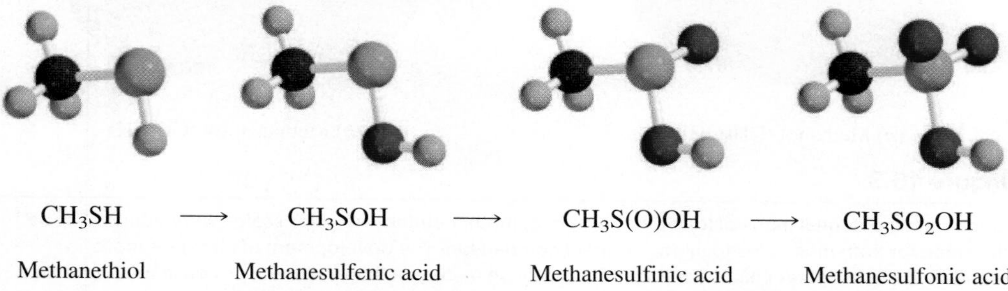

Of these the most important are the sulfonic acids. In general though, sulfonic acids are not prepared by oxidation of thiols. Benzenesulfonic acid ($C_6H_5SO_2OH$), for example, is prepared by sulfonation of benzene (see Section 13.4).

From a biochemical perspective the most important oxidation is the conversion of thiols to **disulfides.**

$$\underset{\text{Thiol}}{2\text{RSH}} \underset{\text{reduction}}{\overset{\text{oxidation}}{\rightleftharpoons}} \underset{\text{Disulfide}}{\text{RSSR}}$$

Although a variety of oxidizing agents are available for this transformation, it occurs so readily that thiols are slowly converted to disulfides by the oxygen in air. Dithiols give cyclic disulfides by intramolecular sulfur–sulfur bond formation. An example of a cyclic disulfide is the coenzyme α-*lipoic acid.* The last step in the laboratory synthesis of α-lipoic acid is an iron(III)-catalyzed oxidation of the dithiol shown:

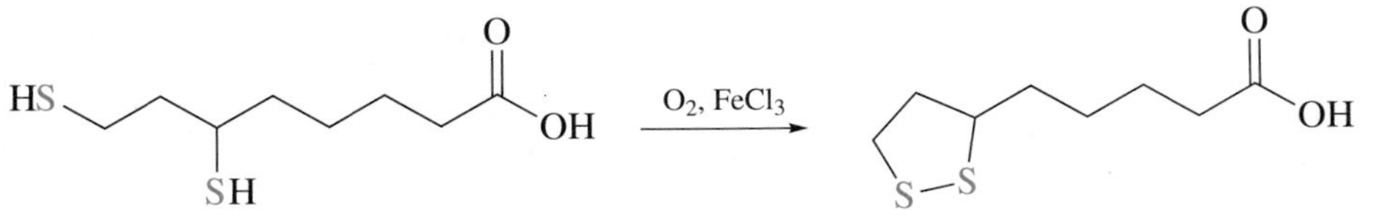

O_2, $FeCl_3$

6,8-Dimercaptooctanoic acid → α-Lipoic acid (78%)

Rapid and reversible making and breaking of the sulfur–sulfur bond is essential to the biological function of α-lipoic acid.

The S—S bonds in disulfides are intermediate in strength between typical covalent bonds and weaker interactions such as hydrogen bonds. Covalent bonds involving C, H, N, and O have bond strengths on the order of 330–420 kJ/mol (79–100 kcal/mol). The S—S bond energy is about 220 kJ/mol (53 kcal/mol), and hydrogen bond strengths are usually less than 30 kJ/mol (7 kcal/mol). Thus, S—S bonds provide more structural stability than a hydrogen bond, but can be broken while leaving the covalent framework intact.

All mammalian cells contain a thiol called *glutathione,* which protects the cell by scavenging harmful oxidants. It reacts with these oxidants by forming a disulfide, which is eventually converted back to glutathione.

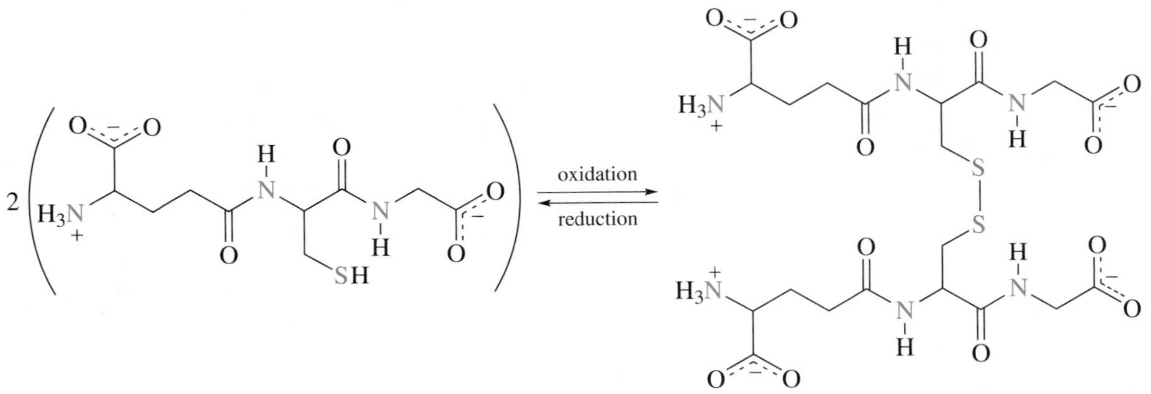

2 Glutathione (reduced form) $\underset{\text{reduction}}{\overset{\text{oxidation}}{\rightleftharpoons}}$ Glutathione (oxidized form)

The three-dimensional shapes of many proteins are governed and stabilized by S—S bonds connecting what would ordinarily be remote segments of the molecule. We'll have more to say about these *disulfide bridges* in Chapter 25.

16.13 Spectroscopic Analysis of Alcohols and Thiols

Infrared: We discussed the most characteristic features of the infrared spectra of *alcohols* earlier (Section 14.21). The O—H stretching vibration is especially easy to identify, appearing in the 3200–3650 cm^{-1} region. As the infrared spectrum of cyclohexanol, presented in Figure 16.4, demonstrates, this peak is seen as a broad absorption of

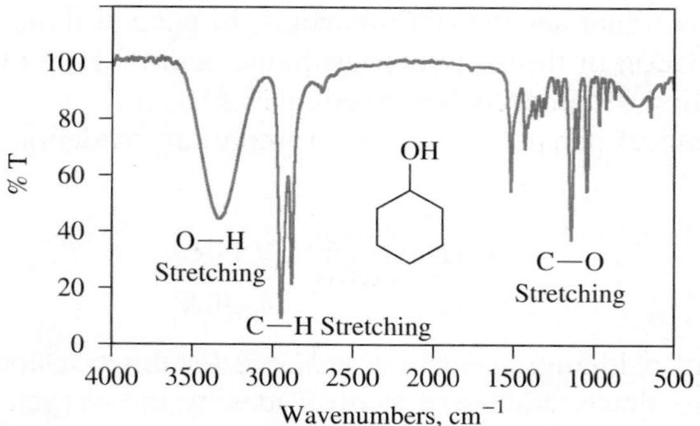

Figure 16.4

The infrared spectrum of cyclohexanol.

moderate intensity. The C—O bond stretching of alcohols gives rise to a moderate to strong absorbance between 1025 and 1200 cm^{-1}. It appears at 1065 cm^{-1} in cyclohexanol, a typical secondary alcohol, but is shifted to slightly higher energy in tertiary alcohols and slightly lower energy in primary alcohols.

The S—H stretching frequency of *thiols* gives rise to a weak band in the range 2550–2700 cm^{-1}.

^{1}H NMR: The most helpful signals in the ^{1}H NMR spectrum of *alcohols* result from the O—**H** proton and the proton in the **H**—C—O unit of primary and secondary alcohols.

H—C—O—H

δ 3.3–4.0 δ 0.5–5

The chemical shift of the hydroxyl proton signal is variable, depending on solvent, temperature, and concentration. Its precise position is not particularly significant in structure determination. Often the signals due to hydroxyl protons are not split by other protons in the molecule and are fairly easy to identify. To illustrate, Figure 16.5 shows the ^{1}H NMR spectrum of 2-phenylethanol, in which the hydroxyl proton signal appears as a singlet at δ 2.2. Of the two triplets in this spectrum, the one at lower field (δ 3.8) corresponds to the protons of the CH_2O unit. The higher-field triplet at δ 2.8 arises from the benzylic CH_2 group. The assignment of a particular signal to the hydroxyl proton can be confirmed by adding D_2O. The hydroxyl proton is replaced by deuterium, and its ^{1}H NMR signal disappears.

Because of its lower electronegativity, sulfur deshields neighboring protons less than oxygen does. Thus, the protons of a CH_2S group appear at higher field than those of a CH_2OH group.

	$CH_3CH_2CH_2$—CH_2—OH	$CH_3CH_2CH_2$—CH_2—SH
^{1}H Chemical shift:	δ 3.6	δ 2.5

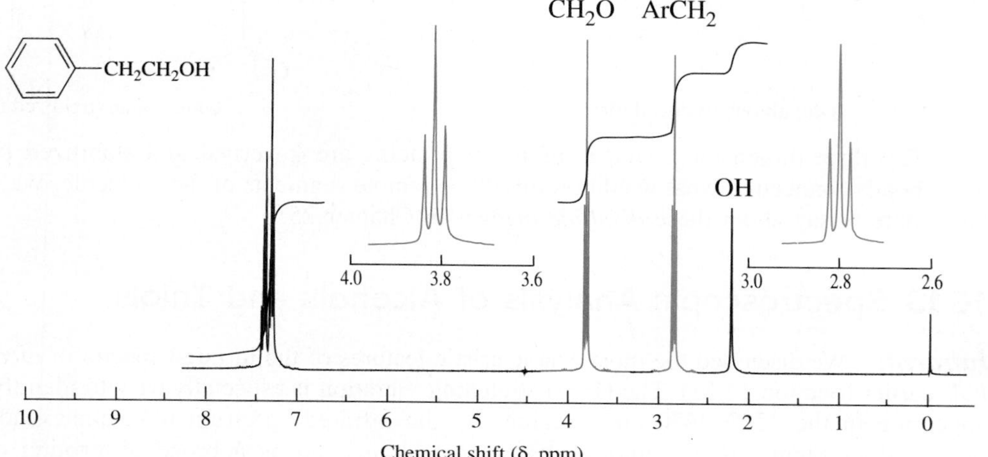

Figure 16.5

The 300-MHz ^{1}H NMR spectrum of 2-phenylethanol ($C_6H_5CH_2CH_2OH$).

^{13}C NMR: The electronegative oxygen of an *alcohol* decreases the shielding of the carbon to which it is attached. The chemical shift for the carbon of the C—OH is 60–75 ppm for most alcohols. Carbon of a C—S group is more shielded than carbon of C—O.

^{13}Chemical shift: δ 14, δ 19, δ 35, δ 62 (OH); δ 13, δ 21, δ 36, δ 24 (SH)

UV-VIS: Unless the molecule has other chromophores, alcohols are transparent above about 200 nm; λ_{max} for methanol, for example, is 177 nm.

Mass Spectrometry: The molecular ion peak is usually quite small in the mass spectrum of an alcohol. A peak corresponding to loss of water is often evident. Alcohols also fragment readily by a pathway in which the molecular ion loses an alkyl group from the hydroxyl-bearing carbon to form a stable cation. Thus, the mass spectra of most primary alcohols exhibit a prominent peak at *m/z* 31.

$$RCH_2\ddot{O}H \longrightarrow R\text{—}CH_2\text{—}\dot{O}H^{+} \longrightarrow R\cdot + H_2C{=}\overset{+}{O}H$$

Primary alcohol — Molecular ion — Alkyl radical — Conjugate acid of formaldehyde, *m/z* 31

Interpreting the mass spectra of sulfur compounds is aided by the observation of an M+2 peak because of the presence of the mass-34 isotope of sulfur. The major cleavage pathway of *thiols* is analogous to that of alcohols.

16.14 SUMMARY

Section 16.1 Functional-group interconversions involving alcohols either as reactants or as products are the focus of this chapter. Alcohols are commonplace natural products. Table 16.1 summarizes reactions discussed in earlier sections that can be used to prepare alcohols.

Section 16.2 Alcohols can be prepared from carbonyl compounds by reduction of aldehydes and ketones. See Table 16.3.

Section 16.3 Alcohols can be prepared from carbonyl compounds by reduction of carboxylic acids. See Table 16.3.

TABLE 16.3 Preparation of Alcohols by Reduction of Carbonyl Functional Groups

	Product of reduction of carbonyl compound by specified reducing agent		
Carbonyl compound	**Lithium aluminum hydride ($LiAlH_4$)**	**Sodium borohydride ($NaBH_4$)**	**Hydrogen (in the presence of a catalyst)**
Aldehyde RC(=O)H (Section 16.2)	Primary alcohol RCH_2OH	Primary alcohol RCH_2OH	Primary alcohol RCH_2OH
Ketone RC(=O)R′ (Section 16.2)	Secondary alcohol RCH(OH)R′	Secondary alcohol RCH(OH)R′	Secondary alcohol RCH(OH)R′
Carboxylic acid RC(=O)OH (Section 16.3)	Primary alcohol RCH_2OH	Not reduced	Not reduced

Section 16.4 Grignard and organolithium reagents react with ethylene oxide to give primary alcohols.

Butylmagnesium bromide (MgBr) + Ethylene oxide (O) $\xrightarrow[\text{2. } H_3O^+]{\text{1. diethyl ether}}$ 1-Hexanol (60–62%) (OH)

Section 16.5 Osmium tetraoxide is a key reagent in the conversion of alkenes to vicinal diols.

2-Phenylpropene $\xrightarrow[\textit{tert}\text{-butyl alcohol, } HO^-]{(CH_3)_3COOH,\ OsO_4\ \text{(cat)}}$ 2-Phenyl-1,2-propanediol (71%) (OH, OH)

This **dihydroxylation** proceeds by syn addition to the double bond. Osmium-based reagents that bear chiral ligands catalyze enantioselective dihydroxylation of alkenes.

Section 16.6 Table 16.4 summarizes reactions of alcohols that were introduced in earlier chapters.

Section 16.7 See Table 16.4.

Section 16.8 See Table 16.4.

Section 16.9 See Table 16.5.

Section 16.10 Oxidation of alcohols to aldehydes and ketones is a common biological reaction. Most require a coenzyme such as the oxidized form of nicotinamide adenine dinucleotide (NAD^+).

Estradiol (CH_3, OH, HO) $\xrightarrow[\text{enzymes}]{NAD^+}$ Estrone (CH_3, O, HO)

Section 16.11 Periodic acid cleaves vicinal diols; two aldehydes, two ketones, or an aldehyde and a ketone are formed.

$$\underset{\text{HO}\quad\ \ \text{OH}}{R_2C-CR_2} \xrightarrow{HIO_4} R_2C{=}O + O{=}CR_2$$

Diol — Two carbonyl-containing compounds

$$\underset{\text{HO}\qquad\quad \text{OH}}{CH_3(CH_2)_7CH-CH(CH_2)_7\overset{O}{\overset{\|}{C}}OH} \xrightarrow{HIO_4} CH_3(CH_2)_7\overset{O}{\overset{\|}{C}}H + H\overset{O}{\overset{\|}{C}}(CH_2)_7\overset{O}{\overset{\|}{C}}OH$$

9,10-Dihydroxyoctadecanoic acid — Nonanal (89%) — 9-Oxononanoic acid (76%)

TABLE 16.4 Reactions of Alcohols Presented in This Chapter

Reaction (section) and comments	General equation and specific example
Conversion to dialkyl ethers (Section 16.7) Heating in the presence of an acid catalyst converts two molecules of a primary alcohol to an ether and water. Diols can undergo an intramolecular condensation if a five- or six-membered cyclic ether results.	$2RCH_2OH \xrightarrow{H^+} RCH_2OCH_2R + H_2O$ Alcohol → Dialkyl ether + Water 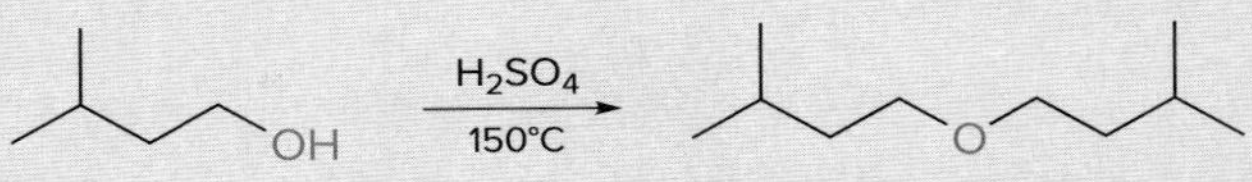 3-Methylbutan-1-ol $\xrightarrow[150°C]{H_2SO_4}$ Di-(3-methylbutyl) ether (27%)
Fischer esterification (Section 16.8) Alcohols react with carboxylic acids in the presence of an acid catalyst to yield an ester and water. The reaction is an equilibrium process that can be driven to completion by using either the alcohol or the acid in excess or by removing the water as it is formed.	$ROH + R'COH \xrightarrow{H^+} R'COR + H_2O$ (R'C(=O)OH; R'C(=O)OR) Alcohol + Carboxylic acid → Ester + Water 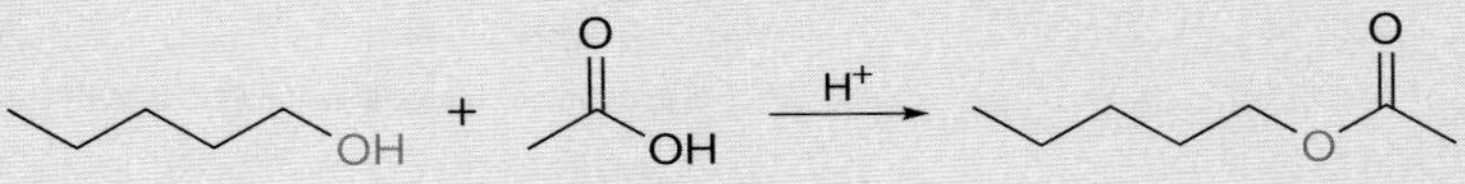 1-Pentanol + Acetic acid $\xrightarrow{H^+}$ Pentyl acetate (71%)
Esterification with acyl chlorides (Section 16.8) Acyl chlorides react with alcohols to give esters. The reaction is usually carried out in the presence of pyridine.	$ROH + R'CCl \longrightarrow R'COR + HCl$ (R'C(=O)Cl; R'C(=O)OR) Alcohol + Acyl chloride → Ester + Hydrogen chloride 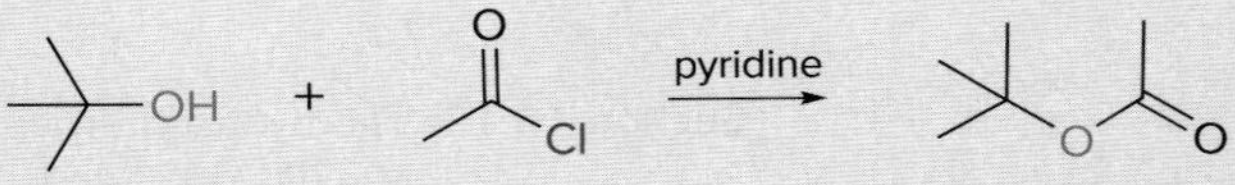 *tert*-Butyl alcohol + Acetyl chloride $\xrightarrow{pyridine}$ *tert*-Butyl acetate (62%)
Esterification with acid anhydrides (Section 16.8) Acid anhydrides react with alcohols to form esters in the same way that acyl chlorides do.	$ROH + R'COCR' \xrightarrow{H^+} R'COR + R'COH$ (R'C(=O)OC(=O)R'; R'C(=O)OR; R'C(=O)OH) Alcohol + Acid anhydride → Ester + Carboxylic acid 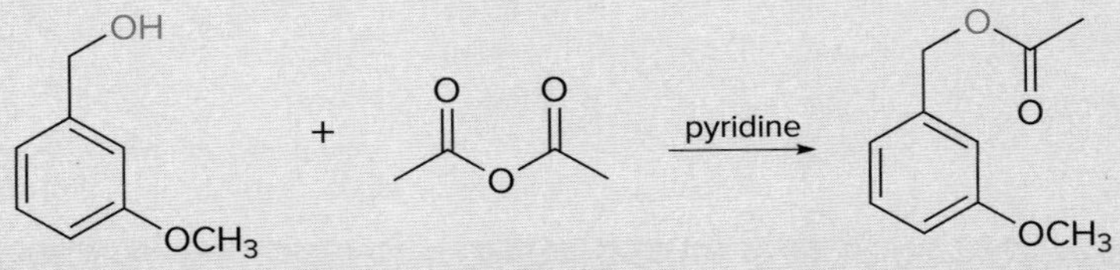 *m*-Methoxybenzyl alcohol (OCH_3) + Acetic anhydride $\xrightarrow{pyridine}$ *m*-Methoxybenzyl acetate (99%)

TABLE 16.5 Oxidation of Alcohols

Class of alcohol	Desired product	Suitable oxidizing agent(s)
Primary, RCH_2OH	Aldehyde RCH (with C=O)	PCC* PDC* $DMSO/(COCl)_2$; $(CH_3CH_2)_3N$
Primary, RCH_2OH	Carboxylic acid $RCOH$ (with C=O)	$Na_2Cr_2O_7$, H_2SO_4, H_2O H_2CrO_4
Secondary, $RCHR'$ (with OH on C)	Ketone RCR' (with C=O)	PCC* PDC* $Na_2Cr_2O_7$, $HsSO_4$, H_2O $DMSO/(COCl)_2$; $(CH_3CH_2)_3N$

*PCC is pyridinium chlorochromate; PDC is pyridinium dichromate. Both are used in dichloromethane.

Section 16.12 **Thiols** are compounds of the type RSH. They are more acidic than alcohols and are readily deprotonated by reaction with aqueous base. Thiols can be oxidized to sulfenic acids (RSOH), sulfinic acids (RSO_2H), and sulfonic acids (RSO_3H). The redox relationship between thiols and disulfides is important in certain biochemical processes.

$$\underset{\text{Thiol}}{2RSH} \underset{\text{reduction}}{\overset{\text{oxidation}}{\rightleftharpoons}} \underset{\text{Disulfide}}{RSSR}$$

Section 16.13 The hydroxyl group of an alcohol has its O—H and C—O stretching vibrations at 3200–3650 and 1025–1200 cm^{-1}, respectively.

The chemical shift of the proton of an O—H group is variable (δ 1–5) and depends on concentration, temperature, and solvent. Oxygen deshields both the proton and the carbon of an H—C—O unit. Typical NMR chemical shifts are δ 3.3–4.0 for ^1H and δ 60–75 for ^{13}C of H—C—O.

The most intense peaks in the mass spectrum of an alcohol correspond to the ion formed according to carbon–carbon cleavage of the type shown:

$$R-\overset{|}{\underset{|}{C}}-\ddot{O}H^{+\cdot} \longrightarrow R\cdot + \;>C=\overset{+}{\ddot{O}}H$$

PROBLEMS

Preparation of Alcohols, Diols, and Thiols

16.16 Write chemical equations, showing all necessary reagents, for the preparation of 1-butanol by each of the following methods:

(a) Hydroboration–oxidation of an alkene
(b) Use of a Grignard reagent
(c) Use of a Grignard reagent in a way different from part (b)
(d) Reduction of a carboxylic acid
(e) Hydrogenation of an aldehyde
(f) Reduction with sodium borohydride

16.17 Write chemical equations, showing all necessary reagents, for the preparation of 2-butanol by each of the following methods:

(a) Hydroboration–oxidation of an alkene

(b) Use of a Grignard reagent

(c) Use of a Grignard reagent different from that used in part (b)

(d–f) Three different methods for reducing a ketone

16.18 Which of the isomeric $C_5H_{12}O$ alcohols can be prepared by lithium aluminum hydride reduction of:

(a) An aldehyde

(b) A ketone

(c) A carboxylic acid

(d) An ester of the type $RCOCH_3$ (with C=O)

16.19 Sorbitol is a sweetener often substituted for cane sugar, because it is better tolerated by diabetics. It is also an intermediate in the commercial synthesis of vitamin C. Sorbitol is prepared by high-pressure hydrogenation of glucose over a nickel catalyst. What is the structure (including stereochemistry) of sorbitol?

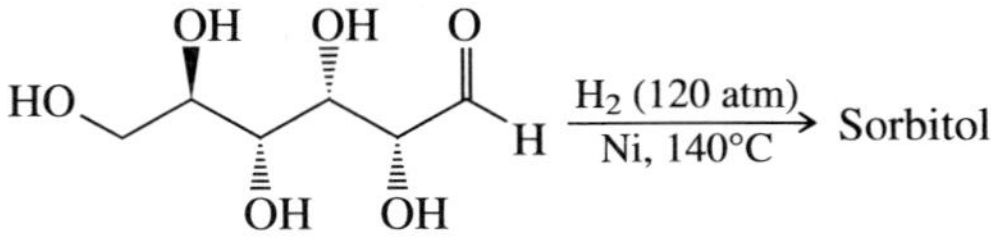

Glucose $\xrightarrow[\text{Ni, 140°C}]{H_2\text{ (120 atm)}}$ Sorbitol

Glucose

16.20 Write equations showing how 1-phenylethanol could be prepared from each of the following starting materials:

(a) Bromobenzene

(b) Benzaldehyde

(c) Benzyl alcohol

(d) Acetophenone

(e) Benzene

16.21 Write equations showing how 2-phenylethanol could be prepared from each of the following starting materials:

(a) Bromobenzene

(b) Styrene

(c) 2-Phenylethanal ($C_6H_5CH_2CHO$)

(d) 2-Phenylethanoic acid ($C_6H_5CH_2CO_2H$)

16.22 Outline a brief synthesis of each of the following compounds from the indicated starting material and any other necessary organic or inorganic reagents.

(a) 2-Propen-1-thiol from propene

(b) 1-Hexanol from 1-bromobutane

(c) 2-Hexanol from 1-bromobutane

(d) 2-Methyl-1,2-propanediol from *tert*-butyl alcohol

(e) 1-Chloro-2-phenylethane from benzene

16.23 Show how each of the following compounds can be synthesized from cyclopentanol and any necessary organic or inorganic reagents. In many cases the desired compound can be made from one prepared in an earlier part of the problem.

(a) 1-Phenylcyclopentanol

(b) 1-Phenylcyclopentene

(c) *trans*-2-Phenylcyclopentanol

(d) 2-C_6H_5-cyclopentanone (structure: C_6H_5, O)

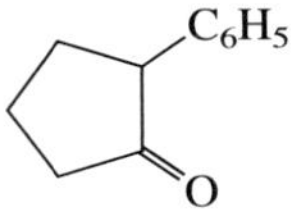

(e) (structure: OH, OH, phenyl on cyclopentane)

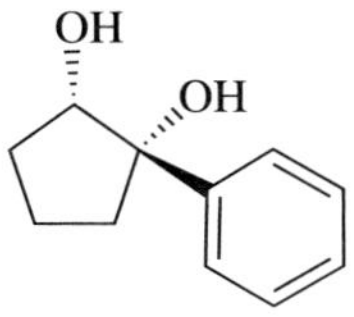

(f) 1-Phenyl-1,5-pentanediol

Reactions

16.24 Several oxidizing reagents for alcohols were described in this chapter. Suggest one for each of the following oxidations.

(a) 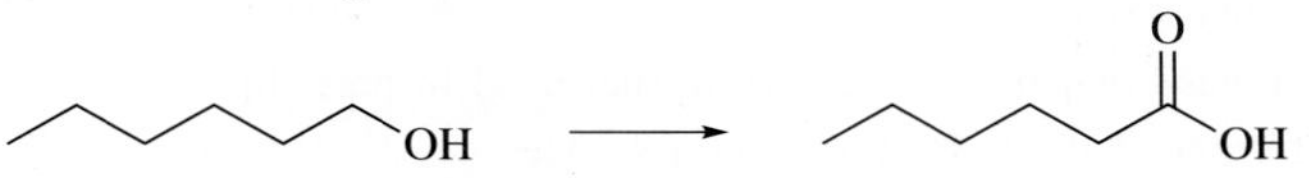 [structure: 1-hexanol] ⟶ [structure: hexanoic acid]

(b)  [structure: 2-hexanol] ⟶ [structure: 2-hexanone]

(c) 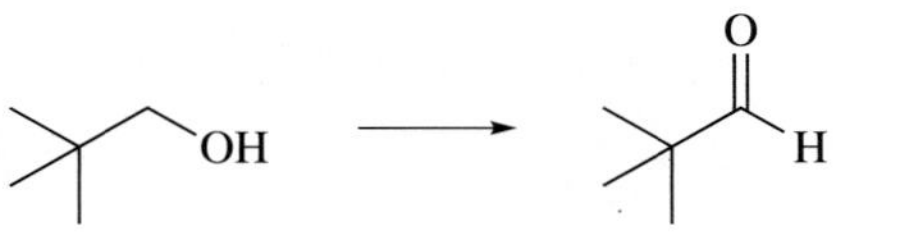 [structure: 2,2-dimethyl-1-propanol] ⟶ [structure: 2,2-dimethylpropanal]

(d) 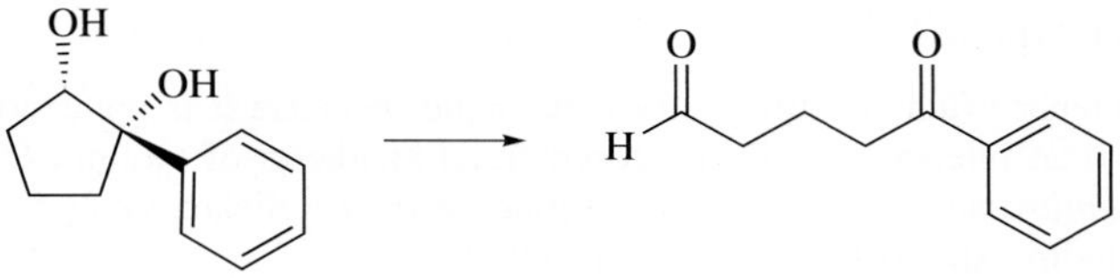 [structure: 1-phenylcyclopentane-1,2-diol] ⟶ [structure: 5-oxo-5-phenylpentanal]

16.25 Write the structure of the principal organic product formed in the reaction of 1-propanol with each of the following reagents:

(a) Sulfuric acid (catalytic amount), heat at 140°C

(b) Sulfuric acid (catalytic amount), heat at 200°C

(c) Dimethyl sulfoxide (DMSO), oxalyl chloride [$(COCl)_2$], triethylamine [$N(CH_2CH_3)_3$]

(d) Pyridinium chlorochromate (PCC) in dichloromethane

(e) Potassium dichromate ($K_2Cr_2O_7$) in aqueous sulfuric acid, heat

(f) Sodium amide ($NaNH_2$)

(g) Acetic acid ($CH_3\overset{O}{\overset{\|}{C}}OH$) in the presence of dissolved hydrogen chloride

(h) $H_3C-C_6H_4-SO_2Cl$ in the presence of pyridine

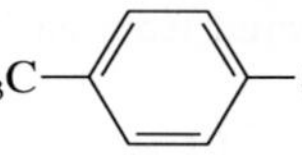

(i) $CH_3O-C_6H_4-\overset{O}{\overset{\|}{C}}Cl$ in the presence of pyridine

(j) $C_6H_5\overset{O}{\overset{\|}{C}}O\overset{O}{\overset{\|}{C}}C_6H_5$ in the presence of pyridine

(k) 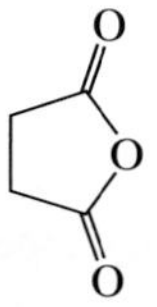 [structure: succinic anhydride] in the presence of pyridine

16.26 Each of the following reactions has been reported in the chemical literature. Predict the product in each case, showing stereochemistry where appropriate.

(a) 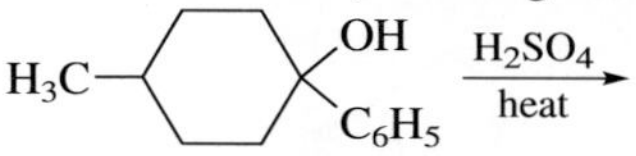 [structure: 4-methyl-1-phenylcyclohexanol, H_3C, OH, C_6H_5] $\xrightarrow[\text{heat}]{H_2SO_4}$

(b) 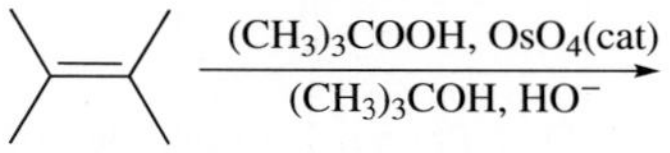 [structure: 2,3-dimethyl-2-butene] $\xrightarrow[(CH_3)_3COH,\ HO^-]{(CH_3)_3COOH,\ OsO_4(\text{cat})}$

(c) [structure: 1-phenylcyclobutene] $\xrightarrow[\text{2. } H_2O_2,\ HO^-]{\text{1. } B_2H_6,\text{ diglyme}}$

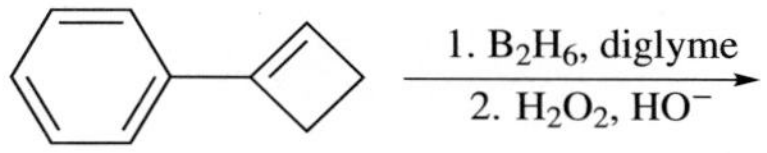

(d) 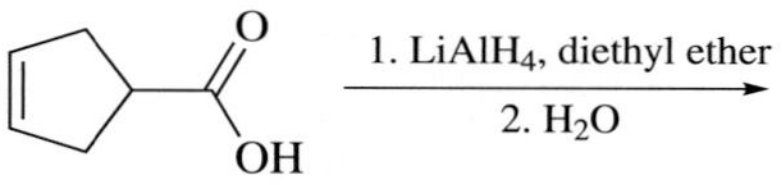 [structure: cyclopent-2-enylacetic acid, O, OH] $\xrightarrow[\text{2. } H_2O]{\text{1. } LiAlH_4,\text{ diethyl ether}}$

(e) $CH_3CHC{\equiv}C(CH_2)_3CH_3$ (with OH on C-2) $\xrightarrow[H_2SO_4,\ H_2O,\ acetone]{H_2CrO_4}$

(f) $CH_3\overset{O}{\overset{\|}{C}}CH_2CH{=}CHCH_2\overset{O}{\overset{\|}{C}}CH_3 \xrightarrow[2.\ H_2O]{1.\ LiAlH_4,\ diethyl\ ether}$

(g) 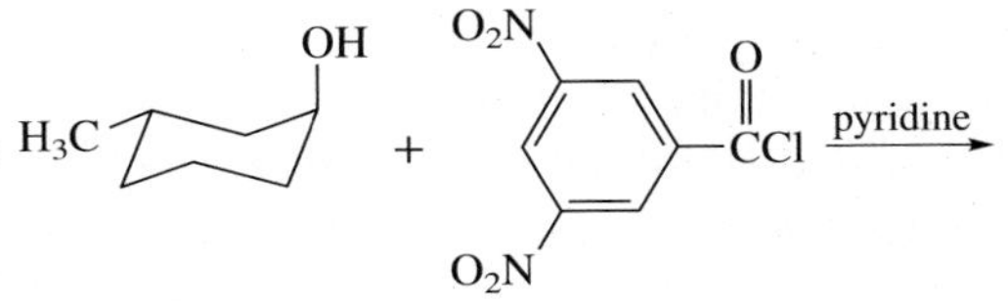 H_3C–cyclohexane–OH (chair) + 3,5-dinitrobenzoyl chloride (O_2N, O_2N, $\overset{O}{\overset{\|}{C}}Cl$) $\xrightarrow{pyridine}$

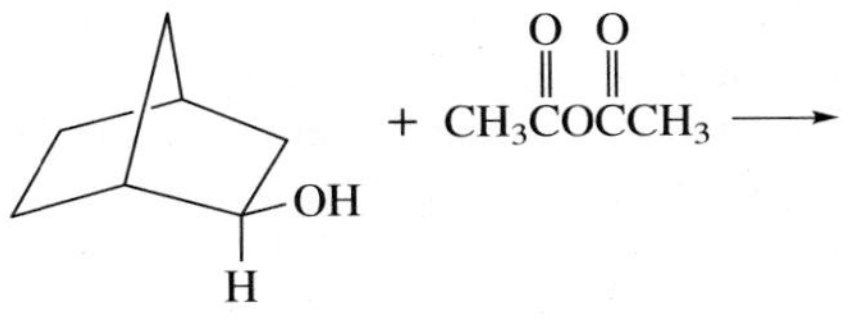

(h) bicyclic alcohol (–OH, H) + $CH_3\overset{O}{\overset{\|}{C}}O\overset{O}{\overset{\|}{C}}CH_3 \longrightarrow$

(i) 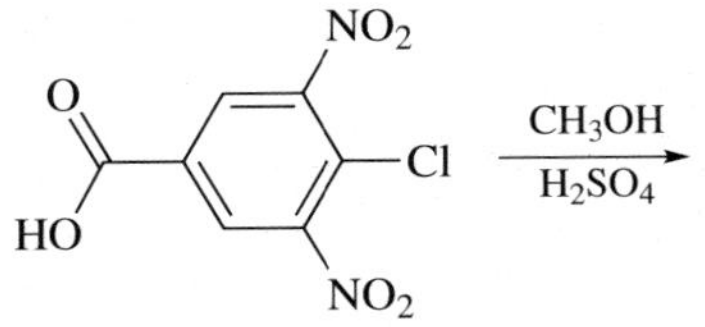 HO–C(=O)–aryl bearing NO_2, Cl, NO_2 $\xrightarrow[H_2SO_4]{CH_3OH}$

16.27 On heating 1,2,4-butanetriol in the presence of an acid catalyst, a cyclic ether of molecular formula $C_4H_8O_2$ was obtained in 81–88% yield. Suggest a reasonable structure for this product.

16.28 The amino acid *cysteine* has the structure

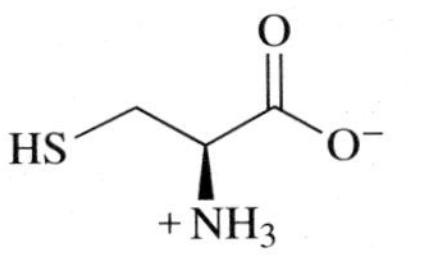

HS–CH₂–CH($\overset{+}{N}H_3$)–C(=O)O⁻

(a) A second sulfur-containing amino acid called *cystine* ($C_6H_{12}N_2O_4S_2$) is formed when cysteine undergoes biological oxidation. Suggest a reasonable structure for cystine.

(b) Another metabolic pathway converts cysteine to *cysteine sulfinic acid* ($C_3H_7NO_4S$), then to *cysteic acid* ($C_3H_7NO_5S$). What are the structures of these two compounds?

Synthesis

16.29 Suggest reaction sequences and reagents suitable for carrying out each of the following conversions. Two synthetic operations are required in each case.

(a) 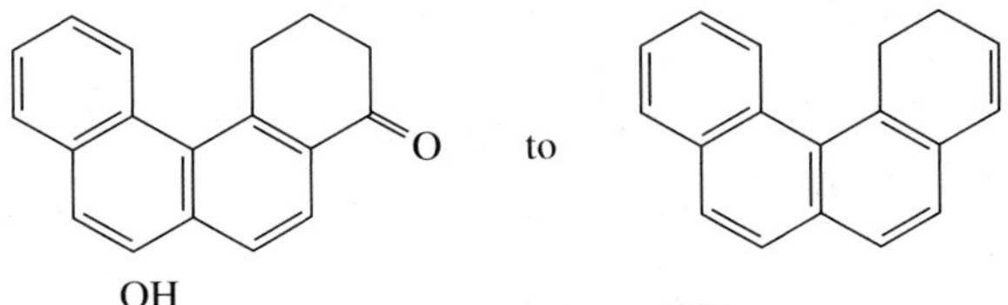 [polycyclic ketone, =O] to [polycyclic alkene]

(b) 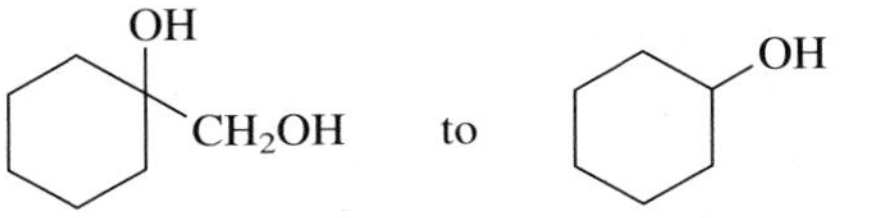 [1-(hydroxymethyl)cyclohexanol: OH, CH_2OH] to [cyclohexane–CH₂? –OH]

(c) 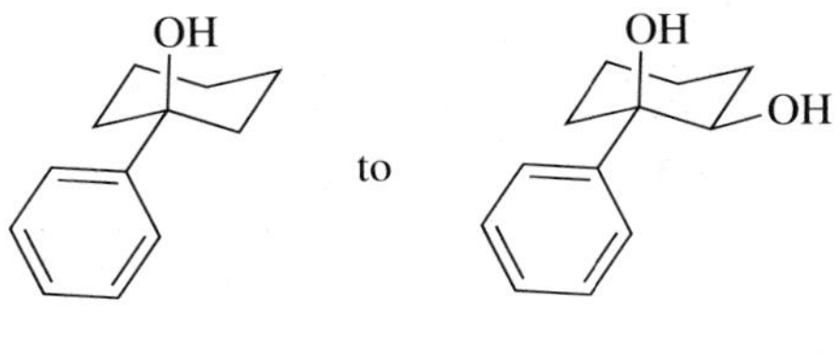 [phenylcyclohexanol, OH] to [phenylcyclohexane-diol, OH, OH]

16.30 The fungus responsible for Dutch elm disease is spread by European bark beetles when they burrow into the tree. Other beetles congregate at the site, attracted by the scent of a mixture of chemicals, some emitted by other beetles and some coming from the tree. One of the compounds given off by female bark beetles is 4-methyl-3-heptanol. Suggest an efficient synthesis of this pheromone from alcohols of five carbon atoms or fewer.

16.31 (a) The cis isomer of 3-hexen-1-ol ($CH_3CH_2CH{=}CHCH_2CH_2OH$) has the characteristic odor of green leaves and grass. Suggest a synthesis for this compound from acetylene and any necessary organic or inorganic reagents.

(b) One of the compounds responsible for the characteristic odor of ripe tomatoes is the cis isomer of $CH_3CH_2CH{=}CHCH_2CH{=}O$. How could you prepare this compound?

16.32 R. B. Woodward was one of the leading organic chemists of the middle part of the twentieth century. Known primarily for his achievements in the synthesis of complex natural products, he was awarded the Nobel Prize in Chemistry in 1965. He entered Massachusetts Institute of Technology as a 16-year-old freshman in 1933 and four years later was awarded the Ph.D. While a student there he carried out a synthesis of *estrone,* a female sex hormone. The early stages of Woodward's estrone synthesis required the conversion of *m*-methoxybenzaldehyde to *m*-methoxybenzyl cyanide, which was accomplished in three steps:

[m-CH_3O–C_6H_4–CH=O] $\xrightarrow{\text{three steps}}$ [m-CH_3O–C_6H_4–CH_2CN] $\xrightarrow{\text{many steps}}$ Estrone

Suggest a reasonable three-step sequence, showing all necessary reagents, for the preparation of *m*-methoxybenzyl cyanide from *m*-methoxybenzaldehyde.

16.33 Complete each of the following equations by writing structural formulas for compounds A through I:

(a) cyclopentene $\xrightarrow{HCl}$ C_5H_7Cl (Compound A) $\xrightarrow[H_2O]{NaHCO_3}$ C_5H_8O (Compound B) $\xrightarrow[H_2SO_4,\ H_2O]{Na_2Cr_2O_7}$ C_5H_6O (Compound C)

(b) $H_2C{=}CHCH_2CH_2CH(OH)CH_3$ $\xrightarrow[pyridine]{SOCl_2}$ $C_6H_{11}Cl$ (Compound D) $\xrightarrow[2.\ Zn,\ H_2O]{1.\ O_3}$ C_5H_9ClO (Compound E) $\xrightarrow{NaBH_4}$ $C_5H_{11}ClO$ (Compound F)

(c) 1-bromo-2-methylnaphthalene $\xrightarrow[\text{benzoyl peroxide, heat}]{NBS}$ Compound G $\xrightarrow[heat]{H_2O,\ CaCO_3}$ Compound H $\xrightarrow[CH_2Cl_2]{PCC}$ ($C_{11}H_7BrO$) Compound I

16.34 Choose the correct enantiomer of 2-butanol that would permit you to prepare (*R*)-2-butanethiol by way of a *p*-toluenesulfonate and write a chemical equation for each step.

Structure Determination

16.35 Suggest a chemical test that would permit you to distinguish between the two glycerol monobenzyl ethers shown.

1-*O*-Benzylglycerol 2-*O*-Benzylglycerol

16.36 A diol ($C_8H_{18}O_2$) does not react with periodic acid. Its ^1H NMR spectrum is shown in Figure 16.6. What is the structure of this diol?

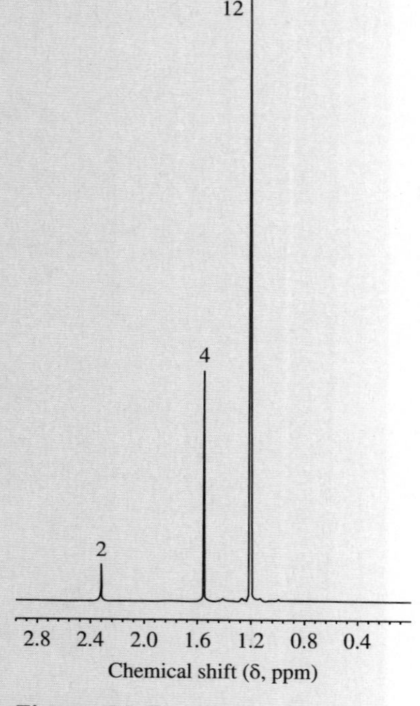

Figure 16.6

^1H NMR spectrum of the diol in Problem 16.36.

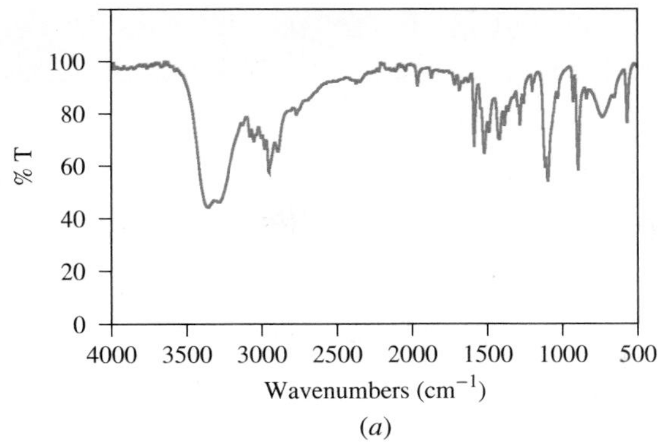

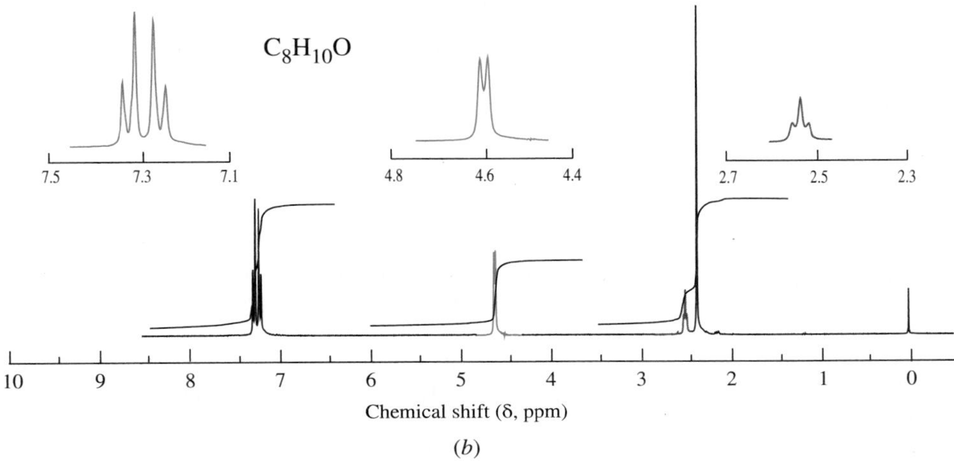

Figure 16.7

The IR (*a*) and 300-MHz ^{1}H NMR (*b*) spectra of a compound $C_8H_{10}O$ (Problem 16.37).

16.37 Identify the compound $C_8H_{10}O$ on the basis of its IR and ^{1}H NMR spectra (Figure 16.7). The broad triplet near δ 2.5 in the NMR spectrum disappears when D_2O is added.

16.38 Identify each of the following $C_4H_{10}O$ isomers on the basis of their ^{13}C NMR spectra:

(a) δ 31.2: CH_3
δ 68.9: C

(b) δ 10.0: CH_3
δ 22.7: CH_3
δ 32.0: CH_2
δ 69.2: CH

(c) δ 18.9: CH_3, area 2
δ 30.8: CH, area 1
δ 69.4: CH_2, area 1

16.39 A compound $C_3H_7ClO_2$ exhibited three peaks in its ^{13}C NMR spectrum at δ 46.8 (CH_2), δ 63.5 (CH_2), and δ 72.0 (CH). What is the structure of this compound?

16.40 A compound $C_6H_{14}O$ has the ^{13}C NMR spectrum shown in Figure 16.8. Its mass spectrum has a prominent peak at *m/z* 31. Suggest a reasonable structure for this compound.

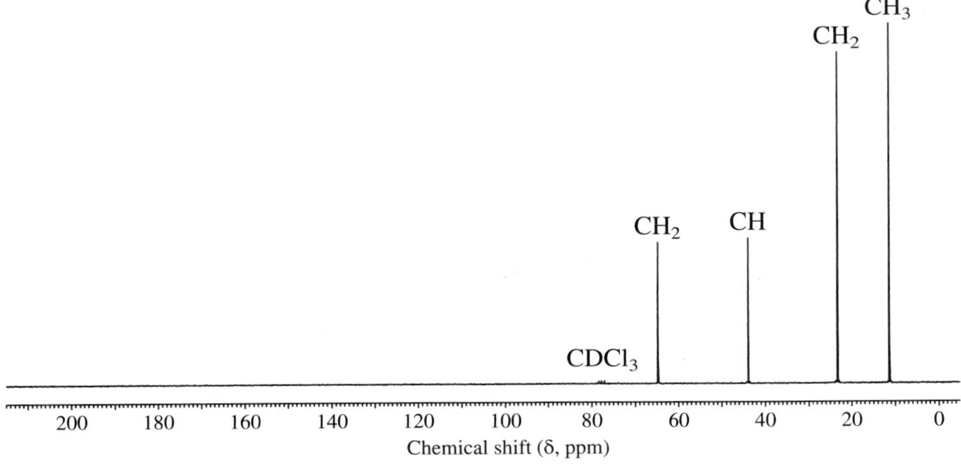

Figure 16.8

The ^{13}C NMR spectrum of the compound $C_6H_{14}O$ (Problem 16.40).

Descriptive Passage and Interpretive Problems 16

The Pinacol Rearrangement

We would expect a vicinal diol such as 2,3-dimethyl-2,3-butanediol to give a conjugated diene by double dehydration on treatment with an acid catalyst.

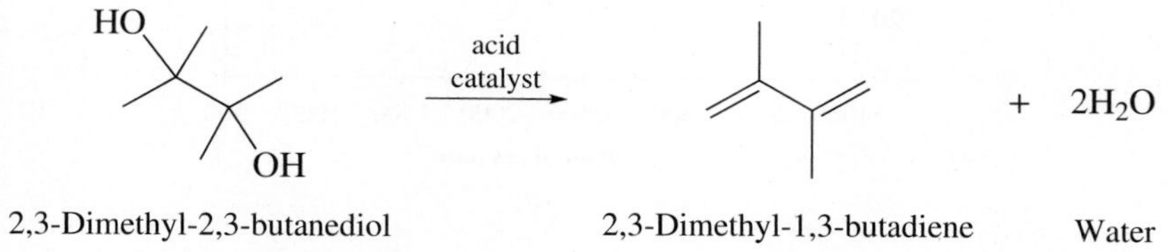

2,3-Dimethyl-2,3-butanediol —(acid catalyst)→ 2,3-Dimethyl-1,3-butadiene + $2H_2O$ (Water)

Although 2,3-dimethyl-1,3-butadiene can be prepared by just such a process, under certain conditions a different reaction occurs.

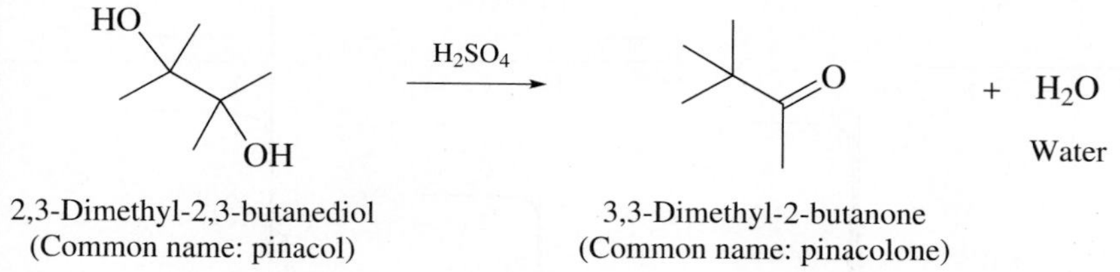

2,3-Dimethyl-2,3-butanediol (Common name: pinacol) —(H_2SO_4)→ 3,3-Dimethyl-2-butanone (Common name: pinacolone) + H_2O (Water)

This reaction is called the *pinacol rearrangement* after the common name of the diol reactant.

The mechanism for conversion of pinacol to pinacolone begins with protonation of one of the OH groups of the vicinal diol.

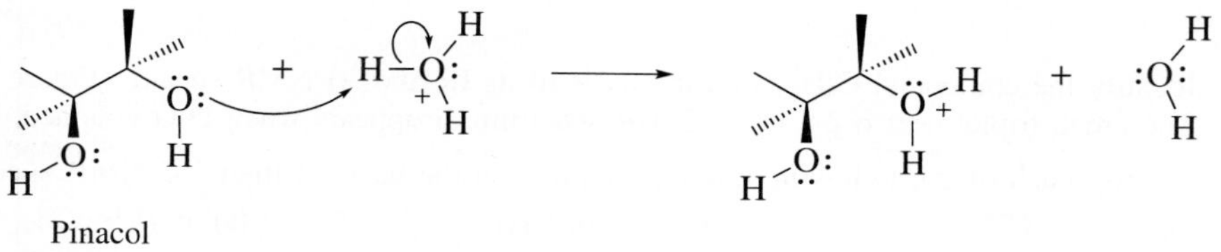

Pinacol

This is followed by loss of water and migration of a methyl group, usually in a single step in which the group anti to the departing water migrates.

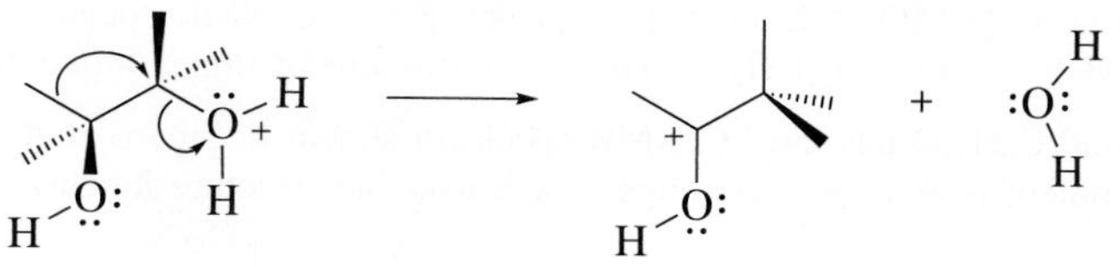

A key to understanding this migration is to recall that carbocations are stabilized by delocalization of a lone pair of an attached oxygen.

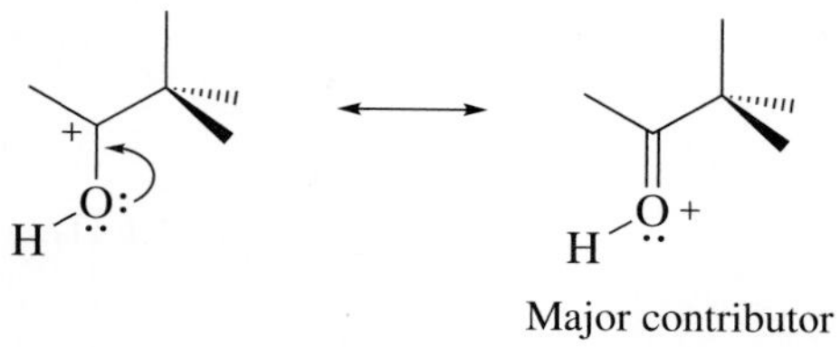

Major contributor

Thus, the rearrangement follows the usual generalization that a less stable carbocation is converted to a more stable one. Deprotonation of oxygen completes the mechanism.

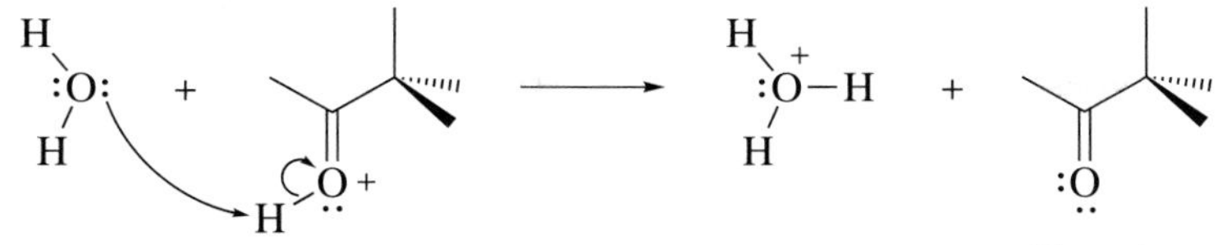

Pinacolone

The term "pinacol rearrangement" is applied in a general way to any rearrangement that transforms a vicinal diol to a ketone.

16.41 Which word or phrase best describes the stereochemistry of the product formed in the pinacol rearrangement of the diol shown?

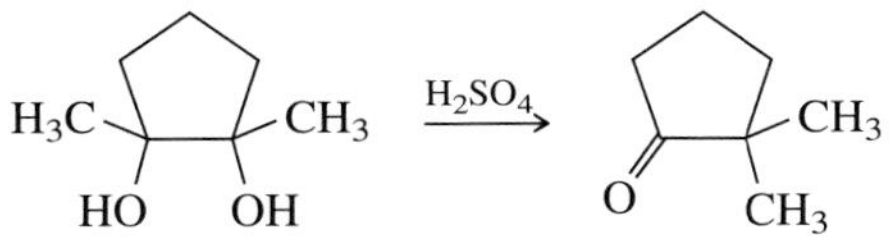

H_2SO_4

A. Achiral

B. A single enantiomer of a chiral molecule

C. Chiral, but racemic

16.42–16.43 Consider the two diols (**1** and **2**) and the two ketones (**3** and **4**).

Vicinal diols

Ketones

A mixture of **3** and **4** is formed by pinacol rearrangement of either **1** or **2.** Given that an ethyl migrates in preference to methyl in pinacol rearrangements, predict the major ketone formed by rearrangement of each diol.

16.42 Diol **1** gives predominantly

A. Ketone **3**

B. Ketone **4**

16.43 Diol **2** gives predominantly

A. Ketone **3**

B. Ketone **4**

16.44 What is the product of the following reaction?

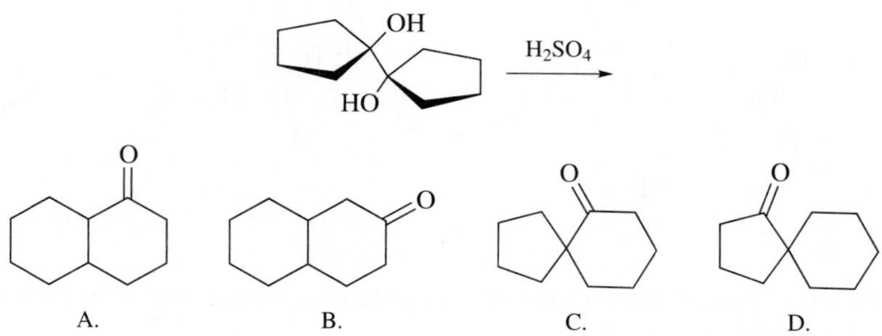

O O O O

A. B. C. D.

16.45 The group that is anti to oxygen migrates in the pinacol rearrangement of the diol shown. What is the product?

CH_3 OH OH CH_3 O CH_3 CH_3 O CH_3 CH_3

A. B.

16.46 Rather than following a mechanism in which a group migrates in the same step as water departs, the pinacol rearrangement of the vicinal diol shown proceeds by way of the more stable of two possible carbocations. A single ketone is formed in 73% yield. What is the structure of this ketone?

OH OH O O

2-Methyl-1,1-diphenyl-1,2-propanediol A. B.

16.47 Like the pinacol rearrangement in the preceding problem, this one also begins with the formation of the more stable of two possible carbocations from a vicinal diol. A 99% yield of a single ketone was isolated. What is this ketone?

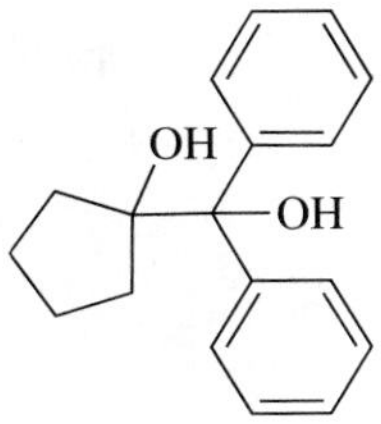

OH

OH

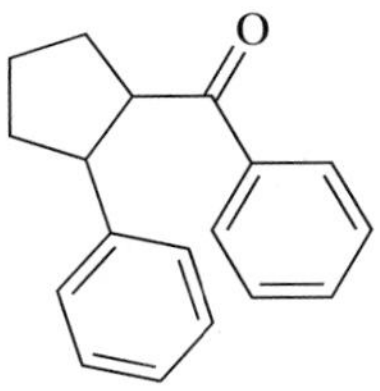

O

O

O

O

A.

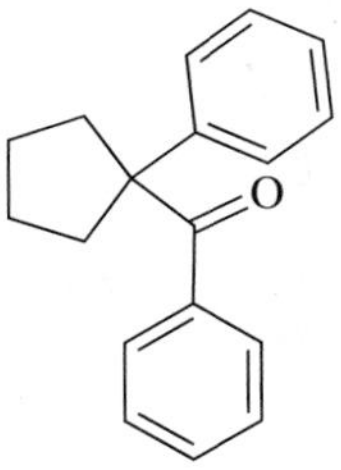

B.

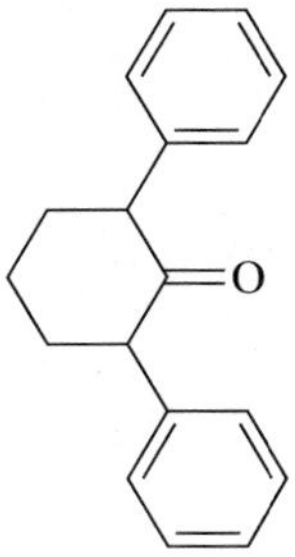

C.

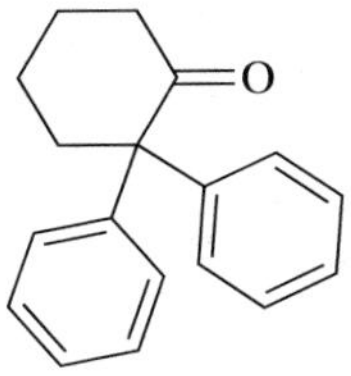

D.

Chapter

17

CHAPTER OUTLINE

Intestinal parasites in cattle are controlled by adding monensin to their feed. Monensin belongs to a class of antibiotics known as *ionophores* ("ion carriers") that act by forming stable complexes of the kind shown in the space-filling model with metal ions. For more, see the boxed essay *Polyether Antibiotics.*
Source: Photo by Jeff Vanuga, USDA Natural Resources Conservation Service

Ethers, Epoxides, and Sulfides

In contrast to alcohols with their rich chemical reactivity, **ethers** (compounds containing a C—O—C unit) undergo relatively few chemical reactions. As you saw when we discussed Grignard reagents in Chapter 15 and lithium aluminum hydride reductions in Chapter 16, this lack of reactivity of ethers makes them valuable as solvents in a number of synthetically important transformations. In the present chapter you will learn of the conditions in which an ether linkage acts as a functional group, as well as the methods by which ethers are prepared.

Unlike most ethers, **epoxides** (compounds in which the C—O—C unit forms a three-membered ring) are very reactive substances. The principles of nucleophilic substitution are important in understanding the preparation and properties of epoxides.

Sulfides (RSR′) are the sulfur analogs of ethers. Just as in Chapter 16, where we saw that the properties of thiols (RSH) are different from those of alcohols, we will explore differences between sulfides and ethers in this chapter.

17.1 Nomenclature of Ethers, Epoxides, and Sulfides

Ethers are named, in substitutive IUPAC nomenclature, as *alkoxy* derivatives of alkanes. Functional class IUPAC names of ethers are derived by listing the two alkyl groups in the general structure

ROR′ in alphabetical order as separate words, and adding the word *ether* at the end. When both alkyl groups are the same, the prefix *di-* precedes the name of the alkyl group.

	$CH_3CH_2OCH_2CH_3$	$CH_3CH_2OCH_3$	$CH_3CH_2OCH_2CH_2CH_2Cl$
Substitutive IUPAC name:	Ethoxyethane	Methoxyethane	1-Chloro-3-ethoxypropane
Functional class IUPAC name:	Diethyl ether	Ethyl methyl ether	3-Chloropropyl ethyl ether

Ethers are described as *symmetrical* or *unsymmetrical* depending on whether the two groups bonded to oxygen are the same or different. Diethyl ether is a symmetrical ether; ethyl methyl ether is an unsymmetrical ether.

Cyclic ethers have their oxygen as part of a ring—they are *heterocyclic compounds* (Section 3.15). Several have specific IUPAC names.

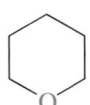

Oxirane Ethylene oxide	Oxetane	Oxolane Tetrahydrofuran	Oxane Tetrahydropyran

In each case the ring is numbered starting at the oxygen. The IUPAC rules also permit oxirane (without substituents) to be called *ethylene oxide. Tetrahydrofuran* and *tetrahydropyran* are acceptable synonyms for oxolane and oxane, respectively.

Problem 17.1

Each of the following ethers has been shown to be or is suspected to be a *mutagen,* which means it can induce mutations in test cells. Write the structure of each of these ethers.

(a) Chloromethyl methyl ether

(b) 2-(Chloromethyl)oxirane (also known as epichlorohydrin)

(c) 3,4-Epoxy-1-butene (2-vinyloxirane)

Sample Solution (a) Chloromethyl methyl ether has a chloromethyl group ($ClCH_2$—) and a methyl group (H_3C—) attached to oxygen. Its structure is $ClCH_2OCH_3$.

Recall from Section 8.11 that epoxides may be named as *epoxy* derivatives of alkanes in substitutive IUPAC nomenclature.

Many substances have more than one ether linkage. Two such compounds, often used as solvents, are the *diethers* 1,2-dimethoxyethane and 1,4-dioxane. Diglyme, also a commonly used solvent, is a *triether.*

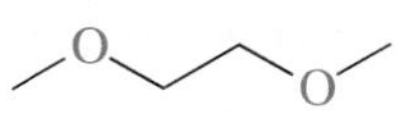
1,2-Dimethoxyethane

1,4-Dioxane

Diethylene glycol dimethyl ether
diglyme

Molecules that contain several ether functions are referred to as *polyethers.* Polyethers have some novel properties and will appear in Section 17.4.

The sulfur analogs (RS—) of alkoxy groups are called *alkylthio* groups. The first two of the following examples illustrate the use of alkylthio prefixes in substitutive nomenclature of sulfides. Functional class IUPAC names of sulfides are derived in exactly the same way as those of ethers but end in the word *sulfide.* Sulfur heterocycles have names analogous to their oxygen relatives, except that *ox-* is replaced by *thi-*. Thus, the sulfur heterocycles containing three-, four-, five-, and six-membered rings are named *thiirane, thietane, thiolane,* and *thiane,* respectively.

Sulfides are sometimes informally referred to as *thioethers,* but this term is not part of systematic IUPAC nomenclature.

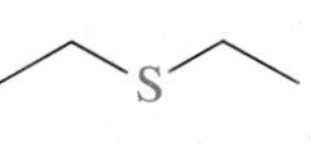
Ethylthioethane
Diethyl sulfide

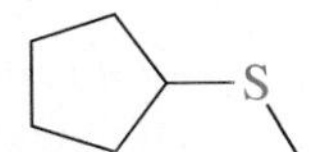
(Methylthio)cyclopentane
Cyclopentyl methyl sulfide

Thiirane

17.2 Structure and Bonding in Ethers and Epoxides

Bonding in ethers is readily understood by comparing ethers with water and alcohols. Van der Waals strain involving alkyl groups causes the bond angle at oxygen to be larger in ethers than in alcohols, and larger in alcohols than in water. An extreme example is di-*tert*-butyl ether, where steric hindrance between the *tert*-butyl groups is responsible for a dramatic increase in the C—O—C bond angle.

H–O–H 105°	H–O–CH_3 108.5°	H_3C–O–CH_3 112°	$(CH_3)_3C$–O–$C(CH_3)_3$ 132°
Water	Methanol	Dimethyl ether	Di-*tert*-butyl ether

Typical carbon–oxygen bond distances in ethers are similar to those of alcohols (≈ 142 pm, 1.42 Å) and are shorter than carbon–carbon bond distances in alkanes (≈ 153 pm, 1.53 Å).

An ether oxygen affects the conformation of a molecule in much the same way that a CH_2 unit does. The most stable conformation of diethyl ether is the all-staggered anti conformation. Tetrahydropyran is most stable in the chair conformation—a fact that has an important bearing on the structures of countless carbohydrates.

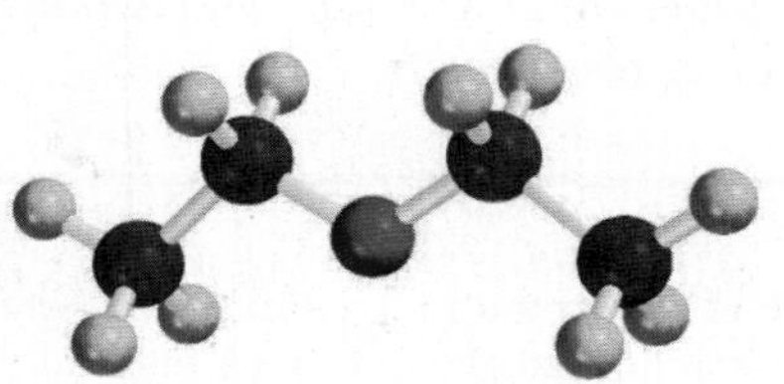
Anti conformation of diethyl ether

Chair conformation of tetrahydropyran

Incorporating an oxygen atom into a three-membered ring requires its bond angle to be seriously distorted from the normal tetrahedral value. In ethylene oxide, for example, the bond angle at oxygen is 61.5°.

H_2C—CH_2 (C—C: 147 pm (1.47 Å)), bridged by O (C—O: 144 pm (1.44 Å))

C—O—C angle 61.5°

C—C—O angle 59.2°

Thus, epoxides, like cyclopropanes, have significant angle strain. They tend to undergo reactions that open the three-membered ring by cleaving one of the carbon–oxygen bonds.

Problem 17.2

The heats of combustion of 1,2-epoxybutane (2-ethyloxirane) and tetrahydrofuran have been measured: one is 2499 kJ/mol; the other is 2546 kJ/mol. Match the heats of combustion with the respective compounds.

17.3 Physical Properties of Ethers

Table 17.1 compares the physical properties of diethyl ether to those of an alkane (pentane) and an alcohol (1-butanol) of similar size and shape. With respect to boiling point, diethyl ether resembles pentane more than 1-butanol. With respect to dipole moment and solubility in water, the reverse is true.

As we have seen before, alcohols have unusually high boiling points because of hydrogen bonding between —OH groups.

$\ddot{O}^{\delta-}$(H):---$H^{\delta+}$—$\ddot{O}$:

Intermolecular hydrogen bonding in 1-butanol

TABLE 17.1 Physical Properties of Diethyl Ether, Pentane, and 1-Butanol

	Compound	Dipole moment, D	Boiling point, °C	Solubility in water, g/100 mL
Diethyl ether	$CH_3CH_2OCH_2CH_3$	1.2	35	7.5
Pentane	$CH_3CH_2CH_2CH_2CH_3$	0	36	≈0
1-Butanol	$CH_3CH_2CH_2CH_2OH$	1.7	117	9

Lacking —OH groups, ethers resemble alkanes in that dispersion forces are the major contributors to intermolecular attractions. Although ethers have significant dipole moments, the fact that their boiling points are closer to alkanes than to alcohols tells us that dipole–dipole attractive forces are minor contributors.

On the other hand, ethers have a negatively polarized oxygen that can hydrogen bond to an —OH proton of water.

Hydrogen bonding between diethyl ether and water

Such hydrogen bonding causes ethers to dissolve in water to approximately the same extent as alcohols of similar size and shape. Alkanes cannot engage in hydrogen bonding to water. Figure 17.1 shows electrostatic potential maps of diethyl ether and water and the hydrogen-bonded complex formed between them.

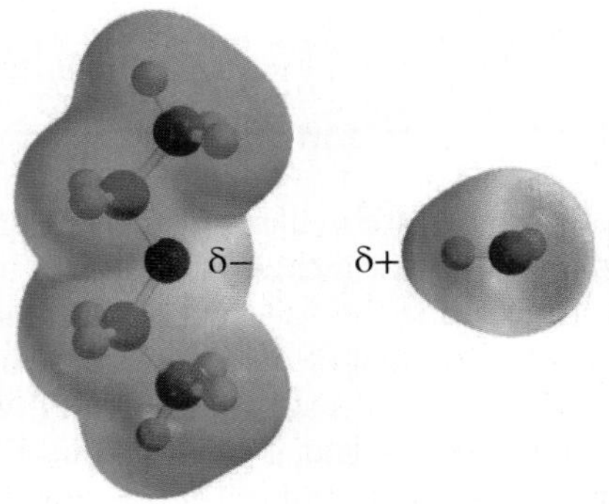

(*a*) Diethyl ether and water as separate molecules

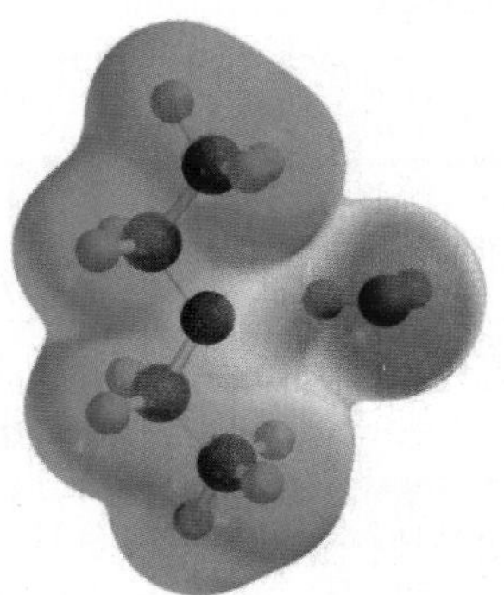

(*b*) Hydrogen-bonded complex

Figure 17.1

Hydrogen bonding between diethyl ether and water results from the attractive force between the negatively polarized oxygen of diethyl ether and the positively polarized hydrogen of water. The color ranges of the electrostatic potential maps are the same.

Problem 17.3

Of the two compounds cyclopentane and tetrahydrofuran, one has a boiling point of 49°C and is insoluble in water; the other has a boiling point of 65°C and is miscible with water in all proportions. Match the properties to the appropriate compound. In which property of which compound is hydrogen bonding important? Sketch the hydrogen-bonding interaction.

17.4 Crown Ethers

Their polar carbon–oxygen bonds and the presence of unshared electron pairs at oxygen contribute to the ability of ethers to form Lewis acid/Lewis base complexes with metal ions.

$$R_2\ddot{O}: \quad + \quad M^+ \quad \rightleftharpoons \quad R_2\overset{+}{\ddot{O}}—M$$

Ether (Lewis base)	Metal ion (Lewis acid)	Ether–metal ion complex

Pedersen was a corecipient of the 1987 Nobel Prize in Chemistry.

The strength of this bonding depends on the kind of ether. Simple ethers form relatively weak complexes with metal ions, but Charles J. Pedersen of DuPont discovered that certain *polyethers* form much more stable complexes with metal ions than do simple ethers.

Pedersen prepared a series of *macrocyclic polyethers,* cyclic compounds containing four or more oxygens in a ring of 12 or more atoms. He called these compounds **crown ethers,** because their molecular models resemble crowns. Their nomenclature is somewhat cumbersome, so Pedersen devised a shorthand description whereby the word *crown* is preceded by the total number of atoms in the ring and is followed by the number of oxygen atoms.

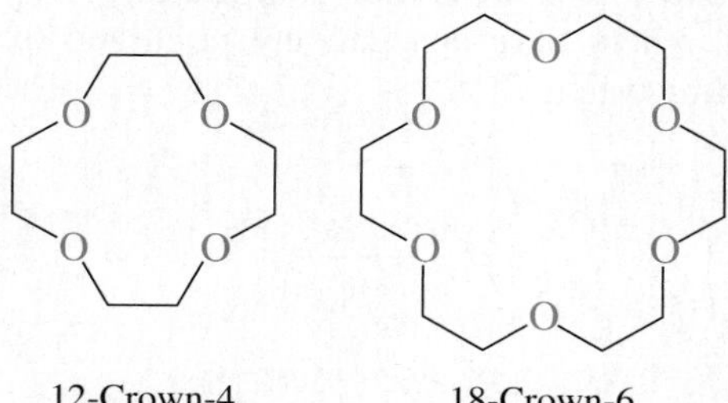

12-Crown-4 18-Crown-6

12-Crown-4 and 18-crown-6 are a cyclic tetramer and hexamer, respectively, of repeating $—OCH_2CH_2—$ units; they are polyethers based on ethylene glycol ($HOCH_2CH_2OH$) as the parent alcohol.

Problem 17.4

What organic compound mentioned earlier in this chapter is a cyclic dimer of $—OCH_2CH_2—$ units?

The metal–ion complexing properties of crown ethers are evident in their effects on the solubility and reactivity of ionic compounds in nonpolar media. Potassium fluoride (KF) is ionic and practically insoluble in benzene alone, but dissolves in it when 18-crown-6 is present. This happens because of the electron distribution of 18-crown-6 as shown in Figure 17.2*a*. The electrostatic potential surface consists of essentially two regions: an electron-rich interior associated with the oxygens and a hydrocarbon-like exterior associated with the CH_2 groups. When KF is added to a solution of 18-crown-6 in benzene, potassium ion (K^+) interacts with the oxygens of the crown ether to form a Lewis acid/Lewis base complex. As can be seen in the space-filling model of this complex (Figure 17.2*b*), K^+, with a diameter of 266 pm (2.66 Å), fits comfortably within the 260–320 pm (2.60–3.20 Å) internal cavity of 18-crown-6. Nonpolar CH_2 groups dominate the outer surface of the complex, mask its polar interior, and permit the

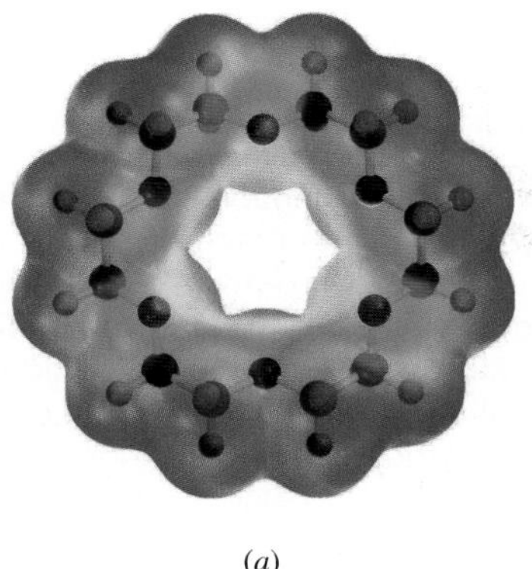

(*a*)

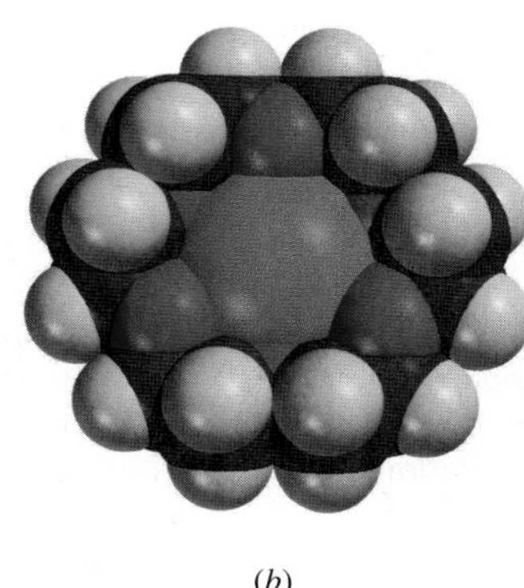

(*b*)

Figure 17.2

(*a*) An electrostatic potential map of 18-crown-6. The region of highest electron density (red) is associated with the negatively polarized oxygens and their lone pairs. The outer periphery of the crown ether (blue) is relatively nonpolar (hydrocarbon-like) and causes the molecule to be soluble in nonpolar solvents such as benzene. (*b*) A space-filling model of the complex formed between 18-crown-6 and potassium ion (K^+). K^+ fits into the cavity of the crown ether where it is bound by a Lewis acid/Lewis base interaction with the oxygens.

complex to dissolve in nonpolar solvents. Every K^+ that is carried into benzene brings a fluoride ion with it, resulting in a solution containing strongly complexed potassium ions and relatively unsolvated fluoride ions.

$$\text{18-Crown-6} + K^+F^- \text{ (solid)} \xrightleftharpoons{\text{benzene}} \text{18-Crown-6/potassium fluoride complex (in solution)}$$

18-Crown-6 | Potassium fluoride (solid) | 18-Crown-6/potassium fluoride complex (in solution)

In solvents such as water and alcohols, fluoride ion is strongly solvated by ion–dipole forces and is neither very basic nor very nucleophilic. On the other hand, the poorly solvated, or "naked," fluoride ions that are present when potassium fluoride dissolves in benzene in the presence of a crown ether are better able to express their anionic reactivity. Thus, alkyl halides react with potassium fluoride in benzene containing 18-crown-6, thereby providing a method for the preparation of otherwise difficultly accessible alkyl fluorides. No reaction is observed in the absence of the crown ether.

$$\text{1-Bromooctane} \xrightarrow[\text{18-crown-6}]{\text{KF, benzene, 90°C}} \text{1-Fluorooctane (92\%)}$$

The reaction proceeds in the direction indicated because a C—F bond is much stronger than a C—Br bond.

Catalysis by crown ethers has been used to advantage to increase the rate of many organic reactions that involve anions as reactants. Just as important, though, is the increased understanding that studies of crown ether catalysis have brought to our knowledge of biological processes in which metal ions, including Na^+ and K^+, are transported through the nonpolar interiors of cell membranes.

17.5 Preparation of Ethers

Because they are widely used as solvents, many simple dialkyl ethers are commercially available. Diethyl ether and dibutyl ether, for example, are prepared by acid-catalyzed condensation of the corresponding alcohols, as described earlier in Section 16.7.

$$\text{1-Butanol} \xrightarrow[\text{130°C}]{H_2SO_4} \text{Dibutyl ether (60\%)} + \underset{\text{Water}}{H_2O}$$

The mechanism for the formation of diethyl ether from ethanol under conditions of acid catalysis was shown in Mechanism 16.1.

In general, this method is limited to the preparation of symmetrical ethers in which both alkyl groups are primary.

Polyether Antibiotics

One way in which pharmaceutical companies search for new drugs is by growing colonies of microorganisms in nutrient broths and assaying the substances produced for their biological activity. This method has yielded thousands of antibiotic substances, of which hundreds have been developed into effective drugs. Antibiotics are, by definition, toxic (*anti* = "against"; *bios* = "life"), and the goal is to find substances that are more toxic to infectious organisms than to their human hosts.

Since 1950, a number of polyether antibiotics have been discovered using fermentation technology. They are characterized by the presence of several cyclic ether structural units, as illustrated for the case of *monensin* in Figure 17.3*a*. Monensin and other naturally occurring polyethers are similar to crown ethers in their ability to form stable complexes with metal ions.

The structure of the sodium salt of monensin is depicted in Figures 17.3*b* and 17.3*c*, where it can be seen that four ether oxygens and two hydroxyl groups surround a sodium ion. The alkyl groups are oriented toward the outside of the complex, and the polar oxygens and the metal ion are on the inside. The hydrocarbon-like surface of the complex permits it to carry its sodium ion through the hydrocarbon-like interior of a cell membrane. This disrupts the normal balance of sodium ions within the cell and interferes with important processes of cellular respiration. Small amounts of monensin are added to animal feed to kill parasites that live in the intestines of chickens, cows, etc. Compounds such as monensin and the crown ethers that affect metal ion transport are referred to as *ionophores* ("ion carriers").

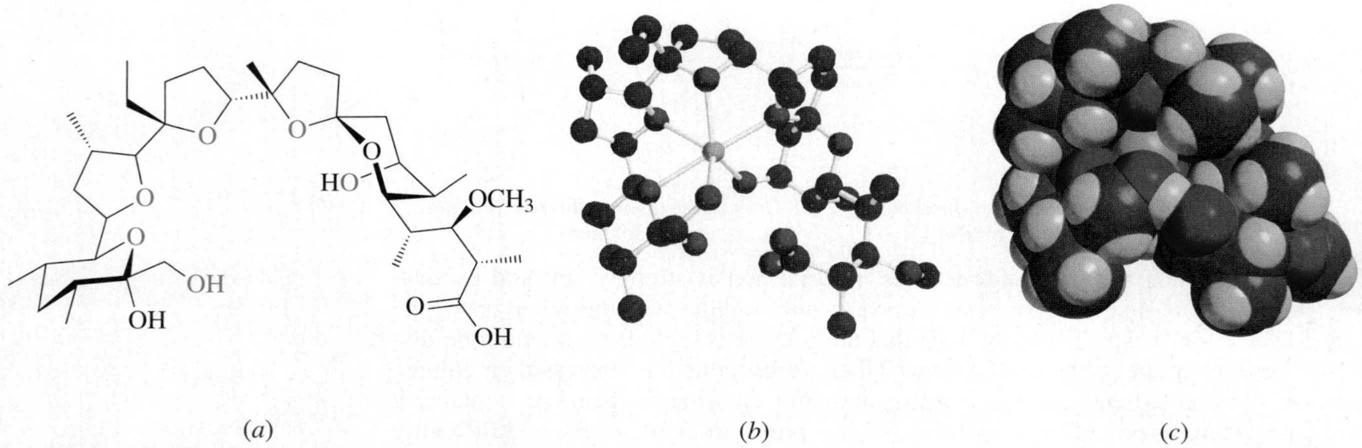

(*a*) (*b*) (*c*)

Figure 17.3

(*a*) The structure of monensin; (*b*) and (*c*) Ball-and-stick and space-filling models, respectively, of the sodium salt of monensin showing close contacts between Na^+ (yellow) and the oxygens in the internal cavity of the complex.

Acid-catalyzed addition of alcohols to alkenes is sometimes used. Billions of pounds of *tert*-butyl methyl ether (MTBE) have been prepared by the reaction:

tert-Butyl methyl ether is often referred to as MTBE, standing for the incorrect name "methyl *tert*-butyl ether." Remember, italicized prefixes are ignored when alphabetizing, and *tert*-butyl precedes methyl.

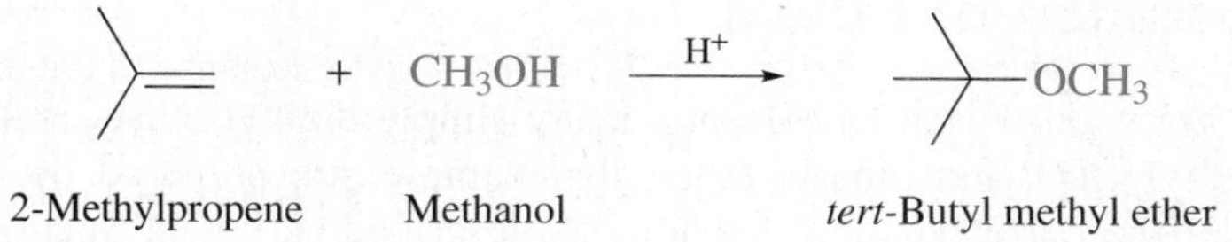

$$\text{2-Methylpropene} + CH_3OH \xrightarrow{H^+} \text{tert-Butyl methyl ether } (-OCH_3)$$

2-Methylpropene + Methanol → *tert*-Butyl methyl ether

Small amounts of *tert*-butyl methyl ether increase the octane rating of gasoline, and before environmental concerns placed limits on its use, the demand for MTBE exceeded the supply.

Problem 17.5

Outline a reasonable mechanism for the formation of *tert*-butyl methyl ether according to the preceding equation and classify it according to its mechanism.

Vicinal haloethers can be prepared by the reaction of an alkene in the presence of an alcohol and a halogen. This reaction is known as *haloetherification* and resembles halohydrin formation (Section 8.10).

1-Methylcyclohexene $\xrightarrow[CH_3CH_2OH]{I_2}$ *trans*-1-Ethoxy-2-iodo-1-methylcyclohexane (67%)

17.6 The Williamson Ether Synthesis

A long-standing method for the preparation of ethers is the **Williamson ether synthesis.** Nucleophilic substitution of an alkyl halide by an alkoxide gives the carbon–oxygen bond of an ether:

The reaction is named for Alexander Williamson, a British chemist who used it to prepare diethyl ether in 1850.

$$RO^- + R'—X \xrightarrow{S_N2} ROR' + X^-$$

Alkoxide ion, Alkyl halide, Ether, Halide ion

The reaction is most successful with methyl and primary alkyl halides.

$$CH_3CH_2CH_2CH_2ONa + CH_3CH_2I \longrightarrow CH_3CH_2CH_2CH_2OCH_2CH_3 + NaI$$

Sodium butoxide, Ethyl iodide, Butyl ethyl ether (71%), Sodium iodide

The alkoxide anion is prepared by the reaction of the alcohol with a base such as sodium hydride. For phenols, which are more acidic than alcohols, weaker bases may be used to generate the phenoxide.

$$CH_3CH_2CH_2CH_2OH + NaH \longrightarrow CH_3CH_2CH_2CH_2O^-Na^+ + H_2$$

$$C_6H_5OH + K_2CO_3 \longrightarrow C_6H_5O^-K^+ + KHCO_3$$

Problem 17.6

(a) Write equations describing two different ways in which benzyl ethyl ether could be prepared by a Williamson ether synthesis.

continued

Sample Solution

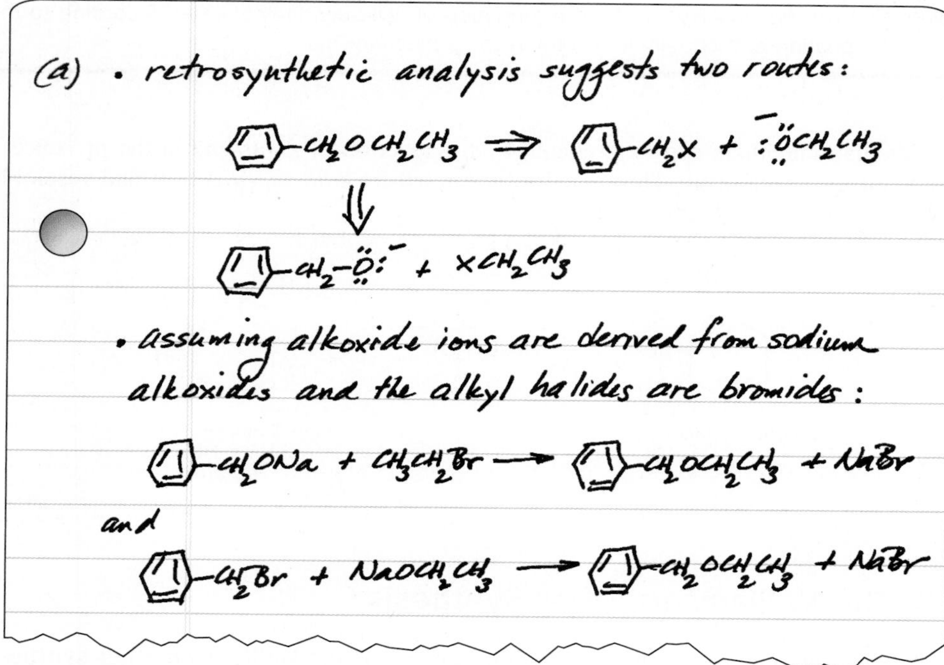

(b) Write an equation showing the most practical synthesis of allyl phenyl ether by the Williamson method.

Secondary and tertiary *alkyl halides* are not suitable, because they react with alkoxide bases by E2 elimination rather than by S_N2 substitution. Whether the *alkoxide base* is primary, secondary, or tertiary is much less important than the nature of the alkyl halide. Thus, benzyl isopropyl ether is prepared in high yield from benzyl chloride, a primary chloride that is incapable of undergoing elimination, and sodium isopropoxide:

Sodium isopropoxide + Benzyl chloride ⟶ Benzyl isopropyl ether (84%) + Sodium chloride (NaCl)

The alternative synthetic route using the sodium salt of benzyl alcohol and an isopropyl halide would be much less effective, because of increased competition from elimination as the alkyl halide becomes more sterically hindered.

Problem 17.7

Only one combination of alkyl halide and alkoxide is appropriate for the preparation of each of the following ethers by the Williamson ether synthesis. What is the correct combination in each case?

(a) CH_3CH_2O–cyclopentyl

(b)

(c)

Sample Solution (a) The ether linkage of cyclopentyl ethyl ether involves a primary carbon and a secondary one. Choose the alkyl halide corresponding to the primary alkyl group, leaving the secondary alkyl group to arise from the alkoxide nucleophile.

CH_3CH_2Br + NaO–(cyclopentyl) $\xrightarrow{S_N2}$ CH_3CH_2O–(cyclopentyl)

Ethyl bromide — Sodium cyclopentanolate — Cyclopentyl ethyl ether

The alternative combination, cyclopentyl bromide and sodium ethoxide, is not appropriate because elimination will be the major reaction:

CH_3CH_2ONa + Br–(cyclopentyl) $\xrightarrow{E2}$ CH_3CH_2OH + (cyclopentene)

Sodium ethoxide — Cyclopentyl bromide — Ethanol — Cyclopentene (major products)

Problem 17.8

Two approaches can be considered for the synthesis of alkyl aryl ethers:

X–C_6H_4–OCH_2CH_3 $\Longrightarrow$ X–C_6H_4–ONa + CH_3CH_2Br

X–C_6H_4–OCH_2CH_3 $\Longrightarrow$ X–C_6H_4–Br + CH_3CH_2ONa

Evaluate the feasibility of both approaches for X = methyl and X = nitro.

17.7 Reactions of Ethers: A Review and a Preview

Up to this point, we haven't seen any reactions of dialkyl ethers. Indeed, ethers are one of the least reactive of the functional groups we shall study. It is this low level of reactivity, along with an ability to dissolve nonpolar substances, that makes ethers so often used as solvents when carrying out organic reactions. Nevertheless, most ethers are hazardous materials, and precautions must be taken when using them. Diethyl ether is extremely flammable and because of its high volatility can form explosive mixtures in air relatively quickly. Open flames must never be present in laboratories where diethyl ether is being used. Other low-molecular-weight ethers must also be treated as fire hazards.

Another dangerous property of ethers is the ease with which they undergo oxidation in air to form explosive peroxides. Air oxidation of diisopropyl ether proceeds according to the equation

$(CH_3)_2CHOCH(CH_3)_2$ + O_2 ⟶ $(CH_3)_2CHOC(CH_3)_2OOH$

Diisopropyl ether — Oxygen — Diisopropyl ether hydroperoxide

The reaction follows a free-radical mechanism and gives a hydroperoxide, a compound of the type ROOH. Hydroperoxides are unstable and shock-sensitive and form related peroxidic derivatives that are prone to violent decomposition. Air oxidation leads to peroxides within a few days if ethers are even briefly exposed to atmospheric oxygen. For this reason, one should never use old bottles of dialkyl ethers, and extreme care must be exercised in their disposal.

17.8 Acid-Catalyzed Cleavage of Ethers

Just as the carbon–oxygen bond of alcohols is cleaved on reaction with hydrogen halides, so too is an ether linkage broken:

$$\underset{\text{Ether}}{ROR'} + \underset{\text{Hydrogen halide}}{HX} \longrightarrow \underset{\text{Alkyl halide}}{RX} + \underset{\text{Alcohol}}{R'OH}$$

The reaction is normally carried out under conditions (excess hydrogen halide, heat) that convert the alcohol formed as one of the original products to an alkyl halide and typically leads to two alkyl halide molecules plus water.

$$\underset{\text{Ether}}{ROR'} + \underset{\text{Hydrogen halide}}{2HX} \xrightarrow{\text{heat}} \underset{\text{Two alkyl halides}}{RX + R'X} + \underset{\text{Water}}{H_2O}$$

sec-Butyl methyl ether $\xrightarrow[\text{heat}]{\text{HBr}}$ 2-Bromobutane (81%) + CH_3Br Bromomethane

The order of hydrogen halide reactivity is HI > HBr >> HCl. Hydrogen fluoride is not effective.

Problem 17.9

A series of dialkyl ethers was allowed to react with excess hydrogen bromide, with the following results. Identify the ether in each case.

(a) One ether gave a mixture of bromocyclopentane and 1-bromobutane.
(b) Another ether gave only benzyl bromide.
(c) A third ether gave one mole of 1,5-dibromopentane per mole of ether.

Sample Solution (a) In the reaction of dialkyl ethers with excess hydrogen bromide, each alkyl group of the ether function is cleaved and forms an alkyl bromide. Because bromocyclopentane and 1-bromobutane are the products, the starting ether must be butyl cyclopentyl ether.

Butyl cyclopentyl ether $\xrightarrow[\text{heat}]{\text{HBr}}$ Bromocyclopentane + 1-Bromobutane

The cleavage of diethyl ether by hydrogen bromide is outlined in Mechanism 17.1. The key step is an S_N2-like attack on a dialkyloxonium ion by bromide (step 2).

Problem 17.10

Adapt Mechanism 17.1 to the reaction:

Tetrahydrofuran $\xrightarrow[150°C]{\text{HI}}$ 1,4-Diiodobutane (65%)

Mechanism 17.1

Cleavage of Ethers by Hydrogen Halides

THE OVERALL REACTION:

$$CH_3CH_2OCH_2CH_3 + HBr \xrightarrow{heat} 2CH_3CH_2Br + H_2O$$

Diethyl ether; Hydrogen bromide; Ethyl bromide; Water

THE MECHANISM:

Step 1: Proton transfer to the oxygen of the ether to give a dialkyloxonium ion.

Diethyl ether + Hydrogen bromide ⇌ (fast) Diethyloxonium ion + Bromide ion

Step 2: Nucleophilic attack of the halide anion on carbon of the dialkyloxonium ion. This step gives one molecule of an alkyl halide and one molecule of an alcohol.

Bromide ion + Diethyloxonium ion → (slow) Ethyl bromide + Ethanol

Steps 3 and 4: These two steps do not involve an ether at all. They correspond to those in which an alcohol is converted to an alkyl halide (Sections 5.8–5.13).

Ethanol + Hydrogen bromide ⇌ (fast) Ethyloxonium ion + Bromide ion → (slow) Ethyl bromide + Water

With ethers of the type ROR′ where R and R′ are different alkyl groups, the question of which carbon–oxygen bond is broken first is not one that we need examine at our level of study. Note also that ethers of tertiary alcohols react with hydrogen halides by an S_N1 mechanism.

Phenols are not converted to aryl halides by reaction with hydrogen halides. Cleavage of an aryl ether by a hydrogen halide is illustrated by the following example.

Guaiacol $\xrightarrow[heat]{HBr}$ Pyrocatechol (85–87%) + CH_3Br Methyl bromide (57–72%)

Guaiacol is obtained by chemical treatment of *lignum vitae*, the wood from a species of tree that grows in warm climates. It is sometimes used as an expectorant to help relieve bronchial congestion.

17.9 Preparation of Epoxides

There are two main methods for the preparation of epoxides:

1. Epoxidation of alkenes
2. Base-promoted ring closure of vicinal halohydrins

Epoxidation of alkenes with peroxy acids was discussed in Section 8.11 and is represented by the general equation

$$R_2C{=}CR_2 + R'\overset{O}{\overset{\|}{C}}OOH \longrightarrow R_2C\underset{O}{—}CR_2 + R'\overset{O}{\overset{\|}{C}}OH$$

Alkene Peroxy acid Epoxide Carboxylic acid

The reaction is a stereospecific syn addition.

Allylic alcohols are converted to epoxides by oxidation with *tert*-butyl hydroperoxide in the presence of certain transition metals. The most significant aspect of this reaction—called the **Sharpless epoxidation**—is its high enantioselectivity when carried out using a combination of *tert*-butyl hydroperoxide, titanium(IV) isopropoxide, and diethyl tartrate.

Diethyl (2*R*,3*R*)-tartrate is the diethyl ester of tartaric acid, a chiral molecule that was discussed in Section 4.13.

$CO_2CH_2CH_3$; H—C—OH; HO—C—H; $CO_2CH_2CH_3$

Diethyl (2*R*,3*R*)-tartrate

tert-Butyl hydroperoxide Titanium(IV) isopropoxide Diethyl (2*R*,3*R*)-tartrate

trans-2-Hexen-1-ol $\xrightarrow[\text{Ti[OCH(CH}_3)_2]_4,\ \text{diethyl }(2R,3R)\text{-tartrate}]{(CH_3)_3COOH}$ (2*S*,3*S*)-2,3-Epoxy-1-hexanol (78% yield, 98% enantiomeric excess)

Oxygen is transferred to the double bond of the allylic alcohol from the hydroperoxy group in a chiral environment and occurs enantioselectively.

The value of this reaction was recognized with the award of the 2001 Nobel Prize in Chemistry to its creator K. Barry Sharpless. Sharpless epoxidation of allylic alcohols can be carried out with catalytic amounts of titanium(IV) isopropoxide and, because both enantiomers of diethyl tartrate are readily available, can be applied to the synthesis of either enantiomer of a desired epoxy alcohol.

Sharpless's work in oxidation also included methods for the enantioselective dihydroxylation of alkenes (see Section 16.5).

Problem 17.11

What would be the absolute configuration of the 2,3-epoxy-1-hexanol produced in the preceding reaction if diethyl (2*S*,3*S*)-tartrate were used instead of (2*R*,3*R*)?

More than a laboratory synthesis, Sharpless epoxidation has been adapted to the large-scale preparation of (+)-disparlure, a sex pheromone used to control gypsy moth infestations, and of (*R*)-glycidol, an intermediate in the synthesis of cardiac drugs known as beta-blockers.

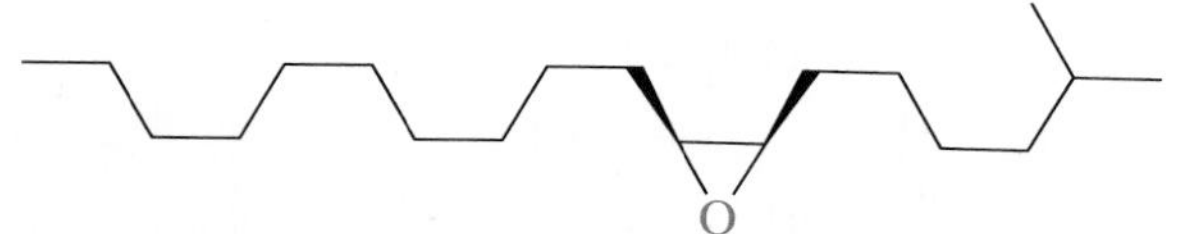

(+)-Disparlure
[(7*R*,8*S*)-7,8-Epoxy-2-methyloctadecane]

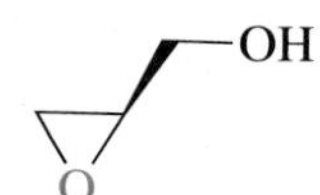

(+)-Glycidol
[(*R*)-2,3-Epoxy-1-propanol]

Examples of naturally occurring epoxides that have interesting biological activity are anticapsin, an amino acid antibiotic, and epothilones A and B, which possess potent anticancer activity.

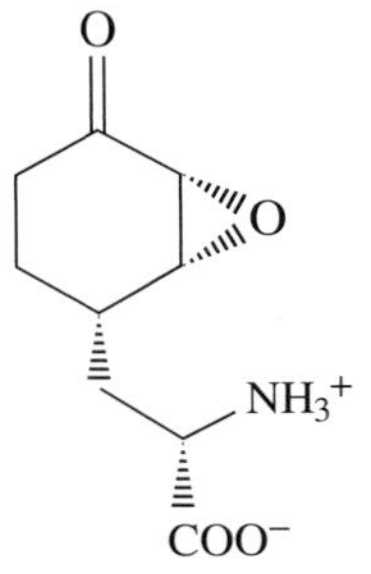

Anticapsin

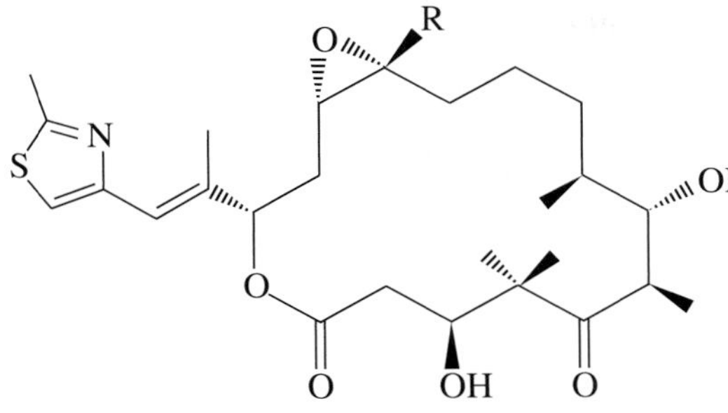

Epothilones: A (R = H) and B (R = CH_3)

The following section describes the preparation of epoxides by the base-promoted ring closure of vicinal halohydrins. Because vicinal halohydrins are customarily prepared from alkenes (Section 8.10), both methods—epoxidation using peroxy acids and ring closure of halohydrins—are based on alkenes as the starting materials for preparing epoxides.

17.10 Conversion of Vicinal Halohydrins to Epoxides

The vicinal halohydrins formed by the reaction of alkenes with aqueous Cl_2, Br_2, or I_2 (Section 8.10) are converted to epoxides in base.

A large portion of the more than 2 billion pounds of 1,2-epoxypropane produced each year in the United States is made from propene by this method. Its main use is in the preparation of polyurethane plastics.

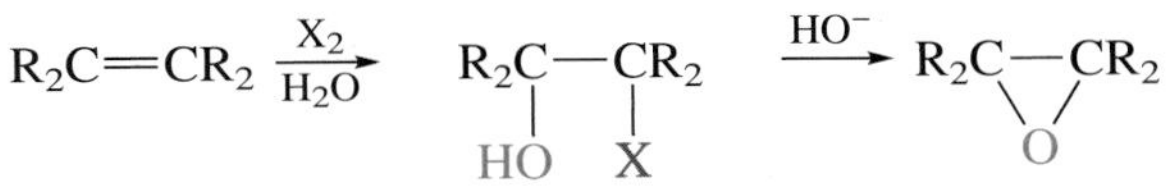

$$R_2C{=}CR_2 \xrightarrow[H_2O]{X_2} R_2C(OH){-}CR_2X \xrightarrow{HO^-} R_2C{-}CR_2 \text{ (epoxide, bridged by O)}$$

Alkene — Vicinal halohydrin — Epoxide

Reaction with base brings the alcohol into equilibrium with its corresponding alkoxide and intramolecular nucleophilic substitution closes the three-membered ring.

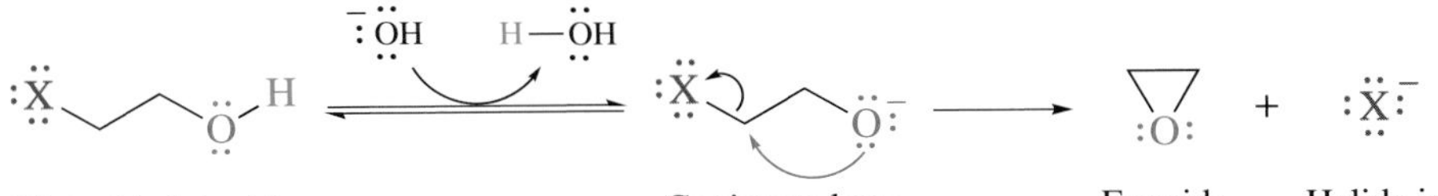

Vicinal halohydrin — Conjugate base — Epoxide + Halide ion

This ring-closing step obeys the usual S_N2 stereochemistry—approach of the nucleophilic oxygen from the side opposite the bond to the leaving group. In cyclohexane rings, this corresponds to a trans-diaxial arrangement of oxygen and halide.

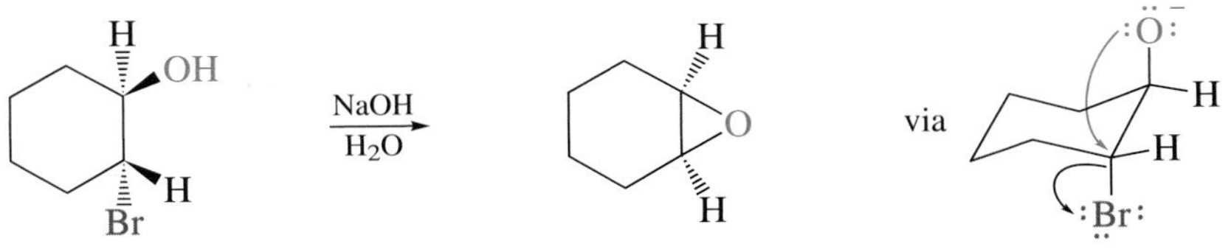

trans-2-Bromocyclohexanol $\xrightarrow[H_2O]{NaOH}$ 1,2-Epoxycyclohexane (81%) via

The overall stereochemistry of the alkene → halohydrin → epoxide sequence is the same as that observed in peroxy acid oxidation of alkenes. Substituents that are cis in the alkene remain cis in the epoxide. The combination of anti addition in forming the bromohydrin, followed by inversion of configuration in conversion of the bromohydrin to the epoxide yields the same stereochemical result as syn epoxidation of an alkene.

cis-2-Butene —(anti addition)→ bromohydrin —(inversion of configuration)→ *cis*-2,3-Epoxybutane

cis-2-Butene

trans-2-Butene —(anti addition)→ bromohydrin —(inversion of configuration)→ *trans*-2,3-Epoxybutane

trans-2-Butene

Problem 17.12

Classify the bromohydrins formed from *cis*- and *trans*-2-butene as erythro or threo. Is either chiral? Is either optically active when formed from the alkene by this method? Classify the epoxides as either chiral or meso. Is either optically active when formed by this method?

17.11 Reactions of Epoxides with Anionic Nucleophiles

Angle strain is the main source of strain in epoxides, but torsional strain that results from the eclipsing of bonds on adjacent carbons is also present. Both kinds of strain are relieved when a ring-opening reaction occurs.

The most striking chemical property of epoxides is their far greater reactivity toward nucleophilic reagents compared with simple ethers. They react rapidly and exothermically with anionic nucleophiles to yield ring-opened products. This enhanced reactivity results from the angle strain of epoxides; ring opening relieves that strain.

We saw an example of nucleophilic ring opening of epoxides in Section 16.4, where the reaction of Grignard and organolithium reagents with ethylene oxide was presented as a synthetic route to primary alcohols:

$$\text{RMgX} + H_2C\overset{O}{—}CH_2 \xrightarrow[\text{2. } H_3O^+]{\text{1. diethyl ether}} RCH_2CH_2OH$$

Grignard reagent | Ethylene oxide | Primary alcohol

Typical anionic nucleophiles react with epoxides in water or alcohols as the solvent to give an alkoxide intermediate that is rapidly converted to an alcohol by proton transfer.

$$Y:^- + R_2C\overset{O}{—}CR_2 \longrightarrow R_2C(Y)—CR_2(\ddot{O}:^-) \xrightarrow{H_2O} R_2C(Y)—CR_2(\ddot{O}H)$$

Nucleophile | Epoxide | Alkoxide ion | Alcohol

Potassium butanethiolate (SK) + Ethylene oxide —(ethanol–water, 0°C)→ 2-(Butylthio)ethanol (99%)

Problem 17.13

What is the principal organic product formed in the reaction of ethylene oxide with each of the following?

(a) Sodium cyanide (NaCN) in aqueous ethanol
(b) Sodium azide (NaN_3) in aqueous ethanol

(c) Sodium hydroxide (NaOH) in water
(d) Phenyllithium (C_6H_5Li) in diethyl ether, followed by addition of dilute sulfuric acid
(e) 1-Butynylsodium ($CH_3CH_2C{\equiv}CNa$) in liquid ammonia

Sample Solution (a) Sodium cyanide is a source of the nucleophilic cyanide anion. Cyanide ion attacks ethylene oxide, opening the ring and forming 2-cyanoethanol:

Ethylene oxide $\xrightarrow[\text{ethanol–water}]{\text{NaCN}}$ 2-Cyanoethanol

The reactions of epoxides with anionic nucleophiles have many of the characteristics of S_N2 reactions. Inversion of configuration occurs at the carbon attacked by the nucleophile:

1,2-Epoxycyclopentane $\xrightarrow[\text{ethanol}]{NaOCH_2CH_3}$ *trans*-2-Ethoxycyclopentanol (67%)

and the nucleophile bonds to the less substituted, less sterically hindered carbon of the ring:

2,3,3-Trimethyloxirane $\xrightarrow[\text{methanol}]{NaOCH_3}$ 3-Methoxy-2,3-dimethyl-2-butanol (53%)

Problem 17.14

Ammonia and amines react with epoxides with the same stereospecificity as anionic nucleophiles. Draw a sawhorse or Newman projection formula for the product of the reaction shown, clearly showing the stereochemistry at both chirality centers. What are the Cahn–Ingold–Prelog *R,S* descriptors for these chirality centers in the reactant and the product?

$\xrightarrow[H_2O]{NH_3}$ 3-Amino-2-butanol

The reactions of Grignard reagents and lithium aluminum hydride with epoxides are regioselective in the same sense as the examples just shown. Substitution occurs at the less substituted carbon of the epoxide ring.

Phenylmagnesium bromide + 1,2-Epoxypropane $\xrightarrow[\text{2. } H_3O^+]{\text{1. diethyl ether}}$ 1-Phenyl-2-propanol (60%)

1,2-Epoxydecane $\xrightarrow[\text{2. } H_2O]{\text{1. } LiAlH_4}$ 2-Decanol (90%)

Epoxidation of an alkene, followed by lithium aluminum hydride reduction of the resulting epoxide, gives the same alcohol as would be obtained by acid-catalyzed hydration of an alkene (Section 8.6).

Experimental observations such as these combine with the principles of nucleophilic substitution to give the picture of epoxide ring opening shown in Mechanism 17.2.

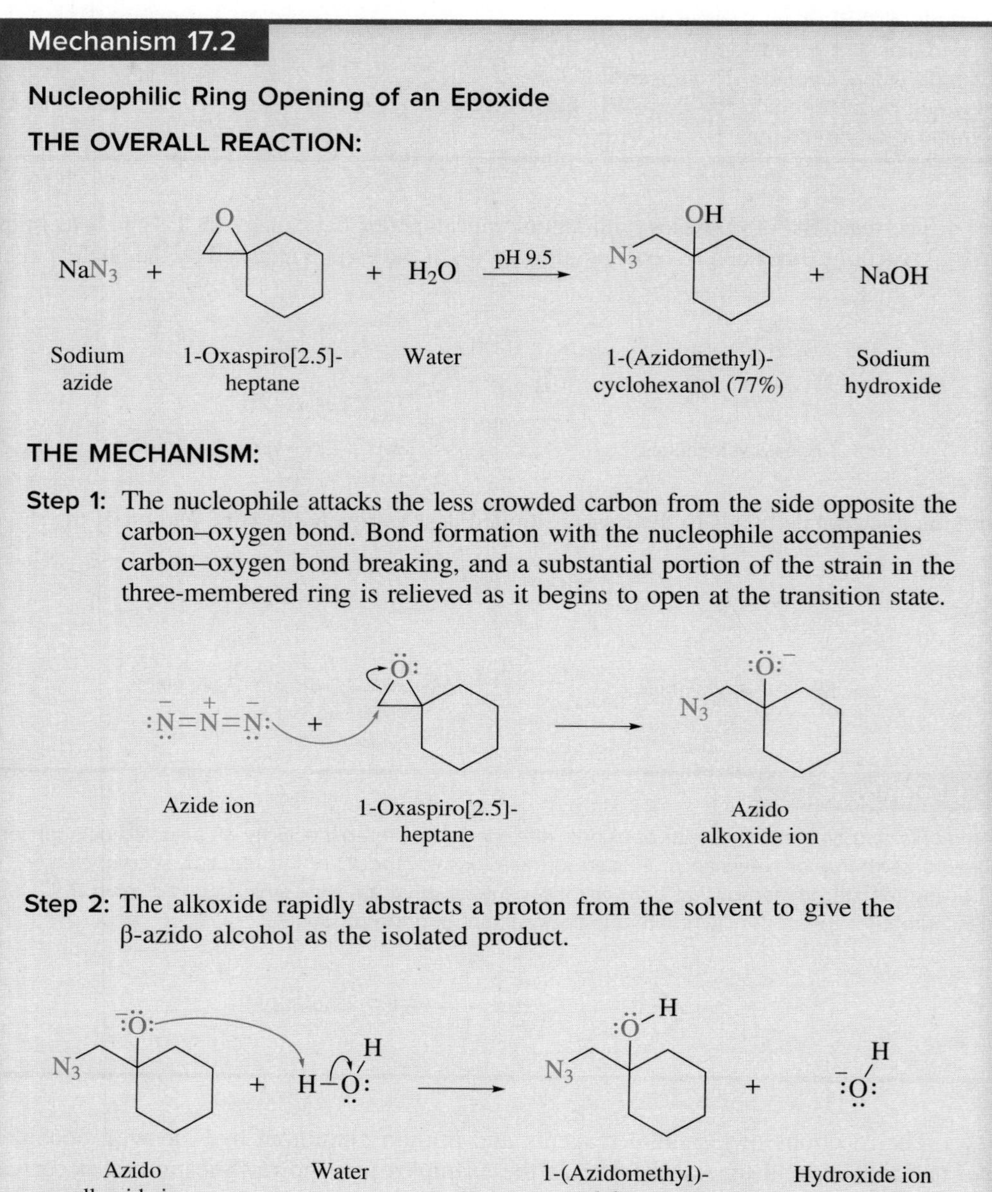

Mechanism 17.2

Nucleophilic Ring Opening of an Epoxide

THE OVERALL REACTION:

THE MECHANISM:

Step 1: The nucleophile attacks the less crowded carbon from the side opposite the carbon–oxygen bond. Bond formation with the nucleophile accompanies carbon–oxygen bond breaking, and a substantial portion of the strain in the three-membered ring is relieved as it begins to open at the transition state.

Step 2: The alkoxide rapidly abstracts a proton from the solvent to give the β-azido alcohol as the isolated product.

17.12 Acid-Catalyzed Ring Opening of Epoxides

Nucleophilic ring openings of epoxides can be catalyzed by acids.

$$\text{Ethylene oxide} + CH_3CH_2OH \xrightarrow[25°C]{H_2SO_4} \text{2-Ethoxyethanol (85\%)}$$

Ethylene oxide   Ethanol   2-Ethoxyethanol (85%)

There is an important difference in the regioselectivity of ring-opening reactions of epoxides in acid solution compared with their counterparts in base. Unsymmetrically substituted epoxides tend to react with anionic nucleophiles at the less hindered carbon of the ring. Under acid-catalyzed conditions, however, reaction occurs at the more substituted carbon.

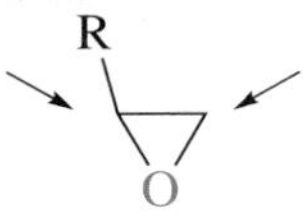

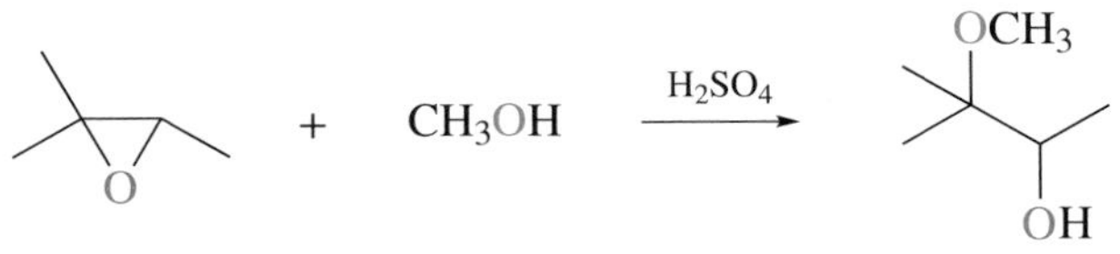

Nucleophiles attack here when reaction is acid-catalyzed.

Anionic nucleophiles attack here.

2,3-Epoxy-2-methylbutane + Methanol (CH_3OH) $\xrightarrow{H_2SO_4}$ 3-Methoxy-3-methyl-2-butanol (76%)

As seen in Mechanism 17.3, the reaction just described involves three steps. Steps 1 and 3 are proton transfers; step 2 involves methanol acting as a nucleophile toward the protonated epoxide. The transition state for this step has a fair amount of carbocation character; breaking the C—O bond of the ring is more advanced than formation of the bond to the nucleophile.

Transition state for reaction of methanol with conjugate acid of 2,3-epoxy-2-methylbutane

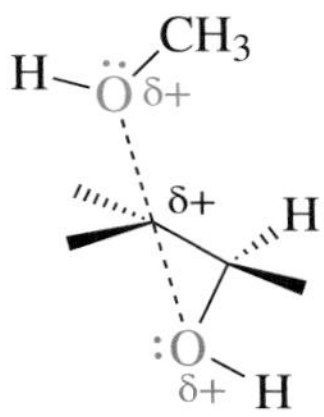

Although nucleophilic participation at the transition state is less than that for reactions between epoxides and anionic nucleophiles, it is enough to ensure that substitution proceeds with inversion of configuration.

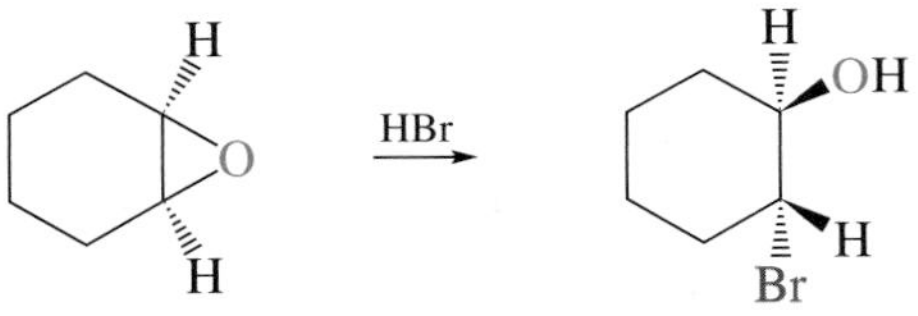

1,2-Epoxycyclohexane $\xrightarrow{HBr}$ *trans*-2-Bromocyclohexanol (73%)

Problem 17.15

The epoxide shown gives a mixture of two azido alcohols on reaction with sodium azide in aqueous acid and in base.

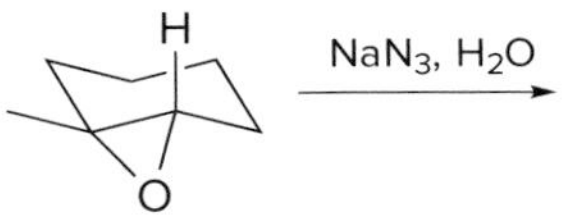

$\xrightarrow{NaN_3,\ H_2O}$

Write a structural formula including stereochemistry for the major product formed at pH 9.5. A different isomer predominates at pH 4.2. What is the structure of this isomer? (Note: The pK_a of HN_3 = 4.2.)

Mechanism 17.3

Acid-Catalyzed Ring Opening of an Epoxide

THE OVERALL REACTION:

2,3-Epoxy-2-methylbutane + Methanol (CH_3OH) $\xrightarrow{H_2SO_4}$ 3-Methoxy-3-methyl-2-butanol (76%)

THE MECHANISM:

Step 1: The most abundant acid in the reaction mixture, the conjugate acid of the solvent methanol, transfers a proton to the oxygen of the epoxide.

2,3-Epoxy-2-methylbutane + Methyloxonium ion $\overset{\text{fast}}{\rightleftharpoons}$ 2,3-Epoxonium-2-methylbutane + Methanol

Step 2: Methanol acts as a nucleophile toward the protonated epoxide and bonds to the more highly substituted (more carbocation-like) carbon of the ring. The carbon–oxygen bond of the ring breaks in this step, and the ring opens.

Methanol + 2,3-Epoxonium-2-methylbutane $\xrightarrow{\text{slow}}$ 3-Methoxonium-3-methyl-2-butanol

Step 3: Proton transfer to methanol completes the reaction and regenerates the acid catalyst.

Methanol + 3-Methoxonium-3-methyl-2-butanol $\overset{\text{fast}}{\rightleftharpoons}$ Methyloxonium ion + 3-Methoxy-3-methyl-2-butanol

A method for achieving net anti hydroxylation of alkenes combines two stereospecific processes: epoxidation of the double bond and hydrolysis of the derived epoxide.

Cyclohexene $\xrightarrow{C_6H_5COOH \text{ (peroxy acid)}}$ 1,2-Epoxycyclohexane $\xrightarrow[HClO_4]{H_2O}$ *trans*-1,2-Cyclohexanediol (80%)

Problem 17.16

Which alkene, *cis*-2-butene or *trans*-2-butene, would you choose in order to prepare *meso*-2,3-butanediol by epoxidation followed by acid-catalyzed hydrolysis? Which alkene would yield *meso*-2,3-butanediol by osmium tetraoxide dihydroxylation?

17.13 Epoxides in Biological Processes

Many naturally occurring substances are epoxides. In most cases, epoxides are biosynthesized by the enzyme-catalyzed transfer of one of the oxygen atoms of an O_2 molecule to an alkene. Because only one of the atoms of O_2 is transferred to the substrate, the enzymes that catalyze such transfers are classified as *monooxygenases*. A biological reducing agent, usually the coenzyme NADH (Section 16.10), is required as well.

$$R_2C{=}CR_2 + O_2 + H^+ + NADH \xrightarrow{\text{enzyme}} R_2C\overset{O}{—}CR_2 + H_2O + NAD^+$$

A prominent example of such a reaction is the biological epoxidation of the polyene squalene. Note the remarkable selectivity of this reaction in which only the terminal alkene undergoes epoxidation.

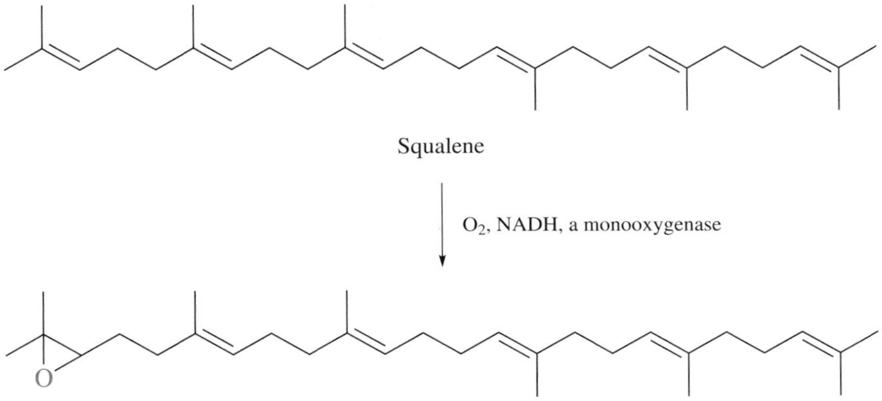

Squalene

O_2, NADH, a monooxygenase

Squalene 2,3-epoxide

The reactivity of epoxides toward nucleophilic ring opening is responsible for one of the biological roles they play. Squalene 2,3-epoxide, for example, is the biological precursor to cholesterol and the steroid hormones, including testosterone, progesterone, estrone, and cortisone. The pathway from squalene 2,3-epoxide to these compounds is triggered by epoxide ring opening and will be described in Chapter 24.

17.14 Preparation of Sulfides

Sulfides, compounds of the type RSR′, are prepared by nucleophilic substitution. Treatment of a primary or secondary alkyl halide with an alkanethiolate ion (RS^-) gives a sulfide:

$$RS^- \; Na^+ + R'—X \xrightarrow{S_N2} RSR' + Na^+ \; X^-$$

Sodium alkanethiolate | Alkyl halide | Sulfide | Sodium halide

$$\text{3-Chloro-1-butene} \xrightarrow[\text{methanol}]{NaSCH_3} \text{Methyl 1-methylallyl sulfide (62\%)}$$

Cl

SCH_3

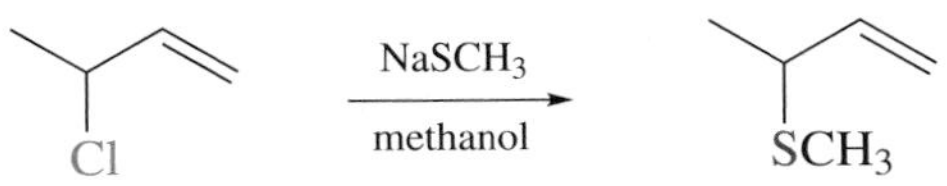

3-Chloro-1-butene

Methyl 1-methylallyl sulfide (62%)

Problem 17.17

The *p*-toluenesulfonate derived from (*R*)-2-octanol and *p*-toluenesulfonyl chloride was allowed to react with sodium benzenethiolate (C_6H_5SNa). Give the structure, including stereochemistry and the appropriate *R* or *S* descriptor, of the product.

17.15 Oxidation of Sulfides: Sulfoxides and Sulfones

We saw in Section 16.12 that thiols differ from alcohols with respect to their behavior toward oxidation. Similarly, sulfides differ from ethers in their behavior toward oxidizing agents. Whereas ethers tend to undergo oxidation at carbon to give hydroperoxides (Section 17.7), sulfides are oxidized at sulfur to give **sulfoxides.** If the oxidizing agent is strong enough and present in excess, oxidation can proceed further to give **sulfones.**

Third-row elements such as sulfur can expand their valence shell beyond eight electrons, and so sulfur–oxygen bonds in sulfoxides and sulfones can be represented as double bonds.

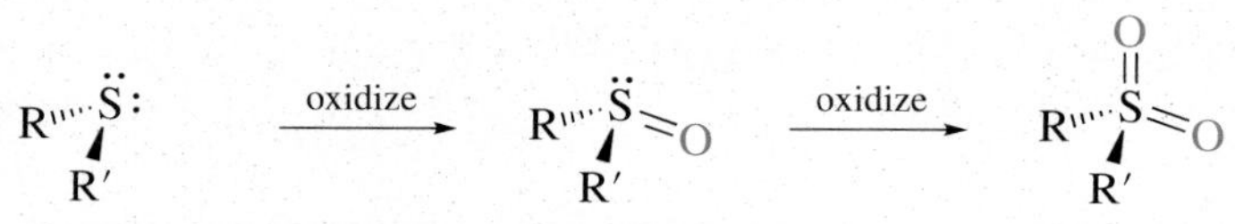

$$RSR' \xrightarrow{\text{oxidize}} RS(=O)R' \xrightarrow{\text{oxidize}} RS(=O)_2R'$$

Sulfide Sulfoxide Sulfone

When the desired product is a sulfoxide, sodium metaperiodate ($NaIO_4$) is an ideal reagent. It oxidizes sulfides to sulfoxides in high yield but shows no tendency to oxidize sulfoxides to sulfones.

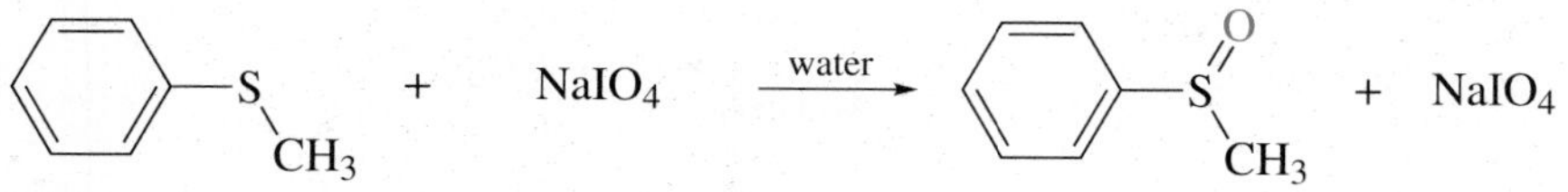

$$C_6H_5SCH_3 + NaIO_4 \xrightarrow{\text{water}} C_6H_5S(=O)CH_3 + NaIO_3$$

Methyl phenyl sulfide; Sodium metaperiodate; Methyl phenyl sulfoxide (91%); Sodium iodate

Peroxy acids, usually in dichloromethane as the solvent, are also reliable reagents for converting sulfides to sulfoxides. One equivalent of a peroxy acid or hydrogen peroxide converts sulfides to sulfoxides; two equivalents gives the corresponding sulfone.

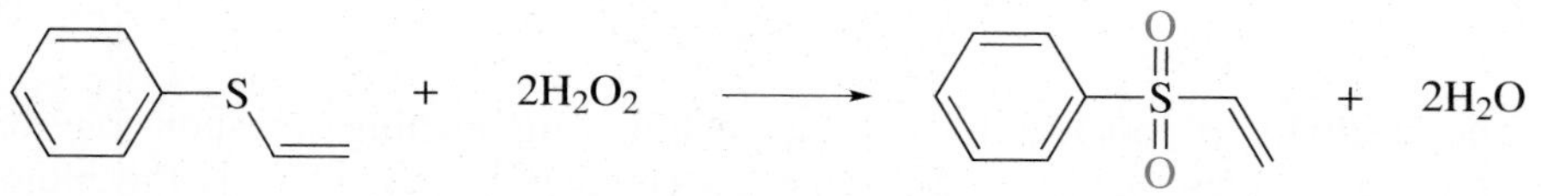

$$C_6H_5SCH{=}CH_2 + 2H_2O_2 \longrightarrow C_6H_5S(=O)_2CH{=}CH_2 + 2H_2O$$

Phenyl vinyl sulfide; Hydrogen peroxide; Phenyl vinyl sulfone (74–78%); Water

Problem 17.18

Prilosec and Nexium are widely used "proton pump" inhibitors for treatment of gastrointestinal reflux disease (GERD). Both have the constitution shown. Prilosec (*omeprazole*) is racemic while Nexium (*esomeprazole*) has the *S* configuration at its chirality center. Revise the drawing so that it represents Nexium.

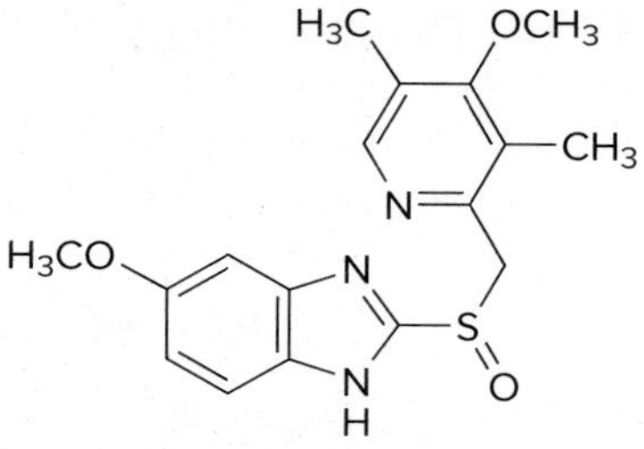

H_3C, OCH_3, CH_3, N, H_3CO, N, S, O, N, H

Omeprazole

Oxidation of sulfides occurs in living systems as well. Among naturally occurring sulfoxides, one that has received recent attention is *sulforaphane,* which is present in broccoli and other vegetables. Sulforaphane holds promise as a potential anticancer agent because, unlike most anticancer drugs, which act by killing rapidly dividing tumor cells faster than they kill normal cells, sulforaphane is nontoxic and may simply inhibit the formation of tumors.

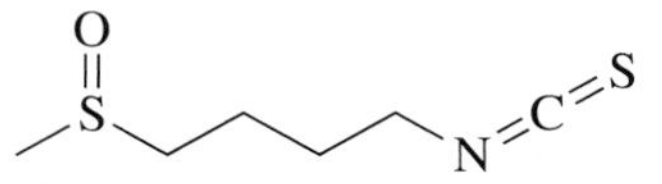

Sulforaphane

The —N=C=S unit in sulforaphane is the *isothiocyanate* group. Isothiocyanates are among the ingredients responsible for the flavor of wasabi.

17.16 Alkylation of Sulfides: Sulfonium Salts

Sulfur is more nucleophilic than oxygen (Section 6.5), and sulfides react with alkyl halides much faster than do ethers. The products of these reactions are called *sulfonium salts.*

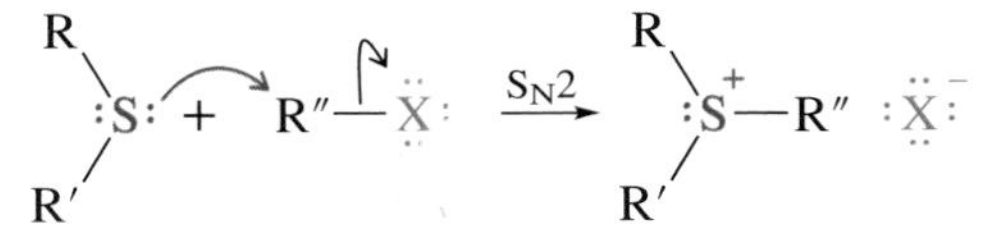

Sulfide + Alkyl halide $\xrightarrow{S_N2}$ Sulfonium salt

$$CH_3(CH_2)_{10}CH_2SCH_3 + CH_3I \longrightarrow CH_3(CH_2)_{10}CH_2\overset{+}{S}(CH_3)CH_3\ I^-$$

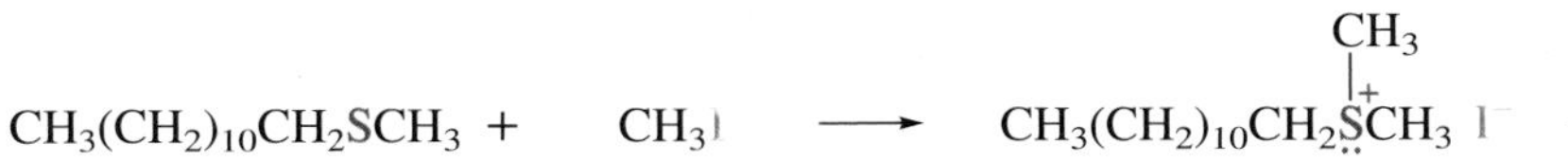

Dodecyl methyl sulfide + Methyl iodide → Dodecyldimethylsulfonium iodide

Problem 17.19

What other combination of alkyl halide and sulfide will yield the same sulfonium salt shown in the preceding example? Predict which combination will yield the sulfonium salt at the faster rate.

A naturally occurring sulfonium salt, *S-adenosylmethionine (SAM),* is a key substance in certain biological processes. It is formed by a nucleophilic substitution in which the sulfur atom of methionine attacks the primary carbon of adenosine triphosphate, displacing the triphosphate leaving group.

The *S* in *S*-adenosylmethionine indicates that the adenosyl group is bonded to sulfur. It does *not* stand for the Cahn–Ingold–Prelog stereochemical descriptor.

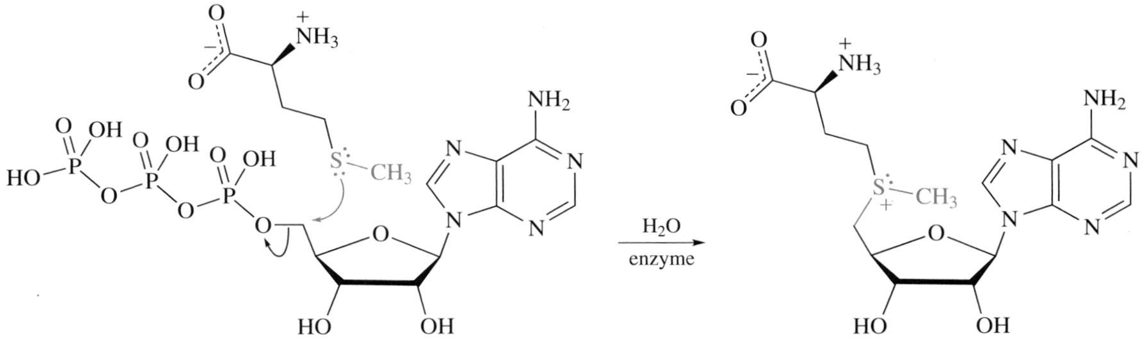

Adenosine triphosphate (ATP) + Methionine $\xrightarrow[\text{enzyme}]{H_2O}$ *S*-Adenosylmethionine (SAM)

S-Adenosylmethionine acts as a biological methyl-transfer agent. Nucleophiles, particularly amines, attack the methyl carbon of SAM, breaking the carbon–sulfur bond. The following equation represents the biological formation of *epinephrine* by methylation of *norepinephrine.* Only the methyl group and the sulfur of SAM are shown explicitly

Epinephrine is also known as *adrenaline* and is a hormone with profound physiological effects designed to prepare the body for "fight or flight."

in the equation to draw attention to the similarity of this reaction to the more familiar S_N2 reactions we have studied.

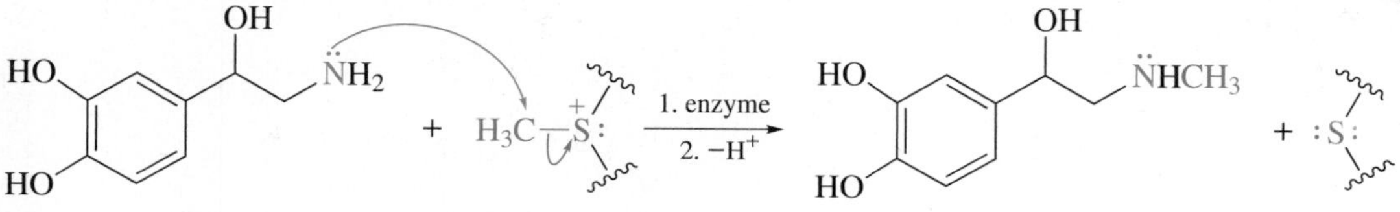

Norepinephrine + SAM $\xrightarrow[2.\ -H^+]{1.\ \text{enzyme}}$ Epinephrine + :S:

Norepinephrine SAM Epinephrine

17.17 Spectroscopic Analysis of Ethers, Epoxides, and Sulfides

The IR, ^{1}H NMR, and ^{13}C NMR spectra of dipropyl ether, which appear in parts *a, b,* and *c,* respectively of Figure 17.4, illustrate some of the spectroscopic features of ethers.

Infrared: The infrared spectra of *ethers* are characterized by a strong, rather broad band due to antisymmetric C—O—C stretching between 1070 and 1150 cm^{-1}. Dialkyl ethers exhibit this band consistently at 1120 cm^{-1}, as shown in the IR spectrum of dipropyl ether.

$CH_3CH_2CH_2OCH_2CH_2CH_3$

Dipropyl ether

C—O—C $\nu = 1118\ cm^{-1}$

The analogous band in alkyl aryl ethers (ROAr) appears at 1200–1275 cm^{-1}.

Epoxides typically exhibit three bands. Two bands, one at 810–950 cm^{-1} and the other near 1250 cm^{-1}, correspond to asymmetric and symmetric stretching of the ring, respectively. The third band appears in the range 750–840 cm^{-1}.

$H_2C—CH(CH_2)_9CH_3$ (epoxide ring, O)

1,2-Epoxydodecane

Epoxide vibrations: $\nu = 837, 917,$ and $1265\ cm^{-1}$

The C—S—C stretching vibration of *sulfides* gives a weak peak in the 600–700 cm^{-1} range. *Sulfoxides* show a strong peak due to S—O stretching at 1030–1070 cm^{-1}. With two oxygens attached to sulfur, *sulfones* exhibit strong bands due to symmetric (1120–1160 cm^{-1}) and asymmetric (1290–1350 cm^{-1}) S—O stretching.

Dimethyl sulfoxide

S=O: $\nu = 1050\ cm^{-1}$

Dimethyl sulfone

S=O: $\nu = 1139$ and $1298\ cm^{-1}$

^{1}H NMR: The chemical shift of the proton in the **H**—C—O—C unit of an *ether* is very similar to that of the proton in the **H**—C—OH unit of an alcohol. A range of δ 3.2–4.0 is typical. The proton in the **H**—C—S—C unit of a *sulfide* appears at higher field than the corresponding proton of an ether because sulfur is less electronegative than oxygen.

$CH_3CH_2CH_2—O—CH_2CH_2CH_3$ $CH_3CH_2CH_2—S—CH_2CH_2CH_3$

^{1}H Chemical shift (δ): 3.2 2.5

Oxidation of a sulfide to a *sulfoxide* or *sulfone* is accompanied by a decrease in shielding of the **H**—C—S—C proton by about 0.3–0.5 ppm for each oxidation.

Epoxides are unusual in that the protons on the ring are more shielded than expected. The protons in ethylene oxide, for example, appear at δ 2.5 instead of the δ 3.2–4.0 range just cited for dialkyl ethers.

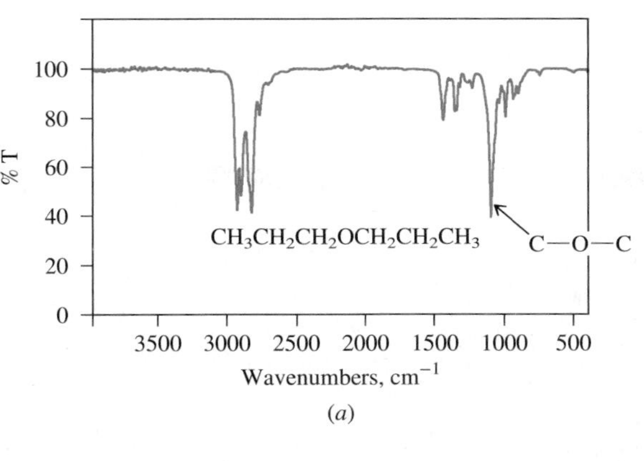

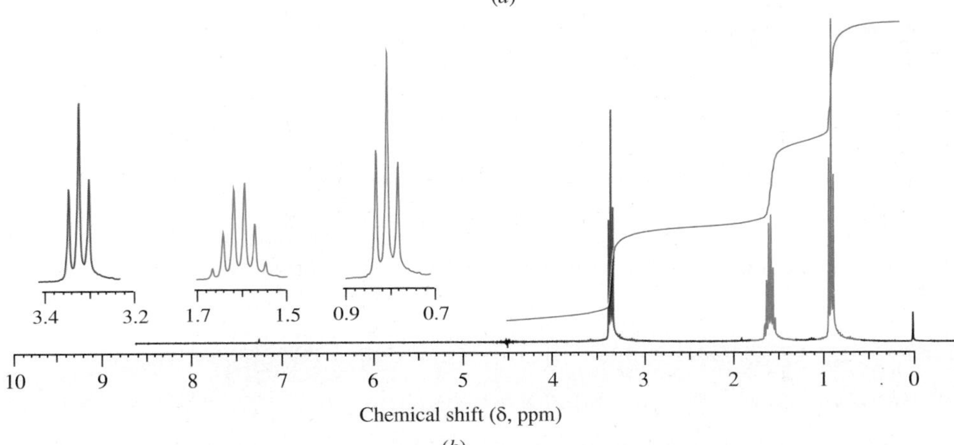

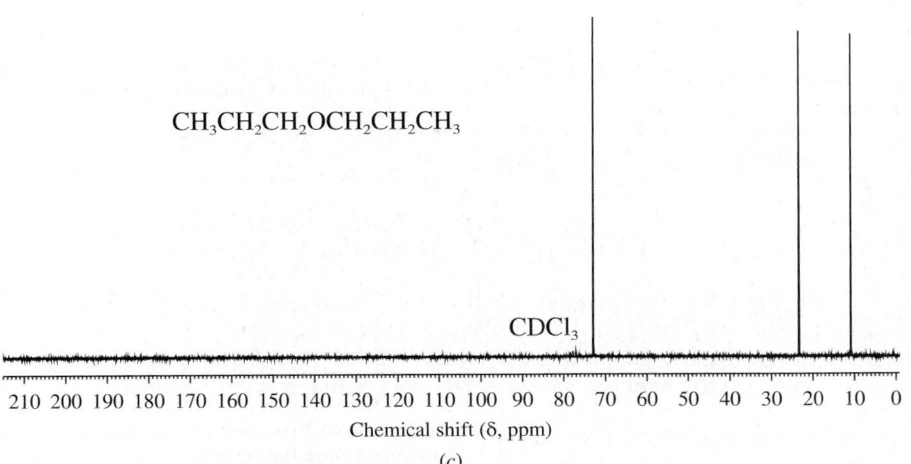

Figure 17.4

The (*a*) infrared, (*b*) 300-MHz ^{1}H NMR, and (*c*) ^{13}C NMR spectra of dipropyl ether ($CH_3CH_2CH_2OCH_2CH_2CH_3$).

^{13}C NMR: The carbons of the C—O—C group in an *ether* are about 10 ppm less shielded than those of an alcohol and appear in the range δ 57–87. The carbons of the C—S—C group in a *sulfide* are significantly more shielded than those of an ether.

	$CH_3CH_2CH_2—O—CH_2CH_2CH_3$	$CH_3CH_2CH_2—S—CH_2CH_2CH_3$
^{13}C Chemical shift (δ):	73	34

The ring carbons of an *epoxide* are somewhat more shielded than the carbons of a C—O—C unit of larger rings or dialkyl ethers.

δ 47 δ 52

$H_2C—CH—CH_2CH_2CH_2CH_3$ (epoxide ring with O)

δ 68 (tetrahydrofuran ring, O)

UV-VIS: Simple ethers have their absorption maximum at about 185 nm and are transparent to ultraviolet radiation above about 220 nm.

Mass Spectrometry: Ethers, like alcohols, lose an alkyl radical from their molecular ion to give an oxygen-stabilized cation. Thus, *m/z* 73 and *m/z* 87 are both more abundant than the molecular ion in the mass spectrum of *sec*-butyl ethyl ether.

+ $\cdot CH_2CH_3$ ← [*m/z* 102] → + $\cdot CH_3$

m/z 73 *m/z* 102 *m/z* 87

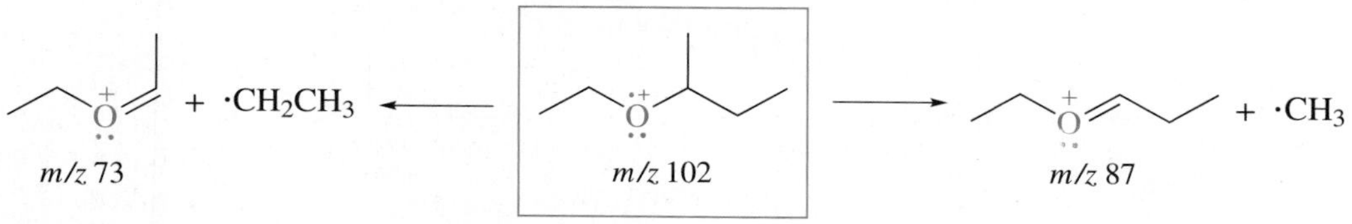

Problem 17.20

There is another oxygen-stabilized cation of *m/z* 87 capable of being formed by fragmentation of the molecular ion in the mass spectrum of *sec*-butyl ethyl ether. Suggest a reasonable structure for this ion.

An analogous fragmentation process occurs in the mass spectra of sulfides. As with other sulfur-containing compounds, the presence of sulfur can be inferred by a peak at *m/z* of M+2.

17.18 SUMMARY

Section 17.1 **Ethers** are compounds that contain a C—O—C linkage. In substitutive IUPAC nomenclature, they are named as *alkoxy* derivatives of alkanes. In functional class IUPAC nomenclature, we name each alkyl group as a separate word (in alphabetical order) followed by the word *ether.*

$CH_3OCH_2CH_2CH_2CH_2CH_2CH_3$ **Substitutive IUPAC name:** 1-Methoxyhexane
Functional class name: Hexyl methyl ether

Epoxides are normally named as *epoxy* derivatives of alkanes or as substituted *oxiranes.*

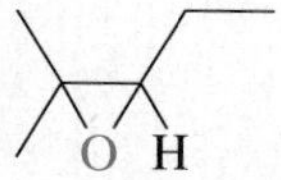

O H

2,3-Epoxy-2-methylpentane
3-Ethyl-2,2-dimethyloxirane

Sulfides are sulfur analogs of ethers: they contain the C—S—C functional group. They are named as *alkylthio* derivatives of alkanes in substitutive IUPAC nomenclature. The functional class IUPAC names of sulfides are derived in the same manner as those of ethers, but the concluding word is *sulfide.*

$CH_3SCH_2CH_2CH_2CH_2CH_2CH_3$ **Substitutive IUPAC name:** 1-(Methylthio)hexane
Functional class name: Hexyl methyl sulfide

Section 17.2 The oxygen atom in an ether or epoxide affects the shape of the molecule in much the same way as an sp^3-hybridized carbon of an alkane or cycloalkane.

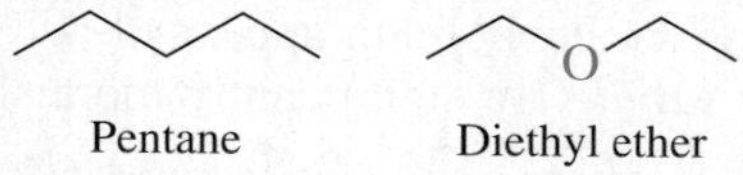

Pentane Diethyl ether

Section 17.3 The carbon–oxygen bond of ethers is polar, and ethers can act as proton *acceptors* in hydrogen bonds with water and alcohols.

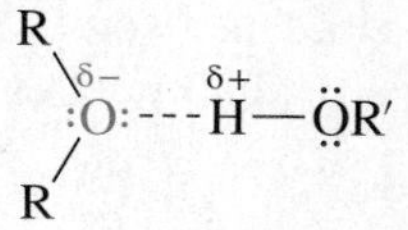

$R_2\overset{\delta-}{O}{:}\cdots\overset{\delta+}{H}—\ddot{O}R'$

But ethers lack OH groups and cannot act as proton *donors* in forming hydrogen bonds.

Section 17.4 Ethers form Lewis acid/Lewis base complexes with metal ions. Certain cyclic polyethers, called **crown ethers,** are particularly effective in coordinating with Na^+ and K^+, and salts of these cations can be dissolved in nonpolar solvents when crown ethers are present. Under these conditions the rates of many reactions that involve anions are accelerated.

1-Bromohexane → (CH_3CO_2K, 18-crown-6; acetonitrile) → Hexyl acetate (96%)

Sections 17.5 and 17.6 The two major methods for preparing ethers are summarized in Table 17.2.

Section 17.7 Dialkyl ethers are useful solvents for organic reactions, but must be used cautiously due to their tendency to form explosive hydroperoxides by air oxidation in opened bottles.

Section 17.8 The only important reaction of ethers is their cleavage by hydrogen halides.

$$\underset{\text{Ether}}{ROR'} + \underset{\text{Hydrogen halide}}{2HX} \longrightarrow \underset{\text{Alkyl halide}}{RX} + \underset{\text{Alkyl halide}}{R'X} + \underset{\text{Water}}{H_2O}$$

The order of hydrogen halide reactivity is HI > HBr > HCl.

Benzyl ethyl ether → (HBr, heat) → Benzyl bromide + Ethyl bromide

Section 17.9 Epoxides are prepared by the methods listed in Table 17.2.

Section 17.10 Epoxides are much more reactive than ethers, especially in reactions that lead to cleavage of their three-membered ring.

Section 17.11 Anionic nucleophiles usually attack the less-substituted carbon of the epoxide in an S_N2-like fashion.

2,2,3-Trimethyloxirane → ($NaOCH_3$, CH_3OH) → 3-Methoxy-2-methyl-2-butanol (53%)

Section 17.12 Under conditions of acid catalysis, nucleophiles attack the carbon that can better support a positive charge. Carbocation character is developed in the transition state.

2,2,3-Trimethyloxirane → (CH_3OH, H_2SO_4) → 3-Methoxy-3-methyl-2-butanol (76%)

Inversion of configuration is observed at the carbon that is attacked by the nucleophile, irrespective of whether the reaction takes place in acidic or basic solution.

Section 17.13 Epoxide functions are present in a great many natural products, and epoxide ring opening is sometimes a key step in the biosynthesis of other substances.

TABLE 17.2 Preparation of Ethers and Epoxides

Reaction (section) and comments	General equation and specific example
Acid-catalyzed condensation of alcohols (Sections 16.7 and 17.5) Two molecules of an alcohol condense in the presence of an acid catalyst to yield a dialkyl ether and water. The reaction is limited to the synthesis of symmetrical ethers from primary alcohols.	$2RCH_2OH \xrightarrow{H^+} RCH_2OCH_2R + H_2O$ Alcohol → Dialkyl ether + Water 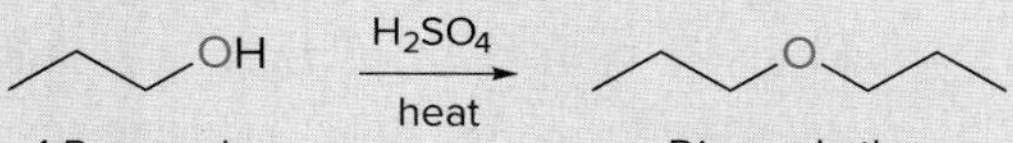 1-Propanol $\xrightarrow[\text{heat}]{H_2SO_4}$ Dipropyl ether
Williamson ether synthesis (Section 17.6) An alkoxide ion displaces a halide or similar leaving group in an S_N2 reaction. The alkyl halide cannot be one that is prone to elimination, and so this reaction is limited to methyl and primary alkyl halides. There is no limitation on the alkoxide ion that can be used.	$RONa + R'X \longrightarrow ROR' + NaX$ Sodium alkoxide + Alkyl halide → Dialkyl ether + Sodium halide 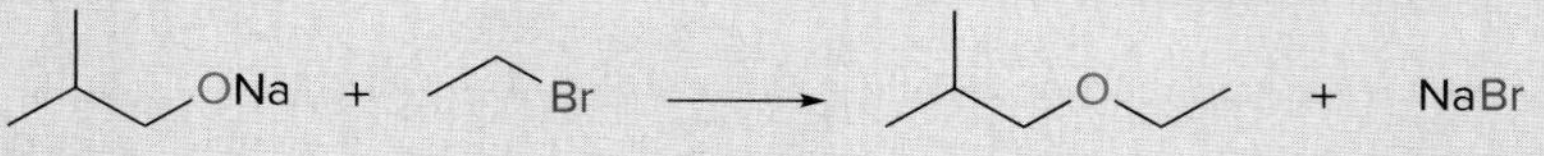 Sodium isobutoxide + Ethyl bromide → Ethyl isobutyl ether (66%) + Sodium bromide (NaBr)
Peroxy acid oxidation of alkenes (Sections 8.11 and 17.9) Peroxy acids transfer oxygen to alkenes to yield epoxides by stereospecific syn addition.	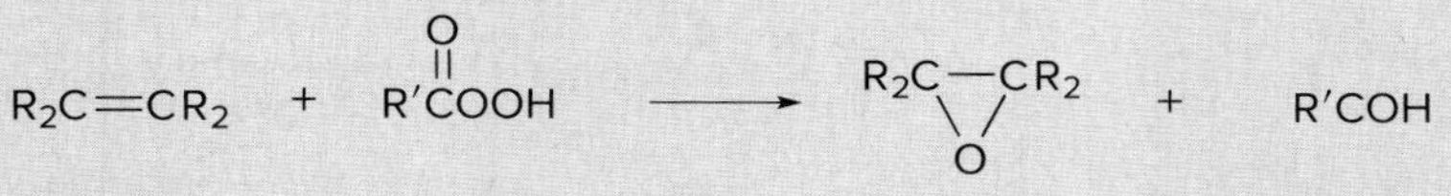 $R_2C{=}CR_2 + R'COOH \longrightarrow R_2C{-}CR_2$ (epoxide, bridged by O) $+ R'COH$ (the peroxy acid and carboxylic acid are drawn with C=O) Alkene + Peroxy acid → Epoxide + Carboxylic acid 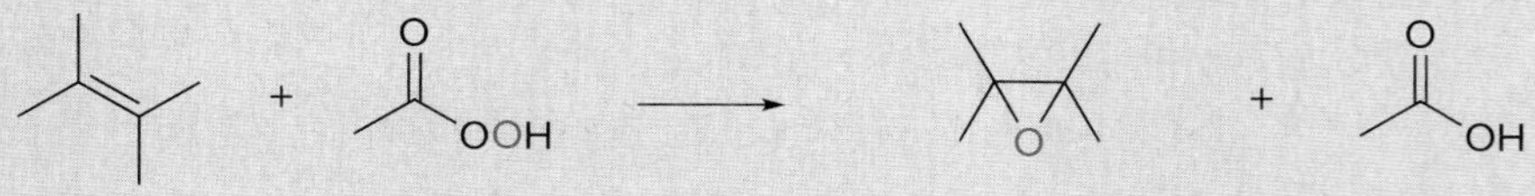 2,3-Dimethyl-2-butene + Peroxyacetic acid → 2,2,3,3-Tetramethyloxirane (70–80%) + Acetic acid
Sharpless epoxidation (Section 17.9) Allylic alcohols are converted to epoxides by treatment with *tert*-butyl hydroperoxide and titanium(IV) alkoxides. The reaction is highly enantioselective in the presence of enantiomerically pure diethyl tartrate.	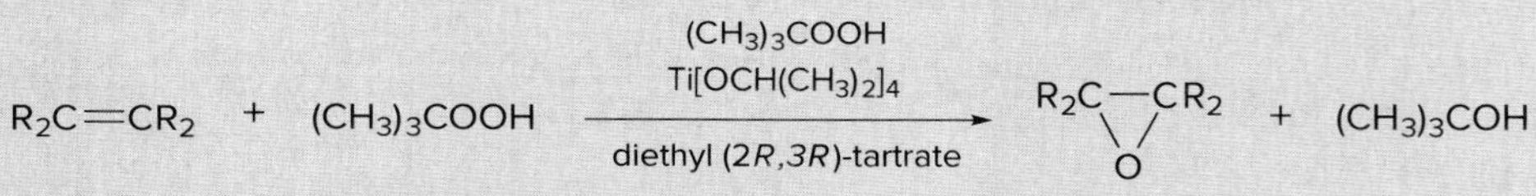 $R_2C{=}CR_2 + (CH_3)_3COOH \xrightarrow[\text{diethyl }(2R,3R)\text{-tartrate}]{(CH_3)_3COOH,\ Ti[OCH(CH_3)_2]_4} R_2C{-}CR_2$ (epoxide, bridged by O) $+ (CH_3)_3COH$ Alkene + *tert*-Butyl hydroperoxide → Epoxide + *tert*-Butyl alcohol 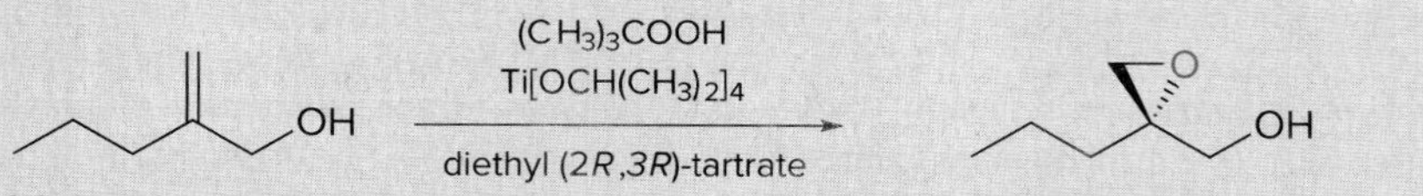 2-Propyl-2-propen-1-ol $\xrightarrow[\text{diethyl }(2R,3R)\text{-tartrate}]{(CH_3)_3COOH,\ Ti[OCH(CH_3)_2]_4}$ (*S*)-2,3-Epoxy-2-propylpropan-1-ol (88% yield; 95% enantiomeric excess)
Cyclization of vicinal halohydrins (Section 17.10) This reaction is an intramolecular version of the Williamson ether synthesis. The alcohol function of a vicinal halohydrin is converted to its conjugate base, which then displaces halide from the adjacent carbon.	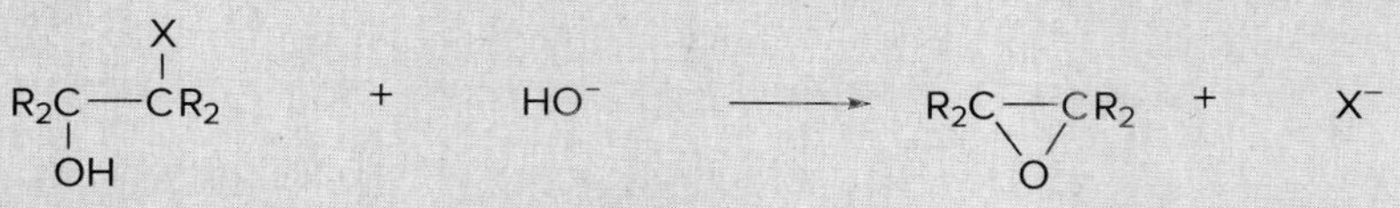 $R_2C(OH){-}CR_2X + HO^- \longrightarrow R_2C{-}CR_2$ (epoxide, bridged by O) $+ X^-$ Vicinal halohydrin + Hydroxide ion → Epoxide + Halide ion 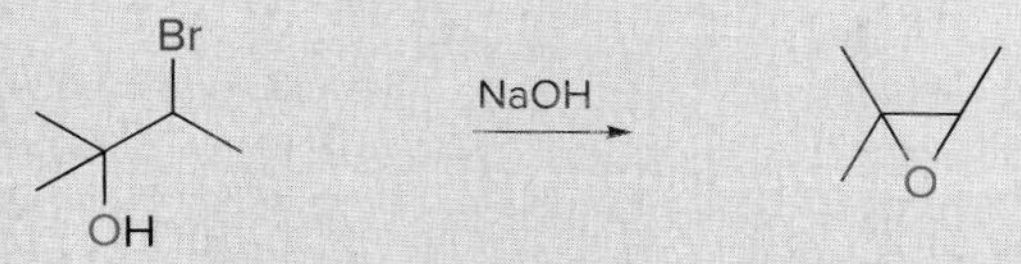 3-Bromo-2-methyl-2-butanol $\xrightarrow{NaOH}$ 2,2,3-Trimethyloxirane (78%)

Section 17.14 Sulfides are prepared by nucleophilic substitution (S_N2) in which an alkanethiolate ion reacts with an alkyl halide.

$$R\ddot{\underset{..}{S}}:^- + R-\ddot{\underset{..}{X}}: \longrightarrow R\ddot{\underset{..}{S}}-R' + :\ddot{\underset{..}{X}}:^-$$

Alkanethiolate Alkyl halide Sulfide Halide

$$C_6H_5SH \xrightarrow[2.\ C_6H_5CH_2Cl]{1.\ NaOCH_2CH_3} C_6H_5SCH_2C_6H_5$$

Benzenethiol Benzyl phenyl sulfide (60%)

Section 17.15 Oxidation of sulfides yields sulfoxides, then sulfones. Sodium metaperiodate is specific for the oxidation of sulfides to sulfoxides, and no further. Hydrogen peroxide or peroxy acids can yield sulfoxides (1 mol of oxidant per mole of sulfide) or sulfones (2 mol of oxidant per mole of sulfide).

$$C_6H_5CH_2SCH_3 \xrightarrow[1\ mol]{H_2O_2} C_6H_5CH_2S(=O)CH_3$$

Benzyl methyl sulfide Benzyl methyl sulfoxide (94%)

Section 17.16 Sulfides react with alkyl halides to give sulfonium salts.

$$RR'\ddot{S}: + R''-\ddot{\underset{..}{X}}: \longrightarrow RR'\overset{+}{S}-R''\ :\ddot{\underset{..}{X}}:^-$$

Sulfide Alkyl halide Sulfonium salt

$$H_3C-S-CH_3 + CH_3I \longrightarrow (CH_3)_3S^+\ I^-$$

Dimethyl sulfide Methyl iodide Trimethylsulfonium iodide (100%)

Section 17.17 An H—C—O—C structural unit in an ether resembles an H—C—O—H unit of an alcohol with respect to the C—O stretching frequency in its infrared spectrum and the H—C chemical shift in its ^1H NMR spectrum. Because sulfur is less electronegative than oxygen, the ^1H and ^{13}C chemical shifts of H—C—S—C units appear at higher field than those of H—C—O—C.

PROBLEMS

Structure and Nomenclature

17.21 Write the structures of all the constitutionally isomeric ethers of molecular formula $C_5H_{12}O$, and give an acceptable name for each.

17.22 Many ethers, including diethyl ether, are effective as general anesthetics. Because simple ethers are quite flammable, their place in medical practice has been taken by highly halogenated nonflammable ethers. Two such general anesthetic agents are *isoflurane* and *enflurane*. These compounds are isomeric; isoflurane is 1-chloro-2,2,2-trifluoroethyl difluoromethyl ether; enflurane is 2-chloro-1,1,2-trifluoroethyl difluoromethyl ether. Write the structural formulas of isoflurane and enflurane.

17.23 Although epoxides are always considered to have their oxygen atom as part of a three-membered ring, the prefix *epoxy* in the IUPAC system of nomenclature can be used to denote a cyclic ether of various sizes. Thus,

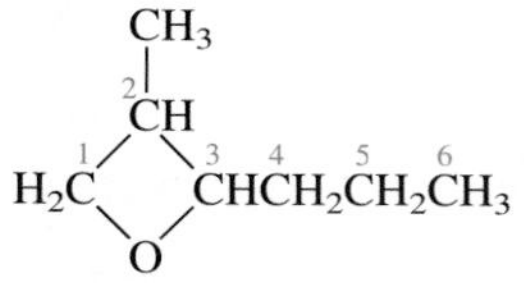

CH_3 / CH / H_2C / $CHCH_2CH_2CH_3$ / O (carbons numbered 1–6)

may be named 1,3-epoxy-2-methylhexane. Using the epoxy prefix in this way, name each of the following compounds:

(a) 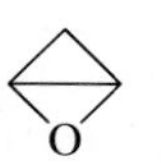 (b) 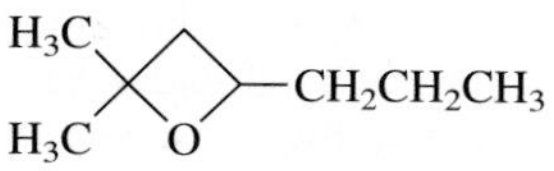 H_3C, H_3C, O, $-CH_2CH_2CH_3$ (c) 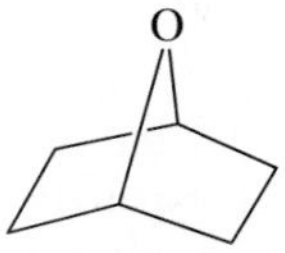 O (d)  O

17.24 The name of the parent six-membered sulfur-containing heterocycle is *thiane.* It is numbered beginning at sulfur. Multiple incorporation of sulfur in the ring is indicated by the prefixes *di-, tri-,* and so on.

(a) How many methyl-substituted thianes are there? Which ones are chiral?

(b) Write structural formulas for 1,4-dithiane and 1,3,5-trithiane.

(c) Which dithiane isomer (1,2-, 1,3-, or 1,4-) is a disulfide?

(d) Draw the two most stable conformations of the sulfoxide derived from thiane.

17.25 Suggest an explanation for the fact that the most stable conformation of *cis*-3-hydroxythiane 1-oxide is the chair in which both substituents are axial.

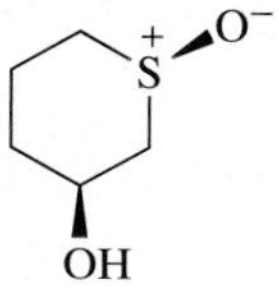

$\overset{+}{S}$, O^-, OH

Reactions and Synthesis

17.26 Predict the principal organic product of each of the following reactions. Specify stereochemistry where appropriate.

(a) 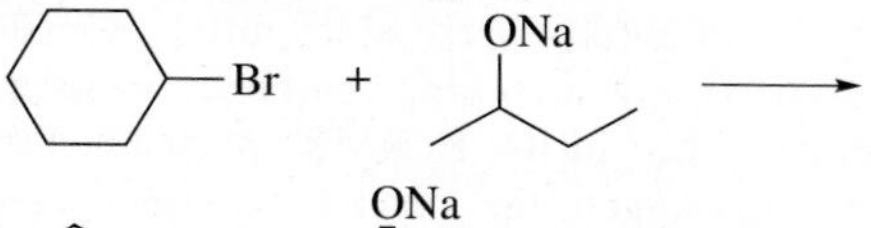 Br + ONa ⟶

(b) 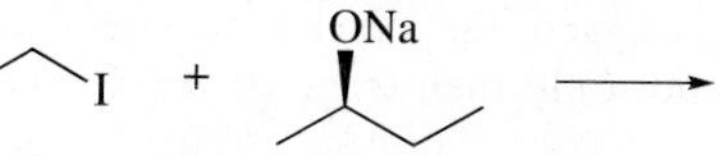 I + ONa ⟶

(c) 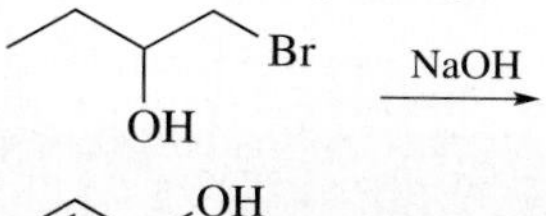 Br, OH $\xrightarrow{NaOH}$

(d) 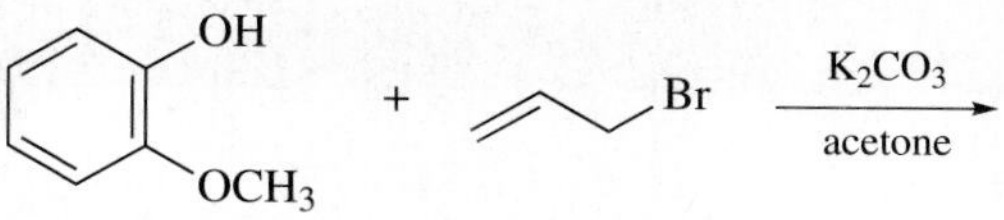 OH, OCH_3 + Br $\xrightarrow[\text{acetone}]{K_2CO_3}$

(e) 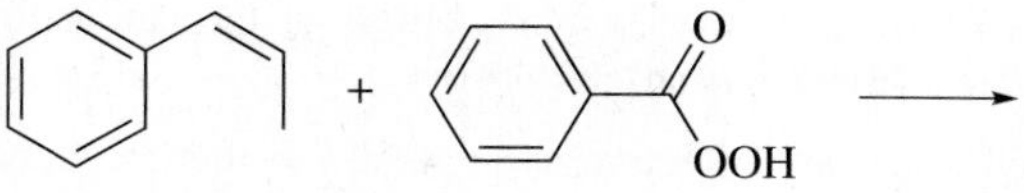 + O, OOH ⟶

(f) 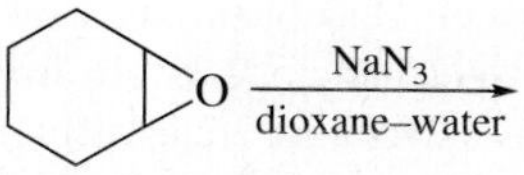 O $\xrightarrow[\text{dioxane–water}]{NaN_3}$

(g) 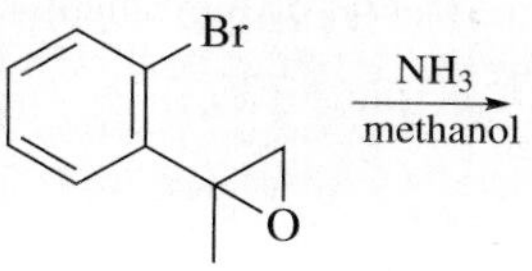 Br, O $\xrightarrow[\text{methanol}]{NH_3}$

(h) 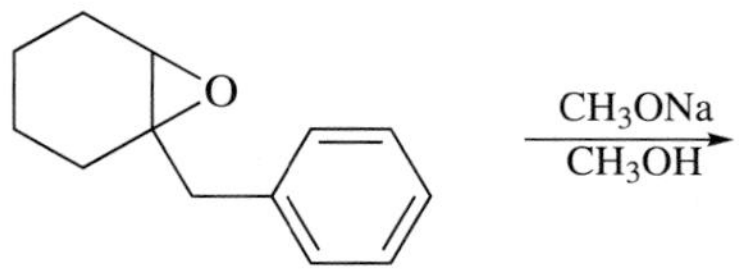 [structure: 1-benzyl-1,2-epoxycyclohexane] $\xrightarrow[CH_3OH]{CH_3ONa}$

(i) 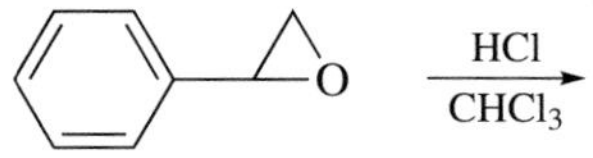 [structure: styrene oxide] $\xrightarrow[CHCl_3]{HCl}$

(j) 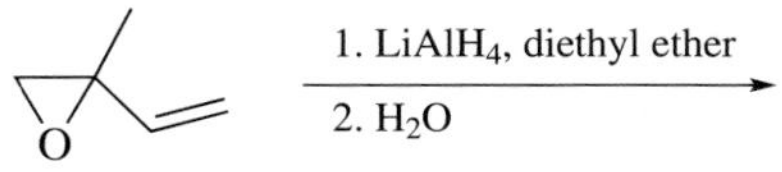 [structure: 2-methyl-2-vinyloxirane] $\xrightarrow[\text{2. } H_2O]{\text{1. } LiAlH_4\text{, diethyl ether}}$

(k) $CH_3(CH_2)_{16}CH_2OTs + CH_3CH_2CH_2CH_2SNa \longrightarrow$

(l) 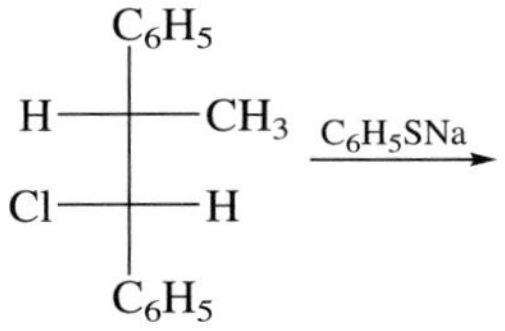 [Fischer projection: top C_6H_5; H–C–CH_3; Cl–C–H; bottom C_6H_5] $\xrightarrow{C_6H_5SNa}$

17.27 The growth of new blood vessels, angiogenesis, is crucial to wound healing and embryonic development. Abnormal angiogenesis is associated with tumor growth, suggesting that inhibition of angiogenesis may be an approach for the treatment of cancer. The diepoxide ovalicin is an angiogenesis inhibitor that was synthesized from compound C, which was in turn prepared from compound A by a two-step sequence. Can you suggest a structure for compound B?

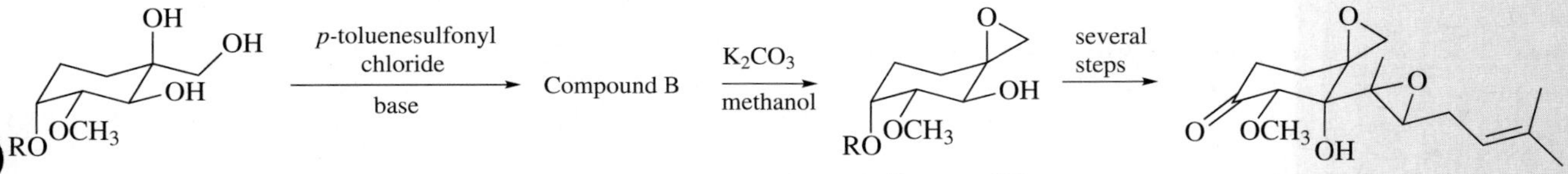

Compound A $\xrightarrow[\text{base}]{p\text{-toluenesulfonyl chloride}}$ Compound B $\xrightarrow[\text{methanol}]{K_2CO_3}$ Compound C $\xrightarrow{\text{several steps}}$ Ovalicin

Compound A
$R = (CH_3CH_2)_3Si—$

Compound C

Ovalicin

17.28 Cineole is the chief component of eucalyptus oil; it has the molecular formula $C_{10}H_{18}O$ and contains no double or triple bonds. It reacts with hydrochloric acid to give the dichloride shown:

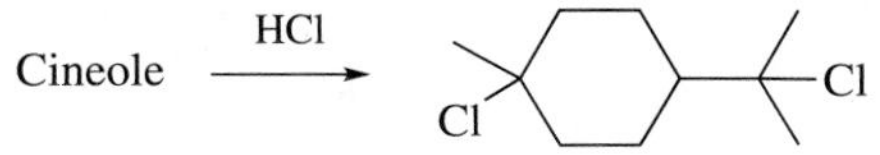

Cineole $\xrightarrow{HCl}$ [structure: cyclohexane bearing Cl/CH₃ and 2-chloro-2-propyl groups]

Deduce the structure of cineole.

17.29 The *p*-toluenesulfonate shown undergoes an intramolecular Williamson reaction on treatment with base to give a spirocyclic ether. Demonstrate your understanding of the terminology used in the preceding sentence by writing the structure, including stereochemistry, of the product.

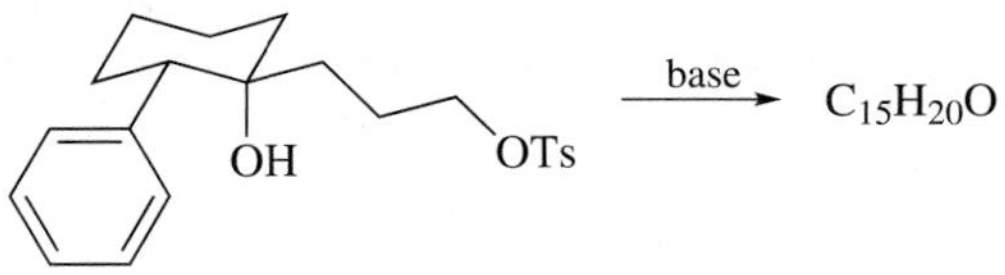

[structure: cyclohexane with phenyl, OH, and $CH_2CH_2CH_2OTs$] $\xrightarrow{\text{base}}$ $C_{15}H_{20}O$

17.30 Given that:

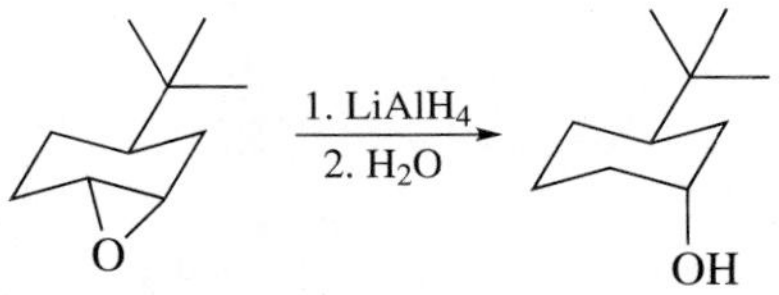

[structure: tert-butylcyclohexene oxide] $\xrightarrow[\text{2. } H_2O]{\text{1. } LiAlH_4}$ [structure: tert-butylcyclohexanol, OH]

does the product of the analogous reaction using $LiAlD_4$ contain an axial or an equatorial deuterium?

17.31 Oxidation of 4-*tert*-butylthiane (see Problem 17.24 for the structure of thiane) with sodium metaperiodate ($NaIO_4$) gives a mixture of two compounds of molecular formula $C_9H_{18}OS$. Both products give the same sulfone on further oxidation with hydrogen peroxide. What is the relationship between the two compounds?

17.32 Deduce the identity of the missing compounds in the following reaction sequences. Show stereochemistry in parts (b) through (d).

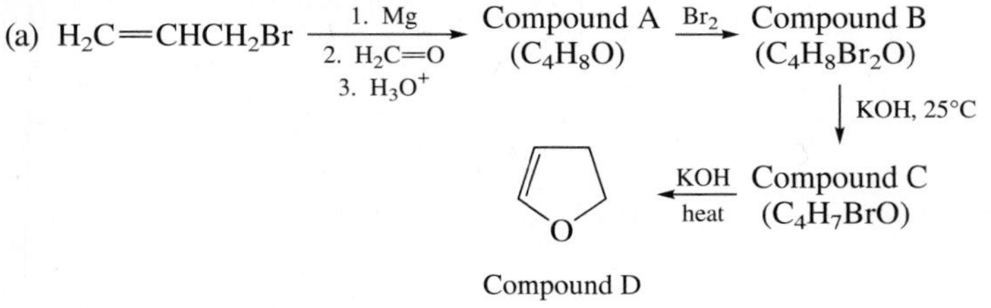

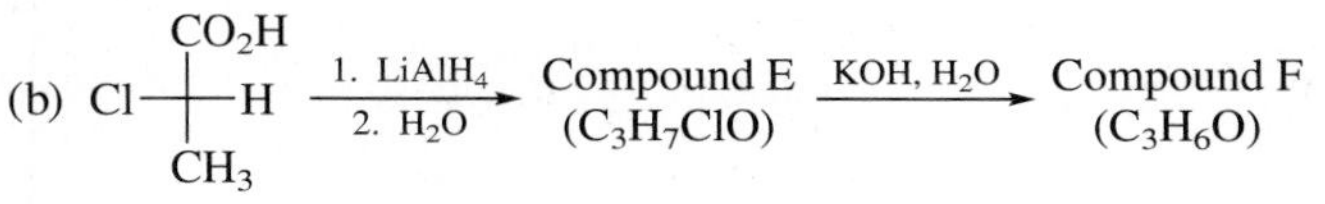

(a) $H_2C{=}CHCH_2Br$ → (1. Mg; 2. $H_2C{=}O$; 3. H_3O^+) Compound A (C_4H_8O) → (Br_2) Compound B ($C_4H_8Br_2O$) → (KOH, 25°C) Compound C (C_4H_7BrO) → (KOH, heat) Compound D

Compound D

(b) Cl—C(CO₂H)(H)(CH₃) → (1. $LiAlH_4$; 2. H_2O) Compound E (C_3H_7ClO) → (KOH, H_2O) Compound F (C_3H_6O)

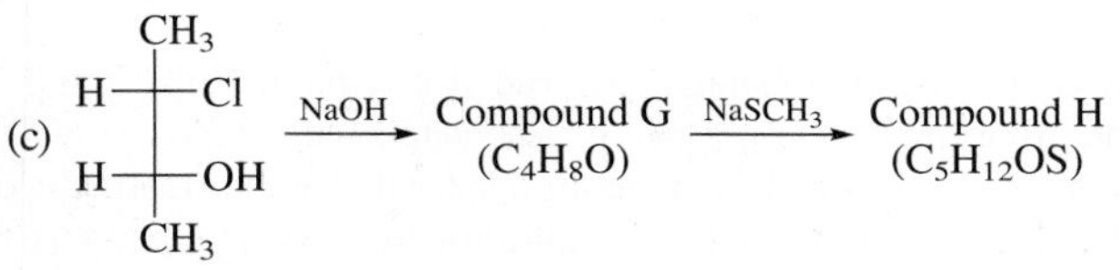

(c) CH_3—CH(H)(Cl)—CH(H)(OH)—CH_3 → (NaOH) Compound G (C_4H_8O) → ($NaSCH_3$) Compound H ($C_5H_{12}OS$)

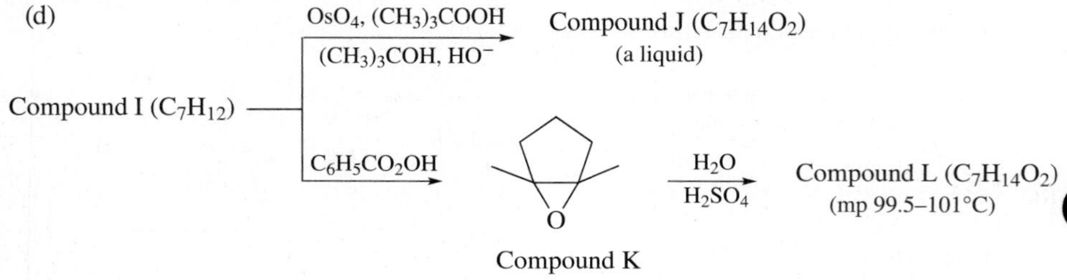

(d) Compound I (C_7H_{12}) → (OsO_4, $(CH_3)_3COOH$; $(CH_3)_3COH$, HO^-) Compound J ($C_7H_{14}O_2$) (a liquid)

Compound I (C_7H_{12}) → ($C_6H_5CO_2OH$) Compound K → (H_2O, H_2SO_4) Compound L ($C_7H_{14}O_2$) (mp 99.5–101°C)

Compound K

17.33 Outline the steps in the preparation of each of the constitutionally isomeric ethers of molecular formula $C_4H_{10}O$, starting with the appropriate alcohols. Use the Williamson ether synthesis as your key reaction.

17.34 Select reaction conditions that would allow you to carry out each of the following stereospecific transformations:

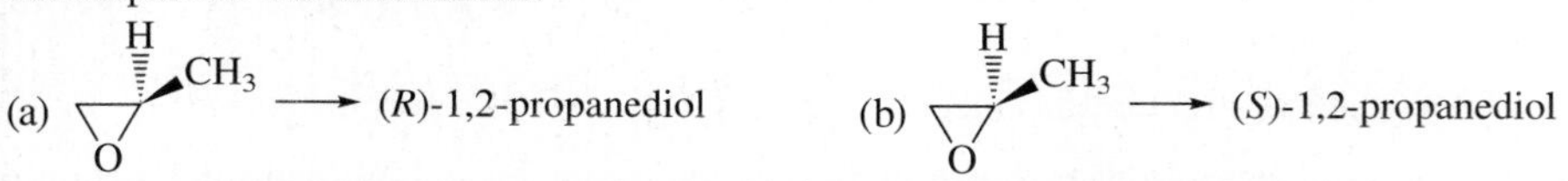

(a) [epoxide with H, CH_3] → (*R*)-1,2-propanediol (b) [epoxide with H, CH_3] → (*S*)-1,2-propanediol

17.35 Propranolol is a drug prescribed to treat cardiac arrhythmia and angina pain and to lower blood pressure. It is chiral, and one enantiomer is responsible for its therapeutic effects. That enantiomer can be synthesized from (*S*)-glycidol as shown. What is the configuration of the propranolol formed by this sequence? (No rearrangements occur.) A bond of the type ⌇ means unspecified stereochemistry.

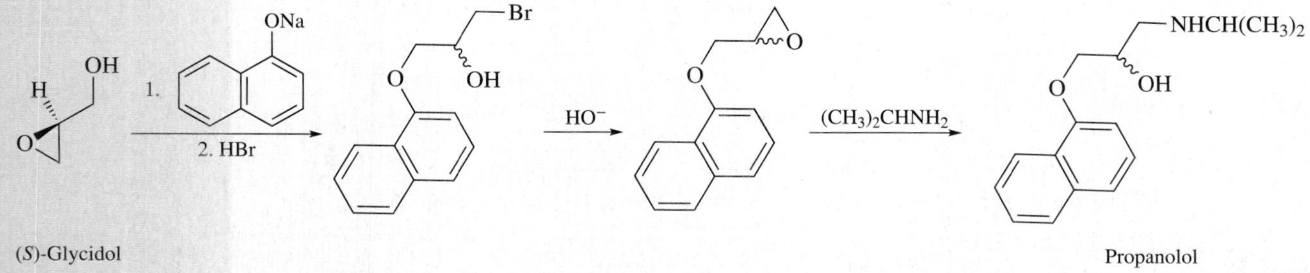

(*S*)-Glycidol → (1. [sodium 1-naphthoxide, ONa]; 2. HBr) [Br, OH compound] → (HO^-) [epoxide] → ($(CH_3)_2CHNH_2$) Propanolol [$NHCH(CH_3)_2$, OH]

(*S*)-Glycidol Propanolol

17.36 *Tamoxifen* is an estrogen receptor modulator that is used in the treatment of breast cancer. Provide the missing reagents and the structure of compound A in the synthesis of tamoxifen.

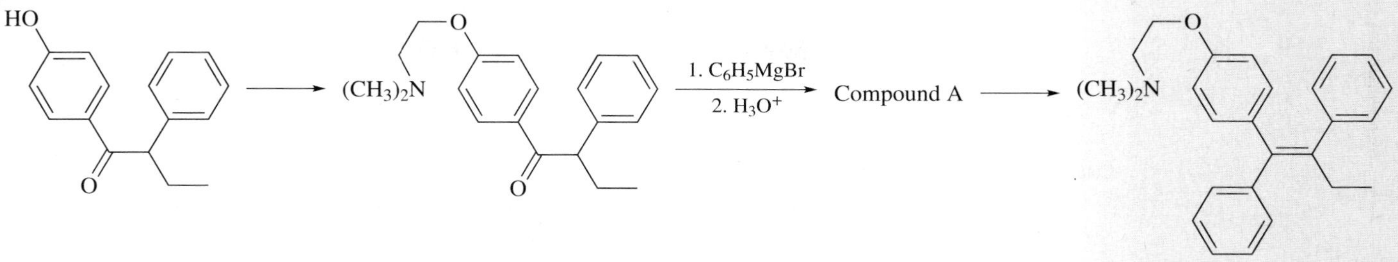

HO ⟶ $(CH_3)_2N$ —— 1. C_6H_5MgBr 2. H_3O^+ ⟶ Compound A ⟶ $(CH_3)_2N$

Tamoxifen

17.37 Suggest short, efficient reaction sequences suitable for preparing each of the following compounds from the given starting materials and any necessary organic or inorganic reagents:

(a) 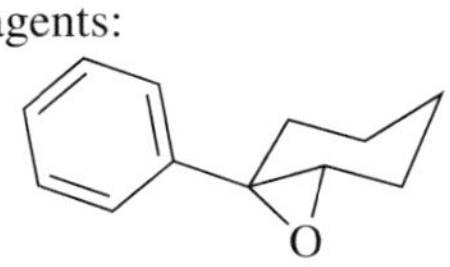 from bromobenzene and cyclohexanol

(b) 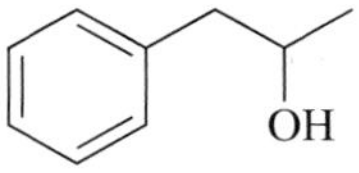 OH from bromobenzene and isopropyl alcohol

(c) 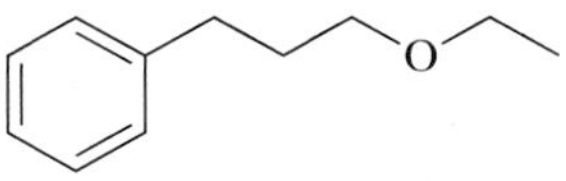 from benzyl alcohol and ethanol

(d) 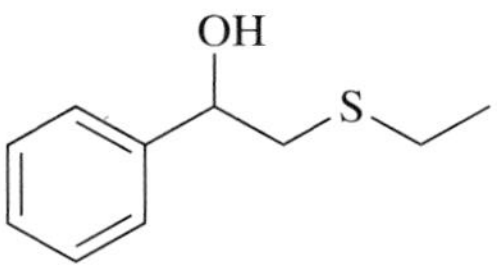 OH S from styrene and ethanol

Mechanisms

17.38 When (*R*)-(+)-2-phenyl-2-butanol is allowed to stand in methanol containing a few drops of sulfuric acid, racemic 2-methoxy-2-phenylbutane is formed. Suggest a reasonable mechanism for this reaction.

17.39 The following reaction has been reported in the chemical literature. Suggest a reasonable mechanism.

O Cl + Li —diethyl ether⟶ O

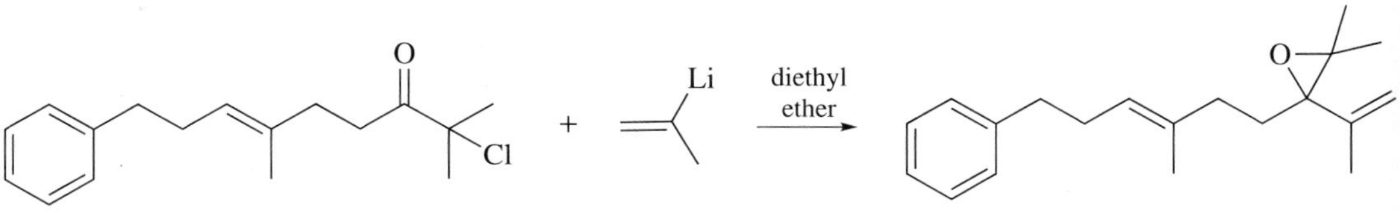

17.40 When bromine is added to a solution of 1-hexene in methanol, the major products of the reaction are as shown:

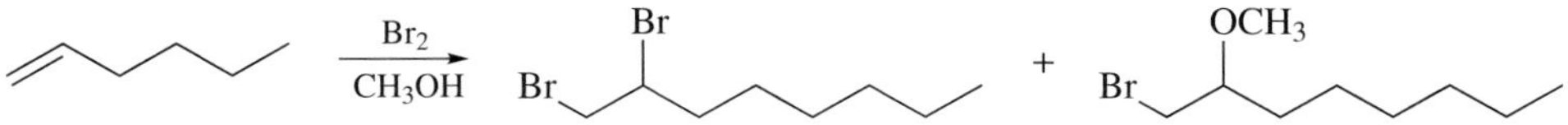

$\xrightarrow[CH_3OH]{Br_2}$ Br Br + OCH_3 Br

1,2-Dibromohexane is not converted to 1-bromo-2-methoxyhexane under the reaction conditions. Suggest a reasonable explanation for the formation of 1-bromo-2-methoxyhexane.

17.41 Write a mechanism for the following reaction.

OH + Br_2 ⟶ O Br + HBr

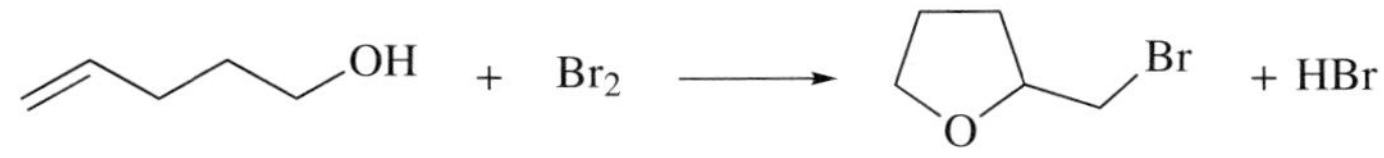

17.42 The antihistamine piperoxan was synthesized using the reaction of catechol and epichlorohydrin in the presence of sodium hydroxide as the first step. Write a mechanism for this reaction.

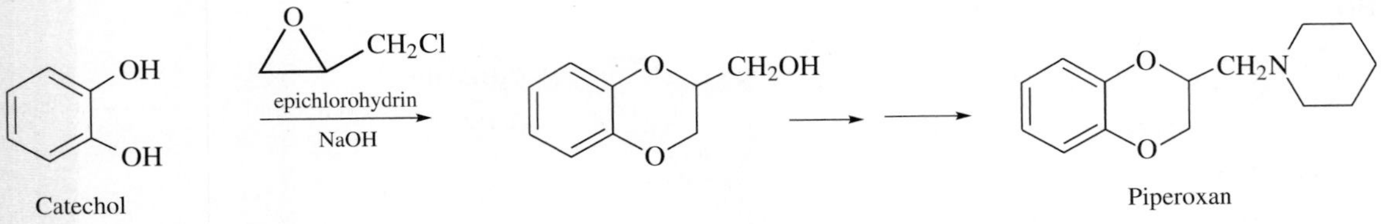

Catechol (OH, OH) $\xrightarrow[\text{NaOH}]{\text{epichlorohydrin}}$ (product with CH_2OH) ⟶ ⟶ Piperoxan (CH_2N)

Spectroscopy and Structure Determination

17.43 This problem is adapted from an experiment designed for undergraduate organic chemistry laboratories.

(a) Reaction of (*E*)-1-(*p*-methoxyphenyl)propene with *m*-chloroperoxybenzoic acid converted the alkene to its corresponding epoxide. Give the structure, including stereochemistry, of this epoxide.

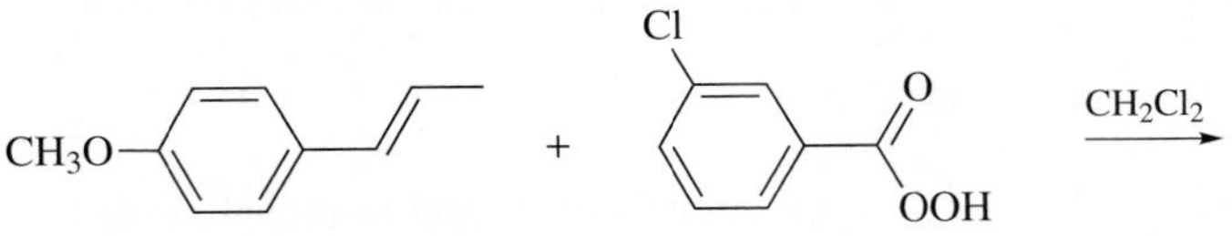

CH_3O–(C₆H₄)–CH=CHCH₃ + (m-ClC₆H₄)C(O)OOH $\xrightarrow{CH_2Cl_2}$

(b) Assign the signals in the ^1H NMR spectrum of the epoxide to the appropriate hydrogens.

δ 1.4 (doublet, 3H)	δ 3.8 (singlet, 3H)
δ 3.0 (quartet of doublets, 1H)	δ 6.9 (doublet, 2H)
δ 3.5 (doublet, 1H)	δ 7.2 (doublet, 2H)

(c) Three signals appear in the range δ 55–60 in the ^{13}C NMR spectrum of the epoxide. To which carbons of the epoxide do these signals correspond?

(d) The epoxide is isolated only when the reaction is carried out under conditions (added Na_2CO_3) that ensure that the reaction mixture does not become acidic. Unless this precaution is taken, the isolated product has the molecular formula $C_{17}H_{17}O_4Cl$. Suggest a reasonable structure for this product and write a reasonable mechanism for its formation.

17.44 A different product is formed in each of the following reactions. Identify the product in each case from their ^1H NMR spectra in Figure 17.5 and suggest an explanation for the observed regioselectivity.

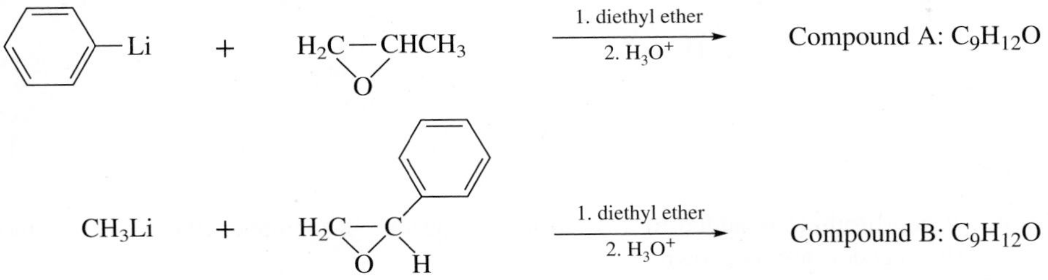

C_6H_5Li + $H_2C{-}CHCH_3$ (epoxide, O) $\xrightarrow[\text{2. } H_3O^+]{\text{1. diethyl ether}}$ Compound A: $C_9H_{12}O$

CH_3Li + $H_2C{-}C(C_6H_5)H$ (epoxide, O) $\xrightarrow[\text{2. } H_3O^+]{\text{1. diethyl ether}}$ Compound B: $C_9H_{12}O$

17.45 The ^1H NMR spectrum of compound A (C_8H_8O) consists of two singlets of equal area at δ 5.1 (sharp) and 7.2 ppm (broad). On treatment with excess hydrogen bromide, compound A is converted to a single dibromide ($C_8H_8Br_2$). The ^1H NMR spectrum of the dibromide is similar to that of A in that it exhibits two singlets of equal area at δ 4.7 (sharp) and 7.3 ppm (broad). Suggest reasonable structures for compound A and the dibromide derived from it.

17.46 The ^1H NMR spectrum of a compound ($C_{10}H_{13}BrO$) is shown in Figure 17.6. The compound gives benzyl bromide, along with a second compound $C_3H_6Br_2$, when heated with HBr. What is the first compound?

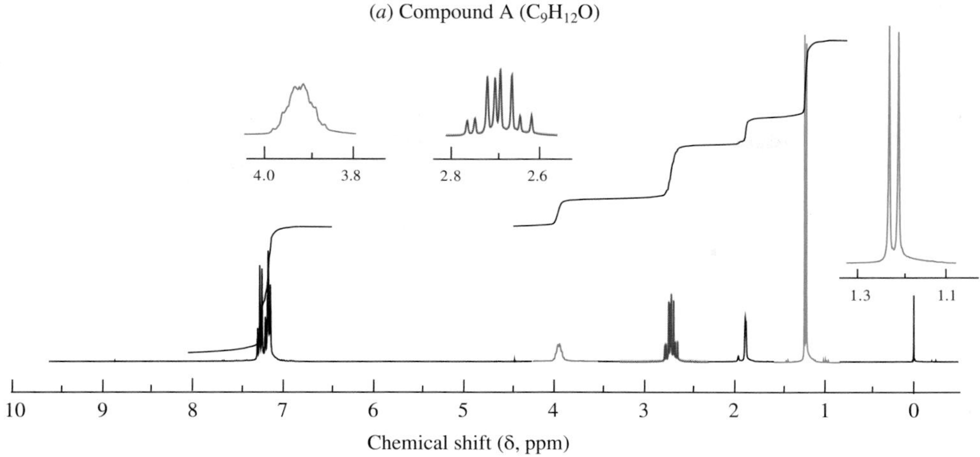

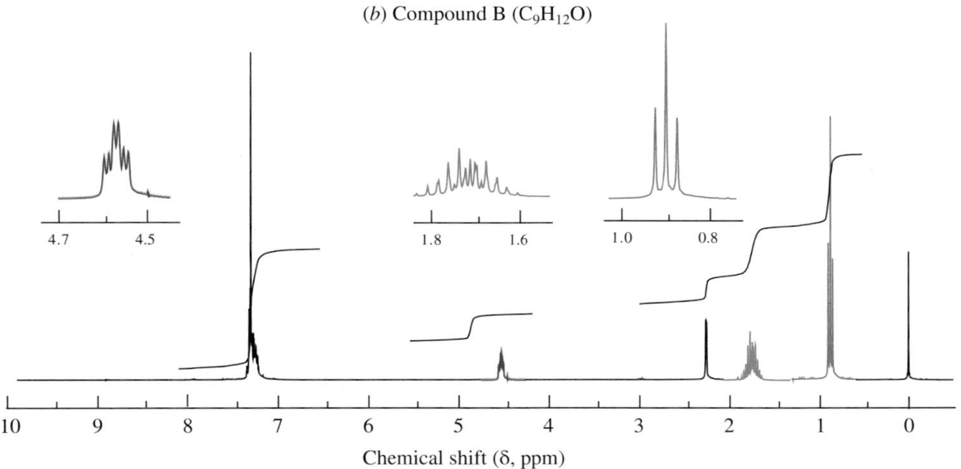

Figure 17.5

The 300-MHz ^{1}H NMR spectra of compounds formed by the reaction of (*a*) phenyllithium with 1,2-epoxypropane and (*b*) methyllithium with styrene oxide (Problem 17.44).

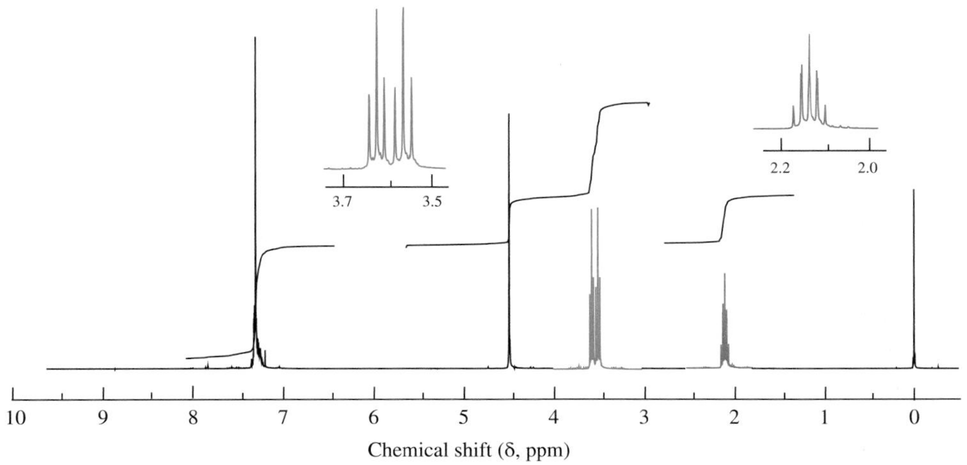

Figure 17.6

The 300-MHz ^{1}H NMR spectrum of a compound, $C_{10}H_{13}BrO$ (Problem 17.46).

17.47 A compound is a cyclic ether of molecular formula $C_9H_{10}O$. Its ^{13}C NMR spectrum is shown in Figure 17.7. Oxidation of the compound with sodium dichromate and sulfuric acid gave 1,2-benzenedicarboxylic acid. What is the compound?

17.48. Identify the compound in Figure 17.8 on the basis of the formula $C_9H_{11}BrO$, IR, and ^{13}C NMR spectra.

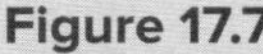

Figure 17.7

The ^{13}C NMR spectrum of a compound, $C_9H_{10}O$ (Problem 17.47).

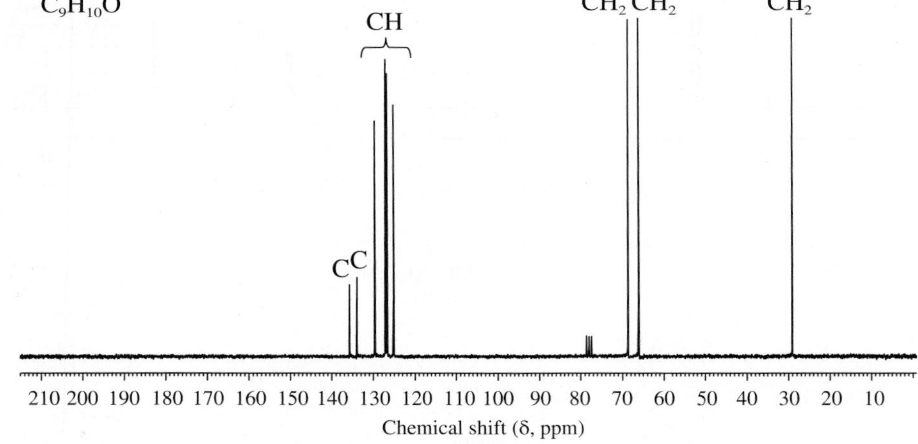

Figure 17.8

(*a*) Infrared and (*b*) ^{13}C NMR spectra of the compound, $C_9H_{11}BrO$ (Problem 17.48).

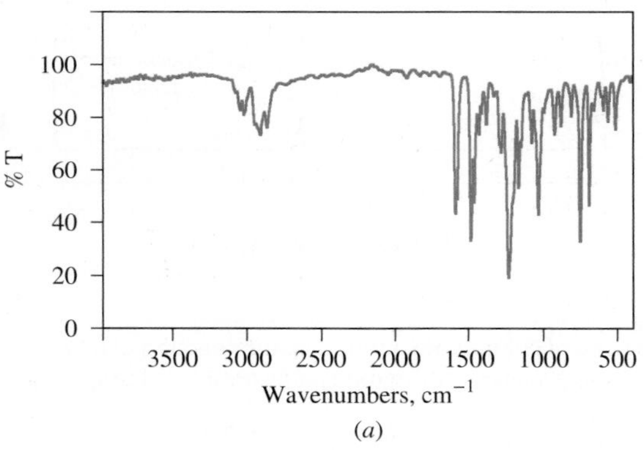

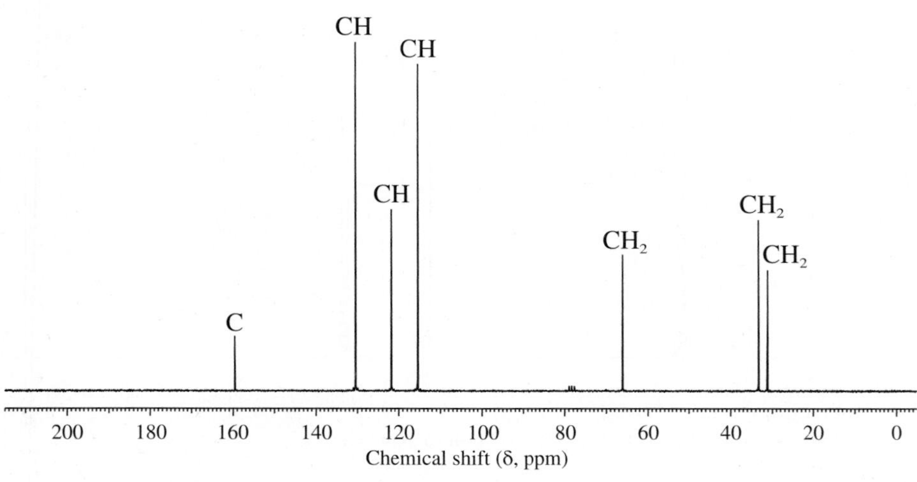

Descriptive Passage and Interpretive Problems 17

Epoxide Rearrangements and the NIH Shift

This passage is about two seemingly unrelated aspects of epoxides:

1. epoxide rearrangements
2. arene oxides

These two topics merge in an important biological transformation in which neither the reactant nor the product is an epoxide—the conversion of the amino acid phenylalanine to tyrosine.

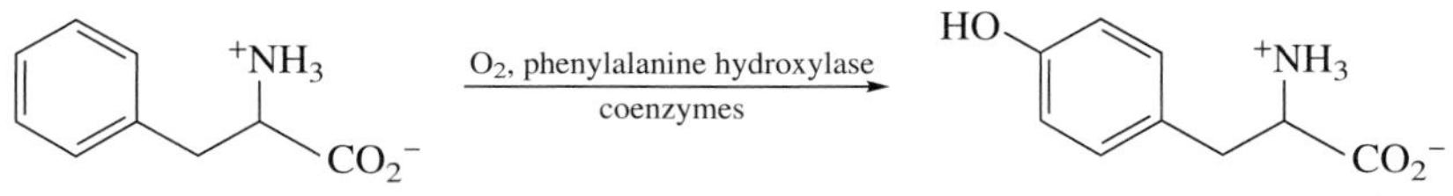

O_2, phenylalanine hydroxylase, coenzymes

Phenylalanine → Tyrosine

Epoxide rearrangements

In some epoxide ring-opening reactions, C—O bond cleavage is accompanied by the development of enough carbocation character at carbon ($^{\delta+}$C—O) to allow rearrangement to occur. These reactions are typically promoted by protonation of the epoxide oxygen or by its coordination to Lewis acids such as boron trifluoride (BF_3) and aluminum chloride ($AlCl_3$).

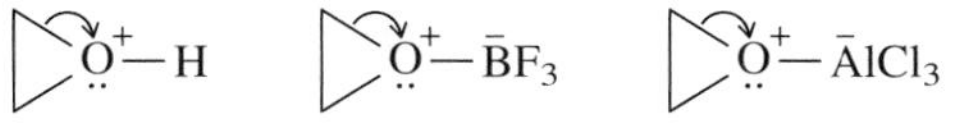

As positive charge develops on the ring carbon, one of the groups on the adjacent carbon migrates to it. This migration is assisted by electron-pair donation from oxygen. It is likely that all of this occurs in the same transition state. Subsequent deprotonation gives an aldehyde or ketone as the isolated product.

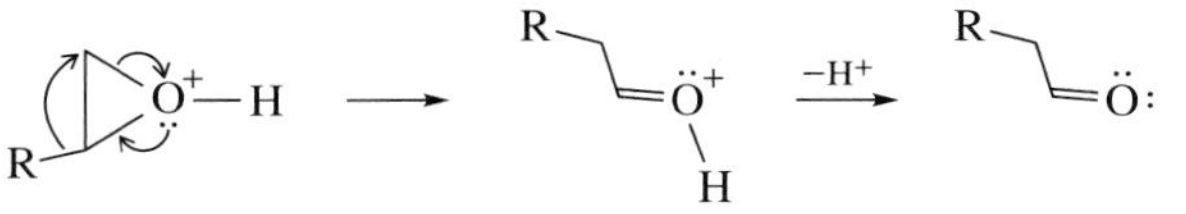

Overall, the reaction resembles the pinacol rearrangement of vicinal diols (see the Chapter 16 Descriptive Passage and Interpretive Problems) and takes place under similar conditions.

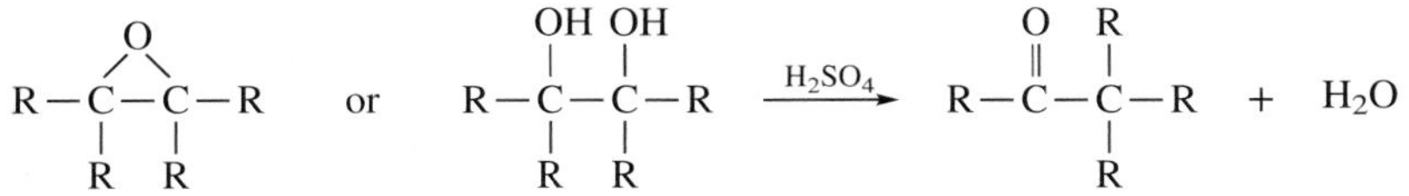

Arene oxides

Aromatic rings are normally inert to the customary reagents that convert alkenes to epoxides, but arene oxides have been synthesized in the laboratory, often by indirect methods. Their chemical reactivity resembles that of other epoxides.

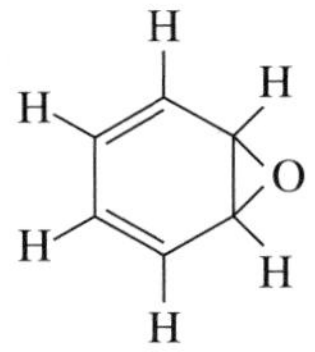

1,2-Epoxycyclohexa-3,5-diene is formally the epoxide of benzene and is the parent of the class of compounds known as arene oxides.

The most striking thing about arene oxides is their involvement in biological processes. Enzymes in the liver oxidize aromatic hydrocarbons to arene oxides, which then react with biological nucleophiles to give compounds used in subsequent reactions or to aid elimination of the arene oxide from the body. Some arene oxides, especially those from polycyclic aromatic hydrocarbons, are carcinogenic and react with nitrogen nucleophiles of DNA to induce mutations (Section 12.6).

The NIH shift

Although hydroxylation of phenylalanine to tyrosine looks like a typical electrophilic aromatic substitution, scientists at the U.S. National Institutes of Health discovered that the biochemical pathway combines epoxidation of the benzene ring followed by epoxide ring opening with rearrangement. This rearrangement, which is the biochemical analog of the pinacol-type reactions described earlier, is known as the "NIH shift."

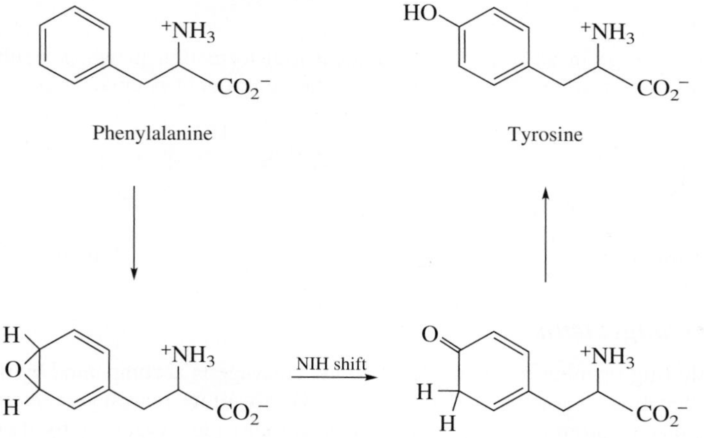

Phenylalanine

Tyrosine

NIH shift

17.49 Epoxides X and Y give the same aldehyde ($C_{14}H_{12}O$) on BF_3-catalyzed rearrangement.

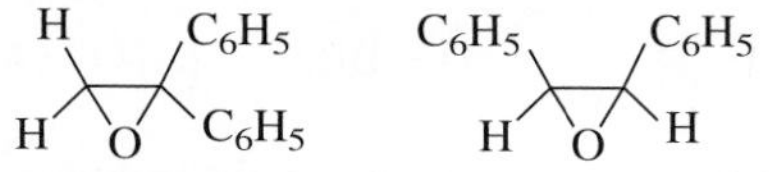

X Y

Which of the following best describes the rearrangement step?

A. H migrates in both X and Y.
B. C_6H_5 migrates in both X and Y.
C. H migrates in X; C_6H_5 migrates in Y.
D. C_6H_5 migrates in X; H migrates in Y.

17.50 Lithium aluminum hydride reduction of 1,2-epoxy-2-methylpropane gives, as expected, predominantly *tert*-butyl alcohol.

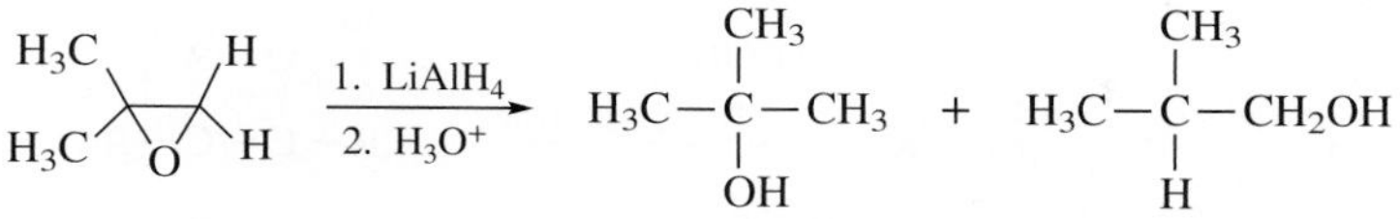

1. $LiAlH_4$ 2. H_3O^+

1,2-Epoxy-2-methylpropane → *tert*-Butyl alcohol (97%) + Isobutyl alcohol (3%)

When the reduction is carried out with an $LiAlH_4/AlCl_3$ mixture, however, epoxide rearrangement precedes reduction and isobutyl alcohol becomes the major product. This rearrangement was confirmed by a deuterium-labeling experiment in which an $LiAlD_4/AlCl_3$ mixture was used. Where was the deuterium located in the isobutyl alcohol product?

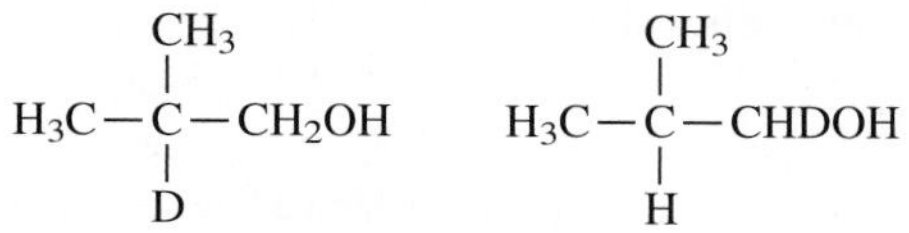

A. $(CH_3)_2CD{-}CH_2OH$

B. $(CH_3)_2CH{-}CHDOH$

17.51 The epoxide derived from benzene, 1,2-epoxycyclohexa-3,5-diene, exists in equilibrium with a monocyclic isomer oxepine.

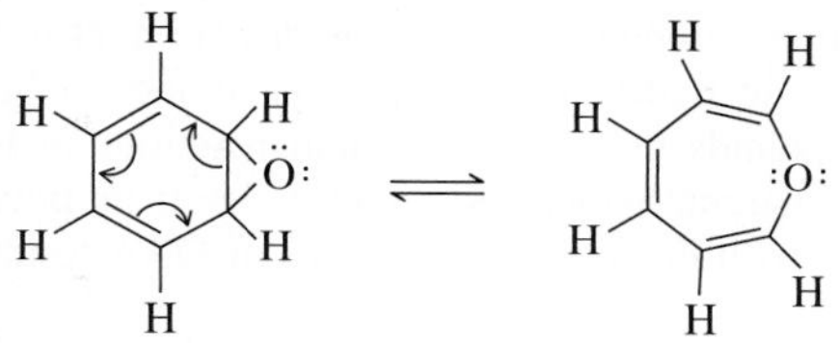

1,2-Epoxycyclohexa-3,5-diene ⇌ Oxepine

Which statement is correct concerning the aromaticity of these two isomers?

A. Both are aromatic.

B. Neither is aromatic.

C. 1,2-Epoxycyclohexa-3,5-diene is aromatic; oxepine is not aromatic.

D. Oxepine is aromatic; 1,2-epoxycyclohexa-3,5-diene is not aromatic.

17.52 Biological oxidation of naphthalene gives a trans vicinal diol by way of an epoxide intermediate. The diol formed is the most stable of the three isomers shown. Which diol is it?

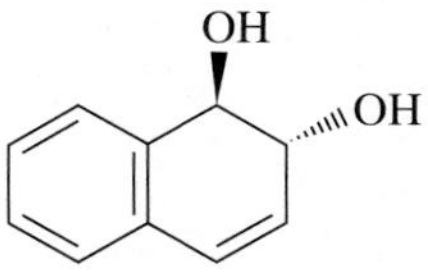

A. 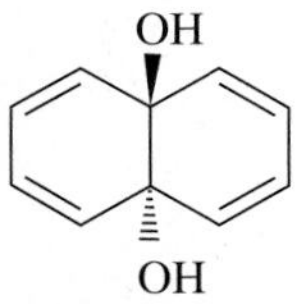 B. 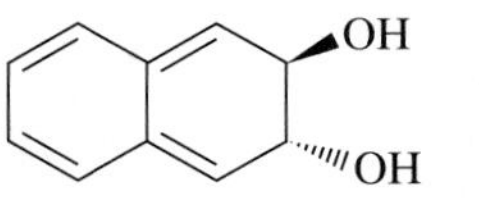 C.

17.53 Acetanilide, which has pain-relieving properties, undergoes a biochemical oxidation similar to that of the NIH shift that occurs with phenylalanine. The product formed from acetanilide is itself a pain reliever. What is the structure of this substance (better known as Tylenol)?

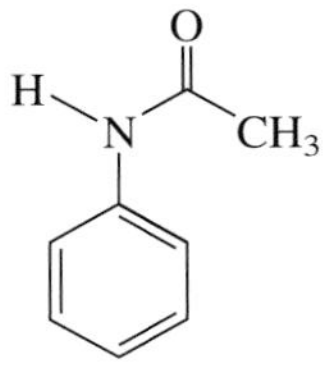

Acetanilide 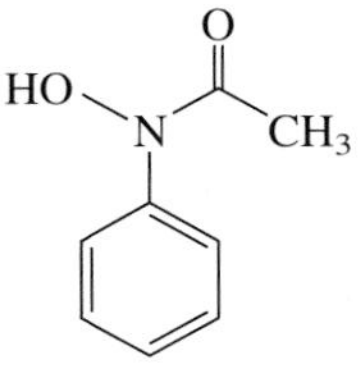 A. 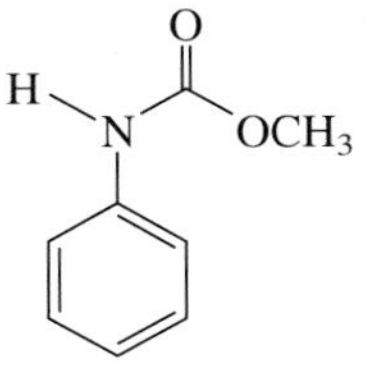 B. 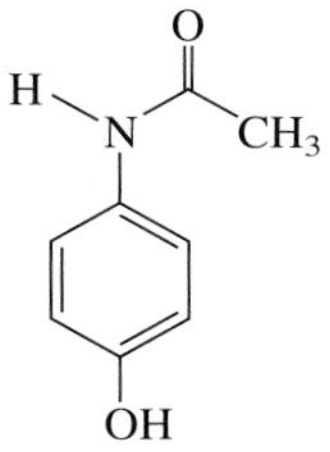 C. 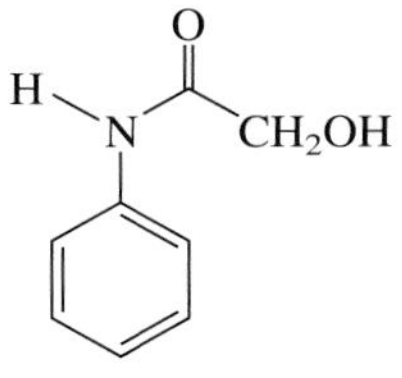 D.

17.54 The hormones serotonin and melatonin are biosynthesized from tryptophan by a series of reactions, including one that involves an NIH shift.

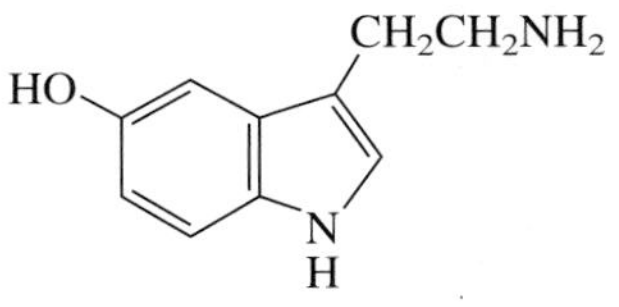

Serotonin 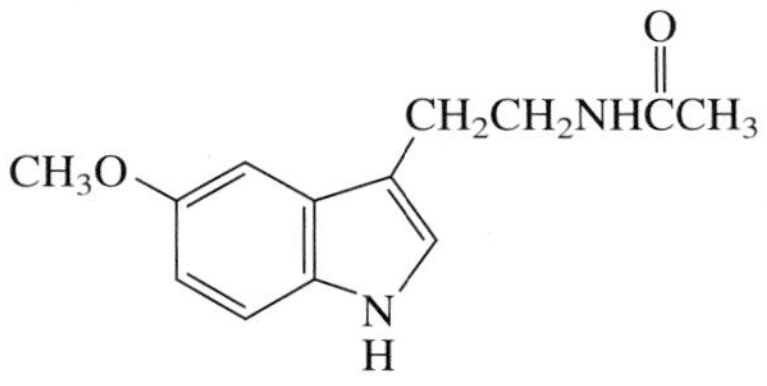 Melatonin

What is the most likely structure for tryptophan?

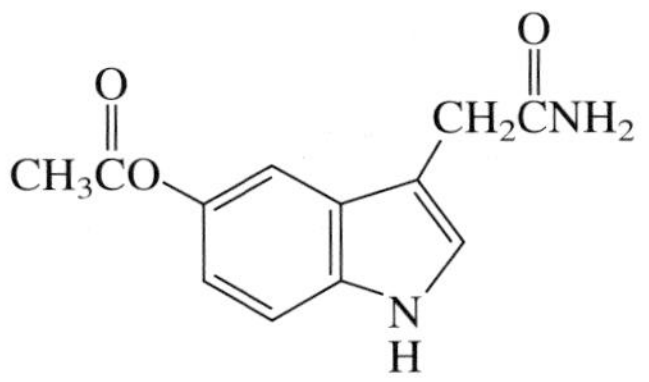

A. 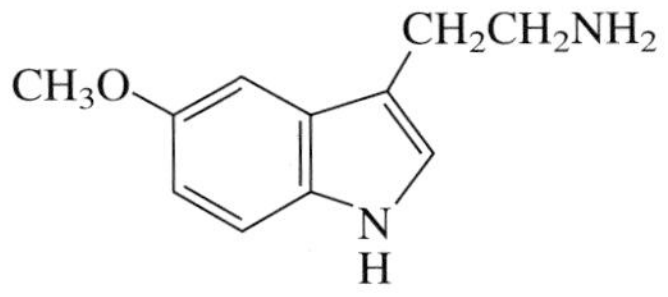 C.

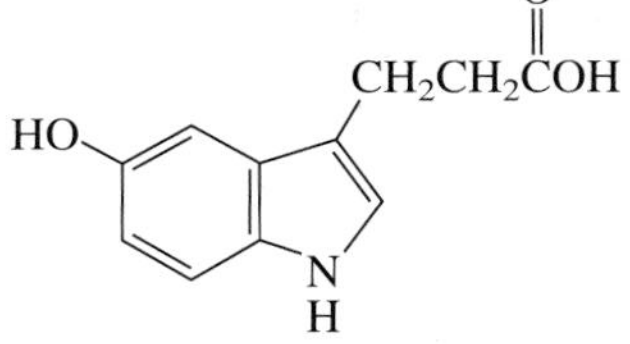

B. 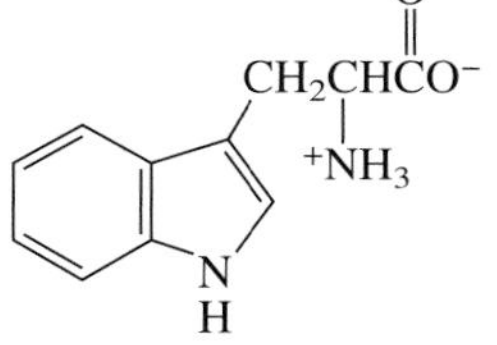 D.

Chapter

18

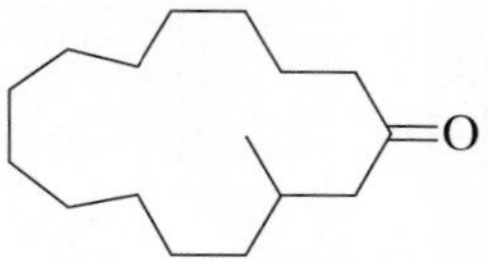

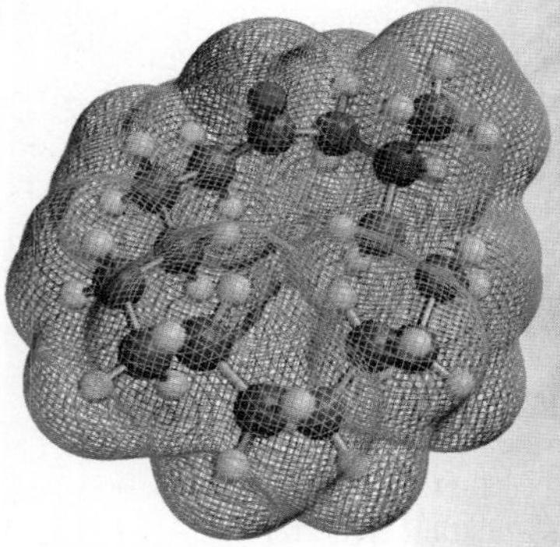

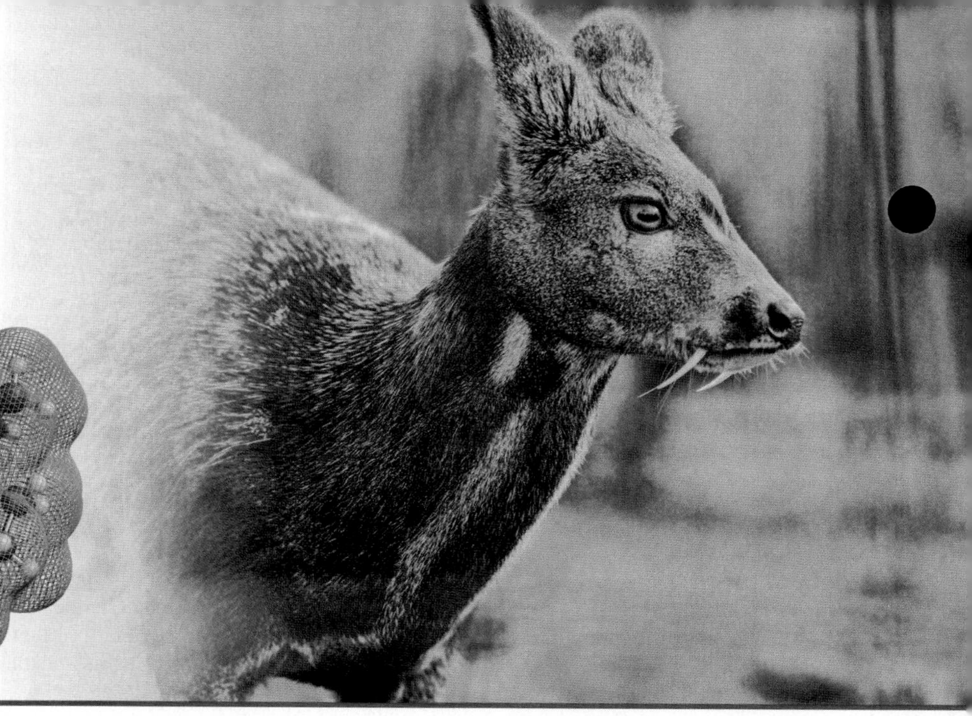

Muscone is a cyclic ketone containing a 15-membered ring. As a product of chemical synthesis, its main application is in perfumery. As obtained from male musk deer native to Asia, it is used in traditional medicine.

CHAPTER OUTLINE

Aldehydes and Ketones: Nucleophilic Addition to the Carbonyl Group

Aldehydes and ketones contain an acyl group $\overset{\text{O}}{\overset{\|}{\text{RC}}}$— bonded either to hydrogen or to another carbon.

$\overset{\text{O}}{\overset{\|}{\text{HCH}}}$	$\overset{\text{O}}{\overset{\|}{\text{RCH}}}$	$\overset{\text{O}}{\overset{\|}{\text{RCR}'}}$
Formaldehyde	Aldehyde	Ketone

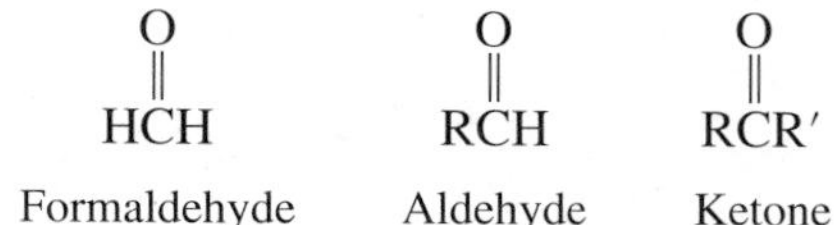

Although the present chapter includes the usual collection of topics designed to acquaint us with a particular class of compounds, its central theme is a fundamental reaction type, *nucleophilic addition to carbonyl groups*. The principles of nucleophilic addition to aldehydes and ketones developed here will be seen to have broad applicability in later chapters when transformations of various derivatives of carboxylic acids are discussed.

18.1 Nomenclature

The longest continuous chain including the —CH=O unit, called a *formyl* group, provides the parent name for aldehydes. The *-e* ending of the corresponding alkane name is replaced by *-al,* and substituents are specified in the usual way. It is not necessary to specify the location of the *formyl* group in the name, because the chain must be numbered starting with this group as C-1. The suffix *-dial* is added to the appropriate alkane name when the compound contains two aldehyde functions.

The -*e* ending of an alkane name is dropped before a suffix beginning with a vowel (-*al*) and retained before one beginning with a consonant (-*dial*).

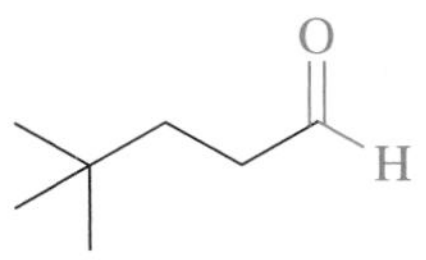

4,4-Dimethylpentanal

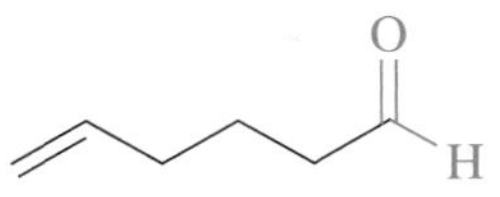

5-Hexenal or Hex-5-enal

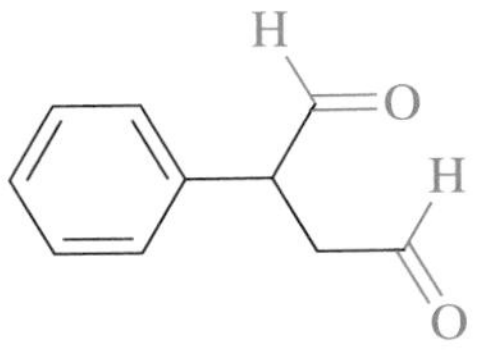

2-Phenylbutanedial

Notice that, because they define the ends of the carbon chain in 2-phenylbutanedial, the aldehyde positions are not designated by numerical locants in the name.

When a formyl group is attached to a ring, the ring name is followed by the suffix *-carbaldehyde*.

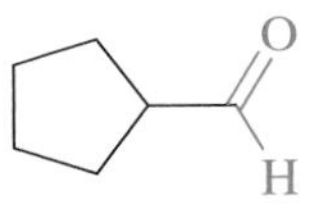

Cyclopentanecarbaldehyde

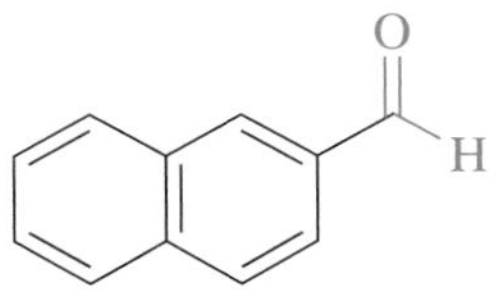

2-Naphthalenecarbaldehyde
or
Naphthalene-2-carbaldehyde

Certain common names of familiar aldehydes are acceptable as IUPAC names. A few examples include

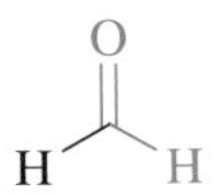

Formaldehyde
(methanal)

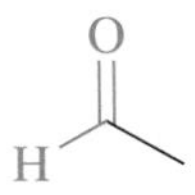

Acetaldehyde
(ethanal)

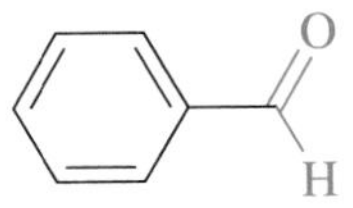

Benzaldehyde
(benzenecarbaldehyde)

Among oxygen-containing groups, a higher oxidation state takes precedence over a lower one in determining the suffix of the substitutive name. Thus, a compound that contains both an alcohol and an aldehyde function is named as an aldehyde.

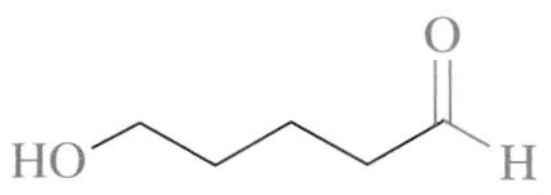

5-Hydroxypentanal

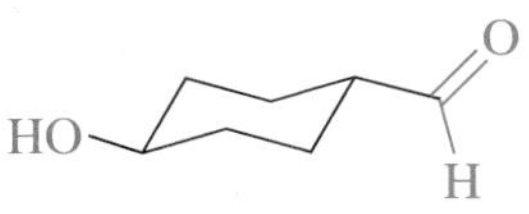

trans-4-Hydroxycyclohexane-carbaldehyde

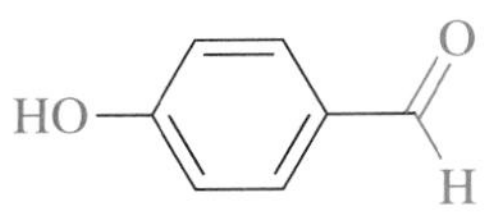

p-Hydroxybenzaldehyde

Problem 18.1

The common names and structural formulas of a few aldehydes follow. Provide an IUPAC name for each.

(a) 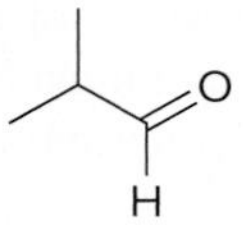 Isobutyraldehyde

(b) 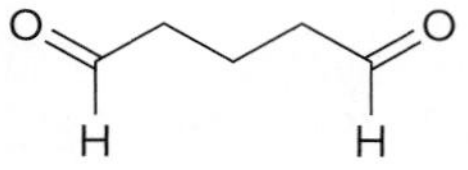 Glutaraldehyde

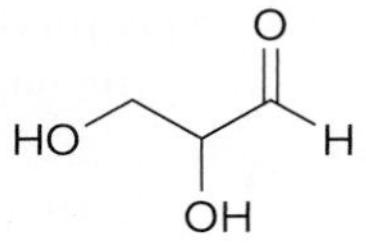

(c) Glyceraldehyde

(d) 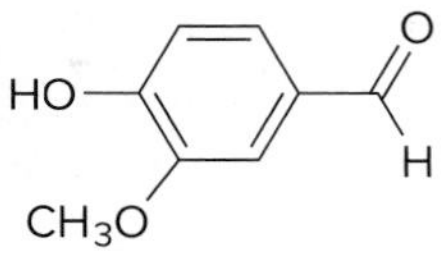 Vanillin

Sample Solution (a) Don't be fooled by the fact that the common name is isobutyraldehyde. The longest continuous chain has three carbons, and so the parent is *propanal*. There is a methyl group at C-2; thus, the compound is 2-methylpropanal.

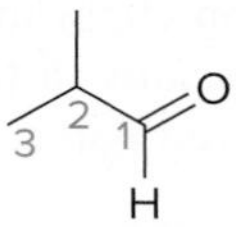

2-Methylpropanal (isobutyraldehyde)

With ketones, the *-e* ending of an alkane is replaced by *-one* in the longest continuous chain containing the carbonyl group. The chain is numbered in the direction that provides the lower number for this group. The carbonyl carbon of a cyclic ketone is C-1 and the number does not appear in the name.

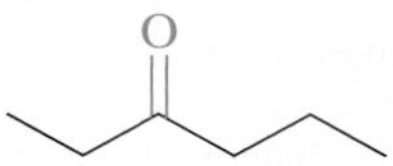

3-Hexanone or Hexan-3-one

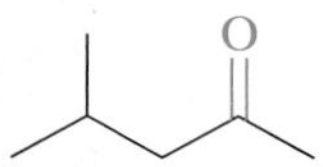

4-Methyl-2-pentanone or 4-Methylpentan-2-one

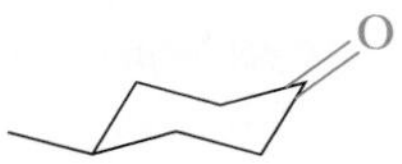

4-Methylcyclohexanone

Like aldehydes, ketone functions take precedence over alcohol functions, halogens, and alkyl groups in determining the parent name and direction of numbering. Aldehydes outrank ketones, however, and a compound that contains both an aldehyde and a ketone carbonyl group is named as an aldehyde. In such cases, the carbonyl oxygen of the ketone is considered an *oxo*-substituent on the main chain.

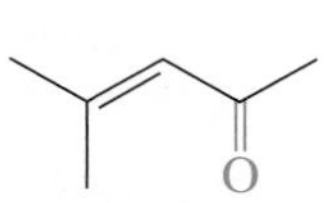

4-Methyl-3-penten-2-one or 4-Methylpent-3-en-2-one

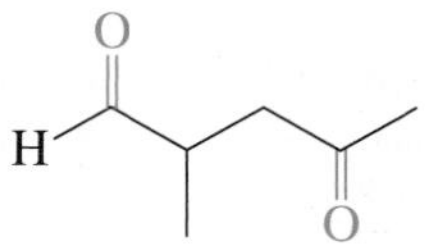

2-Methyl-4-oxopentanal

There are no functional class names for aldehydes in the IUPAC system.

Although substitutive names of the type just described are preferred, the IUPAC rules also permit ketones to be named by functional class nomenclature. The groups attached to the carbonyl group are named as separate words followed by the word *ketone*. They are listed alphabetically.

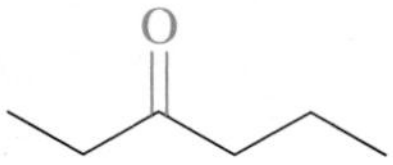

Ethyl propyl ketone

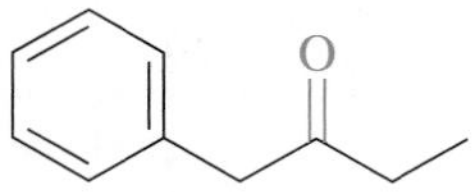

Benzyl ethyl ketone

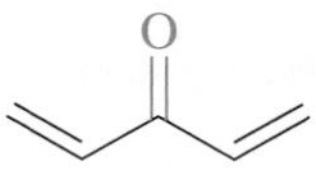

Divinyl ketone

Problem 18.2

Convert each of the following functional class IUPAC names to a substitutive name.

(a) Dibenzyl ketone
(b) Ethyl isopropyl ketone
(c) Methyl 2,2-dimethylpropyl ketone
(d) Allyl methyl ketone

Sample Solution (a) First write the structure corresponding to the name. Dibenzyl ketone has two benzyl groups attached to a carbonyl.

Dibenzyl ketone

The longest continuous chain contains three carbons, and C-2 is the carbon of the carbonyl group. The substitutive IUPAC name for this ketone is *1,3-diphenyl-2-propanone* or *1,3-diphenylpropan-2-one*.

A few of the common names acceptable for ketones in the IUPAC system are

$CH_3\overset{O}{\overset{\|}{C}}CH_3$ — Acetone

$C_6H_5\overset{O}{\overset{\|}{C}}CH_3$ — Acetophenone

$C_6H_5\overset{O}{\overset{\|}{C}}C_6H_5$ — Benzophenone

(The suffix *-phenone* indicates that the acyl group is attached to a benzene ring.)

18.2 Structure and Bonding: The Carbonyl Group

Two notable aspects of the carbonyl group are its *geometry* and *polarity*. The coplanar geometry of the bonds to the carbonyl group is seen in the molecular models of formaldehyde, acetaldehyde, and acetone in Figure 18.1. The bond angles involving the carbonyl group are approximately 120°, but vary somewhat from compound to compound as shown by the examples in Figure 18.1. The C=O bond distance in aldehydes and ketones (122 pm, 1.22 Å) is significantly shorter than the typical C—O bond distance of 141 pm (1.41 Å) seen in alcohols and ethers.

Bonding in formaldehyde can be described according to an sp^2-hybridization model analogous to that of ethylene (Figure 18.2). According to this model, the carbon–oxygen double bond is viewed as one of the $\sigma + \pi$ type. Overlap of half-filled sp^2 hybrid orbitals of carbon and oxygen gives the σ component, whereas side-by-side overlap of half-filled 2*p* orbitals gives the π bond. The oxygen lone pairs occupy sp^2 hybrid orbitals, the axes of which lie in the plane of the molecule. The carbon–oxygen double bond of formaldehyde is both shorter and stronger than the carbon–carbon double bond of ethylene.

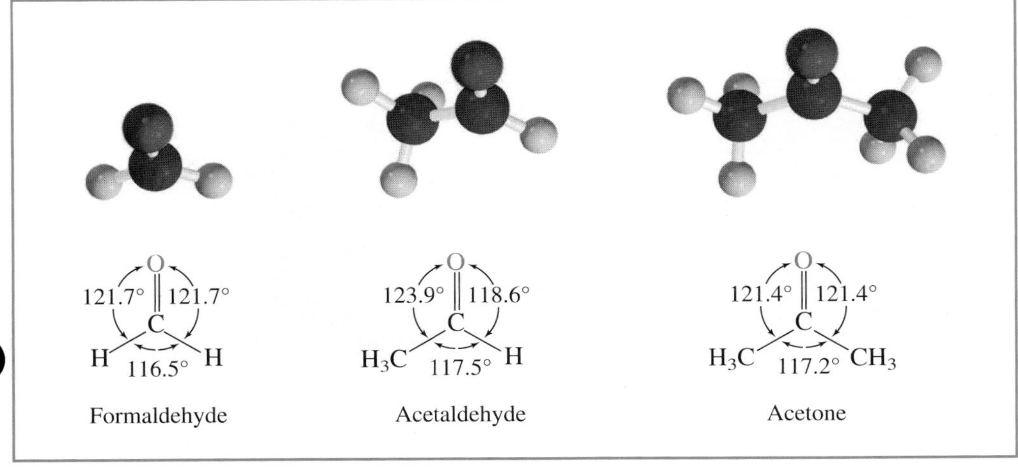

Figure 18.1

The bonds to the carbon of the carbonyl group lie in the same plane, and at angles of approximately 120° with respect to each other.

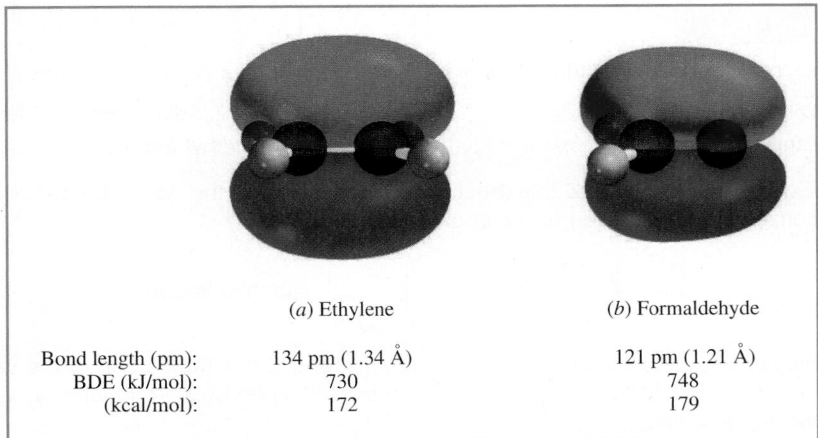

(a) Ethylene
(b) Formaldehyde
Bond length (pm): 134 pm (1.34 Å) 121 pm (1.21 Å)
BDE (kJ/mol): 730 748
(kcal/mol): 172 179

Figure 18.2

Both (*a*) ethylene and (*b*) formaldehyde have the same number of electrons, and carbon is sp^2-hybridized in both. In formaldehyde, one of the carbons is replaced by an sp^2-hybridized oxygen. Like the carbon–carbon double bond of ethylene, the carbon–oxygen double bond of formaldehyde is composed of a σ component and a π component. The values given correspond to the C=C and the C=O units, respectively.

The carbonyl group makes aldehydes and ketones rather polar, with dipole moments that are substantially higher than those of alkenes.

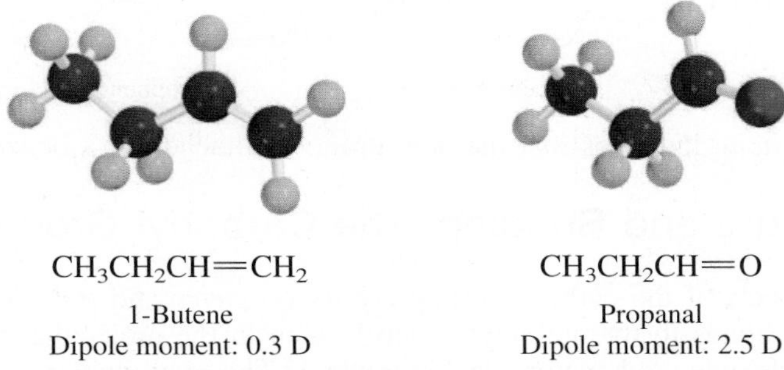

How much a carbonyl group affects the charge distribution in a molecule is apparent in the electrostatic potential maps of 1-butene and propanal (Figure 18.3). The carbonyl carbon of propanal is positively polarized and the oxygen is negatively polarized.

The various ways of representing this polarization include

$$\overset{\delta+}{\text{C}}{=}\overset{\delta-}{\text{O}} \quad \text{or} \quad \overset{+\!\!\longrightarrow}{\text{C}{=}\text{O}} \quad \text{and}$$

$$\text{C}{=}\ddot{\text{O}}\text{:} \longleftrightarrow {}^{+}\text{C}{-}\ddot{\text{O}}\text{:}^{-}$$

The structural features, especially the very polar nature of the carbonyl group, point clearly to the kind of chemistry we will see for aldehydes and ketones in this chapter. The partially positive carbon of C=O has carbocation character and is electrophilic. The

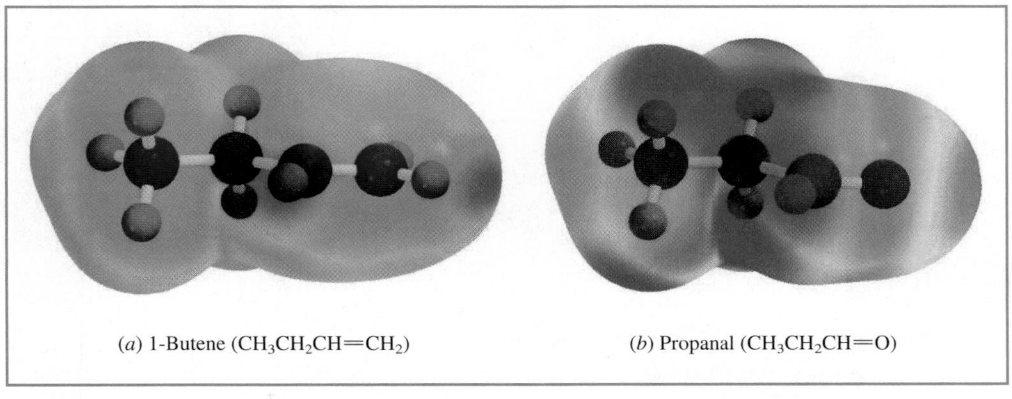

(*a*) 1-Butene ($CH_3CH_2CH{=}CH_2$) (*b*) Propanal ($CH_3CH_2CH{=}O$)

Figure 18.3

Electrostatic potential maps of (*a*) 1-butene and (*b*) propanal. The color ranges are adjusted to a common scale so that the charge distributions in the two compounds can be compared directly. The region of highest negative potential in 1-butene is associated with the electrons of the double bond. The charge separation is greater in propanal. The carbon of the carbonyl group is a site of positive potential. The region of highest negative potential is near oxygen.

planar arrangement of its bonds make this carbon relatively uncrowded and susceptible to attack by nucleophiles. Oxygen is partially negative and weakly basic.

nucleophiles bond to carbon

$C{=}\ddot{O}:$ ⟶ electrophiles, especially protons, bond to oxygen

Alkyl substituents stabilize a carbonyl group in much the same way that they stabilize carbon–carbon double bonds and carbocations—by releasing electrons to sp^2-hybridized carbon. Thus, as their heats of combustion reveal, the ketone 2-butanone is more stable than its aldehyde isomer butanal.

	$CH_3CH_2CH_2\overset{O}{\overset{\|}{C}}H$	$CH_3CH_2\overset{O}{\overset{\|}{C}}CH_3$
	Butanal	2-Butanone
Heat of combustion:	2475 kJ/mol (592 kcal/mol)	2442 kJ/mol (584 kcal/mol)

The carbonyl carbon of a ketone bears two electron-releasing alkyl groups; an aldehyde carbonyl has only one. Just as a disubstituted double bond in an alkene is more stable than a monosubstituted double bond, a ketone carbonyl is more stable than an aldehyde carbonyl. We'll see later in this chapter that structural effects on the relative *stability* of carbonyl groups in aldehydes and ketones are an important factor in their relative *reactivity*.

18.3 Physical Properties

In general, aldehydes and ketones have higher boiling points than alkenes because the dipole–dipole attractive forces between molecules are stronger. But they have lower boiling points than alcohols because, unlike alcohols, two carbonyl groups can't form hydrogen bonds to each other.

	1-Butene	Propanal	1-Propanol
bp (1 atm)	−6°C	49°C	97°C
Solubility in water (g/100 mL)	Negligible	20	Miscible in all proportions

The carbonyl oxygen of aldehydes and ketones can form hydrogen bonds with the protons of OH groups. This makes them more soluble in water than alkenes, but less soluble than alcohols.

Problem 18.3

Sketch the hydrogen bonding between benzaldehyde and water.

18.4 Sources of Aldehydes and Ketones

As we'll see later in this chapter and the next, aldehydes and ketones are involved in many of the most widely used reactions in synthetic organic chemistry. Where do aldehydes and ketones themselves come from?

Figure 18.4

Some naturally occurring aldehydes and ketones.

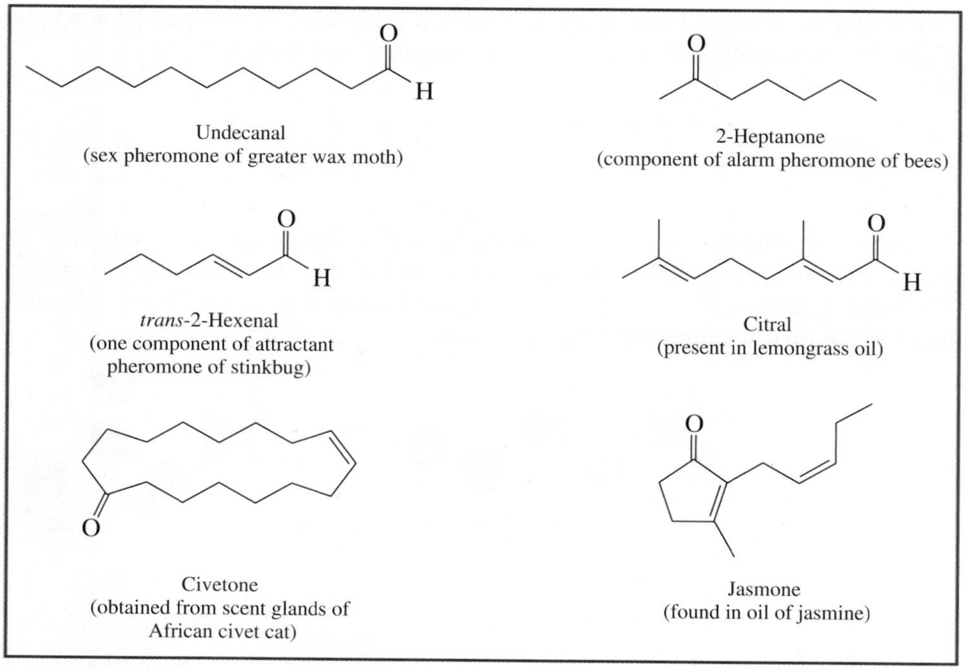

Many occur naturally and arise by the biosynthetic oxidation of alcohols pathway outlined in Section 16.10. In terms of both variety and quantity, aldehydes and ketones rank among the most common and familiar natural products. Several are shown in Figure 18.4.

Many aldehydes and ketones are made in the laboratory by reactions that you already know about, summarized in Table 18.1. To the synthetic chemist, the most important of these are the last two: the oxidation of primary alcohols to aldehydes and secondary alcohols to ketones. *Indeed, when combined with reactions that yield alcohols, the oxidation methods are so versatile that it will not be necessary to introduce any new methods for preparing aldehydes and ketones in this chapter*. A few examples will illustrate this point.

Let's first consider how to prepare an aldehyde from a carboxylic acid. There are no good methods for going from RCO_2H to RCHO directly. Instead, we do it indirectly by first reducing the carboxylic acid to the corresponding primary alcohol, then oxidizing the primary alcohol to the aldehyde.

$$\underset{\text{Carboxylic acid}}{RCO_2H} \xrightarrow{\text{reduce}} \underset{\text{Primary alcohol}}{RCH_2OH} \xrightarrow{\text{oxidize}} \underset{\text{Aldehyde}}{RCH=O}$$

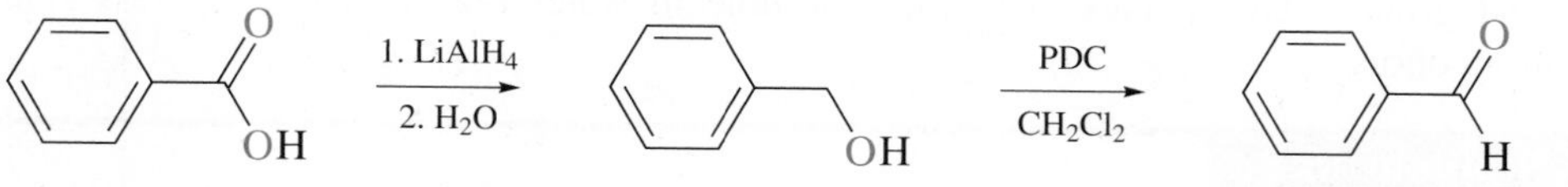

Benzoic acid → Benzyl alcohol (81%) → Benzaldehyde (83%)

Problem 18.4

Can catalytic hydrogenation be used to reduce a carboxylic acid to a primary alcohol in the first step of the sequence $RCO_2H \rightarrow RCH_2OH \rightarrow RCHO$?

TABLE 18.1 Summary of Reactions Discussed in Earlier Chapters That Yield Aldehydes and Ketones

Reaction (section) and comments	General equation and specific example
Ozonolysis of alkenes (Section 8.12) This reaction is used for structure determination and in synthesis. Hydrolysis of the ozonide intermediate in the presence of zinc permits aldehyde products to be isolated without further oxidation. The substituents on a double bond are revealed by identifying the carbonyl-containing compounds in the product.	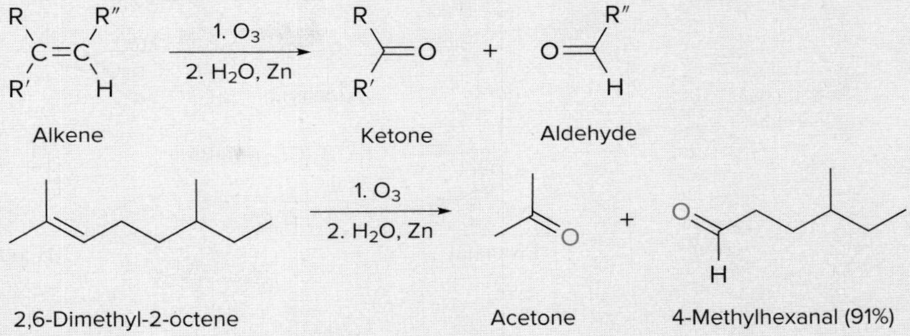 $R(R')C{=}C(R'')H \xrightarrow{\text{1. } O_3;\ \text{2. } H_2O,\ Zn} R(R')C{=}O + O{=}C(R'')H$ Alkene → Ketone + Aldehyde 2,6-Dimethyl-2-octene $\xrightarrow{\text{1. } O_3;\ \text{2. } H_2O,\ Zn}$ Acetone + 4-Methylhexanal (91%)
Hydration of alkynes (Section 9.11) Reaction occurs by way of an enol intermediate formed by Markovnikov addition of water to the triple bond.	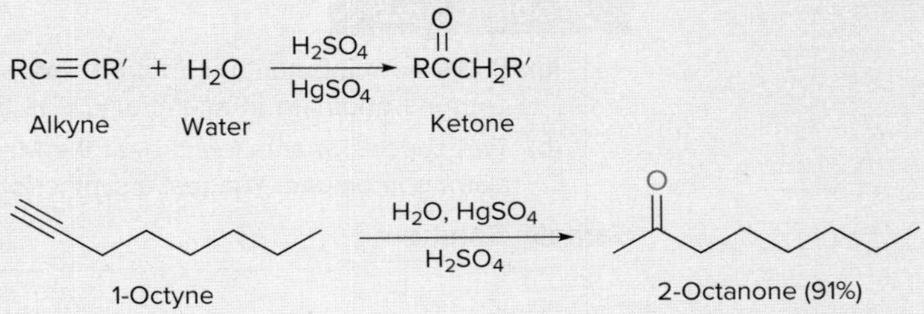 $RC{\equiv}CR' + H_2O \xrightarrow{H_2SO_4,\ HgSO_4} RCCH_2R'$ (C=O) Alkyne + Water → Ketone 1-Octyne $\xrightarrow{H_2O,\ HgSO_4;\ H_2SO_4}$ 2-Octanone (91%)
Friedel–Crafts acylation of aromatic compounds (Section 13.7) Acyl chlorides and acid anhydrides acylate aromatic rings in the presence of aluminum chloride. The reaction is electrophilic aromatic substitution in which acylium ions are generated and attack the ring.	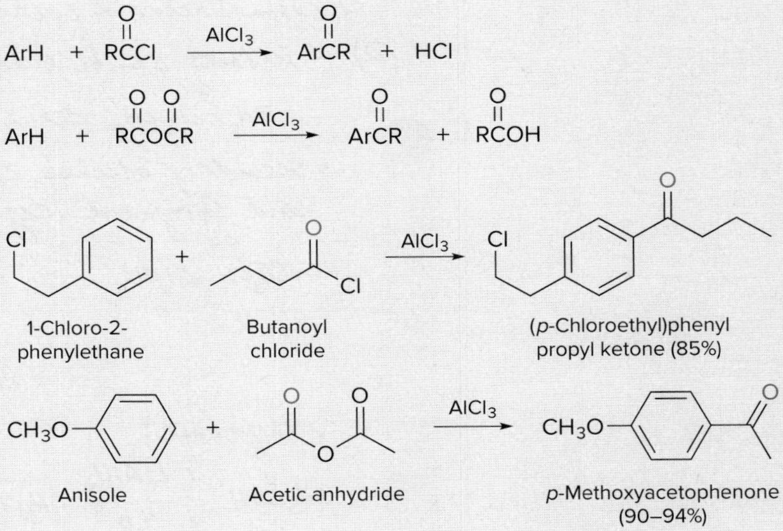 $ArH + RCCl \xrightarrow{AlCl_3} ArCR + HCl$ (C=O) $ArH + RCOCR \xrightarrow{AlCl_3} ArCR + RCOH$ (C=O) 1-Chloro-2-phenylethane + Butanoyl chloride $\xrightarrow{AlCl_3}$ (*p*-Chloroethyl)phenyl propyl ketone (85%) Anisole + Acetic anhydride $\xrightarrow{AlCl_3}$ *p*-Methoxyacetophenone (90–94%)
Oxidation of primary and secondary alcohols to aldehydes and ketones (Section 16.9) Primary alcohols are oxidized to aldehydes and secondary alcohols to ketones by a number of reagents especially pyridinium dichromate (PDC) or pyridinium chlorochromate (PCC) in dichloromethane. Aqueous sources of Cr(VI) are also useful for preparing ketones, but not for aldehydes because of overoxidation to carboxylic acids. Like PDC and PCC, dimethyl sulfoxide in the presence of oxalyl chloride and trimethylamine gives aldehydes from primary alcohols and ketones from secondary alcohols.	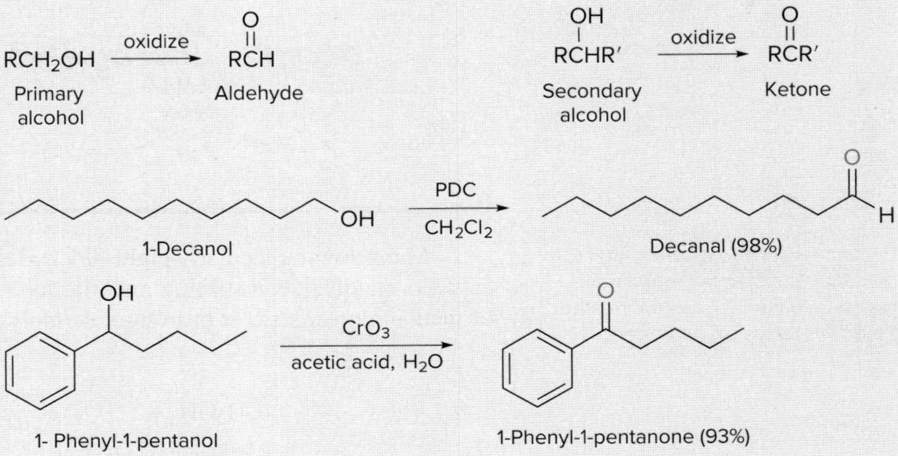 $RCH_2OH \xrightarrow{\text{oxidize}} RCH$ (C=O) Primary alcohol → Aldehyde $RCHR'$ (OH) $\xrightarrow{\text{oxidize}} RCR'$ (C=O) Secondary alcohol → Ketone 1-Decanol $\xrightarrow{PDC,\ CH_2Cl_2}$ Decanal (98%) 1- Phenyl-1-pentanol $\xrightarrow{CrO_3,\ \text{acetic acid, } H_2O}$ 1-Phenyl-1-pentanone (93%)

It is often necessary to prepare ketones by processes involving carbon–carbon bond formation. In such cases the standard method combines addition of a Grignard reagent to an aldehyde with oxidation of the resulting secondary alcohol:

$$\underset{\text{Aldehyde}}{\overset{\overset{\displaystyle O}{\|}}{RCH}} \xrightarrow[2.\ H_3O^+]{1.\ R'MgX,\ \text{diethyl ether}} \underset{\text{Secondary alcohol}}{\overset{\overset{\displaystyle OH}{|}}{RCHR'}} \xrightarrow{\text{oxidize}} \underset{\text{Ketone}}{\overset{\overset{\displaystyle O}{\|}}{RCR'}}$$

$$\underset{\text{Propanal}}{CH_3CH_2CHO} \xrightarrow[2.\ H_3O^+]{1.\ CH_3(CH_2)_3MgBr,\ \text{diethyl ether}} \underset{\text{3-Heptanol}}{CH_3CH_2CH(OH)CH_2CH_2CH_2CH_3} \xrightarrow{H_2CrO_4} \underset{\text{3-Heptanone (57\% from propanal)}}{CH_3CH_2COCH_2CH_2CH_2CH_3}$$

Problem 18.5

(a) Show how 2-butanone could be prepared by a procedure in which all of the carbons originate in acetic acid (CH_3CO_2H).

(b) Two species of ants found near the Mediterranean use 2-methyl-4-heptanone as an alarm pheromone. Suggest a synthesis of this compound from two 4-carbon alcohols.

Sample Solution

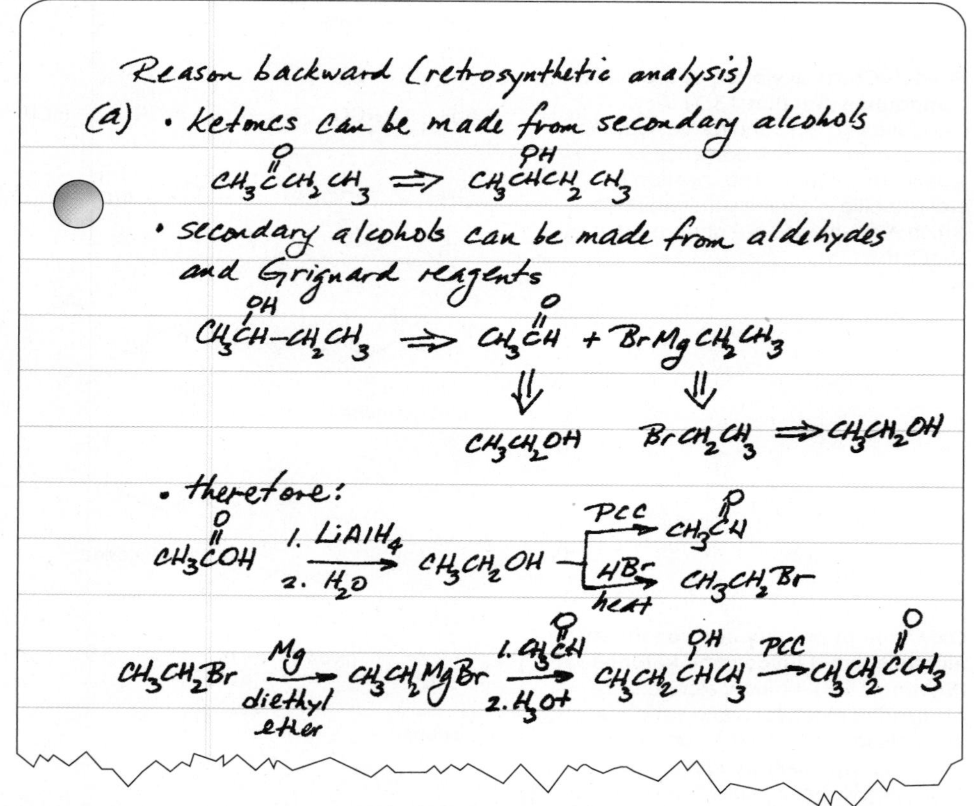

The name *aldehyde* was invented to stand for *al*cohol *dehyd*rogenatum, indicating that aldehydes are related to alcohols by loss of hydrogen.

Many low-molecular-weight aldehydes and ketones are important industrial chemicals. Formaldehyde, a starting material for a number of polymers, is prepared by oxidation of methanol over a silver or iron oxide/molybdenum oxide catalyst at elevated temperature.

$$\underset{\text{Methanol}}{CH_3OH} + \underset{\text{Oxygen}}{\tfrac{1}{2}O_2} \xrightarrow[500^\circ C]{\text{catalyst}} \underset{\text{Formaldehyde}}{\overset{\overset{\displaystyle O}{\|}}{HCH}} + \underset{\text{Water}}{H_2O}$$

Similar processes are used to convert ethanol to acetaldehyde and isopropyl alcohol to acetone.

The "linear α-olefins" described in Section 15.14 are starting materials for the preparation of a variety of aldehydes by reaction with carbon monoxide. The process is called **hydroformylation.**

$$\underset{\text{Alkene}}{RCH{=}CH_2} + \underset{\text{Carbon monoxide}}{CO} + \underset{\text{Hydrogen}}{H_2} \xrightarrow{Co_2(CO)_8} \underset{\text{Aldehyde}}{RCH_2CH_2\overset{O}{\overset{\|}{C}}H}$$

Excess hydrogen brings about the hydrogenation of the aldehyde and allows the process to be adapted to the preparation of primary alcohols. Over 2×10^9 lb/year of a variety of aldehydes and alcohols is prepared in the United States by hydroformylation.

Many aldehydes and ketones are prepared both in industry and in the laboratory by a reaction known as the *aldol condensation,* which will be discussed in detail in Chapter 21.

18.5 Reactions of Aldehydes and Ketones: A Review and a Preview

Table 18.2 summarizes the reactions of aldehydes and ketones that you've seen in earlier chapters. All are valuable tools to the synthetic chemist. Carbonyl groups provide access to hydrocarbons by Clemmensen or Wolff–Kishner reduction, and to alcohols by reduction or by reaction with Grignard or organolithium reagents.

TABLE 18.2 Summary of Reactions of Aldehydes and Ketones Discussed in Earlier Chapters

Reaction (section) and comments	General equation and specific example
Reduction to hydrocarbons (Section 13.8) Two methods for converting carbonyl groups to methylene units are the Clemmensen reduction (zinc amalgam and concentrated hydrochloric acid) and the Wolff–Kishner reduction (heat with hydrazine and potassium hydroxide in a high boiling alcohol).	$R\overset{O}{\overset{\|}{C}}R' \longrightarrow RCH_2R'$ Aldehyde or ketone → Hydrocarbon Citronellal $\xrightarrow[\text{diethylene glycol, heat}]{H_2NNH_2,\ KOH}$ 2,6-Dimethyl-2-octene (80%)
Reduction to alcohols (Section 16.2) Aldehydes are reduced to primary alcohols, and ketones to secondary alcohols, by a variety of reducing agents. Catalytic hydrogenation over a metal catalyst and reduction with sodium borohydride or lithium aluminum hydride are general methods.	$R\overset{O}{\overset{\|}{C}}R' \longrightarrow R\underset{OH}{\underset{\vert}{C}}HR'$ Aldehyde or ketone → Alcohol p-Methoxybenzaldehyde (CH_3O–C₆H₄–CHO) $\xrightarrow[CH_3OH]{NaBH_4}$ p-Methoxybenzyl alcohol (CH_3O–C₆H₄–CH₂OH) (96%)
Addition of Grignard reagents and organolithium compounds (Sections 15.5–15.6) Carbon–carbon bond formation converts aldehydes to secondary alcohols and ketones to tertiary alcohols.	$R\overset{O}{\overset{\|}{C}}R' + R''M \longrightarrow R\overset{O^-M^+}{\underset{R''}{C}}R' \xrightarrow{H_3O^+} R\overset{OH}{\underset{R''}{C}}R'$ Cyclohexanone + Ethylmagnesium bromide $\xrightarrow[2.\ H_3O^+]{1.\ \text{diethyl ether}}$ 1-Ethylcyclohexanol (74%)

The most important chemical property of the carbonyl group is its tendency to undergo **nucleophilic addition** reactions of the type represented in the general equation:

$$\overset{\delta+\ \delta-}{C=O} + \overset{\delta+\ \delta-}{X-Y} \longrightarrow \text{C(O-X)(Y)}$$

Aldehyde or ketone → Product of nucleophilic addition

A negatively polarized atom or group bonds to the positively polarized carbon carbonyl carbon in the rate-determining step of these reactions. Grignard reagents, organolithium reagents, lithium aluminum hydride, and sodium borohydride, for example, all react with carbonyl compounds by nucleophilic addition.

The next section explores the mechanism of nucleophilic addition to aldehydes and ketones. There we'll discuss their *hydration,* a reaction in which water adds to the C=O group. After we use this reaction to develop some general principles, we'll survey a number of related reactions of synthetic, mechanistic, or biological interest.

18.6 Principles of Nucleophilic Addition: Hydration of Aldehydes and Ketones

The convention for writing equilibrium constant expressions without the solvent (water in this case) was discussed in Section 1.12.

Effects of Structure on Equilibrium: Aldehydes and ketones react with water in a rapid equilibrium. The product is a **geminal diol,** also called a "hydrate."

$$\underset{\text{Aldehyde or ketone}}{RC(=O)R'} + \underset{\text{Water}}{H_2O} \overset{\text{fast}}{\rightleftharpoons} \underset{\text{Geminal diol (hydrate)}}{RC(OH)_2R'} \qquad K_{hydr} = \frac{[\text{hydrate}]}{[\text{carbonyl compound}]}$$

TABLE 18.3 Equilibrium Constants (K_{hydr}) and Relative Rates of Hydration of Some Aldehydes and Ketones

Reaction*	K_{hydr}†	Percent hydrate	Relative rate
$HCHO + H_2O \rightleftharpoons H_2C(OH)_2$	2300	>99.9	2200
$CH_3CHO + H_2O \rightleftharpoons CH_3CH(OH)_2$	1.0	50	1.0
$(CH_3)_3CCHO + H_2O \rightleftharpoons (CH_3)_3CCH(OH)_2$	0.2	17	0.09
$CH_3COCH_3 + H_2O \rightleftharpoons (CH_3)_2C(OH)_2$	0.0014	0.14	0.0018

*Neutral solution, 25°C

†$K_{hydr} = \frac{[\text{hydrate}]}{[\text{carbonyl compound}]}$

Overall, the reaction is classified as an *addition*. Water adds to the carbonyl group. Hydrogen becomes bonded to the negatively polarized carbonyl oxygen, hydroxyl to the positively polarized carbon.

Table 18.3 compares the equilibrium constants (K_{hydr}) of some simple aldehydes and ketones. The position of equilibrium depends on what groups are attached to C=O and how they affect its *steric* and *electronic* environment. Both contribute, but the electronic effect controls K_{hydr} more than the steric effect.

Consider first the electronic effect of alkyl groups versus hydrogen atoms attached to C=O. Alkyl substituents stabilize a ketone carbonyl more than the hydrogen of an aldehyde carbonyl. As with all equilibria, factors that stabilize the reactants decrease the equilibrium constant. Thus, the extent of hydration decreases as the number of alkyl groups on the carbonyl increase.

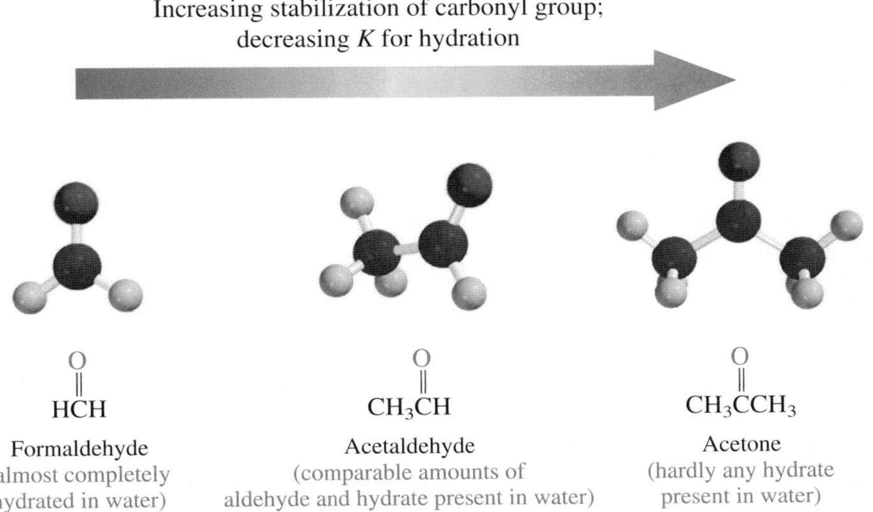

A striking example of an electronic effect on carbonyl group stability and its relation to the equilibrium constant for hydration is seen in the case of hexafluoroacetone. In contrast to the almost negligible hydration of acetone, hexafluoroacetone is completely hydrated.

$$F_3C(C{=}O)CF_3 + H_2O \rightleftharpoons (HO)(OH)C(CF_3)_2 \qquad K = 22{,}000$$

Hexafluoroacetone — Water — 1,1,1,3,3,3-Hexafluoro-2,2-propanediol

Instead of stabilizing the carbonyl group by electron donation as alkyl substituents do, trifluoromethyl groups destabilize it by withdrawing electrons. A less-stabilized carbonyl group is associated with a greater equilibrium constant for addition.

Problem 18.6

Chloral is one of the common names for trichloroethanal. Its hydrate has featured prominently in countless detective stories as the notorious "Mickey Finn" knockout drops. Write a structural formula for chloral hydrate.

Now let's turn our attention to steric effects by looking at how the size of the groups that were attached to C=O affect K_{hydr}. The bond angles at carbon shrink from ≈120° to ≈109.5° as the hybridization changes from sp^2 in the reactant (aldehyde or ketone) to sp^3 in the product (hydrate). The increased crowding this produces in the hydrate is better tolerated, and K_{hydr} is greater when the groups are small (hydrogen) than when they are large (alkyl).

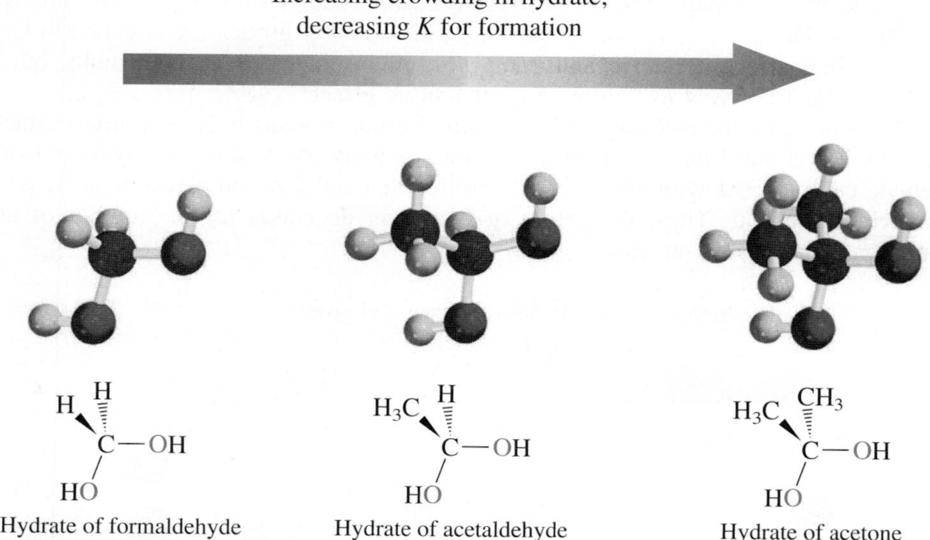

Hydrate of formaldehyde — Hydrate of acetaldehyde — Hydrate of acetone

Electronic and steric effects operate in the same direction. Both cause the equilibrium constants for hydration of aldehydes to be greater than those of ketones.

Effects of Structure on Rate: Electronic and steric effects influence the rate of hydration in the same way that they affect equilibrium. Indeed, the rate and equilibrium data of Table 18.3 parallel each other almost exactly.

Hydration of aldehydes and ketones is a rapid reaction, quickly reaching equilibrium, but faster in acid or base than in neutral solution. Thus, instead of a single mechanism for hydration, we'll look at two mechanisms, one for basic and the other for acidic solution.

Mechanism of Base-Catalyzed Hydration: The base-catalyzed mechanism (Mechanism 18.1) is a two-step process in which the first step is rate-determining. In step 1, the nucleophilic hydroxide ion bonds to the carbon of the carbonyl group. The alkoxide ion formed in step 1 abstracts a proton from water in step 2, yielding the geminal diol. The second step, like all other proton transfers between oxygen that we have seen, is fast.

The role of the basic catalyst (HO^-) is to increase the rate of the nucleophilic addition step. Hydroxide ion, the nucleophile in the base-catalyzed reaction, is much more reactive than a water molecule, the nucleophile in neutral solutions.

Aldehydes react faster than ketones for almost the same reasons that their equilibrium constants for hydration are more favorable. The $sp^2 \rightarrow sp^3$ hybridization change that the carbonyl carbon undergoes on hydration is partially developed in the transition state for the rate-determining nucleophilic addition step (Figure 18.5). Alkyl groups at the reaction site increase the activation energy by simultaneously lowering the energy of the starting state (ketones have a more stabilized carbonyl group than aldehydes) and raising the energy of the transition state (a steric crowding effect).

Mechanism of Acid-Catalyzed Hydration: Three steps are involved in acid-catalyzed hydration (Mechanism 18.2). The first and last are rapid proton transfers between oxygens. The second is a nucleophilic addition and is rate-determining. The acid catalyst activates the carbonyl group toward attack by a weakly nucleophilic water molecule. Protonation of oxygen makes the carbonyl carbon of an aldehyde or a ketone much more electrophilic. Expressed in resonance terms, the protonated carbonyl has a greater degree of carbocation character than an unprotonated carbonyl.

$$\mathrm{C{=}\overset{+}{O}{:}-H \longleftrightarrow {}^{+}C-\ddot{O}{:}-H}$$

Mechanism 18.1

Hydration of an Aldehyde or Ketone in Basic Solution

THE OVERALL REACTION:

$$R'(R)C{=}\ddot{O}: \;+\; H_2\ddot{O}: \;\xrightleftharpoons{HO^-}\; R'R\,C(:\ddot{O}H)_2$$

Aldehyde or ketone — Water — Geminal diol

THE MECHANISM:

Step 1: Nucleophilic addition of hydroxide ion to the carbonyl group

$$H\ddot{O}:^- \;+\; R'(R)C{=}\ddot{O}: \;\xrightleftharpoons{slow}\; R'R\,C(:\ddot{O}:^-)(:\ddot{O}H)$$

Hydroxide — Aldehyde or ketone — Alkoxide ion intermediate

Step 2: Proton transfer from water to the intermediate formed in step 1 gives the geminal diol and regenerates the hydroxide catalyst.

$$R'R\,C(:\ddot{O}:^-)(:\ddot{O}H) \;+\; H{-}\ddot{O}{-}H \;\xrightleftharpoons{fast}\; R'R\,C(:\ddot{O}H)_2 \;+\; {}^-:\ddot{O}{-}H$$

Alkoxide ion intermediate — Water — Geminal diol — Hydroxide

Steric and electronic effects influence the rate of nucleophilic addition to a protonated carbonyl group in much the same way as they do for the case of a neutral one, and protonated aldehydes react faster than protonated ketones.

With this as background, let us now examine how the principles of nucleophilic addition apply to the characteristic reactions of aldehydes and ketones. We'll begin with the addition of hydrogen cyanide.

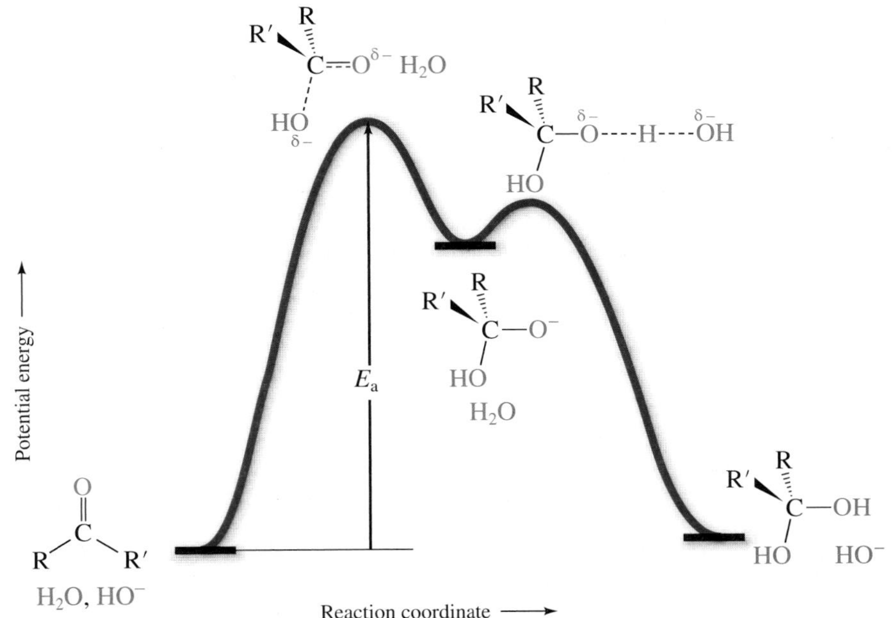

Figure 18.5

Potential energy diagram for base-catalyzed hydration of an aldehyde or ketone.

Mechanism 18.2

Hydration of an Aldehyde or Ketone in Acid Solution

THE OVERALL REACTION:

Aldehyde or ketone + Water $\xrightleftharpoons{H_3O^+}$ Geminal diol

THE MECHANISM:

Step 1: Protonation of the carbonyl oxygen

Aldehyde or ketone + Hydronium ion $\xrightleftharpoons{\text{fast}}$ Conjugate acid of carbonyl compound + Water

Step 2: Nucleophilic addition to the protonated aldehyde or ketone

Water + Conjugate acid of carbonyl compound $\xrightleftharpoons{\text{slow}}$ Conjugate acid of geminal diol

Step 3: Proton transfer from the conjugate acid of the geminal diol to a water molecule

Water + Conjugate acid of geminal diol $\xrightleftharpoons{\text{fast}}$ Hydronium ion + Geminal diol

18.7 Cyanohydrin Formation

The product of addition of hydrogen cyanide to an aldehyde or a ketone contains both a hydroxyl group and a cyano group bonded to the same carbon. Compounds of this type are called **cyanohydrins.**

Aldehyde or ketone + Hydrogen cyanide ($HC{\equiv}N$) $\rightleftharpoons$ Cyanohydrin

Mechanism 18.3 describing cyanohydrin formation is analogous to the mechanism of base-catalyzed hydration. The nucleophile (cyanide ion) bonds to the carbonyl carbon in

Mechanism 18.3

Cyanohydrin Formation

THE OVERALL REACTION:

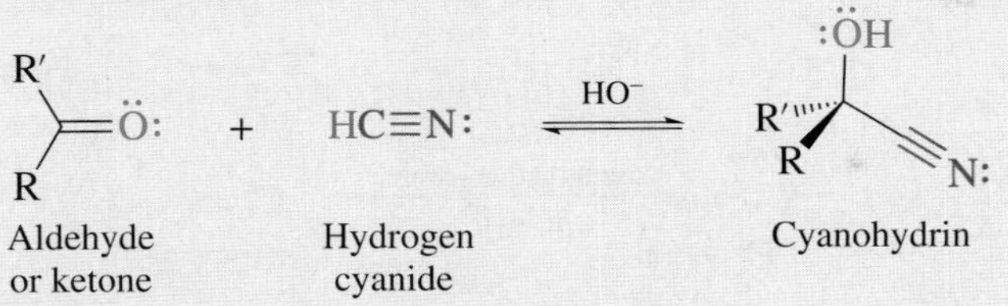

Aldehyde or ketone + Hydrogen cyanide ($HC{\equiv}N$:) $\xrightleftharpoons{HO^-}$ Cyanohydrin

THE MECHANISM:

Step 1: The negatively charged carbon of cyanide ion is nucleophilic and bonds to the carbonyl carbon of the aldehyde or ketone. Hydrogen cyanide itself is not very nucleophilic and does not ionize to form cyanide ion to a significant extent. Thus, a source of cyanide ion such as NaCN or KCN is used.

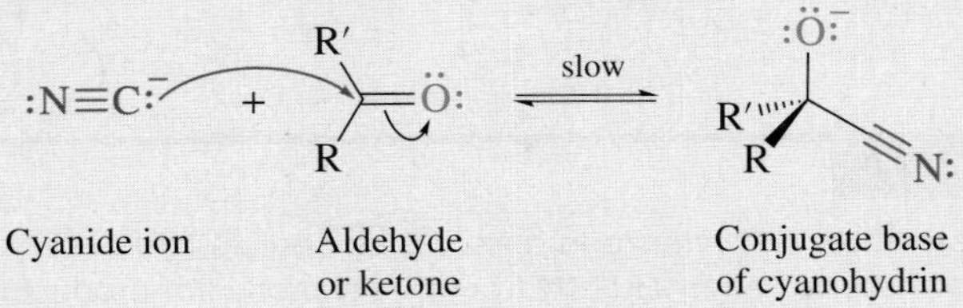

Cyanide ion + Aldehyde or ketone $\xrightleftharpoons{\text{slow}}$ Conjugate base of cyanohydrin

Step 2: The alkoxide ion formed in the first step abstracts a proton from hydrogen cyanide. This step yields the cyanohydrin product and regenerates cyanide ion.

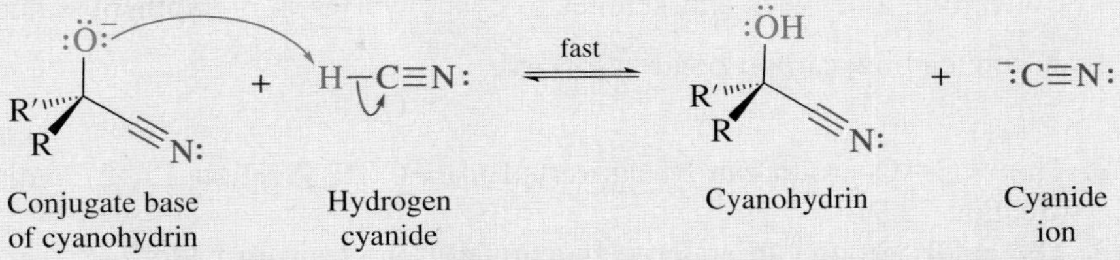

Conjugate base of cyanohydrin + Hydrogen cyanide $\xrightleftharpoons{\text{fast}}$ Cyanohydrin + Cyanide ion

the rate-determining first step, followed by proton transfer to the carbonyl oxygen in the second step.

The addition of hydrogen cyanide is catalyzed by cyanide ion, but HCN is too weak an acid to provide enough $^-{:}C{\equiv}N{:}$ for the reaction to proceed at a reasonable rate. Cyanohydrins are normally prepared by adding an acid to a solution containing the carbonyl compound and sodium or potassium cyanide. This procedure ensures that free cyanide ion is always present in amounts sufficient to increase the rate of the reaction.

Cyanohydrin formation is reversible, and the position of equilibrium depends on the steric and electronic factors governing nucleophilic addition to carbonyl groups described in the preceding section.

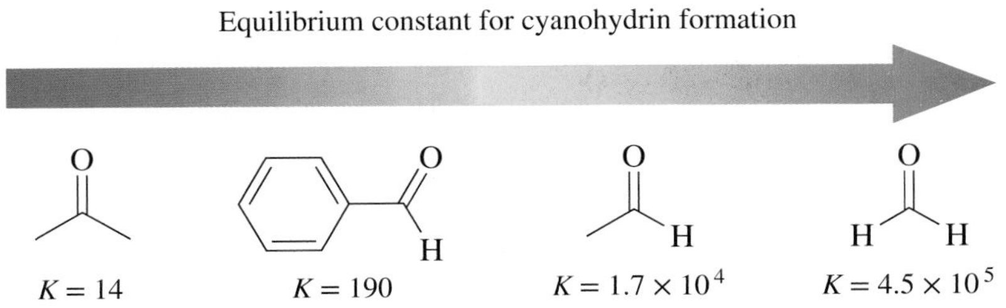

Equilibrium constant for cyanohydrin formation

$K = 14$ $K = 190$ $K = 1.7 \times 10^4$ $K = 4.5 \times 10^5$

Stabilization of the carbonyl group decreases the equilibrium constant for formation of the cyanohydrin. K is greatest for formaldehyde, which has the least stabilized carbonyl, and greater for aldehydes than ketones. Conjugation of the carbonyl and phenyl substituent in benzaldehyde is stabilizing and decreases the value of *K* relative to acetaldehyde.

At the preparative level, aldehydes and unhindered ketones give good yields of cyanohydrins.

In substitutive IUPAC nomenclature, cyanohydrins are named as hydroxy derivatives of nitriles. Because nitrile nomenclature will not be discussed until Section 20.1, we will refer to cyanohydrins as derivatives of the parent aldehyde or ketone as shown in the examples. This conforms to the practice of most chemists.

2,4-Dichlorobenzaldehyde $\xrightarrow[\text{then HCl}]{\text{NaCN, diethyl ether–water}}$ 2,4-Dichlorobenzaldehyde cyanohydrin (100%)

Acetone $\xrightarrow[\text{then } H_2SO_4]{\text{NaCN, } H_2O}$ Acetone cyanohydrin (77–78%)

Problem 18.7

Cyanohydrin formation is reversible in base. Using sodium hydroxide as the base, use curved arrows to show the elimination of HCN from the cyanohydrin product in the presence of sodium hydroxide in step 2 in Mechanism 18.3.

Converting aldehydes and ketones to cyanohydrins is of synthetic value because:

1. A new carbon–carbon bond is formed.
2. The —C≡N group can be converted to —COH (with C=O) (Section 19.12) and —CH_2NH_2 (Section 22.9).
3. The —OH group can undergo functional-group transformations.

Problem 18.8

Methacrylonitrile is an industrial chemical used in the production of plastics and fibers. One method for its preparation is the acid-catalyzed dehydration of acetone cyanohydrin. Deduce the structure of methacrylonitrile.

Cyanohydrins occur naturally, often as derivatives in which the —OH group has been modified to —OR, where R is a carbohydrate unit. These *cyanogenic glycosides* are widespread in plants; one, called *amygdalin,* is found in bitter almonds and in the kernels of peaches, plums, apricots, and related fruits.

Amygdalin

Apricot pits are the most common source of amygdalin.

©lynx/iconotec/Glow Images

Enzyme-catalyzed hydrolysis of amygdalin gives the carbohydrate gentiobiose along with benzaldehyde cyanohydrin, which dissociates to benzaldehyde and hydrogen cyanide.

Gentiobiose

Benzaldehyde cyanohydrin

Depending on the amount of amygdalin present and the manner in which food is prepared from plants containing cyanogenic glycosides, toxic levels of hydrogen cyanide can result.

Problem 18.9

Gynocardin is a naturally occurring cyanogenic glycoside having the structure shown. What cyanohydrin would you expect to be formed on hydrolysis of gynocardin, and to what ketone does this cyanohydrin correspond?

Gynocardin

Cyanogenic compounds are not limited to plants. The defense secretion of many species of millipedes contains the products of cyanohydrin dissociation. These millipedes (Figure 18.6) store either benzaldehyde cyanohydrin or a derivative of it, plus the enzyme that catalyzes its hydrolysis in separate chambers within their bodies. When the millipede is under stress, the contents of the two chambers are mixed and the hydrolysis products—including HCN—are released through the millipede's pores to deter predatory insects and birds.

Figure 18.6

When disturbed, many millipedes protect themselves by converting stored benzaldehyde cyanohydrin to hydrogen cyanide and benzaldehyde.

18.8 Reaction with Alcohols: Acetals and Ketals

Many of the most interesting and useful reactions of aldehydes and ketones involve transformation of the initial product of nucleophilic addition to some other substance under the reaction conditions. An example is the acid-catalyzed addition of alcohols to aldehydes.The expected product, a **hemiacetal,** is not usually isolable, but reacts with an additional mole of the alcohol to give an **acetal.**

$$\text{RCHO} \xrightleftharpoons{R'OH,\ H^+} \text{RCH(OH)OR'} \xrightleftharpoons{R'OH,\ H^+} \text{RCH(OR')}_2 + H_2O$$

Aldehyde — Hemiacetal — Acetal — Water

$$C_6H_5CHO + 2CH_3CH_2OH \xrightleftharpoons{HCl} C_6H_5CH(OCH_2CH_3)_2 + H_2O$$

Benzaldehyde — Ethanol — Benzaldehyde diethyl acetal (66%) — Water

Mechanism 18.4 for formation of benzaldehyde diethyl acetal encompasses two stages. Nucleophilic addition to the carbonyl group characterizes the first stage (steps 1–3), carbocation chemistry the second (steps 4–7). The key carbocation intermediate is stabilized by electron release from oxygen.

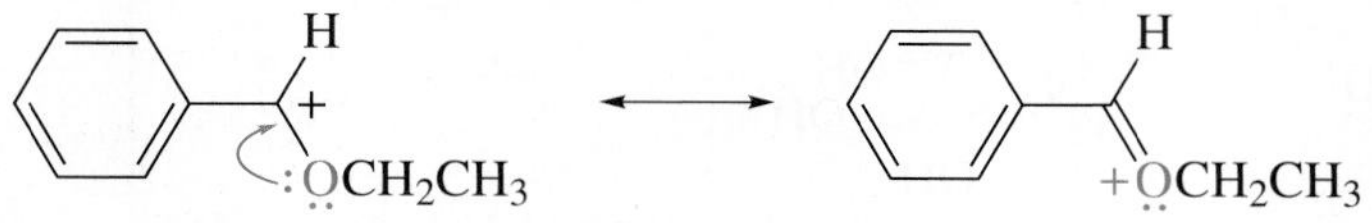

H, +, :ÖCH$_2$CH$_3$ ⟷ H, +ÖCH$_2$CH$_3$

A particularly stable resonance contributor: satisfies the octet rule for both carbon and oxygen

Problem 18.10

Be sure you fully understand Mechanism 18.4 by writing equations for steps 1–3. Use curved arrows to show electron flow.

The position of equilibrium is favorable for acetal formation from most aldehydes, especially when excess alcohol is present as the reaction solvent. For most ketones the position of equilibrium is unfavorable, and other methods must be used for the preparation of acetals from ketones.

Ketal is an acceptable term for acetals formed from ketones. It was once dropped from IUPAC nomenclature, but continued to be so widely used that it was reinstated.

Compounds that contain both carbonyl and alcohol functional groups are often more stable as cyclic hemiacetals or hemiketals than as open-chain structures. An equilibrium mixture of 4-hydroxybutanal contains 11.4% of the open-chain hydroxy aldehyde and 88.6% of the cyclic hemiacetal.

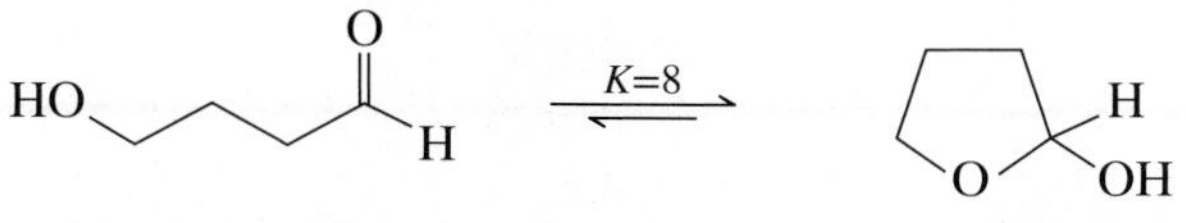

$K=8$

4-Hydroxybutanal ⇌ Tetrahydrofuran-2-ol

Similarly, the carbohydrate D-fructose, which contains a ketone carbonyl, exists almost entirely as a mixture of several cyclic hemiketals, one of which is shown in the equation.

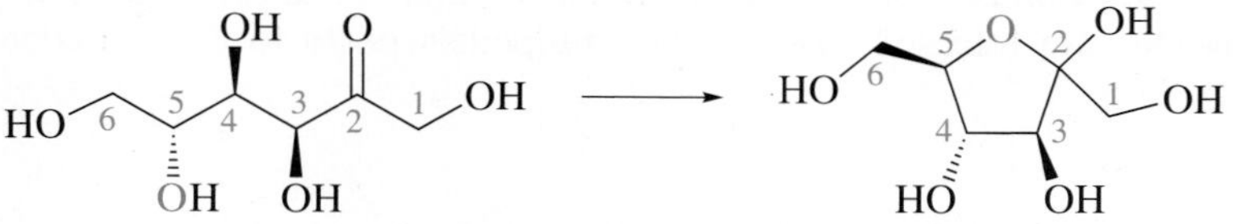

D-Fructose

Diols that bear two hydroxyl groups in a 1,2 or 1,3 relationship to each other yield *cyclic acetals and ketals* with aldehydes and ketones. The five-membered cyclic acetals derived from ethylene glycol are the most commonly encountered examples. Often the position of equilibrium is made more favorable by removing the water formed in the reaction by azeotropic distillation with benzene or toluene:

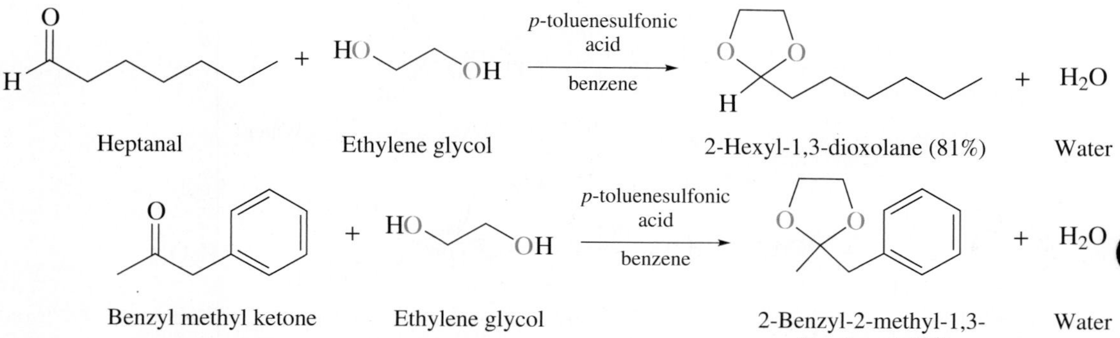

Heptanal + Ethylene glycol $\xrightarrow[\text{benzene}]{p\text{-toluenesulfonic acid}}$ 2-Hexyl-1,3-dioxolane (81%) + H_2O Water

Benzyl methyl ketone + Ethylene glycol $\xrightarrow[\text{benzene}]{p\text{-toluenesulfonic acid}}$ 2-Benzyl-2-methyl-1,3-dioxolane (78%) + H_2O Water

Mechanism 18.4

Acetal Formation from Benzaldehyde and Ethanol

THE OVERALL REACTION:

Benzaldehyde + $2CH_3CH_2OH$ (Ethanol) $\xrightleftharpoons{HCl}$ Benzaldehyde diethyl acetal + H_2O (Water)

THE MECHANISM:

Steps 1–3: Acid-catalyzed nucleophilic addition of 1 mole of ethanol to the carbonyl group. The details of these three steps are analogous to the three steps of acid-catalyzed hydration in Mechanism 18.2. The product of these three steps is a hemiacetal.

Benzaldehyde + CH_3CH_2OH (Ethanol) $\xrightleftharpoons{HCl}$ Benzaldehyde ethyl hemiacetal

Step 4: Steps 4 and 5 are analogous to the two steps in the formation of carbocations in acid-catalyzed reactions of alcohols. Step 4 is proton transfer to the hydroxyl oxygen of the hemiacetal.

Benzaldehyde ethyl hemiacetal + Ethyloxonium ion $\xrightleftharpoons{fast}$ Conjugate acid of benzaldehyde ethyl hemiacetal + Ethanol

Step 5: Loss of water from the protonated hemiacetal gives an oxygen-stabilized carbocation. Of the resonance structures shown, the more stable contributor satisfies the octet rule for both carbon and oxygen.

Conjugate acid of benzaldehyde ethyl hemiacetal $\xrightleftharpoons[H_2O]{-H_2O}$ More stable contributor ⟷ Less stable contributor

Step 6: Nucleophilic addition of ethanol to the oxygen-stabilized carbocation

Oxygen-stabilized carbocation + Ethanol ⇌ Conjugate acid of benzaldehyde diethyl acetal

Step 7: Proton transfer from the conjugate acid of the product to ethanol

Conjugate acid of benzaldehyde diethyl acetal + Ethanol ⇌ Benzaldehyde diethyl acetal + Conjugate acid of ethanol

Problem 18.11

Write the structures of the cyclic acetal or ketal derived from each of the following:

(a) Cyclohexanone and ethylene glycol
(b) Benzaldehyde and 1,3-propanediol
(c) Isobutyl methyl ketone and ethylene glycol
(d) Isobutyl methyl ketone and 2,2-dimethyl-1,3-propanediol

Sample Solution (a) The cyclic acetals derived from ethylene glycol contain a five-membered 1,3-dioxolane ring.

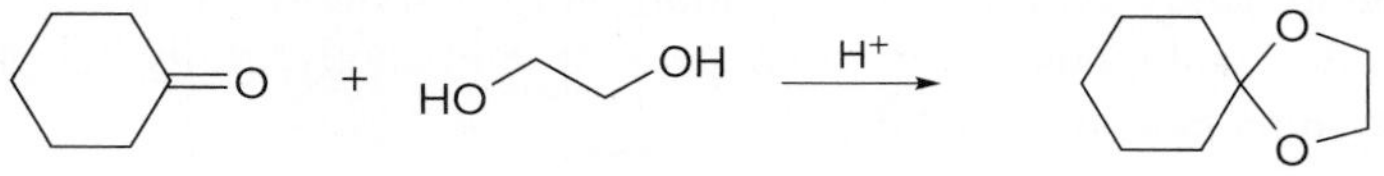

Cyclohexanone + Ethylene glycol $\xrightarrow{H^+}$ Acetal of cyclohexanone and ethylene glycol

Acetals and ketals are susceptible to hydrolysis in aqueous acid:

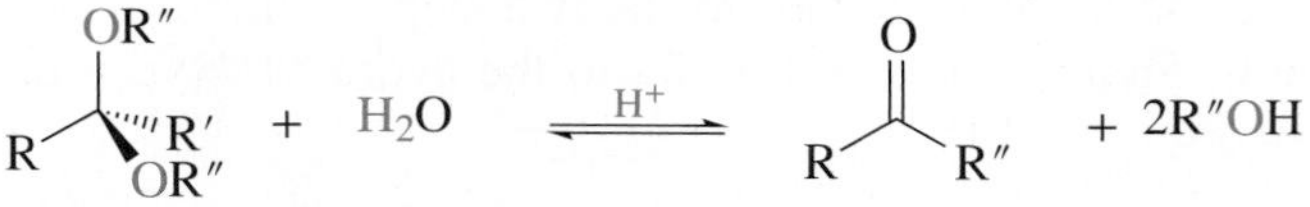

Acetal + H_2O (Water) $\underset{}{\overset{H^+}{\rightleftharpoons}}$ Aldehyde or ketone + $2R''OH$ (Alcohol)

This reaction is simply the reverse of the reaction by which acetals are formed—acetal or ketal formation is favored by excess alcohol, hydrolysis by excess water. The two reactions share the same mechanistic pathway but travel along it in opposite directions. In the following section you'll see how acetal and ketal formation and hydrolysis are applied to synthetic organic chemistry.

Problem 18.12

Problem 18.10 asked you to write details of the mechanism describing formation of benzaldehyde diethyl acetal from benzaldehyde and ethanol. Write a stepwise mechanism for the acid hydrolysis of this acetal.

18.9 Acetals and Ketals as Protecting Groups

In an organic synthesis, it sometimes happens that one of the reactants contains a functional group that is incompatible with the reaction conditions. Consider, for example, the conversion

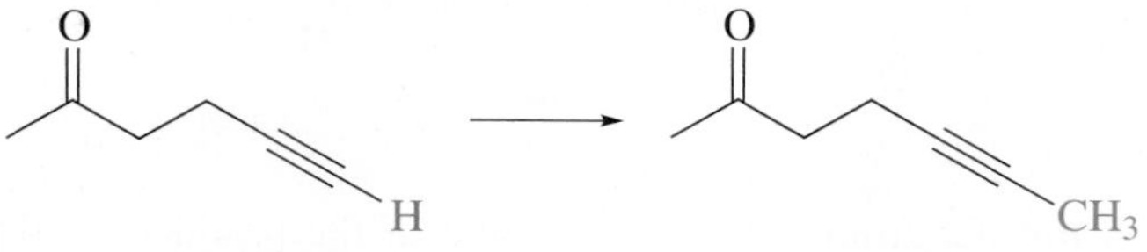

5-Hexyn-2-one ⟶ 5-Heptyn-2-one

It looks as though all that is needed is to prepare the acetylenic anion $CH_3\overset{O}{\overset{\|}{C}}CH_2CH_2C{\equiv}C\!:^-$, then alkylate it with methyl iodide (Section 9.6). There is a complication, however. The carbonyl group in the starting alkyne will neither tolerate the strongly basic conditions required for anion formation nor survive in a solution containing carbanions. Acetylide ions add to carbonyl groups (Section 15.6). Thus, the necessary anion is inaccessible.

The strategy that is routinely followed is to *protect* the carbonyl group during the reactions with which it is incompatible and then to *remove* the protecting group in a subsequent step. Acetals and ketals, especially those derived from ethylene glycol, are the most commonly used groups for carbonyl protection, because they can be introduced and removed readily. They resemble ethers in being inert to many of the reagents, such as hydride reducing agents and organometallic compounds, that react readily with carbonyl groups. The following sequence is the one that was actually used to bring about the desired transformation.

(a) Protection of carbonyl group

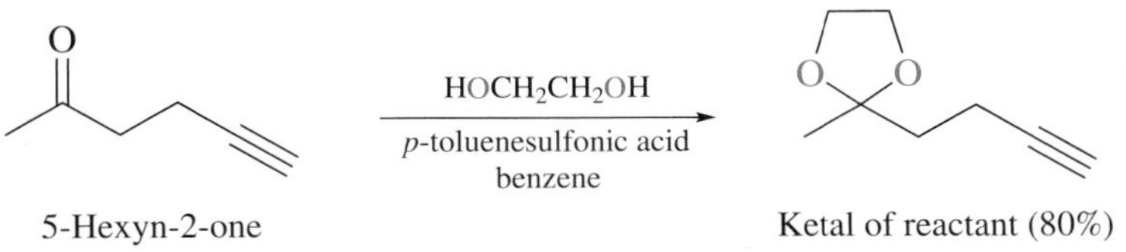

5-Hexyn-2-one $\xrightarrow[\text{benzene}]{\text{HOCH}_2\text{CH}_2\text{OH, } p\text{-toluenesulfonic acid}}$ Ketal of reactant (80%)

(b) Alkylation of alkyne

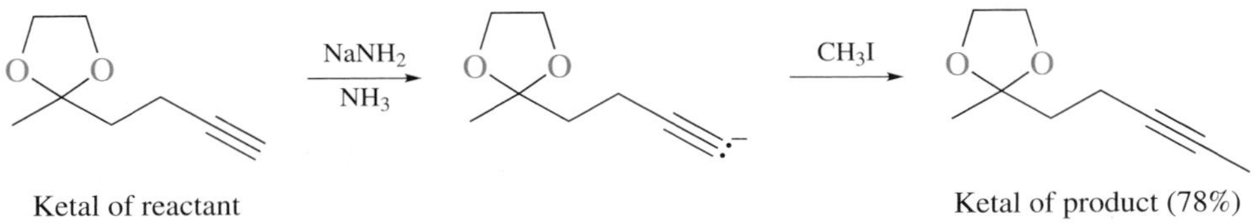

Ketal of reactant $\xrightarrow[\text{NH}_3]{\text{NaNH}_2}$ $\xrightarrow{\text{CH}_3\text{I}}$ Ketal of product (78%)

(c) Removal of the protecting group by hydrolysis

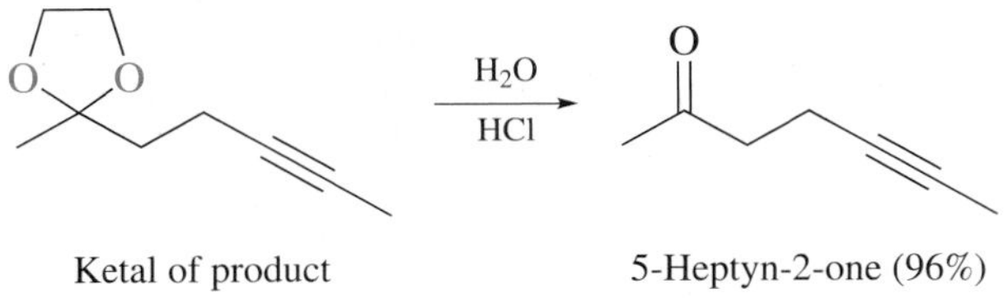

Ketal of product $\xrightarrow[\text{HCl}]{\text{H}_2\text{O}}$ 5-Heptyn-2-one (96%)

Although protecting and deprotecting the carbonyl group adds two steps to the synthetic procedure, both are essential to its success. Functional-group protection is frequently encountered in preparative organic chemistry, and considerable attention has been paid to the design of effective protecting groups for a variety of functionalities.

Problem 18.13

Acetal formation is a characteristic reaction of aldehydes and ketones, but not of carboxylic acids. Use retrosynthetic analysis to show how you could use a cyclic acetal protecting group in the following synthesis, then write equations for the procedure showing the necessary reagents.

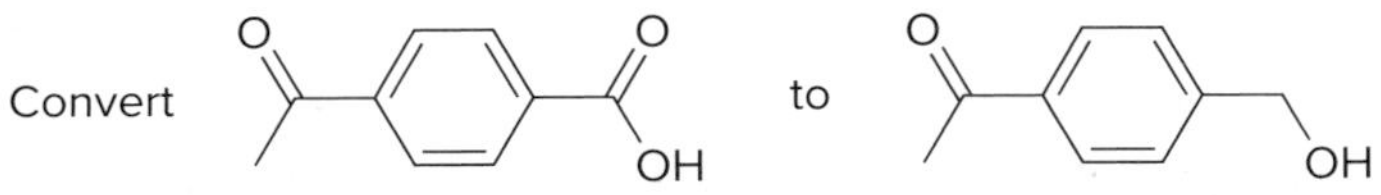

Convert [4-acetylbenzoic acid, OH] to [4-acetylbenzyl alcohol, OH]

18.10 Reaction with Primary Amines: Imines

Like acetal formation, the reaction of aldehydes and ketones with primary amines—compounds of the type RNH_2 and $ArNH_2$—is a two-stage process. Its first stage is nucleophilic addition of the amine to the carbonyl group to give a **hemiaminal.** The

Hemiaminals were formerly known by the now obsolete term *carbinolamine*. Imines are sometimes called **Schiff's bases,** after the nineteenth-century German chemist Hugo Schiff.

second stage is a dehydration and yields an **imine** as the isolated product. Imines from aldehydes are called **aldimines,** those from ketones are **ketimines.**

$$\underset{\text{Aldehyde or ketone}}{RC(=O)R'} + \underset{\text{Primary amine}}{R''\ddot{N}H_2} \xrightleftharpoons{\text{addition}} \underset{\text{Hemiaminal}}{RC(OH)(H\ddot{N}R'')R'} \xrightleftharpoons{\text{elimination}} \underset{N\text{-substituted imine}}{RC(=\ddot{N}R'')R'} + \underset{\text{Water}}{H_2O}$$

$$\underset{\text{Benzaldehyde}}{C_6H_5CHO} + \underset{\text{Methylamine}}{CH_3NH_2} \longrightarrow \underset{N\text{-Benzylidenemethylamine (70\%)}}{C_6H_5CH{=}NCH_3} + \underset{\text{Water}}{H_2O}$$

$$\underset{\text{Cyclohexanone}}{C_6H_{10}{=}O} + \underset{\text{Isobutylamine}}{(CH_3)_2CHCH_2NH_2} \longrightarrow \underset{N\text{-Cyclohexylideneisobutylamine (79\%)}}{C_6H_{10}{=}NCH_2CH(CH_3)_2} + \underset{\text{Water}}{H_2O}$$

Mechanism 18.5 describes the reaction between benzaldehyde and methylamine given in the first example. The first two steps lead to the hemiaminal; the last three show its dehydration to the imine. Step 4, the key step in the dehydration phase, is rate-determining when the reaction is carried out in acid solution. If the solution is too acidic, however, protonation of the amine blocks step 1. Therefore, there is some optimum pH, usually about 5, at which the reaction rate is a maximum. Too basic a solution reduces the rate of step 4; too acidic a solution reduces the rate of step 1.

Imine formation is reversible and can be driven to completion by removing the water that forms. Imines revert to the aldehyde or ketone and amine in the presence of aqueous acid.

Problem 18.14

Write the structure of the aminal or hemiaminal intermediate and the imine product formed in the reaction of each of the following:

(a) Acetaldehyde and benzylamine, $C_6H_5CH_2NH_2$

(b) Benzaldehyde and butylamine, $CH_3CH_2CH_2CH_2NH_2$

(c) Cyclohexanone and *tert*-butylamine, $(CH_3)_3CNH_2$

(d) Acetophenone and cyclohexylamine, $C_6H_{11}NH_2$

Sample Solution (a) A hemiaminal is formed by nucleophilic addition of the amine to the carbonyl group. Its dehydration gives the imine product.

$$\underset{\text{Acetaldehyde}}{CH_3CHO} + \underset{\text{Benzylamine}}{C_6H_5CH_2NH_2} \longrightarrow \underset{\text{Hemiaminal intermediate}}{CH_3CH(OH)NHCH_2C_6H_5} \xrightarrow{-H_2O} \underset{\text{Imine product (}N\text{-ethylidenebenzylamine)}}{CH_3CH{=}NCH_2C_6H_5}$$

A number of compounds of the general type H_2NZ react with aldehydes and ketones in a manner analogous to that of primary amines to form products that are more stable

Mechanism 18.5

Imine Formation from Benzaldehyde and Methylamine

THE OVERALL REACTION:

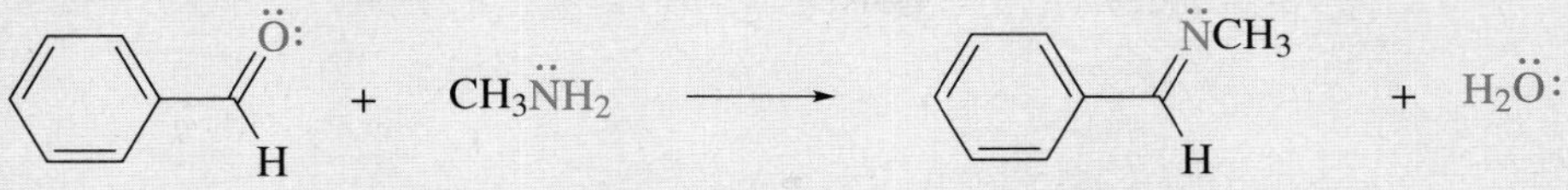

Benzaldehyde + CH_3NH_2 ⟶ N-Benzylidenemethylamine + H_2O

Benzaldehyde Methylamine *N*-Benzylidenemethylamine Water

THE MECHANISM:

Step 1: The amine acts as a nucleophile, adding to the carbonyl group and forming a C—N bond.

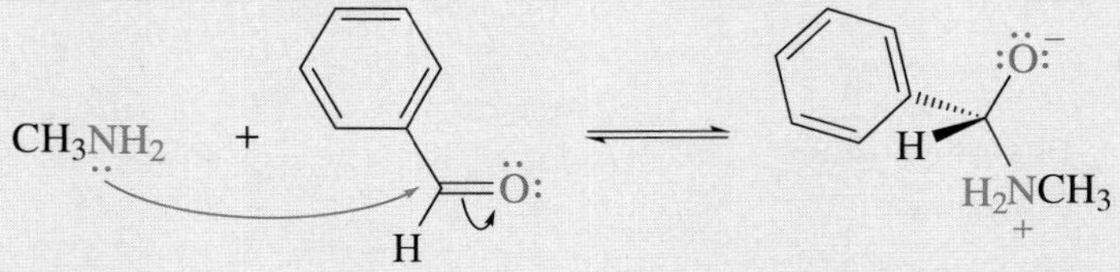

Methylamine Benzaldehyde Dipolar intermediate

Step 2: In a solvent such as water, proton transfers give the hemiaminal.

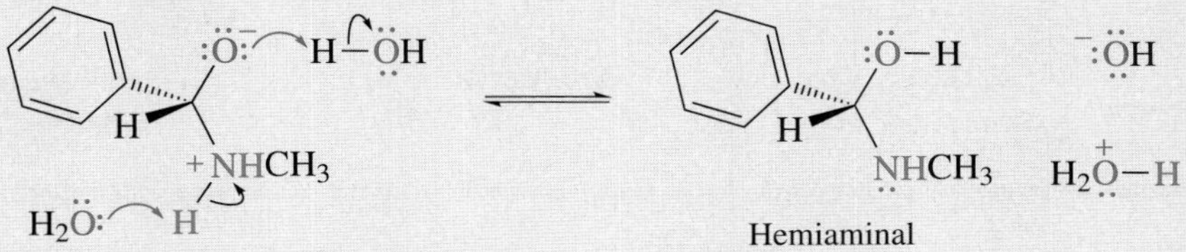

Hemiaminal

Step 3: The dehydration stage begins with protonation of the hemiaminal on oxygen.

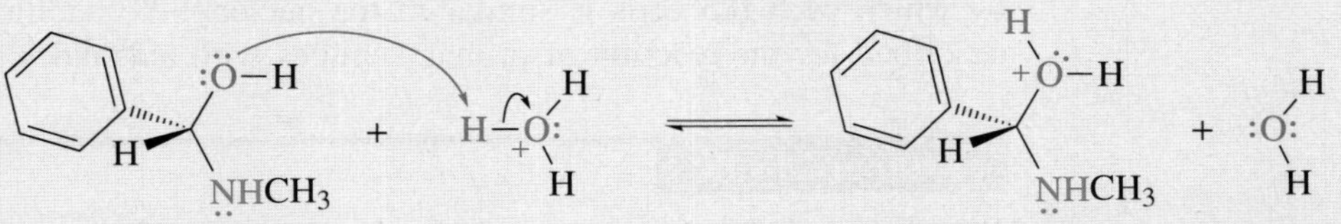

Hemiaminal Hydronium ion *O*-Protonated hemiaminal Water

Step 4: The oxygen-protonated hemiaminal loses water to give a nitrogen-stabilized carbocation. This step is rate-determining at pH = 5.

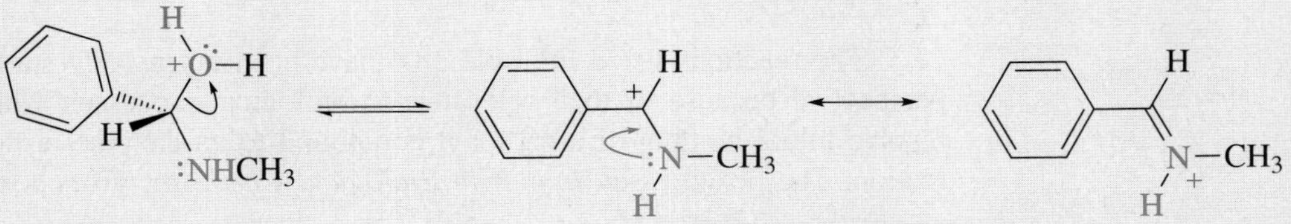

O-Protonated hemiaminal Nitrogen-stabilized carbocation

Step 5: The nitrogen-stabilized carbocation is the conjugate acid of the imine. Proton transfer to water gives the imine.

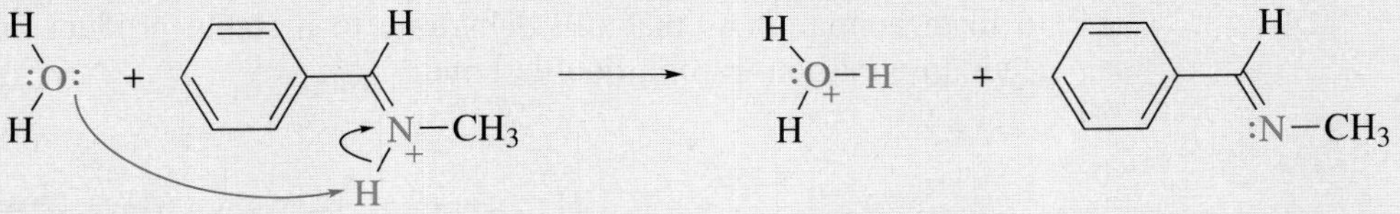

Water Nitrogen-stabilized carbocation Hydronium ion *N*-Benzylidenemethylamine

TABLE 18.4 Reactions of Aldehydes and Ketones with Derivatives of Ammonia

Reagent	Type of product	Specific example of the general reaction: $RC(=O)R' + H_2NZ \longrightarrow RC(=NZ)R' + H_2O$
H_2NOH Hydroxylamine	Oxime	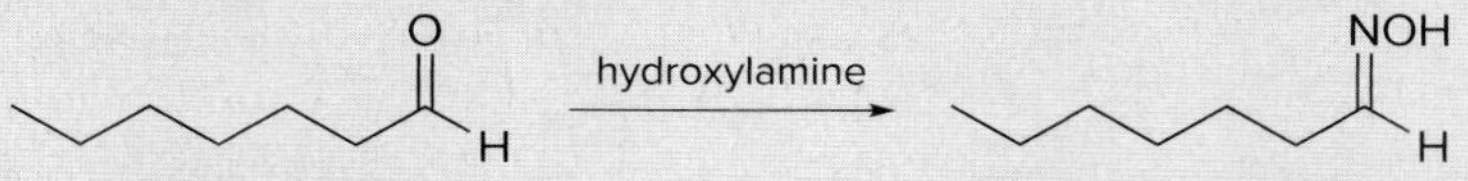 Heptanal —hydroxylamine→ Heptanal oxime (81–93%)
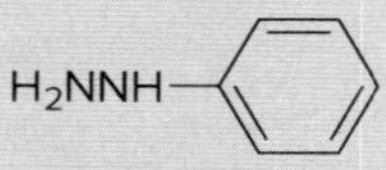 $H_2NNH-C_6H_5$ Phenylhydrazine*	Phenylhydrazone	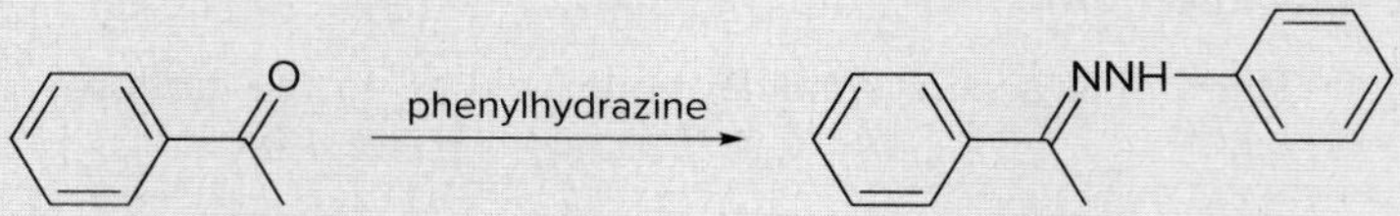 Acetophenone —phenylhydrazine→ Acetophenone phenylhydrazone (87–91%)
$H_2NNHC(=O)NH_2$ Semicarbazide	Semicarbazone	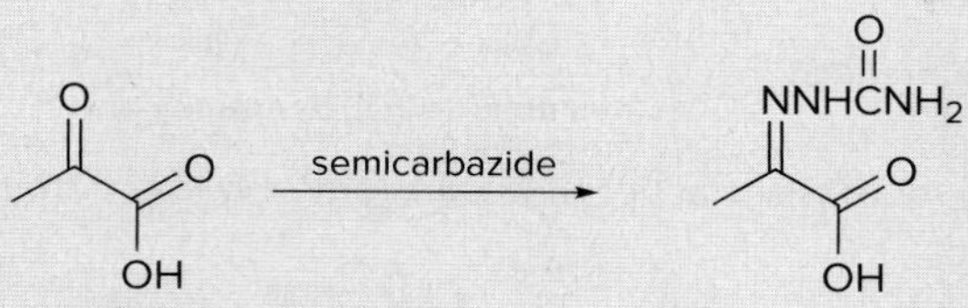 Pyruvic acid —semicarbazide→ Pyruvic acid semicarbazone (94%)

*Compounds related to phenylhydrazine react analogously. *p*-Nitrophenylhydrazine yields *p*-nitrophenylhydrazones; 2,4-dinitrophenylhydrazine yields 2,4-dinitrophenylhydrazones.

than imines. Table 18.4 presents examples of some of these reactions. The mechanism by which each proceeds is similar to the nucleophilic addition–elimination mechanism described for the reaction of primary amines with aldehydes and ketones.

Problem 18.15

The product of the following reaction is a heterocyclic aromatic compound. What is its structure?

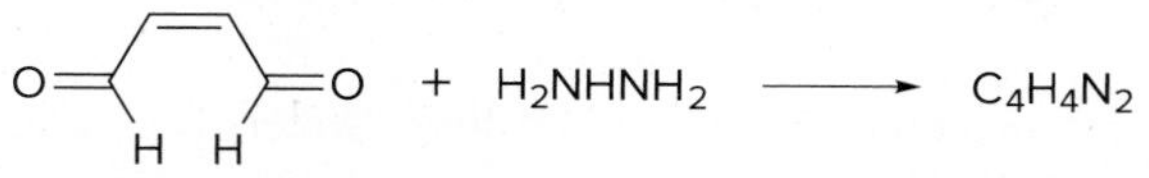

$$O{=}CH{-}CH{=}CH{-}CH{=}O + H_2NNH_2 \longrightarrow C_4H_4N_2$$

The reactions listed in Table 18.4 have been extensively studied from a mechanistic perspective because of their relevance to biological processes. Many biological reactions involve initial binding of a carbonyl compound to an enzyme or coenzyme via imine formation. The boxed essay *Imines in Biological Chemistry* gives some important examples.

18.11 Reaction with Secondary Amines: Enamines

Secondary amines are compounds of the type R_2NH. They add to aldehydes and ketones to form hemiaminals that can dehydrate to a stable product only in the direction that leads to a carbon–carbon double bond:

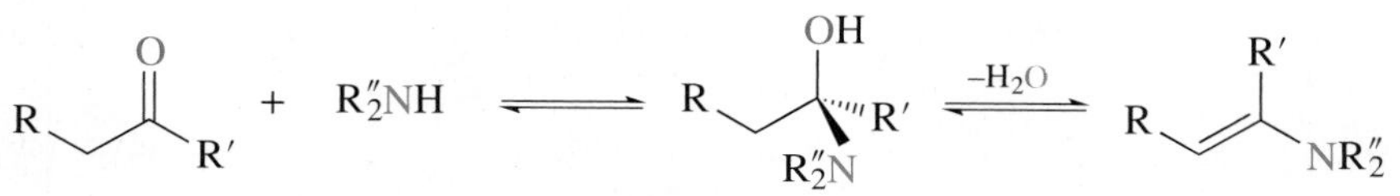

$$RCH_2C(=O)R' + R''_2NH \rightleftharpoons RCH_2C(OH)(R')NR''_2 \overset{-H_2O}{\rightleftharpoons} RCH{=}C(R')NR''_2$$

Aldehyde or ketone; Secondary amine; Hemiaminal; Enamine

Imines in Biological Chemistry

Many biological processes involve an "association" between two species in a step prior to some subsequent transformation. This association can take many forms. It can be a weak association of the attractive van der Waals type, or a stronger interaction such as a hydrogen bond. It can be an electrostatic attraction between a positively charged atom of one molecule and a negatively charged atom of another. Covalent bond formation between two species of complementary chemical reactivity represents an extreme kind of association. It often occurs in biological processes in which aldehydes or ketones react with amines via imine intermediates.

An example of a biologically important aldehyde is *pyridoxal phosphate*, which is the active form of *vitamin B_6* and a coenzyme for many of the reactions of α-amino acids. In these reactions the amino acid binds to the coenzyme by reacting with it to form an imine of the kind shown in the equation. Reactions then take place at the amino acid portion of the imine, modifying the amino acid. In the last step, enzyme-catalyzed hydrolysis cleaves the imine to pyridoxal and the modified amino acid.

In a second example, a key step in the chemistry of vision is binding of an aldehyde to an enzyme via an imine. An outline of the steps involved is presented in Figure 18.7. It starts with *β-carotene*, a pigment that occurs naturally in several fruits and vegetables, including carrots. β-Carotene undergoes oxidative cleavage in the liver to give an alcohol known as *retinol*, or *vitamin A*. Oxidation of vitamin A, followed by isomerization of one of its double bonds, gives the aldehyde 11-*cis*-retinal. In the eye, the aldehyde function of 11-*cis*-retinal combines with an amino group of the protein *opsin* to form an imine called *rhodopsin*. When rhodopsin absorbs a photon of visible light, the cis double bond of the retinal unit undergoes a photochemical cis-to-trans isomerization, which is attended by a dramatic change in its shape and a change in the conformation of rhodopsin. This conformational change is translated into a nerve impulse perceived by the brain as a visual image. Enzyme-promoted hydrolysis of the photochemically isomerized rhodopsin regenerates opsin and a molecule of all-*trans*-retinal. Once all-*trans*-retinal has been enzymatically converted to its 11-cis isomer, it and opsin reenter the cycle.

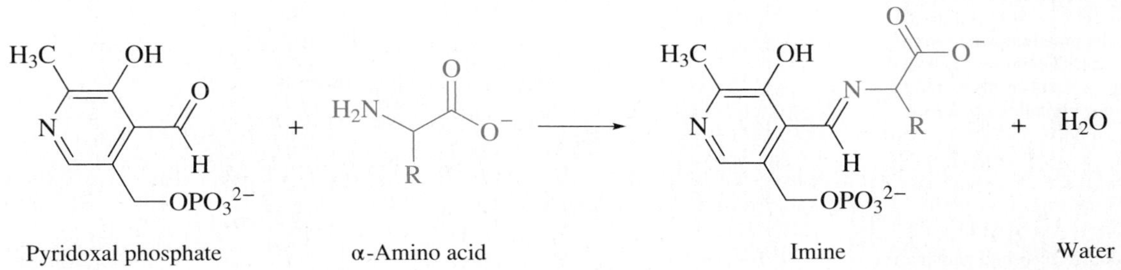

H_3C OH O N H OPO_3^{2-} + H_2N O O^- R ⟶ H_3C OH O N O^- N R H OPO_3^{2-} + H_2O

Pyridoxal phosphate α-Amino acid Imine Water

Problem 18.16

Not all biological reactions of amino acids involving imine intermediates require pyridoxal phosphate. The first step in the conversion of proline to glutamic acid is an oxidation giving the imine shown. Once formed, this imine undergoes hydrolysis to a species having the molecular formula $C_5H_9NO_3$, which then goes on to produce glutamic acid. Suggest a structure for the $C_5H_9NO_3$ species. (*Hint:* There are two reasonable possibilities; one is a hemiaminal, the other is not cyclic.)

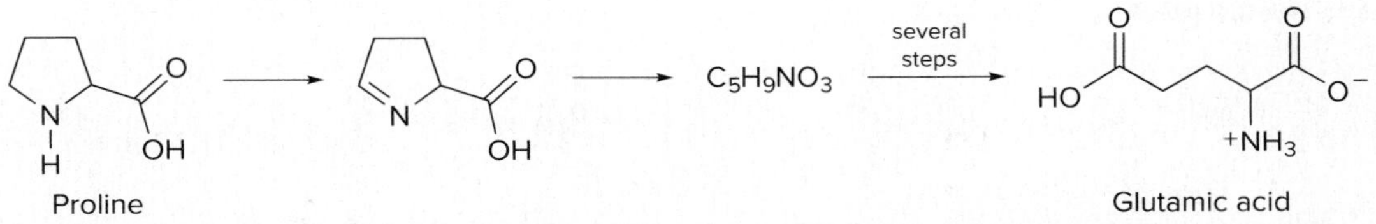

N H O OH ⟶ N O OH ⟶ $C_5H_9NO_3$ $\xrightarrow{\text{several steps}}$ HO O O O^- $^+NH_3$

Proline Glutamic acid

continued

β-Carotene obtained from the diet is cleaved at its central carbon–carbon bond to give vitamin A (retinol).

OH

Oxidation of retinol converts it to the corresponding aldehyde, retinal.

H O

The double bond at C-11 is isomerized from the trans to the cis configuration.

H O

11-*cis*-Retinal is the biologically active stereoisomer and reacts with the protein opsin to form an imine. The covalently bound complex between 11-*cis*-retinal and opsin is called *rhodopsin*.

H_2N-protein

H N-protein

Rhodopsin absorbs a photon of light, causing the cis double-bond at C-11 to undergo a photochemical transformation to trans, which triggers a nerve impulse detected by the brain as a visual image.

$h\nu$

H N-protein

Hydrolysis of the isomerized (inactive) form of rhodopsin liberates opsin and the all-trans isomer of retinal.

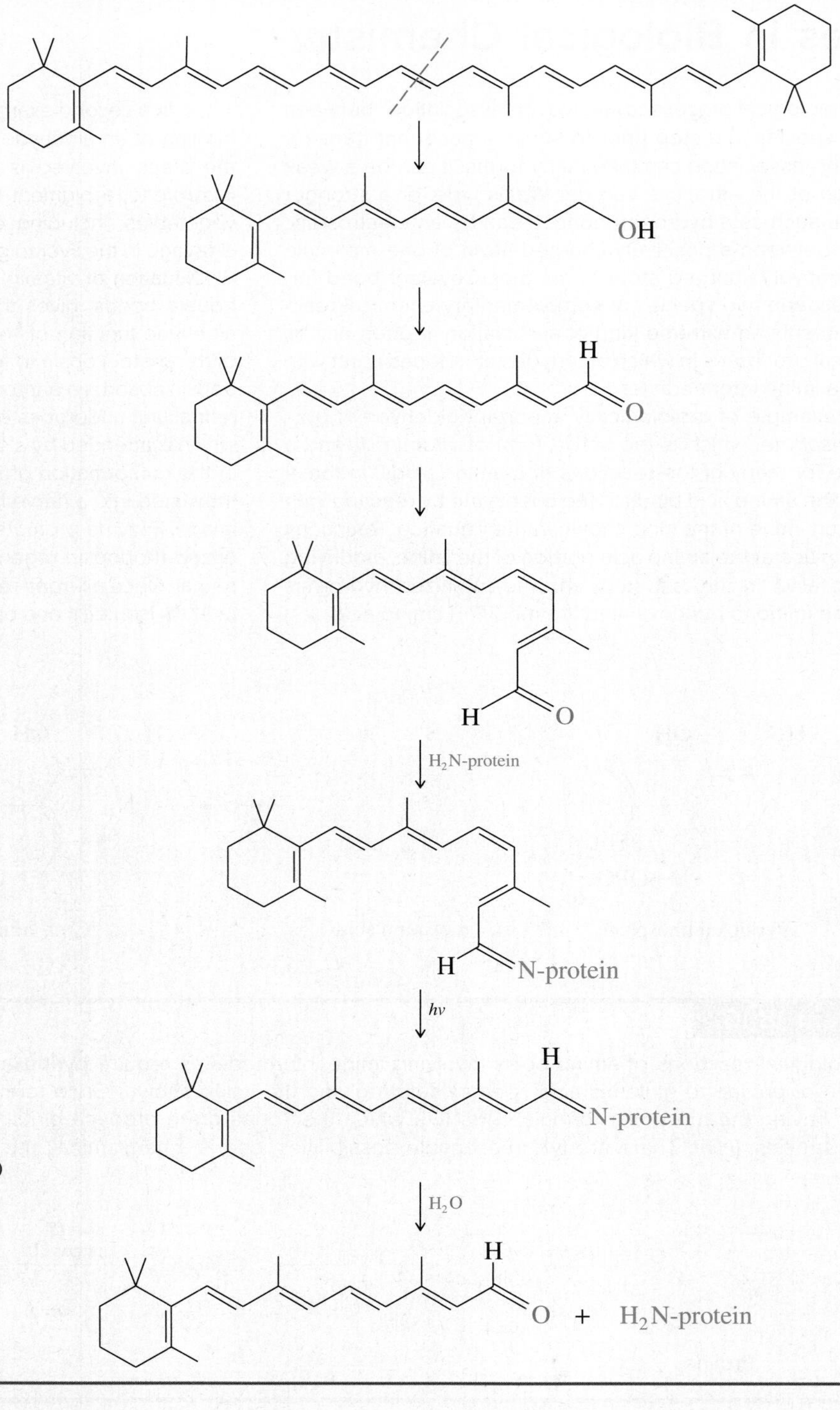

H_2O

H O + H_2N-protein

Figure 18.7

Imine formation between the aldehyde function of 11-*cis*-retinal and an amino group of a protein (opsin) is involved in the chemistry of vision. The numbering scheme in retinal was specifically developed for carotenes and related compounds.

The product is an alkenyl-substituted amine, or **enamine.**

Cyclopentanone + Pyrrolidine $\xrightarrow[\text{heat}]{\text{benzene}}$ *N*-(1-Cyclopentenyl)-pyrrolidine (80–90%) + Water (H_2O)

Mechanism 18.6 outlines the mechanism of the corresponding reaction of pyrrolidine and 2-methylpropanal.

Mechanism 18.6

Enamine Formation

THE OVERALL REACTION:

Pyrrolidine + 2-Methylpropanal $\xrightarrow[\text{heat}]{\text{benzene}}$ 1-(2-Methylpropenyl)-pyrrolidine (94–95%) + Water (H_2O)

THE MECHANISM:

Step 1: Nucleophilic addition of pyrrolidine to 2-methylpropanal gives a protonated hemiaminal. The mechanism is analogous to the addition of primary amines to aldehydes and ketones (Mechanism 18.5).

Pyrrolidine + 2-Methylpropanal ⇌ Protonated hemiaminal intermediate

Step 2: With an assist from the nitrogen lone pair, the protonated hemiaminal expels water to form an iminium ion.

Protonated hemiaminal intermediate ⇌ Iminium ion + Water

Step 3: The iminium ion is then deprotonated in the direction that gives a carbon–carbon double bond.

Iminium ion + Water ⇌ 1-(2-Methylpropenyl)-pyrrolidine + Hydronium ion

Problem 18.17

Write the structure of the hemiaminal intermediate and the enamine product formed in the reaction of each of the following.

(a) Propanal and dimethylamine
(b) 3-Pentanone and pyrrolidine
(c) Acetophenone and [piperidine ring] NH

Sample Solution (a) Nucleophilic addition of dimethylamine to the carbonyl group of propanal gives a hemiaminal that undergoes dehydration to form an enamine.

Propanal + $HN(CH_3)_2$ (Dimethylamine) ⟶ Hemiaminal intermediate $\xrightarrow{-H_2O}$ *N*-(1-Propenyl)-dimethylamine

Enamines are mainly used as reagents for making carbon–carbon bonds, some applications of which are illustrated in the Descriptive Passage and Interpretive Problems accompanying Chapter 22.

18.12 The Wittig Reaction

The reaction is named after Georg Wittig, a German chemist who shared the 1979 Nobel Prize in Chemistry for demonstrating its synthetic potential. Wittig shared the prize with H. C. Brown, who was recognized for developing hydroboration as a synthetic tool.

Wittig reactions, and reactions related to it, are used for the regiospecific synthesis of alkenes from aldehydes and ketones. Their retrosynthetic analysis begins with disconnecting the double bond as shown and introduces a novel structural type called an **ylide.**

$$RR'C{=}CAB \Longrightarrow RR'C{=}O + (C_6H_5)_3\overset{+}{P}{-}\overset{-}{C}AB$$

Alkene target — Aldehyde or ketone — A triphenylphosphonium ylide

An ylide is a neutral molecule having a contributing structure in which two oppositely charged atoms, each with an octet of electrons, are directly bonded to each other. In **Wittig reagents**—ylides of the type shown—the positively charged atom is phosphorus and the negatively charged one is carbon. Most Wittig reagents have three phenyl groups attached to phosphorus and are commonly written as either of two resonance contributors.

$$(C_6H_5)_3\overset{+}{P}{-}\overset{-}{C}AB \longleftrightarrow (C_6H_5)_3P{=}CAB$$

Although second-row elements such as phosphorus can accommodate more than 8 electrons in their valence shell, the dipolar structure is believed to be the major contributing structure for Wittig reagents.

The electrostatic potential map of a very simple ylide, one in which all the atoms other than phosphorus and carbon are hydrogen, is shown in Figure 18.8. The electron distribution is highly polarized in the direction that makes carbon nucleophilic.

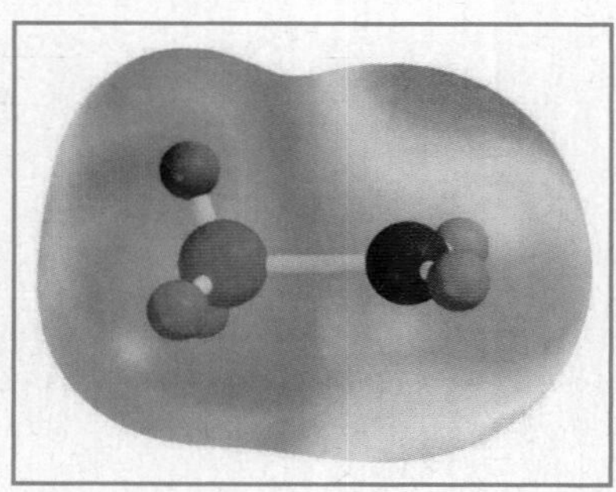

Figure 18.8

An electrostatic potential map of the ylide $H_3\overset{+}{P}{-}\overset{-}{C}H_2$. The region of greatest negative charge is concentrated at carbon.

Ylides are prepared by a two-step procedure. First, an alkyl halide is treated with a phosphine—typically triphenylphosphine—to give a phosphonium salt. The alkyl halide can be methyl, primary, or secondary.

$$(C_6H_5)_3P{:} + RCHR{-}X \xrightarrow{S_N2} (C_6H_5)_3\overset{+}{P}{-}CHR_2 \quad :\ddot{X}:^-$$

Triphenylphosphine — Alkyl halide — An alkyltriphenylphosphonium halide

The phosphonium salt is isolated, then converted to an ylide by an acid–base reaction, using a base such as butyllithium or potassium *tert*-butoxide. The conjugate base of dimethyl sulfoxide in dimethyl sulfoxide as the solvent can also be used, as illustrated in the following example.

The conjugate base of dimethyl sulfoxide is about 10^{20} times more basic than hydroxide.

$$(C_6H_5)_3\overset{+}{P}-CHR_2 + H_2\overset{-}{C}-S(=O)-CH_3 \longrightarrow (C_6H_5)_3\overset{+}{P}-\overset{-}{C}R_2 + H_3C-S(=O)-CH_3$$

An alkyltriphenylphosphonium ion | Conjugate base of dimethyl sulfoxide | Ylide | Dimethyl sulfoxide

Adding an aldehyde or ketone gives the alkene, along with triphenylphospine oxide as a coproduct. The P—O bond strength of the triphenylphosphine oxide coproduct is estimated to be greater than 540 kJ/mol (130 kcal/mol) and contributes to the reaction proceeding in the direction shown.

$$\text{Cyclohexanone} + (C_6H_5)_3\overset{+}{P}-\overset{-}{C}H_2 \xrightarrow{\text{DMSO}} \text{Methylenecyclohexane (86\%)} + (C_6H_5)_3\overset{+}{P}-\overset{-}{O}:$$

Cyclohexanone | Methylenetriphenylphosphorane | Methylenecyclohexane (86%) | Triphenylphosphine oxide

Alternatively, triphenylphosphonium salts may be converted to ylides using an alkyllithium reagent as the base and tetrahydrofuran as the solvent.

Problem 18.18

What other combination of ylide and aldehyde or ketone will give methylenecyclohexane by a Wittig reaction? Write a balanced equation for the reaction.

Problem 18.19

Identify the alkene formed in each of the following reactions:

(a) Benzaldehyde + $(C_6H_5)_3\overset{+}{P}-\overset{-}{C}$(cyclopentylidene) ⟶

(b) Butanal + $(C_6H_5)_3\overset{+}{P}-\overset{-}{C}HCH=CH_2$ ⟶

(c) Cyclohexyl methyl ketone + $(C_6H_5)_3\overset{+}{P}-\overset{-}{C}H_2$ ⟶

Sample Solution (a) In a Wittig reaction the negatively charged substituent on phosphorus is transferred to the aldehyde or ketone, replacing the carbonyl oxygen. Reaction (a) has been used to prepare the indicated alkene in 65% yield.

$$C_6H_5CHO + (C_6H_5)_3\overset{+}{P}-\overset{-}{C}(\text{cyclopentylidene}) \longrightarrow C_6H_5CH{=}(\text{cyclopentane})$$

Benzaldehyde | Cyclopentylidenetriphenylphosphorane | Benzylidenecyclopentane (65%)

Problem 18.20

Write equations outlining two different syntheses of 3-methyl-3-heptene using 1-butanol and 2-butanol as the source of all of the carbons.

3-methyl-3-heptene ⟹ 1-butanol (OH) and 2-butanol (OH)

Extensive mechanistic studies of the Wittig reaction have led to general agreement that a four-membered ring called an oxaphosphetane is an intermediate.

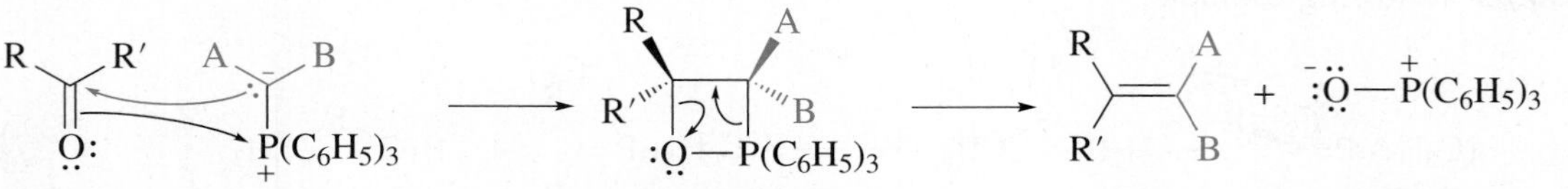

Aldehyde or ketone Ylide Oxaphosphetane Alkene Triphenylphosphine oxide

Less certain is whether the oxaphosphetane is formed from the carbonyl compound and the ylide in a single step or a two-step process. Problem 18.21 explores both possibilities.

Problem 18.21

(a) The product expected from nucleophilic addition of an ylide to an aldehyde or ketone belongs to a class of substances called *betaines*. Like ylides, betaines contain a positively charged and a negatively charged atom and both have an octet of electrons; they differ from ylides in that the two charged atoms are nonadjacent. Write a structural formula for the betaine corresponding to nucleophilic addition of methylenetriphenylphosphorane to cyclohexanone.

(b) Use curved arrows to show the conversion of the betaine in (a) to an oxaphosphetane.

(c) Use curved arrows to show the one-step conversion of the betaine in (a) to methylenecyclohexane and triphenylphosphine oxide.

Sample Solution (a) Nucleophilic addition of the ylide to the carbonyl group leads to C—C bond formation.

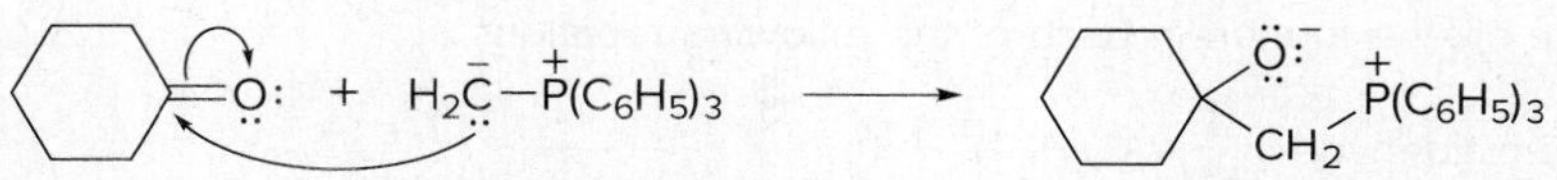

Cyclohexanone Methylenetriphenylphosphorane Betaine intermediate

The stereoselectivity of the Wittig reaction is variable. Simple ylides give a mixture of stereoisomers in which the *Z*-alkene predominates, whereas ylides of the type $(C_6H_5)_3P{=}CHX$, where X is a strongly electron-withdrawing substituent such as —C=O or —C≡N, give mainly the *E*-alkene.

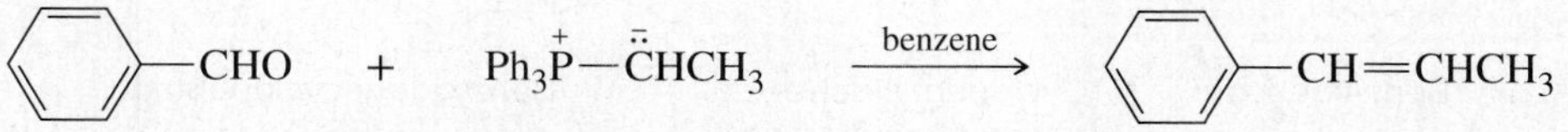

$$C_6H_5CHO + Ph_3\overset{+}{P}-\overset{-}{C}HCH_3 \xrightarrow{\text{benzene}} C_6H_5CH{=}CHCH_3$$

Benzaldehyde Ethylidenetriphenylphosphorane Prop-1-enylbenzene (98% yield; 87% *Z*, 13% *E*)

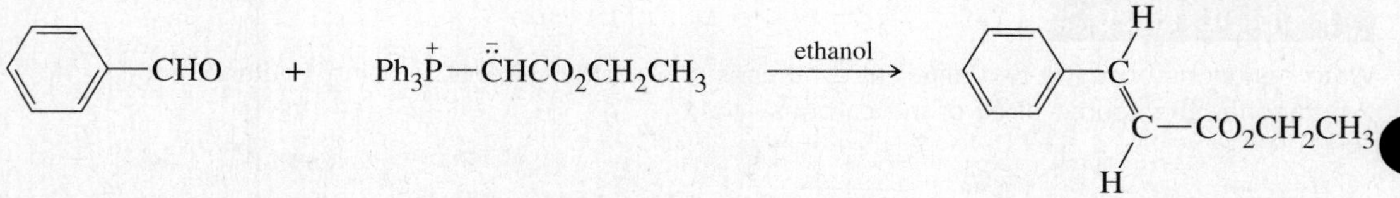

$$C_6H_5CHO + Ph_3\overset{+}{P}-\overset{-}{C}HCO_2CH_2CH_3 \xrightarrow{\text{ethanol}} C_6H_5CH{=}CHCO_2CH_2CH_3$$

Benzaldehyde (Carboethoxymethylidene)-triphenylphosphorane Ethyl cinnamate (77%, only *E*)

18.13 Stereoselective Addition to Carbonyl Groups

Nucleophilic additions to carbonyl groups are often stereoselective with the direction of addition controlled by steric factors. Typically, the nucleophile approaches the less hindered face of the carbonyl. Sodium borohydride reduction of 7,7-dimethylbicyclo-[2.2.1]heptan-2-one illustrates this point:

$NaBH_4$ / isopropyl alcohol, 0°C

7,7-Dimethylbicyclo-[2.2.1]heptan-2-one → *exo*-7,7-Dimethylbicyclo-[2.2.1]heptan-2-ol (80%) + *endo*-7,7-Dimethylbicyclo-[2.2.1]heptan-2-ol (20%)

Approach of borohydride to the top face of the carbonyl group is sterically hindered by one of the methyl groups. The bottom face of the carbonyl group is less congested, and the major product is formed by hydride transfer from this direction.

Approach of borohydride from this direction is hindered by methyl group.

Preferred direction of approach of borohydride is to less hindered face of carbonyl group. $H_3\bar{B}-H$

Problem 18.22

What is the relationship between the major and minor products of the reaction just described? Are they enantiomers or diastereomers?

Enzyme-catalyzed reductions of carbonyl groups are, more often than not, completely stereoselective. Pyruvic acid, for example, is converted exclusively to (*S*)-(+)-lactic acid by the lactate dehydrogenase-NADH system (Section 16.10). The enantiomer (*R*)-(−)-lactic acid is not formed.

Pyruvic acid + NADH (Reduced form of coenzyme) + H^+ —lactate dehydrogenase→ (*S*)-(+)-Lactic acid + NAD^+ (Oxidized form of coenzyme)

The enzyme is a single enantiomer of a chiral molecule and binds the coenzyme and substrate in such a way that hydride is transferred exclusively to the face of the carbonyl group that leads to (*S*)-(+)-lactic acid. Reduction of pyruvic acid in an achiral environment, say with sodium borohydride, also gives lactic acid but as a racemic mixture containing equal quantities of the *R* and *S* enantiomers.

The enantioselectivity of enzyme-catalyzed reactions can be understood on the basis of a relatively simple model. Consider the case of an sp^2-hybridized carbon with prochiral faces as in Figure 18.9*a*. If structural features on the enzyme are complementary in some respect to the groups attached to this carbon, one prochiral face can bind to the enzyme better than the other—there will be a preferred geometry of the enzyme–substrate complex. The binding forces are the usual ones: electrostatic, van der Waals, and so on. If a reaction occurs that converts the sp^2-hybridized carbon to sp^3, there will be a bias toward adding the fourth group from a particular direction as shown in Figure 18.9*b*. As

Prochirality was the subject of the Chapter 4 Descriptive Passage.

Figure 18.9

(*a*) Binding sites of enzyme discriminate between prochiral faces of substrate. One prochiral face can bind to the enzyme better than the other.
(*b*) Reaction attaches fourth group to the top face of the substrate producing only one enantiomer of chiral product.

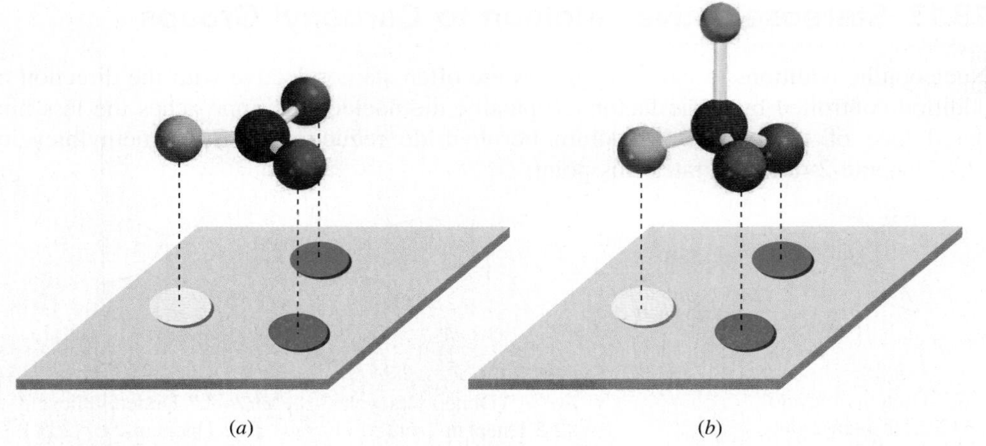

a result, an achiral molecule is converted to a single enantiomer of a chiral one. The reaction is enantioselective because it occurs preferentially at one prochiral face.

18.14 Oxidation of Aldehydes

Aldehydes are readily oxidized to carboxylic acids by a number of reagents, including those based on Cr(VI) in aqueous media.

$$\underset{\text{Aldehyde}}{\mathrm{RCH{=}O}} \xrightarrow{\text{oxidize}} \underset{\text{Carboxylic acid}}{\mathrm{RC(=O)OH}}$$

$$\underset{\text{Furfural}}{\text{Furfural}} \xrightarrow[\mathrm{H_2SO_4,\ H_2O}]{\mathrm{K_2Cr_2O_7}} \underset{\text{Furoic acid (75\%)}}{\text{Furoic acid}}$$

Mechanistically, these reactions probably proceed through the hydrate of the aldehyde and follow a course similar to that of alcohol oxidation.

$$\underset{\text{Aldehyde}}{\mathrm{RCH{=}O}} + \underset{\text{Water}}{\mathrm{H_2O}} \rightleftharpoons \underset{\text{Geminal diol (hydrate)}}{\mathrm{RCH(OH)_2}} \xrightarrow{\text{oxidize}} \underset{\text{Carboxylic acid}}{\mathrm{RC(=O)OH}}$$

Geminal diols are more easily oxidized than alcohols, which is why special reagents such as PCC and PDC (Section 16.9) have been developed for oxidizing primary alcohols to aldehydes and no further. PCC and PDC are effective not only because they are sources of Cr(VI), but also because they are used in nonaqueous media (dichloromethane). By keeping water out of the reaction mixture, the aldehyde is not converted to its hydrate, which is the necessary intermediate that leads to the carboxylic acid.

PCC is pyridinium chlorochromate.
PDC is pyridinium dichromate.

Alcohol oxidation, especially of ethanol, is one of the most common of all biological processes. Two key enzymes, both classified as *dehydrogenases,* are involved. The first catalyzes the oxidation of ethanol to acetaldehyde, the second catalyzes the oxidation of acetaldehyde to acetic acid.

$$\underset{\text{Ethanol}}{\mathrm{CH_3CH_2OH}} \xrightarrow[\text{dehydrogenase}]{\text{alcohol}} \underset{\text{Acetaldehyde}}{\mathrm{CH_3CH{=}O}} \xrightarrow[\text{dehydrogenase}]{\text{aldehyde}} \underset{\text{Acetic acid}}{\mathrm{CH_3C(=O)OH}}$$

Acetaldehyde is toxic and responsible for many of the adverse effects attributed to ethanol. Too much ethanol produces acetaldehyde faster than it can be oxidized to acetic acid and leads to elevated acetaldehyde levels.

18.15 Spectroscopic Analysis of Aldehydes and Ketones

Infrared: Carbonyl groups are among the easiest functional groups to detect by IR spectroscopy. The C=O stretching vibration of aldehydes and ketones gives rise to strong absorption in the region 1710–1750 cm^{-1}, as illustrated for butanal in Figure 18.10. In addition to a peak for C=O stretching, the CH=O group of an aldehyde exhibits two weak bands for C—H stretching near 2720 and 2820 cm^{-1}.

^{1}H NMR: Aldehydes are readily identified by the presence of a signal for the hydrogen of CH=O at δ 9–10. This is a region where very few other protons ever appear. Figure 18.11 shows the 300-MHz ^{1}H NMR spectrum of 2-methylpropanal [$(CH_3)_2CHCH{=}O$)], where the large chemical-shift difference between the aldehyde proton and the other protons in the molecule is clearly evident. As seen in the expanded-scale inset, the aldehyde proton is a doublet, split by the proton at C-2. Coupling between the protons in HC—CH=O is much smaller than typical vicinal couplings, making the multiplicity of the aldehyde peak difficult to see without expanding the scale.

Methyl ketones, such as 2-butanone in Figure 18.12, are characterized by sharp singlets near δ 2 for the protons of $CH_3C{=}O$. Similarly, the deshielding effect of the carbonyl causes the protons of $CH_2C{=}O$ to appear at lower field (2.5) than in a CH_2 group of an alkane.

^{13}C NMR: The signal for the carbon of C=O in aldehydes and ketones appears at very low field, some 190–220 ppm downfield from tetramethylsilane. Figure 18.13 illustrates this for 3-heptanone, in which separate signals appear for each of the seven carbons. The six sp^3-hybridized carbons appear in the range δ 8–42, and the carbon of the C=O group is at δ 210. Note, too, that the intensity of the peak for the C=O carbon is much less than all the others, even though each peak corresponds to a single carbon. This

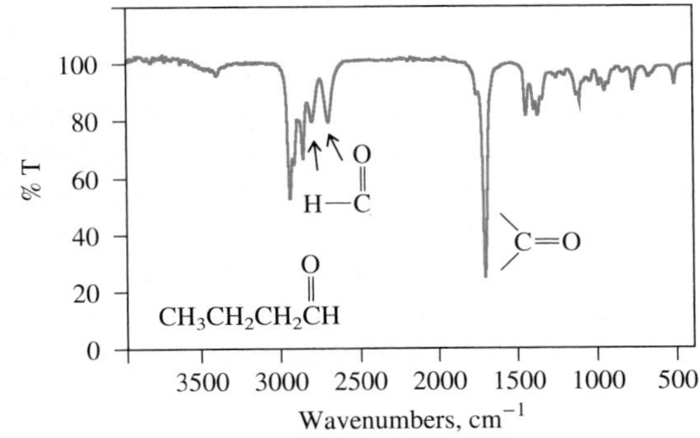

Figure 18.10

IR spectrum of butanal showing peaks characteristic of the CH=O unit at 2700 and 2800 cm^{-1} (C—H) and at 1720 cm^{-1} (C=O).

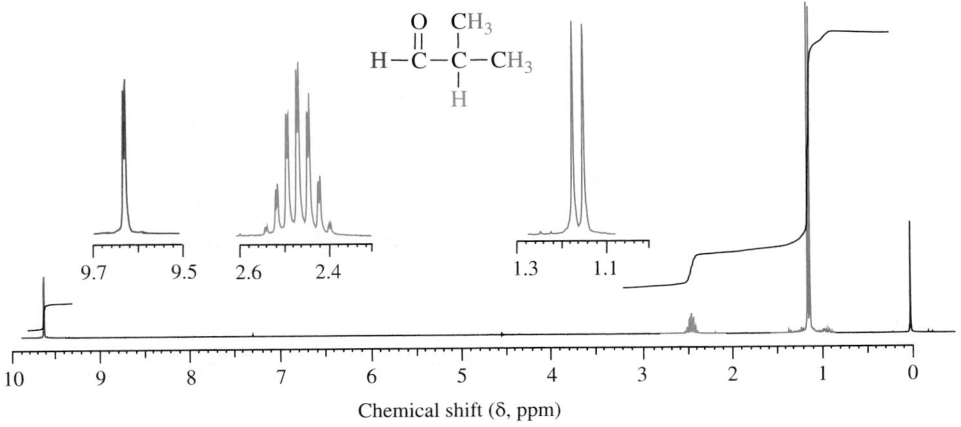

Figure 18.11

The 300-MHz ^{1}H NMR spectrum of 2-methylpropanal, showing the aldehyde proton as a doublet at low field (δ 9.6).

Figure 18.12

The 300-MHz ^{1}H NMR spectrum of 2-butanone.

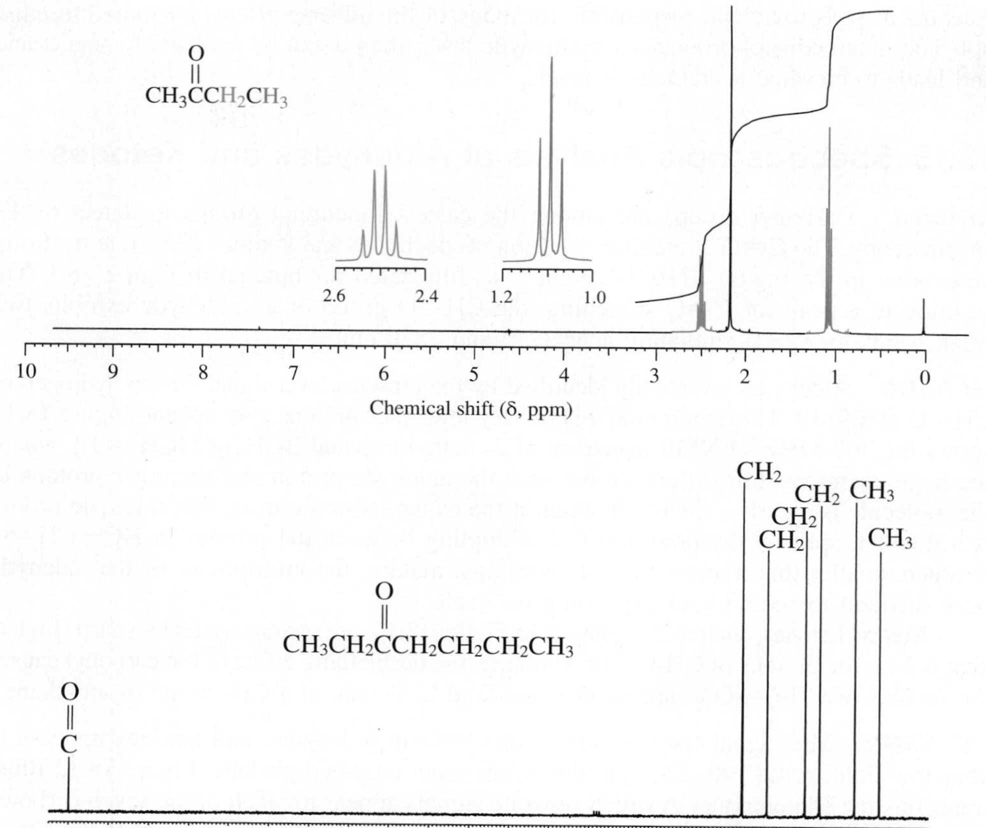

Figure 18.13

The ^{13}C NMR spectrum of 3-heptanone. Each signal corresponds to a single carbon. The carbonyl carbon is the least shielded and appears at δ 210.

decreased intensity is a characteristic of pulsed Fourier transform (FT) spectra for carbons that don't have attached hydrogens.

UV-VIS: Aldehydes and ketones have two absorption bands in the ultraviolet region. Both involve excitation of an electron to an antibonding π^* orbital. In one, called a $\pi \rightarrow \pi^*$ transition, the electron is one of the π electrons of the C═O group. In the other, called an $n \rightarrow \pi^*$ transition, it is one of the oxygen lone-pair electrons. Because the π electrons are more strongly held than the lone-pair electrons, the $\pi \rightarrow \pi^*$ transition is of higher energy and shorter wavelength than the $n \rightarrow \pi^*$ transition. For simple aldehydes and ketones, the $\pi \rightarrow \pi^*$ transition is below 200 nm and of little use in structure determination. The $n \rightarrow \pi^*$ transition, although weak, is of more diagnostic value.

$(H_3C)_2C{=}\ddot{O}{:}$ $\quad \pi \rightarrow \pi^* \quad \lambda_{max}$ 187 nm

$\quad\quad\quad\quad\quad\quad\quad n \rightarrow \pi^* \quad \lambda_{max}$ 270 nm

Acetone

Mass Spectrometry: Aldehydes and ketones typically give a prominent molecular ion peak in their mass spectra. Aldehydes also exhibit an M-1 peak. A major fragmentation pathway for both aldehydes and ketones leads to formation of acyl cations (acylium ions) by cleavage of an alkyl group from the carbonyl. The most intense peak in the mass spectrum of diethyl ketone, for example, is *m/z* 57, corresponding to loss of ethyl radical from the molecular ion.

$$CH_3CH_2C(=\!\!:\!O^{+\cdot})CH_2CH_3 \longrightarrow CH_3CH_2C{\equiv}\ddot{O}^+ + \cdot CH_2CH_3$$

m/z 86 $\quad\quad$ *m/z* 57

18.16 SUMMARY

The chemistry of the carbonyl group is probably the single most important aspect of organic chemical reactivity. Classes of compounds that contain the carbonyl group include many derived from carboxylic acids (acyl chlorides, acid anhydrides, esters, and amides) as well as the two related classes discussed in this chapter: *aldehydes* and *ketones*.

Section 18.1 The substitutive IUPAC names of aldehydes and ketones are developed by identifying the longest continuous chain that contains the carbonyl group and replacing the final *-e* of the corresponding alkane by *-al* for aldehydes and *-one* for ketones. The chain is numbered in the direction that gives the lowest locant to the carbon of the carbonyl group.

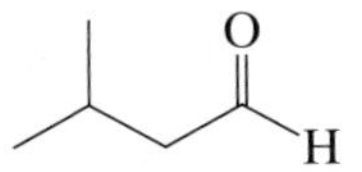

3-Methylbutanal

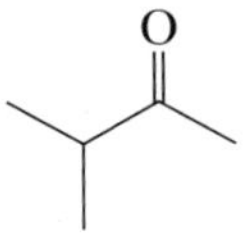

3-Methyl-2-butanone or 3-Methylbutan-2-one

Ketones may also be named using functional class IUPAC nomenclature by citing the two groups attached to the carbonyl in alphabetical order followed by the word *ketone*. Thus, 3-methyl-2-butanone (substitutive) becomes isopropyl methyl ketone (functional class).

Section 18.2 The carbonyl carbon is sp^2-hybridized, and it and the atoms attached to it are coplanar. Aldehydes and ketones are polar molecules. Nucleophiles attack C═O at carbon (positively polarized) and electrophiles, especially protons, attack oxygen (negatively polarized).

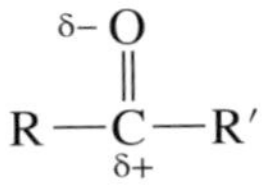

$$\mathrm{R{-}\overset{\overset{\delta-}{O}}{\underset{\delta+}{\overset{\|}{C}}}{-}R'}$$

Section 18.3 Aldehydes and ketones have higher boiling points than hydrocarbons, but have lower boiling points than alcohols.

Section 18.4 The numerous reactions that yield aldehydes and ketones discussed in earlier chapters and reviewed in Table 18.1 are sufficient for most syntheses.

Sections 18.5–18.12 The characteristic reactions of aldehydes and ketones involve *nucleophilic addition* to the carbonyl group and are summarized in Table 18.5. Reagents of the type HY react according to the general equation

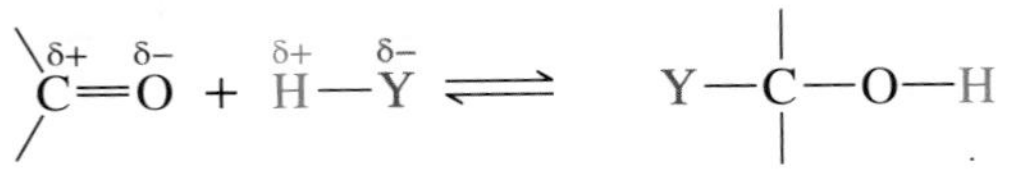

$$\overset{\delta+}{\mathrm{C}}{=}\overset{\delta-}{\mathrm{O}} + \overset{\delta+}{\mathrm{H}}{-}\overset{\delta-}{\mathrm{Y}} \rightleftharpoons \mathrm{Y{-}C{-}O{-}H}$$

Aldehyde or ketone — Product of nucleophilic addition to carbonyl group

Aldehydes undergo nucleophilic addition more readily and have more favorable equilibrium constants for addition than do ketones.

The step in which the nucleophile attacks the carbonyl carbon is rate-determining in both base-catalyzed and acid-catalyzed nucleophilic addition. In the base-catalyzed mechanism this is the first step.

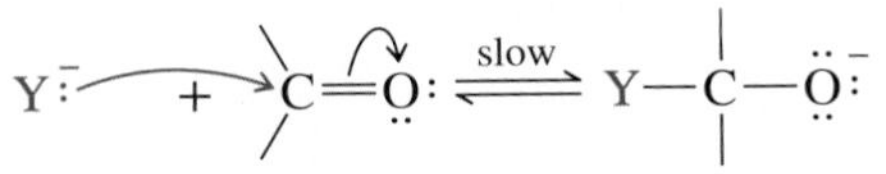

$$\mathrm{Y{:}^-} + \mathrm{C{=}\ddot{O}{:}} \underset{}{\overset{\text{slow}}{\rightleftharpoons}} \mathrm{Y{-}C{-}\ddot{O}{:}^-}$$

Nucleophile — Aldehyde or ketone

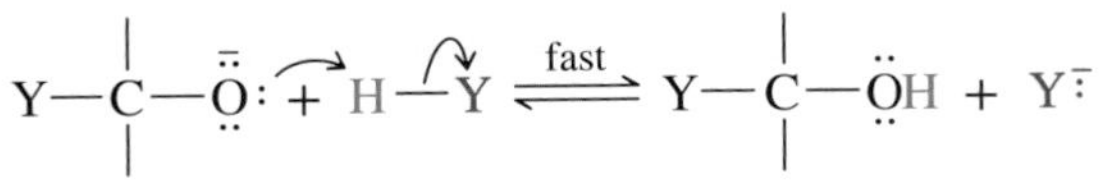

$$\mathrm{Y{-}C{-}\ddot{O}{:}^-} + \mathrm{H{-}Y} \overset{\text{fast}}{\rightleftharpoons} \mathrm{Y{-}C{-}\ddot{O}H} + \mathrm{Y{:}^-}$$

Product of nucleophilic addition

TABLE 18.5 Nucleophilic Addition to Aldehydes and Ketones

Reaction (section) and comments	General equation and specific example
Hydration (Section 18.6) Can be either acid- or base-catalyzed. Equilibrium constant is normally unfavorable for ketones unless R, R′, or both are strongly electron-withdrawing.	RC(=O)R′ (Aldehyde or ketone) + H_2O (Water) ⇌ RC(OH)(OH)R′ (Geminal diol) Chloroacetone (90% at equilibrium) ⇌ (H_2O) Chloroacetone hydrate (10% at equilibrium)
Cyanohydrin formation (Section 18.7) Reaction is catalyzed by cyanide ion. Cyanohydrins are useful synthetic intermediates; cyano group can be hydrolyzed to CO_2H or reduced to CH_2NH_2.	RC(=O)R′ (Aldehyde or ketone) + HCN (Hydrogen cyanide) → RC(OH)(CN)R′ (Cyanohydrin) 3-Pentanone → (KCN, H^+) 3-Pentanone cyanohydrin (75%)
Acetal and ketal formation (Sections 18.8–18.9) Reaction is acid-catalyzed and proceeds by way of a hemiacetal or hemiketal as an intermediate. Equilibrium constant normally favorable for aldehydes, unfavorable for ketones. Cyclic acetals from vicinal diols form readily.	RC(=O)R′ (Aldehyde or ketone) + 2R″OH (Alcohol) ⇌ (H^+) RC(OR″)(OR″)R′ (Acetal when R′ = H; ketal when R′ = alkyl or aryl) + H_2O (Water) *m*-Nitrobenzaldehyde + $2CH_3OH$ (Methanol) → (HCl) *m*-Nitrobenzaldehyde dimethyl acetal (76–85%)
Reaction with primary amines (Section 18.10) Isolated product is an imine, formed by dehydration of a hemiaminal intermediate.	RC(=O)R′ (Aldehyde or ketone) + R″NH_2 (Primary amine) ⇌ RC(=NR″)R′ (Imine) + H_2O (Water) 2-Methylpropanal + *tert*-Butylamine → *N*-(2-Methyl-1-propylidene)-*tert*-butylamine (50%)

TABLE 18.5 Nucleophilic Addition to Aldehydes and Ketones (*Continued*)

Reaction (section) and comments	General equation and specific example
Reaction with secondary amines (Section 18.11) In this case, the hemiaminal intermediate cannot give an imine so dehydrates in the direction that gives a carbon–carbon double bond.	$RCCH_2R'$ (Aldehyde or ketone, C=O) + $R''NHR'''$ (Secondary amine) ⇌ $RC{=}CHR'$ with $NR''R'''$ on C (Enamine) + H_2O (Water) Cyclohexanone + Morpholine —benzene, heat→ 1-Morpholinocyclohexene (85%)
Wittig reaction (Section 18.12) The reaction of a phosphorus ylide with an aldehyde or ketones is a synthetically useful method for the preparation of alkenes.	RCR' (Aldehyde or ketone, C=O) + $(C_6H_5)_3\overset{+}{P}-\overset{-}{C}AB$ (Wittig reagent (an ylide)) → $RR'C{=}CAB$ (Alkene) + $(C_6H_5)_3\overset{+}{P}-O^-$ (Triphenylphosphine oxide) Acetone + 1-Pentylidenetriphenylphosphorane → 2-Methyl-2-heptene (56%)

Under conditions of acid catalysis, the nucleophilic addition step follows protonation of the carbonyl oxygen. Protonation increases the carbocation character of a carbonyl group and makes it more electrophilic.

$$\text{C}{=}\ddot{\text{O}}: + \text{H}{-}\text{Y}: \underset{}{\overset{\text{fast}}{\rightleftharpoons}} \text{C}{=}\overset{+}{\text{O}}\text{H} \longleftrightarrow {}^{+}\text{C}{-}\ddot{\text{O}}\text{H}$$

Aldehyde or ketone — Resonance contributors to protonated aldehyde or ketone

$$\text{HY}: + \text{C}{=}\overset{+}{\text{O}}\text{H} \overset{\text{slow}}{\rightleftharpoons} \text{H}\overset{+}{\text{Y}}{-}\text{C}{-}\ddot{\text{O}}\text{H} \overset{-\text{H}^+}{\rightleftharpoons} \text{Y}{-}\text{C}{-}\ddot{\text{O}}\text{H}$$

Product of nucleophilic addition

Often the product of nucleophilic addition is not isolated but is an intermediate leading to the ultimate product. Most of the reactions in Table 18.5 are of this type.

Section 18.13 Nucleophilic addition to the carbonyl group can be *stereoselective*. When one direction of approach to the carbonyl group is less hindered than the other, the nucleophile normally attacks at the less hindered face.

3,3,5-Trimethylcyclohexanone —1. $LiAlH_4$, diethyl ether; 2. H_2O→ *trans*-3,3,5-Trimethylcyclohexanol (83%) + *cis*-3,3,5-Trimethylcyclohexanol (17%)

Section 18.14 Aldehydes are easily oxidized to carboxylic acids.

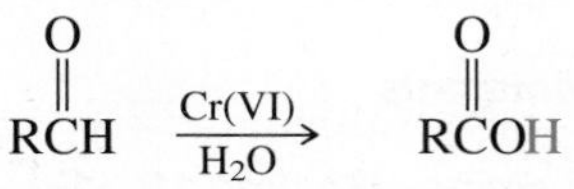

$$\underset{\text{Aldehyde}}{\mathrm{R\overset{O}{\overset{\|}{C}}H}} \xrightarrow[H_2O]{Cr(VI)} \underset{\text{Carboxylic acid}}{\mathrm{R\overset{O}{\overset{\|}{C}}OH}}$$

Section 18.15 A strong peak near 1700 cm^{-1} in the IR spectrum is characteristic of compounds that contain a C═O group. The ^1H and ^{13}C NMR spectra of aldehydes and ketones are affected by the deshielding of a C═O group. The proton of an H—C═O group appears in the δ 8–10 range. The carbon of a C═O group is at δ 190–210.

PROBLEMS

Structure and Nomenclature

18.23 (a) Write structural formulas and provide IUPAC names for all the isomeric aldehydes and ketones that have the molecular formula $C_5H_{10}O$. Include stereoisomers.

(b) Which of the isomers in part (a) yield chiral alcohols on reaction with sodium borohydride?

(c) Which of the isomers in part (a) yield chiral alcohols on reaction with methylmagnesium iodide?

18.24 Each of the following aldehydes or ketones is known by a common name. Its substitutive IUPAC name is provided in parentheses. Write a structural formula for each one.

(a) Chloral (2,2,2-trichloroethanal)

(b) Pivaldehyde (2,2-dimethylpropanal)

(c) Acrolein (2-propenal)

(d) Crotonaldehyde [(*E*)-2-butenal]

(e) Citral [(*E*)-3,7-dimethyl-2,6-octadienal]

(f) Diacetone alcohol (4-hydroxy-4-methyl-2-pentanone)

(g) Carvone (5-isopropenyl-2-methyl-2-cyclohexenone)

(h) Biacetyl (2,3-butanedione)

18.25 The African dwarf crocodile secretes a volatile substance believed to be a sex pheromone. It is a mixture of two stereoisomers, one of which is shown:

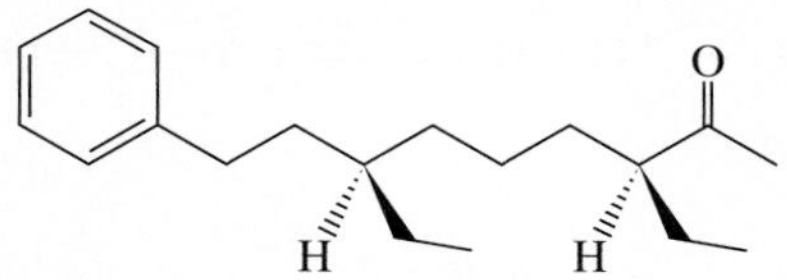

O
H
H

(a) Give the IUPAC name for this compound, including *R* and *S* descriptors for its chirality centers.

(b) One component of the scent substance has the *S* configuration at both chirality centers. How is this compound related to the one shown? Are the compounds enantiomers or diastereomers?

18.26 Compounds that contain a carbon–nitrogen double bond are capable of stereoisomerism much like that seen in alkenes. The structures

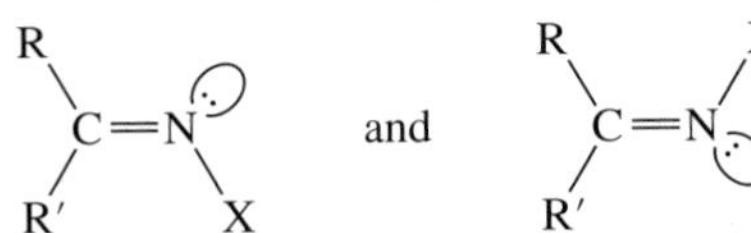

are stereoisomeric. Specifying stereochemistry in these systems is best done by using *E–Z* descriptors and considering the nitrogen lone pair to be the lowest priority group. Write the structures, clearly showing stereochemistry, of the following:

(a) (*Z*)-$CH_3CH{=}NCH_3$

(b) (*E*)-Acetaldehyde oxime

(c) (*Z*)-2-Butanone hydrazone

(d) (*E*)-Acetophenone semicarbazone

Reactions

18.27 Predict the product of the reaction of propanal with each of the following:

(a) Lithium aluminum hydride, followed by water

(b) Sodium borohydride, methanol

(c) Hydrogen (nickel catalyst)

(d) Methylmagnesium iodide, followed by dilute acid

(e) Sodium acetylide, followed by dilute acid

(f) Phenyllithium, followed by dilute acid

(g) Methanol containing dissolved hydrogen chloride

(h) Ethylene glycol, *p*-toluenesulfonic acid, benzene

(i) Aniline ($C_6H_5NH_2$)

(j) Dimethylamine, *p*-toluenesulfonic acid, benzene

(k) Hydroxylamine

(l) Hydrazine

(m) Product of part (l) heated in triethylene glycol with sodium hydroxide

(n) *p*-Nitrophenylhydrazine

(o) Semicarbazide

(p) Ethylidenetriphenylphosphorane [$(C_6H_5)_3\overset{+}{P}{-}\overset{..}{\bar{C}}HCH_3$]

(q) Sodium cyanide with addition of sulfuric acid

(r) Chromic acid

18.28 Repeat the preceding problem for cyclopentanone instead of propanal.

18.29 Hydride reduction (with $LiAlH_4$ or $NaBH_4$) of each of the following ketones has been reported in the chemical literature and gives a mixture of two diastereomeric alcohols in each case. Give the structures of both alcohol products for each ketone.

(a) (*S*)-3-Phenyl-2-butanone

(b) 4-*tert*-Butylcyclohexanone

(c)

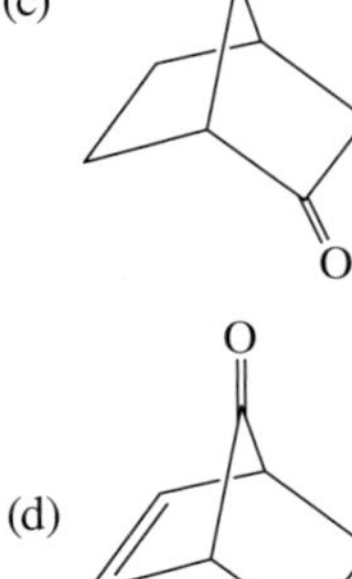

(d)

18.30 Choose which member in each of the following pairs reacts faster or has the more favorable equilibrium constant for reaction with the indicated reagent. Explain your reasoning.

(a) $C_6H_5CH(=O)$ or $C_6H_5C(=O)CH_3$ (rate of reduction with sodium borohydride)

(b) $Cl_3CCH(=O)$ or $CH_3CH(=O)$ (equilibrium constant for hydration)

(c) Acetone or 3,3-dimethyl-2-butanone (equilibrium constant for cyanohydrin formation)

(d) Acetone or 3,3-dimethyl-2-butanone (rate of reduction with sodium borohydride)

(e) $CH_2(OCH_2CH_3)_2$ or $(CH_3)_2C(OCH_2CH_3)_2$ (rate of acid-catalyzed hydrolysis)

18.31 Equilibrium constants for the dissociation (K_{diss}) of cyanohydrins according to the equation

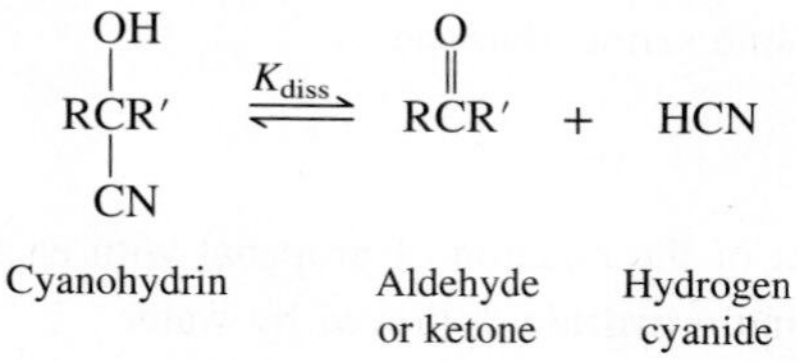

$$RC(OH)(CN)R' \underset{}{\overset{K_{diss}}{\rightleftharpoons}} RC(=O)R' + HCN$$

Cyanohydrin — Aldehyde or ketone — Hydrogen cyanide

have been measured for a number of cyanohydrins. Which cyanohydrin in each of the following pairs has the greater dissociation constant?

(a) 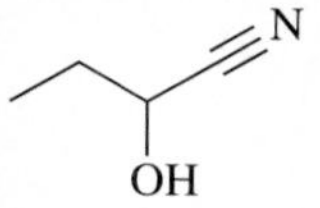 OH, N or OH, N

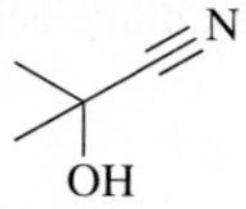

(b) 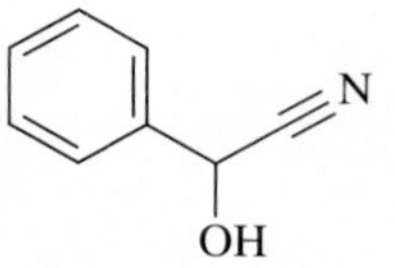 OH, N or OH, N

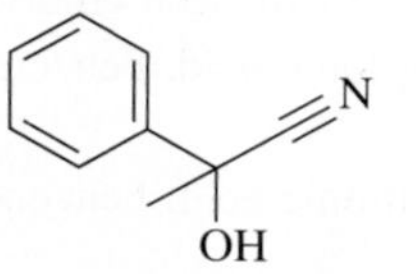

(c) 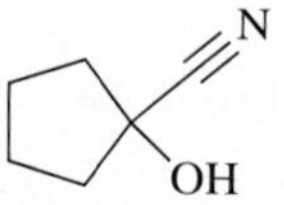 OH, N or OH, N

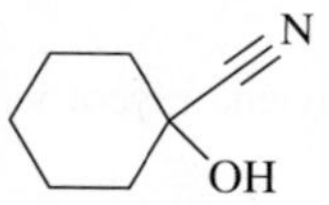

18.32 Each of the following reactions has been reported in the chemical literature and gives a single organic product in good yield. What is the principal product in each reaction?

(a) 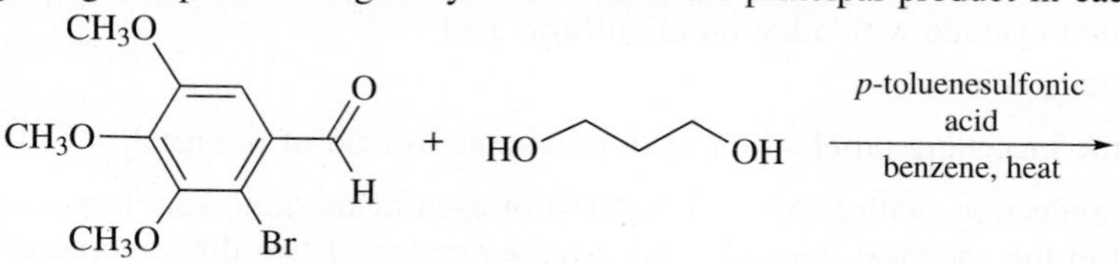 CH_3O, CH_3O, CH_3O, Br, O, H + HO, OH $\xrightarrow[\text{benzene, heat}]{p\text{-toluenesulfonic acid}}$

(b) 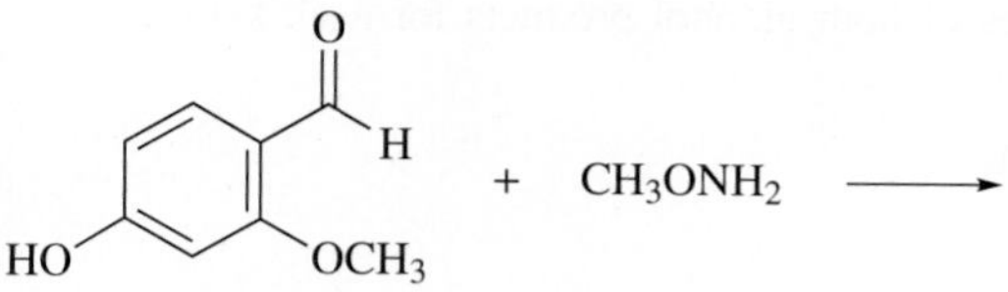 O, H, HO, OCH_3 + CH_3ONH_2 $\longrightarrow$

(c) 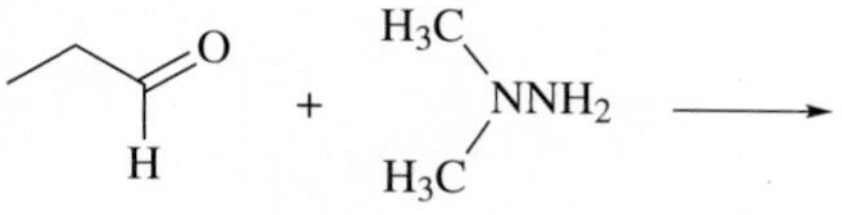 O, H + H_3C, NNH_2, H_3C $\longrightarrow$

(d) 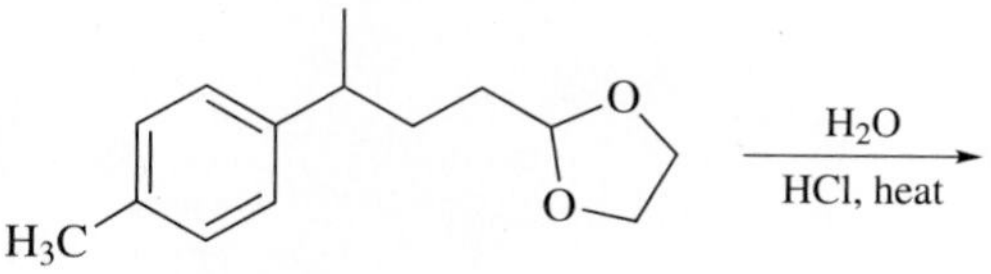 H_3C, O, O $\xrightarrow[\text{HCl, heat}]{H_2O}$

(e) 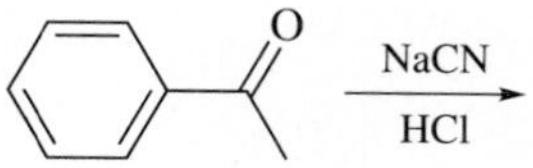 O $\xrightarrow[\text{HCl}]{\text{NaCN}}$

(f) 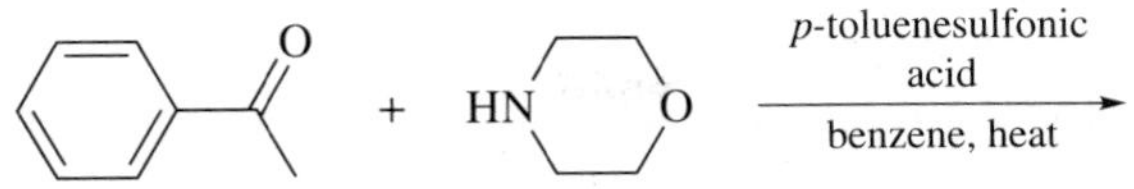 [benzene ring–C(=O)CH₃ structure] + HN(morpholine)O $\xrightarrow[\text{benzene, heat}]{p\text{-toluenesulfonic acid}}$

(g) 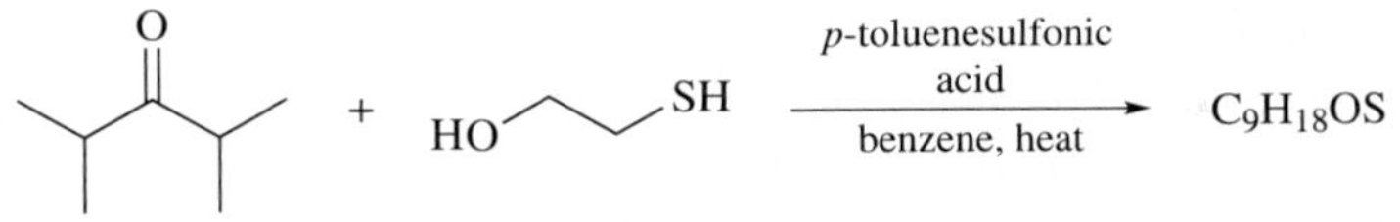 [diisopropyl ketone structure] + HO~~SH $\xrightarrow[\text{benzene, heat}]{p\text{-toluenesulfonic acid}}$ $C_9H_{18}OS$

18.33 Wolff–Kishner reduction (hydrazine, KOH, ethylene glycol, 130°C) of the compound shown gave compound A. Treatment of compound A with *m*-chloroperoxybenzoic acid (MCPBA) gave compound B, which on reduction with lithium aluminum hydride gave compound C. Oxidation of compound C with chromic acid gave compound D ($C_9H_{14}O$). Identify compounds A through D in this sequence.

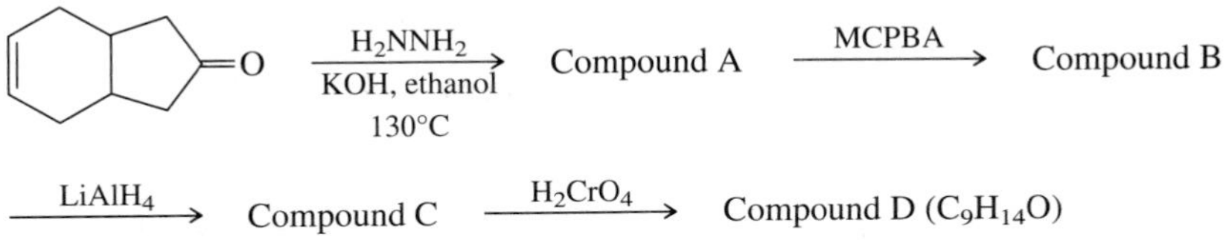

[bicyclic ketone]=O $\xrightarrow[\text{KOH, ethanol, 130°C}]{H_2NNH_2}$ Compound A $\xrightarrow{\text{MCPBA}}$ Compound B

$\xrightarrow{\text{LiAlH}_4}$ Compound C $\xrightarrow{H_2CrO_4}$ Compound D ($C_9H_{14}O$)

18.34 On standing in ^{17}O-labeled water, both formaldehyde and its hydrate are found to have incorporated the ^{17}O isotope of oxygen. Suggest a reasonable explanation for this observation.

18.35 Reaction of benzaldehyde with 1,2-octanediol in benzene containing a small amount of *p*-toluenesulfonic acid yields almost equal quantities of two products in a combined yield of 94%. Both products have the molecular formula $C_{15}H_{22}O_2$. Suggest reasonable structures for these products.

18.36 Compounds that contain both carbonyl and alcohol functional groups are often more stable as cyclic hemiacetals or cyclic acetals than as open-chain compounds. Examples of several of these are shown. Deduce the structure of the open-chain form of each.

(a) 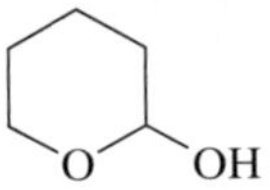 [tetrahydropyran ring] O OH

(b) 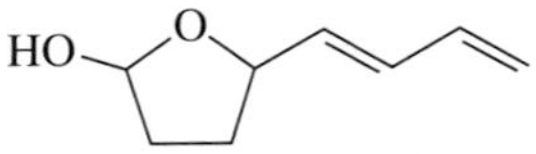 HO–[tetrahydrofuran ring with O]–CH=CH–CH=CH₂

(c) 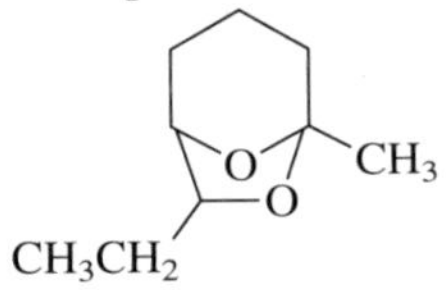 [bicyclic acetal] O O CH_3 CH_3CH_2

Brevicomin (sex attractant of Western pine beetle)

(d) 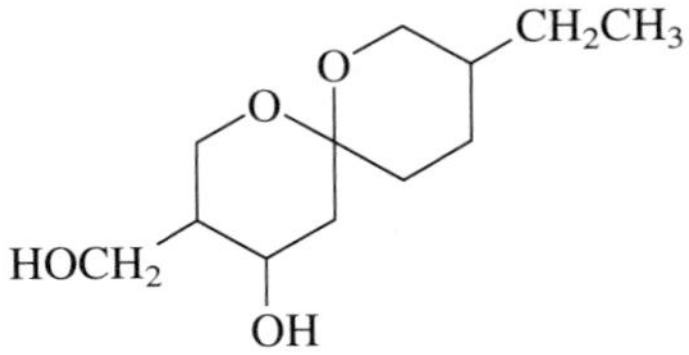 [spiroketal] O O CH_2CH_3 $HOCH_2$ OH

Talaromycin A (a toxic substance produced by a fungus that grows on poultry house litter)

18.37 The OH groups at C-4 and C-6 of methyl α-D-glucopyranoside can be protected by conversion to a benzylidene acetal. What reagents are needed for this conversion?

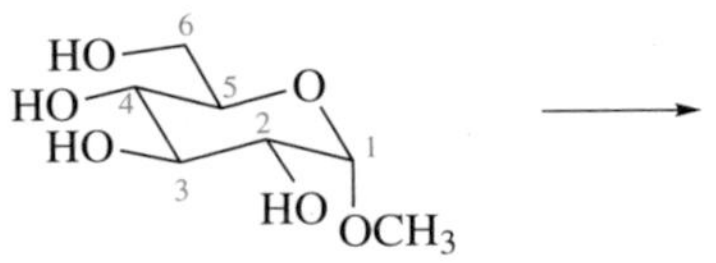

[structure: HO, HO, HO, O, HO, OCH_3; carbons numbered 1–6] ⟶ [structure: benzylidene O, O, O, HO, HO, OCH_3]

Methyl α-D-glucopyranoside

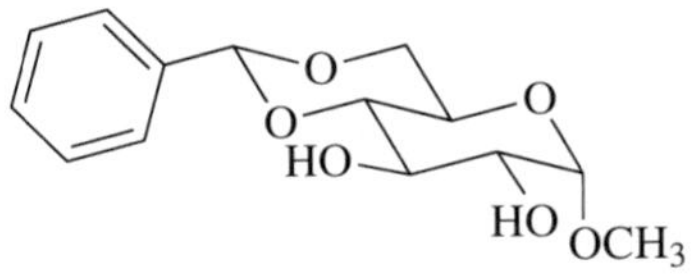

Methyl 4,6-*O*-benzylidene-α-D-glucopyranoside

Synthesis

18.38 Describe reasonable syntheses of benzophenone, $C_6H_5\overset{O}{\overset{\|}{C}}C_6H_5$, from each of the following starting materials and any necessary inorganic reagents.

(a) Benzoyl chloride and benzene

(b) Benzyl alcohol and bromobenzene

(c) Bromodiphenylmethane, $(C_6H_5)_2CHBr$

(d) Dimethoxydiphenylmethane, $(C_6H_5)_2C(OCH_3)_2$

(e) 1,1,2,2-Tetraphenylethene, $(C_6H_5)_2C{=}C(C_6H_5)_2$

18.39 Studies of the sex pheromone of the Douglas fir tussock moth required the synthesis of (*E*)-1,6-henicosadien-11-one. Outline a synthesis of this ketone using (*E*)-5,10-undecadien-1-ol and 1-decanol as sources of all of the carbons.

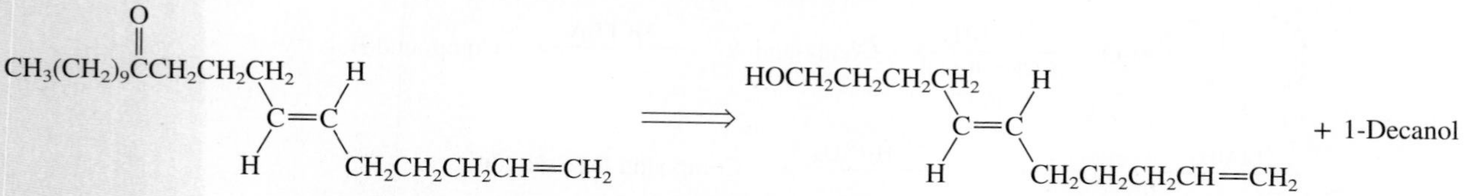

$CH_3(CH_2)_9\overset{O}{\overset{\|}{C}}CH_2CH_2CH_2(H)C{=}C(H)CH_2CH_2CH_2CH{=}CH_2 \Longrightarrow HOCH_2CH_2CH_2CH_2(H)C{=}C(H)CH_2CH_2CH_2CH{=}CH_2$ + 1-Decanol

18.40 The sex attractant of the female winter moth has been identified as the tetraene $CH_3(CH_2)_8CH{=}CHCH_2CH{=}CHCH_2CH{=}CHCH{=}CH_2$. Devise a synthesis of this material from 3,6-hexadecadien-1-ol and allyl alcohol.

18.41 Leukotrienes are substances produced in the body that may be responsible for inflammatory effects. As part of a synthesis of one of these, compound A reacted with the Wittig reagent shown to give B along with some of its *Z* stereoisomer. Explain the origin of these compounds.

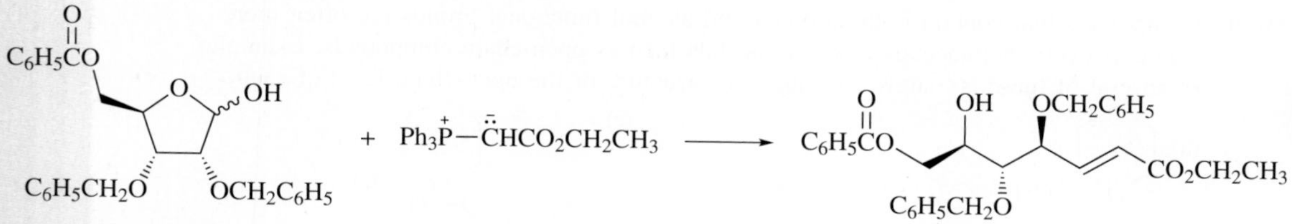

Compound A + $Ph_3\overset{+}{P}-\overset{..}{\bar{C}}HCO_2CH_2CH_3$ ⟶ Compound B

Compound A: $C_6H_5\overset{O}{\overset{\|}{C}}O$, OH, $C_6H_5CH_2O$, $OCH_2C_6H_5$

Compound B: $C_6H_5\overset{O}{\overset{\|}{C}}O$, OH, $OCH_2C_6H_5$, $C_6H_5CH_2O$, $CO_2CH_2CH_3$

18.42 Syntheses of each of the following compounds have been reported in the chemical literature. Using the indicated starting material and any necessary organic or inorganic reagents, describe short sequences of reactions that would be appropriate for each transformation.

(a) 1,1,5-Trimethylcyclononane from 5,5-dimethylcyclononanone

(b) 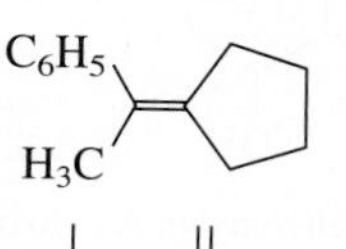 $(C_6H_5)(H_3C)C{=}$cyclopentylidene from $C_6H_5\overset{O}{\overset{\|}{C}}-$cyclopentyl

(c) [structure] from *o*-Bromotoluene and 5-hexenal

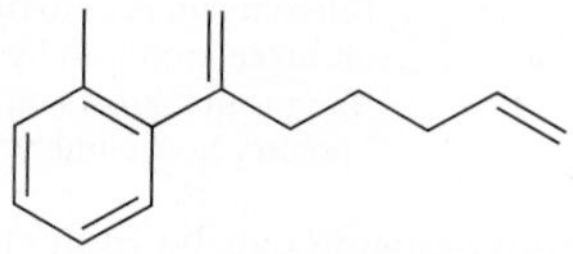

(d) 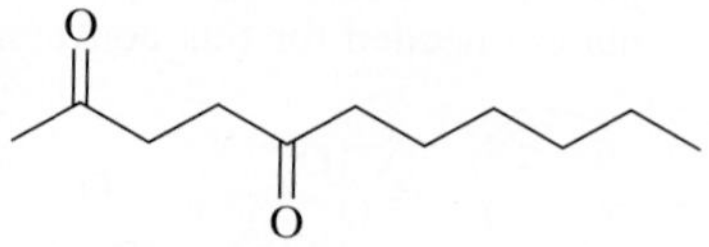 [structure with two C=O groups, O] from [alkyne–OH structure] OH

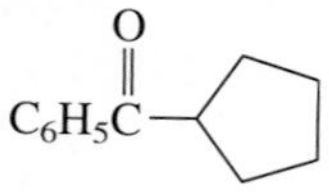

(e) 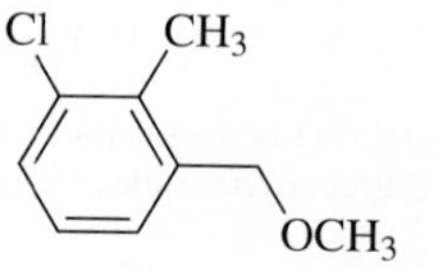 Cl, CH_3, OCH_3 [structure] from 3-Chloro-2-methylbenzaldehyde

(f) 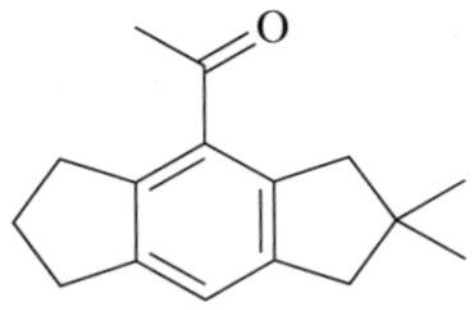 [structure] from [structure]

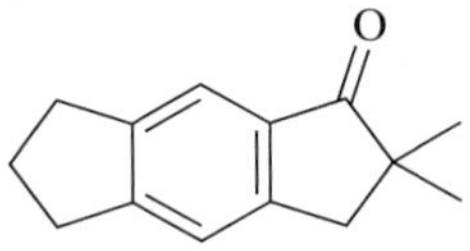

(g) 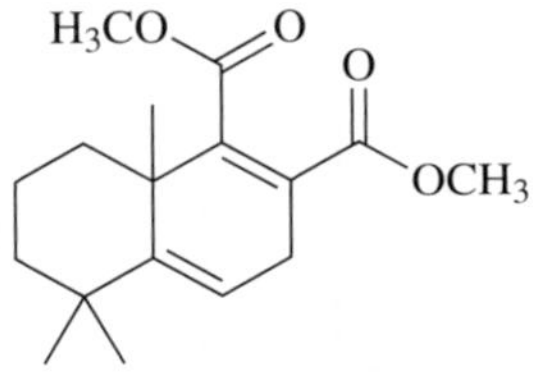 [structure] from [structure]

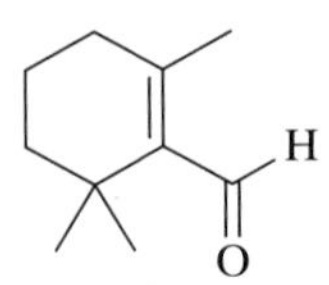

Mechanism

18.43 After heating compound A with a catalytic amount of *p*-toluenesulfonic acid and water in dichloromethane (45C) for 24 hr, compound C was isolated in 79% yield.

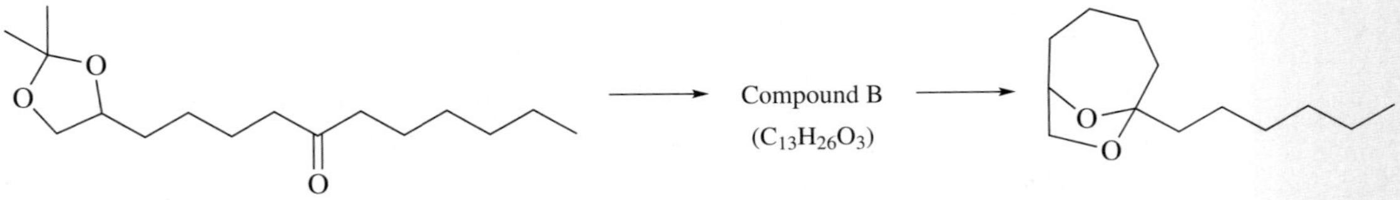

Compound A ($C_{16}H_{30}O_3$) → Compound B ($C_{13}H_{26}O_3$) → Compound C ($C_{13}H_{24}O_2$)

Demonstrate your understanding of the overall reaction by identifying the key intermediate (compound B). What other compound is formed in the reaction?

18.44 Suggest a reasonable mechanism for each of the following reactions:

(a) 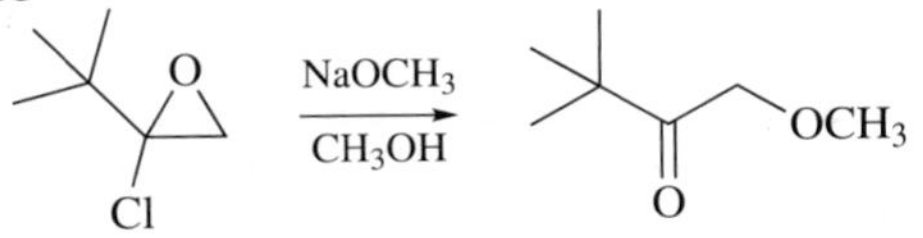 $\xrightarrow[\text{CH}_3\text{OH}]{\text{NaOCH}_3}$ (88%)

(b) 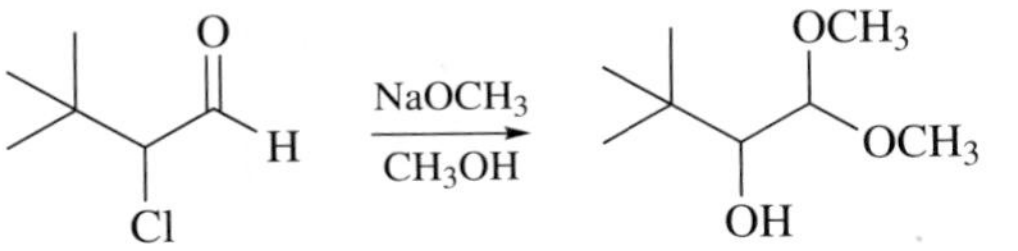 $\xrightarrow[\text{CH}_3\text{OH}]{\text{NaOCH}_3}$ (72%)

18.45 Alcohol functions can be protected as tetrahydropyranyl ethers (THPs) by acid-catalyzed addition to dihydropyran according to the equation:

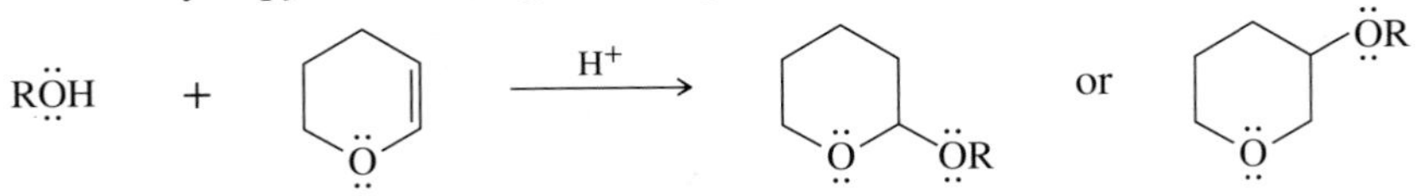

ROH + dihydropyran $\xrightarrow{\text{H}^+}$ I or II

Addition of the alcohol to dihydropyran is regiospecific in that I is formed to the exclusion of II. Suggest a mechanistic reason for the observed regioselectivity.

Spectroscopy

18.46 A compound has the molecular formula C_4H_8O and contains a carbonyl group. Identify the compound on the basis of its ^{1}H NMR spectrum shown in Figure 18.14.

18.47 A compound ($C_7H_{14}O$) has a strong peak in its IR spectrum at 1710 cm^{-1}. Its ^{1}H NMR spectrum consists of three singlets in the ratio 9:3:2 at δ 1.0, 2.1, and 2.3, respectively. Identify the compound.

18.48 Compounds A and B are isomeric diketones of molecular formula $C_6H_{10}O_2$. The ^{1}H NMR spectrum of compound A contains two signals, both singlets, at δ 2.2 (six protons) and 2.8 (four protons). The ^{1}H NMR spectrum of compound B contains two signals, one at δ 1.3 (triplet, six protons) and the other at δ 2.8 (quartet, four protons). What are the structures of compounds A and B?

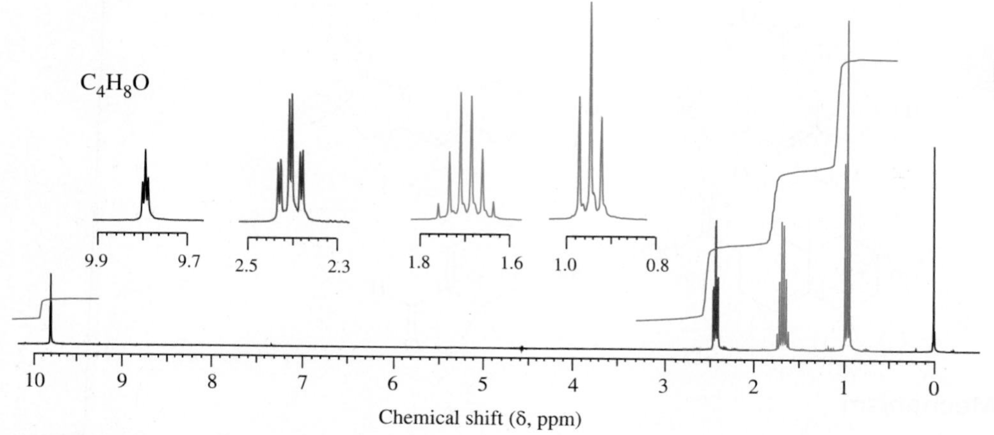

Figure 18.14

The 300-MHz ^1H NMR spectrum of a compound (C_4H_8O) (Problem 18.46).

18.49 A compound ($C_{11}H_{14}O$) has the (*a*) IR and (*b*) 300-MHz ^1H NMR spectra shown in Figure 18.15. What is the structure of this compound?

18.50 A compound is a ketone of molecular formula $C_7H_{14}O$. Its ^{13}C NMR spectrum is shown in Figure 18.16. What is the structure of the compound?

18.51 Compound A and compound B are isomers having the molecular formula $C_{10}H_{12}O$. The mass spectrum of each compound contains an abundant peak at *m/z* 105. The ^{13}C NMR spectra of compound A (Figure 18.17) and compound B (Figure 18.18) are shown. Identify these two isomers.

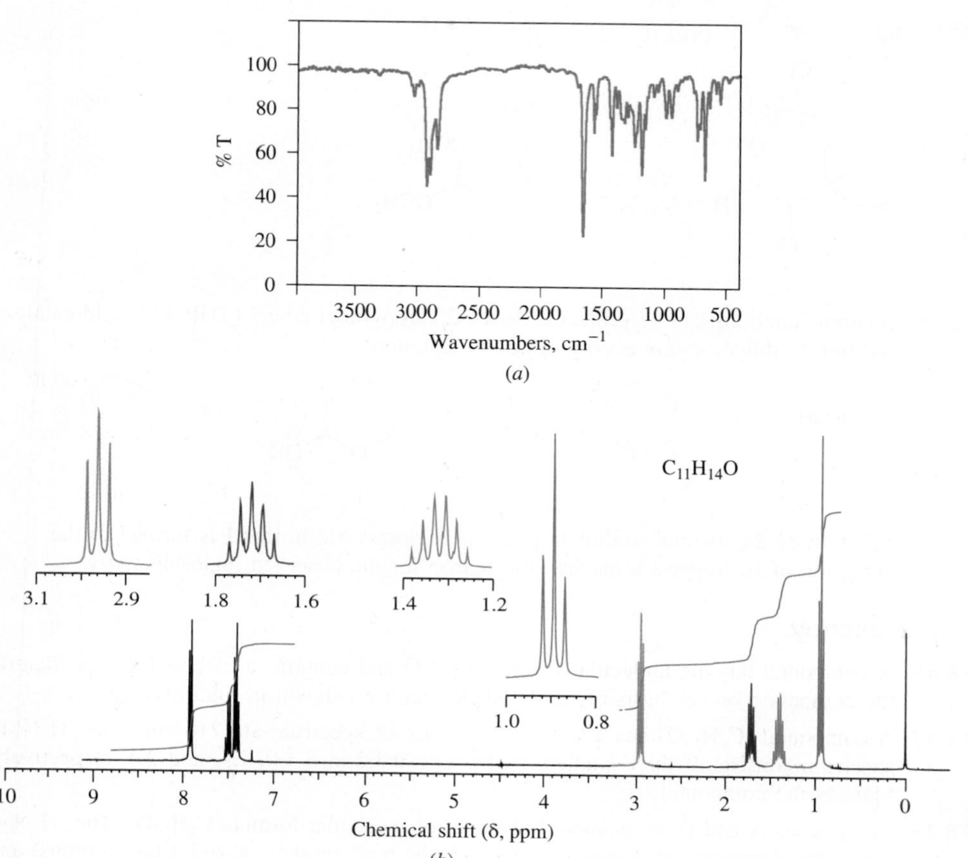

Figure 18.15

The (*a*) IR and (*b*) 300-MHz ^1H NMR spectra of a compound ($C_{11}H_{14}O$) (Problem 18.49).

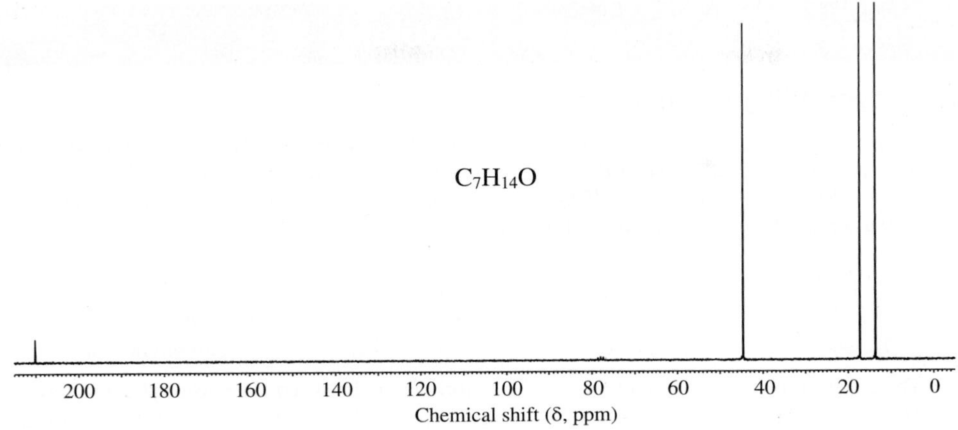

Figure 18.16

The ^{13}C NMR spectrum of a compound ($C_7H_{14}O$) (Problem 18.50).

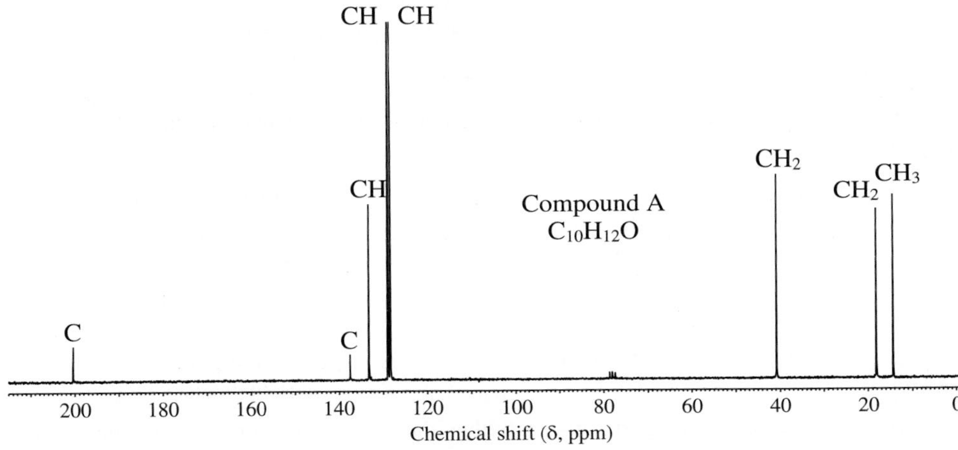

Figure 18.17

The ^{13}C NMR spectrum of compound A ($C_{10}H_{12}O$) (Problem 18.51).

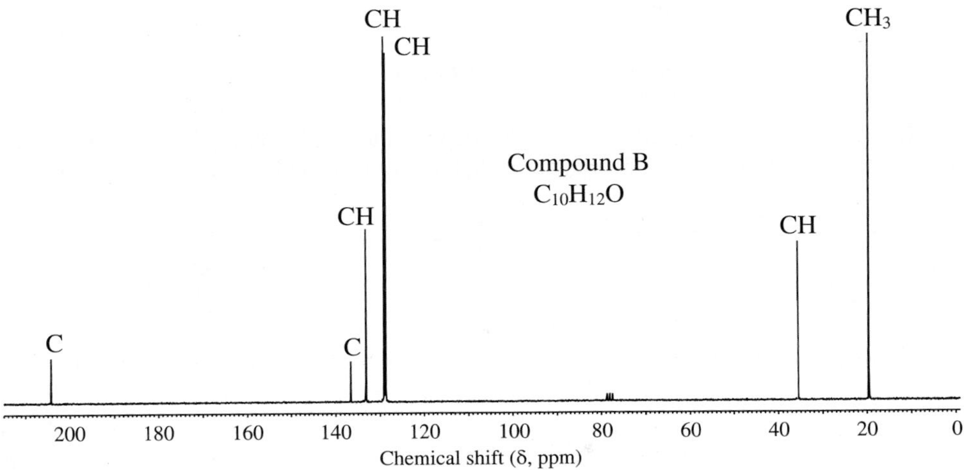

Figure 18.18

The ^{13}C NMR spectrum of compound B ($C_{10}H_{12}O$) (Problem 18.51).

Descriptive Passage and Interpretive Problems 18

The Baeyer–Villiger Oxidation

The oxidation of ketones with peroxy acids is both novel and synthetically useful. An oxygen from the peroxy acid is inserted between the ketone carbonyl group and one of its attached carbons to give an ester. First described by Adolf von Baeyer and Victor Villiger in 1899, reactions of this type are known as **Baeyer–Villiger oxidations.**

$$\underset{\text{Ketone}}{\overset{O}{\overset{\|}{RCR'}}} + \underset{\text{Peroxy acid}}{\overset{O}{\overset{\|}{R''COOH}}} \longrightarrow \underset{\text{Ester}}{\overset{O}{\overset{\|}{RCOR'}}} + \underset{\text{Carboxylic acid}}{\overset{O}{\overset{\|}{R''COH}}}$$

The reaction is regioselective; oxygen insertion occurs between the carbonyl carbon and the larger (R) of the two groups attached to it. Methyl ketones (R = CH_3) give esters of acetic acid:

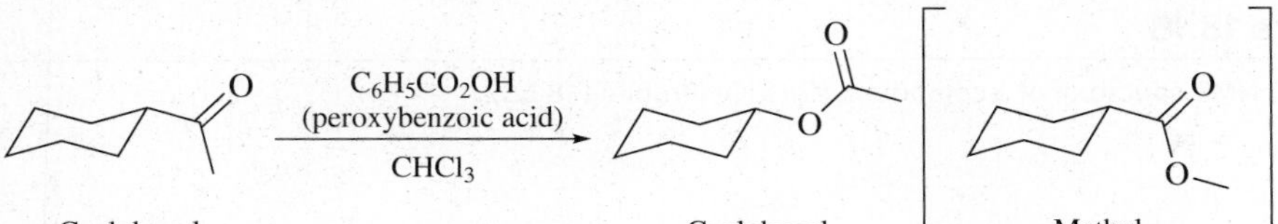

Cyclohexyl methyl ketone —[$C_6H_5CO_2OH$ (peroxybenzoic acid), $CHCl_3$]→ Cyclohexyl acetate (67%) [Methyl cyclohexanecarboxylate (not observed)]

The mechanism of the Baeyer–Villiger reaction begins with nucleophilic addition of the peroxy acid to the carbonyl group.

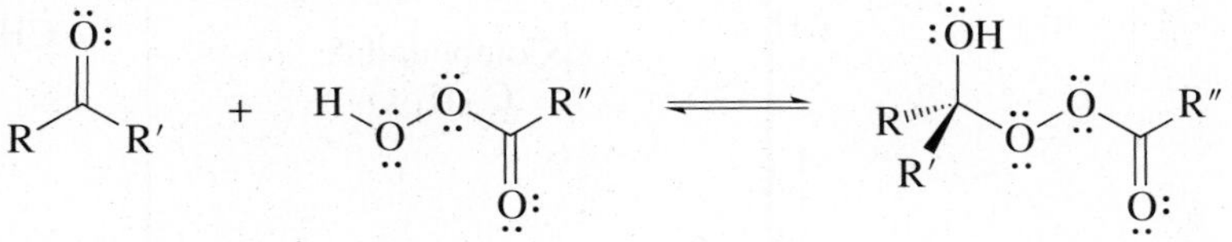

Ketone + Peroxy acid ⇌ Product of nucleophilic addition

After protonation by an acid catalyst (either the peroxy acid or a carboxylic acid), the conjugate acid of the product of the first step rearranges by an alkyl group migration. Normally, it is the larger of the two groups originally bonded to the carbonyl group that migrates.

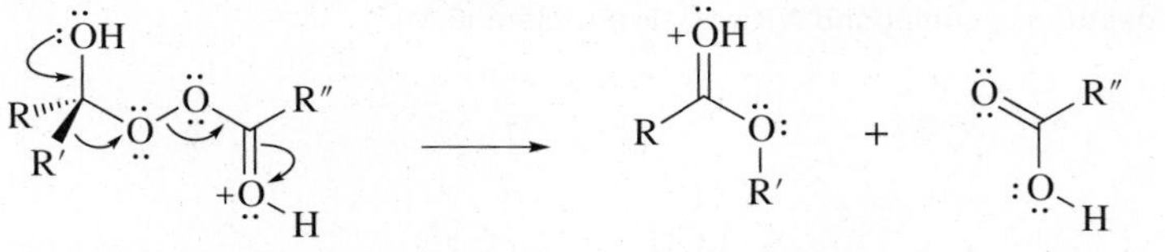

Conjugate acid of product of nucleophilic addition ⟶ Conjugate acid of ester + Carboxylic acid

The reaction is stereospecific; the alkyl group migrates with retention of configuration, as illustrated for the oxidation of *cis*-1-acetyl-2-methylcyclopentane; only the cis product is obtained.

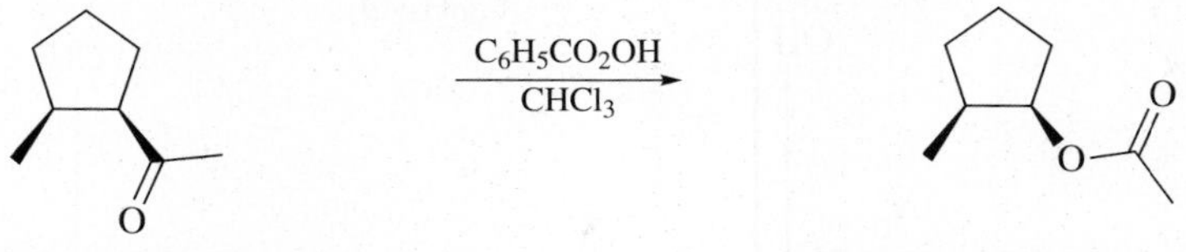

cis-1-Acetyl-2-methylcyclopentane —[$C_6H_5CO_2OH$, $CHCl_3$]→ *cis*-2-Methylcyclopentyl acetate (66%)

When the ketone is cyclic, a cyclic ester, or *lactone,* is formed. Cyclobutanone is oxidized to a lactone by the Baeyer–Villiger reaction.

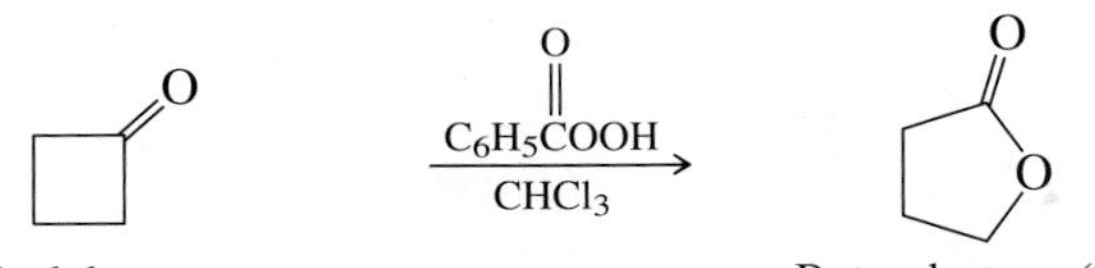

Cyclobutanone γ-Butyrolactone (70%)

18.52 Which of the following are *not* intermediates in the Baeyer–Villiger oxidation of cyclohexyl methyl ketone with peroxybenzoic acid?

A. I and II

B. III and IV

C. I and III

D. II and IV

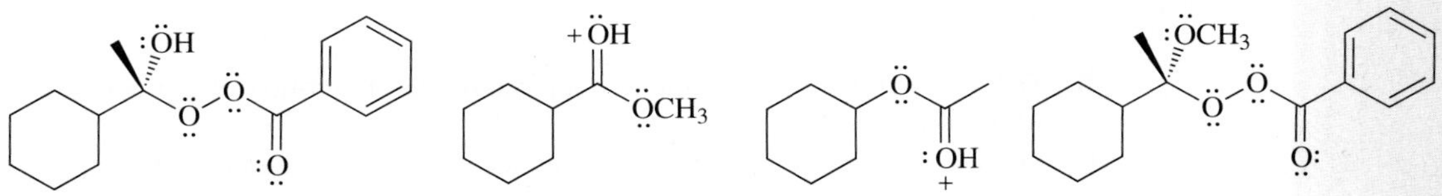

I. II. III. IV.

18.53 Which is the product of the following reaction?

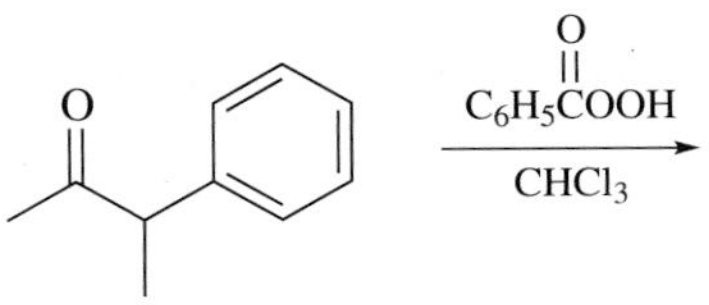

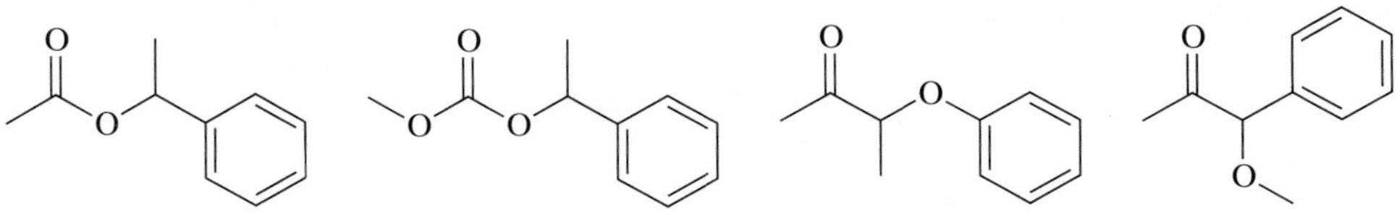

C_6H_5COOH / $CHCl_3$

A. B. C. D.

18.54 If the configuration of the chirality center is *R* in the reactant in Problem 18.53, what will the configuration be at this carbon in the product?

A. *R*

B. *S*

C. an equal mixture of *R* and *S*

D. an unequal mixture of *R* and *S*

18.55 The Baeyer–Villiger oxidations of the substituted diphenyl ketones proceed as indicated because:

$C_6H_5\overset{O}{\overset{\|}{C}}OOH$ / $CHCl_3$

Major product if X = OCH_3

Major product if X = NO_2

A. The electron-withdrawing nitro group retards the migration of the phenyl ring to which it is attached.

B. The electron-releasing methoxy group accelerates the migration of the phenyl ring to which it is attached.

C. Both A and B

D. Neither A nor B. The regioselectivity is due to steric effects.

18.56 A key step in the laboratory synthesis of prostaglandins involves the sequence shown here. What is the identity of compound X?

Compound X → ($C_6H_5\overset{O}{\overset{\|}{C}}OOH$, $CHCl_3$) → [bicyclic lactone with CH_3O, H] → (1. NaOH; 2. H_3O^+) → [cyclopentene with CO_2H, OCH_3, H, OH]

A. [bicyclic ketone with CH_3O]

B. [bicyclic ketone with CH_3O, H]

C. [bicyclic compound with CH_3O, OH, $C(O)CH_3$]

18.57 A reaction analogous to the Baeyer–Villiger reaction occurs in living systems through the action of certain bacterial enzymes. A preparation of the *S* enantiomer of compound Y has been described using a bacterial cyclohexanone monooxygenase enzyme system. What is compound X?

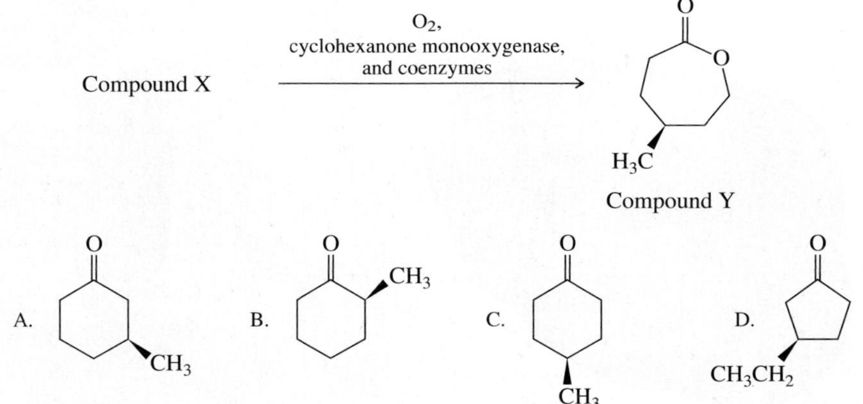

18.58 If compound Y in the preceding problem is prepared by treatment of compound X with peroxybenzoic acid, compound Y would be obtained as:

A. Only the *S* enantiomer
B. Only the *R* enantiomer
C. A racemic mixture
D. An unequal mixture of *R* and *S* enantiomers

Chapter 19

OH, H_3C, O, OH

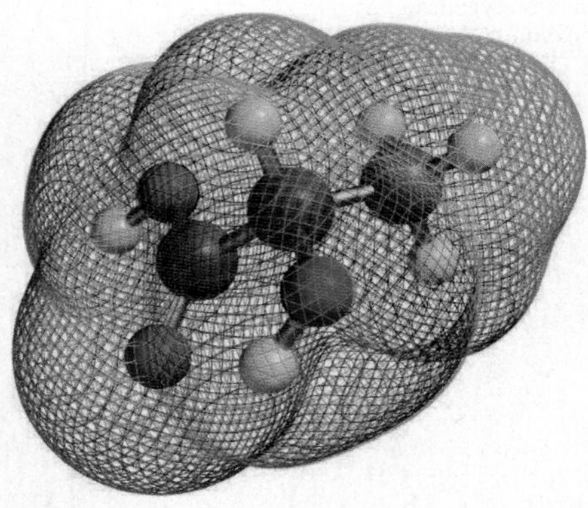

This runner may experience discomfort from the lactic acid that formed in her muscles during her run. The discomfort will be gone in a day or so; the exhilaration lasts much longer.
©Robert Atkins

CHAPTER OUTLINE

Carboxylic Acids

Carboxylic acids, compounds of the type RCOH (with C=O), constitute one of the most frequently encountered classes of organic compounds. Countless natural products are carboxylic acids or are derived from them. Some carboxylic acids, such as acetic acid, have been known for centuries. Others, such as the prostaglandins, which are powerful regulators of numerous biological processes, remained unknown until relatively recently. Still others, aspirin for example, are the products of chemical synthesis. The therapeutic effects of aspirin, known for well over a century, are now understood to result from aspirin's ability to inhibit the biosynthesis of prostaglandins.

$CH_3\overset{O}{\overset{\|}{C}}OH$ — Acetic acid (present in vinegar)

O, $(CH_2)_6CO_2H$, $(CH_2)_4CH_3$, HO, OH — PGE_1 (a prostaglandin; a small amount of PGE_1 lowers blood pressure significantly)

$\overset{O}{\overset{\|}{C}}OH$, $O\overset{\|}{\underset{O}{C}}CH_3$ — Aspirin

The importance of carboxylic acids is magnified when we realize that they are the parent compounds of a large group of derivatives that includes acyl chlorides, acid anhydrides, esters, and amides. Those classes of compounds will be discussed in Chapter 20. Together, this chapter and the next tell the story of some of the most fundamental structural types and functional-group transformations in organic and biological chemistry.

19.1 Carboxylic Acid Nomenclature

It is hard to find a class of compounds in which the common names of its members have influenced organic nomenclature more than carboxylic acids. Not only are the common names of carboxylic acids themselves widely used, but the names of many other compounds are derived from them. Benzene took its name from *benzoic* acid and propane from *propionic* acid, not the other way around. The name butane comes from *butyric* acid, present in rancid butter. The common names of most aldehydes are derived from the common names of carboxylic acids—valeraldehyde from *valeric* acid, for example. Many carboxylic acids are better known by common names than by their systematic ones, and the framers of the IUPAC rules have taken a liberal view toward accepting these common names as permissible alternatives to the systematic ones. Table 19.1 lists both common and systematic names for a number of important carboxylic acids.

Systematic names for carboxylic acids are derived by counting the number of carbons in the longest continuous chain that includes the carboxyl group and replacing the *-e* ending of the corresponding alkane by *-oic acid*. The first four acids in Table 19.1, methanoic (1 carbon), ethanoic (2 carbons), pentanoic (5 carbons), and octadecanoic acid (18 carbons), illustrate this point. When substituents are present, their locations are identified by number; numbering of the carbon chain always begins at the carboxyl group.

Notice that compounds 5 and 6 are named as hydroxy derivatives of carboxylic acids, rather than as carboxyl derivatives of alcohols. This parallels what we saw earlier in Section 18.1 where an aldehyde or ketone function took precedence over a hydroxyl group in defining the main chain. Carboxylic acids take precedence over all the common groups we have encountered to this point in respect to defining the main chain.

Double bonds in the main chain are signaled by the ending *-enoic acid,* and their position is designated by a numerical prefix as shown in entries 7 and 8.

When a carboxyl group is attached to a ring, the parent ring is named (retaining the final *-e*) and the suffix *-carboxylic acid* is added, as shown in entries 9 and 10.

Compounds with two carboxyl groups, as illustrated by entries 11 through 13, are distinguished by the suffix *-dioic acid* or *-dicarboxylic acid* as appropriate. The final *-e* in the name of the parent alkane is retained.

Problem 19.1

The list of carboxylic acids in Table 19.1 is by no means exhaustive insofar as common names are concerned. Many others are known by their common names, a few of which follow. Give a systematic IUPAC name for each.

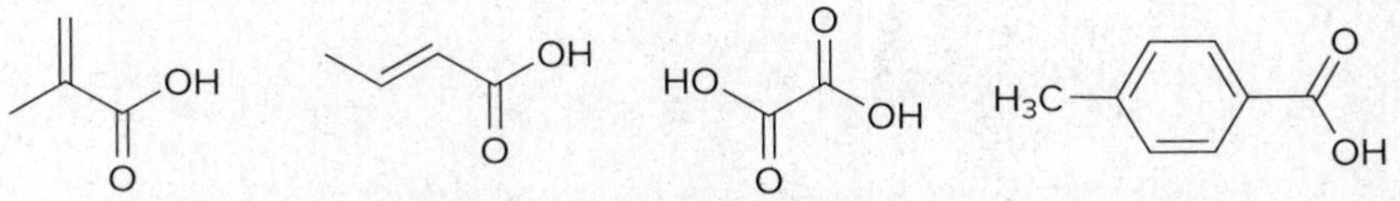

OH O OH O HO O OH O H_3C O OH

Methacrylic acid (a) | Crotonic acid (b) | Oxalic acid (c) | *p*-Toluic acid (d)

Sample Solution (a) Methacrylic acid is an industrial chemical used in the preparation of transparent plastics such as *Lucite* and *Plexiglas*. The carbon chain that includes both the carboxylic acid and the double bond is three carbon atoms in length. The compound is named as a derivative of *propenoic acid*. Both 2-methylpropenoic acid and 2-methylprop-2-enoic acid are acceptable IUPAC names.

TABLE 19.1 Systematic and Common Names of Some Carboxylic Acids

	Structural formula	Systematic name	Common name*
1.	HCOOH	Methanoic acid	Formic acid
2.	CH_3COOH	Ethanoic acid	Acetic acid
3.	$CH_3(CH_2)_3COOH$	Pentanoic acid	Valeric acid
4.	$CH_3(CH_2)_{16}COOH$	Octadecanoic acid	Stearic acid
5.	$CH_3CH(OH)COOH$	2-Hydroxypropanoic acid	Lactic acid
6.	$C_6H_5CH(OH)COOH$	2-Hydroxy-2-phenylethanoic acid	Mandelic acid
7.	$CH_2{=}CHCOOH$	Propenoic acid	Acrylic acid
8.	$CH_3(CH_2)_7CH{=}CH(CH_2)_7COOH$	(*Z*)-9-Octadecenoic acid	Oleic acid
9.	C_6H_5COOH	Benzenecarboxylic acid	Benzoic acid
10.	o-HOC_6H_4COOH	*o*-Hydroxybenzenecarboxylic acid	Salicylic acid
11.	$HOOCCH_2COOH$	Propanedioic acid	Malonic acid
12.	$HOOCCH_2CH_2COOH$	Butanedioic acid	Succinic acid
13.	o-$C_6H_4(COOH)_2$	1,2-Benzenedicarboxylic acid	Phthalic acid

*Except for valeric, mandelic, and salicylic acid, all of the common names in this table are acceptable IUPAC names.

19.2 Structure and Bonding

The structural features of the carboxyl group are most apparent in formic acid, which is planar, with one of its carbon–oxygen bonds shorter than the other, and with bond angles at carbon close to 120°.

Bond Distances	
C=O	120 pm (1.20 Å)
C—O	134 pm (1.34 Å)

Bond Angles	
H—C=O	124°
H—C—O	111°
O—C=O	125°

This suggests sp^2 hybridization at carbon, and a $\sigma + \pi$ carbon–oxygen double bond analogous to that of aldehydes and ketones.

Additionally, sp^2 hybridization of the hydroxyl oxygen allows one of its unshared electron pairs to be delocalized by orbital overlap with the π system of the carbonyl group (Figure 19.1*a*). In resonance terms, this electron delocalization is represented as:

Lone-pair donation from the hydroxyl oxygen makes the carbonyl group less electrophilic than that of an aldehyde or ketone. The electrostatic potential map of formic acid (Figure 19.1*b*) shows the most electron-rich site to be the oxygen of the carbonyl group and the most electron-poor one to be, as expected, the OH hydrogen.

Carboxylic acids are fairly polar, and simple ones such as acetic acid, propanoic acid, and benzoic acid have dipole moments in the range 1.7–1.9 D.

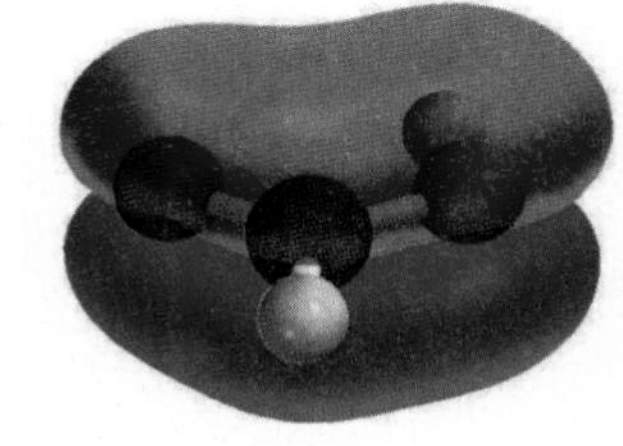

(*a*)

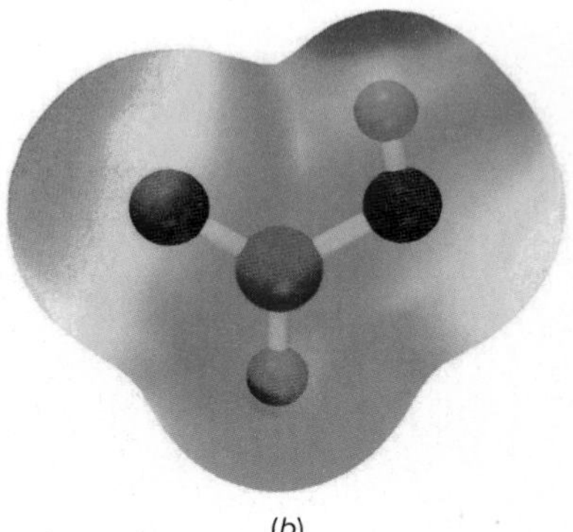

(*b*)

Figure 19.1

(*a*) The *p* orbital of the hydroxyl oxygen of formic acid overlaps with the π component of the double bond of the C=O group to form an extended π system that includes carbon and both oxygens. (*b*) The region of greatest negative charge (red) in formic acid is associated with the oxygen of C=O, and that of positive charge (blue) with the hydrogen of OH.

19.3 Physical Properties

The melting points and boiling points of carboxylic acids are higher than those of hydrocarbons and oxygen-containing organic compounds of comparable size and shape and indicate strong intermolecular attractive forces.

	2-Methyl-1-butene	2-Butanone	2-Butanol	Propanoic acid
bp (1 atm):	31°C	80°C	99°C	141°C

A unique hydrogen-bonding arrangement, shown in Figure 19.2, contributes to these attractive forces. The hydroxyl group of one carboxylic acid molecule acts as a proton donor

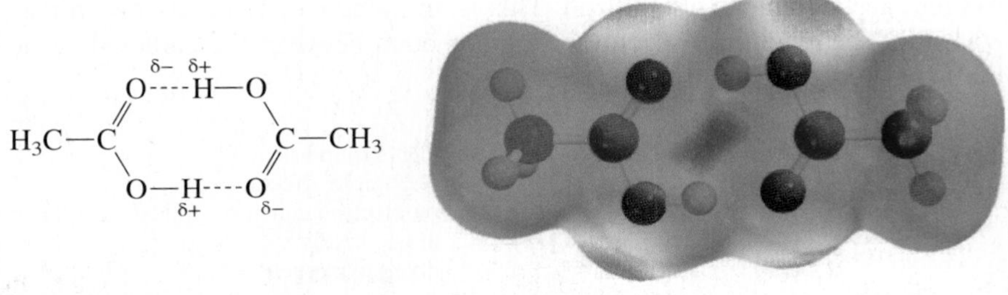

Figure 19.2

Hydrogen bonding between two acetic acid molecules.

toward the carbonyl oxygen of a second. In a reciprocal fashion, the hydroxyl proton of the second carboxyl function interacts with the carbonyl oxygen of the first. The result is that the two carboxylic acid molecules are held together by *two* hydrogen bonds. So efficient is this hydrogen bonding that some carboxylic acids exist as hydrogen-bonded dimers even in the gas phase. In the pure liquid a mixture of hydrogen-bonded dimers and higher aggregates is present.

In aqueous solution, intermolecular association between carboxylic acid molecules is replaced by hydrogen bonding to water. The solubility properties of carboxylic acids are similar to those of alcohols. Carboxylic acids of four carbon atoms or fewer are miscible with water in all proportions.

19.4 Acidity of Carboxylic Acids

Carboxylic acids are the most acidic class of compounds that contain only carbon, hydrogen, and oxygen. With pK_a's of about 5, they are much stronger acids than water and alcohols. The case should not be overstated, however. Carboxylic acids are weak acids; a 0.1 M solution of acetic acid in water, for example, is only 1.3% ionized.

To understand the greater acidity of carboxylic acids compared with water and alcohols, compare the structural changes that accompany the ionization of a representative alcohol (ethanol) and a representative carboxylic acid (acetic acid).

Ionization of ethanol

Ethanol + Water $\xrightleftharpoons{pK_a = 16}$ Ethoxide ion + Hydronium ion $\quad \Delta G^\circ = +91$ kJ/mol (+21.7 kcal/mol)

Ionization of acetic acid

Acetic acid + Water $\xrightleftharpoons{pK_a = 4.7}$ Acetate ion + Hydronium ion $\quad \Delta G^\circ = +27$ kJ/mol (+6.5 kcal/mol)

The large difference in the free energies of ionization of ethanol and acetic acid reflects a greater stabilization of acetate ion relative to ethoxide. Ionization of ethanol yields an alkoxide ion in which the negative charge is localized on oxygen. Solvation forces are the chief means by which ethoxide ion is stabilized. Acetate ion is also stabilized by solvation, but has two additional mechanisms for dispersing its negative charge that are not available to ethoxide:

1. *The inductive effect of the carbonyl group.* The carbonyl group of acetate ion is electron-withdrawing, and by attracting electrons away from the negatively charged oxygen, acetate anion is stabilized. This is an inductive effect, arising in the polarization of the electron distribution in the σ bond between the carbonyl carbon and the negatively charged oxygen.

Positively polarized carbon attracts electrons from negatively charged oxygen. $H_3C{-}C^{\delta+}(=O^{\delta-}){-}O^-$

CH_2 group has negligible effect on electron density at negatively charged oxygen. $H_3C{-}CH_2{-}O^-$

2. *The resonance effect of the carbonyl group.* Electron delocalization, expressed by resonance between the following Lewis structures, causes the negative charge in

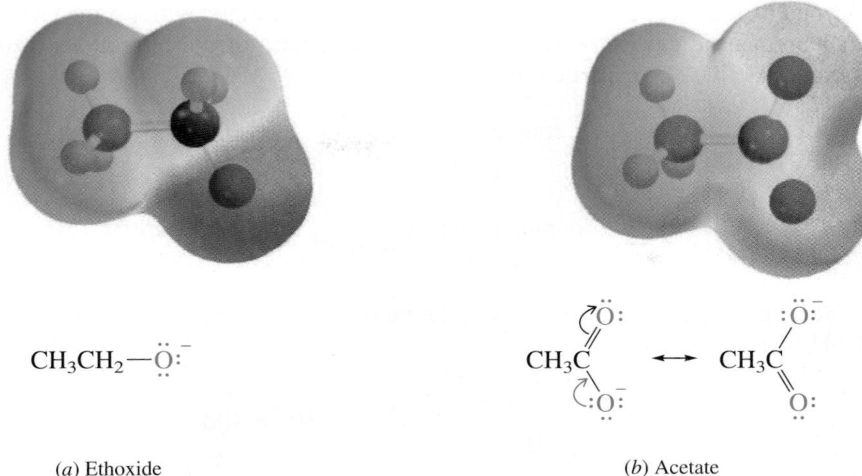

Figure 19.3

The negative charge in ethoxide (*a*) is localized on oxygen. Electron delocalization in acetate (*b*) causes the charge to be shared between two oxygens. The color scale is the same in both electrostatic potential maps.

acetate to be shared equally by both oxygens. Electron delocalization of this type is not available to ethoxide.

$$CH_3C(=O)O^- \longleftrightarrow CH_3C(O^-)=O \quad \text{or} \quad CH_3C(O^{-1/2})_2$$

The electrostatic potential maps in Figure 19.3 contrast the localized negative charge in ethoxide ion with the delocalized charge in acetate.

Measured C—O bond distances also reflect the importance of electron delocalization in acetate ion. Acetic acid's bond distances are consistent with a short double bond (121 pm; 1.21 Å) and a long single bond (136 pm; 1.36 Å), whereas the two carbon–oxygen bond distances in acetate are the same (125 pm; 1.25 Å).

$CH_3C(=O)OH$: C=O 121 pm (1.21Å); C—OH 136 pm (1.36Å)

$CH_3C(O^{-1/2})_2$: 125 pm (1.25 Å); 125 pm (1.25 Å)

Because the electrical properties of a neutral carboxylic acid molecule and a negatively charged carboxylate ion are so different, we need to be aware of which is the major form at the most commonly encountered pH values. For the ionization of a weak acid (HA) in water:

$$H_2\ddot{O}: + H{-}A \rightleftharpoons H_2\overset{+}{O}{-}H + :A^- \qquad K_a = \frac{[H_3O^+][\text{conjugate base}]}{[\text{acid}]}$$

Acid — Conjugate base

and

$$pH = pK_a + \log\frac{[\text{conjugate base}]}{[\text{acid}]}$$

This relationship is known as the **Henderson–Hasselbalch equation.**

Beyond its usual application in calculating the pH of buffer solutions, the Henderson–Hasselbalch equation can be rearranged to tell us the ratio of concentrations of an acid and its conjugate base at a particular pH.

$$\log\frac{[\text{conjugate base}]}{[\text{acid}]} = \text{pH} - \text{p}K_a$$

$$\frac{[\text{conjugate base}]}{[\text{acid}]} = 10^{(\text{pH}-\text{p}K_a)}$$

For a typical carboxylic acid with $\text{p}K_a = 5$, the ratio of the carboxylate ion to the carboxylic acid at pH = 7 is:

$$\frac{[\text{conjugate base}]}{[\text{acid}]} = 10^{(7-5)} = 10^2 = 100$$

Thus, in a solution buffered at a pH of 7, the carboxylate concentration is 100 times greater than the concentration of the undissociated acid.

Notice that this ratio is for a solution at a specified pH, which is not the same as the pH that would result from dissolving a weak acid in pure (unbuffered) water. In the latter instance, ionization of the weak acid proceeds until equilibrium is established at some pH less than 7.

In most biochemical reactions the pH of the medium is close to 7. At this pH, carboxylic acids are nearly completely converted to their conjugate bases. Thus, it is common practice in biological chemistry to specify the derived carboxylate anion rather than the carboxylic acid itself. For example, we say that "glycolysis leads to *lactate* by way of *pyruvate*."

Problem 19.2

(a) Lactic acid has a $\text{p}K_a$ of 3.9. What is the [lactate]/[lactic acid] ratio at the pH of blood (7.4)?

(b) A 0.1 *M* solution of lactic acid in water has a pH of 2.5. What is the [lactate]/[lactic acid] ratio in this solution?

Sample Solution (a) Use the Henderson–Hasselbalch relationship to calculate the ratio of the concentration of the conjugate base (lactate) to the acid (lactic acid).

$$\frac{[\text{conjugate base}]}{[\text{acid}]} = 10^{(\text{pH}-\text{p}K_a)}$$

$$\frac{[\text{lactate}]}{[\text{lactic acid}]} = 10^{(7.4-3.9)}10^{3.5} = 3160$$

19.5 Substituents and Acid Strength

The effect of structure on acidity was introduced in Section 1.13 where we saw that electronegative substituents near an ionizable hydrogen increase its acidity. *Alkyl groups,* on the other hand, have little effect on the ionization constants of carboxylic acids; the K_a values of all acids that have the general formula $C_nH_{2n+1}CO_2H$ are very similar to one another and equal approximately 10^{-5} ($\text{p}K_a = 5$). Table 19.2 gives a few examples.

An electronegative substituent, particularly if it is attached to the α carbon, increases the acidity of a carboxylic acid. All the monohaloacetic acids in Table 19.2 are about 100 times more acidic than acetic acid. Multiple halogen substitution increases the acidity even more; trichloroacetic acid is 7000 times more acidic than acetic acid!

The acid-strengthening effect of electronegative atoms or groups is easily seen as an inductive effect transmitted through the σ bonds of the molecule. According to this model, the σ electrons in the carbon–chlorine bond of chloroacetic acid are drawn toward chlorine, leaving carbon with a slight positive charge. Successive polarization of the electron distribution in the remaining σ bonds increases the positive character of the

TABLE 19.2 Effect of Substituents on Acidity of Carboxylic Acids*

Name of acid	Structure	pK_a
Standard of comparison.		
Acetic acid	CH_3CO_2H	4.7
Alkyl substituents have a negligible effect on acidity.		
Propanoic acid	$CH_3CH_2CO_2H$	4.9
2-Methylpropanoic acid	$(CH_3)_2CHCO_2H$	4.8
2,2-Dimethylpropanoic acid	$(CH_3)_3CCO_2H$	5.1
Heptanoic acid	$CH_3(CH_2)_5CO_2H$	4.9
α-Halogen substituents increase acidity.		
Fluoroacetic acid	FCH_2CO_2H	2.6
Chloroacetic acid	$ClCH_2CO_2H$	2.9
Bromoacetic acid	$BrCH_2CO_2H$	2.9
Dichloroacetic acid	Cl_2CHCO_2H	1.3
Trichloroacetic acid	Cl_3CCO_2H	0.9
Electron-attracting groups increase acidity.		
Methoxyacetic acid	$CH_3OCH_2CO_2H$	3.6
Cyanoacetic acid	$N{\equiv}CCH_2CO_2H$	2.5
Nitroacetic acid	$O_2NCH_2CO_2H$	1.7

*In water at 25°C.

hydroxyl proton, stabilizing the anion. The more stable the anion, the greater the equilibrium constant for its formation.

Electron-withdrawing effect of chlorine is transmitted through σ bonds; increases positive character of OH proton

Inductive effects depend on the electronegativity of the subtituent and the number of σ bonds between it and the affected site. As the number of bonds increases, the inductive effect decreases.

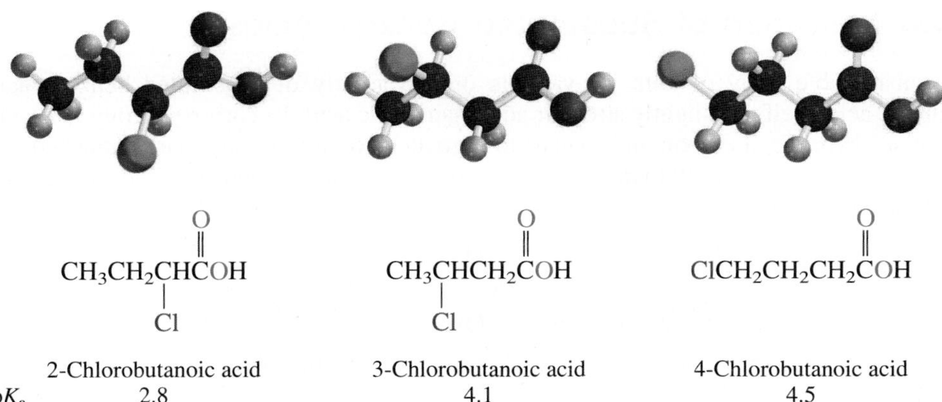

	2-Chlorobutanoic acid	3-Chlorobutanoic acid	4-Chlorobutanoic acid
pK_a	2.8	4.1	4.5

Problem 19.3

Which is the stronger acid in each of the following pairs?

(a) 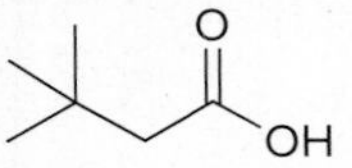 or 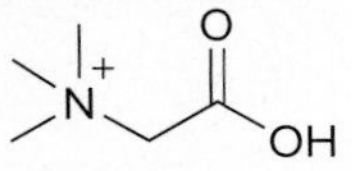

(b) 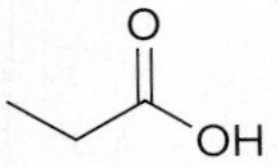 or 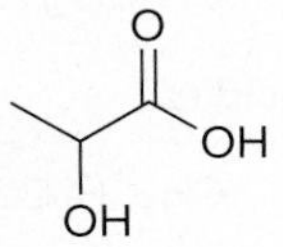

(c) 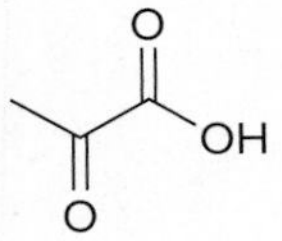 or 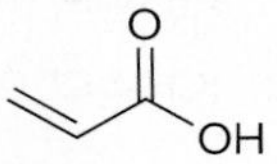

(d) 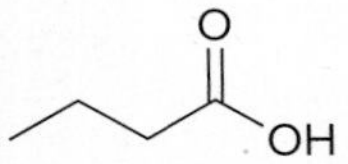 or 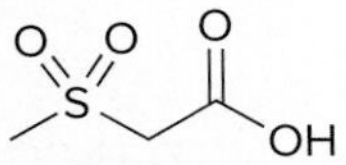

Sample Solution (a) Think of the two compounds as substituted derivatives of acetic acid. A *tert*-butyl group is slightly electron-releasing and has only a modest effect on acidity. The compound $(CH_3)_3CCH_2CO_2H$ is expected to have an acid strength similar to that of acetic acid. A trimethylammonium substituent, on the other hand, is positively charged and is a powerful electron-withdrawing substituent. The compound $(CH_3)_3\overset{+}{N}CH_2CO_2H$ is expected to be a much stronger acid than $(CH_3)_3CCH_2CO_2H$. The measured ionization constants, shown as follows, confirm this prediction.

$(CH_3)_3CCH_2CO_2H$	$(CH_3)_3\overset{+}{N}CH_2CO_2H$
Weaker acid	Stronger acid
$pK_a = 5.3$	$pK_a = 1.8$

Closely related to the inductive effect, and operating in the same direction, is the **field effect.** In the field effect the electronegativity of a substituent is communicated, not by successive polarization of bonds but through the medium, usually the solvent. A substituent in a molecule polarizes surrounding solvent molecules and this polarization is transmitted through other solvent molecules to the remote site.

It is a curious fact that substituents affect the entropy of ionization more than they do the enthalpy term in the expression

$$\Delta G^\circ = \Delta H^\circ - T\Delta S^\circ$$

The enthalpy term ΔH° is close to zero for the ionization of most carboxylic acids, regardless of their strength. The free energy of ionization ΔG° is dominated by the $-T\Delta S^\circ$ term. Ionization is accompanied by an increase in solvation forces, leading to a decrease in the entropy of the system; ΔS° is negative, and $-T\Delta S^\circ$ is positive. Carboxylate ions with substituents capable of dispersing negative charge impose less order on the solvent (water), and less entropy is lost in their production.

19.6 Ionization of Substituted Benzoic Acids

A considerable body of data is available on the acidity of substituted benzoic acids. Benzoic acid itself is a slightly stronger acid than acetic acid. Its carboxyl group is attached to an sp^2-hybridized carbon and ionizes to a greater extent than one that is attached to an sp^3-hybridized carbon. Remember, carbon becomes more electron-withdrawing as its *s* character increases.

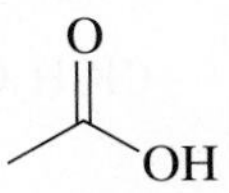

Acetic acid

$pK_a = 4.7$

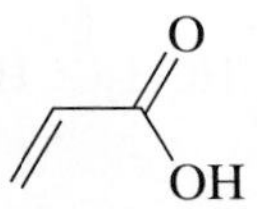

Acrylic acid

$pK_a = 4.3$

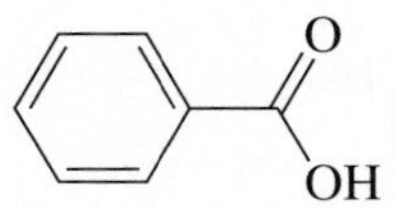

Benzoic acid

$pK_a = 4.2$

TABLE 19.3 Acidity of Some Substituted Benzoic Acids*

	pK_a for different positions of substituent X		
Substituent in $XC_6H_4CO_2H$	Ortho	Meta	Para
H	4.2	4.2	4.2
CH_3	3.9	4.3	4.4
F	3.3	3.9	4.1
Cl	2.9	3.8	4.0
Br	2.8	3.8	4.0
I	2.9	3.9	4.0
CH_3O	4.1	4.1	4.5
O_2N	2.2	3.5	3.4

*In water at 25°C.

Problem 19.4

What is the most acidic neutral molecule characterized by the formula $C_3H_xO_2$?

Table 19.3 lists the ionization constants of some substituted benzoic acids. The largest effects are observed when strongly electron-withdrawing substituents are ortho to the carboxyl group. An *o*-nitro substituent, for example, increases the acidity of benzoic acid 100-fold. Substituent effects are small at positions meta and para to the carboxyl group. In those cases the pK_a values are clustered in the range 3.5–4.5.

19.7 Salts of Carboxylic Acids

In the presence of strong bases such as sodium hydroxide, carboxylic acids are neutralized rapidly and quantitatively:

$$RCO_2H + {}^-OH \xrightleftharpoons{K = 10^{11}} RCO_2^- + H_2O$$

Carboxylic acid (stronger acid) + Hydroxide ion (stronger base) ⇌ Carboxylate ion (weaker base) + Water (weaker acid)

Problem 19.5

Write an ionic equation for the reaction of acetic acid with each of the following, and specify whether the equilibrium favors starting materials or products. What is the value of *K* for each?

(a) Sodium ethoxide
(b) Potassium *tert*-butoxide
(c) Sodium bromide
(d) Sodium acetylide
(e) Potassium nitrate
(f) Lithium amide

Sample Solution (a) This is an acid–base reaction; ethoxide ion is the base.

$$CH_3CO_2H + CH_3CH_2O^- \rightleftharpoons CH_3CO_2^- + CH_3CH_2OH$$

Acetic acid (stronger acid) + Ethoxide ion (stronger base) ⇌ Acetate ion (weaker base) + Ethanol (weaker acid)

The position of equilibrium lies well to the right. Ethanol, with $pK_a = 16$, is a much weaker acid than acetic acid ($pK_a = 4.7$). The equilibrium constant *K* is $10^{(16-4.7)}$, or $10^{11.3}$.

The salts formed on neutralization of carboxylic acids are named by first specifying the metal ion and then adding the name of the acid modified by replacing *-ic acid* by *-ate*. Monocarboxylate salts of diacids are designated by naming both the cation and the hydrogen of the CO_2H group.

$CH_3\overset{O}{\overset{\|}{C}}OLi$ — Lithium acetate

$Cl-C_6H_4-\overset{O}{\overset{\|}{C}}ONa$ — Sodium *p*-chlorobenzoate

$HO\overset{O}{\overset{\|}{C}}(CH_2)_4\overset{O}{\overset{\|}{C}}OK$ — Potassium hydrogen hexanedioate

Metal carboxylates are ionic, and when the molecular weight isn't too high, the sodium and potassium salts of carboxylic acids are soluble in water. Carboxylic acids therefore may be extracted from ether solutions into aqueous sodium or potassium hydroxide.

The solubility behavior of salts of carboxylic acids having 12–18 carbons is unusual and can be illustrated by considering sodium stearate (sodium octadecanoate). Stearate ion contains two very different structural units—a long nonpolar hydrocarbon chain and a polar carboxylate group. The electrostatic potential map of sodium stearate in Figure 19.4 illustrates how different most of the molecule is from its polar carboxylate end.

"Hydrophobic" is often used instead of "lipophilic."

Carboxylate groups are **hydrophilic** ("water-loving") and tend to confer water solubility on species that contain them. Long hydrocarbon chains are **lipophilic** ("fat-loving") and tend to associate with other hydrocarbon chains. Sodium stearate is an example of an **amphiphilic** substance; both hydrophilic and lipophilic groups occur within the same molecule.

When sodium stearate is placed in water, the hydrophilic carboxylate group encourages the formation of a solution; the lipophilic alkyl chain discourages it. The compromise achieved is to form a colloidal dispersion of aggregates called **micelles** (Figure 19.5). Micelles form spontaneously when the carboxylate concentration exceeds a certain minimum value called the **critical micelle concentration.** Each micelle is composed of

Figure 19.4

Structure and electrostatic potential map of sodium stearate.

$O^- Na^+$

lipophilic (hydrophobic)

hydrophilic

Sodium stearate $[CH_3(CH_2)_{16}CO_2Na]$

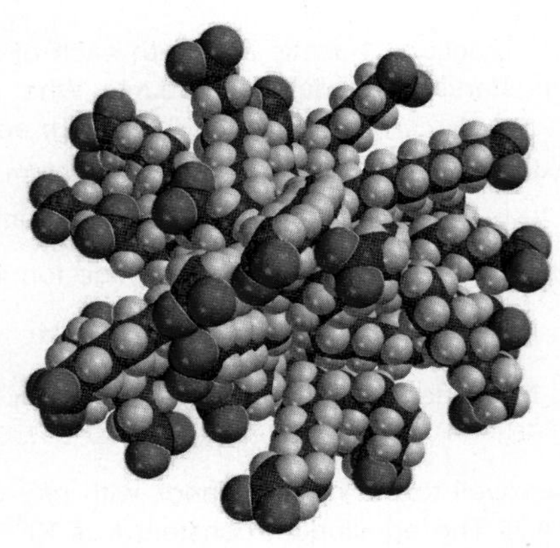

Figure 19.5

Space-filling model of a micelle formed by association of carboxylate ions derived from a long-chain carboxylic acid. The hydrocarbon chains tend to be on the inside and the carboxylate ions on the surface where they are in contact with water molecules and metal cations.

50–100 individual molecules, with the polar carboxylate groups directed toward its outside where they experience attractive forces with water and sodium ions. The nonpolar hydrocarbon chains are directed toward the interior of the micelle, where individually weak but cumulatively significant induced-dipole/induced-dipole forces bind them together. Micelles are approximately spherical because a sphere exposes the minimum surface for a given volume of material and disrupts the water structure least. Because their surfaces are negatively charged, two micelles repel each other rather than clustering to form higher aggregates.

The formation of micelles and their properties are responsible for the cleansing action of soaps. Water that contains sodium stearate removes grease by enclosing it in the hydrocarbon-like interior of the micelles. The grease is washed away with the water, not because it dissolves in the water but because it dissolves in the micelles that are dispersed in the water. Sodium stearate is an example of a soap; sodium and potassium salts of other C_{12}–C_{18} unbranched carboxylic acids possess similar properties.

Detergents are substances, including soaps, that cleanse by micellar action. A large number of synthetic detergents are known. An example is sodium lauryl sulfate [$CH_3(CH_2)_{10}CH_2OSO_3Na$], which has a long hydrocarbon chain terminating in a polar sulfate ion and forms soap-like micelles in water. Detergents are designed to be effective in hard water, meaning water containing calcium salts that form insoluble calcium carboxylates with soaps. These precipitates rob the soap of its cleansing power and form an unpleasant scum. The calcium salts of synthetic detergents such as sodium lauryl sulfate, however, are soluble and retain their micelle-forming ability even in hard water.

19.8 Dicarboxylic Acids

Separate ionization constants, designated K_1 and K_2, respectively, characterize the two successive ionization steps of a dicarboxylic acid.

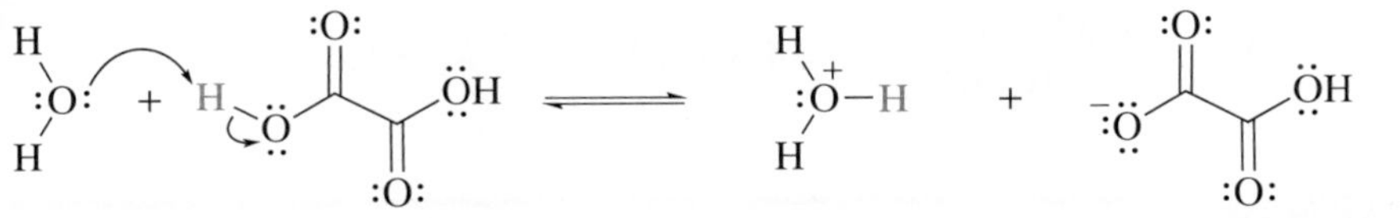

Water + Oxalic acid ⇌ Hydronium ion + Hydrogen oxalate ion $\quad pK_1 = 1.2$

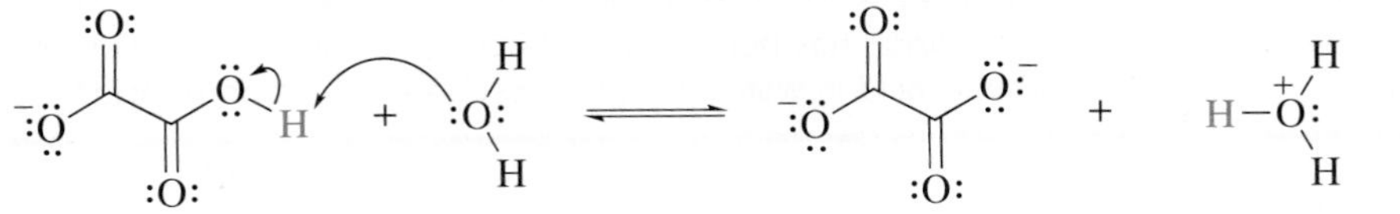

Hydrogen oxalate ion + Water ⇌ Oxalate ion + Hydronium ion $\quad pK_2 = 4.3$

Oxalic acid is poisonous and occurs naturally in a number of plants including sorrel and begonia. It is a good idea to keep houseplants out of the reach of small children, who might be tempted to eat the leaves or berries.

The first ionization constant of dicarboxylic acids is larger than K_a for monocarboxylic analogs. One reason is statistical. There are two potential sites for ionization rather than one, making the effective concentration of carboxyl groups twice as large. Furthermore, one carboxyl group acts as an electron-withdrawing group to facilitate dissociation of the other. This is particularly noticeable when the two carboxyl groups are separated by only a few bonds. Oxalic and malonic acid, for example, are two to three orders of magnitude stronger than simple alkyl derivatives of acetic acid. Heptanedioic acid, in which the carboxyl groups are well separated from each other, is only slightly stronger than acetic acid.

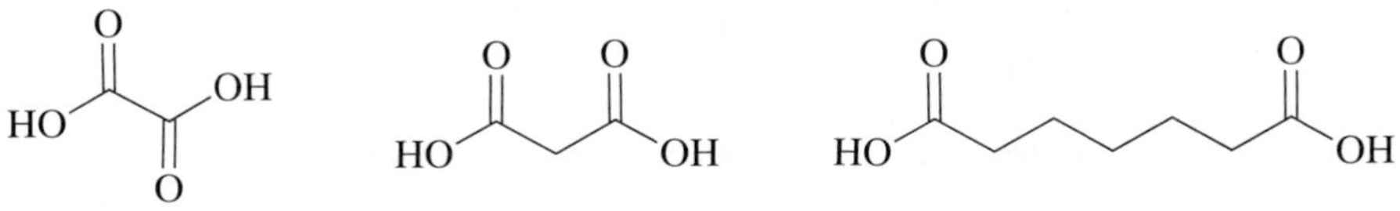

Oxalic acid $pK_1 = 1.2$ | Malonic acid $pK_1 = 2.8$ | Heptanedioic acid $pK_1 = 4.3$

19.9 Carbonic Acid

Through an accident of history, the simplest dicarboxylic acid, carbonic acid, HOC(=O)OH, is not even classified as an organic compound. Because many minerals are carbonate salts, nineteenth-century chemists placed carbonates, bicarbonates, and carbon dioxide in the inorganic realm. Nevertheless, the essential features of carbonic acid and its salts are easily understood on the basis of our knowledge of carboxylic acids.

The systematic name for bicarbonate ion is *hydrogen carbonate*. Thus, the systematic name for sodium bicarbonate ($NaHCO_3$) is *sodium hydrogen carbonate*.

Carbonic acid is formed when carbon dioxide reacts with water. Hydration of carbon dioxide is far from complete, however. Almost all the carbon dioxide that is dissolved in water exists as carbon dioxide; only 0.3% of it is converted to carbonic acid. Carbonic acid is a weak acid and ionizes to a small extent to bicarbonate ion.

:Ö=C=Ö: + H_2O ⇌ HOC(=O)OH

Carbon dioxide Water Carbonic acid

Water + Carbonic acid ⇌ Hydronium ion + Bicarbonate ion

The equilibrium constant for the overall reaction is related to an apparent equilibrium constant K_1 for carbonic acid ionization by the expression

$$K_1 = \frac{[H_3O^+][HCO_3^-]}{[CO_2]} = 4.3 \times 10^{-7} \qquad pK_1 = 6.4$$

Problem 19.6

The value cited for K_1 of carbonic acid, 4.3×10^{-7}, is determined by measuring the pH of water to which a known amount of carbon dioxide has been added. When we recall that only 0.3% of carbon dioxide is converted to carbonic acid in water, what is the "true K_1" of carbonic acid?

Carbonic anhydrase is an enzyme that catalyzes the hydration of carbon dioxide to bicarbonate. The uncatalyzed hydration of carbon dioxide is too slow to be effective in transporting carbon dioxide from the tissues to the lungs, and so animals have developed catalysts to speed this process. The activity of carbonic anhydrase is remarkable; it has been estimated that one molecule of this enzyme can catalyze the hydration of 3.6×10^7 molecules of carbon dioxide per minute.

As with other dicarboxylic acids, the second ionization constant of carbonic acid is far smaller than the first.

Bicarbonate ion + Water ⇌ Carbonate ion + Hydronium ion $\quad pK_2 = 10.2$

Bicarbonate is a weaker acid than carboxylic acids but a stronger acid than water and alcohols.

19.10 Sources of Carboxylic Acids

Many carboxylic acids were first isolated from natural sources and were given common names based on their origin (Figure 19.6). Formic acid (Latin *formica,* meaning "ant") was obtained by distilling ants, but is found in some other insects as well. Since ancient times acetic acid (Latin *acetum,* for "vinegar") has been known to be present in wine that has turned sour. Butyric acid (Latin *butyrum,* meaning "butter") contributes to the odor of both rancid butter and ginkgo berries. Malic acid (Latin *malum,* meaning "apple") occurs in apples. Oleic acid (Latin *oleum,* "oil") takes its name from naturally occurring esters such as those that comprise the major portion of olive oil.

The large-scale preparation of carboxylic acids relies on chemical synthesis. Virtually none of the 3×10^9 lb of acetic acid produced in the United States each year is obtained from vinegar. Most of it comes from the reaction of methanol with carbon monoxide.

$$\underset{\text{Methanol}}{CH_3OH} + \underset{\text{Carbon monoxide}}{CO} \xrightarrow[\text{heat, pressure}]{\text{cobalt or rhodium catalyst}} \underset{\text{Acetic acid}}{CH_3CO_2H}$$

The principal end use of acetic acid is in the production of vinyl acetate for paints and adhesives.

Ants aren't the only insects that use formic acid as a weapon. Some *Galerita* beetles spray attackers with an 80% solution of it.

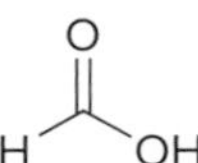

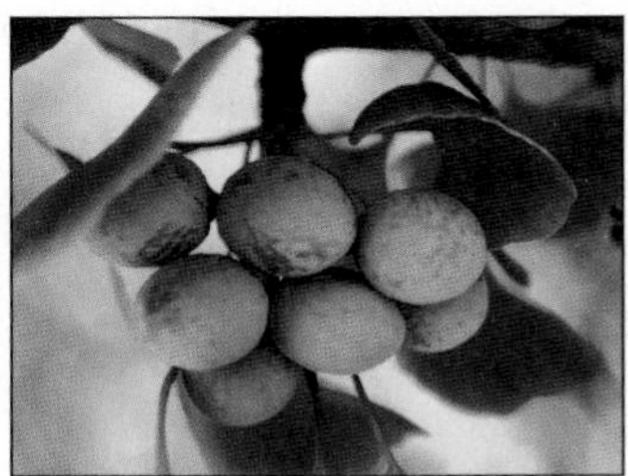

Butanoic and hexanoic acid are responsible for the nasty odor of ginkgo seeds.

Ethanol is oxidized to acetic acid as wine becomes vinegar.

Malic acid and citric acid contribute to the tart taste of many fruits and vegetables.

The oleic acid that forms during decomposition of dead ants is a chemical signal to other ants to carry them from the nest. In an experiment in which live ants had been coated with oleic acid, they were also removed.

Figure 19.6

Some naturally occurring carboxylic acids.

(*Top Left*): ©Nature's Images/Science Source; (*Top Middle*): ©diyanski/123RF; (*Top Right*): ©Iconotec/Glow Images; (*Bottom Left*): ©Photodisc/Getty Images; (*Bottom Right*): ©Gary Vestal/Getty Images

The carboxylic acid produced in the greatest amounts is 1,4-benzenedicarboxylic acid (terephthalic acid). About 5×10^9 lb/year is produced in the United States as a starting material for the preparation of polyester fibers, "PETE" beverage containers, and Mylar film. One important process converts *p*-xylene to terephthalic acid by oxidation with nitric acid:

See Chapter 27 for more on polymers made from terephthalic acid.

$$H_3C-C_6H_4-CH_3 \xrightarrow{HNO_3} HO_2C-C_6H_4-CO_2H$$

p-Xylene → 1,4-Benzenedicarboxylic acid (terephthalic acid)

You will recognize the side-chain oxidation of *p*-xylene to terephthalic acid as a reaction type discussed previously (see Section 12.12). It, and examples of other reactions encountered earlier that can be applied to the synthesis of carboxylic acids, are collected in Table 19.4.

TABLE 19.4 Summary of Reactions Discussed in Earlier Chapters That Yield Carboxylic Acids

Reaction (section) and comments	General equation and specific example
Side-chain oxidation of alkylbenzenes (see Section 12.12) A primary or secondary alkyl side chain on an aromatic ring is converted to a carboxyl group by reaction with a strong oxidizing agent such as potassium permanganate or chromic acid.	$ArCHR_2 \xrightarrow[K_2Cr_2O_7,\ H_2SO_4,\ H_2O]{KMnO_4\ \text{or}} ArCO_2H$ Alkylbenzene → Arenecarboxylic acid 3-Methoxy-4-nitrotoluene $\xrightarrow[2.\ H_3O^+]{1.\ KMnO_4,\ HO^-}$ 3-Methoxy-4-nitrobenzoic acid (100%)
Oxidation of primary alcohols (see Section 16.9) Potassium permanganate and chromic acid convert primary alcohols to carboxylic acids by way of the corresponding aldehyde.	$RCH_2OH \xrightarrow[K_2Cr_2O_7,\ H_2SO_4,\ H_2O]{KMnO_4\ \text{or}} RCO_2H$ Primary alcohol → Carboxylic acid 2-*tert*-Butyl-3,3-dimethyl-1-butanol $\xrightarrow[H_2O,\ H_2SO_4]{H_2CrO_4}$ 2-*tert*-Butyl-3,3-dimethylbutanoic acid (82%)
Oxidation of aldehydes (see Section 18.14) Aldehydes are particularly sensitive to oxidation and are converted to carboxylic acids by a number of oxidizing agents, including potassium permanganate and chromic acid.	$RCH(=O) \xrightarrow{\text{oxidizing agent}} RCO_2H$ Aldehyde → Carboxylic acid Furan-2-carbaldehyde $\xrightarrow{K_2Cr_2O_7,\ H_2SO_4,\ H_2O}$ Furan-2-carboxylic acid (75%)

The examples in the table give carboxylic acids that have the same number of carbon atoms as the starting material. The reactions to be described in the next two sections permit carboxylic acids to be prepared by extending a chain by one carbon atom and are of great value in laboratory syntheses of carboxylic acids.

19.11 Synthesis of Carboxylic Acids by the Carboxylation of Grignard Reagents

We've already seen how Grignard reagents add to carbonyl groups of aldehydes and ketones (Section 15.5). They react in much the same way with carbon dioxide to give magnesium carboxylates, which on acidification yield carboxylic acids.

$$\overset{\delta-}{R}{-}\underset{\delta+}{MgX} + O{=}C{=}O \xrightarrow{\text{diethyl ether}} RCOMgX \xrightarrow{H_3O^+} RCOH$$

Grignard reagent acts as a nucleophile toward carbon dioxide

Halomagnesium carboxylate

Carboxylic acid

Overall, the carboxylation of Grignard reagents transforms an alkyl or aryl halide to a carboxylic acid in which the carbon skeleton has been extended by one carbon atom.

2-Chlorobutane → (1. Mg, diethyl ether; 2. CO_2; 3. H_2O) → 2-Methylbutanoic acid (76–78%)

9-Bromo-10-methylphenanthrene → (1. Mg, diethyl ether; 2. CO_2; 3. H_3O^+) → 10-Methylphenanthrene-9-carboxylic acid (82%)

The major limitation to this procedure is that the alkyl or aryl halide must not bear substituents that are incompatible with Grignard reagents, such as OH, NH, SH, or C=O.

Problem 19.7

2,6-Dimethoxybenzoic acid was needed for a synthesis of the β-lactam antibiotic methicillin. Show how this carboxylic acid could be synthesized from 2-bromo-1,3-benzenediol.

2,6-Dimethoxybenzoic acid ⟹ 2-Bromo-1,3-benzenediol

19.12 Synthesis of Carboxylic Acids by the Preparation and Hydrolysis of Nitriles

Primary and secondary alkyl halides may be converted to the next higher carboxylic acid by a two-step synthetic sequence involving the preparation and hydrolysis of *nitriles.* Nitriles, also known as *alkyl cyanides,* are prepared by nucleophilic substitution.

$$:\ddot{X}-R \; + \; :\bar{C}\equiv N: \xrightarrow{S_N2} RC\equiv N \; + \; :\ddot{X}:^-$$

Primary or secondary alkyl halide; Cyanide ion; Nitrile (alkyl cyanide); Halide ion

The reaction follows an S_N2 mechanism and works best with primary and secondary alkyl halides. Elimination is the only reaction observed with tertiary alkyl halides. Aryl and vinyl halides do not react.

Once the cyano group has been introduced, the nitrile is subjected to hydrolysis. Usually this is carried out by heating in aqueous acid.

The mechanism of nitrile hydrolysis will be described in Section 20.16.

$$RC\equiv N + 2H_2O + H^+ \xrightarrow{heat} RCOH \; (\text{C}=\text{O}) + NH_4^+$$

Nitrile; Water; Carboxylic acid; Ammonium ion

$$C_6H_5CH_2Cl \xrightarrow[DMSO]{NaCN} C_6H_5CH_2CN \xrightarrow[heat]{H_2O,\ H_2SO_4} C_6H_5CH_2COH \; (\text{C}=\text{O})$$

Benzyl chloride; Benzyl cyanide (92%); Phenylacetic acid (77%)

Dicarboxylic acids have been prepared from dihalides by this method:

Applications of dicarboxylic acids in the synthesis of nylon and other polymers are described in Sections 27.11 and 27.12.

$$BrCH_2CH_2CH_2Br \xrightarrow[H_2O]{NaCN} NCCH_2CH_2CH_2CN \xrightarrow[heat]{H_2O,\ HCl} HOC(=O)CH_2CH_2CH_2C(=O)OH$$

1,3-Dibromopropane; 1,5-Pentanedinitrile (77–86%); 1,5-Pentanedioic acid (83–85%)

Problem 19.8

Of the two procedures just described, preparation and carboxylation of a Grignard reagent or formation and hydrolysis of a nitrile, only one is appropriate to each of the following RX → RCO_2H conversions. Identify the correct procedure in each case, and specify why the other will fail.

(a) Bromobenzene → benzoic acid

(b) 2-Chloroethanol → 3-hydroxypropanoic acid

(c) *tert*-Butyl chloride → 2,2-dimethylpropanoic acid

Sample Solution (a) Bromobenzene is an aryl halide and is unreactive toward nucleophilic substitution by cyanide ion. The route $C_6H_5Br \rightarrow C_6H_5CN \rightarrow C_6H_5CO_2H$ fails because the first step fails. The route proceeding through the Grignard reagent is perfectly satisfactory and appears as an experiment in a number of introductory organic chemistry laboratory texts.

$$C_6H_5Br \xrightarrow[\text{diethyl ether}]{Mg} C_6H_5MgBr \xrightarrow[2.\ H_3O^+]{1.\ CO_2} C_6H_5C(=O)OH$$

Bromobenzene; Phenylmagnesium bromide; Benzoic acid

Nitrile groups in cyanohydrins are hydrolyzed under conditions similar to those of alkyl cyanides. Cyanohydrin formation followed by hydrolysis provides a route to the preparation of α-hydroxy carboxylic acids.

Recall the preparation of cyanohydrins in Section 18.7.

2-Pentanone $\xrightarrow[\text{2. } H_3O^+]{\text{1. NaCN}}$ 2-Pentanone cyanohydrin $\xrightarrow[\text{heat}]{H_2O,\ HCl}$ 2-Hydroxy-2-methylpentanoic acid (60% from 2-pentanone)

19.13 Reactions of Carboxylic Acids: A Review and a Preview

The most apparent chemical property of carboxylic acids, their acidity, has already been examined. Three reactions of carboxylic acids—conversion to acyl chlorides, reduction, and esterification—have been encountered in previous chapters and are reviewed in Table 19.5. Acid-catalyzed esterification of carboxylic acids is one of the fundamental reactions of organic chemistry, and this portion of the chapter begins with an examination of the mechanism by which it occurs.

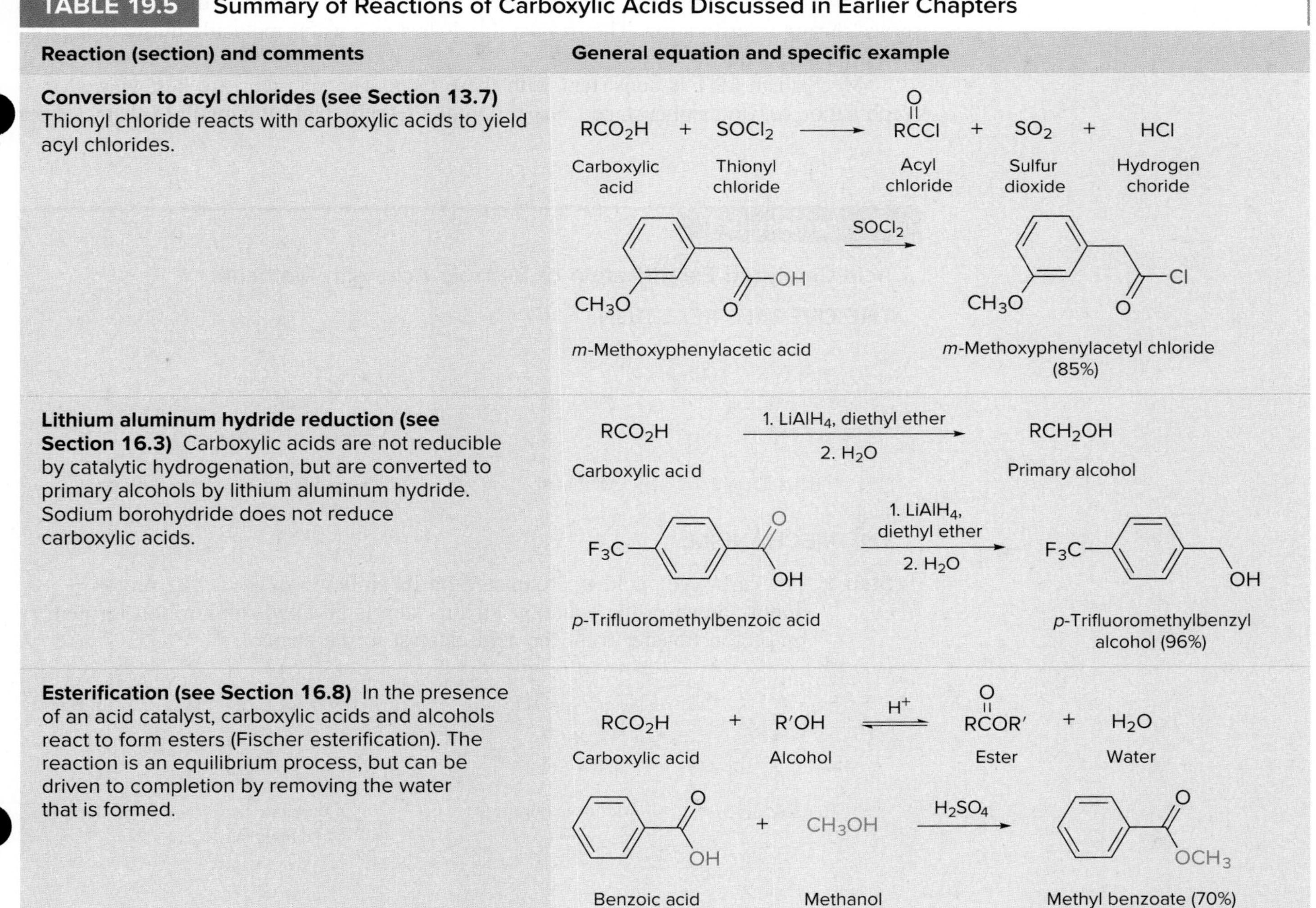

TABLE 19.5 Summary of Reactions of Carboxylic Acids Discussed in Earlier Chapters

Reaction (section) and comments	General equation and specific example
Conversion to acyl chlorides (see Section 13.7) Thionyl chloride reacts with carboxylic acids to yield acyl chlorides.	RCO_2H (Carboxylic acid) + $SOCl_2$ (Thionyl chloride) → RCCl (Acyl chloride) + SO_2 (Sulfur dioxide) + HCl (Hydrogen choride) *m*-Methoxyphenylacetic acid $\xrightarrow{SOCl_2}$ *m*-Methoxyphenylacetyl chloride (85%)
Lithium aluminum hydride reduction (see Section 16.3) Carboxylic acids are not reducible by catalytic hydrogenation, but are converted to primary alcohols by lithium aluminum hydride. Sodium borohydride does not reduce carboxylic acids.	RCO_2H (Carboxylic aci d) $\xrightarrow[\text{2. } H_2O]{\text{1. } LiAlH_4\text{, diethyl ether}}$ RCH_2OH (Primary alcohol) *p*-Trifluoromethylbenzoic acid $\xrightarrow[\text{2. } H_2O]{\text{1. } LiAlH_4\text{, diethyl ether}}$ *p*-Trifluoromethylbenzyl alcohol (96%)
Esterification (see Section 16.8) In the presence of an acid catalyst, carboxylic acids and alcohols react to form esters (Fischer esterification). The reaction is an equilibrium process, but can be driven to completion by removing the water that is formed.	RCO_2H (Carboxylic acid) + R′OH (Alcohol) $\overset{H^+}{\rightleftharpoons}$ RCOR′ (Ester) + H_2O (Water) Benzoic acid + CH_3OH (Methanol) $\xrightarrow{H_2SO_4}$ Methyl benzoate (70%)

19.14 Mechanism of Acid-Catalyzed Esterification

An important question about the mechanism of acid-catalyzed esterification concerns the origin of the alkoxy oxygen. For example, does the methoxy oxygen in methyl benzoate come from methanol, or is it derived from benzoic acid?

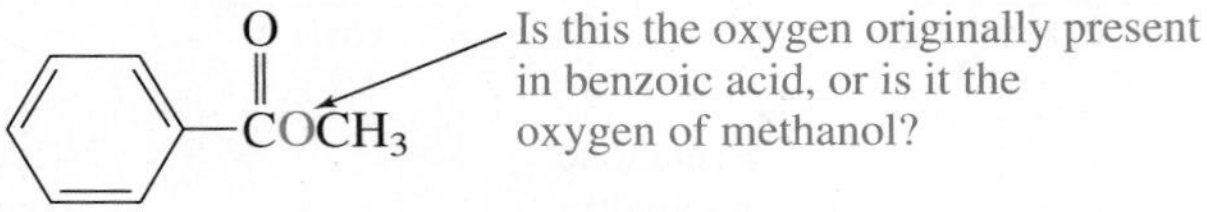

$C_6H_5COCH_3$ (with C=O) — Is this the oxygen originally present in benzoic acid, or is it the oxygen of methanol?

Urey's achievements in chemistry were many and include the discovery of deuterium for which he was awarded the 1934 Nobel Prize in Chemistry.

A clear-cut answer was provided by Irving Roberts and Harold C. Urey of Columbia University in 1938. They prepared methanol that had been enriched in the mass-18 isotope of oxygen and found that when this sample of methanol was esterified with benzoic acid, the methyl benzoate product contained all the ^{18}O label that was originally present in the alcohol.

In this equation, the red O signifies oxygen enriched in its mass-18 isotope; analysis of isotopic enrichment was performed by mass spectrometry.

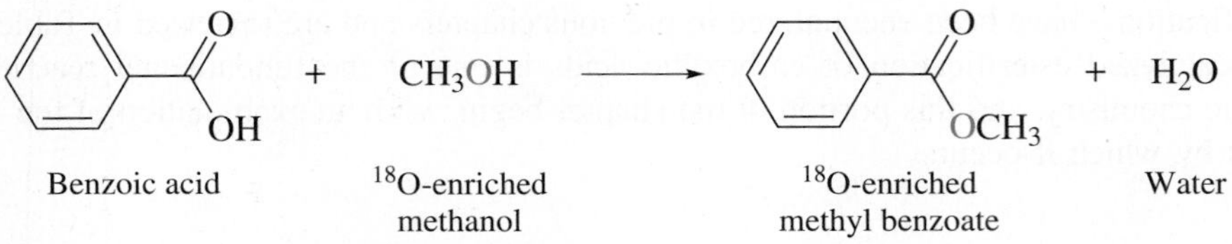

$$C_6H_5COOH + CH_3OH \longrightarrow C_6H_5COOCH_3 + H_2O$$

Benzoic acid + ^{18}O-enriched methanol → ^{18}O-enriched methyl benzoate + Water

The Roberts–Urey experiment tells us that the C—O bond of the alcohol is preserved during esterification. The oxygen that is lost as a water molecule must come from the carboxylic acid.

Mechanism 19.1 is consistent with these facts. The six steps are best viewed as a combination of two distinct stages. *Formation* of a **tetrahedral intermediate** characterizes

Mechanism 19.1

Acid-Catalyzed Esterification of Benzoic Acid with Methanol

THE OVERALL REACTION:

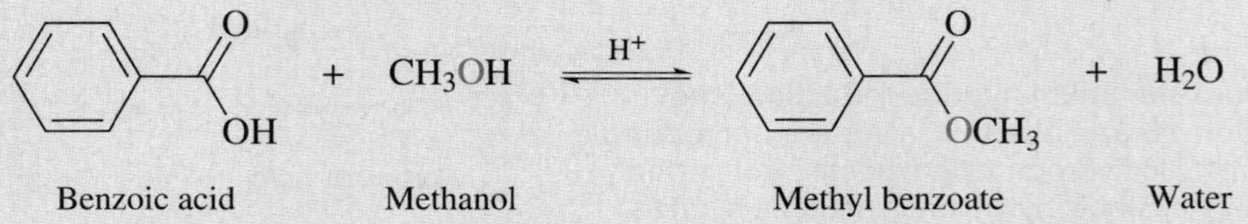

$$C_6H_5COOH + CH_3OH \overset{H^+}{\rightleftharpoons} C_6H_5COOCH_3 + H_2O$$

Benzoic acid + Methanol ⇌ Methyl benzoate + Water

THE MECHANISM:

Step 1: The carboxylic acid is protonated on its carbonyl oxygen. The proton donor shown in the equation for this step is an alkyloxonium ion formed by proton transfer from the acid catalyst to the alcohol.

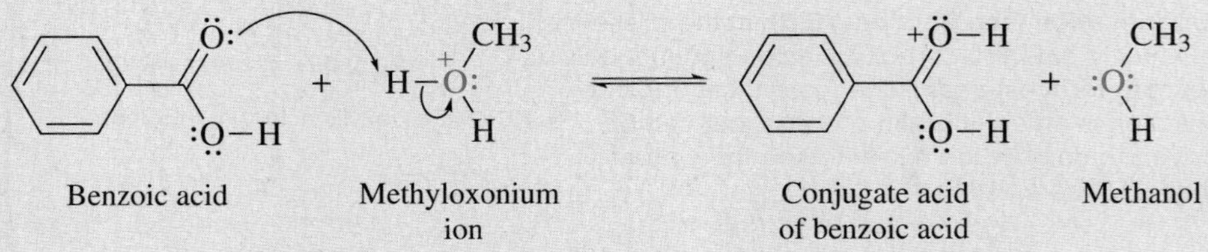

$$C_6H_5C(=\ddot{O}:)\ddot{O}H + H\text{—}\overset{+}{O}(CH_3)H \rightleftharpoons C_6H_5C(=\overset{+}{O}H)\ddot{O}H + :\ddot{O}(CH_3)H$$

Benzoic acid + Methyloxonium ion ⇌ Conjugate acid of benzoic acid + Methanol

Step 2: Protonation of the carboxylic acid increases the positive character of its carbonyl group. A molecule of the alcohol acts as a nucleophile and bonds to the carbonyl carbon.

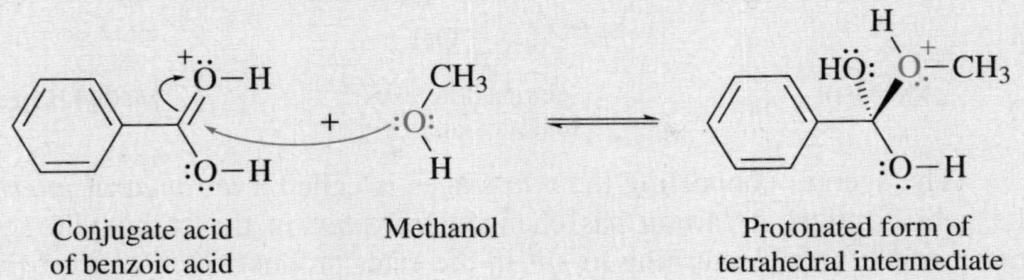

Conjugate acid of benzoic acid + Methanol ⇌ Protonated form of tetrahedral intermediate

Step 3: The oxonium ion formed in step 2 loses a proton to give the tetrahedral intermediate in its neutral form. This concludes the first stage in the mechanism.

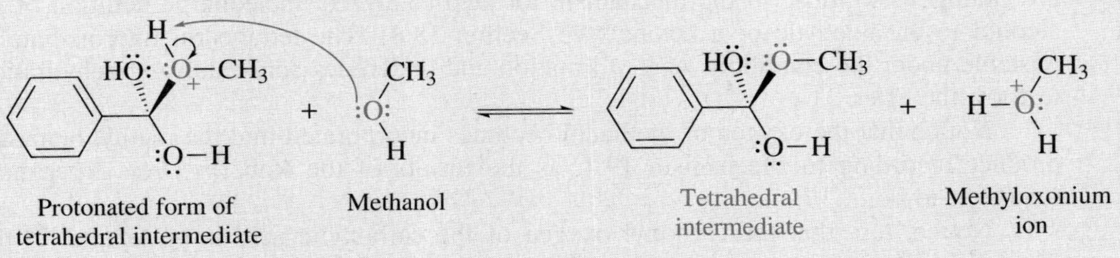

Protonated form of tetrahedral intermediate + Methanol ⇌ Tetrahedral intermediate + Methyloxonium ion

Step 4: The second stage begins with protonation of the tetrahedral intermediate on one of its hydroxyl oxygens.

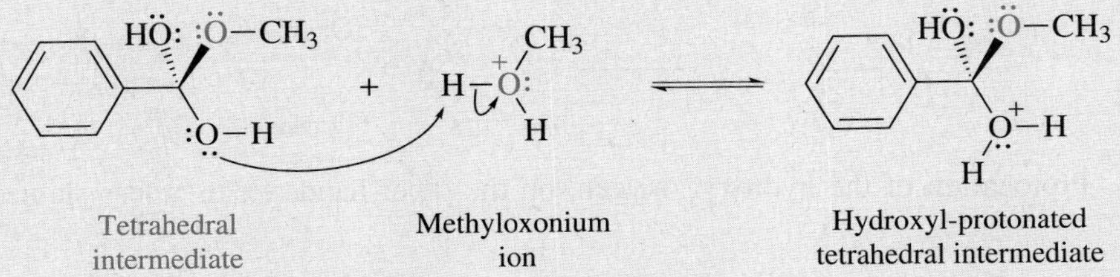

Tetrahedral intermediate + Methyloxonium ion ⇌ Hydroxyl-protonated tetrahedral intermediate

Step 5: The hydroxyl-protonated tetrahedral intermediate dissociates to a molecule of water and gives the protonated form of the ester.

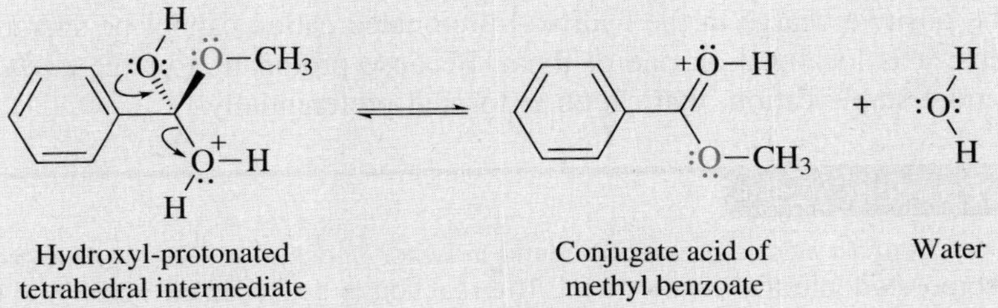

Hydroxyl-protonated tetrahedral intermediate ⇌ Conjugate acid of methyl benzoate + Water

Step 6: Deprotonation of the protonated ester gives the ester.

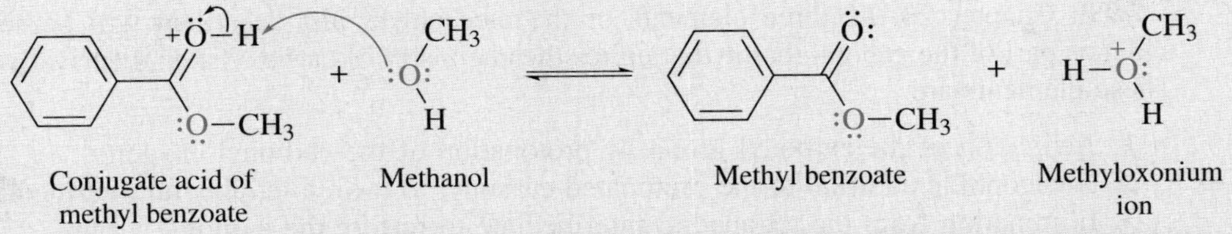

Conjugate acid of methyl benzoate + Methanol ⇌ Methyl benzoate + Methyloxonium ion

the first stage (steps 1–3), and *dissociation* of this tetrahedral intermediate characterizes the second (steps 4–6).

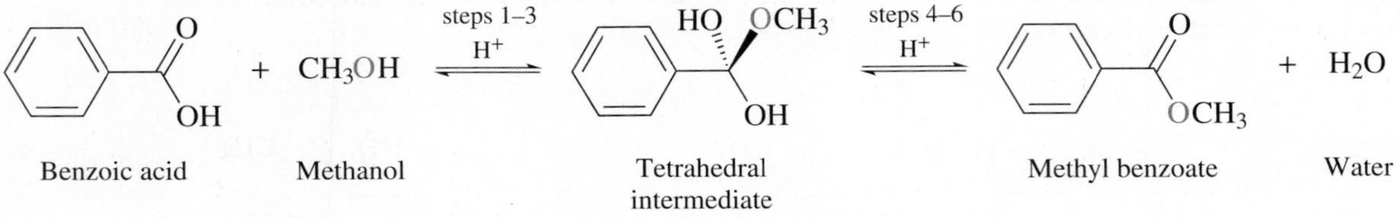

Benzoic acid + CH_3OH (Methanol) $\xrightleftharpoons[]{\text{steps 1–3, } H^+}$ Tetrahedral intermediate $\xrightleftharpoons[]{\text{steps 4–6, } H^+}$ Methyl benzoate + H_2O (Water)

The species connecting the two stages is called a *tetrahedral intermediate* because the hybridization at carbon has changed from sp^2 in the carboxylic acid to sp^3 in the intermediate before returning to sp^2 in the ester product. *The tetrahedral intermediate is formed by nucleophilic addition of an alcohol to a carboxylic acid and is analogous to a hemiacetal formed by nucleophilic addition of an alcohol to an aldehyde or a ketone.* The three steps that lead to the tetrahedral intermediate in the first stage of esterification are analogous to those in the mechanism for acid-catalyzed nucleophilic addition of an alcohol to an aldehyde or a ketone (see Section 18.8). The tetrahedral intermediate is unstable under the conditions of its formation and undergoes acid-catalyzed dehydration to form the ester.

Notice that the oxygen of methanol becomes incorporated into the methyl benzoate product according to Mechanism 19.1, as the results of the Roberts–Urey experiment require it to be.

Notice, too, that the carbonyl oxygen of the carboxylic acid is protonated in the first step and not the hydroxyl oxygen. The species formed by protonation of the carbonyl oxygen is more stable because it is stabilized by electron delocalization. The positive charge is shared equally by both oxygens.

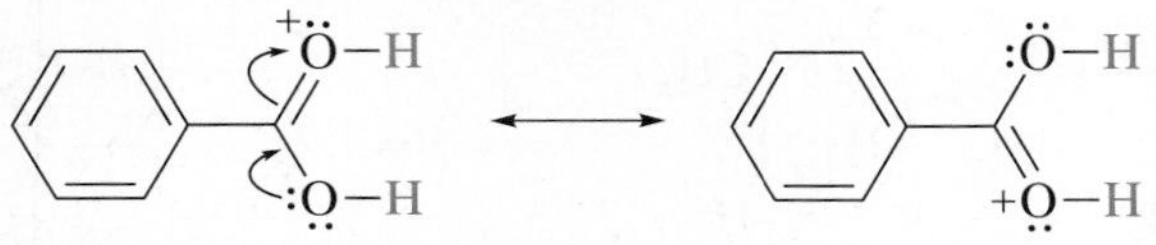

Electron delocalization in carbonyl-protonated benzoic acid

Protonation of the hydroxyl oxygen, on the other hand, yields a less stable cation:

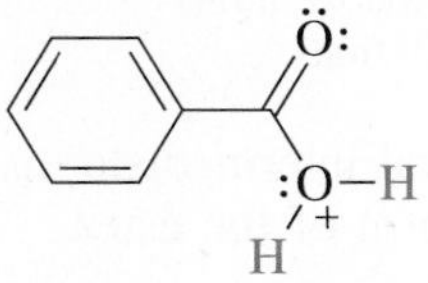

Localized positive charge in hydroxyl-protonated benzoic acid

The positive charge in the hydroxyl-protonated cation cannot be shared by the two oxygens; it is localized on one of them. Because protonation of the *carbonyl oxygen* gives a more stable cation, that cation is formed preferentially.

Problem 19.9

When benzoic acid is allowed to stand in water enriched in ^{18}O, the isotopic label becomes incorporated into the benzoic acid. The reaction is catalyzed by acids. Suggest an explanation for this observation.

In Chapter 20 the three elements of the mechanism just described will be seen again as part of the general theme that unites the chemistry of carboxylic acid derivatives. These elements are

1. Activation of the carbonyl group by protonation of the carbonyl oxygen
2. Nucleophilic addition to the protonated carbonyl to form a tetrahedral intermediate
3. Elimination from the tetrahedral intermediate to restore the carbonyl group

This sequence is fundamental to the carbonyl-group chemistry of carboxylic acids, acyl chlorides, anhydrides, esters, and amides.

19.15 Intramolecular Ester Formation: Lactones

Hydroxy acids, compounds that contain both a hydroxyl and a carboxylic acid function, have the capacity to form cyclic esters called **lactones.** This intramolecular esterification takes place spontaneously when the ring that is formed is five- or six-membered. Lactones that contain a five-membered cyclic ester are referred to as *γ-lactones;* their six-membered analogs are known as *δ-lactones.*

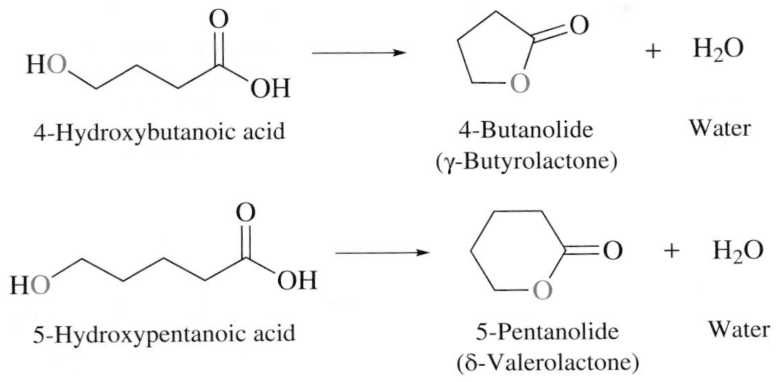

4-Hydroxybutanoic acid → 4-Butanolide (γ-Butyrolactone) + H_2O (Water)

5-Hydroxypentanoic acid → 5-Pentanolide (δ-Valerolactone) + H_2O (Water)

Lactones are named by replacing the *-oic acid* ending of the parent carboxylic acid by *-olide* and identifying its oxygenated carbon by number as illustrated in the preceding equations.

Reactions that are expected to produce hydroxy acids often yield the derived lactones instead if a five- or six-membered ring can be formed.

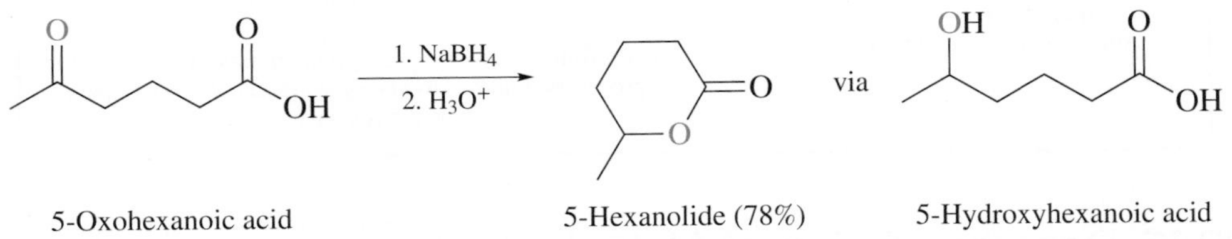

5-Oxohexanoic acid —(1. $NaBH_4$; 2. H_3O^+)→ 5-Hexanolide (78%) via 5-Hydroxyhexanoic acid

Many natural products are lactones, and it is not unusual to find examples in which the ring size is rather large. A few naturally occurring lactones are shown in Figure 19.7. The *macrolide antibiotics,* of which erythromycin is a prominent example, are macrocyclic (large-ring) lactones. The lactone ring of erythromycin is 14-membered.

Problem 19.10

Write the structure of the hydroxy acid corresponding to each of the lactones shown in Figure 19.7.

(a) Mevalonolactone (b) Pentadecanolide (c) Vernolepin

Sample Solution (a) The ring oxygen of the lactone is derived from the OH group of the hydroxy acid. To identify the hydroxy acid, disconnect the O—C(O) bond of the lactone.

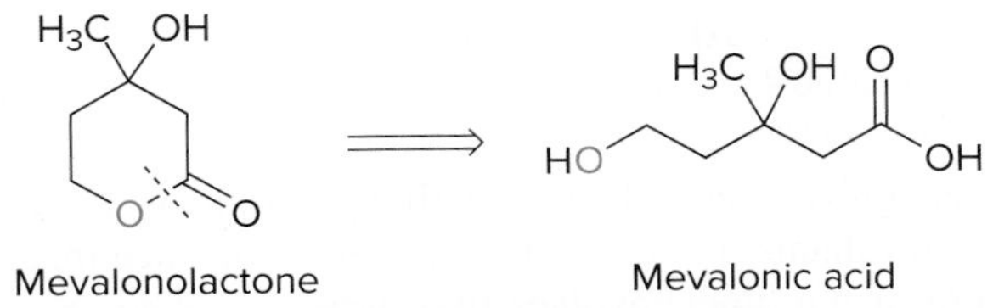

Mevalonolactone ⟹ Mevalonic acid

Lactones with three- or four-membered rings (α-lactones and β-lactones) are very reactive, making their isolation difficult. Special methods are normally required for the laboratory synthesis of small-ring lactones as well as those that contain rings larger than six-membered.

Figure 19.7

Some naturally occurring lactones.

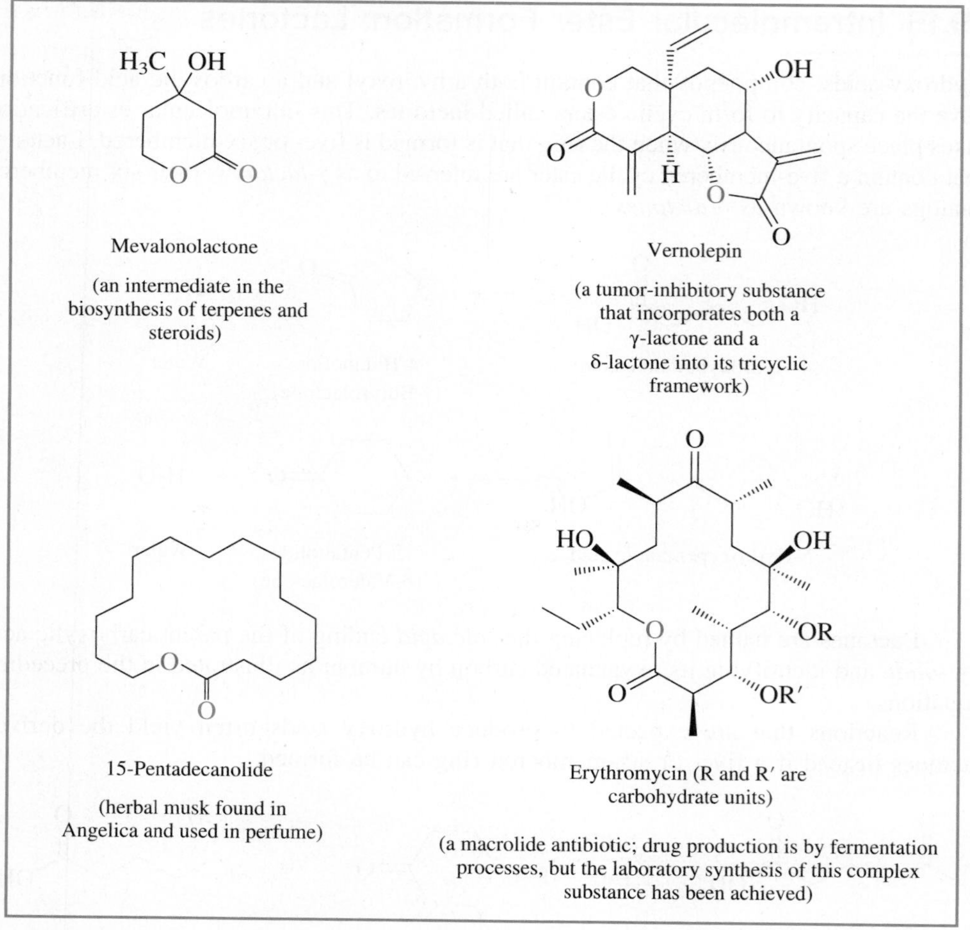

Mevalonolactone

(an intermediate in the biosynthesis of terpenes and steroids)

Vernolepin

(a tumor-inhibitory substance that incorporates both a γ-lactone and a δ-lactone into its tricyclic framework)

15-Pentadecanolide

(herbal musk found in Angelica and used in perfume)

Erythromycin (R and R′ are carbohydrate units)

(a macrolide antibiotic; drug production is by fermentation processes, but the laboratory synthesis of this complex substance has been achieved)

19.16 Decarboxylation of Malonic Acid and Related Compounds

The loss of a molecule of carbon dioxide from a carboxylic acid is known as **decarboxylation.**

$$\underset{\text{Carboxylic acid}}{RCO_2H} \longrightarrow \underset{\text{Alkane}}{RH} + \underset{\text{Carbon dioxide}}{CO_2}$$

Decarboxylation of simple carboxylic acid takes place with great difficulty and is rarely encountered.

Compounds that readily undergo thermal decarboxylation include those related to malonic acid. On being heated, malonic acid is converted to acetic acid and carbon dioxide.

$$\underset{\substack{\text{Malonic acid}\\\text{(propanedioic acid)}}}{HOOCCH_2COOH} \xrightarrow{150^\circ C} \underset{\substack{\text{Carbon}\\\text{dioxide}}}{O{=}C{=}O} + \underset{\text{Acetic acid}}{CH_3COOH}$$

It is important to recognize that only one carboxyl group is lost in this process; the second is retained. A mechanism that accounts for the assistance that one carboxyl group gives to the departure of the other involves two steps.

$$\underset{\text{Malonic acid}}{HOOCCH_2COOH} \xrightarrow{\text{slow}} \underset{\substack{\text{Carbon}\\\text{dioxide}}}{O{=}C{=}O} + \underset{\substack{\text{Enol form}\\\text{of acetic acid}}}{CH_2{=}C(OH)OH} \xrightarrow{\text{fast}} \underset{\text{Acetic acid}}{CH_3COOH}$$

The first step gives carbon dioxide and an enol, which isomerizes to acetic acid in the second step.

Keto–enol tautomerism was introduced in Section 9.11.

The hydrogens attached to C-2 of malonic acid are not directly involved in the decarboxylation step; thus, 1,3-dicarboxylic acids bearing substituents at C-2 undergo an analogous reaction on heating.

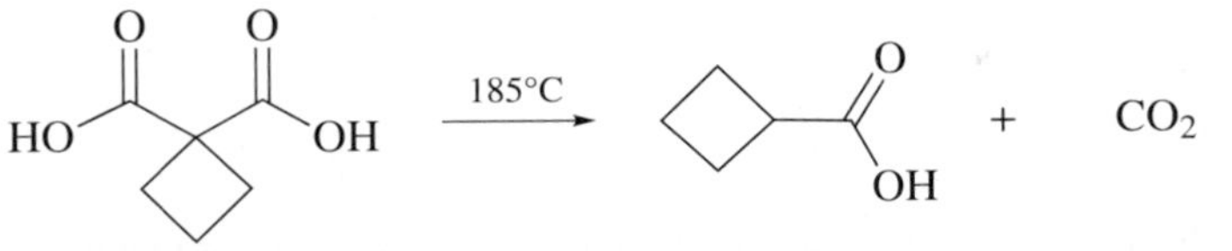

1,1-Cyclobutanedicarboxylic acid —185°C→ Cyclobutanecarboxylic acid (74%) + CO_2 Carbon dioxide

Problem 19.11

What will be the product isolated after thermal decarboxylation of each of the following? Using curved arrows, represent the bond changes that occur in the decarboxylation step.

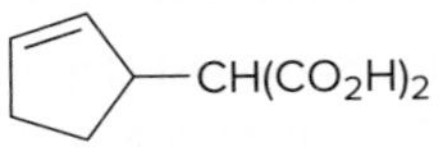

(a) cyclopentenyl–$CH(CO_2H)_2$ (b) 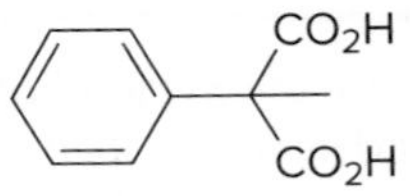 (c)

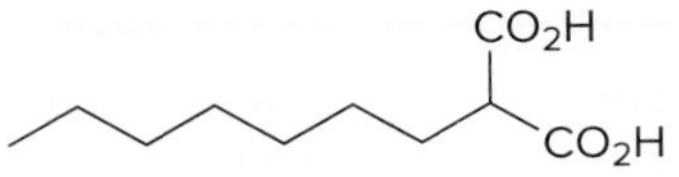

Sample Solution (a) Thermal decarboxylation of malonic acid derivatives leads to the replacement of one of the carboxyl groups by a hydrogen.

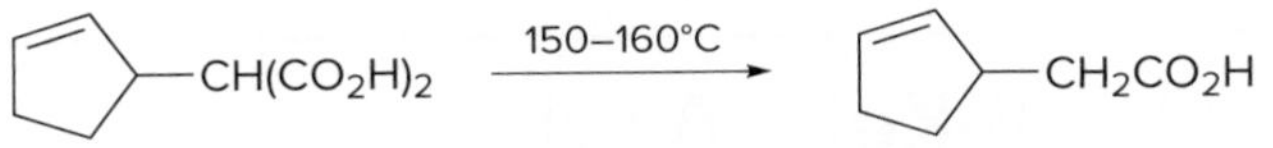

cyclopentenyl–$CH(CO_2H)_2$ —150–160°C→ cyclopentenyl–CH_2CO_2H

In this particular case, the reaction was carried out at the temperature indicated and gave the product shown in 96–99% yield.

The transition state for decarboxylation incorporates a cyclic array of six atoms and gives the enol form of the product as an intermediate:

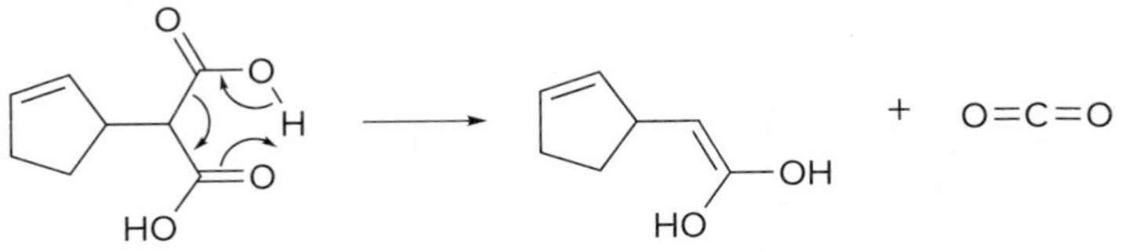

→ + O=C=O

Enol form of product

Notice that the carboxyl group that stays behind during the decarboxylation of malonic acid has a hydroxyl function that is not directly involved. Compounds that have substituents other than OH at this position undergo an analogous decarboxylation. The most frequently encountered ones are β-keto acids—that is, carboxylic acids in which C-3 is C=O.

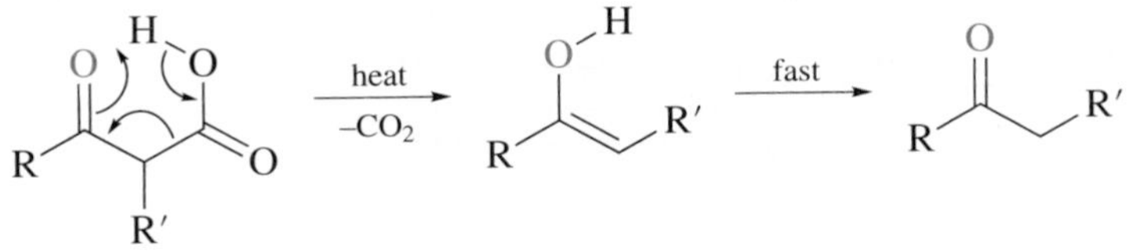

β-Keto acid —heat, $-CO_2$→ Enol form of ketone —fast→ Ketone

Decarboxylation of β-keto acids occurs even at room temperature, giving an enol, which isomerizes to the corresponding ketone as the isolated product.

Problem 19.12

Use curved arrows to show the bonding changes that occur during the decarboxylation of each of the following and give the structure of the resulting enol and the ketone formed from the enol.

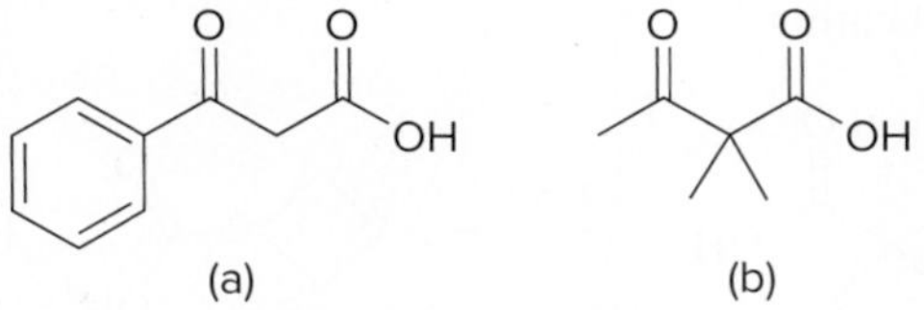

(a) (b)

Sample Solution (a) By analogy to the thermal decarboxylation of malonic acid, we represent the corresponding reaction of benzoylacetic acid as:

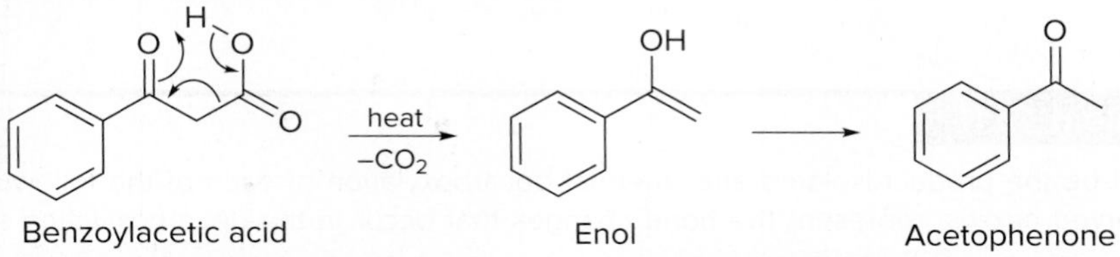

heat, $-CO_2$

Benzoylacetic acid — Enol — Acetophenone

Decarboxylation of derivatives of malonic acid and β-keto acids is the last step in two standard synthetic methods—the *malonic ester synthesis* and the *acetoacetic ester synthesis*—each of which is described in Chapter 21. They also find parallels in certain biochemical reactions. An example of biological decarboxylation is described in the following boxed essay.

Enzymatic Decarboxylation of a β-Keto Acid

In the nineteenth and early twentieth century, simple organic solvents were prepared by fermentation. One of the first industrial-scale processes used the Gram-positive bacterium *Clostridium acetobutylicum* to produce acetone and butanol. This strain of bacterium was discovered by the chemist Chaim Weizmann, whose work in acetone production during World War I indirectly led him to become the first president of Israel. The final step in acetone production is decarboxylation of acetoacetate catalyzed by the enzyme acetoacetate decarboxylase.

acetoacetate decarboxylase

Acetoacetate → Acetone + Carbon dioxide (O=C=O)

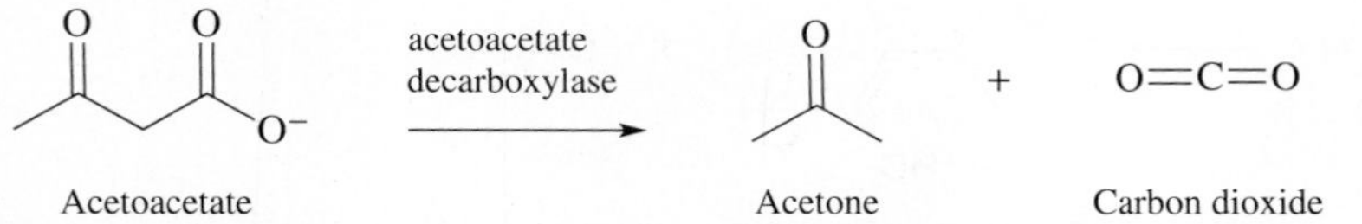

Since the carboxylic acid is not protonated in neutral aqueous solution (pK_a ~ 3.6), the enzymatic reaction was unlikely to mirror the chemical reaction involving a proton transfer from the carboxylic acid to the β-carbonyl. A series of elegant experiments by Frank Westheimer of Harvard University showed that the enzymatic reaction proceeds through an iminium ion intermediate, formed between the ketone of the substrate and an amino group of a lysine residue in the protein (Enz), which lowers the energy of the decarboxylation reaction. Proton transfer from a proton donor (HA) near the active site to the enamine gives another iminium ion intermediate that undergoes hydrolysis to release the desired product, acetone.

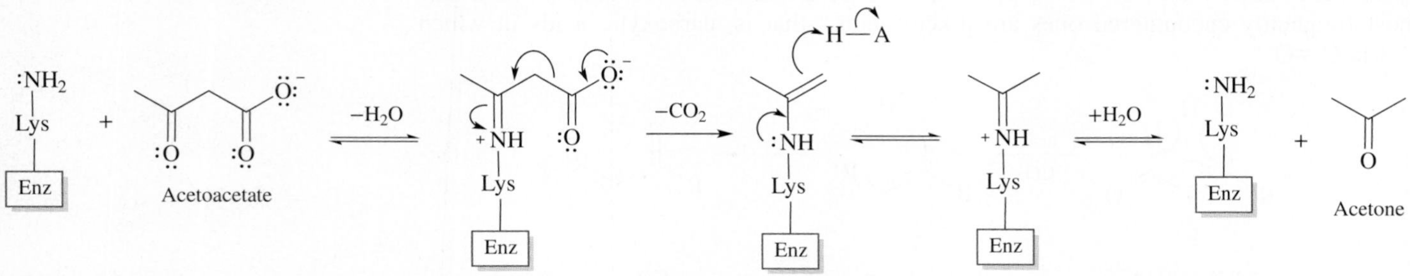

:NH_2–Lys–Enz + Acetoacetate $\xrightarrow{-H_2O}$ $\xrightarrow{-CO_2}$ (H—A) ⇌ $\xrightarrow{+H_2O}$:NH_2–Lys–Enz + Acetone

Fermentation remained an important method for producing acetone until the late 1950s, when solvent production by the petrochemical industry became more economical. Advances in generic engineering and increasing desire to reduce the use of petroleum products are leading to a resurgence of interest in the use of fermentation to produce solvents, biofuels, and industrial and pharmaceutical chemicals.

19.17 Spectroscopic Analysis of Carboxylic Acids

Infrared: The most characteristic peaks in the IR spectra of carboxylic acids are those of the hydroxyl and carbonyl groups. As shown in the IR spectrum of 4-phenylbutanoic acid (Figure 19.8), the O—H and C—H stretching frequencies overlap to produce a broad absorption in the 3500–2500 cm^{-1} region. The carbonyl group gives a strong band for C=O stretching at 1684 cm^{-1}.

^1H NMR: The hydroxyl proton of a CO_2H group is normally the least shielded of all the protons in an NMR spectrum, appearing 10–12 ppm downfield from tetramethylsilane, often as a broad peak. Figure 19.9 illustrates this for 4-phenylbutanoic acid. As with other hydroxyl protons, the proton of a carboxyl group can be identified by adding D_2O to the sample. Hydrogen–deuterium exchange converts $—CO_2H$ to $—CO_2D$, and the signal corresponding to the carboxyl group disappears.

^{13}C NMR: Like other carbonyl groups, the carbon of the $—CO_2H$ group of a carboxylic acid is strongly deshielded (δ 160–185), but not as much as that of an aldehyde or ketone (δ 190–215).

UV-VIS: In the absence of any additional chromophores, carboxylic acids absorb at a wavelength (210 nm) that is not very useful for diagnostic purposes.

Mass Spectrometry: Aside from a peak for the molecular ion, which is normally easy to pick out, aliphatic carboxylic acids undergo a variety of fragmentation processes. The dominant fragmentation in aromatic acids corresponds to loss of OH, then loss of CO.

$$\text{Ar—C(=O)—OH} \xrightarrow{e^-} \underset{M^+}{\text{Ar—C(=}\overset{+}{O}\text{)—OH}} \xrightarrow{-HO\cdot} \underset{[M-17]^+}{\text{Ar—C}\equiv\overset{+}{O}\text{:}} \xrightarrow{-CO} \underset{[M-(17+28)]^+}{Ar^+}$$

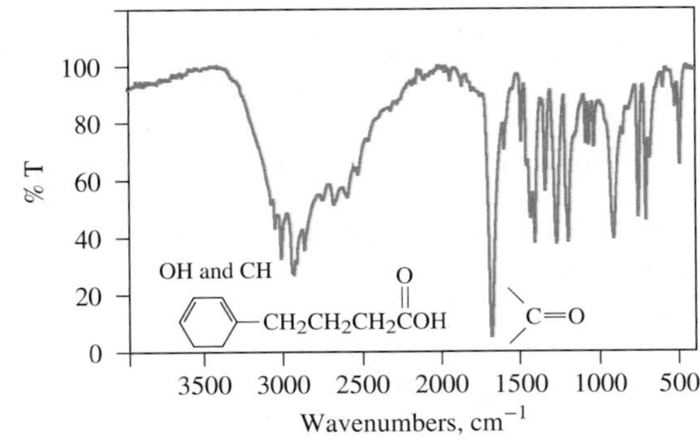

Figure 19.8

The IR spectrum of 4-phenylbutanoic acid.

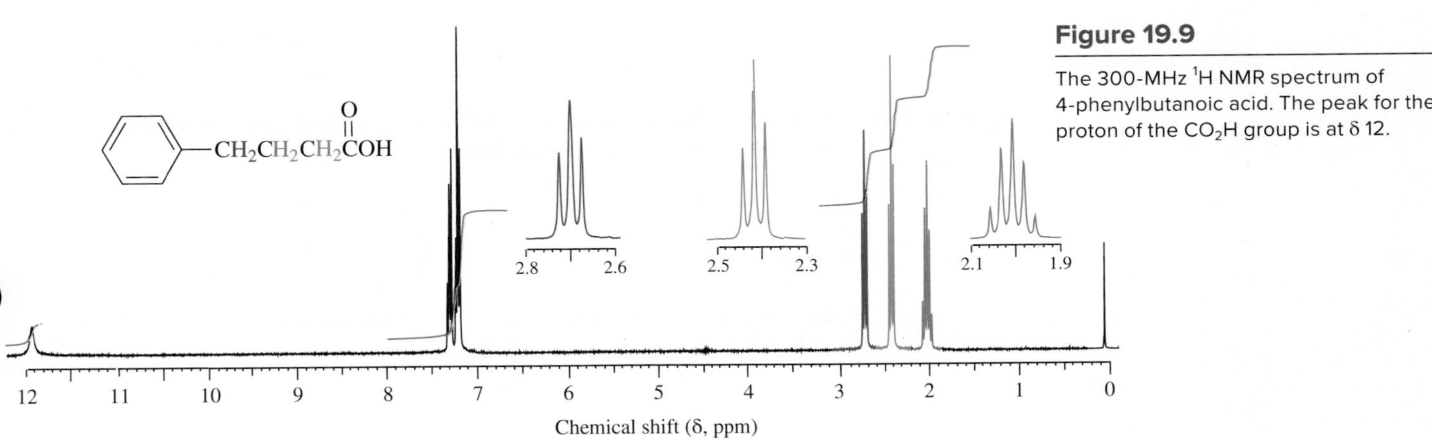

Figure 19.9

The 300-MHz ^1H NMR spectrum of 4-phenylbutanoic acid. The peak for the proton of the CO_2H group is at δ 12.

19.18 SUMMARY

Section 19.1 Carboxylic acids take their names from the alkane that contains the same number of carbons as the longest continuous chain that contains the —CO_2H group. The *-e* ending is replaced by *-oic acid*. Numbering begins at the carbon of the —CO_2H group.

3-Ethylhexane 4-Ethylhexanoic acid

Section 19.2 Like the carbonyl group of aldehydes and ketones, the carbon of a C═O unit in a carboxylic acid is sp^2-hybridized. Compared with the carbonyl group of an aldehyde or ketone, the C═O unit of a carboxylic acid receives an extra degree of stabilization from its attached OH group.

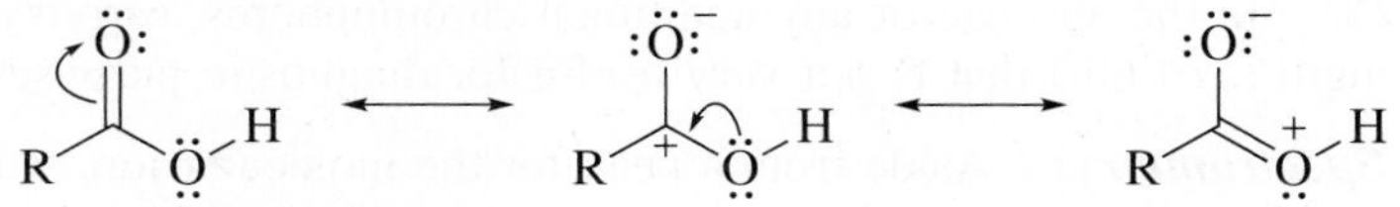

Section 19.3 Hydrogen bonding in carboxylic acids raises their melting points and boiling points above those of comparably constituted alkanes, alcohols, aldehydes, and ketones.

Section 19.4 Carboxylic acids are weak acids and, in the absence of electron-attracting substituents, have pK_a's of approximately 5. Carboxylic acids are much stronger acids than alcohols because of the electron-withdrawing power of the carbonyl group (inductive effect) and its ability to delocalize negative charge in the carboxylate anion (resonance effect).

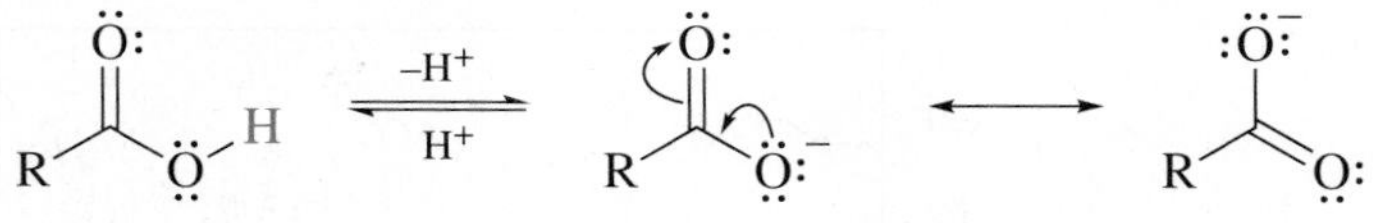

Carboxylic acid Electron delocalization in carboxylate ion

Sections 19.5–19.6 Electronegative substituents, especially those within a few bonds of the carboxyl group, increase the acidity of carboxylic acids.

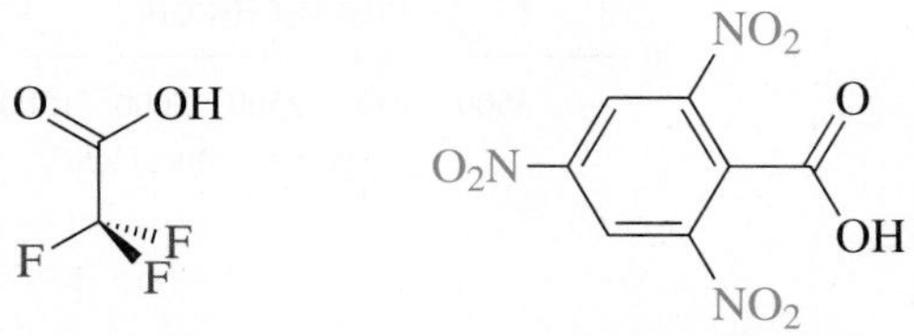

Trifluoroacetic acid
pK_a = 0.2

2,4,6-Trinitrobenzoic acid
pK_a = 0.6

Section 19.7 Although carboxylic acids dissociate to only a small extent in water, they are deprotonated almost completely in basic solution.

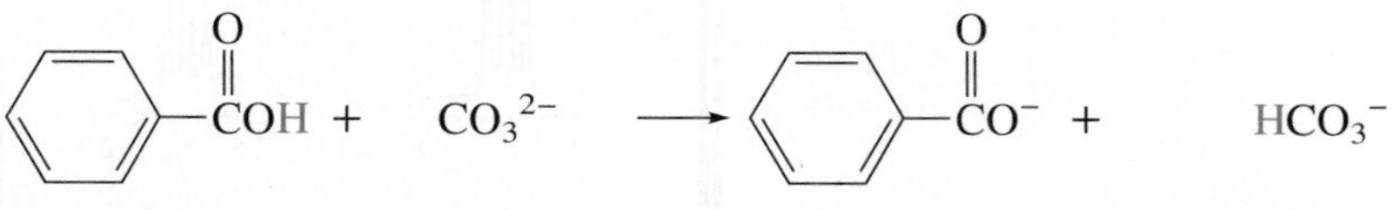

$$C_6H_5COOH + CO_3^{2-} \longrightarrow C_6H_5COO^- + HCO_3^-$$

Benzoic acid pK_a = 4.2 (stronger acid); Carbonate ion; Benzoate ion; Hydrogen carbonate ion pK_a = 10.2 (weaker acid)

Section 19.8 Dicarboxylic acids have separate pK_a values for their first and second ionizations.

Section 19.9 Carbon dioxide and carbonic acid are in equilibrium in water. Carbon dioxide is the major component.

$$O{=}C{=}O + H_2O \underset{99.7\%}{\overset{0.3\%}{\rightleftharpoons}} HO{-}C({=}O){-}OH$$

Section 19.10 Several of the reactions introduced in earlier chapters can be used to prepare carboxylic acids (see Table 19.4).

Section 19.11 Carboxylic acids can be prepared by the reaction of Grignard reagents with carbon dioxide.

4-Bromocyclopentene $\xrightarrow[\text{2. } CO_2 \text{ 3. } H_3O^+]{\text{1. Mg, diethyl ether}}$ Cyclopentene-4-carboxylic acid (66%)

Section 19.12 Nitriles are prepared from primary and secondary alkyl halides by nucleophilic substitution with cyanide ion and can be converted to carboxylic acids by hydrolysis.

2-Phenylpentanenitrile $\xrightarrow[\text{heat}]{H_2O,\ H_2SO_4}$ 2-Phenylpentanoic acid (52%)

Likewise, the cyano group of a cyanohydrin can be hydrolyzed to $-CO_2H$.

Section 19.13 Among the reactions of carboxylic acids, their conversions to acyl chlorides, primary alcohols, and esters were introduced in earlier chapters and were reviewed in Table 19.5.

Section 19.14 The mechanism of acid-catalyzed esterification involves two stages. The first is formation of a tetrahedral intermediate by nucleophilic addition of the alcohol to the carbonyl group and is analogous to acid-catalyzed acetal and ketal formation of aldehydes and ketones. The second is dehydration of the tetrahedral intermediate.

$$RCO_2H + R'OH \overset{H^+}{\rightleftharpoons} R{-}C(OH)_2{-}OR' \overset{H^+}{\rightleftharpoons} RCO_2R' + H_2O$$

Carboxylic acid + Alcohol ⇌ Tetrahedral intermediate ⇌ Ester + Water

Mechanism 19.1 provides details of the six individual steps.

Section 19.15 An intramolecular esterification can occur when a molecule contains both a hydroxyl and a carboxyl group. Cyclic esters are called *lactones* and are most stable when the ring is five- or six-membered.

4-Hydroxy-2-methylpentanoic acid ⟶ 2-Methyl-4-pentanolide

Section 19.16 1,1-Dicarboxylic acids (malonic acids) and β-keto acids undergo thermal decarboxylation by a mechanism in which a β-carbonyl group assists the departure of carbon dioxide.

X = OH: malonic acid derivative

X = alkyl or aryl: β-keto acid

$\xrightarrow{-CO_2}$

Enol form of product

⟶

X = OH: carboxylic acid

X = alkyl or aryl: ketone

Section 19.17 Carboxylic acids are readily identified by the presence of strong IR absorptions near 1700 cm^{-1} (C=O) and between 2500 and 3500 cm^{-1} (OH), a ^{1}H NMR signal for the hydroxyl proton at δ 10–12, and a ^{13}C signal for the carbonyl carbon near δ 180.

PROBLEMS

Structure and Nomenclature

19.13 Many carboxylic acids are much better known by their common names than by their systematic names. Some of these follow. Provide a structural formula for each one on the basis of its systematic name.

(a) 2-Hydroxypropanoic acid (better known as *lactic acid,* it is found in sour milk and is formed in the muscles during exercise)

(b) 2-Hydroxy-2-phenylethanoic acid (also known as *mandelic acid,* it is obtained from plums, peaches, and other fruits)

(c) Tetradecanoic acid (also known as *myristic acid,* it can be obtained from a variety of fats)

(d) 10-Undecenoic acid (also called *undecylenic acid,* it is used, in combination with its zinc salt, to treat fungal infections such as athlete's foot)

(e) 3,5-Dihydroxy-3-methylpentanoic acid (also called *mevalonic acid,* it is a key intermediate in the biosynthesis of terpenes and steroids)

(f) (*E*)-2-Methyl-2-butenoic acid (also known as *tiglic acid,* it is a constituent of various natural oils)

(g) 2-Hydroxybutanedioic acid (also known as *malic acid,* it is found in apples and other fruits)

(h) 2-Hydroxy-1,2,3-propanetricarboxylic acid (better known as *citric acid,* it contributes to the tart taste of citrus fruits)

(i) 2-(*p*-Isobutylphenyl)propanoic acid (an antiinflammatory drug better known as *ibuprofen*)

(j) *o*-Hydroxybenzenecarboxylic acid (better known as *salicylic acid,* it is obtained from willow bark)

19.14 Give an acceptable IUPAC name for each of the following:

(a) $CH_3(CH_2)_6CO_2H$

(b) $CH_3(CH_2)_6CO_2K$

(c) $H_2C{=}CH(CH_2)_5CO_2H$

(d) 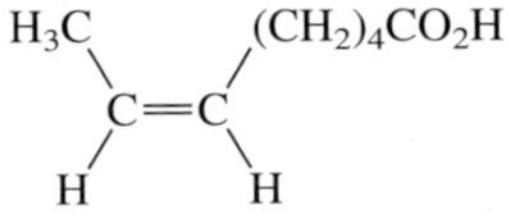

(e) $HO_2C(CH_2)_6CO_2H$

(f) $CH_3(CH_2)_4CH(CO_2H)_2$

(g) 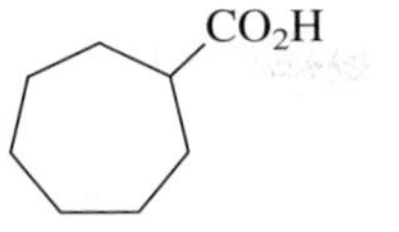

(h) 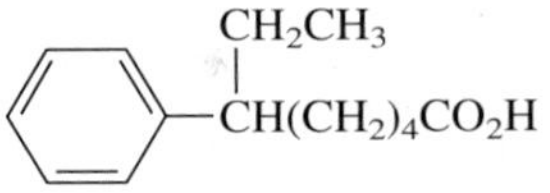

Synthesis

19.15 Propose methods for preparing butanoic acid from each of the following:

(a) 1-Butanol
(b) Butanal
(c) 1-Butene
(d) 1-Propanol
(e) 2-Propanol
(f) $CH_3CH_2CH(CO_2H)_2$

19.16 It is sometimes necessary to prepare isotopically labeled samples of organic substances for probing biological transformations and reaction mechanisms. Various sources of the radioactive mass-14 carbon isotope are available. Describe synthetic procedures by which benzoic acid, labeled with ^{14}C at its carbonyl carbon, could be prepared from benzene and the following ^{14}C-labeled precursors. You may use any necessary organic or inorganic reagents.

(a) CH_3Cl
(b) $H\overset{O}{\overset{\|}{C}}H$
(c) CO_2

19.17 Show how butanoic acid may be converted to each of the following compounds:

(a) 1-Butanol
(b) Butanal
(c) 1-Chlorobutane
(d) Butanoyl chloride
(e) Phenyl propyl ketone
(f) 4-Octanone

19.18 Show by a series of equations how you could synthesize each of the following compounds from the indicated starting material and any necessary organic or inorganic reagents:

(a) 2-Methylpropanoic acid from *tert*-butyl alcohol
(b) 3-Methylbutanoic acid from *tert*-butyl alcohol
(c) 3,3-Dimethylbutanoic acid from *tert*-butyl alcohol
(d) $HO_2C(CH_2)_5CO_2H$ from $HO_2C(CH_2)_3CO_2H$
(e) 3-Phenyl-1-butanol from $CH_3\underset{C_6H_5}{\underset{|}{C}}HCH_2CN$
(f) 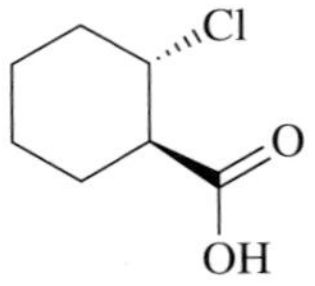

from 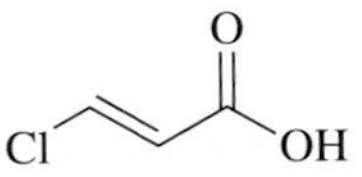

(g) 2,4-Dimethylbenzoic acid from *m*-xylene
(h) 4-Chloro-3-nitrobenzoic acid from *p*-chlorotoluene
(i) (*Z*)-$CH_3CH{=}CHCO_2H$ from propyne

19.19 Devise a synthesis of compound A based on the retrosynthesis shown.

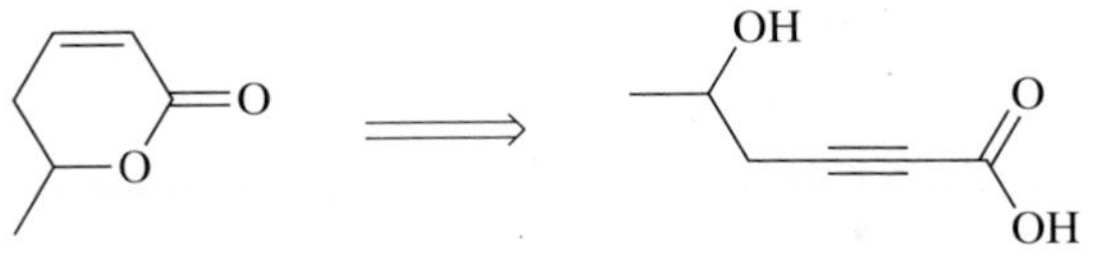

Compound A 5-Hydroxy-2-hexynoic acid

Reactions

19.20 Identify the more acidic compound in each of the following pairs:

(a) $CF_3CH_2CO_2H$ or $CF_3CH_2CH_2CO_2H$

(b) $CH_3CH_2CH_2CO_2H$ or $CH_3C{\equiv}CCO_2H$

(c) Cyclohexane–CO_2H or benzene–CO_2H

(d) 2,3,4,5,6-Pentafluorophenyl–CO_2H or phenyl–CO_2H

(e) F–(C_6F_4)–CO_2H or F–(C_6F_4)–(C_6H_4)–CO_2H

(f) Furan-2-CO_2H or furan-3-CO_2H

(g) Furan-2-CO_2H or pyrrole (N–H)-2-CO_2H

19.21 Rank the compounds in each of the following groups in order of decreasing acidity:

(a) Acetic acid, ethane, ethanol

(b) Benzene, benzoic acid, benzyl alcohol

(c) 1,3-Propanediol, propanedioic acid, propanoic acid

(d) Acetic acid, ethanol, trifluoroacetic acid, 2,2,2-trifluoroethanol, trifluoromethanesulfonic acid (CF_3SO_2OH)

19.22 Give the product of the reaction of pentanoic acid with each of the following reagents:

(a) Sodium hydroxide

(b) Sodium bicarbonate

(c) Thionyl chloride

(d) Phosphorus tribromide

(e) Benzyl alcohol, sulfuric acid (catalytic amount)

(f) Lithium aluminum hydride, then hydrolysis

(g) Phenylmagnesium bromide

19.23 (a) Which stereoisomer of 4-hydroxycyclohexanecarboxylic acid (cis or trans) can form a lactone? What is the conformation of the cyclohexane ring in the starting hydroxy acid? In the lactone?

HO–(cyclohexane)–CO_2H

(b) Repeat part (a) for the case of 3-hydroxycyclohexanecarboxylic acid.

19.24 When compound A is heated, two isomeric products are formed. What are these two products?

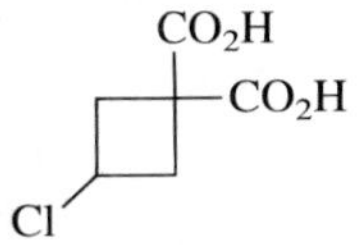

Compound A

19.25 Each of the following reactions has been reported in the chemical literature and gives a single product in good yield. What is the product in each reaction?

(a) 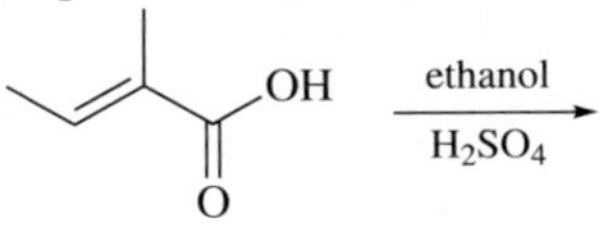 [unsaturated carboxylic acid] $\xrightarrow[H_2SO_4]{\text{ethanol}}$

(b) 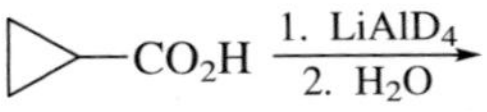 [cyclopropyl]$-CO_2H$ $\xrightarrow[2.\ H_2O]{1.\ LiAlD_4}$

(c) 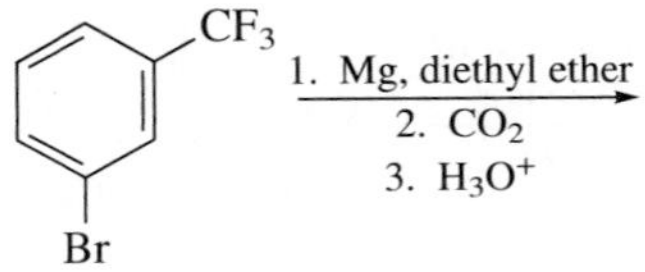 [m-bromo(trifluoromethyl)benzene] $\xrightarrow{1.\ \text{Mg, diethyl ether};\ 2.\ CO_2;\ 3.\ H_3O^+}$

(d) 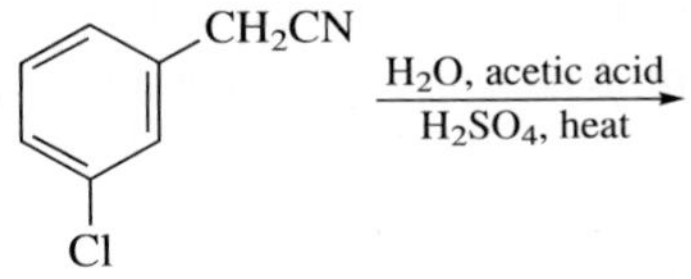 [m-chlorophenyl]CH_2CN $\xrightarrow[H_2SO_4,\ \text{heat}]{H_2O,\ \text{acetic acid}}$

(e) $H_2C{=}CH(CH_2)_8CO_2H$ $\xrightarrow[\text{benzoyl peroxide}]{HBr}$

19.26 The compound shown was subjected to the following series of reactions to give a product having the molecular formula $C_9H_9ClO_3$. What is this product?

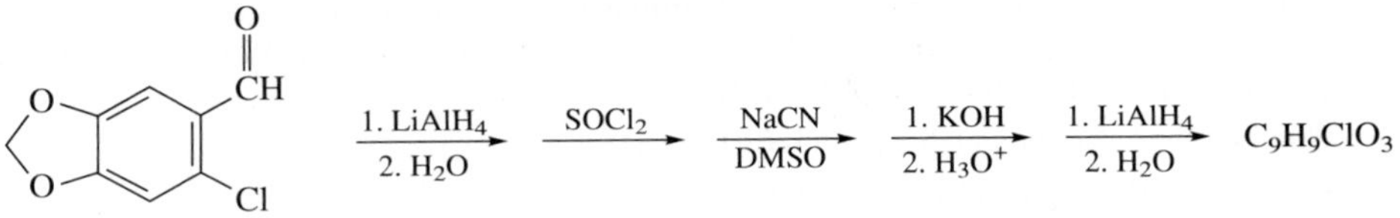

$\xrightarrow[2.\ H_2O]{1.\ LiAlH_4}$ $\xrightarrow{SOCl_2}$ $\xrightarrow[DMSO]{NaCN}$ $\xrightarrow[2.\ H_3O^+]{1.\ KOH}$ $\xrightarrow[2.\ H_2O]{1.\ LiAlH_4}$ $C_9H_9ClO_3$

19.27 A certain carboxylic acid ($C_{14}H_{26}O_2$), which can be isolated from whale blubber or sardine oil, yields nonanal and $O{=}CH(CH_2)_3CO_2H$ on ozonolysis. What is the structure of this acid?

19.28 When levulinic acid ($CH_3\overset{O}{\overset{\|}{C}}CH_2CH_2CO_2H$) was hydrogenated at high pressure over a nickel catalyst at 220°C, a single product, $C_5H_8O_2$, was isolated in 94% yield. This compound lacks hydroxyl absorption in its IR spectrum. What is a reasonable structure for the compound?

19.29 On standing in dilute aqueous acid, compound A is smoothly converted to mevalonolactone.

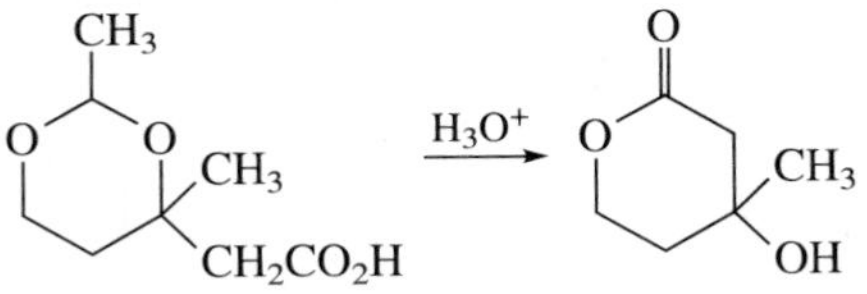

Compound A $\xrightarrow{H_3O^+}$ Mevalonolactone

Suggest a reasonable mechanism for this reaction. What other organic product is also formed?

19.30 In the presence of the enzyme *aconitase,* the double bond of aconitic acid undergoes hydration. The reaction is reversible, and the following equilibrium is established:

Isocitric acid ($C_6H_8O_7$) (6% at equilibrium) $\xrightleftharpoons{H_2O}$ Aconitic acid, $(HO_2C)(H)C{=}C(CO_2H)(CH_2CO_2H)$ (4% at equilibrium) $\xrightleftharpoons{H_2O}$ Citric acid ($C_6H_8O_7$) (90% at equilibrium)

(a) The major tricarboxylic acid present is *citric acid,* the substance responsible for the tart taste of citrus fruits. Citric acid is achiral. What is its structure?

(b) What must be the constitution of isocitric acid? (Assume that no rearrangements accompany hydration.) How many stereoisomers are possible for isocitric acid?

19.31 The enzyme glutaconyl-SCoA decarboxylase catalyzes the decarboxylation of glutaconyl-SCoA to crotonyl-SCoA (CoA = coenzyme A). Write a mechanism for this reaction.

Glutaconyl-SCoA $\xrightarrow[-CO_2]{\text{glutaconyl-SCoA decarboxylase}}$ Crotonyl-SCoA

Spectroscopy

19.32 The ^{1}H NMR spectra of formic acid (HCO_2H), maleic acid (*cis*-$HO_2CCH{=}CHCO_2H$), and malonic acid ($HO_2CCH_2CO_2H$) are similar in that each is characterized by two singlets of equal intensity. Match these compounds with the designations A, B, and C on the basis of the appropriate ^{1}H NMR chemical-shift data.

Compound A: signals at δ 3.2 and 12.1

Compound B: signals at δ 6.3 and 12.4

Compound C: signals at δ 8.0 and 11.4

19.33 Compounds A and B are isomers having the molecular formula $C_4H_8O_3$. Identify A and B on the basis of their ^{1}H NMR spectra.

Compound A: δ 1.3 (3H, triplet); 3.6 (2H, quartet); 4.1 (2H, singlet); 11.1 (1H, broad singlet)

Compound B: δ 2.6 (2H, triplet); 3.4 (3H, singlet); 3.7 (2H triplet); 11.3 (1H, broad singlet)

19.34 Compounds A and B are carboxylic acids. Identify each one on the basis of its ^{1}H NMR spectrum.

(a) Compound A ($C_3H_5ClO_2$) (Figure 19.10).

(b) Compound B ($C_9H_9NO_4$) has a nitro group attached to an aromatic ring (Figure 19.11).

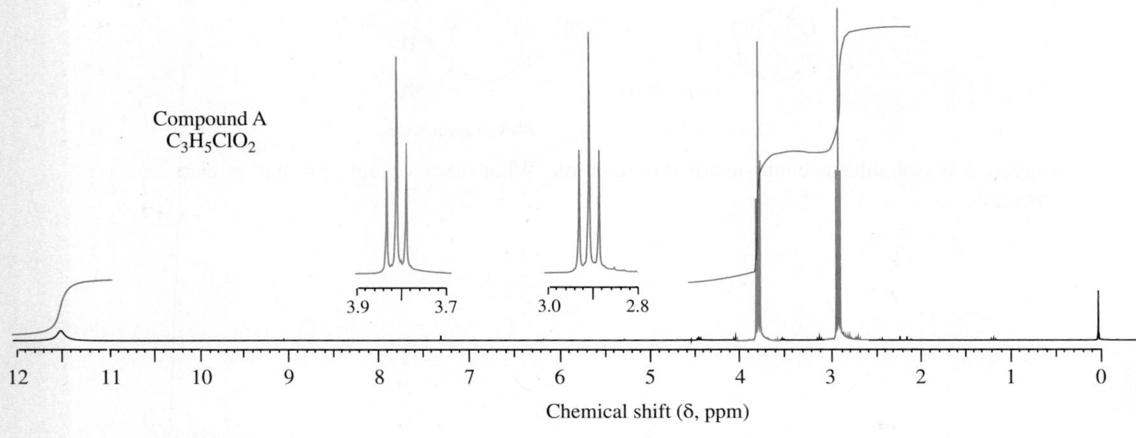

Figure 19.10

The 300-MHz ^{1}H NMR spectrum of compound A ($C_3H_5ClO_2$) (Problem 19.34a).

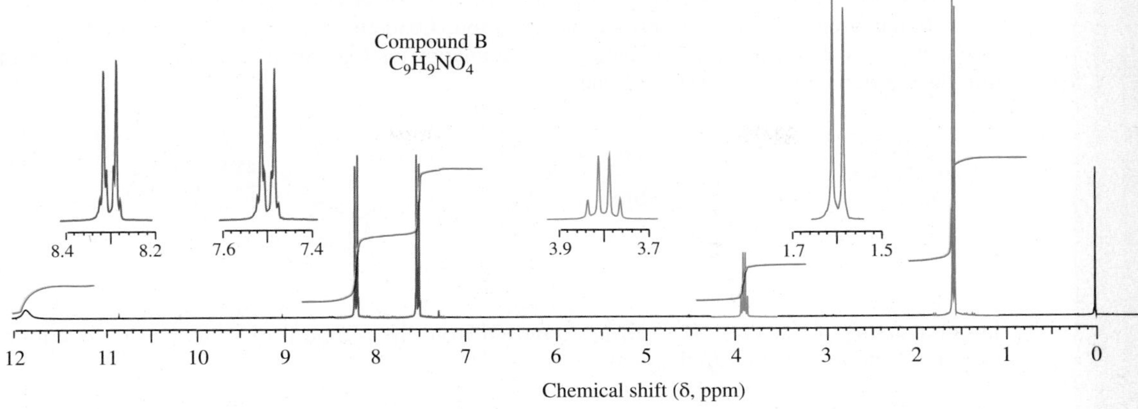

Figure 19.11

The 300-MHz ^{1}H NMR spectrum of compound B ($C_9H_9NO_4$) (Problem 19.34b).

Descriptive Passage and Interpretive Problems 19

Lactonization Methods

In Section 19.15 we saw that hydroxy-substituted carboxylic acids spontaneously cyclize to lactones if a five- or six-membered ring can be formed.

OH OH → + H_2O

Many natural products are lactones, and chemists have directed substantial attention to developing alternative methods for their synthesis. The most successful of these efforts are based on electrophilic addition to the double bond of unsaturated carboxylic acids. For a generalized electrophilic reagent E—Y and 4-pentenoic acid, such reactions give a 5-substituted γ-lactone.

Y—E → E + "HY"

Although the curved arrows show the *overall* electron flow, the mechanism depends on the electrophilic reagent E—Y and normally involves more than one step.

In *iodolactonization* the electrophilic atom E = I, and E—Y represents a source of electrophilic iodine, usually I_2 or *N*-iodosuccinimide. In *phenylselenolactonization,* E = C_6H_5Se and E—Y is benzeneselenenyl chloride (C_6H_5SeCl). Anti addition is observed in both iodo- and phenylselenolactonization.

OH —C_6H_5SeCl / CH_2Cl_2, −78°C→ Se

Both iodo- and phenylselenolactonization offer the advantage of giving a product containing a functional group capable of further modification. Oxidation of the C_6H_5Se substituent, for example, gives a selenoxide that undergoes elimination of C_6H_5SeOH at room temperature to introduce a double bond into the lactone.

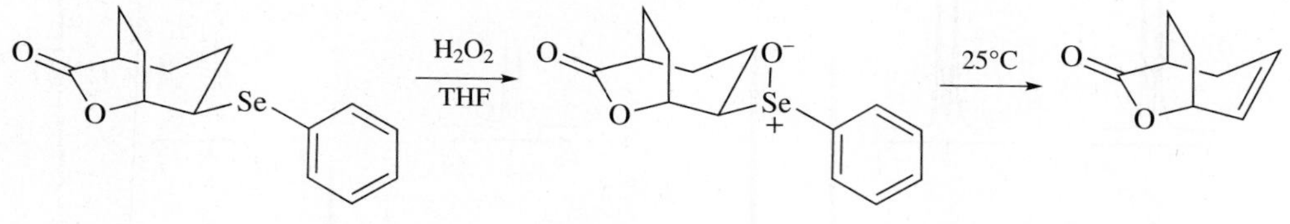

H_2O_2 / THF → 25°C →

In eliminations of this type, H is always removed from the carbon β to selenium that is remote from the lactone oxygen. Elimination is syn.

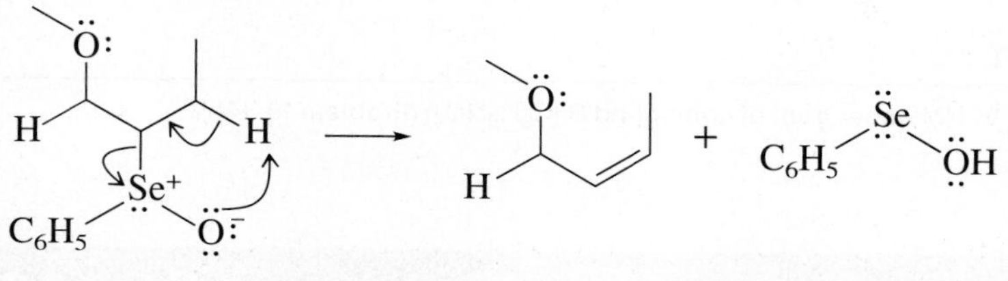

19.35 The dihydroxy acid shown was prepared as a single enantiomer and underwent spontaneous cyclization to give a δ-lactone, What are the *R–S* configurations of the chirality centers in this lactone? (No stereochemistry is implied in the structural drawing.)

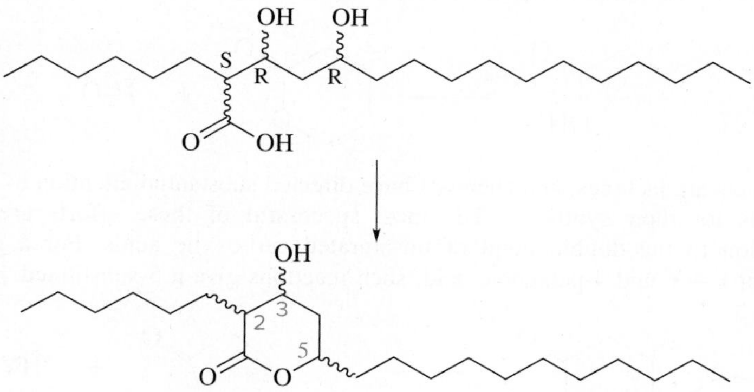

A. 2*R*, 3*R*, 5*R*
B. 2*R*, 3*R*, 5*S*
C. 2*S*, 3*R*, 5*R*
D. 2*S*, 3*R*, 5*S*

19.36 The product of the following reaction has the constitution shown. No stereochemistry is implied. Deduce the stereochemistry on the basis of the fact that iodolactonization is normally an anti addition and it was determined experimentally that the ring junction is cis.

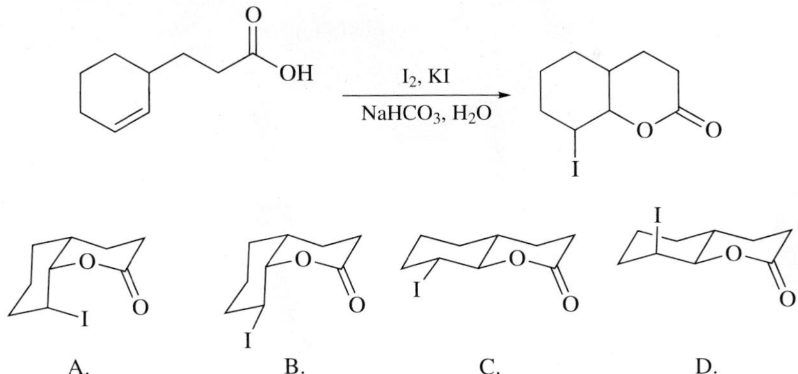

19.37 What is the structure of the γ-lactone formed by iodolactonization of 4-pentynoic acid ($HC{\equiv}CCH_2CH_2CO_2H$)? Anti addition to the triple bond occurs.

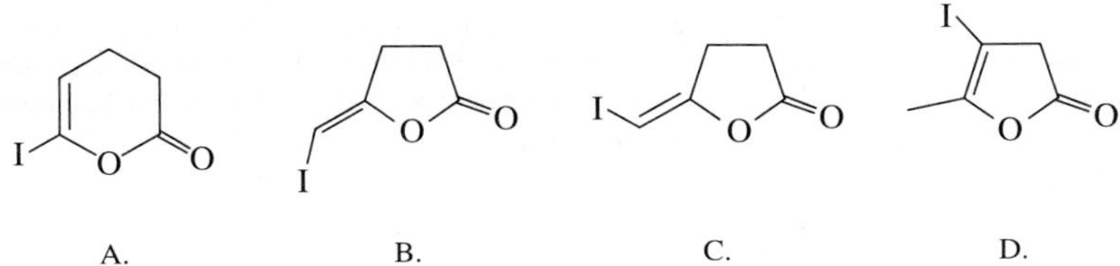

19.38 What is compound X?

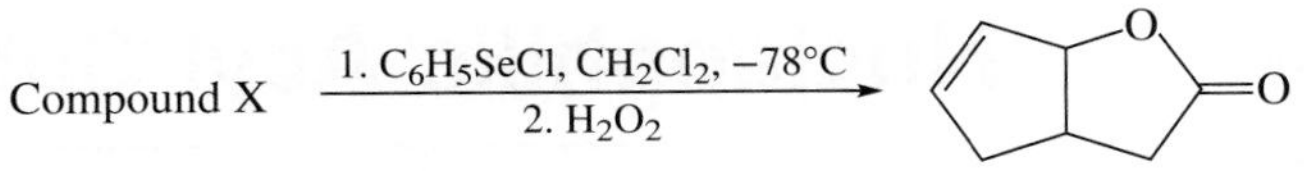

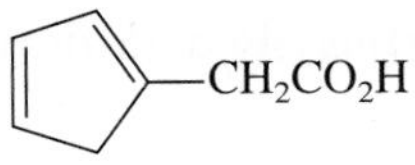

Compound X $\xrightarrow[\text{2. } H_2O_2]{\text{1. } C_6H_5SeCl,\ CH_2Cl_2,\ -78°C}$

A. $-CH_2CO_2H$

B. $-CH_2CO_2H$

C. $-CH_2CO_2H$

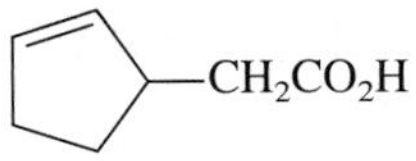

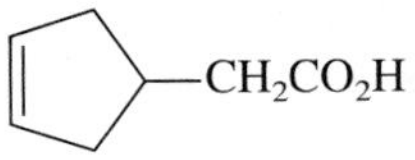

Chapter 20

3-Methylbutyl acetate (isoamyl acetate) is best known for the characteristic odor it gives to bananas. It is also one of the more than 40 compounds in the alarm pheromone a honeybee uses to alert other bees that an intruder has arrived.
©arlindo71/Getty Images

CHAPTER OUTLINE

Carboxylic Acid Derivatives: Nucleophilic Acyl Substitution

This chapter deals with several related classes of compounds collectively referred to as **carboxylic acid derivatives.**

$RCOCl$	$RCOOCOR$	$RCOOR'$	$RCONR'R''$
Acyl chloride	Acid anhydride	Ester	Amide

All have an acyl group bonded to an electronegative element and have **nucleophilic acyl substitution** as their characteristic reaction type.

$$RC(=O)X + Nu{-}H \longrightarrow RC(=O)Nu + H{-}X$$

Nucleophilic acyl substitution includes many useful synthetic methods and is a major participant in numerous biosynthetic processes. Its mechanism, which constitutes a major portion of this chapter, has been thoroughly studied and provides a framework for organizing what might otherwise be a collection of isolated facts, details, and observations. Similar principles apply to the reactions of nitriles and these are included as well.

20.1 Nomenclature of Carboxylic Acid Derivatives

Acyl Chlorides: Although acyl fluorides, bromides, and iodides are all known classes of organic compounds, they are not encountered nearly as often as acyl chlorides. Acyl chlorides, which will be the only acyl halides discussed in this chapter, are named by adding the word "chloride" after the name of the acyl group. To name an acyl group, replace the *-ic acid* ending of the IUPAC name of the corresponding carboxylic acid by *-yl.* The suffix *-carbonyl chloride* is used for attachments to rings other than benzene.

Formyl, acetyl, and benzoyl are preferred over methanoyl, ethanoyl, and benzenecarbonyl, respectively.

Pentanoyl chloride | 3-Butenoyl chloride or But-3-enoyl chloride | *p*-Fluorobenzoyl chloride or 4-Fluorobenzoyl chloride | Cyclopentanecarbonyl chloride

Acid Anhydrides: When both acyl groups are the same, the word "acid" in the corresponding carboxylic acid is replaced by "anhydride." When the two acyl groups are different, their corresponding carboxylic acids are cited in alphabetical order.

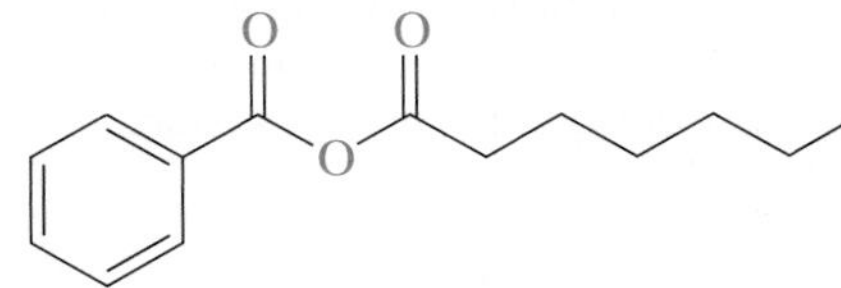

Acetic anhydride | Benzoic anhydride | Benzoic heptanoic anhydride

Esters: The alkyl group and the acyl group of an *ester* are specified independently. Esters are named as *alkyl alkanoates.* The alkyl group R′ of $\mathrm{R\overset{\overset{O}{\|}}{C}OR'}$ is cited first, followed by the acyl portion $\mathrm{R\overset{\overset{O}{\|}}{C}—}$. The acyl portion is named by substituting the suffix *-ate* for the *-ic acid* ending of the corresponding acid.

$\mathrm{CH_3\overset{\overset{O}{\|}}{C}OCH_2CH_3}$ Ethyl acetate

$\mathrm{CH_3CH_2\overset{\overset{O}{\|}}{C}OCH_3}$ Methyl propanoate

$\mathrm{C_6H_5\overset{\overset{O}{\|}}{C}OCH_2CH_2Cl}$ 2-Chloroethyl benzoate

Aryl esters, that is, compounds of the type $\mathrm{R\overset{\overset{O}{\|}}{C}OAr}$, are named in an analogous way.

Amides: When naming amides, replace the *-ic acid* or *-oic acid* of the corresponding carboxylic acid with *-amide.* Substituents, irrespective of whether they are attached to the acyl group or the amide nitrogen, are listed in alphabetical order. Substitution on nitrogen is indicated by the locant *N*-.

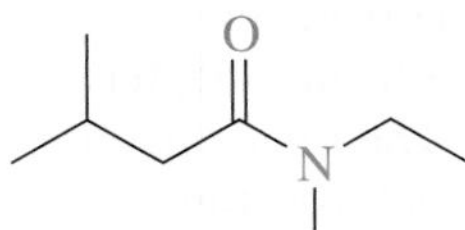

3-Methylbutanamide | *N*-Ethyl-3-methylbutanamide | *N*-Ethyl-*N*,3-dimethylbutanamide

Similar to the *-carbonyl chloride* suffix for acyl chlorides, *-carboxamide* is used when an amide group is attached to a ring.

Nitriles: Substitutive IUPAC names for *nitriles* add the suffix *-nitrile* to the name of the parent hydrocarbon chain that includes the carbon of the cyano group. Nitriles may also be named by replacing the *-ic acid* or *-oic acid* ending of the corresponding carboxylic acid with *-onitrile.* Alternatively, they are sometimes given functional class IUPAC names as alkyl cyanides. The suffix *-carbonitrile* is used when a —CN group is attached to a ring.

$CH_3C{\equiv}N$ — Ethanenitrile or Acetonitrile

5-Methylhexanenitrile or 4-Methylpentyl cyanide

Cyclopentanecarbonitrile or Cyclopentyl cyanide

Problem 20.1

Write a structural formula for each of the following compounds:

(a) 2-Phenylbutanoyl chloride
(b) 2-Phenylbutanoic anhydride
(c) Butyl 2-phenylbutanoate
(d) 2-Phenylbutyl butanoate
(e) 2-Phenylbutanamide
(f) *N*-Ethyl-2-phenylbutanamide
(g) 2-Phenylbutanenitrile

Sample Solution (a) A 2-phenylbutanoyl group is a four-carbon acyl unit that bears a phenyl substituent at C-2. When the name of an acyl group is followed by the name of a halide, it designates an *acyl halide.*

2-Phenylbutanoyl chloride

20.2 Structure and Reactivity of Carboxylic Acid Derivatives

The number of reactions in this chapter is quite large and keeping track of them all can be difficult—or it can be manageable. The key to making it manageable is the same as always: *structure determines properties.*

Figure 20.1 shows the structures of various derivatives of acetic acid (acetyl chloride, acetic anhydride, ethyl acetate, and acetamide) arranged in order of decreasing reactivity toward nucleophilic acyl substitution. Acyl chlorides are the most reactive, amides the least reactive. The reactivity order:

$$\text{acyl chloride} > \text{anhydride} > \text{ester} > \text{amide}$$

is general for nucleophilic acyl substitution and well worth remembering. The range of reactivities is quite large; a factor of about 10^{13} in relative rate separates acyl chlorides from amides.

This difference in reactivity, especially toward hydrolysis, has an important result. We'll see in Chapter 25 that the structure and function of proteins are critical to life itself. The bonds mainly responsible for the structure of proteins are amide bonds, which are about 100 times more stable to hydrolysis than ester bonds. These amide bonds are stable enough to maintain the structural integrity of proteins in an aqueous environment, but susceptible enough to hydrolysis to be broken when the occasion demands.

What structural features are responsible for the reactivity order of carboxylic acid derivatives? Like the other carbonyl-containing compounds that we've studied, they all have a planar arrangement of bonds to the carbonyl group. Thus, all are about the same in offering relatively unhindered access to the approach of a nucleophile. They differ in

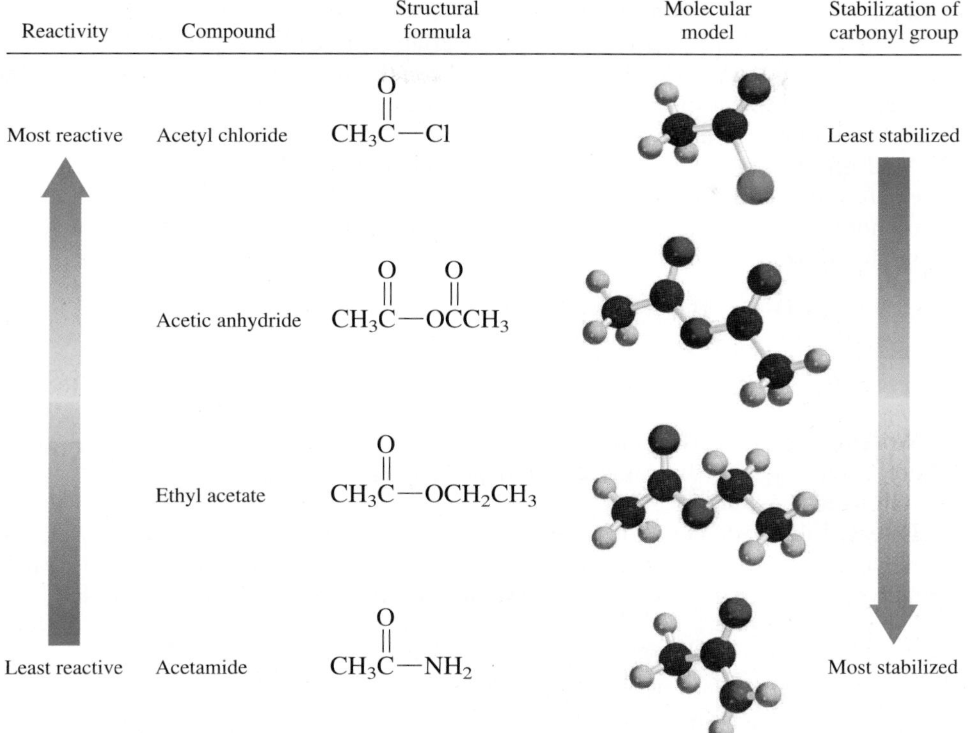

Reactivity
Compound
Structural formula
Molecular model
Stabilization of carbonyl group
Most reactive
Acetyl chloride
$CH_3\overset{O}{\overset{\|}{C}}-Cl$
Least stabilized
Acetic anhydride
$CH_3\overset{O}{\overset{\|}{C}}-O\overset{O}{\overset{\|}{C}}CH_3$
Ethyl acetate
$CH_3\overset{O}{\overset{\|}{C}}-OCH_2CH_3$
Least reactive
Acetamide
$CH_3\overset{O}{\overset{\|}{C}}-NH_2$
Most stabilized

Figure 20.1

Structure, reactivity, and carbonyl-group stabilization in carboxylic acid derivatives. Acyl chlorides are the most reactive, amides the least reactive. Acyl chlorides have the least stabilized carbonyl group, amides the most. Conversion of one class of compounds to another is feasible only in the direction that leads to a more stabilized carbonyl group—that is, from more reactive to less reactive.

the degree to which the atom attached to the carbonyl group can stabilize the carbonyl group by electron donation.

$$R-C(=\ddot{O}:)-\ddot{X}: \longleftrightarrow R-\overset{+}{C}(-\ddot{O}:^-)-\ddot{X}: \longleftrightarrow R-C(-\ddot{O}:^-)=X^+$$

Electron release from the substituent X stabilizes the carbonyl group and makes it less electrophilic.

The order of reactivity of carboxylic acid derivatives toward nucleophilic acyl substitution can be explained on the basis of the electron-donating properties of substituent X. The greater the electron-donating powers of X, the slower the rate.

1. **Acyl chlorides:** Although chlorine has unshared electron pairs, it is a poor electron-pair donor in resonance of the type:

$$R-C(=\ddot{O}:)-\ddot{Cl}: \longleftrightarrow R-C^+(-\ddot{O}:^-)-\ddot{Cl}: \longleftrightarrow R-C(-\ddot{O}:^-)=\ddot{Cl}^+$$

Lone-pair donation ineffective because of poor orbital overlap

Not a significant contributor

Because the C—Cl bond is so long, the lone-pair orbital ($3p$) of chlorine and the π orbital of the carbonyl group do not overlap sufficiently to permit delocalization of a chlorine unshared pair. Not only is the carbonyl group of an acyl chloride not stabilized by electron-pair donation, the electron-withdrawing inductive effect of chlorine makes it more electrophilic and more reactive toward nucleophiles.

2. **Acid anhydrides:** The carbonyl group of an acid anhydride is better stabilized by electron donation than the carbonyl group of an acyl chloride. Even though oxygen is more electronegative than chlorine, it is a far better electron-pair donor toward sp^2-hybridized carbon.

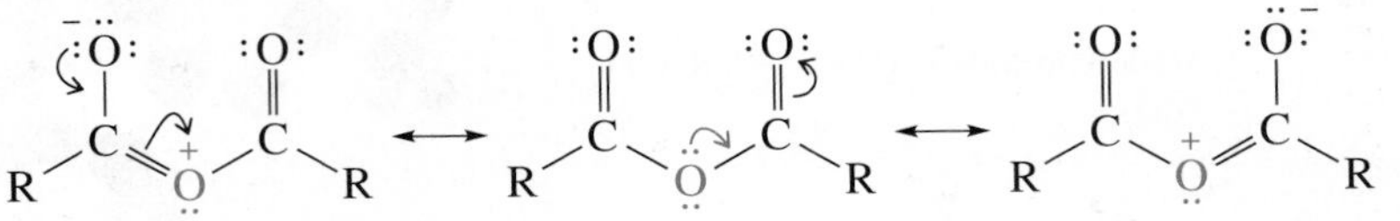

Working against this electron-delocalization is the fact that both carbonyl groups are competing for the same electron pair. Thus, the extent to which each one is stabilized is reduced.

3. **Esters:** Like acid anhydrides, the carbonyl group of an ester is stabilized by electron release from oxygen. Because there is only one carbonyl group, versus two in anhydrides, esters are stabilized more and are less reactive than anhydrides.

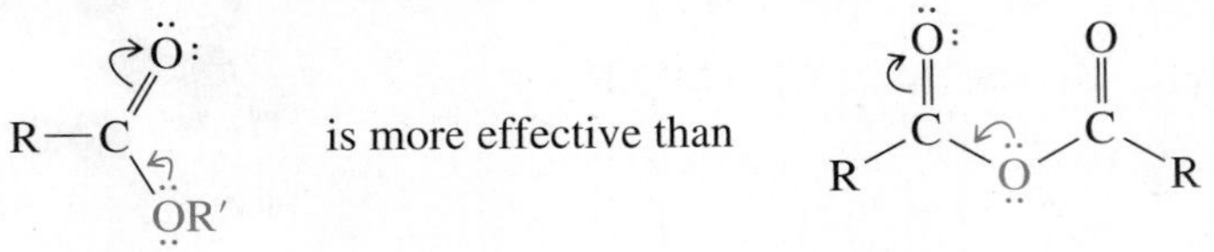

R—C(=O)OR′ is more effective than RC(=O)OC(=O)R

Ester Acid anhydride

4. **Amides:** Nitrogen is less electronegative than oxygen; therefore, the carbonyl group of an amide is stabilized more than that of an ester.

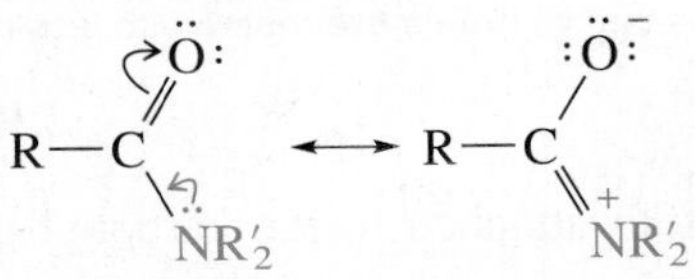

Very effective resonance stabilization

Amide resonance is a powerful stabilizing force and gives rise to a number of structural effects. Unlike the pyramidal arrangement of bonds in ammonia and amines, the bonds to nitrogen in amides lie in the same plane (Figure 20.2*a*). The carbon–nitrogen bond has considerable double-bond character and, at 135 pm (1.35 Å), is substantially shorter than the normal 147 pm (1.47Å) carbon–nitrogen single-bond distance observed in amines.

The barrier to rotation about the carbon–nitrogen bond in amides is 75–85 kJ/mol (18–20 kcal/mol).

E_a = 75–85 kJ/mol (18–20 kcal/mol)

Recall (Section 3.1) that the rotational barrier in ethane is only 12 kJ/mol (3 kcal/mol).

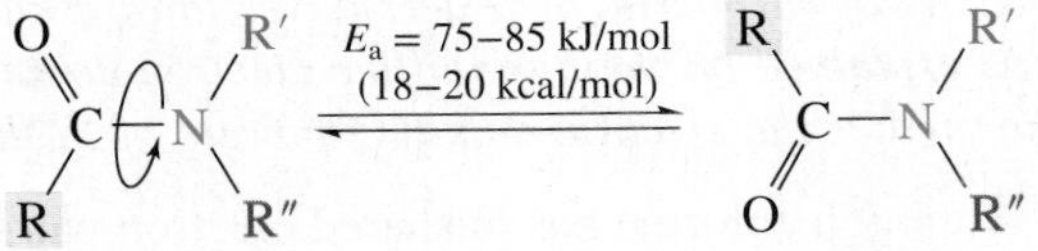

This is an unusually high rotational energy barrier for a single bond and indicates that the carbon–nitrogen bond has significant double-bond character, as the resonance and orbital overlap (Figure 20.2*b*) descriptions suggest.

Problem 20.2

Suggest an explanation for the fact that *N,N*-dimethylformamide [$(CH_3)_2NCH{=}O$] has signals for three nonequivalent carbons (δ 31.3, 36.4, and 162.6) in its ^{13}C NMR spectrum.

Electron release from nitrogen stabilizes the carbonyl group of amides and decreases the rate at which nucleophiles attack the carbonyl carbon.

An extreme example of carbonyl-group stabilization is seen in carboxylate anions:

$$\mathrm{R{-}C({=}\ddot{O}{:})({-}{:}\ddot{O}{:}^{-})} \longleftrightarrow \mathrm{R{-}C({-}{:}\ddot{O}{:}^{-})({=}\ddot{O}{:})}$$

The negatively charged oxygen is a powerful electron donor to the carbonyl group. Resonance in carboxylate anions is more effective than resonance in carboxylic acids, acyl chlorides, anhydrides, esters, and amides. Carboxylate ions do not undergo nucleophilic acyl substitution.

Most methods for their preparation convert one class of carboxylic acid derivative to another by nucleophilic acyl substitution. The order of carbonyl-group stabilization given in Figure 20.1 bears directly on the means by which these transformations may be achieved. A reaction that converts one carboxylic acid derivative to another that lies below it in the figure is practical; a reaction that converts it to one that lies above it is not. This is another way of saying that *one carboxylic acid derivative can be converted to another if the reaction leads to a more stabilized carbonyl group.* Numerous examples of reactions of this type will be presented in the sections that follow.

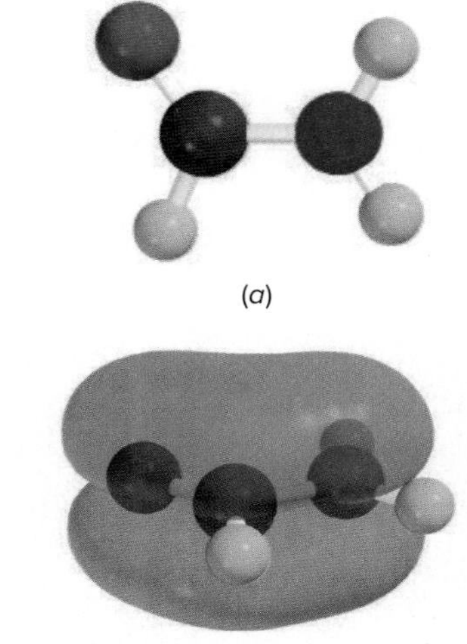

(*a*)

(*b*)

Figure 20.2

(*a*) Formamide ($\mathrm{HCNH_2}$, with C=O) is planar. Carbon and nitrogen are both sp^2-hybridized. (*b*) A π orbital generated by overlap of the 2*p* orbital of nitrogen and the π orbital of the carbonyl group allows delocalization of the nitrogen unshared pair.

20.3 Nucleophilic Acyl Substitution Mechanisms

Experimental support exists for several different mechanisms of nucleophilic acyl substitution, three of which are introduced here.

By far, the most common mechanism is *bimolecular* and involves what is referred to as a *tetrahedral intermediate (TI).* It incorporates two stages, formation of the TI by nucleophilic *addition* to the carbonyl group followed by an *elimination* that restores the carbonyl.

$$\underset{\text{Carboxylic acid derivative}}{\mathrm{RC({=}\ddot{O}{:})X}} + \underset{\text{Nucleophile}}{\mathrm{Nu{-}H}} \xrightarrow[\text{slow}]{\text{addition}} \underset{\text{Tetrahedral intermediate (TI)}}{\mathrm{R{-}C(\ddot{O}{-}H)(X)(Nu)}} \xrightarrow[\text{fast}]{\text{elimination}} \underset{\text{Products}}{\mathrm{RC({=}\ddot{O}{:})Nu} + \mathrm{H{-}X}}$$

The first stage is rate-determining and analogous to nucleophilic addition to the carbonyl group of an aldehyde or ketone. Many of the same nucleophiles that add to the carbonyl groups of aldehydes and ketones—water, alcohols, and amines—add to carbonyl groups of carboxylic acid derivatives and lead to the products of nucleophilic acyl substitution. Each stage can involve several steps, the precise nature of which depends on whether the reaction occurs in neutral, acidic, or basic solution.

The main features that complicate the mechanism involve acid–base chemistry and influence the form in which the reactants, intermediates, and products exist under the reaction conditions. Thus, the rate-determining step can involve formation of the tetrahedral intermediate as a cation ($\mathrm{TI{-}H^+}$), anion ($\mathrm{TI^-}$), or neutral species (TI) depending on the pH at which the reaction is carried out.

$$\underset{\substack{\mathrm{TI{-}H^+}\\ \text{Conjugate acid of}\\ \text{tetrahedral intermediate}}}{\mathrm{R{-}C(\ddot{O}{-}H)(\overset{+}{X}{-}H)(Nu)}} \rightleftharpoons \underset{\substack{\mathrm{TI}\\ \text{Tetrahedral}\\ \text{intermediate}}}{\mathrm{R{-}C(\ddot{O}{-}H)(X)(Nu)}} \rightleftharpoons \underset{\substack{\mathrm{TI^-}\\ \text{Conjugate base of}\\ \text{tetrahedral intermediate}}}{\mathrm{R{-}C({:}\ddot{O}{:}^-)(X)(Nu)}}$$

Once formed, the tetrahedral intermediate can revert to the original carboxylic acid derivative or go on to form the product of nucleophilic acyl substitution. As noted in the preceding section, the reactions generally proceed in the direction that converts a less stabilized carbonyl group to a more stable one.

Alternative mechanisms for nucleophilic acyl substitution are more limited in their scope. One is analogous to the S_N1 mechanism and involves rate-determining formation of an acylium ion.

$$\mathrm{RC(=O)X} \underset{}{\overset{\text{slow}}{\rightleftharpoons}} \ddot{X}^- + R-C\equiv\overset{+}{O}: \xrightarrow[\text{fast}]{:Nu^-} \mathrm{RC(=O)Nu}$$

Carboxylic acid derivative — Acylium ion — Carboxylic acid derivative

Among carboxylic acid derivatives, this mechanism is observed only with acyl chlorides, and even then in only a limited number of cases.

Problem 20.3

The reaction shown is believed to involve an acylium ion intermediate.

$$p\text{-}CH_3OC_6H_4COCl \xrightarrow[F_3CCH_2OH]{H_2O} p\text{-}CH_3OC_6H_4COOH$$

p-Methoxybenzoyl chloride — *p*-Methoxybenzoic acid

Write a structural formula for the acylium ion and describe the structural features that make it a plausible intermediate.

The third mechanism is an S_N2-like process with qualities akin to both the ionization and tetrahedral intermediate pathways.

$$\mathrm{RC(=O)X} + Nu:^- \longrightarrow \left[\mathrm{R}\overset{O^{\delta-}}{C}(X^{\delta-})(Nu^{\delta-})\right] \longrightarrow \mathrm{RC(=O)Nu} + X:^-$$

Carboxylic acid derivative — Nucleophile — Transition state — Product — Anion of leaving group

This mechanism combines features of both the tetrahedral intermediate and acylium ion mechanisms. Like the tetrahedral intermediate mechanism, bond making to the nucleophile contributes to the transition state. Like the ionization mechanism, the bond to the leaving group is partially broken at the transition state. The latter makes it more likely to compete with the tetrahedral intermediate mechanism in reactions of acyl chlorides and less likely in reactions of anhydrides, esters, and amides.

20.4 Nucleophilic Acyl Substitution in Acyl Chlorides

One of the most useful reactions of acyl chlorides was presented in Section 13.7. Friedel–Crafts acylation of aromatic rings takes place when arenes are treated with acyl chlorides in the presence of aluminum chloride.

Among the various carboxylic acid derivatives, acyl chlorides are especially useful because they are readily converted to acid anhydrides, esters, and amides by nucleophilic acyl substitution (Table 20.1). Yields are high and the reaction rates are much greater than the corresponding rates of alkyl halides with the same nucleophiles. Benzoyl chloride, for example, is about 1000 times more reactive than benzyl chloride toward hydrolysis at 25°C.

Problem 20.4

Use Table 20.1 to predict the major organic product obtained by reaction of benzoyl chloride with each of the following:

(a) Acetic acid
(b) Benzoic acid
(c) Ethanol
(d) Methylamine, CH_3NH_2
(e) Dimethylamine, $(CH_3)_2NH$
(f) Water

continued

TABLE 20.1 Conversion of Acyl Chlorides to Other Carboxylic Acid Derivatives

Reaction (section) and comments	General equation and specific example
Reaction with carboxylic acids (see Section 20.4) Acyl chlorides react with carboxylic acids to yield acid anhydrides. When this reaction is used for preparative purposes, pyridine is normally added. Pyridine is a catalyst for the reaction and also acts as a base to neutralize the hydrogen chloride that is formed.	$RCCl$ (=O) + $R'COH$ (=O) ⟶ $RCOCR'$ (O, O) + HCl; Acyl chloride, Carboxylic acid, Acid anhydride, Hydrogen chloride. Heptanoyl chloride + Heptanoic acid —pyridine⟶ Heptanoic anhydride (78–83%)
Reaction with alcohols (see Section 16.8) Acyl chlorides react with alcohols to form esters. The reaction is typically carried out in the presence of pyridine.	$RCCl$ (=O) + $R'OH$ ⟶ $RCOR'$ (=O) + HCl; Acyl chloride, Alcohol, Ester, Hydrogen chloride. Benzoyl chloride + *tert*-Butyl alcohol —pyridine⟶ *tert*-Butyl benzoate (80%)
Reaction with ammonia and amines (see Section 20.12) Acyl chlorides react with ammonia and amines to form amides. A base such as sodium hydroxide is normally added to react with the hydrogen chloride produced.	$RCCl$ (=O) + R'_2NH + HO^- ⟶ $RCNR'_2$ (=O) + H_2O + Cl^-; Acyl chloride, Ammonia or amine, Hydroxide, Amide, Water, Chloride. Benzoyl chloride + Piperidine —NaOH, H_2O⟶ *N*-Benzoylpiperidine (87–91%)
Hydrolysis (see Section 20.4) Acyl chlorides react with water to yield carboxylic acids. In base, the acid is converted to its carboxylate salt. The reaction has little preparative value because the acyl chloride is nearly always prepared from the carboxylic acid itself.	$RCCl$ (=O) + H_2O ⟶ $RCOH$ (=O) + HCl; Acyl chloride, Water, Carboxylic acid, Hydrogen chloride. Phenylacetyl chloride + H_2O (Water) ⟶ Phenylacetic acid

Sample Solution (a) As noted in Table 20.1, the reaction of an acyl chloride with a carboxylic acid yields an acid anhydride.

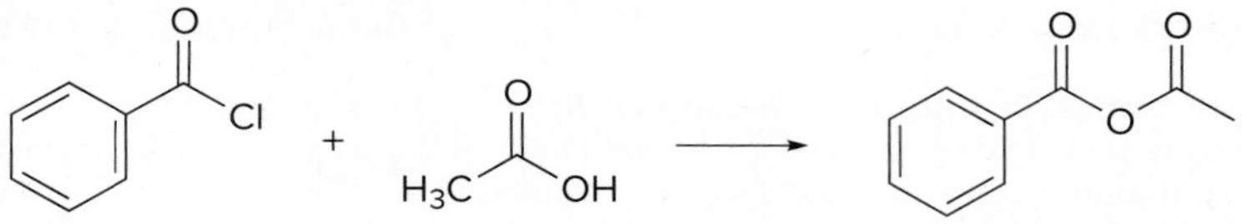

Benzoyl chloride + Acetic acid (H_3C–C(=O)–OH) ⟶ Acetic benzoic anhydride

The product is a mixed anhydride. Acetic acid acts as a nucleophile and substitutes for chloride on the benzoyl group.

On examining the specific examples in Table 20.1, we see that nucleophilic substitutions of acyl chlorides are often carried out in the presence of pyridine. Pyridine is both a catalyst and a weak base. As a catalyst it increases the rate of acylation. As a base it prevents the build-up of HCl.

Table 20.1 concludes with the hydrolysis of acyl chlorides. Because acyl chlorides are themselves prepared by the reaction of carboxylic acids with thionyl chloride (Section 13.7), their hydrolysis is of little synthetic value.

20.5 Nucleophilic Acyl Substitution in Acid Anhydrides

After acyl halides, acid anhydrides are the most reactive carboxylic acid derivatives. Although anhydrides can be prepared by reaction of carboxylic acids with acyl chlorides as was shown in Table 20.1, the three most commonly used anhydrides are industrial chemicals and are prepared by specialized methods. Phthalic anhydride and maleic anhydride, for example, are prepared from naphthalene and butane, respectively.

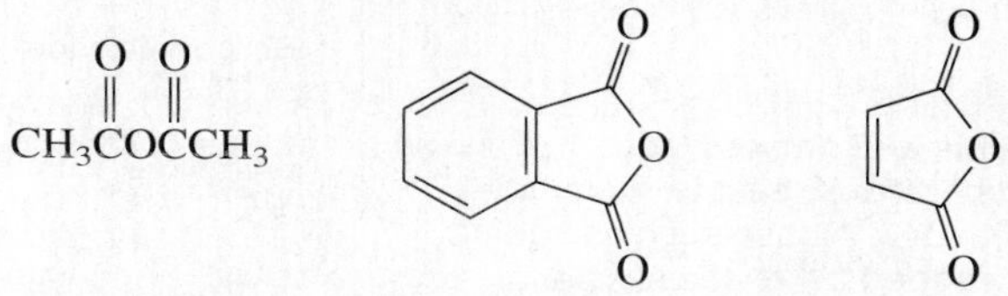

$$CH_3\overset{O}{\overset{\|}{C}}O\overset{O}{\overset{\|}{C}}CH_3$$

Acetic anhydride   Phthalic anhydride   Maleic anhydride

Acid anhydrides contain two acyl groups bonded to the same oxygen. In nucleophilic acyl substitution, one of these acyl groups becomes bonded to the nucleophilic atom. The other acyl group remains on oxygen to become part of a carboxylic acid.

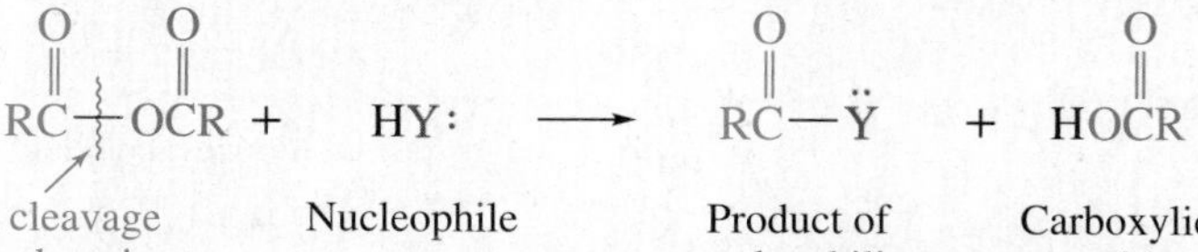

$$R\overset{O}{\overset{\|}{C}}-O\overset{O}{\overset{\|}{C}}R + HY\!: \longrightarrow R\overset{O}{\overset{\|}{C}}-\ddot{Y} + HO\overset{O}{\overset{\|}{C}}R$$

Bond cleavage occurs here in an acid anhydride.   Nucleophile   Product of nucleophilic acyl substitution   Carboxylic acid

Acid anhydrides are more stable and less reactive than acyl chlorides. Acetyl chloride, for example, undergoes hydrolysis about 100,000 times more rapidly than acetic anhydride at 25°C.

Table 20.2 gives examples of two reactions of acetic anhydride. In the first, the anhydride carbonyl is converted to the more stabilized carbonyl group of an ester; in the second, a more stabilized amide carbonyl results. Hydrolysis under neutral or acid-catalyzed conditions yields two moles of a carboxylic acid per mole of anhydride, but, like hydrolysis of acyl chlorides is of little preparative value and is not included in the table. Likewise, hydrolysis in aqueous base according to the following equation is omitted from the table.

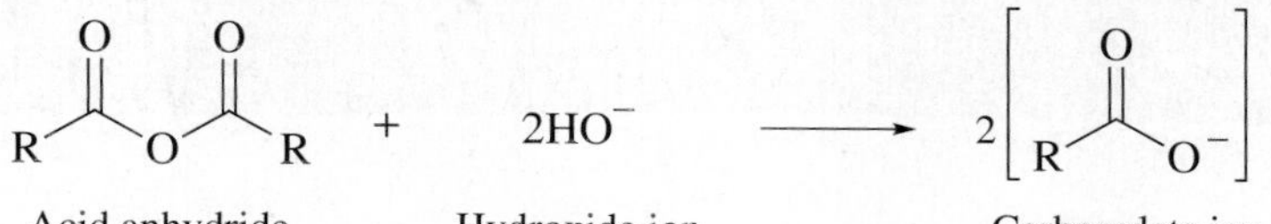

$$RC(=O)OC(=O)R + 2HO^- \longrightarrow 2\left[RC(=O)O^-\right]$$

Acid anhydride   Hydroxide ion   Carboxylate ion

TABLE 20.2 Conversion of Acid Anhydrides to Other Carboxylic Acid Derivatives

Reaction (section) and comments	General equation and specific example
Reaction with alcohols (see Section 16.8) Acid anhydrides react with alcohols to form esters. The reaction may be carried out in the presence of pyridine or it may be catalyzed by acids. Only one acyl group of an anhydride is incorporated into the ester; the other becomes the acyl group of a carboxylic acid.	$\text{RCOCR (Acid anhydride)} + \text{R'OH (Alcohol)} \longrightarrow \text{RCOR' (Ester)} + \text{RCOH (Carboxylic acid)}$ 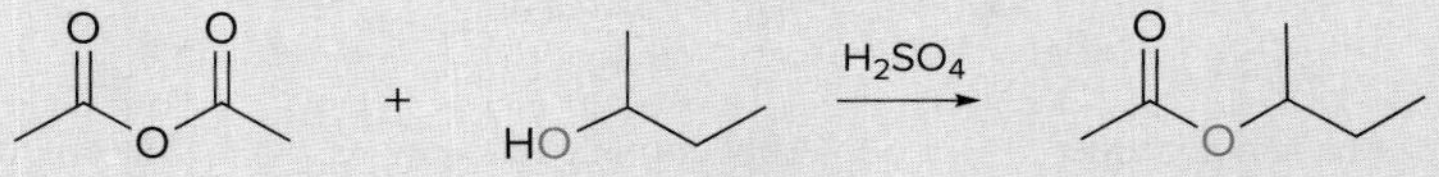 Acetic anhydride + *sec*-Butyl alcohol $\xrightarrow{H_2SO_4}$ *sec*-Butyl acetate (60%)
Reaction with ammonia and amines (see Section 20.12) Acid anhydrides react with ammonia and amines to form amides. Only one acyl group of an anhydride is incorporated into the amide; the other becomes the acyl group of the amine salt of the carboxylic acid.	$\text{RCOCR (Acid anhydride)} + 2\text{R}'_2\text{NH (Amine)} \longrightarrow \text{RCNR}'_2 \text{ (Amide)} + \text{RCO}^- \ \text{H}_2\overset{+}{\text{N}}\text{R}'_2 \text{ (Ammonium carboxylate salt)}$ 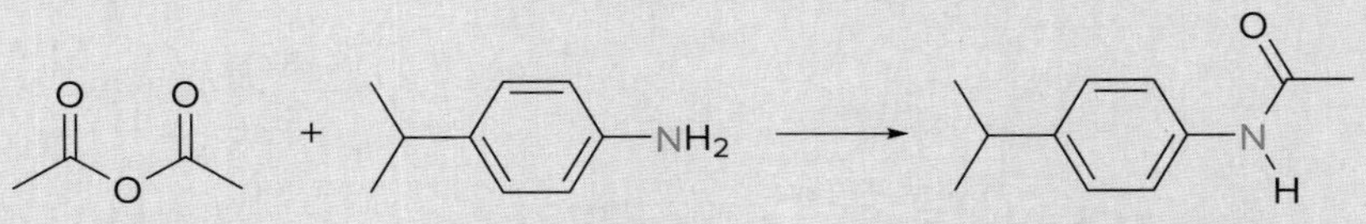 Acetic anhydride + *p*-Isopropylaniline $\longrightarrow$ *p*-Isopropylacetanilide (98%)

Problem 20.5

Predict the major organic product of each of the following reactions.

(a) Benzoic anhydride + methanol $\xrightarrow{H_2SO_4}$

(b) Acetic anhydride + ammonia (2 mol) $\longrightarrow$

(c) Phthalic anhydride + $(CH_3)_2NH$ (2 mol) $\longrightarrow$

(d) Phthalic anhydride + sodium hydroxide (2 mol) $\longrightarrow$

Sample Solution (a) Nucleophilic acyl substitution by an alcohol on an acid anhydride yields an ester.

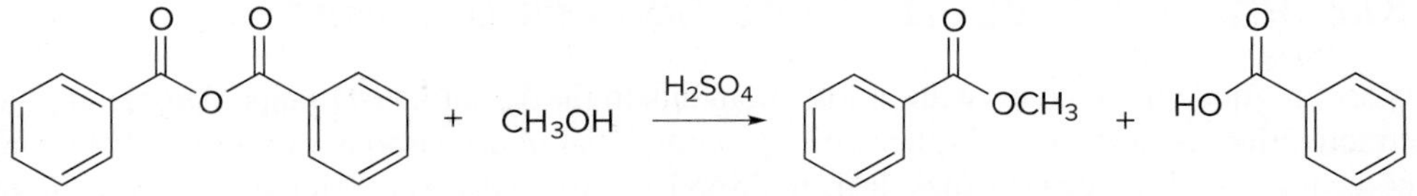

Benzoic anhydride + CH_3OH (Methanol) $\xrightarrow{H_2SO_4}$ Methyl benzoate ($C_6H_5COOCH_3$) + Benzoic acid ($HOCC_6H_5$)

Mechanistically, nucleophilic acyl substitutions of anhydrides normally proceed by way of a tetrahedral intermediate. When the nucleophile is an anion, TI^- is the initial intermediate and its dissociation leads directly to the observed products as shown in Mechanism 20.1 for the reaction:

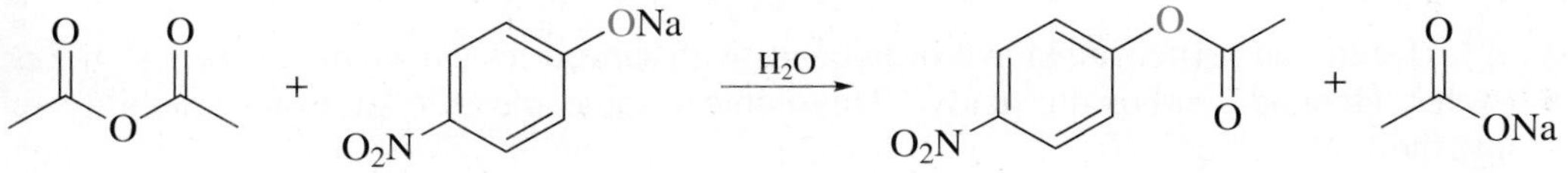

Acetic anhydride + Sodium *p*-nitrophenoxide $\xrightarrow{H_2O}$ *p*-Nitrophenyl acetate (78%) + Sodium acetate

Mechanism 20.1

Nucleophilic Acyl Substitution in an Anhydride

THE OVERALL REACTION:

Acetic anhydride + *p*-Nitrophenoxide ion → *p*-Nitrophenyl acetate + Acetate ion

THE MECHANISM:

Step 1: Nucleophilic addition of *p*-nitrophenoxide to one of the carbonyl groups of the anhydride gives the conjugate base of the tetrahedral intermediate (TI^-).

Acetic anhydride + *p*-Nitrophenoxide ion —slow→ Conjugate base of tetrahedral intermediate (TI^-)

Step 2: Expulsion of acetate from TI^- restores the carbonyl group.

Conjugate base of tetrahedral intermediate (TI^-) —fast→ *p*-Nitrophenyl acetate + Acetate ion

20.6 Physical Properties and Sources of Esters

Esters are moderately polar, with dipole moments in the 1.5- to 2.0-D range. Dipole–dipole attractive forces give esters higher boiling points than hydrocarbons of similar shape and molecular weight. Because they lack hydroxyl groups, however, ester molecules cannot form hydrogen bonds to each other and have lower boiling points than alcohols of comparable molecular weight.

2-Methylbutane: mol wt 72, bp 28°C

Methyl acetate: mol wt 74, bp 57°C

2-Butanol: mol wt 74, bp 99°C

Esters can participate in hydrogen bonds with substances that contain hydroxyl groups (water, alcohols, carboxylic acids). This confers some measure of water solubility on low-molecular-weight esters; methyl acetate, for example, dissolves in water to the extent of 33 g/100 mL. Water solubility decreases as the carbon content of the ester increases.

Many esters occur naturally. Those of low molecular weight are fairly volatile, and many have pleasing odors. Esters often form a significant fraction of the fragrant oil of fruits and flowers. The aroma of oranges, for example, contains 30 different esters along with 10 carboxylic acids, 34 alcohols, 34 aldehydes and ketones, and 36 hydrocarbons.

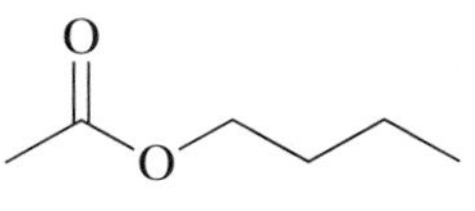

Butyl acetate
(contributes to characteristic pear odor)

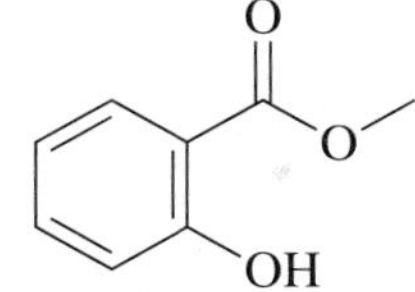

Methyl salicylate
(principal component of oil of wintergreen)

Among the chemicals used by insects to communicate with one another, esters occur frequently.

Ethyl cinnamate
(one of the constituents of the sex pheromone of the male oriental fruit moth)

(*R*)-(*Z*)-5-Tetradecen-4-olide
(sex pheromone of the female Japanese beetle)

Notice that (*R*)-(*Z*)-5-tetradecen-4-olide is a cyclic ester. Recall from Section 19.15 that cyclic esters are called *lactones* and that the suffix *-olide* is characteristic of IUPAC names for lactones.

Esters of glycerol, called *glycerol triesters, triacylglycerols,* or *triglycerides,* are abundant natural products. The most important group of glycerol triesters includes those in which each acyl group is unbranched and has 14 or more carbon atoms. Structurally related phosphatidylcholine is a component of cell membranes (Section 24.4).

Tristearin, a trioctadecanoyl ester of glycerol found in many animal and vegetable fats

Phosphatidylcholine
(R and R′ are carbon chains)

A molecular model of tristearin is shown in Figure 24.2.

Fats and **oils** are naturally occurring mixtures of glycerol triesters. Fats are mixtures that are solids at room temperature; oils are liquids. The long-chain carboxylic acids obtained from fats and oils by hydrolysis are known as **fatty acids.**

The chief methods used to prepare esters are reviewed in Table 20.3.

20.7 Reactions of Esters: A Preview

Nucleophilic acyl substitutions of esters are summarized in Table 20.4. Esters are less reactive than acyl chlorides and acid anhydrides. Nucleophilic acyl substitution in esters, especially ester hydrolysis, has been extensively investigated from a mechanistic perspective. Indeed, much of what we know concerning the general topic of nucleophilic acyl substitution comes from studies carried out on esters. The following sections describe those mechanistic studies.

TABLE 20.3 Preparation of Esters

Reaction (section) and comments	General equation and specific example
From carboxylic acids (see Sections 16.8 and 19.14) In the presence of an acid catalyst, alcohols and carboxylic acids react to form an ester and water. This is the Fischer esterification. The yield of ester can be increased beyond the equilibrium amount by removing the water as it is formed.	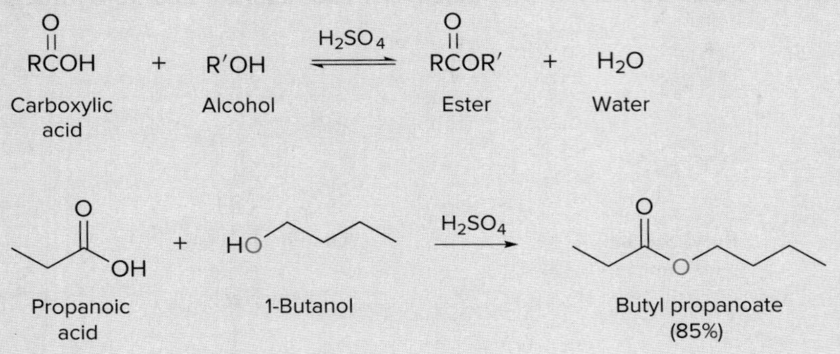 $\underset{\text{Carboxylic acid}}{RCOH} + \underset{\text{Alcohol}}{R'OH} \overset{H_2SO_4}{\rightleftharpoons} \underset{\text{Ester}}{RCOR'} + \underset{\text{Water}}{H_2O}$ Propanoic acid + 1-Butanol $\xrightarrow{H_2SO_4}$ Butyl propanoate (85%)
From acyl chlorides (see Sections 16.8 and 20.4) Alcohols react with acyl chlorides by nucleophilic acyl substitution to yield esters. These reactions are typically performed in the presence of a weak base such as pyridine.	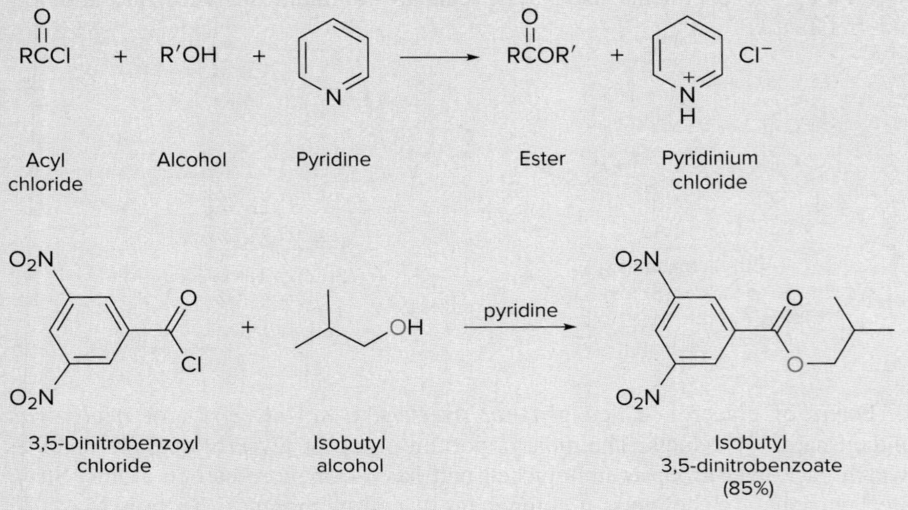 $\underset{\text{Acyl chloride}}{RCCl} + \underset{\text{Alcohol}}{R'OH} + \underset{\text{Pyridine}}{\text{Pyridine}} \longrightarrow \underset{\text{Ester}}{RCOR'} + \underset{\text{Pyridinium chloride}}{\text{Pyridinium } Cl^-}$ 3,5-Dinitrobenzoyl chloride + Isobutyl alcohol $\xrightarrow{\text{pyridine}}$ Isobutyl 3,5-dinitrobenzoate (85%)
From acid anhydrides (see Sections 16.8 and 20.5) Acyl transfer from an acid anhydride to an alcohol is a standard method for preparing esters. The reaction is subject to catalysis by either acids (H_2SO_4) or bases (pyridine).	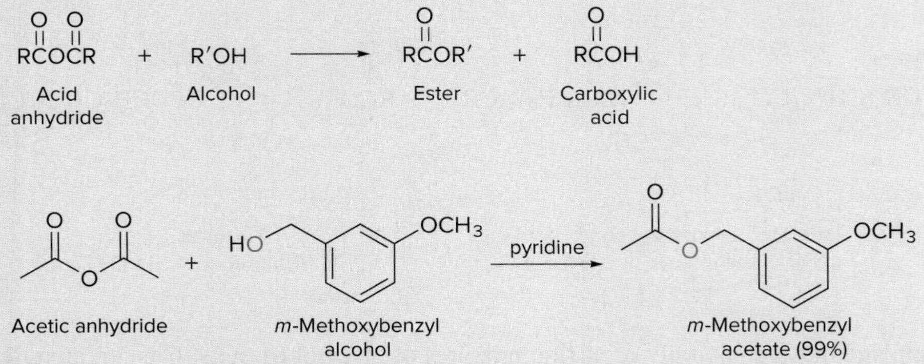 $\underset{\text{Acid anhydride}}{RCOCR} + \underset{\text{Alcohol}}{R'OH} \longrightarrow \underset{\text{Ester}}{RCOR'} + \underset{\text{Carboxylic acid}}{RCOH}$ Acetic anhydride + *m*-Methoxybenzyl alcohol $\xrightarrow{\text{pyridine}}$ *m*-Methoxybenzyl acetate (99%)
Baeyer–Villiger oxidation of ketones (see Descriptive Passage 18) Ketones are converted to esters on treatment with peroxy acids. The reaction proceeds by migration of the group R′ from carbon to oxygen. It is the more highly substituted group that migrates. Methyl ketones give acetate esters.	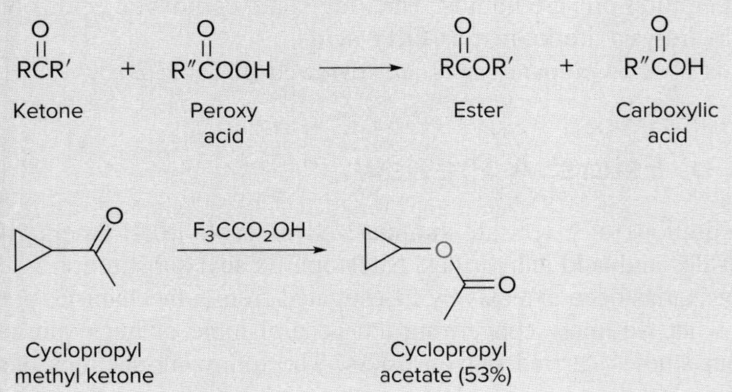 $\underset{\text{Ketone}}{RCR'} + \underset{\text{Peroxy acid}}{R''COOH} \longrightarrow \underset{\text{Ester}}{RCOR'} + \underset{\text{Carboxylic acid}}{R''COH}$ Cyclopropyl methyl ketone $\xrightarrow{F_3CCO_2OH}$ Cyclopropyl acetate (53%)

TABLE 20.4 Conversion of Esters to Other Carboxylic Acid Derivatives

Reaction (section) and comments	General equation and specific example
Reaction with ammonia and amines (see Section 20.10) Esters react with ammonia and amines to form amides. Methyl and ethyl esters are the most reactive.	$RC(=O)OR' + R''_2NH \longrightarrow RC(=O)NR''_2 + R'OH$ Ester, Ammonia or amine, Amide, Alcohol Ethyl fluoroacetate + NH_3 (Ammonia) $\longrightarrow$ Fluoroacetamide (90%) + Ethanol
Hydrolysis (see Sections 20.8 and 20.9) Ester hydrolysis may be catalyzed either by acids or by bases. Acid-catalyzed hydrolysis is an equilibrium-controlled process and the reverse of the Fischer esterification. Hydrolysis in base, called saponification, is irreversible and is the method usually chosen for preparative purposes.	$RC(=O)OR' + H_2O \longrightarrow RC(=O)OH + R'OH$ Ester, Water, Carboxylic acid, Alcohol Methyl *m*-nitrobenzoate $\xrightarrow[2.\ H_3O^+]{1.\ H_2O,\ NaOH}$ *m*-Nitrobenzoic acid (90–96%) + CH_3OH (Methanol)

20.8 Acid-Catalyzed Ester Hydrolysis

Ester hydrolysis is the most studied and best understood of all nucleophilic acyl substitutions. Esters are fairly stable in neutral aqueous media but are cleaved when heated with water in the presence of strong acids or bases. The hydrolysis of esters in dilute aqueous acid is the reverse of the Fischer esterification (Sections 16.8 and 19.14):

$$\underset{\text{Ester}}{RC(=O)OR'} + \underset{\text{Water}}{H_2O} \overset{\text{acid}}{\rightleftharpoons} \underset{\text{Carboxylic acid}}{RC(=O)OH} + \underset{\text{Alcohol}}{R'OH}$$

When esterification is the objective, water is removed from the reaction mixture to encourage ester formation. When ester hydrolysis is the objective, the reaction is carried out in the presence of a generous excess of water. Both reactions illustrate the application of Le Châtelier's principle (Section 8.7) to organic synthesis.

$$\underset{\text{Ethyl 2-chloro-2-phenylacetate}}{C_6H_5CHClCO_2CH_2CH_3} + \underset{\text{Water}}{H_2O} \xrightarrow[\text{heat}]{\text{HCl}} \underset{\text{2-Chloro-2-phenylacetic acid (80–82\%)}}{C_6H_5CHClCO_2H} + \underset{\text{Ethanol}}{CH_3CH_2OH}$$

Problem 20.6

The compound having the structure shown was heated with dilute sulfuric acid to give a product having the molecular formula $C_5H_{12}O_3$ in 63–71% yield. Propose a reasonable structure for this product. What other organic compound is formed in this reaction?

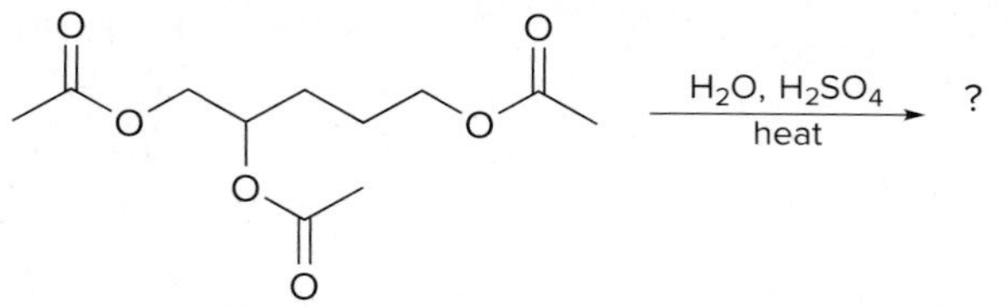

$\xrightarrow[\text{heat}]{H_2O,\ H_2SO_4}$?

The pathway for acid-catalyzed ester hydrolysis is given in Mechanism 20.2. It is precisely the reverse of the mechanism given for acid-catalyzed ester formation in Section 19.14. Like other nucleophilic acyl substitutions, it proceeds in two stages. A tetrahedral intermediate is formed in the first stage, then dissociates to products in the second stage.

A key feature of the first stage (steps 1–3) is the site at which the starting ester is protonated. Protonation of the carbonyl oxygen, as shown in step 1 of Mechanism 20.2, gives a cation that is stabilized by electron delocalization. The alternative site of protonation, the alkoxy oxygen, gives rise to a much less stable cation.

Protonation of carbonyl oxygen ***Protonation of alkoxy oxygen***

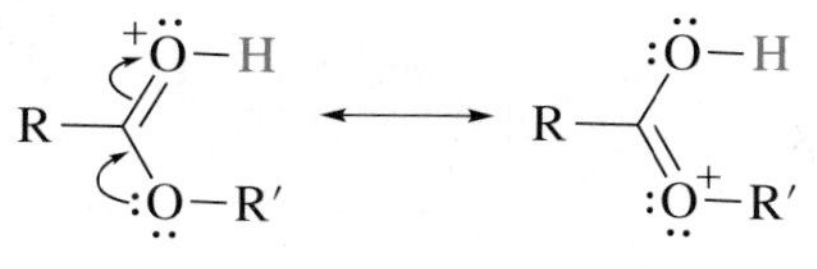

Positive charge is delocalized. 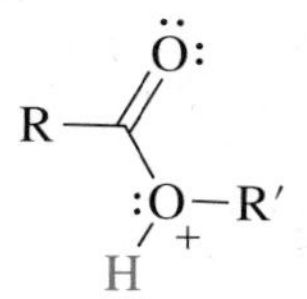 Positive charge is localized on a single oxygen.

Mechanism 20.2

Acid-Catalyzed Ester Hydrolysis

THE OVERALL REACTION:

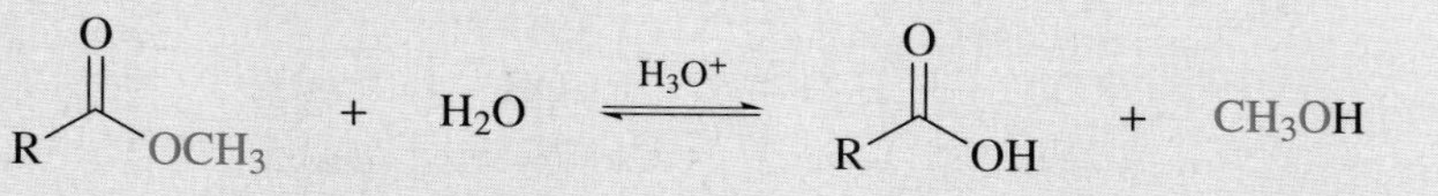

$$RCOOCH_3 + H_2O \underset{}{\overset{H_3O^+}{\rightleftharpoons}} RCOOH + CH_3OH$$

A methyl ester | Water | A carboxylic acid | Methanol

THE MECHANISM:

First Stage: Formation of the tetrahedral intermediate Steps 1–3 are analogous to the mechanism of acid-catalyzed hydration of an aldehyde or ketone.

Step 1: Protonation of the carbonyl oxygen of the ester

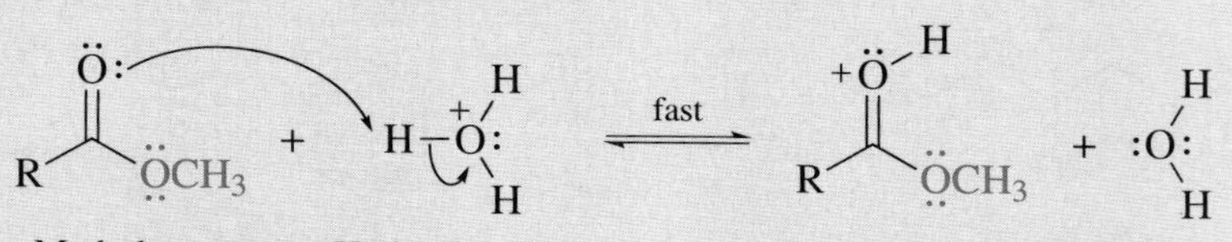

$$RCOOCH_3 + H_3O^+ \overset{\text{fast}}{\rightleftharpoons} RC(=\overset{+}{O}H)OCH_3 + H_2O$$

Methyl ester | Hydronium ion | Protonated ester | Water

Step 2: Nucleophilic addition of water to the protonated ester

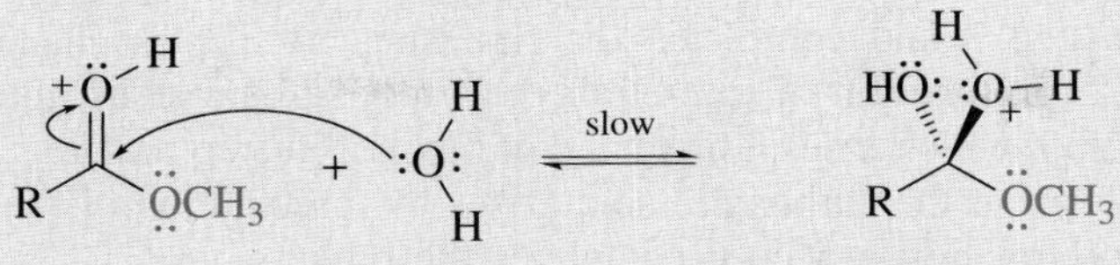

Protonated ester + Water ⇌ (slow) Conjugate acid of tetrahedral intermediate (TI–H$^+$)

Step 3: Deprotonation of TI—H$^+$ to give the neutral form of the tetrahedral intermediate (TI)

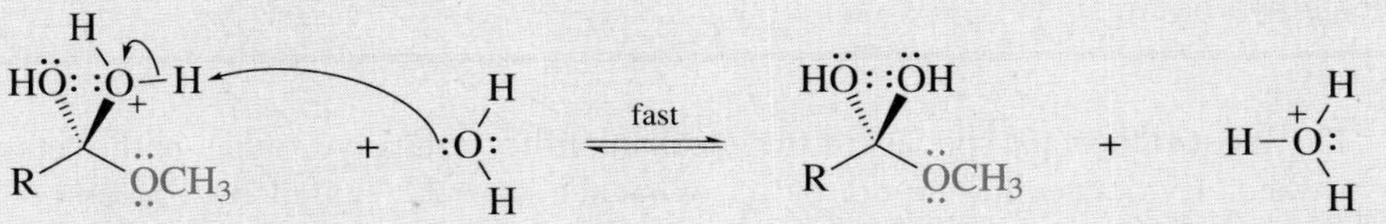

Conjugate acid of tetrahedral intermediate (TI–H$^+$) + Water ⇌ (fast) Tetrahedral intermediate (TI) + Hydronium ion

Second Stage: Dissociation of the tetrahedral intermediate Just as steps 1–3 corresponded to addition of water to the carbonyl group, steps 4–6 correspond to elimination of an alcohol, in this case methanol, from the TI and a restoration of the carbonyl group.

Step 4: Protonation of the alkoxy oxygen of the tetrahedral intermediate

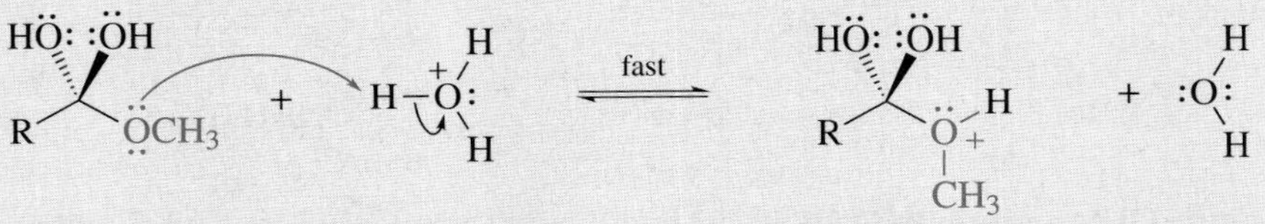

Tetrahedral intermediate (TI) + Hydronium ion ⇌ (fast) Conjugate acid of tetrahedral intermediate (TI–H$^+$) + Water

Step 5: Dissociation of the protonated form of the tetrahedral intermediate gives the alcohol and the protonated form of the carboxylic acid

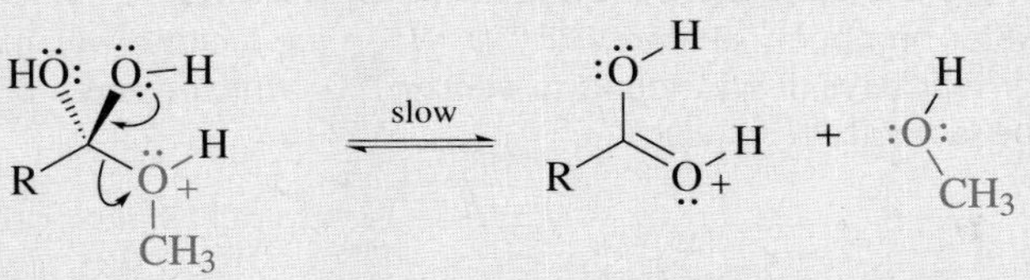

Conjugate acid of tetrahedral intermediate (TI–H$^+$) ⇌ (slow) Protonated carboxylic acid + Methanol

Step 6: Deprotonation of the protonated carboxylic acid completes the process

Protonated carboxylic acid + Water ⇌ (fast) Carboxylic acid + Hydronium ion

Protonation of the carbonyl oxygen makes the carbonyl group more electrophilic. A water molecule adds to the carbonyl group of the protonated ester in step 2. Loss of a proton from the resulting alkyloxonium ion gives the neutral form of the tetrahedral intermediate in step 3 and completes the first stage of the mechanism. In step 4 of Mechanism 20.2, protonation of the tetrahedral intermediate at its alkoxy oxygen gives a new oxonium ion, which loses a molecule of alcohol in step 5. Along with the alcohol, the protonated form of the carboxylic acid arises by dissociation of the tetrahedral intermediate. Its deprotonation in step 6 completes the process.

Problem 20.7

On the basis of the general mechanism for acid-catalyzed ester hydrolysis shown in Mechanism 20.2, write an analogous sequence of steps for the specific case of ethyl benzoate hydrolysis.

The most important species in the mechanism for ester hydrolysis is the tetrahedral intermediate. Evidence in support of its existence was developed by Professor Myron Bender on the basis of labeling experiments he carried out at the University of Chicago. Bender prepared ethyl benzoate, labeled with the mass-18 isotope of the carbonyl oxygen, then subjected it to acid-catalyzed hydrolysis in ordinary (unlabeled) water. He found that ethyl benzoate, recovered from the reaction before hydrolysis was complete, had lost a portion of its isotopic label. This observation is consistent only with the reversible formation of a tetrahedral intermediate under the reaction conditions:

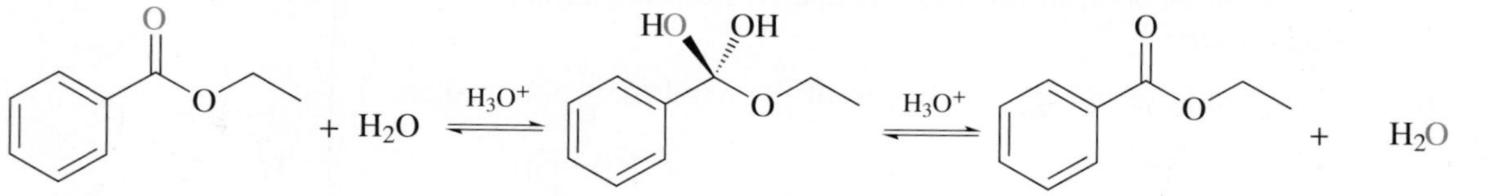

O O HO OH O

+ H_2O $\underset{}{\overset{H_3O^+}{\rightleftharpoons}}$ $\overset{H_3O^+}{\rightleftharpoons}$ + H_2O

^{18}O-Labeled ethyl benzoate | Water | Tetrahedral intermediate | Ethyl benzoate | ^{18}O-Labeled water

The two OH groups in the tetrahedral intermediate are equivalent, and so either the labeled or the unlabeled one can be lost when the tetrahedral intermediate reverts to ethyl benzoate. Both are retained when the tetrahedral intermediate goes on to form benzoic acid.

Problem 20.8

In a similar experiment, unlabeled 4-butanolide was allowed to stand in an acidic solution in which the water had been labeled with ^{18}O. When the lactone was extracted from the solution after four days, it was found to contain ^{18}O. Which oxygen of the lactone do you think became isotopically labeled?

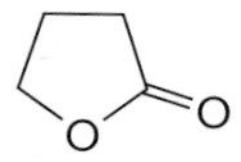

O O

4-Butanolide

20.9 Ester Hydrolysis in Base: Saponification

Unlike its acid-catalyzed counterpart, ester hydrolysis in aqueous base is *irreversible* because carboxylic acids are converted to their corresponding carboxylate anions in aqueous base.

Because it is consumed, hydroxide ion is a reactant, not a catalyst.

$$\underset{\text{Ester}}{\overset{O}{\overset{\|}{RCOR'}}} + \underset{\text{Hydroxide ion}}{HO^-} \longrightarrow \underset{\text{Carboxylate ion}}{\overset{O}{\overset{\|}{RCO^-}}} + \underset{\text{Alcohol}}{R'OH}$$

$$o\text{-Methylbenzyl acetate} + \text{NaOH} \xrightarrow[\text{heat}]{\text{water–methanol}} \text{Sodium acetate} + o\text{-Methylbenzyl alcohol (95–97\%)}$$

o-Methylbenzyl acetate — Sodium hydroxide — Sodium acetate — *o*-Methylbenzyl alcohol (95–97%)

To isolate the carboxylic acid, a separate acidification step following hydrolysis is necessary. Acidification converts the carboxylate salt to the free acid.

$$CH_2{=}C(CH_3)COCH_3 \xrightarrow[2.\ H_2SO_4]{1.\ NaOH,\ H_2O,\ heat} CH_2{=}C(CH_3)COOH + CH_3OH$$

Methyl 2-methylpropenoate (methyl methacrylate) — 2-Methylpropenoic acid (87%) (methacrylic acid) — Methanol

Ester hydrolysis in base is called **saponification,** which means "soap making." Over 2000 years ago, the seafaring peoples bordering the southeastern shores of the Mediterranean generally grouped as *Phoenicians* made soap by heating animal fat with wood ashes. Animal fat is rich in glycerol triesters, and wood ashes are a source of potassium carbonate. Basic hydrolysis of the fats produced a mixture of long-chain carboxylic acids as their potassium salts.

$$CH_3(CH_2)_x\overset{O}{\overset{\|}{C}}O\text{–}CH_2CH(O\overset{O}{\overset{\|}{C}}(CH_2)_yCH_3)CH_2\text{–}O\overset{O}{\overset{\|}{C}}(CH_2)_zCH_3 \xrightarrow[\text{heat}]{K_2CO_3,\ H_2O}$$

$$HOCH_2CH(OH)CH_2OH + KO\overset{O}{\overset{\|}{C}}(CH_2)_xCH_3 + KO\overset{O}{\overset{\|}{C}}(CH_2)_yCH_3 + KO\overset{O}{\overset{\|}{C}}(CH_2)_zCH_3$$

Glycerol — Potassium carboxylate salts

Potassium and sodium salts of long-chain carboxylic acids form micelles that dissolve grease (Section 19.7) and have cleansing properties. The carboxylic acids obtained by saponification of fats and oils are called *fatty acids.*

Problem 20.9

Trimyristin is obtained from coconut oil and has the molecular formula $C_{45}H_{86}O_6$. On being heated with aqueous sodium hydroxide followed by acidification, trimyristin was converted to glycerol and tetradecanoic acid as the only products. What is the structure of trimyristin?

In one of the earliest kinetic studies of an organic reaction, carried out in the nineteenth century, the rate of hydrolysis of ethyl acetate in aqueous sodium hydroxide was found to be first order in ester and first order in base.

$$CH_3\overset{O}{\overset{\|}{C}}OCH_2CH_3 + NaOH \longrightarrow CH_3\overset{O}{\overset{\|}{C}}ONa + CH_3CH_2OH$$

Ethyl acetate — Sodium hydroxide — Sodium acetate — Ethanol

$$\text{Rate} = k[CH_3\overset{O}{\overset{\|}{C}}OCH_2CH_3][NaOH]$$

Overall, the reaction exhibits second-order kinetics. Both the ester and the base are involved in the rate-determining step or in a rapid step that precedes it.

Two processes consistent with second-order kinetics both involve hydroxide ion as a nucleophile but differ in the site of nucleophilic attack. One is an S_N2 reaction, the other is nucleophilic acyl substitution.

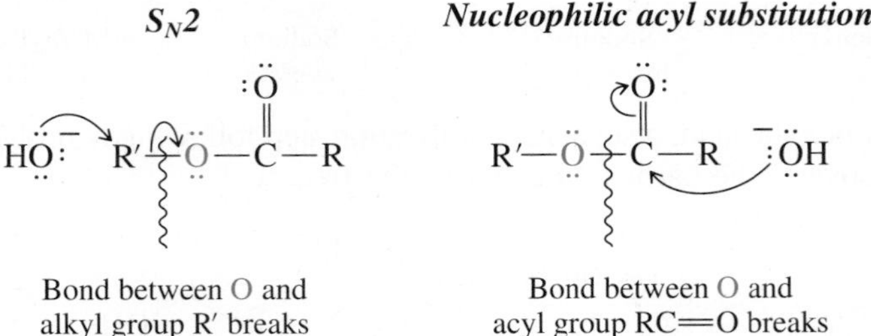

Bond between O and alkyl group R′ breaks

Bond between O and acyl group RC=O breaks

Convincing evidence that ester hydrolysis in base proceeds by a *nucleophilic acyl substitution* mechanism has been obtained from several sources. In one experiment, ethyl propanoate labeled with ^{18}O in the ethoxy group was hydrolyzed. On isolating the products, all the ^{18}O was found in the ethyl alcohol; none was in the sodium propanoate.

^{18}O-labeled ethyl propanoate + NaOH (Sodium hydroxide) ⟶ Sodium propanoate (ONa) + ^{18}O-labeled ethyl alcohol (OH)

The carbon–oxygen bond broken in the process is therefore the one between oxygen and the acyl group. The bond between oxygen and the ethyl group remains intact. An S_N2 reaction at the ethyl group would have broken this bond.

Problem 20.10

In a similar experiment, pentyl acetate was subjected to saponification with ^{18}O-labeled hydroxide in ^{18}O-labeled water. What product do you think became isotopically labeled here, acetate ion or 1-pentanol?

Identical conclusions come from stereochemical studies. Saponification of esters of optically active alcohols proceeds with *retention of configuration.*

(*R*)-(+)-1-Phenylethyl acetate $\xrightarrow[\text{ethanol–water}]{\text{KOH}}$ (*R*)-(+)-1-Phenylethanol (80% yield) + Potassium acetate (KO–C(=O)–CH_3)

None of the bonds to the chirality center is broken when hydroxide adds to the carbonyl group. Had an S_N2 reaction occurred instead, inversion of configuration at the chirality center would have taken place to give (*S*)-(−)-1-phenylethyl alcohol.

In an extension of his work described in the preceding section, Bender showed that ester hydrolysis in base, like acid hydrolysis, takes place by way of a tetrahedral intermediate. The nature of the experiment was the same, and the results were similar to those observed in the acid-catalyzed reaction.

Second-order kinetics, nucleophilic addition to the carbonyl group, and the involvement of a tetrahedral intermediate are accommodated by Mechanism 20.3. Like the

Mechanism 20.3

Ester Hydrolysis in Basic Solution

THE OVERALL REACTION:

$$RC(=O)OCH_3 + HO^- \longrightarrow RC(=O)O^- + CH_3OH$$

A methyl ester — Hydroxide ion — A carboxylate ion — Methanol

THE MECHANISM:

Step 1: Nucleophilic addition of hydroxide to the carbonyl group

$$RC(=O)OCH_3 + {}^-OH \xrightleftharpoons{\text{slow}} R\overset{O^-}{\underset{OCH_3}{C}}OH$$

Methyl ester — Hydroxide ion — Conjugate base of tetrahedral intermediate (TI^-)

Step 2: Dissociation of the anionic tetrahedral intermediate TI^-

$$R\overset{O^-}{\underset{OCH_3}{C}}OH \xrightleftharpoons{\text{fast}} RC(=O)OH + {}^-OCH_3$$

Conjugate base of tetrahedral intermediate (TI^-) — Carboxylic acid — Methoxide ion

Step 3: Proton transfers yield an alcohol and a carboxylate ion

$$CH_3O^- + H{-}OH \rightleftharpoons CH_3OH + {}^-OH$$

Methoxide ion — Water — Methanol — Hydroxide ion

$$RC(=O)OH + {}^-O{-}H \longrightarrow RC(=O)O^- + H{-}O{-}H$$

Carboxylic acid (stronger acid) — Hydroxide ion (stronger base) — Carboxylate ion (weaker base) — Water (weaker acid)

acid-catalyzed mechanism, it has two distinct stages, namely, formation of a tetrahedral intermediate and its subsequent dissociation. Nucleophilic addition to the carbonyl group has a higher activation energy than dissociation of the tetrahedral intermediate; step 1 is rate-determining. All the steps are reversible except the last one. The equilibrium constant for proton abstraction from the carboxylic acid by hydroxide in step 3 is so large that it makes the overall reaction irreversible.

Problem 20.11

On the basis of the general mechanism for basic ester hydrolysis shown in Mechanism 20.3, write an analogous sequence of steps for the saponification of ethyl benzoate.

Problem 20.12

Which ester in each pair would be expected to undergo saponification at the faster rate? Why?

(a) 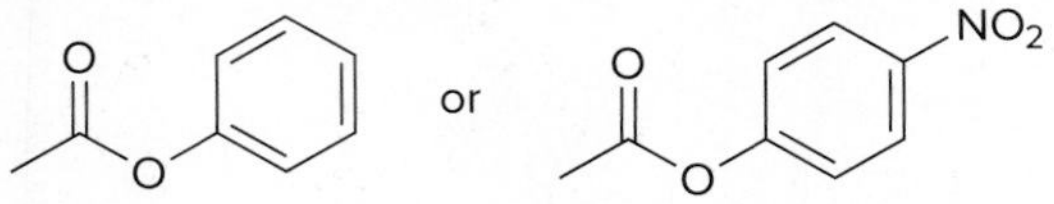 [phenyl acetate] or [p-nitrophenyl acetate]

(b) 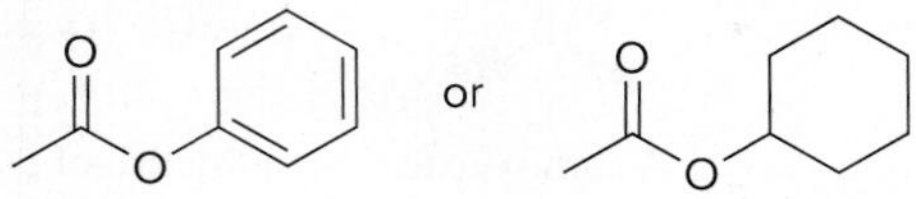 [phenyl acetate] or [cyclohexyl acetate]

(c) 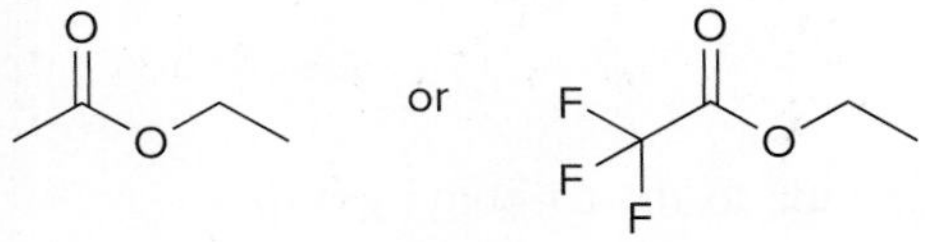 [ethyl acetate] or [ethyl trifluoroacetate]

(d) 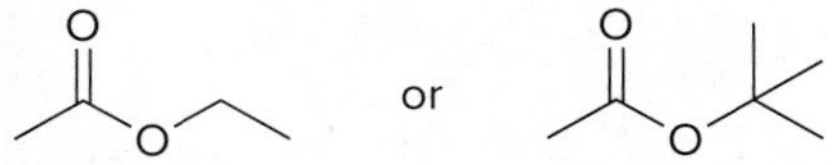 [ethyl acetate] or [tert-butyl acetate]

Sample Solution (a) *p*-Nitrophenyl acetate reacts faster. A *p*-nitrophenyl group withdraws electrons from the ester oxygen which decreases its ability to stabilize the carbonyl group. A less-stabilized carbonyl is more reactive than a more-stabilized one.

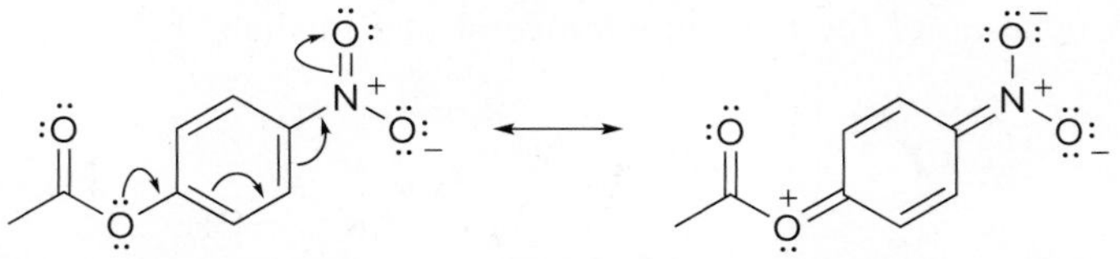

20.10 Reaction of Esters with Ammonia and Amines

Esters react with ammonia to form amides.

$$\underset{\text{Ester}}{\overset{O}{\overset{\|}{RC}}OR'} + \underset{\text{Ammonia}}{NH_3} \longrightarrow \underset{\text{Amide}}{\overset{O}{\overset{\|}{RC}}NH_2} + \underset{\text{Alcohol}}{R'OH}$$

Ammonia is more nucleophilic than water, making it possible to carry out this reaction using aqueous ammonia.

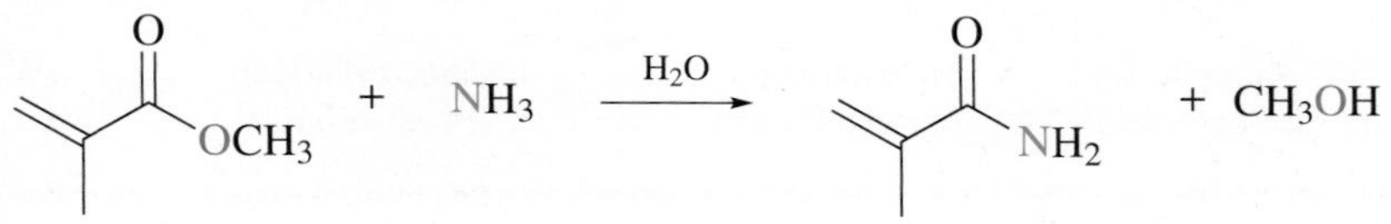

Methyl 2-methylpropenoate + NH_3 (Ammonia) $\xrightarrow{H_2O}$ 2-Methylpropenamide (75%) + CH_3OH (Methanol)

The reaction of amines is analogous to that of ammonia and gives *N*-substituted amides.

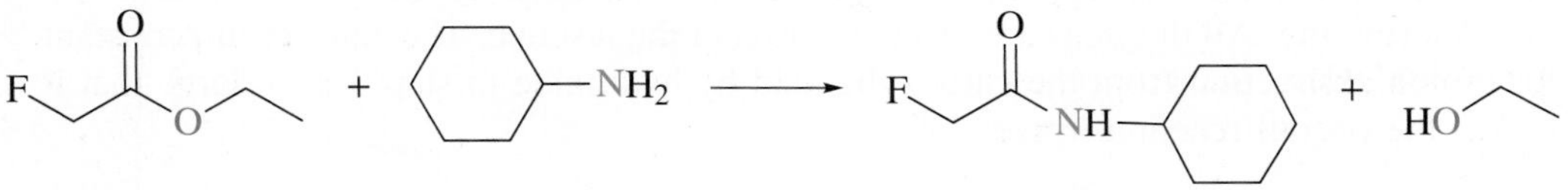

Ethyl fluoroacetate + Cyclohexylamine ($-NH_2$) $\longrightarrow$ *N*-Cyclohexylfluoroacetamide (61%) + Ethanol (HO)

The amine must be primary (RNH_2) or secondary (R_2NH). Tertiary amines (R_3N) cannot form amides because they have no proton on nitrogen that can be replaced by an acyl group.

Problem 20.13

Give the structure of the expected product of the following reaction:

$$\text{(4-methyl lactone)} + CH_3NH_2 \longrightarrow$$

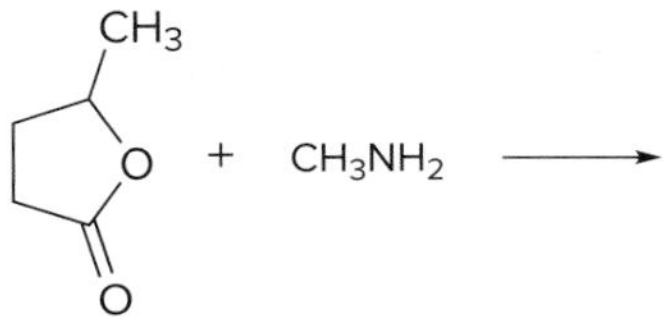

The reaction of ammonia and amines with esters follows the same general mechanistic course as other nucleophilic acyl substitutions. A tetrahedral intermediate is formed in the first stage of the process and dissociates in the second stage.

Problem 20.14

Give the structure of the intermediate formed in the rate-determining step of the following reaction:

$$C_6H_5COOCH_3 + (CH_3)_2CHNH_2 \longrightarrow C_6H_5CONHCH(CH_3)_2 + CH_3OH$$

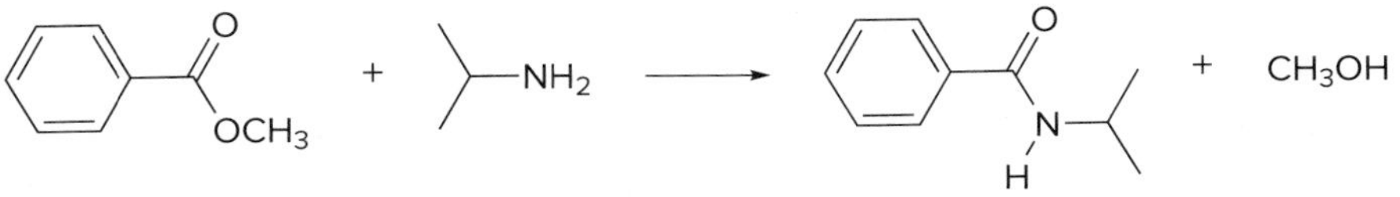

20.11 Reaction of Esters with Grignard and Organolithium Reagents and Lithium Aluminum Hydride

Esters react with two equivalents of a Grignard or organolithium reagent to give tertiary alcohols. Methyl and ethyl esters are normally used.

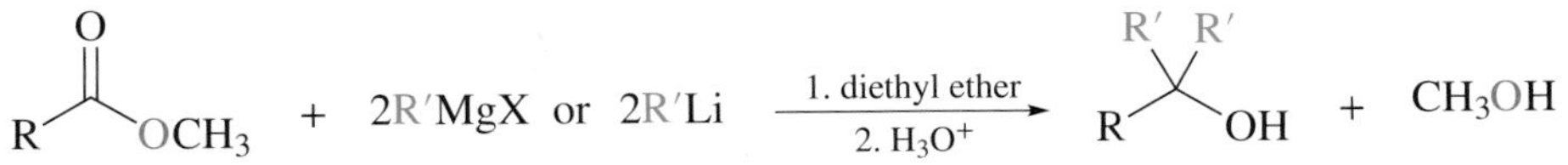

$$\underset{\text{Methyl ester}}{RCOOCH_3} + \underset{\text{Grignard or organolithium reagent}}{2R'MgX \text{ or } 2R'Li} \xrightarrow[\text{2. } H_3O^+]{\text{1. diethyl ether}} \underset{\text{Tertiary alcohol}}{RR'_2COH} + \underset{\text{Methanol}}{CH_3OH}$$

Two of the groups bonded to the hydroxyl-bearing carbon are the same because both are derived from the organometallic reagent.

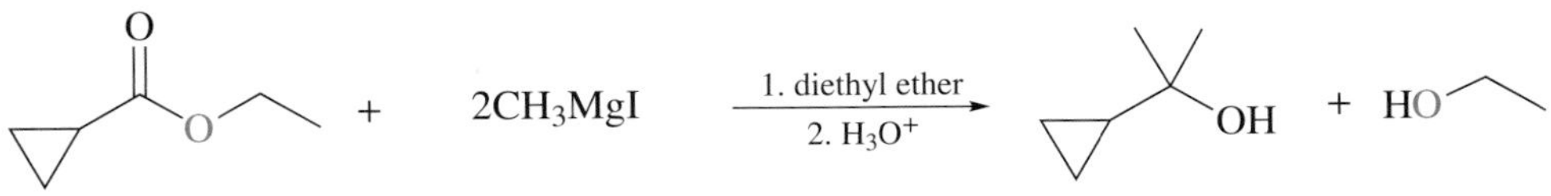

$$\underset{\text{Ethyl cyclopropanecarboxylate}}{C_3H_5COOCH_2CH_3} + \underset{\text{Methylmagnesium iodide}}{2CH_3MgI} \xrightarrow[\text{2. } H_3O^+]{\text{1. diethyl ether}} \underset{\text{2-Cyclopropyl-2-propanol (93\%)}}{C_3H_5C(CH_3)_2OH} + \underset{\text{Ethanol}}{HOCH_2CH_3}$$

The mechanism of the reaction begins with nucleophilic addition of the reagent to the carbonyl group to give a tetrahedral intermediate, which dissociates, giving a ketone.

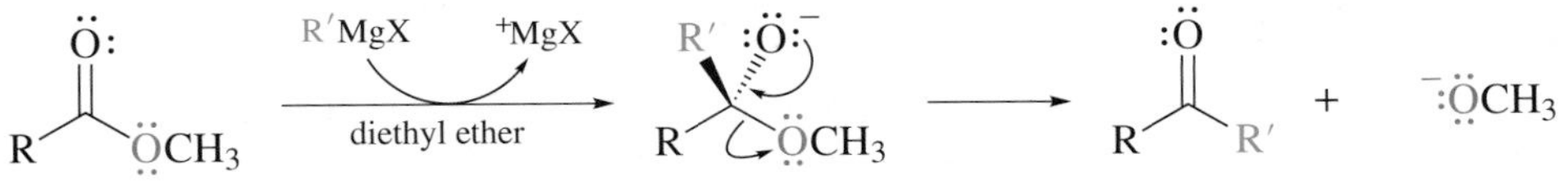

$$\underset{\text{Methyl ester}}{RCOOCH_3} \xrightarrow[\text{diethyl ether}]{R'MgX \quad {}^+MgX} \underset{TI^-}{RR'C(O^-)OCH_3} \longrightarrow \underset{\text{Ketone}}{RCOR'} + \underset{\text{Methoxide ion}}{{}^-OCH_3}$$

Once formed, the ketone intermediate reacts rapidly with the Grignard or organolithium reagent and gives the tertiary alcohol after the usual workup procedure. The ketone is more reactive toward nucleophilic addition than the starting ester; therefore, the reaction cannot be used as a synthesis of ketones by using equimolar amounts of ester and Grignard or organolithium reagent.

Problem 20.15

What combination of ester and Grignard reagent could you use to prepare each of the following tertiary alcohols?

(a) [structure: 3-phenyl-3-pentanol, OH]

(b) $(C_6H_5)_2C(OH)$–cyclopropyl

Sample Solution (a) Tertiary alcohols that have two equivalent groups attached to the C—OH unit are prepared using the Grignard reagent corresponding to those equivalent groups. Retrosynthetically:

[3-phenyl-3-pentanol] ⟹ $C_6H_5C(=O)OR$ + 2 (CH_3CH_2MgX)

An appropriate synthesis is:

$2CH_3CH_2MgBr$ + $C_6H_5COOCH_3$ $\xrightarrow[\text{2. } H_3O^+]{\text{1. diethyl ether}}$ $C_6H_5C(CH_2CH_3)_2OH$

Ethylmagnesium bromide — Methyl benzoate — 3-Phenyl-3-pentanol

Lithium aluminum hydride reduction of esters follows a similar pattern, giving first an aldehyde as an intermediate,

$RC(=O)OCH_3$ $\xrightarrow[\text{diethyl ether}]{H\bar{A}lH_3 \rightarrow AlH_3}$ $RCH(O^-)OCH_3$ (TI⁻) $\longrightarrow$ $RC(=O)H$ + $^-:OCH_3$

Methyl ester — TI⁻ — Aldehyde — Methoxide ion

This aldehyde, however, is itself rapidly reduced under the conditions of its formation. Thus, lithium aluminum hydride reduction of esters gives two alcohols as the isolated products. One is a primary alcohol derived from the aldehyde intermediate, the other corresponds to the alkoxy portion of the original ester.

$C_6H_5COOCH_2CH_3$ $\xrightarrow[\text{2. } H_3O^+]{\text{1. } LiAlH_4\text{, diethyl ether}}$ $C_6H_5CH_2OH$ + CH_3CH_2OH

Ethyl benzoate — Benzyl alcohol (90%) — Ethyl alcohol

Problem 20.16

Which aldehyde is an intermediate in the reduction of ethyl benzoate with lithium aluminum hydride?

Problem 20.17

Give the structure of an ester that will yield a mixture containing equimolar amounts of 1-propanol and 2-propanol on reduction with lithium aluminum hydride.

20.12 Amides

Physical Properties of Amides: Earlier in this chapter (Section 20.2) we noted several of the ways in which electron donation from nitrogen to the carbonyl group affects various structural features of amides. To review, the hybridization of nitrogen in amides is sp^2, and the bonds to nitrogen lie in the same plane. The CN bond is shorter in amides, and the activation energy for rotation about this bond is greater than in amines. According to the resonance picture of formamide, all these properties are consistent with significant CN double-bond character as expressed in contributor C.

A ⟷ B ⟷ C

Of the major classes of organic compounds, amides rank among the most polar. As shown in Table 20.5, acetamide, *N*-methylacetamide, and *N,N*-dimethylacetamide have dipole moments that range from 3.8 to 4.4 D compared with 1.9 for acetic acid. This increased polarity leads to stronger intermolecular attractive forces and causes the boiling points of the amides to be higher. They also contribute to higher melting points for the two amides in the table that contain N—H bonds. *N,N*-Dimethylacetamide does not participate in the hydrogen bonds that characterize the solid phase and has the lowest melting point of the compounds in the table. On the other hand, acetamide has two N—H protons capable of hydrogen bonding and is composed of flat sheets of hydrogen-bonded dimers stacked on top of each other in the crystal.

Problem 20.18

Compare *N*-methylacetamide with its amide isomers propanamide and *N,N*-dimethylformamide. Which do you predict has the highest boiling point? The lowest?

Intermolecular hydrogen bonding in amides, along with the planar geometry of the amide functional group, are the two most important factors governing the conformation of protein chains. We'll learn more about this in Chapter 25.

TABLE 20.5 Intermolecular Forces in Amides

	Acetic acid	*N,N*-Dimethylacetamide	*N*-Methylacetamide	Acetamide
Number of hydrogens available for hydrogen bonds	1	0	1	2
Dipole moment, D	1.9	3.8	4.4	3.9
Melting point, °C	17	−20	31	80
Boiling point, °C	118	165	206	221

Acidity of Amides: Because nitrogen is less electronegative than oxygen, the N—H group of an amide is a weaker acid than the O—H of a carboxylic acid. Typical primary amides have pK_a's near 16, which makes them about as acidic as water. The presence of the carbonyl group makes amides stronger acids and weaker bases than amines. Amides in which two carbonyl groups are bonded to the same nitrogen are called **imides** and have pK_a values near 10.

Increasing acidity →

	$CH_3CH_2NH_2$	$CH_3C(=O)NH_2$	$CH_3C(=O)N(H)C(=O)CH_3$	$CH_3C(=O)OH$
pK_a:	36	15	10	5

Problem 20.19

The pyrimidine thymine, present in DNA, was once thought to be A because A is analogous to benzene. In fact, thymine is B, which is also aromatic. Explain how B satisfies Hückel's rule, and write a contributing resonance structure for B that has a benzene-like ring.

A. (pyrimidine ring with H_3C, OH, N, N, OH)

B. (ring with H_3C, O, N–H, N–H, O)

Sample Solution

- B is planar and monocyclic.
- Completely conjugated. All ring atoms are sp^2-hybridized.

2 π electrons from double bond

2 π electrons from each N

- 6 π electrons satisfies Hückel's rule.
- Resonance

Synthesis of Amides: Tables 20.1, 20.2, and 20.4 included nucleophilic acyl substitutions that are useful for preparing amides by the reaction of amines with acyl chlorides, anhydrides, and esters, respectively. These are the most common methods for the laboratory synthesis of amides.

Because acylation of amines with acyl chlorides and anhydrides yields an acid as one of the products (HCl from acyl chlorides, a carboxylic acid from an anhydride), the efficient synthesis of amides requires some attention to stoichiometry.

Two molar equivalents of amine are frequently used in the reaction with acyl chlorides and acid anhydrides; one molecule of amine acts as a nucleophile, the second as a Brønsted base.

$$\underset{\text{Amine}}{2R_2NH} + \underset{\text{Acyl chloride}}{R'\overset{O}{\overset{\|}{C}}Cl} \longrightarrow \underset{\text{Amide}}{R'\overset{O}{\overset{\|}{C}}NR_2} + \underset{\text{Hydrochloride salt of amine}}{R_2\overset{+}{N}H_2\ Cl^-}$$

$$\underset{\text{Amine}}{2R_2NH} + \underset{\text{Acid anhydride}}{R'\overset{O}{\overset{\|}{C}}O\overset{O}{\overset{\|}{C}}R'} \longrightarrow \underset{\text{Amide}}{R'\overset{O}{\overset{\|}{C}}NR_2} + \underset{\text{Carboxylate salt of amine}}{R_2\overset{+}{N}H_2\ {}^-O\overset{O}{\overset{\|}{C}}R'}$$

It is possible to use only one molar equivalent of amine in these reactions if some other base, such as sodium hydroxide, is present in the reaction mixture to react with the hydrogen chloride or carboxylic acid that is formed. This is a useful procedure in those cases in which the amine is either valuable or available only in small quantities.

Esters and amines react in a 1:1 molar ratio to give amides. No acidic product is formed from the ester, and so no additional base is required.

$$\underset{\text{Amine}}{R_2NH} + \underset{\text{Methyl ester}}{R'\overset{O}{\overset{\|}{C}}OCH_3} \longrightarrow \underset{\text{Amide}}{R'\overset{O}{\overset{\|}{C}}NR_2} + \underset{\text{Methanol}}{CH_3OH}$$

Problem 20.20

Write an equation showing the preparation of the following amides from the indicated carboxylic acid derivative:

(a) $(CH_3)_2CHC(O)NH_2$ from an acyl chloride

(b) $CH_3C(O)NHCH_3$ from an acid anhydride

(c) $HC(O)N(CH_3)_2$ from a methyl ester

Sample Solution (a) Amides of the type $RC(O)NH_2$ are derived by acylation of ammonia.

$$\underset{\text{2-Methylpropanoyl chloride}}{(CH_3)_2CHC(O)Cl} + \underset{\text{Ammonia}}{2NH_3} \longrightarrow \underset{\text{2-Methylpropanamide}}{(CH_3)_2CHC(O)NH_2} + \underset{\text{Ammonium chloride}}{NH_4Cl}$$

Two molecules of ammonia are needed because its acylation produces, in addition to the desired amide, a molecule of hydrogen chloride. Hydrogen chloride (an acid) reacts with ammonia (a base) to give ammonium chloride.

All these reactions proceed by nucleophilic addition of the amine to the carbonyl group. Dissociation of the tetrahedral intermediate proceeds in the direction that leads to an amide.

$$RC(=O)X + R'_2NH \rightleftharpoons RC(OH)(X)NR'_2 \rightleftharpoons RC(=O)NR'_2 + HX$$

Acylating agent · Amine · Tetrahedral intermediate · Amide · Conjugate acid of leaving group

The carbonyl group of an amide is stabilized to a greater extent than that of an acyl chloride, acid anhydride, or ester; amides are formed rapidly and in high yield from each of these carboxylic acid derivatives.

Problem 20.21

Unlike esters, which can be prepared by acid-catalyzed condensation of an alcohol and a carboxylic acid, amides cannot be prepared by an acid-catalyzed condensation of an amine and a carboxylic acid. Why?

20.13 Hydrolysis of Amides

Amides are the least reactive carboxylic acid derivative, and the only nucleophilic acyl substitution reaction they undergo is hydrolysis. Amides are fairly stable in water, but the amide bond is cleaved on heating in the presence of strong acids or bases. Nominally, this cleavage produces an amine and a carboxylic acid. In acid, however, the amine is protonated, giving an ammonium ion:

$$\underset{\text{Amide}}{RCNR'_2} + \underset{\text{Hydronium ion}}{H_3O^+} \longrightarrow \underset{\text{Carboxylic acid}}{RCOH} + \underset{\text{Ammonium ion}}{R'_2\overset{+}{N}H_2}$$

In base the carboxylic acid is deprotonated, giving a carboxylate ion:

$$\underset{\text{Amide}}{RCNR'_2} + \underset{\text{Hydroxide ion}}{HO^-} \longrightarrow \underset{\text{Carboxylate ion}}{RCO^-} + \underset{\text{Amine}}{R'_2NH}$$

The acid–base reactions that occur after the amide bond is broken make the overall hydrolysis irreversible in both cases. The amine product is protonated in acid; the carboxylic acid is deprotonated in base.

$$\text{2-Phenylbutanamide} \xrightarrow[\text{heat}]{H_2O,\ H_2SO_4} \text{2-Phenylbutanoic acid (88–90\%)} + NH_4^+\ HSO_4^-$$

2-Phenylbutanamide · 2-Phenylbutanoic acid (88–90%) · Ammonium hydrogen sulfate

$$\text{N-(4-Bromophenyl)acetamide} \xrightarrow[\text{ethanol–water, heat}]{KOH} CH_3CO^-\ K^+ + H_2N\text{–}C_6H_4\text{–}Br$$

N-(4-Bromophenyl)acetamide · Potassium acetate · *p*-Bromoaniline (95%)

Mechanistically, amide hydrolysis is similar to the hydrolysis of other carboxylic acid derivatives. The mechanism of hydrolysis in acid is presented in Mechanism 20.4. It proceeds in two stages; a tetrahedral intermediate is formed in the first stage and dissociates in the second.

Mechanism 20.4

Amide Hydrolysis in Acid Solution

THE OVERALL REACTION:

$$\mathrm{RC(=O)NH_2} + \mathrm{H_3O^+} \longrightarrow \mathrm{RC(=O)OH} + \mathrm{NH_4^+}$$

An amide | Hydronium ion | A carboxylic acid | Ammonium ion

THE MECHANISM:

First Stage: Formation of the tetrahedral intermediate Steps 1–3 are analogous to the mechanism of acid-catalyzed hydration of aldehydes and ketones and acid-catalyzed hydrolysis of esters.

Step 1: Protonation of the carbonyl oxygen of the amide

$$\mathrm{RC(=\ddot{O}:)\ddot{N}H_2} + \mathrm{H{-}\overset{+}{O}H_2} \underset{}{\overset{\text{fast}}{\rightleftharpoons}} \mathrm{RC(=\overset{+}{\ddot{O}}H)\ddot{N}H_2} + \mathrm{H_2\ddot{O}:}$$

Amide | Hydronium ion | Protonated amide | Water

Step 2: Nucleophilic addition of water to the protonated amide

$$\mathrm{RC(=\overset{+}{\ddot{O}}H)\ddot{N}H_2} + \mathrm{H_2\ddot{O}:} \underset{}{\overset{\text{slow}}{\rightleftharpoons}} \mathrm{RC(H\ddot{O}:)(:\overset{+}{O}H_2)\ddot{N}H_2}$$

Protonated amide | Water | Conjugate acid of tetrahedral intermediate ($TI–H^+$)

Step 3: Deprotonation of $TI—H^+$ to give the neutral form of the tetrahedral intermediate (TI)

$$\mathrm{RC(H\ddot{O}:)(:\overset{+}{O}H_2)\ddot{N}H_2} + \mathrm{H_2\ddot{O}:} \underset{}{\overset{\text{fast}}{\rightleftharpoons}} \mathrm{RC(H\ddot{O}:)(:\ddot{O}H)\ddot{N}H_2} + \mathrm{H{-}\overset{+}{O}H_2}$$

Conjugate acid of tetrahedral intermediate ($TI–H^+$) | Water | Tetrahedral intermediate (TI) | Hydronium ion

Second Stage: Dissociation of the tetrahedral intermediate Just as steps 1–3 corresponded to addition of water to the carbonyl group, steps 4–6 correspond to elimination of ammonia or an amine from TI and restoration of the carbonyl group.

Step 4: Protonation of TI at its amino nitrogen

$$\mathrm{RC(H\ddot{O}:)(:\ddot{O}H)\ddot{N}H_2} + \mathrm{H{-}\overset{+}{O}H_2} \underset{}{\overset{\text{fast}}{\rightleftharpoons}} \mathrm{RC(H\ddot{O}:)(:\ddot{O}H)\overset{+}{N}H_3} + \mathrm{H_2\ddot{O}:}$$

Tetrahedral intermediate (TI) | Hydronium ion | Conjugate acid of tetrahedral intermediate ($TI–H^+$) | Water

continued

Step 5: Dissociation of the *N*-protonated form of the tetrahedral intermediate to give ammonia and the protonated form of the carboxylic acid

Step 6: Proton-transfer processes give the carboxylic acid and ammonium ion

The amide is activated toward nucleophilic acyl substitution by protonation of its carbonyl oxygen. The cation produced in this step is stabilized by resonance involving the nitrogen lone pair and is more stable than the intermediate in which the amide nitrogen is protonated.

Protonation of carbonyl oxygen ***Protonation of amide nitrogen***

Most stable resonance contributors of an *O*-protonated amide

An acylammonium ion; the positive charge is localized on nitrogen

Once formed, the *O*-protonated intermediate is attacked by a water molecule in step 2. The intermediate formed in this step loses a proton in step 3 to give the neutral form of the tetrahedral intermediate. The tetrahedral intermediate has its amino group ($—NH_2$) attached to an sp^3-hybridized carbon, and this amino group is the site at which protonation occurs in step 4. Cleavage of the carbon–nitrogen bond in step 5 yields the protonated form of the carboxylic acid, along with a molecule of ammonia. In acid solution ammonia is immediately protonated to give ammonium ion, as shown in step 6.

The protonation of ammonia in step 6 has such a large equilibrium constant that it makes the overall reaction irreversible.

Problem 20.22

On the basis of the general mechanism for amide hydrolysis in acidic solution shown in Mechanism 20.4, write an analogous sequence of steps for the hydrolysis of acetanilide,

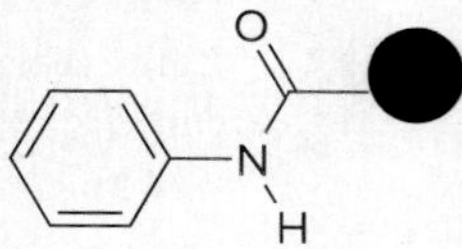

In base the tetrahedral intermediate is formed in a manner analogous to that proposed for ester saponification. Steps 1 and 2 in Mechanism 20.5 show the formation of the tetrahedral intermediate in the basic hydrolysis of amides. In step 3 the basic amino group of the tetrahedral intermediate abstracts a proton from water, and in step 4 the derived ammonium ion dissociates. Conversion of the carboxylic acid to its corresponding carboxylate anion in step 5 completes the process and renders the overall reaction irreversible.

Mechanism 20.5

Amide Hydrolysis in Basic Solution

THE OVERALL REACTION:

$RC(=O)NH_2 + HO^- \longrightarrow RC(=O)O^- + NH_3$

An amide — Hydroxide ion — A carboxylate ion — Ammonia

THE MECHANISM:

Step 1: Nucleophilic addition of hydroxide ion to the carbonyl group

$RC(=O)NH_2 + {}^-OH \underset{}{\overset{slow}{\rightleftharpoons}} RC(OH)(O^-)NH_2$

Amide — Hydroxide ion — Anionic form of tetrahedral intermediate

Step 2: Proton transfer to the anionic form of the tetrahedral intermediate

$RC(OH)(O^-)NH_2 + H{-}OH \longrightarrow RC(OH)(O{-}H)NH_2 + {}^-OH$

Anionic form of tetrahedral intermediate — Water — Tetrahedral intermediate — Hydroxide ion

Step 3: Protonation of the amino nitrogen

$RC(OH)(OH)NH_2 + H{-}OH \longrightarrow RC(OH)(OH)\overset{+}{N}H_3 + {}^-OH$

Tetrahedral intermediate — Water — Ammonium ion — Hydroxide ion

Step 4: Dissociation of the *N*-protonated form of the tetrahedral intermediate

$RC(OH)(O{-}H)\overset{+}{N}H_3 + {}^-O{-}H \longrightarrow RC(=O)OH + :NH_3 + H{-}O{-}H$

Conjugate acid of tetrahedral intermediate — Hydroxide ion — Carboxylic acid — Ammonia — Water

Step 5: Irreversible formation of the carboxylate anion

$RC(=O)O{-}H + {}^-O{-}H \longrightarrow RC(=O)O^- + H{-}O{-}H$

Carboxylic acid — Hydroxide ion — Carboxylate ion — Water

Problem 20.23

On the basis of the general mechanism for basic hydrolysis shown in Mechanism 20.5, write an analogous sequence for the hydrolysis of *N,N*-dimethylformamide,

$HC(=O)N(CH_3)_2$

20.14 Lactams

Lactams are cyclic amides and are analogous to lactones, which are cyclic esters. Most lactams are known by their common names, as the examples shown illustrate.

N-Methylpyrrolidone (a polar aprotic solvent)

ϵ-Caprolactam (industrial chemical used to prepare a type of nylon)

β-Lactam Antibiotics

It may never be known just how spores of *Penicillium notatum* found their way to a Petri dish containing *Staphylococcus* in Alexander Fleming's laboratory at St. Mary's Hospital in London during the summer of 1928. But they did, and the mold they produced made a substance that stopped the *Staphylococcus* colony from growing. His curiosity aroused, Fleming systematically challenged the substance he called "penicillin" with other bacteria and, in addition to *Staphylococcus,* found impressive activity against *Streptococcus* as well as the bacteria that cause diphtheria, meningitis, and pneumonia. Fleming published his findings in 1929, but his efforts to isolate the active substance responsible for penicillin's antibacterial properties were unsuccessful.

By 1938, Fleming had moved on to other research, and Howard Florey and Ernst Chain of the School of Pathology at Oxford were just beginning a program aimed at developing antibacterial agents from natural sources. A candidate that especially appealed to them was Fleming's penicillin.

Their most daunting initial problem was making enough penicillin. Enter Norman Heatley, of whom it has been said: ". . . without Heatley, no penicillin."* Heatley (Figure 20.3), an inventive and careful experimentalist, devised procedures to make and isolate penicillin on a scale sufficient to begin testing. By 1941, Florey, Chain, and Heatley had a drug that was both effective and safe.

England was at war, and the United States soon would be; the need for large amounts of penicillin was obvious. Working with the U.S. Department of Agriculture, Heatley and Andrew J. Moyer of the USDA laboratories in Peoria, Illinois, found better *Penicillium*

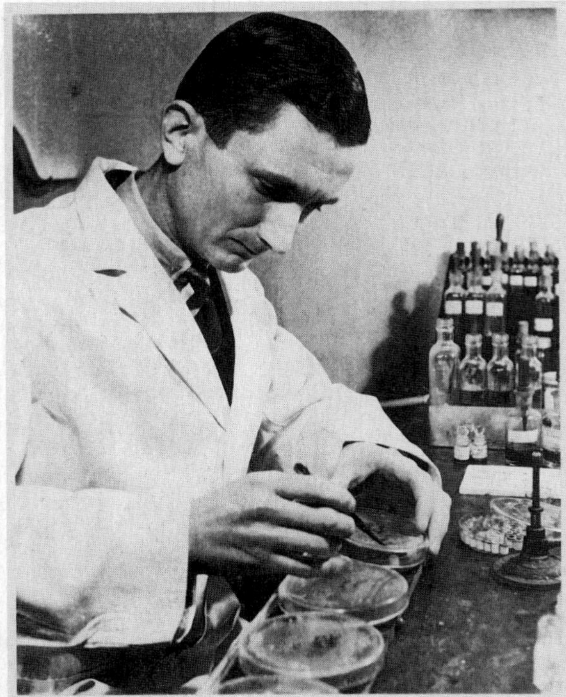

Figure 20.3

Norman Heatley was instrumental in devising methods for obtaining penicillins on a practical scale.

sources and developed novel fermentation methods to produce ever-increasing amounts of penicillin. Treatment of wounded soldiers with penicillin became possible early in 1943 and was widely practiced before the war ended in August 1945. Four months later, Fleming, Florey, and Chain traveled to Stockholm to accept that year's Nobel Prize in Physiology or Medicine. Heatley was not included because custom dictates that a Nobel Prize can be divided among no more than three persons.

The structure of penicillin is unusual because it contains an amide function as part of a four-membered ring (a **β-lactam**). Various penicillins differ in respect to substituent groups and their effectiveness against different strains of bacteria. Penicillin G originated in a strain obtained from a rotting cantaloupe in Peoria and was the first penicillin made on a large scale. Fleming's original penicillin (now called penicillin F) bears a $CH_3CH_2CH{=}CHCH_2{—}$ group in place of the $C_6H_5CH_2{—}$ of penicillin G. A different class of β-lactam antibiotics, the cephalosporins, are similar in structure to the penicillins but have a six-membered instead of a five-membered sulfur-containing ring.

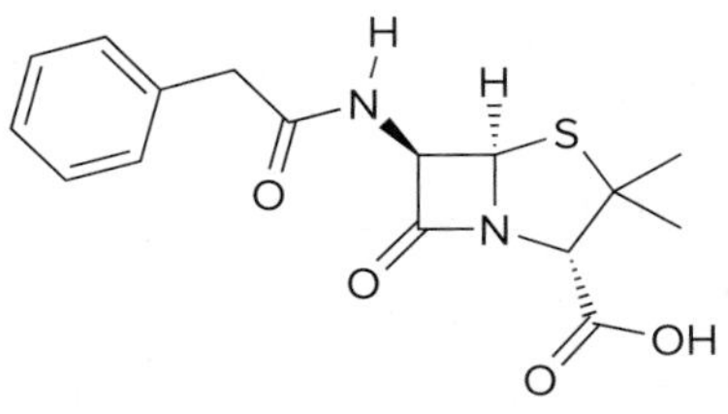

Penicillin G

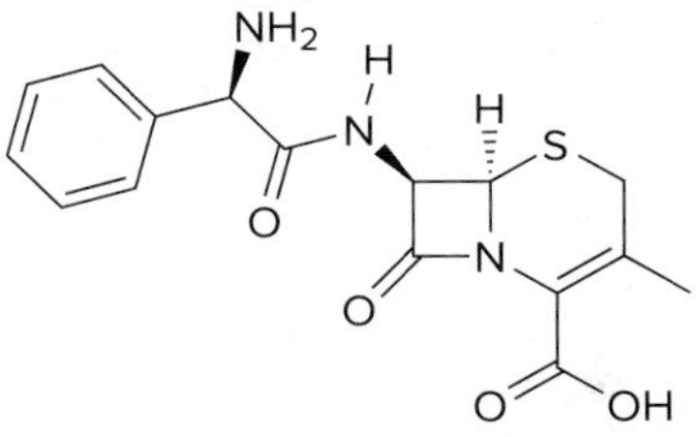

Cephalexin

Although their strained four-membered ring makes β-lactam antibiotics susceptible to hydrolysis, this same elevated reactivity toward nucleophilic acyl substitution is responsible for their antibacterial properties. β-Lactams act by deactivating an enzyme, *transpeptidase,* required for the biosynthesis of bacterial cell walls. The active site of transpeptidase contains a key hydroxyl group, which is converted to an ester by a nucleophilic acyl substitution that cleaves the β-lactam ring. With the acylated form of the enzyme unable to catalyze cell-wall biosynthesis, further bacterial growth is brought under control, and the body's immune system does the rest.

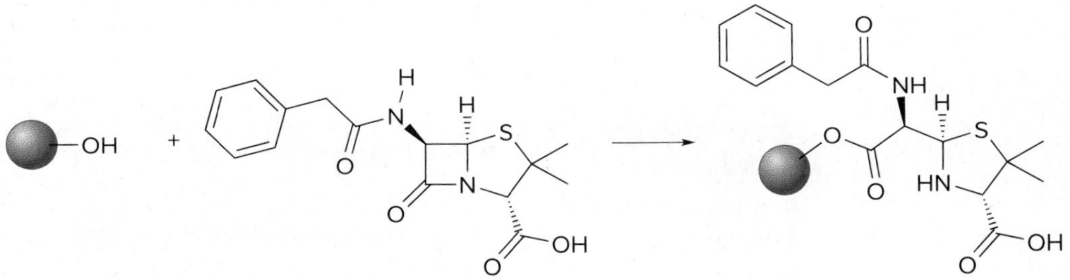

Active form of transpeptidase + Penicillin G → Inactive ester of transpeptidase

Problem 20.24

a. Penicillin-resistant strains of bacteria contain β-*lactamases,* enzymes that catalyze the hydrolysis of a penicillin before the penicillin can acylate *transpeptidase*. Suggest a reasonable structure for the product $C_{16}H_{20}N_2O_5S$ formed by β-lactamase-catalyzed hydrolysis of penicillin G.

b. 6-Aminopenicillanic acid ($C_8H_{12}N_2O_3S$), a key compound in the preparation of "semisynthetic" penicillins, is prepared from penicillin G by *penicillin acyl transferase*-catalyzed hydrolysis. Suggest a reasonable structure for 6-aminopenicillanic acid.

*H. Harris, "The Florey Centenary Lecture and the Development of Penicillin," as quoted in E. Lax, *The Mold in Dr. Florey's Coat,* Henry Holt and Company, New York, 2004, page 89.

Just as amides are more stable than esters, lactams are more stable than lactones. Thus, although β-lactones are rare (Section 19.15), β-lactams are among the best known products of the pharmaceutical industry. The penicillin and cephalosporin antibiotics, which are so useful in treating bacterial infections, are β-lactams and are discussed in the boxed essay accompanying this section.

20.15 Preparation of Nitriles

We have already discussed two procedures by which nitriles are prepared, namely, nucleophilic substitution of alkyl halides by cyanide and conversion of aldehydes and ketones to cyanohydrins. Table 20.6 reviews aspects of these reactions. Neither of the reactions in the table is suitable for aryl nitriles (ArC≡N); these compounds are normally prepared by a reaction to be discussed in Section 22.17.

Both alkyl and aryl nitriles are accessible by dehydration of amides.

$$\underset{\substack{\text{Amide}\\\text{(R may be alkyl}\\\text{or aryl)}}}{\overset{\overset{\displaystyle O}{\|}}{RCNH_2}} \longrightarrow \underset{\substack{\text{Nitrile}\\\text{(R may be alkyl}\\\text{or aryl)}}}{RC\equiv N} + \underset{\text{Water}}{H_2O}$$

P_4O_{10} is comprised of a P_4O_6 core to which four oxygens are attached. The structure of the core is analogous to that of adamantane (Section 3.14). Can you draw a structural formula for P_4O_{10}?

Among the reagents used for this dehydration is P_4O_{10}, known by the common name *phosphorus pentoxide* because it was once thought to have the molecular formula P_2O_5. Phosphorus pentoxide is the anhydride of phosphoric acid and is used in a number of reactions requiring dehydrating agents.

$$\underset{\text{2-Methylpropanamide}}{(CH_3)_2CHCNH_2 \text{ (C=O)}} \xrightarrow[200°C]{P_4O_{10}} \underset{\substack{\text{2-Methylpropanenitrile}\\(69–86\%)}}{(CH_3)_2CHC\equiv N}$$

TABLE 20.6 Preparation of Nitriles

Reaction (section) and comments	General equation and specific example
Nucleophilic substitution by cyanide ion (see Sections 6.1, 6.10) Cyanide ion is a good nucleophile and reacts with alkyl halides to give nitriles. The reaction is of the S_N2 type and is limited to primary and secondary alkyl halides. Tertiary alkyl halides undergo elimination; aryl and vinyl halides do not react.	$\underset{\text{Cyanide ion}}{:N\equiv C:^-} + \underset{\text{Alkyl halide}}{R-X} \longrightarrow \underset{\text{Nitrile}}{R-C\equiv N:} + \underset{\text{Halide ion}}{:X^-}$ $\underset{\text{1-Chlorodecane}}{CH_3(CH_2)_8CH_2Cl} \xrightarrow[\text{ethanol–water}]{KCN} \underset{\text{Undecanenitrile (95\%)}}{CH_3(CH_2)_8CH_2CN}$
Cyanohydrin formation (see Section 18.7) Hydrogen cyanide adds to the carbonyl group of aldehydes and ketones.	$\underset{\text{Aldehyde or ketone}}{RCR'\text{ (C=O)}} + \underset{\text{Hydrogen cyanide}}{HCN} \longrightarrow \underset{\text{Cyanohydrin}}{RC(OH)(CN)R'}$ $\underset{\text{3-Pentanone}}{CH_3CH_2COCH_2CH_3} \xrightarrow[H_3O^+]{KCN} \underset{\text{3-Pentanone cyanohydrin (75\%)}}{(CH_3CH_2)_2C(OH)CN}$

Problem 20.25

Show how ethyl alcohol could be used to prepare (a) CH_3CN and (b) CH_3CH_2CN. Along with ethyl alcohol you may use any necessary inorganic reagents.

20.16 Hydrolysis of Nitriles

Nitriles are classified as carboxylic acid derivatives because they are converted to carboxylic acids on hydrolysis. The conditions required are similar to those for the hydrolysis of amides, namely, heating in aqueous acid or base for several hours. Like the hydrolysis of amides, nitrile hydrolysis is irreversible in the presence of acids or bases. Acid hydrolysis yields ammonium ion and a carboxylic acid.

$$\underset{\text{Nitrile}}{RC{\equiv}N} + \underset{\text{Water}}{H_2O} + \underset{\text{Hydronium ion}}{H_3O^+} \longrightarrow \underset{\text{Carboxylic acid}}{RC(=O)OH} + \underset{\text{Ammonium ion}}{\overset{+}{N}H_4}$$

$$\underset{p\text{-Nitrobenzyl cyanide}}{O_2N{-}C_6H_4{-}CH_2C{\equiv}N} \xrightarrow[\text{heat}]{H_2O,\ H_2SO_4} \underset{p\text{-Nitrophenylacetic acid (92–95\%)}}{O_2N{-}C_6H_4{-}CH_2C(=O)OH}$$

In aqueous base, hydroxide ion abstracts a proton from the carboxylic acid. Isolating the acid requires a subsequent acidification step.

$$\underset{\text{Nitrile}}{RC{\equiv}N} + \underset{\text{Water}}{H_2O} + \underset{\text{Hydroxide ion}}{HO^-} \longrightarrow \underset{\text{Carboxylate ion}}{RC(=O)O^-} + \underset{\text{Ammonia}}{NH_3}$$

$$\underset{\text{Undecanenitrile}}{CH_3(CH_2)_9CN} \xrightarrow[2.\ H_3O^+]{1.\ KOH,\ H_2O,\ heat} \underset{\text{Undecanoic acid (80\%)}}{CH_3(CH_2)_9C(=O)OH}$$

The first four steps of the mechanism for hydrolysis of nitriles in basic solution are given in Mechanism 20.6. These steps convert the nitrile to an amide, which then proceeds to the hydrolysis products according to the mechanism of amide hydrolysis in Mechanism 20.5.

The acid-catalyzed mechanism for nitrile hydrolysis also goes through the amide as an intermediate. Problem 20.26 encourages you to propose a mechanism for that process.

Problem 20.26

Suggest a reasonable mechanism for the conversion of a nitrile (RCN) to the corresponding amide in aqueous acid.

Nucleophiles other than water can also add to the carbon–nitrogen triple bond of nitriles. In the following section we will see a synthetic application of such a nucleophilic addition.

Mechanism 20.6

Nitrile Hydrolysis in Basic Solution

THE OVERALL REACTION: Nitriles are hydrolyzed in base to give ammonia and a carboxylate ion. An amide is an intermediate.

$$RC{\equiv}N \xrightarrow{H_2O} RC(=O)NH_2 \xrightarrow{HO^-} RC(=O)O^- + NH_3$$

Nitrile — Amide — Carboxylate ion — Ammonia

THE MECHANISM:

Step 1: Hydroxide adds to the carbon–nitrogen triple bond. This step is analogous to nucleophilic addition to a carbonyl group.

$$HO^- + RC{\equiv}N: \rightleftharpoons RC(OH){=}N^-$$

Hydroxide ion — Nitrile — An imidate ion

Step 2: The product of step 1 is the conjugate base of an imidic acid to which it is converted by proton abstraction from water.

$$RC(OH){=}N^- + H{-}OH \rightleftharpoons RC(OH){=}NH + {}^-OH$$

An imidate ion — Water — An imidic acid — Hydroxide ion

Step 3: Proton abstraction from oxygen of the imino acid gives the conjugate base of an amide.

$$RC(OH){=}NH + {}^-OH \rightleftharpoons RC(=O)NH^- + H{-}OH$$

An imidic acid — Hydroxide ion — Conjugate base of amide — Water

Step 4: The conjugate base of the amide abstracts a proton from water.

$$RC(=O)NH^- + H{-}OH \rightleftharpoons RC(=O)NH_2 + {}^-OH$$

Conjugate base of amide — Water — Amide — Hydroxide ion

The amide formed in this step then undergoes basic hydrolysis according to the process shown in Mechanism 20.5.

20.17 Addition of Grignard Reagents to Nitriles

The carbon–nitrogen triple bond of nitriles is much less reactive toward nucleophilic addition than the carbon–oxygen double bond of aldehydes and ketones. Strongly basic nucleophiles such as Grignard reagents, however, do react with nitriles in a reaction that is of synthetic value:

$$\underset{\text{Nitrile}}{RC{\equiv}N} + \underset{\text{Grignard reagent}}{R'MgX} \xrightarrow[\text{2. } H_3O^+]{\text{1. diethyl ether}} \underset{\text{Imine}}{RC(=NH)R'} \xrightarrow[\text{heat}]{H_3O^+} \underset{\text{Ketone}}{RC(=O)R'}$$

The imine formed by nucleophilic addition of the Grignard reagent to the nitrile is normally not isolated but is hydrolyzed directly to a ketone.

$$\underset{m\text{-(Trifluoromethyl)benzonitrile}}{m\text{-}F_3CC_6H_4C{\equiv}N} + \underset{\text{Methylmagnesium iodide}}{CH_3MgI} \xrightarrow[\text{2. } H_3O^+\text{, heat}]{\text{1. diethyl ether}} \underset{m\text{-(Trifluoromethyl)acetophenone (79\%)}}{m\text{-}F_3CC_6H_4COCH_3}$$

Problem 20.27

Write an equation showing how you could prepare ethyl phenyl ketone from propanenitrile and a Grignard reagent. What is the structure of the imine intermediate?

Organolithium reagents react in the same way and are often used instead of Grignard reagents.

20.18 Spectroscopic Analysis of Carboxylic Acid Derivatives

Infrared: IR spectroscopy is quite useful in identifying carboxylic acid derivatives. The carbonyl stretching vibration is very strong, and its position is sensitive to the nature of the carbonyl group. In general, electron donation from the substituent decreases the double-bond character of the bond between carbon and oxygen and decreases the stretching frequency. Two distinct absorptions are observed for the symmetric and antisymmetric stretching vibrations of the anhydride function.

$CH_3C(=O)Cl$	$CH_3C(=O)OC(=O)CH_3$	$CH_3C(=O)OCH_3$	$CH_3C(=O)NH_2$
Acetyl chloride	Acetic anhydride	Methyl acetate	Acetamide
$\nu_{C=O} = 1822\ cm^{-1}$	$\nu_{C=O} = 1748\ cm^{-1}$ and $1815\ cm^{-1}$	$\nu_{C=O} = 1736\ cm^{-1}$	$\nu_{C=O} = 1694\ cm^{-1}$

Nitriles are readily identified by absorption due to —C≡N stretching in the 2210–2260 cm^{-1} region.

¹H NMR: Chemical-shift differences in their ¹H NMR spectra aid the structure determination of esters. Consider the two isomeric esters: ethyl acetate and methyl propanoate.

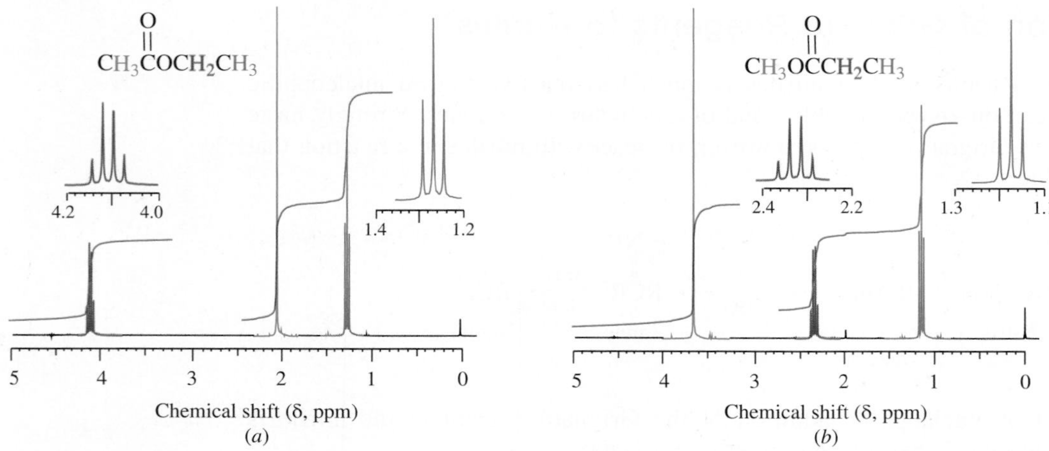

Figure 20.4

The 300-MHz ^{1}H NMR spectra of (*a*) ethyl acetate and (*b*) methyl propanoate.

As Figure 20.4 shows, the number of signals and their multiplicities are the same for both esters. Both have a methyl singlet and a triplet–quartet pattern for the ethyl group.

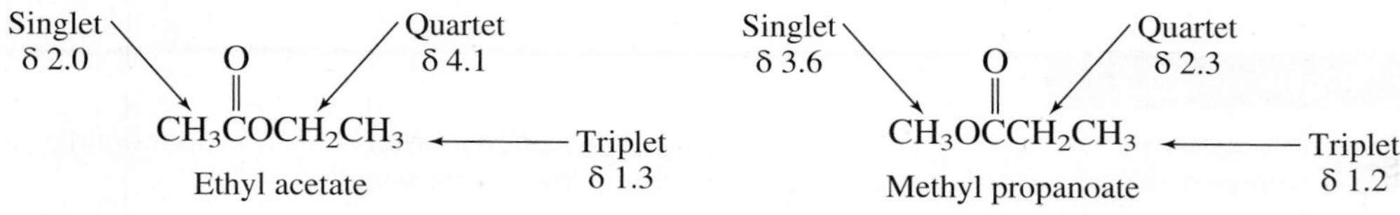

Notice, however, that there is a significant difference in the chemical shifts of the corresponding signals in the two spectra. The methyl singlet is more shielded (δ 2.0) when it is bonded to the carbonyl group of ethyl acetate than when it is bonded to the oxygen of methyl propanoate (δ 3.6). The methylene quartet is more shielded (δ 2.3) when it is bonded to the carbonyl group of methyl propanoate than when it is bonded to the oxygen of ethyl acetate (δ 4.1). Analysis of only the number of peaks and their splitting patterns does not provide an unambiguous answer to structure assignment in esters; chemical-shift data such as that just described must also be considered.

The chemical shift of the N—H proton of amides appears in the range δ 5–8 and is often very broad.

^{13}C NMR: The ^{13}C NMR spectra of carboxylic acid derivatives, like the spectra of carboxylic acids themselves, are characterized by a low-field resonance for the carbonyl carbon in the range δ 160–180. The carbonyl carbons of carboxylic acid derivatives are more shielded than those of aldehydes and ketones, but less shielded than the sp^2-hybridized carbons of alkenes and arenes.

The carbon of a C≡N group appears near δ 120.

UV-VIS: The following values are typical for the $n \rightarrow \pi^*$ absorption associated with the C═O group of carboxylic acid derivatives.

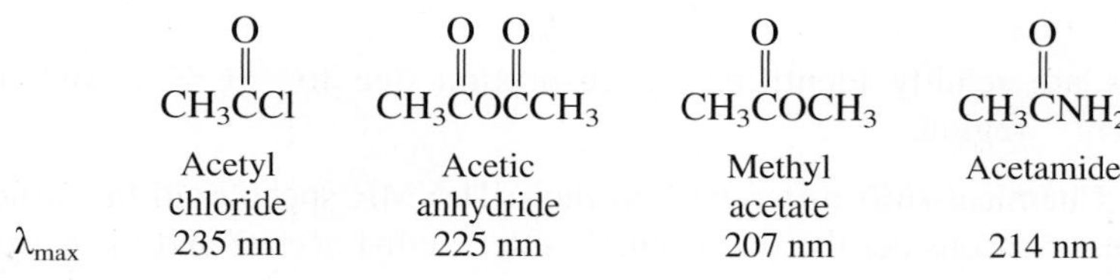

Mass Spectrometry: A prominent peak in the mass spectra of most carboxylic acid derivatives corresponds to an acylium ion derived by cleavage of the bond to the carbonyl group:

$$\mathrm{R{-}C(=\ddot{O}^{+}\cdot)X{:}} \longrightarrow \mathrm{R{-}C{\equiv}\overset{+}{O}{:}} + \cdot\mathrm{X}{:}$$

Amides, however, tend to cleave in the opposite direction to produce a nitrogen-stabilized acylium ion:

$$\mathrm{R{-}C(=\overset{+}{O}{:})\ddot{N}R'_2} \longrightarrow \mathrm{R}\cdot + [{:}\overset{+}{O}{\equiv}C{-}\ddot{N}R'_2 \longleftrightarrow {:}\ddot{O}{=}C{=}\overset{+}{N}R'_2]$$

20.19 SUMMARY

Section 20.1 This chapter concerns the preparation and reactions of *acyl chlorides, acid anhydrides, esters, amides,* and *nitriles.* These compounds are generally classified as carboxylic acid derivatives, and their nomenclature is based on that of carboxylic acids.

Acyl chloride	Acid anhydride	Ester	Amide	Nitrile
RCCl (C=O)	RCOCR (both C=O)	RCOR′ (C=O)	$RCNR'_2$ (C=O)	RC≡N

Section 20.2 The structure and reactivity of carboxylic acid derivatives depend on how well the atom bonded to the carbonyl group donates electrons to it.

$$\mathrm{R{-}C(=\ddot{O}{:}){-}\ddot{X}} \longleftrightarrow \mathrm{R{-}C({-}\ddot{O}{:}^{-}){=}X^{+}}$$

Electron-pair donation stabilizes the carbonyl group and makes it less reactive toward nucleophilic acyl substitution.

Most reactive / Least stabilized carbonyl group: RCCl > RCOCR > RCOR′ > $RCNR'_2$:Least reactive / Most stabilized carbonyl group

Nitrogen is a better electron-pair donor than oxygen, and amides have a more stabilized carbonyl group than esters and anhydrides. Chlorine is the poorest electron-pair donor, and acyl chlorides have the least stabilized carbonyl group and are the most reactive.

Section 20.3 The characteristic reaction of acyl chlorides, acid anhydrides, esters, and amides is **nucleophilic acyl substitution.** In the most common mechanism, addition of a nucleophilic reagent :Nu—H to the carbonyl group leads to a tetrahedral intermediate that dissociates to give the product of substitution:

$$\mathrm{RC(=O)X} + {:}\mathrm{Nu{-}H} \rightleftharpoons \mathrm{RC(OH)(X)\ddot{N}u} \rightleftharpoons \mathrm{RC(=O)Nu{:}} + \mathrm{HX}$$

Carboxylic acid derivative; Nucleophile; Tetrahedral intermediate; Product of nucleophilic acyl substitution; Conjugate acid of leaving group

Section 20.4 Acyl chlorides are converted to acid anhydrides, esters, and amides by nucleophilic acyl substitution.

$$\underset{\text{Acyl chloride}}{\overset{\text{O}}{\overset{\|}{\text{RCCl}}}} + \underset{\text{Carboxylic acid}}{\overset{\text{O}}{\overset{\|}{\text{R}'\text{COH}}}} \longrightarrow \underset{\text{Acid anhydride}}{\overset{\text{O O}}{\overset{\|\ \ \|}{\text{RCOCR}'}}} + \underset{\text{Hydrogen chloride}}{\text{HCl}}$$

$$\underset{\text{Acyl chloride}}{\overset{\text{O}}{\overset{\|}{\text{RCCl}}}} + \underset{\text{Alcohol}}{\text{R}'\text{OH}} \longrightarrow \underset{\text{Ester}}{\overset{\text{O}}{\overset{\|}{\text{RCOR}'}}} + \underset{\text{Hydrogen chloride}}{\text{HCl}}$$

$$\underset{\text{Acyl chloride}}{\overset{\text{O}}{\overset{\|}{\text{RCCl}}}} + \underset{\text{Amine}}{2\text{R}'_2\text{NH}} \longrightarrow \underset{\text{Amide}}{\overset{\text{O}}{\overset{\|}{\text{RCNR}'_2}}} + \underset{\text{Ammonium chloride salt}}{\text{R}'_2\overset{+}{\text{N}}\text{H}_2\ \text{Cl}^-}$$

Examples of each of these reactions may be found in Table 20.1.

Section 20.5 Acid anhydrides are less reactive toward nucleophilic acyl substitution than acyl chlorides, but are useful reagents for preparing esters and amides.

$$\underset{\text{Acid anhydride}}{\overset{\text{O O}}{\overset{\|\ \ \|}{\text{RCOCR}}}} + \underset{\text{Alcohol}}{\text{R}'\text{OH}} \longrightarrow \underset{\text{Ester}}{\overset{\text{O}}{\overset{\|}{\text{RCOR}'}}} + \underset{\text{Carboxylic acid}}{\overset{\text{O}}{\overset{\|}{\text{RCOH}}}}$$

$$\underset{\text{Acid anhydride}}{\overset{\text{O O}}{\overset{\|\ \ \|}{\text{RCOCR}}}} + \underset{\text{Amine}}{2\text{R}'_2\text{NH}} \longrightarrow \underset{\text{Amide}}{\overset{\text{O}}{\overset{\|}{\text{RCNR}'_2}}} + \underset{\text{Ammonium carboxylate salt}}{\text{R}'_2\overset{+}{\text{N}}\text{H}_2\ \ {}^-\overset{\text{O}}{\overset{\|}{\text{OCR}}}}$$

Table 20.2 presents examples of these reactions.

Section 20.6 Esters occur naturally or are prepared from alcohols by Fischer esterification or by acylation with acyl chlorides or acid anhydrides (see Table 20.3). Esters are polar and have higher boiling points than alkanes of comparable size and shape. Esters don't form hydrogen bonds to other ester molecules so have lower boiling points than analogous alcohols. They can form hydrogen bonds to water and so are comparable to alcohols in their solubility in water.

Section 20.7 Esters give amides on reaction with ammonia and amines and are cleaved to a carboxylic acid and an alcohol on hydrolysis (see Table 20.4).

Section 20.8 Ester hydrolysis can be catalyzed by acids and its mechanism (see Mechanism 20.2) is the reverse of the mechanism for Fischer esterification. The reaction proceeds via a tetrahedral intermediate.

$$\text{R}-\text{C}(\text{OH})(\text{OH})(\text{OR}')$$

Tetrahedral intermediate in ester hydrolysis

Section 20.9 Ester hydrolysis in basic solution is called *saponification* and proceeds through the same tetrahedral intermediate (see Mechanism 20.3) as in acid-catalyzed hydrolysis. Unlike acid-catalyzed hydrolysis, saponification is irreversible because the carboxylic acid is deprotonated under the reaction conditions.

$$\underset{\text{Ester}}{\mathrm{RC(=O)OR'}} + \underset{\text{Hydroxide ion}}{\mathrm{HO^-}} \longrightarrow \underset{\text{Carboxylate ion}}{\mathrm{RC(=O)O^-}} + \underset{\text{Alcohol}}{\mathrm{R'OH}}$$

Section 20.10 Esters react with amines to give amides.

$$\underset{\text{Ester}}{\mathrm{RC(=O)OR'}} + \underset{\text{Amine}}{\mathrm{R''_2NH}} \longrightarrow \underset{\text{Amide}}{\mathrm{RC(=O)NR''_2}} + \underset{\text{Alcohol}}{\mathrm{R'OH}}$$

Section 20.11 Esters react with two equivalents of a Grignard or organolithium reagent to form tertiary alcohols.

$$\underset{\text{Methyl ester}}{\mathrm{RC(=O)OCH_3}} + \underset{\text{Grignard reagent}}{\mathrm{2R'MgX}} \xrightarrow[\text{2. } \mathrm{H_3O^+}]{\text{1. diethyl ether}} \underset{\text{Tertiary alcohol}}{\mathrm{RC(OH)(R')-R'}}$$

Lithium aluminum hydride reduces esters to alcohols. Two alcohols are formed; the acyl group is reduced to the primary alcohol.

$$\underset{\text{Ester}}{\mathrm{RC(=O)OR'}} \longrightarrow \underset{\text{Primary alcohol}}{\mathrm{RCH_2OH}} + \underset{\text{Alcohol}}{\mathrm{R'OH}}$$

Section 20.12 Amides having at least one N—H unit can form intermolecular hydrogen bonds with other amide molecules. Compounds of this type have higher melting and boiling points than comparable compounds in which N—H bonds are absent.

Amides are normally prepared by the reaction of amines with acyl chlorides, anhydrides, or esters.

Section 20.13 Like ester hydrolysis, amide hydrolysis can be achieved in either aqueous acid or aqueous base. The process is irreversible in both media. In base, the carboxylic acid is converted to the carboxylate anion; in acid, the amine is protonated to an ammonium ion:

$$\underset{\text{Amide}}{\mathrm{RC(=O)NR'_2}} + \underset{\text{Water}}{\mathrm{H_2O}} \xrightarrow{\mathrm{H_3O^-}} \underset{\text{Carboxylic acid}}{\mathrm{RC(=O)OH}} + \underset{\text{Ammonium ion}}{\mathrm{R'_2\overset{+}{N}H_2}}$$

$$\underset{\text{Amide}}{\mathrm{RC(=O)NR'_2}} + \underset{\text{Water}}{\mathrm{H_2O}} \xrightarrow{\mathrm{HO^-}} \underset{\text{Carboxylate ion}}{\mathrm{RC(=O)O^-}} + \underset{\text{Amine}}{\mathrm{R'_2NH}}$$

Section 20.14 Lactams are cyclic amides.

Section 20.15 Nitriles are prepared by nucleophilic substitution (S_N2) of alkyl halides with cyanide ion, by converting aldehydes or ketones to cyanohydrins (see Table 20.6), or by dehydration of amides.

Section 20.16 The hydrolysis of nitriles to carboxylic acids is irreversible in both acidic and basic solution.

$$\underset{\text{Nitrile}}{RC{\equiv}N} \xrightarrow[\text{or}\;\; 1.\ H_2O,\ HO^-,\ \text{heat};\ 2.\ H_3O^+]{H_3O^+,\ \text{heat}} \underset{\text{Carboxylic acid}}{RCOH\ (\text{C=O})}$$

Section 20.17 Nitriles are useful starting materials for the preparation of ketones by reaction with Grignard reagents.

$$\underset{\text{Nitrile}}{RC{\equiv}N} + \underset{\text{Grignard reagent}}{R'MgX} \xrightarrow[2.\ H_3O^+,\ \text{heat}]{1.\ \text{diethyl ether}} \underset{\text{Ketone}}{RCR'\ (\text{C=O})}$$

Section 20.18 Acyl chlorides, anhydrides, esters, and amides all show a strong band for C=O stretching in the infrared. The range extends from about 1820 cm^{-1} (acyl chlorides) to 1690 cm^{-1} (amides). Their ^{13}C NMR spectra are characterized by a peak near δ 180 for the carbonyl carbon. ^1H NMR spectroscopy is useful for distinguishing between the groups R and R′ in esters (RCO_2R'). The protons on the carbon bonded to O in R′ appear at lower field (less shielded) than those on the carbon bonded to C=O.

PROBLEMS

Structure and Nomenclature

20.28 Write a structural formula for each of the following compounds:

(a) *m*-Chlorobenzoyl chloride
(b) Trifluoroacetic anhydride
(c) *cis*-1,2-Cyclopropanedicarboxylic anhydride
(d) Ethyl cycloheptanecarboxylate
(e) 1-Phenylethyl acetate
(f) 2-Phenylethyl acetate
(g) *p*-Ethylbenzamide
(h) *N*-Ethylbenzamide
(i) 2-Methylhexanenitrile

20.29 Give an acceptable IUPAC name for each of the following compounds:

(a)
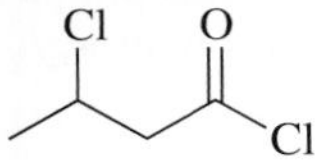

(b)
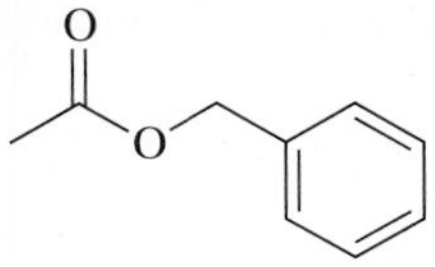

(c)
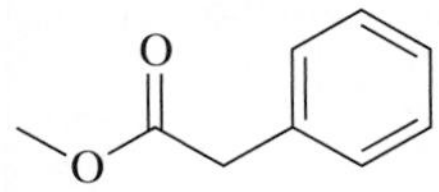

(d)
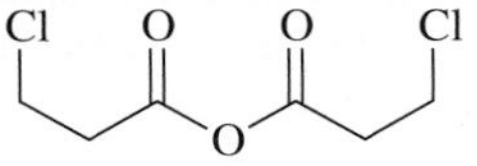

(e)
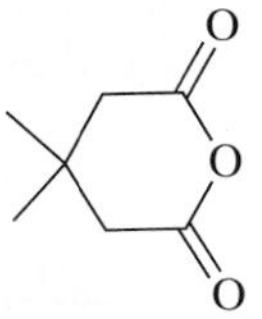

(f)
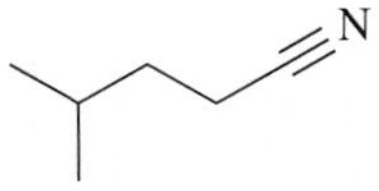

(g)
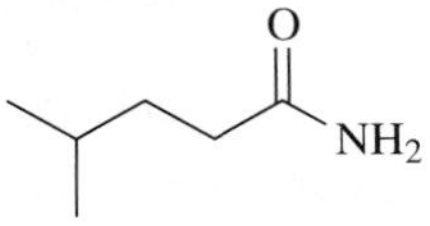

(h)
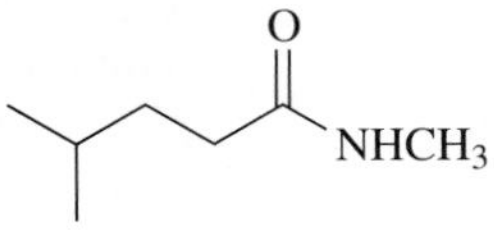

(i)
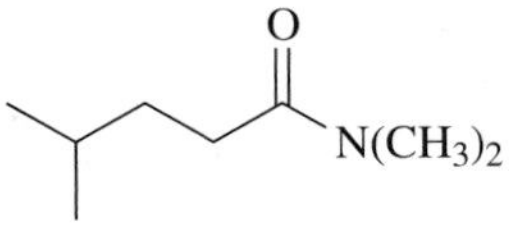

20.30 The serum cholesterol-lowering agent mevinolin (lovastatin) is shown here. Identify the ester and lactone functional groups of lovastatin, and for each, write the structures of the carboxylic acid and alcohol from which the ester and lactone are formed.

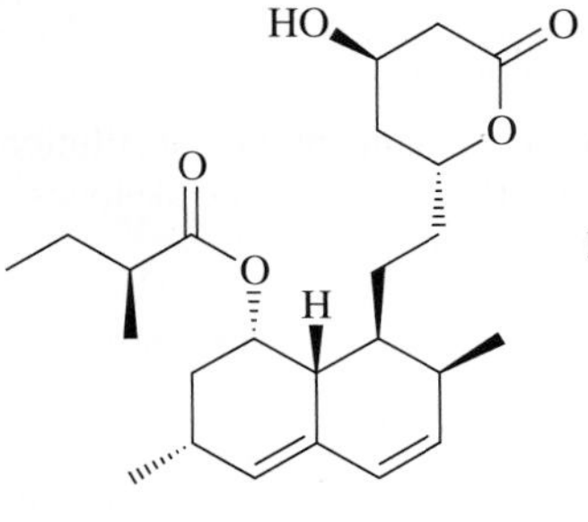

Lovastatin

Reactions

20.31 Write a structural formula for the principal organic product or products of each of the following reactions:

(a) Propanoyl chloride and sodium propanoate

(b) Butanoyl chloride and benzyl alcohol

(c) *p*-Chlorobenzoyl chloride and ammonia

(d) 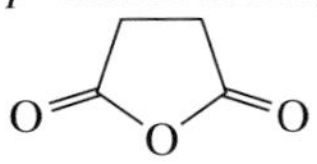 and water

(e) 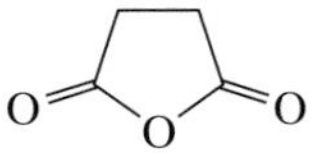 and aqueous sodium hydroxide to give $C_4H_4Na_2O_4$

(f) 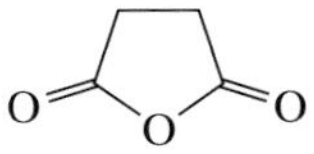 and aqueous ammonia to give $C_4H_{10}N_2O_3$

(g) Methyl benzoate and excess phenylmagnesium bromide, then H_3O^+

(h) Acetic anhydride and 3-pentanol

(i) Ethyl phenylacetate and lithium aluminum hydride, then H_3O^+

(j) 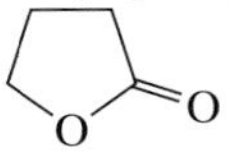 and aqueous sodium hydroxide to give $C_4H_7NaO_3$

(k) 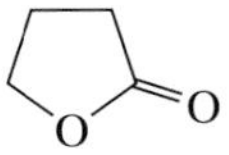 and aqueous ammonia

(l) 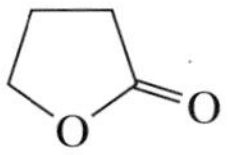 and lithium aluminum hydride, then H_2O

(m) 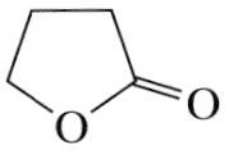 and excess methylmagnesium bromide, then H_3O^+

(n) Ethyl phenylacetate and methylamine (CH_3NH_2)

(o) 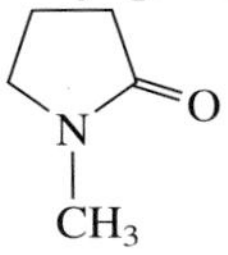 and aqueous sodium hydroxide to give $C_5H_{10}NNaO_2$

(p) 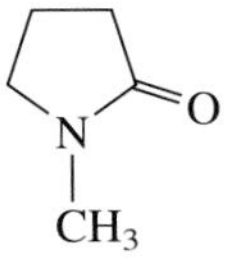 and aqueous hydrochloric acid, heat to give $[C_5H_{12}NO_2]^+$

(q) $C_6H_5NH\overset{O}{\overset{\|}{C}}CH_3$ and aqueous hydrochloric acid, heat to give $C_2H_4O_2 + C_6H_8ClN$

(r) $C_6H_5\overset{O}{\overset{\|}{C}}NHCH_3$ and aqueous sulfuric acid, heat to give $CH_7NO_4S + C_7H_6O_2$

(s) 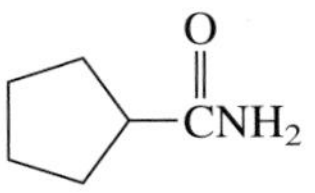 $-\overset{O}{\overset{\|}{C}}NH_2$ and P_4O_{10}

(t) $(CH_3)_2CHCH_2C{\equiv}N$ and aqueous hydrochloric acid, heat

(u) *p*-Methoxybenzonitrile and aqueous sodium hydroxide, heat to give $C_8H_7NaO_3$

(v) Propanenitrile and methylmagnesium bromide, then H_3O^+, heat

20.32 (a) Unlike other esters which react with Grignard reagents to give tertiary alcohols, ethyl formate ($HCOCH_2CH_3$, with C=O on the formyl carbon) yields a different class of alcohols on treatment with Grignard reagents. What kind of alcohol is formed in this case and why?

(b) Diethyl carbonate ($CH_3CH_2OCOCH_2CH_3$, with C=O on the central carbon) reacts with excess Grignard reagent to yield alcohols of a particular type. What is the structural feature that characterizes alcohols prepared in this way?

20.33 Penicillin G is prepared on a large scale by fermentation methods and serves as the starting material for other penicillins. The first step in this process is an enzyme-catalyzed hydrolysis represented by the equation shown. What are the products of this reaction?

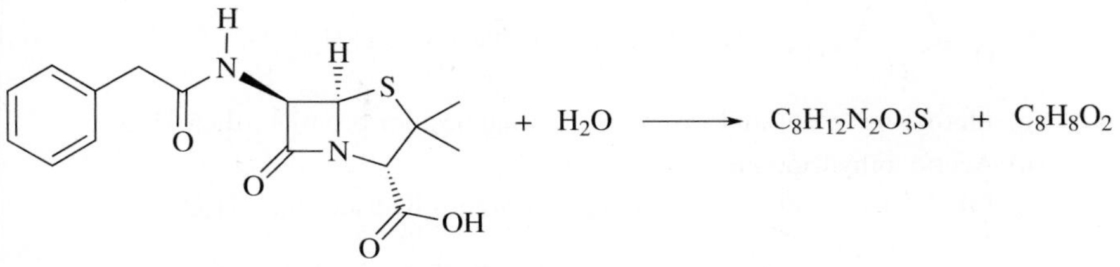

Penicillin G + H_2O ⟶ $C_8H_{12}N_2O_3S$ + $C_8H_8O_2$

20.34 Compound A serves as a prodrug for the analgesic benzocaine. (A prodrug is a pharmacologically inactive compound that is converted in the body to an active drug, usually by a metabolic transformation.) The enzyme *amidase* catalyzes the hydrolysis of compound A into benzocaine. Write the structures of the possible products of A that might be formed by hydrolysis in aqueous HCl.

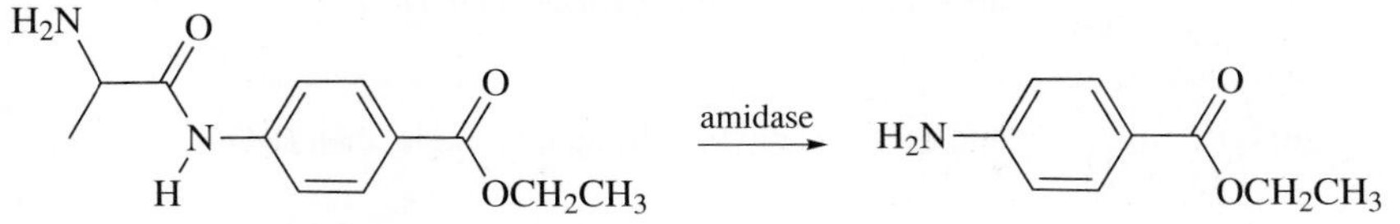

Compound A —(amidase)→ Benzocaine (H_2N–C_6H_4–$CO_2CH_2CH_3$)

20.35 Compound A is a derivative of the carbohydrate perosamine, which is found in the antibiotic perimycin. When A is treated with acetic anhydride in methanol, a monoacyl derivative B ($C_9H_{17}NO_5$) is obtained in 73% yield. What is the structure of compound B?

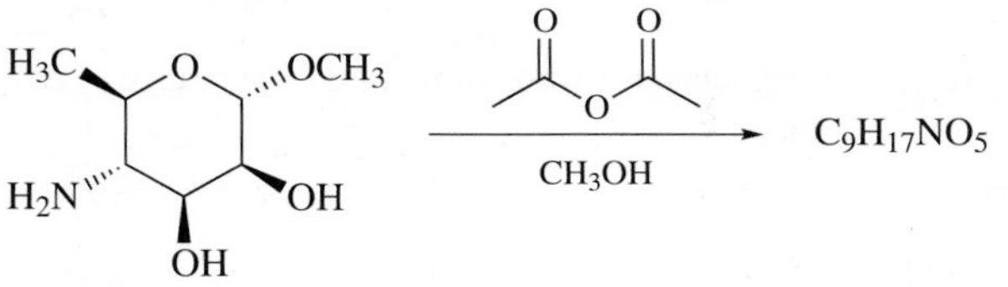

Compound A —(acetic anhydride, CH_3OH)→ $C_9H_{17}NO_5$ (Compound B)

Synthesis

20.36 Using ethanol and sodium or potassium cyanide as the sources of the carbon atoms, along with any necessary inorganic reagents, show how you could prepare each of the following:

(a) Acetyl chloride
(b) Acetic anhydride
(c) Ethyl acetate
(d) Acetamide
(e) 2-Hydroxypropanoic acid

20.37 Using toluene, sodium cyanide, and carbon dioxide as the sources of the carbon atoms, along with any necessary inorganic reagents, show how you could prepare each of the following:

(a) Benzoyl chloride
(b) Benzoic anhydride
(c) Benzyl benzoate
(d) Benzamide
(e) Benzonitrile
(f) Benzyl cyanide
(g) Phenylacetic acid
(h) *p*-Nitrobenzoyl chloride
(i) *m*-Nitrobenzoyl chloride

20.38 Recast the following retrosynthesis in a synthetic format showing all necessary reactants.

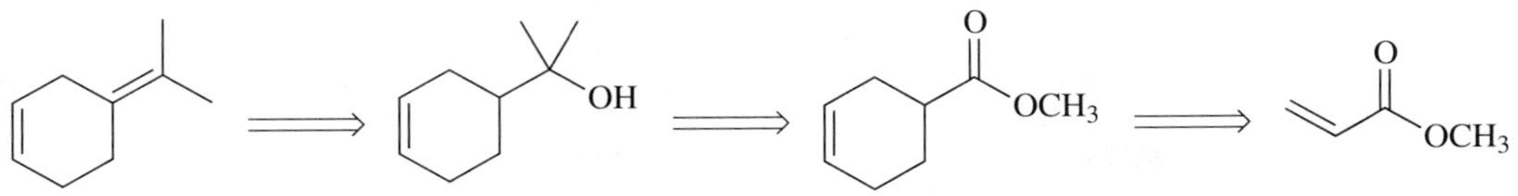

OH OCH$_3$ OCH$_3$

20.39 The saponification of ^{18}O-labeled ethyl propanoate was described in Section 20.9 as one of the significant experiments that demonstrated acyl–oxygen cleavage in ester hydrolysis. The ^{18}O-labeled ethyl propanoate used in this experiment was prepared from ^{18}O-labeled ethyl alcohol, which in turn was obtained from acetaldehyde and ^{18}O-enriched water.

Write a series of equations showing the preparation of $CH_3CH_2\overset{O}{\overset{\|}{C}}OCH_2CH_3$ (where O=^{18}O) from these starting materials.

20.40 The preparation of *cis*-4-*tert*-butylcyclohexanol from its trans stereoisomer was carried out by the following sequence of steps. Write structural formulas, including stereochemistry, for compounds A and B.

Step 1: OH + H_3C–C_6H_4–SO_2Cl $\xrightarrow{\text{pyridine}}$ Compound A ($C_{17}H_{26}O_3S$)

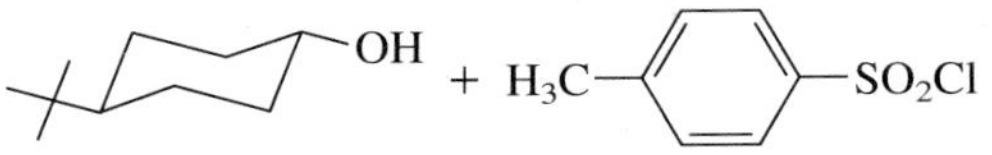

Step 2: Compound A + C_6H_5–$\overset{O}{\overset{\|}{C}}ONa$ $\xrightarrow[\text{heat}]{N,N\text{-dimethylformamide}}$ Compound B ($C_{17}H_{24}O_2$)

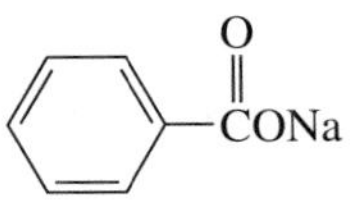

Step 3: Compound B $\xrightarrow[H_2O]{NaOH}$ OH

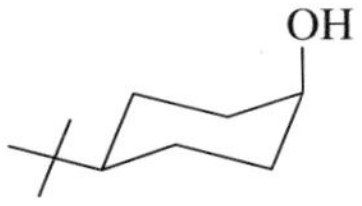

20.41 *Ambrettolide* is obtained from hibiscus and has a musk-like odor. Its preparation from compound A is outlined in the table that follows. Write structural formulas, ignoring stereochemistry, for compounds B through G in this synthesis. (*Hint:* Zinc, as used in step 4, converts vicinal dibromides to alkenes.)

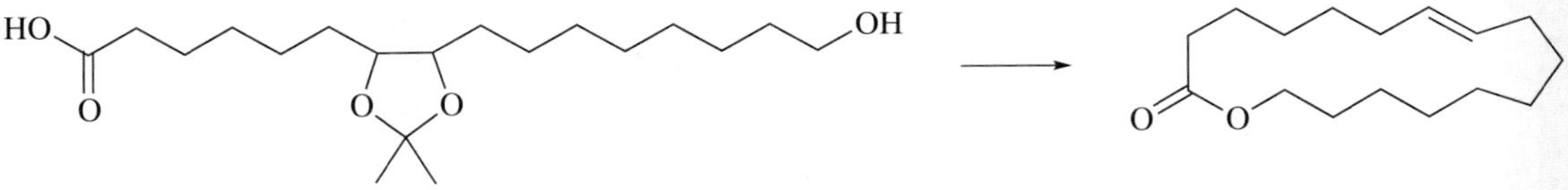

HO O O O OH → O O

Compound A ($C_{19}H_{36}O_5$) Ambrettolide

Step	Reactant	Reagents	Product
1.	Compound A	H_2O, H^+ heat	Compound B ($C_{16}H_{32}O_5$)
2.	Compound B	HBr	Compound C ($C_{16}H_{29}Br_3O_2$)
3.	Compound C	Ethanol, H_2SO_4	Compound D ($C_{18}H_{33}Br_3O_2$)
4.	Compound D	Zinc, ethanol	Compound E ($C_{18}H_{33}BrO_2$)
5.	Compound E	Sodium acetate, acetic acid	Compound F ($C_{20}H_{36}O_4$)
6.	Compound F	KOH, ethanol, then H^+	Compound G ($C_{16}H_{30}O_3$)
7.	Compound G	Heat	Ambrettolide ($C_{16}H_{28}O_2$)

20.42 The ketone shown was prepared in a three-step sequence from ethyl trifluoroacetate. The first step in the sequence involved treating ethyl trifluoroacetate with ammonia to give compound A. Compound A was in turn converted to the desired ketone by way of compound B. Fill in the missing reagents in the sequence shown, and give the structures of compounds A and B.

$$CF_3\overset{O}{\overset{\|}{C}}OCH_2CH_3 \xrightarrow{NH_3} \text{Compound A} \longrightarrow \text{Compound B} \longrightarrow CF_3\overset{O}{\overset{\|}{C}}C(CH_3)_3$$

20.43 The preparation of the sex pheromone of the bollworm moth, (*E*)-9,11-dodecadien-1-yl acetate, from compound A has been described. Suggest suitable reagents for each step in this sequence.

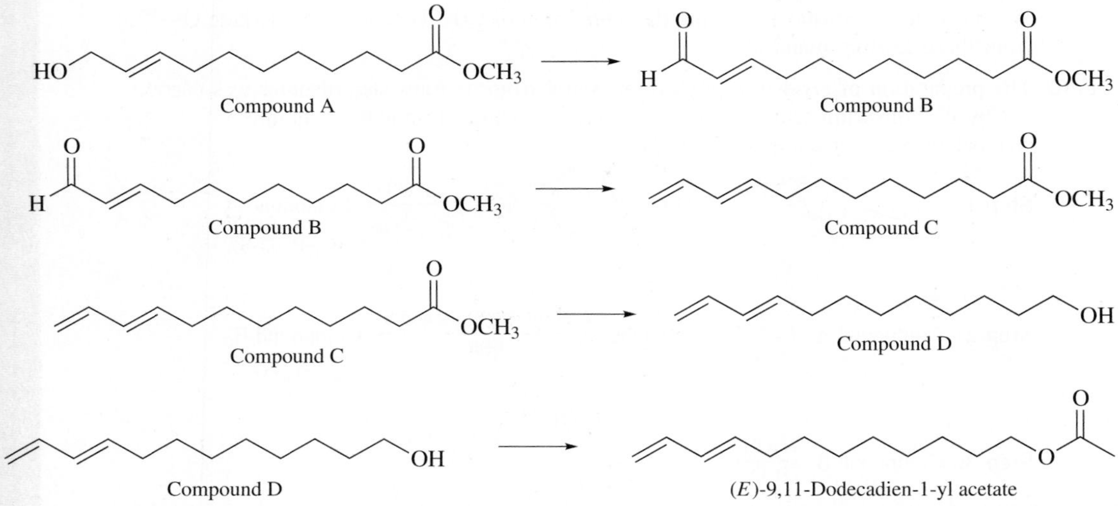

Compound A → Compound B

Compound B → Compound C

Compound C → Compound D

Compound D → (*E*)-9,11-Dodecadien-1-yl acetate

Kinetics and Mechanism

20.44 Acid hydrolysis of *tert*-butyl acetate in ^{18}O-labeled water was found to give *tert*-butyl alcohol having an ^{18}O content nearly identical to that of the solvent. Suggest a mechanism consistent with this observation. (It was shown that incorporation of ^{18}O into *tert*-butyl alcohol after it was formed did not occur.)

20.45 Suggest a reasonable explanation for each of the following observations:

(a) The second-order rate constant k for saponification (basic hydrolysis) of ethyl trifluoroacetate is over 1 million times greater than that for ethyl acetate (25°C).

(b) The second-order rate constant for saponification of ethyl 2,2-dimethylpropanoate, $(CH_3)_3CCO_2CH_2CH_3$, is almost 100 times smaller than that for ethyl acetate (30°C).

(c) The second-order rate constant k for saponification of methyl acetate is 100 times greater than that for *tert*-butyl acetate (25°C).

(d) The second-order rate constant k for saponification of methyl *m*-nitrobenzoate is 40 times greater than that for methyl benzoate (25°C).

(e) The second-order rate constant k for saponification of 5-pentanolide is over 20 times greater than that for 4-butanolide (25°C).

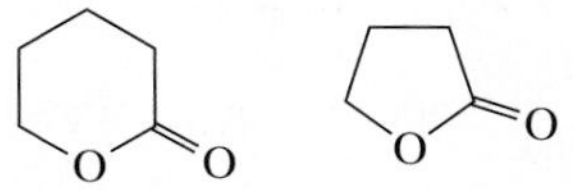

5-Pentanolide 4-Butanolide

(f) The second-order rate constant k for saponification of ethyl *trans*-4-*tert*-butylcyclohexanecarboxylate is 20 times greater than that for its cis diastereomer (25°C).

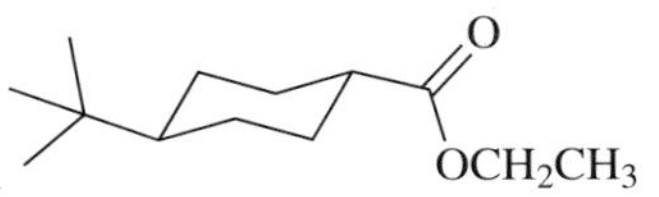

Ethyl *trans*-4-*tert*-butylcyclohexanecarboxylate

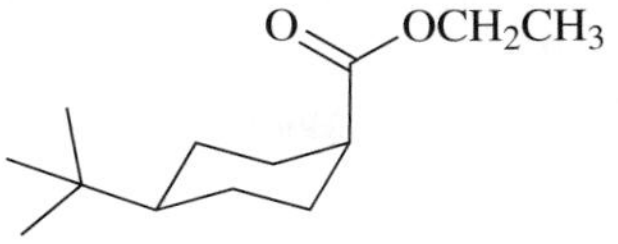

Ethyl *cis*-4-*tert*-butylcyclohexanecarboxylate

20.46 Outline reasonable mechanisms for each of the following reactions:

(a) + BrMg–(CH₂)₄–MgBr $\xrightarrow{\text{1. THF; 2. } H_3O^+}$

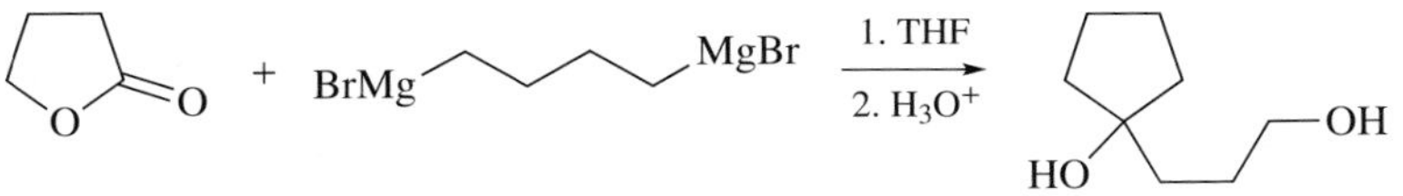

(b) spontaneous

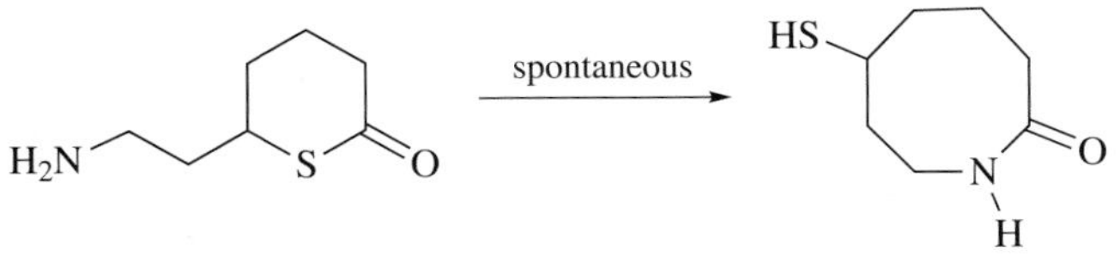

20.47 When compounds of the type represented by A are allowed to stand in pentane, they are converted to a constitutional isomer, compound B.

$RNHCH_2CH_2OC(=O)C_6H_4NO_2 \longrightarrow$ Compound B

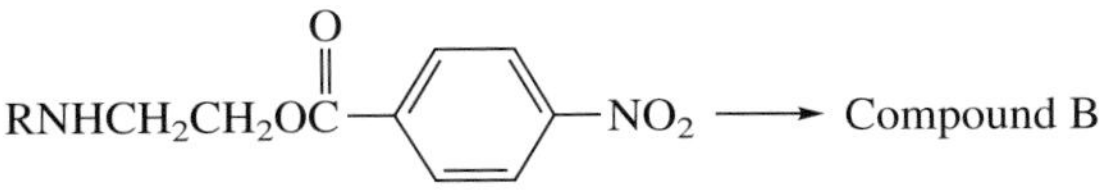

Compound A

Hydrolysis of either A or B yields $RNHCH_2CH_2OH$ and *p*-nitrobenzoic acid. Suggest a reasonable structure for compound B, and demonstrate your understanding of the mechanism of this reaction by writing the structure of the key intermediate in the conversion of compound A to compound B.

20.48 (a) In the presence of dilute hydrochloric acid, compound A is converted to a constitutional isomer, compound B.

$\xrightarrow{H^+}$ Compound B

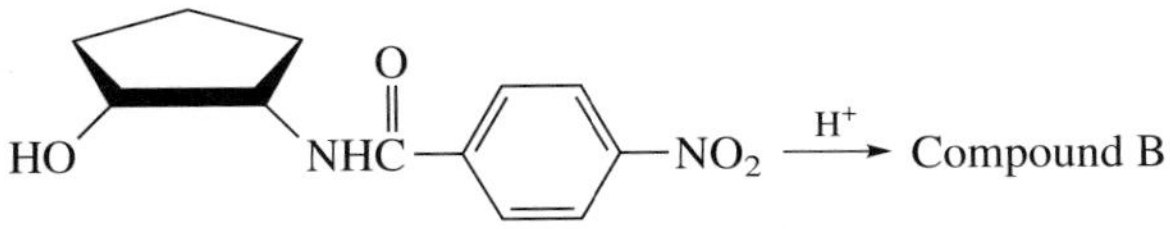

Compound A

Suggest a reasonable structure for compound B.

(b) The trans stereoisomer of compound A is stable under the reaction conditions. Why does it not rearrange?

Spectroscopy

20.49 A certain compound has a molecular weight of 83 and contains nitrogen. Its infrared spectrum contains a moderately strong peak at 2270 cm^{-1}. Its ^1H and ^{13}C NMR spectra are shown in Figure 20.5. What is the structure of this compound?

20.50 A compound has a molecular formula of $C_8H_{14}O_4$, and its IR spectrum contains an intense peak at 1730 cm^{-1}. The ^1H NMR spectrum of the compound is shown in Figure 20.6. What is its structure?

20.51 A compound ($C_4H_6O_2$) has a strong band in the infrared at 1760 cm^{-1}. Its ^{13}C NMR spectrum exhibits signals at δ 20.2 (CH_3), 96.8 (CH_2), 141.8 (CH), and 167.6 (C). The ^1H NMR spectrum of the compound has a three-proton singlet at δ 2.1 along with three other signals, each of which is a doublet of doublets, at δ 4.7, 4.9, and 7.3. What is the structure of the compound?

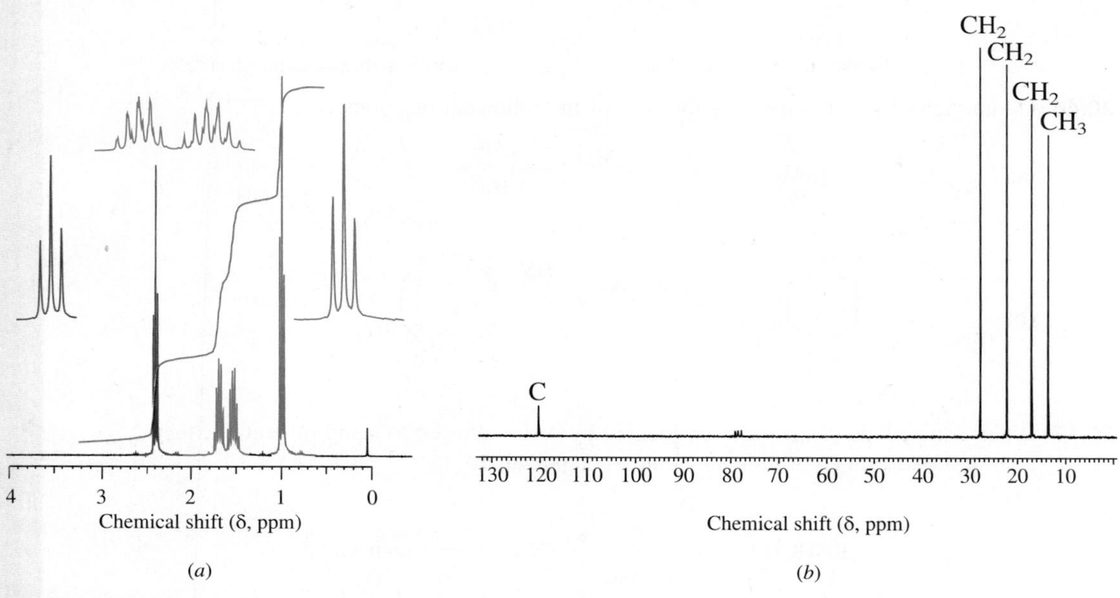

Figure 20.5

The (*a*) 300-MHz ^1H and (*b*) ^{13}C NMR spectra of the compound in Problem 20.49.

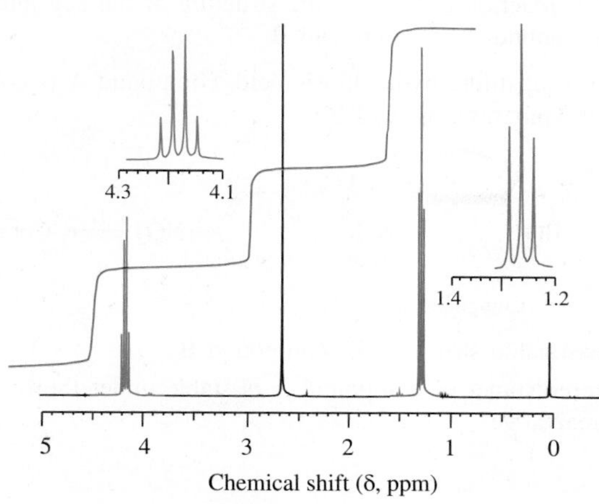

Figure 20.6

The 300-MHz ^1H NMR spectrum of the compound $C_8H_{14}O_4$ in Problem 20.50.

Descriptive Passage and Interpretive Problems 20

Thioesters

Thioesters have the general formula $RCSR'$ (with C=O). They resemble their oxygen counterparts $RCOR'$ (oxoesters) in structure and reactivity more than other carboxylic acid derivatives such as acyl chlorides, acid anhydrides, and amides. Thioesters can be prepared from thiols by reaction with acyl chlorides or acid anhydrides in much the same way as oxoesters are prepared from alcohols.

$$\underset{\text{1-Butanethiol}}{CH_3CH_2CH_2CH_2SH} + \underset{\text{Acetyl chloride}}{CH_3\overset{O}{\overset{\|}{C}}Cl} \longrightarrow \underset{\text{Butyl thioacetate (91\%)}}{CH_3\overset{O}{\overset{\|}{C}}SCH_2CH_2CH_2CH_3}$$

The preparation of thioesters by Fischer esterification is not very effective, however, because the equilibrium is normally unfavorable. Under conditions in which ethanol is converted to ethyl benzoate to the extent of 68%, ethanethiol gives only 15% ethyl thiobenzoate.

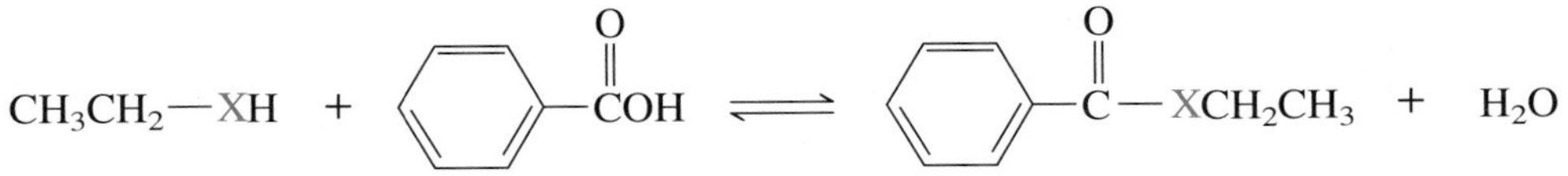

$$CH_3CH_2{-}XH + C_6H_5\overset{O}{\overset{\|}{C}}OH \rightleftharpoons C_6H_5\overset{O}{\overset{\|}{C}}{-}XCH_2CH_3 + H_2O$$

Ethanol: X = O

Ethanethiol: X = S

At equilibrium: X = O; 68%

X = S; 15%

This, and numerous other observations, indicates that S—C═O is less stabilized than O—C═O. Like chlorine, sulfur is a third-row element and does not act as an electron-pair donor to the carbonyl group as well as oxygen.

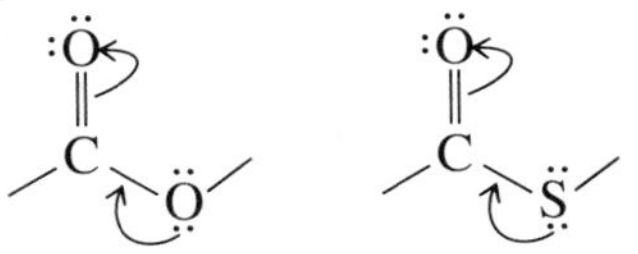

More effective   Less effective

Thioesters and oxoesters are similar in their rates of nucleophilic acyl substitution, except with amine nucleophiles for which thioesters are much more reactive. Many biological reactions involve nucleophilic acyl substitutions referred to as **acyl transfer** reactions. The thioester *acetyl coenzyme A* is an acetyl group donor to alcohols, amines, and assorted other biological nucleophiles.

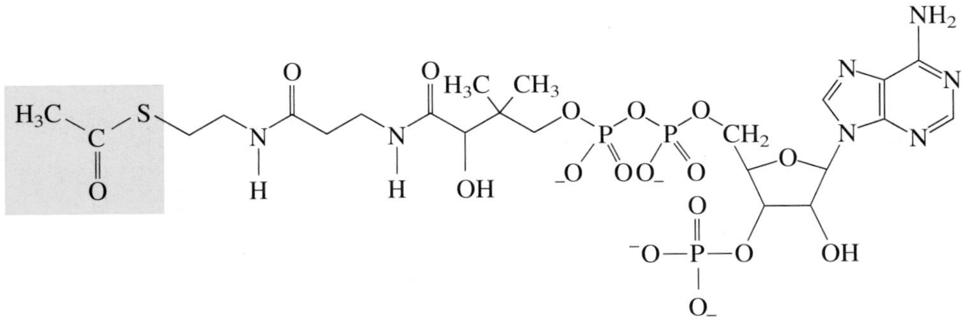

Acetyl coenzyme A ($CH_3\overset{O}{\overset{\|}{C}}SCoA$)

Melatonin, a hormone secreted by the pineal gland that regulates circadian rhythms, including wake–sleep cycles, is biosynthesized by a process in which the first step is an enzyme-catalyzed transfer of the acetyl group from sulfur of acetyl coenzyme A to the $—NH_2$ group of serotonin.

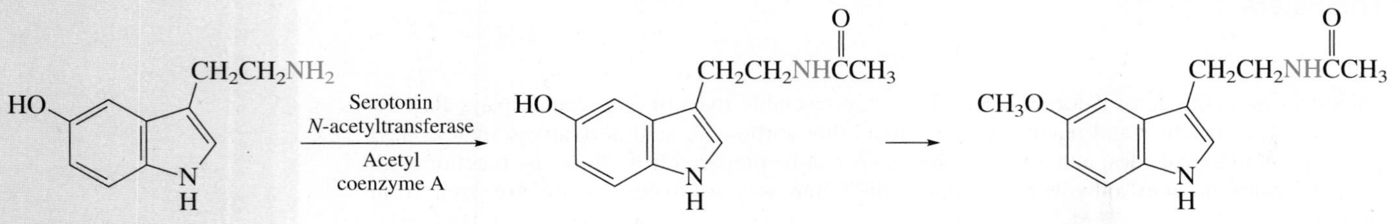

Serotonin → (Serotonin *N*-acetyltransferase; Acetyl coenzyme A) → *N*-Acetylserotonin → Melatonin

20.52 Thioesters react with hydroxylamine by nucleophilic acyl substitution to give hydroxamic acids. What is the structure of the hydroxamic acid formed in the following reaction?

$C_6H_5CSCH_2CH_2CH_3$ (C=O) $\xrightarrow{H_2NOH}$

A. $CH_3CH_2CH_2ONH_2$

B. $CH_3CH_2CH_2NHOH$

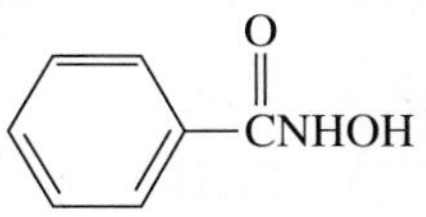

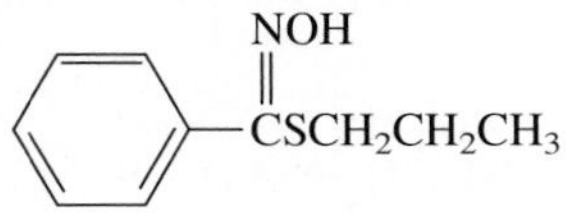

C. C_6H_5CNHOH (C=O)

D. $C_6H_5CSCH_2CH_2CH_3$ (C=NOH)

20.53 The equilibrium constant K equals 56 for the reaction shown.

$$HOCH_2CH_2SCCH_3 \overset{K}{\rightleftharpoons} CH_3COCH_2CH_2SH$$

Complete the following statement so that it correctly describes this reaction.

The sign of:

A. ΔG is + at equilibrium

B. ΔG is – at equilibrium

C. $\Delta G°$ is +

D. $\Delta G°$ is –

20.54 For the reaction shown in Problem 20.53, which of the following better represents the flow of electrons for the step in the mechanism leading to the isomer present in greatest amount at equilibrium?

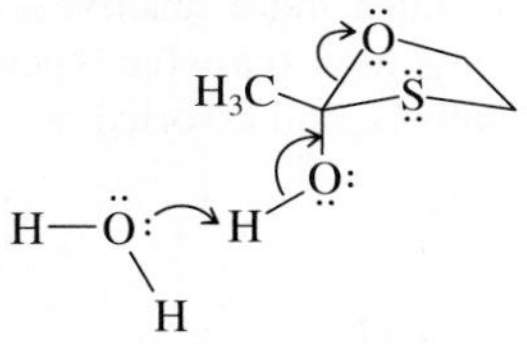

A. [mechanism with curved arrows: H—Ö:(H) ⟶ H ⋯ O:, H_3C, O, S ring]

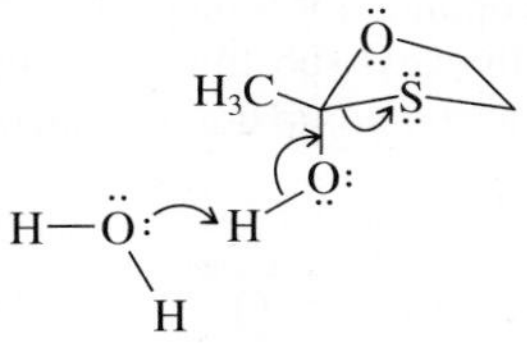

B. [mechanism with curved arrows: H—Ö:(H) ⟶ H ⋯ O:, H_3C, O, S ring]

20.55 Which reaction occurs at the fastest rate?

A. $CH_3\overset{O}{\overset{\|}{C}}SCH_2CH_3 + CH_3NH_2 \longrightarrow CH_3\overset{O}{\overset{\|}{C}}NHCH_3 + CH_3CH_2SH$

B. $CH_3\overset{O}{\overset{\|}{C}}OCH_2CH_3 + CH_3NH_2 \longrightarrow CH_3\overset{O}{\overset{\|}{C}}NHCH_3 + CH_3CH_2OH$

C. $CH_3\overset{O}{\overset{\|}{C}}SCH_2CH_3 + H_2O \longrightarrow CH_3\overset{O}{\overset{\|}{C}}OH + CH_3CH_2SH$

D. $CH_3\overset{O}{\overset{\|}{C}}OCH_2CH_3 + H_2O \longrightarrow CH_3\overset{O}{\overset{\|}{C}}OH + CH_3CH_2OH$

20.56 Which one of the reactions in the preceding problem has the most negative value of $\Delta G°$?

20.57 Acetylcholine is a neurotransmitter formed in nerve cells by the enzyme-catalyzed reaction of choline with acetyl coenzyme A.

Choline + Acetyl coenzyme A $\xrightarrow{\text{choline acetyltransferase}}$ $H_3C-C(=O)-O-CH_2CH_2-\overset{+}{N}(CH_3)_3$

Acetylcholine

What is the most reasonable structure for choline?

A. $HSCH_2CH_2\overset{+}{N}(CH_3)_3$

B. $HOCH_2CH_2\overset{+}{N}(CH_3)_3$

C. $CH_3\overset{O}{\overset{\|}{C}}OCH_2CH_2NH_2$

D. $CH_3\overset{O}{\overset{\|}{C}}SCH_2CH_2\overset{+}{N}(CH_3)_3$

20.58 Thiane was prepared in 76% yield from 5-chloro-1-pentene by the procedure shown. Deduce the structure of compound X in this synthesis.

5-Chloro-1-pentene ($ClCH_2CH_2CH_2CH=CH_2$) + $CH_3\overset{O}{\overset{\|}{C}}SH$ $\xrightarrow{\text{light}}$ Compound X $\xrightarrow[H_2O]{NaOH}$ Thiane

A. $CH_3\overset{O}{\overset{\|}{C}}S-CH_2CH_2CH_2CH=CH_2$

B. $CH_3\overset{S}{\overset{\|}{C}}O-CH_2CH_2CH_2CH=CH_2$

C. $Cl-CH_2CH_2CH_2CH_2CH_2-O\overset{S}{\overset{\|}{C}}CH_3$

D. $Cl-CH_2CH_2CH_2CH_2CH_2-S\overset{O}{\overset{\|}{C}}CH_3$

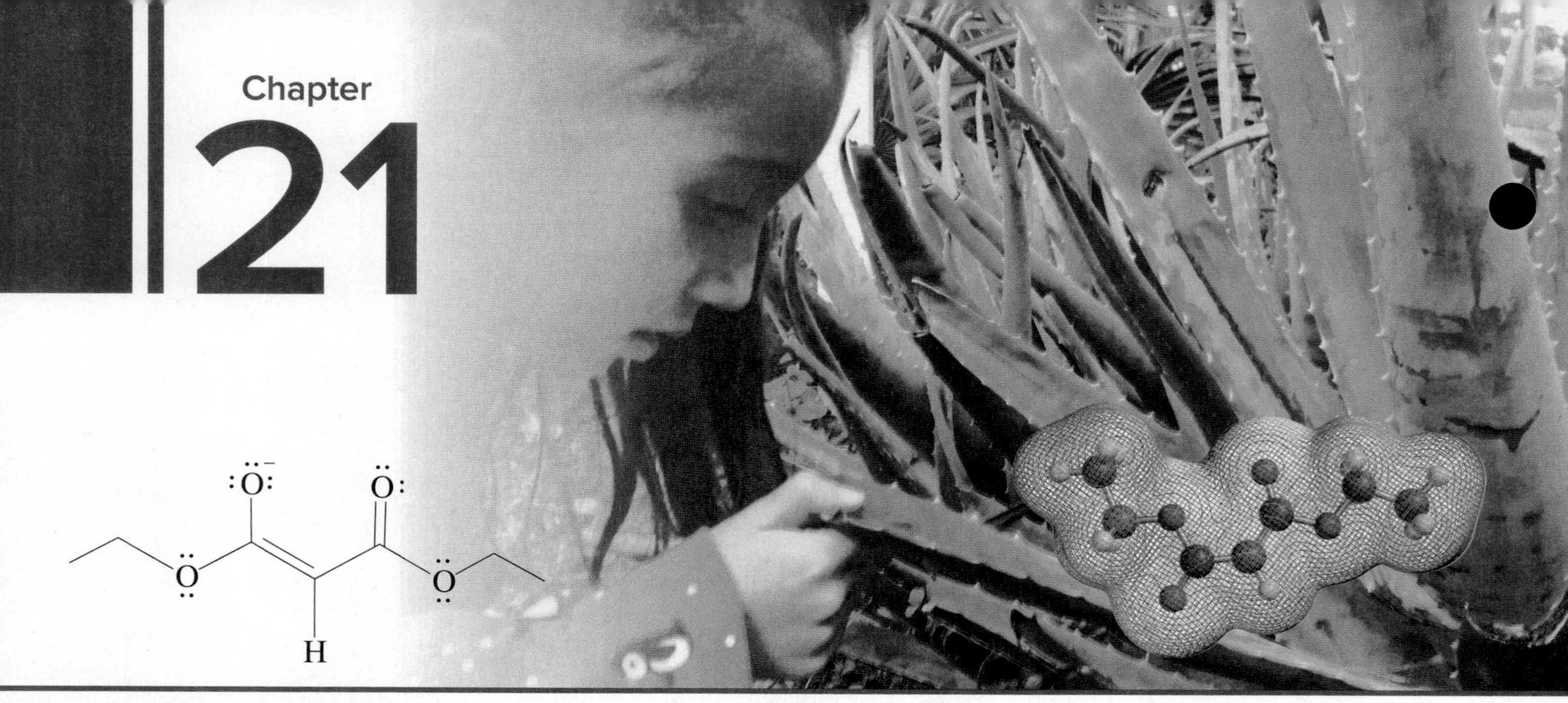

CHAPTER OUTLINE

Coumarins are medicinal compounds found in many plants, including aloe. The electrostatic potential map shows the enolate of diethyl malonate. This enolate has been used in the synthesis of coumarins by a process known as the Knoevenagel condensation.

Enols and Enolates

In Chapter 18 you learned that nucleophilic addition to the carbonyl group of aldehydes and ketones is one of the fundamental types of reactions in organic chemistry. In Chapter 20 you saw how addition to the carbonyl group of esters is a key step in nucleophilic acyl substitution. In the present chapter, you'll encounter yet another pattern of reactivity of these compounds involving **enols** and the **enolates** derived from them.

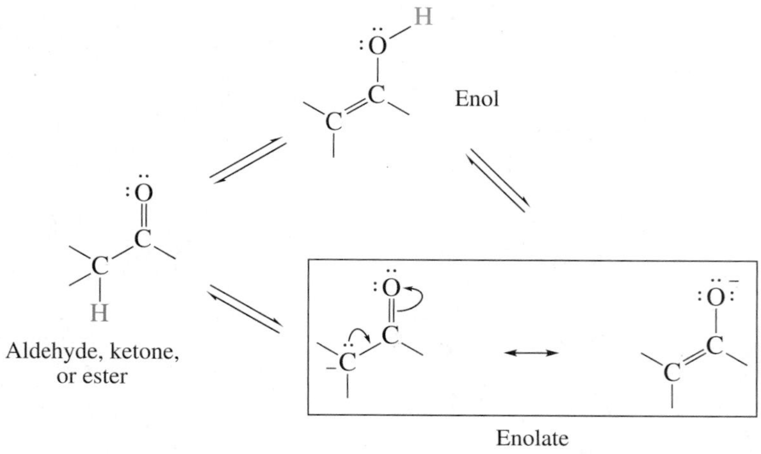

Enol

Aldehyde, ketone, or ester

Enolate

Enolates, represented as a hybrid of the resonance structures shown, are the conjugate bases of enols. The major enolate contributor is the structure with the negative charge on oxygen. It is, however, the carbanionic character of the α carbon that is responsible for the importance of enolates in organic synthesis, and we will sometimes write the enolate in the form that has the negative charge on carbon to emphasize this.

21.1 Aldehyde, Ketone, and Ester Enolates

It is convenient to use Greek letters to designate atoms in relation to the carbonyl group in an aldehyde, ketone, or ester. The carbon adjacent to the carbonyl carbon is the α carbon atom, the next one down the chain is the β carbon, and so on. Butanal, for example, has an α carbon, a β carbon, and a γ carbon.

$$\underset{\gamma}{CH_3}\underset{\beta}{CH_2}\underset{\alpha}{CH_2}\overset{O}{\overset{\|}{C}}H$$

The carbonyl carbon is the reference atom; no Greek letter is assigned to it.

Substituents take the same Greek letter as the carbon atom to which they are attached. A hydrogen connected to the α-carbon atom is an α hydrogen. Butanal has two α hydrogens, two β hydrogens, and three γ hydrogens. No Greek letter is assigned to the hydrogen attached directly to the carbonyl group of an aldehyde.

Our experience to this point has been that C—H bonds are not very acidic. Alkanes, for example, have pK_a's of approximately 60. Compared with them, however, aldehydes, ketones, and esters have relatively acidic hydrogens on their α-carbon atoms (Table 21.1).

TABLE 21.1 pK_a* Values of Some Aldehydes, Ketones, and Esters

Compound	pK_a	Enolate	Compound	pK_a	Enolate
Ethyl acetate	25.6		Diethyl malonate	13	
Acetone	19.1		Ethyl acetoacetate	11	
Acetaldehyde	16.7		2,4-Pentanedione	9	
2-Methylpropanal	15.5				

*The most acidic hydrogens are bonded to the α carbon; one of the α hydrogens is shown in red for each substance.

Two factors—one an electron-withdrawing inductive effect, the other electron delocalization—combine to make an H—C—C=O unit of an aldehyde, ketone, or ester relatively acidic compared with most other C—H bonds. The inductive effect of the carbonyl group increases the positive character of the α hydrogen, and resonance stabilizes the conjugate base.

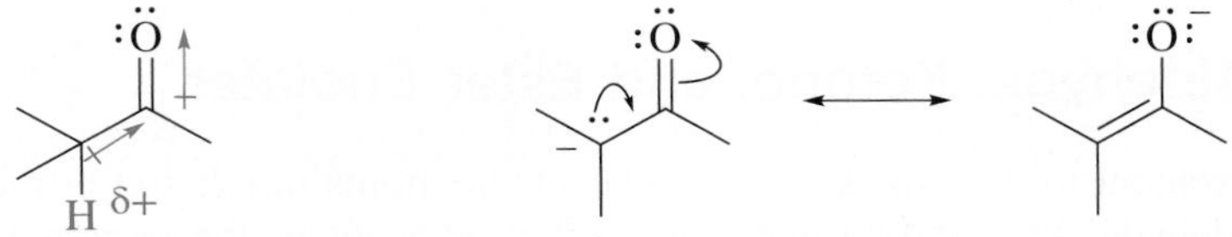

Inductive effect increases positive character of α hydrogen

Electron delocalization stabilizes enolate

The fact that the pK_a's of most simple aldehydes and ketones are about 16–20 means that both the carbonyl compound and its enolate are present when an acid–base equilibrium is established with hydroxide ion. The pK_a's of 2-methylpropanal and water, for example, are so similar that the equilibrium constant for enolate formation is approximately 1.

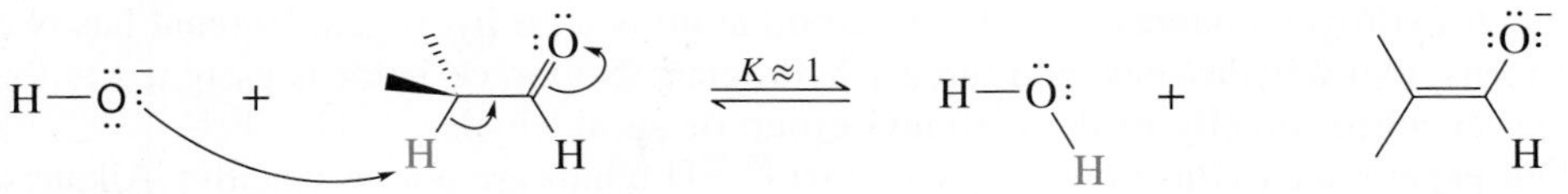

$K \approx 1$

Hydroxide ion | 2-Methylpropanal pK_a = 15.5 | Water pK_a = 15.7 | Enolate of 2-methylpropanal

Aldehydes and ketones can be converted *completely* to their enolates by using very strong bases such as lithium diisopropylamide $[(CH_3)_2CH]_2NLi$.

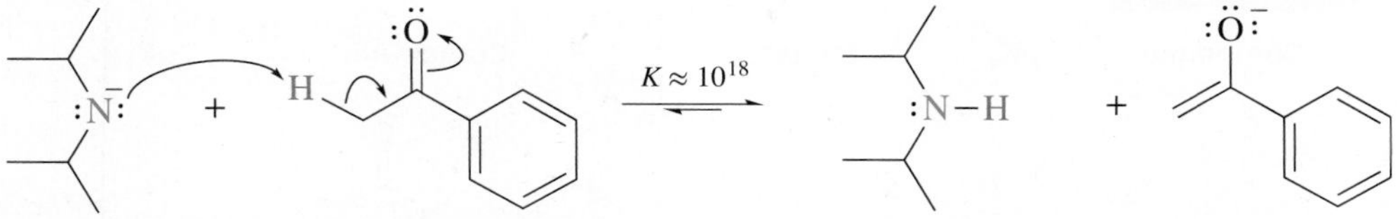

$K \approx 10^{18}$

Diisopropylamide ion | Acetophenone (Stronger acid: pK_a = 18.3) | Diisopropylamine (Weaker acid: pK_a = 36) | Enolate of acetophenone

Problem 21.1

Find the most acidic hydrogen in each of the following and write a chemical equation for the proton-transfer process that occurs on reaction with hydroxide ion. Use curved arrows to show electron flow and label the acid, base, conjugate acid, and conjugate base.

(a) *tert*-Butyl methyl ketone
(b) 3-Methylbutanal
(c) Methyl propanoate

Sample Solution (a) The only α hydrogens in *tert*-butyl methyl ketone are those of the methyl group attached to the carbonyl. Only α hydrogens are acidic enough to be removed by hydroxide. None of the hydrogens of the *tert*-butyl group are α hydrogens.

Hydroxide (base) | *tert*-Butyl methyl ketone (acid) | Water (conjugate acid) | Enolate (conjugate base)

Lithium diisopropylamide (known as **LDA**) is comparable to sodium amide ($NaNH_2$) in basicity, but, unlike $NaNH_2$, is too sterically hindered to undergo competing nucleophilic addition to the carbonyl group.

Problem 21.2

Methyllithium is a stronger base than lithium diisopropylamide but would not be a good choice for converting aldehydes and ketones to their enolates. Why?

The decreased acidity of ester α protons compared with those of aldehydes and ketones reflects the decreased electron-withdrawing ability of an ester carbonyl. Electron delocalization of the type:

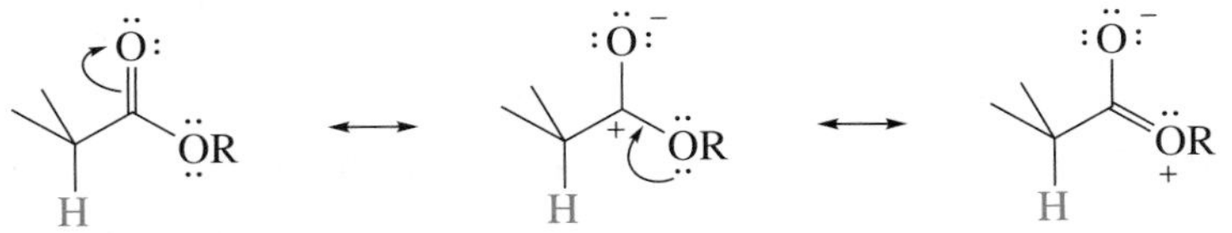

decreases the positive character of an ester carbonyl group and reduces its ability to withdraw electrons away from the α hydrogen.

The conditions for generating enolates from simple esters are similar to those used for aldehydes and ketones except that alkoxide bases are used instead of hydroxide. The alkoxide is chosen to match the ester (sodium ethoxide with ethyl esters, sodium methoxide with methyl esters) to avoid complications due to exchange of alkoxy groups by nucleophilic acyl substitution. An equilibrium is established in which the ester predominates and only a very small amount of enolate is present.

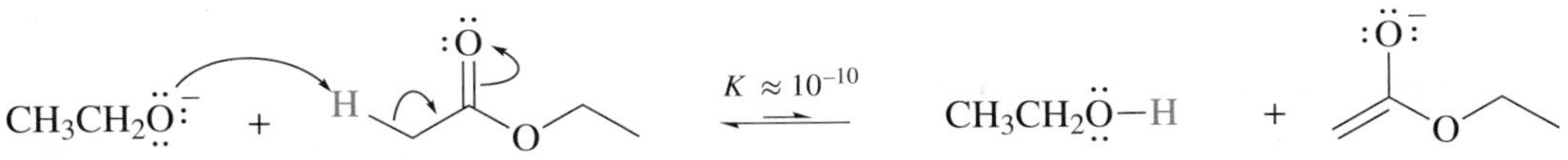

$CH_3CH_2\ddot{O}:^-$ + Ethyl acetate $\underset{}{\overset{K \approx 10^{-10}}{\rightleftharpoons}}$ $CH_3CH_2\ddot{O}-H$ + Enolate of ethyl acetate

Ethoxide ion; Ethyl acetate (Weaker acid: pK_a = 25.6); Ethanol (Stronger acid: pK_a = 16); Enolate of ethyl acetate

Esters can be converted entirely to their enolates by reaction with very strong bases such as lithium diisopropylamide.

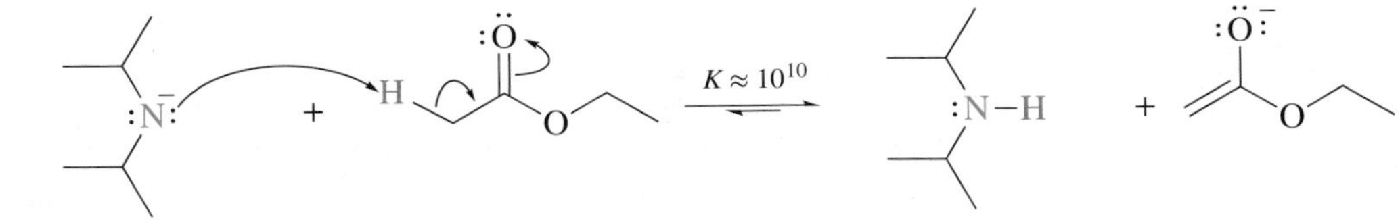

Diisopropylamide ion + Ethyl acetate $\overset{K \approx 10^{10}}{\rightleftharpoons}$ Diisopropylamine + Enolate of ethyl acetate

Diisopropylamide ion; Ethyl acetate (Stronger acid: pK_a = 25.6); Diisopropylamine (Weaker acid: pK_a = 36); Enolate of ethyl acetate

Dicarbonyl compounds such as β-diketones, β-keto esters, and diesters of malonic acid that have two carbonyl groups attached to the same carbon have pK_a's in the 9–13 range and are essentially completely converted to their enolates by hydroxide and alkoxides.

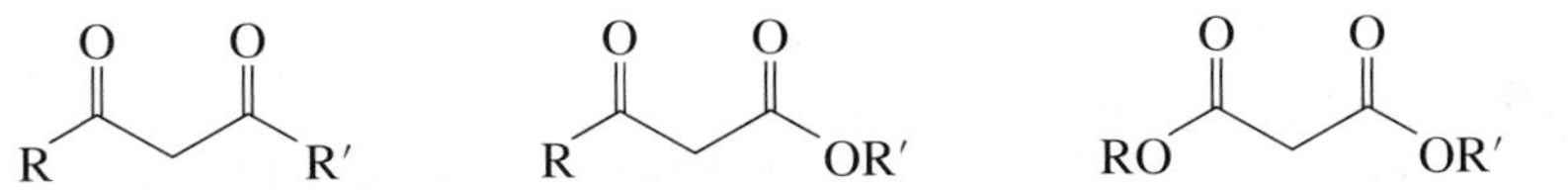

β-Diketone β-Keto ester Diester of malonic acid

The equation shows enolate formation from 2,4-pentanedione, the β-diketone cited in Table 21.1.

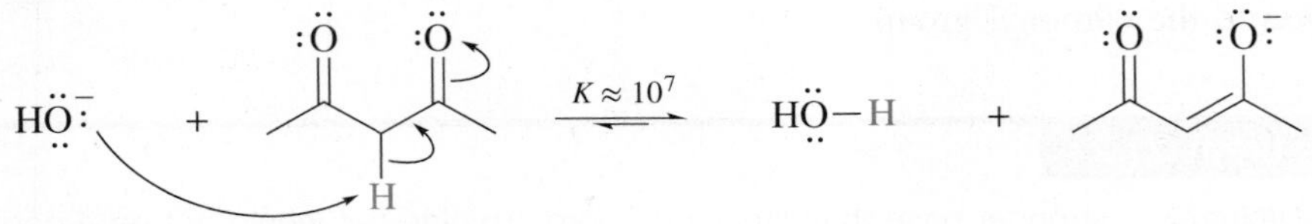

$$\text{HO}^- + \text{2,4-Pentanedione} \underset{}{\overset{K \approx 10^7}{\rightleftharpoons}} \text{HO–H} + \text{Enolate of 2,4-pentanedione}$$

Hydroxide ion | 2,4-Pentanedione (Stronger acid: $pK_a = 9$) | Water (Weaker acid: $pK_a = 15.7$) | Enolate of 2,4-pentanedione

Both carbonyl groups participate in stabilizing the enolate by delocalizing its negative charge.

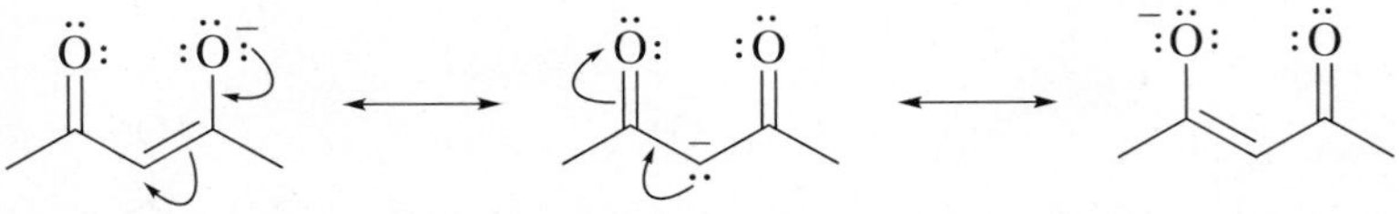

Analogous equations apply to ethyl acetoacetate and diethyl malonate—the β-keto ester and malonate diester, respectively, that are also cited in the table.

Problem 21.3

Write the structure of the most stable enolate derived from each of the following. Give the three major resonance contributors of each enolate.

(a) 2-Methyl-1,3-cyclopentanedione

(b)

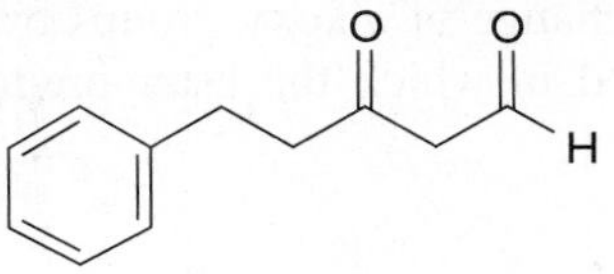

(c)

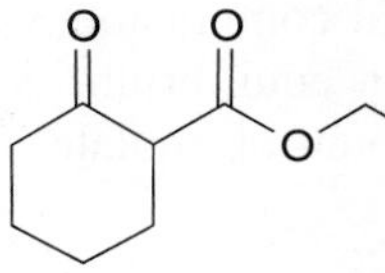

Sample Solution (a)

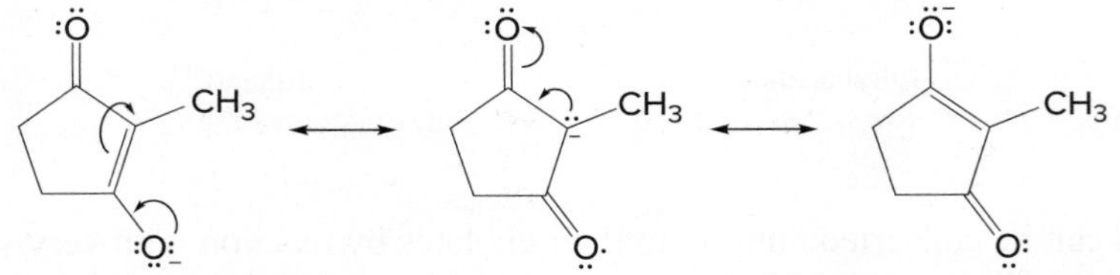

21.2 The Aldol Condensation

We have just seen that treatment of aldehydes and ketones with bases such as hydroxide and alkoxide gives a solution containing both the carbonyl compound and its enolate. Instead of simply maintaining an equilibrium between the two, however, carbon–carbon bond formation occurs.

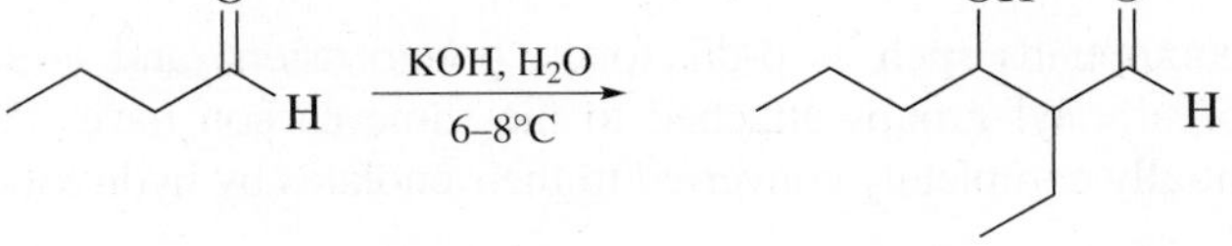

$$\text{Butanal} \xrightarrow[\text{6–8°C}]{\text{KOH, } H_2O} \text{2-Ethyl-3-hydroxyhexanal (75\%)}$$

The β-hydroxy aldehyde product is called an *aldol* because it contains both an *ald*ehyde and an alcoh*ol* function, and the reaction is referred to as **aldol addition.** It proceeds by the series of steps shown in Mechanism 21.1. All the steps in the reaction are reversible,

Mechanism 21.1

Aldol Addition of Butanal

THE OVERALL REACTION:

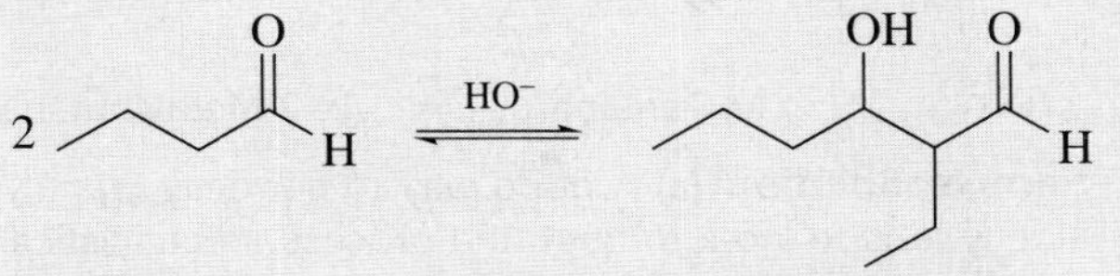

2 Butanal $\xrightleftharpoons{HO^-}$ 2-Ethyl-3-hydroxyhexanal

THE MECHANISM:

Step 1: The base, in this case hydroxide ion, converts a portion of butanal to its enolate by abstracting a proton from the α carbon.

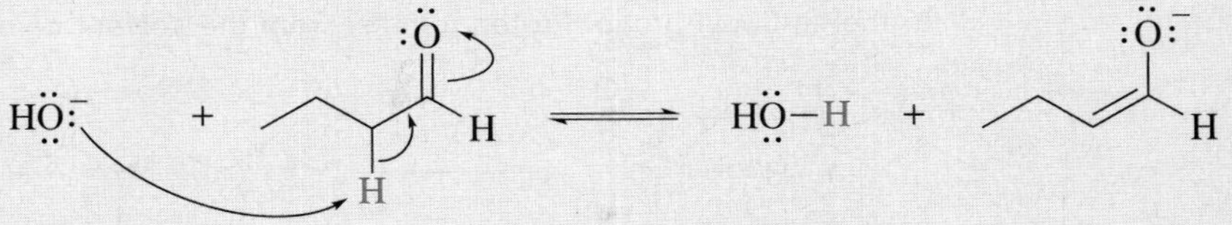

Hydroxide ion + Butanal ⇌ Water + Enolate of butanal

Step 2: The enolate acts as a nucleophile and adds to the carbonyl group.

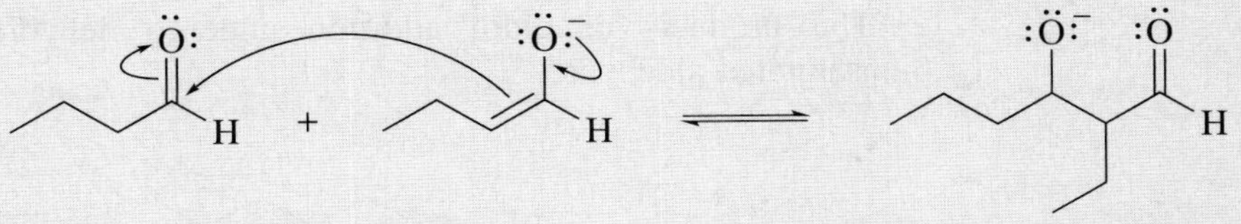

Butanal (electrophile) + Enolate of butanal (nucleophile) ⇌ Alkoxide from nucleophilic addition to the carbonyl

Step 3: The alkoxide product of step 2 abstracts a proton from water to give the aldol and regenerate the hydroxide catalyst.

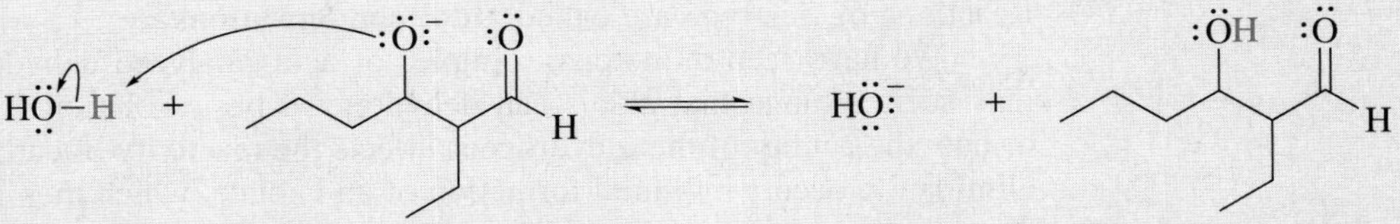

Water + Alkoxide from nucleophilic addition to the carbonyl ⇌ Hydroxide ion + 2-Ethyl-3-hydroxyhexanal

and, like other nucleophilic additions to carbonyl groups, the position of equilibrium is more favorable for aldehydes than for ketones. In the case of acetone, for example, only 2% of the aldol addition product is present at equilibrium.

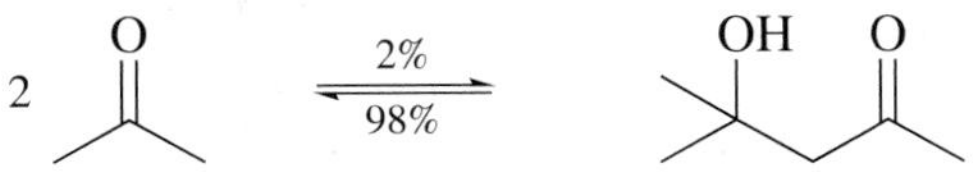

2 Acetone $\underset{98\%}{\overset{2\%}{\rightleftharpoons}}$ 4-Hydroxy-4-methyl-2-pentanone

Problem 21.4

Write the structure of the aldol addition product of each of the following.

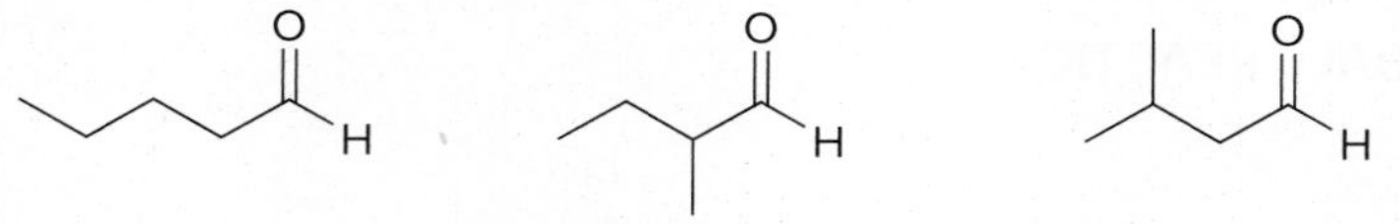

(a) Pentanal (b) 2-Methylbutanal (c) 3-Methylbutanal

Sample Solution (a) A good way to correctly identify the aldol addition product of any aldehyde is to work through the process mechanistically. Remember that the first step is enolate formation and that this *must* involve proton abstraction from the α carbon.

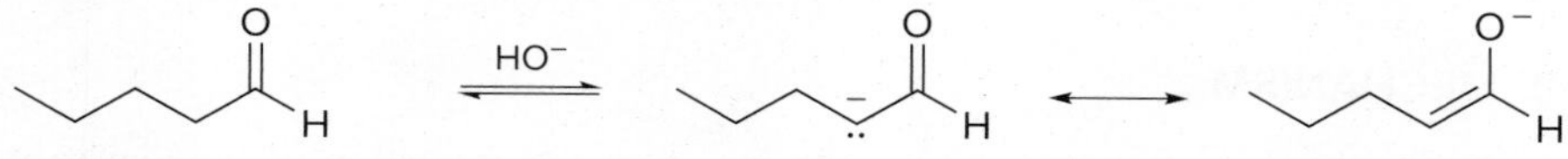

Pentanal $\xrightleftharpoons{HO^-}$ Enolate of pentanal

Now use the negatively charged α carbon of the enolate to form a new carbon–carbon bond to the carbonyl group. Proton transfer from the solvent completes the process.

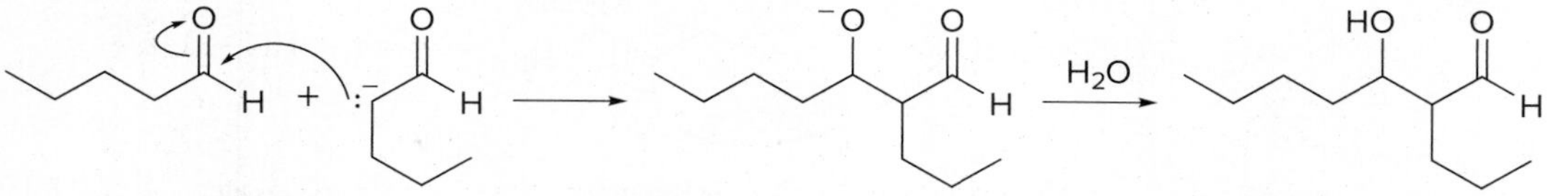

Pentanal + enolate → $\xrightarrow{H_2O}$ Aldol addition product

The products of aldol addition undergo dehydration on heating to yield α,β-unsaturated aldehydes.

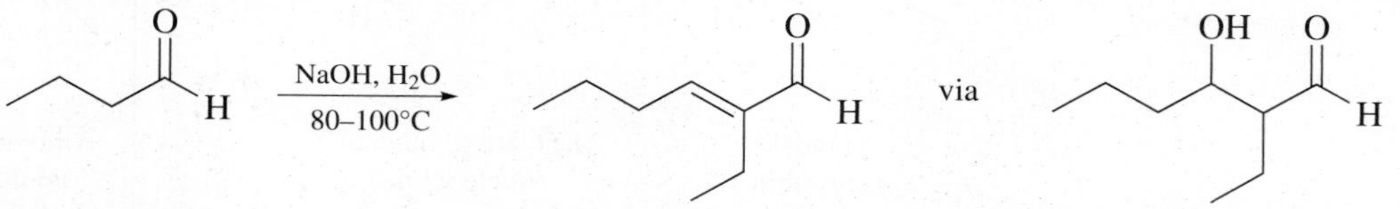

Butanal $\xrightarrow[80\text{–}100°C]{NaOH,\ H_2O}$ (*E*)-2-Ethyl-2-hexenal (86%) via 2-Ethyl-3-hydroxyhexanal (not isolated; dehydrates under reaction conditions)

Reactions of this type are called **aldol condensations.**

We have seen numerous examples of acid-catalyzed dehydration of alcohols, so it may seem strange that aldols can dehydrate in basic solution. This is another example of how the acidity of the α hydrogens affects the reactivity of carbonyl compounds. Here, elimination occurs by initial formation of an enolate, which then loses hydroxide to form the α,β-unsaturated aldehyde.

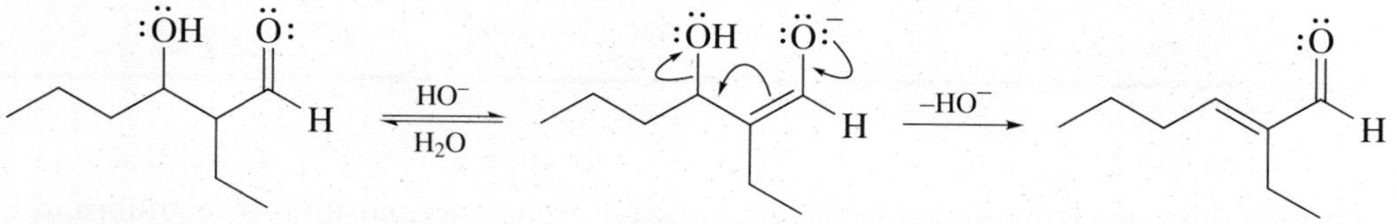

2-Ethyl-3-hydroxyhexanal $\underset{H_2O}{\overset{HO^-}{\rightleftharpoons}}$ Enolate $\xrightarrow{-HO^-}$ (*E*)-2-Ethyl-2-hexenal

Problem 21.5

Write the structure of the aldol condensation product of each of the aldehydes in Problem 21.4. One of these aldehydes can undergo aldol addition, but not aldol condensation. Which one? Why?

Aldol condensations of dicarbonyl compounds—even diketones—occur intramolecularly when five- or six-membered rings are possible.

Na_2CO_3, H_2O, heat

1,6-Cyclodecanedione → Not isolated: dehydrates under reaction conditions → Bicyclo[5.3.0]dec-1(7)-en-2-one (96%)

Problem 21.6

Each of the following can be prepared by an intramolecular aldol condensation of a diketone. Apply retrosynthetic analysis to deduce the structure of the diketone in each case.

(a) (b) (c)

Sample Solution

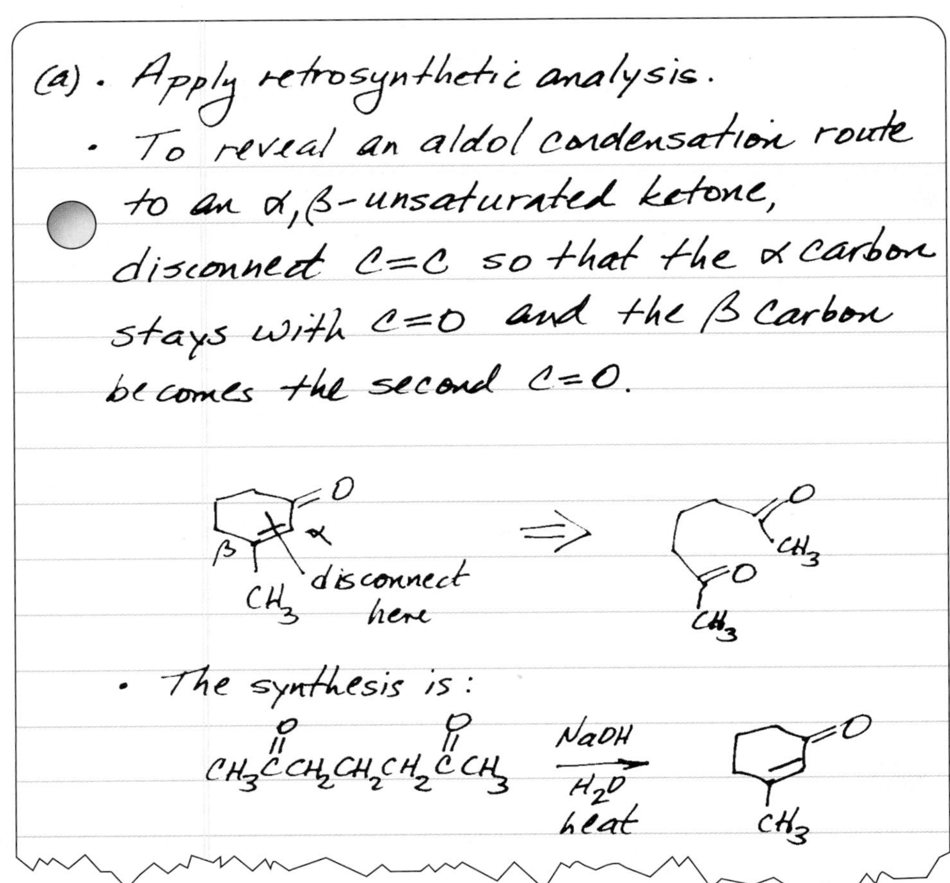

(a) • Apply retrosynthetic analysis.
• To reveal an aldol condensation route to an α,β-unsaturated ketone, disconnect C=C so that the α carbon stays with C=O and the β carbon becomes the second C=O.
α
β
CH3
disconnect here
⇒
CH3
CH3
• The synthesis is:
CH3CCH2CH2CH2CCH3
NaOH
H2O
heat
CH3

21.3 Mixed and Directed Aldol Reactions

Mixed aldol additions and condensations are those involving two different carbonyl compounds. In the most synthetically useful circumstances, only one of the reactants can form an enolate or is more reactive toward nucleophilic addition to its carbonyl group. For example:

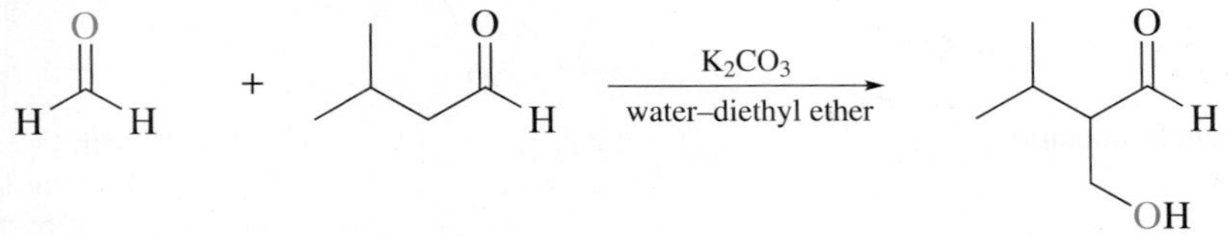

Formaldehyde + 3-Methylbutanal $\xrightarrow[\text{water–diethyl ether}]{K_2CO_3}$ 2-Hydroxymethyl-3-methylbutanal (52%)

Formaldehyde cannot form an enolate and is so reactive toward nucleophilic addition that it suppresses the self-condensation of the other aldehyde.

Aromatic aldehydes are another class of carbonyl compounds that cannot form enolates.

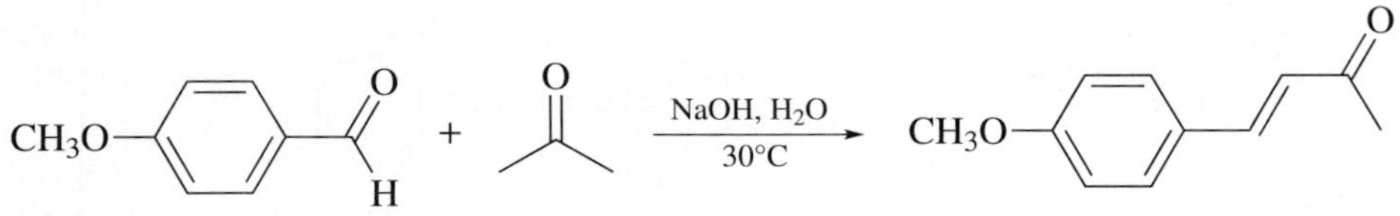

p-Methoxybenzaldehyde + Acetone $\xrightarrow[30°C]{NaOH,\ H_2O}$ 4-*p*-Methoxyphenyl-3-buten-2-one (83%)

Mixed aldol condensations in which a ketone reacts with an aromatic aldehyde are known as **Claisen–Schmidt condensations.**

Here, the mixed condensation is favored by both the reluctance of ketones to self-condense and the readiness with which aldol addition is followed by dehydration to give a product in which the double bond is conjugated to both the carbonyl group and an aromatic ring.

Problem 21.7

Give the structure of the mixed aldol condensation product of benzaldehyde with

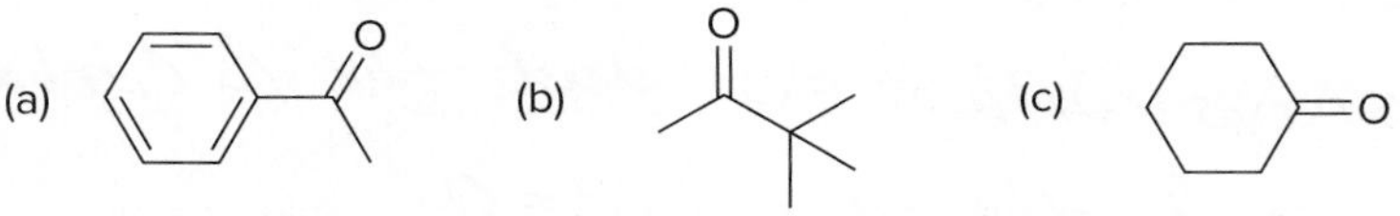

(a) Acetophenone (b) *tert*-Butyl methyl ketone (c) Cyclohexanone

Sample Solution (a) The enolate of acetophenone reacts with benzaldehyde to yield the product of mixed addition. Dehydration of the intermediate occurs, giving the α,β-unsaturated ketone.

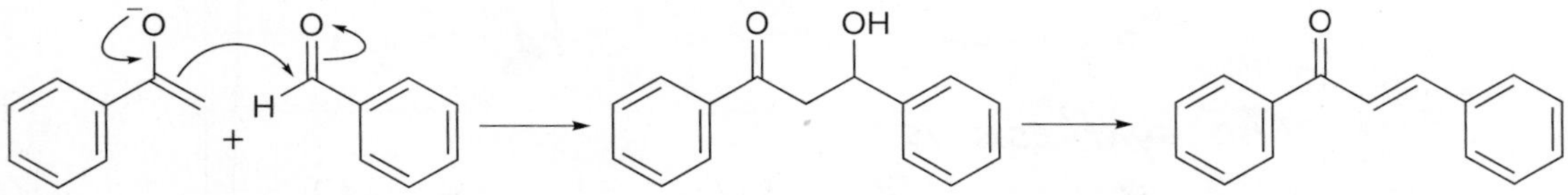

Enolate of acetophenone + benzaldehyde ⟶ ⟶ 1,3-Diphenyl-2-propen-1-one

As actually carried out, the mixed aldol condensation product was isolated in 85% yield on treating benzaldehyde with acetophenone in an aqueous solution of sodium hydroxide at 15–30°C.

From the Mulberry Tree to Cancer Chemotherapy

Estrogens—the primary female sex hormones—are risk factors for developing breast cancer, which, except for certain types of skin cancer, is the most common cancer among women. One approach to breast-cancer treatment is to restrict tumor growth with drugs such as *tamoxifen* that bind to the same cellular receptors as estrogens and block their action by denying them access. Another is to reduce estrogen levels by inhibiting the aromatase enzymes involved in the oxidation of androstenedione to estrone.

$N(CH_3)_2$

Tamoxifen

Androstenedione $\xrightarrow{\text{aromatase}}$ Estrone (an estrogen)

The potential of aromatase inhibitors in the treatment of breast cancer has spurred efforts to find more potent drugs. Bioassays of over 4000 plants led to the discovery of potent aromatase inhibitors from organic extracts of the paper mulberry tree, *Broussonetia papyrifera*. One of these compounds is a chalcone called morachalcone A. Chalcone and its derivatives are present in many plants, have a long history in folk medicine, and continue to be widely used as alternatives to conventional therapies.

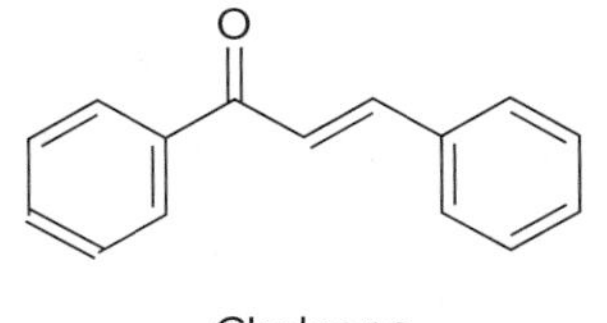

Chalcone

Morachalcone A

The paper mulberry tree, *Broussonetia papyrifera*, is a source of a compound that holds promise as an anticancer drug.

Problem 21.8

Chalcones are based on the parent structure shown above and are conveniently prepared by an aldol condensation. Use retrosynthetic analysis to identify the carbonyl compound and enolate and write an equation for the synthesis of chalcone.

Another way to ensure that only one enolate is present is to use lithium diisopropylamide as the base. LDA is such a strong base that enolate formation is virtually instantaneous and quantitative. The ketone is added to a solution of LDA in a suitable solvent, followed by the compound with which the enolate is to react.

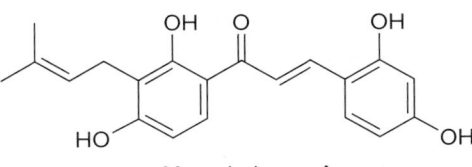

2,2-Dimethyl-3-pentanone $\xrightarrow[\text{THF}]{\text{LDA}}$ Enolate $\xrightarrow[\text{2. } H_2O]{\text{1. } CH_3CH_2CH{=}O}$ 5-Hydroxy-2,2,4-trimethyl-3-heptanone (81%)

Reactions of this type are called *directed aldol additions*.

A number of biochemical reactions resemble aldol addition. One, the first step in the citric acid cycle, is an aldol-type addition of acetyl coenzyme A to the ketone carbonyl of oxaloacetic acid.

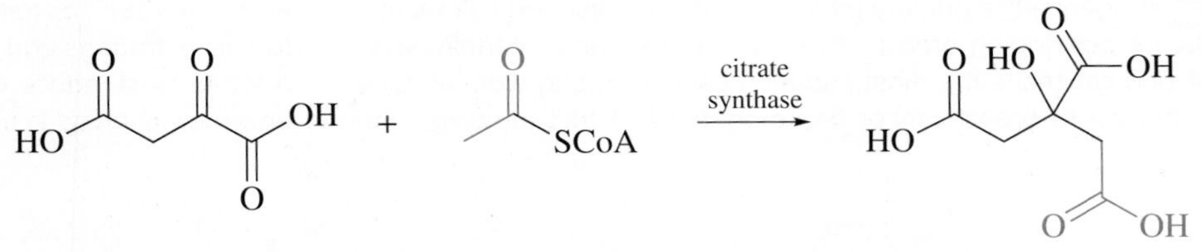

citrate synthase

Oxaloacetic acid + Acetyl coenzyme A → Citric acid

In an early step of *glycolysis,* the energy-producing part of glucose metabolism, the enzyme *aldolase* catalyzes a reverse or *retro-aldol* reaction.

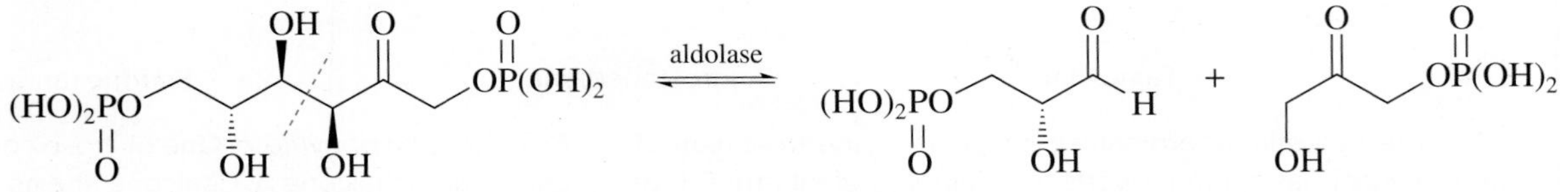

aldolase

D-Fructose 1,6-diphosphate ⇌ D-Glyceraldehyde 3-phosphate + Dihydroxyacetone phosphate

21.4 Acylation of Enolates: The Claisen and Related Condensations

Nucleophilic *addition* to the carbonyl group of an aldehyde or ketone by an enolate is a key step in aldol condensations. Nucleophilic acyl *substitution* at the carbonyl group of an ester by an enolate is a key step in several synthetic procedures applied to making C—C bonds.

In the **Claisen condensation,** one ester is the source of both the acyl group and the enolate and the product is a β-keto ester.

Ludwig Claisen was a German chemist who worked during the last two decades of the nineteenth century and the first two decades of the twentieth. His name is associated with three reactions. The *Claisen–Schmidt reaction* was presented in Section 21.3.

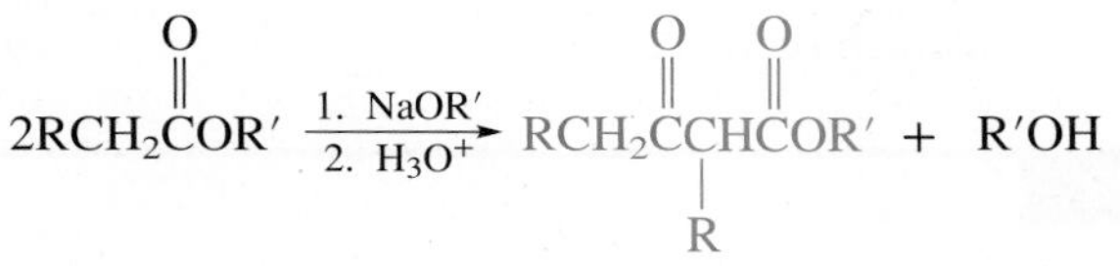

$$2RCH_2\overset{O}{\overset{\|}{C}}OR' \xrightarrow[2.\ H_3O^+]{1.\ NaOR'} RCH_2\overset{O}{\overset{\|}{C}}\underset{R}{\underset{|}{C}}H\overset{O}{\overset{\|}{C}}OR' + R'OH$$

Ester → β-Keto ester + Alcohol

The Claisen condensation is used to prepare β-keto esters and involves bond formation between the α carbon of one molecule and the carbonyl carbon of another. Mechanism 21.2 outlines the steps in the reaction:

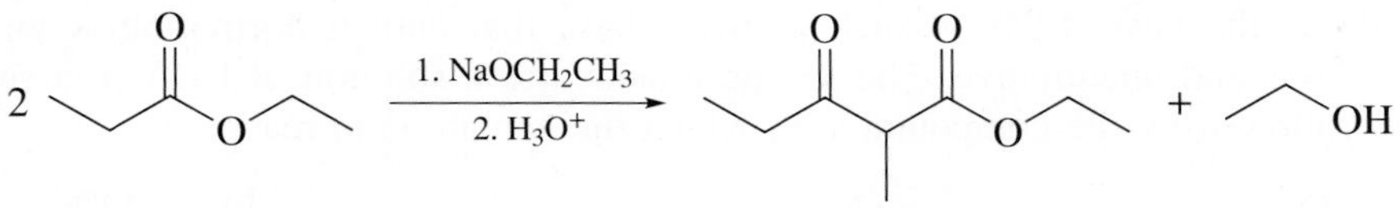

$$2\ \text{Ethyl propanoate} \xrightarrow[2.\ H_3O^+]{1.\ NaOCH_2CH_3} \text{Ethyl 2-methyl-3-oxopentanoate (81\%)} + \text{Ethanol}$$

Ethyl propanoate → Ethyl 2-methyl-3-oxopentanoate (81%) + Ethanol

An ester must have, as in this case, at least two α hydrogens to undergo a synthetically useful Claisen condensation. The first α proton is removed to form the ester enolate (step 1 in Mechanism 21.2). However, the equilibrium concentration of the β-keto ester is unfavorable unless a second α proton is removed as in step 4. Thus, Claisen concentrations succeed for esters of the type RCH_2CO_2R', but not R_2CHCO_2R'.

Mechanism 21.2

Claisen Condensation of Ethyl Propanoate

THE OVERALL REACTION:

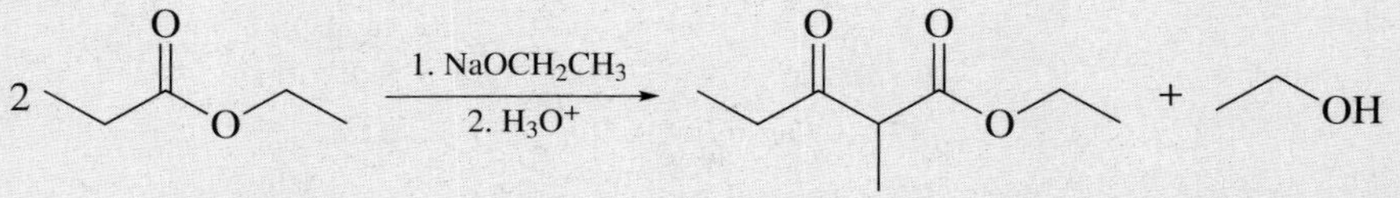

2 Ethyl propanoate → (1. $NaOCH_2CH_3$; 2. H_3O^+) Ethyl 2-methyl-3-oxopentanoate + Ethanol

THE MECHANISM:

Step 1: Proton abstraction from the α carbon of ethyl propanoate gives the corresponding enolate.

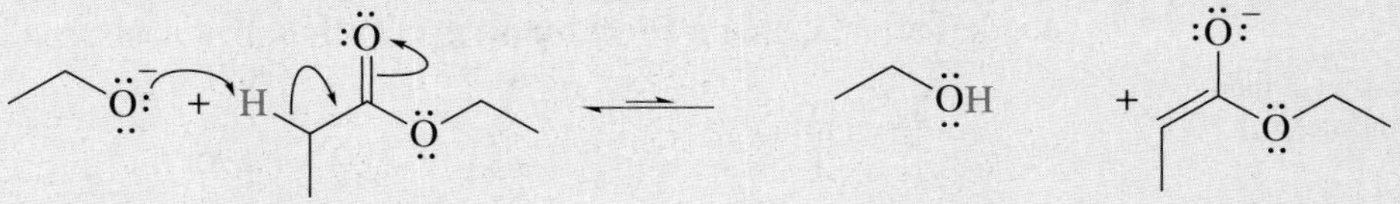

Ethoxide + Ethyl propanoate (weaker acid: pK_a ~ 26) ⇌ Ethanol (stronger acid: pK_a = 16) + Enolate of ethyl propanoate

Step 2: The ester enolate undergoes nucleophilic addition to the carbonyl group of the keto form of the ester. The product of this step is the anionic form of the tetrahedral intermediate.

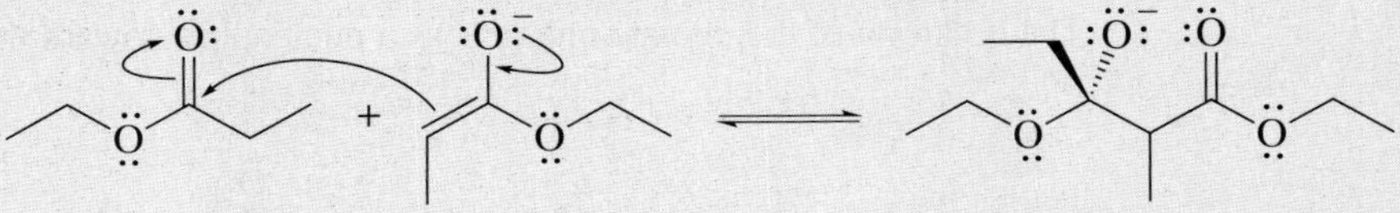

Ethyl propanoate + Enolate of ethyl propanoate ⇌ Anionic form of tetrahedral intermediate

Step 3: The tetrahedral intermediate dissociates by loss of ethoxide and gives the β-keto ester.

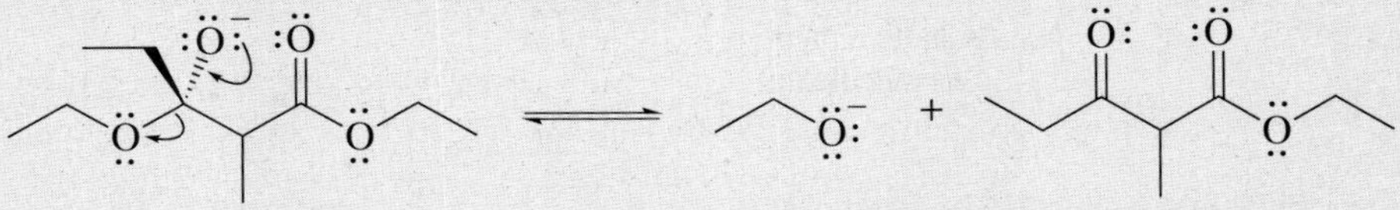

Anionic form of tetrahedral intermediate ⇌ Ethoxide ion + Ethyl 2-methyl-3-oxopentanoate

Step 4: Under the conditions of its formation, the β-keto ester is deprotonated. The equilibrium constant for this step is favorable and drives the equilibrium toward product formation.

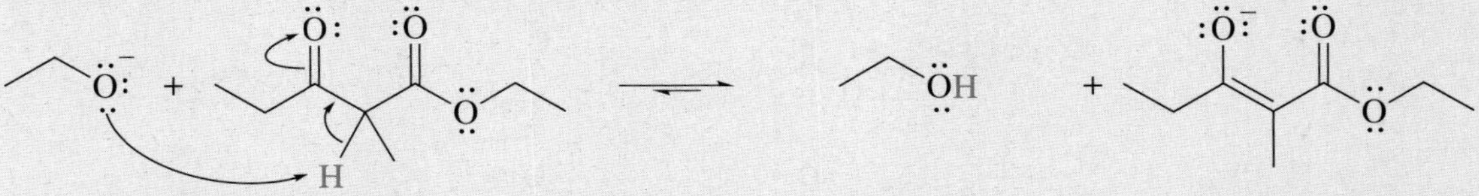

Ethoxide + Ethyl 2-methyl-3-oxopentanoate (stronger acid: pK_a ≈ 11) ⇌ Ethanol (weaker acid: pK_a = 16) + Conjugate base of ethyl 2-methyl-3-oxopentanoate

Step 5: In a separate operation, aqueous acid is added to the reaction mixture to convert the enolate from step 4 to the neutral form of the desired product.

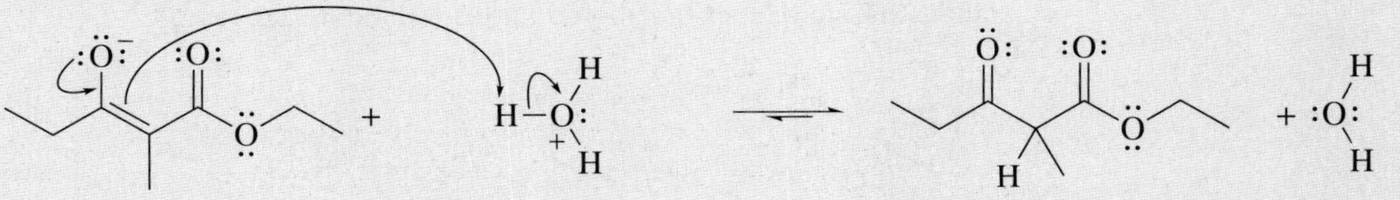

Conjugate base of ethyl 2-methyl-3-oxopentanoate + Hydronium ion (stronger acid: pK_a = −1.7) ⇌ Ethyl 2-methyl-3-oxopentanoate (weaker acid: pK_a ≈ 11) + Water

Problem 21.9

One of the following esters cannot undergo the Claisen condensation. Which one? Write structural formulas for the Claisen condensation products of the other two.

(a) 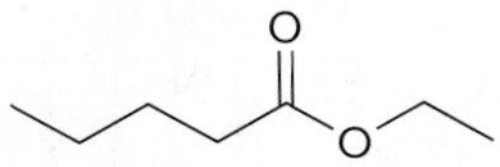 Ethyl pentanoate (b) 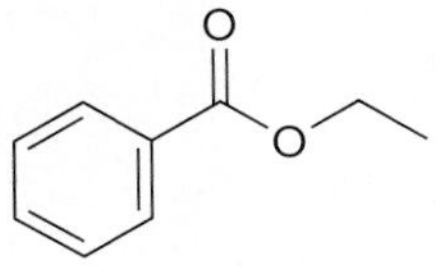 Ethyl benzoate (c) 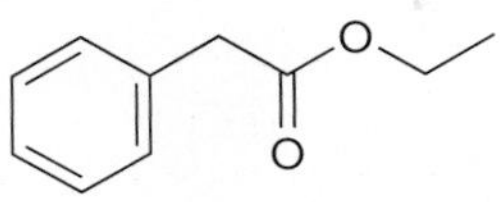 Ethyl phenylacetate

Esters of dicarboxylic acids can undergo an *intramolecular* version of the Claisen condensation, called a **Dieckmann cyclization,** if it leads to a five- or six-membered ring.

Walter Dieckmann was a German chemist and a contemporary of Claisen.

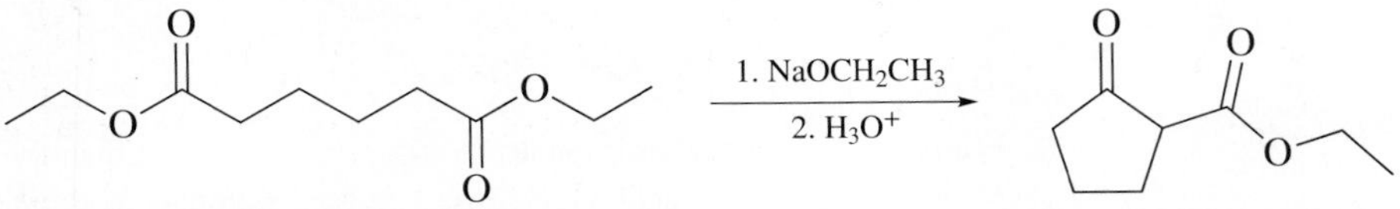

Diethyl hexanedioate → 1. $NaOCH_2CH_3$ 2. H_3O^+ → Ethyl (2-oxocyclopentane)-carboxylate (74–81%)

The α carbon of the enolate unit acts as a nucleophile toward the carbonyl carbon of the other.

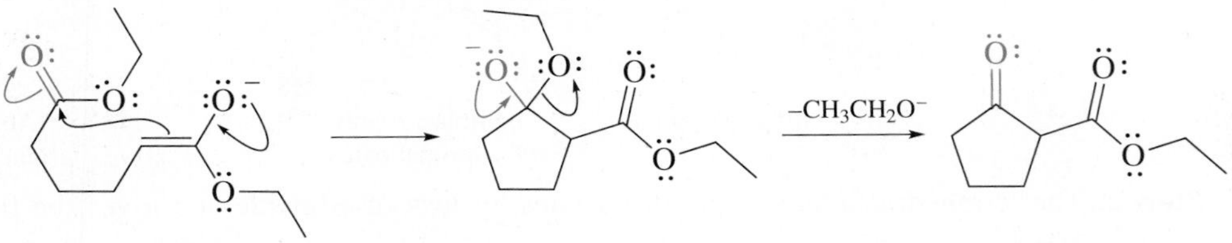

Enolate of diethyl hexanedioate → → $-CH_3CH_2O^-$ → Ethyl (2-oxocyclo-pentane)carboxylate

Problem 21.10

Write the structure of the Dieckmann cyclization product formed on treatment of each of the following diesters with sodium ethoxide, followed by acidification.

(a)

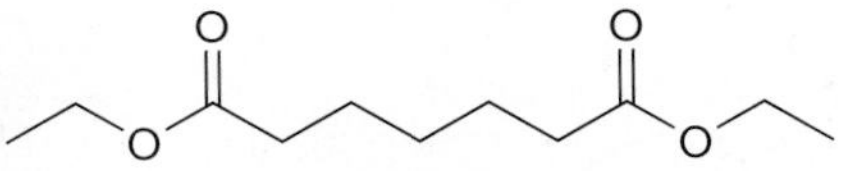

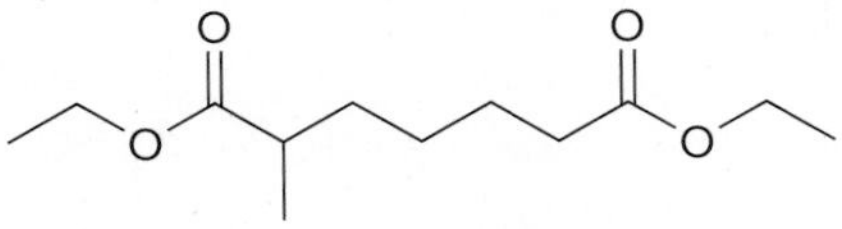

(b)

(c)

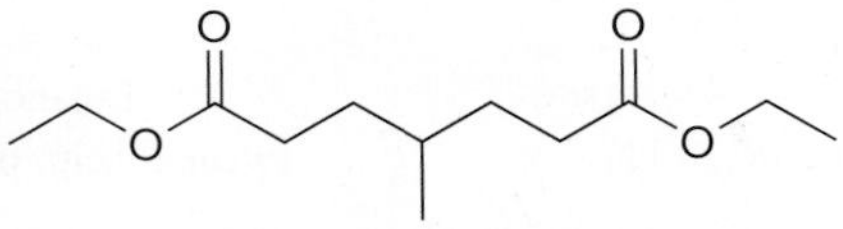

Sample Solution (a) The cyclization involves one C=O unit and the carbon that is α to the other. Therefore, a six-membered ring is formed in this case.

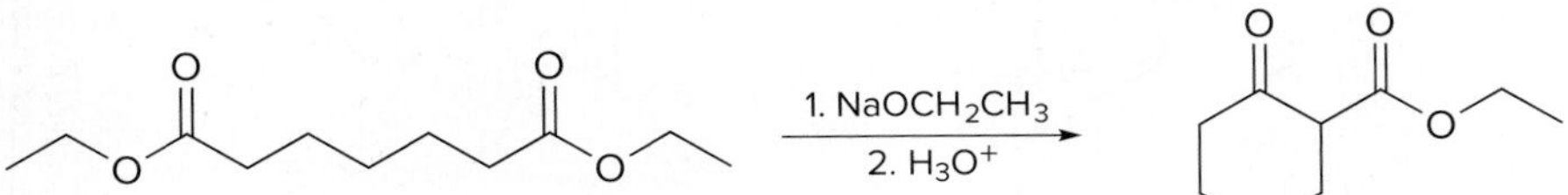

Diethyl heptanedioate → 1. $NaOCH_2CH_3$ 2. H_3O^+ → Ethyl 2-oxo(cyclohexane)-carboxylate)

Mixed Claisen condensations involve C—C bond formation between the α carbon of one ester and the carbonyl carbon of another. The product is a β-keto ester.

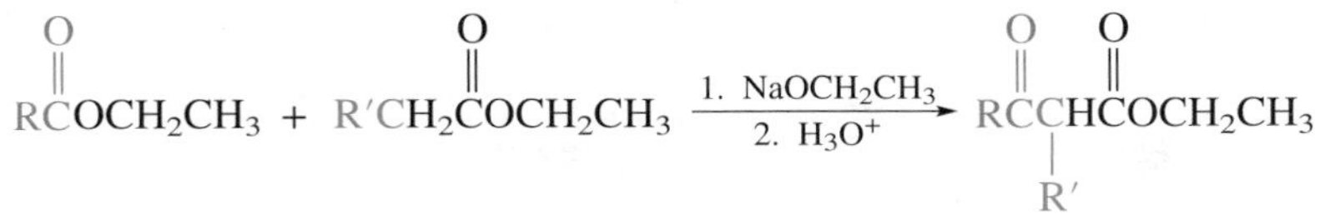

$$\underset{\text{Ester}}{RCOCH_2CH_3} + \underset{\text{Another ester}}{R'CH_2COCH_2CH_3} \xrightarrow[2.\ H_3O^+]{1.\ NaOCH_2CH_3} \underset{\beta\text{-Keto ester}}{RCCHCOCH_2CH_3 \atop |\ R'}$$

Esters that lack α hydrogens, methyl benzoate for example, cannot form enolates so are good candidates for one of the reactants.

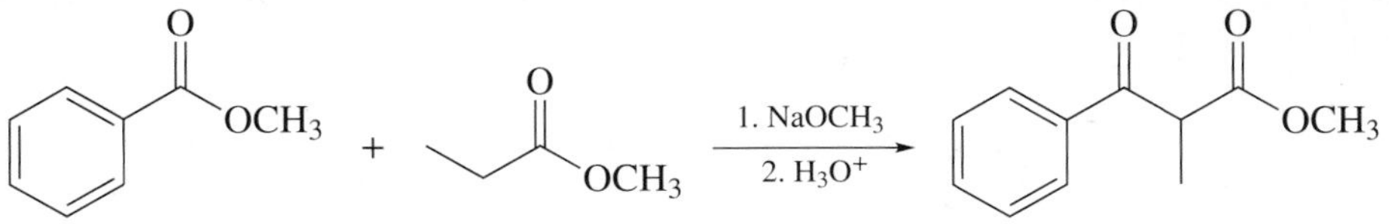

Methyl benzoate (cannot form an enolate) + Methyl propanoate $\xrightarrow[2.\ H_3O^+]{1.\ NaOCH_3}$ Methyl 2-methyl-3-oxo-3-phenylpropanoate (60%)

Problem 21.11

Give the structure of the product obtained when ethyl phenylacetate is treated with each of the following esters under conditions of the mixed Claisen condensation.

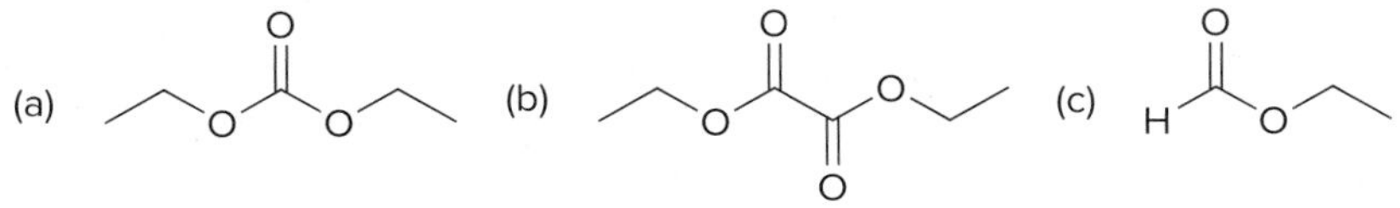

(a) Diethyl carbonate (b) Diethyl oxalate (c) Ethyl formate

Sample Solution (a) Diethyl carbonate cannot form an enolate, but ethyl phenylacetate can. Nucleophilic acyl substitution of diethyl carbonate by the enolate of ethyl phenylacetate yields a diester.

Bond formation is between α carbon of ethyl phenylacetate and carbonyl carbon of diethyl carbonate

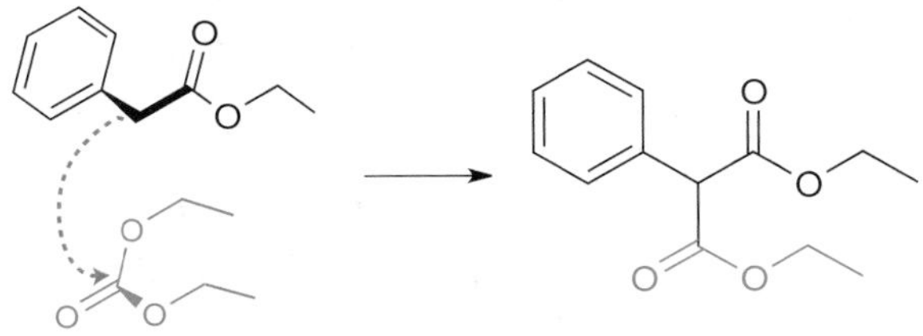

The reaction proceeds in good yield (86%), and the product (diethyl phenylmalonate) is useful in further synthetic transformations of the type to be described in Section 21.5.

In a reaction related to the mixed Claisen condensation, esters that cannot form enolates react with enolates of ketones by nucleophilic acyl substitution to give β-dicarbonyl compounds.

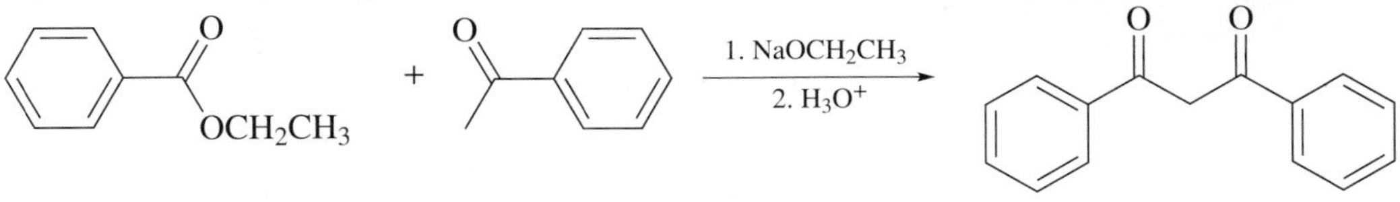

Ethyl benzoate (OCH_2CH_3) + Acetophenone $\xrightarrow[2.\ H_3O^+]{1.\ NaOCH_2CH_3}$ 1,3-Diphenyl-1,3-propanedione (62–71%)

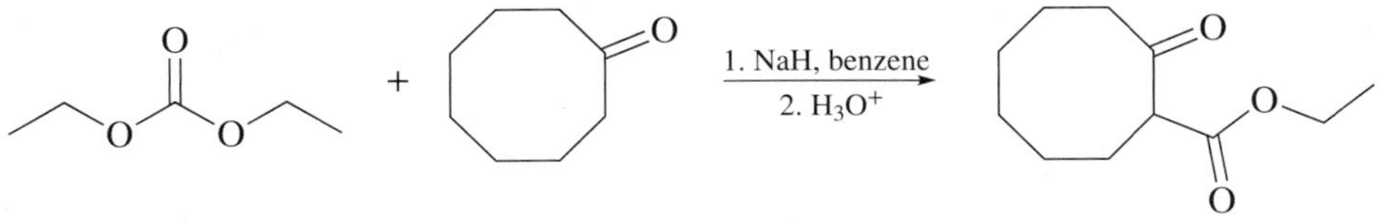

Diethyl carbonate + Cyclooctanone $\xrightarrow[2.\ H_3O^+]{1.\ NaH,\ benzene}$ 2-Carboethoxycyclooctanone (91–94%)

Like the Dieckmann cyclization, intramolecular C—C bond formation can occur if a five- or six-membered ring results.

1. $NaOCH_3$
2. H_3O^+

Ethyl 4-oxohexanoate

2-Methyl-1,3-cyclopentanedione (70–71%)

Problem 21.12

Write an equation for the carbon–carbon bond-forming step in the cyclization just cited. Show clearly the structure of the enolate ion, and use curved arrows to represent its nucleophilic addition to the appropriate carbonyl group. Write a second equation showing dissociation of the tetrahedral intermediate formed in the carbon–carbon bond-forming step.

21.5 Alkylation of Enolates: The Acetoacetic Ester and Malonic Ester Syntheses

Simple aldehyde, ketone, and ester enolates are relatively basic, and their alkylation is limited to methyl and primary alkyl halides; secondary and tertiary alkyl halides undergo elimination. Even when alkylation is possible, other factors intervene that can reduce its effectiveness as a synthetic tool. It is not always possible to limit the reaction to monoalkylation, and aldol addition can compete with alkylation. With unsymmetrical ketones, *regioselectivity* becomes a consideration. For example, a strong, hindered base such as lithium diisopropylamide (LDA) exhibits a preference for abstracting a proton from the less-substituted α carbon of 2-methylcyclohexanone to form the enolate with the less-substituted double bond, termed the kinetic enolate. The isomeric enolate with the more-substituted double bond, termed the thermodynamic enolate, is also formed to some degree. Regioisomeric products are formed on alkylation with benzyl bromide.

1. LDA, 1,2-dimethoxyethane −78°C to 30°C
2. $C_6H_5CH_2Br$, 40°C

2-Methylcyclohexanone

2-Benzyl-6-methylcyclohexanone (≈52% yield) ($CH_2C_6H_5$)

+

2-Benzyl-2-methylcyclohexanone (≈7% yield) ($CH_2C_6H_5$)

Problem 21.13

Write the structures of the kinetic and thermodynamic enolates formed in the alkylation of 2-methylcyclohexanone.

On the other hand, alkylation of the β-diketone 2,4-pentanedione is regiospecific for the position between the two carbonyl groups.

2,4-Pentanedione + Methyl iodide (CH_3I) $\xrightarrow{K_2CO_3}$ 3-Methyl-2,4-pentanedione (only product: 75–77% yield)

With a pK_a of 9, the protons attached to C-3 of 2,4-pentanedione are much more acidic than the protons of C-1 and C-5 (pK_a ≈ 19). This increased acidity reflects stabilization of the enolate by electron delocalization involving both carbonyl groups.

Similar stabilization and reduced basicity of the enolates of β-keto esters and esters of malonic acid makes possible their efficient alkylation as well.

Two procedures called the **acetoacetic ester synthesis** and the **malonic ester synthesis** take advantage of the properties of β-dicarbonyl compounds and are standard methods for making carbon–carbon bonds. Both begin with alkylation of the enolate. Ethyl esters are normally used, with sodium ethoxide as the base.

Acetoacetic ester is a common name for ethyl acetoacetate. Its systematic, but rarely used, name is ethyl 3-oxobutanoate. Malonic ester is a common name for diethyl malonate, which is an acceptable alternative for diethyl propanedioate.

Alkylation phase:

1. $NaOCH_2CH_3$, ethanol
2. $CH_3CH_2CH_2CH_2Br$

Ethyl acetoacetate → Ethyl 2-butyl-3-oxobutanoate (70%)

1. $NaOCH_2CH_3$, ethanol
2. $H_2C{=}CHCH_2CH_2CH_2Br$

Diethyl malonate → Diethyl 2-(4-pentenyl)malonate (85%)

Methyl, primary, and unhindered secondary alkyl halides are satisfactory alkylating agents. Elimination (E2) is the only reaction with tertiary alkyl halides.

After alkylation, hydrolysis in aqueous base and acidification, followed by heating, leads to decarboxylation. The alkylated β-keto ester yields a ketone; the alkylated malonic ester gives a carboxylic acid.

Hydrolysis and decarboxylation phase:

1. NaOH, H_2O
2. H_3O^+

heat, $-CO_2$

Ethyl 2-butyl-3-oxobutanoate → 2-Butyl-3-oxobutanoic acid → 2-Heptanone (60%)

1. NaOH, H_2O
2. H_3O^+

heat, $-CO_2$

Diethyl 2-(4-pentenyl)malonate → 2-(4-Pentenyl)malonic acid → 6-Heptenoic acid (75%)

Decarboxylation of malonic acid and β-keto acids was introduced in Section 19.16.

Problem 21.14

What is the product of each of the following reaction sequences?

(a) [ethyl acetoacetate] 1. $NaOCH_2CH_3$ 2. $C_6H_5CH_2Cl$ 3. NaOH 4. H_3O^+ 5. heat →

(b) [diethyl malonate] 1. $NaOCH_2CH_3$ 2. $(CH_3)_2CHCH_2Br$ 3. NaOH 4. H_3O^+ 5. heat →

A convenient way to determine when to apply the acetoacetic ester or malonic ester approach to a synthetic problem is to incorporate the synthon concept into retrosynthetic analysis. A **synthon** is a structural unit in a molecule that is related to a synthetic operation. A CH_2CO_2H group is a synthon that alerts us to the possibility of preparing a target by a malonic ester synthesis. Likewise, a $CH_2C(O)CH_3$ group is a synthon that suggests an acetoacetic synthesis.

Problem 21.15

Use retrosynthetic analysis to choose the appropriate β-dicarbonyl compound and alkyl halide to prepare each of the following:

(a) 3-Methylpentanoic acid
(b) 1-Phenyl-1,4-pentanedione
(c) 4-Methylhexanoic acid
(d) 5-Hexen-2-one

Sample Solution (a) Locate the appropriate synthon and mentally disconnect the bond to its α carbon. The synthon is derived from diethyl malonate, the remainder of the molecule comes from the alkyl halide.

Disconnect here

3-Methylpentanoic acid ⟹ Alkyl halide (X) + Derived from diethyl malonate (OH)

In this case, a secondary alkyl halide is needed as the alkylating agent. The anion of diethyl malonate is a relatively weak base (the pK_a of its conjugate acid = 11) and reacts with secondary alkyl halides by substitution rather than elimination. Thus, the synthesis begins with the alkylation of the anion of diethyl malonate with 2-bromobutane.

Diethyl malonate + 2-Bromobutane (Br) → 1. $NaOCH_2CH_3$ 2. NaOH 3. H_3O^+ 4. heat → 3-Methylpentanoic acid

As actually carried out, the alkylation phase proceeded in 83–84% yield, and the product of that reaction was converted to 3-methylpentanoic acid by saponification, acidification, and decarboxylation in 62–65% yield.

By carrying out successive alkylations, two different alkyl groups can be added to the α carbon. For the case of diethyl malonate:

Diethyl malonate $\xrightarrow[\text{2. } CH_3Br]{\text{1. } NaOCH_2CH_3\text{, ethanol}}$ Diethyl 2-methylmalonate

then

Diethyl 2-methylmalonate $\xrightarrow{}$ 2-Methyldodecanoic acid (61–74%)

1. $NaOCH_2CH_3$, ethanol
2. $CH_3(CH_2)_8CH_2Br$
3. KOH, ethanol–water
4. H_3O^+
5. heat

HO–C(=O)–CH(CH_3)–$CH_2(CH_2)_8CH_3$

Problem 21.16

Outline a synthesis of the compound shown.

Alkylation of diethyl malonate with dihalides is used to prepare cycloalkanecarboxylic acids when the ring has seven carbons or fewer.

1,3-Dibromopropane + Diethyl malonate $\xrightarrow[\text{ethanol}]{NaOCH_2CH_3}$ Diethyl 1,1-cyclobutanedicarboxylate $\xrightarrow[\text{2. heat}]{\text{1. } H_3O^+}$ Cyclobutanecarboxylic acid

Problem 21.17

Design a synthesis of cyclopentyl methyl ketone based on the partial retrosynthesis:

21.6 Enol Content and Enolization

Keto ⇌ enol equilibration is a property of carbonyl compounds that contain a proton on their α carbon and normally favors the keto form (Table 21.2). Simple aldehydes and ketones exist almost entirely in their keto forms; acetaldehyde contains less than 1 ppm of its enol, and acetone contains 100 times less than that. The enol content of acetic acid and methyl acetate is even smaller because their keto isomers are stabilized by electron release from OH and OCH_3, respectively, to C=O; the enols are not.

Enols were introduced in Section 9.11 as reactive intermediates in the hydration of alkynes.

TABLE 21.2 Enolization Equilibria (keto ⇌ enol) of Some Carbonyl Compounds*

Compound	K
Methyl acetate	$K \approx 10^{-21}$
Acetic acid	$K \approx 10^{-20}$
Acetone	$K = 6 \times 10^{-9}$
Acetaldehyde	$K = 6 \times 10^{-7}$
2-Methylpropanal	$K = 1.4 \times 10^{-4}$
Ethyl acetoacetate	$K = 7 \times 10^{-2}$
2,4-Pentanedione	$K = 2 \times 10^{-1}$

*In water, 25°C.

Problem 21.18

Write structural formulas for the enol isomers of each of the following.

(a) 2,2-Dimethyl-3-pentanone
(b) Acetophenone
(c) 2-Methylcyclohexanone
(d) Methyl vinyl ketone

Sample Solution (a) Only one of the α carbons of 2,2-dimethyl-3-pentanone has an attached hydrogen, so only one constitutional isomer is possible for the enol. *E* and *Z* stereoisomers are possible.

2,2-Dimethyl-3-pentanone (keto) ⇌ (*Z*)-4,4-Dimethyl-2-penten-3-ol (enol) + (*E*)-4,4-Dimethyl-2-penten-3-ol (enol)

The far higher enol content in the β-dicarbonyl compounds of Table 21.2—almost 7% in ethyl acetoacetate and 20% in 2,4-pentanedione—reflects the stabilization of the enol by conjugation of the carbon–carbon double bond with the other carbonyl group plus intramolecular hydrogen bonding of the enolic OH with C=O. Both features are apparent in the structure of the enol of 2,4-pentanedione shown in Figure 21.1. Analogous structural features stabilize the enol of ethyl acetoacetate.

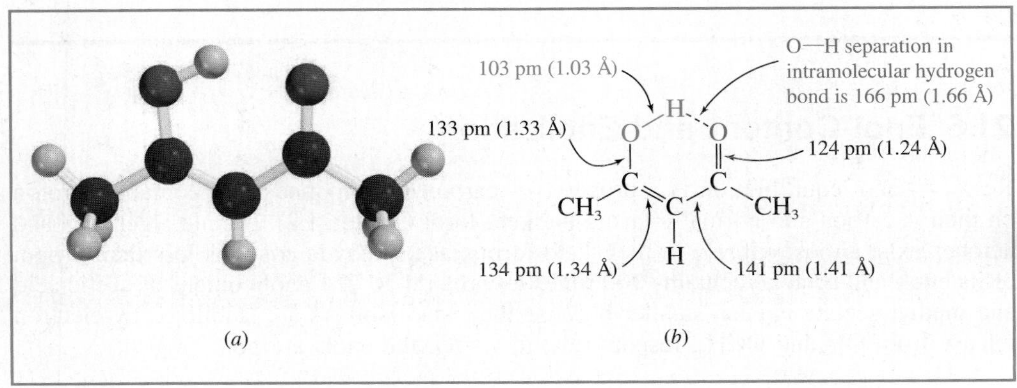

Figure 21.1

(*a*) A molecular model and (*b*) bond distances in the enol of 2,4-pentanedione.

Problem 21.19

Write structural formulas for the two most stable enol isomers of the β-dicarbonyl compounds shown in a form that reflects an intramolecular hydrogen bond.

(a) 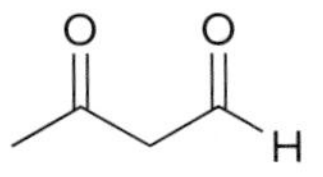 (b) 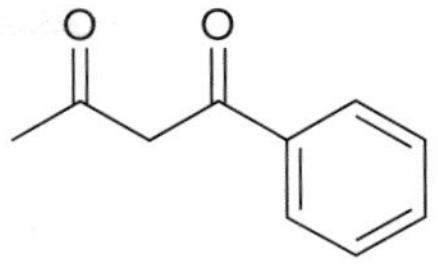 (c)

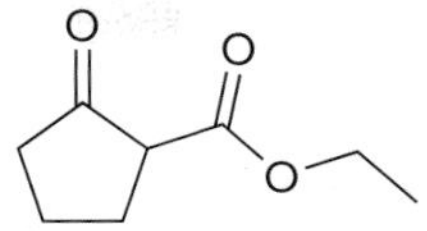

Sample Solution (a) Enolization in β-dicarbonyl compounds involves the carbon between the two carbonyl groups. Hydrogen bonding occurs between OH and C=O.

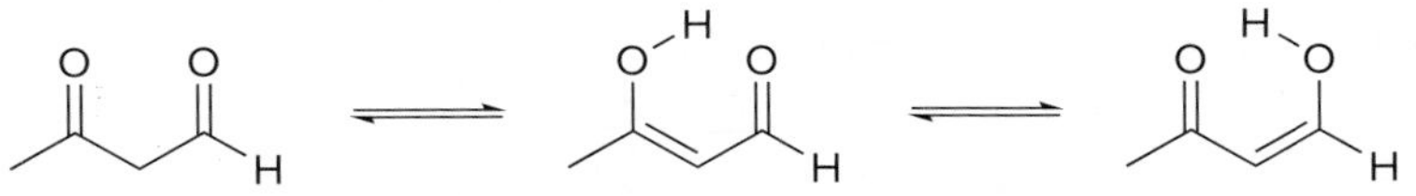

Problem 21.20

K for enolization of 2,4-cyclohexadienone is about 10^{13}. Explain why the enol is so much more stable than the keto tautomer.

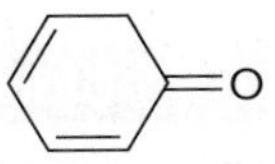

Mechanism 21.3 describes the two steps by which acid-catalyzed keto–enol equilibration occurs; the first is protonation of the carbonyl oxygen, the second is removal of a proton from the α carbon.

Mechanism 21.3

Acid-Catalyzed Enolization of 2-Methylpropanal

THE OVERALL REACTION:

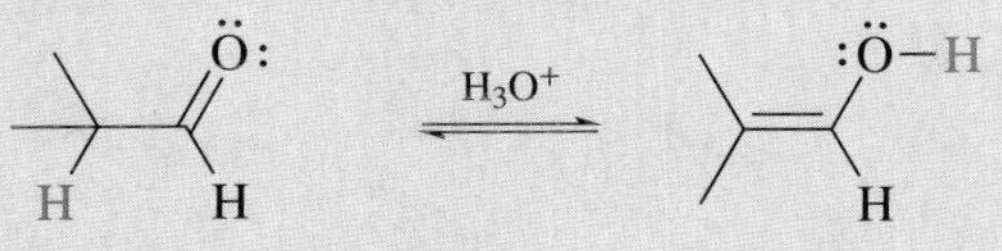

H_3O^+

2-Methylpropanal   Enol

THE MECHANISM:

Step 1: In aqueous solution, hydronium ion transfers a proton to the carbonyl oxygen.

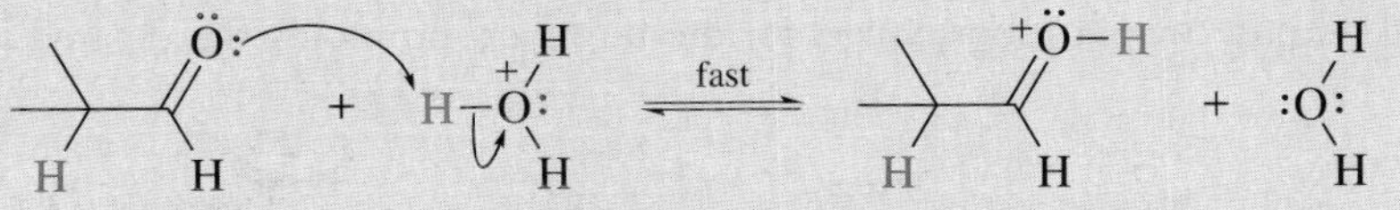

fast

2-Methylpropanal   Hydronium ion   Conjugate acid of 2-methylpropanal   Water

Step 2: A water molecule acts as a Brønsted base to remove a proton from the α carbon of the protonated carbonyl.

slow

Water   Conjugate acid of 2-methylpropanal   Hydronium ion   Enol

The studies that led to this mechanism were carried out in London over a century ago by Arthur Lapworth, who found that the rate of halogenation of acetone is the same regardless of the halogen and its concentration.

Acetone + X_2 (Chlorine, bromine, or iodine) ⟶ Halogen-substituted acetone + HX (Hydrogen halide)

Lapworth reasoned that an enol was a reactive intermediate, that its formation was rate-determining, and that the halogen didn't participate until after the rate-determining step. These early studies provided the basis for the generally accepted mechanism for the α-halogenation of aldehydes and ketones.

slow

fast

fast

Problem 21.21

Chlorination of 2-butanone yields two isomeric products, each having the molecular formula C_4H_7ClO.

(a) What are these two compounds?

(b) Write structural formulas for the enol intermediates that lead to each of these compounds.

(c) Using curved arrows, show the flow of electrons in the reaction of each of the enols with Cl_2.

Synthetically, halogenation of aldehydes and ketones is a useful first step toward introducing other substituents at the α carbon of aldehydes and ketones.

Cyclohexanone $\xrightarrow[H_2O]{Cl_2}$ 2-Chlorocyclohexanone (61–66%) $\xrightarrow[K_2CO_3]{H_2O}$ 2-Hydroxycyclohexanone (76%)

The enol content of carboxylic acids is far smaller than that of aldehydes and ketones and their α halogenation under the usual conditions is not feasible. Adding a small amount of phosphorus trichloride, however, promotes the desired halogenation.

This procedure is called the Hell–Volhard–Zelinsky reaction after its discoverers. In an alternative method, elemental phosphorus is used which reacts with the halogen to generate the phosphorus trihalide.

Phenylacetic acid $\xrightarrow[\text{benzene, 80°C}]{Br_2,\ PCl_3}$ 2-Bromo-2-phenylacetic acid (60–62%)

Phosphorus trichloride converts the carboxylic acid to the corresponding acyl chloride, which has a less stabilized carbonyl group (see Section 20.2) and, therefore, a greater enol content. Bromination of this enol takes place, followed by a process that gives the α-bromo acid plus another molecule of the acyl chloride, which then undergoes enolization, bromination, and so on.

Problem 21.22

α-Halogenation of 3-methylbutanoic acid has been used to prepare the amino acid valine.

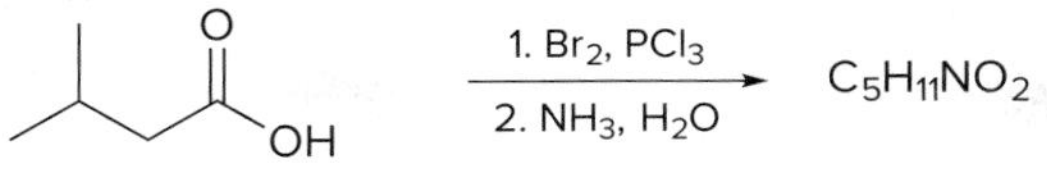

3-Methylbutanoic acid —(1. Br_2, PCl_3; 2. NH_3, H_2O)→ $C_5H_{11}NO_2$ Valine

What is the structure of valine? When isolated from proteins, natural valine has a specific rotation [α] of +26°. What is the rotation of valine prepared via halogenation of 3-methylbutanoic acid?

21.7 The Haloform Reaction

Aldehydes and ketones undergo α halogenation in neutral and acidic media via an enol intermediate. A similar reaction occurs in basic solution except, in this case, the reactive intermediate is an enolate ion.

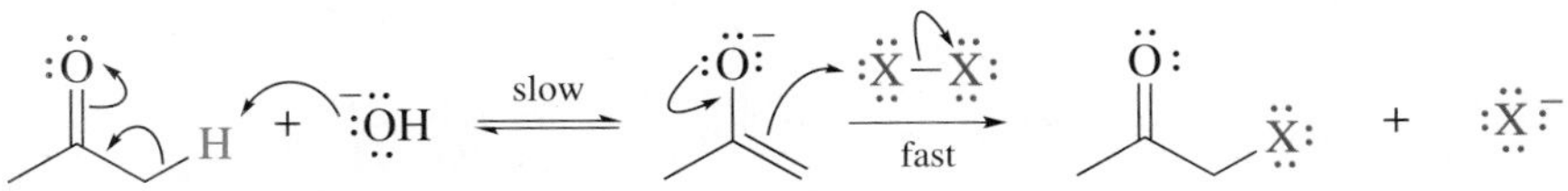

Aldehyde or ketone + $^-$:ÖH ⇌(slow) Enolate —(:X–X:, fast)→ α-Halo aldehyde or ketone + $:X:^-$

Unlike its acid-catalyzed counterpart, this reaction is difficult to limit to monohalogenation.

Methyl ketones undergo a novel C—C cleavage on treatment with halogens in aqueous base that finds some use as a synthesis of carboxylic acids.

$$\underset{\text{Methyl ketone}}{RCCH_3} + \underset{\text{Halogen}}{3X_2} + \underset{\text{Hydroxide ion}}{4HO^-} \longrightarrow \underset{\text{Carboxylate ion}}{RCO^-} + \underset{\text{Trihalomethane}}{CHX_3} + \underset{\text{Halide ion}}{3X^-} + \underset{\text{Water}}{3H_2O}$$

The reaction is called the **haloform reaction** because the trihalomethane produced is chloroform ($CHCl_3$), bromoform ($CHBr_3$), or iodoform (CHI_3), depending on the halogen used.

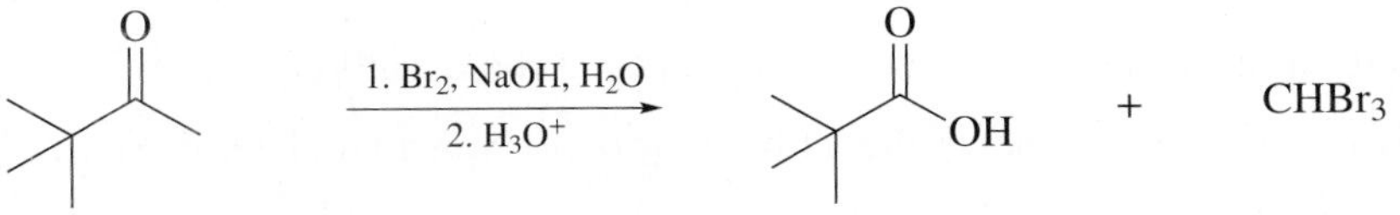

3,3-Dimethyl-2-butanone —(1. Br_2, NaOH, H_2O; 2. H_3O^+)→ 2,2-Dimethylpropanoic acid (71–74%) + $CHBr_3$ Tribromomethane (bromoform)

The methyl ketone shown in the example can enolize in only one direction and typifies the kind of reactant that can be converted to a carboxylic acid in synthetically acceptable yield by the haloform reaction. Mechanism 21.4 describes how this cleavage occurs.

Methyl ketones of the type $RCH_2C(O)CH_3$ and $R_2CHC(O)CH_3$ undergo non-regioselective α halogenation and give a mixture of products.

Problem 21.23

Which of the following is the most suitable for preparing a carboxylic acid by the haloform reaction? Give the structure of the carboxylic acid.

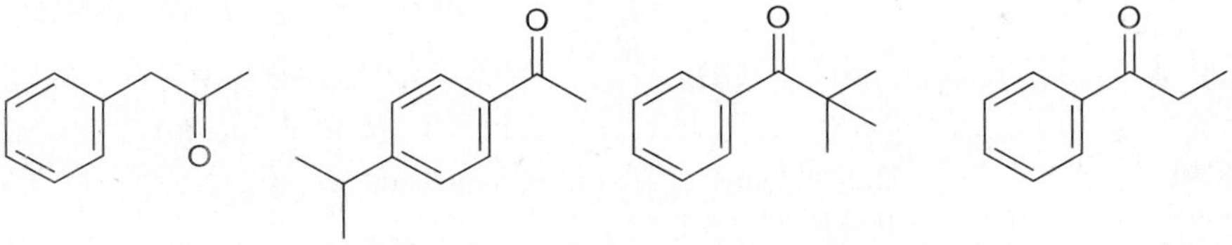

Mechanism 21.4

The Haloform Reaction

THE OVERALL REACTION:

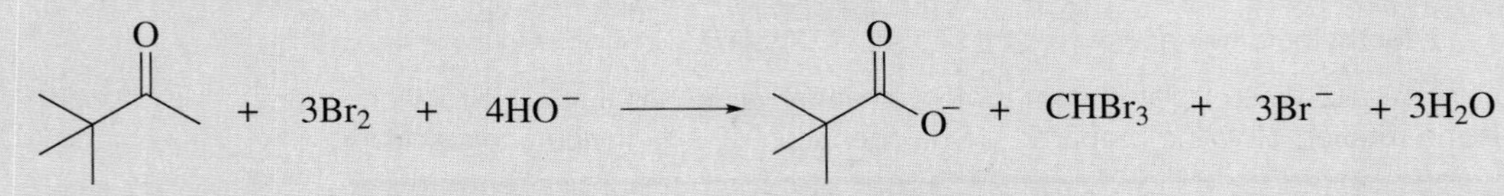

3,3-Dimethyl-2-butanone + $3Br_2$ (Bromine) + $4HO^-$ (Hydroxide ion) ⟶ 2,2-Dimethylpropanoate ion + $CHBr_3$ (Tribromomethane) + $3Br^-$ (Bromide ion) + $3H_2O$ (Water)

THE MECHANISM:

Step 1: The ketone is converted to its enolate by proton abstraction from the α carbon.

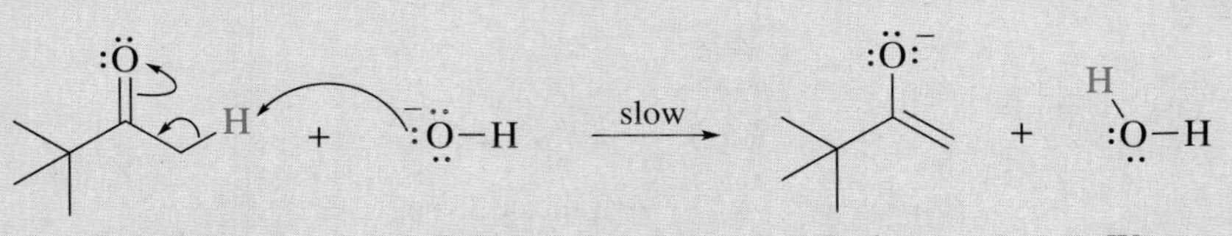

3,3-Dimethyl-2-butanone + Hydroxide ion —slow⟶ Enolate + Water

Step 2: The enolate reacts with bromine to give the α bromoketone.

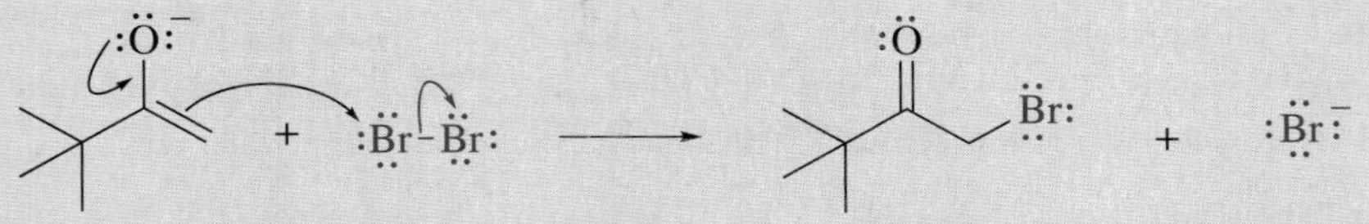

Enolate + Bromine ⟶ 1-Bromo-3,3-dimethyl-2-butanone + Bromide ion

Steps 3–6: Steps 1 and 2 repeat twice more to introduce two additional bromines at the α carbon.

Step 7: Nucleophilic addition of hydroxide to the double bond of the carbonyl group gives a tetrahedral intermediate.

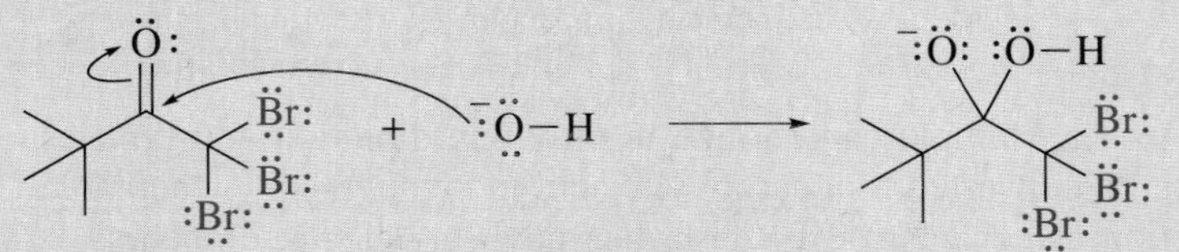

1,1,1-Tribromo-3,3-dimethyl-2-butanone + Hydroxide ion ⟶ Tetrahedral intermediate

Step 8: Expulsion of a tribromomethide ion from the tetrahedral intermediate restores the carbon–oxygen double bond.

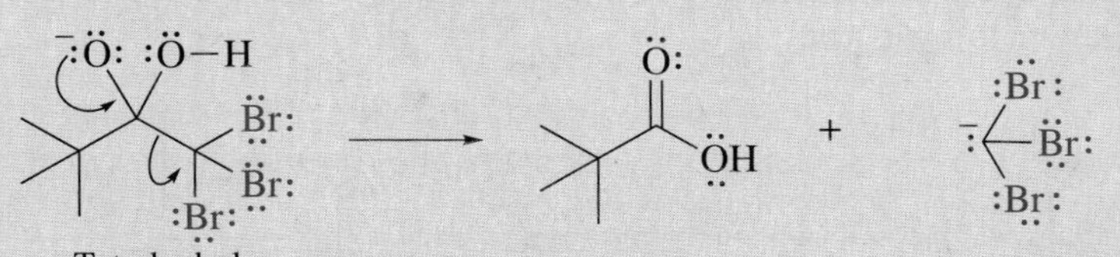

Tetrahedral intermediate ⟶ 2,2-Dimethylpropanoic acid + Tribromomethide ion

Step 9: Both 2,2-dimethylpropanoic acid and tribromomethide ion undergo proton-transfer reactions to give the carboxylate and bromoform.

2,2-Dimethylpropanoic acid + Hydroxide ion ⟶ 2,2-Dimethylpropanoate ion + Water

Water + Tribromomethide ion ⟶ Hydroxide ion + Bromoform

21.8 Some Chemical and Stereochemical Consequences of Enolization

A number of novel reactions involving the α-carbon atom of aldehydes and ketones involve enol and enolate anion intermediates.

Substitution of deuterium for hydrogen at the α-carbon atom of an aldehyde or a ketone is a convenient way to introduce an isotopic label into a molecule and is readily carried out by treating the carbonyl compound with deuterium oxide (D_2O) and base.

$$\text{Cyclopentanone} + 4D_2O \xrightarrow[\text{reflux}]{\text{KOD}} \text{Cyclopentanone-2,2,5,5-}d_4 + 4\text{DOH}$$

Only the α hydrogens are replaced by deuterium in this reaction. The key intermediate is the enolate ion formed by proton abstraction from the α-carbon atom of cyclopentanone. Transfer of deuterium from the solvent D_2O to the enolate gives cyclopentanone containing a deuterium atom in place of one of the hydrogens at the α carbon.

Formation of the enolate

Cyclopentanone + $^-$:ÖD ⇌ Enolate of cyclopentanone + HÖD

Deuterium transfer to the enolate

Enolate of cyclopentanone + D—Ö—D ⇌ Cyclopentanone-2-d_1 + $^-$:ÖD

In excess D_2O the process continues until all four α protons are eventually replaced by deuterium.

Problem 21.24

After the compound shown was heated in D_2O containing K_2CO_3 at 70°C, the only signals that were observed in its ^1H NMR spectrum were at δ 3.9 ppm for the protons in the OCH_3 group and at δ 6.7–6.9 ppm for those on the aromatic ring. What happened?

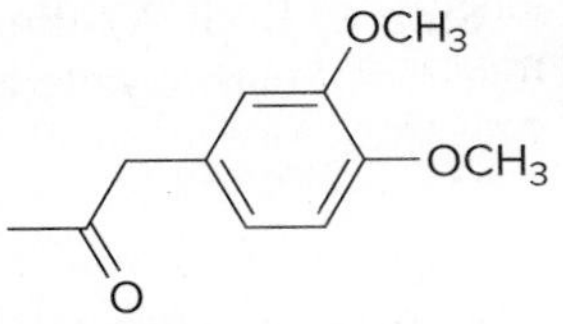

If the α-carbon atom of an aldehyde or a ketone is a chirality center, its stereochemical integrity is lost on enolization. Enolization of optically active *sec*-butyl phenyl ketone leads to its racemization by way of the achiral enol form.

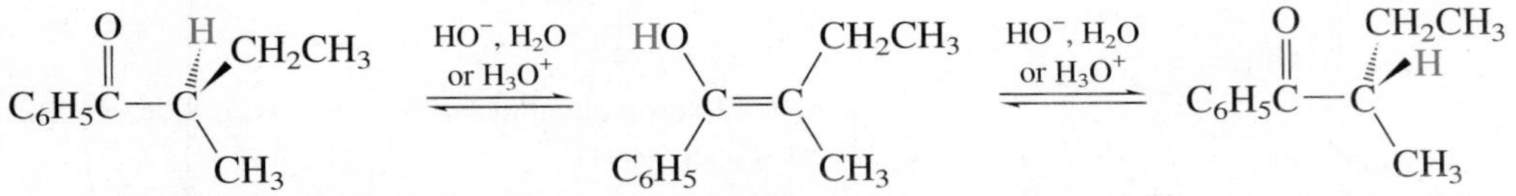

(*R*)-*sec*-Butyl phenyl ketone $\underset{\text{or } H_3O^+}{\overset{HO^-, H_2O}{\rightleftharpoons}}$ Achiral enol $\underset{\text{or } H_3O^+}{\overset{HO^-, H_2O}{\rightleftharpoons}}$ (*S*)-*sec*-Butyl phenyl ketone

21.9 Conjugation Effects in α,β-Unsaturated Aldehydes and Ketones

Aldol condensation (Section 21.3) offers an effective route to α,β-unsaturated aldehydes and ketones, compounds that have interesting and useful properties that result from conjugation of the carbon–carbon double bond with the carbonyl group.

First, in common with other conjugated π-electron systems that we have seen, α,β-unsaturated aldehydes and ketones are more stable than their nonconjugated isomers. Under conditions chosen to bring about their interconversion, for example, the equilibrium between a β,γ-unsaturated carbonyl and its α,β-unsaturated analog favors the conjugated isomer.

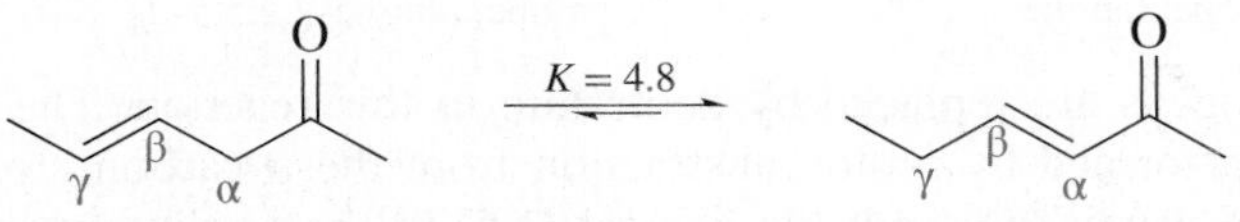

(*E*)-4-Hexen-2-one (17%) $\overset{K = 4.8}{\rightleftharpoons}$ (*E*)-3-Hexen-2-one (83%)

Problem 21.25

Mesityl oxide is an industrial chemical prepared by an aldol condensation. From what organic starting material is mesityl oxide prepared? It often contains about 10% of an isomer with the same carbon skeleton. What is the most likely structure for this contaminant?

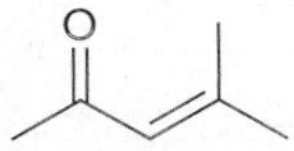

Mesityl oxide

Electron delocalization in α,β-unsaturated carbonyl compounds is represented as resonance among three principal contributors.

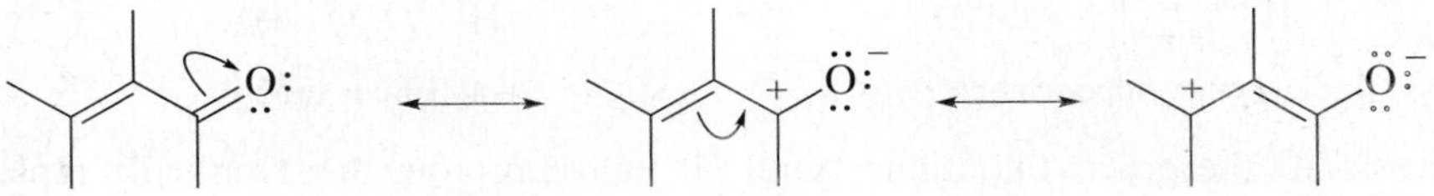

Most stable contributor ⟷ Charge-separated contributors

The positive character of the β carbon suggested by the resonance description is consistent with numerous differences between α,β-unsaturated carbonyl compounds and their nonconjugated relatives. Consistent with its greater separation of positive and negative charge, *trans*-2-butenal has a larger dipole moment than butanal.

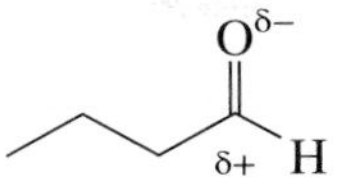

Butanal
$\mu = 2.7$ D

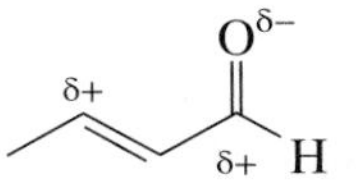

trans-2-Butenal
$\mu = 3.7$ D

Compare also the ^{13}C chemical shifts in the two. In butanal, the electron-withdrawing power of the carbonyl group deshields the α carbon more than the β, and the β more than the γ. Reflecting its increased positive character in 2-butenal, the β carbon is more deshielded than either α or γ.

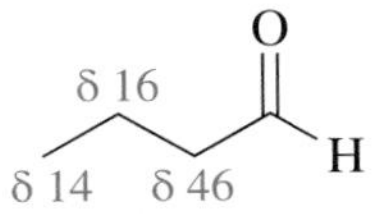

Butanal: δ 14, δ 16, δ 46

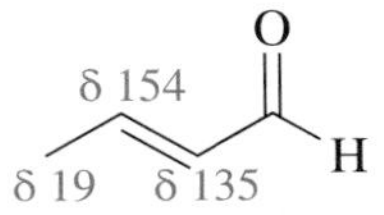

trans-2-Butenal: δ 19, δ 154, δ 135

Chemically, the diminished π-electron density in the double bond makes α,β-unsaturated aldehydes and ketones less reactive than alkenes toward *electrophilic* addition. *Nucleophilic* addition, on the other hand, can involve either the carbonyl group or the α,β double bond.

Strongly basic nucleophiles such as Grignard and organolithium reagents and lithium aluminum hydride tend to react with the carbonyl group by *1,2-addition.*

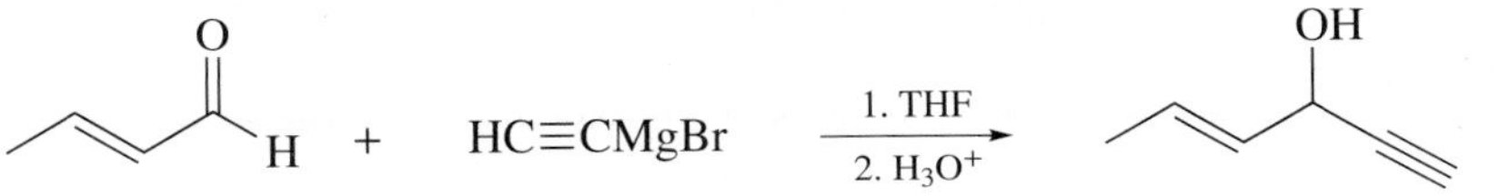

trans-2-Butenal + HC≡CMgBr (Ethynylmagnesium bromide) $\xrightarrow[2.\ H_3O^+]{1.\ THF}$ *trans*-4-Hexen-1-yn-3-ol (84%)

Weakly basic nucleophiles yield products corresponding to reaction at the β carbon of C═C by *conjugate addition* or *1,4-addition.*

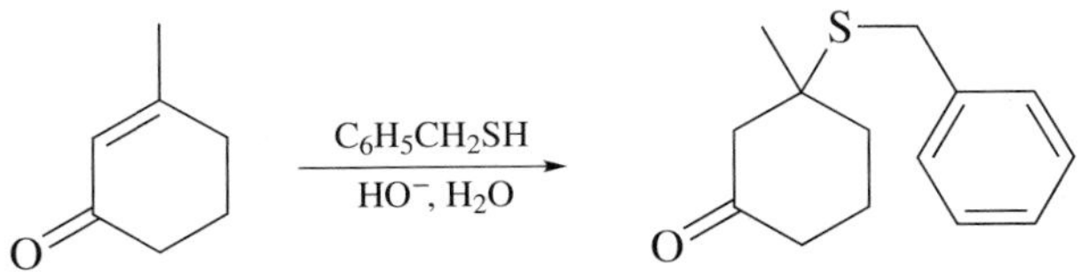

Thiols have pK_a's in the 10–11 range.

3-Methyl-2-cyclohexenone $\xrightarrow[HO^-,\ H_2O]{C_6H_5CH_2SH}$ 3-Benzylthio-3-methylcyclohexanone (58%)

The initial intermediate is an enolate, which is converted to a ketone under the reaction conditions.

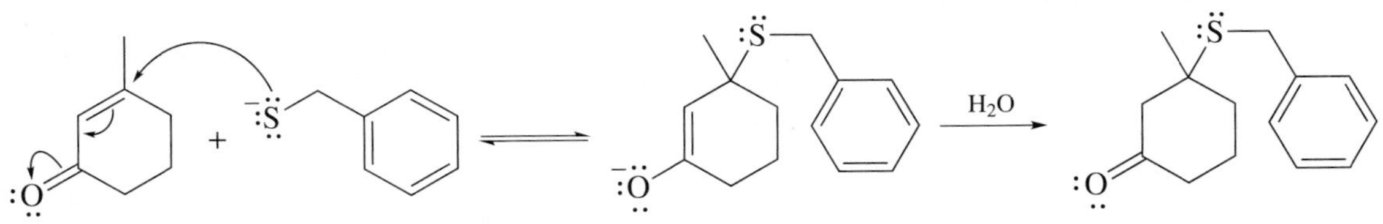

α,β-Unsaturated ketone + Weakly basic nucleophile ⇌ Enolate $\xrightarrow{H_2O}$ Product of 1,4-addition

With strongly basic nucleophiles such as a Grignard or organolithium reagent, nucleophilic addition to C═O is essentially irreversible and 1,2-addition occurs. When the nucleophile is weakly basic, attack at C═O, although rapid, is reversible. The nucleophile goes on and off the carbonyl carbon, thereby allowing the slower, but less reversible,

1,4-addition to compete. The eventual product from 1,4-addition is more stable because it retains the stronger C=O bond at the expense of the weaker C=C bond. 1,2-Addition is kinetically controlled; 1,4-addition is thermodynamically controlled.

Problem 21.26

Acrolein ($H_2C{=}CHCH{=}O$) reacts with sodium azide (NaN_3) in aqueous acetic acid to form a compound, $C_3H_5N_3O$, in 71% yield. Propanal ($CH_3CH_2CH{=}O$), when subjected to the same reaction conditions, is recovered unchanged. Suggest a structure for the product formed from acrolein, and offer an explanation for the difference in reactivity between acrolein and propanal.

The **Michael reaction** is an alkylation in which carbanions, such as the enolates derived from β-diketones, β-keto esters, and diethyl malonate, react with α,β-unsaturated ketones by conjugate addition. The α,β unsaturated ketone serves the same kind of electrophilic role that alkyl halides do toward the enolate.

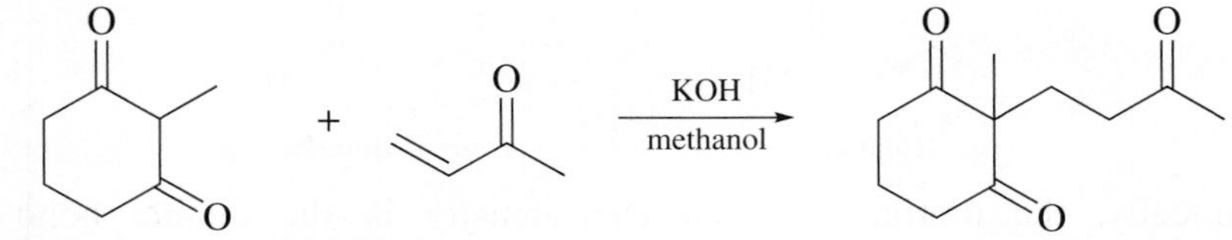

2-Methyl-1,3-cyclohexanedione + Methyl vinyl ketone $\xrightarrow[\text{methanol}]{\text{KOH}}$ 2-Methyl-2-(3′-oxobutyl)-1,3-cyclohexanedione (85%)

Problem 21.27

Outline a synthesis of the compound shown from methyl vinyl ketone and diethyl malonate.

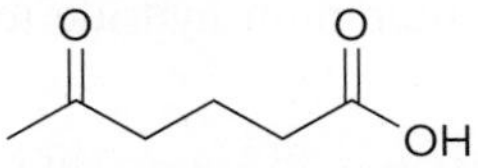

Robinson won the Nobel Prize in Chemistry in 1947 for his studies of biologically important plant products.

Michael reactions of β-diketones have proven especially useful in *annulation*—the grafting of a ring onto some starting molecule. In the **Robinson annulation,** named after Sir Robert Robinson who popularized its use, Michael addition is followed by an intramolecular aldol condensation to give a cyclohexenone.

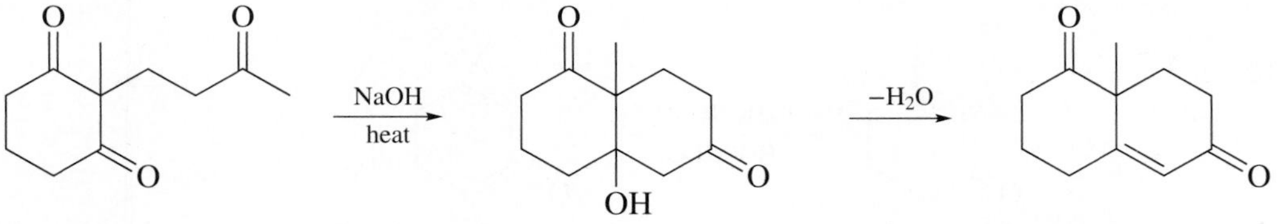

2-Methyl-2-(3′-oxobutyl)-1,3-cyclohexanedione $\xrightarrow[\text{heat}]{\text{NaOH}}$ Intramolecular aldol addition product; not isolated $\xrightarrow{-H_2O}$ 6-Methylbicyclo[4.4.0]dec-1-ene-3,7-dione

Problem 21.28

Both conjugate addition and intramolecular aldol condensation can be carried out in one synthetic operation without isolating any of the intermediates.

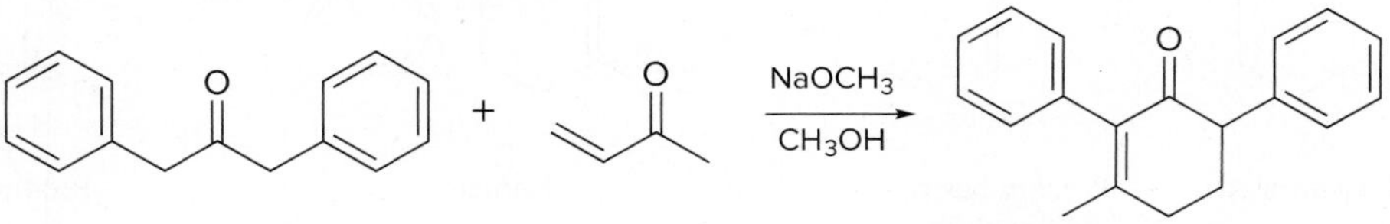

Dibenzyl ketone + Methyl vinyl ketone $\xrightarrow[\text{CH}_3\text{OH}]{\text{NaOCH}_3}$ 3-Methyl-2,6-diphenyl-2-cyclohexenone (55%)

Write structural formulas corresponding to the intermediate formed in the conjugate addition step and in the aldol condensation step.

In many of the preceding synthetic applications, α,β-unsaturated carbonyl compounds resemble alkyl halides in their reactivity toward nucleophiles.

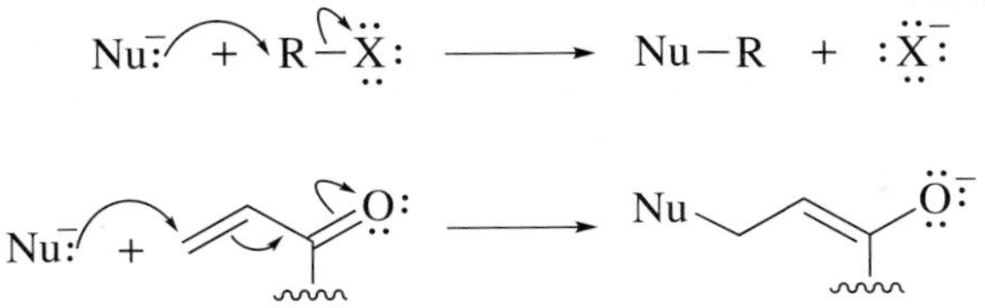

$$\text{Nu:}^- + \text{R}-\ddot{\underset{..}{\text{X}}}: \longrightarrow \text{Nu}-\text{R} + :\ddot{\underset{..}{\text{X}}}:^-$$

The analogy extends toward their reaction with lithium dialkylcuprates. Just as alkyl halides react with lithium dialkylcuprates to form carbon–carbon bonds, so do α,β-unsaturated carbonyl compounds.

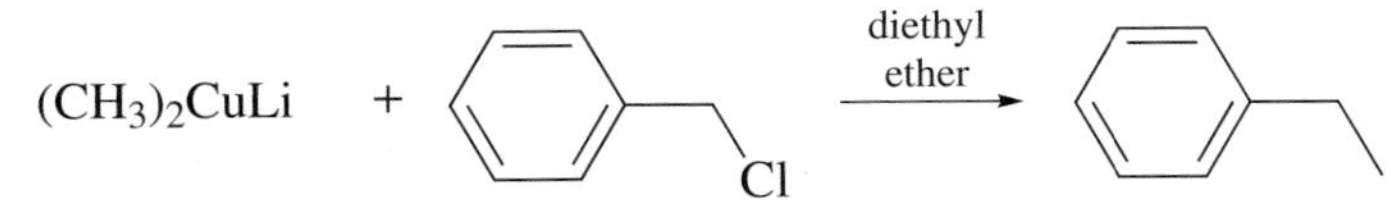

$(CH_3)_2CuLi$ + Benzyl chloride $\xrightarrow{\text{diethyl ether}}$ Ethylbenzene (80%)

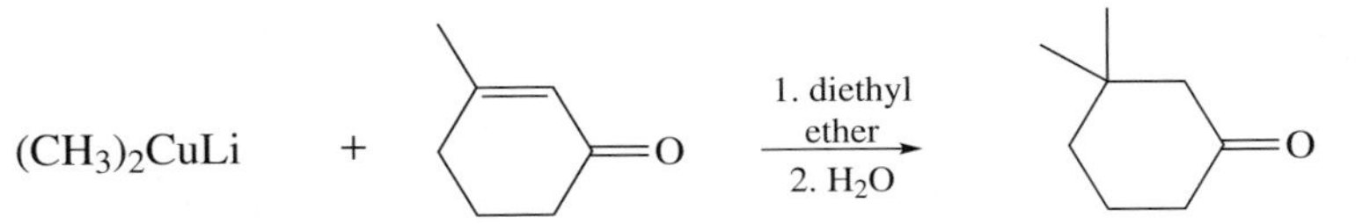

Lithium dimethylcuprate

$(CH_3)_2CuLi$ + 3-Methyl-2-cyclohexenone $\xrightarrow[\text{2. } H_2O]{\text{1. diethyl ether}}$ 3,3-Dimethylcyclohexanone (98%)

Lithium dimethylcuprate

Problem 21.29

Outline two ways in which 4-methyl-2-octanone can be prepared by conjugate addition of an organocuprate to an α,β-unsaturated ketone.

Sample Solution Mentally disconnect one of the bonds to the β carbon so as to identify the group that comes from the lithium dialkylcuprate.

Disconnect this bond

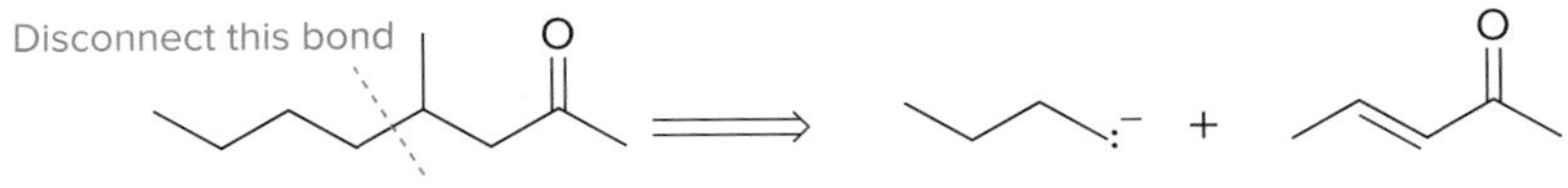

4-Methyl-2-octanone

According to this disconnection, the butyl group is derived from lithium dibutylcuprate. A suitable preparation is

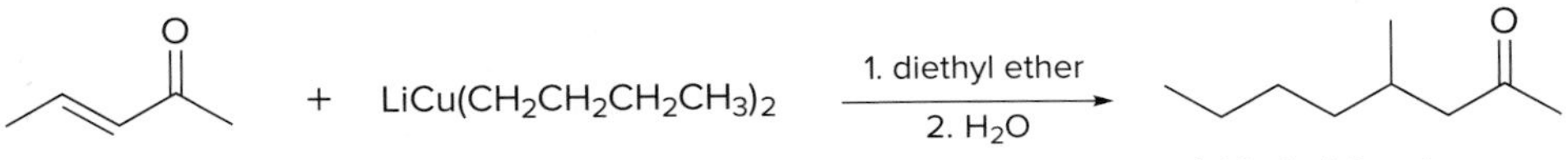

3-Penten-2-one + $LiCu(CH_2CH_2CH_2CH_3)_2$ $\xrightarrow[\text{2. } H_2O]{\text{1. diethyl ether}}$ 4-Methyl-2-octanone

Lithium dibutylcuprate

Now see if you can identify the second possibility.

21.10 SUMMARY

Section 21.1 An α hydrogen of an aldehyde or ketone is more acidic than most other protons bound to carbon (pK_a in the range of 16–20). Their enhanced acidity is due to the electron-withdrawing effect of the carbonyl group and the resonance stabilization of the enolate.

3-Pentanone + Hydroxide ⇌ Enolate + Water

Section 21.2 The aldol condensation is synthetically useful as a method for carbon–carbon bonds. Nucleophilic addition of an enolate to a carbonyl, followed by dehydration, yields an α,β-unsaturated aldehyde or ketone.

Octanal $\xrightarrow[CH_3CH_2OH]{NaOCH_2CH_3}$ 2-Hexyl-2-decenal (79%)

Section 21.3 A mixed aldol condensation between two different carbonyl compounds can be accomplished effectively if only one of them can form an enolate.

Methyl 2-thienyl ketone + *o*-Chlorobenzaldehyde $\xrightarrow[\text{ethanol–water}]{NaOH}$ 3-(*o*-Chlorophenyl-1-(2-thienyl)prop-2-en-1-one (60%)

Section 21.4 Esters of the type RCH_2CO_2R' are converted to β-keto esters on treatment with alkoxide bases. One molecule of an ester is converted to its enolate; a second molecule of the ester acts as an acylating agent toward the enolate.

2 Ethyl butanoate $\xrightarrow[2.\ H_3O^+]{1.\ NaOCH_2CH_3}$ Ethyl 2-ethyl-3-oxohexanoate (76%)

The Dieckmann condensation is an intramolecular version of the Claisen condensation.

Diethyl 1,2-benzenedicarboxylate $\xrightarrow[2.\ H_2O]{1.\ NaOCH_2CH_3}$ Ethyl indan-2-one-1-carboxylate (70%)

Section 21.5 Alkylation of simple aldehydes and ketones via their enolates is difficult. β-Diketones can be converted quantitatively to their enolates, which react efficiently with primary alkyl halides.

2-Benzyl-1,3-cyclohexanedione + Benzyl chloride ($C_6H_5CH_2Cl$) $\xrightarrow[\text{ethanol}]{KOCH_2CH_3}$ 2,2-Dibenzyl-1,3-cyclohexanedione (69%)

In the acetoacetic ester synthesis, a β-keto ester is alkylated as the first step in the preparation of ketones.

Ethyl acetoacetate $\xrightarrow[CH_3CH{=}CHCH_2Br]{NaOCH_2CH_3}$ $\xrightarrow[\text{2. } H_3O^+ \text{ 3. heat}]{\text{1. } HO^-,\ H_2O}$ 5-Hepten-2-one (81%)

Alkyl halides are converted to carboxylic acids by reaction with the enolate derived from diethyl malonate, followed by saponification and decarboxylation.

Diethyl malonate $\xrightarrow[\text{3-chlorocyclopentene}]{NaOCH_2CH_3}$ $\xrightarrow[\text{2. } H_3O^+ \text{ 3. heat}]{\text{1. } HO^-,\ H_2O}$ (3-Cyclopentenyl)acetic acid (66%)

Section 21.6 Aldehydes, ketones, and esters having at least one α hydrogen exist in equilibrium with their enols. The enol content of simple aldehydes, ketones, and esters is small.

Cyclopentanone $\underset{}{\overset{K = 1 \times 10^{-8}}{\rightleftharpoons}}$ Cyclopenten-1-ol

β-Diketones and β-keto esters are more extensively enolized.

1,3-Diphenyl-1,3-dione $\overset{K \approx 10^2}{\rightleftharpoons}$ 1,3-Diphenyl-2-propen-3-ol-1-one

Aldehydes and ketones undergo halogenation at their α carbon via an enol intermediate.

p-Bromoacetophenone $\xrightarrow[\text{acetic acid}]{Br_2}$ *p*-Bromophenacyl bromide (69–72%)

Section 21.7 Methyl ketones are cleaved on reaction with excess halogen in the presence of base. The products are a trihalomethane (haloform) and a carboxylate salt. The reaction is used as a synthesis of carboxylic acids.

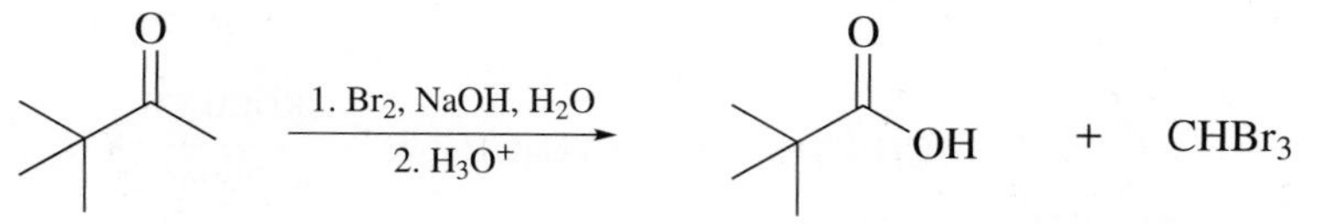
O
1. Br_2, NaOH, H_2O
2. H_3O^+
O
OH
+
$CHBr_3$

3,3-Dimethyl-2-butanone 2,2-Dimethylpropanoic acid (71–74%) Bromoform

Section 21.8 Substitution of deuterium for hydrogen occurs at the α position in aldehydes and ketones in the presence of D_2O and base. If the α carbon of an aldehyde or ketone is a chirality center, racemization can occur upon enolization.

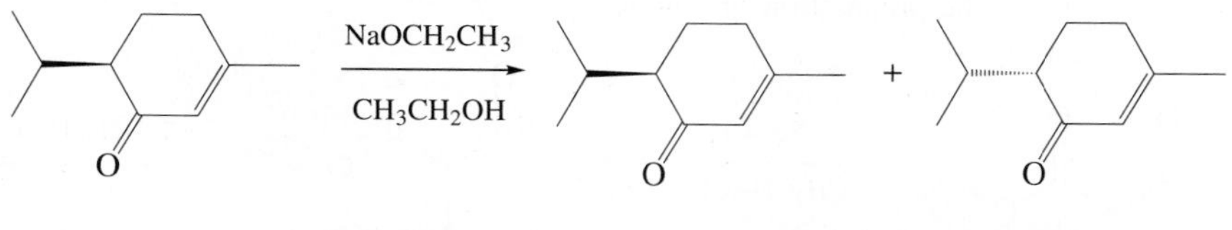
$NaOCH_2CH_3$
CH_3CH_2OH
+
O
O
O

Section 21.9 The β carbon of an α,β-unsaturated carbonyl compound is electrophilic; nucleophiles, especially weakly basic ones, react with α,β-unsaturated aldehydes and ketones by conjugate addition. The nucleophile bonds to the β carbon. Lithium dialkylcuprates react similarly.

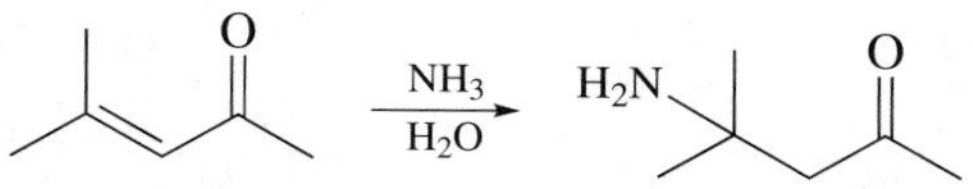
O
NH_3
H_2O
H_2N
O

4-Methyl-3-penten-2-one 4-Amino-4-methyl-2-pentanone (63–70%)

PROBLEMS

Enols and Enolates

21.30 (a) Arrange the following in order of decreasing acidity.

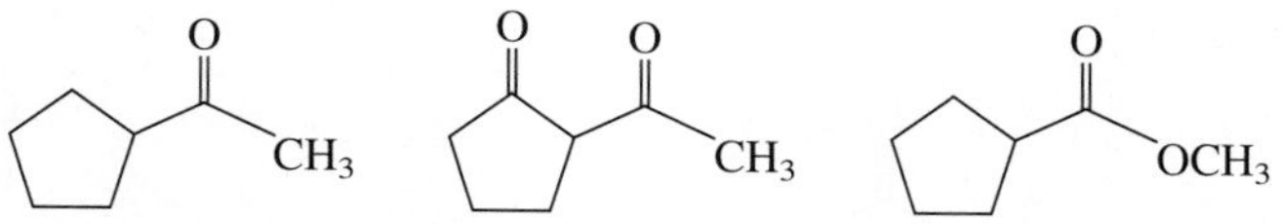
O
CH_3
O
O
CH_3
O
OCH_3

I. II. III.

(b) Write the structures of the kinetic and thermodynamic enolates of compound I.

21.31 Choose the compound in each of the following pairs that has the greater enol content.

(a) 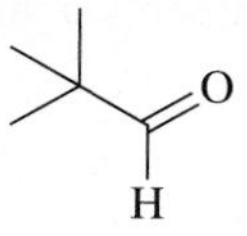
O
H
or 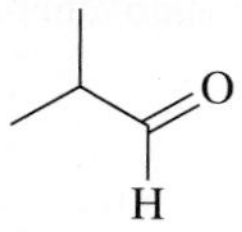
O
H

(b) 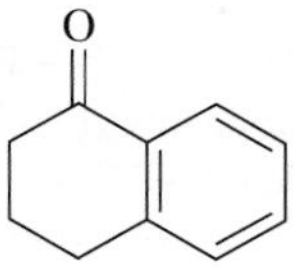
O
or 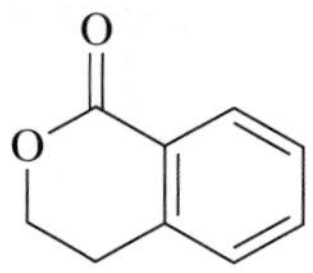
O
O

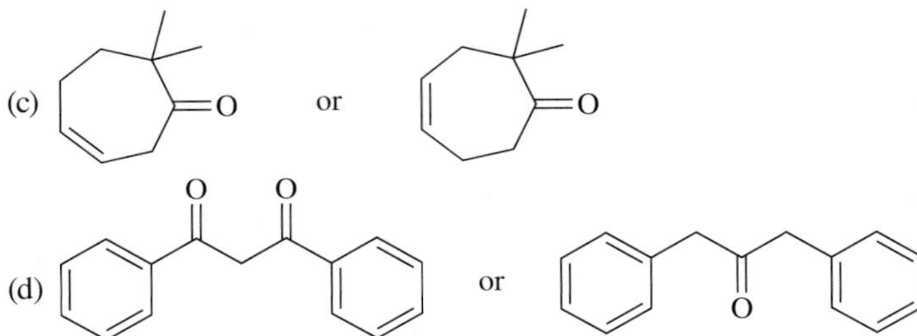

21.32 (a) Only a small amount (less than 0.01%) of the enol form of diethyl malonate is present at equilibrium. Write a structural formula for this enol.

(b) Enol forms are present to the extent of about 8% in ethyl acetoacetate. There are three constitutionally isomeric enols possible. Write structural formulas for these three enols. Which one do you think is the most stable? The least stable? Why?

(c) Bromine reacts rapidly with both diethyl malonate and ethyl acetoacetate. The reaction is acid-catalyzed and liberates hydrogen bromide. What is the product formed in each reaction?

Reactions

21.33 Consider the ketones piperitone, menthone, and isomenthone.

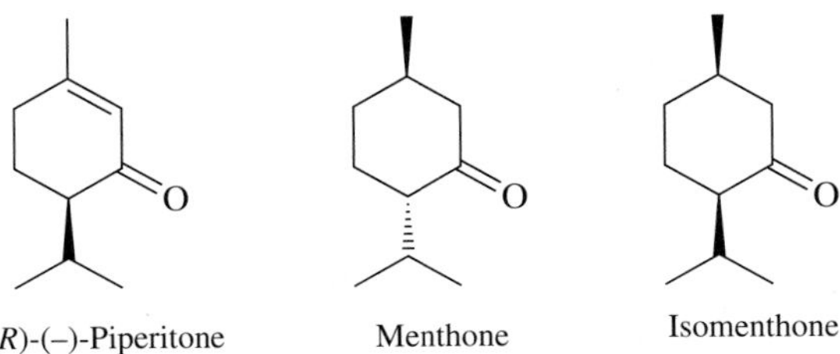

(*R*)-(−)-Piperitone Menthone Isomenthone

Suggest reasonable explanations for each of the following observations.

(a) (*R*)-(−)-Piperitone racemizes on standing in a solution of sodium ethoxide in ethanol.

(b) Menthone is converted to a mixture of menthone and isomenthone on treatment with 90% sulfuric acid.

21.34 In each of the following, the indicated observations were made before any of the starting material was transformed to aldol addition or condensation products:

(a) In aqueous acid, only 17% of $(C_6H_5)_2CHCH{=}O$ is present as the aldehyde; 2% of the enol is present. Some other species accounts for 81% of the material. What is it?

(b) In aqueous base, 97% of $(C_6H_5)_2CHCH{=}O$ is present as a species different from any of those in part (a). What is this species?

21.35 (a) On addition of one equivalent of methylmagnesium iodide to ethyl acetoacetate, the Grignard reagent is consumed, but the only organic product obtained after working up the reaction mixture is ethyl acetoacetate. Why? What happens to the Grignard reagent?

(b) On repeating the reaction but using D_2O and DCl to work up the reaction mixture, it is found that the recovered ethyl acetoacetate contains deuterium. Where is this deuterium located?

21.36 Give the structure of the expected organic product in the reaction of 3-phenylpropanal with each of the following:

(a) Chlorine in acetic acid

(b) Sodium hydroxide in ethanol, 10°C

(c) Sodium hydroxide in ethanol, 70°C

(d) Product of part (c) with lithium aluminum hydride; then H_2O

(e) Product of part (c) with sodium cyanide in acidic ethanol

21.37 Each of the following reactions has been reported in the chemical literature. Write the structure of the product(s) formed in each case.

(a) 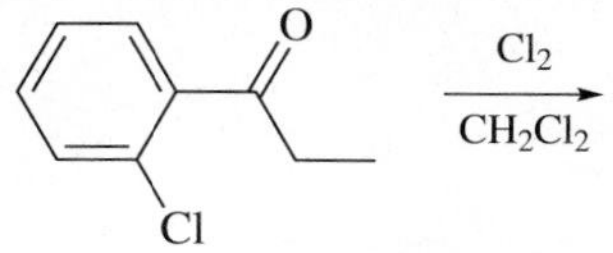 [2-chlorophenyl ethyl ketone] $\xrightarrow[CH_2Cl_2]{Cl_2}$

(b) 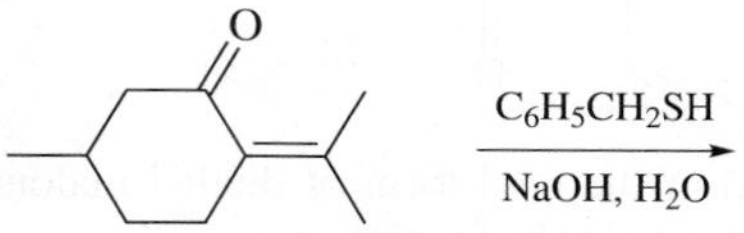 [ketone] $\xrightarrow[NaOH,\ H_2O]{C_6H_5CH_2SH}$

(c) [furan-2-carbaldehyde] + [acetone] $\xrightarrow[water]{NaOH}$

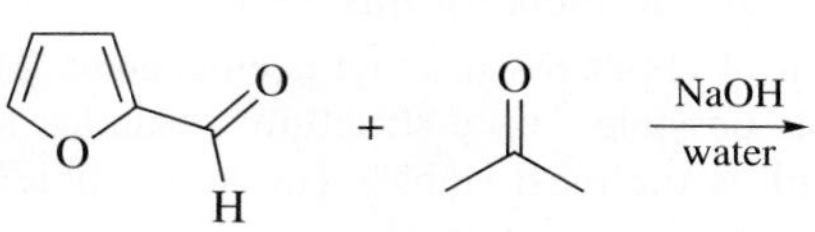

(d) 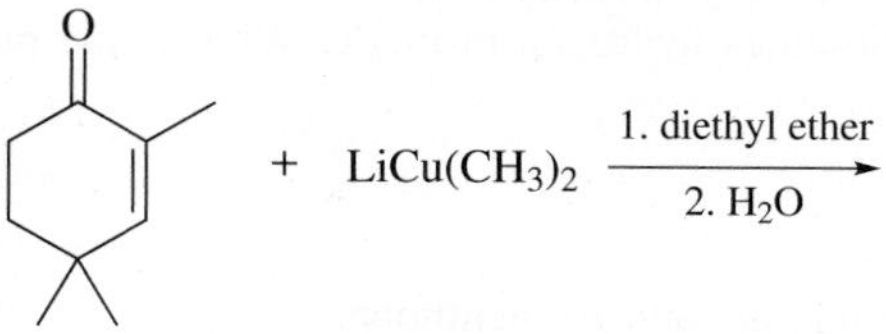 [enone] + $LiCu(CH_3)_2$ $\xrightarrow[2.\ H_2O]{1.\ diethyl\ ether}$

(e) 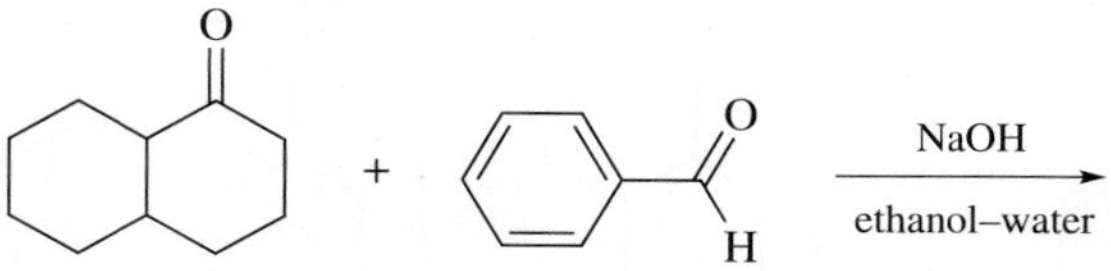 [decalone] + [benzaldehyde] $\xrightarrow[ethanol\text{–}water]{NaOH}$

(f) 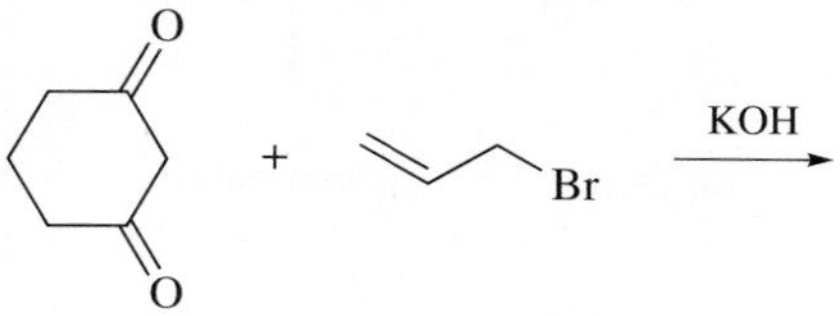 [1,3-cyclohexanedione] + [allyl bromide] $\xrightarrow{KOH}$

21.38 Dibromination of camphor under the conditions shown gave a single product in 99% yield. What is this product?

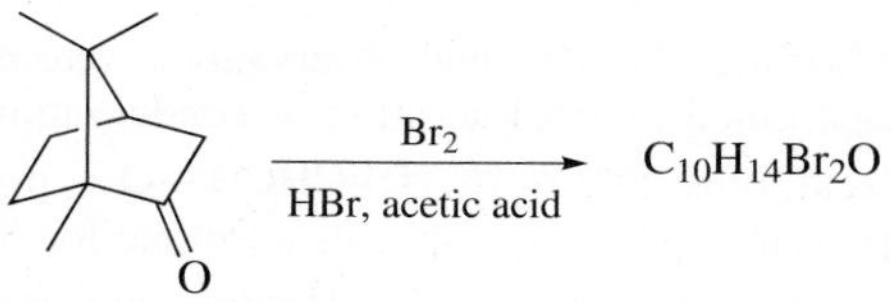

[camphor] $\xrightarrow[HBr,\ acetic\ acid]{Br_2}$ $C_{10}H_{14}Br_2O$

21.39 Bromination of 3-methyl-2-butanone yielded two compounds, each having the molecular formula C_5H_9BrO in a 95:5 ratio. The ^1H NMR spectrum of the major isomer A was characterized by a doublet at δ 1.2 (six protons), a septet at δ 3.0 (one proton), and a singlet at δ 4.1 (two protons). The ^1H NMR spectrum of the minor isomer B exhibited two singlets, one at δ 1.9 and the other at δ 2.5. The lower field singlet had half the area of the higher field one. Suggest reasonable structures for A and B.

21.40 Give the structure of the principal organic product of each of the following reactions:

(a) Ethyl acetoacetate + 1-bromobutane $\xrightarrow{NaOCH_2CH_3,\ ethanol}$

(b) Product of part (a) $\xrightarrow[2.\ H_3O^+\ \ 3.\ heat]{1.\ NaOH,\ H_2O}$

(c) Acetophenone + diethyl carbonate $\xrightarrow[2.\ H_3O^+]{1.\ NaOCH_2CH_3}$

(d) Acetone + diethyl oxalate $\xrightarrow[\text{2. } H_3O^+]{\text{1. } NaOCH_2CH_3}$

(e) Diethyl malonate + 1-bromo-2-methylbutane $\xrightarrow{NaOCH_2CH_3\text{, ethanol}}$

(f) Product of part (e) $\xrightarrow[\text{2. } H_3O^+ \text{ 3. heat}]{\text{1. } NaOH, H_2O}$

(g) Diethyl malonate + 6-methyl-2-cyclohexenone $\xrightarrow{NaOCH_2CH_3\text{, ethanol}}$

(h) Product of part (g) $\xrightarrow{H_2O\text{, HCl, heat}}$

(i) *tert*-Butyl acetate $\xrightarrow[\text{2. benzaldehyde 3. } H_3O^+]{\text{1. } [(CH_3)_2CH]_2NLi\text{, THF}}$

21.41 Give the structure of the product formed on reaction of ethyl acetoacetate with each of the following:

(a) 1-Bromopentane and sodium ethoxide
(b) Saponification (basic hydrolysis) and decarboxylation of the product in part (a)
(c) Methyl iodide and the product in part (a) treated with sodium ethoxide
(d) Saponification and decarboxylation of the product in part (c)
(e) 1-Bromo-3-chloropropane and one equivalent of sodium ethoxide
(f) Product in part (e) treated with a second equivalent of sodium ethoxide
(g) Saponification and decarboxylation of the product in part (f)
(h) Phenyl vinyl ketone and sodium ethoxide
(i) Saponification and decarboxylation of the product in part (h)

21.42 Repeat the preceding problem for diethyl malonate.

21.43 Give the structure of the product ($C_7H_{10}O$) formed by intramolecular aldol condensation of the keto-aldehyde shown.

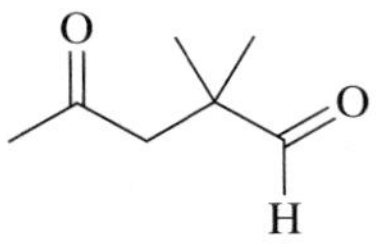

O O H

21.44 The following questions pertain to the esters shown and behavior under conditions of the Claisen condensation.

Ethyl pentanoate | Ethyl 2-methylbutanoate | Ethyl 3-methylbutanoate | Ethyl 2,2-dimethylpropanoate

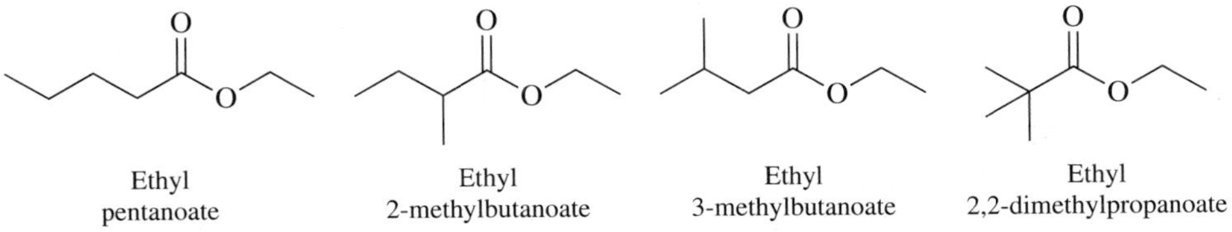

(a) Two of these esters are converted to β-keto esters in good yield on treatment with sodium ethoxide and subsequent acidification of the reaction mixture. Which two are these? Write the structure of the Claisen condensation product of each one.

(b) One ester is capable of being converted to a β-keto ester on treatment with sodium ethoxide, but the amount of β-keto ester that can be isolated after acidification of the reaction mixture is quite small. Which ester is this?

(c) One ester is incapable of reaction under conditions of the Claisen condensation. Which one? Why?

21.45 (a) Give the structure of the Claisen condensation product of ethyl phenylacetate ($C_6H_5CH_2COOCH_2CH_3$).

(b) What ketone would you isolate after saponification and decarboxylation of this Claisen condensation product?

(c) What ketone would you isolate after treatment of the Claisen condensation product of ethyl phenylacetate with sodium ethoxide and allyl bromide, followed by saponification and decarboxylation?

(d) Give the structure of the mixed Claisen condensation product of ethyl phenylacetate and ethyl benzoate.

(e) What ketone would you isolate after saponification and decarboxylation of the product in part (d)?

(f) What ketone would you isolate after treatment of the product in part (d) with sodium ethoxide and allyl bromide, followed by saponification and decarboxylation?

21.46 The following questions concern ethyl (2-oxocyclohexane)carboxylate.

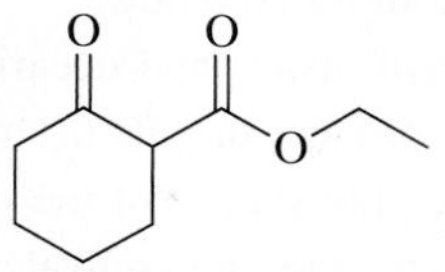

(a) Write a chemical equation showing how you could prepare ethyl (2-oxocyclohexane)-carboxylate by a Dieckmann cyclization.

(b) Write a chemical equation showing how you could prepare ethyl (2-oxocyclohexane)-carboxylate by acylation of a ketone.

(c) Write structural formulas for the two most stable enol forms of ethyl (2-oxocyclohexane)carboxylate.

(d) Write the three most stable resonance contributors to the most stable enolate derived from ethyl (2-oxocyclohexane)carboxylate.

(e) Show how you could use ethyl (2-oxocyclohexane)carboxylate to prepare 2-methylcyclohexanone.

(f) Give the structure of the product formed on treatment of ethyl (2-oxocyclohexane)-carboxylate with acrolein ($H_2C{=}CHCH{=}O$) in ethanol in the presence of sodium ethoxide.

21.47 Each of the reactions shown has been carried out in the course of some synthetic project and gave the compound indicated by the molecular formula. What are those compounds?

(a) $\xrightarrow[\text{heat}]{H_2O,\ H_2SO_4}$ $C_7H_{12}O$

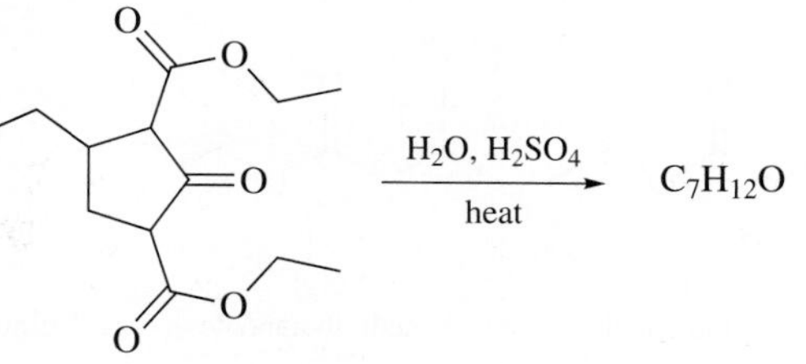

(b) $\xrightarrow[\text{2. } H_3O^+]{\text{1. } NaOCH_2CH_3}$ $C_9H_{12}O_3$ $\xrightarrow[\text{2. } H_3O^+ \text{ 3. heat}]{\text{1. } HO^-,\ H_2O}$ C_6H_8O

(c) CH_3O CH_3O $\xrightarrow[\text{heat}]{HCl,\ H_2O}$ $C_5H_6O_3$

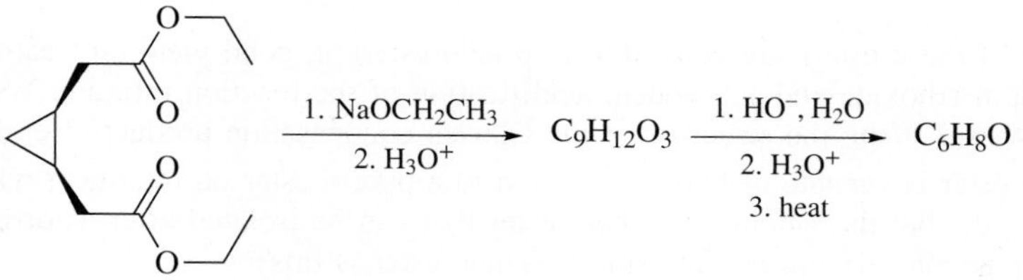

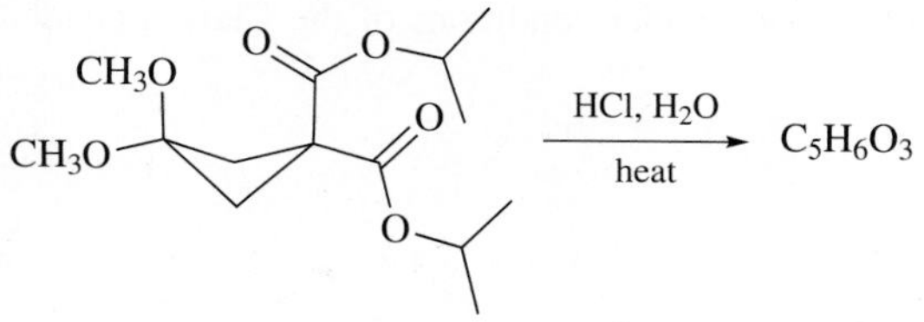

Synthetic Applications

21.48 The use of epoxides as alkylating agents for diethyl malonate provides a useful route to γ-lactones. Write equations illustrating such a sequence for styrene oxide as the starting epoxide. Is the lactone formed by this reaction 3-phenylbutanolide, or is it 4-phenylbutanolide?

3-Phenylbutanolide   4-Phenylbutanolide

21.49 Show how each of the following compounds could be prepared from 3-pentanone. In most cases more than one synthetic transformation will be necessary.

(a) 2-Bromo-3-pentanone
(b) 1-Penten-3-one
(c) 1-Penten-3-ol
(d) 3-Hexanone
(e) 2-Methyl-1-phenyl-1-penten-3-one

21.50 Show how you could prepare each of the following compounds. Use the starting material indicated along with ethyl acetoacetate or diethyl malonate and any necessary inorganic reagents. Assume also that the customary organic solvents are freely available.

(a) 4-Phenyl-2-butanone from benzyl alcohol
(b) 3-Phenylpropanoic acid from benzyl alcohol
(c) 2-Allyl-1,3-propanediol from propene
(d) 4-Penten-1-ol from propene
(e) 5-Hexen-2-ol from propene

21.51 The spicy flavor of cayenne pepper is due mainly to a substance called *capsaicin*. See if you can deduce the structure of capsaicin on the basis of its laboratory synthesis:

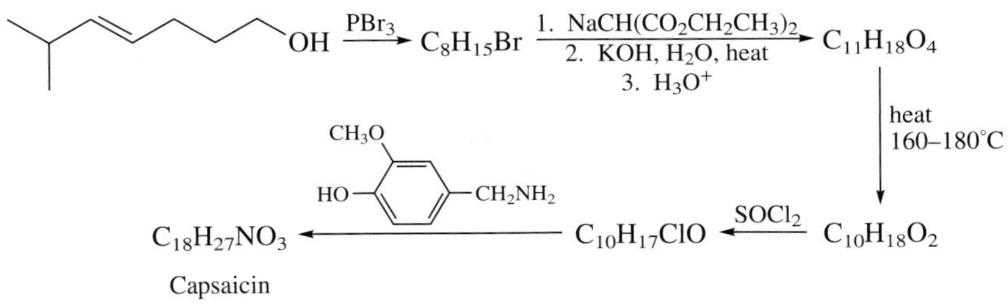

OH $\xrightarrow{PBr_3}$ $C_8H_{15}Br$ $\xrightarrow[\text{2. KOH, } H_2O\text{, heat; 3. } H_3O^+]{\text{1. } NaCH(CO_2CH_2CH_3)_2}$ $C_{11}H_{18}O_4$ $\xrightarrow{\text{heat, 160–180°C}}$ $C_{10}H_{18}O_2$ $\xrightarrow{SOCl_2}$ $C_{10}H_{17}ClO$ $\xrightarrow{\text{(4-HO-3-}CH_3O\text{-}C_6H_3CH_2NH_2)}$ $C_{18}H_{27}NO_3$

CH₃O / HO— —CH₂NH₂

Capsaicin

21.52 Jasmone, which contributes to the odor of jasmine, can be prepared by an intramolecular aldol condensation of a diketone. Use retrosynthetic analysis to deduce the structure of the diketone.

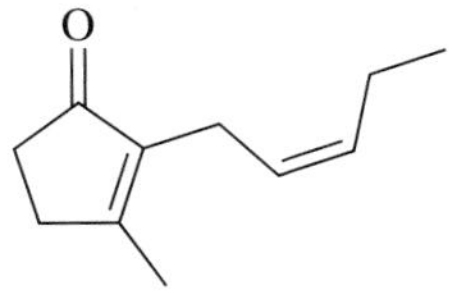

O

21.53 Prepare each of the following target molecules using the compounds shown as the sources of all of the carbons plus any necessary organic or inorganic reagents.

(a) 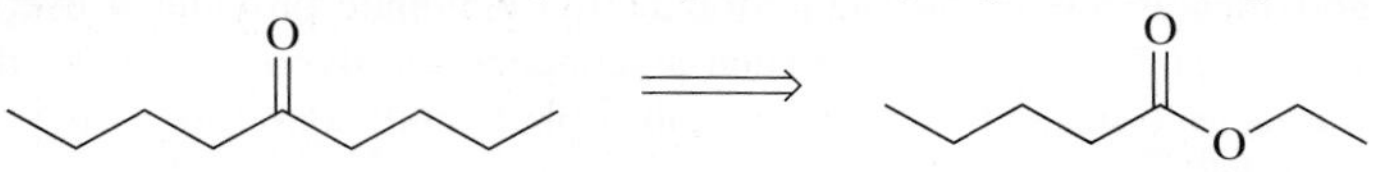 O ⟹ O O

(b) 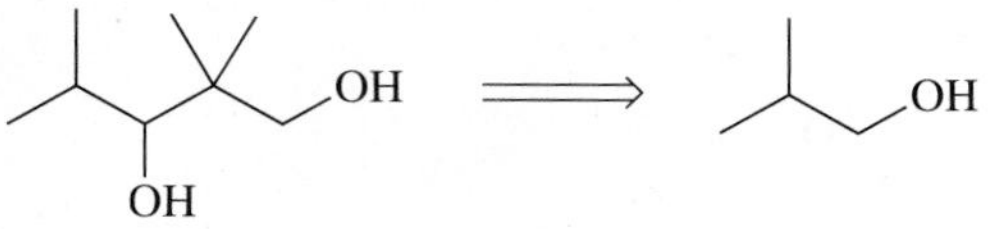 OH OH ⟹ OH

(c) 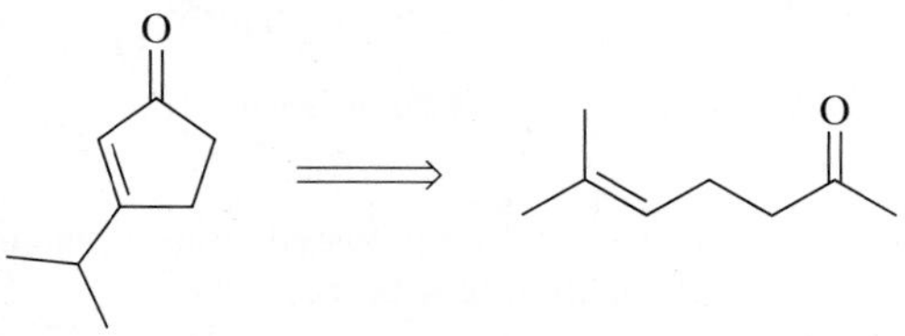 O ⟹ O

(d) 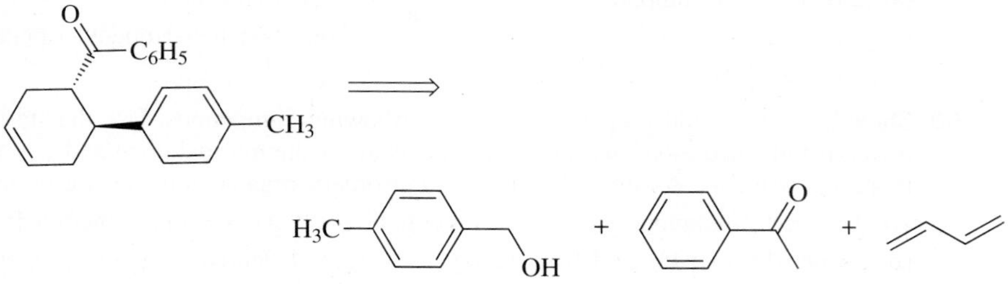 O C_6H_5 CH_3 ⟹

H_3C OH + O +

Mechanisms

21.54 Compound A is difficult to prepare owing to its ready base-catalyzed isomerization to compound B. Write a reasonable mechanism for this isomerization.

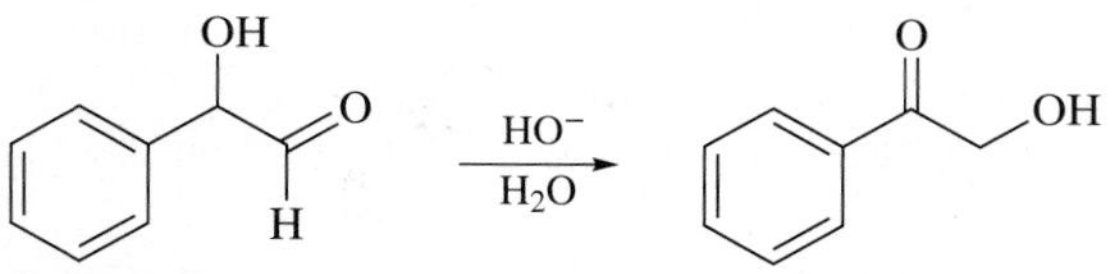

OH O H → (HO^-, H_2O) O OH

Compound A   Compound B

21.55 The α-methylene ketone sarkomycin has an inhibitory effect on certain types of tumors. A key step in the synthesis of sarkomycin is the reaction of lactone ester A with potassium *tert*-butoxide in tetrahydrofuran to give the bicyclic compound B. Write a mechanism for this reaction.

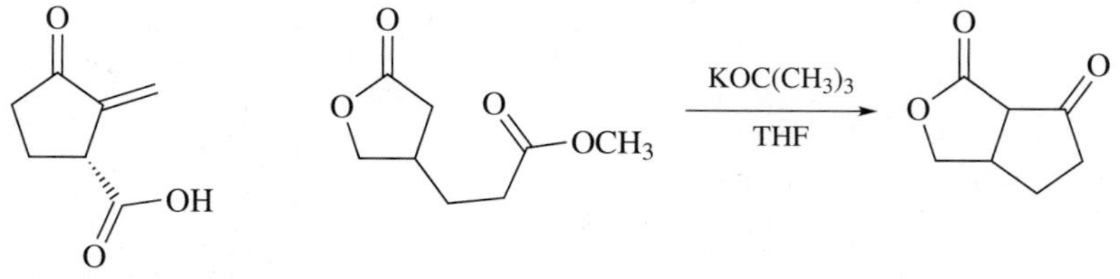

O OH O   O O O OCH_3 → ($KOC(CH_3)_3$, THF) O O O

Sarkomycin   A   B

21.56 β-Lactones can be prepared in good yield from thioester enolates. Suggest a mechanism for the reaction shown.

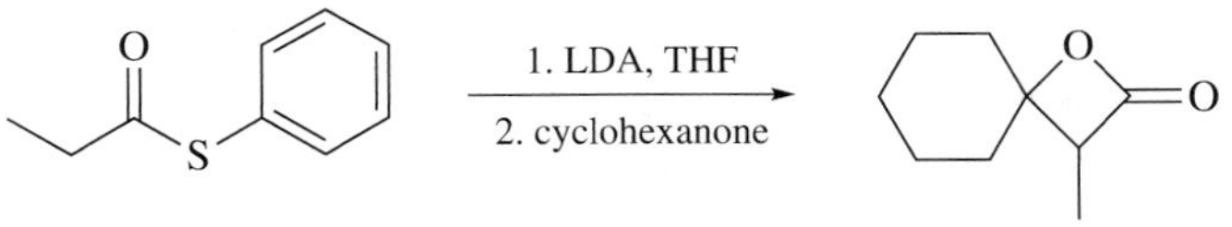
O
S
1. LDA, THF
2. cyclohexanone
O
O

21.57 Outline a reasonable mechanism for each of the following reactions.

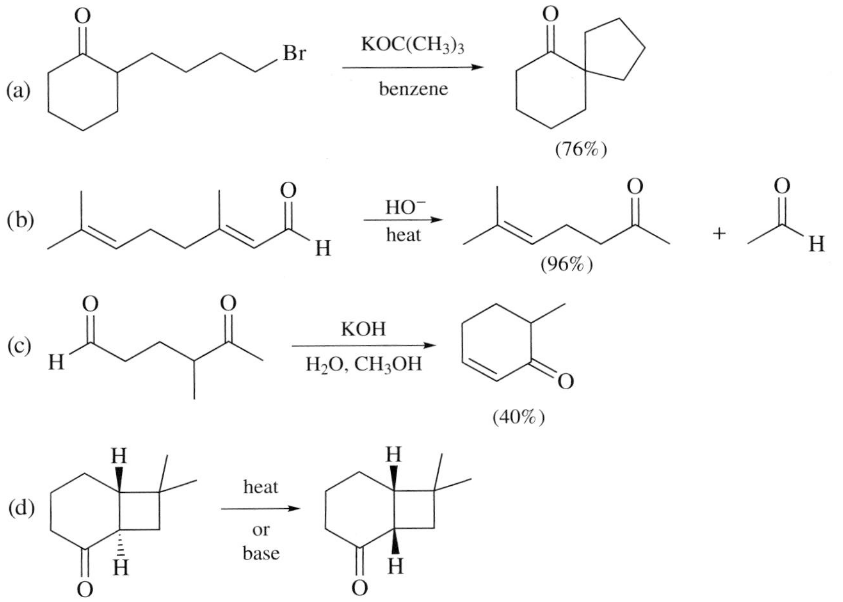
(a)
O
Br
$KOC(CH_3)_3$
benzene
O
(76%)
(b)
O
H
HO^-
heat
O
+
O
H
(96%)
(c)
O
O
H
KOH
H_2O, CH_3OH
O
(40%)
(d)
H
H
O
heat
or
base
H
H
O

21.58 A key step in the synthesis of aspirin, known as the Kolbe–Schmitt reaction, is the preparation of sodium salicylate by the reaction of sodium phenoxide with carbon dioxide under conditions of heat and pressure. Write a mechanism.

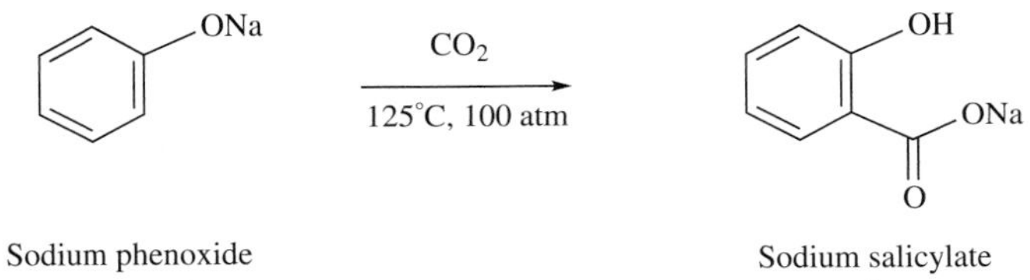
ONa
CO_2
125°C, 100 atm
OH
ONa
O
Sodium phenoxide
Sodium salicylate

Descriptive Passage and Interpretive Problems 21

The Knoevenagel Reaction

Stabilized anions undergo additions to aldehydes and ketones in a reaction that resembles the aldol condensation. The reaction of diethyl malonate with benzaldehyde in the presence of the amine piperidine is an example of this process, which is called the Knoevenagel reaction after its discoverer.

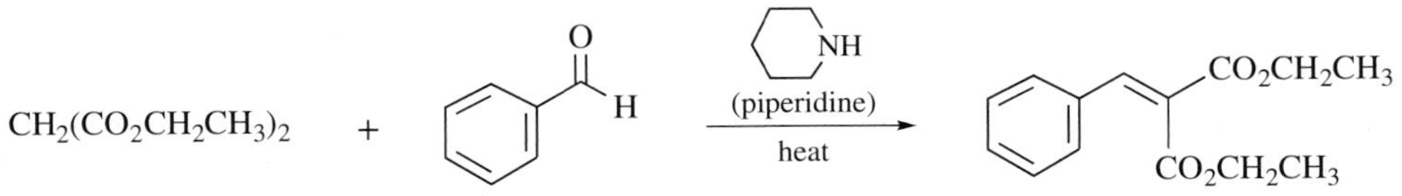
$CH_2(CO_2CH_2CH_3)_2$
+
O
H
NH
(piperidine)
heat
$CO_2CH_2CH_3$
$CO_2CH_2CH_3$

The amine piperidine acts as a basic catalyst in the reaction and also can form an iminium species that reacts with the dimethyl malonate anion.

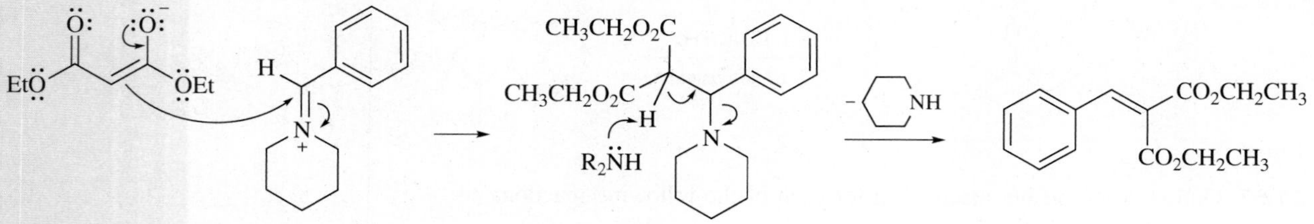

EtÖ ... ÖEt H N+ → $CH_3CH_2O_2C$ $CH_3CH_2O_2C$ H $R_2\ddot{N}H$ N → −(piperidine NH) → $CO_2CH_2CH_3$ $CO_2CH_2CH_3$

Other bases and reaction conditions have been used, and alternative mechanistic pathways may occur. The Knoevenagel reaction has found applications in medicinal chemistry as a method for the preparation of pharmaceuticals.

21.59. What is the product of the following reaction, which was used to synthesize the anticonvulsant drug gabapentin?

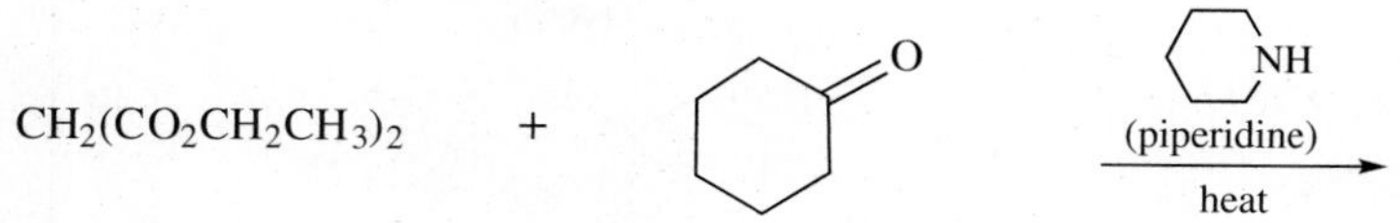

$CH_2(CO_2CH_2CH_3)_2$ + (cyclohexanone) O —NH (piperidine), heat→

A. 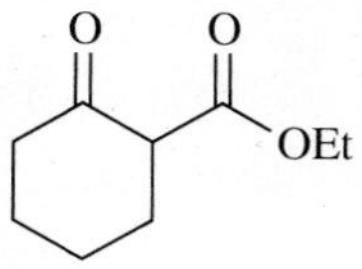 O O OEt

B. 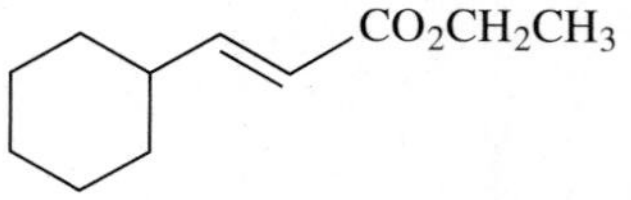 $CO_2CH_2CH_3$

C. 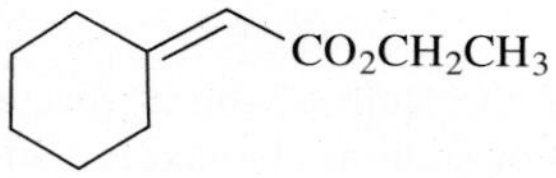 $CO_2CH_2CH_3$

D. 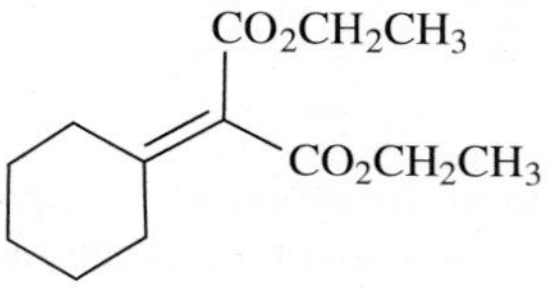 $CO_2CH_2CH_3$ $CO_2CH_2CH_3$

21.60. Which of the following is a possible intermediate in the reaction in Problem 21.59?

A. 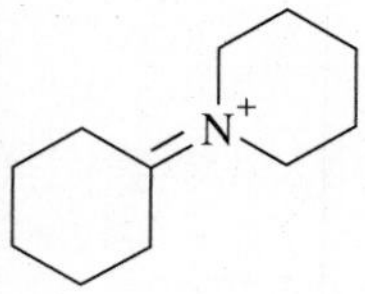 N+

B. 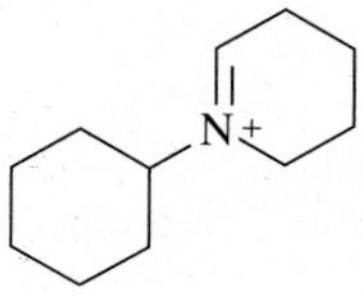 N+

C. 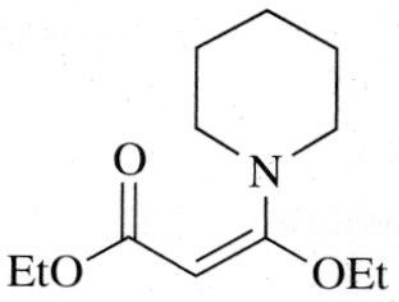 O N EtO OEt

D. 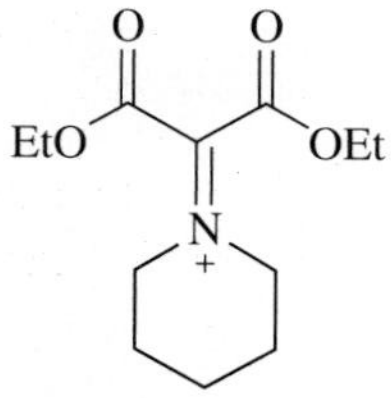 O O EtO OEt N +

21.61. The following Knoevenagel condensation is accompanied by dehydration. Which of the following is *not* a plausible reaction intermediate?

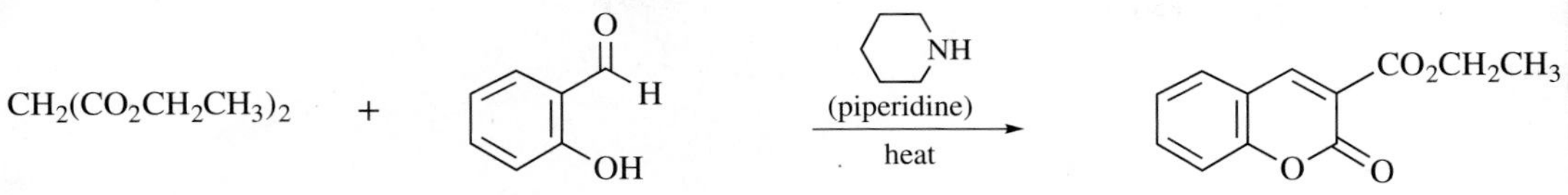

$CH_2(CO_2CH_2CH_3)_2$ + (salicylaldehyde) —(piperidine), heat→ (product with $CO_2CH_2CH_3$)

A. 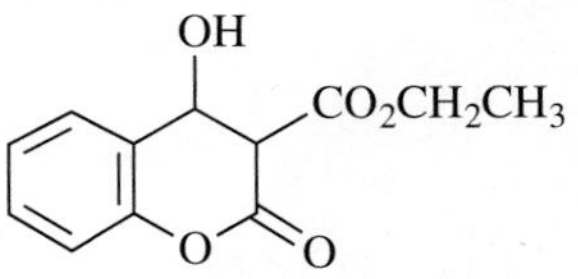 (intermediate with OH, $CO_2CH_2CH_3$)

B. 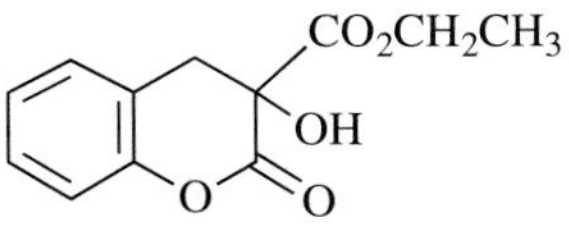 (intermediate with $CO_2CH_2CH_3$, OH)

C. 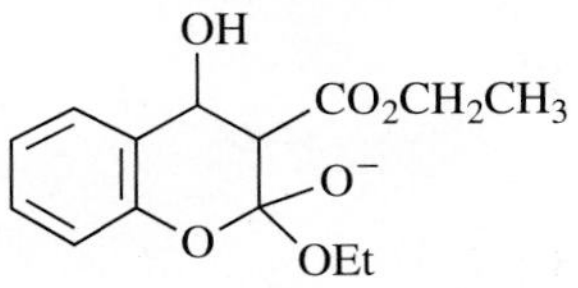 (intermediate with OH, $CO_2CH_2CH_3$, O^-, OEt)

D. 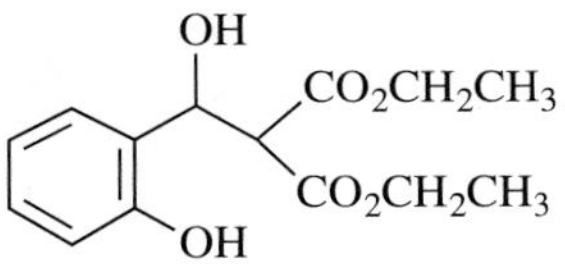 (intermediate with OH, $CO_2CH_2CH_3$, $CO_2CH_2CH_3$, OH)

21.62 The Knoevenagel reaction with malonic acid is often accompanied by decarboxylation. What is the likely product of the following reaction?

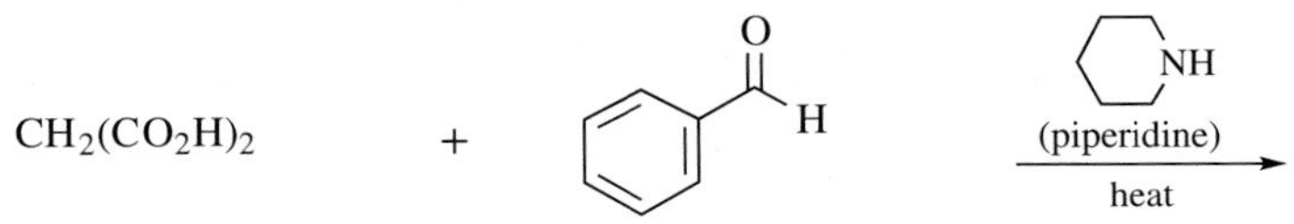

$CH_2(CO_2H)_2$ + (benzaldehyde) —(piperidine), heat→

A. 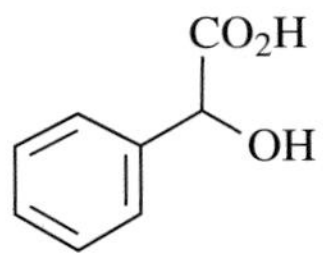 (structure with CO_2H, OH)

B. 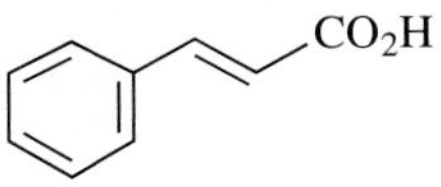 (structure with CO_2H)

C. 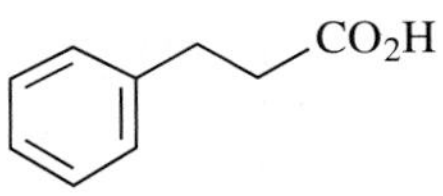 (structure with CO_2H)

D. 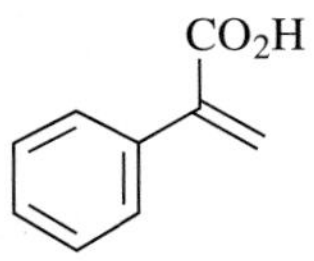 (structure with CO_2H)

Chapter

22

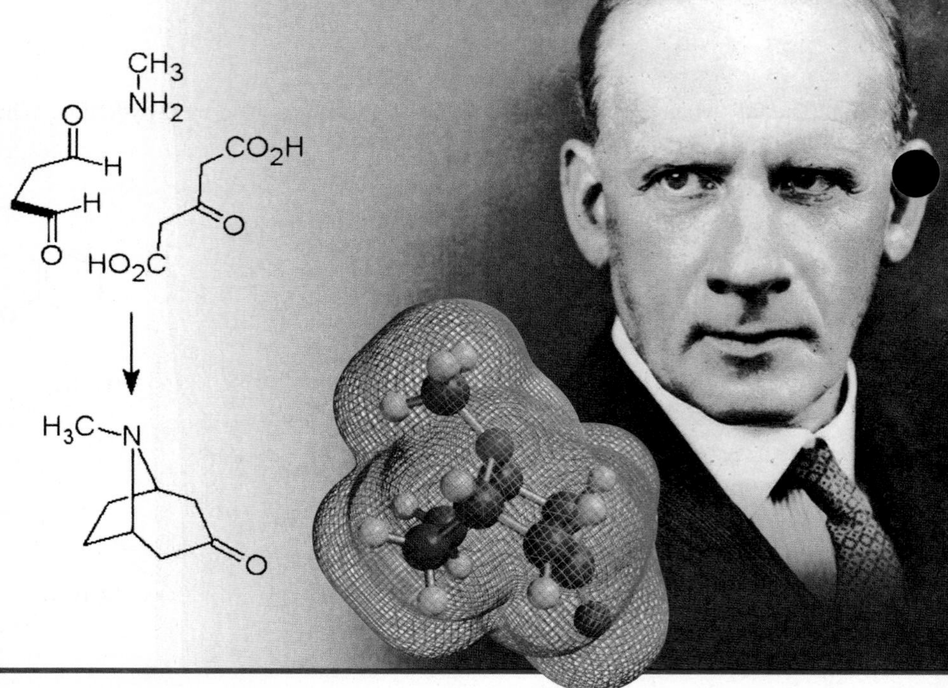

In 1917, Robert Robinson verified his ideas about the biosynthesis of alkaloids by combining methylamine with the compounds shown. The reaction worked as Robinson planned and is recognized as the first chemical synthesis inspired by biochemical thinking.

CHAPTER OUTLINE

Amines

Nitrogen-containing compounds are essential to life. Their ultimate source is atmospheric nitrogen that, by a process known as *nitrogen fixation,* is reduced to ammonia, then converted to organic nitrogen compounds. This chapter describes the chemistry of **amines,** organic derivatives of ammonia. **Alkylamines** have their nitrogen attached to sp^3-hybridized carbon; **arylamines** have their nitrogen attached to an sp^2-hybridized carbon of a benzene or benzene-like ring.

R—N̈<	Ar—N̈<
R = alkyl group: alkylamine	Ar = aryl group: arylamine

Amines, like ammonia, are weak bases. They are, however, the strongest uncharged bases found in significant quantities under physiological conditions. Amines are usually the bases involved in biological acid–base reactions; they are often the nucleophiles in biological nucleophilic substitutions.

Our word *vitamin* was coined in 1912 in the belief that the substances present in the diet that prevented scurvy, pellagra, beriberi, rickets, and other diseases were "vital amines." In many cases, that belief was confirmed; certain vitamins did prove to be amines. In many other cases, however, vitamins were not amines. Nevertheless, the name *vitamin* entered our language and stands as a reminder that early chemists recognized the crucial place occupied by amines in biological processes.

22.1 Amine Nomenclature

Unlike alcohols and alkyl halides, which are classified as primary, secondary, or tertiary according to the degree of substitution at the carbon that bears the functional group, amines are classified according to their *degree of substitution at nitrogen.* An amine with one carbon attached to nitrogen is a **primary amine,** an amine with two is a **secondary amine,** and an amine with three is a **tertiary amine.**

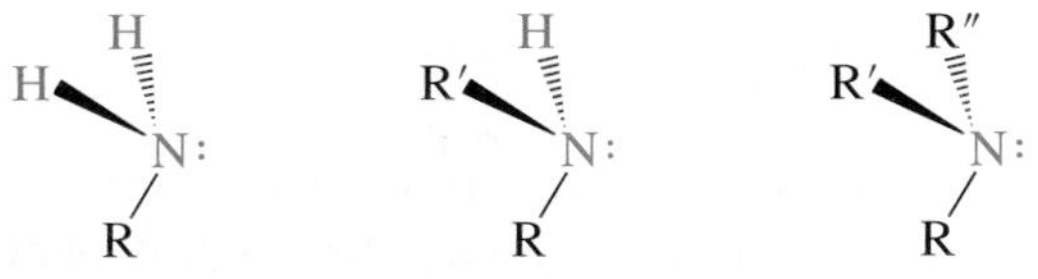

Primary amine Secondary amine Tertiary amine

The groups attached to nitrogen may be any combination of alkyl or aryl groups.

Amines are named in two main ways in the IUPAC system, either as *alkylamines* or as *alkanamines.* When primary amines are named as alkylamines, the ending *-amine* is added to the name of the alkyl group that bears the nitrogen. When named as alkanamines, the alkyl group is named as an alkane and the *-e* ending replaced by *-amine.*

$CH_3CH_2NH_2$

Ethylamine (ethanamine)

Cyclohexylamine (cyclohexanamine)

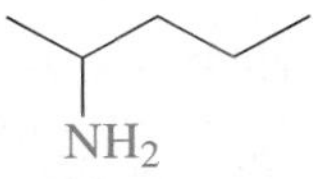

1-Methylbutylamine (2-pentanamine or pentan-2-amine)

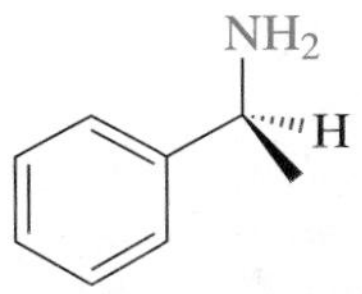

(*S*)-1-Phenylethylamine [(*S*)-1-phenylethanamine]

Problem 22.1

Give an acceptable alkylamine or alkanamine name for each of the following amines:

(a) $C_6H_5CH_2CH_2NH_2$ (b) (c)

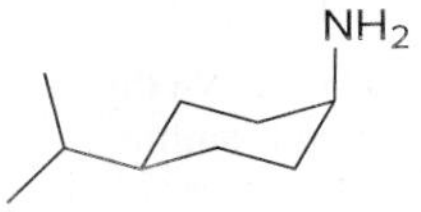

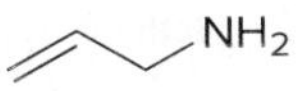

Sample Solution a) The amino substituent is bonded to an ethyl group that bears a phenyl substituent at C-2. The compound $C_6H_5CH_2CH_2NH_2$ may be named as either 2-phenylethylamine or 2-phenylethanamine.

Aniline is the parent IUPAC name for amino-substituted derivatives of benzene. Substituted derivatives of aniline are numbered beginning at the carbon that bears the amino group. Substituents are listed in alphabetical order, and the direction of numbering is governed by the usual "first point of difference" rule.

Aniline was first isolated in 1826 as a degradation product of indigo, a dark blue dye obtained from the West Indian plant *Indigofera anil,* from which the name *aniline* is derived.

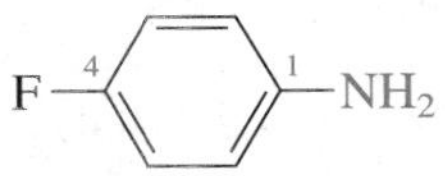

p-Fluoroaniline or 4-Fluoroaniline

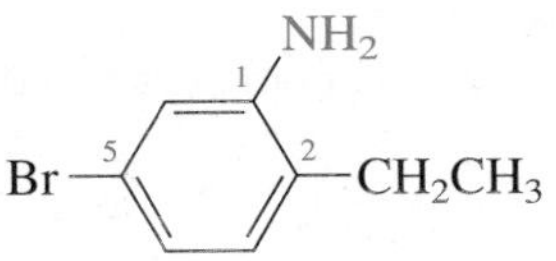

5-Bromo-2-ethylaniline

Arylamines may also be named as *arenamines.* Thus, *benzenamine* is an alternative, but rarely used, name for aniline.

Compounds with two amino groups are named by adding the suffix *-diamine* to the name of the corresponding alkane or arene. The final *-e* of the parent hydrocarbon is retained.

1,2-Propanediamine or Propane-1,2-diamine

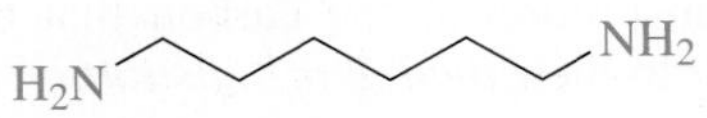

1,6-Hexanediamine or Hexane-1,6-diamine

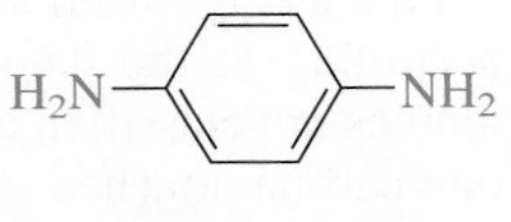

1,4-Benzenediamine or Benzene-1,4-diamine

Amino groups rank rather low in seniority when the parent compound is identified for naming purposes. Hydroxyl groups and carbonyl groups outrank amino groups. In these cases, the amino group is named as a substituent.

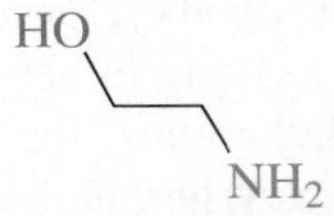

2-Aminoethanol

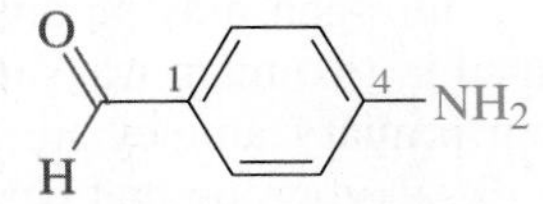

p-Aminobenzaldehyde (4-Aminobenzenecarbaldehyde)

Secondary and tertiary amines are named as *N*-substituted derivatives of primary amines. The parent primary amine is taken to be the one with the longest carbon chain. Rings, however, take precedence over chains. The prefix *N-* is added as a locant to identify substituents on the amine nitrogen.

$CH_3NHCH_2CH_3$

N-Methylethylamine (a secondary amine)

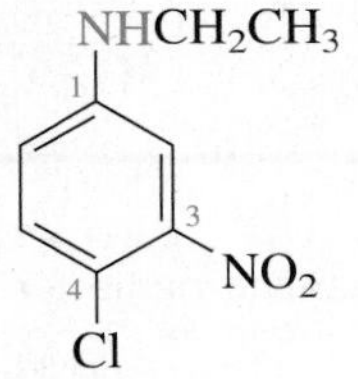

4-Chloro-*N*-ethyl-3-nitroaniline (a secondary amine)

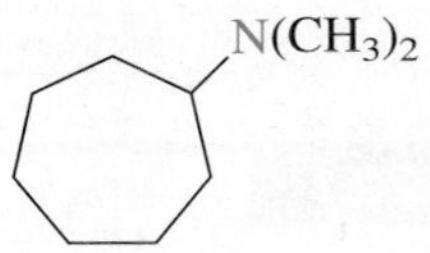

N,N-Dimethylcycloheptylamine (a tertiary amine)

Problem 22.2

Assign alkanamine names to *N*-methylethylamine and to *N,N*-dimethylcycloheptylamine.

Sample Solution *N*-Methylethylamine (given as $CH_3NHCH_2CH_3$ in the preceding example) is an *N*-substituted derivative of ethanamine; it is *N*-methylethanamine.

Problem 22.3

Classify the following amine as primary, secondary, or tertiary, and give it an acceptable IUPAC name.

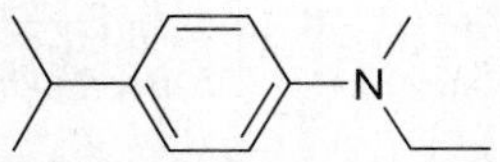

A nitrogen that bears four substituents is positively charged and is named as an *ammonium* ion. The anion that is associated with it is also identified in the name. Ammonium salts that have four alkyl groups bonded to nitrogen are called **quaternary ammonium salts.**

$CH_3\overset{+}{N}H_3\ Cl^-$ — Methylammonium chloride

$(C_5H_9)\overset{+}{N}(CH_3)(H)CH_2CH_3\ CF_3CO_2^-$ — *N*-Ethyl-*N*-methylcyclopentylammonium trifluoroacetate

$C_6H_5CH_2\overset{+}{N}(CH_3)_3\ I^-$ — Benzyltrimethylammonium iodide (a quaternary ammonium salt)

22.2 Structure and Bonding

Alkylamines: As shown in Figure 22.1, methylamine, like ammonia, has a pyramidal arrangement of bonds to nitrogen. Its H—N—H angles (106°) are slightly smaller than the tetrahedral value of 109.5°, whereas the C—N—H angle (112°) is slightly larger. The C—N bond distance of 147 pm (1.47 Å) lies between typical C—C bond distances in alkanes (153 pm; 1.53 Å) and C—O bond distances in alcohols (143 pm; 1.43 Å).

Nitrogen and carbon are both sp^3-hybridized and are joined by a σ bond in methylamine. The unshared electron pair on nitrogen occupies an sp^3-hybridized orbital and is involved in reactions in which amines act as bases or nucleophiles. The electrostatic potential map clearly shows the concentration of electron density at nitrogen in methylamine.

Amines that are substituted with three different groups on the nitrogen are chiral but cannot be resolved into enantiomers because of the low energy barrier for racemization by inversion (see Section 4.14).

Arylamines: Aniline, like alkylamines, has a pyramidal arrangement of bonds around nitrogen, but its pyramid is somewhat shallower. One measure of the extent of this flattening is given by the angle between the carbon–nitrogen bond and the bisector of the H—N—H angle.

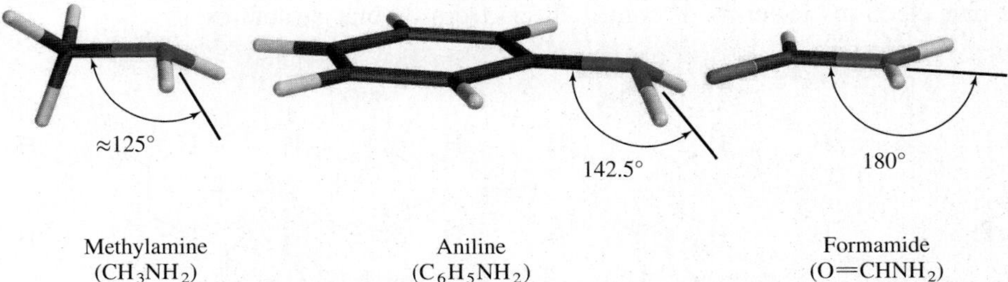

For sp^3-hybridized nitrogen, this angle (not the same as the C—N—H bond angle) is 125°, and the measured angles in simple alkylamines are close to that. The corresponding angle for sp^2 hybridization at nitrogen with a planar arrangement of bonds, as in amides, for example, is 180°. The measured value for this angle in aniline is 142.5°, suggesting a hybridization somewhat closer to sp^3 than to sp^2.

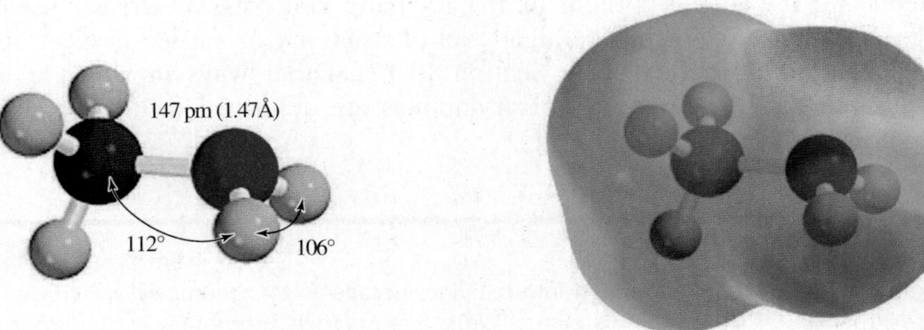

Figure 22.1

Methylamine. (*a*) Bond angles at nitrogen and C—N bond distance. (*b*) The unshared electron pair of nitrogen is a major contributor to the concentration of negative charge indicated by the red region in the electrostatic potential map.

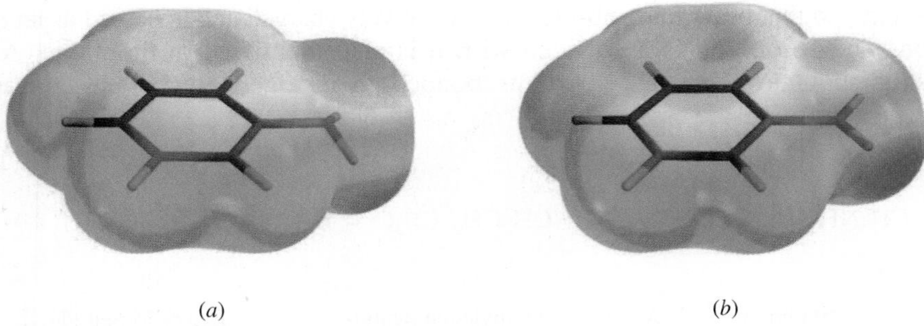

Figure 22.2

Electrostatic potential maps of aniline in which the geometry at nitrogen is (*a*) nonplanar and (*b*) planar. In the nonplanar geometry, the unshared pair occupies an sp^3 hybrid orbital of nitrogen. The region of highest electron density in (*a*) is associated with nitrogen. In the planar geometry, nitrogen is sp^2-hybridized, and the electron pair is delocalized between a *p* orbital of nitrogen and the π system of the ring. The region of highest electron density in (*b*) encompasses both the ring and nitrogen. The actual structure combines features of both; nitrogen adopts a hybridization state between sp^3 and sp^2. The color scale is the same for both models.

The structure of aniline reflects a compromise between two modes of binding the nitrogen lone pair (Figure 22.2). The electrons are more strongly attracted to nitrogen when they are in an orbital with some *s* character—an sp^3-hybridized orbital, for example—than when they are in a *p* orbital. On the other hand, delocalization of these electrons into the aromatic π system is better achieved if they occupy a *p* orbital. A *p* orbital of nitrogen is better aligned for overlap with the *p* orbitals of the benzene ring to form an extended π system than is an sp^3-hybridized orbital. As a result of these two opposing forces, nitrogen adopts an orbital hybridization that is between sp^3 and sp^2.

The corresponding resonance description shows the delocalization of the nitrogen lone-pair electrons in terms of contributions from dipolar structures.

Most stable Lewis structure for aniline

Dipolar resonance contributors of aniline

Delocalization of the nitrogen lone pair decreases the electron density at nitrogen while increasing it in the π system of the aromatic ring. We've already seen one chemical consequence of this in the high level of reactivity of aniline in electrophilic aromatic substitution reactions (see Section 13.12). Other ways in which electron delocalization affects the properties of arylamines are described in later sections of this chapter.

Problem 22.4

As the extent of electron delocalization into the ring increases, the geometry at nitrogen flattens. *p*-Nitroaniline, for example, is planar. Write a resonance contributor for *p*-nitroaniline that shows how the nitro group increases electron delocalization.

22.3 Physical Properties

We have often seen that the polar nature of a substance can affect physical properties such as boiling point. This is true for amines, which are more polar than alkanes but less polar than alcohols. For similarly constituted compounds, alkylamines have boiling points higher than those of alkanes but lower than those of alcohols.

Most commonly encountered alkylamines are liquids with unpleasant, "fishy" odors.

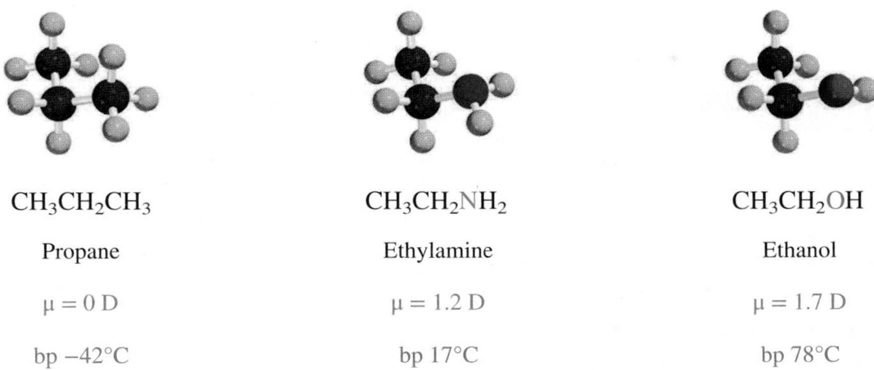

Dipole–dipole interactions, especially hydrogen bonding, are present in amines but absent in alkanes. But because nitrogen is less electronegative than oxygen, an N—H bond is less polar than an O—H bond and hydrogen bonding is weaker in amines than in alcohols.

Among isomeric amines, primary amines have the highest boiling points, and tertiary amines the lowest.

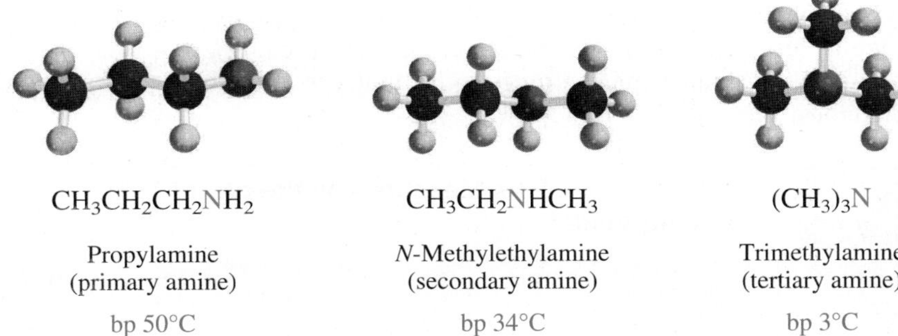

Primary and secondary amines can participate in intermolecular hydrogen bonding, but tertiary amines lack N—H bonds and so cannot.

Amines that have fewer than six or seven carbon atoms are soluble in water. All amines, even tertiary amines, can act as proton acceptors in hydrogen bonding to water molecules.

22.4 Basicity of Amines

When considering the basicity of amines, bear in mind that:

The more basic the amine, the weaker its conjugate acid.

The more basic the amine, the larger the pKa of its conjugate acid.

Citing amine basicity according to pK_a of its conjugate acid makes it possible to analyze acid–base reactions of amines according to the usual Brønsted relationships. For example, we see that amines are converted to ammonium ions by acids even as weak as acetic acid:

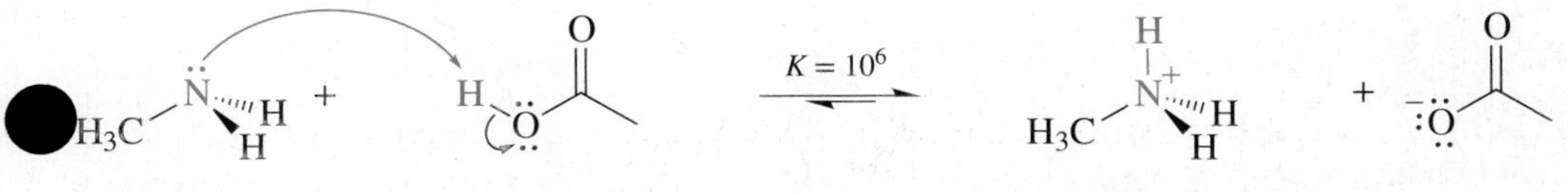

Methylamine | Acetic acid (stronger acid; pK_a = 4.7) | Methylammonium ion (weaker acid; pK_a = 10.7) | Acetate ion

Recall that acid–base reactions are favorable when the stronger acid is on the left and the weaker acid on the right.

Conversely, adding sodium hydroxide to an ammonium salt converts it to the free amine:

$$H_3C{-}\overset{+}{N}H_3 + {:}\ddot{O}{:}^-{-}H \underset{}{\overset{K = 10^5}{\rightleftharpoons}} H_3C{-}\ddot{N}H_2 + H{-}\ddot{O}{:}{-}H$$

Methylammonium ion (stronger acid; $pK_a = 10.7$) — Hydroxide ion — Methylamine — Water (weaker acid; $pK_a = 15.7$)

Problem 22.5

Apply the Henderson–Hasselbalch equation (see Section 19.4) to calculate the $CH_3NH_3^+/CH_3NH_2$ ratio in water buffered at pH 7.

Their basicity provides a means by which amines may be separated from neutral organic compounds. A mixture containing an amine is dissolved in diethyl ether and shaken with dilute hydrochloric acid to convert the amine to an ammonium salt. The ammonium salt, being ionic, dissolves in the aqueous phase, which is separated from the ether layer. Adding sodium hydroxide to the aqueous layer converts the ammonium salt back to the free amine, which is then removed from the aqueous phase by extraction with a fresh portion of ether.

Amines are weak bases, but as a class, *amines are the strongest bases of all neutral molecules.* Table 22.1 lists basicity data for a number of amines. The most important relationships to be drawn from the data are:

1. Alkylamines are slightly stronger bases than ammonia.
2. Alkylamines differ very little among themselves in basicity. Their basicities cover a range of less than 10 in equilibrium constant (1 p*K* unit).
3. Arylamines are about 1 million times (6 p*K* units) weaker bases than ammonia and alkylamines.

TABLE 22.1 Basicity of Amines As Measured by the pK_a of Their Conjugate Acids*

Compound	Structure	pK_a of conjugate acid
Ammonia	NH_3	9.3
Primary amines		
Methylamine	CH_3NH_2	10.6
Ethylamine	$CH_3CH_2NH_2$	10.8
Isopropylamine	$(CH_3)_2CHNH_2$	10.6
tert-Butylamine	$(CH_3)_3CNH_2$	10.4
Aniline	$C_6H_5NH_2$	4.6
Secondary amines		
Dimethylamine	$(CH_3)_2NH$	10.7
Diethylamine	$(CH_3CH_2)_2NH$	11.1
N-Methylaniline	$C_6H_5NHCH_3$	4.8
Tertiary amines		
Trimethylamine	$(CH_3)_3N$	9.7
Triethylamine	$(CH_3CH_2)_3N$	10.8
N,N-Dimethylaniline	$C_6H_5N(CH_3)_2$	5.1

*In water, 25°C.

The small differences in basicity between ammonia and alkylamines, and among the various classes of alkylamines (primary, secondary, tertiary), come from a mix of effects. Replacing hydrogens of ammonia by alkyl groups affects both sides of the acid–base equilibrium in ways that largely cancel.

Replacing hydrogens by aryl groups is a different story, however. An aryl group affects the base much more than the conjugate acid, and the overall effect is large. One way to compare alkylamines and arylamines is by examining the Brønsted equilibrium for proton transfer *to* an alkylamine *from* the conjugate acid of an arylamine.

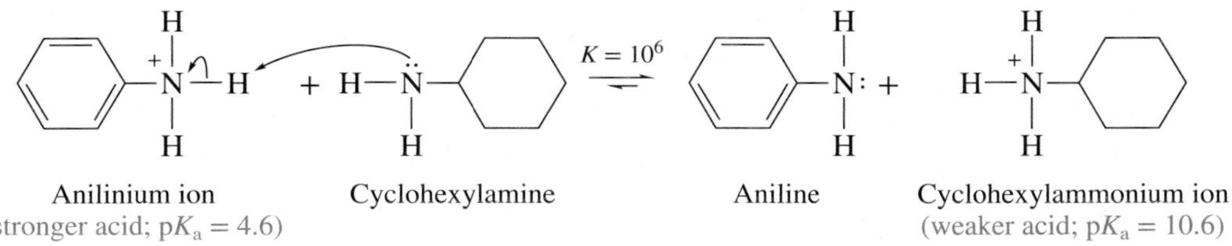

$$C_6H_5NH_3^+ + C_6H_{11}NH_2 \overset{K = 10^6}{\rightleftharpoons} C_6H_5NH_2 + C_6H_{11}NH_3^+$$

Anilinium ion (stronger acid; pK_a = 4.6) | Cyclohexylamine | Aniline | Cyclohexylammonium ion (weaker acid; pK_a = 10.6)

The equilibrium shown in the equation lies far to the right. $K_{eq} = 10^6$ for proton transfer from the conjugate acid of aniline to cyclohexylamine, making cyclohexylamine 1,000,000 times more basic than aniline.

Reading the equation from left to right, we can say that anilinium ion is a stronger acid than cyclohexylammonium ion because loss of a proton from anilinium ion creates a delocalized electron pair of aniline and biases the equilibrium toward the right.

Reading the equation from right to left, we can say that aniline is a weaker base than cyclohexylamine because the electron pair on nitrogen of aniline is strongly held by virtue of being delocalized into the π system of the aromatic ring. The unshared pair in cyclohexylamine is localized on nitrogen, less strongly held, and therefore "more available" in an acid–base reaction.

Problem 22.6

The pK_a's of the conjugate acids of the two amines shown differ by a factor of 40,000. Which amine is the stronger base? Why?

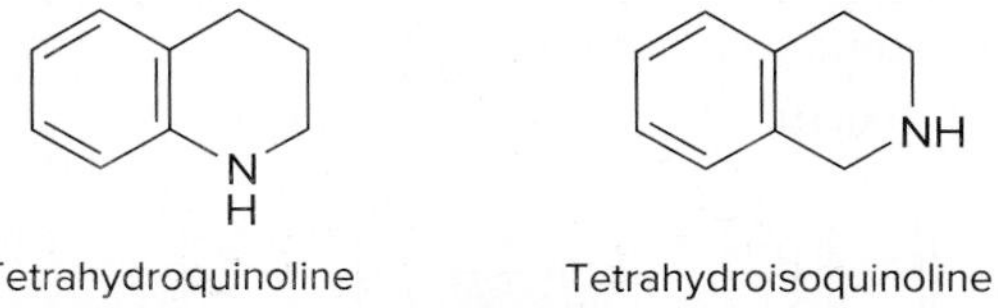

Tetrahydroquinoline | Tetrahydroisoquinoline

Even though they are weaker bases, arylamines, like alkylamines, can be completely protonated by strong acids. Aniline is extracted from an ether solution into 1 M hydrochloric acid by being completely converted to a water-soluble anilinium salt under these conditions.

Conjugation of the amino group of an arylamine with a second aromatic ring, then a third, reduces its basicity even further. Diphenylamine is 6300 times less basic than aniline, whereas triphenylamine is scarcely a base at all, being estimated as 10^{10} times less basic than aniline and 10^{14} times less basic than ammonia.

	$C_6H_5NH_2$	$(C_6H_5)_2NH$	$(C_6H_5)_3N$
	Aniline	Diphenylamine	Triphenylamine
pK_a of conjugate acid:	4.6	0.8	≈ −5

In general, electron-donating substituents on the aromatic ring increase the basicity of arylamines only slightly. Thus, as shown in Table 22.2, an electron-donating methyl group in the para position *increases* the basicity of aniline by less than 1 pK unit. Electron-withdrawing groups are base-weakening and can exert large effects. A *p*-trifluoromethyl group *decreases* the basicity of aniline by a factor of 200 and a *p*-nitro group by a factor

TABLE 22.2 Effect of para Substituents on the Basicity of Aniline

Structure	X	pK_a of conjugate acid
X–C_6H_4–NH_2	H	4.6
	CH_3	5.3
	CF_3	2.5
	O_2N	1.0

of 3800. In the case of *p*-nitroaniline a resonance interaction of the type shown provides for extensive delocalization of the unshared electron pair of the amine group.

Electron delocalization in *p*-nitroaniline

Just as aniline is much less basic than alkylamines because the unshared electron pair of nitrogen is delocalized into the π system of the ring, *p*-nitroaniline is even less basic because the extent of this delocalization is greater and involves the oxygens of the nitro group.

Problem 22.7

Each of the following is a much weaker base than aniline. Present a resonance argument to explain the effect of the substituent in each case.

(a) *o*-Cyanoaniline

(b) 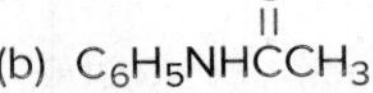
$C_6H_5NHCCH_3$ (with C=O)

(c) *p*-Aminoacetophenone

Sample Solution (a) A cyano substituent is strongly electron-withdrawing. When present at a position ortho to an amino group on an aromatic ring, a cyano substituent increases the delocalization of the amine lone-pair electrons.

This resonance stabilization is lost when the amine group becomes protonated; therefore, *o*-cyanoaniline is a weaker base than aniline.

Multiple substitution by strongly electron-withdrawing groups diminishes the basicity of arylamines still more. Aniline is 3800 times as strong a base as *p*-nitroaniline and 10^9 times more basic than 2,4-dinitroaniline.

Nonaromatic heterocyclic compounds, piperidine, for example, are similar in basicity to alkylamines. When nitrogen is part of an aromatic ring, however, its basicity decreases markedly. Pyridine, for example, is almost 1 million times less basic than piperidine.

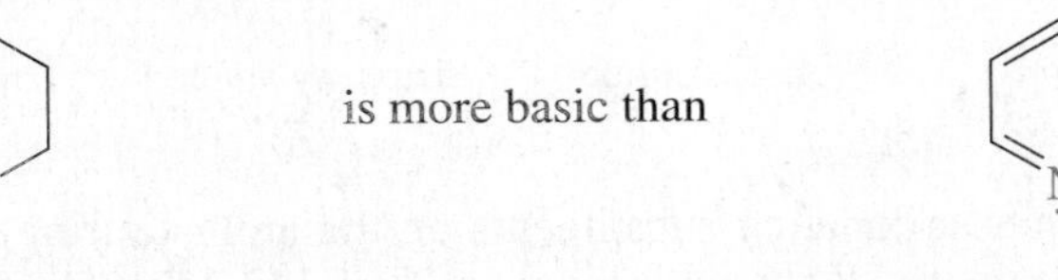

Piperidine
pK_a of conjugate acid = 11.2

Pyridine
pK_a of conjugate acid = 5.2

The difference between the two lies in the fact that the nitrogen lone pair occupies an sp^3-hybridized orbital in piperidine versus an sp^2-hybridized one in pyridine. As we have noted on several occasions, electrons in orbitals with more *s* character are more strongly held than those with less *s* character. For this reason, nitrogen binds its unshared pair more strongly in pyridine than in piperidine and is less basic.

Pyridine and imidazole were two of the heterocyclic aromatic compounds described in Sections 12.21 and 12.22.

Imidazole and its derivatives form an interesting and important class of heterocyclic aromatic amines. Imidazole is approximately 100 times more basic than pyridine. Protonation of imidazole yields an ion that is stabilized by the electron delocalization represented in the resonance structures shown:

H_3O^+

Imidazole
pK_a of conjugate acid = 7

Imidazolium ion

As seen in Figure 22.3, the electrostatic potential map of the conjugate acid of imidazole (imidazolium ion) is consistent with the resonance description that shows both nitrogens as equivalent.

Problem 22.8

Given that the pK_a of imidazolium ion is 7, is a 1 M aqueous solution of imidazolium chloride acidic, basic, or neutral? (b) What about a 1 M solution of imidazole? (c) A solution containing equal molar quantities of imidazole and imidazolium chloride?

Sample Solution

(a) [imidazolium ion] is a stronger acid than H_2O
pK_a = 7   pK_a = 15.7

Therefore, the H_3O^+ concentration is greater in a solution of imidazolium ion in water than it is in pure H_2O—solution of imidazolium chloride is acidic.

An imidazole ring is a structural unit in two biologically important compounds, *histidine* and *histamine.* Histidine is one of the amino acid building blocks of proteins and is directly involved in key proton-transfer processes. The drop in blood pressure

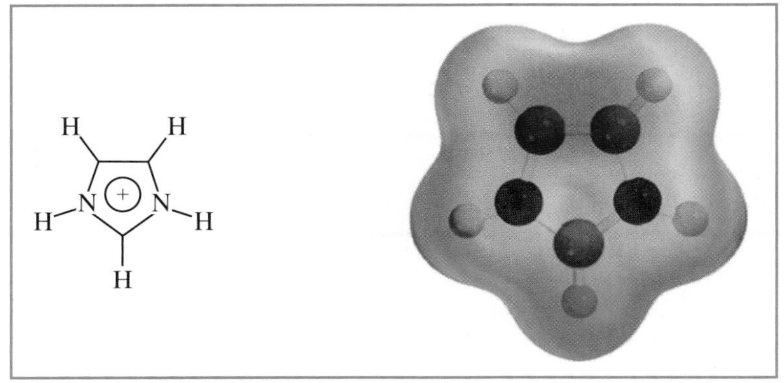

Figure 22.3

Electrostatic potential map of imidazolium ion showing equal distribution of charge between both nitrogens.

associated with shock is a result of the formation of histamine, which stimulates the dilation of blood vessels.

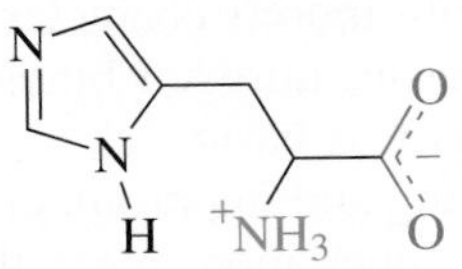

Histidine

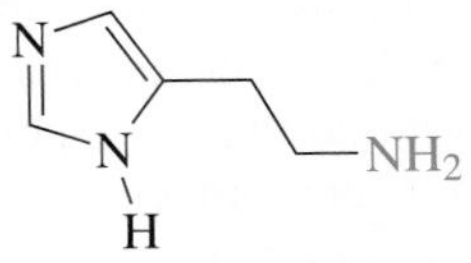

Histamine

Amines as Natural Products

The ease with which amines are extracted into aqueous acid, combined with their regeneration on treatment with base, makes it a simple matter to separate amines from other plant materials, and nitrogen-containing natural products were among the earliest organic compounds to be studied. Their basic (alkaline) properties led amines obtained from plants to be called **alkaloids.** The number of known alkaloids exceeds 5000. They are of special interest because most are characterized by a high level of biological activity. Some examples include *cocaine, coniine,* and *morphine*.

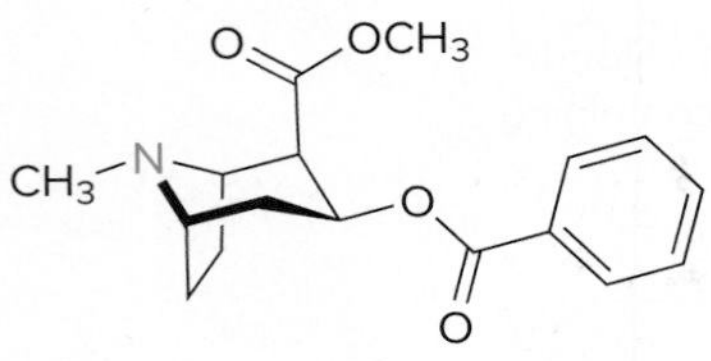

Cocaine

(A central nervous system stimulant obtained from the leaves of the coca plant.)

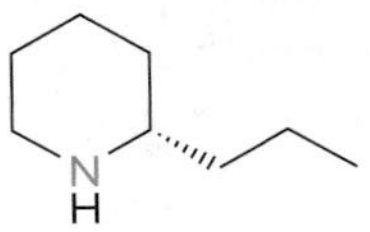

Coniine

(Present among other alkaloids in the hemlock extract used to poison Socrates.)

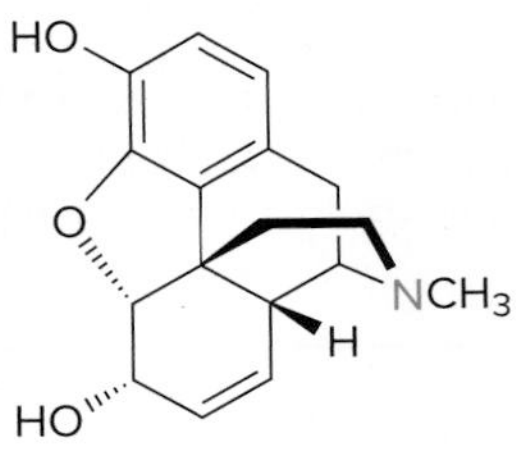

Morphine

(An opium alkaloid. Although an excellent analgesic, its use is restricted because of the potential for addiction. Heroin is the diacetate ester of morphine.)

Many alkaloids, such as *nicotine* and *quinine,* contain two (or more) nitrogen atoms. The nitrogens shown in red in quinine and nicotine are part of a substituted quinoline and pyridine ring, respectively.

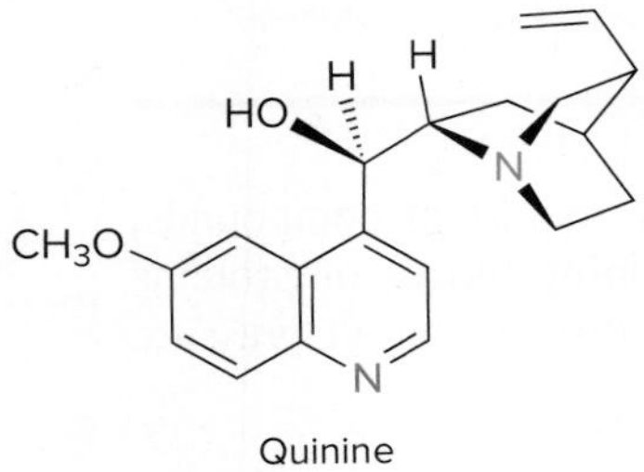

Quinine

(Alkaloid of cinchona bark used to treat malaria.)

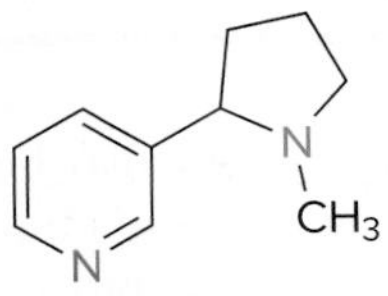

Nicotine

(An alkaloid present in tobacco; a very toxic compound sometimes used as an insecticide.)

Problem 22.9

Write structural formulas for two isomeric conjugate acids of nicotine and use Table 1.8 to estimate their pK_a's.

Several naturally occurring amines mediate the transmission of nerve impulses and are referred to as **neurotransmitters.** Two examples are *epinephrine* and *serotonin.* (Strictly speaking, these compounds are not classified as alkaloids because they are not isolated from plants.)

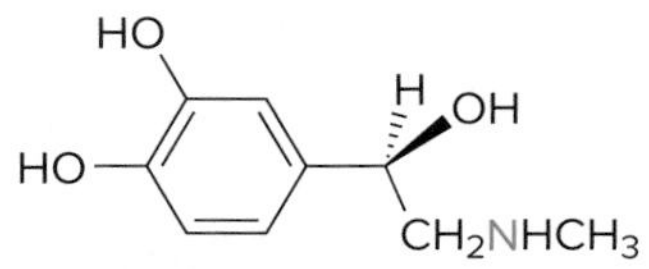

Epinephrine

(Also called *adrenaline;* a hormone secreted by the adrenal gland in preparation for "fight or flight.")

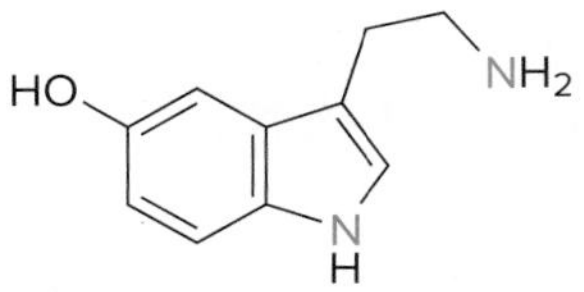

Serotonin

(A hormone synthesized in the pineal gland. Certain mental disorders are believed to be related to serotonin levels in the brain.)

Bioactive amines are also widespread in animals. A variety of structures and properties have been found in substances isolated from frogs, for example. One, called *epibatidine,* is a naturally occurring painkiller isolated from the skin of an Ecuadoran frog (*Epipedobates tricolor*). Another family of frogs produces a toxic mixture of several stereoisomeric amines, called *dendrobines,* on their skin that protects them from attack.

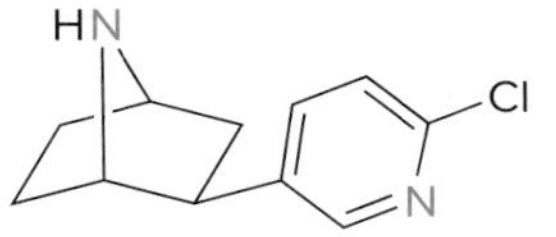

Epibatidine

(Once used as an arrow poison, it is hundreds of times more powerful than morphine in relieving pain, but is too toxic for therapeutic use.)

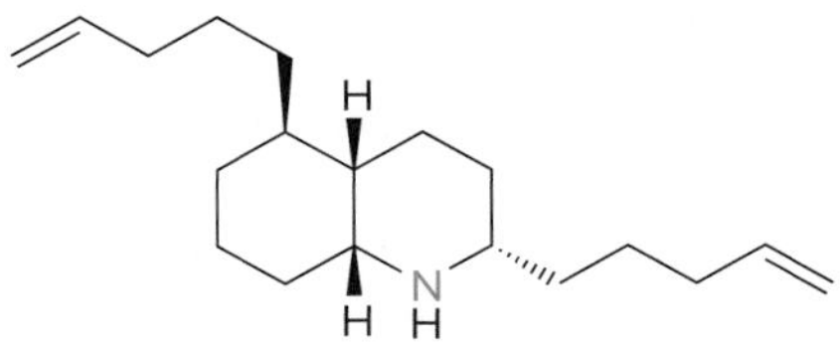

Dendrobine

(Isolated from frogs of the Dendrobatidae family. Related compounds have also been isolated from ants.)

Among the more important amine derivatives found in the body are a group of compounds known as **polyamines,** which contain two to four nitrogen atoms separated by several methylene units:

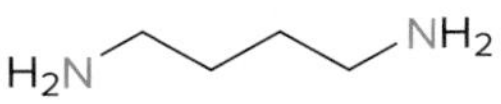

Putrescine

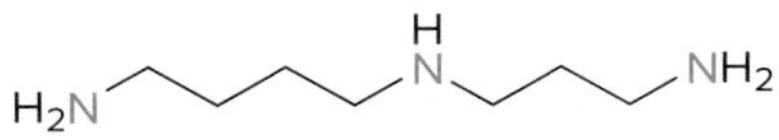

Spermidine

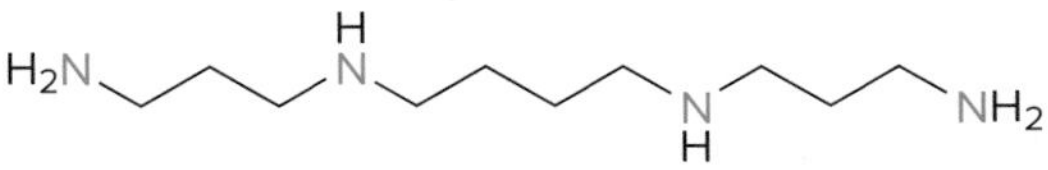

Spermine

These compounds are present in almost all mammalian cells, where they are believed to be involved in cell differentiation and proliferation. Because each nitrogen of a polyamine is protonated at physiological pH (7.4), putrescine, spermidine, and spermine exist as cations with a charge of +2, +3, and +4, respectively, in body fluids. Structural studies suggest that these polyammonium ions affect the conformation of biological macromolecules by electrostatic binding to specific anionic sites—the negatively charged phosphate groups of DNA, for example.

22.5 Tetraalkylammonium Salts as Phase-Transfer Catalysts

In spite of being ionic, many quaternary ammonium salts dissolve in nonpolar media. The four alkyl groups attached to nitrogen shield its positive charge and impart *lipophilic* (hydrophobic) character to the tetraalkylammonium ion. The following two quaternary ammonium salts, for example, are soluble in solvents of low polarity such as benzene, decane, and halogenated hydrocarbons:

$CH_3\overset{+}{N}(CH_2CH_2CH_2CH_2CH_2CH_2CH_2CH_3)_3\ Cl^-$

Methyltrioctylammonium chloride

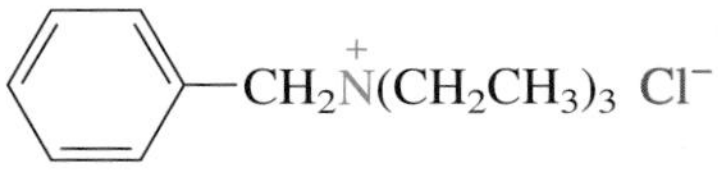

$C_6H_5{-}CH_2\overset{+}{N}(CH_2CH_3)_3\ Cl^-$

Benzyltriethylammonium chloride

$$C_6H_5CH_2\overset{+}{N}(CH_3)_3\ X^- + {}^-CN \rightleftharpoons C_6H_5CH_2\overset{+}{N}(CH_3)_3\ {}^-CN + X^-$$

Benzyltrimethylammonium chloride or bromide; Cyanide ion; Benzyltrimethylammonium cyanide; Halide ion

aqueous phase

butyl bromide phase

$$C_6H_5CH_2\overset{+}{N}(CH_3)_3\ Br^- + CH_3CH_2CH_2CH_2CN \xleftarrow[\text{butyl bromide}]{CH_3CH_2CH_2CH_2Br} C_6H_5CH_2\overset{+}{N}(CH_3)_3\ {}^-CN$$

Benzyltrimethylammonium bromide; Pentanenitrile; Benzyltrimethylammonium cyanide

Figure 22.4

Phase-transfer catalysis. Nucleophilic cyanide ion is transferred from the aqueous to the organic phase as benzyltrimethylammonium cyanide.

This property of quaternary ammonium salts is used to advantage in an experimental technique known as **phase-transfer catalysis.** Imagine that you wish to carry out the reaction

$$\underset{\text{Butyl bromide}}{CH_3CH_2CH_2CH_2Br} + \underset{\text{Sodium cyanide}}{NaCN} \longrightarrow \underset{\text{Pentanenitrile}}{CH_3CH_2CH_2CH_2CN} + \underset{\text{Sodium bromide}}{NaBr}$$

Sodium cyanide does not dissolve in butyl bromide. The two reactants contact each other only at the surface of the solid sodium cyanide, and the rate of reaction under these conditions is too slow to be of synthetic value. Dissolving the sodium cyanide in water is of little help because butyl bromide is not soluble in water and reaction can occur only at the interface between the two phases. Adding a small amount of benzyltrimethylammonium chloride, however, causes pentanenitrile to form rapidly even at room temperature. The quaternary ammonium salt is acting as a *catalyst;* it increases the reaction rate but is not consumed. How?

Quaternary ammonium salts catalyze the reaction between an anion and an organic substrate by transferring the anion from the aqueous phase, where it cannot contact the substrate, to the organic phase. In the cycle shown in Figure 22.4, the first step occurs in the aqueous phase and is an exchange of the anionic partner of the quaternary ammonium salt for cyanide ion. The benzyltrimethylammonium ion migrates to the butyl bromide phase, carrying a cyanide ion along with it. Once in the organic phase, cyanide ion is only weakly solvated and is far more reactive than it is in water or ethanol, where it is strongly solvated by hydrogen bonding. The benzyltrimethylammonium bromide formed in the substitution step returns to the aqueous phase, where it can repeat the cycle. Phase-transfer catalysis incorporates principles of green chemistry. Reactions are faster and proceed in higher yield; fewer side products are produced; the need for excess reagents (cyanide in this case) is minimized; and water can be used as a solvent.

22.6 Reactions That Lead to Amines: A Review and a Preview

Methods for preparing amines address either or both of the following questions:

1. How is the required carbon–nitrogen bond to be formed?
2. Given a nitrogen-containing organic compound such as an amide, a nitrile, or a nitro compound, how is the correct oxidation state of the desired amine to be achieved?

A number of reactions that lead to carbon–nitrogen bond formation were presented in earlier chapters and are summarized in Table 22.3. Among the reactions in the table, the nucleophilic ring opening of epoxides and the reaction of α-halo acids with ammonia give amines directly. The other reactions in Table 22.3 yield products that are converted to amines by some subsequent procedure. As these procedures are described in the following sections, you will see that they are largely applications of principles that you've already learned. You will encounter some new reagents and some new uses for familiar reagents, but very little in the way of new reaction types is involved.

TABLE 22.3 Methods for Carbon–Nitrogen Bond Formation Discussed in Earlier Chapters

Reaction (section) and comments	General equation and specific example
Nucleophilic substitution by azide ion on an alkyl halide (see Sections 6.1, 7.19) Azide ion is a very good nucleophile and reacts with primary and secondary alkyl halides to give alkyl azides. Phase-transfer catalysts accelerate the rate of reaction.	$:\ddot{N}=\overset{+}{N}=\ddot{N}:^- + R-X \longrightarrow :\ddot{N}^- =\overset{+}{N}=\ddot{N}-R + X^-$ Azide ion + Alkyl halide → Alkyl azide + Halide ion Pentyl bromide (1-bromopentane) $\xrightarrow[\text{phase-transfer catalyst}]{NaN_3}$ Pentyl azide (89%) (1-azidopentane)
Nitration of arenes (see Section 13.3) The standard method for introducing a nitrogen atom as a substituent on an aromatic ring is nitration with a mixture of nitric acid and sulfuric acid. The reaction proceeds by electrophilic aromatic substitution.	$ArH + HNO_3 \xrightarrow{H_2SO_4} ArNO_2 + H_2O$ Arene + Nitric acid → Nitroarene + Water Benzaldehyde $\xrightarrow[H_2SO_4]{HNO_3}$ 3-Nitrobenzaldehyde (75–84%)
Nucleophilic ring opening of epoxides by ammonia (see Section 17.11) The strained ring of an epoxide is opened on nucleophilic attack by ammonia and amines to give β-amino alcohols. Azide ion also reacts with epoxides; the products are β-azido alcohols.	$H_3N: +$ Epoxide → β-Amino alcohol Ammonia + Epoxide → β-Amino alcohol (2*R*,3*R*)-2,3-Epoxybutane $\xrightarrow[H_2O]{NH_3}$ (2*R*,3*S*)-3-Amino-2-butanol (70%)
Nucleophilic addition of amines to aldehydes and ketones (see Sections 18.10, 18.11) Primary amines undergo nucleophilic addition to the carbonyl group of aldehydes and ketones to form hemiaminals, which dehydrate under the conditions of their formation to give *N*-substituted imines. Secondary amines yield enamines.	$R\ddot{N}H_2 + R'\overset{O}{\overset{\Vert}{C}}R'' \longrightarrow R'\overset{:NR}{\overset{\Vert}{C}}R'' + H_2O$ Primary amine + Aldehyde or ketone → Imine + Water $CH_3NH_2 + C_6H_5\overset{O}{\overset{\Vert}{C}}H \longrightarrow C_6H_5CH=NCH_3$ Methylamine + Benzaldehyde → *N*-Benzylidenemethylamine (70%)

continued

TABLE 22.3 Methods for Carbon–Nitrogen Bond Formation Discussed in Earlier Chapters (*Continued*)

Reaction (section) and comments	General equation and specific example
Nucleophilic substitution by ammonia on α-halo acids (see Section 21.6) The α-halo acids obtained by halogenation of carboxylic acids under conditions of the Hell–Volhard–Zelinsky reaction are reactive substrates in nucleophilic substitution processes. A standard method for the preparation of α-amino acids is displacement of halide from α-halo acids by nucleophilic substitution using excess aqueous ammonia.	$H_3N{:} + RCH(X)CO_2H \longrightarrow RCH(\overset{+}{N}H_3)CO_2^- + NH_4X$ Ammonia + α-Halo carboxylic acid → α-Amino acid + Ammonium halide 2-Bromo-3-methylbutanoic acid $\xrightarrow[H_2O]{NH_3}$ 2-Amino-3-methylbutanoic acid (47–48%)
Nucleophilic acyl substitution (see Sections 20.4, 20.5, and 20.10) Acylation of ammonia and amines by an acyl chloride, acid anhydride, or ester is an exceptionally effective method for the formation of carbon–nitrogen bonds.	$R_2\ddot{N}H + R'C(O)X \longrightarrow R_2\ddot{N}C(O)R' + HX$ Amine (primary or secondary) or ammonia + Acyl chloride, acid anhydride, or ester → Amide 2 Pyrrolidine + Acetyl chloride → *N*-Acetylpyrrolidine (79%) + Pyrrolidine hydrochloride

22.7 Preparation of Amines by Alkylation of Ammonia

Alkylamines are, in principle, capable of being prepared by nucleophilic substitution reactions of alkyl halides with ammonia.

$$\underset{\text{Alkyl halide}}{RX} + \underset{\text{Ammonia}}{2NH_3} \longrightarrow \underset{\text{Primary amine}}{RNH_2} + \underset{\text{Ammonium halide salt}}{\overset{+}{N}H_4\,X^-}$$

Although this reaction is useful for preparing α-amino acids (see Table 22.3, fifth entry), it is *not* a general method for the synthesis of amines. Its major limitation is that the expected primary amine product is itself a nucleophile and competes with ammonia for the alkyl halide. When 1-bromooctane, for example, is allowed to react with ammonia, both the primary amine and the secondary amine are isolated in comparable amounts.

$$\underset{\substack{\text{1-Bromooctane}\\ \text{(1 mol)}}}{CH_3(CH_2)_6CH_2Br} \xrightarrow{NH_3\text{ (2 mol)}} \underset{\substack{\text{Octylamine}\\ (45\%)}}{CH_3(CH_2)_6CH_2NH_2} + \underset{\substack{\textit{N,N}\text{-Dioctylamine}\\ (43\%)}}{[CH_3(CH_2)_6CH_2]_2NH}$$

Competitive alkylation may continue, resulting in the formation of tertiary amines and quaternary ammonium salts.

Alkylation of ammonia is used to prepare primary amines only when the starting alkyl halide is not particularly expensive and the desired amine can be easily separated from the other components of the reaction mixture.

Problem 22.10

Alkylation of ammonia is sometimes employed in industrial processes such as the preparation of allylamine from propene, chlorine, and ammonia. Write a series of equations summarizing this process. (Allylamine has a number of uses, including the preparation of the diuretic drugs *meralluride* and *mercaptomerin*.)

Aryl halides that are substituted with electron-withdrawing groups react with amines by nucleophilic aromatic substitution (see Section 13.19).

22.8 The Gabriel Synthesis of Primary Alkylamines

A method that achieves the same end result as that of alkylation of ammonia but which avoids the formation of secondary and tertiary amines as byproducts is the **Gabriel synthesis.** Alkyl halides are converted to primary alkylamines without contamination by secondary or tertiary amines. The key reagent is the potassium salt of phthalimide, prepared by the acid–base reaction

Phthalimide ($pK_a = 8.3$) + KOH ⟶ *N*-Potassiophthalimide + H_2O Water ($pK_a = 15.7$)

Phthalimide, with a pK_a of 8.3, can be quantitatively converted to its potassium salt with potassium hydroxide. The potassium salt of phthalimide has a negatively charged nitrogen, which acts as a nucleophile toward primary alkyl halides in a bimolecular nucleophilic substitution (S_N2).

DMF is an abbreviation for *N,N*-dimethylformamide, DMF is a polar aprotic solvent (see Section 6.9) and an excellent medium for S_N2 reactions.

N-Potassiophthalimide + Benzyl chloride $\xrightarrow{\text{DMF}}$ *N*-Benzylphthalimide (74%) + KCl Potassium chloride

The product of this reaction is an imide, a diacyl derivative of an amine. Either aqueous acid or aqueous base can be used to hydrolyze its two amide bonds and liberate the desired primary amine. A more effective method of cleaving the two amide bonds is by reaction with hydrazine:

N-Benzylphthalimide + H_2NNH_2 Hydrazine $\xrightarrow{\text{ethanol}}$ Benzylamine + Phthalhydrazide

Aryl halides cannot be converted to arylamines by the Gabriel synthesis because they do not undergo nucleophilic substitution with *N*-potassiophthalimide in the first step of the procedure.

Among compounds other than simple alkyl halides, α-halo ketones, α-halo esters, and alkyl *p*-toluenesulfonates have also been used. Because phthalimide can undergo only a single alkylation, the formation of secondary and tertiary amines does not occur, and the Gabriel synthesis is a valuable procedure for the laboratory preparation of primary amines.

Problem 22.11

Three of the following amines can be prepared by the Gabriel synthesis; three cannot. Write equations showing the successful applications of this method.

(a) Butylamine
(b) Isobutylamine
(c) *tert*-Butylamine
(d) 2-Phenylethylamine
(e) *N*-Methylbenzylamine
(f) Aniline

Sample Solution

(a) • Gabriel synthesis makes primary amines (RNH_2).

• C—N bond is made by S_N2 reaction. Therefore, works best with primary alkyl halides.

• Use $CH_3CH_2CH_2CH_2Br$ to make $CH_3CH_2CH_2CH_2NH_2$.

$CH_3CH_2CH_2CH_2Br$ + (potassium phthalimide, NK) → (N-butylphthalimide, $NCH_2CH_2CH_2CH_3$)

↓ H_2NNH_2

(phthalhydrazide, NH/NH) + $H_2NCH_2CH_2CH_2CH_3$

22.9 Preparation of Amines by Reduction

Almost any nitrogen-containing organic compound can be reduced to an amine. The synthesis of amines then becomes a question of the availability of suitable precursors and the choice of an appropriate reducing agent.

Alkyl *azides,* prepared by nucleophilic substitution of alkyl halides by sodium azide, as shown in the first entry of Table 22.3, are reduced to alkylamines by a variety of reagents, including lithium aluminum hydride.

$$R{-}\ddot{N}{=}\overset{+}{N}{=}\ddot{N}:^{-} \xrightarrow{\text{reduce}} R\ddot{N}H_2$$

Alkyl azide → Primary amine

$$C_6H_5CH_2CH_2N_3 \xrightarrow[\text{2. } H_2O]{\text{1. } LiAlH_4,\ \text{diethyl ether}} C_6H_5CH_2CH_2NH_2$$

2-Phenylethyl azide → 2-Phenylethylamine (89%)

Catalytic hydrogenation is also effective:

1,2-Epoxycyclohexane $\xrightarrow[\text{dioxane–water}]{NaN_3}$ *trans*-2-Azidocyclohexanol (61%) $\xrightarrow{H_2, Pt}$ *trans*-2-Aminocyclohexanol (81%)

The same reduction methods may be applied to the conversion of *nitriles* to primary amines.

$$RC{\equiv}N \xrightarrow[H_2,\ \text{catalyst}]{LiAlH_4\ \text{or}} RCH_2NH_2$$

Nitrile → Primary amine

p-(Trifluoromethyl)benzyl cyanide $\xrightarrow[\text{2. } H_2O]{\text{1. } LiAlH_4,\ \text{diethyl ether}}$ *p*-(Trifluoromethyl)phenylethylamine (53%)

Pentanenitrile $\xrightarrow[\text{diethyl ether}]{H_2\ (100\ \text{atm}),\ Ni}$ 1-Pentanamine (56%)

The preparation of pentanenitrile under phase-transfer conditions was described in Section 22.5.

Because nitriles can be prepared from alkyl halides by nucleophilic substitution with cyanide ion, the overall process $RX \rightarrow RC{\equiv}N \rightarrow RCH_2NH_2$ leads to primary amines that have one more carbon atom than the starting alkyl halide.

Cyano groups in *cyanohydrins* (see Section 18.7) are reduced under the same reaction conditions.

Nitro groups are readily reduced to primary amines by a variety of methods. Catalytic hydrogenation over platinum, palladium, or nickel is often used, as is reduction by iron or tin in hydrochloric acid. The ease with which nitro groups are reduced is especially useful in the preparation of arylamines, where the sequence $ArH \rightarrow ArNO_2 \rightarrow ArNH_2$ is the standard route to these compounds.

o-Isopropylnitrobenzene $\xrightarrow[\text{methanol}]{H_2,\ Ni}$ *o*-Isopropylaniline (92%)

p-Chloronitrobenzene $\xrightarrow[\text{2. NaOH}]{\text{1. Fe, HCl}}$ *p*-Chloroaniline (95%)

m-Nitroacetophenone $\xrightarrow[\text{2. NaOH}]{\text{1. Sn, HCl}}$ *m*-Aminoacetophenone (82%)

For reductions carried out in acidic media, a pH adjustment with sodium hydroxide is required in the last step in order to convert $ArNH_3^+$ to $ArNH_2$.

Problem 22.12

Outline the synthesis of each of the following arylamines from benzene:

(a) *o*-Isopropylaniline
(b) *p*-Isopropylaniline
(c) 4-Isopropyl-1,3-benzenediamine
(d) *p*-Chloroaniline
(e) *m*-Aminoacetophenone

Sample Solution (a) The last step in the synthesis of *o*-isopropylaniline, the reduction of the corresponding nitro compound by catalytic hydrogenation, is given as one of the three preceding examples. The necessary nitroarene is obtained by fractional distillation of the ortho–para mixture formed during nitration of isopropylbenzene.

Isopropylbenzene $\xrightarrow{HNO_3}$ *o*-Isopropylnitrobenzene (bp 110°C) + *p*-Isopropylnitrobenzene (bp 131°C)

As actually performed, a 62% yield of a mixture of ortho and para nitration products was obtained with an ortho–para ratio of about 1:3.

Isopropylbenzene is prepared by the Friedel–Crafts alkylation of benzene using isopropyl chloride and aluminum chloride (see Section 13.6).

Reduction of an azide, a nitrile, or a nitro compound furnishes a primary amine. A method that provides access to primary, secondary, or tertiary amines is reduction of the carbonyl group of an amide by lithium aluminum hydride.

$$\underset{\text{Amide}}{RC(=O)NR'_2} \xrightarrow[2.\ H_2O]{1.\ LiAlH_4} \underset{\text{Amine}}{RCH_2NR'_2}$$

In this general equation, R and R′ may be either alkyl or aryl groups. When R′ = H, the product is a primary amine:

3-Phenylbutanamide $\xrightarrow[2.\ H_2O]{1.\ LiAlH_4,\ \text{diethyl ether}}$ 3-Phenyl-1-butanamine (59%)

N-Substituted amides yield secondary amines:

Acetanilide $\xrightarrow[2.\ H_2O]{1.\ LiAlH_4,\ \text{diethyl ether}}$ *N*-Ethylaniline (92%)

Acetanilide is an acceptable IUPAC synonym for *N*-phenylethanamide.

N,N-Disubstituted amides yield tertiary amines:

N,N-Dimethylcyclohexanecarboxamide $\xrightarrow[2.\ H_2O]{1.\ LiAlH_4,\ \text{diethyl ether}}$ *N,N*-Dimethyl(cyclohexylmethyl)amine (88%)

Mechanism 22.1 outlines the reduction of amides with lithium aluminum hydride. The reduction of amides follows a similar course to the reduction of esters (see Section 20.11). A tetrahedral intermediate is formed by the addition of hydride (step 1), and undergoes elimination (step 2). In the case of an ester, the alkoxy group is lost to give an intermediate aldehyde. Amides, on the other hand, retain the nitrogen and lose the oxygen from the tetrahedral intermediate. The iminium ion formed undergoes addition of a second hydride and is reduced to the amine in step 3. Because amides are so easy to prepare, this is a versatile method for preparing amines.

Mechanism 22.1

Lithium Aluminum Hydride Reduction of an Amide

THE OVERALL REACTION:

$$2RC(=O)NR'_2 + LiAlH_4 \xrightarrow{\text{diethyl ether}} 2RCH_2NR'_2 + LiAlO_2$$

Amide — Amine

THE MECHANISM:

Step 1: Hydride is transferred from aluminum to the carbonyl carbon of the amide to give a tetrahedral intermediate. The carbonyl oxygen becomes bound to aluminum.

Lithium aluminum hydride + Amide → Tetrahedral intermediate

Step 2: The tetrahedral intermediate undergoes elimination to form an iminium ion.

Tetrahedral intermediate → Iminium ion + H_3Al^{-}—ÖLi

Step 3: The iminium ion undergoes addition of a second hydride. For simplicity, the source of hydride shown in the next equation is the $[LiOAlH_3]^-$ species formed in step 2, but it can be any species present in the reaction mixture that retains an Al—H bond. Thus, the overall stoichiometry corresponds to reduction of two moles of amide per mole of $LiAlH_4$ and the final inorganic product is $LiAlO_2$.

Iminium ion + $[H_3AlOLi]^-$ → Amine + H_2Al—ÖLi

The methods described in this section involve the prior synthesis and isolation of some reducible material that has a carbon–nitrogen bond: an azide, a nitrile, a nitro-substituted arene, or an amide. The following section describes a method that combines the two steps of carbon–nitrogen bond formation and reduction into a single operation. Like the reduction of amides, it offers the possibility of preparing primary, secondary, or tertiary amines by proper choice of starting materials.

22.10 Reductive Amination

A class of nitrogen-containing compounds that was omitted from the section just discussed includes *imines* and their derivatives. Imines are formed by the reaction of aldehydes and ketones with ammonia and primary amines (see Section 18.10). Imines derived from ammonia can be reduced to primary amines by catalytic hydrogenation.

$$\underset{\text{Aldehyde or ketone}}{RC(=O)R'} + \underset{\text{Ammonia}}{NH_3} \longrightarrow \underset{\text{Imine}}{RC(=NH)R'} \xrightarrow[\text{catalyst}]{H_2} \underset{\text{Primary amine}}{RCH(NH_2)R'}$$

The overall reaction converts a carbonyl compound to an amine by carbon–nitrogen bond formation and reduction; it is commonly known as **reductive amination.** What makes it a particularly valuable synthetic procedure is that it can be carried out in a single operation by hydrogenation of a solution containing both ammonia and the carbonyl compound along with a hydrogenation catalyst. The intermediate imine is not isolated but undergoes reduction under the conditions of its formation. Also, the reaction is broader in scope than implied by the preceding equation. All classes of amines—primary, secondary, and tertiary—may be prepared by reductive amination.

When *primary amines* are desired, the reaction is carried out as just described, using ammonia as the nitrogen source.

$$\underset{\text{Cyclohexanone}}{C_6H_{10}{=}O} + \underset{\text{Ammonia}}{NH_3} \xrightarrow[\text{ethanol}]{H_2,\ Ni} \underset{\text{Cyclohexylamine (80\%)}}{C_6H_{11}NH_2} \quad \text{via} \quad C_6H_{10}{=}NH$$

Secondary amines are prepared by hydrogenation of a carbonyl compound in the presence of a primary amine. An *N*-substituted imine, or *Schiff's base,* is an intermediate:

$$\underset{\text{Heptanal}}{CH_3(CH_2)_5CH{=}O} + \underset{\text{Aniline}}{H_2N{-}C_6H_5} \xrightarrow[\text{ethanol}]{H_2,\ Ni} \underset{N\text{-Heptylaniline (65\%)}}{CH_3(CH_2)_5CH_2NH{-}C_6H_5} \quad \text{via} \quad CH_3(CH_2)_5CH{=}N{-}C_6H_5$$

Reductive amination has been successfully applied to the preparation of *tertiary amines* from carbonyl compounds and secondary amines even though a neutral imine is not possible in such cases.

$$\underset{\text{Butanal}}{CH_3CH_2CH_2CH{=}O} + \underset{\text{Piperidine}}{H{-}NC_5H_{10}} \xrightarrow[\text{ethanol}]{H_2,\ Ni} \underset{N\text{-Butylpiperidine (93\%)}}{CH_3CH_2CH_2CH_2{-}NC_5H_{10}}$$

Presumably, the species that undergoes reduction here is a hemiaminal, an iminium ion derived from it, or an enamine.

Hemiaminal ⇌ Iminium ion + $HÖ:^-$

Problem 22.13

Show how you could prepare each of the following amines from benzaldehyde by reductive amination:

(a) Benzylamine
(b) Dibenzylamine
(c) *N,N*-Dimethylbenzylamine
(d) *N*-Benzylpiperidine

Sample Solution (a) Because benzylamine is a primary amine, it is derived from ammonia and benzaldehyde.

$$\text{Benzaldehyde} + NH_3 + H_2 \xrightarrow{Ni} \text{Benzylamine (89\%)} + H_2O$$

Benzaldehyde Ammonia Hydrogen Benzylamine (89%) Water

The reaction proceeds by initial formation of the imine $C_6H_5CH{=}NH$, followed by its hydrogenation.

A variation of the classical reductive amination procedure uses sodium cyanoborohydride ($NaBH_3CN$) instead of hydrogen as the reducing agent and is better suited to amine syntheses in which only a few grams of material are needed. All that is required is to add sodium cyanoborohydride to an alcohol solution of the carbonyl compound and an amine.

$$\text{Benzaldehyde} + H_2N\text{CH}_2\text{CH}_3 \xrightarrow[\text{methanol}]{NaBH_3CN} \text{N-Ethylbenzylamine (91\%)}$$

Benzaldehyde Ethylamine *N*-Ethylbenzylamine (91%)

Sodium cyanoborohydride reduces aldehydes and ketones less rapidly than sodium borohydride, but it reduces iminium ions rapidly. To take advantage of this selectivity, reductive aminations are carried out at mildly acidic pH, where the imines are protonated. Iminium ions are also more reactive than imines toward reduction with hydride.

22.11 Reactions of Amines: A Review and a Preview

The most noteworthy properties of amines are their *basicity* and their *nucleophilicity*. The basicity of amines has been discussed in Section 22.4. Several reactions in which amines act as nucleophiles have already been encountered in earlier chapters. These are summarized in Table 22.4.

Both the basicity and the nucleophilicity of amines originate in the unshared electron pair of nitrogen. When an amine acts as a base, this electron pair abstracts a proton from a Brønsted acid. When an amine undergoes the reactions summarized in Table 22.4, the first step in each case is nucleophilic addition to the positively polarized carbon of a carbonyl group.

$R_3N\!:\ \ H{-}X$ $R_3N\!:\ \ C{=}O$

Amine acting as a base Amine acting as a nucleophile

TABLE 22.4 Reactions of Amines Discussed in Previous Chapters*

Reaction (section) and comments	General equation and specific example
Reaction of primary amines with aldehydes and ketones (see Section 18.10) Imines are formed by nucleophilic addition of a primary amine to the carbonyl group of an aldehyde or a ketone. The key step is formation of a hemiaminal intermediate, which then dehydrates to the imine.	$R\ddot{N}H_2 + R'R''C{=}O \longrightarrow [R\ddot{N}H{-}C(R')(R''){-}OH] \xrightarrow{-H_2O} R\ddot{N}{=}CR'R''$ Primary amine; Aldehyde or ketone; Hemiaminal intermediate; Imine CH_3NH_2 + Benzaldehyde ⟶ N-Benzylidenemethylamine (70%) + H_2O Methylamine; Benzaldehyde; *N*-Benzylidenemethylamine (70%); Water
Reaction of secondary amines with aldehydes and ketones (see Section 18.11) Enamines are formed in the corresponding reaction of secondary amines with aldehydes and ketones.	$R_2\ddot{N}H + R'CH_2(R'')C{=}O \longrightarrow [R_2\ddot{N}{-}C(CH_2R')(R''){-}OH] \xrightarrow{-H_2O} R_2\ddot{N}{-}C(R''){=}CHR'$ Secondary amine; Aldehyde or ketone; Hemiaminal intermediate; Enamine Pyrrolidine + Cyclohexanone $\xrightarrow[\text{heat}]{\text{benzene}}$ N-(1-Cyclohexenyl)pyrrolidine (85–90%) + H_2O Pyrrolidine; Cyclohexanone; *N*-(1-Cyclohexenyl)pyrrolidine (85–90%); Water
Reaction of amines with acyl chlorides (see Section 20.4) Amines are converted to amides on reaction with acyl chlorides. Other acylating agents, such as acid anhydrides and esters, may also be used but are less reactive.	$R_2\ddot{N}H + R'\overset{O}{\overset{\Vert}{C}}Cl \longrightarrow [R_2\ddot{N}{-}C(OH)(R'){-}Cl] \xrightarrow{-HCl} R_2\ddot{N}\overset{O}{\overset{\Vert}{C}}R'$ Primary or secondary amine; Acyl chloride; Tetrahedral intermediate; Amide Butylamine + Pentanoyl chloride ⟶ N-Butylpentanamide (81%) Butylamine; Pentanoyl chloride; *N*-Butylpentanamide (81%)

*Both alkylamines and arylamines undergo these reactions.

In addition to being more basic than arylamines, alkylamines are more nucleophilic. All the reactions in Table 22.4 take place faster with alkylamines than with arylamines.

The sections that follow introduce some additional reactions of amines. In all cases our understanding of how these reactions take place starts with a consideration of the role of the unshared electron pair of nitrogen.

We'll begin with an examination of the reactivity of amines as nucleophiles in S_N2 reactions.

22.12 Reaction of Amines with Alkyl Halides

The reaction of amines with alkyl halides was seen earlier (see Section 22.7) as a complicating factor in the preparation of amines by alkylation of ammonia.

Nucleophilic substitution results when primary alkyl halides are treated with amines.

$$R'CH_2X + RNH_2 \xrightarrow{S_N2} R'CH_2\overset{+}{N}H_2R\ :X^- \xrightarrow{RNH_2 \to RNH_3^+} R'CH_2NHR$$

Primary alkyl halide; Primary amine; Ammonium halide salt; Secondary amine

A second alkylation may follow, converting the secondary amine to a tertiary amine. Alkylation need not stop there; the tertiary amine may itself be alkylated, giving a quaternary ammonium salt.

$$RNH_2 \xrightarrow[S_N2]{R'CH_2X} RNHCH_2R' \xrightarrow[S_N2]{R'CH_2X} RN(CH_2R')_2 \xrightarrow[S_N2]{R'CH_2X} R\overset{+}{N}(CH_2R')_3\ :X^-$$

Primary amine; Secondary amine; Tertiary amine; Quaternary ammonium salt

Thus, if the goal is to convert a primary to a secondary amine, it is customary to carry out the alkylation in the presence of excess primary amine.

$$C_6H_5CH_2Cl + H_2NC_6H_5 \xrightarrow[90°C]{NaHCO_3} C_6H_5CH_2NHC_6H_5$$

Benzyl chloride (1 mol); Aniline (4 mol); *N*-Benzylaniline (85–87%)

Among the family of amine-derived compounds, quaternary ammonium salts are the most different from the others in their properties and applications. We have touched on one of these applications earlier in their role as phase-transfer catalysts (see Section 22.5). In general, the most commonly encountered quaternary ammonium compounds are those derived by alkylation of primary, secondary, or tertiary amines with methyl iodide.

$$C_6H_{11}CH_2NH_2 + 3CH_3I \xrightarrow{\text{methanol}} C_6H_{11}CH_2\overset{+}{N}(CH_3)_3\ :I:^-$$

(Cyclohexylmethyl)amine; Methyl iodide; (Cyclohexylmethyl)-trimethylammonium iodide (99%)

In Section 22.13, we'll see how quaternary ammonium hydroxides are used to prepare alkenes by a novel elimination method.

22.13 The Hofmann Elimination

When quaternary ammonium *hydroxides* are heated, they undergo an E2 β elimination to form an alkene and an amine. The quaternary ammonium hydroxide itself is prepared by treating the corresponding iodide with an aqueous slurry of silver oxide. Silver iodide precipitates, and a solution of the quaternary ammonium hydroxide is formed. Heating the quaternary ammonium hydroxide causes the elimination of trimethylamine.

(Cyclohexylmethyl)trimethylammonium iodide $\xrightarrow[\text{H}_2\text{O, CH}_3\text{OH}]{\text{Ag}_2\text{O}}$ (Cyclohexylmethyl)trimethylammonium hydroxide $\xrightarrow[-\text{H}_2\text{O}]{160°\text{C}}$ Methylenecyclohexane (69%) + Trimethylamine

This reaction is known as the **Hofmann elimination;** it was developed by August W. Hofmann in the middle of the nineteenth century and is both a synthetic method to prepare alkenes and an analytical tool for structure determination.

The most useful aspect of the Hofmann elimination is its *regioselectivity*. Elimination in alkyltrimethylammonium hydroxides proceeds in the direction opposite to that of the Zaitsev rule (Sections 7.10 and 7.15) and favors formation of the *less-substituted* alkene.

sec-Butyltrimethylammonium hydroxide $\xrightarrow[-\text{H}_2\text{O},\ -(\text{CH}_3)_3\text{N}]{\text{heat}}$ 1-Butene (95%) + *trans*-2-Butene + *cis*-2-Butene (5% combined)

The least sterically hindered β hydrogen is removed by the base. Methyl groups are deprotonated in preference to methylene groups, and methylene groups are deprotonated in preference to methines. Elimination reactions of alkyltrimethylammonium hydroxides are said to obey the **Hofmann rule;** they yield the less-substituted alkene.

Problem 22.14

Give the structure of the major alkene formed when the hydroxide of each of the following quaternary ammonium ions is heated.

(a) [1-methylcyclopentyl with $\overset{+}{N}(CH_3)_3$] (b) [$\overset{+}{N}(CH_3)_3$ on 2,4,4-trimethylpentan-2-yl] (c) [ethyl(pentyl)dimethylammonium ion]

Sample Solution (a) Two alkenes are capable of being formed by β elimination: methylenecyclopentane and 1-methylcyclopentane.

(1-Methylcyclopentyl)trimethylammonium hydroxide $\xrightarrow[-\text{H}_2\text{O},\ -(\text{CH}_3)_3\text{N}]{\text{heat}}$ Methylenecyclopentane + 1-Methylcyclopentene

Methylenecyclopentane has the less-substituted double bond and is the major product. The observed isomer distribution is 91% methylenecyclopentane and 9% 1-methylcyclopentene.

We can understand the regioselectivity of the Hofmann elimination by comparing steric effects in the E2 transition states for formation of 1-butene and *trans*-2-butene from *sec*-butyltrimethylammonium hydroxide. In terms of its size, $(CH_3)_3\overset{+}{N}$— (trimethylammonio) is comparable to $(CH_3)_3C$— (*tert*-butyl). As Figure 22.5 illustrates, the E2 transition state requires an anti relationship between the proton that is removed and the trimethylammonio group. No serious van der Waals repulsions are evident in the transition state geometry

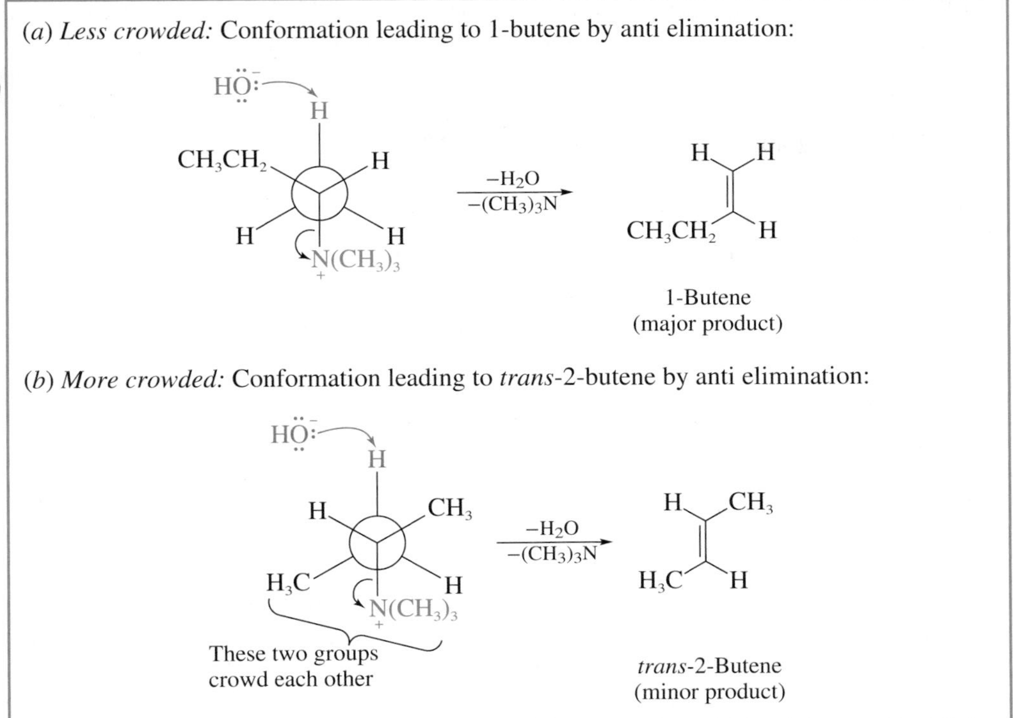

Figure 22.5

Newman projections showing the conformations leading to (*a*) 1-butene, and (*b*) *trans*-2-butene by Hofmann elimination of *sec*-butyltrimethylammonium hydroxide. The major product is 1-butene.

for formation of 1-butene. The conformation leading to *trans*-2-butene, however, is destabilized by van der Waals strain between the trimethylammonio group and a methyl group gauche to it. Thus, the activation energy for formation of *trans*-2-butene exceeds that of 1-butene, which becomes the major product because it is formed faster.

With a regioselectivity opposite to that of the Zaitsev rule, the Hofmann elimination is sometimes used in synthesis to prepare alkenes not accessible by dehydrohalogenation of alkyl halides. This application decreased in importance once the Wittig reaction (see Section 18.12) became established as a synthetic method.

22.14 Electrophilic Aromatic Substitution in Arylamines

Arylamines contain two functional groups, the amine group and the aromatic ring; they are *difunctional compounds*. The reactivity of the amine group is affected by its aryl substituent, and the reactivity of the ring is affected by its amine substituent. The same electron delocalization that reduces the basicity and the nucleophilicity of an arylamine nitrogen increases the electron density in the aromatic ring and makes arylamines extremely reactive toward electrophilic aromatic substitution.

The reactivity of arylamines was noted in Section 13.12, where it was pointed out that $-\ddot{N}H_2$, $-\ddot{N}HR$, and $-\ddot{N}R_2$ are ortho, para-directing and exceedingly powerful activating groups. These substituents are such powerful activators that electrophilic aromatic substitution is only rarely performed directly on arylamines.

Direct nitration of aniline and other arylamines fails because oxidation leads to the formation of dark-colored "tars." As a solution to this problem it is standard practice to first protect the amino group by acylation with either acetyl chloride or acetic anhydride.

$$\underset{\text{Arylamine}}{ArNH_2} \xrightarrow[\text{or } CH_3\overset{\overset{O}{\|}}{C}O\overset{\overset{O}{\|}}{C}CH_3]{CH_3\overset{\overset{O}{\|}}{C}Cl} \underset{N\text{-Acetylarylamine}}{ArNH\overset{\overset{O}{\|}}{C}CH_3}$$

Amide resonance within the *N*-acetyl group competes with delocalization of the nitrogen lone pair into the ring. Protecting the amino group of an arylamine in this way moderates its reactivity and permits nitration of the ring. The acetamido group is activating toward electrophilic aromatic substitution and is ortho, para-directing. After the *N*-acetyl-protecting group has served its purpose, it may be removed by hydrolysis, restoring the amino group:

p-Isopropylaniline → (acetic anhydride; protection step) → *p*-Isopropylacetanilide (98%) → (HNO_3, 20°C; nitration step) → 4-Isopropyl-2-nitroacetanilide (94%) → (KOH, ethanol, heat; deprotection step) → 4-Isopropyl-2-nitroaniline (100%)

The net effect of the sequence *protect–nitrate–deprotect* is the same as if the substrate had been nitrated directly. Because direct nitration is impossible, however, the indirect route is the only practical method.

Problem 22.15

Outline syntheses of each of the following from aniline and any necessary organic or inorganic reagents.

(a) *p*-Nitroaniline (b) 2,4-Dinitroaniline (c) *p*-Aminoacetanilide

Sample Solution (a) Because direct nitration of aniline is not a practical reaction, the amino group must first be protected as its *N*-acetyl derivative.

Aniline → (acetic anhydride) → Acetanilide → (HNO_3, H_2SO_4) → *o*-Nitroacetanilide + *p*-Nitroacetanilide

Nitration of acetanilide yields a mixture of ortho and para substitution products. The para isomer is separated, then subjected to hydrolysis to give *p*-nitroaniline.

p-Nitroacetanilide → (H_2O, HO^- or 1. H_3O^+ 2. HO^-) → *p*-Nitroaniline

Unprotected arylamines are so reactive that it is difficult to limit halogenation to monosubstitution. Generally, halogenation proceeds rapidly to replace all the available hydrogens that are ortho or para to the amino group.

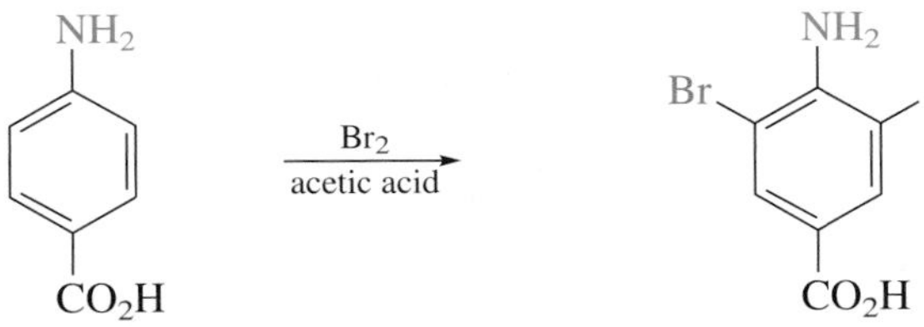

NH_2 / CO_2H —Br_2, acetic acid→ NH_2, Br, Br / CO_2H

p-Aminobenzoic acid 4-Amino-3,5-dibromobenzoic acid (82%)

Decreasing the electron-donating ability of an amino group by acylation makes monohalogenation possible.

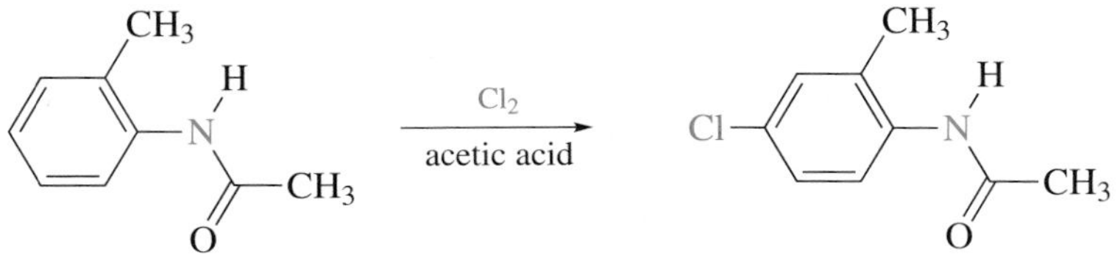

—Cl_2, acetic acid→

2-Methylacetanilide 4-Chloro-2-methylacetanilide (74%)

Friedel–Crafts reactions are normally not successful when attempted on an arylamine, but can be carried out readily once the amino group is protected.

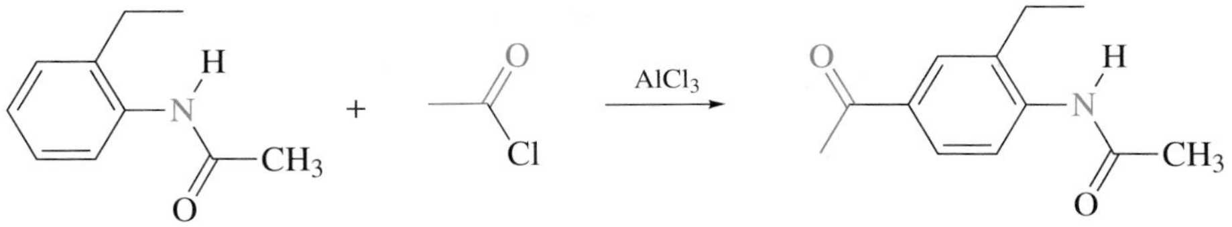

—$AlCl_3$→

2-Ethylacetanilide Acetyl chloride 4-Acetamido-3-ethylacetophenone (57%)

22.15 Nitrosation of Alkylamines

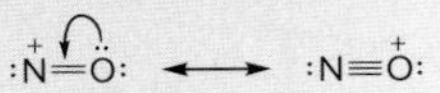

When solutions of sodium nitrite ($NaNO_2$) are acidified, a number of species are formed that act as sources of nitrosyl cation, $:\overset{+}{N}{=}\ddot{O}:$. For simplicity, organic chemists group all these species together and speak of the chemistry of one of them, *nitrous acid,* as a generalized precursor to nitrosyl cation.

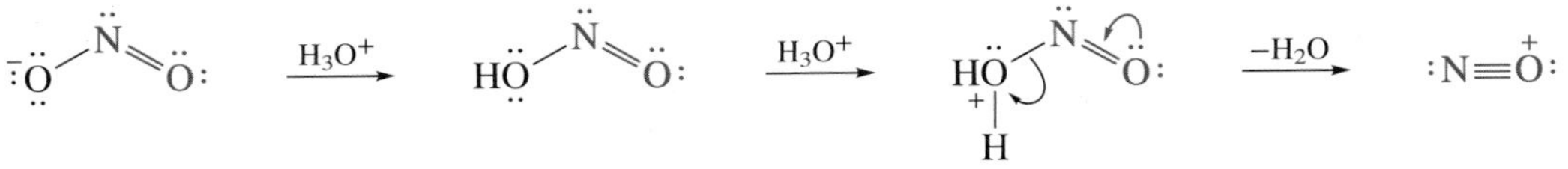

Nitrosyl cation is also called *nitrosonium* ion. It can be represented by the two resonance structures

$:\overset{+}{N}{=}\ddot{O}: \longleftrightarrow :N{\equiv}\overset{+}{O}:$

$^{-}O{-}N{=}O \xrightarrow{H_3O^+} HO{-}N{=}O \xrightarrow{H_3O^+} H_2\overset{+}{O}{-}N{=}O \xrightarrow{-H_2O} :N{\equiv}\overset{+}{O}:$

Nitrite ion (from sodium nitrite) Nitrous acid Nitrosyl cation

Nitrosation of amines is best illustrated by examining what happens when a secondary amine "reacts with nitrous acid." The amine acts as a nucleophile toward the nitrogen of nitrosyl cation. The intermediate that is formed in the first step loses a proton to give an ***N*-nitroso amine** as the isolated product.

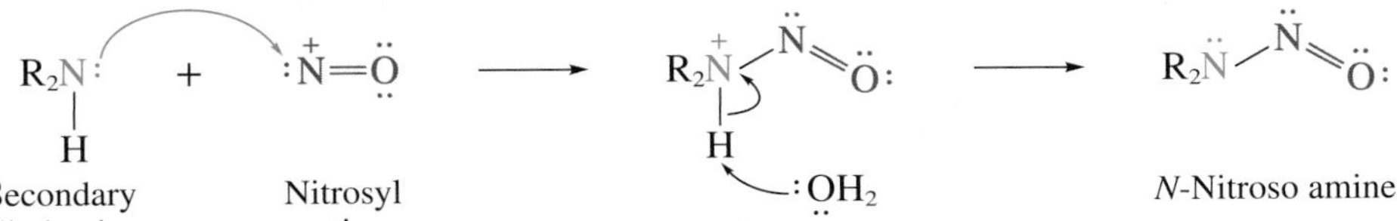

$R_2NH + :\overset{+}{N}{=}\ddot{O} \longrightarrow R_2\overset{+}{N}(H){-}N{=}O \;(:OH_2) \longrightarrow R_2N{-}N{=}O$

Secondary alkylamine Nitrosyl cation *N*-Nitroso amine

For example,

$$(CH_3)_2\ddot{N}H \xrightarrow[H_2O]{NaNO_2,\ HCl} (CH_3)_2\ddot{N}{-}\ddot{N}{=}\ddot{O}:$$

Dimethylamine → *N*-Nitrosodimethylamine (88–90%)

Problem 22.16

N-Nitroso amines are stabilized by electron delocalization. Write the two most stable resonance contributors of *N*-nitrosodimethylamine, $(CH_3)_2NNO$.

N-Nitroso amines are more often called **nitrosamines,** and because many of them are potent carcinogens, they have been the object of much investigation. We encounter nitrosamines in the environment on a daily basis. A few of these, all of which are known carcinogens, are:

N-Nitrosodimethylamine (formed during tanning of leather; also found in beer and herbicides)

N-Nitrosopyrrolidine (formed when bacon that has been cured with sodium nitrite is fried)

N-Nitrosonornicotine (present in tobacco smoke)

Nitrosamines are formed whenever nitrosating agents come in contact with secondary amines, and more are probably synthesized within our body than enter it by environmental contamination. Enzyme-catalyzed reduction of nitrate (NO_3^-) produces nitrite (NO_2^-), which combines with amines present in the body to form *N*-nitroso amines.

When primary amines are nitrosated, their *N*-nitroso compounds can't be isolated because they react further.

$$R\ddot{N}H_2 \xrightarrow[H_3O^+]{NaNO_2} R\ddot{N}H{-}\ddot{N}{=}\ddot{O}: \xrightarrow{H_3O^+} R{-}\ddot{N}H{-}\ddot{N}{=}\overset{+}{\ddot{O}}H \xrightarrow{-H_3O^+} R{-}\ddot{N}{=}\ddot{N}{-}\ddot{O}H \xrightarrow{H_3O^+} R{-}\ddot{N}{=}\ddot{N}{-}\overset{+}{\ddot{O}}H_2 \xrightarrow{-H_2O} R\overset{+}{N}{\equiv}N:$$

Primary alkylamine; (Not isolable); (Not isolable); (Not isolable); (Not isolable); Alkyl diazonium ion

The product of this series of steps is an alkyl **diazonium ion,** and the amine is said to have been **diazotized.** Alkyl diazonium ions are not very stable, decomposing rapidly to form a carbocation and molecular nitrogen when the alkyl group is secondary or tertiary.

Recall from Section 6.10 that decreasing basicity is associated with increasing leaving-group ability. Molecular nitrogen is an exceedingly weak base and an excellent leaving group.

$$R{-}\overset{+}{N}{\equiv}N: \longrightarrow R^+ + :N{\equiv}N:$$

Alkyl diazonium ion — Carbocation — Nitrogen

HONO → − N_2 →

1,1-Dimethylpropylamine 1,1-Dimethylpropyldiazonium ion 1,1-Dimethylpropyl cation

The ultimate products arise by nucleophilic substitution (S_N1) and/or elimination (E1) of the diazonium ion via the carbocation.

HONO, H_2O →

1,1-Dimethylpropylamine 2-Methyl-2-butanol (80%) + 2-Methyl-1-butene (3%) + 2-Methyl-2-butene (2%)

Problem 22.17

Nitrous acid deamination of 2,2-dimethylpropylamine, $(CH_3)_3CCH_2NH_2$, gives the same products as from 1,1-dimethylpropylamine. Suggest a mechanism for their formation from 2,2-dimethylpropylamine.

Aryl diazonium ions, prepared by nitrous acid diazotization of primary arylamines, are substantially more stable than alkyl diazonium ions and are of enormous synthetic value. Their use in the synthesis of substituted aromatic compounds is described in Sections 22.16 and 22.17.

The nitrosation of tertiary alkylamines is rather complicated, and no generally useful chemistry is associated with reactions of this type.

22.16 Nitrosation of Arylamines

We learned in the preceding section that different reactions are observed when the various classes of alkylamines—primary, secondary, and tertiary—react with nitrosating agents. Although no useful chemistry attends the nitrosation of tertiary alkylamines, electrophilic aromatic substitution by nitrosyl cation ($:N\equiv\overset{+}{O}:$) takes place with *N,N*-dialkylarylamines.

1. $NaNO_2$, HCl, H_2O, 8°C
2. HO^-

N,N-Diethylaniline → *N,N*-Diethyl-*p*-nitrosoaniline (95%)

Nitrosyl cation is a relatively weak electrophile and attacks only very strongly activated aromatic rings.

N-Alkylarylamines resemble secondary alkylamines in that they form *N*-nitroso compounds on reaction with nitrous acid.

$NaNO_2$, HCl, H_2O, 10°C

N-Methylaniline → *N*-Methyl-*N*-nitrosoaniline (87–93%)

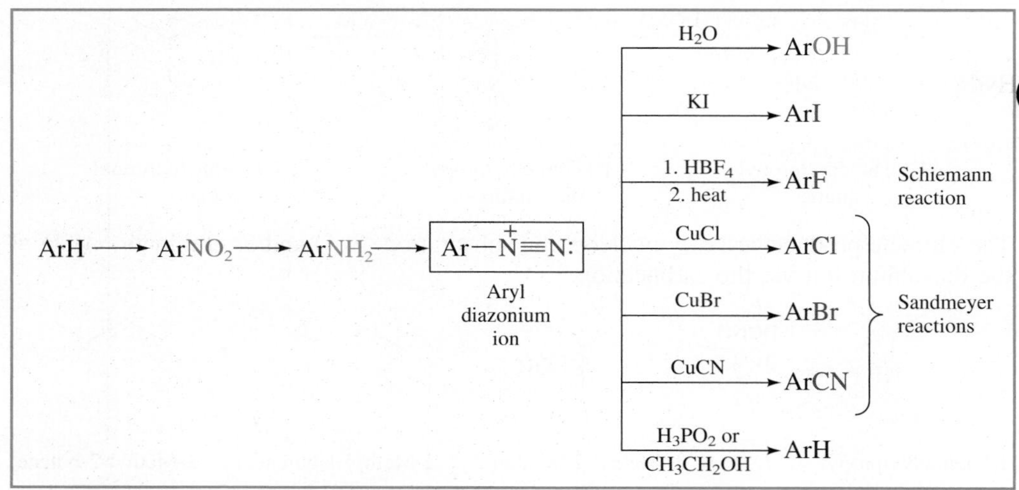

Figure 22.6

The synthetic origin of aryl diazonium ions and their most useful transformations.

Primary arylamines, like primary alkylamines, form diazonium ion salts on nitrosation. Whereas alkyl diazonium ions decompose under the conditions of their formation, aryl diazonium salts are considerably more stable and can be stored in aqueous solution at 0–5°C for a reasonable time. Loss of nitrogen from an aryl diazonium ion generates an unstable aryl cation and is much slower than loss of nitrogen from an alkyl diazonium ion.

$$\text{p-Isopropylaniline} \xrightarrow[H_2O,\ 0\text{–}5°C]{NaNO_2,\ H_2SO_4} \text{p-Isopropylbenzenediazonium hydrogen sulfate } (\overset{+}{N}{\equiv}N\colon\ HSO_4^-)$$

p-Isopropylaniline — *p*-Isopropylbenzenediazonium hydrogen sulfate

Aryl diazonium ions undergo a variety of reactions that make them versatile intermediates for preparing a host of ring-substituted aromatic compounds. In these reactions, summarized in Figure 22.6 and discussed individually in Section 22.17, molecular nitrogen acts as a leaving group and is replaced by another atom or group. All the reactions are regiospecific; the entering group becomes bonded to the same carbon from which nitrogen departs.

22.17 Synthetic Transformations of Aryl Diazonium Salts

An important reaction of aryl diazonium ions is their conversion to *phenols* by hydrolysis:

$$Ar\overset{+}{N}{\equiv}N\colon + 2H_2O \longrightarrow ArOH + H_3O^+ + \colon N{\equiv}N\colon$$

Aryl diazonium ion — Water — A phenol — Nitrogen

This is the most general method for preparing phenols. It is easily performed; the aqueous acidic solution in which the diazonium salt is prepared is heated and gives the phenol directly. An aryl cation is probably generated, which is then captured by water acting as a nucleophile.

$$\text{p-Isopropylaniline} \xrightarrow[2.\ H_2O,\ heat]{1.\ NaNO_2,\ H_2SO_4,\ H_2O} \text{p-Isopropylphenol (73\%)}$$

p-Isopropylaniline — *p*-Isopropylphenol (73%)

Sulfuric acid is normally used instead of hydrochloric acid in the diazotization step so as to minimize the competition with water for capture of the cationic intermediate. Hydrogen sulfate anion (HSO_4^-) is less nucleophilic than chloride.

Problem 22.18

Design a synthesis of *m*-bromophenol from benzene.

The reaction of an aryl diazonium salt with potassium iodide is the standard method for the preparation of *aryl iodides.* The diazonium salt is prepared from a primary aromatic amine in the usual way, a solution of potassium iodide is then added, and the reaction mixture is brought to room temperature or heated to accelerate the reaction.

$$Ar—\overset{+}{N}{\equiv}N: \; + \; I^- \longrightarrow ArI \; + \; :N{\equiv}N:$$

Aryl diazonium ion — Iodide ion — Aryl iodide — Nitrogen

o-Bromoaniline → (1. $NaNO_2$, HCl, H_2O, 0–5°C; 2. KI, room temperature) → *o*-Bromoiodobenzene (72–83%)

o-Bromoaniline

o-Bromoiodobenzene (72–83%)

Problem 22.19

Show how you could prepare *m*-bromoiodobenzene from benzene.

Diazonium salt chemistry provides the principal synthetic method for the preparation of *aryl fluorides* through a process known as the **Schiemann reaction.** In this procedure the aryl diazonium ion is isolated as its fluoroborate salt, which then yields the desired aryl fluoride on being heated.

$$Ar—\overset{+}{N}{\equiv}N: \overset{-}{B}F_4 \xrightarrow{\text{heat}} ArF \; + \; BF_3 \; + \; :N{\equiv}N:$$

Aryl diazonium fluoroborate — Aryl fluoride — Boron trifluoride — Nitrogen

A standard way to form the aryl diazonium fluoroborate salt is to add fluoroboric acid (HBF_4) or a fluoroborate salt to the diazotization medium.

m-Aminophenyl ethyl ketone → (1. $NaNO_2$, H_2O, HCl; 2. HBF_4; 3. heat) → Ethyl *m*-fluorophenyl ketone (68%)

m-Aminophenyl ethyl ketone

Ethyl *m*-fluorophenyl ketone (68%)

Problem 22.20

Show the proper sequence of synthetic transformations in the conversion of benzene to ethyl *m*-fluorophenyl ketone.

Although it is possible to prepare *aryl chlorides* and *aryl bromides* by electrophilic aromatic substitution, it is often necessary to prepare these compounds from an aromatic amine. The amine is converted to the corresponding diazonium salt and then treated with copper(I) chloride or copper(I) bromide as appropriate.

$$\mathrm{Ar{-}\overset{+}{N}{\equiv}N{:}} \xrightarrow{\text{CuX}} \mathrm{ArX} + \mathrm{{:}N{\equiv}N{:}}$$

Aryl diazonium ion — Aryl chloride or bromide — Nitrogen

m-Nitroaniline $\xrightarrow[\text{2. CuCl, heat}]{\text{1. NaNO}_2\text{, HCl, H}_2\text{O, 0–5°C}}$ *m*-Chloronitrobenzene (68–71%)

o-Chloroaniline $\xrightarrow[\text{2. CuBr, heat}]{\text{1. NaNO}_2\text{, HBr, H}_2\text{O, 0–10°C}}$ *o*-Bromochlorobenzene (89–95%)

Reactions that use copper(I) salts to replace nitrogen in diazonium salts are called **Sandmeyer reactions.** The Sandmeyer reaction using copper(I) cyanide is a good method for the preparation of aromatic *nitriles:*

$$\mathrm{Ar{-}\overset{+}{N}{\equiv}N{:}} \xrightarrow{\text{CuCN}} \mathrm{ArCN} + \mathrm{{:}N{\equiv}N{:}}$$

Aryl diazonium ion — Aryl nitrile — Nitrogen

o-Toluidine $\xrightarrow[\text{2. CuCN, heat}]{\text{1. NaNO}_2\text{, H}_2\text{O, HCl, 0°C}}$ *o*-Methylbenzonitrile (64–70%)

The preparation of aryl chlorides, bromides, and nitriles by the Sandmeyer reaction is mechanistically complicated and may involve arylcopper intermediates.

Because cyano groups may be hydrolyzed to carboxylic acids (see Section 20.16), the Sandmeyer preparation of aryl nitriles is a key step in the conversion of arylamines to substituted benzoic acids. In the example just cited, the *o*-methylbenzonitrile that was formed was subsequently subjected to acid-catalyzed hydrolysis to give *o*-methylbenzoic acid in 80–89% yield.

It is possible to replace amino groups on an aromatic ring by hydrogen by reducing a diazonium salt with hypophosphorous acid (H_3PO_2) or with ethanol. These reductions are free-radical reactions in which ethanol or hypophosphorous acid acts as a hydrogen atom donor:

$$\mathrm{Ar{-}\overset{+}{N}{\equiv}N{:}} \xrightarrow[\text{CH}_3\text{CH}_2\text{OH}]{\text{H}_3\text{PO}_2\text{ or}} \mathrm{ArH} + \mathrm{{:}N{\equiv}N{:}}$$

Aryl diazonium ion — Arene — Nitrogen

Reactions of this type are called *reductive deaminations.*

4-Isopropyl-2-nitroaniline $\xrightarrow[\text{CH}_3\text{CH}_2\text{OH}]{\text{NaNO}_2,\ \text{H}_2\text{O, HCl}}$ *m*-Isopropylnitrobenzene (59%)

Sodium borohydride has also been used to reduce aryl diazonium salts in reductive deamination reactions.

Problem 22.21

Cumene (isopropylbenzene) is a relatively inexpensive commercially available starting material. Show how you could prepare *m*-isopropylnitrobenzene from cumene.

The value of diazonium salts in synthetic organic chemistry rests on two main points. Through the use of diazonium salt chemistry:

1. Substituents that are otherwise accessible only with difficulty, such as fluoro, iodo, cyano, and hydroxyl, may be introduced onto a benzene ring.
2. Compounds that have substitution patterns not directly available by electrophilic aromatic substitution can be prepared.

The first of these two features is readily apparent and is illustrated by Problems 22.18 to 22.20. If you have not done these problems yet, you are strongly encouraged to attempt them now.

The second point is somewhat less obvious but is illustrated by the synthesis of 1,3,5-tribromobenzene. This particular substitution pattern cannot be obtained by direct bromination of benzene because bromine is an ortho, para director. Instead, advantage is taken of the powerful activating and ortho, para-directing effects of the amino group in aniline. Bromination of aniline yields 2,4,6-tribromoaniline in quantitative yield. Diazotization of the resulting 2,4,6-tribromoaniline and reduction of the diazonium salt gives the desired 1,3,5-tribromobenzene.

Aniline $\xrightarrow[\text{H}_2\text{O}]{\text{Br}_2}$ 2,4,6-Tribromoaniline (100%) $\xrightarrow[\text{CH}_3\text{CH}_2\text{OH}]{\text{NaNO}_2,\ \text{H}_2\text{SO}_4,\ \text{H}_2\text{O}}$ 1,3,5-Tribromobenzene (74–77%)

To exploit the synthetic versatility of aryl diazonium salts, be prepared to reason backward. When you see a fluorine attached to a benzene ring, for example, realize that it probably will have to be introduced by a Schiemann reaction of an arylamine; realize that the required arylamine is derived from a nitroarene, and that the nitro group is introduced by nitration. Be aware that an unsubstituted position of a benzene ring need not have always been that way. It might once have borne an amino group that was used to control the orientation of electrophilic aromatic substitution reactions before being removed by reductive deamination. The strategy of synthesis is intellectually demanding, and a considerable sharpening of your reasoning power can be gained by attacking the synthesis problems at the end of each chapter. Use retrosynthetic analysis to plan your sequence of accessible intermediates, then fill in the details on how each transformation is to be carried out.

22.18 Azo Coupling

A reaction of aryl diazonium salts that does not involve loss of nitrogen takes place when they react with phenols and arylamines. Aryl diazonium ions are relatively weak electrophiles but have sufficient reactivity to attack strongly activated aromatic rings. The reaction is known as *azo coupling;* two aryl groups are joined together by an azo (—N=N—) function.

(ERG is a powerful electron-releasing group such as —OH or —NR_2) + Aryl diazonium ion ($:N{\equiv}\overset{+}{N}Ar$) ⟶ Intermediate in electrophilic aromatic substitution $\xrightarrow{-H^+}$ Azo compound

N,N-Dimethylaniline + Diazonium ion from o-aminobenzoic acid ⟶ Methyl red (62–66%)

N,N-Dimethylaniline ($(CH_3)_2N$—C_6H_5) + Diazonium ion from *o*-aminobenzoic acid ⟶ Methyl red (62–66%)

The product of this reaction, as with many azo couplings, is highly colored. It is called *methyl red* and was a familiar acid–base indicator before the days of pH meters. It is red in solutions of pH 4 and below, yellow above pH 6.

Soon after azo coupling was discovered in the mid-nineteenth century, the reaction received major attention as a method for preparing dyes. Azo dyes first became commercially available in the 1870s and remain widely used, with more than 50% of the

From Dyes to Sulfa Drugs

The medicine cabinet was virtually bare of antibacterial agents until *sulfa drugs* burst on the scene in the 1930s. Before sulfa drugs became available, bacterial infection might transform a small cut or puncture wound to a life-threatening event. The story of how sulfa drugs were developed is an interesting example of being right for the wrong reasons. It was known that many bacteria absorbed dyes, and staining was a standard method for making bacteria more visible under the microscope. Might there not be some dye that is both absorbed by bacteria and toxic to them? Acting on this hypothesis, scientists at the German dyestuff manufacturer I. G. Farbenindustrie undertook a program to test the thousands of compounds in their collection for their antibacterial properties.

In general, in vitro testing of drugs precedes in vivo testing. The two terms mean, respectively, "in glass" and "in life." In vitro testing of antibiotics is carried out using bacterial cultures in test tubes or Petri dishes. Drugs that are found to be active in vitro progress to the stage of in vivo testing. In vivo testing is carried out in living organisms: laboratory animals or human volunteers. The I. G. Farben scientists found that some dyes did possess antibacterial properties, both in vitro and in vivo. Others were active in vitro but were converted to inactive substances in vivo and therefore of no use as drugs. Unexpectedly, an azo dye called *Prontosil* was inactive in vitro but active in vivo.

In 1932, a member of the I. G. Farben research group, Gerhard Domagk, used Prontosil to treat a young child suffering from a serious, potentially fatal staphylococcal infection. According to many accounts, the child was Domagk's own daughter; her infection was cured and her recovery was rapid and complete. Systematic testing followed and Domagk was awarded the 1939 Nobel Prize in Medicine or Physiology.

In spite of the rationale on which the testing of dyestuffs as antibiotics rested, subsequent research revealed that the antibacterial properties of Prontosil had nothing at all to do with its being a dye! In the body, Prontosil undergoes a reductive cleavage of its azo linkage to form *sulfanilamide,* which is the substance actually responsible for the observed biological activity. This is why Prontosil is active in vivo, but not in vitro.

Prontosil $\xrightarrow{\text{in vivo}}$ Sulfanilamide (H_2N—C_6H_4—SO_2NH_2)

Bacteria require *p*-aminobenzoic acid to biosynthesize *folic acid,* a growth factor. Structurally, sulfanilamide resembles *p*-aminobenzoic acid and is mistaken for it by the bacteria. Folic acid biosynthesis is inhibited and bacterial growth is slowed sufficiently to allow the body's natural defenses to effect a cure. Because animals do not biosynthesize folic acid but obtain it in their food, sulfanilamide halts the growth of bacteria without harm to the host.

Identification of the mechanism by which Prontosil combats bacterial infections was an early triumph of *pharmacology,* a branch of science at the interface of physiology and biochemistry that studies the mechanism of drug action. By recognizing that sulfanilamide was the active agent, the task of preparing structurally modified analogs with potentially superior properties was considerably simplified. Instead of preparing Prontosil analogs, chemists synthesized sulfanilamide analogs. They did this with a vengeance; over 5000 compounds related to sulfanilamide were prepared during the period 1935–1946. Two of the most widely used sulfa drugs are *sulfathiazole* and *sulfadiazine.*

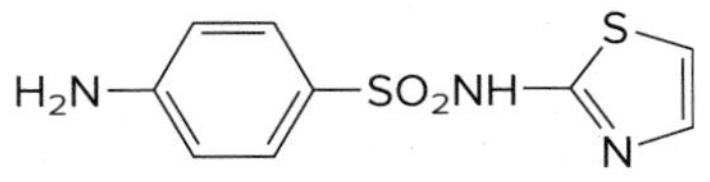

Sulfathiazole

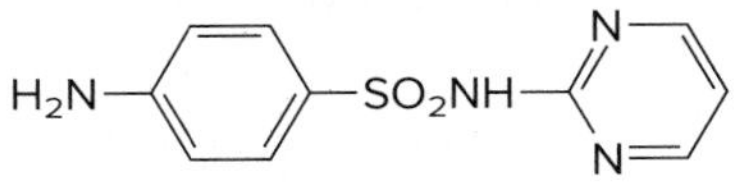

Sulfadiazine

We tend to take the efficacy of modern drugs for granted. One comparison with the not-too-distant past might put this view into better perspective. Once sulfa drugs were introduced in the United States, the number of pneumonia deaths alone decreased by an estimated 25,000 per year. The sulfa drugs are used less now than they were in the mid-twentieth century. Not only are more-effective, less-toxic antibiotics available, such as the penicillins and tetracyclines, but many bacteria that were once susceptible to sulfa drugs have become resistant.

synthetic dye market. Chrysoidine, an azo dye for silk, cotton, and wool, first came on the market in 1876 and remains in use today.

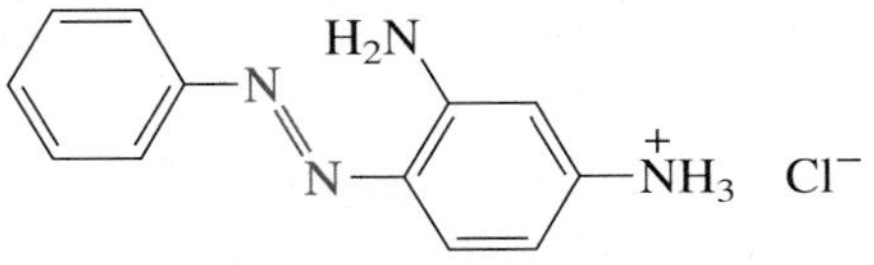

Chrysoidine

Problem 22.22

What amine and what diazonium salt would you use to prepare chrysoidine?

Dyes are regulated in the United States by the Food and Drug Administration (FDA). Over the years the FDA has removed a number of dyes formerly approved for use in food and cosmetics because of concerns about toxicity or cancer-causing potential or because they are skin irritants. Naturally occurring pigments, too numerous to count (saffron, turmeric, fruit colors, for example), are exempt from the approval process.

Of the seven synthetic dyes presently approved for food use, the three shown in Figure 22.7 are azo dyes. Red dye #40, which provides the red color to cherry-flavored

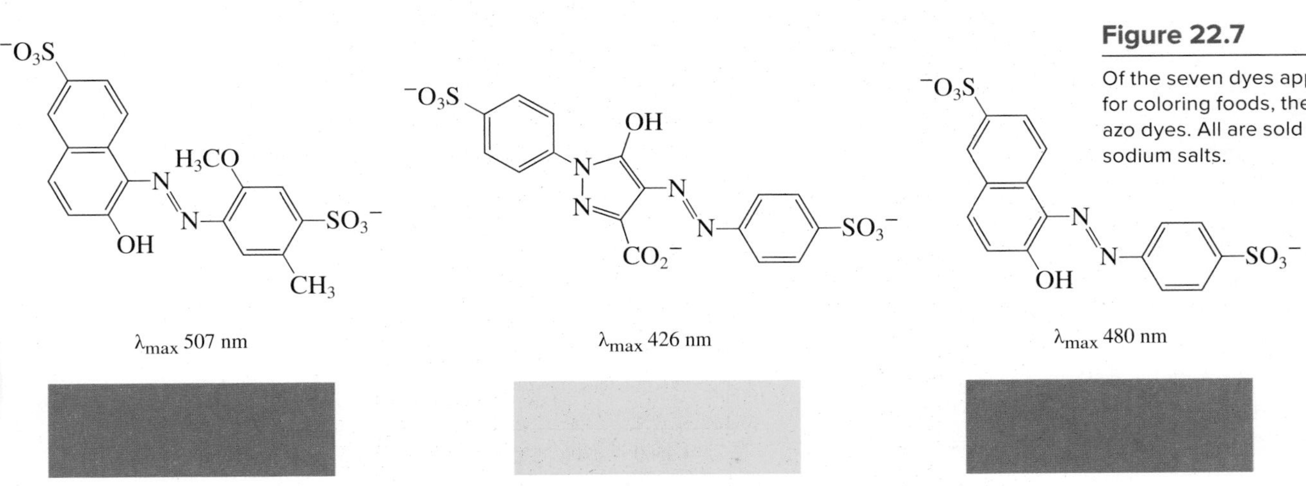

Figure 22.7

Of the seven dyes approved for coloring foods, these three are azo dyes. All are sold as their sodium salts.

The official names of these dyes are FD&C Red No. 40, FD&C Yellow No. 5, and FD&C Yellow No. 6, where FD&C stands for "Food, Drug, and Cosmetic," which is both the name of the law under which these dyes are regulated and the purposes for which they are approved.

foods, is the most popular. Not only is red dye #40 used to color foods, but you may have noticed that almost every over-the-counter cold medicine is a red liquid or comes in a red capsule. The color is red dye #40 and is there by custom more than necessity. Yellow #5 is a lemon color; yellow #6 is orange. The highly conjugated azo linkage and combination of electron-donating and electron-attracting groups are responsible for the intense absorption of visible light by these molecules. Substituents affect the wavelengths absorbed and ultimately the color. Red #40, yellow #5, and yellow #6 all are sodium salts of sulfonic acids, which confers on them the water solubility they need to be effective food colors.

22.19 Spectroscopic Analysis of Amines

Infrared: The absorptions of interest in the IR spectra of amines are those associated with N—H vibrations. Primary alkyl- and arylamines exhibit two peaks in the range 3000–3500 cm^{-1}, which are due to symmetric and antisymmetric N—H stretching modes.

Symmetric N—H stretching of a primary amine — R—N(H)(H) — R—N(H)(H) — Antisymmetric N—H stretching of a primary amine

These two vibrations appear at 3290 and 3370 cm^{-1} in the IR spectrum of butylamine, shown in Figure 22.8. Secondary amines such as diethylamine exhibit only one peak due to N—H stretching. Tertiary amines are transparent in this region because they have no N—H bonds.

^{1}H NMR: Characteristics of the nuclear magnetic resonance spectra of amines may be illustrated by comparing 4-methylbenzylamine (Figure 22.9*a*) with 4-methylbenzyl alcohol (Figure 22.9*b*). Nitrogen is less electronegative than oxygen and so shields neighboring nuclei to a greater extent. The benzylic methylene group attached to nitrogen in 4-methyl-benzylamine appears at higher field (δ 3.8) than the benzylic methylene of 4-methylbenzyl alcohol (δ 4.5). The N—H protons are somewhat more shielded than the O—H protons of an alcohol. In 4-methylbenzylamine the protons of the amino group correspond to the signal at δ 1.4, whereas the hydroxyl proton signal of 4-methylbenzyl alcohol is found at δ 2.5. The chemical shifts and splittings of amino group protons, like those of hydroxyl protons, are variable and are sensitive to solvent, concentration, and temperature.

^{13}C NMR: Similarly, carbons that are bonded to nitrogen are more shielded than those bonded to oxygen, as revealed by comparing the ^{13}C chemical shifts of methylamine and methanol.

δ 26.9 CH_3NH_2	δ 48.0 CH_3OH
Methylamine	Methanol

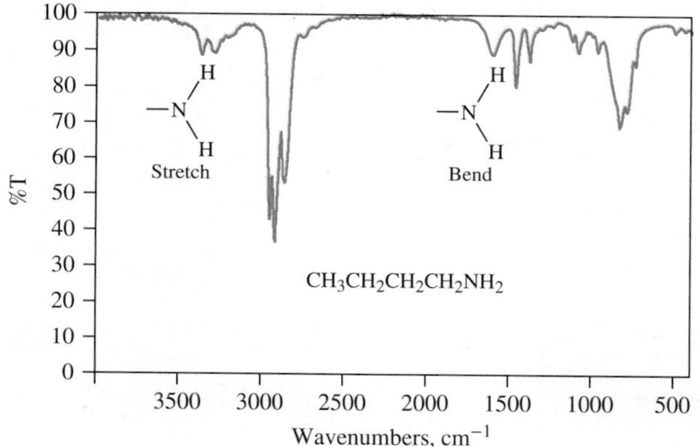

Figure 22.8

The infrared spectrum of butylamine has peaks for N—H stretching at 3290 and 3370 cm^{-1}. One corresponds to a symmetric stretch of the two N—H bonds, the other to an antisymmetric stretch. The peak at 1600 cm^{-1} is for NH_2 bending (scissoring).

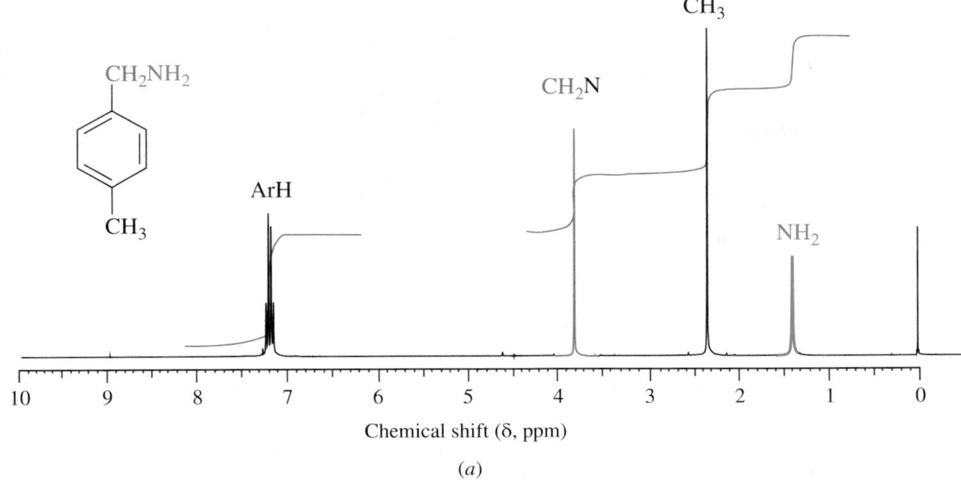

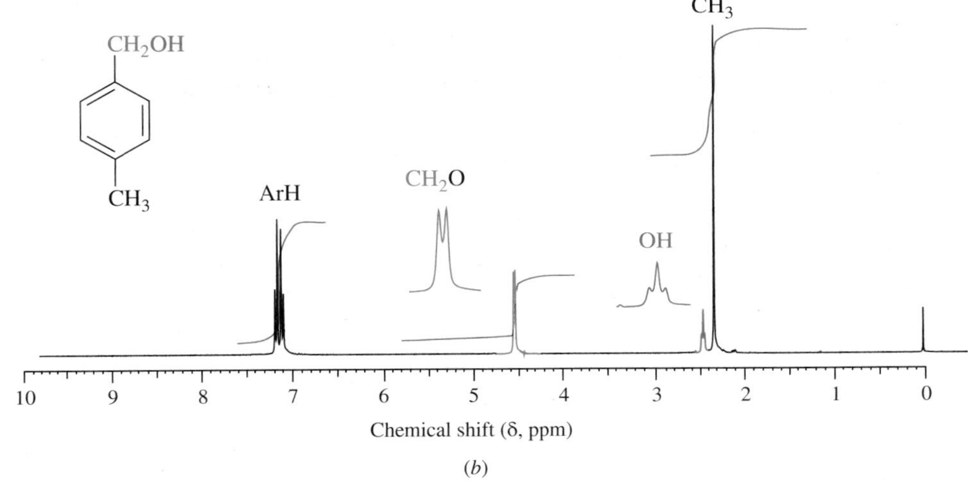

Figure 22.9

The 300-MHz ^1H NMR spectra of (*a*) 4-methylbenzylamine and (*b*) 4-methylbenzyl alcohol. The singlet corresponding to CH_2N in (*a*) is more shielded than that of CH_2O in (*b*).

UV-VIS: In the absence of any other chromophore, the UV-VIS spectrum of an alkylamine is not very informative. The longest wavelength absorption involves promoting one of the unshared electrons of nitrogen to an antibonding σ* orbital (n nσ*) with a λ_{max} in the relatively inaccessible region near 200 nm. In arylamines the interaction of the nitrogen lone pair with the π-electron system of the ring shifts the ring's absorptions to longer wavelength. Tying up the lone pair by protonation causes the UV-VIS spectrum of anilinium ion to resemble benzene.

	X	λ_{max} (*nm*)
Benzene	H	204, 256
Aniline	NH_2	230, 280
Anilinium ion	NH_3^+	203, 254

Mass Spectrometry: A number of features make amines easily identifiable by mass spectrometry.

Recall the "nitrogen rule" from Section 14.25.

First, the peak for the molecular ion M^+ for all compounds that contain only carbon, hydrogen, and oxygen has an *m/z* value that is an even number. The presence of a nitrogen atom in the molecule requires that the m/z value for the molecular ion be odd. An odd number of nitrogens corresponds to an odd value of the molecular weight; an even number of nitrogens corresponds to an even molecular weight.

Second, nitrogen is exceptionally good at stabilizing adjacent carbocation sites. The fragmentation pattern seen in the mass spectra of amines is dominated by cleavage of groups from the carbon atom attached to the nitrogen, as the data for the following pair of constitutionally isomeric amines illustrate:

$$(CH_3)_2\ddot{N}CH_2CH_2CH_2CH_3 \xrightarrow{e^-} (CH_3)_2\overset{+\cdot}{N}-CH_2-CH_2CH_2CH_3 \longrightarrow (CH_3)_2\overset{+}{N}=CH_2 + \cdot CH_2CH_2CH_3$$

N,N-Dimethyl-1-butanamine; M^+ (m/z 101); (m/z 58) (most intense peak)

$$CH_3\ddot{N}HCH_2CH_2CH(CH_3)_2 \xrightarrow{e^-} CH_3\overset{+\cdot}{N}H-CH_2-CH_2CH(CH_3)_2 \longrightarrow CH_3\overset{+}{N}H=CH_2 + \cdot CH_2CH(CH_3)_2$$

N,3-Dimethyl-1-butanamine; M^+ (m/z 101); (m/z 44) (most intense peak)

22.20 SUMMARY

Section 22.1 Alkylamines are compounds of the type shown, where R, R′, and R″ are alkyl groups. One or more of these groups is an aryl group in arylamines.

$$RNH_2 \qquad R'RNH \qquad R''R'RN:$$

Primary amine; Secondary amine; Tertiary amine

Alkylamines are named in two ways. One method adds the ending *-amine* to the name of the alkyl group. The other applies the principles of substitutive nomenclature by replacing the *-e* ending of an alkane name by *-amine* and uses appropriate locants to identify the position of the amino group. Arylamines are named as derivatives of aniline.

Section 22.2 Nitrogen's unshared electron pair is of major importance in understanding the structure and properties of amines. Alkylamines have a pyramidal arrangement of bonds to nitrogen, with an unshared electron pair in an sp^3-hybridized orbital. The geometry at nitrogen in arylamines is somewhat flatter, and the unshared electron pair is delocalized into the π system of the ring. Delocalization binds the electron pair more strongly in arylamines than in alkylamines. Arylamines are less basic and less nucleophilic than alkylamines.

Section 22.3 Amines are less polar than alcohols. Hydrogen bonding in amines is weaker than in alcohols because nitrogen is less electronegative than oxygen. Amines have lower boiling points than alcohols, but higher boiling points than alkanes. Primary amines have higher boiling points than isomeric secondary amines; tertiary amines, which cannot form intermolecular hydrogen bonds, have the lowest boiling points. Amines resemble alcohols in their solubility in water.

Section 22.4 The basicity of amines is conveniently expressed in terms of the pK_a of their conjugate acids.

$$R_3\overset{+}{N}-H + H_2\ddot{O}: \rightleftharpoons R_3N: + H-\overset{+}{O}H_2$$

Conjugate acid of amine; Amine

The stronger base is associated with the weaker conjugate acid. The greater the pK_a of the conjugate acid, the stronger the base. The pK_a's of the conjugate acids of alkylamines lie in the 9–11 range. Arylamines are much weaker bases than alkylamines. The pK_a's of the conjugate acids of arylamines are usually 3–5. Strong electron-withdrawing groups can weaken the basicity of arylamines even more.

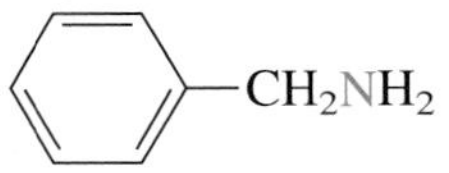

$C_6H_5CH_2NH_2$ — Benzylamine (alkylamine: pK_a of conjugate acid = 9.3)

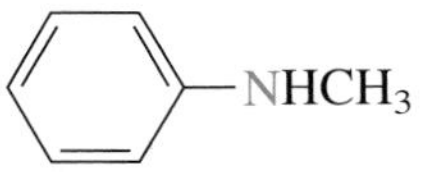

$C_6H_5NHCH_3$ — *N*-Methylaniline (arylamine: pK_a of conjugate acid = 4.8)

Section 22.5 Quaternary ammonium salts, compounds of the type $R_4N^+X^-$, find application as **phase-transfer catalysts.** A small amount of a quaternary ammonium salt promotes the transfer of an anion from aqueous solution, where it is highly solvated, to an organic solvent, where it is much less solvated and much more reactive.

Sections 22.6–22.10 Methods for the preparation of amines are summarized in Table 22.5.

Sections 22.11–22.18 The reactions of amines are summarized in Tables 22.6 and 22.7.

TABLE 22.5 Preparation of Amines

Reaction (section) and comments	General equation and specific example
Alkylation methods	
Alkylation of ammonia (Section 22.7) Ammonia can act as a nucleophile toward primary and some secondary alkyl halides to give primary alkylamines. Yields tend to be modest because the primary amine is itself a nucleophile and undergoes alkylation. Alkylation of ammonia can lead to a mixture containing a primary amine, a secondary amine, a tertiary amine, and a quaternary ammonium salt.	$RX + 2NH_3 \longrightarrow RNH_2 + NH_4X$ Alkyl halide + Ammonia → Alkylamine + Ammonium halide $C_6H_5CH_2Cl \xrightarrow{NH_3\ (8\ mol)} C_6H_5CH_2NH_2 + (C_6H_5CH_2)_2NH$ Benzyl chloride (1 mol) → Benzylamine (53%) + Dibenzylamine (39%)
Alkylation of phthalimide. The Gabriel synthesis (Section 22.8) The potassium salt of phthalimide reacts with alkyl halides to give *N*-alkylphthalimide derivatives. Hydrolysis or hydrazinolysis of this derivative yields a primary alkylamine.	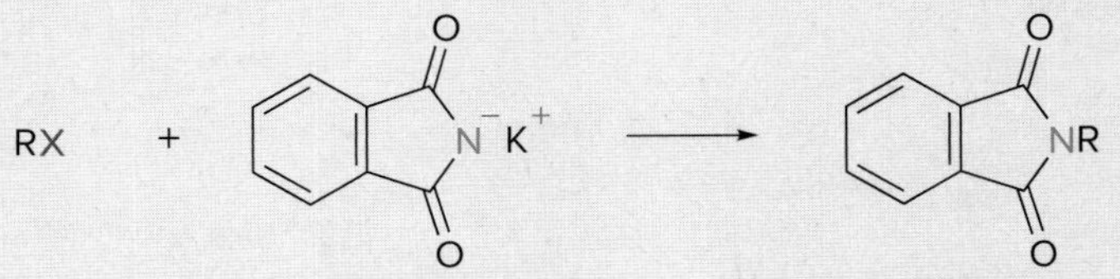 RX + *N*-potassiophthalimide (N^-K^+) → *N*-alkylphthalimide (NR) Alkyl halide, *N*-Potassiophthalimide, *N*-Alkylphthalimide 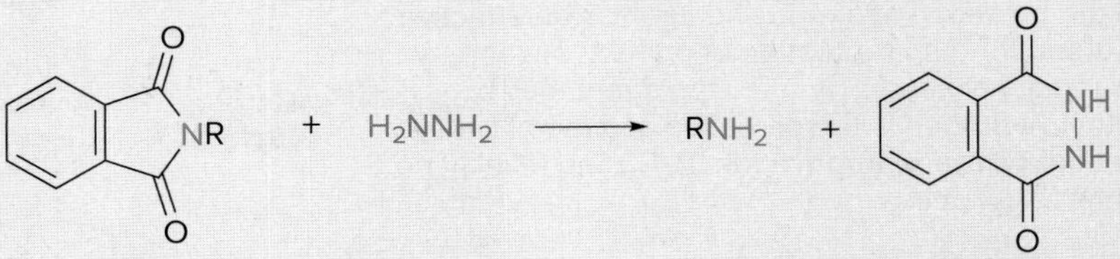 *N*-Alkylphthalimide + H_2NNH_2 → RNH_2 + Phthalhydrazide *N*-Alkylphthalimide, Hydrazine, Primary amine, Phthalhydrazide $CH_3CH{=}CHCH_2Cl \xrightarrow[2.\ H_2NNH_2,\ ethanol]{1.\ N\text{-potassiophthalimide, DMF}} CH_3CH{=}CHCH_2NH_2$ 1-Chloro-2-butene → 2-Buten-1-amine (95%)

continued

TABLE 22.5 Preparation of Amines (*Continued*)

Reaction (section) and comments	General equation and specific example
Reduction methods	
Reduction of alkyl azides (Section 22.9) Alkyl azides, prepared by nucleophilic substitution by azide ion in primary or secondary alkyl halides, are reduced to primary alkylamines by lithium aluminum hydride or by catalytic hydrogenation.	$R\ddot{N}{=}\overset{+}{N}{=}\ddot{\underset{\cdot\cdot}{N}}{:}^{-} \xrightarrow{\text{reduce}} R\ddot{N}H_2$ Alkyl azide → Primary amine 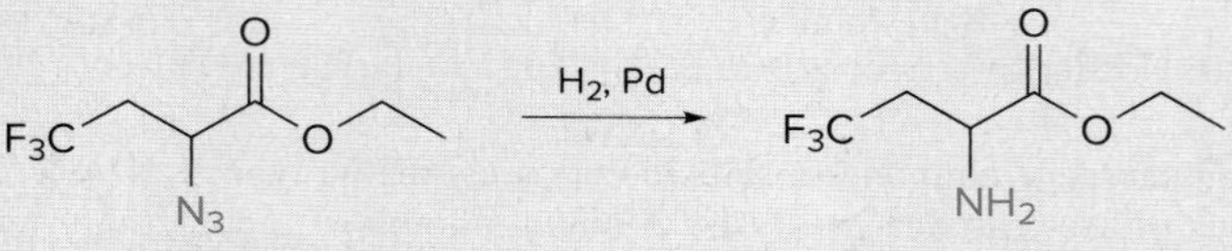 Ethyl 2-azido-4,4,4-trifluorobutanoate $\xrightarrow{H_2,\ Pd}$ Ethyl 2-amino-4,4,4-trifluorobutanoate (96%)
Reduction of nitriles (Section 22.9) Nitriles are reduced to primary amines by lithium aluminum hydride or by catalytic hydrogenation.	$R{-}C{\equiv}N \xrightarrow{\text{reduce}} RCH_2NH_2$ Nitrile → Primary amine 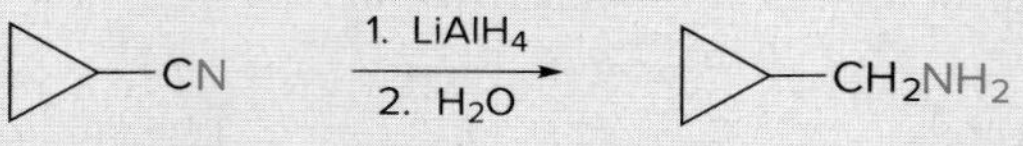 Cyclopropyl cyanide $\xrightarrow[2.\ H_2O]{1.\ LiAlH_4}$ Cyclopropylmethanamine (75%)
Reduction of aryl nitro compounds (Section 22.9) The standard method for the preparation of an arylamine is by nitration of an aromatic ring, followed by reduction of the nitro group. Typical reducing agents include iron or tin in hydrochloric acid or catalytic hydrogenation.	$ArNO_2 \xrightarrow{\text{reduce}} ArNH_2$ Nitroarene → Arylamine $C_6H_5NO_2 \xrightarrow[2.\ HO^-]{1.\ Fe,\ HCl} C_6H_5NH_2$ Nitrobenzene → Aniline (97%)
Reduction of amides (Section 22.9) Lithium aluminum hydride reduces the carbonyl group of an amide to a methylene group. Primary, secondary, or tertiary amines may be prepared by proper choice of the starting amide. R and R′ may be either alkyl or aryl.	$R\overset{O}{\overset{\|}{C}}NR'_2 \xrightarrow{\text{reduce}} RCH_2NR'_2$ Amide → Amine 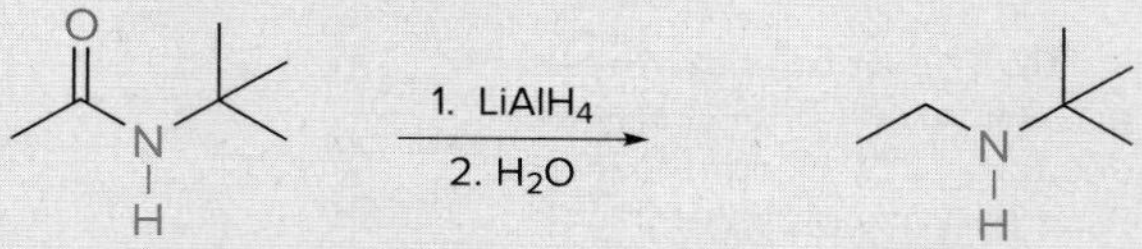 *N-tert*-Butylacetamide $\xrightarrow[2.\ H_2O]{1.\ LiAlH_4}$ *N-tert*-Butyl-*N*-ethylamine (60%)
Reductive amination (Section 22.10) Reaction of ammonia or an amine with an aldehyde or a ketone in the presence of a reducing agent is an effective method for the preparation of primary, secondary, or tertiary amines. The reducing agent may be either hydrogen in the presence of a metal catalyst or sodium cyanoborohydride. R, R′, and R″ may be either alkyl or aryl.	$R\overset{O}{\overset{\|}{C}}R' + R''_2NH \xrightarrow{\text{reducing agent}} R\overset{NR''_2}{\overset{\|}{C}}R'$ (with H on C) Aldehyde or ketone + Ammonia or an amine → Amine 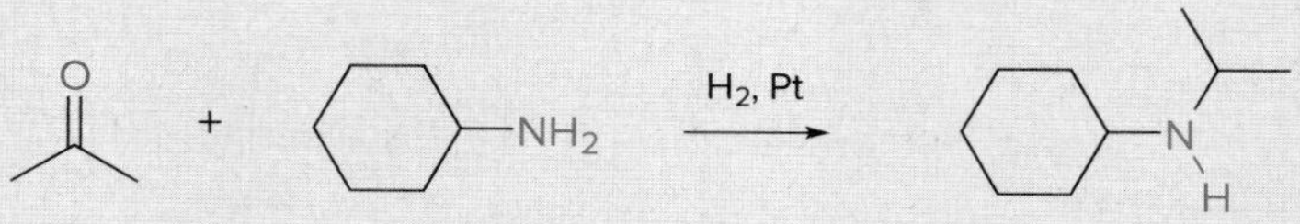 Acetone + Cyclohexylamine $\xrightarrow{H_2,\ Pt}$ *N*-Isopropylcyclohexylamine (79%)

TABLE 22.6 Reactions of Amines Discussed in This Chapter

Reaction (section) and comments	General equation and specific example
Alkylation (Section 22.12) Amines act as nucleophiles toward alkyl halides. Primary amines yield secondary amines, secondary amines yield tertiary amines, and tertiary amines yield quaternary ammonium salts.	$R\ddot{N}H_2 \xrightarrow{R'CH_2X} R\ddot{N}HCH_2R' \xrightarrow{R'CH_2X} R\ddot{N}(CH_2R')_2 \xrightarrow{R'CH_2X} R\overset{+}{N}(CH_2R')_3\,X^-$ Primary amine; Secondary amine; Tertiary amine; Quaternary ammonium salt 2-Chloromethylpyridine + Pyrrolidine $\xrightarrow{heat}$ 2-(Pyrrolidinylmethyl)pyridine (93%)
Hofmann elimination (Section 22.13) Quaternary ammonium hydroxides undergo elimination on being heated. It is an anti elimination of the E2 type. The regioselectivity of the Hofmann elimination is opposite to that of the Zaitsev rule and leads to the less highly substituted alkene.	$RCH_2CHR'(\overset{+}{N}(CH_3)_3)\ HO^- \xrightarrow{heat} RCH{=}CHR' + :N(CH_3)_3 + H_2O$ Alkyltrimethylammonium hydroxide; Alkene; Trimethylamine; Water Cycloheptyltrimethylammonium hydroxide $\xrightarrow{heat}$ Cycloheptene (87%)
Electrophilic aromatic substitution (Section 22.14) Arylamines are very reactive toward electrophilic aromatic substitution. It is customary to protect arylamines as their *N*-acyl derivatives before carrying out ring nitration, chlorination, bromination, sulfonation, or Friedel–Crafts reactions.	$ArH + E^+ \longrightarrow ArE + H^+$ Arylamine; Electrophile; Product of electrophilic aromatic substitution; Proton *p*-Nitroaniline $\xrightarrow[\text{acetic acid}]{2Br_2}$ 2,6-Dibromo-4-nitroaniline (95%)
Nitrosation (Sections 22.15–22.16) Nitrosation of amines occurs when sodium nitrite is added to a solution containing an amine and an acid. *Primary amines* yield alkyl diazonium salts. Alkyl diazonium salts are very unstable and yield carbocation-derived products. Aryl diazonium salts are exceedingly useful synthetic intermediates. Their reactions are described in Table 22.7.	$RNH_2 \xrightarrow[H_3O^+]{NaNO_2} R\overset{+}{N}{\equiv}N:$ Primary amine; Diazonium ion *m*-Nitroaniline $\xrightarrow[H_2O,\ 0\text{–}5°C]{NaNO_2,\ H_2SO_4}$ *m*-Nitrobenzenediazonium hydrogen sulfate (HSO_4^-)

continued

TABLE 22.6 Reactions of Amines Discussed in This Chapter (*Continued*)

Reaction (section) and comments	General equation and specific example
Secondary alkylamines and *secondary arylamines* yield *N*-nitroso amines.	$R_2NH \xrightarrow[H_3O^+]{NaNO_2} R_2N{-}N{=}O$ Secondary amine → *N*-Nitroso amine 2,6-Dimethylpiperidine $\xrightarrow[H_2O]{NaNO_2,\ HCl}$ 2,6-Dimethyl-*N*-nitrosopiperidine (72%)
Tertiary alkylamines illustrate no useful chemistry on nitrosation. *Tertiary arylamines* undergo nitrosation of the ring by electrophilic aromatic substitution.	$(CH_3)_2N{-}C_6H_5 \xrightarrow[H_2O]{NaNO_2,\ HCl} (CH_3)_2N{-}C_6H_4{-}N{=}O$ *N,N*-Dimethylaniline → *N,N*-Dimethyl-4-nitrosoaniline (80–89%)

TABLE 22.7 Synthetically Useful Transformations Involving Aryl Diazonium Ions (Section 22.17)

Reaction and comments	General equation and specific example
Preparation of phenols Heating its aqueous acidic solution converts a diazonium salt to a phenol. This is the most general method for the synthesis of phenols.	$ArNH_2 \xrightarrow[2.\ H_2O,\ heat]{1.\ NaNO_2,\ H_2SO_4,\ H_2O} ArOH$ Primary arylamine → Phenol *m*-Nitroaniline $\xrightarrow[2.\ H_2O,\ heat]{1.\ NaNO_2,\ H_2SO_4,\ H_2O}$ *m*-Nitrophenol (81–86%)
Preparation of aryl fluorides Addition of fluoroboric acid to a solution of a diazonium salt causes the precipitation of an aryl diazonium fluoroborate. When the dry aryl diazonium fluoroborate is heated, an aryl fluoride results. This is the Schiemann reaction; it is the most general method for the preparation of aryl fluorides.	$ArNH_2 \xrightarrow[2.\ HBF_4]{1.\ NaNO_2,\ H_3O^+} Ar\overset{+}{N}{\equiv}N{:}\ \overset{-}{B}F_4 \xrightarrow{heat} ArF$ Primary arylamine → Aryldiazonium fluoroborate → Aryl fluoride *m*-Toluidine $\xrightarrow[2.\ HBF_4]{1.\ NaNO_2,\ HCl,\ H_2O}$ *m*-Methylbenzenediazonium fluoroborate (76–84%) $\xrightarrow{heat}$ *m*-Fluorotoluene (89%)

continued

TABLE 22.7 Synthetically Useful Transformations Involving Aryl Diazonium Ions (Section 22.17) (*Continued*)

Reaction and comments	General equation and specific example
Preparation of aryl iodides Aryl diazonium salts react with sodium or potassium iodide to form aryl iodides. This is the most general method for the synthesis of aryl iodides.	$ArNH_2$ (Primary arylamine) $\xrightarrow[\text{2. NaI or KI}]{\text{1. } NaNO_2,\ H_3O^+}$ ArI (Aryl iodide) 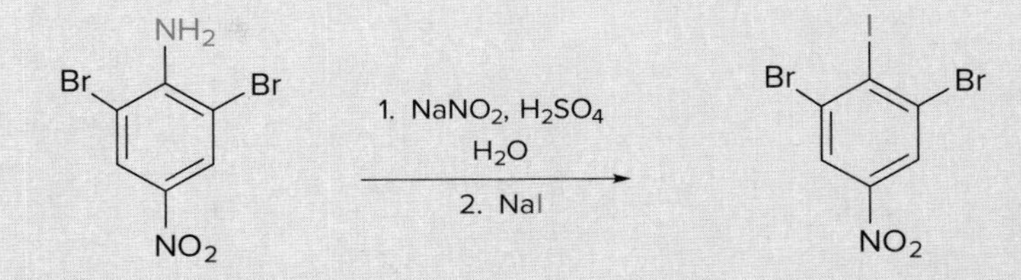 2,6-Dibromo-4-nitroaniline $\xrightarrow[\text{2. NaI}]{\text{1. } NaNO_2,\ H_2SO_4,\ H_2O}$ 1,3-Dibromo-2-iodo-5-nitrobenzene (84–88%)
Preparation of aryl chlorides In the Sandmeyer reaction, a solution containing an aryl diazonium salt is treated with copper(I) chloride to give an aryl chloride.	$ArNH_2$ (Primary arylamine) $\xrightarrow[\text{2. CuCl}]{\text{1. } NaNO_2,\ HCl,\ H_2O}$ ArCl (Aryl chloride) 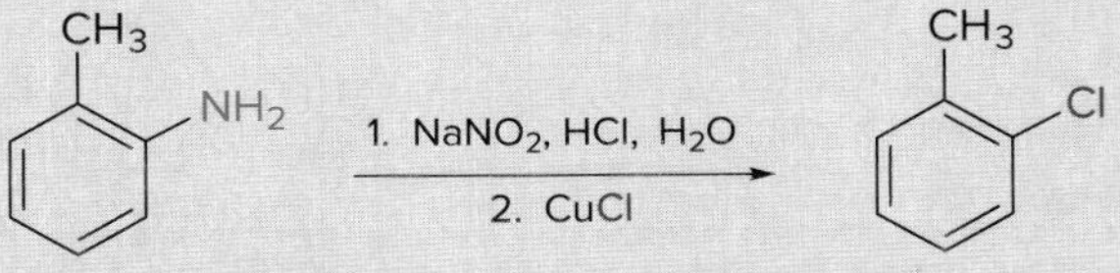 *o*-Toluidine $\xrightarrow[\text{2. CuCl}]{\text{1. } NaNO_2,\ HCl,\ H_2O}$ *o*-Chlorotoluene (74–79%)
Preparation of aryl bromides The Sandmeyer reaction using copper(I) bromide converts primary arylamines to aryl bromides.	$ArNH_2$ (Primary arylamine) $\xrightarrow[\text{2. CuBr}]{\text{1. } NaNO_2,\ HBr,\ H_2O}$ ArBr (Aryl bromide) 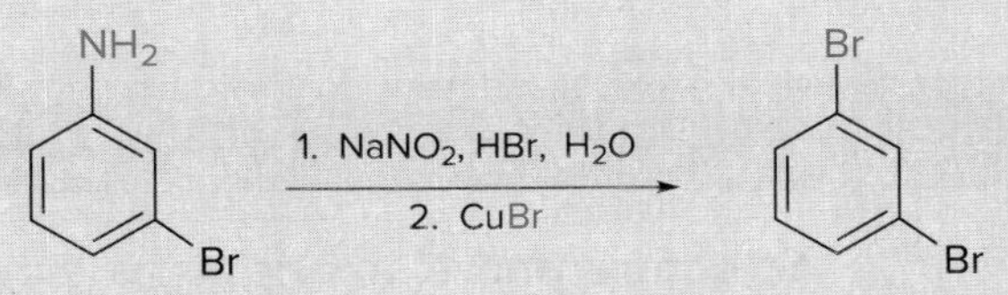 *m*-Bromoaniline $\xrightarrow[\text{2. CuBr}]{\text{1. } NaNO_2,\ HBr,\ H_2O}$ *m*-Dibromobenzene (80–87%)
Preparation of aryl nitriles Copper(I) cyanide converts aryl diazonium salts to aryl nitriles.	$ArNH_2$ (Primary arylamine) $\xrightarrow[\text{2. CuCN}]{\text{1. } NaNO_2,\ H_2O}$ ArCN (Aryl nitrile) 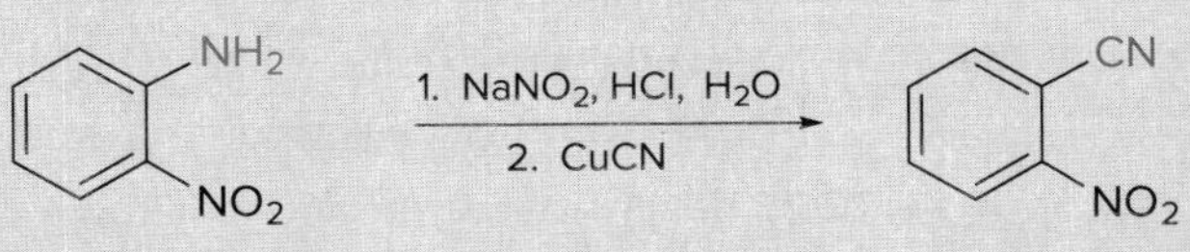 *m*-Nitroaniline $\xrightarrow[\text{2. CuCN}]{\text{1. } NaNO_2,\ HCl,\ H_2O}$ *o*-Nitrobenzonitrile (87%)

continued

TABLE 22.7 Synthetically Useful Transformations Involving Aryl Diazonium Ions (Section 22.17) (*Continued*)

Reaction and comments	General equation and specific example
Reductive deamination of primary arylamines The amino group of an arylamine can be replaced by hydrogen by treatment of its diazonium salt with ethanol or with hypophosphorous acid.	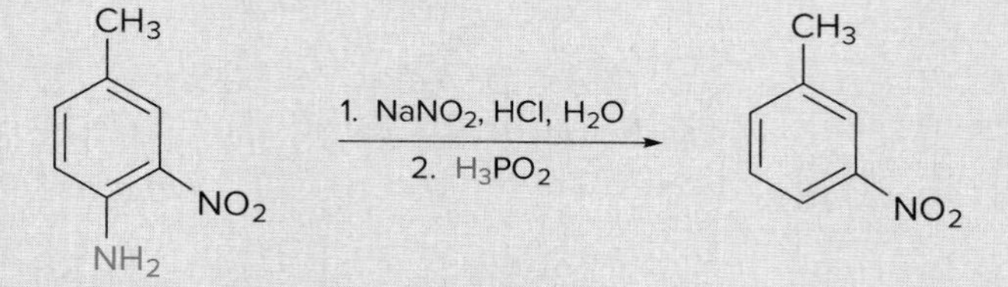 $ArNH_2$ (Primary arylamine) $\xrightarrow[\text{2. } CH_3CH_2OH \text{ or } H_3PO_2]{\text{1. } NaNO_2,\ H_3O^+}$ ArH (Arene) 4-Methyl-2-nitroaniline $\xrightarrow[\text{2. } H_3PO_2]{\text{1. } NaNO_2,\ HCl,\ H_2O}$ *m*-Nitrotoluene (80%)

Section 22.19 The N—H stretching frequency of primary and secondary amines appears in the infrared spectrum in the 3000–3500 cm^{-1} region. In the NMR spectra of amines, protons and carbons of the type H—C—N are more shielded than H—C—O.

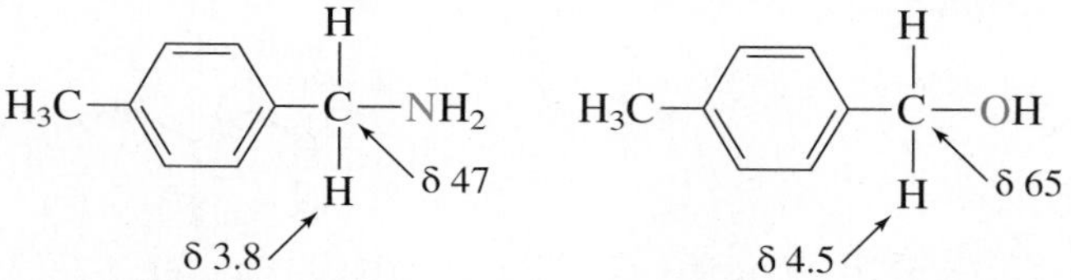

H_3C—C_6H_4—CH_2—NH_2 (H: δ 3.8; C: δ 47) H_3C—C_6H_4—CH_2—OH (H: δ 4.5; C: δ 65)

Amines have odd-numbered molecular weights, which helps identify them by mass spectrometry. Fragmentation tends to be controlled by the formation of a nitrogen-stabilized cation.

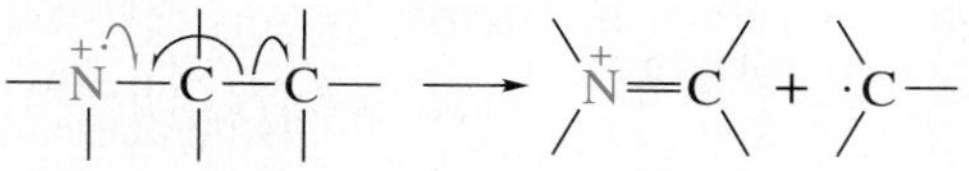

$$-\overset{+\cdot}{N}-C-C- \longrightarrow \overset{+}{N}=C + \cdot C-$$

PROBLEMS

Structure and Nomenclature

22.23 Write structural formulas for all the amines of molecular formula $C_4H_{11}N$. Give an acceptable name for each one, and classify it as a primary, secondary, or tertiary amine.

22.24 Provide a structural formula for each of the following compounds:

(a) 2-Ethyl-1-butanamine
(b) *N*-Ethyl-1-butanamine
(c) Dibenzylamine
(d) Tribenzylamine
(e) Tetraethylammonium hydroxide
(f) *N*-Allylcyclohexylamine
(g) *N*-Allylpiperidine
(h) Benzyl 2-aminopropanoate
(i) 4-(*N,N*-Dimethylamino)cyclohexanone
(j) 2,2-Dimethyl-1,3-propanediamine

22.25 Many naturally occurring nitrogen compounds and many nitrogen-containing drugs are better known by common names than by their systematic names. A few of these follow. Write a structural formula for each one.

(a) *trans*-2-Phenylcyclopropylamine, better known as *tranylcypromine:* an antidepressant drug
(b) *N*-Benzyl-*N*-methyl-2-propynylamine, better known as *pargyline:* a drug used to treat high blood pressure
(c) 1-Phenyl-2-propanamine, better known as *amphetamine:* a stimulant
(d) 1-(*m*-Hydroxyphenyl)-2-(methylamino)ethanol: better known as *phenylephrine:* a nasal decongestant

22.26 Both alkyl- and arylamines have a low barrier for pyramidal inversion at nitrogen, which prevents the separation of chiral amines into their enantiomers. The barrier for inversion at nitrogen in alkylamines is approximately 25 kJ/mol (6 kcal/mol), whereas for arylamines it is much lower, on the order of 6.3 kJ/mol (1.5 kcal/mol). Can you suggest a reason for the difference?

Reactions

22.27 (a) Give the structures and provide an acceptable name for all the isomers of molecular formula C_7H_9N that contain a benzene ring.

(b) Which one of these isomers is the strongest base?

(c) Which, if any, of these isomers yield an *N*-nitroso amine on treatment with sodium nitrite and hydrochloric acid?

(d) Which, if any, of these isomers undergo nitrosation of their benzene ring on treatment with sodium nitrite and hydrochloric acid?

22.28 Arrange the following compounds or anions in each group in order of decreasing basicity:

(a) H_3C^-, H_2N^-, HO^-, F^-

(b) H_2O, NH_3, HO^-, H_2N^-

(c) HO^-, H_2N^-, $:\bar{C}{\equiv}N:$, NO_3^-

(d)

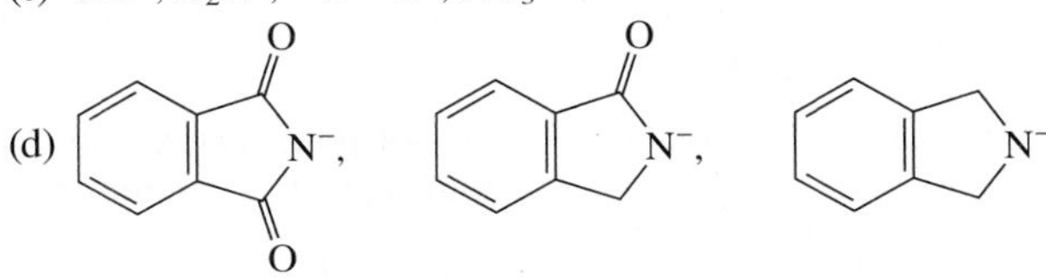

22.29 Arrange the members of each group in order of decreasing basicity:

(a) Ammonia, aniline, methylamine

(b) Acetanilide, aniline, *N*-methylaniline

(c) 2,4-Dichloroaniline, 2,4-dimethylaniline, 2,4-dinitroaniline

(d) 3,4-Dichloroaniline, 4-chloro-2-nitroaniline, 4-chloro-3-nitroaniline

(e) Dimethylamine, diphenylamine, *N*-methylaniline

22.30 *N,N*-Dimethylaniline and pyridine are similar in basicity, whereas 4-(*N,N*-dimethylamino)-pyridine is considerably more basic than either.

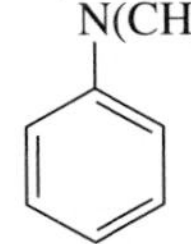

N,N-Dimethylaniline
pK_a of conjugate acid = 5.1

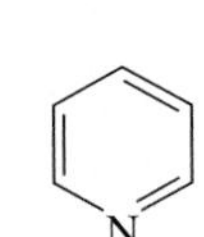

Pyridine
pK_a of conjugate acid = 5.3

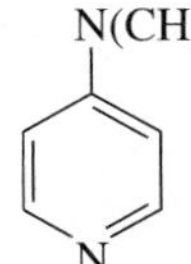

4-(*N,N*-Dimethylamino)pyridine
pK_a of conjugate acid = 9.7

Apply resonance principles to identify the more basic of the two nitrogens of 4-(*N,N*-dimethylamino)pyridine, and suggest an explanation for its enhanced basicity.

22.31 The compound shown is a somewhat stronger base than ammonia. Which nitrogen do you think is protonated when it is treated with an acid? Write a structural formula for the species that results.

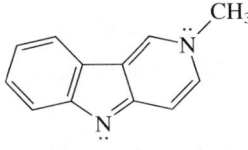

5-Methyl-γ-carboline
pK_a of conjugate acid = 10.5

22.32 Carnosine, found in muscle and brain tissue, acts as a buffer to neutralize small amounts of acid. The pK_a of the conjugate acid of carnosine is close to 7.0. What is its structure?

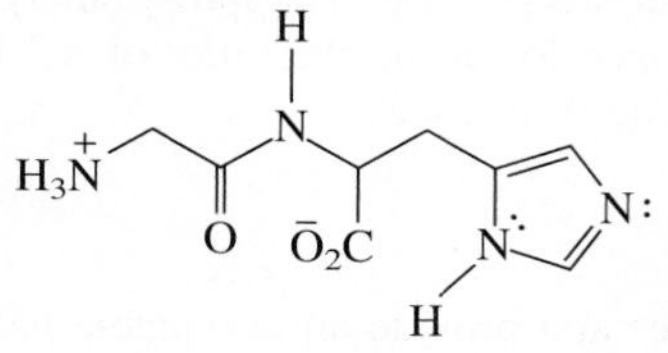

Carnosine

22.33 *Physostigmine,* an alkaloid obtained from a West African plant (*Physotigma venenosum*), is used in the treatment of glaucoma. Treatment of physostigmine with methyl iodide gives a quaternary ammonium salt. What is the structure of this salt?

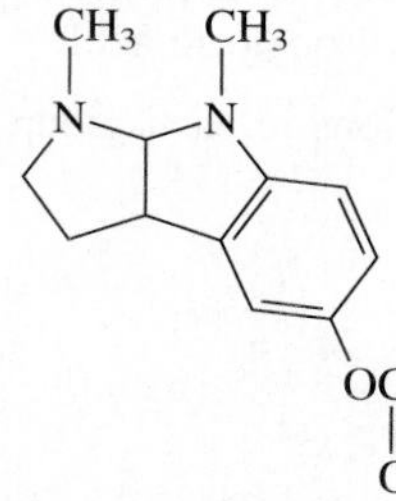

Physostigmine

22.34 9-Aminofluorene has applications in the structural analysis of proteins and carbohydrates. Write a stepwise procedure with equations to show how to separate a mixture of 9-aminofluorene and fluorene in diethyl ether solution.

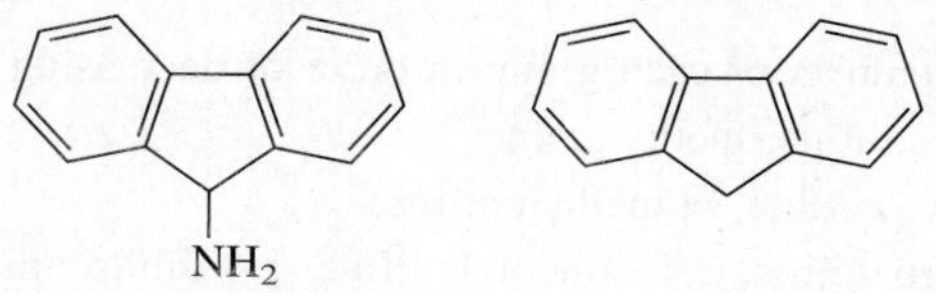

9-Aminofluorene Fluorene

22.35 Give the structure of the expected product formed when benzylamine reacts with each of the following reagents:

(a) Hydrogen bromide
(b) Sulfuric acid
(c) Acetic acid
(d) Acetyl chloride
(e) Acetic anhydride
(f) Acetone
(g) Acetone and hydrogen (nickel catalyst)
(h) Ethylene oxide
(i) 1,2-Epoxypropane
(j) Excess methyl iodide
(k) Sodium nitrite in dilute hydrochloric acid

22.36 Write the structure of the product formed on reaction of aniline with each of the following:

(a) Hydrogen bromide
(b) Excess methyl iodide
(c) Acetaldehyde
(d) Acetaldehyde and hydrogen (nickel catalyst)
(e) Acetic anhydride
(f) Benzoyl chloride
(g) Sodium nitrite, aqueous sulfuric acid, 0–5°C

22.37 Write the structure of the product formed on reaction of acetanilide with each of the following:

(a) Lithium aluminum hydride, followed by water
(b) Nitric acid and sulfuric acid
(c) Sulfur trioxide and sulfuric acid
(d) Bromine in acetic acid
(e) *tert*-Butyl chloride, aluminum chloride
(f) Acetyl chloride, aluminum chloride
(g) 6 M hydrochloric acid, reflux
(h) Aqueous sodium hydroxide, reflux

22.38 Identify the principal organic products of each of the following reactions:

(a) Cyclohexanone + cyclohexylamine $\xrightarrow{H_2,\ Ni}$

(b) 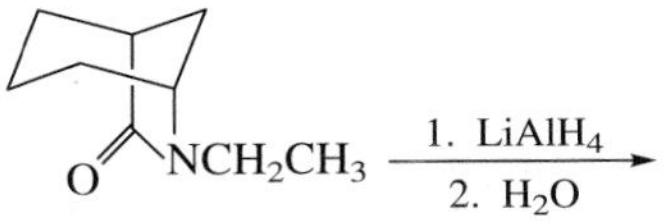 [cyclohexane ring bearing $C(=O)NCH_2CH_3$] $\xrightarrow[2.\ H_2O]{1.\ LiAlH_4}$

(c) $C_6H_5CH_2CH_2CH_2OH \xrightarrow[2.\ (CH_3)_2NH\ (excess)]{1.\ p\text{-toluenesulfonyl chloride, pyridine}}$

(d) 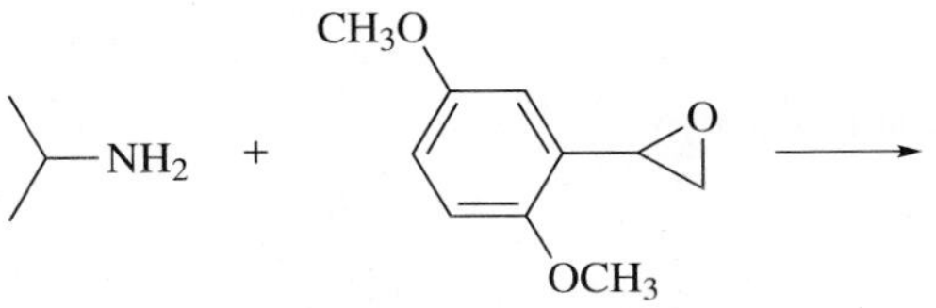 [isopropyl]$-NH_2$ + [benzene ring bearing CH_3O, OCH_3, and epoxide (O)] $\longrightarrow$

(e) $(C_6H_5CH_2)_2NH + CH_3\overset{O}{\overset{\|}{C}}CH_2Cl \xrightarrow[THF]{triethylamine}$

(f)  [cyclohexane ring–$C(CH_3)_2$–$\overset{+}{N}(CH_3)_3$] $HO^- \xrightarrow{heat}$

(g) 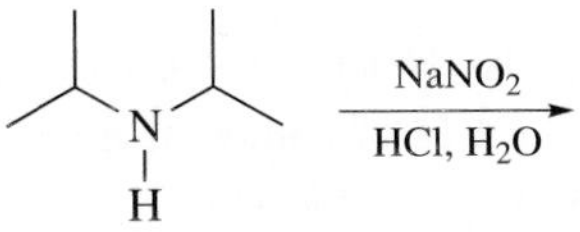 [diisopropyl amine, N–H] $\xrightarrow[HCl,\ H_2O]{NaNO_2}$

22.39 Each of the following reactions has been reported in the chemical literature and proceeds in good yield. Identify the principal organic product of each reaction.

(a) 1,2-Diethyl-4-nitrobenzene $\xrightarrow[ethanol]{H_2,\ Pt}$

(b) 1,3-Dimethyl-2-nitrobenzene $\xrightarrow[2.\ HO^-]{1.\ SnCl_2,\ HCl}$

(c) Product of part (b) + $ClCH_2\overset{O}{\overset{\|}{C}}Cl \longrightarrow$

(d) Product of part (c) + $(CH_3CH_2)_2NH \longrightarrow$

(e) Product of part (d) + HCl $\longrightarrow$

(f) $C_6H_5NH\overset{O}{\overset{\|}{C}}CH_2CH_2CH_3 \xrightarrow[2.\ H_2O]{1.\ LiAlH_4}$

(g) Aniline + heptanal $\xrightarrow{H_2,\ Ni}$

(h) Acetanilide + $ClCH_2\overset{O}{\overset{\|}{C}}Cl \xrightarrow{AlCl_3}$

(i) 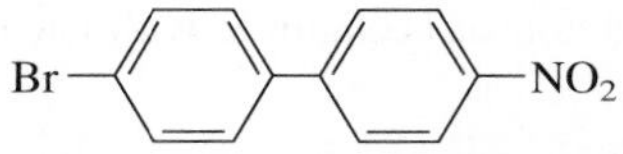 Br–C_6H_4–C_6H_4–NO_2 $\xrightarrow[\text{2. HO}^-]{\text{1. Fe, HCl}}$

(j) Product of part (i) $\xrightarrow[\text{2. H}_2\text{O, heat}]{\text{1. NaNO}_2\text{, H}_2\text{SO}_4\text{, H}_2\text{O}}$

(k) 2,6-Dinitroaniline $\xrightarrow[\text{2. CuCl}]{\text{1. NaNO}_2\text{, H}_2\text{SO}_4\text{, H}_2\text{O}}$

(l) *m*-Bromoaniline $\xrightarrow[\text{2. CuBr}]{\text{1. NaNO}_2\text{, HBr, H}_2\text{O}}$

(m) *o*-Nitroaniline $\xrightarrow[\text{2. CuCN}]{\text{1. NaNO}_2\text{, HCl, H}_2\text{O}}$

(n) 2,6-Diiodo-4-nitroaniline $\xrightarrow[\text{2. KI}]{\text{1. NaNO}_2\text{, H}_2\text{SO}_4\text{, H}_2\text{O}}$

(o) $:N{\equiv}\overset{+}{N}$–C_6H_4–C_6H_4–$\overset{+}{N}{\equiv}N:$ $2\overset{-}{B}F_4$ $\xrightarrow{\text{heat}}$

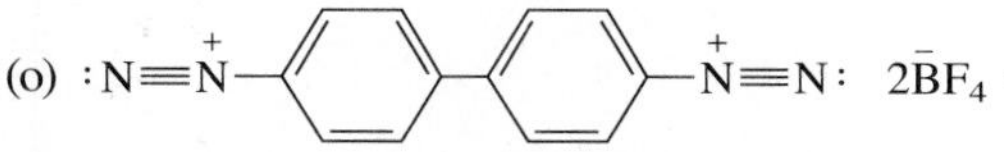

(p) 2,4,6-Trinitroaniline $\xrightarrow[\text{H}_2\text{O, H}_3\text{PO}_2]{\text{NaNO}_2\text{, H}_2\text{SO}_4}$

(q) 2-Amino-5-iodobenzoic acid $\xrightarrow[\text{2. CH}_3\text{CH}_2\text{OH}]{\text{1. NaNO}_2\text{, HCl, H}_2\text{O}}$

(r) Aniline $\xrightarrow[\text{2. 2,3,6-trimethylphenol}]{\text{1. NaNO}_2\text{, H}_2\text{SO}_4\text{, H}_2\text{O}}$

(s) 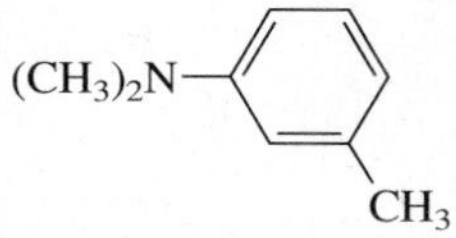 $(CH_3)_2N$–C_6H_4–CH_3 (meta) $\xrightarrow[\text{2. HO}^-]{\text{1. NaNO}_2\text{, HCl, H}_2\text{O}}$

22.40 Provide a reasonable explanation for each of the following observations:

(a) 4-Methylpiperidine has a higher boiling point than *N*-methylpiperidine.

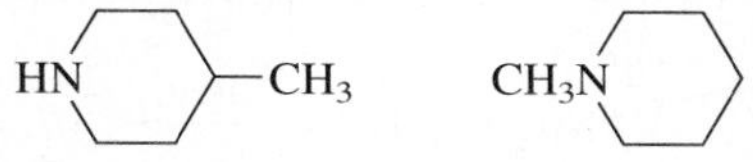

4-Methylpiperidine (bp 129°C) N-Methylpiperidine (bp 106°C)

(b) Two isomeric quaternary ammonium salts are formed in comparable amounts when 4-*tert*-butyl-*N*-methylpiperidine is treated with benzyl chloride.

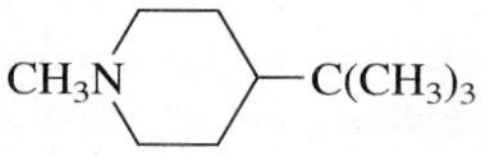

4-*tert*-Butyl-*N*-methylpiperidine

(c) When tetramethylammonium hydroxide is heated at 130°C, trimethylamine and methanol are formed.

(d) The major product formed on treatment of 1-propanamine with sodium nitrite in dilute hydrochloric acid is 2-propanol.

Synthesis

22.41 Describe procedures for preparing each of the following compounds, using ethanol as the source of all their carbon atoms. Once you prepare a compound, you need not repeat its synthesis in a subsequent part of this problem.

(a) Ethylamine
(b) *N*-Ethylacetamide
(c) Diethylamine
(d) *N,N*-Diethylacetamide
(e) Triethylamine
(f) Tetraethylammonium bromide

22.42 Show by writing the appropriate sequence of equations how you could carry out each of the following transformations:

(a) 1-Butanol to 1-pentanamine
(b) *tert*-Butyl chloride to 2,2-dimethyl-1-propanamine
(c) Cyclohexanol to *N*-methylcyclohexylamine

(d) Isopropyl alcohol to 1-amino-2-methyl-2-propanol

(e) Isopropyl alcohol to 1-amino-2-propanol

(f) Isopropyl alcohol to 1-(N,N-dimethylamino)-2-propanol

(g) 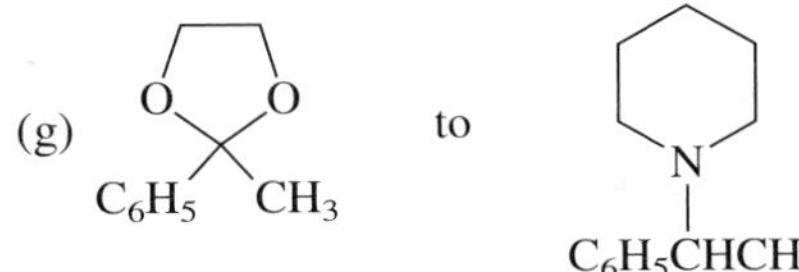

22.43 Each of the following dihaloalkanes gives an N-(haloalkyl)phthalimide on reaction with one equivalent of the potassium salt of phthalimide. Write the structure of the phthalimide derivative formed in each case and explain the basis for your answer.

(a) 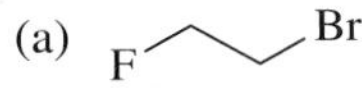

(b) Br, Br (structure)

(c) Br, Br (structure)

22.44 Give the structures, including stereochemistry, of compounds A through C.

$$(S)\text{-2-Octanol} + H_3C\text{—}C_6H_4\text{—}SO_2Cl \xrightarrow{\text{pyridine}} \text{Compound A}$$

$$\text{Compound A} \xrightarrow{NaN_3,\ \text{methanol–water}} \text{Compound B}$$

$$\text{Compound B} \xrightarrow[\text{2. } H_2O]{\text{1. } LiAlH_4} \text{Compound C}$$

22.45 Devise efficient syntheses of each of the following compounds from the designated starting materials. You may also use any necessary organic or inorganic reagents.

(a) 3,3-Dimethyl-1-butanamine from 1-bromo-2,2-dimethylpropane

(b) $H_2C{=}CH(CH_2)_8CH_2$—N(pyrrolidinyl) from 10-undecenoic acid and pyrrolidine

(c) 2-(C_6H_5O)cyclopentyl-NH_2 from 2-(C_6H_5O)cyclopentane-OH

(d) $C_6H_5CH_2N(CH_3)CH_2CH_2CH_2CH_2NH_2$ from $C_6H_5CH_2NHCH_3$ and $BrCH_2CH_2CH_2CN$

(e) $NC\text{—}C_6H_4\text{—}CH_2N(CH_3)_2$ from $NC\text{—}C_6H_4\text{—}CH_3$

22.46 Each of the following compounds has been prepared from p-nitroaniline. Outline a reasonable series of steps leading to each one.

(a) p-Nitrobenzonitrile

(b) 3,4,5-Trichloroaniline

(c) 1,3-Dibromo-5-nitrobenzene

(d) 3,5-Dibromoaniline

(e) p-Acetamidophenol (*acetaminophen*)

22.47 Each of the following compounds has been prepared from o-anisidine (o-methoxyaniline). Outline a series of steps leading to each one.

(a) o-Bromoanisole

(b) o-Fluoroanisole

(c) 3-Fluoro-4-methoxyacetophenone

(d) 3-Fluoro-4-methoxybenzonitrile

(e) 3-Fluoro-4-methoxyphenol

22.48 (a) Outline a synthesis of the following compound from nitrobenzene, p-nitrobenzyl alcohol, and any necessary organic or inorganic reagents.

$$C_6H_5\text{—}N{=}CH\text{—}C_6H_4\text{—}NO_2$$

(b) How would you modify the synthesis if you had to start with p-nitrotoluene instead of p-nitrobenzyl alcohol?

22.49 Design syntheses of each of the following compounds from the indicated starting material and any necessary organic or inorganic reagents:

(a) *p*-Aminobenzoic acid from *p*-methylaniline

(b) *p*-$FC_6H_4\overset{O}{\overset{\|}{C}}CH_2CH_3$ from benzene

(c) 1-Bromo-2-fluoro-3,5-dimethylbenzene from *m*-xylene

(d) 1-Bromo-4-fluoro-2-methylnaphthalene (Br, CH_3, F) from 1-amino-2-methyl-4-nitronaphthalene (NH_2, CH_3, NO_2)

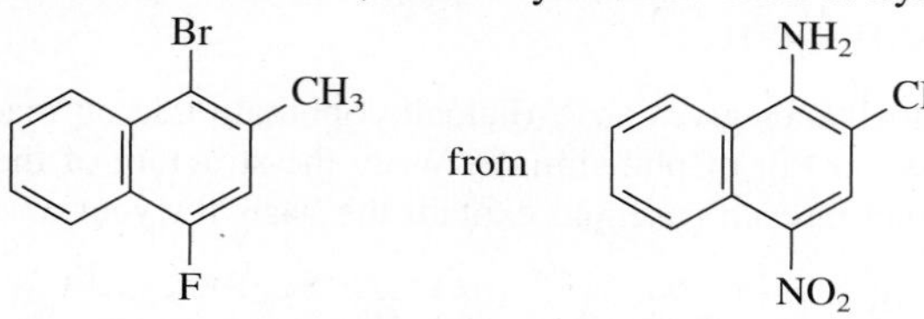

(e) *o*-$BrC_6H_4C(CH_3)_3$ from *p*-$O_2NC_6H_4C(CH_3)_3$

(f) *m*-$ClC_6H_4C(CH_3)_3$ from *p*-$O_2NC_6H_4C(CH_3)_3$

(g) 1-Bromo-3,5-diethylbenzene from *m*-diethylbenzene

(h) Benzene ring bearing CF_3, I, Br from benzene ring bearing CF_3, $NH\overset{O}{\overset{\|}{C}}CH_3$, Br, H_2N

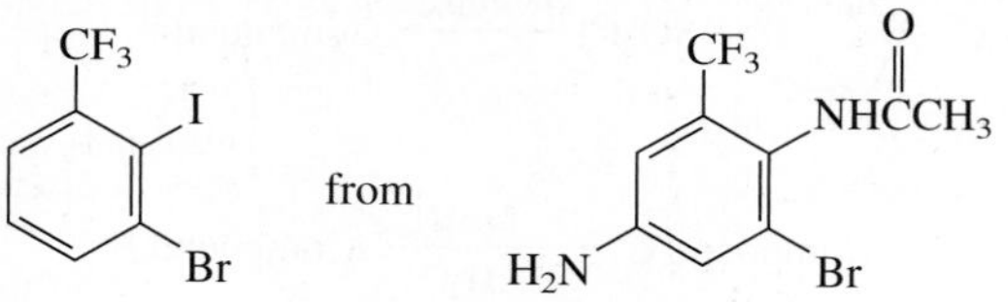

(i) Dibenzazepine bearing NH, CH_3O, CH_3O from biphenyl bearing $CH_2\overset{O}{\overset{\|}{C}}OCH_3$, CH_3O, CH_3O, O_2N

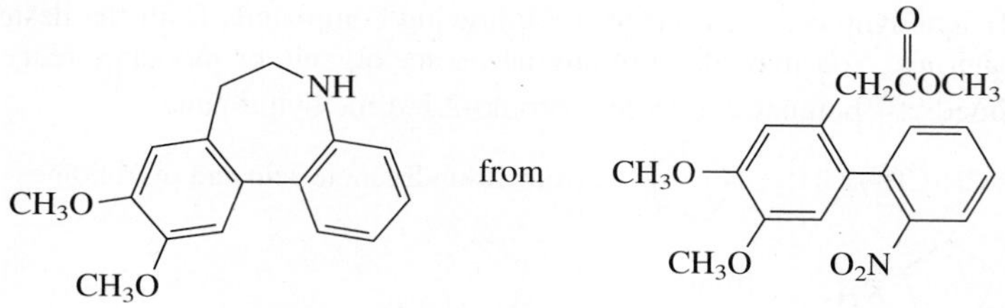

22.50 Show how 2-(2-bromophenyl)ethanamine could be prepared by the Gabriel amine synthesis from *N*-potassiophthalimide and compound A.

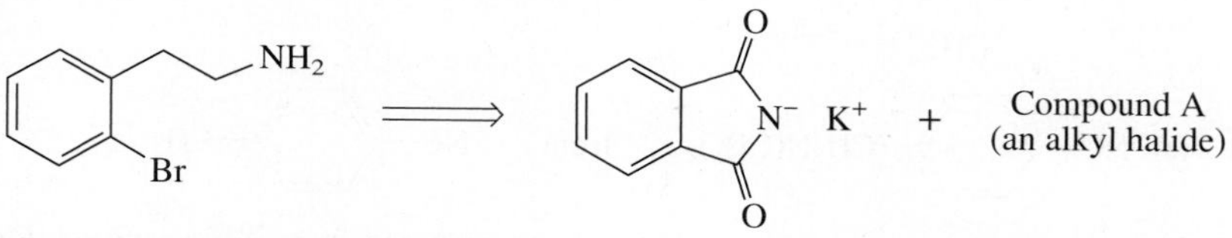

2-(2-Bromophenyl)-ethanamine ⟹ *N*-Potassiophthalimide (N^- K^+) + Compound A (an alkyl halide)

22.51 Ammonia and amines undergo conjugate addition to α,β-unsaturated carbonyl compounds (see Section 21.9). On the basis of this information, predict the principal organic product of each of the following reactions:

(a) 4-Methyl-3-penten-2-one + NH_3 ⟶

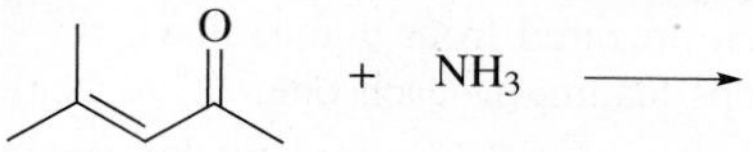

(b) 2-Cyclohexen-1-one + HN(piperidine) ⟶

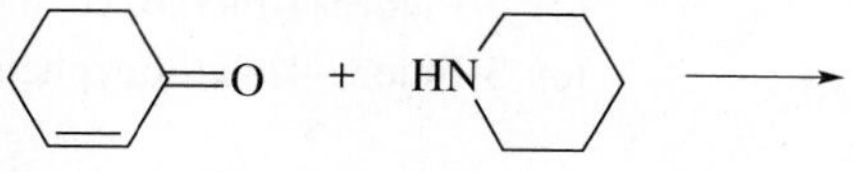

(c) 1,3-Diphenyl-2-propen-1-one + HN(morpholine, O) ⟶

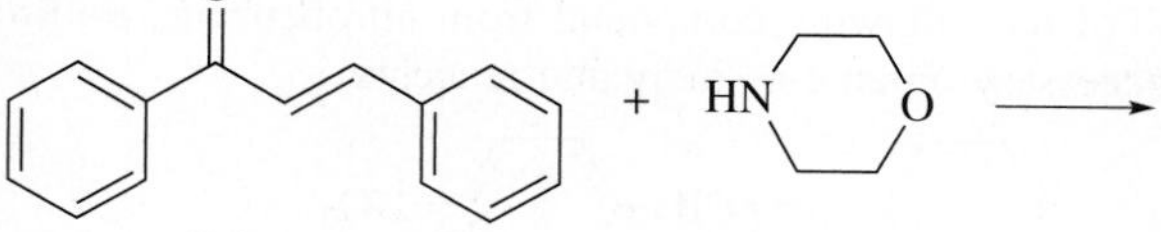

(d) Cyclohexenone bearing a chain with NH_2 —spontaneous→

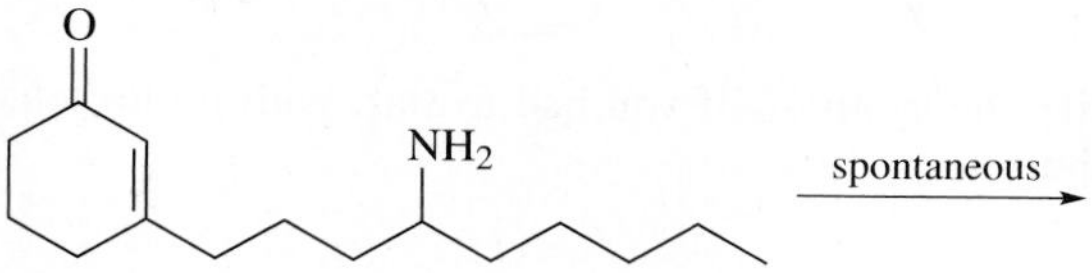

22.52 A number of compounds of the type represented by compound A were prepared for evaluation as potential analgesic drugs. Their preparation is described in a retrosynthetic format as shown.

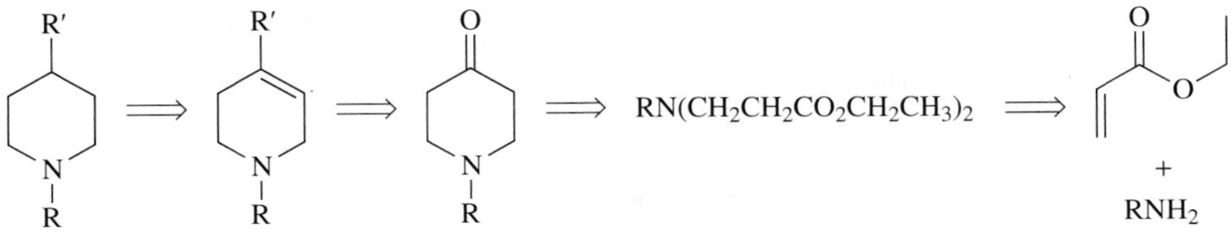

R' R' O O

⟹ ⟹ ⟹ $RN(CH_2CH_2CO_2CH_2CH_3)_2$ ⟹ O

N N N +

R R R RNH_2

On the basis of this retrosynthetic analysis, design a synthesis of *N*-methyl-4-phenylpiperidine (compound A, where R = CH_3, R′ = C_6H_5). Present your answer as a series of equations, showing all necessary reagents and isolated intermediates.

22.53 The reductive amination shown was a key step in the synthesis of a compound for testing as an analgesic. Write a structural formula for this compound.

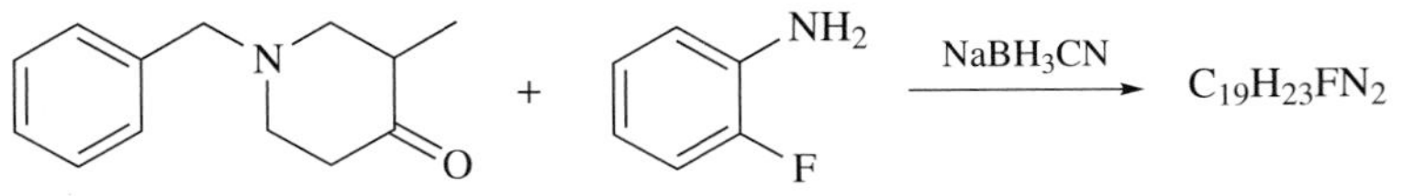

N + NH_2 $\xrightarrow{NaBH_3CN}$ $C_{19}H_{23}FN_2$

O F

22.54 The anticancer drug streptozotocin can be synthesized in two steps from glucosamine. The second step is nitrosation. What is reagent A in the first step?

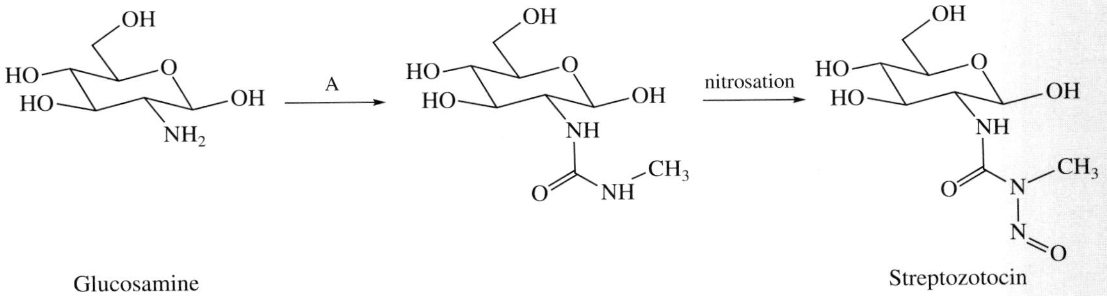

OH OH OH

HO O HO O HO O

HO OH $\xrightarrow{A}$ HO OH $\xrightarrow{\text{nitrosation}}$ HO OH

NH_2 NH NH

O NH CH_3 O N CH_3

N

O

Glucosamine Streptozotocin

22.55 Colchicine, a drug used in the treatment of gout, inhibits mitosis through its interaction with tubulin, a protein that assembles into spindles during cell division. Suggest reagents in the following reaction scheme for the synthesis of colchicine.

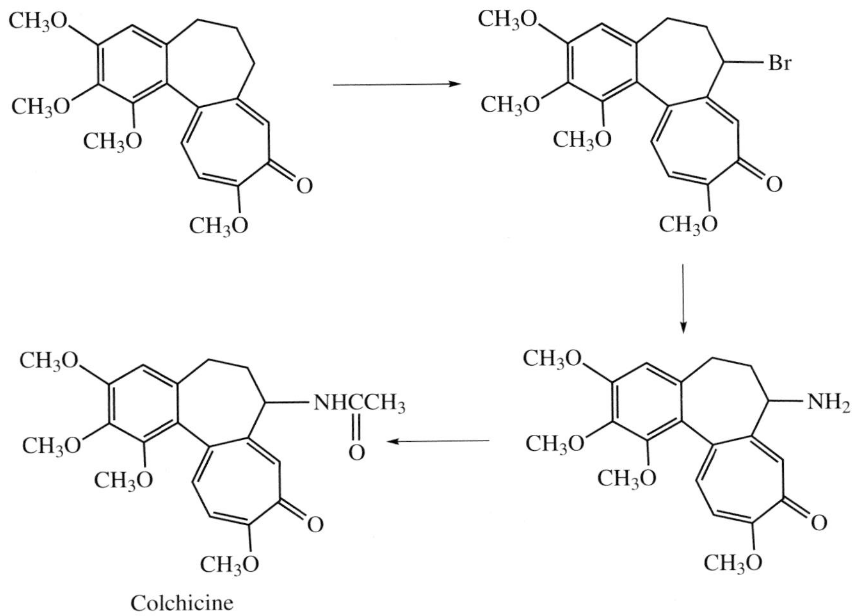

CH_3O CH_3O

⟶ Br

CH_3O CH_3O

CH_3O CH_3O

O O

CH_3O CH_3O

↓

CH_3O CH_3O

NHCCH$_3$ NH_2

CH_3O ‖ ⟵ CH_3O

CH_3O O CH_3O

O O

CH_3O CH_3O

Colchicine

Spectroscopy

22.56 Compounds A and B are isomeric amines of molecular formula $C_8H_{11}N$. Identify each isomer on the basis of the ^1H NMR spectra given in Figure 22.10.

22.57 Does the ^{13}C NMR spectrum shown in Figure 22.11 correspond to that of 1-amino-2-methyl-2-propanol or to 2-amino-2-methyl-1-propanol? Could this compound be prepared by reaction of an epoxide with ammonia?

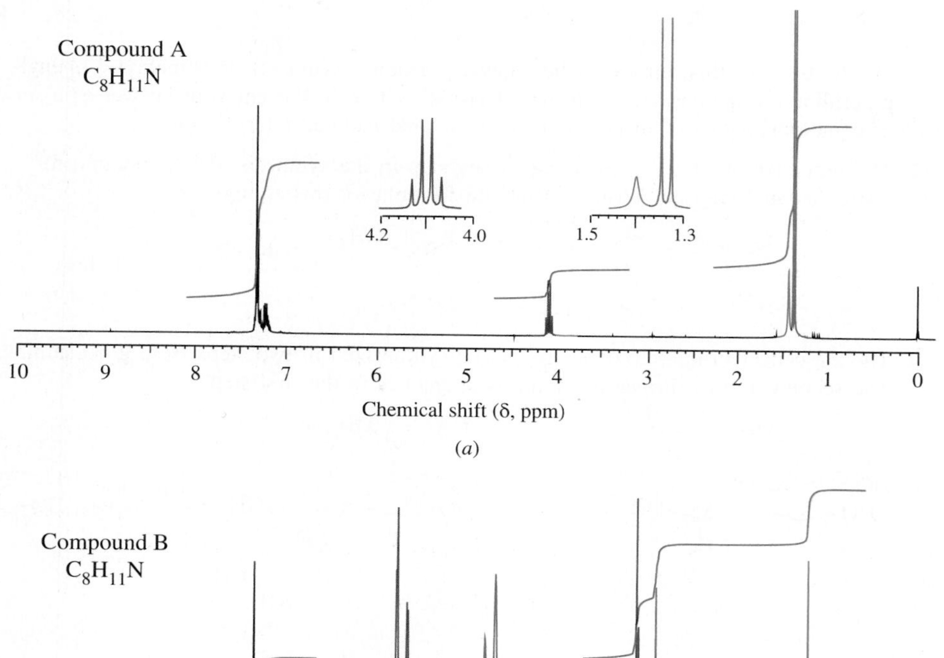

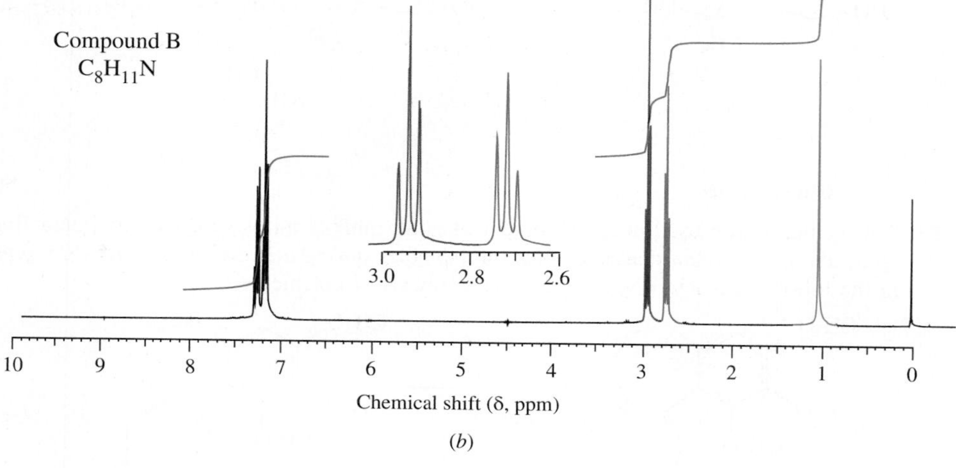

Figure 22.10

The 300-MHz ^1H NMR spectra of (*a*) compound A and (*b*) compound B (Problem 22.56).

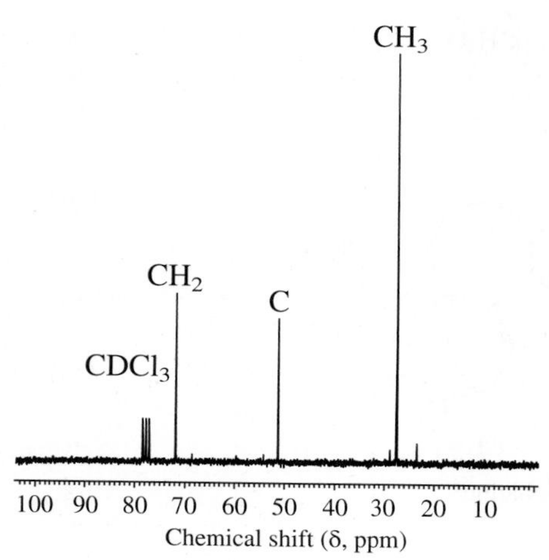

Figure 22.11

The ^{13}C NMR spectrum of the compound described in Problem 22.57.

Descriptive Passage and Interpretive Problems 22

Synthetic Applications of Enamines

The formation of enamines by the reaction of aldehydes and ketones with secondary amines was described in Section 18.11. As the following equation illustrates, the reaction is reversible.

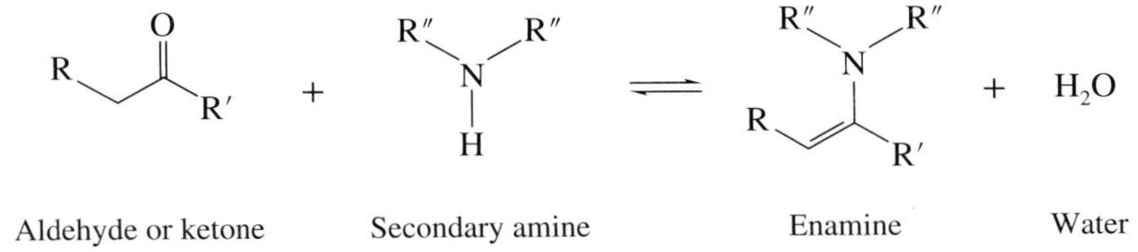

Aldehyde or ketone + Secondary amine ⇌ Enamine + Water (H_2O)

When preparing enamines, the reaction is normally carried out by heating in benzene as the solvent. No catalyst is necessary, but *p*-toluenesulfonic acid is sometimes added. The water formed is removed by distillation of its azeotropic mixture with benzene, which shifts the position of equilibrium to the right to give the enamine in high yield. Conversely, enamines can be hydrolyzed in aqueous acid to aldehydes and ketones.

Enamines resemble enols in that electron-pair donation makes their double bond electron-rich and nucleophilic.

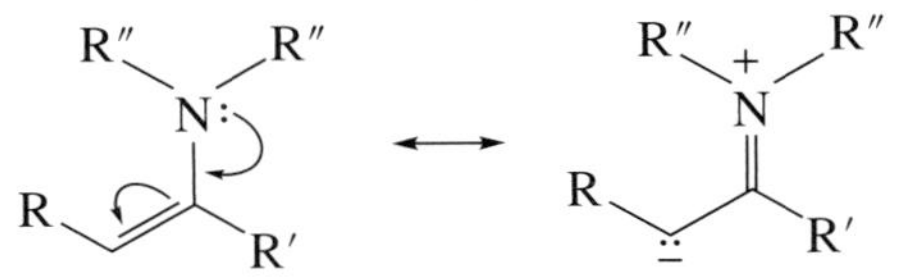

Because nitrogen is a better electron-pair donor than oxygen, an enamine is more nucleophilic than an enol. Enamines, being neutral molecules, are, however, less nucleophilic than enolates, which are anions.

Reactions of enamines with electrophiles (E^+) lead to carbon–carbon bond formation. Subsequent hydrolysis gives an α-substituted derivative of the original aldehyde or ketone.

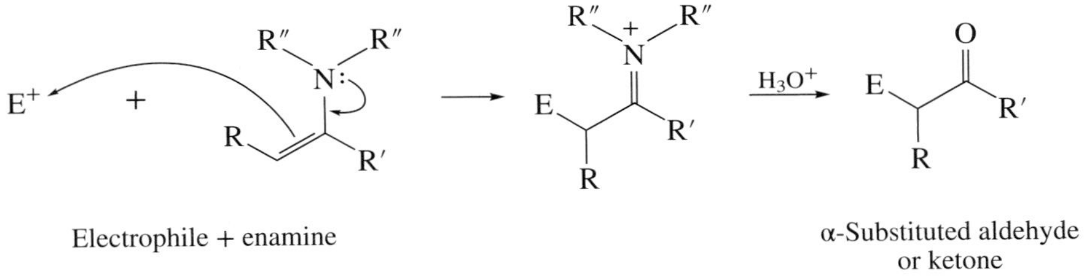

Electrophile + enamine → (via H_3O^+) α-Substituted aldehyde or ketone

Pyrrolidine is the secondary amine used most often for making enamines from aldehydes and ketones.

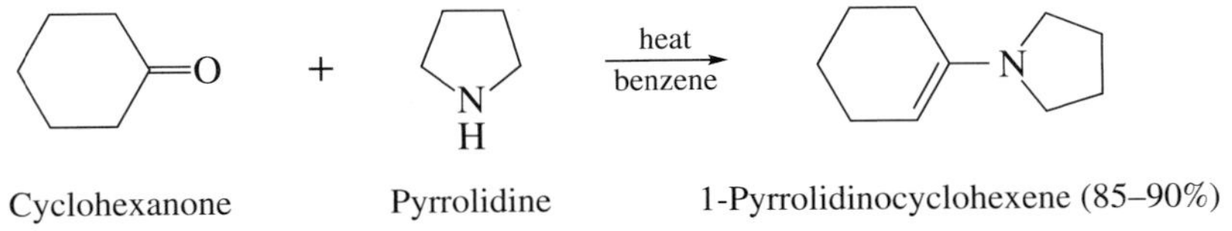

Cyclohexanone + Pyrrolidine —(heat, benzene)→ 1-Pyrrolidinocyclohexene (85–90%)

For synthetic purposes, the electrophilic reagents that give the best yields of α-substituted aldehydes and ketones on reactions with enamines are the following:

1. Alkyl halides that are very reactive in S_N2 reactions such as primary allylic and benzylic halides, α-halo ethers, α-halo esters, and α-halo nitriles.
2. Acyl chlorides and acid anhydrides.
3. Michael acceptors: α,β-unsaturated nitriles, esters, and ketones.

22.58 One of the following is often used to prepare enamines from aldehydes and ketones. The others do not yield enamines. Identify the enamine-forming compound.

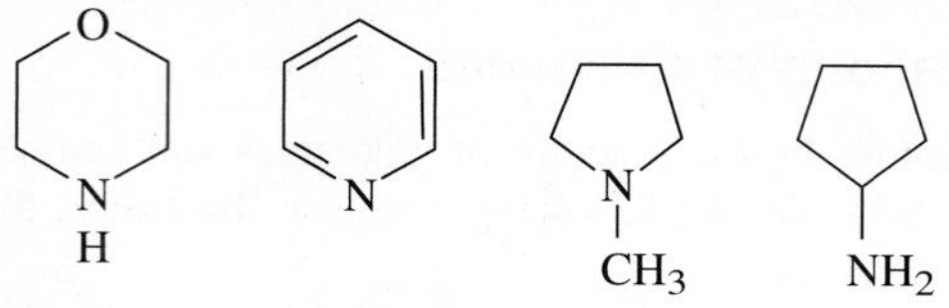

A. (morpholine, N–H) B. (pyridine) C. (N-methylpyrrolidine, N–CH_3) D. (cyclopentylamine, NH_2)

22.59 What is the product of the following reaction?

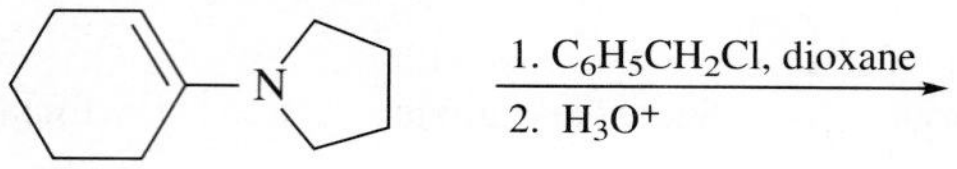

1-(cyclohexen-1-yl)pyrrolidine → 1. $C_6H_5CH_2Cl$, dioxane; 2. H_3O^+

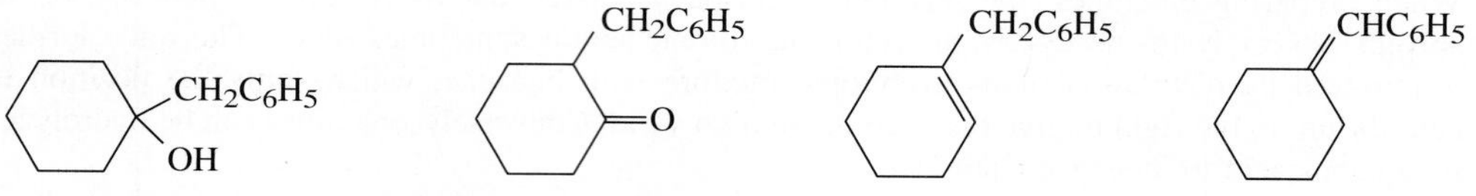

A. (1-benzylcyclohexanol: $CH_2C_6H_5$, OH) B. (2-benzylcyclohexanone: $CH_2C_6H_5$, =O) C. (1-benzylcyclohexene: $CH_2C_6H_5$) D. (benzylidenecyclohexane: CHC_6H_5)

22.60 Unsymmetric ketones give a mixture of two pyrrolidine enamines in which the enamine with the less-substituted double bond predominates. What is the major product of the following reaction sequence?

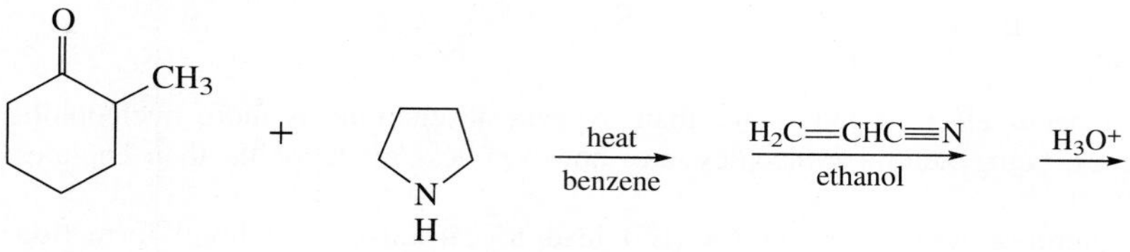

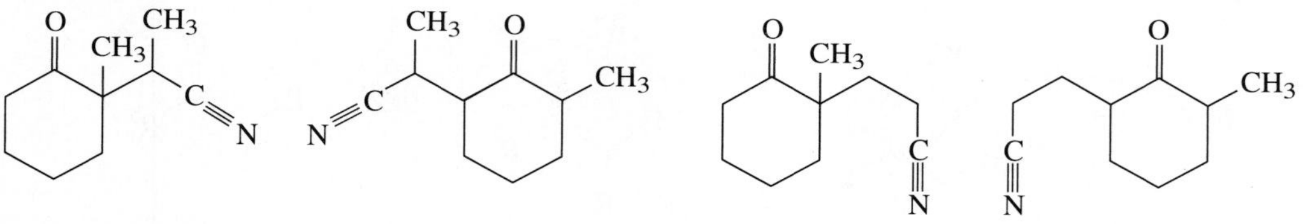

2-methylcyclohexanone (CH_3) + pyrrolidine (N–H) $\xrightarrow[\text{benzene}]{\text{heat}}$ $\xrightarrow[\text{ethanol}]{H_2C{=}CHC{\equiv}N}$ $\xrightarrow{H_3O^+}$

A. (CH_3, CH_3, C≡N) B. (CH_3, O, CH_3, N≡C) (*cis* + *trans*) C. (O, CH_3, C≡N) D. (O, CH_3, C≡N) (*cis* + *trans*)

22.61 What is the product of the following reaction sequence?

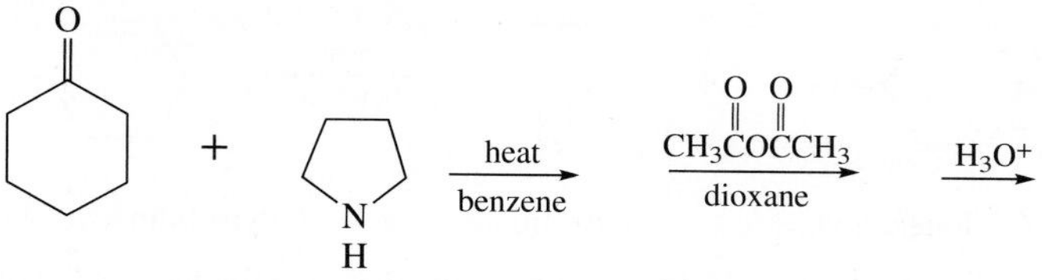

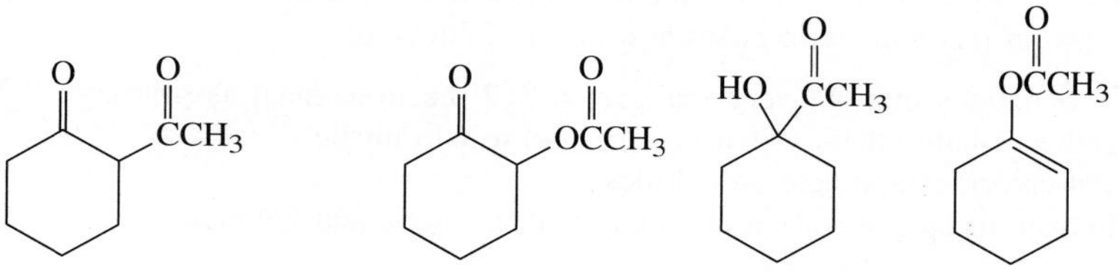

cyclohexanone + pyrrolidine (N–H) $\xrightarrow[\text{benzene}]{\text{heat}}$ $\xrightarrow[\text{dioxane}]{CH_3\overset{O}{\overset{\|}{C}}O\overset{O}{\overset{\|}{C}}CH_3}$ $\xrightarrow{H_3O^+}$

A. (O, $\overset{O}{\overset{\|}{C}}CH_3$) B. (O, $O\overset{O}{\overset{\|}{C}}CH_3$) C. (HO, $\overset{O}{\overset{\|}{C}}CH_3$) D. ($O\overset{O}{\overset{\|}{C}}CH_3$, cyclohexene)

22.62 (+)-2-Allylcyclohexanone has been prepared in 82% enantiomeric excess by alkylation of the optically active enamine prepared from cyclohexanone and an enantiomerically pure pyrrolidine derivative. Of the following, which one is the best pyrrolidine derivative to use in this enantioselective synthesis?

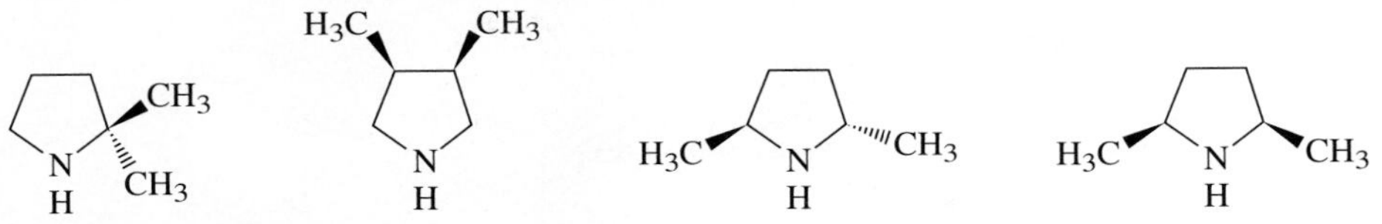

A. B. C. D.

22.63 Cyclooctanecarboxaldehyde was converted to a ketone having the molecular formula $C_{13}H_{20}O$ via its piperidine enamine as shown. What is the structure of the ketone?

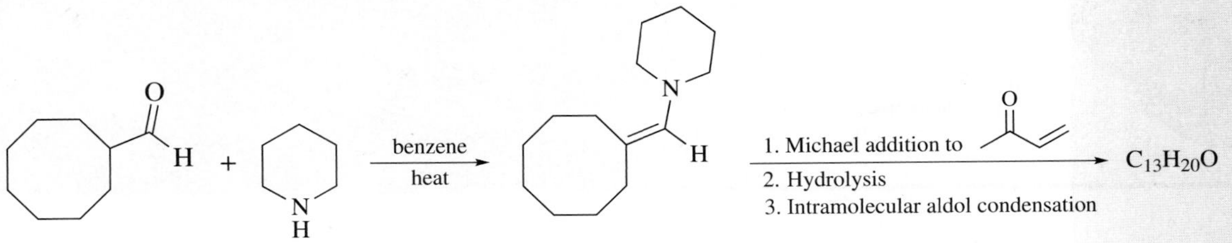

benzene, heat

1. Michael addition to
2. Hydrolysis
3. Intramolecular aldol condensation

$C_{13}H_{20}O$

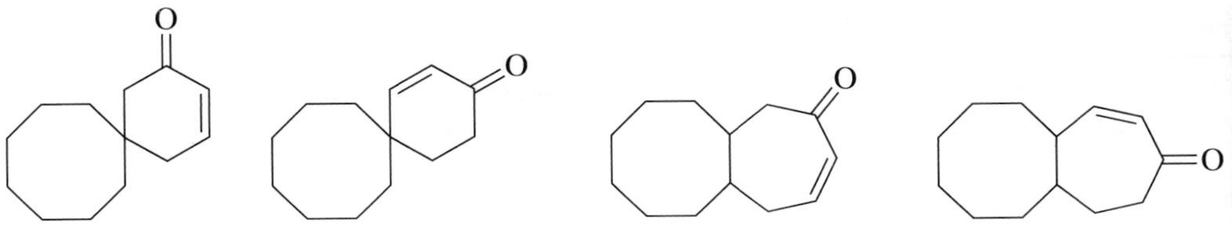

A. B. C. D.

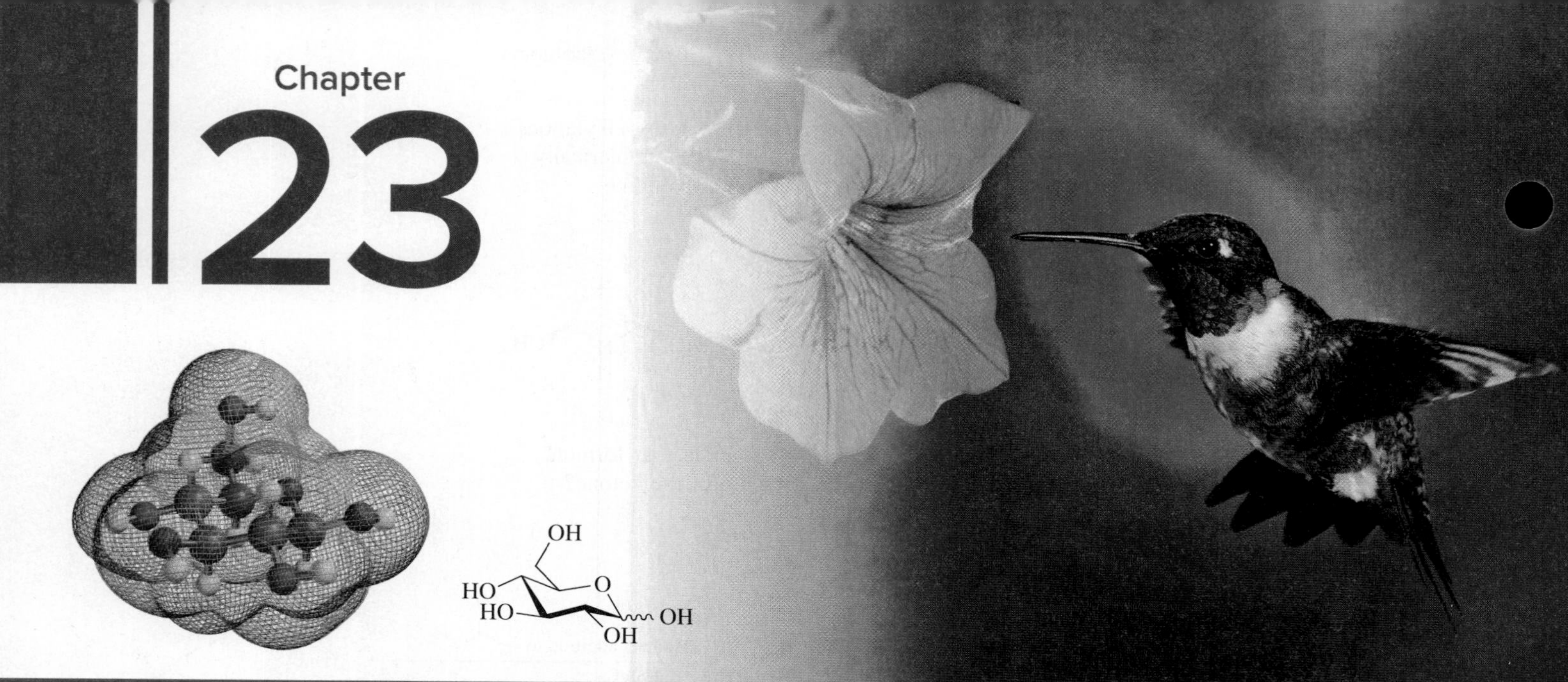

Hummingbirds receive nourishment from flower nectar. Nectar contains glucose (shown), fructose, and sucrose.

CHAPTER OUTLINE

Carbohydrates

The major classes of organic compounds common to living things are *lipids, proteins, nucleic acids,* and *carbohydrates.* Carbohydrates are very familiar to us—we call many of them "sugars." They make up a substantial portion of the food we eat and provide most of the energy that keeps the human engine running. Carbohydrates are structural components of the walls of plant cells and the wood of trees; they are also major components of the exoskeletons of insects, crabs, and lobsters. Carbohydrates are found on every cell surface, where they provide the molecular basis for cell-to-cell communication. Genetic information is stored and transferred by way of nucleic acids, specialized derivatives of carbohydrates, which we'll examine in more detail in Chapter 26.

Historically, carbohydrates were once considered to be "hydrates of carbon" because their molecular formulas in many (but not all) cases correspond to $C_n(H_2O)_m$. It is more realistic to define a carbohydrate as a *polyhydroxy aldehyde* or *polyhydroxy ketone,* a point of view closer to structural reality and more suggestive of chemical reactivity.

This chapter is divided into two parts. The first, and major, portion is devoted to carbohydrate *structure*. You will see how the principles of stereochemistry and conformational analysis combine to aid our understanding of this complex subject.

The second portion describes chemical *reactions* of carbohydrates. Most of these reactions are simply extensions of what you have already learned concerning alcohols, aldehydes, ketones, and acetals. The two areas—structure and reactions—meet in Section 23.20, where we consider the role of carbohydrates in the emerging field of glycobiology.

23.1 Classification of Carbohydrates

Sugar is a combination of the Sanskrit words *su* (sweet) and *gar* (sand). Thus, its literal meaning is "sweet sand."

The Latin word for *sugar* is *saccharum,* and the derived term *saccharide* is the basis of a system of carbohydrate classification. A **monosaccharide** is a simple carbohydrate, one that on hydrolysis is not cleaved to smaller carbohydrates. *Glucose* ($C_6H_{12}O_6$), for example, is a monosaccharide. A **disaccharide** on hydrolysis is cleaved to two monosaccharides, which may be the same or different. *Sucrose*—common table sugar—is a disaccharide that yields one molecule of glucose and one of fructose on hydrolysis.

$$\text{Sucrose } (C_{12}H_{22}O_{11}) + H_2O \longrightarrow \text{glucose } (C_6H_{12}O_6) + \text{fructose } (C_6H_{12}O_6)$$

An **oligosaccharide** (*oligos* is a Greek word that in its plural form means "few") yields two or more monosaccharides on hydrolysis. Thus, the IUPAC classifies disaccharides, trisaccharides, and so on as subcategories of oligosaccharides.

Polysaccharides are hydrolyzed to "many" monosaccharides. The IUPAC has chosen not to specify the number of monosaccharide components that separates oligosaccharides from polysaccharides. The standard is a more practical one; it notes that an oligosaccharide is homogeneous. Each molecule of a particular oligosaccharide has the same number of monosaccharide units joined together in the same order as every other molecule of the same oligosaccharide. Polysaccharides are almost always mixtures of molecules having similar, but not necessarily the same, chain length. *Cellulose,* for example, is a polysaccharide that gives thousands of glucose molecules on hydrolysis, but only a small fraction of the cellulose chains contain exactly the same number of glucose units.

Over 200 different monosaccharides are known. They can be grouped according to the number of carbon atoms they contain and whether they are polyhydroxy aldehydes or polyhydroxy ketones. Monosaccharides that are polyhydroxy aldehydes are called **aldoses;** those that are polyhydroxy ketones are **ketoses.** Aldoses and ketoses are further classified according to the number of carbon atoms in the main chain. Table 23.1 lists the terms applied to monosaccharides having four to eight carbon atoms.

23.2 Fischer Projections and D,L Notation

Fischer determined the structure of glucose in 1900 and won the Nobel Prize in Chemistry in 1902.

Stereochemistry is the key to understanding carbohydrate structure, a fact that was clearly appreciated by the German chemist Emil Fischer. The projection formulas used by Fischer to represent stereochemistry in chiral molecules (see Section 4.7) are particularly well-suited to carbohydrates. Figure 23.1 illustrates their application to the enantiomers of *glyceraldehyde* (2,3-dihydroxypropanal), a fundamental molecule in carbohydrate stereochemistry. When the Fischer projection is oriented as shown in the figure, with the carbon chain

TABLE 23.1 Some Classes of Monosaccharides

Number of carbon atoms	Aldose	Ketose
Four	Aldotetrose	Ketotetrose
Five	Aldopentose	Ketopentose
Six	Aldohexose	Ketohexose
Seven	Aldoheptose	Ketoheptose
Eight	Aldooctose	Ketooctose

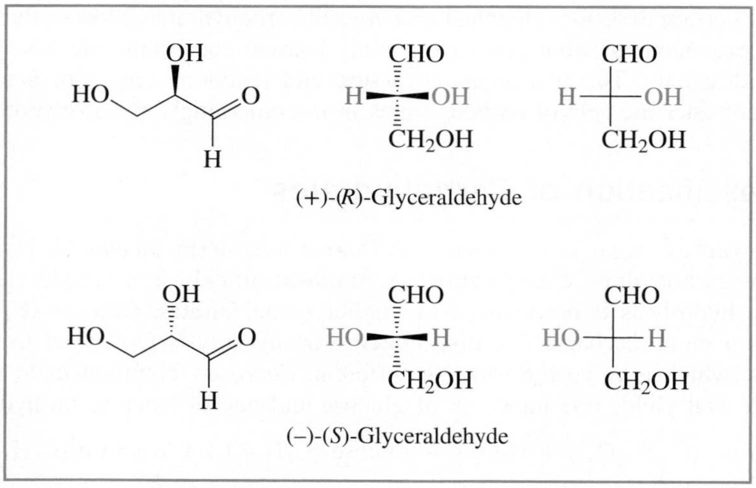

Figure 23.1

Three-dimensional representations and Fischer projections of the enantiomers of glyceraldehyde.

(+)-(*R*)-Glyceraldehyde

(−)-(*S*)-Glyceraldehyde

vertical and the aldehyde carbon at the top, the C-2 hydroxyl group points to the right in (+)-glyceraldehyde and to the left in (−)-glyceraldehyde.

Techniques for determining the absolute configuration of chiral molecules were not developed until the 1950s, and so it was not possible for Fischer and his contemporaries to relate the sign of rotation of any substance to its absolute configuration. A system evolved based on the arbitrary assumption, later shown to be correct, that the enantiomers of glyceraldehyde have the signs of rotation and absolute configurations shown in Figure 23.1. Two stereochemical descriptors were defined: D and L: D from the Latin (*dexter*) for right and L (*laevus*) for left. The absolute configuration of (+)-glyceraldehyde was said to be D and that of its enantiomer, (−)-glyceraldehyde, L, as depicted in Figure 23.1. Compounds that had a spatial arrangement of substituents analogous to (+)-D- or (−)-L-glyceraldehyde were said to have the D or L configurations.

Adopting the enantiomers of glyceraldehyde as stereochemical reference compounds originated with proposals made in 1906 by M. A. Rosanoff, a chemist at New York University.

Problem 23.1

Identify each of the following as either D- or L-glyceraldehyde:

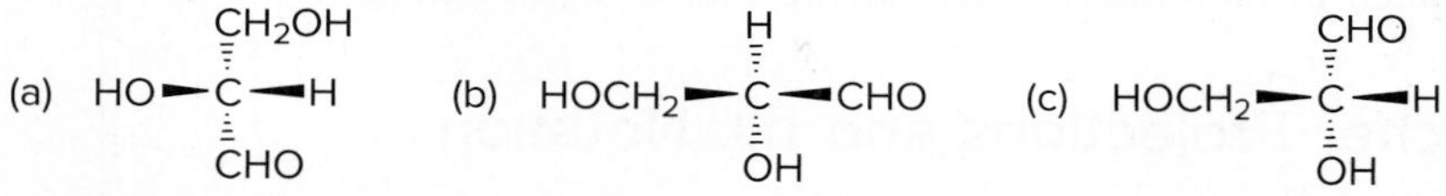

(a) (b) (c)

Sample Solution (a) To compare the structure given to glyceraldehyde most easily, turn it 180° in the plane of the page so that CHO is at the top and CH_2OH is at the bottom. Rotation in this sense keeps the horizontal bonds pointing forward and the vertical bonds pointing back, making it an easy matter to convert the structural drawing to a Fischer projection.

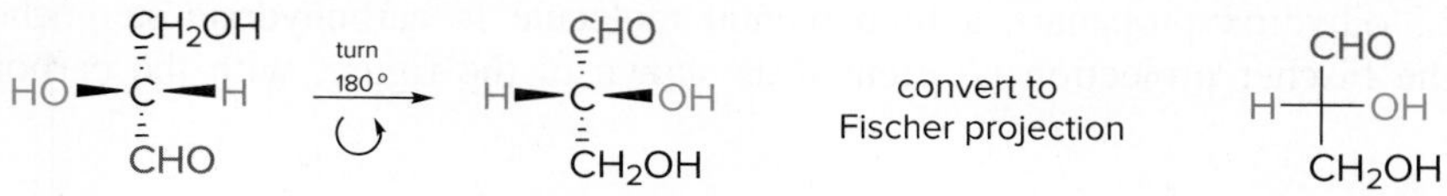

turn 180°

convert to Fischer projection

The structure is the same as that of (+)-glyceraldehyde in Figure 23.1. It is D-glyceraldehyde.

Fischer projections and D, L notation have proved to be so helpful in representing carbohydrate stereochemistry that the chemical and biochemical literature is replete with their use. To read that literature you need to be acquainted with these devices, as well as the more modern Cahn–Ingold–Prelog *R,S* system.

23.3 The Aldotetroses

Glyceraldehyde can be thought of as the simplest chiral carbohydrate. It is an *aldotriose* and, because it contains one chirality center, exists in two stereoisomeric forms: the D and L enantiomers. Moving up the ladder in complexity, next come the *aldotetroses*. Examining their structures illustrates the application of the Fischer system to compounds that contain more than one chirality center.

The aldotetroses are the four stereoisomers of 2,3,4-trihydroxybutanal. Fischer projections are constructed by orienting the molecule in an eclipsed conformation with the aldehyde group at the top. The four carbon atoms define the main chain of the Fischer projection and are arranged vertically. Horizontal bonds point outward, vertical bonds back.

CHO / H—C—OH / H—C—OH / CH_2OH corresponds to the Fischer projection CHO / H—OH / H—OH / CH_2OH

The particular aldotetrose just shown is called D-*erythrose*. The prefix D tells us that the configuration at the *highest numbered chirality center* is analogous to that of (+)-D-glyceraldehyde. Its mirror image is L-erythrose.

Highest numbered chirality center has configuration analogous to that of D-glyceraldehyde

D-Erythrose: CHO (1) / H—(2)—OH / H—(3)—OH / CH_2OH (4)

L-Erythrose: CHO (1) / HO—(2)—H / HO—(3)—H / CH_2OH (4)

Highest numbered chirality center has configuration analogous to that of L-glyceraldehyde

Relative to each other, both hydroxyl groups are on the same side in Fischer projections of the erythrose enantiomers. The remaining two stereoisomers have hydroxyl groups on opposite sides in their Fischer projections. They are diastereomers of D- and L-erythrose and are called D- and L-*threose*. The D and L prefixes again specify the configuration of the highest numbered chirality center. D-Threose and L-threose are enantiomers:

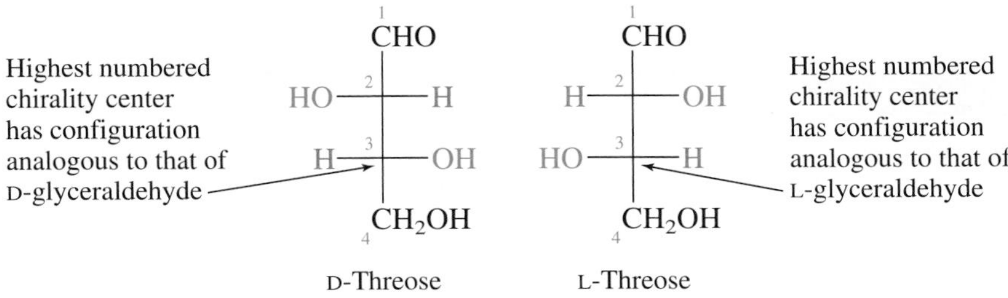

D-Threose L-Threose

Problem 23.2

Which aldotetrose is the structure shown? Is it D-erythrose, D-threose, L-erythrose, or L-threose? (Be careful! The conformation given is not the same as that used to generate a Fischer projection.)

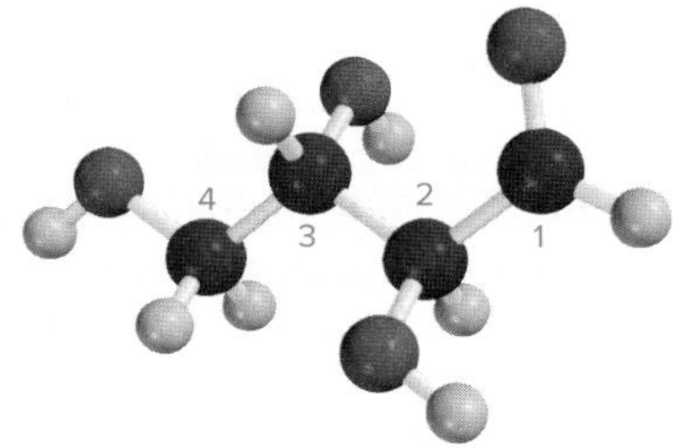

As shown for the aldotetroses, an aldose belongs to the D or the L series according to the configuration of the chirality center farthest removed from the aldehyde function.

Dextrorotatory and levorotatory are older terms for (+) and (−) optical rotation, respectively.

Individual names, such as erythrose and threose, specify the particular arrangement of chirality centers within the molecule relative to each other. Optical activities cannot be determined directly from the D and L prefixes. As it turns out, both D-erythrose and D-threose are levorotatory, but D-glyceraldehyde is dextrorotatory.

23.4 Aldopentoses and Aldohexoses

$2^3 = 8$

Aldopentoses have *three* chirality centers. The *eight stereoisomers* are divided into a set of four D-aldopentoses and an enantiomeric set of four L-aldopentoses. The aldopentoses are named *ribose, arabinose, xylose,* and *lyxose.* Fischer projections of the D stereoisomers of the aldopentoses are given in Figure 23.2. Notice that all these diastereomers have the same configuration at C-4 and that this configuration is analogous to that of (+)-D-glyceraldehyde.

Problem 23.3

(+)-L-Arabinose is a naturally occurring L sugar. It is obtained by acid hydrolysis of the polysaccharide present in mesquite gum. Write a Fischer projection for (+)-L-arabinose.

Among the aldopentoses, D-ribose is a component of many biologically important substances, most notably the ribonucleic acids. D-Xylose is very abundant and is isolated by hydrolysis of the polysaccharides present in corncobs and the wood of trees.

$2^4 = 16$

The aldohexoses include some of the most familiar of the monosaccharides, as well as one of the most abundant organic compounds on Earth, (+)-D-glucose. With *four* chirality centers, *16* stereoisomeric aldohexoses are possible; 8 belong to the D series and 8 to the L series. All are known, either as naturally occurring substances or as the products of synthesis. The eight D-aldohexoses are given in Figure 23.2; the spatial arrangement at C-5, hydrogen to the left in a Fischer projection and hydroxyl to the right, identifies them as carbohydrates of the D series.

Problem 23.4

Use Figure 23.2 as a guide to help you name the aldose shown. What is the D,L configuration at the highest numbered chirality center? The *R,S* configuration? What is its sign of rotation?

CHO
H—OH
H—OH
H—OH
HO—H
CH_2OH

Cellulose is more abundant than glucose, but each cellulose molecule is a polysaccharide composed of thousands of glucose units (Section 23.16). Methane may also be more abundant, but most of the methane comes from glucose.

Of all the monosaccharides, (+)-D-*glucose* is the best known, most important, and most abundant. Its formation from carbon dioxide, water, and sunlight is the central theme of photosynthesis. Carbohydrate formation by photosynthesis is estimated to be on the order of 10^{11} tons per year, a source of stored energy utilized, directly or indirectly, by all higher forms of life on the planet. Glucose was isolated from raisins in 1747 and by hydrolysis of starch in 1811. Its structure was determined, in work culminating in 1900, by Emil Fischer.

(+)-D-*Galactose* is a constituent of numerous polysaccharides. It is best obtained by acid hydrolysis of lactose (milk sugar), a disaccharide of D-glucose and D-galactose. (−)-L-Galactose also occurs naturally and can be prepared by hydrolysis of flaxseed gum

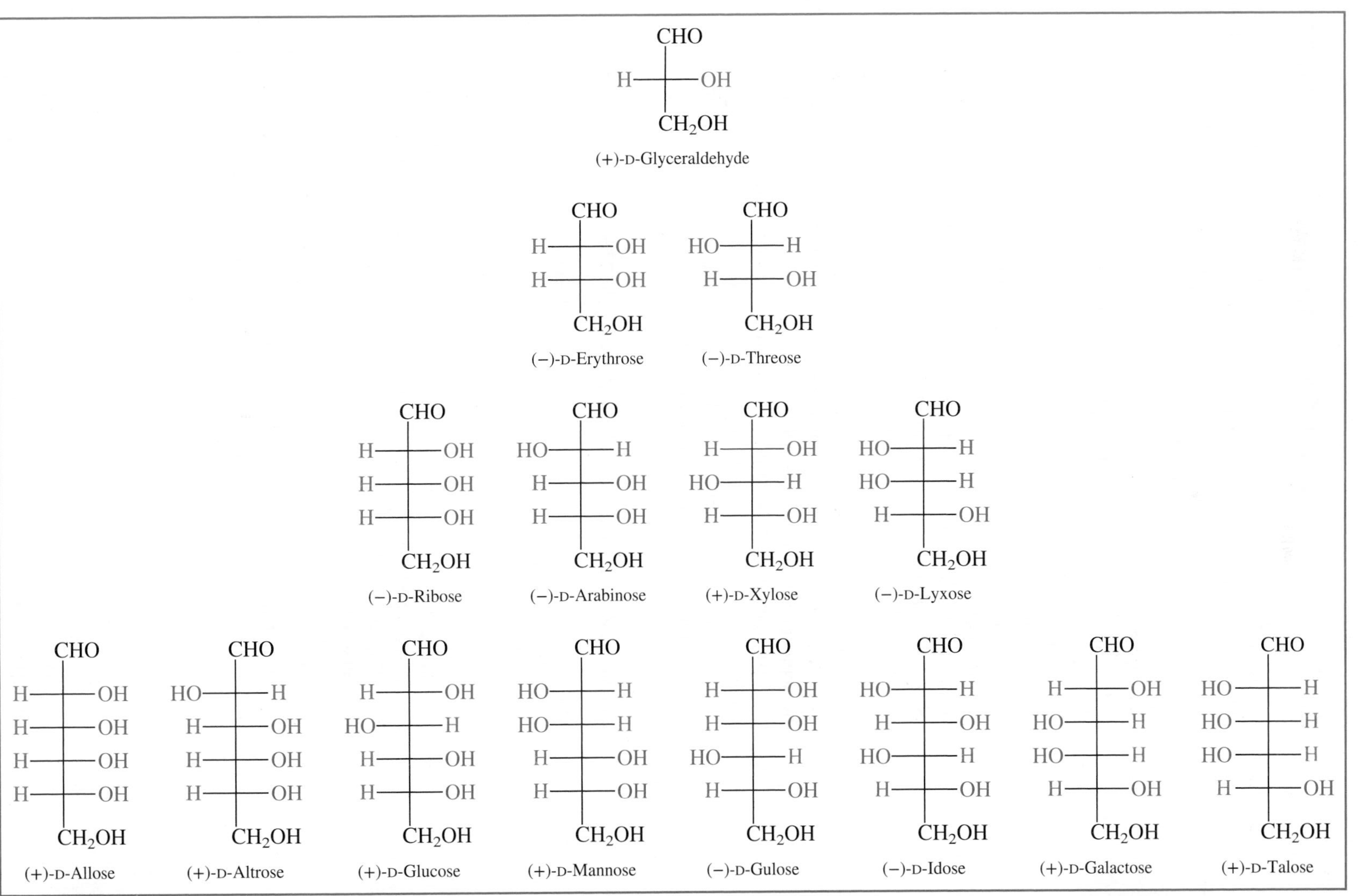

Figure 23.2

Configurations of the D series of aldoses containing three through six carbon atoms.

and agar. The principal source of (+)-D-*mannose* is hydrolysis of the polysaccharide of the ivory nut, a large, nut-like seed obtained from a South American palm.

23.5 A Mnemonic for Carbohydrate Configurations

The task of relating carbohydrate configurations to names requires either a world-class memory or an easily recalled mnemonic. A mnemonic that serves us well here was popularized by Louis F. Fieser and Mary Fieser of Harvard University in their 1956 textbook, *Organic Chemistry.* As with many mnemonics, it's not clear who actually invented it, and references to this particular one appeared in the chemical education literature before publication of the Fiesers' text. The mnemonic has two features: (1) a system for setting down all the stereoisomeric D-aldohexoses in a logical order; and (2) a way to assign the correct name to each one.

A systematic way to set down all the D-aldohexoses (as in Figure 23.2) is to draw skeletons of the necessary eight Fischer projections, placing the C-5 hydroxyl group to the right in each so as to guarantee that they all belong to the D series. Working up the carbon chain, place the C-4 hydroxyl group to the right in the first four structures, and to the left in the next four. In each of these two sets of four, place the C-3 hydroxyl group to the right in the first two and to the left in the next two; in each of the resulting four sets of two, place the C-2 hydroxyl group to the right in the first one and to the left in the second.

Once the eight Fischer projections have been written, they are named in order with the aid of the sentence "All altruists gladly make gum in gallon tanks." The words of the sentence stand for *allose, altrose, glucose, mannose, gulose, idose, galactose, talose*.

An analogous pattern of configurations can be seen in the aldopentoses when they are arranged in the order *ribose, arabinose, xylose, lyxose*. (RAXL is an easily remembered nonsense word that gives the correct sequence.) This pattern is discernible even in the aldotetroses erythrose and threose.

23.6 Cyclic Forms of Carbohydrates: Furanose Forms

Aldoses incorporate two functional groups, C=O and OH, which are capable of reacting with each other. We saw in Section 18.8 that nucleophilic addition of an alcohol function to a carbonyl group gives a hemiacetal. When the hydroxyl and carbonyl groups are part of the same molecule, a *cyclic hemiacetal* results, as illustrated in Figure 23.3.

Cyclic hemiacetal formation is most common when the ring that results is five- or six-membered. Five-membered cyclic hemiacetals of carbohydrates are called **furanose** forms; six-membered ones are called **pyranose** forms. The ring carbon that is derived from the carbonyl group, the one that bears two oxygen substituents, is called the **anomeric carbon.**

Aldoses exist almost exclusively as their cyclic hemiacetals; very little of the open-chain form is present at equilibrium. To understand their structures and chemical reactions, we need to be able to translate Fischer projections of carbohydrates into their cyclic hemiacetal forms. Consider first cyclic hemiacetal formation in D-erythrose. To visualize furanose ring formation more clearly, redraw the Fischer projection in a form more suited to cyclization, being careful to maintain the stereochemistry at each chirality center.

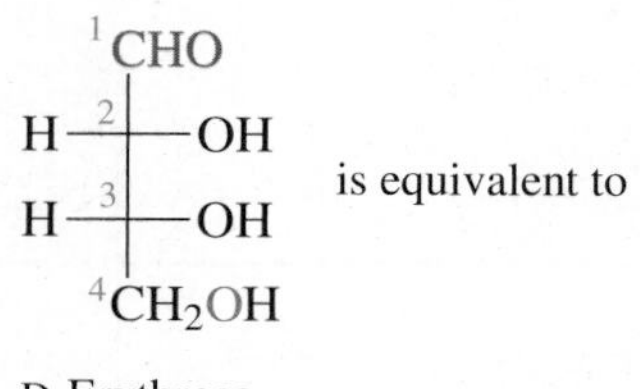

1 CHO
H —2— OH
H —3— OH
4 CH_2OH

D-Erythrose

is equivalent to

H
O
4 1 CH=O
3 2
HO OH

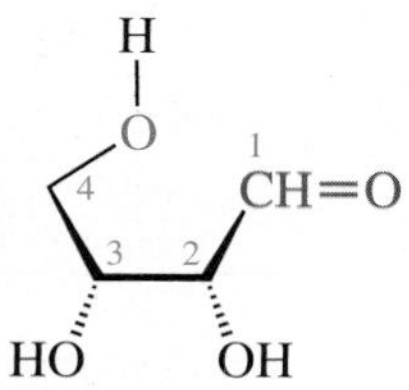

Eclipsed conformation showing C-4 hydroxyl in position to add to carbonyl group

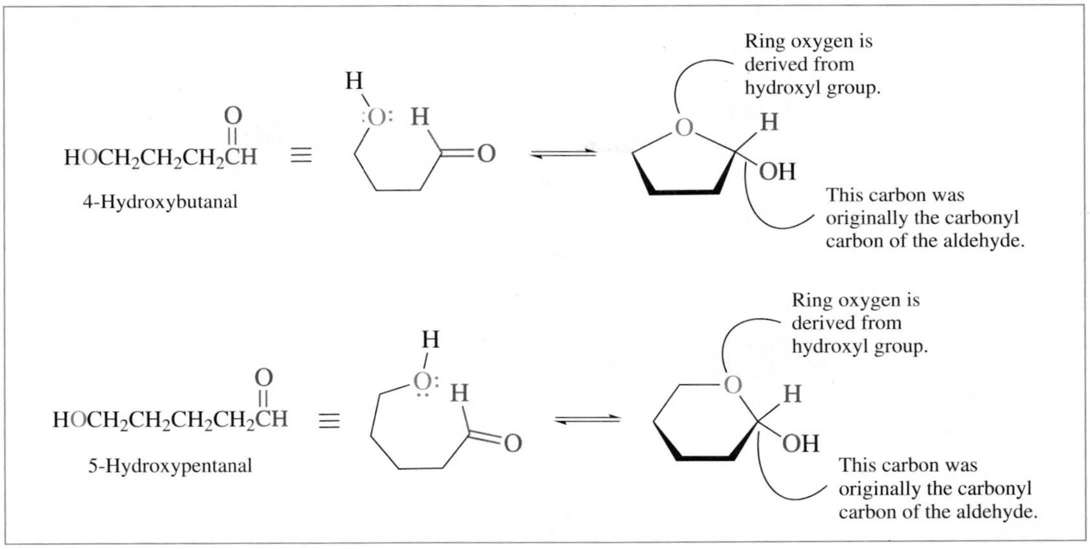

Figure 23.3

Cyclic hemiacetal formation in 4-hydroxybutanal and 5-hydroxypentanal.

Hemiacetal formation between the carbonyl group and the C-4 hydroxyl yields the five-membered furanose ring form. The anomeric carbon is a new chirality center; its hydroxyl group can be either cis or trans to the other hydroxyl groups. The two cyclic forms are diastereomers and are referred to as anomers because they have different configurations at the anomeric carbon.

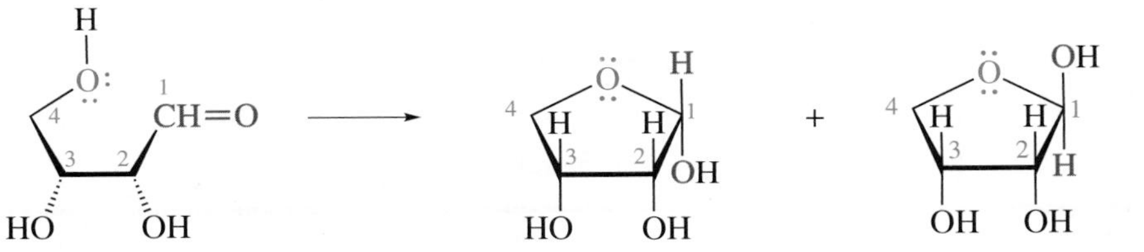

Structural drawings of carbohydrates of this type are called **Haworth formulas,** after the British chemist Sir Walter Norman Haworth (St. Andrew's University and the University of Birmingham). Early in his career Haworth contributed to the discovery that carbohydrates exist as cyclic hemiacetals rather than in open-chain forms. Later he collaborated on an efficient synthesis of vitamin C from carbohydrate precursors. This was the first chemical synthesis of a vitamin and provided an inexpensive route to its preparation on a commercial scale. Haworth was a corecipient of the Nobel Prize in Chemistry in 1937.

The two stereoisomeric furanose forms of D-erythrose are named α-D-erythrofuranose and β-D-erythrofuranose. The prefixes α and β describe the *relative configuration* of the anomeric carbon. The configuration of the anomeric carbon is compared with that of the highest numbered chirality center in the molecule—the one that determines whether the carbohydrate is D or L. Chemists use a simplified, informal version of the IUPAC rules for assigning α and β that holds for carbohydrates up to and including hexoses.

The formal IUPAC rules for α and β notation in carbohydrates are more detailed than our purposes require. These rules can be accessed at http://www.sbcs.qmul.ac.uk/iupac/2carb/06n07.html.

1. Orient the Haworth formula of the carbohydrate with the ring oxygen at the back and the anomeric carbon at the right.

2. For carbohydrates of the D series, the configuration of the anomeric carbon is α if its hydroxyl group is *down,* β if the hydroxyl group at the anomeric carbon is *up*.
3. For carbohydrates of the L series, the configuration of the anomeric carbon is α if its hydroxyl group is *up,* β if the hydroxyl group at the anomeric carbon is *down*. This is exactly the reverse of the rule for the D series.

Substituents that are to the right in a Fischer projection are "down" in the corresponding Haworth formula; those to the left are "up."

Problem 23.5

The structures shown are the four stereoisomeric threofuranoses. Assign the proper D, L and α, β stereochemical descriptors to each.

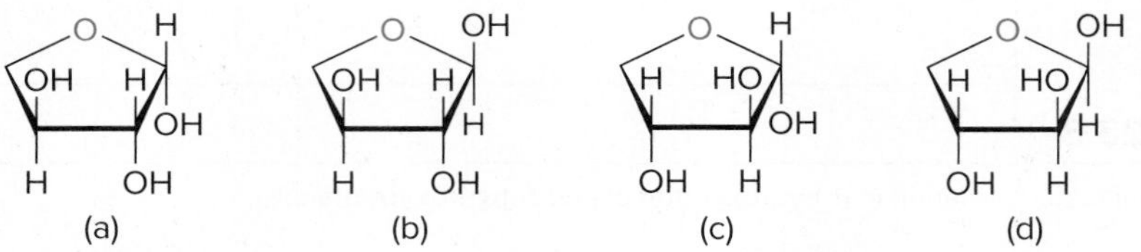

(a) (b) (c) (d)

Sample Solution (a) The —OH group at the highest numbered chirality center (C-3) is up, which places it to the left in the Fischer projection of the open-chain form. The stereoisomer belongs to the L series. The —OH group at the anomeric carbon (C-1) is down, making this the β-furanose form.

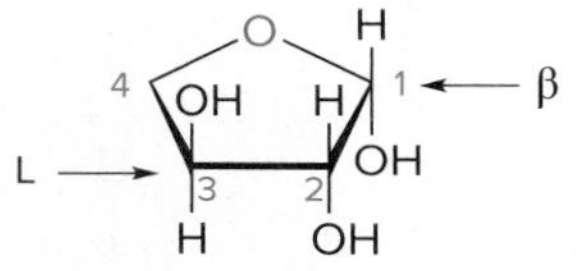

β-L-Threofuranose

Generating Haworth formulas to show stereochemistry in furanose forms of higher aldoses is slightly more complicated and requires an additional operation. Furanose forms of D-ribose are frequently encountered building blocks in biologically important organic molecules. They result from hemiacetal formation between the aldehyde group and the C-4 hydroxyl:

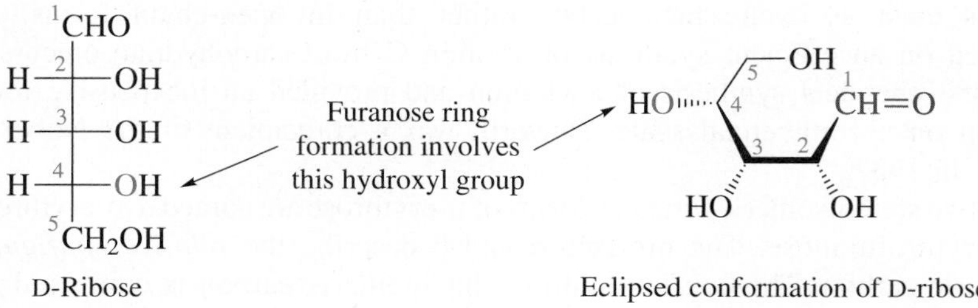

Furanose ring formation involves this hydroxyl group

D-Ribose

Eclipsed conformation of D-ribose

Notice that the eclipsed conformation of D-ribose derived directly from the Fischer projection does not have its C-4 hydroxyl group properly oriented for furanose ring formation. We must redraw it in a conformation that permits the five-membered cyclic hemiacetal to form. This is accomplished by rotation about the C(3)—C(4) bond, taking care that the configuration at C-4 is not changed.

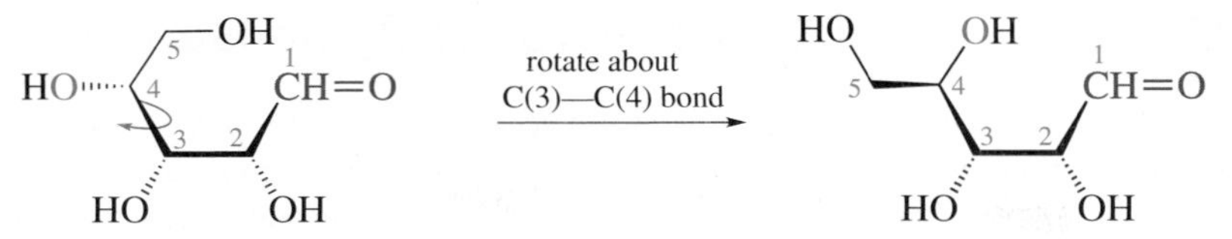

Eclipsed conformation of D-ribose

rotate about C(3)—C(4) bond

Conformation of D-ribose suitable for furanose ring formation

As viewed in the drawing, a 120° counterclockwise rotation of C-4 places its hydroxyl group in the proper position. At the same time, this rotation moves the CH_2OH group to a position such that it will become a substituent that is "up" on the five-membered ring. The hydrogen at C-4 then will be "down" in the furanose form.

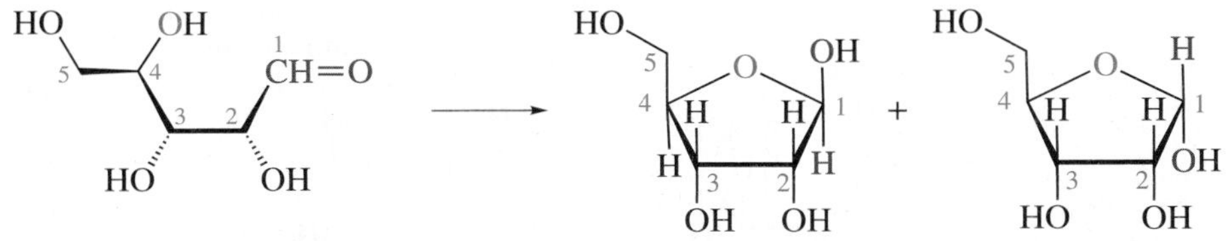

Conformation of D-ribose suitable for furanose ring formation

β-D-Ribofuranose + α-D-Ribofuranose

Problem 23.6

Write Haworth formulas corresponding to the furanose forms of each of the following carbohydrates:

(a) D-Xylose (b) D-Arabinose (c) L-Arabinose

Sample Solution (a) The Fischer projection of D-xylose is given in Figure 23.2.

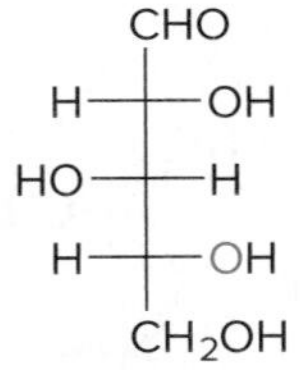

D-Xylose

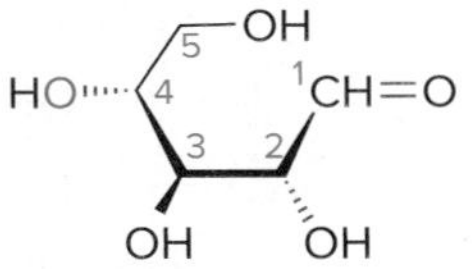

Eclipsed conformation of D-xylose

Carbon-4 of D-xylose must be rotated in a counterclockwise sense to bring its hydroxyl group into the proper orientation for furanose ring formation.

D-Xylose

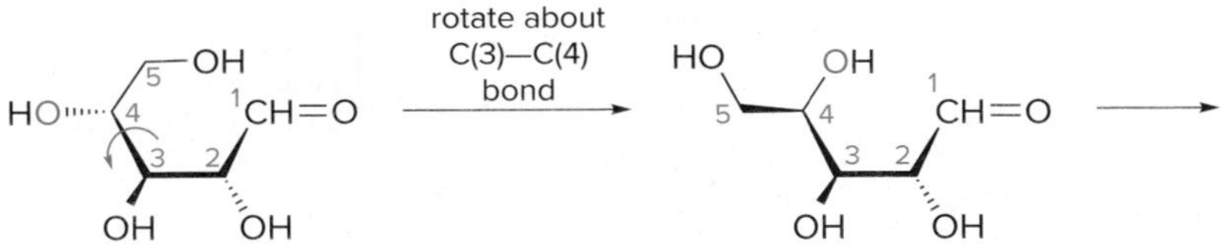

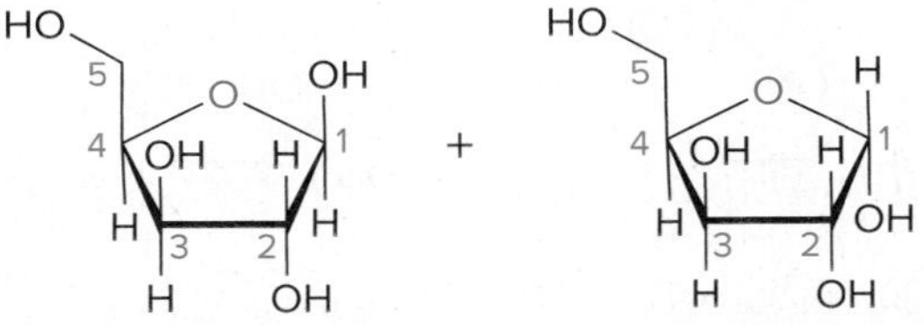

rotate about C(3)—C(4) bond

β-D-Xylofuranose + α-D-Xylofuranose

23.7 Cyclic Forms of Carbohydrates: Pyranose Forms

During the discussion of hemiacetal formation in D-ribose in the preceding section, you may have noticed that aldopentoses can potentially form a six-membered cyclic hemiacetal via addition of the C-5 hydroxyl to the carbonyl group. This mode of ring closure leads to α- and β-*pyranose* forms:

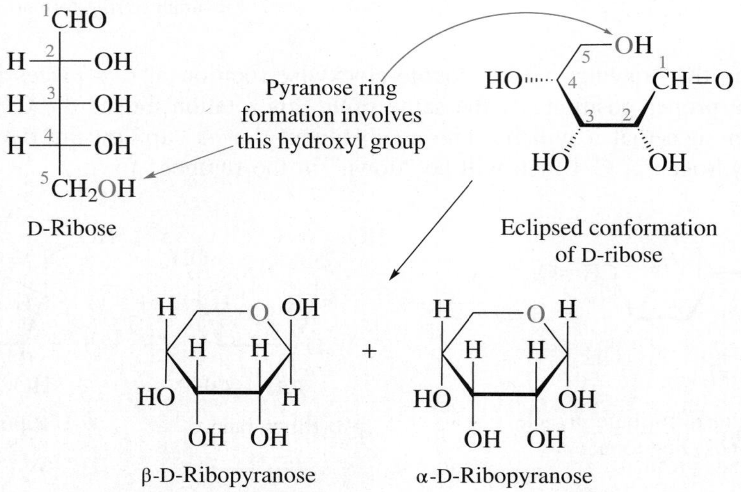

Pyranose ring formation involves this hydroxyl group

D-Ribose

Eclipsed conformation of D-ribose

β-D-Ribopyranose + α-D-Ribopyranose

Like aldopentoses, aldohexoses such as D-glucose are capable of forming two furanose forms (α and β) and two pyranose forms (α and β). The Haworth representations of the pyranose forms of D-glucose are constructed as shown in Figure 23.4; each has a CH_2OH group as a substituent on the six-membered ring.

Haworth formulas are satisfactory for representing *configurational* relationships in pyranose forms but are uninformative as to carbohydrate *conformations*. X-ray

Figure 23.4

Haworth formulas for α- and β-pyranose forms of D-glucose.

D-Glucose (hydroxyl group at C-5 is involved in pyranose ring formation)

turn on side

Eclipsed conformation of D-glucose; hydroxyl at C-5 is not properly oriented for ring formation

rotate about C4–C5 bond in counterclockwise direction

Eclipsed conformation of D-glucose in proper orientation for pyranose ring formation

β-D-Glucopyranose + α-D-Glucopyranose

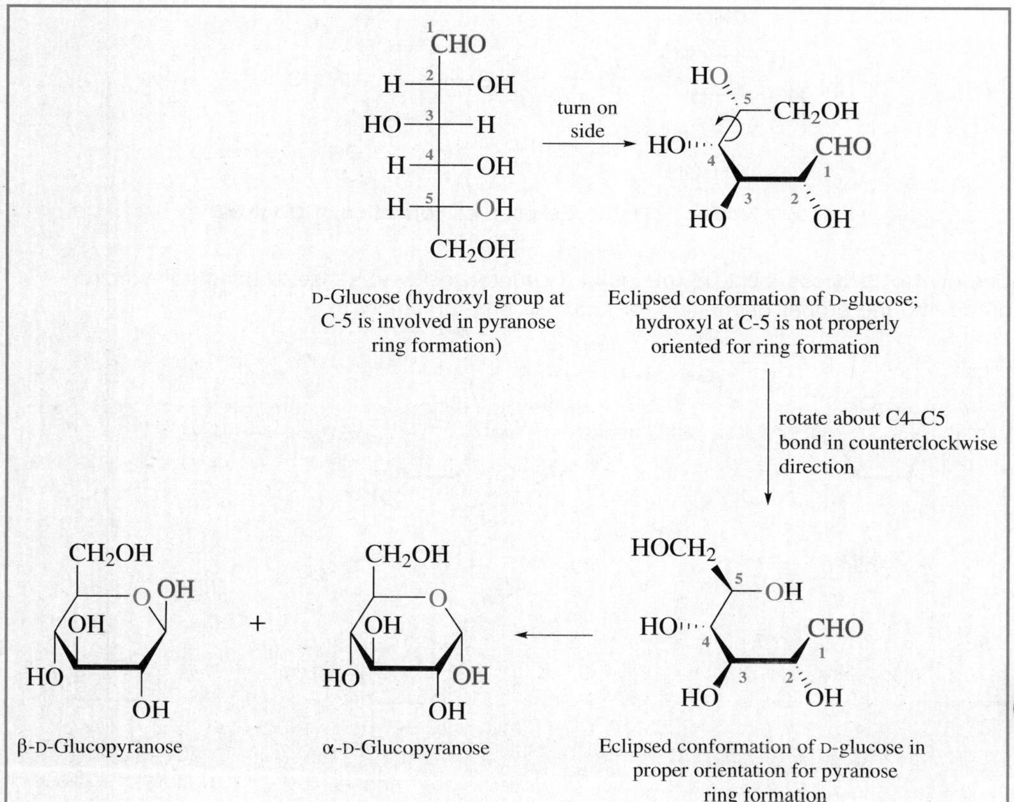

crystallographic studies of a large number of carbohydrates reveal that the six-membered pyranose ring of D-glucose adopts a chair conformation:

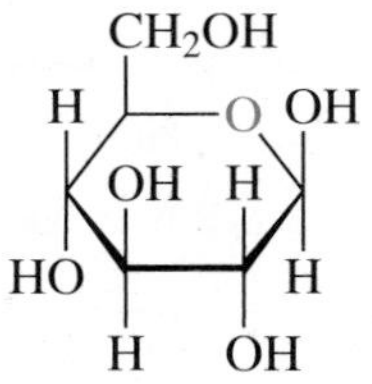

β-D-Glucopyranose

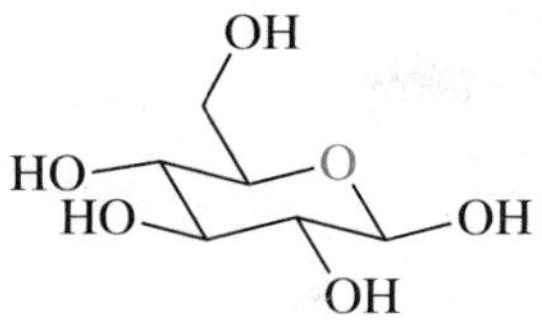

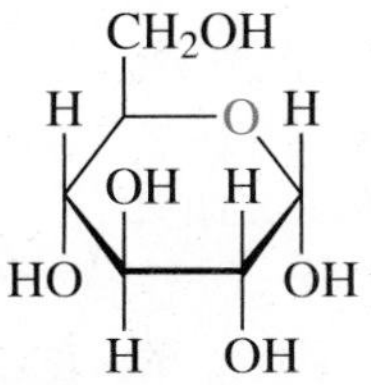

α-D-Glucopyranose

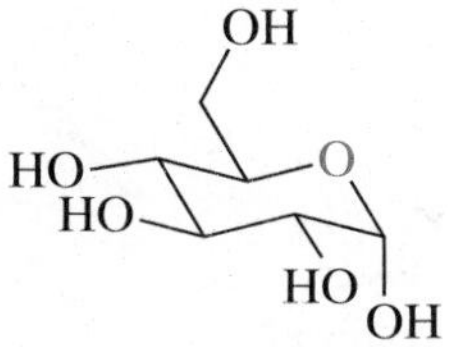

All the ring substituents in β-D-glucopyranose are equatorial in the most stable chair conformation. Only the anomeric hydroxyl group is axial in the α isomer; all the other substituents are equatorial.

Other aldohexoses behave similarly in adopting chair conformations that permit the CH_2OH substituent to occupy an equatorial orientation. Normally the CH_2OH group is the bulkiest, most conformationally demanding substituent in the pyranose form of a hexose.

Problem 23.7

Clearly represent the most stable conformation of the β-pyranose form of each of the following sugars:

(a) D-Galactose (b) D-Mannose (c) L-Mannose (d) L-Ribose

Sample Solution (a) By analogy with the procedure outlined for D-glucose in Figure 23.4, first generate a Haworth formula for β-D-galactopyranose:

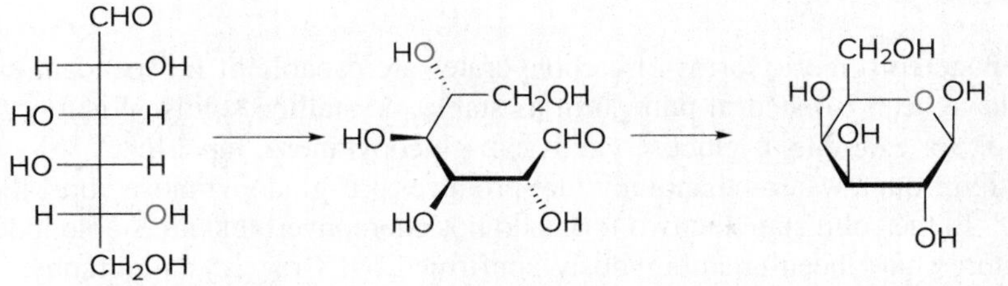

D-Galactose

β-D-Galactopyranose (Haworth formula)

Next, convert the Haworth formula to the chair conformation that has the CH_2OH group equatorial.

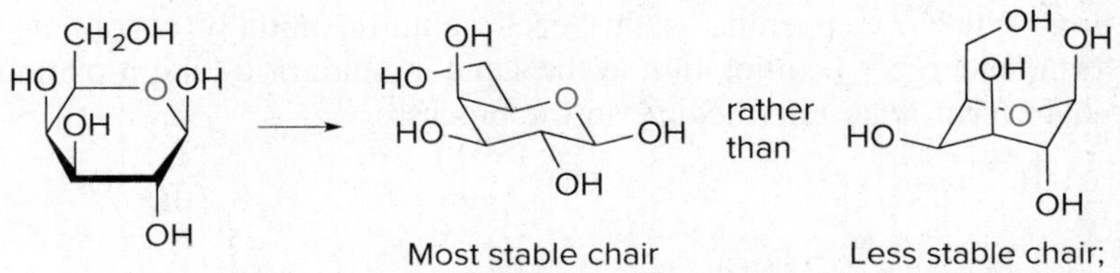

rather than

Most stable chair conformation of β–D-galactopyranose

Less stable chair; CH_2OH group is axial

Galactose differs from glucose in configuration at C-4. The C-4 hydroxyl is axial in β-D-galactopyranose, but it is equatorial in β-D-glucopyranose.

Because six-membered rings are normally less strained than five-membered ones, pyranose forms are usually present in greater amounts than furanose forms at equilibrium, and the concentration of the open-chain form is quite small. The distribution of

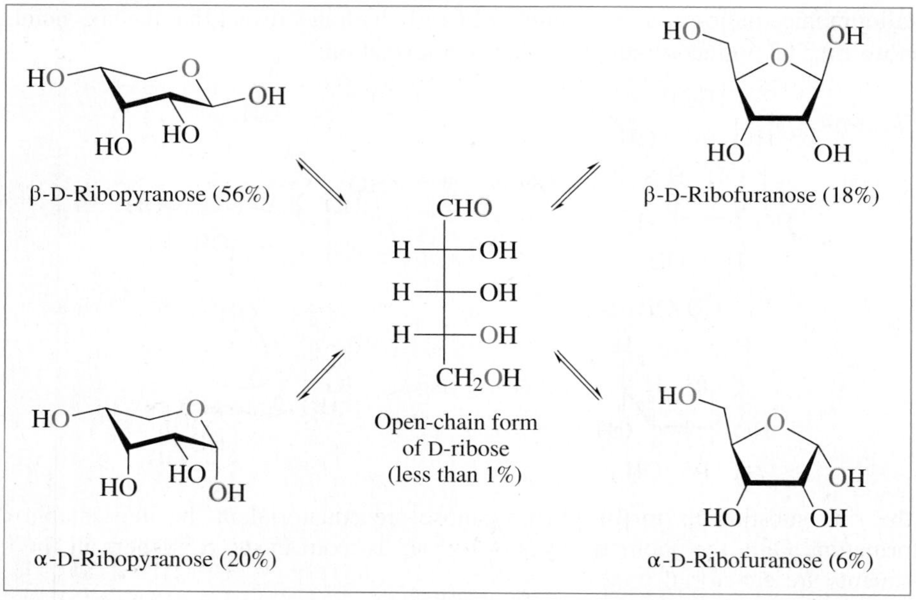

Figure 23.5

Distribution of furanose, pyranose, and open-chain forms of D-ribose in aqueous solution as measured by ^1H and ^{13}C NMR spectroscopy.

carbohydrates among their various hemiacetal forms has been examined by using ^1H and ^{13}C NMR spectroscopy. In aqueous solution, for example, D-ribose is found to contain the various α- and β-furanose and pyranose forms in the amounts shown in Figure 23.5. The concentration of the open-chain form at equilibrium is too small to measure directly. Nevertheless, it occupies a central position, in that interconversions of α and β anomers and furanose and pyranose forms take place by way of the open-chain form as an intermediate. As will be seen later, certain chemical reactions also proceed by way of the open-chain form.

23.8 Mutarotation

The α and β stereoisomeric forms of carbohydrates are capable of independent existence, and many have been isolated in pure form as stable, crystalline solids. When crystallized from ethanol, for example, D-glucose yields α-D-glucopyranose, mp 146°C, $[\alpha]_D$ +112.2°. Crystallization from a water–ethanol mixture produces β-D-glucopyranose, mp 148–155°C, $[\alpha]_D$ +18.7°. In the solid state the two forms do not interconvert and are stable indefinitely. Their structures have been unambiguously confirmed by X-ray crystallography.

The optical rotations just cited for each isomer are those measured immediately after each one is dissolved in water. On standing, the rotation of the solution containing the α isomer decreases from +112.2° to +52.5°; the rotation of the solution of the β isomer increases from +18.7° to the same value of +52.5°. This phenomenon is called **mutarotation.** What is happening is that each solution, initially containing only one anomeric form, undergoes equilibration to the same mixture of α- and β-pyranose forms. The open-chain form is an intermediate in the process.

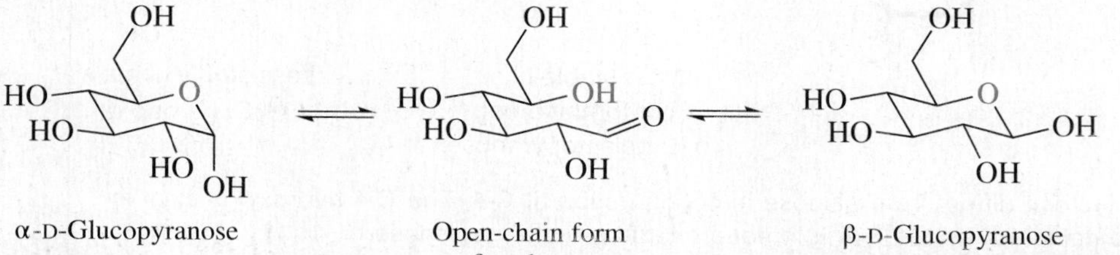

α-D-Glucopyranose (mp 146°C; $[\alpha]_D$ +112.2°) — Open-chain form of D-glucose — β-D-Glucopyranose (mp 148–155°C; $[\alpha]_D$ +18.7°)

Mutarotation occurs slowly in neutral aqueous solution, but can be catalyzed by either acid or base. Mechanism 23.1 shows a four-step, acid-catalyzed mechanism for mutarotation starting with α-D-glucopyranose. Steps 1 and 4 are proton transfers and describe

Mechanism 23.1

Acid-Catalyzed Mutarotation of D-Glucopyranose

THE OVERALL REACTION:

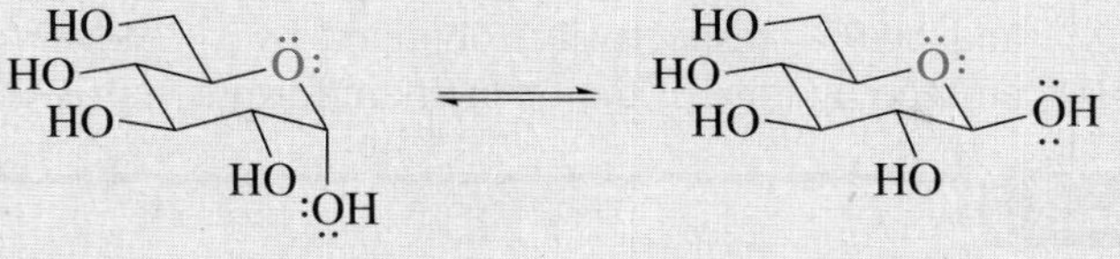

α-D-Glucopyranose ⇌ β-D-Glucopyranose

Step 1: Protonation of the oxygen of the pyranose ring by the acid catalyst. In aqueous solution, the acid catalyst is the hydronium ion.

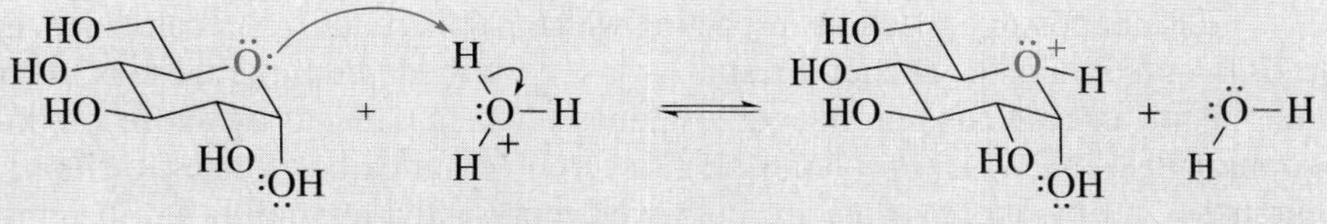

α-D-Glucopyranose + Hydronium ion ⇌ Conjugate acid of α-D-glucopyranose + Water

Step 2: The pyranose ring opens by cleaving the bond between the anomeric carbon and the positively charged oxygen. This ring opening is facilitated by electron release from the OH group at the anomeric carbon and gives the conjugate acid of the open-chain form of D-glucose.

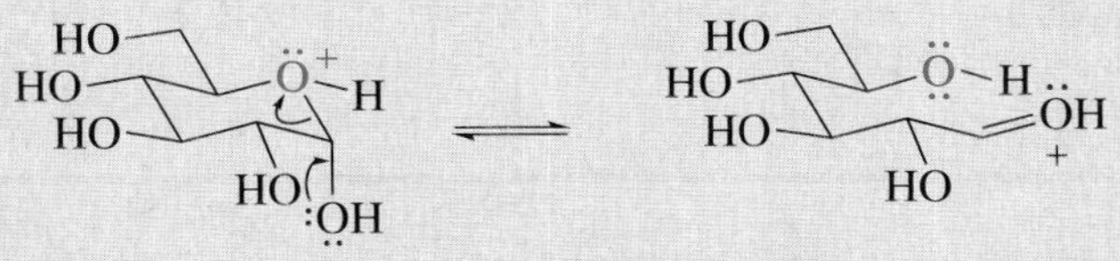

Conjugate acid of α-D-glucopyranose ⇌ Conjugate acid of open-chain form of D-glucose

Step 3: The species formed in the preceding step cyclizes to give the conjugate acid of β-D-glucopyranose. This cyclization is analogous to the acid-catalyzed nucleophilic additions to aldehydes and ketones in Chapter 18.

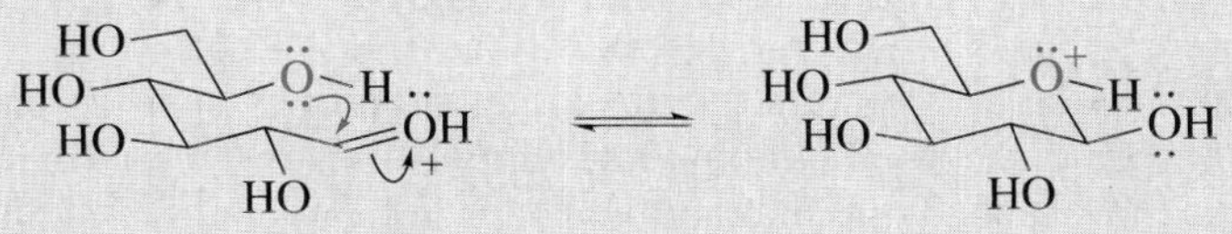

Conjugate acid of open-chain form of D-glucose ⇌ Conjugate acid of β-D-glucopyranose

Step 4: The product of step 3 transfers a proton to water to regenerate the acid catalyst and yield the neutral form of the product.

Conjugate acid of β-D-glucopyranose + Water ⇌ β-D-Glucopyranose + Hydronium ion

the role of the acid catalyst. The combination of step 2 (ring-opening) and step 3 (ring-closing) reverses the configuration at the anomeric carbon. All the steps are reversible and the α/β ratio is governed by the relative energies of the two diastereomers.

The distribution between the α- and β-pyranose forms at equilibrium can be calculated from the optical rotations of the pure isomers and the final optical rotation of the solution. For D-glucose, such a calculation gives 36% α and 64% β. These are close to the values (38.8% and 60.9%, respectively) obtained by ^{13}C NMR measurements. The α- and β-furanoses and the hydrate of the open-chain form comprise the remaining 0.3%.

Problem 23.8

The specific optical rotations of pure α- and β-D-mannopyranose are +29.3° and –17.0°, respectively. When either form is dissolved in water, mutarotation occurs, and the observed rotation of the solution changes until a final rotation of +14.2° is observed. Assuming that only α- and β-pyranose forms are present, calculate the percent of each isomer at equilibrium.

It is not possible to tell by inspection which pyranose form of a particular carbohydrate—α or β—predominates at equilibrium. As just described, the β-pyranose is the major species present in an aqueous solution of D-glucose, whereas the α-pyranose predominates in a solution of D-mannose (Problem 23.8). In certain other carbohydrates, D-ribose for example, furanose and pyranose forms are both well represented at equilibrium (Figure 23.5).

Problem 23.9

Write a four-step mechanism for the mutarotation of D-glucopyranose in aqueous base. Use curved arrows to track electron flow. The first step is:

The factors that control the equilibrium composition of sugars in solution are complex. Although the well-established preference for substituents in six-membered rings to be equatorial rather than axial is important, it is not always the overriding factor. The next section introduces a new structural feature that plays a significant part in determining carbohydrate conformations and α/β anomeric ratios.

23.9 Carbohydrate Conformation: The Anomeric Effect

Not only does carbohydrate structure affect properties such as chemical reactivity, but the structure and shape of carbohydrates are also major factors in a number of biological processes that depend on interactions between molecules—a phenomenon known as *molecular recognition*. In this section, we will consider mainly the conformations of carbohydrates in their pyranose forms. We will return to some familiar concepts of chair conformations and axial versus equatorial groups, but you will see that the presence of an oxygen atom in a six-membered ring leads to some surprising consequences.

The 1969 Nobel Laureate for Chemistry, Odd Hassel, was the first to suggest that the pyranose form of carbohydrates would resemble chair cyclohexane. Replacing a carbon atom in the ring with an oxygen does not change the basic preference for chair forms, even though the pyranose ring has unequal bond lengths. However, in addition to the usual factors that govern the equatorial versus axial orientation of substituents on a six-membered ring, two other factors are important:

1. an equatorial OH is less crowded and better solvated by water than an axial one
2. the anomeric effect

The first of these is straightforward and alerts us to the fact that the relative energies of two species may be different in solution than in the solid state or the gas phase. Hydrogen bonding to water stabilizes equatorial OH groups better than axial ones.

The **anomeric effect,** on the other hand, stabilizes *axial* OH and other electronegative groups at the anomeric carbon in pyranose rings better than equatorial. Consider the mutarotation of glucose just described, which produces an equilibrium mixture containing 36% of the α-anomer and 64% of the β. If we consider only the destabilizing effect of a solvated axial hydroxyl group in the axial position, we would expect only 11% α and 89% β. The presence of more axial hydroxyl than expected results from the contribution of the anomeric effect to the free energy difference between these two *stereoisomers*.

The anomeric effect also influences the *conformational* equilibria in pyranoses with an electronegative atom, usually oxygen or halogen, at C-1. For example, the equilibrium mixture of the β-pyranosyl chloride derived from xylose triacetate contains 98% of the conformer in which chlorine is axial. These two conformations are not interconverted by mutarotation but by chair–chair interconversion. The anomeric effect is sufficiently large so that all four substituents occupy axial positions in the more stable conformer.

$CHCl_3$

2% 98%

2,3,4-Tri-*O*-acetyl-β-D-xylopyranosyl chloride

What is responsible for the anomeric effect? Chemists continue to debate this question, and a number of explanations have emerged, one of which is detailed in Figure 23.6. The structure depicted in (a), in which substituent X is axial, is stabilized by the interaction of an unshared electron pair on the ring oxygen with the antibonding σ* orbital (LUMO) of the C—X bond [(b)]. This interaction is greater when X is axial and anti coplanar to the nonbonding electron pair than it is when X is equatorial (c). The transfer of electron density toward the anomeric carbon is facilitated when X is an electronegative substituent, so the effect is often seen in pyranoses with oxygen or halogen at the anomeric carbon. This model also accounts for the shortening of the O—C-1 bond seen in pyranoses that bear an axial electronegative substituent at the anomeric carbon.

Because five-membered rings are more flexible than six-membered, the anomeric effect is less important in furanose than in pyranose forms.

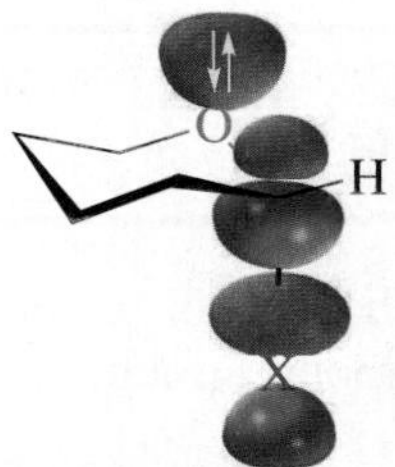

(a) The antibonding C—X orbital σ* (LUMO) can accommodate two electrons.

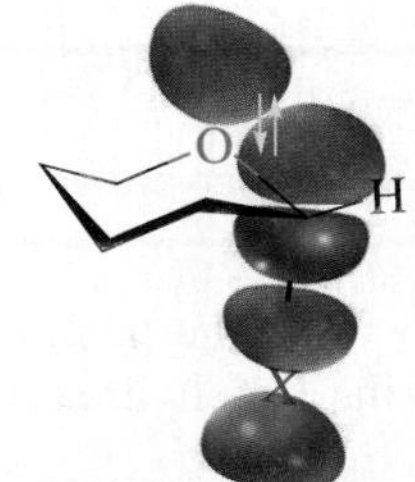

(b) When X is axial, overlap of a nonbonded electron pair of oxygen with the C—X σ* orbital allows oxygen's unshared pair to be delocalized.

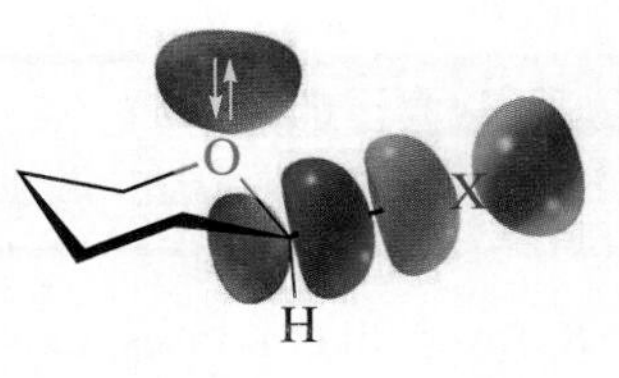

(c) When X is equatorial, the axis of its C—X σ* orbital is gauche to the orbital of the nonbonding electron pair and does not allow oxygen's unshared electron pair to be delocalized as well as in (b).

Figure 23.6

Delocalization of an oxygen lone pair stabilizes the conformation in which an electronegative substituent X at the anomeric carbon is axial.

The anomeric effect is a general property of structural units of the type X—C—Y—R where X and Y are electronegative and Y has at least one unshared pair. Of the conformations about the Y—C bond, gauche is normally more stable than anti.

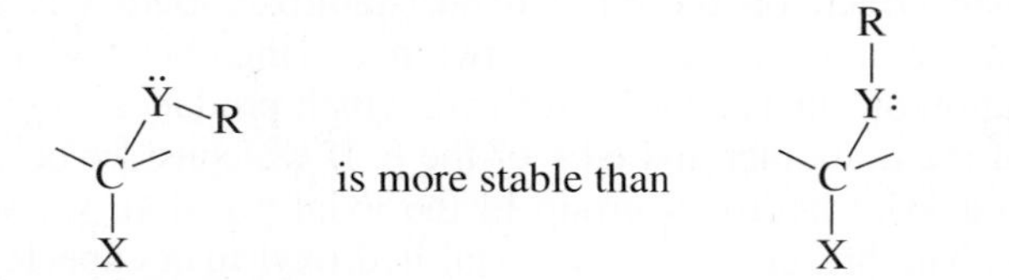

C—X and Y—R bonds are gauche is more stable than C—X and Y—R bonds are anti

For a simple structure such as chloromethyl methyl ether ($ClCH_2OCH_3$), the gauche conformation has been estimated to be about 8 kJ/mol (2 kcal/mol) more stable than the anti.

Problem 23.10

Sketch the most stable conformation of chloromethyl methyl ether.

23.10 Ketoses

To this point our attention has been directed toward aldoses, carbohydrates having an aldehyde function in their open-chain form. Aldoses are more common than ketoses, and their role in biological processes has been more thoroughly studied. Nevertheless, a large number of ketoses are known, and several of them are pivotal intermediates in carbohydrate biosynthesis and metabolism. Examples of some ketoses include D-*ribulose,* L-*xylulose,* and D-*fructose:*

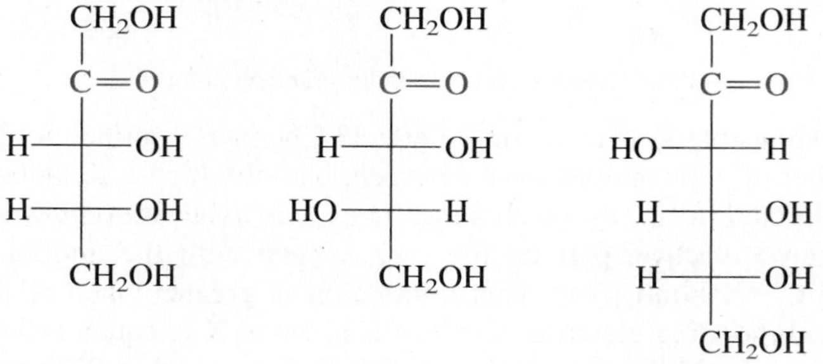

D-Ribulose L-Xylulose D-Fructose

In all three the carbonyl group is at C-2, which is the most common case for naturally occurring ketoses. D-Ribulose is a key intermediate in photosynthesis, the process by which energy from sunlight drives the formation of D-glucose from carbon dioxide and water. L-Xylulose is a product of the abnormal metabolism of xylitol in persons who lack a particular enzyme. D-Fructose is the most familiar ketose; it is present in fruits and honey and is sweeter than sucrose.

Problem 23.11

How many ketotetroses are possible? Write Fischer projections for each.

Ketoses, like aldoses, exist mainly as cyclic hemiacetals. In the case of D-ribulose, furanose forms result from addition of the C-5 hydroxyl to the carbonyl group.

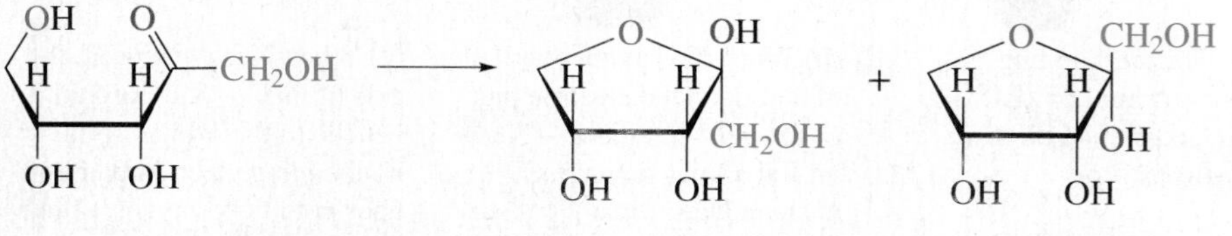

Eclipsed conformation of D-ribulose → β-D-Ribulofuranose + α-D-Ribulofuranose

The anomeric carbon of a furanose or pyranose form of a ketose bears both a hydroxyl group and a carbon substituent. In the case of 2-ketoses, this substituent is a CH_2OH group. As with aldoses, the anomeric carbon of a cyclic hemiacetal is readily identifiable because it is bonded to two oxygens.

Problem 23.12

Use the method outlined in Figure 23.4 to write a Haworth formula for the β-furanose form of D-fructose.

23.11 Deoxy Sugars

A common variation on the general pattern seen in carbohydrate structure is the replacement of one or more of the hydroxyl substituents by some other atom or group. In **deoxy sugars** the hydroxyl group is replaced by hydrogen. Two examples of deoxy sugars are 2-deoxy-D-ribose and L-fucose:

CHO
H—H
H—OH
H—OH
CH_2OH

2-Deoxy-D-ribose

CHO
HO—H
H—OH
H—OH
HO—H
CH_3

L-Fucose
(6-deoxy-L-galactose)

The hydroxyl at C-2 in D-ribose is absent in 2-deoxy-D-ribose. In Chapter 26 we shall see how derivatives of 2-deoxy-D-ribose, called *deoxyribonucleotides,* are the fundamental building blocks of deoxyribonucleic acid (DNA), the material responsible for storing genetic information. L-Fucose, the carbon chain of which terminates in a methyl rather than a CH_2OH group, is often found as one of the carbohydrates in glycoproteins, such as those on the surface of red blood cells that determine blood type (see Section 23.20).

Problem 23.13

Write Fischer projections of

(a) *Cordycepose* (3-deoxy-D-ribose): a deoxy sugar isolated by hydrolysis of the antibiotic substance *cordycepin*

(b) L-*Rhamnose* (6-deoxy-L-mannose): found in plants

Sample Solution (a) The hydroxyl group at C-3 in D-ribose is replaced by hydrogen in 3-deoxy-D-ribose.

CHO
H—OH
H—OH
H—OH
CH_2OH

D-Ribose
(from Figure 23.2)

CHO
H—OH
H—H
H—OH
CH_2OH

3-Deoxy-D-ribose
(cordycepose)

Figure 23.7

The shells of lobsters are mainly chitin, a polymer of *N*-acetyl-D-glucosamine.

23.12 Amino Sugars

Another structural variation is the replacement of a hydroxyl group in a carbohydrate by an amino group to give an **amino sugar.** The most abundant amino sugar is one of the oldest and most abundant organic compounds on Earth. *N*-Acetyl-D-glucosamine is the main component of the polysaccharide in *chitin,* the substance that makes up the tough outer skeleton of arthropods and insects. Chitin has been isolated from a 25-million-year-old beetle fossil, and more than 10^9 tons of chitin is produced in the biosphere each year. Lobster shells, for example, are mainly chitin (Figure 23.7). More than 60 amino sugars are known, many of them having been isolated and identified only recently as components of antibiotics. The anticancer drug doxorubicin hydrochloride (Adriamycin), for example, contains the amino sugar L-daunosamine as one of its structural units.

N-Acetyl-D-glucosamine L-Daunosamine

Sialic acids are a group of carbohydrates that have the interesting structural feature of being amino-substituted derivatives of a nine-carbon ketose. *N*-Acetylneuraminic acid can be considered the parent.

N-Acetylneuraminic acid

More than 50 structurally related sialic acids occur naturally. Sialic acids are often the terminal monosaccharide of cell surface oligosaccharides of glycoproteins and glycolipids. They are involved in many physiological, pathological, and immunological processes.

Problem 23.14

We've included *N*-acetylneuraminic acid in the section on amino sugars and described it as a ketose. We could also call it a deoxy sugar. Locate reasons for these classifications in the structural formula. Number the carbon atoms in the nine-carbon chain. What is the configuration (D or L) at the highest numbered chirality center?

Nitrogen-containing sugars in which nitrogen replaces the ring oxygen are known as imino sugars. An example of an imino sugar is nojirimicin, which occurs naturally. A synthetic derivative, *N*-butyl-1-deoxynojirimicin, is used in the treatment of a lipid metabolism disorder known as Gaucher's disease.

The symbol ∿∿∿ is used to represent a bond of undefined stereochemistry.

Nojirimicin *N*-Butyl-1-deoxynojirimicin

23.13 Branched-Chain Carbohydrates

Carbohydrates that have a carbon substituent attached to the main chain are said to have a **branched chain.** D-Apiose and L-vancosamine are representative branched-chain carbohydrates:

CHO
H—C—OH
HO—C—CH_2OH (Branching group)
CH_2OH

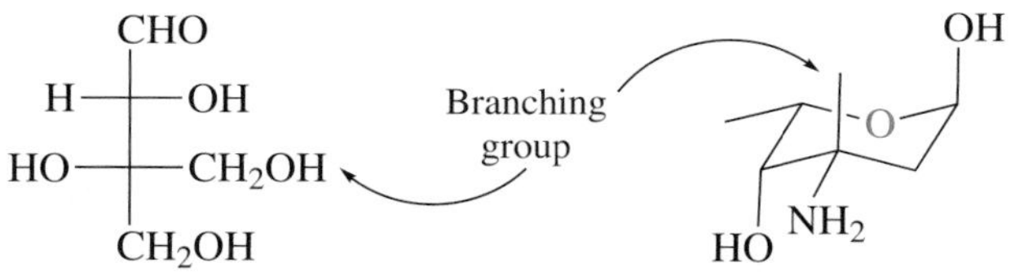

D-Apiose

L-Vancosamine

D-Apiose can be isolated from parsley and is a component of the cell-wall polysaccharide of various marine plants. Among its novel structural features is the presence of only a single chirality center. L-Vancosamine is but one portion of vancomycin, a powerful antibiotic that has emerged as one of only a few that are effective against drug-resistant bacteria. L-Vancosamine is not only a branched-chain carbohydrate, it is a deoxy sugar and an amino sugar as well.

23.14 Glycosides: The Fischer Glycosidation

Glycosides are a large and important class of carbohydrate derivatives characterized by the replacement of the anomeric hydroxyl group by some other substituent. Glycosides are termed *O*-glycosides when the atom attached to the anomeric carbon is oxygen. If the atom is sulfur, the names *S*-glycoside and thioglycoside are both used. Glycosides in which the atom attached to the anomeric carbon is nitrogen are named as glycosylamines.

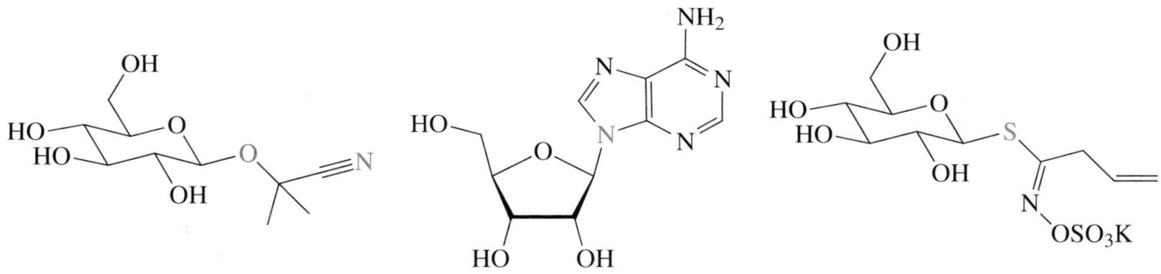

Linamarin Adenosine Sinigrin

Linamarin is an *O*-glycoside of D-glucose and acetone cyanohydrin. It is present in manioc (cassava), a tuberous food plant, and is just one of many cyanogenic glycosides. Nucleosides such as adenosine are glycosylamines of heterocyclic aromatic compounds. The most important ones are those derived from D-ribose and 2-deoxy-D-ribose. Sinigrin is an *S*-glycoside that contributes to the characteristic flavor of mustard and horseradish. All three of the glycosides shown have the β configuration at their anomeric carbon. Many antibiotics occur as glycosides. The most common are *O*-glycosides, such as erythromycin (see Figure 19.7).

The term *glycoside* without a prefix is taken to mean an *O*-glycoside. *O*-Glycosides bear an alkoxy group —OR instead of —OH at the anomeric carbon. Structurally, they are mixed acetals. Recall the sequence of intermediates in acetal formation (see Section 18.8):

$$\underset{\text{Aldehyde or ketone}}{R_2C{=}O} \xrightleftharpoons{R'OH} \underset{\text{Hemiacetal}}{R_2\underset{\displaystyle OH}{\underset{|}{C}}OR'} \xrightleftharpoons{R''OH} \underset{\text{Acetal}}{R_2\underset{\displaystyle OR''}{\underset{|}{C}}OR'}$$

If the aldehyde or ketone bears a γ or δ OH group, the first step takes place intramolecularly to yield a *cyclic* hemiacetal. The second step is intermolecular and requires an alcohol ROH as a reactant.

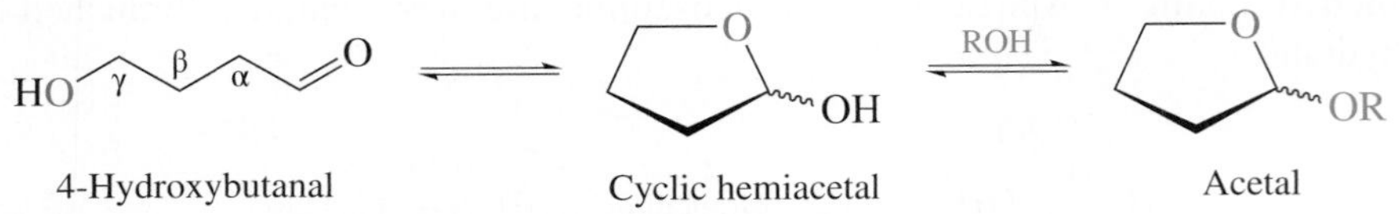

4-Hydroxybutanal ⇌ Cyclic hemiacetal ⇌(ROH) Acetal

In this illustration only a five-membered cyclic hemiacetal, analogous to the furanose form of a carbohydrate, is possible from the γ-hydroxy aldehyde. The final acetal is analogous to a glycoside; in this case, a furanoside. The corresponding products from an aldehyde with a δ —OH group would be a pyranose (cyclic hemiacetal) and a pyranoside (acetal).

In a reaction known as the **Fischer glycosidation,** glycosides are prepared by simply allowing a carbohydrate to react with an alcohol in the presence of an acid catalyst. The reaction is *thermodynamically controlled,* and the major product is the most stable glycoside; for the reaction of D-glucose with methanol this is methyl α-D-glucopyranoside. Six-membered rings are more stable than five-membered ones, and the anomeric effect stabilizes an axial —OCH_3 group.

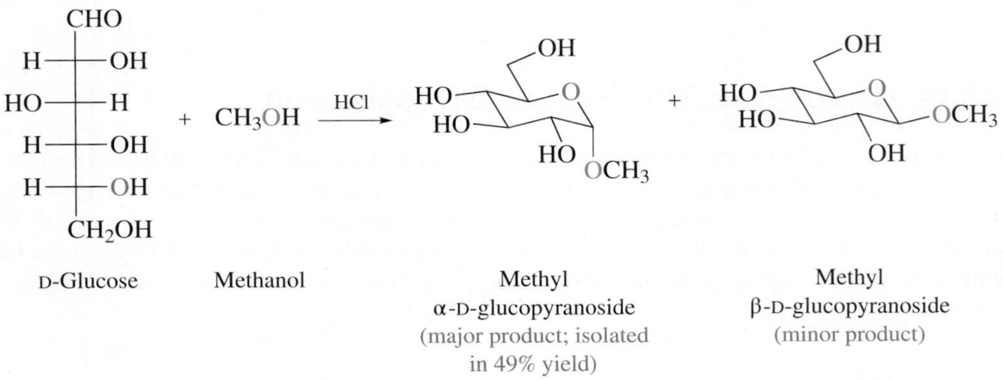

D-Glucose + CH_3OH (Methanol) $\xrightarrow{HCl}$ Methyl α-D-glucopyranoside (major product; isolated in 49% yield) + Methyl β-D-glucopyranoside (minor product)

Problem 23.15

Write structural formulas for the α- and β-methyl pyranosides formed by reaction of D-galactose with methanol in the presence of hydrogen chloride.

Experimental observations suggest that the methyl glycosides are formed by more than one mechanism and can involve formation of a hemiacetal, acetal, or oxonium ion as an intermediate, followed by its cyclization. For the reaction of D-glucose with methanol these key intermediates are:

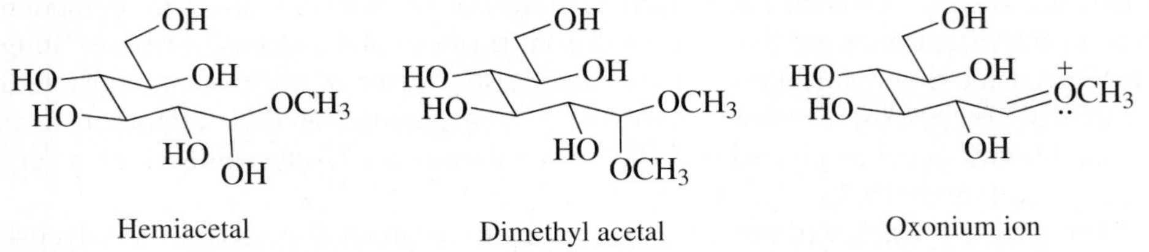

Hemiacetal Dimethyl acetal Oxonium ion

Cyclization can lead to the α- or β-furanoside or α- or β-pyranoside. The furanosides are the kinetic products of the Fischer glycosidation and can be isolated if the reaction is stopped prior to equilibrium. Mechanism 23.2 describes the initial formation of the methyl hemiacetal of D-glucose and its cyclization to a mixture of methyl α-D-glucopyranoside and its β anomer.

A process similar to that of Mechanism 23.2 involving the —OH group at C-4 gives the methyl α- and β-furanosides. These then undergo subsequent conversion to the more stable pyranosides by a mechanism not requiring reversion to D-glucose itself.

Mechanism 23.2

Preparation of Methyl D-Glucopyranosides by Fischer Glycosidation

THE REACTION:

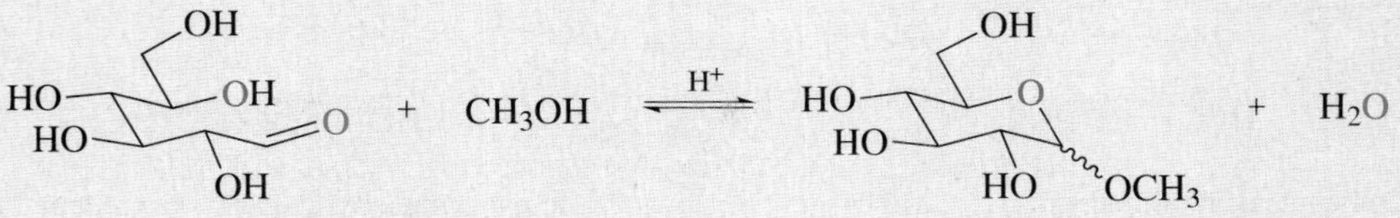

D-Glucose + Methanol $\xrightleftharpoons{H^+}$ Methyl D-glucopyranoside ($\alpha + \beta$) + Water

THE MECHANISM:

Steps 1–3: Acid-catalyzed nucleophilic addition of methanol to the carbonyl group of D-glucose. (See Mechanisms 18.2 and 18.4 for details of acid-catalyzed addition to aldehydes and ketones.)

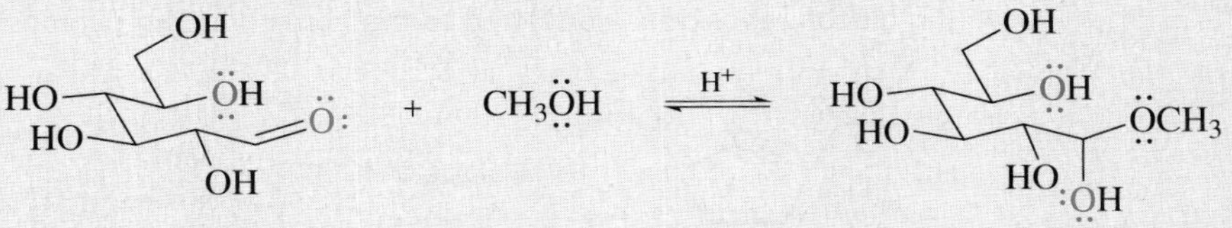

D-Glucose + Methanol $\xrightleftharpoons{H^+}$ D-Glucose methyl hemiacetal

Step 4: Protonation of the —OH group of the hemiacetal unit. The proton donor is shown as the conjugate acid of methanol. It was formed by proton transfer from the acid catalyst to methanol.

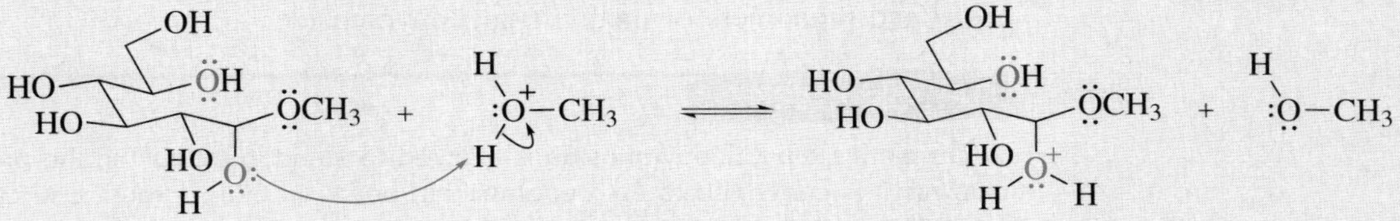

D-Glucose methyl hemiacetal + Methyloxonium ion ⇌ Conjugate acid of D-glucose methyl hemiacetal + Methanol

Step 5: Loss of water from the protonated hemiacetal to give an oxonium ion.

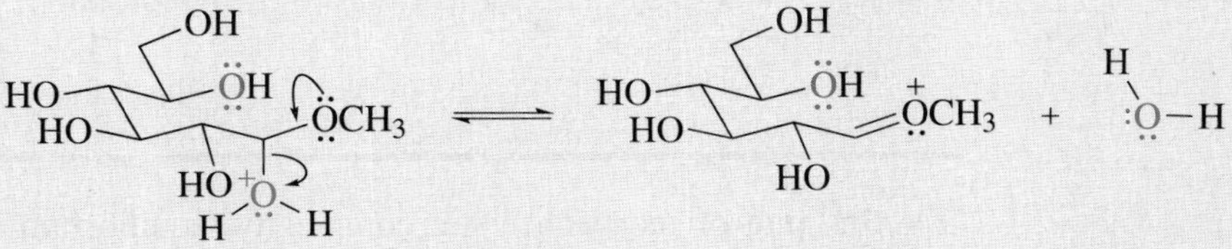

Conjugate acid of D-glucose methyl hemiacetal ⇌ Oxonium ion + Water

Step 6: Cyclization of the oxonium ion. An unshared electron pair of the C-5 oxygen is used to form a bond to C-1, forming the six-membered ring of the glycopyranoside. Both the α and β stereoisomers are formed in this reaction with the α stereoisomer (axial OCH_3) predominating.

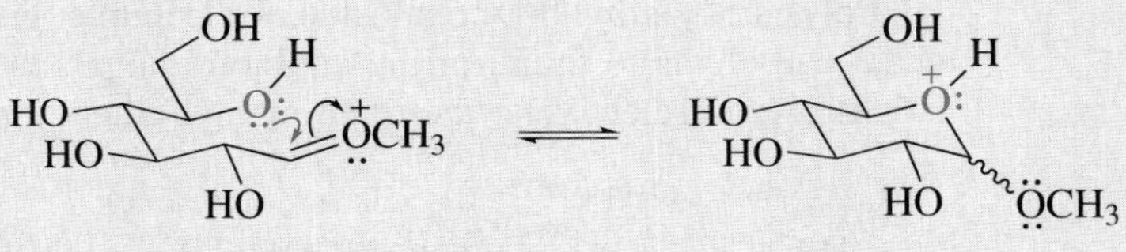

Oxonium ion ⇌ Conjugate acid of methyl D-glucopyranoside ($\alpha + \beta$)

continued

Step 7: Proton transfer from the positively charged ring oxygen to the oxygen of methanol giving a mixture of methyl α- and β-D-glucopyranoside. The acid catalyst is regenerated in this step.

Conjugate acid of methyl D-glucopyranoside (α + β) + Methanol ⇌ Methyl D-glucopyranoside (α + β) + Methyloxonium ion

Problem 23.16

Add curved arrows to the following sequence to show how the conjugate acid of methyl β-D-glucofuranoside is converted to the corresponding pyranoside.

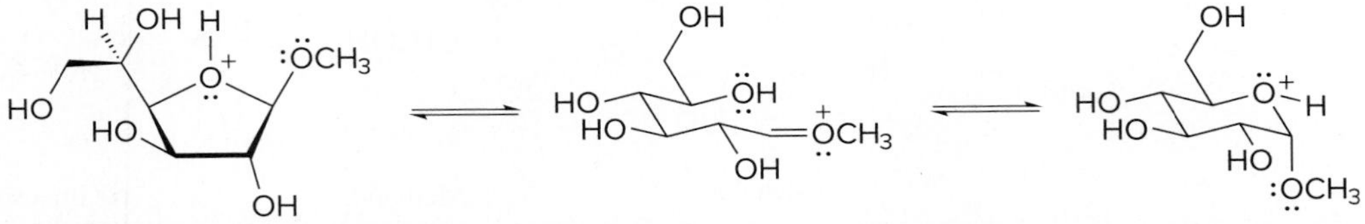

Still another mechanism, one involving a cyclic oxonium ion, is believed to interconvert the α and β anomers of methyl D-glucopyranoside.

Problem 23.17

When methyl β-D-glucopyranoside is allowed to stand in CD_3OH in the presence of an acid catalyst, it is converted to an α anomer that bears an OCD_3 group. Use curved arrows to track the electrons in the reactive intermediates shown.

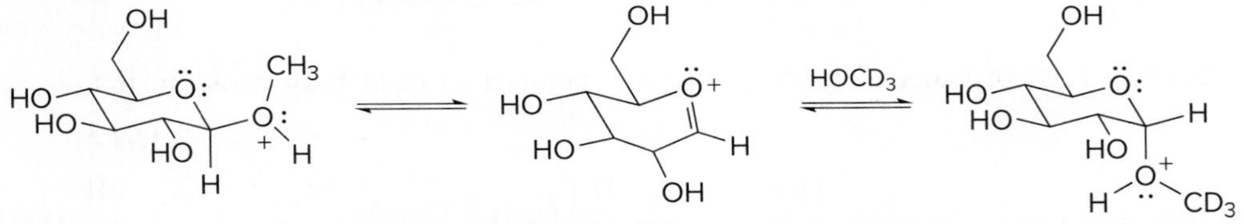

In spite of its mechanistic complexity, equilibrium is established rapidly thereby making the Fischer glycosidation a reliable method for converting a carbohydrate to its *O*-glycoside. Once formed, *O*-glycosides are useful intermediates in the synthesis of a variety of carbohydrate structural types by suitable manipulation of the remaining hydroxyl groups.

Glycosides, like other acetals, are stable in base but undergo hydrolysis in aqueous acid. The products of hydrolysis of an alkyl glycoside are an alcohol and a carbohydrate. Hydrolysis of methyl β-L-nogaloside, for example, gives L-nogalose, which is a branched-chain carbohydrate found in the antibiotic nogalamycin. Note that the methyl glycoside is hydrolyzed selectively; the three methyl ethers remain intact.

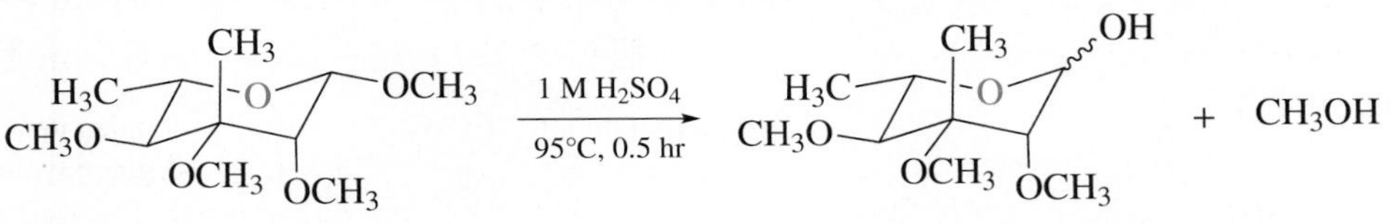

Methyl β-L-nogaloside $\xrightarrow[95°C,\ 0.5\ hr]{1\ M\ H_2SO_4}$ L-Nogalose (88%) + CH_3OH (Methanol)

Problem 23.18

Using Mechanism 23.2 as a guide, write a stepwise mechanism for the acid hydrolysis of methyl α-D-glucopyranoside.

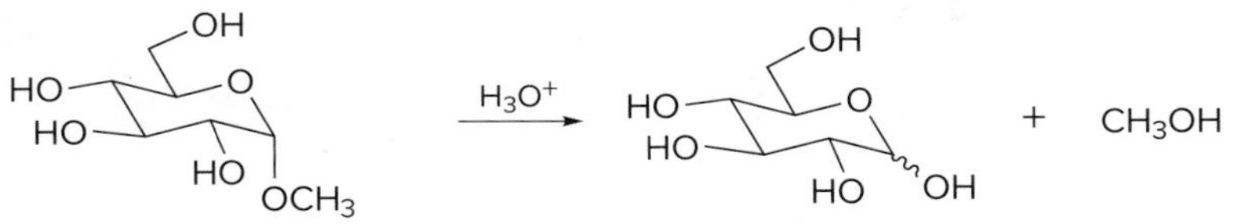

Methyl α-D-glucopyranoside $\xrightarrow{H_3O^+}$ D-Glucose + CH_3OH (Methanol)

23.15 Disaccharides

Disaccharides are carbohydrates that yield two monosaccharide molecules on hydrolysis. Structurally, disaccharides are *glycosides* in which the alkoxy group attached to the anomeric carbon is derived from a second sugar molecule.

Maltose, obtained by the hydrolysis of starch, and *cellobiose,* by the hydrolysis of cellulose, are isomeric disaccharides. In both maltose and cellobiose two D-glucopyranose units are joined by a glycosidic bond between C-1 of one unit and C-4 of the other. The two are diastereomers, differing only in the stereochemistry at the anomeric carbon of the glycosidic bond; maltose is an α-glycoside, cellobiose is a β-glycoside.

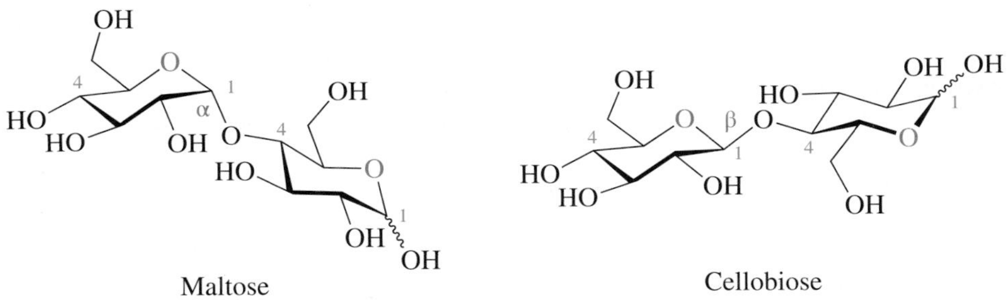

Maltose

Cellobiose

The stereochemistry and points of connection of glycosidic bonds are commonly designated by symbols such as α-(1→4) for maltose and β-(1→4) for cellobiose; α and β designate the stereochemistry at the anomeric position; the numerals specify the ring carbons involved.

Both maltose and cellobiose have a free anomeric hydroxyl group that is not involved in a glycosidic bond. The configuration at the free anomeric center is variable and may be either α or β. Indeed, two stereoisomeric forms of maltose have been isolated: one has its anomeric hydroxyl group in an equatorial orientation; the other has an axial anomeric hydroxyl.

Problem 23.19

The two stereoisomeric forms of maltose just mentioned undergo mutarotation when dissolved in water. What is the structure of the key intermediate in this process?

The single difference in their structures, the stereochemistry of the glycosidic bond, causes maltose and cellobiose to differ significantly in their three-dimensional shape, as the molecular models of Figure 23.8 illustrate. This difference in shape

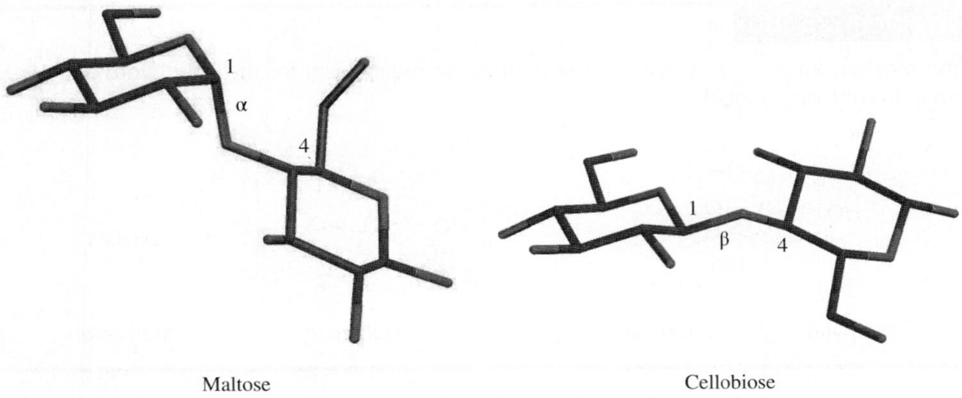

Figure 23.8

Molecular models of the disaccharides maltose and cellobiose. Two D-glucopyranose units are connected by a glycoside linkage between C-1 and C-4. The glycosidic bond has the α orientation in maltose and is β in cellobiose. Maltose and cellobiose are diastereomers.

Maltose

Cellobiose

affects how maltose and cellobiose interact with other chiral molecules such as proteins, causing them to behave much differently toward enzyme-catalyzed hydrolysis. The enzyme *maltase* catalyzes the hydrolytic cleavage of the α-glycosidic bond of maltose but not the β-glycosidic bond of cellobiose. A different enzyme, *emulsin,* produces the opposite result: emulsin catalyzes the hydrolysis of cellobiose but not maltose. The behavior of each enzyme is general for glucosides (glycosides of glucose). Maltase catalyzes the hydrolysis of α-glucosides and is also known as *α-glucosidase,* whereas emulsin catalyzes the hydrolysis of β-glucosides and is known as *β-glucosidase.* The specificity of these enzymes offers a useful tool for structure determination because it allows the stereochemistry of glycosidic linkages to be assigned.

Lactose is a disaccharide constituting 2–6% of milk and is known as *milk sugar.* It differs from maltose and cellobiose in that only one of its monosaccharide units is D-glucose. The other monosaccharide unit, the one that contributes its anomeric carbon to the glycosidic bond, is D-galactose. Like cellobiose, lactose is a β-glycoside.

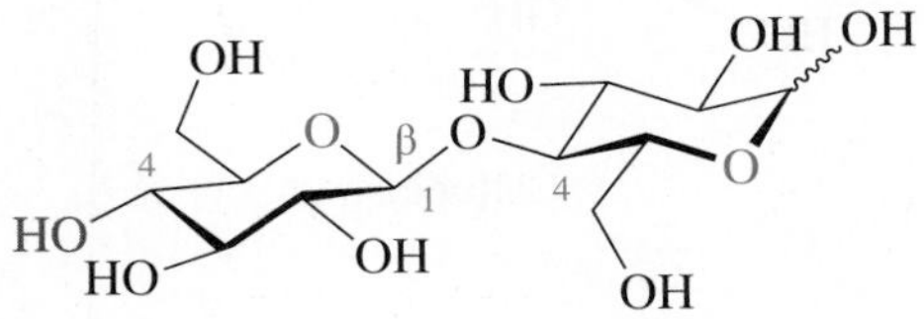

Cellobiose

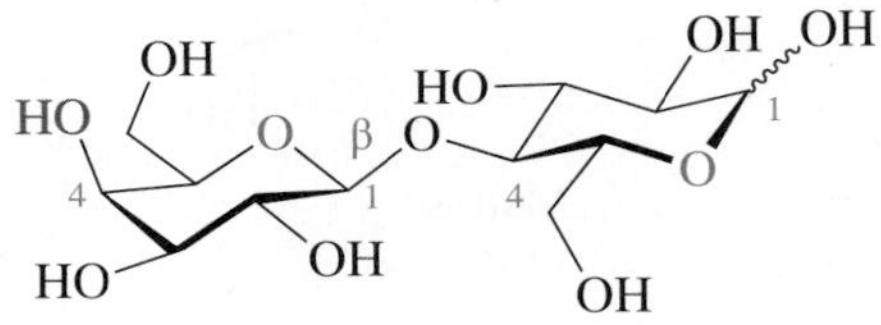

Lactose

Digestion of lactose is facilitated by the β-glycosidase *lactase.* A deficiency of this enzyme makes it difficult to digest lactose and causes abdominal discomfort. Lactose intolerance is a genetic trait; it is treatable through over-the-counter formulations of lactase and by limiting the amount of milk in the diet.

The most familiar of all the carbohydrates is *sucrose*—common table sugar. Sucrose is a disaccharide in which D-glucose and D-fructose are joined at their anomeric carbons by a glycosidic bond.

D-Glucose portion of molecule

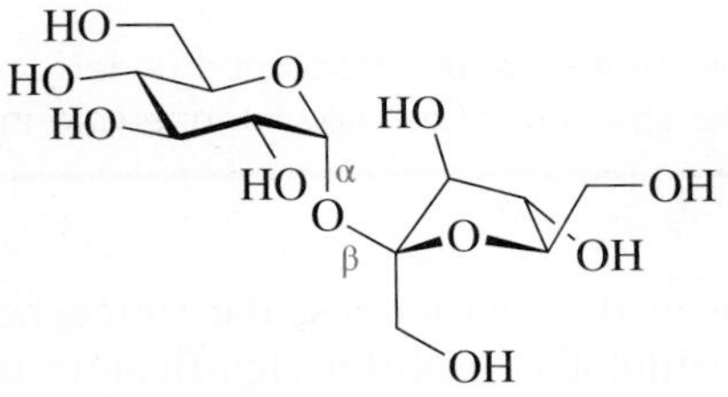

D-Fructose portion of molecule

Sucrose

Its chemical composition is the same irrespective of its source; sucrose from cane and sucrose from sugar beets are identical. Because sucrose does not have a free anomeric hydroxyl group, it does not undergo mutarotation. Hydrolysis of sucrose, catalyzed either by acid or by the enzyme *invertase,* gives a 1:1 mixture of D-glucose and D-fructose, which is sweeter than sucrose. The mixture prepared this way is called "invert sugar" because the sign of rotation of the aqueous solution in which it is carried out "inverts" from + to − as sucrose is converted to the glucose–fructose mixture.

23.16 Polysaccharides

Cellulose is the principal structural component of vegetable matter. Wood is 30–40% cellulose, cotton over 90%. Photosynthesis in plants is responsible for the formation of 10^9 tons per year of cellulose. Structurally, cellulose is a polysaccharide composed of D-glucose units joined by β-(1→4)-glycosidic linkages (Figure 23.9). The average is 7000 glucose units, but can be as many as 12,000. Complete hydrolysis of all the glycosidic bonds of cellulose yields D-glucose. The disaccharide fraction that results from partial hydrolysis is cellobiose.

As Figure 23.9 shows, the glucose units of cellulose are turned with respect to each other. The overall shape of the chain, however, is close to linear. Consequently, neighboring chains can pack together in bundles where networks of hydrogen bonds stabilize the structure and impart strength to cellulose fibers.

Animals lack the enzymes necessary to catalyze the hydrolysis of cellulose and so can't digest it. Cattle and other ruminants use cellulose as a food source indirectly. Colonies of bacteria that live in their digestive tract consume cellulose and in the process convert it to other substances that the animal can digest.

A more direct source of energy for animals is provided by the starches found in many plants. Starch is a mixture containing about 20% of a water-dispersible fraction called *amylose* and 80% of a second component, *amylopectin.*

Like cellulose, amylose is a polysaccharide of D-glucose. However, unlike cellulose in which all of the glycosidic linkages are β, all of the linkages in amylose are α. The small change in stereochemistry between cellulose and amylose creates a large difference in their overall shape and in their properties. Some of this difference can be seen in the structure of a short portion of amylose in Figure 23.10. The presence of the α-glycosidic linkages imparts a twist to the amylose chain. Where the main chain is roughly linear in cellulose, it is helical in amylose. Attractive forces *between* chains are weaker in amylose, and amylose does not form the same kind of strong fibers that cellulose does.

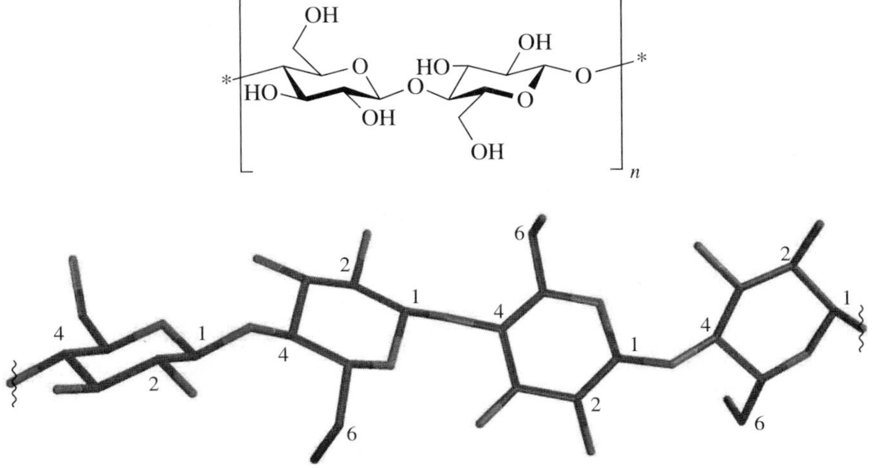

Figure 23.9

Cellulose is a polysaccharide in which D-glucose units are connected by β-(1→4)-glycosidic linkages analogous to cellobiose. Hydrogen bonding, especially between the C-2 and C-6 hydroxyl groups, causes adjacent glucose units to tilt at an angle of 180° with each other.

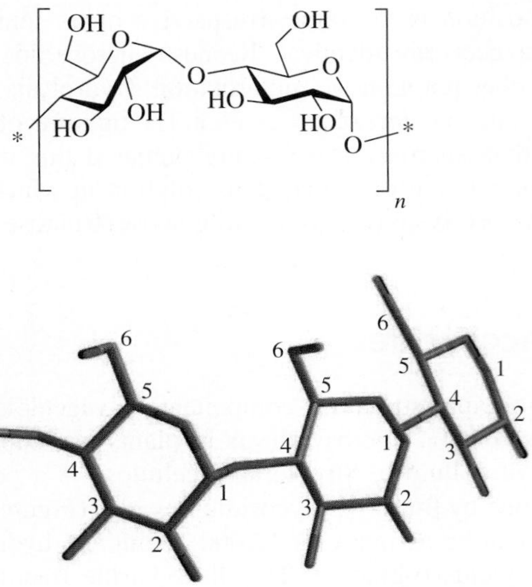

Figure 23.10

Amylose is a polysaccharide in which D-glucose units are connected by α-(1→4)-glycosidic linkages analogous to maltose. The geometry of the glycosidic linkage is responsible for the left-hand helical twist of the chain.

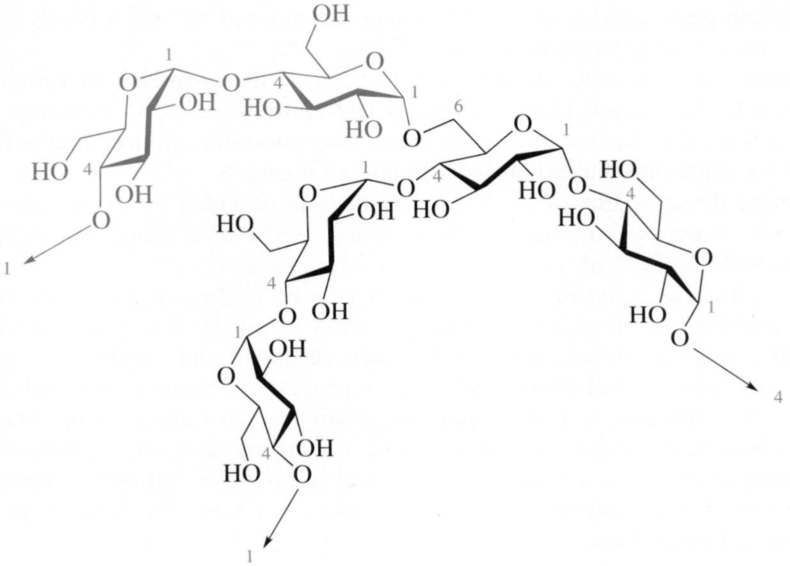

Figure 23.11

Amylopectin. The main chain (black) is the same as in amylose. Amylopectin differs from amylose in having branches (red) linked to the main chain by α-(1→6) glycosidic bonds. Except for the glycosidic bonds connecting the branches to the main chain, all other glycosidic bonds are α-(1→4).

Amylopectin resembles amylose in being a polysaccharide built on a framework of α-(1→4)-linked D-glucose units. In addition to this main framework, however, amylose incorporates polysaccharide branches of 24–30 glucose units joined by α-(1→4)-glycosidic bonds. These branches sprout from C-6 of glucose units at various points along the main framework, connected to it by α-(1→6)-glycosidic bonds (Figure 23.11).

One of the most important differences between cellulose and starch is that animals can digest starch. Because the glycosidic linkages in starch are α, an animal's α-glycosidase enzymes can catalyze their hydrolysis to glucose. When more glucose is available than is needed as fuel, animals store some of it as glycogen. Glycogen resembles amylopectin in that it is a branched polysaccharide of α-(1→4)-linked D-glucose units with branches connected to C-6 of the main chain. The frequency of such branches is greater in glycogen than in amylopectin.

How Sweet It Is!

How sweet is it?

There is no shortage of compounds, natural or synthetic, that taste sweet. The most familiar, sucrose, glucose, and fructose, all occur naturally with worldwide production of sucrose from cane and sugar beets exceeding 100 million tons per year. Glucose is prepared by the enzymatic hydrolysis of starch, and fructose is made by the isomerization of glucose.

Starch + H_2O → (+)-D-Glucose ⇌ (glucose isomerase) (−)-D-Fructose

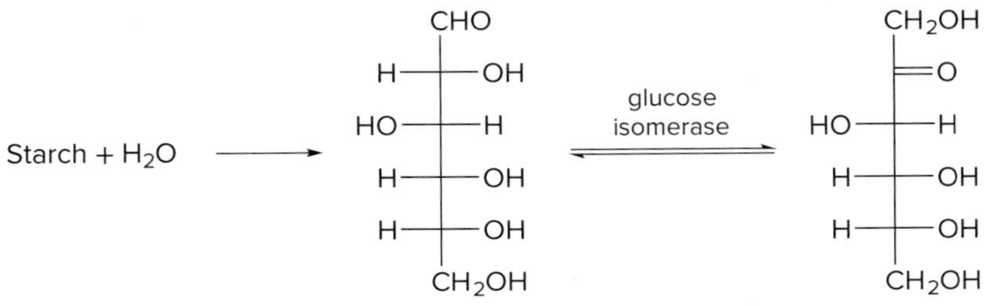

(+)-D-Glucose

(−)-D-Fructose

Among sucrose, glucose, and fructose, fructose is the sweetest. Honey is sweeter than table sugar because it contains fructose formed by the isomerization of glucose as shown in the equation.

You may have noticed that most soft drinks contain "high-fructose corn syrup." Corn starch is hydrolyzed to glucose, which is then treated with glucose isomerase to produce a fructose-rich mixture. The enhanced sweetness permits less to be used, reducing the cost of production. Using less carbohydrate-based sweetener also reduces the number of calories.

Artificial sweeteners are a billion-dollar-per-year industry. The primary goal is to maximize sweetness and minimize calories. We'll look at the following sweeteners to give us an overview of the field.

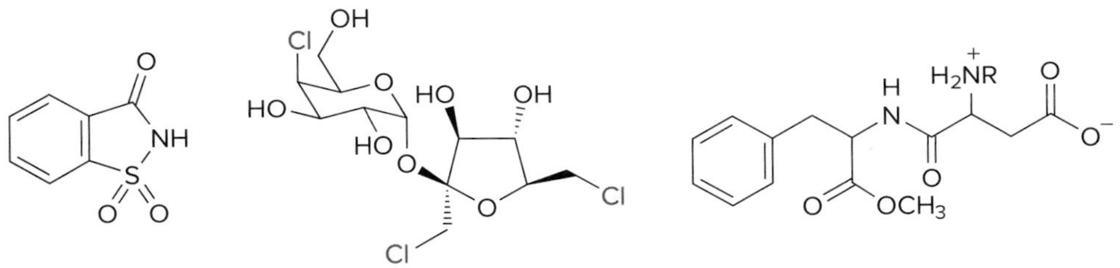

Saccharin

Sucralose

Aspartame, R = H
Neotame, R = $CH_2CH_2C(CH_3)_3$

All of these are hundreds of times sweeter than sucrose and variously described as "low-calorie" or "nonnutritive" sweeteners.

Saccharin was discovered at Johns Hopkins University in 1879 in the course of research on coal-tar derivatives and is the oldest artificial sweetener. In spite of its name, which comes from the Latin word for sugar, saccharin bears no structural relationship to any sugar. Nor is saccharin itself very soluble in water. The proton bonded to nitrogen, however, is fairly acidic and saccharin is normally marketed as its water-soluble sodium or calcium salt. Its earliest applications were not in weight control, but as a replacement for sugar in the diet of diabetics before insulin became widely available.

Sucralose has the structure most similar to sucrose. Galactose replaces the glucose unit of sucrose, and chlorines replace three of the hydroxyl groups. The three chlorine substituents do not diminish sweetness, but do interfere with the ability of the body to metabolize sucralose. It, therefore, has no food value and is "noncaloric."

Aspartame is a methyl ester of a dipeptide, unrelated to any carbohydrate. An aspartame relative, neotame, is even sweeter.

Saccharin, sucralose, and aspartame illustrate the diversity of structural types that taste sweet, and the vitality and continuing development of the industry of which they are a part.

23.17 Application of Familiar Reactions to Monosaccharides

In our discussion of carbohydrate structure to this point, we have already encountered an important reaction—the formation of glycosides under acid-catalyzed conditions (see Section 23.14). Glycoside formation draws our attention to the fact that an OH group on the anomeric carbon of a pyranose or furanose differs in reactivity from the other hydroxyl groups of the carbohydrate. It also demonstrates that what looks like a new reaction is one that we've encountered before—in this case, acetal formation that occurs in the reactions of alcohols with aldehydes and ketones (see Section 18.8). Many other reactions of carbohydrates are also related to familiar functional-group transformations; some of these are recounted in Table 23.2.

The first two entries in Table 23.2 illustrate reactions that involve nucleophilic addition to the carbonyl group of the open-chain form which, although present in small amounts, is continuously replenished as it reacts. Entry 1 is the sodium borohydride reduction of the carbonyl group of the aldose D-galactose. The reaction is a general one; other reducing agents may be used, and the product from reduction of either an aldose or ketose is called an **alditol.** Alditols lack a carbonyl group and exist entirely in open-chain forms.

Problem 23.20

Reduction of the ketose D-fructose gives two alditols on reduction. Which two are they?

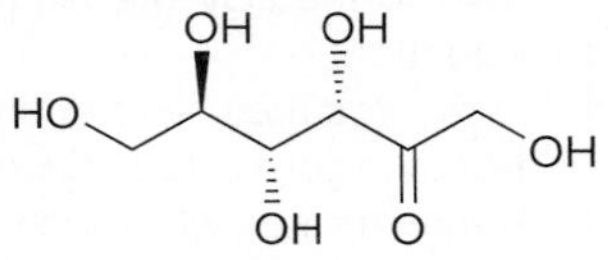

D-Fructose

Entry 2 is cyanohydrin formation by nucleophilic addition of HCN to the carbonyl group. It is the basis of a synthetic method for extending the carbon chain of an aldose. In the example shown in Table 23.2 the two diastereomeric cyanohydrins derived from L-arabinose were separated, and their —C≡N groups converted to —CH═O to yield L-mannose and L-glucose, as shown below for one of the diastereomers. In this conversion, the nitrile group is first reduced to an imine (—C═NH), which is then hydrolyzed to the aldehyde. The sequence extends the chain of L-arabinose, a pentose, to that of L-glucose, a hexose.

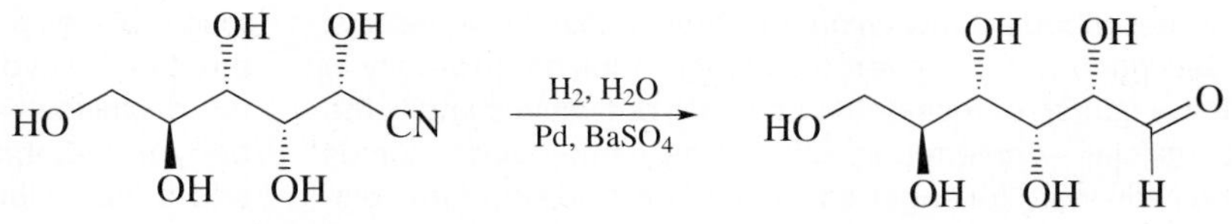

L-Gluconontrile $\xrightarrow[\text{Pd, BaSO}_4]{H_2,\ H_2O}$ L-Glucose (26% from L-arabinose)

Entries 3 and 4 are acylation and alkylation, respectively, of hydroxyl groups. Entry 3 shows the formation of a pentaacetate on reaction of α-D-glucopyranose with acetic anhydride, and entry 4 shows the formation of a tetrabenzyl ether on reaction of methyl α-D-glucopyranoside with benzyl chloride. Benzyl ethers are stable to acid and base hydrolysis, organometallic reagents, and numerous other reaction conditions and are often used to protect OH groups during synthetic operations. They are usually removed by *hydrogenolysis,* a reaction in which catalytic hydrogenation cleaves the $C_6H_5CH_2$—O bond of a benzyl ether.

TABLE 23.2 Familiar Reaction Types of Carbohydrates

Reaction and comments	Example
1. Reduction: Carbonyl groups in carbohydrates are reduced by the same methods used for aldehydes and ketones: reduction with sodium borohydride or lithium aluminum hydride or by catalytic hydrogenation.	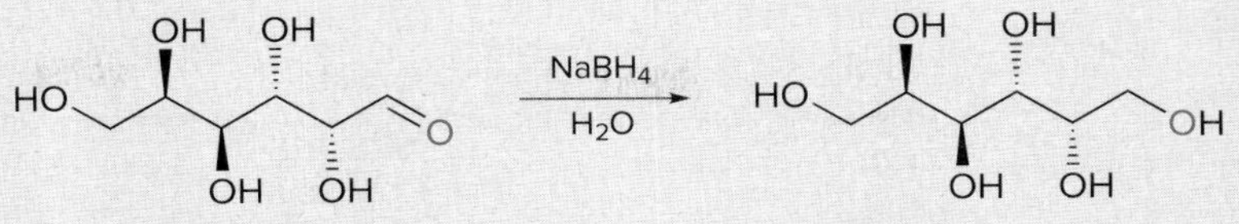 D-Galactose → ($NaBH_4$, H_2O) D-Galactitol (90%)
2. Cyanohydrin formation: Reaction of an aldose with HCN gives a mixture of two diastereomeric cyanohydrins.	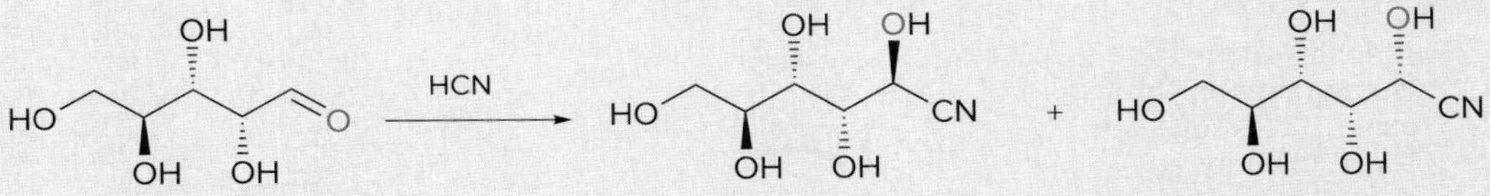 L-Arabinose → (HCN) L-Mannonitrile + L-Gluconitrile
3. Acylation: All available hydroxyl groups of carbohydrates are capable of undergoing acylation to form esters.	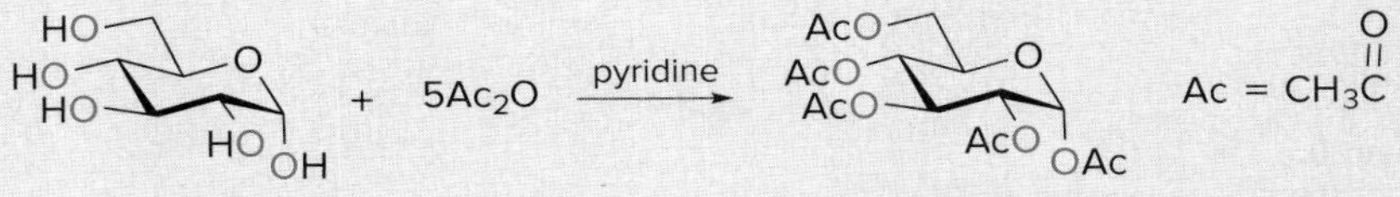 α-D-Glucopyranose + $5Ac_2O$ (Acetic anhydride) → (pyridine) 1,2,3,4,6-Penta-*O*-acetyl-D-glucopyranose (88%); Ac = $CH_3\overset{O}{\overset{\Vert}{C}}$
4. Alkylation: Carbohydrate hydroxyl groups react with alkyl halides, especially methyl and benzyl halides, to give ethers.	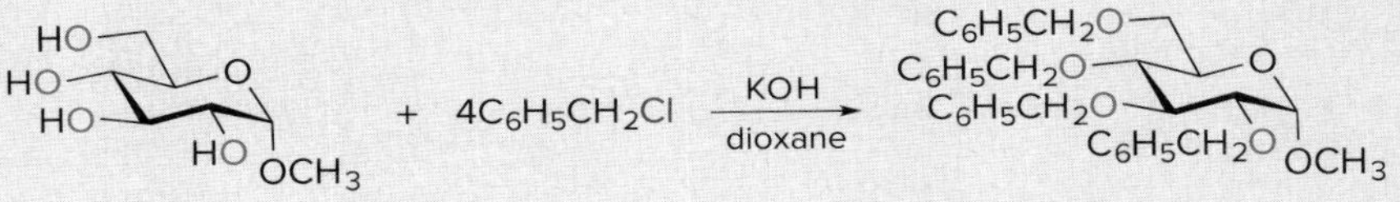 Methyl α-D-glucopyranoside + $4C_6H_5CH_2Cl$ (Benzyl chloride) → (KOH, dioxane) Methyl 2,3,4,6-tetra-*O*-benzyl-α-D-glucopyranoside (95%)
5. Acetal formation: Carbohydrates can serve as the diol component in the formation of cyclic acetals on reaction with aldehydes and ketones in the presence of an acid catalyst. In the example shown, the catalyst is a Lewis acid.	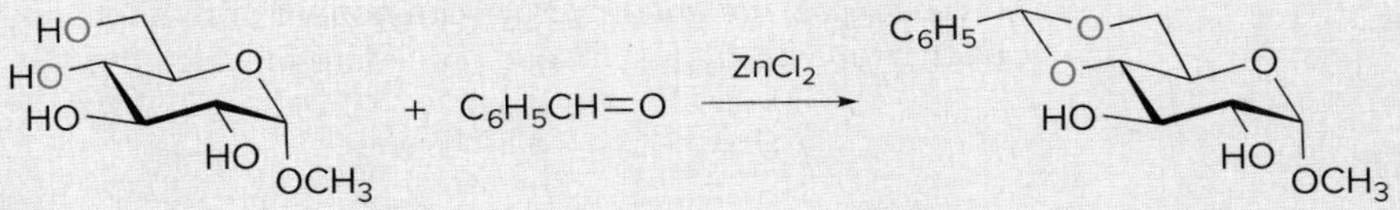 Methyl α-D-glucopyranoside + $C_6H_5CH{=}O$ (Benzaldehyde) → ($ZnCl_2$) Methyl 4,6-*O*-benzylidene-α-D-glucopyranoside (63%)
6. Pyranose–furanose isomerization: The furanose and pyranose forms of a carbohydrate are cyclic hemiacetals and equilibrate by way of their open-chain isomer.	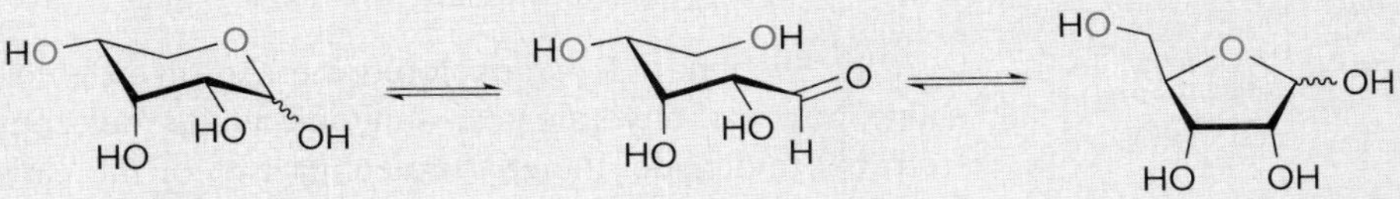 D-Ribopyranose (α and/or β) ⇌ D-Ribose ⇌ D-Ribofuranose (α and/or β)
7. Enolization: Enolization of the open-chain form of a carbohydrate gives an enediol. Carbohydrates that are epimeric at C-2 give the same enediol.	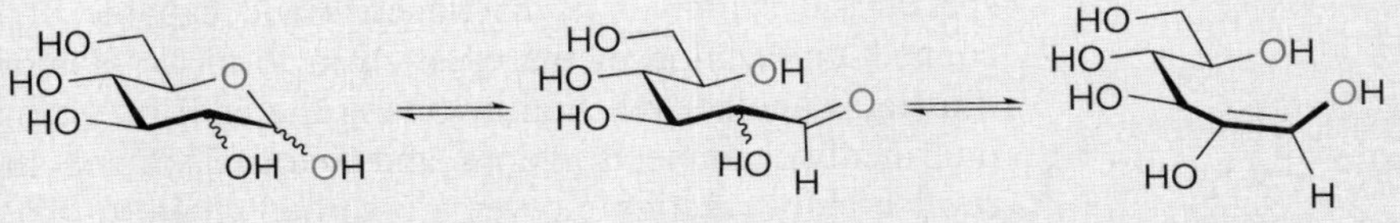 D-Gluco- or D-mannopyranose (α and/or β) ⇌ D-Glucose or D-mannose ⇌ Enediol

continued

TABLE 23.2 Familiar Reaction Types of Carbohydrates (*Continued*)

Reaction and comments	Example
8. Epimerization: Enediol formation provides a pathway for the interconversion of the C-2 epimers. The stereochemistry at C-2 is lost on enolization; reversion to the aldehyde can give either the *R* or *S* configuration at C-2.	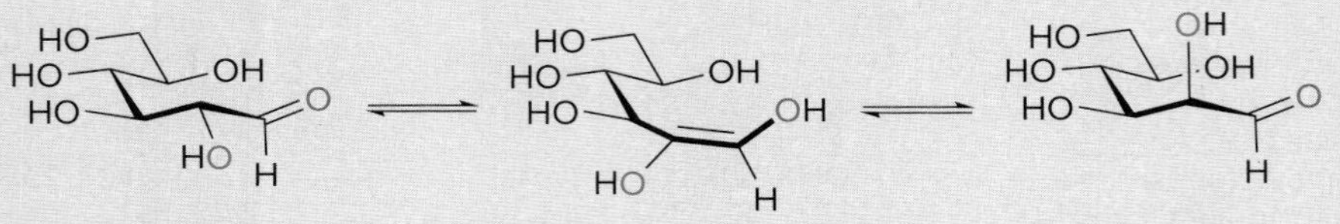 D-Glucose ⇌ Enediol ⇌ D-Mannose
9. Aldose–ketose isomerization: The same 1,2-enediol can revert to either an aldose or a ketose.	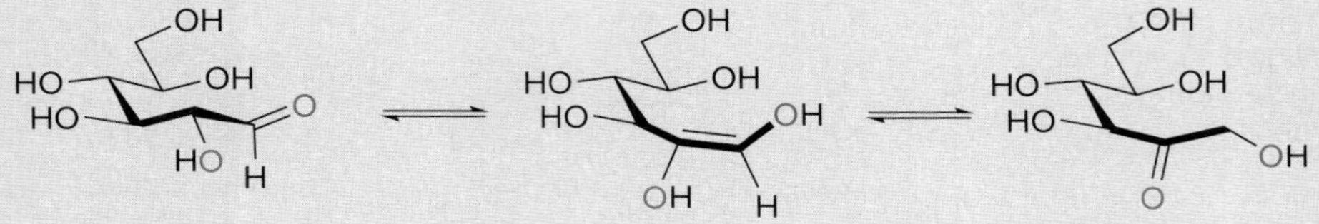 D-Glucose ⇌ Enediol ⇌ D-Fructose
10. Aldol addition and retro-aldol cleavage: The enzyme *aldolase* catalyzes important biochemical reactions related to aldol addition and its reverse.	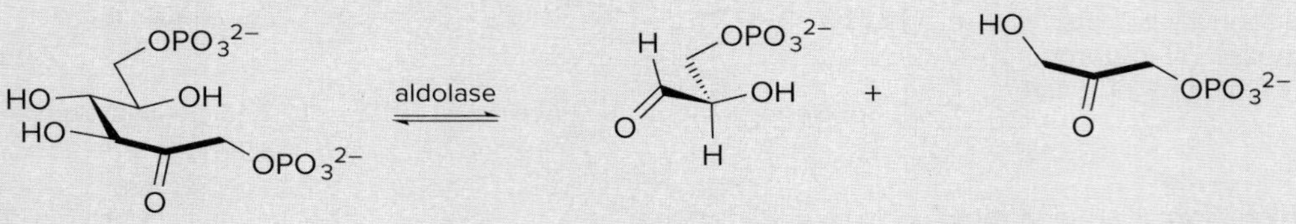 D-Fructose 1,6-diphosphate ⇌ (aldolase) D-Glyceraldehyde 3-phosphate + Dihydroxyacetone phosphate

Problem 23.21

D-Digitoxose, the carbohydrate component of the antiarrhythmic drug *digoxin* (digitalis), has been prepared by a synthetic procedure in which the last step is:

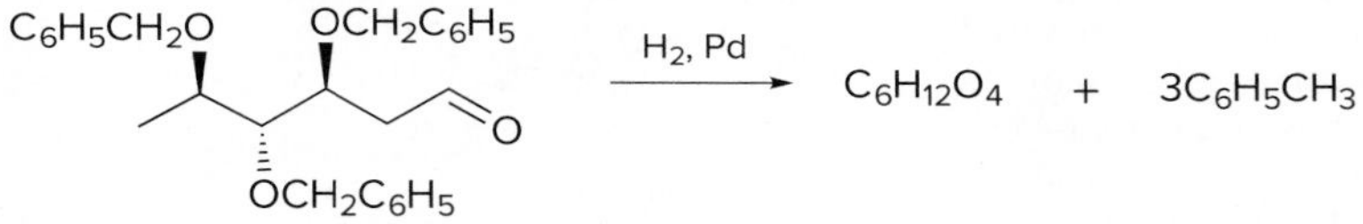

$$\text{(tri-}O\text{-benzyl aldehyde; } C_6H_5CH_2O,\ OCH_2C_6H_5,\ OCH_2C_6H_5) \xrightarrow{H_2,\ Pd} C_6H_{12}O_4 + 3C_6H_5CH_3$$

Give the structure of the α-pyranose form of D-digitoxose.

In Section 18.9 we explored the use of diols to protect carbonyl groups. Entry 5 shows how carbohydrates can function as the diol component in forming a cyclic acetal from benzaldehyde, thereby protecting two of the carbohydrate's hydroxyl groups. This is yet another example of acetal formation, in this case involving the aldehyde group of benzaldehyde and the C-4 and C-6 hydroxyl groups of the carbohydrate.

Entry 6 reminds us that the furanose and pyranose forms of carbohydrates are specialized examples of hemiacetals and capable of interconversion. Entries 7 and 8 connect enolization of the open-chain form to interconversion of C-2 epimers. Entry 9 illustrates another reaction involving the enol of the open-chain form. Here, the enediol intermediate connects aldose and ketose isomers. In entry 10, an enzyme-catalyzed reverse aldol reaction cleaves a 6-carbon chain to two 3-carbon fragments.

Most monosaccharides bear an oxidizable function on every carbon so their oxidation is more complicated than others we have seen and is discussed separately in the next section.

23.18 Oxidation of Carbohydrates

A characteristic property of an aldehyde function is its sensitivity to oxidation. A solution of copper(II) sulfate as its citrate complex (Benedict's reagent) is capable of oxidizing aliphatic aldehydes to the corresponding carboxylic acids.

Benedict's reagent was once used in over-the-counter test kits for monitoring glucose levels in urine.

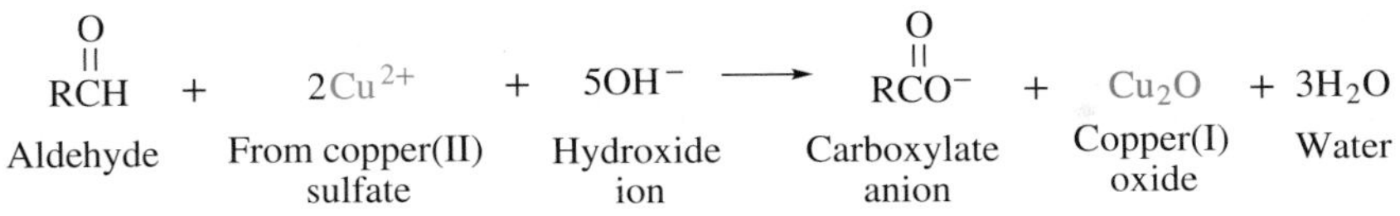

$$\underset{\text{Aldehyde}}{\overset{\text{O}}{\overset{\|}{\text{RCH}}}} + \underset{\text{From copper(II) sulfate}}{2Cu^{2+}} + \underset{\text{Hydroxide ion}}{5OH^-} \longrightarrow \underset{\text{Carboxylate anion}}{\overset{\text{O}}{\overset{\|}{\text{RCO}^-}}} + \underset{\text{Copper(I) oxide}}{Cu_2O} + \underset{\text{Water}}{3H_2O}$$

The formation of a red precipitate of copper(I) oxide by reduction of Cu(II) is taken as a positive test for an aldehyde. Carbohydrates that give positive tests with Benedict's reagent are termed *reducing sugars*.

Aldoses are reducing sugars, since they possess an aldehyde function in their open-chain form. Ketoses are also reducing sugars. Under the conditions of the test, ketoses equilibrate with aldoses by way of enediol intermediates, and the aldoses are oxidized by the reagent.

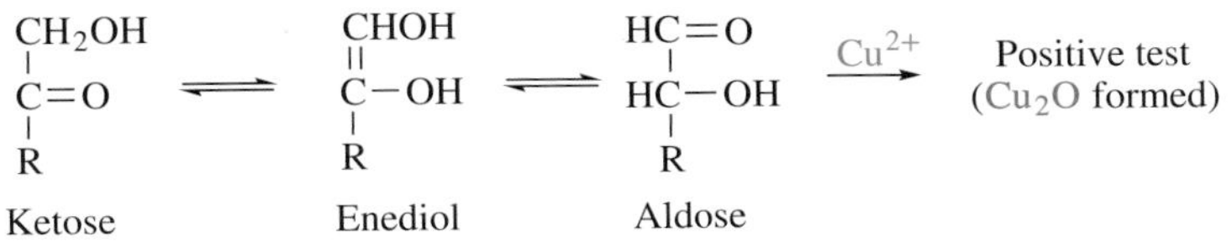

Ketose (CH_2OH–C(=O)–R) ⇌ Enediol (CHOH=C(OH)–R) ⇌ Aldose (HC(=O)–HC(OH)–R) $\xrightarrow{Cu^{2+}}$ Positive test (Cu_2O formed)

The same kind of equilibrium is available to α-hydroxy ketones generally; such compounds give a positive test with Benedict's reagent. Any carbohydrate that contains a free hemiacetal function is a reducing sugar. The free hemiacetal is in equilibrium with the open-chain form, which is susceptible to oxidation. Maltose, for example, gives a positive test with Benedict's reagent.

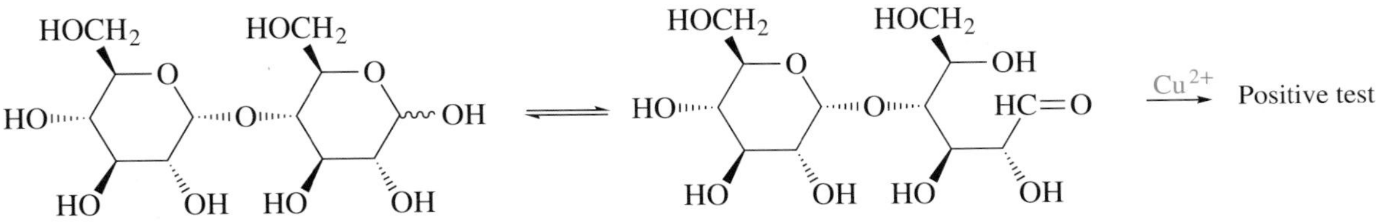

Maltose ⇌ Open-chain form of maltose $\xrightarrow{Cu^{2+}}$ Positive test

Glycosides, in which the anomeric carbon is part of an acetal function, are *not* reducing sugars and do *not* give a positive test.

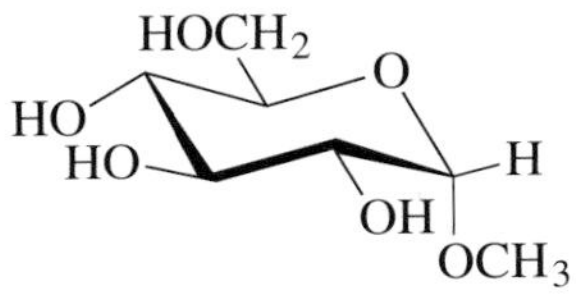

Methyl-α-D-glucopyranoside: not a reducing sugar

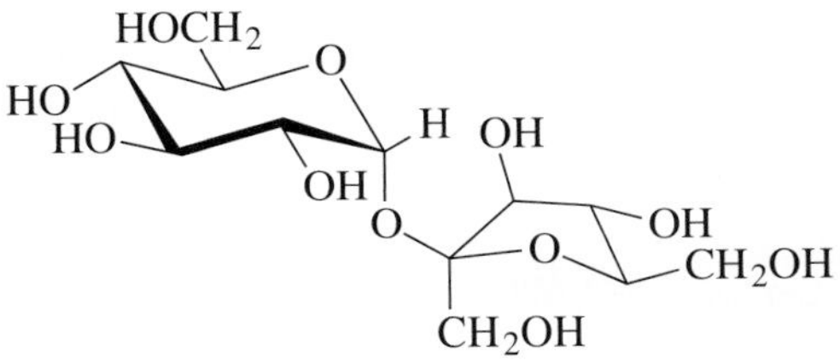

Sucrose: not a reducing sugar

PROBLEM 23.22

Which of the following would be expected to give a positive test with Benedict's reagent? Why?

(a) D-Galactitol (see structure below)
(b) L-Arabinose
(c) 1,3-Dihydroxyacetone
(d) D-Fructose
(e) Lactose
(f) Amylose

Sample Solution (a) D-Galactitol lacks an aldehyde, an α-hydroxy ketone, or a hemiacetal function, so it cannot be oxidized by Cu^{2+} and will not give a positive test with Benedict's reagent.

CH_2OH; H—C—OH; HO—C—H; HO—C—H; H—C—OH; CH_2OH

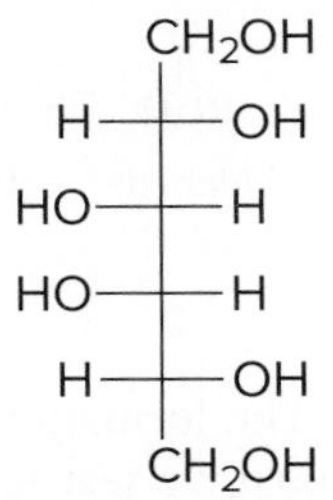

D-Galactitol

The most easily oxidized groups in an aldose are the aldehyde at one end and the primary alcohol at the other. Oxidation of CH═O gives an **aldonic acid;** oxidation of CH_2OH gives a **uronic acid,** and oxidation of both gives an **aldaric acid.**

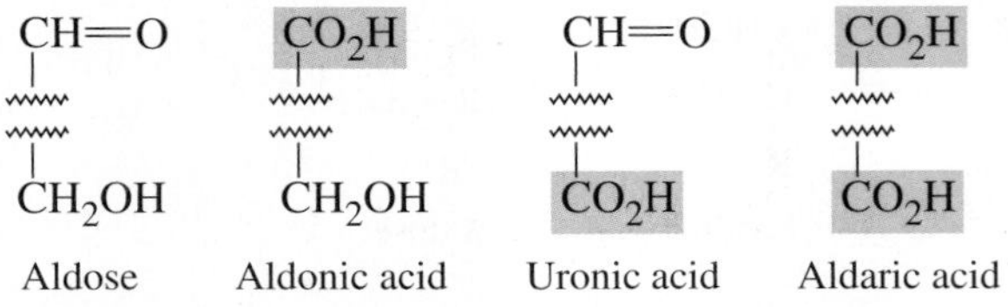

Aldose	Aldonic acid	Uronic acid	Aldaric acid
CH═O	CO_2H	CH═O	CO_2H
CH_2OH	CH_2OH	CO_2H	CO_2H

Aldonic acids are named by replacing the *-ose* ending of the aldose by *-onic acid.* Similarly, the endings *-uronic acid* and *-aric acid* are used in uronic and aldaric acids, respectively.

The most commonly used method for preparing aldonic acids is by oxidation with bromine in aqueous solution. The species that is oxidized is a furanose or pyranose form of the carbohydrate.

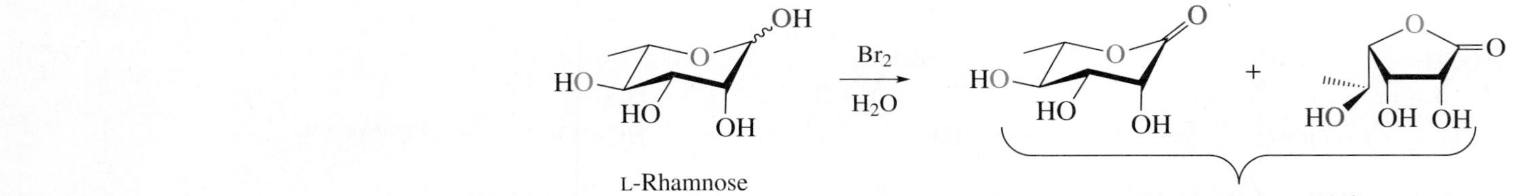

L-Rhamnose $\xrightarrow[H_2O]{Br_2}$ L-Rhamnolactone (80%)

Aldonic acids spontaneously form five-membered (γ) or six-membered (δ) lactones. γ-Lactones are normally more stable than δ-lactones.

Problem 23.23

What is the structure of the aldonic acid that is produced during the oxidation of L-rhamnose?

Direct oxidation of CH_2OH in the presence of CH═O is not practical, so laboratory preparations of uronic acids are limited to processes that include appropriate protection–deprotection strategies.

Problem 23.24

Uronic acids exist as cyclic hemiacetals rather than lactones. Write a structural formulation for the β-pyranose form of D-glucuronic acid—an intermediate in the biosynthesis of vitamin C and in various metabolic pathways.

Aldaric acids are prepared in the laboratory by oxidation of aldoses with nitric acid. Like aldonic acids, aldaric acids exist mainly as lactones when the rings are five-membered or six-membered.

Problem 23.25

Another hexose gives the same aldaric acid on oxidation as does D-glucose. Which one?

Solution

Both D-Glucose and [Fischer projections] → [aldaric acid]; rotate 180° = L-Gulose

Oxidative cleavage of vicinal diol functions in carbohydrates occurs with periodic acid (HIO_4) or sodium metaperiodate ($NaIO_4$). The reaction proceeds through a cyclic intermediate and is similar to what we encountered with diols in Section 16.11.

Vicinal diol —$NaIO_4$→ Cyclic intermediate → Two carbonyl compounds + Iodate ion (IO_3^-) + Water (H_2O)

Once used mainly as an aid in structure determination, periodate oxidation finds its major present use in synthesis as shown in the following example.

1,2:5,6-Di-*O*-isopropylidene-D-mannitol —$NaIO_4$→ 2 2,3-*O*-(Isopropylidene)-D-glyceraldehyde (72%)

23.19 Glycosides: Synthesis of Oligosaccharides

As we saw in Section 23.14, the preparation of glycosides by acid-catalyzed condensation of a carbohydrate with a simple alcohol—the **Fischer glycosidation**—is thermodynamically controlled and favors the formation of pyranose over furanose rings. The anomeric effect causes the α stereoisomer to predominate over the β.

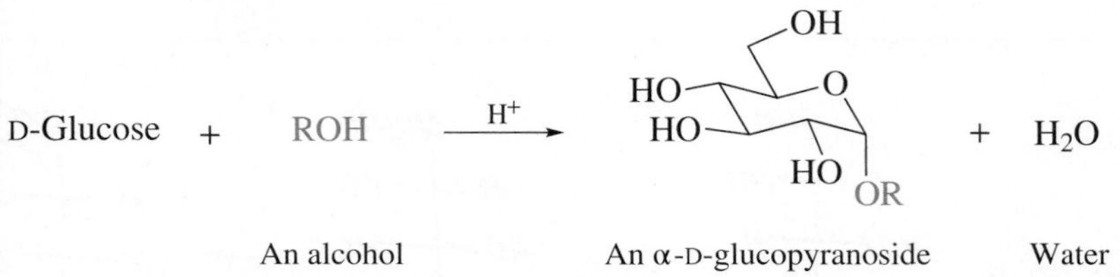

D-Glucose + ROH (An alcohol) $\xrightarrow{H^+}$ An α-D-glucopyranoside + H_2O (Water)

When the desired glycoside is a disaccharide, however, the "alcohol" is no longer simple; it is a carbohydrate with more than one OH group capable of bonding to the anomeric carbon of the other carbohydrate. Thus, constitutionally isomeric as well as stereoisomeric pyranosides are possible.

Consider, for example, a disaccharide such as gentiobiose in which both carbohydrate units are pyranosyl forms of D-glucose. Using the notation introduced in Section 23.15, gentiobiose is a β-(1→6) glycoside. An oxygen atom is bonded to the anomeric carbon of one D-glucopyranose and C-6 of a second D-glucopyranose.

Gentiobiose occurs naturally and has been isolated from gentian root, saffron, and numerous other plant materials. For more, see the boxed essay *Crocuses Make Saffron from Carotenes* in Chapter 24.

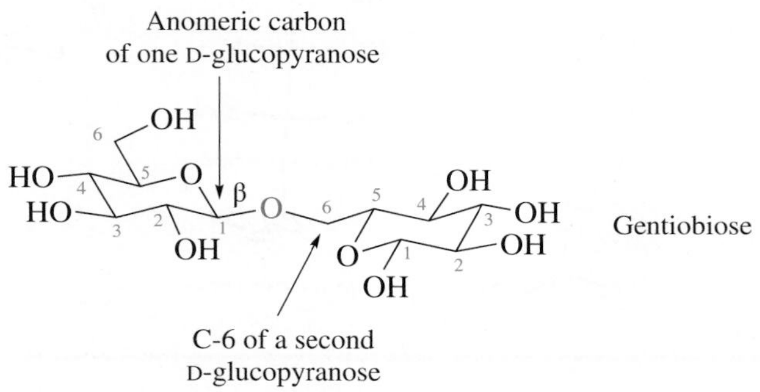

Anomeric carbon of one D-glucopyranose

Gentiobiose

C-6 of a second D-glucopyranose

Although the stereochemistry of the glycosidic linkage is a concern, the first problem to be addressed in a chemical synthesis of a disaccharide such as gentiobiose is achieving the desired 1→6 connectivity. The general strategy involves three stages:

1. Preparation of a suitably protected *glycosyl donor* and *glycosyl acceptor*. A glycosyl donor contains a leaving group at the anomeric carbon. The glycosyl acceptor contains a nucleophilic group, hydroxyl in this case, at the desired carbon.
2. Formation of the glycosidic C—O bond by a nucleophilic substitution in which an OH group of the glycosyl acceptor acts as the nucleophile toward the anomeric carbon of the glycosyl donor.
3. Removal of the protecting groups from the protected disaccharide formed in stage 2.

The glycosyl donor in the synthesis of gentiobiose is 2,3,4,6-tetra-*O*-benzoyl-α-D-glucopyranosyl bromide. It is prepared by treating D-glucose with benzoyl chloride in pyridine, then hydrogen bromide. This reaction places a bromide leaving group on the anomeric carbon and protects the four remaining OH groups as benzoate esters. Like the Fischer glycosylation, the α stereochemistry at the anomeric carbon results from thermodynamic control. The glycosyl acceptor is methyl 2,3,4-tri-*O*-acetyl-β-D-glucopyranoside, in which all of the hydroxyl groups are protected except the one at C-6.

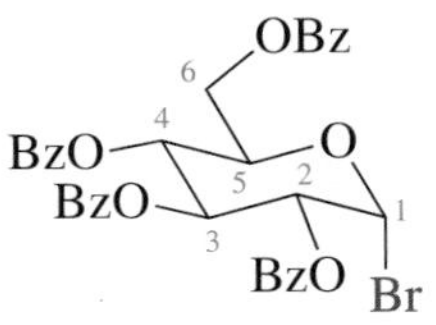

Bz = $C_6H_5\overset{O}{\overset{\|}{C}}$

Ac = $CH_3\overset{O}{\overset{\|}{C}}$

2,3,4,6-Tetra-*O*-benzoyl-α-D-glucopyranosyl bromide

Glycosyl Donor

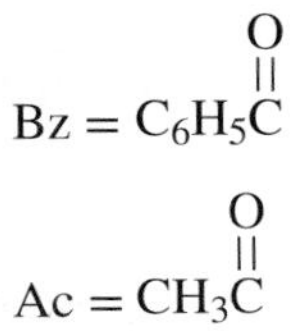

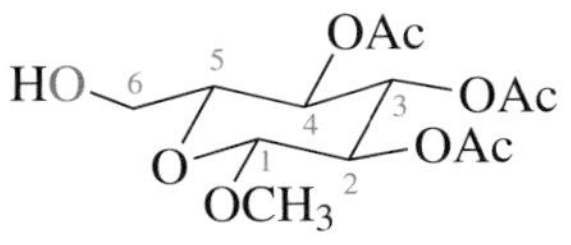

Methyl 2,3,4-tri-*O*-acetyl-β-D-glucopyranoside

Glycosyl Acceptor

Coupling of the glycosyl donor and acceptor takes place in the presence of silver trifluoromethanesulfonate ($AgOSO_2CF_3$) as a source of Ag^+, which activates the pyranosyl bromide toward nucleophilic substitution. A weak base such as 2,4,6-trimethylpyridine is included in order to react with the trifluoromethanesulfonic acid that is produced.

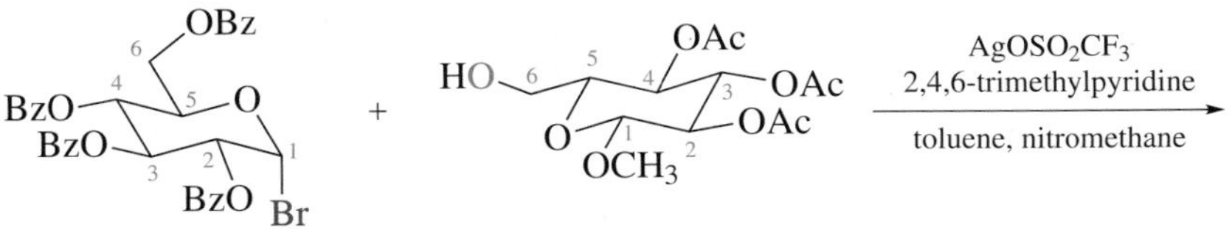

This method of silver-assisted glycoside synthesis is a variation of the *Koenigs–Knorr reaction.* William Koenigs and Edward Knorr first reported their method in 1901.

2,3,4,6-Tetra-*O*-benzoyl-α-D-glucopyranosyl bromide + Methyl 2,3,4-tri-*O*-acetyl-β-D-glucopyranoside

$AgOSO_2CF_3$, 2,4,6-trimethylpyridine; toluene, nitromethane →

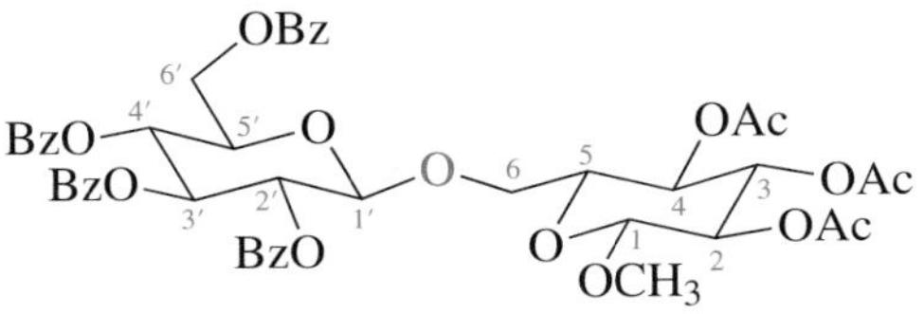

Methyl 2,3,4-tri-*O*-acetyl-2′,3′,4′,6′-tetra-*O*-benzoyl-β-gentiobioside (91%)

The coupling reaction is stereoselective for formation of the β-disaccharide. Removal of the ester protecting groups and hydrolysis of the methyl glycoside are required to complete the synthesis of gentiobiose.

The mechanism of the coupling reaction resembles an S_N1 reaction and begins with a silver-ion-assisted ionization of the carbon–bromine bond of the glycosyl donor to give a carbocation (Mechanism 23.3).

Mechanism 23.3

Silver-Assisted Glycosidation

Step 1: Silver ion acts as a Lewis acid to promote loss of bromide from the anomeric carbon giving an oxygen-stabilized carbocation.

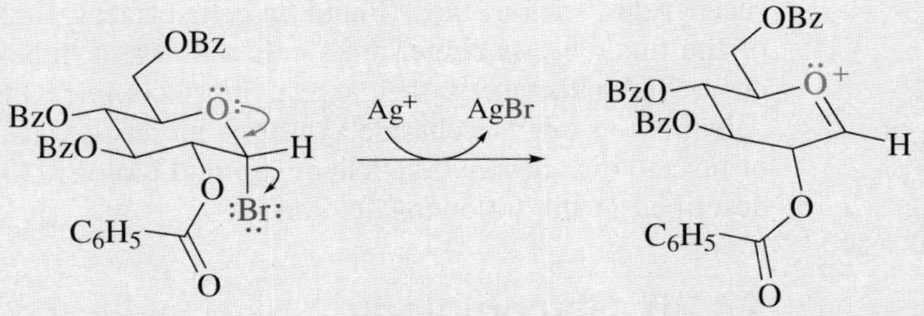

Ag^+ → AgBr

2,3,4,6-Tetra-*O*-benzoyl-α-D-glucopyranosyl bromide

Oxygen-stabilized carbocation

continued

Step 2: Once formed, this carbocation rearranges to a more stable structure, one that is stabilized by electron release from two oxygens.

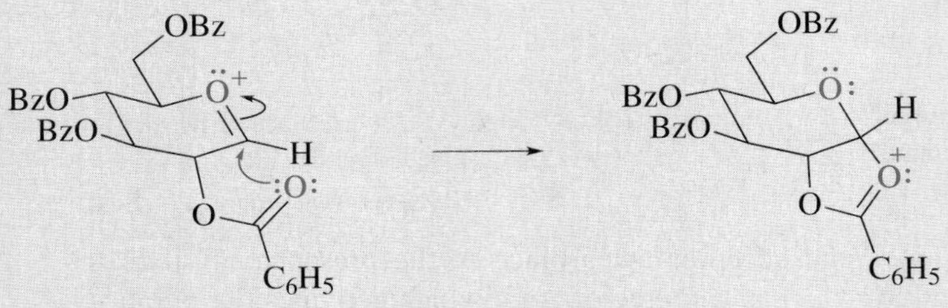

Oxygen-stabilized carbocation

A dioxolenium ion (more stable)

Step 3: The dioxolenium ion reacts with the free hydroxyl group of the glycosyl acceptor to give the conjugate acid of the disaccharide. Attack of the hydroxyl group occurs stereoselectively from the direction opposite the dioxolane ring.

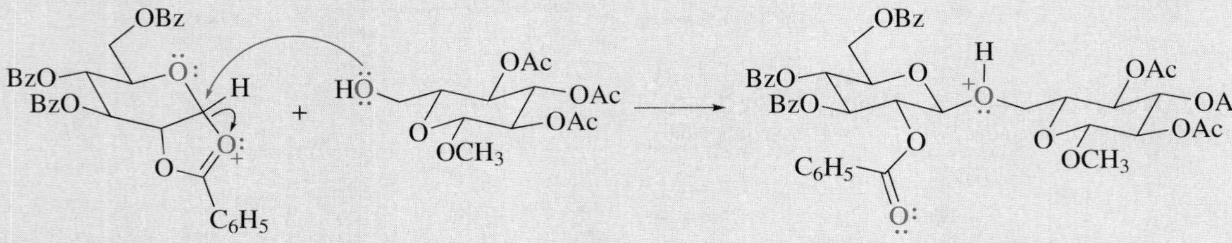

Dioxolenium ion

Methyl 2,3,4-tri-*O*-acetyl-β-D-glucopyranoside

Conjugate acid of β-disaccharide

Step 4: Subsequent to its formation, the conjugate acid is converted to the disaccharide by proton transfer to 2,4,6-trimethylpyridine, which forms a salt with the trifluoromethanesulfonate that is produced in step 1.

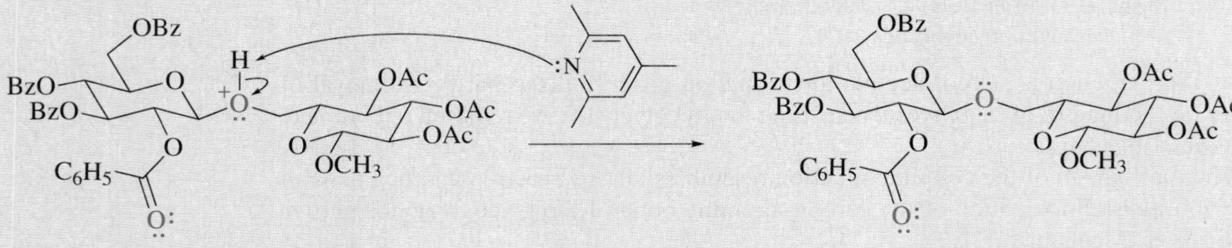

Conjugate acid of β-disaccharide

Methyl 2,3,4-tri-*O*-acetyl-2′,3′,4′,6′-tetra-*O*-benzoyl-β-gentiobioside

The synthesis of di- and trisaccharides was considered a major achievement, until the advent of new synthetic methods made it possible to construct more complex oligosaccharides, such as those found on cell surfaces. In 2017, an oligosaccharide component of the tuberculosis bacterium's cell wall was synthesized. The 92-sugar product, called an arabinogalactan, is the largest, most complex oligosaccharide synthesized to date. Advances in oligosaccharide synthesis are critical to the development of emerging areas of research in carbohydrate chemistry and biology. An exciting new area is **glycobiology,** described in the following section.

In 1947, half of the Nobel Prize in Physiology or Medicine was awarded jointly to Carl Ferdinand Cori and Gerty Theresa (Radnitz) Cori "for their discovery of the course of the catalytic conversion of glycogen." Hans Adolf Krebs was awarded half of the 1953 Nobel Prize in Physiology or Medicine for his discovery of the citric acid cycle.

23.20 Glycobiology

In the early to mid twentieth century, the study of carbohydrates concentrated on their role in energy production. Although a chemical linkage between carbohydrates and proteins was discovered in the late nineteenth century, the term "glycoconjugates" was not coined until the early 1970s. The study of the carbohydrates that occur with these conjugates is the subject of the field of *glycobiology*. A related term, *glycomics,* refers to the study of the complete set of carbohydrates in an organism and their genetic, physiological, and pathological roles. Part of the difficulty in studying these substances is the inherent structural complexity. Two D-glucopyranose molecules, for example, can be joined to form 11 different disaccharides. Although the complexity of oligosaccharides hampers their chemical synthesis, it allows them to serve highly diverse roles in organisms.

Glycoconjugates are widely distributed in living systems. Almost all of human plasma proteins are glycosylated. The outer surfaces of cells and plasma membranes are decorated with oligosaccharides attached to both protein and lipid sources, which are collectively referred to as the "glycocalyx." Bacteria cell walls are particularly dense in glycoproteins and glycolipids. The oligosaccharides of cell walls and membranes are important in cell–cell recognition, cell migration, and interaction with cells of the immune system. There is recent evidence that the composition of the glycocalyx is altered in some cancer cells, which can allow the cancer cells to evade the immune system and make them resistant to immunotherapy. Paradoxically, some glycoproteins, referred to as "tumor-associated antigens," are more highly expressed on the surface of cancer cells and may be exploited for targeted chemotherapy.

One example of the informational role of cell-surface carbohydrates occurs in the distinctions among human blood groups. The structure of the glycoproteins attached to the surface of blood cells determines whether the blood is type A, B, AB, or O. Differences among the carbohydrate components of the various glycoproteins have been identified and are shown in Figure 23.12. Compatibility of blood types is dictated by antigen–antibody interactions. The cell-surface glycoproteins are antigens. Antibodies present in certain blood types can cause the blood cells of certain other types to clump together, and thus set practical limitations on transfusion procedures. The antibodies recognize the antigens they act on by their terminal saccharide units.

Differences in the oligosaccharide composition of bacteria and humans have been exploited in the creation of vaccines. For example, many bacteria are decorated with capsular polysaccharides (CPS), which are long polysaccharide chains from the surface of the virulent bacteria. The CPS need to possess an exposed antigenic determinant that

Edward Jenner, an English physician, is credited with the development of a vaccine for smallpox in the 1790s. Louis Pasteur developed a vaccine against anthrax for use in cattle in the 1870s and coined the term *vaccine* from the Latin *vacca* for cow, in reference to the earlier work of Jenner.

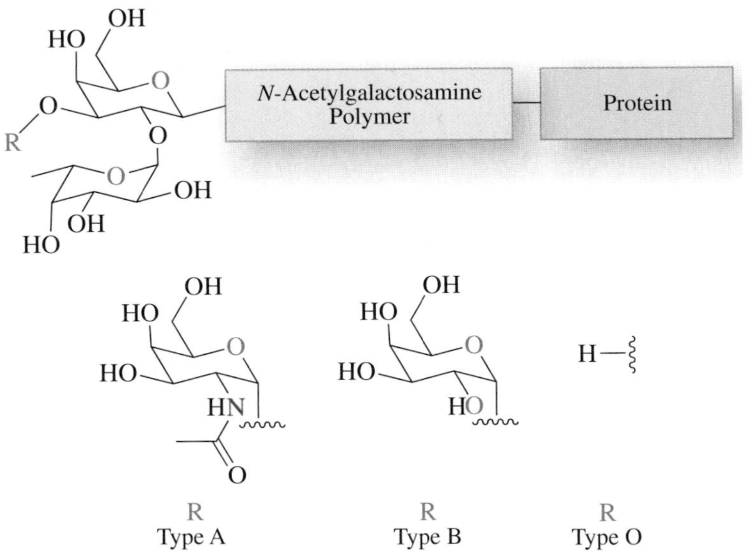

Figure 23.12

Terminal carbohydrate units of human blood-group glycoproteins. The structural difference between the type A, type B, and type O glycoproteins lies in the group designated R.

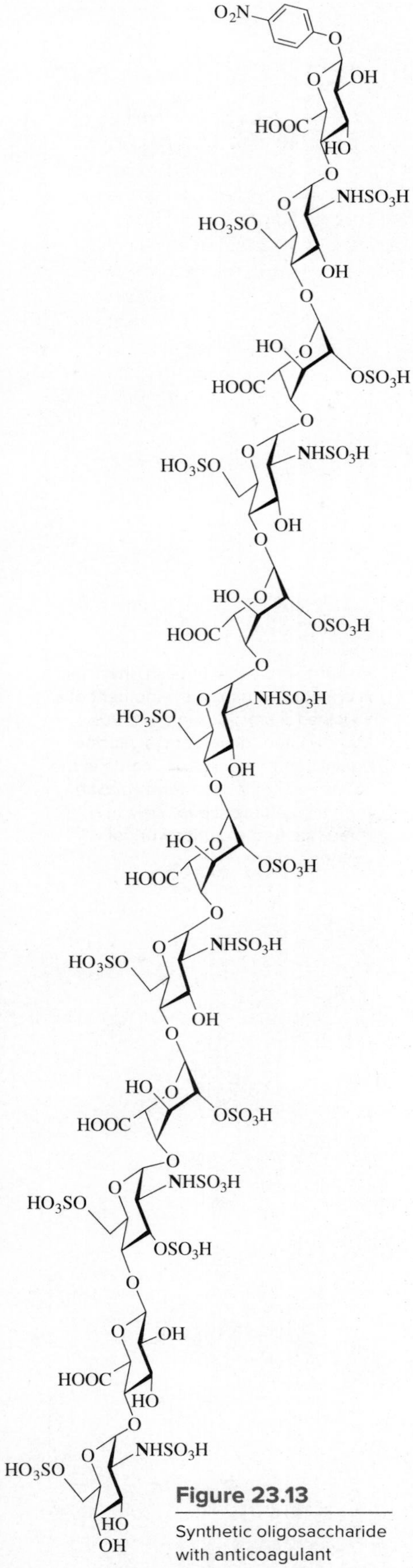

is not found in the corresponding mammalian glycan to promote an immune response. Pneumovax 23, a vaccine against a common bacterium that causes pneumonia, includes 23 different CPS antigens.

An important oligosaccharide used in medicine is the drug heparin, which is a linear polysaccharide of repeating sulfated glucosamine and uronic acid disaccharide units. Heparin is an anticoagulant drug used to prevent blood from clotting. Without heparin, many procedures in modern medicine would be impossible, since it would be impossible to keep blood fluid once it was removed from the human body in procedures like heart–lung oxygenation or kidney dialysis. Heparin is a natural product commonly prepared from animal tissues such as cow lung and pig intestines.

Recently organic and medicinal chemists have begun to prepare heparin oligosaccharides and polysaccharides through chemoenzymatic synthesis. In one such synthesis, a heparin dodecasaccharide (12 sugar residues) was prepared with anticoagulant activity for preclinical evaluation in a primate model. This synthesis was accomplished using a combination of chemical and enzymatic methods and afforded gram quantities of the desired dodecasaccharide shown in Figure 23.13 in 22 steps at 10% overall yield and in 98% purity.

New drugs to treat influenza were developed based on an understanding of the importance of cell-surface oligosaccharides in the life cycle of the viron. After infecting and replicating within a mammalian cell, the influenza virus produces an enzyme called neuraminidase (NA). NA cleaves *N*-acetylneuraminic acid (sialic acid) residues from the glycocalyx of the host cell and in respiratory mucus, which facilitates the release of the virus and propagation of its infection. Inhibitors of this enzyme prevent viral progeny from being released and thus slow the spread of infection. Two drugs that inhibit the activity of NA are oseltamivir (Tamiflu) and zanamivir (Relenza) (Figure 23.14).

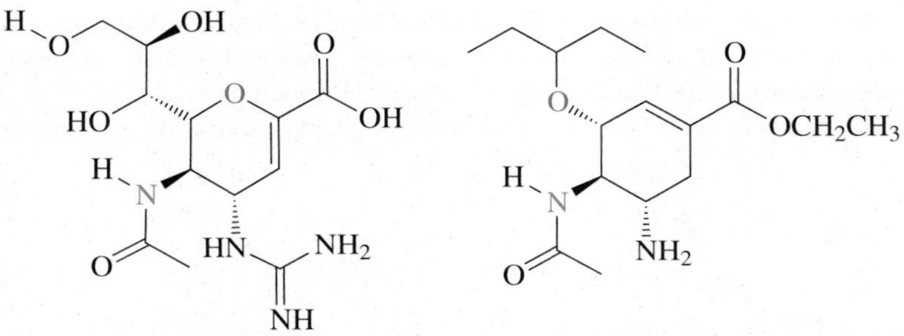

Zanamivir

Oseltamavir

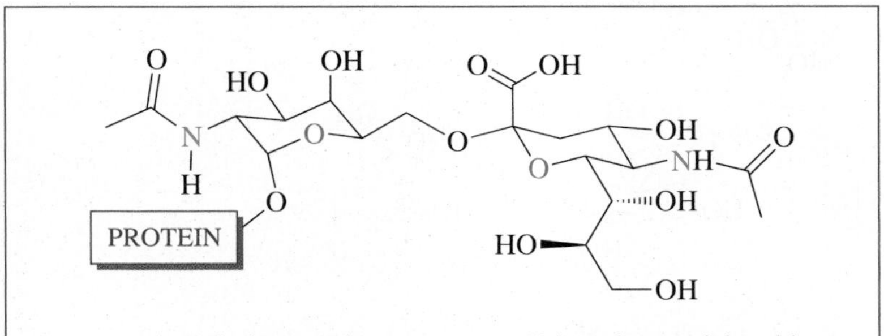

Figure 23.13

Synthetic oligosaccharide with anticoagulant activity.

PROTEIN

N-Acetylgalactosamine

N-Acetylneuraminic acid

Figure 23.14

Diagram of a cell-surface glycoprotein, showing the disaccharide unit that is recognized by the viral neuraminidase enzyme.

Oligosaccharides in Infectious Disease

Staphylococcus aureus is a Gram-positive bacterium that causes both acute and chronic infections in humans. Chronic infections are particularly problematic in host tissues, such as heart valves, and in implanted medical devices, such as prosthetic joints. Chronic *S. aureus* infections are associated with *biofilms,* communities of surface-associated bacteria embedded in a self-produced, hydrated extracellular polymeric matrix. Although bacterial biofilms have been known for over 100 years, their importance in human disease was not recognized until the mid-1970s. Microbiologist J. William Costerton, considered to be the "father of biofilms," proposed that bacterial biofilms are protected from the immune system and the action of antimicrobial agents. Biofilm cells have been shown to be 10–1000-fold less susceptible to various antimicrobial agents than their planktonic (free-floating) counterparts. It is thus not surprising that conventional therapies have proven inadequate in the treatment of many (if not most) chronic biofilm infections due to the extraordinary resistance of biofilms to antimicrobial treatments. Searching for ways to prevent or eradicate biofilms is a very active area of biomedical research. One strategy involves the disaggregation of the bacterial biofilm structure, mainly by attacking the extracellular polymeric matrix that surrounds the biofilm.

Biofilms are encased or embedded in a complex mixture of oligosaccharides, proteins, and extracellular DNA (eDNA). The main oligosaccharide component of the *S. aureus* biofilm matrix is a β-1,6-linked *N*-acetylglucosamine polymer called oligosaccharide intercellular adhesin (PIA). One approach for disruption of the biofilm architecture is to break down the PIA and disperse the individual bacteria, which can then be attacked by the immune system. Dispersion B, a 361-amino-acid protein produced by *Aggregatibacter actinomycetemcomitans*, degrades *S. aureus* biofilms by hydrolyzing the glycosidic linkages of PIA. Clinical uses being explored for Dispersion B include topical application to wounds, coating for medical devices, and cleansing of contact lenses.

β-1,6-linked *N*-acetylglucosamine oligosaccharide

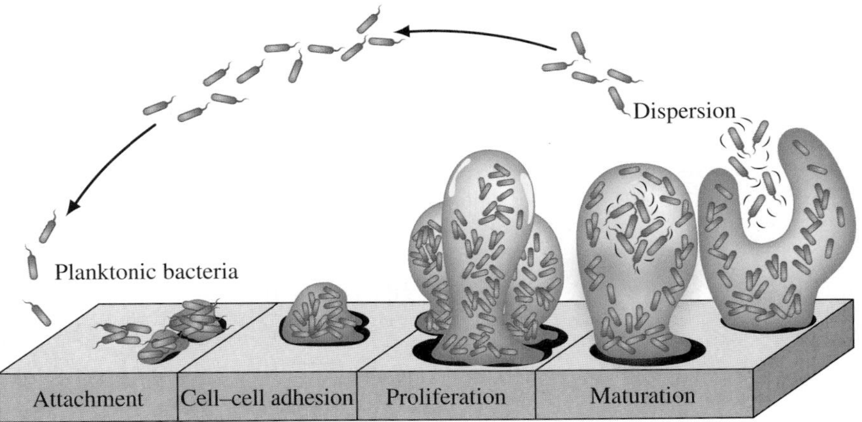

A model of biofilm development. At stage 1, the bacterial cells attach reversibly to the surface. Then, at stage 2, the cells attach irreversibly, a step mediated mainly by exopolymeric substances. At stage 3, the first maturation phase is reached, as indicated by early development of biofilm architecture. The second maturation phase is reached at stage 4 with fully mature biofilms, as indicated by the complex biofilm architecture. At the dispersion stage (stage 5), single motile cells disperse from the microcolonies. Adapted from P. Stoodely et al, "Biofilms as Complex Differentiated Communities," *Annual Review of Microbiology,* 56, 187–209 (2002).

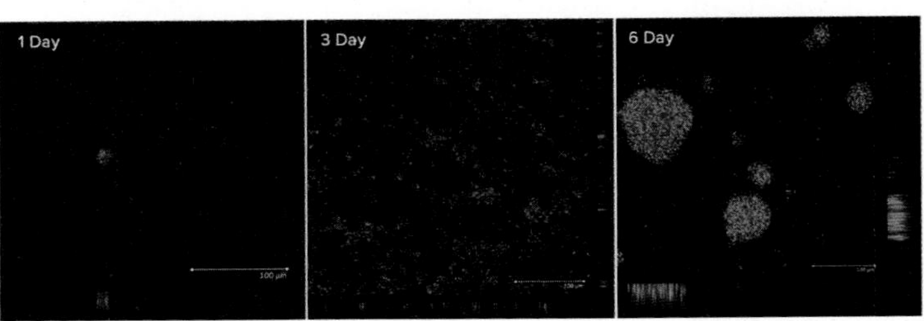

Biofilm formation by *Pseudomonas aeruginosa*. Confocal images were acquired following one, three, and six days of biofilm growth under flowing conditions. White bar = 100 μm.
Images courtesy of Prof. Karin Sauer

23.21 SUMMARY

Section 23.1 Carbohydrates are marvelous molecules! In most of them, every carbon bears a functional group, and the nature of the functional groups changes as the molecule interconverts between open-chain and cyclic hemiacetal forms. Any approach to understanding carbohydrates must begin with structure.

Carbohydrates are polyhydroxy aldehydes and ketones. Those derived from aldehydes are classified as **aldoses;** those derived from ketones are **ketoses.**

Section 23.2 Fischer projections and D,L notation are commonly used to describe carbohydrate stereochemistry. The standards are the enantiomers of glyceraldehyde.

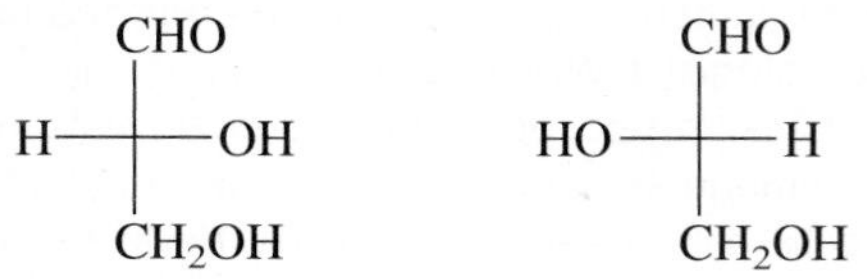

(+)-D-Glyceraldehyde (−)-L-Glyceraldehyde

Section 23.3 Aldotetroses have two chirality centers, so four stereoisomers are possible. They are assigned to the D or the L series according to whether the configuration at their highest numbered chirality center is analogous to D- or L-glyceraldehyde, respectively. Both hydroxyl groups are on the same side of the Fischer projection in erythrose, but on opposite sides in threose. The Fischer projections of D-erythrose and D-threose are shown in Figure 23.2.

Section 23.4 Of the eight stereoisomeric aldopentoses, Figure 23.2 shows the Fischer projections of the D-enantiomers (D-ribose, D-arabinose, D-xylose, and D-lyxose). Likewise, Figure 23.2 gives the Fischer projections of the eight D-aldohexoses.

Section 23.5 The aldohexoses are allose, altrose, glucose, mannose, gulose, idose, galactose, and talose. The mnemonic "All altruists gladly make gum in gallon tanks" is helpful in writing the correct Fischer projection for each one.

Sections 23.6–23.7 Most carbohydrates exist as cyclic hemiacetals. Those with five-membered rings are called **furanose forms;** those with six-membered rings are called **pyranose forms.**

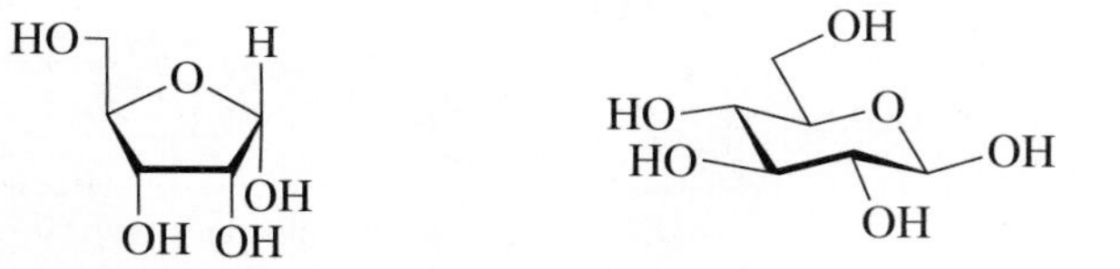

α-D-Ribofuranose β-D-Glucopyranose

The **anomeric carbon** in a cyclic hemiacetal is the one attached to *two* oxygens. It is the carbon that corresponds to the carbonyl carbon in the open-chain form. The symbols α and β refer to the configuration at the anomeric carbon.

Section 23.8 Hemiacetal forms of carbohydrates are interconvertible in water. The equilibrium mixture can contain α and β anomers of furanose and pyranose forms. The change from one form to the equilibrium mixture is accompanied by a change in optical rotation called **mutarotation.** For D-glucose, mutarotation can be described in terms of the interconversion of α-pyranose and β-pyranose forms by way of the open-chain form.

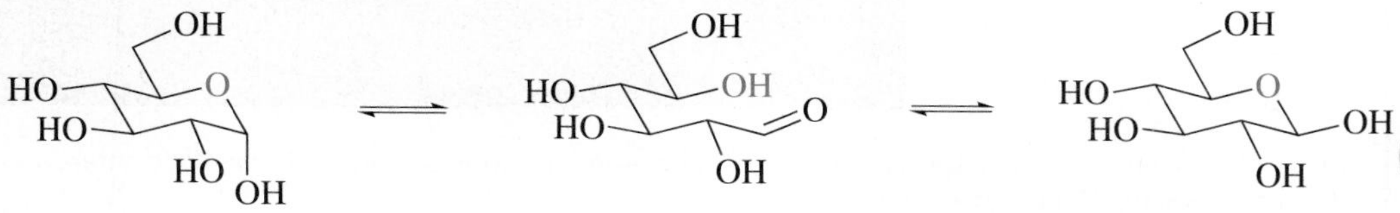

α-D-Glucopyranose ⇌ Open-chain form ⇌ β-D-Glucopyranose

Section 23.9 Pyranose forms of carbohydrates resemble cyclohexane in their conformational preference for chair forms. The **anomeric effect** causes an electronegative substituent at the anomeric (C-1) carbon to be more stable when it is axial than when it is equatorial. The effect is believed to result from the delocalization of an electron pair of the ring oxygen into an antibonding orbital of the anomeric substituent.

Section 23.10 Most naturally occurring ketoses have their carbonyl group located at C-2. Like aldoses, ketoses cyclize to hemiacetals and exist as furanose or pyranose forms.

Sections 23.11–23.13 Structurally modified carbohydrates include **deoxy sugars, amino sugars,** and **branched-chain carbohydrates.**

Section 23.14 Glycosides are acetals, compounds in which the anomeric hydroxyl group has been replaced by an alkoxy group. Glycosides are easily prepared by allowing a carbohydrate and an alcohol to stand in the presence of an acid catalyst.

D-Glucose + ROH $\xrightarrow{H^+}$ [structure: HO, HO, OH, O, HO, OR]

A glycoside

Sections 23.15–23.16 **Disaccharides** are carbohydrates in which two monosaccharides are joined by a glycosidic bond. **Polysaccharides** have many monosaccharide units connected through glycosidic linkages. Complete hydrolysis of disaccharides and polysaccharides cleaves the glycosidic bonds, yielding the free monosaccharide components.

Sections 23.17–23.18 Carbohydrates undergo chemical reactions characteristic of aldehydes and ketones, alcohols, diols, and other classes of compounds, depending on their structure. A review of the reactions described in this chapter is presented in Table 23.2.

Section 23.19 Carbohydrates linked by *O*-glycosidic linkages such as those found in disaccharides are synthesized by the reaction of a glycosyl donor and a glycosyl acceptor.

Section 23.20 Carbohydrates linked to proteins, termed glycoproteins, are one example of glycoconjugates in which a carbohydrate is attached to some other biomolecule. Cell-surface glycoproteins are involved in molecular recognition and are important in the immune system as well as in certain diseases.

PROBLEMS

Structure

23.26 Refer to the Fischer projection of (+)-D-xylose and give structural formulas for

(a) (−)-Xylose (Fischer projection)
(b) Xylitol
(c) β-D-Xylopyranose
(d) α-L-Xylofuranose
(e) Methyl α-L-xylofuranoside
(f) D-Xylonic acid (open-chain Fischer projection)
(g) δ-Lactone of D-xylonic acid
(h) γ-Lactone of D-xylonic acid
(i) Xylaric acid (open-chain Fischer projection)

	CHO	
H		OH
HO		H
H		OH
	CH_2OH	

(+)-D-Xylose

23.27 From among the carbohydrates shown in Problem 23.26, which are reducing sugars?

23.28 What are the *R,S* configurations of the three chirality centers in D-xylose?

23.29 From among the carbohydrates shown in Figure 23.2, choose the D-aldohexoses that yield

(a) An optically inactive product on reduction with sodium borohydride

(b) An optically inactive product on oxidation with bromine

(c) An optically inactive product on oxidation with nitric acid

(d) The same enediol

23.30 Write the Fischer projection of the open-chain form of each of the following:

(a) (c) (b) (d)

23.31 From among the carbohydrates shown in Problem 23.30, choose the one(s) that

(a) Belong to the L series

(b) Are deoxy sugars

(c) Are branched-chain sugars

(d) Are ketoses

(e) Are furanose forms

(f) Have the α configuration at their anomeric carbon

23.32 How many ketopentoses are possible? Write their Fischer projections.

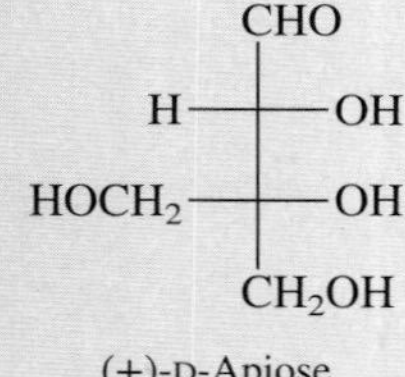

(+)-D-Apiose

23.33 Given the Fischer projection of the branched-chain carbohydrate (+)-D-apiose:

(a) How many chirality centers are in the open-chain form of D-apiose?

(b) Does D-apiose form an optically active alditol on reduction?

(c) How many chirality centers are in the furanose forms of D-apiose?

(d) How many stereoisomeric furanose forms of D-apiose are possible? Write their Haworth formulas.

23.34 The following are the more stable anomers of the pyranose forms of D-glucose, D-mannose, and D-galactose:

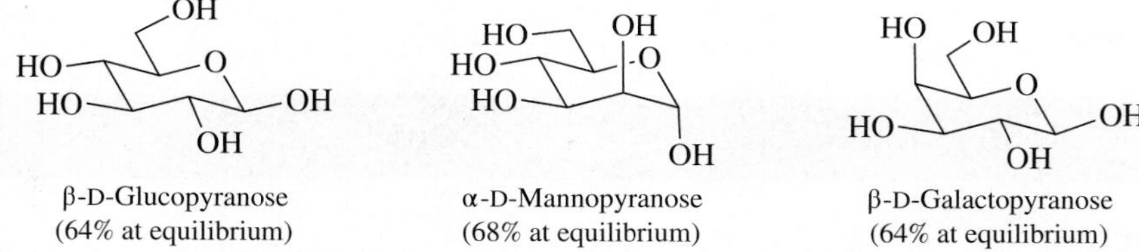

β-D-Glucopyranose (64% at equilibrium) α-D-Mannopyranose (68% at equilibrium) β-D-Galactopyranose (64% at equilibrium)

On the basis of these empirical observations and your own knowledge of steric effects in six-membered rings, predict the preferred form (α- or β-pyranose) at equilibrium in aqueous solution for each of the following:

(a) D-Gulose

(b) D-Talose

(c) D-Xylose

(d) D-Lyxose

23.35 The compound shown is the anticonvulsant drug topiramate. It is a derivative of D-fructopyranose. Identify the acetal carbons in topiramate.

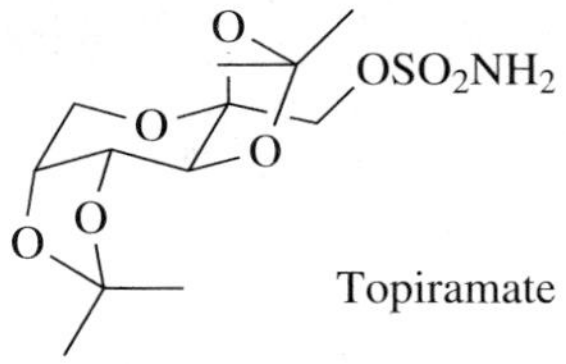

Topiramate

23.36 The γ-lactone of D-gulonic acid was prepared by way of a cyanohydrin derived from an aldopentose.

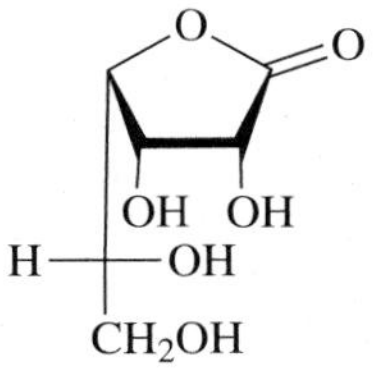

Identify the aldopentose subjected to this chain extension.

23.37 Maltose and cellobiose (Section 23.15) are examples of disaccharides derived from D-glucopyranosyl units joined in an *O*-glycosidic linkage. (a) How many other disaccharides are possible that meet this structural requirement? (b) How many of these are reducing sugars?

Mechanism

23.38 Methyl glycosides of 2-deoxy sugars have been prepared by the acid-catalyzed addition of methanol to unsaturated carbohydrates known as *glycals*.

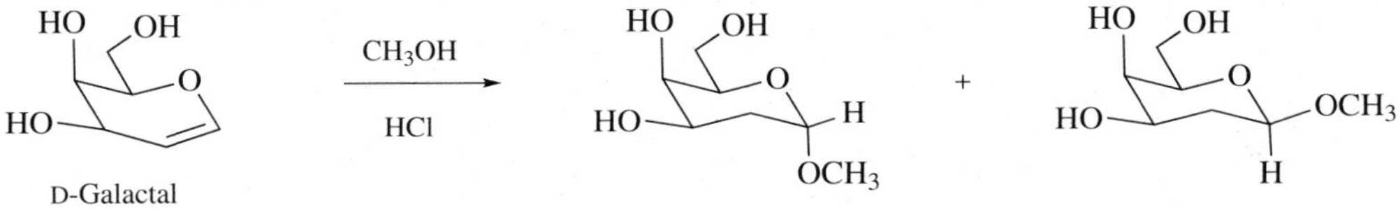

D-Galactal $\xrightarrow[\text{HCl}]{CH_3OH}$ Methyl 2-deoxy-α-D-lyxopyranoside (38%) + Methyl 2-deoxy-β-D-lyxopyranoside (36%)

Suggest a reasonable mechanism for this reaction.

23.39 Basing your answers on the general mechanism for the first stage of acid-catalyzed acetal hydrolysis, suggest reasonable explanations for the following observations:

(a) Methyl α-D-fructofuranoside (compound A) undergoes acid-catalyzed hydrolysis some 10^5 times faster than methyl α-D-glucofuranoside (compound B).

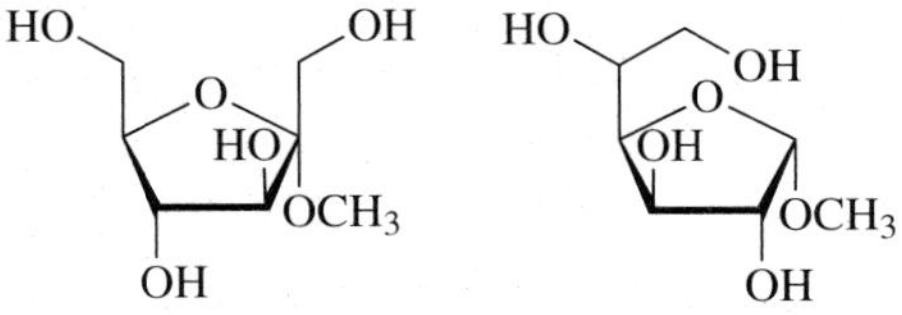

Compound A Compound B

(b) The β-methyl glucopyranoside of 2-deoxy-D-glucose (compound C) undergoes hydrolysis several thousand times faster than that of D-glucose (compound D).

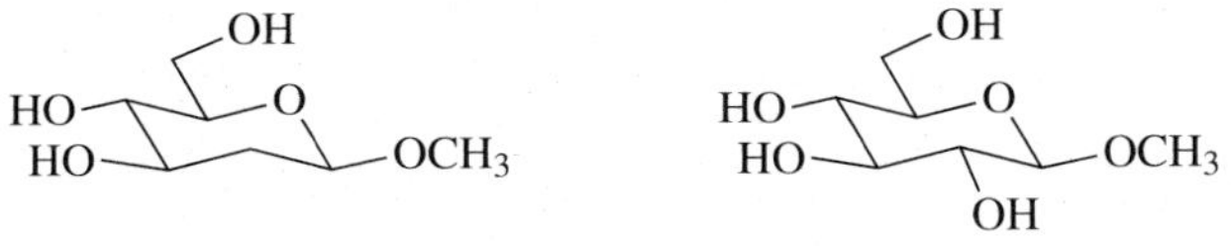

Compound C Compound D

23.40 Acetone reacts with carbohydrates in the presence of an acid catalyst to form products that are commonly referred to as "isopropylidene" or "acetonide" derivatives. The carbohydrate D-ribono-(1,4)-lactone reacts with acetone in the presence of hydrochloric acid to give the acetonide shown here. Write a mechanism for this reaction. (*Hint:* Review Section 18.9 and Problem 18.13.)

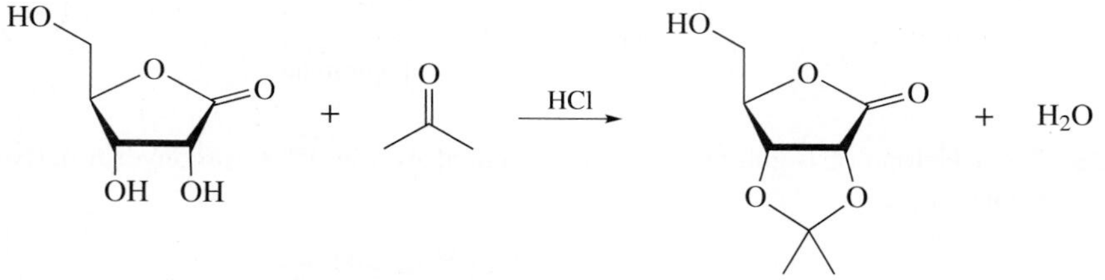

HO, O, O, OH OH + O (acetone) $\xrightarrow{HCl}$ HO, O, O, O, O + H_2O

23.41 In each of the following reactions, the glycosyl donor is activated by reaction with an electrophilic reagent by the general pathway:

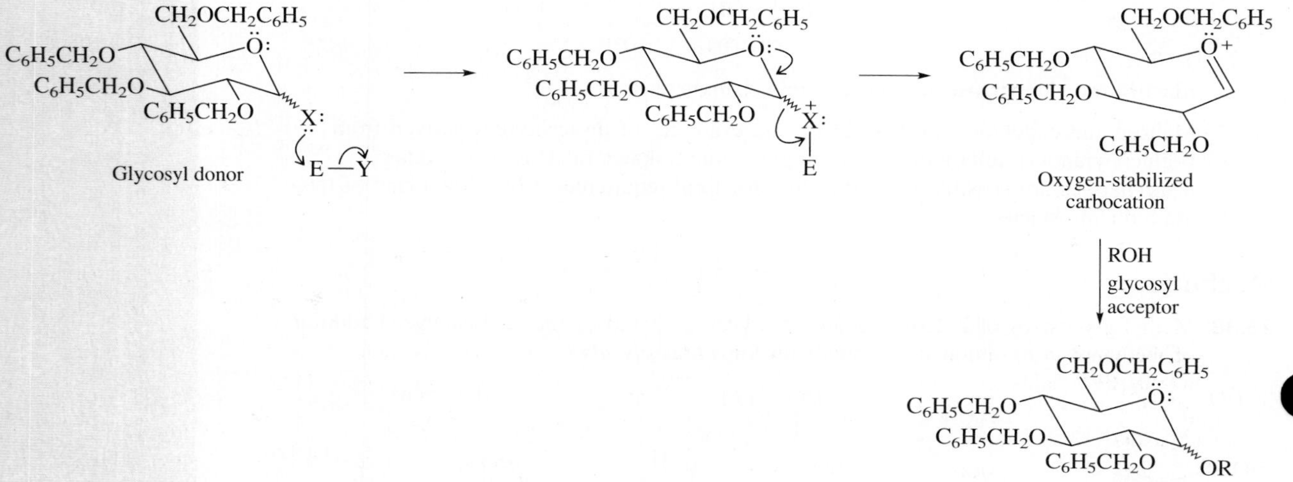

$CH_2OCH_2C_6H_5$, $C_6H_5CH_2O$, $C_6H_5CH_2O$, $C_6H_5CH_2O$, Ö:, X:, E—Y

Glycosyl donor

→ $CH_2OCH_2C_6H_5$, $C_6H_5CH_2O$, $C_6H_5CH_2O$, $C_6H_5CH_2O$, Ö:, $\overset{+}{X}$:, E → $CH_2OCH_2C_6H_5$, $C_6H_5CH_2O$, $C_6H_5CH_2O$, $C_6H_5CH_2O$, Ö+

Oxygen-stabilized carbocation

ROH glycosyl acceptor ↓

$CH_2OCH_2C_6H_5$, $C_6H_5CH_2O$, $C_6H_5CH_2O$, $C_6H_5CH_2O$, Ö:, OR

The oxygen-stabilized carbocation that is formed reacts with the glycosyl acceptor ROH to form an anomeric mixture of *O*-glycosides.

Using curved arrows, write a mechanism to show how each of the glycosyl donors produces the oxygen-stabilized carbocation. For part (a) consult Section 17.16 on sulfides and Section 5.15 on sulfonates; for part (b) examine step 2 in Mechanism 20.6, and assume that silylation occurs at nitrogen in the glycosyl imidate ester.

(a) $OCH_2C_6H_5$, $C_6H_5CH_2O$, $C_6H_5CH_2O$, $C_6H_5CH_2O$, O, SCH_2CH_3 $\xrightarrow[ROH]{F_3CSO_2OCH_3}$ $OCH_2C_6H_5$, $C_6H_5CH_2O$, $C_6H_5CH_2O$, $C_6H_5CH_2O$, O, OR + $CH_3CH_2SCH_3$

Thioglycoside — α/β Glycoside mixture — Ethyl methyl sulfide

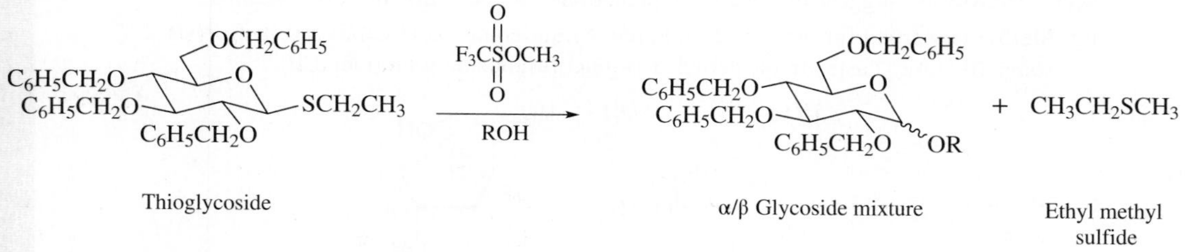

(b) $OCH_2C_6H_5$, $C_6H_5CH_2O$, $C_6H_5CH_2O$, $C_6H_5CH_2O$, O, O–C(=NH)–CCl_3 $\xrightarrow[ROH]{F_3CSO_2OSi(CH_3)_3}$ $OCH_2C_6H_5$, $C_6H_5CH_2O$, $C_6H_5CH_2O$, $C_6H_5CH_2O$, O, OR + Cl_3CCNH_2 (C=O)

Glycosyl imidate ester — α/β Glycoside mixture — Trichloroacetamide

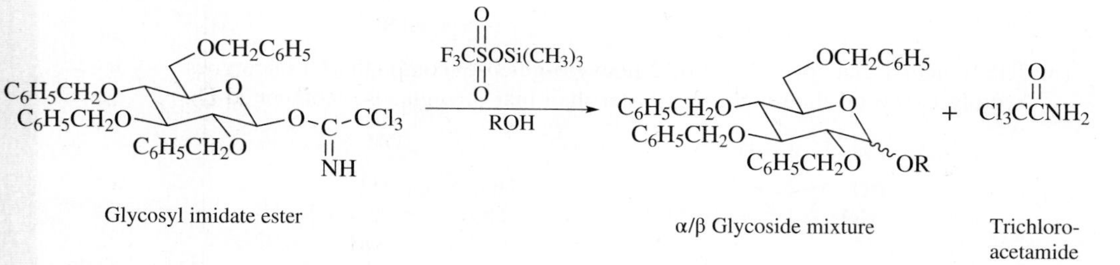

23.42 Pentenyl glycosides are glycosyl donors as illustrated by their conversion to glycosyl bromides with bromine. Write a mechanism for this reaction. Why is the α-anomer formed selectively?

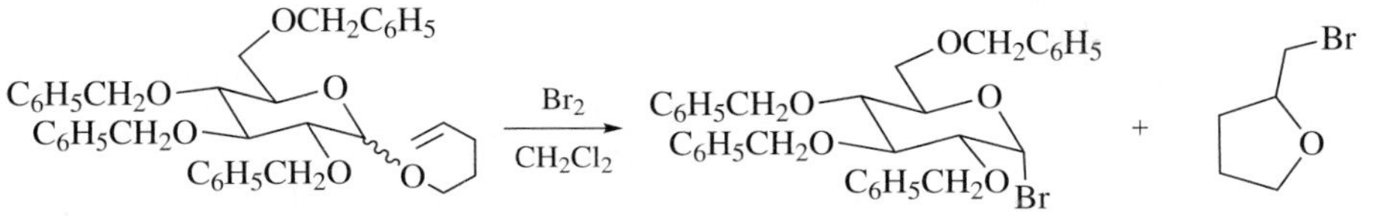

$OCH_2C_6H_5$; $C_6H_5CH_2O$; $C_6H_5CH_2O$; $C_6H_5CH_2O$; O; O — Br_2, CH_2Cl_2 → $OCH_2C_6H_5$; $C_6H_5CH_2O$; $C_6H_5CH_2O$; $C_6H_5CH_2O$; O; Br + Br; O

Pentenyl glycoside — α Glycosyl bromide (90%) — 2-(Bromomethyl)-tetrahydrofuran

Reactions

23.43 Treatment of D-mannose with methanol in the presence of an acid catalyst yields four isomeric products having the molecular formula $C_7H_{14}O_6$. What are these four products?

23.44 Give the products of periodic acid oxidation of each of the following. How many moles of reagent will be consumed per mole of substrate in each case?

(a) D-Arabinose

(b) D-Ribose

(c) Methyl β-D-glucopyranoside

(d) 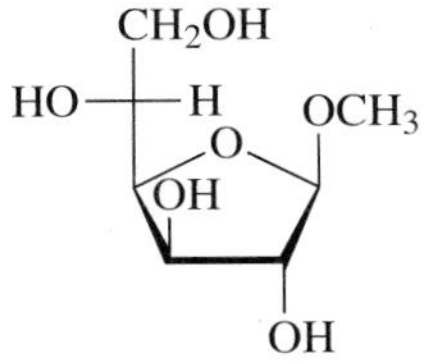 CH_2OH; HO—H; OCH_3; O; OH; OH

23.45 Triphenylmethyl chloride reacts with primary alcohols faster than with secondary ones, making the reaction selective for the hydroxyl group at C-6 in the following case. The ether can be removed by mild acid hydrolysis under conditions that leave the methyl glycoside intact.

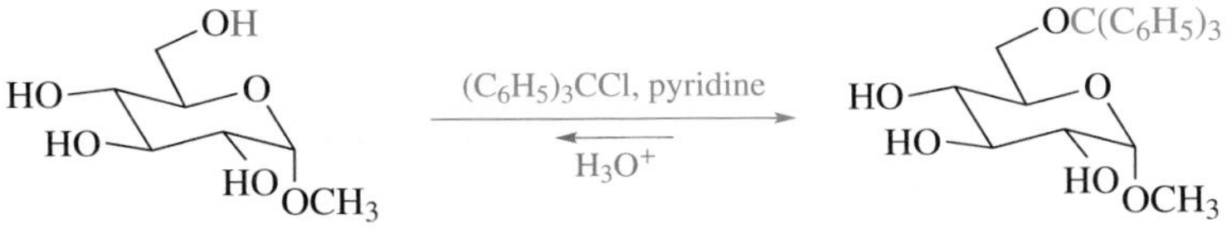

OH; HO; HO; O; HO; OCH_3 — $(C_6H_5)_3CCl$, pyridine → ; ← H_3O^+ — $OC(C_6H_5)_3$; HO; HO; O; HO; OCH_3

Using the above information and any additional information from Table 23.2, show how the compound on the left could be synthesized from the indicated starting material.

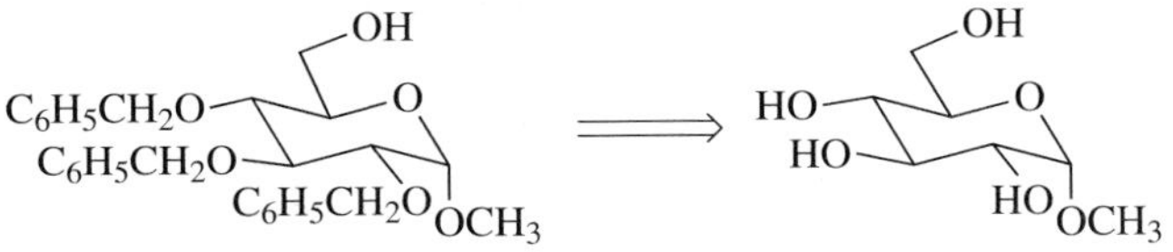

OH; $C_6H_5CH_2O$; $C_6H_5CH_2O$; O; $C_6H_5CH_2O$; OCH_3 ⟹ OH; HO; HO; O; HO; OCH_3

23.46 Compound A was oxidized with periodic acid to give B, which after acid hydrolysis gave C. Bromine oxidation of C gave D. Suggest structural formulas, including stereochemistry, for compounds B, C, and D.

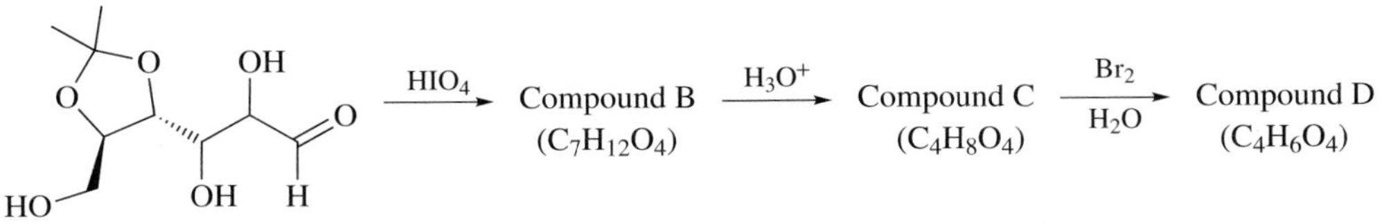

O; O; OH; O; HO; OH; H — HIO_4 → Compound B ($C_7H_{12}O_4$) — H_3O^+ → Compound C ($C_4H_8O_4$) — Br_2, H_2O → Compound D ($C_4H_6O_4$)

Compound A

Descriptive Passage and Interpretive Problems 23

Emil Fischer and the Structure of (+)-Glucose

Emil Fischer's determination of the structure of glucose was carried out as the nineteenth century ended and the twentieth began. The structure of no other sugar was known at that time, and the spectroscopic techniques that now aid organic analysis were not yet available. All Fischer had was information from chemical transformations, polarimetry, and his own intellect.

Fischer knew that (+)-glucose was one of 16 possible stereoisomers having the constitution:

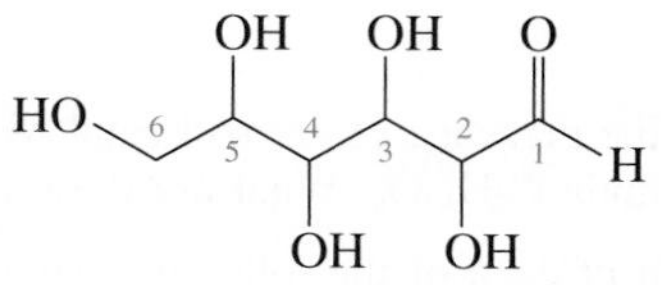

C-2, 3, 4, and 5 are chirality centers of unknown configuration.

By arbitrarily assigning a particular configuration to the chirality center at C-5, Fischer realized that he could determine the configurations of C-2, C-3, and C-4 *relative* to C-5. This reduces the number of structural possibilities to the eight that we now call D-hexoses.

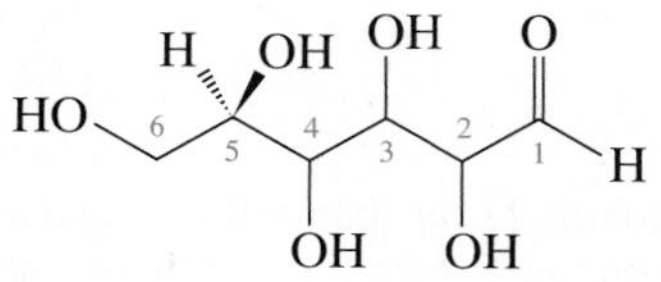

or, as in a Fischer projection

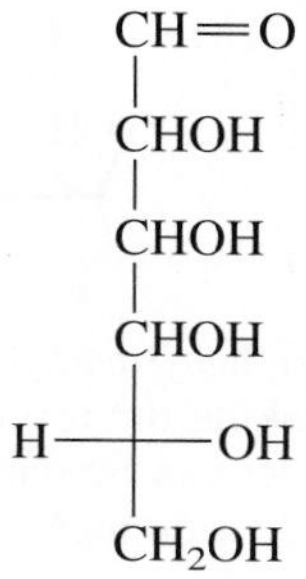

Eventually, Fischer's arbitrary assignment proved correct, which made his stereochemical assignments for all of the chirality centers of (+)-glucose correct in an absolute as well as a relative sense.

The following problems lead you through Fischer's interpretation of the information available to him in determining the structure of (+)-glucose. The order in which the facts are presented is modified slightly from Fischer's, but the logic is the same. We'll begin in Problem 23.47 with (−)-arabinose, a pentose having the same configuration at its highest numbered chirality center as (+)-glucose, a fact that emerges in Problem 23.48.

(−)-Arabinose and (+)-glucose have the same configuration at their highest numbered chirality center. This fact is established in Problem 23.48 but used in Problem 23.47.

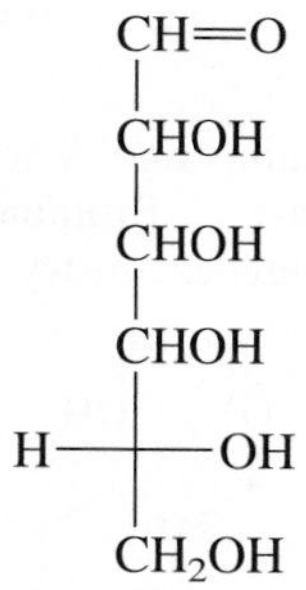

D-Pentoses including (−)-arabinose

D-Hexoses including (+)-glucose

23.47 Oxidation of (−)-arabinose with warm nitric acid gave an optically active aldaric acid. In this reaction, both C-1 and C-5 are oxidized to CO_2H. Assuming the C-4 OH is to the right, which two of the structures shown are possible for (−)-arabinose?

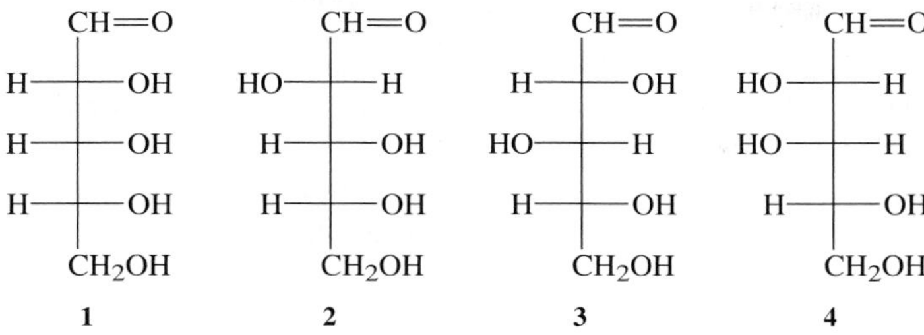

A. **1** and **2**
B. **1** and **3**
C. **1** and **4**
D. **2** and **3**
E. **2** and **4**
F. **3** and **4**

23.48 Chain extension of (−)-arabinose by way of its derived cyanohydrin gave a mixture of (+)-glucose and (+)-mannose. Based on this observation *and* your answer to the preceding problem, which pairs are possible for (+)-mannose and (+)-glucose?

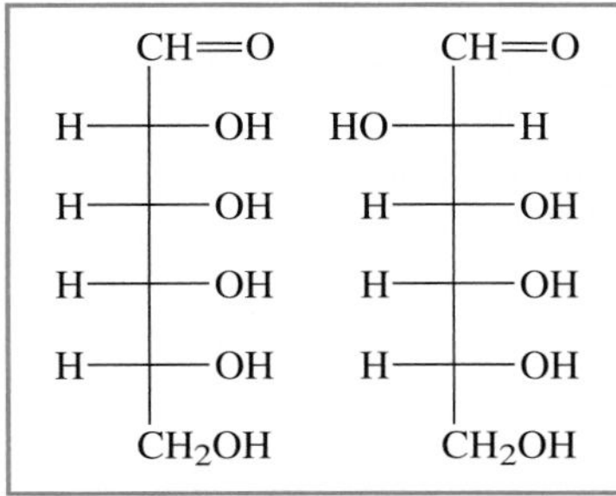

Pair 1

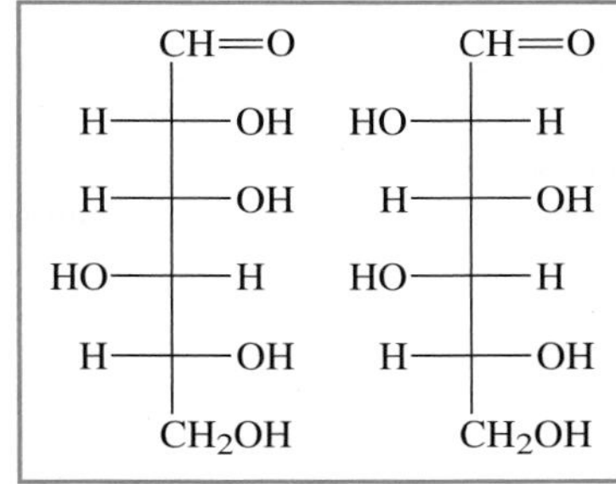

Pair 2

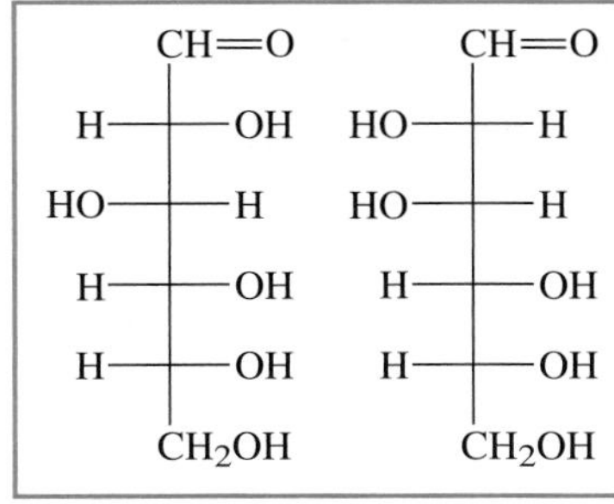

Pair 3

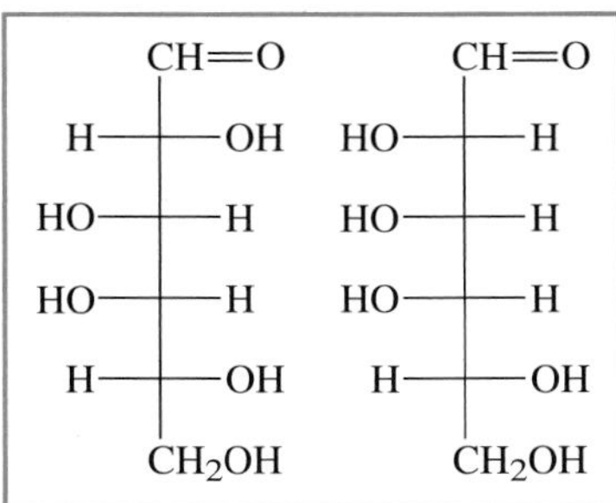

Pair 4

A. Pair 1 and pair 2
B. Pair 1 and pair 3
C. Pair 1 and pair 4
D. Pair 2 and pair 3
E. Pair 2 and pair 4
F. Pair 3 and pair 4

23.49 Both (+)-glucose and (+)-mannose were oxidized to optically active dicarboxylic acids (aldaric acids) ($C_6H_{14}O_4$) with nitric acid. Of the pairs remaining after solving the preceding problem, which one is the (+)-glucose/(+)-mannose pair?

A. Pair 1
B. Pair 2
C. Pair 3
D. Pair 4

In order to do the next problem, you need to know that pair 3 is the correct answer to Problem 23.48. If you are not certain about how this answer is arrived at, it would be a good idea to review the previous questions.

23.50 Because both C-1 and C-6 are oxidized to —CO_2H groups by nitric acid, Fischer recognized that two diastereomeric hexoses could give the same aldaric acid.

CH=O		[6] CH_2OH		CO_2H
CHOH		CHOH		CHOH
CHOH	or	CHOH	$\xrightarrow{HNO_3}$	CHOH
CHOH		CHOH		CHOH
CHOH		CHOH		CHOH
[6] CH_2OH		[1] CH=O		CO_2H

Of the (+)-glucose/(+)-mannose pair (pair 3 in Problem 23.48), only (+)-glucose has a diastereomeric hexose that gives the same aldaric acid. Fischer synthesized that specific diastereomer and found it gave the same aldaric acid as (+)-glucose. Thus, he was able to determine that (+)-glucose and (+)-mannose are:

(+)-Glucose		(+)-Mannose	
CH=O		CH=O	
H	OH	HO	H
HO	H	HO	H
H	OH	H	OH
H	OH	H	OH
CH_2OH		CH_2OH	

Which hexose did Fischer synthesize that gave the same aldaric acid as (+)-glucose?

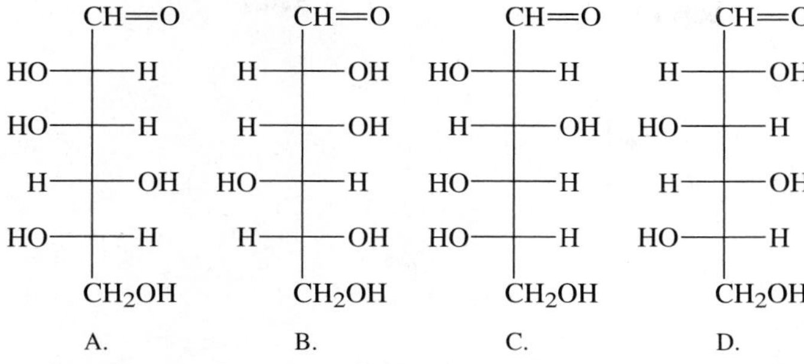

23.51 Refer to Table 23.2 to identify the hexose that is the answer to Problem 23.50.

A. (+)-D-Altrose
B. (+)-D-Galactose
C. (−)-L-Glucose
D. (+)-L-Gulose

Chapter

24

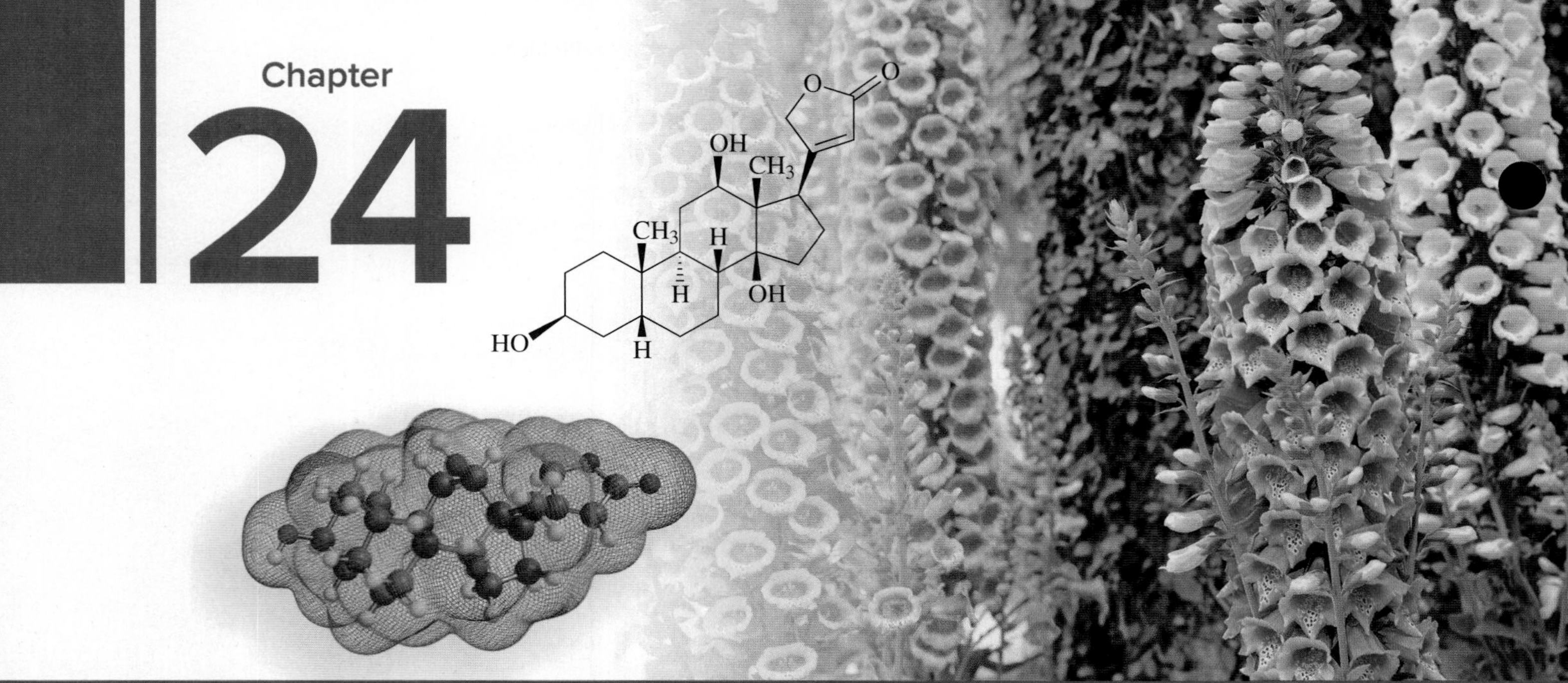

CHAPTER OUTLINE

Digoxigenin is the steroid component of digoxin, a glycoside found in the flowers called foxgloves. Digoxin is used to treat heart arrhythmia and fibrillation. Like all steroids, digoxigenin is a lipid.

Lipids

Among the major classes of naturally occurring biomolecules (lipids, carbohydrates, proteins, and nucleic acids), **lipids** differ in being defined by a physical property—solubility. They are more soluble in nonpolar solvents than in water. Lipids include a variety of structural types, a collection of which is introduced in this chapter. Fatty acids, terpenes, steroids, prostaglandins, and carotenes are all lipids but are very different from one another in both structure and function. They share a common biosynthetic origin in that all are ultimately derived from glucose. During glycolysis, glucose is converted to lactic acid by way of pyruvic acid.

Glucose → Pyruvic acid → (*S*)-Lactic acid

The pathway leading to lactic acid and beyond is concerned with energy storage and production, but some pyruvic acid is converted to acetic acid and used as a starting material in the biosynthesis of more complex substances, especially lipids. This chapter is organized around that theme. We'll begin by looking at the reaction in which acetate (two carbons) is formed from pyruvate (three carbons).

24.1 Acetyl Coenzyme A

Acetate is furnished in most of its important biochemical reactions as its thioester **acetyl coenzyme A** (Figure 24.1*a*). Its formation from pyruvate involves the enzymes of the pyruvate decarboxylase complex summarized as:

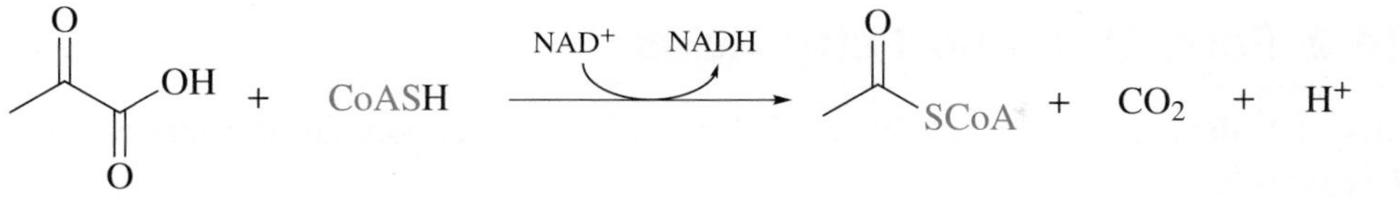

Pyruvic acid + CoASH (Coenzyme A) —($NAD^+ \rightarrow NADH$)→ Acetyl coenzyme A (SCoA) + CO_2 (Carbon dioxide) + H^+ (Proton)

Four cofactors are required, including NAD^+ (see Section 16.10) as an oxidizing agent and coenzyme A (Figure 24.1*b*) as the acetyl group acceptor. Coenzyme A is a thiol; its chain terminates in a sulfhydryl (—SH) group. Acetylation of the sulfhydryl group of coenzyme A gives acetyl coenzyme A.

Thioesters are both more reactive than ordinary esters toward nucleophilic acyl substitution and also contain a greater proportion of enol at equilibrium. Both properties are apparent in the properties of acetyl coenzyme A. In some reactions it is the C=O function that reacts; in others it is the α-carbon atom.

Coenzyme A was isolated and identified by Fritz Lipmann, an American biochemist who shared the 1953 Nobel Prize in Physiology or Medicine for this work.

The Descriptive Passage and Interpretive Problems section at the end of Chapter 20 compares thioesters to their oxygen counterparts and introduces their reactions.

Nucleophilic Acyl Substitution

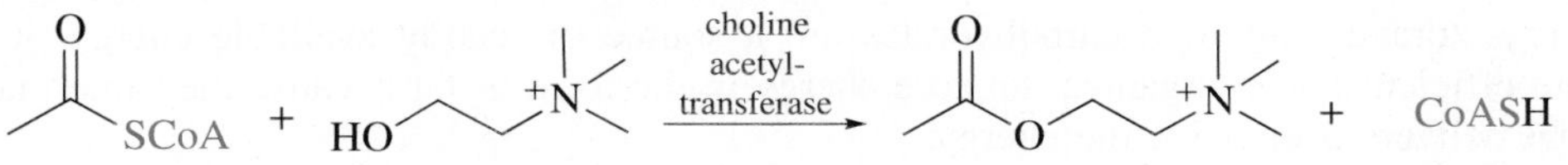

Acetyl coenzyme A (SCoA) + Choline (HO, ^+N) —(choline acetyltransferase)→ Acetylcholine + CoASH (Coenzyme A)

Reaction at the α Carbon

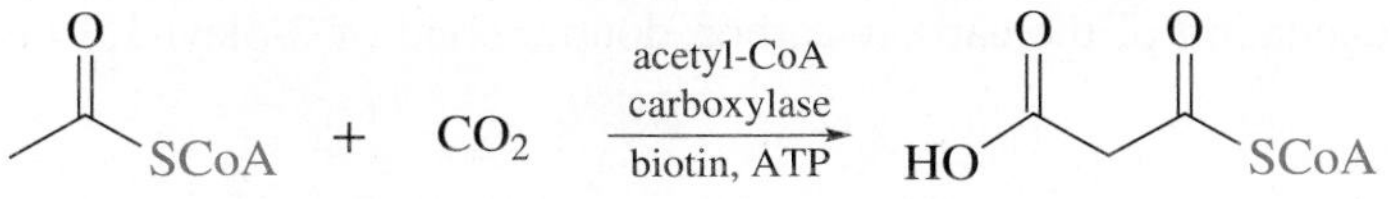

Acetyl coenzyme A (SCoA) + CO_2 (Carbon dioxide) —(acetyl-CoA carboxylase; biotin, ATP)→ Malonyl coenzyme A (HO, SCoA)

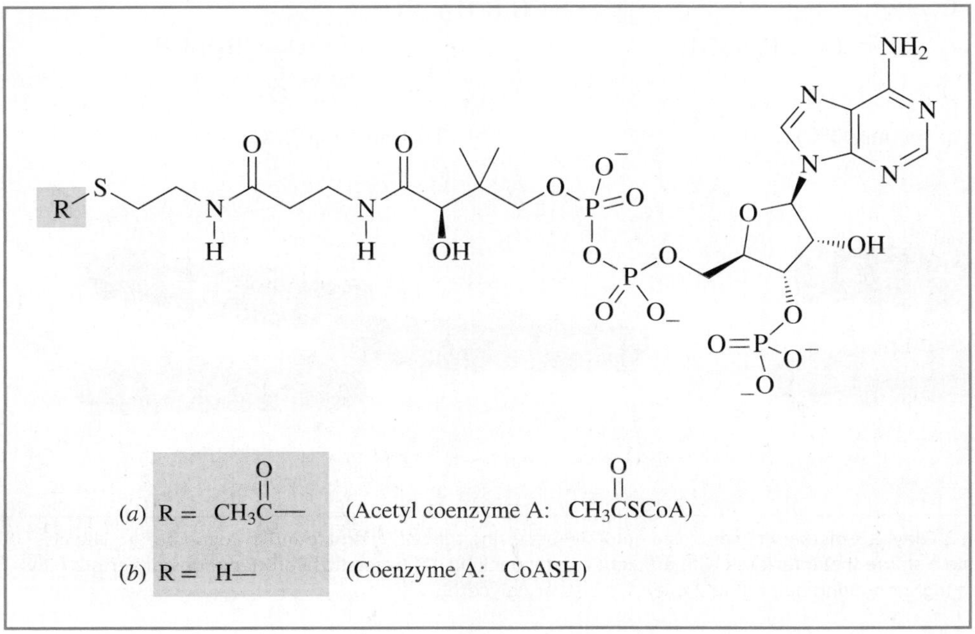

Figure 24.1

Structures of (*a*) acetyl coenzyme A and (*b*) coenzyme A.

We'll see numerous examples of both reaction types in the following sections. Even though these reactions are enzyme-catalyzed and occur at far faster rates than similar transformations carried out in their absence, the types of reactions are essentially the same as the fundamental processes of organic chemistry.

24.2 Fats, Oils, and Fatty Acids

Fats and oils are one class of lipids. In terms of structure, both are triesters of glycerol (*triglycerides*).

Glycerol is the common name for propane-1,2,3-triol.

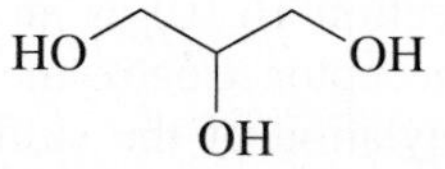

HO–CH₂–CH(OH)–CH₂–OH

Glycerol

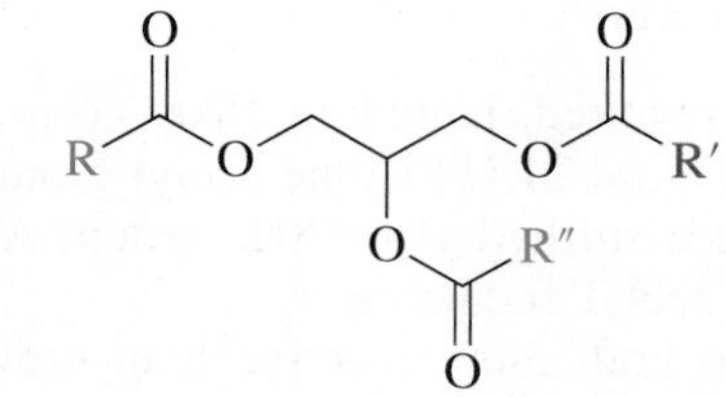

A triacylglycerol: R, R′, and R″ may be the same or different

Historically, fats are solids and oils are liquids, but we refer to both as fats when describing most of their other properties. They serve a number of functions, especially that of energy storage. Although carbohydrates are a source of readily available energy, it is more efficient for an organism to store energy in the form of fat because the same mass of fat delivers over twice the energy.

Figure 24.2 shows the structures of two typical triacylglycerols, 2-oleyl-1,3-distearylglycerol (Figure 24.2*a*) and tristearin (Figure 24.2*b*). Both occur naturally—in cocoa butter, for example. All three acyl groups in tristearin are stearyl (octadecanoyl) groups. In 2-oleyl-1,3-distearylglycerol, two of the acyl groups are stearyl, but the one in the middle is oleyl (*cis*-9-octadecenoyl). As the figure shows, tristearin can be prepared by catalytic hydrogenation of the carbon–carbon double bond of 2-oleyl-1,3-distearylglycerol.

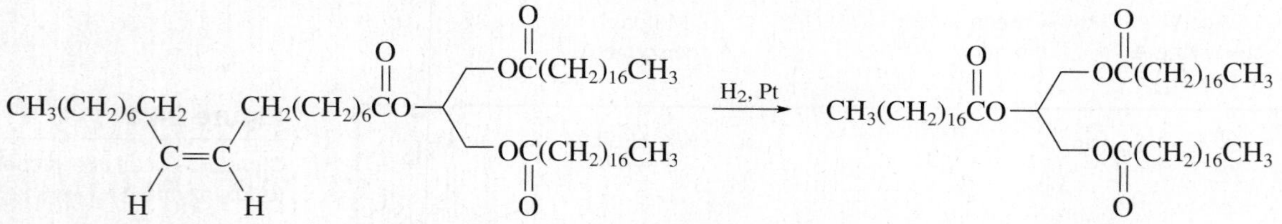

$CH_3(CH_2)_6CH_2$ / $CH_2(CH_2)_6\overset{O}{\overset{\|}{C}}O$ (C=C, H, H) with $O\overset{O}{\overset{\|}{C}}(CH_2)_{16}CH_3$ ×2 $\xrightarrow{H_2,\ Pt}$ $CH_3(CH_2)_{16}\overset{O}{\overset{\|}{C}}O$ with $O\overset{O}{\overset{\|}{C}}(CH_2)_{16}CH_3$ ×2

2-Oleyl-1,3-distearylglycerol (mp 43°C)

Tristearin (mp 72°C)

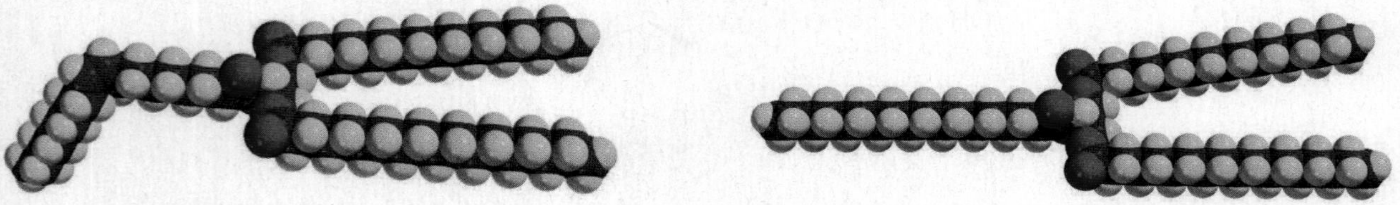

(*a*)

(*b*)

Figure 24.2

The structures of two typical triacylglycerols. (*a*) 2-Oleyl-1,3-distearylglycerol is a naturally occurring triacylglycerol found in cocoa butter. The cis double bond of its oleyl group gives the molecule a shape that interferes with efficient crystal packing. (*b*) Catalytic hydrogenation converts 2-oleyl-1,3-distearylglycerol to tristearin. Tristearin has a higher melting point than 2-oleyl-1,3-distearylglycerol.

Hydrogenation raises the melting point from 43°C in 2-oleyl-1,3-distearylglycerol to 72°C in tristearin and is a standard technique in the food industry for converting liquid vegetable oils to solid "shortenings." The space-filling models of the two show the more compact structure of tristearin, which allows it to pack better in a crystal lattice than the more irregular shape of 2-oleyl-1,3-distearyl-glycerol permits. This irregular shape is a direct result of the cis double bond in the side chain.

Hydrolysis of fats yields glycerol and long-chain **fatty acids.** Thus, tristearin gives glycerol and three molecules of stearic acid on hydrolysis. Table 24.1 lists a few

The term *fatty acid* originally referred to those carboxylic acids that occur naturally in triacylglycerols. Its use has expanded to include all unbranched carboxylic acids, irrespective of their origin and chain length.

TABLE 24.1 Some Representative Fatty Acids

Number of carbons	Common name	Structural formula and systematic name	Melting point, °C
Saturated fatty acids			
12	Lauric acid	Dodecanoic acid	44
14	Myristic acid	Tetradecanoic acid	58.5
16	Palmitic acid	Hexadecanoic acid	63
18	Stearic acid	Octadecanoic acid	69
20	Arachidic acid	Icosanoic acid	75
Unsaturated fatty acids			
18	Oleic acid	*cis*-9-Octadecenoic acid	4
18	Linoleic acid	*cis,cis*-9,12-Octadecadienoic acid	−12

continued

TABLE 24.1 Some Representative Fatty Acids (*Continued*)

Number of carbons	Common name	Structural formula and systematic name	Melting point, °C
Unsaturated fatty acids			
18	Linolenic acid	*cis,cis,cis*-9,12,15-Octadecatrienoic acid	−11
20	Arachidonic acid	*cis,cis,cis,cis*-5,8,11,14-Icosatetraenoic acid	−49

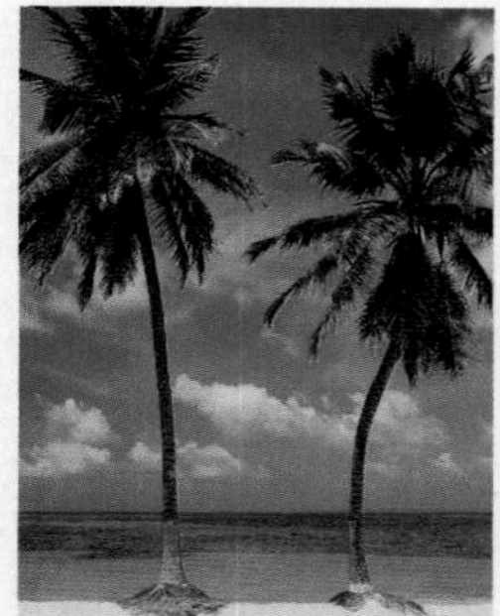

©Glow Images/SuperStock

Palmitic acid is the most abundant naturally occurring fatty acid. It is present in many fats and oils and is best known as the major fatty acid component of palm oil.

representative fatty acids. As these examples indicate, most naturally occurring fatty acids possess an even number of carbon atoms and an unbranched carbon chain. The carbon chain may be saturated or it can contain one or more double bonds. When double bonds are present, they are almost always cis. Acyl groups containing 14–20 carbon atoms are the most abundant in triacylglycerols.

Problem 24.1

What fatty acids are produced on hydrolysis of 2-oleyl-1,3-distearylglycerol? What other triacylglycerol gives the same fatty acids and in the same proportions as 2-oleyl-1,3-distearylglycerol?

A few fatty acids with trans double bonds (trans fatty acids) occur naturally, but the major source of trans fats comes from partial hydrogenation of vegetable oils in, for example, the preparation of margarine. The same catalysts that catalyze the hydrogenation of the double bonds in a triacylglycerol also catalyze double-bond migration and stereoisomerization.

$CH_3(CH_2)_7CH{=}CH(CH_2)_6CO_2R \xrightarrow{H-(M)_n-H} CH_3(CH_2)_7CH_2-CH[(M)_n-H](CH_2)_6CO_2R \xrightarrow{H-(M)_n-H} CH_3(CH_2)_6CH_2CH{=}CH(CH_2)_6CO_2R$

Ester of *cis*-9-octadecenoic acid → Ester of *trans*-8-octadecenoic acid

The intermediate in hydrogenation, formed by reaction of the unsaturated ester with the hydrogenated surface of the metal catalyst M, not only can proceed to the saturated fatty acid ester, but also can dissociate to constitutional and stereoisomers. Unlike polyunsaturated vegetable oils, which tend to reduce serum cholesterol levels, the trans fats produced by stereoisomerization during partial hydrogenation have cholesterol-raising effects similar to those of saturated fats. Increased consumption of trans fats has been linked to higher levels of coronary artery disease.

Fatty acids occur naturally in forms other than as triacylglycerols, and we'll see numerous examples as we go through the chapter. *Anandamide,* for example, is an amide of arachidonic acid (see Table 24.1).

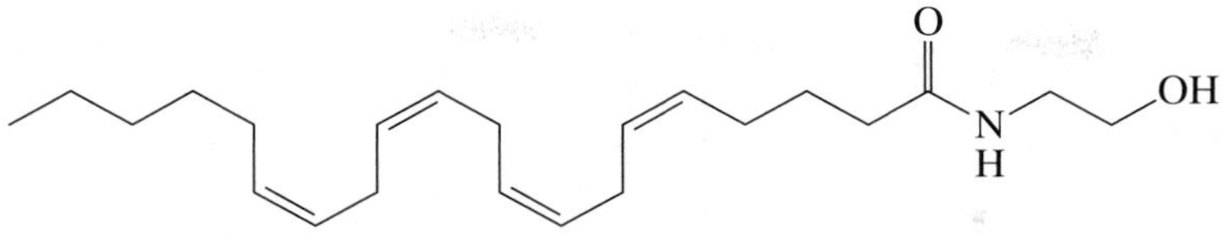

Anandamide

Anandamide was isolated from pig's brain and identified as the substance that normally binds to the "cannabinoid receptor." The active component of marijuana, Δ^9-tetrahydrocannabinol, exerts its effect by binding to a receptor, and scientists had long wondered what compound was the natural substrate for this binding site. Anandamide is that compound and seems to be involved in moderating pain. Once the identity of the "endogenous cannabinoid" was known, scientists looked specifically for it and found it in some surprising places—chocolate, for example.

The cannabinoid receptor belongs to a large family of receptor proteins that span the cell membrane, known as G-coupled protein receptors. Membrane receptor proteins are illustrated later, in Figure 24.4.

24.3 Fatty Acid Biosynthesis

The saturated fatty acids through hexadecanoic acid (palmitic acid) share a biosynthetic pathway that differs among them only in the number of chain elongation events. Four thioesters are involved:

This section outlines fatty acid biosynthesis in animals. The pathway in other organisms such as bacteria is different.

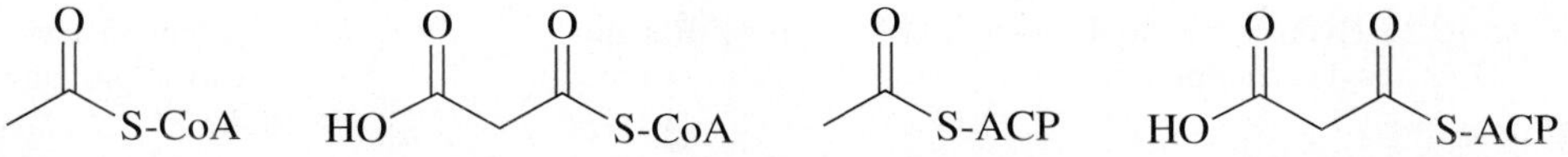

S-CoA — Acetyl coenzyme A; HO … S-CoA — Malonyl coenzyme A; S-ACP — Acetyl acyl carrier protein; HO … S-ACP — Malonyl acyl carrier protein

Of these four, acetyl-CoA and malonyl-ACP deserve special mention: acetyl-CoA because the other three are derived from it, malonyl-ACP because it is the source of all but two of the carbons in the final fatty acid.

Malonyl-CoA is formed by carboxylation of acetyl-CoA. The energy to drive the carboxylation comes from adenosine triphosphate (ATP).

Acetyl coenzyme A + Bicarbonate —(ATP → ADP + PO_4^{3-}; Acetyl-CoA carboxylase)→ Malonyl coenzyme A + H_2O (Water)

ATP is one of the most important organic compounds in biochemistry and appeared earlier (Section 1.8). ATP and ADP are discussed in more detail in Section 26.3.

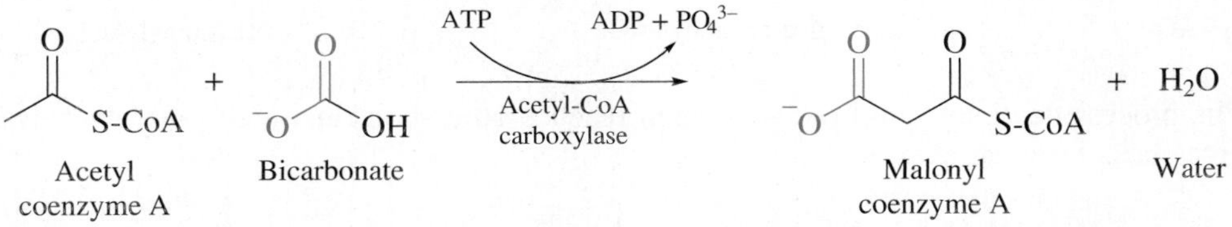

Acetyl and malonyl acyl carrier proteins are formed by acyl transfer to the polypeptide acyl carrier protein.

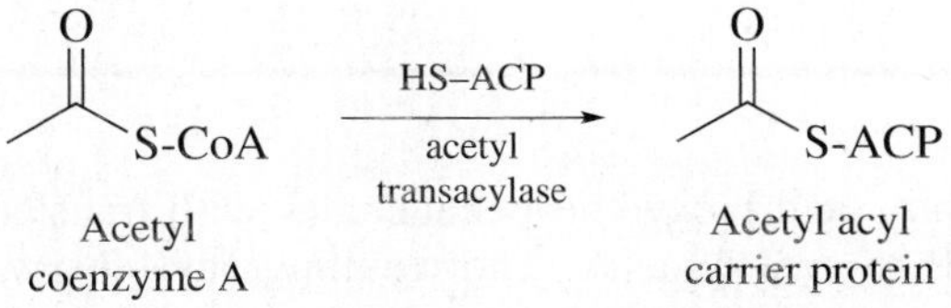

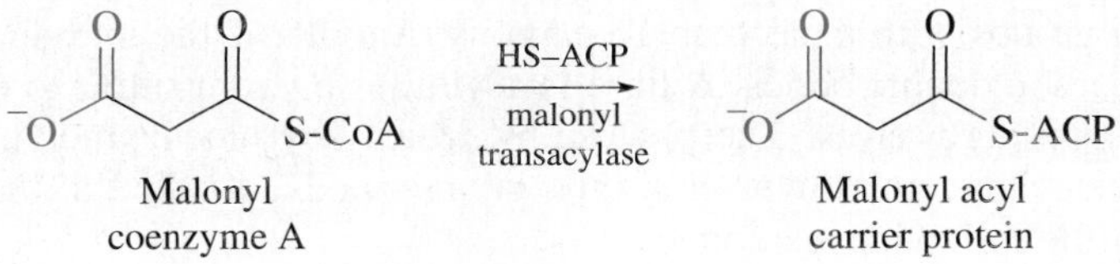

Acetyl coenzyme A —(HS–ACP; acetyl transacylase)→ Acetyl acyl carrier protein

Malonyl coenzyme A —(HS–ACP; malonyl transacylase)→ Malonyl acyl carrier protein

mFAS is a dimer of two polypeptide chains, each of which has a molecular weight of about 270,000.

These attached acyl carrier protein units are handles that bind the growing fatty acid to the multienzyme complex *mammalian fatty acid synthase* (mFAS) and guide it through subsequent transformations. The various enzymes occupy seven domains of mFAS.

One enzyme catalyzes a reaction in which decarboxylation of malonyl-ACP gives an enolate that reacts with acetyl-ACP by nucleophilic acyl substitution to give acetoacetyl-ACP.

Acetyl acyl carrier protein + Malonyl acyl carrier protein → Thiolate leaving group + Acetoacetyl acyl carrier protein + Carbon dioxide

Acetoacetyl-ACP is transported to a second domain of mFAS where it undergoes a sequence of three reactions—reduction, dehydration, reduction—that convert the acetoacetyl group to butanoyl.

Acetoacetyl acyl carrier protein —reduce→ 3-Hydroxybutanoyl acyl carrier protein —dehydrate→ *trans*-2-Butenoyl acyl carrier protein —reduce→ Butanoyl acyl carrier protein

Transfer of an acyl group between the first and second domains is reversible, which allows butanoyl ACP to return to the first domain of mFAS where it reacts with a second malonyl-ACP. The resulting six-carbon β-keto thioester then proceeds through another reduction–dehydration–reduction sequence to give hexanoyl-ACP.

Butanoyl-ACP → 3-Oxohexanoyl-ACP → Hexanoyl-ACP

The process continues until the acyl chain reaches 16 carbons at which point hydrolysis gives hexadecanoic acid.

Problem 24.2

Give the structure of the keto acyl-ACP formed after four cycles of chain extension starting with acetyl-ACP.

This phase of fatty acid biosynthesis concludes with transfer of the newly formed acyl group from ACP to coenzyme A. The resulting acyl coenzyme A molecules can then undergo a number of subsequent biological transformations. One is chain extension, leading to acyl groups with more than 16 carbons. Another is the introduction of one or more carbon–carbon double bonds. A third is acyl transfer from sulfur to oxygen to form esters such as triacylglycerols. The process by which acyl coenzyme A molecules are converted to triacylglycerols involves a type of intermediate called a *phospholipid* and is discussed in the following section.

24.4 Phospholipids

Triacylglycerols arise, not by acylation of glycerol itself, but by a sequence of steps in which the first stage is acyl transfer to L-glycerol 3-phosphate giving a **phosphatidic acid.**

L-Glycerol 3-phosphate + Two acyl coenzyme A molecules (R and R′ may be the same or different) ⟶ Phosphatidic acid + 2HSCoA (Coenzyme A)

Problem 24.3

What is the absolute configuration (*R* or *S*) of L-glycerol 3-phosphate? What must be the absolute configuration of the naturally occurring phosphatidic acids biosynthesized from it?

Hydrolysis of the phosphate ester function of the phosphatidic acid gives a diacylglycerol, which then reacts with a third acyl coenzyme A molecule to produce a triacylglycerol.

Phosphatidic acid $\xrightarrow{H_2O}$ Diacylglycerol $\xrightarrow{R''CS\text{-}CoA}$ Triacylglycerol

Phosphatidic acids not only are intermediates in the biosynthesis of triacylglycerols but also are biosynthetic precursors of other members of a group of compounds called *phosphoglycerides* or *glycerol phosphatides.* Phosphorus-containing derivatives of lipids are known as **phospholipids,** and phosphoglycerides are one type of phospholipid.

One important phospholipid is **phosphatidylcholine,** also called *lecithin.* Phosphatidylcholine is a mixture of diesters of phosphoric acid. One ester function is derived from a diacylglycerol, whereas the other is a choline [$-OCH_2CH_2\overset{+}{N}(CH_3)_3$] unit.

Phosphatidylcholine

Lecithin is added to foods such as mayonnaise as an emulsifying agent to prevent the fat and water from separating into two layers.

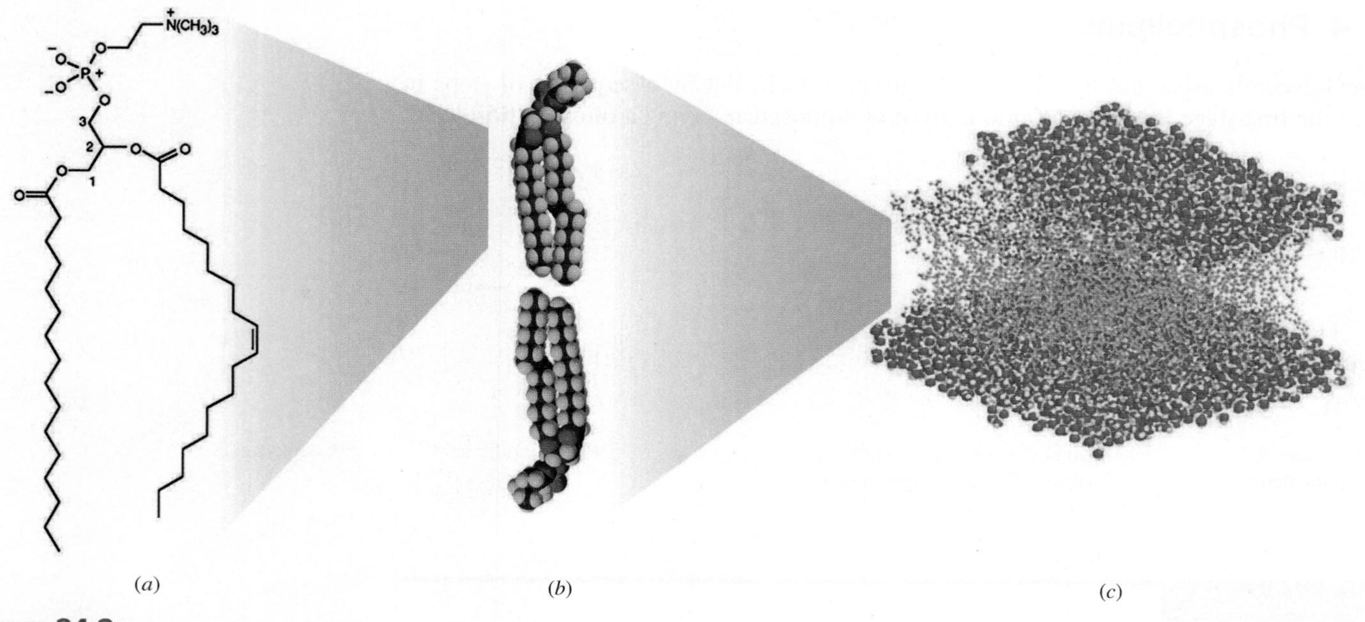

Figure 24.3

(*a*) Phosphatidylcholine. The C-1 and C-2 oxygens of glycerol bear hexadecanoyl (palmityl) and *cis*-9-octadecenoyl (oleyl) groups, respectively; C-3 bears the phosphate ester of choline. (*b*) Two space-filling models of (*a*) oriented so that the polar head group of one points up and the other down. (*c*) A simulation of a phospholipid bilayer. The space-filling models at the top and bottom are water molecules. The polar head groups are in contact with water molecules. The hydrocarbon chains are gray and shown as ball-and-spoke models with the hydrogens omitted. Water molecules are omitted at the upper left corner to make the head groups visible. The simulation is based on the coordinates of H. Heller, M. Schaefer, and K. Schulten, "Molecular Dynamics Simulation of a Bilayer of 200 Lipids in the Gel and in the Liquid-Crystal Phases," *Journal of Physical Chemistry*, 97, 8343–8360 (1993) and taken from an interactive animated tutorial by E. Martz and A. Herráez, "Lipid Bilayers and the Gramicidin Channel" by courtesy of Professor Martz.

Phosphatidylcholine possesses a hydrophilic polar "head group" (the positively charged choline) and two lipophilic (hydrophobic) nonpolar "tails" (the acyl groups). Under certain conditions, such as at the interface of two aqueous phases, phosphatidylcholine forms what is called a *lipid bilayer,* as shown in Figure 24.3. Because there are two long-chain acyl groups in each molecule, the most stable assembly has the polar groups solvated by water molecules at the top and bottom surfaces and the lipophilic acyl groups directed toward the interior of the bilayer.

Phosphatidylcholine is an important component of cell membranes, but cell membranes are more than simply lipid bilayers. Although their composition varies with their source, a typical membrane contains about equal amounts of lipid and protein, and the amount of cholesterol in the lipid fraction can approximate that of phosphatidylcholine. A simplified schematic representation of a cell membrane is shown in Figure 24.4. This diagram originates from the **fluid mosaic model,** which was originally proposed in 1972.

It was known in 1972 that the interior and exterior of the plasma membrane is different with respect to lipid composition, integral membrane protein organization, glycoproteins, and glycolipids. "Flipping" lipids from one side of the membrane to the other does not occur spontaneously, and the overall fluidity of a membrane is affected by the lipid composition. This model has been refined to reflect our increasing understanding of the complexities of biological membranes. It was thought, for example, that lipid composition was fairly homogeneous laterally and that lipids moved freely within the monolayer. Today, more emphasis is placed on the "mosaic" rather than the "fluid" nature of membranes. We now know that organization occurs by lipid-lipid, lipid-protein, and protein-protein interactions within a membrane. For example, "lipid rafts" are ordered subdomains of membranes. In plasma membranes, they contain more cholesterol and sphingolipids. Lipid rafts play important organizational roles in membrane functions such as signal transduction, endocytosis, and cell–cell interactions (Figure 24.5).

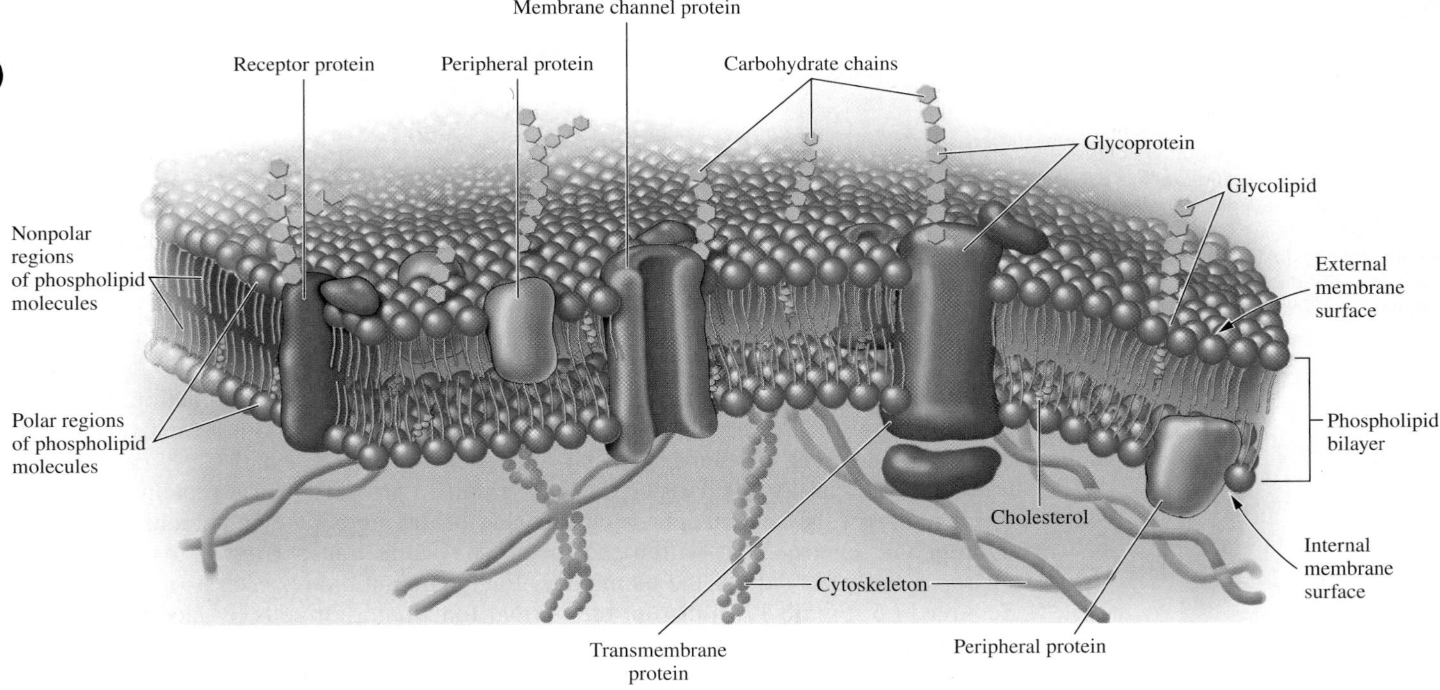

Figure 24.4

Fluid mosaic model of a cell membrane.

Receptor protein: A protein that acts as a receptor toward a hormone, neurotransmitter, or other molecule that can serve as a ligand.

Peripheral protein: A protein that adheres temporarily to the membrane.

Membrane channel protein: A protein that can form a pore through the membrane, through which ions or other solutes may flow.

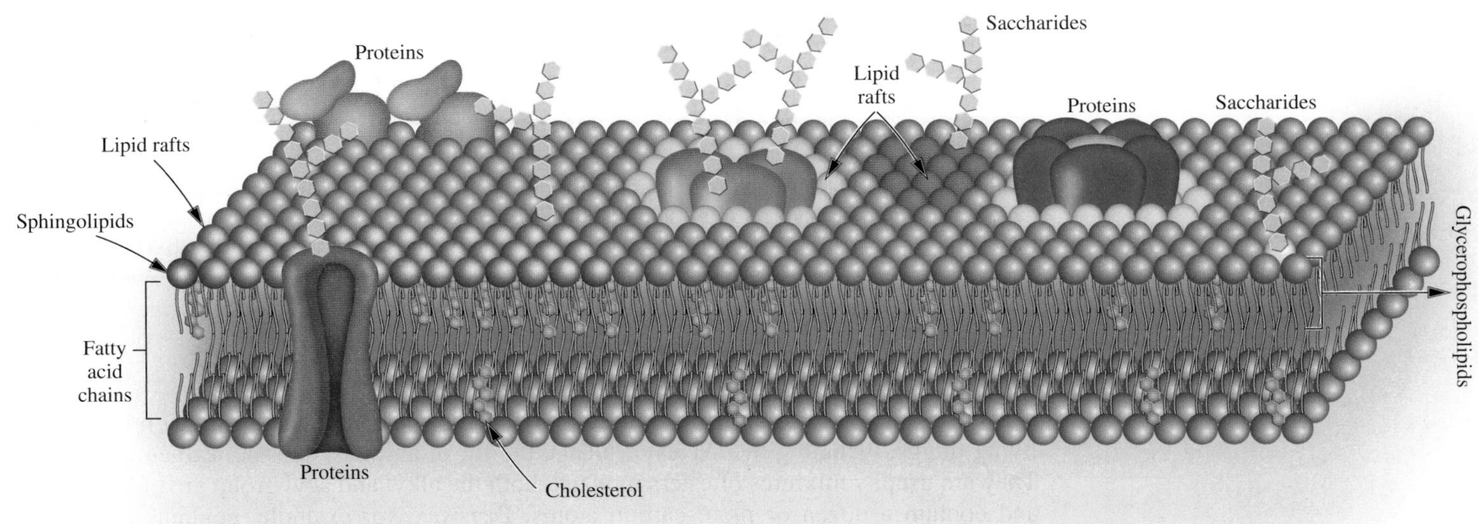

Figure 24.5

Higher-level organization within a membrane. Lateral interactions between membrane components, such as lipid rafts, stabilize different membrane domains.

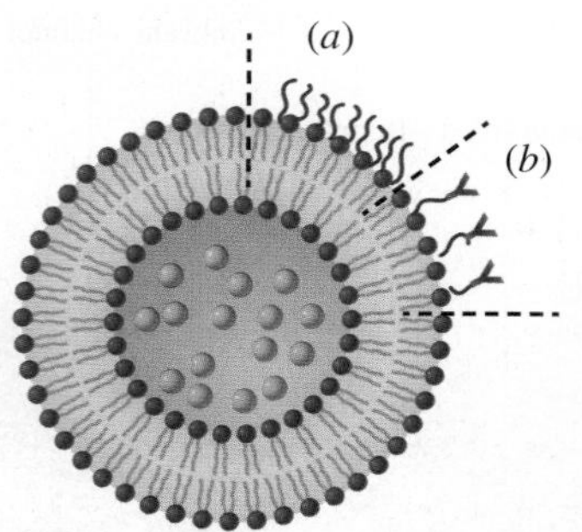

Figure 24.6

A spherical liposome shown in cross section. The membrane is a phospholipid bilayer. The interior is water. The blue spheres represent a water-soluble drug. The outside of the liposome is often modified to improve in vivo behavior. (a) Polyethylene glycol [$-(O-CH_2-CH_2)_n-OH$] can improve serum stability. (b) Targeting molecules such as antibodies can increase cellular specificity.

The protein fraction is responsible for a major part of membrane function. Nonpolar materials can diffuse through the bilayer from one side to the other relatively easily, but polar materials, particularly metal ions such as Na^+, K^+, and Ca^{2+}, cannot. The transport of metal ions is assisted by the membrane proteins. These proteins pick up a metal ion from the aqueous phase on one side of the membrane and shield it from the hydrophobic environment within the membrane while transporting it to the aqueous phase on the other side of the membrane. Ionophore antibiotics such as monensin (see Section 17.5, Figure 17.3) disrupt the normal functioning of cells by facilitating metal ion transport across cell membranes.

Like cells, only much smaller, spherical objects called **liposomes** are enclosed by a phospholipid bilayer that separates a watery interior from an external (also watery) environment. Liposomes occur naturally but can also be prepared from lecithin as a phosphatidylcholine source. Following their chance discovery in 1961, liposomes originally received attention as models for membrane structure. Subsequently, their use as novel vehicles for drug delivery was demonstrated and has led to important applications in medicine.

Generating a liposome in an aqueous solution containing a water-soluble drug yields the species illustrated in Figure 24.6 in which the drug is encapsulated within the interior of the liposome. When given to a patient, the drug-carrying liposome binds to one of the patient's cells and transfers the drug into the cell. Conventional liposomes are rapidly cleared from the bloodstream. Currently, liposomes are modified to increase their stability and to increase their specificity to cellular targets.

Problem 24.4

In addition to bilayers in cell membranes and liposomes, phosphatidylcholine also forms micelles. Compare a typical fatty acid micelle (see Section 19.7) and a phosphatidylcholine micelle, both at pH 7. What is the most significant difference in their surface properties?

24.5 Waxes

Waxes are water-repelling solids that are part of the protective coatings of a number of living things, including the leaves of plants, the fur of animals, and the feathers of birds. They are usually mixtures of esters in which both the alkyl and acyl group are unbranched and contain a dozen or more carbon atoms. Beeswax, for example, contains the ester triacontyl hexadecanoate as one component of a complex mixture of hydrocarbons, alcohols, and esters.

$$CH_3(CH_2)_{14}\overset{\overset{\displaystyle O}{\|}}{C}OCH_2(CH_2)_{28}CH_3$$

Triacontyl hexadecanoate

Problem 24.5

Spermaceti is a wax obtained from the sperm whale. It contains, among other materials, an ester known as *cetyl palmitate,* which is used as an emollient in a number of soaps and cosmetics. The systematic name for cetyl palmitate is *hexadecyl hexadecanoate.* Write a structural formula for this substance.

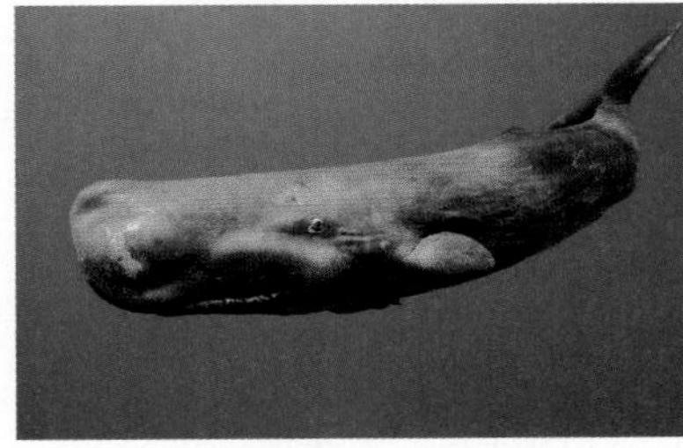

©James R.D. Scott/Getty Images

Fatty acids normally occur naturally as components of esters; fats, oils, phospholipids, and waxes all are unique types of fatty acid esters. There is, however, an important class of fatty acid derivatives that carries out its biological role in the form of the free acid. This class of fatty acid derivatives is described in the following section.

24.6 Prostaglandins

Research in physiology carried out in the 1930s established that the lipid fraction of semen contains small amounts of substances that exert powerful effects on smooth muscle. Sheep prostate glands proved to be a convenient source of this material and yielded a mixture of structurally related substances referred to collectively as **prostaglandins.** We now know that prostaglandins are present in almost all animal tissues, where they carry out a variety of regulatory functions. Among these functions are relaxation or constriction of bronchial muscles, platelet aggregation or disaggregation, induction of labor, and the regulation of inflammation.

Much of the fundamental work on prostaglandins was carried out by Sune Bergström and Bengt Samuelsson of the Karolinska Institute (Sweden) and by Sir John Vane of the Wellcome Foundation (Great Britain). These three shared the Nobel Prize in Physiology or Medicine in 1982.

Prostaglandins are extremely potent substances and exert their physiological effects at very small concentrations. Because of this, their isolation was difficult, and it was not until 1960 that the first members of this class, designated PGE_1 and $PGF_{1\alpha}$ (Figure 24.7), were obtained as pure compounds. More than a dozen structurally related prostaglandins have since been isolated and identified. All the prostaglandins are 20-carbon carboxylic acids and contain a cyclopentane ring. All have hydroxyl groups at C-11 and C-15 (see Figure 24.7). Prostaglandins belonging to the F series have an additional hydroxyl group at C-9, and a carbonyl function is present at this position in the various PGEs. The subscript numerals in their abbreviated names indicate the number of double bonds.

Arachidonic acid gets its name from *arachidic acid,* the saturated C_{20} fatty acid isolated from peanut (*Arachis hypogaea*) oil.

Prostaglandins arise from unsaturated C_{20}-carboxylic acids such as arachidonic acid (see Table 24.1). Animals cannot biosynthesize arachidonic acid directly. They obtain linoleic acid from vegetable oils in their diet and extend the carbon chain of linoleic acid from 18 to 20 carbons while introducing two more double bonds. Linoleic acid is said to be an **essential fatty acid,** forming part of the dietary requirement of animals. Those that feed on diets that are deficient in linoleic acid grow poorly and suffer a number of other disorders, some of which are reversed on feeding them vegetable oils rich in linoleic acid and other *polyunsaturated fatty acids.* One function of these substances is to provide the raw materials for prostaglandin biosynthesis.

Prostaglandin E_1 (PGE_1)   Prostaglandin $F_{1\alpha}$ ($PGF_{1\alpha}$)

Figure 24.7

Structures of two representative prostaglandins. The numbering scheme is illustrated in the structure of PGE_1.

Studies of the biosynthesis of PGE_2 from arachidonic acid have shown that the three new oxygens come from O_2. The enzyme involved, *cyclooxygenase* (COX), catalyzes cycloaddition of O_2 to arachidonic acid to give an *endoperoxide* (PGG_2).

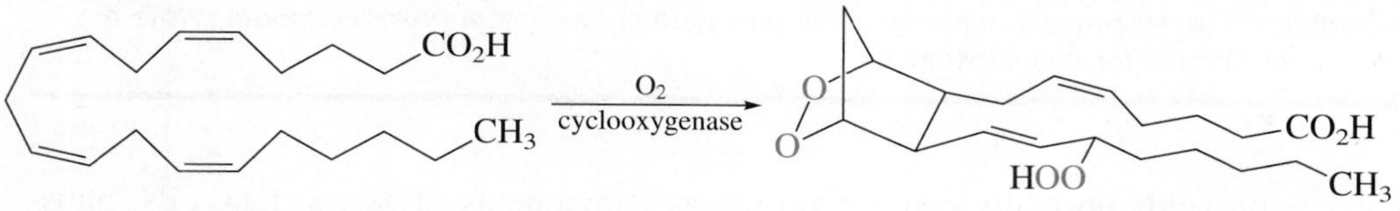

Arachidonic acid $\xrightarrow[\text{cyclooxygenase}]{O_2}$ PGG_2

In the next step, the —OOH group of PGG_2 is reduced to an alcohol function. The product of this step is called PGH_2.

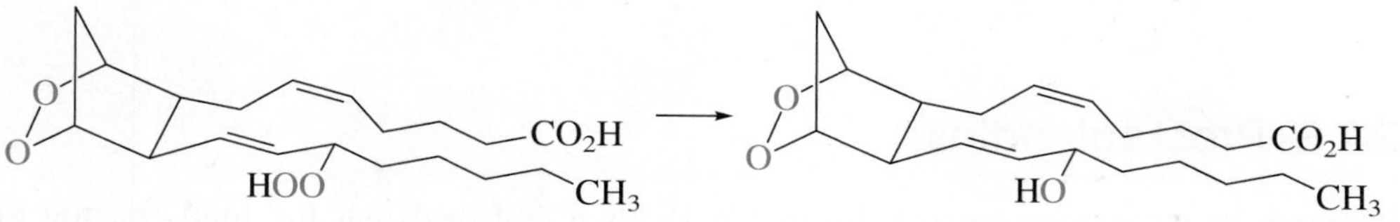

PGG_2 → PGH_2

PGH_2 is the precursor to a number of prostaglandins and related compounds, depending on the enzyme that acts on it. One of these cleaves the O—O bond of the endoperoxide and gives PGE_2.

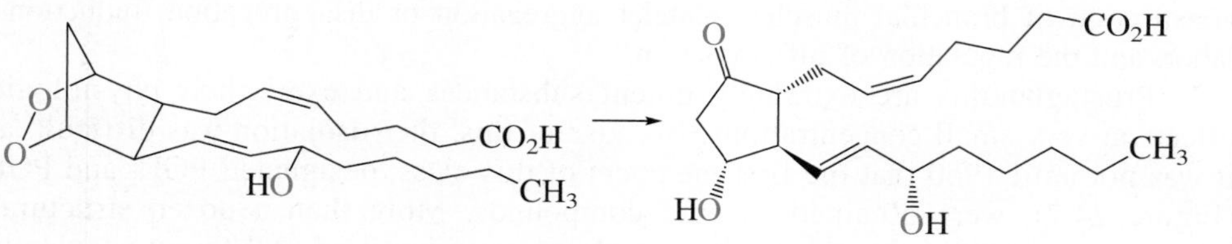

PGH_2 → PGE_2

Before leaving this biosynthetic scheme, notice that PGE_2 has four chirality centers. Even though arachidonic acid is achiral, only the stereoisomer shown as PGE_2 in the equation is formed. Moreover, it is formed as a single enantiomer. Like most enzyme-catalyzed reactions, the transformations in the biosynthesis of PGE_2 are enantioselective.

Problem 24.6

Write the structural formula and give the IUPAC name for the fatty acid from which PGE_1 is biosynthesized. The structure of PGE_1 is shown in Figure 24.7.

Older versions of the IUPAC rules called the unbranched carboxylic acid with 20 carbon atoms *eicosanoic acid*. Consequently, icosanoids are often referred to as *eicosanoids*.

Prostaglandins belong to a group of compounds that, because they are related to icosanoic acid [$CH_3(CH_2)_{18}CO_2H$)], are collectively known as **icosanoids.** The other icosanoids are *thromboxanes, prostacyclins,* and *leukotrienes.*

Thromboxane A_2 (TXA_2) promotes platelet aggregation and blood clotting. The biosynthetic pathway to TXA_2 is the same as that of PGE_2 up to PGH_2. At that point separate pathways lead to PGE_2 and TXA_2.

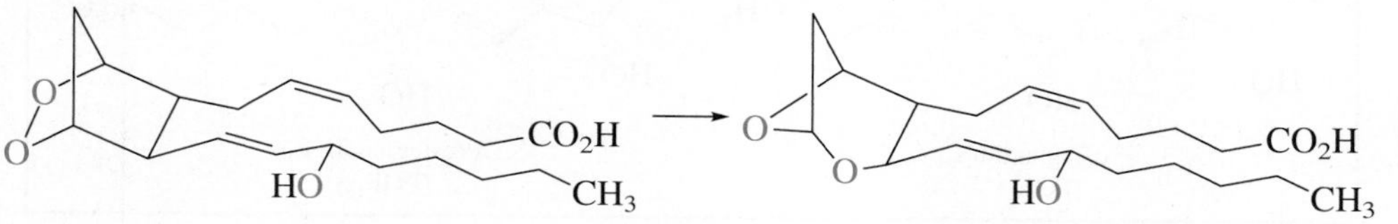

PGH_2 → TXA_2

Nonsteroidal Antiinflammatory Drugs (NSAIDs) and COX-2 Inhibitors

An injection of the steroid cortisone is often effective for reducing the pain and inflammation that comes from an injury. But chronic pain and inflammation, such as occurs with arthritis, is better managed with an orally administered remedy. Enter nonsteroidal antiinflammatory drugs (NSAIDs).

Aspirin is the oldest and best known NSAID. Over the years it has been joined by many others, a few of which are:

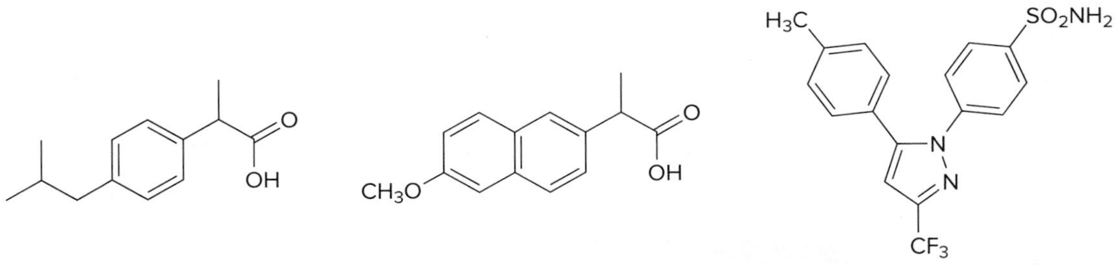

Ibuprofen

Naproxen

Celecoxib (Celebrex)

The long-standing question of how aspirin works has been answered in terms of its effect on prostaglandin biosynthesis. Prostaglandins are made continuously in all mammalian cells and serve a variety of functions. They are biosynthesized in greater amounts at sites of tissue damage, and it is they that cause the pain and inflammation we feel. Cells contain two forms of the cyclooxygenase enzyme, COX-1 and COX-2; both catalyze prostaglandin biosynthesis. Some of the prostaglandins produced with the aid of COX-1 are involved in protecting the stomach and kidneys. COX-2 is concentrated in injured tissue where it works to catalyze the biosynthesis of the prostaglandins responsible for inflammation. Aspirin inhibits prostaglandin biosynthesis by inactivating both COX-1 and COX-2. Although inhibition of COX-2 has the desired effect of relieving pain and inflammation, inhibition of COX-1 causes irritation of the stomach lining.

A good antiinflammatory drug, therefore, will selectively inactivate COX-2 while leaving COX-1 untouched. Aspirin doesn't; in fact, it is about ten times more effective toward inactivating the "wrong" COX.

The classical period of drug development emphasized testing a variety of compounds for biological activity, identifying the features believed to be associated with the desired activity, then synthesizing and testing numerous analogs. Celecoxib, however, was developed with the express goal of inhibiting COX-2. Molecular models of the three-dimensional structures of COX-1 and COX-2 were examined to guide thinking about the kinds of structural units a drug should have in order to selectively inactivate COX-2. This approach not only led to a successful drug, but also validated a structure-based route to new drug development.

Prostacyclin I_2 (PGI_2) inhibits platelet aggregation and relaxes coronary arteries. Like PGE_2 and TXA_2, it is formed from arachidonic acid via PGH_2.

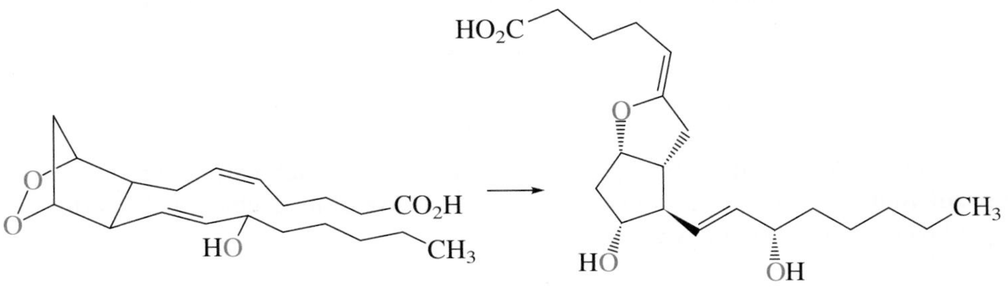

PGH_2 → PGI_2

Leukotrienes are the substances most responsible for the constriction of bronchial passages during asthma attacks. They arise from arachidonic acid by a pathway different from the one that leads to prostaglandins and related compounds. The pathway to leukotrienes does not involve cyclooxygenation. Instead, oxidation simply introduces

—OH groups at specific carbons along the chain. Allylic radicals are involved and some of the double bonds in the product are in different locations than those in arachidonic acid. The enzymes involved are known as *lipoxygenases* and are differentiated according to the carbon of the chain that is oxidized. The biosynthesis of the leukotriene shown begins with a 5-lipoxygenase-catalyzed oxidation of arachidonic acid.

CO_2H CH_3 —several steps→ OH CO_2H S CH_2—$CH\overset{O}{\overset{\|}{C}}NHCH_2CO_2H$ CH_3 NH C=O $^-O_2CCHCH_2CH_2$ $^+NH_3$

Arachidonic acid Leukotriene C_4 (LTC_4)

Problem 24.7

The carbon–sulfur bond in LTC_4 is formed by the reaction of glutathione (see Section 16.12) with leukotriene A_4 (LTA_4). LTA_4 is an epoxide. Suggest a reasonable structure for LTA_4.

Most of the drugs such as epinephrine and albuterol used to treat asthma attacks are *bronchodilators*—substances that expand the bronchial passages. Newer drugs are designed either to inhibit the 5-lipoxygenase enzyme, which acts on arachidonic acid in the first stage of leukotriene biosynthesis, or to block leukotriene receptors.

24.7 Terpenes: The Isoprene Rule

The word *essential* as applied to naturally occurring organic substances can have two different meanings. With respect to fatty acids, *essential* means "necessary." Linoleic acid is an "essential" fatty acid; it must be included in the diet for animals to grow properly because they lack the ability to biosynthesize it directly.

A bayberry (wax myrtle) plant.

Essential is also used as the adjective form of the noun *essence*. The mixtures of substances that make up the fragrant material of plants are called **essential oils** because they contain the essence, that is, the odor, of the plant. The study of the composition of essential oils ranks as one of the oldest areas of organic chemical research. Very often, the principal volatile component of an essential oil belongs to a class of chemical substances called the **terpenes.**

Myrcene, a hydrocarbon isolated from bayberry oil, is a typical terpene:

Myrcene

The structural feature that distinguishes terpenes from other natural products is the **isoprene unit.** The carbon skeleton of myrcene (exclusive of its double bonds) corresponds to the head-to-tail union of two isoprene units.

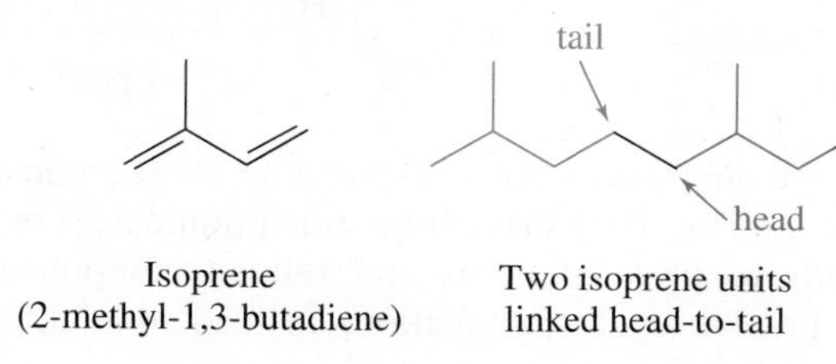

Isoprene (2-methyl-1,3-butadiene) Two isoprene units linked head-to-tail

TABLE 24.2 Classification of Terpenes

Class	Number of isoprene units	Number of carbon atoms
Monoterpene	2	10
Sesquiterpene	3	15
Diterpene	4	20
Sesterpene	5	25
Triterpene	6	30
Tetraterpene	8	40

The German chemist Otto Wallach determined the structures of many terpenes and is credited with setting forth the **isoprene rule:** terpenes are repeating assemblies of isoprene units, normally joined head-to-tail.

Wallach was awarded the 1910 Nobel Prize in Chemistry.

Terpenes are often referred to as *isoprenoid* compounds and are classified according to the number of isoprene units they contain (Table 24.2).

Although the term *terpene* once referred only to hydrocarbons, its use expanded to include functionally substituted derivatives as well, grouped together under the general term *isoprenoids*. Figure 24.8 presents the structural formulas for a number of representative examples. The isoprene units in some of these are relatively easy to identify. The three isoprene units in the sesquiterpene *farnesol,* for example, are indicated as follows in color. They are joined in a head-to-tail fashion.

There are more than 23,000 known isoprenoid compounds.

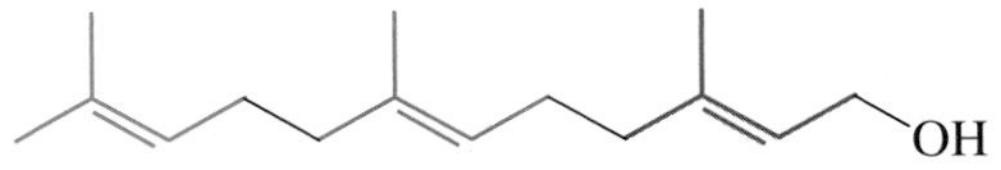

OH

Isoprene units in farnesol ($C_{15}H_{26}O$)

Many terpenes contain one or more rings, but these also can be viewed as collections of isoprene units. An example is α-selinene. Like farnesol, it is made up of three isoprene units linked head-to-tail.

In locating isoprene units within a given carbon skeleton, keep in mind that the double bonds may no longer be present.

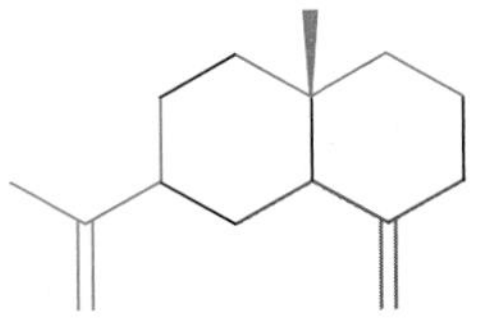

Isoprene units in α-selinene ($C_{15}H_{24}$)

Problem 24.8

Locate the isoprene units in each of the monoterpenes, sesquiterpenes, and diterpenes shown in Figure 24.8. (In some cases there are two equally correct arrangements.)

Tail-to-tail linkages of isoprene units sometimes occur, especially in the higher terpenes. The C(12)—C(13) bond of squalene unites two C_{15} units in a tail-to-tail manner. Notice, however, that isoprene units are joined head-to-tail within each C_{15} unit of squalene.

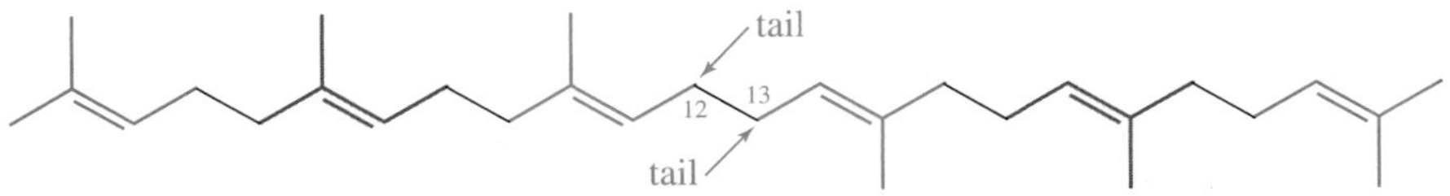

tail
13
12
tail

Isoprene units in squalene ($C_{30}H_{50}$)

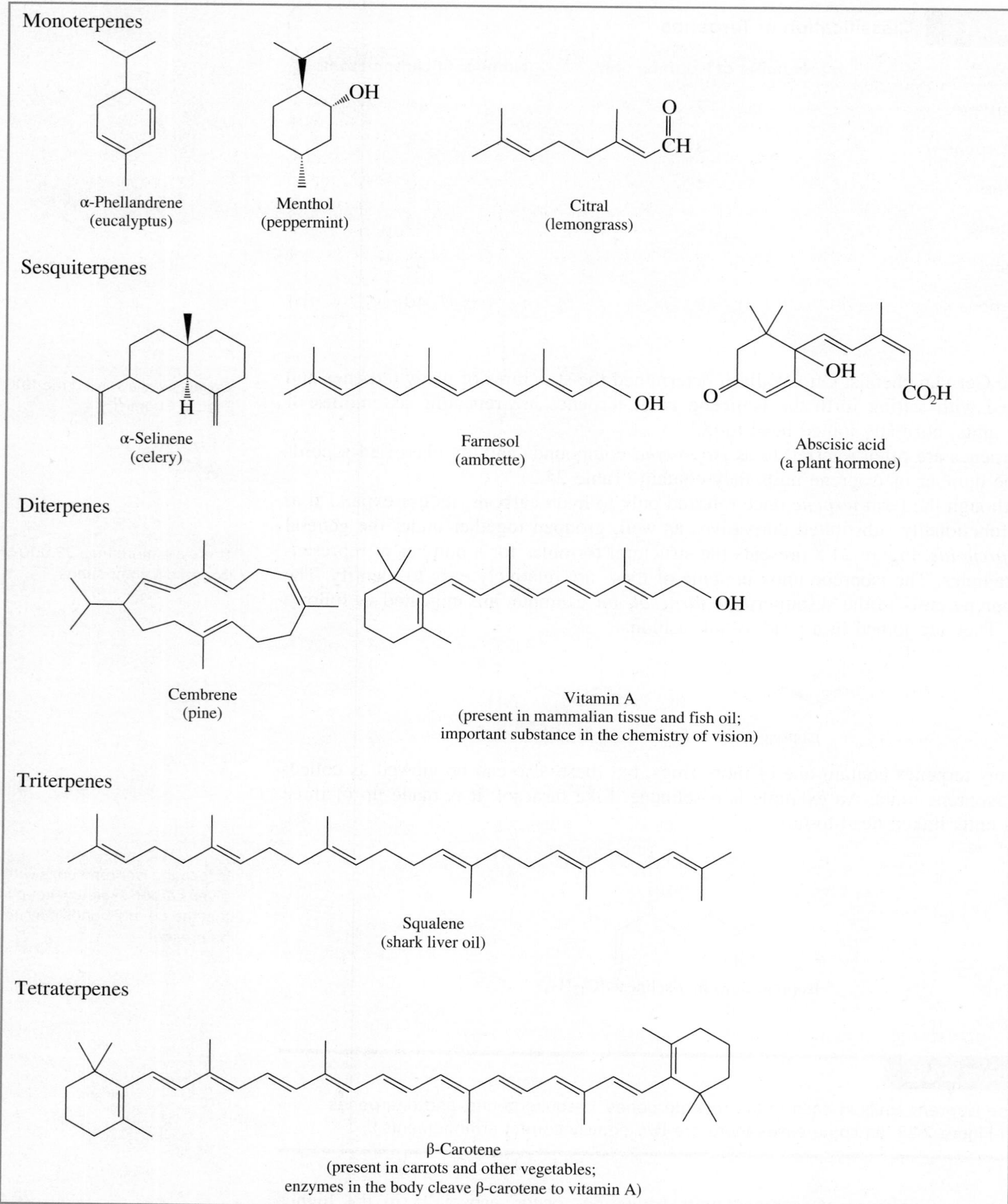

Figure 24.8

Some representative terpenes and related natural products. Structures are customarily depicted as carbon skeleton formulas when describing compounds of isoprenoid origin.

Problem 24.9

Identify the isoprene units in β-carotene (see Figure 24.8). Which carbons are joined by a tail-to-tail link between isoprene units?

Over time, Wallach's original isoprene rule was refined, most notably by Leopold Ruzicka of the Swiss Federal Institute of Technology (Zürich), who put forward a *biological isoprene rule* in which he connected the various classes of terpenes according to their biological precursors. Thus arose the idea of the *biological isoprene unit*. Isoprene is the fundamental structural unit of terpenes and related compounds, but isoprene does not occur naturally—at least in places where biosynthesis is going on. What then is the biological isoprene unit, how is this unit itself biosynthesized, and how do individual isoprene units combine to give terpenes?

Ruzicka was a corecipient of the 1939 Nobel Prize in Chemistry.

24.8 Isopentenyl Diphosphate: The Biological Isoprene Unit

Isoprenoid compounds are biosynthesized from acetate by a process that involves several stages. The first stage is the formation of mevalonic acid from three molecules of acetic acid. In the second stage, mevalonic acid is converted to isopentenyl diphosphate:

3 Acetic acid —several steps→ Mevalonic acid —several steps→ Isopentenyl diphosphate

It is convenient to use the symbol —OPP to represent the diphosphate group. Diphosphate is also known as pyrophosphate.

Isopentenyl diphosphate is the biological isoprene unit; it contains five carbon atoms connected in the same order as in isoprene.

In the presence of the enzyme *isopentenyl diphosphate isomerase,* isopentenyl diphosphate is converted to dimethylallyl diphosphate. The isomerization involves two successive proton transfers: one from an acidic site of the enzyme (Enz—H) to the double bond to give a tertiary carbocation; the other is deprotonation of the carbocation by a basic site of the enzyme to generate the double bond of dimethylallyl diphosphate.

Enz—H + Isopentenyl diphosphate ⇌ Enz:⁻ + Carbocation intermediate ⇌ Enz—H + Dimethylallyl diphosphate

Isopentenyl diphosphate and dimethylallyl diphosphate are structurally similar—both contain a double bond and a diphosphate ester unit—but the chemical reactivity expressed by each is different. The principal site of reaction in dimethylallyl diphosphate is the carbon that bears the diphosphate group. Diphosphate is a reasonably good leaving group in nucleophilic substitution reactions, especially when, as in dimethylallyl diphosphate, it is located at an allylic carbon. Isopentenyl diphosphate, on the other hand, does not have its leaving group attached to an allylic carbon and is far less reactive than dimethylallyl diphosphate toward nucleophilic reagents. The principal site of reaction in isopentenyl diphosphate is the carbon–carbon double bond, which, like the double bonds of simple alkenes, is reactive toward electrophiles.

24.9 Carbon–Carbon Bond Formation in Terpene Biosynthesis

The chemical properties of isopentenyl diphosphate and dimethylallyl diphosphate are complementary in a way that permits them to react with each other to form a carbon–carbon bond that unites two isoprene units. In broad outline, the enzyme-catalyzed process involves bond formation between the allylic CH_2 of dimethylallyl diphosphate and the vinylic CH_2 of isopentenyl diphosphate. Diphosphate is the leaving group and a tertiary carbocation results.

Dimethylallyl diphosphate + Isopentenyl diphosphate ⟶ Ten-carbon carbocation + Diphosphate ion

Alternatively, ionization of dimethylallyl diphosphate could precede carbon–carbon bond formation.

Dimethylallyl diphosphate —(−[⁻ÖPP])⟶ Dimethylallyl cation ⟶ Ten-carbon carbocation

The ten-carbon carbocation that results is the same regardless of whether it is formed in one step or two. Once formed it can react in several different ways, all of which are familiar to us as typical carbocation processes. One is deprotonation to give the carbon–carbon double bond of *geranyl diphosphate.*

Ten-carbon carbocation —(:Enz)⟶ Geranyl diphosphate

Hydrolysis of geranyl diphosphate gives *geraniol,* a pleasant-smelling monoterpene found in rose oil.

Geranyl diphosphate —(H_2O)⟶ Geraniol

Geranyl diphosphate is an allylic diphosphate and, like dimethylallyl diphosphate, can react with isopentenyl diphosphate. A 15-carbon carbocation is formed, which on deprotonation gives *farnesyl diphosphate.* Hydrolysis of farnesyl diphosphate gives the sesquiterpene *farnesol.*

Geranyl diphosphate —(enzyme)⟶ 15-carbon carbocation —(deprotonation)⟶ Farnesyl diphosphate —(hydrolysis)⟶ Farnesol

Repeating the process produces the diterpene geranylgeraniol from farnesyl diphosphate.

Geranylgeraniol

Problem 24.10

Write a sequence of reactions that describes the formation of geranylgeraniol from farnesyl diphosphate.

Geraniol, farnesol, and geranylgeraniol are classified as **prenols,** compounds of the type:

$$H\left[CH_2-\overset{\overset{\displaystyle CH_3}{|}}{C}=CH-CH_2\right]_n OH$$

The group to which the OH (or other substituent) is attached is called a **prenyl** group.

The higher terpenes are formed not by successive addition of C_5 units but by the coupling of simpler terpenes. Thus, the triterpenes (C_{30}) are derived from two molecules of farnesyl diphosphate, and the tetraterpenes (C_{40}) from two molecules of geranylgeranyl diphosphate. These carbon–carbon bond-forming processes involve tail-to-tail couplings and proceed by a more complicated mechanism than that just described.

The reaction of an allylic diphosphate or a carbocation with a source of π electrons is a recurring theme in terpene biosynthesis and is invoked to explain the origin of more complicated structural types. Consider, for example, the formation of cyclic monoterpenes. *Neryl diphosphate,* formed by an enzyme-catalyzed isomerization of the *E* double bond in geranyl diphosphate, has the proper geometry to form a six-membered ring via intramolecular attack of the double bond on the allylic diphosphate unit.

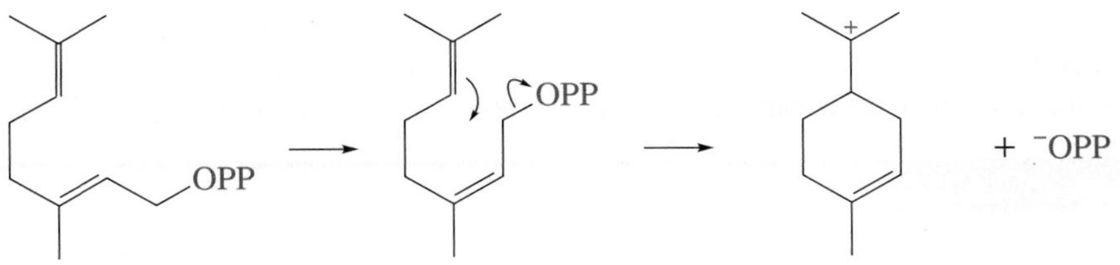

OPP → OPP → + $^-$OPP

Geranyl diphosphate Neryl diphosphate Tertiary carbocation

Loss of a proton from the tertiary carbocation formed in this step gives *limonene,* an abundant natural product found in many citrus fruits. Capture of the carbocation by water gives α-*terpineol,* also a known natural product.

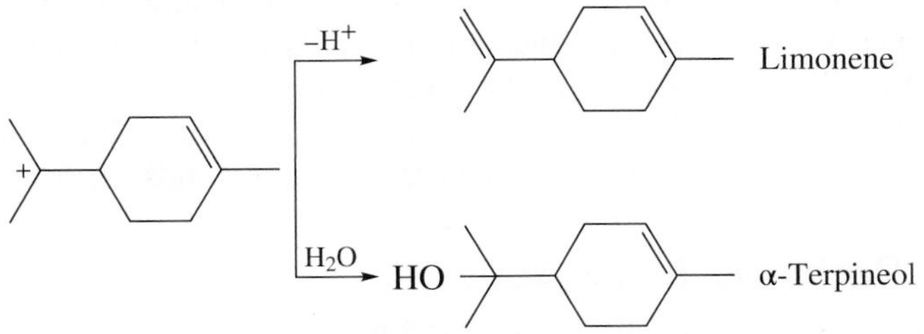

$-H^+$ → Limonene

H_2O → HO α-Terpineol

The same tertiary carbocation serves as the precursor to numerous bicyclic monoterpenes. A carbocation having a bicyclic skeleton is formed by intramolecular attack of the π electrons of the double bond on the positively charged carbon.

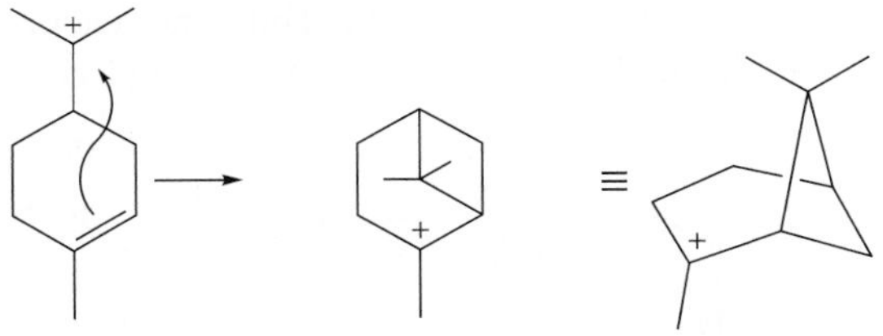

Bicyclic carbocation

Figure 24.9

Two of the reaction pathways available to the C_{10} bicyclic carbocation formed from neryl diphosphate. The same carbocation can lead to monoterpenes based on either the bicyclo[3.1.1] or the bicyclo[2.2.1] carbon skeleton.

A. Loss of a proton from the bicyclic carbocation yields α-pinene and β-pinene. The pinenes are the most abundant of the monoterpenes. They are the main constituents of turpentine.

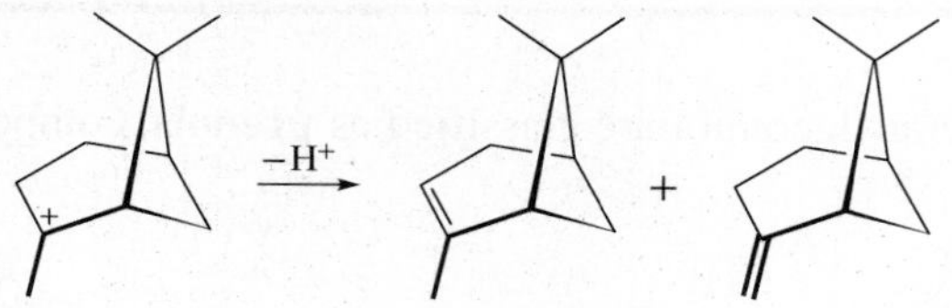

$-H^+$

α-Pinene + β-Pinene

B. Capture of the carbocation by water, accompanied by rearrangement of the bicyclo[3.1.1] carbon skeleton to a bicyclo[2.2.1] unit, yields borneol. Borneol is found in the essential oil of certain trees that grow in Indonesia.

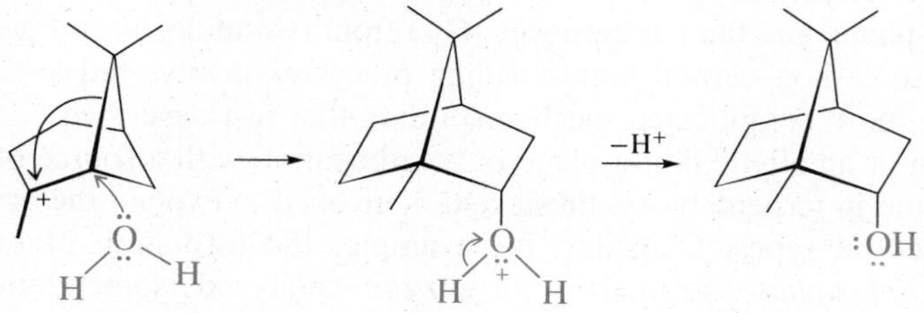

$-H^+$

Borneol

This bicyclic carbocation then undergoes many reactions typical of carbocation intermediates to provide a variety of bicyclic monoterpenes, as outlined in Figure 24.9.

Problem 24.11

The structure of the bicyclic monoterpene borneol is shown in Figure 24.9. Isoborneol, a stereoisomer of borneol, can be prepared in the laboratory by a two-step sequence. In the first step, borneol is oxidized to camphor by treatment with chromic acid. In the second step, camphor is reduced with sodium borohydride to a mixture of 85% isoborneol and 15% borneol. On the basis of these transformations, deduce structural formulas for isoborneol and camphor.

Analogous processes involving cyclizations and rearrangements of carbocations derived from farnesyl diphosphate produce a rich variety of structural types in the sesquiterpene series. We will have more to say about the chemistry of higher terpenes, especially the triterpenes, later in this chapter. For the moment, however, let's return to smaller molecules to complete the picture of how isoprenoid compounds arise from acetate.

24.10 The Pathway from Acetate to Isopentenyl Diphosphate

The introduction to Section 24.8 pointed out that mevalonic acid is the biosynthetic precursor of isopentenyl diphosphate. The early steps in the biosynthesis of mevalonate from three molecules of acetic acid are analogous to those in fatty acid biosynthesis except that they do not involve acyl carrier protein. Thus, the reaction of acetyl coenzyme A with malonyl coenzyme A yields a molecule of acetoacetyl coenzyme A.

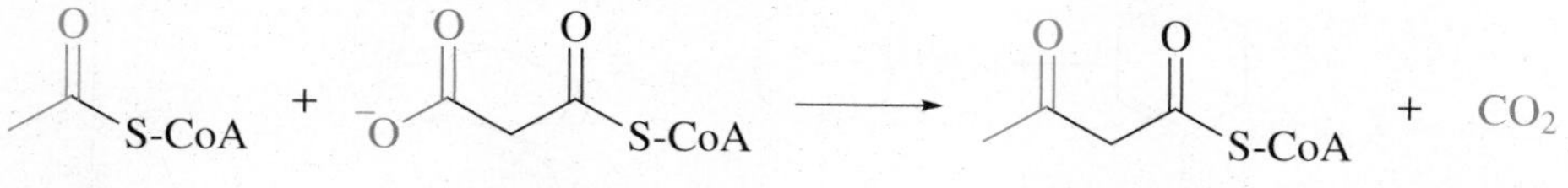

Acetyl coenzyme A + Malonyl coenzyme A ⟶ Acetoacetyl coenzyme A + Carbon dioxide (CO_2)

Carbon–carbon bond formation then occurs between the ketone carbonyl of acetoacetyl coenzyme A and the α carbon of a molecule of acetyl coenzyme A.

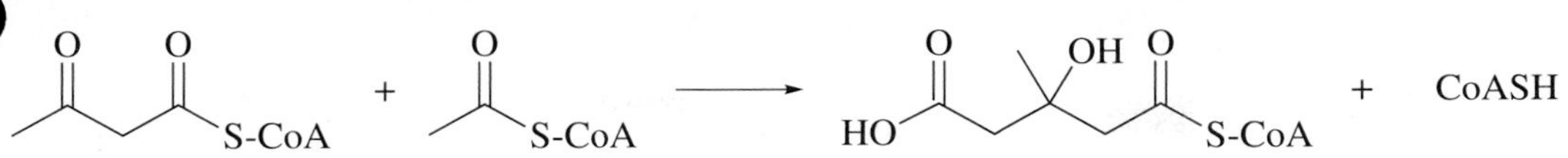

Acetoacetyl coenzyme A + Acetyl coenzyme A → 3-Hydroxy-3-methylglutaryl coenzyme A (HMG-CoA) + CoASH (Coenzyme A)

The product of this reaction, 3-hydroxy-3-methylglutaryl coenzyme A (HMG CoA), has the carbon skeleton of mevalonic acid and is converted to it by enzyme-catalyzed reduction.

3-Hydroxy-3-methylglutaryl coenzyme A (HMG-CoA) → Mevalonic acid

Some of the most effective cholesterol-lowering drugs act by inhibiting the enzyme that catalyzes this reaction.

In keeping with its biogenetic origin in three molecules of acetic acid, mevalonic acid has six carbon atoms. The conversion of mevalonate to isopentenyl diphosphate involves loss of the "extra" carbon as carbon dioxide. First, the alcohol hydroxyl groups of mevalonate are converted to phosphate ester functions—they are enzymatically *phosphorylated,* with introduction of a simple phosphate at the tertiary site and a diphosphate at the primary site. Decarboxylation, in concert with loss of the tertiary phosphate, introduces a carbon–carbon double bond and gives isopentenyl diphosphate, the fundamental building block for formation of isoprenoid natural products.

Some bacteria, algae, and plants make isopentenyl diphosphate by a different route.

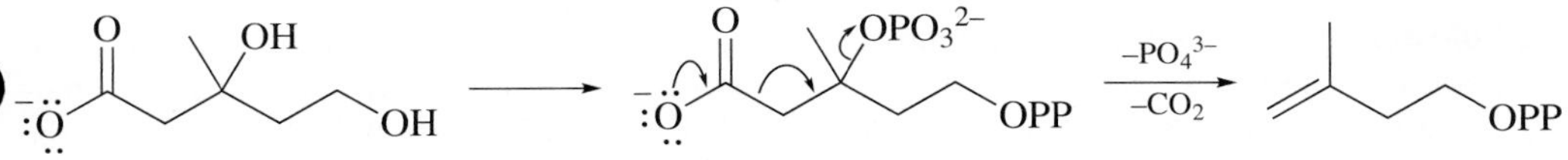

Mevalonate → Unstable; undergoes rapid decarboxylation with loss of phosphate (OPO_3^{2-}, OPP) $\xrightarrow[-CO_2]{-PO_4^{3-}}$ Isopentenyl diphosphate

Much of what we know concerning the pathway from acetate to mevalonate to isopentenyl diphosphate to terpenes comes from "feeding" experiments, in which plants are grown in the presence of radioactively labeled organic substances and the distribution of the radioactive label is determined in the products of biosynthesis. To illustrate, eucalyptus plants were allowed to grow in a medium containing acetic acid enriched with ^{14}C in its methyl group. *Citronellal* was isolated from the mixture of monoterpenes produced by the plants and shown, by a series of chemical degradations, to contain the radioactive ^{14}C label at carbons 2, 4, 6, and 8, as well as at the carbons of both branching methyl groups.

Citronellal occurs naturally as the principal component of citronella oil and is used as an insect repellent.

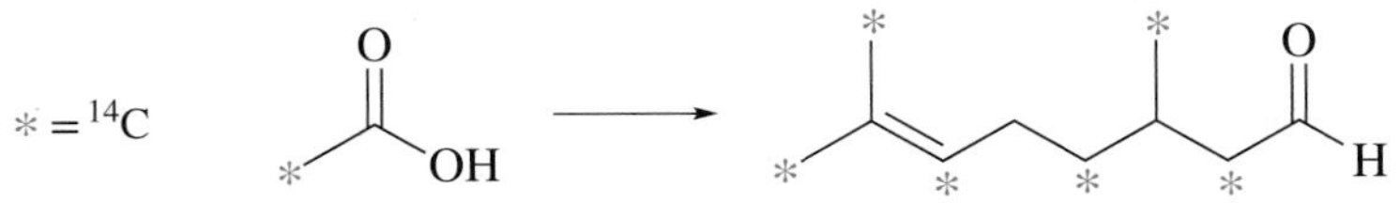

$* = {}^{14}C$

Acetic acid → Citronellal

Figure 24.10 traces the ^{14}C label from its origin in acetic acid to its experimentally determined distribution in citronellal.

Problem 24.12

How many carbon atoms of citronellal would be radioactively labeled if the acetic acid used in the experiment were enriched with ^{14}C at C-1 instead of at C-2? Identify these carbon atoms.

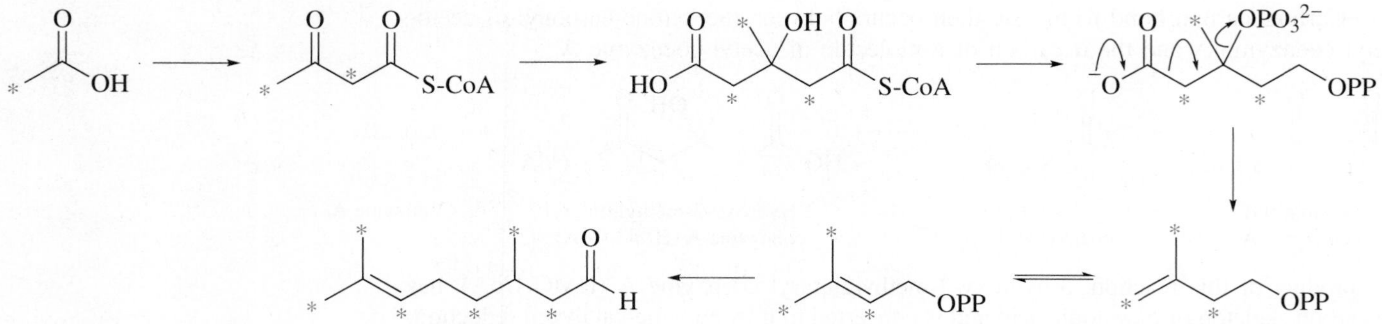

Figure 24.10

The distribution of the ^{14}C label in citronellal biosynthesized from acetate in which the methyl carbon was isotopically enriched with ^{14}C.

The present method employs ^{13}C as the isotopic label. Instead of locating the position of a radioactive ^{14}C label by a laborious degradation procedure, the ^{13}C NMR spectrum of the natural product is recorded. The signals for the carbons that are enriched in ^{13}C are far more intense than those corresponding to carbons in which ^{13}C is present only at the natural abundance level.

Isotope incorporation experiments have demonstrated the essential correctness of the just described scheme for terpene biosynthesis. Considerable effort has been expended toward its detailed elaboration because of the common biosynthetic origin of terpenes and another class of acetate-derived natural products, the steroids.

24.11 Steroids: Cholesterol

Cholesterol is the central compound in any discussion of **steroids.** Its name is a combination of the Greek words for "bile" (*chole*) and "solid" (*stereos*) preceding the characteristic alcohol suffix *-ol.* It is the most abundant steroid present in humans and the most important one as well because all other steroids arise from it. An average adult has over 200 g of cholesterol; it is found in almost all body tissues, with relatively large amounts present in the brain and spinal cord and in gallstones. Cholesterol is the chief constituent of the plaque that builds up on the walls of arteries in atherosclerosis.

Cholesterol was isolated in the eighteenth century, but its structure is so complex that its correct constitution was not determined until 1932 and its stereochemistry not verified until 1955. Steroids are characterized by the tetracyclic ring system shown in Figures 24.11*a* and 24.11*b*, and cholesterol in Figure 24.11*c* modified to include an alcohol function at C-3, a double bond at C-5, methyl groups at C-10 and C-13, and a C_8H_{17} side chain at C-17. Isoprene units may be discerned in various portions of the cholesterol molecule, but the

Figure 24.11

(*a*) The tetracyclic ring system characteristic of steroids and the customary designation of its rings as A, B, C, and D. (*b*) A conformational depiction of a typical steroid showing the stereochemistry of the ring fusions. (*c*) The structure of cholesterol and the steroid numbering system.

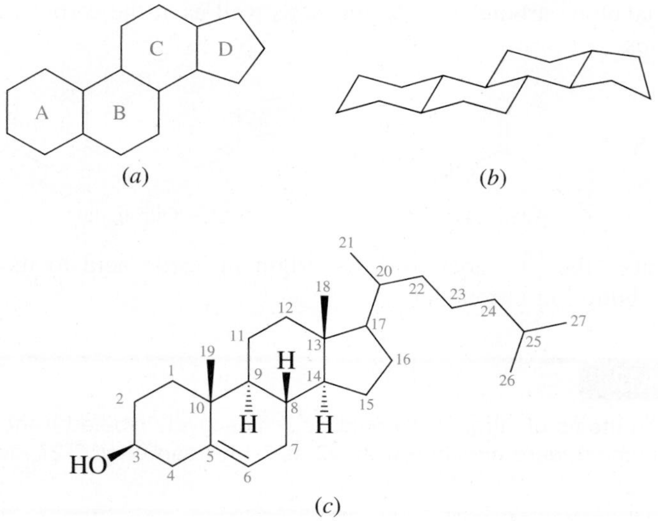

overall correspondence with the isoprene rule is far from perfect. Indeed, cholesterol has only 27 carbon atoms, three too few for it to be classed as a triterpene.

Animals accumulate cholesterol from their diet, but are also able to biosynthesize it from acetate. The pioneering work that identified the key intermediates in the complicated pathway of cholesterol biosynthesis was carried out by Konrad Bloch (Harvard) and Feodor Lynen (Munich). An important discovery was that the triterpene **squalene** (see Figure 24.8) is an intermediate in the formation of cholesterol from acetate. Thus, *the early stages of cholesterol biosynthesis are the same as those of terpene biosynthesis* described in Sections 24.8–24.10. In fact, a significant fraction of our knowledge of terpene biosynthesis is a direct result of experiments carried out in the area of steroid biosynthesis.

Bloch and Lynen shared the 1964 Nobel Prize in Physiology or Medicine.

How does the tetracyclic steroid cholesterol arise from the acyclic triterpene squalene? It begins with the epoxidation of squalene described earlier in Section 17.13 and continues from that point in Mechanism 24.1. Step 1 is an enzyme-catalyzed electrophilic ring opening of squalene 2,3-epoxide. Epoxide ring opening triggers a series of carbocation reactions. These carbocation processes involve cyclization via carbon–carbon bond formation (step 1), ring expansion via a carbocation rearrangement (step 2), another cyclization (step 3), accompanied by a cascade of methyl group migrations and hydride shifts (step 4). The result of all these steps is the tetracyclic triterpene *lanosterol*. Step 5 of Mechanism 24.1 summarizes the numerous remaining transformations by which lanosterol is converted to cholesterol.

Lanosterol is one component of lanolin, a mixture of many substances that coats the wool of sheep.

Problem 24.13

The biosynthesis of cholesterol as outlined in Mechanism 24.1 is admittedly quite complicated. It will aid your understanding of the process if you consider the following questions:

(a) Which carbon atoms of squalene 2,3-epoxide correspond to the doubly bonded carbons of cholesterol?

(b) Which two hydrogen atoms of squalene 2,3-epoxide are the ones that migrate in step 4?

(c) Which methyl group of squalene 2,3-epoxide becomes the methyl group at the C,D ring junction of cholesterol?

(d) What three methyl groups of squalene 2,3-epoxide are lost during the conversion of lanosterol to cholesterol?

Sample Solution (a) As the structural formula in step 5 of Mechanism 24.1 indicates, the double bond of cholesterol unites C-5 and C-6 (steroid numbering). The corresponding carbons in the cyclization reaction of step 1 in the figure may be identified as C-7 and C-8 of squalene 2,3-epoxide (systematic IUPAC numbering).

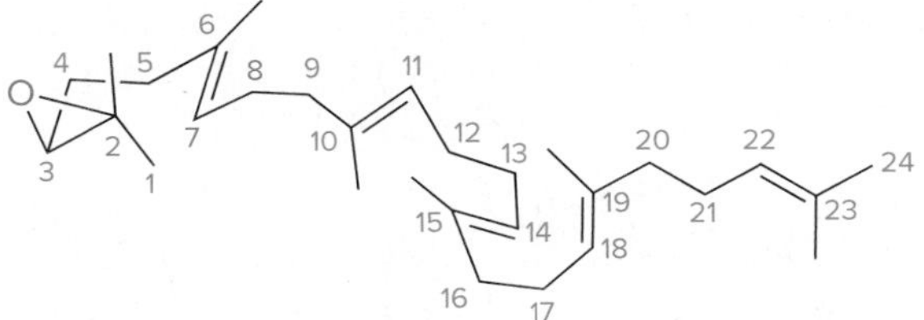

Squalene 2,3-epoxide

Problem 24.14

The biosynthetic pathway shown in Mechanism 24.1 was developed with the aid of isotopic labeling experiments. Which carbon atoms of cholesterol would you expect to be labeled when acetate enriched with ^{14}C in its methyl group ($^{14}CH_3COOH$) is used as the carbon source?

Once formed, cholesterol undergoes a number of biochemical transformations. A very common one is acylation of its C-3 hydroxyl group by reaction with coenzyme A derivatives of fatty acids. Other processes convert cholesterol to the biologically important steroids described in the following sections.

Mechanism 24.1

Biosynthesis of Cholesterol from Squalene

The biosynthetic conversion of squalene to cholesterol proceeds through lanosterol. Lanosterol is formed by enzyme-catalyzed cyclization of the 2,3-epoxide of squalene.

Step 1: An electrophilic species, shown here as $^{+}$Enz—H, catalyzes ring opening of squalene 2,3-epoxide. Ring opening is accompanied by cyclization to give a tricyclic tertiary carbocation. It is not known whether formation of the three new carbon–carbon bonds occurs in a single step or a series of steps.

+Enz—H O → HO H

Squalene 2,3-epoxide   Tricyclic carbocation

Step 2: Ring expansion converts the five-membered ring of the carbocation formed in step 1 to a six-membered ring.

HO H → HO H H

Tricyclic carbocation   Ring-expanded tricyclic carbocation

Step 3: Cyclization of the carbocation formed in step 2 gives a tetracyclic carbocation (*protosteryl cation*).

HO H H → HO H H

Ring-expanded tricyclic carbocation   Protosteryl cation

Step 4: Rearrangement and deprotonation of protosteryl cation gives the tetracyclic triterpene lanosterol.

H Ö H H HO H H → H $\overset{+}{O}$ H H HO H H

Protosteryl cation   Lanosterol

Step 5: A series of enzyme-catalyzed reactions converts lanosterol to cholesterol. The methyl groups at C-4 and C-14 are lost, the C-8 and C-24 double bonds are reduced, and a new double bond is introduced at C-5.

HO 4 5 8 14 H H 24 —many steps→ HO 5 6 H

Lanosterol   Cholesterol

24.12 Vitamin D

A steroid very closely related structurally to cholesterol is its 7-dehydro derivative. 7-Dehydrocholesterol is formed by enzymatic oxidation of cholesterol and has a conjugated diene unit in its B ring. 7-Dehydrocholesterol is present in the tissues of the skin, where it is transformed to vitamin D_3 by a sunlight-induced photochemical reaction.

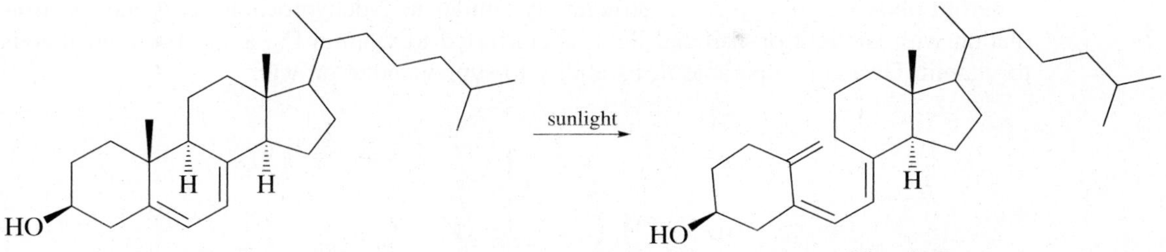

7-Dehydrocholesterol —sunlight→ Vitamin D_3

Vitamin D_3 is a key compound in the process by which Ca^{2+} is absorbed from the intestine. Low levels of vitamin D_3 lead to Ca^{2+} concentrations in the body that are insufficient to support proper bone growth, resulting in rachitis, or "rickets."

Rachitis was once thought to be a dietary deficiency disease because it could be prevented in children by feeding them fish liver oil. Actually, it is an environmental disease brought about by a deficiency of sunlight. Where the winter sun is weak, children may not be exposed to enough of its light to convert the 7-dehydrocholesterol in their skin to

Good Cholesterol? Bad Cholesterol? What's the Difference?

Cholesterol is biosynthesized in the liver, transported throughout the body to be used in a variety of ways, and returned to the liver where it serves as the biosynthetic precursor to other steroids. But cholesterol is a lipid and isn't soluble in water. How can it move through the blood if it doesn't dissolve in it? The answer is that it doesn't dissolve, but is instead carried through the blood and tissues as part of a *lipoprotein* (lipid + protein = lipoprotein).

The proteins that carry cholesterol from the liver are called *low-density lipoproteins*, or LDLs; those that return it to the liver are the *high-density lipoproteins*, or HDLs. If too much cholesterol is being transported by LDL, or too little by HDL, the extra cholesterol builds up on the walls of the arteries, causing atherosclerosis. Blood work done as part of a thorough physical examination measures not only total cholesterol but also the distribution between LDL and HDL cholesterol. An elevated level of LDL cholesterol is a risk factor for heart disease. LDL cholesterol is "bad" cholesterol. HDLs, on the other hand, remove excess cholesterol and are protective. HDL cholesterol is "good" cholesterol.

The distribution between LDL and HDL cholesterol depends mainly on genetic factors, but can be altered. Regular exercise increases HDL and reduces LDL cholesterol, as does limiting the amount of saturated fat in the diet. Much progress has been made in developing new drugs to lower cholesterol. The *statin* class, beginning with lovastatin in 1988, has proven especially effective. The most prescribed cholesterol-lowering drug is atorvastatin (as its calcium salt). A chiral drug, atorvastatin was introduced in 1997 and is sold as a single enantiomer.

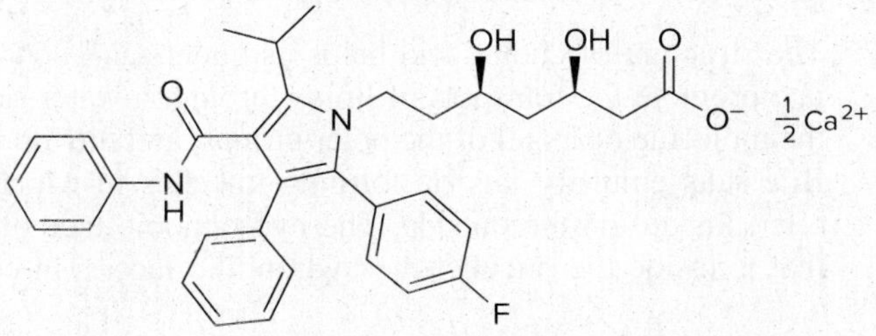

Atorvastatin calcium (Lipitor)

The statins lower cholesterol by inhibiting the enzyme 3-hydroxy-3-methylglutaryl coenzyme A reductase, which is required for the biosynthesis of mevalonic acid (see Section 24.10). Mevalonic acid is an obligatory precursor to cholesterol, so less mevalonic acid translates into less cholesterol.

vitamin D_3 at levels sufficient to promote the growth of strong bones. Fish have adapted to an environment that screens them from sunlight, and so they are not directly dependent on photochemistry for their vitamin D_3 and accumulate it by a different process. Although fish liver oil is a good source of vitamin D_3, it is not very palatable. Synthetic vitamin D_3, prepared from cholesterol, is often added to milk and other foods to ensure that children receive enough of the vitamin for their bones to develop properly. *Irradiated ergosterol* is another dietary supplement added to milk and other foods for the same purpose. Ergosterol, a steroid obtained from yeast, is structurally similar to 7-dehydrocholesterol and, on irradiation with sunlight or artificial light, is converted to vitamin D_2, a substance analogous to vitamin D_3 and comparable in its ability to support bone growth.

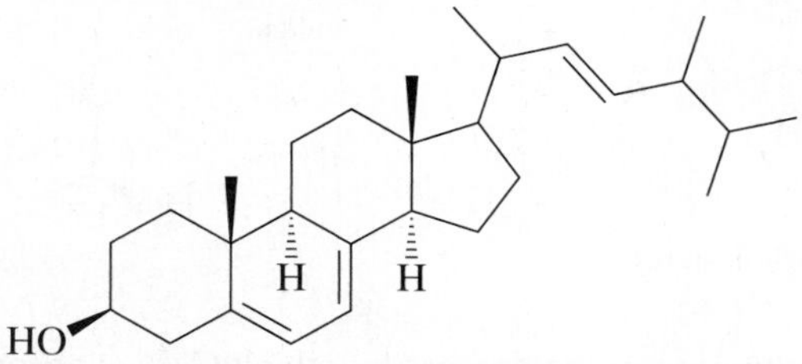

HO H H

Ergosterol

Problem 24.15

Suggest a reasonable structure for vitamin D_2.

24.13 Bile Acids

A significant fraction of the body's cholesterol is used to form **bile acids.** Oxidation in the liver removes a portion of the C_8H_{17} side chain, and additional hydroxyl groups are introduced at various positions on the steroid nucleus. *Cholic acid* is the most abundant of the bile acids. In the form of certain amide derivatives such as *sodium taurocholate,* bile acids act as emulsifying agents to aid the digestion of fats.

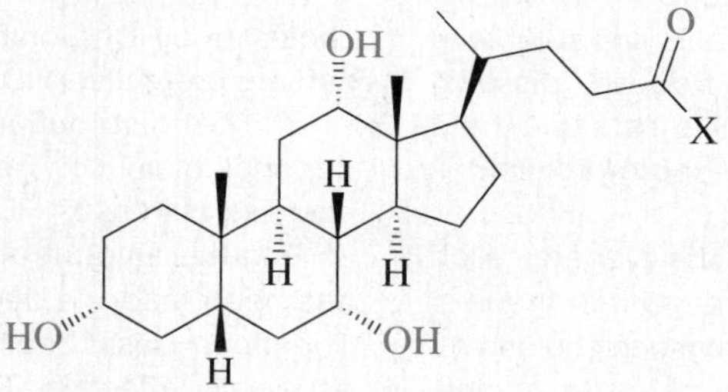

OH O X H H H HO OH H

X = OH; cholic acid
X = $NHCH_2CH_2SO_3Na$;
sodium taurocholate

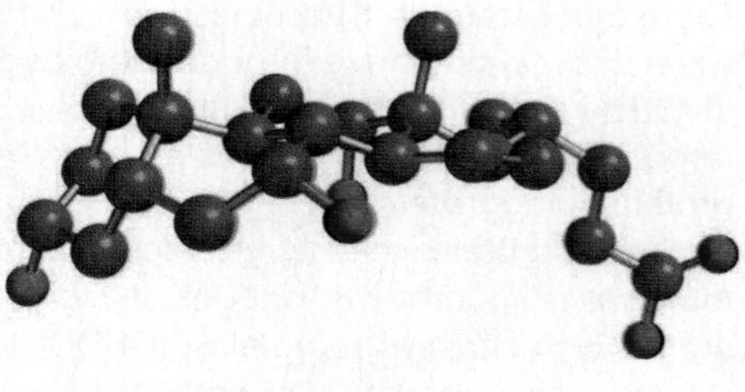

Cholic acid

The structure of cholic acid helps us understand how bile salts such as sodium taurocholate promote the transport of lipids through a water-rich environment. The bottom face of the molecule bears all of the polar groups, and the top face is exclusively hydrocarbon-like. Bile salts emulsify fats by forming micelles in which the fats are on the inside and the bile salts are on the outside. The hydrophobic face of the bile salt associates with the fat that is inside the micelle; the hydrophilic face is in contact with water on the outside.

24.14 Corticosteroids

The outer layer, or *cortex,* of the adrenal gland is the source of a large group of substances known as **corticosteroids.** Like the bile acids, they are derived from cholesterol by oxidation, with cleavage of a portion of the alkyl substituent on the D ring. *Cortisol* is the most abundant of the corticosteroids, but *cortisone* is probably the best known. Cortisone is commonly prescribed as an antiinflammatory drug, especially in the treatment of rheumatoid arthritis.

Cortisol

Cortisone

Corticosteroids exhibit a wide range of physiological effects. One important function is to assist in maintaining the proper electrolyte balance in body fluids. They also play a vital regulatory role in the metabolism of carbohydrates and in mediating the allergic response.

Many antiitch remedies contain dihydrocortisone.

24.15 Sex Hormones

Hormones are the chemical messengers of the body; they are secreted by the endocrine glands and regulate biological processes. Corticosteroids, described in the preceding section, are hormones produced by the adrenal glands. The sex glands—testes in males, ovaries in females—secrete a number of hormones that are involved in sexual development and reproduction. *Testosterone* is the principal male sex hormone; it is an **androgen.** Testosterone promotes muscle growth, deepening of the voice, the growth of body hair, and other male secondary sex characteristics. Testosterone is formed from cholesterol and is the biosynthetic precursor of estradiol, the principal female sex hormone, or **estrogen.** *Estradiol* is a key substance in the regulation of the menstrual cycle and the reproductive process. It is the hormone most responsible for the development of female secondary sex characteristics.

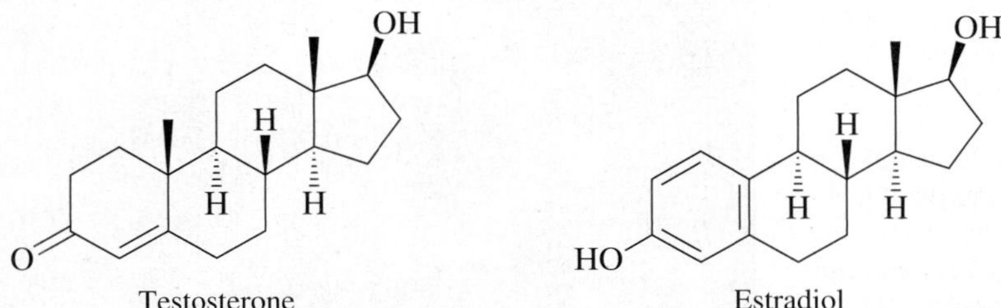

Testosterone Estradiol

Testosterone and estradiol are present in the body in only minute amounts, and their isolation and identification required heroic efforts. In order to obtain 0.012 g of estradiol for study, for example, 4 tons of sow ovaries had to be extracted!

A separate biosynthetic pathway leads from cholesterol to *progesterone,* a female sex hormone. One function of progesterone is to suppress ovulation at certain stages of the menstrual cycle and during pregnancy. Synthetic substances, such as *norethindrone,* have been developed that are superior to progesterone when taken orally to "turn off" ovulation. By inducing temporary infertility, they form the basis of most oral contraceptive agents.

Progesterone

Norethindrone

24.16 Carotenoids

Carotenoids are natural pigments characterized by a tail-to-tail linkage between two C_{20} units and an extended conjugated system of double bonds. They are the most widely distributed of the substances that give color to our world and occur in flowers, fruits, plants, insects, and animals. It has been estimated that biosynthesis from acetate produces approximately a hundred million tons of carotenoids per year. The most familiar carotenoids are lycopene and β-carotene, pigments found in numerous plants and easily isolable from ripe tomatoes and carrots, respectively.

Lycopene (tomatoes)

R = H; β-Carotene (carrots)
R = OH; Zeaxanthyn (yellow corn)

Crocuses Make Saffron from Carotenes

The flowers of *Crocus sativus* are not only pretty, they are valuable. The saffron crocus is cultivated on a large scale because the three gold-colored filaments in each bloom are the source of *saffron*, a dye and a spice that has been used for thousands of years. The amount is small; 75,000 flowers are needed to provide 1 pound of saffron, yet 300 tons of it reach the worldwide market each year.

Saffron is a mixture of substances. Those that make it desirable as a spice and dye are among the ones the plant uses to attract insects. Two of them, *crocetin* and *crocin*, are mainly responsible for its color, another (*safranal*) its odor, and another (*picrocrocin*) its taste. The same 20-carbon conjugated polyene unit is the chromophore that gives crocetin and crocin their yellow color. The difference between the two is that crocin is a glycoside in which both carboxylic acid functions of crocetin are attached to a disaccharide (*gentiobiose*) by ester linkages.

©imageBROKER/Alamy Stock Photo

Crocetin

Crocin

The 20-carbon chromophore originates in biochemical degradation of β-carotene and related carotenoids. Enzyme-catalyzed oxidation cleaves the double bonds at the points indicated to give crocetin.

β-Carotene

Safranal and picrocrocin are both aldehydes. Their structures suggest that they too come from carotenoid precursors. Because it is volatile, safranal contributes to the odor that attracts insects to the flowers. Picrocrocin is a glycoside. Its ability to participate in hydrogen bonding makes it nonvolatile and allows it to remain in place within the flowers where it provides the characteristic saffron flavor.

Safranal

Picrocrocin

Problem 24.16

Can you find the isoprene units in crocetin, crocin, safranal, and picrocrocin?

Not all carotenoids are hydrocarbons. Oxygen-containing carotenes called *xanthophylls,* which are often the pigments responsible for the yellow color of flowers, are especially abundant.

Carotenoids absorb visible light (see Section 14.23) and dissipate its energy as heat, providing protection from the potentially harmful effects of sunlight-induced photochemistry. They are also indirectly involved in the chemistry of vision, owing to the fact that β-carotene is the biosynthetic precursor of vitamin A, also known as retinol, a key substance in the visual process.

The structural chemistry of the visual process, beginning with β-carotene, was described in the boxed essay entitled *Imines in Biological Chemistry* in Chapter 18.

The color of a flamingo's feathers comes from the carotenes in the brine shrimp it eats.

24.17 SUMMARY

Section 24.1 Chemists and biochemists find it convenient to divide the principal organic substances present in cells into four main groups: *carbohydrates, proteins, nucleic acids,* and **lipids.** Structural differences separate carbohydrates from proteins, and both of these are structurally distinct from nucleic acids. Lipids are characterized by a *physical property,* their solubility in nonpolar solvents, rather than by their structure. In this chapter we have examined lipid molecules that share a common biosynthetic origin in that all their carbons are derived from acetic acid (acetate). The form in which acetate occurs in many of these processes is a thioester called acetyl coenzyme A, represented for convenience as:

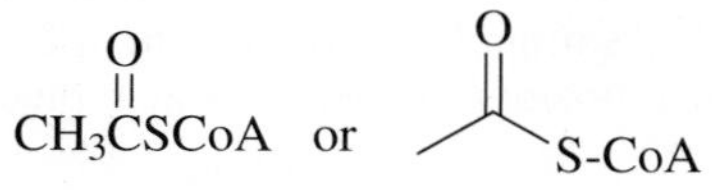

$$CH_3\overset{O}{\overset{\|}{C}}SCoA \quad \text{or} \quad \text{(skeletal structure)-S-CoA}$$

Section 24.2 Acetyl coenzyme A is the biosynthetic precursor to the **fatty acids,** which most often occur naturally as esters. **Fats** and **oils** are glycerol esters of long-chain carboxylic acids. Typically, these chains are unbranched and contain even numbers of carbon atoms.

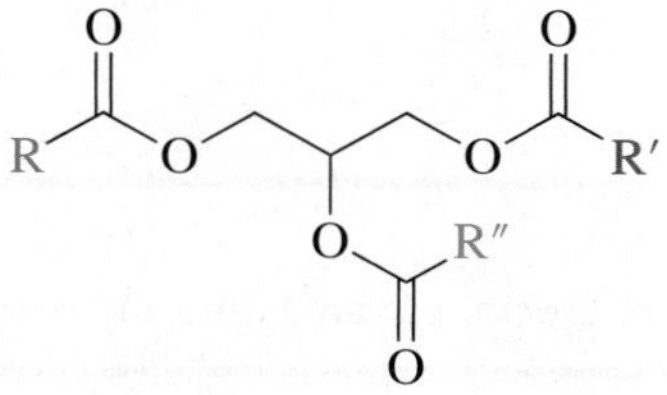

A triacylglycerol: R, R′, and R″ may be the same or different

Section 24.3 The biosynthesis of fatty acids follows the pathway outlined in Section 24.3. Malonyl coenzyme A is a key intermediate.

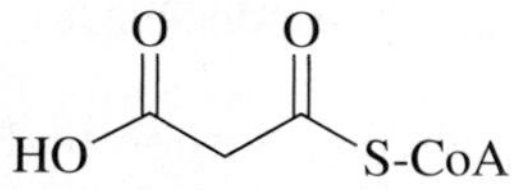

Malonyl coenzyme A

Section 24.4 **Phospholipids** are intermediates in the biosynthesis of triacylglycerols from fatty acids and are the principal constituents of the lipid bilayer component of cell membranes.

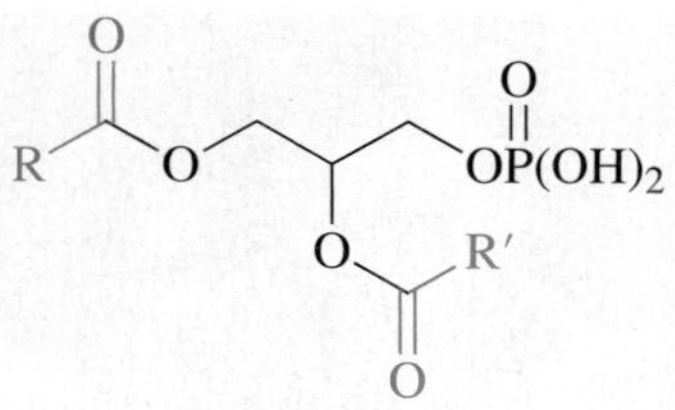

A phospholipid

Section 24.5 **Waxes** are mixtures of substances that usually contain esters of fatty acids and long-chain alcohols.

Section 24.6 **Icosanoids** are a group of naturally occurring compounds derived from unsaturated C_{20} carboxylic acids. Icosanoids include *prostaglandins, prostacyclins, thromboxanes,* and *leukotrienes.* Although present in very small amounts, icosanoids play regulatory roles in a very large number of biological processes.

Section 24.7 **Terpenes** have structures that follow the isoprene rule in that they can be viewed as collections of isoprene units.

β-Thujone

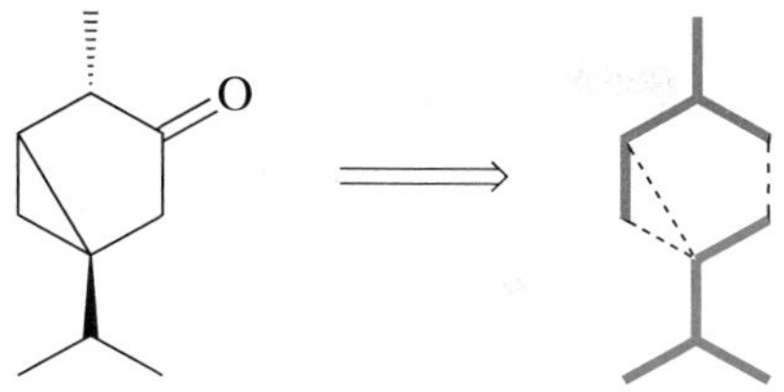

Section 24.8 Terpenes and related *isoprenoid* compounds are biosynthesized from *isopentenyl diphosphate*.

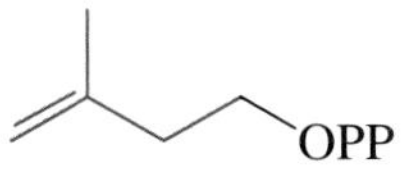

OPP

Isopentenyl diphosphate is the "biological isoprene unit."

Section 24.9 Carbon–carbon bond formation between isoprene units can be understood on the basis of nucleophilic attack of the π electrons of a double bond on a carbocation or an allylic carbon that bears a diphosphate leaving group.

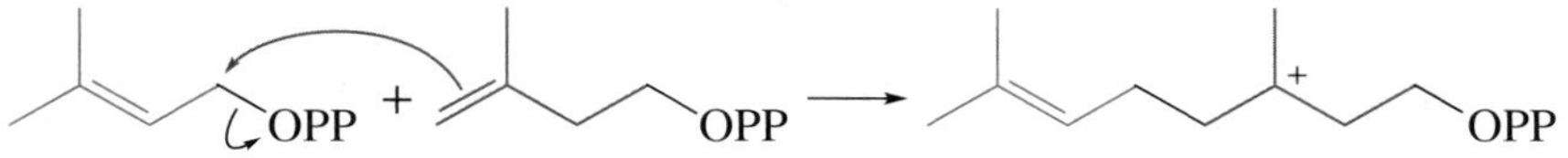

OPP + OPP → OPP

Section 24.10 The biosynthesis of isopentenyl diphosphate begins with acetate and proceeds by way of *mevalonic acid*.

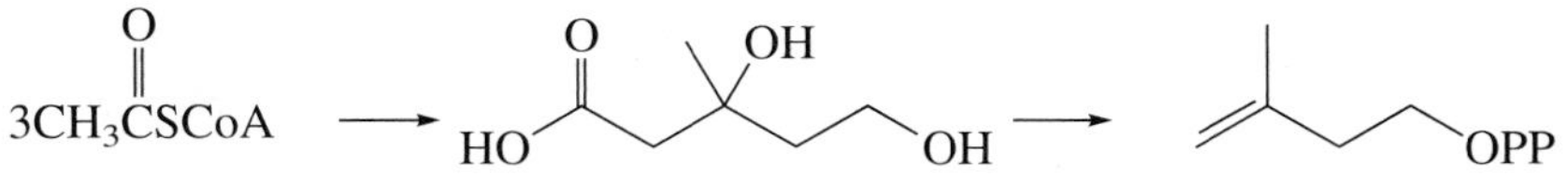

$3CH_3\overset{O}{\overset{\|}{C}}SCoA$ → HO, O, OH, OH → OPP

Acetyl coenzyme A — Mevalonic acid — Isopentenyl diphosphate

Section 24.11 The triterpene *squalene* is the biosynthetic precursor to cholesterol by the pathway shown in Mechanism 24.1.

Sections 24.12–24.15 Most of the steroids in animals are formed by biological transformations of cholesterol.

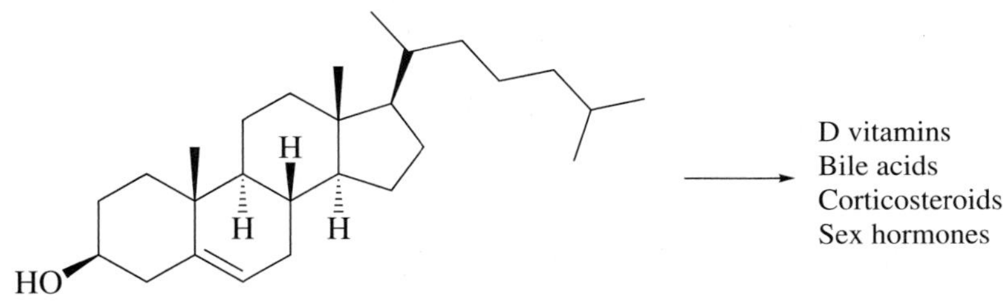

HO, H, H, H →
D vitamins
Bile acids
Corticosteroids
Sex hormones

Section 24.16 **Carotenoids** are tetraterpenes. They have 40 carbons and numerous double bonds. Many of the double bonds are conjugated, causing carotenes to absorb visible light and be brightly colored. They are often plant pigments.

PROBLEMS

Structure

24.17 The structures of each of the following are given within the chapter. Identify the carbon atoms expected to be labeled with ^{14}C when each is biosynthesized from acetate enriched with ^{14}C in its methyl group.

(a) Palmitic acid
(b) PGE_2
(c) PGI_2
(d) Limonene
(e) β-Carotene

24.18 Identify the isoprene units in each of the following naturally occurring substances:

(a) *Ascaridole,* a naturally occurring peroxide present in chenopodium oil:

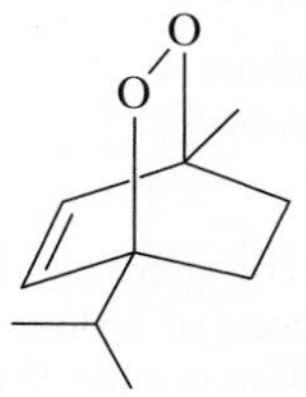

(b) *Dendrolasin,* a constituent of the defense secretion of a species of ant:

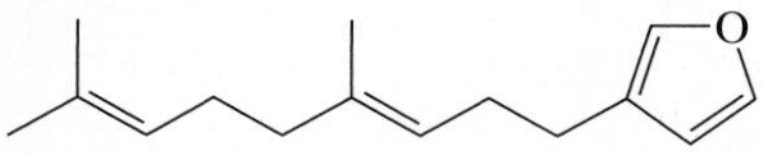

(c) γ-*Bisabolene,* a sesquiterpene found in the essential oils of a large number of plants:

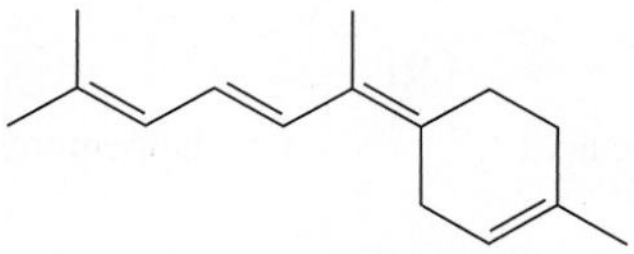

(d) α-*Santonin,* a lactone found in artemisia flowers:

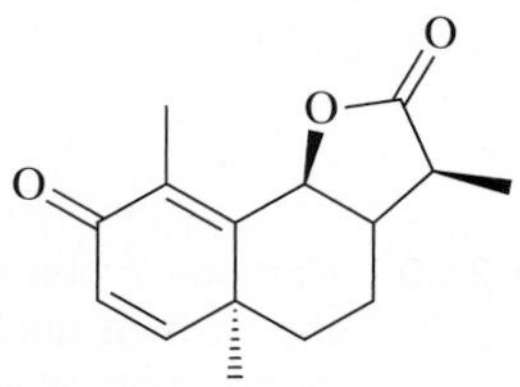

(e) *Tetrahymanol,* a pentacyclic triterpene isolated from a species of protozoa:

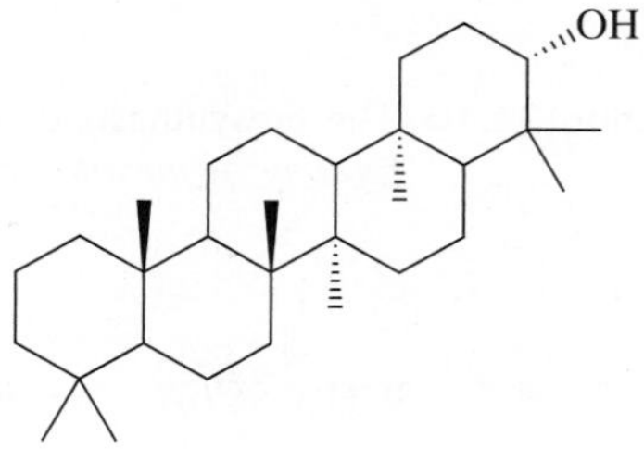

24.19 *Cubitene* is a diterpene present in the defense secretion of a species of African termite. What unusual feature characterizes the joining of isoprene units in cubitene?

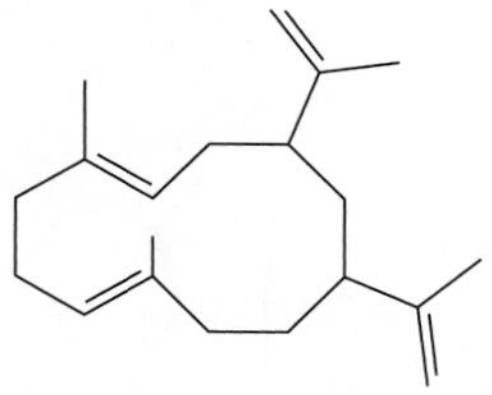

24.20 *Pyrethrins* are a group of naturally occurring insecticidal substances found in the flowers of various plants of the chrysanthemum family. The following is the structure of a typical pyrethrin, *cinerin I* (exclusive of stereochemistry):

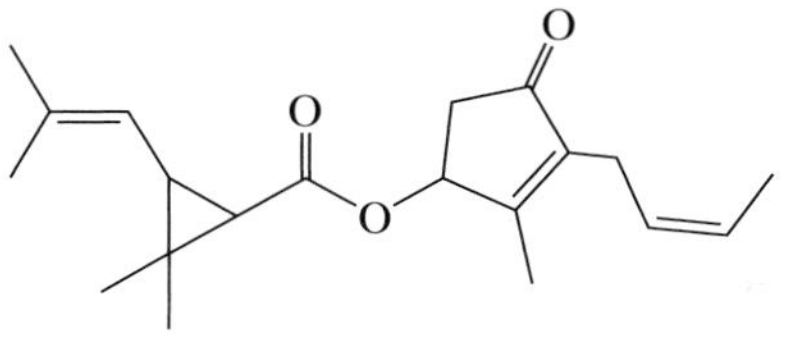

(a) Locate any isoprene units present in cinerin I.

(b) Hydrolysis of cinerin I gives an optically active carboxylic acid, (+)-chrysanthemic acid. Ozonolysis of (+)-chrysanthemic acid, followed by oxidation, gives acetone and an optically active dicarboxylic acid, (−)-caronic acid ($C_7H_{10}O_4$). What is the structure of (−)-caronic acid? Are the two carboxyl groups cis or trans to each other? What does this information tell you about the structure of (+)-chrysanthemic acid?

24.21 *Cerebrosides* are found in the brain and in the myelin sheath of nerve tissue. The structure of the cerebroside *phrenosine* is

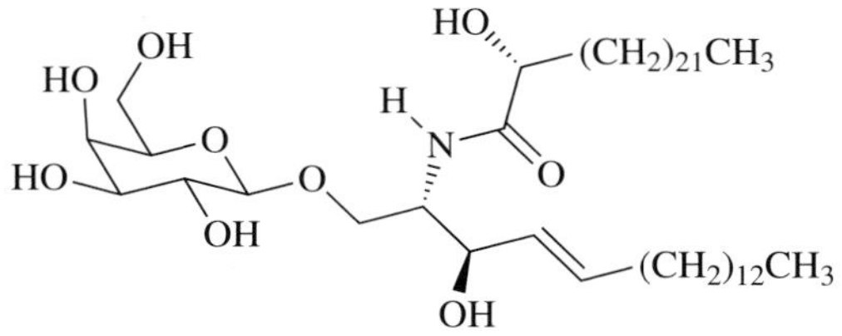

(a) What hexose is formed on hydrolysis of the glycoside bond of phrenosine? Is phrenosine an α- or a β-glycoside?

(b) Hydrolysis of phrenosine gives, in addition to the hexose in part (a), a fatty acid called *cerebronic acid,* along with a third substance called *sphingosine.* Write structural formulas for both cerebronic acid and sphingosine.

Reactions and Mechanism

24.22 Each of the following reactions has been reported in the chemical literature and proceeds in good yield. What are the principal organic products of each reaction? In some of the exercises more than one diastereomer may be possible, but one of them is either the major product or the only product. For those reactions in which one diastereomer is formed preferentially, indicate its expected stereochemistry.

(a) $CH_3(CH_2)_7C{\equiv}C(CH_2)_7COOH + H_2 \xrightarrow{\text{Lindlar Pd}}$

(b) $CH_3(CH_2)_7C{\equiv}C(CH_2)_7COOH \xrightarrow[\text{2. } H^+]{\text{1. Li, } NH_3}$

(c) (Z)-$CH_3(CH_2)_7CH{=}CH(CH_2)_7\overset{O}{\overset{\|}{C}}OCH_2CH_3 + H_2 \xrightarrow{\text{Pt}}$

(d) (Z)-$CH_3(CH_2)_5\underset{OH}{\underset{|}{C}}HCH_2CH{=}CH(CH_2)_7\overset{O}{\overset{\|}{C}}OCH_3 \xrightarrow[\text{2. } H_2O]{\text{1. } LiAlH_4}$

(e) (Z)-$CH_3(CH_2)_7CH{=}CH(CH_2)_7COOH + C_6H_5\overset{O}{\overset{\|}{C}}OOH \longrightarrow$

(f) Product of part (e) + $H_3O^+ \longrightarrow$

(g) (Z)-$CH_3(CH_2)_7CH{=}CH(CH_2)_7COOH \xrightarrow[\text{2. } H^+]{\text{1. } OsO_4\text{, } (CH_3)_3COOH\text{, } HO^-}$

(h) 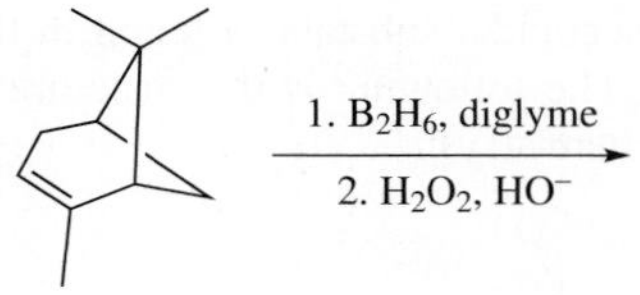 $\xrightarrow[\text{2. } H_2O_2,\ HO^-]{\text{1. } B_2H_6,\ \text{diglyme}}$

(i) 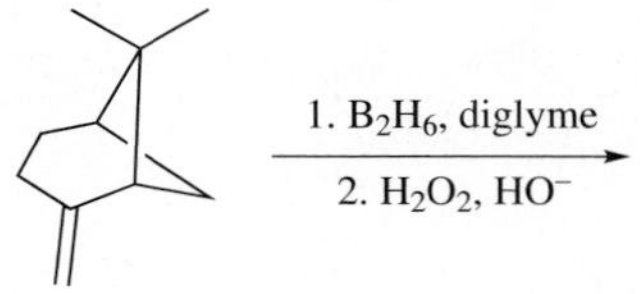 $\xrightarrow[\text{2. } H_2O_2,\ HO^-]{\text{1. } B_2H_6,\ \text{diglyme}}$

(j) 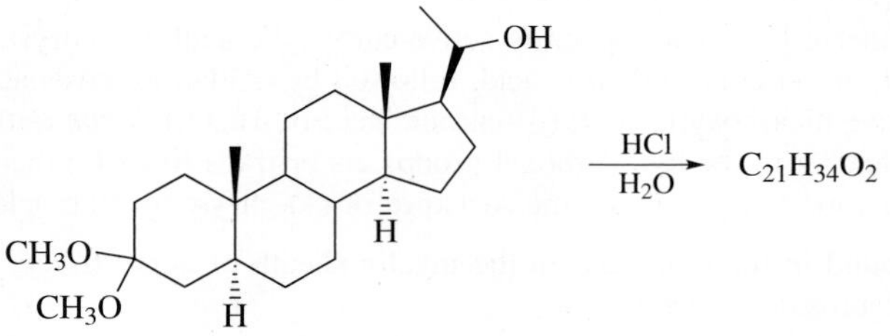 CH_3O, CH_3O, H, H, OH $\xrightarrow[H_2O]{HCl}$ $C_{21}H_{34}O_2$

24.23 Isoprene has sometimes been used as a starting material in the laboratory synthesis of terpenes. In one such synthesis, the first step is the electrophilic addition of 2 mol of hydrogen bromide to isoprene to give 1,3-dibromo-3-methylbutane.

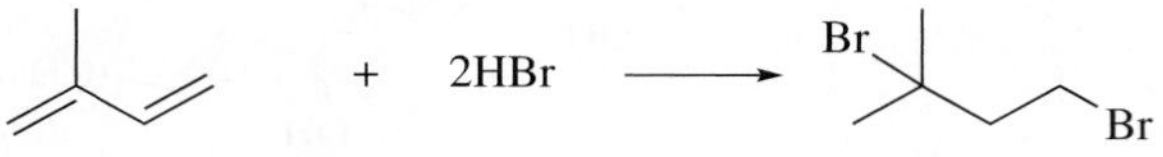

2-Methyl-1,3-butadiene (isoprene) + 2HBr (Hydrogen bromide) ⟶ 1,3-Dibromo-3-methylbutane (Br, Br)

Write a series of equations describing the mechanism of this reaction.

24.24 The ionones are fragrant substances present in the scent of iris and are used in perfume. A mixture of α- and β-ionone can be prepared by treatment of pseudoionone with sulfuric acid.

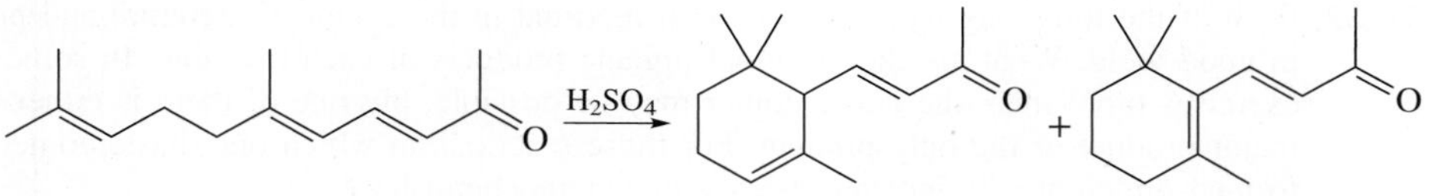

Pseudoionone (O) $\xrightarrow{H_2SO_4}$ α-Ionone (O) + β-Ionone (O)

Write a stepwise mechanism for this reaction.

24.25 β,γ-Unsaturated steroidal ketones represented by the partial structure shown here are readily converted in acid to their α,β-unsaturated isomers. Write a stepwise mechanism for this reaction.

O $\xrightarrow{H_3O^+}$ O

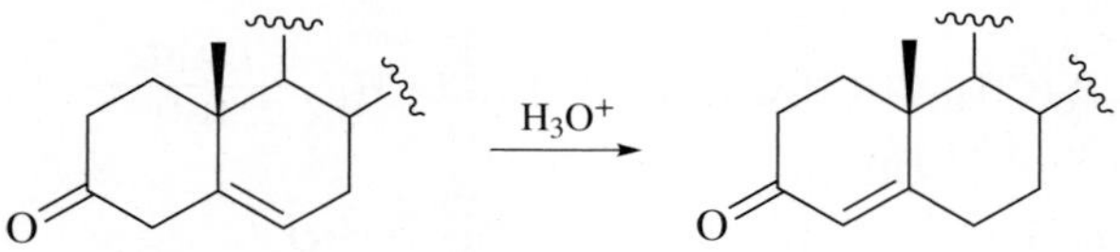

24.26 (a) Suggest a mechanism for the following reaction.

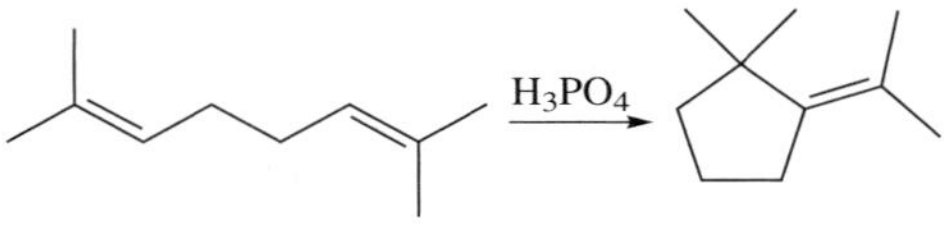

H_3PO_4

(b) The following two compounds are also formed in the reaction given in part (a). How are these two products formed?

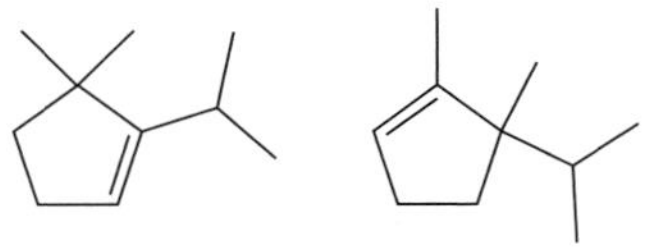

24.27 The following transformation was carried out as part of a multistep synthesis of digitoxigenin. Propose a mechanism.

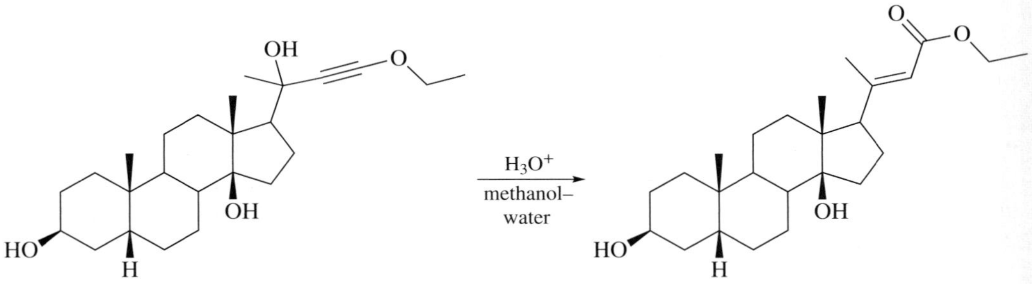

OH O OH HO H $\xrightarrow[\text{methanol–water}]{H_3O^+}$ O O OH HO H

Synthesis

24.28 Describe an efficient synthesis of each of the following compounds from octadecanoic (stearic) acid using any necessary organic or inorganic reagents:

(a) Octadecane
(b) 1-Phenyloctadecane
(c) 3-Ethylicosane
(d) Icosanoic acid
(e) 1-Octadecanamine
(f) 1-Nonadecanamine

24.29 A synthesis of triacylglycerols has been described that begins with the substance shown.

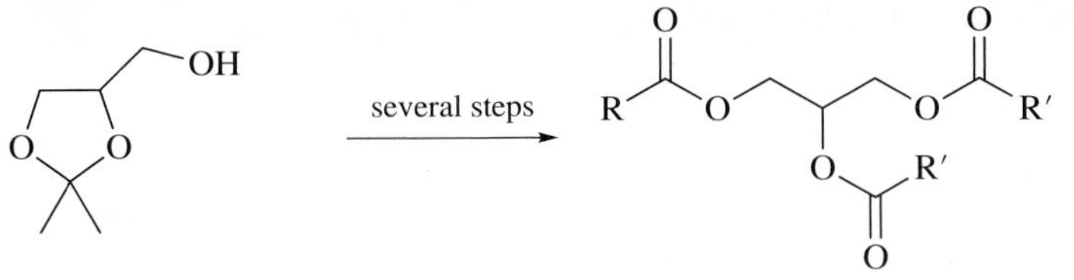

several steps

4-(Hydroxymethyl)-2,2-dimethyl-1,3-dioxolane

Triacylglycerol

Outline a series of reactions suitable for the preparation of a triacylglycerol of the type illustrated in the equation, where R and R′ are different.

24.30 The isoprenoid compound shown is a scent marker present in the urine of the red fox. Suggest a reasonable synthesis for this substance from 3-methyl-3-buten-1-ol and any necessary organic or inorganic reagents.

SCH_3

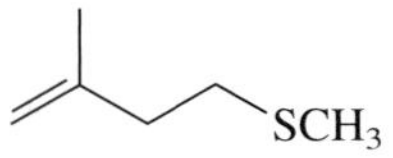

24.31 *Sabinene* is a monoterpene found in the oil of citrus fruits and plants. It has been synthesized from 6-methyl-2,5-heptanedione by the sequence that follows. Suggest reagents suitable for carrying out each of the indicated transformations. (*Hint:* See Section 15.8.)

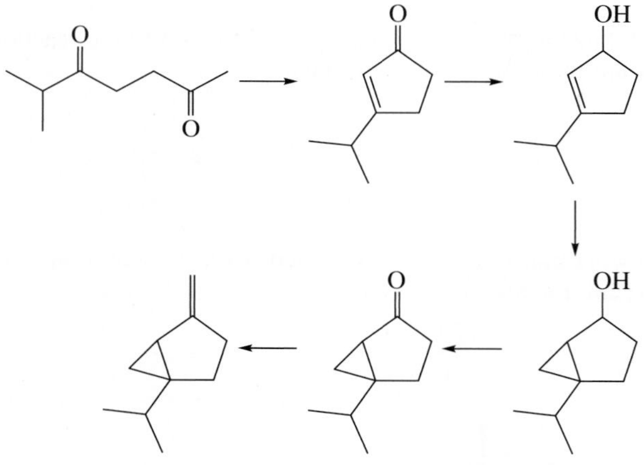

Sabinene

24.32 The glaucoma drug *bimatoprost* is synthesized in two steps from another prostaglandin. Can you suggest a method for this conversion?

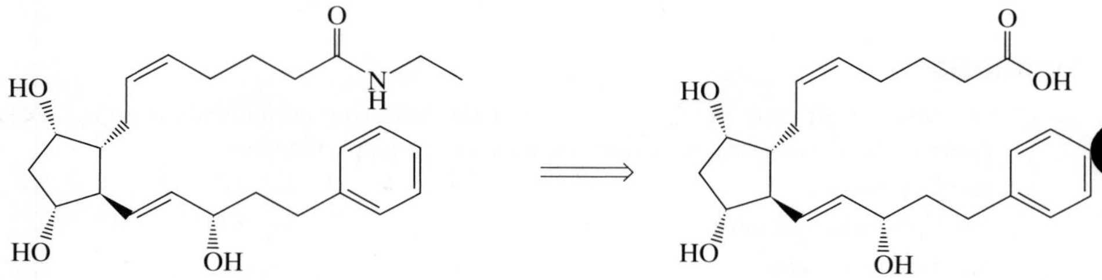

Bimatoprost

Descriptive Passage and Interpretive Problems 24

Polyketides

We have seen in this chapter that acetoacetate is an intermediate in both fatty acid and terpene biosynthesis. It is also an intermediate in the biosynthesis of the **polyketides,** a class of compounds of which more than 7000 are known to occur naturally. Polyketides are composed of alternating C=O and CH_2 groups as well as compounds derived from them. Their biosynthesis resembles fatty acid biosynthesis except that many of the carbonyl groups destined for reduction during fatty acid biosynthesis are retained in polyketide biosynthesis.

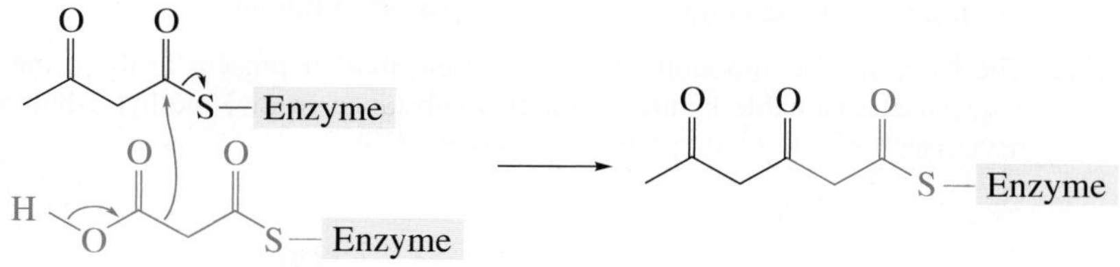

Acetoacetate S— Enzyme

Malonate H O S— Enzyme

S— Enzyme

Many polyketides have one or more methyl substituents on their carbon chain. In some cases *S*-adenosylmethionine is the source of a methyl group; in others methylmalonyl CoA (from propanoic acid) substitutes for acetate during chain assembly.

Acetoacetate
Methylmalonate
S— Enzyme
S— Enzyme
CH_3

1. An enolic OH derived from the β-diketone structural unit can act as the nucleophile in a nucleophilic acyl substitution to give a six-membered oxygen heterocycle known as a pyrone.

2. Intramolecular Claisen condensation gives 1,3,5-trihydroxybenzene.

3. Intramolecular aldol condensation of a slightly longer polyketide chain gives orsellinic acid.

hydrolysis
dehydration
enolization

The number and complexity of structural types that can arise via polyketides is magnified when one realizes that other reactions involving the carbon chain and its carbonyl groups can precede or follow cyclization. Although the number of polyketides for which precise biosynthetic details are known is limited, reasonable suggestions can be made as to their main elements based on a few basic principles of organic reaction mechanisms.

The first enzyme-free intermediate in the biosynthesis of erythromycin is the polyketide 6-deoxyerythronolide B. All of the carbons come from either acetate or propanoate.

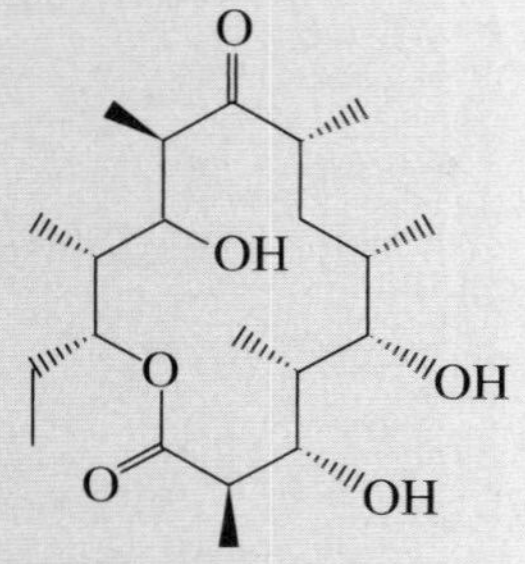

6-Deoxyerythronolide B ($C_{21}H_{38}O_6$)

24.33 How many of the carbons come from acetate? From propanoate?

	Acetate	Propanoate
A.	0	21
B.	6	15
C.	12	9
D.	18	3

24.34 How many methylmalonates are involved in the biosynthesis of 6-deoxyerythronolide B?

(a) 0 (b) 3 (c) 6 (d) 7

24.35 (+)-Discodermolide ($C_{33}H_{55}NO_8$) holds promise as an anticancer drug. Except for its amide carbonyl, all of the carbons of discodermolide are believed to come from acetate or propanoate. How many acetate units? How many propanoate units?

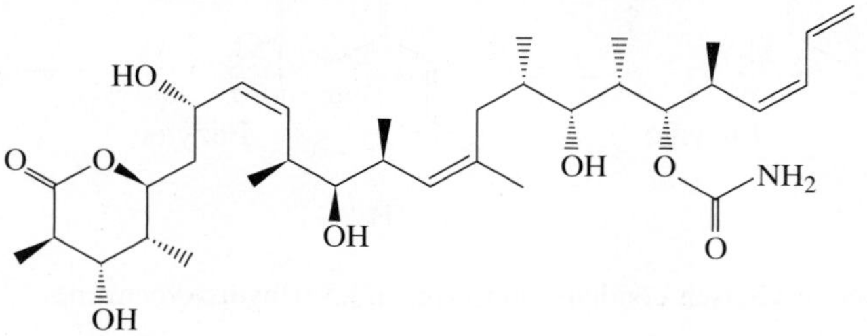

(a) 1 acetate; 10 propanoate
(b) 4 acetate; 8 propanoate
(c) 10 acetate; 4 propanoate
(d) 16 acetate; 0 propanoate

24.36 A key bond-forming step in the biosynthesis of naringenin chalcone is believed to involve an intramolecular Claisen condensation between C-1 and C-6 of the modified polyketide chain shown.

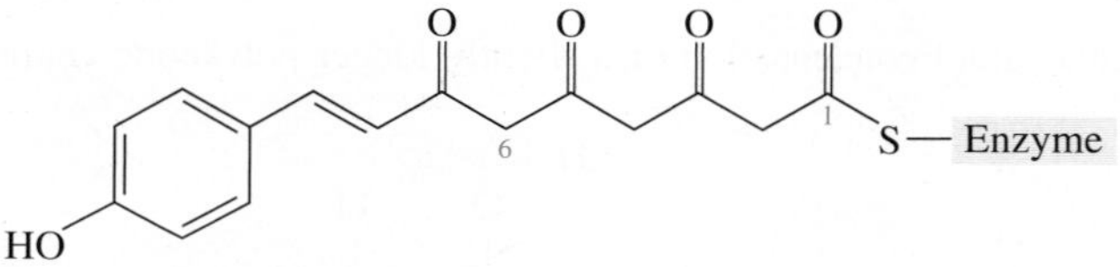

Which of the following is the most reasonable structure of naringenin chalcone based on this hypothesis?

A.

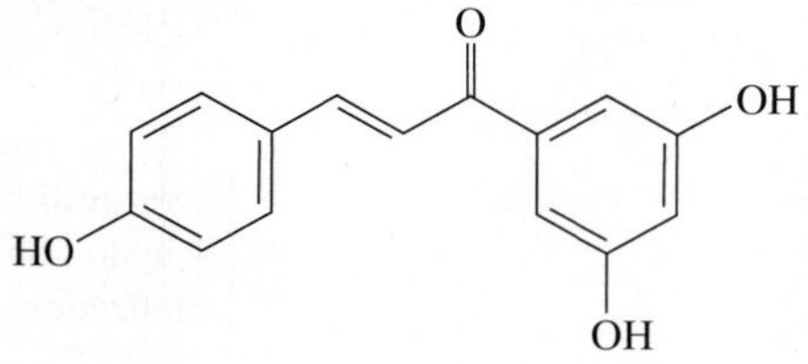

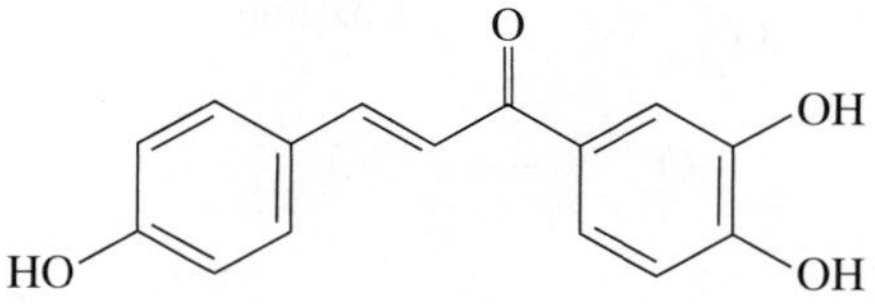

B.

C.

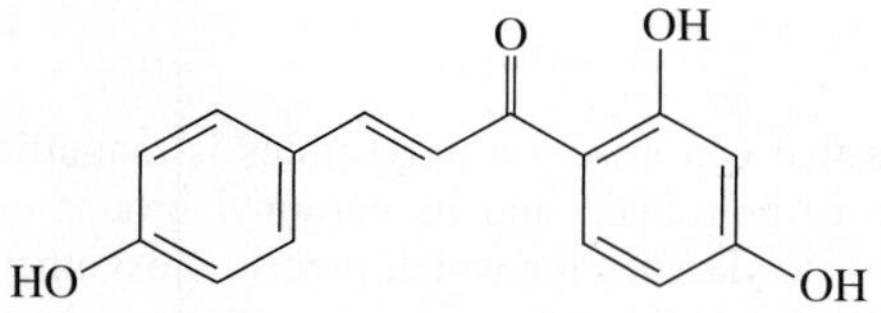

D.

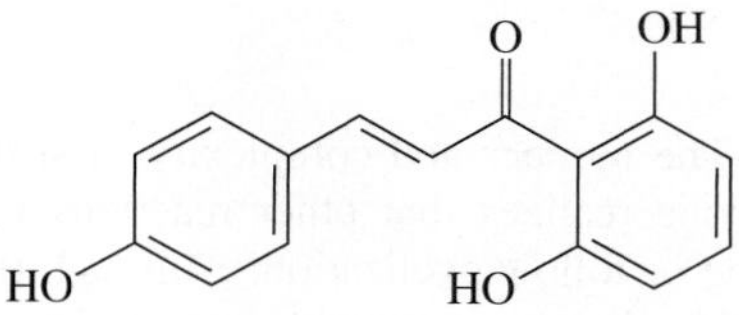

24.37 Carbon–carbon bond formation in the 14-carbon polyketo chain is suggested to be a key biosynthetic step leading to the compound shown.

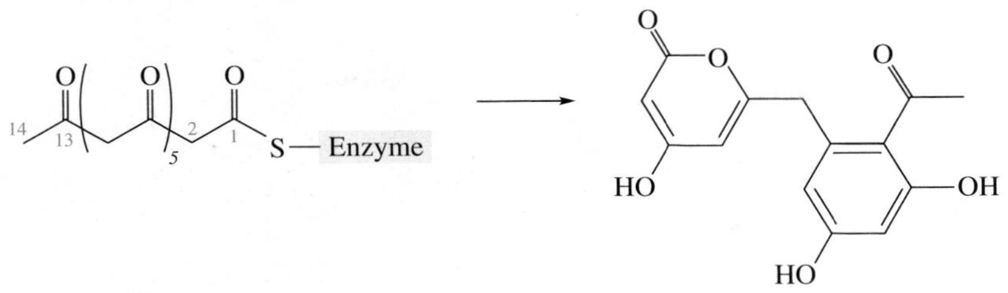

What two carbons are involved in this carbon–carbon bond-forming step?

(a) C-1 and C-5
(b) C-2 and C-14
(c) C-7 and C-12
(d) C-8 and C-13

24.38 Alternariol is a toxin produced by a mold that grows on agricultural products. It is a polyketide derived from seven acetate units. Which of the following is the most reasonable structure for alternariol?

A.

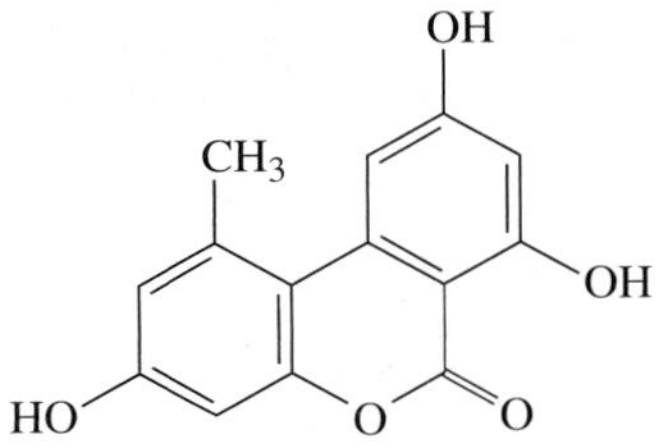

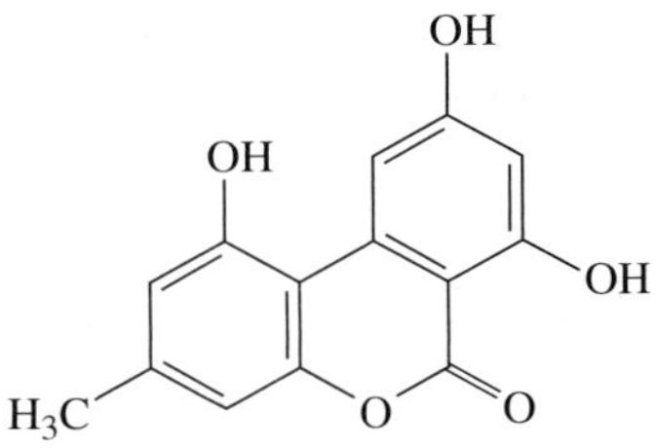

B.

C.

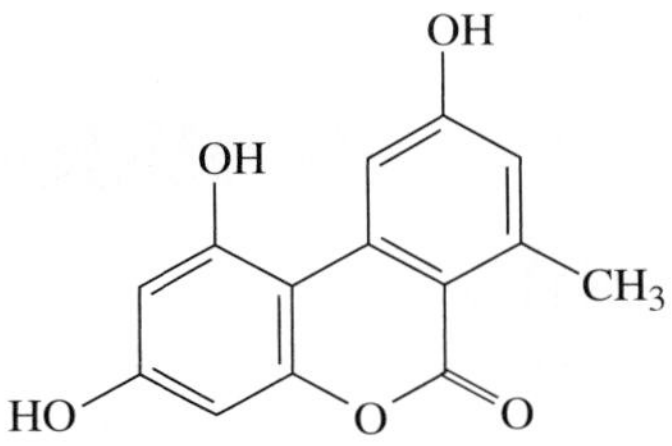

D.

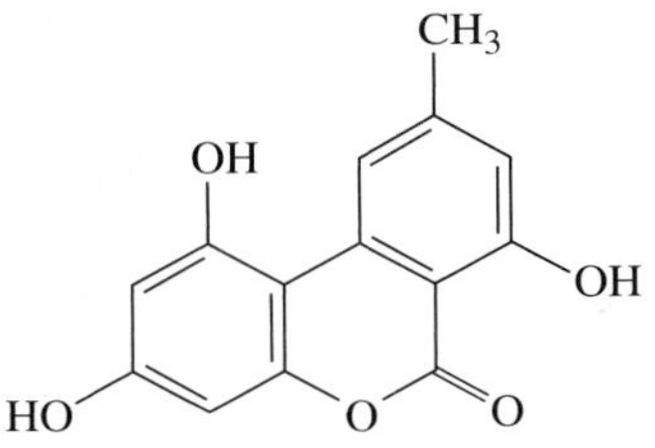

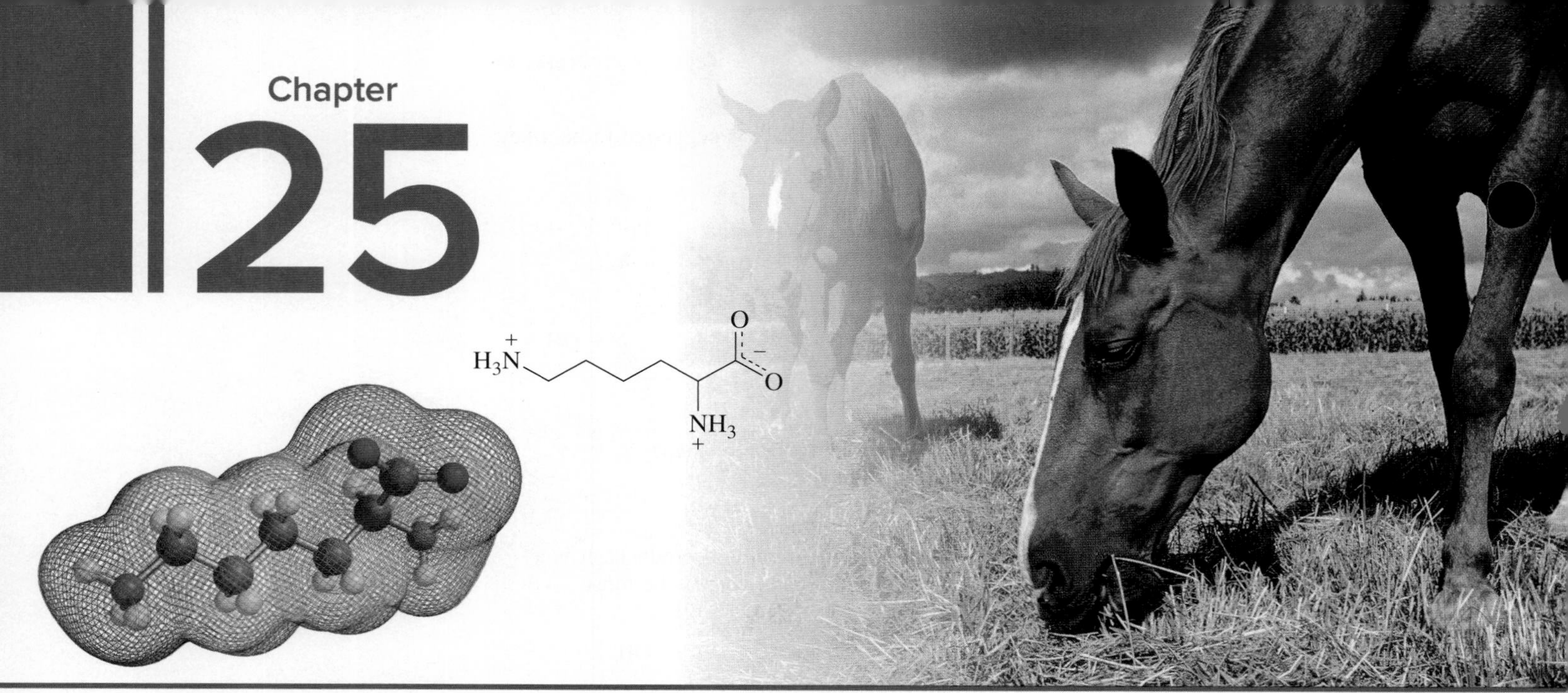

CHAPTER OUTLINE

Horses and many other animals, including humans, must obtain lysine from the food they eat because they can't biosynthesize it; thus, it is one of several "essential" amino acids. Commercial animal feed is often enriched in lysine obtained by methods based in biotechnology.

Amino Acids, Peptides, and Proteins

The relationship between structure and function reaches its ultimate expression in the chemistry of amino acids, peptides, and proteins.

Amino acids are carboxylic acids that contain an amine function. An amide bond between the carboxylic acid function of one amino acid and the amino nitrogen of another is called a **peptide bond.**

peptide bond

$H_3\overset{+}{N}$ R O O^- R′ $H_3\overset{+}{N}$ O O^- $H_3\overset{+}{N}$ R O N H R′ O O^-

Two α-amino acids   Dipeptide

A **dipeptide** is a molecule consisting of two amino acids joined by a peptide bond. A **tripeptide** has three amino acids joined by two peptide bonds, a **tetrapeptide** has four amino acids, and so on. Peptides with more than 30–50 amino acids are **polypeptides. Proteins** are polypeptides that have some biological function.

The most striking thing about proteins is the diversity of their roles in living systems: silk is a protein, skin and hair are mostly proteins, many hormones are proteins, a protein carries oxygen from the lungs to the tissues where it is stored by another protein, and all enzymes are proteins.

As in most aspects of chemistry and biochemistry, structure is the key to function. We'll explore the structure of proteins by first concentrating on their fundamental building-block units, the α-amino acids. Then, after developing the principles of peptide structure, we'll see how the insights gained from these smaller molecules aid our understanding of proteins.

25.1 Classification of Amino Acids

Amino acids are classified as α, β, γ, and so on, according to the location of the amine group on the carbon chain that contains the carboxylic acid function.

1-Aminocyclopropanecarboxylic acid: (an α-amino acid that is the biological precursor to ethylene in plants)

$H_3\overset{+}{N}CH_2CH_2CO_2^-$ (β, α)

3-Aminopropanoic acid: (known as β-alanine, it is a β-amino acid that makes up one of the structural units of coenzyme A)

$H_3\overset{+}{N}CH_2CH_2CH_2CO_2^-$ (γ, β, α)

4-Aminobutanoic acid: (known as γ-aminobutyric acid (GABA), it is a γ-amino acid and is involved in the transmission of nerve impulses)

Although more than 700 different amino acids are known to occur naturally, the group of 20 α-amino acids called the **standard amino acids** listed in Table 25.1 commands special attention. These 20 are coded for in DNA-directed protein synthesis, and all except proline are characterized by the general formula shown where R is a side chain. Proline has a cyclic side chain that incorporates the α-amino function.

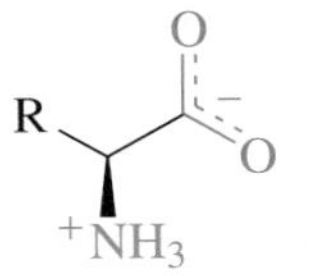

General formula of an α-amino acid

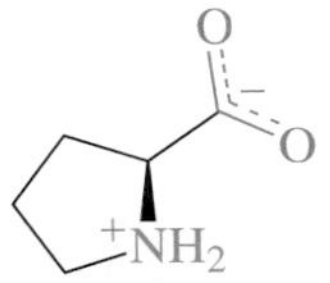

Proline

Two different formats are widely used to abbreviate the names of the standard amino acids. One employs three letters, the other one. Both are given in the table. Our bodies make 12 of the standard amino acids; the other 8 are called **essential amino acids,** which have to be obtained from our diet and are so noted in the table.

The most important aspect of Table 25.1 is that although the 20 standard amino acids share the common feature of being α-amino acids, their side chains differ in respect to their:

1. Size and shape
2. Electronic characteristics; acid–base properties; and ability to engage in ionic bonding, covalent bonding, hydrogen bonding, and van der Waals forces.

Table 25.1 shows the amino acids in the form in which they exist at physiological pH (7.4): amine groups as positively charged ammonium ions, and carboxylic acid groups as negatively charged carboxylates. Their electrostatic potential maps compare their shape and charge distribution.

Amino acids can be grouped by the chemical properties of their side chains. This grouping is a guide, since many side chains have properties common to more than one group.

Aliphatic Side Chains: *Glycine* ($H_3\overset{+}{N}CH_2CO_2^-$) has no side chain and is the smallest amino acid. It adds length and flexibility to a polypeptide chain without sacrificing strength or making spatial demands of its own. After glycine, the next four amino acids

TABLE 25.1 The Standard Amino Acids

Name and abbreviation	Structural formula*	Electrostatic potential map	Name and abbreviation	Structural formula*	Electrostatic potential map
Amino acids with aliphatic side chains					
Glycine Gly (G)	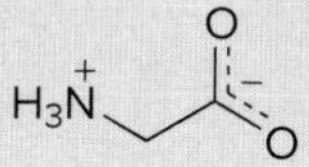	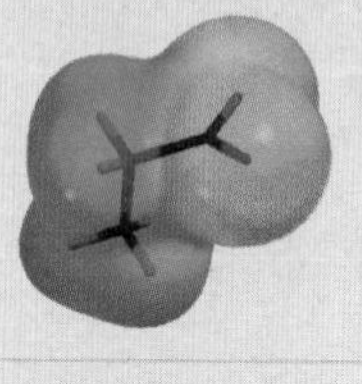	Proline Pro (P)	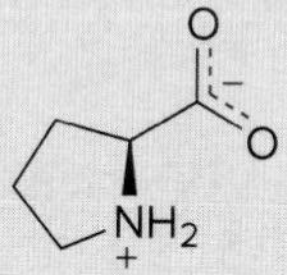	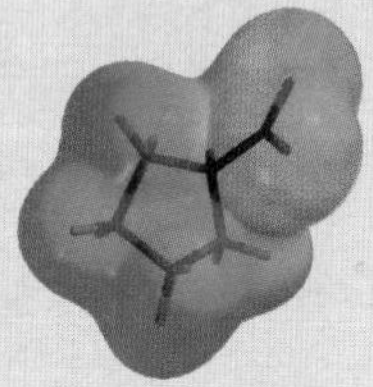
			Amino acids with aromatic side chains		
Alanine Ala (A)	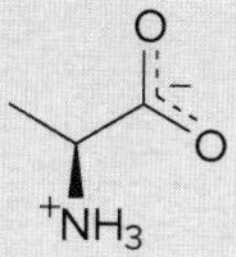	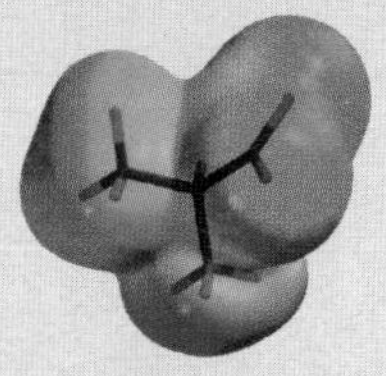	Phenylalanine† Phe (F)	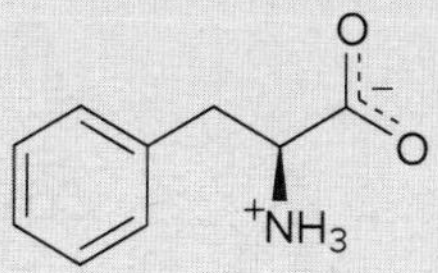	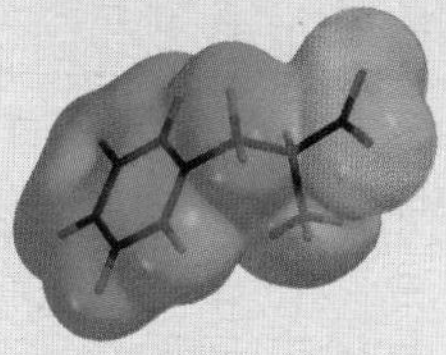
Valine† Val (V)	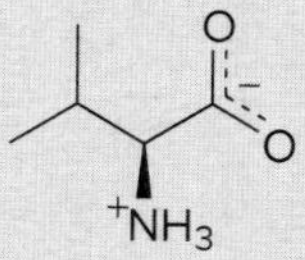	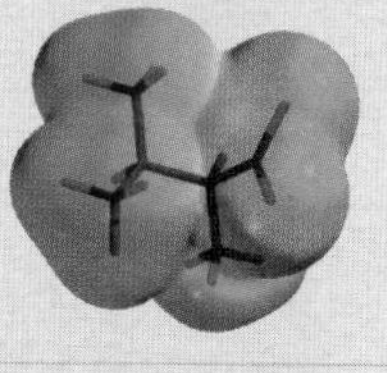	Tyrosine Tyr (Y)	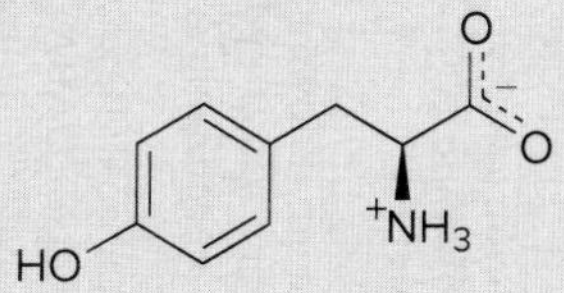	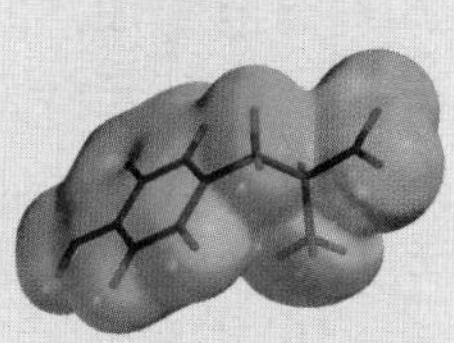
Leucine† Leu (L)	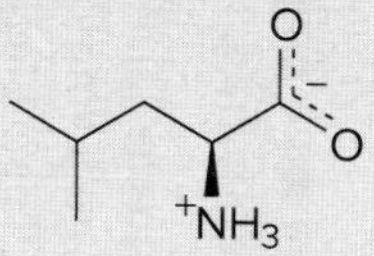	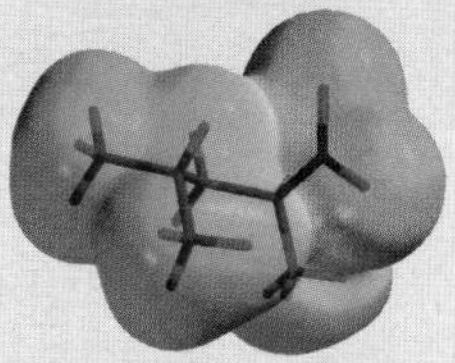	Tryptophan† Trp (W)	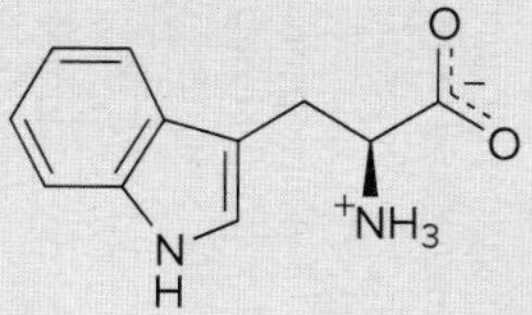	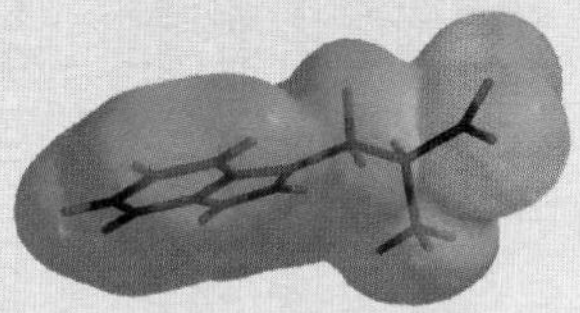
			Amino acids with polar but nonionized side chains		
Isoleucine† Ile (I)	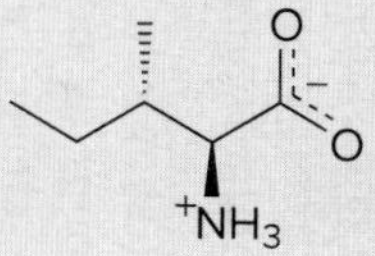	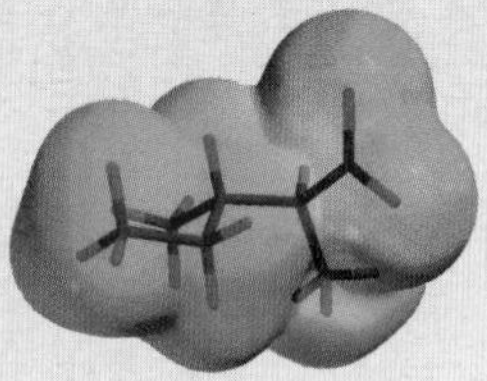	Serine Ser (S)	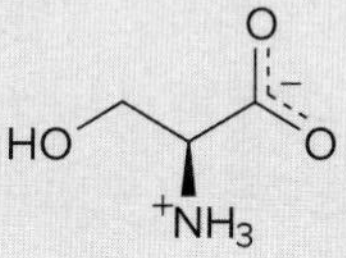	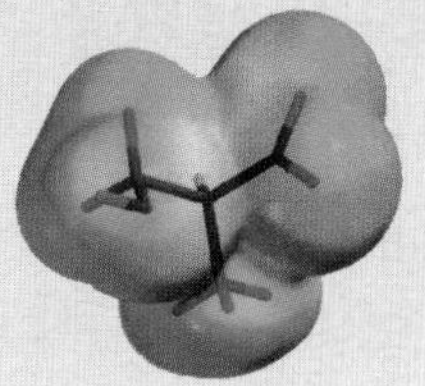

*All amino acids are shown in the form present in greatest concentration at pH 7.

†An essential amino acid, which must be present in the diet of animals to ensure normal growth.

TABLE 25.1 The Standard Amino Acids (*Continued*)

Name and abbreviation	Structural formula*	Electrostatic potential map	Name and abbreviation	Structural formula*	Electrostatic potential map
Amino acids with polar but nonionized side chains (*Continued*)			**Amino acids with acidic side chains**		
Threonine Thr (T)	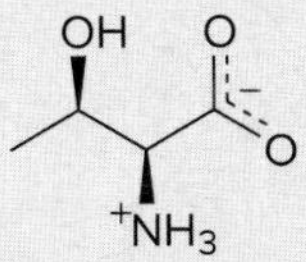	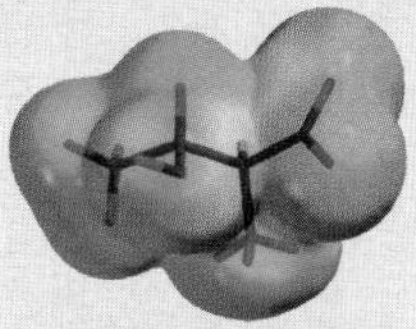	Aspartic acid Asp (D)	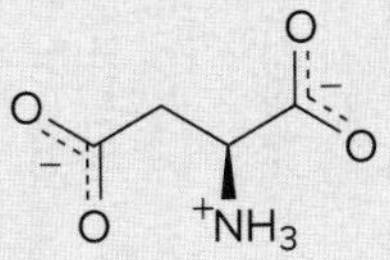	
Cysteine Cys (C)	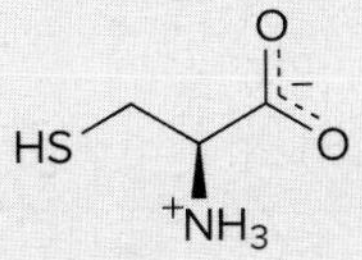		Glutamic acid Glu (E)	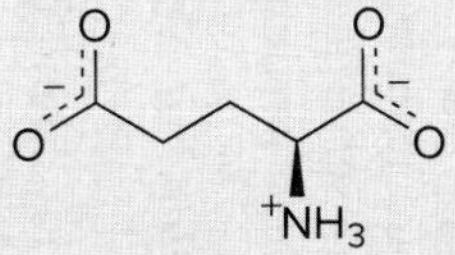	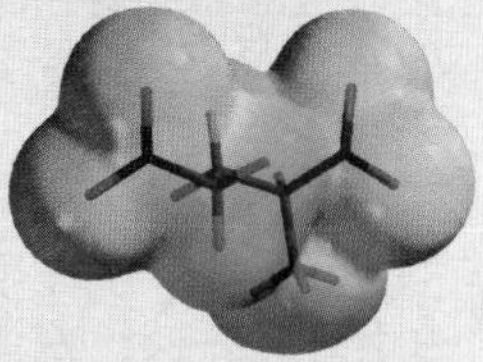
			Amino acids with basic side chains		
Methionine† Met (M)	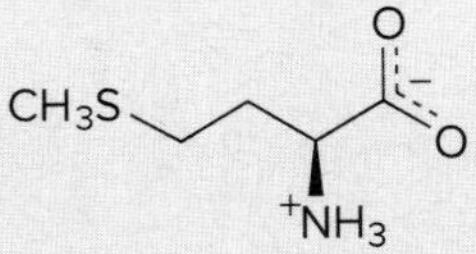	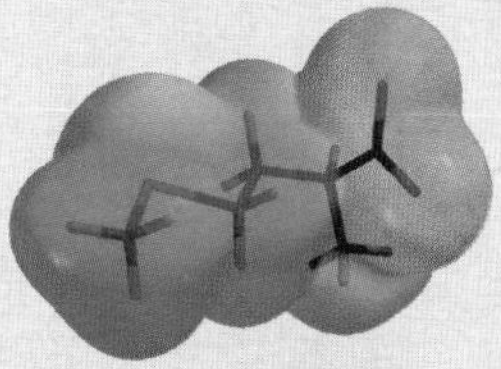	Lysine Lys (K)	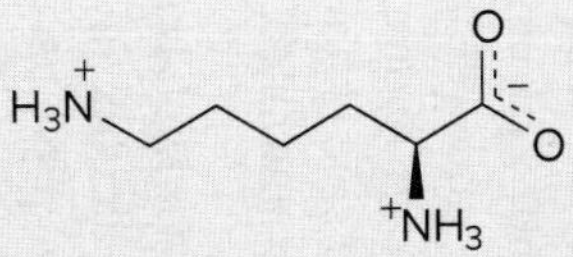	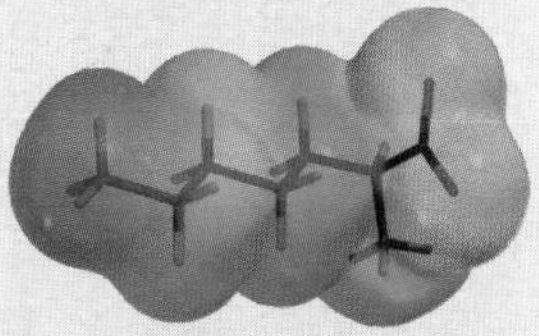
Asparagine Asn (N)	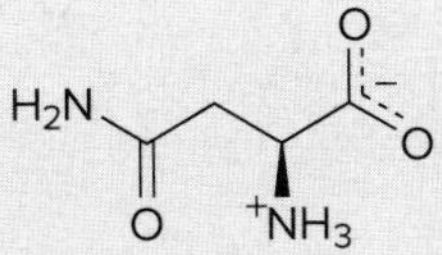	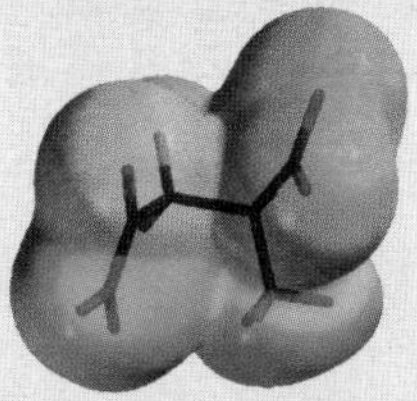	Arginine† Arg (R)	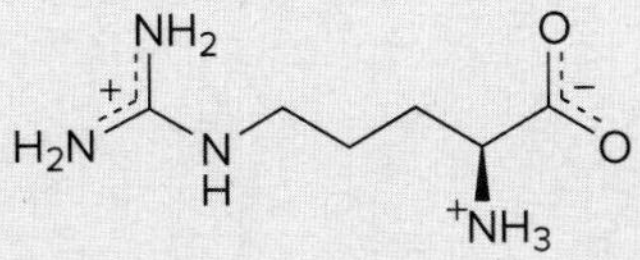	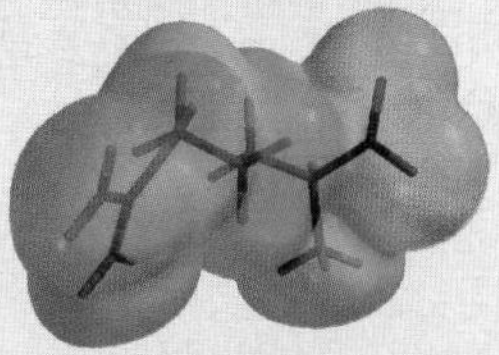
Glutamine Gln (Q)	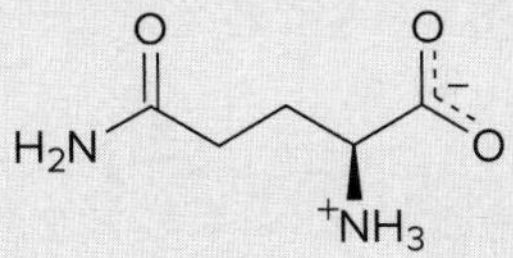	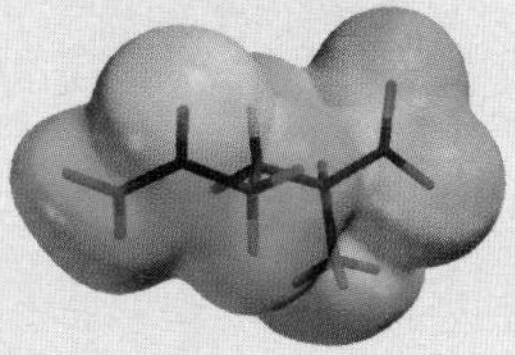	Histidine† His (H)	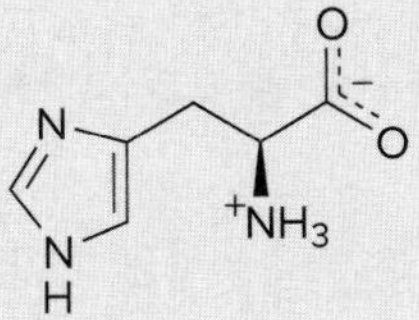	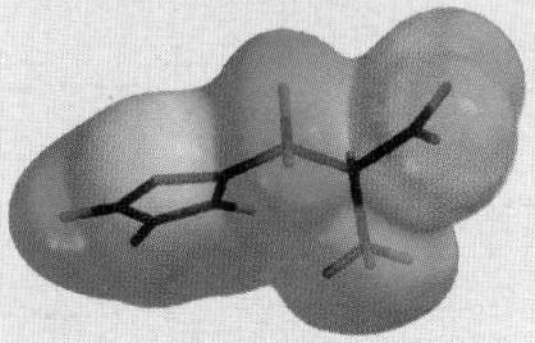

*All amino acids are shown in the form present in greatest concentration at pH 7.

†An essential amino acid, which must be present in the diet of animals to ensure normal growth.

in Table 25.1—*alanine, valine, leucine,* and *isoleucine*—all have simple alkyl groups as side chains: methyl, isopropyl, isobutyl, and *sec*-butyl, respectively. Like aliphatic hydrocarbons, the aliphatic side chains of amino acids are quite unreactive. These side chains are hydrophobic and, although electronically similar, differ in size and shape.

Proline is an aliphatic amino acid that serves important and unique structural roles in proteins. It is sometimes referred to by the archaic term *imino acid* because of the bond between the side chain methylene and the α-amino group. The five-membered ring restricts the conformation of the amino acid and peptide bonds between proline and other amino acids. Connective tissue proteins such as collagen are rich in proline.

Aromatic Amino Acids: Three amino acid side chains contain substituted benzene rings. *Phenylalanine, tyrosine,* and *tryptophan* absorb ultraviolet light (absorption maxima at ~260, 270, and 280 nm, respectively) and are responsible for the lower-energy absorption and fluorescence bands observed in many proteins. Phenylalanine and tryptophan are hydrophobic and are usually found on the interior of proteins. Owing to its phenolic side chain, tyrosine can serve additional roles in a protein, including hydrogen bond donor and acceptor. It can also participate in some enzymatic reactions.

Aliphatic Amino Acids with Polar but Nonionized Side Chains: Two amino acids contain an alcohol in the side chain. *Serine* is a small primary alcohol and serves many functions in proteins. It is frequently found on the surface of proteins and as part of the active site of enzymes as a hydrogen bond donor, a hydrogen bond acceptor, and a nucleophile. *Threonine* has a methyl group in place of one of the hydrogens of the $—CH_2OH$ group of serine, sterically hindering the hydroxyl group. It is less effective than serine in hydrogen bonding and it almost never behaves as a nucleophile in an enzyme.

Cysteine is related to serine except that its side chain is $—CH_2SH$ rather than $—CH_2OH$. This apparently minor difference in structure makes a huge difference in its role in proteins. Recall from Section 16.12 that the —SH group is a weaker acid and a better nucleophile, and it is more easily oxidized than the —OH group. When two cysteines are joined through side chain oxidation ($—CH_2—S—S—CH_2—$), the dimeric amino acid product is referred to as *cystine*. The number of cysteines in a protein is often relatively small, but they often serve important functions in the three-dimensional shape of proteins owing to their ability to undergo reversible oxidation to disulfide bonds. Within a protein, two oxidized cysteines form a disulfide bridge. These covalent bonds can link remote amino acids and are often critically important in the proper folding and maintenance of three-dimensional structures of proteins.

The other sulfur-containing amino acid is *methionine*. In terms of protein structure, the thioether-containing methionine has properties similar to those of aliphatic amino acids. As a free amino acid, however, it plays a vital role in cellular metabolism as the "universal methyl donor," S-adenosylmethionine.

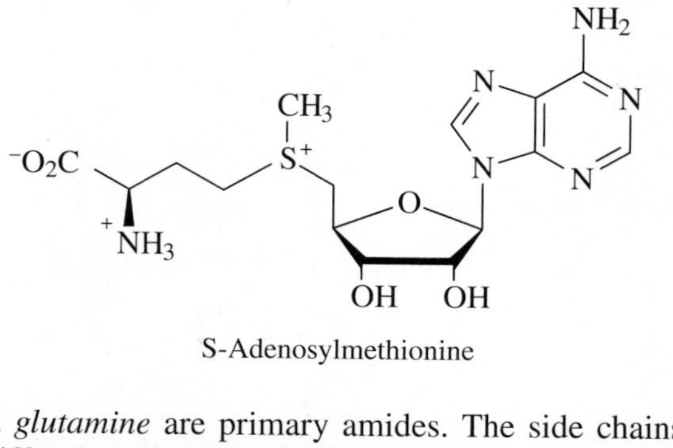

^-O_2C CH_3 S^+ $^+NH_3$ NH_2 N N N N O OH OH

S-Adenosylmethionine

Asparagine and *glutamine* are primary amides. The side chains of both terminate in $—C(O)NH_2$ and differ by only a single CH_2 group. As typical primary amides, they are not nucleophilic and do not form covalent bonds in enzymic reactions. Amide functions are quite polar and interact strongly with water molecules by hydrogen bonding. Like serine, asparagine and glutamine are often found in regions of a peptide that are in contact with water.

Amino Acids with Acidic Side Chains: Two of the standard amino acids have carboxylic acids in their side chains: *aspartic acid* ($—CH_2COOH$) and *glutamic acid* ($—CH_2CH_2COOH$). The electrostatic potential maps of aspartic acid and glutamic acid are two of the most prominent ones in Table 25.1. The $—CO_2H$ side chains of both aspartic and glutamic acid are essentially completely deprotonated to $—CO_2^-$ at biological pH, giving these species the most electron-rich units of all of the common amino acids. Their functions in protein structure include ionic bonding to positively charged species such as metal ions and ammonium ions.

Amino Acids with Basic Side Chains: Three of the standard amino acids are bases. Two of the three are protonated at physiological pH. Lysine, $—CH_2CH_2CH_2CH_2NH_2$, possesses four methylene groups in its side chain. Its acid–base properties are similar to those of an aliphatic amine (Section 22.4). The second basic amino acid, arginine, is the most basic standard amino acid. The traditionally accepted pK_a value of arginine (12.5) has been recently measured using modern spectroscopic techniques to be 13.8. Its pK_a approaches the pK_a of hydroxide, and it is always protonated in proteins. Arginine is so basic because its functional group is not a primary amine, but a four-atom functional group called *guanidine*. Protonation of the imine-like nitrogen of arginine forms a cationic species in which the positive charge is delocalized over all three nitrogens of the functional group. The electrostatic maps of these two amino acids in Table 25.1 illustrate the localized charge on the lysine amine versus the delocalized charge on the arginine guanidine.

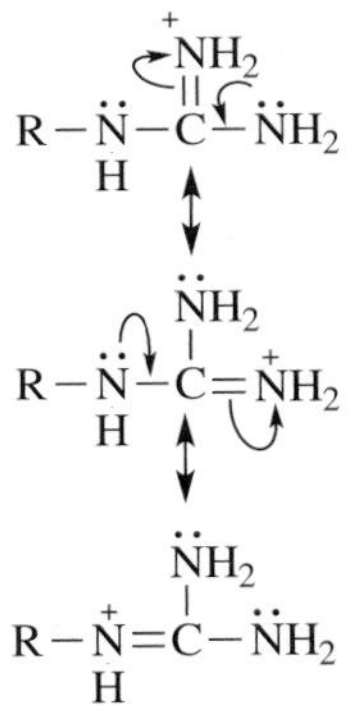

R−NH−C(=$\overset{+}{N}H_2$)−NH_2 ⟷ R−NH−C(−NH_2)=$\overset{+}{N}H_2$ ⟷ R−$\overset{+}{N}$H=C(−NH_2)−NH_2

Resonance structures of the guanidine functional group

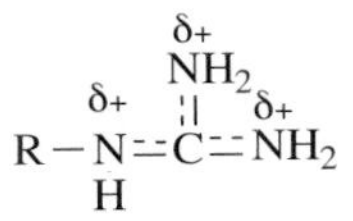

R−NH$\overset{\delta+}{}$−C(−$\overset{\delta+}{N}H_2$)−$\overset{\delta+}{N}H_2$

Resonance hybrid

The side chain of *histidine* has a functional group we saw in Section 22.4, an imidazole. This aromatic heterocycle has a pK_a of approximately 6 in the free amino acid, which means that just a small portion of the side chains of the free amino acid is protonated at neutral pH. Why, then, is it grouped with the basic amino acids? As noted earlier, not all amino acid side chains are neatly categorized by a single chemical property. Histidine is located in many enzyme active sites and often serves a dual role as both a proton donor (in its positively charged form) and a proton acceptor (in its neutral form). Protonation of histidine produces a species in which the positive charge is delocalized over both nitrogens of the side chain. Removal of either proton returns histidine to its neutral state. This property of the amino acid allows it to accomplish a net transfer of protons through space. Its ability to act as part of a "proton relay" system is important in the catalytic activity of a number of enzymes.

Nonstandard Amino Acids: Peptides sometimes contain an amino acid different from the 20 shown in Table 25.1. *Dehydroalanine,* for example, is a component of a toxic substance produced by a strain of cyanogenic bacteria, and *hydroxyproline* is a component of the collagen in connective tissue.

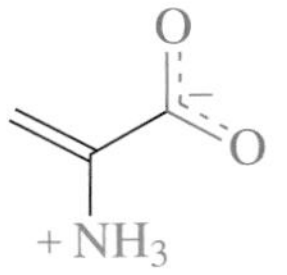

Dehydroalanine

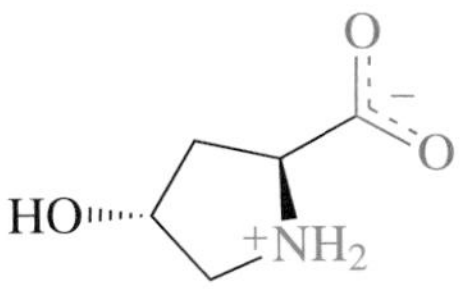

Hydroxyproline

Like most nonstandard amino acids, dehydroalanine and hydroxyproline are formed by modification of one of the standard amino acids that has already been incorporated into a peptide. Two nonstandard amino acids—*selenocysteine* and *pyrrolysine*—the so-called "twenty-first and twenty-second amino acids," however, are coded for by DNA.

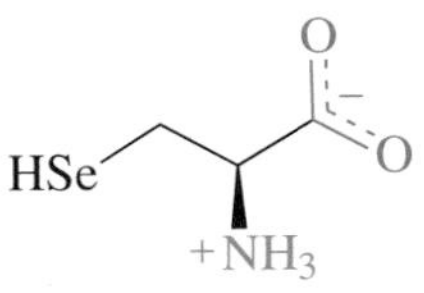

Selenocysteine

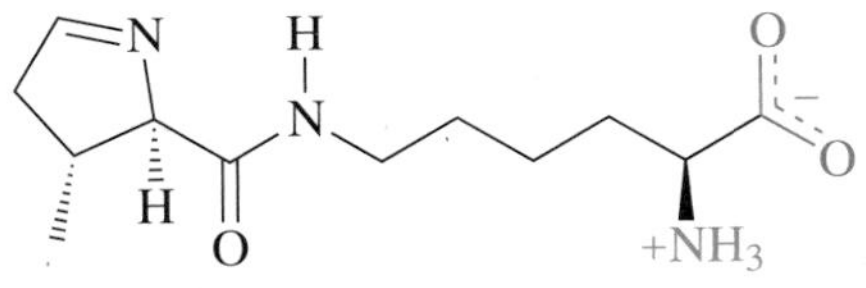

Pyrrolysine

Organic chemists are capable of creating amino acids with functionality not found in nature. New techniques in molecular biology are allowing scientists to use an organism's own ribosomal protein biosynthesis apparati. "Noncanonical" or "unnatural" amino acid mutagenesis has been developed to place synthetic amino acids into the linear sequence of proteins. Just a few of the many noncanonical amino acids that have been inserted into proteins using this technology are shown.

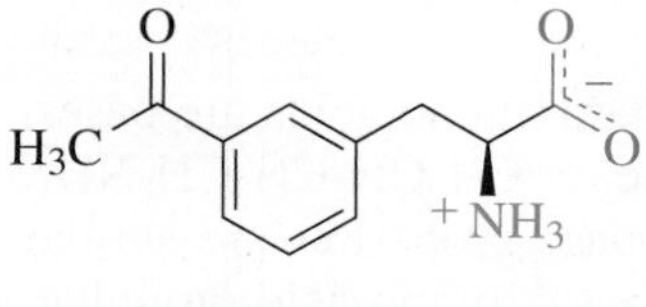

3-Acetylphenylalanine 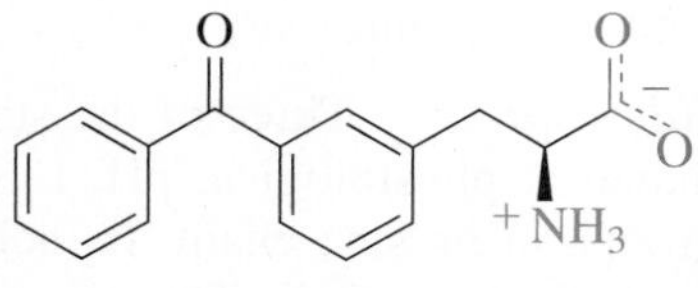 3-Benzoylphenylalanine 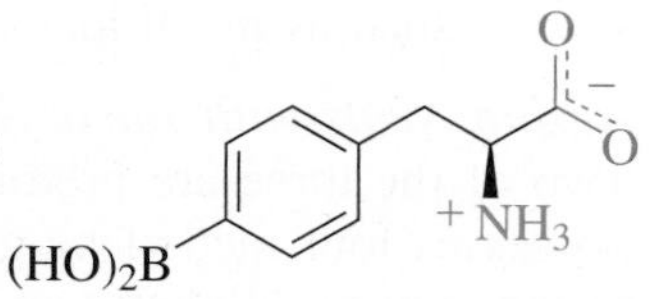 4-Boronophenylalanine

Problem 25.1

The next section deals with amino acid stereochemistry. You can prepare for it by locating all of the chirality centers in the nonstandard amino acids just shown and specifying their configuration using the Cahn–Ingold–Prelog *R,S* notation.

25.2 Stereochemistry of Amino Acids

Glycine is the only amino acid in Table 25.1 that is achiral; the α-carbon atom is a chirality center in all the others. Configurations in amino acids are normally specified by the D,L notational system, which has been described in Section 23.2. Recall that the D,L system describes a three-dimensional structure relative to the stereochemistry of glyceraldehyde. All the chiral amino acids obtained from proteins have the L-configuration at their α-carbon atom, meaning that the amine group is at the left when a Fischer projection is arranged so the carboxyl group is at the top.

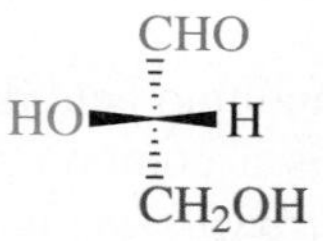

L-Glyceraldehyde 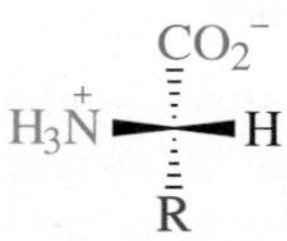 L-Amino acid 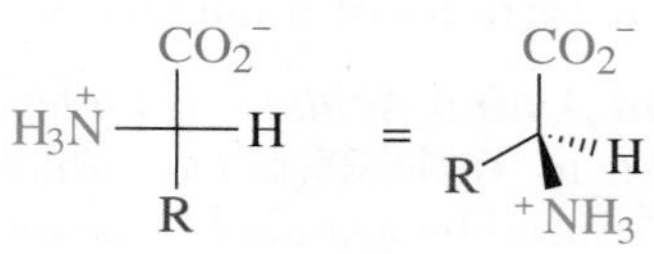 L-Amino acid Fischer projection

Problem 25.2

What is the absolute configuration (*R* or *S*) at the α-carbon atom in each of the following L-amino acids?

(a) 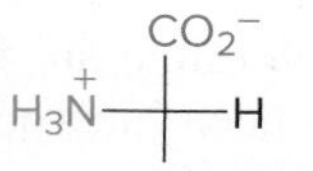 L-Serine (b) 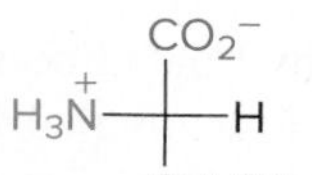 L-Cysteine (c) 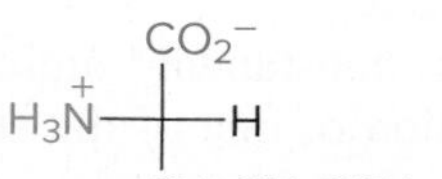 L-Methionine

Sample Solution (a) First identify the four groups attached directly to the chirality center, and rank them in order of decreasing sequence rule precedence. For L-serine these groups are

$$H_3\overset{+}{N}— > —CO_2^- > —CH_2OH > —H$$

Highest ranked Lowest ranked

Next, translate the Fischer projection of L-serine to a three-dimensional representation, and orient it so that the lowest ranked substituent at the chirality center is directed away from you.

CO_2^- ; $H_3\overset{+}{N}$—C—H ; CH_2OH = $\overset{+}{N}H_3$, H, C—CO_2^-, $HOCH_2$ = $HOCH_2$, H, C—CO_2^-, $^+NH_3$

In order of decreasing precedence the three highest ranked groups trace a counterclockwise path.

$HOCH_2$, CO_2^-, $^+NH_3$

The absolute configuration of L-serine is *S*.

Problem 25.3

The amino acid L-threonine is (2*S*,3*R*)-2-amino-3-hydroxybutanoic acid. Draw a Fischer projection and line formula for L-threonine.

Although all the chiral amino acids resulting from ribosomal protein biosynthesis have the L-configuration at their α carbon, that should not be taken to mean that D-amino acids are unknown. In fact, quite a number of D-amino acids occur naturally. It has long been known that bacterial cell walls contain peptides with D-alanine. Antibiotics such as penicillin are toxic to the microbes but not to humans because they target an enzyme necessary in bacterial cell wall biosynthesis that recognizes the sequence D-Ala-D-Ala. D-Ser and D-Asp are involved in mammalian neurological function, although their roles are not well understood. It has been hypothesized that increased levels of D-serine may be involved in conditions such as schizophrenia.

A novel technique for dating archaeological samples called **amino acid racemization (AAR)** is based on the stereochemistry of amino acids. Over time, the configuration at the α-carbon atom of a protein's amino acids is lost in a reaction that follows first-order kinetics. When the α carbon is the only chirality center, this process corresponds to racemization. For an amino acid with two chirality centers, changing the configuration of the α carbon from L to D gives a diastereomer. In the case of isoleucine, for example, the diastereomer is an amino acid not normally present in proteins, called *alloisoleucine.*

L-Isoleucine: CO_2^- (top); $H_3\overset{+}{N}$—C—H; H_3C—C—H; CH_2CH_3 (bottom)

D-Alloisoleucine: CO_2^- (top); H—C—$\overset{+}{N}H_3$; H_3C—C—H; CH_2CH_3 (bottom)

By measuring the L-isoleucine/D-alloisoleucine ratio in the protein isolated from the eggshells of an extinct Australian bird, it was determined that this bird lived approximately 50,000 years ago. Radiocarbon (^{14}C) dating is not accurate for samples older than about 35,000 years, so AAR is a useful addition to the tools available to paleontologists.

25.3 Acid–Base Behavior of Amino Acids

The physical properties of a typical amino acid such as glycine suggest that it is a very polar substance, much more polar than would be expected on the basis of its formulation as $H_2NCH_2CO_2H$. Glycine is a crystalline solid; it does not melt, but on being heated it

Figure 25.1

Titration curve of glycine. At pH values lower than 2.34, $H_3\overset{+}{N}CH_2CO_2H$ is the major species. At pH = 2.34, $[H_3\overset{+}{N}CH_2CO_2H] = [H_3\overset{+}{N}CH_2CO_2^-]$. Between pH = 2.34 and 9.60, $H_3\overset{+}{N}CH_2CO_2^-$ is the major species. Its concentration is a maximum at the isoelectric point (pI = 5.97). At pH = 9.60, $[H_3\overset{+}{N}CH_2CO_2^-] = [H_2NCH_2CO_2^-]$. Above pH = 9.60, $H_2NCH_2CO_2^-$ is the predominant species.

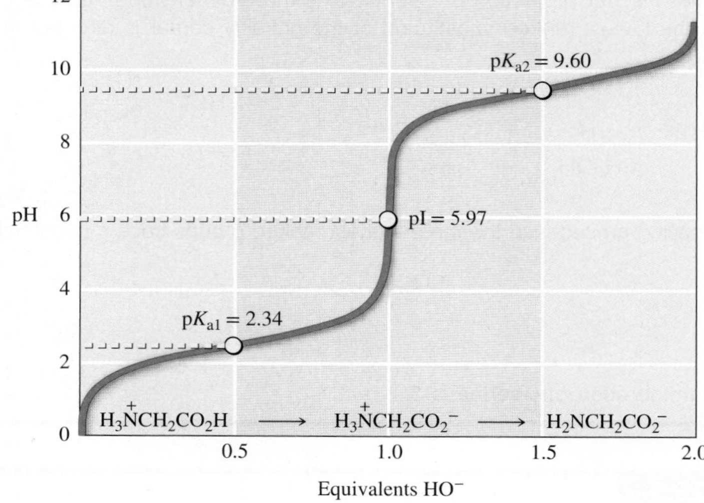

eventually decomposes at 233°C. It is very soluble in water but practically insoluble in nonpolar organic solvents. These properties are attributed to the fact that the stable form of glycine in neutral aqueous solution is a **zwitterion.**

The zwitterion is also often referred to as a *dipolar ion*. Note, however, that it is not an ion, but a neutral molecule.

Glycine

Zwitterionic form of glycine

Glycine, as well as other amino acids, is *amphoteric,* meaning it contains an acidic functional group and a basic functional group. The acidic functional group is the ammonium ion $H_3\overset{+}{N}$—; the basic functional group is the carboxylate ion $—CO_2^-$. How do we know this? Aside from its physical properties, the acid–base properties of glycine, as illustrated by the titration curve in Figure 25.1, require it. In a strongly acidic medium the species present is the cation $H_3\overset{+}{N}CH_2CO_2H$. As the pH is raised, its most acidic proton is removed. Is this proton removed from the positively charged nitrogen or from the carboxyl group? We know what to expect for the relative acid strengths of $R\overset{+}{N}H_3$ and RCO_2H. A typical ammonium ion has $pK_a \approx 9$, and a typical carboxylic acid has $pK_a \approx 5$. The measured pK_a for the conjugate acid of glycine is 2.34, a value closer to that expected for deprotonation of the carboxyl group. As the pH is raised, a second deprotonation step, corresponding to removal of a proton from nitrogen of the zwitterion, is observed. The pK_a associated with this step is 9.60, much like that of typical alkylammonium ions.

$-H^+$ / $+H^+$ $-H^+$ / $+H^+$

Major species at pH < 2.3   Major species at pH 2.3–9.6   Major species at pH > 9.6

Thus, glycine is characterized by two pK_a values: the one corresponding to the more acidic site is designated pK_{a1}, the one corresponding to the less acidic site is designated pK_{a2}. Table 25.2 lists pK_{a1} and pK_{a2} values for the α-amino acids that have neutral side chains, which are the first three groups of amino acids given in Table 25.1. In all cases their pK_a values are similar to those of glycine.

TABLE 25.2 Acid–Base Properties of Amino Acids with Neutral Side Chains

Amino acid	pK_{a1}*	pK_{a2}*	pI
Glycine	2.34	9.60	5.97
Alanine	2.34	9.69	6.00
Valine	2.32	9.62	5.96
Leucine	2.36	9.60	5.98
Isoleucine	2.36	9.60	6.02
Methionine	2.28	9.21	5.74
Proline	1.99	10.60	6.30
Phenylalanine	1.83	9.13	5.48
Tryptophan	2.83	9.39	5.89
Asparagine	2.02	8.80	5.41
Glutamine	2.17	9.13	5.65
Serine	2.21	9.15	5.68
Threonine	2.09	9.10	5.60
Tyrosine	2.20	9.11	5.66

*In all cases, pK_{a1} corresponds to ionization of the carboxyl group; pK_{a2} corresponds to deprotonation of the ammonium ion.

Table 25.2 includes a column labeled pI, which is the *isoelectric point* of the amino acid. The **isoelectric point,** also called the **isoionic point,** is the pH at which the amino acid has no net charge. It is the pH at which the concentration of the zwitterion is a maximum. At a pH lower than pI, the amino acid is positively charged; at a pH higher than pI, the amino acid is negatively charged. For the amino acids in Table 25.2, pI is the average of pK_{a1} and pK_{a2} and lies slightly to the acid side of neutrality.

Some amino acids have side chains that bear acidic or basic groups. As Table 25.3 indicates, these amino acids are characterized by three pK_a values. The third pK_a

TABLE 25.3 Acid–Base Properties of Amino Acids with Ionizable Side Chains

Amino acid	pK_{a1}*	pK_{a2}	pK_a of side chain	pI
Aspartic acid	1.88	9.60	3.65	2.77
Glutamic acid	2.19	9.67	4.25	3.22
Lysine	2.18	8.95	10.53	9.74
Arginine	2.17	9.4**	13.8**	11.6**
Histidine	1.82	9.17	6.00	7.59

*In all cases, pK_{a1} corresponds to ionization of the carboxyl group of $RCH(\overset{+}{N}H_3)CO_2H$ and pK_{a2} to ionization of the ammonium ion.

**Based on pK_a values reported in 2015

Electrophoresis

Electrophoresis is a method for separation and purification that depends on the movement of charged particles in an electric field. Its principles can be introduced by considering some representative amino acids. The medium is a cellulose acetate strip that is moistened with an aqueous solution buffered at a particular pH. The opposite ends of the strip are placed in separate compartments containing the buffer, and each compartment is connected to a source of direct electric current (Figure 25.2*a*). If the buffer solution is more acidic than the isoelectric point (pI) of the amino acid, the amino acid has a net positive charge and migrates toward the negatively charged electrode. Conversely, when the buffer is more basic than the pI of the amino acid, the amino acid has a net negative charge and migrates toward the positively charged electrode. When the pH of the buffer corresponds to the pI, the amino acid has no net charge and does not migrate from the origin.

Thus, if a mixture containing alanine, aspartic acid, and lysine is subjected to electrophoresis in a buffer that matches the isoelectric point of alanine (pH 6.0), aspartic acid (pI = 2.8) migrates toward the positive electrode, alanine remains at the origin, and lysine (pI = 9.7) migrates toward the negative electrode (Figure 25.2*b*).

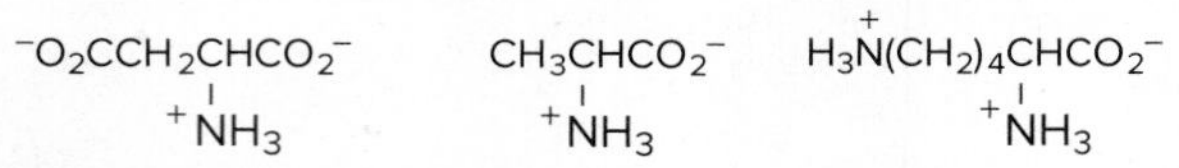

$^{-}O_2CCH_2CHCO_2^{-}$ (with $^{+}NH_3$)	$CH_3CHCO_2^{-}$ (with $^{+}NH_3$)	$H_3\overset{+}{N}(CH_2)_4CHCO_2^{-}$ (with $^{+}NH_3$)
Aspartic acid (−1 ion)	Alanine (neutral)	Lysine (+1 ion)

Electrophoresis is used primarily to analyze mixtures of peptides and proteins, rather than individual amino acids, but analogous principles apply. Because they incorporate different numbers of amino acids and because their side chains are different, two peptides will have slightly different acid–base properties and slightly different net charges at a particular pH. Thus, their mobilities in an electric field will be different, and electrophoresis can be used to separate them. The medium used to separate peptides and proteins is typically a polyacrylamide gel, leading to the term *gel electrophoresis* for this technique.

A second factor that governs the rate of migration during electrophoresis is the size (length and shape) of the peptide or protein. Larger molecules move through the polyacrylamide gel more slowly than smaller ones. In current practice, the experiment is modified to exploit differences in size more than differences in net charge, especially in the *SDS gel electrophoresis* of proteins. Approximately 1.5 g of the detergent *sodium dodecyl sulfate* (SDS) per gram of protein is added to the aqueous buffer. SDS binds to the protein, causing the protein to unfold so that it is roughly rod-shaped with the $CH_3(CH_2)_{10}CH_2$— groups of SDS associated with the lipophilic (hydrophobic) portions of the protein. The negatively charged sulfate groups are exposed to the water. The SDS molecules that they carry ensure that all the protein molecules are negatively charged and migrate toward the positive electrode. Furthermore, all the proteins in the mixture now have similar shapes and tend to travel at rates proportional to their chain length. Thus, when carried out on a preparative scale, SDS gel electrophoresis permits proteins in a mixture to be separated according to their molecular weight. On an analytical scale, it is used to estimate the molecular weight of a protein by comparing its electrophoretic mobility with that of proteins of known molecular weight.

In the next chapter, we'll see how gel electrophoresis is used in nucleic acid chemistry.

Figure 25.2

Application of electrophoresis to the separation of aspartic acid, alanine, and lysine according to their charge type at a pH corresponding to the isoelectric point (pI) of alanine.

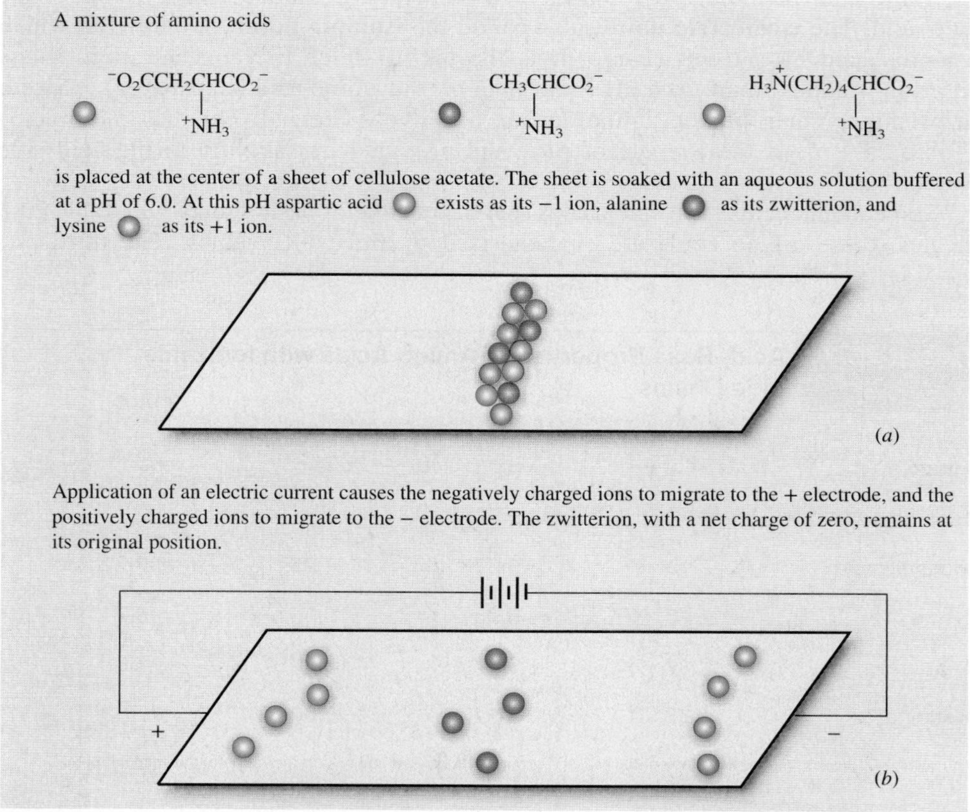

A mixture of amino acids

$^{-}O_2CCH_2CHCO_2^{-}$ ($^{+}NH_3$) $CH_3CHCO_2^{-}$ ($^{+}NH_3$) $H_3\overset{+}{N}(CH_2)_4CHCO_2^{-}$ ($^{+}NH_3$)

is placed at the center of a sheet of cellulose acetate. The sheet is soaked with an aqueous solution buffered at a pH of 6.0. At this pH aspartic acid exists as its −1 ion, alanine as its zwitterion, and lysine as its +1 ion.

(*a*)

Application of an electric current causes the negatively charged ions to migrate to the + electrode, and the positively charged ions to migrate to the − electrode. The zwitterion, with a net charge of zero, remains at its original position.

+ −

(*b*)

reflects the nature of the side chain. Acidic amino acids (aspartic and glutamic acid) have acidic side chains; basic amino acids (lysine, arginine, and histidine) have basic side chains.

The isoelectric points of the amino acids in Table 25.3 are midway between the pK_a values of the zwitterion and its conjugate acid. Take two examples: aspartic acid and lysine. Aspartic acid has an acidic side chain and a pI of 2.77. Lysine has a basic side chain and a pI of 9.74.

Aspartic Acid:

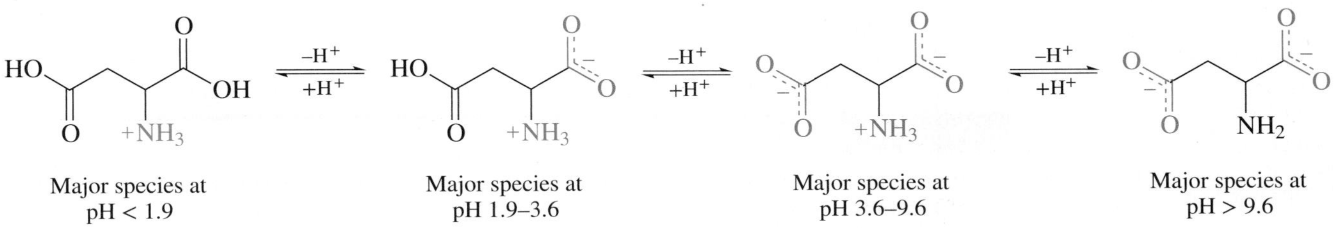

Major species at pH < 1.9 $\underset{+H^+}{\overset{-H^+}{\rightleftharpoons}}$ Major species at pH 1.9–3.6 $\underset{+H^+}{\overset{-H^+}{\rightleftharpoons}}$ Major species at pH 3.6–9.6 $\underset{+H^+}{\overset{-H^+}{\rightleftharpoons}}$ Major species at pH > 9.6

The pI of aspartic acid is the average of pK_{a1} (1.88) and the pK_a of the side chain (3.65), or 2.77.

Lysine:

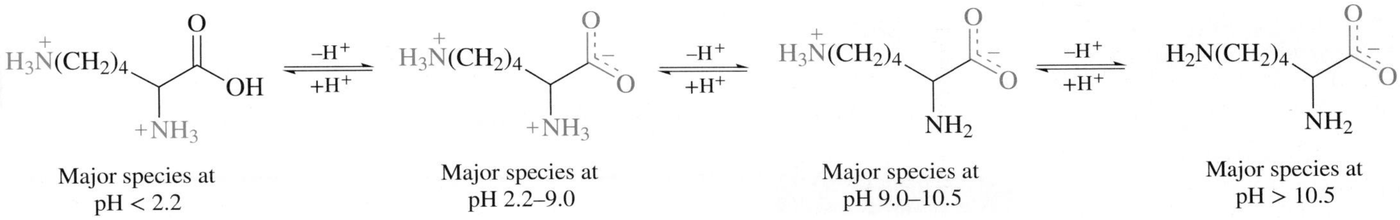

Major species at pH < 2.2 $\underset{+H^+}{\overset{-H^+}{\rightleftharpoons}}$ Major species at pH 2.2–9.0 $\underset{+H^+}{\overset{-H^+}{\rightleftharpoons}}$ Major species at pH 9.0–10.5 $\underset{+H^+}{\overset{-H^+}{\rightleftharpoons}}$ Major species at pH > 10.5

The pI of lysine is the average of pK_{a2} (8.95) and the pK_a of the side chain (10.53), or 9.74.

Problem 25.4

Cysteine has pK_{a1} = 1.96 and pK_{a2} = 10.28. The pK_a for ionization of the —SH group of the side chain is 8.18. What is the isoelectric point of cysteine?

Problem 25.5

Above a pH of about 10, the major species present in a solution of tyrosine has a net charge of −2. Suggest a reasonable structure for this species.

Individual amino acids differ in their acid–base properties. This is important in peptides and proteins, where the properties of the substance depend on its amino acid constituents, especially on the nature of the side chains. It is also important in analyses in which a complex mixture of amino acids is separated into its components by taking advantage of the differences in their proton-donating and accepting power.

25.4 Synthesis of Amino Acids

Two still-used methods for the synthesis of amino acids date from the nineteenth century. One is a nucleophilic substitution in which ammonia reacts with an α-halo carboxylic acid.

2-Bromopropanoic acid + $2NH_3$ $\xrightarrow{H_2O}$ Alanine (65–70%) + NH_4Br

2-Bromopropanoic acid | Ammonia | Alanine (65–70%) | Ammonium bromide

Problem 25.6

Use retrosynthetic analysis to plan a synthesis of valine from 3-methylbutanoic acid and write equations for the synthesis. (*Hint:* See Section 21.6.)

The synthesis of alanine was described by Adolf Strecker of the University of Würzburg (Germany) in a paper published in 1850.

A second method, called the **Strecker synthesis,** combines both functional-group manipulation and chain extension to give an α-amino acid having one more carbon atom than the aldehyde. It begins with two nucleophilic additions that convert the aldehyde to an aminonitrile, followed by hydrolysis of the nitrile function.

Acetaldehyde $\xrightarrow[NaCN]{NH_4Cl}$ 2-Aminopropanenitrile $\xrightarrow[2.\ HO^-]{1.\ H_2O,\ HCl,\ heat}$ Alanine (52–60%)

Problem 25.7

Use retrosynthetic analysis to plan a synthesis of isoleucine from 2-methyl-1-butanol and write equations for the synthesis.

There has been striking success in adapting the Strecker synthesis to the preparation of α-amino acids with greater than 99% enantioselectivity. The numerous methods that have been developed employ specialized chiral reagents or catalysts and feature enantioselective generation of a chirality center by nucleophilic addition to an imine.

Imine (RCH=NR′) $\xrightarrow[\text{chiral catalyst}]{(CH_3)_3SiCN}$ Aminonitrile $\longrightarrow$ Amino acid

The synthesis of the anti-Parkinson drug L-dopa by enantioselective hydrogenation was described earlier in Section 15.13.

These procedures allow chemists to prepare not only L-amino acids, but also their much rarer D-enantiomers.

In an approach related to the malonic ester synthesis (see Section 21.5), carbon–carbon bond formation occurs by nucleophilic substitution as summarized by the disconnection:

$$RCH(\overset{+}{N}H_3)CO_2^- \Longrightarrow R{-}X + {:}\overset{-}{C}H(\overset{+}{N}H_3)CO_2^-$$

The carbanion source is diethyl malonate modified so as to bear an acetamido group at its α carbon.

Diethyl acetamidomalonate $\xrightarrow[\text{2. } C_6H_5CH_2Cl]{\text{1. } NaOCH_2CH_3,\ CH_3CH_2OH}$ Diethyl acetamidobenzylmalonate (90%)

Hydrolysis removes the acetyl group from nitrogen and converts both ester functions to carboxyl groups. Decarboxylation gives the desired product.

Diethyl acetamidobenzylmalonate $\xrightarrow[\text{heat}]{HBr,\ H_2O}$ (not isolated) $\xrightarrow[-CO_2]{\text{heat}}$ Phenylalanine (65%)

Problem 25.8

Outline the steps in the synthesis of valine from diethyl acetamidomalonate. The overall yield of valine by this method is reported to be rather low (31%). Can you think of a reason why this synthesis is not very efficient?

25.5 Reactions of Amino Acids

The reactions that amino acids undergo are those of its two functional groups plus those associated with the side chain. Many, such as the Fischer esterification and amine acylation, are familiar.

Glycine + Acetic anhydride ⟶ *N*-Acetylglycine (89–92%) + Acetic acid

Glycine + Benzyl alcohol $\xrightarrow[\text{heat}]{H_2SO_4}$ Glycine benzyl ester (70–80%)

It's often the case that acylation is carried out for the purpose of protecting the amino group by temporarily suppressing its reactivity. Three widely used protecting groups are *benzyloxycarbonyl, tert-butoxycarbonyl,* and *9-fluorenylmethoxycarbonyl,* which are abbreviated as Z, Boc, and Fmoc, respectively. Various reagents and methods have been developed for introducing these groups, among which are the following examples.

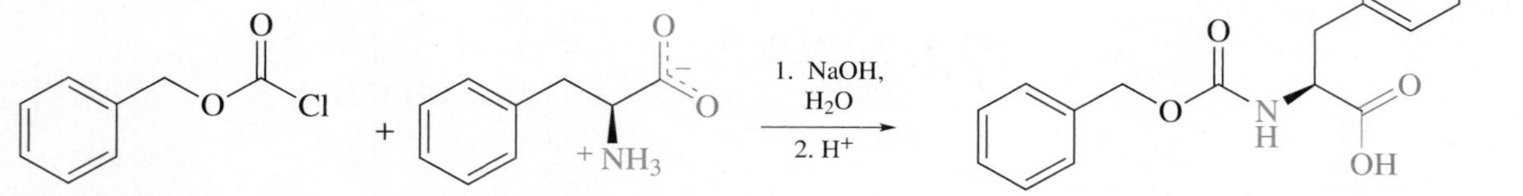

Benzyloxycarbonyl chloride (Z-Cl) + L-Phenylalanine (L-Phe or L-F) → (1. NaOH, H_2O; 2. H^+) *N*-Benzyloxycarbonyl-L-phenylalanine (82–87%) (Z-L-Phe or Z-L-F)

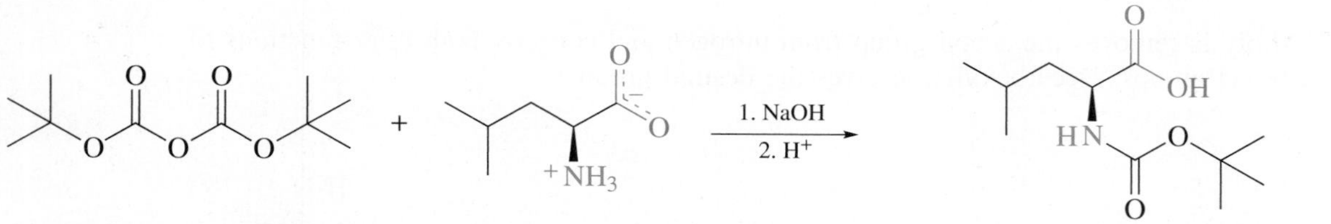

Di-*tert*-Butyl dicarbonate (Boc_2O) + L-Leucine (L-Leu or L-L) → (1. NaOH; 2. H^+) *N*-*tert*-Butoxycarbonyl-L-leucine (96%) (Boc-L-Leu or Boc-L-L)

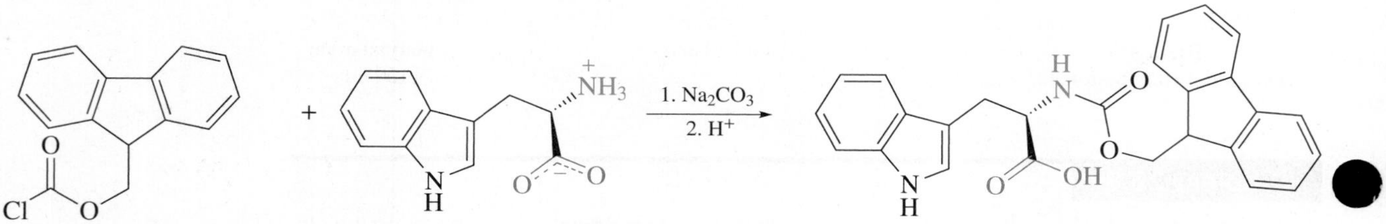

9-Fluorenylmethyloxycarbonyl chloride (Fmoc-Cl) + L-Tryptophan (L-Trp or L-W) → (1. Na_2CO_3; 2. H^+) 9-Fluorenylmethyloxycarbonyl-L-tryptophan (91%) (Fmoc-L-Trp or Fmoc-L-W)

As protecting groups in polypeptide synthesis, Z, Boc, and Fmoc have complementary properties with respect to their removal upon completion of a synthetic step. Boc is removed by acid-cleavage, Fmoc by base, and Z by either acid-cleavage or hydrogenolysis. We'll return to these applications in Section 25.14.

Problem 25.9

Among the typical functional-group reactions of amino acids are the following. Predict the product in each case.

(a) 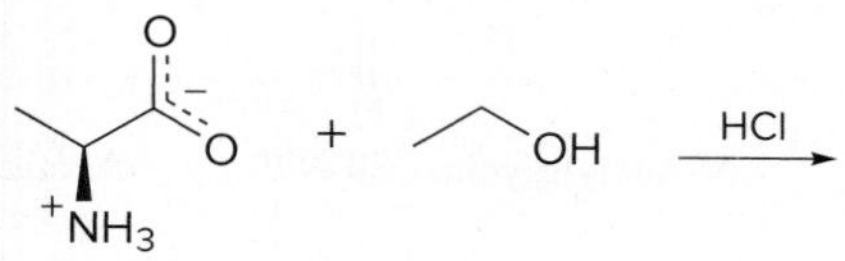 L-alanine (zwitterion, $^+NH_3$) + ethanol (OH) $\xrightarrow{HCl}$

(b) 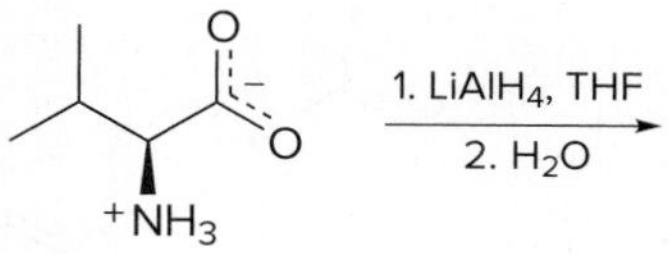 L-valine (zwitterion, $^+NH_3$) $\xrightarrow[2.\ H_2O]{1.\ LiAlH_4,\ THF}$

(c) 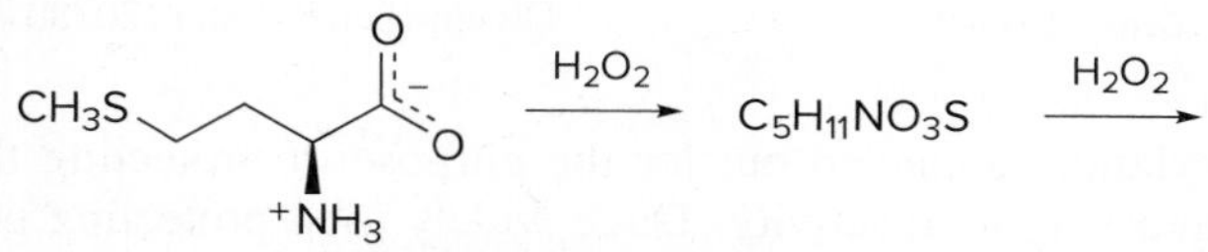 CH_3S–methionine (zwitterion, $^+NH_3$) $\xrightarrow{H_2O_2}$ $C_5H_{11}NO_3S$ $\xrightarrow{H_2O_2}$

25.6 Peptides

A key biochemical reaction of amino acids is their conversion to peptides, polypeptides, and proteins. In all these substances amino acids are linked together by amide bonds. The amide bond between the amino group of one amino acid and the carboxyl of another is called a **peptide bond.** Alanylglycine is a representative dipeptide.

N terminus C terminus N terminus C terminus

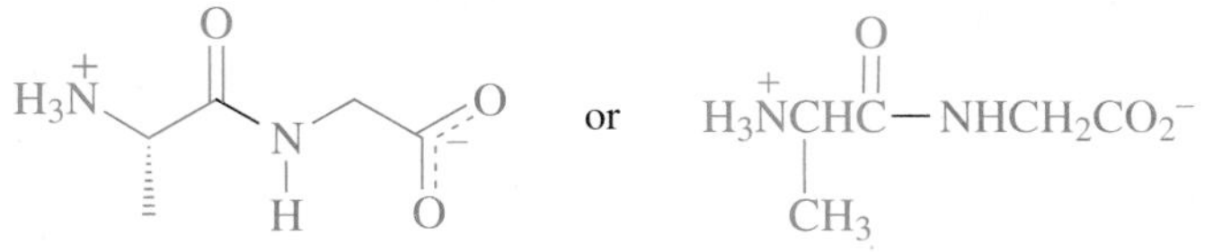

$H_3\overset{+}{N}CHC(=O)—NHCH_2CO_2^-$ (with CH_3 on the α-carbon)

Alanylglycine
(Ala-Gly or AG)

It is understood that α-amino acids occur as their L stereoisomers unless otherwise indicated. The D notation is explicitly shown when a D-amino acid is present, and a racemic amino acid is identified by the prefix DL.

By agreement, peptide structures are written so that the amino group (as $H_3\overset{+}{N}$— or H_2N—) is at the left and the carboxyl group (as CO_2^- or CO_2H) is at the right. The left and right ends of the peptide are referred to as the **N terminus** (or amino terminus) and the **C terminus** (or carboxyl terminus), respectively. Alanine is the N-terminal amino acid in alanylglycine; glycine is the C-terminal amino acid. A dipeptide is named as an acyl derivative of the C-terminal amino acid. The precise order of bonding in a peptide (its amino acid *sequence*) is conveniently specified by using the three-letter amino acid abbreviations for the respective amino acids and connecting them by hyphens. One-letter abbreviations are used without punctuation. Individual amino acid components of peptides are often referred to as **amino acid residues.**

Problem 25.10

Write structural formulas showing the constitution of each of the following dipeptides. Rewrite each sequence using one-letter abbreviations for the amino acids.

(a) Gly-Ala
(b) Ala-Phe
(c) Phe-Ala
(d) Gly-Glu
(e) Lys-Gly
(f) D-Ala-D-Ala

Sample Solution (a) Glycine is the N-terminal amino acid in Gly-Ala; alanine is the C-terminal amino acid.

N terminus C terminus

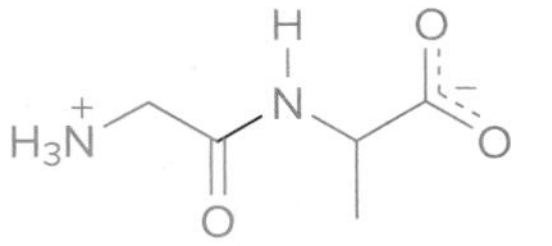

Glycylalanine (GA)

Figure 25.3 shows the structure of Ala-Gly as determined by X-ray crystallography. An important feature is the planar geometry of the peptide bond, and the most stable conformation with respect to this bond has the two α-carbon atoms anti to each other. Rotation about the amide bond is slow because delocalization of the unshared electron pair of nitrogen into the carbonyl group gives partial double-bond character to the carbon–nitrogen bond.

In addition to its planar geometry, the amide bond affects the structure of peptides in another important way. The N—H and the C═O units are candidates for hydrogen bonding with other peptide linkages both within the same and with adjacent polypeptide chains.

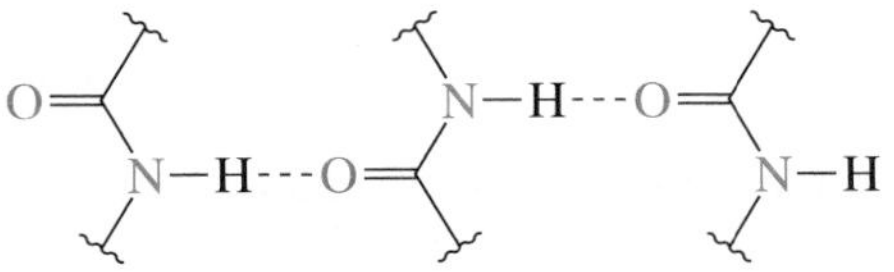

O═C N—H---O═C N—H---O═C N—H

Figure 25.3

Structural features of the dipeptide L-alanylglycine as determined by X-ray crystallography. All of the bonds of the peptide linkage lie in the same plane and both α carbons are anti to each other, as designated by the box.

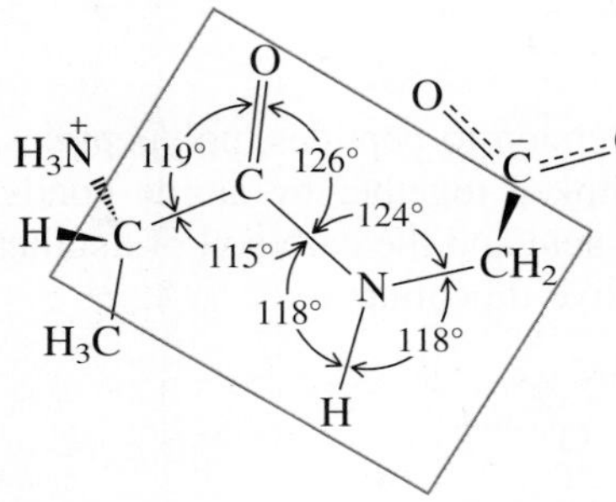

(*a*)

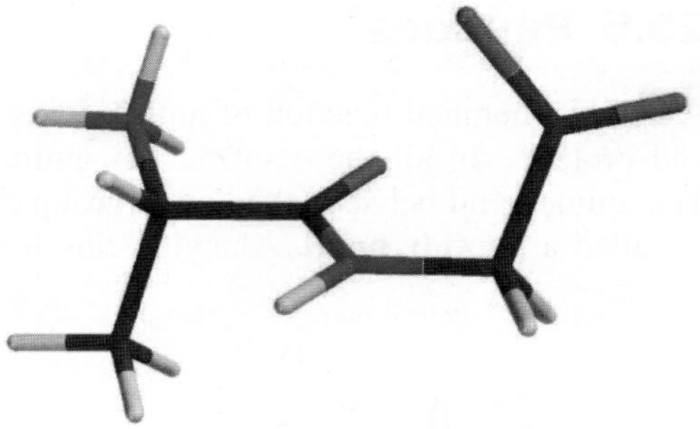

(*b*)

As the only secondary amine among the standard amino acids, L-proline is an exception in that its amides lack an N—H bond.

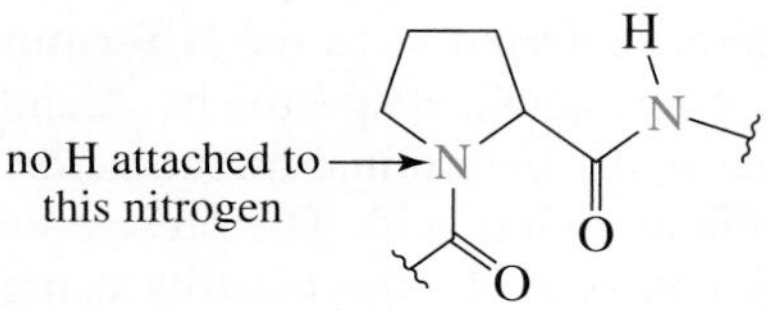

This structural feature of L-proline affects the three-dimensional shape of peptides that contain it by limiting the number of hydrogen-bonding opportunities.

Problem 25.11

Expand your answer to Problem 25.10 by showing the structural formula for each dipeptide in a manner that reveals the stereochemistry at the α-carbon atom. Assume that each chirality center is L.

Sample Solution (a) Glycine is achiral, and so Gly-Ala has only one chirality center, the α-carbon atom of the L-alanine residue. When the carbon chain is drawn in an extended zigzag fashion and L-alanine is the C terminus, its structure is as shown:

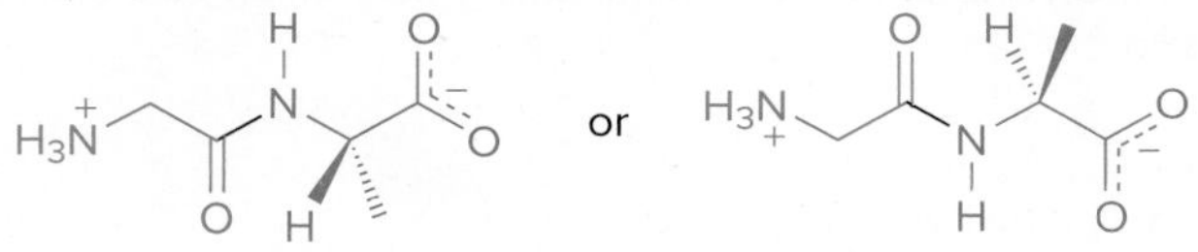

Glycylalanine (GA)

The structures of higher peptides are extensions of the structural features of dipeptides. The neurotransmitter *leucine enkephalin,* for example, has the structure:

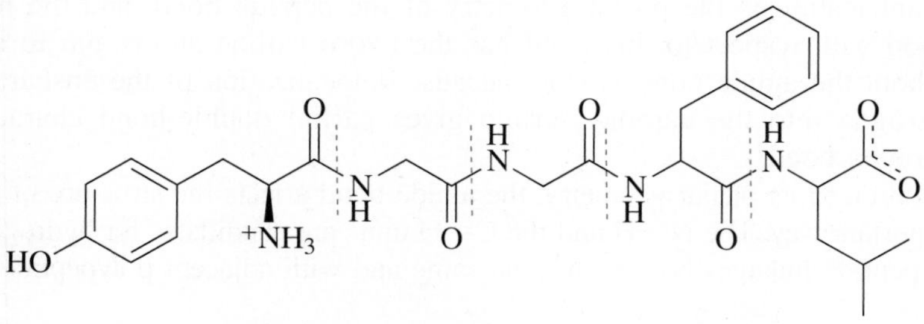

Enkephalins are pentapeptide components of *endorphins,* polypeptides present in the brain that act as the body's own painkillers.

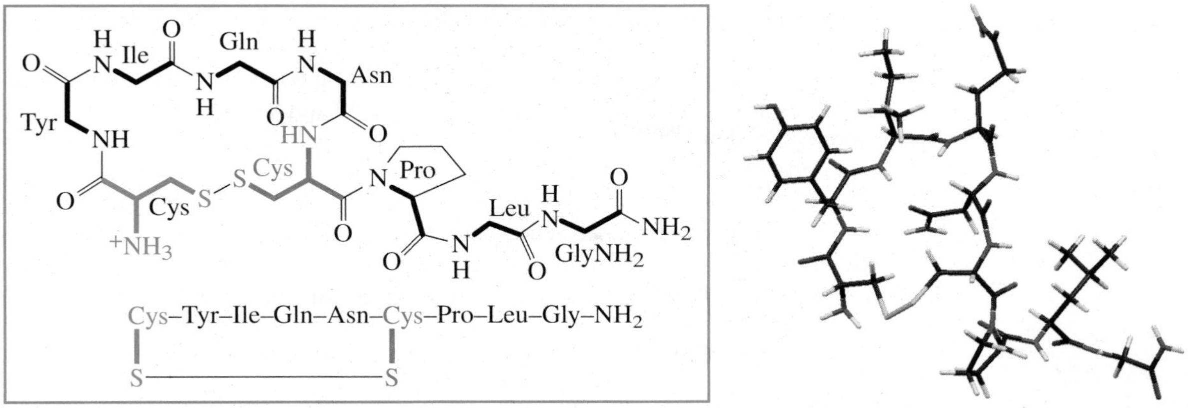

Figure 25.4

(a) Two-dimensional representations of oxytocin. The connectivity of oxytocin with most of the side chains omitted for clarity. A disulfide bond connects the two cysteines. The cysteine shown in blue is the N-terminal amino acid; the one in red is the fourth amino acid beginning at the C terminus. The C terminus is the amide of glycine. (b) Three-dimensional X-ray structure of oxytocin.

Problem 25.12

Methionine enkephalin has the same structure as leucine enkephalin except its C-terminal amino acid is methionine instead of leucine. Using one-letter abbreviations for the amino acids, what is the amino acid sequence of methionine enkephalin?

Peptides having structures slightly different from those described to this point are known. One such variation is seen in the nonapeptide *oxytocin,* shown in Figure 25.4. Oxytocin is a hormone secreted by the pituitary gland that stimulates uterine contractions during childbirth and promotes lactation. Rather than terminating in a carboxyl group, the C-terminal glycine residue in oxytocin has been modified to become its corresponding amide. Two cysteine units, one of them the N-terminal amino acid, are joined by the sulfur–sulfur bond of a large-ring cyclic disulfide unit. This is a common structural modification in polypeptides and proteins that contain cysteine residues. It provides a covalent bond between regions of peptide chains that may be many amino acid residues removed from each other.

Recall from Section 16.12 that compounds of the type RSH are readily oxidized to RSSR.

Problem 25.13

What is the net charge of oxytocin at pH = 7?

Problem 25.14

A certain cyclic peptide bears a side chain that includes two amino acids resembling those in Table 25.1. Which two? How do they differ from those in the table?

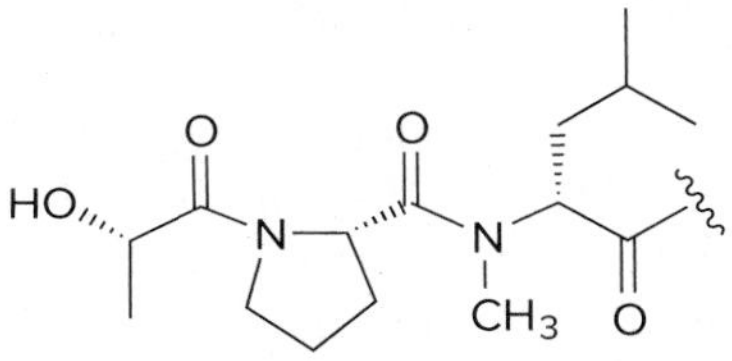

25.7 Introduction to Peptide Structure Determination

There are several levels of peptide structure. The **primary structure** is the amino acid sequence plus any disulfide links. With the 20 amino acids of Table 25.1 as building blocks, 20^2 dipeptides, 20^3 tripeptides, 20^4 tetrapeptides, and so on, are possible. Given a peptide of unknown structure, how do we determine its amino acid sequence?

In this section we will first show how peptide sequencing was first accomplished, which was a monumental achievement in the 1940s. Then we will show the modern chemical method for peptide sequencing, which is still performed today. Finally, we will introduce the most recent advance in peptide sequencing, which is performed by mass spectrometry.

Sanger was a corecipient of a second Nobel Prize in 1980 for devising methods for sequencing nucleic acids. Sanger's strategy for nucleic acid sequencing will be described in Section 26.14.

We'll describe peptide structure determination by first looking at one of the great achievements of biochemistry, the determination of the amino acid sequence of insulin by Frederick Sanger of Cambridge University (England). Sanger was awarded the 1958 Nobel Prize in Chemistry for this work, which he began in 1944 and completed 10 years later. The methods used by Sanger and his coworkers are dated by now, but the overall logic remains important.

Sanger's strategy can be outlined as follows:

1. Determine what amino acids are present and their molar ratios.
2. Cleave the peptide into smaller fragments, separate these fragments, and determine the amino acid composition of the fragments.
3. Identify the N-terminal and the C-terminal amino acid in the original peptide and in each fragment.
4. Organize the information so that the amino acid sequences of small fragments can be overlapped to reveal the full sequence.

25.8 Amino Acid Analysis

Determining what amino acids are present and their ratios begins with hydrolysis of the amide bonds. The peptide is hydrolyzed by heating in 6 M hydrochloric acid to give a solution that contains all of the amino acids. Analysis of the mixture in terms of its components and their relative amounts is typically done by chromatographic methods.

Moore and Stein shared one half of the 1972 Nobel Prize in Chemistry.

Fluorescence is the emission of radiation by a substance after it has absorbed radiation of a higher frequency.

These methods flow from the work of Stanford Moore and William H. Stein of Rockefeller University, who developed automated techniques for separating and identifying amino acids. In their original work, Moore and Stein used ion-exchange chromatography. Modern methods based on high-performance liquid chromatography (HPLC) are both faster and more selective for separating the individual amino acids in a mixture. Either before or after their separation, the amino acids are allowed to react ("tagged") with a substance that bears a group—a naphthalene ring, for example—that fluoresces. The fluorescence is strong enough so that modern analyzers can detect the amino acids obtained from 10^{-5} to 10^{-7}g of peptide.

Problem 25.15

Amino acid analysis of a certain tetrapeptide gave alanine, glycine, phenylalanine, and valine in equimolar amounts. What amino acid sequences are possible for this tetrapeptide?

25.9 Partial Hydrolysis and End Group Analysis

Whereas acid-catalyzed hydrolysis of peptides cleaves amide bonds indiscriminately and eventually breaks all of them, enzymatic hydrolysis is much more selective and is the method used to convert a peptide into smaller fragments.

Papain, the active component of most meat tenderizers, is a peptidase.

The enzymes that catalyze the hydrolysis of peptide bonds are called **peptidases.** *Trypsin,* a digestive enzyme present in the intestine, catalyzes only the hydrolysis of peptide bonds involving the carboxyl group of a lysine or arginine residue. *Chymotrypsin,* another digestive enzyme, is selective for peptide bonds involving the carboxyl group of amino acids with aromatic side chains (phenylalanine, tyrosine, tryptophan). One group of pancreatic enzymes, known as *carboxypeptidases,* catalyzes only the hydrolysis of the peptide bond to the C-terminal amino acid. In addition to these, many other digestive enzymes are known and their selectivity exploited in the selective hydrolysis of peptides.

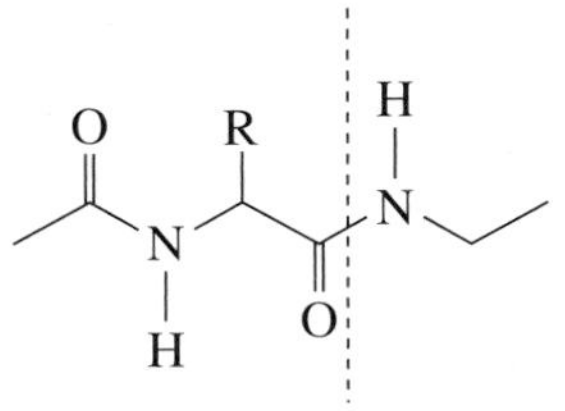

Trypsin cleaves here when
R = side chain of lysine or arginine

Chymotrypsin cleaves here when
R = side chain of phenylalanine, tyrosine, or tryptophan

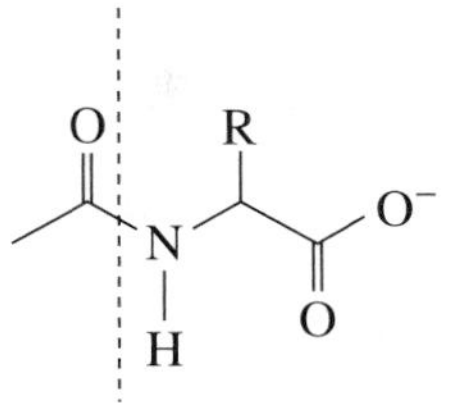

Carboxypeptidase cleaves the peptide bond of the C-terminal amino acid

Problem 25.16

Digestion of the tetrapeptide of Problem 25.15 with chymotrypsin gave a dipeptide that on amino acid analysis gave phenylalanine and valine in equimolar amounts. What amino acid sequences are possible for the tetrapeptide?

An amino acid sequence is ambiguous unless we know the direction in which to read it—left to right, or right to left. We need to know which end is the N terminus and which is the C terminus. As we've just seen, carboxypeptidase-catalyzed hydrolysis cleaves the C-terminal amino acid and so can be used to identify it. What about the N terminus?

Several chemical methods take advantage of the fact that the N-terminal amino group can act as a nucleophile. The α-amino groups of all the other amino acids are part of amide linkages and are much less nucleophilic. Sanger's method for N-terminal residue analysis involves treating a peptide with 1-fluoro-2,4-dinitrobenzene, which is very reactive toward nucleophilic aromatic substitution (see Chapter 13).

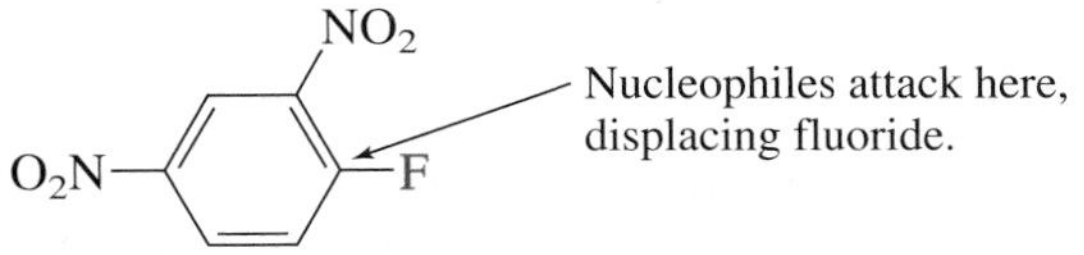

NO_2, O_2N, F

Nucleophiles attack here, displacing fluoride.

1-Fluoro-2,4-dinitrobenzene

1-Fluoro-2,4-dinitrobenzene is commonly referred to as *Sanger's reagent.*

The amino group of the N-terminal amino acid displaces fluoride from 1-fluoro-2,4-dinitrobenzene and gives a peptide in which the N terminus is labeled with a 2,4-dinitrophenyl (DNP) group. This process is shown for Val-Phe-Gly-Ala in Figure 25.5. The 2,4-dinitrophenyl-labeled peptide DNP-Val-Phe-Gly-Ala is isolated and subjected to hydrolysis, after which the 2,4-dinitrophenyl derivative of the N-terminal amino acid is isolated and identified as DNP-Val by comparing its chromatographic behavior with that of standard samples of DNP-labeled amino acids. None of the other amino acid residues bear a 2,4-dinitrophenyl group; they appear in the hydrolysis product as the free amino acids.

Labeling the N-terminal amino acid as its DNP derivative is mainly of historical interest and has been replaced by other methods. We'll discuss one of these—the Edman degradation—in Section 25.11. First, though, we'll complete our review of the general strategy for peptide sequencing by seeing how Sanger tied all of the information together into a structure for insulin.

O_2N
F
+
$H_3\overset{+}{N}$
H
N
O
N
H
O
N
H
O
1-Fluoro-2,4-dinitrobenzene
NO_2
Val-Phe-Gly-Ala (VFGA)
Na_2CO_3
The purpose of Na_2CO_3 is to deprotonate the N-terminal nitrogen. The resulting amino group then reacts with 1-fluoro-2,4-dinitrobenzene by nucleophilic aromatic substitution.
DNP-Val-Phe-Gly-Ala (DNP-VFGA)
H_3O^+
Acid hydrolysis cleaves the amide bonds and gives the 2,4-dinitrophenyl derivative of the N-terminal amino acid and a mixture of unlabeled amino acids.
OH
DNP-Val
Phe (F)
Gly (G)
Ala (A)

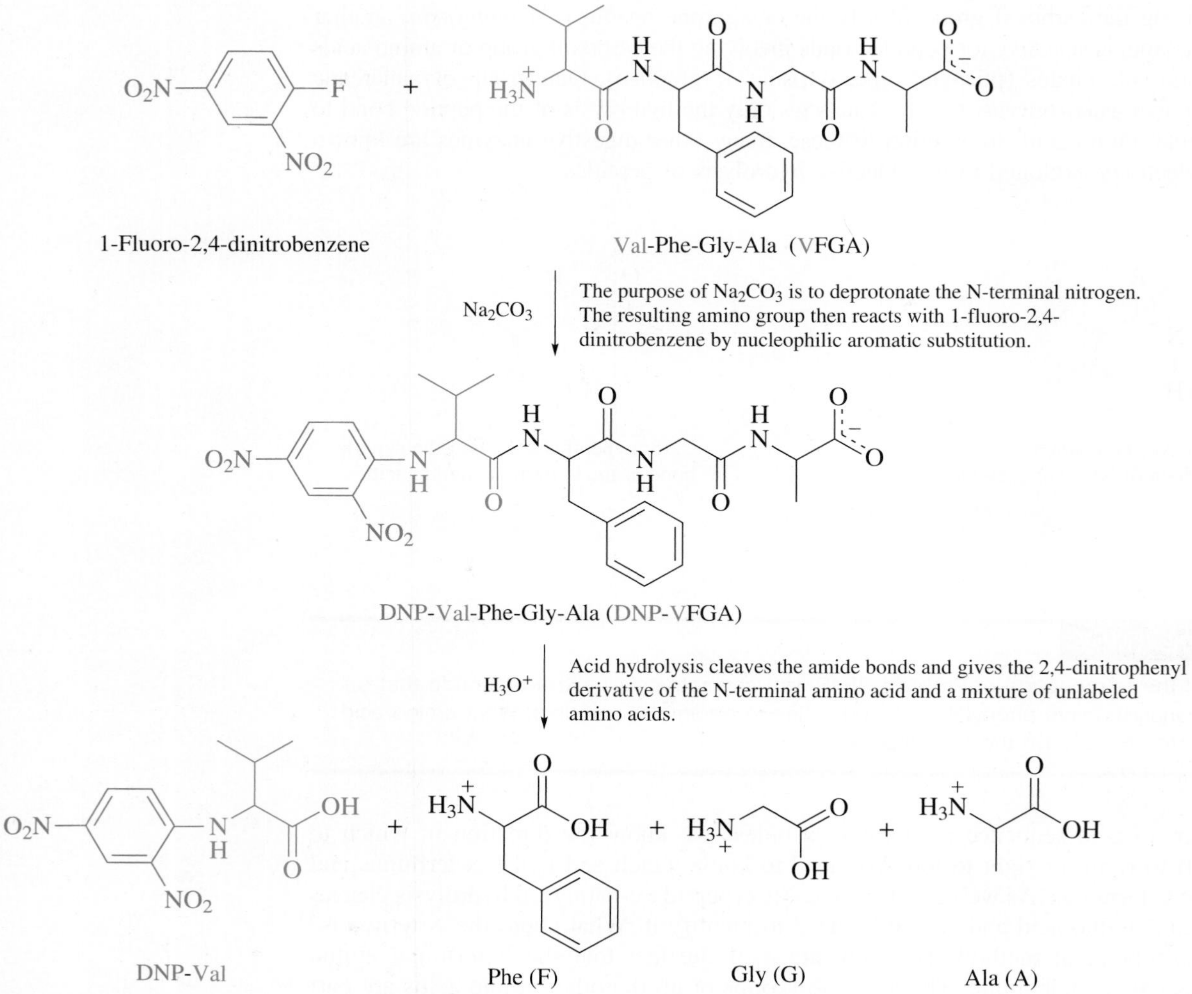

Figure 25.5

Use of 1-fluoro-2,4-dinitrobenzene to identify the N-terminal amino acid of a peptide.

25.10 Insulin

Sanger worked with insulin from cows, which has 51 amino acids, divided between two chains. One of these, the A chain, has 21 amino acids; the other, the B chain, has 30. The A and B chains are joined by disulfide bonds between cysteine residues (Cys-Cys). Figure 25.6 shows some of the information that defines the amino acid sequence of the B chain.

- Reaction of the B chain peptide with 1-fluoro-2,4-dinitrobenzene established that phenylalanine is the N terminus.
- Pepsin-catalyzed hydrolysis gave the four peptides shown in blue in Figure 25.6. (Their sequences were determined in separate experiments.) These four peptides contain 27 of the 30 amino acids in the B chain, but there are no points of overlap between them.
- The sequences of the four tetrapeptides shown in red in Figure 25.6 bridge the gaps between three of the four "blue" peptides to give an unbroken sequence from 1 through 24.
- The peptide shown in green was isolated by trypsin-catalyzed hydrolysis and has an amino acid sequence that completes the remaining overlaps.

The collection of sequenced fragments constitutes the **peptide map** for insulin.

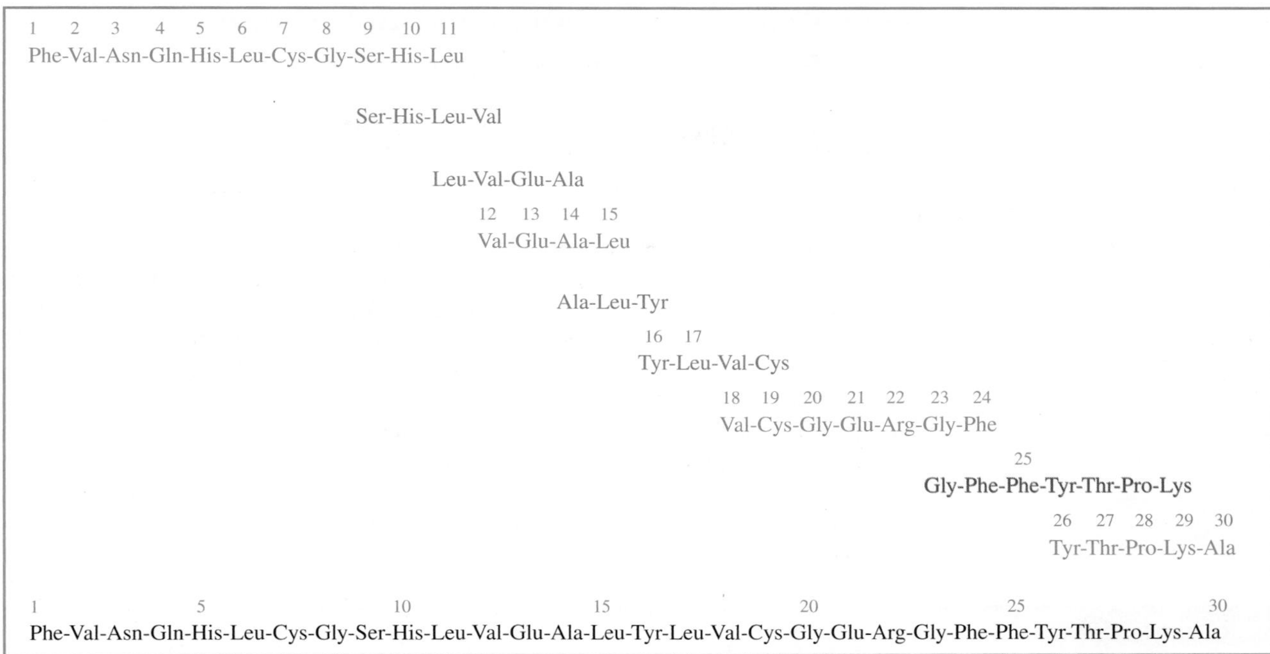

Figure 25.6

Diagram showing how the amino acid sequence of the B chain of bovine insulin can be determined by overlap of peptide fragments. Pepsin-catalyzed hydrolysis produced the fragments shown in blue, trypsin produced the one shown in green, and acid-catalyzed hydrolysis gave many fragments, including the four shown in red. Using one-letter abbreviations, the amino acid sequence is FVNQHLCGSHLVEALYLVCGERGFFYTPKA.

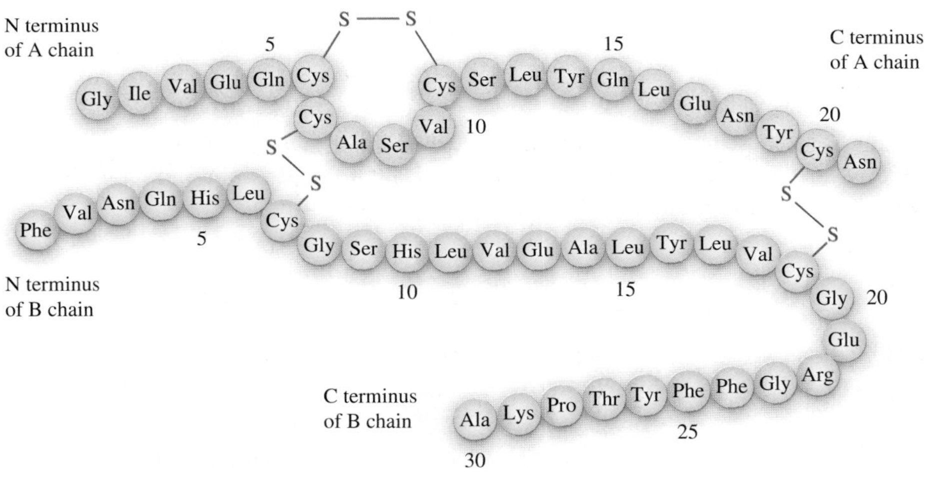

Figure 25.7

The amino acid sequence in bovine insulin. The A chain is joined to the B chain by two disulfide units (shown in green). There is also a disulfide bond linking cysteines 6 and 11 in the A chain. Human insulin has threonine and isoleucine at residues 8 and 10, respectively, in the A′ chain and threonine as the C-terminal amino acid in the B chain.

Sanger also determined the sequence of the A chain and identified the cysteine residues involved in disulfide bonds between the A and B chains as well as in the disulfide linkage within the A chain. The complete insulin structure is shown in Figure 25.7. The structure shown is that of bovine insulin. The A chains of human insulin and bovine insulin differ in only two amino acid residues; their B chains are identical except for the amino acid at the C terminus.

25.11 Edman Degradation and Automated Sequencing of Peptides

When Sanger's method for N-terminal residue analysis was discussed, you may have wondered why it was not done sequentially. Simply start at the N terminus and work steadily back to the C terminus identifying one amino acid after another. The idea is fine, but it just doesn't work well in practice, at least with 1-fluoro-2,4-dinitrobenzene.

A major advance was devised by Pehr Edman (University of Lund, Sweden) that became the standard method for N-terminal residue analysis. The **Edman degradation** is based on the chemistry shown in Mechanism 25.1. A peptide reacts with phenyl isothiocyanate to give a *phenylthiocarbamoyl* (PTC) derivative, as shown in the first step. This PTC derivative is then treated with an acid in an *anhydrous* medium (Edman used nitromethane saturated with hydrogen chloride) to cleave the amide bond between the N-terminal amino acid and the remainder of the peptide. No other peptide bonds are cleaved in this step as amide bond hydrolysis requires water. When the PTC derivative is treated with acid in an anhydrous medium, the sulfur atom of the C=S unit acts as an internal nucleophile, and the only amide bond cleaved under these conditions is the one to the N-terminal amino acid. The

Mechanism 25.1

The Edman Degradation

Step 1: A peptide is treated with phenyl isothiocyanate to give a phenylthiocarbamoyl (PTC) derivative.

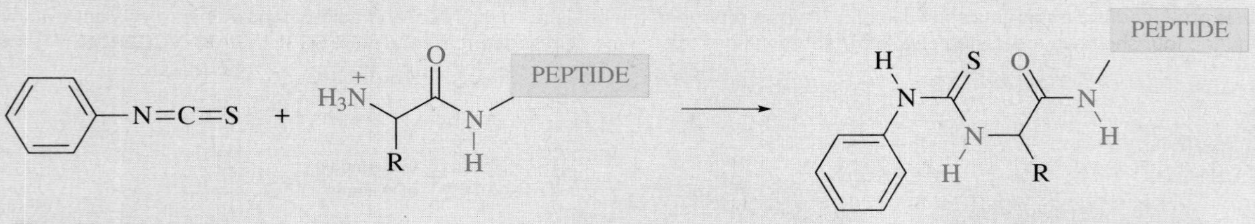

Phenyl isothiocyanate PTC derivative

Step 2: On reaction with hydrogen chloride in an anhydrous solvent, the thiocarbonyl sulfur of the PTC derivative attacks the carbonyl carbon of the N-terminal amino acid. The N-terminal amino acid is cleaved as a thiazolone derivative from the remainder of the molecule.

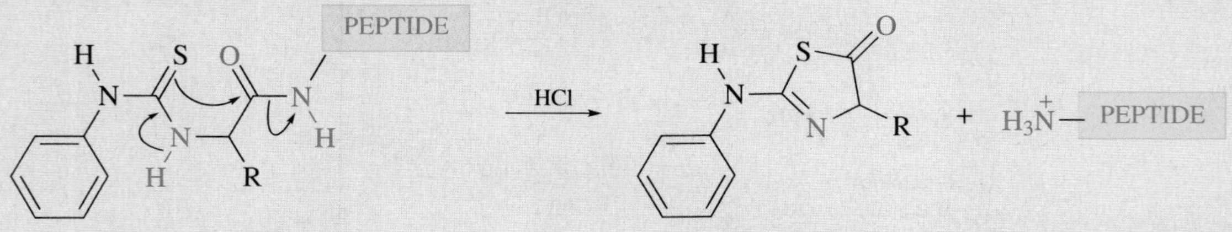

PTC derivative Thiazolone Remainder of peptide

Step 3: Once formed, the thiazolone derivative isomerizes to a more stable phenylthiohydantoin (PTH) derivative, which is isolated and characterized, thereby providing identification of the N-terminal amino acid. The remainder of the peptide (formed in step 2) can be isolated and subjected to a second Edman degradation.

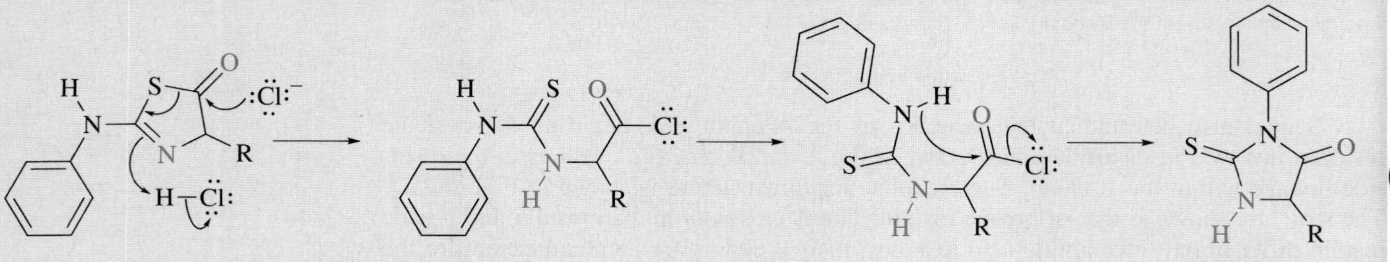

Thiazolone PTH derivative

product of this cleavage, called a *thiazolone,* is unstable under the conditions of its formation and rearranges to a *phenylthiohydantoin* (PTH), which is isolated and identified by comparing it with standard samples of PTH derivatives of known amino acids. This is normally done by chromatographic methods, but mass spectrometry has also been used.

Only the N-terminal amide bond is broken in the Edman degradation; the rest of the peptide chain remains intact. It can be isolated and subjected to a second Edman procedure to determine its new N terminus. We can proceed along a peptide chain by beginning with the N terminus and determining each amino acid in order. The sequence is given directly by the structure of the PTH derivative formed in each successive degradation.

Problem 25.17

Give the structure of the PTH derivative isolated in the second Edman cycle of the tetrapeptide Val-Phe-Gly-Ala.

Ideally, one could determine the primary structure of even the largest protein by repeating the Edman procedure. Because anything less than 100% conversion in any single Edman degradation gives a mixture containing some of the original peptide along with the degraded one, two different PTH derivatives are formed in the

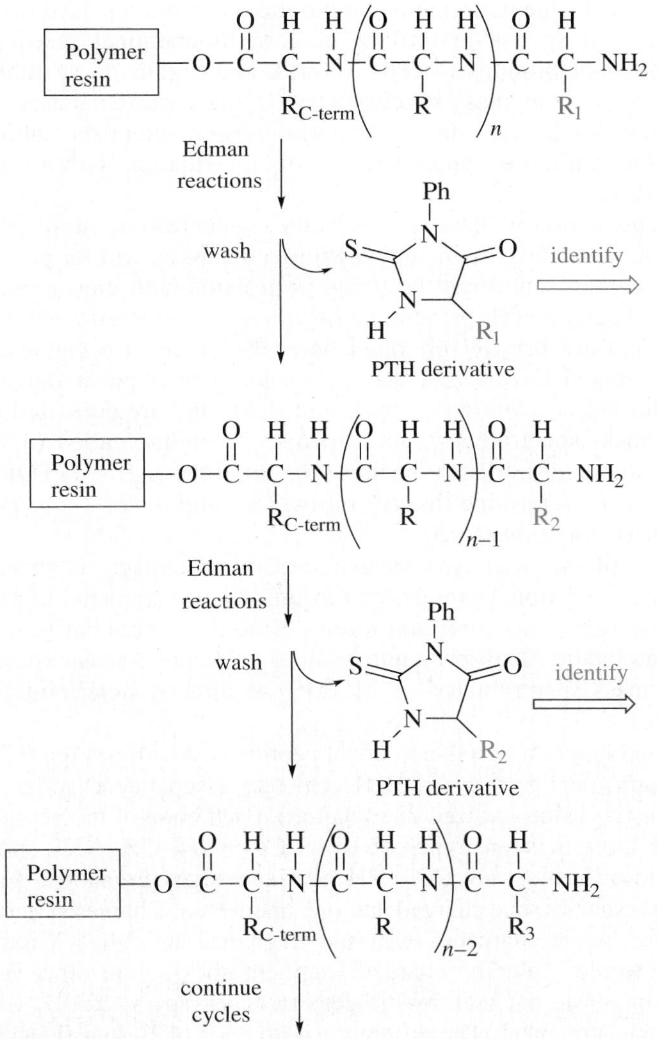

Polymer resin — O—C(=O)—CH($R_{C\text{-term}}$)—N(H)—[C(=O)—CH(R)—N(H)]$_n$—C(=O)—CH(R_1)—NH_2

Edman reactions ↓ wash → PTH derivative (Ph, N, S, O, N—H, R_1) ⟹ identify

Polymer resin — O—C(=O)—CH($R_{C\text{-term}}$)—N(H)—[C(=O)—CH(R)—N(H)]$_{n-1}$—C(=O)—CH(R_2)—NH_2

Edman reactions ↓ wash → PTH derivative (Ph, N, S, O, N—H, R_2) ⟹ identify

Polymer resin — O—C(=O)—CH($R_{C\text{-term}}$)—N(H)—[C(=O)—CH(R)—N(H)]$_{n-2}$—C(=O)—CH(R_3)—NH_2

continue cycles ↓

Figure 25.8

The Edman degradation is performed by attaching the peptide of interest to a resin (solid phase) through its carboxyl terminus. The amino acid PTH derivative is released into solution and separated from the rest of the peptide for identification. The resin bound peptide is then available for the next series of Edman reactions.

next Edman cycle, and the ideal is not realized in practice. After about 20 amino acids, possible mixtures become problematic. Instead, the protein of interest is cleaved into smaller fragments using proteolytic enzymes such as trypsin and chymotrypsin (Section 25.9) and other cleavage techniques. The fragments are then attached to a polymeric support by their carboxyl terminus and the chemical reactions of the Edman degradation are performed on the resin. The released PTH amino acid is then washed from the resin and identified. The remaining peptide is still attached to the resin, exposing the amino terminus of the next amino acid, and the cycle is repeated (Figure 25.8).

As may be evident by the repetitive nature of the processes, the Edman degradation can be automated. Edman protein sequencers are commercially available, and current instrumentation can reliably sequence as little as 10 pmol of a peptide.

25.12 Mass Spectrometry of Peptides and Proteins

The use of mass spectrometry in protein analysis has revolutionized many areas of research and medicine. First developed to sequence unknown peptides, increasing sophistication in instrumentation and computational resources have made protein mass spectrometry an indispensable tool in modern biochemical research. Proteomics, which is the study of proteins in cells, tissues, and organisms, would be impractical if not impossible without mass spectrometry. Using large databases of protein mass spectrometry data and present-day computation power, scientists can identify multiple proteins in complex mixtures and correlate this information with a myriad of cellular processes and diseases.

Protein sequencing by mass spectrometry is generally used for peptides prepared by chemical synthesis, for proteins unknown to a database, and for particularly difficult peptides, such as ones that might be found in proteins with unusual post-translational modifications. The general principles of mass spectrometry were introduced in Section 14.24 and are briefly reiterated here. Molecules are vaporized into the gas phase and are ionized in different ways, depending on a particular application. The charged particles are accelerated by magnetic fields and are deflected to an ion detector. In protein mass spectrometry, an ionization technique called MALDI can create ions from large molecules. Measurement of the ion time-of-flight (TOF) of the charged molecules is used to determine the m/z ratios. (See the boxed essay *Peptide Mapping and MALDI Mass Spectrometry*.)

One utility of mass spectrometry in organic chemistry is that molecular ions containing certain functional groups tend to break into fragments in predictable ways. The m/z of these fragments limit and often define the functional groups contained in the unknown molecule. This principle is used in tandem mass spectrometry (often called tandem mass spectrometry, or MS/MS) to directly determine the sequence of the peptide ion.

A simplified diagram of a hypothetical peptide is shown in Figure 25.10. In tandem mass spectrometry, peptide ions are deflected into a separate chamber and are broken into smaller ions (collision-induced dissociation). Each bond of the peptide backbone can fragment at the three different types of bond: CO—NH, CH—CO, and NH—CH. For simplicity, the locations of only CO—NH bond cleavages are shown. Each bond breakage produces two species, one charged and one uncharged. Charged N-terminal fragments are called B-ions, while charged C-terminal fragments are called Y-ions. For example, cleavage at B_1 would yield the charged fragment shown; the other fragment will be uncharged and therefore not seen by the detector. Cleavage at just B_2 would also give a charged N-terminal fragment. The difference in the m/z of B_1 and B_2 will yield the mass of the amino acid containing R_2. Exact masses of the various amino acids (and frequently common post-translationally modified amino acids) are then consulted to identify the particular amino acid.

Peptide Mapping and MALDI Mass Spectrometry

Biological materials often contain proteins that must be identified. Recent advances in mass spectrometry have made peptide mapping a convenient tool for this purpose. The protein in question is selectively hydrolyzed with a peptidase such as trypsin and the mixture of peptides produced is analyzed by *matrix-assisted laser desorption ionization* (MALDI) as illustrated in Figure 25.9.

MALDI offers two main advantages over traditional mass spectrometric methods.

1. Substances, such as peptides, that lack sufficient volatility to be vaporized for analysis by conventional mass spectrometry can be vaporized by MALDI.
2. The species analyzed by the mass spectrometer is the conjugate acid of a peptide (peptide + H^+). Unlike the highly energetic ions generated by electron impact (see Section 14.24), the cations produced by MALDI have little tendency to fragment. Consequently, a mixture of peptide fragments gives a mass spectrum dominated by peaks with *m/z* values corresponding to those of the individual protonated peptides.

With the aid of freely available Internet tools and databases, the MALDI data set is compared with known proteins to generate a list of potential matches. The analyst inputs the peptidase used to digest the original protein and the *m/z* values of the peptides displayed in the MALDI spectrum. As specified by the search criteria, the search delivers (in a matter seconds!) a list of peptide sequences and the proteins these sequences contain. Next, a different peptidase is used to hydrolyze the protein, to provide a second set of peptides that are also analyzed by MALDI and matched against the database. MALDI mass spectrometry compares the amino acid *composition* of unsequenced peptides with the amino acid *sequence* of known proteins in order to identify an unknown protein. The procedure is repeated until the list of potential matches is narrowed to a single known protein, additional data are needed, or it becomes likely that the protein is new.

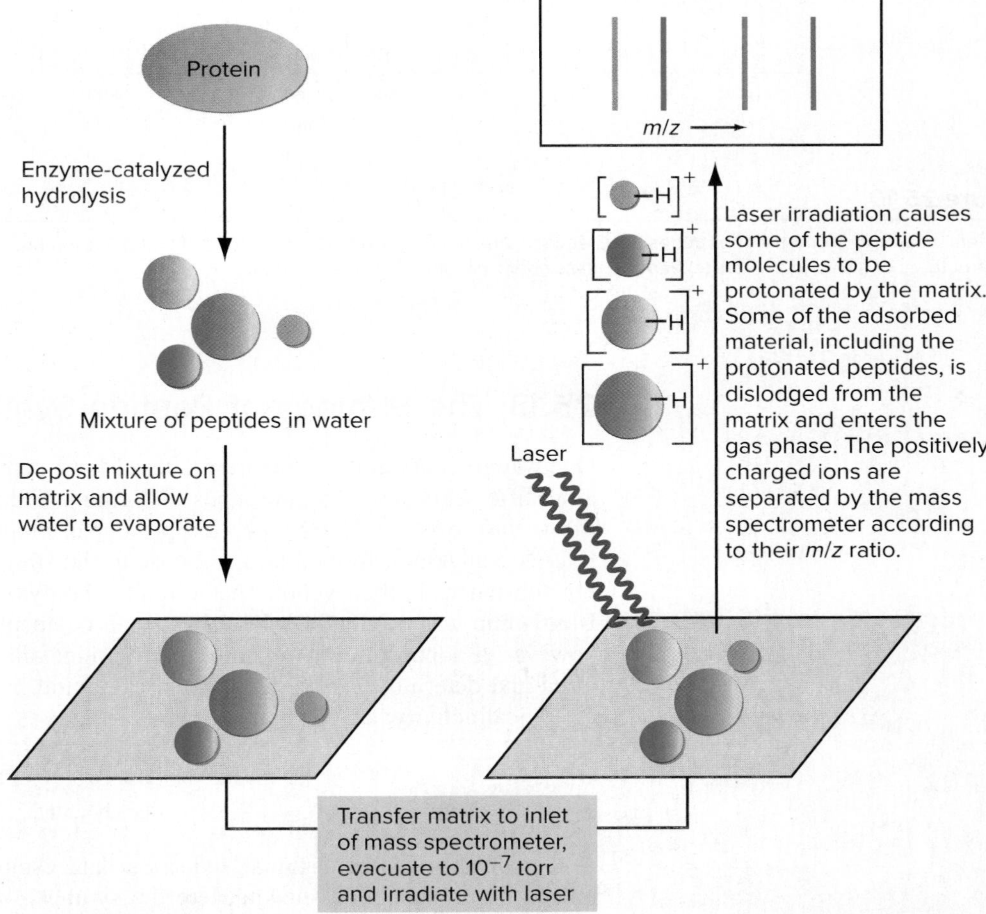
Protein
Enzyme-catalyzed hydrolysis
Mixture of peptides in water
Deposit mixture on matrix and allow water to evaporate
Transfer matrix to inlet of mass spectrometer, evacuate to 10^{-7} torr and irradiate with laser
Laser
H
+
m/z
Laser irradiation causes some of the peptide molecules to be protonated by the matrix. Some of the adsorbed material, including the protonated peptides, is dislodged from the matrix and enters the gas phase. The positively charged ions are separated by the mass spectrometer according to their m/z ratio.

Figure 25.9

The molecular weights of all of the peptides in a mixture obtained by the enzyme-catalyzed hydrolysis of a protein are determined simultaneously by mass spectrometry using matrix-assisted laser desorption ionization (MALDI).

Problem 25.18

Which two of the standard amino acids cannot be differentiated on the basis of their *m/z* ratio?

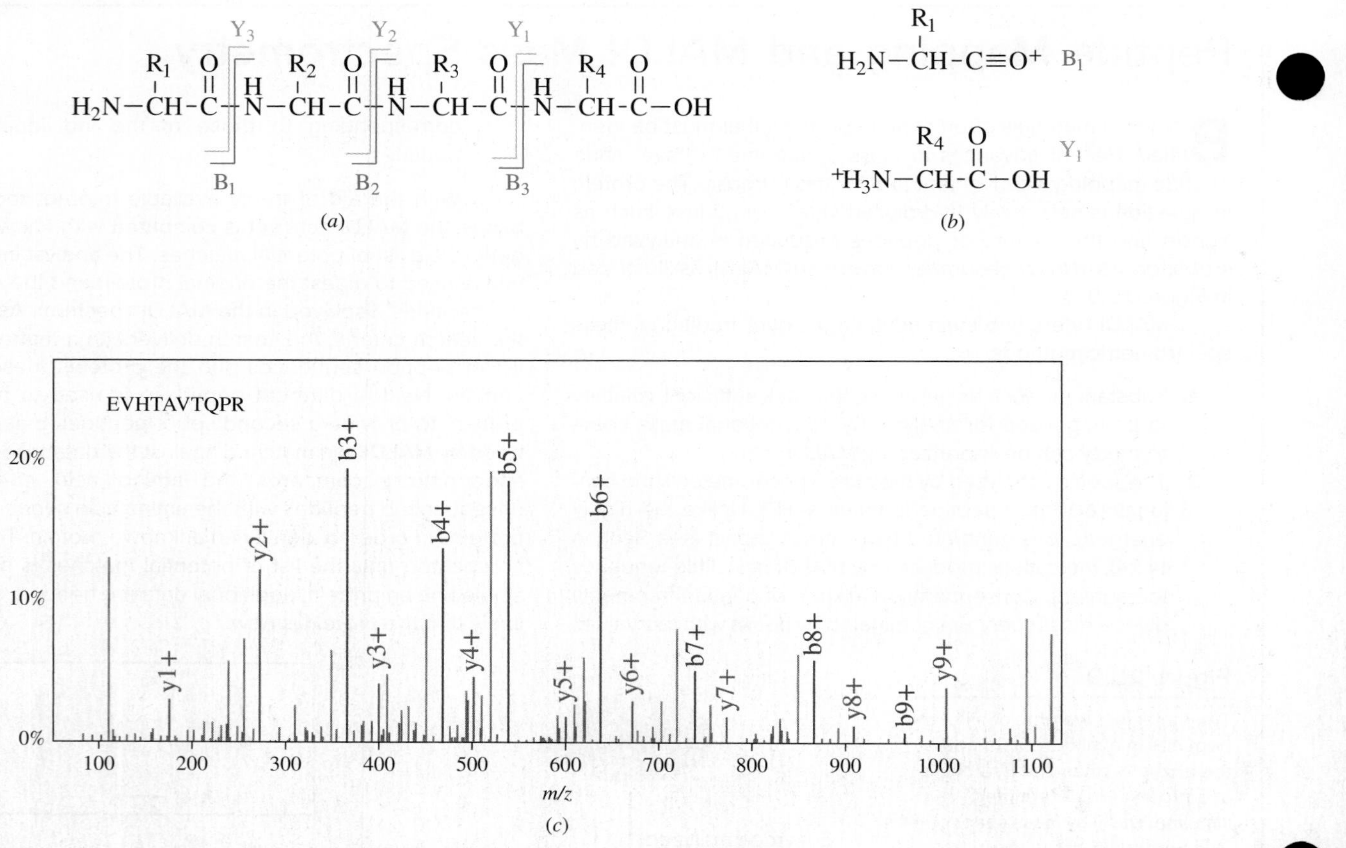

$$H_2N{-}\underset{}{\overset{R_1}{CH}}{-}\overset{O}{\overset{\|}{C}}\,|\,{-}\overset{H}{N}{-}\overset{R_2}{CH}{-}\overset{O}{\overset{\|}{C}}\,|\,{-}\overset{H}{N}{-}\overset{R_3}{CH}{-}\overset{O}{\overset{\|}{C}}\,|\,{-}\overset{H}{N}{-}\overset{R_4}{CH}{-}\overset{O}{\overset{\|}{C}}{-}OH$$

(Cleavage sites: Y_3/B_1, Y_2/B_2, Y_1/B_3)

(a)

$$H_2N{-}\overset{R_1}{CH}{-}C{\equiv}O^+ \quad B_1$$

$$^+H_3N{-}\overset{R_4}{CH}{-}\overset{O}{\overset{\|}{C}}{-}OH \quad Y_1$$

(b)

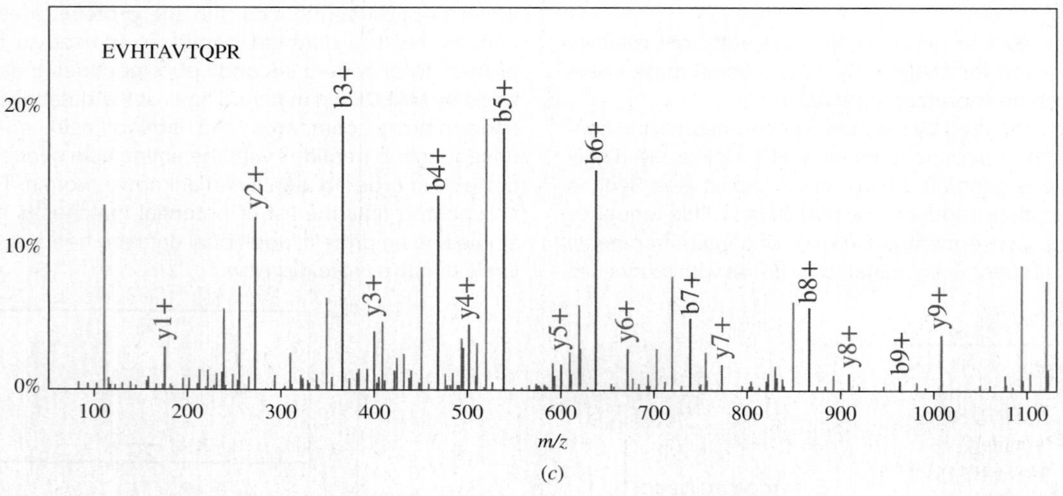

(c)

Figure 25.10

Tandem mass spectrometry of peptides. (a) Cleavage sites that generate B- and Y-ions of a peptide ion are indicated on the generic peptide. (b) Structures of a B- and a Y-ion. (c) Peptide mass spectrum showing B- and Y-ions.

25.13 The Strategy of Peptide Synthesis

One way to confirm the structure proposed for a peptide is to synthesize a peptide having a specific sequence of amino acids and compare the two. This was done, for example, in the case of *bradykinin,* a peptide present in blood that acts to lower blood pressure. Excess bradykinin, formed as a response to the sting of wasps and other insects containing substances in their venom that stimulate bradykinin release, causes severe local pain. Bradykinin was originally believed to be an octapeptide containing two proline residues; however, a nonapeptide containing three prolines in the following sequence was synthesized and determined to be identical with natural bradykinin in every respect, including biological activity:

Arg-Pro-Pro-Gly-Phe-Ser-Pro-Phe-Arg or RPPGFSPFR

Bradykinin

A reevaluation of the original sequence data established that natural bradykinin was indeed the nonapeptide shown. Here the synthesis of a peptide did more than confirm structure; synthesis was instrumental in determining structure.

The synthesis of peptides is an important area of drug development. For many years, the high cost of peptide synthesis and the rapid degradation of peptides when administered in the body hampered the progress of peptide drug discovery. Recent advances in peptide manufacturing and drug delivery have increased the use of peptide-based pharmaceuticals. Two of the most prominent peptide drugs are insulin and calcitonin.

A promising method of drug delivery uses liposomes (see Section 24.6).

The insulin required for the treatment of diabetes used to be obtained by extraction from the pancreas glands of cows and pigs. Since the early 1980s, this "natural" insulin has been replaced by "synthetic" human insulin prepared by recombinant DNA technology.

Recombinant DNA technology is widely used for the synthesis of peptides and proteins and can be exceptionally efficient in producing a desired product. Why, then, is it necessary to use chemical methods to synthesize peptides and proteins? One reason is the nature of protein biosynthesis. It is critical for an organism to make proteins that have the exact sequence of naturally occurring amino acids so it can perform its normal function in an organism. Even one mistake can have catastrophic consequences. For example, sickle cell anemia is the result of just one mutation in the protein hemoglobin, the oxygen-carrying protein in all red blood cells, in which a glutamic acid is replaced by a leucine (see Section 25.19). To avoid errors, the components of in vivo protein synthesis are finely tuned to faithfully translate the genomic message to create the correct protein. If the protein or peptide desired is composed of naturally occurring amino acids, recombinant DNA methods are often the most efficient way of synthesizing these products.

In some cases, though, it is desirable to join amino acids with bonds other than amides, to alter the side chain of an amino acid to a moiety not found in nature, or to change the stereochemistry of one or all of the amino acids in a biopolymer. Pharmaceutical companies use all of these methods to increase the effectiveness, stability, and solubility of peptides used against human diseases. Biochemists use them to understand fundamental science. Chemists can synthesize natural peptides and proteins and variations of these biopolymers not found in nature. We can understand how chemists do so based on what we have learned earlier in this book, but as we will see, there are some complications that we need to take into account to efficiently obtain the desired product.

The objective in peptide synthesis may be simply stated: to connect amino acids in a prescribed sequence by amide bond formation between them. A number of very effective methods and reagents have been designed for peptide bond formation, so that the joining together of amino acids by amide linkages is not difficult. The real difficulty lies in ensuring that the correct sequence is obtained. This can be illustrated by considering the synthesis of a representative dipeptide, Phe-Gly. Random peptide bond formation in a mixture containing phenylalanine and glycine would be expected to lead to four dipeptides:

$$H_3\overset{+}{N}CHCO_2^- \ (CH_2C_6H_5) + H_3\overset{+}{N}CH_2CO_2^- \longrightarrow \text{Phe-Gly} + \text{Phe-Phe} + \text{Gly-Phe} + \text{Gly-Gly}$$

Phenylalanine Glycine

To direct the synthesis so that only Phe-Gly is formed, the amino group of phenylalanine and the carboxyl group of glycine must be protected so that they cannot react under the conditions of peptide bond formation. We can represent the peptide bond formation step by the following equation, where X and Y are amine- and carboxyl-protecting groups, respectively:

N-Protected phenylalanine + C-Protected glycine —couple→ Protected Phe-Gly —deprotect→ Phe-Gly

$$X\text{-NHCH}(CH_2C_6H_5)\text{COOH} + H_2NCH_2C(O)Y \xrightarrow{\text{couple}} X\text{-NHCH}(CH_2C_6H_5)C(O)NHCH_2C(O)Y \xrightarrow{\text{deprotect}} H_3\overset{+}{N}CH(CH_2C_6H_5)C(O)NHCH_2CO_2^-$$

Thus, the synthesis of a dipeptide of prescribed sequence requires at least three operations:

1. *Protect* the amino group of the N-terminal amino acid and the carboxyl group of the C-terminal amino acid.
2. *Couple* the two protected amino acids by amide bond formation between them.
3. *Deprotect* the amino group at the N terminus and the carboxyl group at the C terminus.

Higher peptides are prepared in an analogous way by a direct extension of the logic just outlined for the synthesis of dipeptides.

Sections 25.14 and 25.15 describe the chemistry associated with the protection and deprotection of amino and carboxyl functions, along with methods for peptide bond formation. The focus in those sections is on solution-phase peptide synthesis. Section 25.16 shows how these methods are adapted to solid-phase synthesis.

25.14 Amino and Carboxyl Group Protection and Deprotection

Amino groups are normally protected as their *tert*-butoxycarbonyl (Boc), carbobenzoxy-carbonyl (Z), or 9-fluorenylmethoxycarbonyl (Fmoc) derivative prepared as described in Section 25.5. The Boc group is usually removed by treatment with trifluoroacetic acid in an organic solvent. Other nonaqueous acidic reagents such as hydrogen bromide may also be used.

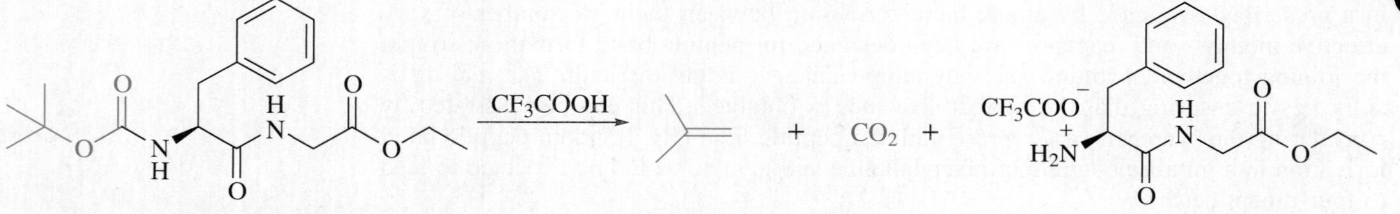

N-tert-Butoxycarbonylphenylalanylglycine ethyl ester — 2-Methylpropene — Carbon dioxide — Phenylalanylglycine ethyl ester trifluoroacetate (100%)

The *tert*-butyl group is cleaved as *tert*-butyl carbocation which, on deprotonation, gives 2-methylpropene. Both of the derived products of the deprotection reaction are gases, which bubble out of the solution and drive the reaction to completion.

A benzyloxycarbonyl-protected amine is more stable than a Boc-protected amine and is not removed by treatment with room-temperature trifluoroacetic acid. It requires a stronger acid such as hydrogen bromide. The derived product from this reaction is benzyl bromide in addition to carbon dioxide.

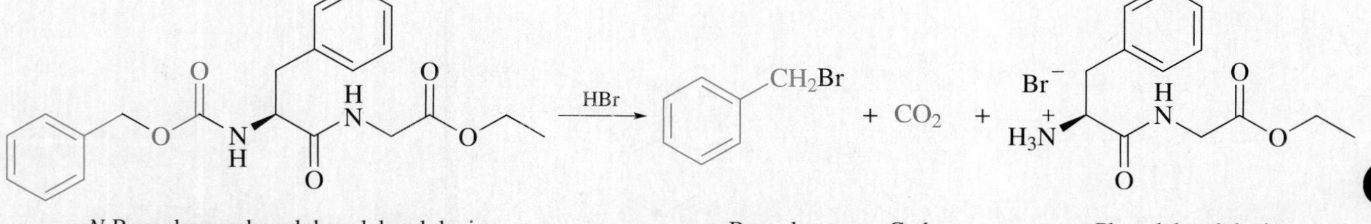

N-Benzyloxycarbonylphenylalanylglycine ethyl ester — Benzyl bromide — Carbon dioxide — Phenylalanylglycine ethyl ester hydrobromide (82%)

Unlike Boc, however, benzyloxycarbonyl can also be removed by *hydrogenolysis* in the presence of palladium. In this case, the derived product is toluene.

H_2, Pd

N-Benzyloxycarbonylphenylalanylglycine ethyl ester → Toluene + Carbon dioxide (CO_2) + Phenylalanylglycine ethyl ester (100%)

9-Fluorenylmethoxycarbonyl (Fmoc) differs from benzyloxycarbonyl and Boc in that it is removable by treatment with base. Primary and some secondary amines are typically used for the deprotection reaction.

9-Fluorenylmethoxycarbonylalanine → Dibenzofulvene + Carbon dioxide (CO_2) + Alanine (98%)

Notice how the conditions for removal of the three amine protecting groups are *orthogonal.* These groups can protect three different amines and can be removed separately. For example, a Boc group may be removed to uncover one amine while other amines bonded to Z and Fmoc would remain protected. This type of strategy is used in peptide synthesis so that side-chain functional groups will not undergo unwanted reactions. In our case, though, we will concern ourselves only with amino acids with unreactive side chains.

Carboxyl groups are normally protected as esters. Methyl and ethyl esters are prepared by Fischer esterification and removed by hydrolysis in base. Benzyl esters are a popular choice because a synthetic peptide, protected at its N terminus with a Z group and at its C terminus as a benzyl ester, can be completely deprotected in a single operation by hydrogenolysis.

H_2, Pd

N-Benzyloxycarbonylphenylalanylglycine benzyl ester → Phenylalanylglycine (87%) + Toluene + Carbon dioxide (CO_2)

Protection of the α-carboxylic acid is not necessary for the most common method used to assemble a larger peptide, the solid-phase method. It will be needed in solution-phase preparation of dipeptides, which we will see next.

25.15 Peptide Bond Formation

To form a peptide bond between two suitably protected amino acids, the free carboxyl group of one of them must be *activated* so that it is a reactive acylating agent. The most familiar acylating agents are acyl chlorides, and they were once extensively used to couple amino acids. Certain drawbacks to this approach, however, led chemists to seek alternative methods.

In one method, treatment of a solution containing the N-protected and the C-protected amino acids with *N,N′*-dicyclohexylcarbodiimide (DCCI) leads directly to peptide bond formation:

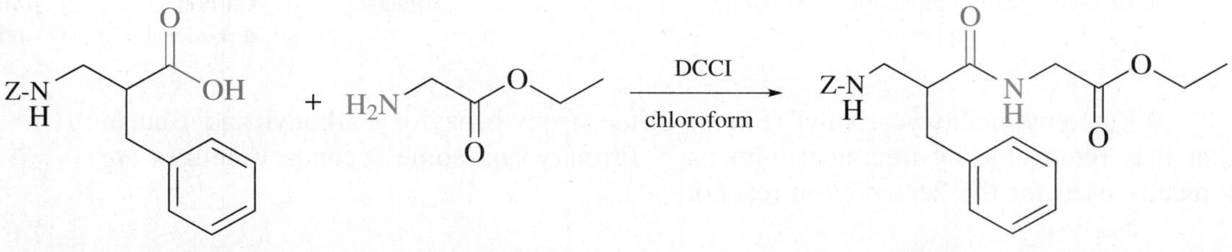

Z-Protected phenylalanine + Glycine ethyl ester $\xrightarrow[\text{chloroform}]{\text{DCCI}}$ Z-Protected Phe-Gly ethyl ester (83%)

N,N′-Dicyclohexylcarbodiimide has the structure:

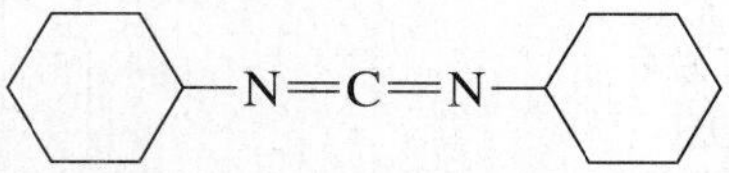

N,N′-Dicyclohexylcarbodiimide (DCCI)

Mechanism 25.2 shows how DCCI promotes the formation of the peptide bond shown in the preceding equation.

Problem 25.19

Show the steps involved in the synthesis of Ala-Leu from alanine and leucine using benzyloxycarbonyl and benzyl ester protecting groups and DCCI-promoted peptide bond formation.

The *N,N′*-dicyclohexylurea that is produced in peptide couplings with DCCI is soluble in the same solvents as the product, making it more difficult to separate the byproducts during purification. An alternative carbodiimide used in peptide synthesis is 1-ethyl-3-(3-dimethylaminopropyl)carbodiimide, known as EDCI. Both EDCI and the urea that is produced from it are water-soluble and can be more easily separated from the peptide product. EDCI can also be used in organic solvents.

$$CH_3CH_2N{=}C{=}NCH_2CH_2CH_2N(CH_3)_2$$

1-Ethyl-3-(3-dimethylaminopropyl)carbodiimide (EDCI)

Problem 25.20

What is the structure of the urea that is produced in the coupling reaction of a protected amino acid and a carboxylic acid with EDCI?

Longer peptides are prepared either by stepwise extension of peptide chains, one amino acid at a time, or by coupling of fragments containing several residues (the *fragment*

Mechanism 25.2

Amide Bond Formation Between a Carboxylic Acid and an Amine Using *N,N′*-Dicyclohexylcarbodiimide

THE OVERALL REACTION:

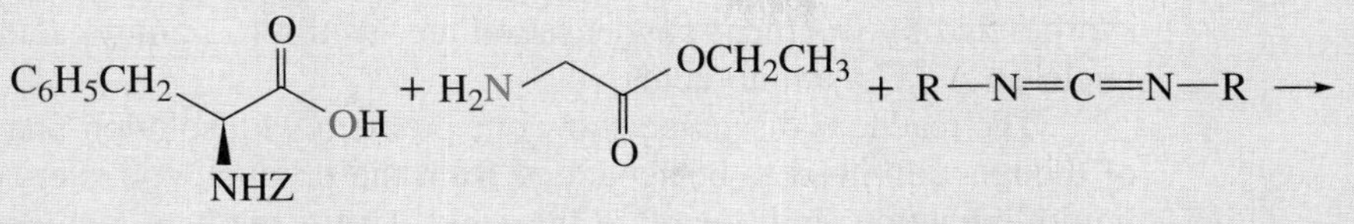

Z-Protected phenylalanine (Z = Benzyloxycarbonyl) + Glycine ethyl ester + *N,N′*-Dicyclohexylcarbodiimide (DCCI) (R = cyclohexyl) ⟶ Z-Protected Phe-Gly ethyl ester + *N,N′*-Dicyclohexylurea

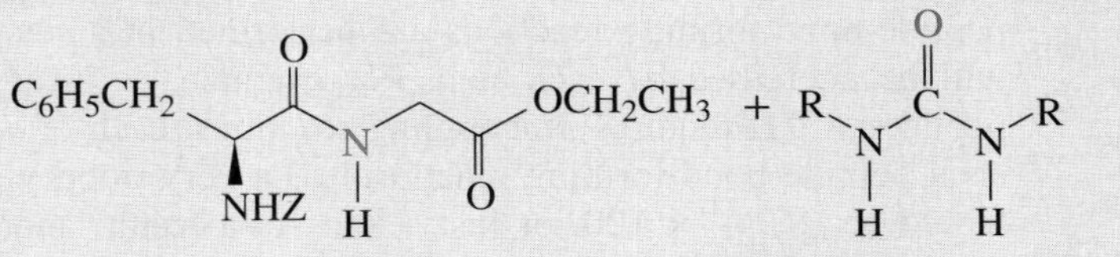

THE MECHANISM:

Step 1: In the first stage of the reaction, the carboxylic acid adds to one of the double bonds of DCCI to give an *O*-acylisourea.

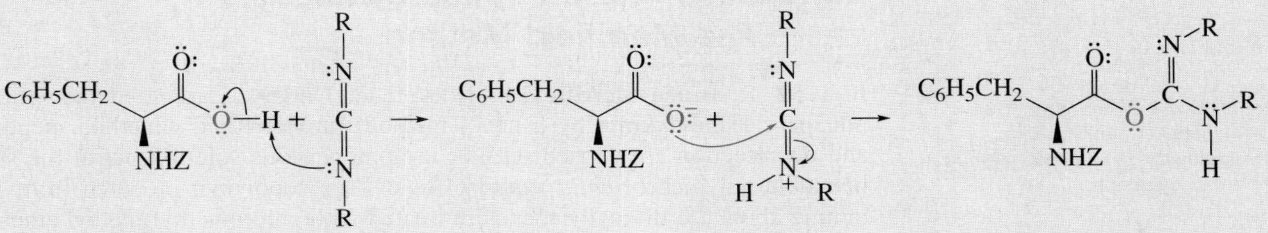

Z-Protected phenylalanine + DCCI ⟶ ⟶ An *O*-Acylisourea

Step 2: Structurally, *O*-acylisoureas resemble acid anhydrides and are powerful acylating agents. In the reaction's second stage the amine adds to the carbonyl group of the *O*-acylisourea to give a tetrahedral intermediate.

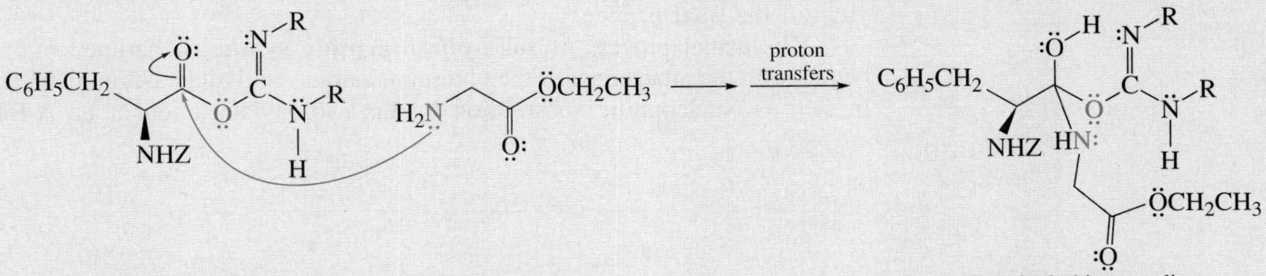

An *O*-acylisourea + Glycine ethyl ester ⟶ (proton transfers) ⟶ Tetrahedral intermediate

Step 3: The tetrahedral intermediate dissociates to an amide and *N,N′*-dicyclohexylurea.

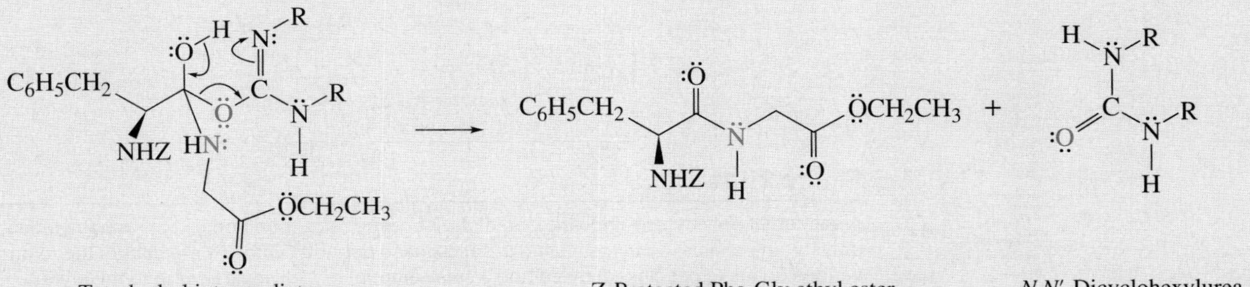

Tetrahedral intermediate ⟶ Z-Protected Phe-Gly ethyl ester + *N,N′*-Dicyclohexylurea

condensation approach). Human pituitary adrenocorticotropic hormone (ACTH), for example, has 39 amino acids and was synthesized by coupling of smaller peptides containing residues 1–10, 11–16, 17–24, and 25–39. An attractive feature of this approach is that the various protected peptide fragments may be individually purified, which simplifies the purification of the final product. Among other substances that have been synthesized by fragment condensation are insulin (51 amino acids) and the protein ribonuclease A (124 amino acids).

The reactions discussed show one problem with solution-phase synthesis. Byproducts of the reaction need to be separated from the product, and every separation step (liquid–liquid extraction) will decrease the yield. Using segment condensation, the overall yield of the reaction is directly affected just by peptide-bond-forming steps. So, if just a few segments are joined together for the product, this approach can be efficient. If many peptide-bond-forming reactions are performed in a sequential manner, the overall yield will be decreased by each step. For example, a linear synthesis of the 39-amino-acid peptide ACTH requires 38 coupling reactions, each of which affects the overall yield. If each peptide-bond-forming reaction had a very good yield (say, 90%), the overall yield would be $(0.9)^{38} \times 100$, or just 1.8%. As a result, another approach is needed for performing sequential peptide synthesis! Fortunately, there is a method to do just that. It is described in the following section.

25.16 Solid-Phase Peptide Synthesis: The Merrifield Method

Merrifield was awarded the 1984 Nobel Prize in Chemistry for developing the solid-phase method of peptide synthesis.

In 1962, R. Bruce Merrifield of Rockefeller University reported the synthesis of the nonapeptide bradykinin by a novel method. In Merrifield's method, peptide coupling and deprotection are carried out not in homogeneous solution but at the surface of an insoluble polymer, or *solid support.* Beads of a copolymer prepared from styrene containing about 2% divinylbenzene are treated with chloromethyl methyl ether and tin(IV) chloride to give a resin in which about 10% of the aromatic rings bear —CH_2Cl groups (Figure 25.11). The growing peptide is anchored to this polymer, and excess reagents, impurities, and byproducts are removed by thorough washing after each operation. This greatly increases the yield of each coupling step (up to 99+%) and therefore the overall yield of the final peptide.

The actual process of **solid-phase peptide synthesis,** outlined in Figure 25.12, begins with the attachment of the C-terminal amino acid to the chloromethylated polymer in step 1. Nucleophilic substitution by the carboxylate anion of an *N*-Boc-protected

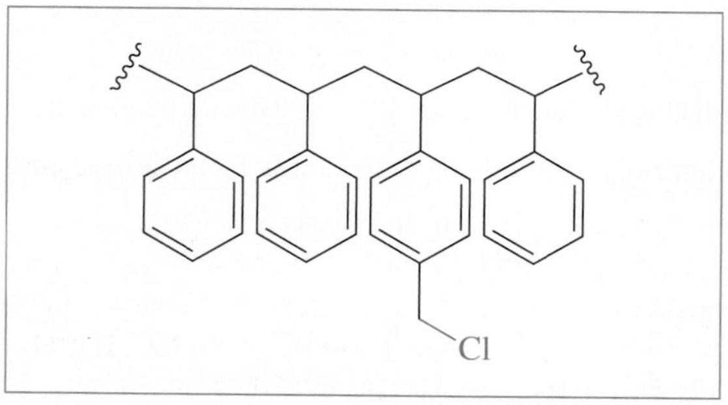

Figure 25.11

A section of polystyrene showing one of the benzene rings modified by chloromethylation. Individual polystyrene chains in the resin used in solid-phase peptide synthesis are connected to one another at various points (cross-linked) by adding a small amount of *p*-divinylbenzene to the styrene monomer. The chloromethylation step is carried out under conditions such that only about 10% of the benzene rings bear —CH_2Cl groups.

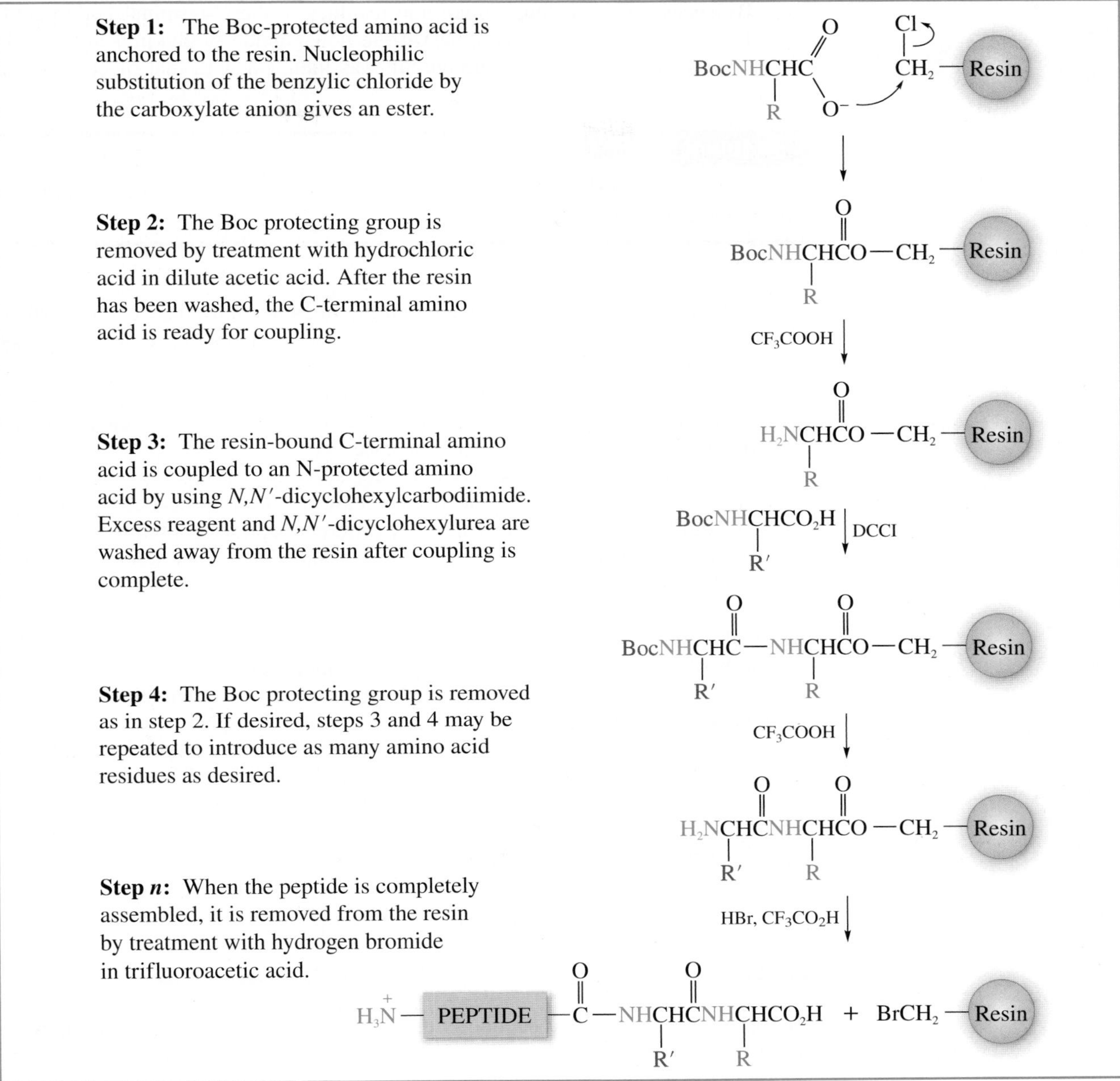

Step 1: The Boc-protected amino acid is anchored to the resin. Nucleophilic substitution of the benzylic chloride by the carboxylate anion gives an ester.
Step 2: The Boc protecting group is removed by treatment with hydrochloric acid in dilute acetic acid. After the resin has been washed, the C-terminal amino acid is ready for coupling.
Step 3: The resin-bound C-terminal amino acid is coupled to an N-protected amino acid by using N,N′-dicyclohexylcarbodiimide. Excess reagent and N,N′-dicyclohexylurea are washed away from the resin after coupling is complete.
Step 4: The Boc protecting group is removed as in step 2. If desired, steps 3 and 4 may be repeated to introduce as many amino acid residues as desired.
Step n: When the peptide is completely assembled, it is removed from the resin by treatment with hydrogen bromide in trifluoroacetic acid.
BocNHCHC
R
O
O⁻
Cl
CH2
Resin
BocNHCHCO—CH2
CF3COOH
H2NCHCO—CH2
BocNHCHCO2H
R′
DCCI
BocNHCHC—NHCHCO—CH2
H2NCHCNHCHCO—CH2
HBr, CF3CO2H
H3N— PEPTIDE —C—NHCHCNHCHCO2H + BrCH2

Figure 25.12

Peptide synthesis by the solid-phase method. Amino acid residues are attached sequentially beginning at the C terminus.

C-terminal amino acid displaces chloride from the chloromethyl group of the polymer to form an ester, protecting the C terminus while anchoring it to a solid support. Next, the Boc group is removed by treatment with acid (step 2), and the polymer containing the unmasked N terminus is washed with a series of organic solvents. Byproducts are removed, and only the polymer and its attached C-terminal amino acid residue remain. Next (step 3), a peptide bond to an *N*-Boc-protected amino acid is formed by condensation in the presence of *N,N′*-dicyclohexylcarbodiimide. Again, the polymer is washed thoroughly. The Boc protecting group is then removed by acid treatment (step 4), and after washing, the polymer is now ready for the addition of another amino acid residue by a repetition of the cycle. When all the amino acids have been added, the synthetic peptide is removed from the polymeric support by treatment with hydrogen bromide in trifluoroacetic acid.

The Merrifield procedure has been adapted to accommodate Fmoc as well as Boc protecting groups.

By successively adding amino acid residues to the C-terminal amino acid, it took Merrifield only eight days to synthesize the nonapeptide bradykinin in 68% yield. The biological activity of synthetic bradykinin was identical with that of natural material.

Problem 25.21

Starting with phenylalanine and glycine, outline the steps in the preparation of Phe-Gly by the Merrifield method.

Merrifield successfully automated all the steps in solid-phase peptide synthesis, and computer-controlled equipment is commercially available to perform this synthesis. Using an early version of his "peptide synthesizer," in collaboration with coworker Bernd Gutte, Merrifield reported the synthesis of the enzyme ribonuclease in 1969. It took them only six weeks to perform the 369 reactions and 11,391 steps necessary to assemble the sequence of 124 amino acids of ribonuclease.

Solid-phase peptide synthesis does not solve all purification problems, however. Even if every coupling step in the ribonuclease synthesis proceeded in 99% yield, the product would be contaminated with many different peptides containing 123 amino acids, 122 amino acids, and so on. Thus, Merrifield and Gutte's six weeks of synthesis was followed by four months spent in purifying the final product. About 12% of the expected mass was obtained after cleaving the protein from the resin. After the purification steps, the final yield was about 3%, which corresponds to a minimum 97% yield for each peptide-bond-forming step. The technique has since been refined to the point that yields at the 99% level and greater are achieved with current instrumentation, and thousands of peptides and peptide analogs have been prepared by the solid-phase method.

Merrifield's concept of a solid-phase method for peptide synthesis and his development of methods for carrying it out set the stage for an entirely new way to do chemical reactions. Solid-phase synthesis has been extended to include numerous other classes of compounds and has helped spawn a whole new field called **combinatorial chemistry.** Combinatorial synthesis allows a chemist, using solid-phase techniques, to prepare hundreds of related compounds (called *libraries*) at a time.

25.17 Secondary Structures of Peptides and Proteins

The primary structure of a peptide is its amino acid sequence. The **secondary structure** is the conformational relationship of nearest-neighbor amino acids with respect to each other. On the basis of X-ray crystallographic studies and careful examination of molecular models, Linus Pauling and Robert B. Corey of the California Institute of Technology showed that certain peptide conformations were more stable than others. Two arrangements, the **α helix** and the **β sheet,** stand out as secondary structural units that are both particularly stable and commonly encountered. Both of these incorporate two important features:

1. The geometry of the peptide bond is planar and the main chain is arranged in an anti conformation (see Section 25.6).
2. Hydrogen bonding can occur when the N—H group of one amino acid unit and the C═O group of another are close in space; conformations that maximize the number of these hydrogen bonds are stabilized by them.

Chains in a β sheet exist in an extended conformation with hydrogen bonds between a carbonyl oxygen of one chain and an amide N—H of another (Figure 25.13). Both the parallel and antiparallel arrangements of chains occur in proteins. Some of the space between peptide chains is occupied by the amino acid side chains, represented by R in Figure 25.13. Van der Waals repulsive forces involving these substituents cause the chains to rotate with respect to one another, giving a rippled effect known as a *β-pleated sheet* (Figure 25.14).

Figure 25.13

Hydrogen bonding between the carbonyl oxygen of one peptide chain and the amide N—H of another in a β-pleated sheet. In the antiparallel arrangement, the N-terminus → C-terminus direction of one chain is opposite to that of the other. In the parallel arrangement, the N-terminus → C-terminus direction is the same for both chains.

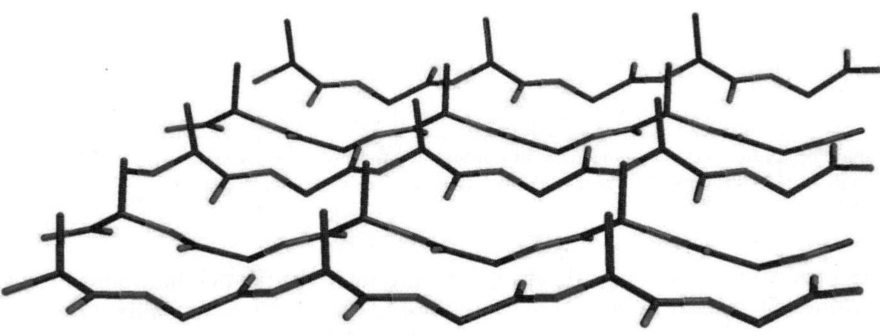

Figure 25.14

The β-pleated sheet secondary structure of a protein, composed of alternating glycine and alanine residues.

The β-pleated sheet is an important secondary structure in proteins that are rich in amino acids with small side chains such as H (glycine), CH_3 (alanine), and CH_2OH (serine). The model in Figure 25.14 is a portion of the calculated structure for a sheet composed of antiparallel strands containing only glycine and alanine in alternating order (-Gly-Ala-Gly-Ala-, etc.). It was designed to resemble *fibroin,* the major protein of silk. Fibroin is almost entirely pleated sheet, and over 80% of it is a repeating sequence of the six-residue unit -Gly-Ser-Gly-Ala-Gly-Ala-. Because the polypeptide backbone adopts an extended zigzag conformation, silk, unlike wool for example, resists stretching.

Problem 25.22

The methyl groups of the alanine residues of the β sheet in Figure 25.14 all point upward. If this pleated sheet were composed of only alanine residues instead of being Gly-Ala-Gly-Ala, etc., what would be the pattern of methyl groups? Would they all point up, alternate up and down, or be random?

The α helix is another commonly encountered secondary structure. Figure 25.15 gives three views of an α-helix model constructed from eight L-alanine residues. Part *a* of the figure is a ball-and-spoke model; part *b* is a view through the center of the helix along its axis. The helix is right-handed with about 3.6 amino acids per turn and is stabilized by hydrogen bonds between the carbonyl oxygens and N—H protons. View *b* shows how the methyl groups of L-alanine project outward from the main chain. This outward orientation of amino acid side chains makes them the points of contact with other amino acids of the same chain, with different protein chains, and with other

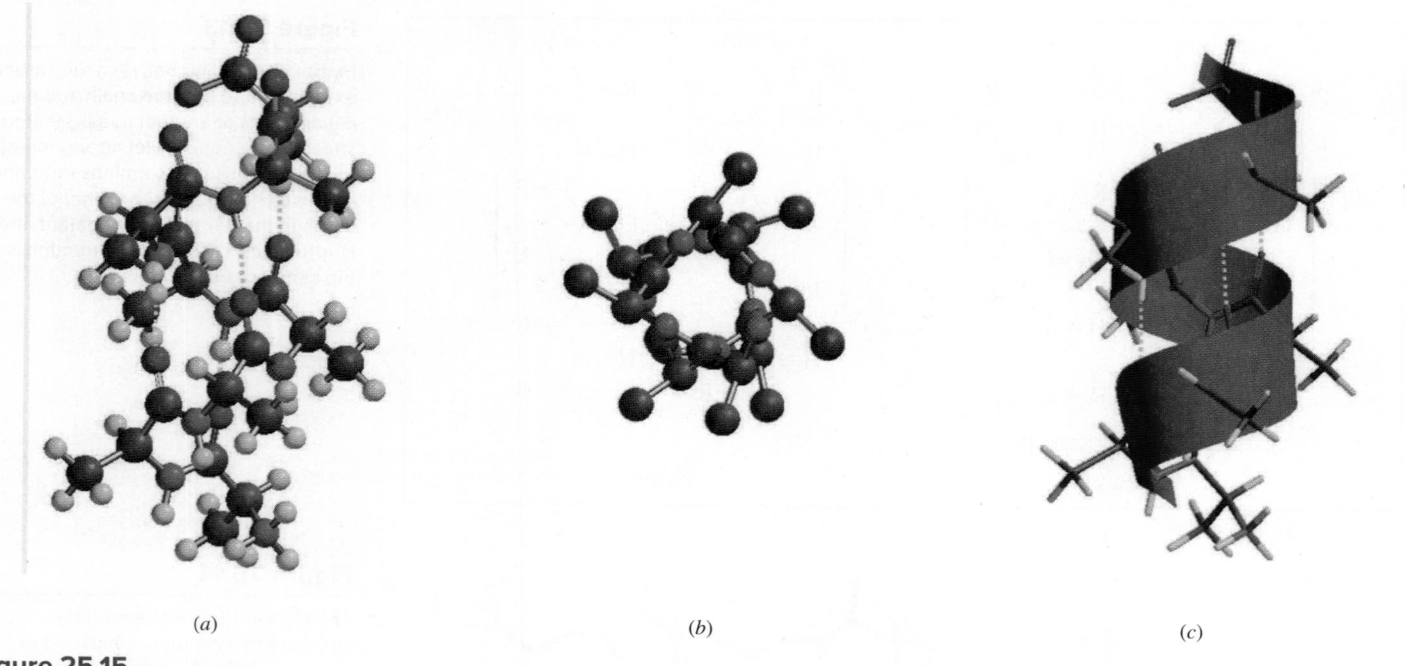

Figure 25.15

Molecular models of an α helix composed of eight alanine residues. The N terminus is at the bottom. (*a*) A ball-and-spoke model. Hydrogen bonds are shown as dashed lines. (*b*) The same model looking up the helical axis from the bottom. Hydrogens have been omitted for clarity. The helix is right-handed, and all of the methyl groups point outward. (*c*) A tube model framed in a ribbon that traces the path of the helix.

biomolecules. Part *c* of the figure uses a ribbon to trace the peptide backbone. The ribbon helps distinguish front from back, makes the right-handedness of the helix more apparent, and is especially useful when looking at how proteins are folded.

The protein components of muscle (*myosin*) and wool (*α-keratin*) contain high percentages of α helix. When wool is stretched, hydrogen bonds break and the peptide chain is elongated. Covalent S—S bonds between L-cysteine residues limit the extent to which the chain can be stretched, however, and once the stretching force is removed the hydrogen bonds re-form spontaneously.

Most proteins cannot be described in terms of a single secondary structure. Rather, most are mixtures of α helix and β sheet, interspersed with regions of **random coils** that have no regular pattern. Figure 25.16 shows a model of ribonuclease, an enzyme that

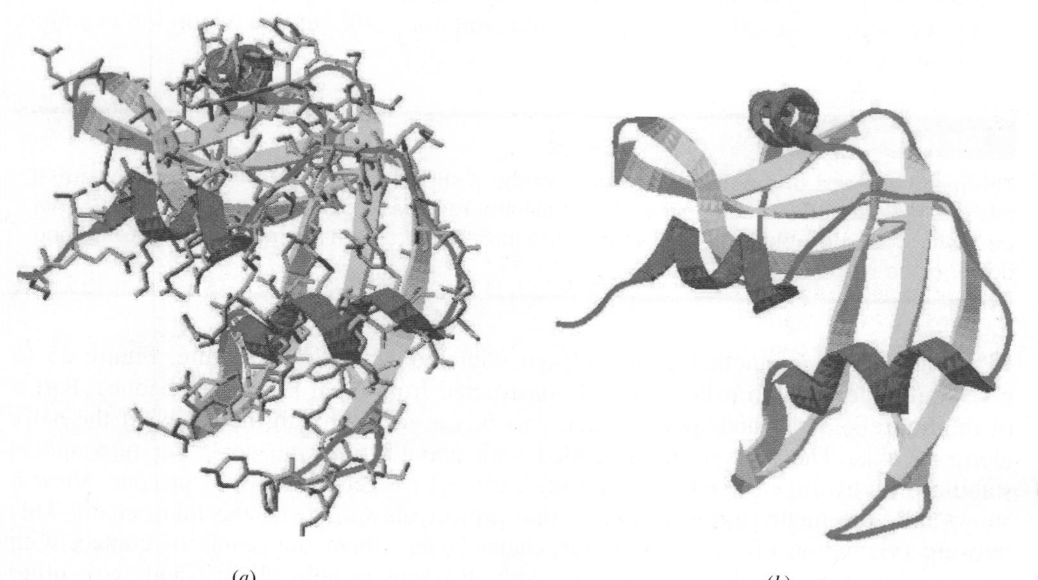

Figure 25.16

Molecular models of ribonuclease. Red ribbons identify sequences where the secondary structure is a helix; yellow ribbons identify strands of β sheet. Arrowheads point in the direction from the N terminus to the C terminus. (*a*) Shows both a molecular model and ribbons. (*b*) Shows only the ribbons.

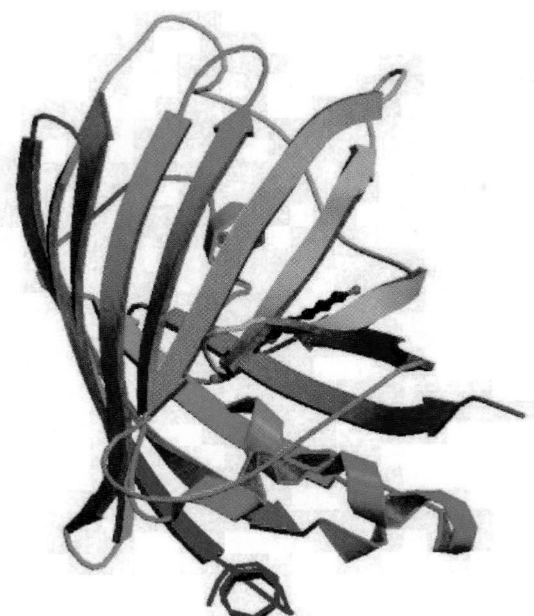

Figure 25.17

Barrel-shaped green fluorescent protein (GFP) has an outer β-sheet structure and an α helix in the inner region.

catalyzes the hydrolysis of RNA. The helical regions are shown in red, the β sheets in yellow. Of the 124 amino acids in this protein, 24 are represented in three sections of α helix. There are two β sheets, one with three strands accounting for 21 amino acids, the other with four strands and 20 amino acids. The strands of each β sheet belong to the same chain and are brought within hydrogen bonding distance because of how the chain is folded. Indeed, the formation of hydrogen bonds such as these is one of the factors that contributes to chain folding.

Another protein which has regions of α helix and β sheet is green fluorescent protein, or GFP. In GFP, the β-sheet structure is barrel-shaped, and the α helix runs along the center of the barrel, where it is shielded from the exterior (Figure 25.17). The α helix in GFP contains an arrangement of amino acids that undergoes fluorescence. Through the use of recombinant DNA technology, GFP can be attached to other proteins that are normally invisible, but now can be imaged using fluorescence microscopy. Using GFP as a biomarker, it is possible to monitor the roles of different proteins in the body, in part because GFP has very low toxicity toward living cells. The development of nerve cells in the brain, the spread of cancer cells, and the damage to neurons that occurs during Alzheimer's disease can be followed by techniques that use GFP.

The 2008 Nobel Prize in Chemistry was awarded to Osamu Shimomura, Martin Chalfie, and Roger Tsien for their work in the development of GFP.

25.18 Tertiary Structure of Polypeptides and Proteins

The way a protein chain is folded, its **tertiary structure,** affects both its physical properties and its biological function. The two main categories of protein tertiary structure are **fibrous** and **globular.**

1. Fibrous proteins are bundles of elongated filaments of protein chains and are insoluble in water.
2. Globular proteins are approximately spherical and either are soluble or form colloidal dispersions in water.

The primary structure of a protein, its amino acid sequence, is the main determinant of tertiary structure. Secondary structure also contributes by limiting the number of conformations available to a polypeptide chain.

Fibrous proteins, being insoluble in water, often have a structural or protective function. The most familiar fibrous proteins are the keratins and collagen. α-Keratin (Figure 25.18) is based on the α-helix secondary structure and is the protein structural

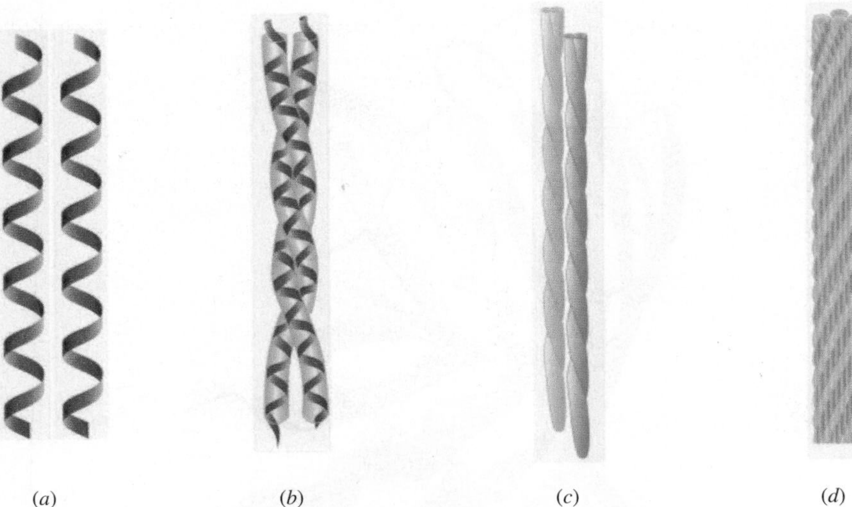

Figure 25.18

α-Keratin. Two α helices (*a*) combine to give a coiled coil (*b*). A pair of coiled coils is a protofilament (*c*). Four protofilaments give a filament (*d*), which is the structural material from which the fibrous protein is assembled.

component of hair, wool, nails, claws, quills, horns, and the outer layer of skin. β-Keratin is based on the β-sheet secondary structure and occurs in silk as fibroin. L-Cysteine is especially abundant in keratins, where it can account for more than 20% of the amino acids present. Collagen occurs mainly in connective tissue (cartilage and tendons) and has a triple helix structure.

Globular proteins include most enzymes and function in aqueous environments. About 65% of the mass of most cells, for example, is water. When placed in water, nonpolar materials, including nonpolar amino acid side chains, cause nearby water molecules to adopt a more ordered arrangement, reducing the entropy of water. This is called the **hydrophobic effect.** The unfavorable negative ΔS is moderated if the protein adopts a spherical shape which places nonpolar side chains inside and polar ones on the surface. Of the various globular arrangements, the one that best offsets the entropy loss with attractive forces among the side chains is the tertiary structure adopted by the protein in its normal, or *native,* state.

Table 25.4 lists the attractive forces that most influence protein tertiary structure. The strongest of these is the covalent S—S bond that unites two cysteine residues. This *disulfide bridge* can form between the $—CH_2SH$ groups of two cysteines, which, although they may be remote from each other in respect to the amino acid sequence, become neighbors when the chain is folded. Formation of the disulfide bond connecting the two stabilizes the local folded arrangement. A typical globular protein normally has only a small number of disulfide bridges. Of the 124 amino acids in ribonuclease (see Figure 25.16 in Section 25.17), 6 are cysteines and each participates in a disulfide bridge; one bridge unites Cys-26 and Cys-84, another Cys-58 and Cys-110, and a third Cys-65 and Cys-72.

The noncovalent interactions are all much weaker than the S—S covalent bond. Among them, the electrostatic attraction between positively and negatively charged side chains, called a salt bridge, is the strongest, followed by hydrogen bonding, then van der Waals forces. Keep in mind though, that the total contribution of the various forces depends not only on the magnitude of an interaction but also on their number. Disulfide bridges may be strong, but there are usually only a few of them. Van der Waals forces are weak, but they outnumber all the other intermolecular attractive forces.

Problem 25.23

Table 25.4 shows a salt bridge between aspartic acid and arginine. Sketch the analogous electrostatic attraction between lysine and an amino acid other than aspartic acid from Table 25.1.

TABLE 25.4 Covalent and Noncovalent Interactions Between Amino Acid Side Chains in Proteins

Description	Type of interaction	Example
Covalent		
Disulfide bridge	S—S bond between two cysteines	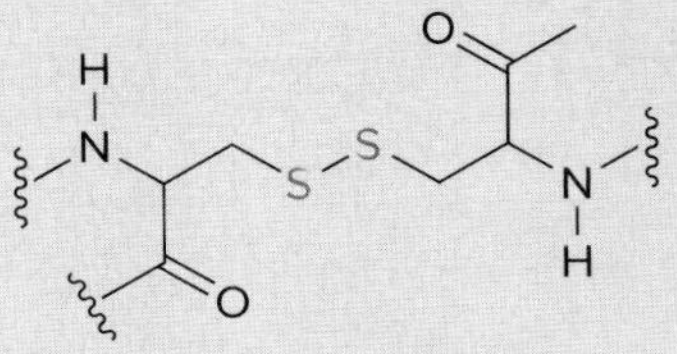 CysS—SCys
Noncovalent		
Salt bridge	Electrostatic attraction between oppositely charged ions	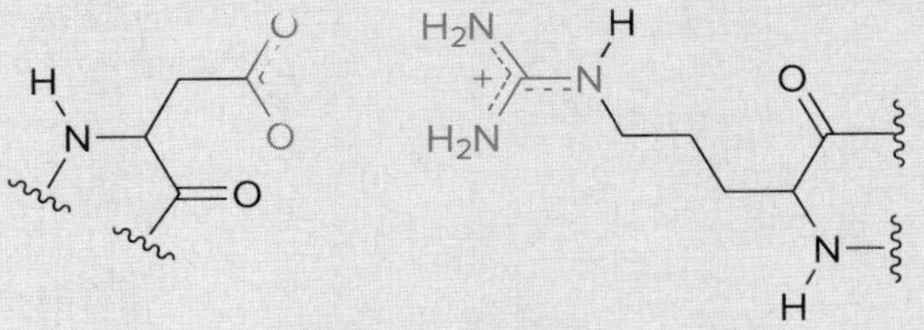 Asp---Arg
Hydrogen bond	Positively polarized H of O—H or N—H group interacts with an electronegative atom (O or N)	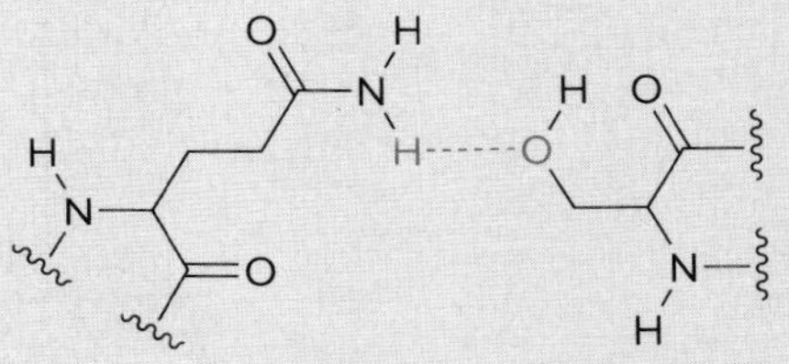 Gln---Ser
Van der Waals	Induced-dipole/induced-dipole attraction (dispersion force) between nonpolar side chains	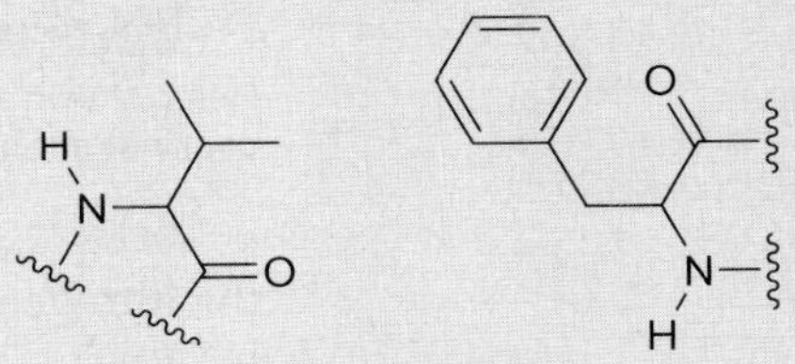 Val---Phe

25.19 Protein Quaternary Structure: Hemoglobin

Rather than existing as a single polypeptide chain, some proteins are assemblies of two or more chains. The manner in which these subunits are organized is called the **quaternary structure** of the protein.

Hemoglobin is an example of a protein that has quaternary structure. Hemoglobin is the oxygen-carrying protein of blood. It binds oxygen at the lungs and transports it to the muscles, where it is stored by myoglobin. Hemoglobin consists of four protein chains, including two identical chains called the *alpha chains* and two identical chains called the *beta chains* (Figure 25.19).

Each chain (*subunit*) contains a *prosthetic group,* heme. Prosthetic groups are nonprotein small molecules that bind tightly to a protein and are required for its function. Heme is an important prosthetic group in which iron(II) is coordinated with the four nitrogen atoms of a type of tetracyclic aromatic substance known as a porphyrin (Figure 25.20). Four of the six available coordination sites of Fe^{2+} are taken up by

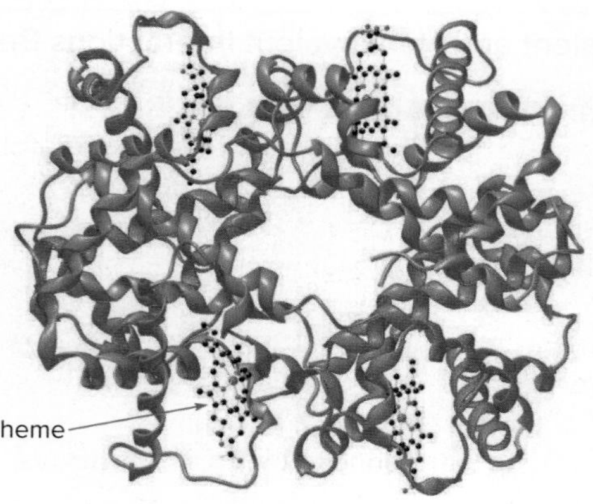

Figure 25.19

Structure of hemoglobin shown as a ribbon model. Each alpha (red) and beta (blue) subunit contains a tightly bound heme (shown in ball-and-stick models), which binds diatomic oxygen.

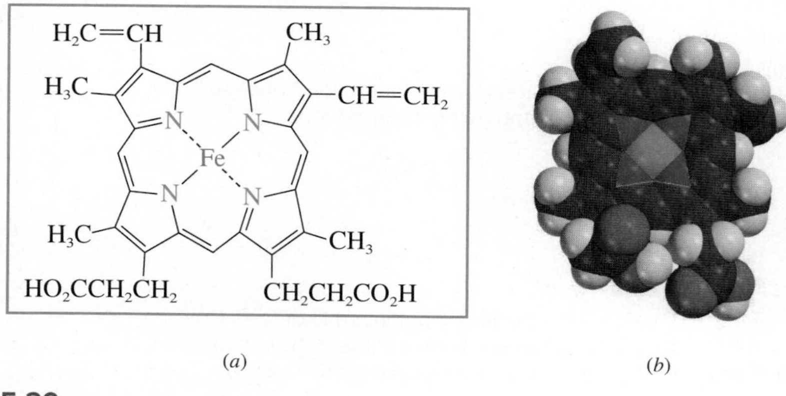

Figure 25.20

Heme shown as (*a*) a structural drawing and (*b*) a space-filling model. The space-filling model shows the coplanar arrangement of the groups surrounding iron.

the nitrogens of the porphyrin, one by a histidine residue of the protein, and the last by a water molecule. Hemoglobin transports oxygen from the lungs by formation of an Fe—O_2 complex. The oxygen displaces water as the sixth ligand on iron and is held there until needed. The protein serves as a container for the heme and prevents oxidation of Fe^{2+} to Fe^{3+}, an oxidation state in which iron lacks the ability to bind oxygen. Separately, neither heme nor the protein binds oxygen in aqueous solution; together, they do it very well.

Some substances, such as CO, form strong bonds to the iron of heme, strong enough to displace O_2 from it. Carbon monoxide binds hundreds of times better than oxygen to hemoglobin. Strong binding of CO at the active site interferes with the ability of heme to perform its biological task of transporting and storing oxygen, with potentially lethal results.

How function depends on structure can be seen in the case of the genetic disorder sickle cell anemia. This is a debilitating, sometimes fatal, disease in which red blood cells become distorted ("sickle-shaped") and interfere with the flow of blood through the capillaries. This condition results from the presence of an abnormal hemoglobin in affected people. The primary structures of the beta chain of normal and sickle cell hemoglobin differ by a single amino acid out of 146; sickle cell hemoglobin has valine in place of glutamic acid as the sixth residue from the N terminus of the β chain. A tiny change in amino acid sequence can produce a life-threatening result! This modification is genetically controlled and probably became established in the gene pool because bearers of the trait have an increased resistance to malaria.

25.20 Enzymes

Knowing how the protein chain is folded is a key element in understanding how an enzyme catalyzes a reaction. *Enzymes* are essential to living systems, which depend on chemical reactions occurring at the right time and in the right location under conditions compatible with life. For mammals, this is generally near neutral pH and at moderate temperature. Protein folding often brings amino acids that are distant from each other in the linear sequence of the protein to precise locations within an enzyme *active site,* the area of the protein in which the catalytic reactions take place.

Enzymes accelerate chemical reactions in a number of ways, some of which are common to all enzymes. For example, all enzymatic reactions begin with a substrate binding to an active site. The energetically unfavorable loss of entropy occurs in this step rather than in the transition state of a catalytic step. This factor alone can account for a substantial increase in the rate of the chemical reaction. In addition, well-placed amino acids dictate the shape, charge, and conformation of substrate(s) that can enter the active site. Amino acid side chains can serve as proton donors or acceptors, nucleophiles, and electrophiles in the catalytic process. Prosthetic groups may be required for the catalytic reaction to take place, which are called *cofactors* when their presence is indispensable.

We saw in Section 25.9 that carboxypeptidase A catalyzes the hydrolysis of the peptide bond to the C-terminal amino acid of polypeptides. Carboxypeptidase A is a metalloprotein; it contains a Zn^{2+} ion as a cofactor, which is essential for catalytic activity. The X-ray crystal structure of carboxypeptidase A (Figure 25.21) locates this Zn^{2+} ion in a hydrophobic cavity near the center of the enzyme, where it is held by coordination to a glutamic acid residue (Glu-72) and two histidines (His-69 and His-196). This region is the active site. The substrate in the case of carboxypeptidase A is a peptide, especially a peptide with a hydrophobic C-terminal amino acid such as phenylalanine or tyrosine. In addition to being hydrophobic, as is the active site, the substrate is bound by an electrostatic attraction between its negatively charged carboxylate and the positively charged side chain of Arg-145. Mechanism 25.3 shows the interactions of the side chains of carboxypeptidase A with Zn^{2+} and a peptide, then describes the mechanism for cleaving the peptide bond to the terminal amino acid. Side chains other than those shown in Mechanism 25.3 have been implicated but have been omitted. The main feature of the mechanism is its relationship to the mechanism of nucleophilic acyl substitution (on which it was patterned). Not only does the enzyme bring the substrate and catalytically active functions together at the active site, but by stabilizing the tetrahedral intermediate it lowers the activation energy for its formation and increases the reaction rate.

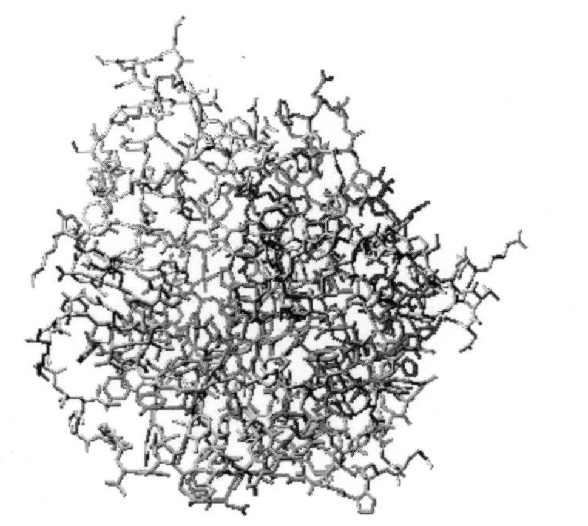

(*a*)

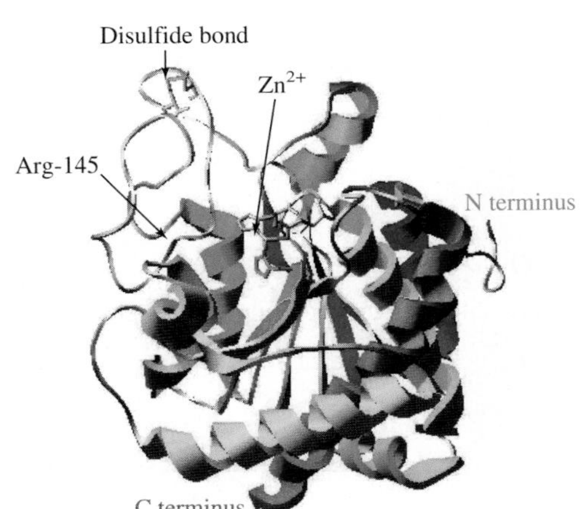

(*b*)

Figure 25.21

The structure of carboxypeptidase A displayed as (*a*) a tube model and (*b*) a ribbon diagram. The most evident feature illustrated by (*a*) is the globular shape of the enzyme. The ribbon diagram emphasizes the folding of the chain.

Mechanism 25.3

Carboxypeptidase-Catalyzed Hydrolysis

THE MECHANISM: The mechanism shown outlines the major stages in carboxypeptidase-catalyzed hydrolysis of a peptide in which the C-terminal amino acid is phenylalanine. Proton transfers accompany stages 2 and 3 but are not shown. Only the major interactions of the substrate with the carboxypeptidase side chains are shown although others may also be involved.

Stage 1: The peptide is positioned in the active site by an electrostatic bond between its negatively charged C-terminal carboxylate and a positively charged arginine side chain of the enzyme. Also at the active site, Zn^{2+} engages in Lewis acid/Lewis base interactions with His-69 and His-196 and an electrostatic attraction with the negatively charged carboxylate of Glu-72. These ligands are shown here but will be omitted in subsequent steps for simplicity.

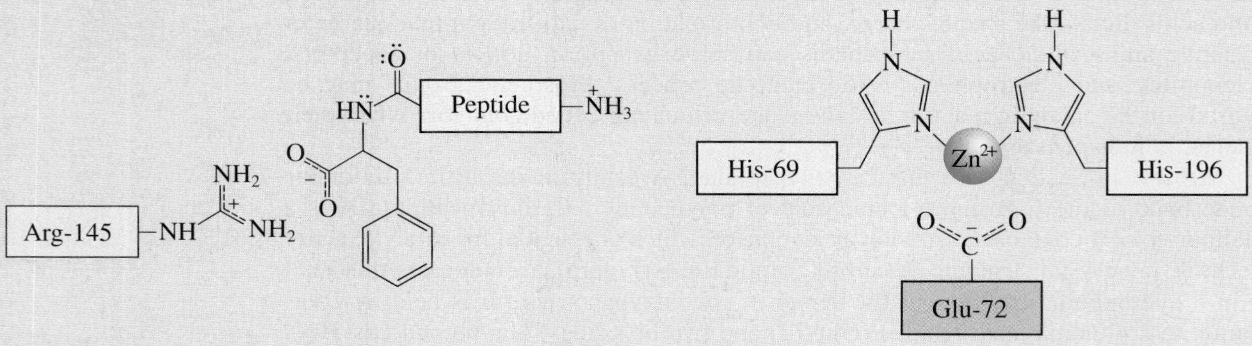

Stage 2: Water adds to the carbonyl group of the peptide bond. The rate of this nucleophilic addition is accelerated by coordination of the carbonyl oxygen to Zn^{2+} and/or to one of the N—H protons of Arg-127 (not shown). The product is a tetrahedral intermediate stabilized by coordination to zinc. Stabilization of the tetrahedral intermediate may be the major factor for the rapid rate of the carboxypeptidase-catalyzed hydrolysis.

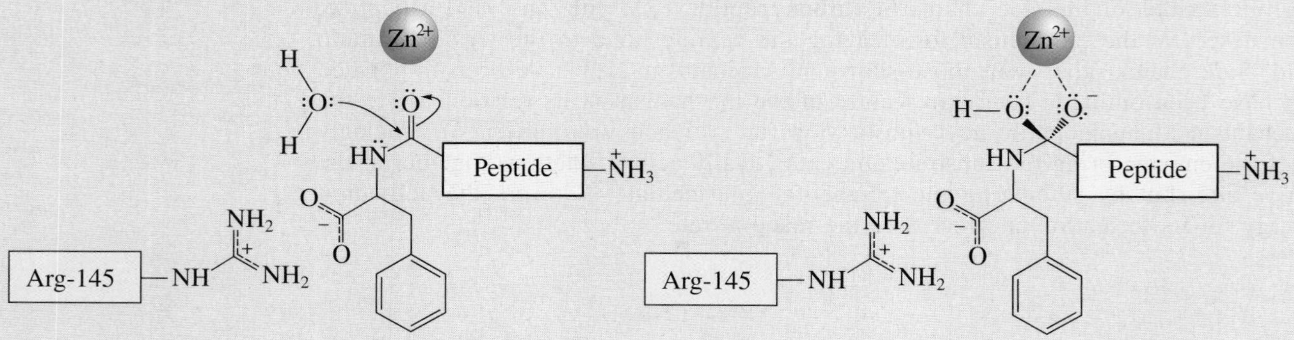

Stage 3: The tetrahedral intermediate dissociates to the C-terminal amino acid (phenylalanine in this case). Subsequent steps restore the active site.

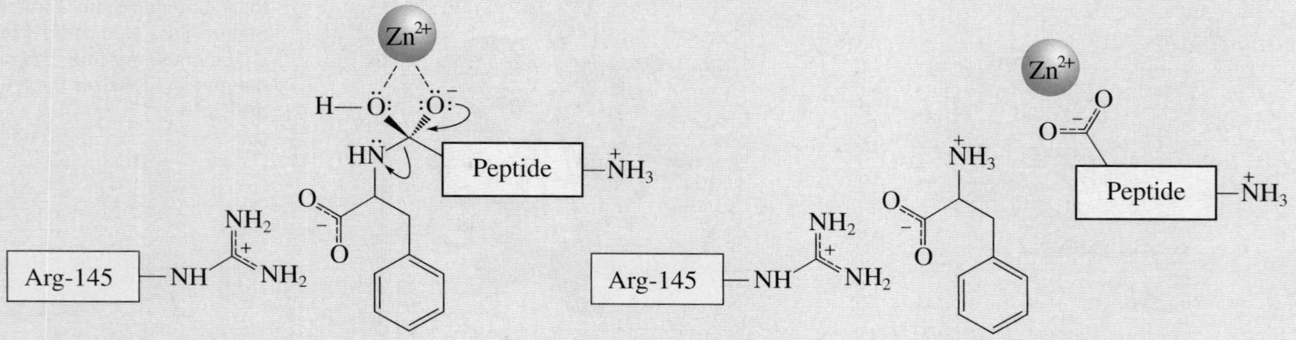

25.21 Coenzymes in Reactions of Amino Acids

The number of chemical processes that protein side chains can engage in is rather limited. Most prominent among them are proton donation, proton abstraction, and nucleophilic addition to carbonyl groups. In many biological processes a richer variety of reactivity is required, and proteins often act in combination with substances other than proteins to carry out the necessary chemistry. These substances are called **prosthetic groups** or **cofactors,** and they can be organic or inorganic. Organic substances that are required for catalytic activity are called **coenzymes.** The protein portion of the enzyme is called the **apoenzyme,** and the combination of the protein and **cofactor** is called the **holoenzyme.** In earlier sections we saw numerous examples of coenzyme-catalyzed reactions that required the cofactor NAD^+ for oxidation reactions and NADH for enzymatic reductions.

Enzymes with particular cofactors tend to catalyze similar transformations. Enzymes that contain **pyridoxal pyrophosphate (PLP),** the active form of vitamin B_6, catalyze reactions that involve carbanionic intermediates. Amino acid metabolism and catabolism pathways frequently involve PLP-containing enzymes.

All PLP-containing enzymes contain an imine bond between the protein and the coenzyme in the resting state. The imine is the result of a chemical reaction between PLP and a lysine residue in the enzyme active site.

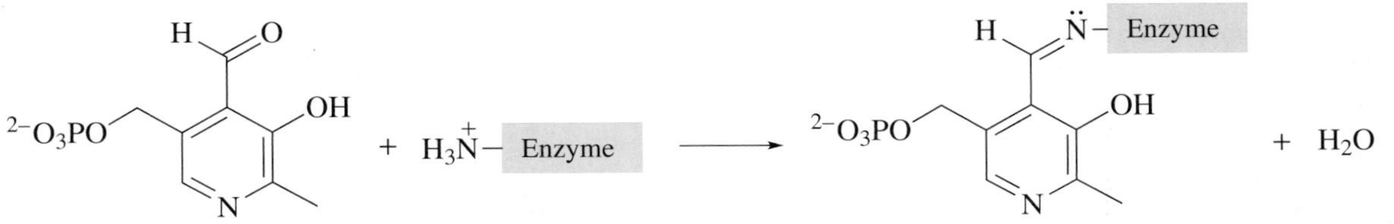

$^{2-}O_3PO$ … OH, H, O, N + $H_3\overset{+}{N}$— Enzyme ⟶ $^{2-}O_3PO$ … H, $\ddot{N}$— Enzyme, OH, N + H_2O

Pyridoxal phosphate (PLP) | Enzyme | Enzyme-bound PLP | Water

Reaction of the enzyme-bound PLP with an amino acid substrate exchanges one imine linkage for another. This is a **transimination** reaction.

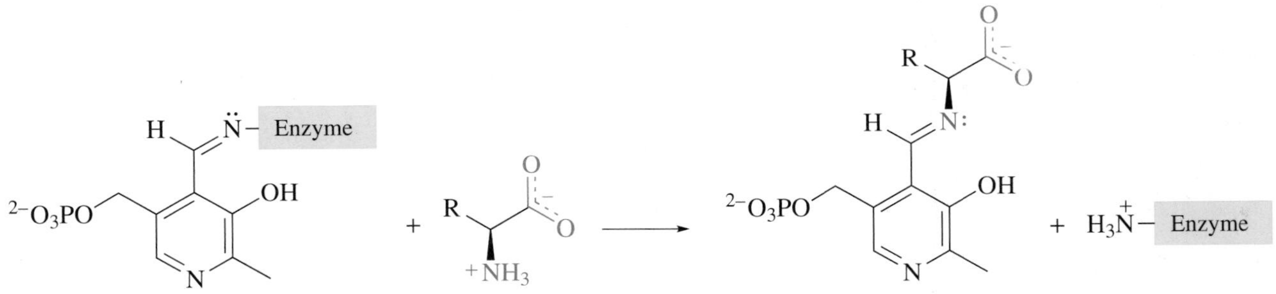

$^{2-}O_3PO$ … H, $\ddot{N}$— Enzyme, OH, N + R, O, O, $^+NH_3$ ⟶ $^{2-}O_3PO$ … H, N:, R, O, O, OH, N + $H_3\overset{+}{N}$— Enzyme

Enzyme-bound PLP | Amino acid | Imine | Enzyme

The pyridine ring of PLP, especially when protonated, facilitates several kinds of reactions at the amino acid's α carbon by acting as an electron-withdrawing group. One reaction is decarboxylation.

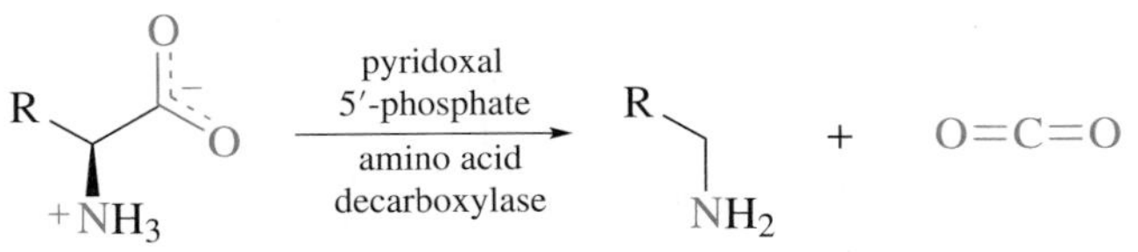

R, O, O, $^+NH_3$ —(pyridoxal 5′-phosphate; amino acid decarboxylase)⟶ R, NH_2 + O=C=O

Amino acid | Amine | Carbon dioxide

Mechanism 25.4 outlines the mechanism of decarboxylation, showing the role played by the coenzyme. A proton transfer from an amino acid residue in the enzyme to the pyridoxal nitrogen occurs prior to stage 2. It was once thought that protonation of the

Mechanism 25.4

Pyridoxal 5′-Phosphate-Mediated Decarboxylation of an α-Amino Acid

THE OVERALL REACTION:

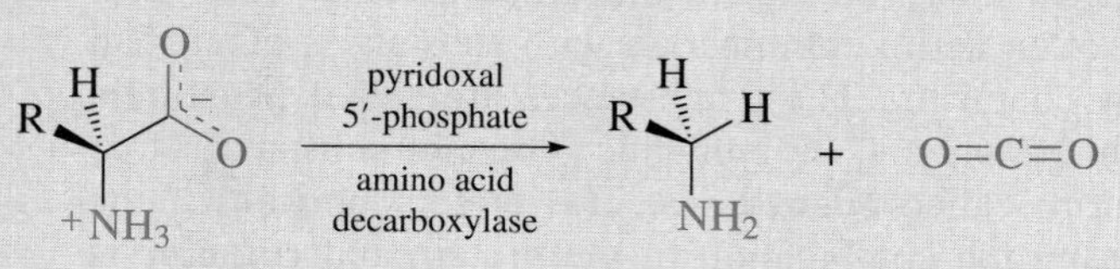

Amino acid —(pyridoxal 5′-phosphate; amino acid decarboxylase)→ Amine + Carbon dioxide ($O{=}C{=}O$)

THE MECHANISM: Each stage is enzyme-catalyzed and can involve more than one elementary step.

Stage 1: The amino acid reacts with enzyme-bound pyridoxal 5′-phosphate (PLP). An imine linkage (C═N) between the amino acid and PLP forms, and the enzyme is displaced.

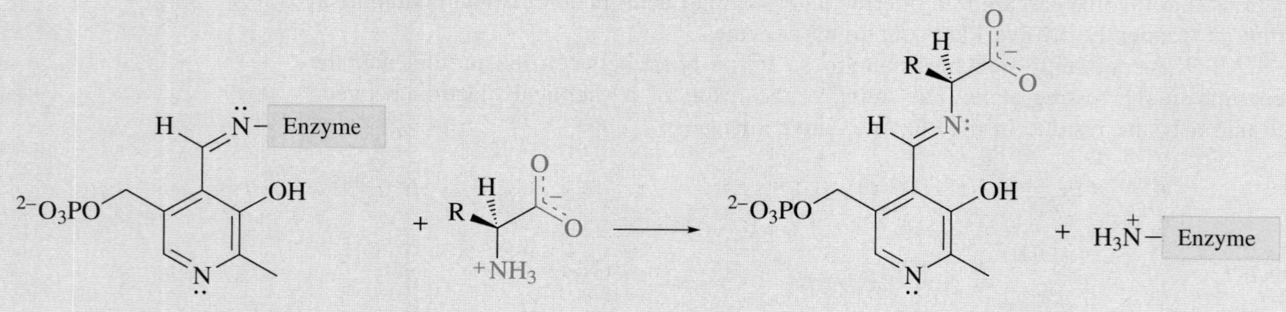

Enzyme-bound PLP + Amino acid → Imine + Enzyme ($H_3\overset{+}{N}$– Enzyme)

Stage 2: When the pyridine ring is protonated on nitrogen, it becomes a stronger electron-withdrawing group, and decarboxylation is facilitated by charge neutralization.

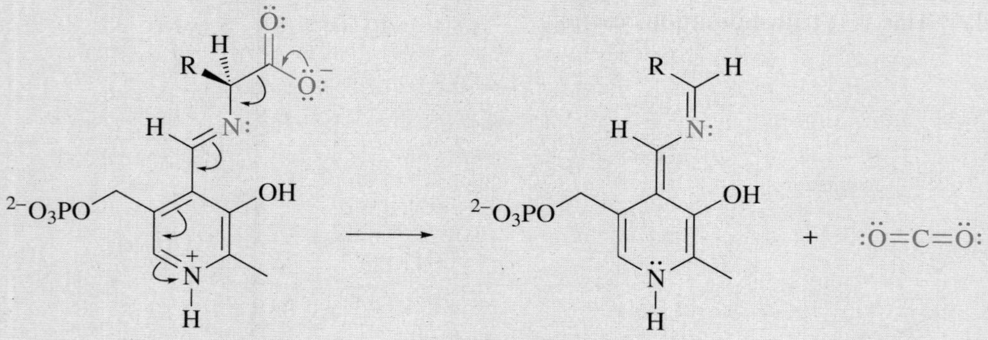

Imine → Decarboxylated imine + Carbon dioxide

Stage 3: Proton transfer to the α carbon and abstraction of a proton from the pyridine nitrogen brings about rearomatization of the pyridine ring.

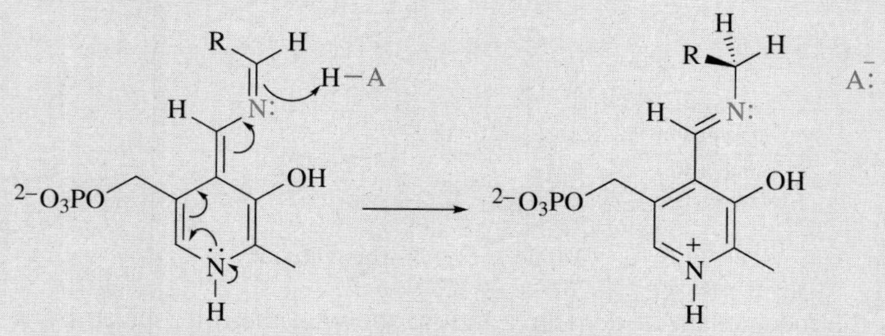

Decarboxylated imine → PLP-bound imine + A^-

Stage 4: Reaction of the PLP-bound imine with the enzyme liberates the amine and restores the enzyme-bound coenzyme.

PLP-bound imine + ^+H_3N–Enzyme → Enzyme-bound PLP + Amine

cofactor was a common feature of all PLP-containing enzyme catalyzed reactions. More recent studies show that this nitrogen may be protonated in some enzymes but it is not in other enzymes.

Many bioactive amines arise by PLP-assisted decarboxylation. Decarboxylation of histidine, for example, gives histamine, a powerful vasodilator normally present in the body but formed in excessive amounts under conditions of anaphylactic shock.

Histidine → (pyridoxal 5′-phosphate, amino acid decarboxylase) → Histamine + $O{=}C{=}O$ (Carbon dioxide)

Histamine is present in various tissues and produces different effects depending on the kind of receptor it binds to. Binding of histamine to H_1 receptors in mast cells triggers, for example, the sneezing and watery eyes of hay fever and the itching of mosquito bites. The H_2 receptors in the cells that line the stomach regulate the secretion of gastric acid. The present generation of antiallergy and antiulcer drugs bind to H_1 and H_2, respectively, and act by denying histamine access to these receptors.

Mast cells reside in tissue rather than blood.

Problem 25.24

One of the amino acids in Table 25.1 is the biological precursor to γ-aminobutyric acid (4-aminobutanoic acid), which it forms by a decarboxylation reaction. Which amino acid is this?

The chemistry of the brain and central nervous system is affected by a group of **neurotransmitters,** substances that carry messages across a synapse from one neuron to another. Several of these neurotransmitters arise from L-tyrosine by structural modification and decarboxylation, as outlined in Figure 25.22.

Problem 25.25

Which of the transformations in Figure 25.22 is catalyzed by an amino acid decarboxylase?

Many of the drugs prescribed to treat anxiety, depression, or attention deficit disorder are "reuptake" inhibitors. They increase the concentration in the brain of a necessary neurotransmitter such as dopamine or epinephrine by slowing the rate at which it is reabsorbed.

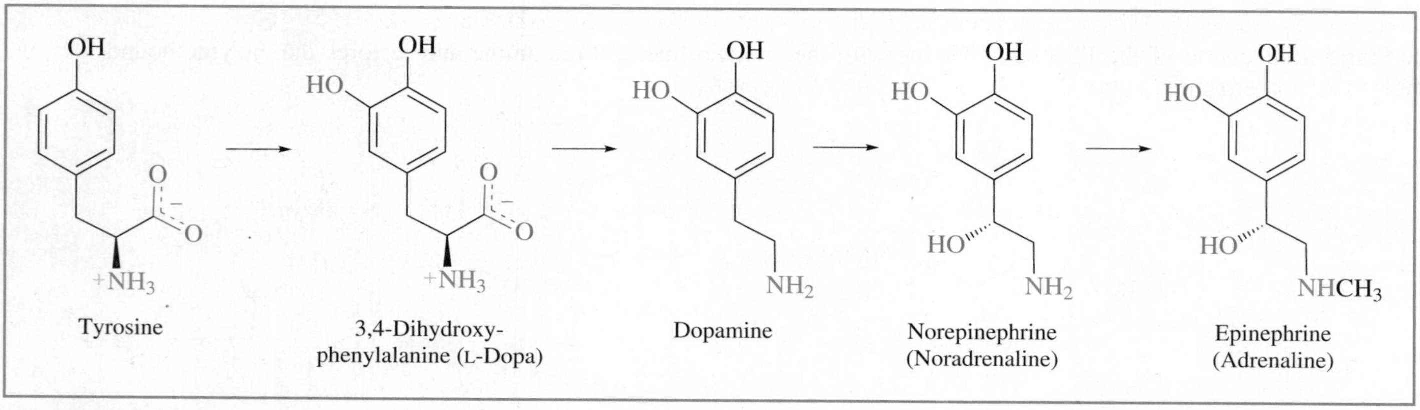

Tyrosine → 3,4-Dihydroxyphenylalanine (L-Dopa) → Dopamine → Norepinephrine (Noradrenaline) → Epinephrine (Adrenaline)

Figure 25.22

Tyrosine is the biosynthetic precursor to a number of neurotransmitters. Each transformation is enzyme-catalyzed.

Pyridoxal 5′-phosphate is also a coenzyme for the enzyme-catalyzed racemization of amino acids. The key reaction is proton abstraction from the α carbon of the amino acid imine of PLP. This step converts the α carbon, which is a chirality center, from sp^3 to sp^2.

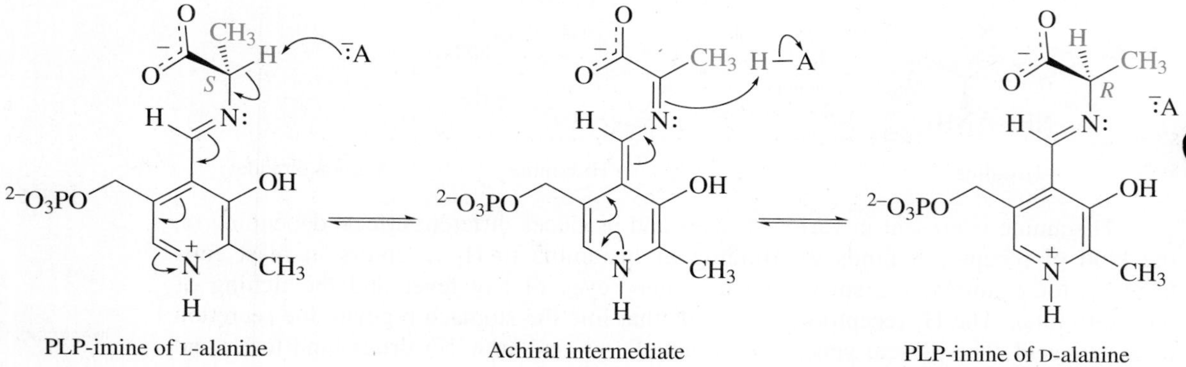

PLP-imine of L-alanine ⇌ Achiral intermediate ⇌ PLP-imine of D-alanine

Proton transfer to the imine carbon of the achiral intermediate gives equal amounts of both enantiomers of the PLP imine. The equation illustrates the racemization of L-alanine, which is catalyzed by the PLP-dependent enzyme *alanine racemase.* Because D-alanine is an essential component of bacterial cell walls, there is considerable interest in designing inhibitors of alanine racemase as potential antibacterial drugs.

In addition to amino acid decarboxylation and racemization, PLP is a coenzyme for **transamination**—the transfer of an amino group from one compound to another. The enzymes that catalyze transaminations are called *aminotransferases* or *transaminases.* Many transaminations involve two compounds: α-ketoglutaric acid and L-glutamic acid—one as a reactant, the other as its product.

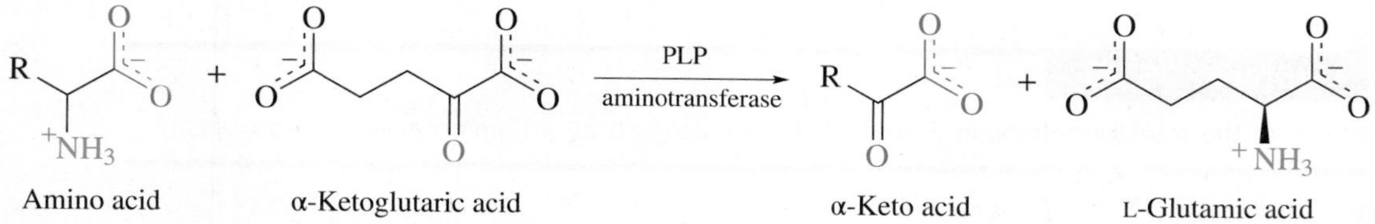

Amino acid + α-Ketoglutaric acid →(PLP, aminotransferase) α-Keto acid + L-Glutamic acid

The reaction shown illustrates a feature of amino acid metabolism, the breaking down of amino acids and using their structural units for other purposes. Written in the other direction, it illustrates a biosynthetic pathway to amino acids. L-Alanine, for example, is not an essential amino acid because we have the capacity to biosynthesize it. One biosynthetic route to L-alanine is the transamination of pyruvic acid.

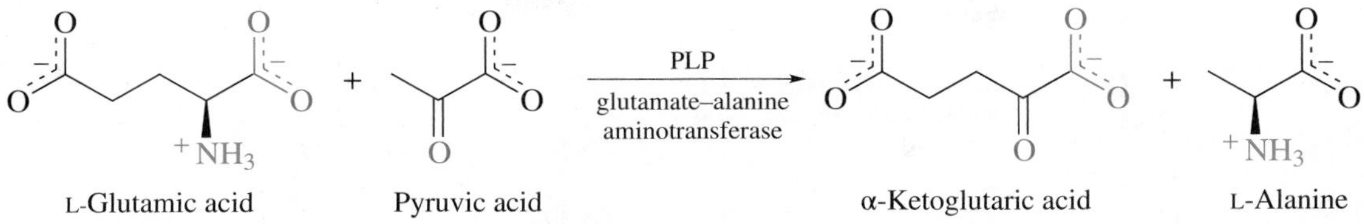

L-Glutamic acid + Pyruvic acid $\xrightarrow[\text{glutamate–alanine aminotransferase}]{\text{PLP}}$ α-Ketoglutaric acid + L-Alanine

As outlined in the first four stages of Mechanism 25.5, the amino group of L-glutamic acid is transferred to the coenzyme PLP to give pyridoxamine 5′-phosphate (PMP). A second transamination, shown in abbreviated form as stage 5, is analogous to the first and gives L-alanine.

Problem 25.26

α-Ketoglutaric acid undergoes a transamination reaction with L-aspartic acid (see Table 25.1), converting it to a compound known as oxaloacetic acid. What is the structure of oxaloacetic acid?

Peptide-bond formation and transamination are the most general reactions of the standard amino acids, but individual amino acids often undergo reactions of more limited scope. One of the biosynthetic pathways to L-tyrosine is oxidation of L-phenylalanine. An *arene oxide* is an intermediate.

For more on this reaction, see Descriptive Passage and Interpretive Problems 17: Epoxide Rearrangements and the NIH Shift.

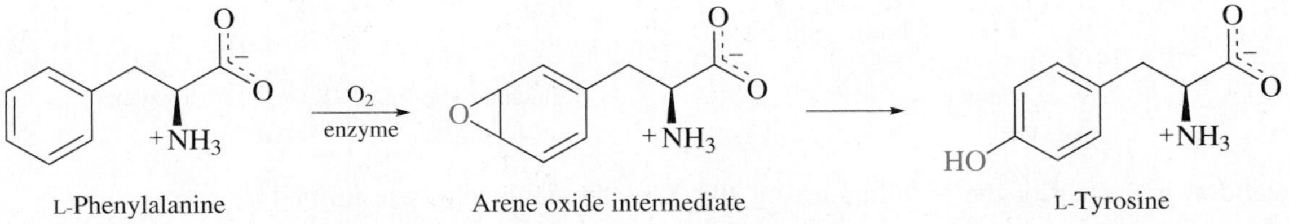

L-Phenylalanine $\xrightarrow[\text{enzyme}]{O_2}$ Arene oxide intermediate $\longrightarrow$ L-Tyrosine

Some individuals lack the enzyme *phenylalanine hydroxylase* required for this conversion, and any L-phenylalanine that would ordinarily be converted to L-tyrosine is converted to phenylpyruvic acid by transamination.

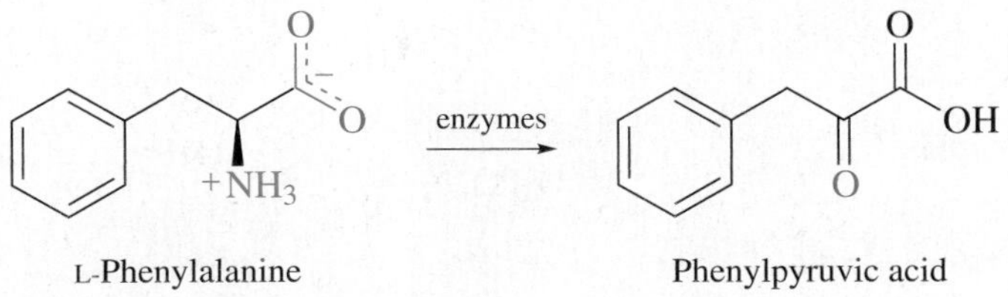

L-Phenylalanine $\xrightarrow{\text{enzymes}}$ Phenylpyruvic acid

Too much phenylpyruvic acid causes *phenylketonuria* (PKU disease), which can lead to mental retardation in growing children. Infants are routinely screened for PKU disease within a few days of birth. PKU disease cannot be cured, but is controlled by restricting the dietary intake of foods, such as meat, that are rich in L-phenylalanine.

Mechanism 25.5

Transamination: Biosynthesis of L-Alanine from L-Glutamic Acid and Pyruvic Acid

THE OVERALL REACTION:

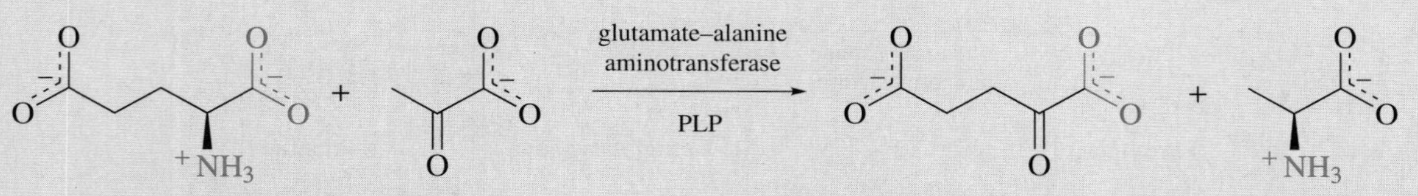

glutamate–alanine aminotransferase

PLP

L-Glutamic acid + Pyruvic acid → α-Ketoglutaric acid + L-Alanine

THE MECHANISM:

Each stage can involve more than one elementary step, and each reaction is enzyme-catalyzed. Stages 1–4 show the transfer of the amino group of L-glutamic acid to pyridoxal 5′-phosphate to give α-ketoglutaric acid and pyridoxamine 5′-phosphate (PMP). PMP reacts with pyruvic acid to give an imine, which then follows stages analogous to 1–4, but in reverse order, to give L-alanine and PLP. These stages are summarized as stage 5.

Stage 1: L-Glutamic acid forms an imine bond to the coenzyme PLP by reaction with the imine formed between PLP and the enzyme.

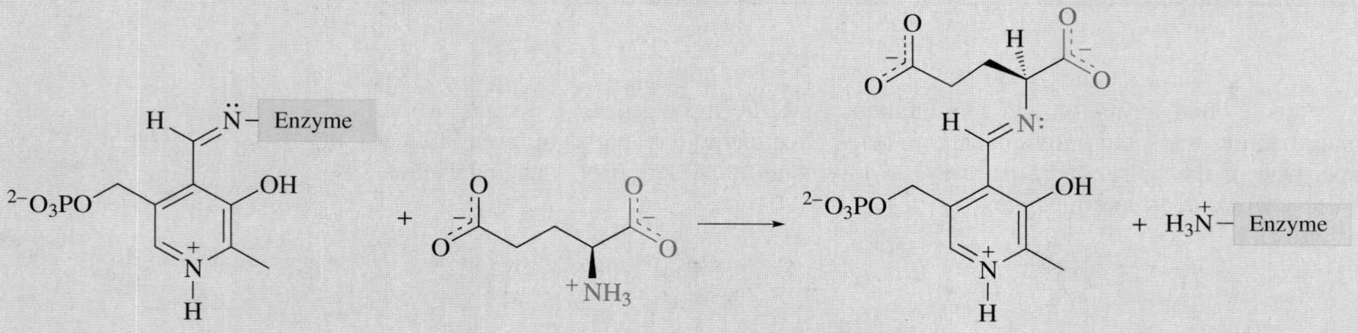

Enzyme-bound PLP + L-Glutamic acid → PLP-glutamic acid imine + $H_3\overset{+}{N}$— Enzyme

Enzyme

Stage 2: The electron-withdrawing effect of the pyridinium ring stabilizes the conjugate base formed by proton abstraction from the α carbon of the imine.

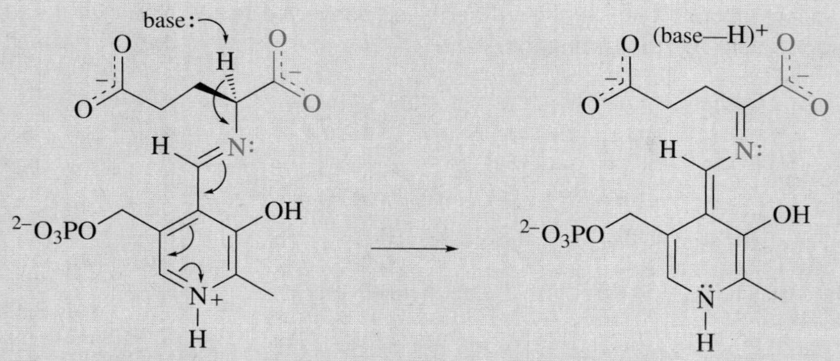

base:

PLP-glutamic acid imine → Conjugate base of PLP-glutamic acid imine + (base—H)$^+$

Stage 3: Electron reorganization and protonation of carbon restores the aromaticity of the pyridine ring while converting a PLP imine to a PMP imine.

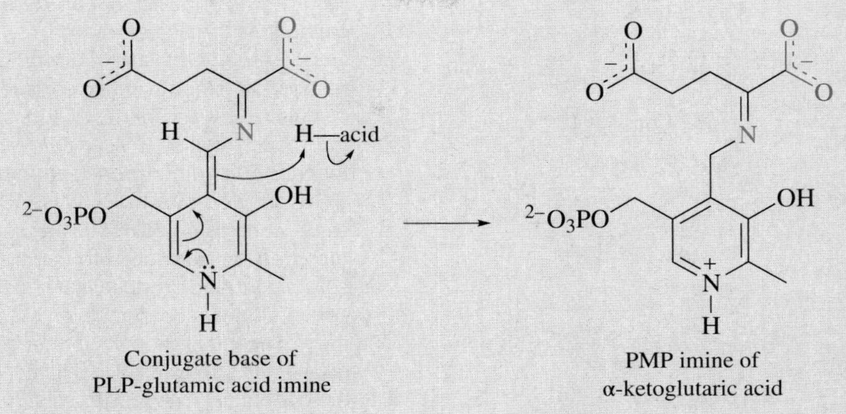

Conjugate base of PLP-glutamic acid imine → PMP imine of α-ketoglutaric acid

Stage 4: Cleavage of the PMP imine, shown here as a hydrolysis, gives pyridoxamine and α-ketoglutaric acid.

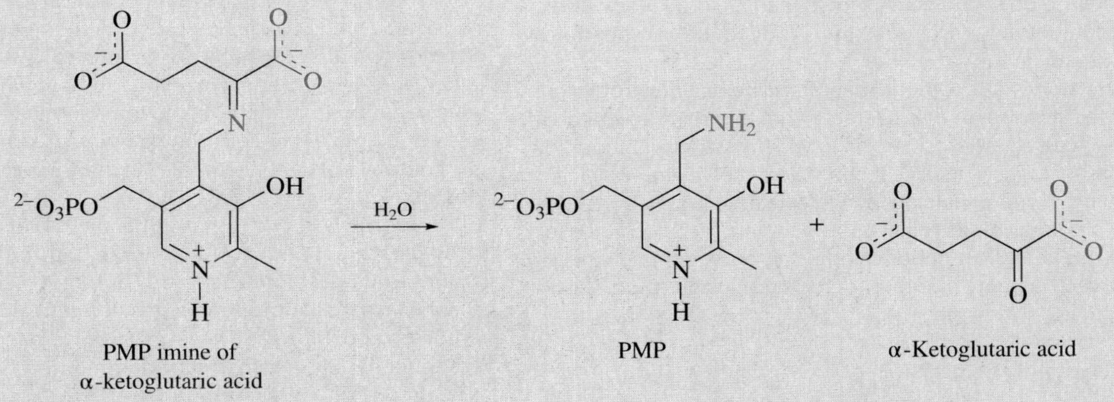

PMP imine of α-ketoglutaric acid $\xrightarrow{H_2O}$ PMP + α-Ketoglutaric acid

Stage 5: Formation of the imine from PMP and pyruvic acid sets the stage for the conversion of pyruvic acid to L-alanine.

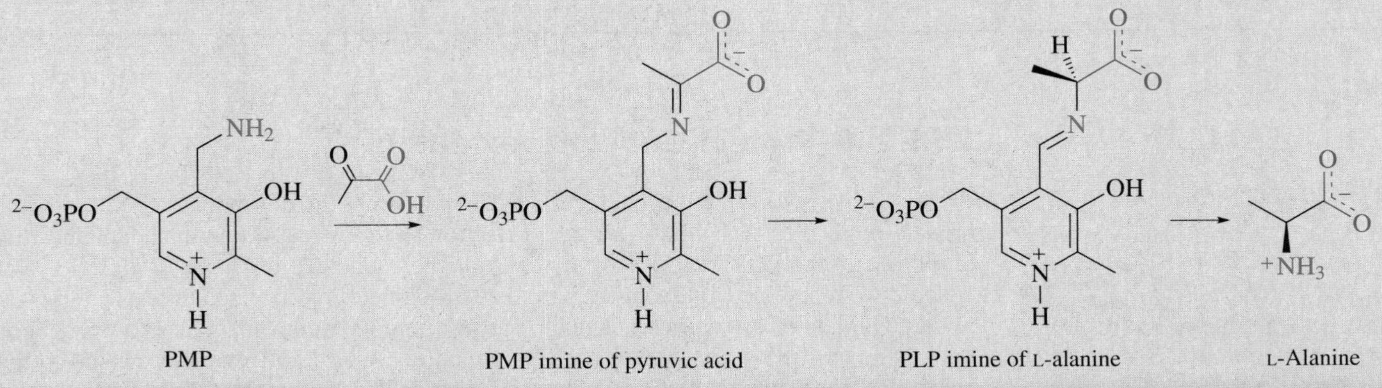

PMP → PMP imine of pyruvic acid → PLP imine of L-alanine → L-Alanine

Oh NO! It's Inorganic!

The amino acid L-arginine undergoes an interesting biochemical conversion.

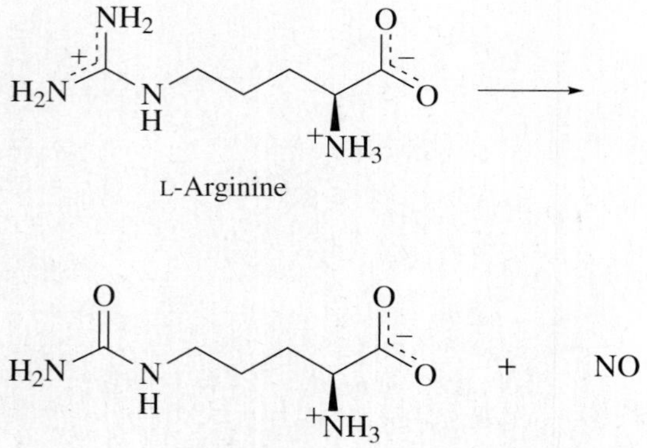

L-Arginine → L-Citrulline + NO (Nitric oxide)

Our experience conditions us to focus on the organic components of the reaction—L-arginine and L-citrulline—and to give less attention to the inorganic one—nitric oxide (nitrogen monoxide, NO). To do so, however, would lead us to overlook one of the most important discoveries in biology in the last quarter of the twentieth century.

Our story starts with the long-standing use of nitroglycerin to treat the chest pain that characterizes angina, a condition in diseases such as atherosclerosis in which restricted blood flow to the heart muscle itself causes it to receive an insufficient amount of oxygen. Placing a nitroglycerin tablet under the tongue provides rapid relief by expanding the blood vessels feeding the heart. A number of other nitrogen-containing compounds such as amyl nitrite and sodium nitroprusside exert a similar effect.

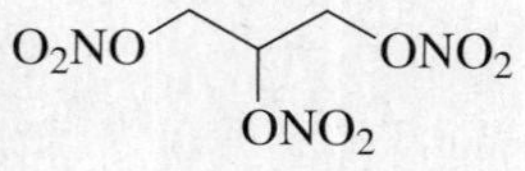

Nitroglycerine

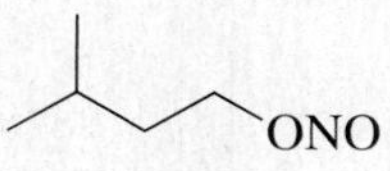

Amyl nitrite

$Na_2[O{=}N{-}Fe(CN)_5]$

Sodium nitroprusside

A chemical basis for their action was proposed in 1977 by Ferid Murad, who showed that all were sources of NO, thereby implicating it as the active agent.

Three years later, Robert F. Furchgott discovered that the relaxing of smooth muscles, such as blood vessel walls, was stimulated by an unknown substance produced in the lining of the blood vessels (the *endothelium*). He called this substance the *endothelium-dependent relaxing factor,* or EDRF and, in 1986, showed that EDRF was NO. Louis J. Ignarro reached the same conclusion at about the same time. Further support was provided by Salvador Moncada, who showed that endothelial cells did indeed produce NO and that the L-arginine-to-L-citrulline conversion was responsible.

The initial skepticism that greeted the idea that NO, which is (a) a gas, (b) toxic, (c) inorganic, and (d) a free radical, could be a biochemical messenger was quickly overcome. An avalanche of results confirmed not only NO's role in smooth-muscle relaxation, but added more and more examples to an ever-expanding list of NO-stimulated biochemical processes. Digestion is facilitated by the action of NO on intestinal muscles. The drug Viagra (*sildenafil citrate*), prescribed to treat erectile dysfunction, works by increasing the concentration of a hormone, the release of which is signaled by NO. A theory that NO is involved in long-term memory receives support from the fact the brain is a rich source of the enzyme *nitric oxide synthase* (NOS), which catalyzes the formation of NO from L-arginine. NO even mediates the glow of fireflies. They glow nonstop when placed in a jar containing NO, but not at all when measures are taken to absorb NO.

Identifying NO as a signaling molecule in biological processes justified a Nobel Prize, but who would get it? Nobel Prizes are often shared, but never among more than three persons. Although four scientists—Murad, Furchgott, Ignarro, and Moncada—made important contributions, the Nobel committee followed tradition and recognized only the first three of them with the 1998 Nobel Prize in Physiology or Medicine.

25.22 G-Protein-Coupled Receptors

Biological receptors have been mentioned a few times in this book, in the context of organic molecules that bind to them. In Section 24.2, we described anandamide, a lipid that binds to the same receptor in the brain as the cannabinoids. In Section 25.13 of this chapter, we mentioned calcitonin, a peptide that regulates blood calcium levels and is used in the treatment of osteoporosis. The biological receptors for anandamide and for calcitonin both belong to a very large class of protein receptors known as **G-protein-coupled receptors,** or GPCRs. The "G" stands for guanine in "guanine nucleotide-binding proteins." GPCRs occur throughout the body and function as "molecular switches" that regulate many physiological processes.

The 2012 Nobel Prize in Chemistry was awarded to Robert J. Lefkowitz (Howard Hughes Medical Institute and Duke University) and Brian K. Kobilka (Stanford University) for their studies of G-protein-coupled receptors.

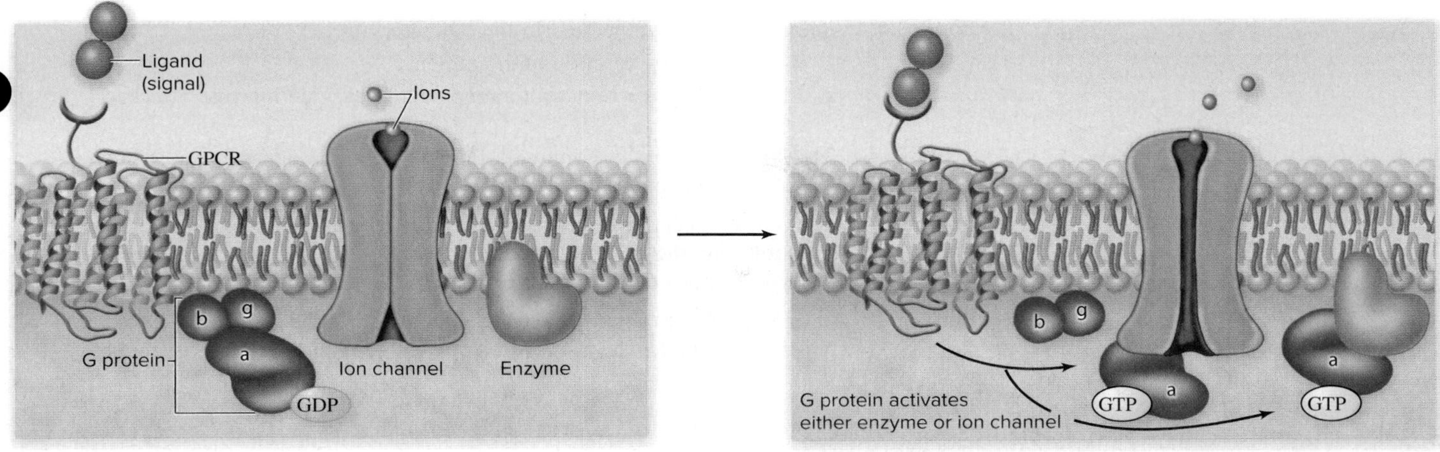

Figure 25.23

Signal transduction is initiated by the binding of a G-protein-coupled receptor (GPCR) to the ligand on the exterior of the cell. Interactions with a nearby G protein inside the cell result in the exchange of bound GDP to GTP and in the release of one of its subunits in a GTP-bound form that activates an ion channel or enzyme.

GPCRs span the cell membrane (see Figure 24.4). When GPCRs bind their specific ligand, such as a lipid, peptide, or ion, they undergo a conformational change, which results in the transduction of a signal across the membrane. The details of how signaling occurs are not completely understood. The conformational change may result in an interaction with a nearby G protein, which in turn can activate enzymes or ion channels (Figure 25.23).

In the biochemistry of vision (see Section 18.11), the interaction of rhodopsin, a GPCR, with light results in the photo-induced isomerization of cis retinal imine to trans retinal imine. This causes conformational changes in rhodopsin, which ultimately result in the *closing* of an ion channel, polarization of the cell membrame, and a nerve impulse that is transmitted to the brain in vision.

G-protein-coupled receptors are involved in many diseases. It is estimated that nearly one half of prescription drugs target GPCRs, in the treatment of cancer, cardiac malfunction, inflammation, pain, and disorders of the central nervous system.

Signal transduction pathways in biochemistry involve a relay of molecular interactions that ultimately produce an intracellular response. These pathways can serve as a connection between events at the cell surface and gene expression in the nucleus.

GDP and GTP are abbreviations for the nucleotides guanosine 5′-diphosphate and guanosine 5′-triphosphate. Nucleotides are discussed in Section 26.3.

25.23 SUMMARY

This chapter revolves around **proteins.** It describes the building blocks of proteins, progressing through **amino acids** and **peptides,** and concludes with proteins themselves.

Section 25.1 The 20 amino acids listed in Table 25.1 are the building blocks of proteins. All are α-amino acids.

Section 25.2 Except for glycine, which is achiral, all of the α-amino acids in Table 25.1 are chiral and have the L-configuration at the α carbon.

Section 25.3 The most stable structure of a neutral amino acid is a **zwitterion.** The pH of an aqueous solution at which the concentration of the zwitterion is a maximum is called the isoelectric point (pI).

CO_2^-

$H_3\overset{+}{N}$—C—H

$CH(CH_3)_2$

Fischer projection of L-valine in its zwitterionic form

Section 25.4 Amino acids are synthesized in the laboratory from

1. α-Halo acids by reaction with ammonia
2. Aldehydes by reaction with ammonia and cyanide ion (the Strecker synthesis)
3. Alkyl halides by reaction with the enolate anion derived from diethyl acetamidomalonate

Section 25.5 Amino acids undergo reactions characteristic of the amino group (e.g., amide formation) and the carboxyl group (e.g., esterification). Amino acid side chains undergo reactions characteristic of the functional groups they contain.

Section 25.6 An amide linkage between two α-amino acids is called a **peptide bond.** By convention, peptides are named and written beginning at the N terminus.

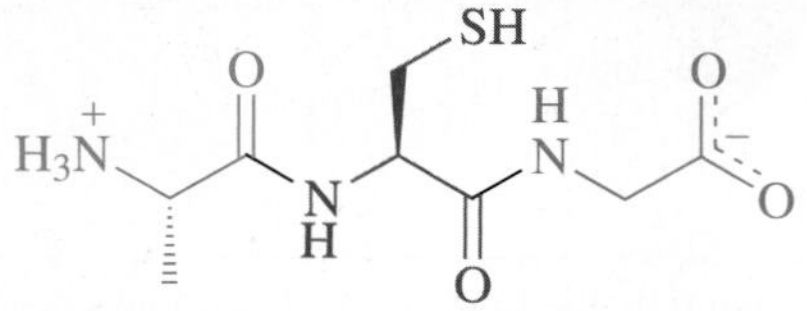

Alanylcysteinylglycine
(Ala-Cys-Gly or ACG)

Section 25.7 The **primary structure** of a peptide is its amino acid sequence plus any disulfide bonds between two cysteine residues. The primary structure is determined by a systematic approach in which the protein is cleaved to smaller fragments, even individual amino acids. The smaller fragments are sequenced and the main sequence deduced by finding regions of overlap among the smaller peptides.

Section 25.8 Complete hydrolysis of a peptide gives a mixture of amino acids. An amino acid analyzer identifies the individual amino acids and determines their molar ratios.

Section 25.9 Selective hydrolysis can be accomplished by using enzymes to catalyze cleavage at specific peptide bonds. Carboxypeptidase-catalyzed hydrolysis can be used to identify the C-terminal amino acid. The N terminus is determined by chemical means. One reagent used for this purpose is Sanger's reagent, 1-fluoro-2,4-dinitrobenzene.

Section 25.10 The procedure described in Sections 25.7–25.10 was used to determine the amino acid sequence of insulin.

Section 25.11 Modern methods of chemoenzymatic peptide sequencing follow a strategy similar to that used to sequence insulin, but are automated and can be carried out on a small scale. An example is repetitive N-terminal amino acid identification using the **Edman degradation.**

Section 25.12 Mass spectrometry is used for peptide sequencing and proteomics.

Section 25.13 Synthesis of a peptide of prescribed sequence requires the use of protecting groups to minimize the number of possible reactions.

Section 25.14 Amino-protecting groups include *benzyloxycarbonyl* (Z), tert-*butoxycarbonyl* (Boc), and *9-fluorenylmethoxycarbonyl* (Fmoc).

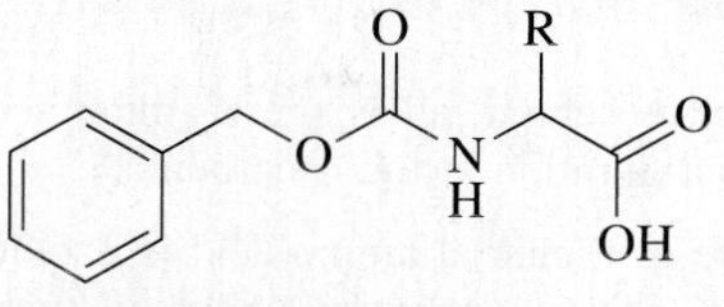

Z-protected amino acid

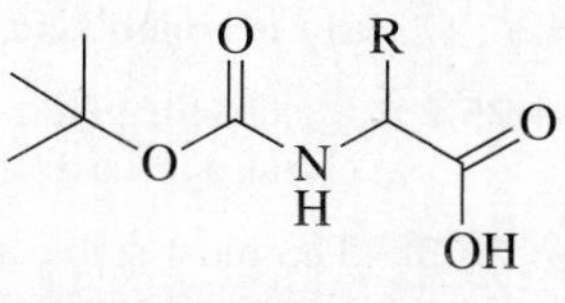

Boc-protected amino acid

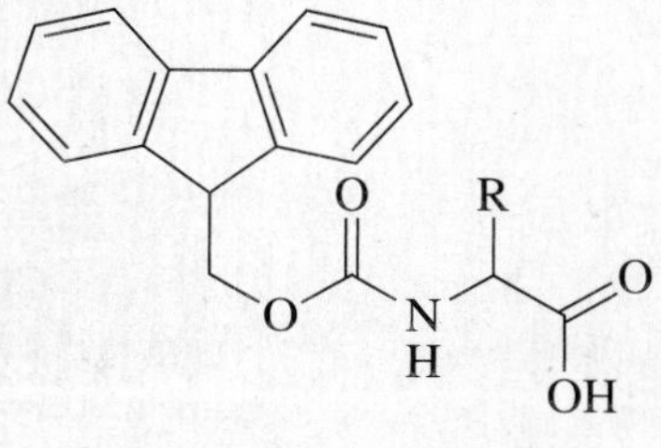

Fmoc-protected amino acid

Trifluoroacetic acid may be used to remove the *tert*-butoxycarbonyl protecting group but not the benzyloxycarbonyl group. The benzyloxycarbonyl protecting group may then be removed using hydrogen bromide or by catalytic hydrogenolysis. Fmoc is removed in base.

Carboxyl groups are normally protected as benzyl, methyl, or ethyl esters. Hydrolysis in dilute base is normally used to deprotect methyl and ethyl esters. Benzyl protecting groups are removed by hydrogenolysis.

Section 25.15 Peptide bond formation between a protected amino acid having a free carboxyl group and a protected amino acid having a free amino group can be accomplished with the aid of *N,N′*-dicyclohexylcarbodiimide (DCCI).

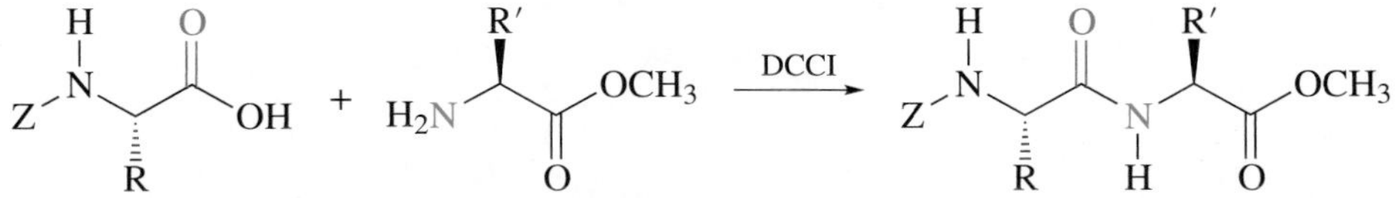

Section 25.16 In the Merrifield method the carboxyl group of an amino acid is anchored to a solid support and the chain extended one amino acid at a time. When all the amino acid residues have been added, the polypeptide is removed from the solid support.

Section 25.17 Two **secondary structures** of proteins are particularly prominent. The *pleated β sheet* is stabilized by hydrogen bonds between N—H and C═O groups of adjacent chains. The *α helix* is stabilized by hydrogen bonds within a single polypeptide chain.

Section 25.18 The folding of a peptide chain is its **tertiary structure.** The tertiary structure has a tremendous influence on the properties of the peptide and the biological role it plays. The tertiary structure is normally determined by X-ray crystallography.

Section 25.19 Many proteins consist of two or more chains, and the way in which the various units are assembled in the native state of the protein is called its **quaternary structure.**

Section 25.20 Many globular proteins are enzymes. They accelerate the rates of chemical reactions in biological systems, but the kinds of reactions that take place are the fundamental reactions of organic chemistry. One way in which enzymes accelerate these reactions is by bringing reactive functions together in the presence of catalytically active functions of the protein.

Section 25.21 Often the catalytically active functions of an enzyme are nothing more than proton donors and proton acceptors. In many cases a protein acts in cooperation with a **cofactor,** a small molecule having the proper functionality to carry out a chemical change not otherwise available to the protein itself. Pyridoxal 5′-phosphate (PLP) is a common cofactor **(coenzyme)** that catalyzes reactions involving amino acids.

Among the biochemical reactions of α-amino acids, several use PLP as a coenzyme. These reactions involve bonds to the α carbon and include transamination, decarboxylation, and racemization.

Section 25.22 G-protein-coupled receptors are transmembrane proteins that function as molecular switches in many physiological processes.

PROBLEMS

Amino Acids (General)

25.27 Which two α-amino acids are the biosynthetic precursors of the penicillins?

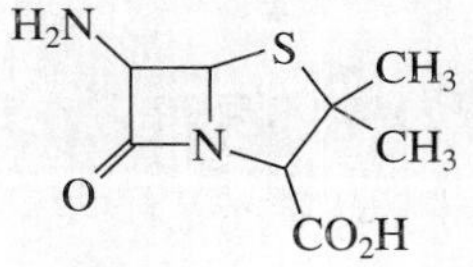

25.28 α-Amino acids are not the only compounds that exist as zwitterions. *p*-Aminobenzenesulfonic acid (sulfanilic acid) is normally written in the form shown but its zwitterionic form is more stable. Write a structural formula for the zwitterion.

$H_2N-C_6H_4-SO_3H$

25.29 Hydrolysis of the following compound in concentrated hydrochloric acid for several hours at 100°C gives one of the amino acids in Table 25.1. Which one? Is it optically active?

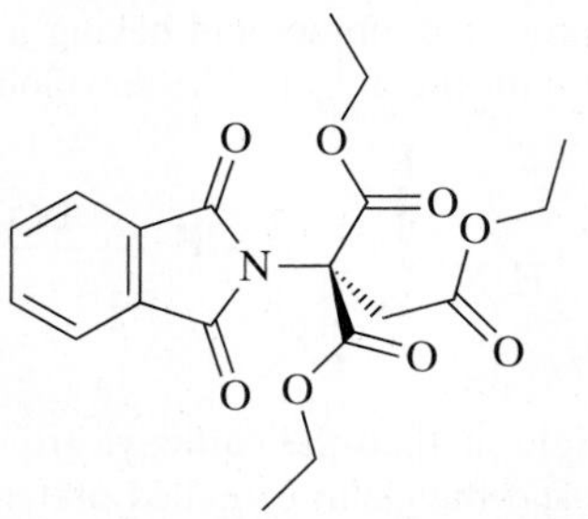

25.30 Acrylonitrile ($H_2C{=}CHC{\equiv}N$) readily undergoes conjugate addition when treated with nucleophilic reagents. Describe a synthesis of β-alanine ($H_3\overset{+}{N}CH_2CH_2CO_2^-$) that takes advantage of this fact.

25.31 (a) Isoleucine has been prepared by the following sequence of reactions. Give the structure of compounds A through D isolated as intermediates in this synthesis.

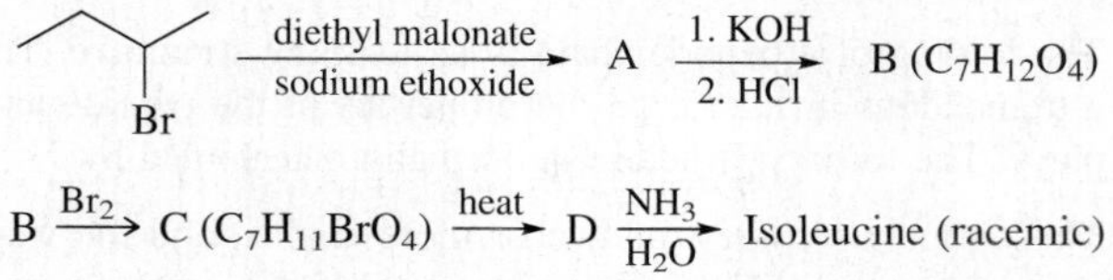

$$\text{(2-bromobutane, Br)} \xrightarrow[\text{sodium ethoxide}]{\text{diethyl malonate}} \text{A} \xrightarrow[\text{2. HCl}]{\text{1. KOH}} \text{B } (C_7H_{12}O_4)$$

$$\text{B} \xrightarrow{Br_2} \text{C } (C_7H_{11}BrO_4) \xrightarrow{\text{heat}} \text{D} \xrightarrow[H_2O]{NH_3} \text{Isoleucine (racemic)}$$

(b) An analogous procedure has been used to prepare phenylalanine. What alkyl halide would you choose as the starting material for this synthesis?

25.32 Identify the major product in each of the following reactions.

(a)

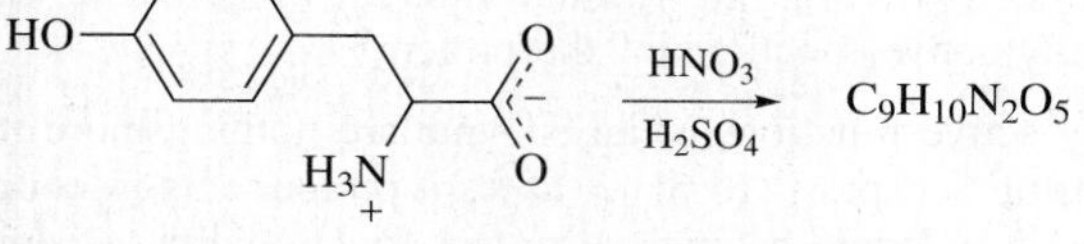

$$\text{HO-C}_6\text{H}_4\text{-CH}_2\text{CH}(\overset{+}{N}H_3)CO_2^- \xrightarrow[H_2SO_4]{HNO_3} C_9H_{10}N_2O_5$$

(b)

$$\text{HO-C}_6\text{H}_4\text{-CH}_2\text{CH(BocNH)C(O)OCH}_3 \xrightarrow[K_2CO_3,\ DMF]{H_2C{=}CHCH_2Br} C_{18}H_{25}NO_5 \xrightarrow[CH_3OH]{NH_3} C_{17}H_{24}N_2O_4$$

(c)

$$H_3\overset{+}{N}CH_2CO_2^- + \text{(}O_2N, O_2N, Cl, Cl, NO_2\text{-substituted benzene)} \xrightarrow[\text{ethanol–water}]{Na_2CO_3} C_{10}H_7N_5Na_2O_{10} \xrightarrow[H_2O]{HCl} C_{10}H_9N_5O_{10}$$

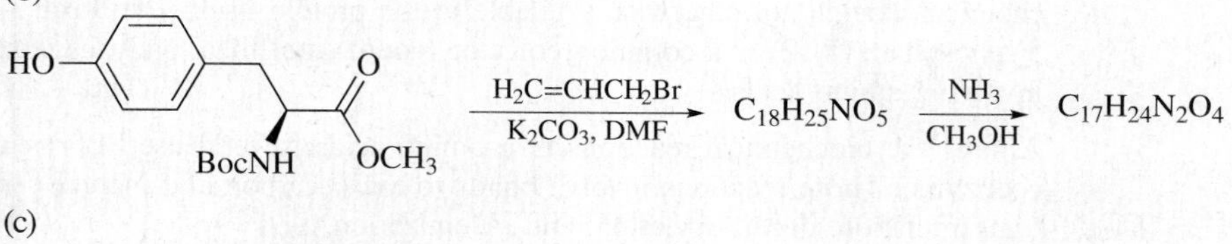

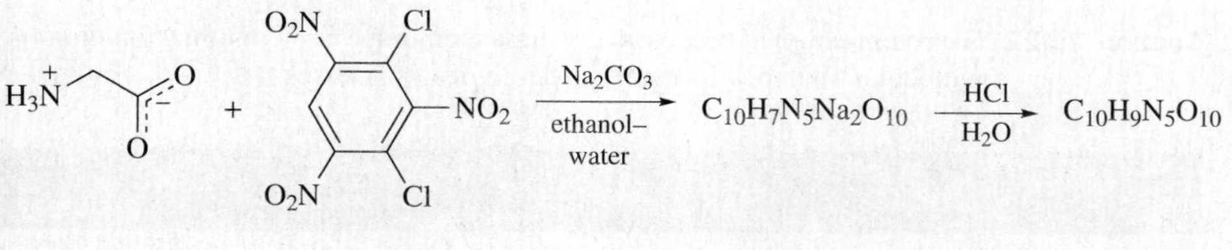

Amino Acids (Acid–Base)

25.33 The imidazole ring of the histidine side chain acts as a proton acceptor in certain enzyme-catalyzed reactions. Which is the more stable protonated form of the histidine residue, A or B? Why?

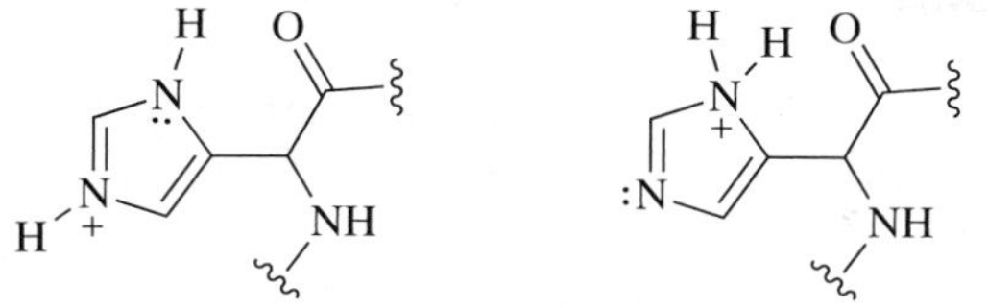

A B

25.34 (a) Use the data in Table 25.2 and the Henderson–Hasselbalch equation (Section 19.4) to calculate the ratio $\frac{[A]}{[B]}$ at pH = 7.

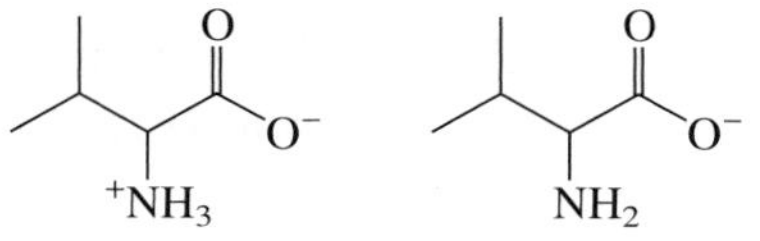

$^+NH_3$ NH_2

A B

(b) At what pH is [A] the largest?

25.35 Putrescine, citrulline, and ornithine are products of arginine metabolism. The molecular formulas are given for the form in which each exists at pH = 7. The net charge corresponding to each formula is zero, +1, and +2. Suggest a reasonable structure for each species.

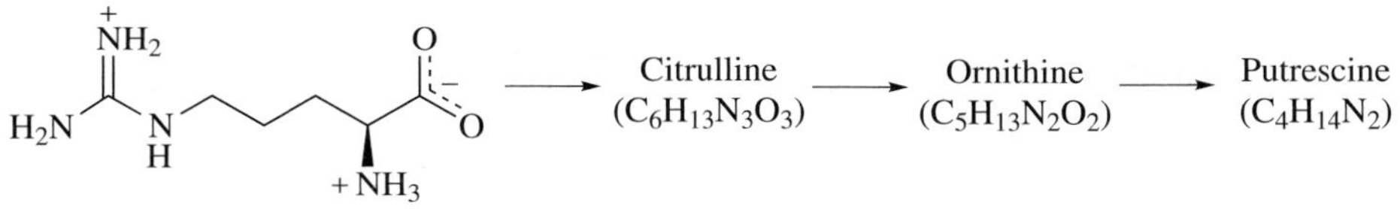

Arginine ($C_6H_{15}N_4O_2$) → Citrulline ($C_6H_{13}N_3O_3$) → Ornithine ($C_5H_{13}N_2O_2$) → Putrescine ($C_4H_{14}N_2$)

Peptides

25.36 Automated amino acid analysis of peptides containing asparagine (Asn) and glutamine (Gln) residues gives a peak corresponding to ammonia. Why?

25.37 The synthetic peptide shown is an inhibitor of the enzyme β-secretase, which plays a role in the development of Alzheimer's disease. It contains five amino acids from Table 25.1. Which ones?

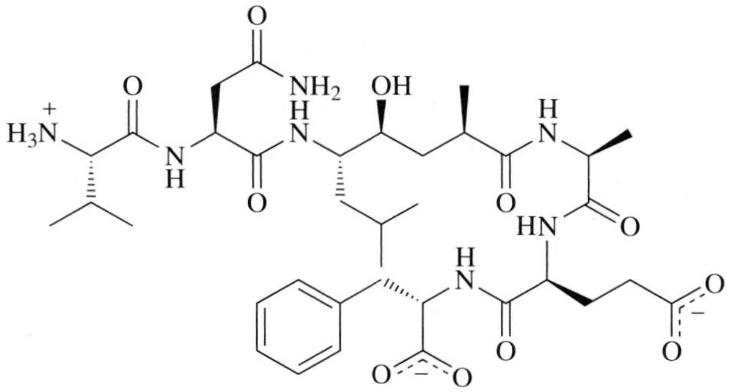

25.38 What are the products of each of the following reactions? Your answer should account for all the amino acid residues in the starting peptides.

(a) Reaction of Leu-Gly-Ser with 1-fluoro-2,4-dinitrobenzene

(b) Hydrolysis of the compound in part (a) in concentrated hydrochloric acid (100°C)

(c) Treatment of Ile-Glu-Phe with $C_6H_5N{=}C{=}S$, followed by hydrogen bromide in nitromethane

(d) Reaction of Asn-Ser-Ala with benzyloxycarbonyl chloride

(e) Reaction of the product of part (d) with *p*-nitrophenol and *N,N′*-dicyclohexylcarbodiimide

(f) Reaction of the product of part (e) with the ethyl ester of valine

(g) Hydrogenolysis of the product of part (f) by reaction with H_2 over palladium

25.39 The first 32 amino acids from the N terminus of the protein *bovine angiogenin* were determined by Edman degradation and have the sequence:

AQDDYRYIHFLTQHYDAKPKGRNDEYCFNMMK

(a) Identify the sites of cleavage during trypsin-catalyzed hydrolysis of this protein.

(b) What are the cleavage sites using chymotrypsin?

25.40 *Somatostatin* is a tetradecapeptide of the hypothalamus that inhibits the release of pituitary growth hormone. Its amino acid sequence has been determined by a combination of Edman degradations and enzymic hydrolysis experiments. On the basis of the following data, deduce the primary structure of somatostatin:

1. Edman degradation gave PTH-Ala.

2. Selective hydrolysis gave peptides having the following indicated sequences:

Phe-Trp
Thr-Ser-Cys
Lys-Thr-Phe
Thr-Phe-Thr-Ser-Cys
Asn-Phe-Phe-Trp-Lys
Ala-Gly-Cys-Lys-Asn-Phe

3. Somatostatin has a disulfide bridge.

25.41 If you synthesized the tripeptide Leu-Phe-Ser from amino acids prepared by the Strecker synthesis, how many stereoisomers would you expect to be formed?

25.42 What protected amino acid would you anchor to the solid support in the first step of a synthesis of oxytocin (see Figure 25.4) by the Merrifield method?

25.43 *Native chemical ligation* (NCL) is a method for coupling peptide chains. It requires that one of the peptides has an N-terminal cysteine, and that the C terminus of the other peptide be a thioester. The thiol of the N-terminal cysteine of peptide 2 reacts with the thioester of peptide 1 to give an initial product that contains a "nonnative" thioester linkage. The initial product rearranges to a more stable, amide-linked (native) peptide. Write mechanisms for the coupling and rearrangement steps in NCL.

Peptide 1–C(=O)SR + HS–CH₂–CH(NH₂)–Peptide 2 —coupling→ Peptide 1–C(=O)–S–CH₂–CH(NH₂)–Peptide 2 + RSH

—rearrangement→ Peptide 1–C(=O)–NH–CH(CH₂SH)–Peptide 2

25.44 Cyanogen bromide ($BrC \equiv N$) cleaves peptides between the carbonyl group of a methionine residue and the next amino acid.

Methionine-containing peptide → (1. $BrC \equiv N$; 2. H_3O^+) → Lactone + C-terminal fragment + $CH_3SC \equiv N$ (Methyl thiocyanate)

(a) The mechanism begins with the reaction of cyanogen bromide with the peptide to give a sulfonium ion:

Methionine-containing peptide → ($BrC \equiv N$) → Sulfonium ion + $:\ddot{Br}:^-$ (Bromide ion)

The sulfonium ion then cyclizes to an iminolactone with loss of methyl thiocyanate.

Sulfonium ion → Conjugate acid of iminolactone + $CH_3\ddot{S}C \equiv N$ (Methyl thiocyanate)

Using curved arrows to show the flow of electrons, suggest a reasonable mechanism for this cyclization.

(b) Hydrolysis of the conjugate acid of the iminolactone to the corresponding lactone cleaves the peptide chain. Write a stepwise mechanism for this reaction.

Enzymes

25.45 Acetoacetate decarboxylase is an enzyme that catalyzes the decarboxylation of acetoacetate to form acetone and carbon dioxide.

$$H_3CCOCH_2COO^- \xrightarrow{\text{enzyme}} H_3CCOCH_3 + CO_2$$

The enzyme contains an active-site lysine residue, which forms a covalent bond with the substrate during the catalytic mechanism.

H_3C, CH_2COO^-, NH^+, Enzyme

a. The covalent intermediate stabilizes a carbanion formed when carbon dioxide is released from the substrate. Show an arrow-pushing mechanism for the formation of this carbanion.

b. Draw a resonance structure of the carbanion that shows how the covalent bond between lysine and the substrate stabilizes the carbanion intermediate.

c. Treatment of the enzyme–product complex with sodium borohydride followed by total acid hydrolysis yields L-amino acids and one modified amino acid. Draw the structure of the modified amino acid.

25.46 The enzyme ornithine decarboxylase is a PLP-containing enzyme that catalyzes the conversion of ornithine to putrescine:

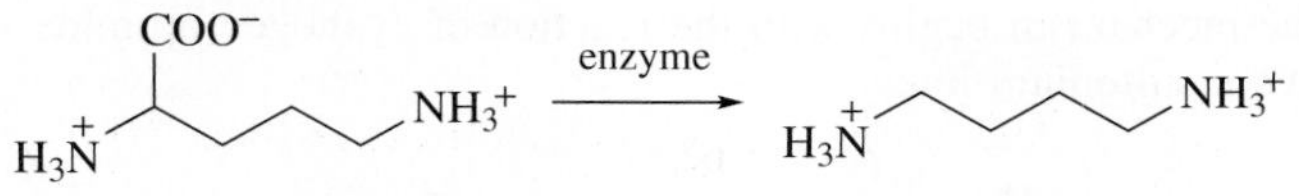

COO^- ; $H_3\overset{+}{N}$; NH_3^+ —enzyme→ $H_3\overset{+}{N}$; NH_3^+

Ornithine   Putrescine

Propose a reasonable mechanism for this reaction.

Descriptive Passage and Interpretive Problems 25

Amino Acids in Enantioselective Synthesis

Organic chemists speak of a "chiral pool," which comprises those naturally occurring compounds that are readily available as a single enantiomer and capable of being used as starting materials for the enantioselective synthesis of other chiral molecules. Amino acids are well represented in the chiral pool. All except glycine have at least one chirality center and, although L-amino acids are more abundant and less expensive than their D-enantiomers, both are available.

Most of the standard amino acids have served as starting materials for enantioselective syntheses. One of the most widely used is L-glutamic acid and its lactam (*S*)-pyroglutamic acid, which is easily prepared by heating an aqueous solution of L-glutamic acid in a sealed container.

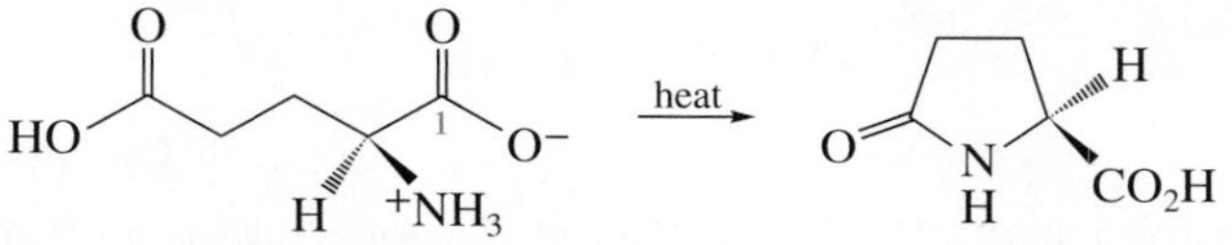

O ; O ; HO ; 1 ; O^- ; H ; $^+NH_3$ —heat→ O ; N ; H ; H ; CO_2H

L-Glutamic acid   (*S*)-Pyroglutamic acid

With three functional groups in a compound with only five carbons, L-glutamic acid and (*S*)-pyroglutamic acid provide access to more complex molecules via functional-group manipulation.

Many of the syntheses are lengthy, and most contain some specialized reactions. The following problems emphasize the planning aspect of amino acid–based enantioselective syntheses starting with either L-glutamic acid or (*S*)-pyroglutamic acid. The few reactions that are included are either familiar or similar to those covered in earlier chapters.

25.47 (*S*)-Pyroglutamic acid was used as the starting material in a synthesis of (+)-ipalbidine, an analgesic alkaloid obtained from the seeds of the white moonflower, *Ipomoea alba*. No bonds are made or broken to the chirality center of (*S*)-pyroglutamic acid in this synthesis. Which of the following is the structure of (+)-ipalbidine?

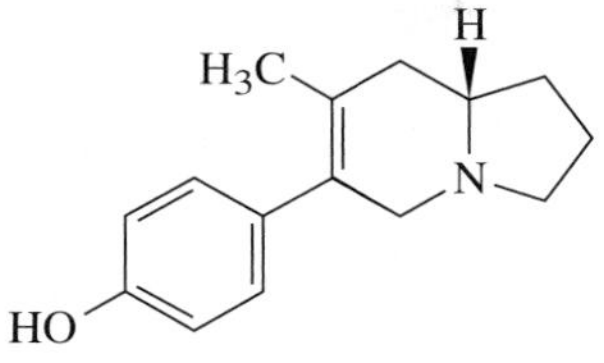

A. 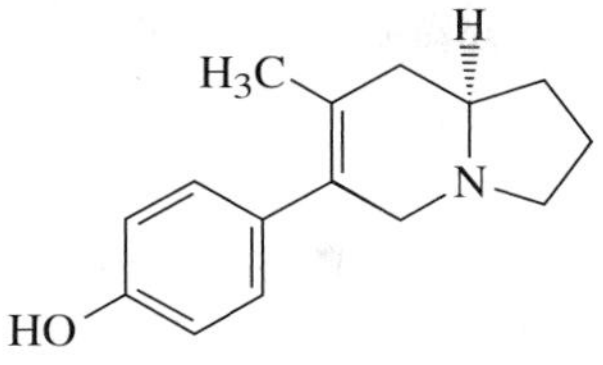 B.

25.48 A synthesis of poison-dart frog toxin has been described that begins with L-glutamic acid.

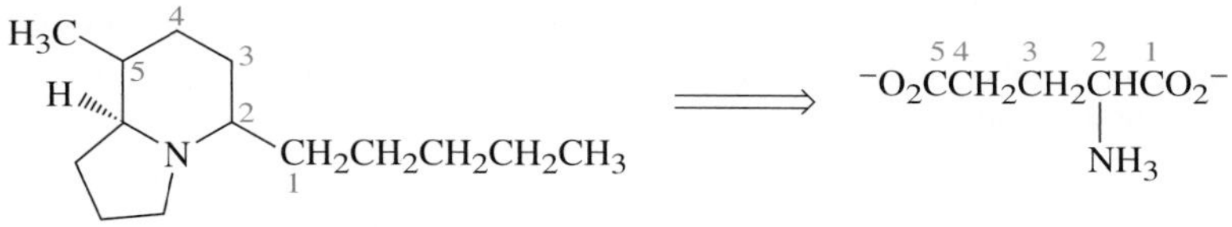

$$^{-}O_2CCH_2CH_2CHCO_2^{-}$$ (with NH_3 on C-2; carbons numbered 5 4 3 2 1)

The numbered carbons in the product correspond to the same numbered carbons in L-glutamic acid. No bonds to carbon-2 are made or broken in the synthesis. If the configuration at carbon-5 in the product is *R*, which of the following best represents the stereochemistry of the frog toxin?

A. H_3C, H, $CH_2CH_2CH_2CH_2CH_3$

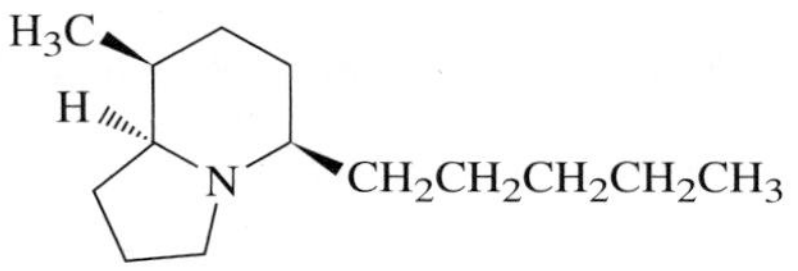

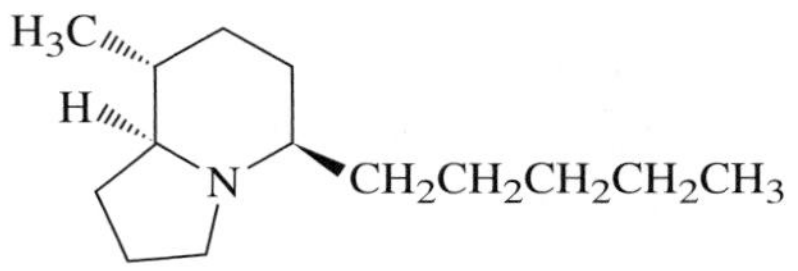

B. H_3C, H, $CH_2CH_2CH_2CH_2CH_3$

C. H_3C, H, $CH_2CH_2CH_2CH_2CH_3$

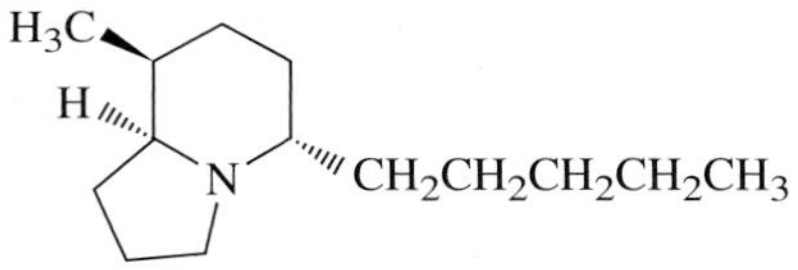

D. H_3C, H, $CH_2CH_2CH_2CH_2CH_3$

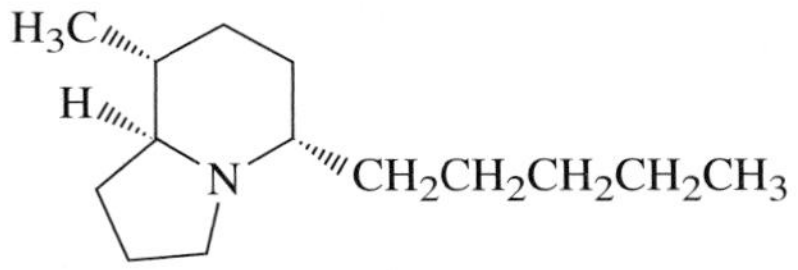

25.49 One synthesis of fosinopril, a drug used to combat high blood pressure, starts with the reduction of (*S*)-pyroglutamic acid to the primary alcohol, followed by protection of the OH and NH groups.

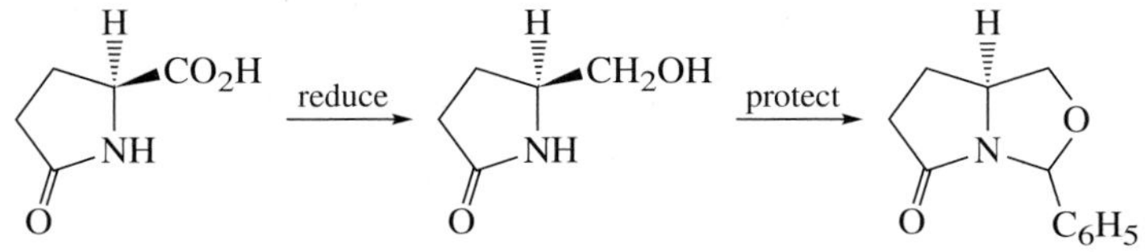

CO_2H —reduce→ CH_2OH —protect→ C_6H_5

What reaction conditions are appropriate for the protection step?

A. $C_6H_5\overset{O}{\overset{\|}{C}}Cl$, pyridine
B. $C_6H_5CH{=}O$, *p*-toluenesulfonic acid, toluene, heat
C. $C_6H_5CO_2CH_3$
D. $C_6H_5CH_2Br$, Na_2CO_3, acetone

25.50 The *N,O*-protected compound formed in the preceding problem was alkylated with 3-bromocyclohexene.

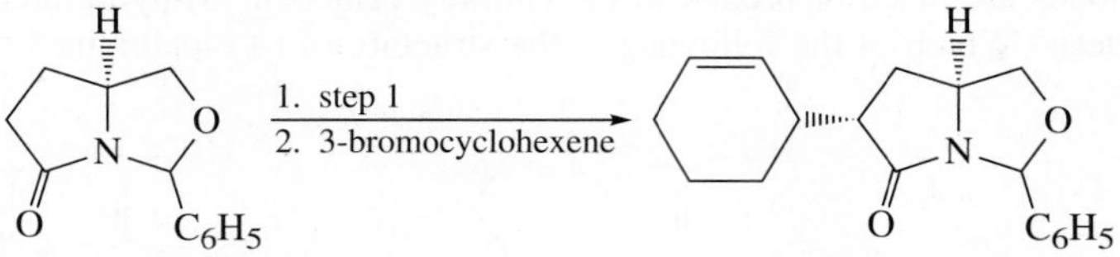

1. step 1
2. 3-bromocyclohexene

What reaction conditions are appropriate for step 1?

A. $LiAlH_4$, diethyl ether
B. $NaOCH_2CH_3$, ethanol
C. Mg, diethyl ether
D. $[(CH_3)_2CH]_2NLi$, tetrahydrofuran

25.51 α-Kainic acid is a neurotoxin produced by certain algae. Several enantioselective syntheses of it have been described, one of which is based on 1-bromo-3-methyl-2-butene and L-glutamic acid.

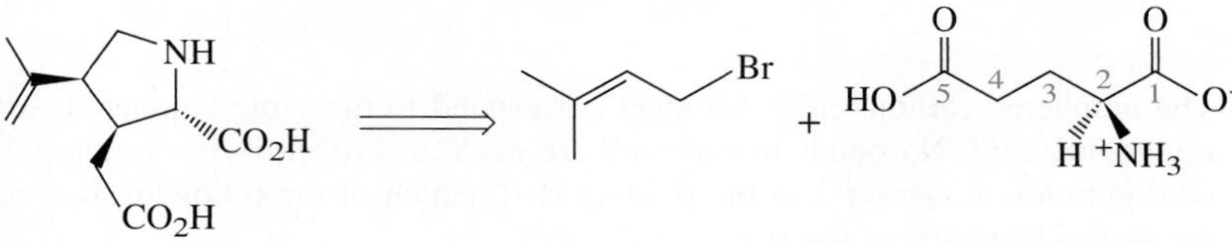

α-Kainic acid   L-Glutamic acid

Which carbon of L-glutamic acid is involved in the only carbon–carbon bond-forming step of the synthesis?

A. C-1
B. C-2
C. C-3
D. C-4
E. C-5

25.52 (*S*)-Tylophorine is an alkaloid isolated from a plant that grows in India and Southeast Asia, which is of interest as a potential antitumor drug. It has been synthesized by a multistep procedure based on L-glutamic acid and the phenanthrene derivative shown.

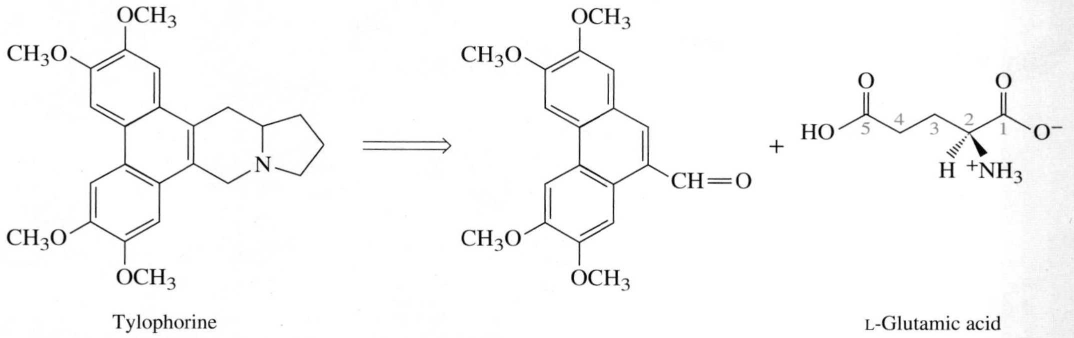

Which carbon of L-glutamic acid is involved in the only carbon–carbon bond-forming step of the synthesis?

A. C-1
B. C-2
C. C-3
D. C-4
E. C-5

Chapter

26

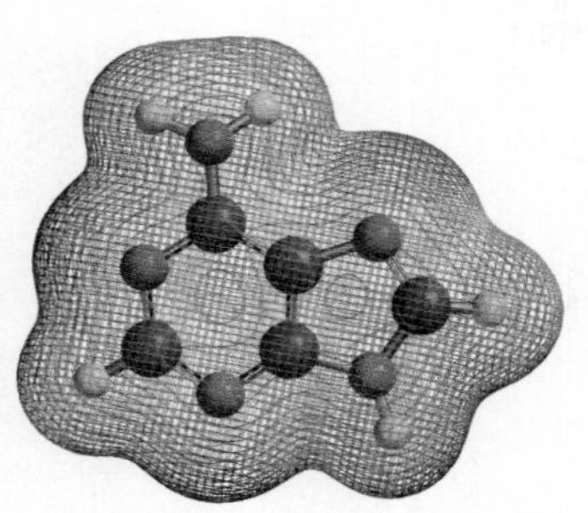

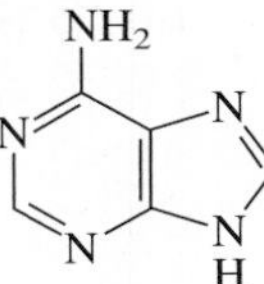

Gregor Mendel's studies of the inherited traits of garden peas received little attention during his lifetime. What he found is now recognized as the beginning of our understanding of genetics. Adenine, shown in the electrostatic potential map, is a "purine base" present in DNA and RNA.
©Robert Giuliano/Villanova University

CHAPTER OUTLINE

Nucleosides, Nucleotides, and Nucleic Acids

In Chapter 1 we saw that a major achievement of the first half of the twentieth century was the picture of atomic and molecular structure revealed by quantum mechanics. In this chapter we examine the major achievement of the second half of that century—a molecular view of genetics based on the structure and biochemistry of nucleic acids.

Nucleic acids are substances present in the nuclei of cells and were known long before anyone suspected they were the primary substances involved in the storage, transmission, and processing of genetic information. There are two kinds of nucleic acids: ribonucleic acid (RNA) and deoxyribonucleic acid (DNA). Both are biopolymers based on three structural units: a carbohydrate, a phosphate ester linkage between carbohydrates, and a heterocyclic aromatic compound. The heterocyclic aromatic compounds are referred to as purine and pyrimidine bases. We'll begin with them and follow the structural thread:

$$\text{Purine and pyrimidine bases} \rightarrow \text{Nucleosides} \rightarrow \text{Nucleotides} \rightarrow \text{Nucleic acids}$$

There will be a few pauses along the way to examine some biochemical roles played by these compounds separate from their genetic one.

26.1 Pyrimidines and Purines

The two types of heterocycles found in DNA and RNA are **pyrimidines** and **purines.** Pyrimidines are monocyclic aromatic compounds that contain two nitrogens in the ring. A purine is a bicyclic molecule in which an imidazole ring is fused to a pyrimidine ring. When derivatives of these molecules are part of DNA or RNA, they are referred to as **nucleic acid bases.**

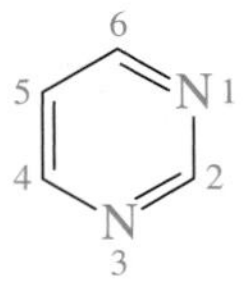

Pyrimidine 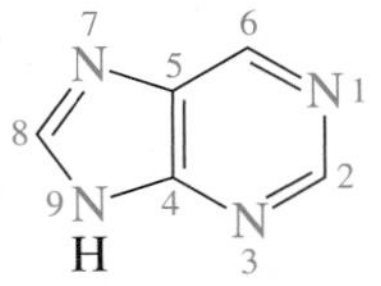 Purine

The five nucleic acid bases of DNA and RNA are shown in Table 26.1. Adenine is a purine with an amine substituent at position 6. The structural resemblance between the parent molecule and the other four nucleic acid bases is not as obvious. Purines and pyrimidines with exocyclic heteroatom substituents can exist in different tautomeric forms. (Recall that tautomers are constitutional isomers that are readily interconvertible. One example seen frequently in this book is the keto–enol pair.) Uracil, for example, can exist in three tautomeric forms. The "keto" **(lactam)** form is the most stable tautomer, and it is the form found in RNA. Tautomerization forms an "enol" **(lactim)** form. The double lactim tautomer most clearly shows why uracil is classified as a pyrimidine.

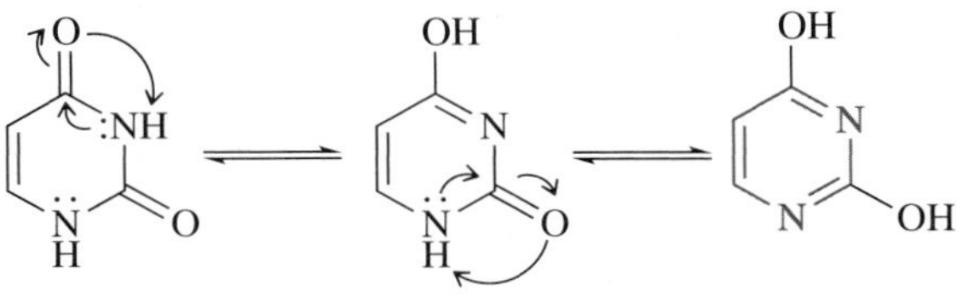

Lactam Lactim Double lactim

There is a pattern in the most stable tautomeric forms of the purines and pyrimidines. In general, derivatives that bear an —NH_2 substituent retain the structure of the parent ring.

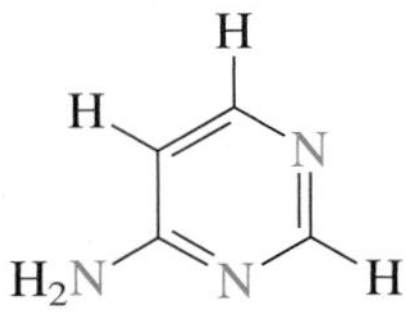

4-Aminopyrimidine 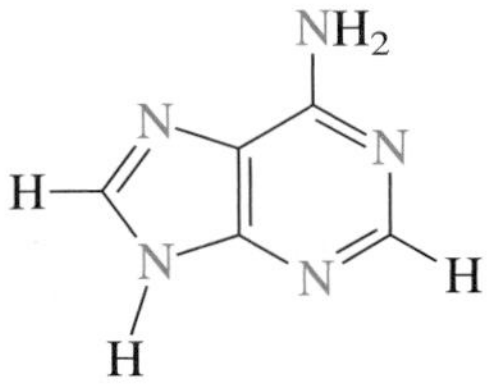 6-Aminopurine

6-Aminopurine is *adenine* and will appear numerous times in this chapter.

Compounds that have an —OH substituent, however, exist almost exclusively in the lactam ("keto") form.

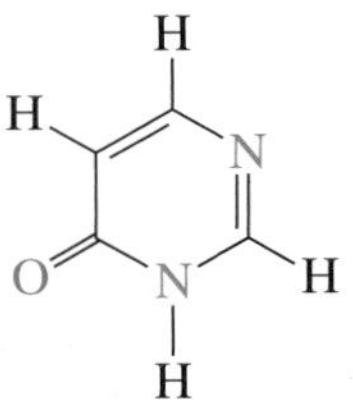

Keto form of 4-hydroxypyrimidine

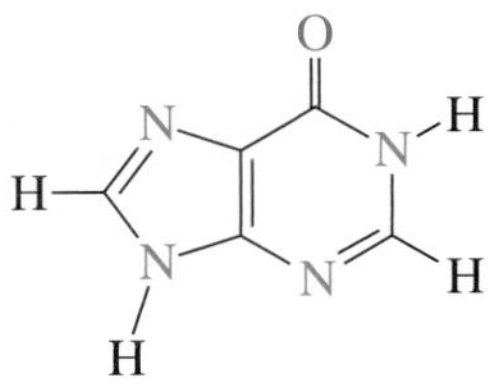

Keto form of 6-hydroxypurine

TABLE 26.1 Pyrimidines and Purines That Occur in DNA and/or RNA

	Name	Structure	Occurrence
Pyrimidines			
	Cytosine	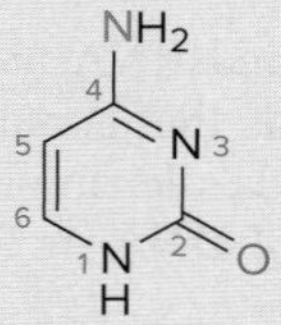	DNA and RNA
	Thymine	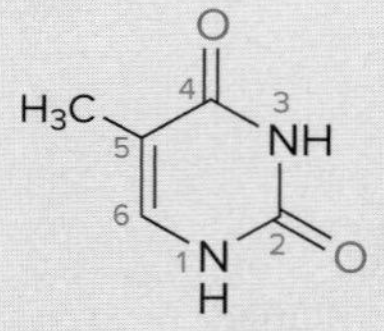	DNA
	Uracil	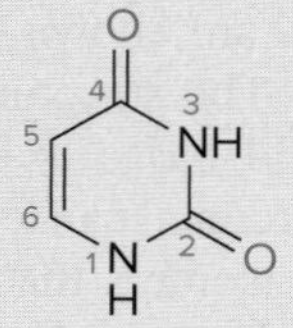	RNA
Purines			
	Adenine	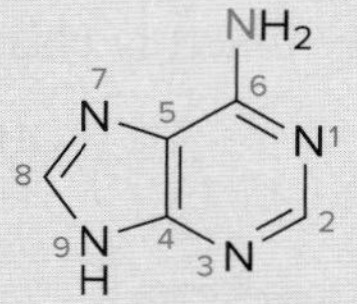	DNA and RNA
	Guanine	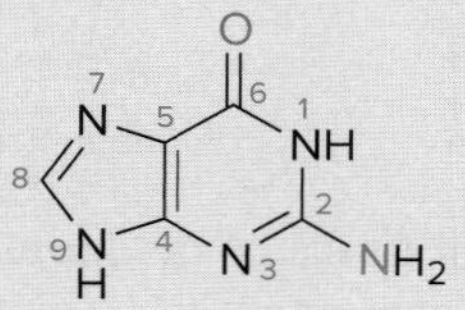	DNA and RNA

The equilibrium for tautomerization lies very far in favor of the familiar forms of the nucleic acid bases. This is fortunate, because tautomerization will change the hydrogen-bonding pattern of the base. For example, the lactim tautomer of thymine possesses the same pattern of hydrogen bond donors and acceptors as cytosine (Figure 26.1).

In the early 1950s, Watson and Crick suggested the possibility that a rare tautomer of a base could lead to spontaneous mutations in the genome. For example, if the lactim tautomer of thymine is present during replication, the newly synthesized DNA strand would possess guanine rather than adenine at the corresponding position. The presence of the "wrong" tautomer at the wrong time can also affect translation of DNA into protein. In fact, it is currently believed that base tautomerization is responsible for a portion of the naturally occurring miscoding errors observed in proteins.

Problem 26.1

Write a structural formula for the enol tautomer of cytosine (Table 26.1).

Thymine
Normal tautomer

Thymine
Lactim tautomer

Cytosine
Normal tautomer

Figure 26.1

Structures of thymine tautomers and of cytosine. The two tautomers of thymine will hydrogen bond to different nucleic acid bases. The minor tautomer's hydrogen-bonding pattern resembles that of cytosine and can base pair with guanosine rather than adenine.

Pyrimidines and purines occur naturally in substances other than nucleic acids. Coffee, for example, is a familiar source of caffeine. Tea contains both caffeine and theobromine.

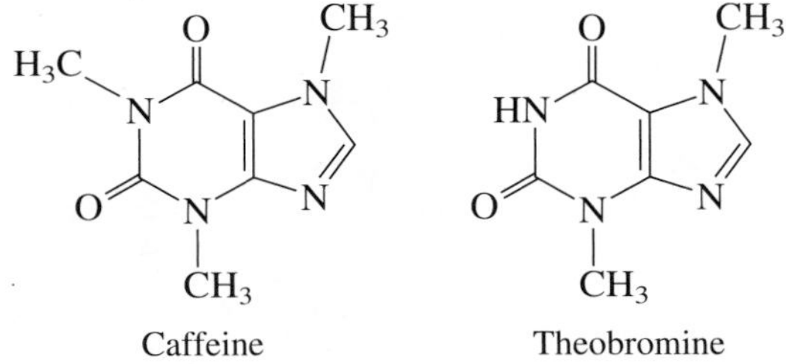

Caffeine Theobromine

The caffeine added to soft drinks is the caffeine that was removed when decaffeinating coffee and tea.

©amnat11/Shutterstock

Problem 26.2

Classify caffeine and theobromine according to whether each is a pyrimidine or a purine. One of these cannot isomerize to an enolic form; two different enols are possible for the other. Explain and write structural formulas for the possible enols.

Several synthetic pyrimidines and purines are useful drugs. *Acyclovir* was the first effective antiviral compound and is used to treat herpes infections. *6-Mercaptopurine* is one of the drugs used to treat childhood leukemia, which has become a very treatable form of cancer with a remission rate approaching 95%.

Acyclovir 6-Mercaptopurine

Acyclovir and 6-mercaptopurine are representative of the kind of drugs for which Gertrude B. Elion and George H. Hitchings of Burroughs Wellcome were awarded a share of the 1988 Nobel Prize in Physiology or Medicine.

26.2 Nucleosides

The most important derivatives of pyrimidines and purines are nucleosides. **Nucleosides** are glycosylamines in which a pyrimidine or purine nitrogen is bonded to the anomeric carbon of a carbohydrate. The nucleosides listed in Table 26.2 are the main building

TABLE 26.2 The Major Pyrimidine and Purine Nucleosides in RNA and DNA

	Pyrimidines			Purines	
Name	**Cytidine**	**Thymidine**	**Uridine**	**Adenosine**	**Guanosine**
Abbreviation*	C	T	U	A	G
Systematic name	1-β-D-Ribo-furanosylcytosine	2′-Deoxy-1-β-D-ribo-furanosylthymine	1-β-D-Ribo-furanosyluracil	9-β-D-Ribo-furanosyladenine	9-β-D-Ribo-furanosylguanine
Structural formula	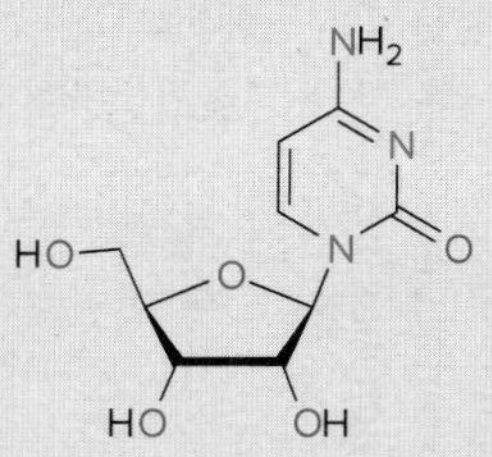	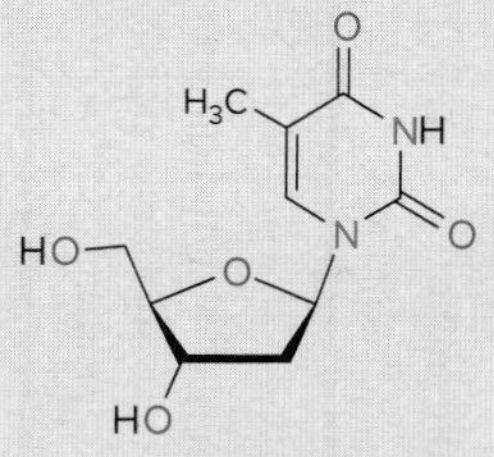	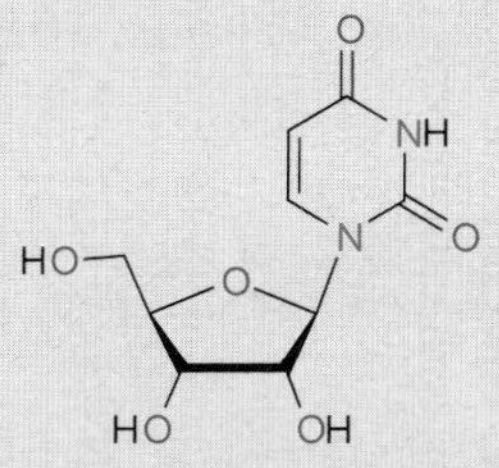	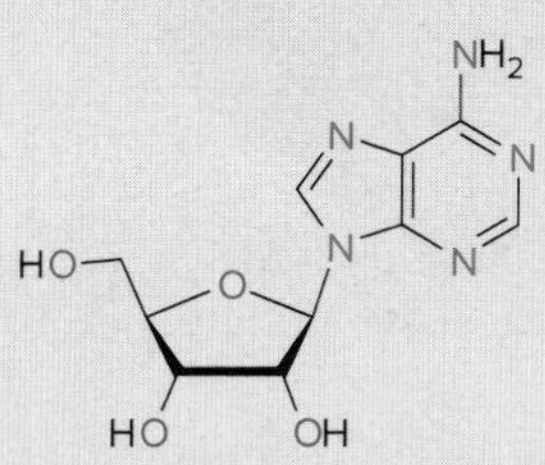	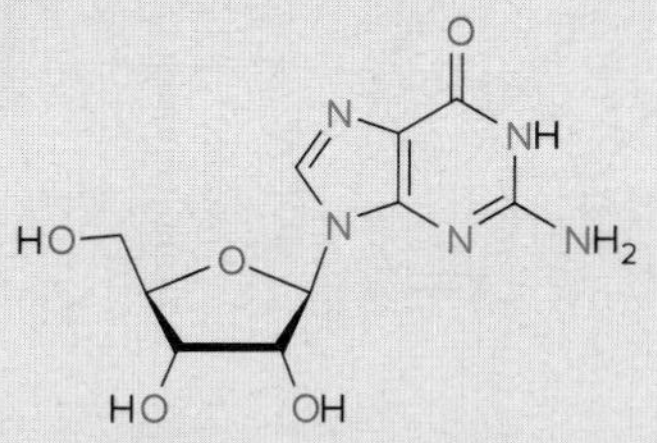
Molecular model	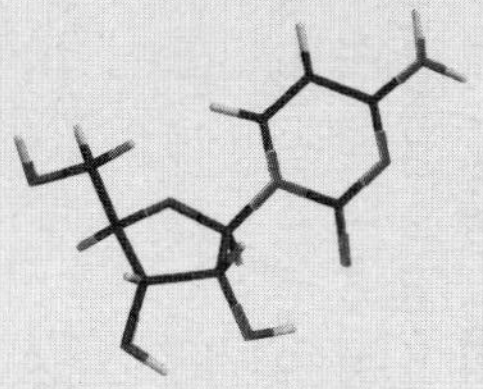	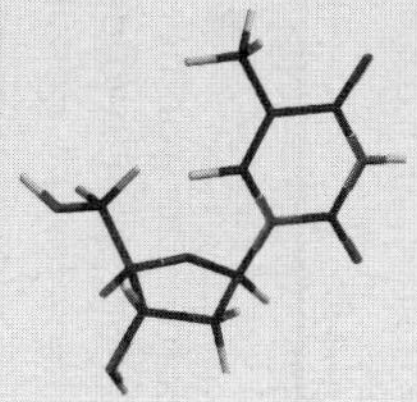	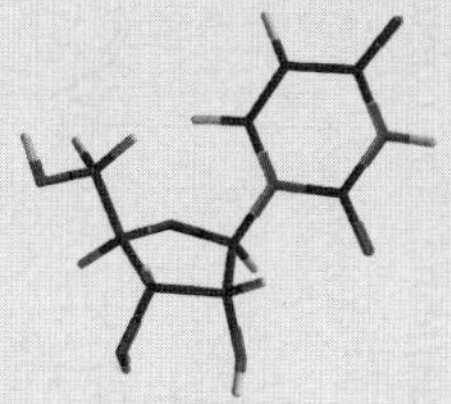	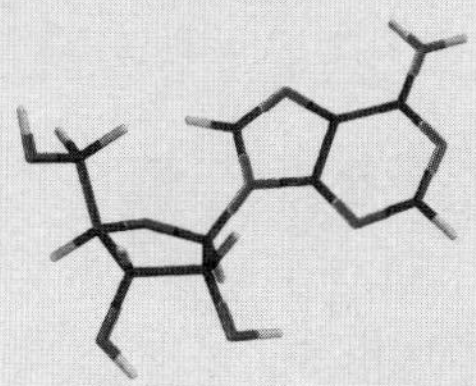	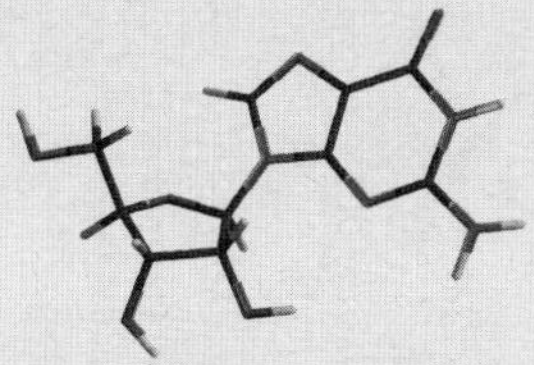
Found in	RNA 2′-Deoxy analog in DNA	DNA	RNA	RNA 2′-Deoxy analog in DNA	RNA 2′-Deoxy analog in DNA

*Sometimes the abbreviation applies to the pyrimidine or purine base, sometimes to the nucleoside. Though this may seem confusing, it is normally clear from the context what is intended and causes no confusion in practice.

blocks of nucleic acids. In RNA the carbohydrate component is D-ribofuranose; in DNA it is 2-deoxy-D-ribofuranose.

Among the points to be made concerning Table 26.2 are the following:

1. Three of the bases (cytosine, adenine, and guanine) occur in both RNA and DNA.
2. Uracil occurs only in RNA; thymine occurs only in DNA.
3. The anomeric carbon of the carbohydrate is attached to N-1 in pyrimidine nucleosides and to N-9 in purines.
4. The pyrimidine and purine bases are cis to the —CH_2OH group of the furanose ring (β stereochemistry).
5. Potential hydrogen-bonding groups (—NH_2 and C=O) point away from the furanose ring.

The numbering scheme used for nucleosides maintains the independence of the two structural units. The pyrimidine or purine is numbered in the usual way. So is the carbohydrate, except that a prime symbol (′) follows each locant. Thus, adenosine is a nucleoside of D-ribose, and 2′-deoxyadenosine is a nucleoside of 2-deoxy-D-ribose.

Problem 26.3

The nucleoside *cordycepin* was isolated from cultures of the fungus *Cordyceps militaris* and found to be 3′-deoxyadenosine. Write its structural formula.

Table 26.2 doesn't include all of the nucleoside components of nucleic acids. The presence of methyl groups on pyrimidine and purine rings is a common, and often important, variation on the general theme.

Although the term *nucleoside* was once limited to the compounds in Table 26.2 and a few others, current use is more permissive. Pyrimidine derivatives of D-arabinose, for example, occur in the free state in certain sponges and are called *spongonucleosides.* The powerful antiviral drug ribavirin, used to treat hepatitis C and Lassa fever, is a synthetic nucleoside analog in which the base, rather than being a pyrimidine or purine, is a *triazole.*

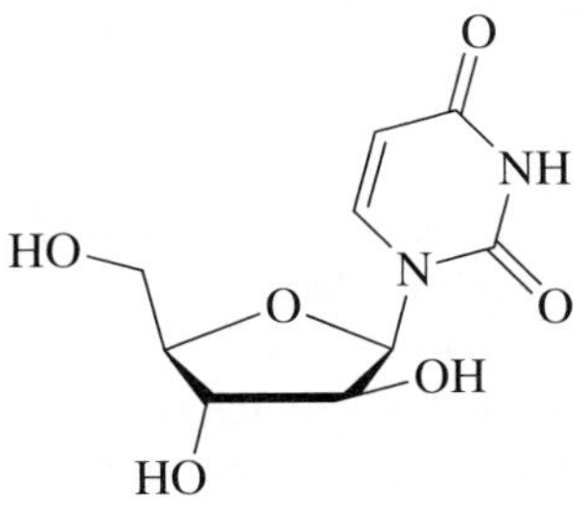

1-β-D-Arabinofuranosyluracil ("spongouridine")

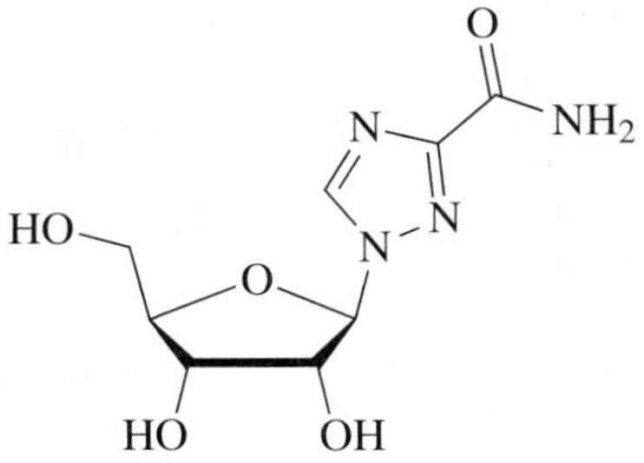

Ribavarin

26.3 Nucleotides

Nucleotides are phosphoric acid esters of nucleosides. Those derived from adenosine, of which *adenosine 5′-monophosphate* (AMP) is but one example, are especially prominent. AMP is a weak diprotic acid with pK_a's for ionization of 3.8 and 6.2, respectively. In aqueous solution at pH 7, both —OH groups of the $P(O)(OH)_2$ unit are ionized.

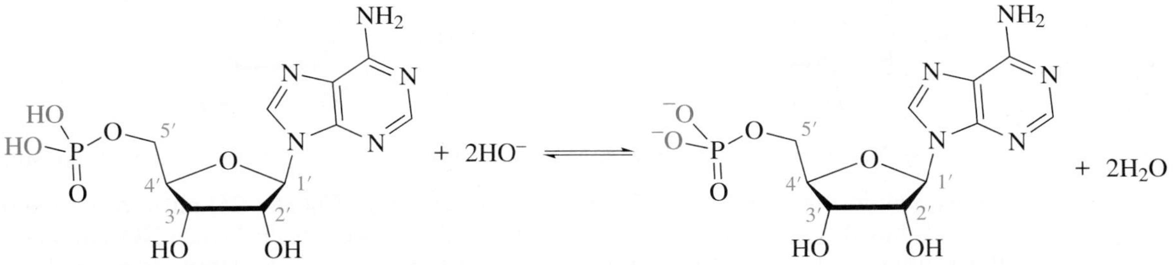

Adenosine 5′-monophosphate (AMP) + $2HO^-$ ⇌ Major species at pH 7 + $2H_2O$

Problem 26.4

Write a structural formula for 2′-deoxycytidine 3′-monophosphate. You may wish to refer to Table 26.2 for the structure of cytidine.

Other important 5′-nucleotides of adenosine include adenosine 5′-diphosphate (ADP) and **adenosine 5′-triphosphate** (ATP):

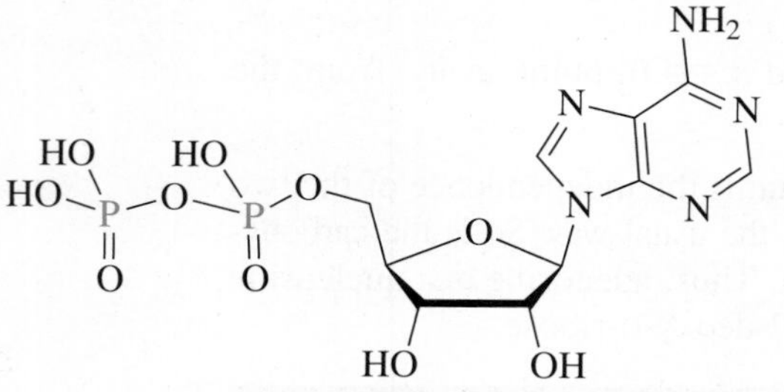

Adenosine 5′-diphosphate (ADP)

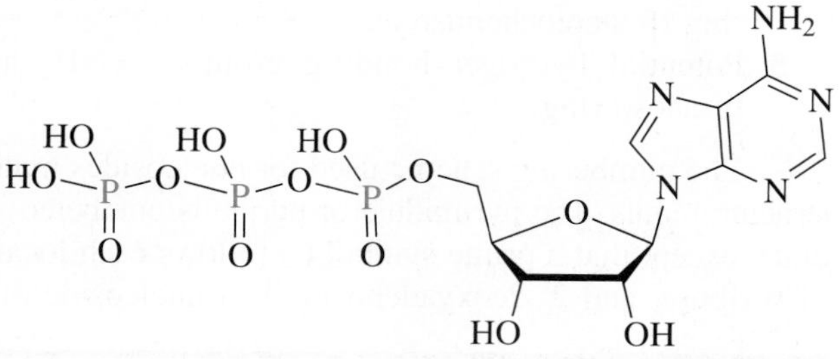

Adenosine 5′-triphosphate (ATP)

ATP is the main energy-storing molecule for practically every form of life on Earth. We often speak of ATP as a "high-energy compound" and its P—O bonds as "high-energy bonds." This topic is discussed in more detail in Sections 26.4 and 26.5.

The biological transformations that involve ATP are both numerous and fundamental. They include, for example, many reactions in which ATP transfers one of its phosphate units to the —OH of another molecule. These *phosphorylations* are catalyzed by enzymes called *kinases.* An example is the first step in the metabolism of glucose:

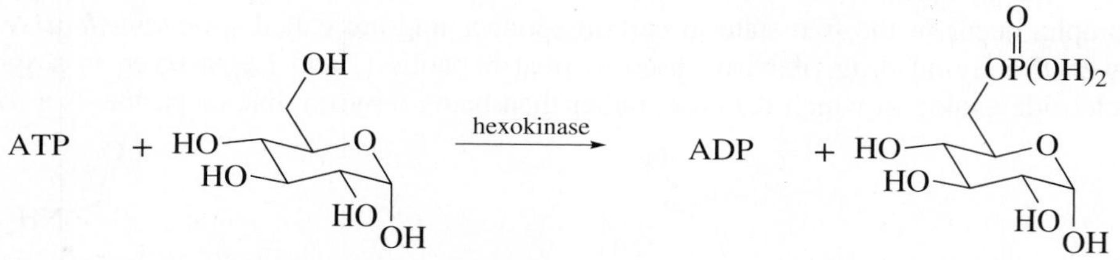

ATP + α-D-Glucopyranose —hexokinase→ ADP + α-D-Glucopyranose 6-phosphate

Adenosine triphosphate; α-D-Glucopyranose; Adenosine diphosphate; α-D-Glucopyranose 6-phosphate

Earl Sutherland of Vanderbilt University won the 1971 Nobel Prize in Physiology or Medicine for uncovering the role of cAMP as a second messenger in connection with his studies of the "fight or flight" hormone epinephrine (see Section 25.21).

Both adenosine and guanosine form cyclic monophosphates (*cyclic-AMP* or *cAMP* and *cyclic-GMP* or *cGMP,* respectively) that are involved in a large number of biological processes as "second messengers." Many hormones (the "first messengers") act by stimulating the formation of cAMP or cGMP on a cell surface, which triggers a series of events characteristic of the organism's response to the hormone. Signalling by cAMP is also involved in the activation of G-protein-coupled receptors (see Section 25.22).

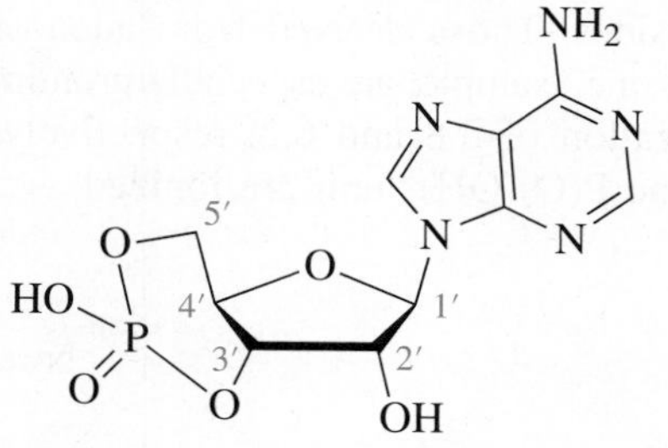

Adenosine 3′,5′-cyclic monophosphate (cAMP)

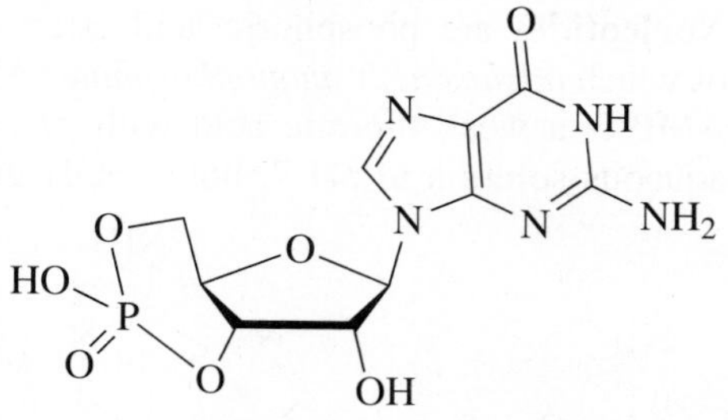

Guanosine 3′,5′-cyclic monophosphate (cGMP)

As we saw in the boxed essay *Oh NO! It's Inorganic!* in Chapter 25, nitric oxide (NO) expands blood vessels and increases blood flow. This process begins when NO stimulates the synthesis of cGMP as a second messenger. Erectile dysfunction drugs such as *sildenafil* (Viagra) increase the concentration of cGMP by inhibiting the enzyme that catalyzes hydrolysis of its cyclic phosphate unit.

Problem 26.5

Cyclic-AMP is formed from ATP in a reaction catalyzed by the enzyme *adenylate cyclase*. Assume that adenylate cyclase acts as a base to remove a proton from the 3′-hydroxyl group of ATP and write a mechanism for the formation of cAMP.

26.4 Bioenergetics

Bioenergetics is the study of the thermodynamics of biological processes, especially those that are important in energy storage and transfer. Some of its conventions are slightly different from those we are accustomed to.

First, it is customary to focus on changes in free energy (ΔG) rather than changes in enthalpy (ΔH). It is instructive to review some of the thermodynamic relationships in chemical reactions that involve free energy. Consider the reaction

Recall that free energy is the energy available to do work. By focusing on free energy, we concern ourselves more directly with what is important to a living organism.

$$mA \rightleftharpoons nB$$

$$K = \frac{[B]^n}{[A]^m}$$

Recall that the **standard free-energy change ($\Delta G°$)** applies to a standard state at constant temperature (298 K) and pressure (1 atm), where initial [A] = initial [B] = 1 M. Reactions are classified as **exergonic** or **endergonic** according to the sign of $\Delta G°$. An exergonic reaction is one in which $\Delta G°$ is negative, whereas an endergonic reaction has a positive value of $\Delta G°$. The sign and magnitude of $\Delta G°$ describes the equilibrium constant for the process at standard state.

$$\Delta G° = -RT \ln K$$

But spontaneity under ***nonstandard state*** depends on the concentrations of reactants and products. If the ratio $[B]^n/[A]^m$ is less than a certain value, the reaction is spontaneous in the forward direction ($\Delta G < 0$); if $[B]^n/[A]^m$ exceeds this value, the reaction is spontaneous in the reverse direction ($\Delta G < 0$). Thus, ΔG tells us about the *reaction* with respect to the substances present and their concentrations. The criterion for spontaneity for a reaction under a given set of conditions is the value of $\mathbf{\Delta G}$**,** which is concentration dependent, **not $\mathbf{\Delta G°}$.**

$$\Delta G = \Delta G° + RT \ln \frac{B^n}{A^m}$$

An important consideration in bioenergetics is that these processes are performed in an aqueous environment in which water and hydronium ions may be both participants in the processes as well as components of the solutions. Standard state in biochemical reactions is additionally defined as occurring in dilute solutions at pH 7 ($[H_3O^+] = 10^{-7}$ M). The activities of water and hydronium ion are defined as 1, and these components do not occur in the equilibrium constant expression. So, for a biochemical reaction such as

$$A + H_2O \rightleftharpoons B + C + H^+$$

The expression for the equilibrium constant, $\mathbf{K'}$**,** is

$$K' = \frac{[B][C]}{[A]}$$

Standard free-energy change ($\mathbf{\Delta G°'}$) for a biological system at equilibrium is therefore

$$\Delta G°' = RT \ln K'$$

To determine if the biochemical reaction is spontaneous under a set of conditions, the free-energy change for the reaction, **$\Delta G'$**, is calculated.

$$\Delta G' = \Delta G^{\circ\prime} + RT \ln \frac{[B][C]}{[A]}$$

This is an important concept for chemical reactions that occur in biochemical systems. Reactions that are not exergonic at the standard state can spontaneously occur if the reaction conditions are such that the sum of **$\Delta G'$** and **$\Delta G^{\circ\prime}$** is < 0.

26.5 ATP and Bioenergetics

The key reaction in bioenergetics is the interconversion of ATP and ADP, usually expressed in terms of the hydrolysis of ATP.

HPO_4^{2-} is often referred to as "inorganic phosphate" and abbreviated P_i.

$$\underset{\text{Adenosine triphosphate}}{\text{ATP}} + \underset{\text{Water}}{H_2O} \longrightarrow \underset{\text{Adenosine diphosphate}}{\text{ADP}} + \underset{\text{Hydrogen phosphate}}{HPO_4^{2-}} \qquad \Delta G^{\circ\prime} = -31 \text{ kJ/mol } (-7.4 \text{ kcal/mol})$$

As written, the reaction is exergonic at pH = 7. The reverse process—conversion of ADP to ATP—is endergonic. ATP is categorized as a "high-energy compound." In biochemistry, these are molecules whose hydrolysis releases more than ~25 kJ/mol (~6 kcal/mol) at standard state. Such molecules are generally short lived in living systems, serving as vehicles for energy transfer rather than energy storage.

When coupled to some other process, the conversion of ATP to ADP can provide the free energy to transform an otherwise endergonic process to an exergonic one. Take, for example, the conversion of glutamic acid to glutamine at pH = 7.

Equation 1:

Glutamic acid + $\overset{+}{N}H_4$ ⟶ Glutamine + H_2O

Equation 1 has $\Delta G^{\circ\prime} = +14$ kJ/mol (3.3 kcal/mol) and is endergonic. The main reason for this is that one of the very stable carboxylate groups of glutamic acid is converted to a less-stable amide function.

Nevertheless, the biosynthesis of glutamine proceeds from glutamic acid. The difference is that the endergonic process in Equation 1 is coupled with the strongly exergonic hydrolysis of ATP.

Equation 2:

Glutamic acid + $\overset{+}{N}H_4$ + ATP ⟶ Glutamine + P_i + ADP

Adding the value of $\Delta G^{\circ\prime}$ for the hydrolysis of ATP (−31 kJ/mol; −7.4 kcal/mol) to that of Equation 1 (+14 kJ/mol; 3.3 kcal/mol) gives $\Delta G^{\circ\prime} = -17$ kJ/mol (−4.1 kcal/mol) for Equation 2. The biosynthesis of glutamine from glutamic acid is exergonic because it is coupled to the hydrolysis of ATP.

Problem 26.6

Verify that Equation 2 is obtained by adding Equation 1 to the equation for the hydrolysis of ATP.

There is an important qualification to the idea that ATP can serve as a free-energy source for otherwise endergonic processes. There must be some mechanism by which ATP reacts with one or more species along the reaction pathway. Simply being present and undergoing independent hydrolysis isn't enough. The mechanism normally involves transfer of a phosphate unit from ATP to some nucleophilic site. In the case of glutamine synthesis, this step is phosphate transfer to glutamic acid to give γ-glutamyl phosphate as a reactive intermediate.

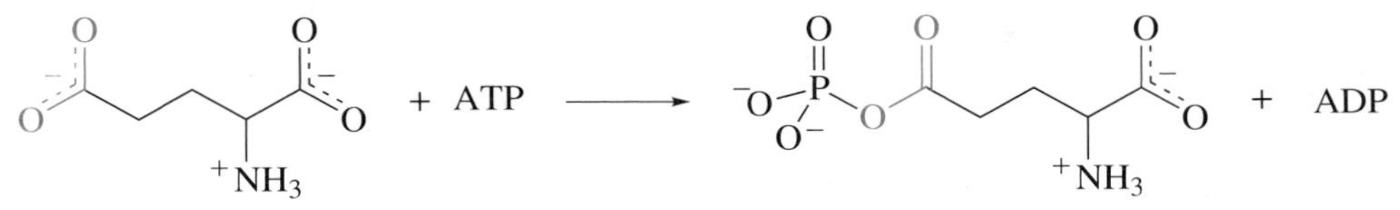

Glutamic acid + ATP ⟶ γ-Glutamyl phosphate + ADP

The γ-glutamyl phosphate formed in this step is a mixed anhydride of glutamic acid and phosphoric acid. It is activated toward nucleophilic acyl substitution and gives glutamine when attacked by ammonia.

Problem 26.7

Write a stepwise mechanism for the formation of glutamine by attack of NH_3 on γ-glutamyl phosphate.

If free energy is stored and transferred by way of ATP, where does the ATP come from? It comes from ADP by the endergonic reaction

$$\underset{\text{Adenosine diphosphate}}{\text{ADP}} + \underset{\text{Hydrogen phosphate}}{HPO_4^{2-}} \longrightarrow \underset{\text{Adenosine triphosphate}}{\text{ATP}} + \underset{\text{Water}}{H_2O} \qquad \Delta G^{\circ\prime} = +31 \text{ kJ/mol } (+7.4 \text{ kcal/mol})$$

which you recognize as the reverse of the exergonic hydrolysis of ATP. The free energy to drive this endergonic reaction comes from the metabolism of energy sources such as fats and carbohydrates. In the metabolism of glucose during glycolysis, for example, about one third of the free energy produced is used to convert ADP to ATP. Glycolysis produces phosphoenolpyruvate, which provides sufficient energy for the conversion of ADP to ATP. Energy-rich compounds are compared in terms of $\Delta G^{\circ\prime}$ for their hydrolysis in Table 26.3.

Problem 26.8

Is $K > 1$ or $K < 1$ for the transfer of a phosphate group from ATP to glucose to give glucose 6-phosphate?

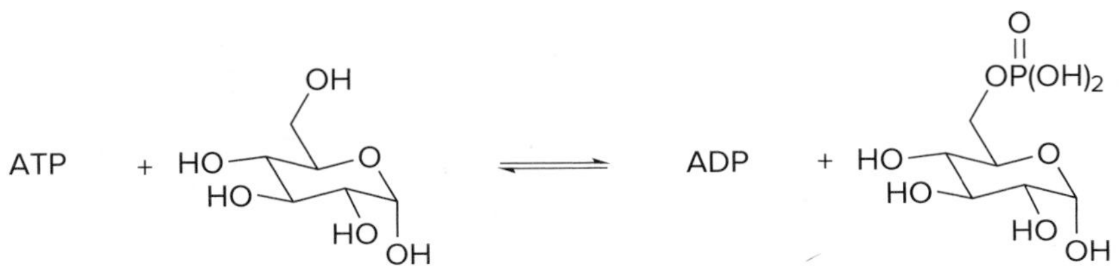

ATP + glucose ⇌ ADP + glucose 6-phosphate ($OP(OH)_2$ with =O)

As important as nucleotides of adenosine are to bioenergetics, that is not the only indispensable part they play in biology. The remainder of this chapter describes how these and related nucleotides are the key compounds in storing and expressing genetic information.

TABLE 26.3 $\Delta G°'$ for the Hydrolysis of Bioenergetically Important Phosphates

Phosphate	$\Delta G°'$ kJ/mol; kcal/mol
Phosphoenolpyruvate ($^{-}O_2C$–C($=CH_2$)–OPO_3^{2-})	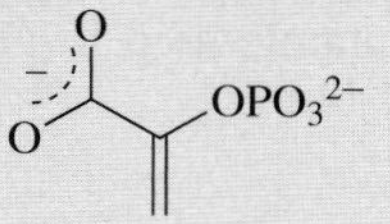 −62 kJ/mol; −15 kcal/mol
ADP	−35 kJ/mol; −8.4 kcal/mol
ATP	−31 kJ/mol; −7.4 kcal/mol
Glucose-1-phosphate	−21 kJ/mol; −5.0 kcal/mol
Glucose-6-phosphate	−14 kJ/mol; −3.3 kcal/mol
AMP	−9.2 kJ/mol; −2.2 kcal/mol

26.6 Phosphodiesters, Oligonucleotides, and Polynucleotides

Just as amino acids can join together to give dipeptides, tripeptides, and so on up to polypeptides and proteins, so too can nucleotides join to form larger molecules. Analogous to the "peptide bond" that connects two amino acids, a **phosphodiester** joins two nucleosides. Figure 26.2 shows the structure and highlights the two phosphodiester units of a trinucleotide of 2′-deoxy-D-ribose in which the bases are adenine (A), thymine (T), and guanine (G). Phosphodiester units connect the 3′-oxygen of one nucleoside to the 5′-oxygen of the next. Nucleotide sequences are written with the free 5′ end at the left and the free 3′ end at the right. Thus, the trinucleotide sequence shown in Figure 26.2 is written as ATG.

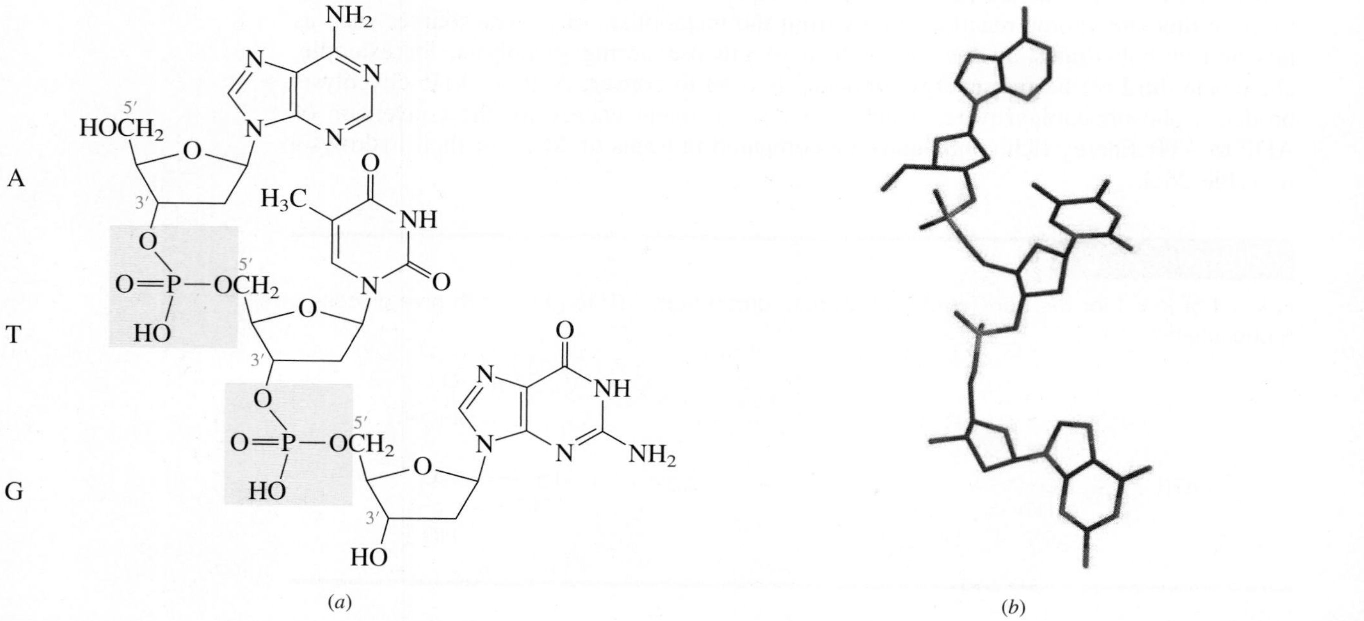

A T G

(a) (b)

Figure 26.2

(*a*) Structural formula and (*b*) molecular model of a trinucleotide ATG. The phosphodiester units highlighted in yellow in (*a*) join the oxygens at 3′ of one nucleoside to 5′ of the next. By convention, the sequence is read in the direction that starts at the free CH_2OH group (5′) and proceeds toward the free 3′-OH group at the other end.

The same kind of 5′→3′ phosphodiester units that join the 2′-deoxy-D-ribose units in Figure 26.2 are also responsible for connecting nucleosides of D-ribose.

Problem 26.9

How would the structures of the trinucleotides AUG and GUA in which all of the pentoses are D-ribose differ from the trinucleotide in Figure 26.2?

Adding nucleotides to the 3′-oxygen of an existing structure is called *elongation* and leads ultimately to a **polynucleotide.** The most important polynucleotides are ribonucleic acid (RNA) and deoxyribonucleic acid (DNA). As we shall see in later sections, the polynucleotide chains of DNA and some RNAs are quite long and contain hundreds of thousands of bases.

Polynucleotides of modest chain length, say 50 or fewer, are called **oligonucleotides.** With the growth of the biotechnology industry, the chemical synthesis of oligonucleotides has become a thriving business with hundreds of companies offering custom syntheses of oligonucleotides of prescribed sequences. Such oligonucleotides are required as "primers" in the polymerase chain reaction (Section 26.16) and as "probes" in DNA cloning and genetic engineering. Their synthesis is modeled after the Merrifield solid-phase method and, like it, is automated. The synthesis of a typical oligonucleotide containing 20–50 bases can be accomplished in a few hours.

Oligonucleotide synthesis is the subject of the Descriptive Passage at the end of this chapter.

26.7 Phosphoric Acid Esters

Phosphate groups are ubiquitous in living systems. In addition to serving as a source of energy, phosphoric acid derivatives are involved in signal transduction, regulation of protein function, and, as described in Section 26.6, they connect the nucleosides in the nucleic acid polymers as phosphate esters. The chemical properties of the phosphate esters in DNA and RNA reflect their role in the cell.

Phosphoric acid possesses three ionizable groups. The pK_a of the parent acid is 2.15, which makes phosphoric acid about 300 times stronger than acetic acid (pK_a = 4.6). The second ionization occurs near neutral pH (pK_a 7.2). The closeness of this pK_a to physiological pH (7.4) is one reason why phosphate buffers are very common in biochemical research laboratories.

$$\mathrm{(HO)_3P{=}O} \underset{pK_a = 2.15}{\overset{H^+}{\rightleftharpoons}} \mathrm{(HO)_2P(O)O^-} \underset{pK_a = 7.20}{\overset{H^+}{\rightleftharpoons}} \mathrm{HOP(O)(O^-)_2} \underset{pK_a = 12.35}{\overset{H^+}{\rightleftharpoons}} \mathrm{P(O)(O^-)_3}$$

Phosphoric acid esters can exist in three different forms: phosphotriester, phosphodiester, or phosphate monoester. Phosphotriesters are not found in living sytems, while phosphodiesters and phosphate monoesters are very common. Phosphotriesters are readily hydrolyzed to phosphodiesters.

$$\mathrm{(H_3CO)_2P(O)OCH_3} + {}^-\mathrm{OH} \xrightarrow[20°C,\ 24\ h]{1\ M\ NaOH} \mathrm{(H_3CO)_2P(O^-)(OH)(OCH_3)} \xrightarrow{-CH_3OH} \mathrm{(H_3CO)(CH_3O)P(O)O^-}$$

Alkyl phosphodiesters, however, are very stable toward hydrolysis. In fact, after 16 days at 100°C, only about half of an alkyl diester is hydrolyzed to its monoester. The phosphate is less electrophilic in the diester anion, and the negative charge on the phosphodiester repulses the incoming hydroxide ion.

Dimethyl phosphate $\xrightarrow[\text{100°C, ~16 days}]{\text{1 M NaOH}}$ Methyl phosphate 50%

The chemical stability of phosphodiesters is reflected in the stability of DNA. Prolonged heating in strong aqueous base does not hydrolyze the DNA phosphodiester bonds.

$$\text{DNA} \xrightarrow[\text{100°C, 1 h}]{\text{1 M NaOH}} \text{No hydrolysis}$$

From an evolutionary point of view, high chemical stability of the genetic material is desirable. The fidelity of the genetic code relies on maintaining the correct sequence. DNA may be cleaved enzymatically, and therefore the location and timing of strand breaking can be controlled. DNA strands are so stable that DNA from ancient animals, plants, and humanoids more than 200,000 years old can be sequenced.

In contrast, RNA tends to serve transient roles in a cell. Chemically, it is highly susceptible to hydrolysis. Contaminating RNA can be eliminated from DNA samples by short treatment with aqueous base.

$$\text{RNA} \xrightarrow[\text{65°C, 1 h}]{\text{0.25 M NaOH}} \text{Complete hydrolysis}$$

The key structural difference in the two polymers is the 2′ hydroxyl group on the ribofuranose of the nucleotide. The hydroxyl group affects the hydrolysis reaction by **anchimeric assistance,** which is defined as a rate increase due to neighboring group participation in a chemical reaction (Figure 26.3). The 2′-alcohol serves as an intramolecular nucleophile, which attacks the adjacent phosphate and forms a new phosphate–oxygen bond. During the process, the bond between the phosphate and the 5′-alcohol in the next ribonucleotide is cleaved and the RNA strand is broken.

The presence of the cyclic intermediate is suggested by the formation of 2′-phosphoesters in addition to the expected 3′-phosphoester. Release of ring strain of the cyclic phosphodiester (~19 kJ/mol; 4.6 kcal/mol) can also contribute to the increased hydrolysis rate for RNA.

−5′-RNA → 2′,3′-phosphodiester → ($^{-}$OH) → 3′-phosphoester + 2′-phosphoester

Figure 26.3

Hydrolysis of RNA by base. The negatively charged hydroxide undergoes reaction with the RNA molecule at the 2′-alcohol, which performs a transesterification reaction to yield a strained phosphodiester while cleaving the RNA polymer. Hydrolysis of the cyclic diester releases the ring strain and produces both 2′- and 3′-monoesters at the cleavage site.

26.8 Deoxyribonucleic Acids

©Villanova University

Gregor Mendel systematically studied and statistically analyzed inherited traits in garden peas.

The nineteenth century saw three things happen that, taken together, prepared the way for our present understanding of genetics. In 1854, an Augustinian monk named Gregor Mendel began growing peas and soon discovered some fundamental relationships about their inherited characteristics. Mendel discovered two laws of heredity: segregation and independent assortment. His work demonstrated the existence of paired elementary units of heredity, and revealed statistical relationships that govern their expression. He described these at a scientific meeting in 1865 and sent copies of a paper describing his work to a number of prominent scientists. At about the same time (1859), Charles Darwin published his book *On the Origin of Species by Means of Natural Selection.* Mendel's work was ignored until it was rediscovered in 1900; Darwin's was widely known and vigorously debated. The third event occurred in 1869 when Johann Miescher isolated a material he called *nuclein* from the nuclei of white blood cells harvested from the pus of surgical bandages. Miescher's nuclein contained both a protein and an acidic, phosphorus-rich substance that, when eventually separated from the protein, was given the name *nucleic acid.*

After 1900, genetic research—but not research on nucleic acids—blossomed. Nucleic acids were difficult to work with, hard to purify, and, even though they were present in all cells, did not seem to be very interesting. Early analyses, later shown to be incorrect, were interpreted to mean that nucleic acids were polymers consisting of repeats of some sequence of adenine (A), thymine (T), guanine (G), and cytosine (C) in a 1:1:1:1 ratio. Nucleic acids didn't seem to offer a rich enough alphabet from which to build a genetic dictionary. Most workers in the field believed proteins to be better candidates.

More attention began to be paid to nucleic acids in 1945 when Oswald Avery of the Rockefeller Institute for Medical Research found that he could cause a nonvirulent strain of a bacterium (*Streptococcus pneumoniae*) to produce virulent offspring by incubating them with a substance isolated from a virulent strain. What was especially important was that this virulence was passed on to succeeding generations and could only result from a permanent change in the genetic makeup—what we now call the **genome**—of the bacterium. Avery established that the substance responsible was DNA and in a letter to his brother speculated that it "may be a gene."

The Mendel Medal, awarded by Villanova University.
©Villanova University

Avery's paper prompted other biochemists to rethink their ideas about DNA. One of them, Erwin Chargaff of Columbia University, soon discovered that the distribution of adenine, thymine, cytosine, and guanine differed from species to species, but was the same within a species and within all the cells of a species. Perhaps DNA did have the capacity to carry genetic information after all. Chargaff also found that regardless of the source of the DNA, half the bases were purines and the other half were pyrimidines. Significantly, the ratio of the purine adenine (A) to the pyrimidine thymine (T) was always close to 1:1. Likewise, the ratio of the purine guanine (G) to the pyrimidine cytosine (C) was also close to 1:1. For human DNA the values are:

Purine	Pyrimidine	Base ratio
Adenine (A) 30.3%	Thymine (T) 30.3%	A/T = 1.00
Guanine (G) 19.5%	Cytosine (C) 19.9%	G/C = 0.98
Total purines 49.8%	Total pyrimidines 50.2%	

Problem 26.10

Estimate the guanine content in turtle DNA if adenine = 28.7% and cytosine = 21.3%.

Avery's studies shed light on the *function* of DNA. Chargaff's touched on *structure* in that knowing the distribution of A, T, G, and C in DNA is analogous to knowing the amino acid composition of a protein, but not its sequence or three-dimensional shape.

The breakthrough came in 1953 when James D. Watson and Francis H. C. Crick proposed a structure for DNA. The Watson–Crick proposal ranks as one of the most important in all of science and has spurred a revolution in our understanding of genetics. The structure of DNA is detailed in the next section. The boxed essay *"It Has Not Escaped Our Notice . . ."* describes how it came about.

Watson and Crick shared the 1962 Nobel Prize in Physiology or Medicine with Maurice Wilkins who, with Rosalind Franklin, was responsible for the X-ray crystallographic work.

26.9 Secondary Structure of DNA: The Double Helix

Watson and Crick relied on molecular modeling to guide their thinking about the structure of DNA. Because X-ray crystallographic evidence suggested that DNA was composed of two polynucleotide chains running in opposite directions, they focused on the forces holding

"It Has Not Escaped Our Notice . . ."

Our text began with an application of physics to chemistry when we described the electronic structure of atoms. We saw then that Erwin Schrödinger's introduction of wave mechanics figured prominently in developing the theories that form the basis for our present understanding. As we near the end of our text, we see applications of chemistry to areas of biology that are fundamental to life itself. Remarkably, Schrödinger appears again, albeit less directly. His 1944 book *What Is Life?* made the case for studying genes, their structure, and function.

Schrödinger's book inspired a number of physicists to change fields and undertake research in biology from a physics perspective. One of these was Francis Crick who, after earning an undergraduate degree in physics from University College, London, and while employed in defense work for the British government, decided that the most interesting scientific questions belonged to biology. Crick entered Cambridge University in 1949 as a 30-year-old graduate student, eventually settling on a research problem involving X-ray crystallography of proteins.

One year later, 22-year-old James Watson completed his Ph.D. studies on bacterial viruses at Indiana University and began postdoctoral research in biochemistry in Copenhagen. After a year at Copenhagen, Watson decided Cambridge was the place to be.

Thus it was that the paths of James Watson and Francis Crick crossed in the fall of 1951. One was a physicist, the other a biologist. Both were ambitious in the sense of wanting to do great things and shared a belief that the chemical structure of DNA was the most important scientific question of the time. At first, Watson and Crick talked about DNA in their spare time because each was working on another project. Soon, however, it became their major effort. Their sense of urgency grew when they learned that Linus Pauling, fresh from his proposal of helical protein structures, had turned his attention to DNA. Indeed, Watson and Crick were using the Pauling approach to structure—take what is known about the structure of small molecules, couple it to structural information about larger ones, and build molecular models consistent with the data.

At the same time, Maurice Wilkins and Rosalind Franklin at King's College, Cambridge, were beginning to obtain high-quality X-ray crystallographic data of DNA. Some of their results were presented in a seminar at King's attended by Watson, and even more were disclosed in a progress report to the Medical Research Council of the U.K. Armed with Chargaff's A = T and G = C relationships and Franklin's X-ray data, Watson and Crick began their model building. A key moment came when Jerry Donohue, a postdoctoral colleague from the United States, noticed that they were using the wrong structures for the pyrimidine and purine bases. Watson and Crick were using models of the enol forms of thymine, cytosine, and guanine, rather than the correct keto forms (recall Section 26.1). Once they fixed this error, the now-familiar model shown in Figure 26.4 emerged fairly quickly and they had the structure of DNA.

Figure 26.4

Molecular modeling—1953 style. James Watson (*left*) and Francis Crick (*right*) with their DNA model.

Watson and Crick published their work in a paper entitled "A Structure for Deoxyribose Nucleic Acid" in the British journal *Nature* on April 25, 1953. In addition to being one of the most important papers of the twentieth century, it is also remembered for one brief sentence appearing near the end.

> "It has not escaped our notice that the specific pairing we have postulated immediately suggests a possible copying mechanism for the genetic material."

True to their word, Watson and Crick followed up their April 25 paper with another on May 30. This second paper, "Genetical Implications of the Structure of Deoxyribonucleic Acid," outlines a mechanism for DNA replication that is still accepted as essentially correct.

the two chains together. Hydrogen bonding between bases seemed the most likely candidate. After exploring a number of possibilities, Watson and Crick hit on the arrangement shown in Figure 26.5 in which adenine and thymine comprise one complementary *base pair* and guanine and cytosine another. This base-pairing scheme has several desirable features.

1. Pairing A with T and G with C gives the proper Chargaff ratios (A = T and G = C).
2. Each pair contains one purine and one pyrimidine base. This makes the A---T and G---C pairs approximately the same size and ensures a consistent distance between the two DNA strands.
3. Complementarity between A and T, and G and C suggests a mechanism for copying DNA. This is called replication and is discussed in Section 26.10.

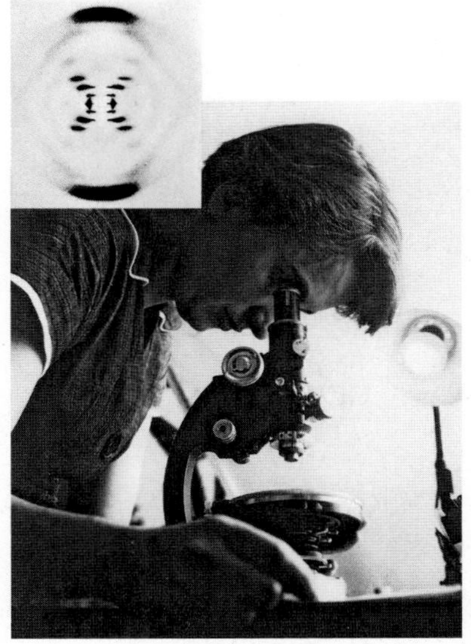

Rosalind Franklin. Her X-ray crystallographic data were used to solve the structure of DNA.
(inset) ©Science Source
©NLM/Science Source

Figure 26.6 supplements Figure 26.5 by showing portions of two DNA strands arranged side by side with the base pairs in the middle.

Hydrogen bonding between complementary bases is responsible for association between the strands, whereas conformational features of its carbohydrate–phosphate backbone and the orientation of the bases with respect to the furanose rings govern the overall shape of each strand. Using the X-ray crystallographic data available to them, Watson and Crick built a molecular model in which each strand took the shape of a right-handed helix. Joining two antiparallel strands by appropriate hydrogen bonds produced the **double helix** shown in the photograph (Figure 26.4). Figure 26.7 shows two modern renderings of DNA models.

In addition to hydrogen bonding between the two polynucleotide chains, the double-helical arrangement is stabilized by having its negatively charged phosphate groups on the outside where they are in contact with water and various cations, Na^+, Mg^{2+}, and ammonium ions, for example. Attractive van der Waals forces between the aromatic pyrimidine and purine rings, called *π-stacking,* stabilize the layered arrangement of the bases on the inside. Even though the bases are on the inside, they are accessible to other substances through two grooves that run along the axis of the double helix. They are more accessible via the *major groove,* which is almost twice as wide as the *minor groove.* The grooves differ in size because of the way the bases are tilted with respect to the furanose ring.

A helical structure for DNA strands had been suggested in 1949 by Sven Furberg in his Ph.D. dissertation at the University of London.

A — T: 2-Deoxyribose; 1080 pm (10.80 Å)

G — C: 2-Deoxyribose; 1080 pm (10.80 Å)

(*a*) (*b*)

Figure 26.5

Hydrogen bonding between DNA bases shown as structural drawings of nucleosides (*top*) and as molecular models (*bottom*) of (*a*) adenine and thymine and (*b*) guanine and cytosine.

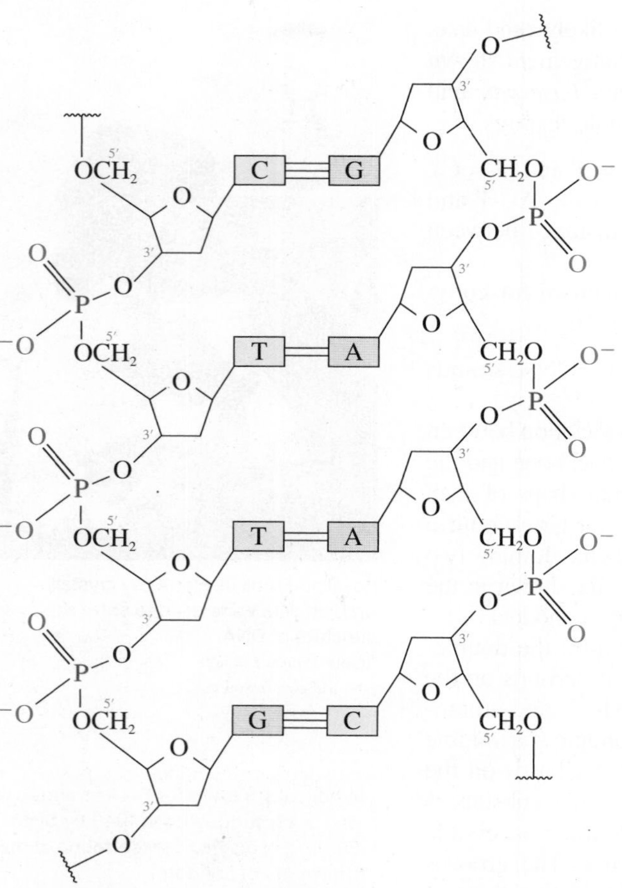

Figure 26.6

Hydrogen bonding between complementary bases (A and T, and G and C) permits pairing of two DNA strands. The strands are antiparallel; the 5′ end of the left strand is at the top, and the 5′ end of the right strand is at the bottom.

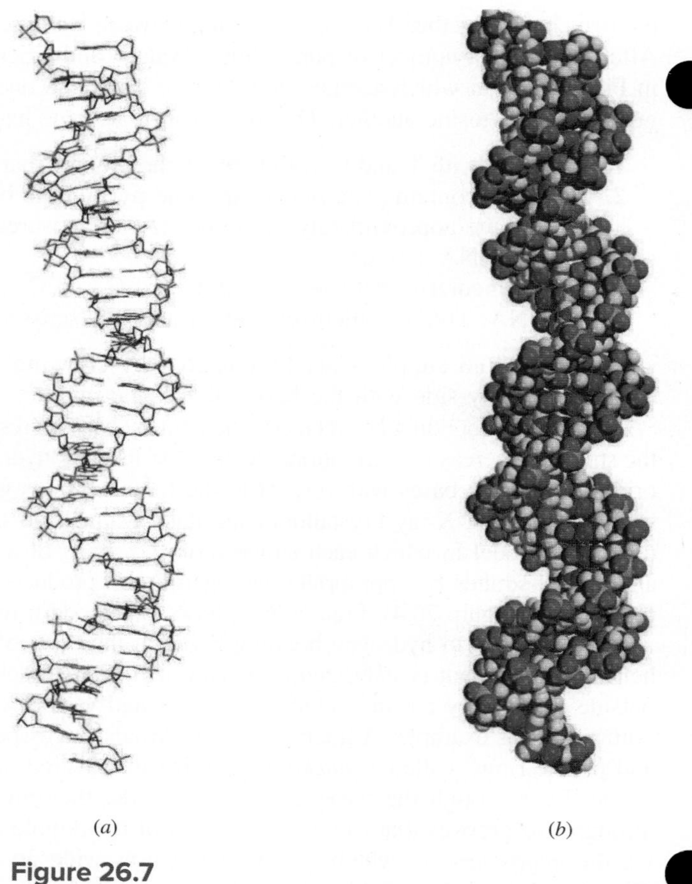

Figure 26.7

(*a*) Tube and (*b*) space-filling models of a DNA double helix. The carbohydrate–phosphate "backbone" is on the outside and can be roughly traced in (*b*) by the red oxygen atoms. The blue atoms belong to the purine and pyrimidine bases and lie on the inside. The base-pairing is more clearly seen in (*a*).

The structure proposed by Watson and Crick was modeled to fit crystallographic data obtained on a sample of the most common form of DNA called B-DNA. Other forms include A-DNA, which is similar to, but more compact than B-DNA, and Z-DNA, which is a left-handed double helix.

By analogy to the levels of structure of proteins, the **primary structure** of DNA is the sequence of bases along the polynucleotide chain, and the A-DNA, B-DNA, and Z-DNA helices are varieties of **secondary structures.**

Not all DNAs are double helices (*duplex DNA*). Some types of viral DNA are single-stranded, and even a few triple and quadruple DNA helices are known.

26.10 Replication of DNA

Every time a cell divides, its DNA is duplicated so that the DNA in the new cell is identical to that in the original one. As Figure 26.8 shows, Watson–Crick base-pairing provides the key to understanding this process of DNA **replication.** During cell division the DNA double helix begins to unwind, generating a **replication fork** separating the two strands. Each strand serves as a template on which a new DNA strand is constructed. The A—T, G—C base-pairing requirement ensures that each new strand is the precise complement of its template strand. Each of the two new duplex DNA molecules contains one original and one new strand.

1. The DNA to be copied is a double helix, shown here as flat for clarity.

2. The two strands begin to unwind. Each strand will become a template for construction of its complement.

3. As the strands unwind, the pyrimidine and purine bases become exposed. Notice that the bases are exposed in the $3' \rightarrow 5'$ direction in one strand, and in the $5' \rightarrow 3'$ direction in the other.

4. Two new strands form as nucleotides that are complementary to those of the original strands are joined by phosphodiester linkages. The sources of the new bases are dATP, dGTP, dCTP, and dTTP already present in the cell.

5. Because nucleotides are added in the $5' \rightarrow 3'$ direction, the processes by which the two new chains grow are different. Chain growth can be continuous in the leading strand, but not in the lagging strand.

6. Two duplex DNA molecules result, each of which is identical to the original DNA.

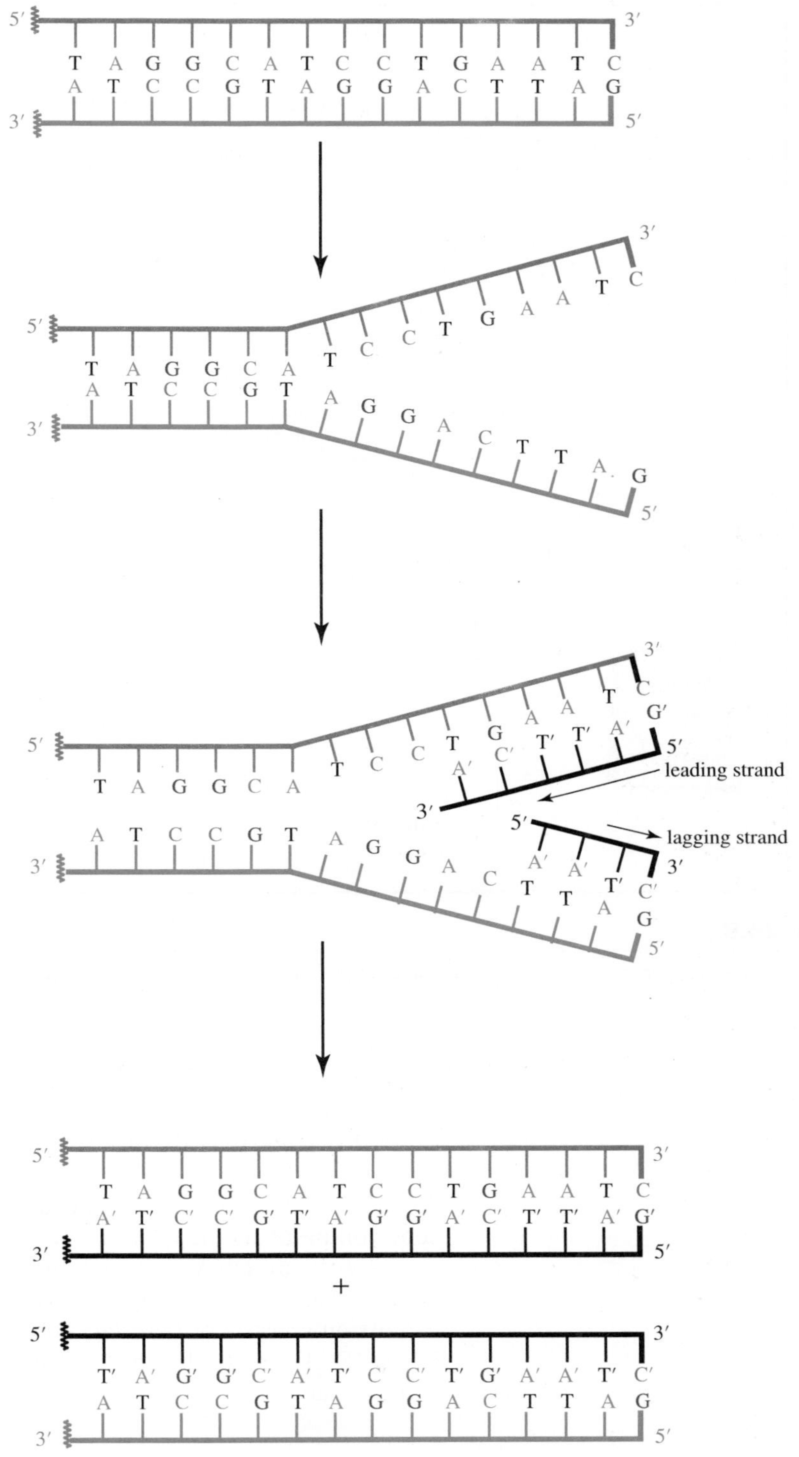

Figure 26.8

Outline of DNA replication. The original strands are shown in red and blue and are the templates from which the new strands, shown in black, are copied.

Both new chains grow in their $5' \rightarrow 3'$ direction. Because of this, one grows toward the replication fork (the **leading strand**) and the other away from it (the **lagging strand**), making the details of chain extension somewhat different for the two. The fundamental chemistry, however, is straightforward (Figure 26.9). The hydroxyl group at the 3′ end of the growing polynucleotide chain acts as a nucleophile, attacking the 5′-triphosphate

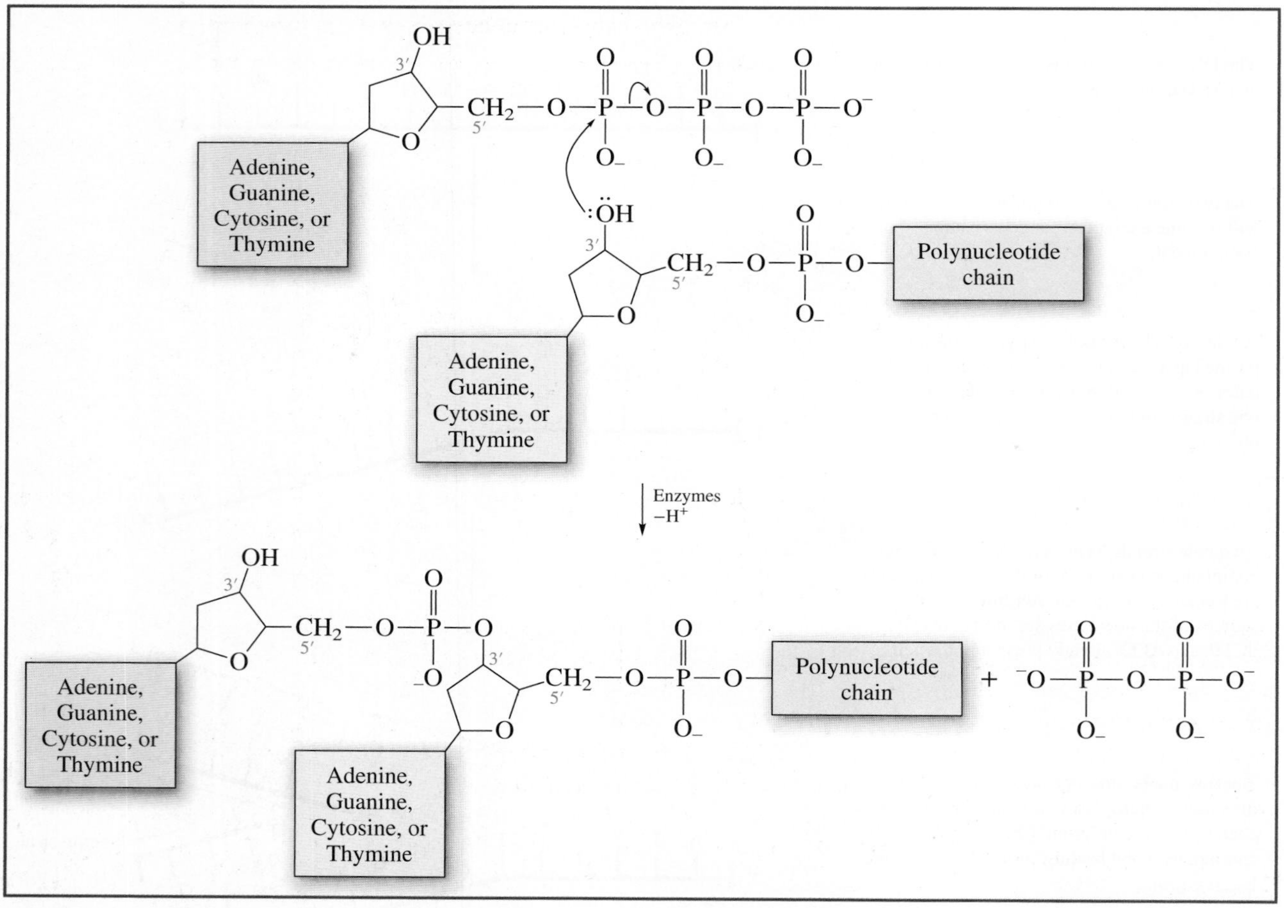

Figure 26.9

The new polynucleotide chain grows by reaction of its free 3′-OH group with the 5′-triphosphate of an appropriate 2′-deoxyribonucleoside.

of 2′-deoxyadenosine, 2′-deoxyguanosine, 2′-deoxycytidine, or thymidine to form the new phosphodiester linkage. The enzyme that catalyzes phosphodiester bond formation is called *DNA polymerase;* different DNA polymerases operate on the leading strand and the lagging strand.

All of the steps, from the unwinding of the original DNA double helix to the supercoiling of the new DNAs, are catalyzed by enzymes.

Genes are DNA and carry the inheritable characteristics of an organism and these characteristics are normally *expressed* at the molecular level via protein synthesis. Gene expression consists of two stages, **transcription** and **translation,** both of which involve RNAs. Sections 26.11 and 26.12 describe these RNAs and their roles in transcription and translation.

26.11 Ribonucleic Acids

The flow of genetic information in an organism starts with DNA, which contains the code for reproducing itself and for the proteins that are required for the life form to function. RNA serves as the conduit through which this information is moved from the nucleus to the cytoplasm, in which the information from DNA is "transferred" (translated) into the RNA sequence, which passes out of the nucleus and is then "read" (transcribed) in the cytosol into the ultimate product, the protein.

The "central dogma" of molecular biology, expressed originally by Francis Crick, is that the flow of genetic information terminates in the protein. That is, DNA and RNA

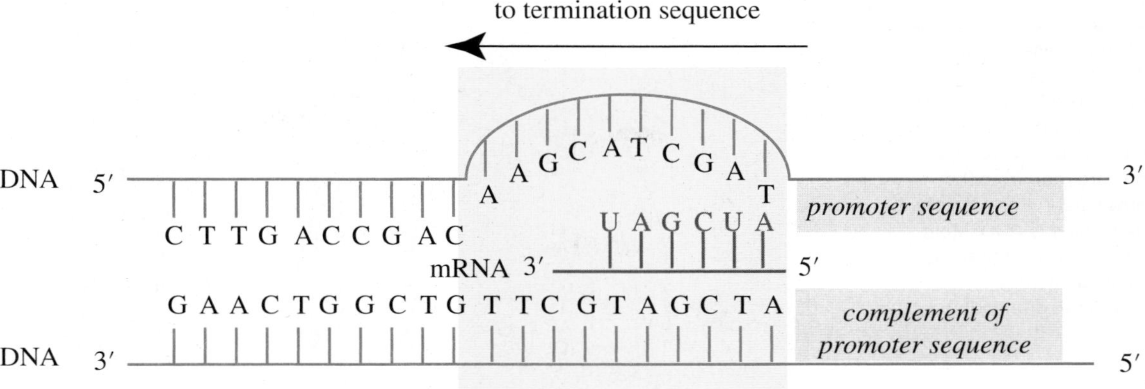

Figure 26.10

During transcription a molecule of mRNA is assembled from a DNA template. Transcription begins at a promoter sequence and proceeds in the 5′→3′ direction of the mRNA until a termination sequence of the DNA is reached. Only a region of about 10 base pairs is unwound at any time.

can both send and receive genetic information. Proteins can only receive genetic information; they cannot replicate or transfer genetic instructions back to nucleic acid polymers. Since that time scientists have discovered that DNA and RNA play more complicated roles than originally envisioned, but the central dogma still holds.

The cellular roles of RNA were originally thought to be restricted to its three major forms, all of which are involved in the direct transfer of genetic information from DNA to protein. These three forms are:

1. Messenger RNA (*mRNA*)
2. Transfer RNA (*tRNA*)
3. Ribosomal RNA (*rRNA*)

Messenger RNA (mRNA): Messenger RNA is synthesized in the cell nucleus during the process of transcription, which is illustrated in Figure 26.10. Transcription resembles DNA replication in that a DNA strand serves as the template for construction of, in this case, a ribonucleic acid. mRNA synthesis begins at its 5′ end, and ribonucleotides complementary to the DNA strand being copied are added. The phosphodiester linkages are formed by reaction of the free 3′-OH group of the growing mRNA with ATP, GTP, CTP, or UTP (recall that uracil, not thymine, is the complement of adenine in RNA). The enzyme that catalyzes this reaction is *RNA polymerase.* Only a small section of about 10 base pairs of the DNA template is exposed at a time. As the synthesis zone moves down the DNA chain, restoration of hydrogen bonds between the two original DNA strands displaces the newly synthesized single-stranded mRNA. The entire DNA molecule is not transcribed as a single mRNA. Transcription begins at a prescribed sequence of bases (the *promoter sequence*) and ends at a *termination sequence.* Thus, one DNA molecule can give rise to many different mRNAs and code for many different proteins. There are thousands of mRNAs and they vary in length from about 500 to 6000 nucleotides.

The **genetic code** (Figure 26.11) is the message carried *by* mRNA. It is made up of triplets of adjacent nucleotide bases called **codons.** Because mRNA has only four different bases and 20 amino acids must be coded for, codes using either one or two nucleotides per amino acid are inadequate. If nucleotides are read in sets of three, however, the four mRNA bases generate 64 possible "words," more than sufficient to code for 20 amino acids.

In addition to codons for amino acids, there are *start* and *stop* codons. Protein biosynthesis begins at a start codon and ends at a stop codon of mRNA. The start codon is the nucleotide triplet AUG, which is also the codon for methionine. The stop codons are UAA, UAG, and UGA. UAG and UGA can also code for pyrrolysine and selenocysteine, respectively. How these two "ambiguous" codons are read depends on the presence of specific genes.

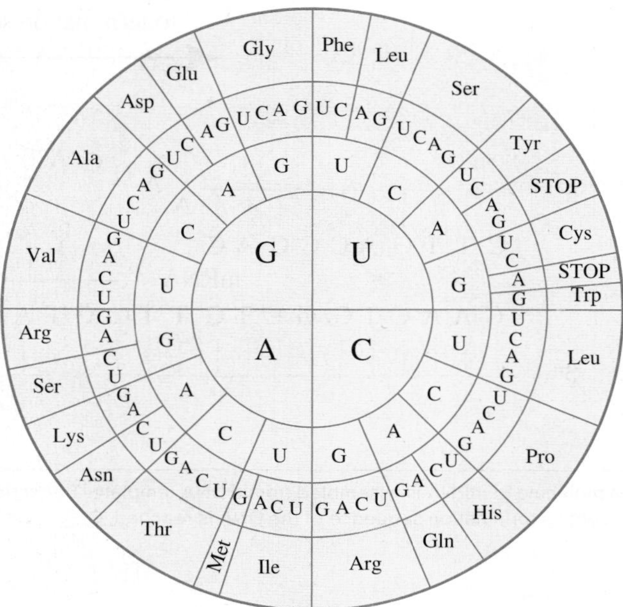

Figure 26.11

The codon wheel can be used to decode a messenger RNA sequence to a peptide sequence. Starting at the 5′-end of the gene, the first position in the codon is found in the center circle. The second and third positions are read from the next two concentric circles. For example, the codon for methionine is AUG. Most of the codons are *redundant,* meaning that an individual amino acid is designated by more than one codon.

Problem 26.11

Write the sequence of the peptide that is coded for by the DNA sequence TACGAAGGACTAAGAATC.

Transfer RNA (tRNA): Transfer RNAs are relatively small nucleic acids, containing only about 70 nucleotides. They get their name because they transfer amino acids to the ribosome for incorporation into a polypeptide. Although 20 amino acids need to be transferred, there are 50–60 tRNAs, some of which transfer the same amino acids. Figure 26.12 shows the structure of phenylalanine tRNA ($tRNA^{Phe}$). Like all tRNAs it is composed of a single strand, with a characteristic shape that results from the presence of paired bases in some regions and their absence in others.

The 1968 Nobel Prize in Physiology or Medicine was shared by Robert W. Holley of Cornell University for determining the nucleotide sequence of phenylalanine transfer RNA.

Among the 76 nucleotides of $tRNA^{Phe}$ are two sets of three that are especially important. The first is a group of three bases called the **anticodon,** which is complementary to the mRNA codon for the amino acid being transferred. Figure 26.11 shows two mRNA codons for phenylalanine, UUU and UUC (reading in the 5′→3′ direction). Because base-pairing requires the mRNA and tRNA to be antiparallel, the two anticodons are read in the 3′→5′ direction as AAA and AAG.

3′ —A—A—G— 5′ tRNA anticodon

5′ —U—U—C— 3′ mRNA codon

The other important sequence is the CCA triplet at the 3′ end. The amino acid that is to be transferred is attached through an ester linkage to the terminal 3′-oxygen of this sequence. *All tRNAs have a CCA sequence at their 3′ end.*

Transfer RNAs normally contain some bases other than A, U, G, and C. Of the 76 bases in $tRNA^{Phe}$, for example, 13 are of the modified variety. One of these, marked G* in Figure 26.12*a*, is a modified guanosine in the anticodon. Many of the modified bases, including G*, are methylated derivatives of the customary RNA bases.

Ribosomal RNA (rRNA): Ribosomes, which are about two-thirds nucleic acid and about one-third protein, constitute about 90% of a cell's RNA. A ribosome is made up of two subunits, referred to as the 60S and 40S subunits in eukaryotes. The larger 60S subunit contains 57 proteins and three strands of RNA: 28S (3354 nucleotides), 5S

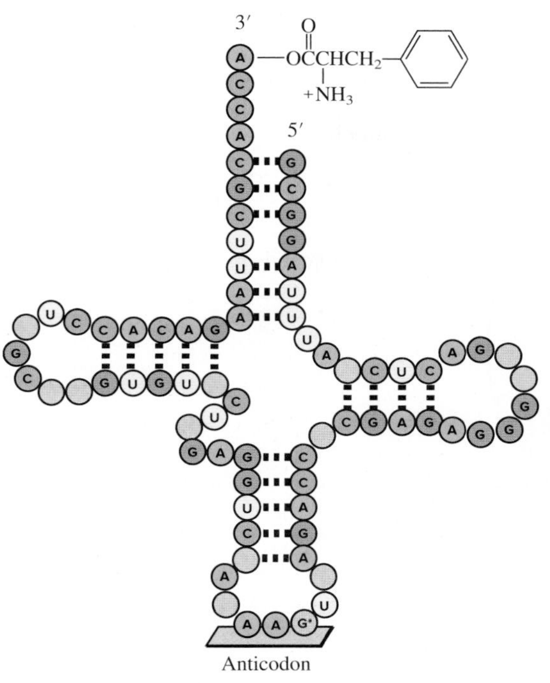

(*a*)

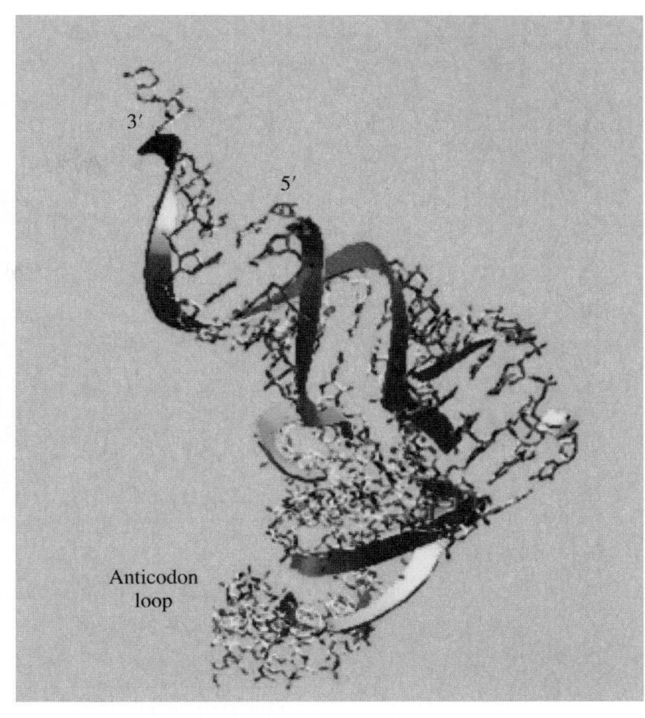

(*b*)

Figure 26.12

Phenylalanine tRNA from yeast. (*a*) A schematic drawing showing the sequence of bases. Transfer RNAs usually contain a number of modified bases (gray circles). One of these is a modified guanosine (G*) in the anticodon. Hydrogen bonds, where present, are shown as dashed lines. (*b*) The structure of yeast $tRNA^{Phe}$ as determined by X-ray crystallography.

(120 nucleotides), and 5.8S (154 nucleotides). The smaller 40S subunit contains 32 proteins and one RNA polymer (18S, 1753 nucleotides). The ribosome is where the message carried by the mRNA is translated into the amino acid sequence of a protein. How it occurs is described in Section 26.12.

An indication of the potential versatility of RNA was the discovery that RNA itself can possess catalytic activity. Until that time, the known catalytic reactions in living systems were performed solely by proteins. In 1982, Thomas Cech and coworkers reported that a strand of RNA from *Tetrahymena thermophila* was capable of breaking and reforming its own phosphodiester bond in the absence of protein. Contemporarily, Sidney Altman of Yale University discovered that the RNA portion of the enzyme RNAse P was sufficient to perform the ribonucleoprotein's catalytic activity in the absence of protein. We now know that RNA catalysts, called **ribozymes,** are found in all kingdoms. One of the functions of RNA in the ribosome is catalytic: an adenine from RNA in the 60S subunit catalyzes the formation of peptide bonds (Figure 26.13).

For their studies on the structure and mode of action of ribosomes, Venkatraman Ramakrishnan (Medical Research Council, U.K.), Thomas Steitz (Yale University, U.S.), and Ada Yonath (Weizmann Institute, Israel) were awarded the 2009 Nobel Prize in Chemistry.

Other Types and Functions of RNA: An important new activity of intracellular RNA emerged from what appeared to be a failed experiment. In the mid 1980s, Richard Jorgensen, a plant scientist at a biotech startup company, was attempting to create a deeper purple color in a petunia. He reasoned that adding an extra gene for the color would produce the dark color. Surprisingly, the genetically modified petunias were white instead of purple. Instead of enhancing the color, the second gene "silenced" the plant's original gene. Observations such as these were explained in 1998. Craig Mello of University of Massachusetts Medical School and Andrew Fire of Stanford University reported that injecting double-stranded RNA into a worm silenced genes complementary to the RNA. It is now known that *RNA interference* (*RNAi*) is a naturally occurring phenomenon for controlling gene expression, and Mello and Fire were awarded the 2006 Nobel Prize in Physiology or Medicine for their discovery.

Sidney Altman (Yale University) and Thomas Cech (University of Colorado) shared the 1989 Nobel Prize in Chemistry for showing that RNAs could function as biological catalysts.

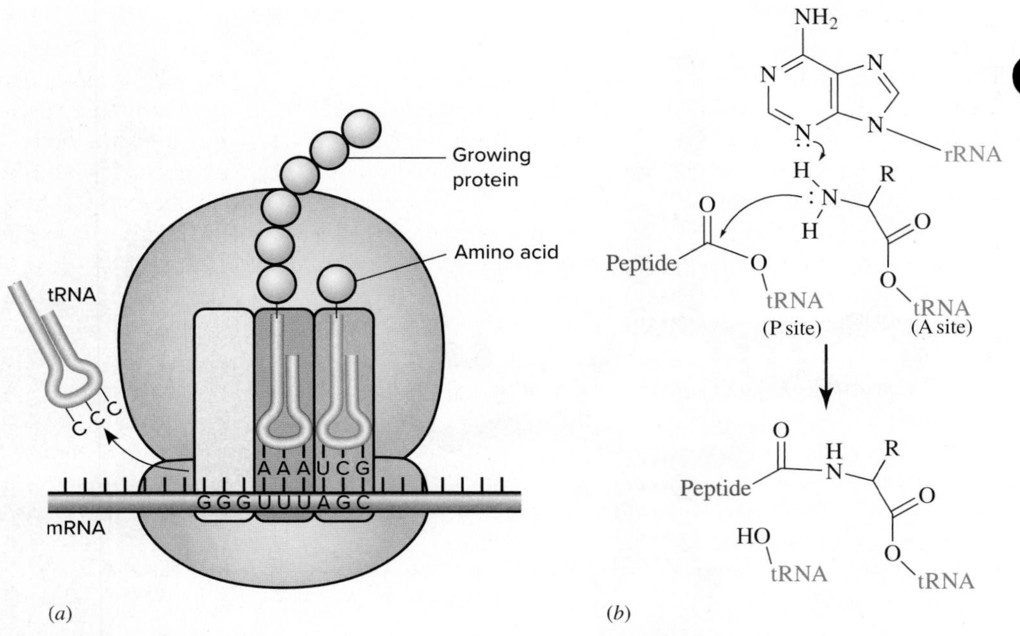

Figure 26.13

Translation of mRNA into an amino acid sequence of a protein involves three types of RNA as well as multiple proteins. (*a*) The ribosome brings together the mRNA and the complementary tRNAs and catalyzes the peptide-bond-forming reaction. (*b*) The growing peptide chain is attached to the tRNA in the P (peptidyl) site. The next amino acid in the polypeptide is attached to a tRNA in the A (aminoacyl) site. Formation of the peptide bond involves a catalytic adenine from rRNA. The process converts an ester to an amide.

New discoveries of RNA structures and functions are occurring at a rapid pace. MicroRNA (miRNA), short hairpin RNA (shRNA), and small interfering RNA (siRNA) are known to be involved in gene regulation in most eukaryotes. Long noncoding RNA (lncRNA) has a wide range of functions, many of which are not yet understood. RNA is more pervasive and versatile in cellular processes than simply serving protein biosynthesis.

26.12 Protein Biosynthesis

As described in the preceding sections, protein synthesis involves transcription of the DNA to mRNA, followed by translation of the mRNA as an amino acid sequence. In addition to outlining the mechanics of transcription, we have described the relationship among mRNA codons, tRNA anticodons, and amino acids.

During translation the protein is synthesized beginning at its N terminus (Figure 26.13). The mRNA is read in its $5' \rightarrow 3'$ direction beginning at the start codon AUG and ending at a stop codon (UAA, UAG, or UGA). Because the start codon is always AUG, the N-terminal amino acid is always methionine (as its *N*-formyl derivative). However, this *N*-formylmethionine residue is normally lost in a subsequent process and the N terminus of the expressed protein is therefore determined by the second mRNA codon. The portion of the mRNA between the start and stop codons is called the coding sequence and is flanked on either side by noncoding regions.

In addition to illustrating the mechanics of translation, Figure 26.13 is important in that it shows the mechanism of peptide bond formation as a nucleophilic acyl substitution. Both methionine and alanine are attached to their respective tRNAs as esters. In a reaction catalyzed by adenine in rRNA, the amino group of new amino acid attacks the ester bond between the growing peptide chain and its tRNA. The process displaces the growing peptide from its tRNA and converts the ester linkage to an amide, forming the peptide bond.

26.13 AIDS

The explosive growth of our knowledge of nucleic acid chemistry and its role in molecular biology in the 1980s coincided with the emergence of AIDS (acquired immune deficiency syndrome) as a major public health threat. In AIDS, a virus devastates the body's defenses to the extent that its victims can die from infections that are normally held in check by a healthy immune system. In the time since its discovery in the early 1980s, AIDS has claimed the lives of over 35 million people, and current estimates place the number of those infected at nearly 78 million. According to the World Health Organization (WHO), AIDS is the fifth leading cause of death in low-income countries. AIDS no longer ranks in the top ten causes of death in high-income countries.

The viruses responsible for AIDS are human immunodeficiency virus 1 and 2 (HIV-1 and HIV-2). Both are **retroviruses,** meaning that their genetic material is RNA rather than DNA. HIVs require a host cell to reproduce, and the hosts in humans are the T4 lymphocytes, which are the cells primarily responsible for inducing the immune system to respond when provoked. The HIV penetrates the cell wall of a T4 lymphocyte and deposits both its RNA and an enzyme called *reverse transcriptase* inside. There, the reverse transcriptase catalyzes the formation of a DNA strand that is complementary to the viral RNA. The transcribed DNA then serves as the template from which the host lymphocyte produces copies of the virus, which then leave the host to infect other T4 cells. In the course of HIV reproduction, the ability of the T4 lymphocyte to reproduce itself is compromised. As the number of T4 cells decreases, so does the body's ability to combat infections.

Problem 26.12

When the RNA of a retrovirus is transcribed, what DNA base is the complement of the uracil in the viral RNA?

Education and effective antiviral medications have decreased AIDS-related deaths by about 50% since the peak in 2005. The first advance in treatment came with drugs such as the nucleoside *zidovudine,* also known as azidothymine, or AZT. During reverse transcription, AZT replaces thymidine in the DNA being copied from the viral RNA. AZT has a 5′-OH group, so can be incorporated into a growing polynucleotide chain. But because it lacks a 3′-OH group, the chain cannot be extended beyond it and synthesis of the viral DNA stops before the chain is complete. Other nucleosides such as 2′,3′-dideoxyinosine (ddI) also block the action of reverse transcriptase.

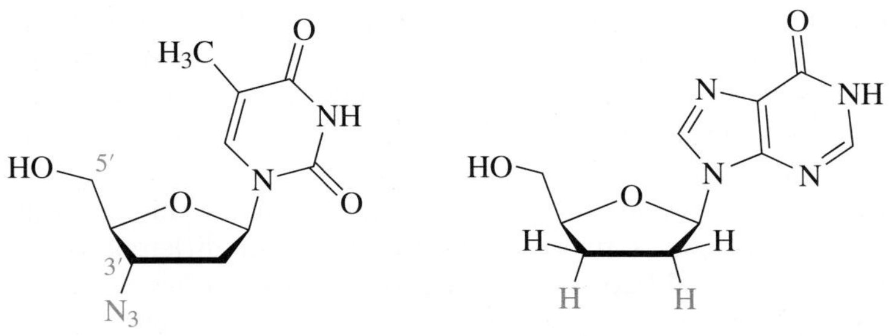

Zidovudine (AZT)

2′,3′-Dideoxyinosine (ddI)

New targets for anti-HIV therapy have been identified and effective drugs against these targets are in clinical use. *Protease inhibitors* block an enzyme necessary for producing the enzymes required for HIV reproduction. *Integrase inhibitors* block the enzyme necessary for insertion of viral DNA into the DNA of the host. Modern AIDS therapy uses combination therapy, a regimen of anti-HIV drugs that includes drugs with at least two different mechanisms of action. Although none of the treatments are considered to be a cure, they can slow or halt the progression of the disease for many years.

Reverse transcriptase inhibitors are also used against certain viruses that, although they are not retroviruses, do require reverse transcriptase to reproduce. The virus that causes hepatitis B is an example.

26.14 DNA Sequencing

Once the Watson–Crick structure was proposed, determining the nucleotide sequence of DNA emerged as an important area of research. Some difficulties were apparent from the beginning, especially if one draws comparisons to protein sequencing. First, most DNAs are much larger biopolymers than proteins. Not only does it take three nucleotides to code for a single amino acid, but vast regions of DNA don't seem to code for anything at all. A less obvious problem is that the DNA alphabet contains only four letters (A, G, C, and T) compared with the 20 amino acids from which proteins are built. Recall too that protein sequencing benefits from having proteases available that cleave the chain at specific amino acids. Not only are there no enzymes that cleave nucleic acids at specific bases but, with only four bases to work with, the resulting fragments would be too small to give useful information. In spite of this, DNA sequencing not only developed very quickly, but also has turned out to be much easier to do than protein sequencing.

To explain how DNA sequencing works, we must first mention **restriction enzymes.** Like all organisms, bacteria are subject to infection by external invaders (e.g.,viruses and other bacteria) and possess defenses in the form of restriction enzymes that destroy the invader by cleaving its DNA. Over 3000 restriction enzymes are known, and hundreds are readily available. Unlike proteases, which recognize a single amino acid, restriction enzymes recognize specific nucleotide *sequences.* Cleavage of the DNA at prescribed sequences gives fragments 100–200 base pairs in length, which are separated, purified, and sequenced independently.

Each sample is used as a template to create complements of itself and placed in a tube containing the materials necessary for DNA synthesis. These materials include the four nucleosides present in DNA, 2′-deoxyadenosine (dA), 2′-deoxythymidine (dT), 2′-deoxyguanosine (dG), and 2′-deoxycytidine (dC), as their triphosphates dATP, dTTP, dGTP, and dCTP. Also present are small amounts of synthetic analogs of ATP, TTP, GTP, and CTP that had been modified in two ways. Their 2′- and 3′-hydroxyl groups have been replaced by hydrogens giving the dideoxy nucleotides 2′,3′-dideoxyadenosine triphosphate (ddATP), 2′,3′-dideoxythymidine triphosphate (ddTTP), 2′,3′-dideoxyguanosine triphosphate (ddGTP), and 2′,3′-dideoxycytidine triphosphate (ddCTP).

Sanger, who had already won a Nobel Prize in 1958 for protein sequencing, shared the 1980 chemistry prize with Walter Gilbert of Harvard University and Paul Berg of Stanford University. Gilbert developed a chemical method for DNA sequencing and Berg was responsible for many important techniques in the study of nucleic acids.

Because their furanose rings lack a 3′ hydroxyl group, incorporation of ddATP, ddTTP, ddGTP, or ddCTP into a growing strand of DNA terminates chain growth (Figure 26.14). In its original formulation by Frederick Sanger the method relied on priming DNA synthesis with ^{32}P-labeled dideoxynucleotides, which were detected by a radioactivity tracing method. Sanger's method has been superceded by using synthetic dideoxy nucleotides identified according to which of four different fluorescent dyes is attached to their purine or pyrimidine base.

As DNA synthesis proceeds, nucleotides from the solution are added to the growing polynucleotide chain. Chain extension takes place without complication as long as the incorporated nucleotides are derived from dATP, dTTP, dGTP, and dCTP. If, however, the incorporated species is derived from a dideoxy analog, chain extension stops. Thus, the sample contains a mixture of DNA fragments of different length, *each of which terminates in a dideoxy nucleotide.*

After separation by electrophoresis, which discriminates among fragments according to their length, the gel is read by irradiation at four different wavelengths. One wavelength causes the modified ddA-containing polynucleotides to fluoresce, another causes modified ddT fluorescence, and so on. The data are stored and analyzed in a computer and printed out as the DNA sequence (Figure 26.14).

In addition to sequencing bits of DNA or individual genes, DNA sequencing has become so powerful a technique that the entire genomes of tens of thousands of organisms have been sequenced. The first and largest number of these organisms were viruses—organisms with relatively small genomes. Then came a bacterium with 1.8 million base pairs, then baker's yeast with 12 million, followed by a roundworm with 97 million. The year 2000 brought announcements of the sequences of the 100-million-base-pair genome of the wild mustard plant and the 180-million-base-pair genome of the fruit fly. On the horizon was the 3-billion-base-pair human genome.

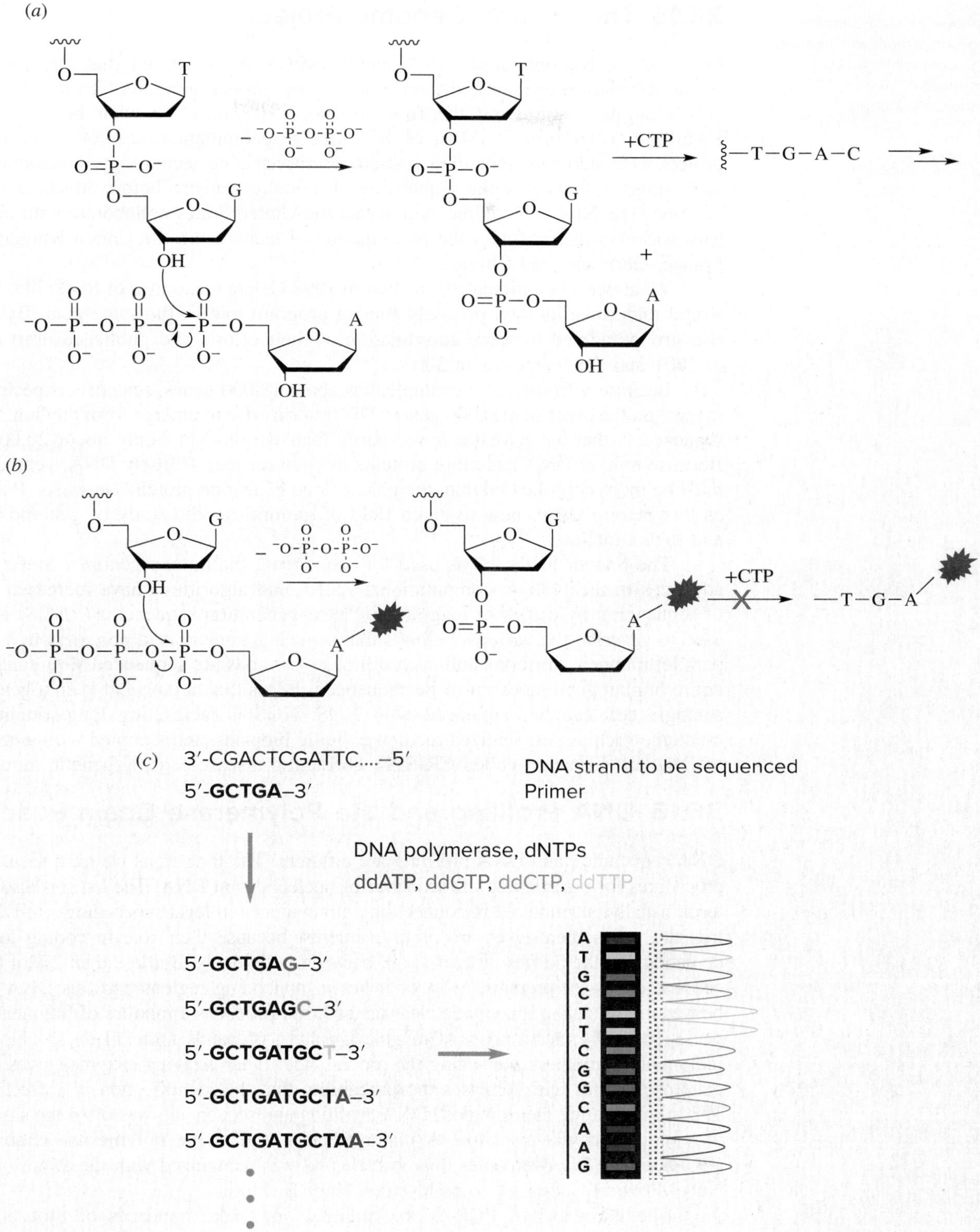

(a)

+CTP

—T—G—A—C

(b)

+CTP

—T—G—A

(c) 3′-CGACTCGATTC....–5′ DNA strand to be sequenced
5′-**GCTGA**–3′ Primer

DNA polymerase, dNTPs
ddATP, ddGTP, ddCTP, ddTTP

5′-**GCTGAG**–3′
5′-**GCTGAGC**–3′
5′-**GCTGATGCT**–3′
5′-**GCTGATGCTA**–3′
5′-**GCTGATGCTAA**–3′

A T G C T T C G G C A A G

Figure 26.14

DNA sequencing. (*a*) DNA synthesis proceeds in the 5′-to-3′ direction, in which the 3′-OH of the growing strand forms a phosphodiester bond with the next nucleotide triphosphate (dNTP), releasing pyrophosphate. The process continues as long as sufficient dNTP is present. (*b*) Adding a dideoxynucleotide triphosphate (ddNTP), which lacks the 3′-OH, terminates synthesis of the complementary strand. If a fluorophore is attached to the ddNTP, the truncated strand will be fluorescent. (*c*) The DNA fragment to be synthesized is annealed to a complementary primer. DNA synthesis proceeds in the presence of dNTPs and ddNTPs, each of which has a different color fluorophore attached. This process generates DNA strands that are different by one base and that contain one fluorophore at the 3′-end of the polymer. The strands are separated by electrophoresis and the identity of the 3′-nucleotide is shown by the fluorophore. An automated DNA sequencer translates this information into the sequence of the desired DNA.

The International Human Genome Sequencing Consortium was headed by Francis S. Collins of the U.S. National Institutes of Health. J. Craig Venter led the Celera effort.

26.15 The Human Genome Project

In 1988, the National Research Council (NRC) recommended that the United States mount a program to map and then sequence the human genome. Shortly thereafter, the U.S. Congress authorized the first allocation of funds for what became a 15-year $3-billion-dollar project. Most of the NRC's recommendations for carrying out the project were adopted, including a strategy emphasizing technology development in the early stages followed by the sequencing of model organisms before attacking the human genome. The NRC's recommendation that the United States collaborate with other countries was also realized with the participation of teams from the United Kingdom, Japan, France, Germany, and China.

What was not anticipated was that in 1998 Celera Genomics of Rockville, Maryland, would undertake its own privately funded program toward the same goal. By 2000, the two groups agreed to some coordination of their efforts and published draft sequences in 2001 and final versions in 2003.

Because a fruit fly, for example, has about 13,000 genes, scientists expected humans to have on the order of 100,000 genes. The first surprise to emerge from the human genome sequence is that we have far fewer genes than we thought—only about 20,000–25,000. Because human DNA has more proteins to code for than fruit-fly DNA, gene expression must be more complicated than the phrase "one gene–one protein" suggests. Puzzles such as this belong to the new research field of **genomics**—the study of genome sequences and their function.

The Sanger method was used for sequencing the human genome. Since that time, advances in technology, computational speed, and algorithms have increased the speed of sequencing by orders of magnitude. "Next-generation sequencing" (NGS) is the term used to describe the various methods that sequence nucleic acid polymers in a massively parallel manner—millions and even billions of strands are sequenced simultaneously. An entire human genome can now be sequenced in less than a day, and even a genome from a single cell can be sequenced with NGS. NGS is facilitating long-sought goals in medicine such as personalized medicine, liquid biopsies, and, coupled with a new method of DNA modification called CRISPR, correcting disease-causing genetic mutations.

26.16 DNA Profiling and the Polymerase Chain Reaction

DNA sequencing and DNA profiling are different. The former, as we have seen, applies to procedures for determining the sequence of nucleotides in DNA. The latter is also a familiar term, usually encountered in connection with evidence in legal proceedings. In DNA profiling, the genes themselves are of little interest because their role in coding for proteins demands that they differ little, if at all, between individuals. But less than 2% of the human genome codes for proteins. Most of it lies in noncoding regions and this DNA does vary between individuals. Enzymatic cleavage of DNA produces a mixture of fragments that can be separated by electrophoresis to give a pattern of bands more likely to belong to one individual than others. Repeating the process with other cleaving enzymes gives a different pattern of bonds and increases the probability that the identification is correct. Until the 1980s, the limiting factor in both DNA profiling and sequencing was often the small amount of sample that was available. A major advance, called the **polymerase chain reaction (PCR),** effectively overcomes this obstacle and was recognized with the award of the 1993 Nobel Prize in Chemistry to its inventor, Kary B. Mullis.

The main use of PCR is to "amplify," or make hundreds of thousands—even millions—of copies of a portion of the polynucleotide sequence in a sample of DNA. Suppose, for example, we wish to copy a 500-base-pair region of a sample of DNA that contains a total of 1 million base pairs. We would begin as described in Section 26.14 by cleaving the DNA into smaller fragments using restriction enzymes, then use PCR to make copies of the desired fragment.

Figure 26.15 illustrates how PCR works. In general, it involves multiple cycles of a three-step sequence. In working through Figure 26.15, be alert to the fact that the material we want does not arise until after the third cycle. After that, its contribution to the mixture

(*a*) Consider double-stranded DNA containing a polynucleotide sequence (the **target region**) that you wish to amplify (make millions of copies of).

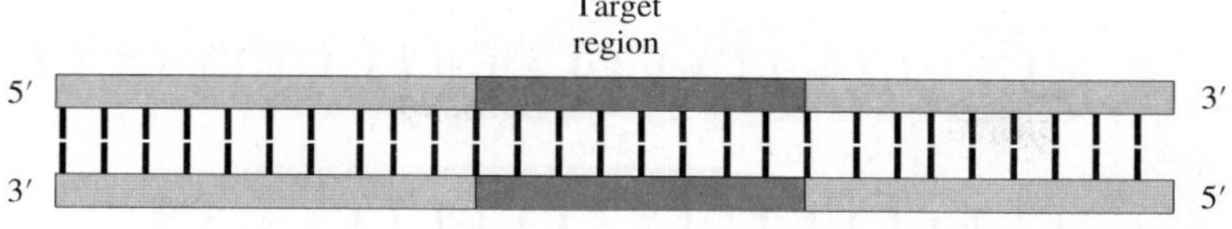

(*b*) Heating the DNA to ≈95°C causes the strands to separate. This is the denaturation step.

(*c*) Cooling the sample to ≈60°C causes one primer oligonucleotide to bind to one strand and the other primer to the other strand. This is the annealing step.

(*d*) In the presence of the four DNA nucleotides and the enzyme DNA polymerase, the primer is extended in its 3′ direction as it adds nucleotides that are complementary to the original DNA strand. This is the synthesis step and is carried out at ≈72°C.

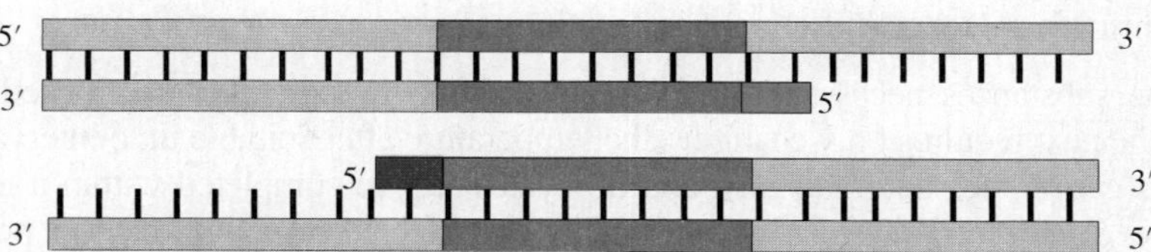

(*e*) Steps (*a*)–(*d*) constitute one cycle of the polymerase chain reaction and produce two double-stranded DNA molecules from one. Denaturing the two DNAs and priming the four strands gives:

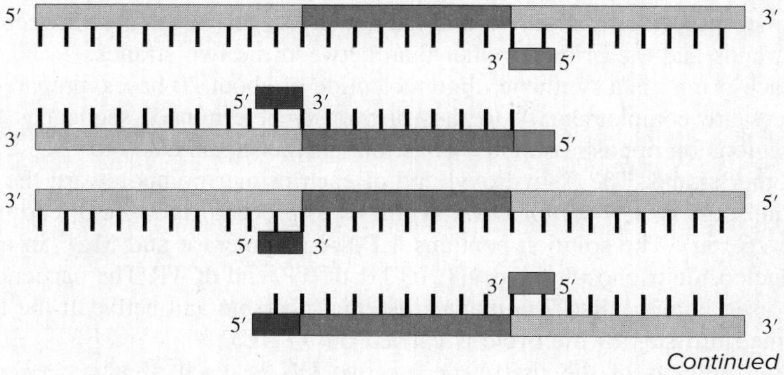

Continued

Figure 26.15

The polymerase chain reaction (PCR). Three cycles are shown; the target region appears after the third cycle. Additional cycles lead to amplification of the target region.

Figure 26.15 *Continued*

(*f*) Elongation of the primed polynucleotide fragments completes the second cycle and gives four DNAs.

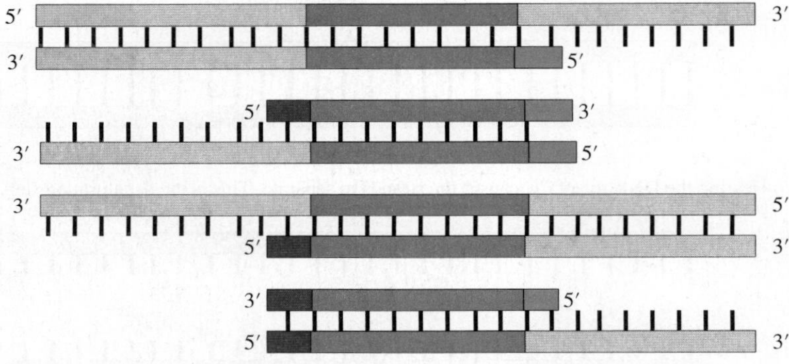

(*g*) Among the eight DNAs formed in the third cycle are two having the structure shown. This is the structure that increases disproportionately in the succeeding cycles.

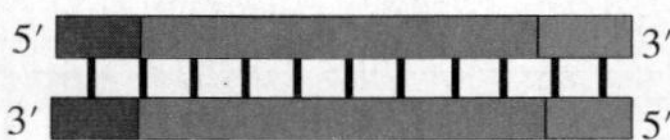

of DNA fragments increases disproportionately. Repetitive PCR cycling increases both the amount of material and its homogeneity (Table 26.4). If every step proceeds in 100% yield, a greater than 1-billionfold amplification is possible after 30 cycles.

Each cycle incorporates three steps:

1. Denaturation
2. Annealing (also called priming)
3. Synthesis (also called extension or elongation)

All of the substances necessary for PCR are present throughout, and proceeding from one cycle to the next requires only changing the temperature after suitable time intervals. The entire process is carried out automatically, and 30 cycles can be completed within a few hours.

The double-stranded DNA shown in Figure 26.15*a* contains the polynucleotide sequence (the target region) we wish to amplify. The DNA is denatured by heating to ≈95°C, which causes the strands to separate by breaking the hydrogen bonds between them (Figure 26.15*b*).

The solution is then cooled to ≈60°C, allowing new hydrogen bonds to form (Figure 26.15*c*). However, the reaction mixture contains much larger concentrations of two primer molecules than DNA, and the new hydrogen bonds are between the separated DNA strands and the primers rather than between the two strands.

Each primer is a synthetic oligonucleotide of about 20 bases, prepared so that their sequences are complementary to the (previously determined) sequences that flank the target regions on opposite strands. Thus, one primer is annealed to one strand, the other to the other strand. The 3′-hydroxyl end of each primer points toward the target region.

The stage is now set for DNA synthesis to proceed from the 3′ end of each primer (Figure 26.15*d*). The solution contains a DNA polymerase and Mg^{2+} in addition to the deoxynucleoside triphosphates dATP, dTTP, dGTP, and dCTP. The particular DNA polymerase used is one called *Taq polymerase* that is stable and active at the temperature at which the third step of the cycle is carried out (72°C).

The products of the first cycle are two DNAs, each of which is composed of a longer and a shorter strand. These products are subjected to a second three-step cycle (Figure 26.15*e–f*) to give four DNAs. Two of these four contain a "strand" that is nothing more than the target region flanked by primers. In the third cycle, these two ultrashort "strands" produce two DNAs of the kind shown in Figure 26.15*g*. This product contains only the target region plus the primers and is the one that increases disproportionately in subsequent cycles.

TABLE 26.4 Distribution of DNAs with Increasing Number of PCR Cycles

Cycle number	Total number of DNAs*	Number of DNAs containing only the target region
0 (start)	1	0
1	2	0
2	4	0
3	8	2
4	16	8
5	32	22
10	1,024	1,004
20	1,048,566	1,048,526
30	1,073,741,824	1,073,741,764

*Total number of DNAs is 2^n, where n = number of cycles.

Since its introduction in 1985, PCR has been applied to practically every type of study that requires samples of DNA. These include screening for genetic traits such as sickle cell anemia, Huntington's disease, and cystic fibrosis. PCR can detect HIV infection when the virus is present in such small concentrations that no AIDS symptoms have as yet appeared. In forensic science, analysis of PCR-amplified DNA from tiny amounts of blood or semen has helped convict the guilty and free the innocent. Anthropologists increasingly use information from DNA analysis to trace the origins of racial and ethnic groups but sometimes find it difficult, for cultural reasons, to convince individuals to volunteer blood samples. Thanks to PCR, the root of a single strand of hair is now sufficient.

26.17 Recombinant DNA Technology

The use of restriction enzymes to cleave DNA at specific sequences was mentioned earlier in this chapter in the context of DNA sequence analysis. These enzymes are also important in the field of **recombinant DNA** technology. We will illustrate this application by describing a method for the production of human insulin.

A plasmid, which is a circular DNA molecule separate from the chromosomal DNA, is obtained from bacterial cells such as *Escherichia coli* and treated with a restriction enzyme to snip the DNA at a specific site (Figure 26.16). The human DNA sequence that codes for the synthesis of insulin is then inserted into the plasmid to give a *recombinant DNA molecule*. The new DNA is the result of the recombination of DNA from the plasmid plus the sequence that codes for human insulin. The new plasmid is termed a *chimeric plasmid* because it contains DNA from two sources, bacterial and human. The plasmid is taken up by growing bacterial cells through a process called *transformation*. The chimeric plasmid serves as a *cloning vector* because it serves as a vehicle to carry the recombinant DNA into *E. coli*. Transcription and translation of the insulin DNA then occur to produce human insulin. When the cells divide, the plasmids are divided between the daughter cells and they continue to produce clones. Insulin produced by recombinant DNA technology is commercially sold as Humulin.

The term *chimera* comes from Greek mythology and refers to a beast composed of the parts of different animals. Homer's *Iliad* describes such a creature: ". . . the Khimaira, of ghastly and inhuman origin, her forepart lionish, her tail a snake's, a she-goat in between. This thing exhaled in jets a rolling fire." (Translation by R. Fitzgerald, Book Six, line 210. Farrar, Straus, and Giroux, 2004, New York.)

The amplification of many other DNA sequences has been carried out by transfection of a cloning vector into a bacterial cell, making it possible to produce quantities of natural proteins that were not previously available, as well as unknown proteins. Green fluorescent protein (see Section 25.17) and its derivatives are produced by recombinant DNA technology. A recombinant human-platelet-derived growth factor (rh-PDGF) *becaplermin* (Regranex) is used clinically in the treatment of diabetic skin ulcers.

Figure 26.16

The production of human insulin by recombinant DNA technology.

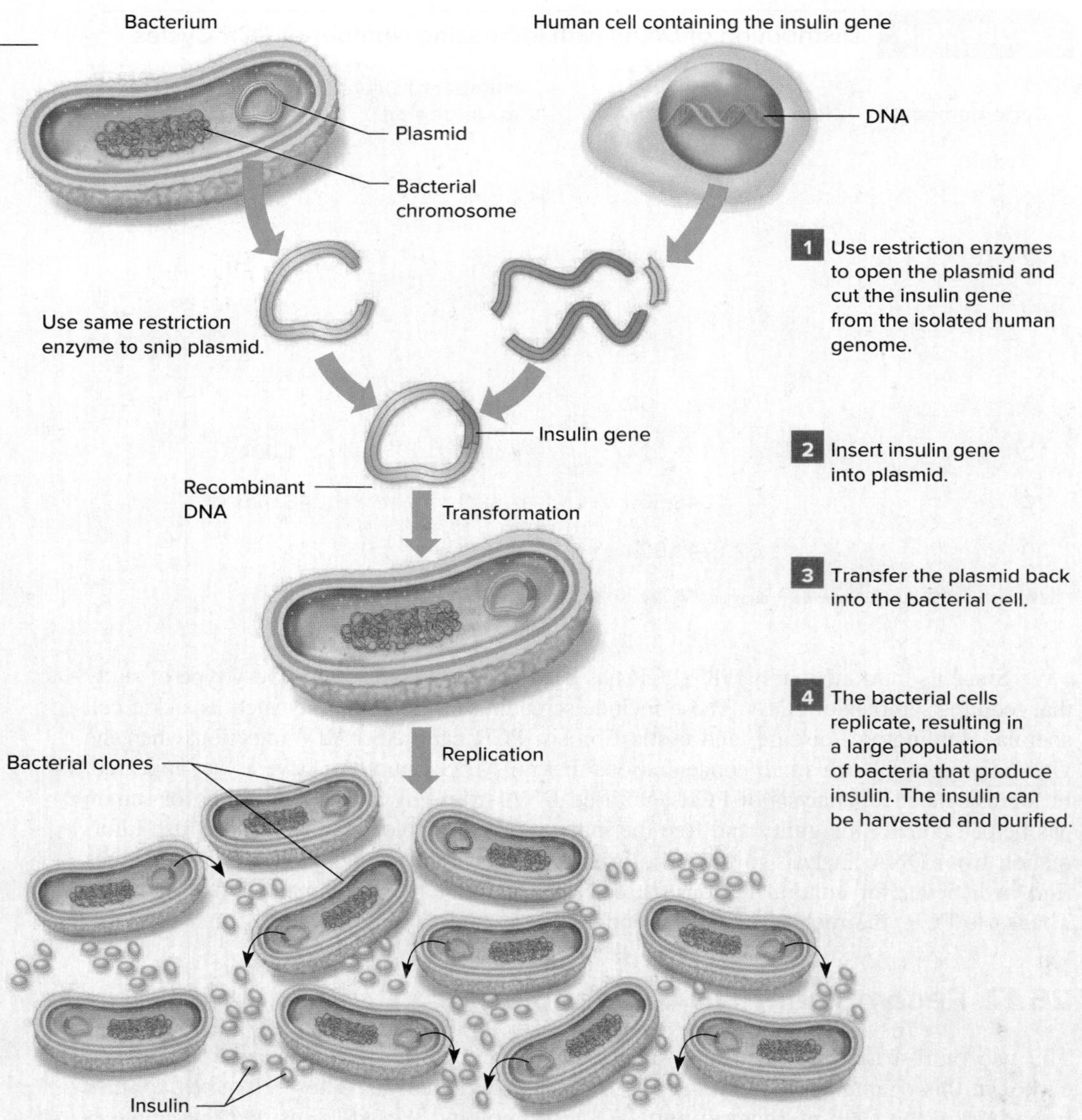

Bacterium

Human cell containing the insulin gene

Plasmid

Bacterial chromosome

DNA

Use same restriction enzyme to snip plasmid.

Insulin gene

Recombinant DNA

Transformation

Replication

Bacterial clones

Insulin

1 Use restriction enzymes to open the plasmid and cut the insulin gene from the isolated human genome.

2 Insert insulin gene into plasmid.

3 Transfer the plasmid back into the bacterial cell.

4 The bacterial cells replicate, resulting in a large population of bacteria that produce insulin. The insulin can be harvested and purified.

26.18 SUMMARY

Section 26.1 Many biologically important compounds are related to the heterocyclic aromatic compounds pyrimidine and purine.

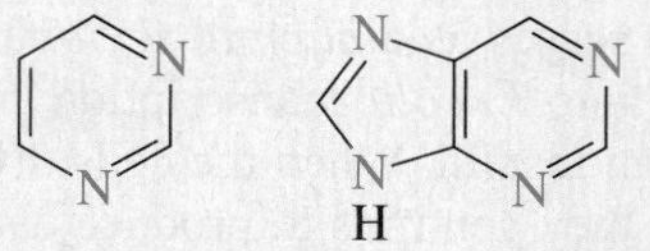

N N N N N H

Pyrimidine Purine

The structure of guanine illustrates an important feature of substituted pyrimidines and purines. Oxygen substitution on the ring favors the keto form rather than the enol. Amino substitution does not.

Guanine

Section 26.2 **Nucleosides** are carbohydrate derivatives of pyrimidine and purine bases. The most important nucleosides are derived from D-ribose and 2′-deoxy-D-ribose.

2′-Deoxy-D-ribose

Thymine

Thymidine

Section 26.3 **Nucleotides** are phosphate esters of nucleosides.

Thymidine 5′-monophosphate

In the example shown, the 5′-OH group is phosphorylated. Nucleotides are also possible in which some other —OH group bears the phosphate ester function. Cyclic phosphates are common and important as biochemical messengers.

Section 26.4 **Bioenergetics** is concerned with the thermodynamics of biological processes. Particular attention is paid to $\Delta G^{\circ\prime}$, the standard free-energy change of reactions at pH = 7. When the sign of $\Delta G^{\circ\prime}$ is +, the reaction is **endergonic;** when the sign of $\Delta G^{\circ\prime}$ is –, the reaction is **exergonic.**

Section 26.5 **Adenosine 5′-triphosphate (ATP)** is a key compound in biological energy storage and delivery.

Adenosine 5′-triphosphate (ATP)

The hydrolysis of ATP to ADP and HPO_4^{2-} is exergonic.

$$\text{ATP} + H_2O \longrightarrow \text{ADP} + HPO_4^{2-} \qquad \Delta G^{\circ\prime} = -31 \text{ kJ } (-7.4 \text{ kcal})$$

Many formally endergonic biochemical processes become exergonic when they are coupled mechanistically to the hydrolysis of ATP.

Section 26.6 Many important compounds contain two or more nucleotides joined together by a **phosphodiester** linkage. The best known are those in which the phosphodiester joins the 5′-oxygen of one nucleotide to the 3′-oxygen of the other.

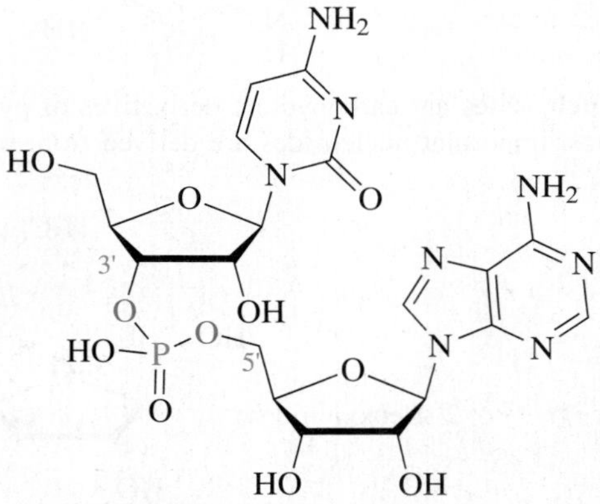

Oligonucleotides contain about 50 or fewer nucleotides held together by phosphodiester links; **polynucleotides** can contain thousands of nucleotides. The carbohydrate component is D-ribose in the polynucleotide ribonucleic acid (RNA) and 2-deoxy-D-ribose in the polynucleotide deoxyribonucleic acid (DNA).

Section 26.7 The phosphodiester bond is very stable toward hydrolysis, but RNA is easily hydrolyzed. The 2′-OH of RNA is responsible for the large difference in hydrolytic stability of DNA and RNA.

Section 26.8 Before the 1940s, the material responsible for genetic transmission was unknown. Research into the structure of genetic material was culminated with Watson and Crick's Nobel Prize–winning DNA structure.

Section 26.9 The most common form of DNA is B-DNA, which exists as a right-handed double helix. The carbohydrate–phosphate backbone lies on the outside, the purine and pyrimidine bases on the inside. The double helix is stabilized by complementary hydrogen bonding (base pairing) between adenine (A) and thymine (T), and guanine (G) and cytosine (C).

Section 26.10 During DNA replication the two strands of the double helix begin to unwind, exposing the pyrimidine and purine bases in the interior. Nucleotides with complementary bases hydrogen bond to the original strands and are joined together by phosphodiester linkages with the aid of DNA polymerase. Each new strand grows in its 5′→3′ direction.

Section 26.11 Three RNAs are involved in gene expression. In the **transcription** phase, a strand of **messenger RNA (mRNA)** is synthesized from a DNA template. The four bases A, G, C, and U, taken three at a time, generate 64 possible combinations called **codons.** These 64 codons comprise the **genetic code** and code for the 20 amino acids found in proteins plus start and stop signals. The mRNA sequence is **translated** into a prescribed protein sequence at the ribosomes. There, small polynucleotides called **transfer RNA (tRNA),** each of which contains an **anticodon** complementary to an mRNA codon, carries the correct amino acid for incorporation into the growing protein. **Ribosomal RNA (rRNA)** is the main constituent of ribosomes and appears to catalyze protein biosynthesis.

Section 26.12 The start codon for protein biosynthesis is AUG, which is the same as the codon for methionine. Thus, all proteins initially have methionine as their N-terminal amino acid, but lose it subsequent to their formation. The reaction responsible for extending the protein chain is nucleophilic acyl substitution.

Section 26.13 HIV, which causes AIDS, is a retrovirus. Its genetic material is RNA instead of DNA. HIV contains an enzyme called reverse transcriptase that allows its RNA to serve as a template for DNA synthesis in the host cell. Some of the successful drugs to treat AIDS are nucleosides that inhibit reverse transcriptase.

Section 26.14 The nucleotide sequence of DNA can be determined by the Sanger method. This is a technique in which a short section of single-stranded DNA is allowed to produce its complement in the presence of dideoxy analogs of ATP, TTP, GTP, and CTP. DNA formation terminates when a dideoxy analog is incorporated into the growing polynucleotide chain. A mixture of polynucleotides differing from one another by an incremental nucleoside is produced and analyzed by electrophoresis. From the observed sequence of the complementary chain, the sequence of the original DNA is deduced.

Section 26.15 The sequence of nucleotides that make up the human genome was completed at the beginning of the twenty-first century. Next-generation sequencing methods are fast enough to sequence a human genome in a day. There is every reason to believe that the increased knowledge of human biology it offers will dramatically affect the practice of medicine.

Section 26.16 In DNA profiling the noncoding regions are cut into smaller fragments using enzymes that recognize specific sequences, and these smaller bits of DNA are then separated by electrophoresis. The observed pattern of DNA fragments is believed to be highly specific for the source of the DNA. Using the **polymerase chain reaction (PCR),** millions of copies of minute amounts of DNA can be produced in a relatively short time.

Section 26.17 DNA sequences that code for the synthesis of a specific protein can be inserted into a bacterial DNA plasmid. The growing bacteria then incorporate the **recombinant DNA** and produce the protein.

PROBLEMS

Purine and Pyrimidine Structural Types

26.13 5-*Fluorouracil* is one component of a mixture of three drugs used in breast-cancer chemotherapy. What is its structure?

26.14 Isoguanine is an isomer of guanine. It was synthesized to create a purine that has the same hydrogen-bonding pattern as the pyrimidine cytosine. In aqueous solution, two forms of the base are observed. About 10% of the isoguanine has a hydrogen-bonding pattern that allows it to base pair with thymine. Provide the structure of the minor form of isoguanine and identify how it is formed from the major isomer.

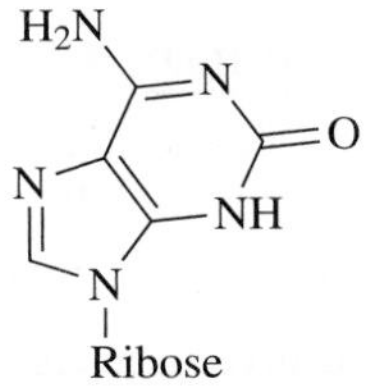

H_2N N O NH N N Ribose

Isoguanine

26.15 (a) Which isomer, the keto or enol form of cytosine, is the stronger acid?

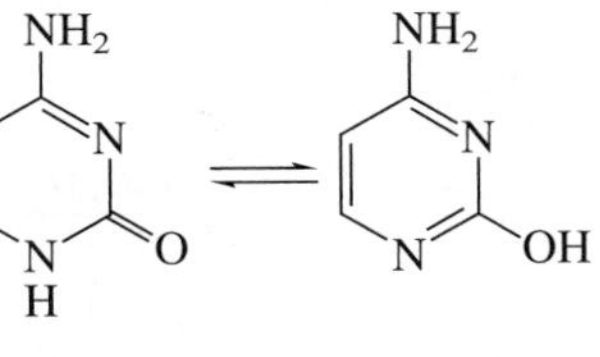

NH_2 N N H O ⇌ NH_2 N N OH

keto enol

(b) What is the relationship between the conjugate base of the keto form and the conjugate base of the enol form?

26.16 Birds excrete nitrogen as *uric acid.* Uric acid is a purine having the molecular formula $C_5H_4N_4O_3$; it has no C—H bonds. Write a structural formula for uric acid.

26.17 *Nebularine* is a toxic nucleoside isolated from a species of mushroom. Its systematic name is 9-β-D-ribofuranosylpurine. Write a structural formula for nebularine.

26.18 The D-arabinose analog of adenosine is an antiviral agent (vidarabine) used to treat conjunctivitis and shingles. Write a structural formula for this compound.

26.19 The 5′-nucleotide of inosine, *inosinic acid* ($C_{10}H_{13}N_4O_8P$), is added to foods as a flavor enhancer. What is the structure of inosinic acid? (The structure of inosine is given in Problem 26.23.)

Reactions of Purines and Pyrimidines

26.20 Adenine is a weak base. Which one of the three nitrogens designated by arrows in the structural formula shown is protonated in acidic solution? Evaluation of the resonance contributors of the three protonated forms will tell you which one is the most stable.

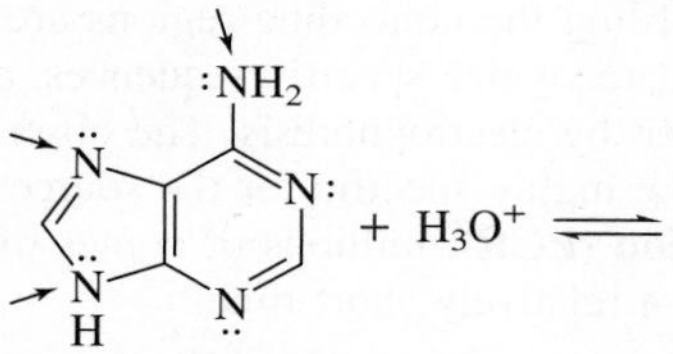

$+ H_3O^+ \rightleftharpoons$

26.21 (a) The two most acidic hydrogens of uracil have pK_a's of 9.5 and 14.2, respectively. Match these pK_a's with the hydrogens in the structural formula and provide structures for the most stable resonance contributors of the monoanion and the dianion.

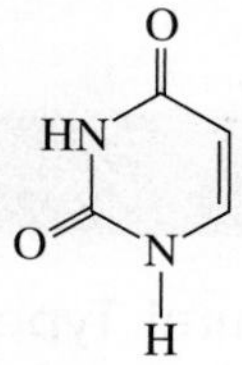

Uracil

(b) The pK_a of the conjugate acid of triethylamine is 10.4. Is triethylamine a strong enough base to convert uracil to its monoanion? To its dianion?

26.22 When 6-chloropurine is heated with aqueous sodium hydroxide, it is quantitatively converted to *hypoxanthine*. Suggest a reasonable mechanism for this reaction.

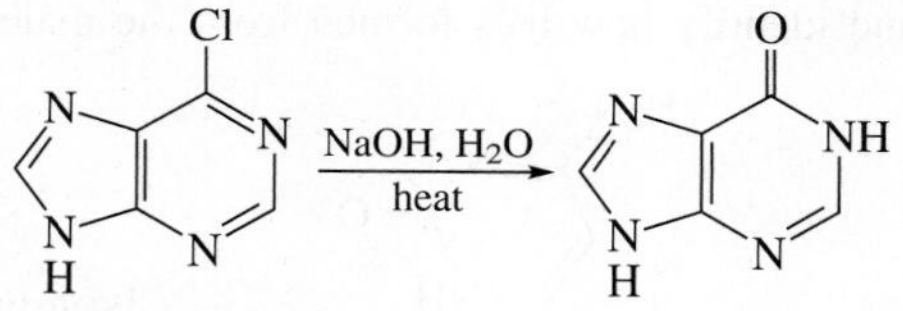

6-Chloropurine $\xrightarrow[\text{heat}]{NaOH,\ H_2O}$ Hypoxanthine

26.23 Treatment of adenosine with nitrous acid gives a nucleoside known as *inosine*. Suggest a reasonable mechanism for this reaction.

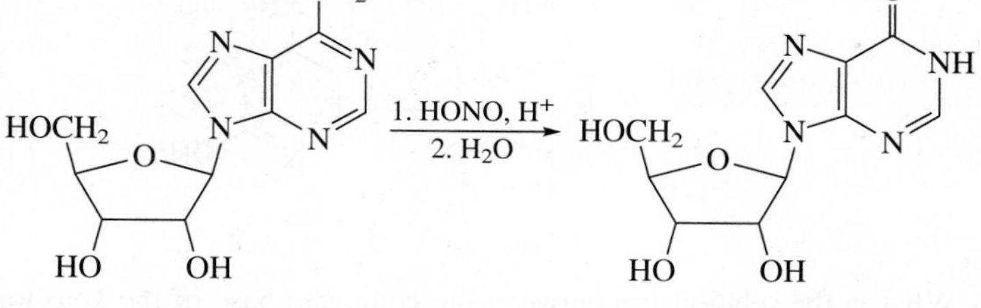

Adenosine $\xrightarrow[2.\ H_2O]{1.\ HONO,\ H^+}$ Inosine

Bioenergetics and Phosphate Esters

26.24 The phosphorylation of α-D-glucopyranose by ATP (Section 26.3) has $\Delta G°' = -23$ kJ/mol (−5.5 kcal/mol) at 298 K.

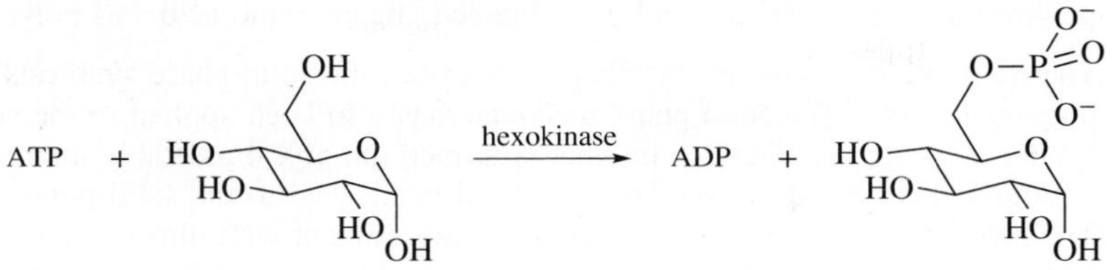

ATP + α-D-glucopyranose —hexokinase→ ADP + α-D-glucopyranose 6-phosphate

(a) Is this reaction exergonic or endergonic?

(b) How would the value of $\Delta G°'$ change in the absence of the enzyme hexokinase? Would it become more positive, more negative, or would it stay the same? Why?

(c) Use the value for the hydrolysis of ATP to ADP (Section 26.5) to calculate $\Delta G°'$ for the reaction of α-D-glucopyranose with inorganic phosphate. Is this reaction exergonic or endergonic?

HPO_4^{2-} + α-D-glucopyranose —hexokinase→ H_2O + α-D-glucopyranose 6-phosphate

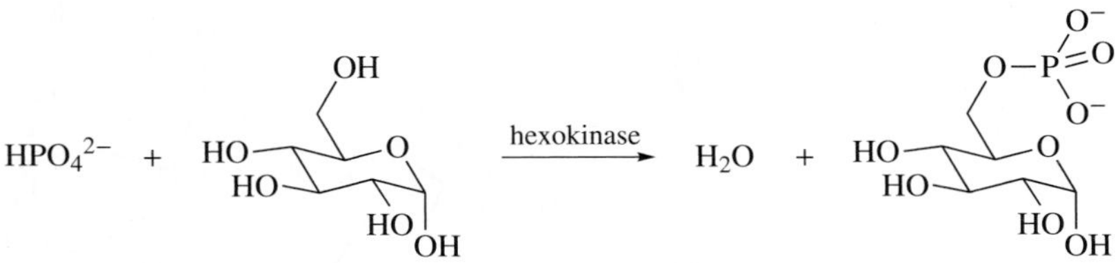

26.25 2′-*O*-Methyl is a naturally occurring modification in RNA. Would you expect this modified RNA to be more stable or less stable than natural RNA toward hydrolysis in base?

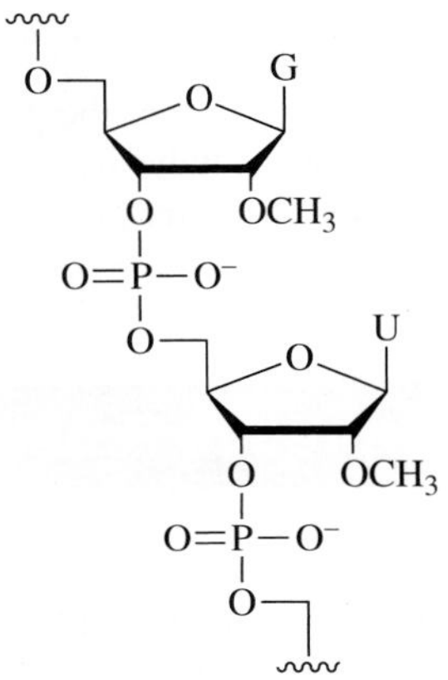

Synthesis of Nucleosides and Nucleotides

26.26 The coupling reaction of 2,6-dichloropurine with 1,2,3,4-tetra-*O*-acetyl-α+β-arabinofuranose takes place when the two are heated in the presence of *p*-toluenesulfonic acid to give the nucleoside in 75% yield. The reaction is stereoselective for the formation of the α-anomer, even though the starting sugar is a mixture of anomers. Can you think of a reason for the stereoselectivity? (*Hint:* See Mechanism 23.3.)

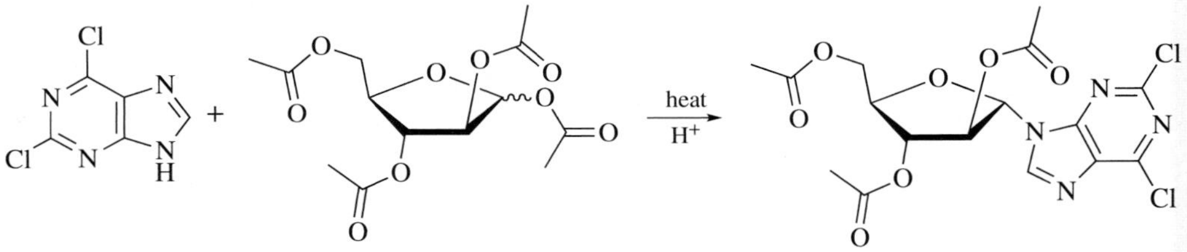

2,6-Dichloropurine + 1,2,3,4-Tetra-*O*-acetyl-α+β-arabinofuranose —heat, H^+→ nucleoside

26.27 In one of the early experiments designed to elucidate the genetic code, Marshall Nirenberg of the U.S. National Institutes of Health (Nobel Prize in Physiology or Medicine, 1968) prepared a synthetic mRNA in which all the bases were uracil. He added this poly(U) to a cell-free system containing all the necessary materials for protein biosynthesis. A polymer of a single amino acid was obtained. What amino acid was polymerized?

26.28 The descriptive passage in this chapter describes the solid-phase synthesis of oligonucleotides. The solid-phase technique has also been applied to the automated synthesis of oligosaccharides in what is termed the glycal assembly method. An unsaturated carbohydrate known as a glycal is attached to the solid polystyrene support. The glycal is then converted to an epoxide by treatment with dimethyldioxirane (DMDO). The epoxide serves as the glycosyl donor and undergoes nucleophilic attack by the hydroxyl group of the glycosyl acceptor. The double bond of the new disaccharide is activated and coupled in the same way to allow extension of the oligosaccharide chain.

resin — $OSiR_2$ — Glycal $\xrightarrow{\text{DMDO}}$ resin — $OSiR_2$ — Glycosyl donor $\xrightarrow[\text{ZnCl}_2]{\text{glycosyl acceptor}}$ resin — $OSiR_2$ — disaccharide (OH)

Glycal

Glycosyl donor

(a) Show a mechanism for the reaction of the glycosyl donor with the glycosyl acceptor. (*Hint:* Zn^{2+} acts as a catalyst.)

(b) Explain the regioselectivity of the reaction in (a).

Descriptive Passage and Interpretive Problems 26

Oligonucleotide Synthesis

In Section 26.6 we noted that synthetic oligonucleotides of defined sequence were commercially available for use as primers for PCR and as probes for cloning DNA. Here we will examine how these oligonucleotides are prepared.

The method bears many similarities to the Merrifield solid-phase synthesis of peptides. A starter unit is attached to a solid support, nucleosides are attached one-by-one until the sequence is complete, whereupon the target oligonucleotide is removed from the support and purified. Like solid-phase peptide synthesis, the preparation of oligonucleotides relies heavily on protecting groups and bond-forming methods.

The starter units are nucleosides in which amine groups on the DNA bases have been protected by acylation.

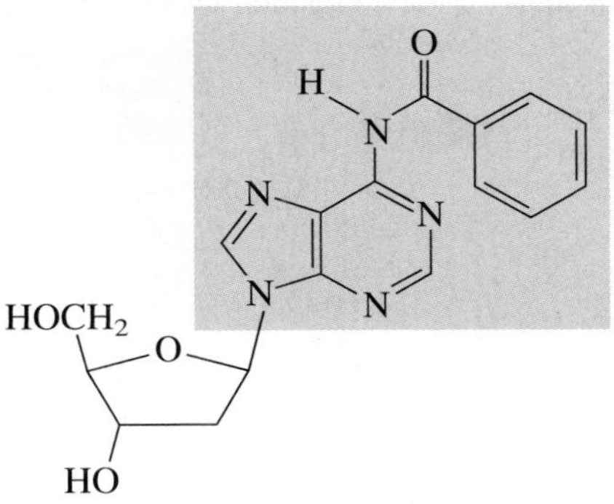

N-Benzoyl-protected 2′-deoxyadenosine

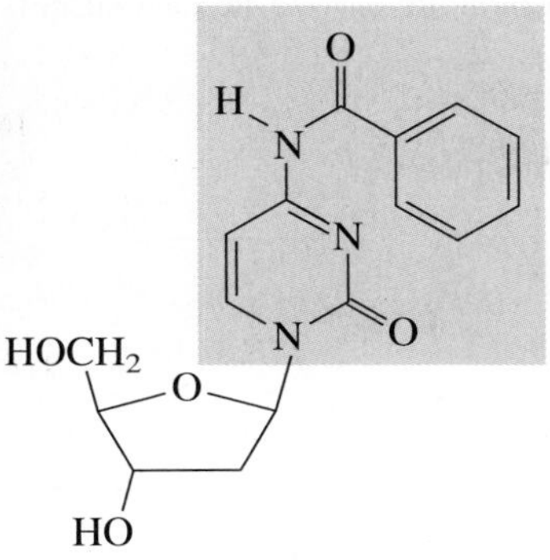

N-Benzoyl-protected 2′-deoxycytidine

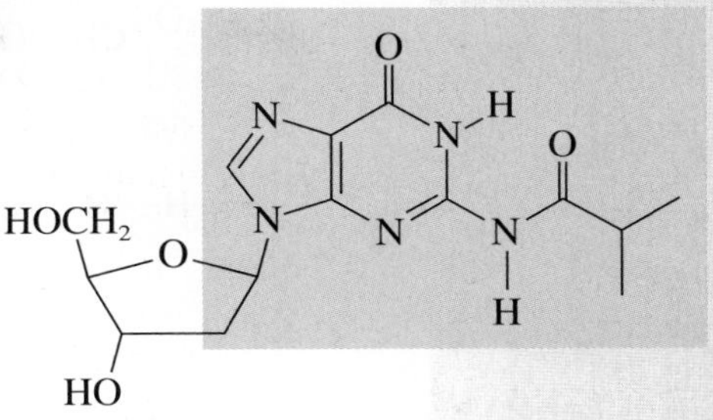

N-2-Methylpropanoyl-protected 2′-deoxyguanosine

Thymidine lacks an —NH_2 group, so needs no protecting group on its pyrimidine base.

These N-protecting groups remain in place throughout the synthesis. They are the first ones added and the last ones removed. None of the further "chemistry" that takes place involves the purine or pyrimidine rings.

The 5′-OH group of the 2′-deoxyribose portion of the nucleosides is primary and more reactive toward ether formation than the 3′-OH group, which is secondary. This difference allows selective protection of the 5′-OH as its 4,4′-dimethoxytriphenylmethyl (DMT) ether.

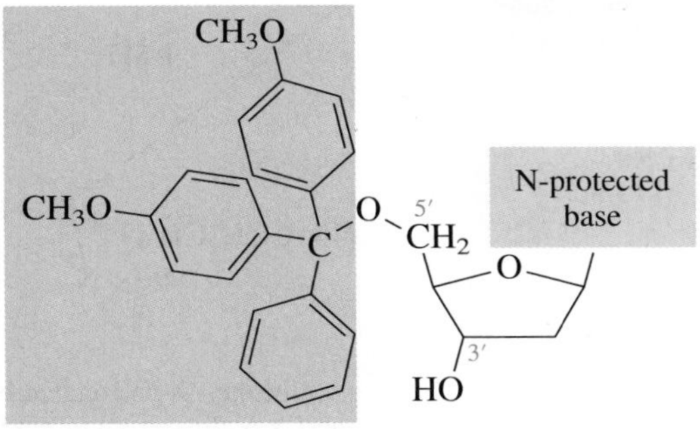

The nucleoside that is to serve as the 3′ end of the final oligonucleotide is attached to a controlled-pore glass (CPG) bead by ester formation between its unprotected 3′-OH and a linker unit already attached to the CPG. In order for chain elongation to proceed in the 3′→5′ direction, the DMT group that protects the 5′-OH of the starter unit is removed by treatment with dichloroacetic acid.

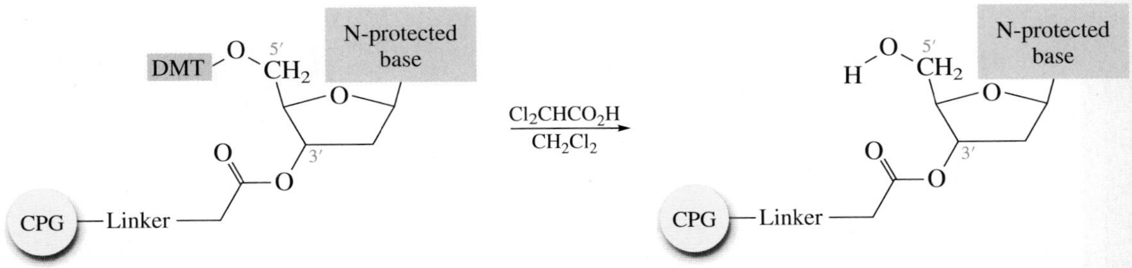

$$\xrightarrow[CH_2Cl_2]{Cl_2CHCO_2H}$$

The stage is now set for adding the second nucleoside. The four blocked nucleosides prepared earlier are converted to their corresponding 3′-phosphoramidite derivatives. An appropriate A, C, T, or G phosphoramidite is used in each successive stage of the elongation cycle.

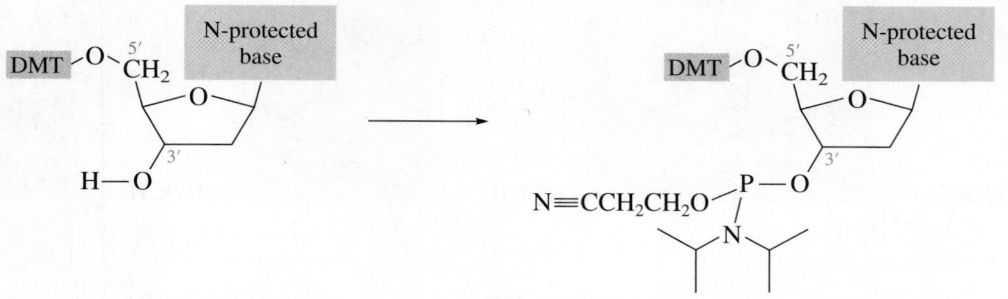

DMT–O–CH_2 (5′) N-protected base; H–O (3′) → DMT–O–CH_2 (5′) N-protected base; $N{\equiv}CCH_2CH_2O$–P(–N(iPr)$_2$)–O (3′)

Each phosphoramidite is coupled to the anchored nucleoside by a reaction in which the free 5′-OH of the anchored nucleoside displaces the diisopropylamino group from phosphorus (Figure 26.17). The coupling is catalyzed by tetrazole, which acts as a weak acid to protonate the diisopropylamino group.

The product of the coupling is a phosphite; it has the general formula $P(OR)_3$. It is oxidized to phosphate $[P(O)(OR)_3]$ in the last step of Figure 26.17.

Figure 26.17

Coupling of a 3′-phosphoramidite derivative of a nucleoside to the starter unit nucleoside in a solid-phase oligonucleotide synthesis. Following the coupling, the resulting phosphite is oxidized to a phosphate.

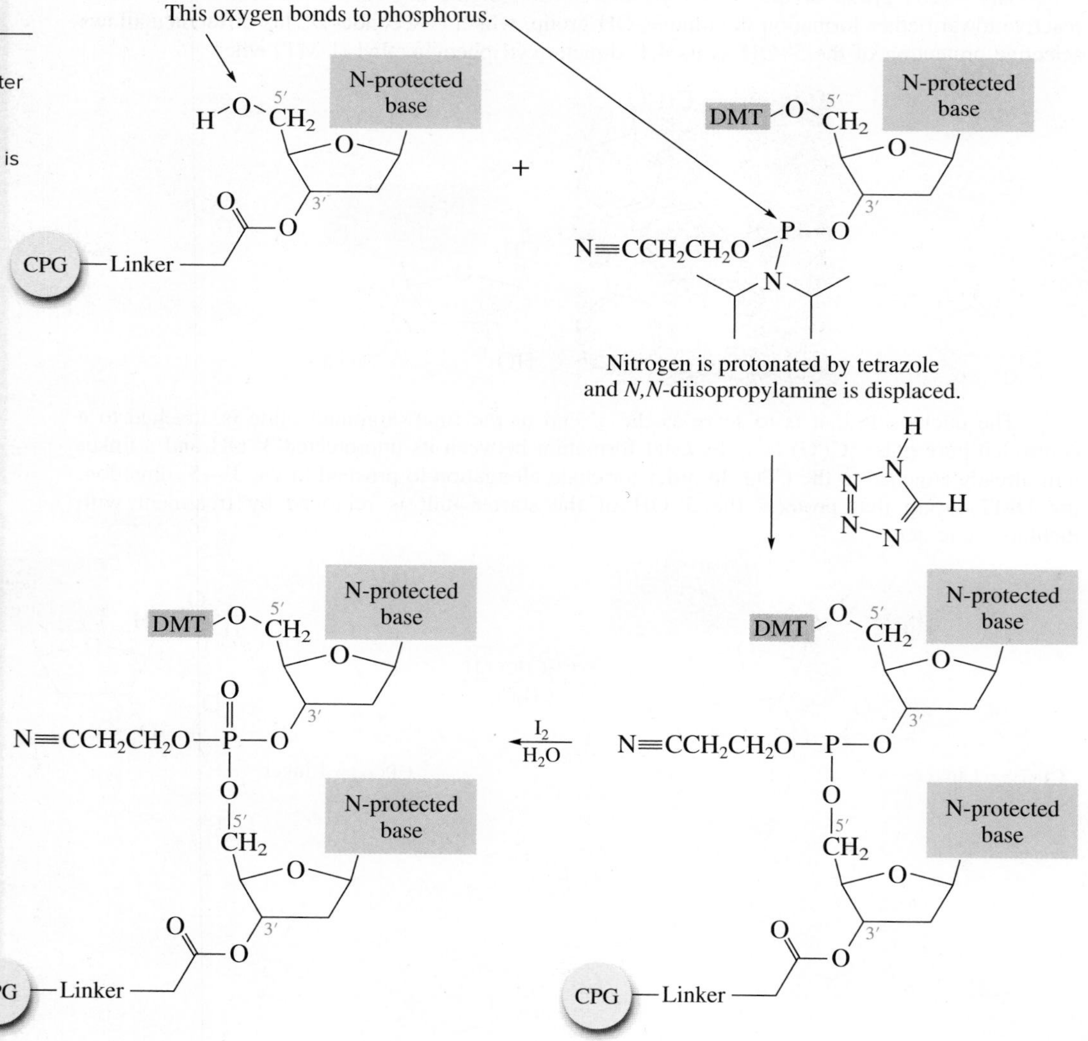

This oxygen bonds to phosphorus.

H–O–CH_2 (5′) N-protected base; CPG–Linker–C(=O)–O (3′) + DMT–O–CH_2 (5′) N-protected base; $N{\equiv}CCH_2CH_2O$–P(–N(iPr)$_2$)–O (3′)

Nitrogen is protonated by tetrazole and *N,N*-diisopropylamine is displaced.

DMT–O–CH_2 (5′) N-protected base; $N{\equiv}CCH_2CH_2O$–P–O (3′); O–CH_2 (5′) N-protected base; CPG–Linker–C(=O)–O (3′)

$\xrightarrow{I_2,\ H_2O}$

DMT–O–CH_2 (5′) N-protected base; $N{\equiv}CCH_2CH_2O$–P(=O)–O (3′); O–CH_2 (5′) N-protected base; CPG–Linker–C(=O)–O (3′)

The 5′-OH of the newly added nucleoside is then deprotected to prepare the bound dinucleotide for the next elongation cycle.

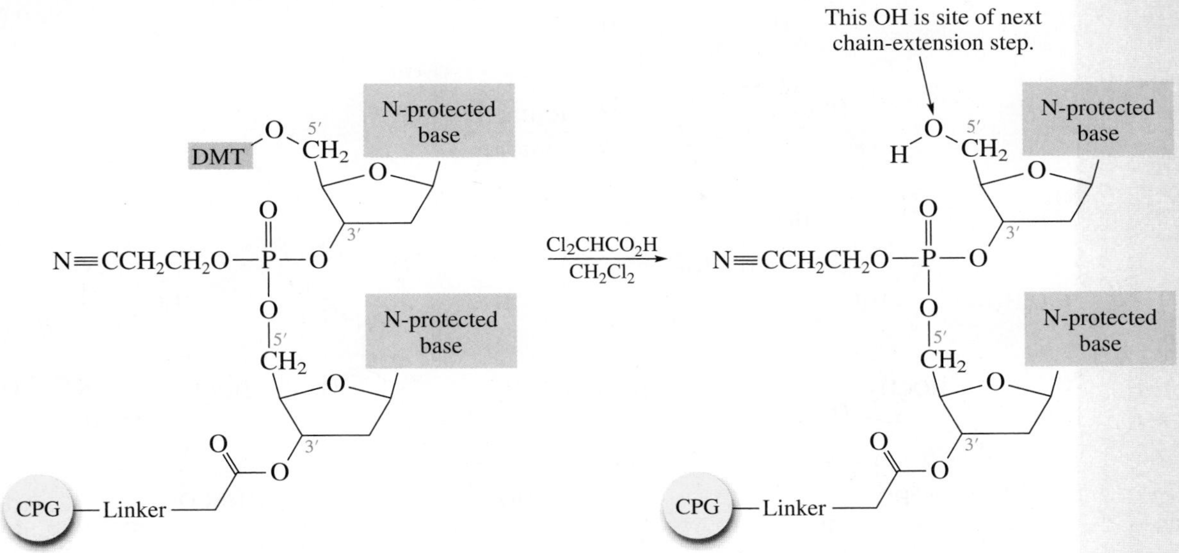

This OH is site of next chain-extension step.

DMT N-protected base $N{\equiv}CCH_2CH_2O$—P—O N-protected base CPG—Linker $\xrightarrow[CH_2Cl_2]{Cl_2CHCO_2H}$ H N-protected base $N{\equiv}CCH_2CH_2O$—P—O N-protected base CPG—Linker

Once all the nucleosides are in place and the last DMT is removed, treatment with aqueous ammonia removes the acyl and cyanoethyl groups and cleaves the oligonucleotide from the CPG support.

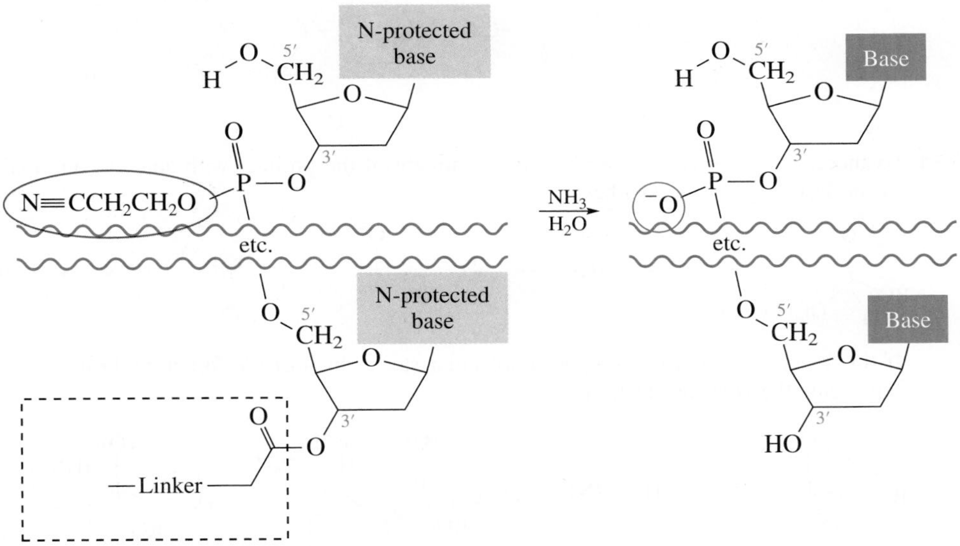

H N-protected base $N{\equiv}CCH_2CH_2O$ P—O etc. N-protected base —Linker $\xrightarrow[H_2O]{NH_3}$ H Base ^-O P—O etc. Base HO

26.29 What is the product of the following reaction?

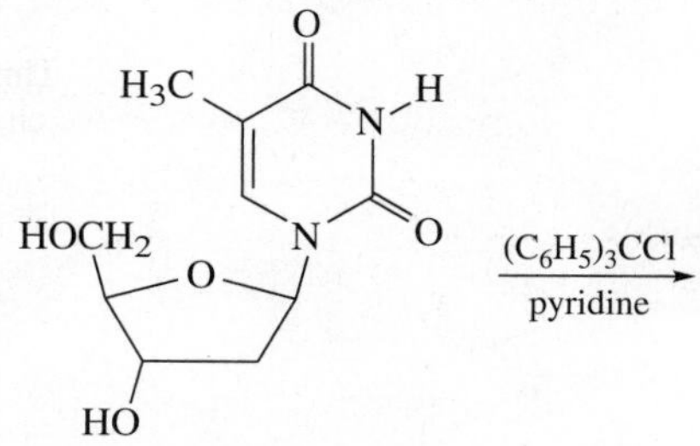

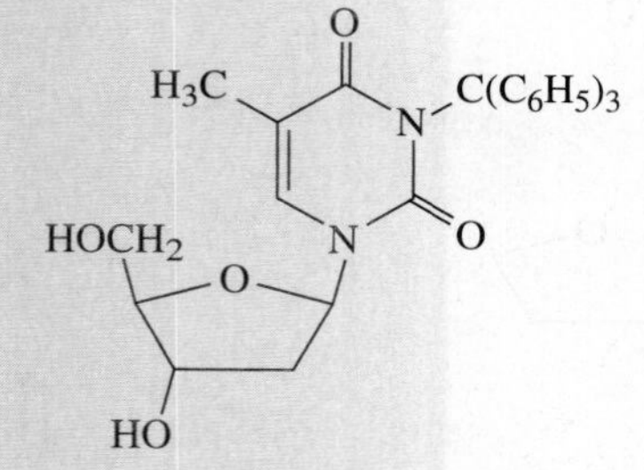

HOCH₂-thymidine (HO at 3′) $\xrightarrow[\text{pyridine}]{(C_6H_5)_3CCl}$

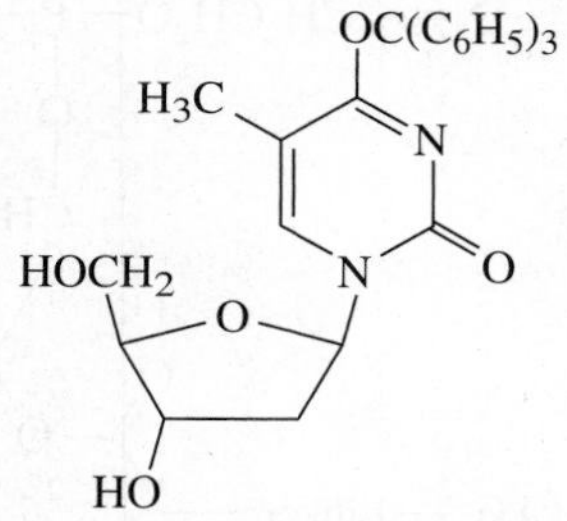

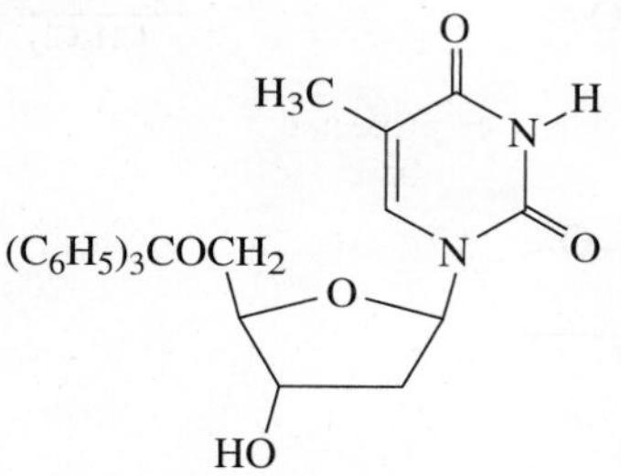

A. N-3 substituted with $C(C_6H_5)_3$; $HOCH_2$, HO

B. 4-$OC(C_6H_5)_3$; $HOCH_2$, HO

C. $(C_6H_5)_3COCH_2$; N–H; HO

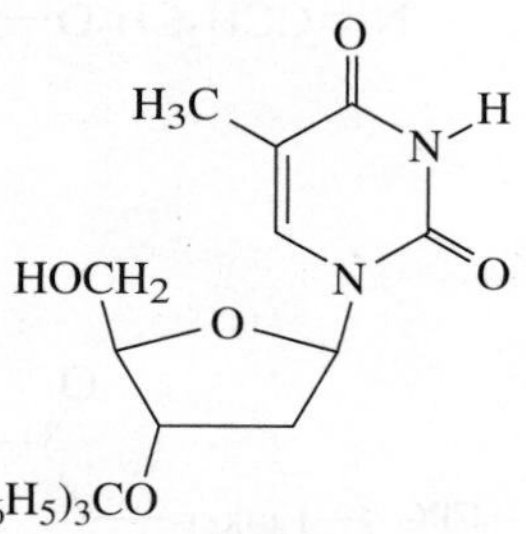

D. $HOCH_2$; N–H; $(C_6H_5)_3CO$

26.30 What species is formed from the DMT-protecting group when it is removed using dichloroacetic acid? (Ar = p-$CH_3OC_6H_4$)

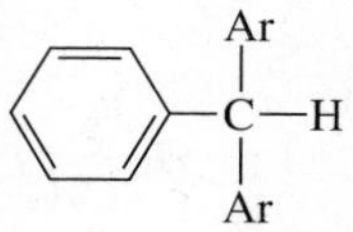

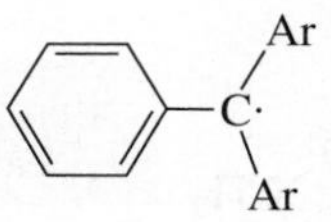

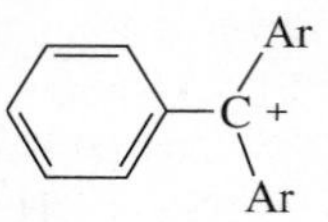

A. $C_6H_5C(Ar)_2H$ B. $C_6H_5C(Ar)_2{:}^-$ C. $C_6H_5C(Ar)_2\cdot$ D. $C_6H_5C(Ar)_2^+$

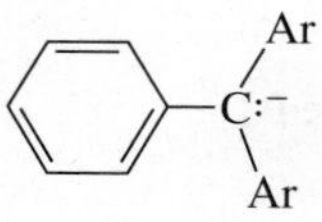

26.31 Cyanoethyl groups are removed during treatment of the product with aqueous ammonia in the last stage of the synthesis.

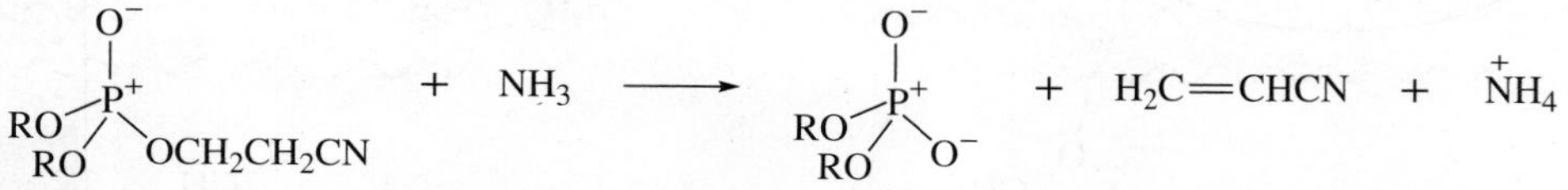

$$(RO)_2P^+(O^-)OCH_2CH_2CN + NH_3 \longrightarrow (RO)_2P^+(O^-)O^- + H_2C{=}CHCN + \overset{+}{N}H_4$$

If this reaction occurs in a single bimolecular step, which of the following best represents the flow of electrons?

A. B. 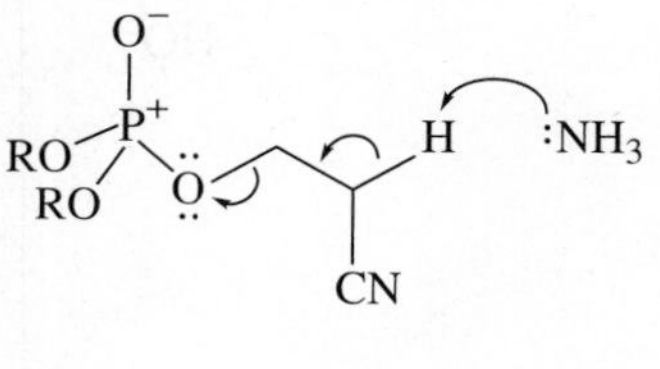 C. 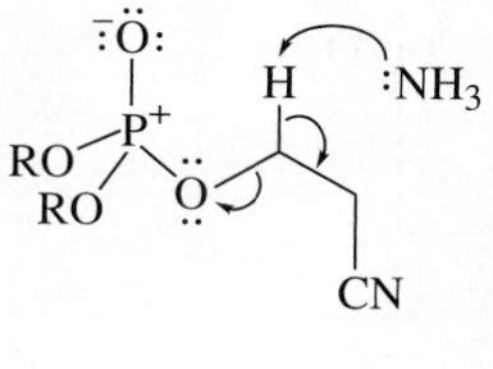 D.

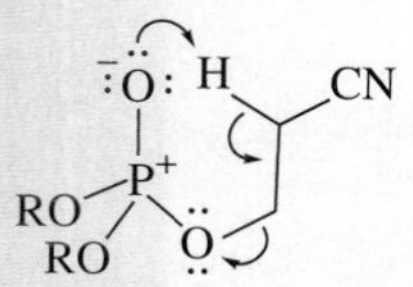

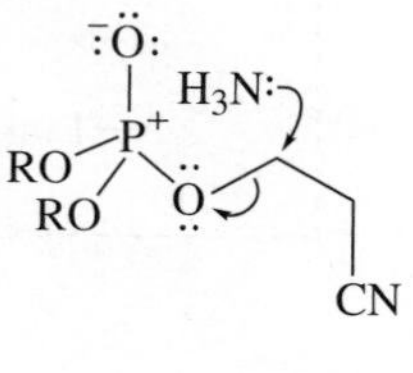

26.32 Structure **1** is the one given for tetrazole in Figure 26.17. Structures **2** and **3** have the same molecular formula (CH_2N_4) and the same number of electrons as **1.** How are these structures related?

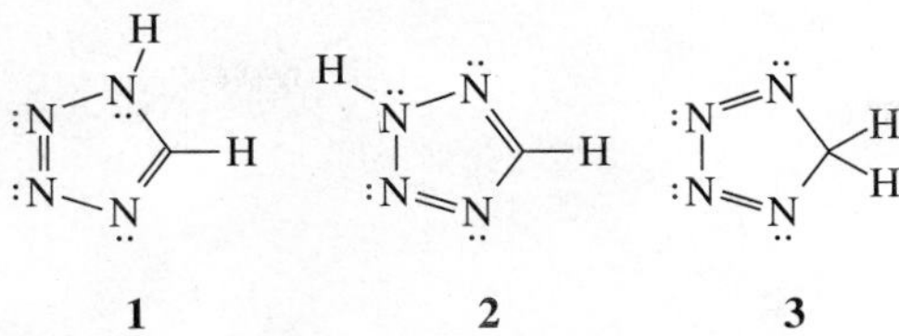

A. **1, 2,** and **3** are constitutional isomers.

B. **1, 2,** and **3** are resonance contributors of the same compound.

C. **1** and **2** are resonance contributors of the same compound; **3** is an isomer of **1** and **2.**

D. **1** and **3** are resonance contributors of the same compound; **2** is an isomer of **1** and **3.**

26.33 Consider the conjugate bases of structures **1, 2,** and **3** in the preceding problem and choose the correct response.

A. **1, 2,** and **3** give different conjugate bases on deprotonation.

B. **1, 2,** and **3** give the same conjugate base on deprotonation.

C. **1** and **2** give the same conjugate base on deprotonation; the conjugate base of **3** is different.

D. **1** and **3** give the same conjugate base on deprotonation; the conjugate base of **2** is different.

26.34 Antisense oligonucleotides are a new class of synthetic drugs, one of which has been approved for use, with numerous others being developed and tested. An antisense drug is designed to have a sequence that is complementary to a portion of a messenger RNA of an organism connected with a disease. The rationale is that the oligonucleotide will bind to the mRNA and interfere with the biosynthesis of a particular protein. An antisense oligonucleotide proposed for treatment of ulcerative colitis has the sequence 5′-GCC CAA GCT GGC ATC GCT CA-3′. In the solid-phase synthesis of this drug, what nucleoside is attached to the controlled-pore glass bead?

A. A

B. T

C. C

D. G

Chapter

27

Aerodynamics of the official 2014 World Cup soccer ball, named the Brazuca ball, were evaluated by NASA. Soccer balls are made of polymeric materials such as polyurethanes, which are prepared from a diol and a diisocyanate such as toluene diisocyanate.

CHAPTER OUTLINE

Synthetic Polymers

A **polymer** is a substance composed of **macromolecules,** molecules that contain a very large number of atoms and have a high molecular weight. Starch, cellulose, silk, and DNA are examples of naturally occurring polymers. Synthetic polymers include nylon, polyethylene, and Bakelite, among countless others. Polymers need not be homogeneous, and most are not. Even one as simple as polyethylene is a mixture of macromolecules with different chain lengths and different degrees of branching.

This chapter is about synthetic polymers, many of which have been introduced in earlier chapters where we emphasized the connection between the reactions used to prepare polymers and the core reactions of organic chemistry. In this chapter, we will add new polymers and methods to those already introduced and expand our understanding of their synthesis, structure, and properties. As we do so, keep in mind that *the reactions used to prepare polymers are the same fundamental reactions that occur with simple organic compounds.*

27.1 Some Background

The earliest applications of polymer chemistry involved chemical modification designed to improve the physical properties of naturally occurring polymers. In 1839, Charles Goodyear transformed natural rubber, which is brittle when cold and tacky

From Bakelite to Nylon

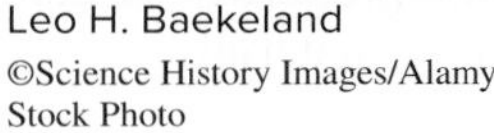

Leo H. Baekeland
©Science History Images/Alamy Stock Photo

Wallace Carothers
©Paul Fearn/Alamy Stock Photo

Polymer science affects our daily lives in ways unimaginable until the early days of the twentieth century. Two men—Leo Baekeland and Wallace Carothers—both chemists by training and innovators by inclination deserve special mention for their roles in bringing it about.

Baekeland was born in Belgium in 1863 and spent his early professional life in Europe. He visited the United States in 1886, became a citizen in 1897, and maintained homes in New York and Florida until his death in 1944. He is best known for introducing polymer chemistry to the world of commerce in a way that affected the daily lives of people everywhere.

Baekeland had already become independently wealthy by inventing one of the earliest photographic films and continued to carry out research designed to produce new materials by chemical methods. By 1907, he and his coworkers had created a novel polymer by heating phenol and formaldehyde in a sealed container. He named the polymer and the company he formed to produce it *Bakelite* (Figure 27.1).

As early as 1912, billiard balls made of Bakelite began to replace ivory ones. Other applications followed and for decades thereafter, it almost always went without saying that anything "plastic" was Bakelite. Now, more than 100 years after Bakelite was invented, numerous products made from it continue to be produced. The company and its name still exist, but as part of an international conglomerate.

Wallace Carothers's story could not be more different, yet his accomplishments rival those of Baekeland. He was born in Iowa in 1896 and, after graduate work and in the early stages of a career in academic teaching and research, was recruited to work at the DuPont Experimental Station in Wilmington, Delaware, in 1928. Carothers was soon given the job of directing a group involved in developing commercially useful polymers. After successfully working out the details for completing an existing project that eventually led to neoprene, Carothers and his group settled on polyamides as attractive candidates for investigation, "nylon 6,6" emerged, and the rest is history.

Nylon 6,6 is a polyamide with each "6" designating the number of carbons in the precursor molecules—adipic acid [$HO_2C(CH_2)_4CO_2H$] and hexamethylenediamine [$H_2N(CH_2)_6NH_2$].

Nylon 6,6 not only proved to be successful in its own right, but inspired DuPont and others to pursue polymer chemistry with increasing vigor, both in designing candidate structures and novel methods for preparing them. Carothers, however, died suddenly in 1937 and one can only speculate what might have been had he not.

Figure 27.1

Many Bakelite items are now sought after as collectibles.
©Comstock/Alamy Stock Photo
©AdShooter/E+/Getty Images

Vulcanization was summarized in the essay *Diene Polymers* in Chapter 11.

when warm, to a substance that maintains its elasticity over a wider temperature range by heating it with sulfur (vulcanization). The first synthetic fibers—called *rayons*—were made by chemical modification of cellulose near the end of the nineteenth century.

Leo Baekeland patented the first totally synthetic polymer, which he called *Bakelite,* in 1910 (Figure 27.1). Bakelite is a versatile, durable material prepared from low-cost materials (phenol and formaldehyde) and was the most successful synthetic material of its kind for many years.

These early successes notwithstanding, knowledge about polymer *structure* was meager. Most chemists believed that rubber, proteins, and the like were colloidal dispersions of small molecules. During the 1920s Hermann Staudinger, beginning at the Swiss Federal Institute of Technology and continuing at the University of Freiburg, argued that polymers were high-molecular-weight compounds held together by normal covalent bonds. Staudinger's views received convincing support in a 1929 paper by Wallace H. Carothers of Du Pont who reached similar conclusions.

Staudinger's studies of polymer structure and Carothers's achievements in polymer synthesis accelerated the development of polymer chemistry, especially its shift from chemical modification of natural polymers to the design and synthesis of new materials. Thousands of synthetic polymers are now known; some mimic the properties of natural materials, others have superior properties and have replaced natural materials.

Staudinger received the 1953 Nobel Prize in Chemistry for his studies of polymers. Were it not for his untimely death in 1937, Carothers would likely have shared in the award.

27.2 Polymer Nomenclature

A monomer is any compound from which a polymer can be prepared.

Although the IUPAC has set forth rules for naming polymers according to structure, an alternative IUPAC *source-based* system that names polymers according to the **monomers** from which they are prepared is more widely used.

Source-based names are, for example, the ones we are accustomed to seeing for polymers such as polyethylene (see Section 10.8) and polystyrene (see Section 12.14). When the name of the monomer is a single word, the polymer derived from it is generated by simply adding the prefix *poly-*. When the name of the monomer consists of two words, both words are enclosed in parentheses immediately following *poly*. Thus, polyacrylonitrile and poly(vinyl chloride) are the polymers of acrylonitrile and vinyl chloride, respectively.

$H_2C{=}CH{-}C{\equiv}N$	$\left[CH_2{-}CH(C{\equiv}N)\right]_n$	$H_2C{=}CH{-}Cl$	$\left[CH_2{-}CH(Cl)\right]_n$
Acrylonitrile	Polyacrylonitrile	Vinyl chloride	Poly(vinyl chloride)

The convention for writing polymer formulas is to enclose the **repeating unit** within brackets, followed by the letter n to indicate that the number of repeating units is not specified. It is, however, assumed to be large.

Problem 27.1

Structural formulas for acrylic and methacrylic acids are as shown. Give the names of the polymers requested in (a) and (b) and represent their structures in the bracketed repeating unit format.

$H_2C{=}C(R){-}C(=O)OH$

R = H; Acrylic acid

R = CH_3; Methacrylic acid

(a) The amide of acrylic acid (acrylamide)
(b) The methyl ester of methacrylic acid (methyl methacrylate)

Sample Solution (a) *Acrylamide* is one word; therefore, its polymer is *polyacrylamide.* The repeating unit follows the pattern illustrated for polyacrylonitrile and poly(vinyl chloride).

$CH_2{=}CHC(O)NH_2$ — Acrylamide

$\left[CH_2{-}CH(CONH_2) \right]_n$ — Polyacrylamide

Source-based nomenclature does not require that a particular polymer actually be made from the "source" monomer. Both poly(ethylene glycol) and poly(ethylene oxide), for example, are made from ethylene oxide and have the same repeating unit.

$$\left[CH_2CH_2O \right]_n$$

The structural difference between the two is that the value of n is larger for poly(ethylene oxide) than for poly(ethylene glycol). Therefore, their physical properties are different and they are known by different source-based names.

Many polymers are routinely referred to by their common names or trade names. The polymer $\left[CF_2CF_2 \right]_n$ is almost always called Teflon rather than polytetrafluoroethylene.

27.3 Classification of Polymers: Reaction Type

Structure, synthesis, production, and applications of polymers span so many disciplines that it is difficult to classify them in a way that serves every interest. Figure 27.2 compares some of the different ways. This section describes how polymers are classified according to the type of reaction—addition or condensation—that occurs.

Addition polymers are formed by reactions of the type:

$$A + B \longrightarrow A{-}B$$

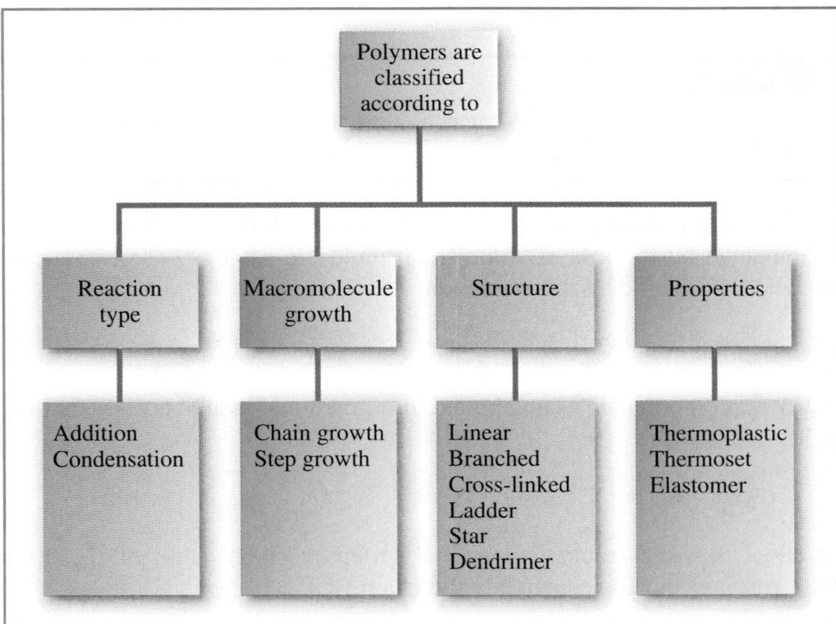

Figure 27.2

Classification of polymers.

where the product (A—B) retains all of the atoms of the reactants (A + B). In the general equation, A and B are monomers that react to give the polymer. When A = B, the resulting polymer is a **homopolymer.** Polystyrene is an example of a homopolymer.

$$C_6H_5-CH{=}CH_2 \longrightarrow \left[-\underset{C_6H_5}{CH}-CH_2- \right]_n$$

Styrene   Polystyrene

When the two monomers are different, the polymer is a **copolymer.** Saran, used as a protective wrap for food, is a copolymer of vinylidene chloride and vinyl chloride.

$$Cl_2C{=}CH_2 + ClCH{=}CH_2 \longrightarrow \left[-\overset{Cl}{\underset{Cl}{C}}-CH_2-\underset{Cl}{CH}-CH_2- \right]_n$$

Vinylidene chloride   Vinyl chloride   Saran

The two components in a copolymer need not be present in equal-molar amounts. In a typical Saran formulation vinylidene chloride is the major monomer (about 85%), and vinyl chloride the minor one.

Polymers prepared from alkenes, regardless of whether they are homopolymers or copolymers, are known as **polyolefins** and are the most familiar addition polymers.

Not all addition polymers are polyolefins. Formaldehyde, for example, polymerizes to give an addition polymer that retains all of the atoms of the monomer.

$$H_2C{=}O \rightleftharpoons \left[CH_2-O \right]_n$$

Formaldehyde   Polyformaldehyde

When monomeric formaldehyde is needed, to react with a Grignard reagent, for example, it is prepared as needed by heating the polymer in order to "depolymerize" it.

Problem 27.2

Under certain conditions formaldehyde forms a cyclic trimer ($C_3H_6O_3$) called *trioxane.* Suggest a structure for this compound.

Condensation polymers are prepared by covalent bond formation between monomers, accompanied by the loss of some small molecule such as water, an alcohol, or a hydrogen halide. The condensation reaction:

$$●-X + Y-○ \longrightarrow ●-○ + X-Y$$

gives a condensation polymer when applied to difunctional reactants. The first condensation step:

$$X-●-X + Y-○-Y \longrightarrow X-●-○-Y + X-Y$$

gives a product that has reactive functional groups. Condensation of these functional groups with reactant molecules extends the chain.

$$Y-○-Y + X-●-○-Y + X-●-X \longrightarrow Y-○-●-○-●-X + 2X-Y$$

The product retains complementary functional groups at both ends and can continue to grow.

The most familiar condensation polymers are polyamides, polyesters, and polycarbonates.

The **aramids,** polyamides in which aromatic rings are joined by amide bonds, are one class of condensation polymer. Heating 1,4-benzenediamine and the acyl chloride of benzene-1,4-dicarboxylic acid (terephthalic acid) gives the aramid *Kevlar* with loss of hydrogen chloride.

$$H_2N{-}C_6H_4{-}NH_2 \;+\; ClC(=O){-}C_6H_4{-}C(=O)Cl \;\longrightarrow\; \left[{-}NH{-}C_6H_4{-}NH{-}C(=O){-}C_6H_4{-}C(=O){-}\right]_n \;+\; nHCl$$

1,4-Benzenediamine | Terephthaloyl chloride | Kevlar | Hydrogen chloride

Kevlar fibers are both strong and stiff and used to make bulletproof vests and protective helmets (Figure 27.3).

Figure 27.3

Police and the military depend on body armor and helmets made of Kevlar fibers. Kevlar protective equipment is more effective than steel, yet far lighter in weight.

Source: U.S. Air Force photo by Tech. Sgt. Jim Varthegyi

Problem 27.3

The amide bond between a molecule of 1,4-benzenediamine and a molecule of terephthaloyl chloride is formed by the usual nucleophilic acyl substitution mechanism. Write a structural formula for the tetrahedral intermediate in this reaction.

27.4 Classification of Polymers: Chain Growth and Step Growth

Addition and *condensation* are familiar to us as reaction types in organic chemistry. The terms we apply to the two different ways that macromolecules arise from lower-molecular-weight units are unique to polymer chemistry and are illustrated in Figure 27.4.

In a **chain-growth** process monomers add one-by-one to the same end of a growing chain (Figure 27.4*a*). Each chain has only one growth point. The concentration of monomer decreases gradually until it is depleted.

The terms *chain growth* and *step growth* are attributed to Paul Flory, who was awarded the 1974 Nobel Prize in Chemistry for his studies on the physical chemistry of polymers.

(*a*) **Chain growth:** Monomers add one-by-one to the same end of a growing chain.

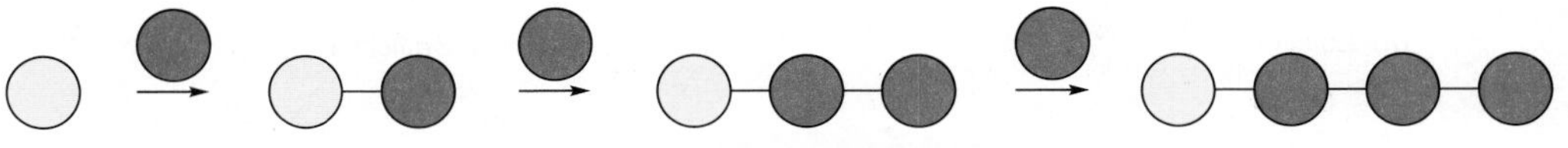

(*b*) **Step growth:** A mixture of polymers of intermediate length (oligomers) form. These oligomers react together to give longer chains.

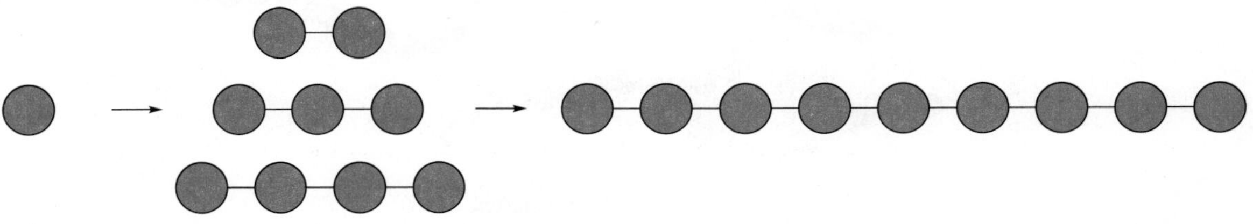

Figure 27.4

Chain-growth (*a*) and step-growth (*b*) polymerization. During chain growth, the amount of monomer remaining decreases gradually. In step growth, most of the monomer is consumed early and the molecular weight of the polymer increases as oligomers combine to form longer chains.

In a **step-growth** process (Figure 27.4*b*), chains have at least two growth points. Most of the monomer molecules are consumed early in the process to give a mixture of compounds of intermediate molecular weight called **oligomers.** These oligomers react with one another to form the polymer. The molecular weight continues to increase even after all the monomer molecules have reacted.

In general, chain growth is associated with addition polymerization and step growth with condensation polymerization. It is not always so, however. We'll see in Section 27.14 that polyurethanes are addition polymers in which step growth, not chain growth, characterizes macromolecule formation.

Problem 27.4

We can anticipate this "later in the chapter" example by examining the reaction:

$$ROH + R'N{=}C{=}O \longrightarrow R{-}O{-}C({=}O){-}N(H){-}R'$$

Is this an addition reaction or a condensation?

27.5 Classification of Polymers: Structure

Polymers made from the same compounds can have different properties depending on how they are made. These differences in physical properties result from differences in the overall *structure* of the polymer chain. The three major structural types—linear, branched, and cross-linked—are illustrated in Figure 27.5. Other, more specialized, structural types—ladders, stars, and dendrimers—have unique properties and are under active investigation.

Linear polymers (Figure 27.5*a*) have a continuous chain of repeating units. The repeating units within the chain are subject to the usual conformational requirements of organic chemistry. The collection of chains can range from *random,* much like a bowl

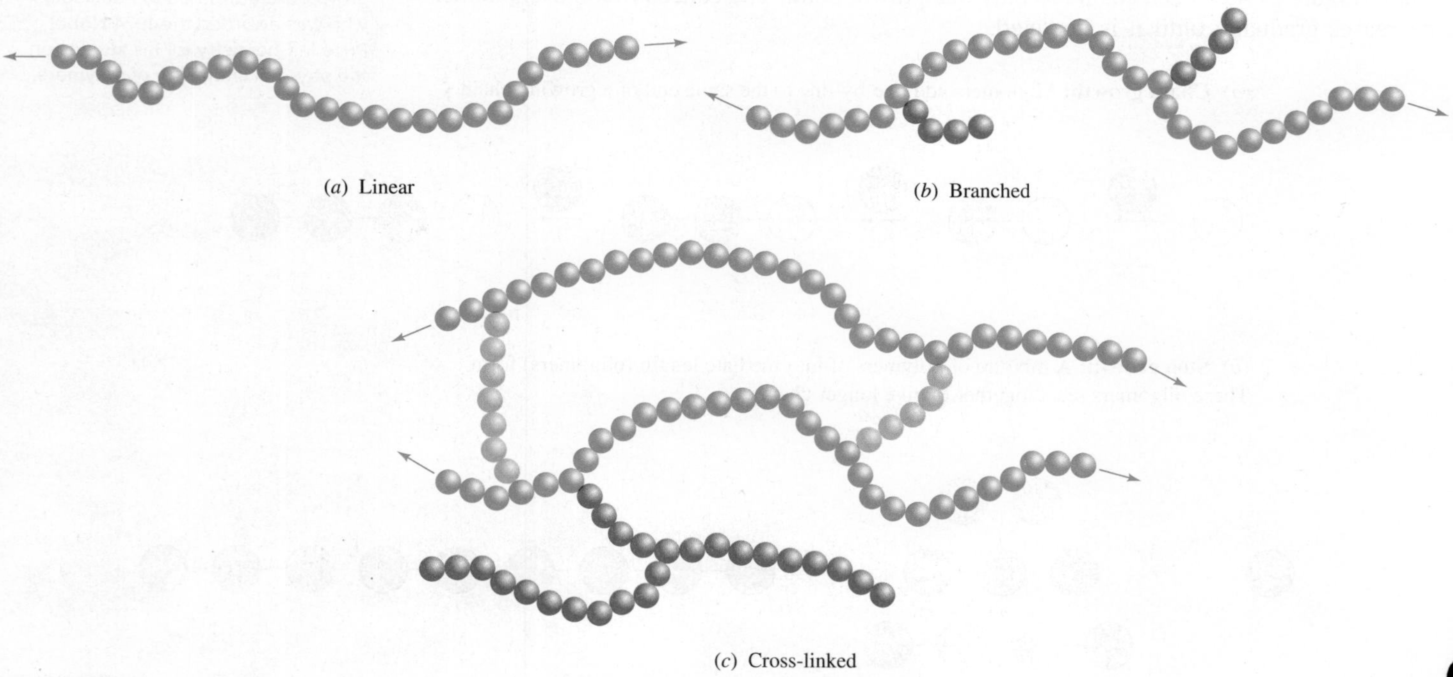

Figure 27.5

(*a*) A linear polymer has a continuous chain. (*b*) A branched polymer has relatively short branches connected to the main chain. (*c*) A cross-linked polymer has covalently bonded linking units between chains. The main chains are shown in blue, the branches in red, and the cross links in yellow.

of spaghetti, to *ordered*. We describe polymers at the random extreme as *amorphous* and those at the ordered extreme as *crystalline*.

Most polymers are a mixture of random tangles interspersed with crystalline domains called **crystallites** (Figure 27.6). The degree of crystallinity of a polymer, that is, the percentage of crystallites, depends on the strength of intermolecular forces between chains. For a particular polymer, density increases with crystallinity because randomly coiled chains consume volume, while closer packing puts the same mass into a smaller volume. The efficiency with which the chains can pack together is strongly affected by the extent to which the chain is branched.

Branched polymers (Figure 27.5*b*) have branches extending from the main chain. In general, increased branching reduces the crystallinity of a polymer and alters properties such as density.

Contrast the properties of low-density polyethylene (LDPE) and high-density (HDPE), two of the six polymers familiar enough to have their own identifying codes for recycling (Table 27.1). Both are homopolymers of ethylene, but are prepared by different methods and have different properties and uses. As their names imply, LDPE has a lower density than HDPE (0.92 g/cm^3 versus 0.96 g/cm^3). LDPE is softer, HDPE more rigid. LDPE has a lower melting point than HDPE. LDPE is the plastic used for grocery store bags; HDPE is stronger and used for water bottles, milk jugs, and gasoline tanks.

The structural difference between the two is that LDPE is more branched, averaging about 20 branches for every thousand carbon atoms compared with about 5 per thousand for HDPE. The greater density of HDPE results from packing more mass into the same volume. Unbranched chains pack more efficiently than branched ones, which translates into stronger intermolecular forces, greater crystallinity, and a tougher, more durable material.

Like HDPE, isotactic polypropylene is highly crystalline with numerous uses, including fibers for rope and carpets. Atactic polypropylene, on the other hand, is much less crystalline and has few applications.

Stereoregular polymers including isotactic polypropylene were described in Section 15.15.

Chains in a **cross-linked** or **network polymer** (Figure 27.5*c*) are connected to one another by linking units, which may be long or short and composed of the same repeating units as the main chain or different ones. Vulcanization, for example, uses sulfur to cross-link the hydrocarbon chains of natural rubber. In general, cross linking increases rigidity by

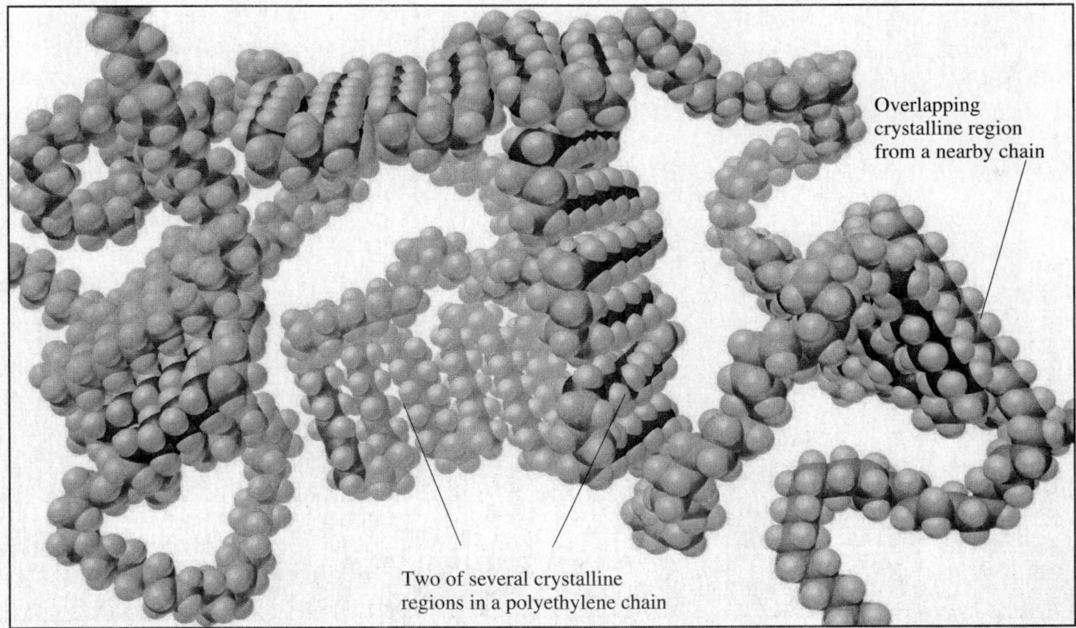

Figure 27.6

Polyethylene contains both randomly coiled (amorphous) and ordered (crystalline) regions. The ordered regions (crystallites) of one chain are shown in a darker color than the random main chain. Crystallites involving the main chain with neighboring ones are in red and yellow.

TABLE 27.1 Recycling of Plastics

Symbol	Polymer	Some uses*	
		New	**Recycled**
1 PETE	Poly(ethylene terephthalate)	Polyester textile fibers, tire cords, photographic film, soft drink and water bottles, food jars	Carpet fibers, detergent bottles, bathtubs, car parts, audio- and videotapes
2 HDPE	High-density polyethylene	Bottles, automobile fuel tanks, milk jugs, bags, cereal box liners	Plastic lumber for exterior uses (picnic tables, mailboxes, decks, trash bins, planters)
3 V	Poly(vinyl chloride)	Floor tiles, vinyl siding, plumbing pipe, gutters and downspouts, garden hoses, shower curtains, window frames, blister packs	Many of the uses of recycled poly(vinyl chloride) are the same as those of new material
4 LDPE	Low-density polyethylene	Trash bags, packaging, squeezable bottles, grocery bags	Packaging film and bags
5 PP	Polypropylene	Indoor–outdoor carpet, rope, medicine bottles, packaging	Indoor–outdoor carpet, rope, fishing nets, tarpaulins, auto parts
6 PS	Polystyrene	Television cabinets, luggage, egg cartons, toys, Styrofoam cups, appliances	Styrofoam insulation and packaging, coat hangers, containers
7 OTHER	Other (acrylics, nylon, polycarbonates, etc.)	5-Gallon reusable water bottles, automobile bumpers and other parts, tires, telephones, safety helmets	

*The uses of new and recycled plastics are often the same, and many products are a mixture of new and recycled material.

restricting the movement of the polymer chains. Vulcanized rubber is a lightly cross-linked elastomer; Bakelite can be so highly cross-linked as to be considered a single molecule.

27.6 Classification of Polymers: Properties

How a polymer responds to changes in temperature is important not only with respect to the conditions under which it can be used, but also in the methods by which it is transformed into a commercial product.

Thermoplastic polymers are the most common and are those that soften when heated. At their *glass transition temperature* (T_g), thermoplastic polymers change from a glass to a flexible, rubbery state. Past this point amorphous polymers are gradually transformed to a liquid as the temperature is raised. Crystalline polymers undergo a second transition, liquefying only when the *melting temperature* (T_m) is reached. Compare the behaviors of atactic, isotactic, and syndiotactic poly(methyl methacrylate) on being heated.

Poly(methyl methacrylate)	T_g(°C)	T_m(°C)
atactic	114	—
isotactic	48	160
syndiotactic	126	200

$$\left[\begin{array}{c} CO_2CH_3 \\ | \\ -C-CH_2- \\ | \\ CH_3 \end{array} \right]_n$$

The atactic form of poly(methyl methacrylate) is amorphous and exhibits only one transition temperature (T_g). The stereoregular isotactic and syndiotactic forms are partially crystalline and undergo both a glass transition and melting.

The process that takes place at T_g is an increase in the conformational mobility of the polymer chains. At T_m, attractive forces in crystallites are broken and individual chains separate.

Melting temperature is an important factor in respect to how polymers are used. The relatively low T_m for low-density polyethylene (115°C) makes it an easy polymer to cast into the desired shape when melted, but at the same time limits its applications. When, for example, a container is required that must be sterilized by heating, the higher T_m of HDPE (137°C) makes it a better choice than LDPE.

Unlike thermoplastic polymers that soften on heating, **thermosetting polymers** (also called *thermosetting resins*) pass through a liquid state then solidify ("cure") on continued heating. The solidified material is a **thermoset.** It is formed by irreversible chemical reactions that create cross links as the thermosetting polymer is heated. *Bakelite,* a highly cross-linked thermoset made from phenol and formaldehyde, is prepared in two stages. In the first stage, condensation between phenol and formaldehyde gives a polymer, which, in its fluid state, is cast in molds and heated, whereupon it solidifies to a hard, rigid mass. The chemical reactions that form the fluid polymer and the solid thermoset are the same kind of condensations; the difference is that there are more cross links in the thermoset. *Melamine* (used in plastic dinnerware) is another example of a thermoset.

Elastomers are flexible polymers that can be stretched but return to their original state when the stretching force is released. Most amorphous polymers become rubbery beyond their glass transition temperature, but not all rubbery polymers are elastic. Cross links in elastomers limit the extent to which elastomers can be deformed then encourage them to return to their original shape when they are relaxed.

27.7 Addition Polymers: A Review and a Preview

Addition polymers are most familiar to us in connection with the polymerization of alkenes.

$$\mathrm{C=C} \longrightarrow \left[-\mathrm{C}-\mathrm{C}- \right]_n$$

Table 27.2 reviews alkene polymerizations that proceed by free radicals and by coordination complexes of the Ziegler–Natta type. Both are chain-growth processes; their propagation steps were outlined in Mechanisms 10.4 and 15.4, respectively. The present section examines two other significant factors in alkene polymerization: initiation and termination.

Initiators of Alkene Polymerization: Whether free-radical or coordination polymerization occurs depends primarily on the substance used to initiate the reaction. Free-radical polymerization occurs when a compound is present that undergoes homolytic bond cleavage when heated. Two examples include

Di-*tert*-butyl peroxide —(100°C)→ Two *tert*-butoxy radicals

Azobisisobutyronitrile (AIBN) —(50°C)→ Two 1-cyano-1-methylethyl radicals + :N≡N: (Nitrogen)

TABLE 27.2 Summary of Alkene Polymerizations Discussed in Earlier Chapters

Reaction (section) and comments	Example
Free-radical polymerization of alkenes (see Section 10.8) Many alkenes polymerize when treated with free-radical initiators. A free-radical chain mechanism is followed and was illustrated for the case of ethylene in Mechanism 10.4.	$H_2C{=}CH_2$ (Ethylene) —(200°C, 2000 atm; O_2 or peroxides)→ Polyethylene
Free-radical polymerization of dienes (see Section 11.9) Conjugated dienes undergo free-radical polymerization under conditions similar to those of alkenes. The major product corresponds to 1,4-addition.	2-Chloro-1,3-butadiene (Chloroprene) —(free-radical initiator)→ Polychloroprene
Free-radical polymerization of styrene (see Section 12.14) Styrene can be polymerized under free-radical, cationic, anionic, and Ziegler–Natta conditions. The mechanism of the free-radical polymerization was shown in Mechanism 12.1.	Styrene —(benzoyl peroxide)→ Polystyrene
Ring-opening metathesis polymerization (see Section 15.14) The double bonds of strained cyclic alkenes are cleaved by certain carbene complexes of tungsten and, in the process, undergo polymerization.	Bicyclo[2.2.1]-2-heptene (Norbornene) —(catalyst; −80°C)→ Polynorbornene
Coordination polymerization (see Section 15.15) Organometallic compounds such as bis(cyclopentadienyl)zirconium dichloride (Cp_2ZrCl_2) catalyze the polymerization of ethylene by the sequence of steps shown in Mechanism 15.4.	$H_2C{=}CH_2$ (Ethylene) —(Cp_2ZrCl_2; methalumoxane)→ Polyethylene

Problem 27.5

(a) Write a chemical equation for the reaction in which *tert*-butoxy radical adds to vinyl chloride to initiate polymerization. Show the flow of electrons with curved arrows.

(b) Repeat part (a) for the polymerization of styrene using AIBN as an initiator.

Sample Solution (a) *tert*-Butoxy radical adds to the CH_2 group of vinyl chloride. The free radical formed in this process has its unpaired electron on the carbon bonded to chlorine.

tert-Butoxy radical + Vinyl chloride → 2-*tert*-Butoxy-1-chloroethyl radical

Coordination polymerization catalysts are complexes of transition metals. The original Ziegler–Natta catalyst, a mixture of titanium tetrachloride and diethylaluminum chloride, has been joined by numerous organometallic complexes such as the widely used bis(cyclopentadienyl)zirconium dichloride.

Bis(cyclopentadienyl)zirconium dichloride

Termination Steps in Alkene Polymerization: The main chain-terminating processes in free-radical polymerization are *combination* and *disproportionation*. In a combination, the pairing of the odd electron of one growing radical chain with that of another gives a stable macromolecule.

$$RO\!\left[CH_2CH_2\right]_x CH_2\dot{C}H_2 \; + \; H_2\dot{C}CH_2\!\left[CH_2CH_2\right]_y OR \longrightarrow$$

Two growing polyethylene chains

$$RO\!\left[CH_2CH_2\right]_x CH_2CH_2{-}CH_2CH_2\!\left[CH_2CH_2\right]_y OR$$

Terminated polyethylene

In disproportionation, two alkyl radicals react by hydrogen-atom transfer. Two stable molecules result; one terminates in a methyl group, the other in a double bond.

$$RO\!\left[CH_2CH_2\right]_x CH_2{-}\dot{C}H_2 \; + \; H_2\dot{C}{-}CH(H)\!\left[CH_2CH_2\right]_y OR \longrightarrow$$

Two growing polyethylene chains

$$RO\!\left[CH_2CH_2\right]_x CH_2{-}CH_2H \; + \; H_2C{=}CH\!\left[CH_2CH_2\right]_y OR$$

Methyl-terminated polyethylene | Double-bond-terminated polyethylene

Both combination and disproportionation consume free radicals and decrease the number of growing chains. Because they require a reaction between two free radicals, each of which is present in low concentration, they have a low probability compared with chain growth, in which a radical reacts with a monomer. Combination involves only bond making and has a low activation energy; disproportionation has a higher activation energy because bond breaking accompanies bond making. Disproportionation has a more adverse effect on chain length and molecular weight than combination.

Problem 27.6

Other than combination, a macromolecule of the type $RO{-}\!\left[CH_2CH_2\right]_x\!{-}CH_2{-}CH_2{-}OR$ can arise by a different process, one which also terminates chain growth. Show a reasonable reaction and represent the flow of electrons by curved arrows.

Among several chain terminating reactions that can occur in coordination polymerization, a common one is an elimination in which a β-hydrogen is transferred to the metal.

$$\text{Cp}_2\overset{+}{Zr}{-}CH_2{-}CH(H)(R) \longrightarrow \text{Cp}_2\overset{+}{Zr}{-}H + H_2C{=}CHR$$

27.8 Chain Branching in Free-Radical Polymerization

Even with the same monomer, the properties of a polymer can vary significantly depending on how it is prepared. Free-radical polymerization of ethylene gives low-density polyethylene; coordination polymerization gives high-density polyethylene. The properties are different because the structures are different, and the difference in the structures comes from the mechanisms by which the polymerizations take place. Free-radical polymerization of ethylene gives a branched polymer, coordination polymerization gives a linear one.

What is the mechanism responsible for the branching that occurs in the free-radical polymerization of ethylene?

By itself, the propagation step in the free-radical polymerization of ethylene cannot produce branches.

$$\boxed{\text{Polymer}}{-}CH_2{-}\dot{C}H_2 \quad H_2C{=}CH_2 \longrightarrow \boxed{\text{Polymer}}{-}CH_2{-}CH_2{-}CH_2{-}\dot{C}H_2$$

In order for the polymer to be branched, an additional process must occur involving a radical site somewhere other than at the end of the chain. The two main ways this can happen both involve hydrogen abstraction from within the polymer chain.

1. Intramolecular hydrogen-atom abstraction
2. Intermolecular hydrogen-atom abstraction (chain transfer)

Intramolecular Hydrogen-Atom Abstraction: Mechanism 27.1 shows how intramolecular hydrogen atom abstraction can lead to the formation of a four-carbon branch. Recall that an intramolecular process takes place *within* a molecule, not *between* molecules. As the mechanism shows, the radical at the end of the growing polymer abstracts a hydrogen atom from the fifth carbon. Five carbons and one hydrogen comprise six atoms of a cyclic transition state. When a hydrogen atom is removed from the fifth carbon, a secondary radical is generated at that site. This, then, is the carbon that becomes the origin for further chain growth. Analogous mechanisms apply to branches shorter or longer than four carbons.

Mechanism 27.1

Branching in Polyethylene Caused by Intramolecular Hydrogen Transfer

THE OVERALL REACTION:

$$\boxed{\text{Polymer}}-CH_2CH_2CH_2CH_2\dot{C}H_2 \xrightarrow{H_2C=CH_2} \boxed{\text{Polymer}}-\underset{\displaystyle CH_2CH_2CH_2CH_3}{CH}CH_2CH_2(CH_2CH_2)_nCH_2\dot{C}H_2$$

THE MECHANISM:

Step 1: The carbon at the end of the chain—the one with the unpaired electron—abstracts a hydrogen atom from the fifth carbon. The transition state is a cyclic arrangement of six atoms.

The resulting radical is secondary and more stable than the original primary radical. Therefore, the hydrogen atom abstraction is exothermic.

$$\boxed{\text{Polymer}}-\dot{C}H-CH_2CH_2CH_2CH_3$$

Step 2: When the radical reacts with ethylene, chain extension takes place at the newly formed radical site. The product of this step has a four-carbon branch attached to the propagating chain.

$$\boxed{\text{Polymer}}-\underset{\displaystyle CH_2CH_2CH_2CH_3}{\dot{C}H} + H_2C=CH_2 \longrightarrow \boxed{\text{Polymer}}-\underset{\displaystyle CH_2CH_2CH_2CH_3}{CH}-CH_2-\dot{C}H_2$$

Step 3: Reaction with additional ethylene molecules extends the growing chain.

$$\boxed{\text{Polymer}}-\underset{\displaystyle CH_2CH_2CH_2CH_3}{CH}CH_2\dot{C}H_2 \xrightarrow{H_2C=CH_2} \boxed{\text{Polymer}}-\underset{\displaystyle CH_2CH_2CH_2CH_3}{CH}CH_2CH_2-(CH_2CH_2)_n-CH_2\dot{C}H_2$$

Problem 27.7

Suggest an explanation for the observation that branches shorter or longer than four carbons are found infrequently in polyethylene. Frame your explanation in terms of how ΔH and ΔS affect the activation energy for intramolecular hydrogen-atom abstraction.

A comparable process cannot occur when Ziegler–Natta catalysts are used because free radicals are not intermediates in coordination polymerization.

Intermolecular Hydrogen-Atom Abstraction (Chain Transfer): Mechanism 27.2 shows how a growing polymer chain abstracts a hydrogen atom from a terminated chain. The original growing chain is now terminated, and the original terminated chain is activated toward further growth. Chain growth, however, occurs at the branch point, not at the end of the chain. An already long chain adds a branch while terminating a (presumably shorter) growing chain. Chain transfer not only leads to branching, but also encourages disparity in chain lengths—more short chains and more long branched chains. Both decrease the crystallinity of the polymer and reduce its strength.

As in the case of intramolecular hydrogen abstraction, branching by chain transfer is not a problem when alkenes are polymerized under Ziegler–Natta conditions because free radicals are not intermediates in coordination polymerization.

Mechanism 27.2

Branching in Polyethylene Caused by Intermolecular Hydrogen Transfer

Step 1: A growing polymer chain abstracts a hydrogen atom from a terminated chain. This step terminates the growing chain and activates the terminated one.

$$\boxed{\text{Polymer}}-CH_2-\dot{C}H_2 + \boxed{\text{Polymer}}-\underset{}{\overset{H}{\overset{|}{C}}H}-\boxed{\text{Polymer}}$$

Growing chain Terminated chain

$$\downarrow$$

$$\boxed{\text{Polymer}}-CH_2-CH_3 + \boxed{\text{Polymer}}-\dot{C}H-\boxed{\text{Polymer}}$$

Terminated chain Growing chain

Step 2: Reaction of the new chain with monomer molecules produces a branch at which future growth occurs.

$$\boxed{\text{Polymer}}-\dot{C}H-\boxed{\text{Polymer}} + H_2C{=}CH_2 \longrightarrow \boxed{\text{Polymer}}-\underset{}{\overset{\dot{C}H_2-CH_2-}{}}CH-\boxed{\text{Polymer}}$$

Growing chain Ethylene Growing branched chain

27.9 Anionic Polymerization: Living Polymers

Anionic polymerization is a useful alternative to free-radical and Ziegler–Natta procedures for certain polymers. Adding butyllithium to a solution of styrene in tetrahydrofuran (THF), for example, gives polystyrene.

$$C_6H_5-CH{=}CH_2 \xrightarrow[\text{THF}]{CH_3CH_2CH_2CH_2Li} \left[\,CH(C_6H_5)-CH_2\,\right]_n$$

Styrene Polystyrene

Mechanism 27.3 shows how addition of butyllithium to the double bond of styrene initiates polymerization. The product of this step is a benzylic carbanion that then adds to a second molecule of styrene to give another benzylic carbanion, and so on by a chain-growth process.

Polystyrene formed under these conditions has a narrower range of molecular weights than provided by other methods. Initiation of polymerization by addition of butyllithium to styrene is much faster than subsequent chain growth. Thus, all the butyllithium is consumed and the number of chains is equal to the number of molecules of butyllithium used. These starter chains then grow at similar rates to produce similar chain lengths.

Problem 27.8

How will the average chain length of polystyrene vary with the amount of butyllithium used to initiate polymerization?

As shown in step 3 of Mechanism 27.3, once all of the monomer is consumed the polymer is present as its organolithium derivative. This material is referred to as

Mechanism 27.3

Anionic Polymerization of Styrene

Step 1: Anionic polymerization of styrene is initiated by addition of butyllithium to the double bond. The regioselectivity of addition is governed by formation of the more stable carbanion, which in this case is benzylic.

$$C_6H_5-CH=CH_2 + :\bar{C}H_2(CH_2)_2CH_3\ Li^+ \longrightarrow C_6H_5-\ddot{C}H-CH_2-CH_2(CH_2)_2CH_3\ Li^+$$

Styrene Butyllithium 1-Phenylhexyllithium

Step 2: The product of the first step adds to a second molecule of styrene.

$$C_6H_5-CH=CH_2 + C_6H_5-\ddot{C}H-CH_2(CH_2)_3CH_3\ Li^+ \longrightarrow C_6H_5-\ddot{C}H(Li^+)-CH_2-CH(C_6H_5)-CH_2(CH_2)_3CH_3$$

Styrene + 1-Phenylhexyllithium 1,3-Diphenyloctyllithium

Step 3: The product of the second step adds to a third molecule of styrene, then a fourth, and so on to give a macromolecule. Reaction continues until all of the styrene is consumed. At this point the polystyrene exists as an organolithium reagent.

$$C_6H_5-\ddot{C}H(Li^+)-CH_2-\left[CH(C_6H_5)-CH_2\right]_n-(CH_2)_3CH_3$$

The organolithium reagent is stable, but easily protonated by water to give polystyrene. Alternatively, another monomer can be added to continue extending the chain.

a **living polymer** because more monomer can be added and anionic polymerization will continue until the added monomer is also consumed. Adding 1,3-butadiene, for example, to a living polymer of styrene gives a new living polymer containing sections ("blocks") of polystyrene and poly(1,3-butadiene).

$$C_6H_5-\ddot{C}H(Li^+)-CH_2-\left[CH(C_6H_5)-CH_2\right]_n-(CH_2)_3CH_3 + H_2C=CH-CH=CH_2$$

"Living" polystyrene 1,3-Butadiene

$$\downarrow$$

$$Li^+ :\bar{C}H_2-CH=CH-CH_2-\left[CH_2-CH=CH-CH_2\right]_m-CH(C_6H_5)-CH_2-\left[CH(C_6H_5)-CH_2\right]_n-(CH_2)_3CH_3$$

"Living" styrene-butadiene copolymer

Living polymerizations are characterized by the absence of efficient termination processes. They are normally terminated by intentionally adding a substance that reacts with carbanions such as an alcohol or carbon dioxide.

The kinds of vinyl monomers that are susceptible to anionic polymerization are those that bear electron-withdrawing groups such as —C≡N and —C(=O)— on the double bond.

Acrylonitrile Methyl acrylate Methyl 2-cyanoacrylate

When a carbonyl and a cyano group are attached to the same carbon as in methyl 2-cyanoacrylate, the monomer that constitutes *Super Glue,* anionic polymerization can be initiated by even weak bases such as atmospheric moisture or normal skin dampness.

Problem 27.9

Write a structural formula for the carbanion formed by addition of hydroxide ion to methyl 2-cyanoacrylate. Accompany this structural formula by a contributing resonance structure that shows delocalization of the negative charge to oxygen, and another to nitrogen.

Sample Solution

• add HO^- to β carbon

• delocalize electron pair from C to O

• delocalize electron pair from C to N

27.10 Cationic Polymerization

Analogous to the initiation of anionic polymerization by addition of nucleophiles to alkenes, cationic polymerization can be initiated by the addition of electrophiles. The alkenes that respond well to cationic polymerization are those that form relatively stable carbocations when protonated. Of these, the one used most often is 2-methylpropene, better known in polymer chemistry by its common name *isobutylene*. Mechanism 27.4 outlines the mechanism of this polymerization as catalyzed by boron trifluoride to which a small amount of water has been added. The active catalyst is believed to be a Lewis acid/Lewis base complex formed from them by the reaction:

$$H_2\ddot{O}: + BF_3 \longrightarrow H_2\overset{+}{O}-\overset{-}{B}F_3$$

Water | Boron trifluoride | Water/Boron trifluoride complex

This complex is a strong Brønsted acid and protonates the double bond of 2-methylpropene in step 1 of the mechanism.

Mechanism 27.4

Cationic Polymerization of 2-Methylpropene

THE OVERALL REACTION:

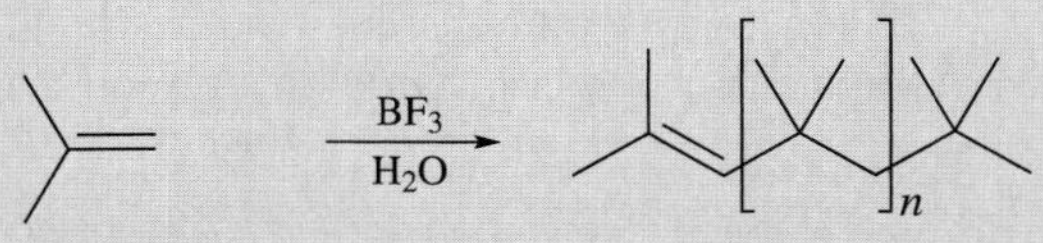

2-Methylpropene $\xrightarrow[H_2O]{BF_3}$ Polyisobutylene

THE MECHANISM:

Step 1: The alkene is protonated, forming a carbocation.

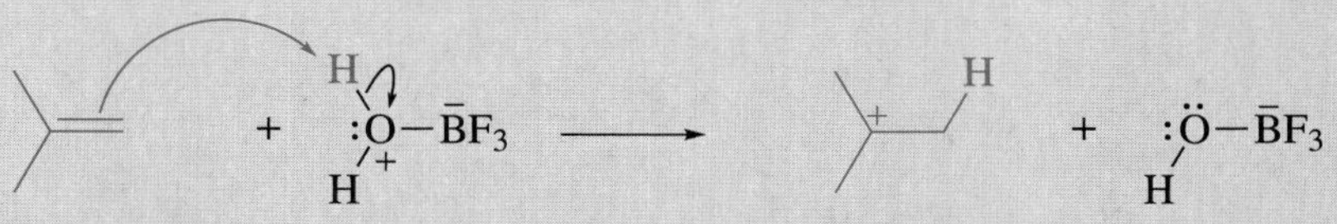

2-Methylpropene + $H_2\overset{+}{O}-\overset{-}{B}F_3$ ⟶ *tert*-Butyl cation + $H\ddot{O}-\overset{-}{B}F_3$

Step 2: The carbocation formed in the preceding step reacts with a molecule of the alkene, forming a new carbocation.

2-Methylpropene + *tert*-Butyl cation ⟶ 1,1,3,3-Tetramethylbutyl cation

Step 3: The process shown in step 2 continues, forming a chain-extended carbocation.

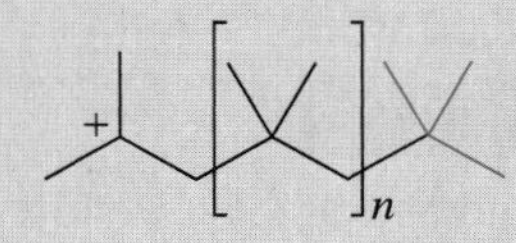

Step 4: One mechanism for chain termination is loss of a proton.

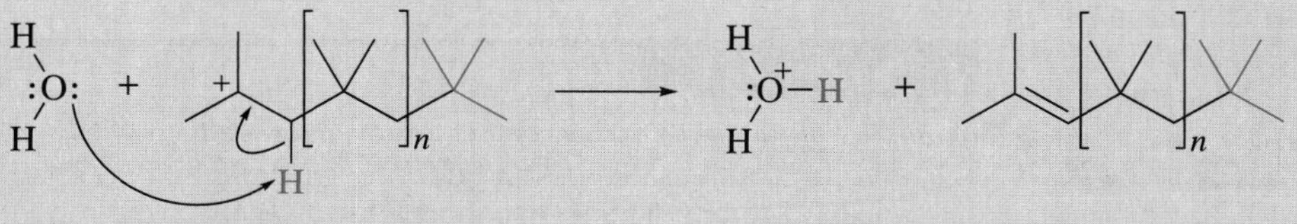

$H_2\ddot{O}:$ + carbocation ⟶ $H_2\overset{+}{O}-H$ + polyisobutylene

Polyisobutylene is the "butyl" in butyl rubber, one of the first synthetic rubber substitutes. Most inner tubes are a copolymer of 2-methylpropene (isobutylene) and 2-methyl-1,3-butadiene (isoprene).

27.11 Polyamides

The polyamide nylon 66 takes its name from the fact that it is prepared from a six-carbon dicarboxylic acid and a six-carbon diamine. The acid–base reaction between adipic acid and hexamethylenediamine gives a salt, which on heating undergoes condensation polymerization in which the two monomers are joined by amide bonds.

The systematic names of adipic acid and hexamethylenediamine are hexanedioic acid and 1,6-hexanediamine, respectively.

$$^{-}\text{O}\overset{\text{O}}{\overset{\|}{\text{C}}}(\text{CH}_2)_4\overset{\text{O}}{\overset{\|}{\text{C}}}\text{O}^{-}\ \ \text{H}_3\overset{+}{\text{N}}(\text{CH}_2)_6\overset{+}{\text{N}}\text{H}_3 \xrightarrow[-\text{H}_2\text{O}]{280\text{–}300^\circ\text{C}} \left[\text{NH}(\text{CH}_2)_6\text{NH}\overset{\text{O}}{\overset{\|}{\text{C}}}(\text{CH}_2)_4\overset{\text{O}}{\overset{\|}{\text{C}}}\right]_n$$

Salt of adipic acid and hexamethylenediamine — Nylon 66

Nylon 66 was the first and remains the most commercially successful synthetic polyamide (Figure 27.7). Others have been developed by varying the number of carbons in the chains of the diamine and the dicarboxylic acid.

Nylon 66 resembles silk in both structure and properties. Both are polyamides in which hydrogen bonds provide an ordered arrangement of adjacent chains.

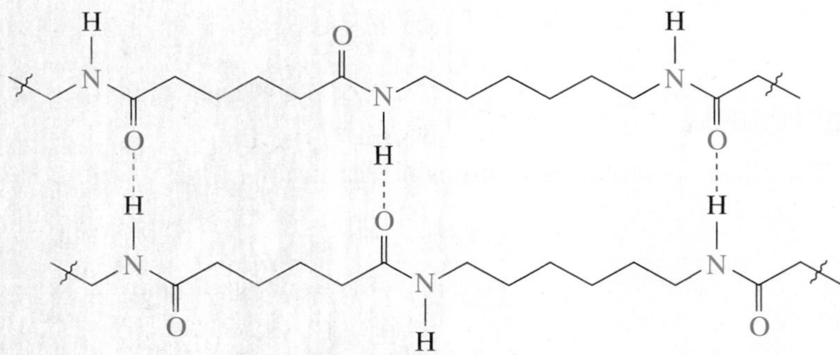

A variation on the diamine/dicarboxylic acid theme is to incorporate the amino and carboxylic acid groups into the same molecule, much as Nature does in amino acids. Nylon 6 is a polyamide derived by heating 6-aminohexanoic acid.

$$\text{H}_3\overset{+}{\text{N}}(\text{CH}_2)_5\text{CO}_2^{-} \xrightarrow{\text{heat}} \left[\text{NH}(\text{CH}_2)_5\overset{\text{O}}{\overset{\|}{\text{C}}}\right]_n + \text{H}_2\text{O}$$

6-Aminohexanoic acid — Nylon 6 — Water

Figure 27.7

Skydivers' parachutes are made of nylon 66.

©Mariusika11/Alamy Stock Photo

Problem 27.10

Nylon 6 is normally prepared from the lactam derived from 6-aminohexanoic acid, called ε-caprolactam. Do you remember what a lactam is? Write the structure of ε-caprolactam.

Problem 27.11

Nomex is an aramid fiber used for fire-resistant protective clothing. It is a polyamide prepared by condensation of 1,3-benzenediamine (*m*-phenylenediamine) and 1,3-benzenedicarboxylic acid (isophthalic acid). What is the repeating unit of Nomex?

27.12 Polyesters

The usual synthetic route to a polyester is by condensation of a dicarboxylic acid with a diol. The best known polyester is poly(ethylene terephthalate) prepared from ethylene glycol and terephthalic acid.

The dimethyl ester of terephthalic acid is used in an analogous method.

$$\text{HO}\overset{\text{O}}{\overset{\|}{\text{C}}}\text{—C}_6\text{H}_4\text{—}\overset{\text{O}}{\overset{\|}{\text{C}}}\text{OH} + \text{HOCH}_2\text{CH}_2\text{OH} \xrightarrow[-\text{H}_2\text{O}]{200\text{–}300°\text{C}} \left[\text{OCH}_2\text{CH}_2\text{O—}\overset{\text{O}}{\overset{\|}{\text{C}}}\text{—C}_6\text{H}_4\text{—}\overset{\text{O}}{\overset{\|}{\text{C}}}\right]_n$$

Terephthalic acid (Benzene-1,4-dicarboxylic acid) — Ethylene glycol — Poly(ethylene terephthalate)

The popularity of clothing made of polyester-cotton blends testifies to the economic impact of this polymer. Poly(ethylene terephthalate) is the PETE referred to in the recycling codes listed in Table 27.1. Plastic bottles for juice, ketchup, and soft drinks are usually made of PETE, as are Mylar film and Dacron fibers (Figure 27.8).

Alkyd resins number in the hundreds and are used in glossy paints and enamels—house, car, and artist's—as illustrated in Figure 27.9. Most are derived from benzene-1,2-dicarboxylic acid (*o*-phthalic acid) and 1,2,3-propanetriol (glycerol). Two of the hydroxyl groups of glycerol are converted to esters of *o*-phthalic acid; the third is esterified with an unsaturated fatty acid that forms cross links to other chains.

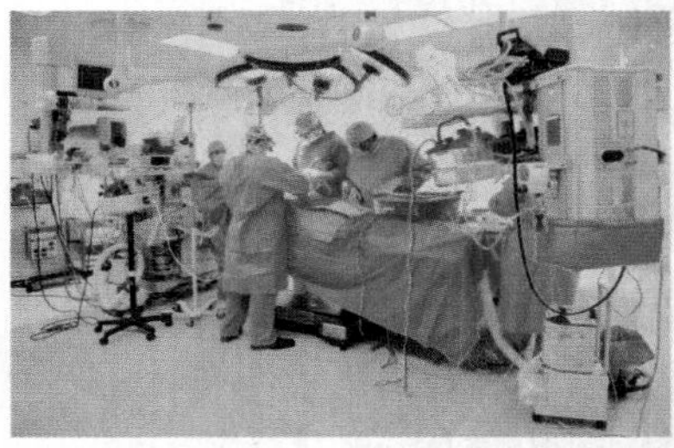

Figure 27.8

Dacron is widely used as the material in surgical sutures.

$$\left[\text{OCH}_2\underset{\underset{\text{O—}\overset{\text{O}}{\overset{\|}{\text{C}}}\text{—R}}{|}}{\text{CH}}\text{CH}_2\text{O—}\overset{\text{O}}{\overset{\|}{\text{C}}}\text{—}(o\text{-C}_6\text{H}_4)\text{—}\overset{\text{O}}{\overset{\|}{\text{C}}}\right]_n$$

An alkyd resin

With both a hydroxyl group and a carboxylic acid function in the same molecule, glycolic acid and lactic acid have the potential to form polyesters. Heating the α-hydroxy acid gives a cyclic diester, which, on treatment with a Lewis acid catalyst ($SnCl_2$ or SbF_3) yields the polymer.

$$\text{HO}\underset{\underset{\text{R}}{|}}{\text{CH}}\overset{\text{O}}{\overset{\|}{\text{C}}}\text{OH} \xrightarrow[-\text{H}_2\text{O}]{\text{heat}} \text{cyclic diester} \xrightarrow{\text{Lewis acid}} \left[\text{O}\underset{\underset{\text{R}}{|}}{\text{CH}}\overset{\text{O}}{\overset{\|}{\text{C}}}\right]_n$$

R = H	Glycolic acid	Glycolide	Poly(glycolic acid)
R = CH_3	Lactic acid	Lactide	Poly(lactic acid)

Surgical sutures made from poly(glycolic acid) and poly(lactic acid), while durable enough to substitute for ordinary stitches, are slowly degraded by ester hydrolysis and don't require a return visit for their removal. Poly(glycolic acid) fibers also hold promise as a scaffold upon which to grow skin cells. This "artificial skin" is then applied to a wound to promote healing.

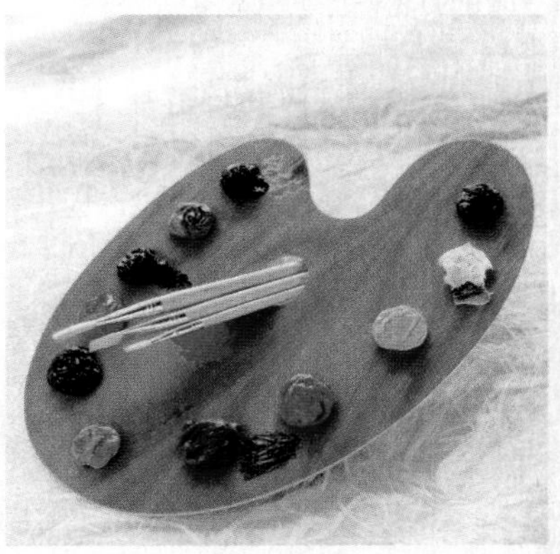

Figure 27.9

Alkyds are used for more than painting rooms. Artists use them too.

Problem 27.12

Another monomer from which surgical sutures are made is ε-caprolactone. What is the repeating unit of poly(ε-caprolactone)?

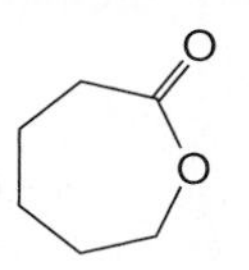

ε-Caprolactone

Polyesters are also used in controlled-release forms of drugs and agricultural products such as fertilizers and herbicides. By coating the active material with a polyester selected so as to degrade over time, the material is released gradually rather than all at once.

27.13 Polycarbonates

Polycarbonates are polyesters of carbonic acid. *Lexan* is the most important of the polycarbonates and is prepared from the diphenolic compound bisphenol A.

$$NaO-C_6H_4-C(CH_3)_2-C_6H_4-ONa + ClC(=O)Cl \xrightarrow{-NaCl} \left[O-C_6H_4-C(CH_3)_2-C_6H_4-OC(=O) \right]_n$$

Disodium salt of bisphenol A   Phosgene   Bisphenol A polycarbonate (Lexan)

Problem 27.13

Write a mechanism for the reaction of one molecule of the disodium salt of bisphenol A with one molecule of phosgene.

Figure 27.10

The polycarbonate lenses in these protective glasses are lightweight, yet shatterproof.

©Getty Images

Lexan is a clear, transparent, strong, and impact-resistant plastic with literally countless applications. It is used in both protective and everyday eyeglasses as illustrated in Figure 27.10. The *Apollo 11* astronauts wore Lexan helmets with Lexan visors on their 1969 trip to the moon. CDs and DVDs are Lexan polycarbonate, as are many cell phones, automobile dashpanels, and headlight and taillight lenses.

27.14 Polyurethanes

A *urethane,* also called a *carbamate,* is a compound that contains the functional group $-OC(=O)NH-$. Urethanes are normally prepared by the reaction of an alcohol and an isocyanate.

$$ROH + R'N{=}C{=}O \longrightarrow ROC(=O)NHR'$$

Alcohol   Isocyanate   Urethane

Polyurethanes are the macromolecules formed from a diol and a diisocyanate. In most cases the diol is polymeric and the diisocyanate is a mixture of the "toluene diisocyanate" isomers.

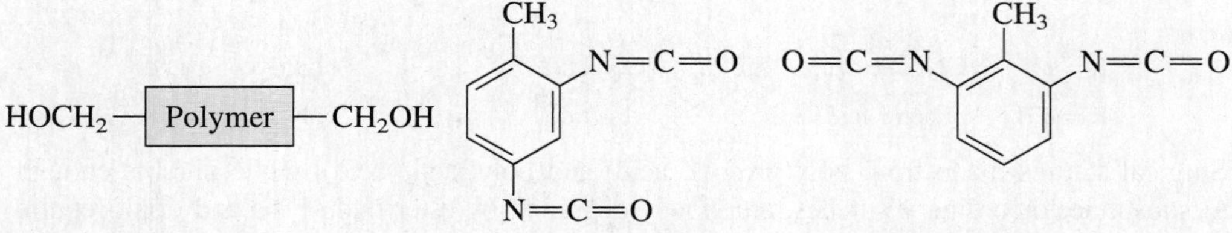

Polymeric diol   Mixture of "toluene diisocyanate" isomers

If, for example, only the 2,6-diisocyanate were present, the repeating unit of the resulting polyurethane would be

$$\left[OCH_2-\boxed{\text{Polymer}}-CH_2OC(=O)NH-C_6H_3(CH_3)-NHC(=O) \right]_n$$

Because a mixture of diisocyanate isomers is actually used, a random mixture of 2,4- and 2,6-substitution patterns results.

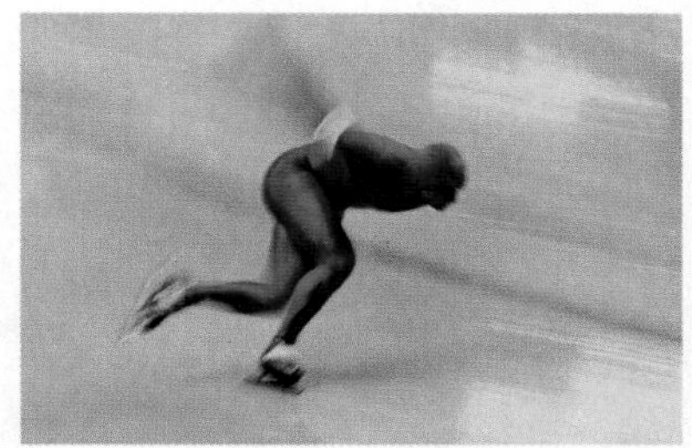

Figure 27.11

Spandex skinsuits make speedskaters more aerodynamic.

Problem 27.14

Write the repeating unit of the "polymeric diol" if it is derived from 1,2-epoxypropane.

The reaction of an alcohol with an isocyanate is addition, not condensation. Therefore, polyurethanes are classified as addition polymers. But because the monomers are difunctional, the molecular weight increases by step growth rather than chain growth.

A major use of polyurethanes is in spandex fibers. Spandex, even when stretched several times its length, has the ability to return to its original state and is a superior substitute for rubber in elastic garments. Its most recognizable application is in athletic wear (swimming, cycling, running) where it is the fabric of choice for high-performance athletes (Figure 27.11).

Polyurethanes have many other applications, especially in paints, adhesives, and foams. Polyurethane foams, which can be rigid (insulation panels) or flexible (pillows, cushions, and mattresses) depending on their degree of cross linking, are prepared by adding foaming agents to the polymerization mixture. One method takes advantage of the reaction between isocyanates and water.

$$\underset{\text{Isocyanate}}{RN{=}C{=}O} + \underset{\text{Water}}{H_2O} \longrightarrow \underset{\text{Carbamic acid}}{RNH{-}\overset{\overset{\displaystyle O}{\|}}{C}{-}OH} \longrightarrow \underset{\text{Amine}}{RNH_2} + \underset{\text{Carbon dioxide}}{CO_2}$$

Although esters of carbamic acid (urethanes) are stable compounds, carbamic acid itself rapidly dissociates to an amine and carbon dioxide. Adding some water to the reactants during polymerization generates carbon dioxide bubbles which are trapped within the polymer.

27.15 Copolymers

Copolymers, polymers made from more than one monomer, are as common as homopolymers. The presence of more than one monomer in a chain makes some control of properties possible. Some structural units stiffen the chain, others make it more flexible. Often a second monomer is added to allow cross linking.

Copolymers are classified according to the distribution of monomers in the macromolecule.

1. Random
2. Block
3. Graft

Random Copolymers: As the name implies, there is no pattern to the distribution of monomer units in a random copolymer.

$$\sim A{-}A{-}B{-}A{-}B{-}B{-}A{-}A{-}B{-}B{-}A{-}B{-}A{-}B \sim$$

Styrene–butadiene rubber (SBR) for automobile tires is a random copolymer. It is prepared by two methods, free-radical and anionic polymerization, both of which are carried out on a mixture of styrene and 1,3-butadiene. Free-radical initiation is essentially nonselective and gives the random copolymer. Anionic initiation is carried out under conditions designed to equalize the reactivity of the two monomers so as to ensure randomness.

Block Copolymers: The main chain contains sections (blocks) of repeating units derived from different monomers. The sequence:

$$\sim A{-}A{-}A{-}A{-}A{-}B{-}B{-}B{-}B{-}B{-}B{-}B{-}B{-}B \sim$$

shows only two blocks, one derived from A and the other from B. A macromolecule derived from A and B can contain many blocks.

The living polymers generated by anionic polymerization are well suited to the preparation of block polymers. Adding 1,3-butadiene to a living polystyrene block sets the stage for attaching a poly(1,3-butadiene) block.

$$\boxed{\text{Polystyrene}}-CH_2-\underset{\cdot\cdot}{C}H(C_6H_5)\ Li^+ \xrightarrow{H_2C=CH-CH=CH_2} \boxed{\text{Polystyrene}}-CH_2-CH(C_6H_5)-CH_2-CH=CH-\overset{-}{\ddot{C}}H_2\ Li^+$$

Further reaction with $H_2C=CHCH=CH_2$

$$\boxed{\text{Polystyrene}}-CH_2-CH(C_6H_5)-\boxed{\text{Poly(1,3-butadiene)}}-CH_2-CH=CH-\overset{-}{\ddot{C}}H_2\ Li^+$$

The properties of the block copolymer prepared by anionic living polymerization are different from the random styrene–butadiene copolymer.

Graft Copolymer: The main chain bears branches (grafts) that are derived from a different monomer.

$$\sim A-A-A-A-A-A-A-A-A-A-A-A-A-A \sim \quad \text{(with B–B branches attached to several A units)}$$

A graft copolymer of styrene and 1,3-butadiene is called "high-impact polystyrene" and is used, for example, in laptop computer cases. It is prepared by free-radical polymerization of styrene in the presence of poly(1,3-butadiene). Instead of reacting with styrene, the free-radical initiator abstracts an allylic hydrogen from poly(1,3-butadiene).

$$\boxed{\text{Poly(1,3-butadiene)}}-CH(H)-CH=CH-CH_2-\boxed{\text{Poly(1,3-butadiene)}} \quad \text{Initiator}\cdot$$

$$\downarrow$$

$$\boxed{\text{Poly(1,3-butadiene)}}-\dot{C}H-CH=CH-CH_2-\boxed{\text{Poly(1,3-butadiene)}}$$

Polystyrene chain growth begins at the allylic radical site and proceeds in the usual way at this and random other allylic carbons of poly(1,3-butadiene).

$$\boxed{\text{Poly(1,3-butadiene)}}-\dot{C}H-CH=CH-CH_2-\boxed{\text{Poly(1,3-butadiene)}} \quad + \quad C_6H_5-CH=CH_2$$

$$\downarrow$$

$$\boxed{\text{Poly(1,3-butadiene)}}-CH(CH_2-\dot{C}H-C_6H_5)-CH=CH-CH_2-\boxed{\text{Poly(1,3-butadiene)}}$$

Polystyrene grafts on a poly(1,3-butadiene) chain are the result.

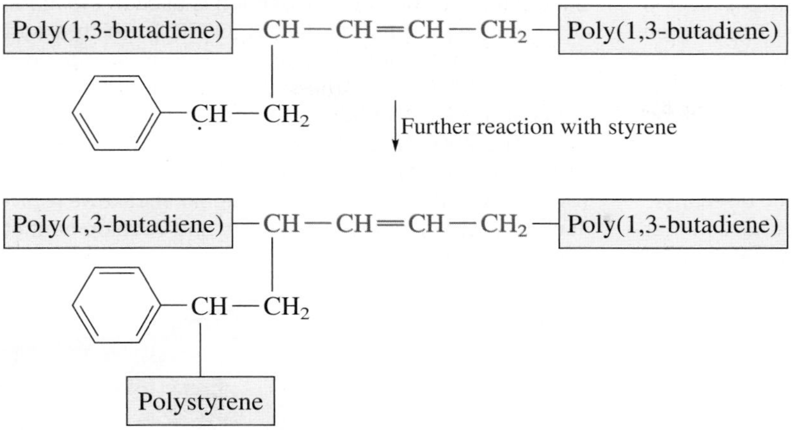

Poly(1,3-butadiene)—CH—CH=CH—CH$_2$—Poly(1,3-butadiene)
(C$_6$H$_5$)ĊH—CH$_2$ (attached to CH)

↓ Further reaction with styrene

Poly(1,3-butadiene)—CH—CH=CH—CH$_2$—Poly(1,3-butadiene)
(C$_6$H$_5$)CH—CH$_2$ (attached to CH), with Polystyrene attached to CH

Polystyrene alone is brittle; poly(1,3-butadiene) alone is rubbery. The graft copolymer is strong, but absorbs shock without cracking because of the elasticity provided by its poly(1,3-butadiene) structural units.

Conducting Polymers

The notion that polymers can conduct electricity seems strange to most of us. After all, the plastic wrapped around the wires in our homes and automobiles serves as insulation. Do polymers exist that can conduct electricity? Even if such materials could be made, why would we be interested in them?

Henry Letheby, a lecturer in chemistry and toxicology at the College of London Hospital, obtained a partially conducting material in 1862 by the anodic oxidation of aniline in sulfuric acid. The material Letheby synthesized was a form of polyaniline. In the 1980s, Alan MacDiarmid of the University of Pennsylvania reinvestigated polyaniline, which is now a widely used conducting polymer. Polyaniline exists in a variety of oxidation states (Figure 27.12), each with different properties. The emaraldine salt is a conductor without the use of additives that enhance conductivity, but its conductivity is enhanced by adding a Brønsted acid that protonates the nitrogen atoms.

Figure 27.12

Polyaniline exists in different forms with varying states of oxidation. One of the forms is a conductor.

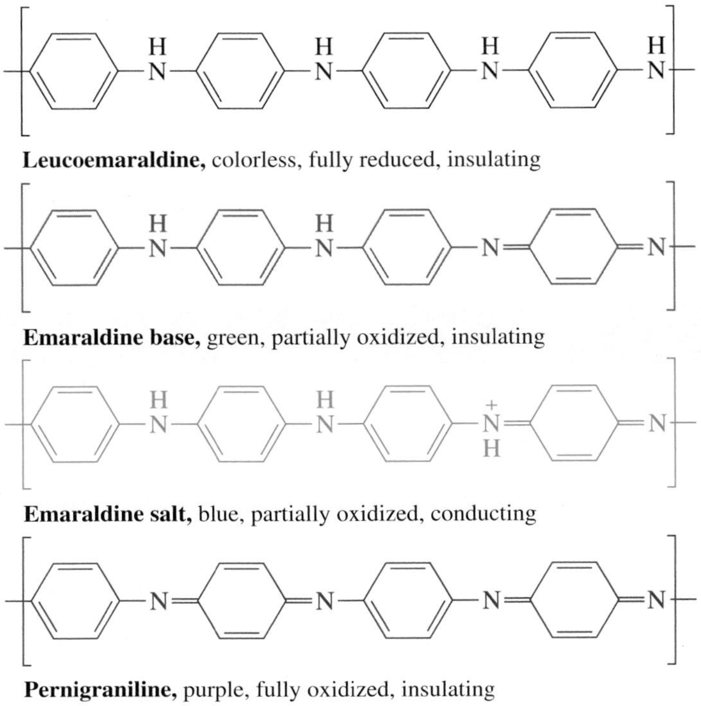

Leucoemaraldine, colorless, fully reduced, insulating

Emaraldine base, green, partially oxidized, insulating

Emaraldine salt, blue, partially oxidized, conducting

Pernigraniline, purple, fully oxidized, insulating

continued

The synthesis of polyaniline can be carried out in aqueous HCl solution, by electrochemical oxidation, or in the presence of a chemical oxidant such as ammonium persulfate. The different forms of polyaniline can then be obtained by altering the current or the pH of the solution. The ability to tailor the process increases the potential for commercial application where the unique properties of a certain polyaniline are desired. Polyanilines are used as corrosion inhibitors and in the electromagnetic shielding of circuits, where they can protect against electrostatic discharge.

Another conducting polymer that has found commercial application is poly(3,4-ethylenedioxythiophene), PEDOT, which is marketed as a dispersion that contains poly(styrene sulfonate). This polymer dispersion is used in the manufacture of organic light-emitting diodes (OLEDs), which are materials that emit light when an electric current is applied to them (Figure 27.13). OLEDs are used for flat panel displays in televisions and cellular telephone displays.

The 2000 Nobel Prize in Chemistry was awarded to Alan Heeger (University of California Santa Barbara), Alan MacDiarmid (University of Pennsylvania), and Hideki Shirakawa (University of Tsukuba, Japan) for their "discovery and development of electrically conductive polymers."

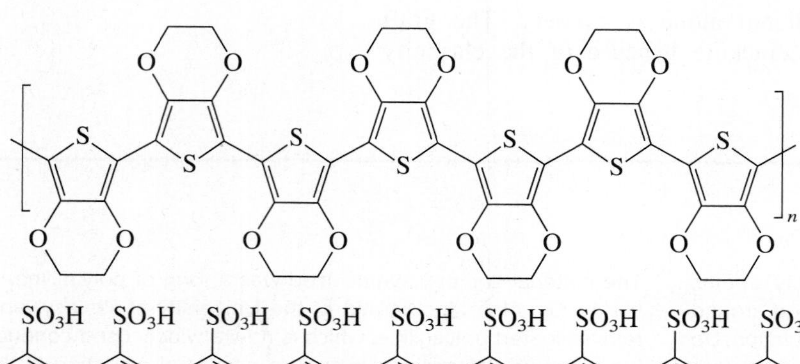

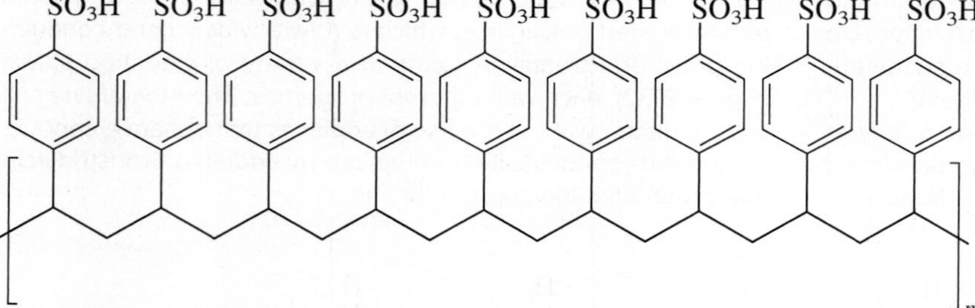

Figure 27.13

A mixture of poly(3,4-ethylenedioxythiophene) and poly(styrene sulfonate) is used in the manufacture of organic light-emitting diodes (OLEDs).

LG markets a television with an OLED screen that is 4.3 mm thick.
©scanrail/123RF

This cellular telephone made by Samsung uses an OLED display.
©Joby Sessions/Future/REX/Shutterstock

27.16 SUMMARY

Section 27.1 Polymer chemistry dates to the nineteenth century with the chemical modification of polymeric natural products. Once the structural features of polymers were determined, polymer synthesis was placed on a rational basis.

Section 27.2 Polymers are usually named according to the monomers from which they are prepared (*source-based nomenclature*). When the name of the monomer is one word, the polymer is named by simply adding the prefix *poly-*. When the name of the monomer is two words, they are enclosed in parentheses and preceded by *poly*.

$$\left[\begin{matrix} CH_3 \\ | \\ CHCH_2 \end{matrix}\right]_n \qquad \left[CH_2CH_2O\right]_n$$

Polypropylene Poly(ethylene oxide)

Sections 27.3–27.6 Polymers may be classified in several different ways:

- Reaction type (addition and condensation)
- Chain-growth or step-growth
- Structure (linear, branched, cross-linked)
- Properties (thermoplastic, thermoset, or elastomer)

Section 27.7 This section emphasizes initiation and termination steps in alkene polymerization. The main terminating reactions in free-radical polymerization are the coupling of two radicals and disproportionation. *Coupling* of two radicals pairs the odd electrons and stops chain growth.

$$\text{Polymer}-\dot{C}H_2 \quad H_2\dot{C}-\text{Polymer} \longrightarrow \text{Polymer}-CH_2-CH_2-\text{Polymer}$$

In *disproportionation,* a hydrogen atom is exchanged between two growing chains, terminating one in a double bond and the other in a new C—H bond.

$$\text{Polymer}-CH_2-\dot{C}H_2 + H_2\dot{C}-CH(H)-\text{Polymer} \longrightarrow \text{Polymer}-CH_2-CH_3 + H_2C{=}CH-\text{Polymer}$$

Section 27.8 Free-radical polymerization of alkenes usually gives branched polymers of low crystallinity. The two main mechanisms by which branches form both involve hydrogen-atom abstraction by the radical site. In one, a growing chain abstracts a hydrogen atom from a terminated polymer.

$$\text{Polymer}-CH(H)-\text{Polymer} + H_2\dot{C}-\text{Polymer} \longrightarrow \text{Polymer}-\dot{C}H-\text{Polymer} + H-CH_2-\text{Polymer}$$

The other is an intramolecular hydrogen-atom abstraction. In most cases this reaction proceeds by a six-center transition state and moves the reactive site from the end of the growing chain to inside it.

$$\text{Polymer}-CH_2-CH(H)-CH_2-CH_2-CH_2-\dot{C}H_2 \longrightarrow \text{Polymer}-CH_2-\dot{C}H-CH_2-CH_2-CH_2-CH_2-H$$

Section 27.9 Anionic polymerization of alkenes that bear a carbanion-stabilizing substituent (X) can be initiated by strong bases such as alkyllithium reagents.

$$R{:}^- \; Li^+ + C{=}C(X) \longrightarrow R-C-\overset{-}{C}{:}(X) \quad Li^+$$

The product of this step is a new organolithium reagent that can react with a second monomer molecule, then a third, and so on. The growing organolithium chain is stable and is called a living polymer.

Section 27.10 Cationic polymerization of alkenes that can form relatively stable carbocations can be initiated by protonation of the double bond or coordination to Lewis acids such as boron trifluoride.

$$\text{C}_6\text{H}_5\text{C(CH}_3\text{)=CH}_2 \xrightarrow{\text{BF}_3} \left[\text{C(CH}_3\text{)(C}_6\text{H}_5\text{)CH}_2\right]_n$$

Section 27.11 The key bond-forming process in many polymerizations is a *condensation* reaction. The most common condensations are those that produce polyamides and polyesters.

Polyamide synthesis is illustrated by the preparation of nylon 66, the most commercially successful synthetic fiber.

$$\text{H}_3\overset{+}{\text{N}}(\text{CH}_2)_6\overset{+}{\text{N}}\text{H}_3 + {}^-\text{O}\overset{\text{O}}{\overset{\|}{\text{C}}}(\text{CH}_2)_4\overset{\text{O}}{\overset{\|}{\text{C}}}\text{O}^- \xrightarrow{\text{heat}} \left[\text{NH}(\text{CH}_2)_6\text{NH}\overset{\text{O}}{\overset{\|}{\text{C}}}(\text{CH}_2)_4\overset{\text{O}}{\overset{\|}{\text{C}}}\right]_n$$

Section 27.12 The condensation of a diol and a dicarboxylic acid produces a *polyester.* Poly(tetramethylene succinate) is a biodegradable polyester derived from butanedioic acid and 1,4-butanediol.

$$\left[\text{O}(\text{CH}_2)_4\text{O}\overset{\text{O}}{\overset{\|}{\text{C}}}(\text{CH}_2)_2\overset{\text{O}}{\overset{\|}{\text{C}}}\right]_n$$

Section 27.13 Most of the applications of *polycarbonates* center on Lexan, a polyester derived from phosgene and bisphenol A.

$$\left[\text{O}-\text{C}_6\text{H}_4-\text{C}(\text{CH}_3)_2-\text{C}_6\text{H}_4-\text{O}\overset{\text{O}}{\overset{\|}{\text{C}}}\right]_n$$

Section 27.14 Like polycarbonates, *polyurethanes* enjoy wide use even though there are relatively few structural types. Most polyurethanes are made from a mixture of the 2,4- and 2,6-diisocyanate derivatives of toluene and a polymeric diol or triol.

$$\left[\text{O}(\text{CH}_2)_2\text{O}(\text{CH}_2)_2\text{O}\overset{\text{O}}{\overset{\|}{\text{C}}}\text{NH}-\text{C}_6\text{H}_3(\text{CH}_3)-\text{NH}\overset{\text{O}}{\overset{\|}{\text{C}}}\right]_n$$

Section 27.15 *Copolymers* are the polymers formed when two or more monomers are present in the mixture to be polymerized. They are classified as random, block, or graft. A *random copolymer* lacks a regular sequence in respect to the appearance of the structural units of the components. A *block copolymer* of monomers A and B is composed of blocks of poly(A) and poly(B). A *graft copolymer* has a main chain of poly(A) to which are grafted branches of poly(B).

PROBLEMS

Monomers

27.15 From what monomer is the polymer with the repeating unit $[\,\text{CH}_2\text{CH}{=}\text{CHCH}_2\,]_n$ prepared? Suggest a source-based name.

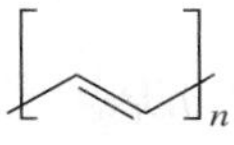

27.16 Give the structure of the lactone from which $\left[OCH_2CH_2\overset{O}{\overset{\|}{C}} \right]_n$ is prepared.

27.17 Kodel fibers are made from the polymer shown. Suggest suitable monomers for its preparation.

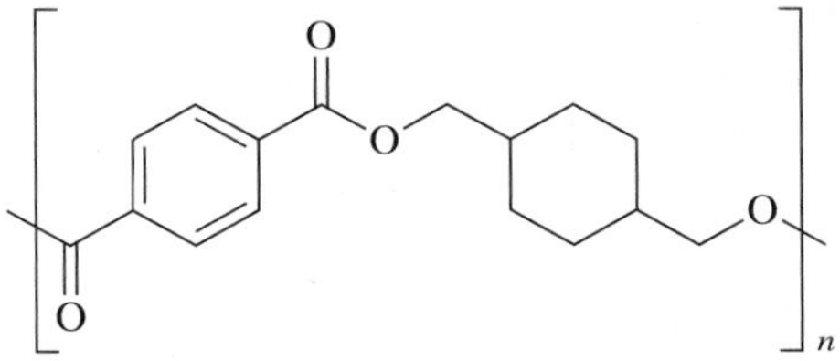

27.18 *Pseudomonas oleovorans* oxidizes nonanoic acid, then stores the 3-hydroxynonanoic acid produced as a homopolymer. Write the formula for the repeating unit of this polyester.

27.19 Nylon 11 is a polyamide used as fishing line and is prepared by heating 11-aminoundecanoic acid [$H_2N(CH_2)_{10}CO_2H$]. What is the repeating unit of nylon 11? Is it a condensation or an addition polymer? Chain growth or step growth?

27.20 Is protein biosynthesis as shown in Figure 26.13 step growth or chain growth? Is the protein that results an addition or a condensation polymer? Why?

Polymerization

27.21 Of the following monomers, which one would undergo cationic polymerization most readily?

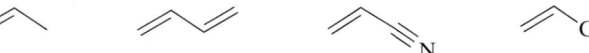

27.22 Of the following monomers, which one would undergo anionic polymerization most readily?

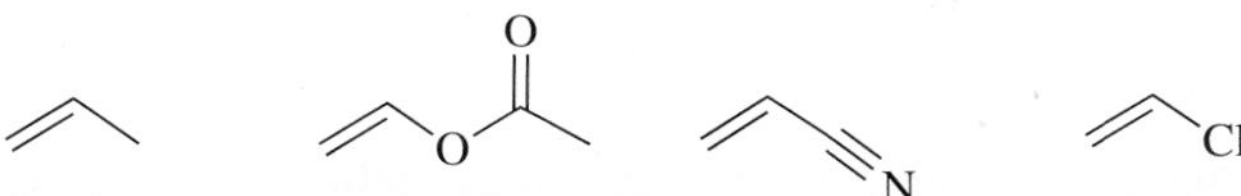

27.23 Polymerization of styrene can occur by a free-radical, cationic, anionic, or coordination mechanism. What mechanism will be followed when each of the compounds shown is used to initiate polymerization?

(a) $TiCl_4$, $(CH_3CH_2)_3Al$ (b) 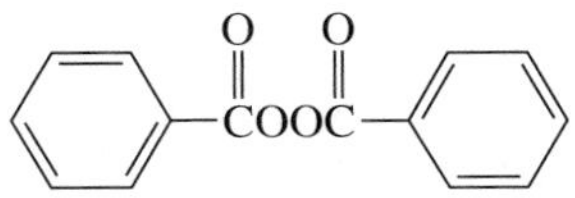 $C_6H_5\overset{O}{\overset{\|}{C}}OO\overset{O}{\overset{\|}{C}}C_6H_5$ (c) BF_3

27.24 Styrene undergoes anionic polymerization at a faster rate than *p*-methoxystyrene. Suggest an explanation for this observation.

27.25 Given that —C≡N stabilizes carbanions better than phenyl, which monomer would you start with to prepare a copolymer of styrene and acrylonitrile?

27.26 *Poly(vinyl butyral)* is the inner liner in safety glass. It is prepared by the reaction shown. What is compound A?

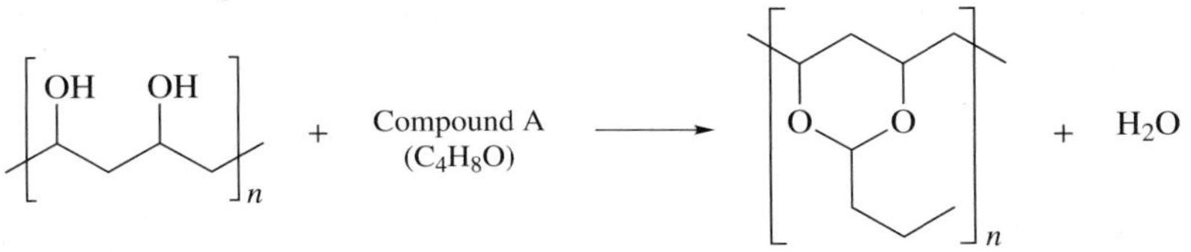

Poly(vinyl alcohol) (OH OH)$_n$ + Compound A (C_4H_8O) ⟶ poly(vinyl butyral)$_n$ + H_2O

27.27 *Linear low-density polyethylene* is a copolymer in which ethylene is polymerized under Ziegler–Natta conditions in the presence of a smaller quantity of a second alkene such as 1-hexene. What structural feature characterizes the resulting polymer?

Mechanisms

27.28 (a) Bisphenol A (shown) is made by the reaction of phenol and acetone. Suggest a mechanism for this reaction. Assume acid (H_3O^+) catalysis.

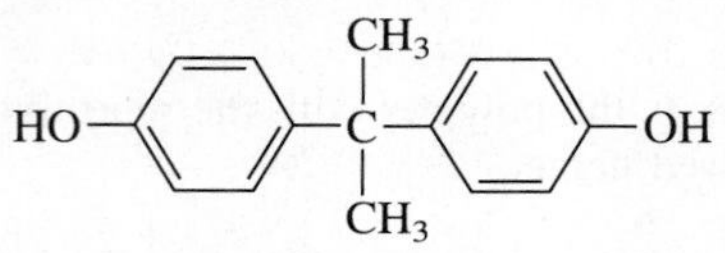

HO–C₆H₄–C(CH_3)₂–C₆H₄–OH

(b) Bisphenol B is made from phenol and 2-butanone. What is its structure?

27.29 Poly(ethylene oxide) can be prepared from ethylene oxide by either anionic or cationic polymerization methods. Write reaction mechanisms for both processes. Use H_3O^+ as the acid and OH^- as the base.

27.30 (a) The first step in the formation of Bakelite from phenol and formaldehyde introduces —CH_2OH groups onto the ring.

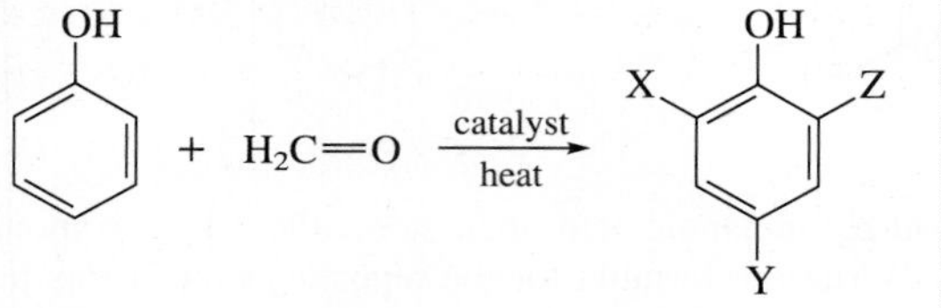

Phenol + $H_2C{=}O$ (Formaldehyde) $\xrightarrow[\text{heat}]{\text{catalyst}}$ substituted phenol (X, Y, Z)

X	Y	Z
H	CH_2OH	H
CH_2OH	H	H
CH_2OH	CH_2OH	H
CH_2OH	H	CH_2OH
CH_2OH	CH_2OH	CH_2OH

Write a mechanism for the formation of *o*-hydroxybenzyl alcohol (X = CH_2OH, Y = Z = H) in this reaction. Assume the catalyst is H_3O^+.

(b) The second step links two of the aromatic rings by a CH_2 group. Write a mechanism for the example shown.

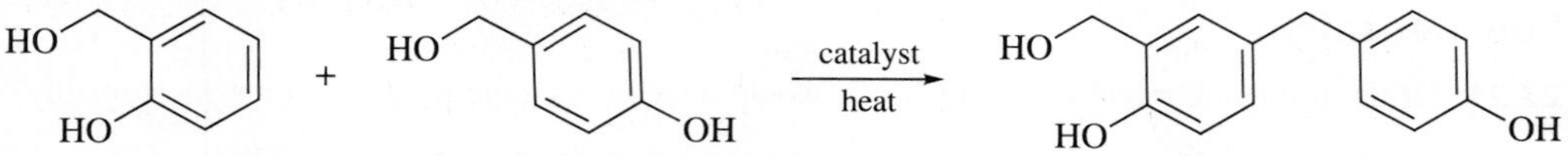

o-hydroxybenzyl alcohol + p-hydroxybenzyl alcohol $\xrightarrow[\text{heat}]{\text{catalyst}}$ CH_2-linked product

27.31 The first step in the mechanism of cationic polymerization of formaldehyde is:

$$H_2C{=}\ddot{O}: + BF_3 \longrightarrow H_2C{=}\overset{+}{\ddot{O}}{-}\overset{-}{B}F_3$$

Write an equation for the second step using curved arrows to track electron movement.

Descriptive Passage and Interpretive Problems 27

Chemically Modified Polymers

Many useful polymers are not themselves the initial products of polymerization but are prepared by chemically modifying the original polymer. Partially fluorinated polyethylene used for protective gloves and to coat automobile gasoline tanks is made by exposing polyethylene to F_2 diluted with nitrogen.

$$\text{polyethylene} + F_2 \longrightarrow \text{partially fluorinated polyethylene (F)} + HF$$

Partial fluorination gives a polymer that, like polyethylene, is easy to cast into films but with a greater resistance to oxidation and water penetration.

The solid support in Merrifield's synthesis of ribonuclease (see Section 25.16) was prepared by incorporating —CH_2Cl groups into a styrene/*p*-divinylbenzene copolymer by electrophilic aromatic substitution.

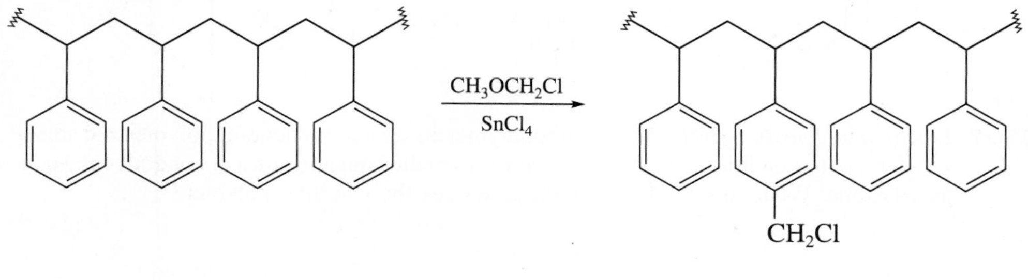

$$\text{polystyrene} \xrightarrow[SnCl_4]{CH_3OCH_2Cl} \text{polymer with } CH_2Cl \text{ on a ring}$$

At the same time that Merrifield was developing his method for the solid-phase synthesis of peptides, Robert Letsinger (Northwestern University) was independently applying the same concept to polynucleotide synthesis. Modern methods for making oligonucleotides are direct descendants of Letsinger's method.

Today's chemists can buy Merrifield-type resins with varying degrees of chloromethyl substitution and cross linking tailored for specific purposes. Because the chlorine atom is primary and benzylic, these resins can be further modified by nucleophilic substitution.

$$\text{PS}-CH_2Cl + Nu:^- \longrightarrow \text{PS}-CH_2Nu + :\ddot{Cl}:^-$$

(In this and succeeding equations, the blue sphere represents a polymer bead and PS stands for polystyrene or a copolymer of polystyrene and *p*-divinylbenzene.)

The products of these reactions form the basis for an entire methodology—*polymer-supported chemical reactions*—wherein the modified polystyrene serves as a reactant, reagent, or catalyst. The reactions are the usual ones of organic chemistry. In the following equation, for example, the modified polystyrene serves as a phase-transfer catalyst (see Section 22.5). The main advantage of using a polymer-supported reagent, or in this case a polymer-supported catalyst, is that it makes isolation of the reaction product easier.

$$\underset{\text{(in toluene)}}{CH_3(CH_2)_6CH_2Br} + \underset{\text{(in water)}}{KCN} \xrightarrow{\text{PS}-CH_2\overset{+}{P}(Bu)_3\ Cl^-} \underset{\text{(in toluene)}}{CH_3(CH_2)_6CH_2CN} + \underset{\text{(in water)}}{KBr}$$

Cyanide ion from aqueous KCN exchanges with Cl^- of the polymer-supported phosphonium chloride and reacts with 1-bromooctane on the surface and within channels of the polymer support. When the reaction is judged to be complete, the polymer (insoluble in both toluene and water) is recovered by filtration and the aqueous layer removed. Distillation of the toluene solution of the product furnishes nonanenitrile, the product of nucleophilic substitution of cyanide for bromide.

The number of applications of chemically modified polymers as materials, reagents, and catalysts is extremely large. The following problems give a few examples.

27.32 Chemical modification of polymers is not always beneficial. Which of the following polymers will be adversely affected by air oxidation the most?

$$\underset{\text{A.}}{\left[CF_2\right]_n} \quad \underset{\text{B.}}{\left[CH_2\right]_n} \quad \underset{\text{C.}}{\left[CH_2C(CH_3){=}CHCH_2\right]_n} \quad \underset{\text{D.}}{\left[CH_2\overset{Cl}{\overset{|}{C}H}\right]_n}$$

27.33 The living polymer formed by reaction of ethylene with butyllithium can be converted to a long-chain alkyldiphenylphosphine by reaction with compound X. The alkyldiphenylphosphine is used in the preparation of phase-transfer catalysts and as a ligand in polymer-supported organometallic compounds. What is compound X?

$$H_2C{=}CH_2 \xrightarrow{\text{BuLi}} Bu\left[CH_2CH_2\right]_nLi \xrightarrow{\text{compound X}} Bu\left[CH_2CH_2\right]_nP(C_6H_5)_2$$

$$\underset{\text{A.}}{(C_6H_5)_2PH} \quad \underset{\text{B.}}{(C_6H_5)_2PCl} \quad \underset{\text{C.}}{(C_6H_5)_2PLi} \quad \underset{\text{D.}}{(C_6H_5)_3P}$$

27.34 The alkyldiphenylphosphine formed in the preceding equation was converted to a dialkyldiphenylphosphonium salt for use as a phase-transfer catalyst. Which of the following is a suitable reactant for such a conversion?

$$\underset{\text{A.}}{CH_3CH_2CH_2CH_2Br} \quad \underset{\text{B.}}{CH_3CH_2CH_2CH_2Li} \quad \underset{\text{C.}}{CH_3CH_2CH_2CH_2OH} \quad \underset{\text{D.}}{CH_3CH_2CH_2CH_2ONa}$$

27.35 A copolymer of styrene and *p*-bromostyrene can be transformed into a living polymer as shown. The aryllithium sites then serve to start chain growth when a suitable monomer is added.

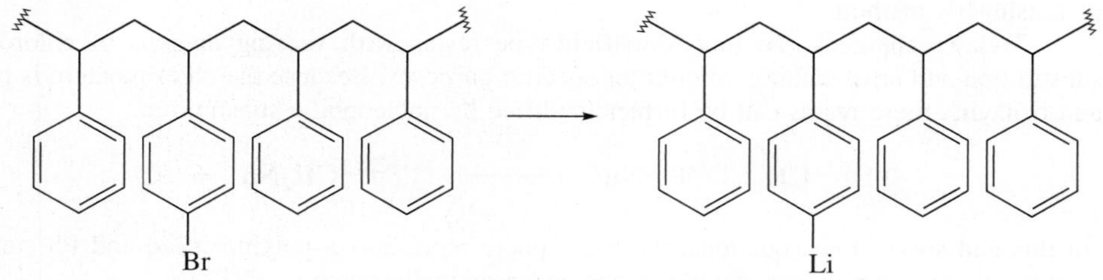

Br → Li

Which of the following is the most suitable for the transformation in the equation?

LiOH	LiCl	$LiCu(CH_3)_2$	Li
A.	B.	C.	D.

27.36 What is the polymer-containing product of the following reaction?

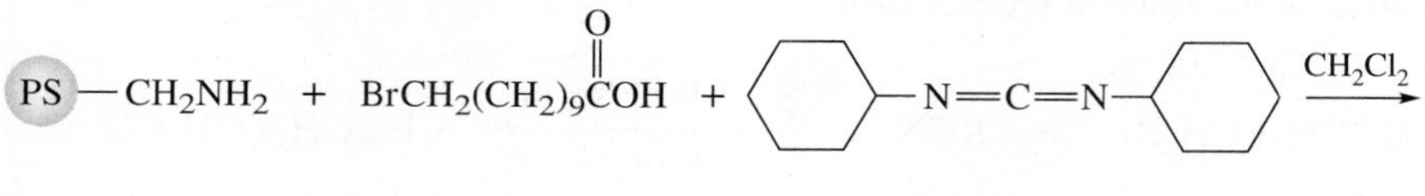

$$\text{PS}-CH_2NH_2 + BrCH_2(CH_2)_9\overset{O}{\overset{\|}{C}}OH + \text{C}_6\text{H}_{11}-N{=}C{=}N-\text{C}_6\text{H}_{11} \xrightarrow{CH_2Cl_2}$$

A. $\text{PS}-CH_2NHCH_2(CH_2)_9\overset{O}{\overset{\|}{C}}OH$

B. $\text{PS}-CH_2NH\overset{O}{\overset{\|}{C}}(CH_2)_9CH_2Br$

27.37 The ethyl ester function in the *R*-BINAP derivative shown was used as the reactive "handle" to bind the chiral unit to polystyrene giving a ligand suitable for ruthenium-catalyzed enantioselective hydrogenation.

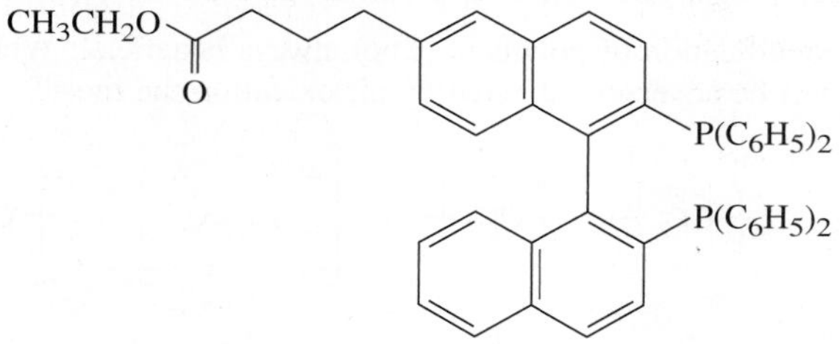

CH_3CH_2O, O, $P(C_6H_5)_2$, $P(C_6H_5)_2$

Which of the following has the proper functionality to react with this ester by nucleophilic acyl substitution to give a polystyrene-supported ligand?

A. $\text{PS}-CH_2Cl$

C. $\text{PS}-CH_2N(CH_3)_2$

B. $\text{PS}-CH_2NH_2$

D. $\text{PS}-CH_2\overset{+}{N}(CH_3)_3\ Cl^-$

27.38 The polystyrene-supported quaternary ammonium chloride shown was treated with aqueous sodium hydroxide, then shaken with a solution of compound X and phenol in toluene at 90°C to give butyl phenyl ether in 97% yield. What is compound X?

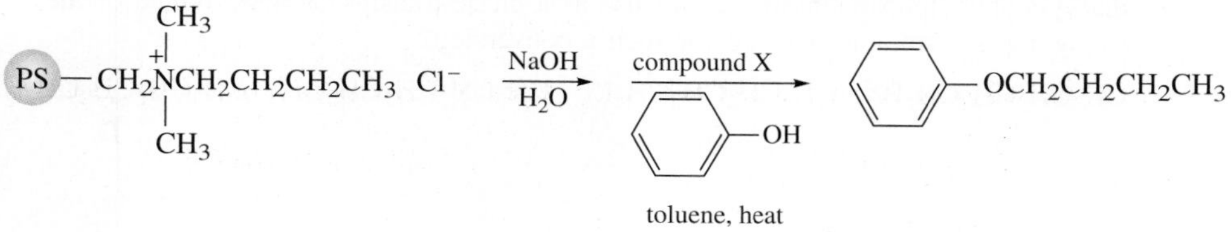

$$\text{PS}-CH_2\overset{+}{N}(CH_3)_2CH_2CH_2CH_2CH_3\ Cl^- \xrightarrow[H_2O]{NaOH} \xrightarrow[\text{toluene, heat}]{\text{compound X},\ C_6H_5OH} C_6H_5-OCH_2CH_2CH_2CH_3$$

A. $CH_3CH_2CH_2CH_2OH$

C. $CH_3CH_2CH_2CH_2Br$

B. $CH_3CH_2CH{=}CH_2$

D. $CH_3CH_2CH_2CH_2NH_2$

Appendix

Summary of Methods Used to Synthesize a Particular Functional Group

Synthesis of Acetals

1. Acid-catalyzed reaction of an aldehyde with two equivalents of an alcohol (18.8)

Synthesis of Acid Anhydrides

1. Reaction of an acyl chloride with a carboxylate ion (20.4)

Synthesis of Alcohols

1. Acid-catalyzed hydration of an alkene (8.6)
2. Hydroboration–oxidation of an alkene (8.8, 8.9)
3. Oxymercuration–demercuration of an alkene (p. 327)
4. Hydrolysis of an alkyl halide (6.5)
5. Hydrolysis of an allylic halide (11.2)
6. Reaction of a Grignard reagent (or RLi) with an aldehyde or ketone (15.5) or ester (20.11)
7. Preparation of an acetylenic alcohol by reaction of sodium alkynides with an aldehyde or ketone (15.6)
8. Reduction of an aldehyde or ketone (16.2)
9. Reduction of a carboxylic acid (16.3)
10. Ring-opening of an epoxide with organometallic reagents (16.4)
11. Acid-catalyzed ester hydrolysis (20.8)
12. Saponification of an ester (20.9)
13. Ring-opening of an epoxide with $LiAlH_4$, then H_2O (17.11)

Synthesis of Aldehydes

1. Ozonolysis of alkenes (8.12)
2. Oxidation of a 1° alcohol (16.9)
3. Cleavage of a 1,2-diol with periodic acid, HIO_4 (16.11)

Synthesis of Alkanes

1. Catalytic hydrogenation of an alkene (8.1, 8.2)
2. Catalytic hydrogenation of an alkyne (9.9)
3. Wolff–Kishner or Clemmensen reduction of an aldehyde or ketone (13.8)
4. Preparation of a cyclopropane by reaction of an alkene with ICH_2ZnI, the Simmons–Smith reagent (15.8–15.9)
5. Reaction of a Gilman reagent (R_2CuLi) with an alkyl halide (15.11)

Synthesis of Alkenes

1. Acid-catalyzed dehydration of an alcohol (7.9–7.13)
2. Dehydrohalogenation of an alkyl halide (7.14–7.16)
3. Hydrogenation of an alkyne with Lindlar catalyst to form a cis alkene (9.9)
4. Reduction of an alkyne with Na (or Li) and liquid NH_3 to form a trans alkene (10.6)
5. Formation of a cyclohexene using a Diels–Alder reaction (11.12–11.14)
6. Formation of a 1,4-cyclohexadiene using a Diels–Alder reaction (Problem 11.41)
7. Formation of a 1,4-cyclohexadiene by the Birch reduction of an arene (12.15)
8. Olefin metathesis (15.14)
9. Wittig reaction: Reaction of an aldehyde or ketone with a phosphonium ylide (18.12)
10. Hofmann reaction: Elimination of a proton and a 3° amine from a quaternary ammonium hydroxide (22.13)

Synthesis of Alkyl Sulfonates (Tosylates)

1. Reaction of *p*-toluenesulfonyl chloride and an alcohol (6.10)

Synthesis of Alkynes

1. Reaction of an acetylide ion (formed by removing a proton from a terminal alkyne) with an alkyl halide (9.6)
2. Two successive eliminations of hydrogen halide from a vicinal or geminal dihalide (9.7)

Synthesis of Amides

1. Reaction of an acyl chloride, acid anhydride, or ester with NH_3, a 1° amine, or a 2° amine (20.10, 20.12)

Synthesis of Amines

1. Treatment of an aryl halide with $NaNH_2/NH_3$ via benzyne intermediate (p. 528)
2. Alkylation of ammonia (22.7, 22.12)
3. Gabriel synthesis of 1° alkyl amines (22.8)
4. Reduction of azides, nitriles, aryl nitro compounds, and amides (22.9)
5. Reductive amination of an aldehyde or ketone (22.10)

Synthesis of Amino Acids

1. Nucleophilic substitution by ammonia on α-halo carboxylic acids (21.6)

Synthesis of Azides

1. Reaction of an alkyl halide with an azide ion (6.1)

Synthesis of Carboxylic Acids

1. Ozonolysis of an alkyne (9.13)
2. Oxidation of alkylbenzenes (12.12)
3. Oxidation of a 1° alcohol (16.9)
4. Oxidation of an aldehyde (18.14)
5. Reaction of a Grignard reagent with CO_2 (19.11)
6. Hydrolysis of a nitrile (19.12, 20.16)
7. Decarboxylation of malonic acids (19.16)
8. Malonic ester synthetic method (21.5)
9. Hydrolysis of an acyl chloride or acid anhydride (20.4, 20.5) or amide (20.13)
10. Acid-catalyzed hydrolysis of an ester (20.8)
11. Saponification of an ester (20.9)
12. Haloform reaction: Reaction of a methyl ketone with excess Br_2 (or Cl_2 or I_2) and HO^- (21.7)

Synthesis of Cyanohydrins

1. Reaction of an aldehyde or ketone with sodium cyanide and HCl (18.7)

Synthesis of Diazonium Ions

1. Nitrosation of 1° amines (22.15, 22.16)

Synthesis of 1,2-Diols

1. Reaction of an alkene with OsO_4 (16.5)
2. Acid-catalyzed hydrolysis of an epoxide (17.12)

Synthesis of Enones

1. The aldol condensation (21.2, 21.3)

Synthesis of Epoxides (Oxiranes)

1. Epoxidation of an alkene with a peracid (8.11)
2. Reaction of a halohydrin with hydroxide ion (17.10)

Synthesis of Esters

1. Reaction of an alkyl halide with a carboxylate ion (6.1)
2. The Fischer esterification: Acid-catalyzed reaction of a carboxylic acid with an alcohol (16.8)
3. Reaction of an acyl chloride or an acid anhydride with an alcohol (16.8, 20.4, 20.5)
4. Baeyer–Villiger oxidation of a ketone (p. 760 and 20.7)

Synthesis of β-Keto Esters

1. The Claisen and Dieckmann condensations (21.4)

Synthesis of Enamines

1. Reaction of an aldehyde or ketone with a 2° amine (18.11)

Synthesis of Ethers

1. Williamson ether synthesis: Reaction of an alkoxide ion with an alkyl halide (6.1, 17.6)
2. Solvolysis of an alkyl halide with an alcohol (6.5)
3. Nucleophilic aromatic substitution of certain activated aryl halides with RO^- (13.19–13.21)
4. Formation of symmetrical ethers by heating an acidic solution of a 1° alcohol (16.7, 17.5)

Synthesis of Halides (Alkyl, Allyl, Dihalides, etc.)

1. Reaction of an alcohol with HX (5.7)
2. Reaction of an alcohol with M^+X^-/H_2SO_4 (5.7)
3. Reaction of an alcohol with $SOCl_2$ or PBr_3 (5.14)
4. Free-radical halogenation of alkanes (10.2–10.4)

5. Addition of HX to alkenes (8.4, 8.5)
6. Addition of Br_2 or Cl_2 to an alkene (8.10)
7. Free-radical addition of HBr/peroxide to an alkene or alkyne (10.5)
8. Reaction of I^- with alkyl chlorides or bromides (6.1)
9. Reaction of alkynes with one or two equivalents of HX (9.10)
10. Reaction of one or two equivalents of X_2 with an alkyne (9.12)
11. Free-radical halogenation of alkenes (11.3)
12. Addition of HX to conjugated dienes (11.10)
13. Addition of Cl_2 or Br_2 to dienes (11.11)
14. Free-radical bromination of alkylbenzenes with NBS (12.10)

Synthesis of Halohydrins

1. Addition of Br_2 or Cl_2 in water to an alkene (8.10)

Synthesis of Hemiacetals

1. Acid-catalyzed reaction of an aldehyde with one equivalent of an alcohol (18.8)

Synthesis of Hemiketals

1. Acid-catalyzed reaction of a ketone with one equivalent of an alcohol (18.8)

Synthesis of Hydrates

1. Hydration of aldehydes and ketones (18.6)

Synthesis of Ketals

1. Acid-catalyzed reaction of a ketone with two equivalents of an alcohol (18.8)

Synthesis of Ketones

1. Ozonolysis of alkenes (8.12)
2. Acid-catalyzed addition of water to an alkyne (9.11)
3. Oxidation of a 2° alcohol (16.9)
4. The pinacol rearrangement of certain vicinal diols (p. 672)
5. Enzymatic decarboxylation of a β-keto acid (19.16)
6. Addition of Grignard reagents to nitriles (20.17)
7. The acetoacetic ester synthetic method (21.5)

Synthesis of Lactones

1. Intramolecular Fischer esterification (19.15)
2. Baeyer–Villiger oxidation of a cyclic ketone (p. 760)

Synthesis of Imines

1. Reaction of an aldehyde or ketone with a 1° amine (18.10)

Synthesis of Nitriles

1. Reaction of an alkyl halide with cyanide ion (6.1)
2. Dehydration of 1° amides with P_4O_{10} (20.15)

Synthesis of Nitrosamines

1. Nitrosation of 2° amines (22.15)

Synthesis of Substituted Benzenes

1. Nitration with $HNO_3 + H_2SO_4$ (13.3)
2. Sulfonation: Reaction with H_2SO_4 (13.4)
3. Halogenation with Br_2 or Cl_2 and a Lewis acid (13.5)
4. Friedel–Crafts alkylation with R–X and $AlCl_3$ (13.6)
5. Friedel–Crafts acylation with an acid chloride (or anhydride) and $AlCl_3$ (13.7)
6. Heck reaction: Couples an aryl halide with an alkene in a basic solution in the presence of $Pd(PPh_3)_4$ (15.12)
7. Stille reaction: Couples an aryl halide with a stannane in the presence of $Pd(PPh_3)_4$ (15.12)
8. Suzuki-Miyaura reaction: Couples an aryl halide with an organoborane in the presence of $Pd(PPh_3)_4$ (15.12)
9. Negishi reaction: Couples an aryl halide with an organozinc in the presence of $Pd(PPh_3)_4$ (15.12)
10. Formation of a phenol by reaction of an aryl diazonium salt with water (22.17)
11. Formation of an aryl fluoride by reaction of an aryl diazonium salt with HBF_4 followed by heat (22.17)
12. Formation of an aryl iodide by reaction of an aryl diazonium salt with KI (22.17)
13. Sandmeyer reaction: Reaction of an aryl diazonium salt with CuCl, CuBr, or CuCN (22.17)
14. Formation of a phenol by cleavage of alkyl aryl ethers with HX (17.8)

Synthesis of Sulfides

1. Reaction of alkanethiolate ions with an alkyl halide (16.12, 17.14)

Synthesis of Sulfonium Salts

1. Alkylation of sulfides (17.16)

Synthesis of Sulfoxides and Sulfones

1. Oxidation of sulfides (17.15)

Synthesis of Thiols

1. Reaction of hydrogen sulfide ion (HS^-) with an alkyl halide (6.1)

Summary of Methods Employed to Form C–C Bonds

1. Reaction of an alkyl halide with cyanide ion (6.1)
2. Reaction of an acetylide ion with an alkyl halide (9.6)
3. Diels–Alder reactions (11.12–11.14; and Problem 11.41)
4. Friedel–Crafts alkylation (13.6)
5. Friedel–Crafts acylation (13.7)
6. Reaction of an aldehyde or ketone with sodium cyanide and HCl (18.7)
7. Reaction of a Grignard reagent (or RLi) with an aldehyde or ketone (15.5) or ester (20.11)
8. Preparation of an acetylenic alcohol by reaction of sodium alkynides with an aldehyde or ketone (15.6)
9. Reaction of a Gilman reagent (R_2CuLi) with an alkyl halide (15.11)
10. Preparation of a cyclopropane by reaction of an alkene with ICH_2ZnI, the Simmons–Smith reagent (15.8)
11. Reaction of a Gilman reagent (R_2CuLi) in conjugate addition with an enone (21.9)
12. Alkylation of enamines (p. 943)
13. Ring-opening of an epoxide with organometallic reagents (16.4)
14. Wittig reaction: Reaction of an aldehyde or ketone with a phosphonium ylide (18.12)
15. Aldol addition (21.2, 21.3)
16. The Claisen and Dieckmann condensations (21.4)
17. Malonic ester synthetic method (21.5)
18. Reaction of a Grignard reagent with CO_2 (19.11)
19. Addition of Grignard reagents to nitriles (20.17)
20. The acetoacetic ester synthetic method (21.5)
21. Heck reaction: Couples an aryl halide with an alkene in a basic solution in the presence of $Pd(PPh_3)_4$ (15.12)
22. Stille reaction: Couples an aryl halide with a stannane in the presence of $Pd(PPh_3)_4$ (15.12)
23. Suzuki-Miyaura reaction: Couples an aryl halide with an organoborane in the presence of $Pd(PPh_3)_4$ (15.12)
24. Negishi reaction: Couples an aryl halide with an organozinc in the presence of $Pd(PPh_3)_4$ (15.12)
25. Knoevenagel reaction (p. 887)

Glossary

A

Absolute configuration: The three-dimensional arrangement of atoms or groups at a chirality center.

Absorbance: In UV-VIS spectroscopy, the value of $\log_{10}(I_0/I)$, where I_0 is the intensity of the incident radiation and I is the intensity of the beam after it has passed through the sample.

Acetal: Product of the reaction of an aldehyde or a ketone with two moles of an alcohol according to the equation

$$RC(=O)R' + 2R''OH \xrightarrow{H^+} RR'C(OR'')_2 + H_2O$$

Acetoacetic ester synthesis: A synthetic method for the preparation of ketones in which alkylation of the enolate of ethyl acetoacetate

$$CH_3\overset{O}{\overset{\|}{C}}CH_2\overset{O}{\overset{\|}{C}}OCH_2CH_3$$

is the key carbon–carbon bond-forming step.

Acetyl coenzyme A: A thioester abbreviated as

$$CH_3\overset{O}{\overset{\|}{C}}SCoA$$

that acts as the source of acetyl groups in biosynthetic processes involving acetate.

Acetylene: The simplest alkyne, $HC{\equiv}CH$.

Achiral: Opposite of *chiral.* An achiral object is superimposable on its mirror image.

Acid: According to the Arrhenius definition, a substance that ionizes in water to produce protons. According to the Brønsted–Lowry definition, a substance that donates a proton to some other substance. According to the Lewis definition, an electron-pair acceptor.

Acid anhydride: Compound of the type

$$R\overset{O}{\overset{\|}{C}}O\overset{O}{\overset{\|}{C}}R$$

Both R groups are usually the same, although they need not always be.

Acidity constant K_a: Equilibrium constant for dissociation of an acid:

$$K_a = \frac{[H^+][A^-]}{[HA]}$$

Activating substituent: A group that when present in place of a hydrogen causes a particular reaction to occur faster. Term is most often applied to substituents that increase the rate of electrophilic aromatic substitution.

Activation energy (E_a): The minimum energy that a reacting system must possess above its most stable state in order to undergo a chemical or structural change.

Active site: The region of an enzyme at which the substrate is bound.

Acylation: Reaction in which an acyl group becomes attached to some structural unit in a molecule. Examples include the Friedel–Crafts acylation and the conversion of amines to amides.

Acyl cation: Synonymous with *acylium ion.*

Acyl chloride: Compound of the type

$$R\overset{O}{\overset{\|}{C}}Cl$$

R may be alkyl or aryl.

Acyl group: The group

$$R\overset{O}{\overset{\|}{C}}—$$

R may be alkyl or aryl.

Acylium ion: The cation $R{-}C{\equiv}\overset{+}{O}:$

Acyl transfer: A nucleophilic acyl substitution. A reaction in which one type of carboxylic acid derivative is converted to another.

Addition: Reaction in which a reagent X–Y adds to a multiple bond so that X becomes attached to one of the carbons of the multiple bond and Y to the other.

1,2 Addition: Addition of reagents of the type X–Y to conjugated dienes in which X and Y add to adjacent doubly bonded carbons:

$$R_2C{=}CH{-}CH{=}CR_2 \xrightarrow{X-Y} R_2CX{-}CHY{-}CH{=}CR_2$$

1,4 Addition: Addition of reagents of the type X—Y to conjugated dienes in which X and Y add to the termini of the diene system (see *conjugate addition*).

$$R_2C{=}CH{-}CH{=}CR_2 \xrightarrow{X-Y} R_2CX{-}CH{=}CH{-}CR_2Y$$

Addition–elimination mechanism: Two-stage mechanism for nucleophilic aromatic substitution. In the addition stage, the nucleophile adds to the carbon that bears the leaving group. In the elimination stage, the leaving group is expelled.

$$C_6H_5X \xrightarrow[\text{addition}]{Y^-} [C_6H_5XY]^- \xrightarrow[\text{elimination}]{-X^-} C_6H_5Y$$

Addition polymer: A polymer formed by addition reactions of monomers.

Adenosine 5′-triphosphate (ATP): The main energy-storing compound in all living organisms.

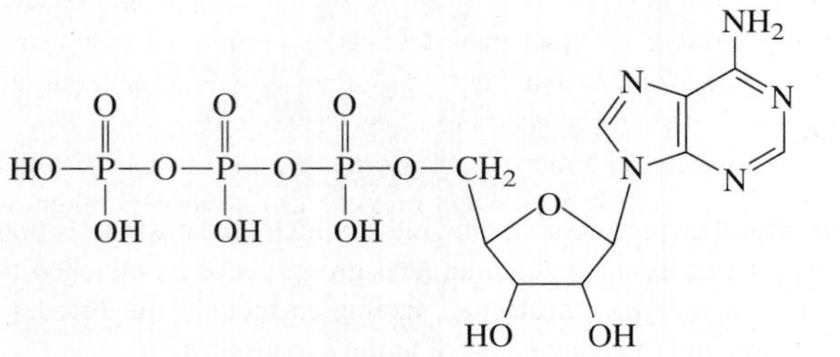

Alcohol: Compound of the type ROH.

Aldaric acid: Carbohydrate in which carboxylic acid functions are present at both ends of the chain. Aldaric acids are typically prepared by oxidation of aldoses with nitric acid.

Aldehyde: Compound of the type

$$\mathrm{RCH{=}O} \quad \text{or} \quad \mathrm{ArCH{=}O}$$

Aldimines: Imines of the type RCH=NHR′ formed by the reaction of aldehydes with primary amines.

Alditol: The polyol obtained on reduction of the carbonyl group of a carbohydrate.

Aldol addition: Nucleophilic addition of an aldehyde or ketone enolate to the carbonyl group of an aldehyde or a ketone. The most typical case involves two molecules of an aldehyde, and is usually catalyzed by bases.

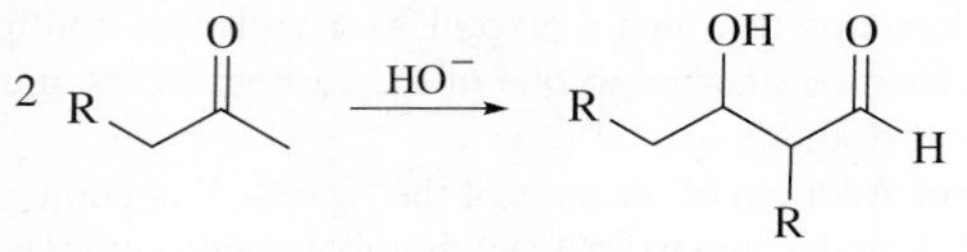

Aldol condensation: When an aldol addition is carried out so that the β-hydroxy aldehyde or ketone dehydrates under the conditions of its formation, the product is described as arising by an aldol condensation.

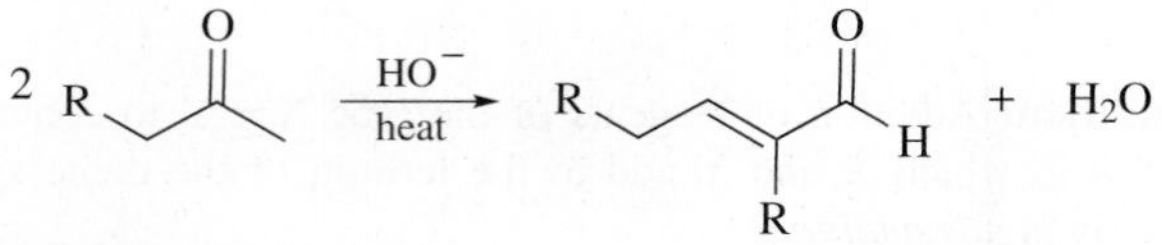

Aldonic acid: Carboxylic acid obtained by oxidation of the aldehyde function of an aldose.

Aldose: Carbohydrate that contains an aldehyde carbonyl group in its open-chain form.

Aliphatic: Term applied to compounds that do not contain benzene or benzene-like rings as structural units. (Historically, *aliphatic* was used to describe compounds derived from fats and oils.)

Alkadiene: Hydrocarbon that contains two carbon–carbon double bonds; commonly referred to as a *diene.*

Alkaloid: Amine that occurs naturally in plants. The name derives from the fact that such compounds are weak bases.

Alkane: Hydrocarbon in which all the bonds are single bonds. Alkanes have the general formula C_nH_{2n+2}.

Alkanethiolate: The conjugate base of a thiol.

Alkene: Hydrocarbon that contains a carbon–carbon double bond (C=C); also known by the older name *olefin.*

Alkoxide ion: Conjugate base of an alcohol; a species of the type $R{-}\ddot{O}{:}^-$.

Alkylamine: Amine in which the organic groups attached to nitrogen are alkyl groups.

Alkylation: Reaction in which an alkyl group is attached to some structural unit in a molecule.

Alkyl group: Structural unit related to an alkane by replacing one of the hydrogens by a potential point of attachment to some other atom or group. The general symbol for an alkyl group is R—.

Alkyl halide: Compound of the type RX, in which X is a halogen substituent (F, Cl, Br, I).

Alkyloxonium ion: Positive ion of the type ROH_2^+.

Alkyne: Hydrocarbon that contains a carbon–carbon triple bond. Also, the class of compounds structurally related to allene as the parent.

Allene: The compound $H_2C{=}C{=}CH_2$.

Allyl group: The group

$$H_2C{=}CHCH_2{-}.$$

Allylic anion: A carbanion in which the negatively charged carbon is allylic.

Allylic carbocation: A carbocation in which the positively charged carbon is allylic.

Allylic carbon: The sp^3-hybridized carbon of a C=C—C unit. Atoms or groups attached to an allylic carbon are termed *allylic substituents.*

Allylic free radical: A free radical in which the unpaired electron is on an allylic carbon.

Allylic rearrangement: Functional-group transformation in which double-bond migration has converted one allylic structural unit to another, as in:

Amide: Compound of the type $\mathrm{RC({=}O)NR'_2}$.

Amine: Molecule in which a nitrogen-containing group of the type $-NH_2$, —NHR, or $-NR_2$ is attached to an alkyl or aryl group.

α-Amino acid: A carboxylic acid that contains an amino group at the α-carbon atom. α-Amino acids are the building blocks of peptides and proteins. An α-amino acid normally exists as a *zwitterion.*

$$\mathrm{RCH(\overset{+}{N}H_3)CO_2^-}$$

L-Amino acid: The Fischer projection of an L-amino acid has the amino group on the left when the carbon chain is vertical with the carboxyl group at the top.

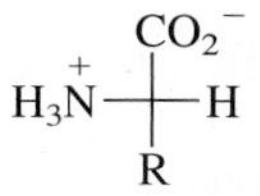

Amino acid racemization: A method for dating archaeological samples based on the rate at which the stereochemistry at the α carbon of amino acid components is randomized. It is useful for samples too old to be reliably dated by ^{14}C decay.

Amino acid residues: Individual amino acid components of a peptide or protein.

Amino sugar: Carbohydrate in which one of the hydroxyl groups has been replaced by an amino group.

Amphiphilic: Possessing both hydrophilic and lipophilic properties within the same species.

Amylopectin: A polysaccharide present in starch. Amylopectin is a polymer of α-(1→4)-linked glucose units, as is amylose (see *amylose*). Unlike amylose, amylopectin contains branches of 24–30 glucose units connected to the main chain by an α-(1→6) linkage.

Amylose: The water-dispersible component of starch. It is a polymer of α-(1→4)-linked glucose units.

Anchimeric assistance: A rate increase due to the direct interaction of a reactive center and a functional group within a molecule. The two components are not directly conjugated to one another. Also called *neighboring group participation.*

Androgen: A male sex hormone.

Angle strain: The strain a molecule possesses because its bond angles are distorted from their normal values.

Anion: Negatively charged ion.

Anionic polymerization: A polymerization in which the reactive intermediates are negatively charged.

Annulene: Monocyclic hydrocarbon characterized by a completely conjugated system of double bonds. Annulenes may or may not be aromatic.

[*x*]Annulene: An annulene in which the ring contains *x* carbons.

Anomeric carbon: The carbon atom in a furanose or pyranose form that is derived from the carbonyl carbon of the open-chain form. It is the ring carbon that is bonded to two oxygens.

Anomeric effect: The preference for an electronegative substituent, especially a hydroxyl group, to occupy an axial orientation when bonded to the anomeric carbon in the pyranose form of a carbohydrate.

Anti: Term describing relative position of two substituents on adjacent atoms when the angle between their bonds is on the order of 180°. Atoms X and Y in the structure shown are anti to each other.

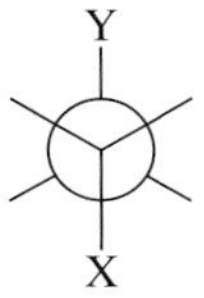

Y

X

Anti addition: Addition reaction in which the two portions of the attacking reagent X—Y add to opposite faces of the double bond.

Antiaromatic: The quality of being destabilized by electron delocalization.

Antibonding orbital: An orbital in a molecule in which an electron is less stable than when localized on an isolated atom.

Anticodon: Sequence of three bases in a molecule of tRNA that is complementary to the codon of mRNA for a particular amino acid.

Aprotic solvent: A solvent that does not have easily exchangeable protons such as those bonded to oxygen of hydroxyl groups.

Aramid: A polyamide of a benzenedicarboxylic acid and a benzenediamine.

Arene: Aromatic hydrocarbon. Often abbreviated ArH.

Arenium ion: The carbocation intermediate formed by attack of an electrophile on an aromatic substrate in electrophilic aromatic substitution. See *cyclohexadienyl cation.*

Aromatic hydrocarbon: An electron-delocalized species that is much more stable than any structure written for it in which all the electrons are localized either in covalent bonds or as unshared electron pairs.

Aromaticity: Special stability associated with aromatic compounds.

Arrhenius equation: The expression $k = Ae^{-E_a/RT}$ relating the rate of a chemical process to temperature.

Arylamine: An amine that has an aryl group attached to the amine nitrogen.

Aryne: A species that contains a triple bond within an aromatic ring (see *benzyne*).

Asymmetric: Lacking all significant symmetry elements; an asymmetric object does not have a plane, axis, or center of symmetry.

Asymmetric synthesis: The stereoselective introduction of a chirality center in a reactant in which the stereoisomeric products are formed in unequal amounts.

Atactic polymer: Polymer characterized by random stereochemistry at its chirality centers. An atactic polymer, unlike an isotactic or a syndiotactic polymer, is not a stereoregular polymer.

Atomic number: The number of protons in the nucleus of a particular atom. The symbol for atomic number is *Z*, and each element has a unique atomic number.

Atropisomers: Stereoisomers that result from restricted rotation about single bonds where the barrier for rotation is sufficient to allow isolation of the isomers.

Axial bond: A bond to a carbon in the chair conformation of cyclohexane oriented like the six "up-and-down" bonds in the following:

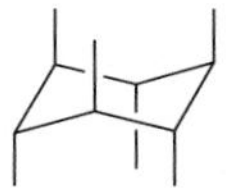

Azo coupling: Formation of a compound of the type ArN=NAr′ by reaction of an aryl diazonium salt with an arene. The arene must be strongly activated toward electrophilic aromatic substitution; that is, it must bear a powerful electron-releasing substituent such as —OH or $—NR_2$.

B

Baeyer–Villiger oxidation: Oxidation of an aldehyde or, more commonly, a ketone with a peroxy acid. The product of Baeyer–Villiger oxidation of a ketone is an ester.

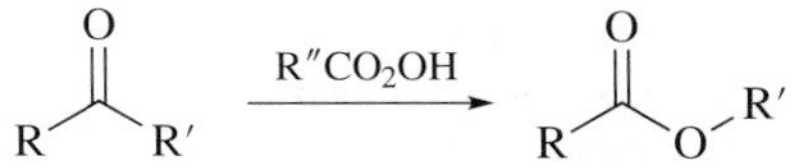

$$RC(=O)R' \xrightarrow{R''CO_2OH} RC(=O)OR'$$

Base: According to the Arrhenius definition, a substance that ionizes in water to produce hydroxide ions. According to the Brønsted–Lowry definition, a substance that accepts a proton from some suitable donor. According to the Lewis definition, an electron-pair donor.

Base pair: Term given to the purine of a nucleotide and its complementary pyrimidine. Adenine (A) is complementary to thymine (T), and guanine (G) is complementary to cytosine (C).

Base peak: The most intense peak in a mass spectrum. The base peak is assigned a relative intensity of 100, and the intensities of all other peaks are cited as a percentage of the base peak.

Bending vibration: The regular, repetitive motion of an atom or a group along an arc the radius of which is the bond connecting the atom or group to the rest of the molecule. Bending vibrations are one type of molecular motion that gives rise to a peak in the infrared spectrum.

Benzyl group: The group $C_6H_5CH_2—$.

Benzylic carbon: A carbon directly attached to a benzene ring. A hydrogen attached to a benzylic carbon is a benzylic hydrogen. A carbocation in which the benzylic carbon is positively charged is a benzylic carbocation. A free radical in which the benzylic carbon bears the unpaired electron is a benzylic radical.

Benzyne: Benzene that lacks two hydrogens.

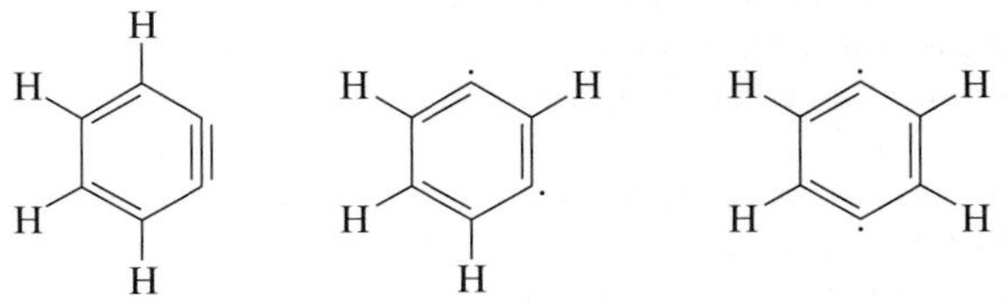

o-Benzyne *m*-Benzyne *p*-Benzyne

Biaryl: A compound in which two aromatic rings are joined by a single bond.

Bile acids: Steroid derivatives biosynthesized in the liver that aid digestion by emulsifying fats.

Bimolecular: A process in which two particles react in the same elementary step.

Bioenergetics: The study of energy transfer in biological processes. The standard state pH = 7 instead of the customary pH = 1.

Biofilms: Communities of microorganisms that grow on surfaces within a matrix of extracellular polymers. Bacteria in biofilms have altered physiology from planktonic bacteria and are usually more resistant to antimicrobial drugs.

Biological isoprene unit: Isopentenyl diphosphate, the biological precursor to terpenes and steroids:

OPP

Birch reduction: Reduction of an aromatic ring to a 1,4-cyclohexa-diene on treatment with a group 1 metal (Li, Na, K) and an alcohol in liquid ammonia.

Block copolymer: A copolymer of monomers A and B in which sections of poly-A and poly-B of variable length alternate.

Boat conformation: An unstable conformation of cyclohexane, depicted as

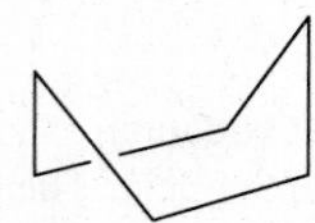

π bond: In alkenes, a bond formed by overlap of *p* orbitals in a side-by-side manner. A π bond is weaker than a σ bond. The carbon–carbon double bond in alkenes consists of two sp^2-hybridized carbons joined by a σ bond and a π bond.

σ bond: A connection between two atoms in which the orbitals involved overlap along the internuclear axis. A cross section perpendicular to the internuclear axis is a circle.

Bond dipole moment: The dipole moment of a bond between two atoms.

Bond dissociation enthalpy: For a substance A:B, the energy required to break the bond between A and B so that each retains one of the electrons in the bond.

Bonding orbital: An orbital in a molecule in which an electron is more stable than when localized on an isolated atom. All the bonding orbitals are normally doubly occupied in stable neutral molecules.

Bond-length distortion: The deviation of the length of a bond between two atoms from its normal value.

Bond-line formula: Formula in which connections between carbons are shown but individual carbons and hydrogens are not. The bond-line formula

represents the compound $(CH_3)_2CHCH_2CH_3$.

Boundary surface: The surface that encloses the region where the probability of finding an electron is high (90–95%).

Branched-chain carbohydrate: Carbohydrate in which the main carbon chain bears a carbon substituent in place of a hydrogen or hydroxyl group.

Branched polymer: A polymer with branches having the same repeating units as the main chain.

Bridged compound: A compound in which two nonadjacent atoms are common to two or more rings.

Broadband decoupling: A technique in ^{13}C NMR spectroscopy that removes the splitting of ^{13}C signals caused by coupling of ^{13}C and ^{1}H nuclei. Thus, all of the ^{13}C signals appear as singlets.

Bromohydrin: A halohydrin in which the halogen is bromine (see *halohydrin*).

Bromonium ion: A halonium ion in which the halogen is bromine (see *halonium ion*).

Brønsted–Lowry approach: A model for acid–base reactions in which acids are proton donors and bases are proton acceptors.

Buckminsterfullerene: Name given to the C_{60} cluster with structure resembling the geodesic domes of R. Buckminster Fuller.

***n*-Butane:** Common name for butane $CH_3CH_2CH_2CH_3$.

***n*-Butyl group:** The group $CH_3CH_2CH_2CH_2—$.

***sec*-Butyl group:** The group

$$CH_3CH_2\underset{|}{C}HCH_3$$

***tert*-Butyl group:** The group $(CH_3)_3C—$.

C

Cahn–Ingold–Prelog system: System for specifying absolute configuration as *R* or *S* on the basis of the order in which atoms or groups are attached to a chirality center. Groups are ranked in order of precedence according to rules based on atomic number.

Carbanion: Anion in which the negative charge is borne by carbon. An example is acetylide ion.

Carbene: A neutral species in which one of the carbon atoms is associated with six valence electrons.

Carbenoid: A compound, usually organometallic, that resembles a carbene in its chemical reactions.

Carbinolamine: An obsolete name for a hemiaminal (see *hemiaminal*).

Carbocation: Positive ion in which the charge resides on carbon. An example is *tert*-butyl cation, $(CH_3)_3C^+$. Carbocations are unstable species that, though they cannot normally be isolated, are believed to be intermediates in certain reactions.

Carbon skeleton diagram: Synonymous with bond-line formula.

Carboxylic acid: Compound of the type $R\overset{\overset{O}{\|}}{C}OH$, also written as RCO_2H.

Carboxylic acid derivative: Compound that yields a carboxylic acid on hydrolysis. Carboxylic acid derivatives include acyl chlorides, acid anhydrides, esters, and amides.

Carcinogen: A cancer-causing substance.

Carotenoids: Naturally occurring tetraterpenoid compounds found in plants and animals.

Catalyst: A substance that increases the rate of a chemical reaction, but is not consumed by it.

Cation: Positively charged ion.

Cationic polymerization: A polymerization in which the reactive intermediates are carbocations.

Cation radical: A positively charged species that has an odd number of electrons.

Cellobiose: A disaccharide in which two glucose units are joined by a β-(1→4) linkage. Cellobiose is obtained by the hydrolysis of cellulose.

Cellulose: A polysaccharide in which thousands of glucose units are joined by β-(1→4) linkages.

Center of symmetry: A point in the center of a structure located so that a line drawn from it to any element of the structure, when extended an equal distance in the opposite direction, encounters an identical element. Benzene, for example, has a center of symmetry.

Chain-growth polymerization: Macromolecule formation by a process in which monomers add sequentially to one end of a chain.

Chain reaction: Reaction mechanism in which a sequence of individual steps repeats itself many times, usually because a reactive intermediate consumed in one step is regenerated in a subsequent step. The halogenation of alkanes is a chain reaction proceeding via free-radical intermediates.

Chain-terminating step: A chemical reaction that stops further growth of a polymer chain.

Chain transfer: A reaction between a growing chain and a terminated chain that terminates the growing chain and activates the previously terminated chain to further growth.

Chair–chair interconversion: Synonymous with ring inversion of cyclohexane and related compounds.

Chair conformation: The most stable conformation of cyclohexane:

Characteristic absorption frequencies: The regions of the infrared (IR) spectrum where peaks characteristic of particular structural units are normally found.

Chemical bond: A connection between atoms.

Chemically nonequivalent: In NMR, synonymous with *chemical-shift-nonequivalent.*

Chemical shift: A measure of how shielded the nucleus of a particular atom is. Nuclei of different atoms have different chemical shifts, and nuclei of the same atom have chemical shifts that are sensitive to their molecular environment. In proton and carbon-13 NMR, chemical shifts are cited as δ, or parts per million (ppm), from the hydrogens or carbons, respectively, of tetramethylsilane.

Chemical-shift-nonequivalent: Nuclei with different chemical shifts in nuclear magnetic resonance (NMR).

Chiral: Term describing an object that is not superimposable on its mirror image.

Chiral drug: A chiral molecule in which the desired therapeutic effect resides in only one of its enantiomers.

Chirality axis: Line drawn through a molecule that is analogous to the long axis of a right-handed or left-handed screw or helix.

Chirality center: An atom that has four nonequivalent atoms or groups attached to it. At various times chirality centers have been called *asymmetric centers* or *stereogenic centers.*

Chlorohydrin: A halohydrin in which the halogen is chlorine (see *halohydrin*).

Chloronium ion: A halonium ion in which the halogen is chlorine (see *halonium ion*).

Cholesterol: The most abundant steroid in animals and the biological precursor to other naturally occurring steroids, including the bile acids, sex hormones, and corticosteroids.

Chromatography: A method for separation and analysis of mixtures based on the different rates at which different compounds are removed from a stationary phase by a moving phase.

Chromophore: The structural unit of a molecule responsible for absorption of radiation of a particular frequency; a term usually applied to ultraviolet-visible spectroscopy.

***cis*-:** Stereochemical prefix indicating that two substituents are on the same side of a ring or double bond. (Contrast with the prefix *trans*-.)

Claisen condensation: Reaction in which a β-keto ester is formed by condensation of two moles of an ester in base:

$$\mathrm{RCH_2\overset{O}{\overset{\|}{C}}OR'} \xrightarrow[\text{2. } H_3O^+]{\text{1. NaOR}'} \mathrm{RCH_2\overset{O}{\overset{\|}{C}}\underset{R}{\underset{|}{C}}H\overset{O}{\overset{\|}{C}}OR'} + \mathrm{R'OH}$$

Claisen rearrangement: Thermal conversion of an allyl phenyl ether to an *o*-allyl phenol. The rearrangement proceeds via a cyclohexadienone intermediate.

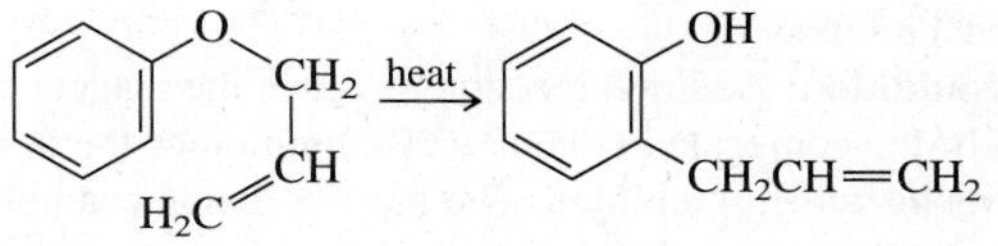

$$\mathrm{C_6H_5OCH_2CH{=}CH_2} \xrightarrow{\text{heat}} o\text{-}\mathrm{HOC_6H_4CH_2CH{=}CH_2}$$

Claisen–Schmidt condensation: A mixed aldol condensation involving a ketone enolate and an aromatic aldehyde or ketone.

Clemmensen reduction: Method for reducing the carbonyl group of aldehydes and ketones to a methylene group ($C{=}O \rightarrow CH_2$) by treatment with zinc amalgam [Zn(Hg)] in concentrated hydrochloric acid.

Closed-shell electron configuration: Stable electron configuration in which all the lowest energy orbitals of an atom (in the case of the noble gases), an ion (e.g., Na^+), or a molecule (e.g., benzene) are filled.

^{13}C NMR: Nuclear magnetic resonance spectroscopy in which the environments of individual carbon atoms are examined via their mass-13 isotope.

Codon: Set of three successive nucleotides in mRNA that is unique for a particular amino acid. The 64 codons possible from combinations of A, T, G, and C code for the 20 amino acids from which proteins are constructed.

Coenzyme: A cofactor that is not chemically bonded to an enzyme.

Coenzyme Q: Naturally occurring group of related quinones involved in the chemistry of cellular respiration. Also known as *ubiquinone*.

Cofactor: A molecule that acts in combination with an enzyme to bring about a reaction. A cofactor may be either a coenzyme or a prosthetic group.

Combinatorial chemistry: A method for carrying out a large number of reactions on a small scale in the solid phase so as to generate a "library" of related compounds for further study, such as biological testing.

Combustion: Burning of a substance in the presence of oxygen. All hydrocarbons yield carbon dioxide and water when they undergo combustion.

Common name: Name given to a compound on some basis other than a systematic set of rules.

σ-Complex: Synonymous with *arenium ion.*

Compound: An assembly of two or more atoms with properties different from the individual atoms.

Concerted reaction: Reaction that occurs in a single elementary step.

Condensation polymer: Polymer in which the bonds that connect the monomers are formed by condensation reactions. Typical condensation polymers include polyesters and polyamides.

Condensation reaction: Reaction in which two molecules combine to give a product accompanied by the expulsion of some small stable molecule (such as water). An example is acid-catalyzed ether formation:

$$2\mathrm{ROH} \xrightarrow{H_2SO_4} \mathrm{ROR} + \mathrm{H_2O}$$

Condensed formula: Structural formula in which subscripts are used to indicate replicated atoms or groups, as in $(CH_3)_2CHCH_2CH_3$.

Configuration: Term describing the spatial arrangement of atoms or groups at a chirality center or double bond.

Conformational analysis: Study of the conformations available to a molecule, their relative stability, and the role they play in defining the properties of the molecule.

Conformational enantiomers: Nonsuperimposable mirror-image conformations of a molecule.

Conformations: Nonidentical representations of a molecule generated by rotation about single bonds.

Conformers: Different conformations of a single molecule.

Conjugate acid: The species formed from a Brønsted base after it has accepted a proton.

Conjugate addition: Addition reaction in which the reagent adds to the termini of the conjugated system with migration of the double bond; synonymous with 1,4 addition. The most common examples include conjugate addition to 1,3-dienes and to α,β-unsaturated carbonyl compounds.

Conjugate base: The species formed from a Brønsted acid after it has donated a proton.

Conjugated diene: System of the type C═C—C═C, in which two pairs of doubly bonded carbons are joined by a single bond. The π electrons are delocalized over the unit of four consecutive sp^2-hybridized carbons.

Conjugated system: A structural arrangement in which electron delocalization permits two groups to interact so that the properties of the conjugated system are different from those of the separate groups.

Conjugation energy: Synonymous with *resonance energy*.

Connectivity: Order in which a molecule's atoms are connected. Synonymous with *constitution*.

Constitutional isomers: Isomers that differ in respect to the order in which the atoms are connected. Butane ($CH_3CH_2CH_2CH_3$) and isobutane [$(CH_3)_3CH$] are constitutional isomers.

Contributing structures: The various resonance structures that can be written for a molecule.

Coordination polymerization: A method of addition polymerization in which monomers are added to the growing chain on an active organometallic catalyst.

Cope rearrangement: The thermal conversion of one 1,5-diene unit to another by way of a cyclic transition state.

Copolymer: Polymer formed from two or more different monomers.

Correlated spectroscopy (COSY): A 2D NMR technique that correlates the chemical shifts of spin-coupled nuclei. COSY stands for correlated spectroscopy.

Corticosteroid: A steroid present in the outer layer, or *cortex,* of the adrenal gland.

Coulombic attraction: The electrical attraction between opposite charges.

Coupling constant *J*: A measure of the extent to which two nuclear spins are coupled. In the simplest cases, it is equal to the distance between adjacent peaks in a split NMR signal.

Covalent bond: Chemical bond between two atoms that results from their sharing of two electrons.

COX-2: Cyclooxygenase-2, an enzyme that catalyzes the biosynthesis of prostaglandins. COX-2 inhibitors reduce pain and inflammation by blocking the activity of this enzyme.

Cracking: A key step in petroleum refining in which high-molecular-weight hydrocarbons are converted to lower-molecular-weight ones by thermal or catalytic carbon–carbon bond cleavage.

Critical micelle concentration: Concentration above which substances such as salts of fatty acids aggregate to form micelles in aqueous solution.

Cross-coupling: A reaction between an organometallic reagent and an alkyl or aryl halide that results in bond formation between the two organic groups.

Cross-linked polymer: A polymer in which two or more chains are covalently bonded.

Crown ether: A cyclic polyether that, via ion–dipole attractive forces, forms stable complexes with metal ions. Such complexes, along with their accompanying anion, are soluble in nonpolar solvents.

Crystallite: An ordered crystalline region within a polymer.

C terminus: The amino acid at the end of a peptide or protein chain that has its carboxyl group intact—that is, in which the carboxyl group is not part of a peptide bond.

Cumulated diene: Diene of the type C═C═C, in which one carbon has double bonds to two others.

Cumulenes: Compounds that contain C═C═C as a structural unit.

Curved arrows: Arrows that show the direction of electron flow in chemical reactions; also used to show differences in electron placement between resonance forms.

Cyanohydrin: Compound of the type

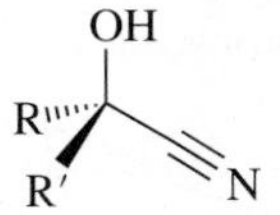

OH
R'''''C—C≡N
R′

Cyanohydrins are formed by nucleophilic addition of HCN to the carbonyl group of an aldehyde or a ketone.

Cycloaddition: Addition, such as the Diels–Alder reaction, in which a ring is formed via a cyclic transition state.

Cycloalkane: An alkane in which a ring of carbon atoms is present.

Cycloalkene: A cyclic hydrocarbon characterized by a double bond between two of the ring carbons.

Cycloalkyne: A cyclic hydrocarbon characterized by a triple bond between two of the ring carbons.

Cyclohexadienyl anion: The key intermediate in nucleophilic aromatic substitution by the addition–elimination mechanism. It is represented by the general structure shown, where Y is the nucleophile and X is the leaving group.

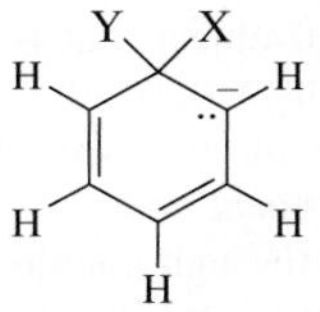

Y X
H H
H H
H

Cyclohexadienyl cation: The key intermediate in electrophilic aromatic substitution reactions. It is represented by the general structure

E H
H H
+
H H
H

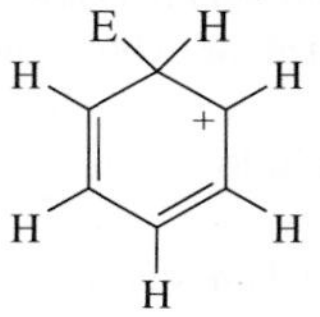

where E is derived from the electrophile that reacts with the ring.

D

d-Block elements: Elements in groups 3–12 of the Periodic Table.

Deactivating substituent: A group that when present in place of hydrogen causes a particular reaction to occur more slowly. The term is most often applied to the effect of substituents on the rate of electrophilic aromatic substitution.

Debye unit (D): Unit customarily used for measuring dipole moments:

$$1\text{D} = 1 \times 10^{-18} \text{ esu·cm}$$

Decarboxylation: Reaction of the type $RCO_2H \rightarrow RH + CO_2$, in which carbon dioxide is lost from a carboxylic acid. Decarboxylation normally occurs readily only when the carboxylic acid is a 1,3-dicarboxylic acid or a β-keto acid.

Decoupling: In NMR spectroscopy, any process that destroys the coupling of nuclear spins between two nuclei. Two types of decoupling are employed in ^{13}C NMR spectroscopy. Broadband decoupling removes all the $^{1}H—^{13}C$ couplings; off-resonance decoupling removes all $^{1}H—^{13}C$ couplings except those between directly bonded atoms.

Dehydration: Removal of H and OH from adjacent atoms. The term is most commonly employed in the preparation of alkenes by heating alcohols in the presence of an acid catalyst.

1,2-, 1,3-, and 1,4-Dehydrobenzene: See *benzyne.*

Dehydrogenation: Elimination in which H_2 is lost from adjacent atoms. The term is most commonly encountered in the industrial preparation of ethylene from ethane, propene from propane, 1,3-butadiene from butane, and styrene from ethylbenzene.

Dehydrohalogenation: Reaction in which an alkyl halide, on being treated with a base such as sodium ethoxide, is converted to an alkene by loss of a proton from one carbon and the halogen from the adjacent carbon.

Delocalization: Association of an electron with more than one atom. The simplest example is the shared electron pair (covalent) bond. Delocalization is important in conjugated π-electron systems, where an electron may be associated with several carbon atoms.

Delocalization energy: Synonymous with *resonance energy.*

Deoxy sugar: A carbohydrate in which one of the hydroxyl groups has been replaced by a hydrogen.

DEPT: Abbreviation for *d*istortionless *e*nhancement of *p*olarization *t*ransfer. DEPT is an NMR technique that reveals the number of hydrogens directly attached to a carbon responsible for a particular signal.

Detergents: Substances that clean by micellar action. Although the term usually refers to a synthetic detergent, soaps are also detergents.

Deuterium isotope effect: The difference in a property, usually reaction rate, that results when one or more atoms of ^{1}H in a compound are replaced by ^{2}H.

Diastereomers: Stereoisomers that are not enantiomers—stereoisomers that are not mirror images of one another.

Diastereoselective reaction: A reaction in which one diastereomeric product is present in excess of the other.

Diastereotopic: Describing two atoms or groups in a molecule that are attached to the same atom but are in stereochemically different environments that are not mirror images of each other. The two protons shown in bold in $\mathbf{H_2}C{=}CHCl$, for example, are diastereotopic. One is cis to chlorine, the other is trans.

1,3-Diaxial repulsion: Repulsive forces between axial substituents on the same side of a cyclohexane ring.

Diazonium ion: Ion of the type $R—\overset{+}{N}{\equiv}N$:. Aryl diazonium ions are formed by treatment of primary aromatic amines with nitrous acid. They are extremely useful in the preparation of aryl halides, phenols, and aryl cyanides.

Diazotization: The reaction by which a primary amine is converted to the corresponding diazonium ion by nitrosation.

Dieckmann cyclization: An intramolecular version of the Claisen condensation.

Dielectric constant: A measure of the ability of a material to disperse the force of attraction between oppositely charged particles. The symbol for dielectric constant is ϵ.

Diels–Alder reaction: Conjugate addition of an alkene to a conjugated diene to give a cyclohexene derivative. Diels–Alder reactions are extremely useful in synthesis.

Dienophile: The alkene that adds to the diene in a Diels–Alder reaction.

Dihydroxylation: Reaction or sequence of reactions in which an alkene is converted to a vicinal diol.

β-Diketone: Compound of the type

R–C(=O)–CH₂–C(=O)–R′

also referred to as a 1,3-diketone.

Dimer: Molecule formed by the combination of two identical molecules.

Diol: A compound with two alcohol functional groups.

Dipeptide: A compound in which two α-amino acids are linked by an amide bond between the amino group of one and the carboxyl group of the other:

$H_3\overset{+}{N}$–CH(R)–C(=O)–NH–CH(R′)–CO_2^-

Dipole–dipole attractive force: A force of attraction between oppositely polarized atoms.

Dipole/induced-dipole force: A force of attraction that results when a species with a permanent dipole induces a complementary dipole in a second species.

Dipole moment: Product of the attractive force between two opposite charges and the distance between them. Dipole moment has the symbol μ and is measured in Debye units (D).

Direct addition: Synonymous with *1,2 addition.*

Disaccharide: A carbohydrate that yields two monosaccharide units (which may be the same or different) on hydrolysis.

Disproportionation: A reaction in which transfer of an atom from one growing polymer chain to another terminates both.

Disubstituted alkene: Alkene of the type $R_2C{=}CH_2$ or $RCH{=}CHR$. The groups R may be the same or different, they may be any length, and they may be branched or unbranched. The significant point is that there are two carbons *directly* bonded to the carbons of the double bond.

Disulfide: A compound of the type RSSR′.

Disulfide bridge: An S—S bond between the sulfur atoms of two cysteine residues in a peptide or protein.

DNA (deoxyribonucleic acid): A polynucleotide of 2′-deoxyribose present in the nuclei of cells that serves to store and replicate genetic information. Genes are DNA.

Double bond: Bond formed by the sharing of four electrons between two atoms.

Double dehydrohalogenation: Reaction in which a geminal dihalide or vicinal dihalide, on being treated with a very strong base such as sodium amide, is converted to an alkyne by loss of two protons and the two halogen substituents.

Double helix: The form in which DNA normally occurs in living systems. Two complementary strands of DNA are associated with each other by hydrogen bonds between their base pairs, and each DNA strand adopts a helical shape.

Downfield: The low-field region of an NMR spectrum. A signal that is downfield with respect to another lies to its left in the spectrum.

E

***E*-:** Stereochemical descriptor used when higher ranked substituents are on opposite sides of a double bond.

E1: See *Elimination unimolecular (E1) mechanism.*

E2: See *Elimination bimolecular (E2) mechanism.*

Eclipsed conformation: Conformation in which bonds on adjacent atoms are aligned with one another. For example, the C—H bonds indicated in the structure shown are eclipsed.

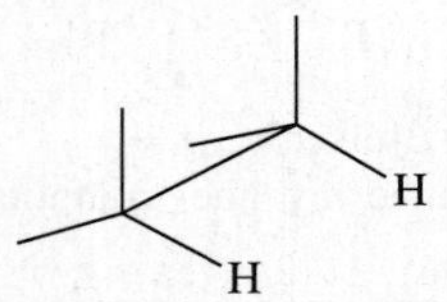

Edman degradation: Method for determining the N-terminal amino acid of a peptide or protein. It involves treating the material with phenyl isothiocyanate ($C_6H_5N{=}C{=}S$), cleaving with acid, and then identifying the phenylthiohydantoin (PTH derivative) produced.

Elastomer: A synthetic polymer that possesses elasticity.

Electromagnetic radiation: Various forms of radiation propagated at the speed of light. Electromagnetic radiation includes (among others) visible light; infrared, ultraviolet, and microwave radiation; and radio waves, cosmic rays, and X-rays.

Electron affinity: Energy change associated with the capture of an electron by an atom.

Electronegativity: A measure of the ability of an atom to attract the electrons in a covalent bond toward itself. Fluorine is the most electronegative element.

Electron configurations: A list of the occupied orbitals of an element or ion including the number of electrons in each. Sodium, for example, has the electron configuration $1s^2 2s^2 2p^6 3s^1$.

Electronic effect: An effect on structure or reactivity that is attributed to the change in electron distribution that a substituent causes in a molecule.

Electron impact: Method for producing positive ions in mass spectrometry whereby a molecule is bombarded by high-energy electrons.

Electron-releasing group: An atom or group that increases the electron density around another atom by an inductive or resonance effect.

18-Electron rule: The number of ligands that can be attached to a transition metal are such that the sum of the electrons brought by the ligands plus the valence electrons of the metal equals 18.

Electron-withdrawing group: An atom or group that decreases the electron density around another atom by an inductive or resonance effect.

Electrophile: A species (ion or compound) that can act as a Lewis acid, or electron pair acceptor; an "electron seeker." Carbocations are one type of electrophile.

Electrophilic addition: Mechanism of addition in which the species that first reacts with the multiple bond is an electrophile ("electron seeker").

Electrophilic aromatic substitution: Fundamental reaction type exhibited by aromatic compounds. An electrophilic species (E^+) replaces one of the hydrogens of an aromatic ring.

$$\text{Ar—H} + \text{E—Y} \longrightarrow \text{Ar—E} + \text{H—Y}$$

Electrophoresis: A method for separation and purification of proteins based on their size.

Electrostatic attraction: Force of attraction between oppositely charged particles.

Electrostatic potential map: The charge distribution in a molecule represented by mapping the interaction energy of a point positive charge with the molecule's electric field on the van der Waals surface.

Elementary step: A step in a reaction mechanism in which each species shown in the equation for this step participates in the same transition state. An elementary step is characterized by a single transition state.

Elements of unsaturation: See *index of hydrogen deficiency.*

β Elimination: Reaction in which a double or triple bond is formed by loss of atoms or groups from adjacent atoms. (See *dehydration, dehydrogenation, dehydrohalogenation,* and *double dehydrohalogenation.*)

Elimination–addition mechanism: Two-stage mechanism for nucleophilic aromatic substitution. In the first stage, an aryl halide undergoes elimination to form an aryne intermediate. In the second stage, nucleophilic addition to the aryne yields the product of the reaction.

Elimination bimolecular (E2) mechanism: Mechanism for elimination of alkyl halides characterized by a transition state in which the attacking base removes a proton at the same time that the bond to the halide leaving group is broken.

Elimination unimolecular (E1) mechanism: Mechanism for elimination characterized by the slow formation of a carbocation intermediate followed by rapid loss of a proton from the carbocation to form the alkene.

Enamine: Product of the reaction of a secondary amine and an aldehyde or a ketone. Enamines are characterized by the general structure

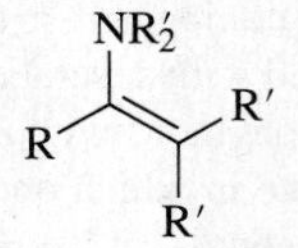

Enantiomeric excess: Difference between the percentage of the major enantiomer present in a mixture and the percentage of its mirror image. An optically pure material has an enantiomeric excess of 100%. A racemic mixture has an enantiomeric excess of zero.

Enantiomeric ratio: The ratio of enantiomers present in a mixture. A mixture with an er = 80:20 has 80% of one enantiomer and 20% of the other.

Enantiomers: Stereoisomers that are related as an object and its nonsuperimposable mirror image.

Enantiopure: An enantiomeric excess of 100%.

Enantioselective reaction: Reaction that converts an achiral or racemic starting material to a chiral product in which one enantiomer is present in excess of the other.

Enantiotopic: Describing two atoms or groups in a molecule whose environments are nonsuperimposable mirror images of each other. The two protons shown in bold in $CH_3\mathbf{CH_2}Cl$, for example, are enantiotopic. Replacement of first one, then the other, by some arbitrary test group yields compounds that are enantiomers of each other.

Endergonic: A process in which $\Delta G°$ is positive.

Endothermic: Term describing a process or reaction that absorbs heat.

Enediyne antibiotics: A family of tumor-inhibiting substances that is characterized by the presence of a C≡C—C=C—C≡C unit as part of a nine- or ten-membered ring.

Enol: Compound of the type

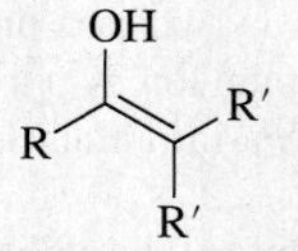

Enols are in equilibrium with an isomeric aldehyde or ketone, but are normally much less stable than aldehydes and ketones.

Enolate ion: The conjugate base of an enol. Enolate ions are stabilized by electron delocalization.

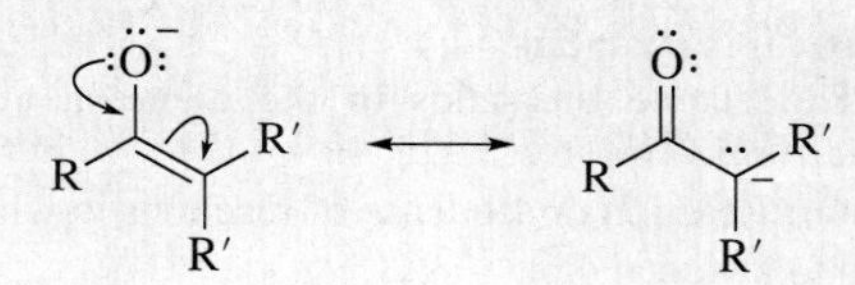

Enolization: A reaction of the type

$$\mathrm{RC(=O)CHR'_2} \longrightarrow \mathrm{RC(OH)=CR'_2}$$

Entgegen: See *E*-.

Enthalpy: The heat content of a substance; symbol, *H*.

Envelope: One of the two most stable conformations of cyclopentane. Four of the carbons in the envelope conformation are coplanar; the fifth carbon lies above or below this plane.

Enzymatic resolution: Resolution of a mixture of enantiomers based on the selective reaction of one of them under conditions of enzyme catalysis.

Enzyme: A protein that catalyzes a chemical reaction in a living system.

Epimers: Diastereomers that differ in configuration at only one of their chirality centers.

Epoxidation: Conversion of an alkene to an epoxide, usually by treatment with a peroxy acid.

Epoxide: Compound of the type

R_2C—CR_2 (bridged by O)

Equatorial bond: A bond to a carbon in the chair conformation of cyclohexane oriented approximately along the equator of the molecule.

Erythro: Term applied to the relative configuration of two chirality centers within a molecule. The erythro stereoisomer has like substituents on the same side of a Fischer projection.

Essential amino acids: Amino acids that must be present in the diet for normal growth and good health.

Essential fatty acids: Fatty acids that must be present in the diet for normal growth and good health.

Essential oils: Pleasant-smelling oils of plants consisting of mixtures of terpenes, esters, alcohols, and other volatile organic substances.

Ester: Compound of the type

RCOR′

Estrogen: A female sex hormone.

Ethene: IUPAC name for $H_2C{=}CH_2$. The common name ethylene, however, is used far more often, and the IUPAC rules permit its use.

Ether: Molecule that contains a C—O—C unit such as ROR′, ROAr, or ArOAr.

Ethylene: $H_2C{=}CH_2$, the simplest alkene and the most important industrial organic chemical.

Ethyl group: The group CH_3CH_2—.

Exergonic: A process in which $\Delta G°$ is negative.

Exothermic: Term describing a reaction or process that gives off heat.

Extinction coefficient: See *molar absorptivity*.

***E–Z* notation for alkenes:** System for specifying double-bond configuration that is an alternative to cis–trans notation. When higher ranked substituents are on the same side of the double bond, the configuration is *Z*. When higher ranked substituents are on opposite sides, the configuration is *E*. Rank is determined by the Cahn–Ingold–Prelog system.

F

Fats and oils: Triesters of glycerol. Fats are solids at room temperature, oils are liquids.

Fatty acid: Carboxylic acids obtained by hydrolysis of fats and oils. Fatty acids typically have unbranched chains and contain an even number of carbon atoms in the range of 12–20 carbons. They may include one or more double bonds.

Fatty acid synthetase: Complex of enzymes that catalyzes the biosynthesis of fatty acids from acetate.

Fibrous protein: A protein consisting of bundled chains of elongated filaments.

Field effect: An electronic effect in a molecule that is transmitted from a substituent to a reaction site via the medium (e.g., solvent).

Fingerprint region: The region 1500–500 cm^{-1} of an infrared spectrum. This region is less characteristic of functional groups than others, but varies so much from one molecule to another that it can be used to determine whether two substances are identical or not.

Fischer esterification: Acid-catalyzed ester formation between an alcohol and a carboxylic acid:

$$\mathrm{RC(=O)OH} + \mathrm{R'OH} \xrightarrow{H^+} \mathrm{RC(=O)OR'} + H_2O$$

Fischer glycosidation: A reaction in which glycosides are formed by treating a carbohydrate with an alcohol in the presence of an acid catalyst.

Fischer projection: Method for representing stereochemical relationships. The four bonds to a tetrahedral carbon are represented by a cross. The horizontal bonds are understood to project toward the viewer and the vertical bonds away from the viewer.

w–C(*x*)(*z*)–*y* is represented in a Fischer projection as a cross with *x* at top, *z* at bottom, *w* at left, *y* at right.

Fluid mosaic model: A schematic representation of a cell membrane.

Formal charge: The charge, either positive or negative, on an atom calculated by subtracting from the number of valence electrons in the neutral atom a number equal to the sum of its unshared electrons plus half the electrons in its covalent bonds.

Fragmentation pattern: In mass spectrometry, the ions produced by dissociation of the molecular ion.

Free energy: The available energy of a system; symbol, *G*. See also *Gibbs energy*.

Free radical: A species in which one of the electrons in the valence shell of carbon is unpaired. An example is methyl radical, CH_3.

Free-radical initiator: A compound present in small amounts that serves as a source of free radicals capable of triggering a free-radical chain reaction involving some other compound.

Free-radical polymerization: An alkene polymerization proceeding via free-radical intermediates.

Frequency: Number of waves per unit time. Although often expressed in hertz (Hz), or cycles per second, the SI unit for frequency is s^{-1}.

Friedel–Crafts acylation: An electrophilic aromatic substitution in which an aromatic compound reacts with an acyl chloride or a carboxylic acid anhydride in the presence of aluminum chloride. An acyl group becomes bonded to the ring.

$$\mathrm{Ar{-}H} + \mathrm{RC(=O){-}Cl} \xrightarrow{AlCl_3} \mathrm{Ar{-}C(=O)R}$$

Friedel–Crafts alkylation: An electrophilic aromatic substitution in which an aromatic compound reacts with an alkyl halide in the presence of aluminum chloride. An alkyl group becomes bonded to the ring.

$$\text{Ar—H} + \text{R—X} \xrightarrow{\text{AlCl}_3} \text{Ar—R}$$

Frontier orbitals: Orbitals involved in a chemical reaction, usually the highest occupied molecular orbital of one reactant and the lowest unoccupied molecular orbital of the other.

Frost circle: A mnemonic that gives the Hückel π MOs for cyclic conjugated molecules and ions.

FT-NMR: Abbreviation for Fourier-transform nuclear magnetic resonance. The most commonly used method for obtaining NMR spectra.

Functional class nomenclature: Type of IUPAC nomenclature in which compounds are named according to functional-group families. The last word in the name identifies the functional group; the first word designates the alkyl or aryl group that bears the functional group. Methyl bromide, ethyl alcohol, and diethyl ether are examples of functional class names.

Functional group: An atom or a group of atoms in a molecule responsible for its reactivity under a given set of conditions.

Furanose form: Five-membered ring arising via cyclic hemiacetal formation between the carbonyl group and a hydroxyl group of a carbohydrate.

G

***G*:** Symbol for Gibbs energy.

Gabriel synthesis: Method for the synthesis of primary alkylamines in which a key step is the formation of a carbon–nitrogen bond by alkylation of the potassium salt of phthalimide.

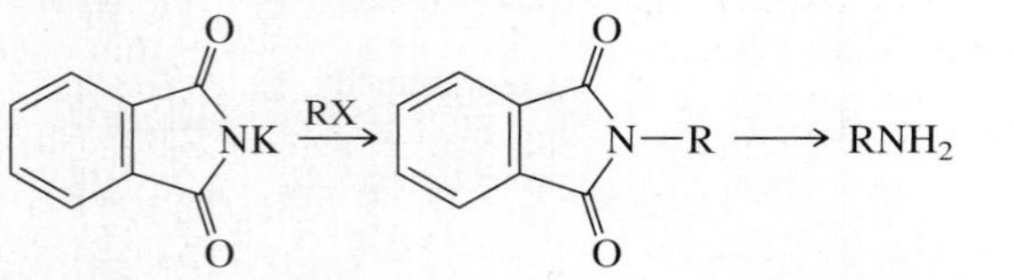

$$\text{(phthalimide)NK} \xrightarrow{\text{RX}} \text{(phthalimide)N—R} \longrightarrow \text{RNH}_2$$

Gauche: Term describing the position relative to each other of two substituents on adjacent atoms when the angle between their bonds is on the order of 60°. Atoms X and Y in the structure shown are gauche to each other.

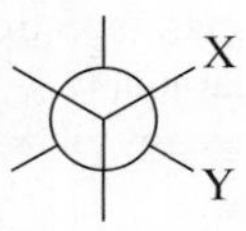

X Y

G-coupled protein receptors: A large family of protein receptors that function as transmembrane molecular switches to regulate many physiological processes.

Geminal dihalide: A dihalide of the form R_2CX_2, in which the two halogen substituents are located on the same carbon.

Geminal diol: The hydrate $R_2C(OH)_2$ of an aldehyde or a ketone.

Generic name: The name of a drug as designated by the U.S. Adopted Names Council.

Genetic code: The relationship between triplets of nucleotide bases in messenger RNA and the amino acids incorporated into a protein in DNA-directed protein biosynthesis.

Genome: The aggregate of all the genes that determine what an organism becomes.

Genomics: The study of genome sequences and their function.

Gibbs energy: The free energy (energy available to do work) of a system.

Gilman reagents: Compounds of the type R_2CuLi used in carbon–carbon bond-forming reactions.

Globular protein: An approximately spherically shaped protein that forms a colloidal dispersion in water. Most enzymes are globular proteins.

Glycobiology: The biochemical study of the structure and function of carbohydrate-containing substances, especially those that occur naturally.

Glycogen: A polysaccharide present in animals that is derived from glucose. Similar in structure to amylopectin.

Glycolysis: Biochemical process in which glucose is converted to pyruvate with release of energy.

Glycoside: A carbohydrate derivative in which the hydroxyl group at the anomeric position has been replaced by some other group. An *O*-glycoside is an ether of a carbohydrate in which the anomeric position bears an alkoxy group.

Graft copolymer: A copolymer of monomers A and B in which branches of poly-A are attached to a poly-B main chain.

Grain alcohol: A common name for ethanol (CH_3CH_2OH).

Graphene: An allotropic form of elemental carbon composed of sheets of planar fused six-membered rings.

Grignard reagent: An organomagnesium compound of the type RMgX formed by the reaction of magnesium with an alkyl or aryl halide.

H

Half-chair: One of the two most stable conformations of cyclopentane. Three consecutive carbons in the half-chair conformation are coplanar. The fourth and fifth carbons lie, respectively, above and below the plane.

Haloform reaction: The formation of CHX_3 (X = Br, Cl, or I) brought about by cleavage of a methyl ketone on treatment with Br_2, Cl_2, or I_2 in aqueous base.

$$\text{R}\overset{\text{O}}{\overset{\|}{\text{C}}}\text{CH}_3 \xrightarrow[\text{HO}^-]{\text{X}_2} \text{R}\overset{\text{O}}{\overset{\|}{\text{C}}}\text{O}^- + \text{CHX}_3$$

Halogenation: Replacement of a hydrogen by a halogen. The most frequently encountered examples are the free-radical halogenation of alkanes and the halogenation of arenes by electrophilic aromatic substitution.

Halohydrin: A compound that contains both a halogen atom and a hydroxyl group. The term is most often used for compounds in which the halogen and the hydroxyl group are on adjacent atoms (vicinal halohydrins). The most commonly encountered halohydrins are chlorohydrins and bromohydrins.

Halonium ion: A species that incorporates a positively charged halogen. Bridged halonium ions are intermediates in the addition of halogens to the double bond of an alkene.

Hammond's postulate: Principle used to deduce the approximate structure of a transition state. If two states, such as a transition state and an unstable intermediate derived from it, are similar in energy, they are believed to be similar in structure.

Haworth formulas: Planar representations of furanose and pyranose forms of carbohydrates.

Heat of combustion: Heat evolved on combustion of a substance. It is the value of $-\Delta H°$ for the combustion reaction.

Heat of hydrogenation: Heat evolved on hydrogenation of a substance. It is the value of $-\Delta H°$ for the addition of H_2 to a multiple bond.

Heck reaction: Palladium-catalyzed cross-coupling of an alkene with an alkyl or aryl halide or sulfonate.

α Helix: One type of protein secondary structure. It is a right-handed helix characterized by hydrogen bonds between NH and C═O groups. It contains approximately 3.6 amino acids per turn.

Hell–Volhard–Zelinsky reaction: The phosphorus trihalide-catalyzed α halogenation of a carboxylic acid:

$$R_2CHCO_2H + X_2 \xrightarrow[\text{or } PX_3]{P} R_2C(X)CO_2H + HX$$

Hemiacetal: Product of nucleophilic addition of one molecule of an alcohol to an aldehyde or a ketone. Hemiacetals are compounds of the type

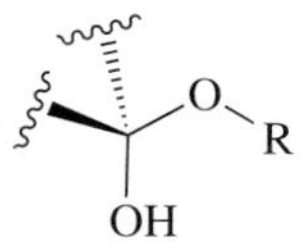

Hemiaminal: A compound of the type shown formed by nucleophilic addition of a primary or secondary amine to an aldehyde or ketone.

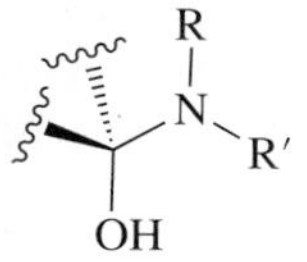

Hemiketal: A hemiacetal derived from a ketone.

Henderson–Hasselbalch equation: An equation that relates degree of dissociation of an acid at a particular pH to its pK_a.

$$pH = pK_a + \log \frac{[\text{conjugate base}]}{[\text{acid}]}$$

Heparin: A naturally occurring anticoagulant, heparin is an oligosaccharide that contains sulfated glycans and aminoglycans.

HETCOR: A 2D NMR technique that correlates the ^1H chemical shift of a proton to the ^{13}C chemical shift of the carbon to which it is attached. HETCOR stands for *heteronuclear chemical shift correlation.*

Heteroatom: An atom in an organic molecule that is neither carbon nor hydrogen.

Heterocyclic aromatic compound: A heterocyclic compound in which the ring that contains the heteroatom is aromatic.

Heterocyclic compound: Cyclic compound in which one or more of the atoms in the ring are elements other than carbon. Heterocyclic compounds may or may not be aromatic.

Heterogeneous reaction: A reaction involving two or more substances present in different phases. Hydrogenation of alkenes is a heterogeneous reaction that takes place on the surface of an insoluble metal catalyst.

Heterolytically: A term used to describe the cleavage of a chemical bond between two atoms in which both electrons are retained by only one of them.

Heterolytic cleavage: Dissociation of a two-electron covalent bond in such a way that both electrons are retained by one of the initially bonded atoms.

Hexose: A carbohydrate with six carbon atoms.

Histones: Proteins that are associated with DNA in nucleosomes.

Hofmann elimination: An elimination reaction that occurs in the direction that gives predominately the less-substituted double bond. Alkyl halides with sterically hindered bases and alkyltrimethylammonium hydroxides follow this reaction path.

$$R_2CH{-}CR'_2(\overset{+}{N}(CH_3)_3)\ HO^- \xrightarrow{\text{heat}} R_2C{=}CR'_2 + N(CH_3)_3 + H_2O$$

Hofmann rule: A β elimination that gives predominately the alkene with the least substituted double bond. Quaternary ammonium hydroxides and alkyl halides with sterically hindered bases follow this rule.

HOMO: Highest occupied molecular orbital (the orbital of highest energy that contains at least one of a molecule's electrons).

Homochiral: A term describing a sample of a chiral molecule in which all of the molecules have the same configuration.

Homogeneous hydrogenation: Hydrogenation of a double bond catalyzed by an organometallic compound that is soluble in the solvent in which the reaction is carried out.

Homologous series: Group of structurally related substances in which successive members differ by a CH_2 group.

Homolytic cleavage: Dissociation of a two-electron covalent bond in such a way that one electron is retained by each of the initially bonded atoms.

Homopolymer: A polymer formed from a single monomer.

Hückel approximation: A type of molecular orbital theory in which π MOs are considered as separate from σ.

Hückel's rule: Completely conjugated planar monocyclic hydrocarbons possess special stability when the number of their π electrons = $4n + 2$, where n is an integer.

Hund's rule: When two orbitals are of equal energy, they are populated by electrons so that each is half-filled before either one is doubly occupied.

Hybrid orbital: An atomic orbital represented as a mixture of various contributions of that atom's *s, p, d,* etc., orbitals.

Hydration: Addition of the elements of water (H, OH) to a multiple bond.

Hydride shift: Migration of a hydrogen with a pair of electrons (H:) from one atom to another. Hydride shifts are most commonly seen in carbocation rearrangements.

Hydroboration–oxidation: Reaction sequence involving a separate hydroboration stage and oxidation stage. In the hydroboration stage, diborane adds to an alkene to give an alkylborane. In the oxidation stage, the alkylborane is oxidized with hydrogen peroxide to give an alcohol. The reaction product is an alcohol corresponding to the anti-Markovnikov, syn hydration of an alkene.

Hydrocarbon: A compound that contains only carbon and hydrogen.

Hydroformylation: An industrial process for preparing aldehydes ($RCH_2CH_2CH{=}O$) by the reaction of terminal alkenes ($RCH{=}CH_2$) with carbon monoxide.

Hydrogenation: Addition of H_2 to a multiple bond.

Hydrogen bonding: Type of dipole–dipole attractive force in which a positively polarized hydrogen of one molecule is weakly bonded to a negatively polarized atom of an adjacent molecule. Hydrogen bonds typically involve the hydrogen of one —OH or —NH group and the oxygen or nitrogen of another.

Hydrolysis: Water-induced cleavage of a bond.

Hydronium ion: The species H_3O^+.

Hydroperoxide: A compound of the type ROOH.

Hydrophilic: Literally, "water-loving"; a term applied to substances that are soluble in water, usually because of their ability to form hydrogen bonds with water.

Hydrophobic: Literally, "water-hating"; a term applied to substances that are not soluble in water, but are soluble in nonpolar, hydrocarbon-like media.

Hydrophobic effect: The excluding of nonpolar molecules from water.

Hyperconjugation: Delocalization of σ electrons.

I

Icosanoids: A group of naturally occurring compounds derived from unsaturated C_{20} carboxylic acids.

Imide: A compound containing the group $-C(=O)-N(H)-C(=O)-$.

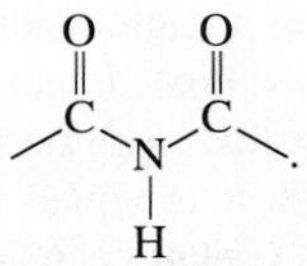

Imine: Compound of the type $R_2C{=}NR'$ formed by the reaction of an aldehyde or a ketone with a primary amine ($R'NH_2$). Imines are sometimes called *Schiff's bases.*

Index of hydrogen deficiency: A measure of the total double bonds and rings a molecule contains. It is determined by comparing the molecular formula C_nH_x of the compound to that of an alkane that has the same number of carbons according to the equation:

$$\text{Index of hydrogen deficiency} = \frac{1}{2}(C_nH_{2n+2} - C_nH_x)$$

Induced-dipole/induced-dipole attractive force: Force of attraction resulting from a mutual and complementary polarization of one molecule by another. Also referred to as *London forces* or *dispersion forces.*

Inductive effect: An electronic effect transmitted by successive polarization of the σ bonds within a molecule or an ion.

Infrared (IR) spectroscopy: Analytical technique based on energy absorbed by a molecule as it vibrates by stretching and bending bonds. Infrared spectroscopy is useful for analyzing the functional groups in a molecule.

Initiation step: A process that causes a reaction, usually a free-radical reaction, to begin but which by itself is not the principal source of products. The initiation step in the halogenation of an alkane is the dissociation of a halogen molecule to two halogen atoms.

Integrase inhibitor: A drug that blocks integrase, an enzyme necessary for the insertion of viral DNA into the DNA of the host.

Integrated area: The relative area of a signal in an NMR spectrum. Areas are proportional to the number of equivalent protons responsible for the peak.

Intermediate: Transient species formed during a chemical reaction. Typically, an intermediate is not stable under the conditions of its formation and proceeds further to form the product. Unlike a transition state, which corresponds to a maximum along a potential energy surface, an intermediate lies at a potential energy minimum.

Intermolecular attractive forces: Forces, either attractive or repulsive, between two atoms or groups in *separate* molecules.

Intramolecular forces: Forces, either attractive or repulsive, between two atoms or groups *within* the same molecule.

Inversion of configuration: Reversal of the three-dimensional arrangement of the four bonds to sp^3-hybridized carbon. The representation shown illustrates inversion of configuration in a nucleophilic substitution where LG is the leaving group and Nu is the nucleophile.

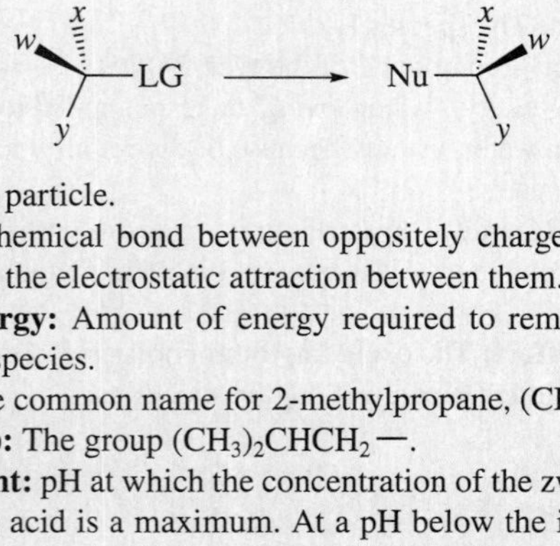

x, *w*, *y* —C—LG ⟶ Nu—C— *x*, *w*, *y*

Ion: A charged particle.

Ionic bond: Chemical bond between oppositely charged particles that results from the electrostatic attraction between them.

Ionization energy: Amount of energy required to remove an electron from some species.

Isobutane: The common name for 2-methylpropane, $(CH_3)_3CH$.

Isobutyl group: The group $(CH_3)_2CHCH_2-$.

Isoelectric point: pH at which the concentration of the zwitterionic form of an amino acid is a maximum. At a pH below the isoelectric point the dominant species is a cation. At higher pH, an anion predominates. At the isoelectric point the amino acid has no net charge.

Isoionic point: Synonymous with *isoelectric point.*

Isolated diene: Diene of the type

$$C{=}C-(C)_x-C{=}C$$

in which the two double bonds are separated by one or more sp^3-hybridized carbons. Isolated dienes are slightly less stable than isomeric conjugated dienes.

Isomers: Different compounds that have the same molecular formula. Isomers may be either constitutional isomers or stereoisomers.

Isoprene rule: Terpenes are composed of repeating head-to-tail-linked isoprene units.

Isoprene unit: The characteristic five-carbon structural unit found in terpenes:

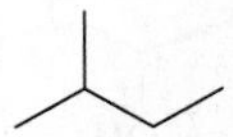

Isopropenyl group: The group $H_2C{=}C(CH_3)-$.

Isopropyl group: The group $(CH_3)_2CH-$.

Isotactic polymer: A stereoregular polymer in which the substituent at each successive chirality center is on the same side of the zigzag carbon chain.

Isotope effect: The difference in a property, usually reaction rate, that is evident when isotopes of the same atom are compared.

Isotopic cluster: In mass spectrometry, a group of peaks that differ in m/z because they incorporate different isotopes of their component elements.

IUPAC rules: The most widely used method of naming organic compounds. It uses a set of rules proposed and periodically revised by the International Union of Pure and Applied Chemistry.

K

Kekulé structure: Structural formula for an aromatic compound that satisfies the customary rules of bonding and is usually characterized by a pattern of alternating single and double bonds. There are two Kekulé formulations for benzene:

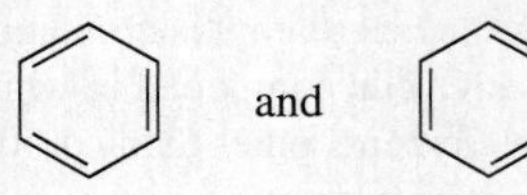

and

A single Kekulé structure does not completely describe the actual bonding in the molecule.

Ketal: An acetal derived from a ketone.

Ketimines: Imines of the type $R_2C{=}NHR$ formed by the reaction of ketones with primary amines. The two groups designated R in the general formula may be the same or different.

Keto-: A tautomeric form that contains a carbonyl group.

Keto–enol tautomerism: Process by which an aldehyde or a ketone and its enol equilibrate:

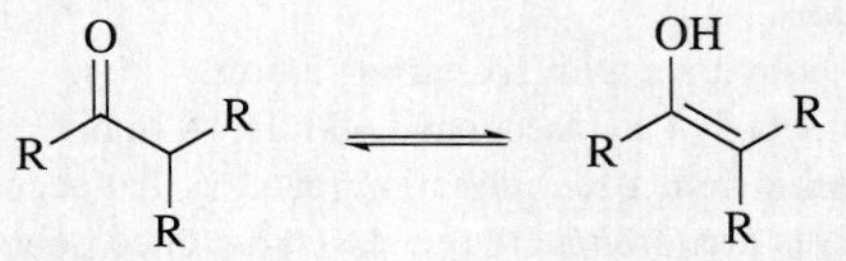

$RC(=O)CHR_2 \rightleftharpoons RC(OH){=}CR_2$

β-Keto ester: A compound of the type

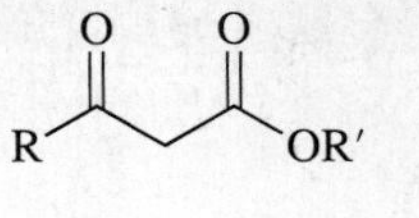

$RC(=O)CH_2C(=O)OR'$

Ketone: A member of the family of compounds in which both atoms attached to a carbonyl group (C=O) are carbon, as in

$$\mathrm{R\overset{\overset{O}{\parallel}}{C}R} \qquad \mathrm{R\overset{\overset{O}{\parallel}}{C}Ar} \qquad \mathrm{Ar\overset{\overset{O}{\parallel}}{C}Ar}$$

Ketose: A carbohydrate that contains a ketone carbonyl group in its open-chain form.

Kiliani–Fischer synthesis: A synthetic method for carbohydrate chain extension. The new carbon–carbon bond is formed by converting an aldose to its cyanohydrin. Reduction of the cyano group to an aldehyde function completes the synthesis.

Kinases: Enzymes that catalyze the transfer of phosphate from ATP to some other molecule.

Kinetic control: Term describing a reaction in which the major product is the one formed at the fastest rate.

Kinetic isotope effect: An effect on reaction rate that depends on isotopic composition.

Kinetic resolution: Separation of enantiomers based on their unequal rates of reaction with a chiral reactant.

Kinetics: The study of reaction rates and the factors that influence them.

Knoevenagel reaction: Stabilized anions undergo additions to aldehydes and ketones in a reaction that resembles the aldol condensation.

$CH_2(CO_2CH_2CH_3)_2$ + C_6H_5CHO $\xrightarrow[\text{heat}]{\text{NH (Piperidine)}}$ $C_6H_5CH{=}C(CO_2CH_2CH_3)_2$

Kolbe–Schmitt reaction: The high-pressure reaction of the sodium salt of a phenol with carbon dioxide to give an *o*-hydroxybenzoic acid. The Kolbe–Schmitt reaction is used to prepare salicylic acid in the synthesis of aspirin.

L

Lactam: A cyclic amide.

β-Lactam: A cyclic amide in which the amide function is part of a four-membered ring. The antibiotic penicillin contains a β-lactam.

Lactone: A cyclic ester.

Lactose: Milk sugar; a disaccharide formed by a β-glycosidic linkage between C-4 of glucose and C-1 of galactose.

Lagging strand: In DNA replication, the strand that grows away from the replication fork.

Leading strand: In DNA replication, the strand that grows toward the replication fork.

Leaving group: The group, normally a halide or sulfonate ion, that is lost from carbon in a nucleophilic substitution or elimination.

Le Châtelier's principle: A reaction at equilibrium responds to any stress imposed on it by shifting the equilibrium in the direction that minimizes the stress.

Lewis acid: A proton donor.

Lewis base: A proton acceptor.

Lewis acid/Lewis base complex: The species that results by covalent bond formation between a Lewis acid and a Lewis base.

Lewis structure: A chemical formula in which electrons are represented by dots. Two dots (or a line) between two atoms represent a covalent bond in a Lewis structure. Unshared electrons are explicitly shown, and stable Lewis structures are those in which the octet rule is satisfied.

Ligand: An atom or group attached to another atom, especially when the other atom is a metal.

Lindlar catalyst: A catalyst for the hydrogenation of alkynes to *cis*-alkenes. It is composed of palladium, which has been "poisoned" with lead(II) acetate and quinoline, supported on calcium carbonate.

Linear polymer: A polymer in which the chain of repeating units is not branched.

Lipid bilayer: Arrangement of two layers of phospholipids that constitutes cell membranes. The polar termini are located at the inner and outer membrane–water interfaces, and the lipophilic hydrocarbon tails cluster on the inside.

Lipids: Biologically important natural products characterized by high solubility in nonpolar organic solvents.

Lipophilic: Literally, "fat-loving"; synonymous in practice with *hydrophobic.*

Liposome: Spherical objects comprised of a phospholipid bilayer.

Lithium diisopropylamide (LDA): $LiN[CH(CH_3)_2]_2$ is a strong, sterically hindered base used in the preparation of enolates.

Living polymer: A polymer that retains active sites capable of further reaction on addition of more monomer.

Localized electrons: Electrons associated with a single atom; that is, not shared with other atoms in a molecule.

Locant: In IUPAC nomenclature, a prefix that designates the atom that is associated with a particular structural unit. The locant is most often a number, and the structural unit is usually an attached substituent as in 2-chlorobutane.

LUMO: The orbital of lowest energy that contains none of a molecule's electrons; the lowest unoccupied molecular orbital.

M

Macromolecule: A substance containing a large number of atoms and having a high molecular weight.

Magnetic resonance imaging (MRI): A diagnostic method in medicine in which tissues are examined by NMR.

Main-group elements: Elements in Groups 1A–8A of the periodic table.

MALDI: Abbreviation for matrix-assisted laser desorption ionization. A mass spectrometric method used in determining the amino acid sequence of peptides and proteins.

Malonic ester synthesis: Synthetic method for the preparation of carboxylic acids involving alkylation of the enolate of diethyl malonate

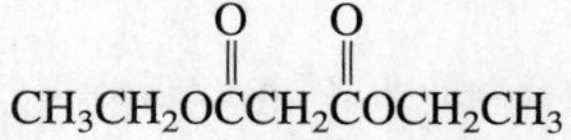

as the key carbon–carbon bond-forming step.

Maltose: A disaccharide obtained from starch in which two glucose units are joined by an α-(1→4)-glycosidic link.

Markovnikov's rule: An unsymmetrical reagent adds to an unsymmetrical double bond in the direction that places the positive part of the reagent on the carbon of the double bond that has the greater number of hydrogens.

Mass spectrometry (MS): Analytical method in which a molecule is ionized and the various ions are examined on the basis of their mass-to-charge ratio.

Mechanism: The sequence of steps that describes how a chemical reaction occurs; a description of the intermediates and transition states that are involved during the transformation of reactants to products.

Mercaptan: An old name for the class of compounds now known as *thiols.*

Merrifield method: See *solid-phase peptide synthesis*.

Meso stereoisomer: An achiral molecule that has chirality centers. The most common kind of meso compound is a molecule with two chirality centers and a plane of symmetry.

Messenger RNA (mRNA): A polynucleotide of ribose that "reads" the sequence of bases in DNA and interacts with tRNAs in the ribosomes to promote protein biosynthesis.

Meta: Term describing a 1,3 relationship between substituents on a benzene ring.

Meta director: A group that when present on a benzene ring directs an incoming electrophile to a position meta to itself.

Metallocene: A transition metal complex that bears a cyclopentadienyl ligand.

Methine group: The group CH.

Methylene group: The group $-CH_2-$.

Methyl group: The group $-CH_3$.

Micelle: A spherical aggregate of species such as carboxylate salts of fatty acids that contain a lipophilic end and a hydrophilic end. Micelles containing 50–100 carboxylate salts of fatty acids are soaps.

Michael reaction: The conjugate addition of a carbanion (usually an enolate) to an α,β-unsaturated carbonyl compound.

Microscopic reversibility: The principle that the intermediates and transition states in the forward and backward stages of a reversible reaction are identical, but are encountered in the reverse order.

Molar absorptivity: A measure of the intensity of a peak, usually in UV-VIS spectroscopy.

Molecular dipole moment: The overall measured dipole moment of a molecule. It can be calculated as the resultant (or vector sum) of all the individual bond dipole moments.

Molecular formula: Chemical formula in which subscripts are used to indicate the number of atoms of each element present in one molecule. In organic compounds, carbon is cited first, hydrogen second, and the remaining elements in alphabetical order.

Molecular ion: In mass spectrometry, the species formed by loss of an electron from a molecule.

Molecularity: The number of species that react together in the same elementary step of a reaction mechanism.

Molecular mechanics: A method for calculating the energy, especially strain energy, of a molecule.

Molecular orbital theory: Theory of chemical bonding in which electrons are assumed to occupy orbitals in molecules much as they occupy orbitals in atoms. The molecular orbitals are described as combinations of the orbitals of all of the atoms that make up the molecule.

Monomer: The simplest stable molecule from which a particular polymer may be prepared.

Monosaccharide: A carbohydrate that cannot be hydrolyzed further to yield a simpler carbohydrate.

Monosubstituted alkene: An alkene of the type $RCH{=}CH_2$, in which there is only one carbon directly bonded to the carbons of the double bond.

Multiplicity: The number of peaks into which a signal is split in nuclear magnetic resonance spectroscopy. Signals are described as singlets, doublets, triplets, and so on, according to the number of peaks into which they are split.

Mutarotation: The change in optical rotation that occurs when a single form of a carbohydrate is allowed to equilibrate to a mixture of isomeric hemiacetals.

N

Nanotube: A form of elemental carbon composed of a cylindrical cluster of carbon atoms.

Negishi reaction: Palladium-catalyzed coupling of an organozinc reagent with an alkyl or aryl halide.

Network polymer: Synonymous with *cross-linked polymer*.

Neurotransmitter: Substance, usually a naturally occurring amine, that mediates the transmission of nerve impulses.

Newman projection: Method for depicting conformations in which one sights down a carbon–carbon bond and represents the front carbon by a point and the back carbon by a circle.

Nitration: Replacement of a hydrogen by an $-NO_2$ group. The term is usually used in connection with electrophilic aromatic substitution.

$$Ar{-}H \xrightarrow[H_2SO_4]{HNO_3} Ar{-}NO_2$$

Nitrile: A compound of the type $R{\equiv}CN$. R may be alkyl or aryl. Also known as alkyl or aryl cyanides.

Nitrogen rule: The molecular weight of a substance that contains C, H, O, and N is odd if the number of nitrogens is odd. The molecular weight is even if the number of nitrogens is even.

Nitrosamine: See N-*nitroso amine*.

Nitrosation: The reaction of a substance, usually an amine, with nitrous acid. Primary amines yield diazonium ions; secondary amines yield *N*-nitroso amines. Tertiary aromatic amines undergo nitrosation of their aromatic ring.

***N*-Nitroso amine:** A compound of the type $R_2N{-}N{=}O$. R may be alkyl or aryl groups, which may be the same or different. *N*-Nitroso amines are formed by nitrosation of secondary amines.

Noble gases: The elements in group 8A of the periodic table (helium, neon, argon, krypton, xenon, radon). Also known as the *rare gases*, they are, with few exceptions, chemically inert.

Nodal surface: A plane drawn through an orbital where the algebraic sign of a wave function changes. The probability of finding an electron at a node is zero.

Nonpolar solvent: A solvent with a low dielectric constant.

N terminus: The amino acid at the end of a peptide or protein chain that has its α-amino group intact; that is, the α-amino group is not part of a peptide bond.

Nuclear magnetic resonance (NMR) spectroscopy: A method for structure determination based on the effect of molecular environment on the energy required to promote a given nucleus from a lower energy spin state to a higher energy state.

Nucleic acid: A polynucleotide present in the nuclei of cells.

Nucleophile: An atom or ion that has an unshared electron pair that can be used to form a bond to carbon. Nucleophiles are Lewis bases.

Nucleophilic acyl substitution: Nucleophilic substitution at the carbon atom of an acyl group.

Nucleophilic addition: The characteristic reaction of an aldehyde or a ketone. An atom possessing an unshared electron pair bonds to the carbon of the $C{=}O$ group, and some other species (normally hydrogen) bonds to the oxygen.

$$RC(=O)R' + H{-}Y{:} \longrightarrow RR'C(OH)Y{:}$$

Nucleophilic aliphatic substitution: Reaction in which a nucleophile replaces a leaving group, usually a halide ion, from sp^3-hybridized carbon. Nucleophilic aliphatic substitution may proceed by either an S_N1 or an S_N2 mechanism.

Nucleophilic aromatic substitution: A reaction in which a nucleophile replaces a leaving group as a substituent on an aromatic ring. Substitution may proceed by an addition–elimination mechanism or an elimination–addition mechanism.

Nucleophilicity: A measure of the reactivity of a Lewis base in a nucleophilic substitution reaction.

Nucleoside: The combination of a purine or pyrimidine base and a carbohydrate, usually ribose or 2-deoxyribose.

Nucleosome: A DNA–protein complex by which DNA is stored in cells.

Nucleotide: The phosphate ester of a nucleoside.

O

Octane rating: The capacity of a sample of gasoline to resist "knocking," expressed as a number equal to the percentage of 2,2,4-trimethylpentane ("isooctane") in an isooctane–heptane mixture that has the same knocking characteristics.

Octet: A filled shell of eight electrons in an atom.

Octet rule: When forming compounds, atoms gain, lose, or share electrons so that the number of their valence electrons is the same as that of the nearest noble gas. For the elements carbon, nitrogen, oxygen, and the halogens, this number is 8.

Olefin metathesis: Exchange of substituents on the double bonds of two alkenes.

$$2R_2C{=}CR'_2 \longrightarrow R_2C{=}CR_2 + R'_2C{=}CR'_2$$

Oligomer: A molecule composed of too few monomer units for it to be classified as a polymer, but more than in a dimer, trimer, tetramer, etc.

Oligonucleotide: A polynucleotide containing a relatively small number of bases.

Oligosaccharide: A carbohydrate that gives three to ten monosaccharides on hydrolysis.

Optical activity: Ability of a substance to rotate the plane of polarized light. To be optically active, a substance must be chiral, and one enantiomer must be present in excess of the other.

Optically pure: Describing a chiral substance in which only a single enantiomer is present.

Optical rotation: The extent to which a chiral substance rotates the plane of plane-polarized light.

Orbital: Strictly speaking, a wave function ψ. It is convenient, however, to think of an orbital in terms of the probability ψ^2 of finding an electron at some point relative to the nucleus, as the volume inside the boundary surface of an atom, or the region in space where the probability of finding an electron is high.

σ Orbital: A bonding orbital characterized by rotational symmetry.

σ* Orbital: An antibonding orbital characterized by rotational symmetry.

Organometallic compound: A compound that contains a carbon-to-metal bond.

Ortho: Term describing a 1,2 relationship between substituents on a benzene ring.

Ortho, para director: A group that when present on a benzene ring directs an incoming electrophile to the positions ortho and para to itself.

Oxidation: A decrease in the number of electrons associated with an atom. In organic chemistry, oxidation of carbon occurs when a bond between carbon and an atom that is less electronegative than carbon is replaced by a bond to an atom that is more electronegative than carbon.

Oxidation number: The formal charge an atom has when the atoms in its covalent bonds are assigned to the more electronegative partner.

Oxidation–reduction: A reaction in which an electron is transferred from one atom to another so that each atom undergoes a change in oxidation number.

Oxidation state: See *oxidation number*.

Oxonium ion: The species H_3O^+ (also called *hydronium ion*).

Oxymercuration–demercuration: A two-stage procedure for alkene hydration.

Ozonide: A compound formed by the reaction of ozone with an alkene.

Ozonolysis: Ozone-induced cleavage of a carbon–carbon double or triple bond.

P

Para: Term describing a 1,4 relationship between substituents on a benzene ring.

Paraffin hydrocarbons: An old name for alkanes and cycloalkanes.

Partial rate factor: In electrophilic aromatic substitution, a number that compares the rate of attack at a particular ring carbon with the rate of attack at a single position of benzene.

Pauli exclusion principle: No two electrons can have the same set of four quantum numbers. An equivalent expression is that only two electrons can occupy the same orbital, and then only when they have opposite spins.

PCC: Abbreviation for pyridinium chlorochromate $C_5H_5NH^+\ ClCrO_3^-$. When used in an anhydrous medium, PCC oxidizes primary alcohols to aldehydes and secondary alcohols to ketones.

PDC: Abbreviation for pyridinium dichromate $(C_5H_5NH)_2^{2+}\ Cr_2O_7^{2-}$. Used in same manner and for same purposes as PCC (see preceding entry).

***n*-Pentane:** The common name for pentane, $CH_3CH_2CH_2CH_2CH_3$.

Pentose: A carbohydrate with five carbon atoms.

Peptidases: Enzymes that catalyze the hydrolysis of peptides and proteins.

Peptide: Structurally, a molecule composed of two or more α-amino acids joined by peptide bonds.

Peptide bond: An amide bond between the carboxyl group of one α-amino acid and the amino group of another.

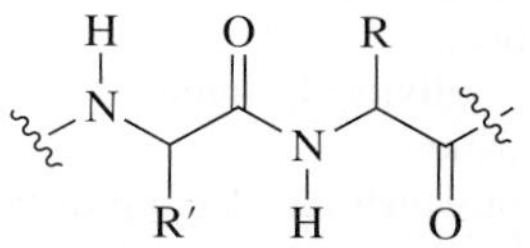

Peptide map: The collection of sequenced fragments of a protein from which its amino acid sequence is determined.

Pericyclic reaction: A reaction that proceeds through a cyclic transition state.

Period: A horizontal row of the periodic table.

Peroxide: A compound of the type ROOR.

Peroxide effect: Reversal of regioselectivity observed in the addition of hydrogen bromide to alkenes brought about by the presence of peroxides in the reaction mixture.

Phase-transfer catalysis: Method for increasing the rate of a chemical reaction by transporting an ionic reactant from an aqueous phase where it is solvated and less reactive to an organic phase where it is not solvated and is more reactive. Typically, the reactant is an anion that is carried to the organic phase as its quaternary ammonium salt.

Phenols: Family of compounds characterized by a hydroxyl substituent on an aromatic ring as in ArOH. *Phenol* is also the name of the parent compound, C_6H_5OH.

Phenyl group: The group

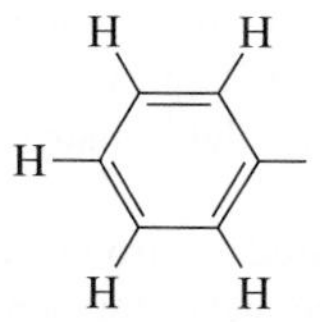

It is often abbreviated C_6H_5—.

Pheromone: A chemical substance used for communication between members of the same species.

Phosphatidic acid: A compound of the type shown, which is an intermediate in the biosynthesis of triacylglycerols.

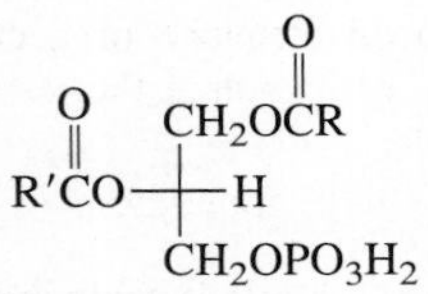

$$\begin{array}{l} \quad\quad\quad\quad\quad\quad \overset{\text{O}}{\overset{\|}{\text{CH}_2\text{OCR}}} \\ \text{R}'\overset{\text{O}}{\overset{\|}{\text{C}}}\text{O}-\!\!\!+\!\!\!-\text{H} \\ \quad\quad\quad\quad \text{CH}_2\text{OPO}_3\text{H}_2 \end{array}$$

Phosphatidylcholine: One of a number of compounds of the type

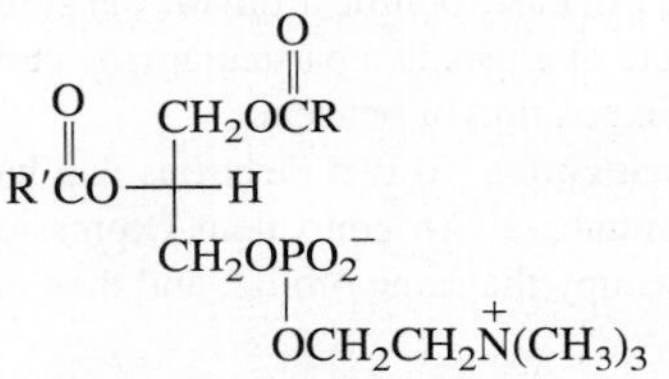

$$\begin{array}{l} \quad\quad\quad\quad\quad\quad \overset{\text{O}}{\overset{\|}{\text{CH}_2\text{OCR}}} \\ \text{R}'\overset{\text{O}}{\overset{\|}{\text{C}}}\text{O}-\!\!\!+\!\!\!-\text{H} \\ \quad\quad\quad\quad \text{CH}_2\text{OPO}_2^- \\ \quad\quad\quad\quad\quad\quad | \\ \quad\quad\quad\quad\quad \text{OCH}_2\text{CH}_2\overset{+}{\text{N}}(\text{CH}_3)_3 \end{array}$$

Phosphodiester: Compound of the type shown, especially when R and R′ are D-ribose or 2-deoxy-D-ribose.

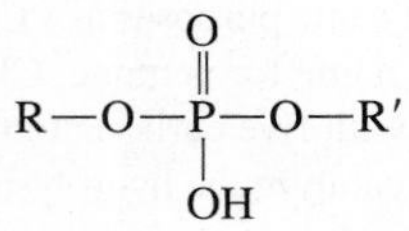

$$\text{R}-\text{O}-\underset{\text{OH}}{\underset{|}{\overset{\text{O}}{\overset{\|}{\text{P}}}}}-\text{O}-\text{R}'$$

Phospholipid: A diacylglycerol bearing a choline-phosphate "head group." Also known as *phosphatidylcholine*.

Photochemical reaction: A chemical reaction that occurs when light is absorbed by a substance.

Photon: Term for an individual "bundle" of energy, or particle, of electromagnetic radiation.

Pi (π) bond: A bond in which the electron distribution is concentrated above and below the internuclear axis, rather than along it as in a σ bond. In organic chemistry π bonds are most often associated with a side-by-side overlap of *p* orbitals on adjacent atoms that are already connected by a σ bond.

Pi (π) electron: Electrons in a π bond or a π orbital.

Pi (π)-electron approximation: A molecular orbital approach in which π orbitals are considered independent from σ.

pK_a: A measure of acid strength defined as $-\log K_a$. The stronger the acid, the smaller the value of pK_a.

Planck's constant: Constant of proportionality (h) in the equation $E = h\nu$, which relates the energy (E) to the frequency (ν) of electromagnetic radiation.

Plane of symmetry: A plane that bisects an object, such as a molecule, into two mirror-image halves; also called a mirror plane. When a line is drawn from any element in the object perpendicular to such a plane and extended an equal distance in the opposite direction, a duplicate of the element is encountered.

Planktonic bacteria: Bacteria that live as individual floating organisms and are not engaged in correlative motions.

Pleated β sheet: Type of protein secondary structure characterized by hydrogen bonds between NH and C=O groups of adjacent parallel peptide chains. The individual chains are in an extended zigzag conformation.

Polar covalent bond: A shared electron pair bond in which the electrons are drawn more closely to one of the bonded atoms than the other.

Polarimeter: An instrument used to measure optical activity.

Polarizability: A measure of the ease of distortion of the electric field associated with an atom or a group. A fluorine atom in a molecule, for example, holds its electrons tightly and is very nonpolarizable. Iodine is very polarizable.

Polar solvent: A solvent with a high dielectric constant.

Polyamide: A polymer in which individual structural units are joined by amide bonds. Nylon is a synthetic polyamide; proteins are naturally occurring polyamides.

Polyamine: A compound that contains many amino groups. The term is usually applied to a group of naturally occurring substances, including spermine, spermidine, and putrescine, that are believed to be involved in cell differentiation and proliferation.

Polycarbonate: A polyester of carbonic acid.

Polycyclic aromatic hydrocarbon: An aromatic hydrocarbon characterized by the presence of two or more fused benzene rings.

Polycyclic hydrocarbon: A hydrocarbon in which two carbons are common to two or more rings.

Polyester: A polymer in which repeating units are joined by ester bonds.

Polyether: A molecule that contains many ether linkages. Polyethers occur naturally in a number of antibiotic substances.

Polyethylene: A polymer of ethylene.

Polymer: Large molecule formed by the repetitive combination of many smaller molecules (monomers).

Polymerase chain reaction (PCR): A laboratory method for making multiple copies of DNA.

Polymerization: Process by which a polymer is prepared. The principal processes include free-radical, cationic, coordination, and condensation polymerization.

Polynucleotide: A polymer in which phosphate ester units join an oxygen of the carbohydrate unit of one nucleoside to that of another.

Polyolefin: An addition polymer prepared from alkene monomers.

Polypeptide: A polymer made up of "many" (more than eight to ten) amino acid residues.

Polypropylene: A polymer of propene.

Polysaccharide: A carbohydrate that yields "many" monosaccharide units on hydrolysis.

Polyurethane: A polymer in which structural units are corrected by a linkage of the type $-\text{NH}\overset{\text{O}}{\overset{\|}{\text{C}}}\text{O}-$.

Potential energy: The energy a system has exclusive of its kinetic energy.

Potential energy diagram: Plot of potential energy versus some arbitrary measure of the degree to which a reaction has proceeded (the reaction coordinate). The point of maximum potential energy is the transition state.

Prenyls: Compounds derived from the group

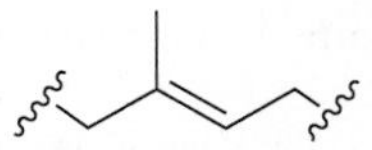

Primary alkyl group: Structural unit of the type RCH_2-, in which the point of attachment is to a primary carbon.

Primary amine: An amine with a single alkyl or aryl substituent and two hydrogens: an amine of the type RNH_2 (primary alkylamine) or $ArNH_2$ (primary arylamine).

Primary carbon: A carbon that is directly attached to only one other carbon.

Primary structure: The sequence of amino acids in a peptide or protein.

Principal quantum number: The quantum number (n) of an electron that describes its energy level. An electron with $n = 1$ must be an *s* electron; one with $n = 2$ has *s* and *p* states available.

Prochiral: The capacity of an achiral molecule to become chiral by replacement of an existing atom or group by a different one.

Prochirality center: An atom of a molecule that becomes a chirality center when one of its attached atoms or groups is replaced by a different atom or group.

Propagation steps: Elementary steps that repeat over and over again in a chain reaction. Almost all of the products in a chain reaction arise from the propagation steps.

Prostaglandin: One of a class of lipid hormones containing 20 carbons, 5 of which belong to an oxygenated cyclopentanoid ring; the remaining 15 carbons are incorporated into two unbranched side chains, adjacent to each other on the ring.

Prosthetic group: A cofactor that is covalently bonded to an enzyme.

Protease inhibitor: A substance that interferes with enzyme-catalyzed hydrolysis of peptide bonds.

Protecting group: A temporary alteration in the nature of a functional group so that it is rendered inert under the conditions in which reaction occurs somewhere else in the molecule. To be synthetically useful, a protecting group must be stable under a prescribed set of reaction conditions, yet be easily introduced and removed.

Protein: A naturally occurring polypeptide that has a biological function.

Protic solvent: A solvent that has easily exchangeable protons, especially protons bonded to oxygen as in hydroxyl groups.

Proton acceptor: A Brønsted base.

Proton donor: A Brønsted acid.

Purine: The heterocyclic aromatic compound

N N N N H

Pyranose form: Six-membered ring arising via cyclic hemiacetal formation between the carbonyl group and a hydroxyl group of a carbohydrate.

Pyrimidine: The heterocyclic aromatic compound

N N

Q

Quantized: Referring to states for which only certain energies are allowed. These states are governed by the relationship $E=nh\nu$, where n is an integer, h is Planck's constant, and ν is the frequency of electromagnetic radiation.

Quantum: The energy associated with a photon.

Quaternary ammonium salt: Salt of the type $R_4N^+ X^-$. The positively charged ion contains a nitrogen with a total of four organic substituents (any combination of alkyl and aryl groups).

Quaternary carbon: A carbon that is directly attached to four other carbons.

Quaternary structure: Description of the way in which two or more protein chains, not connected by chemical bonds, are organized in a larger protein.

Quinone: The product of oxidation of an ortho or para dihydroxybenzene derivative. Examples of quinones include

O O and O O

R

R: Symbol for an alkyl group.

Racemic mixture: Mixture containing equal quantities of enantiomers.

Radical anion: A negatively charged free radical.

Radical cation: A positively charged free radical.

Random coil: A portion of a protein that lacks an ordered secondary structure.

Rare gases: Synonymous with noble gases (helium, neon, argon, krypton, and xenon).

Rate constant k: An experimentally determined proportionality constant that relates the rate of a reaction to the concentrations of the substances present.

Rate-determining step: Slowest step of a multistep reaction mechanism. The overall rate of a reaction can be no faster than its slowest step.

Rearrangement: Intramolecular migration of an atom, a group, or a bond from one atom to another.

Recombinant DNA: DNA molecules that are made by combining nucleotide sequences obtained from different sources. For example, the nucleotide sequence that codes for the synthesis of human insulin can be combined with a bacterial nucleotide sequence to produce recombinant DNA used in the production of insulin.

Reducing sugar: A carbohydrate that possesses an aldehyde or ketone in its linear form or a hydroxide on an anomeric carbon in a cyclic form. Reducing sugars act as mild reducing agents and can be identified by their ability to reduce certain oxidants.

Reduction: Gain in the number of electrons associated with an atom. In organic chemistry, reduction of carbon occurs when a bond between carbon and an atom that is more electronegative than carbon is replaced by a bond to an atom which is less electronegative than carbon.

Reductive amination: Method for the preparation of amines in which an aldehyde or a ketone is treated with ammonia or an amine under conditions of catalytic hydrogenation.

Refining: Conversion of crude oil to useful materials, especially gasoline.

Reforming: Step in oil refining in which the proportion of aromatic and branched-chain hydrocarbons in petroleum is increased so as to improve the octane rating of gasoline.

Regioselective: Term describing a reaction that can produce two (or more) constitutional isomers but gives one of them in greater amounts than the other. A reaction that is 100% regioselective is termed regiospecific.

Relative configuration: Stereochemical configuration on a comparative, rather than an absolute, basis. Terms such as D, L, erythro, threo, α, and β describe relative configuration.

Repeating unit: The structural units that make up a polymer; usually written enclosed in brackets.

Replication: Biosynthetic copying of DNA.

Replication fork: Point at which strands of double-helical DNA separate.

Resolution: Separation of a racemic mixture into its enantiomers.

Resonance: Method by which electron delocalization may be shown using Lewis structures. The true electron distribution in a molecule is regarded as a hybrid of the various Lewis structures that can be written for it.

Resonance energy: Extent to which a substance is stabilized by electron delocalization. It is the difference in energy between the substance and a hypothetical model in which the electrons are localized.

Resonance hybrid: The collection of Lewis structures that, taken together, represent the electron distribution in a molecule.

Restriction enzymes: Enzymes that catalyze the cleavage of DNA at specific sites.

Retention of configuration: Stereochemical pathway observed when a new bond is made that has the same spatial orientation as the bond that was broken.

Retrosynthetic analysis: Technique for synthetic planning based on reasoning backward from the target molecule to appropriate starting materials. An arrow of the type ⇨ designates a retrosynthetic step.

Retrovirus: A virus for which the genetic material is RNA rather than DNA.

Reverse transcriptase inhibitor: A drug that blocks reverse transcriptase, an RNA-directed DNA polymerase that converts viral RNA into DNA for the replication of retroviruses.

Ribosomal RNA (rRNA): The RNA in a cell's ribosomes.

Ribozyme: A polynucleotide that has catalytic activity.

Ring current: Electric field associated with a circulating system of π electrons.

Ring flipping: Synonymous with *ring inversion* of cyclohexane and related compounds.

Ring inversion: Process by which a chair conformation of cyclohexane is converted to a mirror-image chair. All of the equatorial substituents become axial, and vice versa. Also called *ring flipping,* or *chair–chair interconversion.*

RNA (ribonucleic acid): A polynucleotide of ribose.

Robinson annulation: The combination of a Michael addition and an intramolecular aldol condensation used as a synthetic method for ring formation.

Row: Synonymous with *period* in the periodic table.

S

Sandmeyer reaction: Reaction of an aryl diazonium ion with CuCl, CuBr, or CuCN to give, respectively, an aryl chloride, aryl bromide, or aryl cyanide (nitrile).

Saponification: Hydrolysis of esters in basic solution. The products are an alcohol and a carboxylate salt. The term means "soap making" and derives from the process whereby animal fats were converted to soap by heating with wood ashes.

Saturated hydrocarbon: A hydrocarbon in which there are no multiple bonds.

Sawhorse formula: A representation of the three-dimensional arrangement of bonds in a molecule by a drawing of the type shown.

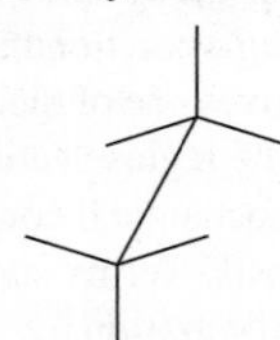

Schiemann reaction: Preparation of an aryl fluoride by heating the diazonium fluoroborate formed by addition of tetrafluoroboric acid (HBF_4) to a diazonium ion.

Schiff's base: Another name for an imine; a compound of the type $R_2C{=}NR'$.

Scientific method: A systematic approach to establishing new knowledge in which observations lead to laws, laws to theories, theories to testable hypotheses, and hypotheses to experiments.

Secondary alkyl group: Structural unit of the type $R_2CH—$, in which the point of attachment is to a secondary carbon.

Secondary amine: An amine with any combination of two alkyl or aryl substituents and one hydrogen on nitrogen; an amine of the type

RNHR′ or RNHAr or ArNHAr′

Secondary carbon: A carbon that is directly attached to two other carbons.

Secondary structure: The conformation with respect to nearest neighbor amino acids in a peptide or protein. The α helix and the pleated β sheet are examples of protein secondary structures.

Sequence rule: Foundation of the Cahn–Ingold–Prelog system. It is a procedure for ranking substituents on the basis of atomic number.

Shared-electron pair: Two electrons shared between two atoms.

Sharpless epoxidation: Epoxidation, especially enantioselective epoxidation, of an allylic alcohol by *tert*-butyl hydroperoxide in the presence of a Ti(IV) catalyst and diethyl tartrate.

β Sheet: A type of protein secondary structure in which the C=O and N—H groups of adjacent chains, or regions of one chain, are hydrogen-bonded in a way that produces a sheet-like structure which may be flat or pleated.

Shell: The group of orbitals that have the same principal quantum number *n.*

Shielding: Effect of a molecule's electrons that decreases the strength of an external magnetic field felt by a proton or another nucleus.

Sigma (σ) bond: In valence bond theory, a bond characterized by overlap of a half-filled orbital of one atom with a half-filled orbital of a second atom along a line connecting the two nuclei.

Sigmatropic rearrangement: Migration of a σ bond from one end of a conjugated π-electron system to the other. The Claisen rearrangement is an example.

Simmons–Smith reaction: Reaction of an alkene with iodomethylzinc iodide to form a cyclopropane derivative.

Skew boat: A conformation of cyclohexane that is less stable than the chair, but slightly more stable than the boat.

Solid-phase peptide synthesis: Method for peptide synthesis in which the C-terminal amino acid is covalently attached to an inert solid support and successive amino acids are attached via peptide bond formation. At the completion of the synthesis the polypeptide is removed from the support.

Solvolysis reaction: Nucleophilic substitution in a medium in which the only nucleophiles present are the solvent and its conjugate base.

Specific rotation: Optical activity of a substance per unit concentration per unit path length:

$$[\alpha] = \frac{100\alpha}{cl}$$

where α is the observed rotation in degrees, c is the concentration in g/100 mL, and l is the path length in decimeters.

Spectrometer: Device designed to measure absorption of electromagnetic radiation by a sample.

Spectroscopy: Study of the interaction of a molecule with electromagnetic radiation so as to learn more about its structure and/or properties.

Spectrum: Output, usually in chart form, of a spectrometer. Analysis of a spectrum provides information about molecular structure.

***sp* Hybridization:** Hybridization state adopted by carbon when it bonds to two other atoms as, for example, in alkynes. The *s* orbital and one of the 2*p* orbitals mix to form two equivalent *sp*-hybridized orbitals. A linear geometry is characteristic of *sp* hybridization.

sp^2 Hybridization: A model to describe the bonding of a carbon attached to three other atoms or groups. The carbon 2*s* orbital and the two 2*p* orbitals are combined to give a set of three equivalent sp^2 orbitals having 33.3% *s* character and 66.7% *p* character. One *p* orbital remains unhybridized. A trigonal planar geometry is characteristic of sp^2 hybridization.

sp^3 Hybridization: A model to describe the bonding of a carbon attached to four other atoms or groups. The carbon 2*s* orbital and the three 2*p* orbitals are combined to give a set of four equivalent orbitals having 25% *s* character and 75% *p* character. These orbitals are directed toward the corners of a tetrahedron.

Spin: Synonymous with *spin quantum number.*

Spin density: A measure of the unpaired electron distribution at the various atoms in a molecule.

Spin quantum number: One of the four quantum numbers that describe an electron. An electron may have either of two different spin quantum numbers, $+\frac{1}{2}$ or $-\frac{1}{2}$.

Spin–spin coupling: The communication of nuclear spin information between two nuclei.

Spin–spin splitting: The splitting of NMR signals caused by the coupling of nuclear spins. Only nonequivalent nuclei (such as protons with different chemical shifts) can split one another's signals.

Spiro compound: A compound in which a single carbon is common to two rings.

Spontaneous reaction: Among several definitions, the one most relevant to the material in this text defines a spontaneous reaction as one that proceeds with a decrease in free energy ($\Delta G < 0$). The "official" definition is that a spontaneous process is one in which the entropy of the universe increases.

Squalene: A naturally occurring triterpene from which steroids are biosynthesized.

Staggered conformation: Conformation of the type shown, in which the bonds on adjacent carbons are as far away from one another as possible.

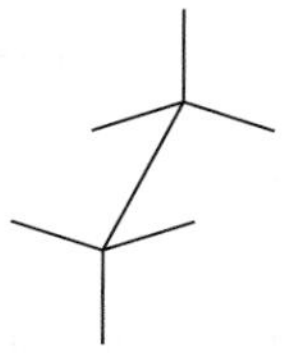

Standard amino acids: The 20 α-amino acids normally present in proteins.

Standard enthalpy of formation ($\Delta H°$): The enthalpy change for formation of one mole of a compound from its elements under standard state conditions. The standard state is the state (solid, liquid, or gas) of a substance at a pressure of 1 atm. The standard state for a solution is 1 M.

Standard free-energy change ($\Delta G°$): The free-energy change ΔG for a reaction occurring under standard state conditions.

Standard free-energy change ($\Delta G°'$): The value of $\Delta G°$ at pH = 7.

Standard heat of formation: The value of $\Delta H°$ for formation of a substance from its elements.

Step-growth polymerization: Polymerization by a process in which monomers are first consumed in oligomer formation followed by subsequent reaction between oligomers to form macromolecules.

Stereochemistry: Chemistry in three dimensions; the relationship of physical and chemical properties to the spatial arrangement of the atoms in a molecule.

Stereoelectronic effect: An electronic effect that depends on the spatial arrangement between the orbitals of the electron donor and acceptor.

Stereoisomers: Isomers with the same constitution but that differ in respect to the arrangement of their atoms in space. Stereoisomers may be either *enantiomers* or *diastereomers.*

Stereoregular polymer: Polymer containing chirality centers according to a regular repeating pattern. Syndiotactic and isotactic polymers are stereoregular.

Stereoselective reaction: Reaction in which a single starting material has the capacity to form two or more stereoisomeric products but forms one of them in greater amounts than any of its stereoisomers. Terms such as "addition to the less hindered side" describe stereoselectivity.

Stereospecific reaction: Reaction in which stereoisomeric starting materials give stereoisomeric products. Terms such as *syn addition, anti elimination,* and *inversion of configuration* describe stereospecific reactions.

Steric effect: An effect on structure or reactivity that depends on van der Waals repulsive forces.

Steric strain: Destabilization of a molecule as a result of van der Waals repulsion, distorted bond distances, bond angles, or torsion angles.

Steroid: Type of lipid present in both plants and animals characterized by a nucleus of four fused rings (three are six-membered, one is five-membered). Cholesterol is the most abundant steroid in animals.

Stille reaction: Palladium-catalyzed coupling of an organotin reagent with an alkyl or aryl halide or sulfonate.

Strain energy: Excess energy possessed by a species because of van der Waals repulsion, distorted bond lengths, bond angles, or torsion angles.

Strecker synthesis: Method for preparing amino acids in which the first step is reaction of an aldehyde with ammonia and hydrogen cyanide to give an amino nitrile, which is then hydrolyzed.

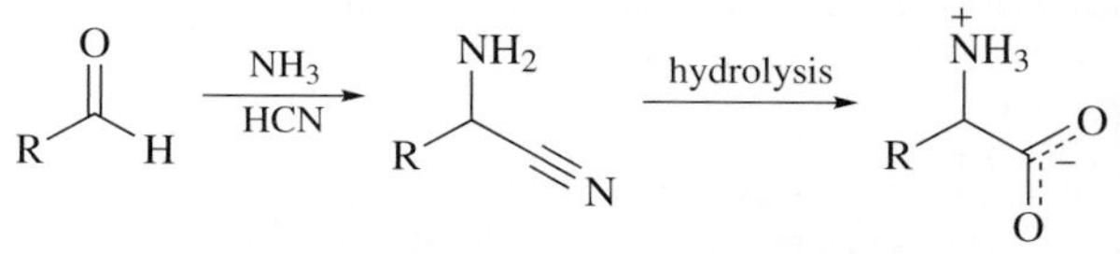

Stretching vibration: A regular, repetitive motion of two atoms or groups along the bond that connects them.

Structural isomer: Synonymous with *constitutional isomer.*

Structure: The sequence of connections that defines a molecule, including the spatial orientation of these connections.

Substitution: The replacement of an atom or group in a molecule by a different atom or group.

Substitution nucleophilic bimolecular (S_N2) mechanism: Concerted mechanism for nucleophilic substitution in which the nucleophile attacks carbon from the side opposite the bond to the leaving group and assists the departure of the leaving group.

Substitution nucleophilic unimolecular (S_N1) mechanism: Mechanism for nucleophilic substitution characterized by a two-step process. The first step is rate-determining and is the ionization of an alkyl halide to a carbocation and a halide ion.

Substitutive nomenclature: Type of IUPAC nomenclature in which a substance is identified by a name ending in a suffix characteristic of the type of compound. 2-Methylbutanol, 3-pentanone, and 2-phenylpropanoic acid are examples of substitutive names.

Sucrose: A disaccharide of glucose and fructose in which the two monosaccharides are joined at their anomeric positions.

Sulfenic acid: Compound of the type RSOH.

Sulfide: A compound of the type RSR′. Sulfides are the sulfur analogs of ethers.

Sulfinic acid: Compound of the type RS(O)OH.

Sulfonation: Replacement of a hydrogen by an $—SO_3H$ group. The term is usually used in connection with electrophilic aromatic substitution.

$$\text{Ar—H} \xrightarrow[\text{H}_2\text{SO}_4]{\text{SO}_3} \text{Ar—SO}_3\text{H}$$

Sulfone: Compound of the type

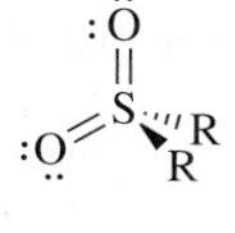

Sulfonic acid: Compound of the type RSO_2OH.

Sulfoxide: Compound of the type

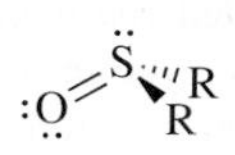

Supercoil: Coiled DNA helices.

Suzuki-Miyaura reaction: Palladium-catalyzed coupling of an organoboron reagent with an alkyl or aryl halide or sulfonate.

Symmetry-allowed reaction: Concerted reaction in which the orbitals involved overlap in phase at all stages of the process.

Symmetry-forbidden reaction: Concerted reaction in which the orbitals involved do not overlap in phase at all stages of the process.

Syn addition: Addition reaction in which the two portions of the reagent that add to a multiple bond add from the same side.

Syndiotactic polymer: Stereoregular polymer in which the configuration of successive chirality centers alternates along the chain.

Synthon: A structural unit in a molecule that is related to a synthetic operation.

Systematic names: Names for chemical compounds that are developed on the basis of a prescribed set of rules. Usually the IUPAC system is meant when the term *systematic nomenclature* is used.

T

Tautomerism: Process by which two isomers are interconverted by the movement of an atom or a group. Enolization is a form of tautomerism.

$$RC(=O)CHR_2 \rightleftharpoons RC(OH)=CR_2$$

Tautomers: Constitutional isomers that interconvert by migration of an atom or group.

Terminal alkyne: Alkyne of the type $RC{\equiv}CH$, in which the triple bond appears at the end of the chain.

Termination steps: Reactions that halt a chain reaction. In a free-radical chain reaction, termination steps consume free radicals without generating new radicals to continue the chain.

Terpenes: Compounds that can be analyzed as clusters of isoprene units. Terpenes with 10 carbons are classified as monoterpenes, those with 15 are sesquiterpenes, those with 20 are diterpenes, and those with 30 are triterpenes.

Tertiary alkyl group: Structural unit of the type $R_3C{-}$, in which the point of attachment is to a tertiary carbon.

Tertiary amine: Amine of the type R_3N with any combination of three alkyl or aryl substituents on nitrogen.

Tertiary carbon: A carbon that is directly attached to three other carbons.

Tertiary structure: A description of how a protein chain is folded.

Tesla: SI unit for magnetic field strength.

Tetrahedral angle: The angle between one line directed from the center of a tetrahedron to a vertex and a second line from the center to a different vertex. This angle is 109° 28′.

Tetrahedral intermediate: The key intermediate in nucleophilic acyl substitution. Formed by nucleophilic addition to the carbonyl group of a carboxylic acid derivative.

Tetramethylsilane (TMS): The molecule $(CH_3)_4Si$, used as a standard to calibrate proton and carbon-13 NMR spectra.

Tetrapeptide: A compound composed of four α-amino acids connected by peptide bonds.

Tetrasubstituted alkene: Alkene of the type $R_2C{=}CR_2$, in which there are four carbons *directly* bonded to the carbons of the double bond. (The R groups may be the same or different.)

Tetrose: A carbohydrate with four carbon atoms.

Thermodynamically controlled reaction: Reaction in which the reaction conditions permit two or more products to equilibrate, giving a predominance of the most stable product.

Thermoplastic polymer: A polymer that softens or melts when heated.

Thermoset: The cross-linked product formed by heating a thermoplastic polymer.

Thermosetting polymer: A polymer that solidifies ("cures") when heated.

Thiol: Compound of the type RSH or ArSH.

Three-bond coupling: Synonymous with *vicinal coupling.*

Threo: Term applied to the relative configuration of two chirality centers within a molecule. The threo stereoisomer has like substituents on opposite sides of a Fischer projection.

Torsional strain: Decreased stability of a molecule associated with eclipsed bonds.

***trans*-:** Stereochemical prefix indicating that two substituents are on opposite sides of a ring or a double bond. (Contrast with the prefix *cis*-.)

Transamination: The transfer (usually biochemical) of an amino group from one compound to another.

Transcription: Construction of a strand of mRNA complementary to a DNA template.

Transfer RNA (tRNA): A polynucleotide of ribose that is bound at one end to a unique amino acid. This amino acid is incorporated into a growing peptide chain.

Transition state: The point of maximum energy in an elementary step of a reaction mechanism.

Translation: The "reading" of mRNA by various tRNAs, each one of which is unique for a particular amino acid.

Transmetalation: Reaction in which the metal atom of an organometallic compound is displaced by a different metal.

Triacylglycerol: A derivative of glycerol (1,2,3-propanetriol) in which the three oxygens bear acyl groups derived from fatty acids.

Tripeptide: A compound in which three α-amino acids are linked by peptide bonds.

Triple bond: Bond formed by the sharing of six electrons between two atoms.

Trisubstituted alkene: Alkene of the type $R_2C{=}CHR$, in which there are three carbons *directly* bonded to the carbons of the double bond. (The R groups may be the same or different.)

Trivial nomenclature: Term synonymous with *common name.*

Twist boat: Synonymous with *skew boat.*

U

Ultraviolet-visible (UV-VIS) spectroscopy: Analytical method based on transitions between electronic energy states in molecules. Useful in studying conjugated systems such as polyenes.

Unimolecular: Describing a step in a reaction mechanism in which only one particle undergoes a chemical change at the transition state.

α,β-Unsaturated aldehyde or ketone: Aldehyde or ketone that bears a double bond between its α and β carbons as in

$$R_2C{=}CH\overset{\overset{\displaystyle O}{\|}}{C}R'$$

Unsaturated hydrocarbon: A hydrocarbon that can undergo addition reactions; that is, one that contains multiple bonds.

Unshared pair: In a Lewis structure, two valence electrons of an atom that are in the same orbital and not shared with any other atom.

Upfield: The high-field region of an NMR spectrum. A signal that is upfield with respect to another lies to its right on the spectrum.

Uronic acids: Carbohydrates that have an aldehyde function at one end of their carbon chain and a carboxylic acid group at the other.

V

Valence bond theory: Theory of chemical bonding based on overlap of half-filled atomic orbitals between two atoms. Orbital hybridization is an important element of valence bond theory.

Valence electrons: The outermost electrons of an atom. For second-row elements these are the 2*s* and 2*p* electrons.

Valence shell: The group of orbitals, filled and unfilled, responsible for the characteristic chemical properties of an atom.

Valence shell electron-pair repulsion (VSEPR) model: Method for predicting the shape of a molecule based on the notion that electron pairs surrounding a central atom repel one another. Four electron pairs will arrange themselves in a tetrahedral geometry, three will assume a trigonal planar geometry, and two electron pairs will adopt a linear arrangement.

Van der Waals forces: Intermolecular forces that do not involve ions (dipole–dipole, dipole/induced-dipole, and induced-dipole/induced-dipole forces).

Van der Waals radius: A measure of the effective size of an atom or a group. The repulsive force between two atoms increases rapidly when they approach each other at distances less than the sum of their van der Waals radii.

Van der Waals strain: Destabilization that results when two atoms or groups approach each other too closely. Also known as van der Waals repulsion.

Vicinal: Describing two atoms or groups attached to adjacent atoms.

Vicinal coupling: Coupling of the nuclear spins of atoms X and Y on adjacent atoms as in X—A—B—Y. Vicinal coupling is the most common cause of spin–spin splitting in ^{1}H NMR spectroscopy.

Vicinal dihalide: A compound containing two halogens on adjacent carbons.

Vicinal diol: Compound that has two hydroxyl (—OH) groups on adjacent sp^3-hybridized carbons.

Vicinal halohydrin: A compound containing a halogen and a hydroxyl group on adjacent carbons.

Vinyl group: The group $H_2C{=}CH—$.

Vinylic carbon: A carbon that is doubly bonded to another carbon. Atoms or groups attached to a vinylic carbon are termed *vinylic substituents.*

W

Wave functions: The solutions to arithmetic expressions that express the energy of an electron in an atom.

Wavelength: Distance between two successive maxima (peaks) or two successive minima (troughs) of a wave.

Wavenumbers: Conventional units in infrared spectroscopy that are proportional to frequency. Wavenumbers are cited in reciprocal centimeters (cm^{-1}).

Wax: A mixture of water-repellent substances that form a protective coating on the leaves of plants, the fur of animals, and the feathers of birds, among other things. A principal component of a wax is often an ester in which both the acyl portion and the alkyl portion are characterized by long carbon chains.

Williamson ether synthesis: Method for the preparation of ethers involving an S_N2 reaction between an alkoxide ion and a primary alkyl halide:

$$RONa + R'CH_2Br \longrightarrow R'CH_2OR + NaBr$$

Wittig reaction: Method for the synthesis of alkenes by the reaction of an aldehyde or a ketone with a phosphorus ylide.

$$R\overset{O}{\overset{\|}{C}}R' + (C_6H_5)_3\overset{+}{P}—\overset{-}{\underset{\cdot\cdot}{C}}R''_2 \longrightarrow (R)(R')C{=}C(R'')_2 + (C_6H_5)_3\overset{+}{P}—O^-$$

Wittig reagents: Name given to the phosphorus ylides used in the Wittig reaction.

Wolff–Kishner reduction: Method for reducing the carbonyl group of aldehydes and ketones to a methylene group ($C{=}O \rightarrow CH_2$) by treatment with hydrazine (H_2NNH_2) and base (KOH) in a high-boiling alcohol solvent.

Wood alcohol: A common name for methanol, CH_3OH.

Ylide: A neutral molecule in which two oppositely charged atoms, each with an octet of electrons, are directly bonded to each other. The compound

$$(C_6H_5)_3\overset{+}{P}—\overset{-}{\underset{\cdot\cdot}{C}}H_2$$

is an example of an ylide.

Z

Z-: Stereochemical descriptor used when higher ranked substituents are on the same side of a double bond.

Zaitsev's rule: When two or more alkenes are capable of being formed by an elimination reaction, the one with the more highly substituted double bond (the more stable alkene) is the major product.

Zusammen: See *Z-*.

Zwitterion: The form in which neutral amino acids actually exist. The amino group is in its protonated form and the carboxyl group is present as a carboxylate

$$\underset{\overset{|}{^{+}NH_3}}{RCHCO_2^-}$$

Index

Note: Page numbers followed by *f* or *t* indicate figures or tables, respectively.

A

C

D

F

I

J

K

L

M

Q

R

S

T

Periodic Table of the Elements

- Metals (main-group)
- Metals (transition)
- Metals (inner transition)
- Metalloids
- Nonmetals

MAIN–GROUP ELEMENTS (groups 1A–2A) | TRANSITION ELEMENTS (groups 3B–2B) | MAIN–GROUP ELEMENTS (groups 3A–8A)

Period	1A (1)	2A (2)	3B (3)	4B (4)	5B (5)	6B (6)	7B (7)	8B (8)	8B (9)	8B (10)	1B (11)	2B (12)	3A (13)	4A (14)	5A (15)	6A (16)	7A (17)	8A (18)
1	1 **H** 1.008																	2 **He** 4.003
2	3 **Li** 6.941	4 **Be** 9.012											5 **B** 10.81	6 **C** 12.01	7 **N** 14.01	8 **O** 16.00	9 **F** 19.00	10 **Ne** 20.18
3	11 **Na** 22.99	12 **Mg** 24.31											13 **Al** 26.98	14 **Si** 28.09	15 **P** 30.97	16 **S** 32.07	17 **Cl** 35.45	18 **Ar** 39.95
4	19 **K** 39.10	20 **Ca** 40.08	21 **Sc** 44.96	22 **Ti** 47.88	23 **V** 50.94	24 **Cr** 52.00	25 **Mn** 54.94	26 **Fe** 55.85	27 **Co** 58.93	28 **Ni** 58.69	29 **Cu** 63.55	30 **Zn** 65.41	31 **Ga** 69.72	32 **Ge** 72.61	33 **As** 74.92	34 **Se** 78.96	35 **Br** 79.90	36 **Kr** 83.80
5	37 **Rb** 85.47	38 **Sr** 87.62	39 **Y** 88.91	40 **Zr** 91.22	41 **Nb** 92.91	42 **Mo** 95.94	43 **Tc** (98)	44 **Ru** 101.1	45 **Rh** 102.9	46 **Pd** 106.4	47 **Ag** 107.9	48 **Cd** 112.4	49 **In** 114.8	50 **Sn** 118.7	51 **Sb** 121.8	52 **Te** 127.6	53 **I** 126.9	54 **Xe** 131.3
6	55 **Cs** 132.9	56 **Ba** 137.3	57 **La** 138.9	72 **Hf** 178.5	73 **Ta** 180.9	74 **W** 183.9	75 **Re** 186.2	76 **Os** 190.2	77 **Ir** 192.2	78 **Pt** 195.1	79 **Au** 197.0	80 **Hg** 200.6	81 **Tl** 204.4	82 **Pb** 207.2	83 **Bi** 209.0	84 **Po** (209)	85 **At** (210)	86 **Rn** (222)
7	87 **Fr** (223)	88 **Ra** (226)	89 **Ac** (227)	104 **Rf** (263)	105 **Db** (262)	106 **Sg** (266)	107 **Bh** (267)	108 **Hs** (277)	109 **Mt** (268)	110 **Ds** (281)	111 **Rg** (272)	112 **Cn** (285)	113 **Nh** (284)	114 **Fl** (289)	115 **Mc** (288)	116 **Lv** (293)	117 **Ts** (293)	118 **Og** (294)

INNER TRANSITION ELEMENTS

6	Lanthanides	58 **Ce** 140.1	59 **Pr** 140.9	60 **Nd** 144.2	61 **Pm** (145)	62 **Sm** 150.4	63 **Eu** 152.0	64 **Gd** 157.3	65 **Tb** 158.9	66 **Dy** 162.5	67 **Ho** 164.9	68 **Er** 167.3	69 **Tm** 168.9	70 **Yb** 173.0	71 **Lu** 175.0
7	Actinides	90 **Th** 232.0	91 **Pa** (231)	92 **U** 238.0	93 **Np** (237)	94 **Pu** (242)	95 **Am** (243)	96 **Cm** (247)	97 **Bk** (247)	98 **Cf** (251)	99 **Es** (252)	100 **Fm** (257)	101 **Md** (258)	102 **No** (259)	103 **Lr** (260)

SOME COMMONLY ENCOUNTERED GROUPS

Group	*Name**	*Group*	*Name**
$CH_3CH_2CH_2—$	Propyl or *n*-propyl	$CH_3C(=O)—$	Ethanoyl or acetyl
$(CH_3)_2CH—$	1-Methylethyl or isopropyl	$C_6H_5—$	Phenyl
$CH_3CH_2CH_2CH_2—$	Butyl or *n*-butyl	$C_6H_5CH_2—$	Phenylmethyl or benzyl
$CH_3CH(—)CH_2CH_3$	1-Methylpropyl or *sec*-butyl	$C_6H_5C(=O)—$	Benzenecarbonyl or benzoyl
$(CH_3)_3C—$	1,1-Dimethylethyl or *tert*-butyl		
$(CH_3)_2CHCH_2—$	2-Methylpropyl or isobutyl		
$(CH_3)_3CCH_2—$	2,2-Dimethylpropyl or neopentyl		
$H_2C{=}CH—$	Ethenyl or vinyl		
$H_2C{=}CHCH_2—$	2-Propenyl or allyl		
$H_2C{=}C(—)CH_3$	1-Methylvinyl or isopropenyl		

*When two names are cited, either one is acceptable in IUPAC nomenclature.

COMMONLY ENCOUNTERED GROUPS LISTED IN ORDER OF INCREASING RANK IN THE CAHN–INGOLD–PRELOG SYSTEM

1. $H—$
2. $CH_3—$
3. $CH_3CH_2—$
4. $CH_3CH(CH_3)CH_2—$
5. $(CH_3)_3CCH_2—$
6. $(CH_3)_2CH—$
7. $H_2C{=}CH—$
8. $(CH_3)_3C—$
9. $HOCH_2—$
10. $HC(=O)—$
11. $CH_3C(=O)—$
12. $HOC(=O)—$
13. $CH_3OC(=O)—$
14. $HSCH_2—$
15. $H_2N—$
16. $HO—$
17. $CH_3O—$
18. $CH_3CH_2O—$
19. $HC(=O)O—$
20. $CH_3C(=O)O—$
21. $F—$
22. $HS—$
23. $Cl—$
24. $Br—$
25. $I—$

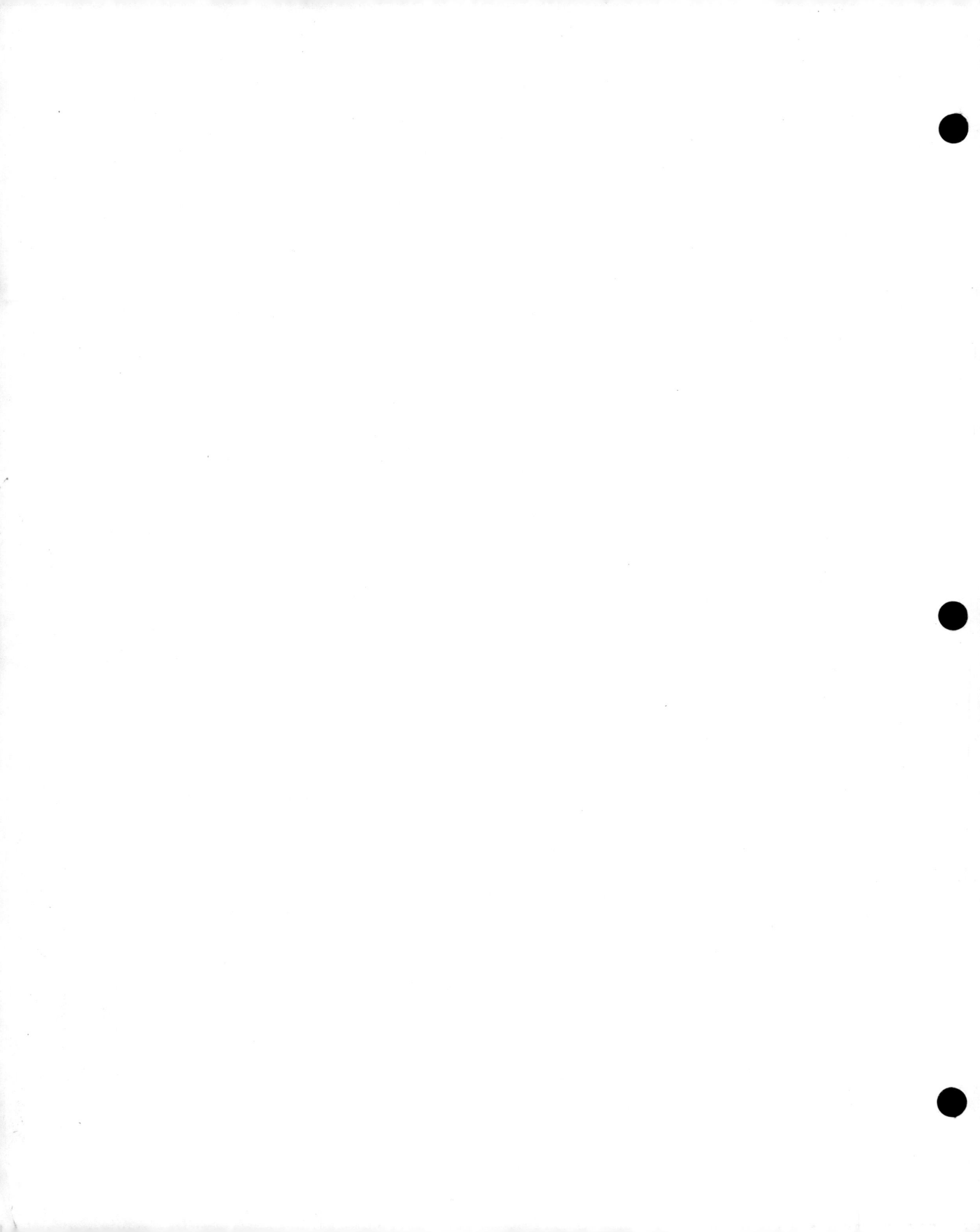